Handbook of Olfaction and Gustation

Handbook of Olfaction and Gustation

Third Edition

Edited by

Richard L. Doty

Copyright © 2015 by Richard L. Doty. All rights reserved

Published by John Wiley & Sons, Inc., Hoboken, New Jersey
Published simultaneously in Canada

No part of this publication may be reproduced, stored in a retrieval system, or transmitted in any form or by any means, electronic, mechanical, photocopying, recording, scanning, or otherwise, except as permitted under Section 107 or 108 of the 1976 United States Copyright Act, without either the prior written permission of the Publisher, or authorization through payment of the appropriate per-copy fee to the Copyright Clearance Center, Inc., 222 Rosewood Drive, Danvers, MA 01923, (978) 750-8400, fax (978) 750-4470, or on the web at www.copyright.com. Requests to the Publisher for permission should be addressed to the Permissions Department, John Wiley & Sons, Inc., 111 River Street, Hoboken, NJ 07030, (201) 748-6011, fax (201) 748-6008, or online at http://www.wiley.com/go/permission.

Limit of Liability/Disclaimer of Warranty: While the publisher and author have used their best efforts in preparing this book, they make no representations or warranties with respect to the accuracy or completeness of the contents of this book and specifically disclaim any implied warranties of merchantability or fitness for a particular purpose. No warranty may be created or extended by sales representatives or written sales materials. The advice and strategies contained herein may not be suitable for your situation. You should consult with a professional where appropriate. Neither the publisher nor author shall be liable for any loss of profit or any other commercial damages, including but not limited to special, incidental, consequential, or other damages.

For general information on our other products and services or for technical support, please contact our Customer Care Department within the United States at (800) 762-2974, outside the United States at (317) 572-3993 or fax (317) 572-4002.

Wiley also publishes its books in a variety of electronic formats. Some content that appears in print may not be available in electronic formats. For more information about Wiley products, visit our web site at www.wiley.com.

Library of Congress Cataloging-in-Publication Data:

Handbook of olfaction and gustation / edited by Richard L. Doty. – Third edition.
   p. ; cm.
  Includes bibliographical references and index.
   ISBN 978-1-118-13922-6 (cloth)
  I. Doty, Richard L., editor.
  [DNLM: 1. Smell–physiology. 2. Chemoreceptor Cells–physiology. 3. Olfaction Disorders. 4. Taste–physiology. 5. Taste Disorders. WV 301]
  QP455
  612.8′6–dc23
               2014037024

Printed in United States of America.

2 2016

Contents

Foreword *Gordon M. Shepherd*   xi

Preface *Richard L. Doty*   xiii

Contributors   xv

Part 1   General Introduction

1 Introduction and Historical Perspective   3

Richard L. Doty

Part 2   Olfaction
Olfactory Anatomy and Neurobiology

2 Anatomy of the Nasal Passages in Mammals   39

Timothy D. Smith, Thomas P. Eiting, and Kunwar P. Bhatnagar

3 Olfactory Mucosa: Composition, Enzymatic Localization, and Metabolism   63

Xinxin Ding and Fang Xie

4 Development, Morphology, and Functional Anatomy of the Olfactory Epithelium   93

John C. Dennis, Shelly Aono, Vitaly J. Vodyanoy, and Edward E. Morrison

5 Olfactory Receptor Function   109

Keiichi Yoshikawa and Kazushige Touhara

6 Odorant Receptor Gene Regulation   123

Akio Tsuboi and Hitoshi Sakano

7 Neurogenesis in the Adult Olfactory Epithelium   133

Alan Mackay-Sim, James St John, and James E. Schwob

Contents

8 Anatomy and Neurobiology of the Main and Accessory Olfactory Bulbs   157

Matthew Ennis and Timothy E. Holy

9 Adult Neurogenesis in the Subventricular Zone and Migration to the Olfactory Bulb   183

John W. Cave and Harriet Baker

10 Cortical Olfactory Anatomy and Physiology   209

Donald A. Wilson, Julie Chapuis, and Regina M. Sullivan

Part 3   Human Olfactory Measurement, Physiology, and Development

11 Psychophysical Measurement of Human Olfactory Function   227

Richard L. Doty and David G. Laing

12 Electrophysiological Measurement of Olfactory Function   261

Allen Osman and Jonathan Silas

13 Structural and Functional Imaging of the Human Olfactory System   279

Jay A. Gottfried

14 Prenatal and Postnatal Human Olfactory Development: Influences on Cognition and Behavior   305

Benoist Schaal

15 Olfactory Memory   337

Theresa L. White, Per Møller, E. P. Köster, Howard Eichenbaum, and Christiane Linster

---

Part 4   Clinical Applications and Perspectives

16 Nasal Patency and the Aerodynamics of Nasal Airflow in Relation to Olfactory Function   355

Kai Zhao and Richard E. Frye

17 Clinical Disorders of Olfaction   375

Richard L. Doty

18 Odor Perception and Neuropathology in Neurodegenerative Diseases and Schizophrenia   403

Richard L. Doty, Christopher H. Hawkes, Kimberley P. Good, and John E. Duda

19 The Olfactory System as a Route of Delivery for Agents to the Brain and Circulation   453

Mary Beth Genter, Mansi Krishan, and Rui Daniel Prediger

20 Influence of Toxins on Olfactory Function and their Potential Association with Neurodegenerative Disease   485

Lilian Calderón-Garcidueñas

Contents

Part 5 Olfaction in Nonhuman Forms

21 Microbial Chemical Sensing   513

Judith Van Houten

22 Olfaction in Insects   531

Paul Szyszka and C. Giovanni Galizia

23 Olfaction in Aquatic Vertebrates   547

Keith B. Tierney

24 The Chemistry of Avian Odors: An Introduction to Best Practices   565

Gabrielle A. Nevitt and Paola A. Prada

25 Olfactory Communication in Rodents in Natural and Semi-Natural Habitats   579

Daniel W. Wesson

26 Olfaction in the Order Carnivora: Family Canidae   591

Peter Hepper and Deborah Wells

27 Olfaction in Nonhuman Primates   605

Matthias Laska and Laura Teresa Hernandez Salazar

Part 6 Gustation
Taste Anatomy and Neurobiology

28 The Role of Saliva in Taste Transduction   625

Ryuji Matsuo and Guy H. Carpenter

29 Anatomy of the Tongue and Taste Buds   637

Martin Witt and Klaus Reutter

30 Chemical Modulators of Taste   665

John A. DeSimone, Grant E. Dubois, and Vijay Lyall

31 The Molecular Basis of Gustatory Transduction   685

Steven D. Munger and Wolfgang Meyerhof

32 Central Taste Anatomy and Physiology of Rodents and Primates   701

Thomas C. Pritchard and Patricia M. Di Lorenzo

33 Development of the Taste System   727

Robin F. Krimm, Shoba Thirumangalathu, and Linda A. Barlow

Contents

Part 7   Human Taste Measurement, Physiology, and Development

34 Psychophysical Measures of Human Oral Sensation   751

Derek J. Snyder, Charles A. Sims, and Linda M. Bartoshuk

35 Mapping Brain Activity in Response to Taste Stimulation   775

Dana M. Small and Annick Faurion

36 The Ontogeny of Taste Perception and Preference Throughout Childhood   795

Catherine A. Forestell and Julie A. Mennella

Part 8   Clinical Applications and Perspectives

37 Nutritional Implications of Taste and Smell Dysfunction   831

Janice Lee, Robin M. Tucker, Sze Yen Tan, Cordelia A. Running, Joshua B. Jones, and Richard D. Mattes

38 Conditioned Taste Aversions   865

Kathleen C. Chambers

39 Clinical Disorders Affecting Taste: An Update   887

Steven M. Bromley and Richard L. Doty

40 Influence of Drugs on Taste Function   911

Susan S. Schiffman

Part 9   Taste in Nonhuman Species

41 Taste Processing in Insects   929

John I. Glendinning

42 Taste in Aquatic Vertebrates   947

Toshiaki J. Hara

43 Comparative Taste Biology with Special Focus on Birds and Reptiles   957

Hannah M. Rowland, M. Rockwell Parker, Peihua Jiang, Danielle R. Reed, and Gary K. Beauchamp

44 Functional Organization of the Gustatory System in Macaques   983

Thomas C. Pritchard and Thomas R. Scott

Contents

Part 10 Central Integration of Olfaction, Taste, and the Other Senses

45 Chemosensory Integration and the Perception of Flavor 1007

John Prescott and Richard Stevenson

46 Neural Integration of Taste, Smell, Oral Texture, and Visual Modalities 1027

Edmund T. Rolls

Part 11 Industrial Applications and Perspectives

47 Olfaction and Taste in the Food and Beverage Industries 1051

Graham A. Bell and Wendy V. Parr

48 Olfaction and Gustation in the Flavor and Fragrance Industries 1067

Benjamin Mattei, Arnaud Montet, and Matthias H. Tabert

49 The Smell and Taste of Public Drinking Water 1079

Gary A. Burlingame and Richard L. Doty

Part 12 Other Chemosensory Systems

50 Trigeminal Chemesthesis 1091

J. Enrique Cometto-Muñiz and Christopher Simons

51 The Vomeronasal Organ 1113

Lisa Stowers and Marc Spehr

52 The Septal Organ, Grueneberg Ganglion, and Terminal Nerve 1133

Minghong Ma, Joerg Fleischer, Heinz Breer, and Heather Eisthen

Author Index 1151

Subject Index 1197

Foreword

It is a special pleasure to provide this Foreword for the lucky readers of the third edition of this classic handbook. The first two editions of the handbook significantly impacted the chemical senses at large, and there is no reason to believe that this will also not be the case for the third edition. Importantly, the third edition has been expanded to cover topics beyond those of its predecessors.

Describing a system in the brain "from molecules to behavior" is one of the grand goals of contemporary neuroscience. This new edition makes clear how the chemosenses – smell and taste – are systems par excellence for achieving this goal. To begin with, the stimuli are themselves molecules, and even in the case of salt and sour, ions. The Handbook covers leading work on how chemosensory receptors are produced by gene expression, and how each type of molecule or ion interacts with its receptor. The combined responses of the receptor cells represent the stimuli as spatial and temporal patterns. These patterned responses are subjected to central processing in the brain centers that transform the neural images created in the sensory domain to smell and taste objects in the central domain. This creates the neural basis of conscious smell and taste perception all the way up to the cerebral cortex, as the Handbook describes.

In addition to acting separately, the taste and smell systems act in concert during eating to give rise to the perception of flavor. The Handbook documents the physiological underpinnings of this process and reveals how flavor and food preferences are shaped from before birth into old age. Clearly, as emphasized in the Handbook, the neural basis of flavor is a rapidly growing field with critical implications for public health policies such as curbing the tendencies of people to eat too much of the foods they crave – tendencies that lead to overeating, obesity, and related disorders.

The editor, Richard Doty, is an expert in both normal and disordered taste and smell, and the Handbook is thus especially strong in disorders to which smell and taste are related. An important example is in neurodegenerative disorders such as Alzheimer's and Parkinson's disease, where the first signs may be decreased smell sensitivity.

Disordered chemosensation may be primary, as well as secondary, elements of disease states elsewhere in the body, and their study have begun to shed light on the etiology of a number of diseases, including some forms of cancer. Altered taste and smell are now recognized as being more prevalent than previously thought and chemosensory testing has begun to be more common in a number of doctors' offices and medical centers. Clearly, chemosensory research has led, and continues to lead, to many practical applications in public health, as the Handbook emphasizes.

The third edition of the Handbook has been expanded to emphasize the important role that the chemical senses play in the national economy. This includes the importance of aromas in cosmetics and perfumes, as well as in determining the flavors of virtually all processed foods. This new edition of the Handbook covers these subjects eloquently, as well as topics such as problems of smell pollution in the environment and chemosensory properties of drinking water.

In addition to these mainstream topics, the Handbook also highlights many little known facts about these fascinating senses. These include the nasal cycle and chemoreception in one-cell organisms, in birds, in dogs, and in insects. Other topics include how taste aversions arise and the role of the frequently overlooked vomeronasal organ in the behavior of a wide range of vertebrates. It provides an excellent introduction into the new field of chemesthesia – a field developing as a result of the discovery of isolated taste cells along the lining of the stomach, esophagus, and intestines that continue to sense our food unconsciously long after its ingestion. Taste and smell thus offer up many surprises about what controls our lives, and the Handbook will continue to be the first place to go to learn about them.

GORDON M. SHEPHERD, M.D., D.Phil.
Professor of Neurobiology
Yale University School of Medicine
New Haven, Connecticut, USA

Preface

This third edition of the *Handbook of Olfaction and Gustation* represents the largest collection of basic, clinical, and applied knowledge on the chemical senses ever compiled in one volume. Since the publication of the second edition a decade ago, many more advances in chemosensory science have occurred, most remarkably in neuroscience, molecular biology, and neurology. A year after the publication of the second edition, the 2004 Nobel Prize in Medicine or Physiology was awarded to Linda Buck and Richard Axel for their discovery of the gene family responsible for expression of the olfactory receptors. This punctuated the fact that the chemical senses had by that time become a key element of modern neuroscience, in part as a result of their unique transduction processes and integral association with stem cell activity and regenerative capacities. Multiple taste receptor proteins and mechanisms have since been identified, particularly for bitter-tasting agents, and taste receptors have now been found to be distributed throughout the alimentary tract, the upper and lower respiratory tracts, and elsewhere. The importance of olfaction in a variety of clinical fields has grown, largely as a consequence of the continued proliferation of commercially available clinical olfactory tests and the discovery that decreased smell function is one of the earliest signs of such neurodegenerative diseases as Alzheimer's disease and Parkinson's disease. The geometric growth of interest in olfaction in neurology and otorhinolaryngology is illustrated in the above figure, a growth that is paralleled in a number of other clinical specialties. Note that in every half decade since 1986, more publications have occurred than in the entire preceding quarter century.

It is important to emphasize that biology, neuroscience, and the clinical specialties are not the only fields where the chemical senses are current centers of focus. The food and beverage industries are increasing,

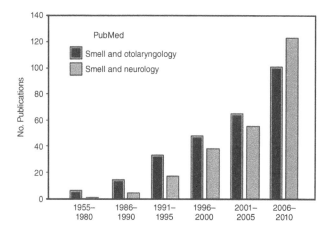

at great cost, research efforts to enhance the flavor of their products and, importantly, to maintain such flavor in light of developing government regulations to minimize the amount of salt, sugar, and other ingredients in their products. The energy industry is now required, by law, to test many of their workers for the ability to smell, making this sensory system of direct interest to this entire industry. Continued interest in olfaction by marketers of perfumes and personal care products goes without saying.

In keeping with previous editions, the third edition of the *Handbook* continues to emphasize history and perspective in its presentations. Also, in keeping with prior editions, in many cases several authors with divergent experience within each of the topic areas have formed collaborations to provide unique and compelling contributions. Compared to earlier editions, more emphasis has been placed on the genetics of olfaction and taste, as well as central nervous system integration of sensory processing. A number of new chapters have been added to incorporate findings from a wider range of species, including birds, fish, and invertebrates, as well as ones that address chemosensory research perspectives within the food, beverage, perfume, personal care, and water industries. To make way for these new chapters and to reduce redundancy and minimize the inevitable increase in the number of volume pages, several previous chapters from the second edition have been combined or omitted.

I am particularly indebted to the contributors for providing such excellent chapters, as well to those who volunteered to be section editors. I thank the staff of Wiley-Liss for their professionalism and suggestions for making the volume the highest quality possible.

RICHARD L. DOTY
Philadelphia, Pennsylvania, USA

Contributors

Editor

Richard L. Doty, Smell & Taste Center, Department of Otorhinolaryngology: Head and Neck Surgery, Perelman School of Medicine, University of Pennsylvania, Philadelphia, Pennsylvania, USA

Authors

Shelly Aono, Department of Anatomy, Physiology and Pharmacology, Auburn University, Auburn, Alabama USA

Harriet Baker, Burke Medical Research Institute, Weill Cornell Medical College, White Plains, New York, USA

Linda A. Barlow, Department of Cell and Developmental Biology, University of Colorado, Aurora, Colorado, USA

Linda M. Bartoshuk, Community Dentistry & Behavioral Science, University of Florida, Gainesville, Florida, USA

Gary K. Beauchamp, Monell Chemical Senses Center, Philadelphia, Pennsylvania, USA

Graham A. Bell, School of Medical Sciences, University of New South Wales, Sydney, Australia

Kunwar P. Bhatnagar, Department of Anatomical Sciences and Neurobiology, University of Louisville School of Medicine, Louisville, Kentucky, USA

Heinz Breer, Institute of Physiology, University of Hohenheim, Stuttgart, Germany

Steven M. Bromley, Outpatient Services, Virtua Neurosciences, Voorhees, New Jersey, USA

Gary A. Burlingame, Bureau of Laboratory Services, Philadelphia Water Department, Philadelphia, Pennsylvania, USA

Lilian Calderón-Garcidueñas, Department of Biomedical and Pharmaceutical Sciences, University of Montana, Missoula, Montana, USA

Guy H. Carpenter, Salivary Research Unit, Dental Institute, King's College, Guy's Hospital, London, United Kingdom

John W. Cave, Burke Medical Research Institute, Weill Cornell Medical College, White Plains, New York, USA

Kathleen C. Chambers, Department of Psychology, University of Southern California, Los Angeles, California, USA

Julie Chapuis, Emotional Brain Institute, Nathan Kline Institute, Orangeburg, New York, USA

J. Enrique Cometto-Muñiz, Chemosensory Perception Laboratory, University of California, San Diego, California, USA

John C. Dennis, Department of Anatomy, Physiology and Pharmacology, Auburn University, Auburn, Alabama, USA

John A. DeSimone, Virginia Commonwealth University, Richmond, Virginia, USA

Patricia M. Di Lorenzo, Department of Psychology, Binghamton University, Binghamton, New York, USA

Xinxin Ding, Department of Environmental Health Sciences, Albany Medical College, Albany, New York, USA

Richard L. Doty, Smell & Taste Center, Department of Otorhinolaryngology: Head and Neck Surgery, Perelman School of Medicine, University of Pennsylvania, Philadelphia, Pennsylvania, USA

Grant E. Dubois, Sweetness Technologies LLC, Rosewell, Georgia, USA

John E. Duda, Department of Neurology, Veteran' Administration Medical Center, Philadelphia, Pennsylvania, USA

Contributors

Thomas P. Eiting, Graduate Program in Organismic and Evolutionary Biology, University of Massachusetts, Amherst, Massachusetts, USA

Howard Eichenbaum, Center for Memory and Brain, Boston University, Boston, Massachusetts, USA

Heather Eisthen, Department of Zoology, Michigan State University, East Lansing, Michigan, USA

Matthew Ennis, Department of Anatomy and Neurobiology, University of Tennessee Health Science Center, Memphis, Tennessee, USA

Annick Faurion, CNRS Institut de Neurobiologie Alfred Fessard (INAF), NeuroBiologie Sensorielle de l'Olfaction et de la Gustation (NBS), Gif-sur-Yvette, France

Joerg Fleischer, Institute of Physiology, University of Hohenheim, Stuttgart, Germany

Catherine A. Forestell, Department of Psychology, The College of William & Mary, Williamsburg, Virginia, USA

Richard E. Frye, Arkansas Children's Hospital Research Institute, University of Arkansas, Little Rock, Arkansas, USA

Giovanni Galizia, Lehrstuhl Neurobiologie, Universität Konstanz, Konstanz, Germany

Mary Beth Genter, Department of Environmental Health, University of Cincinnati, Cincinnati, Ohio, USA

John I. Glendinning, Department of Biological Sciences, Barnard College, Columbia University, New York, New York, USA

Kimberly P. Good, Department of Psychiatry, Dalhousie University, Halifax, Nova Scotia, Canada

Jay A. Gottfried, Department of Neurology, Northwestern University Feinberg School of Medicine, Chicago, Illinois, USA

Toshiaki J. Hara, 12028 Kami-Amakusa, Kumamoto, Japan

Christopher H Hawkes, Neuroscience Centre, Blizard Institute, Barts and The London School of Medicine and Dentistry, London, United Kingdom

Peter Hepper, School of Psychology, Queens University Belfast, Belfast, United Kingdom

Timothy E. Holy, Department of Neurobiology and Developmental Sciences, Washington University, St. Louis, Missouri, USA

Peihua Jiang, Monell Chemical Senses Center, Philadelphia, Pennsylvania, USA

Joshua B. Jones, Department of Nutrition Science, Purdue University, West Lafayette, Indiana, USA

E.P. Köster, Wildforsterweg 4a, 3881 NJ Putten, The Netherlands

Robin F. Krimm, Department of Anatomical Sciences and Neurobiology, University of Louisville School of Medicine, Louisville, Kentucky, USA

Mansi Krishan, Department of Environmental Health, University of Cincinnati, Cincinnati, Ohio, USA

Matthias Laska, Department of Physics, Chemistry and Biology, Linköping University, Linköping, Sweden

David G. Laing, School of Women and Children's Health, University of New South Wales, Randwick, Australia

Janice Lee, Department of Nutrition Science, Purdue University, West Lafayette, Indiana, USA

Christiane Linster, Department of Neurobiology & Behavior, Cornell University, Ithaca, New York, USA

Vijay Lyall, Department of Physiology and Biophysics, Virginia Commonwealth University, Richmond, Virginia, USA

Minghong Ma, Department of Neuroscience, University of Pennsylvania, Philadelphia, Pennsylvania, USA

Alan Mackay-Sim, Eskitis Institute for Cell and Molecular Therapies, Griffith University, Brisbane, Australia

Ryuji Matsuo, Department of Oral Physiology, Okayama University Graduate School of Medicine, Dentistry and Pharmaceutical Sciences, Okayama, Japan

Benjamin Mattei, International Flavors and Fragrances, Hilversum, Netherlands

Richard D. Mattes, Department of Nutrition Science, Purdue University, West Lafayette, Indiana, USA

Julie A. Mennella, Monell Chemical Senses Center, Philadelphia, Pennsylvania, USA

Wolfgang Meyerhof, German Institute of Human Nutrition, Germany

Per Møller, Faculty of Life Sciences, University of Copenhagen, Frederiksberg, Denmark

Arnaud Montet, International Flavors and Fragrances, Neuilly-sur-Seine, France

Edward E. Morrison, College of Veterinary Medicine, Auburn University, Auburn, Alabama, USA

Contributors

Steven D. Munger, Center for Smell and Taste, and Department of Pharmacology and Therapeutics, University of Florida, Gainesville, Florida, USA

Gabrielle A. Nevitt, Department of Neurobiology, Physiology and Behavior, University of California, Davis, California, USA

Allen Osman, Department of Psychology, University of Miami, Coral Gables, Florida, USA

M. Rockwell Parker, Department of Biology, Washington and Lee University, Lexington, Virginia, USA

Wendy V. Parr, Department of Wine, Food, & Molecular Biosciences, Lincoln University, Christchurch, New Zealand

Paola A. Prada, Department of Chemistry and Biochemistry, Florida International University, Miami, Florida, USA

Rui Daniel Prediger, Departamento de Farmacologia, Universidade Federal de Santa Catarina, Florianópolis, SC, Brazil

John Prescott, TasteMatters Research and Consulting, Sydney, Australia

Thomas C. Pritchard, Department of Neural and Behavioral Sciences, The Pennsylvania State University College of Medicine, Hershey, Pennsylvania, USA

Danielle R. Reed, Monell Chemical Senses Center, Philadelphia, Pennsylvania, USA

Klaus Reutter, Department of Anatomy, University of Rostock, Rostock, Germany

Edmund T. Rolls, Oxford Centre for Computational Neuroscience, Oxford, United Kingdom

Hannah M. Rowland, Department of Zoology, University of Cambridge, Cambridge, United Kingdom

Cordelia A. Running, Department of Nutrition Science, Purdue University, West Lafayette, Indiana, USA

Hitoshi Sakano, Department of Biophysics and Biochemistry, The University of Tokyo, Tokyo, Japan

Laura Teresa Hernandez Salazar, Instituto de Neuro-Etologia, Universidad Veracruzana, Xalapa, Veracruz, Mexico

Benoist Schaal, Centre des Sciences du Goût et de l'Alimentation, CNRS, Université de Bourgogne, Dijon, France

Susan S. Schiffman, 18 Heath Place, Durham, North Carolina, USA

James E. Schwob, Department of Anatomy and Cellular Biology, Tufts University, Boston, Massachusetts, USA

Thomas R. Scott, Department of Psychology, San Diego State University, San Diego, California, USA

Jonathan Silas, Department of Psychology, Whitelands College, University of Roehampton, London, United Kingdom

Christopher Simons, Department of Food Science and Technology, The Ohio State University, Columbus, Ohio, USA

Charles A. Sims, Department of Food Science & Human Nutrition, University of Florida, Gainesville, Florida, USA

Dana M. Small, The John B. Pierce Laboratory, New Haven, Connecticut, USA

Timothy D. Smith, School of Physical Therapy, Slippery Rock University, Slippery Rock, Pennsylvania, USA

Derek J. Snyder, Mrs. T. H. Chan Division of Occupational Science and Occupational Therapy, University of Southern California, Los Angeles, California, USA

Marc Spehr, Institute for Biology II / Department of Chemosensation, RWTH-Aachen University, Aachen, Germany

James St. John, National Centre for Adult Stem Cell Research, Griffith University, Brisbane, Australia

Richard Stevenson, Department of Psychology, Macquarie University, Sydney, Australia

Lisa Stowers, Department of Molecular and Cellular Neuroscience, The Scripps Research Institute, La Jolla, California, USA

Regina M. Sullivan, Emotional Brain Institute, Nathan Kline Institute, Orangeburg, New York, USA

Paul Szyszka, Fachbereich Biologie, Universität Konstanz, Konstanz, Germany

Matthias H. Tabert, International Flavors & Fragrances, Union Beach, NJ, USA

Sze Yen Tan, Department of Nutrition Science, Purdue University, West Lafayette, Indiana, USA

Shoba Thirumangalathu, Department of Cell and Developmental Biology, University of Colorado, Aurora, Colorado, USA

Keith B. Tierney, Department of Biological Sciences, University of Alberta, Edmonton, Alberta, Canada

Kazushige Touhara Department of Applied Biological Chemistry, The University of Tokyo, Tokyo, Japan

Robin M. Tucker, Department of Nutrition Science, Purdue University, West Lafayette, Indiana, USA

Akio Tsuboi, Department of Biophysics and Biochemistry, The University of Tokyo, Tokyo, Japan

Judith Van Houten, Department of Biology, University of Vermont, Burlington, Vermont, USA

Vitaly J. Vodyanoy, Department of Anatomy, Physiology and Pharmacology, Auburn University, Auburn, Alabama, USA

Deborah Wells, School of Psychology, Queens University Belfast, Belfast, Northern Ireland, United Kingdom

Daniel W. Wesson, Department of Neurosciences, Case Western Reserve University, Cleveland, Ohio, USA

Theresa L. White, Department of Psychology, Le Moyne College, Syracuse, New York, USA

Donald A. Wilson, Emotional Brain Institute, Nathan Kline Institute, Orangeburg, New York, USA

Martin Witt, Department of Anatomy, University of Rostock, Rostock, Germany

Fang Xie, Department of Environmental Health Sciences, Albany Medical College, Albany, New York, USA

Keiichi Yoshikawa, Department of Applied Biological Chemistry, The University of Tokyo, Tokyo, Japan

Kai Zhao, Monell Chemical Senses Center, Philadelphia, Pennsylvania, USA

Part 1

General Introduction

Chapter 1

Introduction and Historical Perspective

RICHARD L. DOTY

1.1 INTRODUCTION

All environmental nutrients and airborne chemicals required for life enter our bodies by the nose and mouth. The senses of taste and smell monitor the intake of such materials, not only warning us of environmental hazards, but determining, in large part, the flavor of our foods and beverages, largely fulfilling our need for nutrients. These senses are very acute; for example, the human olfactory system can distinguish among thousands of airborne chemicals, often at concentrations below the detection limits of the most sophisticated analytical instruments (Takagi, 1989). Furthermore, these senses are the most ubiquitous in the animal kingdom, being present in one form or another in nearly all air-, water-, and land-dwelling creatures. Even bacteria and protozoa have specialized mechanisms for sensing environmental chemicals – mechanisms whose understanding may be of considerable value in explaining their modes of infection and reproduction (Jennings, 1906; Russo and Koshland, 1983; van Houten, 2000).

While the scientific study of the chemical senses is of relatively recent vintage, the important role of these senses in the everyday life of humans undoubtedly extends far into prehistoric times. For example, some spices and condiments, including salt and pepper, likely date back to the beginnings of rudimentary cooking, and a number of their benefits presumably were noted soon after the discovery of fire. The release of odors from plant products by combustion was likely an early observation, the memory of which is preserved in the modern word *perfume*, which is derived from the Latin *per* meaning "through" and *fumus* meaning "smoke." Fire, with its dangerous and magical connotations, must have become associated early on with religious activities, and pleasant-smelling smoke was likely sent into the heavens in rituals designed to please

or appease the gods, as documented in later civilizations, such as the early Hebrews. Importantly, food and drink became linked to numerous social and religious events, including those that celebrated birth, the attainment of adulthood, graduation to the status of hunter or warrior, and the passing of a soul to a better life.

In this chapter I provide a brief historical overview of the important role that tastes and odors have played in the lives of human beings throughout millennia and key observations from the last four centuries that have helped to form the context of modern chemosensory research. Recent developments, which are described in detail in other contributions to the *Handbook*, are briefly mentioned to whet the reader's appetite for what is to follow. Although an attempt has been made to identify, rather specifically, major milestones in chemosensory science since the Renaissance, some important ones have undoubtedly been left out, and it is not possible to mention, much less discuss, even a small fraction of the many studies of this period that have contributed to our current fund of knowledge. Hopefully the material that is presented provides some insight into the basis for the present Zeitgeist. The interested reader is referred elsewhere for additional perspectives on the history of chemosensory science (e.g., Bartoshuk, 1978, 1988; Beauchamp, 2009; Beidler, 1971a, b; Boring, 1942; Cain, 1978; Cloquet, 1821; Corbin, 1986; Doty, 1976; Douek, 1974; Farb and Armelagos, 1980; Farbman, 1992; Frank, 2000; Garrett, 1998; Gloor, 1997; Harper et al., 1968; Harrington and Rosario, 1992; Jenner, 2011; Johnston et al., 1970; Jones and Jones, 1953; Luciani, 1917; McBurney and Gent, 1979; McCartney, 1968; Miller, 1988; Moulton and Beidler, 1967; Moulton et al., 1975; Mykytowycz, 1986; Nagel, 1905; Ottoson, 1963; Pangborn and Trabue, 1967; Parker, 1922; Pfaff, 1985; Piesse, 1879;

Handbook of Olfaction and Gustation, Third Edition. Edited by Richard L. Doty.
© 2015 Richard L. Doty. Published 2015 by John Wiley & Sons, Inc.

Schiller, 1997; Simon and Nicolelis, 2002; Smith et al., 2000; Takagi, 1989; Temussi, 2006; Vintschgau, 1880; Wright, 1914; von Skramlik, 1926; Zippel, 1993).

1.2 A BRIEF HISTORY OF PERFUME AND SPICE USE

The relatively rich history of a number of ancient civilizations, particularly those of Egypt, Greece, Persia, and the Roman Empire, provides us with examples of how perfumes and spices have been intricately woven into the fabric of various societies. Thousands of years before Christ, fragrant oils were widely used throughout the Middle East to provide skin care and protection from the hot and dry environment, and at least as early as 2000 BCE, spices and fragrances were added to wine, as documented by an inscription on a cuneiform text known as the *Enuma elish* (Heidel, 1949). During the greater part of the 7th to 5th Centuries BCE, Rhodian potters developed vast quantities of perfume bottles in the form of animals, birds, and human heads and busts that were shipped throughout the mediterranian area (Figure 1.1). In Egypt, incense and fragrant substances played a key role in religious rites and ceremonies, including elaborate burial customs, and whole sections of towns were inhabited by men whose sole profession was to embalm the deceased. As revealed in the general body of religious texts collectively termed the "Book of the Dead" (a number of which predate 3000 BCE; Budge, 1960), the Egyptians performed funeral ceremonies at which prayers and recitations of formulae (including ritualistic repeated burning of various types of incense) were made, and where the sharing of meat and drink offerings by the attendees occurred. Such acts were believed to endow the departed with the power to resist corruption from the darkness and from evil spirits that could prevent passage into the next life, as well as to seal the mystic union of the friends and loved ones with the dead and with the chosen god of the deceased. The prayers of the priests were believed to be carried via incense into heaven and to the ears of Osiris and other gods who presided over the worlds of the dead (Budge, 1960).

As noted in detail by Piesse (1879), the ancient Greeks and Romans used perfumes extensively, keeping their clothes in scented chests and incorporating scent bags to add fragrance to the air. Indeed, a different scent was often applied to each part of the body; mint was preferred for the arms; palm oil for the face and breasts; marjoram extract for the hair and eyebrows; and essence of ivy for the knees and neck. At their feasts, Greek and Roman aristocrats adorned themselves with flowers and scented waxes and added the fragrance of violets, roses, and other flowers to their wines. As would be expected, perfume shops were abundant in these societies, serving as meeting places for persons of all walks of life (Morfit, 1847). In Grecian mythology, the invention of perfumes was ascribed to the Immortals. Men learned of them from the indiscretion of Aeone, one of the nymphs of Venus; Helen of Troy acquired her beauty from a secret perfume, whose formula was revealed by Venus. Homer (8th century BCE) reports that whenever the Olympian gods honored mortals by visiting them, an ambrosial odor was left, evidence of their divine nature (Piesse, 1879). Interestingly, bad odors were a key element of a number of myths, including that of Jason and the Argonauts (Burket, 1970). As a result of having been smitten with the wrath of Aphrodite, the women of Lemnos developed a foul odor, which drove their husbands to seek refuge in the arms of Thracian slave girls. The women were so enraged by their husbands' actions that one evening they slew not only their husbands, but all the men of the island. Thereafter, Lemnos was a community of women without men, ruled by the virgin queen Hypsiple, until the day when Jason and the Argo arrived, which ended the period of celibacy and returned the island to heterosexual life.

Perfumes were not universally approved of in ancient Greece. Socrates, for example, objected to them altogether, noting, "There is the same smell in a gentleman and a slave, when both are perfumed," and he believed that the only odors worth cultivating were those that arose from honorable toil and the "smell of gentility" (Morfit, 1847). Nevertheless, the use of perfumes became so prevalent in ancient Greece that laws were passed in Athens in the 6th century BCE to restrain their use. Despite this prohibition, however, their use grew unabated, and the Greeks added greatly to the

Figure 1.1 Example of a horse head perfume bottle manufactured in Rhodes circa 580 BCE. © Heritage Images (Image ID: 2-605-015).

1.2 A Brief History of Perfume and Spice Use 5

stock of fragrant plants from the East that made up the core of the perfume industry.

Perfume and incense had religious significance to the followers of Zoroaster, the Persian religious leader of the 6th century BCE, who offered prayers before altars containing sacred fires to which wood and perfumes were added five times each day (Piesse, 1879). It is noteworthy that, to this day, sandalwood fuels the sacred fires of the Parsees (modem Zoroastrians) in India and that similar rituals were required of the early Hebrews, as indicated by the following instructions from God to Moses (Exodus 30:1, 7–9. 34–38; King James version):

And thou shalt make an altar to burn incense upon: of shittim wood shalt thou make it. And Aaron shall burn thereon sweet incense every morning: when he dresseth the lamps, he shall burn incense upon it [the altar]. And when Aaron lighteth the lamps at even, he shall burn incense upon it, a perpetual incense before the Lord throughout your generations. Ye shall offer no strange incense thereon, nor burnt sacrifice, nor meat offering; neither shall ye pour drink offering thereon.

And the Lord said unto Moses, take unto thee three sweet spices, stacte, and onycha, and galbanum; these sweet spices with pure frankincense; of each shall there be a light weight. And thou shalt make it a perfume <or incense>, a confection after the art of the apothecary, tempered together, pure and holy: And thou shalt beat some of it very small, and put of it before the testimony in the tabernacle of the congregation, where I will meet with thee: it shall be unto you most holy. And as for the perfume which thou shalt make, ye shall not make to yourselves according to the composition thereof: it shall be unto thee holy for the Lord. Whosoever shall make like unto that, to smell thereto, shall even be cut off from his people.

Given such instructions from God and the Christian emphasis on cleansing the soul of evil spirits, as well as the fact that Christ himself, after his crucifixion, had been embalmed in pleasant-smelling myrrh, aloe, and spices (John 19: 39–40), it is perhaps not surprising that bad smells came to signify the unholy at various times in Christian history. Indeed, St. Philip Neri reportedly found the stench emanating from heretics so great that he had to turn his head (Summers, 1926).

One of the more interesting, and tragic, uses of bad smells was to identify witches and warlocks in Europe in the late 1500s. Remy, a distinguished appointee of Charles III to the Provosts of Nancy (a court that judged all criminal cases for some 72 villages in the Nancy region of France), wrote the following in his classic 1595 monograph *Demonolatry*:

In the Holy Scriptures the Devil is constantly referred to as Behemoth, that is to say, "the impure animal and the unclean spirit" (see S. Gregory, in *Memorabilia, Matthew XII, Mark I and V, Job XI*). It is not only because the Devil is, as all his actions and purposes show, impure in his nature and character that we should consider this name to be aptly applied to him; but also because he takes immoderate delight in external filth and uncleanliness. For often he makes his abode in dead bodies; and if he occupies a living body, or even if he forms himself a body out of the air or condensation of vapours, his presence therein is always betrayed by some notable foul and noisome stench. The gifts of the Dcmon are also fashioned from ordure and dung, and his banquets from the flesh of beasts that have died . . . for the most part [he] has for his servants filthy old hags whose age and poverty serve but to enhance their foulness; and these ... he instructs in all impurity and uncleanliness .... Above all he cautions them not to wash their hands, as it is the habit of other men to do in the morning; for he tells them that to do so constitutes a sure obstruction to his incantations. This is the case whether it is the witches themselves who wash their hands, as we learn from the answer freely given to her examiners by Alexia Galaea of Betoncourt at Mirecourt in December 1584, and by countless others whose names I have not now by me; or whether it is the intended victims of their witchcraft who wash their hands, as was stated by Claude Fellet (Mersuay, February 1587) and Catharina Latomia (Haraucourt, February 1587).

In contrast to the detection of witches and warlocks by stench was the verification of sainthood by a pleasant odor, the so-called "odor of sanctity." If a saint had been an impostor, a nauseating smell, rather than a delectable one, was present upon exhumation of his body (Rothkrug, 1981). This concept bears a striking resemblance to the Greek myths of the pleasant odors left by the Olympian gods who visited mortals and may well stem from the same tradition.

It should be noted, however, that cleanliness was not always the vogue for Christianity, as described by McLaughlin (1971) in a series of interesting accounts from the Middle Ages. Thus, in their repudiation of Roman values and their desire to avoid lust and sins of the flesh, early Christians often went unbathed. Every sensation offensive to humans was believed acceptable to God, and the custom of bathing the limbs and anointing them with oil was condemned. Monks shaved their hair, wrapped their heads in cowls to avoid seeing profane objects, and kept legs naked except in the extreme of winter. St. Jerome criticized a number of his followers for being too clean,

6　　　　**Chapter 1　Introduction and Historical Perspective**

and St. Benedict, a key administrator of the early church, pronounced solemnly that "to those that are well, and especially to the young, bathing shall seldom be permitted." St. Agnes reportedly had never washed throughout her life, and a pilgrim to Jerusalem in the 4th century is said to have boasted that her face had gone unwashed for 18 years so as not to disturb the holy water used at her baptism.

During the Middle Ages, perfumery and the widespread use of spices and flavoring agents was little known in Europe, being practiced mainly by Arabs in the East. Marco Polo, visiting the China of Kublai Khan (1216–1295), noted that pleasantly perfumed silk paper money was used for exchange within Khan's kingdom (Boorstin, 1985). The dearth of smell in Europe was to change dramatically, however, as a major element of the Renaissance was the relentless search for perfumes and spices, a number of which were more valuable than silver or gold. The quest was not only for aesthetic enjoyment; some of these agents made it possible, much like cooking itself, to exploit a wider and more diverse range of foodstuffs, including ones that otherwise were unsafe or had little gastronomic appeal. In this regard, it is of interest that at the siege of Rome in 408 BCE, Alaric, the victorious king of the Goths, demanded 3000 pounds of pepper as ransom for the city, and when the Genoese captured Caesarea in 1101 BCE, each soldier received two pounds of pepper as his share of the spoils (Verrill, 1940).

Perfume was introduced, at least in a widespread sense, to medieval Europe by the crusaders. After the downfall of the Roman Empire, the perfume industry moved to the Eastern Roman Empire, and Constantinople became the perfume center of the world. Reportedly, Avicenna (CE 980–1036), the great Arab scientist, philosopher, and physician, discovered a way to extract and maintain the fragrances of plants and invented rose water (Takagi, 1989). In 1190, King Philip II (Philip Augustus, r. 1180–1223) of France granted the first charter to a perfume maker. King Charles V (Charles the Wise, r. 1364–1380) subsequently planted large fields of flowers in France to obtain perfume materials, and Charles VIII (r. 1483–1498) was reportedly the first French monarch to appoint a court perfumer. The soil, climate, and location of southern France made it a natural place for the cultivation of flowers for the perfume industry, which gained world supremacy from the late 1700s – supremacy that has continued to the present time (Vivino, 1960).

According to Piesse (1879), perfumes lost their popularity in England for more than a century prior to the Victorian era, unlike the case in France, Italy, and Spain. Related to this loss of popularity was an act, introduced into the English parliament in 1770, that warned women of the use of scents and other materials in the seduction of men (Piesse, 1879, p. 20):

That all women, of whatever age, rank, profession, or degree, whether virgins, maids, or widows, that shall, from and after such Act, impose upon, seduce or betray into matrimony, any of his Majesty's subjects, by the scents, paints, cosmetic washes, artificial teeth, false hair, Spanish wool, iron stays, hoops, high-heeled shoes, and bolstered hips, shall incur the penalty of the law now in force against witchcraft and the like misdemeanors, and that the marriage, upon conviction, shall stand null and void.

The influences of such attempts to ban perfumes in England were short-lived, as perfume vendors thrived, although the state taxed them and required them, in 1786, to have licenses. By 1800 approximately 40 companies were making perfumes in London. Importantly, in the 19th century the revolution that occurred in organic chemistry ensured the continuance of perfume manufacturing in Britain; the first important successful synthetic odorant, coumarin, was prepared in 1863 by the British chemist Sir William Henry Perkin (Vivino, 1960).

Interestingly, for some time the odor of coffee, when first introduced into London, was viewed as offensive. Thus, a formal complaint was lodged in 1657 by the inhabitants of the parish of St. Dustant's in West London against a barber, James Farr, "for making & Selling of a Drinke called Coffee whereby in making the same he annoyeth his neighbors by evill smells" (Jenner, 2011). As with perfume manufacturing, however, such concerns were relatively short-lived. Thus, in 1708 the English historian Edward Hatton pointed out, in relation to the Farr incident, "who would then have thought London would ever have had nearly 3000 such Nusances, and that Coffee would have been (as now) so much Drank by the best of Quality, and Physicians" (Jenner, p. 30).

1.3　THE CHEMICAL SENSES AND EARLY MEDICINE

The close association between odors, spices, and medicine was undoubtedly forged long before recorded history and was likely fostered not only by stenches associated with plagues and death, but by the utility of essential oils and spices in warding off insects and microbes. Indeed, one reason why perfumes and spices were major objects of international trade in the ancient world was their medicinal properties (van der Veen and Morales, 2014). According to Morris (1984), such properties may have been as important to early civilizations as the development of the X-ray or discovery of penicillin was to our own, as modern studies confirm that numerous essential oils and spices are very effective in controlling pathogens, including *Staphylococcus* and tuberculosis bacilli. Apparently

1.3 The Chemical Senses and Early Medicine

this observation first came to the attention of European scientists in the latter half of the 19th century, when the perfumery workers at Grasse, France, were found to have a much lower rate of cholera and tuberculosis than the rest of the European population. As noted by Morris (1984, p. 15):

Essential oils have shown startling fungitoxic properties. Oil of clove is toxic to specific growths, and oil of geranium is effective against a broad range of fungi. *Cymbopogon* grasse*s*, an Indian genus of aromatic grasses, have been found effective against *Heuninthosporium oryzae*, a source of food poisoning, *Aspergillus niger*, a cause of seborrheic dermatitis of the scalp, *Absidia ramosa*, a cause of otitis, and *Trichoderma viride*, another cause of dermatitis. Man has long guessed that these oils that the plant secreted to protect itself from insect, fungal, and microbial dangers could serve him as well. Thus it is that the story of perfumery is intimately linked to the story of pharmacy. Our ancestors could not formulate the germ theory of disease, but they assumed that whatever smelled clean and healthy must be of use in hygiene.[1]

The history of hygiene and public health is closely associated with the view that bad odors were the source, indeed often the cause, of diseases and pestilence. Places of filth and stench were, in fact, associated with a higher incidence of diseases. The stenches that developed in the cities of Europe during the Middle Ages are unimaginable to us today. Conditions were so bad that, for example, the monks of White Friars in London's Fleet Street complained that the smell from the Fleet River overcame all the frankincense burnt at their altars and killed many of their brethren (McLaughlin, 1971). Such problems were the backdrop of the spread of the plague epidemics that traversed Europe and England in the 12th to the 17th

centuries. As chronicled by Corbin's (1986) fascinating account of the history of hygiene and odors in 18th-century France, health administration of that era was based on a catalog of noxious odors. Indeed, authorities sought to locate the networks of miasmas by "mapping the flux of smells that made up the olfactory texture of the city" (p. 55). The desire to localize odors and to eliminate them in an effort to ward off diseases may well have been one reason why so many odor classification schemes arose during the 18th century, including those of von Haller (1756), Linnaeus (1765), Lorry (1784/85), and Fourcroy (1798).

Throughout this period, as well as in earlier times, infection was believed to be stemmed by wearing a perfume or by burning aromatic pellets in special perfume pans, thereby masking the odors that were considered unhealthy. Lemery's *Pharmacopee universelle* (1697) cataloged the therapeutic value of aromatics and perfumes and suggested the prescription of "apoplectic balms" because "what is pleasing to the nose, being composed of volatile, subtle, and penetrating parts, not only affects the olfactory nerve, but is spread through the whole brain and can deplete its pituita and other overcourse humors, increasing the movement of animal spirits" (Corbin, 1986, p. 62). During outbreaks of the plague, defenses included the burning of incense, juniper, laurel leaves, cypress, pine, balm, rosemary, and lavender, although, if effective, they were only marginally so. Various plague waters, to be poured on handkerchiefs or into pomanders, were invented, including the original eau de cologne. Unpleasant agents were also believed to keep away the plague, and the members of many households crouched over their privies inhaling the fumes in attempts to avert the disaster (McLaughlin, 1971). Specialized plague Physicians wore garmets designed to protect them from the odors emanating from plague victums who were touched by long canes, some of which were hollow to allow for listening to the heart (Figure 1.2). Even in the late 1800s, smells were associated with illnesses, as exemplified by the belief that decaying organic matter in swamps produced malaria (mal = bad, aira = air). This theory, apparently initially proposed by Varro (116-28 BCE) and Palladius (4th century CE), was brought to the more modern stage by Morton (1697) and Lancisi (1717), but was largely abandoned after the French physician Alphonse Laveran (1881) described the responsible parasite and Sir Ronald Ross (1923) demonstrated, a few years later, its transmission by the female anopheline mosquito.

In the history of medicine, both odors and tastes have been used at various times in the diagnosis of diseases (Doty, 1981; Whittle et al., 2007). Even today, diabetes is diagnosed in some areas of the world on the basis of the patient's acetone-like breath and sweet-tasting urine, although, in general, the use of odor and taste in diagnosis has become a lost art. In addition, certain smells and tastes were known to elicit symptoms of some diseases,

---

[1] Billing and Sherman (1998) provide empirical support for the hypothesis that the amount of spice in foodstuffs from various world cuisines is better explained on the basis of their antibacterial than their sensory properties. These investigators quantified the frequency of use of 43 spices in the meat-based cuisines of 136 countries for which traditional cookbooks could be found. A total of 4578 recipes from 93 cookbooks was examined, along with information on the temperature and precipitation in each country, the ranges of spice plants, and the antibacterial properties of each spice. As mean annual temperatures (an index of relative spoilage rates of unrefrigerated foods) increased, the proportion of recipes containing spices, number of spices per recipe, total number of spices used, and use of the most potent antibacterial spices all increased, both within and among countries. The estimated fraction of bacterial species inhibited per recipe in each country was positively correlated with annual temperature. Although alternative hypotheses were considered (e.g., that spices provide macronutrients, disguise the taste and smell of spoiled foods, or increase perspiration and thus evaporative cooling), the data did not support any of these alternatives.

Figure 1.2 Picture of a plague doctor in his protective garb. This costume was invented in France in 1619. It was comprised of a heavy overcoat rubbed in beeswax, glass eye openings, and a beak-like extension which contained scents, such as ambergris, camphor, cloves, mint, and rose petals, to breathe through in an effort to protect the wearer from miasmatic bad air. A cane pointer was used to touch the patient. Courtesy of the Wellcome Library, London.

including epilepsy and hysteria. A classic example is reported by the Roman historian Caius Plinius Caecilius Secundus (Pliny) in his *Historia Naturalis* (circa 50 CE), where sulfur and burning bitumen (asphalt) were noted to induce seizures (Bailey, 1932), a phenomenon that has also been noted in more modern times (West and Doty, 1995). Alum (alumen), which contained aluminum, was used as a deodorant in the Roman empire, predating the use of aluminum salts as deodorants in the United States in the 1880s, as evidenced by the following quotation of Pliny, which extols its values (Bailey, 1932, p. 103):

> ...Liquid alumen has astringent, hardening, and corrosive properties. Mixed with honey, it heals sores in the mouth, pustules, and itchy eruptions. In the latter case, the treatment is applied in a bath to which honey and alumen have been added in the proportion of two to one. Alumen diminishes offensive odours of the axilla, and reduces sweating in general.

To my knowledge, there are no pre-Renaissance treatises on chemosensory dysfunction, *per se*, although descriptions of loss of olfactory function are found in the writings of the ancient Greeks and Romans, as well as in the Bible. In 2 Samuel 19:31–37, a story is told about an 80-year-old man who complained to the king that his ability to taste (or smell) had faded, along with his hearing. Theophrastus noted, in the 3rd century BCE, the following (Stratton, 1917, p. 84):

> ... it is silly to assert that those who have the keenest sense of smell inhale most; for if the organ is not in health or is, for any cause, not unobstructed, more breathing is to no avail. It often happens that man has suffered injury [to the organ] and has no sensation at all.

Although the early Greeks routinely used surgical intervention for the treatment of polyps and other intranasal obstructive problems (for review, see Wright, 1914), the first description of the use of surgery to specifically correct anosmia was apparently made during the Renaissance by Forestus (1591; cited in Lederer, 1959):

> If it [anosmia] is from ethmoidal obstruction, or from the humor discharged from a catarrh, the latter must first be cured. If from the flesh growing from within the nose . . . it is to be cured by the surgeons by operative procedures, either with a cutting instrument, or cautery, or snare.

Claudius Galenus (Galen; 130–200 CE), whose writings had a major impact on Western medicine in general, attributed anosmia to obstruction of the foramina within the cribriform plate (an attribution made by a number of early Greeks, including Plato and Hippocrates). He correctly described the role of the nose in warming and filtering the air and alluded to empirical studies noting the permeability of the dura matter around the cribriform plate to both water and air (Wright, 1914). He believed that the organ for smell was located in the ventricles of the brain and that particles responsible for olfactory sensations passed through the foramina of the cribriform plate during inhalation. As discussed in more detail later in this chapter, this compelling idea continued until the 18th century, when light microscopy revealed that the nasal secretions came from secretory cells within the epithelium. In terms of taste, he posited that the lingual nerve communicated gustatory sensations, in accord with modern perspectives (see Chapters 29 and 32).

1.4 THE RENAISSANCE AND THE BIRTH OF MODERN STUDIES OF TASTE AND SMELL

As is evidenced throughout the volume, major advances have been made in the last quarter century in understanding the senses of taste and smell – advances that follow on the footsteps of a long tradition of scientific observations stemming from treatises written in the 16th century. Indeed, the sense of smell did not escape the attention of Leonardo da Vinci (1452–1519), who, in the *Codex Atlanticus,* presented nine diagrams next to one another in which he compared the behavior of light, the force of a blow, sound, magnetism, and odor (Riti, 1974). Cardinal Gasparao Contarini (1482–1542), an alchemist, wrote about the elements and their combinations in five brief volumes published posthumously in 1548 by Ioannes Gaignaeus. The last of these was dedicated to flavors, odors, and colors. Contarini believed that there were eight flavors or tastes and argued that cooking food or preserving fruit can produce flavors not found in nature. He felt the sense of smell was imperfect and noted that the names of flavors are often employed to explain the variety of odors. Andrea Vesalius devoted one and a half large pages to the sense of smell in his classic anatomy treatise *De Humani Corporis Fabrica* (1543), although he failed to observe the olfactory fila. In 1581, Fernel listed nine types of basic taste qualities, including the seven of Aristotle and Galen (sweet, bitter, sour, salty, astringent, pungent, harsh) and "fatty" and "insipid," the latter apparently reflecting the lack of other taste qualities (Bartoshuk, 1978). During this period, *Two Books on Taste, Sweet and Bitter* was published by Laurence Gyrllus (1566), conceivably being the first work solely devoted to taste. In 1609, Casserius described the detailed structure of the tongue, and Malpighi (1664), Casserius, 1609 and Bellini (1665) associated the sense of taste with lingual papillae. Taste buds were first identified on the barbels and skin of fishes by Leydig (1851), and later were described in mammals (Loven, 1868; Schwalbe, 1868). In 1587, Iohannes Camerarius presented a thesis to the University of Marburg entitled "Themata Physica de Odorum Natura et Affectionibus." In this work, he discussed odor classification, the relationship between taste and smell, a mechanism for explaining the function of olfaction, the ability of smelling in water, and the effect of heat from the sun on odors (Kenneth, 1928).

In 1673 Robert Boyle wrote an article on "Nature, Properties, and Effects of Effluvia" in which he provides vivid and accurate observations on such topics as olfaction in birds, odor tracking in dogs, and the physical nature of the materials released from various odor sources. In his 1675 paper, "Experiments and Observations About the Mechanical Production of Odours," he addresses some simple issues of odorant mixtures and observes that the quality and intensity of odors can be related. He provides, in his "Experiments and Considerations About the Profity of Bodies" (1684), perhaps the first description of intravascular olfaction or taste (i.e., the smelling or tasting of substances that are initially blood-borne):

> One of the notablest instances I ever met with of the porosity of the internal membranes of the human body, was afforded to me by that British nobleman, of whom our famous Harvey tells a memorable, not to say matchless story. This gentleman, having in his youth by an accident, which that doctor relates, had a great and lasting perforation made in his thorax, at which the motion of his heart could be directly perceived, did not only out-live the accident, but grew a strong and somewhat corpulent man; and so robust, as well as gallant, that he afterwards was a soldier, and had the honour to command a body of an army for the King.
>
> This earl of Mount-Alexander . . . gave me the opportunity of looking into his thorax, and of discerning there the motions of the cone, as they call it, or mucro of the heart . . . Having then made several inquiries fit for my purpose, his lordship told me, that, when he did, as he was wont to due from time to time, (though not every day) inject with a syringe some actually warm medicated liquor into his thorax, to cleanse and cherish the parts, he should quickly and plainly find in his mouth the taste and smell of the drugs, wherewith the liquor had been impregnated. And I further learned, that, whereas he constantly wore, upon the unclosed part of his chest, a silken quilt fluffed with aromatic and odoriferous powders, to defend the neighboring parts and keep them warm; when he came, as he used to do after several weeks, to employ a new quilt, the fragrant effluvia of it would mingle with his breath in expiration, and very sensibly perfume it, not, as I declared I suspected, upon the score of the pleasing exhalations, that might get up between his cloathes and his body, but that got into the organs of respiration, and came out with his breath at his mouth, as was confirmed to me by a grave and judicious statesman, that happened to be then present, and knew this general very well.

Another early depiction of what seems to be an example of intravascular olfaction is described by Cloquet (1821). Cloquet notes that Dupuytren, a famous surgeon of the time, performed an ad hoc experiment in which he injected an odorous fluid into a vein of a dog. Soon thereafter, the dog began running around sniffing.

Unfortunately, many of the studies and observations of the period from the 1500s to the mid-1800s confused taste with flavor (which is largely dependent on the sense of smell), thereby obscuring the clear focus needed for

10   Chapter 1   Introduction and Historical Perspective

optimal scientific progress. In these periods, research on gustation was much more limited in scope than that on olfaction, although notable advances were made in taste research, including (1) the discovery that dissimilar metals, when placed on the tongue, produced an "electric taste" sensation (Sulzer, 1752; Volta, 1792), (2) the observation that taste sensations are localized to papillae (Malpighi, 1664; Bell, 1803), (3) the identification of the chorda tympani as the nerve that mediates taste in the anterior tongue (Bellingeri, 1818; see Bartoshuk, 1978), and (4) the demonstration that different regions of the oral cavity are differentially sensitive to different taste qualities (Horn, 1825). As noted above, the observation that taste buds exist within the papillae of the mammalian tongue and depend on an intact nerve supply came in the latter half of the 19th century (see Chapter 29; Loven, 1868; Merkel, 1880; Schwabe, 1867; Vintschgau and Hönigschmied, 1877), as did the painstaking mapping of the sensitivity of individual papillae to stimuli representing the four basic taste qualities (Öhrwall, 1891; Kiesow, 1894).

One reason for the comparatively greater interest in olfaction than taste during the post-Renaissance period stemmed from the compelling, albeit erroneous, conceptual framework in which olfactory functioning, disease, and nasal secretions were viewed. For smelling to occur, odorous bodies had to enter the brain via the foramina of the cribriform plate – the same foramina through which body humors flowed to produce nasal mucus. From this perspective, blockage or alterations in this passageway (e.g., by the changes in the viscosity of the humors) were closely related to diseases that caused (1) anosmia, (2) running noses, (3) high fever, and (4) general ill feeling. There is no doubt that the major conceptual chemosensory advance of this period, indeed perhaps of the entire modern era, was refutation of this ancient concept. The compelling nature of this theory and its adaptation to a more modern era is illustrated by Descartes' (1644) description of how olfaction works (Haldane and Ross, 1955, p. 292):

> . . . two nerves or appendages to the brain, for they do not go beyond the skull, are moved by the corporeal particles separated and flying in the air – not indeed by any particles whatsoever, but only by those which, when drawn into the nostrils, are subtle and lively enough to enter the pores of the bones which we call the spongy, and thus to reach the nerves. And from the diverse motions of these particles, the diverse sensations of smell arise.

Interestingly, convincing evidence for this notion continued to be amassed during this period, as the following quotation from Thomas Willis (1681, p. 100) indicates:

> The Sieve-like Bone in divers Animals is variously perforated for the manifold necessity and difference

of smelling. A Process from the Dura Mater and manifold nervous Fibres pass through every one of its holes, and besmear the inside of the Nostrils. But as the impressions of sensible things, or sensible Species, confined as it were by the undulation or waving of the animal Spirits, ascend through the passages of these bodies stretched out from the Organ towards the Sensory; so the humidities watring the same bodies, for as much as some they may be more superfluous than usual, may distil into the Nostrils through the same ways. For indeed such humors as are perpetually to be sent away from the brain, ought so copiously to be poured upon the Organs of Smelling, as we shall shew hereafter, when we shall speak particularly of the smelling Nerves; in the mean time, that there is such a way of Excretion opening into the Nostrils, some observations, taken of sick people troubled with Cephalick diseases, do further perswade.

> . . . A Virgin living in this City, was afflicted a long time with a most cruel Headach, and in the midst of her pain much and thin yellow Serum daily flowed out from her Nostrils; the last Winter this Excretion stopped for some time, and then the sick party growing worse in the Head, fell into cruel Convulsions, with stupidity; and within three days dyed Apoplectical. Her Head being opened, that kind of yellow Latex overflowed the deeper turnings and windings of the Brain and its interior Cavity or Ventricles . . . I could here bring many other reasons, which might seen to perswade, that the Ventricles of the Brain, of the Cavity made by the complicature or folding up of its border, is a mere sink of the excrementitious Humor; and that the humors there congested, are purged out by the Nose and Palate.

The idea of movement of humors from the brain to the nasal cavity was most likely supported by other types of evidence as well. For example, demonstrations that dyes (e.g., Indian ink), after injection into the subarachnoid space or the cerebral spinal fluid, travel to the nasal mucosa via the cribriform plate were made in the 19th century, and there is no reason to believe that such information was not available in earlier times (see Jackson et al., 1979, for review).

It is not clear who deserves the credit for identifying the olfactory nerves in the upper nasal cavity, although, according to Wright (1914), the 7th-centurey Greek physician Theophilis gave one of the better anatomical accounts of their distribution, despite the potential political ramifications of going against Galen's dictates. Graziadei (1971) credits Massa, in 1536, as having first demonstrated the olfactory nerves in humans, and Scarpa, in 1789, as having shown that the fine fila olfactoria actually end in the regio olfactoria (note, however, Scarpa's 1785 article). Wright

1.4 The Renaissance and the Birth of Modern Studies of Taste and Smell

(1914), on the other hand, notes that the Italian Anatomist Alessandro Achillini, who died in 1512, had described their intranasal distribution.

Regardless of who is responsible for their first description, there was considerable disagreement, which spanned over a century and a half, among authorities as to whether the processes that extended from the olfactory bulbs into the nasal cavity were, in fact, nerves. Indeed, even after they were generally accepted as nerves, debate lasted into the 1840s as to whether they mediated smell sensations. Francois Magendie (1824) was the primary proponent of the idea that such sensations were mediated via the trigeminal nerve, whereas Sir Charles Bell believed that the olfactory nerves subserved such sensations (Shaw, 1833). As late as 1860, experiments appeared in the literature that addressed this point (e.g., Schiff, 1860), although the more authoritative general physiology and medical textbooks from the 1820s to the 1850s correctly noted that the olfactory nerve mediates qualitative odor sensations and the trigeminal nerve somatosensory sensations (e.g., Good, 1822; Kirkes, 1849).

Schneider (1655) and Lower (1670) are generally credited as being the first to show that nasal secretions arise from glands, rather than being secreted through the cribriform plate. However, a century earlier Berenger del Carpi, who taught surgery at Bologna (1502–1527), broke the Hippocratic and Galenic tradition and denied that fluids passed through these foramina, suggesting that they actually passed through the sphenoid sinus (Wright, 1914). The evidence that nasal secretions came from glands, rather than through the cribriform plate, was clearly an important observation in the history of medicine. Collectively, the aforementioned studies placed the first nails in the coffin of the theory propagated largely by Galen that the cribriform plate is pervious to odors and that the sense of smell lies within the ventricles of the brain. Other major studies before 1890, a number of which are now considered classic, contributed the remaining nails to this coffin and include, in chronological order, those by Hunter (1786), Sömmerring (1809), Todd and Bowman (1847), Schultze (1856, 1863), Ecker (1856), Eckhard (1858), Clarke (1861), Hoffman (1866), Martin (1873), Krause (1876), von Brunn (1875, 1880, 1889, 1892), Sidky (1877), Exner (1878), Ehrlich (1886), and Cajal (1889).

An important 17th century discovery was that exocrine glands were responsible for the saliva of the oral cavity, a finding that preceded the aforementioned discovery that similar glands were responsible for nasal secretions. This discovery resulted from the identification of the ducts of the submandibular and parotid glands by Wharton in 1656 and Stensen (Latinized to Nicolaus Stenonis) in 1661, respectively. In 1677, de Graaf employed chronic salivary fistula for collecting saliva from conscious dogs. He and Bordeu (1751) recognized that smell and taste of food, as well as its mastication, increased salivary flow, preceding the classic and defining studies of Pavlov of this phenomenon in the early 1900s. However, investigators at that time did not pay much attention to the neural innervation of the salivary glands, largely due to the belief that saliva arose solely from the small arteries by the opening of pores, being then filtered by the acini (Haller, 1744). This "particle and pore theory" continued until Carl Ludwig discovered, in 1850, that electrical stimulation of the chorda-lingual nerve induced considerable flow of saliva from the submandibular gland in the dog. Ludwig also found that the secretory pressure in the submandibular duct could exceed the arterial pressure, a finding that disproved the particle and pore theory. Claude Bernard followed up Ludwig's discovery by studying reflex salivation and the role of the medulla in altering such salivation, introducing the concept of the salivary reflex arc in 1856.

Since these early studies saliva has been found to serve numerous functions, including ones related to digestion, lubrication, protection of oral tissues, and thermoregulation. In 1963, McBurney and Pfaffmann demonstrated, in a human psychophysical study, that salivary constituents (namely sodium and chloride) could affect taste sensitivity to NaCl. This important study led to the realization that saliva is not only a solvent for tastants, but forms the external milieu of the taste receptors. Their study evolved from previous speculations that lowered sodium content of the blood, such as occurs after adrenalectomy, might directly lower the threshold sensitivity of the salt receptors (e.g., Young 1949). Previously Richter and MacLean (1939) had raised the possibility that the taste threshold might be determined by the salt content of saliva. After McBurney and Pfaffmann's research, the effects of salivary constituents on taste function became the focus of numerous human psychophysical and animal electrophysiological studies. These studies led to the realization that saliva affects not only the perception of salty tasting stimuli, but also the taste of stimuli that induce the other classic taste qualities, as well as fatty tastes and tactile sensations such as astringency (Spielman, 1990; Matsuo, 2000; Bradley and Beidler, 2003).

The studies by Schultze of the olfactory neuroepithelium (1856, 1863) deserve special mention. Despite the lack of modern stains and fixatives, Schultze painstakingly provided the first comprehensive accurate description and drawings of the olfactory receptor cells and supporting cells (Zippel, 1993). He clearly identified the cilia. As noted in his 1856 paper, " … each of the various fiber cells, lying on a refracting knoblet, bears six to ten of these long hairlets which, under resting conditions, appear bristly and extend freely to the nostril air" (Zippel, p. 67). Schultze noted species differences in cilia and identified the non-ciliated supporting cells, as well as basal cells near the lamina propropria. Moreover, he noted the irregularities in the epithelium – irregularities that were further defined by von Brunn in his 1880 and 1882 papers as follows, "… the olfactory region can be compared to a continent which has a sometimes greater, sometimes lesser density of large lakes, extends into peninsulas and splits and contains numerous islands" (Zippel, p. 71). The basal cells were named by Krause (1876). Evidence that the receptor cells arise from the periphery and grow towards the bulb was provided by His (1889).

It is of interest that many of these olfactory studies were performed in an era when the human sense of smell was not highly regarded as important to many notables, a stigma that remains today. For example, Charles Darwin, who might have been expected to champion an exploration of human olfactory studies, indicated the following in 1871 (p. 32):

'… <human olfaction> is of extremely slight service, if any, even to savages, in whom it is generally more highly developed than in the civilised races. It does not warn them of danger, nor guide them to their food; nor does it prevent the Esquimaux from sleeping in the most fetid atmosphere, nor many savages from eating half-putrid meat'.

Darwin's negative perspective on human olfaction was forshadowed nearly a century earlier by Immanuel Kant, the renowned German philosopher, who stated in 1798:

'To which organic sense do we owe the least and which seems to be the most dispensable? The sense of smell. It does not pay us to cultivate it or to refine it in order to gain enjoyment; this sense can pick up more objects of aversion than of pleasure (especially in crowded places) and, besides, the pleasure coming from the sense of smell cannot be other than fleeting or transitory'.

Leading 19th century physicians similarly denigrated the human sense of smell. In 1870, Notta pointed out that smell dysfunction did not represent a life-threatening infliction (Notta, 1870) and three years later Legg indicated, in *The Lancet*, that smell loss "barely amounts to a discomfort" (Legg, 1873). In 1881, Althaus, also writing in *The Lancet* (p. 771), stated the following (p. 771):

if Prince Bismarck and M. Gabetta were to become suddenly blind and deaf, the destinies of Europe would no doubt be changed; while if these two men were to lose their smell and taste, things would probably go on much in the same manner as they do now.

Leading comparative anatomy of the late 19th century fueled, to some degree, the argument that the human sense of smell is not very important and that olfaction is largely

a primitive element of the brain. Broca (1878), using the rhinal fissure as a dividing line, separated the brain into a basal (rhinencephalon) and a superior (pallium) portion. Animals with a large basal region were termed "osmatic" and were assumed to have a keen sense of smell. Those with a small basal brain region were termed "anosmatic" and were assumed to have little or no smell function (Schiller, 1997). Broca extended this concept into human mental development, concluding that the basal, i.e., limbic, part of the brain represented the emotional "brute," whereas the newer cerebral lobes represented the development of intelligence.[2] In 1890, Turner further refined Broca's schemata by dividing animals into three groups, described as follows (p. 106):

1. Macrosmatic, where the organs of smell are largely developed, a condition which is found, for example, in the Ungulata, the proper Carnivora and indeed in the majority of mammals.

2. Microsmatic, where the olfactory apparatus is relatively feeble, as in the Pinnipedia, the Whalebone Whales, Apes, and Man.

3. Anosmatic, where the organs of smell are apparently entirely absent, as in the Dolphins, and it may be in the Toothed Whales generally, though, as regards some genera of Odontoceti, we still require further information.

That being said, considerable efforts were made in the 19th century to quantify human smell and taste function, particularly in Germany where Hermann von Helmholtz's student, Wilhelm Wundt, founded the first experimental psychology laboratory in 1879. Indeed, the dawn of human chemosensory psychophysics occurred in the mid-to-late 19th century, as illustrated by studies in which specific taste qualities were painstakingly mapped on the tongue by Öhrwall (1891) and Kiesow (1894). Although Boring (1942) credits Fischer and Penzoldt (1886) as having measured the first absolute threshold to an odorant, Valentin, in 1848, described a procedure that assessed olfactory sensitivity that predated even Fechner's development of formal threshold methodology by a dozen years. Zwaardemaker (1925), who invented the important draw-tube olfactometer (Chapter 11) and who performed sophisticated studies on a wide range of topics, including adaptation and cross-adaptation, credited Passy (1892) as having made an important step in the development of olfactometry. In essence, Passy dissolved a given amount of odorant in alcohol in a 1:10 ratio. This new solution was then again diluted in such a ratio, and this was repeated

---

[2]This dichotomy was rejuvenated by subsequent investigators, including Maclean's triune brain in which sectors of the brain were divided into the reptilian, paleomammalian (limbic system), and neomammalian (neocortex) complexes (Maclean, 1990).

over and over to provide a series of dilution steps. For testing, a small amount of solution at each concentration was placed in liter bottles which were heated slightly to evaporate the alcohol. Such bottles were then sampled from highest concentration to lowest concentration until no smell was discernible to the subject. Zwaardemaker, however, expressed concern that the alcohol diluent used by Passy might influence the perception of the test odorant. This potential problem was eliminated to a large degree in the successive dilution series described by Toulouse and Vaschide (1899) and Proetz (1924), where water and mineral oil, respectively, served as the diluents.

From the clinical perspective, the more scholarly physicians of the 18th and 19th centuries were very much aware of the major types of olfactory disorders that we recognize today. Good (1822), for example, classified disorders of olfaction into the following categories: *Parosmia acris* (acute smell), *Parosmia obtusa* (obtuse or distorted smell), and *Parosmia expers* (anosmia or lack of smell). Good (1822, pp. 260–261) notes the following regarding *Parosmia obtusa*:

The evil is here so small that a remedy is seldom sought for in idiopathic cases; and in sympathetic affections, as when it proceeds from catarrhs or fevers, it usually, though not always, ceases with the cessation of the primary disease. It is found also as a symptom of hysteria, syncope, and several species of cephalaea, during which the nostrils are capable of inhaling very pungent, aromatic, and volatile errhines, with no other effect than that of a pleasing and refreshing excitement.

Where the sense of smell is naturally weak, or continues so after catarrhs or other acute diseases, many of our cephalic snuffs may be reasonably prescribed, and will often succeed in removing the hebetude. The best are those fond of the natural order verticillatae, as rosemary, lavender, and marjoram; if a little more stimulus be wanted, these may be intermixed with a proportion of the teucrium *Marum*; to which, if necessary, a small quantity of asarum may also be added: but pungent errhines will be sure to increase instead of diminishing the defect.

Good's observations concerning *Parosmia expers* were as follows:

This species is in many instances a sequel of the preceding [Parosmis obtusa]; for whatever causes operate in producing the former, when carried to an extreme or continued for a long period, may also lay a foundation for the latter. But as it often occurs by itself, and without any such introduction, it is entitled to be treated separately. It offers us the two following varieties:

Organica. Organic want of smell. From natural defict, or accidental lesion, injurious to the structure of the organ.

Paralytica. Paralytic want of smell. From local palsy.

The FIRST VARIETY occurs from a connate destitution of olfactory nerves, or other structural defect; or from external injuries of various kinds; and is often found as a sequel in ozaenas, fistula lachrymalis, syphilis, small-pox, and porphyra. The SECOND is produced by neglected and long continued coryzas, and a persevering indulgence in highly acrid sternutatories.

Among the more detailed and vivid descriptions of cases of anosmia in the 19th century literature are those of Ogle (1870). He describes in detail three cases of anosmia due to head injury in which taste function was intact, a case of anosmia associated with facial palsy, a viral-induced case of anosmia, a case of anosmia due to obstruction, and three cases of unilateral olfactory loss that were related to aphasia, agraphia, and seizure attributable to brain lesions. The olfactory losses due to head injury were believed to be caused by the shearing of the olfactory filaments at the level of the cribriform plate from movement of the brain produced by the blow. In this explanation of the problem, Ogle notes (p. 266) that "the anterior brain rests directly upon the bones of the skull, and is not separated from them as is the case elsewhere by the interposition of cerebro-spinal fluid."

1.5 THE MODERN ERA: 20th AND 21st CENTURY ADVANCES

The major progress in the field of chemosensory research that has taken place in the 20th and 21st centuries has largely been due to contemporaneous technical advances in other fields of science. Included among such advances, which are not mutually exclusive, are (1) the invention of sensitive methods for recording minute electrical potentials from the nervous system, including measurements taken from the scalp (e.g., electroencephalography) and recordings from single cells and isolated components of cell membranes, (2) development and refinement of statistical methodologies and experimental designs, such as those employed in epidemiology and pharmacology, (3) development of novel and sensitive psychophysical techniques, (4) the invention of new histological stains and radically novel histological procedures, such as those that utilize autoradiography, immunohistochemistry, and various tracing agents, (5) the development of biochemical techniques for assessing endocrine and neurotransmitter receptor events, including radioimmunochemistry, (6) the

14   **Chapter 1 Introduction and Historical Perspective**

continued development and refinement of microimaging systems, including the electron microscope, (7) advances in tissue preparation procedures that optimize such imaging technology, such as osmium preparations and freeze fracture techniques, (8) the invention of computerized tomography (CT), magnetic resonance imaging (MRI), and other noninvasive imaging tools useful for evaluating the structure of the brain in vivo, (9) the development of functional imaging techniques, such as positron emission tomography (PET), single photon emission computed tomography (SPECT), and functional MRI (fMRI), and (10) innumerable advances in biology, including the development of the fields of animal behavior and, importantly, molecular biology and molecular genetics, where, for example, recombinant DNA techniques and gene manipulations have made it possible to identify and confirm the roles of many proteins involved in olfactory and gustatory function.

Obviously, in this introduction it is only possible to mention a few of the many important observations made in the last hundred and fifty years or so that have contributed to the present research climate. The areas selected for exposition were chosen, in part, on the basis of the amount of chemosensory research they have generated and are continuing to generate. These and other important events or areas of research are mentioned in more detail throughout the *Handbook*.

1.5.1 Electrophysiological Studies

A major 20th-century milestone that had a significant impact on modern chemosensory science was the development of means for electrophysiologically recording nerve impulses from the brain, including the olfactory and gustatory receptors and pathways. Although crude electrical recordings were obtained from the olfactory system in the late 19th century (e.g., Saveliev, 1892; Garten, 1900), the sophisticated equipment necessary for reliable recordings, including sensitive electrodes, was not available until well into the 20th century (e.g., the oscilloscope was invented in 1922; Erlanger and Gasser, 1937). The earliest extracellular recordings of single-cell gustatory primary afferent nerve activity were made by Zotterman (1935) and Pfaffmann (1941), the latter of whom was working in Lord Adrian's laboratory at the time. Kimura and Beidler (1961) and Tateda and Beidler (1964) were the first to record from single cells within the taste bud. Recordings of olfactory nerve fiber bundles were made by Beidler and Tucker (1955); recordings of single-cell olfactory receptor activity from extracellular electrodes were obtained by Hodgson et al. (1955) and Gesteland et al. (1963, 1965). The first evidence for between species differences in single-cell neural firing was presented by Beidler et al. (1955) and Pfaffmann (1955), a point that was later exploited by Frank

(1973) in the use of the hamster as a model for species, such as the human, that exhibit salient responsiveness to sweet-tasting stimuli.

These observations were harbingers for the more recent discoveries of the receptors that largely defined their afferent activity. Other manifestations of this advance in technology include (1) the recording of multicellular summated potentials at the levels of the vertebrate olfactory mucosa [the electro-olfactogram (EOG); Hosoya and Yoshida, 1937; Ottoson, 1956] and insect antenna (the electroantennogram; Schneider, 1957a, b), (2) recording of transduction currents in isolated olfactory receptor cells (e.g., Kurahashi and Shibuya, 1987; Firestein and Werblin, 1989), (3) measurement of ion channel activity in restricted patches of olfactory or taste cell membranes (Nakamura and Gold, 1987; Kinnamon et al.,1988); (4) topographic analysis of responses within the olfactory epithelium and olfactory bulb (e.g., Kauer and Moulton, 1974; Kubie and Moulton, 1980; Leveteau and MacLeod, 1966; Mackay-Sim et al., 1982; Mozell, 1964, 1966; Moulton, 1976), (5) the application of voltage-sensitive dyes for recording electrical changes in chemosensory neural tissue (Kauer, 1988), (6) the recording of olfactory- and taste-evoked potentials in higher brain regions (e.g., Funakoshi and Kawamura, 1968, 1971; Kobal and Hummel, 1988; Kobal and Plattig, 1978; Plattig, 1968/1969), and (7) electrophysiological mapping, in animals, of local, as well as more global, olfactory, and gustatory brain circuits (e.g., Emmers et al., 1962; Getchell and Shepherd, 1975; Komisaruk and Beyer, 1972; Mori and Takagi, 1978; Motokizawa, 1974; Nicoll, 1971; Norgren, 1970; Pfaffmann et al., 1961; Rall et al., 1966; Scott and Pfaffmann, 1972; Shepherd, 1971, 1972; Tanabe et al., 1973, 1975).

The first studies to directly stimulate sectors of the human brain in conscious patients so as to map brain centers associated with taste and smell were those of Foerster (1936) and Penfield and Faulk (1955). These neurosurgeons identified key cortical regions associated with taste and somatosensory perception. Subsequently, numerous anatomical and electrophysiological studies in nonhuman primates confirmed that the primary "gustatory cortex" is located within the insular cortex and parietal operculum and that these regions contain cells sensitive to both gustatory and somatosensory inputs (e.g., Benjamin and Burton, 1968; Ogawa et al., 1989) (see Chapters 32 and 35).

1.5.2 Studies of Receptor Function

Remarkable progress has been made in the last 30 years in identifying the initial events that occur when odorant or tastant molecules activate receptor cells, as is evidenced by the studies reviewed in detail in Chapters 5, 6, 30,

31, and 40–42. A number of these studies have been performed solely at the molecular genetic or biochemical level, but many others have used electrophysiological measures, sometimes in combination with biochemical ones, to address conductance changes that occur in the cell membrane following receptor activation, including activation of single channels. Major technical advances have made it possible to identify the olfactory and gustatory receptors, establish the involved second messengers and ion channels, and to understand phylogenetic associations and likely patterns of evolution. In olfaction, the discovery of a protein termed olfactory marker protein, which is expressed in mature olfactory receptor cells of a range of species, has proved invaluable to the study of olfactory transduction and the anatomy of the olfactory pathways (Margolis, 1972; Farbman and Margolis, 1980). Another protein, alpha-gusducin, has proved similarly useful in understanding taste transduction (McLaughlin et al., 1992a,b).

Soluble proteins have been identified within the olfactory mucus that may play multiple roles, most notably assisting in the transport of hydrophobic molecules to receptor regions or in stimulus removal and/or deactivation (Lee et al., 1987; Pevsner et al., 1988a, b; Schmale et al., 1990). Such "odorant binding proteins" are also found in insects, where they have been said to represent "a major evolutionary adaptation regarding the terrestrialization of the olfactory system, converting hydrophobic odorants into hydrophilic ones by increasing their aqueous solubility" (Vogt et al., 1991, p. 74). The role of mucus and saliva in early elements of olfactory and gustatory transduction are described in Chapters 3 and 28.

It is now appreciated that the interaction of stimulus molecules with the receptor membrane opens or closes, directly or indirectly (i.e., via second-messenger systems), membrane channels, resulting in a change in the flux of ions and alteration of the cell's resting potential (see Chapters 5, 6, 30, 31, 40–42) . Taste receptor cells possess a number of ion channels identical to some of those found in neurons (i.e., voltage-gated Na+, Ca^2+, and K+ channels, as well as Ca+-mediated cation channels and amiloride-sensitive Na+ channels) and release a neurotransmitter that activates the first-order taste neuron (Roper, 1989). In the case of olfaction, the receptor cell is the first-order neuron, so changes in membrane potential, if of sufficient magnitude, produce the action potentials. Bronshtein and Minor (1977) provided the first scientific evidence that these interactions occur on the olfactory cilia, an observation supported by subsequent workers (e.g., Rhein and Cagan, 1980, 1983; Lowe and Gold, 1991; Menco, 1991; see Chapters 3–7). In taste, such interactions occur on microvillae associated with apical portions of the taste receptor cell (Avenet and Lindemann, 1987, 1990; Heck et al., 1984), a region whose importance was apparently first stressed by Renqvist in 1919 (see Kinnamon and Cummings, 1992). Although occluding junctions among cells in the apical region of the taste bud restrict most stimuli to that region, molecules of low molecular weight may permeate these junctions (Holland and Zampighi, 1991). Interestingly, actin filaments are found around the taste pore, suggesting that taste buds may contract or expand, possibly in response to the type of taste stimulus that they encounter (Ohishi et al., 2000). Until the application of electron microscopy to taste buds in the 1950s (de Lorenzo, 1958; Engström and Rytzner, 1956a, b), taste buds were believed to contain cilia, not microvillae.

In the 1980s and 1990s, a number of important findings led to a fuller understanding of the initial events in olfactory transduction. A calcium sensitive enzyme adenylyl cyclase (type III), which is usually coupled to a G-protein, was found to be highly active in olfactory cilia (Kurihara and Koyama,1972; Pace et al., 1985). Several laboratories found that odorants increase, in a dose-related manner, intracellular cyclic adenosine 3′,5′-monophosphate (cAMP) in olfactory receptor cells, triggering the opening of cAMP-gated cation channels (Bruch and Teeter, 1989; Nakamura and Gold, 1987; Pace et al., 1985; Pace and Lancet, 1986; Sklar et al., 1986). In addition to cAMP, cGMP was found to play a likely role in the modulation of the sensitivity of olfactory receptor neurons, such as during adaptation (Leinders-Zufall et al., 1996). A positive correlation was noted between an odorant's ability to activate adenylyl cyclase in a frog ciliary preparation and both the magnitude of the EOG response it produces in frog epithelia (Lowe et al., 1989) and its perceived odor intensity to humans (Doty et al., 1990), suggesting a gross association between the amount of adenylate cyclase activated and the intensity of odor perception.

In 1989, Jones and Reed (1989) isolated a guanine nucleotide-binding protein (G-protein) in olfactory cilia, thereby lending strong support to the view that odorant receptors are G protein-coupled receptors. This protein had 88% sequence identity to conventional G_s and was termed $G\alpha_{olf}$ or G_{olf}. Although G-proteins other than G_{olf} (e.g., G_{i2} and G_o) have been identified in olfactory receptor cells, they appear not to be involved in early transduction events, presumably assisting in such processes as axonal signal propagation, axon sorting, and target innervations (Wekesa and Anholt, 1999).

In 1991, the identification of the primary receptor family responsible for vertebrate olfactory transduction was made by Buck and Axel (1991). This work was the basis for the 2004 Nobel Prize in Medicine or Physiology. Under the assumption that olfactory receptors have elements in common with a large superfamily of surface receptors that evidence seven transmembrane domains and linkage to G-proteins and second-messenger systems, these investigators synthesized oligonucleotides that coded for conserved

16　　　　**Chapter 1　Introduction and Historical Perspective**

(i.e., nearly invariant) amino acid sequences found among receptors from sensory systems other than olfaction. These oligonucleotides were then used as molecular probes. Eighteen clones were found that coded for proteins with seven transmembrane domains within olfactory tissue, but not within brain, retina, or various non-neural tissues. The variability in the amino acid sequences was found to be in regions of the molecule believed to be important in the binding of ligands in other receptor proteins with seven transmembrane domains. Based on this information, it was concluded that a large number of diverse receptors were involved in olfactory transduction and that such receptors comprised a very large multigene family that "encodes seven transmembrane domain proteins whose expression is restricted to the olfactory epithelium."[3]

Subsequent work has found that the number of functional olfactory receptors varies among species, being nearly 2000 in the African elephant, over 1000 in murine rodents, and around 400 in humans and chimpanzees (Glusman et al., 2001; Zozulya et al., 2001; Young and Trask, 2002; Zhang and Firestein, 2002; Niimura et al., 2014). The number of genes appears to be influenced by the environment to which a given species is adapted. In the mid-1990s, it was found that each receptor gene is expressed in only a small percentage of neurons (Ressler et al., 1994) and that a given neuron expresses only one allele of a single receptor gene (Malnic et al., 1999; Lomvardas et al., 2006), although this generalization does not apply to all vertebrates (e.g., fish; see Chapter 23). It was also found that neurons that express the same odorant receptor gene converge on the same glomeruli (Ressler et al., 1994; Mombaerts et al., 1999). Such findings, along with those from other studies (e.g., Johnson and Leon, 2000), have largely clarified the topographic relationships between olfactory receptor cells and the olfactory bulb – relationships that were incompletely and only grossly known from earlier anatomical and physiological studies (e.g., Adrian, 1953; Le Gross Clark, 1951; Døving and Pinching, 1973).

It is important to note that the gene family encoding insect olfactory receptors likely evolved independently from that encoding vertebrate olfactory receptors. This is reflected by the fact that genes of these two families lack sequence homology (Wistrand et al., 2006). One major difference between the two families was discovered in Drosophila, where a "chaperone" receptor (Or83b), unlike other olfactory receptors, was found to be expressed in

all olfactory receptor neurons. Instead of binding volatile ligands, this receptor forms dimers with other olfactory receptors, resulting in ligand-detecting receptor complexes that enhance receptor sensitivity (Benton et al., 2006; Larsson et al., 2004; Neuhaus et al., 2005).

Trace amine-associated receptors (TAARs), a specific family of G-protein coupled receptors, have been identified that function as chemosensory receptors for several volatile amines in mice (Liberles and Buck, 2006) and a range of other vertebrate species (Hashiguchi and Nishida, 2007). TAARs are expressed in the olfactory epithelium, being most abundant in fish and less abundant in tetrapods. Humans express only five putatively functional TAARs, in contrast to zebrafish, who express 109 such TAARs (Hashiguchi and Nishida, 2007). The function of these receptors is yet to be determined, although amines are found among a number of biological secretions in rodents.

As described in detail in Chapter 5, the olfactory-specific receptor guanylyl cyclase D (GC-D, *Gucy2d*) was cloned in the mouse by Fulle et al. (1995). Less than 0.1% of the bipolar olfactory receptor neurons express the GC-D transcript, and these neurons appear to be randomly dispersed among the classic olfactory receptor neurons. Their axons project to unique glomeruli that form a "necklace-like" structure in the caudal olfactory bulb. The function of these unique neurons, which mainly employ cGMP as their second messenger, is poorly understood, although there is evidence that they detect natriuretic peptides from the urine of conspecifics, as well as atmospheric $CO2$ (Hu et al., 2007).

A number of olfactory, taste, and vomeronasal organ receptors have now been deorphened (Glatz and Bailey-Hill, 2011; Kratuwurst et al., 1998; Malnic et al., 1999; Touhara, 2007; Touhara et al., 1999; Wetzel et al., 2001; Zhao et al., 1998; for invertebrates, see Carlson, 2001). For example, Zhao et al. (1998) used an adenovirus-mediated gene transfer procedure to increase the expression of a specific receptor gene in an increased number of receptor neurons in the rat olfactory epithelium, demonstrating ligand-specific increases in EOG amplitude. Krautwurst et al. (1998) employed a polymerase chain reaction (PCR) strategy to generate an olfactory receptor library from which cloned receptors were screened for odorant-induced responsiveness to a panel of odorants, as measured by an assay sensitive to intracellular Ca^{2+} changes. Several receptor types with ligand specificity were found, including one differentially sensitive to the (−) and (+) stereoisomers of citronella. More recently, Saito et al. (2009) deorphanized 10 human and 52 mouse olfactory receptors. Some were broadly tuned to a range of odorants ("generalists") whereas others were narrowly tuned ("specialists") to a small number of structurally related odorants.

---

[3] It is now known that receptor-encoding complement DNA can be expressed in some non-neuronal cells, which, when stimulated with appropriate ligands, generate second-messenger responses (e.g., Raming et al., 1993). The same receptor genes described by Buck and Axel (1991) have now been identified in tissues far removed from the olfactory cilia, including sperm (Parmentier et al., 1992) and the heart (Hillier et al., 1996).

An explosion in the identification of taste receptor genes has occurred in the last decade. Nearly 50 taste receptor genes have been identified and more than 2,000 genes have been associated with primate taste buds (Hevezi et al., 2009). Receptors on three main classes of taste cells have been defined. Type I cells, which are involved in neurotransmitter clearance and ion redistribution and transport, mediate salty taste sensations via specialized Na^+ channels (Chaudhari and Roper, 2010). Type II cells harbor G-protein-coupled receptors that produce sweet, umami (monosodium glutamate-like), and bitter sensations, the latter being mediated by a family of ~30 such receptors. Type III cells have been found to have specialized proton channels for hydrogen ions and mediate sour sensations (Chang, Waters, and Liman, 2010). Of considerable importance is the discovery that sweet receptor genes are compromised in a wide range of vertebrtate species, including bats, horses, cats, chickens, zebra finches, and the western clawed frog (Li et al., 2005; Zhao et al., 2010), as are bitter and umami receptors for dolphins (Feng et al., 2014) (see Chapter 43). During the early part of the 21st century the sequencing and functional expression of umami human taste receptors for glutamate were made, solidifying the concept that umami represents a fifth taste quality (Chaudhari, Landin and Roper, 2000; Nelson et al., 2002; Li et al., 2002). Kikunae Ikeda (1909) was the first to identify glutamic acid as the agent associated with the umami taste, a word derived from the Japanese adjective Umai, meaning delicious (Lindemann, Ogiwara and Ninomiya, 2002).

It has now become apparent that both bitter and sweet taste-related receptors, heretofore believed to occur only in taste buds, are present elsewhere in the body, most notably in the alimentary and respiratory tracts. In pioneering work, Hofer and Drenckhahn (1998), Hofer et al. (1996) discovered that α-gustducin, the taste-specific G-protein α-subunit, is expressed in so-called brush cells within the rat stomach, duodenum, and the pancreatic duct system. Such cells, which were first described in the trachea by Rhodin and Dalham in 1956, have been found in olfactory epithelium (where they are termed microvillar cells; Okano et al., 1967; Moran et al., 1982), lung (Meyrick and Reid, 1968), pancreas (Nakagawa et al., 2009), gall bladder (Luciano and Reale, 1979), and, in nonhumans, the vomeronasal organ (Adams, 1978). Although they have no direct contact with neurons, brush cells are rich in nitric oxide (NO) synthase. NO defends against xenobiotic organisms, protects the mucosa from acid-induced lesions, and, in the case of the gastrointestinal tract, stimulates vagal and splanchnic afferent neurons. NO further acts on nearby cells, including enteroendocrine cells, absorptive or secretory epithelial cells, mucosal blood vessels, and cells of the immune system (Hofer et al., 1998).

Wu et al. (2002) were the first to identify members of the T2R family of bitter receptors within the gastrointestinal tract and in enteroendocrine cell lines. Sweet taste receptors of the T1R family were first found in the intestinal tract and the STC-1 enteroendorine cell line by Dyer et al. (2005). Mace et al. (2007) found that all members of the T1R family were expressed in Paneth and other cells in rat jejunum and were involved in the stimulation of glucose absorption. That same year, Margolskee et al. (2007) and Jang et al. (2007) demonstrated that T1R3 receptors and gustducin play decisive roles in the sensing and transport of dietary sugars from the intestinal lumen into absorptive enterocytes via a sodium-dependent glucose transporter and in regulation of hormone release from gut enteroendocrine cells. Shah et al. (2009) identified a number of T2R bitter receptors in the motile cilia of the human airway that responded to bitter compounds by increasing their beat frequency. Recently, Lee et al. (2012) demonstrated that the T2R38 taste receptor is expressed in human upper respiratory epithelia and responds to acyl-monoserine lactone quorum-sensing molecules secreted by *Pseudomonas aeruginosa* and other gram-negative bacteria. Importantly, they found that differences in T2R38 functionality, as related to TAS2R38 genotype, correlated with susceptibility to upper respiratory infections in humans.

1.5.3 Studies of the Olfactory Pathways in Transport of Agents from the Nose to the Brain

A very important empirical observation, made in the first half of the 20th century, was that the olfactory nerve can serve as a conduit for the movement of viruses and exogenous agents from the nasal cavity into the brain (see Chapters 19 and 20). This route is direct, since the olfactory neurons lack a synapse between the receptive element and the afferent path. The existence of this pathway for viral infection of the brain has been recognized for some time, as evidenced by a number of studies from the 1920s and 1930s (see Clark, 1929; Hurst, 1936). For example, mice intraperitoneally inoculated with louping ill virus showed the first signs of CNS localization of the virus in the olfactory bulbs. Mice whose olfactory mucosa was cauterized with sulfate were partly protected against such infection (Burnet and Lush, 1938). Poliomyelitis virus, placed in the noses of primates, travels to the olfactory bulbs via the axoplasm of the olfactory nerves, rather than along the nerve bundle sheaths (Bodian and Howe, 1941a, b). In a pioneering paper, Armstrong and Harrison (1935) reported that monkeys could be protected against intranasal inoculations of poliomyelitis virus by previous lavage of the nose with solutions of alum or picric acid (or both). Subsequent studies (e.g., Schultz and Gebhardt, 1936) found that zinc sulfate gave a longer-lasting and higher

18   **Chapter 1   Introduction and Historical Perspective**

degree of protection from poliomyelitis, leading to the prophylactic spraying of noses of children with this agent during poliomyelitis outbreaks in the late 1930s (Peet et al., 1937; Schultz and Gebhardt, 1937; Tisdall et al., 1937). Unfortunately, such spraying produced long-lasting, presumably permanent, anosmia in some individuals (Tisdall et al., 1938).

Related to the observation that the olfactory nerves are a major carrier of viruses is the fact that the receptive elements of the olfactory system are exposed, to a large degree, to the vagaries of the external environment, making them susceptible to damage from bacteria, viruses, toxins, and other foreign agents. As reviewed in detail in Chapters 3 and 19, numerous studies have demonstrated that the olfactory mucosa is rich in enzymes that likely minimize the deleterious influences and uptake of most xenobiotic agents into the olfactory receptor cells, including cytochromes P-450, flavin-containing monooxygenase, and aldehyde dehydrogenases and carboxylesterases.

1.5.4   The Discovery that Taste Buds and Olfactory Receptor Cells Regenerate

Another important 20th-century development is the discovery that the gustatory and olfactory receptor cells regenerate periodically. Beidler and Smallman (1965) provided the first scientific demonstration that the sensory cells of the taste bud are in a dynamic state of flux and are constantly being renewed, with the more recently formed cells of the periphery migrating centrally to act as receptors for very limited periods of time. The observation that olfactory receptor cells, which are derived from ectoderm and which serve as the first-order neurons, can regenerate after they are damaged was first noted in mice by Nagahara (1940) and later confirmed in primates by Schultz (1960). This observation is particularly significant, in that it is in conflict with the long held notion that neurons in the adult animal are irreplaceable and suggests that the olfactory system may contain the key to producing neural regeneration in a variety of neural systems (Farbman, 1992). However, questions remain as to why metaplastic respiratory epithelium often invades the region of the damaged olfactory epithelium and, when such metaplasia occurs, the epithelium in that region may never convert to olfactory epithelium. Recent studies, in which the olfactory epithelia of rodents were exposed to airborne or systemically administered toxic agents, may shed some light on this question. Thus, the type of repair seems to correlate with the degree or extent of the initial epithelial damage (Keenan et al., 1990). For example, when the basilar layer of the mucosa is completely damaged, then metaplastic replacement with a respiratory-like epithelium occurs. When the damage is not marked or the toxic insult

is not sustained, regeneration, usually with fewer or irregularly arranged cells, occurs. It is currently believed that horizontal basal cells and globose basal cells near the basement membrane of the epithelium are responsible for the generation of all neuronal and non-neuronal cell types within the epithelium (Chapter 7).

Closely related to the discovery of regeneration within the olfactory epithelium is the important observation made by Andres (1966, 1969) that mitotic cells, young sensory cells, mature sensory cells, and dying cells coexist within the olfactory epithelium (see Farbman, 1992, for a review). This suggested to Andres the hypothesis that the olfactory receptor cells were continually being replaced. The notion that olfactory receptor cells were in a state of flux received subsequent support by others (Chapter 7; Moulton et al., 1970; Thornhill, 1970; Graziadei and Metcalf, 1971) and led to the idea that they are relatively short-lived. Hinds et al. (1984), however, found that a number of the olfactory receptor cells of mice reared in a pathogen-free environment survived for at least 12 months and hypothesized that olfactory nerve cell turnover involves recently formed or immature receptor cells that fail to establish synaptic connections with the olfactory bulb. This hypothesis implies that environmental agents play an important role in dictating which elements of the receptor sheet become replaced and that the rate of regeneration of the olfactory receptor cells is not genetically predetermined, as previously supposed (see Chapter 7). The observation that improvement in olfactory function after cessation of chronic cigarette smoking occurs over a period of years and is dose-related (Frye et al., 1990) suggests, under the assumption that the olfactory epithelium is involved, that either turnover of the olfactory epithelial cell complement takes a much longer time than previously supposed or growth of olfactory epithelium into damaged areas is relatively slow and dependent on the extent of prior trauma, or both.

The study of the regeneration of the olfactory neurons has been greatly enhanced by the ability to culture the olfactory mucosa in vitro (for review, see Mackay-Sim and Chuah, 2000). This was first demonstrated in the culture of olfactory organs from embryonic mice (Farbman, 1977) and used to show the importance of olfactory bulb in promoting differentiation of the olfactory sensory neurons (Chuah and Farbman, 1983). The next major development came with the investigations of dissociated cultures from embryonic and newborn rats (Calof and Chikaraishi, 1989; Pixley, 1992a). This allowed the growth factors regulating olfactory neurogenesis to be explored in the developing olfactory epithelium (DeHamer et al., 1994; Mahanthappa and Schwarting, 1993) and in the adult (MacDonald et al., 1996; Newman et al., 2000).

Interestingly, normal targeting of glomeruli by olfactory receptor axons has been demonstrated in mice lacking functional olfactory cycle nucleotide-gated channels

(Lin et al., 2000) and in mice lacking most intrabulbar GABAergic interneurons (Bulfone et al., 1998). Thus, establishment of the topographical map from the receptor cells to the glomeruli seems to require neither normal neural activity in these pathways nor cues provided by the major neural cell types of the bulb.

More recently, as described in detail in Chapter 7, considerable progress has been made in understanding the regulation of neurogenesis within the olfactory epithelium. Among the more important discoveries was the finding that horizontal basal cells are neural stems cells with the capacity to regenerate both neuronal and non-neuronal elements of the epithelium, including the olfactory receptor nerve precursors, globose basal cells (Leung et al., 2007).

Human olfactory neuronal progenitors have now been grown *in vitro* (Wolozin et al., 1992) and this has been exploited to study biochemical changes in Alzheimer's disease (Wolozin et al., 1993). Primary cultures of human olfactory mucosa (Féron et al., 1998; Murrell et al., 1996) have lead to investigations into the etiology of schizophrenia (Féron et al., 1999), and the vitro growth of olfactory ensheathing cells (Chuah and Au, 1991; Pixley, 1992b; Ramon-Cueto and Nieto-Sampedro, 1992). The latter glial cells assist sensory neuron regeneration (Doucette, 1984) and have, in fact, been employed in cell transplantation therapy for the damaged nervous system (Li et al., 1998; Lu et al., 2001; Ramon-Cueto et al., 2000; Ramon-Cueto and Nieto-Sampedro, 1994).

1.5.5 The Discovery that Some Olfactory Bulb Cells Regenerate

A long-held dogma regarding the nature of the CNS of vertebrates is now known to be false; namely, that the adult brain does not exhibit neurogenesis (for review, see Gross, 2000). Although early studies found mitotic figures within the walls of the lateral ventricle (e.g., Allen, 1912; Globus and Kuhlenbeck, 1944; Öpalski, 1934; Rydberg, 1932), definitive evidence that such cells represented neurogenesis awaited the development of the tritiated thymidine technique, the electron microscope, and immunohistochemistry (Gross, 2000). In the 1960s, Altman and his associates published a series of classic studies based upon tymidine autoradiography that demonstrated neurogenesis in several brain regions of young and adult rats, including the olfactory bulb (Altman, 1969), the neocortex (Altman, 1963, 1966a,b), and the dentate gyrus of the hippocampus (Altman, 1963; Altman and Das, 1965). Regarding the olfactory bulb, proliferating cells were found within the subventricular zone lining segments of the lateral ventricles. These cells were found to reach the core of the olfactory bulb via the rostral migratory stream. Subsequent studies have confirmed and extended these observations (e.g., Luskin, 1993; Lois, Garcia-Verdugo

and Alvarez-Buylla, 1996; O'Rourke, 1996), noting that the precursor cells invade the granule and perioglomerular layers of the bulb, where they differentiate into local interneurons. A major differentiation is into GABAergic granule cells – the most numerous cells of the bulb.

These stem-cell-related phenomena are of considerable significance, as they indicate that the plasticity of the olfactory system goes far beyond simply replacing damaged neuroepithelial cells, and that continual cell replacement may play an integral role in olfactory perception. It is now known, for example, that reducing the numbers of interneurons recently generated via this process impairs the ability of an animal to discriminate among odorants (Gheusi et al., 2000). Moreover, enriching the odorous environment of mice enhances such neurogenesis and improves odor memory (Rochefort et al., 2002). The degree to which such processes influence, or are influenced by, endocrine state and various social processes is not well known, although interestingly glucocorticoids decrease, and estrogens increase, the rate of such neurogenesis within the hippocampus (Gould and Tanapat, 1999; Tanapot et al., 1999).

1.5.6 Functional Imaging Studies

A significant and rapidly evolving modern development in the study of the chemical senses is that of functional imaging. It has long been known or suspected that brain circulation changes selectively with neuronal activity (e.g., Broca, 1879; Mosso, 1881, 1884; Roy and Sherrington, 1890; Fulton, 1928), but was not until the late 1950s and early 1960s, the development of the $[^{131}I]$trifluoroiodomethane ($[^{131}I]CF_3]CF_31$) method provided a potential and novel means for quantitatively examining the influences of sensory, cognitive, and motor processes on local blood flow within regions of the brain (Landau et al., 1955; Freygang and Sakoloff, 1958; Kety, 1960; Sakoloff, 1961). This early work led in the development of the $[^{14}C]$2-deoxy-D-glucose (2-DG) autoradiographic method for determining regional glucose consumption in animals (Reivich et al., 1971; Kennedy et al., 1975; Sakoloff et al., 1977), and set the foundation for modern human functional imaging studies. Reivich et al. (1979) introduced the $[^{18}F]$ flurorodeoxyglucose method for assessing regional glucose metabolism, and Lassen et al. (1963) and Ingvar and Risberg (1965) subsequently developed and applied a procedure in which regional blood flow measurements could be established in humans by using scintillation detectors arrayed over the surface of the scalp. The refinement of such approaches led to the practical development of positron emission tomography (PET) (Ter-Pogossian et al., 1975; Hoffman et al., 1976), which was made possible by the earlier invention of X-ray computed tomography (CT) in 1973 (Hounsfield, 1973). The coincidence of these techniques provided the capability of mapping the regions with increased blood flow or glucose metabolism to specific regions of the brain in three-dimensional coordinates.

Magnetic resonance imaging (MRI) technology emerged contemporaneously with the latter developments (e.g., Lauterbur, 1973). Based upon a set of earlier principles (Block, 1946; Fox and Raichle, 1986; Lauterbur, 1973; Pauling and Coryell, 1936), Ogawa et al. (1990) were able to demonstrate that changes in blood oxygenation could be detected, in vivo, with MRI, setting the stage for the development of functional MRI (fMRI). This phenomenon, known as the blood oxygen level dependent (BOLD) signal, reflects the fact that blood flow changes more than oxygen consumption does in an activated region, reflected by a reciprocal alteration in the amount of local deoxyhemoglobin that is present, thereby altering local magnetic field properties. Details of fMRI, as well as other imaging procedures, are presented in Chapters 13, 35, and 46.

The first study to employ functioning imaging in the chemical senses was that of Sharp et al. (1975). These investigators injected four rats intravenously with 2-DG and immediately placed them in a sealed glass jar containing glass wool saturated with pentyl acetate. After 45 minutes the animals were sacrificed, and sections of the bulbs were appropriately prepared and autoradiographed. Two regions of heightened optical density were noted bilaterally which tended to be centered in the glomerular layer, with variable spread into the external plexiform and olfactory nerve layers. Subsequent studies more clearly defined the regions of apparent activation (e.g., Sharp et al., 1977; Stewart et al., 1979), and resulted in the identification of a unique set of glomeruli in weanling rats responsive to odorants in their mothers' milk (Teicher et al., 1980).

The first published human olfactory PET study was that of Zatorre et al. (1992). These investigators found that odorants increased regional cerebral blood flow (rCBF) bilaterally in the piriform cortex, as well as unilaterally in the right orbitofrontal cortex. The first taste study employing PET was that of Small et al. (1997). Increased rCBF was noted, in response to citric acid, bilaterally within the caudolateral orbitofrontal cortex, and unilaterally within the right anteromedial temporal lobe and the right caudomedial orbitofrontal cortex. The first published fMRI report on olfaction was that of Yousem et al. (1997), who demonstrated (a) odor-induced activation of the orbitofrontal cortex (Brodmann area 11), with a mild right-sided predominance (in accord with the earlier PET study of Zatorre et al., 1992) and (b) unexpected cerebellar activation. Sobel et al. (1998a) noted that olfactory stimulation activated lateral and anterior orbito-frontal gyri of the frontal lobe, and that sniffing behavior, regardless of whether an odor is present, induces piriform cortex activation. These investigators, following up on the unexpected

observation of cerebellar activation by odorants, subsequently demonstrated concentration-dependent odorant activation in the posterior lateral cerebellar hemispheres, and activation from sniffing alone in the anterior cerebellum, most notably the central lobule (Sobel et al., 1998b). More recent advances are summarized in Chapters 13 and 35.

1.5.7 Optogenetics

Optogenetics is an emerging technology with potential for elucidating neural processes associated with smell and taste function. In this technology, which was pioneered by Deisseroth and his associates (e.g., Boyden et al., 2005), photoresponsive microbial proteins are genetically expressed in neurons whose action potentials can then be altered by exposure, in vivo, to various wavelengths of light delivered by fiberoptics to the target neurons. Defined trains of spikes or synaptic events with millisecond-timescale temporal resolution can be induced by brief pulses of light.

To my knowledge, the first application of this technology to an element of the olfactory system, notably the piriform cortex, was by Choi et al. (2011). These investigators found that both appetitive and aversive behaviors could be classically conditioned in mice to light activation of small subsets of piriform neurons that expressed channelrhodopsin. Different sets of neurons could be independently conditioned to elicit distinct behaviors, suggesting that this cortex can produce learned behavioral outputs independent of sensory input. These findings have shed considerable light on the mechanisms by which the piriform cortex associates odors with behavioral responses.

1.5.8 The Animal Behavior Revolution

Another large area of research activity that must be mentioned as having had a profound impact on modern chemosensory research is that of animal behavior (see Chapters 19, 21–27). This field, which grew in geometric proportions after World War II, is a major contributor to chemosensory studies. In addition to providing detailed explications of the many rich and often complicated influences of chemical stimuli on wide range of invertebrate and vertebrate behaviors (including, in mammals, behaviors related to aggression, alarm, suckling and feeding, mating, predator–prey relationships, social status appraisal, territorial marking, and individual and species recognition), this field has provided important methodology for assessing olfactory, gustatory, and vomeronasal function in animals, including preference paradigms (e.g., Richter, 1939; Mainardi et al., 1965), classical conditioning paradigms (Pavlov, 1927), conditioned aversion paradigms (e.g., Garcia et al., 1955), habituation paradigms

1.5 The Modern ERA: 20th and 21st Century Advances

(e.g., Krames, 1970), sniff rate analysis paradigms (e.g., Teichner, 1966), and operant conditioning paradigms using positive or negative reinforcers (Skinner, 1938). Furthermore, behavioral studies have been instrumental in the demonstration of the close association between neuroendocrine and chemoreception systems in both vertebrates and invertebrates, and are critical for demonstrating the effects of various gene manipulations on smell- or taste-mediated behaviors. A number of the chapters of this volume directly relate to this vast literature and, in some cases, provides means for assessing responses of animals to odorants (Chapters 21–27) and tastants (Chapters 38 & 41–44). The reader is referred to the many general reviews of this topic (Albone, 1984; Doty, 1974, 1975, 1976, 1980, 1986; Johnston, 2000; Johnston et al., 1970; Leon, 1983; Meredith, 1983; Mykytowycz, 1970; Slotnick, 1990; Smith, 1970; Stevens, 1975; Vandenbergh, 1983; Verendeev and Riley, 2012; Wysocki, 1979).

According to Stürckow (1970), the studies by Barrows (1907), von Frisch (1919), and Minnich (1921) were seminal for the development of studies of insect chemosensory behavior and physiology, even though earlier, more equivocal, studies had been performed (e.g., Hauser, 1880). Barrows (1907) devised the first insect olfactometer and found, in the pomice fly (*Drosophila ampilophila*), that different degrees of responding were obtained from different concentrations of chemical attractants. Von Frisch (1919) demonstrated that bees could be trained to fly to a fragrant odor using simple reinforcement and later found the location of the olfactory sensilla to be on the eight distal segments of the antennae (von Frisch, 1921, 1922). Minnich (1921, 1926, 1929) explored the responses of various body parts of butterflies, certain muscid flies, and the bee to taste solutions. For example, he found that they extended their proboscises when their tarsi or certain mouth parts were touched with a sugar solution. These and other studies led to electrophysiological studies of the chemoreceptive systems of insects by Dethier (1941), Boistel (1953), Boistel and Coraboeuf (1953), Kaissling and Renner (1968), and Schneider (1955, 1957a, b).

A number of important studies, published in the 1950s, 1960s, and early 1970s, demonstrated a close association between olfaction, social behavior, and reproductive processes in rodents and other mammalian forms. Pioneering reports on this topic include those which showed that odors from male and female mice influence the timing of estrous cycles (Lee and Boot, 1955; Whitten, 1956; Whitten et al., 1968), that urine odor from unfamiliar male mice can block the pregnancy of female mice (Bruce, 1959; Bruce and Parrott, 1960), and that chemical stimuli can accelerate the onset of puberty in mice (Vandenbergh, 1969). Other important studies demonstrated that olfactory bulbectomy, anesthetization, or damage to the olfactory receptor region vomeronasal organ or nervus terminalis,

alone or in combination, can dramatically influence mating behavior, depending on the species involved (e.g., in the male or androgenized female hamster, anesthetization or damage of these systems can eliminate male copulatory behavior; Doty and Anisko, 1973; Doty et al.,1971; Murphy and Schneider, 1970; Powers and Winans, 1973, 1975; Winans and Powers, 1974). Such phenomena have been demonstrated to one degree or another in a wide variety of mammals and have important implications for animal ecology, husbandry, and perhaps even human behavior.

Other studies of this period that had a considerable impact on the field of mammalian social behavior include those that examined, in a systematic manner, sexual odor preferences in rodents. Godfrey (1958), for example, found that estrous female bank voles (*Clethrionomys*) preferred homospecific male odors over heterospecific male odors and that hybrids were discriminated against. Le Magnen (1952) demonstrated that adult male rats (*Rattus norvegicus*) prefer the odor of receptive females to nonreceptive ones, whereas prepubertal or castrated males do not (unless they have been injected with testosterone). Beach and Gilmore (1949) noted that sexually-active male dogs, but not a sexually-inactive male dog, preferred estrous to non-estrous urine. This and other work led to a number of carefully designed studies by Carr and his associates in the 1960s which sought to determine the influences of sexual behavior and gonadal hormones on measures of olfactory function. Carr and Caul (1962) demonstrated that both castrate and noncastrate male rats can be trained to discriminate between the odors of estrous and nonestrous females in a Y-maze test situation, implying that the preference phenomenon observed by Le Magnen (1952) was not due to castration-related influences on olfactory discrimination ability, per se. Carr et al. (1965) subsequently demonstrated the important role of sexual experience in producing strong preferences in male rats for estrous over diestrous odor and in female rats for noncastrate male odors over castrate male odors. These investigators also showed that sexually inexperienced females preferred male noncastrate odors if they were administered gonadal hormones that induced estrus.[4] These general findings have been observed in a wide range of species, although some species differences do exist and castration has been shown to mitigate the increase in detection performance of rats that follows repeated testing (Doty and Ferguson-Segall, 1989; for reviews, see Brown and Macdonald, 1985; Doty, 1974, 1976, 1986).

Animal behavior studies in the 1980s contributed significantly to the understanding of the function of

---

[4] In an unpublished M.A. thesis, Keesey (1962) found that sexually experienced, but not sexually inexperienced, male rats preferred the odor of female urine collected during proestrus than that collected during diestrus.

vomeronasal organ which was described histologically in many species in the 19th century, but whose function was unknown (for review of the early literature, see Wysocki, 1979). In the mouse, removal of the vomeronasal organ eliminates the surge in luteinizing hormone (LH) and subsequent increase in testosterone that ordinarily follows exposure of male mice to an anesthetized novel female mouse or her urine. However, this does not occur following exposure to an awake female mouse, suggesting several sensory cues can produce the LH surge (Coquelin et al., 1984; Wysocki et al., 1983). In both mice and hamsters, vomeronasal organ removal impairs male sexual behavior, particularly in animals that have had no prior adult contact with females (Meredith, 1986; Wysocki et al., 1986). In mice whose vomeronasal organs have been removed soon after birth, long-lasting influences on male sexual behavior in adulthood have been noted (Bean and Wysocki, 1985). Vomeronasal organ removal also greatly decreases aggression in male house mice, particularly those that have not had much fighting experience with other males (Bean, 1982; DaVanzo et al., 1983; Wysocki et al., 1986).

While there is now incontrovertible evidence that most adult humans have a rudimentary vomeronasal organ (VNO) whose opening is present at the base of the nasal chamber, no neural connection exists and the structure is generally considered vestigial (Doty, 2001; Smith and Bhatnagar, 2000; Bhatnagar and Meisami, 1998; Brown, 1979). The presence of a regressive human VNO fits into the general idea that humans have lost much chemosensory capacity, at least relative to many other mammals, over the course of evolution. However, aside from the VNO, the concept of lesser smell function has come under vigorous debate and some animal behavioral studies have suggested that the divide between humans and other mammals is not as great as often believed (Laska et al. 2000) (see Chapter 27). Contributing to this notion is a reassessment of the relationship between nasal cavity structure and both the size and density of the olfactory neuroepithelium of primates, throwing into question the classic distinction between macrosmatic and microsmatic mammals (Smith et al. 2004).

A significant event for the field of odor communication was the coining of the term "pheromone" in insects for "substances which are secreted to the outside by an individual and received by a second individual of the same species, in which they release a specific reaction, for example, a definite behavior or a developmental process" (Karlson and Lüscher, 1959, p. 55). The pheromone concept, which has permeated most areas of biology, has been applied by some workers to nearly any chemical involved in chemosensory communication. The term pheromone replaced an earlier term (ectohormone) and conjures up the idea that the social organization of animals is akin to the endocrine organization of an organism, with disparate

parts being influenced by chemicals that circulate within the social milieu. For many, but not all, insects, this term seems appropriate, given the high degree of stereotypic behavior and evidence for comparatively simple stimuli that induce behavioral or endocrinological changes. However, for many vertebrates, particularly mammals, this concept has questionable utility since it assumes simple associations between stimuli and responses which rarely, if ever, exist. As discussed in *The Great Pheromone Myth* (Doty, 2010), this concept suffers from a range of basic problems when applied to mammals. For example, it leads to the nominal fallacy, i.e., the tendency to confuse naming with explaining, and assumes that a few species-specific molecules of innate origin, largely impervious to learning and distinct from other types of stimuli, are the motive influences. Importantly, it inappropriately dichotomizes complex behaviors and stimuli into simple classes and fails to take into account the complexity of chemical stimuli and the influences of experience in determining their meaning. Despite possibly one or two such claims, no single chemical or set of chemicals has been identified in mammals that can be truly considered analogous to an insect pheromone. A number of phenomena in humans that have been attributed to pheromones, such as menstrual synchrony, have been found to be based upon statistical artifact, further denigrating the usefulness of this popular but flawed concept (Schank, 2006; Strassmann, 1997; Wilson, 1992).

Pioneering behavioral studies of mammalian taste function began in the 1930s, heralded by experiments that sought to explain so-called specific hungers, e.g., salt craving in patients with adrenal gland hypofunction. In seeking to determine whether alterations in taste function are responsible for increased NaCl intake of adrenalectomized rats, Richter (1936, 1939) developed the two-bottle taste test (see also Richter and Campbell, 1940). In this test, differential fluid intake from two bottles, one of which contains a tastant (e.g., a NaCl solution) and the other water alone, is recorded over a period of time. The lowest concentration of the tastant that produces a differential intake is taken as the threshold measure.

Although this behavioral procedure provided a means for measuring a preference threshold, postingestional factors may alter the behavioral response and such a threshold is conceptually different from a sensory threshold. Thus, a lack of preference between two solutions need not reflect an inability to discriminate between them [see Stevens (1975) for reviews of analogous procedures for olfaction]. Subsequent workers, including Carr (1952), Harriman and MacLeod (1953), Morrison (1967), and Morrison and Norrison (1966), utilized shock avoidance paradigms or operant conditioning paradigms that provided positive reinforcement to establish NaCl threshold values – values that were much lower than those obtained using Richter's procedure and which corresponded more closely to neural thresholds. Numerous modifications of behavioral procedures for assessing taste function in mammals have since been developed which incorporate general principles that evolved from these pioneering behavioral studies (e.g., Brosvic et al., 1985, 1989; Spector et al., 1990). Analogous procedures have been developed in olfaction (e.g., Bowers and Alexander, 1967; Braun et al., 1967; Braun and Marcus, 1969; Eayrs and Moulton, 1960; Goff, 1961; Henton, 1969; Moulton, 1960; Moulton and Eayrs, 1960; Pfaffmann et al., 1958; Slotnick and Katz, 1974; Slotnick and Ptak, 1977; Slotnick and Schellinck, 2002). Another noteworthy development in behavioral testing was that of the conditioned aversion paradigm (Garcia et al., 1955, 1974). In one variant of this technique, an animal is allowed to drink or smell a novel tastant or odorant and is then injected with an agent that produces nausea (e.g., lithium chloride). The animal quickly learns to avoid the novel stimulus as a result of a single aversive conditioning experience, even if the aversion occurs long after the presentation of the sensory stimulus. This procedure can be used to establish whether detection of a given stimulus is present and is particularly useful for assessing cross-reactivity of stimuli (i.e., the extent to which a stimulus has elements in common with other stimuli). One of the more novel applications of this technique was by Smotherman (1982), who demonstrated that the olfactory system of rats is functional in utero. In this study, unborn rat pups (gestation day 20) received in utero injections of apple juice and lithium chloride. After birth, these individuals showed evidence of having developed a conditioned aversion to the odor of apple juice (see Chapter 14).

1.5.9 Clinical Chemosensory Studies

Considerable progress in understanding chemosensory disorders has been made in the last few decades, as reviewed in detail in Chapters 16–20, 28, and 36–40. The proliferation of clinical studies has been fueled, in large part, by the development wide-spread commercial availability of standardized psychophysical tests of olfactory function that first occurred in the early 1980s (e.g., Doty et al., 1984a, Doty, 2000, 2001; Kobal et al., 1996; Hummel et al., 1997; for review, see Doty, 2007). It is now widely appreciated that smell loss is markedly depressed in elderly persons (Doty et al., 1984b), and that most common causes of *permanent* smell loss are (1) upper respiratory viral infections, (2) head trauma, and (3) nasal and sinus disease (e.g., Deems et al., 1991). Moreover, it appears that these disorders largely reflect damage to the olfactory neuroepithelium, as revealed by autopsy and biopsy studies (Douek et al., 1975; Hasegawa, Yamagishi and Nakano, 1986; Jafek et al., 1989, 1990; Moran et al., 1992). Most complaints of

taste loss reflect the loss of olfactory function, and flavor sensations are largely derived from retronasal stimulation of the olfactory system during active deglutition (Mozell et al., 1969; Burdach and Doty, 1987).

We now know that the olfactory system seems more susceptible to damage than the taste system, although damage to regional lingual afferents is particularly striking in old age (Matsuda and Doty, 1995), and taste sensitivity is directly related to the number of taste buds or papillae stimulated, regardless of whether stimulation is by chemicals or by electrical current (Doty et al., 2001; Miller et al., 2002; Zuniga et al., 1993). Moreover, it has become increasingly apparent that many medicines, including a number of antibiotics, antidepressants, antihypertensives, antilipid agents, and psychotropic drugs, can produce alterations of the taste system (e.g., severe dysgeusia), alone or in combination with alterations in the smell system (Schiffman, 1983; Schiffman et al., 1998, 1999a,b, 2000). Importantly, recent studies suggest that damage to one of the major taste nerves (e.g., one chorda tympani) may release inhibition on other taste nerves (e.g., the contralateral glossopharyngeal nerve), resulting in hypersensitivity to some tastants and the production of phantom dysgeusias (Lehman et al., 1995; Yanagisawa et al., 1998).

A major advance in the last few years is the discovery that smell loss is among the first, if not the first, signs of such common neurodegenerative diseases as Alzheimer's disease (AD) and idiopathic Parkinson's disease (PD), and that disorders sharing similar motor signs, such as progressive supranuclear palsy (PSP) and MPTP-induced Parkinsonism (MPTP-P), are largely unaccompanied by such loss (see Chapter 18). Such observations imply that olfactory testing can be of value not only in the detection of some neurodegenerative disorders early in their development, but in differential diagnosis. Indeed, odor identification testing accurately differentiates between patients with AD and those with major affective disorder (i.e., depression) (Solomon et al., 1998; McCaffrey et al., 2000). Interestingly, longitudinal studies have now appeared indicating that olfactory dysfunction can be predictive of AD in individuals who are at risk for this disorder, particularly when considered in relation to other risk factors (Bacon et al., 1998; Graves et al., 1999; Devanand et al., 2000). The only neurodegenerative disorder for which a definitive physiological basis has been found to date, however, is multiple sclerosis, where a -0.94 correlation has been observed between odor identification test scores and the number of plaques, as measured by MRI, in the subtemporal and subfrontal regions of the brain (Doty et al., 1997, 1998, 1999).

Among the many diseases or disorders in addition to those noted above that are associated with smell dysfunction include severe alcoholism, amyotrophic lateral sclerosis (ALS), chronic obstructive pulmonary disease, cystic fibrosis, epilepsy, the Guam ALS/PD complex, head trauma, Huntington's disease, Kallmann's syndrome, Korsakoff's psychosis, myasthenia gravis, pseudohypoparathyroidism, psychopathy, restless leg syndrome, schizophrenia, seasonal affective disorder, and Sjogren's syndrome. Neurological disorders in which olfactory seems to be spared in addition to PSP and MPTP-induced PD are corticobasal degeneration, depression, panic disorder, essential tremor, and multiple chemical hypersensitivity (for review, see Doty, 2012). In addition to traditional medical means for treating or managing diseases responsible for decreased olfactory function, surgical intervention at the level of the olfactory neuroepithelium (e.g., by selectively ablating or stripping away the diseased tissue) or the olfactory bulb (e.g., by removal of one or both olfactory bulbs in an anterior cranial approach) has successfully eliminated or markedly reduced the symptoms of some forms of chronic dysosmia or phantosmia (see Chapter 27). Recent advances in understanding the deleterious influences of oxygen radicals on neural tissue, as well as changes that occur in olfactory tissue at menopause, have led to ongoing studies of the prophylactic potential of antioxidants, hormones, and other agents in mitigating toxin-induced damage to the olfactory (e.g., Dhong et al., 1999).

1.6 CONCLUSIONS

In this introduction, a brief description of the significant role that tastes and odors have played throughout the course of human history has been presented. In addition, a number of key studies, events, and trends have been identified which form the backdrop of much of today's chemosensory research enterprise, providing perspective for the chapters that follow. The chapters of this volume provide a detailed contemporary information related to most of these trends and address the important role of chemosensory science in both basic and applied (e.g., clinical) situations. Until recently, the chemical senses have engendered, relative to the other major senses, comparatively little attention on the part of the scientific and medical communities. This has clearly changed, as reflected in the chapters that follow.

REFERENCES

Adams, D. R. (1978). The bovine vomeronasal organ. *Arch. Histol. Jpn.*, 49: 211–225.

Adrian, E. D. (1953). Sensory messages and sensation; the response of the olfactory organ to different smells. *Acta Physiol. Scand.*, 26: 5–14.

Allen, E. (1912). The cessation of mitosis in the central nervous system of the albino rat. *J. Comp. Neurol.* 22: 547–569.

Albone, E. S. (1984). *Mammalian Semiochemistry*. Wiley, New York.

Althaus, J. (1881). A lecture on the physiology and pathology of the olfactory nerve. *Lancet*, 1: 771–773; 813–815.

Altman, J. (1963). Autoradiographic investigation of cell proliferation in the brains of rats and cats. Postnatal growth and differentiation of the mammalian brtain, with implications for a morphological study of memory. *Anat. Rec.* 145: 573–591.

Altman, J. (1966a). Autoradiographic and histological studies of postnatal neurogenesis. II. A longitudinal investigation of the kinetics, migration and transformation of cells incorporating tritiated thymidine in infant rats, with special reference to neurogenesis in some brain regions. *J. Comp. Neurol.* 128: 431–473.

Altman, J. (1966b). Proliferation and migration of undifferentiated precursor cells in the rat during postnatal gliogenesis. *Exp. Neurol.* 16: 263–278.

Altman, J. (1969). Autoradiogeraphic and histological studies of postnatal neurogenesis: IV. Cell proliferation and migration in the anterior forebrain, with special reference to persisting neurogenesis in the olfactory bulb. *J. Comp. Neurol.* 137: 433–458.

Altman, J. and Das, G. D. (1965). Autoradiographic and histological evidence of postnatal hypocampal neuroregenesis in rats. *J. Comp. Neurol.* 124: 319–335.

Altman, J. and Das, G. D. (1966a). Autoradiographic and histological studies of postnatal neurogenesis. I. A longitudinal investigation of the kinetics, migration and transformation of cells incorporating tritiated thymidine in neonate rats, with special reference to postnatal neurogenesis in some brain regions. *J. Comp. Neurol.* 126: 337–390.

Andres. K. H. (1966). Der Feinbau der Regio olfactoria von Makrosmatikem. *Z. Zellforsch.* 69: 140–154.

Andres, K. H. (1969). Der olfaktorische Saum der Katze. *Z. Zellforsch.* 96: 250–274.

Armstrong, C., and Harrison, W. T. (1935). Prevention of intranasally inoculated poliomyelitis of monkeys by instillation of alum into nostrils. *Public Health Rep.* 50: 725–730.

Avenet, P., and Lindemann, B. (1987). Patch–clamp study of isolated taste receptor cells in the frog. *J. Membr. Biol.* 97: 223–240.

Avenet, P., and Lindemann, B. (1990). Fluctuation analysis in the amiloride-blockable currents in membrane patches excised from salt-taste receptor cells. *J. Bas. Gin. Physiol. Pharmacol.* l: 383–391.

Bacon, A. W., Bondi, M. W., Salmon, D. P., and Murphy, C. (1998). Very early changes in olfactory functioning due to Alzheimer's disease and the role of apolipoprotein E in olfaction. *Ann. NY Acad. Sci.*, 855: 723–731.

Bailey, K. C. (1932). *The Elder Pliny's Chapters on Chemical Subjects*. Vol. II. E. Arnold and Company, London.

Barrows, W. M. (1907). The reactions of the pomace fly, *Drosophila ampelophila* Loew, to odorous substances. *J. Exp. Zool.* 4: 515–537.

Bartoshuk, L. M. (1978). History of taste research. In *Handbook of Perception*. Vol. *VI A. Tasting and Smelling*, E. C. Carterette and M. P. Friedman (Eds.). Academic Press, New York, pp. 3–17.

Bartoshuk, L. M. (1988). Taste. In Stevens' Handbook of Experimental Psychology. Vol. 1. *Perception and Motivation*, R. C. Atkinson, R. J. Herrnstein, G. Lindzey, and R. D. Luce (Eds.). Wiley, New York, pp. 461–499.

Beach, F. A. and Gilmore, R. W. (1949). Response of male dogs to urine from females in heat. *J. Mammal.* 30: 391–392.

Bean, N. J. (1982). Modulation of agonistic behavior by the dual olfactory system in male mice. *Physiol. Behav.* 29: 433–437.

Bean, N. J., and Wysocki, C. J. (1985). Behavioral effects of removal of the vomeronasal organ in neonatal mice. *Chem. Senses* 10: 421–422.

Beauchamp, G. K. (2009). Sensory and receptor responses to Umami: an overview of pioneering work. *Am. J. Clin. Nutr.* 90: 7235–7275.

Beidler, L. M., Fishman, I. Y. and Hardiman. Species differences in taste responses. *Am. J. Physiol.* 181: 235–239.

References

Beidler, L. M. (1971a) (Ed.), *Handbook of Sensory Physiology Vol. IV. Chemical Senses. Part 1 Olfaction.* Springer-Verlag, New York.

Beidler, L. M. (1971b) (Ed.), *Handbook of Sensory Physiology. Vol. IV. Chemical Senses. Part 2. Taste.* Springer-Verlag, New York.

Beidler, L. M., and Tucker, D. (1955). Response of nasal epithelium to odor stimulation. *Science* 122: 76.

Beidler, L. M., and Smallman, R. L. (1965). Renewal of cells within taste buds. *J. Cell Biol.* 27: 263–272.

Bell, C. (1803). On the sense of tasting. In: The anatomy of the human body. London: A. Strahan.

Bellingeri, C. F. (1818). Dissertatio inauguralis, quam publice defendebat in reio Atheneo Anno 1818 die IX maji Augustae Taurinorum.

Bellini, L. (1665). Gustus organum novissime deprehensum praemissis ad faciliorem intelligentiam quibusdam de saporibus [Taste organs newly observed; with certain premises about the senses given for easier understanding (translation from Latin into German by Jurisch (1922)]. *Mangetus Bibliotheca Anat. 2*, Bologna.

Benjamin, R. M. and Burton, H. (1968). Projection of taste nerve afferents to anterior opercular-insular cortex in squirrel monkey (Saimiri sciureus). *Brain Res.* 7 : 221–231.

Benton, R., Sachse, S., Michnick, S. W., and Vosshall, L. B. (2006). Atypical membrane topology and heteromeric function of Drosophila odorant receptors in vivo. *PLoS Biol.* 4: 240–257.

Bernard, C. (1856). *Leçons de physiologie expérimentale. Appliwuée à la Médecine.* Vol II. Bailliére, Paris.

Bhatnagar, K. P. and Meisami, E. (1998) Vomeronasal organ in bats and primates: extremes of structural variability and its phylogenetic implications. *Microsc. Res. Tech.* 43: 465–475.

Billing, J. and Sherman, P. W. (1998). Antimicrobial functions of spices: why some like it hot. *Quart. Rev. Biol.* 73: 3–49.

Block, F. (1946). Nuclear introduction. *Physiol. Rev.* 70: 460–474.

Bodian, D. and Howe, H. A. (1941a). Experimntal studies on intraneural spread of poliomyelitis virus. *Bull. John Hopkins Hosp.* 68: 248–267.

Bodian, D. and Howe, H. A. (1941b). The rate of progression of poliomyelitis virus in nerves. *Bull. Johns Hopkins Hosp.* 69: 79–85.

Boistel, J. (1953). Etude fonctionelle des terminaisons sensorielles des antennes d'Hyménoptères. *C. R. Soc. Biol.* 147: 1683-1688.

Boistel, J., and Coraboeuf, E. (1953). L'activité électique dans I'antenne isolée de Léidoptè au cours de l'étude de l'olfaction. *C. R. Soc. Biol.* 147: 1172-1175.

Boorstin, D. J. (1985). *The discoverers.* N.Y.: Vintage Books.

Bordeu, T. D. (1751). *Recherches anatomiques sur la position des glandes et sur leur action.* Brosson, Paris.

Boring, E. G. (1942). *Sensation and Perception in the History of Experimental Psychology.* Appleton Century-Crofts, New York.

Bowers, J. M., and Alexander, B. K. (1967). Mice: individual recognition by olfactory cues. *Science* 158: 1208–1210.

Boyden, E. S., Zhang, F., Bamberg, E., et al. (2005). Millisecond-timescale, genetically targeted optical control of neural activity. *Nat. Neurosci.* 8: 1263–1268.

Bradley, R. M., and Beidler, L. M. (2003). Saliva: Its role in taste function. In *Handbook of Olfaction and Gustation, 2nd,* R. L. Doty (Ed.). Marcel Dekker, New York, pp. 639–650.

Braun, J. J. and Marcus, J. (1969). Stimulus generalization among odorants by rats. *Physiol. Behav.* 4: 245.

Braun, J. J., Wermuth, B. and Haberly, L. (1967). An olfactory discrimination apparatus: modification of the basic wind-tunnel design. *Psychonomic Sci.* 9: 515.

Broca, P. (1878). Anatomie comparée des circonvolutions cérebrales. Le grand lobe linbique et la scissure dans la série des mammifères. *Rev. Anthropol. Ser.* 2. I: 385–498.

Broca, P. (1879). Sur les temperatures morbides locales. *Bull. Acad. Med.* (Paris). 28: 1331–1347.

Bronshtein, A. A., and Minor, A. V. (1977). The regeneration of olfactory flagella and restoration of electro-olfactogram after the treatment of the olfactory mucosa with Triton X–100. *Cytology* 19: 33–39.

Brosvic, G. M., Slotnick, B. M., and Tandeciarz, S. (1985). A computer-controlled automated system for gustatory psychophysics. *Chem. Senses* 10: 447.

Brosvic, G. M., Risser, J. M., and Doty, R. L. (1989). No influence of adrenalectomy on measures of taste sensitivity in the rat. *Physiol. Behav.* 46: 699–705.

Brown, K. (1979). Chemical communication between animals. In *Chemical Influences on Behaviour*, K. Brown, and S. J. Cooper (Eds.). Academic Press, London, pp. 599–649.

Brown, R. E. and Macdonald, D. W. (1985). *Social Odours in Mammals* (2 vol.). Clarendon Press, Oxford, 1985.

Bruce, H. M. (1959). An exteroceptive block to pregnancy in the mouse. *Nature* (Lond.) 814: 105.

Bruce, H. M., and Parrott, D. V. M. (1960). Role of olfactory sense in pregnancy block by strange males. *Science* 131: 1526.

Bruch, R. C., and Teeter, J. H. (1989). Second messenger signalling mechanisms in olfaction. In *Chemical Senses: Receptor Events and Transduction in Taste and Olfaction*, J. G. Brand, J. H. Tetter, M. R. Kam, and R. H. Cagan (Eds.). Marcel Dekker, New York, pp. 283–298.

Buck, L., and Axel, R. (1991). A novel multigene family may encode odorant receptors: a molecular basis for odor recognition. *Cell* 65: 175–187.

Budge, E.A. (1960). *The Book of the Dead.* University Books, New Hyde Park, NY.

Bulfone, A., Wang, F., Hevner, R., et al. (1998). An olfactory sensory map develops in the absence of normal projection neurons or GABAergic interneurons. *Neuron* 21: 1273–1282.

Burdach, K. and Doty, R. L. (1987). Retronasal flavor perception: Influences of mouth movements, swallowing and spitting. *Physiol. Behav.* 41: 353–356.

Burket, W. (1970). Jason, hypsipyle, and new fire at Lemnos. A study in myth and ritual. *Class. Q.* 20: 1–16.

Burnet, F. M., and Lush, D. (1938). Infection of the central nervous system by louping ill virus. *Aust. J. Exp. Biol. Med. Sci.* 16: 233–240.

Cain, W. S. (1978). History of research on smell. In *Handbook of Perception. Vol. VIA. Tasting and Smelling*, E. C. Carterette and M. P. Friedman (Eds.). Academic Press, New York, pp. 197–229.

Cajal, R. S. (1889). *Nuevos aplicaciones del metodo de Golgi: terminaciones del nervio olfactorio en la mucosa nasal.* Barcelona.

Calof, A. L. and Chikaraishi, D. M. (1989) Analysis of neurogenesis in a mammalian neuroepithelium: proliferation and differentiation of an olfactory neuron precursor in vitro. *Neuron* 3: 115–127.

Carlson, J. R. (2001). Viewing odors in the mushroom body of the fly. *Trends Neurosci.* 24: 497–498.

Carr, W. J. (1952). The effect of adrenalectomy upon the NaCl taste threshold in rat. *J. Comp. Physiol. Psychol.* 45: 377–380.

Carr, W. J. and Caul, W. F. (1962). The effect of castration in rat upon the discrimination of sex odours. *Anim. Behav.* 10: 20–27.

Carr, W. J., Solberg, B., and Pfaffmann, C. (1962). The olfactory threshold for estrous female urine in normal and castrated male rats. *J. Comp. Physiol. Psychol.* 55: 415–417.

Carr, W. J., Loeb, L. S., and Dissinger, M. E. (1965). Responses of rats to sex odors. *J. Comp. Physiol. Psychol.* 59: 370–377.

Casserius (1609). *Penthaesteseion* (cited after Jurisch, 1922).

Chang, R.B., Waters, H., and Liman, E.R. (2010). A proton current drives action potentials in genetically identified sour taste cells. *Proc. Natl. Acad. Sci. U.S.A.* 107: 22320–22325.

Chaudhari, N., and Roper, S. D. (2010). The cell biology of taste. *J. Cell Biol.* 190: 285–296.

Choi, G. B., Stettler, D. D., Kallman, B. R., et al. (2011). Driving opposite behaviors with ensembles of piriform neurons. *Cell* 146: 1004–1015.

Chuah, M. I. and Au, C. (1991), Olfactory schwann cells are derived from precursor cells in the olfactory epithelium. *J. Neurosci. Res.* 29: 172–180.

Chuah, M. I. and Farbman, A. I. (1983). Olfactory bulb increases marker protein in olfactory receptor cells. *J. Neurosci.* 3: 2197–2205.

Chaudhari, N., Landin, A. M. and Roper, S. D. (2000) A novel metabotropic glutamate receptor functions as a taste receptor. *Nat. Neurosci.*, 3: 113–119.

Clarke, J. L. (1861). Über den Bau des Bulbus olfactorius and der Geruchss-chleimhaut. *Z. wiss. Zool.* 11: 31–42.

Clark, W. E. L. (1929). Anatomical investigation into the routes by which infections may pass from the nasal cavities into the brain. *Rep. Public Health Med. Subjects No. 54.* London, pp. 1–27.

Cloquet, H. (1821). *Osphrèsiologie, ou Traitè,4 des Odeurs, du sens et des organes de l'olfaction.* 2nd ed. Mèquignon-Marvis, Paris.

Contarini, G. (1548). *De elementis [et] eorum mixtionibus libri quinque.* Jean de Gaigny, Paris.

Coquelin, A., Clancy, A. N., Macrides, F., Nobel, E. P., and Gorski, R. A. (1984). Pheromonally induced release of luteinizing hormone in male mice: involvement of the vomeronasal system. *J. Neurosci.* 4: 2230–2236.

Corbin, A. (1986). *The Foul and the Fragrant.* Harvard University Press, Cambridge, MA.

Darwin, C. R. (1871) *The Descent of Man and Selection in Relation to Sex.* N.Y.: Appleton & Co., p. 23.

DaVanzo, J. P., Sydow, M., and Garris, D. R. (1983). Influence of isolation and training on fighting in mice with olfactory bulb lesions. *Physiol. Behav.* 31: 857–860.

Deems, D. A., Doty, R. L., Settle, R. G., et al. (1991). Smell and taste disorders: a study of 750 patients from the University of Pennsylvania Smell and Taste Center. *Arch. Otolaryngol. Head. Neck Surg.* 117: 519–528.

de Graaf, R. (1677). De succo pancreatico. In *Opera Omnia.* Hackiana, Amsterdam.

DeHamer, M. K., Guevara, J. L., Hannon, K., et al. (1994). Genesis of olfactory receptor neurons in vitro: regulation of progenitor cell divisions by fibroblast growth factors. *Neuron* 13, 1083–1097.

de Lorenzo, A. J. (1958). Electron microscopic observations on the taste buds of the rabbit. *J. Biophys. Biochem. Cytol.* 4: 143–150.

Dethier, V. G. (1941). The function of the antennal receptors in lepidopterous larvae. *Biol. Bull.* 80: 403–414.

Devanand, D. P., Michaels-Marston, K. S., Liu, X., et al. (2000). Olfactory deficits in patients with mild cognitive impairment predict Alzheimer's disease at follow-up. *Amer. J. Psychiat.* 157: 1399–1405.

Dhong, H. J., Chung, S. K., and Doty, R. L. (1999). Estrogen protects against 3-methylindole-induced olfactory loss. *Brain Res.* 824: 312–315.

Doty, R. L. (1974). A cry for the liberation of the female rodent. *Psychol. Bull.* 81: 159–172.

Doty, R. L. (1975). Determination of odour preferences in rodents: a methodological review. In *Methods in Olfactory Research*, D. G. Moulton, A. Turk, and J. W. Johnson, Jr., (Eds.). Academic Press, London, pp. 395–406.

Doty, R. L., Ed. (1976). *Mammalian Olfaction, Reproductive Processes, and Behavior.* Academic Press, New York.

Doty, R. L. (1980). Scent marking in mammals. In *Comparative Psychology: Research in Animal Behavior*, M. R. Denny (Ed.). Wiley, New York, pp. 385–399.

Doty, R. L. (1981). Olfactory communication in humans. *Chem. Senses* 6: 351–376.

Doty, R. L. (1986). Odor-guided behavior in mammals. *Experientia* 42: 257–271.

Doty, R. L. (1998). Cranial Nerve I: Olfaction. In: C. G. Goetz and E. J. Pappert (Eds.), *Textbook of Clinical Neurology.* Philadelphia: W.B. Saunders, pp. 90–101.

Doty, R. L. (2001). Olfaction. *Annu. Rev. Psychol.* 52: 423–452.

Doty, R. L. (2007). Office procedures for quantitative assessment of olfactory function. *Am. J. Rhinol.* 21: 460–473.

Doty, R. L. (2010). *The Great Pheromone Myth.* Baltimore: Johns Hopkins University Press.

Doty, R. L. (2012). Olfaction in Parkinson's disease and related disorders. *Neurobiol. Dis.* 46: 527–552.

Doty, R. L., and Anisko, J. J. (1973). Procaine hydrochloride olfactory block eliminates mounting in the male golden hamster. *Physiol. Behav.* 10: 395-397.

Doty, R. L., Bagla, R., Morgenson, M., and Mirza, N. (2001). NaCl thresholds: relationship to anterior tongue locus, area of stimulation, and number of fungiform papillae. *Physiol. Behav.* 72: 373–378.

Doty, R. L, Carter, C. S., and Clemens, L. G. (1971). Olfactory control of sexual behavior in male and early-androgenized female hamsters. *Horm. Behav.* 2: 325-333.

Doty, R. L. and Ferguson-Segall, M. (1989). Influence of adult castration on the olfactory sensitivity of the male rat: a signal detection analysis. *Behav. Neurosci.* 103: 691-694.

Doty, R. L., Kreiss, D. S., and Frye, R. E. (1990). Human odor intensity perception: correlation with frog epithelial adenylate cyclase activity and transepithelial voltage response. *Brain Res.* 527: 130-134.

Doty, R. L., Li, C., Mannon, L. J., and Yousem, D. M. (1997). Olfactory dysfunction in multiple sclerosis. *New Engl. J. Med.* 336: 1918–1919.

Doty, R. L., Li, C., Mannon, L. J., and Yousem, D. M. (1998). Olfactory dysfunction in multiple sclerosis: Relation to plaque load in inferior frontal and temporal lobes. *Ann. N. Y. Acad. Sci.* 855: 781–786.

Doty, R. L., Li, C., Mannon, L. J., and Yousem, D. M. (1999). Olfactory dysfunction in multiple sclerosis: relation to longitudinal changes in plaque numbers in central olfactory structures. *Neurology*, 53: 880–882.

Doty, R. L., Shaman, P. and Dann, M. (1984a). Development of the University of Pennsylvania Smell Identification Test: A standardized microencapsulated test of olfactory function. *Physiol. Behav.* (Monograph) 32: 489–502.

Doty, R. L., Shaman, P., Applebaum, S. L., et al. (1984b). Smell identification ability: Changes with age. *Science* 226: 1441–1443.

Doucette, J. R. (1984) The glial cells in the nerve fibre layer of the rat olfactory bulb. *Anat. Rec.* 210: 385–391.

Douek, E. (1974). *The Sense of Smell and its Abnormalities.* Livingstone, Edinburgh.

Douek, E., Banniester, L. H. and Dodson, H. C. (1975). Recent advances in the pathology of olfaction. *Proc. Roy. Soc. Med.* 68: 467–470.

Dving, K.B. and Pinching, A.J. (1973). Selective degeneration of neurons in the olfactory bulb following prolonged odour exposure. *Brain Res.* 52: 115–129.

Dyer, J., Salmon, K. S., Zibrik, L., and Shirazi-Beechey, S. P. (2005). Expression of sweet taste receptors of the T1R family in the intestinal tract and enteroendocrine cells. *Biochem. Soc. Trans.* 33: 302–305.

Eayrs, J. T., and Moulton, D. G. (1960). Studies in olfactory acuity. 1. Measurement of olfactory thresholds in the rat. *Q. J. Exp. Psychol.* 12: 90–98.

References

Ecker, A. (1856). Über die Geruchschleimhaut des Menschen. Über das Epithelium der Riechschleimhaut und die wahrscheinliche Endigung des Geruchnerven. *Z. wiss. Zool.* 8: 303-306.

Eckhard, C. (1858). Über die Endigungsweise des Geruchsnerven. *Beitr. Anat. Physiol. (Eckhard) Geissen* 1: 77-84.

Ehrlich, P. (1886). Uber die Methylenblaureaction der lebenden Nerven substanz. *Deutsche Med. Wochenschr.* 12: 49-52.

Emmers, R., Benjamin, R. M., and Blomquist, A. J. (1962). Thalamic localization of afferents from the tongue in albino rat. *J. Comp. Neurol.* 118: 43-48.

Engström, H., and Rytzner, C. (1956a). The structure of taste buds. *Ann. Otolaryngol.* 46: 361-367.

Engström, H., and Rytzner, C. (1956b). The fine structure of taste buds and taste fibres. *Ann Oto. Rhinol. Laryngol.* 65: 361-375.

Erlanger, J. and Gasser, H. S. (1937). *Electrical Signs of Nervous Activity*. University of Pennsylvania Press, Philadelphia.

Exner, S. (1878). Fortgesetzte Studien über die Endigungsweise des Geruchsnerven. *Sitzber. Akad. Wiss. Wien Math.-naturw. Klasse* 76-III: 171-221.

Farb, P., and Armelagos, G. (1980). *Consuming Passions: The Anthropology of Eating*. Houghton Mifflin, Boston.

Farbman, A. I. (1977). Differentiation of olfactory receptor cells in organ culture. *Anat. Rec.* 189: 187–200.

Farbman, A. I. (1992). *Cell Biology of Olfaction*. Cambridge University Press, Cambridge.

Farbman, A. I. and Margolis, F. L. (1980). Olfactory marker protein during ontogeny: Immunohistochemical localization. *Dev. Biol.* 74: 205–215.

Feng, P., Zheng, J., Rossiter, S.J. et al. (2014). Massive losses of taste receptor genes in toothed and baleen whales. *Genome Biol. Evol.* 6: 1254–1265.

Féron, F., Perry, C., Hirning, M., et al. (1999). Altered adhesion, proliferation and death in neural cultures from adults with schizophrenia. *Schiz. Res.* 40: 211–218.

Féron, F., Perry, C., McGrath, J. and Mackay-Sim, A. (1998). New techniques for biopsy and culture of human olfactory epithelial neurons. *Arch. Otolaryngol. Head Neck Surg.* 124: 861–866.

Firestein, S., and Werblin, F. (1989). Odor-induced membrane currents in vertebrate olfactory receptor neurons. *Science* 244: 79-82.

Firestein, S., Darrow, B., and Shepherd, G. M. (1991). Activation of the sensory current in salamander olfactory receptor neurons depends on a G protein-mediated cAMP second messenger system. *Neuron*, 6: 825–835.

Fischer. E., and Penzoldt, F. (1886). Über die Empfindlichkeit des Geruchssinnes. *Sitzber. physik.-med. Sozietät Erlangen* 18: 7–10.

Foerster, O. (1936). Sensible corticale Felder. *Bumke und Foersters Handb. Neurol.*, 6, 358–448.

Fourcroy, A. F. (1798). Mèmoire sur l'esprit recteur de Boerrhaave, l'arome des chimistes francais, ou le principe de l'odeu des vègètaux. *Ann. Chim.* 26: 232-250.

Fox, P.T. and Raichle, M. E. (1986). Focal physiological uncoupling of cerebral blood flow and oxidative metabolism during somatosensory stimulation in human subjects. *Proc. Natl. Acad. Sci.* USA 83: 1140–1144.

Frank, M. E. (1973). An analysis of hamster afferent taste nerve response functions. *J. Gen. Physiol.* 61: 588–623.

Frank, M. E. (2000). Neuron types, receptors, behavior, and taste quality. *Physiol. Behav.* 69: 53–62.

Freygang, W. H. and Sakoloff, L. (1958). Quantitative measurements of regional circulation in the central nervous system by the use of radioactive inert gas. *Adv. Biol. Med. Physics* 6: 263–279.

Frye, R. E., Schwartz, B., and Doty, R. L. (1990). Dose-related effects of cigarette smoking on olfactory function. *JAMA* 263: 2133-2136.

Fulle, H. J., Vassar, R., and Foster, D. C., et al. (1995). A receptor guanylyl cyclase expressed specifically in olfactory sensory neurons. *Proc. Natl. Acad. Sci. USA* 92: 3571–3575.

Fulton, J. F. (1928). Observations upon the vascularity of the human occipital lobe during visual activity. *Brain* 51: 310–320.

Funakoshi, M., and Kawamura, Y. (1968). Summated cortical responses to taste stimulation in man. *J. Physiol. Soc. Jpn.* 30: 282-283.

Funakoshi, M., and Kawamura, Y. (1971). Summated cerebral evoked response to taste stimuli in man. *Electroenceph. Clin. Neurophysiol.* 30: 205–209.

Garcia, J., Kimeldorf, D., and Koelling, R. A. (1955). A conditioned aversion towards saccharin resulting from exposure to gamma radiation. *Science* 122: 157-159.

Garcia, J., Hankins, W. G., and Rusiniak, K. W. (1974). Behavioral regulation of the milieu interne in man and rat. *Science* 185: 824-831.

Garrett, J. R. (1998). Historical introduction to salivary secretion. In *Glandular Mechanisms of Salivary Secretion*, J. R. Garrett, J. Ekström, and L. C. Anderson (Eds.). Karger, Basel, pp. 1–20.

Garten, S. (1900). *Physiologie der marklosen Nerven*. G. Fischer, Jena.

Gesteland, R. C., Leavin, J. Y., Pitts, W. H., and Rojas, A. (1963). Odor specificities of the frog's olfactory receptors. In *Olfaction and Taste,* Y. Zotterman (Ed.), Pergamon Press, Oxford, pp. 19–44.

Gesteland, R. C., Lettvin, J. Y., and Pitts, W. H. (1965). Chemical transmission in the nose of the frog. *J. Physiol.* 181: 525–559.

Getchell, T. V., and Shepherd, G. M. (1975). Short-axon cells in the olfactory bulb: dendrodendritic synaptic interactions. *J. Physiol.* 251: 523–548.

Gheusi, G., Cremer, H., McLean, H., et al. (2000). Importance of newly generated neurons in the adult olfactory bulb for odor discrimination. *Proc. Nat. Acad. Sci. USA* 97: 1823–1828.

Glatz, R., and Bailey-Hill, K. (2011). Mimicking nature's noses: from receptor deorphaning to olfactory biosensing. *Prog. Neurobiol.* 93: 270–296.

Globus, J. H. and Kuhlenbeck, H. (1944). The subependymal cell plate (matrix) and its relationship to brain tumors of he ependymal type. *J. Neuropath. Exp. Neurol.* 3: 1–35.

Gloor, P. (1997). *The Temporal Lobe and Limbic System*. New York, Oxford University Press.

Glusman, G., Yanai, I., Rubin, I. and Lancet, D. (2001). The complete human olfactory subgenome. *Genome Res.* 11: 685–702.

Godfrey, J. (1958). The origin of sexual isolation between bank voles. *Proc. Roy. Phys. Soc. Edinburgh* 27: 47–55.

Goff, W. R (1961). Measurement of absolute olfactory sensitivity in rats. *Am. J. Psychol.* 74: 384–393.

Good, J. M. (1822). *The Study of Medicine*. Vol. III. Baldwin, Cadock and Joy: London.

Gould, E. and Tanapat, P. (1999). Stress and hippocampal neurogenesis. *Biol. Psychiat.* 46: 1472–1479.

Graves, A. B., Bowen, J. D., Rajaram, L., et al. (1999). Impaired olfaction as a marker for cognitive decline: interaction with apolipoprotein E epsilon4 status. *Neurology*, 53: 1480–1487.

Graziadei, P. P. C. (1971). The olfactory mucosa of vertebrates. In *Handbook of Sensory Physiology, Vol. IV. Chemical Senses. Sect. 2. Taste*. L. M. Beidler (Ed.). Springer-Verlag, New York, pp. 27–58.

Graziadei, P. P. C. and Metcalf, J. E. (1971). Autoradiographic and ultrastructural observations on the frog's olfactory mucosa. *Z. Zellforsch.* 116: 305–318.

Gross, C. G. (2000). Neurogenesis in the adult brain: death of a dogma. *Nat. Rev. Neurosci.* 1: 67–73.

Gryllus, L. (1566). *De sapore dulci et amaro.* Prague: Georgium Melantrichum ab Auentino.

Haldane, E. S., and Ross, G. R. T. (1955). *Philosophical Works Rendered into English.* Dover, New York.

Haller, A. V. (1744). *Physiology (trans. S. Mihles).* Innys and Richardson, London.

Harper, R., Bate Smith, E. C., and Land, D. G. (1968). *Odour Description and Odour Classification.* American Elsevier, New York.

Harriman, A. E., and MacLeod, R. B. (1953). Discriminative thresholds of salt for normal and adrenalectomized rats. *Am. J. Psychol.* 66: 465-471.

Harrington, A., and Rosario, V. (1992). Olfaction and the primative: nineteeth-century medical thinking on olfaction. In *Science of Olfaction,* M. J. Serby and K. L. Chobor (Eds.). Springer-Verlag, New York, pp. 3-27.

Hasegawa, S., Yamagishi, M., and Nakano, Y. (1986). Microscopic studies of human olfactory epithelia following traumatic anosmia. *Arch. Otolaryngol.* 243: 112–116.

Hashiguchi, Y. and Nishida, M. (2007). Evolution of trace amine-associated receptor (TAAR) gene family in vertebrates: Lineage-specific expansions and degradations of a second class of vertebrate chemosensory receptors expressed in the olfactory epithelium. *Mol. Biol. Evol.* 24: 2099–2107.

Hatton, E. (1708). *A New View of London; or, An Ample Account of That City, in … Eight Sections,* 2 vols. London, 1: 30. Cited in Jenner, M. S. (2011). Follow your nose? Smell, smelling, and their histories. *Am. Hist. Rev.* 116: 335–351.

Hauser, G. (1880). Physiologische und histologische Untersuchungen über das Geruchsorgan der Insekten. *Z. Wiss. Zool.* 34: 367–403.

Heck, G. L., Mierson, S., and DeSimone, J. (1984). Salt taste tranaduction occurs through an amiloride-sensitive sodium transport pathway. *Science,* 223: 403-405.

Heidel, A. (1949). *The Gilgamesh Epic and Old Testament Parallels,* 2nd ed. University of Chicago Press, Chicago.

Henton, W. W. (1969). Conditioned suppression to odorous stimuli in pigeons. *J. Exp. Anal. Behav.* 12: 175–185.

Hevezi, P., Moyer, B.D., Lu, M. et al. (2009). Genome-wide analysis of gene expression in primate taste buds reveals links to diverse processes. *PLoS One.* Jul 28;4(7):e6395. doi: 10.1371/journal.pone.0006395.

Hillier, L., Lennon, G., Becker, M., et al. (1996) Generation and analysis of 280,000 human expressed sequence tags. *Genome Res.* 1996; 6: 807–828.

Hinds, J. W., Hinds, P. L., and McNelly, N. A. (1984). An autoradiographic study of the mouse olfactory epithelium: evidence for long-lived receptors. *Anat. Rec.* 210: 375–383.

His, W. (1889) Über die Entwicklung des Riechlappens und des Riechganglions und diejenige des verlängerten Markes. *Verh. Anat. Ges.* 3: 63–66.

Hodgson, E. S., Lettvin, J. Y., and Roeder, K. D. (1955). Physiology of a primary chemoreceptor unit. *Science* 122: 417–418.

Hofer, D., Puschel, B., and Drenckhahn, D. (1996). Taste receptor-like cells in the rat gut identified by expression of alpha-gustducin. *Proc. Natl. Acad. Sci. U.S.A,* 93: 6631–6634.

Hofer, D. and Drenckhahn, D. (1998). Identification of the taste cell G-protein, α-gustducin, in brush cells of the rat pancreatic duct system. *Histochem. Cell Biol.* 110: 303–309.

Hofer, D., Jons, T., Kraemer, J., and Drenckhahn, D. (1998). From cytoskeleton to polarity and chemoreception in the gut epithelium. *Ann. N.Y. Acad. Sci.,* 859: 75–84.

Hoffman, C. K. (1866). *Onderzoekingen over den anatomischen Bouw wart de Membranaolfactoria en het peripherische Uitende van den Nervus olfactorius.* Amsterdam, 58 pp.

Hoffman, E. J., Phelps, M. E., Mullani, N. A., et al. (1976). Design and performance characteristics of a whole-body positron transaxial tomography. *J. Nucl. Med.* 17: 493–502.

Holland, V. F., and Zampighi, G. A. (1991). Tight junctions in taste buds: possible role in perception of intravascular gustatory stimuli. *Chem. Senses* 16: 69–80.

Horn, W. (1825). *Über den Geschmackssinn des Menschen.* Karl Groos, Heidelberg.

Hosoya, Y., and Yoshida, H. (1937). Über die bioekktrischen Erscheinungen an der Riechschleimhaut. *Jpn. J. Med. Sci. III Biophys.* 5: 22.

Hounsfield, G. N. (1973). Computerized transverse axial scanning (tomography): Part I. Description of system. *Brit. J. Radiol.* 46: 1016–1022.

Hu, J., Zhong, C., and Ding, C., et al. (2007). Detection of near-atmospheric concentrations of CO_2 by an olfactory subsystem in the mouse. *Science* 317: 953–957.

Hummel, T., Sekinger, B., Wolf, S. R., et al. (1997) 'Sniffin' sticks': olfactory performance assessed by the combined testing of odor identification, odor discrimination and olfactory threshold. *Chem Senses.* 22: 39–52.

Hunter, J. (1786). Description of the nerves which supply the organs of smelling. In *Observations on Certain Parts of the Animal Economy,* J. Hunter. London, pp. 213–219.

Hurst, E. W. (1936). Newer knowledge of virus diseases of nervous system: review and interpretation. *Brain,* 59: 1–34.

Ikeda, K. (1909) New seasonings. *J. Tokyo Chem. Soc.* 30: 820–836 [in Japanese].

Ingvar, G. H. and Risberg, J. (1965). Influence of mental activity upon regional cerebral blood flow in man. *Acta Neurol. Scand. Supp.* 14: 183–186.

Jackson, R. T., Tigges, J., and Arnold, W. (1979). Subarachnoid space of the CNS, nasal mucosa, and lymphatic system. *Arch. Otolaryngol.* 105: 180–184.

Jafek, B. W., Eller, P. M., Esses, B. A. and Moran, D. T. (1989). Posttraumatic anosmia. *Arch. Neurol.* 46:300–304.

Jafek, B. W., Hartman, D., Eller, P. M., et al. (1990). Post-viral olfactory dysfunction. *Am. J. Rhinol.* 4: 91–100.

Jang, H. J., Kokrashvili, Z., Theodorakis, M. J., et al. (2007). Gut-expressed gustducin and taste receptors regulate secretion of glucagon-like peptide-1. *Proc. Natl. Acad. Sci. U.S.A.,* 104: 15069–15074.

Jenner, M. S. (2011). Follow your nose? Smell, smelling, and their histories. *Am. Hist. Rev.* 116: 335–351.

Jennings, H. S. (1906). *Behaviour of the Lower Organisms.* Columbia University Press, New York.

Johnson, B.A. and Leon, M. (2000). Modular representations of odorants in the glomerular layer of the rat olfactory bulb and the effects of stimulus concentration. *J. Comp.Neurol.* 10: 496–509.

Johnston, R. E. (2000). Chemical communication and pheromones: the types of chemical signals and the role of the vomeronasal system. In: T. E. Finger, W. L. Silver and D. Restrepo (Eds.), *The Neurobiology of Taste and Smell.* 2nd edition. New York: Wiley-Liss, Inc., pp. 101–127.

Johnston, J. W., Jr., Moulton, D. G., and Turk, A., Eds. (1970). *Advances in Chemoreception. Vol. 1. Communication by Chemical Signals.* Appleton-Century-Crofts, New York.

Jones. D. T., and Reed, R. R. (1989). G_{olf}: an olfactory neuron specific-G protein involved in odorant signal transduction. *Science,* 244: 790–795.

Jones, F. N., and Jones, M. H. (1953). Modern theories of olfaction: a critical review. *J. Psychol.* 36: 207–241.

Jurisch, A. (1922). Studien über die Papillae vallatae beim Menschen. *Z. Anat. Entwicklungsgesch.* 66: 1–149.

Kaissling. K. E., and Renner, M. (1968). Antennale Rezeptoren für Queen Substance und Sterzelduft bei der Honigbiene. *Z. Vergleich. Physiol.* 59: 357–361.

Kant, E. (translated by Dowdell, V.L.) (1798) *Anthropology from a Pragmatic Point of View*, Southern Illinois University Press

Karlson, P., and Lüscher, M. (1959). "Phermones": a new term for a class of biologically active substances. *Nature* 183: 55–56.

Kauer, J. S. (1988). Real-time imaging of evoked activity in local circuits of the salamander olfactory bulb. *Nature* 331: 166–168.

Kauer, J. S., and Moulton, D. G. (1974). Response patterns of olfactory bulb neurons using odor stimulation and small nasal areas in the salamander. *J. Physiol. (Lond.)* 243: 717–737.

Keenan, C. M., Kelly, D. P., and Bogdanffy, M. S. (1990). Degeneration and recovery of rat olfactory epithelium following inhalation of dibasic esters. *Fund. App. Toxicol.* 15: 381–393.

Kennedy, C., DesRosiers, M. H., Jehl, J. W., et al. (1975). Mapping of functional neural pathways by autoradiographic survey of local metabolic rate with ^{14}C deoxyglucose. *Science* 187: 850–853.

Kenneth, J. H. (1928). A note on a forgotten sixteenth century disputation on smell. *J. Laryngol. Otol.* 43: 103–104.

Keesey, J. C. (1962). *Olfactory preference by heterosexually naïve and experienced male rats for estrus and diestrus female urine.* Unpublished M.A. Thesis. San Jose State College. Cited in J. B. Cooper, *Comparative Psychology.* N.Y.: Ronald Press, 1971, p. 321.

Kety, S. S. (1960). Measurement of local blood flow by the exchange of an inert, diffusible substance. In: H. D. Bruner (Ed.), *Methods in Medical Research,* Chicago: Year Book Publ., VIII: 228–236.

Kiesow, F. (1894). Beiträge zur Physiologischen Psychologie des Geschmackssinnes. *Philos. Stud.* 10: 523–561. Kimura, K., and Beidler, L. M. (1961). Microelectrode study of taste receptors of rat and hanister. *J. Cell Comp. Physiol.* 58: 131–140.

Kimura, K. and Beidler, L.M. (1961). Microelectrode study of taste receptors of rat and hamster. *J. Cell. Comp. Physiol.* 58: 715–731.

Kinnamon, S. C., and Cummings, T. A. (1992). Chemosensory transduction mechanisms in taste. *Annu. Rev. Physiol.* 54: 715–731.

Kinnamon, S. C., Dionne, V. E., and Beam, K. G. (1988). Apical location of K$^+$ channels in taste cells provides a basis for sour taste transduction. *Proc. Natl. Acad. Sci. USA* 85: 7023–7027.

Kirkes. W. S. (1849). *Manual of Physiology.* Lea and Blanchard, Philadelphia, pp. 400–401.

Kobal, G., and Hummel, C. (1988). Cerebral chemosensory evoked potentials elicited by chemical stimulation of the human olfactory and respiratory nasal mucosa. *Electroencephalogr. Clin. Neurophysiol.* 71: 241–250.

Kobal, G., and Plattig, K. H. (1978). Methodische Anmerkungen zur Gewinnung olfaktorischer EEG Antworten des wachen Menschen (objektive Olfaktometrie). *Z. EEM-EMG* 9: 135–145.

Kobal, G., Hummel, T., Sekinger, B., et al. (1996). "Sniffin' sticks": screening of olfactory performance. *Rhinology.* 34: 222–226.

Komisaruk, B. R. and Beyer, C. (1972). Responses of diencephalic neurons to olfactory.bulb stimulation, odor and arousal. *Brain Res.* 36: 153–170.

Krames, L. (1970). Responses of female rats to the individual body odors of male rats. *Psychonomic Sci* 20: 274.

Krause, W. (1876). Allgemeine und microscopische Anatomie. In *Handbuch der menschlichen Anatomie.* Bd. 1. C. F. Th. Krause, Hrsg., Hannover.

Krautwurst, D., Yau. K. W., Reed. R. R. (1998). ldentificarion of ligands for olfactory receptors by functional expression of a receptor library. *Cell* 95: 917–26.

Kubie, J. L., and Moulton, D. G. (1980). Odorant specific patterns of differential sensitivity inherent in the salamander olfactory epithelium. *Soc. Neurosci. Abstr.* 6: 243.

Kurahashi, T., and Shibuya, T. (1987). The odor responses and odor-induced current in the solitary olfactory receptor cells isolated from newts. *Chem. Senses* 12: 508.

Kurihara, K., and Koyama, N. (1972). High activity of adenyl cyclase in olfactory and gustatory organs. *Biochem. Biophys. Res. Commun.* 48: 30–33.

Lancisi, J. M. (1717). *De Noxiis Paludun Effluviis,* Eorumque Remediis. Rome.

Landau, W. H., Freygang, W. H., Rowland, L. P., et al. (1955). The local circulation of the living brain: values in the unanesthetized and anesthetized cat. *Trans. Am. Neurol. Assoc.* 80: 125–129.

Larsson, M. C., Domingos, A. I., Jones, W. D., Chiappe, M. E., Amrein, H., and Vosshall, L .B. (2004). Or83b encodes a broadly expressed odorant receptor essential for Drosophila olfaction. *Neuron* 43, 703–714.

Laska, M., Seibt, A. and Weber, A. (2000) 'Microsmatic' primates revisited: olfactory sensitivity in the squirrel monkey. *Chem. Senses* 25: 47–53.

Lassen, N. A., Hoedt-Rasmussen, K., Sorensen, S. C., Skinhoj, E., Cronquist, B., Bodforss, E., and Ingvar, D. H. (1963). Regional cerebral blood flow in man determined by Krypton-85. *Neurology* 13: 719–727.

Lauterbur, P. (1973). Image formation by induced local interactions: Examples employing nuclear magnetic resonance. *Nature* 242: 190–191.

Laveran, C. L. A. (1881). *Nature parasitaire ties accidents de l'irnpaludisme: description dun Nouvean parasite troue dans le sang ties malades atteints de fievre palustre.* Bailliere, Paris.

Lederer, F. L. (1959). The problem of nasal polyps. *J. Allergy.* 30: 420–432.

Lee, K., Wells, R. G., and Reed, R. R. (1987). Isolation of an olfactory cDNA; similarity to retinolbinding protein suggests a role in olfaction. *Science* 235: 1053–1056.

Lee, R.J., Xiong, G., Kofonow, J.M. et al. (2012). T2R38 taste receptor polymorphisms underlie susceptibility to upper respiratory infection. *J. Clin. Invest.* 122: 4145–4159.

Lee, S. van der, and Boot, L. M. (1955). Spontaneous pseudopregnancy in mice. *Acta Physiol. Pharmacol. Neer.* 4: 442–444.

Legg, J. W. (1873). A case of anosmia following a blow. *Lancet* 2: 659–600.

Le Gross Clark, W.E. (1951). The projection of the olfactory epithelium on the olfactory bulb in the rabbit. *J. Neurol. Neurosurg.* Psychiat. 14: 1–10.

Lehman, C. D., Bartoshuk, L. M., Catalanotto, et al. (1995). Effect of anesthesia of the chorda tympani nereve on taste perception in humans. *Physiol. Behav.* 57: 943–951.

Leinders-Zufall, T., Shepherd, G. M., and Zufall, F. (1996). Modulation by cyclic GMP of the odour sensitivity of vertebrate olfactory receptor cells. *Proc. Roy. Soc. Lond. B.* 263: 803–811.

Le Magnen, J. (1952). Les phènomenes olfacto-sexuells chez le rat blanc. *Arch. Sci. Physiol.* 6: 295–332.

Lemery, N. (1697). Pharmacopee universelle. Cited in *The Foul and the Fragrant,* A. Corbin (1986). Harvard University Press, Cambridge, MA.

Leon, M. (1983). Chemical communication in mother-young interactions. In *Pheromones and Reproduction in Mammals,* J. G. Vandenbergh (Ed.). Academic Press, New York, pp. 39–77.

Leung, C. T., Coulombe, P. A., and Reed, R. R. (2007). Contribution of olfactory neural stem cells to tissue maintenance and regeneration. *Nature Neurosci.* 10: 720–726.

Leveteau, J., and MacLeod, P. (1966). Olfactory discrimination in the rabbit olfactory glomerulus. *Science* 153: 175–176.

Leydig, F. (1851). Über die Haut einiger Süßwasserfische. *Z. wiss. Zool.* 3: 1–12.

Li, X., Staszewski, L., Xu, H., et al. (2002) Human receptors for sweet and umami taste. *Proc. Natl Acad. Sci. USA,* 99: 4692–4696.

Li, X., Li, W., Wang, H. et al. (2005). Pseudogenization of a sweet-receptor gene accounts for cats' indifference toward sugar. *PLoS Genet.* Jul;1(1): 27–35.

Li, Y., Field, P. and Raisman, G. (1998). Regeneration of adult corticospinal axons induced by transplanted olfactory ensheathing cells. *J. Neurosci.* 18: 10514–10524.

Liberles, S.D. and Buck, L.B. (2006). A second class of chemosensory receptors in the olfactory epithelium. *Nature.* 442(7103): 645–650.

Lin, D. M., Wang, F., Lowe, G., et al. (2000). Formation of precise connections in the olfactory bulb occurs in the absence of odorant-evoked neuronal activity. *Neuron* 26: 69–80.

Lindemann, B., Ogiwara, Y. and Ninomiya, Y. (2002). The discovery of umami. *Chem. Senses* 27: 843–844.

Linnaeus, C. (1765). Odores medicamentorum. *Amoenit. Acad.* 3: 183–201.

Lois, C., Garcia-Verdugo, J. M., and Alvarez-Buylla, A. (1966). Chain migration of neuronal precursors. *Science* 271: 978–981.

Lomvardas, S., Barnea, G., Pisapia, D. J., et al. (2006) Interchromosomal interactions and olfactory receptor choice. *Cell* 126: 403–413.

Lorry, D. (1784/1785). Observations sur les parties volatiles et odorantes des médicaments tirés des substances vègètales et animales. *Hist. Mèm. Soc. Roy. Mèd.* 7: 306–318.

Loven, C. (1868). Beiträge zur Kenntnis vom Bau der Geschmackswärzchen der Zunge. *Arch. Mikrosk. Anat.* 4: 96–109.

Lowe, G., and Gold, G. H. (1991). The spatial distribution of odorant-sensitive and odorant-induced currents in salamander olfactory receptor cells. *J. Physiol. (Lond.)* 442: 147–168.

Lowe, G. and Gold, G. H. (1993). Contribution of the ciliary cyclic nucleotide-gated conductance to olfactory transduction in the salamander. *J. Physiol.* 462: 175–196.

Lowe, G., Nakamura, T., and Gold, G. H. (1989). Adenylate cyclase mediates transduction for a wide variety of odorants. *Proc. Natl. Acad. Sci. USA* 86: 5641–5645.

Lower, R. (1670). Dissertatio de origine catarrhi in qua ostenditur non provenire a cerebro. In: *Tractato de corde.* J. Redmayne, London, pp. 221–239.

Lu, J., Feron, F., Ho, S., et al. (2001). Transplantation of nasal olfactory tissue promotes partial recovery in paraplegic rates. *Brain Res.* 889: 344–357.

Luciani, L. (1917). Human Physiology. London: Macmillan and Co.

Luciano, L. and Reale, E. (1979). A new morphological aspect of the brush cells of the mouse gallbladder epithelium. *Cell Tissue Res.* 201: 37–44.

Ludwig, C. (1850). Neue Versuche über die Beihilfe der Nerven zu der Speichelsekretion. *Naturforsh Ges Zürich* 53/54: 210–239.

Luskin, M. B. (1993). Restricted proliferation and migration of postnatally generated neurons derived from the forebrain subventricular zone. *Neuron* 11: 173–189.

MacDonald, K. P. A., Murrell, W. G., Bartlett, P. F., et al. (1996). FGF2 promotes neuronal differentiation in explant cultures of adult and embryonic mouse olfactory epithelium. *J. Neurosci. Res.* 44: 27–39.

Mackay-Sim, A. and Chuah, M. (2000). Neurotrophic growth factors in the primary olfactory pathway. *Prog. Neurobiol.* 62: 527–559.

Mackay-Sim, A., Shaman, P., and Moulton, D. G. (1982). Topographic coding of olfactory quality: odorant specific patterns of epithelial responsivity in the salamander. *J. Neurophysiol.* 48: 584–596.

MacLean, P. D. (1990). *The Triune Brain in Evolution: Role in Paleocerebral Functions.* New York: Plenum Press.

Mace, O.J., Affleck, J., Patel, N. et al. Sweet taste receptors in rat small intestine stimulate glucose absorption through apical GLUT2. *J. Physiol.* 582(Pt 1): 379–392.

Magendie, F. (1824). Le nerf olfactif, est-il l'organe de l'odorat? Experiences sur cette question. *J. Physiol. Exp. Pathol.* 4: 169–176.

Mahanthappa, N. K. and Schwarting, G. A. (1993). Peptide growth factor control of olfactory neurogenesis and neuron survival in vitro: roles of EGF and TGF-bs. *Neuron* 10: 293–305.

Mainardi, D., Marsan, M., and Pasquali, A. (1965). Causation of sexual preferences of the house mouse. The behaviour of mice reared by parents whose odour was artifically altered. *Atti Soc. Ital. Sci. Natur. Museo Civ. Stor. Natur. Milano* 104: 325–338.

Malnic, B., Hirono, J., Sato, T., and Buck, L. B. (1999). Combinatorial receptor codes for odors. *Cell* 96: 713–723.

Malpighi, M. (1664). Exercitatio epistolica de lingua. (1686; Jo. Alphonso Borellio). In: *Opera omnia*, Malpighi, M. (ed.), R. Scott and G. Wells, Londini (London), pp. 13–20.

Margolis, F. L. (1972). A brain protein unique to the olfactory bulb. *Proc. Nat. Acad. Sci.* 69: 1221–1224.

Margolskee, R. F., Dyer, J., Kokrashvili, Z., et al. (2007). T1R3 and gustducin in gut sense sugars to regulate expression of Na+-glucose cotransporter 1. *Proc. Natl. Acad. Sci. U.S.A.*, 104: 15075–15080.

Martin, H. N. (1873). Notes on the structure of the olfactory mucous membrane. *J. Anat. Physiol. (Lond.)* 8: 39–44.

Matsuda, T. and Doty, R. L. (1976). Regional taste sensitivity to NaCl: relationship to subject age, tongue locus and area of stimulation. *Chem. Senses* 20: 283–290.

Matsuo, R. (2000). Role of saliva in the maintenance of taste sensitivity. *Crit. Rev. Oral Biol.* 11: 216–229.

McBurney, D. H., and Gent, J. F. (1979). On the nature of taste qualities. *Psychol. Bull.* 86: 151–167.

McBurney, D. H., and Pfaffmann, C. (1963). Gustatory adaptation to saliva and sodium chloride. *J. Exp. Psychol.* 65: 523–529.

McCaffrey, R. J., Duff, K., and Solomon, G. S. (2000). Olfactory dysfunction discriminates probable Alzheimer's dementia from major depression: a cross-validation and extension. *J. Neuropsychiat. Clin. Neurosci.* 12: 29–33.

McCartney, W. (1968). *Olfaction and Odours.* Springer-Verlag, Berlin.

McLaughlin, T. (1971). *Dirt: A Social History as Seen Through the Uses and Abuses of Dirt.* Dorset Press, New York.

McLaughlin, S. K., McKinnon, P. J., and Margolskee, R. F. (1992a). Gustducin is a taste-cell-specific G protein closely related to the transducins. *Nature*, 357: 563–569.

McLaughlin, S. K., McKinnon, P. J., and Margolskee, R. F. (1992b). α Gustducin: a taste cell specific G protein subunit closely related to the α transducins. In *Chemical Signals in Vertebrates 6*, R. L. Doty and D. Müller-Schwarze (Eds.) Plenum Press, New York, pp. 9–14.

Menco, B. P. M. (1991). Ultrastructual localization of the transduction apparatus in the rat's olfactory epithelium. *Chem. Senses* 15: 555.

Meredith, M. (1983). Sensory physiology of pheromone communication. In *Pheromones and Reproduction in Mammals*, J. G. Vandenbergh (Ed.). Academic Press, New York, pp. 199–252.

Meredith, M. (1986). Vomeronasal organ removal before sexual experience impairs male hamster mating behavior. *Physiol. Behav.* 36: 737–743.

Merkel, F. (1880). Über die Endigungen der sensiblen Nerven in der Haut der Wirbelthiere. Rostock, 214 pp.

Meyrick, B. and Reid, L. (1968). The alveolar brush cell in rat lung–a third pneumonocyte. *J. Ultrastruct. Res.* 23: 71–80.

Miller, I. J. Jr., Ed. (1988). *The Beidler Symposium on Taste and Smell.* Book Service Associates, WinstonSalem, NC.

References

Miller, S. L., Mirza, N., and Doty, R. L. (2002). Electrogustometric thresholds: Relationship to anterior tongue locus, area of stimulation, and number of fungiform papillae. *Physiol. Behav.* 75: 753–757.

Minnich, D. E. (1921). An experimental study of the tarsal chemoreceptors of two nymphalid butterflies. *J. Exp. Zool.* 33: 173–203.

Minnich, D. E. (1926). The chemical sensitivity of the tarsi of certain muscid flies. (*Phormia regina* Meigen, *Phormia terrae-novae* R. D. and *Lucilia sericate* Meigen). *Biol. Bull.* 51: 166–178.

Minnich, D. E. (1929). The chemical sensitivity of the legs of the blow-fly, *Calliphora vomitoria* Linn., to various sugars. *Z. Vergl. Physiol.* 11: 1–55.

Mombaerts, P. (1999). Molecular biology of odorant receptors in vertebrates. *Ann. Rev. Neurosci.* 22: 487–509.

Moran, D. T., Jafek, B. W., Eller, P. M. and Rowley, J. C. III,. (1992). The ultrastructural histopathology of human olfactory dysfunction. *Microsc. Res. Tech.* 23: 103–110.

Moran, D. T., Jafek, B. W., and Rowley, J. C., III, (1982). Electron microscopy of the olfactory epithelium reveals a new cell type: the microvillar cell. *Brain Res*, 253: 39–46.

Morfit, C. (1847). *Chemistry Applied to the Manufacture of Soap and Candles*. Carey and Hart, Philadelphia.

Mori, K., and Takagi, S. F. (1978). An intracellular study of dendrodendritic inhibitory synapses on mitral cells in the rabbit olfactory bulb. *J. Physiol.* 719: 589–604.

Morris, E. T. (1984). *Fragrance: The Story of Perfume from Cleopatra to Chanel*. Scribner, New York.

Morrison, G. R. (1967). Behavioral response patterns to salt stimuli in the rat. *Can. J. Psychol.* 21: 141–152.

Morrison, G. R., and Norrison, W. (1966). Taste detection in the rat. *Can. J. Psychol.* 20: 208–217.

Morton, R. (1697). *Opera Medica*. Sumptibus Anisson and Posuel, Lugduni.

Mosso, A. (1881). *Über den Kreislauf des Blutes im menschlichen Gehirn*. Leipzig: Verlag von Veit.

Mosso, A. (1884). *La temperatura del cervèllo*. Milan.

Motokizawa, F. (1974). Olfactory input to the thalamus: electrophysiological evidence. *Brain Res.* 67: 334–337.

Moulton, D. G. (1960). Studies in olfactory acuity. III. Relative detectability of n-alphatic acetates by the rat. *Q. J. Exp. Psychol.* 12: 203–213.

Moulton, D. G. (1976). Spatial patterning of response to odors in the peripheral olfactory system. *Physiol. Rev.* 56: 578–593.

Moulton, D. G., and Beidler, L. M. (1967). Structure and function in the peripheral olfactory system. *Physiol. Rev.* 47: 1–52.

Moulton, D. G., and Eayrs, J. T. (1960). Studies in olfactory acuity. II. Relative detectability of n-aliphatic alcohols by the rat. *Q. J. Exp. Psychol.* 12: 99–109.

Moulton, D. G., Celebi, G., and Fink, R. P. (1970). Olfaction in mammals – two aspects; proliferation of cells in the olfactory epithelium and sensitivity to odours. In *Ciba Foundation Symposium on Taste and Smell in Vertebrates,* G. E. W. Wolstenholme and J. Knight (Eds.), Churchill, London, pp. 227–250.

Moulton, D. G., Turk, A., and Johnson, J. W., Jr., (1975). *Methods in Olfactory Research*. Academic Press, London.

Mozell, M. M. (1964). Olfactory discrimination: electrophysiological spatiotemporal basis. *Science* 143: 1336–1337.

Mozell, M. M. (1966). The spatiotemporal analysis of odorants at the level of the olfactory receptor sheet. *J. Gen. Physiol.* 50: 25–i1.

Mozell, M. M., Smith, B. P., Smith, P. E., et al. (1969). Human chemoreception in flavor identification. *Arch. Otolaryngol.* 90: 131–137.

Murphy, M. R., and Schneider, G. E. (1970). Olfactory bulb removal eliminates mating behavior in the male golden hamster. *Science* 167: 302–304.

Murrell, W., Bushell, G. R., Livesey, J., et al. (1996) Neurogenesis in adult human. *Neuroreport* 7: 1189–1194.

Mykytowycz, R. (1970). The role of skin glands in mammalian communication. In *Advances in Chemoreception. Vol. 1. Communication by Ctemical Signals,* J. W. Johnston, Jr., D. G. Moulton, and A. Turk (Eds.). Appleton-Century-Crofts, New York, pp. 327–360.

Mykytowycz, R. (1986). A quarter of a century of studies of chemical communication in vertebrates. In *Chemical Signals in Vertebrates 4*, D. Duvall, D. Müller-Schwarze, and R. M. Silverstein (Eds.). Plenum Press, New York, pp. 1–11.

Nagahara, Y. (1940). Experimentelle Studien über die histologischen Veränderungen des Geruchssorgans nach der Olfactoriusdurchschneidung. *Jpn. J. Med. Sci. Pt. V. Pathol.* 6: 165–199.

Nagel, W.A. (1905). Handbuch der Physiologie des Menschen. Braunschwieg: Vieweg & Sohn.

Nakagawa, Y., Nagasawa, M., Yamada, S., et al. (2009). Sweet taste receptor expressed in pancreatic beta-cells activates the calcium and cyclic AMP signaling systems and stimulates insulin secretion. *PLoS.One.*, 4, e5106.

Nakamura, T., and Gold, G. H. (1987). A cyclic nucleotide-gated conductance in olfactory receptor cilia. *Nature* 325: 442–444.

Nelson, G., Chandrashekar, J., Hoon, M. A., et al. (2002). An amino-acid taste receptor. *Nature*, 416: 199–120.

Neuhaus, E. M., Gisselmann, G., and Zhang, W., et al. (2005). Odorant receptor heterodimerization in the olfactory system of *Drosophila melanogaster*. *Nat. Neurosci.* 8: 15-17.

Newman, M., Féron, F. and Mackay-Sim, A. (2000). Growth factor control of olfactory neurogenesis. *Neuroscience* 99: 343–350.

Nicoll, R. A. (1971). Recurrent excitation of secondary olfactory neurons: a possible mechanism for signal amplification. *Science* 171: 825–825.

Niimura, Y., Matsui, A., and Touhara, K. (2014). Extreme expansion of the olfactory receptor gene repertoire in African elephants and evolutionary dynamics of orthologous gene groups in 13 placental mammals. *Genome Res.* 24: 1485–1496.

Norgren, R. (1970). Gustatory responses in the hypothalamus. *Brain Res.* 21: 63–77.

Notta, A. (1870). Recherches sur la perte de l'odorat. *Arch. Gen. Med.* 15: 385–407.

Ogawa, H., Ito, S., and Nomura, T. (1989). Oral cavity representation at the frontal operculum of macaque monkeys. *Neurosci. Res.*, 6: 283–298.

Ogawa, S., Lee, T. M., Kay, A. R. and Tank, D. W. (1990). Brain magnetic resonance imaging with contrast dependent on blood oxygenation. *Proc. Nat. Acad. Sci. USA* 87: 9868–9872.

Ogle, W. (1870). Anosmia (or cases illustrating the physiology and pathology of the sense of smell). *Med. Chir. Trans.* 35: 263–290.

Ohishi, Y., Komiyama, S., Wakida, K., Uchida, T., and Shiba, Y. (2000). Immunohistochemical observation of actin filaments in epithelial cells encircling the taste pore cavity of rat fungiform papillae. *Eur. J. Histochem.* 44: 353–358.

Öhrwall, H. (1891). Untersuchungen über den Geschmackssinn. *Skand. Arch. Physiol.* 2: 1–69.

Okano, M., Weber, A. F., and Frommes, S. P. (1967). Electron microscopic studies on the distal border of the canine olfactory epithelium. *J. Ultrastruct. Res.*, 17: 487–502.

Öpalski, A. (1934). Über locale Unterschiede im Bau der Ventrikelwände beim Menschen. *Z. Ges. Neurol. Psychiat.* 149: 221–254.

O'Rourke, N. A. (1996). Neuronal chain gangs: homotypic contacts support migration into the olfactory bulb. *Neuron* 16: 1061–1064.

Ottoson, D. (1956). Analysis of the electrical activity of the olfactory epithelium. *Acta Physiol. Scand.* 35 (Suppl. 122): 1–83.

Ottoson, D. (1963). Some aspects of the function of the olfactory system. *Pharmacol. Rev.* 15: 1–42.

Pace, U., and Lancet, D. (1986). Olfactory GTP-binding protein: signal transducing polypeptide of vertebrate chemosensory neurons. *Proc. Natl. Acad. Sci. USA* 83: 4947–4951.

Pace, U., Hanski, E., Salomon, Y., and Lancet, D. (1985). Odorant-sensitive adenylate cyclase may mediate olfactory reception. *Nature* 316: 255–258.

Palladius, R. T. A. (4th century A.D.), *DeRostica*, Lib. 1.

Pangborn, R. M. and Trabue, I. M. (1967). Bibliography on the sense of taste (1566–1966). In M. R. Kare and O. Maller (Eds.), *The Chemical Senses and Nutrition*. Baltimore: Johns Hopkins Press, pp. 355–471.

Parker, G. H. (1922). *Smell, Taste, and Allied Senses in the Vertebrates.* Lippincott, Philadelphia.

Parmentier, M., Libert, F., Schurmans, S., et al. (1992). Expression of members of the putative olfactory receptor gene family in mammalian germ cells. *Nature* 355: 453–455.

Passy, J. (1892). Sur les minimums perceptibles de quelques odeurs. *Comp. Rend. Soc. Biol. (Paris)* 44: 84–88, 239–243.

Pauling, L. and Coryell, C. D. (1936). The magnetic properties and structure of hemoglobin, oxyhemoglobin and carbonmonoxyhemoglobin. *Proc. Nat. Acad. Sci. USA* 89: 5951–5955.

Pavlov, I. P. (1927). *Conditioned Reflexes*. Oxford University Press, Oxford.

Peet, M. M., Echols, D. H., and Richter, H. J. (1937). The chemical prophylaxis for poliomyelitus: the technique of applying zinc sulfate intranasally. *JAMA* 108: 2184–2187.

Penfield, W. and Faulk, M. E., Jr., (1955). The insula; further observations on its function. *Brain*, 78: 445–470.

Pevsner, J., Hwang, P. M., Sklar, P. B., et al. (1988a). Odorant-binding protein and its mRNA are localized to lateral nasal gland implying a carrier function. *Proc. Natl. Acad. Sci.USA* 85: 2382–2387.

Pevsner, J., Reed, R. R., Feinstein, P. G., and Snyder, S. H. (1988b). Molecular cloning of odorantbinding protein: member of a ligand carrier family. *Science* 241: 336–339.

Pfaff, D. W. (1985). *Taste, Olfaction, and the Central Nervous System: A Festschrift in Honor of Carl Pfaffmann*. Rockefeller University Press, New York.

Pfaffmann, C. (1941). Gustatory afferent impulses. *J. Cell. Comp. Physiol.* 17: 243–258.

Pfaffmann, C. (1955). Gustatory nerve impulses in rat, cat and rabbit. *J. Neurophysiol.* 23: 429–440.

Pfaffmann, C., Goff, W. R., and Bare, J. K. (1958). An olfactometer for the rat. *Science* 128: 1007–1008.

Pfaffmann, C., Erickson, R., Frommer, G., and Halpern, B. (1961). Gustatory discharges in the rat medulla and thalamus. In *Sensory Communication*, W. A. Rosenblith (Ed.). MIT Press, Cambridge, MA, pp. 455–473.

Piesse, G. W. H. (1879). *Art of Perfumery*. Longmans, London.

Pixley, S. K. (1992a). CNS glial cells support in vitro survival, division, and differentiation of dissociated olfactory neuronal progenitor cells. *Neuron* 8: 1191–1204.

Pixley, S. K. (1992b) The olfactory nerve contains two populations of glia, identified both in vivo and in vitro. *Glia* 5: 269–284.

Plattig, K. H. (1968/1969). Über den elektrischen Geschmack, Reizstärkeabhängige evozierte Hirnpotentiale nach elektrischer Reizung der Zunge beim Menschen. Habil-Schr. Erlangen 1968. Z. *Biol.* 116: 161–211.

Powers, J. B., and Winans, S. S. (1973). Sexual behavior in periphally anosmic male hamsters. *Physiol. Behav.* 10: 361–368.

Powers, J. B., and Winans, S. S. (1975). Vomeronasal organ: critical role in mediating sexual behavior of the male hamster. *Science* 187: 961–963.

Proetz, A. W. (1924). Exact olfactometry. *Ann. Otol. Rhinol. Laryngol.* 33: 275–278.

Rall, W., Shepherd, G. M., Reese. T. S., and Brightman, M. W. (1966). Dendrodendritic synaptic pathway for inhibition in the olfactory bulb. *Exp. Neurol.* 14: 44–56.

Ramon-Cueto, A., Cordero, M., Santos-Benito, F. and Avila, J. (2000). Functional recovery of paraplegic rats and motor axon regeneration in their spinal cords by olfactory ensheathing glia. *Neuron* 25: 425–435.

Ramon-Cueto, A. and Nieto-Sampedro, M. (1992). Glial cells from adult rat olfactory bulb: immunocytochemical properties of pure cultures of ensheathing cells. *Neuroscience* 47: 213–220.

Ramon-Cueto, A. and Nieto-Sampedro, M. (1994). Regeneration into the spinal cord of transected dorsal root axons is promoted by ensheathing glia transplants. *Exp. Neurol.* 27: 232–244.

Raming, K., Krieger, J., Strotmann, J., et al. (1993). Cloning and expression of odorant receptors. *Nature* 361: 353–356.

Remy, N. (1595). *Demonolatry*. (Translation to English by E. A. Ashwin.) John Rodker, London, 1930.

Reivich, M., Sano, N. and Sakoloff, L. (1971). Development of an autoradiographic method for the determination of regional glucose consumption. In: R. W. Ross-Russell (Ed.), *Brain and Blood Flow*. London: Pitman, 397–400.

Reivich, M., Kuhl, D., Wolf, A., et al. (1979). The [^{18}F]fluorodeoxyglucose method for the measurement of local cerebral glucose utilization in man. *Circ. Res.* 44: 127–137.

Renqvist, Y. (1919). Über den Geschmack. *Scand. Arch. Physiol.* 38: 7–201.

Ressler, K. J., Sullivan, S. L. and Buck, L. B. (1994). Information coding in the olfactory system: evidence for a stereotyped and highly organized epitope map in the olfactory bulb. *Cell* 79: 1245–1255.

Rhein, L. D., and Cagan, R. H. (1980). Biochemical studies of olfaction: isolation, characterization, and odorant binding activity of cilia from rainbow trout olfactory rosettes. *Proc. Natl. Acad. Sci. USA* 77: 4412–4416.

Rhein, L. D., and Cagan, R. H. (1983). Biochemical studies of olfaction: binding specificity to odorants to a ciliar preparation from rainbow trout olfactory rosettes. *J. Neurochem.* 41: 569–577.

Rhodin, J. and Dalham, T. (1956). Electron microscopy of the tracheal ciliated mucosa in rat. *Z. Zellforsch.* 44: 345–412.

Richter, C. P. (1936). Increased salt appetitie in adrenalectomized rats. *Amer. J. Physiol.* 115: 115–161.

Richter, C. P. (1939). Salt taste thresholds of normal and adrenalectomized rats. *Endocrinology* 24: 367–371.

Richter, C. P., and Campbell, K. H. (1940). Taste thresholds and taste preference of rats for five common sugars. *J. Nutr.* 20: 31–46.

Richter, C. P., and MacLean, A. (1939). Salt taste thresholds of humans. *Amer. J. Physiol.* 126: 1–6.

Riti, L. (1974). *The Unknown Leonardo*. McGraw-Hill, New York.

Rochefort, C., Gheusi, G., Vincet, J-D. and Liedo, P-M. (2002). Enriched odor exposure increases the number of newborn neurons in the adult olfactory bulb and improves odor memory. *J. Neurosci.* 22: 2679–2689.

Roper, S. D. (1989). The cell biology of vertebrate taste receptors. *Annu. Rev. Neurosci.* 12: 329–353.

Ross, R. (1923). *Memoirs*. E. P. Dutton, New York.

Rothkrug, L. (1981). The "odour of sanctity," and the Hebrew origins of Christian relic veneration. *Hist. Reflect. Refex. Hist.* 8: 95–137.

Roy, C. S., and Sherrington, C. S. (1890). On the regulation of the blood supply of the brain. *J. Physiol. (London)* 11: 85–108.

References

Rydberg, E. (1932). Cerebral injury in newborn children consequent on birth trauma. *Acta Path. Microbiol. Scand.*, Suppl. 10: 1–247.

Russo, A. F., and Koshland, D. E., Jr., (1983). Separation of signal transduction and adaptation functions of the asparate receptor in bacterial sensing. *Science* 220: 1016–1020.

Saito, H., Chi, Q., Zhuang, H., Matsunami, H., and Mainland, J. D. (2009). Odor coding by a mammalian receptor repertoire. Sci. Signal 2:ra9.

Saveliev, N. A. (1892). *Fisiologiya nervi olfactorii Istoricheskiya i Eksperimentalniya Izsliedovaniya.* Univ. Tipograf, Noskva.

Scarpa, A. (1785). *Anatomicarum annotationum liber secundus. De organo olfactus praecipuo, deque nervis nasalibus interioribus e pari quinto nervorum cerebri.* Regis, Ticini.

Scarpa, A. (1789). *Anatomicae disquisitiones de auditu et olfacto.* Regis, Ticini.

Schank, J. C. (2006). Do human menstrual-cycle pheromones exist? *Hum. Nature* 17: 448–470.

Schiff, J. M. (1860). Der erste Hirnnerv ist der Geruchsnerv. *Untersuchungen zur Naturlehre des Menschen und der Thiere* 6: 254–267.

Schiffman, S. S. (1983). Taste and smell in disease. *New Eng. J. Med.* 308: 1275–1279, 1337–1343.

Schiffman, S. S., Zervakis, J., Suggs, M. S., et al. (1999a). Effect of medications on taste: example of amitriptyline HCl. *Physiol. Behav.* 66: 183–191.

Schiffman, S. S., Zervakis, J., Shaio, E. and Heald, A. E. (1999b). Effect of the nucleoside analogs zidovudine, didanosine, stavudine, and lamivudine on the sense of taste. *Nutrition* 15: 854–859.

Schiffman, S. S., Zervakis, J., Suggs, M. S., et al. (2000). Effect of tricyclic antidepressants on taste responses in humans and gerbils. *Pharmacol. Biochem. Behav.* 65: 599–609.

Schiller, F. (1997). A memoir of olfaction. *J. Hist. Neurosci.* 6: 133–146.

Schmale, H., Holtgreve-Grez, H. and Christiansen, H. (1990). Possible role for salivary gland protein in taste reception indicated by homology to lipophilic-ligand carrier proteins. *Nature* 343: 366–369.

Schneider, C. V. (1655). *Liber de osse cribriformi, et sensu ac organo odoratus, et morbis ad utrumque spectantibus, de coryza, haemorrhagia Narium, polypo, steruntatione, a missione odoratus*; Mebius et Schumacher, Wittebergae.

Schneider, D. (1955). Mikro-Elektroden registrieren die elektrischen Impulse einzelner Sinneszellen der Schmetterlingsantenne. *Ind.-Elektron.* 3: 3–7.

Schneider, D. (1957a). Electrophysiological investigation on the antennal receptors of the silk moth during chemical and mechanical stimulation. *Experientia* 13: 89–91.

Schneider, D. (1957b). Elektrophysiologische Untersuchungen von Chemo- and Menschanorezeptoren der Antenne des Seidenspinners *Bombyx mori* L. *Z. Vergl. Physiol.* 40: 8–41.

Schultz, E. W. (1960). Repair of olfactory mucosa, with special reference to regeneration of olfactory cells (sensory neurones). *Am. J. Pathol.* 37: 1–19.

Schultz, E. W., and Gebhardt, L. P. (1936). Prevention of intranasally inoculated poliomyelitis in Monkeys by previous intranasal irrigation with chemical agents. *Proc. Soc. Exp. Biol. Med.* 34: 133–135.

Schultz, E. W., and Gebhardt, L. P. (1937). Zinc sulfate prophylaxis in poliomyelitis. *JAMA.* 108: 2182–2184.

Schultze, M. (1856). Über die Endigungsweise des Geruchsnerven und der Epithelialgebilde der Nasenschleimhaut. *Monatsber. Deutsche Akad. Wiss. Berlin* 21: 505–515.

Schultze, M. (1863). Untersuchungen über den Bau der Nasenschleimhaut, namentlich die Structur und Endigungsweise der Geruchsnerven bei dem Menschen und den Wirbelthieren. *Abh. Naturforsch. Ges. Halle* 7: 1–100.

Schwalbe, G. (1868). Über die Geschmacksorgane der Säugethiere und des Menschen. *Arch. mikr. Anat.* 4: 154–187.

Schwabe, G. (1867). Das Epithel der Papillae vallatae. *Arch. Mikrosk. Anat.* 3: 504–508.

Scott, J. W., and Pfaffmann, C. P. (1972). Characteristics of responses of lateral hypothalamic neurons to stimulation of the olfactory bulb. *Brain Res.* 48: 251–264.

Sharp, F. R., Kauer, J. S., and Shepherd, G. M. (1975). Local sites of activity-related glucose metabolism in rat olfactory bulb during olfactory stimulation. *Brain Res.* 98: 596–600.

Sharp, F. R., Kauer, J. S., and Shepherd, G. M. (1977). Laminar analysis of 2-deoxyglucose uptake in olfactory bulb and olfactory cortex of rabbit and rat. *J Neurophysiol,* 40: 800–813.

Shah, A.S., Ben-Shahar, Y., Moninger, T.O. et al. (2009). Motile cilia of human airway epithelia are chemosensory. *Science.* 325: 1131–1134.

Shaw, A. (1833). *Narrative of the Discoveries of Sir Charles Bell in the Nervous System.* Longman, Orme, Brown, Green and Longmans. London.

Shepherd, G. M. (1971). Physiological evidence for dendrodendritic interactions in the rabbit's olfactory glomerulus. *Brain Res.* 32: 212–217.

Shepherd, G. M. (1972). Synaptic organization of the mammalian olfactory bulb. *Physiol. Rev.* 52: 864–917.

Sidky, M. (1877). *Recherches anatomo-microscopiques sur la muqueuse olfactive.* Paris, 74 pp.

Simon, S. A. and Nicolelis, M. A. L. (2002). *Methods in Chemosensory Research.* Boca Raton: CRC Press.

Skinner, B. F. (1938). *The Behavior of Organisms.* Appeton-Century, New York.

Sklar, P. B., Anholt, R. R. H., and Snyder, S. H. (1986). The odorant-sensitive adenylate cyclase of olfactory receptor cells. *J. Biol. Chem.* 261: 15538–15543.

Slotnick, B. M. (1990). Olfactory perception. In *Comparative Perception, Vol.* I. Basic Mechanisms, M. A. Berkeley and W. C. Stebbins (Eds.). Wiley, New York, pp. 155–214.

Slotnick, B. M., and Katz, H. (1974). Olfactory learning-set in rats. *Science* 185: 796–798.

Slotnick, B. M., and Ptak, J. E. (1977). Olfactory intensity-difference thresholds in rats and humans. *Physiol. Behav.* 19: 795–802.

Small, D. M., Jones-Gotman, M., Zatorre, R. J., et al. (1997). A role for the right anterior temporal lobe in taste quality recognition. *J. Neurosci.* 17: 5136–5142.

Smith, D. V., St. John, S. J. and Boughter, J. D. Jr. (2000). Neuronal cell types and taste quality coding. *Physiol. Behav.* 69: 77–85.

Smith, J. (1970). Conditioned suppression as a animal psychophysical technique. In *Animal Psychophysics,* W. C. Stebbins (Ed.). Appleton-Century-Crofts, New York, pp. 125–159.

Smith, T. D. and Bhatnagar, K. P. (2000). The human vomeronasal organ. Part II. prenatal development. *J. Anat.* 197: 421–436.

Smith, T. D., Bhatnagar, K. P., Tuladhar, P., et al. (2004). Distribution of olfactory epithelium in the primate nasal cavity: are microsmia and macrosmia valid morphological concepts? *Anat. Rec.,* 281, 1173–1181.

Smotherman, W. P. (1982). Odor aversion learning by the rat fetus. *Physiol. Behav.* 29: 769–777.

Sobel, N., Prabhakaran, V., Desmond, J. E., et al. (1998a). Sniffing and smelling: separate subsystems in the human olfactory cortex. *Nature,* 392, 282–286.

Sobel, N., Prabhakaran, V., Hartley, C. A., et al. (1998b). Odorant-induced and sniff-induced activation in the cerebellum of the human. *J. Neurosci.* 18: 8990–9001.

Sakoloff, L. (1961). Local cerebral circulation at rest and during altered cerebral activity induced by anesthesia or visual stimulation. In: S. S. Kety and J. Elkes (Ed.), *The Regional Chemistry, Physiology, and Pharmacology of the Nervous System.* Oxford, Pergamon Press, pp. 107–117.

Sakoloff, L., Reivich, M., Kennedy, C., et al. (1977). The [^{14}C]deoxyglucose method for the meaurement of local cerebral glucose utilization: theory, procedure, and normal values in the conscious and anesthetized albino rat. *J. Neurochem.* 28: 897–916.

Solomon, G. S., Petrie, W. M., Hart, J. R., and Brackin, H. B. Jr., (1998). Olfactory dysfunction discriminates Alzheimer's dementia from major depression. *J. Neuropsychiat. Clin. Neurosci.* 10: 64–67.

Sömmerring, S. T. (1809). *Abbildungen der menschlichen Organe des Geruches.* Frankfurt am Main, Varrentrapp and Wenner.

Spector, A. C., Andrews-Labenski, J. and Letterio, F. C. (1990). A new gustometer for psychophysical taste testing in the rat. *Physiol. Behav.* 47: 795–803.

Spielman, A. I. (1990). Interaction of saliva and taste. *J. Dent. Res.* 69: 838–843.

Stenonis, N. (1661). *Glandulis oris & novis earundem vasis.* Leiden.

Stevens, D. A. (1975). Laboratory methods for obtaining olfactory discrimination in rodents. In *Methods in Olfactory Research,* D. G. Moulton. A. Turk, and J. W. Johnson, Jr., (Eds.), Academic Press, London, pp. 375–394.

Stewart, W. B., Kauer, J. S., and Shepherd, G. M. (1979). Functional organization of rat olfactory bulb analysed by the 2-deoxyglucose method. *J. Comp. Neurol.* 185: 715–734.

Strassmann, B. I. (1997). The biology of menstruation in *Homo sapiens*: total lifetime menses, fecundity, and nonsynchrony in a natural-fertility population. *Curr. Anthro.* 38: 123–129.

Stratton, G. M. (1917). *Theophrastus and the Greek Physiological Psychology before Aristotle.* George Allen & Unwin, London.

Stürckow, B. (1970). Responses of olfactory and gustatory receptor cells in insects. In *Advances in Chemoreception. Vol. I. Communication by Chemical Signals,* J. W. Johnston, Jr., D. G. Moulton, and A. Turk (Eds.). Appleton-Century-Crofts, New York, pp. 107–159.

Sulzer, M. (1752). Resherches sur l'origine des sentiments agreables et desagreables. Troisieme partie: Des plaisirs des sens. Historic de l'academie des sciences et belle lettrns de Berlin. Cited in Bujas, Z., Electrical Taste. In *Handbook of Sensory Physiology. Vol. IV. Chemical Senses. Sect. 2. Taste,* L. M. Beidler (Ed.). Springer-Verlag, New York, pp. 180–199.

Summers, M. (1926). *The History of Witchcraft and Demonology.* Knopf, New York.

Takagi, S. F. (1989). *Human Olfaction.* University of Tokyo Press, Tokyo.

Tanabe, T., Iino, M., Ooshima, Y., and Takagi, S. F. (1973). The olfactory center in the frontal lobe of the monkey. *J. Physiol. Soc. Jpn.* 35: 550.

Tanabe, T., Yarita, H., Iino, M., et al. (1975). An olfactory projection area in orbitofrontal cortex of the monkey. *J. Neurophysiol.* 38: 1269–1283.

Tanapat, P., Hastings, N. B., Reeves, A. J., and Gould, E. (1999). Estrogen stimulates a transient increase in the number of new neurons in the dentate gyrus of the adult female rat. *J. Neurosci.* 19: 5792–5801.

Tateda, H., and Beidler, L. M. (1964). The receptor potential of the taste cell of the rat. *J. Gen. Physiol.* 47: 479–486.

Teichner, W. H. (1966). A method for studying olfaction in the unrestrained rat. *J. Psychol.* 63: 291–297.

Teicher, M. H., Stewart, W. B., Kauer, J. S., and Shepherd, G. M. (1980). Suckling pheromone stimulation of a modified glomerular region in the developing rat olfactory bulb revealed by the 2-deoxyglucose method. *Brain Res.* 194: 530–535.

Temussi, P. (2006). The history of sweet taste: not exactly a piece of cake. *J. Mol. Recognit.* 19: 188–199.

Ter-Pogossian, M. M., Phelps, M. E., Hoffman, E. J. and Mullani, N. A. (1975). A positron-emission tomography for nuclear imaging (PET). *Radiology* 114: 89–98.

Thornhill, R. A. (1970). Cell division in the olfactory epithelium of the lamprey, *Lampetra fluviatilis. Z. Zellforsch.* 109: 147–157.

Tisdall, F. F., Brown, A., Defries, R. D., et al. (1937). Nasal spraying as preventive of poliomylitis. *Can. Public Health J.* 28: 431–434.

Tisdall, F. F., Brown, A., and Defries, R. D. (1938). Persistent anosmia following zinc sulfate nasal spraying. *J. Pediatr.* 13: 60–62.

Todd, R. B., and Bowman, W. (1847). *The Physiological Anatomy and Physiology of Man,* Vol. 11. Parker, London.

Touhara, K. (2007). Deorphanizing vertebrate olfactory receptors: recent advances in odorant-response assays. *Neurochem. Int.* 51, 132–139.

Touhara, K., Sengoku, S., Inaki, K., et al. (1999). Functional indentification and reconstitution of an odorant recptor in single olfactory neurons. *Proc. Natl. Acad Sci. USA* 96: 4040–4045.

Toulouse, E., and Vaschide, N. (1899). Mesure de l'odorat chez l'homme et chez la femme. *Comp. Rend. Soc. Biol.* 51: 381–383.

Valentin, G. (1848). *Lehrbuch der Physiologie des Menschen.* Braunschweig.

Vandenbergh, J. G. (1969). Male odor accelerates female sexual maturation in mice. *Endocrinology* 84: 658–660.

Vandenbergh, J. G. (1983). *Phermones and Reproduction in Mammals.* Academic Press, New York.

Van der Lee, S. and Boot. L. M. (1955). Spontaneous pseudopregnancy in mice. *Acta Physiol. Pharm. Neerl.* 4: 442–444.

van der Veen, M., and Morales, J. (2014). The Roman and Islamic spice trade: New archaeological evidence. *J. Ethnopharmacol.* http://dex.doi.org/10.1016/j.jep.2014.09.036.

Van Houten, J. (2000). Chemoreception in microorganisms. In: T. E. Finger, W. L. Silver and D. Restrepo (Eds.), *The Neurobiology of Taste and Smell,* 2nd edition. New York: Wiley-Liss, Inc., pp. 11–40.

Varro, M. T. (116–28 BCE). *Rerum Rusticarium,* Lib. 1.

Verendeev, A. L., and Riley, A. (2012). Conditioned taste aversion and drugs of abuse: history and interpretation. *Neurosci. Biobehav. Rev.* 36: 2193–2205.

Verrill, A. H. (1940). *Perfumes and Spices.* L. C. Page, Boston.

Vesalius, Andrea. (1543). De Humani Corporis Fabrica. *Basileae.*

Vintschgau, M. and von Hönigschmied, J. (1877). Nervus glossopharyngeus und Schmeckbecher. *Arch. Ges. Physiol.* 14: 443–448.

Vintschgau, M. von (1880). Physiologie des Geschmacksines. In L. Hermann (ed), *Handbuch der Physiologie der Sinnesorgane.* Leipzig: Vogel, pp. 145–224.

Vivino, A. E. (1960). Perfumes and perfumery. In *The Encyclopedia Americana.* Americana Corporation, New York, pp. 577–582.

Vogt, R. G., Prestwich, G. D., and Lerner, M. R. (1991). Odorant-binding-protein subfamilies associate with distinct classes of olfactory receptor neurons in insects. *J. Neurobiol.* 22: 74–84.

Volta, A. (1792). Briefe über thierische Electricität. In *Ostwald's Klassiker der exakten Wissenschaten,* A. J. Oettingen (Ed.). Engelmann, Leipzig, 1900. Cited in Bujas, Z., Chapter 10, Electrical Taste. In *Handbook of Sensory Physiology.* Vol. IV. Chemical Senses. Sect. 2. Taste, L. M. Beidler (Ed.). Springer-Verhag, New York, pp. 180–199. [Original: Volta, A., 1816 Sull elettricitá animale, In: Collezione dell' Opere. G.Piatti, Firenze, 268pp. Vol. 2, Part1 (see pp. 55–118)].

von Brunn, A. (1875). Untersuchungen über das Riechepithel. *Arch. Mikroskop. Anat.* 11: 468–477.

von Brunn, A. (1880). Weitere Untersuchungen über das Riechepithel und sein Verhalten zum Nervus olfactorius. *Arch. Mikroskop. Anat.* 17: 141–151.

References

von Brunn, A. (1889). Zwei Mikroskopische Präparate vom Riechepithel eines Hingerichteten. *Verh. Anat. Ges.* 3: 133–134.

von Brunn, A. (1892). Beiträge zur mikroskopischen Anatomie der menschlichen Nasenhöhle. *Arch. Mikroskop. Anat.* 39: 632–651.

von Frisch, K. (1919). Über den Geruchssinn den Biene und seine blütenbiologische Bedeutung. *Zool. Jahrb. Physiol.* 37: 1–238.

von Frisch, K. (1921). Über den Sitz des Geruchssinnes bei Insekten. *Zool. Jahrb. Physiol.* 38: 449–516.

von Frisch, K. (1922). Morphologische and biologische Untersuchungen der Putzapparate der Hymenopteren. *Arch. Naturgeschichte Abt. A* 88: 1–63.

von Haller, A. (1756). *Olfactus, Elementa physiologiae corporis humani.* Liber XlV Tomus Quintus, Francisci Grasset, Lausanne, pp. 125–185.

von Skramlik, E. (1926). *Handbuch der Physiologie der niederen Sinne.* Georg Thieme-Verlag, Leipzig.

Wekesa, K. S. and Anholt, R. R. H. (1999). Differential expression of G proteins in the mouse olfactory system. *Brain Res,* 837: 117–126.

West, S. E. and Doty, R. L. (1995). Olfactory function in epilepsy and temporal lobe resection lobectomy: A review. *Epilepsia* 36: 531–542.

Wetzel, C. H., Behrendt, H. J., Gisselmann, G., et al. (2001). Functional expression and characterization of a Drosophila odorant receptor in a heterologous cell system. *Proc. Natl. Acad. Sci. USA* 98: 9317–9380.

Wharton, T. (1656). Adenographia. London.

Whitten, W. K. (1956). Modification of the oestrous cycle of the mouse by external stimuli associated with the male. *J. Endocrinol.* 13: 399–404.

Whitten, W. K., Bronson, F. H., and Greenstein, J. A. (1968). Estrus-inducing phermone of male mice: transport by movement of air. *Science* 161: 584–585.

Whittle, C. E., Fakharzadeh, S., Eades, J., and Preti, G. (2007). Human breath odors and their use in diagnosis. *Ann. N. Y. Acad. Sci.* 1098: 252–266.

Winans, S. S., and Powers, J. B. (1974). Neonatal and two-stage olfactory bulbectomy: effects on male sexual behavior. *Behav. Biol.* 10: 461–471.

Willis, T. (1681). *Five Treatises.* Dring, Harper, Leight and Martin, London.

Wilson, H. C. (1992). A critical review of menstrual synchrony research. *Psychoneuroendocrinology* 17: 565–591.

Wistrand, M., Kall, L., and Sonnehammer, E. L. (2006). A general model of G protein coupled receptor sequences and its application to detect remote homologs. *Protein Sci.* 15: 509–521.

Wolozin, B., Lesch, P., Lebovics, R. and Sunderland, T. (1993). Olfactory neuroblasts from Alzheimer donors: studies on APP processing and cell regulation. *Biol. Psychiat.* 34: 8241–8380.

Wolozin, B., Sunderland, T., Zheng, B., et al. (1992). Continuous culture of neuronal cells from adult human olfactory epithelium. *J. Mol. Neurosci.* 3:137–146.

Wong, S. T., Trinh, K., Hacker, R., et al. (2000). Disruption of the type III adenylyl cyclase gene leads to peripheral and behavioral anosmia in transgenic mice. *Neuron* 27: 487–497.

Wright, J. (1914). *A History of Laryngology and Rhinology.* Lea & Febiger, Philadelphia.

Wu, S. V., Rozengurt, N., Yang, M., et al. (2002). Expression of bitter taste receptors of the T2R family in the gastrointestinal tract and enteroendocrine STC-1 cells. *Proc. Natl. Acad. Sci.U.S.A.,* 99: 2392–2397.

Wysocki, C. J. (1979). Neurobehavioral evidence for the involvement of the vomeronasal system in mammalian reproduction. *Neurosci. Biobehav. Rev.* 3: 301–341.

Wysocki, C. J., Katz, Y., and Bernard, R. (1983). The male vomeronasal organ mediates female-induced testosterone surges. *Biol. Reprod.* 28: 917–922.

Wysocki, C. J., Bean, N. J., and Beauchamp, G. K. (1986). The mammalian vomeronasal system: its role in learning and social behaviors. In *Chemical Signals in Vertebrates 4,* D. Duvall, D. Müller-Schwarze and R. M. Silverstein (Eds.) New York, Plenum Press, pp. 471–485.

Yanagisawa, K., Bartoshuk, L. M., Catalanotto, F. A., et al. (1998). Anesthesia of the chorda tympani nerve and taste phantoms. *Physiol. Behav.* 63: 329–335.

Young, J. M. and Trask, B. J. (2002). The sense of smell: genomics of vertebrate olfactory receptors. *Hum. Molec. Genet.* 11: 1153–1160.

Young, P. T. (1949). Studies of food preference, appetite and dietary habit: V. Techniques for testing food preference and the significance of results obtained with different methods. *Comp. Psychol. Monogr.* 19: 1–58.

Yousem, D. M., Williams, S. C., Howard, R. O., et al. (1997). Functional MR imaging during odor stimulation: preliminary data. *Radiology,* 204: 833–838.

Zatorre, R. J., Jones-Gotman, M., Evans, A. C., and Meyer, E. (1992). Functional localization and lateralization of human olfactory cortex. *Nature* 360: 339–340.

Zhang, X. and Firestein, S. (2002). The olfactory receptor gene superfamily of the mouse. *Nat. Neurosci.* 5:124–133.

Zhao. H., Ivic, L, Otaki, J. M., et al. (1998). Functional expression of a mammalian odorant receptor. *Science* 279: 237–242.

Zhao H, Zhou Y, Pinto CM, et al. Evolution of the sweet taste receptor gene Tas1r2 in bats. *Mol. Biol. Evol.* 2010 27: 2642–2650.

Zippel, H. P. (1993). Historical aspects of research on the vertebrate olfactory system. *Naturwissenschaften* 80: 65–76.

Zotterman, Y. (1935). Action potentials in the glossopharyngeal nerve and in the chorda tympani. *Scand. Arch. Physiol.* 72: 73–77.

Zozulya, S., Echeverri, F. and Nguyen, T. (2001). The human olfactory receptor repertoire. *Genome Biol.* 2(6): RESEARCH0018. Epub 2001 Jun 1.

Zuniga, J. R., Davis, S. H., Englehardt, R. A., et al. (1993). Taste performance on the anterior human tongue varies with fungiform taste bud density. *Chem. Senses* 18: 449–460.

Zwaardemaker, H. (1925). *L'Odorat.* Doin, Paris.

Part 2

Olfaction
Olfactory Anatomy and Neurobiology

Chapter 2

Anatomy of the Nasal Passages in Mammals

TIMOTHY D. SMITH, THOMAS P. EITING, and KUNWAR P. BHATNAGAR

2.1 INTRODUCTION

Human nasal anatomy has been well-documented for centuries. The human nasal passages represent the proximal (upper, or facial) part of the respiratory tract, and communicate posteriorly with the pharynx, a shared conduit of the respiratory and digestive tracts (Figure 2.1a). In skeletal anatomy, the piriform aperture is the portal into the nasal cavity (Figure 2.1b). This opening is elaborated by cartilaginous support structures that house the downwardly-oriented, fleshy anterior nares (Figure 2.1c). The anterior nares open to a dilated vestibule on each side. Between the right and left vestibules is the columella, a fleshy structure right under the tip of the nose (Figure 2.1c). The columella is supported internally by the nasal septal cartilage, which extends posteriorly to join the midline portions of the ethmoid and vomer bones. The nasal cavity is thus separated into right and left halves, or nasal fossae, beginning at the anterior nares. The end of the nasal passages is commonly located with reference to paired exit points, the choanae (or posterior nares), that transmit air from the nasal fossae into the nasopharynx (Figures 2.1a,d). The nasal choanae are separated by the posterior end of the nasal septum, which is supported by the vomer bone.

The nasal septum is composed of three elements: the septal cartilage, the perpendicular plate of the ethmoid, and the vomer bone. These three are all related in that the nasal septal cartilage articulates posteriorly with the perpendicular plate of the ethmoid and the vomer bone (Figure 2.1e). In the nasal cavity, this median (septal) component forms the medial boundary for right and left nasal fossae. The lateral wall represents the most complex boundary of the nasal fossa (Figures 2.1f,g). Formed by portions of the maxillary, ethmoid, lacrimal, and palatine bones, this wall has both invaginations that extend into the nasal airway and ostia that communicate to paranasal spaces (Figure 2.1f). The cribriform plate of the anterior cranial base is the roof of each nasal fossa, while the palate forms the floor (Figures 2.1f,g).

Once inspired air is drawn beyond the vestibule, it enters the nasal cavity proper, in which the air passages become more complex. For this reason, the lateral wall of the nasal fossa has received much attention. As reviewed by Wright (1914), the human nasal conchae (turbinates, turbinals) that arise from the lateral walls of the ethmoid bone (Figures 2.1f,g) extend into the nasal passages. These were described to some degree by such early anatomists as Galen, Colombo, and Ingrassias. Their function is now understood to be in both air-conditioning and olfaction (see review by Clerico et al., 2003). In addition, they divide the nasal airways into compartments, called meatuses. The inferior nasal concha articulates with the maxillary bone (a basis for its alternate name – maxilloturbinal; Figure 2.1f). The middle and superior nasal conchae are part of the lateral mass of the ethmoid bone.

Interestingly, Galen had recognized the role of turbinals in altering, cleansing, and warming the inspired air, although he apparently did not clearly differentiate between the different turbinals. An exacting description of the turbinals was made by von Haller in his "Icones anatomicae" (1756). The nasal turbinals received their name from Casserius, who had described the turbinals not only of humans, but also of other species, including

---

Handbook of Olfaction and Gustation, Third Edition. Edited by Richard L. Doty.
© 2015 Richard L. Doty. Published 2015 by John Wiley & Sons, Inc.

Chapter 2 Anatomy of the Nasal Passages in Mammals

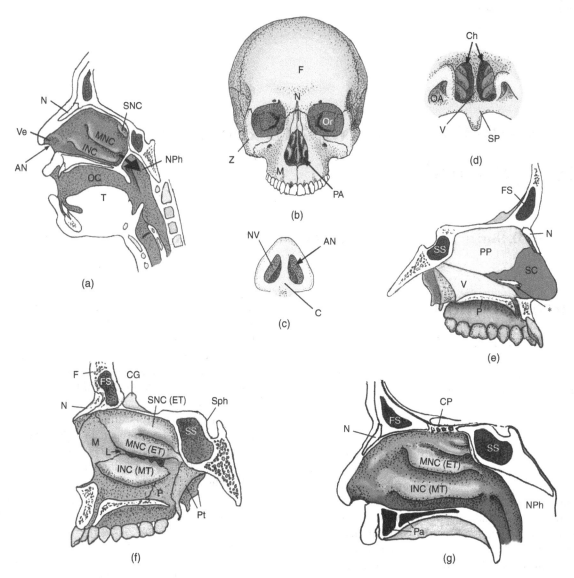

Figure 2.1 Anatomy of human nasal passages. (a) The nasal passages are the upper part of the respiratory tract. On each side, air enters the nasal passage through the anterior nares (AN) and joins the pharynx by passing through the choana (route indicated by arrow). In nasal breathing (as opposed to breathing through the oral cavity – OC), inspired air passes across the complex surfaces of the nasal cavity. First, it is drawn into a small hairy chamber, the vestibule (Ve), in which vibrissa filter relatively large airborne items. Subsequently, the air is drawn across the inferior, middle, and superior nasal concha (INC, MNC, SNC) that extend into the airways with mucosa that functions for air conditioning or olfaction (see text for more details). (b) The piriform aperture is anterior-most osseous opening of the nasal airways, shown here in a frontal view of the skull. (c) In the flesh, nasal cartilages and the integument extend from the piriform aperture to form downwardly-oriented anterior nares, shown here from an inferior perspective. The columella (C) is the fleshy base of the nasal septum, which divides the nasal cavity into right and left sides. (d) The choanae (Ch), seen here in a posterior view, transmit air to the nasopharynx. These channels are separated into right and left sides by the posterior end of the nasal septum, formed by the vomer bone (V). (e) Structural support for the nasal septum is formed by the nasal septal cartilage (SC), vomer bone, and the perpendicular plate of the ethmoid bone (PP). This forms the median boundary for both halves of the nasal cavity. The vestigial vomeronasal organ (*) is found within the mucosa of the nasal septum. (f) A lateral view reveals the numerous bones that form the lateral wall of the nasal fossa, including the maxilla (M), the inferior nasal concha (INC), the ethmoid (the MNC and SNC are parts of this bone), the palatine (P), and lacrimal (L; most of this bone is hidden by the MNC. These bones comprise the lateral wall (g). The "floor" of the nasal cavity is formed in part by the hard palate (Pa) and the "roof" of the nasal cavity is formed by the cribriform plate of the ethmoid (CP), the nasal bone (N), and the sphenoid bone (Sph), shown surrounding the sphenoid sinus (SS). CG, crista galli; ET, ethmoturbinal (alternate name for MNC, SNC); F, frontal bone; FS, frontal sinus; M, maxillary bone; MT (maxilloturbinal, alternate name of INC)); NPh, nasopharynx; NV, "nasal valve"; Pt, pterygoid plates; T, tongue; Z, zygomatic bone.

the cow, horse, sheep, hare, cat, and dog. He stated in his 1610 work, *Pentaesthesion*, that hidden in the depth of the nostrils are

> oblong little bones which may be called spongy, and seem like the steps of a ladder, because one is placed above the other. ... Hippocrates not inaptly calls them sleeves. Turbines I would call them from their form and function. They are bones, not cartilages. Turbinated bones (Turbinata Ossa) they are rightly called. They are usually three in number, indeed this many at least always.

While knowledge of the engorgement of the erectile tissue covering the turbinals was known by Ruppert as early as 1754, the classic description of the vascular erectile tissue within the nasal mucosa was made in 1853 by Kohlrausch. The so-called nasal cycle – the fluctuating side-to-side variations in nasal engorgement – was quantified by Kayser in 1895. This discovery resulted from his invention of an ingenious device employing calibrated bellows to withdraw a known volume of air through the nose from the oral cavity. By determining the time required to draw the air, he was able to establish an index of nasal resistance for each side of the nose.

The literature on human nasal anatomy, spanning centuries, has thus revealed great complexity. By comparison, a broader understanding of mammalian nasal anatomy has been somewhat hampered because our picture of evolutionary history of nasal anatomy relies mostly on comparisons of living vertebrates, since nasal anatomy fossilizes quite poorly. Moreover, destructive methods and delicate dissection was required to study other vertebrates until quite recently. Modern methods such as computed tomography (CT) have greatly increased our breadth of knowledge (e.g., Rae and Koppe, 2003; Rossie, 2006). In this chapter we provide a comparative overview of vertebrate nasal anatomy, with a primary focus on mammals.

2.2 EVOLUTION OF COMPLEXITY IN THE VERTEBRATE NASAL CAVITY

2.2.1 Comparative Nasal Anatomy of Terrestrial Vertebrates

The complexity of the nasal cavity of land dwelling vertebrates is intertwined with the evolution of air filtration, breathing, water recovery, and, in some species, the transmission of vocalizations, including echolocation calls of bats.[1] The vertebrate nasal cavity begins as a bilateral invagination of the nasal pits, paired depressions that form

at the juncture of the maxillary and medial and lateral nasal prominences (Dieulafe, 1906). These pits invaginate into deep nasal sacs, which are bilateral in most extant vertebrates.[2] In postnatal morphology, the respiratory and digestive tracts of air breathing vertebrates often share the same lumina, but this is not the case for more primitive forms. In most species of fish, for example, the nasal sac is entirely separated from the oropharyngeal passageways (Negus, 1958). This changed as vertebrates began to invade terrestrial habitats. While amphibians typically retained relatively simple nasal cavities in the form of nasal sacs, most amniotes – vertebrates fully adapted to terrestrial life – developed complex nasal cavities containing secondary and tertiary turbinals distinct from the ridges found in amphibians (Negus, 1958).

Extant reptiles have a wide range of complexity in internal nasal anatomy. Turtles have a nasal cavity similar to that of amphibians, whereas crocodilians have a somewhat more complex nasal anatomy, possessing three turbinals (Parsons, 1967). Birds and mammals have more complex turbinals than reptiles – complexity that is often related to endothermic regulation of body temperature (Hillenius, 1994; Geist, 2000). In some mammals, the turbinals create a complex labyrinth of recesses, thereby increasing internal surface area and compartmentalizing the internal nose into functional regions. Because the nasal mucosa is specialized for more than one function (Negus, 1958; Moore, 1981; Hillenius, 1992), variations of the mammalian nasal cavity reveal respiratory and special sensory adaptations.

2.2.2 Development of the Mammalian Nasal Cavity

The complexity of the nasal cavity in mammals arises from the embryonic and early fetal development of a series of mesenchymal condensations. In some cases, the larger turbinals[3] can be considered primary structures, since they chondrify at or near the same time as the outer walls of the nasal cavity (Vidic, 1971; Moore, 1981). In other cases, the turbinals are clearly secondary, forming from mesenchymal condensations that develop after most of the nasal capsule has chondrified (Smith and Rossie, 2008; Smith et al., 2012). The complexity of individual turbinals in adults appears to arise progressively during development. This developmental progression in mammals that have complex turbinals has been described as

---

[1] The effects of some of the more specialized functions on the form and functional anatomy of the nasal passages (e.g., sound transmission and echolocation) are interesting and perhaps understudied, but will not be considered further in this chapter.

[2] These pits form a single cavity in some chordates, such as the lamprey (Dieulafe, 1906). However, it forms via fusion of the two growing nasal pits, and the lamprey larvae preserve evidence of this fusion in their bilobed nasal sac (Weichert, 1958).

[3] The term "turbinal" is preferred here since it is relatively more commonly used, in recent decades, than "turbinate" in work on evolution and development of the mammalian nasal cavity. However, these terms are entirely synonymous when used with prefixes (endo, fronto, maxillo, etc.).

42   **Chapter 2   Anatomy of the Nasal Passages in Mammals**

"splitting" (Parker, 1874), although this may more appropriately be described as a "budding" of accessory lamellae via mesenchymal proliferation.

The largest turbinals within the mammalian nasal cavity project closest to the midline, and are often provided with a collective positional name, the "endoturbinals" (Figure 2.2; Table 2.1). Smaller turbinals, hidden from view deep in the midline row of turbinals are sometimes referred to as "ectoturbinals." However, due to inconsistent usage and developmental factors (see Smith et al., 2011, and Macrini, 2012 for recent critical evaluations), we use other terms here that relate turbinals to the bones with which they ultimately articulate: maxilloturbinals, nasoturbinals, and ethmoturbinals (Figure 2.2; Table 2.1). Smaller turbinals include those that reside in recesses (e.g., the fronto-turbinals) and those that exist between ethmoturbinals (interturbinals) (Maier, 1993a, b). The interturbinals are secondary structures, beginning as mesenchymal condensations after other adjacent regions have started to chondrify (Smith and Rossie, 2008; Smith et al., 2012).

The turbinals develop as inward projections of the nasal capsule cartilage that surrounds and supports the nasal airways during fetal development. The largest space within the nasal airway, bordered by the largest proportion of internal nasal surface area, is termed the "nasal cavity proper." In humans, this space corresponds to the entire nasal cavity (excluding paranasal spaces and vestibule). Most mammals have a more extensive nasal cavity proper compared to humans, including a posterior dead-ended space, the olfactory recess. The olfactory recess will be discussed functionally below. Developmentally, inwardly-projecting shelves from the inferior margin of the pars posterior form part of the floor of the olfactory recess. On each side, these shelves are formed by cartilaginous elements (lamina transversalis posterior) that join choanal (lateral) processes of the vomer, thus sealing off a more superior space (the olfactory recess), from the remainder of the nasal fossa (the nasopharyngeal duct). Once ossified, this horizontal plate is referred to as the transverse lamina (Figures 2.2, 2.3).

A final consideration on the fetal template is the organization of accessory (paranasal) spaces within the nasal cavities. Most paranasal spaces form where embryonic parts of the nasal capsule intersect, due to folding of the middle portion of the nasal capsule (Rossie, 2006). The paranasal spaces are accessory chambers, or recesses (Figure 2.3). Although humans do possess paranasal spaces, which can pneumatize into fairly large paranasal sinuses postnatally, many mammals have more complex paranasal spaces, or recesses, compared to humans and other "higher" primates.[4] The difference in complexity may be related to a greater amount of encroachment of the nasal cavity by the orbits in primates. In most mammals, the space between the orbits is wide-set enough to make room for portions of the paranasal recesses (especially, the frontal recess). Another consideration pertains to development of the paranasal space after the cartilaginous capsule is mostly ossified or resorbed. If a paranasal recess is enlarged by the process of secondary pneumatization, the term paranasal "sinus" is instead used. In recent literature, the process of secondary pneumatization denotes a critical distinction between paranasal spaces (see Witmer, 1999; Rossie, 2006; Smith et al., 2008, 2010; Macrini, 2012). For example, while the maxillary recess is ubiquitous among extant mammals, it is not clear how commonly the space is pneumatized beyond the limits established by the cartilaginous nasal capsule. Therefore, possession of a maxillary "sinus" cannot presently be considered a primitive feature of mammals.

2.3 MAMMALIAN NASAL ANATOMY

2.3.1 General Organization of the Mammalian Nasal Region

Structures within the mammalian nose are bilateral. The exceptions are the columella and the septal core (cartilaginous and/or bony), these being the median and single structures. The nasal septum is a unique structure which has both median as well as bilateral components. While the bony or cartilaginous septum is unpaired, all the mucosal elements, such as the vomeronasal organ (VNO), septal olfactory organ (SOO), paraseptal cartilages, the various nerves, blood vessels, glands and ducts, are bilateral structures within the septum. Other structures, such as the paravomeronasal ganglion that bears connection to vomeronasal nerves, have been observed in the septal lamina propria of some species of bats and primates (Bhatnagar and Kallen, 1974a; Bhatnagar and Smith, 2007; Smith and Bhatnagar, 2009). This structure may merit a more in-depth investigation for its link, if any, to the terminal nerve.

The prototypical mammalian nasal cavity is divided into right and left nasal fossae by the midline nasal septum. As in human anatomy, the nasal fossa may be divided into a small, anterior, vestibule, and the nasal fossa proper (Figure 2.3). The vestibule may be extremely abbreviated or a small "chamber" of its own (as in humans). Its transition to the nasal fossa proper is indicated by a shift in mucosal type, and may be marked by a small ridge (e.g., the limen

---

[4]"Higher" primates refers to anthropoid primates (monkeys, apes, and humans). However, whenever discussing reduction of nasal anatomy we could equally be referring to Southeast asian tarsiers – primates that possess some primitive traits. Tarsiers, like anthropoids, have marked reduction of the nasal cavity. This may be especially related to the large eye size in these nocturnal species.

2.3 Mammalian Nasal Anatomy

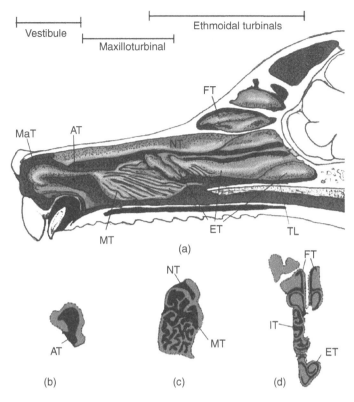

Figure 2.2 (a) Lateral nasal wall of the dog, showing major cartilaginous and bony structures in mammals. Descriptions of the lateral nasal wall are reflection of its great intricacy. For example, five regions of the nasal spaces are described in canine anatomy: vestibular, maxilloturbinate, nasomaxillary, ethmoid, and frontal sinus regions (Craven et al., 2007). These regions correspond almost neatly to physiological compartments, such as the highly branched canine maxilloturbinal that warms and humidifies inspired air, or the ethmoid that supports most of the olfactory neuroepithelium. The regions of the internal nose, which overlap to a varying degree in mammals, are indicated above. Cross-sections of these regions are shown below, (b) vestibule, showing the atrioturbinal (AT), (c) maxilloturbinal, (d) ethmoturbinals in the olfactory recess, with frontal recesses above. Cross sections are illustrated after (Craven et al., 2007). In B-D, turbinals are emphasized in black and the airway is shown in gray. ET, ethmoturbinal; FT, frontoturbinal; IT, intertrubinal; MaT, marginoturbinal; MT, maxilloturbinal; NPD, nasopharyngeal duct; NT, nasoturbinal; TL, transverse lamina.

nasi of humans). The osseous and cartilaginous boundaries of the nasal cavities are essentially the same as described for humans above, except the premaxillary bone is a separate element that forms lateral and inferior boundaries at the anterior limit (this bone is completely merged with the maxilla in humans). The extent to which certain bones (e.g., nasal) border the nasal cavity varies greatly among mammals, and the recesses and turbinals are more elaborate in many mammals compared to humans (Smith et al., 2007a; Stößel et al., 2010). The cribriform plate (essentially a bilateral structure divided by a bony ridge, the crista galli) stands between the posterior nasal fossa and the anterior cranial fossa. The various nerves (olfactory, vomeronasal, trigeminal branches, autonomic, nervus terminalis) pass through the openings in the plate to the olfactory bulb and other central brain regions. Blood vessels also enter or leave through their own foramina (see Bhatnagar and Kallen, 1974b).

Some nasal regions appear to bear physiological significance in a broad range of mammals. For instance, the nasal fossa may be divided into nasal and paranasal component spaces (Figure 2.3b). In addition, the more midline nasal spaces are sometimes divided physiologically into different spaces (meatuses), associated with respiratory or special sensory roles (Figure 2.3). In each of these spatial constructs, the turbinals play an important role in directing airflow to compartments of the nasal airway (see below).

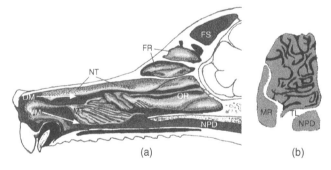

Figure 2.3 (a) Lateral nasal wall of the dog, showing major internal nasal spaces in mammals. Two major airflow routes are also indicated. The dorsal meatus (DM) is beneath the nasoturbinal (NT); the ventral meatus (VM) is beneath the maxilloturbinal (MT). A cross-section just anterior to the transverse lamina (b) shows lateral spaces including the maxillary recess (hidden from view in (a). Cross section is illustrated after Craven et al. (2007). FR, frontal recess; FS, frontal sinus; FT, frontoturbinal; MT, maxilloturbinal; NPD, nasopharyngeal duct; NT, nasoturbinal; OR, olfactory recess; TL, transverse lamina.

The organization of the canine nasal cavity provides an excellent example of a mammal in which turbinals and spaces can be discussed according to relatively discrete functions (Craven et al., 2007). However, there are departures from this regional specialization in species with proportionally reduced nasal dimensions (Figure 2.1, and

Table 2.1 Terminology for the bony and cartilaginous structures and recesses of the nasal fossa (based on Smith and Rossie, 2008)

Structure	Homologous Human Structure[1]	Synonyms and Comments
Structures		
nasoturbinal (entire)	see below	concha nasalis dorsalis[2]; endoturbinal I (Moore, 1981)
nasoturbinal, mucosal part	aggar nasi	pars rostralis[2]; nasoturbinal (Moore, 1981);
(soft tissue structure)		aggar nasi region of nasoturbinal (Kollmann and Papin, 1925)
nasoturbinal, osseous part	semicircular crest and	pars caudalis[2]; ethmoturbinal I (Martin, 1990);
	uncinate process are	endoturbinal I (Paulli, 1901); nasoturbinal, region of uncinate process
	remnants	(Kollmann and Papin, 1925)
maxilloturbinal	inferior nasal concha	concha ventralis[2]; maxilloturbinate; ventral concha
ethmoturbinal I[3]	middle nasal concha	concha media[2]; ethmoturbinal II (Martin, 1990); endoturbinal II, upper lamella (Paulli, 1901; Moore, 1981); cardinal concha II (Pedziwiatr, 1972); endoturbinal I (Allen, 1882)
ethmoturbinal II	?	superior nasal concha concha ethmoidalis[2]; ethmoturbinal III (Martin, 1990); endoturbinal II, lower lamella (Paulli, 1901; Moore, 1981)
ethmoturbinal III	?	concha ethmoidalis[2]; ethmoturbinal IV (Martin, 1990); endoturbinal II (Paulli, 1900a,b,c; Moore, 1981)
ethmoturbinal IV	?	concha ethmoidalis[2]; ethmoturbinal V (Martin, 1990); endoturbinal III (Moore, 1981)
ethmoturbinal V[4]	?	
semicircular crest	same	continuous with, and part of, nasoturbinal, osseous part
frontoturbinal	ethmoid bulla	ectoturbinal (Moore, 1981)
frontomaxillary septum	-	lateral root of ethmoturbinal I (Rossie, 2006); anterior root of ethmoturbinal I (de Beer, 1937); horizontal lamina (Maier, 1993a, b)
interturbinal (after Maier, 1993b)		ectoturbinal (Paulli, 1900a, b, c; Moore, 1981); accessory turbinal (Dieulafe, 1906)
transverse lamina	ossiculum Bertini?	lamina terminalis (Kollmann and Papin, 1925; Hill, 1953) posterior transverse lamina (Macrini, 2012); subethmoidal shelf (Negus, 1958)
vomeronasal cartilage	paraseptal cartilage (vestige)	anterior paraseptal cartilage

Spaces

lateral recess	maxillary recess	properly refers to a fetal paranasal space which results in an elongated in fetal human cavity in adults of many mammals (anterolateral and posterolateral recesses); see Rae and Koppe (2003) regarding confusion of this term regarding Old World primates
anterolateral recess		recessus anterior (de Beer, 1937); maxillary sinus (Rowe et al., 2005, see Figure 2.5 therein)
posterolateral recess		(this is equivalent to both the frontal and maxillary recesses, together - Smith and Rossie, 2008)
frontal recess		superior maxillary recess (Negus, 1958, p. 313)
maxillary recess	fetus: maxillary recess postnatal: maxillary sinus[5]	inferior maxillary recess (Negus, 1958, p. 313)
olfactory recess	sphenoethmoidal recess (vestige)	sphenoethmoidal recess, ethmoturbinal recess (Maier, 1993a, b); cupular recess (Van Gilse, 1927); sphenoidal recess (Loo, 1974)

[1] See Moore (1981) and Smith and Rossie (2006, 2008) for further discussion.

[2] NAV (Nomina Anatomica Veterinaria, 2005).

[3] Terms for these evaginations within the nose of terrestrial vertebrates are most commonly termed turbinates or turbinals. But nomenclature is complex and, in some cases, different turbinals have been referred to using the same term (for more detailed accounts, see Moore, 1981; Novacek, 1993; Rowe et al., 2005; Smith and Rossie, 2006, 2008; Smith et al., 2007a; Smith et al., 2011; Macrini, 2012).

[4] Macrini (2012), in a thorough reconsideration of previous studies, notes as many as 18 ethmoturbinals (p. 73, termed "endoturbinals") have been reported in the monotreme *Tachyglossus*.

[5] The maxillary sinus, technically, results from pneumatic expansion beyond the limits of the fetal maxillary recess in humans.

46   Chapter 2   Anatomy of the Nasal Passages in Mammals

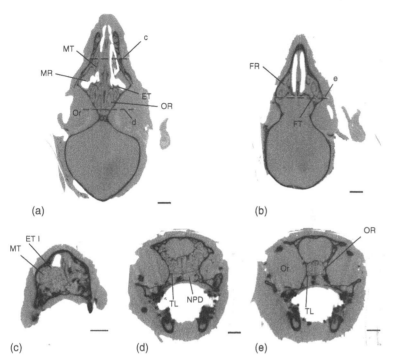

Figure 2.4 Micro-CT slices through the nasal region of adult tree shrews (*Tupaia glis*). Horizontal slices are shown in (a) and (b), at more ventral and dorsal levels of the nasal cavity, respectively. (c) a coronal slice through nasal cavity is shown (approximately at the plane indicated in (a). Here it can be seen that the first ethmoturbinal (ET I) extends far anteriorly, overlapping the maxilloturbinal (MT). (d) a coronal slice showing the nasal cavity at a more posterior level, where the olfactory recess (OR, in (a) is separated from the nasopharyngeal duct (NPD) by the transverse lamina (TL). (e) a coronal slice showing the olfactory recess at a more anterior level, where the frontal recess and olfactory recess overlap. Scale bars: 2 mm.

see below), and many other mammals have more overlap of these regions than seen in the dog (Figure 2.4).

Major variations in the nasal passages have been documented from the anterior nares to the choanae. For some, adaptive explanations have been proposed. The size and shape of the anterior narial opening correlates with the ecological niche for the species (e.g., slit-like in the camel and in the mouse-tailed bat, *Rhinopoma*, which are denizens of arid environments, and for which respiratory water conservation is crucial). In animals which have a snout or trunk (e.g., elephant, aardvark, ant-eaters, male proboscis monkey, tube-nosed bat), the narial openings are situated at the snout/trunk tip. Posteriorly, the extent to which the vomer divides the posterior nasopharyngeal passages varies. In some mammals there are no choanae; instead a single common nasopharygeal duct connect the nasal airways to the nasopharynx (Smith et al., 2012). The significance of this anatomy (e.g., to airflow) has yet to be clarified. Structural variation of the lateral nasal wall and paranasal spaces is well-studied, which will be considered further below.

2.3.2 Anatomical and Physiological Roles of the Turbinals

The inferolateral margin of the nasal fossa is framed by inwardly scrolled edges of the nasal capsule, the most anterior of which do not ossify. Forming an inferolateral ridge along the anterior naris is the marginoturbinal (Maier, 2000). This turbinal is mobile, owing to insertions of some muscles of facial expression, and thus can modify airflow. In most mammals, the marginoturbinal forms a continuous ridge with the cartilaginous atrioturbinal in the vestibule. The latter is continuous with the osseous maxilloturbinal in the nasal fossa proper.

Articulating mostly with the maxillary bone, the maxilloturbinal varies from a simple, single scroll (as in the human inferior nasal concha) to a highly branched network of lamellae, with a correspondingly enormous surface area (Negus, 1958; Van Valkenburgh et al., 2004; Figure 2.2c). The maxilloturbinal plays important roles in heat and water conservation and in saturating inspired air before it reaches the lungs. Species with very large maxilloturbinal surface areas (e.g., pinnipeds) often have a great need to conserve both heat and moisture.

Ethmoturbinals are part of the lateral mass of the ethmoid bone. These are rarely branched to the same degree as the maxilloturbinal, but they can be complex nonetheless (e.g., see Macrini, 2012). Ethmoturbinals may vary more in overall number than individual complexity (Cave, 1973; Smith et al., 2007a). The nasoturbinal articulates with the inferior margin of the nasal bone, but also extends posteriorly into the ethmoid complex. In some mammals, the nasoturbinal is a ridge that runs continuously from the anterior nasal fossa to merge posteriorly with ethmoturbinal I. In such cases, the most anterior part may be a glandular swelling, while more posterior parts may ossify from portions of the cartilaginous nasal capsule (see Smith and Rossie, 2008, for more details). To the extent that the nasoturbinal is osseous, the bone itself is part of the ethmoid.

2.3 Mammalian Nasal Anatomy

Smaller turbinals are exceedingly variable in number. The frontoturbinals are located within the frontal recess. A thorough reassessment of the number of frontoturbinals is arguably important, because previous work has collectively combined frontoturbinals and other small turbinals according to spatial position away from the midline (i.e., ectoturbinals). Since these are distinguished as elements of the paranasal as opposed to more midline spaces, the frontoturbinals have been argued to receive different air streams, and possibly different odorant types (see below), during inspiration. Interturbinals occur closer to midline spaces. Specifically, one or more interturbinals may be found in the posterior nasal fossa between two adjacent ethmoturbinals (Smith et al., 2011; 2012).

2.3.3 Microanatomy

Variations in internal nasal morphology have served as the basis for inferring functional processes, including olfaction and both the warming and humidifying of inspired air. Although the skeletal morphology may be used to infer some physiological specializations (see below), the actual functional anatomy is mostly dictated by the distribution of specific types of nasal epithelia, as well as solitary chemosensory cells that are innervated by the trigeminal nerve (Finger et al., 2003; Ogura et al., 2010). Broadly, the nasal fossa is lined with four types of epithelia: stratified, respiratory, transitional, and olfactory (Harkema et al., 1987; 2006). Stratified epithelium typically lines the vestibule, the floor of the nasal cavity, or the walls of the nasopalatine ducts. Respiratory epithelium is characterized by numerous motile cilia (kinocilia) that propel mucous toward the pharynx. This epithelium also possesses at least four other cell types, such as non-ciliated, goblet (mucous-producing), secretory (small granule), and basal cells (Harkema et al., 1987). Transitional epithelium is a non-stratified type that occurs between stratified squamous and respiratory epithelium, typically as a narrow strip (Harkema et al., 1987; 2006). Additionally, a non-descript, simple epithelium, may be found between the respiratory and olfactory epithelium and in other locations. These epithelia could be termed "junctional epithelia," as these small patches may only represent transitions, rather than functional types of epithelia. The olfactory epithelium comprises, collectively, the epithelia of the main olfactory system (see Chapter 3), the vomeronasal system (see Chapter 51), and the septal organ of Masera (see Chapter 52). The majority of our discussion concerns the distribution of respiratory and olfactory epithelia, which are the most abundant types that line the nasal cavity proper (Harkema et al., 1987).

Most mammals have an anteriorly situated vestibule just inside the anterior nares. The facial skin (with hair and glands) is reflected within the vestibule. In the nasal fossa proper, the stratified squamous epithelium lines the floor of the nasal fossa, but respiratory mucosa appears laterally, superiorly, and medially by the level of the maxilloturbinal. Olfactory neuroepithelium may appear first along the roof of the nasal fossa or on the superior margin of ethmoturbinal I (Figure 2.5a). Figure 2.5 shows olfactory and non-olfactory epithelia that are distributed along the boundaries of the nasal fossa at different anteroposterior levels, in the coronal plane. Although all ethmoturbinals bear olfactory neuroepithelium (even the human middle nasal concha may have some – Read, 1908; Leopold et al., 2000; and see discussion in Clerico et al., 2003, regarding variations), the first ethmoturbinal bears a substantial amount of ciliated, non-olfactory neuroepithelium (Smith et al., 2004, 2007b; Rowe et al., 2005; Smith and Rossie, 2008). Thus, to generalize all ethmoturbinals as strictly "olfactory" turbinals (Dieulafe, 1908; Moore, 1981) may be ill advised. Indeed, in some species the first ethmoturbinal and the maxilloturbinal appear to have closely opposed surfaces of respiratory epithelia, implying these structures are collaboratively involved in air conditioning (Figure 2.6). More posterior ethmoturbinals do bear more olfactory surface area, proportionally, compared to ethmoturbinal I (Adams, 1972; Bhatnagar and Kallen, 1975; Smith and Rossie, 2008). In many mammals, one or more of the more posterior ethmoturbinals is sequestered in the olfactory recess, where olfactory neuroepithelium predominates. Below the transverse lamina, the nasopharyngeal ducts are lined with respiratory epithelium.

The septal mucosa matches the lateral wall in terms of superoinferior distribution of mucosa generally (Figure 2.6e). Internally, a specialized chemosensory epithelial organ is found near the base of the nasal septum anteriorly. This structure, the vomeronasal organ (VNO), is present bilaterally and often encapsulated by cartilage or bone (Figure 2.5). In most mammals, the VNO is an epithelial tube lined with two types of epithelia: chemosensory and receptor-free ciliated epithelia (Bhatnagar and Smith, 2007). Teleost fishes do not have a VNO. Their VNO receptors (V1Rs, V2Rs) are expressed in the main olfactory neuroepithelium (Hashiguchi et al., 2008), as may be some of the V2R genes in amphibians (Syed et al., 2013). Amphibians, reptiles and mammals have VNO in numerous stages of development, whereas birds lack it altogether (Halpern and Martínez-Marcos, 2003). The VNO is frequently considered a pheromone-receptor organ, but recent discussion has questioned whether it is truly evolved as such (Baxi et al., 2006) or whether mammals have the equivalent of true pheromonal communication (Doty, 2010). Vomeronasal chemoreception may be multipurpose for sociosexual communication (Powers and Winans, 1975; Wysocki et al., 1991), prey detection (Cooper, 1997), and perhaps even dietary preference (Bhatnagar,

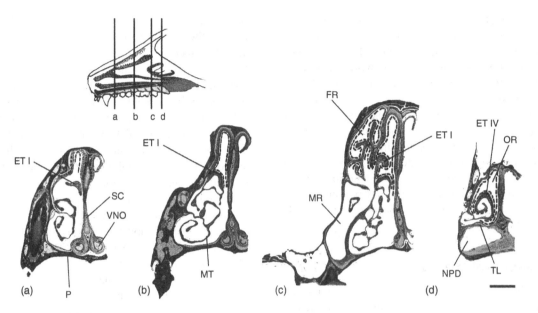

Figure 2.5 Distribution of respiratory and olfactory epithelia (dashed lines) at four different coronal cross-sectional levels (a through d) in the mouse lemur (*Microcebus murinus*). Modified from Smith et al. (2007a). ET I, ethmoturbinal I; ET IV, ethmoturbinal IV; FR, frontal recess; MR, maxillary recess; MT, maxilloturbinal; NPD, nasopharyngeal duct; OR, olfactory recess; P, osseous palate; SC, septal cartilage; TL, transverse lamina; VNO, vomeronasal organ. Scale bar: 1 mm.

1980; Cooper and Bhatnagar, 1976; Bhatnagar and Kallen, 1974b; Halpern et al., 2005). The VNO receptor epithelium is unlike the olfactory neuroepithelium which presents olfactory cilia, olfactory glands, and other morphological specialties such as a very large number of receptors (which are several thousand times greater in number compared to the VN receptors). The VNO neuroepithelium forms a rather diminutive sensory patch (compared to olfactory neuroepithelium) and has sensory neurons that only project microvilli. Goblet cells may be found in the receptor-free epithelium of the VNO; glands for the VNO are from more distant sources than observed for the olfactory epithelium (the latter having intimately nested Bowman's glands). Vomeronasal nerve bundles (the fila from the vomeronasal (VN) receptors) collect into the VN nerves which reach and ramify in the accessory olfactory bulb. In mammals, ventral septal glands drain into the lumen of the VNO; Smith and Bhatnagar (2009) suggest these glands may not be homologous with "vomeronasal glands" of other vertebrates. The lumen of the VNO itself, in turn, drains into the nasal cavity or the nasopalatine duct through the vomeronasal duct. The VN duct exits anteriorly, quite close to the vestibule in many mammals. In the last several decades, numerous reports have scrutinized a vestige of the human VNO that persists postnatally in humans and perhaps great apes (Johnson et al., 1985; Trotier et al., 2000; Bhatnagar and Smith, 2001; Smith et al., 2002). The most thorough histological studies have categorically demonstrated the absence of a functional VNO neuroepithelium in humans after early fetal stages (Smith and Bhatnagar, 2000; Trotier et al., 2000; Bhatnagar and Smith, 2001, 2010; Witt et al., 2002).

The SOO has been described as an island of olfactory neuroepithelium in some species (e.g., rodents – Menco and Morrison, 2003; Ma, 2007). This structure has received far less attention than the VNO and MOE, and its distribution throughout mammals is unclear (e.g., it has not been described in primates). This patch of epithelium, spatially anterior to much of the MOE, expresses Olfactory Marker Protein (Weiler and Benali, 2005). In this sense, it possesses olfactory sensory receptor cells, like the VNO and MOE. Overall, Weiler and Benali (2005) consider the SOO more similar to the MOE than VNO in terms of expression of various neuronal markers. This patch of olfactory neurons likely detects chemostimuli that overlap with those perceived by other chemosensory systems (Ma, 2007). More research is needed to establish the extent to which the SOO is distinguished from other olfactory tissues.

Lastly, the nasal mucosa of mammals is traversed by numerous nerves and blood vessels. Nerves include the nervus terminalis, vomeronasal nerve bundles, trigeminal nerve branches, autonomic nerves, and olfactory nerve bundles. Blood vessels to the nasal (and paranasal) mucosae include branches of the sphenopalatine artery (a terminal branch of the maxillary artery), ethmoidal branches of the ophthalmic artery, and the infraorbital artery. Vessels in nasal mucosa also include large venous sinuses. As one of the few examples of erectile tissues in the body, this complex system of arteries and veins

2.3 Mammalian Nasal Anatomy

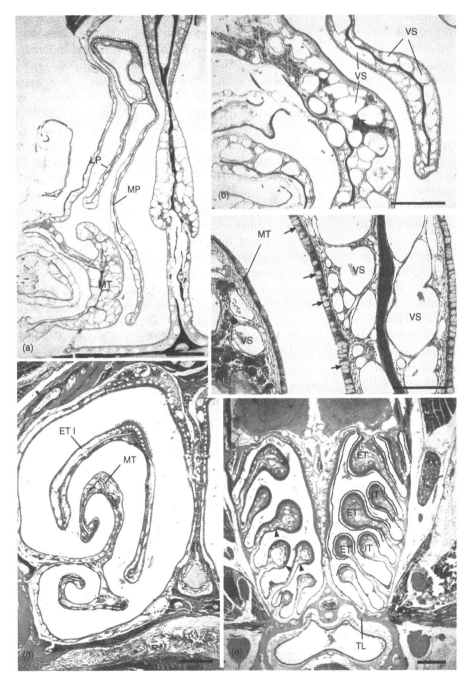

Figure 2.6 Complementary surfaces of the nasal mucosae on the first ethmoturbinal (ET I) and the maxilloturbinal (MT). (a) The folded ET I of an adult primate (bushbaby) has a descending lateral plate (LP) and medial plate (MP) that closely associate with the MT. Only the dorsal rim ET I has olfactory neuroepithelium (dashed line), while the MP and LP are covered with respiratory epithelium. (b) and (c) Higher magnifications of the MP reveals the MP has a dense array of venous sinuses (VS) in the lamina propria. (d) A similar close association of ET I and MT is seen in a megachiropteran bat (*Megaerops ecaudatus*). (e) The olfactory recess of a bat (*Megaderma lyra*) showing multiple rows of ethmoturbinals (ET) with interturbinals (IT) between them. The transition between olfactory and non-olfactory neuroepithelium is clear even at low magnification (arrow heads, on left). Olfactory neuroepithelium (dashed line) covers the turbinals and septum to approximately the same level ventrally. small arrows, goblet cells; TL, transverse lamina. Scale bars: a, 1 mm; b, 600 μm; c, 150 μm; d, 500 μm; e, 500 μm.

significantly influence the diameter of major sections of the nasal airway (see Chapter 16).

2.3.4 Phylogenetic Variations of Internal Nasal Anatomy

Certain specifics about the early form of the nasal fossa in the class Mammalia are difficult to reconstruct, primarily because the delicate anatomy is not pristinely preserved in fossils. It is possible to infer some generalities about the primitive condition by surveying the common nasal features of extant mammals. Representatives of most extant mammals possess a transverse lamina (Negus, 1958). An extinct mammal from the late Triassic, *Morganucodon*, reportedly possessed a transverse lamina, but earlier mammals likely did not (Kielan-Jaworowska et al., 2004). From this it may be inferred that a likely primitive characteristic of "crown" mammals (the common ancestor of living mammals, and their descendants) is to possess an olfactory recess, in which the posterior-most ethmoturbinals are sequestered.

Among extant species, marsupials have a great complexity of the lateral wall, e.g., with three to seven ethmoturbinals in addition to smaller turbinals (Rowe et al., 2005; Macrini, 2012, 2014). A notable characteristic of the lateral wall in many marsupials is the elaborate folding and branching of the maxilloturbinal and ethmoturbinals, and an obliquely sloping transverse lamina (Macrini, 2012). Extant monotremes have contrasting patterns of olfactory morphology, with the aquatic platypus possessing a relatively reduced complement of turbinals and echidnas possessing a far more complex lateral nasal wall (Ashwell, 2012) and perhaps as many as 18 ethmoturbinals (Paulli, 1900a). Macrini (2012) cautions that as serial homologues, it is difficult to establish equivalency of ethmoturbinals between taxa.

Most placental mammals possess a similar arrangement of turbinals, although the total number of turbinals of the ethmoid bone (including smaller turbinals) may vary greatly (Paulli, 1900a,b,c). Paulli analyzed nasal anatomy in a broad array of mammals and set some of the practices still used in nomenclature, but use of terminology has varied (see discussion by Rowe et al., 2005; Smith and Rossie, 2008; Macrini, 2012, 2014). With recent efforts to clarify (if not standardize) terminology in mind, as well as the incorporation of CT data, a thorough reanalysis of internal nasal anatomy in extant placental mammals is sorely needed. Reviewing the literature, Macrini (2012) notes that placental mammals possess between two and ten ethmoturbinals. Ethmoturbinal I can be tracked developmentally as an important boundary between the nasal cavity proper and paranasal spaces (Maier, 1993a, b; Smith and Rossie, 2008). However, for the remainder of turbinals it is difficult to assess which turbinals are lost in species that have the greatest reduction in turbinals. Conversely, species that possess numerous ethmoturbinals that reach the midline may have a similar or greater number of interturbinals that lie hidden lateral to them (Smith et al., 2012). These interturbinals reflect the multiplicity of the ethmoidal scrolls. The main difference between the interturbinals and the larger ethmoturbinals is the extent of growth. This iterative theme of development in the ethmoid renders species differences much more understandable. The difference between the echidna (18 ethmoturbinals) on the one hand and most other mammals (one to four ethmoturbinals) on the other may be related to the timing and extent of growth of the smaller mesenchymal condensations that form at the lower margins of the nasal capsule (see Macrini, 2014, for a discussion of the implications on turbinal homology).

The extent to which the variation in the number of turbinals relates to olfactory performance is clear in certain comparisons. For example, a comparison of extremes in internal nasal complexity on the one hand (e.g., dogs) and simplicity on the other (e.g., humans) appears to reveal profoundly different special sensory adaptations; this is largely borne out by other lines of evidence (e.g., olfactory gene repertoire, Quignon et al., 2003). Yet, such stark comparisons do little to reveal how variations in internal nasal form are adaptive. Recent work reveals that even for carnivorans, mammals that rely on turbinals both for being olfactory specialists and for regulating body temperature, variation in turbinal surface area may relate to diet and other ecological variables (Van Valkenburgh et al., 2011, 2014).

Many mammals exhibit some degree of reduction of the turbinals. There are two suggested reasons for such reductions. The first is ecological. Like the platypus, many aquatic forms of placental mammals appear to have reduced olfactory structures relative to terrestrial relatives (Gittleman, 1991; Van Valkenburgh et al., 2011). It is postulated that, for mammals, olfaction is of less utility in aquatic environments (Van Valkenburgh et al., 2011). Certainly the reduction is extreme in cetaceans, mammals that are fully readapted to aquatic life, although some evidence suggests some whales may retain olfactory abilities (Thewissen et al., 2011). Conversely, the maxilloturbinal is relatively enlarged in aquatic mammals (Huntley et al., 1984; Van Valkenburgh et al., 2011). A second idea invokes "competing" functions of nasal structures to explain reductions. The maxilloturbinal, especially important to body temperature regulation in carnivorans (Van Valkenburgh et al., 2004), is enlarged in aquatic environments where body temperature regulation is more challenging. Furthermore, because the nasal surface area is finite within the nasal cavity, selection cannot enlarge one turbinal (e.g., the larger maxilloturbinal in aquatic vertebrates) without taking out potential space for another turbinal (Van Valkenburgh et al., 2011). Van Valkenburgh et al. (2011) present evidence suggesting that the ratios of maxilloturbinals to ethmoturbinals are interdependent, and postulate the existence of an ecological trade-off of turbinal size in aquatic forms.

Two ecological explanations have been put forth to explain olfactory reduction in primate ancestors. One idea, that olfaction is of limited importance in an arboreal habitat (Elliot Smith. 1927; Le Gros Clark, 1959), has been refuted (Cartmill, 1970). A second basis for reduction of nasal structures has also been postulated: an evolutionary trade-off of special senses (see below).

2.3.5 Turbinals: Their Mucosal Properties and Role in Airflow

Certain generalizations regarding the roles of the turbinals in nasal physiology, derived mainly from rodents and canines, may apply broadly, albeit imprecisely, to mammals as a whole. They provide an example of physiological roles of turbinals that are potentially dichotomous. The maxilloturbinal resides in the ventral air space between the anterior nares and the nasopharyngeal ducts posteriorly. Most inhaled air passes through the maxilloturbinal (Figure 2.3), which in canines is quite complex and elaborately branched. This air is humidified and warmed before passing to the lower respiratory airways. Mucosal properties in the lower part of the nasal fossa can dictate the extent to which the maxilloturbinal is exposed to airflow. Dilation of blood vessels in the mucosa at the base of the septum (septal swell bodies) blocks air from passing directly through the unobstructed lower part of the inferior meatus, forcing more air to traverse the complicated spaces around the maxilloturbinal (Craven et al., 2007).

During normal respiration, approximately 85–90% of inspired air bypasses the olfactory region and flows ventrally though the nasal cavity of rodents and canids (Kimbell et al. 1997; Craven et al., 2010). However, these animals can increase the percentage of air that reaches the olfactory region by forcefully inhaling air ("sniffing") (Craven et al. 2010; Kimbell et al., 1997; Yang et al., 2007). Sniffing may allow animals to sample more odorant molecules with each breath, thus aiding olfactory discrimination (Yang et al., 2007). Craven et al. (2010) suggest that this airflow pattern, in which air travels superior to the complex maxilloturbinal, through the dorsal meatus, and to the olfactory recess, may be a characteristic of all macrosmatic species. The arrangement in rodents, canines, and possibly other mammals also leads to prolonged exposure of odorants to the nasal mucosa, for example, during exhalation (Eiting et al., 2014). At least some odorants seem to become entrapped in the olfactory recess and are not washed out during exhalation (Craven et al., 2010; Yang et al., 2007; Youngentob et al., 1987). In contrast, the airflow patterns of humans (and presumably other higher primates) differ significantly from carnivores, ungulates, rodents, and marsupials, all of which possess an olfactory recess.

Another directional influence on inspired air is found in the first ethmoturbinal. Ethmoturbinal I passes diagonally upward from its root, toward the nasal septum. Posteriorly, it typically merges with the nasoturbinal. In doing so, this turbinal divides the nasal fossa into more centralized airflow channels (leading to both the olfactory recess and nasopharyngeal ducts) and lateral (paranasal) spaces (Clancy et al., 1994; Smith et al., 2011). Primitive extant primates, rodents, and many other mammals share this anatomical arrangement. However, airflow in paranasal spaces is poorly understood compared to the main airway, even in well-studied species (Craven et al., 2007). Evidence on the nature of receptivity of the olfactory neuroepithelium indicates these two regions may not detect an identical range of odorant types. In rodents, the olfactory receptor neurons located along more central ethmoturbinals project axons to spatially distinct portions of the main olfactory bulb compared to those along frontoturbinals (Miyamichi et al., 2005). These regions correspond roughly to regions

hypothesized to have differences in odorant absorption properties as predicted by computational studies, suggesting a structure-function relationship between airway anatomy and odorant absorption (Schoenfeld and Cleland, 2005; Zhao et al., 2006).

2.4 PRIMATE INTERNAL NASAL ANATOMY

Olfactory anatomy has been described as reduced in the order Primates by some authors (e.g., Elliot Smith, 1927; Cave, 1973). Yet, there is currently no consensus that primates lack a keen sense of smell (Schaal and Porter, 1991; Heymann, 2006; Laska et al., 2000). Behavioral studies, for example, show many "lower" (e.g., lemurs and lorsises) and some "higher" primates (e.g., New World monkeys) extensively scent mark (Lewis, 2006; Mertl-Millhollen, 2006; Epple et al., 1993). Still, a combination of evidence suggests some primates have a relatively diminished complement of genes as well as morphological structures for olfaction (Liman and Innan, 2003; Gilad et al., 2003; Cave, 1973; Loo, 1974).

The dichotomous arrangement of nasal anatomy in primates is shown in Figures 2.7–2.10. Lemurs and lorises resemble other mammals in possessing an olfactory recess, yet there are also certain distinctions. First, the overlap of ethmoturbinals with other turbinals is quite different in the large lemurids compared to the mouse lemur. Large diurnal or cathemeral lemurids have a somewhat more dog-like arrangement of turbinals compared to nocturnal lemurs and lorises. The ethmoturbinals are more sequestered posterosuperiorly, with less spatial overlap of the maxilloturbinal and nasoturbinal.

In contrast, the tree shrew and mouse lemur illustrate a somewhat more common arrangement. The first ethmoturbinal is extremely elongated anteriorly, and thus overlaps the maxilloturbinal and nasoturbinal to a great extent (Figures 2.7a,b). The olfactory neuroepithelium projects along the dorsal edge of the first ethmoturbinal, although not to the anterior tip of the turbinal. The portions of the turbinals that descend ventrally are covered, in part, with respiratory epithelium. These laminae overlap the maxilloturbinal, and can be seen to provide complementary surfaces with venous sinuses in the mucosa (Figures 2.6a-c). Based on the available literature, this describes the morphology in all mouse and dwarf lemurs, bushbabies, and lorises (Kollmann and Papin, 1925; Smith et al., 2007b; Figures 2.7b, 2.8a, 2.9a,b). The broad overlap of the ethmoturbinal with the surfaces of other turbinals is therefore common in the most primitive living primates, and could represent a primitive condition. Given that this overlap occurs in tree shrews, at least some megabats (Figures 2.6d,e), and at least some cats (Van Valkenburgh

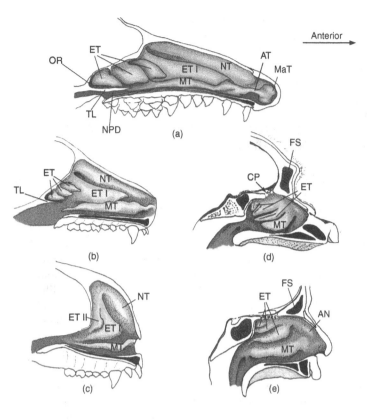

Figure 2.7 Lateral nasal wall in (a) tree shrew (*Ptilocercus lowii*), (b) mouse lemur (*Microcebus murinus*), (c) monkey (common marmoset, *Callithrix jacchus*), (d) chimpanzee (*Pan troglodytes*), and (e) humans (*Homo sapiens*). A primitive feature retained in lemurs is the presence of a transverse lamina (TL) which separates the olfactory recess (OR) from the nasopharyngeal duct (NPD). However, the transverse lamina of many lemurs (b) does not extend as far anteriorly as in tree shrews (a), which are extant models for many aspects of ancestral primate anatomy. All anthropoids c–e and tarsiers (not shown) lack the transverse lamina and olfactory recess, except as a vestige of the latter. ET, ethmoturbinals; AN, agger, nasi; AT, atrioturbinal; ET I, ethmoturbinal I; ET II, ethmoturbinal II; FS, frontal sinus; MaT, marginoturbinal; MT, maxilloturbinal; NT, nasoturbinal. Figure 2.7a is modified after Le Gros Clark (1959) and Wible (2011).

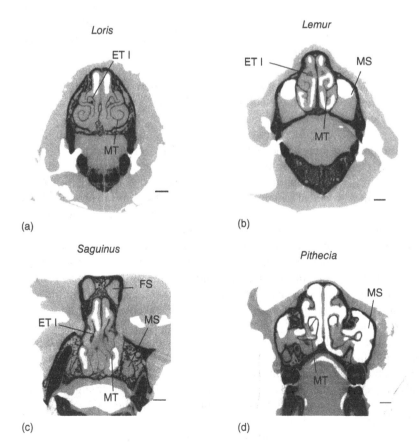

Figure 2.8 Micro-CT slices of the maxilloturbinal region in primates, showing more primitive nasal form in a loris and lemur (a) and (b) compared to two species of anthropoids (c) and (d). Variation exists in both groups. Nocturnal lemurs and lorises have more elongated first ethmoturbinals (ET I) which extend anterior to the paranasal spaces, and extensively overlap the dorsal surfaces of the maxilloturbinal (MT). Some anthropoids have extremely attenuated midfaces, as in tamarins (e.g., *Saguinus* spp) and many others (including humans). In such cases, the turbinal and paranasal regions are largely coextensive in anteroposterior space. In some other species, especially those with somewhat more pronounced midfaces (or snouts) there is less spatial overlap of maxilloturbinal and ethmoturbinal regions (e.g., *Pithecia*). *Saguinus* has a relatively simple MT with two lamellae. In *Pithecia*, the MT is more extensively folded. Scale bars: a, 4 mm; b; c, 2 mm; d, 3 mm.

2.4 Primate Internal Nasal Anatomy

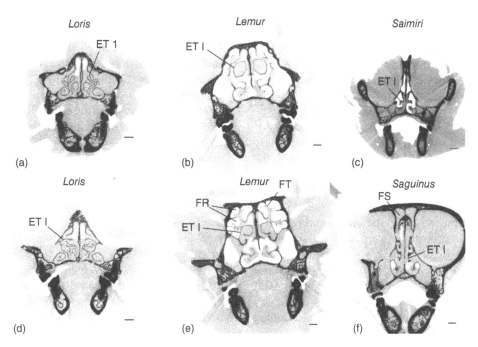

Figure 2.9 Micro-CT slices of the ethmoturbinal region in primates, showing more primitive nasal form in a loris and lemur (a, b, d) and (e) compared to two species of anthropoids (c) and (f). The loris shows continued overlap of first ethmoturbinal (ET I) with the maxilloturbinal (MT). Overall, the central nasal airway is highly packed with turbinals, while the paranasal spaces are markedly reduced compared to the lemur. In the lemur, the interorbital width is greatly expanded and the turbinals are less packed. In addition, the frontal recess (FR) is large, with three frontoturbinals (FT). In some anthropoids the ethmoturbinal region is greatly reduced posteriorly due to enlarged orbits (e.g., *Saimiri,* 8c). Reduction is far less pronounced in the smaller-eyed *Saguinus* (8F). FS, frontal sinus. Scale bars: a, d, 4 mm; b, e, 2 mm; c, f, 3 mm.

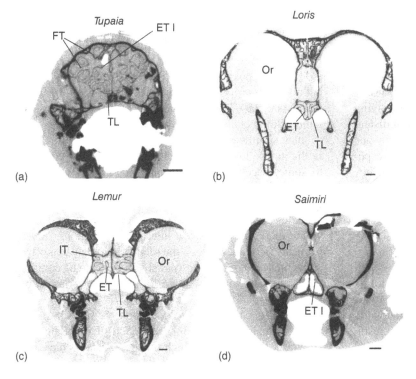

Figure 2.10 Micro-CT slices of the posterior ethmoturbinal region, showing more primitive nasal form in (a) tree shrew, (b) loris and (c) lemur compared to (d) a monkey. In the first three species, the CT slice is at the anterior end of the olfactory recess. The olfactory recess extends farther anteriorly in the tree shrew, enclosing even the posterior end of the first ethmoturbinal (ET I), and overlapping the frontal recess (FT, frontoturbinals). In the loris and lemur only the most posterior ethmoturbinal (ET) is enclosed in the olfactory recess. The ethmoturbinal region may be complicated by the presence of smaller interturbinals (IT) between the larger turbinals. The orbit (Or) of *Saimiri* is so enlarged and convergent (facing forward) that the interorbital region is devoid of bone. A membrane-covered interorbital fenestra (*) is found between the orbits. Not even a vestige of the olfactory recess is seen. TL, transverse lamina. Scale bars: a, c, d, 4 mm; b, 4 mm.

54   **Chapter 2   Anatomy of the Nasal Passages in Mammals**

et al., 2014) the possibility exists that in many mammals the turbinals function collaboratively (see discussion above in Section 2.3.3). This arrangement blurs the roles of ethmoturbinals to a greater degree than is commonly known for so-called "macrosmatic" mammals (e.g., rodents and canines).

The anthropoid pattern, which also applies to the large-eyed Southeast Asian tarsiers, involves a loss of posterosuperior nasal structures. Instead, a more vertically arranged set of turbinals is observed. In all anthropoids, the maxilloturbinal is found most inferiorly, and is the most anteroposteriorly extensive of the turbinals in primates (Figure 2.7). One or two ethmoturbinals are found superiorly and somewhat posteriorly to the maxilloturbinal. The frontoturbinal and the posterior part of the nasoturbinal are absent in many higher primates (some species may possess a nasoturbinal – e.g., *Pithecia*, Figure 2.8). These characteristics describe human nasal anatomy as well (Figure 2.7e). The lateral nasal wall of humans has a very simple, single fold. At least some anthropoids have folded turbinals, or even double scrolls (Figures 2.8c,d).

Overall, the nasal anatomical characteristics of primates appear, at least in gross anatomical terms, to support the notion that there are some taxa that are "microsmatic" and others that are "macrosmatic" (relatively reduced and relatively heightened olfactory abilities, respectively). However, recent literature includes great debate on the topic (see Schaal and Porter, 1991; Laska et al., 2000; Smith and Bhatnagar, 2004; Heymann, 2006). In light of continued disagreement on the importance of olfaction in primates, it is worth discussing existing anatomical evidence that bears on two particular questions regarding extant primates.

2.4.1   Is Olfactory Reduction a Characteristic of All Primates?

Epple and Moulton (1978) contrasted olfactory anatomy of various primates to the tree shrew (*Tupaia glis*), a living model that may bear some resemblance to the most distant ancestors of all primates. Their description reflects one common view: that all primates have a reduced nasal (especially olfactory) anatomy; some more than others. They describe the tree shrew as being distinct from primates in having olfactory neuroepithelium covering all of the ethmoturbinals as well as the olfactory recess. Primates are then described to display reductions, with lorises having some reduction of ethmoturbinals (compared to *Tupaia*) and macaques and apes having still more reduction as well as absence of an olfactory recess. Such statements imply a "gradistic" reduction across the order Primates. At the same time, these are generalizations that rely perhaps too strongly on isolated observations of a few taxa; as a result some details may be inaccurate as generalizations whilst others may remain overlooked.

Some evidence supports the claim that even the most primitive living primates (lemurs and lorises) have some degree of nasal reduction. While it has been long known that lemurs and lorises retain more ethmoturbinals than higher primates (Kollmann and Papin, 1925), they are also acknowledged to have some degree of crowding of the interorbital space. This is because in all extant primates, the orbits converge further toward the midline than in other living mammals (Figures 2.10, 2.11). Thus, progressive encroachment of the potential space for nasal structures is postulated as a basis for olfactory reduction (Cartmill, 1970, 1972). Looking very broadly, this is true in a graded fashion: More primitive extant primates have a lesser degree of orbital convergence compared to more specialized anthropoids and tarsiers (Ross, 1995). This explains in great part (though not entirely) why nasal cavities are more capacious in the former.

Quantitative data known for extant mammals do not clearly indicate that lemurs have a lesser percentage of olfactory surface area (SA) compared to other mammals. The lesser mouse lemur (*Microcebus murinus*) has a similar or greater percentage of olfactory mucosa compared to some bats of similar or smaller body size, and to the domestic cat (see review in Smith et al., 2012, 2014). Insectivores, rodents, and the opossum have a considerably greater percentage of olfactory SA, while humans have considerably less (Negus, 1958; Smith et al., 2012). Still, reduction of olfactory anatomy may be subtle. Two regions that are known to be relatively small in some lemurs are the frontal and olfactory recesses (Smith and Rossie, 2008; Smith et al., 2011). For example, the olfactory recess SA comprises only 7% of the total internal nasal SA in *Microcebus*, compared to 31% in a bat of nearly the same body mass (*Megaderma lyra*, Smith et al., 2011, 2012). However, total ethmoturbinal SA is similar in these species, indicating a possible relationship with airflow dynamics but not necessarily the amount of olfactory SA. The frontal recess is 6.5% of the total nasal surface area in the mouse lemur (Smith et al., 2012). Although the frontal recess contributes significantly to total olfactory surface area (20% in the mouse lemur), it appears relatively small compared to some insectivores. Smith and Rossie (2008) attribute this to its closer adjacency to the orbits in mouse lemurs compared to insectivores (Figure 2.11). An analysis of all primates to determine the relative size of the frontal recess would therefore be of great interest. Even among "lower" primates (lemurs, lorises), the relationship between orbital convergence and frontal recess size is likely to be complex, because eye position relative to the snout varies among taxa (Cartmill, 1972).

Thus, existing evidence supports the hypothesis that living primates exhibit some reduction of nasal anatomy. However, the extent to which mucosal surfaces have been reduced as opposed to reorganized is not entirely clear.

2.4 Primate Internal Nasal Anatomy

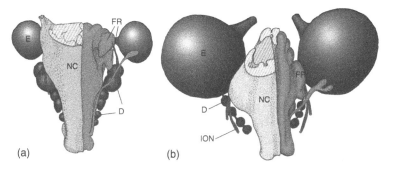

Figure 2.11 Three dimensional reconstruction of the soft tissue structures of the face in a fetal non-primate (a), *Tenrec ecaudatus* and a mouse lemur (b), *Microcebus murinus*, anterior and slightly dorsal view. By stripping away bone, the association of the eyes (E) with interorbital structures is clearer. On the left side of each image, the as-yet unossified nasal capsule (NC) is shown. On the right side, the nasal capsule is stripped away to reveal nasal mucosa. Note the eyes of the lemur are relatively large and somewhat more forward facing (convergent) compared to the non-primate. This has an effect to reorganize the nasal capsule and cavity, shifting the frontal recess (FR) anteriorly. The extent to which primate nasal capsules are reorganized or possibly reduced might now be investigated using micro-CT methods. D, deciduous teeth; ION, infraorbital nerve.

Moreover, Epple and Moulton's (1978) specific suggestion that primates bear less olfactory neuroepithelium in the olfactory recess than tree shrews has not been well tested. Indeed, the contrast in internal architecture is more profound when contrasting the frontal recess anatomy of tree shrews to lorises and lemurs (Figures 2.9–2.10) – which suggests a link between either eye size or convergence and paranasal reduction. Moreover, detailed "maps" and quantitative information of the internal microanatomy, including olfactory surface area, olfactory sensory neuron density, olfactory cilia length, and other parameters are known only for very few primate species (Smith and Bhatnagar, 2004, 2014). Therefore, our ability to know whether some primate species, let alone entire suborders, are microsmatic is limited at this time.

2.4.2 Is Olfactory Reduction a Characteristic of Higher Primates?

In part, the categorization of some mammals as macrosmatic, or "keen-scented" (Negus, 1958; Turner, 1891), has been made as a contrast to our own species – a manner of better understanding our own sensory specializations. We also recognize some similar morphological traits of humans and our closest relatives. "Higher" primates (i.e., anthropoids) are often discussed as being microsmatic, or having reduced olfactory abilities (Elliot Smith, 1927; Cave, 1973; Baron et al., 1983; Martin, 1990; Wako et al., 1999; Gilad et al., 2003). This is based on morphological, physiological, and/or genetic contrasts to relatively "macrosmatic" species (lemurs and lorises). The validity of this view need not be debated again here, as a rather large base of literature from recent years can provide the reader with differing opinions (e.g., Schaal and Porter, 1991; Laska et al., 2000; Gilad et al., 2003; Smith and Bhatnagar, 2004; Heymann, 2006; Smith et al., 2007a; Hepper et al., 2008). However, recent work clearly establishes that some morphological concepts related to olfactory anatomy are in error. Specifically, the persistently greater "snout" length, on average, in more primitive living primates (lemurs and lorises) compared to higher primates is not by itself an indicator of olfactory reduction. Much of the snout region is lined with respiratory mucosa (Smith et al., 2004; 2007a,b; 2011). Accordingly, a recent text on primate anatomy draws no direct parallel between snout length and olfactory abilities (Ankel-Simons, 2007). Here, we revisit the extent to which a true dichotomy exists between lemurs and lorises compared to higher primates as it pertains to olfactory and frontal recesses.

All anthropoids and tarsiers have extreme reduction of the posterosuperior nasal fossa, the location of the olfactory recess in other mammals. In some anthropoids the interorbital region is so highly approximated that there is only a single midline bony plate, the interorbital septum, or even a partial opening (interorbital fenestra) in which only a membrane separates the orbits (Maier, 1983). Such extreme orbital approximation also occurs in some large-eyed lemurs and lorises, but the resulting interorbital septum is found above the olfactory recess (Cartmill, 1972). In lieu of an olfactory recess, anthropoids may have a small vestige, the *recessus cupularis*, which is devoid of turbinals (Rossie, 2006), but in all cases the result of this extreme approximation is a reduction in ethmoturbinal number. Anthropoids also lack frontoturbinals (Maier and Ruf, 2014).

Few studies have quantitatively studied the nasal mucosa of primates, so it is still unclear to what extent the distribution of olfactory mucosa is reduced (or perhaps redistributed) as a result of this nasal cavity reduction. Qualitatively, several studies suggest that the distribution of olfactory mucosa covers a portion of each of the ethmoturbinals in anthropoids (Wako et al., 1999; Harkema et al., 1987, 2006), including the middle and superior nasal

56 Chapter 2 Anatomy of the Nasal Passages in Mammals

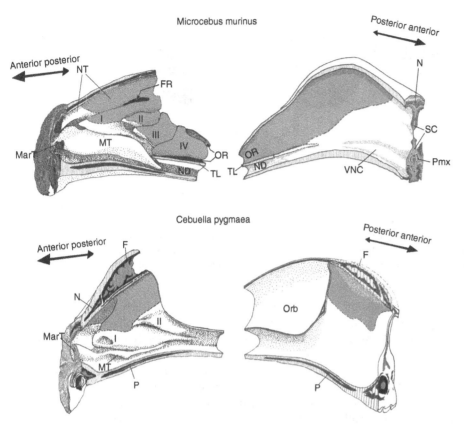

Figure 2.12 Three dimensional reconstruction of the nasal mucosa in the smallest lemur (*Microcebus murinus*) and smallest monkey (*Cebuella pygmaea*). Olfactory (shaded) versus non-olfactory (unshaded) surface areas of the lateral nasal wall (left) and nasal septum (right) are shown (*Microcebus* modified from Smith and Rossie, 2008; *Cebuella* from Smith et al., 2014). Note that the absence of the more posterior ethmoturbinals in the monkey corresponds to the close approximation of the orbits. The orbital surface of the ethmoid bone (Orb) is included to emphasize this. I-IV, ethmoturbinals I through IV; F, frontal bone; FR, frontal recess; MarT, marginoturbinal; MT, maxilloturbinal; N, nasal bone; ND, nasopharyngeal duct; NT, nasoturbinal; OR, olfactory recess; SC, septal cartilage; TL, transverse lamina; VNC, bulge of vomeronasal cartilage.

concha of humans (Read, 1908). Only recently have detailed "maps" of mucosa distribution been produced in a lemur. Figure 2.12 shows the distribution of olfactory mucosa in the world's smallest lemur and monkey. The absence of an olfactory recess in the monkey is readily related to the approximation of the eyes toward the midline. The orbit wall appears posterosuperiorly along the septum, and on the contralateral side it may be inferred that the orbit obliterates the space occupied by the olfactory recess in the mouse lemur. Although the morphological differences are extreme and loss of more posterior ethmoturbinals in the monkey is clear, there are two respects in which these differences reflect reorganization (rather than loss) of mucosal SA. First, while the lemur has a dedicated respiratory passageway beneath the olfactory recess (the nasopharyngeal ducts), in the monkey the ethmoturbinals simply end in the nasopharyngeal meatus (albeit with non-olfactory SA). Second, olfactory SA appears to be distributed somewhat more anteriorly in the monkey compared to the mouse lemur (see also Smith et al., 2004, 2014).

In summary, it is beyond dispute that the humans and other anthropoids have reduced or eliminated certain olfactory regions of the nasal airway (olfactory recess, frontal recess), probably in connection with encroachment by the orbits (Maier and Ruf, 2014). The functional consequences of this reduction are unclear. Indeed, since our understanding of how morphology relates directly to airflow is based on very few species of mammals (e.g., see Schoenfeld and Cleland, 2005; Zhao and Dalton, 2007), the significance of regional reduction or reorganization are presently hard to assess.

2.5 CONCLUSIONS

The internal nasal airway of mammals is a morphologically complex space that reflects the developmental constraints and functional necessities of the mid-facial region. This chapter has reviewed mammalian nasal anatomy to provide a context for understanding our own, rather unusual nasal form. In brief, our major differences in nasal anatomy compared to other mammals may be related, in part, to 1) some degree of olfactory reduction in our own Infraorder (Anthropoidea) compared to most other mammals and some other primates; 2) different nasal adaptations to climate, perhaps due to an alternative means of body temperature regulation (e.g., skin cooling).

A broad comparative understanding of internal nasal anatomy has remained elusive, in part due to the difficulty in exposing and gathering data on the delicate internal nasal region. In addition, our understanding of quantitative aspects of the nasal mucosae in different mammals has progressed slowly over the past five decades, which have for a long time been limited to a comparison of humans to several domestic animals (Negus, 1958; Moulton and Beidler, 1967). Still, recent work shows that modern methods such as computed tomography (CT) hold immense promise as tools to uncover adaptive morphological patterns in internal nasal structure (Van Valkenburgh et al., 2004, 2011). The combination of CT with knowledge of microanatomical variations (e.g., epithelial surface areas) across mammals may prove an especially powerful research method (Rowe et al., 2005, Smith et al., 2011). One pattern is already clear: not all "macrosmatic" mammals are alike (Stößel et al., 2010). If that term is used too broadly, we may lose sight of variations among mammals that may pertain to ecology, either directly (e.g., dietary correlates of turbinal surface area) or indirectly (e.g., airflow patterns) (Van Valkenburgh et al., 2011; Gross et al., 1982).

Some authors use the term macrosmatic in a very narrow sense, to refer to carnivores, ungulates, and rodents (Craven et al., 2007); a very restrictive use of this term is wise. These and perhaps a few other mammals may be the original basis for some morphological generalizations (e.g., "olfactory" and "respiratory" turbinals), which are not equally applicable to all mammals. This knowledge offers insight into our own species' nasal evolutionary history. None of our closest extant relatives bear a very close similarity to noted olfactory specialists, such as the canids. Many nocturnal lemurs and lorises have vastly overlapping surfaces of ethmoturbinals, nasoturbinal, and maxilloturbinal. This condition is common in other mammals, including many bats (Smith, unpublished observations) and the tree shrew (Wible, 2011; note that Scandentia (tree shrews) and Primates are two of the three extant orders comprising the clade Euarchonta). As such, this is quite possibly the ancestral appearance from which anthropoids (such as ourselves) and their allies evolved. Much more work is needed to understand how turbinals may work collaboratively, via overlapping complementary mucosal types. In tandem with the need for a more detailed appreciation of comparative microanatomy of the mammalian nasal boundaries, a broader understanding of airflow patterns in mammals with differently organized nasal airways is essential. In this broad context, we may more fully understand our own nasal airflow dynamics (see Chapter 16), olfactory anatomy, and olfactory abilities.

A2.1 SOME TECHNIQUES FOR STUDYING THE NASAL PASSAGEWAYS

A full appreciation of this topic can be reached by consulting bibliographies of scholarly writings of Charles Evans, Richard L. Doty, Charles Wysocki, Mimi Halpern and others. See also Simon and Nicolelis (2001) *Methods in Chemosensory Research*. CRC Press, Boca Raton, 328 pp.

1. **GROSS EXAMINATION:** The entire head of the specimen is partially decalcified; the anterior nasal segment (posterior to the cribriform plate) is then bisected a bit lateral to the nasal septum, thus separating the right and left nasal fossae. Using a dissecting microscope, the minute internal anatomy of the structures therein (such as various turbinals) is examined. We have found formic acid decalcifier (see Bhatnagar and Kallen, 1974a for details) to be very safe for the delicate nasal parts.

2. **ENDOCASTS:** Some delicate nasal and related cranial anatomy has been successfully studied by creating endocasts using elastic molding material. The cribriform plate was studied in this way by Pihlström et al. (2005), although computerized reconstructions are a more recent option (Bird et al., 2014).

3. *IN VITRO* **STAINING** of the bones and cartilages using Alizarin Red S, toluidine blue, and other procedures.

4. **PREPARATION OF THE ENTIRE NASAL SEPTUM** to study the Vomeronasal organ, VN nerves, Nervus terminalis, ganglionic structures, and glands (see, Bojsen-Moller, 1975).

5. **EMBRYONIC (DEVELOPMENTAL) studies in staged embryos** (Müller and O'Rahilly, 2004).

6. **HISTO-MORPHOLOGY:** The entire head, specially the anterior nasal segment, is prepared for 10 μm thick serial sections. Depending on the size, the head is generally first decalcified, then divided into three parts: anterior, middle, and posterior cranium. This is to allow maximum paraffin infiltration. Further decalcification, processing through paraffin (Cellosolve or other media) under vacuum leading to complete serial section series. Ten-micron thick serial sections are recommended to allow morphometric studies of the neuronal nuclei. Generally every fifth serial section is arranged on glass slides; the remaining sections are saved. A trichrome stain (e.g., Gomori one-step) works brilliantly in well-preserved (embalmed) specimens (see Bhatnagar and Kallen, 1974a; Bhatnagar and Smith, 2007). Results from a study from such carefully prepared serial series cannot be overemphasized.

7. **IMAGE-ADVANCED, MICRO-CT studies, computer reconstruction techniques.** Comparative anatomy of the internal nasal region has been notoriously difficult to study, except through destructive means (e.g., Allen, 1882; Paulli, 1900a,b,c; Cave, 1973). The availability of non-destructive imaging methods, such as micro-computed tomography (CT),

has made nasal anatomy far more accessible (e.g., Van Valkenburgh et al., 2004, 2011; Rowe et al., 2005; Rossie, 2006; Macrini, 2012). Various software programs are used in conjunction with micro-CT to reconstruct and quantify delicate nasal structures (e.g., see Rossie, 2006; Van Valkenburgh et al., 2014). Histology and micro-CT can be co-registered to enhance imaging of bone with soft tissue information (Rehorek and Smith, 2007; DeLeon and Smith, 2014).

8. **SCANNING, TRANSMISSION, AND CONFOCAL MICROSCOPIC STUDY** of tissues obtained by dissection (after intracardiac perfusion with EM fixatives).

9. **ELECTRO-OLFACTOGRAMS, VOMERONASOGRAMS, AND ELECTRODE IMPLANTING STUDIES.** See review by Scott and Scott-Johnson, 2002).

10. **ETHOLOGICAL STUDIES OF FLEHMEN** (see Estes, 1972).

11. **PARASITOLOGICAL EXAMINATION OF THE NASAL EPITHELIA** (see, Lichtenfels et al., 1981).

12. **MOLECULAR GENE RECEPTOR STUDIES, UPTAKE OF RECEPTORS** (Bannister and Dodson, 1992, and **autoradiographic** studies.

13. **EXPERIMENTAL PROTOCOLS:** An understanding of the "division of labor" among olfactory systems was achieved in great part through nerve ablation studies (e.g., Powers and Winans, 1975).

14. **IMMUNOREACTIVE STUDIES:** Our understanding of functional capacity of the human vomeronasal organ was achieved, in part, through immunohistochemical studies that showed only few cells that express neuronal markers (e.g., Witt et al., 2002), in contrast to other primate (Dennis et al., 2004). More broadly, all olfactory epithelia express a Olfactory Marker Protein (Weiler and Benali, 2005).

15. **ARTIFICIAL (ELECTRONIC) NOSES:** The first artificial-nose system was reported by Persaud and Dodd (1982). Since then sensory designs have been progressively updated and refined. The e-noses generally have three main components: a vapor-delivery system, a sensory array, and a pattern-recognition algorithm (also see Walt et al., 2012).

16. **AIRFLOW MEASUREMENTS THROUGH THE NASAL PASSAGES.** The olfactory and respiratory epithelium directly and intimately react with the passing air currents serving olfaction, warming of the internal *milieu* and other functions (see discussion in Craven et al., 2007, 2010; Hillenius, 1992). Note: the VNO is not directly involved with the incoming or outgoing air currents (see Brunner, 1914; Hillenius and Rehorek, 2005).

REFERENCES

Adams, D. R. (1972). Olfactory and non-olfactory epithelia in the nasal cavity of the mouse, *Peromyscus. Am. J. Anat.* 133: 37–49.

Allen, H. (1882). On a revision of the ethmoid bone in Mammalia, with special reference to the description of this bone in and of the sense of smell in the Chiroptera. *Bull. Mus. Comp. Zool., Harvard* 10: 135–164.

Ankel-Simons, F. (2007). *Primate Anatomy: An Introduction*. New York: Academic Press.

Ashwell, K. W. S. (2012). Development of the olfactory pathways in platypus and echidna. *Brain, Behav. Evol* 79: 45–56.

Bannister, L. H. and Dodson, H. C. (1992). Endocytic pathways in the olfactory and vomeronasal epithelia of the mouse: ultrastructure and uptake of tracers. *Microsc. Res. Tech.* 23: 128–141.

Baron, G., Frahm, H. D., Bhatnagar, K. P., and Stephan, H. (1983). Comparison of brain structure volumes in Insectivora and Primates: III, main olfactory bulb (MOB). *J. Hirnforsch.* 24: 551–568.

Baxi, K. N., Dorries, K. M., and Eisthen, H. L. (2006). Is the vomeronasal system really specialized for detecting pheromones? *Trends Neurosci.* 29: 1–7.

Bhatnagar, K. P. (1980). The chiropteran vomeronasal organ: its relevance to the phylogeny of bats. In *Proceedings of the 5th International Bat Research Conference*, Wilson, D. E. and Gardner, A. L. (Eds.). Lubbock: Texas Tech, pp. 289–316.

Bhatnagar, K. P. and Kallen, F. C. (1974a) Morphology of the nasal cavities and associated structures in *Artibeus jamaicensis* and *Myotis lucifugus. Am. J. Anat.* 139: 67–190.

Bhatnagar, K. P. and Kallen, F. C. (1974b). Cribriform plate of ethmoid, olfactory bulb and olfactory acuity in forty species of bats. *J. Morphol.* 142: 71–89.

Bhatnagar, K. P. and Kallen, F. C. (1975) Quantitative observations on the nasal epithelia and olfactory innervation in bats. *Acta Anat.* 91: 272–282.

Bhatnagar, K. P. and Smith, T. D. (2001). The human vomeronasal organ-Part III: postnatal development from infancy through the ninth decade. *J. Anat.* 199: 289–302.

Bhatnagar, K. P. and Smith, T. D. (2007). Light microscopic and ultrastructural observations on the vomeronasal organ of *Anoura* (Chiroptera: Phyllostomidae). *Anat. Rec.* 290: 1341–1354.

Bhatnagar, K. P. and Smith, T. D. (2010). The human vomeronasal organ. Part VI. A nonchemosensory vestige in the context of major variations of the mammalian vomeronasal organ. *Curr. Neurobiol.* 1: 1–9.

Bird, D., Amirkhanian, A., Pang, B., and Van Valkenburgh, B. (2014). Quantifying the cribriform plate: Influences of allometry, function and phylogeny in carnivora. *Anat. Rec.* 297:2080–2092.

Bojsen-Moller, F. (1975). Demonstration of terminalis, olfactory, trigeminal and perivascular nerves in the rat nasal septum. *J. Comp. Neurol.* 159: 245–256.

Brunner, H. L. (1914). Jacobson's organ and the respiratory mechanism of amphibians. *Morphol. Jb.* 48: 157–165.

Cartmill, M. (1970). *The orbits of arboreal mammals: a reassessment of the arboreal theory of primate evolution.* Ph.D. Dissertation. Duke University.

Cartmill, M. (1972). Arboreal adaptations and the origin of the order Primates. In *The Functional and Evolutionary Biology of Primates*, Tuttle, R. (Ed.). Chicago: Aldine Atherton, pp. 97–122.

Cave, A. J. E. (1973). The primate nasal fossa, *Biol. J. Linn. Soc., Lond.* 5: 377–387.

Clancy, A. N. Schoenfeld, T. A., Forbes, W. B., and Macrides, F. (1994). The spatial organization of the peripheral olfactory system of the hamster. II: Receptor surfaces and odorant passageways within the nasal cavity. *Brain Res. Bull.* 34: 211–241.

References

Clerico, D. M., To, W. C., and Lanza, D. C. (2003). Anatomy of the human nasal passages. In *Handbook of Olfaction and Gustation*. 2nd Ed. R. L. Doty Ed. New York: Marcell Dekker, pp. 1–16.

Cooper, J. G. and Bhatnagar, K. P. (1976). Comparative anatomy of the vomeronasal organ complex in bats. *J. Anat.* 122: 571–601.

Cooper, W. E., Jr., (1997). Correlated evolution of prey chemical discrimination with foraging, lingual morphology, and vomeronasal chemoreceptor abundance in lizards. *Behav. Ecol. Sociobiol.* 41: 257–265.

Craven, B. A., Neuberger, T., Paterson, E. G., et al. (2007). Reconstruction and morphometric analysis of the nasal airway of the dog (*Canis familiaris*) and implications regarding olfactory airflow. *Anat. Rec.* 290: 1325–1340.

Craven, B. A., Paterson, E. G., and Settles, G. S. (2010). The fluid dynamics of canine olfaction: unique nasal airflow patterns as an explanation of macrosmia. *J. Roy. Soc. Interface* 7: 933–943.

de Beer, G. R. (1937). *The Development of the Vertebrate Skull*. Chicago: Chicago University Press.

Deleon, V.B., and Smith, T.D. (2014). Mapping the nasal airways: using histology to enhance CT-based three-dimensional reconstruction. *Anat. Rec.* 297:2113–2120.

Dennis, J. C., Smith, T. D., Bhatnagar, K. P. et al. (2004). Expression of neuron-specific markers by the vomeronasal neuroepithelium in six species of primates. *Anat. Rec.* 281: 1190–1199.

Dieulafe, L. (1906). Morphology and embryology of the nasal fossae of vertebrates. *Ann. Otol. Rhinol. Laryngol.* 15: 1–584.

Doty, R. L. (2010). *The great pheromone myth*. Baltimore: The Johns Hopkins University Press.

Eiting, T.P., Smith, T.D., Perot, J.B., and Dumont, E.R. (2014). The role of the olfactory recess in olfactory airflow. *J. Exp. Biol.* 217:1799–803.

Elliot Smith, G. (1927). *The Evolution of Man*. London: Oxford University Press.

Epple, G. and Moulton, D. (1978). Structural organization and communicatory function of olfaction in nonhuman primates. In *Sensory Systems of Primates*, Noback, C. R. (Ed.). New York: Plenum Press, pp. 1–22.

Epple, G., Belcher, A., Küderling, I. et al. (1993). Making sense out of scents: species differences in scent gland, scent-marking behaviour, and scent-mark composition in the Callitrichidae. In *Marmosets and tamarins: systematics, behaviour, and ecology*, A. B. Rylands (Ed.). Oxford: Oxford University Press, pp. 123–151.

Estes, R. D. (1972). The role of the vomeronasal organ in mammalian reproduction. *Mammalia* 36: 315–34l.

Finger, T. E., Böttger, B., Hansen, A. et al. (2003). Solitary chemoreceptor cells in the nasal cavity serve as sentinels of respiration. *Proc. Natl. Acad. Sci. USA* 100: 8981–8986.

Geist, N. R. (2000). Nasal respiratory turbinate function in birds. *Physiol. Biochem. Zool.* 73: 581–589.

Gilad, Y., Man, O., Pääbo, S., and Lancey, D. (2003). Human specific loss of olfactory receptor genes. *Proc. Natl. Acad. Sci. USA* 100: 3324–3327.

Gittleman, J. L. (1991) Carnivore olfactory bulb size, allometry, phylogeny, and ecology. *J. Zool.* 225: 253–272.

Gross, E. A., Swenberg, J. A., Fields, A., and Popp, J. A. (1982). Comparative morphometry of the nasal cavity in rats and mice. *J. Anat.* 135: 83–88.

Halpern, M. and Martínez-Marcos, A. (2003). Structure and function of the vomeronasal system: an update. *Progr. Neurobiol.* 70: 245–318.

Halpern, M., Daniels, Y., and Zuri, I. (2005). The role of the vomeronasal system in food preferences of the gray short-tailed opossum, *Monodelphis domestica*. *Nutrit. Metab.* 2: 6 doi:10.1186/1743-7075-2-6.

Hansen, T. H. (1997). Stabilizing selection and the comparative analysis of adaptation. *Evolution* 51: 1341–1351.

Harkema, J. R., Popper, C. G., Hyde, D. M. et al. (1987). Nonolfactory surface epithelium of the nasal cavity of the bonnet monkey: A morphologic and morphometric study of the transitional and respiratory epithelium. *Am. J. Anat.* 180: 266–279.

Harkema, J. R., Carey, S. A., and Wagner, J. G. (2006). The nose revisited: A brief review of the comparative structure, function, and toxicologic pathology of the nasal epithelium. *Toxicol. Pathol.* 34: 253–269.

Hashiguchi, Y., Furuta, Y., and Nishida, M. (2008). Evolutionary patterns and selective pressures of odorant/pheromone receptor gene families in teleost fishes. *PLoS ONE* 3(12): e4083.

Hepper, P., Wells, W., McArdle, P. et al. (2008). Olfaction in the gorilla. In *Chemical Signals in Vertebrates 11*, Hurst, J., Beynon, R. J., Roberts, S. C., and Wyatt, T. (Eds.). New York: Springer, pp. 103–110.

Heymann, E. W. (2006). The neglected sense – olfaction in primate behavior, ecology, and evolution. *Am. J. Primatol.* 68: 519–524.

Hill, W. C. O. (1953). *Primates, Comparative Anatomy and Taxonomy. I. Strepsirrhini*. Edinburgh: Edinburgh University Press.

Hillenius, W. J. (1992). The evolution of nasal turbinates and mammalian endothermy. *Paleobiology* 18: 17–29.

Hillenius, W. J. (1994). Turbinates in therapsids: evidence for endothermy in mammal-like reptiles. *Evolution* 48: 207–229.

Hillenius, W. J. and Rehorek, S. J. (2005). From the eye to the nose: Ancient orbital to vomeronasal communication in tetrapods? In *Chemical Signals in Vertebrates 10*, Mason, R. T., LeMaster, M. P., and Muller-Schwarze, D. (Eds.). New York: Springer Verlag. pp. 228–241.

Huntley, A. C., Costa, D. P., and Rubin, R. D. (1984). The contribution of nasal counter-current heat exchange to water balance in the Northern elephant seal, *Mirounga angustirostris*. *J. Exp. Biol.* 113: 447–454.

Johnson, A., Josephson, R., Hawke, M. (1985). Clinical and histological evidence for the presence of the vomeronasal (Jacobson's) organ in adult humans. J. Otolaryngol. 14:71–79.

Kayser, R. (1895) Die exacte Messung der Luftdurchgängikeit der Nase. *Arch Laryngol* 3: 101–210.

Kielan-Jaworowska, Z., Cifelli, R., and Luo, Z-X. (2004). *Mammals from the age of dinosaurs: origins, evolution, and structure*. New York: Columbia University Press.

Kimbell, J. S., Godo, M. N., Gross, E. A. et al. (1997). Computer simulation of inspiratory airflow in all regions of the F344 rat nasal passages. *Toxicol. Appl. Pharmacol.* 145: 388–398.

Kohlrausch, O. (1853). Über das Schwellgewebe an den Muscheln der Nasenschleimhaut. *Müller Arch. Anat. Physiol. Wiss. Med.*, 149–150.

Kollmann, M. and Papin, L. (1925). Etudes sur lémuriens. Anatomie compareé des fosses nasales et de leurs annexes. *Archs. Morph. gé. exp.* 22, 1–60.

Laska, M., Seibt, A., and Weber, A. (2000). 'Microsmatic' primates revisited: olfactory sensitivity in the squirrel monkey. *Chem. Senses* 25: 47–53.

Le Gros Clark, W. E. (1959). *The Antecedents of Man*. Edinburgh: Edinburgh University Press.

Leopold, D. A., Hummel, T., Schwob, J. E. et al. (2000). Anterior distribution of human olfactory epithelium. *Laryngoscope* 110: 417–421.

Lewis, R. J. (2006). Scent marking in sifakas: No one function explains it all. *Am. J. Primatol.* 68: 622–636.

Lichtenfels, J. R., Bhatnagar, K. P., Whittaker, F. H., and Frahm, H. D. (1981). Filarioid nematodes in olfactory mucosa, olfactory bulb, and brain ventricular system of bats. *Trans. Amer. Micros. Soc.* 100: 216–219.

Liman, E. R. and Innan, H. (2003). Relaxed selective pressure on an essential component of pheromone transduction in primate evolution. *Proc. Natl. Acad. Sci. USA* 100: 3328–3332.

Loo, S. K. (1974). Comparative study of the histology of the nasal fossa in four primates. *Folia Primatol.* 21: 290–303.

Ma, M. (2007). Encoding olfactory signals via multiple chemosensory systems. *Crit. Rev. Biochem. Mol. Biol.* 42: 463–480.

Macrini, T. E. (2012). Comparative morphology of the internal nasal skeleton of adult marsupials based on x-ray computed tomography. *Bull. Amer. Mus. Nat. Hist.* 365: 1–91.

Macrini. T.E. (2014). Development of the ethmoid in *Caluromys philander* (Didelphidae, Marsupialia) with a discussion on the homology of the turbinal elements in marsupials. *Anat. Rec.* 297:2007–2017.

Maier, W. (1983). Morphology of the interorbital region of *Saimiri sciureus*. *Folia Primatol.* 41: 277–303.

Maier, W. (1993). Cranial morphology of the therian common ancestor, as suggested by adaptations of neonate marsupials. In *Mammal Phylogeny*, Szalay, F., Novacek, M. J., and McKenna, M. C. (Eds.). New York: Springer-Verlag. pp. 165–181.

Maier, W. (1993b). Zur evolutiven und funktionellen Morpholgie des Gesichtsschädels der Primaten. *Z. Morph. Anthrop.* 79: 279–299.

Maier, W. (2000). Ontogeny of the nasal capsule in cercopithecoids: a contribution to the comparative and evolutionary morphology of catarrhines. In *Old World Monkeys*, Whitehead, P. F. and Jolly, C. (Eds.). Cambridge: Cambridge University Press, pp. 99–132.

Maier, W., and Ruf, I. (2014). Morphology of the nasal capsule of Primates – with special reference to Daubentonia and Homo. *Anat. Rec.* 297:1985–2006.

Martin, R. D. (1990). *Primate Origins and Evolution. A Phylogenetic Reconstruction*. Princeton, NJ: Princeton University Press.

Menco, B. P. M. and Morrison, E. E. (2003). Morphology of the mammalian olfactory epithelium: Form, fine structure, function, and pathology. In *Handbook of Olfaction and Gustation*. 2nd Ed, Doty, R. L. (Ed.). New York: Marcell Dekker, pp. 17–49.

Mertl-Millhollen, A. S. (2006). Scent marking as resource defense by female *Lemur catta*. *Am. J. Primatol.* 68: 605–621.

Miyamichi, K., Serizawa, S., Kimura, H., and Sakano, H. (2005). Continuous and overlapping expression domains of odorant receptor genes in the olfactory epithelium determine the dorsal/ventral positioning of glomeruli in the olfactory bulb. *J. Neurosci.* 25: 3586–3592.

Moore, W. J. (1981). *The Mammalian Skull*. Cambridge: Cambridge University Press.

Moulton, D. G. and Beidler, L. M. (1967). Structure and function of the peripheral olfactory system. *Physiol. Rev.* 47: 1–51.

Müller, F. and O'Rahilly, R. (2004). Olfactory structures in staged human embryos. *Cells Tiss. Org.* 178: 93–116.

Negus, V. (1958). *The Comparative Anatomy and Physiology of the Nose and Paranasal Sinuses*. Livingston: Edinburgh and London.

Nomina Anatomica Veterinaria. (2005). www.wava-amav.org/nav.htm.

Novacek, M. J. (1993). Patterns of diversity in the mammalian skull. In *The Skull, vol. 2*. Hanken, J. and Hall, B. K. (Eds.). Chicago: Chicago University Press. pp. 438–545.

Ogura, T., Krosnowski, K., Zhang, L. et al. (2010). Chemoreception regulates chemical access to mouse vomeronasal organ: role of solitary chemosensory cells. *PLoS ONE* 5(7): e11924. doi:10.1371/journal.pone.0011924.

Parker, W. K. (1874) On the structure and development of the skull in the pig. *Phil. Trans. Royal Soc. Lond.* 164: 289–336.

Parsons, T. S. (1967) Evolution of the nasal structure in the lower tetrapods. *Am. Zool.*, 7: 397–413.

Paulli, S. (1900a). Über die Pneumaticität des Schädels bei den Säugethieren. III. Über die Morphologie des Siebbeins der Pneumaticität bei den Insectivoran, Hyracoideen, Chiropteren, Carnivoren, Pinnepedien, Edentates, Rodentiern, Prosimien und Primaten. *Gegenb. Morpholog. Jahrb.* 28: 483–564.

Paulli, S. (1900b). Über die Pneumaticität des Schädels bei den Säugethieren. II. Über die Morphologie des Siebbeins der Pneumaticität bei den Ungulaten und Probosciden. *Gegenb. Morpholog. Jahrb.* 28: 179–251.

Paulli, S. (1900c) Über die Pneumaticität des Schädels bei den Säugethieren. I. Über den Bau des Siebbiens. Über die Morphologie des Siebbeins der Pneumaticität bei den Monotremen und den Marsupialiern. *Gegenb. Morpholog. Jahrb.* 28: 147–178.

Pedziwiatr, Z.F. (1972). Das Siebbeinlabyrinth. II. Differenzierung und Systematik der Hauptmuschel bein einigen Gattungen der Säugetiere. *Anat. Anz.* 131: 378–390.

Persaud, K. and Dodd, G. (1982). Analysis of discrimination mechanisms in the mammalian olfactory system using a model nose. *Nature* 299: 352–355.

Pihlström, H., Fortelius, M., Hemilä, S., et al. (2005). Scaling of mammalian ethmoid bones can predict olfactory organ size and performance. *Proc. Biol. Sci.* 272: 957–962.

Powers, J. B. and Winans, S. S. (1975). Vomeronasal organ: critical role in mediating sexual behavior of the male hamster. *Science* 187: 961–963.

Quignon, P., Kirkness, E., Cadieu, E., et al. (2003). Comparison of the canine and human olfactory receptor gene repertoires. *Genome Biol.* 4: R80. (doi:10.1186/gb-2003-4-12-r80).

Rae, T. C. and Koppe, T. (2003). The term "lateral recess" and craniofacial pneumatization in Old World monkeys (Mammalia, Primates, Cercopithecoidea). *J. Morphol.* 258: 193–199.

Read, E. A. (1908) A contribution to the knowledge of the olfactory apparatus in dog, cat and man. *Am. J. Anat.* 8: 17–47.

Rehorek, S. J. and Smith, T. D. (2007). Concurrent 3-D visualization of multiple microscopic structures. In: *Modern Research and Educational Topics on Microscopy. No.3*. Volume 2: 917–923.

Ross, C. F. (1995). Allometric and functional influences on primate orbit orientation and the origins of Anthropoidea. *J. Hum. Evol.* 29: 201–227.

Rossie, J. B. (2006). Ontogeny and homology of the paranasal sinuses in Platyrrhini (Mammalia: Primates). *J. Morphol.* 267: 1–40.

Rowe, T., Eiting, T. P., Macrini, T. E. and Ketcham, R. A. (2005) Organization of the olfactory and respiratory skeleton in the nose of the gray short-tailed opossum *Monodelphis domestica*. *J. Mamm. Evol.* 12: 303–336.

Ruppert, B. (1754). Dissertatio inauguralis medica de tunica pituitaria – exponens ajust anatomiam, physiologiam, et pathologiam. Prague.

Schaal, B., and Porter, R. H. (1991). "Microsmatic humans" revisited: The generation and perception of chemical signals. *Adv. Study Behav.* 20: 474–482.

Schoenfeld, T. A. and Cleland, T. A. (2005). The anatomical logic of smell. *Trends Neurosci.* 28: 620–627.

Scott, J. W. and Scott-Johnson, P. (2002). The electroolfactogram: A review of its history and uses. *Microsc. Res. Techn.* 58: 152–160.

Simon, S.A., and Nicolelis, M.A.L. (2001). Methods in Chemosensory Research. CRC Press, Boca Raton, 328 pp.

Smith, T.D., Bhatnagar, K.P., Shimp. K.L., Kinzinger, J.H., Bonar, C.J., Burrows, A.M., Mooney, M.P., Siegel, M.I. (2002). Histological definition of the vomeronasal organ in humans and chimpanzees with a comparison to other primates. Anat. Rec. 267:166–176.

Smith, T. D. and Bhatnagar, K. P. (2004). "Microsmatic" primates: Reconsidering how and when size matters. *Anat. Rec.* 279: 24–31.

Smith, T. D. and Bhatnagar, K. P. (2009). Vomeronasal system evolution. In: L. Squire (Ed.), *New Encyclopedia of Neuroscience, Vol. 10*. Oxford: Academic Press, pp. 461–470.

Smith, T. D., and Rossie, J. B. (2006). Primate olfaction: Anatomy and evolution. In *Olfaction and the Brain: Window to the Mind*, Brewer W., Castle, D. and Pantelis, C. (Eds.). Cambridge: Cambridge University Press, pp. 135–166.

References

Smith, T.D., Rossie, J.B., Docherty, et al. 2008. Fate of the nasal capsular cartilages in prenatal and perinatal tamarins (Saguinus geoffroyi) and extent of secondary pneumatization of maxillary and frontal sinuses. *Anat. Rec.* 291:1397–1413.

Smith, T. D., and Rossie J. B. (2008). The nasal fossa of mouse and dwarf lemurs (Primates, Cheirogaleidae). *Anat. Rec.* 291: 895–915.

Smith, T. D., Bhatnagar, K. P., Tuladhar, P., and Burrows, A. M. (2004). Distribution of olfactory epithelium in the primate nasal cavity: are "microsmia" and "macrosmia" valid morphological concepts? *Anat. Rec.* 281: 1173–1181.

Smith, T. D., Rossie, J. B., and Bhatnagar, K. P. (2007a). Evolution of the nose and nasal skeleton in primates. *Evol. Anthropol.* 16: 132–146.

Smith, T. D., Bhatnagar, K. P., Rossie, J. B., et al. (2007b). Scaling of the first ethmoturbinal in nocturnal strepsirrhines: olfactory and respiratory surfaces. *Anat. Rec.* 290: 215–237.

Smith, T.D., Rossie, J.B., Cooper, G.M., et al. (2010). The maxillary sinus in three genera of New World monkeys: Factors that constrain secondary pneumatization. *Anat. Rec.*, 203:91–107.

Smith, T. D., Eiting, T. P., and Rossie, J. B. (2011) Distribution of olfactory and nonolfactory surface area in the nasal fossa of *Microcebus murinus*: implications for microcomputed tomography and airflow studies. *Anat. Rec.* 294: 1217–1225.

Smith, T. D., Eiting, T. P., and Bhatnagar, K. P. (2012). A quantitative study of olfactory, non-olfactory, and vomeronasal epithelia in the nasal fossa of the bat *Megaderma lyra. J. Mamm. Evol.* 19: 27–41.

Smith, T.D., Eiting, T.P., Bonar, C.J., and Craven, B.A. (2014). Nasal morphometry in marmosets: loss and redistribution of olfactory surface area. *Anat. Rec.*, 297:2093–2104.

Stößel, A., Junold, A., and Fischer, M. S. (2010). The morphology of the eutherian ethmoidal region and its implications for higher-order phylogeny. *J. Zool. Syst. Evol. Res.* 48: 167–180.

Syed, A. S., Sansone, A., Nadler, W. et al. (2013). Ancestral amphibian v2rs are expressed in the main olfactory epithelium. *Proc. Natl. Acad. Sci. USA* 110: 7714–7719.

Thewissen, J. G. M., George, J., Rosa, C., and Kishida, T. (2011). Olfaction and brain size in the bowhead whale (*Balaena mysticetus*). *Mar. Mamm. Sci.* 27: 282–294.

Trotier, D., Eloit, C., Wassef, M., et al. (2000). The vomeronasal cavity in adult humans. *Chem. Senses* 25: 369–380.

Turner, W. (1891). The convolutions of the brain: A study in comparative anatomy. *J. Anat. Physiol.* 25: 105–153.

Van Gilse, P. H. G. (1927). Development of the sphenoidal sinus in man and its homology in mammals. *J. Anat.* 61: 153–166.

Van Valkenburgh, B., Theodor, J., Friscia, A., and Rowe, T. (2004). Respiratory turbinates of canids and felids: a quantitative comparison. *J. Zool. Lond.* 264: 281–293.

Van Valkenburgh, B., Curtis, A., Samuels, J. X., et al. (2011). Aquatic adaptations in the nose of carnivorans: evidence from the turbinates. *J. Anat.* 218: 298–310.

Valkenburgh, B., Pang, B., Bird, D., Curtis, A., Yee, K., Wysocki, C., and Craven, B.A. (2014). Respiratory and Olfactory Turbinals in Feliform and Caniform Carnivorans: the influence of snout length. *Anat. Rec.* 297:2065–2079.

Von Haller, A. (1756). *Icones Anatomicae*. Gottingen.

Vidic, B. (1971). The prenatal morphogenesis of the lateral nasal wall in the rat (*Mus rattus*). *J. Morph.* 133: 303–317.

Wako, K., Hiratsuka, H., Katsuta, O., and Tsuchitani, M. (1999). Anatomical structure and surface epithelial distribution in the nasal cavity of the common cotton-eared marmoset (*Callithrix jacchus*). *Exp. Anim.* 48: 31–36.

Walt, D. R., Stitzel, S. E., and Aernecke, M. J. (2012). Artificial noses. *Amer. Sci.* 100: 38–45.

Weichert, C. K. (1958) *Anatomy of the Chordates* 2nd ed. McGraw Hill.

Weiler, E. and Benali, A. (2005). Olfactory epithelia differentially express neuronal markers. *J. Neurocytol.* 34: 217–240.

Wible, J. R. (2011) On the tree shrew skull (Mammalia, Placentalia, Scandentia). *Ann. Carnegie Mus.* 79: 149–230.

Witmer, L. M. (1999). The phylogenetic history of the paranasal air sinuses. In *The Paranasal Sinuses of Higher Primates*, Koppe, T., Nagai, H., and Alt, K. W. (Eds.). Berlin: Quintessence. pp. 21–34.

Witt, M., Georgiewa, B., Knecht, M., and Hummel, T. (2002). On the chemosensory nature of the vomeronasal epithelium in adult humans. *Histochem. Cell Biol.* 117: 493–509.

Wright, J. (1914). A history of laryngology and rhinology. 2nd ed. Lea and Febiger, New York, New York.

Wysocki, C. J., Kruczek, M., Wysocki, L. M., and Lepri, J. J. (1991). Activation of reproduction in nulliparous and primiparous voles is blocked by vomeronasal organ removal. *Biol. Reprod.* 45: 611–616.

Yang, G. C., Scherer, P. W., Zhao, K., and Mozell, M. M. (2007). Numerical modeling of odorant uptake in the rat nasal cavity. *Chem. Senses* 32: 273–284.

Youngentob, S. L., Mozell, M. M., Sheehe, P. R., and Hornung, D. E. (1987). A quantitative analysis of sniffing strategies in rats performing odorant discrimination tasks. *Physiol. Behav.* 41: 59–69.

Zhao, K., Dalton, P., Yang, G.C., Scherer, P.W. (2006). Numerical modeling of turbulent and laminar airflow and odorant transport during sniffing in the human and rat nose. *Chem. Senses* 31:107–118.

Zhao, K. and Dalton, P. (2007). The way the wind blows: Implications of modeling nasal airflow. *Curr. All. Asthma Rep.* 7: 117–125.

Chapter 3

Olfactory Mucosa: Composition, Enzymatic Localization, and Metabolism

XINXIN DING and FANG XIE

3.1 INTRODUCTION

The olfactory mucosa, as well as the nasal respiratory mucosa, has a very high metabolic capacity for endogenous and exogenous, or xenobiotic, substrates. Olfactory tissue also has a high degree of inflammatory and immune responsiveness stimulated by contact with foreign substances, exfoliates in response to toxic insults, and regenerates to varying degrees following this exfoliation. The olfactory epithelium (OE) is distinctive in containing the only recognized mammalian neurons that regenerate from precursor basal cells. In addition, these neurons are unique in contacting the external environment with their dendritic processes while the axonal processes of the same cells synapse within the central nervous system (CNS) in the olfactory bulbs. The olfactory mucosa, therefore, represents a tissue where interactions are continually occurring between external stimuli and various physiological processes, including biotransformation, mucus secretion, immune responses, neural signaling, and cell differentiation, development, and death. This chapter examines the basic structure and cell types of the olfactory mucosa and then focuses primarily on the enzymatic capacity of this tissue. The anatomic characteristics generally common to all species are outlined and followed by a brief discussion of interspecies variability in the magnitude, localization, or occurrence of these characteristics. The identity and localization of nasal enzymes and their metabolic capacities are compared across several species. Finally, the potential for these enzymes to modulate the toxicity of xenobiotics and to influence tissue function, including homeostasis and odor signal detection, is discussed.

This chapter is not a review of all enzymes detected in nasal tissues. A number of review articles on various biotransformation enzymes detected in nasal tissues, their localization, regulation, and substrates, or the specific toxic effects that they are thought to mediate, have been published, including Dahl and Hadley (1991), Ding and Coon (1993), Thornton-Manning and Dahl (1997), Ding and Kaminsky (2003), Jeffrey et al. (2006), Harkema et al. (2006); and, Watelet et al. (2009). This chapter describes the complexity in the composition and regulation of the major known biotransformation enzymes in the olfactory mucosa. Additionally, it discusses various physiological and pathological processes in which nasal metabolic activity is thought to play a role, providing the reader with a framework in which to incorporate enzyme-specific information and relevant sources for a more detailed examination of specific questions. In many cases, data on nasal metabolism have been obtained from whole tissue homogenates, making it impossible to determine the relative contributions of epithelial or subepithelial enzymes, or olfactory versus respiratory mucosa. Because the olfactory mucosa occupies the most caudal region of the nose, metabolism in other nasal regions or in nasal glands can affect olfactory function as well. Therefore, data from nasal homogenates are also discussed and noted as such.

Handbook of Olfaction and Gustation, Third Edition. Edited by Richard L. Doty.
© 2015 Richard L. Doty. Published 2015 by John Wiley & Sons, Inc.

3.2 ANATOMY OF THE NASAL CAVITY

To understand the contribution of metabolism in the olfactory mucosa, it is necessary to understand its relationship to other nasal tissues. Before reaching the olfactory mucosa, inhaled air comes in contact with three other epithelial types in the nasal cavity: squamous, transitional, and respiratory. These epithelial regions differ in their metabolic capacities, but metabolism in these tissues can alter the chemical composition of inhaled toxicants before they reach the olfactory mucosa.

The anterior vestibule of the nasal cavity is lined with a stratified squamous epithelium. Although enzymes primarily involved in metabolism of endogenous substrates such as alkaline phosphatase and gamma glutamyl peptidase have been localized to squamous epithelium (Randall et al., 1987), this epithelium has not been reported to have significant xenobiotic-metabolizing capacity. The squamous epithelium has neither secretory capacity nor cilia. The squamous epithelium gives way, in some species, to a narrow region of transitional epithelium, which is a cuboidal, nonciliated epithelium that has a high metabolic capacity for substrates of specific cytochrome P450 (P450 or CYP) enzymes (Bond et al., 1988). This transitional epithelium often displays metaplastic changes in response to toxicants. For example, chronic exposure to ozone results in a metaplastic change of the transitional epithelium to secretory, respiratory epithelium, and cigarette smoke exposure produces squamous metaplasia in this region (Harkema, 1990).

Continuing in a caudal direction, the nasal cavity is lined by a respiratory mucosa consisting of an epithelium made up of ciliated cells and mucous-secreting goblet cells. Underlying this epithelium are subepithelial glands that produce the majority of serous secretions in the nose. The respiratory mucosa plays the major role in both production of nasal secretions and clearance of inhaled materials. This epithelium contains ciliated cells that are responsible for movement of the mucous layer through the nasal cavity as well as secretory goblet cells that produce and secrete mucus. The respiratory epithelium is underlaid by subepithelial glands, which also secrete mucus to the mucous layer. High metabolic capacity is found in the respiratory mucosa as well.

The most dorsal and caudal region of the nasal cavity is lined by the olfactory mucosa. The surface area of this mucosa is greatly enhanced by a convoluted turbinate structure, which varies greatly across species (Harkema et al., 2006). The olfactory mucosa is composed of the OE lining the nasal cavity and separated from the underlying lamina propria by the basal lamina (Figure 3.1).

3.3 COMPOSITION OF THE OLFACTORY MUCOSA

The following is a brief, general description of the olfactory mucosa discussing primarily those elements common to most species. For more details, the reader is referred to the following reviews: for general nasal and olfactory tissue anatomy–Sorokin (1988) and Uraih and Maronpot (1990); for comparative anatomy – Reznik (1990); and for *human* olfactory anatomy – Chapters 2 and 4 in this volume, as well as a recent immunohistochemical characterization of human olfactory mucosa using a comprehensive series of marker antibodies (Holbrook et al., 2011).

3.3.1 Epithelial Cell Types

The OE is made up of four primary cell types: the olfactory receptor cells; the sustentacular or supporting cells; the basal cells; and the duct cells of Bowman's glands (Figure 3.1). Cilia containing the olfactory receptors project from the receptor cells into the mucous layer lining the nasal cavity. These cells are unique in two respects: (1) they project directly into the brain before their first synapse, which makes them the only cells directly contacting both the CNS and the external environment; and (2) in contrast to almost all other neuronal cells, olfactory receptor cells regenerate from basal cells after damage (Graziadei and Monti-Graziadei, 1983; Huard et al., 1998; Jang et al., 2003; Leung et al., 2007; Iwai et al., 2008). The cilia on the receptor cells are nonmotile. The ciliary membrane has unique properties; the contents of cardiolipin and phosphatidylethanolamine in an enriched porcine cilia preparation are lower than in lipid extracted from the whole olfactory mucosa, and sulfoglycosphingolipids are detected only in the ciliary membrane but not in the whole olfactory mucosa lipid extract (Lobasso et al., 2010). It remains to be determined whether these unique membrane properties are required for optimal functioning of the odorant receptors.

The olfactory receptor neurons (ORNs) generally have very little xenobiotic-metabolizing capacity. The majority of xenobiotic-metabolizing enzymes in the OE have been localized to the sustentacular cells, the duct cells of Bowman's glands, and the progenitor basal cells. Sustentacular cells have secretory functions in some species (Getchell et al., 1988; Zielinski et al., 1988), but generally are not the primary source of the seromucous secretions covering the olfactory epithelium. The microvilli of sustentacular cells contain a transporter protein named low density lipoprotein receptor (also named megalin), which seems to provide a local mechanism for rapid clearance of hydrophobic odorants, by mediating internalization of complexes of odorant-binding proteins (OBP) and odorants/xenobiotics into lysosomes within sustentacular cells (Strotmann and Breer, 2011).

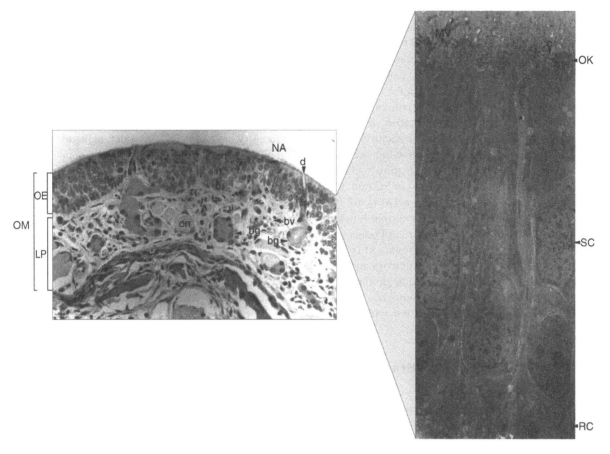

Figure 3.1 Cellular anatomy of the olfactory mucosa. The left panel shows a transverse section in the region of the ethmoturbinates of an adult rat. Tissues on the lumenal side of the basal lamina compose the olfactory epithelium (OE), and tissue inferior to the basal lamina forms the lamina propria (LP). The two layers are included in the olfactory mucosa (OM). Structures identified include sustentacular cell nuclei (*sn*), olfactory neuronal cells (*n*), basal cells (*b*), olfactory nerve bundles (*on*), Bowman's glands (*bg*), Bowman's gland ducts (*d*), blood vessels (*bv*), and nasal airway (NA). [5 μm paraffin embedded sections stained with hematoxylin and eosin; approximately x 250 (Modified from Gu et al., 1997)]. The right panel shows a transmission electron micrograph of the olfactory epithelial surface of an adult rat (Approximately x 3500). The structures identified are olfactory receptor cells (RC); olfactory dendritic knobs (OK); cilia of receptor neurons (C); sustentacular (or supporting) cells (SC), and microvilli (MV) at the lumenal surface of supporting cells.

3.3.2 Subepithelial Structure

The lamina propria consists of the acinar cells of Bowman's glands, olfactory nerves and their associated Schwann cells, blood vessels, lymphatic vessels, and connective tissue. Because this tissue is so often exposed to inhaled foreign substances, cells associated with inflammation and immunity – including neutrophils, plasma cells, monocytes, and macrophages – often are present within the submucosa and epithelium.

3.3.2.1 Bowman's Glands.
These subepithelial glands are the primary source of mucous and serous secretions in the olfactory mucosa. The acinar and duct cells of Bowman's glands contain many xenobiotic-metabolizing enzymes, although this localization is species dependent. In some cases, there is evidence that these enzymes are secreted, again depending on the species (Bogdanffy et al., 1987; Chen et al., 1992; Lewis et al., 1992a). However, whether the secretion is from Bowman's glands or from sustentacular cells has not been determined. Secretory glands from the more anterior part of the nasal cavity may also contribute to the mucus content in the olfactory region; the complex structure and cellular compositions of these glands in the lamina propria of the human anterior middle nasal turbinate have been described (Tandler et al., 2000).

3.3.2.2 Blood Vessels and Lymphatic Vessels.
The highly vascularized lamina propria in the olfactory mucosa is supplied by the ethmoidal artery, a source distinct from the sphenopalatine supply of the respiratory mucosa. In mice, the blood flow through the total nasal mucosa has been estimated at 0.87% of cardiac output

66   **Chapter 3   Olfactory Mucosa: Composition, Enzymatic Localization, and Metabolism**

(Stott et al., 1983); and in rats, 0.32 ml/min (Morris and Cavanagh, 1986), or 0.53% of cardiac output (Stott et al., 1983). The high perfusion rates in nasal tissues allow for rapid absorption and systemic distribution of substances that penetrate the olfactory epithelium. Conversely, this high perfusion can also allow toxicants in the bloodstream to come in contact with xenobiotic-metabolizing enzymes in the olfactory mucosa. Therefore, because of the high metabolic capacity for xenobiotics, the olfactory mucosa can show significant tissue damage following even systemic administration of toxicants that require metabolic activation. Examples of this are discussed in Section 3.5.

The nasal mucosa also contains a network of lymphatic vessels, which, in addition to their normal functions in transporting fluids, lymphocytes, and antibodies, may also contribute to the transport of xenobiotics to or from the nasal cavity. In normal human nasal mucosa and sinus mucosa, the lymphatic vessels are located in the superficial layer of the submucosa, arranged in a network consisting of antler-like branches and blind ends, but not lymphangion chains (Kim et al., 2007).

3.3.3 Secretions of the Olfactory Mucosa

Acinar cells of Bowman's glands and, in some species, sustentacular cells secrete acidic, sulfated, or neutral mucopolysaccharides, the percentage of each depending on the species and specific physiological, neuronal, and environmental conditions. The distribution of different carbohydrate residues in the mucociliary complex is not homogeneous (Getchell et al., 1993b; Berger et al., 2005). Human nasal secretions contain immune factors including IgA, IgM, and IgG (Kaliner, 1991). Secretory component and J chain have been localized to acinar and duct cells of Bowman's glands, as well as the mucociliary apparatus in the human olfactory mucosa (Mellert et al., 1992). Other components of mucus thought to play a defensive role include the antimicrobial proteins lysozyme and lactoferrin (Mellert et al., 1992; Mullol et al., 1992), enzymatic constituents, including aminopeptidases, endopeptidases, carboxypeptidases, angiotensin-converting enzyme, peroxidase, and kallikrein (Kaliner, 1991; Ohkubo et al., 1998), and a number of antioxidants (Cross et al., 1994), such as reduced glutathione (GSH), mucin, and an abundant, thiol-specific antioxidant protein belonging to the mono-cysteine subfamily of peroxiredoxins (Novoselov et al., 1999). Nasal mucus also contains regulatory proteins and peptides, including secretory leukoprotease inhibitor (Lee et al., 1993), substance P, vasoactive intestinal peptide (Chaen et al., 1993), adrenomedullin (Fujikura and Okubo, 2011), antimicrobial peptides such as psoriasin (Laudien et al., 2011), and insulin-like growth factor-I and its binding proteins (Federico et al., 1999), as well as transport

proteins such as OBP (Pelosi, 1996; Badonnel et al., 2009), salivary androgen binding protein (sABP) (Zhou et al., 2011), and vomeromodulin (Krishna et al., 1995b).

The concentrations of ions and proteins in the nasal mucus are influenced by rates of water loss to evaporation and rates of replenishment; the latter are partly controlled by aquaporins, which are water channel proteins, expressed in the supporting cells and basal cells in the OE and in secretory acinar cells and duct cells in the Bowman's glands (Ablimit et al., 2006; Solbu and Holen, 2012). Nasal secretion may be controlled by nerve stimulation (Revington et al., 1997) and by corticosteroids (Fong et al., 1999). The latter, via mineralocorticoid receptors in supporting cells and Bowman's glands (Robinson et al., 1999), may modulate olfactory Na^+, K^+-ATPase and active ion transport, which results in hyperosmolarity of mucus with respect to serum, as well as secretion of water. In mouse anterior nasal glands and lateral nasal glands (LNG), mucus secretion is mainly mediated by receptors for inositol 1,4,5-trisphosphate (IP3R2 and IP3R3); genetic ablation of these receptors leads to large decreases in secretion of fluid and protein from LNG acinar cells, nasal inflammation, and increases in detection threshold for odorants (Fukuda et al., 2008). Nasal secretion can also be regulated by nutritional cues. The secretion of OBPs, which function as scavengers of reactive oxidants (Grolli et al., 2006), as well as odorant transporters and possibly facilitator for odorant binding to their olfactory receptors (Hajjar et al., 2006), can be upregulated in the olfactory mucosa by food deprivation or by exposure to leptin, a satiety hormone (Badonnel et al., 2009).

The composition as well as quantity of nasal secretions can change dramatically with inflammation, disease, or toxicant exposure (Harkema et al., 1988; Corps et al., 2010). For example, cigarette smoke, antigens, and diethyl ether cause leakage of macromolecules and small ions into the mucous layer (Morgan et al., 1986). The levels of growth factors and neuropeptides in the mucus also change in pathological conditions (Chaen et al., 1993; Federico et al., 1999). The concentration of adrenomedullin, a vasodilatory and anti-inflammatory peptide produced by mast cells in the nasal mucosa, is lower in the nasal discharge of patients with allergic rhinitis than in the nasal discharge of non-allergic patients (Fujikura and Okubo, 2011); whereas the levels of lipids with antimicrobial properties were higher in maxillary sinus secretions from patients with chronic rhinosinusitis when compared with those without the disease (Lee et al., 2010). Finally, the levels of signaling cyclic nucleotides, cAMP and cGMP, are decreased in nasal mucus from patients with impaired taste and smell functions (Henkin and Velicu, 2008).

The cilia on respiratory epithelial cells move mucus over the surface of the epithelium to produce mucociliary clearance of environmental airway contaminants. However, the OE does not contain beating cilia and therefore must rely on the movement and flow created by the cilia in the respiratory epithelium for clearance. As is the case with mucous composition, the efficacy of ciliary beating can be altered by toxicant exposure or disease (e.g., Morgan et al., 1986; Min et al., 1999; Iravani and Melville, 1974). In humans, clearance rates have been described as characteristic of a given individual, which may vary from 1 to 20 mm/min (Proctor et al., 1978), and nasal ciliary beat frequency appears to be age independent (Jorissen et al., 1998).

The presence in nasal secretions of macromolecules suggests that the protective function of secretions is not simply related to clearance, but includes reactions such as bacterial destruction by lysozyme, proinflammatory peptide degradation by peptidases, viral inactivation by IgA interaction, and possibly metabolism of toxicants prior to tissue contact or absorption into systemic circulation. For example, carbonic anhydrase VI is secreted to nasal mucus, where it may play a role in CO_2 sensation and acid-base balance (Kimoto et al., 2004). Plunc, a member of the secretory gland protein family, is found in nasal respiratory epithelium and nasal mucus, where it may offer protection against bacterial infection (Sung et al., 2002). The precise activity and in vivo function of these macromolecules in mucous secretions have not been well studied to date. However, it is clear that transport, binding, and clearance of xenobiotics occurring in the mucociliary apparatus will influence deposition of these substrates in nasal tissues. Therefore, alterations in mucous constituents and ciliary function may alter metabolism as well.

3.3.4 Comparative Aspects of Mucosal Composition

The primary interspecies anatomical differences in olfactory mucosa result from differences in turbinate structure and related proportion of the nasal cavity lined with olfactory mucosa. In general, the surface area to nasal cavity volume ratio reflects the reliance on olfaction of a given species. For example, the rat has a surface area to lumenal volume ratio of 3350 mm^2/cm^3; macaque monkey, 775 mm^2/cm^3; and human, 820 mm^2/cm^3 with comparative lumenal volumes of 0.4 cm^3 for rat and 25 cm^3 for human (Harkema, 1991). Increased surface area results from an increase in the complexity of turbinates in the nasal cavity; generally the greatest difference occurs in the number of olfactory turbinates. For example, in the rat, the percentage of the nasal surface area covered by OE is nearly 50%, a much greater percentage than in the human (Harkema, 1991). Increased infolding in the turbinates also results in alteration in airflow patterns, and therefore in intranasal deposition patterns. However, studies on airflow indicate that the percentage of inspired air reaching the olfactory mucosa is roughly 15% in rat, monkey, and human (Kimbell et al., 1993; Jaillardon et al., 1992; Hahn et al., 1993). Because of the differences in relative proportion of olfactory tissue, however, the percentage of inhaled dose deposited in olfactory tissue may still be quite different across these species. These differences are therefore important considerations in extrapolating data derived from laboratory animal research to the human population.

Although the cytoarchitecture of the OE is remarkably similar across mammalian species, there are species-related differences in the topographical organization of immature and mature ORNs. In human OE, immature and mature ORNs are dispersed, in striking difference from the highly compact, laminar organization found in rodents (Hahn et al., 2005). In addition, there are species differences in basal cell morphology. In rodent OE, the horizontal basal cells and globose basal cells are morphologically distinct; whereas, in human OE, almost all basal cells appear similar to the globose basal cells found in rodents (Hahn et al., 2005). The capacity and localization of xenobiotic metabolism can also be markedly different among different species (Dahl and Lewis, 1993; Hadley and Dahl, 1983; Gervasi et al., 1989; Morris, 1997; Sheng et al., 2000).

Because biopsy samples of human olfactory mucosa are difficult to obtain, most human nasal enzyme activity to date has been studied in respiratory mucosa. Notably, human fetal olfactory mucosa is readily available, and has been shown to contain efficient metabolic activity toward a xenobiotic compound (Wong et al., 2005).

The cells containing xenobiotic metabolic activity in the olfactory mucosa are relatively consistent across species. The primary localization for xenobiotic metabolizing enzymes is within the sustentacular, basal, and duct cells of the epithelium and within the acinar and duct cells of Bowman's glands in the lamina propria. Much like squamous epithelial cells, olfactory receptor cells contain enzymes having primarily a homeostatic function, such as alkaline phosphatase (Bourne, 1948) and carbonic anhydrase (Brown et al., 1984); however, xenobiotic-metabolizing enzymes have generally not been localized to these cells.

Although the cell types identified in the previous paragraph are consistent sites of enzyme localization across species, the specific distribution of a given enzyme, such as the cyanide-metabolizing enzyme rhodanese (Lewis et al., 1991), within these cell types can vary across species; whereas other enzymes, such as carboxylesterase (Lewis et al., 1994b) and the P450 CYP2A isoforms (Su and Ding, 2004), show remarkable similarity in distribution across species. These features can be important determinants of toxicant-induced damage in different species, and must be kept in mind when generalizing from one species to another.

3.4 IDENTITY, TISSUE- AND CELL-SELECTIVE EXPRESSION, AND DEVELOPMENTAL REGULATION OF NASAL BIOTRANSFORMATION ENZYMES

The dramatic capacity of mammalian nasal mucosa to metabolize inhaled substances is now well recognized. Since the first reports that P450 activity in rat nasal mucosa sometimes exceeded activity in liver when normalized to tissue protein content (Hadley and Dahl, 1982), numerous laboratories have reported activity in the nasal mucosa for families of xenobiotic-metabolizing enzymes, including flavin-containing monooxygenases, aldehyde dehydrogenases, alcohol dehydrogenase, carboxylesterases, epoxide hydrolases, UDP glucuronosyltransferase (UGT), glutathione transferase (GST), and rhodanese (Dahl and Hadley, 1991). In addition, xenobiotic-metabolizing capacity has been demonstrated in olfactory and other nasal tissues from a broad range of species, including Drosophila melanogaster (Wang et al., 1999), lobsters (Trapido-Rosenthal et al., 1990), rainbow trout (Starcevic and Zielinski, 1995), rabbits (Ding and Coon, 1988, 1990a; Shehin-Johnson et al., 1995), rodents (Hadley and Dahl, 1982; Genter et al., 1995b), dogs (Dahl et al., 1982), pigs (Marini et al., 1998), sheep (Larsson et al., 1994), cows (Longo et al., 1997), and humans (Lewis et al., 1991, 1994b; Gervasi et al., 1989; Getchell et al., 1993a; Gu et al., 2000; Wong et al., 2005).

Steadfast progress has been made in the identification and characterization of nasal biotransformation enzymes. While studies on P450 enzymes have moved forward to the determination of in vivo functions, the nasal expression of many other phase-I biotransformation enzymes as well as phase II conjugation enzymes and transporters has been reported. The biological model systems ranged from insects to fish and to humans. Although most of the xenobiotic-metabolizing enzymes localized in the nose are also found in other tissues, some are uniquely or preferentially expressed in the olfactory mucosa in a number of species.

3.4.1 Cytochrome P450

The P450 gene superfamily encodes a large number of structurally similar monooxygenases (Nelson et al., 1996). All P450s contain a heme prosthetic group ligated to a highly conserved cysteine residue in the carboxyl terminal portion of the proteins. In a single species, for example, the humans, the total number of P450 genes can be more than 50, and individual genes are expressed more or less in tissue- and cell-selective fashions. Within a cell, the majority of P450s are located in the endoplasmic reticulum (microsomal fraction) while some are specifically located in the mitochondria.

The substrates for microsomal P450s include physiologically important substances such as steroid hormones, eicosanoids, and retinoids, and xenobiotics such as drugs, procarcinogens, antibiotics, organic solvents, anesthetics, pesticides, and odorants. P450-catalyzed biotransformations lead to the formation of more polar compounds that are more readily excreted directly or after conjugation with water-soluble agents such as glucuronic acid and GSH (Porter and Coon, 1991).

P450 and NADPH-cytochrome P450 reductase (CPR), a flavoprotein required for microsomal P450-catalyzed monooxygenase reactions, have been found in relatively high concentration in the olfactory mucosa of rodents, rabbits, cows, dogs, pigs, monkeys (Dahl and Hadley, 1991; Ding and Coon, 1993; Hua et al., 1997; Longo et al., 1997; Marini et al., 1998), and humans (Getchell et al., 1993a; Su et al., 1996). Both P450 and CPR have also been identified in the olfactory organ of Drosophila melanogaster (Hovemann et al., 1997; Wang et al., 1999). On a per mg microsomal protein basis, the level of total microsomal P450 in olfactory mucosa is second only to liver among all tissues examined in rodents and rabbits; the level of CPR in olfactory mucosa microsomes is even higher than in liver (Ding et al., 1986; Reed et al., 1986). The evolutionarily conserved presence of the P450 enzymes supports their functional importance in olfaction.

Many different P450s have been identified in mammalian olfactory mucosa, including members of the CYP1A, 2A, 2B, 2C, 2D, 2E, 2G, 2F, 2J, 2S, 3A, 4A, and 4B subfamilies (Dahl and Hadley, 1991; Deshpande et al., 1999; Ding and Coon, 1993; Gu et al., 1998; Zhang et al., 1997; Minn et al., 2005; Zhang et al., 2005; Saarikoski et al., 2005; Tyden et al., 2008; Li et al., 2011). Of these, CYP1A2, CYP2A, and CYP2G1 are the major forms in rabbits and mice (Ding and Coon, 1990a; Genter et al., 1998; Gu et al., 1998). Multiple genes are present in the CYP2A subfamily, which were named sequentially according to the time of discovery. The functional CYP2A genes expressed in the olfactory mucosa include CYP2A3 in rats, CYP2A5 in mice, CYP2A6 and CYP2A13 in humans, and CYP2A10 and CYP2A11 in rabbits (Koskela et al., 1999; Peng et al., 1993; Su et al., 1996, 2000). There is only a single CYP2G gene in rats, mice, and rabbits, all named CYP2G1 (Nef et al., 1990; Ding et al., 1991, Hua et al., 1997). In humans, there may be two copies of the CYP2G gene, but both contain loss-of-function mutations in the majority of individuals, and a functional cDNA has not been identified (Sheng et al., 2000). Monkeys also have two copies of CYP2G genes; the second copy, CYP2G2, is expressed in the nasal mucosa in cynomolgus monkey, and is active toward coumarin, an odorant (Uno et al., 2011).

3.4 Identity, Tissue- and Cell-Selective Expression, and Developmental Regulation

The human CYP2A13 gene is preferentially expressed in the respiratory tract, with highest levels found in the nasal mucosa, and it is the most efficient P450 enzyme in the in vitro metabolic activation of 4-(methylnitrosamino)-1-(3-pyridyl)-1-butanone (NNK), a tobacco-specific nitrosamine and potent nasal and lung carcinogen (Su et al., 2000; Jalas et al., 2005). CYP2A13 is also active toward many other toxicants and carcinogens, including aflatoxin B_1 (He et al., 2006), 4-aminobiphenyl (Nakajima et al., 2006), naphthalene, styrene, and toluene (Fukami et al., 2008), and 3-methylindole (3-MI) (D'Agostino et al., 2009), as well as odorants, such as nicotine (Bao et al., 2005). CYP2F1 is another human P450 that is primarily expressed in the nasal mucosa and the lung (Zhang and Ding 2008; Carlson, 2008; Weems et al., 2010), with little expression in other tissues (Carr et al., 2003). CYP2F1 appears to be active toward several pulmonary toxicants, including naphthalene, styrene, 3-MI, and benzene (Lanza et al., 1999; Nakajima et al., 1994; Powley and Carlson, 2000; Cruzan et al., 2002); however, as a result of difficulties encountered in attempts to heterologously express CYP2F1, the substrate specificity and biochemical properties of this enzyme are still unclear (e.g., Baldwin et al., 2005). A CYP2A13/2F1-transgenic mouse has been produced (Wei et al., 2012), in which both CYP2A13 and CYP2F1 proteins are abundantly expressed in the olfactory mucosa. Several P450 knockout mouse strains targeting CYPs preferentially expressed in the nasal mucosa have also been generated, including *Cyp2a5*-null (Zhou et al., 2010), *Cyp2f2*-null (Li et al., 2011), *Cyp2g1*-null (Zhuo et al., 2004; Zhou et al., 2010), and *Cyp2s1*-null (Wei et al., 2013), as well as a *Cyp2a(4/5)bgs*-null mouse model, in which nine *Cyp* genes on mouse chromosome 7 (including, sequentially, *Cyp2a5, 2g1, 2b19, 2b23, 2a4, 2b9, 2b13, 2b10,* and *2s1*) are deleted (Wei et al., 2013). Additionally, a mouse model (named Cpr-low) that targets CPR, thus affecting all microsomal P450 enzymes in all cells, has been described (Wu et al., 2005; Wei et al., 2010). These mouse models are being utilized for *in vivo* functional studies of the various P450 enzymes in the olfactory mucosa.

In addition to the P450 forms in gene families 1–4, which are often referred to as the xenobiotic-metabolizing P450s, there are also several microsomal P450 gene families specifically involved in steroid biosynthetic pathways in the endocrine and reproductive organs or bile acid metabolism in liver, such as *CYP7*, 17, 19, 21, and 51 (Nelson et al., 1996). Of these, *CYP19*, also known as the aromatase, which is responsible for estrogen biosynthesis, is expressed in olfactory mucosa in mice (Zhou et al., 2009).

Several olfactory mucosal P450s are specifically or preferentially expressed in this tissue. For example, *CYP2G1* is only expressed in the olfactory mucosa (Ding and Coon, 1990a; Hua et al., 1997; Nef et al., 1989) and, at much lower levels, in the vomeronasal organ (VNO) (Gu et al., 1999). Several *CYP2A*s are expressed in olfactory mucosa at levels much higher than in other tissues (Ding and Coon, 1990a; Su et al., 1996). Preferential expression of a P450 in the olfactory organ was also found in Drosophila (Wang et al., 1999). Although the specific biological roles of these tissue-selective P450s have not been identified, their unique or preferential presence in the olfactory mucosa strongly suggests functional importance in the chemosensory organ. In mice, *CYP2A5* has been found to play a role in the maintenance of androgen homeostasis in the LNG (Zhou et al., 2011).

Immunohistochemical studies of several olfactory mucosa microsomal P450s, including *CYP1A, 2A, 2B, 2G,* and *4B*, indicated that they are expressed in non-neuronal cells, particularly in the sustentacular cells in the epithelium and in the Bowman's glands in the submucosa (Adams et al., 1991; Chen et al., 1992; Getchell et al., 1993a; Thornton-Manning et al., 1997; Voigt et al., 1985; 1993; Zupko et al., 1991; Chen et al., 2003; Piras et al., 2003; Genter et al., 2006). Distribution of CPR was found to resemble that of the P450s (Adams et al., 1991; Baron et al., 1986; Voigt et al., 1985); however, CPR expression in ORNs has also been reported (Voigt et al., 1985; Verma et al., 1993). The lack of known microsomal P450 expression in the neuronal cells is also supported by toxicological studies implicating the Bowman's glands and the supporting cells as the initial targets following chemical treatment (Brittebo, 1997). The localization of P450s to the mucus-producing cells in the Bowman's glands and the detection of P450 immunoreactivity in the mucociliary complex at the epithelial surface led to suggestions that they may be secreted to the mucus layer where they may directly act on inhaled chemicals (Adams et al., 1991; Chen et al., 1992; Genter et al., 2006).

Little is known about the molecular mechanisms that regulate the tissue- and cell-selective expression of P450s and other transformation enzymes in the olfactory mucosa. Early studies identified nuclear factor I (NFI)-like cis-acting elements in the proximal promoter region of rat *CYP2A3* and *CYP1A2* genes (Zhang and Ding, 1998; Zhang et al., 2000). These highly conserved DNA sequences, which are critical for transcriptional activity of the cognate P450 promoters in vitro, appear to interact with olfactory mucosa-restricted nuclear proteins. A similar regulatory element is not identified in the promoter of mouse *Cyp2g1* gene, but a 3.6-kilobase *Cyp2g1* 5′-flanking sequence was sufficient to direct olfactory mucosa-specific and proper developmental expression of a beta-galactosidase reporter gene in transgenic mice (Zhuo et al., 2001). However, the transgene expression pattern in the olfactory mucosa did not exactly match that of the *Cyp2g1* gene, which suggested that additional regulatory sequences are required for correct cell type-selectivity.

70   **Chapter 3   Olfactory Mucosa: Composition, Enzymatic Localization, and Metabolism**

In vitro studies showed that the transcriptional regulation of human *CYP2A13* gene involves C/EBP transcription factors, instead of NFI. In vivo studies using chromatin immunoprecipitation assays indicated that C/EBP is associated with *CYP2A13* promoter in the olfactory mucosa of *CYP2A13*-transgenic mice (Ling et al., 2007). It also appears that the tissue specific expression of CYP2A3 and CYP2A13 may involve epigenetic repression of the genes' expression in non-permissive tissues (Ling et al., 2004; 2007). Notably, genetic polymorphisms in CYP2A13 that are associated with decreased CYP2A13 expression in lung and, presumably, also nasal mucosa have been identified (D'Agostino et al., 2008; Wu et al., 2009; Zhang et al., 2004; 2007); more in-depth studies of the relationships between these genetic variations with expression of CYP2A13 in the nasal tissues may not only reveal the underlining regulatory mechanisms, but also provide opportunities to study relationships between variations in CYP2A13 expression and olfactory function or disease susceptibility. A *CYP2A13*-null allele has also been identified (Zhang et al., 2003).

The developmental expression of P450s and CPR in olfactory mucosa has also been examined. In rabbits, CYP2G1 was detected at 2 days before birth (Ding et al., 1992). In rats, CYP2G1 expression was detected at E20, which was suggested to coincide with the appearance of Bowman's glands (Margalit and Lancet, 1993). Prenatal expression of several P450s and CPR has also been found in humans (Gu et al., 2000). The earlier onset of P450 expression in olfactory mucosa than in other tissues may indicate a functional significance in the perinatal period when olfactory ability is important for the survival of the newborn. In humans, a number of biotransformation genes are expressed in the fetal nasal mucosa at gestational ages of ~3 months, including aldehyde dehydrogenase 6 (ALDH6), ALDH7, CYP1B1, CY2E1, CYP2F1, CYP4B1, UGT2A1, flavin-containing monooxygenase 1, and GSTP1 (Zhang et al., 2005).

3.4.2   Other Enzymes

GSTs catalyze the conjugation of GSH with numerous electrophilic substrates, including reactive intermediates formed in P450-catalyzed reactions, which decrease their reactivity with proteins and other cellular macromolecules (Armstrong, 1997; Eaton and Bammler, 1999), as well as unaltered odorants (Ben-Arie et al., 1993). Most GSTs are located in the cytosol, although some have also been found in microsomes and mitochondria (Eaton and Bammler, 1999). At least five cytosolic GST gene families are known in humans. Multiple GSTs have been detected in the olfactory mucosa in a number of species (Aceto et al., 1993; Ben-Arie et al., 1993, Banger et al., 1993, Krishna et al., 1995a, Starcevic and Zielinski, 1995, Rogers et al., 1999).

A tissue-specific GST has not been found in mammals, although one has been found in the sphinx moth *Manduca sexta* (Rogers et al., 1999). In rats, GSTA and GSTM immunoreactivity was detected in sustentacular cells and Bowman's glands in the olfactory mucosa. In humans, GSTA immunoreactivity was detected mainly in the acinar cells of the Bowman's glands, as well as in the supranuclear region of supporting cells, but GSTP immunoreactivity was detected only in the supporting cells in the olfactory mucosa (Krishna et al., 1994; 1995a). In mice, GSTA and GSTM, but not GSTP, show a preferential expression in the main olfactory tissue, in comparison to the VNO (Green et al., 2005), and GSTM expression as well as GST activity displays a zonal distribution pattern in the olfactory mucosa, favoring the lateral olfactory turbinates over the dorsal or septal regions (Whitby-Logan et al., 2004).

UDP glucuronosyltransferases (UDPGT or UGTs), also named UDP glycosyltransferases, catalyze the conjugation of UDP glucuronic acid with a variety of substrates (Mackenzie et al., 1997). In mammals, the UGTs are found in microsomal fractions and belong to two different gene families, each having multiple genes (Mackenzie et al., 1997). An olfactory mucosa-specific UGT has been identified in rats, cows, and humans, which are all named UGT2A1, and are active toward numerous compounds, including many odorants (Lazard et al., 1990, 1991; Mackenzie et al., 1997; Jedlitschky et al., 1999). A tissue-specific UGT (DmeUgt35a) has also been identified in the olfactory organ of Drosophila melanogaster (Wang et al., 1999). Multiple UGTs are believed to be expressed in mammalian olfactory mucosa (Marini et al., 1998). In rats, UGT2A1 mRNA is found in sustentacular cells and Bowman's gland, as well as in the olfactory sensory neuron and olfactory bulb (Heydel et al., 2001). UGT2A1 is also preferentially expressed in the nasal mucosa than in liver of human fetus (Zhang et al., 2005). A second member of the human UGT2A subfamily, UGT2A2, has been studied. It is also preferentially expressed in the nasal mucosa and, as UGT2A1, it has broad glucuronidation activities toward different endobiotic and xenobiotic substrates, including odorants (Sneitz et al., 2009). A comprehensive survey of the expression profile of 19 human UGT1A, 2A and 2B genes showed that, in both adult and fetal olfactory mucosa, the composite UGT mRNA expression level is also high, ~20% of the adult hepatic level, and that there is a large developmental shift from high UGT2A1 and UGT2A2 expression in the fetus to predominantly UGT1A6 expression in the adult (Court et al., 2012).

Sulfotransferases (ST), which include phenol ST (PST), hydroxysteroid ST (HSST), and, in plants, flavonol ST (FST) gene families, catalyze the transfer of a sulfonate group from 3'-phosphoadenosine 5'-phosphosulfate to both endogenous and xenobiotic compounds (Weinshilboum et al., 1997). Mouse nasal cytosol had high activity for a

3.4 Identity, Tissue- and Cell-Selective Expression, and Developmental Regulation

number of phenolic aromatic odorants (Miyawaki et al., 1996; Tamura et al., 1997). Mouse PSTG immunoreactivity, which is detectable prenatally, is localized mainly in the sustentacular cells in the most dorsal and medial zone of mouse olfactory mucosa (Miyawaki et al., 1996).

Microsomal epoxide hydrolase often plays a protective role against xenobiotic toxicity by detoxifying reactive epoxides formed by monooxygenase reactions. Microsomal epoxide hydrolase is abundantly expressed in the olfactory mucosa (Genter et al., 1995a; Faller et al., 2001); however, it is intriguingly absent in the epithelium lining the dorsal medial meatus in rats, making that region vulnerable to toxicity induced by epoxide-forming compounds such as 2,6-dichlorobenzonitrile (DCBN) (Genter et al., 1995a). The NADPH:quinone oxidoreductase also displays a zone-specific expression in olfactory epithelium, which may underlie zone-specific toxicity of certain quinoidal drugs and nasal toxicants (Gussing and Bohm, 2004).

Carboxylesterases are abundantly expressed in the olfactory mucosa in a number of mammalian species; human nasal mucosal carboxylesterases actively participate in the metabolism of drugs and other xenobiotics, such as methyl methacrylate, dexamethasone cipecilate and ciclesonide (Mainwaring et al., 2001; Sato et al., 2007a,b,c; Sasagawa et al., 2011). A molybdo-flavoenzyme named aldehyde oxidase homologue 3 is expressed in Bowman's glands of mouse olfactory mucosa and has aldehyde oxidase activity toward exogenous substrates (Kurosaki et al., 2004). Other nasal biotransformation enzymes include acetyltransferases, which show strikingly higher activity in olfactory mucosa than in liver (Genter, 2004), as well as aminopeptidase and dipeptidyldipeptidase, which may influence stability of nasally applied peptide drugs (Agu et al., 2009).

The nasal expression and function of a number of transporters have been studied. These include peptide transporters (Agu et al., 2011; Quarcoo et al., 2009), drug/xenobiotic transporters, such as the multidrug-resistance related protein 1 (MRP1) (Kudo et al., 2010), and metal transporters, such as the divalent metal transporters (DMTs) (Thompson et al., 2007). MRP1, which is expressed in supporting cells in olfactory mucosa, but not in cells of the respiratory mucosa, has been suggested to participate in olfactory signal termination (Kudo et al., 2010). The metal transporters DMT1, which is found on both the apical microvilli and the end feet of sustentacular cells, is believed to mediate absorption of inhaled manganese into the CNS (Thompson et al., 2007).

The expression of enzymes and transporters mainly involved in the metabolism or transport of endogenous signaling compounds has also been studied in the olfactory mucosa. For example, monoamine oxidase (MAO), active in dopamine inactivation, and dopamine transporter, responsible for dopamine uptake, are both expressed and functional in the olfactory mucosa, where they may influence bioavailability of intranasally administered dopamine in replacement therapy (Chemuturi et al., 2006; Chemuturi and Donovan, 2006). Enzymes important for arachidonic acid metabolism, including cyclooxygenases, lipoxygenases, and phospholipase A2, are abundant in nasal glands, sinonasal mucosa and respiratory mucosa, and may play important roles in nasal inflammatory responses associated with rhinitis and cystic fibrosis (Gosepath et al., 2004; Owens et al., 2008; Ishige et al., 2008). On the other hand, leukotriene A4 hydrolase, a key enzyme in leukotriene B-4 biosynthesis, is widely distributed in neurons of the OE and VNO, and proposed to play a neuromodulatory function (Chiba et al., 2006).

The expression of many genes that are crucial for various cellular "house-keeping" functions has been detected in the olfactory mucosa. For example, heparanase, which degrades proteoglycans associated with cell surface and extracellular matrix, is possibly involved in neural differentiation (Moretti et al., 2006). Peptidylglycine alpha-amidating monooxygenase, responsible for amidation of neuropeptides, such as the pituitary adenylyl cyclase-activating polypeptide, may play important roles in proliferation of the OE (Hansel et al., 2001). Glucose transporters and monocarboxylate transporters, some of which show a polarized pattern of expression in different cell types, are believed to regulate energy supply throughout olfactory mucosa (Nunez-Parra et al., 2011); whereas the vacuolar proton-pumping ATPase is thought to regulate the pH in the mucous layer (Paunescu et al., 2008). Both vacuolar proton-pumping ATPase and carbonic anhydrases, which catalyze the reversible hydration of CO_2, may contribute to interindividual differences in the nasal ability to detect CO_2 (Tarun et al., 2003; Paunescu et al., 2008). Finally, several anti-oxidative enzymes, including superoxide dismutase, catalase, glutathione peroxidase, and DT-diaphorase, are present in both nasal respiratory and olfactory mucosa (Reed et al., 2003), and likely play important roles in maintaining cellular redox balance.

A number of studies have examined global gene or protein expression profiles in nasal tissues from various sources, including olfactory, transitional, and respiratory mucosa, or olfactory sensory cilia, or olfactory cleft mucus, of rats, mice, or humans (Barbour et al., 2008; Debat et al., 2007; DeStefano-Shields et al., 2010; Hester et al., 2002; 2003; Mayer et al., 2008, 2009; Poon et al., 2005; Roberts et al., 2007; Simoes et al., 2011). Results of these and other similar studies provide a rich resource for further identification of additional biotransformation genes that are expressed in the nasal tissues, and/or regulated by pathophysiological events, such as aging and xenobiotic exposure.

72   Chapter 3   Olfactory Mucosa: Composition, Enzymatic Localization, and Metabolism

3.5 FUNCTIONS OF NASAL BIOTRANSFORMATION ENZYMES

Nasal xenobiotic metabolism likely serves multiple functions. Four possibilities, discussed in more detail below, are: (1) detoxication of inhaled and systemically-derived xenobiotics; (2) protection of other tissues, such as the lung and CNS, from inhaled toxicants; (3) modification of inhaled odorants, including the special case of steroids as reproductive stimuli; and (4) modulation of endogenous signaling molecules. In addition, the roles of nasal biotransformation enzymes in the metabolic activation and toxicity of inhaled or systemically-derived xenobiotics are also considered.

3.5.1 Detoxication of Inhaled Toxicants

The nose is the portal of entry for inhaled chemicals and, as such, is continually exposed to toxic insults. Therefore, one function of xenobiotic metabolism could be detoxication of inhaled toxicants. Because of the small mass of nasal mucosa, the protective function of nasal metabolism is probably more a form of local tissue protection than protection of downstream tissues such as lung. It can be said for the vast majority of lipophilic compounds that would normally build up in the nasal tissue that combined P450 and phase II metabolism, or metabolism by other routes, decreases toxicity either by increasing solubility and subsequent clearance or by other chemical modification to less toxic forms. The presence of relatively high amounts of anti-oxidative enzymes may also serve to protect the nasal tissues against damage.

In some instances, nasal metabolic systems may work in tandem to provide local protection (Dahl and Hadley, 1983; Morgan and Monticello, 1990; Bogdanffy, 1990). Therefore, the regional distribution of damage will be influenced by the relationships among airflow, deposition, and chemical solubility and reactivity, as well as localization of enzymes involved in the sequential metabolic steps. For example, the enzyme rhodanese, which metabolizes cyanide to the less toxic metabolite thiocyanate (Lewis et al., 1991), may also metabolize cyanide produced from the P450 metabolism of inhaled organonitrile compounds such as acetonitrile (Dahl and Waruszewski, 1990).

With respect to the secondary detoxication of toxic metabolites, the cellular localization of specific enzymes may be important when generalizing across species. The organonitrile β,β'-iminodipropionitrile (IDPN) is toxic to the acinar cells of Bowman's glands following systemic administration in rats (Genter et al., 1992). P450 metabolism of IDPN would yield cyanide. The cells of Bowman's glands in the rat contain several isoforms of P450 (Dahl and Hadley, 1991); however, they do not

contain rhodanese (Lewis et al., 1992b). It is therefore likely that the toxicity of IDPN in these cells is caused by a buildup of the metabolite, cyanide. Because the cellular distribution of rhodanese differs across species, the target cells for IDPN toxicity may also differ.

It should be remembered that nasal xenobiotic enzymes act not only on inhaled substrates, but on substrates in the systemic circulation as well, as evidenced by metabolite-induced toxicity to the olfactory mucosa seen following intravenous administration of toxicants such as NNK (Belinsky et al., 1990), 3-MI (Turk et al., 1986), and acetaminophen (Jeffery and Haschek, 1988). Teleologically, this metabolism of systemic compounds could serve to reduce stimulation of olfactory receptors by circulating odorants, thereby eliminating possible interference with or masking of inhaled odorants. In addition, such metabolism could serve to protect this important sensory tissue from damage induced by circulating toxicants. Notably, as will be discussed later, the toxic metabolites formed are usually removed by phase II enzymes or other further biotransformation mechanisms, thus avoiding tissue damage.

3.5.2 Protection of Other Tissues from Inhaled Toxicants

3.5.2.1 Lung. Based on activity, most nasal enzymes probably have little effect on reducing concentrations of toxicants entering systemic circulation unless inhaled concentrations are very low. Two exceptions to this might be toxic substrates of carboxylesterases and cyanogenic compounds detoxified by the cyanide-metabolizing enzyme rhodanese (Dahl, 1988; Lewis et al., 1991). Across several species, the capacity of nasal carboxylesterases is sufficient to detoxify inhaled concentrations of esters such as ethyleneglycol monomethyl ether acetate in the 1000 to 3000-ppm range (Dahl, 1988), a concentration in excess of occupational exposure limits. However, in many cases, such as with dibasic esters, the metabolites of esters are themselves toxic to olfactory tissues (Bogdanffy, 1990). Likewise, nasal rhodanese activity is sufficient to detoxify inhaled concentrations of hydrogen cyanide as high as 2800 ppm in the rat (Lewis et al., 1991).

This capacity to significantly alter systemic toxicant exposure through nasal metabolism does not hold for all enzyme families. The capacity for nasal metabolism of some P450 substrates is considerably lower: 0.1-5 ppm for *p*-nitroanisole and 0.1-3 ppm for aniline. At these levels of activity, significant systemic protection from inhaled toxic substrates would probably not result from nasal metabolism.

3.5.2.2 Central Nervous System. Xenobiotic metabolism in the OE as well as in the olfactory bulbs may

3.5 Functions of Nasal Biotransformation Enzymes

be a component of a "nose-brain barrier." The OE has a unique anatomy wherein a single receptor cell contacts the external environment in the nasal lumen and projects directly to its synapse within the CNS in the glomeruli of the olfactory bulbs. These cells, then, provide direct access for inhalants to the CNS. Several studies have demonstrated that a variety of materials instilled or surgically implanted into the nasal cavity can be transported to the olfactory bulbs (Shipley, 1985; Schultz and Gebhardt, 1934; Tomlinson and Esiri, 1983; McLean et al., 1989; Henriksson and Tjälve, 2000; Larsson and Tjälve, 2000). Generally, these studies have used concentrations of material far in excess of those encountered environmentally. Therefore, the importance of this phenomenon for *inhalation* of environmentally relevant concentrations of toxicants is not yet clear. The factors involved in a nose-brain barrier that might protect the brain from toxicant exposure have not yet been elucidated. However, they most likely include xenobiotic metabolism and transport in the olfactory mucosa, as well as in the olfactory bulb, where biotransformation enzymes such as UGT2A1 (Heydel et al., 2001) and xenobiotic efflux transporters such as the multidrug resistance P-glycoprotein are also expressed (Thiebaud et al., 2011). In addition, mucociliary clearance, immune responses in the olfactory or other nasal mucosa, tight junctions between epithelial cells, and the rapid death of epithelial cells following toxicant exposure are also likely to play a role (Lewis et al., 1994a).

3.5.3 Modification of Olfactory Stimuli

3.5.3.1 *Odorants.*

A third possible function for nasal metabolism is either activation of inhaled nonodorants to odorants or, conversely, clearance of odorants from the olfactory receptor cells to allow reactivation of receptors (Dahl, 1988; Getchell and Getchell, 1977; Lazard et al., 1991). Although most biotransformation enzymes are located in the non-neuronal cells, the lipophilic substrates can quickly diffuse to all cells in the olfactory mucosa. This phenomenon of receptor reactivation has also been demonstrated in lobsters (Trapido-Rosenthal et al., 1990) and in silk moths (Vogt and Riddiford, 1981).

Odorant metabolites may contribute to potency and odor quality. Many odorant metabolites are more water soluble than the parent odorants, and they may reach very high concentrations in the mucus bathing ORNs (Dahl, 1988; Price, 1984). If olfactory stimulation includes summation of signals from parent odorant and its metabolites (Kashiwayanagi et al., 1987; Price, 1984), such metabolites could be important to the sensitivity, intensity, and quality perception of an odor. Thus, it was hypothesized that odor quality and intensity may reflect effects of the odorant and its metabolites on ORNs (Dahl, 1988; Price, 1984, 1986).

In a recent study in mice, it was convincingly demonstrated that pharmacological inhibition of enzymatic conversion of odorants that are aldehydes or esters to corresponding acids and alcohols in the nasal mucus is associated with changes in odorant-induced glomerular activation patterns in the olfactory bulb, as well as failure of the animals to identify the tested odorants, which they normally can discriminate, in olfactory discrimination tests (Nagashima and Touhara, 2010). These results, which suggested that biotransformation of odorants in the nasal mucus is rapid enough to affect odorant detection and identification, support the previously proposed role of nasal biotransformation enzymes in odorant signal modification.

Lipophilic odorants can partition favorably into the membranous structure of the neuroepithelium. Their accumulation may adversely affect many aspects of cellular function. They may saturate the odorant clearance mechanism, disturb mitochondrial energy production, and suppress or sensitize local immune systems. They may also change the electrophysiological properties of the plasma membrane and thus the functional capacity of ion channels and other signal transduction components. Thus, in addition to the role in odorant signal modification, biotransformation reactions which convert these lipophilic compounds into more water-soluble metabolites may also be indispensable for maintaining the homeostasis of the chemosensory tissue.

3.5.3.2 *Steroids.*

Steroids are likely to represent a special case of metabolic modulation of olfactory stimuli. Inhaled steroids can serve as primary olfactory cues in the regulation of reproductive function in a number of species, as well as modulators of olfactory function. Androstenone is a steroid found in the urine and saliva of pigs and humans, and also in human sweat. In pigs, androstenone excreted by boars has been shown to initiate mating behavior in estrus sows with intact olfactory function (Beauchamp et al., 1976). Sensitivity to the odor of androstenone varies widely in humans (Dorries et al., 1989). Interaction of exogenous steroids with the olfactory system has also been demonstrated through modification of serum testosterone, testicular size, and spermatogenesis in rhesus monkeys by intranasal administration of estradiol and progesterone (Anand-Kumar et al., 1980). Estradiol present in the urine of male mice can cause failure of blastocyst implantation in females (Guzzo et al., 2012).

It is likely that the ability of the olfactory mucosa to metabolize steroids will influence responses to exogenous steroids. Mammalian olfactory mucosa has very high activities in the metabolism of all three major sex steroids (Brittebo and Rafter, 1984; Brittebo, 1985; Ding and Coon, 1990a, 1994; Hua et al., 1997, Longo et al., 1997; Marini et al., 1998). The VNO is also capable of metabolizing sex steroids (Gu et al., 1999). Accumulation of sex steroids and other endogenous or exogenous compounds that are normally removed by P450 metabolism could affect signal transduction by competing for receptors. To that end, Rosenblum et al. (1991) reported that receptor binding of 17α,20β-dihydroxy-4-pregnen-3-one to goldfish olfactory mucosa is competitively inhibited by progesterone and other sex steroids. Androstenone is metabolized by the nasal mucosa in pigs (Gennings et al., 1974) and is a competitive inhibitor of steroid metabolism by CYP2G1 (Ding and Coon, 1994), which suggests that it may be metabolized by this or other P450 enzymes.

The LNG, which is a non-steroidogenic organ, may play an important role as a storage organ for androgens (Zhou et al., 2009). In adult male mice, testosterone levels in the LNG are substantially higher than those in the olfactory mucosa and other non-reproductive or non-endocrine tissues examined (Figure 3.2). The high levels in the LNG are accompanied by high levels of sABP, low microsomal testosterone-hydroxylase activities, and absence of steroid 5α-reductases and the aromatase, enzymes that can convert testosterone to other forms of steroid hormones. The levels of testosterone in olfactory mucosa and LNG are influenced by circulating androgen levels, by rates of steroid metabolism in the tissue, and by the presence of steroid binding proteins, such as sABP, which may impede testosterone metabolism, and sequester steroids inside cells. Thus, the unusually high testosterone level in the LNG can be explained by the very high sABP levels and very low testosterone metabolism activities in this tissue. The role of P450-mediated testosterone hydroxylation in maintaining LNG testosterone levels are demonstrated by the finding that LNG testosterone levels are further increased in $Cyp2a5$-null mice (Zhou et al., 2011); CYP2A5 encodes an avid testosterone hydroxylase.

The functional significance of the unusually high testosterone level in the LNG, a gland known for its secretion of OBPs (Snyder et al., 1988) and IgA (Adams et al., 1981) to the nasal cavity, is intriguing. It has been proposed that the stored androgen can be secreted, together with sABP, from LNG to nasal mucus, thus potentially functioning in a paracrine fashion in other parts of the male olfactory system, including the olfactory mucosa and VNO. Testosterone in the nasal mucus may activate membrane androgen receptors; interact with odorant/pheromone receptors (either as an agonist or an antagonist); or inhibit putative pheromone-metabolizing enzymes in the mucus (Zhou et al., 2009).

The sABPs are members of the secretoglobin family (Laukaitis and Karn, 2005). They are expressed in a number of organs in the head and neck region, including the VNO, olfactory bulb, salivary glands, lacrimal gland, and Harderian gland (Laukaitis et al., 2005). They can bind testosterone, progesterone, but not estradiol or cholesterol, with high affinity (Karn, 1998). Their biological functions have not been determined, though, in addition to the proposed roles in concentrating testosterone in the LNG and delivery of testosterone to the nasal mucus, they are also believed to function as pheromones, in modulation of odorant detection, or in reproductive behavior (Emes et al., 2004; Laukaitis et al., 1997; Talley et al., 2001). Interestingly, CYP2A5, testosterone and sABP are integral parts of a defensive mechanism against xenobiotic toxicity in the LNG and possibly other nasal tissues (Zhou et al., 2011). The $Cyp2a5$ gene knockout not only led to increases in testosterone levels, but also induction of sABP, the latter can protect glutathione and cellular proteins by quenching reactive metabolites of acetaminophen, a drug and model nasal toxicant, and presumably other xenobiotic compounds. This mechanism (Figure 3.3) explains the finding that $Cyp2a5$-null mice are resistant to the toxicity of acetaminophen in the LNG even though CYP2A5 is not responsible for the bioactivation of acetaminophen in LNG (Zhuo et al., 2004; Zhou et al., 2011). Earlier studies on the liver-Cpr-null mice also showed that, while the toxicity of acetaminophen in the olfactory mucosa was not dependent on bioactivation by hepatic P450 enzymes, its toxicity in the LNG was at least partly dependent on hepatic bioactivation, thus supporting the idea that liver produced acetaminophen metabolites can be transported to the LNG to cause toxicity (Gu et al., 2005).

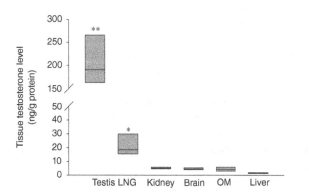

Figure 3.2 Serum and tissue testosterone level in male mice (From Zhou et al., 2009, with permission). Testosterone levels were determined for 8-week old male mice (n = 8). The values shown are medians, together with the 25% (lower bar) and 75% (upper bar) percentile marks. **, levels in testis were significantly higher than those in other tissues (P < 0.05). *, levels in the LNG were significantly higher than levels in all other tissues examined, except for testis (P < 0.05).

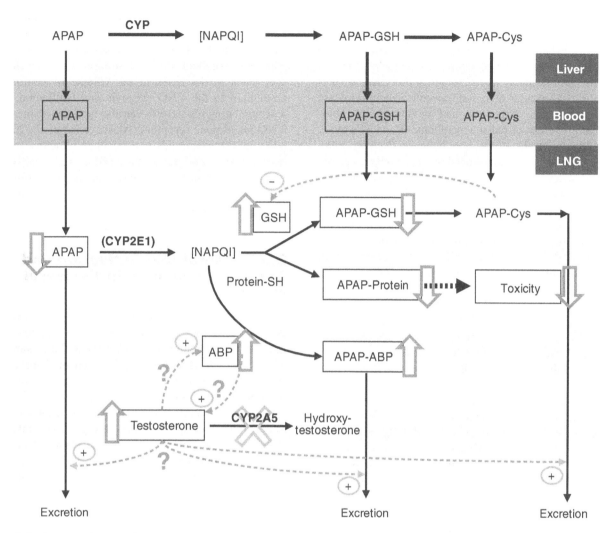

Figure 3.3 A proposed novel defensive mechanism against acetaminophen (APAP) toxicity in the male mouse LNG (From Zhou et al., 2011, with permission). Boxed items represent entities measured in APAP-treated mice. Boxed arrows indicate direction of changes observed in *Cyp2a5*-null mice. *In situ* metabolic activation of APAP in the LNG (presumably mediated by CYP2E1) leads to the formation of the reactive N-acetyl-*p*-benzoquinone imine (NAPQI), which forms conjugates with GSH or protein (including ABP), and eventually leads to cytotoxicity. APAP-GSH and APAP-CYS, either locally formed, or absorbed from systemic circulation, cause further depletion of GSH (through the γ-glutamyl cycle), and thus increase vulnerability of cellular proteins to attack by N-acetyl-*p*-benzoquinone imine. Testosterone is speculated as a positive regulator of ABP expression and possibly mucus secretion. LNG testosterone level is regulated negatively by CYP2A5 and positively by ABP. Genetic ablation (or xenobiotic inhibition) of CYP2A5 leads to suppression of testosterone disposition in the LNG, consequent increases in tissue testosterone levels, up-regulation of ABP, and enhanced protection against APAP toxicity, through increased formation of APAP-ABP adducts, reduced depletion of cellular GSH, reduced formation of APAP conjugates with other cellular proteins, and (presumably) increased excretion of APAP and APAP-GSH/APAP-CYS to nasal mucus.

3.5.4 Modulation of Endogenous Signaling Molecules

The P450 isoforms identified in the olfactory mucosa are all active in the metabolism of endogenous compounds, although they are also involved in metabolizing foreign chemicals. For example, CYP1B, 2A, 2B, 2G, and 3A are active in the hydroxylation of sex steroids (Ding and Coon, 1994; Hayes et al., 1996; Liu et al., 1996; Waxman et al., 1991), CYP1A, 2B, 2C, 2E, 2J, and 4A are active in the hydroxylation or epoxygenation of arachidonic acid (Laethem et al., 1992; Luo et al., 1998; Scarborough et al., 1999), CYP1A2 and CYP2J4 are active in converting retinals to retinoic acids (RAs), and CYP1A and 2B are active in the hydroxylation of RA (Roberts et al., 1992; Zhang et al., 1998). The consequences of microsomal

76   Chapter 3   Olfactory Mucosa: Composition, Enzymatic Localization, and Metabolism

P450 catalyzed metabolism of endogenous compounds are usually inactivation of the bioactive substance, as with hydroxylation of testosterone and RAs. However, the epoxygenated or hydroxylated products of arachidonic acid have been implicated in many biological processes, such as regulation of vascular tone, ion transport, calcium release from endoplasmic reticulum, and modification of biophysical properties of plasma membrane (Capdevila et al., 1992; Makita et al., 1996). It is believed (Nebert, 1990, 1991) that the xenobiotic-metabolizing P450s regulate steady state levels of endogenous compounds important for growth, homeostasis, differentiation, and neuroendocrine functions.

P450-catalyzed formation of arachidonic acid epoxide and hydroxides can regulate vascular tone and thus rate of blood flow (Capdevila et al., 1992; Makita et al., 1996). Decreased ability to produce these regulatory molecules may lead to congestion and restrictions in airflow in the nasal cavity, which could affect threshold sensitivity in odor detection. On the other hand, accumulation of arachidonic acid may lead to increased production of leukotrienes and other mediators through the lipoxygenase pathway and potentially induce airway hypersensitivity (Pinto et al., 1997).

ORNs are one of the few vertebrate neuronal populations that undergo continuous turnover and replacement throughout the life of an animal and following injury (Graziadei and Monti-Graziadei, 1983; Goldstein et al., 1998). Such remarkable regenerative capacity may be at least partly related to the presence of the highly active biotransformation enzymes, which control the availability and level of various endogenous bioactive substances capable of regulating growth and differentiation in the target tissue. The relatively high efficiency and broad substrate specificity of the P450 enzymes toward steroid hormones (Ding and Coon, 1994) and retinoids (Roberts et al., 1992) suggest that these compounds may accumulate in olfactory mucosa or other nasal tissues when the P450s are inhibited or down-regulated, as has been demonstrated for testosterone in the LNG (Zhou et al., 2011).

RA plays essential regulatory roles in the differentiation and regeneration of the olfactory system (for a review see Rawson and LaMantia, 2006). RAs are degraded in target tissues by microsomal P450s (Duester, 1996). RA inactivation catalyzed by an embryonic P450 isoform (P450RA) results in RA hyposensitivity in cultured cells (Fujii et al., 1997). RA biosynthetic activity and retinoid receptors are expressed in the olfactory mucosa of adult mice (Zhang, 1999). Metabolism of RA by CYP26B1 in the olfactory neurons impacts neuronal survival and formation of neural circuit map (Hagglund et al., 2006).

Olfactory mucosa is a known target tissue for steroid hormone action (Balboni, 1967; Balboni and Vannelli, 1982; Fong et al., 1999; Vannelli and Balboni, 1982). In male rats, olfactory mucosa morphology is altered by castration, and testosterone replacement counteracts these alterations (Balboni, 1967). Castration during development also leads to decreases in the volume of neurosensory epithelium in the VNO (Segovia and Guillamon, 1982), and circulating sex steroids can also modulate expression of VNO pheromone receptors (Alekseyenko et al., 2006). In addition, corticosteroids may regulate olfactory secretion by modulating Na^+,K^+-ATPase (Fong et al., 1999). Thus, prolonged accumulation of these endogenous compounds may lead to changes in olfactory mucosa structure, cell biology, and functional capacity.

3.5.5 Metabolic Activation and Xenobiotic Toxicity in the Nasal Mucosa

The powerful biotransformation enzymes, particularly the P450 enzymes, generate reactive intermediates from inhaled or systemically-derived xenobiotic substrates, which could lead to toxicity. The activated metabolites, such as the proposed epoxide intermediates from DCBN or coumarin (Ding et al., 1996; Zhuo et al., 1999), can usually be efficiently removed by phase II enzymes such as GST. However, when the dose is high or when the phase II enzymes are compromised due to chemical inhibition or genetic deficiency, the reactive metabolites would accumulate and cause cytotoxicity in the olfactory mucosa. In cases where a reactive intermediate with relatively long half-life is generated, such as the benzo(a)pyrene epoxides (Dahl et al., 1985), it may be transported to nearby organs, such as the pharynx, the esophagus, the anterior nasal cavity, and the olfactory bulb, where local biotransformation activities are much lower compared to the olfactory mucosa, and potentially cause toxicity (Dahl et al., 1985; Ghantous et al., 1990; Persson et al., 2002).

Numerous compounds, such as ferrocene (Sun et al., 1991), 3-trifluoromethylpyridine (Gaskell, 1990), acetaminophen (Jeffery and Haschek, 1988, Genter et al., 1998), NNK (Belinsky et al., 1990), DCBN (Brittebo, 1997), styrene (Green et al., 2001), methimazole (Bergstrom et al., 2003), 2,6-dichlorophenyl methylsulfone (Franzen et al., 2006), naphthalene (Lee et al., 2005), and coumarin (Gu et al., 1997) are metabolized to toxicants that produce necrosis of the OE. This toxicity can occur following not only inhalation, but systemic administration as well. Recent studies with P450 and Cpr knockout (global or tissue-specific) mouse models indicate that, for DCBN and methimazole, bioactivation in the target tissue plays an essential role, and that CYP2A5 is the major enzyme responsible for the bioactivation (Xie et al., 2010; Xie et al., 2011).

3.5 Functions of Nasal Biotransformation Enzymes

For some compounds, which induce toxicities both in nasal mucosa and elsewhere, the specific enzymes responsible for the nasal toxicity may differ from those responsible in other tissues. For instance, 3-MI is toxic to both olfactory mucosa and lung. However, 3-MI bioactivation is mediated mainly by CYP2F2 in the lungs, but by both CYP2A5 and CYP2F2 in the olfactory mucosa (Zhou et al., 2012). Furthermore, the reactive metabolites produced by the two enzymes through common dehydrogenation and epoxidation pathways have different fates, with CYP2A5 supporting direct conversion to stable metabolites and CYP2F2 supporting further formation of reactive iminium ions (Zhou et al., 2012) (Figure 3.4). Similarly, whereas CYP2F2 is essential for naphthalene induced lung toxicity, it only had a minor role in naphthalene-induced olfactory mucosal toxicity (Li et al., 2011). Therefore, knowledge of the specific enzymes involved in the biotransformation of a given toxicant in olfactory mucosa is important for understanding mechanisms of toxicity.

Often, the relative toxicity of a compound in different species or in tissues within a given species is affected by levels of activating or detoxicating enzymes. However, when compounds are administered systemically, the capacity for other organs to clear the compound must be considered, as it may reduce the amounts reaching the nose. An example is the finding that although rat and mouse olfactory P450s are equally active in metabolic activation of coumarin, rats are much more sensitive to the nasal toxicity of coumarin than mice are, because of a lower hepatic clearance of the parent compound (Zhuo et al., 1999). The observation that nasal toxicity of DCBN is augmented in a mouse model with ablated hepatic DCBN metabolic clearance (the liver-specific Cpr-null mouse) further supports the idea that hepatic clearance can impact nasal exposure to a parent toxicant (Xie et al., 2010). Additionally, when a toxicant is inhaled, differences in nasal airflow patterns, mucociliary clearance, or epithelial status may also affect the toxicity. For example, with inhaled naphthalene, cellular injury is confined to the medial meatus, regions characterized as having relatively slow airflow but high P450 activity, whereas with systemically administered naphthalene, injury occurs throughout the olfactory mucosa (Lee et al., 2005).

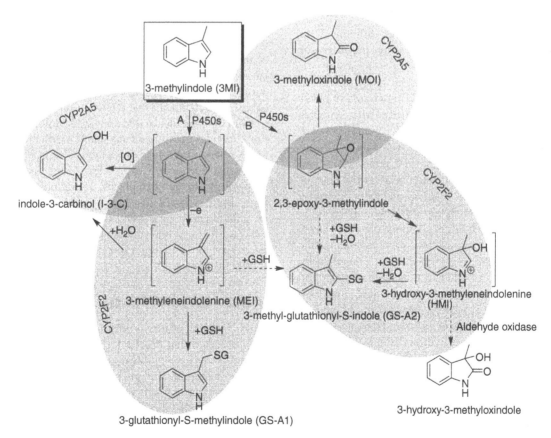

Figure 3.4 Scheme for the metabolic activation of 3MI by mouse olfactory mucosa and lung microsomal P450s (From Zhou et al., 2012, with permission). Pathway A represents dehydrogenation and pathway B represents epoxidation. Shaded areas show metabolites preferentially formed by CYP2A5 or CYP2F2.

78   **Chapter 3   Olfactory Mucosa: Composition, Enzymatic Localization, and Metabolism**

Nasal cancers are relatively uncommon in humans, although in certain populations, notably Chinese males, the rate of occurrence is quite high (Tricker and Preussman, 1991). The occurrence of nasal tumors in laboratory animals exposed by inhalation to toxic materials, on the other hand, is a common finding. Often, the toxic materials are procarcinogens requiring metabolic activation, suggesting that differences in nasal xenobiotic metabolism between humans and laboratory animals may underlie the observed differences in nasal tumor formation.

The ability of rodent nasal mucosa to bioactivate procarcinogens is well documented, for example, for the herbicide acetochlor (Green et al., 2000) and the tobacco-specific carcinogene N'-nitrosonornicotine (Murphy et al., 2000). Few studies have been able to demonstrate bioactivation activities for nasal procarcinogens in human nasal mucosa (e.g., Green et al., 2001). However, the human nasal tissues studied are rarely of olfactory origin. A study with human fetal olfactory mucosa demonstrated relatively high efficiency for bioactivation of NNK, another tobacco-specific procarcinogen, although the rates were low, as a result of the very low levels of P450 expression in the human tissue, compared to rodents (Wong et al., 2005). More studies on the capability of human nasal tissues to bioactivate procarcinogens are needed.

Among laboratory animal species, differences in rates of carcinogen bioactivation are often predictive of species sensitivity and tissue specificity in nasal tumorigenesis. For example, alachlor induces nasal tumors in rats in regions with high CYP2A3 expression (Genter et al., 2000). Rat nasal mucosa preferentially activates N-nitrosopiperidine, a potent rat nasal carcinogen, over N-nitrosopyrrolidine, a weak nasal carcinogen (Wong et al., 2003). However, although such relationships are compelling as explanations of toxicity, as in the case of noncarcinogenic responses, relative metabolic capacity is not always sufficient to explain differences in toxicity. In the case of alachlor-induced olfactory carcinogenesis, differential rates of hepatic alachlor elimination in mice and rats may explain the resistance in mice (Genter et al., 2004), and differences in other factors, such as induction of olfactory mucosal matrix metalloproteinase, may also be relevant (Genter et al., 2005). The nasal carcinogens NNK and N'-nitrosonornicotine produce DNA adducts in nasal tissue via the reactive α-hydroxylated N-nitrosamine metabolites. Although metabolic capacity of the tissues might lead to the prediction that the olfactory mucosa would produce comparatively more adducts than the respiratory mucosa, more adducts were actually found in the respiratory mucosa (Belinsky et al., 1990; Zhang et al., 2009a; Zhang et al., 2009b). Again, other factors, perhaps in this case route of exposure and DNA repair rates, must be taken into account to explain or predict toxicity.

3.6   MODIFICATION OF OLFACTORY XENOBIOTIC METABOLISM

As is the case of other xenobiotic-metabolizing enzymes, nasal enzymes are susceptible to modification in their levels of activity. Specific chemical inhibitors, and in some cases inducers, can alter nasal metabolic capacity. In at least some instances, such an alteration has been demonstrated to subsequently alter toxicity as well. Induction of P450 activity by administration of ß-naphthoflavone decreased the severity of 3-MI-induced olfactory lesions (Turk et al., 1986), possibly as a result of lower blood levels due to enhanced liver metabolism. Conversely, treatment with dexamethasone potentiates the 3-MI olfactory toxicity, which could be partly due to the inducing action of dexamethasone on the P450 responsible for metabolic bioactivation of 3-MI in the olfactory mucosa (Kratskin et al., 1999). In addition, changes in endogenous steroid hormones may modify the biotransformation capacity of the olfactory mucosa, as demonstrated by the effects of castration on nasal metabolism of testosterone. Castrated male rats have a reduced ability to metabolize testosterone, while testosterone replacement restores metabolic capacity for the steroid (Lupo et al., 1986). However, little else is known about this potentially very interesting subject.

As the following section will detail, many common environmental exposure scenarios can result in alterations of nasal enzymatic activity, thereby enhancing individual variations in responses to toxicant exposures. These alterations can result from either direct inhalation or systemic exposure to chemicals that induce or inhibit nasal enzymes, or from toxicant insults that alter the histology of the tissue. For example, many toxicants cause exfoliation of the OE and a concomitant loss of metabolic capacity from those lost cells. Enzymatic expression also appears sensitive to hyperplastic or metaplastic alterations in the epithelium that can result from toxicant exposures, infections, or inflammatory processes. Consideration of a patients' exposure history may therefore be helpful in diagnosing what appears to be an atypical response to a subsequent toxicant exposure.

3.6.1   Enzyme Induction

Early reports, primarily from studies on rat nasal tissue, indicated that nasal P450s were relatively refractory to induction by a wide range of inducers effective for hepatic enzymes. Either no induction or mild induction of rat nasal P450 activity was observed following treatment with the classic inducers: phenobarbital, benzo(a)pyrene, 2,3,7,8-tetrachlorodibenzo-p-dioxin, or 3-methylcholanthrene (Hadley and Dahl, 1982; Bond, 1983; Baron et al., 1988; Longo et al., 1988). Although one study reported the induction of mouse nasal P450 activity by phenobarbital (Brittebo, 1982), the apparently increased activity may have resulted from induction of phase II enzymes occurring downstream from the P450 breakdown step, as only increased $^{14}CO_2$ production was reported (Dahl and Hadley, 1991). Induction of phase II enzymes would be consistent with reports of induction of these enzymes in rat nasal tissue by phenobarbital (Guengerich et al., 1982; Longo et al., 1988). The phase II enzyme UGT is also induced by both Arochlor 1254 and 3-methylcholanthrene (Bond, 1983; Longo et al., 1988). Nevertheless, not all phase II enzymes are readily inducible in the olfactory mucosa. Olfactory GSTs were not induced in rats by trans-stilbene oxide, which caused a 2-fold induction in the liver (Banger et al., 1996); only a marginal induction (1.3-fold) by phenobarbital was achieved in the olfactory mucosa while a 2.8-fold induction was found in liver (Banger et al., 1996).

Rabbit nasal CYP2E1 (involved in the metabolism of ethanol and other alcohols, acetone, acetaminophen, nitrosamines, and diethyl ether) can be increased 2-fold by treatment with ethanol and 6-fold following acetone treatment (Ding and Coon, 1990b). These data represent the first evidence of an increase in nasal xenobiotic-metabolizing capacity of a magnitude that can be considered important physiologically. Induction of CYP2E1 in the olfactory mucosa has been confirmed in other species (Longo and Ingelman-Sundberg, 1993; Genter et al., 1994; Gu et al., 1998; Wang et al., 2002; Nannelli et al., 2005). Nasal CYP2E1 and CYP1A2 can also be induced in rats by fasting (Longo et al., 2000). Interestingly, induction of rat CYP2E1 by dioxane occurred only at the protein level in the liver, but was at both protein and mRNA levels in the nasal mucosa (Nannelli et al., 2005). CYP1A1, which is active in the metabolic activation of polycyclic aromatic hydrocarbons, was not induced in the olfactory mucosa by 3-methylcholanthrene, but was significantly induced in Bowman's glands and in the olfactory and respiratory epithelia following a single intraperitoneal injection of Arochlor 1254 in rats (Voigt et al., 1993). CYP1A1 protein is also induced in olfactory mucosa of mainstream cigarette smoke-exposed rats, but a corresponding increase in CYP1A1 activity was not observed (Wardlaw et al., 1998). Induction of nasal P450 enzymes by tobacco smoke has been proposed as a possible mechanism for developing resistance to the environmental toxins implicated in Parkinsonism and other neurologic diseases (Gresham et al., 1993).

CYP2As represent major P450 isoforms in the olfactory mucosa of a number of species. A study by Beréziat et al. (1995) suggested that a CYP2A-like P450 may be induced in rats by treatment with coumarin in drinking water. However, the same results were not obtained in another study with a different strain of rats (Gu et al., 1997), and no induction of CYP2A was found following

treatment of mice with several chemicals known to induce the same enzyme in the liver (Su et al., 1998). Expression of CYP2A3 mRNA in rat nasal mucosa was not induced by 3-methylcholanthrene, pyrazole, or beta-ionone (Robottom-Ferreira et al., 2003). Interestingly, one of the known CYP2E1 and CYP2A5 inducers, pyrazole, was found to induce CYP2J4 in rat olfactory mucosa as well as in other tissues (Zhang et al., 1999). Pyrazole also induces CYP2J proteins in mouse olfactory mucosa (Xie et al., 2000). Expression of CYP2B and 2C, as well as CYP1A and microsomal epoxide hydrolase, is induced in rat olfactory mucosa by the environmental contaminant 2,3,7,8-tetrachlorodibenzo-p-dioxin (Genter et al., 2002). A recent study in rats showed that mRNAs for P450 2A3, CYP3A9, UGT2A1, multidrug resistance-related protein type 1, and sulfotransferase 1C1 was increased by dexamethasone in the olfactory mucosa (Thiebaud et al., 2010).

The tissue-differential inducibility of some, but not all, P450 enzymes may be related to the unique (yet still unknown) function of each P450 enzyme and the need for the olfactory mucosa to maintain certain enzymes at a relatively constant level. Alternatively, it is possible that nasal biotransformation enzymes may respond preferentially to inhaled odorants. For example, carboxylesterase is induced in the olfactory mucosa following inhalation exposure to the common solvent pyridine (Nikula et al., 1995), which is not a substrate for this enzyme.

3.6.2 Inhibition of Nasal Xenobiotic Metabolism

Unlike the case for nasal enzyme induction, inhibition of nasal xenobiotic metabolism occurs in a wide range of enzyme families. Several P450 isoforms have been inhibited in homogenates of nasal mucosa by hepatic P450 inhibitors such as metyrapone, α-naphthoflavone, piperonyl butoxide, and a number of odorants, including 5-α-androstenone. Because these inhibitors are common ingredients in many products in everyday usage such as perfumes, cosmetics, and household insecticides, exposure to these compounds may alter nasal metabolic capacity from that observed in controlled laboratory situations (Dahl and Brezinski, 1985; Dahl, 1982; Laethem et al., 1992; Ding and Coon, 1994; Marini et al., 2001).

Cigarette smoke is another common environmental pollutant that is known to modify olfaction (Frye et al., 1989) and to alter the capacity for nasal xenobiotic metabolism (Wardlaw et al., 1998). Alterations in nasal metabolism may be the direct result of exposure to the myriad of components of cigarette smoke known to be metabolized in the nasal epithelium including benzo(a)pyrene, N-nitrosonornicotine, and cyanide. Rhodanese, the primary enzyme of cyanide metabolism, shows nearly a 50% reduction in activity in respiratory mucosa from human smokers compared to nonsmokers (Lewis et al., 1991).

Inhibitors of biotransformation enzymes have been used in vivo to demonstrate the role of local metabolism in xenobiotic toxicity. For example, treatment with metyrapone reduced or abolished cytotoxicity caused by a number of toxic chemicals, such as methimazole (Bergman and Brittebo, 1999), 2,6-dichlorothiobenzamide (Eriksson and Brittebo, 1995), DCBN (Walters et al., 1993), and IDPN (Genter et al., 1994). Metyrapone has also been used to demonstrate that inspired styrene is metabolized in nasal tissues in the rat and mouse (Morris, 2000). Other P450 inhibitors used for in vivo studies include diethyldithiocarbamate (Eriksson and Brittebo, 1995; Deamer and Genter, 1995), carbon tetrachloride (Genter et al., 1994), disulfiram (Deamer and Genter, 1995), 3-aminobenzamide (Eriksson et al., 1996), and cobalt protoporphyrin IX (Chamberlain et al., 1998); the latter depletes P450 by interfering with heme synthesis. Some inhibitors cause inactivation of a subset of P450s, such as xylene and xylene metabolites (Blanchard and Morris, 1994; Vaidyanathan et al., 2003; Foy and Schatz, 2004) and chlormethiazole (Longo et al., 2000). Inhibition of nasal GST-dependent conjugation activity has been achieved by depleting GSH with phorone (Larsson and Tjälve, 1995) or phorone plus L-buthionine sulphoximine (Chamberlain et al., 1998). But, GST expression and activity in mouse olfactory mucosa are unchanged at 24 hr after treatment with polyinosinic:polycytidylic acid to induce acute phase response, while the expression of CYP2E1 was decreased (Weech et al., 2003). In addition, the role of aldehyde dehydrogenase on nasal uptake of inspired acetaldehyde has been examined using cyanamide as an inhibitor. While these inhibitors have been useful for the initial identification of biotransformation enzymes or pathways involved in the metabolism and toxicity of a compound, they are generally not specific for any single enzyme or even a single family of enzymes (e.g., Eriksson et al., 1996; Von Weymarn et al., 2005; D'Agostino et al., 2009). Furthermore, these inhibitors are most likely also toxicants. Therefore, caution should be exercised in interpreting the results obtained using chemical inhibitors as tools. Alternatively, mouse models with targeted gene deletion of specific P450 and other biotransformation enzymes are more preferable.

3.6.3 Effects of Mucosal Damage on Nasal Metabolism

Expression of olfactory mucosa P450s and CPR is suppressed when ORNs undergo degeneration either as a consequence of chemical toxicity, unilateral naris closure, or olfactory bulbectomy (Gu et al., 1997; Schwob et al., 1995; Walters et al., 1992, 1993; Sammeta and McClintock,

2010). In the cases of tissue damage caused by exposure to DCBN and methimazole in mice, CYP2A5 expression disappeared in damaged Bowman's glands, but persisted in damaged OE (Piras et al., 2003). In rats treated with methyl bromide, another olfactory toxicant, P450 expression returns to normal following successful regeneration of ORNs, but not when degenerated olfactory mucosa was replaced by respiratory type of epithelium (Schwob et al., 1995). In alachlor-induced tumors in rat olfactory mucosa, neither olfactory marker protein nor CYP2A3 was detected (Genter et al., 2000). The suppressed expression of P450 following olfactory bulbectomy is particularly intriguing since the P450-expressing cells were apparently intact after the operation (Walters et al., 1992). The latter result contrasts with report that the expression of PSTg protein is not affected in the olfactory mucosa following olfactory bulbectomy (Miyawaki et al., 1996). Decreases in GSH and GST levels have also been found in the peripheral olfactory organ of rainbow trout during retrograde olfactory nerve degeneration, which are followed by widespread recovery as the ORNs begin to repopulate the olfactory mucosa (Starcevic and Zielinski, 1997).

Tissue damage resulting from chemically-induced nasal toxicity may underlie some of the in vivo inhibitory effects of enzyme inhibitors described in the previous section, as well as the apparent resistance of nasal biotransformation enzymes to xenobiotic induction (Su et al., 1996). Furthermore, although it may appear logical to assume parallel alterations in histopathology and metabolic activity in nasal tissue, interpretation of data from these correlative studies can be complicated by the multifaceted nature of the toxic response. For example, tissue damage may lead to increased influx of immune cells, which may contribute to local metabolic activity.

Another complexity in interpretation of data indicating altered metabolism following exposure to specific toxicants lies in the parallel pathological alterations to the nasal epithelium. Biochemical data are often normalized per mg protein, per mg tissue, per mg mitochondrial or microsomal protein, and so forth. If cellularity has decreased in the tissue (as is often the case in the OE), the normalized data may show no alteration in metabolism, but the total capacity of the tissue to metabolize inhalants may be severely reduced. Conversely, hyperplastic responses may greatly increase the metabolic capacity without altering the normalized biochemical data. Although this problem exists in other tissues as well, the structure of the nasal epithelium and the close association with cartilage and bone make it difficult to control the problem by normalizing to total tissue weight, as can readily be done in most other organs. In addition, toxicant exposure and age are both known to produce metaplastic alteration in the OE. Because these alterations in cell type can also affect metabolic capacity, close evaluation of both biochemical and histopathological alterations in the interpretation of data from nasal epithelium is necessary for valid extrapolations. Finally, because inhalants can contact and be metabolized in the respiratory or transitional mucosa before reaching the olfactory mucosa, metabolic processes occurring in these nasal mucosa will also affect olfactory processes.

3.7 CONCLUSIONS

The olfactory system has several unique properties, which make it interesting as well as important to study. These include its continuous exposures to inhaled environmental toxicants, its vulnerability to cell loss resulting from toxicant insult, and its capacity to regenerate neuronal cells following this loss. In addition, its histological structure, with neuronal cells contacting the external environment at the nasal lumen and projecting directly to the olfactory bulb, makes it a viable portal of entry for inhaled environmental toxicants, as well as a potential route of entry for therapeutic drugs, into the CNS. Although olfaction has traditionally been thought of as a sensory system of minor importance in humans, evidence is accumulating that olfaction plays an important role in learning and memory, hedonic responses, and reproductive function in humans as well as in other species.

Current knowledge on nasal biotransformation enzymes indicate that they are very likely to play important roles in many cellular processes in the olfactory mucosa, as supported by their high metabolic capacity, their diverse substrates of both endogenous and exogenous origins, and their tissue and cell-type specific expression. The precise nature of these roles, which will likely be enzyme specific, is not yet fully understood; however, as reviewed in this chapter, with the increasing availability of knockout mouse and other genetically modified animal models, there are already exciting results.

Nasal biotransformation enzymes can alter inhaled toxicants either by converting them to less toxic metabolites or by activating them to reactive chemicals that result in local damage and in some cases damage to other tissues as well. As such, the role of nasal biotransformation enzymes has historically received most attention in the field of toxicology. The biotransformation capacity in the olfactory mucosa is vulnerable to modification by a variety of toxicant exposures, histological changes, and disease or inflammatory processes. Further, the nasal xenobiotic-metabolism activities will be influenced by genetic polymorphisms of the participating biotransformation enzymes. Thus, the capacity to detoxify or activate inhaled toxicants is likely to be a fluid system best understood with respect to an individual case history and genetic makeup. Future studies on the interactions of various components of the olfactory system, the activities of individual biotransformation enzymes and their genetic polymorphisms, the link between genetic polymorphisms in biotransformation enzymes and various diseases that involve the nasal tissues, and the impact of nasal xenobiotic-metabolism on other systems such as the CNS will make it possible to identify situations, periods, or individuals of increased vulnerability to inhaled xenobiotics, and to reduce the risk of toxicity through targeted prevention.

ACKNOWLEDGMENTS

The authors acknowledge Drs. Bert Menco and Virginia Carr for the transmission electron micrograph in Figure 3.1. Research in the authors' laboratory was supported in part by NIH Grants ES07462 and ES020867 from the National Institute of Environmental Health Sciences and CA092596 from the National Cancer Institute.

REFERENCES

Ablimit, A., Matsuzaki, T., Tajika, Y., et al. (2006). Immunolocalization of water channel aquaporins in the nasal olfactory mucosa. *Arch. Histol. Cytol.* 69: 1–12.

Aceto, A., Sacchetta, P., Dragani, B. et al. (1993). Glutathione transferase isoenzymes in olfactory and respiratory epithelium of cattle. *Biochem. Pharmacol.* 46: 2127–2133.

Adams, D. R., Deyoung, D. W., and Griffith, R. (1981). The lateral nasal gland of dog: its structure and secretory content. *J. Anat.* 132: 29–37.

Adams, D. R., Jones, A. M., Plopper, C. G., et al. (1991). Distribution of cytochrome P-450 monoxygenase enzymes in the nasal mucosa of hamster and rat. *Am. J. Anat.* 190: 291–298.

Agu, R. U., Obimah, D. U., Lyzenga, W. J., et al. (2009). Specific aminopeptidases of excised human nasal epithelium and primary culture: a comparison of functional characteristics and gene transcripts expression. *J. Pharm. Pharmacol.* 61: 599–606.

Agu, R., Cowley, E., Shao, D., et al. (2011). Proton-coupled Oligopeptide Transporter (POT) family expression in human nasal epithelium and their drug transport potential. *Mol. Pharmaceut.* 8: 664–672.

Alekseyenko, O. V., Baum, M. J., and Cherry, J. A. (2006). Sex and gonadal steroid modulation of pheromone receptor gene expression in the mouse vomeronasal organ. *Neuroscience* 140: 1349–1357.

Anand-Kumar, T. C., Sehgal, A., David, G. F. X., et al. (1980). Effect of intranasal administration of hormonal steroids on serum testosterone and spermatogenesis in rhesus monkey (Macaca mulatta). *Biol. Reprod.* 22: 935–940.

Armstrong, R. N. (1997). Structure, catalytic mechanism, and evolution of the glutathione transferases. *Chem. Res. Toxicol.* 10: 2–18.

Badonnel, K., Durieux, D., Monnerie, R., et al. (2009). Leptin-sensitive OBP-expressing mucous cells in rat olfactory epithelium: a novel target for olfaction-nutrition crosstalk? *Cell Tissue Res.* 338: 53–66.

Balboni, G. C. (1967). L'ultrastuttura dell'epitelio olfattiva nel ratto e sue modificazioni in seguito a castrazione e alla somministrazione, a ratti castrati, di testosterone. *Arch. Ital. Anat. Embriol.* 52: 203–223.

Balboni, G. C., and Vannelli, G. B. (1982). Morphological features of the olfactory epithelium in prepubertal and postpubertal rats, in *Olfaction and Endocrine Regulation* (Breiphol, ed.). IRL Press, London, pp. 285–295.

Baldwin, R. M., Shultz, M. A., and Buckpitt, A. R. (2005). Bioactivation of the pulmonary toxicants naphthalene and 1-nitronaphthalene by rat CYP2F4. *J. Pharmacol. Exp. Ther.* 312: 857–865.

Banger, K. K., Lock, E. A., and Reed, C. J. (1993). The characterization of glutathione S-transferases from rat olfactory epithelium. *Biochem. J.* 290: 199–204.

Banger, K. K., Lock, E. A., and Reed, C. J. (1996). Regulation of rat olfactory glutathione S-transferase expression. Investigation of sex differences, induction, and ontogenesis. *Biochem. Pharmacol.* 52: 801–808.

Bao, Z. P., He, X. Y., Ding, X.X., et al. (2005). Metabolism of nicotine and cotinine by human cytochrome P450 2A13. *Drug Metab.* 33: 258–261.

Barbour, J., Neuhaus, E.M., Piechura, H., et al. (2008). New insight into stimulus-induced plasticity of the olfactory epithelium in Mus musculus by quantitative proteomics. *J. Proteome. Res.* 7: 1594–1605.

Baron, J., Burke, J. P., Guengerich, F. P., et al. (1988). Sites for xenobiotic activation and detoxication within the respiratory tract: implications for chemically induced toxicity. *Toxicol. Appl. Pharmacol.* 93: 493–505.

Baron, J., Voigt, J. M., Whitter, T. B., et al. (1986). Identification of intratissue sites for xenobiotic activation and detoxication. *Adv. Expt. Med. Biol.* 197: 119–144.

Beauchamp, G. K., Doty, R. L., Moulton, D. G., and Mugford, R. A. (1976). The pheromone concept in mammalian chemical communication: a critique. In *Mammalian Olfaction, Reproductive Processes, and Behavior*, R. L. Doty (Ed.). Academic Press, New York, pp. 147–153.

Belinsky, S. A., Foley, J. F., White, C. M., et al. (1990). Dose-response relationship between O^6-methylguanine formation in Clara cells and induction of pulmonary neoplasia in the rat by 4-(methylnitrosamino)-1-(3-pyridyl)-1-butanone. *Cancer Res.* 50: 3772–3780.

Ben-Arie, N., Khen, M., and Lancet, D. (1993). Glutathione S-transferases in rat olfactory epithelium: purification, molecular properties and odorant biotransformation. *Biochem. J.* 292: 379–384.

Beréziat, J. C., Raffalli, F., Schmezer, P., et al. (1995). Cytochrome P450 2A of nasal epithelium: regulation and role in carcinogen metabolism. *Mol. Carcinog.* 14: 130–139.

Berger, G., Kogan, T., Skutelsky, E., and Ophir, D. (2005). Glycoconjugate expression in normal human inferior turbinate mucosa: A lectin histochemical study. *Am. J. Rhinol.* 19: 97–103.

Bergman, U. and Brittebo, E. B. (1999). Methimazole toxicity in rodents: covalent binding in the olfactory mucosa and detection of glial fibrillary acidic protein in the olfactory bulb. *Toxicol. App. Pharmacol.* 155: 190–200.

Bergstrom, U., Giovanetti, A., Piras, E., and Brittebo, E. B. (2003). Methimazole-induced damage in the olfactory mucosa: Effects on ultrastructure and glutathione levels. *Toxicol. Pathol.* 31: 379–387.

Blanchard, K. T. and Morris, J. B. (1994). Effects of m-xylene on rat nasal cytochrome P450 mixed function oxidase activities. *Toxicol. Lett.* 70: 253–259.

Bogdanffy, M. S. (1990). Biotransformation enzymes in the rodent nasal mucosa: the value of a histochemical approach. *Environ. Health Perspect.* 85: 177–186.

Bogdanffy, M. S., Randall, H. W., and Morgan, K. T. (1987). Biochemical quantitation and histochemical localization of carboxylesterase in the nasal passages of the Fischer-344 rat and B6C3F1 mouse. *Toxicol. Appl. Pharmacol.* 88: 183–194.

Bond, J. A. (1983). Some biotransformation enzymes responsible for polycyclic aromatic hydrocarbon metabolism in rat nasal turbinates: effects on enzyme activities of in vitro modifiers and intraperitoneal and inhalation exposures of rats to inducing agents. *Cancer Res.* 43: 4804–4811.

Bond, J. A., Harkema, J. R., and Russell, V. I. (1988). Regional distribution of xenobiotic metabolizing enzymes in respiratory airways of dogs. *Drug Metab. Dispos.* 16: 116–124.

Bourne, G. H. (1948). Alkaline phosphatase in taste buds and nasal mucosa. *Nature* 161: 445–446.

Brittebo, E. B. (1982). Demethylation of aminopyrine by the nasal mucosa in mice and rats. *Acta Pharmacol. Toxicol.* 51: 227–232.

Brittebo, E. B. (1985). Localization of oestradiol in the rat nasal mucosa. *Acta Pharmacol. Toxicol.* 57: 285–290.

Brittebo, E. B. (1997). Metabolism-dependent activation and toxicity of chemicals in nasal glands. *Mut. Res.* 380: 61–75.

Brittebo, E. B. and Rafter, J. J. (1984). Steroid metabolism by rat nasal mucosa: studies on progesterone and testosterone. *J. Ster. Biochem.* 20: 1147–1151.

Brown, D., Garcia-Segura, L. M., and Orci, L. (1984). Carbonic anhydrase is present in olfactory receptor cells. *Histochemistry* 80: 307–309.

Capdevila, J. H., Falck, J. R., and Estabrook, R. W. (1992). Cytochrome P450 and the arachidonate cascade. *FASEB J.* 6: 731–736.

Carlson, G. P. (2008). Critical appraisal of the expression of cytochrome P450 enzymes in human lung and evaluation of the possibility that such expression provides evidence of potential styrene tumorigenicity in humans. *Toxicol.* 254: 1–10.

Carr, B. A., Wan, J., Hines, R. N., and Yost, G. S. (2003). Characterization of the human lung CYP2F1 gene and identification of a novel lung-specific binding motif. *J. Biol. Chem.* 278: 15473–15483.

Chaen, T., Watanabe, N., Mogi, G., et al. (1993). Substance P and vasoactive intestinal peptide in nasal secretions and plasma from patients with nasal allergy. *Ann. Otol. Rhinol. Laryngol.* 102: 16–21.

Chamberlain, M. P., Lock, E. A., Gaskell, B. A., and Reed, C. J. (1998). The role of glutathione S-transferase- and cytochrome P450-dependent metabolism in the olfactory toxicity of methyl iodide in the rat. *Arch. Toxicol.* 72: 420–428.

Chemuturi, N. V. and Donovan, M. D. (2006). Metabolism of dopamine by the nasal mucosa. *J. Pharm. Sci.* 95: 2507–2515.

Chemuturi, N. V., Haraldsson, J. E., Prisinzano, T., and Donovan, M. (2006). Role of dopamine transporter (DAT) in dopamine transport across the nasal mucosa. *Life Sci.* 79: 1391–1398.

Chemuturi, N. V., Hayden, P., Klausner, M., and Donovan, M. D. (2005). Comparison of human tracheal/bronchial epithelial cell culture and bovine nasal respiratory explants for nasal drug transport studies. *J. Pharm. Sci.* 94: 1976–1985.

Chen, P. H. and Fang, S. Y. (2004). Expression of human beta-defensin 2 in human nasal mucosa. *Eur. Arch. Oto-Rhino-Laryngol* 261: 238–241.

Chen, Y., Getchell, M. L., Ding, X., and Getchell, T. V. (1992). Immunolocalization of two cytochrome P450 isozymes in rat nasal chemosensory tissue. *NeuroReport* 3: 749–752.

Chen, Y., Liu, Y., Su, T., et al. (2003). Immunoblot analysis and immunohistochemical characterization of CYP2A expression in human olfactory mucosa. *Biochem. Pharmacol.* 66: 1245–1251.

Chiba, Y., Shimada, A., Satoh, M., et al. (2006). Sensory system-predominant distribution of leukotriene A(4) hydrolase and its colocalization with calretinin in the mouse nervous system. *Neuroscience* 141: 917–927.

Corps, K. N., Islam, Z., Pestka, J. J., and Harkema, J. R. (2010). Neurotoxic, inflammatory, and mucosecretory responses in the nasal airways of mice repeatedly exposed to the macrocyclic trichothecene mycotoxin roridin a: dose-response and persistence of injury. *Toxicol. Pathol.* 38: 429–451.

Court, M. H., Zhang, X. L., Ding, X. X., et al. (2012). Quantitative distribution of mRNAs encoding the 19 human UDP-glucuronosyltransferase enzymes in 26 adult and 3 fetal tissues. *Xenobiotica* 42: 266–277.

Cross, C. E., van, d., V., O'Neill, C. A., Louie, S., and Halliwell, B. (1994). Oxidants, antioxidants, and respiratory tract lining fluids. *Environ. Health Persp.* 102 Suppl 10: 185–191.

Cruzan, G., Carlson, G. P., Johnson, K. A., et al. (2002). Styrene respiratory tract toxicity and mouse lung tumors are mediated by CYP2F-generated metabolites. *Reg. Toxicol. Pharmacol.* 35: 308–319.

D'Agostino, J., Zhang, X., Wu, H., et al. (2008). Characterization of CYP2A13*2, a variant cytochrome P450 allele previously found to be associated with decreased incidences of lung adenocarcinoma in smokers. *Drug Metab. Dispos.* 36: 2316–2323.

D'Agostino, J., Zhuo, X., Shadid, M., et al. (2009). The pneumotoxin 3-methylindole is a substrate and a mechanism-based inactivator of CYP2A13, a human cytochrome P450 enzyme preferentially expressed in the respiratory tract. *Drug Metab. Dispos.* 37: 2018–2027.

Dahl, A. R. (1982). The inhibition of rat nasal cytochrome P-450-dependent monooxygenase by the essence heliotropin (piperonal). *Drug Metab. Dispos.* 10: 553–554.

Dahl, A. R. (1988). The effect of cytochrome P-450-dependent metabolism and other enzyme activities on olfaction. In *Molecular Neurobiology of the Olfactory System: Molecular, Membranous, and Cytological Studies*, F. L. Margolis and T. V. Getchell (Eds.). Plenum Press, New York, pp. 51–70.

Dahl, A. R., and Brezinski, D. A. (1985). The inhibition of rabbit nasal and hepatic cytochrome P-450-dependent hexamethylphosphoramide (HMPA) N-demethylase by methylenedioxphenyl compounds. *Biochem. Pharmacol.* 34: 631–636.

Dahl, A. R., and Hadley, W. M. (1983). Formaldehyde production promoted by rat nasal cytochrome P450-dependent monooxygenases with nasal decongestants, essences, solvents, air pollutants, nicotine, and cocaine as substrates. *Toxicol. Appl. Pharmacol.* 67: 200–205.

Dahl, A. R., and Hadley, W. M. (1991). Nasal cavity enzymes involved in xenobiotic metabolism: effects on the toxicity of inhalants. *Crit. Rev. Toxicol.* 21: 345–372.

Dahl, A. R., and Lewis, J. L. (1993). Respiratory tract uptake of inhalants and metabolism of xenobiotics. *Annu. Rev. Pharmacol. Toxicol.* 32: 383–407.

Dahl, A. R., and Waruszewski, B. A. (1990). Metabolism of organonitriles to cyanide by rat nasal tissue enzymes. *Xenobiotica* 19: 1201–1205.

Dahl, A. R., Coslett, D. S., Bond, J. A., and Hesseltine, G. R. (1985). Metabolism of benzo(a)pyrene on the nasal mucosa of Syrian hamsters: comparison to metabolism by other extrahepatic tissues and possible role of nasally produced metabolites in carcinogenesis. *J. Natl. Cancer Inst.* 75: 135–139.

Dahl, A. R., Hadley, W. M., Hahn, F. F., et al. (1982). Cytochrome P-450-dependent monooxygenases in olfactory epithelium in dogs: possible role in tumorigenicity. *Science* 216: 57–59.

Deamer, N. J. and Genter, M. B. (1995). Olfactory toxicity of diethyldithiocarbamate (DDTC) and disulfiram and the protective effect of DDTC against the olfactory toxicity of dichlobenil. *Chem. Biol. Interact.* 95: 215–226.

Debat, H., Eloit, C., Blon, F., et al. (2007). Identification of human olfactory cleft mucus proteins using proteomic analysis. *J. Proteome. Res.* 6: 1985–1996.

Deshpande, V. S., Genter, M. B., Jung, C., and Desai, P. B. (1999). Characterization of lidocaine metabolism by rat nasal microsomes: implications for nasal drug delivery. *Eur. J. Drug Metab. Pharmacokinetics* 24: 177–182.

DeStefano-Shields, C., Morin, D., and Buckpitt, A. (2010). Formation of Covalently Bound Protein Adducts from the Cytotoxicant Naphthalene in Nasal Epithelium: Species Comparisons. *Environ. Health Perspec.* 118: 647–652.

Diaz, D., Krejsa, C. M., and Kavanagh, T. J. (2002). Expression of glutamate-cysteine ligase during mouse development. *Mol. Reprod. Dev.* 62: 83–91.

Dimova, S., Brewster, M. E., Noppe, M., et al. (2005). The use of human nasal in vitro cell systems during drug discovery and development. *Toxicol. in Vitro.* 19: 107–122.

Ding, X. and Kaminsky, L. S. (2003). Human extrahepatic cytochromes P450: function in xenobiotic metabolism and tissue-selective chemical toxicity in the respiratory and gastrointestinal tracts. *Annu. Rev. Pharmacol. Toxicol.* 43: 149–173.

Ding, X., and Coon, M. J. (1988). Purification and characterization of two unique forms of cytochrome P-450 from rabbit nasal microsomes. *Biochemistry.* 27: 8330–8337.

Ding, X., and Coon, M. J. (1990a). Immunochemical characterization of multiple forms of cytochrome P-450 in rabbit nasal microsomes and evidence for tissue-specific expression of P450s NMa and NMb. *Mol. Pharmacol.* 37: 489–496.

Ding, X., and Coon, M. J. (1990b). Induction of cytochrome P-450 isozyme 3a (P-450IIE1) in rabbit olfactory mucosa by ethanol and acetone. *Drug Metab. Dispos.* 18: 742–745.

Ding, X., and Coon, M. J. (1993). Olfactory cytochrome P450. In: Schenkman, J. B., Greim, H. (eds) *Cytochrome P450, Handbook of Experimental Pharmacology*, Volume 105, Springer-Verlag, New York, pp. 351–361.

Ding, X., and Coon, M. J. (1994). Steroid metabolism by rabbit olfactory-specific P450 2G1. *Arch. Biochem. Biophys.* 315: 454–459.

Ding, X., Koop, D. R., Crump, B. L., and Coon, M. J. (1986). Immunochemical identification of cytochrome P-450 isozyme 3a (P-450alc) in rabbit nasal and kidney microsomes and evidence for differential induction by alcohol. *Mol. Pharmacol.* 30: 370–378.

Ding, X., Peng, H. M., and Coon, M. J. (1992). P450 cytochromes NMa, NMb (2G1), and LM4 (1A2) are differentially expressed during development in rabbit olfactory mucosa and liver. *Mol. Pharmacol.* 42: 1027–1032.

Ding, X., Porter, T. D., Peng, H. M., and Coon, M. J. (1991). cDNA and derived amino acid sequence of rabbit nasal cytochrome P450NMb (P450IIG1), a unique isozyme possibly involved in olfaction. *Arch. Biochem. Biophysics* 285: 120–125.

Ding, X., Spink, D. C., Bhama, J. K., et al. (1996). Metabolic activation of 2,6-dichlorobenzonitrile, an olfactory-specific toxicant, by rat, rabbit, and human cytochromes P450. *Mol. Pharmacol.* 49: 1113–1121.

Dorries, K. M., Schmidt, H. J., Beauchamp, G. K., and Wysocki, C. J. (1989). Changes in sensitivity to the odor of androstenone during adolescence. *Dev. Psychobiol.* 22: 423–435.

Duester, G. (1996). Involvement of alcohol dehydrogenase, short-chain dehydrogenase/reductase, aldehyde dehydrogenase, and cytochrome P450 in the control of retinoid signaling by activation of retinoic acid synthesis. *Biochemistry.* 35: 12221–12227.

Eaton, D. L. and Bammler, T. K. (1999). Concise review of the glutathione S-transferases and their significance to toxicology. *Toxicol. Sciences* 49: 156–164.

Emes, R. D., Riley, M. C., Laukaitis, C. M., et al. (2004). Comparative evolutionary genomics of androgen-binding protein genes. *Genome Res.* 14: 1516–1529.

Eriksson, C. and Brittebo, E. B. (1995). Effects of the herbicide chlorthiamid on the olfactory mucosa. *Toxicol. Letts.* 76: 203–208.

Eriksson, C., Busk, L., and Brittebo, E. B. (1996). 3-Aminobenzamide: effects on cytochrome P450-dependent metabolism of chemicals and on the toxicity of dichlobenil in the olfactory mucosa. *Toxicol. Appl. Pharm.* 136: 324–331.

Faller, T. H., Csanady, G. A., Kreuzer, P. E., et al. (2001). Kinetics of propylene oxide metabolism in microsomes and cytosol of different organs from mouse, rat, and humans. *Toxicol. Applied Pharmacol.* 172: 62–74.

Federico, G., Maremmani, C., Cinquanta, L., et al. (1999). Mucus of the human olfactory epithelium contains the insulin-like growth factor-I system which is altered in some neurodegenerative diseases. *Brain Res.* 835: 306–314.

Fong, K. J., Kern, R. C., Foster, J. D., et al. (1999). Olfactory secretion and sodium, potassium-adenosine triphosphatase: regulation by corticosteroids. *Laryngoscope* 109: 383–388.

Foy, J. W. D. and Schatz, R. A. (2004). Inhibition of rat respiratory-tract cytochrome P-450 activity after acute low-level m-xylene inhalation: Role in 1-nitronaphthalene toxicity. *Inhalat. Toxicol.* 16: 125–132.

Franzen, A., Carlsson, C., Hermansson, V., et al. (2006). CYP2A5-mediated activation and early ultrastructural changes in the olfactory mucosa: Studies on 2,6-dichlorophenyl methylsulfone. *Drug Metab.* 34: 61–68.

Frye, R. E., Doty, R. L., and Schwartz, B. (1989). Influence of cigarette smoking on olfaction: evidence for a dose-response relationship. *JAMA* 263: 1233–1236.

Fujii, H., Sato, T., Kaneko, S., et al. (1997). Metabolic inactivation of retinoic acid by a novel P450 differentially expressed in developing mouse embryos. *EMBO J.* 16: 4163–4173.

Fujikura, T. and Okubo, K. (2011). Adrenomedullin level in the nasal discharge from allergic rhinitis cohort. *Peptides* 32: 368–373.

Fukami, T., Katoh, M., Yamazaki, H., et al. (2008). Human cytochrome P450 2A13 efficiently metabolizes chemicals in air pollutants: naphthalene, styrene, and toluene. *Chem Res Toxicol* 21: 720–725.

Fukuda, N., Shirasu, M., Sato, K., et al. (2008). Decreased olfactory mucus secretion and nasal abnormality in mice lacking type 2 and type 3 IP3 receptors. *Eur. J. Neurosci.* 27: 2665–2675.

Gaskell, B. A. (1990). Nonneoplastic changes in the olfactory epithelium – experimental studies. *Environ. Health Perspect.* 85: 275–289.

Gennings, J. N., Gower, D. B., and Bannister, L. H. (1974). Studies on the metabolism of the odoriferous ketones, 5-androst-16-en-3-one and 4,16-androstadien-3-one by the nasal epithelium of the mature and immature sow. *Biochem. Biophys. Acta.* 369: 294–303.

Genter, M. B. (2004). Update on olfactory mucosal metabolic enzymes: Age-related changes and N-acetyltransferase activities. *J. Biochem. Mol. Toxicol.* 18: 239–244.

Genter, M. B., Apparaju, S., and Desai, P. B. (2002). Induction of olfactory mucosal and liver metabolism of lidocaine by 2,3,7,8-tetrachlorodibenzo-p-dioxin. *J. Biochem. Mol. Toxicol.* 16: 128–134.

Genter, M. B., Burman, D. M., Dingeldein, M. W., et al. (2000). Evolution of alachlor-induced nasal neoplasms in the Long-Evans rat. *Toxicol. Pathol.* 28: 770–781.

Genter, M. B., Owens, D. M., and Deamer, N. J. (1995a). Distribution of microsomal epoxide hydrolase and glutathione S-transferase in the rat olfactory mucosa: relevance to distribution of lesions caused by systemically-administered olfactory toxicants. *Chem. Senses* 20: 385–392.

Genter, M. B., Deamer, N. J., Blake, B. L., et al. (1995b). Olfactory toxicity of methimazole: dose-response and structure-activity studies and characterization of flavin-containing monooxygenase activity in the Long-Evans rat olfactory mucosa. *Toxicol. Path.* 23: 477–486.

Genter, M. B., Deamer, N. J., Cao, Y., and Levi, P. E. (1994). Effects of P450 inhibition and induction on the olfactory toxicity of beta,beta′-iminodipropionitrile (IDPN) in the rat. *J. Biochem. Toxicol.* 9: 31–39.

Genter, M. B., Goss, K. H., and Groden, J. (2004). Strain-specific effects of alachlor on murine olfactory mucosal responses. *Toxicol. Path.* 32: 719–725.

Genter, M. B., Liang, H. C., Gu, J., et al. (1998). Role of CYP2A5 and 2G1 in acetaminophen metabolism and toxicity in the olfactory mucosa of the Cyp1a2(−/−) mouse. *Biochem. Pharmacol.* 55: 1819–1826.

Genter, M. B., Llorens, J., O'Callaghan, J. P., et al. (1992). Olfactory toxicity of β,β′-iminodipropionitrile in the rat. *J. Pharmacol. Exp. Ther.* 263: 1432–1439.

Genter, M. B., Warner, B. M., Krell, H. W., and Bolon, B. (2005). Reduction of alachlor-induced olfactory mucosal neoplasms by the matrix metalloproteinase inhibitor Ro 28–2653. *Toxicol. Pathol.* 33: 593–599.

Genter, M. B., Yost, G. S., and Rettie, A. E. (2006). Localization of CYP4B1 in the rat nasal cavity and analysis of CYPs as secreted proteins. *J. Biochem. Mol. Toxicol.* 20: 139–141.

Gervasi, P. G., Longo, V., Ursino, F., and Panattoni, G. (1989). Drug metabolizing enzymes in respiratory mucosa of humans. Comparison with rats. In *Cytochrome P-450: Biochemistry and Biophysics.* I. Schuster (Ed.). Taylor & Francis, New York, pp. 97–100.

Getchell, M. L., Chen, Y., Ding, X. et al. (1993a). Immunohistochemical localization of a cytochrome P-450 isozyme in human nasal mucosa: age-related trends. *Ann. Otol. Rhinol Laryngol* 102: 368–374.

Getchell, T. V., Su, Z., and Getchell, M. L. (1993b). Mucous domains: microchemical heterogeneity in the mucociliary complex of the olfactory epithelium. *Ciba Foundation Symposium* 179: 27–40.

Getchell, M. L., Zielinski, B., and Getchell, T. V. (1988). Odorant and autonomic regulation of secretion in the olfactory mucosa. In *Molecular Neurobiology of the Olfactory System: Molecular, Membranous, and Cytological Studies.* F. L. Margolis and T. V. Getchell (Eds.). Plenum Press, New York, pp. 71–98.

Getchell, T. V. and Getchell, M. L. (1977) *Early events in vertebrate olfaction. Chem. Senses* 2: 313–326.

Ghantous, H., Dencker, L., Gabrielsson, J., et al. (1990). Accumulation and turnover of metabolites of toluene and xylene in nasal mucosa and olfactory bulb in the mouse. *Pharmacol. Toxicol.* 66: 87–92.

Goldstein, B. J., Fang, H. S., Youngentob, S. L., and Schwob, J. E. (1998). Transplantation of multipotent progenitors from the adult olfactory epithelium. *Neuroreport* 9: 1611–1617.

Gosepath, J., Brieger, J., Gletsou, E., and Mann, W. J. (2004). Expression and localization of cyclooxygenases (Cox-1 and Cox-2) in nasal respiratory mucosa. Does Cox-2 play a key role in the immunology of nasal polyps? *J. Investig. Allergol. Clin Immunol.* 14(2):114–118.

Graziadei, P. P., and Monti-Graziadei, A. G. (1983). Regeneration in the olfactory system of vertebrates. *Am. J. Otolaryngol.* 4: 228–233.

Green, N., Weech, M., and Walters, E. (2005). Localization and characterization of glutathione-s-transferase isozymes alpha, mu, and pi within the mouse vomeronasal organ. *Neurosci. Lett.* 375: 198–202.

Green, T., Lee, R., Moore, R. B., et al. (2000). Acetochlor-induced rat nasal tumors: Further studies on the mode of action and relevance to humans. *Regulatory Toxicol. Pharmacol.* 32: 127–133.

Green, T., Lee, R., Toghill, A., et al. (2001). The toxicity of styrene to the nasal epithelium of mice and rats: studies on the mode of action and relevance to humans. *Chemico-Biolog Interactions* 137: 185–202.

Gresham, L. S., Molgaard, C. A., and Smith, R. A. (1993). Induction of cytochrome P-450 enzymes via tobacco smoke: a potential mechanism for developing resistance to environmental toxins as related to Parkinsonism and other neurologic diseases. *Neuroepidemiology* 12: 114–116.

Grolli, S., Merli, E., Conti, V., et al. (2006). Odorant binding protein has the biochemical properties of a scavenger for 4-hydroxy-2-nonenal in mammalian nasal mucosa. *FEBS Journal* 273: 5131–5142.

Gu, J., Cui, H. D., Behr, M., et al. (2005). In vivo mechanisms of tissue-selective drug toxicity: Effects of liver-specific knockout of the NADPH-cytochrome P450 reductase gene on acetaminophen toxicity in kidney, lung, and nasal mucosa. *Mol. Pharmacol.* 67: 623–630.

Gu, J., Dudley, C., Su, T., et al. (1999). Cytochrome P450 and steroid hydroxylase activity in mouse olfactory and vomeronasal mucosa. *Biochem.Biophys.Res.Commun.* 266: 262–267.

Gu, J., Su, T., Chen, Y., et al. (2000). Expression of biotransformation enzymes in human fetal olfactory mucosa: potential roles in developmental toxicity. *Toxicol. Appl. Pharmacol.* 165: 158–162.

Gu, J., Walker, V. E., Lipinskas, T. W., et al. (1997). Intraperitoneal administration of coumarin causes tissue-selective depletion of cytochromes P450 and cytotoxicity in the olfactory mucosa. *Toxicol. Applied Pharmacol.* 146: 134–143.

Gu, J., Zhang, Q. Y., Genter, M. B., et al. (1998). Purification and characterization of heterologously expressed mouse CYP2A5 and CYP2G1: role in metabolic activation of acetaminophen and 2,6-dichlorobenzonitrile in mouse olfactory mucosal microsomes. *J. Pharmacol. Exp. Ther.* 285: 1287–1295.

Guengerich, F. P., Dannan, G. A., Wright, S. T., et al. (1982). Purification and characterization of liver microsomal cytochromes P-450: electrophoretic, spectral, catalytic, and immunochemical properties and inducibility of eight isozymes isolated from rats treated with phenobarbital or -naphthoflavone. *Biochem.* 21: 6019–6030.

Gussing, F. and Bohm, S. (2004). NQO1 activity in the main and the accessory olfactory systems correlates with the zonal topography of projection maps. *Eur. J. Neurosci.* 19: 2511–2518.

Guzzo, A. C., Jheon, J., Imtiaz, F., and Decatanzaro, D. (2012). Oestradiol transmission from males to females in the context of the Bruce and Vandenbergh effects in mice (Mus musculus). *Reproduction* 143: 539–548.

Hadley, W. M., and Dahl, A. R. (1982). Cytochrome P-450 dependent monooxygenase activity in rat nasal epithelial membranes. *Toxicol. Lett.* 10: 417–422.

Hadley, W. M., and Dahl, A. R. (1983). Cytochrome P-450-dependent monooxygenase activity in nasal membranes of six species. *Drug Metab. Dispos.* 11: 275–276.

Hagglund, M., Berghard, A., Strotmann, J., and Bohm, S. (2006). Retinoic acid receptor-dependent survival of olfactory sensory neurons in postnatal and adult mice. *J. Neurosci.* 26: 3281–3291.

Hahn, C. G., Han, L. Y., Rawson, N. E., et al. (2005). In vivo and in vitro neurogenesis in human olfactory epithelium. *J. Comp. Neurol.* 483: 154–163.

Hahn, I., Scherer, P. W., and Mozell, M. M. (1993). Velocity profiles measured for airflow through a large scale model of the human nasal cavity. *J. Appl. Physiol.* 75: 2273–2287.

Hajjar, E., Perahia, D., Debat, H., et al. (2006). Odorant binding and conformational dynamics in the odorant-binding protein. *J. Biol. Chem.* 281: 29929–29937.

Hansel, D. E., May, V., Eipper, B. A., and Ronnett, G. V. (2001). Pituitary adenylyl cyclase-activating peptides and alpha-amidation in olfactory neurogenesis and neuronal survival in vitro. *J. Neurosci.* 21: 4625–4636.

Harkema, J. R. (1990). Comparative pathology of the nasal mucosa in laboratory animals exposed to inhaled irritants. *Environ. Health Perspect.* 85: 231–238.

Harkema, J. R. (1991). Comparative aspects of nasal airway anatomy: relevance to inhalation toxicology. *Toxicol. Pathol.* 19: 321–336.

Harkema, J. R., Hotchkiss, J. A., Harlnsen, A. G., and Henderson, R. F. (1988). In vivo effects of transient neutrophil influx on nasal respiratory epithelial mucosubstances: quantitative histochemistry. *Am. J. Pathol.* 130: 605–615.

Hayes, C. L., Spink, D. C., Spink, B. C., et al. (1996). 17 beta-estradiol hydroxylation catalyzed by human cytochrome P450 1B1. *Proc. Natl. Acad. Sci. U.S.A.* 93: 9776–9781.

He, X. Y., Tang, L., Wang, S. L., et al. (2006). Efficient activation of aflatoxin B1 by cytochrome P450 2A13, an enzyme predominantly expressed in human respiratory tract. *Int. J. Cancer* 118: 2665–2671.

Henkin, R. I. and Velicu, I. (2008). cAMP and cGMP in nasal mucus: relationships to taste and smell dysfunction, gender and age. *Clin. Investigative Med.* 31: E71–E77.

Henriksson, J. and Tjälve, H. (2000). Manganese taken up into the CNS via the olfactory pathway in rats affects astrocytes. *Toxicol. Sci.* 55: 392–398.

Hester, S. D., Benavides, G. B., Sartor, M., et al. (2002). Normal gene expression in male F344 rat nasal transitional and respiratory epithelium. *Gene* 285: 301–310.

Hester, S. D., Benavides, G. B., Yoon, L., et al. (2003). Formaldehyde-induced gene expression in F344 rat nasal respiratory epithelium. *Toxicology.* 187: 13–24.

Heydel, J. M., Leclerc, S., Bernard, P., et al. (2001). Rat olfactory bulb and epithelium UDP-glucuronosyltransferase 2A1 (UGT2A1) expression: in situ mRNA localization and quantitative analysis. *Mol. Brain Res.* 90: 83–92.

Holbrook, E. H., Wu, E. M., Curry, W. T., et al. (2011). Immunohistochemical Characterization of Human Olfactory Tissue. *Laryngoscope* 121: 1687–1701.

Hovemann, B. T., Sehlmeyer, F., and Malz, J. (1997). Drosophila melanogaster NADPH-cytochrome P450 oxidoreductase: pronounced expression in antennae may be related to odorant clearance. *Gene* 189: 213–219.

Hua, Z., Zhang, Q. Y., Su, T., et al. (1997). cDNA cloning, heterologous expression, and characterization of mouse CYP2G1, an olfactory-specific steroid hydroxylase. *Arch. Biochem. Biophysics* 340: 208–214.

Huard, J. M. T., Youngentob, S. L., Goldstein, B. J., et al. (1998). Adult olfactory epithelium contains multipotent progenitors that give rise to neurons and non-neural cells. *J. Comp. Neurol.* 400: 469–486.

Iravani, J., and Melville, G. N. (1974). Long-term effect of cigarette smoke on mucociliary function in animals. *Respiration* 31: 358–366.

Ishige, T., Okamura, H. O., and Kitamura, K. (2008). Immunohistochemical localization of phospholipase A2 in the guinea pig nasal mucosa. *J. Med. Dental Sci.* 55(1): 29–32.

Iwai, N., Zhou, Z., Roop, D. R., et al. (2008). Horizontal basal cells are multipotent progenitors in normal and injured adult olfactory epithelium. *Stem Cells* 26: 1298–1306.

Jaillardon, E., Astic, L., Cattarelli, M., and Gay, B. (1992). A numerical model of the flow in a rat nasal cavity. *Chem. Senses* 17(6): 844.

Jalas, J. R., Hecht, S. S., and Murphy, S. E. (2005). Cytochrome P450 enzymes as catalysts of metabolism of 4-(methylnitrosamino)-1-(3-pyridyl)-1-butanone, a tobacco specific carcinogen. *Chem.Res.Toxicol.* 18: 95–110.

Jang, W., Youngentob, S. L., Schwob, J. E. (2003). Globose basal cells are required for reconstitution of olfactory epithelium after methyl bromide lesion. *J. Comp. Neurol.* 460: 123–140.

Jedlitschky, G., Cassidy, A. J., Sales, M., et al. (1999). Cloning and characterization of a novel human olfactory UDP-glucuronosyltransferase. *Biochem. J.* 340: 837–843.

Jeffrey, A. M., Iatropoulos, M. J., and Williams, G. M. (2006). Nasal cytotoxic and carcinogenic activities of systemically distributed organic chemicals [Review]. *Toxicol. Pathol.* 34: 827–852.

Jeffery, E. H., and Haschek, W. M. (1988). Protection by dimethyl sulfoxide against acetaminophen-induced hepatic, but not respiratory toxicity in the mouse. *Toxicol. Appl. Pharmacol.* 93: 452–461.

Jorissen, M., Willems, T., and Van der, S. B. (1998). Nasal ciliary beat frequency is age independent. *Laryngoscope* 108: 1042–1047.

Kaliner, M. A. (1991). Human nasal respiratory secretions and host defense. *Am. Rev. Respir. Vis.* 144: 552–556.

Karn, R. C. (1998). Steroid binding by mouse salivary proteins. *Biochem. Genet.* 36: 105–117.

Kashiwayanagi, M. Sai, K. and Kurihara, K. (1987) Cell suspensions from porcine olfactory mucosa. Changes in membrane potential and membrane fluidity in response to various odorants. *J. Gen. Physiol.* 89: 443–457.

Kim, T. H., Lee, S. H., Moon, J. H., et al. (2007). Distributional characteristics of lymphatic vessels in normal human nasal mucosa and sinus mucosa. *Cell Tissue Res.* 327: 493–498.

Kimoto, M., Iwai, S., Maeda, T., et al. (2004). Carbonic anhydrase VI in the mouse nasal gland. *J. Histochem. Cytochem.* 52: 1057–1062.

Kimbell, J. S., Gross, E. A., Joyner, D. R., et al. (1993). Application of computational fluid dynamics to regional dosimetry of inhaled chemicals in the upper respiratory tract of the rat. *Toxicol. Appl. Pharmacol.* 121: 253–263.

Koskela, S., Hakkola, J., Hukkanen, J., et al. (1999). Expression of CYP2A genes in human liver and extrahepatic tissues. *Biochem. Pharmacol.* 57: 1407–1413.

Kratskin, I. L., Kimura, Y., Hastings, L., and Doty, R. L. (1999). Chronic dexamethasone treatment potentiates insult to olfactory receptor cells produced by 3-methylindole. *Brain Res.* 847: 240–246.

Krishna, N. S., Getchell, T. V., and Getchell, M. L. (1994). Differential expression of alpha, mu, and pi classes of glutathione S-transferases in chemosensory mucosae of rats during development. *Cell Tissue Res.* 275: 435–450.

Krishna, N. S., Getchell, T. V., Dhooper, N., et al. (1995a). Age- and gender-related trends in the expression of glutathione S-transferases in human nasal mucosa. *Ann. Ontol. Rhinol. Laryngol.* 104: 812–822.

Krishna, N. S., Getchell, M. L., Margolis, F. L., and Getchell, T. V. (1995b). Differential expression of vomeromodulin and odorant-binding protein, putative pheromone and odorant transporters, in the developing rat nasal chemosensory mucosae. *J. Neuro. Res.* 40: 54–71.

Kudo, H., Doi, Y., and Fujimoto, S. (2010). Expressions of the multidrug resistance-related proteins in the rat olfactory epithelium: A possible role in the phase III xenobiotic metabolizing function. *Neurosci. Lett.* 468: 98–101.

Kurosaki, M., Terao, M., Barzago, M. M., et al. (2004). The aldehyde oxidase gene cluster in mice and rats - Aldehyde oxidase homologue 3, a novel member of the molybdo-flavoenzyme family with selective expression in the olfactory mucosa. *J. Biol. Chem.* 279: 50482–50498.

Laethem, R. M., Laethem, C. L., Ding, X., and Koop, D. R. (1992). P-450-dependent arachidonic acid metabolism in rabbit olfactory microsomes. *J. Pharmacol. Exp. Ther.* 262: 433–438.

Lanza, D. L., Code, E., Crespi, C. L., et al. (1999). Specific dehydrogenation of 3-methylindole and epoxidation of naphthalene by recombinant human CYP2F1 expressed in lymphoblastoid cells. *Drug Metab. Dispos.* 27: 798–803.

Larsson, P. and Tjälve, H. (1995). Extrahepatic bioactivation of aflatoxin B1 in fetal, infant and adult rats. *Chem. Biol. Interact.* 94: 1–19.

Larsson, P. and Tjälve, H. (2000). Intranasal instillation of aflatoxin B-1 in rats: Bioactivation in the nasal mucosa and neuronal transport to the olfactory bulb. *Toxicol. Sci.* 55: 383–391.

Larsson, P., Busk, L., and Tjälve, H. (1994). Hepatic and extrahepatic bioactivation and GSH conjugation of aflatoxin B1 in sheep. *Carcinogenesis* 15: 947–955.

Laudien, M., Dressel, S., Harder, J., and Glaser, R. (2011). Differential expression pattern of antimicrobial peptides in nasal mucosa and secretion. *Rhinology* 49: 107–111.

Laukaitis, C. M., Critser, E. S., and Karn, R. C. (1997) Salivary androgen-binding protein (ABP) mediates sexual isolation in Mus musculus. *Evolution* 51: 2000–2005.

Laukaitis, C. M., and Karn, R.C. (2005) Evolution of the secretoglobins: a genomic and proteomic view. *Biol. J. Linnean Soc.* 84: 493–501.

Laukaitis, C. M., Dlouhy, S. R., Emes, R. D., et al. (2005). Diverse spatial, temporal, and sexual expression of recently duplicated androgen-binding protein genes in Mus musculus. *BMC. Evol. Biol.* 5: 40.

Lazard, D., Tal, N., Rubinstein, M., et al. (1990). Identification and biochemical analysis of novel olfactory-specific cytochrome P-450IIA and UDP-glucuronyl transferase. *Biochemistry.* 29: 7433–7440.

Lazard, D., Zupko, K., Poria, Y., et al. (1991). Odorant signal termination by olfactory UDP glucuronosyl transferase. *Nature* 349: 790–793.

Lee, C. H., Igarashi, Y., Hohman, R. J., et al. (1993). Distribution of secretory leukoprotease inhibitor in the human nasal airway. *Am. Rev. Resp. Dis.* 147: 710–716.

Lee, J. T., Jansen, M., Yilma, A. N., et al. (2010). Antimicrobial lipids: Novel innate defense molecules are elevated in sinus secretions of patients with chronic rhinosinusitis. *Am. J. Rhinol. Allergy* 24: 99–104.

Lee, M. G., Phimister, A., Morin, D., et al. (2005). In situ naphthalene bioactivation and nasal airflow cause region-specific injury patterns in the nasal mucosa of rats exposed to naphthalene by inhalation. *J. Pharmacol. Exp. Therapeut.* 314: 103–110.

Leung, C. T., Coulombe, P. A., Reed, R. R. (2007). Contribution of olfactory neural stem cells to tissue maintenance and regeneration. *Nat. Neurosci.* 10: 720–726.

Lewis, J. L., Hahn, F. F., and Dahl, A. R. (1994a). Transport of inhaled toxicants to the central nervous system: characteristics of a nose-brain barrier. In *The Vulnerable Brain and Environmental Risks. Vol. 3. Toxins in Air and Water*, R. L. Isaacson and K. F. Jensen (Eds.). Plenum Press, New York, pp. 77–103.

Lewis, J. L., Nikula, K. J., Novak, R., and Dahl, A. R. (1994b). Comparative localization of carboxylesterase in F344 rat, beagle dog, and human nasal tissue. *Anat. Rec.* 239: 55–64.

Lewis, J. L., Nikula, K. J., and Dahl, A. R. (1992a). Comparative analysis of activity and distribution of nasal carboxylesterases (CE). *Toxicologist* 12: 398.

Lewis, J. L., Rhoades, C. E., Bice, D. E., et al. (1992b). Interspecies comparison of cellular localization of the cyanide metabolizing enzyme rhodanese within olfactory mucosa. *Anat. Rec.* 232: 620–627.

Lewis, J. L., Rhoades, C. E., Gervasi, P. G., et al. (1991). The cyanide-metabolizing enzyme rhodanese in human nasal respiratory mucosa. *Toxicol. Appl. Pharmacol.* 108: 114–120.

Li, L., Wei, Y., Van Winkle, L., et al. (2011). Generation and Characterization of a Cyp2f2-Null Mouse and Studies on the Role of CYP2F2 in Naphthalene-Induced Toxicity in the Lung and Nasal Olfactory Mucosa. *J. Pharm. Exp. Ther.* 339: 62–71.

Ling, G., Hauer, C. R., Gronostajski, R. M., et al. (2004). Transcriptional regulation of rat CYP2A3 by nuclear factor 1: identification of a novel NFI-A isoform, and evidence for tissue-selective interaction of NFI with the CYP2A3 promoter in vivo. *J. Biol. Chem.* 279: 27888–27895.

Liu, C., Zhuo, X., Gonzalez, F. J., and Ding, X. (1996). Baculovirus-mediated expression and characterization of rat CYP2A3 and human CYP2A6: role in metabolic activation of nasal toxicants. *Mol. Pharmacol.* 50: 781–788.

Lobasso, S., Lopalco, P., Angelini, R., et al. (2010). Lipidomic Analysis of Porcine Olfactory Epithelial Membranes and Cilia. *Lipids* 45: 593–602.

Longo, V. and Ingelman-Sundberg, M. (1993). Acetone-dependent regulation of cytochromes P4502E1 and P4502B1 in rat nasal mucosa. *Biochem. Pharmacol.* 46: 1945–1951.

Longo, V., Amato, G., Santucci, A., and Gervasi, P. G. (1997). Purification and characterization of three constitutive cytochrome P-450 isoforms from bovine olfactory epithelium. *Biochem. J.* 323: 65–70.

Longo, V., Citti, L., and Gervasi, P. G. (1988). Biotransformation enzymes in nasal mucosa and liver of Sprague–Dawley rats. *Toxicol. Lett.* 44: 289–297.

Longo, V., Ingelman-Sundberg, M., Amato, G., et al. (2000). Effect of starvation and chlormethiazole on cytochrome P450s of rat nasal mucosa. *Biochem. Pharmacol.* 59: 1425–1432.

Luo, G., Zeldin, D. C., Blaisdell, J. A., et al. (1998). Cloning and expression of murine CYP2CS and their ability to metabolize arachidonic acid. *Arch. Biochem. Biophysics* 357: 45–57.

Lupo, D., Lodi, L., Canonaco, M., et al. (1986). Testosterone metabolism in the olfactory epithelium of intact and castrated male rats. *Neurosci. Lett.* 69: 259–262.

Mackenzie, P. I., Owens, I. S., Burchell, B., et al. (1997). The UDP glycosyltransferase gene superfamily: recommended nomenclature update based on evolutionary divergence. *Pharmacology.* 7: 255–269.

Mainwaring, G., Foster, J. R., Lund, V., and Green, T. (2001). Methyl methacrylate toxicity in rat nasal epithelium: studies of the mechanism of action and comparisons between species. *Toxicology.* 158: 109–118.

Makita, K., Falck, J. R., and Capdevila, J. H. (1996). Cytochrome P450, the arachidonic acid cascade, and hypertension: new vistas for an old enzyme system. *FASEB J.* 10: 1456–1463.

Margalit, T., and Lancet, D. (1993). Expression of olfactory receptor and transduction genes during rat development. *Dev. Brain Res.* 73: 7–16.

Marini, S., Longo, V., Mazzaccaro, A., and Gervasi, P. G. (1998). Xenobiotic-metabolizing enzymes in pig nasal and hepatic tissues. *Xenobiotica* 28: 923–935.

Marini, S., Longo, V., Zaccaro, C., et al. (2001). Selective inactivation of rat and bovine olfactory cytochrome P450 by three haloethanes. *Toxicol. Letts.* 124: 83–90.

Mayer, U., Kuller, A., Daiber, P. C., et al. (2009). The proteome of rat olfactory sensory cilia. *Proteomics* 9: 322–334.

Mayer, U., Ungerer, N., Klimmeck, D., et al. (2008). Proteomic analysis of a membrane preparation from rat olfactory sensory cilia. *Chem. Senses* 33: 145–162.

McLean, J. H., Shipley, M. T., and Bernstein, D. I. (1989). Golgi-like, transneuronal retrograde labeling with CNS injections of herpes simplex virus type 1. *Brain Res. Bull.* 22: 867–881.

Mellert, T. K., Getchell, M. L., Sparks, L., and Getchell, T. V. (1992). Characterization of the immune barrier in human olfactory mucosa. *Otolaryngol. Head Neck Surg.* 106: 181–188.

Min, Y. G., Ohyama, M., Lee, K. S., et al. (1999). Effects of free radicals on ciliary movement in the human nasal epithelial cells. *Auris, Nasus, Larynx* 26: 159–163.

Minn, A. L., Pelczar, H., Denizot, C., et al. (2005). Characterization of microsomal cytochrome P450-dependent monooxygenases in the rat olfactory mucosa. *Drug Metab.* 33: 1229–1237.

Miyawaki, A., Homma, H., Tamura, H., et al. (1996). Zonal distribution of sulfotransferase for phenol in olfactory sustentacular cells. *EMBO J.* 15: 2050–2055.

Moretti, M., Sinnappah-Kang, N. D., Toller, M., et al. (2006). HPSE-1 expression and functionality in differentiating neural cells. *J. Neurosci. Res.* 83: 694–701.

Morgan, K. T., and Monticello, T. M. (1990). Airflow, gas deposition, and lesion distribution in the nasal passages. *Environ. Health Perspect.* 88: 209–218.

Morgan, K. T., Patterson, D. L., and Gross, E. A. (1986). Responses of the nasal mucociliary apparatus to airborne irritants. In *Toxicology of the Nasal Passages,* Chapter 8, C. S. Barrow (Ed.). Hemisphere Publishing Corp., New York, pp. 123–134.

Morris, J. B. (1997). Uptake of acetaldehyde vapor and aldehyde dehydrogenase levels in the upper respiratory tracts of the mouse, rat, hamster, and guinea pig. *Fund. Appl. Toxicol.* 35: 91–100.

Morris, J. B. (2000). Uptake of styrene in the upper respiratory tract of the CD mouse and Sprague–Dawley rat. *Toxicol. Sci.* 54: 222–228.

Morris, J. B., and Cavanagh, D. G. (1986). Deposition of ethanol and acetone vapors in the upper respiratory tract of the rat. *Fund. Appl. Toxicol.* 6: 78–88.

Mullol, J., Raphael, G. D., Lundgren, J. D., et al. (1992). Comparison of human nasal mucosal secretion in vivo and in vitro. *J. Allergy Clin. Immunol.* 89: 584–592.

Murphy, S. E., Isaac, I. S., Ding, X. X., and McIntee, E. J. (2000). Specificity of cytochrome P450 2A3-catalyzed alpha-hydroxylation of N′-nitrosonornicotine enantiomers. *Drug Metab.* 28: 1263–1266.

Nagashima, A. and Touhara, K. (2010). Enzymatic conversion of odorants in nasal mucus affects olfactory glomerular activation patterns and odor perception. *J. Neurosci.* 30: 16391–16398.

Nakajima, M., Itoh, M., Sakai, H., et al. (2006). CYP2A13 expressed in human bladder metabolically activates 4-aminobiphenyl. *Int. J. Cancer* 119: 2520–2526.

Nakajima, T., Elovaara, E., Gonzalez, F. J., et al. (1994). Styrene metabolism by cDNA-expressed human hepatic and pulmonary cytochromes P450. *Chem. Res. Toxicol.* 7: 891–896.

Nannelli, A., De Rubertis, A., Longo, V., and Gervasi, P. (2005). Effects of dioxane on cytochrome P450 enzymes in liver, kidney, lung and nasal mucosa of rat. *Arch. Toxicol.* 79: 74–82.

Nebert, D. W. (1990). Drug metabolism. *Growth signal pathways. Nature* 347: 709–710.

Nebert, D. W. (1991). Proposed role of drug-metabolizing enzymes: regulation of steady state levels of the ligands that effect growth, homeostasis, differentiation, and neuroendocrine functions. *Mol. Endocrinol.* 5: 1203–1214.

Nef, P., Heldmann, J., Lazard, D., et al. (1989). Olfactory-specific cytochrome P-450: cDNA cloning of a novel neuroepithelial enzyme possibly involved in chemoreception. *J. Biol. Chem.* 264: 6780–6785.

Nef, P., Larabee, T. M., Kagimoto, K., and Meyer, U. A. (1990). Olfactory-specific cytochrome P-450 (P-4500lfl; IIGl): gene structure and developmental regulation. *J. Biol. Chem.* 265: 2903–2907.

Nelson, D. R., Koymans, L., Kamataki, T., et al. (1996). P450 superfamily: update on new sequences, gene mapping, accession numbers and nomenclature. *Pharmacogen* 6: 1–42.

Nikula, K. J., Novak, R. F., Chang, I. Y., et al. (1995). Induction of nasal carboxylesterase in F344 rats following inhalation exposure to pyridine. *Drug Metab. Disposition* 23: 529–535.

Novoselov, S. V., Peshenko, I. V., Popov, V. I., et al. (1999). Localization of 28-kDa peroxiredoxin in rat epithelial tissues and its antioxidant properties. *Cell Tissue Res.* 298: 471–480.

Nunez-Parra, A., Cortes-Campos, C., Bacigalupo, J., et al. (2011). Expression and Distribution of Facilitative Glucose (GLUTs) and Monocarboxylate/H+ (MCTs) Transporters in Rat Olfactory Epithelia. *Chem. Senses* 36: 771–780.

Ohkubo, K., Baraniuk, J. N., Hohman, R., et al. (1998). Aminopeptidase activity in human nasal mucosa. *J. Allergy Clin. Immunol.* 102: 741–750.

Owens, J. M., Shroyer, K. R., and Kingdom, T. T. (2008). Expression of cyclooxygenase and lipoxygenase enzymes in sinonasal mucosa of patients with cystic fibrosis. *Arch. Otolaryngol Head Neck Surg.* 134: 825–831.

Paunescu, T. G., Jones, A. C., Tyszkowski, R., and Brown, D. (2008). V-ATPase expression in the mouse olfactory epithelium. *Am. J. Physiol. - Cell Physiol.* 295: C923–C930.

Pelosi, P. (1996). Perireceptor events in olfaction. *J. Neurobiology.* 30: 3–19.

Peng, H. M., Ding, X., and Coon, M. J. (1993). Isolation and heterologous expression of cloned cDNAs for two rabbit nasal microsomal proteins, CYP2A10 and CYP2A11, that are related to nasal microsomal cytochrome P450 form a. *J. Biol. Chem.* 268: 17253–17260.

Persson, E., Larsson, P., and Tjalve, H. (2002). Cellular activation and neuronal transport of intranasally instilled benzo(a)pyrene in the olfactory system of rats. *Toxicol. Lett.* 133: 211–219.

Pinto, S., Gallo, O., Polli, G., et al. (1997). Cyclooxygenase and lipoxygenase metabolite generation in nasal polyps. *Prostaglan. Leukotr. Ess.* 57: 533–537.

Piras, E., Franzen, A., Fernandez, E. L., et al. (2003). Cell-specific expression of CYP2A5 in the mouse respiratory tract: Effects of olfactory toxicants. *J. Histochem. Cytochem.* 51: 1545–1555.

Poon, H. F., Vaishnav, R. A., Butterfield, D. A., et al. (2005). Proteomic identification of differentially expressed proteins in the aging murine olfactory system and transcriptional analysis of the associated genes. *J. Neurochem.* 94: 380–392.

Porter, T. D. and Coon, M. J. (1991). Cytochrome P-450. Multiplicity of isoforms, substrates, and catalytic and regulatory mechanisms. *J. Biol. Chem.* 266: 13469–13472.

Powley, M. W. and Carlson, G. P. (2000). Cytochromes P450 involved with benzene metabolism in hepatic and pulmonary microsomes. *J. Biochem. Mol. Toxicol.* 14: 303–309.

Price, S. (1984). Mechanisms of stimulation of olfactory neurons: An essay. *Chem. Senses* 8: 341–354.

Price, S. (1986). Effects of odorant mixtures on olfactory receptor cells. *Ann. N.Y. Acad. Sci.* 512: 55–60.

Proctor, D. F., Adams, G. K., Andersen, I., and Man, S. F. (1978). Nasal mucociliary clearance in man. *Ciba Found. Symp.* 54: 219–234.

Quarcoo, D., Fischer, T. C., Heppt, W., et al.(2009). Expression, localisation and functional implications of the transporter protein PEPT2 in the upper respiratory tract. *Respiration* 77: 440–446.

Randall, H. W., Bogdanffy, M. S., and Morgan, K. T. (1987). Enzyme histochemistry of the rat nasal mucosa embedded in cold glycol methacrylate. *Am. J. Anat.* 179: 10–17.

Rawson, N. E. and LaMantia, A. S. (2006). Once and again: Retinoic acid signaling in the developing and regenerating olfactory pathway. *J. Neurobiol.* 66: 653–676.

Reed, C. J., Lock, E. A., and De, M. F. (1986). NADPH: cytochrome P-450 reductase in olfactory epithelium. Relevance to cytochrome P-450-dependent reactions. *Biochem. J.* 240: 585–592.

Reed, C. J., Robinson, D. A., and Lock, E. A. (2003). Antioxidant status of the rat nasal cavity. *Free Radical Biol. Med.* 34: 607–615.

Revington, M., Lacroix, J. S., and Potter, E. K. (1997). Sympathetic and parasympathetic interaction in vascular and secretory control of the nasal mucosa in anaesthetized dogs. *J. Physiol.* 505: 823–831.

Reznik, G. (1990). Comparative anatomy, physiology, and function of the upper respiratory tract. *Environ. Health. Perspect.* 85: 171–184.

Roberts, E. S., Soucy, N. V., Bonner, A. M., et al. (2007). Basal gene expression in male and female Sprague–Dawley rat nasal respiratory and olfactory epithelium. *Inhal. Toxicol.* 19: 941–949.

Roberts, E. S., Vaz, A. D., and Coon, M. J. (1992). Role of isozymes of rabbit microsomal cytochrome P-450 in the metabolism of retinoic acid, retinol, and retinal. *Mol. Pharmacol.* 41: 427–433.

Robinson, A. M., Kern, R. C., Foster, J. D., et al. (1999). Mineralocorticoid receptors in the mammalian olfactory mucosa. *Ann. Otol. Rhinol. Laryngol.* 108: 974–981.

Robottom-Ferreira, A. B., Aquino, S. R., Queiroga, R., et al. (2003). Expression of CYP2A3 mRNA and its regulation by 3-methylcholanthrene, pyrazole, and beta-ionone in rat tissues. *Brazilian J. Med. Biol. Res.* 36: 839–844.

Rogers, M. E., Jani, M. K., and Vogt, R. G. (1999). An olfactory-specific glutathione-S-transferase in the sphinx moth Manduca sexta. *J. Exp. Biol.* 202: 1625–1637.

Rosenblum, P. M., Sorensen, P. W., Stacey, N. E., and Peter, R. E. (1991). Binding of the steroidal pheromone 17a,20b-dihydroxy-4-pregnen-3-one to goldfish (Carassius auratus) olfactory epithelium membrane preparations. *Chem. Senses* 16: 143–154.

Saarikoski, S. T., Wikman, H. A., Smith, G., Wolff, C. H., Husgafvel-Pursiainen, K. (2005). Localization of cytochrome P450 CYP2S1 expression in human tissues by in situ hybridization and immunohistochemistry. *J. Histochem. Cytochem.* 53:549–56.

Sammeta, N. and McClintock, T. S. (2010). Chemical Stress Induces the Unfolded Protein Response in Olfactory Sensory Neurons. *J. Comp. Neurol.* 518: 1825–1836.

Sasagawa, T., Yamada, T., Nakagawa, T., et al. (2011). In vitro metabolism of dexamethasone cipecilate, a novel synthetic corticosteroid, in human liver and nasal mucosa. *Xenobiotica* 41: 874–884.

Sato, H., Nave, R., Nonaka, T., et al. (2007). In vitro activation of the corticosteroid ciclesonide in animal nasal mucosal homogenates. *Biopharm. Drug Disposition.* 28: 59–64.

Sato, H., Nave, R., Nonaka, T., et al. (2007). In vitro metabolism of ciclesonide in human nasal epithelial cells. *Biopharm. Drug Disposition.* 28(1): 43–50.

Sato, H., Nave, R., Nonaka, T., et al. (2007). Uptake and metabolism of ciclesonide and retention of desisobutyryl-ciclesonide for up to 24 hours in rabbit nasal mucosa. *BMC Pharmacol.* 7:7.

Scarborough, P. E., Ma, J. X., Qu, W., and Zeldin, D. C. (1999). P450 subfamily CYP2J and their role in the bioactivation of arachidonic acid in extrahepatic tissues. *Drug Metab. Rev.* 31: 205–234.

Schultz, E. W., and Gebhardt, L. P. (1934). Olfactory tract and poliomyelitis. *Proc. Soc. Exp. Biol. Med.* 31: 728–730.

Schwob, J. E., Youngentob, S. L., and Mezza, R. C. (1995). Reconstitution of the rat olfactory epithelium after methyl bromide-induced lesion. *J. Comp. Neurol.* 359: 15–37.

Segovia, S. and Guillamon, A. (1982). Effects of sex steroids on the development of the vomeronasal organ in the rat. *Brain Res.* 281: 209–212.

Shehin-Johnson, S. E., Williams, D. E., Larsen-Su, S., et al. (1995). Tissue-specific expression of flavin-containing monooxygenase (FMO) forms 1 and 2 in the rabbit. *J. Pharmacol. Exp. Ther.* 272: 1293–1299.

Sheng, J., Guo, J., Hua, Z., et al. (2000). Characterization of human CYP2G genes: widespread loss-of-function mutations and genetic polymorphism. *Pharmacogenetics*, 10: 667–678.

Shipley, M. T. (1985). Transport of molecules from nose to brain: transneuronal anterograde and retrograde labeling in the rat olfactory system by wheat germ agglutinin-horseradish peroxidase applied to the nasal epithelium. *Brain Res. Bull.* 15: 129–142.

Simoes, T., Charro, N., Blonder, J., et al. (2011). Molecular profiling of the human nasal epithelium: A proteomics approach. *J. Proteomics.* 75: 56–69.

Sneitz, N., Court, M. H., Zhang, X., et al. (2009). Human UDP-glucuronosyltransferase UGT2A2: cDNA construction, expression, and functional characterization in comparison with UGT2A1 and UGT2A3. *Pharmacogenetics Genomics* 19: 923–934.

Snyder, S. H., Sklar, P. B., and Pevsner, J. (1988). Molecular mechanisms of olfaction. *J. Biol. Chem.* 263: 13971–13974.

Solbu, T. T. and Holen, T. (2012). Aquaporin pathways and mucin secretion of Bowman's glands might protect the olfactory mucosa. *Chem. Senses* 37: 35–46.

Sorokin, S. P. (1988). The respiratory system. In *Cell and Tissue Biology: A Textbook of Histology*. Chapter 25, L. Weiss (Ed.). Uban & Schwarzenberg, Baltimore, pp. 753–814.

Starcevic, S. L. and Zielinski, B. S. (1995). Immunohistochemical localization of glutathione S-transferase pi in rainbow trout olfactory receptor neurons. *Neurosci. Lett.* 183: 175–178.

Starcevic, S. L. and Zielinski, B. S. (1997). Glutathione and glutathione S-transferase in the rainbow trout olfactory mucosa during retrograde degeneration and regeneration of the olfactory nerve. *Exp. Neurol.* 146: 331–340.

Stott, W. T., Dryzga, M. D., and Ramsey, J. C. (1983). Blood-flow distribution in the mouse. *J. Appl. Toxicol.* 3: 310–312.

Strotmann, J. and Breer, H. (2011). Internalization of odorant-binding proteins into the mouse olfactory epithelium. *Histochem. Cell Biol.* 136: 357–369.

Su, T. and Ding, X. (2004). Regulation of the cytochrome P450 2A genes. *Toxicol. Appl. Pharmacol.* 199: 285–294.

Su, T., Bao, Z., Zhang, Q. Y., et al. (2000). Human cytochrome P450 CYP2A13: predominant expression in the respiratory tract and its high efficiency metabolic activation of a tobacco-specific carcinogen, 4-(methylnitrosamino)-1-(3-pyridyl)-1-butanone. *Cancer Res.* 60: 5074–5079.

Su, T., He, W., Gu, J. et al. (1998). Differential xenobiotic induction of CYP2A5 in mouse liver, kidney, lung, and olfactory mucosa. *Drug Metabol. Disposition* 26: 822–824.

Su, T., Sheng, J. J., Lipinskas, T. W., and Ding, X. (1996). Expression of CYP2A genes in rodent and human nasal mucosa. *Drug Metabol. Disposition* 24: 884–890.

Sun, J. D., Dahl, A. R., Gillett, N. A., et al. (1991). Two-week repeated inhalation exposure of F344/N rats and B6C3F1 mice to ferrocene. *Fund. Appl. Toxicol.* 17: 150–158.

Sung, Y. K., Moon, C., Yoo, J. Y., et al. (2002). Plunc, a member of the secretory gland protein family, is up-regulated in nasal respiratory epithelium after olfactory bulbectomy. *J. Biol. Chem.* 277: 12762–12769.

Talley, H. M., Laukaitis, C. M., and Karn, R. C. (2001). Female preference for male saliva: implications for sexual isolation of Mus musculus subspecies. *Evolution* 55: 631–634.

Tamura, H., Miyawaki, A., Inoh, N., et al. (1997). High sulfotransferase activity for phenolic aromatic odorants present in the mouse olfactory organ. *Chem. Biol. Interact.* 104: 1–9.

Tandler, B., Edelstein, D. R., and Erlandson, R. A. (2000). Ultrastructure of submucosal glands in human anterior middle nasal turbinates. *J. Anat.* 197: 229–237.

Tarun, A. S., Bryant, B., Zhai, W. W., et al. (2003). Gene expression for carbonic anhydrase isoenzymes in human nasal mucosa. *Chem. Senses* 28: 621–629.

Thiebaud, N., Menetrier, F., Belloir, C., et al. (2011). Expression and differential localization of xenobiotic transporters in the rat olfactory neuro-epithelium. *Neurosci. Lett.* 505: 180–185.

Thiebaud, N., Sigoillot, M., Chevalier, J., et al. (2010). Effects of typical inducers on olfactory xenobiotic-metabolizing enzyme, transporter, and transcription factor expression in rats. *Drug Metab* 38: 1865–1875.

Thompson, K., Molina, R. M., Donaghey, T., et al. (2007). Olfactory uptake of manganese requires DMT1 and is enhanced by anemia. *FASEB J.* 21: 223–230.

Thornton-Manning, J. R. and Dahl, A. R. (1997). Metabolic capacity of nasal tissue interspecies comparisons of xenobiotic-metabolizing enzymes. *Mut. Res.* 380: 43–59.

Thornton-Manning, J. R., Nikula, K. J., Hotchkiss, J. A., et al. (1997). Nasal cytochrome P450 2A: identification, regional localization, and metabolic activity toward hexamethylphosphoramide, a known nasal carcinogen. *Toxicol. Appl. Pharmacol.* 142: 22–30.

Tomlinson, A. H., and Esiri, M. M. (1983). Herpes simplex encephalitis. Immunohistological demonstration of spread of virus via olfactory pathways in mice. *J. Neurol. Sci.* 60: 473–484.

Trapido-Rosenthal, H. G., Carr, W. E. S., and Gleeson, R. A. (1990). Ectonucleotidase activities associated with the olfactory organ of the spiny lobster. *J. Neurochem.* 55: 88–96.

Tricker, A. R., and Preussman, R. (1991). Carcinogenic N-nitrosamines in the diet: occurrence, formation, mechanisms and carcinogenic potential. *Mutat. Res.* 259: 277–289.

Turk, M. A. M., Flory, W., and Henk, W. G. (1986). Chemical modulation of 3-methylindole toxicosis in mice: effect of bronchiolar and olfactory mucosal injury. *Vet. Pathol.* 23: 563–570.

Tyden, E., Olsen, L., Tallkvist, J., et al. (2008). Cytochrome P450 3A, NADPH cytochrome P450 reductase and cytochrome b(5) in the upper airways in horse. *Res. Vet. Sci.* 85: 80–85.

Uno, Y., Uehara, S., Murayama, N., and Yamazaki, H. (2011). CYP2G2, Pseudogenized in Human, Is Expressed in Nasal Mucosa of Cynomolgus Monkey and Encodes a Functional Drug-Metabolizing Enzyme. *Drug Metab.* 39: 717–723.

Uraih, L. C., and Maronpot, R. R. (1990). Normal histology of the nasal cavity and application of special techniques. *Environ. Health Perspect.* 85: 187–208.

Vaidyanathan, A., Foy, J. W. D., and Schatz, R. A. (2003). Inhibition of rat respiratory-tract cytochrome P-450 isozymes following inhalation of m-xylene: Possible role of metabolites. *J. Toxicol. Environment Health* 66: 1133–1143.

Vannelli, G. B., and Balboni, G. C. (1982). On the presence of estrogen receptors in the olfactory epithelium of the rat, in *Olfaction and Endocrine Regulation* (Breiphol, ed.). IRL Press, London, pp. 279–282.

Verma, A., Hirsch, D. J., Glatt, C. E., et al. (1993). Carbon monoxide: a putative neural messenger. *Science* 259: 381–384.

Vogt, R. G., and Riddiford, L. M. (1981). Pheromone binding and inactivation by moth antennae. *Nature* 293: 161–163.

Voigt, J. M., Guengerich, F. P., and Baron, J. (1985). Localization of a cytochrome P-450 isozyme (cytochrome P-450 PB-B) and NADPH-cytochrome P-450 reductase in rat nasal mucosa. *Cancer Lett.* 27: 241–247.

Voigt, J. M., Guengerich, F. P., and Baron, J. (1993). Localization and induction of cytochrome P450 1A1 and aryl hydrocarbon hydroxylase activity in rat nasal mucosa. *J. Histochem. Cytochem.* 41: 877–885.

von Weymarn, L. B., Zhang, Q. Y., Ding, X., and Hollenberg, P. F. (2005). Effects of 8-methoxypsoralen on cytochrome P450 2A13. *Carcinogen* 26: 621–629.

Walters, E., Buchheit, K., and Maruniak, J. A. (1992). Receptor neuron losses result in decreased cytochrome P450 immunoreactivity in associated non-neuronal cells of mouse olfactory mucosa. *J. Neurosci. Res.* 33: 103–111.

Walters, E., Buchheit, K., and Maruniak, J. A. (1993). Olfactory cytochrome P-450 immunoreactivity in mice is altered by dichlobenil but preserved by metyrapone. *Toxicology.* 81: 113–122.

Wang, H. B., Chanas, B., and Ghanayem, B. I. (2002). Effect of methacrylonitrile on cytochrome P-450 2E1 (CYP2E1) expression in male F344 rats. *J. Toxicol. Environ. Health Part A* 65: 523–537.

Wang, Q., Hasan, G., and Pikielny, C. W. (1999). Preferential expression of biotransformation enzymes in the olfactory organs of Drosophila melanogaster, the antennae. *J. Biol. Chem.* 274: 10309–10315.

Wardlaw, S. A., Nikula, K. J., Kracko, D. A., et al. (1998). Effect of cigarette smoke on CYP1A1, CYP1A2 and CYP2B1/2 of nasal mucosae in F344 rats. *Carcinogenesis* 19: 655–662.

Watelet, J. B., Strolin-Benedetti, M., and Whomsley, R. (2009). Defence mechanisms of olfactory neuro-epithelium: mucosa regeneration, metabolising enzymes and transporters. *B-ENT*: 21–37.

Waxman, D. J., Lapenson, D. P., Aoyama, T., et al. (1991). Steroid hormone hydroxylase specificities of eleven cDNA-expressed human cytochrome P450s. *Arch. Biochem. Biophys.* 290: 160–166.

Weech, M., Quash, M., and Walters, E. (2003). Characterization of the mouse olfactory glutathione s-transferases during the acute phase response. *J. Neurosci. Res.* 73: 679–685.

Weems, J. M., Lamb, J. G., D'Agostino, J., et al. (2010). Potent mutagenicity of 3-methylindole requires pulmonary cytochrome P450-mediated bioactivation: a comparison to the prototype cigarette smoke mutagens B(a)P and NNK. *Chem. Res. Toxicol.* 23: 1682–1690.

Wei, Y., Li, L., Zhou, X., et al. (2013). Generation and Characterization of a Novel Cyp2a(4/5)bgs-null Mouse Model. *Drug Metab. Dispos.* 41: 132–140.

Wei, Y., Wu, H., Li, L., et al. (2012). Generation and characterization of a CYP2A13/2B6/2F1-transgenic mouse model. *Drug Metab. Dispos.* 40: 1144–1150.

Wei, Y., Zhou, X., Fang, C., et al. (2010). Generation of a mouse model with a reversible hypomorphic cytochrome P450 reductase gene: utility for tissue-specific rescue of the reductase expression, and insights from a resultant mouse model with global suppression of P450 reductase expression in extrahepatic tissues. *J. Pharmacol. Exp. Ther.* 334: 69–77.

Weinshilboum, R. M., Otterness, D. M., Aksoy, I. A., et al. (1997). Sulfation and sulfotransferases 1: Sulfotransferase molecular biology: cDNAs and genes. *FASEB J* 11: 3–14.

Whitby-Logan, G. K., Weech, M., and Walters, E. (2004). Zonal expression and activity of glutathione S-transferase enzymes in the mouse olfactory mucosa. *Brain Res.* 995: 151–157.

Wong, H. L., Murphy, S. E., and Hecht, S. S. (2003). Preferential metabolic activation of N-nitrosopiperidine as compared to its structural homologue N-nitrosopyrrolidine by rat nasal mucosal microsomes. *Chem. Res. Toxicol.* 16: 1298–1305.

Wong, H. L., Zhang, X. L., Zhang, Q. Y., et al. (2005). Metabolic activation of the tobacco carcinogen 4-(methylnitrosamino)-(3-pyridyl)-1-butanone by cytochrome P450 2A13 in human fetal nasal microsomes. *Chem. Res. Toxicol.* 18: 913–918.

Wu, H., Zhang, X., Ling, G., et al. (2009). Mechanisms of differential expression of the CYP2A13 7520C and 7520G alleles in human lung: allelic expression analysis for CYP2A13 heterogeneous nuclear RNA, and evidence for the involvement of multiple cis-regulatory single nucleotide polymorphisms. *Pharmacogenet. Genom.* 19: 852–863.

Wu, L., Gu, J., Cui, H. D., et al. (2005). Transgenic mice with a hypomorphic NADPH-Cytochrome P450 reductase gene: Effects on development, reproduction, and microsomal cytochrome P450. *J. Pharmacol. Exp. Therapeut.* 312: 35–43.

Xie, F., Zhou, X., Behr, M., et al. (2010). Mechanisms of olfactory toxicity of the herbicide 2,6-dichlorobenzonitrile: essential roles of CYP2A5 and target-tissue metabolic activation. *Toxicol. Appl. Pharmacol.* 249(1): 101–106.

Xie, F., Zhou, X., Genter, M. B., et al. (2011). The tissue-specific toxicity of methimazole in the mouse olfactory mucosa is partly mediated through target-tissue metabolic activation by CYP2A5. *Drug Metab.* 39: 947–951.

Xie, Q., Zhang, Q. Y., Zhang, Y., et al. (2000). Induction of mouse CYP2J by pyrazole in the eye, kidney, liver, lung, olfactory mucosa, and small intestine, but not in the heart. *Drug Metab.* 28: 1311–1316.

Zhang, J. H. and Ding, X. (1998). Identification and characterization of a novel tissue-specific transcriptional activating element in the 5′-flanking region of the CYP2A3 gene predominantly expressed in rat olfactory mucosa. *J. Biol. Chem.* 273: 23454–23462.

Zhang, J. H., Zhang, Q.-Y., Guo, J. C., et al. (2000). Identification and functional characterization of a conserved, nuclear factor 1-like element in the proximal promoter region of CYP1A2 gene specifically expressed in the liver and olfactory mucosa. *J. Biol. Chem.* 275: 8895–8902.

Zhang, Q.-Y. (1999). Retinoic acid biosynthetic activity and retinoid receptors in the olfactory mucosa of adult mice. *Biochem. Biophys. Res. Comms.* 256: 346–351.

Zhang, Q-Y. and Ding, X. (2008) The CYP2F, CYP2G and CYP2J Subfamilies. In *Cytochrome P450: Role in the Metabolism and Toxicity of Drugs and Other Xenobiotics*. (Ioannides C, Ed), Chapter 10, RSC Publishing, Cambridge, UK, pp. 309–353.

Zhang, Q.-Y., Ding, X., and Kaminsky, L. S. (1997). cDNA cloning, heterologous expression, and characterization of rat intestinal CYP2J4. *Arch. Biochem. Biophysics* 340: 270–278.

Zhang, Q.-Y., Ding, X., Dunbar, D., et al. (1999). Induction of rat small intestinal cytochrome P-450 2J4. *Drug Metab. Dispos.* 27: 1123–1127.

Zhang, Q.-Y., Raner, G., Ding, X., et al. (1998). Characterization of the cytochrome P450 CYP2J4 - expression in rat small intestine and role in retinoic acid biotransformation from retinal. *Arch. Biochem. Biophys.* 353: 257–264.

Zhang, S. Y., Wang, M. Y., Villalta, P. W., et al. (2009a). Quantitation of Pyridyloxobutyl DNA Adducts in Nasal and Oral Mucosa of Rats Treated Chronically with Enantiomers of N′-Nitrosonornicotine. *Chem. Res. Toxicol.* 22: 949–956.

Zhang, S. Y., Wang, M. Y., Villalta, P. W., et al. (2009b). Analysis of Pyridyloxobutyl and Pyridylhydroxybutyl DNA Adducts in Extrahepatic Tissues of F344 Rats Treated Chronically with 4-(Methylnitrosamino)-1-(3-pyridyl)-1-butanone and Enantiomers of 4-(Methylnitrosamino)-1-(3-pyridyl)-1-butanol. *Chem. Res. Toxicol.* 22: 926–936.

Zhang, X. L., Chen, Y., Liu, Y. Q., et al. (2003). Single nucleotide polymorphisms of the human CYP2A13 gene: Evidence for a null allele. *Drug Metab.* 31: 1081–1085.

Zhang, X. L., Zhang, Q. Y., Liu, D. Z., et al. (2005). Expression of cytochrome P450 and other biotransformation genes in fetal and adult human nasal mucosa. *Drug Metab.* 33: 1423–1428.

Zhang, X., Caggana, M., Cutler, T. L., and Ding, X. (2004). Development of a real-time polymerase chain reaction-based method for the measurement of relative allelic expression and identification of CYP2A13 alleles with decreased expression in human lung. *J. Pharmacol. Exp. Ther.* 311: 373–381.

Zhang, X., D'Agostino, J., Wu, H., et al. (2007). CYP2A13: variable expression and role in human lung microsomal metabolic activation of the tobacco-specific carcinogen 4-(methylnitrosamino)-1-(3-pyridyl)-1-butanone. *J. Pharmacol. Exp. Ther.* 323: 570–578.

Zhou, X., D'Agostino, J., Li, L., et al. (2012). Respective Roles of CYP2A5 and CYP2F2 in the bioactivation of 3-methylindole in mouse olfactory mucosa and lung: studies using Cyp2a5-null and Cyp2f2-null mouse models. *Drug Metab.* 40: 642–647.

References

Zhou, X., Wei, Y. A., Xie, F., et al. (2011). A novel defensive mechanism against acetaminophen toxicity in the mouse lateral nasal gland: role of Cyp2a5-mediated regulation of testosterone homeostasis and salivary androgen-binding protein expression. *Mol. Pharmacol.* 79: 710–723.

Zhuo, X., Gu, J., Zhang, Q-Y., et al. (1999). Biotransformation of coumarin by rodent and human cytochrome P450: metabolic basis of tissue-selective toxicity in the olfactory mucosa of rats and mice. *J. Pharmacol. Expt. Ther.* 288: 463–471.

Zhou, X., Zhang, X. L., Weng, Y., et al. (2009). High abundance of testosterone and salivary androgen-binding protein in the lateral nasal gland of male mice. *J. Steroid Biochem.* 117: 81–86.

Zhou, X., Zhuo, X. L., Xie, F., et al. (2010). Role of CYP2A5 in the Clearance of Nicotine and Cotinine: Insights from Studies on a Cyp2a5-null Mouse Model. *J. Pharmacol. Exp. Ther.* 332: 578–587.

Zhuo, X. L., Gu, J., Behr, M. J., et al. (2004). Targeted disruption of the olfactory mucosa-specific Cyp2g1 gene: Impact on acetaminophen toxicity in the lateral nasal gland, and tissue-selective effects on Cyp2a5 expression. *J. Pharmacol. Exp. Ther.* 308: 719–728.

Zhuo, X. L., Schwob, J. E., Swiatek, P. J., and Ding, X. X. (2001). Mouse Cyp2g1 gene: Promoter structure and tissue-specific expression of a Cyp2g1-LacZ fusion gene in transgenic mice. *Arch. Biochem. Biophys.* 391: 127–136.

Zielinski, B. S., Getchell, M. L., and Getchell, T. V. (1988). Ultrastructural characteristics of sustentacular cells in control and odorant-treated olfactory mucosa of the salamander. *Anal. Rec.* 221: 769–779.

Zupko, K., Poria, Y., Lancet, D. (1991). Immunolocalization of cytochrome P-450olf1 and P-450olf2 in rat olfactory mucosa. *Eur J Biochem* 196: 51–58.

Zhuo, X. L., Zhao, W. P., Zheng, J. N., et al. (2009). Bioactivation of coumarin in rat olfactory mucosal microsomes: Detection of protein covalent binding and identification of reactive intermediates through analysis of glutathione adducts. *Chem-Biol. Interact.* 181: 227–235.

Chapter 4

Development, Morphology, and Functional Anatomy of the Olfactory Epithelium

JOHN C. DENNIS, SHELLY AONO, VITALY J. VODYANOY, and EDWARD E. MORRISON

4.1 INTRODUCTION

The main olfactory epithelium (MOE) has provided an excellent model for studying several aspects of neural development for more than a century (Bedford, 1904; Ramon y Cajal, 1909). Examination of, and experimentation on, the MOE has produced important insights into adult neurogenesis, remodeling of central synaptic circuits and neuroendocrine functions, to mention a few. Key advances in understanding the MOE were made employing classical histological techniques that described the structural MOE components and a neuroblast population that uniquely separates the MOE from other sensory epithelia (Breipohl et al., 1974; Cuschieru and Bannister, 1975; Graziadei and Monti-Graziadei, 1978; Morrison and Costanzo, 1989). Powerful modern techniques like gene identification and manipulation have allowed further definition of the MOE (Buiakova et al., 1996; Youngentob and Margolis, 1999; Lee et al., 2011).

The MOE has been extensively examined in a number of vertebrates including humans (Costanzo and Morrison, 1989; Engstrom and Bloom, 1953; Graziadei, 1973a; Menco, 1977; Moran et al., 1982; Morrison and Costanzo, 1990) and, in all tetrapod species, the sensory epithelium consists primarily of three cell types: columnar, nonsensory supporting cells, bipolar olfactory receptor neurons (ORNs), and a neuroblast-containing population of basal cells. The underlying lamina propria contains vascular structures, olfactory axon fascicles, and secretory Bowman's glands that are present in all terrestrial vertebrates (Figure 4.1). This chapter reviews the development and cytoarchitecture of the MOE, a recently discovered defense mechanism that protects the olfactory sensory apparatus and the nasal cavity in general, and a novel functional property of ORN signal transduction.

4.1.1 Development of the Olfactory Epithelium

The olfactory placodes develop in the rostrolateral regions of the embryonic head (Cuschieru and Bannister, 1975; Klein and Graziadei, 1983). The placodal epithelium and adjacent mesenchyme will give rise to the nasal cavity. These ectodermal regions proliferate, thicken, and then form the olfactory pits, which are invaginations that extend into the underlying mesenchyme (Figure 4.2). The nasal pit epithelial organization shows a remarkable similarity to the developing neural tube, namely a proliferating ependymal/ventricular layer and a cellular mantle layer (Figures 4.3 and 4.4). The subsequent development of the nasal cavity involves a concert of molecular signals involving, among others, retinoic acid (RA), fibroblast growth factor, and bone morphogenetic factors from the developing MOE and surrounding mesenchyme (see Beites et al., 2005 for review). Studies have shown that RA signaling

---

Handbook of Olfaction and Gustation, Third Edition. Edited by Richard L. Doty.
© 2015 Richard L. Doty. Published 2015 by John Wiley & Sons, Inc.

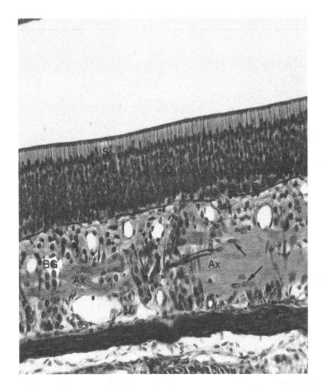

Figure 4.1 Silver stain of rodent olfactory epithelium illustrates cells of epithelium and underlying structures of the lamina propria. Supporting cell (S); Olfactory receptor neurons (O); Basal cells (B); Olfactory axon fascicles (Ax); Bowman glands (BG); Olfactory ensheathing cells (arrows). (*See plate section for color version.*)

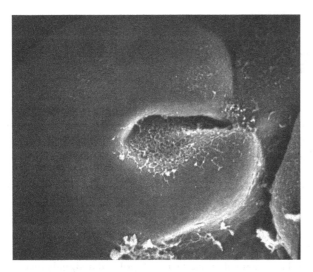

Figure 4.2 Scanning electron micrograph of the head region rodent embryo. The olfactory pit forms a groove bounded by lateral and medial nasal processes.

between placode ectoderm and mesenchyme is critical to driving molecular and cellular diversity within the MOE (LaMantia et al., 2000; Rawson and LaManita, 2006). The general process of molecular signaling between

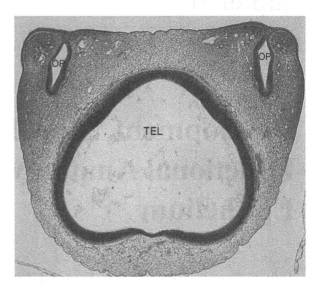

Figure 4.3 Horizontal section through telencephalon (TEL) and olfactory pit (OP) region of pig embryo. Olfactory pits form a groove as it extends into underlying mesenchyme.

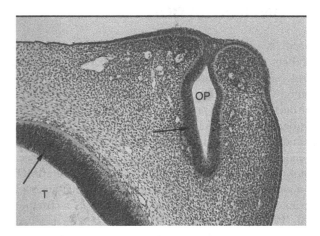

Figure 4.4 Higher magnification of olfactory pit groove bounded by lateral and medial nasal processes. The epithelium of the developing telencephalon and nasal epithelium are similar in structure with a proliferative region and a cellular mantle (arrows).

mesenchyme and epithelia is not specific to the MOE but is common in the development of complex systems such as limb bud, heart; brachial arches and lower motor neuron development (see Balmer and Lamantia, 2005 for review).

MOE development gives rise to a pseudostratified neuroepithelium composed of several cell types. Mature and immature ORN cell bodies typically occupy the lower two thirds of the neuroepithelium, whereas cell bodies of the non-neuronal supporting cells are located in the upper third. There are two basal cell populations: horizontal basal cells (HBC) that reside along the basal lamina and globose basal cells (GBC) that are located just above the HBC layer.

4.1.2 Olfactory Receptor Neuron

The ORN is a true bipolar neuron. A single dendritic process extends from the 5–7 μm cell bodies to the epithelial surface. The dendrite in most vertebrate species terminates in a knob-like swelling with 5–30 sensory cilia, which originate from basal bodies within the dendritic knob. In some fish, reptile, and bird species the dendrite is covered with microvilli similar to those of vomeronasal neurons. Proximally, olfactory cilia resemble typical motile nonsensory cilia in that microtubules are present in a 9 + 2 arrangement. As they extend to form a dense blanket on the epithelial surface, the cilia taper and may contain only a few microtubules (Figure 4.5) (Graziadei, 1973b; Menco, 1977; Morrison and Costanzo, 1990). Although motile respiratory cilia and olfactory sensory cilia have a similar morphology, olfactory cilia typically extend relatively long distances from their dendrites, forming a dense network that greatly increases their sensory surface area (Figure 4.6). Individual transmembrane odorant receptors (OR) are located on the olfactory cilia. The classic work of Buck and Axel (1991) led to the understanding that each olfactory receptor neuron expresses only one of ~1000 olfactory receptors on the ciliary processes (Mombaerts, 2004). The OR is a seven transmembrane G-protein coupled receptor that, following the binding of an odorant, activates the signal cascade that results in olfactory receptor neuron depolarization (see below).

4.1.3 Microvillar Cells

Microvillar cells have neuron-like morphology and are observed in the apical neuroepithelium. These cells have a thin unmyelinated axon arising from the basal cell body which courses through the basal lamina (Figure 4.7). These cells have been observed in humans,

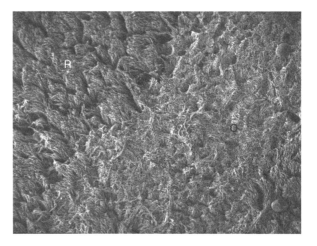

Figure 4.6 Surface view of human nasal mucosa illustrating the morphological differences between the respiratory (R) and sensory olfactory cilia (O). The olfactory cilia extend over the mucosa surface forming a dense cilia network.

canines, rodents, and turtles. The function and cell lineage of the microvillar cell is unknown (Okano et al., 1967; Graziadei, 1973a; Moran et al., 1982; Morrison and Costanzo, 1990).

4.1.4 Basal Cells

The structure of the MOE has been compared to that of the developing neural tube of the embryonic CNS but the MOE only gives rise to one neuron type. Like the neural tube, the dividing MOE neuroblast stem cell layer is located basally. Olfactory basal cells are located in the lower epithelial compartment near the basal lamina (Figure 4.1). Two types of basal cells have been identified in vertebrates: horizontal and globose. Horizontal basal

Figure 4.5 (A) Human olfactory epithelium illustrating a dendritic knob located between microvillar supporting cells (M). Olfactory dendrites can have 5–30 tapering ciliary processes (arrow). (B) Transmission electron micrograph of rodent olfactory dendritic knob. Gradually tapering sensory cilia (arrows) arise from basal bodies in the cytoplasm. Supporting cell microvilli (M).

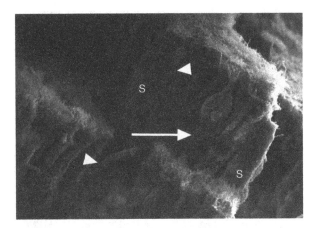

Figure 4.7 Fractured canine olfactory epithelium illustrates a microvillar cell near the apical surface. An axon process extends from the basal region of the cell body extending towards the basal lamina (arrow). Olfactory dendrites (arrow heads). Supporting cells (S).

cells (HBCs) contain tonofilaments and are immunoreactive to anti-cytokeratin antisera. Globose basal cells typically reside above the HBCs and are cytokeratin negative. Early studies by Graziadei and Monti-Graziadei and others showed that the basal cells contain a neuroblast population that constitutively divides to replace olfactory receptor neurons normally and following injury (Figure 4.8) (Graziadei and Monti-Graziadei, 1978; Graziadei and Monti-Graziadei, 1979; Costanzo, 1985; Huard et al., 1998; Schwob, 2002). Studies from several groups showed that the HBC is a relatively quiescent population that lacks neuronal markers and divides slowly, similar to the ependymal cells lining the post natal ventricles (Holbrook et al., 1995). The GBCs differ morphologically and biochemically: they are round, do not express cytokeratin and give rise to a line of MASH-1 positive, transit amplifying cells. These in turn produce a second line of intermediate precursors, which subsequently divide and produce terminally differentiated ORNs (Caggiano et al., 1994; Beites et al., 2005; Nicolay et al., 2006). Thus, the GBCs are the proliferative cell population within the MOE and the HBCs may be regarded as the tissue stem cells (Mackay-Sim, 2010; Suzuki et al., 2013). The HBC lineage is thought to be of placode origin. Recently, it was demonstrated that the MOE contains neural crest derived cells (NCDC) that comingle with the placode and olfactory pit (Forni et al., 2011). Using a transgenic mouse strain that expresses EGFP-tagged NCDCs, Suzuki et al. (2013) observed labeled HBCs in adult MOE that were of both NCDC and placode origin. These observations, along with NCDCs recently discovered in post-natal inner ear (Freyer et al., 2011) and eye (Yoshida et al., 2006), suggest that NCDCs may have a common "back up" function as tissue stem cells in sensory organs. The findings that the MOE contains both placode and pluripotent neural crest cells is exciting and opens research avenues via which to examine NCDC potential in regenerative medicine.

4.1.5 Supporting Cells

The supporting cell is a columnar cell that spans the epithelium. Supporting cell nuclei form the most superficial row of epithelial cell nuclei. Their apical surfaces project short microvilli into the overlying mucus (see Figure 4.5). Throughout their length, supporting cells are intimately associated with the olfactory neurons via intercellular bridges/appendages connecting the two cell types. In the basal region the supporting cell has a foot process that anchors it to the basal lamina (Figure 4.9) (Rafols and Getchell, 1983; Costanzo and Morrison, 1989; Morrison and Costanzo, 1989). Supporting cells express enzymes involved in metabolism of foreign compounds, suggesting a glia-like protective role of the neuronal environment. Recent studies demonstrated that defensins, a family of antiviral and antibacterial proteins, are expressed in the MOE (see below). Although additional studies are necessary to identify the cellular origin of defensins, it is tempting to speculate that supporting cells secrete these peptides in the MOE. Supporting cells are also phagocytic and, like CNS glia, provide structural support. The lineage of supporting cells is still unclear. Transplantation of GBCs into an experimentally degenerated MOE in mice gave rise to both olfactory neurons and supporting cells. However, some supporting cells labeled as neural crest-derived have been observed in the MOE, suggesting both placodal and neural crest origin (Chen et al., 2004; Forni et al., 2011).

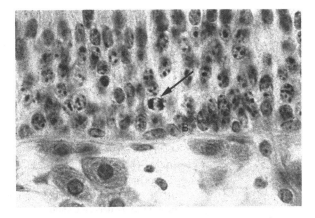

Figure 4.8 Light microscopy of olfactory epithelium 5 days following injury. Mitotic activity is located in the basal cell (globose) region of the neuroepithelium (arrow).

4.1 Introduction

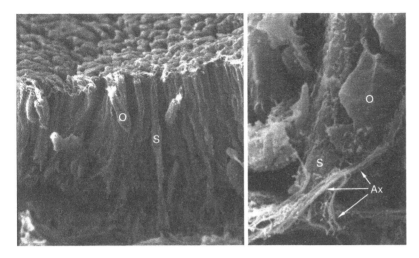

Figure 4.9 Left: Scanning electron micrograph of rodent olfactory epithelium illustrates the columnar shaped supporting cells (S) spanning the epithelium. Olfactory receptor neuron (O). Right: Higher magnification of basal region of the olfactory epithelium. Supporting cell (S) foot process attaches to basal lamina. Olfactory receptor neuron (O); Small olfactory axon fascicles (Ax).

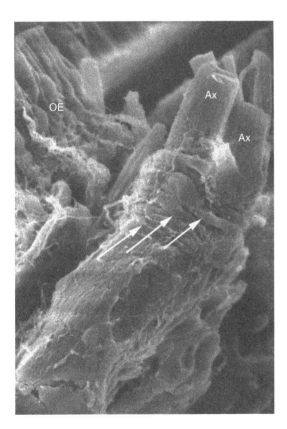

Figure 4.10 Scanning electron micrograph of the lamina propria region. Olfactory axons (Ax) are surrounded by the cytoplasmic processes of the (arrows). Olfactory epithelium (OE).

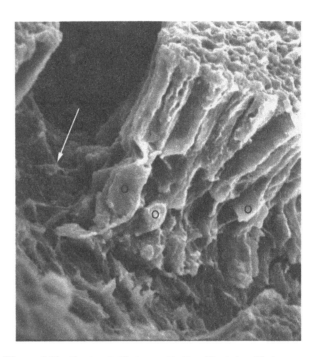

Figure 4.11 Fractured olfactory epithelium illustrates olfactory neurons (O) and basal lamina. Growing olfactory axon with a growth cone on the leading edge can be observed (arrow). From Morrison and Costanzo, 1989.

4.1.6 Olfactory Ensheathing Cells

The olfactory ensheathing cell (OEC) is a glia type cell associated with axons (Figure 4.10). Axon fascicles are surrounded by OEC cytoplasmic processes, which form a mesaxon for each fascicle (Figure 4.11). OECs express GFAP and S-100, cytoplasmic markers for CNS glia (Au and Roskams, 2003). For many years the origin of OECs and Schwann cells has puzzled investigators (Chuah and Au, 1991). Indirect evidence suggested similarities between these two cell types based on morphology and molecular markers. The recent studies by two separate groups (Forni et al., 2011; Katoh et al., 2011) showed that pluripotent neural crest cells are present in the olfactory placode and give rise to OECs. The neural crest also gives rise to Schwann cells. These results help explain the high level of biological and molecular similarity between OECs and Schwann cells.

OECs have been shown to promote axon growth and recovery in several nerve lesion models, suggesting that their presence is responsible for the permissive and dramatic olfactory axogenesis (Mackay-Sim and St. John, 2011). A recent clinical study of canine spinal cord injury transplants showed unequivocally that transplanted OECs enhanced locomotor improvement (Granger et al., 2012). These exciting results suggest that OECs can mediate substantial change in function in local intraspinal tracts.

4.1.7 Migrating Cells

As mentioned above, the olfactory placode is capable of generating a number of different cell types. There are also indications that migratory cell types present in the developing MOE are directed toward the developing telencephalon. Early observations by Bedford (1904) showed cells leaving the placode and forming cords within the adjacent mesenchyme. Later investigators described this mass of cells and axons as the "migrating mass" (Valverde et al., 1992). Some of these migrating cells are related to olfactory functions. Among these are OECs that play an important role in guidance of the migrating mass (Doucette, 1990; Chuah and West, 2002). Another cell type not directly related to olfaction but which migrates along MOE/VNO axons is the gonadotropin releasing hormone (GnRH1) cell. These cells are dependent on the guidance cues of developing peripherin-positive axons to transverse several complex terrains to ultimately reach the developing diencephalon. Gonadotropin releasing hormone neurons are a population of cells with a key role in vertebrate sexual development and behavior (Schwanzel-Fukuda and Pfaff, 1989; Wray et al., 1989).

Although these migrating cells are observed leaving the placodal epithelium, their origin has been unclear (Mendoza et al., 1982). Studies using a transgenic mouse model system examined the possible contribution of cranial neural crest derived cells. The olfactory placode and cranial neural crest are closely positioned to each other and mixing of the tissues has long been suggested (Couley and Le Douarin, 1985). Forni et al. (2011) used two complementary mouse systems to label NC and placode ectoderm contributions. They showed that pluripotent NC cells mingled with placode ectoderm and gave rise to olfactory ensheathing cells and GnRH1 expressing neurons.

The developing vertebrate head relies a great deal on the interactions between the two ectodermal derivatives NC and cranial placodes. Previous studies have proven NC origin and cellular contributions to ectodermally derived cranial placodes (D'Amico-Martel and Noden, 1983). These findings not only offer new insights into olfactory development but also a rich field of investigation into GnRH1 neurons and the olfactory system, and could provide insight into NC defects and anosmia associated with normal and pathological development of the olfactory system.

4.1.8 Olfactory Nerve and Olfactory Bulb Formation

An axon process arises from the basal aspect of the olfactory receptor neuron. At its tip is a growth cone. The growing axon and growth cone must navigate through the mesenchyme to reach the telencephalon (Figure 4.12). The growth cone guides the axon by providing the motile force for their extension and by acting as exquisite sensors that detect and subsequently respond to a variety of environmental guidance cues. Growth cone associated cell surface receptors interpret these signals as attractive or repulsive forces (Mueller, 1999). The definitive cues that guide olfactory axon development and growth towards the telencephalon are not fully understood. No doubt guidance cues and signpost (e.g., semaphorins, slit, ephrin, and netrins) within the mesenchyme and possible chemotaxic cues from the CNS direct and attract axon growth cones. One unique feature of the olfactory growth cone is the expression of a seven transmembrane olfactory receptor that plays a critical role for the axon to target the olfactory bulb (Feinstein and Mombaerts, 2004).

Pearson (1941), in his classic study on the development of the human olfactory nerve, noted that "the development of the olfactory bulb reminds one of Kipling's story *The Elephant's Child*." Pearson was not suggesting that the axons were "pulling" part of the telencephalon to form the olfactory bulb, but that the olfactory axons influence bulb development. When the axons reach the telencephalon there is no evidence of olfactory bulb formation. After the axons grow into the brain, a protrusion from the telencephalon begins to develop, forming the rudimentary olfactory bulb. The olfactory axons increase in number, forming the presumptive olfactory nerve layer. They enter

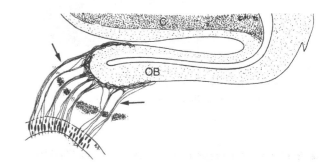

Figure 4.12 Diagram of olfactory axons growing from the nasal cavity reaching the developing olfactory bulb (OB). At the time olfactory axons reach the neural tube the olfactory bulb arises out of the ventral aspect of the telencephalon.

4.1 Introduction

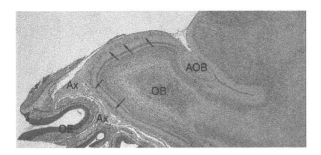

Figure 4.13 Low power light micrograph of rodent olfactory primary neuraxis. Olfactory epithelium (OE). Olfactory axons (Ax). Olfactory bulb (OB). Accessory olfactory bulb (AOB). Glomeruli (arrows). (*See plate section for color version.*)

the developing olfactory bulb, ultimately forming the glomerular layer (Figures 4.12 and 4.13). A small population of axons penetrates the ventricular zone where they are thought to stimulate genesis of the olfactory bulb (Gong and Shiply, 1995). However, recent studies in mutant mice that lack an olfactory epithelium revealed that an olfactory bulb-like structure develops in the rostral telencephalon. The exact nature of the signal(s) that induce olfactory bulb formation are not fully understood and warrant further studies.

4.1.9 Olfactory Marker Protein

Olfactory marker protein (OMP) is a widely recognized molecular marker of mature olfactory neurons (Margolis, 1972). It is the first powerful marker for fully differentiated olfactory receptor neurons and provides a key tool to investigate the dynamic relations between peripheral sensory neurons and central connections with the olfactory bulb. Olfactory marker protein is a low molecular weight soluble protein found almost exclusively in the mature olfactory neuron cell body, dendrites, and axon processes and is absent from the other cell components of the olfactory epithelium (Figure 4.14). OMP expression is developmentally regulated. Beginning at embryonic day 14 its expression reaches the adult level at postnatal day 30 (P30) and coincides with synapse formation on second-order olfactory bulb neurons (Farbman and Margolis, 1980; Graziadei et al., 1980).

Although OMP is present in a wide range of vertebrates and is highly conserved, its full function is still unclear. Genetic manipulations and deletion of OMP results in slower odorant response kinetics in EOG, single cell recordings, and calcium imaging (Lee et al., 2011; Kwon et al., 2009). Behaviorally, OMP $^{-/-}$ mice exhibit reduced odorant sensitivity and altered odor discrimination (Youngentob and Margolis, 1999; Youngentob et al., 2001). Developmentally, patch clamp recordings from the first postnatal month show functional changes with P30

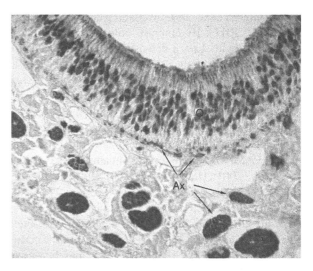

Figure 4.14 Olfactory marker protein immunohistochemistry of the rodent MOE. Mature olfactory receptor neurons (O) and axon fascicles of varying size (arrows) are OMP+. (*See plate section for color version.*)

showing greater odor sensitivity, indicating OMP may play a role in ORN maturity. Taken together, these observations illustrate that OMP exerts its molecular and cellular actions at multiple sites (Buiakova et al., 1996; Lee et al., 2011).

4.1.10 Innate Immunity in the Main Olfactory Epithelium

Epithelia and leukocytes contribute to a physical and biochemical barrier called innate immunity in mammals, the first line of defense against pathogenic insult. Defensins are a family of small cationic peptides and have a wide range of antibacterial, antiviral, antifungal, and antiparasitic activity, being part of the biochemical barrier. Originally identified in the 1980s as part of the antibiotic arsenal of neutrophils, defensin and defensin-like peptides have since been found throughout the Metazoa, as well as in plants. The amino acid sequence among the 40 human defensins so far identified varies greatly but it is the highly conserved tertiary structure upon which the classification of these peptides is based. In the amino acid sequence, six cysteine residues are conserved and the pairing of these, which forms three disulfide bridges, is the basis for identifying α and β subfamilies (reviewed in Ganz, 2003; Lehrer and Lu, 2012; Weinberg, 2012; Carvalho and Gomes, 2009; Gachomo et al., 2012). Marsupial and placental mammal expression of both α- and β-defensins is highly compartmentalized. For example, Paneth cells express defensins that regulate intestinal flora (Cunliffe, 2003; Wehkamp et al., 2006). Down regulation of paneth cell α-defensin expression and secretion is associated with inflammatory bowel disease (Crohn's disease) in the ileum (Fahlgren et al., 2003; Wehkamp et al., 2005; Gersemann et al., 2011). By contrast, colonic Crohn's disease is correlated with low β-defensins 2 and 3 (Wehkamp et al., 2002; Fellermann et al., 2006; Gersemann et al., 2008). More β-defensins than α-defensins have been identified and β-defensins are thought to be evolutionarily more ancient. Primarily secreted by epithelial cells, some β-defensins are constitutively expressed, while others are induced by cytokines, pathogen associated molecular patterns (PAMPs), and pro- or anti-inflammatory cytokines (Hao et al., 2001; Harder et al., 2001; Joly et al., 2005; Batoni et al., 2006).

Defensins exert their antibiotic activity through disruption of the target's lipid bilayer and binding glycoproteins in the case of anti-viral activity (Kagan et al., 1990; Klotman and Chang 2006). Secondary effects may be derived from interactions with intracellular components such as liposomes, RNA, and DNA (Zasloff, 2002; Sahl et al., 2005) and chemoattractant properties, which recruit cells of the adaptive immunity system (Taylor et al., 2008). Antibiotic potency varies between defensins and depends upon environmental factors such as oxygen levels, salt concentrations, serum proteins, and proteases (Harder et al., 2001; Maisetta et al., 2008; Schroeder et al., 2011; Goldman et al., 1997). Unlike other innate immunity genes, some defensins show high copy number variation which, coupled with tissue specific expression and large variations in interindividual expression, may contribute to disease susceptibility and tissue specific pathogen resistance (Linzmeier and Ganz 2005; Vankeerberghen et al., 2005; Linzmeier and Ganz 2006; Wehkamp et al., 2006; Kisich et al., 2007; Milanese et al., 2009; Zanger et al., 2011).

The MOE is an avenue via which pathogens enter the CNS and defensins are variably expressed in all parts of the nasal cavity (Figures 4.15A and B) (Zanusso et al., 2003; Mori et al., 2005; Sjölinder and Jonsson 2010). Human β-defensins (hBD) 1, 2, and 3 all show variable expression in rostral nasal tissues and secretions (Cole et al., 1999; Laudien et al., 2011). Of these three defensins, hBD3 is thought to be the most potent against a wide variety of pathogens and works synergistically with other antibiotic agents (Lee et al., 2002; Matthews et al., 1999). hBD3 is critical for oral keratinocyte-mediated killing of *Staphylococcus aureus*, a common opportunistic pathogen in both the oral and nasal cavities (Maisetta et al., 2003; Kisich et al., 2007). In the nasal cavity, hBD3 is mainly expressed in the *vestibulum nasi*, which is the colonization site of *S. aureus* (Laudien et al., 2011). In vivo, reduced cutaneous hBD3 expression is coincident with persistently elevated carriage of *S. aureus* in the nares, a risk factor for hospital-acquired, as well as recurrent, infections (Huang and Platt, 2003; Zanger et al., 2010; Zanger et al., 2011). Canine and rat hBD3 orthologues show similar expression

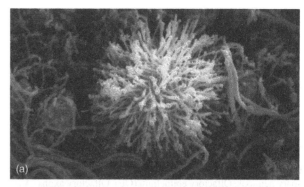
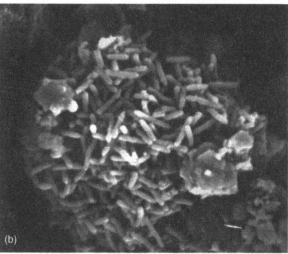

Figure 4.15 Foreign particles and pathogens can reach the olfactory region and cause irritation/infection causing an inflammatory responses. (a) Pollen grain embedded on the surface of the olfactory mucosa. (b) Bacteria (*Edswarsiella ictaluri*) infection of olfactory mucosa.

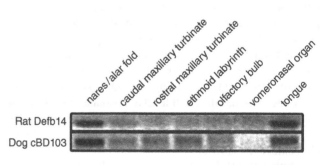

Figure 4.16 RNA expression of rat β-defensin 14 and canine β-defensin 103 (human β-defensin 3 paralogs) in oronasal tissue.

patterns (Figure 4.16), with the highest hBD3 mRNA levels stratified squamous epithelia. This suggests that hBD3 in nasal secretions may originate from the rostral tissues in the nasal cavity (Figure 4.17), offering limited protection to caudal tissues, including the olfactory neurons.

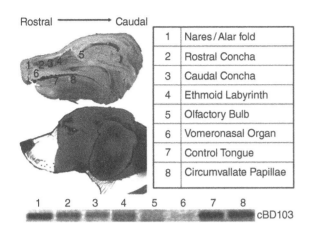

Figure 4.17 β-defensin 103 (cBD103) RNA expression in canine olfactory bulb (brain) and oronasal tissues and their anatomic locations. cBD103 expression is representative of 15 dogs. However, some interindividual variation was visible, especially in the rostral and caudal concha samples.

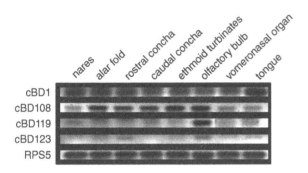

Figure 4.18 Canine β-defensin 1, 108, 119 and 123 RNA expression in the oronasal cavity and olfactory bulb (brain) of a single dog. Antibiotic activity and function in the innate immune system of the nasal cavity is not yet known. RPS5 is a reference gene.

Canine β-defensin (cBD) 1, 108, 119, and 123 transcripts were detected in nasal cavity respiratory epithelium, the MOE, and the olfactory bulb (Figure 4.18), indicating they may protect the caudal portions of the nasal cavity. Previous work has shown these cBDs are variably expressed in the trachea, lung, and skin (Erles and Brownlie, 2010; Leonard et al., 2012) but most of these β-defensins are not well characterized, so their role in the innate immunity of the nasal cavity and central nervous system remains to be elucidated.

4.1.11 Zinc Nanoparticles and Olfactory Receptor Neuron Signaling

Zinc is essential for normal physiological function and, in humans, is the second most abundant trace element (see Takeda, 2001 and references therein). The CNS contains zinc and its local concentrations and subcellular segregation are tightly regulated (Howell and Frederickson, 1990; Christensen et al., 1992; Jo et al., 2000a; Danscher et al., 1985). Throughout the CNS, zinc ions are largely sequestered in synaptic vesicles and released into the synaptic cleft during signal transmission. These zinc-containing neurons are called zinc-enriched neurons (ZENs) and, in the forebrain, constitute a subpopulation of glutamatergic neurons. Changes in zinc concentration are associated with pathologies that include Alzheimer's disease, amyotrophic lateral sclerosis, depression, and epilepsy (reviewed in Bitanihirwe and Cunningham, 2009; Takeda and Tamano, 2009). In particular, perturbations in zinc concentrations caused by various agencies result in a decline of olfactory acuity or inappropriate responses to olfactory cues (Mackay-Sim and Dreosti, 1989; Takeda et al., 1999). The OB contains one of the largest populations of ZEN terminals in the brain and these are concentrated in the granule cell and glomerular layers (see Figure 4.13) (Jo et al., 2000b; Friedman and Price, 1984). ZENs synapsing on granule cells are centrifugal axons, just over half of which come from the anterior olfactory nucleus (Carson, 1984). The zinc positive axons in the glomerular layer are the terminals of main olfactory sensory neurons (Jo et al., 2000b). In both regions, zinc is a modulator of OB neuron activity (Ferraris et al., 1997; Trombley and Shepherd, 1996).

In the context of zinc's functions and spatial sequestration in the CNS generally, and the olfactory system specifically, Vodyanoy and coworkers began experiments with metallic (nonionic) zinc nanoparticles. Samoylov et al. (2005) isolated metallic nanoparticles from the blood of several vertebrate species. Subsequent analysis showed that zinc constituted a subset of these serum borne metals leading Vodyanoy (2010) to hypothesize that, given zinc's significant functions in the olfactory system, metallic zinc nanoparticles would affect signal transduction across the main olfactory sensory neuron membrane.

Odorants containing zinc clusters in nanomolar concentrations produced increased responses in both

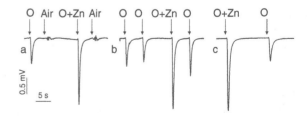

Figure 4.19 EOG recordings from rat olfactory epithelium. Responses were induced by a mixture of ordorants (O) each at 16 mM; air; or odorants with zinc nanoparticles at 10 pM (O+Zn). Representative traces were obtained from one tissue sample with 10 contacts and 39 traces.

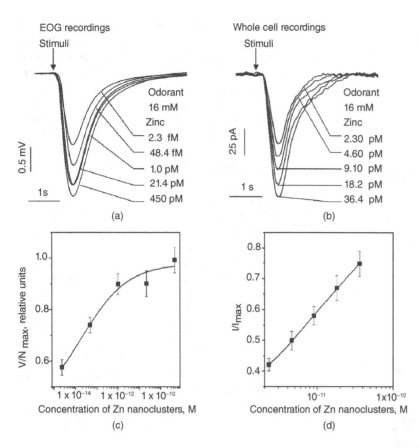

Figure 4.20 EOG and single cells recordings from rat MOE exposed to odorant + zinc nanoparticles at constant odorant concentration. The stimulus was a 0.25 second (s) air pulse or odorant + zinc nanocluster. (a) Representative EOG traces in 4 tissues, 10 contacts, and 34 EOG recordings. (b) Whole cell current traces from olfactory neurons with holding potentials at −70 mV. Downward direction represents inward current. Representative traces were obtained from 23 recordings and a total of 198 traces. (c) Normalized peak EOG voltage (V/V_{max}) plot of odorant + zinc particles and zinc nanocluster concentration. (d) Normalized peak negative current (I/I_{max}) plot of odorants + zinc particles and zinc nanocluster concentration.

EOG (Figure 4.19) and single-cell recordings made from freshly dissected MOE (Figure 4.20) (Moore et al., 2012; Viswaprakash et al., 2009) and in EOG recordings made from tissue-culture preparations (Viswaprakash et al., 2010). The increased responses were reversible (Figure 4.19) and dose dependent (Figure 4.20). Zinc particles alone elicited no response.

The same experiments were repeated with gold, silver, or copper nanoparticles at 5 nM in 16 mM odorant mixtures (Figure 4.21). Both gold and silver elevated EOG responses, although the relative amplitudes were smaller and their effective durations were attenuated relative to zinc. The metals' effects peaked at about 20 minutes and then began to decline, whereas zinc's effects continued to increase EOG amplitude. Silver nanoparticles in solution also increase EOG responses in a fish species at low concentration but at higher concentrations silver nanoparticles reduced EOG response amplitudes (Bilberg et al., 2011).

Tissues exposed to 5 nM copper nanoparticles with odorants responded at control levels to about 40 minutes. At that time, activity evoked by odorants alone began to decline but the tissue's response in the presence of copper nanoparticles continued at a constant level. This observation indicates that copper can modulate sensory neuron activity. However, we cannot conclude that copper is a functioning modulator in the odorant/receptor signal transduction complex in vivo.

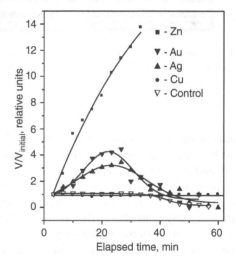

Figure 4.21 Relative EOG amplitudes showing metal effects on responses to a 16 mM odorant mixture delivered in 0.25 s puffs. The recording pipette contained zinc, gold, silver or copper nanoclusters at 5 nM or no metal as a control. The first trace for each metal was made immediately following contact. Subsequent traces were recorded at 200 s intervals. Representative data were recorded as follows. Zinc: 3 tissue samples, 4 contacts, 60 recordings, 450 traces; Gold: 1 tissue sample, 2 contacts, 28 recordings, 221 traces; silver: 2 tissues, 4 contacts, 58 recordings, 460 traces; copper: 1 tissue, 2 contacts, 23 recordings, 172 traces; control: 2 tissues, 37 recordings, 340 traces.

4.1 Introduction

The supposition that copper is active at the MOE apical surface is not baseless since copper is involved in modulating OB neuron signaling activity (Trombley and Shepherd, 1996) and zinc and copper would have modulating functions throughout the olfactory system's primary neuraxis. It must be noted that the observations enumerated above concern the effects of metallic nanoparticles. Ionic forms of these metals inhibit olfactory responses: zinc ions in rats (Viswaprakash et al., 2009); silver ions in a carp species (Bilberg et al., 2011); and copper ions in a salmon species (Kennedy et al., 2012). This modulating behavior of copper nanoparticles at the MOE apical surface in vitro suggests that zinc and copper have functions throughout the primary neuraxis of the olfactory system and, in the MOE, mirrors the two metals' modulating effects in the OB reported by Trombley and Shepherd (1996).

Using a straightforward theoretical model, the binding stoichiometry of odorant and zinc displayed in single cell recordings can be calculated (Vodyanoy, 2010). Patch clamp data derived from a number of experiments yields a Hill coefficient suggesting that two odorant molecules are needed to activate one receptor/G_{olf} complex. The Hill coefficient for zinc nanoclusters was approximately 0.5, which means that one zinc particle binds two receptors in a given receptor activation event. These values are consonant with theoretical and empirical observations reported by other investigators (Giraldo, 2008; Connors, 1987; reviewed in Kleene, 2008). Taken together, the best explanation for these data is that signal transduction across the sensory neuron membrane results from the dimerization of receptor molecules. The entire transduction complex would be composed of two receptor molecules, a zinc particle, and one G_{olf} heterotrimer.

These findings corroborate reports of G protein-coupled receptor (GPCR) dimerization in general (reviewed by Milligan, 2004) and, more specifically, sensory system GPCR dimerization like that of rhodopsin in the vertebrate retina (Fotiadis et al., 2003). An early study of MOE proteins characterized a 19 kD band by sodium dodecyl sulfide polyacrylamide gel electrophoresis. Gel filtration produced a native 38 kD homodimer and binding kinetics with the pepper-derived odorant 2-isobutyl-3-[^{3}H]methoxypyrazine. Such observations suggested that the polypeptide was an odorant receptor with two odorant binding sites when dimerized (Pevsner et al., 1985). A second study by Dean and co-workers (2001) used a second method, the evolutionary trace, to analyze and model the pertinent helices in a number of GPCRs, including olfactory receptor homodimerization. More recently, insect receptors were shown to form heterodimers (Sato et al., 2008; Neuhaus et al., 2005). Finally, Wade et al. (2011) was able to establish stable heterologous human QR1740 receptor G-protein complex expression in the yeast *Saccharomyces cerevisiae*. Using bioluminescence resonance energy transfer together with immunoprecipitation and immunoblotting to analyze membrane fractions, Wade and co-workers demonstrated that, in the heterologous system, functional olfactory receptor units were homodimers. But contrary to Vodyanoy's (2010) calculated stoichiometry, as well as previous analyses, Wade et al. (2011) report a single odorant binding event necessary to precipitate signal transduction.

In vitro EOG signal amplification and single cell recordings facilitated by zinc nanoparticles is documented. The question then becomes where in the signal transduction cascade zinc has its effect? In general, odorant binding causes adenylate cyclase activation and subsequent synthesis of the second messenger cAMP (Figure 4.22). In turn, cAMP synthesis leads to the opening of nucleotide and calcium gated channels, which activate the sensory neuron. The signal is terminated by the phosphodiesterase catalyzed degradation of cAMP (reviewed by Kleen, 2008).

EOG responses are elicited without odorant from MOE tissue following application of the phosphodiesterase inhibitor 3-isobutyl-1-methylxanthine (IBMX). Addition of zinc nanoparticles does not affect IBMX-induced EOG responses indicating that zinc does not act on the ion channels (Figure 4.23) (Moore et al., 2012).

Since metallic zinc alone does not evoke a response, the effect is not at the level of cAMP production or activation

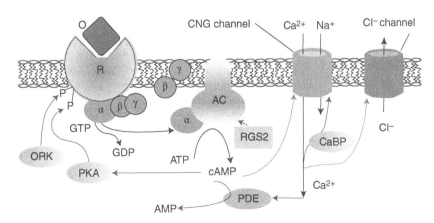

Figure 4.22 In a generalized schematic of signal transduction, an odorant (O) binds a receptor molecule and the G-alpha (α) subunit activates adenylate cyclase (AC). AC activation produces cAMP the production of which opens the cyclic nucleotide gated calcium channels.

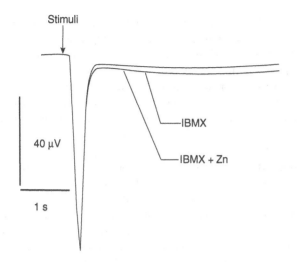

Figure 4.23 EOG response induced by a pulse of 400 μM IBMX with and without 4 nM zinc nanoparticles. The figure shows representative traces from 3 tissues and 25 recordings.

of adenylate cyclase. Taken together, these data indicate that zinc's site of action is the olfactory receptor/G-protein complex. Although zinc ions inhibited MOE response to odorants in the experiments outlined above, these observations do not contradict Turin's proposed model (1996) of odorant induced signal transduction at the MOE surface. One possibility is that the nanoparticles bind at the apical surface and release zinc ions (Vodyanoy, 2010). These ions would be the source of electrons that are involved in the tunneling proposed by Turin.

ACKNOWLEDGMENTS

The authors thank Dr. Jishu Shi for his help editing portions of this manuscript and Mr. Mark Ambery for the beagle profile illustration. The work described here was supported in part by grants DHS-01-G-022-A10 and Auburn University AHDR Morrison (101002 139269 2055) to EEM; and Fetzer Institute Inc., Grant 2231 to VV.

REFERENCES

Au, E. and Roskams, A. J. (2003). Olfactory ensheathing cells of the lamina propria in vivo and in vitro. *Glia* 41: 224–236.

Balmer, C. W. and LaMantia, A. S. (2005). Noses and neurons: induction, morphogenesis, and neuronal differentiation in the peripheral olfactory pathway. *Dev. Dyn.* 234(3): 464–481.

Batoni, G., Maisetta, G., Esin, S., and Campa, M. (2006). Human beta-dfensin-3: a promising antimicrobial peptide. *Mini-Rev. Med. Chem.* 6: 1063–1073.

Bedford, E. A. (1904). The early history of the olfactory nerve in swine. *J. Comp. Neurol.* 14: 390–410.

Beites, C. L., Kawauchi, S., Crocker, C. E., and Calof, A. L. (2005). Identification and molecular regulation of neural stem cells in the olfactory epithelium. *Exp. Cell Res.* 306: 309–316.

Bilberg, K., Doving, K. B., Breedholm, K. and Baatrup, E. (2011). Silver nanoparticles disrupt olfaction in Crusian carp (*Carassius carassius*) and Eurasian perch (*Perca fluviatilis*). *Aquat. Toxicol.* 104: 145–152.

Bitanihirwe, B. K. Y. and Cunningham, M. G. (2009). Zinc: the brain's dark horse. *Synapse* 63: 1029–1049.

Breipohl, W., Laugwitz, H. J. and Bornfeld, N. (1974). Topographical relations between the dendrites of olfactory sensory cells and sustentacular cells in different verterates. *J. Anat.* 117: 89–94.

Buck, L. and Axel, R. (1991). A novel multigene family may encode odorant receptors: a molecular basis of odor recognition. *Cell* 65: 175–187.

Buiakova, O. I., Baker, H., Scott, J. W., et al. (1996). Olfactory marker protein (OMP) gene deletion causes altered physiological activity of olfactory sensory neurons. *Proc. Natl. Acad. Sci. USA* 93: 9858–9863.

Caggiano, M., Kauer, J. S., and Hunter, D. D. (1994). Globase basal cells are neuronal progenitors in the olfactory epithelium: A lineage analysis using a replication-incompetent retrovirus. *Neuron* 13: 339–352.

Carson, K. A. (1984). Quantitative localization of neurons projecting to the mouse main olfactory bubl. *Brain Res. Bull.* 12: 629–634.

Carvalho, A. de O., and Gomes, V. M. (2009). Plant defensins—prospects for the biological functions and biothechnological properties. *Peptides* 30: 1007–1020.

Chen, X., Fang, H. and Schwob, J. E. (2004). Multipotency of purified, transplanted globose basal cells in olfactory epithelium. *J. Comp. Neurol.* 469: 457–474.

Christensen, M. K., Frederickson, C. J., and Danscher, G. (1992). Retrograde tracing of zinc-containing neurons by selenide ions: a survey of seven selenium compounds. *J. Histochem. Cytochem.* 40: 575–579.

Chuah, M. I. and Au, C. (1991). Olfactory Schwann cells are derived from precursor cells in the olfactory epithelium. *J. Neurosci.* 29: 172–180.

Chuah, M. I. and West, A. K. (2002). Cellular and molecular biology of ensheathing cells. *Microsc. Res. Tech.* 58: 216–227.

Cole, A. M., Dewan, P., and Ganz, T. (1999). Innate antimicrobial activity of nasal secretions. *Infect. Immun.* 67: 3267.

Connors, K. A. (1987). Binding Constants. The Measurements of Molecular Complex Stability. New York: Wiley.

Costanzo, R. M. (1985). Neural regeneration and functional connection following olfactory nerve transection in hamster. *Brain Res.* 361: 258–266.

Costanzo, R. M. and Morrison, E. E. (1989). Three dimensional scanning electron microscopic study of the normal hamster olfactory epithelium. *J. Neurocytol.* 18: 381–391.

Couley, G. F. and Le Douarin, N. M. (1985) Mapping of the early neural primordium in quail-chick chimeras. I. Developmental relationships between placodes, facial ectoderm, and prosencephalon. *Dev. Biol.* 110: 422–439.

Cunliffe, R. (2003). A-defensins in the gastrointestinal tract. *Mol. Immunol.* 40: 463–467.

Cuschieru, A. and Bannister, L. H. (1975). The development of the olfactory mucosa in the mouse: light microcopy. *J. Anat.* 119: 277–286.

D'Amico-Martel, A. and Noden, D. M. (1983). Contributions of placodal and neural crest cells to avian cranial peripheral ganglia. *Am. J. Anat.* 166: 445–468.

Danscher, G., Howell, G., Perex-Clausell, J., and Hertel, N. (1985). The dithizone Timm's sulphide silver and the selenium methods demonstrate a chelatable pool of zinc in CNS. A proton activation (PIXE) analysis of carbon tetrachloride extracts from rat brains and spinal cords intravitally treated with dithizone. *Histochemistry* 83: 419–422.

Dean, M. K., Higgs, C., Smith, R. E., et al. (2001). Dimerization of G-protein-coupled receptors. *J. Med. Chem.* 44: 4595–4614.

References

Doucette, R. (1990). Glial influences on axonal growth in the primary olfactory system. *Glia* 3: 433–449.

Engstrom, H. and Bloom, G. (1953). The structure of the olfactory region in man. *Acta Otolaryngol.* 43: 11–21.

Erles, K. and Brownlie, J. (2010) Expression of beta-defensins in the canine respiratory tract and antimicrobial activity against Bordetella bronchiseptica. *Vet. Immunol. Immunopathol.* 135: 12–19.

Fahlgren, A., Hammarström, S., Danielsson, Å. and Hammarstrom, M. L. (2003). Increased expression of antimicrobial peptides and lysozyme in colonic epithelial cells of patients with ulcerative colitis. *Clin. Exp. Immunol.* 131: 90–101.

Farbman, A. I. and Margolis, F. L. (1980). Olfactory marker protein during ontogeny: immunohistochemical localization. *Dev. Biol.* 74: 205–215.

Feinstein, P. and Mombaerts, P. (2004). Contextual model for axonal sorting into glomeruli in the mouse olfactory system. *Cell.* 117: 817–831.

Fellermann, K., Strange, D. E., Schaeffeler, E. et al. (2006). A chromosome 8 gene-cluster polymorphism with low human β-defensin 2 gene copy number predisposes to Crohn's disease of the colon. *Am. J. Hum. Genet.* 79: 439–752.

Ferraris, N., Perroteau, I., De Marchis, S., et al. (1997). Glutamatergic deafferentation of olfactory bulb modulates the expression of mGluR1a mRNA. *NeuroReport* 8: 1949–1953.

Forni, P. E., Taylor-Burds, C., Melvin, V.S., et al. (2011). Neural crest and ectodermal cells intermix in the nasal placode to give rise to GnRH-1 neurons, sensory neurons, and olfactory ensheathing cells. *J. Neurosci.* 31(18): 6915–6927.

Fotiadis, D., Liang, Y., Filipek, S., et al. (2003). Atomic-force microscopy: rhodopsin dimers in native disc membranes. *Nature* (London, UK). 421: 127–128.

Freyer, L., Aggarwal, V. and Morrow, B. E. (2011). Dual embryonic origin of the mammalian otic vesicle forming the inner ear. *Development* 138: 5403–5414.

Friedman, B. and Price, J. L. (1984). Fiber systems in the olfactory bulb and cortex: a study in adult and developing rats, using the Timm method with the light and electron microscope. *J. Comp. Neurol.* 223: 88–109.

Gachomo, E. W., Jimenez-Lopex, J.C., Kayode, A.P.P. et al. (2012). Structural characterization of plant defensin protein superfamily. *Mol. Biol. Rep.* 39: 4461–4469.

Ganz, T. (2003). Defensins: Antimicrobial peptides of innate immunity. *Nat. Rev. Immunol.* 3(9): 710–720.

Gersemann, M., Wehkamp, J., Fellermann, K., and Strange, E. F. (2008). Crohn's disease-defect in innate defence. *World J. Gastroenterol.* 14: 5499–5503.

Gersemann, M., Wehkamp, J., and Stange, E. (2011). Innate immune dysfunction in inflammatory bowel disease. *J. Intern. Med.* 271: 421–428.

Giraldo, J. (2008). On the fitting of binding data when receptor dimerization is suspected. *Br. J. Pharmacol.* 155: 17–23.

Goldman, M. J., Anderson, G. M., Stolzenberg, E. D. et al. (1997). Human β-defensin-1 is a salt-sensitive antibiotic in lung that is inactivated in cystic fibrosis. *Cell* 88(4): 553–560.

Gong, Q. and Shipley, M. T. (1995). Evidence that pioneer olfactory axons regulate telencephalon cell cycle kinetics to induce the formation of the olfactory bulb. *Neuron* 14: 91–101.

Graham, B. J. (2001). Integration between the epibranchial placodes and the hindbrain. *Science* 294: 595–598.

Granger, N., Blamires, H., Franklin, R. J. M., and Jeffery, N. D. (2012). Autologous olfactory mucosal cell transplants in clinical spinal cord injury: a randomized double-blinded trial in a canine translational model. *Brain* 135: 3227–3237.

Graziadei, P. P. C. (1973a). The ultrastructure of vertebrates olfactory mucosa. In: The Ultrastructure of Sensory Organs, Friedman, I. (Ed). Oxford: Elsevier, pp. 267–305.

Graziadei, P. P. C. (1973b). Cell dynamics in the olfactory mucosa. *Tissue Cell* 5: 113–131.

Graziadei, P. P. C. and Monti-Graziadei, G. A. (1978). The olfactory system: a model for the study of neurogenesis and axon regeneration in mammals. In: Neuronal Plasticity, Wotman, C. W. (Ed). Raven Press: New York, pp. 113–153.

Graziadei, P. P. C. and Monti-Graziadei, G. A. (1979). Neurogenesis and neuron regeneration in the olfactory system of mammals. I. Morphological aspects of differentiation and structural organization of the olfactory sensory neurons. *J. Neurocytol.* 8: 1–18.

Graziadei, G. A., Stanley, R. S., and Grazadei, P. P. (1980). The olfactory marker protein in the olfactory system of the mouse during development. *Neuroscience* 5: 1239–1252.

Hao, H. N., Zhao, J., Lotoczky, G., et al. (2001). Induction of human β-defensin 2 expression in human astrocytes by lipopolysaccharide and cytokines. *J. Neurochem.* 77(4): 1027–1035.

Harder, J., Bartels, J., Christophers, E., and Schröder, J. (2001). Isolation and characterization of human -defensin-3, a novel human inducible peptide antibiotic. *J. Biol. Chem.* 276(8): 5707.

Holbrook, E. H., Szumowski, K. E. and Schwob, J. E. (1995). An immunochemical, ultrastructural and development characterization of the horizontal basal cells of rat olfactory epithelium. *J. Comp. Neural.* 363: 129–146.

Howell, G. A. and Frederickson, C. J. (1990). A retrograde transport method for mapping zinc-containing fiber systems in the brain. *Brain Res.* 515: 277–186.

Huang, S. S. and Platt, R. (2003) Risk of methicillin-resistant Staphylococcus aureus infection after previous infection or colonization. *Clin. Infect. Dis.* 36: 281–285.

Huard, J. M., Youngentob, S. L., Goldstein, B. J., et al. (1998). Adult olfactory epithelium contains multipotent progenitors that give rise to neurons and non-neural cells. *J. Comp. Neurol.* 400: 469–586.

Jo, S. M., Danscher, G., Schroder, H. D., et al. (2000a). Zinc-enriched (ZEN) terminals in mouse spinal cord: immunohistochemistry and autometallography. *Brain Res.* 870: 1623–1629.

Jo, S. M., Won, M. H., Cole, T. B. et al. (2000b). Zinc-enriched (ZEN) terminals in mouse olfactory bulb. *Brain Res.* 865: 227–236.

Joly, S., Organ, C. C., Johnson, G. K. et al. (2005). Correlation between beta-defensin expression and induction profiles in gingival keratinocytes. *Mol. Immunol.* 42(9): 1073.

Kagan, B. L., Selsted, M. E., Ganz, T. and Lehrer, R. I. (1990). Antimicrobial defensin peptides form voltage-dependent ion-permeable channels in planar lipid bilayer membranes. *Proc. Natl. Acad. Sci. USA* 87(1): 210–214.

Katoh, H., Shibata, S., Fukuda, K., et al. (2011). The duel origin of the peripheral olfactory system: placode and neural crest. *Mol. Brain* 4: 34.

Kennedy, C. J., Stecko, P., Truelson, B. and Petkovich, D. (2012). Dissolved organic carbon modulates the effects of copper on olfactory-mediated behaviors of chinook slamon. *Environ. Toxicol. Chem.* 31: 2281–2288.

Kisich, K. O., Howell, M. D., Boguniewicz, M., et al. (2007). The constitutive capacity of human keratinocytes to kill staphylococcus aureus is dependent on beta-defensin 3. *J. Investig. Dermatol.* 127(10): 2368–2380.

Kleene, S. J. (2008). The electrochemical basis of odor transduction in vertebrate olfactory cilia. *Chem. Senses.* 33: 839–859.

Klein, S. L. and Graziadei, P. P. C. (1983). The differentiation of the olfactory placode in Xenopus laevis: a light and electron microscope study. *J. Comp. Neurol.* 217: 17–30.

Chapter 4 Development, Morphology, and Functional Anatomy of the Olfactory Epithelium

Klotman, M. E. and Chang, T. L. (2006). Defensins in innate antiviral immunity. *Nat. Rev. Immunol.* 6: 447–456.

Kwon, H. J., Koo, J. H., Zufall, F., et al. (2009). Ca extrusion by NCX is compromised in olfactory sensory neurons of OMP mice. *PLoS One* 4: e4260.

LaMantia, A. S., Bhasin, N., Rhodes, K., and Heemskerk, J. (2000). Mesechymal/epithelial induction mediates olfactory pathway formation. *Neuron* 28: 411–425.

Laudien, M., Dressel, S., Harder, J. and Glaser, R. (2011). Differential expression pattern of antimicrobial peptides in nasal mucosa and secretion. *Rhinology* 49: 107–111.

Lee, A. C., He, J., and Ma, M. (2011). Olfactory marker protein is critical for functional maturation of olfactory sensory neurons and development of mother preference. *J. Neurosci.* 31(8): 2974–2982.

Lee, S. H. A. G., Kim, J. E. U. N., Lee, H. M. A. N., et al. (2002). Antimicrobial defensin peptides of the human nasal mucosa. *Ann. Otol. Rhinol. Laryn.* 111(2): 135–141.

Lehrer, R. I. and Lu, W. (2012). A-defensins in human innate immunity. *Immunol. Rev.* 245: 84–112.

Leonard, B. C., Marks, S. L., Outerbridge, C. A., et al. (2012). Activity, expression and genetic variation of canine β-defensin 103: A multifunctional antimicrobial peptide in the skin of domestic dogs. *J. Innate Immun.* 4(3): 248–259.

Linzmeier, R. M. and Ganz, T. (2005). Human defensin gene copy number polymorphisms: Comprehensive analysis of independent variation in alpha-and beta-defensin regions at 8p22–p23. *Genomics* 86(4): 423.

Linzmeier, R. M. and Ganz, T. (2006). Copy number polymorphisms are not a common feature of innate immune genes. *Genomics* 88: 122–126.

Mackay-Sim, A. (2010). Stem cells and their niche in the adult olfactory mucosa. *Arch. Ital. Biol.* 148: 47–58.

Mackay-Sim, A. and Dreosti, I. E. (1989). Olfactory function in zinc-deficient adult mice. *Exp. Brain Res.* 76: 207–212.

Mackay-Sim, A. and St. John, J. A. (2011). Olfactory ensheathing cells from the nose: clinical application in human spinal cord injuries. *Exp. Neuro.* 229: 174–180.

Maisetta, G., Batoni, G., Esin, S., et al. (2003). Activity of human beta-defensin 3 alone or combined with other antimicrobial agents against oral bacteria. *Antimicrob. Agents Chemother.* 47: 33–49.

Maisetta, G., Di Luca, M., Esin, S., et al. (2008). Evaluation of the inhibitory effects of human serum components on bactericidal activity of human beta defensin 3. *Peptides* 29: 1–6.

Margolis, F. L. (1972). A brain protein unique to the olfactory bulb. *Proc. Natl. Acad. Sci., USA* 69: 1221–1224.

Mathews, M., Jia, H. P., Guthmiller, J. M., et al. (1999). Production of β-defensin antimicrobial peptides by the oral mucosa and salivary glands. *Infect. Immunol.* 67(6): 2740–2745.

Menco, B. Ph. M. (1977). A qualitative and quantitative investigation of olfactory and respiratory nasal surfaces of cow and sheep based on various ultrastructural and biochemical methods. *Comm. Agri. Univ. Wageningen* 77(13): 1–157.

Mendoza, A. S., Breipohl, W., and Miragall, F. (1982). Cell migration from the chick olfactory placode: a light and electron microscopic study. *J. Embryol. Exp. Morphol.* 69: 47–59.

Milanese, M., Segat, L., Arraes, L.C., et al. (2009). Copy number variation of defensin genes and hiv infection in brazilian children. *JAIDS J. Acquir. Immune Defic. Syndr.* 50: 331–333.

Milligan, G. (2004). G protein-coupled receptor dimerization: function and ligand pharmacology. *Mol. Pharmacol.* 66: 1–7.

Mombaerts, P. (2004). Odorant receptor gene choice in olfactory sensory neurons: the one receptor-one neuron hypothesis revisited. *Curr. Opin. Neurobiol.* 14: 31–36.

Moore, C. H., Pustovyy, O., Dennis, J. C., et al. (2012). Olfactory responses to explosives associated odorants are enhanced by zinc nanoparticles. *Talanta* 88: 730–733.

Moran, D. T., Rowley, J. C., and Jafek, B. W. (1982). Electron microscopy of human olfactory epithelium reveals a new cell type: the microvillar cell. *Brain Res.* 253: 39–46.

Mori, I., Nishiyama, Y., Yokochi, T. and Kimura, Y. (2005). Olfactory transmission of neurotropic viruses. *J. Neurovirol.* 11: 129–137.

Morrison, E. E. and Costanzo, R. M. (1989). Scanning electron microscopic study of degeneration and regeneration in the olfactory epithelium after axotomy. *J. Neurocytol.* 18: 393–405.

Morrison, E. E. and Costanzo, R. M. (1990). Morphology of the human olfactory epithelium. *J. Comp. Neurol.* 297: 1–14.

Mueller, B. K. (1999). Growth cone guidance: first steps towards a deeper understanding. *Ann. Rev. Neurosci.* 22: 351–388.

Neuhaus, E. M., Gisselmann, G., Zhang, W., et al. (2005). Odorant receptor heterodimerization in the olfactory system of *Drosophilia melanogaster*. *Nat. Neurosci.* 8: 15–17.

Nicolay, D. J., Doucette, J. R., and Nazarali, A. J. (2006). Transcriptional regulation of neurogenesis in the olfactory epithelium. *Cell Mol. Neurobiol.* 26: 803–821.

Okano, M., Weber, A. F., and Frommes, S. P. (1967). Electron microscopic studies of the distan border of the canine olfactory epithelium. *J. Ultrastruct. Res.* 17: 487–502.

Pearson, A. A. (1941). The development of the olfactory nerve in man. *J. Comp. Neurol.* 75: 199–217.

Pevsner, J., Trifiletti, R. R., Strittmatter, S. M., and Snyder, S. H. (1985). Isolation and characterization of an olfactory receptor protein for odorant pyrazines. *Proc. Natl. Acad. Sci.USA* 82: 3050–3054.

Rafols, J. A. and Getchell, T. V. (1983). Morphological relations between the receptor neurons, sustentacular cells and Schwann cells in the olfactory mucosa of the salamander. *Anat. Rec.* 206: 87–101.

Ramon Y Cajal, S. (1909–1911). Histologie du systeme neveux de l'homme et des vertebres, 2 vols. Translated by L. Azoulay. Reprinted by Instituto Ramon Y Cajal del C.S.T.S. Madrid, 1952–1955.

Rawson, N. E. and LaMantia, A. S. (2006). Once and again: retinoic acid signaling in the developing and regenerating olfactory pathway. *J. Neurobiol.* 66: 653–676.

Sahl, H. G., Pag, U., Bonness, S., et al. (2005). Mammalian defensins: Structures and mechanism of antibiotic activity. *J. Leukoc. Biol.* 77(4): 466–475.

Samoylov, A. M., Samoylova, T. I., Pustovyy, O. M., et al. (2005). Novel metal clusters isolated from blood are lethal to cancer cells. *Cells Tissues Organs.* 179: 115–124.

Sato, K., Pellegrino, M., Nakagawa, T., et al. (2008). Insect olfactory receptors are heteromeric ligand-gated ion channels. *Nature* 452: 1002–1006.

Schroeder, B. O., Wu, Z., Nuding, S., et al. (2011). Reduction of disulphide bonds unmasks potent antimicrobial activity of human beta-defensin 1. *Nature* 469: 419–423.

Schwanzel-Fukuda, M. and Pfaff, D. W. (1989). Origin of luteinizing hormone-releasing hormone neurons. *Nature* 338: 161–164.

Schwob, J. E. (2002). Neural regeneration and the peripheral olfactory system. *Anat. Rec.* 269: 33–49.

Sjölinder, H. and Jonsson, A. B. (2010). Olfactory nerve—a novel invasion route of neisseria meningitidis to reach the meninges. *PloS one* 5(11): e14034.

Suzuki, J., Yoshizaki, K., Kobayashi, T., and Osumi, N. (2013). Neural crest derived horizontal basal cells as a tissue stem cells in the adult olfactory epithelium. *Neurosci. Res.* 75: 112–120.

References

Takeda, A. (2001). Zinc homeostasis and functions of zinc in the brain. *Biometals* 14: 343–351.

Takeda, A., Sawashita, J., Takefuta, S., et al. (1999). Role of zinc released by stimulation in rat amygdala. *J. Neurosci. Res.* 57: 405–410.

Takeda, A., and Tamano, H. (2009). Insight into zinc signaling from dietary zinc deficiency. *Brain Res. Rev.* 62: 33–44.

Taylor, K., Clarke, D. J., McCullough, B., et al. (2008). Analysis and separation of residues important for the chemoattractant and antimicrobial activities of beta-defensin 3. *J. Biol. Chem.* 283: 6631–6639.

Trombley, P. Q. and Shepherd, G. M. (1996). Differential modulation by zinc and copper of amino acid receptors from rat olfactory bulb neurons. *J. Neurophysiol.* 76: 2536–2546.

Turin, L. (1996). A spectroscopic mechanism for primary olfactory reception. *Chem. Senses* 21: 773–791.

Valverde, F., Santacana, M. and Heredia, M. (1992). Formation of an olfactory glomerulus: morphological aspects of development and organization. *Neuroscience* 49: 255–275.

Vankeerberghen, A., Nuytten, H., Dierickx, K., et al. (2005). Differential induction of human beta-defensin expression by periodontal commensals and pathogens in periodontal pocket epithelial cells. *J. Periodontol.* 76: 1293–1303.

Viswaprakash, N., Dennis, J. C., Globa, L., et al. (2009). Enhancement of odorant-induced responses in olfactory receptor neurons by zinc nanoparticles. *Chem. Senses* 34: 547–557.

Viswaprakash, N., Josephson, E. M., Dennis, J. C., et al. (2010). Odorant response kinetics from cultured mouse olfactory epithelium at different ages in vitro. *Cells Tissues Organs* 192: 361–373.

Vodyanoy, V. (2010). Zinc nanoparticles interact with olfactory receptor neurons. *Biometals* 23: 1097–1103.

Wade, F., Espagne, A., Persuy, M-A., et al. (2011). Relationship between homo-oligomerization of a mammalian olfactory receptor and its activation state demonstrated by bioluminescence resonance energy transfer. *J. Biol. Chem.* 286: 15252–15259.

Wehkamp, J., Fellermann, K., Herrlinger, K. R., et al. (2002). Human β-defensin 2 but not β-defensin 1 is expressed preferentially in colonic mucosa of inflammatory blowel disease. *Eur. J. Gastroenterol. Hepatol.* 14: 745–752.

Wehkamp, J., Harder, J., Weichenthal, M., et al. (2006). Inducible and constitutive β-defensins are differentially expressed in crohn's disease and ulcerative colitis. *Inflamm. Bowel Dis.* 9(4): 215–223.

Wehkamp, J., Salzman, N. H., Porter, E., et al. (2005). Reduced paneth cell α-defensins in ileal crohn's disease. *Proc. Natl. Acad. Sci. USA.* 102: 18129–18134.

Weinberg, A., Jin, G., Sieg, S., and McCormick, T. S. (2012). The yin and yang of human beta-defensins in health and disease. *Front. Immun.* 3: 294. doi: 10.3389/fimmu.2012.00294.

Wray, S., Grant, P., and Gainer, H. (1989). Evidence that cells expressing luteinizing hormone-releasing hormone mRNA in the mouse are derived from progenitor cells in the olfactory placode. *Proc. Natl. Acad. Sci. USA* 86: 8132–8136.

Yoshida, S., Shimmura, S., Nagoshi, N., et al. (2006). Isolation of multipotent neural crest-derived stem cells from the adult mouse cornea. *Stem Cells* 24: 2714–2722.

Youngentob, S. L. and Margolis, F. L. (1999). OMP gene deletion causes an elevation in behavioral threshold sensitivity. *Neuroreport* 10: 15–19.

Youngentob, S. L., Margolis, F. L., and Youngentob, L. M. (2001). OMP gene deletion results in an alteration in odorant quality perception. *Behav. Neurosci.* 115: 626–631.

Zanger, P., Holzer, J., Schleucher, R., et al. (2010). Severity of staphylococcus aureus infection of the skin is associated with inducibility of human β-defensin 3 but not human β-defensin 2. *Infect. Immun.* 78: 3112–3117.

Zanger, P., Nurjadi, D., Vath, B., and Kremsner, P. G. (2011). Persistent nasal carriage of staphylococcus aureus is associated with deficient induction of human β-defensin 3 after sterile wounding of healthy skin in vivo. *Infect. Immun.* 79: 2658–2662.

Zanusso, G., Ferrari, S., Cardone, F., et al. (2003). Detection of pathologic prion protein in the olfactory epithelium in sporadic creutzfeldt–jakob disease. *N. Engl. J. Med.* 348: 711–719.

Zasloff, M. (2002). Antimicrobial peptides of multicellular organisms. *Nature* 415: 389–395.

Chapter 5

Olfactory Receptor Function

KEIICHI YOSHIKAWA and KAZUSHIGE TOUHARA

5.1 INTRODUCTION

In the natural environment, a complex chemical space is constructed by a variety of compounds that are emitted from organisms and other sources. Such compounds include small volatile organic compounds, inorganic gases, and peptidic molecules, and are produced via multiple metabolic processes. These environmental compounds provide critical information for the survival of an animal, as they reflect the existence of food, predators, and conspecifics. Environmental compounds are specifically detected and processed by the mammalian olfactory system, a system that has a remarkable capacity to recognize and discriminate numerous compounds. The olfactory neural system begins within a large neuroepithelium termed the olfactory epithelium (OE) located in the roof, superior conchae, and septum of the nasal cavity (Figure 5.1). The OE contains millions of olfactory sensory neurons (OSNs) which extend ciliated dendrites to its surface. In addition, the OSNs project axons that synapse with second order neurons in the olfactory bulb (OB). The fundamental role of the OSNs is to detect environmental information and convey it to secondary neurons.

The underlying molecular basis of olfactory capacity is the large variety and number of olfactory receptors expressed on the sensory neurons. Olfactory receptors are defined on the basis that they are: (1) expressed in OSNs, (2) function to detect specific environmental chemical cues, and (3) initiate intracellular signal transduction which ultimately drives action potential generation in neurons. The discovery of the first mammalian olfactory receptor family, the odorant receptor (OR) family, initiated the molecular era of olfactory research. This achievement has become the template for the identification of other types of mammalian olfactory receptors. Currently, three types of olfactory receptors have been shown to be expressed in a distinct subpopulation of OSNs. Each olfactory receptor detects specific cues and contributes to the remarkable capacity of olfaction. Detection of these specific cues provides animals not only with odor perception, but also several physiological and behavioral effects. This chapter discusses the expression, structure, signal transduction pathways, and ligand identity of each type of olfactory receptors, with an emphasis on their physiological function.

5.2 VERTEBRATE ODORANT RECEPTORS

5.2.1 Discovery of the Odorant Receptor Multigene Family

In 1991, Buck and Axel isolated and cloned OR genes based on three criteria (Buck and Axel, 1991). First, they assumed that ORs belong to the G protein-coupled receptor (GPCR) family. This was due to the finding that the activation of OSN by odorants resulted in the production of the G protein-mediated second messenger cAMP, which subsequently induced action potentials in the OSN (Nakamura and Gold, 1987; Pace et al., 1985). Second, ORs should be encoded by a multigene family with a capacity to distinguish numerous distinct odors. Third, ORs should be specifically expressed in the OE. The investigators then used a series of degenerate oligo-nucleotides designed according to conserved amino acid sequences in a variety of GPCRs to amplify the sequence of GPCRs expressed in rat OE. Resulting PCR products were then selected using the

---

Handbook of Olfaction and Gustation, Third Edition. Edited by Richard L. Doty.
© 2015 Richard L. Doty. Published 2015 by John Wiley & Sons, Inc.

110 Chapter 5 Olfactory Receptor Function

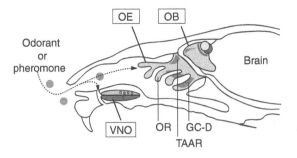

Figure 5.1 Schematic representation of the rodent olfactory system anatomy. Sagittal view of the rodent head showing the olfactory system. Stimulus molecules are transported into the nasal cavity, and reach two neuroepithelia known as the olfactory epithelium (OE) and the vomeronasal organ (VNO) epithelium. The vast majority of OSNs express ORs (green), whereas a distinct subpopulation of sensory neurons expresses TAARs (magenta) or GC-D (grey). The axons of OSNs project to a distinct area on the OB, termed glomeruli. Necklace glomeruli, where GC-D neurons project, are shown in gray, and axons of TAAR-expressing OSNs target to a distinct dorsal area of the OB (magenta). (*See plate section for color version.*)

second criterion of whether the product consisted of multiple related genes. Finally, the OE-specific expression of the candidate genes was verified, and described as a novel multigene family encoding putative OR. The identified genes encoded proteins that possessed common structural features in GPCRs. That is, the proteins contained seven putative trans-membrane regions, a glycosylation site in the extracellular N-terminal region, and putative disulfide bonds between the conserved cystein molecules in the extracellular loops (Figure 5.2).

ORs are commonly distinguished from other GPCRs by several consensus motifs including a LHTPMY motif within the first intracellular loop, a MAYDRYVAIC motif at the end of trans-membrane segment 3 (TM3), FSTCSSH at the beginning of trans-membrane segment 6 (TM6) and PMLNPF in trans-membrane segment 7 (TM7) (Fleischer et al., 2009). These conserved motifs are thought to contribute to the maintenance of the normal conformational states of nascent OR proteins. ORs are also known to possess relatively variable amino acid residues within trans-membrane segments that are involved in ligand binding (Abaffy et al., 2007; Katada et al., 2005).

Using genome analysis, all tetrapod animals have been shown to express between 400 and 4,200 OR genes, including 20–60% pseudogenes (Nei et al., 2008; Niimura et al., 2014). In humans and the mouse, the number of intact OR genes is 396 and 1130, respectively (Figure 5.3). These numbers indicate that the OR gene family represents the largest gene family present in the entire genome. In each animal, OR genes are distributed throughout the genome and are generally found as small clusters. Phylogenetically, OR genes are categorized into nine classes known as α, β, γ, δ, ε, ζ, η, θ, and κ, each of which originated from distinct ancestral genes. Among the nine classes, the ORs identified in the genome of tetrapods are largely of the α and γ classes. These classes are widely known as class I and class II ORs, respectively (Niimura and Nei, 2005). The remaining seven classes of ORs are expressed mainly in fish and amphibians. These findings suggest that these seven classes of OR are able to detect water-soluble compounds, whereas ORs of the α and γ classes likely function to detect terrestrial volatile odorants (Nei et al., 2008). Two forms of OR nomenclature have been reported and include the Zhang and Firestein nomenclature (Zhang and Firestein, 2002), and the mouse genome informatics (MGI) nomenclature. The latter being the one most currently used.

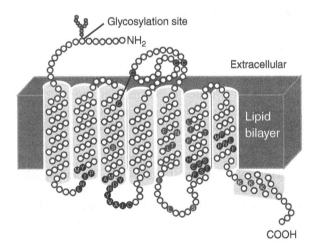

Figure 5.2 Molecular structure of the odorant receptor. The membrane topology of the OR is shown. Common amino acid sequences among ORs are shaded black. Key amino acid residues involved in odorant binding and G protein-coupling are shown in red and blue, respectively. (*See plate section for color version.*)

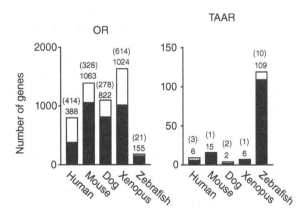

Figure 5.3 Total number of olfactory receptor genes in vertebrates. The black and white bars represent the number of intact genes and pseudogenes, respectively. The number shown above the bar indicates the number of intact genes and pseudogenes (parentheses). These findings were taken from Nei et al., 2008 and Niimura et al., 2014.

5.2 Vertebrate Odorant Receptors

It took approximately seven years following the seminal discovery of the putative OR gene family for evidence of OR mediation of OSN response to odorants to be obtained. This long delay was largely due to difficulties in functionally expressing ORs in cell lines. Thus, the first verification of OR function was achieved using an in vivo-based approach where adenovirus-mediated over-expression of cloned OR I7, a rat orthologue of mouse Olfr2, in rat OSNs (Zhao et al., 1998). The infected rat OE showed a greater response to odorous aliphatic aldehydes. This suggested that the OR I7 gene encoded a receptor for aldehydes. A separate approach was based on the logic that a single OSN specifically expressed only one of a thousand OR genes, and that the OR should be responsible for responsiveness of the OSN to particular odorants. Thus, from an odorant responsive OSN, an OR gene was cloned using degenerate primers that recognized conserved sequence motifs (Malnic et al., 1999; Touhara et al., 1999). Subsequent demonstration that functionally cloned OR genes actually functioned as receptors for odorants was performed using adenovirus-mediated over-expression of OR in OSNs to verify their responsiveness (Touhara et al., 1999). The hallmark of these achievements was taking advantage of in vivo OSN machinery. These achievements also highlighted a lack of knowledge of the components responsible for functional expression of ORs in heterologous cells.

5.2.2 Development of Functional Assay Systems

The in vivo-based analysis of specific ORs was often complicated by the presence of endogenous ORs. In most cases, heterologous expression allows for a clear analysis of single receptor function. However, when an OR gene was exogenously introduced into cell lines, nascent OR protein was retained in the endoplasmic reticulum (ER) and subsequently degraded (Lu et al., 2003; McClintock et al., 1997). Consequently, the OR protein was not expressed on the cell surface membrane. Therefore, it was presumed that OSN contained specific molecular machinery that allowed for the expression of nascent OR protein at the cell surface, however, corresponding components did not exist in cell lines.

The first major success in heterologous expression of an OR was achieved in 1998. Krautwurst et al. (1998) designed a chimeric OR where the N-terminal sequence of the OR was extended with N-terminal 20 amino acids of bovine rhodopsin (Rho-tag). This strategy was based on a previous finding that rhodopsin could be expressed on the membrane of HEK293 cells, and that the N-terminal motif was required for membrane transport (Sung et al., 1991). These Rho-tagged OR proteins were successfully expressed on the cell surface of HEK293 cells and were activated by cognate ligands, resulting in intracellular signaling (Krautwurst et al., 1998). The activity of the

N-terminal epitope tag underlying the promotion of membrane expression was attributed to an increase in the glycosylation sites of the OR protein (Katada et al., 2004). A later study suggested that the ligand specificity and sensitivity of ORs were not modulated by the additional sequence (Zhuang and Matsunami, 2007). This approach paved the way for the characterization of numerous ORs. However, difficulties still remained in achieving heterologous expression for the large majority of ORs.

In 2004, trans-membrane proteins called receptor transporting proteins (RTPs) were discovered to act as chaperones of ORs from RNAs expressed in mouse OSNs (Saito et al., 2004). Receptor expression enhancing proteins (REEPs) were also identified in the OSNs, but were shown to exhibit lesser activity (Saito et al., 2004). Among the four members of RTP genes, RTP1 and 2 were found to be expressed in OSNs, and promoted membrane expression of ORs in heterologous cells. A more recent report demonstrated that RTP1 exhibited multiple roles in the expression of functional ORs (Wu et al., 2012). These functions included permitting the exit of ORs from the ER, trafficking ORs from Golgi to the cell membrane, and formation of co-receptors required for the modulation of OR activity. Collectively, RTPs are known to greatly enhance OR-mediated signaling in a heterologous cell.

Resistant to inhibitors of cholinesterase 8 homologue B (Ric-8B) also functions as a modulator of functional OR expression. Ric-8B is expressed in the OE and, when introduced into heterologous cells, amplifies the OR-mediated intracellular signaling (Von Dannecker et al., 2005; Von Dannecker et al., 2006; Yoshikawa and Touhara, 2009). Two distinct molecular mechanisms underlying Ric-8B activity have been proposed. The first mechanism involves the maintenance of G protein expression levels, while the second mechanism involves GPCR-dependent guanine nucleotide exchange factor (GEF) (Chan et al., 2011; Kerr et al., 2008; Nagai et al., 2010). Due to the short ligand-dwelling time of OR, an OR-ligand complex has a low probability of coupling with a cognate G protein in order to catalyze GDP-GTP exchange (Bhandawat et al., 2005). Ric-8B may work to increase the probability of coupling by ensuring adequate levels of G proteins on the cell surface, and to re-activate the small number of G proteins that have been previously activated by ORs. Although the precise physiological relevance of RTPs and Ric-8B in OSNs remains to be determined, the introduction of these factors in combination with the fusion of N-terminal epitope tags proved pivotal in overcoming the difficulties in the functional expression of OR in heterologous cells. To date, approximately 10% of murine ORs have been functionally expressed in heterologous cells and paired with their cognate ligands (Glatz and Bailey-Hill, 2011; Saito et al., 2009).

5.2.3 OR-Mediated Signal Transduction and Modulation of OR Activation

A fundamental role of ORs is to initiate intracellular signal transduction, a role that ultimately produces action potentials in the OSNs (Mombaerts, 2004). At the ciliary membrane of OSNs, an agonist-bound OR is coupled to the olfactory specific G protein Gαolf, which subsequently activates adenylyl cyclase III (ACIII) OSNs, resulting in an intracellular increase in cAMP levels (Figure 5.4) (Breer et al., 1990; Kleene, 2008). The cAMP positively regulates the open probability of cAMP-sensitive cyclic nucleotide-gated (CNG) channels. These channels are heterotetramers comprised of four subunits termed CNGA2, CNGA4, and CNGB1b, with a stoichiometry of 2:1:1 (Nakamura and Gold, 1987; Zheng and Zagotta, 2004). The CNG channel causes calcium influx and subsequent opening of calcium-gated chloride ion channels (Kurahashi and Yau, 1993; Leinders-Zufall et al., 1997; Lowe and Gold, 1993). The membrane depolarization of OSNs caused by these ion permeations is the precipitating event for generation of action potentials in the OSNs. This signaling pathway is widely accepted as a canonical pathway, due to the anosmic phenotype of mice lacking the Gαolf, ACIII, or CNG channel (Belluscio et al., 1998; Brunet et al., 1996; Wong et al., 2000). More recent studies argue that anoctamin 2, also known as TMEM16B, contributes to OR-mediated signaling as a calcium-gated chloride ion channel in the OSNs (Reisert et al., 2003; Reisert et al., 2005; Stephan et al., 2009).

In addition to the established OR signaling pathways, recent studies have revealed several mechanisms thought to modulate OR activation and signal transduction in OSNs. In order to initiate signal transduction, ORs are localized on the cell surface of OSNs, most likely in conjunction with RTPs. Co-localization of OR with RTP1 on lipid rafts of a heterologous cell membrane is known to be essential for normal OR activation (Wu et al., 2012). RTPs may also form complexes with some ORs on the surface of the ciliated membrane and may modulate their signaling efficacy through orthosteric or allosteric mechanisms. A Gq-coupled receptor termed M3 muscarinic acetylcholine receptor (M3-R) has been shown to be expressed on the cilia of OSNs, and has the potential to modulate OR-mediated signaling (Li and Matsunami, 2011). When co-expressed with ORs in heterologous cells, M3-R may physically interact with various ORs in order to amplify their signal, but does not affect the expression level of OR proteins. Interestingly, the active state of M3-R further enhances OR-mediated signals, whereas inactive M3-R reverses this effect. Although the precise molecular mechanisms underlying these events remains unclear, the hypothesis that non-olfactory GPCRs modulate OR activity is supported by further observations, including the finding that the purinergic receptor P2Y, the G protein-coupled 2 (P2Y2) receptor, and the β2 adrenergic receptor (β2AR) enhance the function of some ORs (Bush et al., 2007; Hague et al., 2004). Ache and co-workers found that phosphoinositide-3-kinase (PI3K) inhibited OR-mediated signal transduction in OSNs (Ukhanov et al., 2011). Blockade of PI3K with inhibitors resulted in a significant enhancement of odorant-mediated calcium response of OSNs. This result is in line with previous observations that PIP$_3$, which is generated by PI3K, negatively regulates olfactory CNG channels (Brady et al., 2006; Zhainazarov et al., 2004). It would be interesting to determine whether acetylcholine or a hormonal compound that activates PI3K in OSNs is secreted in the OE under distinct physiological conditions such as nutritional changes and the estrus cycle.

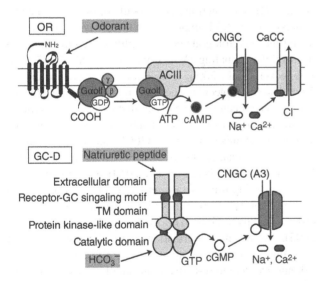

Figure 5.4 Signal transduction mediated by olfactory receptors. (Upper panel) In OSNs, the OR activates signal components in order of heterotrimeric Gαolf and ACIII. CNG channel (CNGC) is composed of CNGA2, CNGA4 and CNGB1b. Opening of the two types of ion channels, CNGC and a Ca^{2+}-dependent Cl$^-$ channel, leads to membrane depolarization. (Lower panel) The GC-D exhibits the catalytic activity in response to extracellular binding natriuretic peptide or intracellular binding HCO$_3^-$. This results in an intracellular increase in cGMP that stimulates gating of the CNG channel complex, including the CNGA3 subunit. The opening of the CNG channel results in membrane depolarization via influx of Ca^{2+} and Na$^+$.

5.2.4 Structural Basis of the Ligand Binding Mode and G Protein Coupling

The development of *in vitro* functional assay systems has allowed investigators to address various questions regarding how ORs bind to their cognate agonist and activate Gαolf in OSNs. ORs are generally classified into class-A

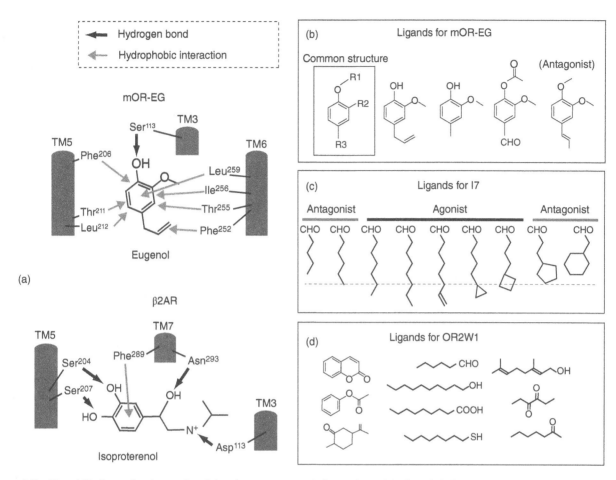

Figure 5.5 Ligand-binding and tuning modes of the odorant receptor. (a) Comparison of the ligand-binding mode of mOR-EG (upper schematic) versus that of β2AR (lower schematic). For mOR-EG, binding of eugenol is achieved via numerous hydrophobic interactions and only one hydrogen bond with Ser113, whereas β2AR forms four hydrogen bonds with isoproterenol. (b) Representative ligand repertoire of the mouse OR mOR-EG. (c) Ligand repertoire of I7. Dashed line shows length requirement for receptor activation. Short chain aldehydes are able to bind to I7 but are not able to activate. (d) An agonist repertoire of the relatively non-specific OR OR2W1.

family GPCRs which contain GPCRs including rhodopisn and β2AR (Jacoby et al., 2006). GPCRs in the class-A family are activated by small ligands whose ligand binding sites are contained within a cavity formed by the seven trans-membrane domains. The agonist-bound activated GPCRs couple to cognate G proteins with their intracellular loops and C-terminus region (Liggett, 2002; Liggett et al., 1991; Nikiforovich et al., 2007). The structural basis of ORs shares similarity with that of the other members of class-A family GPCRs.

Computational docking simulation and rational receptor design in combination with site-directed mutagenesis resulted in the successful definition of odorant binding sites in the murine OR mOR-EG, also known as MOR174-9 or Olfr73 (Katada et al., 2005). Consistent with other class-A type GPCRs, ligand binding sites in the mOR-EG are mainly formed by several amino acid residues that reside in TM3, TM5, and TM6 (Figure 5.2).

A characteristic feature of the OR in the ligand binding mode is that the vast majority of the critical amino acid residues involved in odorant binding are hydrophobic (Katada et al., 2005). Thus, ORs associate with cognate odorants via weak hydrophobic interactions, in contrast to other GPCRs including β2AR, that bind ligands mainly via electrostatic interaction (Figure 5.5a, (Klabunde and Hessler, 2002). This weak ligand binding may in turn result in a relatively low affinity (μM order) and short life span of the OR-ligand complex (<1 ms) (Bhandawat et al., 2005). Despite the low affinity of ORs, mammalian species show extremely high sensitivity to many odorants (pM-nM order). This discrepancy may be attributed to the presence of the nasal mucus that covers the OSNs layer (Oka et al., 2006). The nasal mucus may play a role in trapping odorants and presenting them to the OSNs in a timely and spatially concentrated manner.

It has been proposed that a specific molecular basis of ORs may exist for the recognition of divalent sulfur-containing odorants. A recent study has shown that the presence of copper ion at a physiological level (30 μM) enhances the response of MOR244-3 (Olfr1509)-expressing HEK293 cells to a series of sulfur-containing odorants (Duan et al., 2012). This copper ion-induced enhancement was abolished when the H105 residue in MOR244-3 contained a mutation. Molecular modeling of MOR244-3 has predicted that H105 is located between TM2 and TM5 facing the extracellular side, thus representing a potential copper-associating amino acid residue. The authors of the study proposed the hypothesis that the complex of copper ions and divalent sulfur compounds exhibits a more appropriate size to match the binding site of ORs and present a more stable interaction with the ORs. This metal ion-mediated OR-ligand recognition may serve as the molecular basis which accounts for our high sensitivity to sulfur compounds such as the warning agent added to natural gas, 2-methyl-2-propanethiol.

When an OR binds to an agonist, the OR undergoes conformational change and consequently couples to Gαolf (Kato and Touhara, 2009). Mutagenesis studies have revealed that OR-coupling to Gαolf is achieved via a series of conserved amino acid residues located in the third intracellular loop and C-terminal domain (Figure 5.2), (Kato et al., 2008). The crystal structure of the OR is yet to be determined. Recent isolation of large quantities of ORs may aid in the construction of high-resolution crystal structures of ORs, and provide further insight into the structural basis of ORs in the future (Cook et al., 2009).

5.2.5 Ligand Selectivity

A variety of organic compounds that have a molecular weight of less than M.W. 300 and that contain a relatively hydrophobic motif are odorous to humans. A recent study suggests that the human nose has the potential to distinguish a trillion different odorants (Bushdid et al., 2014). Against the numerous odorants, each of the hundreds of ORs show a distinct breadth of molecular receptive range, from narrowly tuned to broadly tuned. Most ORs recognize a variety of compounds that share a common structural backbone or a functional group. For example, the mouse OR mOR-EG is activated by more than 25 compounds, most of which contain a phenol moiety in their chemical structure (Figure 5.5b), (Katada et al., 2005). These relatively selective ORs may also recognize compounds that are structurally similar to agonists as competitive antagonists (Figure 5.5c), (Oka et al., 2004; Peterlin et al., 2008). In contrast, a relatively non-specific OR, such as human OR2W1, is activated by some 37 compounds of 63 odorants tested that do not share a structural backbone or functional groups (Figure 5.5d) (Saito et al., 2009).

Generally, ORs exhibit a relatively broad ligand spectra compared with other GPCRs for hormones and neurotransmitters. This is most likely due to the finding that the ligand binding mode of ORs is based on weak hydrophobic interactions.

On the other hand, ORs also show high selectivity in the discrimination of closely related structures. MOR139-3 discriminates meta-cresol from the ortho and para geometric isomers (Yoshikawa and Touhara, 2009). MOR107-1 (Olfr323) recognizes the (−) enantiomer of fenchone, but not the (+) enantiomer, while MOR271-1 (Olfr168) is activated by the (+) enantiomer only (Saito et al., 2009). These broad but selective molecular receptive ranges appear to allow humans to recognize and discriminate between numerous odorants. It is worth noting that the ligand selectivity of ORs observed in the screening of synthetic odorant collections does not necessarily reflect the physiological context. Olfr288, which is able to recognize at least eight different synthetic lactone-derivatives, appears to be specifically tuned to (Z)-5-tetradecen-1-ol in the urine of male mice against a background of several exocrine gland secretions (Yoshikawa et al., 2013). Thus, although ORs apparently exhibit a relatively broad ligand spectrum against synthetic odorants, they may demonstrate significant selectivity in a natural environment.

An important question in olfactory research is whether odor quality is encoded in the periphery. It may be hypothesized that each OR is responsible for a distinct odor quality and that distinct OR-expressing OSNs are innervated by an individually tuned area of the brain. However, it has been demonstrated that a pair of odorants that represent different odor qualities can activate the same OR, whereas the same odorant can activate multiple ORs. These observations, and the fact that humans are able to recognize numerous odorants, led to the proposition of the combinatorial coding model. This model suggests that among the hundreds of ORs, an activated OR pattern is transmitted to higher brain areas to define an odor quality. Nevertheless, there are a few examples in which an odorant appears to be encoded by an extremely small number of ORs (Keller et al., 2007). OR7D4 is identified as a human OR for androstenone (5α-androst-16-en-3-one). Large scale OR screening using a heterologous cell line suggested that detection of androstenone is achieved mainly by OR7D4. Muscone, a macrocyclic ketone with fascinating fragrance, is recognized mainly by MOR215-1 in mice and probably by OR5AN1 in humans (Shirasu et al., 2014). In the case of such a specific OR-ligand pair, the efficacy of the single OR-mediated signaling can directly affect the odor perception of individuals. Genetic variation of human OR7D4 includes two amino acid substitutions which impaired responsiveness to androstenone *in vitro* and led to reduced sensitivity in human subjects.

5.3 A NOVEL MEMBER OF THE VERTEBRATE OLFACTORY RECEPTORS IN THE OLFACTORY EPITHELIUM

The possibility that additional classes of olfactory receptors exist in the OE has also been suggested. This suggestion was based on previous findings. First, minor populations of OSNs do not possess the OR signaling component Gαolf, implying that these OSNs contain a different molecular mechanism for the recognition of environmental cues (Meyer et al., 2000). Second, the OE shows a remarkable ability to recognize diverse compounds beyond the small odorous molecules. It has been reported that major histocompatibility complex peptides induce calcium response in some OSNs of the OE (Spehr et al., 2006). In addition, the detection of a murine pheromone was thought to be achieved not only in the vomeronasal organ (VNO), an accessory olfactory system, but also in the OE (Boehm et al., 2005; Restrepo et al., 2004; Yoon et al., 2005). These observations suggest that some OSNs express novel types of receptors for peptides and/or pheromones.

5.3.1 Trace Amine-Associated Receptors

In 2006, a second class of olfactory GPCRs, termed trace amine-associated receptors (TAARs), was identified as the receptor for volatile amines (Liberles and Buck, 2006). An enriched population of OSNs was isolated from the OE using cell sorting for endogenous β-galactosidase activity. The candidate genes encoding novel chemosensory GPCRs were then screened using real-time quantitative PCR with primers designed against GPCRs that did not include previously known chemosensory receptors. The identified TAAR genes encoded proteins that were classed into the class-A GPCR family, but were unrelated to ORs. Rather, TAARs were found to be distantly related to biogenic amine receptors such as the serotonin and dopamine receptors. The TAAR genes were identified in several vertebrate species including fish and human (Figure 5.3), (Gloriam et al., 2005; Nei et al., 2008). The total number of TAAR genes is thought to be smaller than that for the OR. One exception to this rule is the zebrafish, which has a similar number of intact genes encoding ORs (155 genes) and TAARs (109 genes). The human and mouse genome has been shown to contain 6 and 15 intact TAAR genes, respectively.

In the murine OE, TAARs (excluding TAAR1) are expressed in a minor population of OSNs in a mutually exclusive manner (Figure 5.1; Liberles and Buck, 2006). That is, an OSN expressing one particular TAAR gene does not co-express other members of TAARs. In addition, TAARs-expressing OSNs are unlikely to co-express any ORs. Thus, it is conceivable that TAARs define the response specificity of a subset of OSNs. A recent study demonstrated that the TAAR-expressing OSNs innervate a distinct population of glomeruli from those connected with ORs-expressing OSNs (Johnson et al., 2012; Pacifico et al., 2012). In order to convert volatile amine-signal to an electrical signal in OSNs, TAARs may utilize a similar signal transduction mechanism to ORs. In OSNs, TAARs are co-expressed with Gαolf, suggesting active TAARs couple to Gαolf and initiate cAMP signaling (Liberles and Buck, 2006). This is supported by the finding that TAARs increase intracellular cAMP when expressed in a heterologous cell (Liberles and Buck, 2006). Furthermore, the electrical response in the OE to volatile amines was shown to disappear following disruption of genes encoding ACIII or the CNG channel (Brunet et al., 1996; Wong et al., 2000).

It had been presumed that TAARs are able to recognize volatile amines in the OE, as TAARs were originally shown to act as receptors for amines such as tyramine in the brain (Borowsky et al., 2001). Indeed, functional assays using a heterologous cell line provided evidence that TAARs recognize a series of odorous amines with unique selectivity and sensitivity (Liberles and Buck, 2006). Until recently, 7 of the 15 mouse TAARs (mTAAR1, mTAAR3, mTAAR4, mTAAR5, mTAAR7b, mTAAR7e and mTAAR7f) have been paired with 1–4 of the cognate amine agonists (Ferrero et al., 2012; Liberles and Buck, 2006). One of human TAARs has also been shown to recognize volatile amines (Wallrabenstein et al., 2013). TAAR-expressing OSNs can respond to volatile amines at nanomolar range (Pacifico et al., 2012; Zhang et al., 2013). Thus, TAAR repertoire mediates high sensitivity detection of volatile amines.

To address where TAARs bind to ligands, mutagenesis analysis on TAAR1 was conducted that focused on the potent endogenous agonist 3-iodothyronamine, which is a derivative of the thyroid hormone (Tan et al., 2009). This study suggested that the ligand-binding site of TAAR is located in a trans-membrane region, a finding that is consistent with recent works indicating that a key amino acid residue that defines ligand specificity is located in TM3 of TAARs (Ferrero et al., 2012). The ligand-binding site of a TAAR contains the charged amino acid residue aspartic acid that forms a salt bridge with an amine group located in the ligand. In addition, TAARs also contain several amino acid residues that may be involved in hydrogen bonding to the ligand, which is similar to the epinephrine-binding mode of β2AR (Figure 5.5A).

In contrast to ORs, TAARs demonstrate a relatively low copy number throughout vertebrate species and a specific projection pattern to the OB. These findings imply that TAARs play an important role in olfaction. Some TAARs appear to be involved in the detection of chemosignals that trigger an innate behavioral response. The volatile amine 2-phenylethylamine, which is highly abundant in the urine of carnivore species, acts as a key chemosignal for the avoidance of predators in rodents (Ferrero et al., 2011). 2-Phenylethylamine present in carnivore urine is finely recognized and discriminated by the mouse TAAR4. When mice smell 2-phenylethylamine in carnivore urine, they exhibit an avoidance behavior. TAAR4-decicient mice did not show avoidance behavior to 2-phenylethylamine (Dewan et al., 2013). Thus, TAAR4-expressing OSNs may contribute to the formation of a hard-wired aversion circuit. Thus, TAARs appear to play an important role in the detection of chemosignals utilized in intra-species chemical communication.

5.3.2 Guanylyl Cyclase D

In 1995, the olfactory-specific receptor guanylyl cyclase D (GC-D, *Gucy2d*) was cloned from the rat OE using degenerate primers that recognized conserved motifs within the guanylyl cyclase receptor (Fulle et al., 1995). The expression of GC-D transcripts was restricted to a small (less than 0.1%), randomly dispersed, population of OSNs located among the OR-expressing OSNs (Figure 5.1). GC-D-expressing OSNs demonstrated a typical bipolar morphology (Fulle et al., 1995; Zufall and Munger, 2010). That is, the cell body extended a single dendrite containing sensory cilia and a single axon. The axons of the GC-D neuron specifically projected to a set of interspersing glomeruli that form a "necklace" structure at the caudal area of the OB (Juilfs et al., 1997). In addition, the GC-D neuron utilized a distinct signal transduction pathway when compared to the majority of OSNs that expressed ORs or TAARs. The GC-D neurons were also shown to lack canonical signaling components of ORs including Gαolf, ACIII, and the cAMP-sensitive CNG channel subunits CNGA2 and CNGB1b (Juilfs et al., 1997; Meyer et al., 2000). Instead, these neurons were found to express the cGMP-specific CNG channel subunit CNGA3, cGMP-stimulated phosphodiesterase PDE2, and GC-D. This expression of signaling components suggests that GC-D neurons mainly utilize cGMP as a second messenger to mediate conversion of environmental signals to a neuronal response. We now know that GC-D neurons form part of the molecular basis of environmental cue detection, and transmit information to secondary neurons through intracellular cGMP signaling (Figure 5.4) (Zufall and Munger, 2010). The gene encoding GC-D is thought to have become a pseudogene early in primate evolution, indicating that the underlying molecular and cellular mechanisms do not exist in primates.

Today three chemosensory functions of GC-D neurons have been suggested through studies identifying different types of agonists including natriuretic peptides, carbon dioxide, and carbon disulfide. Early research in this field demonstrated that mammalian receptor guanylyl cyclases

functioned as receptors for hormonal natriuretic peptides, or functioned as a component of GPCR-dependent signaling via photoreceptor-specific GC-E and GC-F (Bennett et al., 1991; Kuhn, 2009; Potter et al., 2006). The natriuretic peptides were originally shown to be secreted from the atrium and ventricle of the heart in response to the expansion of blood volume (Dietz, 2005; Field et al., 1991; John et al., 1995). The hormonal activities of natriuretic peptides include the stimulation of salt excretion and the dilation of blood vessels to lower blood pressure. The natriuretic peptide target receptor guanylyl cyclase expressed in the vascular smooth muscle stimulates guanylyl cyclase activity (Potter et al., 2006). As a result, intracellular cGMP levels are elevated, eliciting physiological response via cGMP-regulated protein kinases, phosphodiesterases, and ion channels. One research group has demonstrated that olfactory GC-D neurons in the OE respond to two kinds of natriuretic peptides, uroguanylin and guanylyn, but not urodilatin (a GC-A agonist), and the heat stable enterotoxin STp (a GC-C agonist) (Leinders-Zufall et al., 2007). The response to the natriuretic peptides was found to disappear when GC-D neurons lacked an intact GC-D or CNGA3 gene, but not CNGA2, indicating that the response requires intracellular cGMP but not cAMP signaling. Both agonistic natriuretic peptides may be present within the excreted urine. Consistent with these findings, diluted urine was able to activate GC-D neurons, although it remains unclear as to whether the activity of urine was actually attributed to natriuretic peptides (Leinders-Zufall et al., 2007). Given the fact that nonvolatile compounds were able to gain access to a region where OSNs are scattered in the nasal cavity (Spehr et al., 2006), urinary natriuretic peptides were thought to most likely act as a physiological stimulant for GC-D neurons. Therefore, one function of the GC-D neurons may be to detect natriuretic peptides present within the urine of other individuals, and to obtain information about their metabolic status, for example water and salt balance.

It has also been suggested that GC-D neurons function as a sensor for atmospheric carbon dioxide concentration (Hu et al., 2007). Carbon dioxide is thought to serve as an important environmental cue for animals, as its levels are often associated with the presence of food, predators, and environmental stress. GC-D neurons exhibit intracellular calcium response to carbon dioxide at a threshold near 0.1%, with mice also able to recognize this concentration. Carbon dioxide-exposed GC-D neurons demonstrated an increased frequency of action potential firing, most likely due to the opening of CNG channels following intracellular cGMP increase. The responsiveness of GC-D neurons to carbon dioxide requires enzyme carbonic anhydrase II (CAII) expressed in GC-D neurons (Hu et al., 2007). Intracellular CAII is thought to metabolize carbon dioxide via the reversible reaction of $CO_2 + H_2O \rightleftharpoons HCO_3^- + H^+$ (Khalifah, 1971). It was previously demonstrated that

5.4 Olfactory Receptors in Non-Olfactory Tissues

highly diffusible intracellular HCO_3^- in turn activated GC-D, resulting in an increase in intracellular cGMP levels. Heterologous expression of GC-D in HEK293T cells resulted in activation following the addition of $NaHCO_3$ or $KHCO_3$, but not following pH change, resulting in intracellular cGMP increase (Sun et al., 2009).

More recently, it has been indicated that GC-D neurons exhibit sensitivity to carbon disulfide (CS_2), an isoelectronic isomer of CO_2 (Munger et al., 2010). The CS_2-responsiveness of GC-D neurons also requires CAII activity, suggesting that only CS_2 metabolites may activate GC-D. It has been previously shown that the rats emit CS_2 in their outward breath at a concentration of approximately 1 ppm (13 µM; Galef et al., 1988), and that this physiological concentration of CS_2 is capable of inducing a response in GC-D neurons (Galef et al., 1988; Munger et al., 2010). It has been suggested that CS_2 acts as a semiochemical that mediates social influences on food selection in the rodent (Galef et al., 1988). That is, rodents will exhibit an enhanced preference for a food after smelling food-derived odorants and CS_2 from living conspecifics that have eaten that food. This response is termed the socially transmitted food preference (STFP) and is particularly important for rodents that are continuously faced with the problem of selecting appropriate foods. Although the underlying molecular mechanisms that mediate and form STFP are yet to be elucidated, a recent study has provided evidence that GC-D neurons mediate STFP via the detection of CS_2 (Munger et al., 2010). Munger et al. investigated mice lacking GC-D and CNGA3. Disruption of these receptor components resulted in a reduced sensitivity of GC-D neurons to CS_2 and a failure to form STFP. Thus, GC-D neurons appear essential for normal sensitivity to CS_2 and acquisition of the STFP.

GC-D neurons exhibit sensitivity to CS_2 at a sub-micromolar range, whereas the response threshold of carbon dioxide is more than 6.8 mM, a 10,000-fold lesser sensitivity than CS_2 (Hu et al., 2007; Munger et al., 2010). Although sensitivity to both stimulants appears to be at a physiologically relevant level for when animals may be exposed via conspecific breathing or atmospheric air, the affinity of the GC-D neurons appears more highly tuned to CS_2 than to CO_2. GC-D neurons also act as sensitive chemosensors that respond to the pM of natriuretic peptides (EC_{50} = 60–770 pM) (Leinders-Zufall et al., 2007). This finding is consistent with reports for other members of the guanylyl cyclase receptors that exhibit a dissociation constant at the pM range of natriuretic peptides (Bennett et al., 1991). The higher sensitivity to natriuretic peptides raises the question of whether natriuretic peptides act as a conspecific cue that GC-D neurons detect for the acquisition of STFP in a physiological context. Mice genetically or pharmacologically lacking CAII demonstrate a severely reduced sensitivity to CS_2 and an impaired acquisition

of an artificial 'CS_2-mediated' food preference, but show normal sensitivity to natriuretic peptides (Munger et al., 2010). It would be interesting to examine whether pre-treatment with a pair of food odor and natriuretic peptides confers murine food preference, and whether CAII-deficient mice are able to detect social cues from living conspecifics and exhibit the food preference.

The structural basis of receptor guanylyl cyclase has also been extensively studied for GC-A, also known as natriuretic receptor-A (Kuhn, 2003). Receptor guanylyl cyclases including GC-D are single trans-membrane proteins that contain a ligand-binding extracellular domain, a trans-membrane domain and an intracellular catalytic domain (Figure 5.4). Two highly conserved motifs have been demonstrated in the extracellular juxtamembrane region, termed the receptor-GC signaling motif, and the intracellular protein kinase-like domain (Chinkers and Garbers, 1989; Liu et al., 1997; Wilson and Chinkers, 1995). The crystal structure of the extracellular domain indicates that GC-A binds to one natriuretic peptide as a homodimer (Ogawa et al., 2004). The dimeric extracellular domain consists of several key amino acid residues that interact with ligands via hydrogen bonds and hydrophobic interactions (He et al., 2006; Parat et al., 2008). The protein backbone of GC-A extracellular domains form a short parallel β-sheet with the C-terminal peptide backbone of natriuretic peptides. When a pair of extracellular domains binds to an agonist, receptor-GC signaling motifs are rotated (Ogawa et al., 2004). This extracellular conformational change may transmit to intracellular domains via trans-membrane helices and induce active conformation of two catalytic domains. Each intracellular protein kinase-like domain binds to ATP as a positive allosteric modulator which augments cyclase activity (Duda et al., 2009). For olfactory GC-D, a similar mechanism may function to detect natriuretic peptides and initiate intracellular signaling. In contrast, GC-D binds to the bicarbonate ion derived from carbon dioxide with the intracellular catalytic domain. This is indicated by the observation that targeted deletion of the extracellular domain or intracellular protein kinase-like domain of GC-D does not affect sensitivity to the bicarbonate ion, and that the recombinant catalytic domain alone shows an enhanced cyclase activity in the presence of this ion in HEK293T cells (Sun et al., 2009).

5.4 OLFACTORY RECEPTORS IN NON-OLFACTORY TISSUES

In 1992, one year after the discovery of the OR gene family, it was reported that mammalian ORs were also expressed in the testis (Parmentier et al., 1992). This finding raised the hypothesis that the largest family of chemoreceptors may serve some biological functions in tissues other than in the OE. Similarly, TAARs have been shown to function in the recognition of volatile amine in the nose, while in the brain functions as a receptor for trace-amine. Further evidence has suggested that ORs are expressed in a variety of tissues including the brain, heart, kidney, and muscle (Feldmesser et al., 2006; Ferrand et al., 1999; Otaki et al., 2004; Pluznick et al., 2009; Zhang et al., 2004). A GPCR identified as a prostate specific G protein-coupled receptor (PSGR) was later classed into the OR family (OR51E2), (Neuhaus et al., 2009). In many cases, olfactory signaling components such as Gαolf, ACIII, and CNG channel are co-expressed in non-olfactory tissues. The function of these ectopic ORs has been investigated. The human OR known as OR1D2 was found to be expressed in the testis, where it is used to aid in the recognition of human sperm to an odorant ligand and induce intracellular signaling that regulates sperm chemotaxis (Spehr et al., 2003). Similar findings have been observed in the mouse, where a ligand for MOR23 (Olfr16) was expressed in testis and found to regulate sperm motility (Fukuda et al., 2004). It is noteworthy that a recent study suggests that odorant recognition by sperm is mediated by a sperm-specific CatSper channel (Brenker et al., 2012). In murine muscle, ORs are expressed in myocytes during unique stages of muscle regeneration and regulate cellular migration (Griffin et al., 2009). PSGR is also known to regulate proliferation of human prostate cancer cells (Neuhaus et al., 2009). In the kidney, Olfr78 modulates blood pressure in response to short chain fatty acids produced by gut microbiota (Pluznick et al., 2013). Although the existence of endogenous ligands for the ectopic ORs has not been determined, the ORs appear to function in the detection of not only environmental volatile cues but also endocrine cues, and as a consequence regulate cell- or tissue-level function within the body. Intriguingly, one report has revealed that ORs expressed in a non-olfactory tissue were more highly conserved in animals than ORs expressed exclusively in the OE (De la Cruz et al., 2009). The acquisition of particular ligand selectivity may serve a selective advantage in the function of body tissues. These non-olfactory ORs are future targets of gene deletion studies that will help determine their physiological function. Understanding the precise function of the non-olfactory ORs may help pave the way for use of the OR family as novel targets for therapeutic agents.

5.5 CONCLUSIONS

Recent studies have significantly enhanced our understanding of the molecular mechanisms underlying the functional organization of the primary olfactory system. In the murine OE, multiple types of olfactory receptor have been identified including two GPCR families and one receptor enzyme. The pairing of olfactory receptors with cognate ligands has revealed a wide range of olfactory cues including small organic compounds, peptidic molecules, and inorganic gases. The emerging large scale relationship between environmental chemical cues and olfactory receptor repertoires has also revealed not only a greater role for olfaction, but also a considerably complex fashion of odor coding. For example, mice recognize safe food through CS_2 detection at multiple neuronal inputs including GC-D, while ORs are required for food-related odor detection (Munger et al., 2010). Future studies that dissect the function of each olfactory receptor will aid in the clarification of the activation pattern of olfactory receptors against a chemical repertoire in a specific natural environment, and will help us to further understand olfactory coding strategies and the neuronal mechanisms underlying adaptive behaviors.

REFERENCES

Abaffy, T., Malhotra, A., and Luetje, C. W. (2007). The molecular basis for ligand specificity in a mouse olfactory receptor: a network of functionally important residues. *J. Biol. Chem.* 282: 1216–1224.

Belluscio, L., Gold, G. H., Nemes, A., and Axel, R. (1998). Mice deficient in G(olf) are anosmic. *Neuron* 20: 69–81.

Bennett, B. D., Bennett, G. L., Vitangcol, R. V., et al. (1991). Extracellular domain-IgG fusion proteins for three human natriuretic peptide receptors. Hormone pharmacology and application to solid phase screening of synthetic peptide antisera. *J. Biol. Chem.* 266: 23060–23067.

Bhandawat, V., Reisert, J., and Yau, K. W. (2005). Elementary response of olfactory receptor neurons to odorants. *Science* 308: 1931–1934.

Boehm, U., Zou, Z., and Buck, L. B. (2005). Feedback loops link odor and pheromone signaling with reproduction. *Cell* 123: 683–695.

Borowsky, B., Adham, N., Jones, K. A., et al. (2001). Trace amines: identification of a family of mammalian G protein-coupled receptors. *Proc. Natl. Acad. Sci. USA* 98: 8966–8971.

Brady, J. D., Rich, E. D., Martens, J. R., et al. (2006). Interplay between PIP3 and calmodulin regulation of olfactory cyclic nucleotide-gated channels. *Proc. Natl. Acad. Sci. USA* 103: 15635–15640.

Breer, H., Boekhoff, I., and Tareilus, E. (1990). Rapid kinetics of second messenger formation in olfactory transduction. *Nature* 345: 65–68.

Brenker, C., Goodwin, N., Weyand, I., et al. (2012). The CatSper channel: a polymodal chemosensor in human sperm. *EMBO J.* 31: 1654–1665.

Brunet, L. J., Gold, G. H., and Ngai, J. (1996). General anosmia caused by a targeted disruption of the mouse olfactory cyclic nucleotide-gated cation channel. *Neuron* 17: 681–693.

Buck, L. and Axel, R. (1991). A novel multigene family may encode odorant receptors: a molecular basis for odor recognition. *Cell* 65: 175–187.

Bush, C. F., Jones, S. V., Lyle, A. N., et al. (2007). Specificity of olfactory receptor interactions with other G protein-coupled receptors. *J. Biol. Chem.* 282: 19042–19051.

Bushdid, C., Magnasco, M. O., Vosshall, L. B., et al. *Science* 343: 1370–1372.

Chan, P., Gabay, M., Wright, F. A., and Tall, G. G. (2011). Ric-8B is a GTP-dependent G protein alphas guanine nucleotide exchange factor. *J. Biol. Chem.* 286: 19932–19942.

Chinkers, M., and Garbers, D. L. (1989). The protein kinase domain of the ANP receptor is required for signaling. *Science* 245: 1392–1394.

Cook, B. L., Steuerwald, D., Kaiser, L., et al. (2009). Large-scale production and study of a synthetic G protein-coupled receptor: human olfactory receptor 17–4. *Proc. Natl. Acad. Sci. USA* 106, 11925–11930.

De la Cruz, O., Blekhman, R., Zhang, X., et al. (2009). A signature of evolutionary constraint on a subset of ectopically expressed olfactory receptor genes. *Mol. Biol. Evol.* 26: 491–494.

Dewan, A., Pacifico, R., Zhan, R., et al. (2013). Non-redundant coding of aversive odours in the main olfactory pathway. *Nature* 497: 486–489.

Dietz, J. R. (2005). Mechanisms of atrial natriuretic peptide secretion from the atrium. *Cardiovasc Res.* 68: 8–17.

Duan, X., Block, E., Li, Z., et al. (2012). Crucial role of copper in detection of metal-coordinating odorants. *Proc. Natl. Acad. Sci. U S A* 109: 3492–3497.

Duda, T., Bharill, S., Wojtas, I., et al. (2009). Atrial natriuretic factor receptor guanylate cyclase signaling: new ATP-regulated transduction motif. *Mol. Cell Biochem.* 324: 39–53.

Feldmesser, E., Olender, T., Khen, M., et al. (2006). Widespread ectopic expression of olfactory receptor genes. *BMC Genomics* 7: 121.

Ferrand, N., Pessah, M., Frayon, S., et al. (1999). Olfactory receptors, Golf alpha and adenylyl cyclase mRNA expressions in the rat heart during ontogenic development. *J. Mol. Cell Cardiol.* 31: 1137–1142.

Ferrero, D. M., Lemon, J. K., Fluegge, D., et al. (2011). Detection and avoidance of a carnivore odor by prey. *Proc. Natl. Acad. Sci. USA* 108: 11235–11240.

Ferrero, D. M., Wacker, D., Roque, M. A., et al. (2012). Agonists for 13 trace amine-associated receptors provide insight into the molecular basis of odor selectivity. *ACS Chem. Biol.* 7: 1184–1189.

Field, L. J., Veress, A. T., Steinhelper, M. E., et al. (1991). Kidney function in ANF-transgenic mice: effect of blood volume expansion. *Am. J. Physiol.* 260: R1–5.

Fleischer, J., Breer, H., and Strotmann, J. (2009). Mammalian olfactory receptors. *Front Cell Neurosci* 3: 9.

Fukuda, N., Yomogida, K., Okabe, M., and Touhara, K. (2004). Functional characterization of a mouse testicular olfactory receptor and its role in chemosensing and in regulation of sperm motility. *J. Cell Sci.* 117: 5835–5845.

Fulle, H. J., Vassar, R., Foster, D. C., et al. (1995). A receptor guanylyl cyclase expressed specifically in olfactory sensory neurons. *Proc. Natl. Acad. Sci. USA* 92: 3571–3575.

Galef, B. G., Jr., Mason, J. R., Preti, G., and Bean, N. J. (1988). Carbon disulfide: a semiochemical mediating socially-induced diet choice in rats. *Physiol. Behav.* 42: 119–124.

Glatz, R., and Bailey-Hill, K. (2011). Mimicking nature's noses: from receptor deorphaning to olfactory biosensing. *Prog. Neurobiol.* 93: 270–296.

Gloriam, D. E., Bjarnadottir, T. K., Yan, Y. L., et al. (2005). The repertoire of trace amine G-protein-coupled receptors: large expansion in zebrafish. *Mol. Phylogenet Evol.* 35, 470–482.

Griffin, C. A., Kafadar, K. A., and Pavlath, G. K. (2009). MOR23 promotes muscle regeneration and regulates cell adhesion and migration. *Dev. Cell* 17: 649–661.

Hague, C., Uberti, M. A., Chen, Z., et al. (2004). Olfactory receptor surface expression is driven by association with the beta2-adrenergic receptor. *Proc. Natl. Acad. Sci. USA* 101: 13672–13676.

He, X. L., Dukkipati, A., and Garcia, K. C. (2006). Structural determinants of natriuretic peptide receptor specificity and degeneracy. *J. Mol. Biol.* 361: 698–714.

Hu, J., Zhong, C., Ding, C., et al. (2007). Detection of near-atmospheric concentrations of CO_2 by an olfactory subsystem in the mouse. *Science* 317: 953–957.

Jacoby, E., Bouhelal, R., Gerspacher, M., and Seuwen, K. (2006). The 7 TM G-protein-coupled receptor target family. *ChemMedChem* 1: 761–782.

John, S. W., Krege, J. H., Oliver, P. M., et al. (1995). Genetic decreases in atrial natriuretic peptide and salt-sensitive hypertension. *Science* 267: 679–681.

Johnson, M. A., Tsai, L., Roy, D. S., et al. (2012). Neurons expressing trace amine-associated receptors project to discrete glomeruli and constitute an olfactory subsystem. *Proc. Natl. Acad. Sci. U S A* 109: 13410–13415.

Juilfs, D. M., Fulle, H. J., Zhao, A. Z., et al. (1997). A subset of olfactory neurons that selectively express cGMP-stimulated phosphodiesterase (PDE2) and guanylyl cyclase-D define a unique olfactory signal transduction pathway. *Proc. Natl. Acad. Sci. U S A* 94: 3388–3395.

Katada, S., Hirokawa, T., Oka, Y., et al. (2005). Structural basis for a broad but selective ligand spectrum of a mouse olfactory receptor: mapping the odorant-binding site. *J. Neurosci.* 25: 1806–1815.

Katada, S., Tanaka, M., and Touhara, K. (2004). Structural determinants for membrane trafficking and G protein selectivity of a mouse olfactory receptor. *J. Neurochem.* 90: 1453–1463.

Kato, A., Katada, S., and Touhara, K. (2008). Amino acids involved in conformational dynamics and G protein coupling of an odorant receptor: targeting gain-of-function mutation. *J. Neurochem.* 107: 1261–1270.

Kato, A. and Touhara, K. (2009). Mammalian olfactory receptors: pharmacology, G protein coupling and desensitization. *Cell Mol. Life Sci.* 66: 3743–3753.

Keller, A., Zhuang, H., Chi, Q., et al. (2007). Genetic variation in a human odorant receptor alters odour perception. *Nature* 449: 468–472.

Kerr, D. S., Von Dannecker, L. E., Davalos, M., et al. (2008). Ric-8B interacts with G alpha olf and G gamma 13 and co-localizes with G alpha olf, G beta 1 and G gamma 13 in the cilia of olfactory sensory neurons. *Mol. Cell. Neurosci.* 38: 341–348.

Khalifah, R. G. (1971). The carbon dioxide hydration activity of carbonic anhydrase. I. Stop-flow kinetic studies on the native human isoenzymes B and C. *J. Biol. Chem.* 246: 2561–2573.

Klabunde, T. and Hessler, G. (2002). Drug design strategies for targeting G-protein-coupled receptors. *Chembiochem* 3: 928–944.

Kleene, S. J. (2008). The electrochemical basis of odor transduction in vertebrate olfactory cilia. *Chem. Senses* 33: 839–859.

Krautwurst, D., Yau, K. W., and Reed, R. R. (1998). Identification of ligands for olfactory receptors by functional expression of a receptor library. *Cell* 95: 917–926.

Kuhn, M. (2003). Structure, regulation, and function of mammalian membrane guanylyl cyclase receptors, with a focus on guanylyl cyclase-A. *Circ. Res.* 93: 700–709.

Kuhn, M. (2009). Function and dysfunction of mammalian membrane guanylyl cyclase receptors: lessons from genetic mouse models and implications for human diseases. *Handb. Exp. Pharmacol.*, 47–69.

Kurahashi, T., and Yau, K. W. (1993). Co-existence of cationic and chloride components in odorant-induced current of vertebrate olfactory receptor cells. *Nature* 363: 71–74.

Leinders-Zufall, T., Cockerham, R. E., Michalakis, S., et al. (2007). Contribution of the receptor guanylyl cyclase GC-D to chemosensory function in the olfactory epithelium. *Proc. Natl. Acad. Sci. U S A* 104: 14507–14512.

Leinders-Zufall, T., Rand, M. N., Shepherd, G. M., et al. (1997). Calcium entry through cyclic nucleotide-gated channels in individual cilia of olfactory receptor cells: spatiotemporal dynamics. *J. Neurosci.* 17: 4136–4148.

Li, Y. R., and Matsunami, H. (2011). Activation state of the M3 muscarinic acetylcholine receptor modulates mammalian odorant receptor signaling. *Sci. Signal 4: ra1.*

Liberles, S. D., and Buck, L. B. (2006). A second class of chemosensory receptors in the olfactory epithelium. *Nature* 442: 645–650.

Liggett, S. B. (2002). Update on current concepts of the molecular basis of beta2-adrenergic receptor signaling. *J. Allergy. Clin. Immunol.* 110: S223–227.

Liggett, S. B., Caron, M. G., Lefkowitz, R. J., and Hnatowich, M. (1991). Coupling of a mutated form of the human beta 2-adrenergic receptor to Gi and Gs. Requirement for multiple cytoplasmic domains in the coupling process. *J. Biol. Chem.* 266: 4816–4821.

Liu, Y., Ruoho, A. E., Rao, V. D., and Hurley, J. H. (1997). Catalytic mechanism of the adenylyl and guanylyl cyclases: modeling and mutational analysis. *Proc. Natl. Acad. Sci. U S A* 94: 13414–13419.

Lowe, G., and Gold, G. H. (1993). Nonlinear amplification by calcium-dependent chloride channels in olfactory receptor cells. *Nature* 366: 283–286.

Lu, M., Echeverri, F., and Moyer, B. D. (2003). Endoplasmic reticulum retention, degradation, and aggregation of olfactory G-protein coupled receptors. *Traffic* 4: 416–433.

Malnic, B., Hirono, J., Sato, T., and Buck, L. B. (1999). Combinatorial receptor codes for odors. *Cell* 96: 713–723.

McClintock, T. S., Landers, T. M., Gimelbrant, A. A., et al. (1997). Functional expression of olfactory-adrenergic receptor chimeras and intracellular retention of heterologously expressed olfactory receptors. *Brain Res. Mol. Brain Res.* 48: 270–278.

Meyer, M. R., Angele, A., Kremmer, E., et al. (2000). A cGMP-signaling pathway in a subset of olfactory sensory neurons. *Proc. Natl. Acad. Sci. U S A* 97: 10595–10600.

Mombaerts, P. (2004). Genes and ligands for odorant, vomeronasal and taste receptors. *Nat. Rev. Neurosci.* 5: 263–278.

Munger, S. D., Leinders-Zufall, T., McDougall, L. M., et al. (2010). An olfactory subsystem that detects carbon disulfide and mediates food-related social learning. *Curr. Biol.* 20: 1438–1444.

Nagai, Y., Nishimura, A., Tago, K., et al. (2010). Ric-8B stabilizes the alpha subunit of stimulatory G protein by inhibiting its ubiquitination. *J. Biol. Chem.* 285: 11114–11120.

Nakamura, T. and Gold, G. H. (1987). A cyclic nucleotide-gated conductance in olfactory receptor cilia. *Nature* 325: 442–444.

Nei, M., Niimura, Y., and Nozawa, M. (2008). The evolution of animal chemosensory receptor gene repertoires: roles of chance and necessity. *Nat. Rev. Genet.* 9: 951–963.

Neuhaus, E. M., Zhang, W., Gelis, L., et al. (2009). Activation of an olfactory receptor inhibits proliferation of prostate cancer cells. *J. Biol. Chem.* 284: 16218–16225.

Niimura, Y. and Nei, M. (2005). Evolutionary dynamics of olfactory receptorgenes in fishes and tetrapods. *Proc. Natl. Acad. Sci. U S A* 102: 6039–6044.

Niimura, Y., Matsui, A., and Touhara, K. (2014). Extreme expansion of the olfactory receptor gene repertoire in African elephants and evolutionary dynamics of orthologous gene groups in 13 placental mammals. *Genome Res.* 24: 1485–1496.

Nikiforovich, G. V., Taylor, C. M., and Marshall, G. R. (2007). Modeling of the complex between transducin and photoactivated rhodopsin, a prototypical G-protein-coupled receptor. *Biochemistry* 46: 4734–4744.

Ogawa, H., Qiu, Y., Ogata, C. M., and Misono, K. S. (2004). Crystal structure of hormone-bound atrial natriuretic peptide receptor extracellular domain: rotation mechanism for transmembrane signal transduction. *J. Biol. Chem.* 279: 28625–28631.

Oka, Y., Katada, S., Omura, M., et al. (2006). Odorant receptor map in the mouse olfactory bulb: in vivo sensitivity and specificity of receptor-defined glomeruli. *Neuron* 52: 857–869.

Oka, Y., Omura, M., Kataoka, H., and Touhara, K. (2004). Olfactory receptor antagonism between odorants. *Embo. J.* 23: 120–126.

Otaki, J. M., Yamamoto, H., and Firestein, S. (2004). Odorant receptor expression in the mouse cerebral cortex. *J. Neurobiol.* 58: 315–327.

Pace, U., Hanski, E., Salomon, Y., and Lancet, D. (1985). Odorant-sensitive adenylate cyclase may mediate olfactory reception. *Nature* 316: 255–258.

Pacifico, R., Dewan, A., Cawley, D., et al. (2012). An olfactory subsystem that mediates high-sensitivity detection of volatile amines. *Cell reports* 2: 76–88.

Parat, M., McNicoll, N., Wilkes, B., et al. (2008). Role of extracellular domain dimerization in agonist-induced activation of natriuretic peptide receptor A. *Mol. Pharmacol.* 73: 431–440.

Parmentier, M., Libert, F., Schurmans, S., et al. (1992). Expression of members of the putative olfactory receptor gene family in mammalian germ cells. *Nature* 355: 453–455.

Peterlin, Z., Li, Y., Sun, G., et al. (2008). The importance of odorant conformation to the binding and activation of a representative olfactory receptor. *Chem. Biol.* 15: 1317–1327.

Pluznick, J. L., Zou, D. J., Zhang, X., et al. (2009). Functional expression of the olfactory signaling system in the kidney. *Proc. Natl. Acad. Sci. U S A* 106: 2059–2064.

Pluznick, J. L., Protzko, R. J., Gevorgyan, H., et al. (2013). Olfactory receptor responding to gut microbiota-derived signals plays a role in renin secretion and blood pressure regulation. *Proc. Natl. Acad. Sci. U S A* 110: 4410–4415.

Potter, L. R., Abbey-Hosch, S., and Dickey, D. M. (2006). Natriuretic peptides, their receptors, and cyclic guanosine monophosphate-dependent signaling functions. *Endocr. Rev.* 27: 47–72.

Reisert, J., Bauer, P. J., Yau, K. W., and Frings, S. (2003). The Ca-activated Cl channel and its control in rat olfactory receptor neurons. *J. Gen. Physiol.* 122: 349–363.

Reisert, J., Lai, J., Yau, K. W., and Bradley, J. (2005). Mechanism of the excitatory Cl- response in mouse olfactory receptor neurons. *Neuron* 45: 553–561.

Restrepo, D., Arellano, J., Oliva, A. M., et al. (2004). Emerging views on the distinct but related roles of the main and accessory olfactory systems in responsiveness to chemosensory signals in mice. *Horm. Behav.* 46: 247–256.

Saito, H., Chi, Q., Zhuang, H., et al. (2009). Odor coding by a Mammalian receptor repertoire. *Sci. Signal* 2: ra9.

Saito, H., Kubota, M., Roberts, R. W., et al. (2004). RTP family members induce functional expression of mammalian odorant receptors. *Cell* 119: 679–691.

Shirasu, M., Yoshikawa, K., Takai, Y., et al. (2014). Olfactory receptor and neural pathway responsible for highly selective sensing of musk odors. *Neuron* 81: 165–178.

Spehr, M., Gisselmann, G., Poplawski, A.et al. (2003). Identification of a testicular odorant receptor mediating human sperm chemotaxis. *Science* 299: 2054–2058.

Spehr, M., Kelliher, K. R., Li, X. H.et al. (2006). Essential role of the main olfactory system in social recognition of major histocompatibility complex peptide ligands. *J. Neurosci.* 26: 1961–1970.

Stephan, A. B., Shum, E. Y., Hirsh, S.et al. (2009). ANO2 is the cilial calcium-activated chloride channel that may mediate olfactory amplification. *Proc. Natl. Acad. Sci. U S A* 106: 11776–11781.

Sun, L., Wang, H., Hu, J.et al. (2009). Guanylyl cyclase-D in the olfactory CO_2 neurons is activated by bicarbonate. *Proc. Natl. Acad. Sci. U S A* 106: 2041–2046.

References

Sung, C. H., Schneider, B. G., Agarwal, N. et al. (1991). Functional heterogeneity of mutant rhodopsins responsible for autosomal dominant retinitis pigmentosa. *Proc. Natl. Acad. Sci. U S A* 88: 8840–8844.

Tan, E. S., Naylor, J. C., Groban, E. S.et al. (2009). The molecular basis of species-specific ligand activation of trace amine-associated receptor 1 (TAAR(1)). *ACS Chem. Biol.* 4: 209–220.

Touhara, K., Sengoku, S., Inaki, K.et al. (1999). Functional identification and reconstitution of an odorant receptor in single olfactory neurons. *Proc. Natl. Acad. Sci. U S A* 96: 4040–4045.

Ukhanov, K., Brunert, D., Corey, E. A., and Ache, B. W. (2011). Phosphoinositide 3-kinase-dependent antagonism in mammalian olfactory receptor neurons. *J. Neurosci.* 31: 273–280.

Von Dannecker, L. E., Mercadante, A. F., and Malnic, B. (2005). Ric-8B, an olfactory putative GTP exchange factor, amplifies signal transduction through the olfactory-specific G-protein Galphaolf. *J. Neurosci.* 25: 3793–3800.

Von Dannecker, L. E., Mercadante, A. F., and Malnic, B. (2006). Ric-8B promotes functional expression of odorant receptors. *Proc. Natl. Acad. Sci. U S A* 103: 9310–9314.

Wallrabenstein, I., Kuklan, J., Weber, L., Zborala, S., Werner, M., et al. (2013). Human Trace Amine-Associated Receptor TAAR5 Can Be Activated by Trimethylamine. PLoS ONE 8(2): e54950. doi:10.1371/journal.pone.0054950.

Wilson, E. M., and Chinkers, M. (1995). Identification of sequences mediating guanylyl cyclase dimerization. *Biochemistry* 34: 4696–4701.

Wong, S. T., Trinh, K., Hacker, B., et al. (2000). Disruption of the type III adenylyl cyclase gene leads to peripheral and behavioral anosmia in transgenic mice. *Neuron* 27: 487–497.

Wu, L., Pan, Y., Chen, G. Q.et al. (2012). Receptor-transporting Protein 1 Short (RTP1S) Mediates Translocation and Activation of Odorant Receptors by Acting through Multiple Steps. *J. Biol. Chem.* 287: 22287–22294.

Yoon, H., Enquist, L. W., and Dulac, C. (2005). Olfactory inputs to hypothalamic neurons controlling reproduction and fertility. *Cell* 123: 669–682.

Yoshikawa, K. and Touhara, K. (2009). Myr-Ric-8A enhances G(alpha15)-mediated Ca^{2+} response of vertebrate olfactory receptors. *Chem. Senses* 34: 15–23.

Yoshikawa, K., Nakagawa, H., Mori, N., et al. (2013). *Nature Chem. Biol.* 9: 160–162.

Zhainazarov, A. B., Spehr, M., Wetzel, C. H.et al. (2004). Modulation of the olfactory CNG channel by Ptdlns(3,4,5)P3. *J. Membr. Biol.* 201: 51–57.

Zhang, J., Pacifico, R., Cawley, D., et al. (2013). Ultrasensitive detection of amines by a trace amine-associated receptor. *J Neurosci.* 33: 3228–3239.

Zhang, X. and Firestein, S. (2002). The olfactory receptor gene superfamily of the mouse. *Nat. Neurosci.* 5: 124–133.

Zhang, X., Rogers, M., Tian, H.et al. (2004). High-throughput microarray detection of olfactory receptor gene expression in the mouse. *Proc. Natl. Acad. Sci. U S A* 101: 14168–14173.

Zhao, H., Ivic, L., Otaki, J. M.et al. (1998). Functional expression of a mammalian odorant receptor. *Science* 279: 237–242.

Zheng, J. and Zagotta, W. N. (2004). Stoichiometry and assembly of olfactory cyclic nucleotide-gated channels. *Neuron* 42: 411–421.

Zhuang, H. and Matsunami, H. (2007). Synergism of accessory factors in functional expression of mammalian odorant receptors. *J. Biol. Chem.* 282: 15284–15293.

Zufall, F. and Munger, S. D. (2010). Receptor guanylyl cyclases in mammalian olfactory function. *Mol. Cell Biochem.* 334: 191–197.

Chapter 6

Odorant Receptor Gene Regulation

AKIO TSUBOI and HITOSHI SAKANO

6.1 INTRODUCTION AND OVERVIEW

The main olfactory system recognizes a vast range of molecules that represent vital information about an animal's environment, including the location of prey and predators. Sensory information is specially encoded in the brain, forming neural maps that are fundamental for higher-order processing of sensory information. Molecular mechanisms of sensory map formation in the visual system in particular have been extensively studied. It is well established that axonal projection of retinal ganglion cells is instructed by multiple pairs of axon-guidance molecules that demonstrate graded expression in the retina and tectum (McLaughlin and O'Leary, 2005). The spatial organization of the projecting neurons in the retina is maintained and projected onto the target, preserving the nearest-neighbor relationship. By contrast, for olfactory map formation instructed by gene expression, our beginning of understanding such maps stemmed from the ground-breaking discovery of the odorant receptor gene family by Buck and Axel (1991).

In the *Mus musculus* olfactory system, there are ~1,000 functional odorant receptor (OR) genes comprising the largest multigene family of this species (Buck and Axel, 1991). These OR genes are clustered at about 50 different loci that are scattered among most of the chromosomes. Each olfactory sensory neuron (OSN) in the olfactory epithelium (OE) expresses only one member of the OR gene family in a mutually-exclusive and monoallelic manner. Because OSN axons of spatially dispersed cell bodies with the same neuronal identity converge on a pair of glomerular locations in the target olfactory bulb (OB) (Mombaerts et al., 1996), the odorant stimuli received in the OE are converted to an olfactory map of activated glomeruli (Figure 6.1). Thus, the one neuron – one receptor rule forms the genetic basis for OR-instructed axonal projection. Much of the olfactory map formation appears to occur autonomously by axon-axon interaction, without involving target-derived cues (Sakano, 2010). Within the OB, the odor information coded in the olfactory map is processed in the local neuronal circuits and then conveyed by mitral and tufted cells to various areas of the olfactory cortex (Mori et al., 1999: Shepherd et al., 2004; Mori and Sakano, 2011). In this chapter, we review recent progress on the study of OR gene regulation and OR-instructed axonal projection of OSNs in rodents.

6.2 ORGANIZATION AND STRUCTURE OF THE ODORANT RECEPTOR GENES

A multigene family encoding OR molecules was first identified in rat, which consists of hundreds of related OR genes (Buck and Axel, 1993). In humans, only ~380 of the 800 OR genes have an intact open reading frame and are assumed to be functional (Glusman et al., 2001; Malnic et al., 2004). In the mouse, of ~1,400 OR sequences, some 20% are pseudogenes (Zhang and Firestein, 2002; Young et al., 2002; Zhang et al., 2004a; Godfrey et al., 2004). The mouse repertoire of ~1,100 functional OR genes is the largest multigene family that encompasses ~5% of all genes. Mouse ORs are grouped into ~230 subfamilies, based on the amino-acid identities of >40% (Zhang and Firestein, 2002). Vertebrate OR genes are divided phylogenetically into two different classes, I and II. The class I genes resemble the fish OR genes, while the class II genes

---

Handbook of Olfaction and Gustation, Third Edition. Edited by Richard L. Doty.
© 2015 Richard L. Doty. Published 2015 by John Wiley & Sons, Inc.

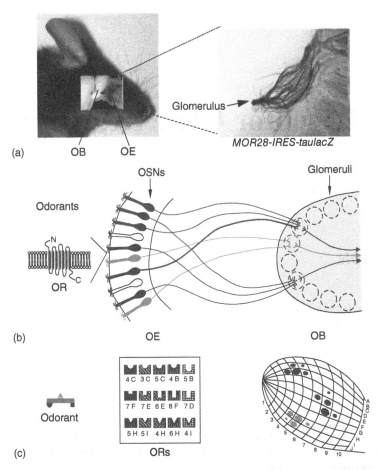

Figure 6.1 Olfactory map in the mouse olfactory bulb. (a) Odorant receptor (OR)-instructed axonal projection of olfactory sensory neurons (OSNs). *MOR28*-expressing axons are stained blue with X-gal in the transgenic mouse, *MOR28-ires-tau-lacZ*. OSNs expressing the transgene converge their axons to a specific site, forming a glomerulus (*arrow*) in the olfactory bulb (OB). (b) Axonal segregation and olfactory map formation. OSN axons are guided to approximate destinations in the OB by a combination of dorsal-ventral (D-V) patterning and anterior-posterior (A-P) patterning. D-V projection occurs based on anatomical locations of OSNs in the olfactory epithelium (OE). A-P projection is regulated by OR-derived cAMP signals. The map is further refined in an activity-dependent manner during the early neonatal period. (c) Conversion of olfactory signals. In the OE, each OSN expresses only one functional OR gene in a monoallelic manner. Furthermore, OSN axons expressing the same OR species converge to a specific glomerulus in the OB. Thus each glomerulus represents one OR species. Because a given odorant typically activates multiple OR species and a given OR responds to multiple odorants, odorant signals received by OSNs in the OE are converted into a two-dimensional odor map of activated glomeruli with varying magnitudes of activity. (*See plate section for color version.*)

are unique to terrestrial vertebrates. The mouse OR genes are clustered at ~50 different loci spread over almost all chromosomes. Each cluster contains 1–189 OR genes, with an average intergenic distance of ~25 kilobases. Most OR genes are composed of 2–5 exons, where usually one exon encodes the entire protein. At least two-thirds of OR genes exhibit multiple transcriptional variants with alternative isoforms of noncoding exons (Young et al., 2003). Comparative analysis of the mouse and human OR genes revealed that coding homologies within the clusters are accounted for by recent gene duplication as well as gene conversion among the coding sequences (Nagawa et al., 2002). To maintain the integrity of the domain structure of the olfactory sensory map, gene conversion may have played an active role in having the sequence similarities between the paralogs of the OR genes (Tsuboi et al., 2011).

6.2.1 Zonal Expression of Odorant Receptor Genes in the Olfactory Epithelium

In rodents, the OE can be divided into four distinct zones, based on the expression patterns of OR genes (Ressler et al., 1993; Vassar et al., 1993; Strotmann et al., 1994a, b; Tsuboi et al., 1999). The most dorso-medial zone of the OE is defined as zone 1 (Sullivan et al., 1996), also as dorsal (D) zone (Kobayakawa et al., 2007). D-zone OSNs are positive for O-MACS (Oka et al., 2003), but negative

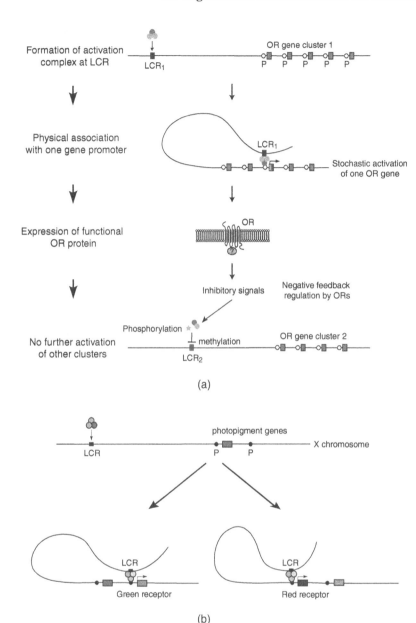

Figure 6.2 Mutually-exclusive expression of odorant receptor and photopigment genes. (a) In the odorant receptor (OR) gene system, the locus control region (LCR)-promoter (P) interaction alone would not preclude the activation of a second OR gene located in the other allele or in other OR gene clusters. Once a functional gene is expressed, the OR molecules transmit inhibitory signals to block the further activation of additional OR gene clusters. (b) In the human visual system, an LCR plays an important role in choosing either the red or green photopigment gene in a mutually exclusive manner in the cone cells of the retina. Stochastic interaction between the LCR and either promoter (P) ensures the mutually exclusive expression of the human photopigment genes encoded on the X chromosome. From Smallwood et al., 2002. (*See plate section for color version.*)

for OCAM (Yoshihara et al., 1997). Conversely, ventral (V) zone OSNs are negative for O-MACS, but positive for OCAM. The D zone is further divided into D_I and D_{II} zones based on the expressing OR gene classes described below (Mori and Sakano, 2011).

6.2.1.1 Class I Odorant Receptors.
The class I ORs were first identified in fish (Ngai et al., 1993) and later found in amphibians (Freitag et al., 1995). They are assumed to recognize water-soluble odorous ligands. Class I ORs are also found in mammals, including human (Glusman et al., 2001; Malnic et al., 2004) and mouse (Zhang and Firestein, 2002; Young et al., 2002; Zhang et al., 2004a; Godfrey et al., 2004). In the mouse, there are over 150 class I-OR genes, classified into 42 subfamilies (Zhang and Firestein, 2002; Zhang et al., 2004a). They are all clustered at a single locus on the mouse chromosome 7. Comprehensive *in situ* hybridization (ISH) revealed that almost all class I-OR genes are expressed specifically in D zone (Tsuboi et al., 2006), confirming the data from a high-throughput microarray analysis (Zhang et al., 2004b). It appears that they are coordinately regulated by a common locus control region (LCR) and/or promoter sequences (Figure 6.2). In the mutant mouse lacking the homeodomain protein Lhx2, class I expression becomes spared, while class II expression is abolished (Hirota et al., 2007).

6.2.1.2 Class II Odorant Receptors.

In contrast to class I, class II OR genes account for a large share of the mammalian OR repertoire. ISH analyses with various OR gene probes demonstrated that each OR gene is expressed in a restricted area in the OE. It has been reported that OSNs expressing a given OR gene are confined to one of the four OE zones but are randomly distributed within the zone (Ressler et al., 1993; Vassar et al., 1993; Strotmann et al., 1994a, b Tsuboi et al., 1999). This appears to be true for D-zone specific class I genes. However, for most class II OR gene, the expression area is not always confined to one of the four conventional zones (Norlin et al., 2001; Iwema et al., 2004; Miyamichi et al., 2005). Expression areas appear to be specific to each OR gene and are arranged in an overlapping and continuous manner in the OE. It was demonstrated by DiI retrograde staining experiments that the dorsal/ventral arrangement of glomeruli in the OB is well correlated with the expression areas of corresponding ORs along the dorsomedial/ventrolateral (DM/VL) axis in the OE (Miyamichi et al., 2005). Thus, OR gene choice may not be totally stochastic, but rather it may be restricted by the OSN location in the OE. How does this positional information within the OE regulate OR gene choice? It is possible that this regulation is determined by cell lineage, resulting in the use of zone-specific transcription factors, e.g., Lhx2, Msx1 and Foxg1 (Hirota et al., 2007; Norlin et al., 2001; Duggan et al., 2008).

6.2.2 Single Odorant Receptor Gene Expression

In the mouse olfactory system, each OSN chooses for expression only one functional OR gene in a mutually-exclusive and monoallelic manner (Chess et al., 1994; Serizawa et al., 2000; Ishii et al., 2001). Such unique expression forms the genetic basis for the OR-instructed axonal projection of OSNs to the OB. How is the singular OR gene choice regulated and how is the one neuron – one receptor rule maintained in OSNs? Based on previous studies of other multigene families, three mechanisms have been considered for the choice and activation of OR genes: (a) DNA recombination, which brings a promoter and the enhancer region into close proximity; (b) gene conversion, which transfers a copy of the gene into the expression cassette; and (c) a regulatory DNA region, which interacts with only one promoter site. Irreversible DNA changes, such as recombination and gene conversion, had been attractive explanations for single OR gene expression because of the many parallels between the immune system and the olfactory system. However, these theories were dismissed after two groups independently cloned mice from postmitotic OSN nuclei and determined that the mice showed no irreversible DNA rearrangement in the OR genes (Eggan et al., 2004; Li et al., 2004).

6.2.2.1 Odorant Receptor Gene Choice.

Because the genetic translocation model appeared unlikely, another possibility was explored, namely a locus control region (LCR) that might regulate the single OR gene choice (Figure 6.2a). LCR is defined as a *cis*-acting regulatory region that controls multiple genes clustered at a specific genetic locus. The first example of an LCR was identified in the globin gene locus containing developmentally regulated and related genes (Grosveld et al., 1987). It has been assumed that transcription-activating factors bind to the LCR (holocomplex) that physically interacts with the remote promoter site by looping out the intervening DNA. In the globin gene system, an active chromatin hub structure has been reported, in which the LCR is in close proximity to the gene to be expressed (Carter et al., 2002; Tolhuis et al., 2002). One example of such an LCR in the mouse olfactory system, named *H* for the homology between the mouse and human DNA sequences, was identified upstream of the *MOR28* cluster containing seven murine OR genes (Nagawa et al., 2002; Serizawa et al., 2003). Deletion and mutation analyses of the *H* region further revealed that the 124-bp core-*H* region, which contains two homeodomain sequences and one O/E-like sequence, is sufficient to achieve the enhancer activity (Nishizumi et al., 2007). Based on the deletion analysis, another example of OR-gene LCR, named *P*, was identified between the *P3* and *P4* genes (Bozza et al., 2009; Vassalli et al., 2011, Khan et al., 2011). Both the homeodomain and O/E-like sequences are found in the *H* and *P* regions as well as in many OR-gene promoters (Vassalli et al., 2002; Hirota and Mombaerts, 2004; Rothman et al., 2005; Michaloski et al., 2006; Bozza et al., 2009; Vassalli et al., 2011). Homeodomain factors, Lhx2 and Emx2, and O/E family proteins are known to bind to their motifs in the OR gene promoter (Wang et al., 2004; Hirota et al., 2007; McIntyre et al., 2008). It is possible that these nuclear factors bind to the OR-gene LCRs and form a complex that remodels the chromatin structure near the cluster, thereby activating one OR promoter site at a time by physical interaction (Serizawa et al., 2003) (Figure 6.2a). This model is attractive because it reduces the likelihood of the simultaneous activation of multiple OR genes, from a probability of ~1,000 individual genes to ~50 LCR loci.

In the immune system, it has been reported that the LCR of the *IL-4* gene in naïve T helper cells interacts not only with the adjacent genes on chromosome 11, but also with the alternatively expressed *IFN-γ* gene on chromosome 10 (Spilianakis et al., 2005). Such an inter-chromosomal association of multiple gene loci may be a common feature of chromosomal organization for the coordinately regulated genes. Does the *H* region act on other OR gene clusters, not only in *cis* but also in *trans*,

6.2 Organization and Structure of the Odorant Receptor Genes

like the *IL-4* LCR in T cells? Based on fluorescent *in situ* hybridization and chromosome conformation capture analyses of OSN nuclei, Lomvardas et al. (2006) reported that the single *trans*-acting LCR *H* may allow the stochastic activation of only one OR gene in each OSN. However, knockout studies of *H* contradict the single *trans*-acting LCR model for the OR gene choice (Fuss et al., 2007; Nishizumi et al., 2007). Targeted deletion of *H* abolished the expression of only three proximal OR genes in *cis*, indicating the presence of another LCR in the downstream region to regulate the four distal OR genes in the same *MOR28* cluster. Furthermore, in heterozygous (H^+/H^-) knockout mice, the wild-type H^+ allele could not rescue the H^- mutant allele in *trans*, indicating that *H* can act only in *cis* and not in *trans* (Nishizumi et al., 2007).

6.2.2.2 Negative Feedback Regulation to Maintain the One Neuron – One Receptor Rule.

In the human visual system, an LCR plays an important role in choosing either the red or green photopigment gene in a mutually exclusive manner in the cone cells of the retina (Wang et al., 1999). It has been shown that stochastic interaction between the LCR and either of the two promoters ensures the mutually exclusive expression of the human photopigment genes that are encoded on the X chromosome (Smallwood et al., 2002) (Figure 6.2b). In contrast, in the OR gene system, the LCR-promoter interaction alone would not preclude the activation of a second OR gene located in the other allele or in other OR gene clusters. Therefore, it has been postulated that the functional OR proteins have an inhibitory role to prevent further activation of other OR genes (Serizawa et al., 2003; Lewcock and Reed, 2004). Transgenic experiments demonstrated that the mutant OR genes lacking either the entire coding sequence or the start codon can permit a second OR gene to be expressed (Serizawa et al., 2003; Feinstein et al., 2004; Lewcock and Reed, 2004; Shykind et al., 2004). Naturally occurring frameshift mutants of OR genes also allow the coexpression of a functional OR gene (Serizawa et al., 2003). It is known that a substantial number of pseudogenes are present in the mammalian OR gene families (~30% of total OR genes in mice and ~60% in humans). If an activated LCR has selected a pseudogene and has been trapped by its promoter, other clusters must undergo a similar process to ensure the activation of a functional OR gene (Roppolo et al., 2007). A pseudogene may help slow the process of OR gene activation and further reduce the likelihood of activating two functional OR genes. Thus rate-limited activation of an OR gene by *cis*-acting LCRs and negative feedback regulation by the OR gene product together appear to

ensure the maintenance of the one neuron – one receptor rule (Serizawa et al., 2003, 2004) (Figure 6.2a). However, the exact nature of the negative feedback signals has not been explored. Targets of the feedback signals are also issues for future studies. Promoters of OR genes, enhancers of OR gene clusters, and protein factors binding them could be silenced by OR-derived negative feedback signals. Histone modification could be another target of the feedback regulation.

It has been reported that forced expression of an OR gene with the Tet transactivation system can suppress the expression of endogenous ORs (Nguyen et al., 2007; Fleischmann et al., 2008). As the forced expression was found to be inefficient, researchers have proposed that the OR coding sequence has an inhibitory effect on transcriptional regulation. However, it has not been demonstrated that the suppressive effect would result from the DNA sequence and not the potentially toxic effects of OR overexpression. It is important to identify the essential nucleotide sequences responsible for the suppressive effects of functional ORs. In neonatal mice, a small fraction of OSNs express two functional OR species simultaneously, which are eventually eliminated in an activity-dependent manner (Tian and Ma, 2008). This form of negative selection has also been considered to be a fail-safe mechanism to ensure monogenic OR expression (Mombaerts, 2006).

OR-mediated negative feedback does not require G protein signaling (Imai et al., 2006; Nguyen et al., 2007). The involvement of G proteins in negative feedback regulation was tested using transgenic mice expressing a mutant-type OR, in which the DRY motif, essential for G protein signaling, was changed to RDY. This mutation completely abolished odor-evoked calcium responses in OSNs, indicating the loss of G protein signaling. However, OSNs expressing the mutant OR maintained the one neuron–one receptor rule. Seven-transmembrane receptors often utilize G protein-independent signaling pathways, such as those involving β-arrestin (Shenoy and Lefkowitz, 2011). Notably, the DRY motif is dispensable for β-arrestin-mediated seven-transmembrane-receptor signaling (Seta et al., 2002). Knockout studies will clarify whether β-arrestin-mediated signaling is involved in negative feedback regulation.

Recently, Magklara et al. (2011) reported that chromatin-mediated silencing is important to prevent multiple expression of OR genes in each OSN. They found that heterochromatic compaction of OR-gene clusters occurs before OR transcription via two trimethyl marks of H3K9me3 and H4K20me3. The enrichment for these silent marks is significantly reduced in an activated OR-gene region, which is marked instead with H3K4me3. It is assumed that although all OR-gene loci become silenced before OR-gene transcription, at a later stage, a limited histone demethylase Lsd1 activity removes methylation of H3K9me3 from a stochastically chosen locus, activating its transcription (Lyons et al., 2013; Magklara et al., 2011). Once a functional gene is expressed, negative feedback signals may prevent the enzyme/selector from activating other OR-gene loci.

6.2.3 Regulation of Odorant Receptor-Instructed Axonal Projection

The olfactory map in the OB comprises discrete glomeruli, each representing a single OR species (Ressler et al., 1994; Vassar et al., 1994; Mombaerts et al., 1996) (Figure 6.3b). Coding-swap experiments of OR genes indicated the instructive role of the OR protein in OSN projection (Mombaerts et al., 1996; Wang et al., 1998; Feinstein and Mombaerts, 2004). Because OSNs expressing the same OR are scattered in the OE for anterior-posterior projection, topographic organization must occur during the process of axonal projection of OSNs. Unlike axonal projection in other sensory systems in which relative positional information is preserved between the periphery and the brain (Lemke and Reber, 2005; McLaughlin and O'Leary, 2005; Petersen, 2007), there is no such correlation for the projection along the anterior-posterior axis in the mouse olfactory system. Although OR molecules have been known to play an instructive role in projecting OSN axons to form the glomerular map, it has remained entirely unclear how this occurs at the molecular level. Intriguingly, OR molecules are detected in axon termini by tagging with green fluorescent protein (Feinstein and Mombaerts, 2004) or by immunostaining with anti-OR antibodies (Barnea et al., 2004; Strotmann et al., 2004). On the basis of these observations, it was suggested that the OR protein itself may recognize positional cues in the OB and also may mediate homophilic interaction of similar axons (Feinstein et al., 2004; Mombaerts, 2006). Although these models were attractive, recent studies argue against them. Instead of directly acting as guidance receptors or adhesion molecules, ORs regulate transcription levels of axon-guidance and axon-sorting molecules by OR-derived cAMP signals with levels uniquely determined by the OR species (Imai et al., 2006; Serizawa et al., 2006). In the mouse olfactory system, axon guidance/sorting molecules whose expression levels are regulated by ORs can be categorized into two different types (Sakano, 2010). One is type I, including Nrp1 and Plexin-A1, which is expressed at axon termini in a graded manner along the anterior-posterior axis in the OB. The other is type II, including Kirrel2 and Kirrel3, which is expressed at axon termini showing a mosaic pattern in the OB. In the knockout mouse for adenylyl-cyclase III, Nrp1 and Kirrel2 are downregulated, whereas Plexin-A1 and Kirrel3

are upregulated, suggesting that both type I and type II genes are under the control of cAMP signals. However, type II expression is regulated in an activity-dependent manner, whereas type I expression is independent from the neuronal activity. It is interesting to study how the same second messenger, cAMP, is capable of regulating two types of genes differently.

Recently, Nakashima et al. (2013) reported that the agonist-independent OR activity determines transcription levels of type I molecules, e.g., Neuropilin-1 and Plexin-A1, thereby regulating the anterior-posterior targeting of OSN axons. They demonstrated that G_s, but not G_{olf}, is responsible for mediating the agonist-independent OR activity. In contrast, expression of type II molecules, e.g., Kirrel2 and Kirrel3, is regulated by the ligand-induced OR activity using Golf. Nakashima et al. (2013) conclude that the equilibrium of conformational transitions set by each OR is the major determinant of expression levels of anterior-posterior targeting molecules.

6.2.3.1 Positional Information of Olfactory Sensory Neurons in the Olfactory Epithelium Regulates Dorsal-Ventral Projection in the Olfactory Bulb.

In the dorsal region of the OE, OSNs expressing a given OR gene are distributed throughout D zone (Tsuboi et al., 2006). However, in the ventral region, OR genes exhibit spatially limited expression patterns along the DM-VL axis of the OE (Miyamichi et al., 2005). The relationship between the dorsal-ventral positioning of glomeruli and the locations of OSNs in the OE has been demonstrated by retrograde DiI staining of OSN axons (Astic et al., 1987; Miyamichi et al., 2005). These observations suggest that the anatomical locations of OSNs in the OE contribute to the dorsal-ventral positioning of glomeruli in the OB. In the OE, an axon-guidance receptor, Nrp2, and its repulsive ligand, Sema3F, are both expressed by OSN axons in a complementary manner to regulate dorsal-ventral projection (Takahashi et al., 2010; Takeuchi et al., 2010) (Figure 6.3a). Another set of repulsive signaling molecules, Robo2 and Slit1, are needed to restrict the first wave of OSN projection to the anterodorsal OB (Cho et al., 2007; Nguyen-Ba-Charvet et al., 2008). Expression levels of dorsal-ventral guidance molecules, such as Nrp2 and Sema3F, are closely correlated with the expressed OR species. However, unlike Nrp1 and Sema3A, which are involved in anterior-posterior positioning, the transcription of *Nrp2* and *Sema3F* is not downstream of OR signaling. If dorsal-ventral guidance molecules are not regulated by OR-derived signals, how are their expression levels determined and correlated with the expressed OR species? It appears that both OR gene choice and expression levels of *Nrp2* and *Sema3F* are commonly regulated by positional information within the OE. This regulation

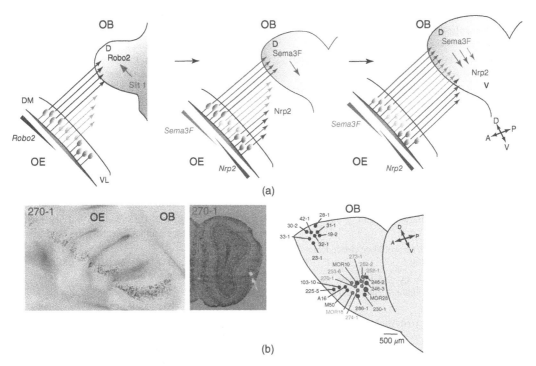

Figure 6.3 OR-gene expression in the OE contributes to the dorsal-ventral positioning of glomeruli in the OB. (a) Axonal extension of OSNs occurs sequentially along the dorsomedial-ventrolateral (DM-VL) axis of the OE as the OB grows from dorsal to ventral during development (Takeuchi et al. 2010). (Left panel) Dorsal-zone OSNs express Robo2 and project their axons to the prospective dorsal (D) domain of the embryonic OB. Repulsive interactions between Robo2 and Slit1 are needed to restrict early OSN projection to the embryonic OB. (Middle panel) In the OE, *Nrp2* and *Sema3F* genes are expressed in a complementary manner. (Right panel) Sema3F is deposited at the anterodorsal region of the OB by early arriving dorsal-zone axons and prevents the late-arriving Nrp2$^+$ axons from invading the dorsal region of the OB, resulting in projection to the ventral (V) domain. (b) A glomerular map generated by *in situ* hybridization (ISH) of OB sections. Axonal projection of OSNs occurs sequentially from the DM to the VL area in the OE, leading to establishment of the glomerular order in the OB along the D-V axis. Representative data for MOR270-1 were shown in the whole-mount ISH of the OE (left) and the radioisotope ISH of OB section (center). The glomerular map is schematically illustrated in the right on the basis of OB-ISH data. From Takahashi et al., 2010. (*See plate section for color version.*)

is likely determined by cell lineage, resulting in the use of specific sets of transcription factors, which can explain the anatomical correlation along the DM-VL axis of the OE.

6.3 CONCLUSION

In the mammalian olfactory system, each OSN activates only one functional OR gene in a mutually-exclusive and monoallelic manner. Such unique expression forms the genetic basis for the OR-instructed axonal projection of OSNs to the OB. Thus, the one neuron – one receptor rule is essential for the conversion of olfactory signals received in the OE into the olfactory map in the OB. Two examples of OR-gene LCRs, *H* and *P*, have been reported for the *MOR28* cluster (Serizawa et al., 2003) and P3 cluster (Khan et al., 2011). Like the LCR in other gene systems, physical interaction between the LCR and promoter probably enables the expression of only one OR gene within the cluster. Double *in situ* hybridization experiments demonstrated that the mutant OR genes, which contain the entire coding deletion or frameshift sequence, can permit a second OR gene to be expressed. Although it was recently reported that chromatin-mediated silencing is important for the single OR expression (Magklara et al., 2011), an inhibitory role has been postulated for the functional OR protein that may prevent a histone-demethylation enzyme/selector from activating another allele (Lyons et al., 2013). Stochastic activation of an OR gene by LCR and negative-feedback regulation by the OR gene product, together, may ensure the maintenance of the one neuron – one receptor rule in the mammalian olfactory system.

REFERENCES

Astic, L., Saucier, D., and Holley, A. (1987). Topographical relationships between olfactory receptor cells and glomerular foci in the rat olfactory bulb. *Brain Res.* 424: 144–152.

Barnea, G., O'Donnell, S., Mancia, F. et al. (2004). Odorant receptors on axon termini in the brain. *Science* 304: 1468.

Bozza, T., Vassalli, A., Fuss, S. et al. (2009). Mapping of class I and class II odorant receptors to glomerular domains by two distinct types of olfactory sensory neurons in the mouse. *Neuron* 61: 220–233.

Buck, L. and Axel, R. (1991). A novel multigene family may encode odorant receptors: a molecular basis for odor recognition. *Cell* 65: 175–187.

Carter, D., Chakalova, L., Osborne, C. S., Dai, Y. F., Fraser, P. (2002). Long-range chromatin regulatory interaction in vivo. *Nat. Genet.* 32: 623–626.

Chess, A., Simon, I., Cedar, H., and Axel, R. (1994). Allelic inactivation regulates olfactory receptor gene expression. *Cell* 78: 823–834.

Cho, J. H., Lepine, M., Andrews, W., et al. (2007). Requirement for Slit-1 and Robo-2 in zonal segregation of olfactory sensory neuron axons in the main olfactory bulb. *J. Neurosci.* 27: 9094–9104.

Duggan, C. D., Demaria, S., Baudhuin, A., et al. (2008). Foxg1 is required for development of the vertebrate olfactory system. *J. Neurosci.* 28: 5229–5239.

Eggan, K., Baldwin, K., Tackett, M., et al. (2004). Mice cloned from olfactory sensory neurons. *Nature* 428: 44–49.

Feinstein, P., Bozza, T., Rodriguez, I., et al. (2004). Axon guidance of mouse olfactory sensory neurons by odorant receptors and the β2 adrenergic receptor. *Cell* 117: 833–846.

Feinstein, P. and Mombaerts, P. (2004). A contextual model for axonal sorting into glomeruli in the mouse olfactory system. *Cell* 117: 817–831.

Fleischmann, A., Shykind, B. M., Sosulski, D. L., et al. (2008). Mice with a "monoclonal nose": perturbations in an olfactory map impair odor discrimination. *Neuron* 60: 1068–1081.

Freitag, J., Krieger, J., Strotmann, J., and Breer, H. (1995). Two classes of olfactory receptors in *Xenopus laevis*. *Neuron* 15: 1383–1392.

Fuss, S. H., Omura, M., and Mombaerts, P. (2007). Local and *cis* effects of the H element on expression of odorant receptor genes in mouse. *Cell* 130: 373–384.

Glusman, G., Yanai, I., Rubin, I., and Lancet, D. (2001). The complete human olfactory subgenome. *Genome Res.* 11: 685–702.

Godfrey, P. A., Malnic, B., and Buck, L. B. (2004). The mouse olfactory receptor gene family. *Proc. Natl. Acad. Sci. U.S.A.* 101: 2156–2161.

Grosveld, F., van Assendelft, D. B., Greaves, D. R., and Kollias, G. (1987). Position-independent, high-level expression of the human β-globin gene in transgenic mice. *Cell* 51: 975–985.

Hirota, J. and Mombaerts, P. (2004). The LIM-homeodomain protein Lhx2 is required for complete development of mouse olfactory sensory neurons. *Proc. Natl. Acad. Sci. USA* 101: 8751–8755.

Hirota, J., Omura, M., and Mombaerts, P. (2007). Differential impact of Lhx2 deficiency on expression of class I and class II odorant receptor genes in mouse. *Mol. Cell. Neurosci.* 34: 679–688.

Imai, T., Suzuki, M., and Sakano, H. (2006). Odorant receptor-derived cAMP signals direct axonal targeting. *Science* 314: 657–661.

Ishii, T., Serizawa, S., Kohda, A., et al. (2001). Monoallelic expression of the odourant receptor gene and axonal projection of olfactory sensory neurons. *Genes Cells* 6: 71–78.

Iwema, C. L., Fang, H., Kurtz, D. B., et al. (2004). Odorant receptor expression patterns are restored in lesion-recovered rat olfactory epithelium. *J. Neurosci.* 24: 356–369.

Khan, M., Vaes, E., and Mombaerts, P. (2011). Regulation of the probability of mouse odorant receptor gene choice. *Cell* 147: 907–921.

Kobayakawa, K., Kobayakawa, R., Matsumoto, H., et al. (2007). Innate versus learned odour processing in the mouse olfactory bulb. *Nature* 450: 503–508.

Lemke, G. and Reber, M. (2005). Retinotectal mapping: new insights from molecular genetics. *Annu. Rev. Cell Dev. Biol.* 21: 551–580.

Lewcock, J. W. and Reed, R. R. (2004). A feedback mechanism regulates monoallelic odorant receptor expression. *Proc. Natl. Acad. Sci. USA* 101: 1069–1074.

Li, J., Ishii, T., Feinstein, P., and Mombaerts, P. (2004). Odorant receptor gene choice is reset by nuclear transfer from mouse olfactory sensory neurons. *Nature* 428: 393–399.

Lomvardas, S., Barnea, G., Pisapia, D. J., et al. (2006). Interchromosomal interactions and olfactory receptor choice. *Cell* 126: 403–413.

Lyons, D.B., Allen, W.E., Goh, T., et al. (2013). An epigenetic trap stabilizes singular olfactory receptor expression. *Cell* 154: 325–336.

Magklara, A., Yen, A., Colquitt, B. M., et al. (2011). An epigenetic signature for monoallelic olfactory receptor expression. *Cell* 145: 555–570.

Malnic, B., Godfrey, P. A., and Buck, L. B. (2004). The human olfactory receptor gene family. *Proc. Natl. Acad. Sci. U.S.A.* 101: 2584–2589.

McIntyre, J. C., Bose, S. C., Stromberg, A. J., and McClintock, T. S. (2008). Emx2 stimulates odorant receptor gene expression. *Chem. Senses.* 33: 825–837.

McLaughlin, T. and O'Leary, D. D. (2005). Molecular gradients and development of retinotopic maps. *Annu. Rev. Neurosci.* 28: 327–355.

Michaloski, J. S., Galante, P. A., and Malnic, B. (2006). Identification of potential regulatory motifs in odorant receptor genes by analysis of promoter sequences. *Genome Res.* 16: 1091–1098.

Miyamichi, K., Serizawa, S., Kimura, H. M., and Sakano, H. (2005). Continuous and overlapping expression domains of odorant receptor genes in the olfactory epithelium determine the dorsal/ventral positioning of glomeruli in the olfactory bulb. *J. Neurosci.* 25: 3586–3592.

Mombaerts, P. (2006). Axonal wiring in the mouse olfactory system. *Annu. Rev. Cell Dev. Biol.* 22: 713–737.

Mombaerts, P., Wang, F., Dulac, C., et al. (1996). Visualizing an olfactory sensory map. *Cell* 87: 675–686.

Mori, K., Nagao, H., and Yoshihara, Y. (1999). The olfactory bulb: coding and processing of odor molecule information. *Science* 286: 711–715.

Mori, K. and Sakano, H. (2011). How is the olfactory map formed and interpreted in the mammalian brain? *Annu. Rev. Neurosci.* 34: 467–499.

Nagawa, F., Yoshihara, S., Tsuboi, A., et al. (2002). Genomic analysis of the murine odorant receptor *MOR28* cluster: a possible role of gene conversion in maintaining the olfactory map. *Gene* 292: 73–80.

Nakashima, A., Takeuchi, H., Imai, T., et al. (2013). Agonist-independent GPCR activity regulates anterior-posterior targeting of olfactory sensory neurons. *Cell* 154: 1314–1325.

Ngai, J., Dowling, M. M., Buck, L., et al. (1993). The family of genes encoding odorant receptors in the channel catfish. *Cell* 72: 657–666.

Nguyen-Ba-Charvet, K. T., DiMeglio, T., Fouquet, C., and Chedotal, A. (2008). Robos and slits control the pathfinding and targeting of mouse olfactory sensory axons. *J. Neurosci.* 28: 4244–4249.

Nguyen, M. Q., Zhou, Z., Marks, C. A., et al. (2007). Prominent roles for odorant receptor coding sequences in allelic exclusion. *Cell* 131: 1009–1017.

Nishizumi, H., Kumasaka, K., Inoue, N., et al. (2007). Deletion of the core-H region in mice abolishes the expression of three proximal odorant receptor genes in *cis*. *Proc. Natl. Acad. Sci. USA* 104: 20067–20072.

Norlin, E. M., Alenius, M., Gussing, F., et al. (2001). Evidence for gradients of gene expression correlating with zonal topography of the olfactory sensory map. *Mol. Cell. Neurosci.* 18: 283–295.

Oka, Y., Kobayakawa, K., Nishizumi, H., et al. (2003). O-MACS, a novel member of the medium-chain acyl-CoA synthetase family, specifically expressed in the olfactory epithelium in a zone-specific manner. *Eur. J. Biochem.* 270: 1995–2004.

Petersen, C. C. (2007). The functional organization of the barrel cortex. *Neuron* 56: 339–355.

References

Ressler, K. J., Sullivan, S. L., and Buck, L. B. (1993). A zonal organization of odorant receptor gene expression in the olfactory epithelium. *Cell* 73: 597–609.

Ressler, K. J., Sullivan, S. L., and Buck, L. B. (1994). Information coding in the olfactory system: evidence for a stereotyped and highly organized epitope map in the olfactory bulb. *Cell* 79: 1245–1255.

Roppolo, D., Vollery, S., Kan, C. D., et al. (2007). Gene cluster lock after pheromone receptor gene choice. *EMBO J.* 26: 3423–3430.

Rothman, A., Feinstein, P., Hirota, J., and Mombaerts, P. (2005). The promoter of the mouse odorant receptor gene M71. *Mol. Cell. Neurosci.* 28: 535–546.

Sakano, H. (2010). Neural map formation in the mouse olfactory system. *Neuron* 67: 530–542.

Serizawa, S., Ishii, T., Nakatani, H., et al. (2000). Mutually exclusive expression of odorant receptor transgenes. *Nat. Neurosci.* 3: 687–693.

Serizawa, S., Miyamichi, K., Nakatani, H., et al. (2003). Negative feedback regulation ensures the one receptor–one olfactory neuron rule in mouse. *Science* 302: 2088–2094.

Serizawa, S., Miyamichi, K., and Sakano, H. (2004). One neuron–one receptor rule in the mouse olfactory system. *Trends Genet.* 20: 648–653.

Serizawa, S., Miyamichi, K., Takeuchi, H., et al. (2006). A neuronal identity code for the odorant receptor-specific and activity-dependent axon sorting. *Cell* 127: 1057–1069.

Seta, K., Nanamori, M., Modrall, J. G., et al. (2002). AT1 receptor mutant lacking heterotrimeric G protein coupling activates the Src-Ras-ERK pathway without nuclear translocation of ERKs. *J. Biol. Chem.* 277: 9268–9277.

Shenoy, S. K. and Lefkowitz, R. J. (2011). β-Arrestin-mediated receptor trafficking and signal transduction. *Trends Pharmacol. Sci.* 32: 521–533.

Shepherd, G., Chen, W. R., and Greer, C. A. (2004). Olfactory bulb. In The Synaptic Organization of the Brain, ed. GM Shepherd, pp. 165–216. New York: Oxford University Press.

Shykind, B. M., Rohani, S. C., O'Donnell, S., et al. (2004). Gene switching and the stability of odorant receptor gene choice. *Cell* 117: 801–815.

Smallwood, P. M., Wang, Y., and Nathans, J. (2002). Role of a locus control region in the mutually exclusive expression of human red and green cone pigment genes. *Proc. Natl. Acad. Sci. USA* 99: 1008–1011.

Spilianakis, C. G., Lalioti, M. D., Town, T., et al. (2005). Interchromosomal associations between alternatively expressed loci. *Nature* 435: 637–645.

Strotmann, J., Levai, O., Fleischer, J., et al. (2004). Olfactory receptor proteins in axonal processes of chemosensory neurons. *J. Neurosci.* 24: 7754–7761.

Strotmann, J., Wanner, I., Helfrich, T., et al. (1994a). Rostro-caudal patterning of receptor-expressing olfactory neurones in the rat nasal cavity. *Cell Tissue Res.* 278: 11–20.

Strotmann, J., Wanner, I., Helfrich, T., et al. (1994b). Olfactory neurons expressing distinct odorant receptor subtypes are spatially segregated in the nasal epithelium. *Cell Tissue Res.* 276: 429–438.

Sullivan, S. L., Adamson, M. C., Ressler, K. J., et al. (1996). The chromosomal distribution of mouse odorant receptor genes. *Proc. Natl. Acad. Sci. U.S.A.* 93: 884–888.

Takahashi, H., Yoshihara, S., Nishizumi, H., and Tsuboi, A. (2010). Neuropilin-2 is required for the proper targeting of ventral glomeruli in the mouse olfactory bulb. *Mol. Cell. Neurosci.* 44: 233–245.

Takeuchi, H., Inokuchi, K., Aoki, M., et al. (2010). Sequential arrival and graded secretion ofSema3F by olfactory neuron axons specify map topography at the bulb. *Cell* 141: 1056–1067.

Tian, H. and Ma, M. (2008). Activity plays a role in eliminating olfactory sensory neurons expressing multiple odorant receptors in the mouse septal organ. *Mol. Cell. Neurosci.* 38: 484–488.

Tolhuis, B., Palstra, R. J., Splinter, E., et al. (2002). Looping and interaction between hypersensitive sites in the active β-globin locus. *Mol. Cell* 10: 1453–1465.

Tsuboi, A., Imai, T., Kato, H. K., et al. (2011). Two highly homologous mouse odorant receptors encoded by tandemly-linked MOR29A and MOR29B genes respond differently to phenyl ethers. *Eur. J. Neurosci.* 33: 205–213.

Tsuboi, A., Miyazaki, T., Imai, T., and Sakano, H. (2006). Olfactory sensory neurons expressing class I odorant receptors converge their axons on an antero-dorsal domain of the olfactory bulb in the mouse. *Eur. J. Neurosci.* 23: 1436–1444.

Tsuboi, A., Yoshihara, S., Yamazaki, N., et al. (1999). Olfactory neurons expressing closely linked and homologous odorant receptor genes tend to project their axons to neighboring glomeruli on the olfactory bulb. *J. Neurosci.* 19: 8409–8418.

Vassalli, A., Feinstein, P., and Mombaerts, P. (2011). Homeodomain binding motifs modulate the probability of odorant receptor gene choice in transgenic mice. *Mol. Cell. Neurosci.* 46: 381–396.

Vassalli, A., Rothman, A., Feinstein, P., et al. (2002). Minigenes impart odorant receptorspecific axon guidance in the olfactory bulb. *Neuron* 35: 681–696.

Vassar, R., Chao, S. K., Sitcheran, R., et al. (1994). Topographic organization of sensory projections to the olfactory bulb. *Cell* 79: 981–991.

Vassar, R., Ngai, J., and Axel, R. (1993). Spatial segregation of odorant receptor expression in the mammalian olfactory epithelium. *Cell* 74: 309–318.

Wang, F., Nemes, A., Mendelsohn, M., and Axel, R. (1998). Odorant receptors govern the formation of a precise topographic map. *Cell* 93: 47–60.

Wang, S. S., Lewcock, J. W., Feinstein, P., et al. (2004). Genetic disruptions of O/E2 and O/E3 genes reveal involvement in olfactory receptor neuron projection. *Development* 131: 1377–1388.

Wang, Y., Smallwood, P. M., Cowan, M., et al. (1999). Mutually exclusive expression of human red and green visual pigment-reporter transgenes occurs at high frequency in murine cone photoreceptors. *Proc. Natl. Acad. Sci. USA* 96: 5251–5256.

Young, J. M., Friedman, C., Williams, E. M., et al. (2002). Different evolutionary processes shaped the mouse and human olfactory receptor gene families. *Hum. Mol. Genet.* 11: 535–546.

Young, J. M., Shykind, B. M., Lane, R. P., et al. (2003). Odorant receptor expressed sequence tags demonstrate olfactory expression of over 400 genes, extensive alternative splicing and unequal expression levels. *Genome Biol.* 4: R71.

Yoshihara, Y., Kawasaki, M., Tamada, A., et al. (1997). OCAM: a new member of the neural cell adhesion molecule family related to zone-to-zone projection of olfactory and vomeronasal axons. *J. Neurosci.* 17: 5830–5842.

Zhang, X. and Firestein, S. (2002). The olfactory receptor gene superfamily of the mouse. *Nat. Neurosci.* 5: 124–133.

Zhang, X., Rodriguez, I., Mombaerts, P., and Firestein, S. (2004a). Odorant and vomeronasal receptor genes in two mouse genome assemblies. *Genomics* 83: 802–811.

Zhang, X., Rogers, M., Tian, H., et al. (2004b). High-throughput microarray detection of olfactory receptor gene expression in the mouse. *Proc. Natl. Acad. Sci. U.S.A.* 101: 14168–14173.

Chapter 7

Neurogenesis in the Adult Olfactory Epithelium

ALAN MACKAY-SIM, JAMES ST JOHN, and JAMES E. SCHWOB

7.1 INTRODUCTION

It is now recognized that neurogenesis occurs in a number of sites within the adult brain, even in the brain of aged humans (Eriksson et al., 1998). Neurogenesis has long been recognized as a property of the adult olfactory epithelium. Indeed, neurogenesis in this tissue is robust enough to replenish piecemeal or replace wholesale the neuronal population to such an extent that structure and function are restored after injury severe enough to compromise or destroy the sense of smell. Olfactory sensory neurons regenerate in humans (Wolozin et al., 1992; Murrell et al., 1996), even into old age (Holbrook et al., 2011), making the olfactory system one of the most continually variable regions of the human nervous system.

7.2 NEUROGENESIS IN ADULT OLFACTORY EPITHELIUM

7.2.1 Overview of Cell Types in Olfactory Mucosa

The olfactory epithelium is a pseudo-stratified, columnar epithelium containing three cell compartments: the superficial supporting cells, the medially located sensory neurons, and the basal cells (see Chapter 4). The supporting cells, including sustentacular cells and several variants of microvillar cells, can be identified by a number of antigens including the anonymous monoclonal antibodies SUS1 (Hempstead and Morgan, 1983) and SUS4 (Goldstein and Schwob, 1996), as well as more defined markers, such as keratin 18, E-cadherin, and transcription factors (more on this below) (Guo et al., 2010; Packard et al., 2011a, b). The mature sensory neurons are identified by their expression of olfactory marker protein (OMP; Danciger et al., 1989). Below the sensory neurons and blending in with the basal cells below are the immature sensory neurons, which are identified by expression of GAP43 and β-tubulin type III (Verhaagen et al., 1989; Roskams et al., 1998). The basal cells of the olfactory epithelium have traditionally been divided into two categories on the basis of morphology: horizontal basal cells (HBCs) and globose basal cells (GBCs). Under the electron microscope HBCs resemble the basal cells of the adjacent respiratory epithelium, appearing highly differentiated, with a dark cytoplasm, forming a flattened monolayer along the basal lamina with which they make hemidesmosomal connections (Holbrook et al., 1995). HBCs envelop the olfactory axons as they exit the epithelium and are identified immunologically by their expression of a variety of antigens including keratins 5 and 14, ICAM-1, and a surface glycoprotein that binds to the lectin BS-I (Holbrook et al., 1995). GBCs, more superficially located than HBCs, are round cells with large nucleus and sparse cytoplasm and are not seen in respiratory epithelium (Goldstein and Schwob, 1996). All GBCs (and some immature neurons) are immunologically labelled with the monoclonal antibodies GBC1-3 (Goldstein and Schwob, 1996); the antigen recognized by GBC3 was identified as the immature form of the non-integrin laminin receptor (Jang et al., 2007). The GBC population is heterogeneous and can be subdivided further based on expression of transcription factors, as described below. The GBC population also includes a

---

Handbook of Olfaction and Gustation, Third Edition. Edited by Richard L. Doty.
© 2015 Richard L. Doty. Published 2015 by John Wiley & Sons, Inc.

134 Chapter 7 Neurogenesis in the Adult Olfactory Epithelium

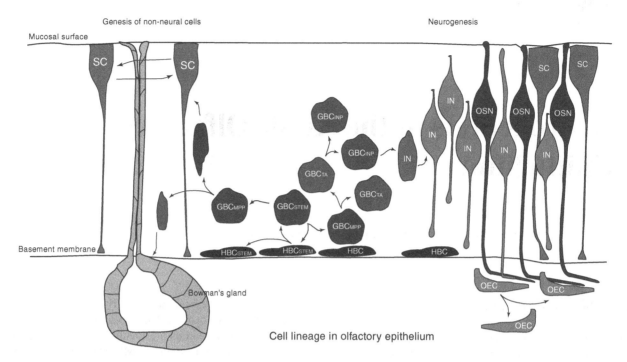

Figure 7.1 Cell lineage in olfactory epithelium. The globose basal cell population (GBC) contains a multipotent stem cell (GBC$_{STEM}$) that can proliferate and produce all neural and non-neural cells of the olfactory epithelium, including the Bowman's gland cells (GC) and duct cells, supporting cells (SC) and olfactory sensory neurons (OSN). Supporting cells can divide in situ. Globose basal cells may form a hierarchy from multipotent proliferating progenitors (GBC$_{MPP}$) through transit amplifying cells (GBC$_{TA}$) to immediate neuronal precursors (GBC$_{INP}$) which leave the cell cycle after dividing, to differentiate into immature neurons (IN). The horizontal basal cell (HBC$_{STEM}$) is a quiescent stem cell whose asymmetric division provides another horizontal basal stem cell and a multipotent globose basal cell stem cell (GBC$_{STEM}$). The horizontal basal cell may only be active after massive epithelial damage to maintain their population. Olfactory ensheathing cells (OEC) in the lamina propria can divide in situ. They may be derived from an epithelial stem cell.

"third-type" of basal cell located above the HBC layer but extending a slender process to contact the basal lamina at an adherens-type junction (Holbrook et al., 1995). These have been interpreted as an intermediate form between HBCs and GBCs, and their existence has been cited as evidence that HBCs give rise to GBCs (Graziadei and Monti Graziadei, 1979) but the same histological data could be interpreted as a transition in the opposite direction. As discussed below, in the normal adult olfactory epithelium these transitions are probably rare events. Figure 7.1 shows an overview of the cell types in the olfactory epithelium and the lineage relations between them.

7.2.2 Replacement of Sensory Neurons in the Adult Olfactory Epithelium

After the early reports of basal cell mitosis in mouse olfactory epithelium (Nagahara, 1940) and regeneration of olfactory sensory neurons after zinc sulfate lesion in monkey (Schultz, 1941), there followed numerous reports confirming these observations in a variety of vertebrates: frog (Smith, 1951), fish (Westerman and von Baumgarten, 1964), cat and dog (Andres, 1966), lamprey (Thornhill, 1970), and mouse (Smart, 1971). Proliferating cells and their progeny can be identified at different times after an animal is injected with thymidine analogs that label cells synthesizing DNA during the cell cycle. At early survival periods after injection of ^{3}[H]-thymidine in the mouse, the dividing cells are located in two layers of the epithelium – most are among the basal cells, close to the basement membrane, with a few located apically among the supporting cells (Moulton et al., 1970; Graziadei and Monti Graziadei, 1979). With increasing periods after injection of ^{3}[H]-thymidine, the labelled basal cells migrate away from the basement membrane until their nuclei lie in the mid-zone of the epithelium in the region of the sensory neuron nuclei (Moulton et al., 1970; Graziadei and Monti Graziadei, 1979). These observations are consistent with basal cells giving rise to neurons and, under the electron microscope, there appear to be transitional cell types whose morphology suggests that they are immature neurons (Graziadei, 1973; Graziadei and Monti Graziadei, 1979). The rate of basal cell mitosis is inversely proportional to epithelial thickness, indicating regulatory mechanisms at work within the epithelium (Mackay-Sim and Patel, 1984). Quantitative analyses show that 70–80% of the labelled cells are lost from the epithelium between 14 and 21 days after migrating into the neuronal layer (Moulton et al., 1970; Mackay-Sim and Kittel, 1991a), but labelled cells in the neuronal layer survive for 3 months (Moulton et al., 1970; Mackay-Sim and Kittel, 1991b) or longer (Hinds et al., 1984). Retrograde labelling by injection of colloidal gold provided direct evidence that olfactory neurons remain connected to the olfactory bulb for at least 3 months (Mackay-Sim and Kittel, 1991b). To survive, neurons must find synaptic space at the bulb and dendritic space at the epithelial surface (Breipohl et al., 1986; Mackay-Sim and Kittel, 1991b). Olfactory neuron survival is dependent on contact with, and trophic support from, the olfactory bulb (Schwob et al., 1992). Cell death is an integral part of neurogenesis during embryonic development and analysis of cell death in the adult olfactory epithelium indicates that all cell types undergo apoptosis, not just mature sensory neurons (Mahalik, 1996). Cell death can occur within one day of birth (Carr and Farbman, 1993), indicating that apoptosis in the olfactory epithelium is an integral and early part of neurogenesis in the adult olfactory epithelium. Olfactory neurogenesis is regulated by endocrine, autocrine, and paracrine factors similar to those operating during embryonic organogenesis (Breipohl et al., 1986).

7.2.3 Multipotent Stem Cells in the Olfactory Epithelium

Tissues capable of repair and regeneration are thought to contain a proliferating "stem" cell whose progeny become the functionally mature cells of the tissue. "Tissue-resident" or "adult" stem cells are usually defined as a "self-renewing" population (they proliferate to maintain their number) which is "multipotent" (their progeny can make all the cell types of the tissue). Tissue-resident stem cells are thought to divide "asymmetrically" to produce another stem cell and a proliferating daughter cell often called a "multipotent progenitor" or "transit amplifying" cell, to indicate that this daughter cell population has lost the ability to self-renew and, with each cell division, the progeny become more committed along a certain lineage of differentiation. In the olfactory neuron lineage, the final dividing cell is known as the "immediate neuronal precursor" (Figure 7.1). In the olfactory epithelium, advances are being made in discriminating among the basal cells, all the cells in lineage from the "true" stem cell (the self-renewing cell giving rise to multipotent progeny) through to mature sensory neuron.

The fact that neurons are born in normal olfactory epithelium and can be restored following epithelial injury provides prima facie evidence for the existence

of neural stem cell(s) in the olfactory epithelium. Certainly, the GBC population contains cells committed to making neurons: direct analysis of genetically labelled GBCs and their progeny using a replication-incompetent retrovirally-derived vector (RRV), demonstrated that the dividing GBCs cells are only capable of producing other GBCs or neurons (Caggiano et al., 1994; Schwob et al., 1994a). This and evidence in vitro makes it clear that the immediate neuronal precursor is a proliferating GBC (Graziadei and Monti Graziadei, 1979; Calof and Chikaraishi, 1989; Goldstein and Schwob, 1996; Newman et al., 2000). This then establishes a lineage from GBC to mature sensory neuron but does not identify the stem cell, nor shed light on whether it is capable of making other cell types of the epithelium (Figure 7.1). The multipotent capacity of the stem cells is demonstrated by the use of toxic agents that cause the death of most cell types in the olfactory epithelium, not just the sensory neurons (e.g., zinc sulphate; Harding et al., 1978). Methyl bromide gas is selectively toxic to the olfactory epithelium. It is administered at small doses by passive inhalation for 6–8 hours and causes widespread destruction of neurons, supporting cells, glands, and duct cells, but spares many or even most GBCs and HBCs (Schwob et al., 1995). Despite this widespread destruction, the epithelium regenerates all cell types including neurons, supporting cells, basal cells, and Bowman's duct/gland cells (Schwob et al., 1995). Use of this model and lineage analysis to trace the genetically labelled progeny of the remaining basal cells indicated that there may be two multipotent progenitors, one that gives rise to non-neuronal cells only (supporting cells, Bowman's gland cells and duct cells) and another that gives rise to basal cells, neurons, and supporting cells (Huard et al., 1998). Additionally, it was concluded that the multipotent cell giving rise to the non-neuronal cells is a Bowman's gland duct cell and the other multipotent, neural lineage cell is a GBC (Huard et al., 1998). In these studies the dividing cells are genetically labelled by transduction with the RRV, applied intranasally. In rats, GBCs are the most superficial and also the most highly proliferative cell type that survives MeBr lesion and are infectable 1 day after exposure (Schwob et al., 1995), making them the most likely target for RRV transduction after intranasal infusion. These experiments take us further down the lineage from immediate neuronal precursor to multipotent progenitors but do not identify whether there is a self-renewing stem cell giving rise to these two multipotent progenitors.

HBCs were proposed as stem cells based on histological analysis after epithelial reconstitution (Rehn et al., 1981). Quantitative analysis of the undamaged epithelium suggested that the stem cell resides on the basement membrane in the location of the horizontal basal cell (Mackay-Sim and Kittel, 1991a), but this hypothesis remained inconclusive. This is because the asymmetrically dividing cells were slowly dividing and located on the basement membrane, where they were not identified as horizontal basal cells, immunologically or morphologically (Mackay-Sim and Kittel, 1991a). Additional experiments in vitro also suggested that neurons can arise from horizontal basal cells (Mahanthappa and Schwarting, 1993; Satoh and Takeuchi, 1995), but the progenitors were identified with a single basal cell antibody (keratin) and the progeny with a single neuronal antibody (NCAM) (Mahanthappa and Schwarting, 1993; Satoh and Takeuchi, 1995). Without accounting for other potential cells in the cultures (GBCs, supporting cells), the neurons may have arisen from unidentified cells but ascribed to the keratin-positive HBCs. The role of HBCs remained moot even after the retroviral labeling experiments of undisturbed epithelium (Caggiano et al., 1994; Schwob et al., 1994a) because the animals were killed too early to observe the division of a horizontal basal cell with a cell cycle period of about 50 days (Mackay-Sim and Kittel, 1991). The contribution of HBCs as multipotent progenitors was demonstrated unequivocally using lineage tracing with genetically labelled cells (Leung et al., 2007). HBCs express keratin-5 (K5). Transgenic mice carried a reporter gene (*LacZ*) that was expressed only by cells expressing K5 and only when they were administered the drug Tamoxifen, such that the reporter gene was transcribed into protein (β-galactosidase) in the K5-expressing cells and their progeny. The K5 cells and progeny could be identified in tissue sections through a color reaction catalyzed by β-galactosidase. By administering the Tamoxifen (to switch on the labeling) after methyl bromide exposure to damage the olfactory epithelium, it was shown that K5-expressing HBCs could reconstitute all the cell types, both neuronal and non-neuronal, of the olfactory epithelium (Leung et al., 2007). A second study used a similar reporter gene in K5-positive HBCs with sensory neuron damage by olfactory bulbectomy and came to essentially the same conclusions (Iwai et al., 2008).

These studies showed that HBCs were *sufficient* to give rise to all cells. After epithelial damage not all the epithelial cells were labeled (Leung et al., 2007; Iwai et al., 2008) and in the undisturbed epithelium, despite extensive labeling of HBCs, there were only a few patches of labeled cells to indicate that HBCs had contributed to the cells above them in the epithelium (Leung et al., 2007; Iwai et al., 2008). These observations indicate that HBCs, although sufficient, were *not necessary* for regeneration of the epithelium, either undisturbed or after olfactory bulbectomy (Iwai et al., 2008) or chemical damage (Leung et al., 2007). This leaves open the question as to whether there are two multipotent progenitors, the HBC and a GBC. The stem cell potential of GBCs was specifically explored by a series of experiments using cell transplantation techniques in combination with genetic lineage tracing. In the first experiment, proliferating cells of the rat olfactory mucosa (i.e., mainly GBCs; Schwartz Levey et al., 1991) were genetically labeled by infection with RRV after olfactory bulbectomy (Goldstein et al., 1998). These cells were then transplanted into the olfactory epithelium of MeBr-lesioned hosts and their fate was followed histologically. The descendants of these presumed GBCs included HBCs, GBCs, neurons and supporting cells (Goldstein et al., 1998). In a second study, GBC-2 antibody immunoreactivity was used to select GBCs and immature neurons from the normal, undamaged mouse epithelium by fluorescence-activated cell sorting (Chen et al., 2004b). These cells were infused intranasally into MeBr-lesioned hosts and their fate was followed by their expression of GFP or β-galactosidase (Chen et al., 2004b). The descendants of these GBCs included GBCs, neurons, sustentacular cells, duct/gland cells, and even respiratory columnar epithelial cells, i.e., every epithelial cell type *except* HBCs (Chen et al., 2004b). In contrast to this multipotency, sustentacular, and duct/gland cells sorted from the normal mouse epithelium using the antibody SUS-4 gave rise only to themselves (Chen et al., 2004b) and HBCs sorted from normal mouse epithelium using the lectin BS-I failed to engraft, suggesting that HBCs isolated from normal epithelium lacked the potency expected of stem cells (Chen et al., 2004b). Recent experiments show that some GBCs are very slow cycling (Guo et al., 2010), a feature taken as ancillary evidence for "stemness" (Cotsarelis et al., 1989; Mackay-Sim and Kittel, 1991; Potten, 1998). These experiments demonstrate that GBCs, like HBCs, satisfy the criteria for stemness, namely multipotency and self-renewal. As discussed above, each cell type can give rise to the other, indicating that both can be considered "stem cells" to all the cell types in the olfactory epithelium, although the transplantation experiments demonstrate that the GBCs are a mixed population of stem cells, multipotent proliferating progenitors, and immediate neuronal precursors (Figure 7.1).

In summary, the transplantation experiments above demonstrate that GBC stem cells are active in the normal epithelium and during sensory neuron replacement after damage to the olfactory epithelium. The lineage tracing experiments confirm this: in the normal epithelium the HBCs were labeled, with little contribution to other cell types; after bulbectomy (Iwai et al., 2008) or MeBr lesion (Leung et al., 2007), the HBCs contributed to all epithelial cell types. Even so, much of the epithelium remained unlabeled, indicating that GBCs continue to contribute to repair, as confirmed by combined lineage tracing and transplantation (Chen et al., 2004b). From these studies it seems probable that the GBC stem cell is constitutively active and the HBC stem cell is quiescent, requiring specific activation. This hypothesis was tested directly using transgenic mice similar to those used previously (Leung et al., 2007) allowing us to identify HBCs and their progeny after transplantation into unlabeled hosts (Schnittke, Herrick, and Schwob, manuscript in preparation). It was found that transplanted HBCs did not show multipotency if they were derived from normal epithelium, in accordance with previous transplantation results (Chen et al., 2004b). However, if the donor olfactory epithelium was exposed to MeBr lesion 18–48 hr prior to HBC harvest, the HBCs contributed to all cell types in the host epithelium, consistent with previous lineage-tracing results (Leung et al., 2007). These observations demonstrate directly that HBCs must undergo a kind of "activation" *in situ* by some type of lesion-dependant signal in order to proliferate and become multipotent. GBCs do not require prior activation by a lesioned environment, but immediately function as multipotent following transplantation and engraftment (Chen et al., 2004b).

Taken together, the best current formulation for understanding the nature of stem cells in the olfactory epithelium is the existence of two kinds of tissue stem cell – a quiescent, reserve population (HBCs), and a more facultative population (GBCs). Each of these stem cells can generate the other, as well as all other neuronal and non-neuronal cell types in the adult olfactory epithelium. It is significant that genetic deletion of the transcription factor *p63* eliminated HBCs but did not prevent the development of GBCs, and all the other cells of the olfactory epithelium, including neurons, and their projection onto the olfactory bulb (Packard et al., 2011a). Furthermore, considering that HBCs emerge after GBCs during development (Packard et al., 2011), it is possible that HBCs are originally a GBC-derived population of quiescent stem cell that is removed from the cell cycle until re-activated by signals released during severe damage to the olfactory epithelium. Recent data indicate that down-regulation of the transcription factor *p63* is necessary and sufficient to activate HBCs. Within 18 hours of unilateral MeBr exposure, *p63* levels in HBCs are reduced in the lesioned side compared to the unlesioned side (Schnittke, Packard, and Schwob, manuscript in preparation). During the next day, some HBCs proliferate and lose all detectable *p63* expression (while remaining identifiable by labeling for K5, K14, and ICAM-1/CD54). When harvested at this time point, the HBCs are functionally multipotent following transplantation, indicating that loss of *p63* activates return to multipotency. Conversely, RRV-driven expression of *p63* prevents activation and multipotency (Schnittke and Schwob, manuscript in preparation). It is proposed that *p63* acts to promote HBC cell renewal and inhibits its differentiation (Fletcher et al., 2011), which is consistent with the finding that downregulation of *p63* releases this inhibition to activate a multipotent state. Specific molecular signals sustaining quiescence of HBCs, or causing their activation during epithelial damage, have yet to be identified but they may include those described below.

7.2.4 Globose Basal Cell Lineage in Olfactory Neurogenesis

It should be emphasized that GBC cells form a heterogeneous population in terms of function within a hierarchy of cells from stem cell through to immediate neuronal precursor (Figure 7.1). GBCs are histologically very similar but can be distinguished from each other on the basis of the expression of a group of transcription factors, including well-known members of the basic helix-loop-helix (bHLH) family of transcription factors. Figure 7.2 is a heuristically useful model of the flow of cells during normal neurogenesis and during the reconstitution of the epithelium following injury. Needless to say, there is much that is not understood about the details of the process.

According to this model, the stages include: (1) multipotent stem cells (GBC_{STEM}) (Chen et al., 2004b), (2) multipotent progenitors (GBC_{MPP}) with the capacity to generate multiple epithelial cell types in response to direct epithelial injury (Goldstein et al., 1998; Huard et al., 1998; Chen et al., 2004b), (3) transit amplifying cells, which are apparently committed to making neurons (GBC_{TA}; Guillemot et al., 1993; Gordon et al., 1995), and (4) immediate neuronal precursors, which undergo a final division and differentiate into neurons (GBC_{INP}; Calof and Chikaraishi, 1989; Cau et al., 1997). Based on expression analysis and genetic epitasis, the transcription factors that can be used to separate the groups are Sox2 (a HMG transcription factor; Kawauchi et al., 2009; Guo et al., 2010), Pax6 (paired box), Ascl1 (bHLH; formerly known as Mash1; Guillemot et al., 1993; Cau et al., 1997), Neurog1 (bHLH; Cau et al., 1997; Cau et al., 2002), and NeuroD1 (bHLH; Cau et al., 1997; Packard et al., 2011a, b). Sox2 and Pax6 appear earliest in development and during regeneration; they are universally co-expressed by the residual GBCs 1 day after MeBr lesion, when the progenitor cells are multipotent (Guo et al., 2010). Ascl1 and then Neurog1 and NeuroD1 appear sequentially, followed by the differentiation of post-mitotic daughters into immature olfactory neurons (Cau et al., 1997; Manglapus et al., 2004). Ascl1 was identified as a marker for the transit amplifying stage based on the proliferation of that subpopulation of GBCs in advance of the enhanced proliferation caused by neuronal death and on the reduction of olfactory neurogenesis in Ascl1 knockout animals (Gordon et al., 1995). Epistasis in Ascl1 knockout epithelium and the timing of expression in normal embryos and lesioned-recovering adult epithelium indicate that Neurog1 and NeuroD1 expression are expressed downstream of Ascl1 (Cau et al., 1997; Manglapus et al., 2004). Further analysis suggests that Neurog1

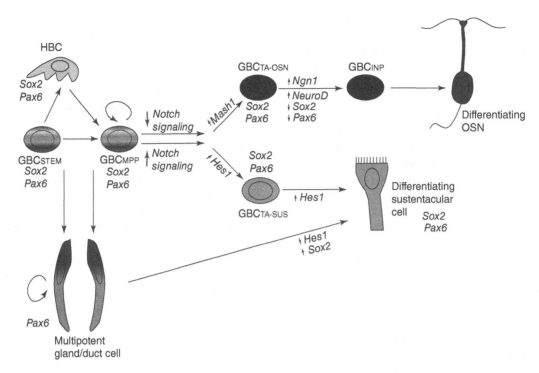

Figure 7.2 Stem cell-progenitor cell relationships in the adult olfactory epithelium. In this model, the heterogeneous population of GBCs includes stem cells (GBC$_{STEM}$), committed multipotent precursors (GBC$_{MPP}$), neuron-committed transit amplifying cells (GBC$_{TA-OSN}$) and immediate neuronal precursors (GBC$_{INP}$) that give rise to olfactory sensory neurons (OSN). Different GBCs are distinguished by the expression of a limited number of transcription factors (*Sox2*, *Pax6*, *Ngn1*, *NeuroD*, *Mash1*, *Hes1*, *p63*). After damage to the epithelium, *Notch* signaling, and the upregulation of *Hes1*, directs some of the multipotent GBCs toward a sustentacular fate (GBC$_{TA-SUS}$). Multipotent gland/duct cells are also capable of making sustentacular cells, although the signals required are not yet clear. The HBCs, a reserve stem cell population, are set aside when the transcription factor *p63* is expressed and are activated to function in recover after epithelial injury when *p63* is suppressed. At several points along the pathway, the stem and progenitor cells have a capacity to give rise to and renew themselves, at least to a limited extent.

is slightly upstream of NeuroD1, but shows substantial overlap with it (Cau et al., 2002; Manglapus et al., 2004; Packard et al., 2011a). All olfactory neurons apparently derive from a NeuroD1-expressing stage in the progenitor progression, based on genetic lineage tracing and expression of NeuroD1 by proliferating GBCs (Packard et al., 2011a). Sox2 and Pax6 perdure through the Ascl1 (+), transit-amplifying stage, but both are undetectable in immediate neuronal precursor-stage GBCs marked by either NeuroD1 or Neurog1 (Guo et al., 2010). Interestingly, the patterns of gene expression by the GFP-labelled immediate neuronal precursors and immature neurons in mice expressing a Neurog1-GFP transgene are virtually identical between normal epithelium and post-bulbectomy epithelium when proliferation is accelerated (Krolewski et al., 2012). This finding suggests that the Neurog1$^+$ GBCs are downstream of the cell that is responsible for the acceleration of proliferation following bulb ablation.

Additional evidence can be adduced for the identification of Sox2$^+$/Pax6$^+$/Ascl1$^-$ cell as the multipotent progenitor (GBC$_{MPP}$). One day after MeBr lesion, some identified GBCs labeled with RRV (Huard et al., 1998) express the neuroinhibitory bHLH transcription factor, Hes1, and then shift apically and differentiate into Hes1-expressing sustentacular cells (Manglapus et al., 2004). This finding accounts for the generation of sustentacular cell-only clones following RRV-LacZ infection or GBC transplantation. Hes1 is one of the canonical targets for up-regulation by Notch signaling in both vertebrates and invertebrates, a pathway that is commonly used to switch progenitor cells between fates, especially neuronal versus non-neuronal (Kageyama and Ohtsuka, 1999). Indeed, Hes1 up-regulation is often used as an indirect measure of Notch activation. To test whether Notch signaling was responsible for the switch between generating neurons versus sustentacular cells, two vectors were used to modify Notch signaling: RRV encoding a constitutively active form of Notch1 (the Notch1 intracellular domain or NICD) and another encoding a dominant negative version of Mastermind-like (dnMAML), a co-factor required for activation of transcription following Notch signaling. Infection with RRV-dnMAML prevented the generation of sustentacular cells, such that infected clones were composed of neurons and, on occasion, non-HBC basal cells (Guo, Schnittke, and Schwob, manuscript in preparation). In contrast, RRV-NICD prevented the generation of neurons and produced clones that were extremely large and uniformly composed of sustentacular-like cells of a primitive, K18-expressing epithelial cell (*ibid*). Taken together, the data suggest strongly that Notch signalling is involved in the regulation of neuron production versus generation of sustentacular and other non-neuronal cells by multipotent GBCs (GBC$_{MPP}$). Although firmly based on available data, the model in Figure 7.2 is still incomplete in describing the lineage relationships during neurogenesis and the production of non-neuronal cell types in the adult olfactory epithelium. There may also be differences during development. For example, some Ascl1$^+$ GBCs or early GBC-like progenitors on the way to becoming GBCs express *p63* and may go on to make HBCs (Packard et al., 2011b).

7.2.5 Ectomesenchymal Stem Cell in Olfactory Mucosa

In addition to the basal cells of the olfactory epithelium, the human olfactory mucosa contains a stem cell that is described as an "ectomesenchymal" stem cell, with properties of both mesenchymal and neural stem cells (Delorme et al., 2010). This stem cell is grown in vitro from the whole mucosa, including cells from both epithelium and the underlying lamina propria (Murrell et al., 2005; Delorme et al., 2010). It can be cultured in "neurospheres," in common with other neural stem cells, but is capable of differentiating into neurons, astrocytes, and oligodendrocytes, as well as a wide range of non-neural cell types, including cardiac and skeletal muscle, chondrocytes, osteocytes, liver cells, and fat cells (Murrell et al., 2005; Murrell et al., 2008; Murrell et al., 2009; Delorme et al., 2010; Wetzig et al., 2011). After transplantation into the chick embryo, olfactory stem cells from mouse and human olfactory mucosa contributed to heart, liver, kidney, and nervous system (Murrell et al., 2005). When transplanted into the bone marrow of irradiated adult rat, olfactory stem cells contributed to all blood lineages (Murrell et al., 2005). A "mesenchymal-like" stem cell was isolated from embryonic rat olfactory mucosa with cellular markers that are also present in the lamina propria (Tome et al., 2009). Adult mouse and human lamina propria cultures yield multipotent stem cells in the absence of cells from the olfactory epithelium (unpublished observations), but definitive identity of a lamina propria stem cell is still lacking. It is possible that the "ectomesenchymal" stem cell in adult olfactory mucosa is derived from the ectomesenchymal stem cell

that migrates out from the cranial neural crest during embryogenesis. These multipotent cells give rise to multiple neural and non-neural tissues of the head, including neurons and glia of the peripheral nervous system, smooth muscle, bone, cartilage, corneal endothelium, and dental pulp. The olfactory ensheathing cells in the olfactory mucosa are derived from neural crest (Barraud et al., 2010; Forni et al., 2011) and there is some evidence for contributions to the olfactory placode by the neural crest in mouse (Katoh et al., 2011). A neural crest origin of stem cells in adult olfactory mucosa may help explain their multipotent capacity, although definitive evidence for a neural crest origin of olfactory mucosa stem cells is still lacking.

7.2.6 Summary

Life-long neurogenesis, now accepted to be integral to the adult brain, was first demonstrated in the adult olfactory epithelium. As shown in Figures 7.1 and 7.2, there is now a deep understanding of the cell lineage of olfactory neurogenesis from stem cell to differentiated sensory neuron and the other cell types of the olfactory epithelium. It is apparent that both HBCs and GBCs have the capacity of multipotent stem cells, with HBCs likely to be a quiescent population requiring activation by epithelial damage to re-enter the multipotent lineage. In the lamina propria below the epithelium there is another multipotent ectomesenchymal stem cell. A major outstanding question is whether there is a direct lineage relationship between the neural stem cell of the olfactory epithelium and the multipotent ectomesenchymal stem cell in the lamina propria.

7.3 REGULATION OF OLFACTORY NEUROGENESIS

Early analyses of olfactory neurogenesis were interpreted to mean that the neurons "remain in the epithelium as mature functional elements for approximately 25 days" (Moulton, 1975; Graziadei and Monti Graziadei, 1979; Samanen and Forbes, 1984). Along with this concept of a short-lived sensory neuron, there came to be an assumption that turnover of sensory neurons from basal cell mitosis was a "predetermined … genetic characteristic" (Graziadei and Monti Graziadei, 1978). Contrary to this oft-cited view, there is now overwhelming evidence that olfactory neurogenesis is a highly regulated and homeostatic process that in some respects reflects aspects of embryonic development in other parts of the nervous system. Under this hypothesis neurogenesis in the adult olfactory epithelium follows rules that govern brain development (Breipohl et al., 1986): (1) cell death occurs at all stages of neuronal development; (2) developing neurons are over-produced;

140 Chapter 7 Neurogenesis in the Adult Olfactory Epithelium

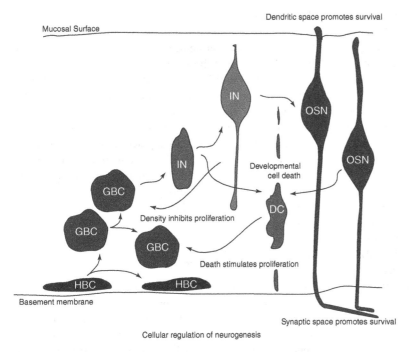

Figure 7.3 Cellular regulation of neurogenesis in vivo. Neurogenesis is regulated by extrinsic factors (epithelial damage, toxins, olfactory bulb ablation) causing cell death which, in turn, stimulates proliferation of the basal cell progenitors and neuronal precursors (HBC, GBC). Cell death (DC, dying cell) is also caused by intrinsic developmental factors (hormones, growth factors, dendritic and synaptic space) that affect maturation and survival of developing neurons (IN). Basal cell proliferation is inhibited by the density of immature neurons. These dynamic processes involve many paracrine and autocrine signals which eventually select a few neurons to undergo final maturation. Once mature, the olfactory sensory neuron (OSN) lives until damaged by extrinsic factors.

Table 7.1 Regulation of basal cell proliferation in vivo.

Experiment	Stimulus	Effect on Basal Cell Proliferation	Possible Regulatory Pathway
Olfactory nerve cut	Death of neurons	Increased	Proliferating factor released from dying cells
Chemical destruction of neurons	Death of neurons	Increased	Proliferating factor released from dying cells
Destruction of olfactory bulb mitral cells	Death of neurons	Increased	Lack of trophic factor from synaptic target
Naris occlusion, open side	Increased usage, loss of neurons	Increased	Proliferating factor released from dying cells
Naris occlusion, closed side	Reduced usage, no cell death	Decreased	Lack of proliferating factor, increase in inhibitory factor
Epithelial thickness	Increase in immature neuron density	Decreased	Increase in inhibitory factor

(3) developing neurons pass through a "critical period" during which they must find synaptic space at their target, or die; (4) successful neurons are dependent on their target for survival; and (5) mature neurons are not programmed to die but may die from external influences. Figure 7.3 and Table 7.1 summarize the cycle of neurogenesis and some of the regulating factors.

During neurogenesis in the adult olfactory epithelium autocrine and paracrine signals and cell–cell contact maintain or induce different cells types. Signals potentially arise within the epithelium, lamina propria, and olfactory bulb. Furthermore, because the surface density of sensory neuron dendrites is controlled and stable (Mackay-Sim et al., 1988; Mackay-Sim and Kittel, 1991a), it is possible that signals may be present in the mucus, acting as target-derived factors for the dendrites.

Neurogenesis is stimulated by death of the sensory neurons (Nagahara, 1940; Graziadei, 1973) followed by restoration of olfactory function (Harding et al., 1978). Long-term sensory neuron survival requires synaptic contact with the olfactory bulb (Schwob et al., 1994b) and loss of the synaptic targets in the olfactory bulb stimulates basal proliferation for more than a year (Weiler and Farbman, 1999). Neurogenesis is affected by usage, as shown by experiments in which the naris is sealed unilaterally in early development (Farbman et al., 1988) or in adulthood

(Maruniak et al., 1989). Neurogenesis is regulated by the density of immature neurons, as shown by an inverse correlation between basal cell proliferation and epithelial thickness in the salamander. Quantitative analysis of the cell types in this epithelium demonstrated that the only cell type whose numbers increased with epithelial thickness was the immature neuron (Mackay-Sim et al., 1988). This led to the conclusion that the developing neurons exert an inhibitory influence on basal cell proliferation (Mackay-Sim et al., 1988). In agreement is evidence that the rate of proliferation in vitro was reduced when precursor cells were co-cultured with sensory neurons (Mumm et al., 1996). Cell death in the normal undisturbed epithelium is apoptotic (Magrassi and Graziadei, 1995) and can occur as early as one day after cell birth (Carr and Farbman, 1993), such that apoptotic cells can be horizontal basal cells, globose basal cells, immature neurons, and mature neurons (Holcomb et al., 1995; Mahalik, 1996). The apoptotic cascade can be induced by activation of the cell surface receptors Fas and TNF receptor-1 by their ligands FasL and TNF-α (Farbman et al., 1999). Neurogenesis is reduced with age in the human olfactory epithelium (Naessen, 1971; Nakashima et al., 1984; Nakashima et al., 1985).

7.4 MOLECULAR REGULATION OF OLFACTORY NEUROGENESIS

Regulation of proliferation, differentiation, survival, and death during olfactory neurogenesis requires multiple factors operating in multiple signaling pathways. All cells in epithelium and lamina propria may play roles in maintaining sensory neuron number and survival. For example, the supporting cells could monitor the local density of sensory neurons because they surround the dendrites of the sensory neurons (Breipohl et al., 1974; Graziadei and Monti Graziadei, 1979) and may release factors to regulate proliferation of basal cells or differentiation of neuronal precursors and immature neurons supporting cells. Similarly horizontal basal cells wrap around the axons as they leave the epithelium (Holbrook et al., 1995) and the ensheathing cells do so when they enter the lamina propria and guide them to the olfactory bulb (Doucette, 1984; Gong et al., 1994). Either of these cell types could regulate neuronal survival and differentiation. Broadly speaking, there are two types of signals that could regulate olfactory neurogenesis at the local or cellular level: diffusible and fixed. Growth factors are diffusible signals that can have paracrine or autocrine actions. Fixed signals include physical interactions via direct cell surface contacts and indirect contacts through the extracellular matrix. Extracellular signals have not been extensively investigated in olfactory neurogenesis—much more is known about growth factors.

7.4.1 Growth Factors and Receptors Present in Olfactory Epithelium

A long list of growth factors and their receptors have been identified in the olfactory epithelium (Table 7.2), although the cell types that express them are not always known. The exceptions are CNTF (basal cells and neurons; Buckland and Cunningham, 1999), dopamine (mucus; Lucero and Squires, 1998), dopamine D2 receptor (basal cells and neurons; Feron et al., 1999; Koster et al., 1999), EGF receptor and TGFα (horizontal basal cells and supporting cells; Farbman et al., 1994; Holbrook et al., 1995; Farbman and Buchholz, 1996; Krishna et al., 1996), FGF2 (supporting cells, neurons, basal cells; Matsuyama et al.,1992; Gall et al., 1994; Goldstein et al., 1997; Chuah and Teague, 1999; Hsu et al., 2001), GDNF (neurons; Buckland and Cunningham, 1999), IGF binding proteins 2–4 (mucus; Federico et al., 1999), p75^{NTR} (supporting cells and horizontal basal cells in the human; Feron et al., 2008; Hahn et al., 2005; Minovi et al., 2010), TrkA (supporting cells and horizontal basal cells; Feron et al., 2008; Roskams et al., 1996; Miwa et al., 1998), TrkB and TrkC (neurons; Roskams et al., 1996; Feron et al., 2008), NGF (neurons; Williams and Rush, 1988; Aiba et al., 1993; Roskams et al., 1996; supporting cells, neurons, immature neurons, globose basal cells; Feron et al., 2008), BDNF (horizontal basal cells; Buckland and Cunningham, 1999; supporting cells, neurons, immature neurons Feron et al., 2008), NT3 (supporting cells, neurons, immature neurons; Feron et al., 2008), NT4 (all cell types in epithelium; Feron et al., 2008), Neuregulin1 isoforms (lamina propria) and ErbB receptors (GBCs and immature neurons; Salehi-Ashtiani and Farbman, 1996; Donovan and Schwob, manuscript in preparation). The variety of growth factors and receptors present in the olfactory epithelium and the variation in expression by the different olfactory cell types suggests a rich complexity in the regulation of olfactory neurogenesis.

7.4.2 Growth Factor Functions in Olfactory Neurogenesis

Defining the actions of growth factors can be very difficult. For example, does the growth factor act directly or via a neighboring cell? Does the putative target cell have the appropriate receptors? Is an increase in sensory neurons in vitro the result of a growth factor action on progenitor proliferation or differentiation, or neuronal survival or death? Generalizing from the literature is further complicated by the variety of culture systems, cell types, species, and culture media. Consequently the conclusions drawn from most studies about growth factor actions should be considered as "working hypotheses" until the various complexities are explored. Nonetheless converging data suggest distinct and

Table 7.2 Growth factors, receptors, binding proteins and synthetic enzymes present in olfactory epithelial cells, identified immunologically, using in situ hybridisation or RT-PCR. Papers include adult and developing tissues and some primary cultures of adult epithelial cells. ATP, dopamine, nitric oxide and vitamin A are not peptide growth factors but have growth factor-like actions in olfactory cell cultures. ActR, Activin receptor; BDNF, brain derived growth factor; BMP, bone morphogenic protein; BMPR, bone morphogenic protein receptor; CNTF, ciliary neurotrophic factor; D2, dopamine D2 receptor; DA, dopamine; EGF, epidermal growth factor; EGFR, epidermal growth factor receptor; FGF, fibroblast growth factor; FGFR, fibroblast growth factor receptor; GDNF, glial cell line-derived growth factor; GFR, GDNF receptor; GGF, glial growth factor; IGF, insulin-like growth factor; LIF, leukocyte inhibitory factor; NDF, Neu differentiation factor; NGF, nerve growth factor; NO, nitric oxide; NOS, nitric oxide synthase; NPY, neuronal protein Y; NRG, neuregulin; NT3, neurotrophin 3; p75[NTR], p75 neurotrophin receptor; PDGF, platelet-derived growth factor; PDGFR, platelet-derived growth factor receptor; P2XR, purine receptor type 2X; PYY, protein YY; RA, retinoic acid; RAR, retinoic acid receptor; RXR, retinoid X receptor; TGF, transforming growth factor; Trk, tyrosine kinase.

Growth Factor Family	Ligands	Reference	Receptors, Binding Proteins, Enzymes	Reference
Purines	ATP		P2XR	Gao et al., 2010
Cytokines	CNTF	Buckland and Cunningham 1999		
	LIF	Getchell et al., 2002b; Bauer et al., 2003; Lopez et al., 2012	LIFR	Getchell et al., 2002a, b; Lopez et al., 2012
Chemokines	MIP1α	Getchell et al., 2002c		
	CX3CL1	Ruitenberg et al., 2008	CX3CR1	Vukovic et al., 2010; Blomster et al., 2011
Dopamine	DA	Lucero and Squires 1998	D2	Coronas et al., 1997b; Feron et al., 1999a; Koster et al., 1999
EGF family	TGFα	Farbman and Buchholz 1996	EGFR	Farbman et al., 1994; Holbrook et al., 1995; Krishna et al., 1996; Salehi-Ashtiani and Farbman 1996
	NRG			
	NDF	Salehi-Ashtiani and Farbman 1996	ErbB2	Salehi-Ashtiani and Farbman 1996
			ErbB3	Perroteau et al., 1998
			ErbB4	Perroteau et al., 1998
FGF family	FGF2	Goldstein et al., 1997; Chuah and Teague 1999; Hsu et al., 2001	FGFR1	DeHamer et al., 1994
			FGFR1b,c	Hsu et al., 2001
			FGFR2	DeHamer et al., 1994
			FGFR2b,c	Hsu et al., 2001
			FGFR3b,c	Hsu et al., 2001
GDNF family	GDNF	Nosrat et al., 1996; Buckland and Cunningham 1999; Woodhall et al., 2001	Ret	Nosrat et al., 1996
			GFRα1	Nosrat et al., 1996; Woodhall et al., 2001
			GFRα2	Woodhall et al., 2001
IGF family	IGF-I	Ayer-le Lievre et al., 1991	IGFR-I	Pixley et al., 1998
	IGF-II	Ayer-le Lievre et al., 1991	IGFBP-2	Bondy and Lee 1993; Federico et al., 1999
			IGFBP-3	Federico et al., 1999
			IGFBP-4	Federico et al., 1999
			IGFBP-5	Bondy and Lee 1993
Neurotrophins	NGF	Ayer-Lelievre et al., 1983; Williams and Rush 1988; Woodhall et al., 2001	TrkA	Roskams et al., 1996; Miwa et al., 1998; Feron et al., 2008
	BDNF	Buckland and Cunningham 1999; Woodhall et al., 2001; Feron et al., 2008; Uranagase et al., 2012	TrkB	Roskams et al., 1996; Woodhall et al., 2001; Feron et al., 2008
	NT3	Feron et al., 2008	TrkC	Roskams et al., 1996; Feron et al., 2008
			p75[NTR]	Feron et al., 2008

7.4 Molecular Regulation of Olfactory Neurogenesis

Table 7.2 (*Continued*)

Growth Factor Family	Ligands	Reference	Receptors, Binding Proteins, Enzymes	Reference
Nitric oxide	NO	Sulz et al., 2009	eNOS	Sulz et al., 2009
			iNOS	Sulz et al., 2009; Lopez et al., 2012
			nNOS	Chen et al., 2004a; Sulz et al., 2009
NPY family	NPY	Kanekar et al., 2009; Doyle et al., 2012		
	PYY	Doyle et al., 2012		
PDGF family	PDGFA	Orr-Urtreger and Lonai 1992	PDGFRα	Lee et al., 1990; Orr-Urtreger and Lonai 1992
TGFβ family	BMP2,4,7	Shou et al., 2000	BMPR-Ib	Zhang et al., 1998
			ActR-Ib	Verschueren et al., 1995
			TGFβR2	Getchell et al., 2002b
Vitamin A	RA	Asson-Batres et al., 2009	RARα	Yee and Rawson 2005; Oztokatli et al., 2012
	Retinol	Asson-Batres et al., 2009	RARβ	Zhang 1999
			RXRα	Zhang 1999
			RXRβ	Zhang 1999
			RXRγ	Zhang 1999
			RA synthetic enzymes	Asson-Batres and Smith 2006; Peluso et al., 2012
Wnt family	Wnt2	Wang et al., 2011	Fzd9	Wang et al., 2011
	Wnt3A	Wang et al., 2011		
	Wnt7b	Wang et al., 2011		
	Sfrp1	Wang et al., 2011		

definable roles in olfactory neurogenesis for EGF family factors, FGF2, TGFβ family factors, LIF, NO, and ATP (Figure 7.4).

EGF and TGFα stimulate proliferation of horizontal basal cells (Satoh and Takeuchi, 1995; Farbman and Buchholz, 1996; Feron et al., 1999c). TGFα, but not EGF, is expressed in the olfactory epithelium and their cognate receptor (EGFR) is also present on horizontal basal cells and supporting cells (Farbman et al., 1994; Holbrook et al., 1995; Farbman and Buchholz, 1996; Krishna et al., 1996). Neuregulins, other members of the EGF family, are implicated in olfactory neurogenesis. The neu differentiation factors (NDFβ-1, -2, and -3) stimulate proliferation of olfactory ensheathing cells and the ensheathing cells, GBCs and immature neurons express ErbB2 receptors (Pollock et al., 1999). Another neuregulin, glial growth factor 2 (GGF2), is weakly proliferative and induces differentiation of ensheathing cells and is expressed by them (Chuah et al., 2000).

In vitro, FGF2 stimulates proliferation of a globose basal cell neuronal precursor (DeHamer et al., 1994; Newman et al., 2000), a globose basal cell-like cell line (Goldstein et al., 1997), and a human olfactory progenitor cell line (Ensoli et al., 1998). FGF2 induced differentiation in explant cultures of mouse and human olfactory epithelium (MacDonald et al., 1996; Murrell et al., 1996), but this is an indirect effect of stimulation of globose basal

cell proliferation (Newman et al., 2000). These observations are consistent with in vivo delivery of FGF2 both intraperitoneally (Nakamura et al., 2002) and intranasally (Nishikawa et al., 2009). FGF2 also stimulated proliferation of ensheathing cells (Chuah and Teague, 1999), and the related FGF1 stimulates ensheathing cell differentiation (Key et al., 1996).

Several members of the TGFβ family of growth factors are implicated in olfactory neurogenesis. TGFβ2 induces differentiation of neuronal precursors (Mahanthappa and Schwarting, 1993), a keratin-positive basal cell line (Satoh and Takeuchi, 1995) and the globose basal cell (Newman et al., 2000). In vivo experiments (Mackay-Sim and Patel, 1984) and in vitro experiments (Mumm et al., 1996) indicated that neurons or immature neurons exert an inhibitory effect on basal cell proliferation. It is possible that this inhibition is mediated via BMPs and their receptors. BMPs 2, 4, and 7 can inhibit proliferation of neuronal precursors in vitro (Shou et al., 1999). BMP receptor subtype Ib is present in embryonic olfactory epithelium (Zhang et al., 1998), mRNA for BMP receptor subtypes Ia, Ib, and II are present in adult olfactory epithelium (Hsu and Mackay-Sim, unpublished). In the embryo, BMPs 2, 4, and 7 are expressed by cells in the lamina propria beneath the olfactory epithelium and noggin, a BMP antagonist, inhibited olfactory neurogenesis in embryonic cultures and low concentrations of BMP4 promoted survival of newly

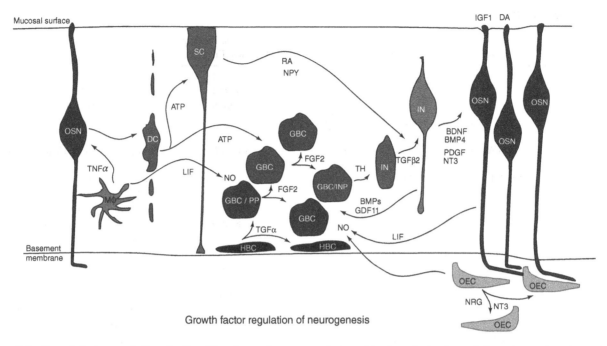

Figure 7.4 Growth factor regulation of cell proliferation and neurogenesis. A working hypothesis of growth factor interactions that maintain neurogenesis in olfactory epithelium. See text for details.

generated olfactory receptor neurons (Shou et al., 2000). These results suggest both anti-proliferative and neuronal survival roles for BMPs, at least during embryogenesis. Another member of the TGFβ family, GDF11, also arrests the cell cycle of olfactory progenitors (Wu et al., 2003). *GDF11*-deficient mice have more olfactory progenitors and neurons (Wu et al., 2003), whereas mice deficient in a GDF11 antagonist, follistatin, have decreased neurogenesis (Gokoffski et al., 2011). *Foxg1* is a transcription factor that negatively regulates TGFβ family signaling. *Foxg1*-deficient mice have deficient follistatin signaling, and like the follistatin deficient mouse have fewer olfactory progenitors and sensory neurons (Kawauchi et al., 2009a). This phenotype is rescued by mutations in *GDF11*, leading to the conclusion that *Foxg1* inhibits GDF11 signaling via follistatin (Kawauchi et al., 2009a). It is proposed that GDF11 inhibits and follistatin stimulates expansion of the immediate neuronal precursors, but a second follistatin binding factor, activin betaB, inhibits expansion of earlier stages in the lineage, namely stem cells and early progenitors (Gokoffski et al., 2011).

The cytokine, LIF, regulates proliferation of olfactory progenitors during regeneration after injury to the olfactory epithelium. Olfactory bulbectomy transiently induces *LIF* mRNA in injured sensory neurons (Bauer et al., 2003) and globose basal cells (Nan et al., 2001). Bulbectomy also induces expression of mRNA for the LIF receptor, *LIFR*, and downstream signaling molecules STAT3 and Cyclin D1 (Getchell et al., 2002b). Bulbectomy normally stimulates progenitor proliferation in the olfactory epithelium, but this fails to occur in the LIF-deficient mouse (Bauer et al., 2003). In vitro, LIF stimulates proliferation neuronal precursors (Satoh and Yoshida, 1997) and inhibits sensory neuron maturation (Moon et al., 2002). Recent evidence demonstrates that the effects of LIF are mediated via the diffusible factor, nitric oxide (NO) (Sulz et al., 2009; Lopez et al., 2012). Nitric oxide synthase (NOS) is induced in olfactory epithelium after bulbectomy (Roskams et al., 1994). In vitro, NO induced proliferation of basal cells and blockers of inducible-NOS inhibited proliferation and promoted differentiation into sensory neurons, effects that were reversed by a NO donor (Sulz et al., 2009). LIF induced expression of LIFR and iNOS in horizontal basal cells and globose basal cells and stimulated proliferation of neural progenitors (Lopez et al., 2012). The proliferative effect of LIF was prevented by a LIFR blocking antibody and by an inhibitor of iNOS, with both effects reversed by a NO donor (Lopez et al., 2012). Taken together these experiments suggest that damage to the olfactory sensory neurons induces expression of LIF, which then stimulates proliferation of globose basal cells, the neuronal progenitors, and inhibits differentiation of sensory neurons via induction of iNOS and production of NO. Because NO diffuses readily through tissues, this mechanism would spatially enhance the transient effect of LIF release from affected sensory neurons. After damage to the olfactory epithelium, LIF is released by macrophages in the olfactory

7.4 Molecular Regulation of Olfactory Neurogenesis

mucosa (Getchell et al., 2002b) and iNOS is induced in olfactory ensheathing cells (Harris et al., 2009), further amplifying this signaling pathway.

Purines such as ATP are involved in regeneration of the sensory neurons after epithelial injury. The purinergic P2X receptor is found in basal cells, and blocking PTX receptors inhibits the injury-induced proliferation of basal cells (Gao et al., 2010). Intraperitoneal and intranasal ATP and purinergic agonists induced proliferation of progenitors in the olfactory epithelium, an effect that was reversed by purinergic antagonists and similar results were obtained in vitro (Jia et al., 2009). The mechanism of ATP-induced proliferation may act via several growth factor signaling pathways. ATP-induced expression in olfactory epithelium of $TGF\alpha$, FGF2, and NPY via purinergic receptors (Jia et al., 2011). The ATP-induced proliferation was blocked by inhibitors of EGFR and FGFR, indicating that ATP signaling acts via the well documented $TGF\alpha$ and FGF2 actions in the olfactory epithelium. NPY is expressed by supporting cells, sensory neurons and some olfactory ensheathing cells (Doyle et al., 2012). NPY-deficient mice have fewer olfactory neurons but more MASH1-positive globose basal cells (Doyle et al., 2012). The number of neurospheres generated from olfactory mucosa is reduced compared to wild-type mice in NPY-deficient and Y1 receptor-deficient mice (Doyle et al., 2008). In the neonatal olfactory epithelium, purinergic receptor agonists induced release of NPY in slice cultures (Kanekar et al., 2009) and in the adult olfactory epithelium, ATP promoted NPY synthesis and increased the number of NPY-positive cells (Jia and Hegg, 2010). This was blocked with purinergic receptor antagonist, which also blocked the effect of ATP on basal cell proliferation (Jia and Hegg, 2010). The emerging picture from these results is that cellular damage to the olfactory epithelium releases ATP into the extracellular milieu and this then induces proliferation of neuronal progenitors via the EGF, FGF2, and NPY signaling pathways. PYY, a molecule in the NPY family may also play a role in neurogenesis (Doyle et al., 2008; Doyle et al., 2012).

Other peptide growth factors are implicated in olfactory neurogenesis, although their functions are less well defined. IGF-I is present in human olfactory mucus (Federico et al., 1999) and infusion of IGF-I into the external naris increased the thickness of the olfactory epithelium and increased the number of proliferating cells (Pixley et al., 1998). In vitro, IGF-I reduced GBC proliferation and promoted morphological differentiation of neurons (McCurdy et al., 2005). Of the neurotrophins, BDNF and NT-3, but not NGF, increased the numbers of immature neurons in primary cultures of olfactory neurons (Holcomb et al., 1995; Roskams et al., 1996; Liu et al., 1998). BDNF induces proliferation and survival of neuronal precursors,

effects that were blocked by atrial C-type natriuretic peptide (Simpson et al., 2002). Given the distribution of the neurotrophin receptors (see above), it is not surprising that sensory neurons were not affected by the presence of NGF because they lack its receptors TrkA and $p75^{NTR}$ (Feron et al., 2008). It is interesting to note that in co-cultures of neurons and ensheathing cells withdrawal of NGF resulted in a dramatic decrease in neuron number (Bakardjiev, 1997). This may have been an indirect effect via loss of the ensheathing cells, which express the low affinity NGF receptor (see below). The Wnt signaling pathway is a major regulator of embryonic brain development. Several Wnt signaling molecules were expressed in the adult olfactory epithelium, including agonists (Wnt2, Wnt3a, Wnt7b), an antagonist (Sfrp1), a receptor (Fzd), and signaling pathway molecules (Axin1, Tcf7l2; Wang et al., 2011). Wnt signaling inhibitors disrupted neurogenesis after epithelial damage with significant reductions in olfactory sensory neurons and neuronal progenitors, indicating that Wnt signaling may be necessary for progenitor proliferation (Wang et al., 2011). In vitro, Wnt activators increased the number of immature neurons and reduced the number of non-neural cell types, leading to the suggestion that Wnt promotes neuronal differentiation (Wang et al., 2011).

Dopamine induced apoptosis and differentiation of an olfactory cell line (Coronas et al., 1997a) and promoted differentiation in explant culture of olfactory epithelium of the adult mouse (Feron et al., 1999c). In human explant cultures, dopamine inhibited mitosis and induced apoptosis (Feron et al., 1999c). These effects were mediated via the dopamine D2 receptor (Feron et al., 1999c), which is present in the neuronal layer of the olfactory epithelium (Feron et al., 1999c; Koster et al., 1999), Dopamine is present in the mucus above the epithelium (Lucero and Squires, 1998). Dopamine modulates electrical properties of sensory neurons (Vargas and Lucero, 1999) via adenylyl cyclase (Mania-Farnell et al., 1993; Coronas et al., 1999) and hence could signal that the developing dendrite has reached the epithelial surface. Dopamine is also present in the olfactory bulb at the site of axon termination (Halasz et al., 1977; Davis and Macrides, 1983), and it could act there as a signal to the developing neuron that its axon has reached its target. It is possible, therefore, that dopamine may act continually as a trophic factor at both ends of the active sensory neuron.

Retinoic acid (vitamin A) is involved in many aspects of neural development. The many components of retinoic signaling are present in the olfactory epithelium: receptors ($RAR\alpha$, $RAR\beta$, $RXR\alpha$, $RXR\beta$, and $RXR\gamma$; Zhang, 1999; Yee and Rawson, 2005), synthetic enzymes (RALDH 1–3; Peluso et al., 2012), binding proteins (CRABP I and II; Asson-Batres et al., 2003) and degrading enzyme (Cyp26B1; Hagglund et al., 2006). Vitamin A deficiency increases proliferation of basal cells (Asson-Batres et al., 2003a, b) and retinoic acid induced differentiation of an immortalized olfactory progenitor cell line into OMP-positive neurons (Lakard et al., 2007) and induced neurite outgrowth of olfactory neurons (Whitesides et al., 1998). The direction of retinoic acid signaling appears to be from supporting cells and lamina propria fibroblasts onto sensory neurons and basal cells because the synthetic enzymes are found in the former (Asson-Batres and Smith, 2006; Peluso et al., 2012) and the receptors are found in all cell types (Yee and Rawson, 2005; Oztokatli et al., 2012). Cyp26B1, the retinoic acid degrading enzyme, was found in some sensory neurons (Hagglund et al., 2006). Expression of a dominant-negative retinoic acid receptor (RARα) in sensory neurons during development led to their premature death (Hagglund et al., 2006; Oztokatli et al., 2012) and after epithelial damage, retroviral transduction with the same dominant-negative retinoic acid receptor after epithelial damage blocked the reappearance of mature OMP-positive sensory neurons (Peluso et al., 2012). Epithelial damage and treatment with retinoic acid increased the number of RARα – positive supporting cells and basal cells (Yee and Rawson, 2005). These experiments suggest that retinoic acid signaling regulates differentiation of neuronal progenitors into sensory neurons and that this signaling operates via supporting cells.

7.4.3 Macrophages and Olfactory Neurogenesis

The olfactory mucosa, including the olfactory epithelium, is populated by macrophages and dendritic cells, the surveillance cells of the immune system (Getchell et al., 2002b; Ruitenberg et al., 2008). Macrophages respond dynamically to tissue damage. Recruitment of macrophages to the epithelium after damage is essential for regeneration. When macrophage recruitment was reduced after bulbectomy, apoptosis was increased whereas epithelial thickness, proliferation, and sensory neuron numbers were all reduced (Borders et al., 2007). Bulbectomy increased levels of macrophage attractant chemokines MIP-1α and MCP-1 and their receptors CCR1 and CCR2 – all found on resident and infiltrating macrophages (Getchell et al., 2002c). *MIP-1α*-deficient mice had reduced recruitment of macrophages to the epithelium after bulbectomy that was reversed by MIP-1α treatment (Kwong et al., 2004). Macrophage infiltration and activation and neuronal progenitor proliferation were all reduced after bulbectomy in mice deficient of the macrophage scavenger receptor, which mediates binding of macrophages to apoptotic cells (Getchell et al., 2006). Macrophages and dendritic cells in the olfactory epithelium express another chemokine receptor, CX3CR1 and they depend on its ligand, fractalkine/CX3CL1, for intraepithelial migration (Ruitenberg et al., 2008). CX3CR1-deficient

mice have increased death of sensory neurons after bulbectomy (Blomster et al., 2011). This was associated with increased recruitment of macrophages into the damaged epithelium and increased levels of proinflammatory cytokines (TNFα and IL6), despite diminished macrophage phagocytic activity (Blomster et al., 2011). Taken together these studies demonstrate the importance of macrophages in regulating regeneration of the olfactory epithelium after damage to the sensory neurons.

7.4.4 Olfactory Ensheathing Cells and Neurogenesis

The remarkable capacity for neurogenesis is partly dependent on the olfactory ensheathing cells (OECs), the glia that surround the sensory axons to form the olfactory nerve fascicles from the epithelium to the olfactory bulb (Doucette, 1984; Doucette, 1989; Marin-Padilla and Amieva, 1989; Doucette, 1990; Valverde and Lopez-Mascaraque, 1991; Franceschini and Barnett, 1996). In the lamina propria they surround local fascicles of olfactory axons that become increasingly larger towards the olfactory bulb (Doucette, 1990). Upon reaching the bulb, the olfactory nerve defasciculates; presumably with the aid of the OECs of the outer nerve fibre layer (Doucette, 1989). The axons then refasciculate according to their odorant receptor type, prior to targeting to their topographically appropriate glomerulus and a process in which the OECs of the inner nerve fibre layer are thought to contribute (Doucette, 1989; Mombaerts et al., 1996; Au et al., 2002).

OECs are heterogeneous in morphology, gene expression, protein expression, and functions associated with their location in the olfactory nerve or olfactory bulb. In the olfactory nerve, two types are distinguished by distinct cytoplasmic densities under the electron microscope (Cuschieri and Bannister, 1975; Doucette, 1989; Valverde et al., 1993) and immunologically, whereby some express S-100 protein and glial acidic fibrillary protein (GFAP) and others express only GFAP (Pixley, 1992). In the nerve and outer nerve fibre layer of olfactory bulb, OECs express p75^{NTR} (Vickland et al., 1991; Pixley, 1992; Turner and Perez-Polo, 1992; Gong et al., 1994; Mombaerts et al., 1996; Roskams et al., 1996), S100 (Pixley, 1992), brain lipid binding protein (BLBP; Vincent et al., 2008), and polysialated neural cell adhesion molecule (E-NCAM) (Franceschini and Barnett, 1996). OECs in the inner nerve fibre layer of olfactory bulb express low levels of S100B, but not p75^{NTR}, and a subset expresses TROY and NPY (Ubink and Hokfelt, 2000; Hisaoka et al., 2004; Morikawa et al., 2008). Microarray studies demonstrate that OECs from olfactory nerve outside the olfactory bulb expressed higher levels of genes involved in modulating the extracellular matrix; OECs from the outer nerve layer of the olfactory bulb expressed higher levels of genes contributing to modulating the extracellular matrix, cell sorting and neural development; and OECs from the inner nerve layer of the olfactory bulb expressed higher levels of genes that regulate axon guidance (Guerout et al., 2010; Honore et al., 2012).

In the healthy adult, OEC proliferation is limited but after olfactory bulbectomy OECs rapidly proliferate throughout the olfactory nerve pathway and from precursors within the basal layer of the olfactory epithelium (Chehrehasa et al., 2012). Newly formed OECs migrate into the lesion cavity in the skull and enable the regenerating axons to grow and extend on the pre-formed surface of OECs (Chehrehasa et al., 2010). Time-lapse imaging analysis in vitro reveals that OECs from olfactory nerve predominantly adhere to each, other whereas OECs of the olfactory bulb display a mix of repulsion and adhesion (Windus et al., 2007; Windus et al., 2010). OEC behaviour is regulated by extracellular matrix (Tisay and Key, 1999), affecting sensory neurons via soluble factors and by direct contact. Olfactory sensory neurons preferentially grow on OECs (Chuah and Au, 1994; Tisay and Key, 1999; Windus

et al., 2011), which form a substrate for neuronal migration (Windus et al., 2010; Windus et al., 2011). OECs produce numerous growth factors that would assist olfactory neurogenesis including NGF (Woodhall et al., 2001; Lipson et al., 2003), BDNF (Woodhall et al., 2001; Lipson et al., 2003), FGF1 (Key et al., 1996), FGF2 (Matsuyama et al., 1992; Gall et al., 1994; Chuah and Teague 1999), GDNF (Woodhall et al., 2001; Lipson et al., 2003), CNTF (Guthrie et al., 1997; Lipson et al., 2003), Neurotrophin (NT)-4/5, and VEGF (Mackay-Sim and Chuah, 2000; Boruch et al., 2001; Woodhall et al., 2001; Wewetzer et al., 2002; Au and Roskams, 2003).

7.4.5 Summary

There are clearly many molecular signals regulating proliferation, differentiation, survival and death in neurogenesis in the adult olfactory epithelium (Table 7.3, Figure 7.4). The picture is still incomplete because there are still many growth factors and receptors in the olfactory epithelium for

Table 7.3 Growth factors that have proven active in vitro and in vivo that act on different cell types in olfactory mucosa. "Growth factor" is used here to encompass all extracellular signalling molecule affecting neurogenesis.

Growth Factor	Cell Type	Action	Reference
ATP	Neuronal precursors	Stimulates proliferation	Jia et al., 2011
BDNF	Mature neurons	Promotes survival	Newman et al., 2000
BMP 2/4/7	Neuronal precursors	Inhibits proliferation	Shou et al., 1999; Shou et al., 2000
BMP4	Immature neurons	Promotes survival	Shou et al., 2000
CX3CL1	Macrophages	Promotes recruitment	Blomster et al., 2011
Dopamine	Immature neurons	Promotes differentiation	Feron et al., 1999c
EGF / TGFα	Horizontal basal cells	Stimulates proliferation	Farbman et al., 1994; Farbman and Buchholz 1996; Newman et al., 2000
	Supporting cells	Stimulates proliferation	Farbman and Buchholz 1996; Newman et al., 2000
FGF2	Globose basal cell neuronal precursors		DeHamer et al., 1994; Newman et al., 2000; Nishikawa et al., 2009
	Olfactory ensheathing cells	Stimulates proliferation	Chuah and Teague 1999
GDF11	Neuronal precursors	Inhibits proliferation	Wu et al., 2003
GGF2	Olfactory ensheathing cells	Stimulates proliferation	Chuah et al., 2000
IGF-I	Immature neurons	Promotes differentiation	McCurdy et al., 2005
LIF	Neuronal precursors	Promotes proliferation	Lopez et al., 2012
MIP1α	Macrophages	Promotes recruitment	Kwong et al., 2004
NO	Neuronal precursors	Stimulates proliferation	Sulz et al., 2009; Lopez et al., 2012
NRG	Olfactory ensheathing cells	Stimulates proliferation	Pollock et al., 1999
NT3	Olfactory ensheathing cells	Stimulates proliferation	Bianco et al., 2004
PDGF	Mature neurons	Promotes survival	Newman et al., 2000
RA	Immature neurons	Promotes differentiation	Whitesides et al., 1998; Lakard et al., 2007
	Neuronal precursors	Inhibits proliferation	Asson-Batres et al., 2003b
TGFβ2	Globose basal cell neuronal precursors	Promotes differentiation	Mahanthappa and Schwarting 1993; Newman et al., 2000
TH	Immature neurons	Promotes survival	Mackay-Sim and Beard 1987
Wnt	Neuronal precursors	Stimulates proliferation	Wang et al., 2011
	Immature neurons	Promotes differentiation	Wang et al., 2011

which functional roles are not yet defined (Table 7.2). Regulation of cell lineage and neurogenesis would occur locally in a cell-type and time-dependent manner, involving cells within the epithelium, lamina propria, and olfactory bulb.

7.5 CLINICAL APPLICATIONS FROM OLFACTORY NEUROGENESIS

Olfactory mucosa can be sampled in humans without affecting the sense of smell (Lanza et al., 1993; Feron et al., 1998), thus offering access to stem cells, neural progenitors, neurons, and olfactory ensheathing cells. This convergence of biological properties and practical accessibility are leading to clinical applications in transplantation repair of the injured spinal cord (Mackay-Sim and St John, 2011) and in drug discovery for brain diseases (Mackay-Sim, 2012). Deeper understanding of olfactory neurogenesis may help to unravel the cause of esthesio-(also known as olfactory) neuroblastomas, which originate from the olfactory epithelium. These tumors present a highly heterogeneous cellular composition indicating that they may derive from a multipotent GBC stem cells with evidence for HBC-like differentiation, sustentacular-like cells, and other cells with a phenotype intermediate between olfactory neurons and Bowman's duct/gland cells. Such cells have their own sub-territory within the tumor, indicating that they may derive from multipotent GBC stem cells (Holbrook et al., 2011).

7.5.1 Transplantation Repair of the Spinal Cord

Olfactory ensheathing cells were chosen as candidates for repair of the injured spinal cord because of their role in assisting neurogenesis in the olfactory mucosa and their ability to integrate with the central nervous system (Doucette, 1990; Ramon-Cueto and Nieto-Sampedro, 1994; Li et al., 1997; Li et al., 1998; Ramon-Cueto et al., 1998). There are now numerous studies demonstrating the efficacy of olfactory ensheathing cell transplantation for the repair of the injured rodent spinal cord (Mackay-Sim and St John, 2011). Importantly for clinical translation, olfactory ensheathing cells from the nose of adult rats demonstrated that autologous transplantation could assist functional recovery after complete spinal cord transaction (Lu et al., 2001; Lu et al., 2002). Additionally, olfactory ensheathing cells from the human nose induced functional recovery after transplantation into the contused spinal cord of the rat (Gorrie et al., 2010). There are numerous ways in which olfactory ensheathing cells contribute after transplantation into the injured spinal cord: provision of growth factors (Woodhall et al., 2001; Chung et al., 2004); direct physical contact (Li et al., 2004); stimulation of axon sprouting (Chung et al., 2004); and myelination (Imaizumi et al., 1998).

A Phase I clinical trial demonstrated that intra-spinal cord transplantation of autologous olfactory ensheathing cells is feasible, with no adverse outcomes over an observation period of three years (Feron et al., 2005; Mackay-Sim et al., 2008). Another clinical approach has been to transplant whole pieces of autologous olfactory mucosa into the injured spinal cord after surgical removal of the injured cord tissue (Lima et al., 2006). This method was safe in most patients, although significant adverse advents were reported (Lima et al., 2006; Chhabra et al., 2009). Some patients noted improvements in locomotor function after olfactory mucosa transplantation, but such improvements were not universal (Lima et al., 2006; Chhabra et al., 2009). In these studies there was no control group, as expected in an efficacy trial. Obviously olfactory mucosa pieces contain many cells types other than olfactory ensheathing cells, including stem cells, which may contribute to recovery of the injured spinal cord.

7.5.2 Olfactory Stem Cell Models of Brain Diseases

Autopsy specimens of olfactory mucosa from patients show immunohistological evidence of pathological changes in Alzheimer's disease (Talamo et al., 1989) including increased frequency of β-amyloid aggregates and dystrophic tau-positive neurites (Lee et al., 1993; Crino et al., 1995; Arnold et al., 2010) and markers of oxidative stress (Perry et al., 2003). "Neuroblasts" can be grown in continuous culture from autopsy specimens of human olfactory mucosa (Wolozin et al., 1992) and, in Alzheimer's disease, these show elevated levels of the precursor to β-amyloid protein (Wolozin et al., 1993) and increased oxidative stress (Ghanbari et al., 2004) compared to neuroblasts from people without the disease. Olfactory mucosa from schizophrenia patients has different proportions of cells staining for immature and mature neuronal markers at autopsy (Arnold et al., 2001) and olfactory mucosa cultures have altered cell proliferation and altered tissue adhesion, not seen in tissue from controls or patients with bipolar disorder (Feron et al., 1999b; McCurdy et al., 2006). Neurodevelopmental abnormalities are also reported in biopsy specimens from patients with Rett's syndrome (Ronnett et al., 2003).

Olfactory stem cells can be grown from human biopsy tissue either as neurally restricted progenitors (Roisen et al., 2001) or as more broadly multipotent stem cells (Murrell et al., 2005; Delorme et al., 2010). Olfactory stem cells are grown as self-renewing "neurospheres" that can differentiate into neurons, astrocytes, and oligodendrocytes (Roisen et al., 2001; Murrell et al., 2005). Olfactory neurosphere-derived cells from patients and controls can be grown in continuous culture to investigate cellular and molecular aspects of the effects of disease on the nervous system (Mackay-Sim, 2012). In Parkinson's disease, olfactory neurosphere-derived cells show evidence of oxidative stress in their gene and protein expression, in several metabolic assays, and in reduced activity of the *NRF2* signaling pathway (Matigian et al., 2010; Cook et al., 2011). Olfactory neurosphere-derived cells from patients with schizophrenia have altered gene and protein expression and metabolic function in cell signaling pathways associated with neurodevelopment (Matigian et al., 2010). Closer analysis of schizophrenia patient-derived cells reveals dysregulation of the cell cycle including increased levels of cyclin D1, a shorter cell cycle period and an increased proliferation rate (Matigian et al., 2010; Fan et al., 2012). Gene expression differences from control cells of patient-derived cells were disease-specific for schizophrenia and Parkinson's disease (Matigian et al., 2010) and additionally, there were disease-specific differences in the variability in gene expression, indicating an additional level of dysregulation as a consequence of schizophrenia and Parkinson's disease (Mar et al., 2011). Olfactory cell cultures are also instructive for understanding monogenic diseases. Neuroblasts from olfactory epithelium of two patients with fragile X mental retardation had no FMR1 protein, consistent with their genetic mutation and protein expression in their leukocytes (Abrams et al., 1999). Similarly, olfactory ectomesenchymal stem cells derived from olfactory mucosa of patients with familial dysautonomia show the expected reduction in protein levels of the affected gene (*IPKAPB*) and have promise in testing small molecules to correct the condition (Boone et al., 2010).

7.5.3 Future Clinical Applications

Olfactory function is commonly lost with age (Doty et al., 1984), although it may be due to ageing-associated disease, rather than ageing itself (Mackay-Sim et al., 2006). This is associated with olfactory nerve fibrosis, replacement of olfactory by respiratory epithelium, and olfactory epithelium that lacks neurons and GBCs, but retains HBCs (Holbrook et al., 2005). Potential therapeutic strategies include activation of HBCs to proliferate and re-enter the multipotent cell lineage and even transplanting autologous HBCs or GBCs after generating them in vitro from small biopses. A deeper understanding of the stem cells in the olfactory mucosa may also lead to clinical applications for transplantation outside the olfactory system. For example, human olfactory neurosphere-derived cells can promote functional recovery after transplantation into the injured rat spinal cord (Xiao et al., 2005), transplantation into the parkinsonian rat (Murrell et al., 2008) and after

transplantation into the injured intervertebral disc (Murrell et al., 2009).

7.6 CONCLUSIONS

Olfactory neurogenesis is a continuing, regulated process that is similar in many respects to neurogenesis in the embryonic nervous system. It is noteworthy that although neurogenesis continues throughout adult life, the rate of basal cell proliferation declines in old age and there is a significant decrease in olfactory function in aging humans. The regulation of olfactory neurogenesis is not completely understood; there are a large number of candidate factors whose function is unknown but whose presence is inferred by the expression of appropriate receptors in the epithelium. Apart from its intrinsic interest in scientific terms, understanding olfactory neurogenesis is becoming important now for its potential clinical applications in diseases of brain development and in nervous system repair.

ABBREVIATIONS

ActR	Activin receptor
Ascl-1	Achaete-scute like 1 (formerly known as Mash1)
BDNF	brain-derived growth factor
BMP	bone morphogenic protein
BMPR	BMP receptor
CNTF	ciliary neurotrophic factor
Cre-ERT2	Cre recombinase fused with the selectively tamoxifen-sensitive estrogen receptor
D2	dopamine D2 receptor
DA	dopamine
EC	ensheathing cell
EGF	epidermal growth factor
EGFR	epidermal growth factor receptor
FACS	fluorescence-activated cell sorting
FGF	fibroblast growth factor
FGFR	fibroblast growth factor receptor
GAP43	growth associated protein 43
GBC	globose basal cell
GDNF	glial cell line-derived growth factor
GFAP	glial fibrilliary acidic protein
GFP	enhanced green fluorescent protein
GFR	GDNF receptor
GGF	glial growth factor HBC horizontal basal cell
HBC	horizontal basal cell
IGF	insulin-like growth factor
IGFBP	insulin-like growth factor binding protein

IGFR	insulin-like growth factor receptor
IN	immature neuron
LIF	leukocyte inhibitory factor
MeBr	methyl bromide
MN	mature neuron
NCAM	neural cell adhesion molecule
NDF	Neu differentiation factor
Neurog1	Neurogenin 1
NGF	nerve growth factor
NO	nitric oxide
NOS	nitric oxide synthase
eNOS	endothelial NOS
iNOS	inducible NOS
nNOS	neural NOS
NPY	neuronal protein Y
NT3	neurotrophin 3
OMP	olfactory marker protein
p75^{NTR}	p75 neurotrophin receptor
PDGF	platelet-derived growth factor
PDGFR	platelet-derived growth factor receptor
P2XR	purine receptor, type 2X
PYY	protein YY
RA	retinoic acid
RAR	retinoic acid receptor
RXR	retinoid X receptor
RRV	replication-incompetent retrovirus
TGF	transforming growth factor
TNF	tumour necrosis factor
Trk	tyrosine kinase

ACKNOWLEDGMENTS

Alan Mackay-Sim is supported by grants from the National Health and Medical Research Council of Australia. James E. Schwob is supported by grants R01 DC002167 and R01 DC010242 from the NIH.

REFERENCES

Abrams, M. T., Kaufmann, W. E., et al. (1999). FMR1 gene expression in olfactory neuroblasts from two males with fragile X syndrome. *Am. J. Med. Genet.* 82(1): 25–30.

Aiba, T., J. Mori, et al. (1993). Nerve growth factor (NGF) and its receptor in rat olfactory epithelium. *Acta. Otolaryngol. Suppl.* 506: 37–40.

Andres, K. (1966). Der Feinbau der Regio olfactoria von Makrosmatikern. *Z Zellforsch. Mikrosk. Anat.* 69: 140–154.

Arnold, S. E., Han, L. Y., et al. (2001). Dysregulation of olfactory receptor neuron lineage in schizophrenia. *Arch. Gen. Psychiatry.* 58(9): 829–835.

Arnold, S. E., Lee, E. B., et al. (2010). Olfactory epithelium amyloid-beta and paired helical filament-tau pathology in Alzheimer disease. *Ann. Neurol.* 67(4): 462–469.

Asson-Batres, M. A., Ahmad, O., et al. (2003). Expression of the cellular retinoic acid binding proteins, type II and type I, in mature rat olfactory epithelium. *Cell Tissue Res.* 312(1): 9–19.

Asson-Batres, M. A. and Smith, W. B. (2006). Localization of retinaldehyde dehydrogenases and retinoid binding proteins to sustentacular cells, glia, Bowman's gland cells, and stroma: potential sites of retinoic acid synthesis in the postnatal rat olfactory organ. *J. Comp. Neurol.* 496(2): 149–171.

Asson-Batres, M. A., Smith, W. B., et al. (2009). Retinoic acid is present in the postnatal rat olfactory organ and persists in vitamin A–depleted neural tissue. *J. Nutr.* 139(6): 1067–1072.

Asson-Batres, M. A., Zeng, M. S., et al. (2003b). Vitamin A deficiency leads to increased cell proliferation in olfactory epithelium of mature rats. *J. Neurobiol.* 54(4): 539–554.

Au, E. and Roskams, A. J. (2003). Olfactory ensheathing cells of the lamina propria in vivo and in vitro. *Glia* 41(3): 224–236.

Au, W. W., Treloar, H. B., et al. (2002). Sublaminar organization of the mouse olfactory bulb nerve layer. *J. Comp. Neurol.* 446(1): 68–80.

Ayer-le Lievre, C., Stahlbom, P. A., et al. (1991). Expression of IGF-I and -II mRNA in the brain and craniofacial region of the rat fetus. *Development* 111(1): 105–115.

Ayer-Lelievre, C. S., Ebendal, T., et al. (1983). Localization of nerve growth factor-like immunoreactivity in rat nervous tissue. *Med. Biol.* 61(6): 296–304.

Bakardjiev, A. (1997). Biosynthesis of carnosine in primary cultures of rat olfactory bulb. *Neurosci. Lett.* 227(2): 115–118.

Barraud, P., Seferiadis, A. A., et al. (2010). Neural. crest origin of olfactory ensheathing glia. *Proc. Natl. Acad. Sci. U. S. A.* 107(49): 21040–21045.

Bauer, S., Rasika, S., et al. (2003). Leukemia inhibitory factor is a key signal for injury-induced neurogenesis in the adult mouse olfactory epithelium. *J. Neurosci.* 23(5): 1792–1803.

Bianco, J. I., Perry, C., et al. (2004). Neurotrophin 3 promotes purification and proliferation of olfactory ensheathing cells from human nose. *Glia* 45(2): 111–123.

Blomster, L. V., Vukovic, J., et al. (2011). CX(3)CR1 deficiency exacerbates neuronal loss and impairs early regenerative responses in the target-ablated olfactory epithelium. *Mol. Cell. Neurosci.* 48(3): 236–245.

Bondy, C. and Lee, W. H. (1993). Correlation between insulin-like growth factor (IGF)-binding protein 5 and IGF-I gene expression during brain development. *J. Neurosci.* 13(12): 5092–5104.

Boone, N., Loriod, B., et al. (2010). Olfactory stem cells, a new cellular model for studying molecular mechanisms underlying familial dysautonomia. *PLoS One* 5(12): e15590.

Borders, A. S., Getchell, M. L., et al. (2007). Macrophage depletion in the murine olfactory epithelium leads to increased neuronal death and decreased neurogenesis. *J. Comp. Neurol.* 501(2): 206–218.

Boruch, A. V., Conners, J. J., et al. (2001). Neurotrophic and migratory properties of an olfactory ensheathing cell line. *Glia* 33(3): 225–229.

Breipohl, W., Laugwitz, H. J., et al. (1974). Topological relations between the dendrites of olfactory sensory cells and sustentacular cells in different vertebrates. *An ultrastructural study. J. Anat.* 117(Pt 1): 89–94.

Breipohl, W., Mackay-Sim, A., et al. (1986). Neurogenesis in the vertebrate main olfactory epithelium. *Ontogeny of Olfaction.* W. Breipohl (Eds). Berlin, Springer-Verlag, pp. 21–33.

Buckland, M. E. and Cunningham, A. M. (1999). Alterations in expression of the neurotrophic factors glial cell line-derived neurotrophic factor, ciliary neurotrophic factor and brain-derived neurotrophic factor, in the target-deprived olfactory neuroepithelium. *Neuroscience* 90(1): 333–347.

References

Caggiano, M., Kauer, J. S., et al. (1994). Globose basal cells are neuronal progenitors in the olfactory epithelium: a lineage analysis using a replication-incompetent retrovirus. *Neuron* 13(2): 339–352.

Calof, A. L. and Chikaraishi, D. M. (1989). Analysis of neurogenesis in a mammalian neuroepithelium: proliferation and differentiation of an olfactory neuron precursor in vitro. *Neuron* 3(1): 115–127.

Carr, V. M. and Farbman, A. I. (1993). The dynamics of cell death in the olfactory epithelium. *Exp. Neurol.* 124(2): 308–314.

Cau, E., Casarosa, S., et al. (2002). Mash1 and Ngn1 control distinct steps of determination and differentiation in the olfactory sensory neuron lineage. *Development* 129(8): 1871–1880.

Cau, E., Gradwohl, G., et al. (1997). Mash1 activates a cascade of bHLH regulators in olfactory neuron progenitors. *Development* 124(8): 1611–1621.

Chehrehasa, F., Ekberg, J. A., et al. (2012). Two phases of replacement replenish the olfactory ensheathing cell population after injury in postnatal mice. *Glia* 60(2): 322–332.

Chehrehasa, F., Windus, L. C., et al. (2010). Olfactory glia enhance neonatal axon regeneration. *Mol. Cell. Neurosci.* 45(3): 277–288.

Chen, J., Tu, Y., et al. (2004a). The localization of neuronal nitric oxide synthase may influence its role in neuronal precursor proliferation and synaptic maintenance. *Dev. Biol.* 269(1): 165–182.

Chen, X., Fang, H., et al. (2004b). Multipotency of purified, transplanted globose basal cells in olfactory epithelium. *J. Comp. Neurol.* 469(4): 457–474.

Chhabra, H. S., Lima, C., et al. (2009). Autologous olfactory [corrected] mucosal transplant in chronic spinal cord injury: an Indian Pilot Study. *Spinal Cord* 47(12): 887–895.

Chuah, M. I. and Au, C. (1994). Olfactory cell cultures on ensheathing cell monolayers. *Chem. Senses* 19(1): 25–34.

Chuah, M. I., Cossins, J., et al. (2000). Glial growth factor 2 induces proliferation and structural changes in ensheathing cells. *Brain Res.* 857(1–2): 265–274.

Chuah, M. I. and Teague, R. (1999). Basic fibroblast growth factor in the primary olfactory pathway: mitogenic effect on ensheathing cells. *Neuroscience* 88(4): 1043–1050.

Chung, R. S., Woodhouse, A., et al. (2004). Olfactory ensheathing cells promote neurite sprouting of injured axons in vitro by direct cellular contact and secretion of soluble factors. *Cell. Mol. Life Sci.* 61(10): 1238–1245.

Cook, A. L., Vitale, A. M., et al. (2011). NRF2 activation restores disease related metabolic deficiencies in olfactory neurosphere-derived cells from patients with sporadic Parkinson's disease. *PLoS One* 6(7): e21907.

Coronas, V., Feron, F., et al. (1997a). In vitro induction of apoptosis or differentiation by dopamine in an immortalized olfactory neuronal cell line. *J. Neurochem.* 69(5): 1870–1881.

Coronas, V., Krantic, S., et al. (1999). Dopamine receptor coupling to adenylyl cyclase in rat olfactory pathway: a combined pharmacological-radioautographic approach. *Neuroscience* 90(1): 69–78.

Coronas, V., Srivastava, L. K., et al. (1997b). Identification and localization of dopamine receptor subtypes in rat olfactory mucosa and bulb: a combined in situ hybridization and ligand binding radioautographic approach. *J. Chem. Neuroanat.* 12(4): 243–257.

Cotsarelis, G., Cheng, S. Z., et al. (1989). Existence of slow-cycling limbal epithelial basal cells that can be preferentially stimulated to proliferate: implications on epithelial stem cells. *Cell* 57(2): 201–209.

Crino, P. B., Martin, J. A., et al. (1995). Beta-Amyloid peptide and amyloid precursor proteins in olfactory mucosa of patients with Alzheimer's disease, Parkinson's disease, and Down syndrome. *Ann. Otol. Rhinol. Laryngol.* 104(8): 655–661.

Cuschieri, A. and Bannister, L. H. (1975). The development of the olfactory mucosa in the mouse: electron microscopy. *J. Anat.* 119(Pt 3): 471–498.

Danciger, E., Mettling, C., et al. (1989). Olfactory marker protein gene: its structure and olfactory neuron-specific expression in transgenic mice. *Proc. Natl. Acad. Sci. U. S. A.* 86(21): 8565–8569.

Davis, B. and Macrides, F. (1983). Tyrosine hydroxylase immunoreactive neurons and fibers in the olfactory system of the hamster. *Journal of Comp. Neuro.* 214: 427–440.

DeHamer, M. K., Guevara, J. L., et al. (1994). Genesis of olfactory receptor neurons in vitro: regulation of progenitor cell divisions by fibroblast growth factors. *Neuron* 13(5): 1083–1097.

Delorme, B., Nivet, E., et al. (2010). The human nose harbors a niche of olfactory ectomesenchymal stem cells displaying neurogenic and osteogenic properties. *Stem Cells Dev.* 19(6): 853–866.

Doty, R. L., Shaman, P., et al. (1984). Smell identification ability: changes with age. *Science* 226(4681): 1441–1443.

Doucette, J. R. (1984). The glial cells in the nerve fiber layer of the rat olfactory bulb. *Anat. Rec.* 210(2): 385–391.

Doucette, R. (1989). Development of the nerve fiber layer in the olfactory bulb of mouse embryos. *J. Comp. Neurol.* 285(4): 514–527.

Doucette, R. (1990). Glial influences on axonal growth in the primary olfactory system. *Glia* 3(6): 433–449.

Doyle, K. L., Hort, Y. J., et al. (2012). Neuropeptide Y and peptide YY have distinct roles in adult mouse olfactory neurogenesis. *J. Neurosci. Res.* 90(6): 1126–1135.

Doyle, K. L., Karl, T., et al. (2008). Y1 receptors are critical for the proliferation of adult mouse precursor cells in the olfactory neuroepithelium. *J. Neurochem.* 105(3): 641–652.

Ensoli, F., Fiorelli, V., et al. (1998). Basic fibroblast growth factor supports human olfactory neurogenesis by autocrine/paracrine mechanisms. *Neuroscience* 86(3): 881–893.

Eriksson, P. S., Perfilieva, E., et al. (1998). Neurogenesis in the adult human hippocampus. *Nat. Med.* 4(11): 1313–1317.

Fan, Y., Abrahamsen, G., et al. (2012). Altered cell cycle dynamics in schizophrenia. *Biol. Psychiat.* 71(2): 129–135.

Farbman, A. I., Brunjes, P. C., et al. (1988). The effect of unilateral naris occlusion on cell dynamics in the developing rat olfactory epithelium. *J. Neurosci.* 8(9): 3290–3295.

Farbman, A. I., Buchholz, J. A., et al. (1994). Growth factor regulation of olfactory cell proliferation. *Olfaction and Taste XI*. K. Kurihara, N. Suzuki and H. Ogawa (Eds). Tokyo, Springer-Verlag: 45–48.

Farbman, A. I. and Buchholz, J. A. (1996). Transforming growth factor-alpha and other growth factors stimulate cell division in olfactory epithelium in vitro. *J. Neurobiol.* 30(2): 267–280.

Farbman, A. I., Buchholz, J. A., et al. (1999). A molecular basis of cell death in olfactory epithelium. *J. Comp. Neurol.* 414(3): 306–314.

Federico, G., Maremmani, C., et al. (1999). Mucus of the human olfactory epithelium contains the insulin-like growth factor-I system which is altered in some neurodegenerative diseases. *Brain Res.* 835(2): 306–314.

Feron, F., Bianco, J., et al. (2008). Neurotrophin expression in the adult olfactory epithelium. *Brain Res.* 1196: 13–21.

Feron, F., Mackay-Sim, A., et al. (1999a). Stress induces neurogenesis in non-neuronal cell cultures of adult olfactory epithelium. *Neuroscience* 88(2): 571–583.

Feron, F., Perry, C., et al. (2005). Autologous olfactory ensheathing cell transplantation in human spinal cord injury. *Brain* 128(Pt 12): 2951–2960.

Feron, F., Perry, C., et al. (1999b). Altered adhesion, proliferation and death in neural cultures from adults with schizophrenia. *Schizophr. Res.* 40(3): 211–218.

Feron, F., Perry, C., et al. (1998). New techniques for biopsy and culture of human olfactory epithelial neurons. *Arch. Otolaryngol. Head Neck Surg.* 124(8): 861–866.

Feron, F., Vincent, A., et al. (1999c). Dopamine promotes differentiation of olfactory neuron in vitro. *Brain Res.* 845(2): 252–259.

Fletcher, R. B., Prasol, M. S., et al. (2011). *p63* regulates olfactory stem cell self-renewal and differentiation. *Neuron* 72(5): 748–759.

Forni, P. E., Taylor-Burds, C., et al. (2011). Neural crest and ectodermal cells intermix in the nasal placode to give rise to GnRH-1 neurons, sensory neurons, and olfactory ensheathing cells. *J. Neurosci.* 31(18): 6915–6927.

Franceschini, I. A. and Barnett, S. C. (1996). Low-affinity NGF-receptor and E-N-CAM expression define two types of olfactory nerve ensheathing cells that share a common lineage. *Dev. Biol.* 173(1): 327–343.

Gall, C. M., Berschauer, R., et al. (1994). Seizures increase basic fibroblast growth factor mRNA in adult rat forebrain neurons and glia. *Brain Res. Mol. Brain Res.* 21(3–4): 190–205.

Gao, L., Cao, L., et al. (2010). Blocking P2X receptors can inhibit the injury-induced proliferation of olfactory epithelium progenitor cells in adult mouse. *Int. J. Pediatr. Otorhinolaryngol.* 74(7): 747–751.

Getchell, M. L., Boggess, M. A., et al. (2002a). Expression of TGF-beta type II receptors in the olfactory epithelium and their regulation in TGF-alpha transgenic mice. *Brain Res.* 945(2): 232–241.

Getchell, M. L., Li, H., et al. (2006). Temporal gene expression profiles of target-ablated olfactory epithelium in mice with disrupted expression of scavenger receptor A: impact on macrophages. *Physiol. Genomics* 27(3): 245–263.

Getchell, T. V., Shah, D. S., et al. (2002b). Leukemia inhibitory factor mRNA expression is upregulated in macrophages and olfactory receptor neurons after target ablation. *J. Neurosci. Res.* 67(2): 246–254.

Getchell, T. V., Subhedar, N. K., et al. (2002c). Chemokine regulation of macrophage recruitment into the olfactory epithelium following target ablation: involvement of macrophage inflammatory protein-1alpha and monocyte chemoattractant protein-1. *J. Neurosci. Res.* 70(6): 784–793.

Ghanbari, H. A., Ghanbari, K., et al. (2004). Oxidative damage in cultured human olfactory neurons from Alzheimer's disease patients. *Aging Cell* 3(1): 41–44.

Gokoffski, K. K., Wu, H. H., et al. (2011). Activin and GDF11 collaborate in feedback control of neuroepithelial stem cell proliferation and fate. *Development* 138(19): 4131–4142.

Goldstein, B. J., Fang, H., et al. (1998). Transplantation of multipotent progenitors from the adult olfactory epithelium. *Neuroreport* 9(7): 1611–1617.

Goldstein, B. J. and Schwob, J. E. (1996). Analysis of the globose basal cell compartment in rat olfactory epithelium using GBC-1, a new monoclonal antibody against globose basal cells. *J. Neurosci.* 16(12): 4005–4016.

Goldstein, B. J., Wolozin, B. L., et al. (1997). FGF2 suppresses neuronogenesis of a cell line derived from rat olfactory epithelium. *J. Neurobiol.* 33(4): 411–428.

Gong, Q., Bailey, M. S., et al. (1994). Localization and regulation of low affinity nerve growth factor receptor expression in the rat olfactory system during development and regeneration. *J. Comp. Neurol.* 344(3): 336–348.

Gordon, M. K., Mumm, J. S., et al. (1995). Dynamics of MASH1 expression in vitro and in vivo suggest a non-stem cell site of MASH1 action in the olfactory receptor neuron lineage. *Mol. Cell. Neurosci.* 6(4): 363–379.

Gorrie, C. A., Hayward, I., et al. (2010). Effects of human OEC-derived cell transplants in rodent spinal cord contusion injury. *Brain Res.* 1337: 8–20.

Graziadei, P. P., and G. A. Monti, Graziadei (1978). Continuous nerve cell renewal in the olfactory system. *Handbook of Sensory Physiology.* M. Jacobson (Eds). New York, Springer-Verlag. IX. Development of Sensory Systems: 55–83.

Graziadei, P. P. (1973). Cell dynamics in the olfactory mucosa. *Tissue Cell* 5(1): 113–131.

Graziadei, P. P. and Monti Graziadei, G. A. (1979). Neurogenesis and neuron regeneration in the olfactory system of mammals. I. *Morphological aspects of differentiation and structural organization of the olfactory sensory neurons. J. Neurocytol.* 8(1): 1–18.

Guerout, N., Derambure, C., et al. (2010). Comparative gene expression profiling of olfactory ensheathing cells from olfactory bulb and olfactory mucosa. *Glia* 58(13): 1570–1580.

Guillemot, F., Lo, L. C., et al. (1993). Mammalian achaete-scute homolog 1 is required for the early development of olfactory and autonomic neurons. *Cell* 75(3): 463–476.

Guo, Z., Packard, A., et al. (2010). Expression of pax6 and sox2 in adult olfactory epithelium. *J. Comp. Neurol.* 518(21): 4395–4418.

Guthrie, K. M., Woods, A. G., et al. (1997). Astroglial ciliary neurotrophic factor mRNA expression is increased in fields of axonal sprouting in deafferented hippocampus. *J. Comp. Neurol.* 386(1): 137–148.

Hagglund, M., Berghard, A., et al. (2006). Retinoic acid receptor-dependent survival of olfactory sensory neurons in postnatal and adult mice. *J. Neurosci.* 26(12): 3281–3291.

Hahn, C. G., Han, L. Y., et al. (2005). In vivo and in vitro neurogenesis in human olfactory epithelium. *J. Comp. Neurol.* 483(2): 154–163.

Halasz, N., Ljungdahl, A., et al. (1977). Transmitter histochemistry of the rat olfactory bulb. I. Immunohistochemical localization of monoamine synthesizing enzymes. Support for intrabulbar, periglomerular dopamine neurons. *Brain. Res.* 126(3): 455–474.

Harding, J. W., Getchell, T. V., et al. (1978). Denervation of the primary olfactory pathway in mice. V. Long-term effect of intranasal $ZnSO_4$ irrigation on behavior, biochemistry and morphology. *Brain Res.* 140(2): 271–285.

Harris, J. A., West, A. K., et al. (2009). Olfactory ensheathing cells: nitric oxide production and innate immunity. *Glia* 57(16): 1848–1857.

Hempstead, J. L. and Morgan, J. I. (1983). Monoclonal antibodies to the rat olfactory sustentacular cell. *Brain Res.* 288(1–2): 289–295.

Hinds, J. W., Hinds, P. L., et al. (1984). An autoradiographic study of the mouse olfactory epithelium: evidence for long-lived receptors. *Anat. Rec.* 210(2): 375–383.

Hisaoka, T., Morikawa, Y., et al. (2004). Expression of a member of tumor necrosis factor receptor superfamily, TROY, in the developing olfactory system. *Glia* 45(4): 313–324.

Holbrook, E. H., Leopold, D. A., et al. (2005). Abnormalities of axon growth in human olfactory mucosa. *Laryngoscope* 115(12): 2144–2154.

Holbrook, E. H., Szumowski, K. E., et al. (1995). An immunochemical, ultrastructural, and developmental characterization of the horizontal basal cells of rat olfactory epithelium. *J. Comp. Neurol.* 363(1): 129–146.

Holbrook, E. H., Wu, E., et al. (2011). Immunohistochemical characterization of human olfactory tissue. *Laryngoscope* 121(8): 1687–1701.

Holcomb, J. D., Mumm, J. S., et al. (1995). Apoptosis in the neuronal lineage of the mouse olfactory epithelium: regulation in vivo and in vitro. *Dev. Biol.* 172(1): 307–323.

Honore, A., Le Corre, S., et al. (2012). Isolation, characterization, and genetic profiling of subpopulations of olfactory ensheathing cells from the olfactory bulb. *Glia* 60(3): 404–413.

Hsu, P., Yu, F., et al. (2001). Basic fibroblast growth factor and fibroblast growth factor receptors in adult olfactory epithelium. *Brain Res.* 896(1–2): 188–197.

Huard, J. M., Youngentob, S. L., et al. (1998). Adult olfactory epithelium contains multipotent progenitors that give rise to neurons and non-neural cells. *J. Comp. Neurol.* 400(4): 469–486.

Imaizumi, T., Lankford, K. L., et al. (1998). Transplanted olfactory ensheathing cells remyelinate and enhance axonal conduction in the demyelinated dorsal columns of the rat spinal cord. *J. Neurosci.* 18(16): 6176–6185.

Iwai, N., Zhou, Z., et al. (2008). Horizontal basal cells are multipotent progenitors in normal and injured adult olfactory epithelium. *Stem Cells* 26(5): 1298–1306.

Jang, W., Kim, K. P., et al. (2007). Nonintegrin laminin receptor precursor protein is expressed on olfactory stem and progenitor cells. *J. Comp. Neurol.* 502(3): 367–381.

Jia, C., Cussen, A. R., et al. (2011). ATP differentially upregulates fibroblast growth factor 2 and transforming growth factor alpha in neonatal and adult mice: effect on neuroproliferation. *Neuroscience* 177: 335–346.

Jia, C., Doherty, J. P., et al. (2009). Activation of purinergic receptors induces proliferation and neuronal differentiation in Swiss Webster mouse olfactory epithelium. *Neuroscience* 163(1): 120–128.

Jia, C. and Hegg, C. C. (2010). NPY mediates ATP-induced neuroproliferation in adult mouse olfactory epithelium. *Neurobiol. Dis.* 38(3): 405–413.

Kageyama, R. and Ohtsuka, T. (1999). The Notch-Hes pathway in mammalian neural development. *Cell Res.* 9(3): 179–188.

Kanekar, S., Jia, C., et al. (2009). Purinergic receptor activation evokes neurotrophic factor neuropeptide Y release from neonatal mouse olfactory epithelial slices. *J. Neurosci. Res.* 87(6): 1424–1434.

Katoh, H., Shibata, S., et al. (2011). The dual origin of the peripheral olfactory system: placode and neural crest. *Mol. Brain.* 4: 34.

Kawauchi, S., Kim, J., et al. (2009a). Foxg1 promotes olfactory neurogenesis by antagonizing Gdf11. *Development* 136(9): 1453–1464.

Kawauchi, S., Santos, R., et al. (2009b). The role of foxg1 in the development of neural stem cells of the olfactory epithelium. *Ann. N. Y. Acad. Sci.* 1170: 21–27.

Key, B., Treloar, H. B., et al. (1996). Expression and localization of FGF-1 in the developing rat olfactory system. *J. Comp. Neurol.* 366(2): 197–206.

Koster, N. L., Norman, A. B., et al. (1999). Olfactory receptor neurons express D2 dopamine receptors. *J. Comp. Neurol.* 411(4): 666–673.

Krishna, N. S., Little, S. S., et al. (1996). Epidermal growth factor receptor mRNA and protein are expressed in progenitor cells of the olfactory epithelium. *J. Comp. Neurol.* 373(2): 297–307.

Krolewski, R. C., Packard, A., et al. (2012). Global expression profiling of globose basal cells and neurogenic progression within the olfactory epithelium. *J. Comp. Neurol.* 20(3): 14–34.

Kwong, K., Vaishnav, R. A., et al. (2004). Target ablation-induced regulation of macrophage recruitment into the olfactory epithelium of Mip-1alpha-/- mice and restoration of function by exogenous MIP-1alpha. *Physiol. Genomics* 20(1): 73–86.

Lakard, S., Lesniewska, E., et al. (2007). In vitro induction of differentiation by retinoic acid in an immortalized olfactory neuronal cell line. *Acta Histochem.* 109(2): 111–121.

Lanza, D. C., Moran, D. T., et al. (1993). Endoscopic human olfactory biopsy technique: a preliminary report. *Laryngoscope* 103(7): 815–819.

Lee, J. H., Goedert, M., et al. (1993). Tau proteins are abnormally expressed in olfactory epithelium of Alzheimer patients and developmentally regulated in human fetal spinal cord. *Exp. Neurol.* 121(1): 93–105.

Lee, K. H., Bowen-Pope, D. F., et al. (1990). Isolation and characterization of the alpha platelet-derived growth factor receptor from rat olfactory epithelium. *Mol. Cell. Biol.* 10(5): 2237–2246.

Leung, C. T., Coulombe, P. A., et al. (2007). Contribution of olfactory neural stem cells to tissue maintenance and regeneration. *Nat. Neurosci.* 10(6): 720–726.

Li, Y., Carlstedt, T., et al. (2004). Interaction of transplanted olfactory-ensheathing cells and host astrocytic processes provides a bridge for axons to regenerate across the dorsal root entry zone. *Exp. Neurol.* 188(2): 300–308.

Li, Y., Field, P. M., et al. (1997). Repair of adult rat corticospinal tract by transplants of olfactory ensheathing cells. *Science* 277(5334): 2000–2002.

Li, Y., Field, P. M., et al. (1998). Regeneration of adult rat corticospinal axons induced by transplanted olfactory ensheathing cells. *J. Neurosci.* 18(24): 10514–10524.

Lima, C., Pratas-Vital, J., et al. (2006). Olfactory mucosa autografts in human spinal cord injury: a pilot clinical study. *J. Spinal. Cord. Med.* 29(3): 191–203; discussion 204–206.

Lipson, A. C., Widenfalk, J., et al. (2003). Neurotrophic properties of olfactory ensheathing glia. *Exp. Neurol.* 180(2): 167–171.

Liu, N., Shields, C. B., et al. (1998). Primary culture of adult mouse olfactory receptor neurons. *Exp. Neurol.* 151(2): 173–183.

Lopez, E., Mackay-Sim, A., et al. (2012). Leukaemia inhibitory factor stimulates proliferation of olfactory neuronal progenitors via inducible nitric oxide synthase. *PLoS One.* 7(9):e45018.

Lu, J., Feron, F., et al. (2001). Transplantation of nasal olfactory tissue promotes partial recovery in paraplegic adult rats. *Brain Res.* 889(1–2): 344–357.

Lu, J., Feron, F., et al. (2002). Olfactory ensheathing cells promote locomotor recovery after delayed transplantation into transected spinal cord. *Brain* 125(Pt 1): 14–21.

Lucero, M. T. and Squires, A. (1998). Catecholamine concentrations in rat nasal mucus are modulated by trigeminal stimulation of the nasal cavity. *Brain Res.* 807(1–2): 234–236.

MacDonald, K. P., Murrell, W. G., et al. (1996). FGF2 promotes neuronal differentiation in explant cultures of adult and embryonic mouse olfactory epithelium. *J. Neurosci. Res.* 44(1): 27–39.

Mackay-Sim, A. (2012). Patient-derived olfactory stem cells: new models for brain diseases. *Stem Cells* 2:36–47.

Mackay-Sim, A. and Beard, M. D. (1987). Hypothyroidism disrupts neural development in the olfactory epithelium of adult mice. *Brain Res.* 433(2): 190–198.

Mackay-Sim, A., Breipohl, W., et al. (1988). Cell dynamics in the olfactory epithelium of the tiger salamander: a morphometric analysis. *Exp. Brain Res.* 71(1): 189–198.

Mackay-Sim, A. and Chuah, M. I. (2000). Neurotrophic factors in the primary olfactory pathway. *Prog. Neurobiol.* 62(5): 527–559.

Mackay-Sim, A., Feron, F., et al. (2008). Autologous olfactory ensheathing cell transplantation in human paraplegia: a 3-year clinical trial. *Brain* 131(Pt 9): 2376–2386.

Mackay-Sim, A., Johnston, A. N., et al. (2006). Olfactory ability in the healthy population: reassessing presbyosmia. *Chem. Senses* 31(8): 763–771.

Mackay-Sim, A. and Kittel, P. (1991a). Cell dynamics in the adult mouse olfactory epithelium: a quantitative autoradiographic study. *J. Neurosci.* 11(4): 979–984.

Mackay-Sim, A. and Kittel, P. W. (1991b). On the Life Span of Olfactory Receptor Neurons. *Eur. J. Neurosci.* 3(3): 209–215.

Mackay-Sim, A. and Patel, U. (1984). Regional differences in cell density and cell genesis in the olfactory epithelium of the salamander, Ambystoma tigrinum. *Exp. Brain. Res.* 57(1): 99–106.

Mackay-Sim, A. and St John, J. A. (2011). Olfactory ensheathing cells from the nose: clinical application in human spinal cord injuries. *Exp. Neurol.* 229(1): 174–180.

Magrassi, L. and Graziadei, P. P. (1995). Cell death in the olfactory epithelium. *Anat. Embryol. (Berl.)* 192(1): 77–87.

Mahalik, T. J. (1996). Apparent apoptotic cell death in the olfactory epithelium of adult rodents: death occurs at different developmental stages. *J. Comp. Neurol.* 372(3): 457–464.

Mahanthappa, N. K. and Schwarting, G. A. (1993). Peptide growth factor control of olfactory neurogenesis and neuron survival in vitro: roles of EGF and TGF-beta s. *Neuron* 10(2): 293–305.

Manglapus, G. L., Youngentob, S. L., et al. (2004). Expression patterns of basic helix-loop-helix transcription factors define subsets of olfactory progenitor cells. *J. Comp. Neurol.* 479(2): 216–233.

Mania-Farnell, B. L., Farbman, A. I., et al. (1993). Bromocriptine, a dopamine D2 receptor agonist, inhibits adenylyl cyclase activity in rat olfactory epithelium. *Neuroscience* 57(1): 173–180.

Mar, J. C., Matigian, N. A., et al. (2011). Variance of gene expression identifies altered network constraints in neurological disease. *PLoS Genet.* 7(8): e1002207.

Marin-Padilla, M. and Amieva, M. R. (1989). Early neurogenesis of the mouse olfactory nerve: Golgi and electron microscopic studies. *J. Comp. Neurol.* 288(2): 339–352.

Maruniak, J. A., Lin, P. J., et al. (1989). Effects of unilateral naris closure on the olfactory epithelia of adult mice. *Brain Res.* 490(2): 212–218.

Matigian, N., Abrahamsen, G., et al. (2010). Disease-specific, neurosphere-derived cells as models for brain disorders. *Dis. Model Mech.* 3(11–12): 785–798.

Matsuyama, A., Iwata, H., et al. (1992). Localization of basic fibroblast growth factor-like immunoreactivity in the rat brain. *Brain Res.* 587(1): 49–65.

McCurdy, R. D., Feron, F., et al. (2005). Regulation of adult olfactory neurogenesis by insulin-like growth factor-I. *Eur. J. Neurosci.* 22(7): 1581–1588.

McCurdy, R. D., Feron, F., et al. (2006). Cell cycle alterations in biopsied olfactory neuroepithelium in schizophrenia and bipolar I disorder using cell culture and gene expression analyses. *Schizophr. Res.* 82(2–3): 163–173.

Minovi, A., Witt, M., et al. (2010). Expression and distribution of the intermediate filament protein nestin and other stem cell related molecules in the human olfactory epithelium. *Histol. Histopathol.* 25(2): 177–187.

Miwa, T., Horikawa, I., et al. (1998). TrkA expression in mouse olfactory tract following axotomy of olfactory nerves. *Acta. Otolaryngol. Suppl.* 539: 79–82.

Mombaerts, P., Wang, F., et al. (1996). Visualizing an olfactory sensory map. *Cell* 87(4): 675–686.

Moon, C., Yoo, J. Y., et al. (2002). Leukemia inhibitory factor inhibits neuronal terminal differentiation through STAT3 activation. *Proc. Natl. Acad. Sci. U. S. A.* 99(13): 9015–9020.

Morikawa, Y., Hisaoka, T., et al. (2008). TROY, a novel member of the tumor necrosis factor receptor superfamily in the central nervous system. *Ann. N. Y. Acad. Sci.* 1126: A1–10.

Moulton, D. (1975). Cell renewal in the olfactory epithelium of the mouse. *Olfaction and Taste V*. D. Denton and J. Coughlan (Eds). New York, Academic Press: 111–114.

Moulton, D., Çelebi, G., et al. (1970). Olfaction in mammals - two aspects: proliferation of cells in the olfactory epithelium and sensitivity to odours. *Ciba Foundation on Taste and Smell in Vertebrates*. G. Wolstenholme and J. Knight (Eds). London, J. & A. Churchill: 227–250.

Mumm, J. S., Shou, J., et al. (1996). Colony-forming progenitors from mouse olfactory epithelium: evidence for feedback regulation of neuron production. *Proc. Natl. Acad. Sci. U. S. A.* 93(20): 11167–11172.

Murrell, W., Bushell, G. R., et al. (1996). Neurogenesis in adult human. *Neuroreport* 7(6): 1189–1194.

Murrell, W., Feron, F., et al. (2005). Multipotent stem cells from adult olfactory mucosa. *Dev. Dyn.* 233(2): 496–515.

Murrell, W., Sanford, E., et al. (2009). Olfactory stem cells can be induced to express chondrogenic phenotype in a rat intervertebral disc injury model. *Spine J.* 9(7): 585–594.

Murrell, W., Wetzig, A., et al. (2008). Olfactory mucosa is a potential source for autologous stem cell therapy for Parkinson's disease. *Stem Cells* 26(8): 2183–2192.

Naessen, R. (1971). An enquiry on the morphological characteristics and possible changes with age in the olfactory region of man. *Acta Otolaryngol.* 71(1): 49–62.

Nagahara, Y. (1940). Experimentelle Studien uber die histologischen Veranderungen des Geruchsorgans nach der Olfactorius durchschneidung. Beutrage zur Kenntnis des feineren Baus des Geruchsorgans. *Jap. J. Med. Sci. V, Path.* 5: 46–63.

Nakamura, H., Higuchi, Y., et al. (2002). The effect of basic fibroblast growth factor on the regeneration of guinea pig olfactory epithelium. *Eur. Arch. Otorhinolaryngol.* 259(3): 166–169.

Nakashima, T., Kimmelman, C. P., et al. (1984). Structure of human fetal and adult olfactory neuroepithelium. *Arch. Otolaryngol.* 110(10): 641–646.

Nakashima, T., Kimmelman, C. P., et al. (1985). Immunohistopathology of human olfactory epithelium, nerve and bulb. *Laryngoscope* 95(4): 391–396.

Nan, B., Getchell, M. L., et al. (2001). Leukemia inhibitory factor, interleukin-6, and their receptors are expressed transiently in the olfactory mucosa after target ablation. *J. Comp. Neurol.* 435(1): 60–77.

Newman, M. P., Feron, F., et al. (2000). Growth factor regulation of neurogenesis in adult olfactory epithelium. *Neuroscience* 99(2): 343–350.

Nishikawa, T., Doi, K., et al. (2009). Effect of intranasal administration of basic fibroblast growth factor on olfactory epithelium. *Neuroreport* 20(8): 764–769.

Nosrat, C. A., Tomac, A., et al. (1996). Cellular expression of GDNF mRNA suggests multiple functions inside and outside the nervous system. *Cell Tissue Res.* 286(2): 191–207.

Orr-Urtreger, A. and Lonai, P. (1992). Platelet-derived growth factor-A and its receptor are expressed in separate, but adjacent cell layers of the mouse embryo. *Development* 115(4): 1045–1058.

Oztokatli, H., Hornberg, M., et al. (2012). Retinoic acid receptor and CNGA2 channel signaling are part of a regulatory feedback loop controlling axonal convergence and survival of olfactory sensory neurons. *FASEB J.* 26(2): 617–627.

Packard, A., Giel-Moloney, M., et al. (2011a). Progenitor cell capacity of NeuroD1-expressing globose basal cells in the mouse olfactory epithelium. *J. Comp. Neurol.* 519(17): 3580–3596.

Packard, A., Schnittke, N., et al. (2011b). DeltaNp63 regulates stem cell dynamics in the mammalian olfactory epithelium. *J. Neurosci.* 31(24): 8748–8759.

Peluso, C. E., Jang, W., et al. (2012). Differential expression of components of the retinoic acid signaling pathway in the adult mouse olfactory epithelium. *J. Comp. Neurol.* 520(16): 3707–3726.

Perroteau, I., Oberto, M., et al. (1998). ErbB-3 and ErbB-4 expression in the mouse olfactory system. *Ann. N. Y. Acad. Sci.* 855: 255–259.

References

Perry, G., Castellani, R. J., et al. (2003). Oxidative damage in the olfactory system in Alzheimer's disease. *Acta Neuropathol.* 106(6): 552–556.

Pixley, S. K. (1992). The olfactory nerve contains two populations of glia, identified both in vivo and in vitro. *Glia* 5(4): 269–284.

Pixley, S. K., Dangoria, N. S., et al. (1998). Effects of insulin-like growth factor 1 on olfactory neurogenesis in vivo and in vitro. *Ann. N. Y. Acad. Sci.* 855: 244–247.

Pollock, G. S., Franceschini, I. A., et al. (1999). Neuregulin is a mitogen and survival factor for olfactory bulb ensheathing cells and an isoform is produced by astrocytes. *Eur. J. Neurosci.* 11(3): 769–780.

Potten, C. S. (1998). Stem cells in gastrointestinal epithelium: numbers, characteristics and death. *Philos. Trans. R. Soc. Lond. B. Biol. Sci.* 353(1370): 821–830.

Ramon-Cueto, A. and Nieto-Sampedro, M. (1994). Regeneration into the spinal cord of transected dorsal root axons is promoted by ensheathing glia transplants. *Exp. Neurol.* 127(2): 232–244.

Ramon-Cueto, A., Plant, G. W., et al. (1998). Long-distance axonal regeneration in the transected adult rat spinal cord is promoted by olfactory ensheathing glia transplants. *J. Neurosci.* 18(10): 3803–3815.

Rehn, B., Breipohl, W., et al. (1981). Chemical blockade of olfactory perception by N-methyl-formimino-methylester in albino mice. II. light microscopical investigations. *Chem. Senses* 6: 317–328.

Roisen, F. J., Klueber, K. M., et al. (2001). Adult human olfactory stem cells. *Brain Res.* 890(1): 11–22.

Ronnett, G. V., Leopold, D., et al. (2003). Olfactory biopsies demonstrate a defect in neuronal development in Rett's syndrome. *Ann. Neurol.* 54(2): 206–218.

Roskams, A. J., Bethel, M. A., et al. (1996). Sequential expression of Trks A, B, and C in the regenerating olfactory neuroepithelium. *J. Neurosci.* 16(4): 1294–1307.

Roskams, A. J., Bredt, D. S., et al. (1994). Nitric oxide mediates the formation of synaptic connections in developing and regenerating olfactory receptor neurons. *Neuron* 13(2): 289–299.

Roskams, A. J., Cai, X., et al. (1998). Expression of neuron-specific beta-III tubulin during olfactory neurogenesis in the embryonic and adult rat. *Neuroscience* 83(1): 191–200.

Ruitenberg, M. J., Vukovic, J., et al. (2008). CX3CL1/fractalkine regulates branching and migration of monocyte-derived cells in the mouse olfactory epithelium. *J. Neuroimmunol.* 205(1–2): 80–85.

Salehi-Ashtiani, K. and Farbman, A. I. (1996). Expression of neu and Neu differentiation factor in the olfactory mucosa of rat. *Int. J. Dev. Neurosci.* 14(7–8): 801–811.

Samanen, D. W. and Forbes, W. B. (1984). Replication and differentiation of olfactory receptor neurons following axotomy in the adult hamster: a morphometric analysis of postnatal neurogenesis. *J. Comp. Neurol.* 225(2): 201–211.

Satoh, M. and Takeuchi, M. (1995). Induction of NCAM expression in mouse olfactory keratin-positive basal cells in vitro. *Brain Res. Dev. Brain Res.* 87(2): 111–119.

Satoh, M. and Yoshida, T. (1997). Promotion of neurogenesis in mouse olfactory neuronal progenitor cells by leukemia inhibitory factor in vitro. *Neurosci. Lett.* 225(3): 165–168.

Schultz, E. (1941). Regeneration of olfactory cells. *Proc. Soc. Exp. Biol. Med.* 46: 41–43.

Schwartz Levey, M., Chikaraishi, D. M., et al. (1991). Characterization of potential precursor populations in the mouse olfactory epithelium using immunocytochemistry and autoradiography. *J. Neurosci.* 11(11): 3556–3564.

Schwob, J. E., Huard, J. M., et al. (1994a). Retroviral lineage studies of the rat olfactory epithelium. *Chem. Senses.* 19(6): 671–682.

Schwob, J. E., Szumowski, K. E., et al. (1992). Olfactory sensory neurons are trophically dependent on the olfactory bulb for their prolonged survival. *J. Neurosci.* 12(10): 3896–3919.

Schwob, J. E., Youngentob, S. L., et al. (1994). On the formation of neuromata in the primary olfactory projection. *J. Comp. Neurol.* 340(3): 361–380.

Schwob, J. E., Youngentob, S. L., et al. (1995). Reconstitution of the rat olfactory epithelium after methyl bromide-induced lesion. *J. Comp. Neurol.* 359(1): 15–37.

Shou, J., Murray, R. C., et al. (2000). Opposing effects of bone morphogenetic proteins on neuron production and survival in the olfactory receptor neuron lineage. *Development* 127(24): 5403–5413.

Shou, J., Rim, P. C., et al. (1999). BMPs inhibit neurogenesis by a mechanism involving degradation of a transcription factor. *Nat. Neurosci.* 2(4): 339–345.

Simpson, P. J., Miller, I., et al. (2002). Atrial natriuretic peptide type C induces a cell-cycle switch from proliferation to differentiation in brain-derived neurotrophic factor- or nerve growth factor-primed olfactory receptor neurons. *J. Neurosci.* 22(13): 5536–5551.

Smart, I. H. (1971). Location and orientation of mitotic figures in the developing mouse olfactory epithelium. *J. Anat.* 109(Pt 2): 243–251.

Smith, C. G. (1951). Regeneration of sensory olfactory epithelium and nerves in adult frogs. *Anat. Rec.* 109(4): 661–671.

Sulz, L., Astorga, G., et al. (2009). Nitric oxide regulates neurogenesis in adult olfactory epithelium in vitro. *Nitric Oxide* 20(4): 238–252.

Talamo, B. R., Rudel, R., et al. (1989). Pathological changes in olfactory neurons in patients with Alzheimer's disease. *Nature* 337(6209): 736–739.

Thornhill, R. A. (1970). Cell division in the olfactory epithelium of the lamprey, Lampetra fluviatilis. *Z. Zellforsch. Mikrosk. Anat.* 109(2): 147–157.

Tisay, K. T. and Key, B. (1999). The extracellular matrix modulates olfactory neurite outgrowth on ensheathing cells. *J. Neurosci.* 19(22): 9890–9899.

Tome, M., Lindsay, S. L., et al. (2009). Identification of nonepithelial multipotent cells in the embryonic olfactory mucosa. *Stem Cells* 27(9): 2196–2208.

Turner, C. P. and Perez-Polo, J. R. (1992). Regulation of the low affinity receptor for nerve growth factor, p75NGFR, in the olfactory system of neonatal and adult rat. *Int. J. Dev. Neurosci.* 10(5): 343–359.

Ubink, R. and Hokfelt, T. (2000). Expression of neuropeptide Y in olfactory ensheathing cells during prenatal development. *J. Comp. Neurol.* 423(1): 13–25.

Uranagase, A., Katsunuma, S., et al. (2012). BDNF expression in olfactory bulb and epithelium during regeneration of olfactory epithelium. *Neurosci. Lett.* 516(1): 45–49.

Valverde, F., Heredia, M., et al. (1993). Characterization of neuronal cell varieties migrating from the olfactory epithelium during prenatal development in the rat. Immunocytochemical study using antibodies against olfactory marker protein (OMP) and luteinizing hormone-releasing hormone (LH-RH). *Brain Res. Dev. Brain Res.* 71(2): 209–220.

Valverde, F. and Lopez-Mascaraque, L. (1991). Neuroglial arrangements in the olfactory glomeruli of the hedgehog. *J. Comp. Neurol.* 307(4): 658–674.

Vargas, G. and Lucero, M. T. (1999). Dopamine modulates inwardly rectifying hyperpolarization-activated current (Ih) in cultured rat olfactory receptor neurons. *J. Neurophysiol.* 81(1): 149–158.

Verhaagen, J., Oestreicher, A. B., et al. (1989). The expression of the growth associated protein B50/GAP43 in the olfactory system of neonatal and adult rats. *J. Neurosci.* 9(2): 683–691.

Verschueren, K., Dewulf, N., et al. (1995). Expression of type I and type IB receptors for activin in midgestation mouse embryos suggests distinct functions in organogenesis. *Mech. Dev.* 52(1): 109–123.

Vickland, H., Westrum, L. E., et al. (1991). Nerve growth factor receptor expression in the young and adult rat olfactory system. *Brain Res.* 565(2): 269–279.

Vincent, A. J., Lau, P. W., et al. (2008). SPARC is expressed by macroglia and microglia in the developing and mature nervous system. *Dev. Dyn.* 237(5): 1449–1462.

Vukovic, J., Blomster, L. V., et al. (2010). Bone marrow chimeric mice reveal a role for CX(3)CR1 in maintenance of the monocyte-derived cell population in the olfactory neuroepithelium. *J. Leukoc. Biol.* 88(4): 645–654.

Wang, Y. Z., Yamagami, T., et al. (2011). Canonical Wnt signaling promotes the proliferation and neurogenesis of peripheral olfactory stem cells during postnatal development and adult regeneration. *J. Cell. Sci.* 124(Pt 9): 1553–1563.

Weiler, E. and Farbman, A. I. (1999). Mitral cell loss following lateral olfactory tract transection increases proliferation density in rat olfactory epithelium. *Eur. J. Neurosci.* 11(9): 3265–3275.

Westerman, R. A. and von Baumgarten, R. (1964). Regeneration of olfactory paths in carp (Cyprinus carpio L.). *Experientia* 20(9): 519–520.

Wetzig, A., Mackay-Sim, A., et al. (2011). Characterization of olfactory stem cells. *Cell Transplant.* 20(11–12): 1673–1691.

Wewetzer, K., Verdu, E., et al. (2002). Olfactory ensheathing glia and Schwann cells: two of a kind? *Cell Tissue Res.* 309(3): 337–345.

Whitesides, J., Hall, M., et al. (1998). Retinoid signaling distinguishes a subpopulation of olfactory receptor neurons in the developing and adult mouse. *J. Comp. Neurol.* 394(4): 445–461.

Williams, R. and Rush, R. A. (1988). Electron microscopic immunocytochemical localization of nerve growth factor in developing mouse olfactory neurons. *Brain Res.* 463(1): 21–27.

Windus, L. C., Chehrehasa, F., et al. (2011). Stimulation of olfactory ensheathing cell motility enhances olfactory axon growth. *Cell. Mol. Life Sci.* 68(19): 3233–3247.

Windus, L. C., Claxton, C., et al. (2007). Motile membrane protrusions regulate cell-cell adhesion and migration of olfactory ensheathing glia. *Glia* 55(16): 1708–1719.

Windus, L. C., Lineburg, K. E., et al. (2010). Lamellipodia mediate the heterogeneity of central olfactory ensheathing cell interactions. *Cell. Mol. Life Sci.* 67(10): 1735–1750.

Wolozin, B., Lesch, P., et al. (1993). A.E. Bennett Research Award 1993. Olfactory neuroblasts from Alzheimer donors: studies on APP processing and cell regulation. *Biol. Psychia.* 34(12): 824–838.

Wolozin, B., Sunderland, T., et al. (1992). Continuous culture of neuronal cells from adult human olfactory epithelium. *J. Mol. Neurosci.* 3(3): 137–146.

Woodhall, E., West, A. K., et al. (2001). Cultured olfactory ensheathing cells express nerve growth factor, brain-derived neurotrophic factor, glia cell line-derived neurotrophic factor and their receptors. *Brain Res. Mol. Brain Res.* 88(1–2): 203–213.

Wu, H. H., Ivkovic, S., et al. (2003). Autoregulation of neurogenesis by GDF11. *Neuron* 37(2): 197–207.

Xiao, M., Klueber, K. M., et al. (2005). Human adult olfactory neural progenitors rescue axotomized rodent rubrospinal neurons and promote functional recovery. *Exp. Neurol.* 194(1): 12–30.

Yee, K. K. and Rawson, N. E. (2005). Immunolocalization of retinoic acid receptors in the mammalian olfactory system and the effects of olfactory denervation on receptor distribution. *Neuroscience* 131(3): 733–743.

Zhang, D., Mehler, M. F., et al. (1998). Development of bone morphogenetic protein receptors in the nervous system and possible roles in regulating trkC expression. *J. Neurosci.* 18(9): 3314–3326.

Zhang, Q. Y. (1999). Retinoic acid biosynthetic activity and retinoid receptors in the olfactory mucosa of adult mice. *Biochem. Biophys. Res. Commun.* 256(2): 346–351.

Chapter 8

Anatomy and Neurobiology of the Main and Accessory Olfactory Bulbs

MATTHEW ENNIS and TIMOTHY E. HOLY[1]

8.1 INTRODUCTION

The olfactory system of most mammals, excluding humans, consists of two parallel components, the main and accessory olfactory systems. They operate in parallel and are largely separate anatomically and functionally. The main olfactory system primarily processes volatile odors that enter the nasal cavity during respiration. Odors are transduced by nasal epithelium olfactory receptor neurons (ORNs) and are then relayed to, and synapse in, the main olfactory bulb (MOB). In contrast, the accessory olfactory system processes non-volatile odors, transduced by vomeronasal receptor neurons in the vomeronasal organ and transmitted to the accessory olfactory bulb (AOB). In tandem, the two systems underlie our sense of "smell" which is critical for flavor perception, food selection, and identification of conspecifics, predators, and prey. Odors produce strong and lasting memories in humans, and exert potent control of animal behavior. This chapter focuses on olfactory processing in the mammalian MOB and AOB. Knowledge in both structures has undergone explosive growth over the last two decades. We focus primarily on these advances; readers are referred to other sources that review earlier primary literature (Mori, 1987; Shepherd, 1972; Shipley et al., 1996; Ennis et al., 2007). Due to late developmental maturation in the olfactory system we emphasize findings from mature or post-weaning ages.

[1]We acknowledge grant support for portions of this chapter that refer to research of the authors: PHS Grants DC003195, DC008702, DC005964, DC010381, and NS068409.

8.2 MAIN OLFACTORY BULB (MOB)

The rodent MOB is a cylindrical structure connected by a slender peduncle to the anterior pole of the brain. The MOB is laminated with the layers forming concentric circles (Figures 8.1, 8.2); from superficial to deep these are the: olfactory nerve layer (ONL), glomerular layer (GL), external plexiform layer (EPL), mitral cell layer (MCL), internal plexiform layer (IPL), granule cell layer (GCL), and the subependymal layer/subventricular zone. Odors are transduced by ORNs (Chapters 5 and 6), giving rise to action potentials that travel along olfactory nerve (ON) fibers and synapse in spherical shaped glomeruli in the GL. ORNs expressing the same odorant receptor project to one or two glomeruli located on the medial and/or lateral side of each MOB (Figure 8.2; Ressler et al., 1993, 1994; Vassar et al., 1994; Mombaerts et al., 1996; Wang et al., 1998; Potter et al., 2001; Treloar et al., 2002; Wang et al., 1998). Within the glomeruli, ORN axons synapse with mitral and tufted cells (M/TCs) – the MOB output neurons – and with intrinsic neurons, juxtaglomerular cells (JGCs) (Pinching and Powell, 1971a; Shepherd, 1972; Najac et al., 2011). ORNs utilize glutamate as their primary neurotransmitter (Sassoè-Pognetto et al., 1993) and the initial olfactory sensory input to MOB is excitatory. ORN terminals also contain carnosine, zinc, and copper that may function as neuromodulators (Sassoè-Pognetto et al., 1993; Ennis et al. 2007). The relatively cell sparse EPL contains tufted cells (TCs), interneurons, and an extensive network of dendrodendritic synapses formed

---

Handbook of Olfaction and Gustation, Third Edition. Edited by Richard L. Doty.
© 2015 Richard L. Doty. Published 2015 by John Wiley & Sons, Inc.

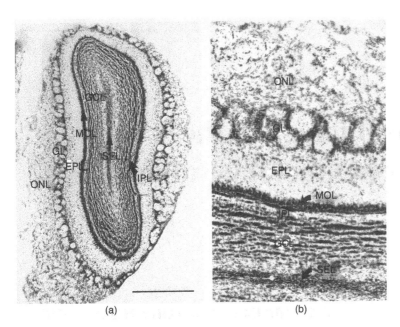

Figure 8.1 Architecture of the MOB. (a) and (b): Coronal sections (Nissl stain) of the rat MOB at low (a) and high (b) magnifications. Abbreviations: SEL, subependymal layer. Scale bar = 1 mm.

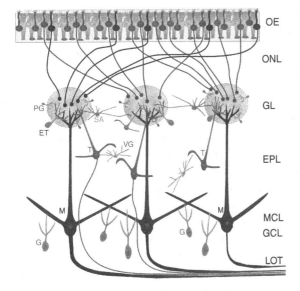

Figure 8.2 Basic circuitry of the MOB. Abbreviations: G, granule cell; M, mitral cell; T, tufted cell, OE, olfactory epithelium; VG, van Gehuchten cell. (*See plate section for color version.*)

between M/TC and GCs. The thin MCL contains MC somata as well as GCs. The IPL is relatively thin and cell sparse, while the GCL contains the largest number of cells, primarily granule cells using the inhibitory transmitter gamma-aminobutryric acid (GABA).

8.2.1 Glomerular Layer (GL)

Glomeruli have long been considered to represent fundamental units of olfactory processing. In line with this, each glomerulus: (1) receives input from a homogeneous ORN population expressing the same odorant receptor; (2) contains a dedicated population of intrinsic and output neurons organized in a columnar like arrangement; and (3) exhibits unique odor input response properties (Wachowiak and Shipley, 2006; Willhite et al., 2006).

8.2.1.1 GL Neurons.
GL neurons comprise three primary classes collectively referred to as JGCs: external tufted cells (ETCs), periglomerular cells (PGCs) cells, and (3) short axon cells (SACs) (Pinching and Powell, 1971a). The vast majority are interneurons with local connections in the GL, although some ETCs project outside of MOB (Ennis et al., 2007).

EXTERNAL TUFTED CELLS (ETCs) Like their deeper M/TC cousins, ETCs extend a single apical dendrite that arborizes in one glomerulus and receive monosynaptic ORN input (Pinching and Powell, 1971b, c; Hayar et al., 2004a, b; Liu and Shipley, 2008a, b); only ~6% of ETCs have apical dendrites in two glomeruli. ET/M/TCs contain glutamate and form excitatory dendrodendritic synapses with PG and SACs, and in turn receive inhibitory synapses from PGCs Pinching and Powell, 1971b, c). The apical dendrites of ET/M/TCs located within the same glomerulus are coupled by connexin protein-based gap junctions that function as bidirectional electrical synapses important in synchronizing intraglomerular activity (Ennis et al., 2007). Some ETCs cells have secondary dendrites that extend into the EPL, and some also appear to have axons that may synapse with PG or SACs (Hayar et al., 2004a). ETCs recorded in slices exhibit spontaneous rhythmic spike bursts at frequencies ranging from ~0.5-11 Hz

8.2 Main Olfactory Bulb (MOB)

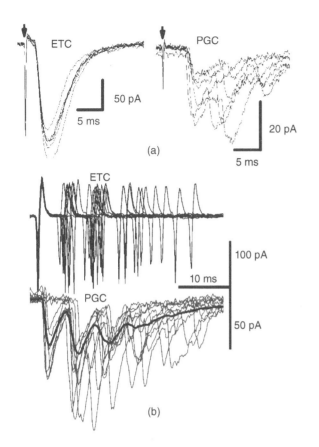

Figure 8.3 (a) Voltage-clamp traces showing of ON-evoked synaptic responses (EPSCs) in ET and PGCs. Right, ON stimulation (arrow) produced short, constant latency EPSCs in an ETC, indicative of monosynaptic input. Left, ON stimulation evoked longer, variable latency EPSCs in a PGC, indicative of di- or poly-synaptic responses. (b) ETCs monosynaptically excite PGCs. Dual recordings from an ETC (cell-attached recording mode, upper trace) traces) and a PGC (whole-cell voltage clamp mode, lower traces). Spontaneous ETC spike bursts (10 superimposed bursts) are accompanied by phase-locked busts of EPSCs in the PGC (black line is the average of 135 traces).

(Hayar et al., 2004a, b, 2005; Liu and Shipley, 2008a, b) (Figures 8.3, 8.4), a range overlapping the 2–12 Hz theta frequencies of rodent sniffing (see Section 8.7.2). ETCs with secondary dendrites in the EPL do not appear to rhythmically burst, indicating that there may be at least two ETC subtypes (Antal et al.; 2006). ETC rhythmic bursting is generated intrinsically and involves a number of voltage-dependent currents, but especially important in burst initiation is a TTX-sensitive slowly inactivating or persistent Na$^+$ current (Hayar et al., 2004a, b; Liu and Shipley, 2008a, b). ETCs respond monosynaptically to ON stimulation with a short latency excitatory postsynaptic potential or current (EPSP/EPSC) or a spike burst (Figure 8.3; Hayar et al., 2004a, b, 2005; Liu and Shipley, 2008a, 2008b).

PERIGLOMERULAR CELLS (PGCs) PGCs, the largest GL neuronal population, comprise multiple subtypes that are all thought to be inhibitory. Their dendrites are typically restricted to a small subregion in 1–2 glomeruli (Hayar et al., 2004a, b; Shao et al., 2009; Kiyokage et al., 2010). Only a relatively small subpopulation (30%) receive ORN synapses (Shao et al., 2009; Kiyokage et al., 2010). Their dendrites receive synapses from ET/M/TC dendrites and in turn make reciprocal inhibitory synapses back onto the parent ET/M/TC dendrite or onto those of other ET/M/TCs (Kasowski et al., 1999; Toida et al., 1998, 2000; Hayar et al., 2004b). Most PGC cells are GABAergic (Ennis et al., 2007; Shao et al., 2009). The GL contains a population of dopamine (DA) cells that were classically described as a subset of GABAergic PGCs (Halász et al., 1977; Kosaka et al., 1985; Gall et al., 1987; Ennis et al., 2007). Recent studies (see below) however, indicate that most if not all DA-GABA cells in the mouse are SACs. PGCs also stain for other neurochemical markers or neuropeptides (Ennis et al., 2007; Kosaka and Kosaka, 2011). PGCs are thus neurochemically heterogeneous and not surprisingly exhibit varying electrophysiological properties (Puopolo and Belluzzi, 1998; McQuiston and Katz, 2001; Pignatelli et al., 2005; Puopolo et al., 2005). However, most PG and SACs recorded in slices typically have low spontaneous spike activity even though they receive barrages of spontaneous glutamatergic EPSP/EPCS (Figures 8.3, 8.4) (Hayar et al., 2004a, b; Shao et al., 2009; Kiyokage et al., 2010). In vivo, PGCs exhibit broader odor tuning (respond to more odorants) than ET/M/TCs (Tan et al., 2010).

SHORT AXON CELLS (SACs) SACs are distinguished by dendrites that seem to harvest information from multiple glomeruli (Aungst et al., 2003; Hayar et al., 2004a). Like PGCs, most (70%) do not receive direct ORN synaptic input but instead receive excitatory input from ET/M/TCs dendrites (Hayar et al., 2004b; Kiyokage et al., 2010). They extend axons up to 1–2 mm in the GL providing an anatomical substrate for interglomerular interactions (Aungst et al., 2003). SAC axons were initially thought to synapse onto PGCs in other glomeruli, while more recent work indicates that they may also target ETCs (see below; Aungst et al., 2003). Mouse SACs were recently shown to contain DA and GABA (Kiyokage et al., 2010; Kosaka and Kosaka, 2011). These DA-GABA SACs exhibit marked heterogeneity with regard to glomerular contacts. Those that receive direct ORN input have processes in ~5 glomeruli, but 50% of the process ramify in one "preferred" glomerulus. SACs lacking direct ORN input fall into two categories associated with ~15 vs. ~39 glomeruli respectively. Some cells in a later group had processes in hundreds of glomeruli, but it is unclear if the process of these and the other subtypes are dendrites, axons or both.

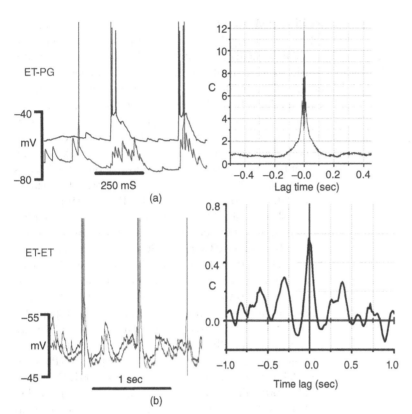

Figure 8.4 Synchronous activity among ET and PGCs associated with the same glomerulus. (a) Simultaneous current clamp recordings show that ETC spike bursts (blue trace) are associated with PGC EPSP bursts (red trace). Cross-correlogram at left shows that ETC spikes are highly synchronous with PGC EPSPs. (b) Simultaneous current clamp recordings from two ETCs shows synchronous membrane potential and spike activity, further evidenced in the membrane potential cross-correlogram at left.

8.2.1.2 Excitatory Intraglomerular Processing.

ORN INPUT Each glomerulus contains the apical dendritic tufts of about 20 MCs, 50 TCs and 1500–2000 JGCs (Ennis et al., 2007). Glutamate release from ORN terminals, controlled by N- and P/Q-type Ca^{2+} channels, excites these populations mainly via AMPA and NMDA receptors that mediate respectively fast and slow synaptic response components (Ennis et al., 1996, 2001; Aroniadou-Anderjaska et al., 1997, 1999a; Chen and Shepherd, 1997; Keller et al., 1998). ET/M/TCs also exhibit smaller, slow synaptic responses to ORN input or prolonged odor responses mediated by the metabotropic glutamate receptor1 (mGluR1) subtype (De Saint Jan and Westbrook, 2005; Ennis et al., 2006; Yuan and Knopfel, 2006; Matsumoto et al., 2009; Dong et al. 2009).

ORN TO PG/SAC VS. ORN TO ET/M/TC TO PG/SAC CIRCUITS PG and SACs fall into two populations based on spontaneous and ON-evoked synaptic activity (i.e., EPSPs/EPSCs.) Approximately 30% of PG/SACs exhibit a non-bursting spontaneous EPSC pattern and respond monosynaptically to ON stimulation (Hayar et al., 2004b; Shao et al., 2009; Kiyokage et al., 2010). By contrast, ~70% of PG/SACs exhibit spontaneous EPSC bursts and respond polysynaptically to ON stimulation (Figure 8.3). Since ETC spontaneously burst, receive monosynaptic ORN input, and form excitatory synapses with PG/SAC dendrites, it is thought that the ETC population largely mediates ON responses in this subset of PG/SACs cells – an ORN-ETC-PG/SAC circuit. Consistent with this, ETCs monosynaptically drive EPSC bursts in PG/SACs (Figure 8.3; Hayar et al., 2004b) with estimates that a given ETC activates ~5 PG/SACs (Murphy et al., 2005). Spontaneous and ON-evoked activity in these ETC-driven PG/SACs is similar when the GL is isolated, consistent with control by ETCs, although this does not exclude a more minor degree of excitation by M/TC apical dendrites. As discussed below, these functionally distinct PG/SAC populations exert differential roles in regulating spontaneous and ON-evoked inhibition. The fact that some PG/SACs are devoid of ORN input is consistent with the bi-compartmental organization of glomeruli into zones rich in vs. devoid of ORN terminals (Kosaka et al., 1997; Kasowski et al., 1999). Cabinda-positive JGCs extend dendrites into compartments devoid of ORN terminals and may correspond to PG/SACs shown electrophysiologically to lack ORN input (Toida et al., 1998, 2000).

8.2.1.3 Dendritic Glutamate Spillover Among ET/M//TCs.

Although ET/M/TC dendrites do not form conventional anatomical synapses, dendritic glutamate release (i.e., spillover) can excite α-amino-3-

8.2 Main Olfactory Bulb (MOB)

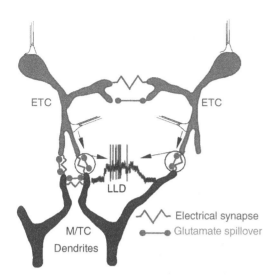

Figure 8.5 Glutamate spillover and electrical synapses mediate intraglomerular synchrony among ET/M/TCs. Glutamate release from, and spillover among, the apical dendrites of ET/M/TCs mediates excitatory interactions among these cells that strengthen intraglomerular synchrony. The apical dendrites of these cells also form bidirectional electrical synapses that facilitate glutamate release and intraglomerular synchrony. Convergent spike bursts from multiple ETCs engage long-lasting depolarizations (LLDs) in M/TCs via glutamate spillover and electrical synapses. (*See plate section for color version.*)

hydroxy-5-methyl-4-isoxazolepropionic acid (AMPA), N-methyl-D-asparate (NMDA) and metabotropic glutamate receptor 1 (mGluR1) receptors on the parent dendrite (self- or autoexcitation) or neighboring dendrites (lateral or feedforward excitation) (Figure 8.5; Aroniadou-Anderjaska et al., 1999b; Isaacson, 1999; Carlson et al., 2000; Friedman and Strowbridge, 2000; Salin et al., 2001; Schoppa and Westbrook, 2001, 2002; Hayar at al. 2005; Murphy et al., 2005; Christie et al., 2005; Christie and Westbrook, 2006; Pimentel and Margrie, 2008). Spillover excitatory responses typically have a slow time course and chiefly occur among dendrites in the same glomerulus. However, fast monosynaptic-like AMPA receptor-mediated lateral excitation between M/TCs or from ET to M/TCs also occurs (Urban and Sakmann, 2002; Pimentel and Margrie, 2008; De Saint Jan et al., 2009; Najac et al., 2011). Spillover excitation is enhanced by low Mg^{2+} or $GABA_A$ receptor block, which can engage/amplify an NMDA receptor component and trigger autoexcitation of M/TC lateral dendrites. Dendritic glutamate release is facilitated by gap junction electrotonic coupling among ET/M//TC dendrites (Figure 8.5; Schoppa and Westbrook, 2002; Christie et al., 2005; Christie and Westbrook, 2006). Lateral excitation was abolished by genetic deletion of electrical synapses and its strength was correlated with electrotonic coupling strength between MC pairs (Christie and Westbrook, 2006; but see Pimentel and Margrie, 2008). Lateral excitation is thought to comprise a specialized electrochemical form of communication that synchronizes glomerular activity.

Lateral glutamate spillover is responsible for unusually large and prolonged (1–2 sec) depolarizations (LLDs) in MCs spontaneously, or in response to ON stimulation or antidromic activation of multiple M/TCs, but never by activation of single MCs (Carlson et al., 2000). They are all-or-none in nature, require AMPA receptor activation, originate in the glomeruli and are synchronous among MCs of the same glomerulus (Carlson et al., 2000; Karnup et al., 2006; De Saint Jan et al., 2009). LLDs are generated by network mechanism involving intraglomerular glutamate release from an ensemble of ET/M//TC dendrites and appear to involve gap junctional-coupled amplification of glutamate release (Figure 8.5; Christie and Westbrook, 2006). ETCs appear to play a critical role as their bursts often precede spontaneous MC LLDs and evoked ETC bursts can elicit LLD-like responses in MCs but not vice-versa (De Saint Jan et al., 2009). LLDs clearly amplify ON input onto M/TCs, but there is controversy about their role in driving entirely, in an all-or-none fashion, ON responses in MCs via a disynaptically ORN-ET-MC circuit. In support: (1) ON stimulation produced either no synaptic or spiking response in MCs or an all-or-none disynaptic LLD/spiking response, and (2) monosynaptic ON responses in ETCs preceded LLDs (Gire and Schoppa, 2009; Gire et al., 2012). At odds are observations of (1) graded MC spiking responses to graded intensity ON stimuli or odors (Jiang et al., 1996; Ciombor et al., 1999; Margrie and Schaefer, 2003; Cang and Isaacson, 2003; Davison and Katz, 2007 Cury and Uchida, 2010; Tan et al., 2010), (2) differences in the magnitude or direction of odor-evoked spiking among MCs of the same glomerulus recorded simultaneously (Dhawale et al., 2010), and (3) LLDs producing graded spiking levels or that occur without MC spikes (Karnup et al., 2006). Although ETCs have a lower ON response threshold than MCs, there is clear electrophysiological and anatomical evidence for monosynaptic ORN input to MCs (De Saint Jan et al., 2009; Najac et al., 2011; Shao et al., 2012). Thus, it appears that weak ON input preferentially engaging the more excitable ETCs leads to disynaptic MC LLDs while stronger input engages monosynaptic responses and typically the LLD (Najac et al., 2011; Shao et al., 2012).

8.2.1.4 Inhibitory Intraglomerular Processing.
PRESYNAPTIC INHIBITION OF ORN TERMINALS
The strength of olfactory input at the first stage of synaptic transfer in the MOB is regulated by presynaptic control of ORN terminal excitability. ORN terminals express $GABA_B$ and DA D2 receptors (Ennis et al., 2007). GABA and DA

released from PGCs can, despite lack of anatomical PGC to ORN synapses, bind to these presynaptic receptors to inhibit glutamate release. In the case of $GABA_B$ receptors, presynaptic inhibition is mediated by suppression of Ca^{2+} influx into ORN terminals (Wachowiak et al., 2005). $GABA_B$ or D2 receptor activation reduces spontaneous and ON-evoked activity in all cells tested (PG, ET, MCs), as well as ON- or odor-evoked ORN terminal activity (Keller et al., 1998; Hsia et al., 1999; Aroniadou-Anderjaska et al., 2000; Berkowicz and Trombley, 2000; Ennis et al., 2001; Palouzier-Paulignan et al., 2002; McGann et al., 2005; Wachowiak et al., 2005; Vucinić et al., 2006; Pírez and Wachowiak, 2008; Maher and Westbrook, 2008; Shao et al., 2009). PGCs can, via $GABA_B$ receptors, presynaptically inhibit their own ORN monosynaptic input (Murphy et al., 2005; Shao et al., 2009). $GABA_B$ receptor presynaptic inhibition is tonically active in vitro and in vivo and thus $GABA_B$ receptor antagonists uniformly increase ON- or odor-evoked responses (Aroniadou-Anderjaska et al., 2000; Pírez and Wachowiak, 2008; Shao et al., 2009); this tonic inhibition appears to mainly be driven by the ETC to PGC circuit. D2 receptor presynaptic inhibition in vitro does not appear to have a strong tonic component (Maher and Westbrook, 2008). Presynaptic inhibition increases with repetitive ON stimulation from ~2-10 Hz in slices, suggesting it may provide a sniff frequency dependent negative feedback mechanism (Aroniadou-Anderjaska et al., 2000; Wachowiak et al., 2005; McGann et al., 2005). However, presynaptic inhibition of ORN terminals was constant for odors presented at 1–5 Hz sniffs (Pírez and Wachowiak, 2008), perhaps because robust tonic inhibition in vivo masks frequency-dependent effects. Presynaptic inhibition of ORN terminals may provide a negative gain control mechanism for adjusting the scale or dynamic range of odor concentration that can be processed by MOB neurons or to prevent saturation of input. Thus, presynaptic inhibition magnitude may scale with the odor intensity conveyed to PGCs. Alternatively, presynaptic inhibition may increase contrast (i.e., sharpen the spatial pattern) of odorant representations and facilitate detection of strong odors by preferentially suppressing weakly responsive glomeruli. However, $GABA_B$ presynaptic inhibition does not appear to alter the relative spatial pattern of weakly and strongly activated glomeruli as determined by imaging ORN terminal activity (Pírez and Wachowiak, 2008; see also Vucinić et al., 2006). It remains to be established if this holds for the influence of presynaptic inhibition on postsynaptic odor responses and perception. In this regard, it is noteworthy that D2 blockade increased the number of M/TCs responding to single or multiple odorants and improved odor discrimination (Wilson and Sullivan, 1995; Wei et al., 2006; Escanilla et al., 2009). The later effect seems counter to what one would expect in coding strategies where increases in the overlap of odor

responsive M/TCs would impair discrimination (Linster and Hasselmo, 1997; Linster and Cleland, 2002). Alternatively, a D2 antagonist increase in odor-evoked ORN glutamate release would strengthen perceived odor intensity to enhance discriminability (Wei et al., 2006). D2 receptor agonists by contrast impaired odorant specific glomerular response patterns (2-dexyglucose), reduced odorant detectability, or impaired discrimination of low concentration odorants (Doty and Risser, 1989; Sallaz and Jourdan, 1992; Wei et al., 2006).

PGC TO ET/M/TC INHIBITION While it has been known for decades that M/TCs receive inhibitory input spontaneously or in response to ON input, the relative contributions of PGC vs. GCs to such inhibition has only recently come to light. PGCs are driven by rhythmically bursting ETCs and ETC stimulation readily elicits PGC mediated feedback inhibition (Murphy et al., 2005). This circuit is responsible for a significant portion of intraglomerular inhibition. Spontaneous $GABA_A$ receptor inhibition (i.e., IPSPs/IPSCs) is observed in ET/M/TCs; rhythmic IPSC bursts are prominent in ETCs and are also observed in M/TCs (Hayar et al., 2004b, 2005; Dong et al., 2007). Most MC spontaneous IPSC bursts are eliminated, and overall spontaneous IPSCs are reduced by ~50%, when the GL is excised to remove PGC inhibition. Focal GL $GABA_A$ receptor block reduces MC spontaneous IPSCs by ~50% (Shao et al., 2012). ON-evoked excitation of ET/M/TCs is sculpted by temporally overlapping inhibition (Carlson et al., 2000; Gire and Schoppa, 2009; Najac et al., 2011). Such inhibition has early and longer duration delayed phases (Shao et al., 2012). The early onset and ~50% of the late inhibition in MCs is abolished by GL $GABA_A$ receptor block; this also eliminated all spontaneous and ON-evoked IPSCs in ETCs lacking secondary dendrites in the EPL. These and other findings indicate the PGCs mediate most if not all inhibition in ETCs, and the early and half of the late inhibition in M/TCs. Block of PGC mediated intraglomerular inhibition increases the magnitude and duration of ON-evoked LLDs and spiking in ET/M/TCs (Gire and Schoppa, 2009; Shao et al., 2012). Intraglomerular postsynaptic inhibition, along with ORN presynaptic inhibition, are thought to be critical in regulating the gain of glomerular input–output function, and also to curtail prolonged odor responses to allow precise temporal modulation of ET/M/TC activity by the sniff cycle (Wachowiak and Shipley, 2006; Shao et al., 2012). PGC GABA release has also been reported to lead to $GABA_A$ receptor-mediated self-inhibition or inhibition of adjacent PGCs (Smith and Jahr, 2002; Murphy et al., 2005).

8.2.1.5 Interglomerular Processing. As noted earlier, SACs receive input from, and project to, multiple glomeruli, thus forming a circuit for interglomerular

8.2 Main Olfactory Bulb (MOB)

communication. The direct synaptic effect of the SAC interglomerular projection seems to be glutamatergic excitation of ET and PGCs in neighboring glomeruli; MCs by contrast are polysynaptically inhibited (Aungst et al., 2003). Pre-stimulation of one glomerulus suppressed ON-evoked excitation of MCs (i.e., LLDs) in adjacent glomeruli, an effect eliminated by GABA$_A$ blockade. Based on this, it was proposed that SACs excite PGCs in neighboring glomeruli directly, or disynaptically via ETCs, to increase GABA release and inhibit MCs. This interglomerular circuit may function in a center-on, surround-off manner whereby glomeruli receiving strong odor input inhibit input to weakly responsive surround glomeruli (Aungst et al., 2003). If SACs contain GABA and DA as recently reported (Kiyokage et al., 2010), their excitatory cellular effect would require co-expression of another excitatory transmitter such as glutamate. Also, if SACs excite ETCs, and if ETCs play a critical role in LLDs, one would expect to see excitation in some M/TCs via this interglomerular circuit.

8.2.2 External Plexiform Layer (EPL)

8.2.2.1 *Tufted Cells (TCs).* TCs are the most numerous cells of the EPL and three TC subclasses are recognized based on location in the EPL: superficial, middle, and deep TCs. All share several features in common with ET/MCs, including a typical single apical dendrite ending in one glomerulus that receives monosynaptic ORN input and forms dendrodendritic synapses with PG/SACs (Ennis et al., 2007). They also utilize glutamate as their principle transmitter (Christie et al., 2001), although superficial TCs also contain CCK (Liu and Shipley, 1994). Their lateral dendrites are relatively short compared to MCs, but like MCs ramify in the EPL forming dendrodendritic synapses with GCs and other inhibitory EPL interneurons. Indeed, much of the EPL neuropil is devoted by dendrodendritic processing as discussed below. TC axons, together with those of MCs, terminate locally in the IPL/GCL on GCs or exit the MOB to innervate primary olfactory cortical (POC) structures including olfactory or piriform cortex. There are differences in M vs. TC projections to POC, with TCs projecting preferentially to the anterior olfactory nucleus, olfactory tubercle, and rostral olfactory cortex (Ennis et al., 2007; Nagayama et al., 2010). Axons of CCK-containing superficial TCs form the intrabulbar association system (IAS) that connects lateral and medial regions of each MOB (Schoenfeld et al., 1985; Liu and Shipley, 1994). IAS axons travel to the opposite side of the bulb and synapse on GCs apical dendrites. The IAS exhibits a high degree of topographic specificity, such that TCs associated with one glomerulus preferentially target the IPL below the opposite side isofunctional glomerulus and vice-versa

(Belluscio et al., 2002; Lodovichi et al., 2003). Thus, the IAS is configured to modulate M/TCs, via GCs, receiving input from ORNs expressing the same receptor, on opposite sides of the MOB. TCs are excitatory, as is CCK, and therefore the IAS should depolarize GCs to inhibit M/TCs, although this has yet to be tested. TC physiology is discussed below with MCs, but there are some notable differences between the two output classes and among TC subtypes. TCs in vivo tend to fire at higher spontaneous rates than MCs (Nagayama et al., 2004). In slices, some TCs exhibit spontaneous and ON-evoked bursting activity somewhat similar to ETCs, although there is significant variability (Hamilton et al., 2005; Ma and Lowe 2010). Different TC classes have differential sensitivity to ORN input, with superficial cells more easily excited by ORN stimulation (Schneider and Scott, 1983; Ezeh et al., 1993; Wellis et al., 1989).

8.2.2.2 *EPL Interneurons.* The EPL contains morphologically and neurochemically heterogeneous interneurons, such as van Gehuchten cells and other subtypes. Many are GABAergic and presumed to be inhibitory, but others express Substance P, enkephalin, neurotensin, somatostatin, neuropeptide Y, and other neuropepetides (Ennis et al., 2007). The role of these transmitters is unknown. EPL interneurons respond to ON input polysynaptically, probably via glutamate release from M/TC lateral dendrites (Hamilton et al., 2005). EPL interneurons are thought to inhibit M/TC, and possibly GC, dendrites. Many EPL interneurons have dendrites spanning several adjacent glomeruli and may provide localized inhibition of M/TCs that are topographically related to overlying glomeruli (Hamilton et al., 2005).

8.2.3 Mitral Cell Layer (MCL)

8.2.3.1 *Anatomical Features of MCs.* The thin MCL contains MC somata as well as numerous GCs. Indeed, it is often overlooked that ~90% of the cells in the MCL are GABAergic, corresponding to superficially located GCs (Frazier and Brunjes, 1988; Parrish-Aungst et al., 2007). MC features are similar to those of TCs although MC lateral dendrites are more extensive and may extend up to 1–2 mm. MCs have been subdivided into two classes, Type I and II, based on extension of dendrites into the deep or superficial parts of the EPL, respectively (Orona et al., 1984). Type II MCs are more readily activated by ORN stimulation than Type I cells (Schneider and Scott, 1983; Ezeh et al., 1993; Wellis et al., 1989). M/TC lateral dendrites receive axonal input from certain classes of centrifugal neuromodulatory transmitter systems (see Section 8.9), and dendrodendritic inputs from EPL interneurons and GCs. M/TC axon collaterals synapse with

164 Chapter 8 Anatomy and Neurobiology of the Main and Accessory Olfactory Bulbs

GCs, or exit the MOB as the lateral olfactory tract (LOT) to innervate POC. In addition to glutamate, some M/TCs contain neuropeptides such as corticotropin-releasing factor.

8.2.3.2 Spontaneous Firing and Intrinsic Membrane Properties.
In anesthetized rodents, M/TCs fire spontaneously over a wide ~1–40 Hz range with mean rates generally 14–18 Hz and faster instantaneous rates during bursts or odors (Chaput and Holley, 1979; Chaput, 1983; Yu et al., 1993; Jiang et al., 1996; Davison and Katz, 2007; Nica et al., 2010. Spontaneous firing in awake rodents is generally similar to that in anesthetized preparations; range, 1–33 Hz, mean range, 12–17 Hz (Kay and Laurent, 1999; Fuentes et al., 2008; Shusterman et al., 2011). However, in one study recordings of M/TCs before and after anesthesia exhibited a large drop in firing from ~27 to 9 Hz (Rinberg et al., 2006). In slices, M/TC discharge is more modest (~0.5–20 Hz range; 3–6 Hz mean), presumably because tonic sensory input is absent (Ennis et al., 1996; Ciombor et al., 1999; Heyward et al., 2001; Ma and Lowe, 2010). Spontaneous firing in M/TCs is intrinsically generated and persists when their synaptic inputs are inactivated (Ennis et al., 1996; Heyward et al., 2001; Ma and Lowe, 2010). MCs, but not other MOB cell types, exhibit intrinsic membrane bistability characterized by spontaneous alternation between a perithreshold upstate membrane potential and a more hyperpolarized (~−10 mV) downstate potential (Heyward et al., 2001; Heinbockel et al., 2004). Spontaneous or ON-evoked spikes are readily launched from the upstate characterized by high frequency (10–50 Hz) subthreshold membrane potential oscillations (Desmaisons et al., 1999; Heyward et al., 2001; Balu et al., 2004). MC intrinsic membrane properties support temporally precise spiking responses to brief sniff-like phasic inputs (Balu et al., 2004).

A hallmark of spontaneous and odor-evoked activity in M/TCs, and of nearly all MOB subtypes, is phasic rhythmic modulation as a function of the 1–12 Hz theta frequency range of the respiratory cycle (Buonviso et al., 2006; Scott, 2006). Respiratory- or sniff-coupled spiking and membrane potential oscillations occur in the presence and absence of odors and depend upon rhythmic ORN glutamate input in response to airflow across the nasal epithelium (Philpot et al., 1997; Chalansonnet and Chaput, 1998; Kay and Laurent, 1999; Luo and Katz, 2001; Charpak et al., 2001; Margrie and Schaefer, 2003; Debarbieux et al., 2003). Thus, rhythmic activity persists when the MOB is isolated from the brain by olfactory peduncle trans section, but is eliminated when nasal respiration is bypassed by tracheal respiration (Sobel and Tank, 1993; Philpot et al., 1997; Margrie and Schaefer, 2003; Phillips et al., 2012). Peak

depolarization of the MC membrane potential, as well as EPSPs and spikes, occur during or after a short lag from the inhalation/inspiration transition (Charpak et al., 2001; Margrie and Schaefer, 2003; Cang and Isaacson, 2003; Debarbieux et al., 2003). Other aspects of M/TC odor responses are discussed below in Section 8.8.

8.2.3.3 Dendritic Spike Propagation.
Subthreshold ON-evoked EPSPs are decremental and decrease by ~30% over a 300 µm distance in the apical dendrite (Djurisic et al., 2004). Suprathreshold input typically elicits spikes that are initiated at the soma, but stronger input can trigger actively propagating apical dendritic Na⁺ spikes originally described in vivo as fast prepotentials (Mori, 1987; Chen et al., 1997; Isaacson and Strowbridge, 1998; Debarbieux et al., 2003). Somatic spikes also back-propagate non-decrementally along the apical dendrite and thus could engage M/TC glutamate spillover and dendro-dendritic input to PG/SACs. Somatic or lateral dendritic inhibitory input (IPSPs) can block somatic spike initiation providing the cellular basis for modulation of odor responses by GC-mediated inhibition (Chen and Shepherd, 1997; Chen et al., 1997). Na⁺ spikes actively propagate in M/TC lateral dendrites, suggesting that M/TC spikes, via GCs, could inhibit other M/TCs over considerable spatial distances. (Charpak et al., 2001; Xiong and Chen, 2002; Debarbieux et al., 2003). However, other studies report that lateral dendritic spike propagation is attenuating (Margrie et al., 2001; Lowe, 2002; Christie and Westbrook, 2003).

8.2.4 Granule Cell Layer (GCL)

8.2.4.1 Granule Cells (GCs).
GCs are small axon-less cells packed into row-like aggregates in the GCL (Price and Powell, 1970b; Reyher et al., 1991). As noted above, superficially-located GCs are also present in the MCL. GCs have an apical dendrite that ramifies in the EPL and shorter basal dendrites in the GCL. The apical dendrites of superficial GCs arborize in the superficial and deep EPL whereas those of deeper GCs ramify in the deep EPL (Orona et al., 1983; Greer, 1987). GCs contain GABA (Ribak et al., 1977), and some contain enkephalin and other neuropeptides (Bogan et al., 1982; Davis et al., 1982). GC apical dendrites receive synapses from, and make synapses onto, M/TCs. GCs also receive synapses from a variety of centrifugal afferents, including inputs from primary olfactory cortex (POC) (Section 8.6) and neuromodulatory transmitter systems (Section 8.9). Input from POC (e.g., piriform cortex, anterior olfactory nucleus) comprise the bulk of synaptic contacts onto GCs (Price and Powell, 1970a,c). GCs also receive synapses from M/TC axons, as well as inputs from other GCL interneurons. Anatomically, gap junctions have been reported among GCs (Reyher

et al., 1991), although electrophysiological studies found no evidence for electrotonic coupling (Schoppa, 2006). In vivo and in vitro, GCs spontaneous spiking is relatively infrequent, probably due to their hyperpolarized resting potential (−65 to −75 mV), and is driven primarily by glutamatergic input (Wellis and Scott, 1990; Cang and Isaacson, 2003; Zelles et al., 2006; Heinbockel et al., 2007; Pressler et al., 2007). Synaptic input or direct depolarization can elicit Na^+ and Ca^{2+} spikes in GCs (Halabisky et al., 2000; Egger et al., 2005). GC activity is strongly regulated by a transient A-type K^+ current, I_A, preferentially expressed in dendrites where it counters brief depolarizing synaptic inputs (Schoppa and Westbrook, 1999). Blockade of I_A with 4-AP decreases the lag or delay for evoked spikes. GCs in the MOB and AOB are potently activated by mGluR1 agonists while mGluR1 antagonists reduce M/TC-evoked excitation (Castro et al., 2007; Heinbockel et al., 2007). GCs exhibit broader odor tuning (respond to more odorants) than ET/M/TCs (Tan et al., 2010) and seem to lack inhibitory odor responses (Luo and Katz, 2001). In contrast to M/TCs, GC odor responses attenuate rapidly after the first respiratory cycle (Cang and Isaacson, 2003).

8.2.4.2 Deep SACs.
The IPL and GCL contain a population of SACs, collectively referred to as deep SACs, that form several morphologically distinct subtypes (i.e., Golgi, Cajal, horizontal and Blanes cells) (see Eyre et al., 2008 for review). Most deep SACs have dendrites restricted to the IPL/GCL, while their axons tend to terminate and ramify either in the GL, EPL or GCL. Although long considered interneurons, a subpopulation was found to project to POC. These cells are GABAergic, but may also express neuropeptides such as vasoactive intestinal peptide, neuropeptide-Y, enkephalin, and somatostatin (Eyre et al., 2008). Deep SAC axons seem to exclusively target inhibitory GABA neurons, including PG and GCs. Consistent with this, activation of deep SACs evokes $GABA_A$ receptor mediated inhibition of GCs (Eyre et al., 2008; Pressler and Strowbridge, 2006). Thus, deep SACs comprise a major intrabulbar inhibitory circuit and provide the only known local regulatory system from inhibitory neurons below the MCL to the sensory input layer. Since the targets of deep SACs are inhibitory interneurons, functionally deep SAC activation should lead to ET/M/TC disinhibition. One deep SAC type, Blanes cells, appears to exert prolonged modulation of MOB excitability. Brief excitation elicits prolonged (up to 44 min) Blanes cell spiking and in turn, prolonged barrages of IPSCs in GCs (Pressler and Strowbridge, 2006). This would be expected to yield long-lasting excitation (i.e., disinhibition) of M/TCs. Deep SACs cells receive excitatory glutamatergic input from axon terminals, which may include M/TC axons and centrifugal feedback input from POC (Eyre et al., 2008).

8.2.5 Dendrodendritic Transmission Between M/T and Gcs

8.2.5.1 Overview Of M/T - GC Dendrodendritic Interactions.
The majority of synapses of the EPL are between M/TC lateral dendrites and GC apical dendrites. Most of the synapses are: (1) asymmetrical synapses from M/TC dendrites onto GC dendrites, and (2) symmetrical synapses from the spines ("gemmules") of GC dendrites to M/TC dendrites (Price and Powell, 1970b-d). These synapses are mostly reciprocal and occur in roughly equal proportion (Hirata, 1964; Rall et al., 1966; Price and Powell, 1970d; Jackowski et al., 1978; Woolf et al., 1991). A M/TC can therefore generate self- or feedback inhibition and also lateral inhibition of other M/TCs.

8.2.5.2 Excitatory M/T to GC Transmission.
M/TC glutamate release requires activation of Ca^{2+} influx via high voltage activated (HVA) channels located close to active release sites (Isaacson and Strowbridge, 1998). M/TC stimulation evokes dual component EPSCs in GCs, consisting of a fast AMPA and slow NMDA receptor components (Isaacson and Strowbridge, 1998; Schoppa et al., 1998; Aroniadou-Anderjaska et al., 1999a; Chen et al., 2000; Isaacson, 2001). The AMPA receptor component is less effective than the NMDA receptor component in evoking spikes in GCs, especially in Mg^{2+} free conditions (Schoppa et al., 1998). Block of GC I_A enhances AMPA receptor-dependent spiking (Schoppa and Westbrook, 1999). An mGluR1 component is involved in M/TC to GC transmission in the MOB and AOB (Heinbockel et al., 2007; Castro et al., 2007).

8.2.5.3 Inhibitory GC to M/TC Transmission.
GC activation evokes $GABA_A$ receptor-mediated IPSPs/IPSCs in M/TCs (Chen et al., 2000; Isaacson and Vitten, 2003; Dietz and Murthy, 2005). Ca^{2+} influx through P/Q- and N-type channels and through NMDA receptor channels near dendrodendritic synapses, presumably on GC spines, directly couple to GABA release (Isaacson and Strowbridge, 1998; Chen et al., 2000; Halabisky et al., 2000; Isaacson, 2001; Egger et al., 2003, 2005). T-type Ca^{2+} currents activated by subthreshold depolarization and store Ca^{2+} release also contribute to GABA release (Egger et al., 2003, 2005).

8.2.5.4 Self and Lateral Dendrodendritic Inhibition.
M/TC depolarization results in a dendrodentrically-mediated feedback IPSP/IPSC; that is, self-inhibition (Isaacson and Strowbridge, 1998; Schoppa et al., 1998; Chen et al., 2000; Halabisky et al., 2000; Dietz and Murthy, 2005). Such inhibition is of long duration (1–2 sec) and consists of a flurry of individual IPSPs/IPSCs suggesting

asynchronous GABA release from multiple GCs (Schoppa et al., 1998). In normal extracellular Mg^{2+}, self-inhibition is reduced to similar levels by AMPA or NMDA receptor antagonism, and is abolished when both receptors are blocked (Isaacson and Strowbridge, 1998). Mg^{2+}-free conditions enhance self-inhibition and cause it to be dominated by NMDA receptor activation (Isaacson and Strowbridge, 1998; Schoppa et al., 1998; Isaacson, 2001; Chen et al., 2000; Christie et al., 2001; Halabisky and Strowbridge, 2003). Self-inhibition is reduced by mGluR1 blockade (Heinbockel et al., 2007; Castro et al., 2007). Self-inhibition has been reported to be unaffected (Halabisky et al., 2000) or enhanced (Schoppa and Westbrook, 1999; Isaacson, 2001) by blockers of I_A. Lateral or feedforward inhibition is mediated through the M/TC-GC-M/TC dendrodendritic circuit (Isaacson and Strowbridge, 1998; Urban and Sakmann, 2002). Like self-inhibition, lateral inhibition in Mg^{2+}-free conditions is dominated by the NMDA receptor synaptic response (Isaacson and Strowbridge, 1998; Schoppa et al., 1998). I_A currents, which oppose brief AMPA depolarizations, strongly regulate lateral inhibition. Thus, block of I_A by 4-AP reduces the duration of lateral inhibition by increasing the ability of AMPA receptors to elicit more rapid and synchronous GABA release (Schoppa and Westbrook, 1999).

LOCAL VS. GLOBAL MODES OF DENDRODENDRITIC INHIBITION Current views are that weak stimulation producing small or subthreshold synaptic responses in GCs, and thus Ca^{2+} influx into isolated spines, triggers GABA release and inhibition of M/TC that are synaptically-coupled, and in close proximity, to those spines. Such inhibition would tend to be spatially restricted. NMDA receptors, voltage-dependent Ca^{2+} channels and store Ca^{2+} can trigger GABA release. By contrast, stronger input that elicits Na^+ or Ca^{2+} spikes would produce more global or widespread inhibition (Chen et al., 2000; Egger et al., 2003, 2005). This mechanism would tend to dominate long-range lateral inhibition as well as inhibition following excitatory input to GC somata or basal dendrites. The spatial extent of lateral inhibition following TC activation is less than that for MCs, presumably due to the shorter TC lateral dendrites (Christie et al., 2001).

TEMPORAL MODULATION OF DENDRODENDRITIC INHIBITION M/TC self-inhibition and GC-M/TC inhibition exhibit paired pulse depression with a slow recovery time constant (Isaacson and Vitten, 2003; Dietz and Murthy, 2005). This may be due to intrinsic GC properties leading to spike adaptation across repetitive sniff frequency input (Pressler et al., 2007). These factors may account for the decrement in GABAergic inhibition of M/TCs across sniffs of an odor (Cang and Isaacson,

2003). Thus, the strength of dendrodendritic inhibition is likely to be temporally modulated by respiration (see also Phillips et al., 2012). Other studies indicate that higher frequency local gamma oscillatory activity generated by M/TC-GC interactions modulates GABAergic inhibition. Gamma frequency stimulation in the GCL or ELP can facilitate M/TC feedback and lateral inhibition by relieving the Mg^{2+} block of GC NMDA receptors (Halabisky and Strowbridge, 2003). Lateral inhibition is strengthened in an activity dependent manner among M/TCs that have correlated spike activity (Arevian et al., 2008).

8.2.6 Primary Olfactory Cortical Inputs

Extrinsic afferent input to the MOB, also referred to as centrifugal fibers (CFFs), can be subdivided into two classes: (1) inputs arising from non-olfactory, so called neuromodulatory transmitter systems (see Section 8.9), and (2) feedback inputs arising from olfactory-related structures, in particular those arising from POC. CFF projections from POC arise from glutamatergic pyramidal neurons in layer II and III of piriform cortex, as well as other POC structures (Shipley et al., 1996; Ennis et al., 2007). These projections massively target GCs and terminate mostly heavily on somata/basal dendrites and proximal apical dendrites. CFF activation excites GCs and produces IPSPs in M/TCs, that is, dendrodendritic inhibition (Ennis and Hayar, 2008). CFF-evoked synaptic responses in GC proximal dendrites have AMPA and NMDA receptor components, but have a much faster time course than M/TC synaptic input to GC distal dendrites in the EPL (Balu et al., 2007; Laaris et al., 2007). Activation of proximal synaptic input can reverse the Mg block of NMDA responses at distal EPL inputs, facilitating the expression of M/TC dendrodendritic inhibition (Balu et al., 2007). CFF, but not M/TC synaptic input, exhibits NMDA receptor-dependent LTP that amplifies M/TC inhibition by GCs (Guo and Strowbridge, 2009).

8.2.7 Adult-Born Neurons and Sensory/Experience-Dependent Plasticity

In adult animals MOB interneurons (GCs and PGCs) are continually generated from progenitor cells of the rostral migratory stream (see Chapter 9). These new neurons functionally integrate into MOB circuits in a sensory experience-dependent manner and contribute to odor learning and discrimination. This topic is beyond the scope of this chapter and readers are referred to Chapter 9 and other studies on this topic (Gheusi et al., 2000; Rochefort et al., 2002; Carleton et al., 2003; Lledo et al., 2005; Lazarini and Lledo, 2011; Moreno et al.,

2009; Sawada et al., 2011; Sultan et al., 2011; Yokoyama et al., 2011). Similarly, reviewers are referred elsewhere to the effects of odor activity, experience, and learning on ORN input to the glomeruli, MOB circuitry, odor responses, and oscillations and olfactory-guided behaviors (Fletcher and Wilson, 2002, 2003; Gervais et al., 2007; Jones et al., 2008; Mandairon and Linster, 2009; Cummings and Belluscio, 2010; Parrish-Aungst et al., 2011).

8.2.8 Odor Responses, Oscillations and Synchrony

8.2.8.1 Odor Responses.
A comprehensive review of odor coding is beyond the scope of this chapter and we refer the reader to more extensive reviews of this topic (Wachowiak and Shipley, 2006; Mori et al., 2006; Wilson and Mainen, 2006; Johnson and Leon, 2007). ORNs expressing a given odorant receptor are promiscuous or broadly tuned such that odors tend to activate ORN populations expressing different receptors and ORNs expressing the same receptor respond to multiple odorants. ORNs expressing the same receptor are thought to have a receptive field or molecular receptive range (MRR) that encompasses all of the odors they respond to. Optical imaging techniques have revealed that these features are reflected in odor response profiles of ORN terminals in the glomeruli. Thus, glomerular activity mirrors ORN activity and each glomerulus has a unique odor response profile or MRR predicated by the MRR of the ORN population by which it is innervated. An odorant may activate multiple glomeruli and a glomerulus may respond to multiple odorants in a manner that varies with stimulus concentration. ORNs responsive to structurally similar odorants or odorants sharing similar molecular features tend to map onto adjacent glomeruli, and thus similar odorants tend to activate spatially adjacent glomerular clusters. In general, increasing odorant concentration increases the response magnitude of a glomerulus and increases the number of activated glomeruli (Meister and Bonhoeffer, 2001; Wachowiak and Cohen, 2001; Fried et al., 2002; Spors and Grinvald, 2002; Spors et al., 2006. Additionally, glomerular activation patterns have strong temporal components (kinetics, respiratory sniff modulation) that differ across the population of odor responsive glomeruli and vary within a glomerulus in an odorant specific manner (Spors et al., 2006; Carey et al., 2009). Therefore, odor information is conveyed by the pattern of activated glomeruli, their relative activation magnitude, and temporal response dynamics. Each of these response features, at the level of glomerular input and M/TC output, may contribute to the odor code but their relative contribution has yet to be determined.

M/TC odor responses are complex and variable and include inhibitory responses alone or in combination with excitation (Luo and Katz, 2001; Margrie et al., 2001; Cang and Isaacson, 2003; David et al., 2009; Tan et al., 2010). The type (excitation vs. inhibition), magnitude and direction of response can change dramatically with odor concentration. In general, odor-evoked EPSP amplitudes vary proportionately with stimulus strength and spikes are often launched from the rising phase of EPSPs (Margrie and Schaefer, 2003; Cang and Isaacson, 2003). Odor-evoked IPSP strength does not seem to vary with stimulus concentration, although the IPSP amplitude declines across sniffs (Cang and Isaacson, 2003). It has long been known that closely adjacent M/TCs exhibit similar activity profiles and odor responses as one would expect from cells receiving sensory input from the same glomerulus (Buonviso et al., 2006). This has been directly confirmed by the high similarity of odor responses of M/TCs of the same glomerulus; however, odor response dynamics can vary among "sister" M/TCs of the same glomerulus (Dhawale et al., 2010). Nonetheless, it is thought sister M/TCs have a similar molecular receptive range (MRR) to a series of structurally or molecularly similar odorants, with a preferential sensitivity (most vigorous response) to one odorant in the family; more dissimilar members of the family evoke weaker responses with inhibition occurring to even more dissimilar members (Yokoi et al., 1995; Kashiwadani et al., 1999; Fletcher and Wilson, 2003). In addition to a defined MRR, M/TCs may also respond to very different, structurally unrelated odorants. Major differences in the breath of M/TC receptive fields have been reported, perhaps due to the range or concentration of odorants tested and/or the influences of anesthesia (Kay and Laurent, 1999; Rinberg et al., 2006; Davison and Katz, 2007; Fantana et al., 2008; Tan et al., 2010). There are conflicting reports on the breadth of odor tuning in M vs TCs with some studies reporting similar tuning curves (Tan et al., 2010) and others that TCs exhibit more robust and widely tuned excitatory responses (Nagayama et al., 2004); the later study also reported that TCs exhibit weaker and narrower inhibitory odor response tuning than MCs. Superficial TCs exhibit more prolonged odor responses than MCs (Luo and Katz, 2001).

Inhibitory circuits in the glomerular and deeper layers modulate M/TC odor responses. In some models, inhibitory odor responses are thought to arise from M/TCs in neighboring glomeruli with slightly different, but overlapping MRRs (Yokoi et al., 1995; Nagayama et al., 2004). Thus, neighboring cells that are highly tuned to an odorant suppress activity of M/TC populations that are weakly tuned or unresponsive to an odorant via lateral inhibitory circuits. Observations that M/TCs associated with different but nearby glomeruli interact with a common population of GCs is consistent with this model (Arevian et al., 2008). Also partially consistent are findings that M/TCs below odor responsive glomeruli are excited, but are inhibited or show no response to odors activating adjacent glomeruli (Luo and Katz, 2001). Inhibitory circuits in the glomerular layer and at the deeper M/TC-GC network are both thought to contribute to inhibitory odor responses. In theory, inhibition is thought to play a key role in population response odor coding strategies via contrast enhancement – sharpening odor representations – by inhibiting weakly responsive M/TCs and/or reducing overlapping populations that respond in common to odors (Yokoi et al., 1995; Linster and Hasselmo, 1997; Linster and Cleland, 2002; Cleland and Sethupathy 2006; Urban and Arevian, 2009). Consistent with this concept, blockade of inhibition or activity at M/TC-GC synapses, enhances excitatory and reduces inhibitory odor responses, and also broadened breadth of tuning (Yokoi et al., 1995; Margrie et al., 2001; Tan et al., 2010). Inhibition is also thought to play a key role in discrimination. Odors that produce overlapping activation of M/TCs will be perceived as more similar (i.e., more difficult to discriminate) than odors that activate non-overlapping populations (Linster and Hasselmo, 1997; Linster and Cleland, 2002); see also Section 8.8.2.

In addition to the magnitude or strength of the odor response (Bathellier et al., 2008), the temporal dynamics of responses are thought to contribute to odor coding. Temporal components of odor responses or phase relative to respiration are also thought to be important to the odor code. The latency of the first spike following stimulus onset is inversely proportional to the number of spikes per sniff, such that shorter latencies occur during bursts (Margrie and Schaefer, 2003; Cang and Isaacson, 2003). This suggests that the initial spike latency provides information about sensory input strength, i.e., odor concentration. It has long been known that M/TC odor responses generally synchronize to the sniff cycle, but also that the number of spikes and their timing evolve over multiple sniffs (Buonviso et al., 2006; Scott, 2006; Schaefer and Margrie, 2007). However, recent studies indicate that the timing of odor responses relative to the sniff phase are relatively invariant but odor specific, suggesting that sub-sniff spike timing contains sufficient information for odor discrimination (Cury and Uchida, 2010; Carey and Wachowiak, 2011; Shusterman et al., 2011). Rodent sniff frequency varies with behavior and this is also thought to modulate odor responses. Imaging studies indicate that the sniff-locked pattern of presynaptic input to the glomeruli attenuates and decouples from respiration as sniff frequency increases (Spors et al., 2006; Verhagen et al. 2007; Wesson et al., 2009; Carey et al., 2009; but see also Oka et al., 2009). How this impacts the magnitude and temporal dynamics of M/TC odor responses is unclear.

Odor response magnitude seems to decrease with increases in sniff rate (Bathellier et al., 2008; Carey and Wachowiak, 2011). Some studies indicate that odor response timing is relatively invariant to sniff frequency, at least up to ~4–5 Hz (Cury and Uchida, 2010; Carey and Wachowiak, 2011; Shusterman et al., 2011). This is at odds with other reports that temporal variability increases or that spike phase locking decouples from the respiratory cycle as sniff frequency increases (Bhalla and Bower, 1997; Kay and Laurent, 1999; Rinberg et al., 2006; Courtiol et al., 2011). Just as inhibition is thought to provide contrast enhancement of odorant representations, it is also thought to regulate gain control of the frequency or amplitude of M/TC odor responses, and to modulate temporal and synchronization dynamics of odor responses. Blockade of GABA inhibition increases the number, rate and temporal pattern of spikes elicited by ON or odor stimuli and this can occur when inhibition is blocked in the glomeruli or in the EPL (Yokoi et al., 1995; Lagier et al., 2004; Gire and Schoppa, 2009; Tan et al., 2010, Najac et al., 2011; Shao et al., 2012).

8.2.8.2 Oscillations. Oscillations and synchronous activity are characteristic features of spontaneous and odor-driven activity in the MOB and other olfactory structures. Temporal firing patterns, including rhythmic oscillations and neuronal synchronization are thought to be important in odor processing and coding at intra- and interglomerular levels within the MOB, and also in POC (Wehr and Laurent, 1996; Kauer, 1998; Kay and Laurent, 1999; Friedrich, 2002; Lledo et al., 2005; Gelperin, 2006; Kepecs et al., 2006; Schoppa, 2006; Spors et al., 2006; Beshel et al., 2007; Kay et al., 2009; Wachowiak, 2011). Prominent oscillatory activity is present in MOB field potentials and membrane potential/spike activity of individual MOB neurons. Olfactory sensory input occurs as a result of respiration which imparts slow rhythmical temporal structure to olfactory activity in the theta band (1–12 Hz) (Wachowiak, 2011). Higher frequency oscillations, superimposed on the theta, occur in the beta (15–40 Hz) and gamma (40–80 Hz) bands. Rodent MOB oscillations in the 1–12 Hz range are referred to as "theta" principally by similarity to hippocampal theta oscillations (Kay and Laurent, 1999; Kay, 2003, 2005). Since theta oscillations are generated entirely by respiratory-driven sensory input their prevailing frequency varies with sniff frequency; this includes components related to slow (1–3 Hz) "passive" sniffing as well as a faster components (5–12 Hz) characteristic of "active" investigative sniffing (Adrian, 1950; Welker, 1964; Macrides et al., 1982; Eeckman and Freeman, 1990; Kay and Laurent, 1999; Kay, 2003; Kay et al., 2009). Sniff frequency varies with task demands and increases in response to novel odors or

8.2 Main Olfactory Bulb (MOB)

discriminations tasks, suggesting that rapid sniffing may improve odor processing and perception (Kepecs et al., 2006; Wesson et al., 2008a, b; Wachowiak, 2011).

Beta oscillations do not have a prominent spontaneous component, but are evoked by odorants and are amplified or emerge as animals learn to correctly recognize an odorant (Martin et al., 2004a, 2006; Fletcher et al., 2005; Kay et al., 2009). Sufficiently strong odorants trigger spike bursts occurring within the beta to gamma frequency range (20–60 Hz (Debarbieux et al., 2003). Although beta oscillations vary with molecular features of odorants, they do not appear to be generated within the MOB but instead involve interactions with the piriform cortex (Neville and Haberly, 2003; Martin et al., 2004b, 2006; David et al., 2009). Gamma oscillations, evoked by odors or ON stimulation, occur on the crest of the slower respiratory driven theta cycle, and are apparent in the EEG, MOB field potentials, and M/TC subthreshold membrane potential and spiking activity (Adrain, 1950; Eeckman and Freeman, 1990; Kay and Freeman, 1998; Kashiwadani et al., 1999; Debarbieux et al., 2003; Friedman and Strowbridge, 2003; Neville and Haberly, 2003; Fletcher et al., 2005; Lagier et al., 2004; Martin et al., 2004b; David et al., 2009; Courtiol et al., 2011). Gamma is generated in the MOB and persists when centrifugal input is disrupted or when the MOB is isolated by peduncle trans section (Gray and Skinner, 1988; Neville and Haberly, 2003; Martin et al., 2004b). Gamma oscillations are generated by reverberatory bidirectional interactions at MT-GC synapses. Manipulations that impair M/TC to GC or GC to M/TC interactions suppress gamma oscillations (Friedman and Strowbridge, 2003; Neville and Haberly, 2003; Lagier et al., 2004, 2007). Gamma oscillations are also strengthened by electrical synapses, and disrupted by gap junction inhibition or in connexin36-knockout mice (Hormuzdi et al., 2001; Friedman and Strowbridge, 2003). Inhibition and gamma oscillations via the M/TC-GC system are thought to regulate spike timing and synchronization to enhance odor discrimination and learning (Linster and Hasselmo, 1997; Desmaisons et al., 1999; Linster and Cleland, 2001; Lagier et al., 2004; Beshel et al. 2007; Kay et al. 2009). Gamma oscillations are enhanced as animals learn to discriminate similar odorants (Kay et al., 2009). Disruption of inhibition through the M/TC-GC system or disruption of gamma oscillations impairs odor discrimination, while strengthening inhibition or gamma oscillations improves discrimination (Nusser et al., 2001; Abraham et al., 2010).

8.2.8.3 Synchrony. Temporally correlated or synchronized activity among the elements of the odor responsive neuronal population is thought to be a key element in odor coding (Linster and Cleland, 2001;

Laurent, 2002). Synchronous activity and similar odor response patterns among M/TCs in close proximity have been observed for decades (see Buonviso et al., 2006 for review). Studies over the last decade have revealed that synchronous activity is prominent among MOB neurons associated with the same glomerulus, consistent with the idea that glomeruli comprise fundamental units of olfactory processing. Spontaneous (spikes, subthreshold membrane potential oscillations, EPSPs, IPSPs) and ON-evoked activity is highly synchronous among sister ETCs (Figure 8.4; Hayar et al., 2004b, 2005). Synaptic input can synchronize ETCs, but synchrony persists when glutamate and GABA receptors are blocked; synchronous activity in this condition is mediated by electrical synapses between these cells (Hayar et al., 2004b, 2005). ET and PG/SACs of the same glomerulus exhibit synchronous activity which depends on glutamate synapses from ET to PG/SACs (Figure 8.4; Hayar et al., 2004b). LLDs synchronize sister MCs on a slow (~1 Hz) timescale (Carlson et al., 2000). LLDs seem to be driven by ETCs and consistent with this sister ET and MCs exhibit synchronous activity (Figure 8.5; De Saint Jan et al., 2009). Slow synchronous oscillations (2 Hz) among MCs can be elicited by ON stimulation or NMDA application (Schoppa and Westbrook, 2001, 2002; Christie and Westbrook, 2006). Such oscillations involve AMPA and mGlu receptor mechanisms and are absent in mice lacking electrical synapses. Induced spiking can also elicit faster timescale synchrony among sister MCs that depends on electrical synapses (Schoppa and Westbrook, 2002; Christie et al., 2005; Christie and Westbrook, 2006). These and other electrophysiological studies show that bi-directional electrical synapses exist between ET/M/TCs and also play a role in fast and slow timescale synchrony among sister TCs and among TCs and MCs (Figure 8.5; Pimentel and Margrie, 2008; De Saint Jan et al., 2009; Ma and Lowe, 2010; Gire et al., 2012). GABAergic inhibition is also thought to play an important role in synchronizing M/TC activity, especially over long distances. Such inhibition would coordinate the activity of neurons associated with different glomeruli (Desmaisons et al., 1999; Linster and Cleland, 2001; Lagier et al., 2004; Schoppa, 2006; Arevian et al., 2008).

8.2.9 Neuromodulatory Inputs

Context- and behavioral state dependent regulation of neural networks involved in sensory processing are mediated by neuromodulatory transmitter systems, such as the monoamines and acetylcholine (ACh). Context and behavioral state modify network oscillations, neuronal excitability and odor responses in the olfactory system (Kay and Laurent, 1999; Murakami et al., 2005; Tsuno et al., 2008; Linster et al., 2011). MOB receives inputs from cholinergic, noradrenergic, and serotonergic cell groups in the basal forebrain and brainstem. These systems terminate diffusely in multiple MOB layers and modulate multiple neuronal subtypes. The older literature pertaining to these neuromodulatory inputs is reviewed elsewhere (Ennis et al., 2007; Ennis and Hayar, 2008).

8.2.9.1 Cholinergic Input.
Cholinergic projections to the MOB and AOB originate mainly in the nucleus of the horizontal limb of the diagonal band (NDB) with a small contribution from the vertical limb (Shipley and Adamek, 1984). About 20% of the NDB neurons that project to the bulb are cholinergic but many NDB-MOB projection neurons are GABAergic (Zaborszky et al., 1986). Cholinergic fibers are present in most layers of the MOB, except the ONL, and are especially dense in the GL and IPL. Indeed, muscarinic and nicotinic receptors are found in most layers of MOB (Ennis et al., 2007). Nicotinic receptor agonists were reported to excite MCs and a bipolar PGC subpopulation (Castillo et al., 1999), while ACh or muscarinic agonists inhibited GL DA cells but had no effect on other PGCs (Pignatelli and Belluzzi, 2008). NBD stimulation was reported to depress (Nickell and Shipley, 1988) or to increase (Kunze et al., 1991, 1992) M/TC activity indirectly via primary actions on GCs. Consistent with the later studies, MOB ACh infusion reduced paired-pulse depression of LOT-evoked GCL field potentials via muscarinic receptor-mediated inhibition of GC GABA release (Elaagouby et al., 1991). Comparable findings indicated that ACh-muscarinic receptor activation reduces GC to M/TC inhibition in a behavioral state dependent manner (Tsuno et al., 2008). Thus, GC to M/TC inhibition is low with higher levels of ACh release during waking or fast wave sleep, and higher during slow wave sleep when ACh release declines. This ACh suppression of GC to M/TC inhibition is at odds with conclusions that muscarinic receptor activation presynaptically enhances GC GABA release (Castillo et al., 1999; Ghatpande et al., 2006), and separately, increases GC excitability in the MOB and AOB by enhancing afterdepolarizations (Pressler et al., 2007; Smith and Araneda, 2010). The later action decreased GC spike adaptation across repetitive sniff-like depolarizations or M/TC synaptic input, and in turn, prevented the reduction of MC GABAergic inhibition during repetitive stimulation. Muscarinic receptor excitation of GCs and enhanced lateral inhibition would be expected to sharpen or reduce overlap of odor responses among M/TCs. Consistent with this, NDB lesions or systemic or MOB injection of muscarinic receptor antagonists impairs odor habituation and discrimination (Linster and Cleland, 2002; Wilson et al., 2004). Other studies show that increasing ACh in the MOB sharpened M/TC odorant receptive fields and enhanced odor discrimination; both effects involved nicotinic and muscarinic receptor components (Mandairon et al., 2006; Chaudhury et al., 2009).

8.2.9.2 Norepinephrine (NE) Input.
NE input to the bulb arises from the pontine nucleus locus coeruleus (LC) and terminates in nearly all layers but is especially dense in the IPL and GCL (Shipley et al., 1985; McLean et al., 1989). NE receptors occur in multiple layers of the MOB and are expressed by multiple cell types, congruent with the pattern of NE fiber innervation (Ennis et al., 2007). LC neuronal activity varies in a state dependent manner and the activity of these cells increases in conditions when sniff frequency increases; for example, novel odors or environments (see Linster et al., 2011 for review). Odors excite LC neurons and trigger rapid increases in MOB NE levels (Linster et al., 2011). Over the years, a number of often discrepant effects have been reported in the MOB that appear to be due to species or age, mode or site of NE application, or NE concentration (Linster et al., 2011). In particular, NE can elicit concentration dependent effects since NE receptor subtypes have different affinities and some MOB cells express multiple subtypes. LC-evoked NE release in vivo or NE activation of α1 receptors in slices enhances MC spiking in response to weak or perithreshold intensity ON stimulation, suggesting that NE may function to enhance detection of weak odors (Jiang et al., 1996; Ciombor et al., 1999; Hayar et al. 2001). In slices, NE evokes a moderate direct depolarization of MCs via α1 receptors due to closure of K^+ channels; this effect, together with increased input resistance, appears to mediate enhanced responses to weak ON stimuli. GC to M/TC inhibition is regulated in a non-linear or bidirectional manner by NE concentration. Low NE levels engaging α2 receptors suppress GC excitability and decrease MC inhibition (i.e., disinhibition), while higher levels acting at α1 receptors excite GCs and inhibit MCs (Nai et al., 2009; 2010); similar α1 effects occur in the AOB (Araneda and Firestein 2006; Smith et al., 2009). These studies indicate that the differential affinities of NE receptor subtypes allow for differential modulation of GABA release and olfactory processing as a function of the level of NE release, which in turn, is regulated by behavioral state (Linster et al., 2011). In preweaning rodents, the disinhibitory α2 actions are predominant (Pandipati et al., 2010).

NE modulation of odor responses has been less well studied. Repetitive LC stimulation-odor pairings induced prolonged (4 hr) suppression of M/TC responses to the paired odor that was prevented by blockade of α and β receptors (Shea et al., 2008). Behaviorally, such pairing decreased investigation of the paired odor, suggesting that NE release facilitates a habituation-like memory (i.e., decreased interest), consistent with other odor habituation memory results (Veyrac et al., 2007; Guerin et al., 2008). NE release appears to be important in excitatory-rewarded odor and inhibitory-non rewarded odor responses in pairs of synchronous M/TCs (Doucette et al., 2011). In spontaneous habituation-dishabituation tests, discrimination between chemically related odorants was impaired when MOB NE receptors, and in particular α1 receptors, were blocked (Mandairon et al., 2008). A similar testing paradigm showed that MOB NE infusion improved odor detection and discrimination in a non-linear manner that varied as a function of NE and odor concentration (Escanilla et al., 2010). Computational modeling studies suggested that NE excitation of MCs and GCs via the α1 receptor improved odor signal-to-noise ratios, gamma oscillatory dynamics, and M/TC spike synchronization to improve odor detection and discrimination (Escanilla et al., 2010). By contrast, odorant discrimination in reward-motivated tasks was impaired only when all NE receptors in MOB were blocked (Doucette et al., 2007; Mandairon et al., 2008). The well documented role of LC projections to the MOB and AOB in neonatal conditioned odor preference learning, chemosensory regulation of pregnancy, and in post-partum maternal behavior has been reviewed elsewhere (Ennis and Hayar, 2008).

8.2.9.3 Serotonergic (5-HT) Input to MOB.

Serotonergic inputs to the MOB and AOB arise from midbrain dorsal and median raphe neurons (McLean and Shipley, 1987). Serotonergic fibers are present in most layers of MOB, but are especially dense in the GL where they target GABAergic PGC and perhaps ET/M/TC dendrites (Gracia-Llanes et al., 2010). 5-HT receptors are localized in most layers of the MOB (Ennis et al., 2007). 5-HT depolarized 34% of JGCs in vitro via of $5HT_{2C}$ receptor activation of a nonselective cation current (Hardy et al., 2005); the heterogeneous properties of 5-HT-responsive cells suggested that several JGC types are targeted by 5-HT fibers. More recent work showed that 5-HT depolarized and increased the bursting frequency of ETCs via engaging a transient receptor potential channel-mediated cation current linked to $5HT_{2A}$ receptor activation (Liu et al., 2012). Some MCs are depolarized via $5\text{-}HT_{2A}$ receptors, while others are hyperpolarized by an indirect action mediated by GCs (Hardy et al., 2005); the inhibition was mediated by $5\text{-}HT_{2A}$ receptors. Subsequently, stimulation of dorsal raphe 5-HT neurons or topical $5\text{-}HT_{2A}$ agonists reduced odor-evoked input to the glomeruli as visualized by imaging activity in ORN terminals (Petzold et al., 2009). This was due to $GABA_B$ receptor presynaptic inhibition of ON terminals subsequent to 5-HT excitation of GABAergic PGCs via the $5\text{-}HT_{2C}$ receptor. Behavioral studies indicate that lesion of serotonergic fibers reversed or impaired conditioned olfactory learning, while 5-HT release or $5\text{-}HT_2$ receptor agonists promoted odor conditioning (Morizumi et al., 1994; McLean et al., 1996; Yuan et al., 2003; McLean and Harley, 2004).

8.3 ACCESSORY OLFACTORY BULB (AOB)

The AOB is a distinct region of the olfactory bulb. In mice, it is located at the posterior face, separated from forebrain by the rhinal fissure. Unlike the MOB, the AOB does not receive input from the nasal epithelium: its sensory information comes from the vomeronasal organ (VNO). The VNO detects non-volatile (liquid-phase) stimuli which are delivered by a pumping mechanism (Meredith and O'Connell, 1979), so that operationally the accessory olfactory system is reminiscent of the taste system.

8.3.1 Glomerular Layer (GL)

8.3.1.1 Major Divisions of the GL.

Like the MOB, the AOB is a laminar structure, and sensory input from the nose arrives in the glomerular layer. Anatomically, one of the most striking features of the rodent AOB is its regional organization, which is closely linked to the molecular organization of the VNO. Vomeronasal sensory neurons (VSNs) express G protein-coupled receptors from different gene families: the *V1R*s (Dulac and Axel, 1995), the *V2R*s (Herrada and Dulac, 1997; Matsunami and Buck, 1997; Ryba and Tirindelli, 1997), and the *FPR*s (Liberles et al., 2009; Rivière et al., 2009; Roberts and Matsunami, personal communication). The *V2R*-expressing cells can be subdivided further by whether they express members of a family of non-classical class I major histocompatibility genes denoted *H2-Mv* (Loconto et al., 2003; Ishii et al., 2003; Ishii and Mombaerts, 2008).

The anterior portion of the AOB glomerular layer receives input from VSNs expressing members of the *V1R* family (Halpern et al., 1995; Jia and Halpern, 1996). The posterior AOB receives inputs from *V2R*-expressing neurons, with the *H2-Mv*-positive VSNs projecting to the most posterior AOB, and the *H2-Mv*-negative VSNs projecting to a "middle" (anterior part of the posterior) region (Ishii and Mombaerts, 2008). The division between anterior and posterior, called the *linea alba* (Larriva-Sahd, 2008), is often readily visible in sections and three-dimensional reconstructions. It is worth noting that many mammals, including ruminants and carnivores, do not express V2R receptors. Consequently, their AOBs consist of just the anterior compartment (Takigami et al., 2000; Tirindelli et al., 2009).

Anatomically, AOB glomeruli are much less distinct than their counterparts in the MOB (Ramón y Cajal, 1901; Yokosuka, 2012), largely because they are not as clearly delineated by JGCs (Section 8.10.3). Glomeruli are quite variable in size, many as small as 10 μm with a few reaching approximately 100 μm in diameter. While MOB glomeruli form a (mostly) two-dimensional sheet, in the AOB the glomeruli are stacked one atop the other

(Ramón y Cajal, 1901). This physical arrangement necessitates the use of physical or optical sectioning to visualize the three-dimensional structure of the glomerular layer.

8.3.1.2 *Spatial Organization of Glomeruli.*

As in the MOB, the VSNs expressing the same receptor type send their axons to common targets. However, while the typical main olfactory receptor type targets two glomeruli per bulb (the usual range being 1–4, Zou et al., 2004), in the AOB they typically target more. While only a few receptor types have been examined, the anterior (V1R) receptor types have targeted an average of 4–30 glomeruli, depending on the receptor (Belluscio et al., 1999; Rodriguez et al., 1999; Wagner et al., 2006), and posterior types have targeted three to ten glomeruli (Del Punta et al., 2002; Haga et al., 2010). These studies also revealed that the "coarse" features – but often not the details – of the projection pattern were reproducible across animals.

Physiological mapping via calcium imaging has shown that neurons responding to sulfated steroids target both the anterior (*V1R*) AOB and the "middle" region innervated by the *H2-Mv*-negative *V2R*-expressing neurons (Hammen et al., 2014). The latter projections appear to be largely restricted to cells responding to a particular subfamily of sulfated steroids, the pregnanolones, known as metabolites of progesterone and as potent neurosteroids (Shu et al., 2004). Many individual glomeruli have seemingly-identical stimulus responses, in accord with the finding that single receptor types target multiple glomeruli. A battery of 11 sulfated steroids allowed at least 10 distinct patterns to be identified in individual animals, each pattern encompassing an average of 12 glomerular regions of interest. Collectively, these steroid-responsive glomeruli account for approximately 25% of the volume of the anterior AOB, and thus sample more comprehensively than molecular labeling.

There are indications that individual glomeruli in the AOB may receive input from more than one receptor type (Belluscio et al., 1999); however, not all glomeruli show evidence for this phenomenon (Del Punta et al., 2002). When tested physiologically using sulfated steroids, virtually all glomeruli showed similar patterns of responsiveness as observed in the VNO (Hammen et al., 2014), suggesting a high degree of functional "purity."

While the sensory map varies significantly from animal to animal, reproducible features of the projections have been identified by examining multiple receptor types simultaneously. Using mice expressing three different reporter genes in cells expressing three different receptor types, it was shown that two receptor types within the same V1R subfamily targeted neighboring glomeruli, even when the precise position of these targets varied among animals (Wagner et al., 2006). Physiologically, the most striking form of chemotopic organization is "heterophylic": reproducible clumps of glomeruli that respond to very disparate stimuli, or glomeruli responding to similar stimuli that tend to be systematically far apart. It seems possible that these two apparently-opposing forms of organization may coexist at different levels, one operating within receptor subfamilies and the other operating between receptor subfamilies. Individual ligands can activate multiple VSN types, and individual VSNs can respond to multiple ligands (Meeks et al., 2010; Turaga and Holy, 2012). Consequently, just as with the main olfactory system, sensory coding in the accessory system is "combinatorial" (Revial et al., 1982; Friedrich and Korsching, 1997). It is possible that the physical arrangement of glomeruli plays a role in processing these combinatorial responses, through the actions of local interneurons.

8.3.1.3 *GL Neurons.*

By comparison with the MOB, the neurons of AOB glomerular layer are more sparse (Larriva-Sahd, 2008; Yokosuka, 2012). The most abundant neurons are GAD67-positive, and their density seems approximately commensurate with the MOB. In contrast, there are far fewer TH-positive neurons, and their cell bodies tend to reside at the edge of the external cellular layer (see below). There are some indications that the morphology of AOB JGCs is quite diverse (Yokosuka, 2012). Superficial SACs have also been identified, with somewhat larger somata than is typical in the MOB (Larriva-Sahd, 2008).

TCs are not universally recognized in the AOB, but studies have identified a class of neurons, sometimes called "superficial MCs," that may correspond to the ETCs of the MOB (Larriva-Sahd, 2008; Yonekura and Yokoi, 2008). These neurons are protocadherin-21 positive, reside near the boundary of the glomerular and external cellular layers, possess a single dendritic tuft, and do not project outside the bulb. They are much less numerous than their MOB counterparts: by Golgi and protocadherin-21 staining, these cells are seen only 10–15% as frequently as the AOB MCs (discussed below). At present, we are not aware of any studies of the physiology and/or function of AOB glomerular layer neurons. Some AOB MCs are powerfully inhibited by particular stimuli (Hendrickson et al., 2008), but the cellular origin of this inhibition has not been studied in detail.

8.3.2 External Cellular Layer (ECL)

While the AOB is often divided into the same layers as the MOB, there is no dramatic separation between "EPL" and "MCL." For this reason, here we adopt a suggestion (Larriva-Sahd, 2008) to use the term "external cellular layer" (ECL) for the combined region.

8.3.2.1 Mitral Cells (MCs). The most numerous class of cells in the ECL are the so-called MCs, the main projection neurons of the AOB. These differ in several important respects from their counterparts in the MOB. First, their appearance is quite different, and not at all mitral, leading some authors to prefer the term large principal cells (Larriva-Sahd, 2008). Second, they are not organized in a thin layer as in the MOB, but are found throughout the ECL. Third, AOB MCs possess multiple dendrites that innervate glomeruli in tufts, with a reported range of 1–9 and a median of three tufts per cell (Takami and Graziadei, 1991; Larriva-Sahd, 2008; Yonekura and Yokoi, 2008). All of the innervated glomeruli reside on one side of the *linea alba*, indicating that these neurons receive primary sensory input from only one of the two major families of *VR*s. While the somata of the MCs are typically found on the same side of the *linea alba* as their glomerular tufts (Jia et al., 1997; Larriva-Sahd, 2008), this is not guaranteed (Yonekura and Yokoi, 2008). Moreover, their secondary dendrites are frequently observed to span the gap between anterior and posterior bulb (Larriva-Sahd, 2008).

Because AOB MCs have multiple dendritic tufts, in principle they may be able to integrate, in a direct excitatory sense, the outputs of multiple receptor types (Takami and Graziadei, 1991). Current evidence suggests some heterogeneity in the degree to which this actually happens in practice. A study of two genetically-labeled receptor types, V1rb2 and V2r1b (Del Punta et al., 2002), showed that the MCs receiving input from these receptor types formed tufts exclusively in glomeruli innervated by these receptor types – consequently, these neurons did not integrate directly from multiple receptor types. However, another study of three different receptor types, V1ra1, V1ra3, and V1rb1, did identify MCs receiving input from both labeled and unlabeled glomeruli (Wagner et al., 2006). Determining the "typical" rule by genetic labeling appears to be a difficult challenge, since it requires a specific mouse for each receptor.

More broadly, several lines of evidence suggest that direct integration happens, but perhaps in only a minority of MCs. In the anterior AOB, approximately two-thirds of the MCs have tufts that span a relatively narrow range of the depth of the glomerular layer (Yonekura and Yokoi, 2008). As the different subfamilies of V1Rs form glomeruli in layers in the anterior AOB (Wagner et al., 2006), this is consistent with the notion that these neurons sample from a restricted set of inputs. Moreover, a physiological study of responses of AOB neurons to dilute urine and sulfated steroids showed that approximately three-quarters had excitatory responses consistent with the profile of known single VSNs (Meeks et al., 2010). However, both studies also suggest that integration may happen in a subset of cells: approximately one third of MCs had tufts at multiple depths within the anterior AOB (Yonekura and Yokoi, 2008), and approximately one quarter of MCs had physiological responses which appeared to directly integrate multiple sensory streams (Meeks et al., 2010).

When stimulated with either complex mixtures or pure compounds, AOB MCs show both excitatory and inhibitory responses (Luo et al., 2003; Hendrickson et al., 2008; Meeks et al., 2010; Ben-Shaul et al., 2010). Even in some neurons lacking obvious inhibitory responses, the prevalence of such input was indicated by the observation of strong mixture suppression (Hendrickson et al., 2008). Consequently, inhibitory processing may substantially alter MCs' output.

8.3.2.2 Other Neurons of the ECL. Several other cell types in the ECL have been identified (Larriva-Sahd, 2008). The "round projecting cell" lacks a glomerular tuft, and its dendrites are confined to the ECL, but (as suggested by its name) it projects outside of the AOB. Some TCs are displaced from the glomerular/ECL boundary, and hence may be a distinct type. External GCs are relatively abundant spiny neurons that tend to reside deeper within the ECL. The dwarf cells and polygonal neurons are two additional, rarer, classes of interneurons within this layer.

8.3.3 Internal Cellular Layer (ICL)

The internal cellular layer (ICL) is deep in the lateral olfactory tract. The most abundant neurons of the internal cellular are the internal GCs, spiny neurons whose morphology is similar to their counterparts of the MOB. More rarely, one finds cells which appear to be the analog of Cajal cells (Larriva-Sahd, 2008). These cells have elaborate dendrites, some of which are entangled with the lateral olfactory tract, and it has been speculated that they might integrate information between the main and accessory olfactory bulbs (Larriva-Sahd, 2008). Finally, the "interstitial neurons of the bulbi" lie at the junction between main and accessory olfactory bulbs, and consist of recognizably-distinct subtypes.

8.4 EXTRINSIC INPUTS

Several distinct sources of afferents are known, of which the most extensively investigated is the noradrenergic input from locus coeruleus (Broadwell and Jacobowitz, 1976). This neuromodulatory input has attracted attention as a trigger for the Bruce effect, a form of olfactory learning that occurs when a recently-mated female mouse is exposed to the scent of a male (Brennan and Keverne, 1997; Brennan, 2009).

Briefly (see Brennan (2009) for an excellent in-depth review), a wide range of behavioral and pharmacological

174   Chapter 8   Anatomy and Neurobiology of the Main and Accessory Olfactory Bulbs

evidence suggests that mating increases NE levels in the AOB, and this has been hypothesized to disinhibit AOB MCs and thus lead to potentiation of their outputs onto GCs. While this model received support from in vivo neuronal recordings employing vagino-cervical stimulation presumably triggering NE release in the AOB (Otsuka et al., 2001), *in vitro* studies indicated that NE decreases output of MCs (Kaba and Huang, 2005) and increases GC inhibition (Araneda and Firestein, 2006; Smith et al., 2009). The differences among these findings, and delicate balance between excitation and inhibition mediated by NE in the MOB (see Section 8.9.2), suggest the need for additional study.

AOB neuronal excitability is also modulated by ACh, using both nicotinic and muscarinic receptors (Smith and Araneda, 2010). ACh increased the excitability of both MCs and GCs, through very different mechanisms. A direct behavioral role for cholinergic modulation of the AOB does not yet seem to have been documented, but afferent projections from the NDB have recently been identified (Mohedano-Moriano et al., 2012).

In addition, the AOB receives projections from regions that are not known to be directly neuromodulatory. Two of the AOB's principal output targets, the bed nucleus of the stria terminalis (BST) and the vomeronasal amygdala, provide feedback projections (Raisman, 1972; Broadwell and Jacobowitz, 1976; Davis et al., 1978; Kevetter and Winans, 1981) with laminar specificity (Barber, 1982; Fan and Luo, 2009). The ECL is targeted exclusively by the BST, and these projections are GABAergic (Fan and Luo, 2009). Conversely, the ICL receives efferent input from the medial amygdala posteroventral nucleus and the posterior medial cortical amygdala. This projection is probably glutamatergic, and likely involves considerably more neurons than the one from the BST (Fan and Luo, 2009). Interestingly, more than half of the neurons providing this feedback, in both the BST and the vomeronasal amygdala, express estrogen receptor α, and hence are candidates for incorporating aspects of the animal's hormonal status (Fan and Luo, 2009). Another major projection from the bed nucleus of the accessory olfactory tract, another target of the AOB, has also been identified (Mohedano-Moriano et al., 2012).

Two studies (Cenquizca and Swanson, 2007; de la Rosa-Prieto et al., 2009) have identified feedback projections to the AOB ICL from the CA1 region of the hippocampus and ventral subiculum. These regions also project to the MOB, but the projection to the AOB is more substantial. Consequently, all areas that receive projections from the AOB, and even some that do not, provide feedback to the AOB.

Clues as to the circuit role of the feedback connections from the medial amygdala to the AOB were found by combining retrograde tracing with c-*fos* labeling after exposure

to odors (Martel and Baum, 2009). A significant increase in *fos*-positive retrogradely-labeled neurons was seen in those mice (all female) which were exposed to male mouse urine volatiles; using other odors produced little or no increase. This result may help explain the increase in c-*fos* seen in the AOB when the mouse is exposed only to volatiles (Martel and Baum, 2007), which presumably do not gain access to the VNO. Since the medial amygdala receives projections from both major olfactory systems (Kang et al., 2009), this feedback connection may incorporate signals from the main olfactory system to modulate the function of the accessory olfactory system.

REFERENCES

Abraham, N. M., Egger, V., Shimshek, D. R., et al. (2010). Synaptic inhibition in the olfactory bulb accelerates odor discrimination in mice. *Neuron.* 65: 399–411.

Adrian, E. D. (1950). The electrical activity of the olfactory bulb. *Electroenceph. Clin. Neurophysiol.* 2: 377–388.

Antal, M., Eyre, M., Finklea, B., and Nusser, Z. (2006). External tufted cells in the main olfactory bulb form two distinct subpopulations. *Eur. J. Neurosci.* 24: 1124–1136.

Araneda, R. C. and Firestein, S. (2006). Adrenergic enhancement of inhibitory transmission in the accessory olfactory bulb. *J. Neurosci.* 26: 3292–3298.

Arevian, A. C., Kapoor, V., and Urban, N. N. (2008). Activity-dependent gating of lateral inhibition in the mouse olfactory bulb. *Nat. Neurosci.* 11: 80–87.

Aroniadou-Anderjaska, V., Ennis, M., and Shipley, M. T. (1997). Glomerular synaptic responses to olfactory nerve input in rat olfactory bulb slices. *Neurosci.* 79: 425–434.

Aroniadou-Anderjaska, V., Ennis, M., and Shipley, M. T. (1999a). Current-source density analysis in the rat olfactory bulb: laminar distribution of kainate/AMPA and NMDA receptor-mediated currents. *J. Neurophysiol.* 81: 15–28.

Aroniadou-Anderjaska, V., Ennis, M., and Shipley, M. T. (1999b). Dendrodendritic recurrent excitation in mitral cells of the rat olfactory bulb. *J. Neurophysiol.* 82: 489–494.

Aroniadou-Anderjaska, V., Zhou, F. M., Priest, C. A., et al. (2000). Tonic and synaptically evoked presynaptic inhibition of sensory input to the rat olfactory bulb via GABA(B) heteroreceptors. *J. Neurophysiol.* 84: 1194–1203.

Aungst, J. L., Heyward, P. M., Puche, A. C., et al. (2003). Center-surround inhibition among olfactory bulb glomeruli. *Nature.* 426: 623–629.

Balu, R., Larimer, P., and Strowbridge, B. W. (2004). Phasic stimuli evoke precisely timed spikes in intermittently discharging mitral cells. *J. Neurophysiol.* 92: 743–753.

Balu, R., Pressler, R. T., and Strowbridge, B. W. (2007). Multiple modes of synaptic excitation of olfactory bulb granule cells. *J. Neurosci.* 27: 5621–5632.

Barber, P. C. (1982). Adjacent laminar terminations of two centrifugal afferent pathways to the accessory olfactory bulb in the mouse. *Brain Res.* 245: 215–221.

Bathellier, B., Buhl, D. L., Accolla, R. and Carleton, A. (2008). Dynamic ensemble odor coding in the mammalian olfactory bulb: sensory information at different timescales. *Neuron.* 57: 586–598.

References

Belluscio, L., Koentges, G., Axel, R., and Dulac, C. (1999). A map of pheromone receptor activation in the mammalian brain. *Cell* 97: 209–220.

Belluscio, L., Lodovichi, C., Feinstein, P., et al. (2002). Odorant receptors instruct functional circuitry in the mouse olfactory bulb. *Nature.* 419: 296–300.

Ben-Shaul, Y., Katz, L. C., Mooney, R., and Dulac, C. (2010). In vivo vomeronasal stimulation reveals sensory encoding of conspecific and allospecific cues by the mouse accessory olfactory bulb. *Proc. Natl. Acad. Sci. U.S.A.* 107: 5172–5177.

Berkowicz, D. A. and Trombley, P. Q. (2000). Dopaminergic modulation at the olfactory nerve synapse. *Brain Res.* 855: 90–99.

Beshel, J., Kopell, N., and Kay, L. M. (2007). Olfactory bulb gamma oscillations are enhanced with task demands. *J. Neurosci.* 27: 8358–8365.

Bhalla, U. S. and Bower, J. M. (1997). Multiday recordings from olfactory bulb neurons in awake freely moving rats: spatially and temporally organized variability in odorant response properties. *J. Comput. Neurosci.* 4: 221–256.

Bogan, N., Brecha, N., Gall, C. and Karten, H. J. (1982). Distribution of enkephalin-like immunoreactivity in the rat main olfactory bulb. *Neurosci.* 7: 895–906.

Brennan P. A. (2009). Outstanding issues surrounding vomeronasal mechanisms of pregnancy block and individual recognition in mice. *Behav. Brain Res.* 200: 287–294.

Brennan P. A. and Keverne E. B. (1997). Neural mechanisms of mammalian olfactory learning. *Prog. Neurobiol.* 51: 457–481.

Broadwell R. D. and Jacobowitz D. M. (1976). Olfactory relationships of the telencephalon and diencephalon in the rabbit. *III. The ipsilateral centrifugal fibers to the olfactory bulbar and retrobulbar formations. J. Comp. Neurol.* 170: 321–345.

Buonviso, N., Amat, C., and Litaudon, P. (2006). Respiratory modulation of olfactory neurons in the rodent brain. *Chem. Senses* 31: 145–154.

Cang, J. and Isaacson, J. S. (2003). In vivo whole-cell recording of odor-evoked synaptic transmission in the rat olfactory bulb. *J. Neurosci.* 23: 4108–4116.

Carey, R. M., Verhagen, J. V., Wesson, D. W., et al. (2009). Temporal structure of receptor neuron input to the olfactory bulb imaged in behaving rats. *J. Neurophysiol.* 101: 1073–1088.

Carey, R. M. and Wachowiak, M. (2011). Effect of sniffing on the temporal structure of mitral/tufted cell output from the olfactory bulb. *J. Neurosci.* 31: 10615–10626.

Carleton, A., Petreanu, L. T., Lansford, R., et al. (2003). Becoming a new neuron in the adult olfactory bulb. *Nat. Neurosci.* 6: 607–618.

Carlson, G. C., Shipley, M. T., and Keller, A. (2000). Long-lasting depolarizations in mitral cells of the rat olfactory bulb. *J. Neurosci.* 20: 2011–2021.

Castro, J. B., Hovis, K. R., and Urban, N. N. (2007). Recurrent dendrodendritic inhibition of accessory olfactory bulb mitral cells requires activation of group I metabotropic glutamate receptors. *J. Neurosci.* 27: 5664–5671.

Castillo, P. E., Carleton, A., Vincent, J. D., and Lledo, P. M. (1999). Multiple and opposing roles of cholinergic transmission in the main olfactory bulb. *J. Neurosci.* 19: 9180–9191.

Cenquizca L. A., Swanson L. W. (2007). Spatial organization of direct hippocampal field CA1 axonal projections to the rest of the cerebral cortex. *Brain Res. Rev.* 56: 1–26.

Chalansonnet, M. and Chaput, M. A. (1998). Olfactory bulb output cell temporal response patterns to increasing odor concentrations in freely breathing rats. *Chem. Senses* 23: 1–9.

Chaput, M. (1983). Effects of olfactory peduncle sectioning on the single unit responses of olfactory bulb neurons to odor presentation in awake rabbits. *Chem. Senses* 8: 161–177.

Chaput, M. and Holley, A. (1979). Spontaneous activity of olfactory bulb neurons in awake rabbits, with some observations on the effects of pentobarbital anaesthesia. *J. Physiol. (Paris).* 75: 939–948.

Charpak, S., Mertz, J., Beaurepaire, E., et al. (2001). Odor-evoked calcium signals in dendrites of rat mitral cells. *Proc Natl Acad Sci.* 98: 1230–1234.

Chaudhury, D., Escanilla, O., and Linster, C. (2009). Bulbar acetylcholine enhances neural and perceptual odor discrimination. *J. Neurosci.* 29: 52–60.

Chen, W. R., Midtgaard, J., and Shepherd, G. M. (1997). Forward and backward propagation of dendritic impulses and their synaptic control in mitral cells. *Science* 278: 463–467.

Chen, W. R. and Shepherd, G. M. (1997). Membrane and synaptic properties of mitral cells in slices of rat olfactory bulb. *Brain Res.* 745: 189–196.

Chen, W. R., Xiong, W., and Shepherd, G. M. (2000). Analysis of relations between NMDA receptors and GABA release at olfactory bulb reciprocal synapses. *Neuron* 25: 625–633.

Christie, J. M. and Westbrook, G. L. (2003). Regulation of backpropagating action potential in mitral cell lateral dendrites by A-type potassium currents. *J. Neurophysiol.* 89: 2466–2472.

Christie, J. M. and Westbrook, G. L. (2006). Lateral excitation within the olfactory bulb. *J. Neurosci.* 26: 2269–2277.

Christie, J. M., Bark, C., Hormuzdi, S. G., et al. (2005). Connexin36 mediates spike synchrony in olfactory bulb glomeruli. *Neuron* 46: 761–772.

Christie, J. M., Schoppa, N. E., and Westbrook, G. L. (2001). Tufted cell dendrodendritic inhibition in the olfactory bulb is dependent on NMDA receptor activity. *J. Neurophysiol.* 85: 169–173.

Ciombor, K. J., Ennis, M., and Shipley, M. T. (1999). Norepinephrine increases rat mitral cell excitatory responses to weak olfactory nerve input via alpha-1 receptors in vitro. *Neuroscience* 90: 595–606.

Cleland, T. A. and Sethupathy, P. (2006). Non-topographical contrast enhancement in the olfactory bulb. *BMC Neurosci.* 7: 7.

Courtiol, E., Hegoburu, C., Litaudon, P., et al. (2011). Individual and synergistic effects of sniffing frequency and flow rate on olfactory bulb activity. *J. Neurophysiol.* 106: 2813–2824.

Cummings, D. M. and Belluscio L. (2010). Continuous neural plasticity in the olfactory intrabulbar circuitry. *J. Neurosci.* 30: 9172–9180.

Cury, K. M. and Uchida, N. (2010). Robust odor coding via inhalation-coupled transient activity in the mammalian olfactory bulb. *Neuron* 68: 570–585.

David, F. O., Hugues, E., Cenier, T., et al. (2009). Specific entrainment of mitral cells during gamma oscillation in the rat olfactory bulb. *PLoS Comput. Biol.* Oct; 5(10): e1000551.

Davis, B. J., Burd, G. D., and Macrides, F. (1982). Localization of methionine-enkephalin, substance P, and somatostatin immunoreactivities in the main olfactory bulb of the hamster. *J. Comp. Neurol.* 204: 377–383.

Davis, B. J., Macrides, F., Youngs, W. M., et al. (1978). Efferents and centrifugal afferents of the main and accessory olfactory bulbs in the hamster. *Brain Res. Bull.* 3: 59–72.

Davison, I. G. and Katz, L. C. (2007). Sparse and selective odor coding by mitral/tufted neurons in the main olfactory bulb. *J. Neurosci.* 27: 2091–2101.

Debarbieux, F., Audinat, E., and Charpak S. (2003). Action potential propagation in dendrites of rat mitral cells in vivo. *J. Neurosci.* 23: 5553–5560.

de la Rosa-Prieto, C., Ubeda-Banon, I., Mohedano-Moriano, A., et al. (2009). Subicular and CA1 hippocampal projections to the accessory olfactory bulb. *Hippocampus* 19: 124–129.

Del Punta, K., Puche, A., Adams, N., et al. (2002). A divergent pattern of sensory axonal projections is rendered convergent by second-order neurons in the accessory olfactory bulb. *Neuron* 35: 1057–1066.

Desmaisons, D., Vincent, J-D., and Lledo, P-M. (1999). Control of action potential timing by intrinsic subthreshold oscillations in olfactory bulb neurons. *J. Neurosci.* 19: 10727–10737.

De Saint Jan, D., Hirnet, D., Westbrook, G. L. and Charpak, S. (2009). External tufted cells drive the output of olfactory bulb glomeruli. *J. Neurosci.* 18: 2043–2052.

De Saint Jan, D. and Westbrook, G. L. (2005). Detecting activity in olfactory bulb glomeruli with astrocyte recording. *J. Neurosci.* 25: 2917–2924.

Dhawale, A. K., Hagiwara, A., Bhalla, U. S., et al. (2010). Non-redundant odor coding by sister mitral cells revealed by light addressable glomeruli in the mouse. *Nat. Neurosci.* 13: 1404–1412.

Dietz, S. B. and Murthy, V. N. (2005). Contrasting short-term plasticity at two sides of the mitral-granule reciprocal synapse in the mammalian olfactory bulb. *J. Physiol.* 569: 475–488.

Djurisic, M., Antic, S., Chen, W. R., and Zecevic, D. (2004). Voltage imaging from dendrites of mitral cells: EPSP attenuation and spike trigger zones. *J. Neurosci.* 24: 6703–6714.

Dong, H-W., Hayar, A., Callaway, J., et al. (2009). I mGluR activation enhances Ca^{2+}-dependent nonselective cation currents and rhythmic bursting in main olfactory bulb external tufted cells. *J. Neurosci.* 29: 11943–11953.

Dong, H-W., Hayar, A. and Ennis, M. (2007). Activation of group I metabotropic glutamate receptors on main olfactory bulb granule cells and periglomerular cells enhances synaptic inhibition of mitral cells. *J. Neurosci.* 27: 5654–5663.

Doty, R. L. and Risser, J. M. (1989). Influence of the D-2 dopamine receptor agonist quinpirole on the odor detection performance of rats before and after spiperone administration. *Psychopharmacology* 98: 310–315.

Doucette, W., Gire, D. H., Whitsell, J., et al. (2011). Associative cortex features in the first olfactory brain relay station. *Neuron.* 69: 1176–1187.

Doucette, W., Milder, J., and Restrepo, D. (2007). Adrenergic modulation of olfactory bulb circuitry affects odor discrimination. *Learn. Mem.* 14: 539–547.

Dulac, C. and Axel, R. (1995). A novel family of genes encoding putative pheromone receptors in mammals. *Cell* 83: 195–206.

Eeckman, F. H. and Freeman, W. J. (1990). Correlations between unit firing and EEG in the rat olfactory system. *Brain Res.* 528: 238–244.

Egger, V., Svoboda, K., and Mainen, Z. F. (2003). Mechanisms of lateral inhibition in the olfactory bulb: efficiency and modulation of spike-evoked calcium influx into granule cells. *J. Neurosci.* 23: 7551–7558.

Egger, V., Svoboda, K., and Mainen, Z. F. (2005). Dendrodendritic synaptic signals in olfactory bulb granule cells: local spine boost and global low-threshold spike. *J. Neurosci.* 25: 3521–3530.

Elaagouby, A., Ravel, N., and Gervais, R. (1991). Cholinergic modulation of excitability in the rat olfactory bulb: effect of local application of cholinergic agents on evoked field potentials. *Neurosci.* 45: 653–662.

Ennis, M., Hamilton, K. A., and Hayar A. (2007). Neurochemistry of the main olfactory system. In *Handbook of Neurochemistry and Molecular Neurobiology*, 3rd Edition, Lajtha, A. (Ed), Vol. 20, *Sensory Neurochemistry*, Johnson, D. (Ed.). Heidelberg, Germany: Springer, pp. 137–204.

Ennis, M. and Hayar, A. (2008). Physiology of the Olfactory Bulb, In *The Senses: A Comprehensive Reference*, Basbaum, A., Kaneko, A., and Shepherd, G. (Eds.-In Chief) Vol. 4, *Olfaction and Taste*, Firestein, S. and Beauchamp, G. (Eds). San Diego: Academic Press, pp. 641–686.

Ennis, M., Zhou, F. M., Ciombor, K. J., et al. (2001). Dopamine D2 receptor-mediated presynaptic inhibition of olfactory nerve terminals. *J. Neurophysiol.* 86: 2986–2997.

Ennis, M., Zhu, M., Heinbockel, T., and Hayar, A. (2006). Olfactory nerve-evoked metabotropic glutamate receptor -mediated responses in olfactory bulb mitral cells. *J. Neurophysiol.* 95: 2233–2241.

Ennis, M., Zimmer, L. A., and Shipley, M. T. (1996). Olfactory nerve stimulation activates rat mitral cells via NMDA and non- NMDA receptors in vitro. *NeuroReport.* 7: 989–992.

Escanilla, O., Arrellanos, A., Karnow, A., et al. (2010). Noradrenergic modulation of behavioral odor detection and discrimination thresholds in the olfactory bulb. *Eur. J. Neurosci.* 32: 458–468.

Escanilla, O., Yuhas, C., Marzan, D., and Linster, C. (2009). Dopaminergic modulation of olfactory bulb processing affects odor discrimination learning in rats. *Behav. Neurosci.* 123: 828–833.

Eyre, M. D., Antal, M., and Nusser, Z. (2008). Distinct deep short-axon cell subtypes of the main olfactory bulb provide novel intrabulbar and extrabulbar GABAergic connections. *J. Neurosci.* 28: 8217–8229.

Ezeh, P. I., Wellis, D. P., and Scott, J. W. (1993). Organization of inhibition in the rat olfactory bulb external plexiform layer. *J. Neurophysiol.* 70: 263–274.

Fan, S. and Luo, M. (2009). The organization of feedback projections in a pathway important for processing pheromonal signals. *Neuroscience* 161: 489–500.

Fantana, A. L., Soucy, E. R., and Meister, M. (2008). Rat olfactory bulb mitral cells receive sparse glomerular inputs. *Neuron.* 59: 802–814.

Fletcher, M. L. and Wilson, D. A. (2002). Experience modifies olfactory acuity: acetylcholine-dependent learning decreases behavioral generalization between similar odorants. *J. Neurosci.* 22: RC201.

Fletcher, M. L. and Wilson, D. A. (2003). Olfactory bulb mitral-tufted cell plasticity: odorant-specific tuning reflects previous odorant exposure. *J. Neurosci.* 23: 6946–55.

Fletcher, M. L., Smith, A. M., Best, A. R., and Wilson, D. A. (2005). High-frequency oscillations are not necessary for simple olfactory discriminations in young rats. *J. Neurosci.* 25: 792–798.

Frazier, L. L. and Brunjes, P. C. (1988). Unilateral odor deprivation: early postnatal changes in olfactory bulb cell density and number. *J. Comp. Neurol.* 269: 355–370.

Fried, H. U., Fuss, S. H., and Korsching, S. I. (2002). Selective imaging of presynaptic activity in the mouse olfactory bulb shows concentration and structure dependence of odor responses in identified glomeruli. *Proc. Natl. Acad. Sci. USA.* 99: 3222–3227.

Friedman, D. and Strowbridge, B. W. (2000). Functional role of NMDA autoreceptors in olfactory mitral cells. *J. Neurophysiol.* 84: 39–50.

Friedman, D. and Strowbridge, B. W. (2003). Both electrical and chemical synapses mediate fast network oscillations in the olfactory bulb. *J. Neurophysiol.* 89: 2601–2610.

Friedrich, R. W. (2002. Real time odor representations. *Trends Neurosci.* 25: 487–489.

Friedrich, R. and Korsching, S. (1997). Combinatorial and chemotopic odorant coding in the zebrafish olfactory bulb visualized by optical imaging. *Neuron* 18: 737–752.

Fuentes, R. A., Aguilar, M. I., Aylwin, M. L., and Maldonado, P. E. (2008). Neuronal activity of mitral-tufted cells in awake rats during passive and active odorant stimulation. *J. Neurophysiol.* 100: 422–430.

Gall, C. M., Hendry, S. H. C., Seroogy, K. B., et al. (1987). Evidence for coexistence of GABA and dopamine in neurons of the rat olfactory bulb. *J. Comp. Neurol.* 266: 307–318.

Gelperin, A. (2006). Olfactory computations and network oscillation. *J. Neurosci.* 26: 1663–1668.

Gervais, R., Buonviso, N., Martin, C., and Ravel, N. (2007). What do electrophysiological studies tell us about processing at the olfactory bulb level? *J. Physiol. Paris.* 101: 40–45.

References

Ghatpande, A. S., Sivaraaman, K., and Vijayaraghavan, S. (2006). Store calcium mediates cholinergic effects on mIPSCs in the rat main olfactory bulb. *J. Neurophysiol.* 95: 1345–1355.

Gheusi, G., Cremer, H., McLean, H., et al. (2000). Importance of newly generated neurons in the adult olfactory bulb for odor discrimination. *Proc Natl Acad Sci USA* 97: 1823–1828.

Gire, D. H. and Schoppa, N. E. (2009). Control of on/off glomerular signaling by a local GABAergic microcircuit in the olfactory bulb. *J. Neurosci.* 29: 13454–13464.

Gire, D. H., Franks, K. M., Zak, J. D., et al. (2012). Mitral cells in the olfactory bulb are mainly excited through a multistep signaling path. *J. Neurosci.* 32: 2964–2975.

Gracia-Llanes, F. J., Blasco-Ibáñez, J. M., Nácher, J., et al. (2010). Synaptic connectivity of serotonergic axons in the olfactory glomeruli of the rat olfactory bulb. *Neurosci.* 169: 770–780.

Gray, C. M. and Skinner, J. E. (1988). Centrifugal regulation of olfactory bulb of the waking rabbit as revealed by reversible cryogenic blockade. *Exp. Brain Res.* 69: 378–386.

Greer, C. A. (1987). Golgi analyses of dendritic organization among denervated olfactory bulb granule cells. *J. Comp. Neurol.* 256: 442–452.

Guerin, D., Peace, S. T., Didier, A., et al. (2008). Noradrenergic neuromodulation in the olfactory bulb modulates odor habituation and spontaneous discrimination. *Behav. Neurosci.* 122: 816–826.

Guo, Y. and Strowbridge, B. W. (2009). Long-term plasticity of excitatory inputs to granule cells in the rat olfactory bulb. *Nature Neurosci.* 12: 731–733.

Haga, S., Hattori, T., Sato, T., et al. (2010). The male mouse pheromone ESP1 enhances female sexual receptive behaviour through a specific vomeronasal receptor. *Nature* 466: 118–122.

Halabisky, B. and Strowbridge, B. W. (2003). γ-Frequency excitatory input to granule cells facilitates dendrodendritic inhibition in the rat olfactory bulb. *J. Neurophysiol.* 90: 644–654.

Halabisky, B., Friedman, D., Radojicic, M., and Strowbridge, B. W. (2000). Calcium influx through NMDA receptors directly evokes GABA release in olfactory bulb granule cells. *J. Neurosci.* 20: 5124–5134.

Halász, N., Ljungdahl, A., Hokfelt, T., et al. (1977). Transmitter histochemistry of the rat olfactory bulb. I. Immunohistochemical localization of monoamine synthesizing enzymes. Support for intrabulbar, peri-glomerular dopamine neurons. *Brain Res.* 126: 455–474.

Halpern, M., Shapiro, L.S., and Jia, C. (1995). Differential localization of G proteins in the opossum vomeronasal system. *Brain Res.* 677: 157–161.

Hamilton, K. A., Heinbockel, T., Ennis, M., et al. (2005). Properties of external plexiform layer interneurons in mouse olfactory bulb slices. *Neurosci.* 133: 819–829.

Hammen, G. F., Turaga, D., Holy, T. E. and Meeks, J. P. (2014). Functional organization of glomerular maps in the mouse accessory olfactory bulb. *Nat. Neurosci.* 17: 953–961.

Hardy, A., Palouzier-Paulignan, B., Duchamp, A., et al. (2005). 5-hydroxytryptamine action in the rat olfactory bulb: In vitro electrophysiological patch-clamp recordings of juxtaglomerular and mitral cells. *Neuroscience* 131: 717–731.

Hayar, A., Heyward, P. M., Heinbockel, T., et al. (2001). Direct excitation of mitral cells via activation of alpha1-noradrenergic receptors in rat olfactory bulb slices. *J. Neurophysiol.* 86: 2173–2182.

Hayar, A., Karnup, S., Shipley, M. T., and Ennis, M. (2004a). Olfactory bulb glomeruli: external tufted cells intrinsically burst at theta frequency and are entrained by patterned olfactory input. *J. Neurosci.* 24: 1190–1199.

Hayar, A., Karnup, S., Ennis, M., and Shipley, M. T. (2004b). External tufted cells: a major excitatory element that coordinates glomerular activity. *J. Neurosci.* 24: 6676–6685.

Hayar, A., Shipley, M. T., and Ennis, M. (2005). Olfactory bulb external tufted cells are synchronized by multiple intraglomerular mechanisms. *J. Neurosci.* 25: 8197–8208.

Heinbockel, T., Laaris, N., and Ennis M. (2007). Activation of metabotropic glutamate receptors in the main olfactory bulb drive granule cell-mediated inhibition. *J. Neurophysiol.* 97: 858–870.

Heinbockel, T., Heyward, P., Conquet, F., and Ennis, M. (2004). Regulation of main olfactory bulb mitral cell excitability by metabotropic glutamate receptor mGluR1. *J. Neurophysiol.* 92: 3085–3096.

Hendrickson, R., Krauthamer, S., Essenberg, J., and Holy, T.E. (2008). Inhibition shapes sex-selectivity in the mouse accessory olfactory bulb. *J. Neurosci.* 28: 12523–12534.

Herrada, G. and Dulac, C. (1997). A novel family of putative pheromone receptors in mammals with a topographically organized and sexually dimorphic distribution. *Cell* 90: 763–773.

Heyward, P. M., Ennis, M., Keller, A., and Shipley, M. T. (2001). Membrane bistability in olfactory bulb mitral cells. *J. Neurosci.* 21: 5311–5320.

Hirata, Y. (1964). Some observations on the fine structure of the synapses in the olfactory bulb of the mouse, with particular reference to the atypical synaptic configurations. *Arch. Histol. Jap.* 24: 293–302.

Hormuzdi, S. G., Pais, I., LeBleu, F. E. Net al. (2001). Impaired electrical signaling disrupts gamma frequency oscillations in connexin 36-deficient mice. *Neuron.* 31: 487–495.

Hsia, A. Y., Vincent, J. D., and Lledo, P. M. (1999). Dopamine depresses synaptic inputs into the olfactory bulb. *J. Neurophysiol.* 82: 1082–1085.

Isaacson, J. S. (1999). Glutamate spillover mediates excitatory transmission in the rat olfactory bulb. *Neuron.* 23: 377–384.

Isaacson, J. S. (2001). Mechanisms governing dendritic gamma-aminobutyric acid (GABA) release in the rat olfactory bulb. *Proc. Natl. Acad. Sci.* 98: 337–342.

Isaacson, J. S. and Strowbridge, B. W. (1998). Olfactory reciprocal synapses: dendritic signaling in the CNS. *Neuron.* 20: 749–761.

Isaacson, J. S. and Vitten, H. (2003. GABAB receptors inhibit dendrodendritic transmission in the rat olfactory bulb. *J. Neurosci.* 23: 2032–2039.

Ishii, T., Hirota, J., and Mombaerts, P. (2003). Combinatorial coexpression of neural and immune multigene families in mouse vomeronasal sensory neurons. *Curr. Biol.* 13: 394–400.

Ishii, T. and Mombaerts, P. (2008). Expression of nonclassical class I major histocompatibility genes defines a tripartite organization of the mouse vomeronasal system. *J. Neurosci.* 28: 2332–2341.

Jackowski, A., Parnavelas, J. G., and Lieberman, A. R. (1978). The reciprocal synapse in the external plexiform layer of the mammalian olfactory bulb. *Brain Res.* 159: 17–28.

Jia, C., Goldman, G., and Halpern, M. (1997). Development of vomeronasal receptor neuron subclasses and establishment of topographic projections to the accessory olfactory bulb. *Brain Res. Dev. Brain Res.* 102: 209–216.

Jia, C. and Halpern, M. (1996). Subclasses of vomeronasal receptor neurons: differential expression of G proteins (Gi alpha 2 and G(o alpha)) and segregated projections to the accessory olfactory bulb. *Brain Res.* 719: 117–128.

Jiang, M. R., Griff, E. R., Ennis, M., Zimmer, L. A. and Shipley, M. T. (1996). Activation of locus coeruleus enhances the responses of olfactory bulb mitral cells to weak olfactory nerve input. *J. Neurosci.* 16: 6319–6329.

Johnson, B. A. and Leon, M. (2007). Chemotopic odorant coding in a mammalian olfactory system. *J. Comp. Neurol.* 503: 1–34.

Jones, S. V., Choi, D. C., Davis, M., and Ressler, K. J. (2008). Learning-dependent structural plasticity in the adult olfactory pathway. *J. Neurosci.* 28: 13106–13211.

Kaba, H. and Huang, G. Z. (2005). Long-term potentiation in the accessory olfactory bulb: a mechanism for olfactory learning. *Chem. Senses* 30 Suppl 1: i150–151.

Kang, N., Baum, M.J., and Cherry, J.A. (2009). A direct main olfactory bulb projection to the 'vomeronasal' amygdala in female mice selectively responds to volatile pheromones from males. *Eur. J. Neurosci.* 29: 624–634.

Karnup, S. V., Hayar, A., Shipley, M. T., and Kurnikova, M. G. (2006). Spontaneous field potentials in the glomeruli of the olfactory bulb: the leading role of juxtaglomerular cells. *Neuroscience* 142: 203–221.

Kashiwadani, H., Sasaki, Y. F., Uchida, N., and Mori, K. (1999). Synchronized oscillatory discharges of mitral/tufted cells with different molecular receptive ranges in the rabbit olfactory bulb. *J. Neurophysiol.* 182: 1786–1792.

Kasowski, H. J., Kim, H., and Greer, C. A. (1999). Compartmental organization of the olfactory bulb glomerulus. *J. Comp. Neurol.* 407: 261–274.

Kauer, J. S. (1998). Olfactory processing: a time and place for everything. *Curr. Biol.* 8: R282–R283.

Kay, L. M. (2003). A challenge to chaotic intinerancy from brain dynamics. *Chaos* 13: 1057–1066.

Kay, L. M. (2005). Theta oscillations and sensorimotor performance. *Proc. Natl. Acad. Sci. USA.* 102: 3863–3868.

Kay, L. M., Beshel, J., Brea, J. et al. (2009). Olfactory oscillations: the what, how and what for. *Trends. Neurosci.* 32: 207–214.

Kay, L. M. and Freeman, W. J. (1998). Bidirectional processing in the olfactory-limbic axis during olfactory behavior. *Behav. Neurosci.* 112: 541–553.

Kay, L. M. and Laurent, G. (1999). Odor- and context-dependent modulation of mitral cell activity in behaving rats. *Nat. Neurosci.* 2: 1003–1009.

Keller, A., Yagodin, S., Aroniadou-Anderjaska, V., et al. (1998). Functional organization of rat olfactory bulb glomeruli revealed by optical imaging. *J. Neurosci.* 18: 2602–2612.

Kepecs, A., Uchida, N., and Mainen, Z. F. (2006). The sniff as a unit of olfactory processing. *Chem. Senses* 31: 167–179.

Kevetter, G. A. and Winans, S. S. (1981). Connections of the corticomedial amygdala in the golden hamster. I. Efferents of the "vomeronasal amygdala". *J. Comp. Neurol.* 197: 81–98.

Kiyokage, E., Pan, Y. Z., Shao, Z., et al. (2010). Molecular identity of periglomerular and short axon cells. *J. Neurosci.* 30: 1185–1196.

Kosaka, T., Hataguchi, Y., Hama, K., et al. (1985). Coexistence of immunoreactivities for glutamate decarboxylase and tyrosine hydroxylase in some neurons in the periglomerular region of the rat main olfactory bulb: possible coexistence of gamma-aminobutyric acid (GABA) and dopamine. *Brain Res.* 343: 166–171.

Kosaka, T. and Kosaka, K. (2011). "Interneurons" in the olfactory bulb revisited. *Neurosci. Res.* 69: 93–99.

Kosaka, K., Toida, K., Margolis, F. L., and Kosaka, T. (1997). Chemically defined neuron groups and their subpopulations in the glomerular layer of the rat main olfactory bulb–II. Prominent differences in the intraglomerular dendritic arborization and their relationship to olfactory nerve terminals. *Neuroscience* 76: 775–786.

Kunze, W. A. A., Shafton, A. D., Kemm, R. E., and McKenzie, J. S. (1991). Effect of stimulating the nucleus of the horizontal limb of the diagonal band on single unit activity in the olfactory bulb. *Neurosci.* 40: 21–27.

Kunze, W. A. A., Shafton, A. D., Kemm, R. E., and McKenzie, J. S. (1992). Intracellular responses of olfactory bulb granule cells to stimulating the horizontal diagonal band nucleus. *Neuroscience* 48: 63–369.

Laaris, N., Heonbockel, T., and Ennis, M. (2007). Complementary postsynaptic activity patterns elicited in olfactory bulb by stimulation of mitral/tufted and centrifugal fiber inputs to granule cells. *J. Neurophysiol.* 97: 296–306.

Lagier, S., Carleton, A., and Lledo, P. M. (2004). Interplay between local GABAergic interneurons and relay neurons generates gamma oscillations in the rat olfactory bulb. *J. Neurosci.* 24: 4382–4392.

Lagier, S., Panzanelli, P., Russo, R. E., et al. (2007). GABAergic inhibition at dendrodendritic synapses tunes gamma oscillations in the olfactory bulb. *Proc. Natl. Acad. Sci. USA.* 104: 7259–7264.

Larriva-Sahd, J. (2008). The accessory olfactory bulb in the adult rat: a cytological study of its cell types, neuropil, neuronal modules, and interactions with the main olfactory system. *J. Comp. Neurol.* 510: 309–350.

Laurent, G. (2002). Olfactory network dynamics and the coding of multidimensional signals. *Nat. Rev. Neurosci.* 3: 884–895.

Lazarini, F. and Lledo, P. M. (2011). Is adult neurogenesis essential for olfaction? *Trends Neurosci.* 34: 20–30.

Liberles, S. D., Horowitz, L. F., Kuang, D.et al. (2009). Formyl peptide receptors are candidate chemosensory receptors in the vomeronasal organ. *Proc. Natl. Acad. Sci. U.S.A.* 106: 9842–9847.

Linster, C. and Cleland, T. A. (2001). How spike synchronization among olfactory neurons can contribute to sensory discrimination. *J. Comput. Neurosci.* 10: 187–193.

Linster, C. and Cleland, T. A. (2002). Cholinergic modulation of sensory representations in the olfactory bulb. *Neural Netw.* 15: 709–717.

Linster, C. and Hasselmo, M. (1997). Modulation of inhibition in a model of olfactory bulb reduces overlap in the neural representation of olfactory stimuli. *Behav. Brain Res.* 84: 117–127.

Linster, C., Nai, Q., and Ennis, M. (2011). Nonlinear effects of noradrenergic modulation of olfactory bulb function in adult rodents. *J. Neurophysiol.* 105: 1432–1443.

Liu, S., Aungst, J. L., Puche, A. C., and Shipley, M. T. (2012). Serotonin modulates the population activity profile of olfactory bulb external tufted cells. *J. Neurophysiol.* 107: 473–483.

Liu, S. and Shipley, M. T. (2008a). Multiple conductances cooperatively regulate spontaneous bursting in mouse olfactory bulb external tufted cells. *J. Neurosci.* 28: 1628–1635.

Liu, S. and Shipley, M. T. (2008b). Intrinsic conductances actively shape excitatory and inhibitory postsynaptic responses in olfactory bulb external tufted cells. *J. Neurosci.* 28: 10311–10322.

Liu, W. L. and Shipley, M. T. (1994). Intrabulbar associational system in the rat olfactory bulb comprises cholecystokinin-containing tufted cells that synapse onto the dendrites of GABAergic granule cells. *J. Comp. Neurol.* 346: 541–558.

Lledo, P. M., Gheusi, G., and Vincent, J. D. (2005). Information processing in the mammalian olfactory system. *Physiol. Rev.* 85: 281–317.

Loconto, J., Papes, F., Chang, E. et al. (2003). Functional expression of murine V2R pheromone receptors involves selective association with the M10 and M1 families of MHC class Ib molecules. *Cell* 112: 607–618.

Lodovichi, C., Belluscio, L., and Katz, L. C. (2003). Functional topography of connections linking mirror-symmetric maps in the mouse olfactory bulb. *Neuron.* 38: 265–276.

Lowe, G. (2002). Inhibition of backpropagating action potentials in mitral cell secondary dendrites. *J. Neurophysiol.* 88: 64–85.

Luo, M., Fee, M., and Katz, L. (2003). Encoding pheromonal signals in the accessory olfactory bulb of behaving mice. *Science* 299: 1196–1201.

Luo, M. and Katz, L. (2001). Response correlation maps of neurons in the mammalian olfactory bulb. *Neuron.* 32: 1165–1179.

Ma, J. and Lowe, G. (2010). Correlated firing in tufted cells of mouse olfactory bulb. *Neurosci.* 169: 1715–1738.

Macrides, F., Eichenbaum, H. B., and Forbes, W. B. (1982). Temporal relarionship between sniffing and the limbic θ rhythm during odor discrimination and reversal learning. *J. Neurosci.* 2: 1705–1717.

Maher, B. J. and Westbrook, G. L. (2008). Co-transmission of dopamine and GABA in periglomerular cells. *J. Neurophysiol.* 99: 1559–1564.

References

Mandairon, N., Ferretti, C. J., Stack, C. M., et al. (2006). Cholinergic modulation in the olfactory bulb influences spontaneous olfactory discrimination in adult rats. *Eur. J. Neurosci.* 24: 3234–3244.

Mandairon, N. and Linster, C. (2009). Odor perception and olfactory bulb plasticity in adult mammals. *J. Neurophysiol.* 101: 2204–2209.

Mandairon, N., Peace, S., Karnow, et al. (2008). Noradrenergic modulation in the olfactory bulb influences spontaneous and reward-motivated discrimination, but not the formation of habituation memory. *Eur. J. Neurosci.* 27: 1210–1219.

Margrie, T. W., Sakmann, B., and Urban, N. N. (2001). Action potential propagation in mitral cell lateral dendrites is decremental and controls recurrent and lateral inhibition in the mammalian olfactory bulb. *Proc. Natl. Acad. Sci. USA.* 98: 319–324.

Margrie, T. W. and Schaefer, A. T. (2003). Theta oscillation coupled spike latencies yield computational vigour in a mammalian sensory system. *J. Physiol.* 546: 363–374.

Martel, K. L. and Baum, M. J. (2007). Sexually dimorphic activation of the accessory, but not the main, olfactory bulb in mice by urinary volatiles. *Eur. J. Neurosci.* 26: 463–475.

Martel, K. L. and Baum, M. J. (2009). A centrifugal pathway to the mouse accessory olfactory bulb from the medial amygdala conveys gender-specific volatile pheromonal signals. *Eur. J. Neurosci.* 29: 368–376.

Martin, C., Gervais, R., Hugues, E., et al. (2004a). Learning modulation of odor-induced oscillatory responses in the rat olfactory bulb: a correlate of odor recognition? *J. Neurosci.* 24: 389–397.

Martin, C., Gervais, R., Hugues, E., et al. (2004b). Learning-induced modulation of oscillatory activities in the mammalian olfactory system: the role of centrifugal fibers: a correlate of odor recognition? *J. Physiol. (Paris).* 98: 467–478.

Martin, C., Gervais, R., Hugues, E., et al. (2006). Learning-induced oscillatory activities correlated to odour recognition: a network activity. *Eur. J. Neurosci.* 23: 1801–1810.

Matsumoto, H., Kashiwadani, H., Nagao, H., et al. (2009). Odor-induced persistent discharge of mitral cells in the mouse olfactory bulb. *J. Neurophysiol.* 101: 1890–1900.

Matsunami, H. and Buck, L. (1997). A multigene family encoding a diverse array of putative pheromone receptors in mammals. *Cell* 90: 775–784.

McGann, J. P., Pírez, N., Gainey, M. A., et al. (2005). Odorant representations are modulated by intra- but not interglomerular presynaptic inhibition of olfactory sensory neurons. *Neuron* 48: 1039–1053.

McLean, J. H., Darby-King, A., and Hodge, E. (1996). 5-HT2 receptor involvement in conditioned olfactory learning in the neonatal rat pup. *Behav. Neurosci.* 110: 1426–1434.

McLean, J. H. and Harley, C. W. (2004). Olfactory learning in the rat pup: a model that may permit visualization of a mammalian memory trace. *NeuroReport* 15: 1691–1697.

McLean, J. H. and Shipley, M. T. (1987). Serotonergic afferents to the rat olfactory bulb: I. Origins and laminar specificity of serotonergic inputs in the adult rat. J. Neurosci. 7: 3016–3028.

McLean, J. H., Shipley, M. T., Nickell, W.T., et al. (1989). Chemoanatomical organization of the noradrenergic input from locus coeruleus to the olfactory bulb of the adult rat. *J. Comp. Neurol.* 285: 339–349.

McQuiston, A. R. and Katz, L. C. (2001). Electrophysiology of interneurons in the glomerular layer of the rat olfactory bulb. *J. Neurophysiol.* 86: 1899–1907.

Meeks, J.P., Arnson, H.A., and Holy, T.E. (2010). Representation and transformation of sensory information in the mouse accessory olfactory system. *Nat. Neurosci.* 13: 723–730.

Meister, M. and Bonhoeffer, T. (2001). Tuning and topography in an odor map on the rat olfactory bulb. *J. Neurosci.* 21: 1351–1360.

Meredith, M. and O'Connell, R. (1979). Efferent control of stimulus access to the hamster vomeronasal organ. *J. Physiol.* 286: 301–316.

Mohedano-Moriano, A., de la Rosa-Prieto, C., Saiz-Sanchez, D., et al. (2012). Centrifugal telencephalic afferent connections to the main and accessory olfactory bulbs. *Front. Neuroanat.* 6: 19.

Mombaerts, P., Wang, F., Dulac, C., et al. (1996). Visualizing an olfactory sensory map. *Cell.* 87: 675–686.

Moreno, M. M., Linster, C., Escanilla, O., et al. (2009). Olfactory perceptual learning requires adult neurogenesis. *Proc. Natl. Acad. Sci. USA.* 106: 17980–17985.

Mori, K. (1987). Membrane and synaptic properties of identified neurons in the olfactory bulb. *Prog. Neurobiol.* 29: 275–320.

Mori, K., Takahashi, Y. K., Igarashi, K. M., and Yamaguchi, M. (2006). Maps of odorant molecular features in the mammalian olfactory bulb. *Physiol. Rev.* 86: 409–433.

Morizumi, T., Tsukatani, H., and Miwa, T. (1994). Olfactory disturbance induced by deafferentation of serotonergic fibers in the olfactory bulb. *Neuroscience* 61: 733–738.

Murakami, M., Kashiwadani, H., Kirino, Y., and Mori, K. (2005). State-dependent sensory gating in olfactory cortex. *Neuron.* 46: 285–296.

Murphy, G. J., Darcy, D. P., and Isaacson, J. S. (2005). Intraglomerular inhibition: signaling mechanisms of an olfactory microcircuit. *Nat. Neurosci.* 8: 354–364.

Nagayama, S., Enerva, A., Fletcher, M. L., et al. (2010). Differential axonal projection of mitral and tufted cells in the mouse main olfactory system. *Front. Neural Circuits.* 4: 120.

Nagayama, S., Takahashi, Y. K., Yoshihara, Y., and Mori, K. (2004). Mitral and tufted cells differ in the decoding manner of odor maps in the rat olfactory bulb. *J. Neurophysiol.* 91: 2532–2540.

Nai, Q., Dong, H. W., Hayar, A., et al. (2009). Noradrenergic regulation of GABAergic inhibition of main olfactory bulb mitral cells varies as a function of concentration and receptor subtype. *J. Neurophysiol.* 101: 2472–2484.

Nai, Q., Dong, H. W., Linster, C., and Ennis, M. (2010). Activation of alpha1 and alpha2 noradrenergic receptors exert opposing effects on excitability of main olfactory bulb granule cells. *Neuroscience* 169: 882–892.

Neville, K. R. and Haberly, L. B. (2003). Beta and gamma oscillations in the olfactory system of the urethane-anesthetized rat. *J. Neurophysiol.* 90: 3921–3930.

Najac, M., De Saint Jan, D., Reguero, L., et al. (2011). Monosynaptic and polysynaptic feed-forward inputs to mitral cells from olfactory sensory neurons. *J. Neurosci.* 15: 8722–8729.

Nica, R., Matter, S. F., and Griff, E. R. (2010). Physiological evidence for two classes of mitral cells in the rat olfactory bulb. *Brain Res.* 1358: 81–88.

Nickell, W. T. and Shipley, M. T. (1988). Neurophysiology of magnocellular forebrain inputs to the olfactory bulb in the rat: frequency potentiation of field potentials and inhibition of output neurons. *J. Neurosci.* 8: 4492–4502.

Nusser, Z., Kay, L. M., Laurent, G., et al. (2001). Disruption of GABAA receptors on GABAergic interneurons leads to increased oscillatory power in the olfactory bulb network. *J. Neurophysiol.* 86: 2823–2833.

Oka, Y., Takai, Y., and Touhara, K. (2009). Nasal airflow rate affects the sensitivity and pattern of glomerular odorant responses in the mouse olfactory bulb. *J. Neurosci.* 29: 12070–12078.

Orona, E., Rainer, E. C., and Scott, J. W. (1984). Dendritic and axonal organization of mitral and tufted cells in the rat olfactory bulb. *J. Comp. Neurol.* 226: 346–356.

Orona, E., Scott, J. W., and Rainer, E. C. (1983). Different granule cell populations innervate superficial and deep regions of the external plexiform layer in rat olfactory bulb. *J. Comp. Neurol.* 217: 227–237.

Otsuka, T., Ishii, K., Osako, Y., et al. (2001). Modulation of dendrodendritic interactions and mitral cell excitability in the mouse accessory olfactory bulb by vaginocervical stimulation. *Eur J. Neurosci.* 13: 1833–1838.

Palouzier-Paulignan, B., Duchamp-Viret, P., Hardy, A.B., and Duchamp, A. (2002). GABA(B) receptor-mediated inhibition of mitral/tufted cell activity in the rat olfactory bulb: a whole-cell patch-clamp study in vitro. *Neurosci.* 111: 241–250.

Pandipati, S., Gire, D.H., and Schoppa NE. (2010). Adrenergic receptor-mediated disinhibition of mitral cells triggers long-term enhancement of synchronized oscillations in the olfactory bulb. *J. Neurophysiol.* 104: 665–674.

Parrish-Aungst, S., Kiyokage, E., Szabo, G., et al. (2011). Sensory experience selectively regulates transmitter synthesis enzymes in interglomerular circuits. *Brain Res.* 1382: 70–76.

Parrish-Aungst, S., Shipley, M. T., Erdelyi, F., et al. (2007). Quantitative analysis of neuronal diversity in the mouse olfactory bulb. *J. Comp. Neurol.* 501: 825–836.

Petzold, G. C., Hagiwara, A., and Murthy, V. N. (2009). Serotonergic modulation of odor input to the mammalian olfactory bulb. *Nat. Neurosci.* 12: 784–791.

Phillips, M. E., Sachdev, R. N., Willhite, D. C., and Shepherd, G. M. (2012). Respiration drives network activity and modulates synaptic and circuit processing of lateral inhibition in the olfactory bulb. *J. Neurosci.* 32: 85–98.

Philpot, B. D., Foster, T. C., and Brunjes, P. C. (1997). Mitral/tufted cell activity is attenuated and becomes uncoupled from respiration following naris closure. *J. Neurobiol.* 33: 374–86.

Pignatelli, A. and Belluzzi, O. (2008). Cholinergic modulation of dopaminergic neurons in the mouse olfactory bulb. *Chem. Senses* 33: 331–338.

Pignatelli, A., Kobayashi, K., Okano, H., and Belluzzi, O. (2005). Functional properties of dopaminergic neurones in the mouse olfactory bulb. *J. Physiol.* 564: 501–514.

Pinching, A. J. and Powell, T. P. (1971a). The neuron types of the glomerular layer of the olfactory bulb. *J. Cell. Sci.* 9: 305–345.

Pinching, A. J. and Powell, T. P. (1971b). The neuropil of the glomeruli of the olfactory bulb. *J. Cell. Sci.* 9: 347–377.

Pinching, A. J. and Powell, T. P. (1971c). The neuropil of the periglomerular region of the olfactory bulb. *J. Cell. Sci.* 9: 379–409.

Pimentel, D. O. and Margrie, T. W. (2008). Glutamatergic transmission and plasticity between olfactory bulb mitral cells. *J. Physiol.* 586: 2107–2119.

Pírez, N. and Wachowiak, M. (2008). In vivo modulation of sensory input to the olfactory bulb by tonic and activity-dependent presynaptic inhibition of receptor neurons. *J. Neurosci.* 28: 6360–6371.

Potter, S. M., Zheng, C., Koos, D. S., et al. (2001). Structure and emergence of specific olfactory glomeruli in the mouse. *J. Neurosci.* 21: 9713–9723.

Pressler, R. T. and Strowbridge, B. W. (2006). Blanes cells mediate persistent feedforward inhibition onto granule cells in the olfactory bulb. *Neuron.* 49: 889–904.

Pressler, R. T., Inoue, T., and Strowbridge, B. W. (2007). Muscarinic receptor activation modulates granule cell excitability and potentiates inhibition onto mitral cells in the rat olfactory bulb. *J. Neurosci.* 27: 10969–10981.

Price, J. L. and Powell, T. P. S. (1970a). An electron-microscopic study of the termination of the afferent fibres to the olfactory bulb from the cerebral hemisphere. *J. Cell. Sci.* 7: 157–187.

Price, J. L. and Powell, T. P. S. (1970b). The mitral and short axon cells of the olfactory bulb. *J. Cell. Sci.* 7: 631–651.

Price, J. L. and Powell, T. P. S. (1970a,c). The morphology of the granule cells of the olfactory bulb. *J. Cell. Sci.* 9: 91–123.

Price, J. L. and Powell, T. P. S. (1970d). The synaptology of the granule cells of the olfactory bulb. *J. Cell. Sci.* 7: 125–155.

Puopolo, M., Bean, B. P., and Raviola, E. (2005). Spontaneous activity of isolated dopaminergic periglomerular cells of the main olfactory bulb. *J. Neurophysiol.* 94: 3618–3627.

Puopolo, M. and Belluzzi, O. (1998). Functional heterogeneity of periglomerular cells in the rat olfactory bulb. *Eur. J. Neurosci.* 10: 1073–1083.

Raisman, G. (1972). An experimental study of the projection of the amygdala to the accessory olfactory bulb and its relationship to the concept of a dual olfactory system. *Exp. Brain Res.* 14: 395–408.

Rall, W., Shepherd, G. M., Reese, T. S., and Brightman, M. W. (1966). Dendrodendritic synaptic pathway for inhibition in the olfactory bulb. *Exp. Neurol.* 14: 44–56.

Ramón y Cajal, S. (1901). Textura del lóbulo olfativo accesorio. *Trab. Lab. Invest. Biol.* 1: 141–149.

Ressler, K. J., Sullivan, S. L., and Buck, L. B. (1993). A zonal organization of odorant receptor gene expression in the olfactory epithelium. *Cell.* 73: 597–609.

Ressler, K. J., Sullivan, S. L. and Buck, L. B. (1994). Information coding in the olfactory system: evidence for a stereotyped and highly organized epitope map in the olfactory bulb. *Cell.* 79: 1245–1255.

Revial, M. F., Sicard, G., Duchamp, A., and Holley, A. (1982). New studies on odour discrimination in the frog's olfactory receptor cells. i. experimental results. *Chem. Senses* 7: 175–190.

Reyher, C. K., Lubke, J., Larsen, W. J., et al. (1991). Olfactory bulb granule cell aggregates: morphological evidence for interperikaryal electrotonic coupling via gap junctions. *J. Neurosci.* 11: 1485–1495.

Ribak, C. E., Vaughn, J. E., Saito, K., et al. (1977). Glutamate decarboxylase localization in neurons of the olfactory bulb. *Brain Res.* 126: 1–18.

Rinberg, D., Koulakov, A., and Gelperin, A. (2006). Sparse odor coding in awake behaving mice. *J. Neurosci.* 26: 8857–8865.

Rivière, S., Challet, L., Fluegge, D., et al. (2009). Formyl peptide receptor-like proteins are a novel family of vomeronasal chemosensors. *Nature* 459: 574–577.

Rochefort, C., Gheusi, G., Vincent, J.-D., and Lledo, P.-M. (2002). Enriched odor exposure increases the number of newborn neurons in the adult olfactory bulb and improves odor memory. *J. Neurosci.* 22: 2679–2689.

Rodriguez, I., Feinstein, P., and Mombaerts, P. (1999). Variable patterns of axonal projections of sensory neurons in the mouse vomeronasal system. *Cell* 97: 199–208.

Ryba, N. and Tirindelli, R. (1997). A new multigene family of putative pheromone receptors. *Neuron* 19: 371–379.

Salin, P. A., Lledo, P. M. Vincent, J. D., and Charpak, S. (2001). Dendritic glutamate autoreceptors modulate signal processing in rat mitral cells. *J. Neurophysiol.* 85: 1275–1282.

Sallaz, M. and Jourdan, F. (1992). Apomorphine disrupts odour-induced patterns of glomerular activation in the olfactory bulb. *NeuroReport.* 3: 833–836.

Sassoè-Pognetto, M., Cantino, D., and Panzanelli, P. (1993). Presynaptic co-localization of carnosine and glutamate in olfactory neurons. *NeuroReport.* 5: 7–10.

Sawada, M., Kaneko, N., Inada, H., et al. (2011). Sensory input regulates spatial and subtype-specific patterns of neuronal turnover in the adult olfactory bulb. *J. Neurosci.* 31: 11587–11596.

Schaefer, A. T. and Margrie, T. W. (2007). Spatiotemporal representations in the olfactory system. *Trends Neurosci.* 30: 92–100.

References

Schneider, S. P. and Scott, J. W. (1983). Orthodromic response properties of rat olfactory mitral and tufted cells correlate with their projection patterns. *J. Neurophysiol.* 50: 358–378.

Schoenfeld, T. A., Marchand, J. E., and Macrides, F. (1985). Topographic organization of tufted cell axonal projections in the hamster main olfactory bulb: an intrabulbar associational system. *J. Comp. Neurol.* 235: 503–518.

Schoppa, N. E. (2006). Synchronization of olfactory bulb mitral cells by precisely timed inhibitory inputs. *Neuron.* 49: 271–283.

Schoppa, N. E., Kinzie, J. M., Sahara, Y., et al. (1998). Dendrodendritic inhibition in the olfactory bulb is driven by NMDA receptors. *J. Neurosci.* 18: 6790–6802.

Schoppa, N. E. and Westbrook, G. L. (1999). Regulation of synaptic timing in the olfactory bulb by an A-type potassium current. *Nat. Neurosci.* 2: 1106–1113.

Schoppa, N. E. and Westbrook, G. L. (2001). Glomerulus-specific synchronization of mitral cells in the olfactory bulb. *Neuron.* 31: 639–651.

Schoppa, N. E. and Westbrook, G. L. (2002). AMPA receptors drive correlated spiking in olfactory bulb glomeruli. *Nat. Neurosci.* 5: 1194–1202.

Scott, J. W. (2006). Sniffing and spatiotemporal coding in olfaction. *Chem. Senses* 31: 119–130.

Shao, Z., Puche, A. C., Kiyokage, E., et al. (2009). Two GABAergic intraglomerular circuits differentially regulate tonic and phasic presynaptic inhibition of olfactory nerve terminals. *J. Neurophysiol.* 101: 1988–2001.

Shao, Z., Puche, A. C., Liu, S., and Shipley, M. T. (2012). Intraglomerular inhibition shapes the strength and temporal structure of glomerular output. *J. Neurophysiol.* 108: 782–793.

Shea, S. D., Katz, L. C., and Mooney, R. (2008). Noradrenergic induction of odor-specific neural habituation and olfactory memories. *J. Neurosci.* 28: 10711–10719.

Shepherd, G. M. (1972). Synaptic organization of the mammalian olfactory bulb. *Physiol. Rev.* 52: 864–917.

Shipley, M. T. and Adamek, G. D. (1984). The connections of the mouse olfactory bulb: a study using orthograde and retrograde transport of wheat germ agglutinin conjugated to horseradish peroxidase. *Brain Res. Bull.* 12: 669–688.

Shipley, M. T., Halloran, F. J., and de la Torre, J. (1985). Surprisingly rich projection from locus coeruleus to the olfactory bulb in the rat. *Brain Res.* 329: 294–299.

Shipley, M. T., McLean, J. H., Ennis, M., and Zimmer, L. A. (1996). The olfactory system, in *Handbook of Chemical Neuroanatomy, Vol. 12: Integrated systems of the CNS, Part III*, Bjorklund, A., Hokfelt, T., and Swanson, L. (Eds). Amsterdam:Elsevier, pp. 467–571.

Shu, H. J., Eisenman, L. N., Jinadasa, D., et al. (2004). Slow actions of neuroactive steroids at GABAA receptors. *J. Neurosci.* 24: 6667–6675.

Shusterman, R., Smear, M. C., Koulakov, A. A., and Rinberg, D. (2011). Precise olfactory responses tile the sniff cycle. *Nat. Neurosci.* 14: 1039–1044.

Smith, R. S. and Araneda, R. C. (2010). Cholinergic modulation of neuronal excitability in the accessory olfactory bulb. *J. Neurophysiol.* 104: 2963–2974.

Smith, R. S., Weitz, C. J., and Araneda, R. C. (2009). Excitatory actions of noradrenaline and metabotropic glutamate receptor activation in granule cells of the accessory olfactory bulb. *J. Neurophysiol.* 102: 1103–1114.

Smith, T. C. and Jahr, C. E. (2002). Self-inhibition of olfactory bulb neurons. *Nat. Neurosci.* 5: 760–766.

Sobel, E. C. and Tank, D. W. (1993). Timing of odor stimulation does not alter patterning of olfactory bulb unit activity in freely breathing rats. *J. Neurophysiol.* 69: 1331–1337.

Spors, H. and Grinvald, A. (2002). Spatio-temporal dynamics of odor representations in the mammalian olfactory bulb. *Neuron* 34: 301–315.

Spors, H., Wachowiak, M., Cohen, L. B., and Friedrich, R. W. (2006). Temporal dynamics and latency patterns of receptor neuron input to the olfactory bulb. *J. Neurosci.* 26: 1247–1259.

Sultan, S., Rey, N., Sacquet, J., et al. (2011). Newborn neurons in the olfactory bulb selected for long-term survival through olfactory learning are prematurely suppressed when the olfactory memory is erased. *J. Neurosci.* 31: 14893–14898.

Takami, S. and Graziadei, P. P. (1991). Light microscopic Golgi study of mitral/tufted cells in the accessory olfactory bulb of the adult rat. *J. Comp. Neurol.* 311: 65–83.

Takigami, S., Mori, Y., and Ichikawa, M. (2000). Projection pattern of vomeronasal neurons to the accessory olfactory bulb in goats. *Chem. Senses* 25: 387–393.

Tan, J., Savigner, A., Ma, M., and Luo, M. (2010). Odor information processing by the olfactory bulb analyzed in gene-targeted mice. *Neuron* 65: 912–926.

Tirindelli, R., Dibattista, M., Pifferi, S., and Menini, A. (2009). From pheromones to behavior. *Physiol. Rev.* 89: 921–956.

Toida, K., Kosaka, K., Aika, Y., and Kosaka, T. (2000). Chemically defined neuron groups and their subpopulations in the glomerular layer of the rat main olfactory bulb–IV. Intraglomerular synapses of tyrosine hydroxylase-immunoreactive neurons. *Neuroscience* 101: 11–17.

Toida, K., Kosaka, K., Heizmann, C.W., and Kosaka, T. (1998). Chemically defined neuron groups and their subpopulations in the glomerular layer of the rat main olfactory bulb: III. Structural features of calbindin D28K-immunoreactive neurons. *J. Comp. Neurol.* 392: 179–198.

Treloar, H. B., Feinstein, P., Mombaerts, P., and Greer, C. A. (2002). Specificity of glomerular targeting by olfactory sensory axons. *J. Neurosci.* 22: 2469–2477.

Tsuno, Y., Kashiwadani, H., and Mori, K. (2008). Behavioral state regulation of dendrodendritic synaptic inhibition in the olfactory bulb. *J. Neurosci.* 28: 9227–9238.

Turaga, D. and Holy, T. E. (2012). Organization of vomeronasal sensory coding revealed by fast volumetric calcium imaging. *J. Neurosci.* 32: 1612–1621.

Urban, N. N. and Arevian, A. C. (2009). Computing with dendrodendritic synapses in the olfactory bulb. *Ann. NY Acad. Sci.* 1170: 264–269.

Urban, N. N. and Sakmann, B. (2002). Reciprocal intraglomerular excitation and intra- and interglomerular lateral inhibition between mouse olfactory bulb mitral cells. *J. Physiol. (Lond.)* 542: 355–367.

Vassar, R., Chao, S. K., Sitcheran, R., et al. (1994). Topographic organization of sensory projections to the olfactory bulb. *Cell.* 79: 981–991.

Verhagen, J. V., Wesson, D. W., Netoff, T. I., et al. (2007). Sniffing controls an adaptive filter of sensory input to the olfactory bulb. *Nat. Neurosci.* 10: 631–639.

Veyrac, A., Nguyen, V., Marien, M., et al. (2007). Noradrenergic control of odor recognition in a nonassociative olfactory learning task in the mouse. *Learn. Mem.* 14: 847–854.

Vucinić, D., Cohen, L. B. and Kosmidis, E. K. (2006). Interglomerular center-surround inhibition shapes odorant-evoked input to the mouse olfactory bulb in vivo. *J. Neurophysiol.* 95: 1881–1887.

Wachowiak, M. (2011). All in a sniff: Olfaction as a model for active sensing. *Neuron.* 71: 962–973.

Wachowiak, M. and Cohen, L. B. (2001). Representation of odorants by receptor neuron input to the mouse olfactory bulb. *Neuron.* 32: 723–735.

Wachowiak, M., McGann, J. P., Heyward, P. M., et al. (2005). Inhibition of olfactory receptor neuron input to olfactory bulb glomeruli mediated by suppression of presynaptic calcium influx. *J. Neurophysiol.* 94: 2700–2712.

Wachowiak, M. and Shipley, M. T. (2006). Coding and synaptic processing of sensory information in the glomerular layer of the olfactory bulb. *Semin. Cell. Dev. Biol.* 17: 411–423.

Wagner, S., Gresser, A. L., Torello, A.T., and Dulac, C. (2006). A multireceptor genetic approach uncovers an ordered integration of VNO sensory inputs in the accessory olfactory bulb. *Neuron* 50: 697–709.

Wang, F., Nemes, A., Mendelsohn, M., and Axel, R. (1998). Odorant receptors govern the formation of a precise topographic map. *Cell.* 93: 47–60.

Wehr, M. and Laurent, G. (1996). Odour encoding by temporal sequences of firing in oscillating neural assemblies. *Nature.* 384: 162–166.

Wei, C. J., Linster, C., and Cleland, T. A. (2006). Dopamine D(2) receptor activation modulates perceived odor intensity. *Behav. Neurosci.* 120: 393–400.

Welker, W. I. (1964). Analysis of sniffing of the albino rat. *Behaviour* 22: 223–244.

Wellis, D. P. and Scott, J. W. (1990). Intracellular responses of identified rat olfactory bulb interneurons to electrical and odor stimulation. *J. Neurophysiol.* 64: 932–947.

Wellis, D. P., Scott, J. W., and Harrison, T. A. (1989). Discrimination among odorants by single neurons of the rat olfactory bulb. *J. Neurophysiol.* 61: 1161–1177.

Wesson, D. W., Carey, R. M., Verhagen, J. V., and Wachowiak, M. (2008a). Rapid encoding and perception of novel odors in the rat. *PLoS Biol.* 6: e82.

Wesson, D. W., Donahou, T. N., Johnson, M. O., and Wachowiak, M. (2008b). Sniffing behavior of mice during performance in odor-guided tasks. *Chem Senses* 33: 581–596.

Wesson, D. W., Verhagen, J. V., and Wachowiak, M. (2009). Why sniff fast? The relationship between sniff frequency, odor discrimination, and receptor neuron activation in the rat. *J. Neurophysiol.* 101:1089–1102.

Willhite, D. C., Nguyen, K. T., Masurkar, A. V., et al. (2006). Viral tracing identifies distributed columnar organization in the olfactory bulb. *Proc. Natl. Acad. Sci. USA.* 103: 12592–12597.

Wilson, D. A., Fletcher, M. L., and Sullivan, R. M. (2004). Acetylcholine and olfactory perceptual learning. *Learn. Mem.* 11: 28–34.

Wilson, D. A. and Sullivan, R. M. (1995). The D2 antagonist spiperone mimics the effects of olfactory deprivation on mitral/tufted cell odor response patterns. *J. Neurosci.* 15: 5574–5581.

Wilson, R. I. and Mainen, Z. F. (2006). Early events in olfactory processing. *Ann. Rev. Neurosci.* 29: 163–201.

Woolf, T. B., Shepherd, G. M., and Greer, C. A. (1991). Serial reconstruction of granule cell spines in the mammalian olfactory bulb. *Synapse* 7: 181–191.

Xiong, W. and Chen, W. R. (2002). Dynamic gating of spike propagation in the mitral cell lateral dendrites. *Neuron* 34: 115–126.

Yokoi, M., Mori, K., and Nakanishi, S. (1995). Refinement of odor molecule tuning by dendrodendritic synaptic inhibition in the olfactory bulb. *Proc. Natl. Acad. Sci. USA.* 92: 3371–3375.

Yokosuka, M. (2012). Histological properties of the glomerular layer in the mouse accessory olfactory bulb. *Exp. Anim.* 61: 13–24.

Yokoyama, T. K., Mochimaru, D., Murata, K., et al. (2011). Elimination of adult-born neurons in the olfactory bulb is promoted during the postprandial period. *Neuron.* 71: 883–897.

Yonekura, J. and Yokoi, M. (2008). Conditional genetic labeling of mitral cells of the mouse accessory olfactory bulb to visualize the organization of their apical dendritic tufts. *Mol. Cell. Neurosci.* 37: 708–718.

Yu, G-Z., Kaba, H., Saito, H., and Seto, K. (1993). Heterogeneous characteristics of mitral cells in the rat olfactory bulb. *Brain Res. Bull* 31: 701–706.

Yuan, Q., Harley, C. W., and McLean, J. H. (2003). Mitral cell $\alpha 1$ and 5-HT2A receptor colocalization and cAMP coregulation: a new model of norepinephrine-induced learning in the olfactory bulb. *Learn. Mem.* 10: 5–15.

Yuan, Q. and Knopfel, T. (2006). Olfactory nerve stimulation-evoked mGluR1 slow potentials, oscillations and calcium signaling in mouse olfactory bulb mitral cells. *J. Neurophysiol.* 95: 3097–3104.

Zaborszky, L., Carlsen, J., Brashear, H. R., and Heimer, L. (1986). Cholinergic and GABAergic afferents to the olfactory bulb in the rat with special emphasis on the projection neurons in the nucleus of the horizontal limb of the diagonal band. *J. Comp. Neurol.* 243: 488–509.

Zelles, T., Boyd, J. D., Hardy, A. B., and Delaney, K. R. (2006). Branch-specific Ca^{2+} influx from Na^+–dependent dendritic spikes in olfactory granule cells. *J. Neurosci.* 26: 30–40.

Chapter 9

Adult Neurogenesis in the Subventricular Zone and Migration to the Olfactory Bulb

JOHN W. CAVE and HARRIET BAKER

9.1 INTRODUCTION

The past two decades have witnessed an explosive growth of research investigating neurogenesis in adult vertebrate brains. Prior to this intense exploration, prevailing opinion was that adult brains did not generate new neurons beyond peri-natal development, despite pioneering studies showing evidence of proliferation and the generation of new neurons in the adult rat brain (Altman, 1962; Kaplan and Hinds, 1977). Consensus began to change, however, with the conclusive demonstration of neurogenesis in adult brains of birds (Alvarez-Buylla and Nottebohm, 1988; Goldman and Nottebohm, 1983; Paton and Nottebohm, 1984). Within a decade of these findings in avians, several studies in adult rodents confirmed that both the subventricular zone (SVZ) of the lateral ventricles and dentate gyrus produce progenitors that migrate and become neurons in the olfactory bulb (OB) and hippocampus, respectively (Altman and Bayer, 1990; Cameron et al., 1993; Lois and Alvarez-Buylla, 1993, 1994; Luskin, 1993). The current consensus is that adult neurogenesis occurs in several brain regions of non-mammalian vertebrates (Chapouton et al., 2007; Kaslin et al., 2008), but in mammals it is limited to the SVZ/OB and hippocampus (Bonfanti and Peretto, 2011; Lledo et al., 2006; Ming and Song, 2011). Neurogenesis in other regions of the adult mammalian brain has been reported, but these findings are either species-specific or controversial (Bonfanti and Peretto, 2011; Gould, 2007).

This chapter will focus on both neurogenesis in the SVZ and the migration of neuroblasts through the rostral migratory stream (RMS) in the adult mammalian brain. Much of the research in this field has been conducted with rodents, specifically with mice. A growing number of studies with animals other than rodents, however, indicate that some significant differences exist between mammalian species (Bonfanti and Peretto, 2011). Although the discussion in this chapter is based largely on studies performed with rodents, documented differences in humans are noted.

9.2 SUBVENTRICULAR ZONE AND ROSTRAL MIGRATORY STREAM AS A NEUROGENIC NICHE

9.2.1 Cytoarchitecture of SVZ

The SVZ of the adult lateral ventricles is derived from several regions of the embryonic neuroepithelium, including the lateral and medial ganglionic eminences as well as cortex and septum (Merkle et al., 2007; Stenman et al., 2003; Willaime-Morawek et al., 2006; Young et al., 2007). In rodents, OB interneuron progenitors are generated from neural stem cells that span the entire rostro-caudal length of the ventricle (Merkle et al., 2007). By contrast, generation of human progenitors is limited to only the ventral region of the anterior horn in the lateral ventricle (Curtis et al., 2007; Quinones-Hinojosa et al., 2006; Sanai et al., 2011; Sanai et al., 2004; Wang et al., 2011a). Neuronal and glial progenitors also can be generated *in vitro* from multi-potent stem cells isolated from RMS and OB core in both rodents and humans (Gritti et al., 2002; Liu and

---

Handbook of Olfaction and Gustation, Third Edition. Edited by Richard L. Doty.
© 2015 Richard L. Doty. Published 2015 by John Wiley & Sons, Inc.

Martin, 2003; Pagano et al., 2000). Only in rodents, however, have multi-potent neural stem cells outside the SVZ been demonstrated to generate OB interneurons in vivo (Alonso et al., 2008; Hack et al., 2005; Mendoza-Torreblanca et al., 2008; Merkle et al., 2007).

The rodent SVZ lies adjacent to ependymal cells lining the ventricular wall (Figure 9.1) (Doetsch et al., 1997). Most ependymal cells are multi-ciliated (type E1), but there is a small subset with two partially invaginated cilia (type E2) (Mirzadeh et al., 2008). The cilia of E1 cells have been suggested to direct cerebrospinal fluid (CSF) flow in the ventricle and establish gradients of various signaling molecules (Sawamoto et al., 2006). Alternatively, these cilia may serve as mechanical or chemical sensors of the CSF (Gotz and Stricker, 2006), which is analogous to the proposed function of the E2 cells (Mirzadeh et al., 2008).

Adjacent to the ependymal cells are astrocytes (B1 cells), which have an apical process that contacts the CSF and a basal process that terminates on blood vessels (Mirzadeh et al., 2008; Shen et al., 2008). Together, the B1 and ependymal cells have a pinwheel-like organization on the ventricular wall with the apical processes of several B1 cells at the center and the ependymal cells forming a rosette around this center (Mirzadeh et al., 2008). Vascular cell adhesion molecule 1 (VCAM1) is critical for maintaining this organization (Kokovay et al., 2012). The B1 cells are derived from embryonic radial glia and function as slowly dividing neural stem cells in the SVZ (Doetsch et al., 1999; Laywell et al., 2000; Merkle et al., 2004; Ventura and Goldman, 2007). These cells are typically identified by the expression of glial fibrillary acidic protein (GFAP), epidermal growth factor receptor (EGFR), and the filament protein Nestin.

Slowly dividing B1 cells generate rapidly proliferating transit amplifying cells (type C cells), which can be identified by the expression of EGFR and Nestin as well as the transcription factors Distal-less homeobox 2 (Dlx2) and Achaete-Scute Complex-like 1 (ASCL1). Transit amplifying cells produce neuroblasts (type A cells), which are the progenitor cells that migrate to the OB and become interneurons (Doetsch et al., 1999). Neuroblasts characteristically express neuron-specific class III beta-tubulin (βTubIII/Tuj1), Doublecortin (Dcx) and polysialylated-neural cell adhesion molecule (PSA-NCAM). The collection of A, B1, and C cells are separated from the neighboring striatum by a second set of GFAP-containing astrocytes that have a more stellate morphology (B2 cells) (Doetsch et al., 1997). Mitotically active cells in the SVZ and RMS typically express markers such as the transcription factor SRY related HMG box 2 (Sox2), proliferating cell nuclear antigen (PCNA), nuclear protein Ki-67, and phospho-histone H3.

The cytoarchitecture of adult human SVZ is significantly different from rodents (Figure 9.1) (Quinones-Hinojosa et al., 2006; Sanai et al., 2004). The border with the ventricular space is formed by a single row of multi-ciliated ependymal cells, but unlike rodents, the layer adjacent to these ependymal cells is a region with few cell bodies and has an extensive network of astrocytic processes. Many ependymal cells have basal processes that also project into this region. On the other side of this hypocellular gap is a dense ribbon of GFAP-expressing astrocytic cell bodies. Within this ribbon, some astrocytes express markers of mitotic activity. These proliferative cells may be the multi-potent progenitors observed with in vitro cultured explants, but a rigorous determination as to whether they act as neural stem cells in vivo remains to be established. Furthest from the ventricle is a transitional region that starts to resemble the neighboring striatal region and contains the first neuronal cell bodies.

Also unlike rodents, the adult human SVZ has noticeably very few cells that express markers of migrating neuroblasts, such as PSA-NCAM or Dcx (Quinones-Hinojosa et al., 2006; Sanai et al., 2011; Sanai et al., 2004; Wang et al., 2011a). The few cells that are present typically are found as individuals or pairs of cells in either the hypocellular layer or astrocytic ribbon on the ventral surface of the anterior horn. There is no dense network of migrating neuroblasts, as in adult rodents. The SVZ in both the human fetus (22–24 weeks post-conception) and infant (less than 18 months), however, has large numbers of PSA-NCAM expressing cells (presumably neuroblasts) adjacent to the ependymal layer that resemble the migrating chains of neuroblasts observed in rodents (Figure 9.1) (Guerrero-Cazares et al., 2011; Sanai et al., 2011; Wang et al., 2011a; Weickert et al., 2000). There are also several EGFR-expressing cells in the infant human SVZ that are either transit amplifying cells or neural stem cells. Unlike rodents, however, neither these EGFR-expressing cells nor the GFAP-expressing cells form tubes around the putative neuroblasts. Rather, the GFAP-expressing cells that line the ventricle wall have radial glia-like processes. Approximately eighteen months after birth, however, the population of PSA-NCAM expressing cells is depleted and the apparent migratory route adjacent to the ependymal cells transforms into the hypocellular space observed in adults. Although blood vessels have a major role in the organization of the rodent SVZ and RMS, their role, if any, in humans has not been described.

9.2.2 Neuroblast Migration Through the SVZ and RMS

In the adult rodent SVZ, migrating neuroblasts form chains that are ensheathed by astrocytes and transit-amplifying cells and tangentially migrate along the rostro-caudal axis of the ventricle (Doetsch and Alvarez-Buylla, 1996;

9.2 Subventricular Zone and Rostral Migratory Stream as a Neurogenic Niche

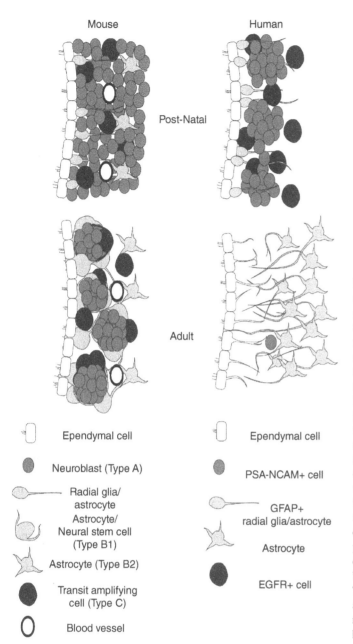

Figure 9.1 Cytoarchitecture of the post-natal and adult SVZ in mice and humans. These cartoons represent coronal views of either the anterior SVZ of the murine lateral ventricle or the ventral SVZ in the anterior horn of the human lateral ventricle. In the post-natal mouse, neuroblasts (Type A cells) migrate in a mass tangentially along the rostro-caudal axis of the ventricle. Transit amplifying cells (Type C cells) are interspersed among the neuroblasts and GFAP-expressing glia have radial glia-like projection. In adult mice, the astrocytes (Type B1 cells), including the neural stem cells, have organized their processes into a mesh that form tubes surrounding the tangentially migrating chains of neuroblasts. Transit amplifying cells (Type C cells) are also part of these glial tubes. This cellular network is separated from the adjacent brain parenchyma by astrocytes (Type B2 cells). By contrast, the human post-natal SVZ contains chains of tangentially oriented PSA-NCAM expressing cells that are adjacent to the ependymal cells. Interspersed among these presumptive neuroblasts are EGFR-expressing cells (likely either transit amplifying cells or neural stem cells) and GFAP-expressing glia with radial processes. In the adult, the populations of PSA-NCAM and EGFR expressing cells are drastically depleted, leaving a hypocellular gap between ependymal cells and a thick ribbon of astrocyte cell bodies. This gap region is filled largely with processes from the neighboring astrocytes and ependymal cells. Although blood vessels have a major role in the organization of the murine SVZ, a similar role in humans has not been be documented.

Doetsch et al., 1997; Lois et al., 1996). This progenitor migration parallels CSF flow in the lateral ventricle, which is suggested to contain gradients of guidance molecules that provide direction for migrating neuroblasts (Sawamoto et al., 2006). The SVZ also has a rich plexus of blood vessels that are closely associated with astrocytes in the SVZ and serve as a scaffold for migrating of neuroblast chains (Shen et al., 2008; Snapyan et al., 2009; Tavazoie et al., 2008).

Migrating neuroblasts in adults reach the OB by traversing the RMS, which is a rostral extension of the SVZ formed by closing of the olfactory ventricle that extends from main lateral ventricle during embryonic development. The RMS in adult rodents contains a network of interconnected glial tubes through which neuroblasts migrate and blood vessels serve as important scaffolds for these glial tubes (Whitman et al., 2009). This cytoarchitecture of the adult RMS, however, does not develop until about the third postnatal week of life (Alves et al., 2002; Pencea and Luskin, 2003; Peretto et al., 2005). At birth, astrocytes in the SVZ and RMS still have radial glial-like fibers and neuroblasts migrate as a mass with little organization. During the first several weeks of postnatal development, the astrocytic processes start to form a mesh that becomes glial tubes and the migrating neuroblasts organize into chains ensheathed by this meshing. The organization of the astrocytic processes in the glial tubes is actively modulated by neuroblasts in order to form and maintain their migration routes (Kaneko et al., 2010).

Astrocytes have a prominent role in RMS and SVZ cytoarchitecture, but they are not required for migrating neuroblasts to form chains (Peretto et al., 2005; Wichterle et al., 1997). Rather, glia may have other critical organizational functions. Since astrocytes express several molecules that regulate both neural precursor proliferation and neuroblast migration (discussed further in Section 9.2.3), the mature glial cytoarchitecture of the SVZ and RMS may be necessary to effectively regulate neurogenesis in adults. Alternatively or in addition, the mature glial cytoarchitecture may ensure that neuroblasts are shunted towards the OB. In early postnatal rodent brains, SVZ-derived progenitors also migrate to cortical and sub-cortical regions in addition to the OB (De Marchis et al., 2004a; Inta et al., 2008). In human infants, SVZ-derived progenitors migrate along a medial migratory stream to the ventromedial prefrontal cortex in addition to the OB via the RMS (Sanai et al., 2011). In both adult rodents and humans, however, these alternative post-natal migratory pathways are absent.

Neuroblasts migrate through the SVZ and RMS in a saltatory manner with rapid changes in their morphology (Kakita and Goldman, 1999). Although neuroblasts appear organized in chains, their individual movements are complex and include turning motions between 30 to 90 degrees as well as direction reversals (Martinez-Molina et al., 2011; Nam et al., 2007; Suzuki and Goldman, 2003). This non-linear migration makes it difficult to estimate the speed at which individual progenitors migrate, but studies using confocal microscopy of slice cultures have estimated average migration speeds between 50–105 μm/hr (De Marchis et al., 2001; Nam et al., 2007; Platel et al., 2008b). Neuroblast migration rates are inherently variable, however, and they are regulated by several mechanisms (discussed further in Section 9.2.3).

Once neuroblasts reach the OB, they switch from tangential to radial migration. The cellular motion remains saltatory, but radial migration rates tend to be slower than chain migration and range between 5 to 65 μm/hr (Bovetti et al., 2007b; De Marchis et al., 2001). Similar to the SVZ and RMS, blood vessel networks guide radially migrating neuroblasts towards their final positions in the OB (Bovetti et al., 2007b). Adult olfactory radial glia also assist in this migration (Emsley et al., 2012). These glia are distinct from those in either the SVZ or RMS and have cell bodies in the OB granule cell layer that send non-arborized radial processes into the mitral cell layer. Olfactory radial glia are born in either the RMS or OB of both embryos and adults, but how these cells direct radial migrating neuroblasts remains to be established.

The nature of the RMS in adult humans has been a source of controversy. The human RMS descends ventrally and caudally from the ventral surface of the anterior horn in the lateral ventricle before turning in a rostral direction through the olfactory tract towards the OB (Curtis et al., 2007; Sanai et al., 2011; Wang et al., 2011a). A similar laminar organization and shared expression of Neogenin (Netrin and RGMa receptor) indicates that the human RMS is a rostral extension of SVZ as it is in rodents (Bradford et al., 2010; Kam et al., 2009). Some studies have reported that the adult human RMS contains either an open or partially closed ventricle that extends the length of RMS (Curtis et al., 2007; Kam et al., 2009), but others have failed to observe this ventricular space (Sanai et al., 2011; Wang et al., 2011a). The human SVZ and RMS have been consistently reported to lack an intricate network of glial tubes containing migrating chains of neuroblasts found in rodents, but the number of progenitors present in the SVZ and RMS has been debated. Some investigators have suggested that both the SVZ in the ventral surface of anterior horn and descending limb of the RMS contain many proliferating cells and neuroblasts expressing PSA-NCAM (Curtis et al., 2007; Kam et al., 2009). By contrast, others have reported that the entire SVZ/RMS axis is only sparsely populated with proliferating cells and PSA-NCAM expressing neuroblasts (Sanai et al., 2011; Sanai et al., 2004; Wang et al., 2011a). Differences in both the experimental methodology and patient populations used for the human studies may account for some discrepancies between these two sets of studies, but further investigation is necessary to resolve these fundamental issues regarding the cytoarchitecture of the adult human SVZ and RMS.

9.2.3 Factors that Control Proliferation and Migration

The cytoarchitecture of the SVZ and RMS places neural stem cells, transit amplifying cells and neuroblasts in close proximity to blood vessels, CSF and axon terminals originating from other brain regions. The micro-environmental niche of the SVZ and RMS integrates many different developmental and physiological signals to control neurogenesis (Ihrie and Alvarez-Buylla, 2011; Riquelme et al., 2008). Some of the factors that modulate proliferation and migration in the adult SVZ and RMS are listed in Table 9.1. The list is extensive, but not exhaustive, and is intended to

9.2 Subventricular Zone and Rostral Migratory Stream as a Neurogenic Niche

Table 9.1 Factors that regulate neural progenitor proliferation and migration in the adult SVZ and RMS

Factor	Reported Function in either Proliferation or Migration	Reference
Cell Cycle		
Cyclin dependent kinase 2 (Cdk2)	Promotes proliferation of transit amplifying cells	Jablonska et al., 2007
Cyclin dependent kinase 6 (Cdk6)	Promotes proliferation in transit amplifying cells and neuroblasts	Beukelaers et al., 2011
p16[INK4a]	Reduces proliferative capacity of progenitors in an age-dependent manner	Molofsky et al., 2006
p19[INK4d]	Promotes cell cycle arrest in migrating neuroblasts	Coskun and Luskin, 2001
p21[Cip1]	Controls size of neural stem cell pool by promoting mitotic quiescence	Kippin et al., 2005
p27[Kip1]	Reduces proliferation of transit amplifying cells	Doetsch et al., 2002b; Li et al., 2009
p53	Reduces proliferation of neural stem cells and promotes differentiation of transit amplifying cells into neuroblasts	Gil-Perotin et al., 2011; Gil-Perotin et al., 2006
p107	Regulates neuroblast production by controlling self-renewal of neural stem cells and transit amplifying cells	Vanderluit et al., 2004; Vanderluit et al., 2007
Phosphatase and tensin homolog (PTEN)	Inhibits self-renewal of neural stem cells	Gregorian et al., 2009
Cytoskeleton		
Doublecortin (Dcx)	Microtubule-associated protein that is necessary for proper tangential and radial migration of neuroblasts	Koizumi et al., 2006
Glial fibrillary acidic protein (GFAP)	Delta (δ) splice-variant of this intermediate filament protein is expressed in human SVZ astrocytes and promotes perinuclear collapse	Roelofs et al., 2005
Pericentrin (Pcnt)	Centrosomal protein that promotes neuroblast migration along RMS	Endoh-Yamagami et al., 2010
Extracellular Matrix/ Guidance Molecules		
A disintegrin and metalloprotease 2 (ADAM2)	Facilitates migration of neuroblasts	Murase et al., 2008
Anosmin-1	Chemoattractant guiding neuroblasts towards OB	Garcia-Gonzalez et al., 2010
ErbB4/Neuregulins	Promote proliferation of neural stem cells and facilitate chain formation of migrating neuroblasts	Anton et al., 2004; Ghashghaei et al., 2006
Integrins	Chemoattractants guiding neuroblasts towards OB; promote chain formation of neuroblasts and their association with blood vessels; maintain integrity of glial tubes in RMS	Belvindrah et al., 2007; Emsley and Hagg, 2003a; Kokovay et al., 2010; Mobley and McCarty, 2011; Mobley et al., 2009; Shen et al., 2008
Matrix metalloproteases	Facilitate migration of neuroblasts	Bovetti et al., 2007a; Lee et al., 2006; Wang et al., 2006

(continued)

Table 9.1 (*Continued*)

Factor	Reported Function in either Proliferation or Migration	Reference
Neogenin/ Deleted in Colorectal Cancer (DCC)/Netrin	Chemoattractant that guides migrating neuroblasts towards OB	Hakanen et al., 2011; Murase and Horwitz, 2002
Neural cell adhesion molecule (NCAM)	Facilitates chain formation of migrating neuroblasts	Battista and Rutishauser, 2010; Chazal et al., 2000; Cremer et al., 1994; Hu, 2000; Hu et al., 1996; Ono et al., 1994
Prokinecticin 2 (PK2)	Chemoattractant for migrating neuroblasts and facilitates switch from tangential to radial migration in OB	Ng et al., 2005; Prosser et al., 2007; Puverel et al., 2009
Reelin	Facilitates switch from tangential to radial migration in OB	Courtes et al., 2011; Hack et al., 2002; Simo et al., 2007
Robo/Slit	Chemorepellant guiding neuroblasts towards OB and maintain orgnanization of glial surrounding migrating neuroblasts	Chen et al., 2001; Hu, 1999; Kaneko et al., 2010; Nguyen-Ba-Charvet et al., 2004; Wu et al., 1999
Stromal derived factor 1 (SDF1/CXCL12)/ Chemokine receptor 4(CXCR4)	Maintains progenitor cells in quiescent state in absence of activated EGFR; stimulates migration to blood vessels with EGFR is activated	Kokovay et al., 2010
Tenascin-R	Facilitates switch from tangential to radial migration in OB	Saghatelyan et al., 2004
Vascular Cell Adhesion Molecule 1(VCAM1)	Maintains neural stem cell and ependymal cell pinwheel structure in SVZ	Kokovay et al., 2012
Transcription Factors and miRNA		
Aristaless related homeobox (Arx)	Promotes proliferation and migration of progenitors	Yoshihara et al., 2005
Arsenite-resistance protein 2 (Ars2)	Promotes neural stem cell self-renewal	Andreu-Agullo et al., 2012
Bmi1	Promotes neural stem cell self-renewal	Fasano et al., 2009; Molofsky et al., 2003; Zencak et al., 2005
Cyclic AMP response element binding protein (CREB)	Promotes proliferation of neural precursor cells	Iguchi et al., 2011
Distal-less related homeobox (Dlx) family	Promotes proliferation of transit amplifying cells as well as their transition from neural stem cells	Suh et al., 2009
Empty spiracles homeobox 2 (Emx2)	Negative regulator of neural precursor proliferation	Galli et al., 2002
Forkhead box protein O3 (FoxO3)	Promotes maintenance of precursor cells in a mitotically quiescence and undifferentiated state	Renault et al., 2009
Genome screened homodomain 2 (Gsx2)	Promotes maintenance of precursor cells in a mitotically quiescence and undifferentiated state	Mendez-Gomez and Vicario-Abejon, 2012
High mobility group A protein 2 (Hmg2a)	Promotes neural stem cell self-renewal	Nishino et al., 2008
Let-7b	Promotes neural stem cell proliferation and self-renewal	Nishino et al., 2008; Zhao et al., 2010

9.2 Subventricular Zone and Rostral Migratory Stream as a Neurogenic Niche

Table 9.1 (*Continued*)

Factor	Reported Function in either Proliferation or Migration	Reference
miR-124	Promotes differentiation of progenitors	Akerblom et al., 2012; Cheng et al., 2009
miR-9	Promotes neural stem cell proliferation and self-renewal	Zhao et al., 2009
Querkopf (Qkf/Myst4)	Promotes neural stem cell self-renewal	Merson et al., 2006; Sheikh et al., 2012
SRY related HMG box 2 (Sox2)	Promotes proliferation of neural precursor cells	Ferri et al., 2004
SRY related HMG box 9 (Sox9)	Inhibits transition of transit amplifying cells to neuroblasts	Cheng et al., 2009
Serum Response Factor (SRF)	Promotes neuroblasts migration by modulating regulating structure and dynamics of actin microfilaments	Alberti et al., 2005
Smad interacting protein 1 (Sip1)	Regulates production of astrocytes associated with glial tubes	Nityanandam et al., 2012
Specificity protein (Sp8)	Promotes survival in a subset of neuroblasts	Waclaw et al., 2006
Tailless (Tlx)	Promotes proliferation and self-renewal of neural stem cells	Liu et al., 2010; Qu et al., 2010
Ventral anterior homeobox 1 (Vax1)	Negative regulator of neural precursor proliferation	Soria et al., 2004

Growth Factors

Factor	Reported Function in either Proliferation or Migration	Reference
Brain derived neurotrophic factor (BDNF)	Increases neuroblast migration rate; its role in regulating proliferation of either neural stem cell or transit amplifying cell is controversial	Bath et al., 2012
Ciliary neurotrophic factor (CNTF)	Increases proliferation of neural stem cells and transit amplifying cells	Emsley and Hagg, 2003b; Shimazaki et al., 2001
Epidermal growth factor (EGF)	Increases proliferation in a subpopulation of neural stem cells and transit amplifying cells	Doetsch et al., 2002a; Gonzalez-Perez and Alvarez-Buylla, 2011
Fibroblast growth factor (FGF)	Increases proliferation of neural stem cells and transit amplifying cells	Mudo et al., 2009
Glial derived neurotrophic factor (GDNF)	Chemoattractant for RMS-derived neuroblasts	Paratcha et al., 2006
Hepatocyte growth factor (HGF)	Increases transit amplifying cell proliferation and acts as a chemoattractant for neuroblast migration to OB	Garzotto et al., 2008; Nicoleau et al., 2009; Wang et al., 2011b
Insulin growth factor (IGF)	Promotes neuroblast migration transition from SVZ to RMS	Hurtado-Chong et al., 2009
Leukemia inducible factor (LIF)	Promotes self-renewal of neural stem cells	Shimazaki et al., 2001
Oncostatin M	Decreases proliferation of neural stem cells and transit amplifying cells	Beatus et al., 2011
Platelet-derived growth factor (PDGF)	Promotes proliferation of a subset of neural stem cells	Jackson et al., 2006
Transforming growth factor-alpha (TGF-α)	Promotes proliferation of transit amplifying cells	Tropepe et al., 1997

(*continued*)

Table 9.1 (*Continued*)

Factor	Reported Function in either Proliferation or Migration	Reference
Vascular endothelial growth factor (VEGF)	Increases production of migrating neuroblasts and facilitates development of vascular scaffold surrounding RMS; its role as a chemoattractant for migrating neuroblasts is controversial	Bozoyan et al., 2012; Jin et al., 2002; Licht et al., 2010; Wittko et al., 2009; Zhang et al., 2003
Signaling Pathways		
Eph tyrosine kinase receptor family /Ephrin ligands	Reduce proliferation of neural stem cells and, in some cases, facilitates migration of neuroblasts	Conover et al., 2000; Holmberg et al., 2005; Katakowski et al., 2005; Ricard et al., 2006
Hedgehog pathways	Promote proliferation of neural stem and transit amplifying cells and promotes migration of neuroblasts; Sonic hedgehog (Shh) ligand is also a chemoattractant for migrating neuroblasts	Ahn and Joyner, 2005; Angot et al., 2008; Balordi and Fishell, 2007a; Machold et al., 2003; Palma et al., 2005
Notch pathway	Maintains pool size of proliferating neural stem and transit amplifying cells by promoting self-renewal	Ables et al., 2011; Alexson et al., 2006; Basak et al., 2012; Hitoshi et al., 2002
Nucleotide	Increases proliferation of transit amplifying cells	Mishra et al., 2006; Suyama et al., 2012
Wingless/Integration 1 (Wnt) pathway	Promotes proliferation of neural stem and transit amplifying cells, but co-expression of homeodomain interacting protein kinase (Hipk1) in some neural stem cells may induce Wnt signaling to promote cell cycle arrest	Adachi et al., 2007; Marinaro et al., 2012
Neurotransmitters		
Dopamine	Promotes proliferation of transit amplifying cells	Borta and Hoglinger, 2007; O'Keeffe et al., 2009a
GABA	Decreases proliferaton of both neural stem cells and neuroblasts, also reduces neuroblast migration rates	Bolteus and Bordey, 2004; Gascon et al., 2006; Liu et al., 2005; Nguyen et al., 2003
Glutamate	Promotes neuroblast proliferation and survival, but decreases migration rates	Brazel et al., 2005; Di Giorgi Gerevini et al., 2004; Platel et al., 2008b
Serotonin	Increases markers of proliferation	Banasr et al., 2004; Hitoshi et al., 2007; Soumier et al., 2010

provide a sense for the wide range of mechanisms that regulate neurogenesis and migration in the adult SVZ and RMS.

9.2.3.1 *Cell Cycle Regulators.*

The finding that cell cycle gene disruption affects neural precursor proliferation is not surprising, but the cell type specificity for some of these genes during adult neurogenesis is unexpected (Beukelaers et al., 2012). For example, loss of p21[Cip1] preferentially increases proliferation in neural stem cells in the SVZ (Kippin et al., 2005). When compared to wild-type controls, loss of p21[Cip1] increases the number of mitotically active neural stem cells in adult mice. This increase is not sustained throughout adult life, however, since the number of mitotically active neural stem cells in aged adults is significantly reduced. These findings suggest that SVZ neural stem cells have a limited capacity for self-renewal and p21^{Cip1} is required to maintain these stem cells in a quiescent state so that only a limited number of neural stem cells are mitotically active at any one time.

In contrast to p21^{Cip1}, loss of p27^{Kip1} selectively increases proliferation of transit amplifying cells in the SVZ without noticeably affecting either the stem cells or neuroblasts (Doetsch et al., 2002b). Although the proliferation rate of mitotically active neuroblasts is not disturbed, there is a significant reduction in the number of neuroblasts present in the SVZ and RMS that is likely due to transit amplifying cells undergoing rounds of self-renewal instead of generating neuroblasts.

9.2.3.2 Cytoskeletal Proteins.

Neuroblast migration through the SVZ and RMS is initiated by the extension of a leading process that is followed by movement of the centrosomes and translocation of the nucleus in the same direction (Kakita and Goldman, 1999). The cell soma moves forward when the trailing process detaches and repositions itself. Thus, migration requires the movement of centrosomes coordinated with the rearrangement of cytoskeletal structures necessary for the extension and repositioning of processes. Consistent with a pivotal role for this coordination, neuroblast migration is impaired by disruptions in either microtubule-associated proteins, like Dcx and Microtubule affinity-regulating kinase 2 (Mark2), or centrosome-interacting proteins, like Pericentrin (Pcnt) (Endoh-Yamagami et al., 2010; Koizumi et al., 2006; Mejia-Gervacio et al., 2012). The human-specific GFAP-δ splice variant promotes collapse of cytoskeletal intermediate filaments near the nucleus and its expression in the SVZ astrocytic ribbon suggests that these astrocytes have motility requirements distinct from other regions (Roelofs et al., 2005).

9.2.3.3 Extra-Cellular Matrix and Guidance Molecules.

The extra-cellular matrix environment in the SVZ and RMS contains chondroitin sulfate, heparin sulfate, proteoglycans, Tenascin-C, and laminin (Cayre et al., 2009; Leong and Turnley, 2011; Sun et al., 2010b). These extra-cellular matrix molecules regulate neuroblast migration by serving as ligands for integrin receptors, which are heterodimeric cell surface receptors composed of one α and β subunit. Inhibition of α6β1 integrin disrupts both the association of neuroblasts with blood vessels and tangential chain migration of neuroblasts (Belvindrah et al., 2007; Emsley and Hagg, 2003a; Kokovay et al., 2010; Shen et al., 2008). Similar disruptions are also observed with inhibition of the β8 subunit (Mobley and McCarty, 2011;

Mobley et al., 2009). Interactions between α6β1 integrin and laminin are also critical for guiding the neuroblasts towards the OB and infusion of exogenous laminin can misroute progenitor towards sites of infusion (Emsley and Hagg, 2003a).

Before chain migration and glial tubes are established in the postnatal SVZ and RMS, neuroblast migration is dependent on matrix metalloproteases (Bovetti et al., 2007a). Once the mature glial cytoarchitecture is established, however, metalloprotease enzymes are not required for tangential chain migration. Rather, both integrin and ErbB4 receptors as well as PSA-NCAM are necessary for neuroblast chain organization (Anton et al., 2004; Belvindrah et al., 2007; Hu et al., 1996; Mobley and McCarty, 2011; Ono et al., 1994). In the case of PSA-NCAM, genetic deletion of NCAM disrupts chain migration, in part, by disrupting glial tube formation (Chazal et al., 2000). Since NCAM is only expressed by neuroblasts, this perturbation of glial tube formation indicates that interactions between neuroblasts and the surrounding astrocytes are critical for maintenance of the glial cytoarchitecture.

Further evidence that neuroblasts are necessary for the maintenance of the adult glial cytoarchitecture in the SVZ and RMS comes from studies with Slit ligands and their cognate Robo receptors. The canonical roles for Slit 1/2 ligands are as chemorepellants secreted by the choroid plexus and septum that guide migrating neuroblasts towards the OB (Hu, 1999; Nguyen-Ba-Charvet et al., 2004; Wu et al., 1999). Recent studies, however, show that migrating neuroblasts also secrete Slit1 and activate Robo receptors on astrocytic processes (Kaneko et al., 2010). This signaling alters the distribution and extension of astrocytic processes forming the glial tube. The expression of Slit1 in the RMS and SVZ may be regulated by Sonic hedgehog (Shh), which can partially account for neuroblast migration defects observed when Hedgehog signaling is disrupted (Balordi and Fishell, 2007a).

Some of the factors required for tangential migration, such as ErbB4 receptors and Dcx, are also necessary for radial migration (Anton et al., 2004; Koizumi et al., 2006), whereas others, like PSA-NCAM, are dispensable (Hu et al., 1996; Ono et al., 1994). The transition to radial migration also requires Reelin (Courtes et al., 2011; Hack et al., 2002; Simo et al., 2007), Tenascin-R (Saghatelyan et al., 2004) and Prokinecticin 2 (Ng et al., 2005; Prosser et al., 2007; Puverel et al., 2009).

The extra-cellular signals and guidance cues that determine the final location of radially migrating neuroblasts are not established. Since OB interneurons in different OB lamina have distinct sets of characteristic synaptic connectivities (Shepherd, 1972), elucidating the cues that terminate radial migration is critical for understanding how OB circuits are established during development and maintained by adult neurogenesis.

9.2.3.4 Transcription Factors and miRNA.

Each successive stage of neurogenesis requires a specific set of transcription factors to establish the proper gene expression profile of the cell. MicroRNAs (miRNA) are important modifiers of transcription factor gene expression during both embryonic and adult neurogenesis. These non-coding RNAs target complementary regions in transcription factor mRNA transcripts, which promotes transcript degradation and suppresses protein production. Thus, regulation of both transcription factor and miRNA expression is critical for controlling both the timing and progression of neurogenesis.

Proliferation and self-renewal of neural stem cells in the adult SVZ is tightly controlled by several of transcription factors and miRNA. Several of these factors, including a pivotal role by the transcription factor Tail-less homeobox (Tlx), are shown in Figure 9.2. Tlx is a transcription repressor protein and its expression is directly activated by Sox2 (Shimozaki et al., 2012), a common marker of proliferation in the SVZ. Tlx expression is antagonized by negative auto-feedback and both miR-9 and Let-7b miRNAs, which also target other transcription factors that regulate neural stem cell proliferation and self-renewal (Nishino et al., 2008; Shibata et al., 2011; Zhao et al., 2010; Zhao et al., 2009). Expression of miR-9, however, is directly repressed by Tlx (Zhao et al., 2009). Thus, Tlx participates in multiple feedback loops to regulate its own expression levels. Tlx promotes neural stem cell proliferation and self-renewal by directly repressing the transcription of cell cycle inhibitors, such as Pten and p21^{Cip1} (Sun et al., 2007).

The differentiation of transit amplifying cells into neuroblasts is controlled, in part, by Dlx2 and Sox9 transcription factors and miR-124. Dlx2 cooperates with EGFR signaling to promote self-renewal of cells at the transition between transit amplifying cells and migrating neuroblasts (Suh et al., 2009). This function of Dlx2 may be mediated by its activation of Arista-less homeobox (Arx) expression (Cobos et al., 2005; Colasante et al., 2008), which is critical for the proliferation and migration of progenitors (Yoshihara et al., 2005). Sox9 impedes the transition from transit amplifying cells to neuroblasts, but its target genes are not identified (Cheng et al., 2009). In contrast to both Dlx2 and Sox9, miR-124 promotes the differentiation of neuronal progenitors. This miRNA is expressed at low levels in neuroblasts, but progressively increases as they differentiate into mature neurons (Akerblom et al., 2012; Cheng et al., 2009). miR-124 advances differentiation of transit amplifying cells to mitotically quiescent neuroblasts by down regulating both Dlx2 and Sox9. Since miR-124 expression is maintained in post-mitotic mature OB interneurons, it likely has many additional and stage-specific targets that are essential for promoting neuronal differentiation.

9.2.3.5 Growth Factors.

Most growth factors stimulate proliferation and/or facilitate migration, but their actions are not always uniform throughout the SVZ. Transforming growth factor-α (TGF-α), for example, is required for the proliferation of transit amplifying cells in a specific region of SVZ (Tropepe et al., 1997). Genetic loss of TGF-α diminishes neural precursor proliferation in only the dorsolateral corner, but intraventricular infusion of TGF-α can stimulate proliferation throughout the entire SVZ, suggesting that other growth factors are functionally redundant to TGF-α outside of the dorsolateral corner.

This partial functional redundancy within the SVZ indicates that there is heterogeneity in the expression of different growth factor receptors. Such heterogeneity in

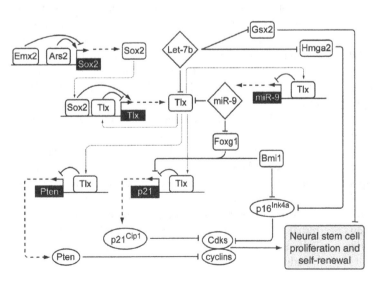

Figure 9.2 A network of select transcription factors and miRNAs that regulate neural stem cell proliferation and self-renewal. The Tlx transcription factor has a pivotal role in maintaining the neural stem cell pool size. Tlx promotes proliferation and self-renewal by directly repressing transcription of cyclin and cyclin dependent kinase (Cdk) inhibitor genes, such as Pten and p21^{Cip1}. The expression of Tlx itself in neural stem cells is tightly controlled by several transcription factors and miRNA, including feedback loops that are either positive (repressing miR-9 expression) or negative (binding its own promoter).

growth factor sensitivity suggests that different factors can target specific subsets of neural stem cells, transit amplifying cells and neuroblasts. As discussed further in Section 9.4, the SVZ is organized in spatially distinct regions that preferentially generate specific OB interneuron phenotypes. Heterogeneity of growth factor receptors in neural stem cells and transit amplifying cells may be integral for regulating the production of specific interneurons phenotypes. This heterogeneity may also be an important determinant of neuronal vs. glial cell-fate production in the SVZ. For example, EGF and Platelet-derived growth factor (PDGF) promote proliferation of specific subsets of neural stem cells and transit amplifying cells that preferentially generate oligodendrocytes (Doetsch et al., 2002a; Gonzalez-Perez and Alvarez-Buylla, 2011; Jackson et al., 2006; Menn et al., 2006).

Although many growth factors promote proliferation and migration in the SVZ, Oncastatin M is an exception. The receptor protein for this member of the gp130-dependent family of cytokines is expressed in a subset of neural stem cells and transit amplifying cells (Beatus et al., 2011). Both *in vitro* studies with neurospheres and intraventricular infusion with Oncastatin M in vivo demonstrate that it is a potent inhibitor of neural precursor proliferation. This inhibitory role contrasts with other members of the gp130-dependent family of cytokines, including ciliary neurotrophic factor (CNTF) and leukemia inducible factor (LIF), which promote progenitor proliferation and contribute to the long-term maintenance of proliferating neural precursors pools (Emsley and Hagg, 2003b; Shimazaki et al., 2001). The differential mitogenic responses to these factors may be mediated by the precise subunit composition of the receptor complexes receiving either the Oncostatin M, CNTF, or LIF signals (Beatus et al., 2011).

9.2.3.6 Intercellular Signaling Pathways.
Several inter-cellular signaling pathways are necessary for maintaining the pool size of proliferating neural precursor cells. As discussed with the role of cell cycle components (Section 9.2.3.1), the transition from mitotic quiescence to active proliferation in neural stem cells as well as self-renewal rate of both neural stem cells and transit amplifying cells needs to be tightly controlled so that a reservoir of proliferating precursor cells can be maintained through life. Disruption of either Notch or Hedgehog signaling pathways, results in the precocious differentiation of precursors cells and readily depletes the number of neural stem cells and transit amplifying cells in the SVZ (Alexson et al., 2006; Balordi and Fishell, 2007b; Basak et al., 2012; Hitoshi et al., 2002). Consistent with these roles in regulating proliferation, both Notch and Hedgehog pathways have been shown in other cellular systems to

directly target genes encoding cell cycle regulatory proteins. The Notch pathway regulates the transcription of Cyclin D1, Cyclin dependent kinase 2 (Cdk2) and p21^{Cip1} (Castella et al., 2000; Ronchini and Capobianco, 2001), whereas the Hedgehog pathway directly targets Cdk6 and p16Ink4a (Bishop et al., 2010; Lopez-Rios et al., 2012).

Also consistent with the role that growth factors have in stimulating proliferation in the SVZ, there are extensive interactive networks between growth factors and intercellular signaling pathways. Activated EGFR signaling in neural stem cells, for example, promotes Numb-mediated ubiquitination of the Notch-1 receptor (Aguirre et al., 2010). This degradation of Notch-1 suppresses Notch pathway activity and reduces the number, proliferation, and self-renewal of neural stem cells. Several other interactions between growth factor and intercellular signaling pathways are reported in the adult SVZ, including: Notch and CNTF (Chojnacki et al., 2003), Notch and pigment epithelium-derived factor (PEDF) (Andreu-Agullo et al., 2009), Notch and soluble factors, including VEGF, released by endothelial cells (Shen et al., 2004; Sun et al., 2010a), Hedgehog and FGF (Goncalves et al., 2009), as well as purinergic signaling pathways and EGF (Grimm et al., 2009).

9.2.3.7 Neurotransmitters.
Neurotransmitters are essential regulators of cell proliferation and migration during adult neurogenesis. The role of GABA is the most extensively studied. In the SVZ, GABA is synthesized and released by neuroblasts (De Marchis et al., 2004b; Plachez and Puche, 2012; Wang et al., 2003). Functional GABA$_A$ receptors are on both neuroblasts and neural stem cells, but not transit amplifying cells (Gascon et al., 2006; Nguyen et al., 2003; Stewart et al., 2002; Wang et al., 2003). For neuroblasts, GABA functions as an autocrine signal that reduces both their migration speed and proliferation rate (Bolteus and Bordey, 2004; Gascon et al., 2006; Nguyen et al., 2003). GABA depolarizes neuroblasts and increases intracellular Ca^{+2} concentrations, in part, by activating L-type calcium channels (Darcy and Isaacson, 2009; Nguyen et al., 2003; Wang et al., 2003). GABA depolarizes neuroblasts because of their high intracellular chloride concentrations generated by the expression of sodium-potassium-chloride co-transporter 1 (NKCC1), and pharmacological disruption of NKCC1 reverses the effect of GABA on neuroblasts (Gascon et al., 2006; Mejia-Gervacio et al., 2011; Sun et al., 2012). In neural stem cells, GABA is a paracrine signal that reduces proliferation (Liu et al., 2005). GABA stimulates opening of L- and T-type calcium channels in neural stem cells (Young et al., 2010) and this calcium influx may be communicated throughout the SVZ by an extensive network

of connexin-43 gap junctions between astrocytes (Lacar et al., 2011).

Glutamate synthesized and released by astrocytes can also regulate proliferation and migration in the SVZ (Platel et al., 2010; Platel et al., 2007). Functional glutamate receptors have not been identified on either neural stem cells or transit amplifying cells in the SVZ, but neuroblasts express metabotropic mGluR5 (GRM5) and kainate GluK5-7 (GRIK 1-3) receptors (Platel et al., 2008b). Stimulation of mGluR5 increases neuroblast proliferation, but activation of the GluK5 receptors reduces neuroblast migration rates (Di Giorgi Gerevini et al., 2004; Platel et al., 2008b). As neuroblasts migrate towards the OB, they also progressively increase in their responsiveness to NMDA (Platel et al., 2010). The specific sub-unit composition of the NMDA receptors is not established, but stimulation of these receptors blocks apoptosis and promotes survival of neuroblasts. Studies with SVZ-derived neurospheres indicate that metabotropic receptors can also block apoptosis as well as increase proliferation (Brazel et al., 2005; Castiglione et al., 2008), but whether these receptors serve this function in vivo remains to be established.

In addition to locally generated GABA and glutamate, the SVZ receives serotonergic input from axon terminals originating in the raphe nuclei (Azmitia and Segal, 1978). Studies using either lesions to the raphe nuclei or treatment with serotonin receptor agonists indicate the activation of serotonin receptors in the SVZ stimulates proliferation (Banasr et al., 2004; Brezun and Daszuta, 1999; Soumier et al., 2010). The precise cell-types that express these receptors, however, are not established. Analysis of transgenic mice show that neuroblasts express the 5-HT3 receptors (Inta et al., 2008), and both pharmacological studies and reverse-transcription/polymerase chain reaction analyses indicate that 5-HT1 and 5-HT2 classes of serotonin receptors are also expressed (Banasr et al., 2004; Councill et al., 2006; Hitoshi et al., 2007; Soumier et al., 2010). Some 5-HT1 receptors, however, may have species-specific functions (Arnold and Hagg, 2012).

Dopaminergic terminals originating in the substantia nigra also modulate neurogenesis in the SVZ (Freundlieb et al., 2006). D2-like dopamine receptors are expressed by transit amplifying cells and both D2-like and D1-like receptors are expressed by neuroblasts (Hoglinger et al., 2004). Loss of midbrain dopaminergic input either by deafferentation or chemical lesions indicate that dopamine stimulates proliferation in the SVZ (Borta and Hoglinger, 2007). The mechanism of action for dopamine may be integrated with the role of growth factors, such as CNTF and EGF (O'Keeffe et al., 2009b; Yang et al., 2008). *In vitro* studies, however, have produced some conflicting results regarding the action of dopamine on SVZ neurogenesis, but technical experimental differences, such as the specific antagonists or agonists used, may underlie some of these discrepancies (O'Keeffe et al., 2009a).

The regulation of neurogenesis by the locally generated GABA and glutamate make a convenient reciprocal feedback pathway between neuroblasts and neural stem cells to modulate progenitor production (Platel et al., 2008a). The input from either the raphe nuclei or substantia nigra, however, appears to be one-way since there is no evidence that the SVZ sends any direct signal back to these regions. The physiological significance of regulating adult SVZ neurogenesis by either of these distant regions is uncertain, but suggests that olfaction in adults can be modified by the control of either alertness/mood by the raphe nuclei or movement by the midbrain.

9.3 INTEGRATION OF ADULT-GENERATED NEURONS IN THE OB

The integration of adult-born interneurons into the pre-existing OB circuitry begins when progenitors reach their final position in either the periglomerular or granule cell layer. The temporal dynamics of morphological maturation and functional integration, however, are different for progenitors in each of these layers. Adult-generated periglomerular interneurons take up to 6 weeks from progenitor formation in the SVZ to develop mature dendritic arbors and spine densities (Whitman and Greer, 2007a). They develop voltage-dependent sodium currents and the capacity to fire action potentials before the formation of synaptic connections (Belluzzi et al., 2003). The first synapses made on maturing periglomerular interneurons are GABAergic, whereas glutamatergic synaptic input from olfactory receptor neurons and/or mitral/tufted cells requires an additional two weeks. As these interneurons mature, their glutamatergic synaptic inputs evolve as both the AMPA/NMDA ratio increases and the contribution of the NR2B (GRIN2B) subunit to the NMDA response decreases (Grubb et al., 2008). Thus, the functional character of the excitatory input received by periglomerular interneurons changes during the course of maturation.

In contrast to the periglomerular interneurons, granule cells receive synaptic input before they generate action potentials (Carleton et al., 2003; Whitman and Greer, 2007b). Almost as soon as granule cell progenitors stop radial migration and begin to extend dendritic processes, both GABAergic and glutamatergic synaptic inputs form on the basal and apical dendrites as well as the soma (Katagiri et al., 2011; Kelsch et al., 2008; Panzanelli et al., 2009; Whitman and Greer, 2007b). The glutamatergic input is from either centrifugal or mitral/tufted cell axon collaterals (Kelsch et al., 2008; Whitman and Greer, 2007b). The initial density of GABAergic and glutamatergic synapses

9.3 Integration of Adult-Generated Neurons in the OB

are about the same, but the glutamatergic synaptic density substantially increases as maturation proceeds (Kelsch et al., 2008; Panzanelli et al., 2009; Whitman and Greer, 2007b). During this maturation, there is a transient increase in both the number of functional glutamatergic input sites and the NMDA/AMPA ratio at these sites (Katagiri et al., 2011). Dendritic arbors and spine densities mature within 4 weeks of progenitor formation in the SVZ (Whitman and Greer, 2007b). Maturation of dendritic branching and spine density as well as glutamatergic synapses requires GABAergic input mediated by the $GABA_A$ receptor $\alpha 2$ subunit (Pallotto et al., 2012). The ability to generate action potentials develops between 1–2 weeks after granule cell progenitors are generated in the SVZ, but the large majority of output synapses are not formed until a month after the cell is able to generate action potentials (Bardy et al., 2010). Thus, granule cells "silently integrate" and "listen" to the pre-existing circuitry before they "speak" (Kelsch et al., 2008).

Approximately 10,000 to 30,000 neuroblasts per day migrate into the OB from the SVZ and RMS in adult rodents (Lois and Alvarez-Buylla, 1994), but many of these new neurons do not stably integrate into the OB and are eliminated. In the murine periglomerular layer, approximately 50% of the new interneurons are gone within 1–2 months of their formation in the SVZ (Bovetti et al., 2009; Kato et al., 2000; Whitman and Greer, 2007a). The surviving neurons are stable and remain in the periglomerular layer for up to another 19 months (Winner et al., 2002). The turnover rate in this layer is estimated to be about 3% per month, which would require almost 35 months, essentially the maximum life span of mice, to completely replace all interneurons (Lagace et al., 2007; Mizrahi et al., 2006; Ninkovic et al., 2007). In the granule cell layer, approximately 50% of new interneurons are also lost within 1–2 months of their generation and the surviving neurons are stable for up to another 19 months (Lemasson et al., 2005; Petreanu and Alvarez-Buylla, 2002; Winner et al., 2002). The turnover rate for granule cells, however, is controversial. Some have suggested that adult born interneurons constitute only a minor population of the granule cell interneuron population (Ninkovic et al., 2007). Several other studies, however, have reported substantially higher turnover rates and that adult-born granule cell progenitors predominantly target the deep granule cell region (Imayoshi et al., 2008; Lagace et al., 2007; Lemasson et al., 2005; Mandairon et al., 2006). In one study, almost the entire deep granule cell layer was reported to turn over within a year, and only about 50% of the superficial granule cells were replaced during the same time (Imayoshi et al., 2008).

In both periglomerular and granule cell layers, the window between 15–45 days after progenitor formation in the SVZ is a critical period for establishing which adult-born neurons survive (Kelsch et al., 2009; Mandairon et al., 2006; Petreanu and Alvarez-Buylla, 2002; Yamaguchi and Mori, 2005). During this critical period, odorant-induced synaptic activity is an important determinant of survival (Bovetti et al., 2009; Mandairon et al., 2006; Petreanu and Alvarez-Buylla, 2002; Rochefort and Lledo, 2005; Saghatelyan et al., 2005; Yamaguchi and Mori, 2005). Modulation of odorant-induced synaptic activity before this critical period does not affect progenitor maturation, likely because effective synaptic connections on the progenitors have not developed. Sensory input is critical for modifying the distribution and density of dendritic spine densities as well as the complexity of dendritic branching (Dahlen et al., 2011; Kelsch et al., 2009; Livneh et al., 2009; Saghatelyan et al., 2005). In addition to glutamatergic input from mitral/tufted cells (and olfactory receptor neurons in the glomerular layer), centrifugal input from other brain regions is suspected to be an essential component of the activity-dependent maturation and survival of adult-born OB interneurons (Katagiri et al., 2011; Whitman and Greer, 2007b). Furthermore, both feeding and sleeping behaviors strongly influence the temporal dynamics of survival (Yokoyama et al., 2011). The molecular mechanisms that integrate behavioral and sensory input to promote survival of adult-born interneurons has not been elucidated, but the transcription factor cAMP Response Element-Binding (CREB) protein has an established and pivotal role for the maturation and survival of newborn adult OB neurons (Giachino et al., 2005; Herold et al., 2011).

For interneurons that achieve long-term survival, their dendritic morphology and spine density retain some plasticity. Spine densities on mature periglomerular interneurons are dynamic in response to changes to sensory input (Livneh et al., 2009; Mizrahi, 2007). The age of the animal at which time neurons are born also affects the mature dendritic structure. Periglomerular interneurons born in older adults have a reduced total dendritic length and increased spine density relative to those generated in younger animals (Livneh and Mizrahi, 2011). By contrast, mature granule cell interneurons born in older animals have lower spine densities relative to those generated in younger adults. The spine densities in adult granule cell interneurons, however, are more stable when compared to those born during neo-natal development (Kelsch et al., 2012). The regulation of dendritic maturation and plasticity in adult OB interneurons is regulated by several molecular factors, including 5T4 (Yoshihara et al., 2012), VEGF (Licht et al., 2010), Agrin signaling (Burk et al., 2012), Wnt signaling (Pino et al., 2011) and miR-132 (Pathania et al., 2012).

9.4 PHENOTYPE SPECIFICATION AND MAINTENANCE

OB interneurons contain a wide range of mature interneuron phenotypes. In the periglomerular layer, there are numerous phenotypic subsets based on distinct combinatorial expression patterns of neurotransmitters, neuroactive peptides, transmembrane receptors, and calcium binding proteins (Brill et al., 2009; Panzanelli et al., 2007; Parrish-Aungst et al., 2007). Most interneurons in this layer are GABAergic, but there is also a small subset of excitatory glutamatergic cells (Brill et al., 2009). Different subsets of periglomerular layer interneurons also make distinct synaptic connections within the OB circuitry (Kosaka and Kosaka, 2005). In the granule cell layer, all interneurons are GABAergic, but the superficial and deep granule cells layers preferentially co-express 5T4 and calcium/calmodulin-dependent protein kinase type IV (CamKIV), respectively (Saino-Saito et al., 2007; Yoshihara et al., 2012). Superficial and deep granule cells also preferentially synapse with tufted and mitral cells, respectively (Mori et al., 1983; Orona et al., 1983). Although there is greater phenotypic diversity in the periglomerular layer, adult neurogenesis produces more granule cells relative to periglomerular interneurons at approximately a 3:1 ratio (Luskin, 1993; Zigova et al., 1996).

Nearly all OB interneuron phenotypes in both periglomerular and granule cells layers are regenerated by adult neurogenesis, but they turn over at different rates. Deep granule cells turn over and are regenerated at a faster rate than the superficial granule cells (Imayoshi et al., 2008; Lagace et al., 2007; Lemasson et al., 2005; Mandairon et al., 2006). In the periglomerular layer, interneurons co-expressing both GABA and dopamine turn over faster than those that co-express GABA with either Calbindin or Calretinin (Kohwi et al., 2007; Whitman and Greer, 2007a). The turnover rate for mature OB dopaminergic neurons is not constant, however, and is inversely proportional to the levels of odorant-induced synaptic activity (Bastien-Dionne et al., 2010; Bovetti et al., 2009; Sawada et al., 2011).

In addition to differential turnover rates, production rates for specific phenotypes are age-dependent. For example, the production of superficial granule cells peaks during the first postnatal week of development (Lemasson et al., 2005), but the 5T4-expressing subset of superficial granule cells are generated at a constant rate through postnatal and adult life (Batista-Brito et al., 2008). By contrast, production of Calbindin-expressing periglomerular interneurons peak during a late embryonic and neo-natal time window before dropping to a lower rate in adults (Batista-Brito et al., 2008; De Marchis et al.,

2007). Periglomerular layer calretinin-expressing interneurons display the opposite temporal relationship, where their production is lowest during embryonic and neo-natal development and greatest in adults (Batista-Brito et al., 2008; De Marchis et al., 2007).

The molecular mechanisms that regulate which phenotype is adopted by SVZ-derived progenitors is not well understood, but combinatorial transcription factor codes are proposed to direct progenitors towards distinct differentiation pathways (Allen et al., 2007; Brill et al., 2008). For example, the co-expression of Dlx2, Pax6, Er81, and Meis2 direct OB neuronal progenitors towards the combined GABAergic and dopaminergic neurotransmitter phenotype. By contrast, co-expression of Sp8 with either Er81 or Meis2 is associated with periglomerular neurons expressing Calretinin.

The diversity of transcription factor codes specifying OB interneuron phenotypes reflects the intrinsic heterogeneity of neural stem cells in the SVZ. As noted in Section 9.2.1, the SVZ in the adult lateral ventricles is derived from several regions of the embryonic neuroepithelium, including the lateral and medial ganglionic eminences as well as cortex (Merkle et al., 2007; Stenman et al., 2003; Willaime-Morawek et al., 2006; Young et al., 2007). During embryonic development, each of these regions express distinct sets of transcription factors that specify different neuronal phenotypes (Wonders and Anderson, 2006). In the adult SVZ, these regions generate distinct domains that can be visualized by lineage-tracing studies (Allen et al., 2007; Kohwi et al., 2007; Young et al., 2007). Coupled with spatially restricted expression patterns of signaling molecules, such as Shh (Ihrie et al., 2011), these distinct lineage domains establish unique dorso-ventral and rostro-caudal origins within the adult SVZ for different interneuron phenotypes (Figure 9.3) (Brill et al., 2009; Kelsch et al., 2007; Merkle et al., 2007; Young et al., 2007). For example, cells derived from Emx1 and Pax6 lineages are limited to dorso-lateral edges of the SVZ and preferentially generate dopaminergic periglomerular layer interneurons and superficial granule cells (Merkle et al., 2007; Young et al., 2007). Within this region, a subdomain Tbr2-expressing cells produce glutamatergic juxtaglomerular neurons (Brill et al., 2009). By contrast, Gsx2 and Nkx2.1 lineages populate the lateral and ventral portions of the SVZ, respectively. Both Calbindin-expressing periglomerular layer interneurons and deep granule cells are preferentially generated in the ventral region of the adult SVZ (Merkle et al., 2007; Young et al., 2007). Thus, neural stem cells are temporally and spatially heterogeneous in their potential to produce progenitors of specific OB interneuron phenotypes (Bovetti et al., 2007c; Lledo et al., 2008; Weinandy et al., 2011).

Adult neural stem cells in the SVZ also produce glial progenitors that migrate separately from neuroblasts and

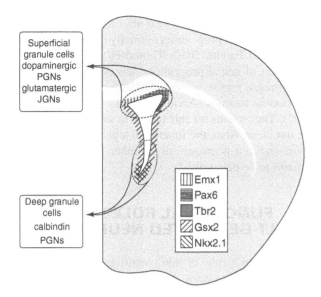

Figure 9.3 Heterogeneity in the origin of specific OB interneuron phenotypes derived from the adult SVZ. A cartoon of a coronal section through an adult mouse brain showing the partial overlap in the lineage of regions comprising the SVZ. During embryonic development, transcription factors (such as Emx1, Pax6, Tbr2, Gsx2 and Nkx2.1) are expressed in distinct regions of the neuroepithelium. These regions eventually fuse and form the adult SVZ. The neural stem cells derived from these different lineages generate specific OB interneuron phenotypes in the adult SVZ. Superficial granule cells, dopaminergic periglomerular layer neurons (PGNs) and glutamatergic juxtaglomerular neurons (JGNs) are preferentially generated in the dorsal SVZ regions. By contrast, deep granule cells and Calbindin-expressing PGNs are produced in the ventral regions of the SVZ.

populate several regions, including the corpus callosum and striatum (Menn et al., 2006; Suzuki and Goldman, 2003). Peak production of glial progenitors occurs in early post-natal development before tapering off to low levels in adults, but either demyelination in the corpus callosum or seizures increases production of adult-generated glia (Marshall et al., 2003; Menn et al., 2006; Nait-Oumesmar et al., 1999; Parent et al., 2006). Oligodendrocyte lineage transcription factor 2 (Olig2) is a key regulator of the neuronal vs. glial cell fate decision. The Olig2 transcription factor is an early marker of the glial fate that is expressed in a subset of neural stem cells and transit-amplifying cells and is both necessary and sufficient for preventing neuronal differentiation (Hack et al., 2005; Marshall et al., 2005; Menn et al., 2006). Glial cell fates are also promoted by stimulation of EGF and PDGF receptors, which preferentially enhance oligodendritic progenitor production (Aguirre et al., 2007; Gonzalez-Perez and Alvarez-Buylla, 2011; Gonzalez-Perez and Quinones-Hinojosa, 2010; Jackson et al., 2006). In addition to EGFR and PDGFR, activation of bone morphogenic protein (BMP) receptors by BMP ligands also promotes glial differentiation of SVZ-derived progenitors (Lim et al., 2000). BMP receptors are expressed by neural stem cells, transit amplifying cells, and neuroblasts, whereas BMP ligands are highly expressed in only the neural stem cells and transit amplifying cells. Ependymal cells lining the lateral ventricle, however, do not express either BMP ligands or receptors; rather they strongly express the BMP antagonist, Noggin. The blockade of BMP signaling by Noggin released from ependymal cells represses gliogenesis and enables the formation of neurogenic microenvironments in the SVZ. Continued elucidation of the molecular mechanisms controlling the neuronal vs. glial cell fate in SVZ-derived progenitors may foster the development of novel therapeutic strategies to treat demyelination disorders and brain tumors (Jackson and Alvarez-Buylla, 2008; Nait-Oumesmar et al., 2008).

9.5 ADULT NEUROGENESIS IN THE SVZ/RMS OF THE AGING AND PATHOLOGICAL BRAIN

As discussed above in Section 9.2, neuroblast production in the human SVZ precipitously declines after the first 18 months after birth (Sanai et al., 2011). The precise size of the proliferative neural progenitor population present in the adult SVZ is controversial (Curtis et al., 2007; Kam et al., 2009; Quinones-Hinojosa et al., 2006; Sanai et al., 2004), but the comparative size of the adult proliferative population consistently has been found smaller than in the infant SVZ (Sanai et al., 2011; Wang et al., 2011a; Weickert et al., 2000). Despite the presence of some multi-potent neural stem cells in the adult SVZ, post-mortem C^{14} analysis of human OBs confirms that most, if not all, OB interneurons are generated at the time of birth and not during adolescence or adulthood (Bergmann et al., 2012).

Adult rodents maintain neurogenesis in the SVZ, but neurogenic output does progressively decline during aging (Conover and Shook, 2011). There is an approximately 50–70% decrease in the number of both proliferative cells in the SVZ and progenitors migrating into the OB of aged mice (22–26 months old) when compared to younger mice (2–5 months) (Ahlenius et al., 2009; Tropepe et al., 1997). This reduction in neurogenic capacity is caused, at least in part, by an increase in the number of neural stem cells becoming mitotically quiescent (Luo et al., 2006). Several cell cycle regulators are critical for regulating the pool size of proliferating neural stem cells in the SVZ (discussed in Section 9.3.1), and the expression levels of Cdk inhibitors, such as $p27^{Kip1}$ and $p16^{INKa}$, in the SVZ progressively increase with age (Ahlenius et al., 2009; Molofsky et al., 2006). Cell cycle inhibitor expression levels are also likely critical for regulating neural stem cell pool size in the human SVZ, but the mechanistic differences responsible for the rapid reduction in the number of proliferative human progenitors during postnatal development remains to be established.

Aging-related pathological conditions in the brain also modulate SVZ neurogenesis. Parkinson's disease, for example, is associated with a decrease in the rate of SVZ neurogenesis (Marxreiter et al., 2013; Winner et al., 2011). Several rodent lesion models report decreases in both proliferation in the SVZ and survival of new neurons in the OB. Transgenic mice over-expressing variants of alpha-synuclein have produced mixed results regarding progenitor proliferation, but they have consistently displayed a diminished survival of new neurons in the OB. Some studies of human Parkinson's disease patients have also indicated that there is a disease-related reduction in markers of SVZ progenitor proliferation (Hoglinger et al., 2004; O'Keeffe et al., 2009b), but others have challenged these findings (van den Berge et al., 2011). Analysis of the SVZ in Parkinson's patients is difficult, however, because proliferation levels in the SVZ may vary significantly due to confounding factors that are unique to each patient, such as the presence of other neurological conditions and differences in medication histories.

In contrast to the diminution of adult SVZ neurogenesis associated with Parkinson's disease, human patients with Huntington's disease show an increase in markers of proliferation in the SVZ (Curtis et al., 2003; Curtis et al., 2005a; Curtis et al., 2005b). The SVZ in Huntington's disease patients is thicker and contains substantially more neural stem cells, as well as additional transit amplifying cells and neuroblasts. Consistent with these findings, rodent models of Huntington's disease show that progenitor proliferation in the adult SVZ does not decrease, rather it remains unchanged or is increased (Batista et al., 2006; Kohl et al., 2010; Phillips et al., 2005; Tattersfield et al., 2004). Rodent studies also show that neuroblasts migrate into the damaged striatal regions, which comes at the expense of new neurons migrating into the OB (Kohl et al., 2010).

In addition to neurodegenerative disease, ischemia also alters adult neurogenesis in the SVZ (Cayre et al., 2009; Conover and Shook, 2011; Kernie and Parent, 2010). Studies with rodent models have consistently reported that focal ischemia stimulates adult SVZ neurogenesis (Arvidsson et al., 2002; Jin et al., 2001; Parent et al., 2006; Zhang et al., 2001). Although the neurogenesis rates peak one or two weeks post-stroke before declining, significantly elevated levels can persist for up to 4 months (Thored et al., 2006). Consistent with the rodent models, human stroke victims also show evidence for increased cell proliferation in the SVZ (Jin et al., 2006; Macas et al., 2006; Marti-Fabregas et al., 2010). Analyses of rodent stroke models reveal that the increase in SVZ neurogenesis is accompanied by a redirection of migrating neuroblasts towards the infarct area. This redirection of neuroblasts

can be detected for up to a year after the stroke and comes at the expense of progenitors normally destined for the OB (Kernie and Parent, 2010; Thored et al., 2007). Despite this influx of neural progenitors, few progenitors reaching the damaged regions survive and differentiate into mature functioning neurons (Arvidsson et al., 2002; Parent et al., 2002). The reasons for this failure to survive and integrate are not clear. Also, the functional significance, if any, for stroke-induced neurogenesis in either rodents or humans remains to be determined.

9.6 FUNCTIONAL ROLES OF ADULT-GENERATED NEURONS

The functional role of adult neurogenesis in the SVZ, RMS, and OB has been challenging to assess. In rodents, conflicting conclusions have been reported. Typical rodent studies have disrupted adult neurogenesis (e.g., through irradiation (Lazarini et al., 2009) or by genetically driven toxin expression (Imayoshi et al., 2008)) then compared olfactory-based functional behavior between the experimental and control groups. Differences in the experimental methods used to disrupt neurogenesis may account for some discrepancies in the reported findings, but the specific behavioral testing methods used are a critical factor in the outcome of functional testing (Lazarini and Lledo, 2011). Studies using operant learning paradigms show that adult neurogenesis is required for long-term odor associative memory (Lazarini et al., 2009; Sultan et al., 2010). By contrast, those using non-operant learning indicate that adult neurogenesis is not necessary (Breton-Provencher et al., 2009; Imayoshi et al., 2008). The dependence of newborn adult OB neuron survival on operant, but not non-operant, learning is consistent with the importance of selecting appropriate testing methods to assess the functional analysis of adult neurogenesis (Mandairon et al., 2011). Even within operant learning paradigms, however, the difficulty of an odor discrimination task is a critical parameter. For easy discrimination challenges (i.e., presentation of dissimilar odors), adult neurogenesis is dispensable. Difficult discrimination tasks (presentation of similar odors), however, require input specifically from adult-born and not early post-natal born neurons (Alonso et al., 2012).

The importance of olfaction in rodents underlies the significance that adult neurogenesis has in shaping odor discrimination and associative learning and memory, as well as odor-based fear discrimination (Valley et al., 2009), mate preference (Mak et al., 2007), and maternal behavior (Shingo et al., 2003). In humans, olfaction is integral for many behaviors, but it does not have the same prominent role as in rodents. Furthermore, the precipitous decline in neuroblast production by the SVZ in infants suggests that humans do not rely on adult neurogenesis for odor discrimination and associative learning. Human olfactory function does decline with senescence (Doty et al., 1984), but age-related decreases in the number of glomeruli and mitral cells (Meisami et al., 1998), as well as an increased prevalence of neurodegenerative disease or injury (as discussed in Section 9.5), make it difficult to assess whether the negligible addition of adult-born OB interneurons underlies this diminution of olfactory function. Thus, the functional role of neural proliferation in the adult human SVZ is unknown and elucidating this role is a critical challenge for future studies to address.

REFERENCES

Ables, J. L., Breunig, J. J., Eisch, A. J., and Rakic, P. (2011) Not(ch) just development: Notch signalling in the adult brain. *Nat. Rev. Neurosci.* 12: 269–283.

Adachi, K., Mirzadeh, Z., Sakaguchi, M., et al. (2007) Beta-catenin signaling promotes proliferation of progenitor cells in the adult mouse subventricular zone. *Stem Cells* 25: 2827–2836.

Aguirre, A., Dupree, J. L., Mangin, J. M., and Gallo, V. (2007) A functional role for EGFR signaling in myelination and remyelination. *Nat. Neurosci.* 10: 990–1002.

Aguirre, A., Rubio, M. E., and Gallo, V. (2010) Notch and EGFR pathway interaction regulates neural stem cell number and self-renewal. *Nature* 467: 323–327.

Ahlenius, H., Visan, V., Kokaia, M., et al. (2009) Neural stem and progenitor cells retain their potential for proliferation and differentiation into functional neurons despite lower number in aged brain. *J. Neurosci.* 29: 4408–4419.

Ahn, S. and Joyner, A. L. (2005) In vivo analysis of quiescent adult neural stem cells responding to Sonic hedgehog. *Nature* 437: 894–897.

Akerblom, M., Sachdeva, R., Barde, I., et al. (2012) MicroRNA-124 Is a Subventricular Zone Neuronal Fate Determinant. *J. Neurosci.* 32: 8879–8889.

Alberti, S., Krause, S. M., Kretz, O., et al. (2005) Neuronal migration in the murine rostral migratory stream requires serum response factor. *Proc. Natl. Acad. Sci. U S A.* 102: 6148–6153.

Alexson, T. O., Hitoshi, S., Coles, B. L., et al. (2006) Notch signaling is required to maintain all neural stem cell populations–irrespective of spatial or temporal niche. *Dev. Neurosci.* 28: 34–48.

Allen, Z. J., Waclaw, R. R., Colbert, M. C., and Campbell, K. (2007) Molecular identity of olfactory bulb interneurons: transcriptional codes of periglomerular neuron subtypes. *J. Mol. Histol.* 38: 517–525.

Alonso, M., Lepousez, G., Wagner, S., et al. (2012) Activation of adult-born neurons facilitates learning and memory. *Nat. Neurosci.* 15(6)897–904.

Alonso, M., Ortega-Perez, I., Grubb, M. S., et al. (2008) Turning astrocytes from the rostral migratory stream into neurons: a role for the olfactory sensory organ. *J. Neurosci.* 28: 11089–11102.

Altman, J. (1962) Are new neurons formed in the brains of adult mammals? *Science* 135: 1127–1128.

Altman, J. and Bayer, S. A. (1990) Migration and distribution of two populations of hippocampal granule cell precursors during the perinatal and postnatal periods. *J. Comp. Neurol.* 301: 365–381.

Alvarez-Buylla, A. and Nottebohm, F. (1988) Migration of young neurons in adult avian brain. *Nature* 335: 353–354.

Alves, J. A., Barone, P., Engelender, S., et al. (2002) Initial stages of radial glia astrocytic transformation in the early postnatal anterior subventricular zone. *J. Neurobiol.* 52: 251–265.

Andreu-Agullo, C., Maurin, T., Thompson, C. B., and Lai, E. C. (2012) Ars2 maintains neural stem-cell identity through direct transcriptional activation of Sox2. *Nature* 481: 195–198.

Andreu-Agullo, C., Morante-Redolat, J. M., Delgado, A. C., and Farinas, I. (2009) Vascular niche factor PEDF modulates Notch-dependent stemness in the adult subependymal zone. *Nat. Neurosci.* 12: 1514–1523.

Angot, E., Loulier, K., Nguyen-Ba-Charvet, K. T., et al. (2008) Chemoattractive activity of sonic hedgehog in the adult subventricular zone modulates the number of neural precursors reaching the olfactory bulb. *Stem Cells* 26: 2311–2320.

Anton, E. S., Ghashghaei, H. T., Weber, J. L., et al. (2004) Receptor tyrosine kinase ErbB4 modulates neuroblast migration and placement in the adult forebrain. *Nat. Neurosci.* 7: 1319–1328.

Arnold, S. A., and Hagg, T. (2012) Serotonin 1A receptor agonist increases species- and region-selective adult CNS proliferation, but not through CNTF. *Neuropharmacology* 63(7): 1238–1247.

Arvidsson, A., Collin, T., Kirik, D., et al. (2002) Neuronal replac(7)ement from endogenous precursors in the adult brain after stroke. *Nat. Med.* 8: 963–970.

Azmitia, E. C. and Segal, M. (1978) An autoradiographic analysis of the differential ascending projections of the dorsal and median raphe nuclei in the rat. *J. Comp. Neurol.* 179: 641–667.

Balordi, F. and Fishell, G. (2007a) Hedgehog signaling in the subventricular zone is required for both the maintenance of stem cells and the migration of newborn neurons. *J. Neurosci.* 27: 5936–5947.

Balordi, F. and Fishell, G. (2007b) Mosaic removal of hedgehog signaling in the adult SVZ reveals that the residual wild-type stem cells have a limited capacity for self-renewal. *J. Neurosci.* 27: 14248–14259.

Banasr, M., Hery, M., Printemps, R., and Daszuta, A. (2004) Serotonin-induced increases in adult cell proliferation and neurogenesis are mediated through different and common 5-HT receptor subtypes in the dentate gyrus and the subventricular zone. *Neuropsychopharmacology* 29: 450–460.

Bardy, C., Alonso, M., Bouthour, W., and Lledo, P. M. (2010) How, when, and where new inhibitory neurons release neurotransmitters in the adult olfactory bulb. *J. Neurosci.* 30: 17023–17034.

Basak, O., Giachino, C., Fiorini, E., et al. (2012) Neurogenic subventricular zone stem/progenitor cells are Notch1-dependent in their active but not quiescent state. *J. Neurosci.* 32: 5654–5666.

Bastien-Dionne, P. O., David, L. S., Parent, A., and Saghatelyan, A. (2010) Role of sensory activity on chemospecific populations of interneurons in the adult olfactory bulb. *J. Comp. Neurol.* 518: 1847–1861.

Bath, K. G., Akins, M. R., and Lee, F. S. (2012) BDNF control of adult SVZ neurogenesis. *Dev. Psychobiol.* 54: 578–589.

Batista, C. M., Kippin, T. E., Willaime-Morawek, S., et al. (2006) A progressive and cell non-autonomous increase in striatal neural stem cells in the Huntington's disease R6/2 mouse. *J. Neurosci.* 26: 10452–10460.

Batista-Brito, R., Close, J., Machold, R., and Fishell, G. (2008) The distinct temporal origins of olfactory bulb interneuron subtypes. *J. Neurosci.* 28: 3966–3975.

Battista, D. and Rutishauser, U. (2010) Removal of polysialic acid triggers dispersion of subventricularly derived neuroblasts into surrounding CNS tissues. *J. Neurosci.* 30: 3995–4003.

Beatus, P., Jhaveri, D. J., Walker, T. L., et al. (2011) Oncostatin M regulates neural precursor activity in the adult brain. *Dev. Neurobiol.* 71: 619–633.

Belluzzi, O., Benedusi, M., Ackman, J., and LoTurco, J. J. (2003) Electrophysiological differentiation of new neurons in the olfactory bulb. *J. Neurosci.* 23: 10411–10418.

Belvindrah, R., Hankel, S., Walker, J., et al. (2007) Beta1 integrins control the formation of cell chains in the adult rostral migratory stream. *J. Neurosci.* 27: 2704–2717.

Bergmann, O., Liebl, J., Bernard, S., et al. (2012) The age of olfactory bulb neurons in humans. *Neuron* 74: 634–639.

Beukelaers, P., Vandenbosch, R., Caron, N., et al. (2011) Cdk6-dependent regulation of G(1) length controls adult neurogenesis. *Stem Cells* 29: 713–724.

Beukelaers, P., Vandenbosch, R., Caron, N., et al. (2012) Cycling or not cycling: cell cycle regulatory molecules and adult neurogenesis. *Cell Mol. Life Sci.* 69: 1493–1503.

Bishop, C. L., Bergin, A. M., Fessart, D., et al. (2010) Primary cilium-dependent and -independent Hedgehog signaling inhibits p16(INK4A). *Mol. Cell* 40: 533–547.

Bolteus, A. J. and Bordey, A. (2004) GABA release and uptake regulate neuronal precursor migration in the postnatal subventricular zone. *J. Neurosci.* 24: 7623–7631.

Bonfanti, L. and Peretto, P. (2011) Adult neurogenesis in mammals–a theme with many variations. *Eur. J. Neurosci.* 34: 930–950.

Borta, A. and Hoglinger, G. U. (2007) Dopamine and adult neurogenesis. *J. Neurochem.* 100: 587–595.

Bovetti, S., Bovolin, P., Perroteau, I., and Puche, A. C. (2007a) Subventricular zone-derived neuroblast migration to the olfactory bulb is modulated by matrix remodelling. *Eur. J. Neurosci.* 25: 2021–2033.

Bovetti, S., Hsieh, Y. C., Bovolin, P., et al. (2007b) Blood vessels form a scaffold for neuroblast migration in the adult olfactory bulb. *J. Neurosci.* 27: 5976–5980.

Bovetti, S., Peretto, P., Fasolo, A., and De Marchis, S. (2007c) Spatio-temporal specification of olfactory bulb interneurons. *J. Mol. Histol.* 38: 563–569.

Bovetti, S., Veyrac, A., Peretto, P., et al. (2009) Olfactory enrichment influences adult neurogenesis modulating GAD67 and plasticity-related molecules expression in newborn cells of the olfactory bulb. *PLoS One* 4: e6359.

Bozoyan, L., Khlghatyan, J., and Saghatelyan, A. (2012) Astrocytes control the development of the migration-promoting vasculature scaffold in the postnatal brain via VEGF signaling. *J. Neurosci.* 32: 1687–1704.

Bradford, D., Faull, R. L., Curtis, M. A., and Cooper, H. M. (2010) Characterization of the netrin/RGMa receptor neogenin in neurogenic regions of the mouse and human adult forebrain. *J. Comp. Neurol.* 518: 3237–3253.

Brazel, C. Y., Nunez, J. L., Yang, Z., Levison, S. W. (2005) Glutamate enhances survival and proliferation of neural progenitors derived from the subventricular zone. *Neuroscience* 131: 55–65.

Breton-Provencher, V., Lemasson, M., Peralta, M. R., and Saghatelyan, A. (2009) Interneurons produced in adulthood are required for the normal functioning of the olfactory bulb network and for the execution of selected olfactory behaviors. *J. Neurosci.* 29: 15245–15257.

Brezun, J. M. and Daszuta, A. (1999) Depletion in serotonin decreases neurogenesis in the dentate gyrus and the subventricular zone of adult rats. *Neuroscience* 89: 999–1002.

Brill, M. S., Ninkovic, J., Winpenny, E., et al. (2009) Adult generation of glutamatergic olfactory bulb interneurons. *Nat. Neurosci.* 12: 1524–1533.

Brill, M. S., Snapyan, M., Wohlfrom, H., et al. (2008) A dlx2- and pax6-dependent transcriptional code for periglomerular neuron specification in the adult olfactory bulb. *J. Neurosci.* 28: 6439–6452.

Burk, K., Desoeuvre, A., Boutin, C., et al. (2012) Agrin-signaling is necessary for the integration of newly generated neurons in the adult olfactory bulb. *J. Neurosci.* 32: 3759–3764.

Cameron, H. A., Woolley, C. S., McEwen, B. S., and Gould, E. (1993) Differentiation of newly born neurons and glia in the dentate gyrus of the adult rat. *Neuroscience* 56: 337–344.

Carleton, A., Petreanu, L. T., Lansford, R., et al. (2003) Becoming a new neuron in the adult olfactory bulb. *Nat. Neurosci.* 6: 507–518.

Castella, P., Sawai, S., Nakao, K., et al. (2000) HES-1 repression of differentiation and proliferation in PC12 cells: role for the helix 3-helix 4 domain in transcription repression. *Mol. Cell. Biol.* 20: 6170–6183.

Castiglione, M., Calafiore, M., Costa, L., et al. (2008) Group I metabotropic glutamate receptors control proliferation, survival and differentiation of cultured neural progenitor cells isolated from the subventricular zone of adult mice. *Neuropharmacology* 55: 560–567.

Cayre, M., Canoll, P., and Goldman, J. E. (2009) Cell migration in the normal and pathological postnatal mammalian brain. *Prog. Neurobiol.* 88: 41–63.

Chapouton, P., Jagasia, R., and Bally-Cuif, L. (2007) Adult neurogenesis in non-mammalian vertebrates. *Bioessays* 29: 745–757.

Chazal, G., Durbec, P., Jankovski, A., et al. (2000) Consequences of neural cell adhesion molecule deficiency on cell migration in the rostral migratory stream of the mouse. *J. Neurosci.* 20: 1446–1457.

Chen, J. H., Wen, L., Dupuis, S., et al. (2001) The N-terminal leucine-rich regions in Slit are sufficient to repel olfactory bulb axons and subventricular zone neurons. *J. Neurosci.* 21: 1548–1556.

Cheng, L. C., Pastrana, E., Tavazoie, M., and Doetsch, F. (2009) miR-124 regulates adult neurogenesis in the subventricular zone stem cell niche. *Nat. Neurosci.* 12: 399–408.

Chojnacki, A., Shimazaki, T., Gregg, C., et al. (2003) Glycoprotein 130 signaling regulates Notch1 expression and activation in the self-renewal of mammalian forebrain neural stem cells. *J. Neurosci.* 23: 1730–1741.

Cobos, I., Broccoli, V., and Rubenstein, J. L. (2005) The vertebrate ortholog of Aristaless is regulated by Dlx genes in the developing forebrain. *J. Comp. Neurol.* 483: 292–303.

Colasante, G., Collombat, P., Raimondi, V., et al. (2008) Arx is a direct target of Dlx2 and thereby contributes to the tangential migration of GABAergic interneurons. *J. Neurosci.* 28: 10674–10686.

Conover, J. C., Doetsch, F., Garcia-Verdugo, J. M., et al. (2000) Disruption of Eph/ephrin signaling affects migration and proliferation in the adult subventricular zone. *Nat. Neurosci.* 3: 1091–1097.

Conover, J. C., and Shook, B. A. (2011) Aging of the subventricular zone neural stem cell niche. *Aging. Dis.* 2: 49–63.

Coskun, V. and Luskin, M. B. (2001) The expression pattern of the cell cycle inhibitor p19(INK4d) by progenitor cells of the rat embryonic telencephalon and neonatal anterior subventricular zone. *J. Neurosci.* 21: 3092–3103.

Councill, J. H., Tucker, E. S., Haskell, G. T., et al. (2006) Limited influence of olanzapine on adult forebrain neural precursors in vitro. *Neuroscience* 140: 111–122.

Courtes, S., Vernerey, J., Pujadas, L., et al. (2011) Reelin controls progenitor cell migration in the healthy and pathological adult mouse brain. *PLoS One* 6: e20430.

Cremer, H., Lange, R., Christoph, A., et al. (1994) Inactivation of the N-CAM gene in mice results in size reduction of the olfactory bulb and deficits in spatial learning. *Nature* 367: 455–459.

Curtis, M. A., Kam, M., Nannmark, U., et al. (2007) Human neuroblasts migrate to the olfactory bulb via a lateral ventricular extension. *Science* 315: 1243–1249.

Curtis, M. A., Penney, E. B., Pearson, A. G., et al. (2003) Increased cell proliferation and neurogenesis in the adult human Huntington's disease brain. *Proc. Natl. Acad. Sci. U S A* 100: 9023–9027.

References

Curtis, M. A., Penney, E. B., Pearson, J., et al. (2005a) The distribution of progenitor cells in the subependymal layer of the lateral ventricle in the normal and Huntington's disease human brain. *Neuroscience* 132: 777–788.

Curtis, M. A., Waldvogel, H. J., Synek, B., and Faull, R. L. (2005b) A histochemical and immunohistochemical analysis of the subependymal layer in the normal and Huntington's disease brain. *J. Chem. Neuroanat.* 30: 55–66.

Dahlen, J. E., Jimenez, D. A., Gerkin, R. C., and Urban, N. N. (2011) Morphological analysis of activity-reduced adult-born neurons in the mouse olfactory bulb. *Front. Neurosci.* 5: 66.

Darcy, D. P. and Isaacson, J. S. (2009) L-type calcium channels govern calcium signaling in migrating newborn neurons in the postnatal olfactory bulb. *J. Neurosci.* 29: 2510–2518.

De Marchis, S., Bovetti, S., Carletti, B., et al. (2007) Generation of distinct types of periglomerular olfactory bulb interneurons during development and in adult mice: implication for intrinsic properties of the subventricular zone progenitor population. *J. Neurosci.* 27: 657–664.

De Marchis, S., Fasolo, A., and Puche, A. C. (2004a) Subventricular zone-derived neuronal progenitors migrate into the subcortical forebrain of postnatal mice. *J. Comp. Neurol.* 476: 290–300.

De Marchis, S., Fasolo, A., Shipley, M., and Puche, A. (2001) Unique neuronal tracers show migration and differentiation of SVZ progenitors in organotypic slices. *J. Neurobiol.* 49: 326–338.

De Marchis, S., Temoney, S., Erdelyi, F., et al. (2004b) GABAergic phenotypic differentiation of a subpopulation of subventricular derived migrating progenitors. *Eur. J. Neurosci.* 20: 1307–1317.

Di Giorgi Gerevini, V. D., Caruso, A., Cappuccio, I., et al. (2004) The mGlu5 metabotropic glutamate receptor is expressed in zones of active neurogenesis of the embryonic and postnatal brain. *Brain Res. Dev. Brain Res.* 150: 17–22.

Doetsch, F. and Alvarez-Buylla, A. (1996) Network of tangential pathways for neuronal migration in adult mammalian brain. *Proc. Natl. Acad. Sci. U S A* 93: 14895–14900.

Doetsch, F., Caille, I., Lim, D. A., et al. (1999) Subventricular zone astrocytes are neural stem cells in the adult mammalian brain. *Cell* 97: 703–716.

Doetsch, F., Garcia-Verdugo, J. M., and Alvarez-Buylla, A. (1997) Cellular composition and three-dimensional organization of the subventricular germinal zone in the adult mammalian brain. *J. Neurosci.* 17: 5046–5061.

Doetsch, F., Petreanu, L., Caille, I., et al. (2002a) EGF converts transit-amplifying neurogenic precursors in the adult brain into multipotent stem cells. *Neuron* 36: 1021–1034.

Doetsch, F., Verdugo, J. M., Caille, I., et al. (2002b) Lack of the cell-cycle inhibitor p27Kip1 results in selective increase of transit-amplifying cells for adult neurogenesis. *J. Neurosci.* 22: 2255–2264.

Doty, R. L., Shaman, P., Applebaum, S. L., et al. (1984) Smell identification ability: changes with age. *Science* 226: 1441–1443.

Emsley, J. G. andHagg, T. (2003) alpha6beta1 integrin directs migration of neuronal precursors in adult mouse forebrain. *Exp. Neurol.* 183: 273–285.

Emsley, J. G. and Hagg, T. (2003) Endogenous and exogenous ciliary neurotrophic factor enhances forebrain neurogenesis in adult mice. *Exp. Neurol.* 183: 298–310.

Emsley, J. G., Menezes, J. R., Madeiro Da Costa, R. F., et al. (2012) Identification of radial glia-like cells in the adult mouse olfactory bulb. *Exp. Neurol.* 236: 283–297.

Endoh-Yamagami, S., Karkar, K. M., May, S. R., et al. (2010) A mutation in the pericentrin gene causes abnormal interneuron migration to the olfactory bulb in mice. *Dev. Biol.* 340: 41–53.

Fasano, C. A., Phoenix, T. N., Kokovay, E., et al. (2009) Bmi-1 cooperates with Foxg1 to maintain neural stem cell self-renewal in the forebrain. *Genes. Dev.* 23: 561–574.

Ferri, A. L., Cavallaro, M., Braida, D., et al. (2004) Sox2 deficiency causes neurodegeneration and impaired neurogenesis in the adult mouse brain. *Development* 131: 3805–3819.

Freundlieb, N., Francois, C., Tande, D., et al. (2006) Dopaminergic substantia nigra neurons project topographically organized to the subventricular zone and stimulate precursor cell proliferation in aged primates. *J. Neurosci.* 26: 2321–2325.

Galli, R., Fiocco, R., De Filippis, L., et al. (2002) Emx2 regulates the proliferation of stem cells of the adult mammalian central nervous system. *Development* 129: 1633–1644.

Garcia-Gonzalez, D., Clemente, D., Coelho, M., et al. (2010) Dynamic roles of FGF-2 and Anosmin-1 in the migration of neuronal precursors from the subventricular zone during pre- and postnatal development. *Exp. Neurol.* 222: 285–295.

Garzotto, D., Giacobini, P., Crepaldi, T., et al. (2008) Hepatocyte growth factor regulates migration of olfactory interneuron precursors in the rostral migratory stream through Met-Grb2 coupling. *J. Neurosci.* 28: 5901–5909.

Gascon, E., Dayer, A. G., Sauvain, M. O., et al. (2006) GABA regulates dendritic growth by stabilizing lamellipodia in newly generated interneurons of the olfactory bulb. *J. Neurosci.* 26: 12956–12966.

Ghashghaei, H. T., Weber, J., Pevny, L., et al. (2006) The role of neuregulin-ErbB4 interactions on the proliferation and organization of cells in the subventricular zone. *Proc. Natl. Acad. Sci. U S A* 103: 1930–1935.

Giachino, C., De Marchis, S., Giampietro, C., et al. (2005) cAMP response element-binding protein regulates differentiation and survival of newborn neurons in the olfactory bulb. *J. Neurosci.* 25: 10105–10118.

Gil-Perotin, S., Haines, J. D., Kaur, J., aet al. (2011) Roles of p53 and p27(Kip1) in the regulation of neurogenesis in the murine adult subventricular zone. *Eur. J. Neurosci.* 34: 1040–1052.

Gil-Perotin, S., Marin-Husstege, M., Li, J., et al. (2006) Loss of p53 induces changes in the behavior of subventricular zone cells: implication for the genesis of glial tumors. *J. Neurosci.* 26: 1107–1116.

Goldman, S. A. and Nottebohm, F. (1983) Neuronal production, migration, and differentiation in a vocal control nucleus of the adult female canary brain. *Proc. Natl. Acad. Sci. U. S. A.* 80: 2390–2394.

Goncalves, M. B., Agudo, M., Connor, S., et al. (2009) Sequential RARbeta and alpha signalling in vivo can induce adult forebrain neural progenitor cells to differentiate into neurons through Shh and FGF signalling pathways. *Dev. Biol.* 326: 305–313.

Gonzalez-Perez, O. and Alvarez-Buylla, A. (2011) Oligodendrogenesis in the subventricular zone and the role of epidermal growth factor. *Brain. Res. Rev.* 67: 147–156.

Gonzalez-Perez, O. and Quinones-Hinojosa, A. (2010) Dose-dependent effect of EGF on migration and differentiation of adult subventricular zone astrocytes. *Glia* 58: 975–983.

Gotz, M. and Stricker, S. H. (2006) Go with the flow: signaling from the ventricle directs neuroblast migration. *Nat. Neurosci.* 9: 470–472.

Gould, E. (2007) How widespread is adult neurogenesis in mammals? *Nat. Rev. Neurosci.* 8: 481–488.

Gregorian, C., Nakashima, J., Le Belle, J., et al. (2009) Pten deletion in adult neural stem/progenitor cells enhances constitutive neurogenesis. *J. Neurosci.* 29: 1874–1886.

Grimm, I., Messemer, N., Stanke, M., Gachet, C., and Zimmermann, H. (2009) Coordinate pathways for nucleotide and EGF signaling in cultured adult neural progenitor cells. *J. Cell. Sci.* 122: 2524–2533.

Gritti, A., Bonfanti, L., Doetsch, F., et al. (2002) Multipotent neural stem cells reside into the rostral extension and olfactory bulb of adult rodents. *J. Neurosci.* 22: 437–445.

Grubb, M. S., Nissant, A., Murray, K., and Lledo, P. M. (2008) Functional maturation of the first synapse in olfaction: development and adult neurogenesis. *J. Neurosci.* 28: 2919–2932.

Guerrero-Cazares, H., Gonzalez-Perez, O., Soriano-Navarro, M., et al. (2011) Cytoarchitecture of the lateral ganglionic eminence and rostral extension of the lateral ventricle in the human fetal brain. *J. Comp. Neurol.* 519: 1165–1180.

Hack, I., Bancila, M., Loulier, K., et al. (2002) Reelin is a detachment signal in tangential chain-migration during postnatal neurogenesis. *Nat. Neurosci.* 5: 939–945.

Hack, M. A., Saghatelyan, A., de Chevigny, A., et al. (2005) Neuronal fate determinants of adult olfactory bulb neurogenesis. *Nat. Neurosci.* 8: 865–872.

Hakanen, J., Duprat, S., and Salminen, M. (2011) Netrin1 is required for neural and glial precursor migrations into the olfactory bulb. *Dev. Biol.* 355: 101–114.

Herold, S., Jagasia, R., Merz, K., et al. (2011) CREB signalling regulates early survival, neuronal gene expression and morphological development in adult subventricular zone neurogenesis. *Mol. Cell. Neurosci.* 46: 79–88.

Hitoshi, S., Alexson, T., Tropepe, V., et al. (2002) Notch pathway molecules are essential for the maintenance, but not the generation, of mammalian neural stem cells. *Genes Dev.* 16: 846–858.

Hitoshi, S., Maruta, N., Higashi, M., et al. (2007) Antidepressant drugs reverse the loss of adult neural stem cells following chronic stress. *J. Neurosci. Res.* 85: 3574–3585.

Hoglinger, G. U., Rizk, P., Muriel, M. P., et al. (2004) Dopamine depletion impairs precursor cell proliferation in Parkinson disease. *Nat. Neurosci.* 7: 726–735.

Holmberg, J., Armulik, A., Senti, K. A., et al. (2005) Ephrin-A2 reverse signaling negatively regulates neural progenitor proliferation and neurogenesis. *Genes. Dev.* 19: 462–471.

Hu, H. (1999) Chemorepulsion of neuronal migration by Slit2 in the developing mammalian forebrain. *Neuron* 23: 703–711.

Hu, H. (2000) Polysialic acid regulates chain formation by migrating olfactory interneuron precursors. *J. Neurosci. Res.* 61: 480–492.

Hu, H., Tomasiewicz, H., Magnuson, T., and Rutishauser, U. (1996) The role of polysialic acid in migration of olfactory bulb interneuron precursors in the subventricular zone. *Neuron* 16: 735–743.

Hurtado-Chong, A., Yusta-Boyo, M. J., Vergano-Vera, E., et al. (2009) IGF-I promotes neuronal migration and positioning in the olfactory bulb and the exit of neuroblasts from the subventricular zone. *Eur. J. Neurosci.* 30: 742–755.

Iguchi, H., Mitsui, T., Ishida, M., et al. (2011) cAMP response element-binding protein (CREB) is required for epidermal growth factor (EGF)-induced cell proliferation and serum response element activation in neural stem cells isolated from the forebrain subventricular zone of adult mice. *Endocr. J.* 58: 747–759.

Ihrie, R. A. and Alvarez-Buylla, A. (2011) Lake-front property: a unique germinal niche by the lateral ventricles of the adult brain. *Neuron* 70: 674–686.

Ihrie, R. A., Shah, J. K., Harwell, C. C., et al. (2011) Persistent sonic hedgehog signaling in adult brain determines neural stem cell positional identity. *Neuron* 71: 250–262.

Imayoshi, I., Sakamoto, M., Ohtsuka, T., et al. (2008) Roles of continuous neurogenesis in the structural and functional integrity of the adult forebrain. *Nat. Neurosci.* 11: 1153–1161.

Inta, D., Alfonso, J., von Engelhardt, J., et al. (2008) Neurogenesis and widespread forebrain migration of distinct GABAergic neurons from the postnatal subventricular zone. *Proc. Natl. Acad. Sci. U. S. A.* 105: 20994–20999.

Jablonska, B., Aguirre, A., Vandenbosch, R., et al. (2007) Cdk2 is critical for proliferation and self-renewal of neural progenitor cells in the adult subventricular zone. *J. Cell Biol.* 179: 1231–1245.

Jackson, E. L. and Alvarez-Buylla, A. (2008) Characterization of adult neural stem cells and their relation to brain tumors. *Cells Tissues Organs* 188: 212–224.

Jackson, E. L., Garcia-Verdugo, J. M., Gil-Perotin, S., et al. (2006) PDGFR alpha-positive B cells are neural stem cells in the adult SVZ that form glioma-like growths in response to increased PDGF signaling. *Neuron* 51: 187–199.

Jin, K., Minami, M., Lan, J. Q., et al. (2001) Neurogenesis in dentate subgranular zone and rostral subventricular zone after focal cerebral ischemia in the rat. *Proc. Natl. Acad. Sci. U S A* 98: 4710–4715.

Jin, K., Wang, X., Xie, L., et al. (2006) Evidence for stroke-induced neurogenesis in the human brain. *Proc. Natl. Acad. Sci. U S A* 103: 13198–13202.

Jin, K., Zhu, Y., Sun, Y., et al. (2002) Vascular endothelial growth factor (VEGF) stimulates neurogenesis in vitro and in vivo. *Proc. Natl. Acad. Sci. U S A* 99: 11946–11950.

Kakita, A. and Goldman, J. E. (1999) Patterns and dynamics of SVZ cell migration in the postnatal forebrain: monitoring living progenitors in slice preparations. *Neuron* 23: 461–472.

Kam, M., Curtis, M. A., McGlashan, S. R., et al. (2009) The cellular composition and morphological organization of the rostral migratory stream in the adult human brain. *J. Chem. Neuroanat.* 37: 196–205.

Kaneko, N., Marin, O., Koike, M., et al. (2010) New neurons clear the path of astrocytic processes for their rapid migration in the adult brain. *Neuron* 67: 213–223.

Kaplan, M. S. and Hinds, J. W. (1977) Neurogenesis in the adult rat: electron microscopic analysis of light radioautographs. *Science* 197: 1092–1094.

Kaslin, J., Ganz, J., and Brand, M. (2008) Proliferation, neurogenesis and regeneration in the non-mammalian vertebrate brain. *Philos. Trans. R. Soc. Lond. B. Biol. Sci.* 363: 101–122.

Katagiri, H., Pallotto, M., Nissant, A., et al. (2011) Dynamic development of the first synapse impinging on adult-born neurons in the olfactory bulb circuit. *Neural. Syst. Circuits.* 1: 6.

Katakowski, M., Zhang, Z., deCarvalho, A. C., and Chopp, M. (2005) EphB2 induces proliferation and promotes a neuronal fate in adult subventricular neural precursor cells. *Neurosci. Lett.* 385: 204–209.

Kato, T., Yokouchi, K., Kawagishi, K., et al. (2000) Fate of newly formed periglomerular cells in the olfactory bulb. *Acta. Otolaryngol* 120: 876–879.

Kelsch, W., Lin, C. W., and Lois, C. (2008) Sequential development of synapses in dendritic domains during adult neurogenesis. *Proc. Natl. Acad. Sci. U S A* 105: 16803–16808.

Kelsch, W., Lin, C. W., Mosley, C. P., and Lois, C. (2009) A critical period for activity-dependent synaptic development during olfactory bulb adult neurogenesis. *J. Neurosci.* 29: 11852–11858.

Kelsch, W., Mosley, C. P., Lin, C. W., and Lois, C. (2007) Distinct mammalian precursors are committed to generate neurons with defined dendritic projection patterns. *PLoS Biol.* 5: e300.

Kelsch, W., Sim, S., and Lois, C. (2012) Increasing heterogeneity in the organization of synaptic inputs of mature olfactory bulb neurons generated in newborn rats. *J. Comp. Neurol.* 520: 1327–1338.

Kernie, S. G. and Parent, J. M. (2010) Forebrain neurogenesis after focal Ischemic and traumatic brain injury. *Neurobiol. Dis.* 37: 267–274.

Kippin, T. E., Martens, D. J., and van der Kooy, D. (2005) p21 loss compromises the relative quiescence of forebrain stem cell proliferation leading to exhaustion of their proliferation capacity. *Genes Dev.* 19: 756–767.

Kohl, Z., Regensburger, M., Aigner, R., et al. (2010) Impaired adult olfactory bulb neurogenesis in the R6/2 mouse model of Huntington's disease. *BMC Neurosci.* 11: 114.

Kohwi, M., Petryniak, M. A., Long, J. E., et al. (2007) A subpopulation of olfactory bulb GABAergic interneurons is derived from Emx1- and Dlx5/6-expressing progenitors. *J. Neurosci.* 27: 6878–6891.

Koizumi, H., Higginbotham, H., Poon, T., et al. (2006) Doublecortin maintains bipolar shape and nuclear translocation during migration in the adult forebrain. *Nat. Neurosci.* 9: 779–786.

Kokovay, E., Goderie, S., Wang, Y., et al. (2010) Adult SVZ lineage cells home to and leave the vascular niche via differential responses to SDF1/CXCR4 signaling. *Cell Stem Cell* 7: 163–173.

Kokovay, E., Wang, Y., Kusek, G., et al. (2012) VCAM1 Is essential to maintain the structure of the SVZ niche and acts as an environmental sensor to regulate SVZ lineage progression. *Cell Stem Cell* 11: 220–230.

Kosaka, K. and Kosaka, T. (2005) synaptic organization of the glomerulus in the main olfactory bulb: compartments of the glomerulus and heterogeneity of the periglomerular cells. *Anat. Sci. Int.* 80: 80–90.

Lacar, B., Young, S. Z., Platel, J. C., and Bordey, A. (2011) Gap junction-mediated calcium waves define communication networks among murine postnatal neural progenitor cells. *Eur. J. Neurosci.* 34: 1895–1905.

Lagace, D. C., Whitman, M. C., Noonan, M. A., et al. (2007) Dynamic contribution of nestin-expressing stem cells to adult neurogenesis. *J. Neurosci.* 27: 12623–12629.

Laywell, E. D., Rakic, P., Kukekov, V. G., et al. (2000) Identification of a multipotent astrocytic stem cell in the immature and adult mouse brain. *Proc. Natl. Acad. Sci. U. S. A.* 97: 13883–13888.

Lazarini, F. and Lledo, P. M. (2011) Is adult neurogenesis essential for olfaction? *Trends Neurosci.* 34: 20–30.

Lazarini, F., Mouthon, M. A., Gheusi, G., et al. (2009) Cellular and behavioral effects of cranial irradiation of the subventricular zone in adult mice. *PLoS One* 4: e7017.

Lee, S. R., Kim, H. Y., Rogowska, J., et al. (2006) Involvement of matrix metalloproteinase in neuroblast cell migration from the subventricular zone after stroke. *J. Neurosci.* 26: 3491–3495.

Lemasson, M., Saghatelyan, A., Olivo-Marin, J. C., and Lledo, P. M. (2005) Neonatal and adult neurogenesis provide two distinct populations of newborn neurons to the mouse olfactory bulb. *J. Neurosci.* 25: 6816–6825.

Leong, S. Y. and Turnley, A. M. (2011) Regulation of adult neural precursor cell migration. *Neurochem. Int.* 59: 382–393.

Li, X., Tang, X., Jablonska, B., et al. (2009) p27(KIP1) regulates neurogenesis in the rostral migratory stream and olfactory bulb of the postnatal mouse. *J. Neurosci.* 29: 2902–2914.

Licht, T., Eavri, R., Goshen, I., et al. (2010) VEGF is required for dendritogenesis of newly born olfactory bulb interneurons. *Development* 137: 261–271.

Lim, D. A., Tramontin, A. D., Trevejo, J. M., et al. (2000) Noggin antagonizes BMP signaling to create a niche for adult neurogenesis. *Neuron* 28: 713–726.

Liu, H. K., Wang, Y., Belz, T., et al. (2010) The nuclear receptor tailless induces long-term neural stem cell expansion and brain tumor initiation. *Genes. Dev.* 24: 683–695.

Liu, X., Wang, Q., Haydar, T. F., and Bordey, A. (2005) Nonsynaptic GABA signaling in postnatal subventricular zone controls proliferation of GFAP-expressing progenitors. *Nat. Neurosci.* 8: 1179–1187.

Liu, Z. and Martin, L. J. (2003) Olfactory bulb core is a rich source of neural progenitor and stem cells in adult rodent and human. *J. Comp. Neurol.* 459: 368–391.

Livneh, Y., Feinstein, N., Klein, M., and Mizrahi, A. (2009) Sensory input enhances synaptogenesis of adult-born neurons. *J. Neurosci.* 29: 86–97.

Livneh, Y. and Mizrahi, A. (2011) Long-term changes in the morphology and synaptic distributions of adult-born neurons. *J. Comp. Neurol.* 519: 2212–2224.

Lledo, P. M., Alonso, M., and Grubb, M. S. (2006) Adult neurogenesis and functional plasticity in neuronal circuits. *Nat. Rev. Neurosci.* 7: 179–193.

Lledo, P. M., Merkle, F. T., and Alvarez-Buylla, A. (2008) Origin and function of olfactory bulb interneuron diversity. *Trends Neurosci.* 31: 392–400.

Lois, C. and Alvarez-Buylla, A. (1993) Proliferating subventricular zone cells in the adult mammalian forebrain can differentiate into neurons and glia. *Proc. Natl. Acad. Sci. U. S. A.* 90: 2074–2077.

Lois, C. and Alvarez-Buylla, A. (1994) Long-distance neuronal migration in the adult mammalian brain. *Science* 264: 1145–1148.

Lois, C., Garcia-Verdugo, J. M., and Alvarez-Buylla, A. (1996) Chain migration of neuronal precursors. *Science* 271: 978–981.

Lopez-Rios, J., Speziale, D., Robay, D., et al. (2012) GLI3 constrains digit number by controlling both progenitor proliferation and BMP-dependent exit to chondrogenesis. *Dev. Cell* 22: 837–848.

Luo, J., Daniels, S. B., Lennington, J. B., et al. (2006) The aging neurogenic subventricular zone. *Aging Cell* 5: 139–152.

Luskin, M.B. (1993) Restricted proliferation and migration of postnatally generated neurons derived from the forebrain subventricular zone. *Neuron* 11: 173–189.

Macas, J., Nern, C., Plate, K. H., and Momma, S. (2006) Increased generation of neuronal progenitors after ischemic injury in the aged adult human forebrain. *J. Neurosci.* 26: 13114–13119.

Machold, R., Hayashi, S., Rutlin, M., et al. (2003) Sonic hedgehog is required for progenitor cell maintenance in telencephalic stem cell niches. *Neuron* 39: 937–950.

Mak, G. K., Enwere, E. K., Gregg, C., et al. (2007) Male pheromone-stimulated neurogenesis in the adult female brain: possible role in mating behavior. *Nat. Neurosci.* 10: 1003–1011.

Mandairon, N., Sacquet, J., Jourdan, F., and Didier, A. (2006) Long-term fate and distribution of newborn cells in the adult mouse olfactory bulb: Influences of olfactory deprivation. *Neuroscience* 141: 443–451.

Mandairon, N., Sultan, S., Nouvian, M., et al. (2011) Involvement of newborn neurons in olfactory associative learning? The operant or non-operant component of the task makes all the difference. *J. Neurosci.* 31: 12455–12460.

Marinaro, C., Pannese, M., Weinandy, F., et al. (2012) Wnt signaling has opposing roles in the developing and the adult brain that are modulated by Hipk1. *Cereb Cortex* 22: 2415–2427.

Marshall, C. A., Novitch, B. G., and Goldman, J. E. (2005) Olig2 directs astrocyte and oligodendrocyte formation in postnatal subventricular zone cells. *J. Neurosci.* 25: 7289–7298.

Marshall, C. A., Suzuki, S. O., and Goldman, J. E. (2003) Gliogenic and neurogenic progenitors of the subventricular zone: who are they, where did they come from, and where are they going? *Glia* 43: 52–61.

Marti-Fabregas, J., Romaguera-Ros, M., Gomez-Pinedo, U., et al. (2010) Proliferation in the human ipsilateral subventricular zone after ischemic stroke. *Neurology* 74: 357–365.

Martinez-Molina, N., Kim, Y., Hockberger, P., and Szele, F. G. (2011) Rostral migratory stream neuroblasts turn and change directions in stereotypic patterns. *Cell Adh. Migr.* 5: 83–95.

Marxreiter, F., Regensburger, M., and Winkler, J. (2013) Adult neurogenesis in Parkinson's disease. *Cell. Mol. Life Sci.* 70: 459–473.

Meisami, E., Mikhail, L., Baim, D., and Bhatnagar, K. P. (1998) Human olfactory bulb: aging of glomeruli and mitral cells and a search for the accessory olfactory bulb. *Ann. N. Y. Acad. Sci.* 855: 708–715.

Mejia-Gervacio, S., Murray, K., and Lledo, P. M. (2011) NKCC1 controls GABAergic signaling and neuroblast migration in the postnatal forebrain. *Neural. Dev.* 6: 4.

Mejia-Gervacio, S., Murray, K., Sapir, T., et al. (2012) MARK2/Par-1 guides the directionality of neuroblasts migrating to the olfactory bulb. *Mol. Cell. Neurosci.* 49: 97–103.

Mendez-Gomez, H. R. and Vicario-Abejon, C. (2012) The homeobox gene Gsx2 regulates the self-renewal and differentiation of neural stem cells and the cell fate of postnatal progenitors. *PLoS One* 7: e29799.

Mendoza-Torreblanca, J. G., Martinez-Martinez, E., Tapia-Rodriguez, M., et al. (2008) The rostral migratory stream is a neurogenic niche that predominantly engenders periglomerular cells: in vivo evidence in the adult rat brain. *Neurosci. Res.* 60: 289–299.

Menn, B., Garcia-Verdugo, J. M., Yaschine, C., et al. (2006) Origin of oligodendrocytes in the subventricular zone of the adult brain. *J. Neurosci.* 26: 7907–7918.

Merkle, F. T., Mirzadeh, Z., and Alvarez-Buylla, A. (2007) Mosaic organization of neural stem cells in the adult brain. *Science* 317: 381–384.

Merkle, F. T., Tramontin, A. D., Garcia-Verdugo, J. M., and Alvarez-Buylla, A. (2004) Radial glia give rise to adult neural stem cells in the subventricular zone. *Proc. Natl. Acad. Sci. U. S. A.* 101: 17528–17532.

Merson, T. D., Dixon, M. P., Collin, C., et al. (2006) The transcriptional coactivator Querkopf controls adult neurogenesis. *J. Neurosci.* 26: 11359–11370.

Ming, G. L. and Song, H. (2011) Adult neurogenesis in the mammalian brain: significant answers and significant questions. *Neuron* 70: 687–702.

Mirzadeh, Z., Merkle, F. T., Soriano-Navarro, M., et al. (2008) Neural stem cells confer unique pinwheel architecture to the ventricular surface in neurogenic regions of the adult brain. *Cell Stem Cell* 3: 265–278.

Mishra, S. K., Braun, N., Shukla, V., et al. (2006) Extracellular nucleotide signaling in adult neural stem cells: synergism with growth factor-mediated cellular proliferation. *Development* 133: 675–684.

Mizrahi, A. (2007) Dendritic development and plasticity of adult-born neurons in the mouse olfactory bulb. *Nat. Neurosci.* 10: 444–452.

Mizrahi, A., Lu, J., Irving, R., et al. (2006) In vivo imaging of juxtaglomerular neuron turnover in the mouse olfactory bulb. *Proc. Natl. Acad. Sci. U. S. A.* 103: 1912–1917.

Mobley, A. K. and McCarty, J. H. (2011) beta8 integrin is essential for neuroblast migration in the rostral migratory stream. *Glia* 59: 1579–1587.

Mobley, A. K., Tchaicha, J. H., Shin, J., et al. (2009) Beta8 integrin regulates neurogenesis and neurovascular homeostasis in the adult brain. *J. Cell Sci.* 122: 1842–1851.

Molofsky, A. V., Pardal, R., Iwashita, T., et al. (2003) Bmi-1 dependence distinguishes neural stem cell self-renewal from progenitor proliferation. *Nature* 425: 962–967.

Molofsky, A. V., Slutsky, S. G., Joseph, N. M., et al. (2006) Increasing p16INK4a expression decreases forebrain progenitors and neurogenesis during ageing. *Nature* 443: 448–452.

Mori, K., Kishi, K., and Ojima, H. (1983) Distribution of dendrites of mitral, displaced mitral, tufted, and granule cells in the rabbit olfactory bulb. *J. Comp. Neurol.* 219: 339–355.

Mudo, G., Bonomo, A., Di Liberto, V., et al. (2009) The FGF-2/FGFRs neurotrophic system promotes neurogenesis in the adult brain. *J. Neural Transm.* 116: 995–1005.

Murase, S., Cho, C., White, J. M., and Horwitz, A. F. (2008) ADAM2 promotes migration of neuroblasts in the rostral migratory stream to the olfactory bulb. *Eur. J. Neurosci.* 27: 1585–1595.

Murase, S. and Horwitz, A. F. (2002) Deleted in colorectal carcinoma and differentially expressed integrins mediate the directional migration of neural precursors in the rostral migratory stream. *J. Neurosci.* 22: 3568–3579.

Nait-Oumesmar, B., Decker, L., Lachapelle, F., et al. (1999) Progenitor cells of the adult mouse subventricular zone proliferate, migrate and differentiate into oligodendrocytes after demyelination. *Eur. J. Neurosci.* 11: 4357–4366.

Nait-Oumesmar, B., Picard-Riera, N., Kerninon, C., and Baron-Van Evercooren, A. (2008) The role of SVZ-derived neural precursors in demyelinating diseases: from animal models to multiple sclerosis. *J. Neurol Sci.* 265: 26–31.

Nam, S. C., Kim, Y., Dryanovski, D., et al. (2007) Dynamic features of postnatal subventricular zone cell motility: a two-photon time-lapse study. *J. Comp. Neurol.* 505: 190–208.

Ng, K. L., Li, J. D., Cheng, M. Y., et al. (2005) Dependence of olfactory bulb neurogenesis on prokineticin 2 signaling. *Science* 308: 1923–1927.

Nguyen, L., Malgrange, B., Breuskin, I., et al. (2003) Autocrine/paracrine activation of the GABA(A) receptor inhibits the proliferation of neurogenic polysialylated neural cell adhesion molecule-positive (PSA-NCAM+) precursor cells from postnatal striatum. *J. Neurosci.* 23: 3278–3294.

Nguyen-Ba-Charvet, K. T., Picard-Riera, N., Tessier-Lavigne, M., et al. (2004) Multiple roles for slits in the control of cell migration in the rostral migratory stream. *J. Neurosci.* 24: 1497–1506.

Nicoleau, C., Benzakour, O., Agasse, F., et al. (2009) Endogenous hepatocyte growth factor is a niche signal for subventricular zone neural stem cell amplification and self-renewal. *Stem Cells* 27: 408–419.

Ninkovic, J., Mori, T., and Gotz, M. (2007) Distinct modes of neuron addition in adult mouse neurogenesis. *J. Neurosci.* 27: 10906–10911.

Nishino, J., Kim, I., Chada, K., and Morrison, S. J. (2008) Hmga2 promotes neural stem cell self-renewal in young but not old mice by reducing p16Ink4a and p19Arf Expression. *Cell* 135: 227–239.

Nityanandam, A., Parthasarathy, S., and Tarabykin, V. (2012) Postnatal subventricular zone of the neocortex contributes GFAP+ cells to the rostral migratory stream under the control of Sip1. *Dev. Biol.* 366: 341–356.

O'Keeffe, G. C., Barker, R. A., and Caldwell, M. A. (2009a) Dopaminergic modulation of neurogenesis in the subventricular zone of the adult brain. *Cell Cycle* 8: 2888–2894.

O'Keeffe, G. C., Tyers, P., Aarsland, D., et al. (2009b) Dopamine-induced proliferation of adult neural precursor cells in the mammalian subventricular zone is mediated through EGF. *Proc. Natl. Acad. Sci. U. S. A.* 106: 8754–8759.

Ono, K., Tomasiewicz, H., Magnuson, T., and Rutishauser, U. (1994) N-CAM mutation inhibits tangential neuronal migration and is phenocopied by enzymatic removal of polysialic acid. *Neuron* 13: 595–609.

Orona, E., Scott, J. W., and Rainer, E. C. (1983) Different granule cell populations innervate superficial and deep regions of the external plexiform layer in rat olfactory bulb. *J. Comp. Neurol.* 217: 227–237.

Pagano, S. F., Impagnatiello, F., Girelli, M., et al. (2000) Isolation and characterization of neural stem cells from the adult human olfactory bulb. *Stem Cells* 18: 295–300.

Pallotto, M., Nissant, A., Fritschy, J. M., et al. (2012) Early Formation of GABAergic Synapses Governs the Development of Adult-Born Neurons in the Olfactory Bulb. *J. Neurosci.* 32: 9103–9115.

Palma, V., Lim, D. A., Dahmane, N., et al. (2005) Sonic hedgehog controls stem cell behavior in the postnatal and adult brain. *Development* 132: 335–344.

Panzanelli, P., Bardy, C., Nissant, A., et al. (2009) Early synapse formation in developing interneurons of the adult olfactory bulb. *J. Neurosci.* 29: 15039–15052.

Panzanelli, P., Fritschy, J. M., Yanagawa, Y., et al. (2007) GABAergic phenotype of periglomerular cells in the rodent olfactory bulb. *J. Comp. Neurol.* 502: 990–1002.

Paratcha, G., Ibanez, C. F., and Ledda, F. (2006) GDNF is a chemoattractant factor for neuronal precursor cells in the rostral migratory stream. *Mol. Cell. Neurosci.* 31: 505–514.

Parent, J. M., Vexler, Z. S., Gong, C., et al. (2002) Rat forebrain neurogenesis and striatal neuron replacement after focal stroke. *Ann. Neurol.* 52: 802–813.

Parent, J. M., von dem Bussche, N., and Lowenstein, D. H. (2006) Prolonged seizures recruit caudal subventricular zone glial progenitors into the injured hippocampus. *Hippocampus* 16: 321–328.

Parrish-Aungst, S., Shipley, M. T., Erdelyi, F., et al. (2007) Quantitative analysis of neuronal diversity in the mouse olfactory bulb. *J. Comp. Neurol.* 501: 825–836.

Pathania, M., Torres-Reveron, J., Yan, L., et al. (2012) miR-132 enhances dendritic morphogenesis, spine density, synaptic integration, and survival of newborn olfactory bulb neurons. *PLoS One* 7: e38174.

Paton, J. A. and Nottebohm, F. N. (1984) Neurons generated in the adult brain are recruited into functional circuits. *Science* 225: 1046–1048.

Pencea, V. and Luskin, M. B. (2003) Prenatal development of the rodent rostral migratory stream. *J. Comp. Neurol.* 463: 402–418.

Peretto, P., Giachino, C., Aimar, P., et al. (2005) Chain formation and glial tube assembly in the shift from neonatal to adult subventricular zone of the rodent forebrain. *J. Comp. Neurol.* 487: 407–427.

Petreanu, L. and Alvarez-Buylla, A. (2002) Maturation and death of adult-born olfactory bulb granule neurons: role of olfaction. *J. Neurosci.* 22: 6106–6113.

Phillips, W., Morton, A. J., and Barker, R. A. (2005) Abnormalities of neurogenesis in the R6/2 mouse model of Huntington's disease are attributable to the in vivo microenvironment. *J. Neurosci.* 25: 11564–11576.

Pino, D., Choe, Y., and Pleasure, S. J. (2011) Wnt5a controls neurite development in olfactory bulb interneurons. *ASN Neuro.* 3: e00059.

Plachez, C., and Puche, A. C. (2012) Early specification of GAD67 subventricular derived olfactory interneurons. *J. Mol. Histol.* 43: 215–221.

Platel, J. C., Dave, K. A., and Bordey, A. (2008a) Control of neuroblast production and migration by converging GABA and glutamate signals in the postnatal forebrain. *J. Physiol.* 586: 3739–3743.

Platel, J. C., Dave, K. A., Gordon, V., et al. (2010) NMDA receptors activated by subventricular zone astrocytic glutamate are critical for neuroblast survival prior to entering a synaptic network. *Neuron* 65: 859–872.

Platel, J. C., Heintz, T., Young, S., et al. (2008b) Tonic activation of GLUK5 kainate receptors decreases neuroblast migration in whole-mounts of the subventricular zone. *J. Physiol.* 586: 3783–3793.

Platel, J. C., Lacar, B., and Bordey, A. (2007) GABA and glutamate signaling: homeostatic control of adult forebrain neurogenesis. *J. Mol. Histol.* 38: 602–610.

Prosser, H. M., Bradley, A., and Caldwell, M. A. (2007) Olfactory bulb hypoplasia in Prokr2 null mice stems from defective neuronal progenitor migration and differentiation. *Eur. J. Neurosci.* 26: 3339–3344.

Puverel, S., Nakatani, H., Parras, C., and Soussi-Yanicostas, N. (2009) Prokineticin receptor 2 expression identifies migrating neuroblasts and their subventricular zone transient-amplifying progenitors in adult mice. *J. Comp. Neurol.* 512: 232–242.

Qu, Q., Sun, G., Li, W., et al. (2010) Orphan nuclear receptor TLX activates Wnt/beta-catenin signalling to stimulate neural stem cell proliferation and self-renewal. *Nat. Cell Biol.* 12: 31–40; sup pp 31–39.

Quinones-Hinojosa, A., Sanai, N., Soriano-Navarro, M., et al. (2006) Cellular composition and cytoarchitecture of the adult human subventricular zone: a niche of neural stem cells. *J. Comp. Neurol.* 494: 415–434.

Renault, V. M., Rafalski, V. A., Morgan, A. A., et al. (2009) FoxO3 regulates neural stem cell homeostasis. *Cell Stem Cell* 5: 527–539.

Ricard, J., Salinas, J., Garcia, L., and Liebl, D. J. (2006) EphrinB3 regulates cell proliferation and survival in adult neurogenesis. *Mol. Cell. Neurosci.* 31: 713–722.

Riquelme, P. A., Drapeau, E., and Doetsch, F. (2008) Brain micro-ecologies: neural stem cell niches in the adult mammalian brain. *Philos. Trans. R. Soc. Lond. B. Biol. Sci.* 363: 123–137.

Rochefort, C. and Lledo, P. M. (2005) Short-term survival of newborn neurons in the adult olfactory bulb after exposure to a complex odor environment. *Eur. J. Neurosci.* 22: 2863–2870.

Roelofs, R. F., Fischer, D. F., Houtman, S. H., et al. (2005) Adult human subventricular, subgranular, and subpial zones contain astrocytes with a specialized intermediate filament cytoskeleton. *Glia* 52: 289–300.

Ronchini, C. and Capobianco, A. J. (2001) Induction of cyclin D1 transcription and CDK2 activity by Notch(ic): implication for cell cycle disruption in transformation by Notch(ic). *Mol. Cell. Biol.* 21: 5925–5934.

Saghatelyan, A., de Chevigny, A., Schachner, M., and Lledo, P. M. (2004) Tenascin-R mediates activity-dependent recruitment of neuroblasts in the adult mouse forebrain. *Nat. Neurosci.* 7: 347–356.

Saghatelyan, A., Roux, P., Migliore, M., et al. (2005) Activity-dependent adjustments of the inhibitory network in the olfactory bulb following early postnatal deprivation. *Neuron* 46: 103–116.

Saino-Saito, S., Cave, J. W., Akiba, Y., et al. (2007) ER81 and CaMKIV identify anatomically and phenotypically defined subsets of mouse olfactory bulb interneurons. *J. Comp. Neurol.* 502: 485–496.

Sanai, N., Nguyen, T., Ihrie, R. A., et al. (2011) Corridors of migrating neurons in the human brain and their decline during infancy. *Nature* 478: 382–386.

Sanai, N., Tramontin, A. D., Quinones-Hinojosa, A., et al. (2004) Unique astrocyte ribbon in adult human brain contains neural stem cells but lacks chain migration. *Nature* 427: 740–744.

Sawada, M., Kaneko, N., Inada, H., et al. (2011) Sensory input regulates spatial and subtype-specific patterns of neuronal turnover in the adult olfactory bulb. *J. Neurosci.* 31: 11587–11596.

Sawamoto, K., Wichterle, H., Gonzalez-Perez, O., et al. (2006) New neurons follow the flow of cerebrospinal fluid in the adult brain. *Science* 311: 629–632.

Sheikh, B. N., Dixon, M. P., Thomas, T., and Voss, A. K. (2012) Querkopf is a key marker of self-renewal and multipotency of adult neural stem cells. *J. Cell. Sci.* 125: 295–309.

Shen, Q., Goderie, S. K., Jin, L., et al. (2004) Endothelial cells stimulate self-renewal and expand neurogenesis of neural stem cells. *Science* 304: 1338–1340.

Shen, Q., Wang, Y., Kokovay, E., et al. (2008) Adult SVZ stem cells lie in a vascular niche: a quantitative analysis of niche cell-cell interactions. *Cell Stem Cell* 3: 289–300.

Shepherd, G. M. (1972) Synaptic organization of the mammalian olfactory bulb. *Physiol. Rev.* 52: 864–917.

Shibata, M., Nakao, H., Kiyonari, H., et al. (2011) MicroRNA-9 regulates neurogenesis in mouse telencephalon by targeting multiple transcription factors. *J. Neurosci.* 31: 3407–3422.

Shimazaki, T., Shingo, T., and Weiss, S. (2001) The ciliary neurotrophic factor/leukemia inhibitory factor/gp130 receptor complex operates in the maintenance of mammalian forebrain neural stem cells. *J. Neurosci.* 21: 7642–7653.

Shimozaki, K., Zhang, C. L., Suh, H., et al. (2012) SRY-box-containing gene 2 regulation of nuclear receptor tailless (Tlx) transcription in adult neural stem cells. *J. Biol. Chem.* 287: 5969–5978.

Shingo, T., Gregg, C., Enwere, E., et al. (2003) Pregnancy-stimulated neurogenesis in the adult female forebrain mediated by prolactin. *Science* 299: 117–120.

Simo, S., Pujadas, L., Segura, M. F., et al. (2007) Reelin induces the detachment of postnatal subventricular zone cells and the expression of the Egr-1 through Erk1/2 activation. *Cereb. Cortex* 17: 294–303.

Snapyan, M., Lemasson, M., Brill, M. S., et al. (2009) Vasculature guides migrating neuronal precursors in the adult mammalian forebrain via brain-derived neurotrophic factor signaling. *J. Neurosci.* 29: 4172–4188.

Soria, J. M., Taglialatela, P., Gil-Perotin, S., et al. (2004) Defective postnatal neurogenesis and disorganization of the rostral migratory stream in absence of the Vax1 homeobox gene. *J. Neurosci.* 24: 11171–11181.

Soumier, A., Banasr, M., Goff, L. K., and Daszuta, A. (2010) Region- and phase-dependent effects of 5-HT(1A) and 5-HT(2C) receptor activation on adult neurogenesis. *Eur. Neuropsychopharmacol.* 20: 336–345.

Stenman, J., Toresson, H., and Campbell, K. (2003) Identification of two distinct progenitor populations in the lateral ganglionic eminence: implications for striatal and olfactory bulb neurogenesis. *J. Neurosci.* 23: 167–174.

Stewart, R. R., Hoge, G. J., Zigova, T., and Luskin, M. B. (2002) Neural progenitor cells of the neonatal rat anterior subventricular zone express functional GABA(A) receptors. *J. Neurobiol.* 50: 305–322.

Suh, Y., Obernier, K., Holzl-Wenig, G., et al. (2009) Interaction between DLX2 and EGFR regulates proliferation and neurogenesis of SVZ precursors. *Mol. Cell. Neurosci.* 42: 308–314.

Sultan, S., Mandairon, N., Kermen, F., et al. (2010) Learning-dependent neurogenesis in the olfactory bulb determines long-term olfactory memory. *FASEB J* 24: 2355–2363.

Sun, G., Yu, R. T., Evans, R. M., and Shi, Y. (2007) Orphan nuclear receptor TLX recruits histone deacetylases to repress transcription and regulate neural stem cell proliferation. *Proc. Natl. Acad. Sci. U. S. A.* 104: 15282–15287.

Sun, J., Zhou, W., Ma, D., and Yang, Y. (2010a) Endothelial cells promote neural stem cell proliferation and differentiation associated with VEGF activated Notch and Pten signaling. *Dev Dyn* 239: 2345–2353.

Sun, L., Yu, Z., Wang, W., and Liu, X. (2012) Both NKCC1 and anion exchangers contribute to Cl(−) accumulation in postnatal forebrain neuronal progenitors. *Eur. J. Neurosci.* 35: 661–672.

Sun, W., Kim, H., and Moon, Y. (2010b) Control of neuronal migration through rostral migration stream in mice. *Anat. Cell Biol.* 43: 269–279.

Suyama, S., Sunabori, T., Kanki, H., et al. (2012) Purinergic signaling promotes proliferation of adult mouse subventricular zone cells. *J. Neurosci.* 32: 9238–9247.

Suzuki, S. O. and Goldman, J. E. (2003) Multiple cell populations in the early postnatal subventricular zone take distinct migratory pathways: a dynamic study of glial and neuronal progenitor migration. *J. Neurosci.* 23: 4240–4250.

Tattersfield, A. S., Croon, R. J., Liu, Y. W., et al. (2004) Neurogenesis in the striatum of the quinolinic acid lesion model of Huntington's disease. *Neuroscience* 127: 319–332.

Tavazoie, M., Van der Veken, L., Silva-Vargas, V., et al. (2008) A specialized vascular niche for adult neural stem cells. *Cell Stem Cell* 3: 279–288.

Thored, P., Arvidsson, A., Cacci, E., et al. (2006) Persistent production of neurons from adult brain stem cells during recovery after stroke. *Stem Cells* 24: 739–747.

Thored, P., Wood, J., Arvidsson, A., et al. (2007) Long-term neuroblast migration along blood vessels in an area with transient angiogenesis and increased vascularization after stroke. *Stroke* 38: 3032–3039.

Tropepe, V., Craig, C. G., Morshead, C. M., and van der Kooy, D. (1997) Transforming growth factor-alpha null and senescent mice show decreased neural progenitor cell proliferation in the forebrain subependyma. *J. Neurosci.* 17: 7850–7859.

Valley, M. T., Mullen, T. R., Schultz, L. C., et al. (2009) Ablation of mouse adult neurogenesis alters olfactory bulb structure and olfactory fear conditioning. *Front. Neurosci.* 3: 51.

van den Berge, S. A., van Strien, M. E., Korecka, J. A., et al. (2011) The proliferative capacity of the subventricular zone is maintained in the parkinsonian brain. *Brain* 134: 3249–3263.

Vanderluit, J. L., Ferguson, K. L., Nikoletopoulou, V., et al. (2004) p107 regulates neural precursor cells in the mammalian brain. *J. Cell. Biol.* 166: 853–863.

Vanderluit, J. L., Wylie, C. A., McClellan, K. A., et al. (2007) The Retinoblastoma family member p107 regulates the rate of progenitor commitment to a neuronal fate. *J. Cell. Biol.* 178: 129–139.

Ventura, R. E. and Goldman, J. E. (2007) Dorsal radial glia generate olfactory bulb interneurons in the postnatal murine brain. *J. Neurosci.* 27: 4297–4302.

Waclaw, R.R., Allen, Z.J., Bell, S.M., et al. (2006) The zinc finger transcription factor Sp8 regulates the generation and diversity of olfactory bulb interneurons. *Neuron* 49: 503–516.

Wang, C., Liu, F., Liu, Y. Y., et al. (2011a) Identification and characterization of neuroblasts in the subventricular zone and rostral migratory stream of the adult human brain. *Cell Res.* 21: 1534–1550.

Wang, D. D., Krueger, D. D., and Bordey, A. (2003) GABA depolarizes neuronal progenitors of the postnatal subventricular zone via GABAA receptor activation. *J. Physiol.* 550: 785–800.

Wang, L., Zhang, Z. G., Zhang, R. L., et al. (2006) Matrix metalloproteinase 2 (MMP2) and MMP9 secreted by erythropoietin-activated endothelial cells promote neural progenitor cell migration. *J. Neurosci.* 26: 5996–6003.

Wang, T. W., Zhang, H., Gyetko, M. R., and Parent, J. M. (2011b) Hepatocyte growth factor acts as a mitogen and chemoattractant for postnatal subventricular zone-olfactory bulb neurogenesis. *Mol. Cell. Neurosci.* 48: 38–50.

Weickert, C. S., Webster, M. J., Colvin, S. M., et al. (2000) Localization of epidermal growth factor receptors and putative neuroblasts in human subependymal zone. *J. Comp. Neurol.* 423: 359–372.

Weinandy, F., Ninkovic, J., and Gotz, M. (2011) Restrictions in time and space–new insights into generation of specific neuronal subtypes in the adult mammalian brain. *Eur. J. Neurosci.* 33: 1045–1054.

Whitman, M. C., Fan, W., Rela, L., et al. (2009) Blood vessels form a migratory scaffold in the rostral migratory stream. *J. Comp. Neurol.* 516: 94–104.

Whitman, M. C. and Greer, C. A. (2007a) Adult-generated neurons exhibit diverse developmental fates. *Dev. Neurobiol.* 67: 1079–1093.

Whitman, M. C. and Greer, C. A. (2007b) Synaptic integration of adult-generated olfactory bulb granule cells: basal axodendritic centrifugal input precedes apical dendrodendritic local circuits. *J. Neurosci.* 27: 9951–9961.

Wichterle, H., Garcia-Verdugo, J. M., and Alvarez-Buylla, A. (1997) Direct evidence for homotypic, glia-independent neuronal migration. *Neuron* 18: 779–791.

Willaime-Morawek, S., Seaberg, R. M., Batista, C., et al. (2006) Embryonic cortical neural stem cells migrate ventrally and persist as postnatal striatal stem cells. *J. Cell. Biol.* 175: 159–168.

Winner, B., Cooper-Kuhn, C. M., Aigner, R., et al. (2002) Long-term survival and cell death of newly generated neurons in the adult rat olfactory bulb. *Eur. J. Neurosci.* 16: 1681–1689.

References

Winner, B., Kohl, Z., and Gage, F. H. (2011) Neurodegenerative disease and adult neurogenesis. *Eur. J. Neurosci.* 33: 1139–1151.

Wittko, I. M., Schanzer, A., Kuzmichev, A., et al. (2009) VEGFR-1 regulates adult olfactory bulb neurogenesis and migration of neural progenitors in the rostral migratory stream in vivo. *J. Neurosci.* 29: 8704–8714.

Wonders, C. P. and Anderson, S. A. (2006) The origin and specification of cortical interneurons. *Nat. Rev. Neurosci.* 7: 687–696.

Wu, W., Wong, K., Chen, J., et al. (1999) Directional guidance of neuronal migration in the olfactory system by the protein Slit. *Nature* 400: 331–336.

Yamaguchi, M. and Mori, K. (2005) Critical period for sensory experience-dependent survival of newly generated granule cells in the adult mouse olfactory bulb. *Proc. Natl. Acad. Sci. U. S. A.* 102: 9697–9702.

Yang, P., Arnold, S. A., Habas, A., et al. (2008) Ciliary neurotrophic factor mediates dopamine D2 receptor-induced CNS neurogenesis in adult mice. *J. Neurosci.* 28: 2231–2241.

Yokoyama, T. K., Mochimaru, D., Murata, K., et al. (2011) Elimination of adult-born neurons in the olfactory bulb is promoted during the postprandial period. *Neuron* 71: 883–897.

Yoshihara, S., Omichi, K., Yanazawa, M., et al. (2005) Arx homeobox gene is essential for development of mouse olfactory system. *Development* 132: 751–762.

Yoshihara, S., Takahashi, H., Nishimura, N., et al. (2012) 5T4 glycoprotein regulates the sensory input-dependent development of a specific subtype of newborn interneurons in the mouse olfactory bulb. *J. Neurosci.* 32: 2217–2226.

Young, K. M., Fogarty, M., Kessaris, N., and Richardson, W. D. (2007) Subventricular zone stem cells are heterogeneous with respect to their embryonic origins and neurogenic fates in the adult olfactory bulb. *J. Neurosci.* 27: 8286–8296.

Young, S. Z., Platel, J. C., Nielsen, J. V., et al. (2010) GABA(A) Increases Calcium in Subventricular Zone Astrocyte-Like Cells Through L- and T-Type Voltage-Gated Calcium Channels. *Front. Cell Neurosci.* 4: 8.

Zencak, D., Lingbeek, M., Kostic, C., et al. (2005) Bmi1 loss produces an increase in astroglial cells and a decrease in neural stem cell population and proliferation. *J. Neurosci.* 25: 5774–5783.

Zhang, H., Vutskits, L., Pepper, M. S., and Kiss, J. Z. (2003) VEGF is a chemoattractant for FGF-2-stimulated neural progenitors. *J. Cell. Biol.* 163: 1375–1384.

Zhang, R. L., Zhang, Z. G., Zhang, L., and Chopp, M. (2001) Proliferation and differentiation of progenitor cells in the cortex and the subventricular zone in the adult rat after focal cerebral ischemia. *Neuroscience* 105: 33–41.

Zhao, C., Sun, G., Li, S., et al. (2010) MicroRNA let-7b regulates neural stem cell proliferation and differentiation by targeting nuclear receptor TLX signaling. *Proc. Natl. Acad. Sci. U. S. A.* 107: 1876–1881.

Zhao, C., Sun, G., Li, S., and Shi, Y. (2009) A feedback regulatory loop involving microRNA-9 and nuclear receptor TLX in neural stem cell fate determination. *Nat. Struct. Mol. Biol.* 16: 365–371.

Zigova, T., Betarbet, R., Soteres, B. J., et al. (1996) A comparison of the patterns of migration and the destinations of homotopically transplanted neonatal subventricular zone cells and heterotopically transplanted telencephalic ventricular zone cells. *Dev. Biol.* 173: 459–474.

Chapter 10

Cortical Olfactory Anatomy and Physiology

DONALD A. WILSON, JULIE CHAPUIS, and REGINA M. SULLIVAN

10.1 INTRODUCTION

The olfactory cortex is defined as those regions receiving direct monosynaptic input from the olfactory bulb and includes areas such as the olfactory tubercle, piriform cortex, and lateral entorhinal cortex, among others. Over the past 5–10 years (since the previous edition of this handbook), there has been an exponential growth in research on the anatomy, physiology, and function of olfactory cortex. This work has substantially enhanced understanding of what the cortex contributes to odor perception, and how disruption of olfactory cortical function in disease and aging may contribute to olfactory perceptual deficits. In this chapter we review the current knowledge of olfactory cortical anatomy and physiology, and describe current ideas about how this functional anatomy shapes odor perception. In addition, we touch on the contributions of other cortical regions, for example the orbitofrontal cortex, to odor perception.

There is a long history of neuroanatomical and electrophysiological interest in the olfactory cortex which created a rich groundwork for ideas about how olfactory cortex functions and what it contributes to odor perception. Comparative anatomical studies noted the evolutionary conservation of piriform cortex throughout vertebrates relative to other cortical areas (Herrick, 1933). Herrick argued that given that olfactory cortex is the most evolutionarily primitive cortex, it has unique associative impact on other regions, both providing subtle modulation of those other regions and receiving multi-sensory feedback from them (Herrick, 1933). "At all stages of cortical elaboration an important function of the olfactory cortex, in addition to participation in its own specific way in cortical

associations, is to serve as a non-specific activator for all cortical activities" (1933: 14). Electrophysiological analyses of odor-evoked piriform cortical activity was also explored very early on, first with the pioneering use of local field potentials (Adrian, 1942), and subsequently with single-unit recordings (Haberly, 1969). The piriform cortex was the first cortical region successfully used for in vitro slice electrophysiology, pre-dating development of hippocampal slices (Yamamoto and McIlwain, 1966). This has led to a wealth of information on synaptic physiology and connectivity within olfactory cortex that rivals that of any other brain region (e.g., Suzuki and Bekkers, 2010; Poo and Isaacson, 2011a; Suzuki and Bekkers, 2011). As experimental data mounted, hypotheses of olfactory cortical function and the role of piriform cortex in odor perception matured, especially in the mid-1980s (Haberly, 1985; Lynch, 1986; Wilson and Bower, 1988; Hopfield, 1991). These early models included basic concepts still held today, though current models are now more strongly supported by experimental data.

Currently, the olfactory cortex is hypothesized to function as both the primary olfactory sensory cortex and as association cortex (Haberly, 2001; Gottfried, 2010; Wilson and Sullivan, 2011). Activity in the olfactory cortex merges odorant feature information extracted at the sensory epithelium and refined in the olfactory bulb into odor objects which form the basis of perception. However, these odor object representations include not only odorant information, but also information conveying hedonics, context, and expectation (Gottfried, 2009). For example, odor evoked activity in the olfactory cortex can be shaped

---

Handbook of Olfaction and Gustation, Third Edition. Edited by Richard L. Doty.
© 2015 Richard L. Doty. Published 2015 by John Wiley & Sons, Inc.

by a visual image associated with the smell (Gottfried and Dolan, 2003), just as the color of the source of an odorant can influence the perception of its odor (Zellner and Kautz, 1990; Zellner et al., 1991). The olfactory cortex is the target of descending, multimodal, and modulatory inputs that may contribute to these rich perceptual outcomes. Furthermore, based on anatomical and physiological differences, the olfactory cortex can be divided into many sub-regions. Anatomically, these sub-regions differ in both local circuitry and afferent and efferent connectivity. While still under examined, there is increasing evidence for specialization of function among these sub-regions, perhaps forming a parallel processing stream of information leading to different aspects of the overall perception. It is increasingly apparent that experience and cortical plasticity play an important role in olfactory cortical function and odor perception.

10.2 OLFACTORY CORTEX: STRUCTURE AND FUNCTION

The olfactory cortex (Figure 10.1) is unique among mammalian sensory cortices in that: 1) it is primarily a 3-layered paleocortical structure in contrast to the 6-layered neocortex of other systems; 2) despite remarkable spatial patterning of odor-evoked activity in the olfactory bulb, the olfactory cortex shows no detectable topography, in contrast to neocortical sensory regions; and 3) in contrast to thalamocortical sensory systems, there is no thalamic nucleus between the peripheral receptors and olfactory cortex, though functions similar to those of the thalamus may occur in the olfactory bulb (Kay and Sherman, 2007) or cortex itself (Shepherd, 2005).

The olfactory cortex, that is, the termination zone of olfactory bulb mitral and tufted cells (Figure 10.2), includes the anterior olfactory nucleus (AON), tenia tecta, olfactory tubercle (OT), anterior and posterior piriform cortices (aPCX and pPCX respectively), the cortical nucleus of the amygdala, and the lateral entorhinal cortex. In rodents, these structures lie within the olfactory peduncle and along the ventral lateral edge of the forebrain, extending the full length of the forebrain with the lateral entorhinal cortex at the caudal end. In primates, the majority of these structures have moved medially with the expansion of the neocortex, with AON and piriform cortex lying anterior medial and the base of the olfactory peduncle in the same anterior-posterior plane as the amygdala. All regions of the olfactory cortex send efferents back to the olfactory bulb, as well as to a number of ipsilateral and contralateral targets.

Many of the sub-regions of olfactory cortex have received relatively little experimental attention, and will be only briefly described here. The majority of the review will focus on piriform cortex. Although evidence is beginning to emerge for specialization of function among cortical sub-regions (e.g., (Sosulski et al., 2011; Wesson and Wilson, 2011)), perhaps suggestive of parallel processing of olfactory bulb output, the data are currently sparse.

10.2.1 Anterior Olfactory Nucleus

The anterior olfactory nucleus lies in the olfactory peduncle and is the most anterior region of the olfactory cortex. It is a laminated structure, with clear molecular and cell body layers (Brunjes et al., 2005), and the argument has been made that it should more properly be named anterior olfactory cortex (Haberly, 2001). The AON can be divided into at least two major sub-regions, the pars external and the pars principalis. AON neurons receive direct input from mitral and tufted cell axons as they leave the olfactory bulb along the lateral surface of the olfactory peduncle forming the lateral olfactory tract. A definitive review of AON structure and cell types is published elsewhere (Brunjes et al., 2005).

The pars externa appears to receive a topographic input from the olfactory bulb, and in turn projects in a highly precise topographic manner to the contralateral olfactory bulb (Yan et al., 2008). In accord with this topographical anatomy, odor-evoked activity within the AON pars externa displays odor-specific spatial patterning, similar to that observed in the olfactory bulb (Kay et al., 2011). Odor-evoked activity in pars externa is also bilaterally asymmetrical, with ipsilateral inputs driving primarily excitatory activity and contralateral inputs primarily inducing activity suppression (Kikuta et al., 2008; Kikuta et al., 2010). This odor-specific, bilateral convergence suggests that the AON pars externa could play a role in odor localization (Rajan et al., 2006; Porter et al., 2007).

Similar to the pars externa, the AON pars principalis is a major source of commissural fibers to the contralateral olfactory bulb, AON, and piriform cortex (Brunjes et al., 2005; Illig and Eudy, 2009). However, in contrast to the AON pars externa, the pars principalis does not show spatial patterning in odor-evoked activity, but instead shows highly distributed odor-evoked activity (Meyer et al., 2006) similar to that observed in the piriform cortex (Illig and Haberly, 2003; Rennaker et al., 2007; Stettler and Axel, 2009).

In addition to serving as an important point of bilateral convergence and comparison, the AON may also begin the process of synthesizing molecular features extracted at the periphery and olfactory bulb into odor objects (Haberly, 2001). In fact, AON single-units are driven most effectively by odor mixtures, with subthreshold presentation of components in a mixture adding non-linearly to evoke a robust mixture response (Lei et al., 2006).

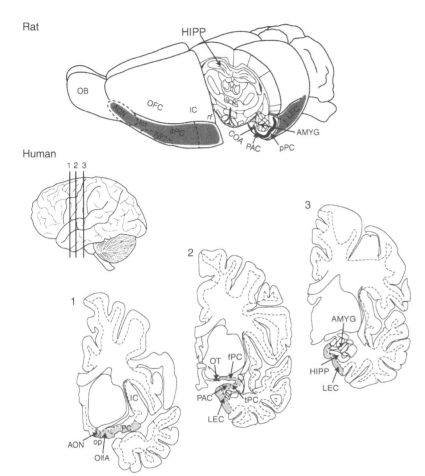

Figure 10.1 General anatomy of the olfactory system in rodent (top) and human (bottom). The olfactory cortex is labeled in gray. Abbreviations: AMYG: amygdaloid complex; AON: anterior olfactory nucleus; COA: cortical amygdala; HIPP: hippocampus; LEC: lateral entorhinal cortex; IC: insular cortex; lot: lateral olfactory tract; OB: olfactory bulb; OFC: orbitofrontal cortex; OlfA: olfactory area; OT: olfactory tubercle; op: olfactory peduncle; PAC: periamygdaloid cortex; aPC: piriform cortex, anterior part; pPC: piriform cortex, posterior part; fPC: piriform cortex, frontal area; tPC: piriform cortex, temporal area; rf: rhinal fissure.

10.2.2 Olfactory Tubercle

The olfactory tubercle is a relatively large region of the rodent olfactory cortex, lying along the ventral surface of the brain, medial to the piriform cortex. The human OT is located in a similar location, though appears to be much smaller compared to other olfactory cortical areas. Developmentally, the OT is ventral striatal in origin, and maintains strong reciprocal connections with the nucleus accumbens, ventral tegmentum and ventral pallidum. In fact, as might be expected from this connectivity, the largest body of research on the OT relates to its role in reward and motivation (Heimer, 2003; Ikemoto, 2007).

The OT receives its strongest input from the ventrolateral olfactory bulb (Imamura et al., 2011), with tufted cells more dominant than mitral cells (Nagayama et al., 2010; Igarashi et al., 2012). The ventrolateral olfactory bulb is robustly responsive to conspecific urine odor in rodents (Schaefer et al., 2001), and thus may be important in certain social odor-motivated social behaviors. Single-units in OT respond to a variety of odors including biologically meaningful ones (e.g., estrus odors, urine, food odors) (Payton et al., 2012; Rampin et al., 2012), as well as monomolecular odorants (Wesson and Wilson, 2010). Simple odor coding in OT single-units appears relatively similar to that in anterior piriform cortex (Payton et al., 2012), though such comparative studies have only just begun. Given the differences in both inputs and outputs between the OT and other olfactory cortical areas, it has been suggested that olfactory bulb output may display characteristics consistent with a parallel processing system (Wesson and Wilson, 2011; Payton et al., 2012). Regions of the bulb, or specific output cell types from individual glomeruli, may target specific cortical regions for localized information processing and transmittal to specialized downstream circuits.

In addition to odor responses, the rodent OT displays multimodal convergence. OT single-units respond to both odor and sound (Wesson and Wilson, 2010). The behavioral significance of such convergence is unclear, though coincident sound and odor may contribute to localization of moving objects in nocturnal rodents.

Finally, data from both rodents and humans suggests a potential role for the OT in odor-evoked attention or arousal. Lesions of the OT in rats impair the normal

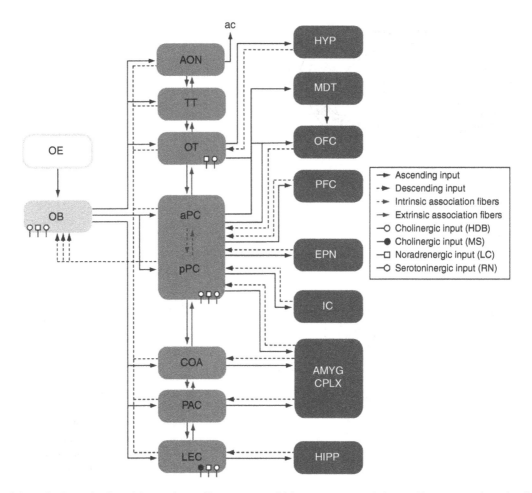

Figure 10.2 Schematized organization of the vertebrate olfactory system. Light gray structures (primary olfactory cortex) receive direct input from the main olfactory bulb. Most areas receiving direct input from the main olfactory bulb project back to the bulb. The intrinsic association fibers interconnect the anterior and posterior region of the piriform cortex. The extrinsic association fibers bind the different structures composing the primary olfactory cortex. Dark gray structures receive direct inputs from the primary olfactory cortex. Modulatory inputs project broadly to all primary olfactory structures Abbreviations: ac: anterior commissure; AMYG CPLX: amygdaloid complex; AON: anterior olfactory nucleus; COA: cortical amygdala; EPN: endopiriform nucleus; HDB: horizontal limb of the diagonal band of Broca; HIPP: hippocampus; HYP: hypothalamus; IC: insular cortex; LEC: lateral entorhinal cortex; LC: locus coeruleus ; MDT: mediodorsal thalamus; MS: medial septum; OB:olfactory bulb; OE: olfactory epithelium; OFC: orbitofrontal cortex; OT: olfactory tubercle; PAC: periamygdaloid cortex; aPC: piriform cortex, anterior part; pPC: piriform cortex, posterior part; PFC: prefrontal cortex; RN: raphe nucleus; TT: tenia tecta.

sleep-wake cycle modulation of odor-evoked arousal in cortical EEG (Gervais, 1979). In humans, active sniffing of clean air in an attempt to detect an odor evokes more activity in the anterior piriform/OT than sniffing clean air when no odor is expected (Zelano et al., 2005). This elevated activity in the OT/anterior piriform has been interpreted to reflect attention to sniff and sampled air. No such attention effects were observed in the more posterior temporal piriform cortex.

10.2.3 Piriform Cortex

The piriform cortex is unlike other mammalian sensory cortices in several respects (Figure 10.3). First, its primary sensory input does not originate in the thalamus, but rather from second order sensory neurons in the olfactory bulb, mitral, and tufted cells. Second, it is not neocortical in architecture, but rather is a 'simpler' 3-layered cortex, with afferents directed at superficial layer I and primary cell bodies located in layers II and III. Finally, rather than being organized topographically relative to its primary afferent structure and the sensory world, the piriform cortex is highly non-topographic and instead highly associative.

The primary input to the piriform cortex is from the ipsilateral olfactory bulb's mitral and tufted cells. Mitral cells project throughout both the anterior and posterior piriform cortices, while tufted cells appear to terminate rostrally, primarily in the anterior olfactory nucleus and

10.2 Olfactory Cortex: Structure and Function

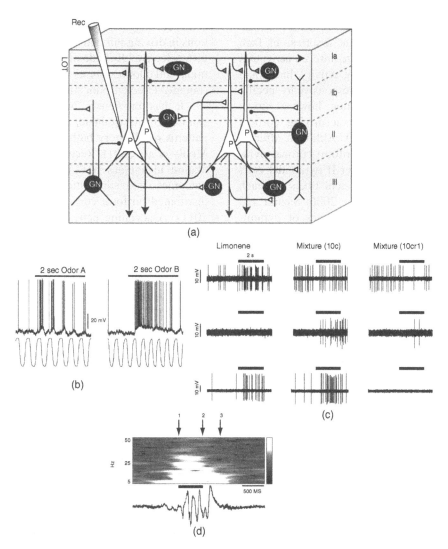

Figure 10.3 (a) Cellular organization of the piriform cortex. Layers II and III contain the cell bodies of pyramidal (P) neurons; the bulbar afferents contact their apical dendrites in superficial layer Ia (gray synapses). Layer Ib contains the intrinsic and extrinsic association fibers (white synapses). The different classes of GABAergic neurons (GN) described by Suzuki & Bekkers (2010, *Cereb. Cortex*, 20:2971–2984) are also represented. Example of cortical intracellular (b), extracellular (c) and population (d) recordings are illustrated below. (a) Intracellular activity and respiration over the course of 2 s odor stimulation in urethane-anesthetized rat (b) Odor-evoked response of three piriform cortical neurons to a monomolecular odorant (limonene) and to two different complex mixtures, each containing 10 components and differing by just one. (c) Representative piriform local field potentials (bottom: raw signal; top: time-frequency representation) obtained from a rat performing an olfactory discrimination task (1: nose poke in the odor port; 2: end of odor sampling; 3: access to the water port). The odor sampling induced the emergence of a powerful oscillation in the beta (15–40 Hz) frequency band.

olfactory tubercle, and much less so in the piriform cortex (Nagayama et al., 2004; Igarashi et al., 2012). Given the differences in mitral and tufted cell odor-evoked activity (Nagayama et al., 2004; Igarashi et al., 2012), these different projection patterns may underlie a parallel processing structure in olfaction (Igarashi et al., 2012; Payton et al., 2012), analogous to that observed in vision (Nassi and Callaway, 2009). Despite the remarkable spatial patterning of olfactory sensory neurons to the olfactory bulb glomerular layer and thus, spatial patterns of odor-evoked activity in the glomeruli (Rubin and Katz, 2001; Johnson and Leon, 2007) and output neurons (Imamura et al., 1992; Luo and Katz, 2001; Fletcher, 2012), the projection to the piriform cortex loses all detectable topography. Output neurons from individual glomeruli project widely throughout piriform cortex, with no detectable spatial patterning (Ghosh et al., 2011; Mitsui et al., 2011; Sosulski et al., 2011). Furthermore, individual cortical neurons receive input from broadly distributed glomeruli, producing a classic divergent-convergent pattern (Miyamichi et al., 2011).

This input pattern creates an anatomical substrate for synthesis of co-occurring odorant features. In fact, piriform cortical neurons appear to require co-activation of multiple glomeruli to drive spiking activity. Individual piriform cortical pyramidal cells are responsive to specific spatial patterns of glomerular activation, however single glomerulus activation is ineffective at driving cortical neurons (Davison and Ehlers, 2011). Although afferent fiber glutamatergic synapses onto piriform cortical pyramidal cells are relatively strong, layer II pyramidal cells require co-activation of multiple afferent fibers to reach spike threshold (Franks and Isaacson, 2006; Suzuki and Bekkers, 2006, 2011). However, subclasses of pyramidal cells show differential sensitivity to afferent input. Semi-lunar cells, which have apical dendrites with large spines located selectively within Layer Ia and thus anatomically appear highly sensitive to afferent input, are in fact more strongly depolarized by afferent input rather than superficial pyramidal cells in Layer II (Suzuki and Bekkers, 2011). In addition, semi-lunar cells have no basal dendrites (Neville and Haberly, 2004), and thus appear to be primarily tuned to afferent input and only minimally responsive to association fiber input (Suzuki and Bekkers, 2011). Thus, these cells may have unique contributions to the intra-cortical association fiber system described below. For example, semilunar cells form a major component of the association fiber input to superficial pyramidal cells, forming in essence a second layer of processing in piriform cortex (Suzuki and Bekkers, 2011).

The divergent/convergent projections from olfactory bulb to piriform cortex mean that there are no obvious odor-specific spatial patterns of odor-evoked activity within the piriform cortex (Cattarelli et al., 1988; Rennaker et al., 2007; Stettler and Axel, 2009). Neighboring neurons are as likely to respond to different odors as they are to respond to the same odors (Rennaker et al., 2007; Stettler and Axel, 2009). Although intracortical association fiber activity contributes to this odor-evoked pyramidal cell activity (Franks et al., 2011; Poo and Isaacson, 2011b), the results strongly suggest convergence of multiple glomerular input onto individual pyramidal cells. This allows individual, distributed piriform cortical neurons to synthesize odorant features extracted by the periphery and refined in the olfactory bulb, beginning the process of odor object formation.

As noted, however, a critical component of the synthesis of odorant features is the extensive intracortical association fiber system (Haberly, 2001). This intrinsic network helps bind distributed co-active neurons into an ensemble unique to a given input (Figure 10.4). These association fiber connections are an important component in driving odor-evoked activity. In some cases, pyramidal cells that do not respond directly to stimulation of individual glomeruli, do respond when specific combinations of glomeruli are activated, suggesting a role for intrinsic excitatory connections in driving this activity (Davison and Ehlers, 2011). In fact, selective blockade of association fibers reduces pyramidal cell odor responses and narrows receptive field width (range of effective odor stimuli) (Poo and Isaacson, 2011b).

It has previously been demonstrated that rodents can detect, discriminate, and learn about different spatial patterns of olfactory bulb activation (Roman et al., 1987; Mouly et al., 2001). Recent work using optogenetic stimulation techniques has demonstrated similar behavioral outcomes with stimulation of a relatively small set of distributed piriform cortical pyramidal cells (Choi et al., 2011). Associating activation of the distributed pyramidal cells with aversive (foot shock) or appetitive (food) rewards can produce conditioned learned approaches or avoidance behaviors, similar to natural odor conditioning. Activation of around 500 cells was sufficient to mediate this behavior (Choi et al., 2011). Such a small ensemble of neurons (0.5% of the piriform cortical population) capable of driving behavior allows for high capacity storage of many odor objects and is consistent with Marr's model of archicortex (Marr, 1971).

In addition to allowing convergence of multiple odorant features, the association fibers also play an important role in odor memory (Haberly, 2001). Association fiber synapses exhibit robust NMDA-dependent long-term potentiation (Kanter and Haberly, 1990; Poo and Isaacson, 2007). Associative learning with odors can increase synaptic currents evoked by association fiber stimulation (Saar et al., 2002), as well as dendritic spine density in regions of the apical dendritic where association fibers terminate (Knafo et al., 2001).

This plasticity of association fiber synapses is believed to help bind members of a co-active cortical ensemble. With this binding of spatially distributed neurons, discrimination and odor acuity improve. A second hypothesized consequence of this network effect is pattern completion. Computational models of piriform cortex have demonstrated that optimal associative plasticity in association fiber synapses helps store a template of familiar odor patterns which allow "filling-in" features of degraded inputs and full response to an odor object (Hasselmo et al., 1992; Barkai et al., 1994). Either too much or too little plasticity can result in excessive or impaired pattern completion and thus, impaired recognition and discrimination (Hasselmo and McGaughy, 2004). Recent work has directly tested the pattern completion ability of piriform cortical circuits (Barnes et al., 2008; Wilson, 2009). Complex mixtures of monomolecular odorants were "morphed" by either removing individual components (10 component mix, 10 component mix with 1 missing, 10 component mix with 2 missing, etc.) or by replacing individual components with a novel contaminant. Ensembles of mitral/tufted cells de-correlated (responded significantly differently between)

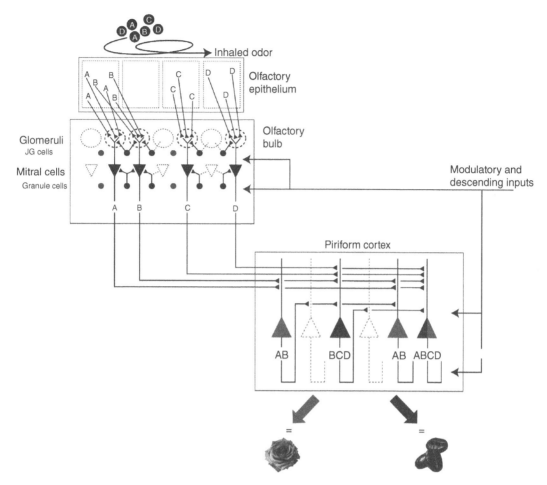

Figure 10.4 Neurobiological steps leading to the synthetic perception of odors. Individual receptors localized in the olfactory epithelium express one of 1000 different receptor proteins and are randomly scattered within one of four zones. Receptors expressing the same receptor protein converge on to a small number of exclusive glomeruli (four receptor types are labeled A, B, C and D in this example). The receptors are hypothesized to be responsive to individual odorant features, rather than odorant molecules as a whole. Mitral cells receive receptor input from a single glomerulus (and thus convergent receptor input; e.g., A or B), and project to the piriform cortex. Within the olfactory bulb, inter-glomerular and inter-output neuron lateral inhibition is mediated by juxtaglomerular and granule cells, respectively, heightening contrast between similar odorant features. Neurons in the piriform cortex form a combinatorial array, allowing convergence of multiple odorant features and/or behavioral state/non-olfactory inputs to occur on single neurons. Thus, distributed ensembles of pyramidal cells can synthesize the bulbar messages into unified percepts like the smell of "rose" (here synthesis of A and B) or "coffee" (synthesis of B, C and D) and store these representations in memory. Both the olfactory bulb and piriform cortex receive extensive input from neuromodulatory and non-olfactory inputs.

all the various mixture morphs from the standard 10 component mixture. This is consistent with a pattern separation role for the olfactory bulb, similar to that of the hippocampal dentate gyrus (Sahay et al., 2011). In contrast, piriform cortical single-unit ensembles failed to de-correlate the 10 component mixture from that missing a single component, despite the fact that sufficient information was available in the olfactory bulb. This suggests that the piriform cortical ensembles completed the slightly degraded input, and responded as if the entire odor object was present. As more components are removed, or novel contaminants added, piriform cortical ensembles de-correlate the mixtures even more strongly than the olfactory bulb (Barnes et al., 2008). Behavioral discrimination performance in a two-alternative choice task mirrored the cortical ensemble de-correlation—mixtures not de-correlated by the cortex are difficult for the animals to discriminate (Barnes et al., 2008). These results suggest that, as originally hypothesized (Haberly, 2001), the piriform cortex can perform pattern completion which contributes to perceptual stability. Interestingly, new data suggest that the boundary between cortical pattern completion and separation is experience dependent. Cortical pattern completion can be enhanced in tasks requiring odor generalization and pattern separation can be enhanced in tasks requiring fine odor acuity (Chapuis and Wilson, 2011). These changes in olfactory cortical processing lead to changes in perceptual acuity in both rodents (Fletcher and Wilson, 2002; Chapuis and Wilson, 2011; Chen et al., 2011) and humans (Li et al., 2006; Li et al., 2008).

Piriform cortical based pattern completion may contribute to perceptual stability in the face of changing odor intensity or mixture content, which will affect feature input to the cortex. If such changes are minimal, the cortex can "ignore" them by filling in the missing components through the activation of previously strengthened synapses, thus allowing perceptual stability. More significant changes, or previous odor training, can enhance a shift toward pattern separation and odor discrimination.

A second form of synaptic plasticity within the piriform cortex contributes to odor background segmentation. The piriform cortex rapidly adapts to stable odor input (Wilson, 1998a), and this cortical adaptation to odor is associated with afferent (mitral/tufted cell) synaptic depression recorded intracellularly, in vivo (Wilson, 1998b). The recovery of odor responses occurs within about 2 min, as does the synaptic depression (Best and Wilson, 2004). This cortical adaptation is mediated by pre-synaptic metabotropic receptors (group III) which reduce glutamate release from mitral/tufted cell axons during repetitive stimulation (Best and Wilson, 2004). Blockade of mGluRIII receptors within the piriform cortex prevents the synaptic depression, cortical odor adaptation, and short-term behavioral habituation (Best et al., 2005; Yadon and Wilson, 2005; Bell et al., 2008).

The synaptic depression is homo-synaptic, leaving afferent inputs conveying information from other non-active mitral/tufted cells (and glomeruli) intact (Best and Wilson, 2004). Thus, homosynaptic depression may contribute to the fact that short-term cortical adaptation (Wilson, 2000) and short-term behavioral habituation are highly odor specific (Fletcher and Wilson, 2002; McNamara et al., 2008). However, homosynaptic depression is not sufficient to account for habituation specificity between highly overlapping input patterns (Linster et al., 2009). Potentiation of association fiber synapses also plays a major role in this odor-specificity. In a computational model of the olfactory system which incudes olfactory sensory neurons, olfactory bulb neurons, and piriform cortex (Linster et al., 2007), cortical odor adaptation was induced if afferent homosynaptic depression was included in the model, although this cortical adaptation was only minimally odor specific. In contrast, if long-term potentiation was included in association fiber synapses, and odor exposure was sufficiently long enough to induce familiarization, then cortical adaptation was highly odor specific (Linster et al., 2009). The same constraints hold true in vivo. The specificity of cortical odor adaptation and of behavioral odor habituation is dependent on how familiar the odors are (e.g., duration of exposure (Fletcher

and Wilson, 2002; Wilson, 2003), and this specificity can be disrupted by disruption of normal synaptic plasticity in association fiber synapses, for example with modulation of piriform cortical acetylcholine muscarinic receptors (Wilson, 2001; Fletcher and Wilson, 2002).

Having a central mechanism of short-term odor adaptation (rather than at the olfactory sensory neuron) allows the system to adjust to contingency changes. For example, using odors to guide performance in a discrimination task would be severely impacted if the cortex stopped responding to repeated cues across trials. Importantly, noradrenergic inputs to the piriform cortex can modulate synaptic depression within the piriform cortex (Best and Wilson, 2004), potentially via pre-synaptic beta receptors on mitral cell axons. Activation of noradrenergic beta receptors can inhibit mGluRIII receptor function via a protein kinase A dependent phosphorylation (Cai et al., 2001). Loud sounds which elevate norepinephrine within the piriform cortex (Smith et al., 2009) can induce dishabituation of odor-evoked behavioral responses (Smith et al., 2009). The behavioral dishabituation is blocked by intra-cortical infusion of the noradrenergic beta receptor antagonist propranolol (Smith et al., 2009). A similar reduction in cortical adaptation may occur under conditions of arousal or attention, allowing sustained cortical odor responses.

Together these results suggest that while the olfactory epithelium and olfactory bulb are obviously required for normal olfaction, the piriform cortex moves the system from an odorant feature oriented process towards an odor object oriented system, capable of synthetic or configural processing, and capable of treating odors as distinct from background. Both of these processes—odor object formation and odor-background segmentation—require synaptic plasticity and experience (Wilson and Stevenson, 2006; Gottfried, 2010; Wilson and Sullivan, 2011).

Experience also shapes hedonic responses to odors (Davis, 2004; Moriceau and Sullivan, 2006; Li et al., 2008; Hegoburu et al., 2009; Fletcher, 2012). Simple associative learning through pairing of an odor with either an aversive or appetitive stimulus modifies both the behavioral response to the odor and odor-evoked activity within the olfactory cortex (Sevelinges et al., 2004; Barkai, 2005; Martin et al., 2006; Chapuis and Wilson, 2011; Chen et al., 2011). These changes appear to contribute to both learned changes in acuity as described above, and to the memory of the hedonic association. For example, odor-fear conditioning can result in either generalized fear to many odors or relatively odor-specific to the conditioned stimulus (Chen et al., 2011). Although the basolateral amygdala is required for the fear conditioning itself (Johansen et al., 2011), the stimulus control over the learned fear (generalized or specific) appears to be related to changes in receptive fields of piriform cortical single-units. In animals trained to show generalized odor fear, chronically recorded piriform cortex single-units were more broadly tuned than those in pseudo-conditioned animals. In contrast, animals trained to show highly odor-specific fear had piriform single-units that were more narrowly tuned than those in pseudo-conditioned animals. Similar changes in 'tuning' of piriform cortical odor-evoked activity have been observed in humans undergoing odor-aversion conditioning (Li et al., 2008).

However, in addition to these learned changes in cortical acuity, learned pleasant and aversive odors may differentially activate the piriform cortex (Gottfried et al., 2002a). For example, in young rats, the anterior piriform cortex appears more robustly activated by learned appetitive/preferred odors while the posterior piriform cortex is more strongly activated by learned aversive odors (Moriceau et al., 2006). Importantly however, it should be noted that how an odor is experienced may shape the circuit activated in response to odors with learned hedonic significance. Food odors, which may be experienced either orthonasally before they enter the mouth or retronasally during chewing can activate piriform cortex (Chabaud et al., 2000; Small et al., 2005; Yoshida and Mori, 2007) Ortho- and retronasal routes of stimulation can activate different neural circuits during recall of a learned aversion odor (Chapuis et al., 2009).

The piriform cortex is also a major source of top-down control of the olfactory bulb (Restrepo et al., 2009; Manabe et al., 2011). This top-down control provides another source of either immediate or long-lasting fine-tuning of odor processing, potentially similar to that produced by neocortical feedback to thalamic nuclei in other sensory systems. Piriform cortical input to the olfactory bulb primarily targets inhibitory granule cells (Price and Powell, 1970), and thus could help shape contrast or gain of bulb output via modulation of this inhibitory circuit (Lledo et al., 2004; Strowbridge, 2009). Descending input to olfactory bulb granule cells shows long-term potentiation (Gao and Strowbridge, 2009) [perhaps most robustly on newborn granule cells (Nissant et al., 2009)], and thus experience-dependent changes in olfactory bulb activity mediated by this pathway could be long lasting (Manabe et al., 2011).

While plasticity is necessary for the piriform cortex to process odors, create odor objects, and adapt to background odors, this reliance on plasticity may contribute to the vulnerability of the olfactory system across a wide range of pathologies (Mesholam et al., 1998), including Alzheimer's disease (Doty et al., 1987; Nordin and Murphy, 1998; Morgan and Murphy, 2002; Olichney et al., 2005), Parkinson's disease (Sobel et al., 2001; Siderowf et al., 2005; Wattendorf et al., 2009; Doty, 2012), and schizophrenia (Hudry et al., 2002; Rupp, 2010; Malaspina et al., 2012). For example, in both animal models (Wesson et al., 2010; Cramer et al., 2012) and human functional

imaging studies (Li et al., 2010), olfactory deficits are strongly correlated with piriform cortex dysfunction, though damage (e.g., amyloid-beta deposition or tauopathy) can be expressed throughout the olfactory pathway from the olfactory sensory neurons to the entorhinal cortex (Macknin et al., 2004; Braak et al., 2011; Rawson et al., 2011). It should be noted that functional disruption of a single region within a pathway may reflect damage localized to that region, damage to other areas targeting that region, or abnormal connectivity between regions. In fact, normal functional connectivity within the primary olfactory pathway is impaired by amyloid deposition in mouse models of Alzheimer's disease (Wesson et al., 2011)

10.2.4 Entorhinal Cortex

The lateral entorhinal cortex is located caudal to the piriform cortex in rodents and anteromedially along the temporal lobe in humans (Brodmann areas 28 and 34). The rodent entorhinal cortex is divided into medial and lateral subdivisions, which together form the primary cortical input to the hippocampal formation. The entorhinal cortex is highly multisensory, with cortical inputs from other perirhinal and parahippocampal cortices and return projections to these structures (Kerr et al., 2007; Agster and Burwell, 2009). Its primary thalamic input is primarily from the dorsal midline thalamic nuclei (Kerr et al., 2007). However, relevant to the current discussion, in both rodents (Schwob and Price, 1978; Sosulski et al., 2011; Igarashi et al., 2012) and primates (Carmichael et al., 1994) the lateral entorhinal cortex receives direct input from olfactory bulb mitral cells and thus fits the definition of olfactory cortex. The lateral entorhinal also receives input from the piriform cortex. Given these inputs, it is not surprising that neurons in the lateral entorhinal cortex respond to odors (Deshmukh and Knierim, 2011; Xu and Wilson, 2012). Although very few sensory physiology studies have examined the lateral entorhinal, single units there appear to show more selectivity in their responses than cells in piriform cortex (Xu and Wilson, 2012).

In addition to receiving input from the olfactory bulb and piriform cortex for transmission to the hippocampal formation, the lateral entorhinal is also a major source of top-down feedback to more peripheral parts of the olfactory system. This pathway has not been explored in much detail, though both the piriform cortex and olfactory bulb are targets (Wyss, 1981; Insausti et al., 1997). Aspiration lesions of the entorhinal cortex enhance odor-evoked c-fos in piriform cortex (Bernabeu et al., 2006), suggesting entorhinal cortex-mediated a top-down suppression of olfactory cortical activity. In fact, electrical stimulation of the lateral entorhinal cortex can inhibit piriform cortical response to olfactory bulb stimulation, though the use of electrical stimulation for these studies and the resulting antidromic activation of the circuits complicates interpretation. Nonetheless, these studies suggest a powerful role of the lateral entorhinal cortex in modulating olfactory function. Given the nature of other inputs to the entorhinal noted above, including the amygdala, hippocampus, and perirhinal cortices, this feedback could be rich in memory, hedonic, and multimodal information.

Interestingly, aspiration lesions of the entorhinal cortex disrupt some forms of odor memory but enhance others. Odor-based tasks that require the hippocampus are disrupted by entorhinal lesions, as might be expected (Staubli et al., 1984). However, hippocampal independent odor-based memory tasks, such as conditioned odor aversion are enhanced by entorhinal lesions (Ferry et al., 1996; Wirth et al., 1998). This enhancement could reflect a release from inhibition that the lateral entorhinal cortex imposes on the olfactory cortex (Bernabeu et al., 2006; Mouly and Di Scala, 2006).

Finally, it should be mentioned that the entorhinal cortex is heavily impacted by Alzheimer's disease and associated deposition of amyloid beta and phosphorylated tau pathologies. The trans-entorhinal area appears to be the earliest forebrain region impacted by Alzheimer's disease (Braak and Braak, 1992). The role of entorhinal pathology in the early emergence of olfactory deficits in mild cognitive impairment and early Alzheimer's disease (Doty et al., 1987; Murphy, 1999; Murphy et al., 1999) is not clear. Declines in odor identification test scores in mild cognitive impairment do not correlate with entorhinal cortex volume (Devanand et al., 2010). However, entorhinal dysfunction could affect both the transmission of olfactory information towards memory circuits in the hippocampus, and also impact descending modulation of the olfactory bulb and cortex. In both humans (Li et al., 2010) and rodents (Wesson et al., 2010; Wesson et al., 2011), pathology and dysfunction in the piriform cortex are highly correlated with olfactory perceptual and memory breakdown. The contribution of disrupted top-down modulation of piriform cortical function in this disorder deserves further examination.

10.2.5 Orbitofrontal Cortex

Although orbitofrontal cortex does not fit the classic definition of olfactory cortex since it does not receive direct input from the olfactory bulb, it has an important role in olfaction and chemosensation. The orbitofrontal cortex is located dorsal to the piriform cortex in rodents and at the anterior base of the frontal lobes in humans (Brodmann areas 10, 11, and 47). The orbitofrontal cortex is the target of the dorsomedial nucleus of the thalamus, and also receives inputs from sensory cortices

of all modalities. Both the ventral tegmentum and the nucleus accumbens project to the orbitofrontal cortex, which may provide important information regarding relative reward value and its relationship to sensory objects. There are reciprocal connections between the orbitofrontal cortex and the insular and parahippocampal cortices, amygdala nuclei, the hippocampus and hypothalamus. The orbitofrontal cortex even extends projections to the brainstem including the locus coeruleus. Directly relevant to the present discussion, the orbitofrontal cortex also has a strong, reciprocal connection with the piriform cortex, with the strongest projections to and from the anterior piriform cortex, and less to the posterior piriform cortex (Johnson et al., 2000; Ekstrand et al., 2001; Illig, 2005). The piriform cortex also projects indirectly to the orbitofrontal cortex via the dorsomedial thalamus.

The orbitofrontal cortex is broadly involved in reward or outcome value, decision making, memory, and emotion (Schoenbaum et al., 2011). Similar to the entorhinal cortex, the orbitofrontal cortex can be divided into anatomically and functionally based sub-regions along both the medial-lateral and anterior-posterior axes (Carmichael and Price, 1994; Rudebeck and Murray, 2011).

Single units in the orbitofrontal cortex can be highly multimodal (Schoenbaum and Eichenbaum, 1995; Rolls et al., 2003; Kadohisa et al., 2004), including strong responses to chemosensory stimuli (Tanabe et al., 1975; Schoenbaum and Eichenbaum, 1995; Critchley and Rolls, 1996b; Rolls et al., 1996). For example, orbitofrontal single units can respond to the smell, taste, and sight or mouth feel of food items. Similar odor-evoked activity can be observed in human functional imaging analyses of the orbitofrontal cortex (Jones-Gotman and Zatorre, 1993; Royet et al., 2001; Gottfried et al., 2002b; Small et al., 2005), often displaying a lateralization (Zatorre et al., 1992; Royet and Plailly, 2004).

The orbitofrontal cortex, particularly ventrolateral regions, appear to play several roles in odor perception. This area may play an important role in flavor perception (Schul et al., 1996), though separation of pure sensory information from hedonic or reward value is difficult. For example, responses to food odors are reduced by feeding to satiety or previous negative/aversive associative learning (Critchley and Rolls, 1996a; O'Doherty et al., 2000; Gottfried et al., 2002a). In fact, odor pleasantness is a strong modulator of orbitofrontal cortical odor responses in humans (Royet et al., 2000; Rolls et al., 2008). The orbitofrontal cortex may also be involved in attention to odors. For example, attention to odors selectively enhances the functional connectivity between the dorsomedial nucleus of the thalamus and the orbitofrontal cortex compared to the direct piriform-orbitofrontal cortex pathway (Plailly et al., 2008). These findings suggest a potential thalamocortical contribution to sensory attention in olfaction, similar to that seen in other sensory systems (McAlonan et al., 2008).

Finally, as with the entorhinal cortex, the orbitofrontal cortex provides top-down input to the piriform cortex. This top-down feedback presumably provides information about stimulus value or motivational significance (Schoenbaum and Eichenbaum, 1995; Schoenbaum et al., 1999; Schoenbaum et al., 2011). Given the nature of associative odor coding in piriform cortex, this feedback could become directly incorporated into the cortical representation of the odor itself. Furthermore, this top-down input is plastic, showing robust enhancement induced by olfactory learning (Cohen et al., 2008).

10.3 SUMMARY

The olfactory cortex is a structurally and functionally heterogeneous cortical region targeted by olfactory bulb output neurons. Olfactory cortical neuroarchitecture includes ventral striatal (olfactory tubercle), paleocortical (piriform cortex), and neocortical (lateral entorhinal cortex) components. There is no classic thalamic relay between the peripheral receptors and olfactory cortex, though the dorsomedial nucleus of the thalamus receives input from the piriform cortex and projects on the orbitofrontal neocortex—an important higher order olfactory and chemosensory structure.

Data currently support a role for the piriform cortex in pattern recognition of olfactory bulb odor-specific spatial-temporal output, with an experience-dependent synthesis of those feature-driven patterns into perceptual odor objects. The anterior piriform cortex may be specialized for encoding odor identity while posterior regions more specialized for encoding odor categories and adding hedonic content. The piriform cortex is also importantly involved in short-term odor habituation, which contributes to odor-background segmentation.

Less is known about the other olfactory cortex components. The olfactory tubercle displays multi-sensory sensory characteristics, and based on its anatomical circuitry could be an important mediator between odors and motivated behavior. Initial investigations of the entorhinal cortex suggest it may play a role not only as a gateway for olfactory input to the hippocampal formation, but also as an important top-down modulator of olfactory cortical and olfactory bulb function. The orbitofrontal cortex appears important in odor hedonics and reward/decision-making consequences of odor stimulation. The orbitofrontal cortex is also an important multisensory site for convergence of taste, olfaction, mouth feel, and vision that together contribute to flavor.

Finally, olfaction is increasingly recognized to express vulnerability to early stages of several neuropathologies including Parkinson's and Alzheimer's disease. Dysfunction within the piriform cortex has been strongly associated with olfactory dysfunction in both humans suffering from Alzheimer's disease and in rodent models of the disease. Understanding how the primary olfactory system functions, and how it fits within larger cortical and subcortical networks will help elucidate why this seemingly simple system is so susceptible to such a wide variety of disorders.

REFERENCES

Adrian, E. D. (1942) Olfactory reactions in the brain of the hedgehog. *J. Physiol.* 100: 459–473.

Agster, K. L., and Burwell, R. D. (2009) Cortical efferents of the perirhinal, postrhinal, and entorhinal cortices of the rat. *Hippocampus* 19: 1159–1186.

Barkai, E. (2005) Dynamics of learning-induced cellular modifications in the cortex. *Biol. Cybern.* 92: 360–366.

Barkai, E., Bergman, R. E., Horwitz, G., and Hasselmo, M. E. (1994) Modulation of associative memory function in a biophysical simulation of rat piriform cortex. *J. Neurophysiol.* 72: 659–677.

Barnes, D. C., Hofacer, R. D., Zaman, A. R., et al. (2008) Olfactory perceptual stability and discrimination. *Nat. Neurosci.* 11: 1378–1380.

Bell, H., Chenoweth, B., and Wilson, D. A. (2008) Neurobehavioral consequences of cortical adaptation disruption during ontogeny. *Neurosci. Lett.* 445: 47–52.

Bernabeu, R., Thiriet, N., Zwiller, J., and Di Scala, G. (2006) Lesion of the lateral entorhinal cortex amplifies odor-induced expression of c-fos, junB, and zif 268 mRNA in rat brain. *Synapse* 59: 135–143.

Best, A. R., and Wilson, D. A. (2004) Coordinate synaptic mechanisms contributing to olfactory cortical adaptation. *J. Neurosci.* 24: 652–660.

Best, A. R., Thompson, J. V., Fletcher, M. L., and Wilson, D. A. (2005) Cortical metabotropic glutamate receptors contribute to habituation of a simple odor-evoked behavior. *J. Neurosci.* 25: 2513–2517.

Braak, H. and Braak, E. (1992) The human entorhinal cortex: normal morphology and lamina-specific pathology in various diseases. *Neurosci. Res.* 15: 6–31.

Braak, H., Thal, D.R., Ghebremedhin, E., and Del Tredici, K. (2011) Stages of the pathologic process in Alzheimer disease: age categories from 1 to 100 years. *J. Neuropathol. Exp. Neurol.* 70: 960–969.

Brunjes, P. C., Illig, K. R., and Meyer, E. A. (2005) A field guide to the anterior olfactory nucleus (cortex). *Brain Res. Rev.* 50: 305–335.

Cai, Z., Saugstad, J. A., Sorensen, S. D., et al. (2001) Cyclic AMP-dependent protein kinase phosphorylates group III metabotropic glutamate receptors and inhibits their function as presynaptic receptors. *J. Neurochem.* 78: 756–766.

Carmichael, S. T. and Price, J. L. (1994) Architectonic subdivision of the orbital and medial prefrontal cortex in the macaque monkey. *J. Comp. Neurol.* 346: 366–402.

Carmichael, S. T., Clugnet, M. C., and Price, J. L. (1994) Central olfactory connections in the macaque monkey. *J. Comp. Neurol.* 346: 403–434.

Cattarelli, M., Astic, L., and Kauer, J. S. (1988) Metabolic mapping of 2-deoxyglucose uptake in the rat piriform cortex using computerized image processing. *Brain Res.* 442: 180–184.

Chabaud, P., Ravel, N., Wilson, D. A., et al. (2000) Exposure to behaviourally relevant odour reveals differential characteristics in rat central olfactory pathways as studied through oscillatory activities. *Chem. Senses* 25: 561–573.

Chapuis, J. and Wilson, D. A. (2011) Bidirectional plasticity of cortical pattern recognition and behavioral sensory acuity. *Nat. Neurosci.* 15: 155–163.

Chapuis, J., Garcia, S., Messaoudi, B., et al. (2009) The way an odor is experienced during aversive conditioning determines the extent of the network recruited during retrieval: a multisite electrophysiological study in rats. *J. Neurosci.* 29: 10287–10298.

Chen, C. F., Barnes, D. C., and Wilson, D. A. (2011) Generalized versus stimulus-specific learned fear differentially modifies stimulus encoding in primary sensory cortex of awake rats. *J. Neurophysiol.* 106: 3136–3144.

Choi, G. B., Stettler, D. D., Kallman, B. R., et al. (2011) Driving opposing behaviors with ensembles of piriform neurons. *Cell* 146: 1004–1015.

Cohen, Y., Reuveni, I., Barkai, E., and Maroun, M. (2008) Olfactory learning-induced long-lasting enhancement of descending and ascending synaptic transmission to the piriform cortex. *J. Neurosci.* 28: 6664–6669.

Cramer, P. E., Cirrito, J. R., Wesson, D. W., et al. (2012) ApoE-directed therapeutics rapidly clear beta-Amyloid and reverse deficits in AD mouse models. *Science* 335: 1503–1506.

Critchley, H. D. and Rolls, E. T. (1996a) Hunger and satiety modify the responses of olfactory and visual neurons in the primate orbitofrontal cortex. *J. Neurophysiol.* 75: 1673–1686.

Critchley, H. D. and Rolls, E. T. (1996b) Olfactory neuronal responses in the primate orbitofrontal cortex: analysis in an olfactory discrimination task. *J. Neurophysiol.* 75: 1659–1672.

Davis, R. L. (2004) Olfactory learning. *Neuron* 44: 31–48.

Davison, I. G. and Ehlers, M. D. (2011) Neural circuit mechanisms for pattern detection and feature combination in olfactory cortex. *Neuron* 70: 82–94.

Deshmukh, S. S. and Knierim, J. J. (2011) Representation of non-spatial and spatial information in the lateral entorhinal cortex. *Front Behav. Neurosci.* 5: 69.

Devanand, D. P., Tabert, M. H., Cuasay, K., et al. (2010) Olfactory identification deficits and MCI in a multi-ethnic elderly community sample. *Neurobiol. Aging* 31: 1593–1600.

Doty, R. L. (2012) Olfactory dysfunction in Parkinson disease. *Nat. Rev. Neurol.* 8: 329–339.

Doty, R. L., Reyes, P. F., and Gregor, T. (1987) Presence of both odor identification and detection deficits in Alzheimer's disease. *Brain Res. Bull.* 18: 597–600.

Ekstrand, J. J., Domroese, M. E., Johnson, D. M.et al. (2001) A new subdivision of anterior piriform cortex and associated deep nucleus with novel features of interest for olfaction and epilepsy. *J. Comp. Neurol.* 434: 289–307.

Ferry, B., Oberling, P., Jarrard, L. E., and Di Scala, G. (1996) Facilitation of conditioned odor aversion by entorhinal cortex lesions in the rat. *Behav. Neurosci.* 110: 443–450.

Fletcher, M. L. (2012) Olfactory aversive conditioning alters olfactory bulb mitral/tufted cell glomerular odor responses. *Front Syst. Neurosci.* 6: 16.

Fletcher, M. L. and Wilson, D. A. (2002) Experience modifies olfactory acuity: acetylcholine-dependent learning decreases behavioral generalization between similar odorants. *J. Neurosci.* 22: RC201.

Franks, K. M. and Isaacson, J. S. (2006) Strong single-fiber sensory inputs to olfactory cortex: implications for olfactory coding. *Neuron* 49: 357–363.

Franks, K. M., Russo, M. J., Sosulski, D. L.et al. (2011) Recurrent circuitry dynamically shapes the activation of piriform cortex. *Neuron* 72: 49–56.

Gao, Y. and Strowbridge, B. W. (2009) Long-term plasticity of excitatory inputs to granule cells in the rat olfactory bulb. *Nat. Neurosci.* 12: 731–733.

Gervais, R. (1979) Unilateral lesions of the olfactory tubercle modifying general arousal effects in the rat olfactory bulb. *Electroencephalogr Clin. Neurophysiol.* 46: 665–674.

Ghosh, S., Larson, S. D., Hefzi, H., et al. (2011) Sensory maps in the olfactory cortex defined by long-range viral tracing of single neurons. *Nature* 472: 217–220.

Gottfried, J. A. (2009) Function follows form: ecological constraints on odor codes and olfactory percepts. *Curr. Opin. Neurobiol.* 19: 422–429.

Gottfried, J. A. (2010) Central mechanisms of odour object perception. *Nat. Rev. Neurosci.* 11: 628–641.

Gottfried, J. A. and Dolan, R. J. (2003) The nose smells what the eye sees: crossmodal visual facilitation of human olfactory perception. *Neuron* 39: 375–386.

Gottfried, J. A., O'Doherty, J., and Dolan, R. J. (2002a) Appetitive and aversive olfactory learning in humans studied using event-related functional magnetic resonance imaging. *J. Neurosci.* 22: 10829–10837.

Gottfried, J. A., Deichmann, R., Winston, J. S., and Dolan, R. J. (2002b) Functional heterogeneity in human olfactory cortex: an event-related functional magnetic resonance imaging study. *J. Neurosci.* 22: 10819–10828.

Haberly, L. B. (1969) Single unit responses to odor in the prepyriform cortex of the rat. *Brain Res.* 12: 481–484.

Haberly, L. B. (1985) Neuronal circuitry in olfactory cortex: anatomy and functional implications. *Chem. Senses* 10: 219–238.

Haberly, L. B. (2001) Parallel-distributed processing in olfactory cortex: new insights from morphological and physiological analysis of neuronal circuitry. *Chem. Senses* 26: 551–576.

Hasselmo, M. E. and McGaughy, J. (2004) High acetylcholine levels set circuit dynamics for attention and encoding and low acetylcholine levels set dynamics for consolidation. *Prog. Brain Res.* 145: 207–231.

Hasselmo, M. E., Anderson, B. P., and Bower, J. M. (1992) Cholinergic modulation of cortical associative memory function. *J. Neurophysiol.* 67: 1230–1246.

Hegoburu, C., Sevelinges, Y., Thevenet, M., et al. (2009) Differential dynamics of amino acid release in the amygdala and olfactory cortex during odor fear acquisition as revealed with simultaneous high temporal resolution microdialysis. *Learn Mem.* 16: 687–697.

Heimer, L. (2003) A new anatomical framework for neuropsychiatric disorders and drug abuse. *Am. J. Psychiatry* 160: 1726–1739.

Herrick, C. J. (1933) The functions of the olfactory parts of the cerebral cortex. *Proc. Natl. Acad. Sci. U. S. A.* 19: 7–14.

Hopfield, J. J. (1991) Olfactory computation and object perception. *Proc. Natl. Acad. Sci. U. S. A.* 88: 6462–6466.

Hudry, J., Saoud, M., D'Amato, T., et al. (2002) Ratings of different olfactory judgements in schizophrenia. *Chem. Senses* 27: 407–416.

Igarashi, K. M., Ieki, N., An, M., et al. (2012) Parallel mitral and tufted cell pathways route distinct odor information to different targets in the olfactory cortex. *J. Neurosci.* 32: 7970–7985.

Ikemoto, S. (2007) Dopamine reward circuitry: two projection systems from the ventral midbrain to the nucleus accumbens-olfactory tubercle complex. *Brain Res. Rev.* 56: 27–78.

Illig, K. R. (2005) Projections from orbitofrontal cortex to anterior piriform cortex in the rat suggest a role in olfactory information processing. *J. Comp. Neurol.* 488: 224–231.

Illig, K. R. and Haberly, L. B. (2003) Odor-evoked activity is spatially distributed in piriform cortex. *J. Comp. Neurol.* 457: 361–373.

Illig, K. R. and Eudy, J. D. (2009) Contralateral projections of the rat anterior olfactory nucleus. *J. Comp. Neurol.* 512: 115–123.

References

Imamura, F., Ayoub, A. E., Rakic, P., and Greer, C. A. (2011) Timing of neurogenesis is a determinant of olfactory circuitry. *Nat. Neurosci.* 14: 331–337.

Imamura, K., Mataga, N., and Mori, K. (1992) Coding of odor molecules by mitral/tufted cells in rabbit olfactory bulb. I. Aliphatic compounds. *J. Neurophysiol.* 68: 1986–2002.

Insausti, R., Herrero, M. T., and Witter, M. P. (1997) Entorhinal cortex of the rat: cytoarchitectonic subdivisions and the origin and distribution of cortical efferents. *Hippocampus* 7: 146–183.

Johansen, J. P., Cain, C. K., Ostroff, L. E., and LeDoux, J. E. (2011) Molecular mechanisms of fear learning and memory. *Cell* 147: 509–524.

Johnson, B. A. and Leon, M. (2007) Chemotopic odorant coding in a mammalian olfactory system. *J. Comp. Neurol.* 503: 1–34.

Johnson, D. M., Illig, K. R., Behan, M., and Haberly, L. B. (2000) New features of connectivity in piriform cortex visualized by intracellular injection of pyramidal cells suggest that "primary" olfactory cortex functions like "association" cortex in other sensory systems. *J. Neurosci.* 20: 6974–6982.

Jones-Gotman, M. and Zatorre, R. J. (1993) Odor recognition memory in humans: Role of right temporal and orbitofrontal regions. *Brain Cog.* 22: 182–198.

Kadohisa, M., Rolls, E. T., and Verhagen, J. V. (2004) Orbitofrontal cortex: neuronal representation of oral temperature and capsaicin in addition to taste and texture. *Neuroscience* 127: 207–221.

Kanter, E. D. and Haberly, L. B. (1990) NMDA-dependent induction of long-term potentiation in afferent and association fiber systems of piriform cortex in vitro. *Brain Res.* 525: 175–179.

Kay, L. M. and Sherman, S. M. (2007) An argument for an olfactory thalamus. *Trends Neurosci.* 30: 47–53.

Kay, R. B., Meyer, E. A., Illig, K. R., and Brunjes, P. C. (2011) Spatial distribution of neural activity in the anterior olfactory nucleus evoked by odor and electrical stimulation. *J. Comp. Neurol.* 519: 277–289.

Kerr, K. M., Agster, K. L., Furtak, S. C., and Burwell, R. D. (2007) Functional neuroanatomy of the parahippocampal region: the lateral and medial entorhinal areas. *Hippocampus* 17: 697–708.

Kikuta, S., Kashiwadani, H., and Mori, K. (2008) Compensatory rapid switching of binasal inputs in the olfactory cortex. *J. Neurosci.* 28: 11989–11997.

Kikuta, S., Sato, K., Kashiwadani, H., et al. (2010) Neurons in the anterior olfactory nucleus pars externa detect right or left localization of odor sources. *Proc. Natl. Acad. Sci. U. S. A.* 107: 12363–12368.

Knafo, S., Grossman, Y., Barkai, E., and Benshalom, G. (2001) Olfactory learning is associated with increased spine density along apical dendrites of pyramidal neurons in the rat piriform cortex. *Eur. J. Neurosci.* 13: 633–638.

Lei, H., Mooney, R., and Katz, L. C. (2006) Synaptic integration of olfactory information in mouse anterior olfactory nucleus. *J. Neurosci.* 26: 12023–12032.

Li, W., Howard, J. D., and Gottfried, J. A. (2010) Disruption of odour quality coding in piriform cortex mediates olfactory deficits in Alzheimer's disease. *Brain* 133: 2714–2726.

Li, W., Luxenberg, E., Parrish, T., and Gottfried, J. A. (2006) Learning to smell the roses: experience-dependent neural plasticity in human piriform and orbitofrontal cortices. *Neuron* 52: 1097–1108.

Li, W., Howard, J. D., Parrish, T. B., and Gottfried, J. A. (2008) Aversive learning enhances perceptual and cortical discrimination of indiscriminable odor cues. *Science* 319: 1842–1845.

Linster, C., Henry, L., Kadohisa, M., and Wilson, D. A. (2007) Synaptic adaptation and odor-background segmentation. *Neurobiol. Learn Mem.* 87: 352–360.

Linster, C., Menon, A. V., Singh, C. Y., and Wilson, D. A. (2009) Odor-specific habituation arises from interaction of afferent synaptic adaptation and intrinsic synaptic potentiation in olfactory cortex. *Learn. Mem.* 16: 452–459.

Lledo, P. M., Saghatelyan, A., and Lemasson, M. (2004) Inhibitory interneurons in the olfactory bulb: from development to function. *Neuroscientist* 10: 292–303.

Luo, M. and Katz, L. C. (2001) Response correlation maps of neurons in the mammalian olfactory bulb. *Neuron* 32: 1165–1179.

Lynch, G. (1986) *Synapses, Circuits and the Beginnings of Memory.* Cambridge, MA: MIT Press.

Macknin, J. B., Higuchi, M., Lee, V. M., et al. (2004) Olfactory dysfunction occurs in transgenic mice overexpressing human tau protein. *Brain Res.* 1000: 174–178.

Malaspina, D., Goetz, R., Keller, A., et al. (2012) Olfactory processing, sex effects and heterogeneity in schizophrenia. *Schizophr. Res.* 135: 144–151.

Manabe, H., Kusumoto-Yoshida, I., Ota, M., and Mori, K. (2011) Olfactory cortex generates synchronized top-down inputs to the olfactory bulb during slow-wave sleep. *J. Neurosci.* 31: 8123–8133.

Marr, D. (1971) Simple memory: a theory for archicortex. *Philos. Trans R. Soc. Lond. B. Biol. Sci.* 262: 23–81.

Martin, C., Gervais, R., Messaoudi, B., and Ravel, N. (2006) Learning-induced oscillatory activities correlated to odour recognition: a network activity. *Eur. J. Neurosci.* 23: 1801–1810.

McAlonan, K., Cavanaugh, J., and Wurtz, R. H. (2008) Guarding the gateway to cortex with attention in visual thalamus. *Nature* 456: 391–394.

McNamara, A. M., Magidson, P. D., Linster, C., et al. (2008) Distinct neural mechanisms mediate olfactory memory formation at different timescales. *Learn. Mem.* 15: 117–125.

Mesholam, R. I., Moberg, P. J., Mahr, R. N., and Doty, R. L. (1998) Olfaction in neurodegenerative disease. *Arch. Neurol.* 55: 84–90.

Meyer, E. A., Illig, K. R., and Brunjes, P. C. (2006) Differences in chemo- and cytoarchitectural features within pars principalis of the rat anterior olfactory nucleus suggest functional specialization. *J. Comp. Neurol.* 498: 786–795.

Mitsui, S., Igarashi, K. M., Mori, K., and Yoshihara, Y. (2011) Genetic visualization of the secondary olfactory pathway in Tbx21 transgenic mice. *Neural Syst. Circuits* 1: 5.

Miyamichi, K., Amat, F., Moussavi, F., et al. (2011) Cortical representations of olfactory input by trans-synaptic tracing. *Nature* 472: 191–196.

Morgan, C. D. and Murphy, C. (2002) Olfactory event-related potentials in Alzheimer's disease. *J. Int. Neuropsychol. Soc.* 8: 753–763.

Moriceau, S. and Sullivan, R. M. (2006) Maternal presence serves as a switch between learning fear and attraction in infancy. *Nat. Neurosci.* 9: 1004–1006.

Moriceau, S., Wilson, D. A., Levine, S., and Sullivan, R. M. (2006) Dual circuitry for odor-shock conditioning during infancy: corticosterone switches between fear and attraction via amygdala. *J. Neurosci.* 26: 6737–6748.

Mouly, A. M. and Di Scala, G. (2006) Entorhinal cortex stimulation modulates amygdala and piriform cortex responses to olfactory bulb inputs in the rat. *Neuroscience* 137: 1131–1141.

Mouly, A. M., Fort, A., Ben-Boutayab, N., and Gervais, R. (2001) Olfactory learning induces differential long-lasting changes in rat central olfactory pathways. *Neuroscience* 102: 11–21.

Murphy, C. (1999) Loss of olfactory function in dementing disease. *Physiol. Behav.* 66: 177–182.

Murphy, C., Nordin, S., and Jinich, S. (1999) Very early decline in recognition memory for odors in Alzheimer's disease. *Aging Neuropsychol. Cog.* 6: 229–240.

Nagayama, S., Takahashi, Y. K., Yoshihara, Y., and Mori, K. (2004) Mitral and tufted cells differ in the decoding manner of odor maps in the rat olfactory bulb. *J. Neurophysiol.* 91: 2532–2540.

Nagayama, S., Enerva, A., Fletcher, M. L., et al. (2010) Differential axonal projection of mitral and tufted cells in the mouse main olfactory system. *Front. Neural Circuits* 4.

Nassi, J. J. and Callaway, E. M. (2009) Parallel processing strategies of the primate visual system. *Nat. Rev. Neurosci.* 10: 360–372.

Neville, K. R. and Haberly, L. (2004) Olfactory cortex. In: *The Synaptic Organization of the Brain*, 5th Edition (Shepherd, G. M., ed.), pp. 415–454. New York: Oxford University Press.

Nissant, A., Bardy, C., Katagiri, H., et al. (2009) Adult neurogenesis promotes synaptic plasticity in the olfactory bulb. *Nat. Neurosci.* 12: 728–730.

Nordin, S. and Murphy, C. (1998) Odor memory in normal aging and Alzheimer's disease. *Ann. N. Y. Acad Sci.* 855: 686–693.

O'Doherty, J., Rolls, E. T., Francis, S., et al. (2000) Sensory-specific satiety-related olfactory activation of the human orbitofrontal cortex. *Neuroreport* 11: 893–897.

Olichney, J. M., Murphy, C., Hofstetter, C. R., et al. (2005) Anosmia is very common in the Lewy body variant of Alzheimer's disease. *J. Neurol. Neurosurg. Psychiatry* 76: 1342–1347.

Payton, C. A., Wilson, D. A., and Wesson, D. W. (2012) Parallel odor processing by two anatomically distinct olfactory bulb target structures. *PLoS One* 7: e34926.

Plailly, J., Howard, J. D., Gitelman, D. R., and Gottfried, J. A. (2008) Attention to odor modulates thalamocortical connectivity in the human brain. *J. Neurosci.* 28: 5257–5267.

Poo, C. and Isaacson, J. S. (2007) An early critical period for long-term plasticity and structural modification of sensory synapses in olfactory cortex. *J. Neurosci.* 27: 7553–7558.

Poo, C. and Isaacson, J. S. (2011a) A major role for intracortical circuits in the strength and tuning of odor-evoked excitation in olfactory cortex. *Neuron* 72: 41–48.

Poo, C. and Isaacson, J. S. (2011b) A major role for intracortical circuits in the strength and tuning of odor-evoked excitation in olfactory cortex. *Neuron* 72: 41–48.

Porter, J., Craven, B., Khan, R. M., et al. (2007) Mechanisms of scent-tracking in humans. *Nat. Neurosci.* 10: 27–29.

Price, J. L. and Powell, T. P. (1970) An electron-microscopic study of the termination of the afferent fibers to the olfactory bulb from the cerebral hemisphere. *J. Cell Sci.* 7: 157–187.

Rajan, R., Clement, J. P., and Bhalla, U. S. (2006) Rats smell in stereo. *Science* 311: 666–670.

Rampin, O., Bellier, C., and Maurin, Y. (2012) Electrophysiological responses of rat olfactory tubercle neurons to biologically relevant odours. *Eur. J. Neurosci.* 35: 97–105.

Rawson, N. E., Gomez, G., Cowart, B. J., et al. (2011) Age-associated loss of selectivity in human olfactory sensory neurons. *Neurobiol. Aging* 33:1913–1919.

Rennaker, R. L., Chen, C. F., Ruyle, A. M., et al. (2007) Spatial and temporal distribution of odorant-evoked activity in the piriform cortex. *J. Neurosci.* 27: 1534–1542.

Restrepo, D., Doucette, W., Whitesell, J. D., et al. (2009) From the top down: flexible reading of a fragmented odor map. *TINS* 32: 525–531.

Rolls, E. T., Critchley, H. D., and Treves, A. (1996) Representation of olfactory information in the primate orbitofrontal cortex. *J. Neurophysiol.* 75: 1982–1996.

Rolls, E. T., Verhagen, J. V., and Kadohisa, M. (2003) Representations of the texture of food in the primate orbitofrontal cortex: neurons responding to viscosity, grittiness, and capsaicin. *J. Neurophysiol.* 90: 3711–3724.

Rolls, E. T., Grabenhorst, F., Margot, C., et al. (2008) Selective attention to affective value alters how the brain processes olfactory stimuli. *J. Cogn. Neurosci.* 20: 1815–1826.

Roman, F., Staubli, U., and Lynch, G. (1987) Evidence for synaptic potentiation in a cortical network during learning. *Brain Res.* 418: 221–226.

Royet, J. P. and Plailly, J. (2004) Lateralization of olfactory processes. *Chem. Senses* 29: 731–745.

Royet, J. P., Zald, D., Versace, R., et al. (2000) Emotional responses to pleasant and unpleasant olfactory, visual, and auditory stimuli: a positron emission tomography study. *J. Neurosci.* 20: 7752–7759.

Royet, J. P., Hudry, J., Zald, D. H., et al. (2001) Functional neuroanatomy of different olfactory judgments. *Neuroimage* 13: 506–519.

Rubin, B. D. and Katz, L. C. (2001) Spatial coding of enantiomers in the rat olfactory bulb. *Nat. Neurosci.* 4: 355–356.

Rudebeck, P. H. and Murray, E. A. (2011) Dissociable effects of subtotal lesions within the macaque orbital prefrontal cortex on reward-guided behavior. *J. Neurosci.* 31: 10569–10578.

Rupp, C. I. (2010) Olfactory function and schizophrenia: an update. *Curr. Opin. Psychiatry* 23: 97–102.

Saar, D., Grossman, Y., and Barkai, E. (2002) Learning-induced enhancement of postsynaptic potentials in pyramidal neurons. *J. Neurophysiol.* 87: 2358–2363.

Sahay, A., Wilson, D. A., and Hen, R. (2011) Pattern separation: a common function for new neurons in hippocampus and olfactory bulb. *Neuron* 70: 582–588.

Schaefer, M. L., Young, D. A., and Restrepo, D. (2001) Olfactory fingerprints for major histocompatibility complex-determined body odors. *J. Neurosci.* 21: 2481–2487.

Schoenbaum, G. and Eichenbaum, H. (1995) Information coding in the rodent prefrontal cortex. I. Single-neuron activity in orbitofrontal cortex compared with that in pyriform cortex. *J. Neurophysiol.* 74: 733–750.

Schoenbaum, G., Chiba, A. A., and Gallagher, M. (1999) Neural encoding in orbitofrontal cortex and basolateral amygdala during olfactory discrimination learning. *J. Neurosci.* 19: 1876–1884.

Schoenbaum, G., Gottfried, J. A., Murray, E. A., and Ramus, S. J., eds (2011) *Critical Contributions of the Orbitofrontal Cortex to Behavior*. New York: New York Academy of Sciences.

Schul, R., Slotnick, B. M., and Dudai, Y. (1996) Flavor and the frontal cortex. *Behav. Neurosci.* 110: 760–765.

Schwob, J. E. and Price, J. L. (1978) The cortical projection of the olfactory bulb: development in fetal and neonatal rats correlated with quantitative variations in adult rats. *Brain Res.* 151: 369–374.

Sevelinges, Y., Gervais, R., Messaoudi, B., et al. (2004) Olfactory fear conditioning induces field potential potentiation in rat olfactory cortex and amygdala. *Learn Mem.* 11: 761–769.

Shepherd, G. M. (2005) Perception without a thalamus how does olfaction do it? *Neuron* 46: 166–168.

Siderowf, A., Newberg, A., Chou, K. L., et al. (2005) [99mTc]TRODAT-1 SPECT imaging correlates with odor identification in early Parkinson disease. *Neurology* 64: 1716–1720.

Small, D. M., Gerber, J. C., Mak, Y. E., and Hummel, T. (2005) Differential neural responses evoked by orthonasal versus retronasal odorant perception in humans. *Neuron* 47: 593–605.

Smith, J. J., Shionoya, K., Sullivan, R. M., and Wilson, D. A. (2009) Auditory stimulation dishabituates olfactory responses via noradrenergic cortical modulation. *Neural Plast.* 2009: 754014.

Sobel, N., Thomason, M. E., Stappen, I., et al. (2001) An impairment in sniffing contributes to the olfactory impairment in Parkinson's disease. *Proc. Natl. Acad Sci. U. S. A.* 98: 4154–4159.

References

Sosulski, D. L., Lissitsyna Bloom, M., Cutforth, T., et al. (2011) Distinct representations of olfactory information in different cortical centres. *Nature* 472: 213–216.

Staubli, U., Ivy, G., and Lynch, G. (1984) Hippocampal denervation causes rapid forgetting of olfactory information in rats. *Proc. Natl. Acad Sci. U. S. A.* 81: 5885–5887.

Stettler, D. D. and Axel, R. (2009) Representations of odor in the piriform cortex. *Neuron* 63: 854–864.

Strowbridge, B. W. (2009) Role of cortical feedback in regulating inhibitory microcircuits. *Ann. N. Y. Acad Sci.* 1170: 270–274.

Suzuki, N. and Bekkers, J. M. (2006) Neural coding by two classes of principal cells in the mouse piriform cortex. *J. Neurosci.* 26: 11938–11947.

Suzuki, N. and Bekkers, J. M. (2010) Inhibitory neurons in the anterior piriform cortex of the mouse: classification using molecular markers. *J. Comp. Neurol.* 518: 1670–1687.

Suzuki, N. and Bekkers, J. M. (2011) Two layers of synaptic processing by principal neurons in piriform cortex. *J. Neurosci.* 31: 2156–2166.

Tanabe, T., Iino, M., and Takagi, S. F. (1975) Discrimination of odors in olfactory bulb, pyriform-amygdaloid areas, and orbitofrontal cortex of the monkey. *J. Neurophysiol.* 38: 1284–1296.

Wattendorf, E., Welge-Lussen, A., Fiedler, K., et al. (2009) Olfactory impairment predicts brain atrophy in Parkinson's disease. *J. Neurosci.* 29: 15410–15413.

Wesson, D. W. and Wilson, D. A. (2010) Smelling sounds: olfactory-auditory sensory convergence in the olfactory tubercle. *J. Neurosci.* 30: 3013–3021.

Wesson, D. W. and Wilson, D. A. (2011) Sniffing out the contributions of the olfactory tubercle to the sense of smell: hedonics, sensory integration, and more? *Neurosci. Biobehav. Rev.* 35: 655–668.

Wesson, D. W., Levy, E., Nixon, R. A., and Wilson, D. A. (2010) Olfactory dysfunction correlates with amyloid-beta burden in an Alzheimer's disease mouse model. *J. Neurosci.* 30: 505–514.

Wesson, D. W., Borkowski, A. H., Landreth, G. E., et al. (2011) Sensory network dysfunction, behavioral impairments, and their reversibility in an Alzheimer's beta-amyloidosis mouse model. *J. Neurosci.* 31: 15962–15971.

Wilson, D. A. (1998a) Habituation of odor responses in the rat anterior piriform cortex. *J. Neurophysiol.* 79: 1425–1440.

Wilson, D. A. (1998b) Synaptic correlates of odor habituation in the rat anterior piriform cortex. *J. Neurophysiol.* 80: 998–1001.

Wilson, D. A. (2000) Odor specificity of habituation in the rat anterior piriform cortex. *J. Neurophysiol.* 83: 139–145.

Wilson, D. A. (2001) Scopolamine enhances generalization between odor representations in rat olfactory cortex. *Learn Mem.* 8: 279–285.

Wilson, D. A. (2003) Rapid, experience-induced enhancement in odorant discrimination by anterior piriform cortex neurons. *J. Neurophysiol.* 90: 65–72.

Wilson, D. A. (2009) Pattern separation and completion in olfaction. *Ann. N. Y. Acad Sci.* 1170: 306–312.

Wilson, D. A. and Stevenson, R. J. (2006) *Learning to Smell: Olfactory Perception from Neurobiology to Behavior.* Baltimore: Johns Hopkins University Press.

Wilson, D. A. and Sullivan, R. M. (2011) Cortical processing of odor objects. *Neuron* 72: 506–519.

Wilson, M. A. and Bower, J. M. (1988) A computer simulation of olfactory cortex with functional implications for storage and retrieval of olfactory information. In: *Neural Information Processing Systems* (Anderson, D. Z., ed.), pp. 114–126: American Institute of Physics.

Wirth, S., Ferry, B., and Di Scala, G. (1998) Facilitation of olfactory recognition by lateral entorhinal cortex lesion in rats. *Behav. Brain Res.* 91: 49–59.

Wyss, J. M. (1981) An autoradiographic study of the efferent connections of the entorhinal cortex in the rat. *J. Comp. Neurol.* 199: 495–512.

Xu, W. and Wilson, D. A. (2012) Odor-evoked activity in the mouse lateral entorhinal cortex. *Neuroscience* 223: 12–20.

Yadon, C. A. and Wilson, D. A. (2005) The role of metabotropic glutamate receptors and cortical adaptation in habituation of odor-guided behavior. *Learn. Mem.* 12: 601–605.

Yamamoto, C. and McIlwain, H. (1966) Potentials evoked in vitro in preparations from the mammalian brain. *Nature* 210: 1055–1056.

Yan, Z., Tan, J., Qin, C., et al. (2008) Precise circuitry links bilaterally symmetric olfactory maps. *Neuron* 58: 613–624.

Yoshida, I. and Mori, K. (2007) Odorant category profile selectivity of olfactory cortex neurons. *J. Neurosci.* 27: 9105–9114.

Zatorre, R. J., Jones-Gotman, M., Evans, A. C., and Meyer, E. (1992) Functional localization and lateralization of human olfactory cortex. *Nature* 360: 339–340.

Zelano, C., Bensafi, M., Porter, J., et al. (2005) Attentional modulation in human primary olfactory cortex. *Nat. Neurosci.* 8: 114–120.

Zellner, D. A. and Kautz, M. A. (1990) Color affects perceived odor intensity. *J. Exp. Psychol. Hum. Percept. Perform* 16: 391–397.

Zellner, D. A., Bartoli, A. M., and Eckard, R. (1991) Influence of color on odor identification and liking ratings. *Am. J. Psychol.* 104: 547–561.

Part 3

Human Olfactory Measurement, Physiology, and Development

Chapter 11

Psychophysical Measurement of Human Olfactory Function

RICHARD L. DOTY and DAVID G. LAING

11.1 INTRODUCTION

During the last two centuries, numerous tests have been devised to assess the function of the human sense of smell. Many of these tests have been modeled on procedural and mathematical concepts developed in the mid-nineteenth century by Weber (1834) and Fechner (1860), and by Thurstone, Stevens, and others in the twentieth century (e.g., Anderson, 1970; Thurstone, 1927a,b; Stevens, 1961). Tests derived from these traditions include absolute detection thresholds (the lowest odorant concentration that can be perceived), differential thresholds (the smallest difference in concentration of a given chemical that can be perceived), and various indices of suprathreshold sensation magnitude. Most of these tests were developed within the theoretical framework of establishing mathematical rules or laws that govern the build-up of suprathreshold sensation relative to stimulus intensity. To achieve all of these ends, well-defined stimuli (e.g., single chemicals of known chemical purity) were usually employed, allowing for straightforward stimulus specification.

In addition to these developments, other trends resulted in the development or application of tests more useful in applied settings. For example, eighteenth and nineteenth-century physicians simply presented familiar odorants to patients to see if they could be identified, usually without insight into prior psychophysical developments. In the twentieth century relatively sophisticated procedures were developed within the food industry (e.g., the forced-choice triangle test; Helm and Trolle, 1946; Peryman and Swartz, 1950), where the need exists to quantify the acceptability of various product formulations in relation to perceived qualitative attributes. Unlike the academic psychophysical traditions, and akin to the clinical traditions, the stimuli were generally chemically complex, that is, not made up of single chemicals. Although quantitative, the metrics employed in these paradigms were more operational and rarely linked to physicochemical properties such as odorant concentration.

More recently, techniques have been developed to study odor perception arising from stimuli within the oral cavity, i.e., so-called retronasal odor perception (Burdach and Doty, 1987; Bender et al., 2009; Chen and Halpern, 2000; Croy et al., 2014; Heilmann et al., 2002). The measurement of such perception has practical consequences for both clinicians and food manufacturers, given the critical role of olfaction in determining the flavor of foods and beverages.

The present chapter has two major goals. The first is to provide the reader with an overview of the quantitative methods currently available for assessing the sense of smell, regardless of the historical traditions that led to their development. Emphasis is placed on the relative utility of various approaches for achieving this end. The second goal is to examine elements of odor mixture perception, including how well individual components can be discerned. An understanding of odor mixture processing is of value in elucidating how the olfactory system works, as its neural architecture seems to be designed to filter or collapse complex arrays of chemical information into distinct, interpretable, and manageable percepts. Although many of the examples described in this chapter come from clinical studies, the tenants of the chapter are broadly applicable to settings outside the clinic, including industrial and regulatory ones.

Handbook of Olfaction and Gustation, Third Edition. Edited by Richard L. Doty.
© 2015 Richard L. Doty. Published 2015 by John Wiley & Sons, Inc.

11.2 STIMULUS CONTROL AND PRESENTATION

In some chemosensory paradigms, extremely accurate stimulus specification is required, and elaborate olfactometers and other devices for presenting known concentrations of odorants in specific quantities for various durations have been devised (for review, see Prah et al., 1995). This is particularly true for devices employed in human chemosensory electrophysiology (Chapter 12) and in research that requires precise measurement, such as that involved in establishing regulatory guidelines or relationships among physicochemical parameters of odorants and perceptual processes (Schmidt and Cain, 2010). In other paradigms, including those related to assessing olfactory function in patients, such precision is not necessary. The requirement is only that the odorants are presented in a reliable manner and that norms are available to establish whether a patient's responses are normal or abnormal. Thus, accurate clinical assessment of chemosensory function can be made using surprisingly simple stimulus presentation equipment.

As illustrated in Figure 11.1, devices that have been used to present odorants to humans include (a) the draw tube olfactometer of Zwaardemaker (1925, 1927), (b) glass sniff bottles and specialized cannisters (Cheesman and Townsend, 1956; Doty et al., 1986; Frank et al., 2003; Nordin et al., 1998), (c) odorized glass rods, wooden sticks, felt-tipped pens, alcohol pads, plastics embedded with odorants, or strips of blotter paper (Semb, 1968; Toyota et al., 1978; Davidson and Murphy, 1997; Hummel et al., 1997; Saito, 2006; Simmen et al., 1999), (d) plastic squeeze bottles (Amoore and Ollman, 1983; Cain et al., 1988; Doty, 2000; Guadagni et al., 1963), (e) bottles from which blasts of odorized air are presented (Elsberg and Levy, 1935; Ikeda et al., 1999), (f) microencapsulated "scratch and sniff" odorized strips (Doty, Agrawal and Frye, 1989; Cameron and Doty, 2013; Doty et al., 1984a; Richman, 1992; Doty, 1995), (g) laboratory based air-dilution olfactometers (Cheesman and Kirkby, 1959; Cook et al., 2014; Doty et al., 1988; Hayes et al., 2013; Kobal and Plattig, 1978; Johnson and Sobel, 2007; Lorig et al., 1999; Lundström et al., 2010; Philpott et al., 2009; Punter, 1983; Schmidt and Cain, 2010; Schriever, 2011; Sezille et al., 2013; Walker et al., 1990; Wenzel, 1948) and (h) exposure chambers, including mobile units with analytical equipment and subject waiting rooms (e.g., Berglund et al., 1984; Dalton et al., 1997; Springer, 1974).

Stimulus presentation devices that increase the speed and reliability of olfactory testing for both adults and children have recently been developed. These include a game-like rotating disk test for measuring function in children (Figure 11.2), an olfactometer that allows for computerized self-administration of threshold, discrimination, memory, and identification tests, among others (Figure 11.3), and stimulus presentation wands that allow

Figure 11.1 Procedures for presenting odorants to subjects for assessment. a. Early draw-tube olfactometer of Zwaardemaker. In this apparatus, an outer tube made of rubber or another odorous material slides along a calibrated inner tube, one end of which is inserted into the subject's nostril. When the odorized tube is slid toward the subject, less of its internal surface is exposed to the inspired airstream, resulting in a weaker olfactory sensation. b. Sniff bottle. c. Perfumer's strip. d. Squeeze bottle. e. Blast injection device. The experimenter injects a given volume of odor into the bottle and releases the pressure by squeezing a clamp on the tube leading to the nostril, producing a stimulus pulse. f. Microencapsulated "scratch and sniff" test. g. Sniff ports on a rotating table connected to one of the University of Pennsylvania's dynamic air-dilution olfactometers. h. Odor evaluation room of mobile odor evaluation laboratory designed to evaluate responses of panel members to diesel exhaust. Copyright © 2006, Richard L. Doty.

11.2 Stimulus Control and Presentation

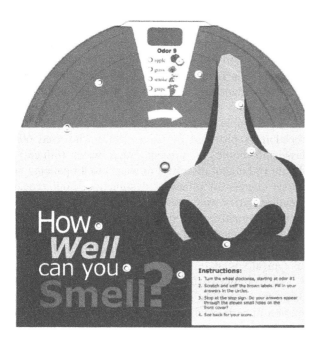

Figure 11.2 Photograph of the front of the Smell Wheel, a game-like rotating odor identification test for assessing smell function of children. A disk containing odors rotates within a cardboard jacket, exposing one odor at a time in the top window. The child releases the odor from the microencapsulated brown taste strip and makes an answer and moves to the next odor. When all 11 items are marked, the correct answers appear in the holes around the front of the test. The test score is the number of holes that are filled in. This forced-choice test was designed to capture the child's imagination and to provide a standardized test measure with odors known to young people and a minimum number of trials. Copyright © 2012 Sensonics International, Haddon Heights, NJ USA.

for rapid presentation of odors to subjects without removing the tops from sniff bottles or pen-like dispensers (Figure 11.4). These wands have the advantage of being able to use existing normative data and eliminate the touching of an odor source to the nose.

In addition to these approaches to stimulus presentation, intravenous administration of odorants has been used to produce chemosensory sensations. This approach, which is only rarely used today, was employed primarily by Japanese otolaryngologists in an attempt to assess whether the olfactory receptors are working when nasal congestion or blockage eliminates or mitigates airflow to the receptor region. The assumption underlying this technique is that the stimulus makes its way to the olfactory receptors via the bloodstream. Most commonly thiamine propyldisulfide (Alinamin) is injected into the median cubital vein, and recordings of the duration and latency of the onset of a garlic-like sensation experienced by the patient are made (see Takagi, 1989, for review). Although this procedure may be of value in some cases, there is controversy regarding its physiological basis (i.e., whether the stimulus reaches the receptors via diffusion from nasal capillaries,

from lung air, or both; see Maruniak et al., 1983). Furthermore, such testing is invasive, highly variable, not readily adaptable to forced-choice paradigms, and lacks normative referents.

Figure 11.3 The self-administered computerized olfactory test system (SCOTS). The dome, which can be readily exchanged for other domes with different sets of odors, contains up to 40 odorants that can be individually released, or released in combinations, to the sniffing port. This system eliminates subject error in the presentation of test stimuli and allows for exact control of stimulus duration, interstimulus intervals, and other factors. Photo courtesy of Sensonics International, Haddon Heights, NJ, USA. Copyright © 2013 Sensonics International.

Figure 11.4 A modern pen-like device, termed the Snap & Sniff™ wand, by which odorants can be presented for sampling. Sets of these devices come in multiple odorant configurations, some of which can employ normative data available from published sources. Photo courtesy of Sensonics International, Haddon Heights, NJ, USA. Copyright © 2014 Sensonics International.

11.3 PSYCHOPHYSICAL TEST PROCEDURES

Any procedure that provides a quantitative measure of sensory function and requires a verbal or conscious overt response on the part of the examinee is generally considered a psychophysical procedure. In this section, the basic psychophysical paradigms available for measuring olfactory function are discussed and examples of their application are provided. The interested reader is referred to other sources for more detailed information about psychophysical methods, including their mathematical foundations (Ekman and Sjöberg, 1965; Gescheider, 1997; Guilford, 1954; Köster, 1975; Marks, 1974; Stevens, 1961; Tanner and Swets, 1954; Wise et al., 2000).

11.3.1 Detection and Recognition Threshold Tests

A popular means for assessing chemosensory function is to establish, operationally, a measure of the lowest concentration of a stimulus that can be detected. A qualitative odor sensation (e.g., "banana-like") is rarely perceived at very low odorant concentrations, where only the faint presence of something is noted. The absolute or detection threshold is the lowest odorant concentration where such a presence is reliably detected, whereas the recognition threshold is the lowest concentration where odor quality is discerned. In modern olfactory detection threshold testing, a subject is asked to indicate, on a given trial, which of two or more stimuli (e.g., a low concentration odorant and one or more non-odorous blanks) smells strongest, rather than to report whether an odor is perceived or not. Recognition thresholds are obtained in a similar manner, but the requirement is to report which one has the target quality. Such "forced-choice" procedures are less susceptible than non-forced choice procedures to contamination by response biases (i.e., the conservatism or liberalism in reporting the presence of an odor under uncertain conditions). In addition, they are typically more reliable and produce lower threshold values (Blackwell, 1953; Doty et al., 1995).

Two types of threshold procedures that have received the most clinical and industrial use are the ascending method of limits (AML) and the single staircase (SS) procedures. In the AML procedure, odorants are presented sequentially from low to high concentrations and the point of transition between detection and no detection is estimated. Forced-choice responses are required on each trial. Such a paradigm has been adopted by the water industry worldwide and codified by the American Society for Testing and Materials as E679-04 (ASTM, 2008; see also Lawless, 2010). Unfortunately, as noted in Chapter 49, this approach, as practiced following ASTM guidelines, is fraught with difficulties. In the SS method (a variant of the method of limits technique; see Cornsweet, 1962), the concentration of the stimulus is increased following trials on which a subject fails to detect the stimulus and decreased following trials where correct detection occurs. In both these procedures, the direction of initial stimulus presentation is made from weak to strong in an effort to reduce adaptation effects of prior stimulation (see Pangborn et al., 1964).

An example of a clinical application of the AML procedure is provided by Cain (1982) who used 60-ml glass sniff bottles to present either water (diluent) or odorant (n-butanol dissolved in water) to 43 patients with various degrees of olfactory dysfunction. Four repeated ascending series were presented to each side of the nose in a two-alternative, forced-choice format. This test, which took approximately half an hour per patient to administer, demonstrated that the olfactory dysfunction in these cases was typically bilateral.

An example of the clinical use of a SS procedure comes from a study that demonstrates loss of olfactory function in early Alzheimer's disease (Doty et al., 1987). In this experiment, a trial consisted of the presentation of two 100-ml glass sniff bottles to the patient in rapid succession. One bottle contained 20 ml of a given concentration of phenyl ethyl alcohol dissolved in odorless propylene glycol, whereas the other contained mineral oil alone. The patient was asked to report which of the two bottles in a pair produced the strongest sensation. The first trial was presented at a -6.50 log (liquid volume/volume) concentration step. If a miss occurred on any trial before five were correctly completed at that concentration, the process was repeated at one log concentration step higher. When five consecutive correct trials occurred at a given concentration level, the staircase was "reversed" and the next pair of trials was presented at a 0.5 log concentration step lower. From this point on, only one or two trials were presented at each step (i.e., if the first trial was missed, the second was not given and the staircase was moved to the next higher 0.5 log step concentration). When correct performance occurred on both trials, the concentration of the next trial was given at one half-log unit step lower. The average of the last four of seven staircase reversal points served as the threshold estimate. Examples of individual data obtained using the SS procedure are shown in Figure 11.5.

In general, threshold values are relative and dependent upon such factors as the method of stimulus dilution, volume of inhalation, species of molecule, type of psychophysical task, and number of trials presented (Pierce et al., 1996; Tsukatani et al., 2003). A number of investigators have noted that threshold measures often exhibit considerable intra- and intersubject variability. For example, in one study of 60 subjects, intersubject variation as great as 5 log units was reported (Brown et al., 1968). In another, in which a non-forced-choice ascending threshold procedure was used (the Japanese "T&T Olfactometer"), variation on the order of 16 log units was present among groups of 430–1000 young subjects (Yoshida, 1984).

11.3 Psychophysical Test Procedures

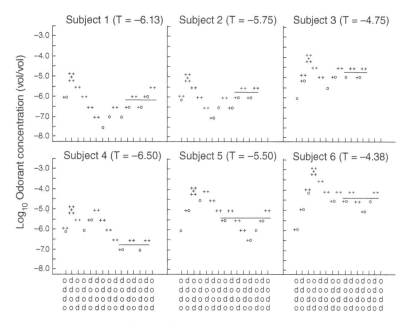

Figure 11.5 Data illustrating single-staircase detection threshold determinations. Each plus (+) indicates a correct detection when an odorant versus a blank is presented. Each minus (−) indicates an incorrect report of an odorant. Threshold value (T; phenyl ethyl alcohol vol/vol in light USP grade mineral oil) is calculated as the mean of the last four of seven staircase reversals. Although the geometric mean is often employed to assess such measures, given that odorant concentration is already on a log scale, the arithmetic mean is preferred. The arithmetic means are indicated on the figure. The corresponding geometric means for the six respective subjects are: −6.11, −5.74, −4.74, −5.47 and −4.38. The o's and d's on the abcissa indicate the counterbalancing order of the presentation sequences for each trial and are read downward (o—odorant presented first, then diluent; d—diluent presented first, then odorant). In the first reversal point (where five correct sets of pairs occur at the same concentration), the fifth order sequence is determined by the first o or d of the subsequent column of four order sequences. Modified from Doty (1991) with permission. Copyright © 1991, Raven Press.

Stevens et al. (1988) obtained 60 threshold values over the course of 30 days from three subjects (20 for butanol, 20 for pyridine, and 20 for β-phenylethylmethylethylcarbinol). These investigators found that the within-subject variability across test days was as great as the between-subject variability on a given test day, suggesting to these authors that the large individual differences observed in threshold values are not a reflection of big differences among stable threshold values of subjects, but reflected large day-to-day fluctuations in the test measures. Unfortunately, much of this fluctuation likely reflects the use of the single ascending detection threshold technique in which the apparent limen is traversed only once. Test procedures incorporating more trials, such as the SS procedure, produce less variable measures and, when employed, do not exhibit as marked day-to-day fluctuations. The reliability of such procedures is discussed later in this chapter.

11.3.2 Difference Threshold Tests

In classical psychophysics, the smallest amount by which a stimulus must be changed to make it perceptibly stronger or weaker is termed a "just noticeable difference" or JND. This value is also called a difference or differential threshold (in contrast to an absolute threshold, as described above). The size of the increment in odorant concentration (ΔI) required to produce a JND increases as the comparison concentration (I) increases, with the ratio approximating a constant; i.e., $\Delta I / I = K$ (Weber's law; Weber, 1834). K is a rough index of the sensory system's sensitivity (i.e., the smaller the K value, the more sensitive the system is to fine changes in stimulation). However, numerous studies suggest that K is not a constant, being influenced by the size of I, particularly at the extremes of the sensory continuum (Doty, 1991; Hidalgo et al., 2011).

An example of a brief clinical test used to establish a difference threshold is described by Eichenbaum et al. (1983). In this test, 10 binary dilutions (in water) of acetone, ethanol, almond extract, and lemon extract were presented. Initially, the highest and lowest concentrations of a given odorant were presented and the subject was required to choose the stronger stimulus. Successively stronger stimuli were then paired with the strongest stimulus until, on the last of the 10 trials, the two samples were identical. Eichenbaum operationally defined the difference threshold as the lowest concentration for which discrimination up to and including the dilution was effortless.

11.3.3 Signal Detection Tests

Signal detection theory (SDT) differs fundamentally from the approach of sensory measurement inherent in classical threshold theory. Thus, SDT rejects the notion of a threshold (whether absolute or differential), and focuses on (a) noise and signal plus noise as the milieu of the detection situation and (b) the influences of subject expectancies and rewards on the detection decision. Signal detection procedures provide both a measure of sensory sensitivity and the subject's response criterion or bias (Tanner and Swets, 1954). In effect, the response criterion is the internal rule used by a subject in deciding whether or not to report detecting a stimulus (e.g., the liberalism or conservatism in reporting a sensation under uncertain circumstances). For example, two subjects may experience the same subtle degree of sensation from a very weak stimulus. One, however, may report that no sensation was perceived (e.g., perhaps because of lack of self-confidence), whereas the other may report the presence of the sensation. In both cases, the stimulus was perceived to the same degree. However, the two subjects had different criteria for reporting its presence. In a traditional non-forced-choice detection threshold paradigm, the investigator would conclude that these two subjects differed in sensitivity to the stimulus, when, in fact, they only differed in regards to their response biases.

Expectation is not a trivial matter, as it can significantly increase the variability of threshold measures. A classical example of the role of expectation, as well as perhaps group pressure, was noted in class demonstration by Slosson in 1899 (p. 407):

> I had prepared a bottle filled with distilled water carefully wrapped in cotton and packed in a box. After some other experiments I stated that I wished to see how rapidly an odor would be diffused through the air, and requested that as soon as anyone perceived the odor he should raise his hand. I then unpacked the bottle in the front of the hall, poured the water over the cotton, holding my head away during the operation and started a stopwatch. While awaiting the results I explained that I was quite sure that no one in the audience had ever smelled the chemical compound which I had poured out, and expressed the hope that, while they might find the odor strong and peculiar, it would not be too disagreeable to anyone. In fifteen seconds most of those in the front row had raised their hands, and in forty seconds the "odor" had spread to the back of the hall, keeping a pretty regular "wave front" as it passed on. About three-fourths of the audience claimed to perceive the smell … More would probably have succumbed to the suggestion, but that

at the end of a minute I was obliged to stop the experiment, for some on the front seats were being unpleasantly affected and were about to leave the room …

Signal detection theory was designed not only to overcome such problems, but, as noted above, to provide a quantitative measure of the response criterion. SDT assumes that a stimulus is imbedded within a background of noise. Noise can arise from a variety of sources and can be conceptualized at a number of levels (e.g., variations in attention, stimulus fidelity, neural firing unrelated to the stimulus, fluctuations in distracting physiological processes). In most cases noise is assumed to be normally distributed (as is done here to simplify discussion). Whenever a signal is added to the "noise" (N) distribution, a "signal plus noise" (SN) distribution results. Both the N and SN distributions can be placed on the same set of axes, as shown in Figure 11.6. The measure of the subject's sensitivity is the distance between the means of these distributions, signified as d'.

The concept of the response criterion is also illustrated in this figure (Doty, 1991). Given constant sensitivity (d'), on any given trial a low-concentration odorant (SN) or a blank stimulus (N) is presented, and the subject's task is to report whether or not an odor was presented. Reports of "yes" are represented by the areas under the N and SN curves to the right of the vertical line depicting the subject's response criterion, whereas reports of "no" are indicated by the areas to the left of this line. In case 1, the subject exhibits a very liberal criterion, reporting the presence of an odor on the majority of the SN trials (ß) and on half of the N trials (α). Thus, although correct detection of the odorant occurred nearly all of the time (ß), many false alarms (α) were present. Perhaps in this instance the subject was rewarded for reporting the detection of an odor and not admonished for making false alarms. In case 2, the subject chose a less liberal response criterion. Although fewer correct detections of the odor were made (ß), fewer false alarms were also made (α). In case 3, the observer chose a very conservative response criterion, making few false alarms but similarly making fewer correct detections. This would tend to result, for example, when a subject is penalized for making false positives and given few rewards for successful detection of the odor. In all three of these hypothetical cases, the sensitivity (i.e., d') was equivalent, as indicated by the constant distance between the N and SN distributions.

In a typical olfactory experiment employing SDT, the subject is presented with a large number of trials of a single low concentration of odorant interspersed with blank trials (Doty et al, 1981; Semb, 1968). Even though the number of blank and odorant trials need not be equivalent, this is commonly the case. The proportion or percent of the total odor trials (S) on which a subject reports detecting an odor (the hit rate) is calculated, as is the percent of blank trials (N) on

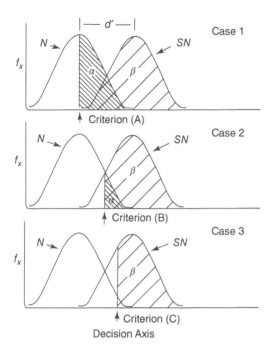

Figure 11.6 Hypothetical distributions of signal plus noise (SN) and noise alone (N) plotted on the same axes. When the strength of the perceived signal increases, the SN distribution moves to the right, increasing d', the measure of the distance between the two distributions in standard deviation units (z scores). The response criterion can vary when perceptual sensitivity (d') remains constant. In case 1, a liberal criterion was chosen in which a relatively large number of false positives occurred (i.e., α, the reports of the presence of odor when the blank (N) is presented). In cases 2 and 3 more conservative criteria were chosen, decreasing both the number of false positives (α) and hits (β). Traditional threshold measures confound the influences of perceptual sensitivity and the setting of the response criterion. Reprinted from Doty (1976) with permission. Copyright © 1976, Academic Press.

which an odor is reported (the false alarm rate). The parametric sensitivity measure, d', can then be computed by converting the proportions to normal distribution standard deviation values (z-scores) via a normal probability table; d' equals the z-score for hits minus the z-score for false alarms. More convenient procedures for determining d' for any combination of hit and false alarm proportions include the use the table provided by Elliot (1964) and various free online calculators, such as that of ComputerPsych LLC: http://www.computerpsych.com/Research_Software/Norm Dist/Online/ Detection_Theory?hits=10&fa=5&misses=5 &cr=5. In addition, nonparametric signal detection measures are also available (Brown, 1974; Frey and Colliver, 1973; Grier, 1971; Hodos, 1970; but see Macmillan and Creelman, 1996), as are methods for testing the parametric assumptions of traditional signal detection analyses (Gescheider, 1995; Green and Swets, 1966).

The classical parametric measure of response bias is termed ß. Not to be confused with the ß in Figure 11.6, ß represents the ratio, at the criterion point, of the ordinate of the SN distribution to that of the N distribution. This value can be easily calculated from the hit and false-alarm rates by use of ordinate values from the normal curve, as discussed by Gescheider (1995). The aforementioned online calculator computes both d' and ß.

Despite the fact that hundreds of trials have been traditionally used in signal detection studies, some chemical senses studies have employed far fewer trials, largely due to practicality. For example, Potter and Butters (1980) and Eichenbaum et al. (1983) computed d' using only 30 test trials. Even though such estimates are somewhat unstable (because a test's reliability is usually a function of its length), they may be less so than typically assumed, and there is at least some empirical rationale for the use of abbreviated signal detection tests. Thus, O'Mahony et al. (1979b), in a study of gustatory sensitivity to sodium chloride, found that Brown's (1974) nonparametric R index fell, after 40 trials, within 5% of the values obtained after 200 trials in slightly over half the subjects tested. However, an analogous olfactory study has not been performed, and ideally all of the subjects should evidence such response stability. For these reasons, it is prudent to use as many trials as possible in signal detection tasks.

11.3.4 Suprathreshold Scaling Procedures

A number of psychological attributes can be assigned to odors, including strength, hedonics (i.e., pleasantness/unpleasantness), and quality. These attributes change as a function of concentration, although odor quality is more stable and hedonics is more variable and idiosyncratic (Doty, 1975). Importantly, associations between hedonics and intensity are not always linear and are dependent upon odorant concentration (Figure 11.7). Because the intensity of a stimulus is related to the number of neurons that are recruited and the frequency at which they fire, suprathreshold measures of intensity may relate to the extent of neural damage present in the afferent pathway (Drake et al., 1969). However, suprathreshold rating or scaling methods may be less sensitive, at least in some cases, to olfactory dysfunction than a number of other methods (e.g., detection threshold tests and tests of odor identification). They do, however, have the advantage of being relatively brief, easy to administer, and less susceptible than threshold tests to subtle stimulus contamination. Moreover, negative findings must be conservatively interpreted. While it is true that suprathreshold rating scales have completely missed major changes in olfaction observed by other methods (e.g., the influences of age on olfactory function; see Rovee et al., 1975), it is also possible that they may be

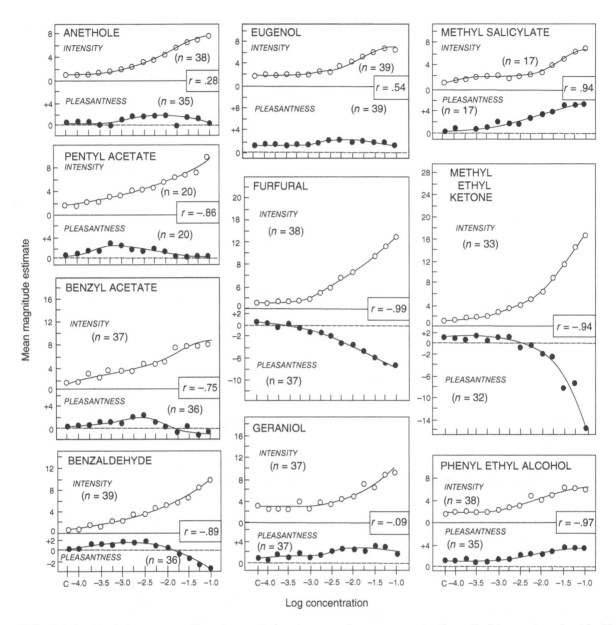

Figure 11.7 Relationship of pleasantness and intensity magnitude estimates to odorant concentration (log vol/vol) in propylene glycol for 10 odorants. C = control (propylene glycol alone). r = Pearson product moment correlation between intensity and pleasantness estimates across data points differing significantly in intensity from control. All correlations significant beyond the 0.01 level with the exception of those for anethole (p > 0.20), eugenol (p > 0.06), and geraniol (p > 0.20). Lines fitted to data points by visual inspection. Modified from Doty (1975).

measuring processes quite different from those measured by thresholds and most other tests and that such processes are, in fact, preserved in aging and some age-related diseases. In other words, they could be measuring aspects of perceptual activity that go undetected by many other methods (Doty et al., 1994; see, however, Doty et al., 2014).

Rating scales are a popular means for assessing the relative amount of a suprathreshold sensory stimulus or psychological attribute perceived by a subject. Such scales have been used for more than a thousand years (for review, see McReynolds and Ludwig, 1987). Galen, for example, employed a rudimentary hot-cold scale that later led to hot and cold body temperature scales employed by physicians until the nineteenth-century development of practical clinical thermometers. Christian Thomasius (1655–1728) used rating scales to measure such psychological attributes as acquisitiveness, sensuousness, and social ambition and, in 1807, the British navy adopted a 12-point rating scale for assessing wind strength (0 = calm, 12 = hurricane). The first use of a rating scale in the chemical senses appears to be that of Fernberger (1932), who had subjects categorize the taste of a bitter tasting agent, PTC, into the categories of "tasteless," "slightly bitter," "bitter," "very bitter," and "extremely bitter."

In olfactory assessment, two types of scales are most popular: category scales, where the relative amount of a sensation is signified by indicating which of a series of discrete categories best describes the sensation, and line scales (also termed visual analog or graphic scales), where the subject or patient indicates the strength of the sensation by placing a mark along a line that has descriptors (termed anchors) located at its extremes (e.g., very weak-very strong). For discussions of the general properties of these and other scales, including the influence of category number on their psychometric properties, see Anderson (1970), Guilford (1954), Hjermstad et al. (2011), Markon et al. (2011), and Preston and Colman (2000).

As pointed out by Bartoshuk and her associates in Chapter 34, across-group comparisons using traditional rating scales can be problematic for a number of reasons. One problem is that labels at the ends of scales, such as "very strong," do not necessarily mean the same to all persons and are context dependent. As she noted elsewhere (Bartoshuk et al., 2005, p. 122),

A woman giving birth might describe her pain as "very strong," but she might also describe a cup of tea as "very strong." Clearly the sensory intensities of the experiences are different; a listener understands that "very strong" depends on the sensory context to which it is applied. It is useful to think of sensory descriptors as printed on an elastic ruler; their relative positions on the ruler are essentially fixed, but the entire ruler can be stretched or compressed to fit the domain of interest.

Classic rating scales also suffer from the tendency of subjects to group responses at the extreme end of the scale (i.e., a ceiling effect). In an attempt to overcome this problem, a number of scales have been developed in which logarithmic-like elements have been incorporated into their design in an effort to more closely mimic ratio-like properties of magnitude estimation procedures (see below) (e.g., Borg, 1982; Neely et al., 1992; Green et al., 1996). In some such scales, verbal anchors are positioned along strategic points of the scale where they seem most appropriate.

A more sophisticated means for understanding the build-up of odor intensity as a function of stimulus concentration involves magnitude estimation. In this procedure, the relative magnitude of each member of a stimulus set is estimated by employing some other sensory modality or cognitive domain in making the assessment. A key difference between this procedure and rating scale procedures is that the *ratio relations among the intensities (or other attributes) of the stimuli are estimated*, and a subject's responses are not confined to categories or a short response

line. Continua commonly used in magnitude estimation tasks for intensity include number (e.g., assigning numbers proportionate to an odor's intensity) and distance (e.g., pulling a tape measure a distance proportional to an odor's intensity) (Berglund et al., 1971; Stevens, 1961).

In the most common magnitude estimation procedure, the subject assigns numbers relative to the magnitude of the sensations. For example, if the number 60 is used to indicate the intensity of one concentration of an odorant, a concentration that smells four times as intense would be assigned the number 240. If another concentration is perceived to be half as strong as the initial stimulus, it would be assigned the value 30. In some cases, a standard for which a number has been pre-assigned (often the middle stimulus of the series) is presented to the subject in an effort to make his or her responses more reliable. In other cases, the individual is free to choose any desired number system, so long as the numbers are made proportional to the magnitude of the attribute (the "free modulus method"). The important point is that, in this specific paradigm, the absolute values of the numbers are not the key element of interest, only the relations between the numbers.

To obtain an index of suprathreshold function, magnitude estimation data are most commonly plotted on log-log coordinates (log magnitude estimates on the ordinate and log odorant concentrations on the abscissa) and the best line of fit determined using linear regression. The resulting function, $\log P = n \log \Phi + \log k$, where P = perceived intensity, k = the Y intercept, Φ = stimulus concentration, and n = the slope, can be represented in its exponential form as a power function, $P = k\Phi^n$, where the exponent n is the slope of the function on the log-log plot. In olfaction, n varies in magnitude from odor to odor, but is generally less than 1, reflecting a negatively accelerated function on linear-linear coordinates (Figure 11.8). As noted elsewhere, various investigators have made modifications in these equations in an attempt to take into account differences in such factors as threshold sensitivity, vapor pressures, and adaptation (Doty, 1991; Jones, 1958; Overbosch, 1986).

It is noteworthy that magnitude estimation, like most other sensory procedures, can be biased or influenced in systematic ways by procedural and subject factors (Doty, 1991; Jones, 1966; Marks, 1974). The magnitude estimation task is relatively complex in that accurate responses to a stimulus require a good memory for the prior stimulus. If too much time lapses between the presentations of stimuli, the memory of the prior stimulus fades. On the other hand, if the trials are spaced too closely together, adaptation can distort the relationship. Not all subjects consistently provide ratio estimates of stimuli, and a number do not understand the concept of producing ratios (Baird et al, 1970; Moskowitz, 1977). Comparison stimuli can alter the stimulus function; for example, moderately intense odor is reported to be more intense when presented with weak com-

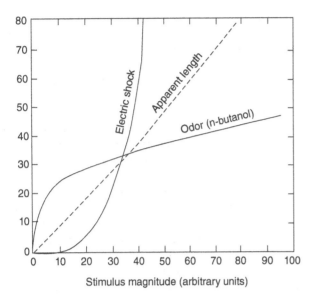

Figure 11.8 Relationship between perceived magnitude of three types of stimuli, as measured by magnitude estimation, and stimulus magnitude. Note that the perceived intensity of the example odorant increases in a negatively accelerated fashion, indicating a power function exponent less than 1 (i.e., in this case 0.33). Adapted and modified, with permission, from Stevens (1961). Copyright © 1961, MIT Press.

parison stimuli than with strong comparison stimuli) (Helson, 1964; Eyman et al., 1975). Moreover, the magnitude of the exponent is dependent on the choice of the stimulus scale (i.e., the units in which the stimulus concentration is expressed), although in olfaction this is probably of minor consequence (Myers, 1982). As in other sensory systems, the assignment of a standard will influence the obtained slope as will the range of presented stimuli. In general, subjects have a preferred range of numbers they assign to whatever range of stimuli that is presented. Sets of stimuli of a restricted range generally yield larger exponents than those of a more expansive range; that is, spreading the same set of numbers over a smaller stimulus range yields a steeper slope (Jones and Woskow, 1966).

The degree to which these and other potential shortcomings hinder the use of magnitude estimation procedures in applied settings, such as the clinic, is not known; however, it is likely that such problems can be minimized by ensuring that the instructions, test procedures, and test stimuli are carefully standardized and monitored. Comparative assessments of nine-point rating scales, line scales, magnitude estimation scales, and a hybrid of category and line scales suggest that, for untrained or mathematically unsophisticated subjects, category scales and line scales may be superior to magnitude estimation when such factors as variability, reliability, and ease of use are considered (Lawless and Malone, 1986a,b). For such reasons, the use of labeled scales that take into account the exponential aspects of sensory perception, such as the scale developed by Green et al. (1996), would seem preferable.

Since the magnitude estimation function's intercept and height above the origin depend on idiosyncratic differences in the use of numbers and the specific magnitude estimation method employed (e.g., fixed vs. free modulus), only its slope has traditionally been used as an index of sensory function. In an attempt to gain additional information from the function's ordinate position, investigators have employed the method of cross-modal magnitude matching, which provides, at least theoretically, information about the perceived intensity of stimuli from the absolute position of the magnitude estimation function and corrects, to some degree, for differences among subjects in number usage or other idiosyncratic responses (for a detailed discussion of this procedure, see Marks et al., 1988). In a common application of this method, judgments of the intensity of sensations from two sensory modalities (e.g., loudness and odor intensity) are made on a common magnitude estimation scale (Marks et al., 1986). Under the assumption that subjects experience stimuli on one of the continua (i.e., loudness) in a similar manner (an assumption that some question), differences among their loudness ratings would be expected to reflect differences in number usage. The odor intensity continuum can then be adjusted accordingly. Such normalization allows, theoretically, for a direct comparison of scale values across subjects; thus, if the adjusted odor intensity magnitude value for one subject is 10 and for another subject is 20 at the same concentration level, the second subject is presumed to experience twice the odor intensity as the first subject.

11.3.5 Quality Discrimination Tests

The most straightforward chemosensory quality discrimination test requires individuals to decide whether two stimuli have the same or different quality. In one scenario, a series of same-odorant and different-odorant pairs is presented, and the proportion of pairs that are correctly differentiated is taken as the measure of discrimination (Potter and Butters, 1980; O'Mahony, 1979a,b). Variants on this theme include picking the "odd" stimulus from a set from which only the "odd" stimulus differs, either at a constant delay interval (Frijters et al., 1980; Hummel et al., 1997) or at varying delay intervals (Choudhury et al., 2003), the latter allowing for the assessment of short-term memory in addition to basic discrimination. A recently developed discrimination test employs a set of 15 stimuli representing pairs of different ratios of two odorants from a set of six odorants (Weierstall and Pause, 2012).

Another form of discrimination test is based on a procedure called multidimensional scaling (MDS). In one variant of this procedure, ratings are made for all possible pairs of stimuli (or selected subsets of pairs) on a line scale anchored with descriptors like "completely different vs. exactly the same," and the correlation matrix among these ratings is subjected to an algorithm that places the stimuli in two or more dimensional space relative to their perceived similarities (e.g., Schiffman et al., 1981). The process is akin to constructing a map of a country from a list of distances available between the cities of that country. Persons with poor discrimination abilities fail to discern differences and similarities among stimuli, as illustrated by multidimensional spaces that have no distinct or reliable groupings. Because of its time-consuming nature and the fact that statistical procedures for comparing one person's MDS space to another's (or to a norm) are poorly worked out, MDS has not been used routinely in the clinic. Interestingly, when subjects are asked to rate the similarity of stimuli that are only indicated to them by name (i.e., the odorants, per se, are never presented), stimulus spaces derived by MDS are analogous to those obtained by the actual use of the odorants (Carrasco and Ridout, 1993; Ueno, 1992). This implies that well-defined imagery, or at least conceptual representations, exist for odorous stimuli.

Wise and Cain (2000) used a response latency approach to determine the discriminability of unmixed odors and mixed odors. A clear monotonic relationship was found between latency and accuracy, with latency decreasing with accuracy. In addition, subjects required more time and made more errors in discriminations between binary mixtures and their unmixed components than between the unmixed components. It was concluded that this approach may provide a novel measure of differences in odor quality, since latency provides information about discriminability.

11.3.6 Quality Recognition Tests

Two general classes of quality recognition tests can be defined. In the first class, the subject is asked whether each stimulus of a presented set is recognized. Identification is not always required. As indicated at the beginning of the chapter, this procedure is relatively crude, despite the fact that it is perhaps the most common means used by neurologists to measure olfactory function (e.g., Sumner, 1962). In the second class, a patient is presented with a "target" stimulus and subsequently asked to select the target from a larger set of stimuli. The number of correct responses of a series serves as the test score.

A variant on this theme is the stimulus matching task, in which a set of stimuli are provided and the subject is required to match the stimuli, one by one, to those of a set of identical stimuli. As an example, Abraham and Matha (1983) presented patients with eight vials that contained four odorants (two vials per odor). The task was to pair up the equivalent two-vial containers. The number of pairs correctly matched on each of two administrations of the test was the test score.

11.3.7 Odor Identification Tests

Among the most popular procedures for assessing taste and smell function are those that require stimulus quality identification. Such tests can be divided into three groups: naming tests, yes/no identification tests, and multiple-choice identification tests. The respective responses required, on a given trial, in these three classes of tests are: (1) to provide a name for the stimulus, (2) to signify whether the stimulus smells like an object named by the examiner (e.g., does this smell like a rose?), and (3) to identify the stimulus from a list of names or pictures.

Odor naming tests in which no response alternatives are provided have been used clinically (e.g., Gregson et al., 1981), but are of limited value since many normal individuals have difficulty in naming or identifying even familiar odors without cues. Yes/no identification tests are much more useful, since they require a patient to report whether or not each of a set of stimuli smells like a particular substance named by the experimenter. Two trials with each stimulus are usually given, with the correct alternative provided on one trial and an incorrect one on the other (e.g., orange odor is presented and the subject is asked on one trial whether the odor smells like orange and on another trial whether the odor smells like peppermint). Although such a test requires the patient to keep the percept in memory long enough to compare it with the target word (which, of course, must also be recalled from memory), some of its proponents argue that it is less influenced by cognitive and memory demands than multiple-choice identification tests (see below). Since chance performance on this type of test is 50% compared to 25% on a four-alternative multiple-choice identification test, its range of discriminability is lower and therefore more trials are needed to obtain the same statistical power as the multiple-choice odor identification test.

Numerous multiple-choice odor identification tests have been described in the clinical literature (Bensafi et al., 2003; Cain et al., 1983; Doty et al., 1984a; Gregson et al., 1981; Wood and Harkins, 1987; Hummel et al., 1997; Simmen et al., 1999; Nordin et al., 1998; Wright, 1987). These tests are conceptually similar and, in the few cases that have been examined, strongly correlated with one another (Cain and Rabin, 1989; Doty et al., 1994; Wright, 1987). The most widely used of these tests — the University of Pennsylvania Smell Identification Test (UPSIT) — examines the ability of subjects to identify, from sets of four descriptors, each of 40 "scratch and sniff" odorants (Doty, 1995; Doty et al., 1984a,b) (Figure 11.9). The number of correct items out of 40 serves as the test measure. This value is compared to norms and a percentile rank is determined, depending on the age and gender of the subject (Doty, 1995). This

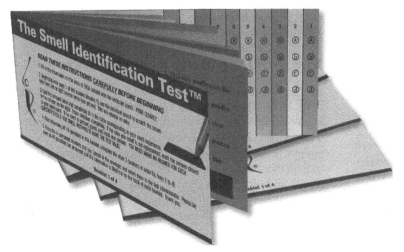

Figure 11.9 The University of Pennsylvania Smell Identification Test (UPSIT; known commercially as the Smell Test) (Doty et al., 1984a). This test, developed in the early 1980's, is comprised of 40 microencapsulated odorants located next to forced-choice questions on each page of 10-page booklets. A subject's test score can be compared to peers employing norms based upon ~4,000 persons, allowing for a determination of absolute loss (mild, moderate, severe, total) as well as a percentile rank relative to the age and gender of the subject. Copyright © 2004, Sensonics International, Haddon Heights, New Jersey.

test has several unique features, including amenability to self-administration and a means for detecting malingering (see Section 11.6). Furthermore, it is available in 15 different language versions. The popularity of this test is attested to by the fact that hundreds of scientific publications have arisen from its use by investigators from many laboratories and clinics.

Several odor identification confusion matrix tests have been described that are applicable to clinical settings (Köster, 1975; Wright, 1987). The test that has been most widely applied is that of Wright (1987). In Wright's test, each of 10 suprathreshold stimuli are presented to a patient in counterbalanced order 10 times (100 total trials). The response alternatives are the names of the 10 stimuli: ammonia, chlorine bleach, licorice, mothballs, peppermint, roses, turpentine, vanilla, Vicks vapor rub, and vinegar. No feedback as to the correctness of the subjects' responses is given. The percentage of responses given to each alternative for each odorant is determined and displayed in a rectangular matrix (stimuli making up rows and response alternatives making up equivalently ordered columns). Responses along the negative diagonal therefore represent correct responses, whereas those that fall away from the diagonal represent "confusions." The percentage of correct responses is used as the main test measure, although some of its proponents argue that the confusions (off-diagonal responses) may provide meaningful clinical information.

The main limitations of Wright's confusion matrix are its long administration time (approximately 45 min) and the lack of evidence that the off-diagonal responses provide any meaningful clinical information (although such responses may be of value in detecting malingering; see Kurtz et al., 1999). It would seem that if off-diagonal responses are to be sensitive to aberrations or distortions seen in most clinical cases, more subtle differences in the response alternatives need to be employed within the matrix. Should subtle aberrations be reliably categorized, this approach would have considerable clinical value.

11.3.8 Memory Tests

As described in detail in Chapter 15, the measurement of odor memory is not always straightforward, particularly in relatively short tests applicable to clinic settings. In a typical odor recognition memory test, a subject is required to smell an odorant or a small set of odorants (termed the target or inspection stimulus or stimulus set) and to select, after an interval of time ranging from less than a minute to hours or even days, that odorant or set of odorants from foils (distracters). Repeated trials may be performed at one or more retention intervals for each of several stimuli or sets of stimuli.

Unfortunately, despite attempts to minimize labeling of the inspection odor with a familiar word or item on the part of a subject, such labeling often occurs and, thus, what is being measured across intervals is the memory of the label, not the memory of the odor. In other words, once an individual recognizes an odor as that of an orange, for example, all that has to be remembered over time is the concept "orange," not the specific smell of the orange. Later, when given stimuli from which to select the earlier perceived odor, the subject simply looks for the smell of an orange (which has been known and stored in long-term memory for much of his or her life). In effect, the odor is not what is being uniquely remembered over the retention interval, only its name or concept and remembrance of having smelled the substance.

For this reason, investigators have attempted to employ novel, nondescript, and unfamiliar odorants in such tasks (Møller et al., 2004). Unfortunately, it is difficult to find target odors that are not labeled by subjects as pleasant or unpleasant, fruity or non-fruity, medicine-like or non-medicine like, chemical-like or non-chemical like, etc. Importantly, it is important to recognize that performance across the delay intervals is what is assessing the "memory" component of the task, not necessarily the overall test score. A number of odor memory tests are, in fact, essentially odor discrimination tests with varying inspection (delay) intervals. When scores on a nominal odor memory task differ between two groups (as evidenced by a main group effect in an analysis of variance), then a significant interaction term between delay interval and group would indicate an effect on memory, per se. Without an interaction with delay interval, the difference could simply reflect discrimination, not memory.

A number of examples of clinical applications of odor memory tests are available from the literature, although convincing evidence for a true odor memory deficit is lacking in most cases.

Campbell and Gregson (1972) developed a test of short-term odor memory in which four odors in a row were presented and the patient was asked if the fourth, which was the same as one of the first three, was equivalent to the first, second, or third odorant. No delay interval, per se, was defined between the presentation of the stimuli, but presumably the trials were presented closely after one another. Seven three-odor combinations of 12 inspection stimuli were administered. Patients who had difficulty with this task were subsequently given two-odor combinations. The test score was the number of odors that were consistently recognized by the subject. This test was shown to be sensitive to olfactory deficits due to schizophrenia (Campbell and Gregson, 1972), Kallmann's syndrome (Gregson and Smith, 1981), and Korsakoff psychosis (Gregson et al., 1981).

Jones et al. (1975) presented 20 pairs of odorants at 0- and 30-sec delay intervals to 14 alcoholic Korsakoff psychosis patients, 14 alcoholic controls, and 14 nonalcoholic controls. On a given trial, the subject's task was to report whether the second stimulus was the same as or different from the first. In the 30-sec delay interval, the subjects counted backward by threes in an effort to minimize semantic labeling. Since the Korsakoff psychosis patients performed significantly more poorly than did the control groups at *both* the 0- and 30-sec retention intervals, it is questionable whether odor memory, per se, is the main trait being influenced in this case.

Jones-Gotman and Zatorre (1993) reported that—in patients having undergone surgical cerebral extirpation for control of epilepsy—odor memory impairment was evident between the controls and two of the eight surgical groups evaluated; namely, those who had received excision from the right temporal or right orbitofrontal cortices. The memory task consisted of eight target odors and eight new foils, and the yes-no recognition testing was performed twice after the initial testing—20 minutes later and 24 hours later. The authors interpret their findings as evidence of a "right hemisphere predominance in odor memory."

11.3.9 Sniff Magnitude Test

In 2003, a novel test was developed which measures the change in respiration that occurs when an odorant, usually an unpleasant one, is smelled for the first time (Frank et al., 2003; Figure 11.10). Upon sniffing the canister, either fresh air or a malodor is released. Upon perceiving a malodor, subjects with a normal sense of smell inhibit their inhalation, with the timing and degree of the inhibition being related to the intensity and unpleasantness of the odor. The negative pressure is monitored by a tube at the base of the nose that is connected to a piezoelectric pressure transducer. Sniff pressure measurements are calculated every 10 ms once a sniff is detected and continue until return to ambient air pressure occurs. The *sniff magnitude ratio* is the ratio of the integrated sniff magnitude value given to a malodor to that obtained when sniffing non-odorized air. When suppression occurs more to an odorant than to air, this ratio is less than one.

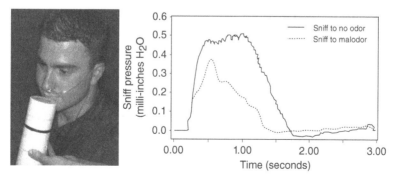

Figure 11.10 Left: Sniff Magnitude Test canister positioned for a sniff. Upon sniffing the top of the canister bad smelling odor or odorless air is released. The pressure changes during inhalation are signaled from the nasal cannula at the base of the nares to a pressure transducer. The digitized signal allows for the dynamics of the entire sniff epoch to be recorded by computer. Right: Top tracing represents a sniff to no odor (air); Bottom tracing represents a sniff to malodor. Note abupt attenuation of sniff response upon encountering the malodor. Reprinted from Frank et al. (2004). Copyright © 2004 Elsevier Inc.

This type of test may be particularly useful in detecting malingering and in assessing persons who have difficulty verbalizing their responses. It does require, however, that an individual perceives and reacts to odor quality. Correlations ranging from 0.33 to 0.57 have been found between sniff magnitude ratios and measures of odor identification, memory and threshold detection (Tourbier and Doty, 2007), although, like some psychophysical tests (see Tsukatani et al., 2005), it does not differentiate well between various degrees of hyposmia. Principal component analyses suggest that the SMT may evaluate aspects of olfactory function somewhat distinct from those measured by most other tests, probably reflecting suprathreshold hedonics. A hood and well ventilated room is required for this test, given the very unpleasant nature of its stimuli.

11.4 TEST RELIABILITY

The utility of an olfactory test reflects the degree to which it is reliable (consistent, dependable, or stable) and valid (accurately measures what it portends to measure). Related to a test's validity are its sensitivity (ability to detect abnormalities when present) and specificity (ability to detect abnormalities with a minimum number of false positives). Although a test cannot be valid without being reliable, the reverse is not the case; that is, a test can be reliable but not valid. Despite the fact that measures of test reliability and validity are available for many medical and psychological tests, this is not the case for most olfactory tests. Indeed, measures of validity (other than a few intercorrelations among different tests) are extremely rare; hence, in this section studies of reliability are emphasized (for more discussion on this point, see Schwartz, 1991).

The reliability of a test can be determined in several ways. First, the test can be administered on two occasions to each member of a group of subjects and a correlation coefficient computed between the test scores on the two occasions (termed the *test-retest reliability coefficient* or the *coefficient of stability*). Second, when parallel forms of a test are available, the two forms can be administered to the same set of subjects and a correlation coefficient computed between the two forms. Third, subsections of some types of tests (e.g., multiple-item odor identification tests) can be correlated with one another to provide an estimate of test stability. The test is viewed, in this case, as consisting of parallel forms and the resulting coefficient, when based upon the correlation of half of the items with the other half of the items, is termed the *split-half reliability coefficient*. Since reliability is related to test length, as will be noted below, a statistical correction for test length must be applied to the correlation coefficient obtained in this way to provide the estimated reliability coefficient for the full test

(Guilford, 1954). A widely employed statistic, Cronbach's coefficient alpha, assesses the internal consistency of a test and is mathematically equivalent to the average of all possible split-half reliability coefficents of the test's items, given certain assumptions (e.g., that the item standard deviations are equal) (Cortina, 1993).

The magnitude of reliability coefficients depends, to a large degree, on the variation of the test scores of the group upon which they are computed. If all members of a group score exactly the same on a test administered on two test occasions, the reliability coefficient would be suspect, since there is no inter-individual variability. If only a small variation occurs among the subjects, then the reliability coefficient may be spuriously low. Importantly, if test scores are included from normal and patient groups, the reliability coefficient can be inflated. Thus, in assessing reliability one must have some understanding of the variation among the test scores. Also, it should be noted that while a high reliability coefficient indicates that a group of individuals scored similarly *relative to one another* on a test from one test occasion to the other, all of the individual's test scores still may be lower (or higher) on the second than on the first test occasion. In other words, systematic changes in the test values can occur which are not reflected in the reliability coefficient. In such a case, a high reliability coefficient is misleading to the extent that the overall stability of the test may vary systematically over time.

Since it has been reported that olfactory thresholds vary considerably among individuals and evidence considerable day-to-day fluctuations within the same individuals (e.g., Stevens et al., 1988), one might expect their reliability to be suspect. Indeed, reliability coefficients for various threshold tests do vary considerable from study to study, and extremely low reliability coefficients have been reported in some cases (e.g., Heywood and Costanzo, 1986; Punter, 1983). Nonetheless, particularly in cases where repeated estimates of the threshold are obtained, respectable reliability coefficients are evident. In general, forced-choice odor identification tests with a relatively large number of items exhibit a high degree of reliability, whereas shorter identification tests evidence lower reliability. Very short tests (e.g., three items) may exhibit spuriously high reliability coefficients, since smell abilities are forced into three broad categories. In general, longer tests exhibit higher reliabilities. The relationship between test length and reliability is shown for a number of olfactory tests in Figure 11.11. A listing of a number of olfactory test procedures applicable to clinical situations is provided in Table 11.1, along with reported reliabilities when indicated in the literature. Note that many of these

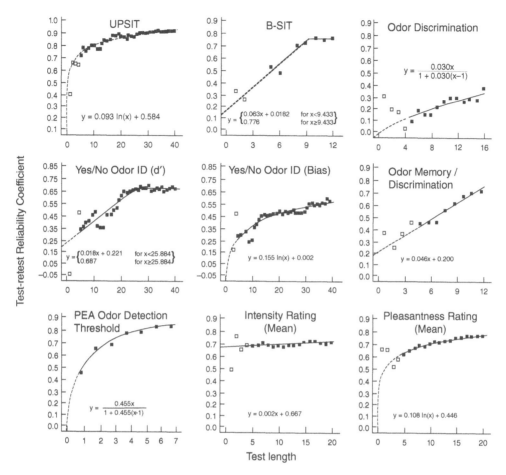

Figure 11.11 Relationship of reliability to cumulative test length for test measures amenable to such an evaluation. Best fit formulae are indicated. Modified from Doty et al. (1995). Copyright © 1995 Oxford University Press.

reliabilities are unlikely to be statistically different from one another.

11.5 OTHER CONSIDERATIONS

11.5.1 Comparing Test Results from Nominally Distinct Olfactory Tests

Many studies make comparisons across olfactory tests having different names and operational constraints, for example, tests of odor identification, detection, discrimination, and memory. One must keep in mind, however, that despite their different names, such tests are not mutually exclusive and are usually correlated with one another (Doty et al., 1994). Thus, a test of odor identification requires the ability to detect an odor, as well as to remember the odor. A forced-choice threshold test requires a comparison of an odor or a weak sensation with alternatives, a task that again requires memory as well as discrimination. Differing degrees of semantic labeling may occur in odor discrimination tasks, just as occurs in odor identification tasks, again pointing to communalities among the procedures.

When differences are observed in findings among different "types" of olfactory tests, caution is warranted in their interpretation, not only for the reasons noted above, but because of test-related differences in reliability or sensitivity, employment of different odorants, and variations in non-olfactory operational elements of testing (e.g., ability to discriminate or identify independent of the ability to smell) (Hedner et al., 2010). Ideally, non-olfactory controls (e.g., tests in another modality to require the same operational tasks) should be employed to be certain that any deficits that are observed on a given test are not due to such non-olfactory factors. This is particularly true when mentation may be altered, as in head trauma or dementia. It is for this reason that we often employ the Picture Identification Test, a test essentially identical to the UPSIT but which uses pictures rather than odors as test stimuli (Vollmecke and Doty, 1985). If persons do well on this test, we know that the putative olfactory deficits are not due to lack of understanding the concepts involved in the test or to the non-olfactory operational procedures involved in taking the test.

242   Chapter 11   Psychophysical Measurement of Human Olfactory Function

Table 11.1 Basic information for modern olfactory tests amenable to clinical settings. Abbreviations: ID, identification; DT, detection threshold; RT, recognition threshold; IR, intensity rating; DIS, discrimination; MEM, memory

Test	Type of Test	# of Odors or Items	Test-Retest Reliability Coefficient	Estimated Test Duration	Availability of Norms
Alberta Smell Test (Green & Iverson, 1998)	ID	8	Not Reported	10 min	NO
Alcohol Sniff Test (Davidson et al., 1998; Davidson & Murphy, 1997)	DT	1	0.80	<5 min	NO
Amoore Threshold Test (Amoore & Ollman, 1983)	DT	1	0.70	<10 min	NO
Barcelona Smell Test (Cardesin et al., 2006)	DT & RT	24	Not Reported	~30 min	NO
Biolfa® Olfactory Test (Bonfils et al., 2004)	DT & RT	3 & 8	Not Reported	~30 min	NO
Brief Smell ID Test™ (aka Cross-Cultural Smell ID Test) (Doty et al., 1996; Liu et al., 1995)	ID	12	0.73	<5 min	YES
Candy Smell Test (Renner et al., 2009)	ID	23	0.75	~20 min	Limited
Connecticut Chemosensory Clinical Research Center (CCRC) Test (Cain et al., 1983)	ID & DT	10 ID; 1 DT	DT: 0.36 ID: 0.60[1]	~30 min	Limited
Combined Olfactory Test (COT) (Robson et al., 1996)	ID & DT	9 ID; 1 DT	0.87	10 min	NO
Dusseldort Odour Discrimination Test (Weierstall & Pause, 2012)	DIS	15[2]	0.66	15 min	NO
European Test of Olfactory Capabilities (ETOC) (Thomas-Danguin et al., 2003)	ID & DT	16	0.90	< 20 min	NO
Italian Olfactory Identification Test (IOIT) (Maremmani et al., 2012)	ID	33	0.96 & 0.99[3]	10–15 min	Limited
Jet Stream Olfactometer (Ikeda et al., 1999)	ID	8	Not Reported	<5 min	NO
Jones' 6 single Ascending Series Threshold Tests (Jones, 1955)	DT (6 repeats)	3	0.61, 0.77, 0.80	15–20 minutes/ DT	NO
Koelega threshold test (Koelega, 1979)	DT	1	0.65 bilateral; 0.51 right side; 0.59 left side	15–20 minutes/ DT	NO
Kremer Olfactory Test (Kremer et al., 1998)	ID	6	Not Reported	<5 min	NO

11.5 Other Considerations

Table 11.1 (*Continued*)

Test	Type of Test	# of Odors or Items	Test-Retest Reliability Coefficient	Estimated Test Duration	Availability of Norms
Le Nez du Vin (McMahon & Scadding, 1996)	ID	6	Not Reported	<5 min	NO
Monell Extended Sniffin' Sticks Identification Test (MONEX-40) Freiherr et al., 2012)	ID	40	0.68	~15 min	NO
Odor Confusion Matrix OCM) (Kurtz et al., 2001; Wright, 1987)	ID	10	0.94	~1 hr	NO
Odor Identification Test for Children (Laing et al., 2008; Armsstrong et al., 2008).	ID	16	0.98	~5 min	Limited
Odor Discrimination/ Memory Test™ (ODMT) (Bromley & Doty, 1995; Choudhury et al., 2003)	OM & DIS	12	0.68	15 min	Limited
Odor Stick ID Test (Hashimoto et al., 2004; Kobayashi et al., 2004).	ID	13	0.77	<15 min	Limited
Pediatric Smell Wheel (Cameron & Doty, 2013)	ID	11	0.70	5 min	Limited
Pocket Smell Test™ (PST) (Duff et al., 2002; Makowska et al., 2011; Solomon et al., 1998)	ID	3	0.80 & .85[4]	<1 min	NO
Quick Smell Test (Q-SIT) (Jackman & Doty, 2005)	ID	3	0.87[4]	< 1 min	NO
Quick Sniff Test (Gilbert et al., 2002)	ID & IR	1	ID Not Reported; IR: 0.51	<1 min	NO
q-Sticks (Hummel et al., 2010)	ID	3	Not Reported	<1 min	NO
Random Threshold Test (Kobal et al., 2001)	RT	2; 16 concentrations	0.71	~10 min	NO
San Diego Odor ID Test (Anderson et al., 1992; Krantz et al., 2009)	ID	6	0.86	~10 min	NO
Scandinavian Odor ID Test (SOIT) (Nordin et al., 1998)	ID	16	0.79	10–15 min	Limited
Smell Diskettes (Simmen et al., 1999)	ID	8	Not Reported	<5 min	NO
Smell Threshold Test™ (Doty 2000)	DT	1	0.88	~20 min	YES
Sniffin' Sticks Test (Kobal et al., 1996)	ID, DT, Disc	12	ID:0.73, DT: 0.54; Comb: 0.72[5]	~60 min for whole test	YES
T&T Olfactometer Test (Takagi, 1989)	DT & RT	5	0.22–0.45[6]	15–30 min	Limited

(*continued*)

Table 11.1 (*Continued*)

Test	Type of Test	# of Odors or Items	Test-Retest Reliability Coefficient	Estimated Test Duration	Availability of Norms
University of Pennsylvania Smell Identification Test (UPSIT; aka Smell Identification Test) (Doty, 1995; Doty et al., 1984a)	ID	40	0.94	10–15 min	YES
Utrecht Odour ID Test (Geur Identificatie Test Utrecht (GITU) (Hendriks 1988)	ID	18 or 36	0.68–0.77	30–45 min	NO
Viennese Odor Test (Lehrner & Deecke, 2000)	ID	20	0.75	15 min	NO

[1]Cain and Gent (1991) computed correlations between left:right sides of the nose (which some might consider an index of reliability) were much higher—butanol, 0.68; phenyl ethyl methyl ethyl carbinol, 0.96; isoamyl butyrate, 0.86; pyridine, 0.83.

[2]Reflects 15 pairs of stimuli, each pair being a mixture of two odorants selected from a total of six chemicals.

[3]Based upon sample containing normal subjects and patients with Parkinson's disease.

[4]These values are inflated given that only three response scores (1,2, or 3) were possible.

[5]When computed in a study population containing normosmics and patients with olfactory dysfunction, these values increased to 0.80 (discrimination), 0.88 (identification) & 0.92 (threshold) (Haehner et al., 2009).

[6]Doty et al. (1995) found test-retest reliability coefficients for DTs of the five odorants of this non-forced choice test (skatole, isovaleric acid, γ-undecalactone, β-phenyl ethanol, cyclotene) to range from 0.56 to 0.71; RT coefficients ranged from 0.22 to 0.45.

This brings up the question of whether one should add the results of tests of nominally different domains together, such as commonly done with Sniffin' Sticks in producing a "TDI" score (Hummel et al., 1997). There are, in fact, numerous problems with this simple additive approach. The most obvious one is that if one assumes that the elements to be combined are measuring different entities (e.g., identification, detection, discrimination), one person may do well on identification but poorly on detection, for example, and another may do well on detection but poorly on identification, with both ending up with the same combined score for quite different reasons. Thus, simply adding test scores together results in differential arbitrary weightings of the test components for each subject—weightings that are differentially influenced by each test's response scale or range of test scores. Importantly, the tests within the composite have different non-olfactory test demands, employ different stimuli, and have different reliabilities. The proper statistical solution, if one feels the need to combine such data, would be to convert the scores of each test to standardized z-scores and then perform the addition or averaging of the scores. One is still left, however, with the issue of knowing the relative contributions of the various tests for a given composite score, making the composite difficult to conceptually interpret.

11.5.2 Unilateral Versus Bilateral Testing

Most individuals with chemosensory dysfunction evidence the dysfunction bilaterally (Cain and Rabin, 1989; Gudziol et al., 2007). In cases where unilateral losses are present, they often go unnoticed. When time is at a premium, bilateral testing is preferable to unilateral testing since it reflects clinically meaningful deficits. However, there are a number of occasions when unilateral olfactory testing is of considerable value (e.g., in the detection of some types of tumors and in differentiating between some disease subtypes) (Doty, 1979; Good et al., 2002; Mainland et al., 2005), and the ideal assessment of a patient includes unilateral, as well as bilateral, testing. In normosmic subjects, directionally lateralized left:right differences in threshold and identification test scores are not present (Betchen and Doty, 1998; Brand et al., 2004; Good et al., 2003; Gudziol et al., 2007). Interestingly, odor recognition memory is better under bilateral than unilateral test conditions (Bromley and Doty, 1995) and, in the menopause, estrogen replacement therapy appears to influence asymmetry noted on such memory tasks (Doty et al., 2008).

Unilateral testing is straightforward. Although it is possible to present a stimulus to one naris and obtain mainly unilateral stimulation, the possibility of the crossing of odorant to the contralateral side within the rear of the nasopharynx upon exhalation cannot be excluded. Thus, it is prudent to close the contralateral naris without distorting the septum [e.g., by using a piece of Microfoam™ tape (3M Corporation, Minneapolis, MN) cut to fit tightly over the borders of the naris] and have the patient exhale through the mouth after inhaling through the nose (Doty et al., 1992). As in the case when both nares are blocked, this precaution decreases the likelihood for air to enter the blocked nasal chamber via the nasopharynx.

Furukawa et al. (1988) noted that seven of 94 patients (7%) they examined, all of whom evidenced no bilateral threshold deficits, evidenced significant unilateral threshold deficits. They reported a similar phenomenon in six of 12 patients who had had brain surgery. Of 82 consecutive nonanosmic patients presenting to the University of Pennsylvania Smell and Taste Center with chemosensory dysfunction, 14 (i.e., 17%) were observed whose unilateral detection thresholds were discrepant from one another by at least three orders of magnitude (Doty, unpublished). Interestingly, nine of these 14 individuals were anosmic on one side of the nose, even though only three had bilateral detection threshold values that were obviously abnormal. Larger more recent studies have suggested that up to 25% of patients presenting with chemosensory disturbance exhibit meaningful left:right differences in olfactory test scores (Welge-Lüssen et al., 2010) which, interestingly, may precede the onset of bilateral dysfunction (Gudziol et al., 2010) and may relate to lateralized differences in nasal chamber and olfactory bulb volumes (Altundag et al., 2014).

11.5.3 Detection of Malingering

Because considerable compensation can be involved in accident cases that alter the ability to smell, malingering on chemosensory tasks is not uncommon. It is frequently suggested in the medical literature that if a patient cannot readily perceive the vapors from an irritating substance presented to the nose, he or she is malingering (e.g., Griffith, 1976). However, this is not a definitive method for detecting malingering. Thus, individuals who, on other grounds, are believed to be feigning anosmia usually have difficulty in denying experiencing the effects of NH_4 or other irritants, particularly since these stimuli often produce eye watering, coughing, and other reflexes that are manifested overtly. Furthermore, there appears to be considerable variability among normal individuals in trigeminal responsiveness to such stimulants and persons with smell loss may have less trigeminal sensitivity (Frasnelli and Hummel, 2007).

A more valid approach for detecting cheating on the basis of psychophysical testing is to examine response strategies of patients on forced-choice tests, since malingerers often avoid the correct response more often than expected on the basis of chance. This is well illustrated by responses to the UPSIT. Since the UPSIT is a four-alternative forced-choice test, approximately 25% of the test items (i.e., 10) are correctly answered, on average, by an anosmic, although obviously there is a sampling distribution around this number. The probability of scoring 5 or less on the UPSIT and not having at least some ability to smell is less than 5 in 100. The probability of scoring zero on the UPSIT and having no sense of smell is less than 1 in 100,000.

Electrophysiological testing, such as described in Chapter 12, can be useful, in some cases, in detecting malingering. However, it must be kept in mind that electrophysiological measures are not immune to malingering (e.g., patients can move or provide less cooperation during testing) and care must be made to differentiate between subtle EEG changes, such as event-related potential amplitudes or latencies, that arise from the trigeminal nerve and perhaps other sources. Considerable variability can be present in such measurements. The Sniff Magnitude Test, described earlier in this chapter, may also be useful in detecting malingering, since the presented odorants are so vile that it is difficult not to suppress a sniff once they are perceived. To date, no one has explored the usefulness of this novel test in such detection.

It should be noted that chemosensory malingerers differ in many ways from psychiatric malingerers. While the latter typically exaggerate their general health problems, chemosensory malingerers do the opposite. Thus, in a study which compared responses to an intake questionnaire of 22 chemosensory malingerers (based on improbable responses in forced-choice testing) to 66 non-malingerers matched on etiology, the malingerers significantly exaggerated symptom severity, including weight loss, appetite change, and interference with everyday activities, and underreported factors that might be construed as contributing to their dysfunction, including smoking behavior, medication use, nasosinus problems, dental problems, nasal or oral surgical operations, and allergies (Doty and Crastnopol, 2010). Interestingly, the two groups did not differ on sex, education, and how many words were used in an intake questionnaire to describe the purported chemosensory problem.

11.5.4 Subject Variables

The reader should be aware that numerous factors influence olfactory function in "normal" individuals and that these factors can significantly alter the ability to smell. Among variables that meaningfully alter such ability are age, gender, and smoking habits. Of these three variables, age is the most important (for review, see Doty and Kamath, 2014). Indeed, over the age of 80 years nearly three out of four persons exhibit marked olfactory dysfunction; half of those between the ages of 65 and 80 years evidence such dysfunction (Doty et al., 1984b; see also Murphy et al., 2002).

Estimates of heritability of olfactory function are less in old persons than in young persons (Doty et al., 2011), likely reflecting, among other things, cumulative damage to the olfactory epithelium from environmental insults, such as from viruses and pollution. Age-related declines in olfactory performance are observed for a variety of olfactory tests, including tests of odor detection threshold, identification, discrimination, adaptation, and suprathreshold odor intensity perception (Doty and Kamath, 2014). In addition, age influences the responsiveness of the nasal mucosa to volatile chemicals that produce irritation and other skin sensations (Stevens and Cain, 1986). In general, (a) large individual differences are present in the test scores of older individuals, (b) olfactory dysfunction is most evident after the sixth decade of life, and (c) women, on average, exhibit age-related declines in odor perception later in life then do men. The age-related decline in the ability to smell is not inconsequential. Thus, in older persons, an olfactory deficit, particularly in combination with the ApoE-e4 allele, has been associated with greater subsequent cognitive decline (e.g., Graves et al., 1999; Olofsson et al., 2009). Importantly, a disproportionate number of older persons die from accidental gas poisoning (Chalke et al., 1958), and many complain that their food has no flavor (Doty et al. 1984b). The latter phenomenon, which can lead to decreased interest in food, may explain some cases of age-related nutritional deficiencies and weight loss. As documented clinically (e.g., Deems et al., 1991), decreased "taste" perception during deglutition largely reflects the loss of stimulation of the olfactory receptors via the retronasal route (Burdach and Doty, 1987; Mozell et al., 1969).

In general, women of all ages outperform men on tests of odor identification, detection, discrimination, and suprathreshold intensity and pleasantness perception (for review, see Doty and Cameron, 2009). Such differences are present for a wide variety of odorants, including human breath and bodily secretions (Doty et al., 1975, 1978b, 1982), and are observed as early as such testing can be reliably performed (Doty, 1986). The fact that female babies more readily show a preference for odors from their own mothers than do male babies suggests that such sex differences are present at birth and are either inborn or due to early sexually dimorphic developmental influences (Makin and Porter, 1989; see Chapter 14).

The influence of tobacco smoking on olfactory function is less marked, on average, than that of age or gender (e.g., Doty et al., 1984b). This influence, however, is dose-related and present in both previous and past smokers (Frye et al., 1990). Moreover, second-hand smoke can adversely influence the ability to smell, even in children (Nageris et al., 2002). Interestingly, cessation from smoking results in some improvement of olfactory function over time—improvement that is related to the amount of previous smoking and the duration of such cessation (Frye et al., 1990).

Both reversible and irreversible changes in smell function have been observed following exposure to a wide variety of environmental agents, including industrial chemicals and dusts (see Chapter 20). In the most extensive study on this point, the olfactory function of 731 workers at a chemical plant that manufactures acrylates and methacrylates was tested (Schwartz et al., 1989). Decrements in odor identification test scores proportionate to the estimated dose exposure levels of these acrylates were found. Interestingly, individuals who had never smoked cigarettes but who had been exposed to acrylates were six times more likely than their non-exposed counterparts to evidence olfactory decrements. Long-term decrements in smell function of workers exposed the airborne toxins during the cleanup of the World Trade Center debris in 2001 have been documented (Altman et al., 2011; Dalton et al., 2010).

Prior experience with odors, particularly that obtained on taste and smell organoleptic panels, clearly influences measures of the ability to smell. For example, repeated testing within the perithreshold odorant concentration range results in decreased thresholds or enhancement of signal detection sensitivity measures (Dalton et al., 2002; Doty et al., 1981; Engen, 1960; Rabin and Cain, 1986; Wysocki et al., 1989); practice with feedback influences the ability to name odors (Desor and Beauchamp, 1974; Engen and Ross, 1973). Interestingly, the hedonic quality of odorants can be influenced by repeated exposure, making unpleasant odors less unpleasant and pleasant odors less pleasant (Cain and Johnson, 1978). Assuming that adaptation is not the primary basis for this phenomenon, affective components of odors appear to habituate somewhat independently of odor intensity.

Bodily conditions such as hunger or satiety also have been reported to influence measures of olfactory function. For example, Albrecht et al. (2009) reported the paradoxical finding that thresholds for the food-related odor amyl acetate were actually lower under conditions of satiety than hunger, an effect not observed for the non-food related odorant n-butanol. In contrast, Cameron et al. (2012) found that fasting for 24 hours enhanced performance on tests of odor identification, detection, and discrimination. However, Janowitz and Grossman (1949) found no evidence that hunger or satiety altered recognition thresholds for coffee odor, as measured using the blast injection method.

11.5.5 Adaptation

Exposure to an odorant, if recent, of sufficient strength, and relatively continuous, can produce a temporary decrease in its ability to be perceived, empirically reflected by heightened detection threshold values or decreased intensity ratings (for a review, see Cometto-Muñiz and Cain, 1995).

Some chemicals produce a decrement in the perception of other chemicals (termed cross-adaptation). Fortunately, most modern clinical olfactory tests are either little influenced by adaptation or operationally are standardized in such a way that any adaptation that occurs is unlikely to meaningfully influence the test results. For example, the UPSIT was designed to minimize adaptation by (a) employing largely multicomponent "natural" odorants, (b) requiring minimal sampling of each odorant, (c) having verbal, rather than odorous, response alternatives, (d) ordering the presentation of odorants such that dissimilar odorants follow one another (thereby minimizing cross-adaptation), and (e) allowing adequate time between the smelling of each odorant item (Doty et al., 1984a).

Several general rules have emerged from studies of adaptation that are worthy of note (Cometto-Muñiz and Cain, 1995; Frank et al., 2010; Goyert et al., 2007; Jacob et al., 2003; Köster and De Wijk, 1991; Stuiver, 1958). First, the amount of adaptation induced is a function of the duration of exposure and the concentration of the adapting stimulus. Second, the subject's attention level influences the degree of adaptation. Third, pleasant odors may adapt less than unpleasant ones, although intensity is often confounded with pleasantness. Fourth, the rate and degree of recovery from adaptation are a function of the magnitude and duration of the adapting stimulus. Fifth, cross-adaptation is most commonly asymmetrical. For example, while exposure to odorant A decreases the perceived intensity of odorant B, exposure to odorant B may not decrease the exposure to odorant A to the same degree. Sixth, the sensitivity to a given odorant is reduced more by the exposure to that odorant than to any other odorant. Seventh, an odorant may have a larger adapting effect on the sensitivity to another odorant than it does on itself. Eighth, the sensitivity to an odorant which self-adapts strongly is usually also reduced strongly by other odorants. Ninth, adaptation of one side of the nose produces adaptation, albeit less, in the other side of the nose. Tenth, adaptation to complex odorants (i.e., odorants made up of more than one chemical) is generally less than adaptation to single-component odorants. Eleventh, adaptation to a component odorant of a mixture can "unmask" the perception of other odorants within the mixture, a process which varies as a function of exposure time to the adapting stimulus. Finally, adaptation to odorants can be relatively rapid. For example, Aronsohn (1886) found that subjects continuously exposed to the vapors of lemon or orange oil reported complete loss of olfactory sensations, on average, in three minutes (range: 2.5 to 11 min). Recovery occurred in about the same time as that required to produce the loss.

The issue of adaptation comes up in practical matters regarding the direction of trial sequences in threshold tasks, that is, ascending vs. descending trials. In one study it was found that better performance, that is, more frequent stimulus detection, occurred at lower perithreshold concentrations for the odorant phenyl ethyl alcohol relative to a mineral oil blank when the trials began at higher concentrations and subsequently descended to lower concentrations (Doty et al., 2003). Conversely, better performance occurred at higher perithreshold odorant concentrations when trials began at lower concentrations and ascended to higher concentrations. Importantly, no meaningful differences between ascending and descending trials were evident at concentrations in the range that would typically be used in the calculation of a threshold, that is, at 75% correct performance in a two-alternative forced-choice task. This suggests that threshold estimates are little influenced by the direction of trial sequences, at least for phenyl ethyl alcohol.

11.5.6 Responsivity Versus Sensitivity

A number of persons report being hypersensitivity to odors, such as women in the first stages of pregnancy (Cameron, 2014) and persons who report hypersensitivity to environmental odors (Doty et al., 1988). However, such hypersensitivity is not generally detected by odor identification or odor detection threshold tests (Caccappolo et al., 2000; Doty et al., 1988; Nordin et al., 2005). To date, such persons have not been evaluated using signal detection analysis, but it is conceivable that, in some cases, their "sensitivity" to odors reflects response factors and personality. Karnekull et al. (2011) recently demonstrated, in a non-clinical population, a positive correlation between neuroticism and both negative responses to environmental odors and noise. Those highest on the measure of neuroticism also rated odors presented to them as more intense. Other work suggests that beliefs about odors can determine the amount of attention paid to them and how strongly they are rated. For example, when subjects are told they are being exposed to an unhealthy odor, their ratings of its intensity are greater, and the time required for adaptation is longer, than if they are told the odor is healthy (Dalton, 1996, 1999; Zucco et al., 2008). Such findings demonstrate how attitudes towards odors can influence their perception.

11.6 THE PERCEPTION OF ODORANT MIXTURES

As noted above, a number of modern olfactory tests, including the UPSIT, employ stimuli that, for the most part, are complex mixtures of chemicals, mimicking stimuli encountered in everyday life. More often than not such stimuli are perceived as a unitary gestalt, and given a name associated with the object or source from which they are known to emanate; for example, cinnamon, pizza, cheese, gasoline, orange, lemon, walnut, etc. (see Livermore and Laing, 1998b). There is considerable clinical utility in using such tests, since many receptor types are activated. This is in contrast to threshold tests employing single odorants, as they presumably examine the responses of the olfactory system to a much smaller subset of the ~450 receptors. It has been shown that rodents who have sustained damage to 80–90% of their olfactory receptor cells still retain their ability to detect some single odorants. Similarly, odor sensitivity is retained unchanged when large lesions have been made in the bulb. Therefore, from at least a theoretical standpoint, major changes in the olfactory system can occur and not be detectable by the use of some single odorants. In contrast, the perception of mixtures invariably involves inhibitory interactions at the bulb (and possibly other olfactory centers) that occur through complex neural circuitry. Lesions that disrupt the circuitry are likely to alter the characteristic suppression effects observed between odors in mixtures. Rat data indicate that lesions involving much of the bulb can result in the failure to re-learn a mixture analysis task, compared to their successful retention of odor sensitivity and ability to discriminate between odor qualities (Slotnick et al., 1997).

How is it that mixtures of chemicals end up providing a largely unitary perceptual gestalt? How much information, in terms of discriminating individual components, can humans obtain from complex mixtures? If one odorant suppresses the odor of another (as is seen in the case of deodorants or room fresheners), how does this relate to the relative concentrations of the odorants within the mixture? Are there psychophysical rules or laws explaining mixture relationships? These and other questions related to odorant mixtures are the basis of the remainder of the chapter.

11.6.1 Effects of Mixing Odorants on their Perceived Intensity

Usually when two single compound odorants are mixed together, the perceived intensity of one or both is altered substantially, the net result being a lowering of the intensity of the components. However, on rare occasions enhancement may occur. In early mixture studies, Aronsohn (1886) reported that the odor of camphor was neutralized by such odors as gasoline, cologne water, and oil of juniper, and Nagel (1897) found that counteraction between two odorants could result in both being rendered almost odorless. Zwaardemaker (1900), the most famous of early olfactory scientists, confirmed these observations for a number of mixtures using an olfactometer (Figure 11.1), and demonstrated that the extent of perceptual interactions between two odorants was more dependent on their concentrations than on their qualities. Similar results have been reported

by others, including Moncrieff (1959) and Jones and Woskow (1964), the latter reporting that the perceived intensity of a binary mixture, although less than the sum of its component intensities, is more than a simple average of the two.

Zwaardemaker (1930) conceptualized the mutual weakening of the perceived intensity of a mixture of two components as follows: "The two sensations can be imagined as two vectors representing two forces counteracting each other in our intellect." The interaction between two odorants was later formalized by Berglund et al. (1973) in a mathematical model that incorporated the application of vector addition to odor mixtures for the prediction of the overall intensity of mixtures. Although the vector model has received widespread attention (e.g., Berglund, 1974; Berglund et al. 1976; Cain and Drexler, 1974; Cain, 1975; Moskowitz and Barbe, 1977; Laing et al. 1984; Berglund and Olsson, 1993a; Berglund and Olsson, 1993b; Olsson, 1994), after decades of investigation its best predictions have been for simple binary mixtures. Other models for predicting the perceived intensity of simple mixtures include the Strongest Component Model and the U (Patte and Laffort, 1979) and UPL (Laffort and Dravnieks, 1982) models. Both represent modifications of the vector model but did not widen the application of the original model to multi-component mixtures. The most recent model in this series was the UPL2 model (Laffort, 1989) which incorporated the power function that normally relates perceived odor intensity to concentration. The ERM model of Schiet and Frijters (1988) was also based on a power function relating these factors and, although applied with some success to simple gustatory mixtures, was not an improvement in the models just described for olfactory mixtures. As summarized by Cain et al. (1995), "the principle by which psychophysical information on single components reflects itself in a model of interaction seems to evade the psychophysical models presented here" (all the above).

Clearly, none of the aforementioned models adequately describes the changes in perceived intensity for all pairs of odorants examined, and none have been demonstrated to reliably predict the intensity of mixtures containing more than two odorants. Booth (1995) provides an interesting critique on the modeling of odor interactions, but provides no firm ground for future studies to proceed. Amongst a number of shortcomings, none of the above models have been based upon the receptive and neural processes that underlie the perception of mixtures, nor has attention been given to choosing odors that have physicochemical features that might provide some basis for antagonistic interactions. Furthermore, these models have provided no insight as to the nature of the sensory processes, and none adequately predicts the intensity of multi-component mixtures. Present evidence suggests that addition or partial

11.6 The Perception of Odorant Mixtures

addition of the perceived intensities of the components of mixtures occurs with binary and ternary mixtures; beyond this number of components neural processes limit intensity addition (Berglund et al., 1976; Moskowitz and Barbe, 1977; Laing et al., 1994).

The interactions noted above concern suprathreshold concentrations of odorants and provide examples of where the sense of smell compresses rather than adds intensity information. In contrast, additivity of neural input appears to be inherent in mixtures containing only sub-threshold quantities of odorants (Laska et al., 1990; Laska and Hudson, 1991). Indeed, in mixtures with only three odorants, the magnitude of the addition was noted by Laska et al. to be substantial and to often exceed that obtained from simple summation. Patterson et al. (1993) reports instances of near-true additivity of sub-threshold components, and suggested additivity may function to enhance sensitivity to the typically complex (and often sub-threshold) odor stimuli encountered in everyday life. They noted that the number of chemicals activating the system could be as important as the strength of any one of the odorants, providing a type of "biological economy" of the input.

11.6.2 Discrimination of Components in Odorant Mixtures

Since, as mentioned earlier, odors are commonly encountered as mixtures in our environment, an important characteristic of the human sense of smell is to discriminate differences between mixtures. Discriminating the odors of fresh and "off" milk, ripe and overripe fruit, cork taint in wine, and various perfumes are examples. In the area of pollution control, changes in the complex odor of sewage provide engineers with an insight as to the part of the treatment process that is not functioning properly; sulfides emanate if the anaerobic process is malfunctioning, and sour, rancid, and acid odors appear if the sludge treatment is inappropriate. In studies with binary mixtures, Rabin (1988) and Rabin and Cain (1989) showed that humans are particularly sensitive to the presence of small amounts of odorants that are not normally found in a stimulus. They reported that (a) high familiarity with a major component and the ability to label it consistently facilitates the detection of a minor component, (b) the minor component is not detected as readily if it is unfamiliar, and (c) unpleasant stimuli are more detectable than pleasant ones, although the effect was not as large as the effect of familiarity. Experience, therefore, and to a lesser extent pleasantness, improves discrimination between two single odorants or two mixtures, suggesting that similar cognitive processes operate with the two types of stimuli.

Although the Rabin studies suggested humans are very sensitive to small changes in an olfactory stimulus, Laska and Hudson (1992) reported that relatively large changes in the composition of mixtures are sometimes required for discrimination to occur. Thus, discrimination of 3-, 6-, or 12-component mixtures from the same mixtures minus one component produced error levels of 20–40%, with the level depending on the type of odorant that was removed. Accordingly, the dependence on the type of odor removed precluded defining a limit in the ability of humans to discriminate between two complex mixtures.

11.6.3 Identification of Components in Odorant Mixtures

Prior to studies of the abilities of humans to analyze mixtures, informal information from perfumers and flavorists suggested that between 5 and 30 components may be identified in mixtures (Laing, 1994b). Over the past decade, it has become clear that these numbers are an overestimate, as most individuals, including perfumers, are only able to identify up to three, or rarely, four components. An early hint that only a small number of odorants can be identified in mixtures was apparent in the report by Berglund (1974), who suggested from studies of the addition of the perceived intensities of components, that an analytic or additive process occurred up to three components, whereupon above this number an interactive process predominated. The latter was apparent as an asymptote in the total perceived intensity of a mixture, with little change occurring as the number of components increased. In accord with this notion, Moskowitz and Barbe (1977) found that in some instances the overall intensity of five-component mixtures was less than that of mixtures with fewer components. More recently it was shown that mixtures with 30 or more equal intensity components smell alike, suggesting that the single non-descript odor represents an 'olfactory white' that is equivalent to white noise in addition (Weiss et al., 2012).

In perhaps the first formal scientific studies on this topic, Laing and Francis (1989) and Livermore and Laing (1996) reported that training and experience did not increase the number of components identified with subjects that had been trained for a few minutes, three weeks, or who were perfumers and flavorists, the maximum still being three to four. Varying the task or the odorants resulted in no improvement in the number identified; thus, in another study a selective attention procedure was not more efficient than a procedure which required subjects to identify as many components as possible during an an lib sampling method (Laing and Glemarec, 1992). Futhermore, this maxima was not altered if the odorants used were those classified by perfumers as 'poor blenders,' that is, odorants that they used to 'stand out' in mixtures (Livermore and Laing, 1998a). Schiet and Frijters (1988), using another approach to this problem, reported that subjects invariably underestimate the number of components in mixtures containing up to four components. A similar result was obtained by Jellinek and Köster (1979), whose subjects found the odor of single chemicals to be as complex as that of mixtures.

Clearly, the data of the aforementioned studies indicate that there is a significant limitation in the olfactory system in the processing of information from more than about three odorants. Mixtures of complex odors tend to behave like mixtures of single odorants, with a maximum of about three being identified in stimuli containing up to eight complex odors. The possibility that the entry of hundreds, perhaps thousands, of odorants into the nose would produce a non-identifiable smell sensation has not yet eventuated (Livermore and Laing, 1998b).

11.6.4 Mechanisms Involved in Odor Mixture Perception

The limited ability of humans to discriminate and identify odorants in mixtures is likely due to a number of mechanisms. Changes in spatial processing arising from competition for receptor sites and cells at the periphery, and inhibition in the bulb and at other olfactory centers, would be expected to reduce the information within the activated receptor cell arrays, making it difficult to recognize the patterns of activation due to different odorants. Since temporal processing favors the first processed odorant, the initial odorant has the opportunity to act as an antagonist towards other odorants at the periphery and to inhibit neural activity arising from other odorants in the bulb. However, if the delivery of different odorants in ternary or more complex mixtures to the nose is less than the time for processing two odorants in working memory, the latter becomes the ultimate limiting factor as regards the number of odorants identified. These mechanisms are discussed in detail below.

11.6.4.1 Spatial Processing. As noted elsewhere in this volume, a given odorant activates unique arrays of receptor cells in the nose (Mackay-Sim et al. 1982; Kauer, 1991) which, in turn, are reflected by patterns of activation of glomeruli and mitral/tufted cells in the olfactory bulb. Different odors produce different arrays that represent the spatial codes of identification. However, when a mixture of two odorants is sensed and the perception of one or both are suppressed to some degree, the arrays representing the two stimuli in the bulb show a reduction in the number of glomeruli that are activated (Bell et al., 1987; Joerges et al., 1997). If the suppression of one of the odorants is such that it cannot be perceived, little of the normal array of activated glomeruli is seen (Bell et al., 1987). The suppression may be due to fewer receptor cells being activated (Kurahashi et al., 1994; Ache et al., 1988; Simon and Derby, 1995) because of competition by the odorants for the same receptor sites, resulting in less input to the bulb. Suppression can

also be caused by lateral inhibition between glomeruli or mitral cells in the bulb (Pinching and Powell, 1971; White, 1979; Shepherd and Greer, 1990). The loss of identity of up to five odorants in eight-component mixtures (Livermore and Laing, 1998a) prompted Jinks and Laing (1999a) to propose that the competition and inhibition between odorants could result in no odorant being identified in mixtures containing double this number of components. Their psychophysical study showed that one and zero components were identified in 12- and 15-component stimuli, respectively. The fact that the 15-component stimulus had an odor, albeit not one that could be associated with any of the components or an object or source, indicates that neural input from some or all of the arrays characterizing the components was registered.

In light of such observations, it is interesting to note that, in the rat, neural images of complex olfactory stimuli, including rat nest odors comprised of volatiles from urine, feces, and bodies (Stewart et al., 1979), have shown that the number of activated glomeruli is similar to that found with a simple single odorant such as limonene (Bell et al., 1987). Therefore, spatial processing of single and complex odorants involves both peripheral and bulbar interactions that reduce and simplify identification. Accordingly, the olfactory system uses spatial coding to analyze and identify single odorants when presented alone and in simple mixtures, and simplifies identification of complex mixtures by combining the remaining parts of the arrays into a single characteristic array that is associated with the object or source of the stimulus. This interpretation is in agreement with the finding of Jellinek and Köster (1979) noted above that single odors are perceived to be as complex as that of mixtures.

But there is another aspect to spatial processing. An intriguing feature of single odorants and odor mixtures is that they can be characterized by several qualities or 'notes' (Moskowitz and Barbe, 1977; Laing and Willcox, 1983). Hexenal, for example, is described as having 'green' and 'fatty' qualities, and ethyl butyrate as 'sweet' and 'fruity'. However, when single odorants are components of mixtures but cannot be identified, often one or more of their qualities can be discerned. Jinks and Laing (2001) investigated the qualities of binary, ternary, and quaternary mixtures of four dissimilar odorants to determine the information about odor quality that needs to be retained for identification of the odorant. The data indicated that failure to identify an odorant could occur with the loss of some but not all of the qualities. However, failure could also occur when the major qualities were present but the ratios of their perceived intensities were substantially altered. This suggested that a different smell could be produced using the same qualities but in different ratios. Identification, therefore, was affected by the type and/or the perceived intensity of the qualities of an odorant. These results were interpreted in

11.6 The Perception of Odorant Mixtures

terms of a Configurational Hypothesis of Olfaction, in analogy with the Configurational Hypothesis of Facial Recognition (Enns and Shore, 1997; Rakover and Teucher, 1997). In brief, in the case of a face, identification of a person requires not only certain features to be present in a drawing or photograph, but these features must be in the correct proportion to each other. Similarly, identification of an odorant or a complex mixture requires some of the characteristic qualities in the correct proportions to be perceived.

But what is the neural basis of a quality or 'odor note,' and how is it represented in the spatial code? Thanks to advances in molecular biological studies of the chemoreceptive process, an insight to this problem is possible. For example, as noted throughout this volume, it is commonly accepted that each human receptor cell has only one type of receptor (Rawson et al., 1997) and that there are nearly 500 functional receptor types in humans, as indicated by the number of receptor genes. Stimulation of receptor cells by a single odorant most commonly results in a variety of cells being activated in accordance with the degree or 'ease of fit' of the odorant to each receptor site type (Chapter 4 and 5). If the fit is predominantly to two or three receptor types they will be the main inputs to the array of glomeruli and mitral/tufted cells activated in the bulb. However, the conformations adopted by an odorant to fit the two to three receptor types will be dictated by the structural features of the odorant and receptor molecule. In one conformation a molecule may be aligned within a receptor site according to its length and functional group, e.g., the I7 receptor for octanal (Araneda et al., 2000), in another it may sense a structural feature common to a number of odorants, e.g., an 8 carbon chain containing a terminal carbonyl group common to aliphatic aldehydes, acids, esters, and ketone (Imamura et al. 1992). Since the overall odors of these latter aliphatic carbonyl substances are easily discriminable (Laska et al. 2000), each odorant must require at least two receptor types to be occupied for this to occur. Accordingly, it is tempting to suggest that activation of the cells with the common receptor for these odorants results in an odor quality common to each odorant, whilst activation of the cells unique to each odorant produces a quality unique to each odorant. In addition, the spatial map of each odorant should show glomeruli or mitral/tufted cells that are activated by all four odorants, and others that only one of the odorants will activate. From the limited data available, it is suggested that the conformations an odorant can adopt in different types of receptors defines the important structural features that provide the qualities perceived. This interpretation suggests the spatial code for an odorant contains information about molecular structure and odor qualities. In contrast, the spatial map of complex mixtures such as chocolate aroma, where none of dozens of odorants can be identified, will be composed of input from receptor cells representing features of many odorants

and it may be the location and magnitude of the input to the bulb rather than molecular features that define its identity. Nevertheless, several qualities can usually be discerned in complex aromas, and these are likely to be those that remain from individual odorants in the mixture which are insufficient to identify the latter but contribute to the overall aroma of the complex mixture.

11.6.4.2 *Temporal Processing.*

During the 1980s, Getchell et al. (1984) reported that odorants can differ by hundreds of milliseconds in the time they take to activate receptor cells, whilst Kuznicki and Turner (1986) showed that humans require different reaction times to recognize the four common tastants. These findings prompted Laing (1987) to propose that if the time differences between the activating times of odorants at the periphery were maintained as the neural message traveled through the bulb and other olfactory processing centers in the brain that dealt with memory and identification, then a "fast" odorant would have a number of advantages if presented in a mixture with a "slow" odorant. For example, the faster odorant may be more successful in competition for receptor sites and cells. Being the first to activate the bulb, it could trigger lateral inhibition between glomeruli or between mitral cells to further reduce neural input from the slower odorant. Accordingly, it was predicted that the faster odorant would be the first odorant identified in a mixture, the slower odorant would incur the greatest suppression of intensity, and the number of cells and glomeruli in spatial arrays activated by the latter odorant would be reduced.

To investigate the first two of the above predictions, Laing et al. (1994) used a specially designed computer-controlled olfactometer which allowed odorants to be delivered together in a mixture or in series separated by intervals as small as 50 ms. By asking subjects which of two odorants was perceived first during a trial, and varying the time between delivery of both odorants from 100–600 ms, the processing time difference between them was established as that which produced a chance response, i.e., 50% for the forced-choice yes/no task. The magnitude of the differences varied from zero to more than a second, and was dependent on both the quality and perceived intensity of the odorants, with the latter being more important. Perceived intensity was also reduced more for the slower odorant. With both predictions upheld, the existence of temporal processing and its implications for mixture perception were demonstrated. A later study (Jinks and Laing, 1999b) confirmed that knowledge of processing time differences allowed predictions of which odor would be perceived first in other mixtures. Thus, they showed that when odor A was perceived before B and B was perceived before C, that A was perceived before C, demonstrating that transitivity had occurred (Figure 11.12). However,

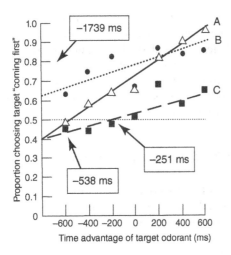

Figure 11.12 Regression lines representing the proportion of trials an odor in a binary mixture was perceived 'first' when presented with a time advantage (+ values on X axis), disadvantage (- values on X axis), or as true mixture (0 value on X axis). Arrows and times in boxes indicate when both odorants were perceived first on 50% of trials. A = stimulus of coniferan and triethylamine with coniferan being perceived 538 ms before triethylamine; B = a stimulus of carvone/triethylamine with carvone perceived first 1739 ms before triethylamine; C = a stimulus of carvone/coniferan with carvone perceived first 251 ms before coniferan.

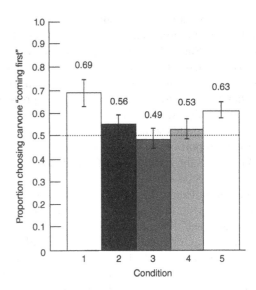

Figure 11.13 Proportion of trials (numbers above bars) in which subjects selected an odor 'coming first' in binary and ternary mixtures, and mixtures where the presentation of triethylamine was delayed. Conditions are; 1 - binary mixture of carvone/coniferan; 2 - ternary mixture of carvone/coniferan/triethylamine; 3,4 and 5 are the ternary mixture with the presentation of triethylamine delayed by 300, 600 and 900 ms, respectively. Open and shaded bars indicate that the means were significantly/not significantly different from 0.5 (chance), respectively.

investigation of temporal processing in ternary mixtures revealed a substantial limitation in the ability of humans to indicate which odor is perceived first, and the existence of a third mechanism that affects perception of components in odor mixtures (Jinks and Laing, 1999b). Temporal processing of ternary mixtures and the third mechanism, which is postulated to involve olfactory working memory, are discussed below.

11.6.4.3 The Role of Memory.

The perception of the order of processing odorants in ternary mixtures, however, has proved to be very difficult (Jinks and Laing, 1999b). Initial experiments indicated that subjects recorded chance level responses when asked to indicate which odorant was perceived first or last. To investigate whether the chance results were due to a limitation in the capacity of olfactory working memory to process both order and identity of the odorants, presentation of the third odorant was delayed by 300, 600, and 900 ms. With one of the two sets of three odorants studied, the results indicated that a delay of between 600 and 900 ms was needed before the usually faster odorant was perceived first (Figure 11.13). With the other set, subjects recorded chance responses even with the 900 ms delay. The mean responses of subjects, when asked to identify the odorants in the mixture or delay conditions, showed that this was at chance level. The result is in agreement with the earlier studies of Laing and colleagues who found that few subjects could identify all the components of ternary mixtures. Overall, the results with ternary mixtures indicated that a mechanism related to the speed of information retrieval about the identity or temporal order of the components was the cause. The most likely candidate appears to be the inability of olfactory working memory to process the information about the identity and order of the first two components before neural input from the third began to be processed. Although it is not fully understood, working memory is defined as the "system responsible for the temporary storage and manipulation of information, forming an important link between perception and controlled action" (Baddeley, 1998). The process of identifying an odor within working memory is likely to involve several steps; encoding of the odor by neurons, recalling of the coded representation of the odor from long term memory, comparison of the two representations, and the judging and responding to the representations. This type of process has been proposed for visual information (Eskandar et al., 1992). Indeed the interference of a third odorant with the perception of others is reminiscent of that reported for visualspatial memory where it was proposed that an irrelevant visual stimulus may have obligatory access to a visual store and interfere with the storage and processing of other visualspatial information in working memory (Toms et al. 1994). Limitations in the capacity of olfactory working memory to process more than two odorants within 600–900 ms appears to be a major factor limiting the discrimination and identification of odorants in multi-component mixtures. Such a finding has implications for the perception of odorants released during an eating episode where many can be released within the processing time differences cited here, but only a few may be identified. Controlled release of odorants from different food media could, however, allow products to be developed with high flavor impact.

11.7 CONCLUSIONS

The present chapter has provided an up-to-date review of the psychophysical means for testing the human sense of smell, and has examined how the human olfactory system likely integrates information from complex arrays of odorant chemicals which, individually, would seem to produce conflicting odorous sensations. It is of interest that relatively high correlations exist among the scores derived from nominally distinct olfactory tests, regardless of whether they are based upon single- or multi-component stimuli. Test reliability has been shown to be largely a result of test length, irrespective of the nature of the stimuli included in the tests. To what extent tests employing multicomponent odors are superior to ones employing single odorants is an empirical issue, although it would seem that by sampling more elements of the system, a test should be more sensitive. Continued efforts to refine the procedural elements of olfactory tests should help in the development of test batteries sensitive to wider ranges of olfactory deficits than those that are currently available.

ACKNOWLEDGMENTS

This chapter was supported, in part, by USAMRAA W81XWH-09-1-0467 from the Department of Defense and Grants PO1 DC 00161, RO1 DC 04278, RO1 DC 02974, RO1 AG 27496, and RO1 AG 08148 from the National Institutes of Health, Bethesda, MD USA, a contribution from the Baker Foundation, and an Australian Research Council Large Grant.

REFERENCES

Abraham, A. and Matha, K. V. (1983). The effect of right temporal lobe lesions on matching of smells. *Neuropsychologia* 21: 277–281.

Ache, B. W., Gleeson, R. A., and Thompson, H. A. (1988). Mechanisms for mixture suppression in olfactory receptors of the spiny lobster. *Chem. Senses* 13: 425–434.

Albrecht, J., Schreder, T., Kleemann, A.M. et al. (2009). Olfactory detection thresholds and pleasantness of a food-related and a non-food odour in hunger and satiety. *Rhinology.* 47: 160–165.

Altman, K. W., Desai, S. C., Moline, J., et al. (2011). Odor identification ability and self-reported upper respiratory symptoms in workers at the post-9/11 World Trade Center site. *Int. Arch. Occup. Environ. Health.* 84: 131–137.

Altundag, A., Salihoglu, M., Tekeli, H., et al. (2014). Lateralized differences in olfactory funcction and olfactory bulb volume relate to nasal septum deviation. *J. Craniofac. Surg.* 25: 359–362.

Amoore, J. E. and Ollman, B. G. (1983). Practical test kits for quantitatively evaluating the sense of smell. *Rhinology* 21: 49–54.

Anderson, N. H. (1970). Functional measurement and psychophysical judgment. *Psychol. Rev.* 77: 153–170.

Anderson, J., Maxwell, L., and Murphy, C. (1992). Odorant identification testing in the young child. *Chem. Senses* 17: 590.

Araneda, R. C., Kini, A. D., and Firestein, S. (2000). The molecular receptive range of an odorant receptor. *Nature Neurosci.* 3: 1248–1255.

Aronsohn, E. (1886). Experimentalle untersuchungen zur physiologie des geruchs. *Dies. Archiv. Physiol. Abthlg.* 321–357.

Armstrong, J. E., Laing, D. G., Wilkes, F. J. and Laing, O. N. (2008). Olfaction function in Australian Aboriginal children and chronic otitis media. *Chem. Senses* 33: 503–507.

ASTM (2008).Standard practice for determining odor and taste thresholds by a forced-choice ascending concentration series method of limits, E-679-04, Annual Book of Standards, Vol. 15.08. ASTM International, West Conshohocken, PA, pp. 36–41.

Baddeley, A. (1998). Working memory. *C.R. Acad. Sci. Paris, Life Sci.* 321: 167–173.

Baird, J. C., Lewis, C., and Romer, D. (1970). Relative frequencies of numerical responses in ratio estimation. *Percept. Psychophys.* 8: 358–362.

Bartoshuk, L. M., Fast, K., and Snyder, D. J. (2005). Differences in our sensory worlds: invalid comparisons with labeled scales. *Curr. Dir. Psychol. Sci.* 14: 122–125.

Bell, G. A., Laing, D. G., and Panhuber, H. (1987). Odor mixture suppression: evidence for a peripheral mechanism in human and rat. *Brain Res.* 426: 8–18.

Bender, G., Hummel, T., Negoias, S., and Small, D. M. (2009). Separate signals for orthonasal vs. retronasal perception of food but not nonfood odors. *Behav. Neurosci.* 123: 481–489.

Bensafi, M., Rouby, C., Farget, V. et al. (2003). Perceptual, affective, and cognitive judgments of odors: pleasantness and handedness effects. *Brain Cogn.* 51: 270–275.

Berglund, B. (1974). Quantitative and qualitative analysis of industrial odors with human observers. *Ann. NY. Acad. Sci.* 237: 35–51.

Berglund, B., and Olsson, M. J. (1993a). Odor-intensity interaction in binary mixtures. *J. Exp. Psychol. Human Percept. Perform.* 19: 302–314.

Berglund, B. and Olsson, M. J. (1993b). Odor-intensity interaction in binary and ternary mixtures. *Percept. Psychophys.* 53: 475–482.

Berglund, B., Berglund, U., and Lindvall, T. (1976). Psychological processing of odor mixtures. *Psychol. Rev.* 83: 432–441.

Berglund, B., Berglund, U., Lindvall, T., and Svensson, L. T. (1973). A quantitative principle of perceived intensity summation in odor mixtures. *J. Exp. Psychol.* 100: 29–38.

Berglund, B., Berglund, U., Ekman, G., and Engen, T. (1971). Individual psychophysical functions for 28 odorants. *Percept. Psychophys.* 9: 379–384.

Berglund, B., Berglund, U., Johansson, I., and Lindvall, T. (1984). Mobile laboratory for sensory air-industrial environments. In Indoor Air, *Vol. 3*, B. Berglund, T. Lindvall, and J. Sumndeel (Eds.). Swedish Council for Building Research, Stockholm, pp. 467–472.

Betchen, S. A., and Doty, R. L. (1998). Bilateral detection thresholds in dextrals and sinistrals reflect the more sensitive side of the nose, which is not lateralized. *Chem. Senses* 23: 453–457.

Blackwell, H. R. (1953). Psychophysical. thresholds: experimental studies of methods of measurement. *Bull. Engin. Res. Inst.* No. 36, Ann Arbor, University of Michigan Press.

Bonfils, P., Faulcon, P., and Avan, P. (2004). Screening of olfactory function using the Biolfa olfactory test: investigations in patients with dysosmia. *Acta Otolaryngol. (Stockh)* 124: 1063–1071.

Booth, D. A. (1995). Cognitive processes in odorant mixture assessment. *Chem. Senses* 20: 639–643.

Borg, G. (1982). A category scale with ratio properties for intermodal and interindividual comparisons. In: H. G. Geissler and P. Penzold (Eds.), Psychophysical Judgment and the Process of Perception. Berlin: VEG Deutscher Verlag Der Wissenschaften, pp. 25–34.

Brand, G., Millot, J. L., Jacquot, L., Thomas, S., and Wetzel, S. (2004). Left:right differences in psychophysical and electrodermal measures of olfactory thresholds and their relation to electrodermal indices of hemispheric asymmetries. *Percept. Mot. Skills* 98: 759–769.

Bromley, S. M., and Doty, R. L. (1995). Odor recognition memory is better under bilateral than unilateral test conditions. *Cortex* 31: 25–40.

Brown, J. (1974). Recognition assessed by rating and ranking. *Brit. J. Psychol.* 65: 13–22.

Brown, K. S., Maclean, C. M., and Robinette, R. R. (1968). The distribution of the sensitivity to chemical odors in man. *Hum. Biol.* 40: 456–472.

Burdach, K. J. and Doty R. L. (1987). The effects of mouth movements, swallowing, and spitting on retronasal odor perception. *Physiol. Behav.* 41: 353–356.

Caccappolo, E., Kipen, H., Kelly-McNeil, K., et al. (2000). Odor perception: multiple chemical sensitivities, chronic fatigue, and asthma. *J. Occup. Environ. Med.* 42: 629–638.

Cain, W. S. (1975). Odor intensity, mixtures and masking. *Chem. Sens. Flav.* 1: 339–352.

Cain, W. S. (1982). Sumner's "on testing the sense of smell" revisited. *Yale J. Biol. Med.* 55: 515–519.

Cain, W. S. and Drexler, M. (1974). Scope and evaluation of odor counteraction and masking. *Ann. N. Y. Acad. Sci.* 237: 427–439.

Cain, W. S. and Gent, J. F. (1991). Olfactory sensitivity: reliability, generality, and association with aging. *J. Exp. Psychol. Hum. Percept. Perform.* 17: 382–391.

Cain, W, S. and Johnson, F., Jr. (1978). Lability of odor pleasantness: influence of mere exposure. *Perception* 7: 459–465.

Cain, W. S. and Rabin, R. D. (1989). Comparability of two tests of olfactory functioning. *Chem. Senses* 14: 479–485.

Cain, W. S., Gent, J., Catalanotto, F. A., and Goodspeed, R. B. (1983). Clinical evaluation of olfaction. *Am. J. Otolaryngol.* 4: 252–256.

Cain, W. S., Gent, J. P., Goodspeed, R. B., and Leonard, G. (1988). Evaluation of olfactory dysfunction in the Connecticut Chemosensory Clinical Research Center. *Laryngoscope* 98: 83–88.

Cain, W. S. Schiet, F. T., Olsson, M. J., and de Wijk, R. A. (1995). Comparison of models of odor interaction. *Chem. Senses* 20: 625–637.

Cameron, E. L. (2014). Pregnancy and olfaction: a review. *Front. Psychol.* Feb 6;5:67. doi: 10.3389/fpsyg.2014.00067.

Cameron, E. L. and Doty, R. L. (2013). Odor identification testing in children and young adults using the smell wheel. *Int. J. Pediatr. Otorhinolaryngol.* 77: 346–350.

Cameron, J. D., Goldfield, G. S., and Doucet, E. (2012). Fasting for 24 h improves nasal chemosensory performance and food palatability in a related manner. *Appetite* 58: 978–981.

Campbell, L. M., and Gregson, R. A. M. (1972). Olfactory short term memory in normal, schizophrenic, and brain damaged cases. *Aust. J. Psychol.* 24: 179–185.

Cardesin, A., Alobid, I., Benitez, P., et al. (2006). Barcelona Smell Test - 24 (BAST-24): validation and smell characteristics in the healthy Spanish population. *Rhinology* 44: 83–9.

Carrasco, M. and Ridout J. B. (1993). Olfactory perception and olfactory imagery: a multidimensional scaling analysis. *J. Exp. Psychol. Hum. Percept. Perform.* 19: 287–301.

Chalke, H. D., Dewhurst, J. R., and Ward, C. W. (1958). Loss of sense of smell in old people. *Public Health* 72: 223–230.

Cheesman, G. H. and Townsend, M. J. (1956). Further experiments on the olfactory thresholds of pure chemical substances, using the "sniff-bottle method." *Q. J. Exp. Psychol.* 8: 8–14.

Cheesman, G. H. and Kirkby, H. M. (1959). An air dilution olfactometer suitable for group threshold measurements. *Quart. J. Exp. Psychol.* 11: 115–123.

Chen, V., and Halpern, B. P. Retronasal but not oral-cavity-only identification of "purely olfactory" odorants. *Chem. Senses* 33: 107–118.

Choudhury, E. S., Moberg, P. and Doty, R. L. (2003). Influences of age and sex on a microencapsulated odor memory test. *Chem. Senses* 28: 799–805.

Cometto-Muñiz, J. E, and Cain, W. S. (1995). Olfactory adaptation. In R. L. Doty (Ed.), Handbook of Olfaction and Gustation. N.Y.: Marcel Dekker, pp. 257–281.

Cook, G. R., Krithika, S., Edwards, M., et al. (2014). Quantitative measurement of odor detection thresholds using an air dilution olfactometer, and association with genetic variants in a sample of diverse ancestry. *PeerJ.* 2:e643. doi:10.7717/peerj.643.

Cornsweet, T. N. (1962). The staircase-method in psychophysics. *Am. J. Psychol.* 75: 485–491.

Cortina, J. M. (1993). What is coefficient alpha? An examination of theory and applications. *J. Appl. Psychol.* 78: 98–104.

Croy, I., Hoffman, H., Philpott, C., et al. (2014). Retronasal testing of olfactory function: an investigation and comparison in seven countries. *Eur. Arch. Otorhinolaryngol.* 271: 1087–1095.

Dalton, P. (1996). Odor perception and beliefs about risk. *Chem. Senses* 21: 447–458.

Dalton, P. (1999). Cognitive influences on health symptoms from acute chemical exposure. *Health Psychol.* 18: 579–590.

Dalton, P. H., Opiekun, R. E., Gould, M., et al. (2010). Chemosensory loss: functional consequences of the world trade center disaster. *Environ. Health Perspect.* 118: 1251–1256.

Dalton, P., Wysocki, C. J., Brody, M. J., and Lawley, H. J. (1997). Perceived odor, irritation, and health symptoms following short-term exposure to acetone. *Amer. J. Indust. Med.* 31: 558–569.

Dalton, P., Doolittle, N., and Breslin, P. A. (2002). Gender-specific induction of enhanced sensitivity to odors. *Nature Neurosci.* 5: 199–200.

Davidson, T.M., Freed, C., Healy, M.P. et al. (1998). Rapid clinical evaluation of anosmia in children: the Alcohol Sniff Test. *Ann. N.Y. Acad. Sci.* 855: 787–792.

Davidson, T. M., and Murphy, C. (1997). Rapid clinical evaluation of anosmia. *The alcohol sniff test. Arch. Otolaryngol. Head Neck Surg.* 123: 591–594.

Deems, D. A., Doty, R. L., Settle, R. G., Moore-Gillon, V., Shaman, P., Mester, A. F., Kimmelman, C. P., Brightman, V. J., and Snow, J. B. Jr. (1991). Smell and taste disorders, a study of 750 patients from the University of Pennsylvania Smell and Taste Center. *Arch. Otolaryngol. Head Neck Surg.* 117: 519–528.

Desor, J. A. and Beauchamp, G. K. (1974). The human capacity to transmit olfactory information. *Percept. Psychophys.* 16: 551–556.

References

Doty, R. L. (1975). An examination of relationships between the pleasantness, intensity and concentration of 10 odorous stimuli. *Percept. Psychophys.* 17: 492–496.

Doty, R. L. (1976). Reproductive endocrine influences upon human nasal chemoreception: A review. In Mammalian Olfaction, Reproductive Processes, and Behavior, Academic Press, New York, pp. 295–321.

Doty, R. L. (1979). A review of olfactory dysfunctions in man. *Am. J. Otolaryngol.* 1: 57–79.

Doty, R.L. (1986). Ontogeny of human olfactory function. In W. Breipohl (Ed.), *Ontogeny of Olfaction in Vertebrates.* Berlin: Springer-Verlag, pp. 3-17.

Doty, R. L. (1991). Olfactory system. In Smell and Taste in Health and Disease, T. V. Getchell, R. L. Doty, L. M. Bartoshuk and J. B. Snow, Jr. (Eds.). Raven Press, New York, pp. 175–203.

Doty, R. L. (1995). The Smell Identification Test™ Administration Manual, 3rd ed. Haddon Hts., NJ: Sensonics, Inc.

Doty, R. L. (2000). The Smell Threshold Test™ Administration Manual. Haddon Hts., NJ: Sensonics, Inc.

Doty, R. L., Agrawal, U., and Frye, R. (1989). Evaluation of the internal consistency reliability of the fractionated and whole University of Pennsylvania Smell Identification Test. *Percept. Psychophys.* 45: 381–384.

Doty, R. L., Beals, E., Osman, A., et al. (2014). Suprathreshold odor intensity perception in early stage Parkinson's disease. *Mov. Disord.* 29: 1208–1212.

Doty, R. L. and Cameron, E. L. (2009). Sex differences and reproductive hormone influences on human odor perception. *Physiol. Behav.* 97: 213–228.

Doty, R. L. and Crastnopol, B. (2010). Correlates of chemosensory malingering. *Laryngoscope* 120: 707–711.

Doty, R. L., Deems, D. A., Frye, R., Pelberg, R., and Shapiro, A. (1988). Olfactory sensitivity, nasal resistance, and autonomic function in the multiple chemical sensitivities (MCS) syndrome. *Arch. Otolaryngol. Head Neck Surg.* 114: 1422–1427.

Doty, R. L., Diez, J. M., Turnacioglu, S., et al. (2003). Influences of feedback and ascending and descending trial presentations on perithreshold odor detection performance. *Chem. Senses* 28: 523–526.

Doty, R. L., Ford, M., Preti, G., and Huggins, G. (1975). Human vaginal odors change in pleasantness and intensity during the menstrual cycle. *Science* 190: 1316–1318.

Doty, R. L., Gregor, T., and Settle, R. G. (1986). Influences of intertrial interval and sniff bottle volume on the phenyl ethyl alcohol olfactory detection threshold. *Chem. Senses* 11: 259–264.

Doty, R. L., and Kamath, V. (2014). The influences of age on olfaction: a review. *Frontiers in Psychology – Cognitive Science*, 5:20. doi:10.3389/fpsyg.2014.00020.

Doty, R. L., Kisat, M., and Tourbier, I. (2008). Estrogen replacement therapy induces functional asymmetry on an odor memory/discrimination test. *Brain Res.* 1214: 35–39.

Doty, R. L., Kligman, A., Leyden, J., and Orndorff, M. M. (1978b). Communication of gender from human axillary odors: relationship to perceived intensity and hedonicity. *Behav. Biol.* 23: 373–380.

Doty, R. L., Marcus, A., and Lee, W. W. (1996). Development of the 12-item cross-cultural smell identification test (CC-SIT). *Laryngoscope* 106: 353–356.

Doty, R. L., McKeown, D., Lee W. W. and Shaman, P. (1995). Test-retest reliability of 10 olfactory tests. *Chem. Senses* 20: 645–656.

Doty, R. L., Petersen, I., Mensah, N., and Christensen, K. (2011). Genetic and environmental influences on odor identification ability in the very old. *Psychol. Aging* 26: 864–871.

Doty, R. L., Reyes, P., and Gregor, T. (1987). Presence of both odor identification and detection deficits in Alzheimer's disease. *Brain Res. Bull.* 18: 597–600.

Doty, R. L., Shaman, P., and Dann, M. (1984a). Development of the University of Pennsylvania Smell Identification Test: a standardized microencapsulated test of olfactory function. *Physiol. Behav.* 32: 489–502.

Doty, R. L., Shaman, P., Applebaum, S. L., et al. (1984b). Smell identification ability: changes with age. *Science* 226: 1441–1443.

Doty, R. L., Snyder, P., Huggins, G., and Lowry, L. D. (1981). Endocrine, cardiovascular, and psychological correlates of olfactory sensitivity changes during the human menstrual cycle. *J. Comp. Physiol. Psychol.* 95: 45–60.

Doty, R. L., Smith, R., McKeown, D. and Raj, J. (1994). Tests of human olfactory function: Principal components analysis suggests that most measure a common source of variance. *Percept. Psychophys.* 56: 701–707.

Doty, R. L., Stem, M. B., Pfeiffer, C., et al. (1992). Bilateral olfactory dysfunction in early stage treated and untreated idiopathic Parkinson's disease. *J. Neurol. Neurosurg. Psychiat.* 15(2): 35–47.

Drake, B., Johansson, B., von Sydow, D., and Döving, K. B. (1969). Quantitative psychophysical and electrophysiological data on some odorous compounds. *Scand. J. Psychol.* 10: 89–96.

Duff, K., McCaffrey, R. J., and Solomon, G. S. (2002). The Pocket Smell Test: successfully discriminating probable Alzheimer's dementia from vascular dementia and major depression. *J. Neuropsychiat. Clin. Neurosci.* 14: 197–201.

Eichenbaum, H., Morton, T. H., Potter, H., and Corkin, C. (1983). Selective olfactory deficits in case H. M. *Brain* 106: 459–472.

Ekman, G. and Sjöberg, L. (1965). Scaling. *Ann. Rev. Psychol.* 16: 451–474.

Elliot, P. B. (1964). Tables of d'. In: Swets, J. A., ed. Signal Detection and Recognition by Human Observers. New York: Wiley, 1964.

Elsberg, C. A. and Levy, L. (1935). The sense of smell: I. A new and simple method of quantitative olfactometry. *Bull. Neurol. Inst.* NY 4: 4–19.

Engen, T. (1960). Effect of practice and instruction on olfactory thresholds. *Percept. Motor Skills* 10: 195–198.

Engen, T. and Ross, B. M. (1973). Long-term memory of odors with and without verbal descriptions. *J. Exp. Psychol.* 100: 221–227.

Enns, J. T., and Shore, D. I. (1997). Separate influences of orientation and lighting in the inverted-face effect. *Percept. Psychophys.* 59: 23–31.

Eskandar, E. N., Optican, L. M., and Richmond, B. R.1992). Role of inferior temporal neurons in visual memory: II. Multiplying temporal waveforms related to vision and memory. *J. Neurophysiol.* 68: 1296–1306.

Eyman, R. K., Kim, P. J., and Call, T. (1975). Judgment error in category vs magnitude scales. *Percept. Motor Skills* 40: 415–423.

Fechner, G. T. (1860). Elemente der Psychophysik. Leipzig, Breitkopf and Hartrel.

Fernberger, S. H. (1932). A preliminary study of taste deficiency. *Am. J. Psychol.* 44: 322–326.

Frank, M. E., Goyert, H. F., and Hettinger, T. P. (2010). Time and intensity factors in identification of components of odor mixtures. *Chem. Senses* 35: 777–787.

Frank, R. A., Dulay, M. F., Niergarth, K. A., and Gesteland, R. C. (2004). A comparison of the sniff magnitude test and the University of Pennsylvania Smell Identification Test in children and nonnative English speakers. *Physiol. Behav.* 81: 475–480.

Freiherr, J., Gordon, A. R., Alden, E. C., et al. (2012). The 40-item Monell Extended Sniffin' Sticks Identification Test (MONEX-40). *J. Neurosci. Methods* 205: 10–16.

Frasnelli, J., and Hummel, T. (2007). Interactions between the chemical senses: trigeminal function in patients with olfactory loss. *Internat. J. Psychophysiol.* 65: 177–181.

Frijters, J. E. R., Kooistra, A., and Vereijken, P. F. G. (1980). Tables of d' for the triangular method and the 3-AFC signal detection performance. *Percept. Psychophys.* 27: 176–178.

Frey, P. W. and Colliver, J. A. (1973). Sensitivity and responsivity measures for discrimination learning. *Learn. Motiv.* 4: 327–342.

Frye, R. E., Schwartz, B., and Doty, R. L. (1990). Dose-related effects of cigarette smoking on olfactory function. *JAMA* 263: 1233–1236.

Furukawa, M., Kamide, M., Miwa, T., and Umeda, R. (1988). Importance of unilateral examination in olfactometry. *Auris Nasus Larynx (Tokyo)* 15: 113–116.

Gescheider, G. A. (1997). Psychophysics: The fundamentals. Hillsdale, N.J.: Lawrence Erlbaum Associates.

Gescheider, G. A. (1988). *Psychophysical scaling. Ann. Rev. Psychol.* 39: 169–200.

Getchell, T. V., Margolis, F. L., and Getchell, M. L. (1984). Perireceptor and receptor events in vertebrate olfaction. *Progr. Neurobiol.* 23: 317–345.

Gilbert, A. N., Popper, R., Kroll, J. J., Nicklin, L., and Zellner, D. A. (2002). The Cranial 1 Quick Sniff®: a new screening test for olfactory function. *Chem. Senses* 27: A23.

Good, K. P., Martzke, J. S., Daoud, M. A., and Kopala, L. C. (2003). Unirhinal norms for the University of Pennsylvania Smell Identification Test. *Clin. Neuropsychol.* 17: 226–234.

Good, K. P., Martzke, J. S., Milliken, H. I., et al. (2002). Unirhinal olfactory identification deficits in young male patients with schizophrenia and related disorders: association with impaired memory function. *Schizophr Res.* 56: 211–223.

Goyert, H. F., Frank, M. E., Gent, J. F. and Hettinger, T. P. (2007). Characteristic component odors emerge from mixtures after selective adaptation. *Brain Res. Bull.* 72: 1–9.

Graves, A. B., Bowen, J. D., Rajaram, L., et al. (1999). Impaired olfaction as a marker for cognitive decline: interaction with apolipoprotein E epsilon4 status. *Neurology* 53: 1480–1487.

Green, B. G., Dalton, P., Cowart, B., et al. (1996). Evaluating the 'Labeled Magnitude Scale' for measuring sensations of taste and smell. *Chem. Senses* 21: 323–334.

Green, D. M. and Swets, J. A. (1966). Signal detection theory and psychophysics. New York: Wiley.

Green, P. and Iverson, G. (1998). Exaggeration of anosmia in 80 litigating head injury cases. *Arch. Clin. Neuropsychol.* 13: 138.

Gregson, R. A. M., and Smith, D. A. R. (1981). The clinical assessment of olfaction: differential diagnoses including Kallmann's syndrome. *J. Psychosomat. Res.* 25: 165–174.

Gregson, R. A. M., Free, M. L., and Abbott, M. W. (1981). Olfaction in Korsakoffs, alcoholics and normals. *Br. J. Clin. Psychol.* 20: 3–10.

Grier, J. B. (1971). Nonparametric indexes for sensitivity and bias: computing formulas. *Psychol. Bull.* 75: 424–429.

Griffith, L. P. (1976). Abnormalities of smell and taste. *Practitioner* 217: 907–913.

Guadagni, D. G., Buttery, R. G., and Okano, S. (1963). Odour thresholds of some organic compounds associated with food flavours. *J. Sci. Food Agr.* 14: 761–765.

Gudziol, V., Hummel, C., Negoias, S., Ishimaru, T., and Hummel, T. (2007). Lateralized differences in olfactory function. *Laryngoscope* 117: 808–811.

Gudziol, V., Paech, I. and Hummel, T. (2010). Unilateral reduced sense of smell is an early indicator for global olfactory loss. *J. Neurol.* 257: 959–963.

Guilford, J. P. (1954). Psychometric Methods. McGraw-Hill, New York.

Haehner, A., Mayer, A. M., Landis, B. N., et al. (2009). High test-retest reliability of the extended version of the "Sniffin' Sticks" test. *Chem. Senses* 34: 705–711.

Hashimoto, Y., Fukazawa, K., Fujii, M., et al. (2004). Usefulness of the odor stick identification test for Japanese patients with olfactory dysfunction. *Chem Senses* 29: 565–571.

Hayes, J. E., Jinks, A. L, and Stevenson, R. J. (2013). A comparison of sniff bottle staircase and olfactometer-based threshold tests. *Behav. Res. Methods* 45: 178–182.

Hedner, M., Larsson, M., Arnold, N., et al. (2010). Cognitive factors in odor detection, odor discrimination, and odor identification tasks. *J. Clin. Exp. Neuropsychol.* 32: 1062–1067.

Heilmann, S., Strehle, G., Rosenheim, K., Damm, M., and Hummel, T. (2002). Clinical assessment of retronasal olfactory function. *Arch. Otolaryngol. Head Neck Surg.* 128: 414–418.

Helm, E. and Trolle, B. (1946). Selection of a taste panel. *Wallerstein Lab. Communications.* 9: 181–194.

Helson, H. (1964). Adaptation-level Theory: An experimental and systematic approach to behavior. Harper and Row: New York.

Hendriks, A. P. (1988). Olfactory dysfunction. *Rhinology* 26: 229–251.

Heywood, P. G. and Costanzo, R. M. (1986). Identifying normosmics: a comparison of two populations. *Am. J. Otolaryngol.* 7: 194–199.

Hidalgo, J., Chopard, G., Galmiche, J., et al. (2011). Just noticeable difference in olfaction: a discriminative tool between healthy elderly and patients with cognitive disorders associated with dementia. *Rhinology* 49: 513–518.

Hjermstad, M. J., Fayers, P. M., Haugen, D. F., et al. (2011). Studies comparing numerical rating scales, verbal rating scales, and visual analogue scales for assessment of pain intensity in adults: A systematic literature review. *J. Pain. Symptom. Manage* 41: 1073–1093.

Hodos, W. (1970). Nonparametric index of response bias for use in detection and recognition experiments. *Psychol. Bull.* 74: 351–354.

Hummel, T., Pfetzing, U., and Lotsch, J. (2010). A short olfactory test based on the identification of three odors. *J. Neurol.* 257: 1316–1321.

Hummel, T., Sekinger, B., Wolf, S. R., et al. (1997). 'Sniffin' sticks': olfactory performance assessed by the combined testing of odor identification, odor discrimination and olfactory threshold. *Chem. Senses* 22: 39–52.

Ikeda, K., Tabata, K., Oshima, T., et al. (1999). Unilateral examination of olfactory threshold using the Jet Stream Olfactometer. *Auris Nasus Larynx* 26: 435–439.

Imamura, K., Mataga, N., and Mori, K. (1992). Coding of odor molecules by mitral/tufted cells in rabbit olfactory bulb. *I. Aliphatic Compounds. J. Neurophysiol.* 68: 1986–2002.

Jackman, A. H., and Doty, R. L. (2005). Utility of a three-item smell identification test in detecting olfactory dysfunction. *Laryngoscope* 115: 2209–2212.

Jacob, T. J. C., Fraser, C., Wang, L., et al. (2003). Psychophysical evaluation of responses to pleasant and mal-odour stimulation in human subjects: adaptation, dose response and gender differences. *Internat. J. Psychophysiol.* 48: 67–80.

Janowitz, H.D. and Grossman, M.I. (1949). Gustoolfactory thresholds in relation to appetite and hunger sensations. *J.Appl. Physiol.* 2: 217–222.

Jellinek, J. S., and Köster, E. P. (1979). Perceived fragrance complexity and its relation to familiarity and pleasantness. *J. Soc. Cosmet. Chem.* 30: 253–262.

Jinks, A. and Laing, D. G. (1999a). A limit in the processing of components in odour mixtures. *Perception* 28: 395–404.

Jinks, A., and Laing, D. G. (1999b). Temporal processing reveals a mechanism for limiting the capacity of humans to analyze odor mixtures. *Cogn. Brain Res.* 8: 311–325.

Jinks, A., and Laing, D. G. (2001). The analysis of odor mixtures by humans: Evidence for a configurational process. *Physiol. Behav.* 72: 51–63.

Johnson, B. N., and Sobel, N. (2007). Methods for building an olfactometer with known concentration outcomes. *J. Neurosci. Meth.* 160: 231–245.

Joerges, J., Kuttner, A., Galizia, C. G., and Menzel, R. (1997). Representations of odours and odour mixtures visualized in the honeybee brain. *Nature* 387: 285–288.

Jones, B. P., Moskowitz, H. R., and Butters, N. (1975). Olfactory discrimination in alcoholic Korsakoff patients. *Neuropsychologia* 13: 173–179.

Jones, F. N. (1955). Olfactory absolute thresholds and their implications for the nature of the receptor process. *J. Psychol.* 40: 223–227.

Jones, F. N. (1958). Scales of subjective intensity for odors of diverse chemical nature. *Am. J. Psychol.* 71: 305–310.

Jones, F. N. (1966). Some effects of context on the slope in magnitude estimation. *J. Exp. Psychol.* 71: 177–180.

Jones, F. N. and Woskow, M. H. (1964). On the intensity of odor mixtures. *Ann. NY. Acad. Sci.* 116: 484–494.

Jones-Gotman, M., and Zatorre, R. J. (1993). Odor recognition memory in humans: role of right temporal and orbitofrontal regions. *Brain Cognition* 22: 182–198.

Karnekull, S. C., Jonsson, F. U., Larsson, M., and Olofsson, J. K. (2011). Affected by smells? Environmental chemical responsivity predicts odor perception. *Chem. Senses* 36: 641–648.

Kauer, J. S. (1991). Contributions of topography and parallel processing to odor coding in the vertebrate olfactory pathway. *TINS* 14: 79–85.

Kobal, G., Hummel, T., Sekinger, B., et al. (1996). "Sniffin' sticks": screening of olfactory performance. *Rhinology* 34: 222–226.

Kobal, G., Palisch, K., Wolf, S. R., et al. (2001) A threshold-like measure for the assessment of olfactory sensitivity: the "random" procedure. *Eur. Arch. Otorhinolaryngol.* 258: 168–172.

Kobal, G. and Plattig, K.–H. (1978). Methodische Anmerkungen zur Gewinnung olfaktorischer EEG-Antworten des wachen Menschen (objektive Olfaktometrie). *Z. EEG-EMG* 9: 135–415.

Kobayashi, M., Nishida, K., Nakamura, S., et al. (2004) Suitability of the odor stick identification test for the Japanese in patients suffering from olfactory disturbance. *Acta Otolaryngol. Suppl.* 553: 74–79.

Koelega, H. S. (1979). Olfaction and sensory asymmetry. *Chem. Senses Flav.* 4: 89–95.

Köster, E. P. (1975). Human psychophysics in olfaction. In Methods in Olfactory Research, D. G. Moulton, A. Turk, and J. W. Johnston, Jr. (Eds.). Academic Press: New York, pp. 345–374.

Köster, E. P. and de Wijk, R. A. (1991). Olfactory adaptation. In D. G. Laing, R. L. Doty, and W. Breipohl (Eds.), The Human Sense of Smell. Berlin: Springer-Verlag, pp. 199–215.

Krantz, E. M., Schubert, C. R., Dalton, D. S., et al. (2009). Test-retest reliability of the San Diego Odor Identification Test and comparison with the brief smell identification test. *Chem. Senses* 34: 435–440.

Kremer, B., Klimek, L., and Mosges, R. (1998). Clinical validation of a new olfactory test. *Eur. Arch. Otorhinolaryngol.* 255: 355–358.

Kurahashi, T., Lowe, G., and Gold, G. H. (1994). Suppression of odorant responses by odorants in olfactory receptor cells. *Science* 265: 118–120.

Kurtz, D. B., White, T. L., Hornung, D. E., and Belknap, E. (1999). What a tangled web we weave: discriminating between malingering and anosmia. *Chem. Senses* 24: 697–700.

Kurtz, D. B., White, T. L., Sheehe, P. R., et al. (2001). Odorant confusion matrix: the influence of patient history on patterns of odorant identification and misidentification in hyposmia. *Physiol. Behav.* 72: 595–602.

Kuznicki, J. T. and Turner, L. S. (1986). Reaction time in the perceptual processing of taste quality. *Chem. Senses* 11: 183–201.

Laffort, P. (1989). Models for describing intensity interactions in odor mixtures: a reappraisal. In Perception of Complex Smells and Tastes, D. G. Laing, W. S. Cain, R. L. McBride, and B. W. Ache (Eds.). Academic Press: Sydney, pp. 205–223.

Laffort, P. and Dravnieks, A. (1982). Several models of suprathreshold quantitative olfactory interaction in humans applied to binary, ternary and quaternary mixtures. *Chem. Senses* 7: 153–174.

Laing, D. G. (1987). Coding of chemosensory stimulus mixtures. *Ann. N.Y. Acad. Sci.* 510: 61–66.

Laing, D. G. and Francis, G. W. (1989). The capacity of humans to identify odors in mixtures. *Physiol. Behav.* 46: 809–814.

Laing, D. G. and Glemarec, A. (1992). Selective attention and the perceptual analysis of odor mixtures. *Physiol. Behav.* 52: 1047–1053.

Laing, D. G., Segovia, C., Fark, T., et al. (2008). Tests for screening olfactory and gustatory function in school-age children. *Otolaryngol. Head Neck Surg.* 139: 74–82.

Laing, D. G. and Willcox, M. E. (1983). Perception of components in binary odor mixtures. *Chem. Senses* 7: 249–264.

Laing, D. G., Eddy, A., and Best, D. J. (1994). Perceptual characteristics of binary, trinary and quaternary odor mixtures consisting of unpleasant constituents. *Physiol. Behav.* 56: 81–93.

Laing, D. G., Eddy, A., Francis, G. W., and Stephens, L. (1994). Evidence for temporal processing of odor mixtures in humans. *Brain Res.* 651: 317–328.

Laing, D. G., Panhuber, H., Willcox, M. E., and Pittman, E. A. (1984). Quality and intensity of binary odor mixtures. *Physiol. Behav.* 33: 309–319.

Laska, M. and Hudson, R. (1991). A comparison of the detection thresholds of odor mixtures and their components. *Chem. Senses* 16: 651–662.

Laska, M. and Hudson, R. (1992). Ability to discriminate between related mixtures. *Chem. Senses* 17: 403–415.

Laska, M., Hudson, R., and Distel, H. (1990). Olfactory sensitivity to biologically relevant odors may exceed the sum of component thresholds. *Chemoecology* 1: 139–141.

Laska, M., Ayabe-Kanamura, S., Hubener, F., and Saito, S. (2000). Olfactory discrimination ability for aliphatic odorants as a function of oxygen moiety. *Chem. Senses* 25: 189–197.

Lawless, H. T. (2010). A simple alternative analysis for threshold data determined by ascending forced-choice methods of limits. *J. Sens. Stud.* 25: 332–346.

Lawless, H. T., and Malone, G. T. (1986a). The discrimination efficiency of common scaling methods. *J. Sens. Stud.* 1: 85–98.

Lawless, H. T., and Malone, G. T. (1986b). A comparison of rating scales: sensitivity, replicates and relative measurement. *J. Sens. Stud.* 1: 155–174.

Lehrner, J., and Deecke, L. (2000). The Viennese olfactory test battery - A new method for assessing human olfactory functions. *Aktuelle Neurologie* 27: 170–177.

Liu, H. C., Wang, S. J., Lin, K. P., et al. (1995). Performance on a smell screening test (the MODSIT): a study of 510 predominantly illiterate Chinese subjects. *Physiol Behav* 58: 1251–1255.

Livermore, A. and Laing, D. G. (1996). Influence of training and experience on the perception of multicomponent odor mixtures. *J. Exp. Psychol. Human Percept. Perform.* 22: 267–277.

Livermore, A. and Laing, D. G. (1998a). The influence of odor type on the discrimination and identification of odorants in multicomponent odor mixtures. *Physiol. Behav.* 65: 311–320.

Livermore, A. and Laing, D. G. (1998b). The influence of chemical complexity on the perception of multicomponent odor mixtures. *Percept. Psychophys.* 60: 650–661.

Lorig, T. S., Elmes, D. G., Zald, D. H., and Pardo, J. V. (1999). A computer-controlled olfactometer for fMRI and electrophysiological studies of olfaction. *Behav. Res. Meth. Inst. Comput.* 31: 370–375.

Lundström, J. N., Gordon, A. R., Alden, E. C., et al. (2010). Methods for building an inexpensive computer-controlled olfactometer for temporally-precise experiments. *Int. J. Psychophysiol.* 78: 179–189.

Mackay-Sim, A., Shaman, P., and Moulton, D. G. (1982). Topographic coding of olfactory quality: odorant-specific patterns of epithelial responsivity in the salamander. *J. Neurophysiol.* 48: 584–596.

Macmillan, N. and Creelman, C. (1996). Triangles in ROC space: History and theory of "nonparametric" measures of sensitivity and response bias. *Psychonomic Bull. Rev.* 3: 164–170.

Mainland, J. D., Johnson, B. N., Khan, R., et al. (2005). Olfactory impairments in patients with unilateral cerebellar lesions are selective to inputs from the contralesional nostril. *J. Neurosci.* 25: 6362–6371.

Makin, J. W. and Porter, R. H. (1989). Attractiveness of lactating females' breast odors to neonates. *Child Dev.* 60: 803–810.

Makowska, I., Kloszewska, I., Grabowska, A., et al. (2011). Olfactory deficits in normal aging and Alzheimer's disease in the Polish elderly population. *Arch. Clin. Neuropsychol.* 26: 270–279.

Markon, K. E., Chmielewski, M., and Miller, C.J. (2011). The reliability and validity of discrete and continuous measures of psychopathology: a quantitative review. *Psychol Bull.* 137:856–79.

Maremmani, C., Rossi, G., Tambasco, N., et al. (2012). The validity and reliability of the Italian Olfactory Identification Test (IOIT) in healthy subjects and in Parkinson's disease patients. *Parkinsonism. Relat. Disord.* 18: 788–793.

Marks, L. E. (1974). Sensory Processes. Academic Press: New York.

Marks, L. E., Szczesiul, R., and Ohlott, P. (1986). On the cross-modal perception of intensity. *J. Exp. Psychol. Hum. Percept. Perform.* 12: 517–534.

Marks, L. E., Stevens, J. C., Bartoshuk, L. M., et al. (1988). Magnitude matching: the measurement of taste and smell. *Chem. Senses* 13: 63–87.

Maruniak, J. A., Silver, W. L., and Moulton, D. G. (1983). Olfactory receptors respond to blood-borne odorants. *Brain Res.* 265: 312–316.

McMahon, C., and Scadding, G. K. (1996). Le Nez du Vin--a quick test of olfaction. *Clin. Otolaryngol. Allied Sci.* 21: 278–280.

McReynolds, P. and Ludwig, K. (1987). On the history of rating scales. *Pers. Ind. Diff.* 8: 281–283.

Møller, P., Wulff, C., and Köster, E. P. (2004). Do age differences in odour memory depend on differences in verbal memory?. *Learn. Mem.* 15: 915–917.

Moncrieff, R. W. (1959). The counteraction of odors. *Chem. Canada.* 11: 66–72.

Moskowitz, H. (1977). Magnitude estimation: notes on what, how, when, and why to use it. *J. Food Qual.* 3: 195–227.

Moskowitz, H. R. and Barbe, C. D. (1977). Profiling of odor components and their mixtures. *Sens. Process.* 1: 212–226.

Mozell, M. M., Smith, B. P., Smith, P. E., et al. (1969). Nasal chemoreception in flavor identification. *Arch. Otolaryngol.* 90: 131–137.

Murphy, C., Schubert, C. R., Cruickshanks, K. J., et al. (2002). Prevalence of olfactory impairment in older adults. *JAMA* 288: 2307–2312.

Myers, A. L. (1982). Psychophysical scaling and scales of physical stimulus measurement. *Psychol. Bull.* 92: 203–214.

Nagel, W. A. (1897). Über Mischgerüche und die Komponentengliederung des Geruchssines. *Z. Psychol. Physiol. Sinnesorg.* 15: 82–101.

Nageris, B., Hadar, T., and Hansen, M. C. (2002). The effects of passive smoking on olfaction in children. *Rev. Laryngol. Otol. Rhinol. (Bord.)* 123: 89–91.

Neely, G., Ljunggren, G., Sylven, C., and Borg, G. (1992). Comparison between the Visual Analogue Scale (VAS) and the Category Ratio Scale (CR-10) for the evaluation of leg exertion. *Internat. J. Sports Med.* 13: 133–136.

Nordin, S., Brämerson, Lidén E., and Bende, M. (1998). The Scandinavian odor-identification test: development, reliability, validity, and normative data. *Acta Otolaryngol.* 118: 226–234.

Nordin, S., Martinkauppi, M., Olofsson, J. et al. (2005). Chemosensory perception and event-related potentials in self-reported chemical hypersensitivity. *Int. J. Psychophysiol.* 55: 243–255.

Olofsson, J. K., Rönnlund, M., Nordin, S., et al. (2009). Odor identification deficit as a predictor of five-year global cognitive change: interactive effects with age and ApoE-epsilon4. *Behav. Genet.* 39: 496–503.

Olsson, M. J. (1994). An interaction model for odor quality and intensity. *Percept. Psychophys.* 55: 363–372.

O'Mahony, M. (1979a). Short-cut signal detection measurements for sensory analysis. *J. Food Sci.* 44: 302–303.

O'Mahony, M., Gardner, L., Long, D., et al. (1979b) Salt taste detection: an R-index approach to signal-detection measurements. *Perception* 8: 497–506.

Overbosch, P. (1986). A theoretical model for perceived intensity in human taste and smell as a function of time. *Chem. Senses* 11: 315–329.

Pangborn, R. M., Berg, H. W., Roessler, E. B., and Webb, A. D. (1964) Influence of methodology on olfactory response. *Percept. Motor Skills* 18: 91–103.

Patte, R. and Laffort, P. (1979). An alternative model of olfactory quantitative interaction in binary mixtures. *Chem. Sens. Flav.* 4: 267–274.

Patterson, M. Q., Stevens, J. C., Cain, W. S., and Cometto-Muniz, J. E. (1993). Detection thresholds for an olfactory mixture and its three constituent compounds. *Chem. Senses* 18: 723–734.

Peryman, D. R. and Swartz, V.W. (1950). Measurement of sensory differences. *Food Tech.* 4: 390–395.

Philpott, C. M., Gaskin, J. A., McClelland, L., et al. (2009). The Leicester semi-automated olfactory threshold test—a psychophysical olfactory test for the 21st century. *Rhinology* 47: 248–253.

Pierce, J. D., Jr., Doty, R. L., and Amoore, J. E. (1996). Analysis of position of trial sequence and type of diluent on the detection threshold for phenyl ethyl alcohol using a single staircase method. *Percept. Motor Skills* 82: 451–458.

Pinching, A. J. and Powell, T. P. S. (1971). The neuropil of the glomeruli of the olfactory bulb. *J. Cell Sci.* 9: 347–377.

Potter, H. and Butters, N. (1980). An assessment of olfactory deficits in patients with damage to prefrontal cortex. *Neuropsychologia* 18: 621–628.

Prah, J. D., Sears, S. B., and Walker, J. C. (1995). Modern approaches to air dilution olfactometry. In: R. L. Doty (Ed.), Handbook of Olfaction and Gustation. New York: Marcel Dekker, pp. 227–255.

Preston, C. C. and Colman, A. M. (2000). Optimal number of response categories in rating scales: reliability, validity, discriminating power, and respondent preferences. *Acta Psychologica* 104: 1–15.

Punter, P. H. (1983). Measurement of human olfactory thresholds for several groups of structurally related compounds. *Chem. Senses* 7: 215–235.

Rabin, M. D. and Cain, W. S. (1986). Determinants of measured olfactory sensitivity. *Percept. Psychophys.* 39: 281–286.

Rabin, M. D. (1988). Experience facilitates olfactory quality discrimination. *Percept. Psychophys.* 44: 532–540.

Rabin, M. D. and Cain, W. S. (1989). Attention and learning in the perception of odor mixtures. In Perception of Complex Smells and Tastes, D. G. Laing, W. S. Cain, R. L. McBride, and B. W. Ache (Eds.). Academic Press: Sydney, pp. 173–188.

References

Rakover, S. S. and Teucher, B. (1997). Facial inversion effects - parts and whole relationship. *Percept. Psychophys.* 59: 752–761.

Rawson, N. E., Gomez, G., Cowart, B., et al. (1997). Selectivity and response characteristics of human olfactory neurons. *J. Neurophysiol.* 77: 1606–1613.

Renner, B., Mueller, C. A., Dreier, J., Faulhaber, S., Rascher, W., and Kobal, G. (2009). The candy smell test: a new test for retronasal olfactory performance. *Laryngoscope* 119: 487–495.

Richman, R. A., Post, E. M., Sheehe, P. R., and Wright, H. N. (1992). Olfactory performance during childhood. I. Development of an odorant identification test for children. *J. Pediatr.* 121: 908–911.

Robson, A. K., Woollons, A. C., Ryan, J., et al. (1996). Validation of the combined olfactory test. *Clin. Otolaryngol.* 21: 512–518.

Rovee, C. K., Cohen, R. Y. and Shlapack, W. (1975). Life-span stability in olfactory sensitivity. *Dev. Psychol* 11: 311–318.

Saito, S., Ayabe-Kanamura, S., Takashima, Y., et al. (2006). Development of a smell identification test using a novel stick-type odor presentation kit. *Chem Senses* 31: 379–391.

Schiet, F. T. and Frijters, J. E. R. (1988). An investigation of the equiratio-mixture model in olfactory psychophysics: a case study. *Percept. Psychophys.* 44: 304–308.

Schiffman, S. S., Reynolds, M. L., and Young, F. W. (1981). Introduction to Multidimensional Scaling: Theory, Methods, and Applications. Academic Press: Orlando, FL.

Schmidt, R. and Cain, W. S. (2010). Making scents: Dynamic olfactometry for threshold measurement. *Chem. Senses* 35: 109–120.

Schriever, V. A., Körner, J., Beyer, R., et al. (2011). A computer-controlled olfactometer for a self-administered odor identification test. *Eur. Arch. Otorhinolaryngol.* 268: 1293–1297.

Schwartz, B. S. (1991). The epidemiology of olfactory dysfunction. In The Human Sense of Smell, D. G. Laing, R. L. Doty, and W. Breipohl (Eds.). Springer-Verlag: Berlin, pp. 307–334.

Schwartz, B. S., Doty, R. L., Monroe, C., et al. (1989). The evaluation of olfactory function in chemical workers exposed to acrylic acid and acrylate vapors. *Am. J. Public Health* 79: 613–618.

Semb, G. (1968). The detectability of the odor of butanol. *Percept. Psychophys.* 4: 335–340.

Sezille, C., Messaoudi, B., Bertrand, A., et al. (2013). A portable experimental apparatus for human olfactory fMRI experiments. *J. Neurosci. Methods* 218(1): 29–38.

Shepherd, G. M., and Greer, C. A. (1990). Olfactory bulb. In The Synaptic Organization of the Brain, G.M. Shepherd (Ed.). Oxford University Press: New York, pp. 133–169.

Simmen, D., Briner, H. R., and Hess, K. (1999). Screeningtest des Geruchssinnes mit Riechdisketten. *Laryngorhinootologie* 78: 125–130.

Simon, T. W. and Derby, C. D. (1995). Mixture suppression without inhibition for binary mixtures from whole cell patch clamp studies of in situ olfactory receptor neurons of the spiny lobster. *Brain Res.* 678: 213–224.

Slosson, E. E. (1899). A lecture experiment in hallucinations. *Psychol. Rev.* 6: 407–408.

Slotnick, B. M., Bell, G. A., Panhuber, H., and Laing, D. G. (1997). Detection and discrimination of propionic acid after removal of its 2-DG identified major focus in the olfactory bulb: a psychophysical analysis. *Brain Res.* 762: 89–96.

Solomon, G. S., Petrie, W. M., Hart, J. R., and Brackin, H. B. Jr. (1998). Olfactory dysfunction discriminates Alzheimer's dementia from major depression. *J. Neuropsychiat. Clin. Neurosci.* 10: 64–67.

Springer, K. (1974). Combustion odors-a case study. In Human Responses to Environmental Odors, A. Turk, J. W. Johnston, Jr., and D. G. Moulton (Eds.). Academic Press: New York, pp. 227–262.

Stevens, J. C., and Cain, W. S. (1986). Smelling via the mouth: effect of aging. *Percept. Psychophys.* 40: 142–146.

Stevens, J. C., Cain, W. S., and Burke, R. J. (1988). Variability of olfactory thresholds. *Chem. Senses* 13: 643–653.

Stevens, S. S. (1961). The psychophysics of sensory function. In W. A. Rosenblith (Ed.), Sensory Communication. Cambridge: MIT Press.

Stewart, W. B., Kauer, J. S., and Shepherd, G. M. (1979). Functional organization of rat olfactory bulb analyzed by the 2-deoxyglucose technique. *J. Comp. Neurol.* 185: 715–734.

Stuiver, M. (1958). Biophysics of the Sense of Smell. Doctoral thesis. University of Groningen, The Netherlands.

Sumner, D. (1962). On testing the sense of smell. *Lancet* 2: 895–897.

Takagi, S. F. (1989). Human Olfaction. Tokyo Press, Tokyo, 1989.

Tanner, W. P., Jr. and Swets, J. A. (1954). A decision-making theory of visual detection. *Psychol. Rev.* 61: 401–409.

Thomas-Danguin, T., Rouby, C., Sicard, G., et al. (2003). Development of the ETOC: A European test of olfactory capabilities. *Rhinology* 41: 142–151.

Thurstone, L. L. (1927a). A law of comparative judgment. *Psychol Rev.* 34: 273–286.

Thurstone, L. L. (1927b). Psychophysical analysis. *Amer. J. Psychol.* 38: 368–369.

Toms, M., Morris, N., and Foley, P. (1994). Characteristics of visual interference with visuospatial working memory. *Br. J. Psychol.* 85: 131–144.

Tourbier, I. A. and Doty, R. L. (2007). Sniff magnitude test: Relationship to odor identification, detection, and memory tests in a clinic population. *Chem. Senses* 32, 515–523.

Toyota, B., Kitamura, T., and Takagi, S. F. (1978). Olfactory Disorders - Olfactometry and Therapy. Igaku-Shoin, Tokyo, 1978.

Tsukatani, T., Miwa, T., Furukawa, M., and Costanzo, R. M. (2003). Detection thresholds for phenyl ethyl alcohol using serial dilutions in different solvents. *Chem. Senses* 28: 25–32.

Tsukatani, T., Reiter, E. R. Miwa, T., and Costanzo, R. M. (2005). Comparison of diagnostic findings using different olfactory test methods. *Laryngoscope* 115: 1114–1117.

Ueno, Y. (1992). Perception of odor quality by free image-association test. *Jpn. J. Psychol.* 63: 256–261.

Vollmecke, T. and Doty, R. L. (1985). Development of the Picture Identification Test (PIT): A research companion to the University of Pennsylvania Smell Identification Test. *Chem. Senses* 10: 413–414.

Walker, J. C., Kurtz, D. B., Shore, F. M., et al. (1990). Apparatus for the automated measurement of the responses of humans to odorants. *Chem. Senses* 15: 165–177.

Weber, E. H. (1834). De pulsu, resorptione, auditu et tactu: Annotationes anatomicae et physiologicae. Koehler: Leipzig.

Wenzel, B. (1948). Techniques in olfactometry. *Psychol. Bull.* 45: 231–246.

Weiss, T., Snitz, K., Yablonka, A., et al. (2012). Perceptual convergence of multi-component mixtures in olfaction implies an olfactory white. *Proc. Natl. Acad. Sci. USA* 109: 19959–19964.

Weierstall, R. and Pause, B. M. (2012). Development of a 15-item odour discrimination test (Düsseldorf Odour Discrimination Test). *Perception* 41: 193–203.

Welge-Lüssen, A., Gudziol, V., Wolfensberger, M. and Hummel, T. (2010). Olfactory testing in clinical settings—is there additional benefit from unilateral testing? *Rhinology* 48: 156–159.

Wise, P. M., Olsson, M. J., and Cain, W. S. (2000). Quantification of odor quality. *Chem. Senses* 25: 429–443.

White, E. L. (1979). Synaptic organization of the mammalian olfactory glomerulus: new findings including an intraspecific variation. *Brain Res.* 60: 299–313.

Wise, P. M. and Cain, W. S. (2000). Latency and accuracy of discriminations of odor quality between binary mixtures and their components. *Chem. Senses* 25: 247–265.

Wood, J. B. and Harkins, S. W. (1987). Effects of age, stimulus selection, and retrieval environment on odor identification. *J. Gerontol.* 42: 584–588.

Wright, H. N. (1987). Characterization of olfactory dysfunction. *Arch. Otolaryngol. Head Neck Surg.* 113: 163–168.

Wysocki, C. J., Dorries, K. M., and Beauchamp, G. K. (1989). Ability to perceive androstenone can be acquired by ostensibly anosmic people. *Proc. Natl. Acad. Sci.* 4(86): 7976–7978.

Yoshida, M. (1984). Correlation analysis of detection threshold data for "standard test" odors. *Bull. Facul. Sci. Eng. Chuo Univ.* 27: 343–353.

Zucco, G. M., Militello, C., and Doty, R. L. (2008). Discriminating between organic and psychological determinants of multiple chemical sensitivity: A case study. *Neurocase*, 14: 485–493.

Zwaardemaker, H. C. (1900). Die compensation von Geruchsempfindungen. Arch. Physiol. Leipzig, pp. 423–432, as translated in *Perf. Ess. Oil Rec.* (1959). 50: 217–221.

Zwaardemaker, H. (1925). L'Odorat. Doin: Paris.

Zwaardemaker, H. (1927). The sense of smell. *Acta Oto-Laryngol.* 11: 3–15.

Zwaardemaker, H. (1930). An intellectual history of a physiologist with psychophysical aspirations. In A History of Psychology in Autobiography, Vol. 1, C. Murchison (Ed.). Clarke University Press: Worcester, MA, p. 491.

Chapter 12

Electrophysiological Measurement of Olfactory Function

ALLEN OSMAN and JONATHAN SILAS

12.1 INTRODUCTION

In this chapter, we introduce the fundamentals of electroencephalography (EEG) and survey its many applications in basic and applied research on olfactory function. Both classic and modern approaches to data collection and analysis are reviewed, providing a comprehensive overview of this emerging field.

In 1875, Richard Caton discovered that electrical potentials could be measured directly from the exposed surface of the cerebral cortex in animals. This paved the way for the first EEG recordings from the human scalp by Hans Berger (1929). Both Caton and Berger noted that brain electrical activity could be influenced by external sensory events. Such effects were first reported for odorants in 1890 by Fleischl von Marxow in response to ammonia presented to a rabbit's nose. It was not until the late 1960s, however, that modulation of scalp-recorded EEG by odorants was demonstrated in humans (Finkenzeller, 1965; Allison and Goff, 1967). This has now been documented in numerous studies and shown to be mediated by both the olfactory and trigeminal systems. Before turning to these studies, let us consider first the neurophysiological basis of EEG.

EEG is a time series of voltages produced by electrical fields in the brain. This chapter will focus exclusively on EEG potentials recorded from the human scalp. These electrical signals arise from the activity of a large number of neurons which summates and is conducted to scalp electrodes through brain, bone, and other intervening tissue. For summation and conduction to occur, two requirements must be met. First, electrical activity from many neurons must overlap in time, that is, occur synchronously. Because action potentials are so brief, they are unlikely to occur synchronously. Consequently, EEG is thought to arise mainly from graded post-synaptic potentials, which have a longer duration. Second, the neurons must have similar orientations. When neurons all point in the same direction, they produce "open" electrical fields that can be detected at a distance. But, when arranged randomly or radiating from a common center, they produce "closed" fields that are not conducted far and cannot be detected at the scalp.

Thus, scalp-recorded EEG is merely the tip of an electrical iceberg, comprising mostly synchronous postsynaptic potentials from populations of neurons that produce open fields. Open fields are obtained when neurons are organized in layers, as is the case in the cerebral cortex, which is also proximal to the scalp. Indeed, cortical pyramidal cells are especially well suited to generate measurable electric fields, given that their apical dendrites are aligned in common perpendicular to the cortical sheet. It is important to recognize, however, that the synchrony of cortical neurons is modulated by afferents from subcortical sources.

When recorded from the scalp, EEG provides a noninvasive tool for studying neural states and responses in humans. In comparison to other neuroimaging measures, it has a relatively direct connection to neural activity, albeit a subset. Perhaps one of its most useful features is its good temporal resolution. Modern signal acquisition allows voltage to be sampled at intervals of less than 1 msec. Though temporal resolution depends on other factors as well, including inherent temporal variability of

---

Handbook of Olfaction and Gustation, Third Edition. Edited by Richard L. Doty.
© 2015 Richard L. Doty. Published 2015 by John Wiley & Sons, Inc.

the measured processes, it is greater for EEG than most neuroimaging techniques.

EEG time series have been analyzed quantitatively in at least three ways. A separate section of this chapter is devoted to each as employed in modern studies of olfaction. The oldest is to compute power spectra based on extended recordings lasting several seconds or minutes. Like light, sound, and other wave phenomena, "brain waves" contain periodic components at specific frequencies whose power can be represented as a spectrum (see Section 12.2). Another form of analysis computes event-related potentials (ERPs) by averaging multiple short EEG segments that proceed or follow repeated "events" (stimuli or responses). ERPs, which reflect that portion of the EEG with a constant time-relation to the event, often have a temporal resolution on the order of msecs (see Section 12.3). These two types of analysis can be thought of as emphasizing, respectively, EEG in the "frequency domain" (power as a function of frequency) and "time domain" (amplitude as a function of time). The most recent type of analysis ("time-frequency") examines EEG changes simultaneously in both the time and frequency domains with a high degree of resolution in both (see Section 12.4).

12.2 OSCILLATORY CHANGES IN EEG DURING OLFACTION

Berger and other early researchers observed changes in on-going EEG rhythms induced by non-olfactory stimuli, for example, visual ones. Modern studies have shown that odorants can induce analogous effects. In this section of the chapter we turn to these latter studies and the type of EEG analyses they employ.

12.2.1 Neuronal Mechanisms

Decreases or increases in power, at any particular frequency, caused by a specific event of interest are thought to reflect the synchronicity of the firing of an underlying neuronal population. Absolute power is a measure of the intensity of energy (in μV) squared and calculated in a series of frequency bands (termed the power spectrum) for selected periods of time. A desynchronization of neuronal firing results in a decreased measure of power in the EEG. Conversely, a synchronization of the underlying neuronal population results in increased power. In most cases, more synchronization and greater power usually indicates less cortical processing and vice versa (Pfurtscheller and da Silva, 1999).

Changes of power, at any particular frequency, measured while olfactory processing takes place do not necessarily reflect low-level perceptual components of olfactory processing. Rather, these frequency changes have

been commonly viewed as an indication of higher-order cognitive or conscious processes associated with the "olfactory percept" and not necessarily activities germane to activity within the olfactory pathways, per se (Kobal, 2003). Nevertheless, some research has demonstrated a relation between cortical frequency changes and olfactory processing that are relevant here, even if not associated with so-called "sensory" mechanisms.

As research in the field of continuous changes in the on-going EEG develops, it now appears that these measures may, in fact, be more useful or specific then other electrophysiological changes classically associated with olfactory sensory processing. Furthermore, to what degree so-called "sensory" and cognitive components are dissociable in the formation of any percept, including an olfactory one, is questionable (see Doty et al., 1994).

12.2.2 Early Research

One of the early documentations of quantitative odorant-induced changes in the on-going EEG was that by Lorig et al. (1988). Using a method of analyzing changes in power at different frequencies for discrete time periods, these investigators noted distinct differences in EEG power when comparing the inhalation of room air through the nose to inhalation through the mouth. Specifically, they documented a more widely distributed decrease in frequency power across the scalp during nasal than during mouth inhalation in the beta (13–64 Hz) frequency range as well as a greater decrease in the alpha (8–13 Hz) frequency bandwidth over the left hemisphere. This suggested that nasal inhalation recruits an olfactory processing network, whereas oral inhalation does not. This explanation likely holds under the constraints of their paradigm, although under other circumstances retronasal olfactory processing via the oral cavity could also be in play (Halpern, 2009).

Alpha and beta rhythms are often associated with attentional, motoric, and higher-order perceptual cortical processes that are believed to be largely independent of sensory processing (Pfurtscheller and da Silva, 1999). Lorig et al. (1988) suggested, however, that odor-related cortical changes demonstrate the effect of unaware or unconscious olfactory processing in a distributed system in the central nervous system (CNS). However, other explanations of this phenomenon are possible. For example, such changes could simply reflect changes in oxygen intake. That being said, subsequent research has shown that different odorants can be differentiated by a decrease in alpha frequency power at different electrode locations on the scalp (van Toller and Reed, 1989). Indeed, alpha and beta decreases in power seem to reflect the involvement of cortical processes involved in olfactory perception.

Early investigations also revealed some interesting findings in relation to modulations of power within the theta frequency range (~4–8 Hz). Sensory stimulation, regardless of the modality, tends to decrease theta activity (Morruzzi and Magoun, 1949). This is thought to be reliant upon a direct link between the ascending reticular formation and clusters of neurons involved in sensory processing (Motokizawa and Furuya, 1973). However, despite the connection from primary olfactory cortical structures to the ascending reticular formation, some evidence seems to suggest an *increase* in power in the theta frequency bandwidth after odorant stimulation (see Lorig, 1989). Theta increases are most prominent when measured at posterior areas on the scalp and occur across different types of odors, whereas inhalation of odorless air seems to produce a decrease in theta activity (Klemm et al., 1992).

Some have suggested that an increase in theta activity is associated with decreased arousal perhaps due to some kind of cognitive associative function that olfaction has with food (Lorig, 1989). However, given that these increases tend to occur across different odors, even non-food ones, this seems somewhat unlikely.

That being said, one study examining a modulation of the theta frequency bandwidth found a consistent *decrease* in theta activity during the perception of food-related olfactory stimuli (Martin, 1998). This discrepant finding in the theta range is difficult to reconcile with studies finding an increase in theta activity and could reflect some mediation by subjective state or mood of the individual (Lorig and Schwartz, 1988). It is also possible that such an inconsistency could reflect noise associated with the analytical technique that was employed, that is, a "discrete" method of frequency analysis. Later in the chapter (Section 12.4) we discuss in greater detail why discrete methods of frequency analysis likely confer a decreased signal to noise ratio.

12.2.3 Coherence

In addition to power, it is possible to examine the *coherence* of EEG oscillations. Coherence, in this context, is the correlation between measurements of the EEG signal at two sites in phase at a specific frequency. Harada et al. (1996), in a pilot study, presented two different odors to participants, methyl-cyclopentenolone (a nutty maple smell) and scatol (a feces-like smell) and measured coherence changes relative to rest. For methyl-cyclopentenolone olfactory perception, coherence in the delta frequency bandwidth decreased in frontal regions, whereas alpha coherence increased bilaterally in temporal regions. For stimulation with scatol, delta coherence decreased in the frontal region but alpha coherence increased in frontal and occipital regions.

Coherence is thought to represent the information associated with a state or stimulus in a distributed cortical network. Increased coherence in the alpha frequency range may reflect a distributed processing of the subjective and psychophysical aspect of the olfactory percept. Harada

et al. (1996) give little in the way of interpretation for their findings but suggest further work may build on these results; unfortunately little exists. To our knowledge only one other study had examined coherence during olfactory processing (Cherninskii et al., 2009). This more recent account of coherence reports an increase in coherence in the beta frequency bandwidth measured over left temporal regions. Cherninskii et al. (2009) suggest that this may be indicative of higher-order cognitive functions related to the formation of semantic concepts associated with olfactory perception. Although more research is needed, analysis of frequency coherence during olfaction may prove to be a fruitful method for better understanding the role of distributed cortical processes.

12.2.4 Clinical and Hedonic Significance

Changes in the EEG as a result of olfactory stimulation have also been used to better define or diagnose a range of diseases. It is well established, for example, that olfactory deficits are an early sign of a number of neurodegenerative diseases, including Alzheimer's disease (AD) and Parkinson's disease (PD) (Doty et al., 1991; Doty, 2012). Seal et al. (1998) report differences in resting state EEG power between patients with AD and patients with generalized vasculature dementia (VaSD). Furthermore, changes in EEG power during an odor detection task were better than resting state EEG activity at identifying those with AD. Specifically, greater increases in theta wave power and lesser decreases in beta and alpha wave power were found among the AD participants, compared to those with VaSD, during an odor detection task. The authors suggest the changes in the theta frequency bandwidth may reflect a general slowing of the EEG signal common among those with AD (Berg et al., 1984). However, during the odor detection task the decreased power of the beta frequency bandwidth was best at differentiating between those with AD and vascular dementia, conceivably reflecting cortical processes specifically related to odor detection.

Some studies import a primacy to the hedonic, or "pleasantness," perceptual component of olfactory stimuli (e.g., Khan et al., 2007). Proponents of this concept assume that the perceptual hedonic component can be mapped onto the physiochemical properties of an odorant, thereby suggesting a partially innate aspect of olfactory liking (Khan et al., 2007). This indicates that the liking of olfactory stimuli is fundamental to forming an olfactory percept.

Unfortunately, only a few studies have examined modulations of frequency components of the EEG in relation to the hedonic aspect of odor perception. Brauchli et al. (1995) recorded EEG activity during the suprathreshold presentation of phenylethyl alcohol (PEA; a pleasant "floral" smell) and valeric acid (VA; an unpleasant "cheesy" smell). Overall, perception of VA was associated with a relatively greater increase in an upper alpha frequency bandwidth (i.e., 9.78–12.5 Hz) than PEA at frontal and parietal electrode locations. These authors opined that the increased alpha power reflected a decrease in cortical activation since "olfactory processing occupies large subcortical areas" (p. 513). This suggested to them that unpleasant stimuli recruit a primary subcortical olfactory system to a greater degree than pleasant ones and that this resulted in a suppression of cortical activation. Their conclusion, based only on five participants and two odorants, was supported by Owen and Patterson (2002), who noted that an odor rated as unpleasant by one group of participants tended to increase alpha activity in frontal locations compared to those who rated the same odor as pleasant. It should be noted, however, that unlike PEA, VA is a relatively strong stimulus for intranasal trigeminal (CN V) afferents, implying a potential confound of the involved neural activity (Doty et al., 1978).

Observations of cortical suppression during olfactory processing contrast with observations of decreased alpha activity during the presentation of non-olfactory suprathreshold stimuli thought to reflect attentional orienting towards the stimuli. However, more recent research has found an olfaction-related decrease in alpha power over the occipital cortex, similar to the traditional decrease seen in other sensory modalities. Interestingly, in one study approximately 66% of patients who reported being anosmic from head injury still displayed a definite decrease in power of the alpha frequency during olfactory stimulation (Bonanni et al., 2006). This so-called "olfactory stop response" of the alpha frequency suggests that, despite self-reports of olfactory perceptual deficits, cortical processing of the stimuli may in some cases still be detectable. Further research, including control groups and an examination of the relation between alpha modulation and standardized smell tests, is required to further understand this finding. It is possible that residual cortical functioning in the olfactory system is detectable without conscious perception of a given odorant. Moreover, most persons with head trauma related olfactory deficits do not have complete loss of olfactory function, despite beliefs to the contrary (Doty et al., 1997).

12.2.5 Temporal Resolution

The research reviewed in Section 12.2 is representative of the majority of work that has examined gradual changes in on-going EEG activity during olfactory stimulation. The measures employed all involve basic techniques commonly used to analyze changes in the on-going EEG over an extended period of time. Sections 12.3 and 12.4 review

research on olfaction employing EEG techniques that allow for much greater temporal resolution. We now turn to work involving what is perhaps the most well-established approach to precise measurement of the time-course of EEG changes, the use of Event-Related Potentials (ERPs).

12.3 CHEMOSENSORY EVENT-RELATED POTENTIALS

ERPs are the portion of the EEG elicited by a specific event, for example, the presentation of a stimulus or the start of an overt movement. The ERP associated with a single event is usually very difficult to observe in an EEG recording because, like the ripples from a stone tossed into a strong current, it is dwarfed by the background activity in which it is embedded. The most common method for extracting ERPs from the background EEG involves a process of averaging. An event is repeated many times while the EEG is recorded and the EEG is then divided into segments, each "time-locked" to the event. An average segment is calculated by averaging the voltages at corresponding time-points from individual segments, where "corresponding" means preceding or following the start of an event by the same amount of time. As more individual segments are added to the average, variations in amplitude of the background EEG diminish. This is because these variations are random fluctuations around an expected mean voltage that is the same at any given moment relative to the event. The ERP, in contrast, is a systematic pattern of voltage changes that occur at the same time relative to each repetition of the event. Thus, it remains (relatively) unscathed by the averaging process while fluctuations in the background EEG are cancelled out.

12.3.1 Measurement

ERPs elicited by odorants are most often referred to as chemosensory event-related potentials or CSERPs (Evans et al., 1993). Two factors have made the measurement of CSERPs especially challenging. First, as highlighted above, the measurement of ERPs is crucially dependent on the occurrence of brief events with abrupt onsets. Punctate stimulation is more difficult to achieve for the chemical senses than for vision, audition, or touch. Second, it is difficult to stimulate the olfactory nerve without also stimulating the trigeminal nerve, whose intranasal free nerve endings signal such sensations as warm/cool, irritation, and pungency. To do so, requires (a) the presentation of olfactory stimuli without concomitant changes in pressure, temperature, or humidity, and (b) the use of odorants that stimulate only the olfactory nerve. These two problems were a central concern of early work involving CSERPs and caused the field to have a somewhat rocky start.

CSERPs were first reported by Finkenzeller (1965) and Allison and Goff (1967). However, enthusiasm for CSERPs as a means to study olfaction soon diminished after Smith et al. (1971) found them to be solely the result of trigeminal stimulation. We now know that CSERPs can arise from either the trigeminal or olfactory systems. Most odorants stimulate both (Doty et al., 1978) and thus elicit composite CSERPs (Kobal and Hummel, 1988). A few, such as vanillin and PEA, are not generally detectable by anosmics with preserved trigeminal function, particularly at low to moderate concentrations. Thus, they mainly stimulate the olfactory nerve alone (Doty et al., 1978). Such odorants elicit CSERPs in individuals with normal olfactory function (Kobal and Hummel, 1988; Kobal and Hummel, 1998), but not in anosmics (Kobal and Hummel, 1998). In contrast, odorless carbon dioxide is used to selectively stimulate the trigeminal nerve (Kobal and Hummel, 1988; 1998). The disparity between Smith et al. (1971) and later studies was probably due to the subsequent development of more effective methodologies for adequately stimulating the olfactory system and preventing trigeminal stimulation.

Chief among these advances were specialized olfactometers (e.g., Benignus and Prah, 1980; Evans and Starr, 1992; Kobal and Plattig, 1978; Kobal, 1981; Kobal and Hummel, 1988; Lorig et al., 1999). The first, originally developed by Kobal and Plattig (1978), became commercially available and added considerable impetus to the study of CSERPs. In their device, one of two concurrent air streams is directed to the olfactometer outlet, which consists of a teflon tube inserted into one of the subject's nostrils. Both streams have the same flow rate, temperature, and humidity. They differ only in that one contains a mixture of air and odorant, while the other contains air alone. Switching which stream exits the olfactometer into the nose, which can occur in less than 20 msec, is used to present a series of brief odorants embedded in a background of odorless air without producing any concomitant changes in stimulation.

Other foundational methodological developments concern parameters of odorant presentation, such as duration, inter-stimulus interval, and flow rate. Highlighting the need for abrupt onsets, CSERPs result primarily from the initial portion (e.g., 100 msec.) of odorant presentation (Kobal, 2003). The duration of inter-stimulus intervals is important because of habituation at short intervals. Though the degree of habituation depends on the odorant type and concentration, CSERP amplitude can be reduced substantially at intervals of less than 30 sec (Kobal and Hummel, 1991). Odorant flow rate is important because it influences the number of molecules reaching and being absorbed into the olfactory mucosa. Another factor that has received considerable attention is breathing. Subjects are often asked to mouth breath, while odorants are presented asynchronously with their respiratory cycle. Under such conditions, CSERPs appear to be larger in response to odorants presented during inhalation than exhalation (Haehner et al., 2011; but see Pause et al., 1999). To prevent mouth breathing from influencing airflow in the nasal cavity, and thus odorant presentation, subjects are sometimes taught a technique involving velopharyngeal closure (Kobal, 1981). Reviews of these and other methodological issues can be found in Kobal and Hummel (1991), Lorig (2000), and Kobal (2003).

In contrast to the methodological hurdles posed by the presentation of olfactory stimuli, the recording of CSERPs is straightforward relative to that of other ERPs. In many cases (see Section 12.3.2 below), it is sufficient to record EEG from a small number of electrodes along the midline near the vertex. The range of frequencies recorded (online filter settings) is usually similar to what is typically employed for recordings of late cortical ERPs (e.g., 0.01–100 Hz), which easily includes all the frequencies that carry information examined in current CSERP analyses. The scalp recordings are typically referenced to recordings at one of the mastoid bones or the algebraic average of recordings at both. Earlobes are also popular. Following the session, segments of the recordings containing electrical contamination from blinks, eye-movements, or other artifacts are removed or corrected. After some additional offline filtering, the EEG recordings are ready to be averaged to form CSERPs.

12.3.2 Representation

CSERPs have been represented in a number of ways. The most common form to date is as a time-series of voltages recorded at one or a few locations on the scalp. These time-series are often conceived of as comprising a series of distinct "components," each arising from a specific neural population during a specific type of information processing. Components are operationally defined by their polarity, latency, shape, topography over the scalp, and sensitivity (or lack thereof) to various experimental manipulations. Though not always the case, each component is often identified with a specific peak or trough that occurs (more-or-less) consistently in the voltage time-series.

Figure 12.1 provides an example of voltage time-series with labeled components from a study by Hawkes et al. (1997). Each is the average of recordings from an individual subject in response to odorants stimulating either the trigeminal (CO_2, Panel A) or olfactory (Hydrogen Sulfide, H_2S, Panel B) nerve and was obtained at the location on the scalp where each type of CSERP is maximal (Panel C). They are representative of a general pattern in which the first of a series of large deflections is positive and occurs at about 250 msec. after stimulus onset, the second is negative, and the third is positive and may be a complex of several deflections (Kobal, 2003). The three labeled

266 Chapter 12 Electrophysiological Measurement of Olfactory Function

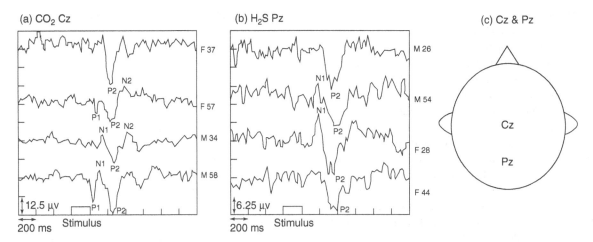

Figure 12.1 (a) CSERPs from four healthy subjects in response to trigeminal simulation (200 msec. pulses of CO_2) recorded at electrode site CZ. (b) CSERPs from four healthy subjects in response to olfactory stimulation (200 msec. pulses of H_2S) recorded at electrode site Pz. Time and voltage (negative up) are shown respectively on the horizontal and vertical axes. Each subject's age and sex is shown to the right of their CSERP. (c) Scalp locations of electrode sites Cz and Pz. Adapted with permission from "Olfactory dysfunction in Parkinson's disease," by C.H. Hawkes, B.C. Shephard, and S.E. Daniel, 1997. *Journal of Neurology, Neurosurgery, and Psychiatry*, 62(5), p. 438. Copyright 1997 by BMJ Group.

components—P1, N1, and P2—are identified with these deflections and named on the basis of their polarity and order (Evans et al., 1993). Most often measured are the peak amplitude and latency of N1 and P2, which have test-retest reliabilities comparable to those of cortical components elicited by visual, auditory, and tactile stimuli (Thesen and Murphy, 2002; Welge-Lussen et al., 2003; Nordin et al., 2011). These CSERP components arise from cortical areas involved in early stages of sensory processing (Pause and Krauel, 2000) but occur later than analogous cortical components in other sensory modalities. This delay may be due, at least in part, to the time required after presentation for odorants to reach and stimulate receptors in the nose (Kobal, 2003).

Figure 12.2 provides examples of several alternative ways to represent CSERPs. The figure comes from a study by Lascano et al. (2010) and shows the grand average of responses from 12 healthy subjects to H_2S. Panel A shows a voltage time-series which, though in the same format as those in Figure 12.1, displays broader N1 and P2 components. This is to be expected in a grand average, given the variability in their peak latencies between subjects. Panel B shows a butterfly plot displaying the voltage time series for each of 64 recording sites across the scalp. As can be seen, the variability of voltages across scalp sites at a given time point (vertical spread) increases at times when components are evident in the display above (A). This is because the difference in voltage between locations across the scalp at a given moment is proportional to the overall strength of the neural response. Panel C displays a measure of this spatial variability in voltage called the Global Field Power (GFP; Lehmann and Skrandies, 1980). GFP is represented by the contour of the plot. The colored segments under the contour indicate "micro-states" (Lehmann and Skrandies, 1980), which are likewise based on the topographic distribution of voltages. Changes in ERP topography indicate changes in the identity, orientation, and/or relative strengths of its neural generators. Microstates correspond to intervals of time during which the topography, and hence the configuration of neural generators, remains relatively constant. The ones shown here were inferred using a clustering algorithm. The average topography corresponding to each microstate is shown in Panel D. Those for States 2 and 4 provide a good idea of the voltage distribution across the entire scalp around the time of N1 and P2.

A final example shown in Figure 12.3 is based on work by Kayser et al. (2010) and involves grand average CSERPs from 35 healthy subjects in response to H_2S. The approach employed here was motivated in part by two problems associated with ERPs in general. First, they are very much influenced by the choice of reference electrode. Second, recordings from multiple electrode sites need somehow to be integrated. Both problems are in fact solved by GFP, which is a single measure based on the entire voltage topography and independent of reference. Kayser et al. chose a different approach that provided detailed inferences about components. First, they transformed their recordings at each electrode site from voltage to current source density (CSD), a reference-free measure. A set of components underlying the resulting CSD time-series at all electrode sites was then derived by Principal Component Analysis (PCA). These components are displayed in Panel A, which shows their time-courses and the percentage of the variance accounted for by each. Kayser et al. interpreted the earliest two (peaks at 305 and 630 msec) as corresponding to the

12.3 Chemosensory Event-Related Potentials

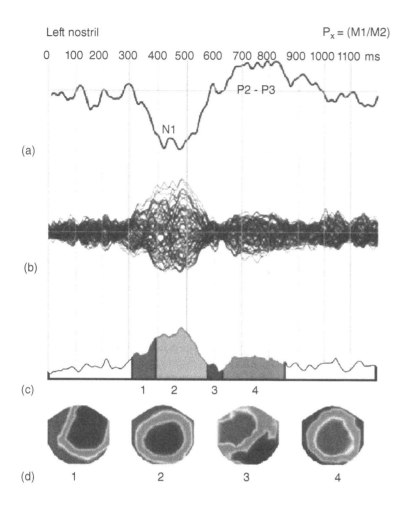

Figure 12.2 Olfactory CSERPs 0–1200 msec. following left-nostril stimulation from 12 healthy subjects. (a) Grand average of time-series recorded at electrode site Pz. (b) Butterfly plot of overlaid grand average time-series at all 64 electrode sites. (c) Temporal extent (colored segments) and global field power (contour) of the microstates identified by a cluster analysis of the 64 grand average time-series. (d) Scalp topographies of the microstates (red, positive voltage; blue, negative voltage). Adapted from "Spatio-temporal dynamics of olfactory processing in the human brain: An event-related source imaging study," by A.M. Lascano et al., 2010. *Neuroscience*, 167(3), p. 703. Copyright 2010 by Elsevier LTD. Adapted with permission. (*See plate section for color version.*)

traditional N1 and P2 components and the two later components as related to a motor response performed by subjects. Panel B displays the topographic distribution of the N1 and P2 CSD components for weak and strong odorants, as well as the difference between the two (intensity effect). Of particular interest are the locations of maximal current flow for each component, which are consistent with the putative neural sources discussed in the next section.

12.3.3 Neural Sources

The neural generators of CSERPs are not yet well known. This is in part because few studies have so far attempted to locate them, and also because of difficulties inherent in ERP source localization. One prerequisite for inferring the neural generators of ERPs from electrical measurements on the scalp is a "head model," that is, an estimate of the resistance and geometry of each type of tissue comprising the head. Electric fields, however, are accompanied by magnetic ones that arise from the same sources and are uninfluenced as they pass through these tissues. Thus, because it minimizes the need for a head model, it is somewhat more tractable to infer sources from magnetic than electrical measurements. In most cases, however, neither type of measurement at the scalp provides sufficient information to uniquely specify a set of underlying sources. It is therefore necessary to constrain the range of possibilities by making a number of physiologically plausible assumptions, including adoption of a model concerning the nature and/or number of sources. The earliest and perhaps best known "source models" involve a small number of current dipoles (source and sink), each of which can vary in location, orientation, and strength.

For these reasons, most extant studies providing information germane to the sources of CSERPs have employed magnetoencephalography (MEG) and dipole modeling along with concurrent recording of EEG. Two of these studies, including the earliest, employed CO_2 stimulation of the trigeminal nerve in an effort to elucidate the neural basis of pain (Huttunen et al., 1986; Hari et al., 1997). Both found a source that was most active around the time of N1 and located bilaterally in the upper bank of the Sylvian fissure near the lateral end of the central sulcus in the vicinity of the secondary somatosensory area. The later study (Hari et al., 1997) found this source to have a right-hemispheric dominance and detected a second source located in the

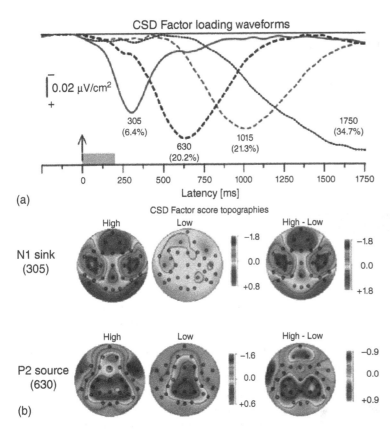

Figure 12.3 Principal Component Analyses (PCA) of current-source density (CSD) measures of olfactory CSERPs to H_2S. (a) Factor loadings of the first four PCA factors (labels indicate peak latency and variance explained) extracted from olfactory CSD waveforms. (b) PCA-CSD factor score topographies corresponding to N1 sink (top) and P2 source (bottom) for 35 healthy subjects. Topographies are shown for responses to low- and high-intensity stimuli, as well as the difference (high minus low). Adapted from "Neuronal generator patterns of olfactory event-related brain potentials in schizophrenia," by J. Kayser et al., 2010. *Psychophysiology*, 47, p 1082. Copyright 2010 by Wiley Periodicals, Inc. Adapted with permission. (*See plate section for color version.*)

Rolandic area, that is, around primary sensorimotor cortex. Other studies have involved odorants affecting primarily the olfactory nerve. Most have found dipole sources during the P1-N1-P2 interval located bilaterally in the Sylvian and superior temporal fissures, as well as the intervening superior temporal lobe (Kettenmann et al., 1996, 1997; Sakuma et al., 1997). These all lie in cortical areas with hypothesized connections to the primary olfactory cortex, including anterior-central parts of the insula and parainsular cortex (Kettenmann et al., 1997). Bilateral sources have been reported also in inferior parietal cortex (Sakuma et al., 1997) and near the orbitofrontal sulcus (Tonoike et al., 1998).

Though magnetic fields are generally preferred for source localization, electric fields do have some advantages. Magnetic fields measurable over the scalp arise from only a subset of the sources that produce measureable electric fields, specifically ones closer to the scalp involving current flow tangential to the scalp, for example, from the walls of sulci. Electric fields can be used to detect these sources as well as somewhat deeper ones and radial ones in gyrii. So far, the only attempt to model the sources of CSERPs based directly on the ERP signals has been the study discussed above by Lascano et al. (2010). Besides taking into account the electrical resistance of tissues, there were two other noteworthy methodological differences from the above MEG studies. First, Lascono et al. used a distributed source model. Such models involve a 3-D volume comprised of a large number of voxels, in which each voxel "contains" a dipole with a specific strength and orientation. The volume used here corresponded to grey matter of the cerebral cortex and limbic structures segmented from an average MRI (Montreal Neurological Institute). Second, they separately modeled the voxels active during each of the four microstates shown in Figure 12.2. As mentioned, each microstate corresponds to a time interval with a stable ERP topography and hence a stable configuration of generators. During the first microstate, activation ipsilateral to the stimulated nostril was found in the medial temporal lobe (parahippocampal gyrus and amygdala) and lateral temporal lobe (middle and superior temporal gyrus, inferior insular cortex). During the second and third microstates, activation spread to the same structures in the contralateral hemisphere. The inferior frontal gyrus became activated bilaterally during fourth state.

12.3.4 Functional Significance

A rough idea has begun to emerge as well about the relation between CSERPs and olfaction at the functional level. Relevant information has come from the examination of effects on CSERPs of factors and manipulations that influence the efficacy or quality of olfactory perception, as well as correlations between CSERPs and other olfactory measures.

Included are the effects of stimulus properties, such as concentration. When odorants stimulating the olfactory and/or trigeminal nerve are strong enough to be perceived, further increases in intensity decrease the latency and/or increase the amplitude of the P1 component, N1 component, and at least part of the late positive complex (Kobal and Hummel, 1991; Pause et al., 1997; Tateyama et al., 1998; Wang et al., 2002). Consistent with the stimulus-intensity effects, these components have been found to be larger (Murphy et al., 1994) or earlier (Tateyama et al., 1998) in individuals with lower olfactory thresholds, that is, a more sensitive sense of smell. The amplitude of CSERP components is influenced also by the state of olfactory adaptation or habituation, as manipulated by the rate of odorant presentation. Wang et al. (2002) reported such effects, as well as an impressive correlation between component amplitudes and detection performance across experimental conditions defined by different combinations of stimulus intensity and rate of presentation.

Olfactory perception is known to vary with traits that vary among individuals, including age and gender. The decline of olfactory function throughout the lifespan of healthy adults is well established, including increased thresholds for detection and diminished ability to identify odorants (e.g., Doty et al., 1984; Murphy et al., 2004). Likewise, olfactory CSERP components become progressively smaller (P2 and N1-P2 difference) and slower (N1) over much of the course of normal adult aging (Murphy et al., 1994; Evans et al., 1995; Hummel et al., 1998; Covington et al., 1999; Murphy et al., 2000). Within each age group, women have been found to have lower thresholds and better identification than men (for a review, see Doty and Cameron, 2009). So, it is perhaps not surprising that women have larger CSERPs (Becker et al., 1993; Evans et al., 1995; Morgan et al., 1997; Olofsson and Nordin, 2004; Stuck et al., 2006).

CSERPs are also influenced by psychological states. Especially susceptible to such effects is a constituent of the late positive complex thought to be a P300.[1] The P300 is not exclusively an olfactory ERP component; it can be elicited by stimuli in any sensory modality and is thought to arise from brain regions, such as the hippocampus and amygdala (Halgren et al., 1980) and/or cortical association areas (Polich and Squire, 1993), not devoted solely to any particular modality. The P300 is a response to unexpected stimuli and is studied most often using the "oddball" paradigm, which involves sequences comprised of frequently and infrequently repeated stimuli. P300 amplitude is larger in response to the infrequent stimuli, even when "frequent" and "infrequent" are defined by

membership in an abstract category based on a semantic feature, for example, gender in a list of names (Kutas et al., 1977). A number of studies employing oddball paradigms have demonstrated robust P300s in response to infrequent odorants (e.g., Durand-Lagarde and Kobal, 1991; Pause et al., 1996; Pause and Krauel, 2000). Perhaps because it is related to "surprise," P300s can be elicited also by a single repeated odorant when the intervals between repetitions are as long as those typically employed to avoid habituation (e.g., Morgan et al., 1999; Geisler and Murphy, 2000).

P300s elicited by olfactory stimuli have the same characteristics as those elicited by stimuli in other sensory modalities. They require attention to the stimuli (Krauel et al., 1998; Geisler and Murphy, 2000) and are sensitive to their subjective significance (Pause et al., 1996; Laudien et al., 2008; Bulsing et al., 2010) and emotional valence (Pause and Krauel, 2000). These effects are consistent with the idea that the P300 reflects cognitive states that can be induced through any sensory channel. They also indicate that the contribution of P300s to the CSERP is likely to be reduced under conditions with equally probable odorants of little intrinsic significance to which subjects need not attend. It is noteworthy, however, that the N1 and P2 are also sensitive to manipulations of attention and stimulus significance (Krauel et al., 1998; Geisler and Murphy, 2000; Laudien et al., 2008; Bulsing et al., 2010). This would suggest that the neural processes that directly produce these olfactory ERP components either perform cognitive functions or receive feedback from other (more central) processes that do. Further clues concerning the functional significance of CSERPs are contained in the next two sections, which detail some of their applications.

12.3.5 In the Olfactory Clinic

There has been considerable interest in applying CSERPs in clinical settings devoted to olfactory disorders. This interest stems in part from their potential to detect olfactory responses in individuals who might be unwilling (e.g., malingerers) or unable (e.g., infants or those with cognitive disabilities) to provide accurate responses on behavioral tests. Moreover, as a psychophysiological measure, they hold the promise of detecting subtle responses that may evade behavioral tests. CSERPs have been obtained from patients with olfactory disorders due to a variety of causes, including nasal disease, upper respiratory infection, poisoning or adverse reactions to medications, traumatic head injury, neurological problems, and congenital or idiopathic causes. So far, the feature of CSERPs most often employed has been their presence vs. absence as judged by trained observers. As mentioned earlier, the absence of CSERP responses to olfactory stimuli has been demonstrated in anosmics (Cui and Evans, 1997; Kobal and Hummel, 1998). However, as illustrated in Figure 12.4, they can

---

[1] Despite its name, this component often follows the stimulus by more than 300 msec. For reviews see Donchin (1981), Donchin and Coles (1988), and Verleger (1988).

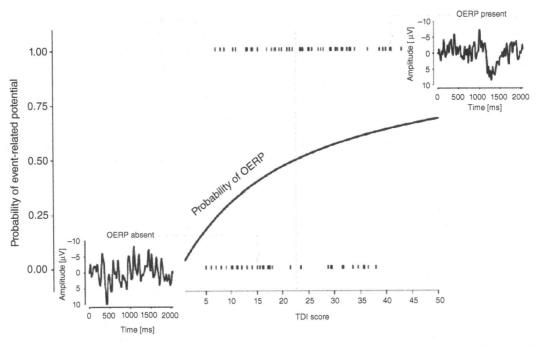

Figure 12.4 Probability of detecting an olfactory CSERP as a function of TDI score (sum of a patient's test scores for odor threshold, discrimination, and identification). Dots indicate the presence or absence of a CSERP from individual patients, and the vertical dotted line indicates a 0.5 probability at a TDI of 22.6. TDIs of less than 15.5, between 15.5 and 30.5, and greater than 30.5 indicate respectively functional anosmia, hyposmia, and normal function. The two inserts depict examples of the absence of a CSERP from a hyposmic patient (left bottom, TDI score indicated with an arrow) and presence of a CSERP from a normosmic patient (right top, TDI score indicated with an arrow). Reprinted from "The clinical significance of electrophysiological measures of olfactory function," by J. Lotsch and T. Hummel, 2006. *Behavioral Brain Research*, 170, p. 80. Copyright 2006 by Elsevier B.V. Reprinted with permission.

be undetectable as well in patients for whom behavioral measures have demonstrated some preserved olfactory function.

Figure 12.4, from Lotsch and Hummel (2006), shows the probability of detecting olfactory CSERPs from patients as a function of their combined threshold, discrimination, and identification (TDI) score on a behavioral test (Kobal et al., 2000). The patients consisted of 59 men and 64 women (ages 19–89) who visited a smell and taste clinic with chemosensory complaints due to all of the causes mentioned in the previous paragraph. As can be seen, those with TDI scores in the "functional anosmia" range (0–15) sometimes produced detectable CSERPs, those with scores in the middle (around 23) of the hyposmic range (16–30) had a 50% chance of producing detectable CSERPs, and those with scores in the normosmic range (31–50) did not always produce detectable CSERPs. Overall CSERPs were detected in 20%, 48%, and 70% of patients with scores in the three TDI ranges. A similar relation between TDI score and detection of olfactory CSERPs was observed at an outpatient rhinology clinic for 107 males and 122 females (ages 15–75+) with a similar variety of problems (Rombaux et al., 2009). Overall, though trigeminal CSERPs were detectable in 99% of 221 recordings, olfactory CSERPs were detectable in only 29% of 224 recordings. Moreover, while TDI scores differed significantly between patients with problems due to different etiologies, the probability of detecting olfactory CSERPs did not.

The interpretation of CSERPs as currently used in the olfactory clinic would appear to be as follows. An olfactory CSERP implies the presence of at least some olfactory function. Thus, when detected with a high degree of confidence, it should call into question a diagnosis of complete anosmia. A detectable olfactory CSERP in hyposmic patients may even be associated with a greater chance of improvement (Rombaux et al., 2010). In contrast, a failure to detect an olfactory CSERP does not necessarily imply diminished olfactory function, though alternative explanations involving methodological problems can be ruled out to some extent by detection of trigeminal CSERPs in the same recording session. In sum, though still somewhat limited, the utility of CSERPs in the olfactory clinic is likely to increase as a result of promising developments on the horizon. For example, given the functional interactions between the olfactory and trigeminal systems and robust presence of trigeminal CSERPs, these latter CSERPs might be used to assess olfactory function (Rombaux et al., 2009). Perhaps most importantly, methodological advances may enhance the detection of olfactory responses in the EEG (Rombaux, et al., 2012), including one to be discussed in Section 12.4 (Huart et al., 2012).

12.3.6 Effects of Brain Disease

An increasing number of neurological and neuropsychiatric diseases have been found to be associated with early olfactory dysfunction. Moreover, such olfactory signs often antedate more classic symptoms. The search for early biomarkers of these diseases, as well as insight into their development, has thus spurred considerable research on their relation to olfaction. Given the potential of CSERPs to index specific neural and functional components of the olfactory system, it is only natural that some of this work has involved their measurement. Chief among the brain diseases studied so far with CSERPs are Alzheimer's disease (AD), Parkinson's disease (PD), and schizophrenia.

Both AD and PD have been reported to be associated with increases in the latency of olfactory CSERPs but with little change in their amplitude (Barz et al., 1997; Hawkes et al., 1997; Wetter and Murphy, 2001; Morgan and Murphy, 2002; Corby et al., 2012). Figure 12.5, which compares component latencies from 12 AD patients and 12 age- and gender-matched controls (Morgan and Murphy, 2002), shows just how robust these differences in AD can be. As can be seen, N1, P2, and P3 latencies are longer for AD patients by approximately 150–300 msec. These components have likewise been found to occur later for individuals at risk for AD, that is, those who test positive for the E4 allele of the apolipoprotein gene (Wetter and Murphy, 2001; Corby et al., 2012). In PD, both the N1 and P2 have been found to be delayed for olfactory but not trigeminal CSERPs (Barz et al., 1997; Hawkes et al., 1997). Delays in N1 associated with AD and PD would seem to indicate a slowing of early (possibly subcortical) olfactory processing. Slowing of early olfactory processes would likewise be expected to delay the P2 and P3 components, though additional slowing of later olfactory and/or cognitive processes could also contribute to their observed delay in AD and PD.

In contrast to AD and PD, olfactory CSERP amplitudes have been found to differ from normal in schizophrenia. Reduced N1 and P2 components have been recorded from both schizophrenics (Turetsky et al., 2003; Kayser et al., 2010) and unaffected first-degree relatives of schizophrenics (Turetsky et al., 2008). Latency differences have also been reported. Turetsky et al. (2003) found P2 latency for strong stimuli to be prolonged in schizophrenics, while Pause et al. (2008) found N1 and P2 latencies to be reduced. This latter finding, which was most evident in responses to an unpleasant odor (rotten-butter-like isobutyraldehyde), was interpreted by Pause et al. as reflecting increased sensitivity to emotionally negative odors. Other neurological

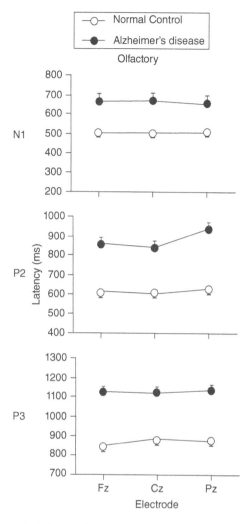

Figure 12.5 Olfactory CSERP latencies for N1, P2, and P3 components as a function of electrode site for normal controls and Alzheimer's disease patients. Adapted from "Olfactory event-related potentials in Alzheimer's disease," by C.D. Morgan and C. Murphy, 2002. *Journal of the International Neuropsychological Society*, 8(6), p. 760. Copyright 2002 by Cambridge University Press.

diseases studied so far with CSERPs include Down's syndrome (Wetter and Murphy, 1999), Huntington's disease (Wetter et al., 2005), multiple sclerosis (Hawkes, 1996), motor neuron disease (Hawkes et al., 1998), and epilepsy (Hummel et al., 1995).

The influence of brain pathology on CSERPs has provided insight as well into the functional anatomy of the olfactory and trigeminal systems. One example involves unilateral epileptic foci in the temporal lobes. Hummel et al. (1995) found foci in each hemisphere to be associated with later N1 and P2 components when olfactory stimuli were presented to the ipsilateral than to the contralateral nostril. This is congruent with an ipsilateral dominance of early olfactory processing that extends to cortical areas in the temporal lobe. In contrast, N1 and P2 components occurred later when trigeminal stimuli were presented to the left than to the right nostril, regardless of the side of the foci. Another example concerns evidence for right-hemispheric dominance in the olfactory system obtained by Daniels et al. (2001) from subjects with unilateral tumors located mainly in the temporal and frontal lobes. Right-hemisphere tumors were associated with diminished P2 and P3 amplitudes at parietal sites and a frontal shift in topography regardless of which nostril was stimulated. With left-hemisphere tumors, these effects were less pronounced and associated mainly with left-nostril stimulation. Finally, grouping patients based on their CSERPs can help identify structures in which olfactory or trigeminal responses are compromised by brain pathology. For example, Welge-Lussen et al. (2009) compared fMRI responses between hyposmic Parkinson patients with and without detectable olfactory CSERPs. Stronger responses were found for the group with detectable CSERPs in the amygdala, parahippocampal cortex, inferior frontal gyrus, insula, cingulate gyrus, striatum, and inferior temporal gyrus.

12.4 RECENT ADVANCES IN ANALYSIS OF EEG ACTIVITY

In Section 12.2, we reviewed research using older methodologies that examined changes in the on-going EEG at each of several frequency ranges over an extended period of time. Section 12.3 presented research involving ERPs which, though they provide much better temporal resolution, do not distinguish between different frequencies. Moreover, as discussed below, ERPs are sensitive only to a portion of neural responses to stimulation. In this section, we discuss recent advances in analyzing EEG activity that combine the capabilities of both traditional EEG and ERP methodologies. As we shall see, it is now possible to examine changes in power induced by a stimulus at individual frequencies with a high degree of temporal precision.

12.4.1 Discrete Frequency and Continuous Time-Frequency Measures

The processing and analysis of *induced* power modulations at any given frequency is unlike the processing of *evoked* ERPs (see Section 12.3). A change in the on-going EEG activity can be time locked to a specific event but not necessarily *phase* locked to that event. As the induced signal is not phase-locked and the phase varies randomly in relation to the event, a process of averaging, as is conducted in ERP analysis, would eliminate the signal of

interest (Pfurtscheller and da Silva, 1999). Therefore, a more complex form of signal processing must be implemented in order to measure power modulations, at any given frequency range, during a specific event; this is known as *frequency analysis*.

There are several analytical techniques under the umbrella of frequency analysis that are used to measure specific modulations of power largely divisible into two broad categories: discrete and continuous. The most common form of discrete frequency measures is the Fast Fourier Transform (FFT). This process effectively expresses a complex signal (such as that measured with EEG) as a summation of a series of simple sine and cosine terms. These distinct sinusoidal components of a signal can be converted into continuous time-specific data in the frequency domain. The FFT therefore produces measures of power, of any specified frequency, for a given *discrete* time period. These power values are then usually made relative to a baseline measure. This is by far the most common method used in the literature, as reviewed in Section 12.2.

There are specific limitations in using an FFT to reliably measure power. The better the signal is located in the time domain the less accurate the location of the signal is in the frequency domain and vice versa (formalized in the "uncertainty principle"; Heisenberg, 1927). More modern "continuous time-frequency" analysis techniques allow for more accurate measurements in the time domain as well as in the frequency domain, the most common of which is known as the Continuous Wavelet Transform (CWT).

The CWT is unlike the Fourier Transform in that it passes a series of small frequency windows, or "wavelets," over every sample that is subject to spectral calculation. Effectively, the resulting CWT measurements describe the evolution of power for specific frequencies over the time period of interest. Moreover, the spectral calculation is relative to a specified time period, much like the baseline period used in ERP analysis.

The ability to observe an evolution of frequency power over time confers several advantages over discrete methods (i.e., FFTs). For example, psychological questions pertaining to the speed or timing of cognitive processes can be better answered as power is described over short time intervals. Further, when no modulation power is found using an FFT calculation, this may be due to the lengthy time period incorporated by the FFT, obscuring modulation that may only occur over shorter time periods. Wavelet analysis is better at minimizing this problem. These methods also incorporate non-phase locked information that is eradicated in the averaging process of ERPs.

Discrete methods of frequency analysis are not temporally specific enough to examine event related changes. Rather, a coarse time period is taken and either one, or several odors in the same experimental condition, is presented to the participant. This method increases the likelihood of the modulation of brain frequencies by higher-order cognitive processes and decreases the signal to-noise ratio. Indeed, this is one reason why event-related potentials are more widely used in olfactory research than discrete methods of frequency analysis.

12.4.2 Continuous Time-Frequency EEG

To our knowledge, only two studies have used continuous time frequency measures at the onset of an olfactory event (Huart et al., 2012; Lorig and Randol, 1996). The study by Lorig and Randol (1996), only available as a conference presentation abstract, describes a modulation of power across a wide range of frequency bandwidths only 1.5 seconds after olfactory stimulation. Further, only low frequency modulations were found to show a variable response in relation to the odor presented. This research suggests that the temporal evolution of the frequency power is somewhat slower than ERP modulations, implying that EEG and ERP modulations may be tapping into different components of the olfactory network. However, this is only a preliminary report of findings; calculations of both EEG and ERP modulations of the stimuli to the same EEG signal would allow for a better examination of the association between the two components.

Exploring EEG changes in response to olfactory and trigeminal stimulation in 11 individuals with normal smell function, Huart et al. (2012) characterized both the phase locked component of the EEG (the traditional ERP; discussed in greater detail above) and the non-phase locked continuous changes in a frequency spectrum. Via an olfactometer, participants were presented with CO_2 for trigeminal stimulation and with PEA for olfactory stimulation.

Huart et al. (2012) used a wavelet analysis, described above, to assess the evolution of EEG power across time during olfactory stimulation for a frequency spectrum ranging from 0.3 to 30 Hz (see Figure 12.6). The results showed a long lasting increase in power around the theta frequency range (~5 Hz), followed by a shorter decrease in alpha power (~10 Hz). Modulations in power were most prominent over the central electrode Cz in a similar fashion to the olfactory ERP. The authors suggested that this common topography indicates that both the ERP and the changes in time-frequency reflect the same stimulus-evoked cortical activity.

In their study, trigeminal nerve stimulation resulted in an increase in power in a high frequency range (~10–15 Hz) and, similar to olfactory stimulation, was followed by a long lasting decrease in alpha power (See Figure 12.7). The early high frequency component was thought, tentatively, to suggest specific processing of trigeminal stimulation and the alpha to reflect a decrease in neuronal idling related to cortical processing of the trigeminal stimulus. The similarity

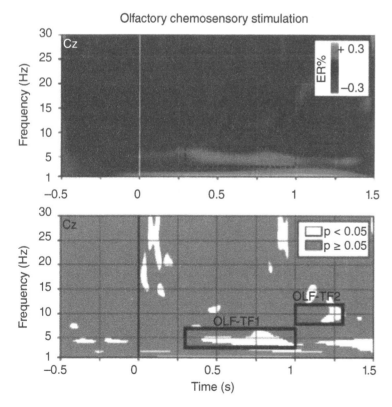

Figure 12.6 Modulation of EEG power across time during olfactory stimulation. Top panel: continuous power changes in color, bottom panel: only those power changes that differed significantly from zero. Adapted from "Time-frequency analysis of chemosensory event-related potentials to characterize the cortical representation of odors in humans," by C. Huart et al., 2012. *PloS ONE*, 7(3), p. 6. (*See plate section for color version.*)

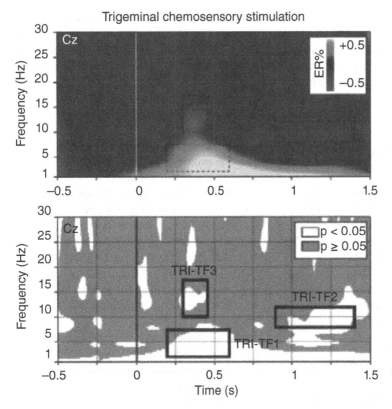

Figure 12.7 Modulation of EEG power across time during trigeminal stimulation. Top panel: continuous power changes in color. Bottom panel: only those power changes that differed significantly from zero. Adapted from "Time-frequency analysis of chemosensory event-related potentials to characterize the cortical representation of odors in humans," by C. Huart et al., 2012. *PloS ONE*, 7(3), p. 6. (*See plate section for color version.*)

between olfactory and trigeminal nerve stimulation in cortical functioning, as indicated by alpha power modulations, may reflect a functional overlap in trigeminal and olfactory processing. This functional overlap may suggest that the processing of nasal somatosensory and olfactory information is non-independent. This is in line with recent research and theoretical positions that suggest that "the sniff is part of the olfactory percept" (Mainland and Sobel, 2006).

Huart et al. (2012) also compared the time-frequency modulations induced by olfactory and trigeminal stimulation to time-frequency information taken from background EEG noise to determine the "discriminability" of the signal. The ability to discriminate between the presence and absence of olfactory cortical processing is obviously useful for clinical examinations of olfactory functioning, as well as for the validity of the method as a whole. Importantly, the time-frequency signal was able to discriminate significantly better than the olfactory ERPs between background noise and olfactory induced modulations. This peaked with a sensitivity of 81.8% and a specificity of 90.9%.

Perhaps most importantly, Huart et al. (2012) report that above all other EEG and ERP measures, increases in power at the theta frequency range were positively ($r = 0.7$, $p = 0.02$) correlated with a behavioral measure of olfactory function. This suggests that, in fact, time-frequency EEG modulations in response to olfactory stimulation reflect olfactory related cortical processing of functional relevance.

REFERENCES

Allison, T. and Goff, W. R. (1967). Human cerebral evoked responses to odorous stimuli. *Electroencephalogr. Clin. Neurophysiol.* 23(6): 558–560.

Barz, S., Hummel, T., Pauli, E., et al. (1997). Chemosensory event-related potentials in response to trigeminal and olfactory stimulation in idiopathic Parkinson's disease. *Neurology* 49(5): 1424–1431.

Becker, E., Hummel, T., Piel, E., et al. (1993). Olfactory event-related potentials in psychosis-prone subjects. *Int. J. Psychophysiol.* 15(1): 51–58.

Benignus, V. A., and Prah, J. D. (1980). A computer-controlled vapor-dilution olfactometer. *Behav. Res. Methods Instrum.* 12: 535–540.

Berg, L., Danziger, W. L., Storandt, M., et al. (1984). Predictive features in mild senile dementia of the Alzheimer's type. *Neurology* 34: 563–569.

Berger, H. (1929) On the electroencephalogram of man. (English translation). *Electroencephalogr. Clin. Neurophysiol.* 1969: Suppl 28: 37–73.

Bonanni, E., Borghetti, D., Fabbrini, M., et al. (2006). Quantitative EEG analysis in post-traumatic anosmia. *Brain Res. Bull.* 71: 69–75.

Brauchli, P., Rüegg, P. B., Etzweiler, F., and Zeier, H. (1995). Electrocortical and autonomic alteration by administration of a pleasant and an unpleasant odor. *Chem. Senses* 20: 505–515.

Bulsing, P. J., Smeets, M. A. M., Gemainhardt, C., et al. (2010). Irritancy expectancy alters odor perception: Evidence from olfactory event-related potential research. *J. Neurophysiol.* 104: 2749–2756.

References

Caton, R. (1875). The electrical currents of the brain. *Br. Med. J.* 2: 278 (abstract).

Cherninskii, A. A., Zima, I. G., Makarchouk, N. Y., et al. (2009). Modifications of EEG related to directed perception and analysis of olfactory information in humans. *Neurophysiology* 41: 63–70.

Corby, K., Morgan, C. D., and Murphy, C. (2012). Abnormal event-related potentials in young and middle-aged adults with the ApoE ϵ4 allele. *Int. J. Psychophysiol.* 83: 276–281.

Covington, J. W., Geisler, M. W., Polich, J., and Murphy, C. (1999). Normal aging and odor intensity effects on the olfactory event-related potential. *Int. J. Psychophysiol.* 32(3): 205–214.

Cui, L., and Evans, W. J. (1997). Olfactory event-related potentials to acetate in congenital anosmia. *Electroencephalogr. Clin. Neurophysiol.* 102: 303–306.

Daniels, C., Gottwald, B., Pause, B. M., et al. (2001). Olfactory event-related potentials in patients with brain tumors. *Clin. Neurophysiol.* 112: 1523–1530.

Donchin, E. (1981). Surprise! ... Surprise? *Psychophysiology* 18: 493–511.

Donchin, E. and Coles, M. G. H. (1988). Is the P300 a manifestion of context updating? *Behav. Brain Sci.* 11: 357–428.

Doty, R. L. (2012). Olfaction in Parkinson's disease and related disorders. *Neurobiol. Dis.* 46: 527–552, 2012.

Doty, R. L., Brugger, W. E., Jurs, P. C., et al. (1978). Intranasal trigeminal stimulation from odorous volatiles: Psychometric responses from anosmic and normal humans. *Physiol. Behav.* 20(2): 175–185.

Doty, R. L. and Cameron, E. L. (2009). Sex differences and reproductive hormone influences on human odor perception. *Physiol. Behav.* 97(2): 213–228.

Doty, R. L., Perl, D. P., Steele, J. C., et al. (1991). Olfactory dysfunction in three neurodegenerative diseases. *Geriatrics* 43: 47–51.

Doty, R. L., Shaman, P., Applebaum, S. L., et al. (1984). Smell identification ability—Changes with age. *Science* 226: 1441–1443.

Doty, R. L., Smith, R., McKeown, D. A., and Raj, J. (1994). Tests of human olfactory function: Principal components analysis suggests that most measure a common source of variance. *Percept. Psychophys.* 56: 701–707.

Doty, R. L., Yousem, D. M., Pham, L. T., et al. (1997). Olfactory dysfunction in patients with head trauma. *Arch. Neurol.* 54: 1131–1140.

Durand-Lagarde, M. and Kobal, G. (1991). P300: a new technique of recording a cognitive component in the evoked potentials. *Chem. Senses* 16: 379 (abstract).

Evans, W. J., Cui, L., and Starr, A. (1995). Olfactory event- related potentials in normal human subjects: Effects of age and gender. *Electroencephalogr. Clin. Neurophysiol.* 95(4): 293–301.

Evans, W. J., Kobal, G., Lorig, T. S., and Prah, J. (1993). Suggestions for collection and reporting of chemosensory olfactory event-related potentials. *Chem. Senses* 18: 751–756.

Evans, W. J. and Starr, A. (1992). Stimulation parameters and temporal evolution of the olfactory evoked potentials in rats. *Chem. Senses* 17: 61–78.

Finkenzeller, P. (1965). Gemittelte EEG-Potentiale bei olfactorischer Reizung. *Pflügers Arch.* 292: 76–85.

Fleischl von Marxow, E. (1890). Mitteilung betreffend der Physiologie der Hirnrinde. *Zbl. Physiol.* 4: 537–540.

Geisler, M. W. and Murphy, C. (2000). Event-related brain potentials to attended and ignored olfactory and trigeminal stimuli. *Int. J. Psychophysiol.* 37(3): 309–315.

Haehner, A., Gruenewald, G., Dibenedetto, M., and Hummel, T. (2011). Responses to olfactory and intranasal trigeminal stimuli: Relation to the respiratory cycle. *Neuroscience* 175: 178–183.

Halgren, E., Squires, N. K., Wilson, C. L., et al. (1980). Endogenous potentials generated in the human hippocampal formation and amygdala by infrequent events. *Science* 210: 803–805.

Halpern, B. P. (2009). Retronasal olfaction. In *Encyclopedia of Neuroscience*, L. R. Squire (Ed.), Elsevier, Amsterdam, Online.

Harada, H., Shiraishi, K., Kato, T., and Soda, T. (1996). Coherence analysis of EEG changes during odour stimulation in humans. *J. Laryngol Otol.* 110: 652–656.

Hari, R., Portin, K., Kettenmann, B., et al. (1997). Right hemisphere preponderance of responses to painful CO_2 stimulation of the human nasal mucosa. *Pain* 72: 145–151.

Hawkes, C. H. (1996). Assessment of olfaction in multiple sclerosis. *Chem. Senses* 21: 486 (abstract).

Hawkes, C. H., Shephard, B. C., and Daniel, S. E. (1997). Olfactory dysfunction in Parkinson's disease. *J. Neural. Neurosurg. Psychiatry* 62(5): 436–446.

Hawkes, C. H., Shephard, B. C., Geddes, J. F., et al. (1998). Olfactory disorder in motor neuron disease. *Exp. Neurol.* 150(2): 248–253.

Heisenberg, W. (1927). *Über quantentheoretische Umdeutung kinematischer und mechanischer Beziehungen' Zeitschrift für Physik.* English translation in *Quantum Theory and Measurement*, A. Wheeler and W. H. Zurek (Eds.), Princeton University Press: Princeton, 1983, pp. 62–84.

Huart, C., Legrain, V., Hummel, T., et al. (2012). Time-frequency analysis of chemosensory event-related potentials to characterize the cortical representation of odors in humans. *PloS One* 7: 1–11.

Hummel, T., Pauli, E., Schuler, P., et al. (1995). Chemosensory event-related potentials in patients with temporal lobe epilepsy. *Epilepsia* 36(1): 79–85.

Hummel, T., Barz, S., Pauli, E., and Kobal, G. (1998). Chemosensory event-related potentials change with age. *Electroencephalogr. Clin. Neurophysiol.* 108(2): 208–217.

Huttunen, J., Kobal, G., Kaukoranta, E., and Hari, R. (1986). Cortical responses to painful CO_2 stimulation of the nasal mucosa: A magnetoencephalographic study in man. *Electroencephalogr. Clin. Neurophysiol.* 64: 347–349.

Kayser, J., Tenke, C. E., Malaspina, D., et al. (2010). Neuronal generator patterns of olfactory event-related brain potentials in schizophrenia. *Psychophysiology* 47: 1075–1086.

Kettenmann, B., Jousmaki, V., Portin, K., et al. (1996). Odorants activate the human superior temporal sulcus. *Neurosci. Lett.* 203(2): 143–145.

Kettenmann, B., Hummel, C., Stefan, H., and Kobal, G. (1997). Multiple olfactory activity in the human neocortex identified by magnetic source imaging. *Chem. Senses* 22(5): 493–502.

Khan, R. M., Luk, C. H., Flinker, A., et al. (2007). Predicting odor pleasantness from odorant structure: Pleasantness as a reflection of the physical world. *J. Neurosci.* 27: 10015–10023.

Klemm, W. R., Lutes, S. D., Hendrix, D. V., and Warrenburg, S. (1992). Topographical EEG maps of human responses to odors. *Chem. Senses* 17: 347–361.

Kobal, G. (1981). *Elektrophysiologische Untersuchungen des menschlichen Geruchssinns.* Thieme Verlag, Stuttgart.

Kobal, G. (2003). Electrophysiological measurement of olfactory function. In *Handbook of Olfaction and Gustation*, 2nd edition, R. L. Doty (Ed.), Marcel Dekker: N.Y., pp. 229–249.

Kobal, G. and Hummel, C. (1988). Cerebral chemosensory evoked potentials elicited by chemical stimulation of the human olfactory and respiratory nasal mucosa. *Electroencephalogr. Clin. Neurophysiol.* 71(4): 241–250.

Kobal, G. and Hummel, T. (1991). Olfactory evoked potentials in humans. In *Smell and Taste in Health and Disease*, T. V. Getchell (Ed.), Raven Press, N.Y., pp. 255–275.

Kobal, G. and Hummel, T. (1998). Olfactory and intranasal trigeminal event-related potentials in anosmic patients. *Laryngoscope* 108(7): 1033–1035.

Kobal, G., Klimek, L., Wolfensberger, M., et al. (2000). Multicenter investigation of 1,036 subjects using a standardized method for the assessment of olfactory function combining tests of odor identification, odor discrimination, and olfactory thresholds. *Eur. Arch. Otorhinolaryngol.* 257(4): 205–211.

Kobal, G., and Plattig, K. H. (1978). Methodische Anmerkungen zur Gewinnung olfaktorischer EEG-Antworten des wachen Menschen (objektive Olfaktometrie). *Z. EEG-EMG* 9(3): 135–415.

Krauel, K., Pause, B. M., Sojka, B., et al. (1998). Attentional modulation of central odor processing. *Chem. Senses* 23(4): 423–432.

Kutas, M., McCarthy, G., and Donchin, E. (1977). Augmenting mental chronometry: The P300 as a measure of stimulus evaluation time. *Science* 197: 792–795.

Lascano, A. M., Hummel, T., Lacroix, J. S., et al. (2010). Spatio-temporal dynamics of olfactory processing in the human brain: An event-related source imaging study. *Neuroscience* 167(3): 700–708.

Laudien, J. H., Wencker, S., Ferstl, R., and Pause, B. M. (2008). Context effects on odor processing: An event-related potential study. *NeuroImage* 41: 1426–1436.

Lehmann, D. and Skrandies, W. (1980). Reference-free identification of components of checkerboard-evoked multichannel potential fields. *Electroencephalogr. Clin. Neurophysiol.* 48(6): 609–621.

Lorig, T. S. (1989). Human EEG and odor response. *Prog. Neurobiol.* 33: 387–398.

Lorig, T. S. (2000). The application of electroencephalographic techniques to the study of human olfaction: A review and tutorial. *Int. J. Psychophysiol.* 36: 91–104.

Lorig, T. S., Elmes, D. G., Zald, D. H., and Pardo, J. V. (1999). A computer-controlled olfactometer for fMRI and electrophysiological studies of olfaction. *Behav. Res. Methods Instrum. Comput.* 31(2): 370–375.

Lorig, T. S. and Randol, M. (1996). Evaluation of event-related synchronous brain activity following chemosensory stimulation. *Chem. Senses* 21: 636 (abstract).

Lorig, T. S. and Schwartz, G. E. (1988). Brain and odor: I. Alteration of human EEG by odor administration. *Psychobiology* 16: 281–284.

Lorig, T. S., Schwartz, G. E., Herman, G. E., and Lane, R. D. (1988). Brain and Odor: II. EEG activity during nose and mouth breathing. *Psychobiology* 16: 285–287.

Lotsch, J. and Hummel, T. (2006). The clinical significance of electrophysiological measures of olfactory function. *Behav. Brain Res.* 170: 78–83.

Mainland, J., and Sobel, N. (2006). The sniff is part of the olfactory percept. *Chem. Senses* 31: 181–196.

Martin, G. N. (1998). Human electroencephalographic (EEG) response to olfactory stimulation: Two experiments using the aroma of food. *Int. J. Psychophysiol.* 30: 287–302.

Morgan, C. D., Covington, J. W., Geisler, M.W., et al. (1997). Olfactory event-related potentials: Older males demonstrate the greatest deficits. *Electroencephalogr. Clin. Neurophysiol.* 104(4): 351–358.

Morgan, C. D., Geisler, M. W., Covington, J. W., et al. (1999). Olfactory P3 in young and older adults. *Psychophysiology* 36(3): 281–287.

Morgan, C. D. and Murphy, C. (2002). Olfactory event-related potentials in Alzheimer's disease. *J. Int. Neuropsychol. Soc.* 8(6): 753–763.

Morruzzi, G. and Magoun, H. W. (1949). Brainstem reticular formation and activation of the EEG. *Electroencephalogr. Clin. Neurophysiol.* 1: 455–473.

Motokizawa, F. and Furuya, N. (1973). Neural pathway associated with the EEG arousal response by olfactory stimulation. *Electroencephalogr. Clin. Neurophysiol.* 35: 83–91.

Murphy, C., Nordin, S., de Wijk, R. A., et al. (1994). Olfactory-evoked potentials: Assessment of young and elderly, and comparison to psychophysical threshold. *Chem. Senses* 19(1): 47–56.

Murphy, C., Morgan, C. D., Geisler, M. W., et al. (2000). Olfactory event-related potentials and aging: normative data. *Int. J. Psychophysiol.* 36(2): 133–145.

Nordin, S., Andersson, L., Olofsson, J. K., et al. (2011). Evaluation of auditory, visual and olfactory event-related potentials for comparing interspersed- and single-stimulus paradigms. *Int. J. Psychophysiol.* 81: 252–262.

Olofsson, J. K. and Nordin, S. (2004). Gender differences in chemosensory perception and event-related potentials. *Chem. Senses* 29: 629–637.

Owen, C. and Patterson, J. (2002). Odour liking physiological indices: A correlation of sensory and electrophysiological responses to odour. *Food Qual. Prefer.* 13: 307–316.

Pause, B. M., Hellmann, G., Goder, R., et al. (2008). Increased processing speed for emotionally negative odors in schizophrenia. *Int. J. Psychophysiol.* 70: 16–22.

Pause, B. M. and Krauel, K. (2000). Chemosensory event-related potentials (CSERP) as a key to the psychology of odors. *Int. J. Psychophysiol.* 36(2): 105–122.

Pause, B. M., Sojka, B., Krauel, K., and Ferstl, R. (1996). The nature of the late positive complex within the olfactory event-related potential (OERP). *Psychophysiology* 33(4): 376–384.

Pause, B. M., Sojka, B., and Ferstl, R. (1997). Central processing of odor concentration is a temporal phenomenon as revealed by chemosensory event-related potentials (CSERP). *Chem. Senses* 22(1): 9–26.

Pause, B. M., Krauel, K., Sojka, B., and Ferstl, R. (1999). Is odor processing related to oral breathing? *Int. J. Psychophysiol.* 32(3): 251–260.

Pfurtscheller, G. and da Silva, F. H. L. (1999). Event-related EEG/MEG synchronization and desynchronization: Basic principles. *Clin. Neurophysiol.* 110: 1842–1857.

Polich, J. and Squire, L. R. (1993). P300 from amnestic patients with bilateral hippocampal lesions. *Electroencephalogr. Clin. Neurophysiol.* 86: 408–417.

Rombaux, P., Huart, C., Collet, S., et al. (2010). Presence of olfactory event-related potentials predicts recovery in patients with olfactory loss following upper respiratory tract infection. *Laryngoscope* 120: 2115–2118.

Rombaux, P., Huart, C., and Mouraux, A. (2012). Assessment of chemosensory function using electroencephalographic techniques. *Rhinology* 50(1): 13–21.

Rombaux, P., Mouraux, A., Collet, S., et al. (2009). Usefulness and feasibility of psychophysical and electrophysiological olfactory testing in the rhinology clinic. *Rhinology* 47: 28–35.

Sakuma, K., Kakigi, R., Kaneoke, Y., et al. (1997). Odorant evoked magnetic fields in humans. *Neurosci. Res.* 27(2): 115–122.

Seal, E. C. J., Hintum, C. J. A. V., Pierson, J. M., and Helme, R. D. (1998). Quantitative electroencephalography, with serial subtraction and odour detection in the differentiation of Alzheimer's disease and vascular dementia. *Arch. Gerontol. Geriatr.* 27: 115–126.

Smith, D. B., Allison, T., Goff, W. R., and Principato, J. J. (1971). Human odorant evoked responses: effects of trigeminal or olfactory deficit. *Electroenceph. Clin. Neurophysiol.* 30(4): 313–317.

Stuck, B. A., Frey, S., Freiburg, C., et al. (2006). Chemosensory event-related potentials in relation to side of stimulation, age, sex, and stimulus concentration. *Clin. Neurophysiol.* 117: 1367–1375.

Tateyama, T., Hummel, T., Roscher, S., et al. (1998). Relation of olfactory event-related potentials to changes in stimulus concentration. *Electroenceph. Clin. Neurophysiol.* 108(5): 449–455.

Thesen, T. and Murphy, C. (2002). Reliability analysis of event-related brain potentials to olfactory stimuli. *Psychophysiology* 39: 733–738.

Tonoike, M., Yamaguchi, M., Kaetsu, I., et al. (1998). Ipsilateral dominance of human olfactory activated centers estimated from event-related magnetic fields measured by 122-channel whole-head neuromagnetometer using odorant stimuli synchronized with respirations. *Ann. N.Y. Acad. Sci.* 855: 579–590.

Turetsky, B. I., Kohler, C. G., Gur, R. E., and Moberg, P. J. (2008). Olfactory physiological impairment in first–degree relatives of schizophrenia patients. *Schizophr. Res.* 102: 220–229.

Turetsky, B. I., Moberg, P. J., Owzar, K., et al. (2003). Physiologic impairment of olfactory stimulus processing in schizophrenia. *Biol. Psychiatry* 53: 403–411.

Van Toller, S. and Reed, M. K. (1989). Brain electrical activity topographical maps produced in response to olfactory and chemosensory stimulation. *Psychiatry Res.* 29: 429–430.

Verleger, R. (1988). Event-related potentials and cognition: A critique of the context updating hypothesis and an alternative interpretation of P3. *Behav. Brain Sci.* 11: 343–356.

Wang, L., Walker, V. E., Sardi, H., et al. (2002). The correlation between physiological and psychological responses to odour stimulation in human subjects. *Clin. Neurophysiol.* 113: 542–551.

Welge-Lussen, A., Wattendorf, E., Schwerdtfeger, U., et al. (2009). Olfactory-induced brain activity in Parkinson's disease relates to the expression of event-related potentials: A functional magnetic resonance imaging study. *Neuroscience* 162: 537–543.

Welge-Lussen, A., Wille, C., Renner, B., and Kobal, G. (2003). Test-retest reliability of chemosensory evoked potentials. *J. Clin. Neurophysiol.* 20: 135–142.

Wetter, S. and Murphy, C. (1999). Individuals with Down's syndrome demonstrate abnormal olfactory event-related potentials. *Clin. Neurophysiol.* 110(9): 1563–1569.

Wetter, S. and Murphy, C. (2001). Apolipoprotein E epsilon 4 positive individuals demonstrate delayed olfactory event-related potentials. *Neurobiol. Aging* 22(3): 439–447.

Wetter, S., Peavy, G., Jacobson, M., et al. (2005). Olfactory and auditory event-related potentials in Huntington's disease. *Neurophysiology* 19(4): 428–436.

Chapter 13

Structural and Functional Imaging of the Human Olfactory System

JAY A. GOTTFRIED

13.1 INTRODUCTION

An image is a visual reproduction or rendition of an object externally presented to the eye. (In the case of Art, an image might be the visual reproduction of an object *internally* presented to the mind's eye.) Imaging is thus the process of generating these images. From prehistoric cave pigments to extraterrestrial Mars rovers, technological advances and human curiosity have shaped our ability to generate images of the very near and very far, the very big and very small, and everything in between. Imaging of the human nervous system has moved closely in step with scientific progress to make the unobservable observable: from the classical techniques of pen-and-ink, camera lucida, dry-plate photography, silver staining, and light microscopy, to the incipient clinical milieu of roentgenograms (x-ray) and pneumoencephalography, and onward to the modern era of computed axial tomography (CAT scan), single photon emission computed tomography (SPECT), positron emission tomography (PET), and magnetic resonance imaging (MRI). These last two methods have revolutionized our understanding of the human brain, particularly its functional aspects, and constitute the primary focus of this chapter.

Since the Second Edition of the *Handbook* appeared approximately ten years ago, the field of human olfactory imaging has grown by sniffs and bounds. Technical and methodological innovations have made it possible to test scientific hypotheses that could not have been previously tested. New imaging approaches have induced a paradigm shift from the "where" to the "how" of odor information processing, providing novel mechanistic insights into the structural and functional organization of human olfaction.

Work in patient populations has helped inform basic understanding of the human olfactory system, and translational studies have identified putative imaging biomarkers to optimize medical diagnostic decisions. The main goal of this chapter is to review these recent research developments and advances. In line with the *Handbook*'s important message that the research present (and future) are inextricably linked to the research past, this chapter will briefly discuss human olfactory imaging in a historical framework. To begin we must retreat to the wood-paneled theaters of the 19th-century German anatomists, where the first images of the human olfactory system were generated.

One of the first scientific treatises dedicated to human olfactory imaging was published in the early 1800s by the German physician, anatomist, and inventor, Samuel Thomas Sömmerring (1755–1830). Among his more notable achievements was the first systematic characterization of the cranial nerves, development of a protective inoculation against smallpox, and an influential though ultimately misguided description of the pterodactyl (stating it was a batlike mammal rather than a reptile). Sömmerring's 1809 monograph, *Abbildungen der menschlichen Organe des Geruches* (i.e., *Images of the Human Smell Organs*), provided one of the first faithful anatomical renderings of the human peripheral olfactory system, including the distribution of olfactory nerve fibers along the superior turbinate, their penetration through the cribriform plate, the position of the olfactory bulb in the anterior cranial fossa, and the posterior course of the olfactory tract (Figure 13.1a). These macroscopic images remained the definitive source of information regarding human olfaction, up until the advent of microscopic and histological studies half a century later,

---

Handbook of Olfaction and Gustation, Third Edition. Edited by Richard L. Doty.
© 2015 Richard L. Doty. Published 2015 by John Wiley & Sons, Inc.

280   **Chapter 13   Structural and Functional Imaging of the Human Olfactory System**

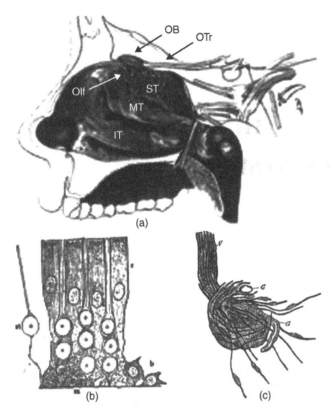

Figure 13.1 Pre-Golgi images of the human olfactory system, 19th century. (a) Anatomical rendering of the right lateral wall of the nasal cavity (septum removed), viewed in sagittal cross-section. Olf, olfactory nerves, resembling a lace-like sheet of fibers concentrated along the surfaces of the superior turbinate and upper middle turbinate, en route to the olfactory bulb (OB) and olfactory tract (OTr). IT, inferior turbinate; MT, middle turbinate; ST, superior turbinate. From Sömmerring, 1809. (b) Vertical cross-section through the neuroepithelium of the human olfactory region, after exposure in 1% osmic acid solution for 15 minutes. Note three different cell types, including basal cells (at the bottom edge of the image), olfactory sensory neurons (with cell bodies in the lower third of the image, tapering towards the top), and supporting cells (with cell bodies in the upper third of the image). From Krause and Krause, 1876. (c) Microscopic image of the human olfactory bulb, in which a spherical aggregation of cells and fibers (glomerulus) is joined by olfactory nerve fibers coursing from the upper left of the image, and by dendrites of mitral cells approaching from the bottom right and right side of the image. From Meynert, 1872.

again from the German school of anatomy, including the pre-Golgi work of Schultze (1856), Meynert (1872), and Krause (1876) that depicted the cellular organization of the human olfactory epithelium and olfactory bulb glomerulus (Figure 13.1b,c). This work first identified three of the fundamental cell types in the olfactory epithelium, including ciliated receptor neurons, columnar supporting cells, and basal cells, and described the spherical masses in the olfactory bulb (glomeruli) where afferent inputs and mitral cell fibers converged (Meynert, 1872; Krause and Krause, 1876).

Only sporadic progress on human olfactory imaging was made over the next 100 years, again of a gross macroscopic or histological disposition. In 1954 A. C. Allison had a rare opportunity to examine the brain of a 52-year-old man whose olfactory peduncle had been accidentally transected on one side during a prefrontal leucotomy two years prior to death (Allison, 1954). Using different staining methods to evaluate neurons, axons, and myelin, he was able to delineate the secondary projection areas of the human olfactory bulb, based on regional patterns of fiber degeneration and trans-synaptic cellular atrophy (Figure 13.2a,b). These findings provided the most systematic anatomical images of the human olfactory brain yet available, with insightful descriptions of the anterior olfactory nucleus, olfactory tubercle, frontal and temporal piriform cortex, and medial and cortical nuclei of the amygdala. Around the same time that Allison was toiling with tissue stains, Engstrom and Bloom (1953) used electron microscopy (EM) to obtain high-resolution images of the human olfactory mucosa. In an ingenious step, Morrison and Costanzo (1990, 1992) used freeze-fracture scanning EM to describe the fine microstructure of the human olfactory epithelium in three dimensions (Figure 13.2c,d), considerably extending prior EM studies of olfactory epithelial surface features (Engstrom and Bloom, 1953; Naessen, 1971; Nakashima et al., 1984). Their breathtaking images revealed the ultrastructural details of the dendrites, knobs, and cilia originating from the cell bodies of human olfactory sensory neurons, and captured the olfactory fibers of passage coursing from the epithelium to the olfactory bulb. One important take-home message from this work was that the distribution of olfactory epithelium within the nasal mucosa was highly patchy and irregularly interspersed with respiratory epithelial tissue.

These pioneering forays into olfactory imaging held one distinct disadvantage for investigators and subjects alike: the reliance on post-mortem material. The advent of noninvasive imaging techniques, arguably beginning with the first CT scanner in the early 1970s, has provided new ways to probe the structure and function of the human brain, both in health and disease. However, it was also apparent from the earlier anatomical data that most olfactory-related structures in the human brain were small, implying that successful imaging of these regions would require a technique with sufficient spatial resolution. The development of magnetic resonance imaging (MRI), and its many cousins of image "contrast" methods—most famously the blood oxygen level-dependent [BOLD] contrast utilized for functional MRI—have revolutionized our understanding of the human olfactory brain. The next section of this chapter will briefly review the principles and methodology of MRI scanning, given its domineering

13.2 A Brief Primer on MRI Scanning

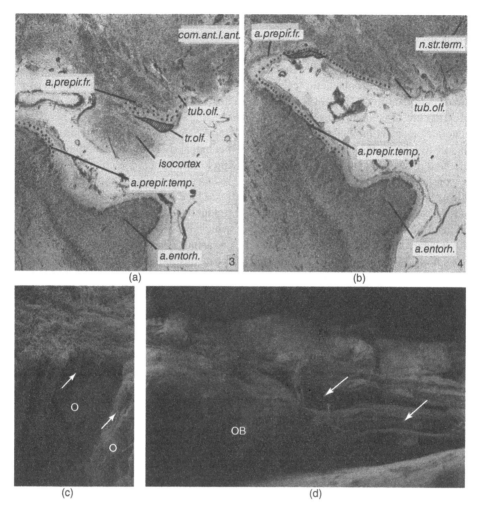

Figure 13.2 Pre-MRI images of the human olfactory system, 20th century. (a) and (b) Coronal cross-sections through the human basal forebrain and dorsomedial temporal lobe (cresyl violet stain, x7.5). Cross-hatching represents the olfactory tract, with dots indicating presumptive terminations of the olfactory tract. Key: *a.entorh*, entorhinal area; *a.prepir.fr.*, frontal portion of the piriform area; *a.prepir.temp.*, temporal portion of the piriform area; *com.ant.l.ant.*, anterior limb of the anterior commissure; *n.str.term.*, nucleus of the stria terminalis; *tr.olf.*, olfactory tract; *tub.olf.*, olfactory tubercle. From Allison, 1954. (c) and (d) Freeze-fracture electron microscopy images of the human olfactory epithelium (c) and bulb (d). Panel (c) (x4,400 magnification) shows olfactory sensory neurons (O) in the epithelial layer, with their dendrites (arrows) projecting up toward the epithelial surface, which is densely covered with a mossy mat of olfactory cilia. Panel (d) (x44 magnification) shows a view of the dorsal cribriform plate in the floor of the anterior cranial fossa, with the olfactory bulb (OB) lying in place, with axon bundles of the olfactory nerves (arrows) from the nasal cavity projecting to and enveloping the anterior end of the OB.

contribution to modern imaging techniques. Subsequent chapters will focus on key advances in structural and functional imaging of human olfaction, with an emphasis on recent data that have emerged over the last 10 years.

13.2 A BRIEF PRIMER ON MRI SCANNING

This short overview of MRI scanning will administer a minimal dose of physics to answer three questions. First, how does an MRI scanner work? Second, how is the MRI signal generated? Third, how is the MR signal transformed into a spatial image? For detailed discussion of these topics, the reader is referred to many excellent textbooks (Toga and Mazziotta, 1996; Jezzard et al., 2002; Huettel et al., 2008; Buxton, 2009; Poldrack et al., 2011); much of this section is specifically drawn from the book, *Functional Magnetic Resonance Imaging*, 2nd Edition (Huettel et al., 2008).

13.2.1 How Does an MRI Scanner Work?

The historical antecedent to modern MRI was nuclear magnetic resonance, that is, NMR. This technique was

282　**Chapter 13**　**Structural and Functional Imaging of the Human Olfactory System**

founded on the idea that atomic nuclei resonate (oscillate) at a specific magnetic frequency, and that if placed within an external magnetic field resonating at that same frequency, the nuclei would absorb energy. Subsequent emission, or radiation, of this electromagnetic energy could then be measured and quantified. Thus, at the heart of NMR, and also of MRI, are three components: a static magnetic field to generate an external magnetic perturbation; a transmitter coil to convey the magnetic field to the sample of interest; and a receiver coil to read out the electromagnetic emission from the sample. It is important that the magnetic field is spatially homogeneous, because local and unexpected variations in the field can induce signal distortion and measurement error. In addition, a larger magnetic field is generally preferable because it increases the signal-to-noise ratio (though this can come at the expense of increased signal artifact, an early confound that plagued olfactory functional MRI). Achieving very stable and high-strength magnetic fields (1.5-Tesla and higher) requires the use of superconducting electromagnets that are cooled to temperatures near absolute zero, typically using liquid helium, reducing electrical resistance within the electromagnet wires to optimize current flow while minimizing electrical power requirements.

13.2.2 How is the MRI Signal Generated?

Generation of the MRI signal follows from several basic physics principles. Thermal energy causes atomic nuclei (being a composite of protons and neutrons) to spin. If these nuclei have a net positive charge, as in the case of odd numbers of protons (e.g., ^{1}H or ^{31}P), then the spinning motion creates an electrical current, which in turn creates a small internal magnetic field. Given that humans are approximately 70% water, MRI scanning of human biological tissue is most commonly tuned to imaging of hydrogen nuclei (a single proton, i.e., ^{1}H).

Normally, the spin axes of the nuclei are randomly oriented with regard to each other, so that the aggregate of nuclei within a tissue sample does not generate a net magnetic field. However, in the presence of an MRI scanner's strong magnetic field, the spin axes of all of the atomic nuclei become aligned, resulting in net magnetization longitudinal to the scanner's magnetic field. Interestingly, the spin axes are not rigidly aligned with the magnetic field, but instead exhibit rotation, or "precession," around the main magnetic field, similar to the wobbling motion of a gyroscope or spinning top. The resonance frequency of precession, that is, the frequency at which the nucleus will absorb energy, is specific to the type of nucleus (for protons it is 42.58 MHz/Tesla) and is known as the Larmor frequency.

Note that the process of magnetization does not itself constitute the MRI signal. In order to measure the signal, a radiofrequency "excitation pulse" is delivered from the transmission coil to inject energy and perturb the system. This pulse consists of a burst of electromagnetic energy waves, set to oscillate at the same frequency as the resonant (Larmor) frequency of the nucleus of interest. The immediate effect is to tip the spins of the atomic nuclei from the longitudinal axis (a low-energy state) into the transverse plane (a high-energy state), shifting the orientation of net magnetization. With ongoing excitation there is increasing loss of spin phase coherence among nuclei, resulting in reduced net magnetization in the transverse plane.

Once radiofrequency excitation ceases, the nuclei gradually transition back to their low-energy state, during which time they emit energy (also at the same Larmor frequency). An MRI receiver coil tuned to the same frequency is used to detect changes in electrical current between excitation and recovery. This is the essence of the recorded MR signal. Recovery, or "relaxation," of magnetization along the longitudinal axis follows the time-constant, T_1, typically a few hundred milliseconds, whereas recovery within the transverse plane follows the time-constant, T_2, typically a few tens of milliseconds. Thus, by adjusting the timing parameters of signal acquisition within the relaxation phase, one can control the relative contributions of T_1 signal and T_2 signal to the MR image. In T_1-weighted brain scans, gray matter appears bright, whereas white matter, water, and cerebrospinal fluid (CSF) appear dark, providing excellent structural anatomical information. In T_2-weighted scans, water and CSF appear bright, making this imaging sequence sensitive to tissue edema and other pathological changes. A related sequence protocol, T_2*-weighted imaging, is the basis for acquisition of BOLD signal during functional MRI, which will be described later in the chapter.

13.2.3 How is the MR Signal Transformed into a Spatial Image?

A critical early innovation of MR technology was to convert single-point measurements of bulk matter into two-dimensional spatial images. Although static magnetic fields could be used to excite an entire tissue slice in two dimensions, the fact that all points within the slice were excited at the same resonant frequency meant that all points within the slice *would also emit an MR signal at the same frequency*. Thus, there was no way to resolve different spatial locations using this basic technique. An initial attempt to overcome this issue was to collect MR data from one voxel in space at a time, then stitch together the data from all voxels to create an MR image. This approach was laborious, time-consuming, and impractical.

In the early 1970s Dr. Paul Lauterbur devised a solution to the spatial encoding problem. He determined that by spatially varying the strength of the external magnetic

13.3 STRUCTURAL IMAGING

Structural noninvasive imaging of the human olfactory system has particular technical, almost existential, constraints due to the fact that the anatomical boundaries of most olfactory areas in the human brain (especially the piriform cortex and OFC) are just not evident. Adding to this problem is the fact that even the best structural MRI sequences cannot achieve a spatial resolution much below 1 mm³, so that morphological descriptions remain fairly coarse, without finer characterization of laminar organization or cellular or architectural features. Indeed, the use of functional MRI techniques (as discussed in the next section) has played a more important role in defining the "structural" organization of human olfaction, insofar as these approaches have identified brain areas that are consistently involved in processing smells. There is one brain area that structural imaging has excelled over functional imaging, and that is the olfactory bulb. The small slender size of the olfactory bulb, and its artifact-prone location near an air-tissue interface, make this region extremely challenging to visualize using fMRI. Olfactory structural imaging has also found a role in clinical diagnostics. These topics are presented below.

13.3.1 Structural Imaging of the Human Olfactory Bulb

Efforts in the 1980s to image the human olfactory system (Schellinger et al., 1983) using computed tomography (CT) were laudatory, but sub-optimal, because skull-based artifacts substantially reduced visualization of the olfactory bulbs and adjacent structures. The first systematic MRI investigation of the human olfactory bulb was conducted using a T_1-weighted sequence, though identification of the olfactory bulbs and tracts was inconsistent across participants (Suzuki et al., 1989). In subsequent work by Yousem, Doty, and colleagues (Yousem et al., 1996a; Yousem et al., 1996b; Yousem et al., 1998), a high-resolution MR surface coil was placed over the nasion (where the frontal bone meets the two nasal bones in the midline) to help improve signal recovery. In healthy subjects, coronal T_1-weighted images clearly demonstrated olfactory bulbs and tracts, in position between the ventromedial prefrontal cortex and the ethmoid sinuses (Figure 13.3a,b). By comparison, in patients with congenital or posttraumatic anosmia (smell loss), the anatomical volumes of the bulbs and tracts were reduced, with relative preservation of central structures, suggesting a primarily peripheral basis for their olfactory perceptual impairments.

In related work examining volumetric changes across the life span in healthy subjects (Yousem et al., 1998), olfactory bulb volumes were observed to peak in the fourth decade of life ($\sim$150 mm³), then progressively declined into the eighth decade ($\sim$120 mm³), hinting at one possible explanation for why the sense of smell diminishes with age. However, the lack of a significant correlation between MRI bulb volumes and odor identification scores failed to support this hypothesis. One other cautionary note from this study is that considerable variation was seen across individuals and age groups, implying that volumetric data of the olfactory bulb might have limited diagnostic applicability at the level of single subjects.

More recent imaging studies from the Hummel laboratory have in fact shown that the volume of the olfactory bulb correlates with odor perceptual performance, not only in healthy subjects (Rombaux et al., 2006; Buschhuter et al., 2008) but also in patients with postinfectious and posttraumatic olfactory impairment (Mueller et al., 2005; Rombaux et al., 2006). Again, prominent interindividual variation in the healthy population was evident, with over two-fold changes between different subjects (e.g., a range of 37 mm³ to 98 mm³ for the left olfactory bulb (Buschhuter et al., 2008). A longitudinal study of patients with hyposmia and anosmia (Haehner et al., 2008) demonstrated that over a mean time interval of 15 months, changes in odor detection thresholds correlated with changes in olfactory bulb volume (interestingly, both increases and decreases were observed). To the extent that odor thresholds are a reflection of peripheral olfactory processing, these findings suggest a potential causal link between peripheral afferent input and structural integrity of the olfactory bulb. Similarly, in patients before and after surgical treatment for chronic rhinosinusitis with polyps, increased olfactory bulb volumes (up to 19%) were correlated to improvements in odor detection, but not to odor discrimination or identification, following a three-month interval (Gudziol et al., 2009). That these changes were observed in patients with an average age of 53 underscores the point that even the aging human olfactory system has a capacity to regenerate.

An unanticipated finding to emerge from MR volumetric imaging is that the depth of the olfactory sulcus in healthy subjects is predictive of odor perceptual performance, summed over tests of olfactory threshold, discrimination, and identification (Hummel et al., 2003). The olfactory sulcus represents the cleft between the lateral border of the gyrus rectus and the medial border of the medial orbitofrontal gyrus, forming a groove where the olfactory tract resides. Examination of patients with isolated anosmia (smell loss) indicates that the depth of this sulcus is a good indicator of the presence of the olfactory tract (Abolmaali et al., 2002).

One technological development for imaging the olfactory bulb has been a shift from using T_1-weighted sequences in favor of T_2-weighted sequences. Although T_1-weighted MRI strictly provides better anatomical resolution, T_2-weighted MRI holds several advantages. Chief among these is the fact that CSF, which appears bright on T_2, essentially encompasses the olfactory bulb and tract. As a result, these structures become sharply delineated by an outer bright ring, making it easier to trace their full extent. With improved contrast between the olfactory bulb, the CSF, and nearby blood vessels, another impressive benefit is that penetration of the olfactory nerves through the bony cribriform plate en route to the bulb can be visualized (Duprez and Rombaux, 2010) (Figure 13.3c,d). These specialized T_2-weighted MR scans rely on an imaging protocol known as a fast-spin-echo two-dimensional sequence, collected in the coronal plane, with 0.5 mm² in-plane (within-slice) resolution and 2 mm thick slices (Gudziol et al., 2009; Duprez and Rombaux, 2010). Notably, the recent introduction of a "constructive interference in steady state" (CISS) three-dimensional sequence, with an isotropic voxel resolution of 0.5 mm³, promises even greater signal resolution and reproducibility for olfactory bulb volumetry (Burmeister et al., 2011).

In exciting new work, Burmeister and collaborators (Burmeister et al., 2012) had an opportunity to obtain MR structural data of the human olfactory bulbs and tracts at high-resolution (0.2 mm³ voxels at 3-Tesla and 0.1 mm³ voxels at 9.4-Tesla) from embalmed post-mortem material that had been fixed in ethanol-formaldehyde (Figure 13.4). These MRI scans were compared to Nissl-stained 20-μm tissue sections from the same patient donors. Cross-sectional analysis of MR intensity through the olfactory bulb revealed four discrete intensity bands when imaged at 3-Tesla, appearing to correspond to major cortical lamina of the bulb on the histological sections,

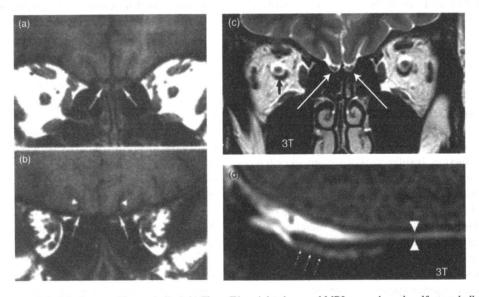

Figure 13.3 Structural MRI of the human olfactory bulb. (a,b) These T1-weighted coronal MRI scans show the olfactory bulbs (A, arrows) lying beneath the gyrus rectus bilaterally, and the olfactory tracts (b, arrows) lying at the base of the olfactory sulcus (arrowheads) bilaterally. Note that on these T1-weighted images, the olfactory bulb signal is brighter than the adjacent cerebrospinal fluid (CSF). (c) A T2-weighted coronal section from a 3-T MRI scan shows that the MR signal in the olfactory bulbs (white arrows) is darker than surrounding CSF and in fact encircles the bulbs, making it easy to identify the structure. Black arrow points to the dural sheath surrounding the optic nerve. (d) A sagittal T2-weighted image (where anterior is left and dorsal is up) highlights thin streaks of dark MR signal (small dotted arrows), which represent bony perforations in the cribriform plate through which the olfactory nerve filia pass on their way to the bulb. Arrowheads depict the CSF space between the ethmoid bone and the basal frontal lobe. a and b from Yousem et al., 1998; c and d from Duprez and Rombaux, 2010.

13.3 Structural Imaging

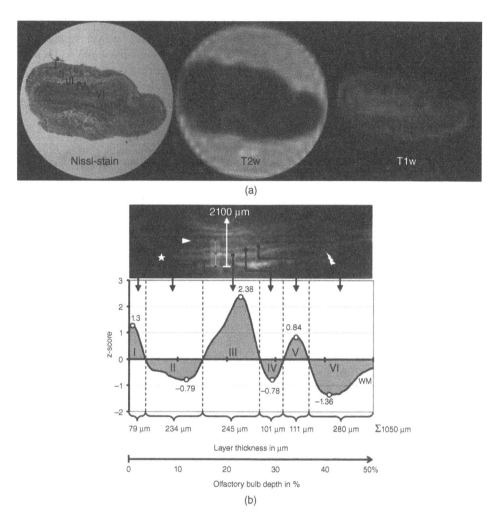

Figure 13.4 High-resolution MRI of the human olfactory bulb. (a) A histological section of the human olfactory bulb (Nissl-stained; 20 μm coronal sections) is compared to T2-weighted (T2w) and T1-weighted (T2w) MRI scans acquired at high-resolution (3-T MRI; 200 μm sections). Both the histological and MRI images were obtained from human post-mortem samples. Six discrete layers can be identified in the Nissl-stained section of the bulb, including the olfactory nerve layer (I), glomerular layer (II), external plexiform layer (III), mitral cell layer (IV), internal plexiform layer (V), and granule cell layer (VI), as well as central white matter (wm). At least three layers can be discerned in the MR images, with alternating bands of hypointense (dark) and hyperintense (bright) signal. (b) Structural MRI scans at even higher resolution (9.4 T MRI; 100 μm) reveals identification of all six lamina in the olfactory bulb (upper half of panel). MRI intensity profile analysis was conducted in serial coronal sections across an area of the bulb delimited by the white asterisk and lightning bolt symbols. White arrow indicates diameter of the olfactory bulb; white arrowhead indicates central white matter. Intensity quantification was conducted from the inferior surface (i.e., 0% depth) to the mid-white matter (i.e., 50% depth). Normalized signal profiles are shown in the lower half of the panel. Modified from Burmeister et al., 2012.

including (from the outside in) the olfactory nerve layer, the glomerular layer, the external plexiform layer, and a portion spanning the mitral and granule cell layers. Intensity profiles were even more impressive at 9-Tesla, with intensity profile analysis capturing all cytoarchitectonic layers of the bulb. Additional 3-Tesla high-resolution image analyses indicated that the olfactory tracts could adopt four unique morphological configurations, including a tripartite shape (akin to a Mexican hat) in which a dorsal zone of MR signal hypointensity was suggested to represent the anterior olfactory nucleus. These new data accentuate the remarkable potential of MRI to characterize microstructural details of the human brain. That being said, the imaging techniques described here would be extremely difficult to implement in living humans, if for no other reason that subjects would need to lie perfectly immobile during the 75–100 minutes of high-resolution 3-Tesla scanning, and that most subjects' heads would not fit within the dedicated rat head coil (inner diameter, 3.8 cm) used to transmit and receive the radiofrequency signals. Thus, at the present time, laminar characterization of the olfactory bulb is out of reach of conventional 3-Tesla

MRI, but this situation will hopefully change as technology continues to move forward.

13.3.2 Olfactory Structural Imaging in Clinical Diagnostics

Perceptual deficits in the sense of smell occur in a wide number of neurological and psychiatric conditions, likely reflecting the fact that the neuropathology in many of these disorders involves, or even arises in, olfactory-related limbic brain regions of the medial temporal lobe and orbitofrontal cortex (OFC). As a result, a main goal of structural imaging of the human olfactory system has been to develop potential biomarkers of disease, perhaps even for pre-clinical diagnosis prior to onset of overt symptoms.

Some of the original work along these lines examined patients with schizophrenia, identifying a 23% reduction in olfactory bulb volume compared to control subjects (Turetsky et al., 2000), as well as volumetric reductions in perirhinal cortices that correlated with odor detection deficits (Turetsky et al., 2003b). The same investigators went on to show that the volume of the olfactory bulb (on the right side only) was also reduced in first-degree relatives of patients with schizophrenia (Turetsky et al., 2003a), suggesting a genetic predisposition to structural derangements in the olfactory system, irrespective of health status. Volumetric changes in the amygdala and hippocampus have also been implicated in schizophrenia, with hippocampal volumes correlating with odor discrimination performance (Rupp et al., 2005). Olfactory bulb volumes have recently been found to be reduced in another psychiatric disorder, major depression (Negoias et al., 2010). Interestingly, olfactory bulb volume inversely correlated with the magnitude of depression, that is, the most depressed patients had the largest olfactory bulbs. This somewhat paradoxical finding raises the intriguing idea that individuals with a more acute sense of smell are more prone to developing depression.

Patients with Alzheimer's disease develop smell deficits early in the course of their illnesses (Doty et al., 1987; Koss et al., 1988; Murphy et al., 1990). Such observations have motivated studies coupling structural MRI with olfactory perceptual testing. Murphy and colleagues (Murphy et al., 2003) found a significant association in Alzheimer's disease patients between the size of the left hippocampus and performance on an olfactory identification task where patients had to match an odor stimulus to line drawings. To the extent that successful performance on this task required binding together information across olfactory and visual modalities, this finding suggested key involvement of the hippocampus in olfactory relational memory. Subsequent work has shown that patients in the early stages of Alzheimer's disease have reduced olfactory bulb volumes, which correlated with a measure of global

cognitive decline (Thomann et al., 2009b). Similar volumetric changes were seen in patients with "age-associated cognitive decline," presumably in a pre-clinical phase of Alzheimer's disease (Thomann et al., 2009a), though because these data were pooled alongside patients with clinically established Alzheimer's disease, these findings remain to be confirmed.

A variety of structural imaging techniques have been applied to patients with Parkinson's disease, for whom olfactory impairments are an early non-motor manifestation. Using an imaging software platform known as volumetric brain mapping (VBM) to detect alterations in gray matter, Wattendorf et al. (2009) evaluated structural brain changes in mild and moderately advanced Parkinson's disease. Compared to control participants, test scores combining olfactory threshold, discrimination, and identification tasks were significantly correlated with piriform cortex atrophy in mild-stage patients and with amygdala atrophy in moderate-advanced patients, in spite of preservation of total gray matter volume. These findings suggest that at least a portion of olfactory perceptual dysfunction in Parkinson's disease has a central basis.

An alternative MRI technique, diffusion tensor imaging (DTI), has also been used to characterize structural changes in Parkinson's disease. DTI is based on the idea that water molecules diffuse more quickly along the longitudinal axis of white matter tracts than perpendicular to those tracts (Mori and Zhang, 2006; Hagmann et al., 2007; Assaf and Pasternak, 2008). By measuring diffusion in multiple directions, DTI provides a way to quantify the difference between parallel and perpendicular diffusion, which comprises the DTI signal.

In a group of 12 Parkinson's patients, increased diffusivity was observed in the base of the frontal lobes, in the vicinity of the olfactory tracts, when compared to 12 age-matched controls (Scherfler et al., 2006). Moreover, these data correctly classified 16 of 17 independent subjects into the Parkinson's group (9/9 patients) and the control group (8/9 controls), attesting to the robustness of the group differences in this brain region. However, these results should be interpreted with caution. The volume of increased diffusivity was ~1300 mm^3, which far exceeds the volumes reported in the literature (the very upper estimate of olfactory bulb volume, based on structural MRI, is ~100 mm^3; Buschhuter et al., 2008). Furthermore, the scanning protocols were not optimized for imaging of the olfactory tract or bulb, where susceptibility artifact can introduce signal distortions. A separate DTI study identified changes in fractional anisotropy (an index of water diffusivity in the brain) in the gyrus rectus and primary olfactory cortex (near entorhinal cortex) in Parkinson's patients with smell loss, but not in control subjects, and fractional anisotropy measures from olfactory cortex correlated with odor identification scores (Ibarretxe-Bilbao et al., 2010). Given that the control subjects had an intact sense of smell, it remains unclear whether these DTI changes simply reflect group differences in olfactory perception, or something unique about olfactory processing in Parkinson's disease.

13.4 FUNCTIONAL IMAGING

Functional imaging of the human brain is based on a simple principle: it takes work to think. It even takes work not to think (perhaps applicable to both resting brain states and certain neuroscientists). Neuronal activity is an energy-demanding process: for every ion-gated action potential and every ligand-gated post-synaptic potential, chemical concentration gradients between the extracellular and intracellular spaces need to be restored, and neurotransmitters need to be repackaged into vesicles. These cellular 'housekeeping' operations require energy, which for the brain equals glucose, along with oxygen to drive efficient conversion of glucose to ATP (this is known as aerobic respiration). It thus follows that a given brain region with increased activity will need more glucose and consume more oxygen. Depending on available local concentrations of glucose and oxygen, the brain may also need increased blood flow to deliver these substances and thereby sustain ongoing activity. The important concept is that all human functional imaging techniques provide only surrogate measures of brain activity.

13.4.1 The First Human Olfactory Functional Imaging Study

Initial approaches to identify odor-dependent activity in the human brain (Zatorre et al., 1992) used positron emission tomography (PET). This modestly invasive imaging technique involves introduction of a radioactive tracer into the bloodstream, and was the leading functional imaging method through the late 1980s and early 1990s. The idea is that as the radioactive signal decays, it generates a positron, which quickly collides with an electron to generate two gamma rays that travel in opposite directions and are detected by an array of "coincidence detectors" positioned around the subject's head. After sufficient time, the number of events accumulated at each detector can be transformed into a three-dimensional spatial map, with hot spots corresponding to brain areas of increased metabolism or uptake of the tracer. Note that PET scanning has nothing to do with MRI, though the PET images are aligned to a T1-weighted anatomical MRI scan for establishing which brain regions are activated.

In 1992 Zatorre and colleagues conducted the first olfactory imaging study, using PET with H_2O^{15} as the radioactive tracer (Zatorre et al., 1992), providing measures of oxygen consumption and regional cerebral blood flow (rCBF) in the human brain. (Another commonly used PET tracer is a fluorine isotope, ^{18}F, conjugated to glucose, that is, fluoro-2-deoxy-D-glucose, or FDG, providing a measure of glucose uptake.) During scanning subjects were presented with either an odorless cotton wand (control condition) or a cotton wand saturated with one of eight different odorants. Comparison of these two conditions revealed odor-specific activations in the piriform cortex bilaterally and the OFC on the right side (Figure 13.5a). These findings nicely complemented data from patient lesion models of odor perception (Rausch et al., 1977; Potter and Butters, 1980; Eskenazi et al., 1983; Jones-Gotman and Zatorre, 1988) and suggested that imaging approaches could provide meaningful functional

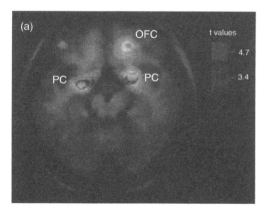
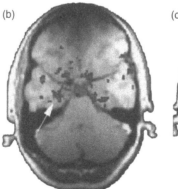
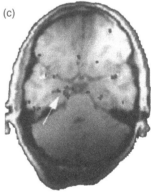

Figure 13.5 Functional imaging correlates of olfactory processing in the human brain. (a) Odor-evoked increases in PET regional cerebral blood flow (rCBF) are superimposed onto an axial section of an MRI scan (shown 17 mm below the intercommissural plane). Activations are depicted as a *t*-statistic image. PC, piriform cortex; OFC, orbitofrontal cortex. Modified from Zatorre et al., 1992. (b, c) Sniffing (in the absence of odor) evoked fMRI activity in the temporal piriform and entorhinal cortices (b, white arrow), but following topical anesthesia to the nasal cavity, fMRI was markedly reduced in this same participant (c, white arrow). Modified from Sobel et al., 1998. (*See plate section for color version.*)

information about the human olfactory system. By identifying some of the critical brain regions involved in human olfaction, this landmark work set the stage for the olfactory imaging work to follow.

Note that although PET imaging offers direct insights into cerebral metabolic states, there are several obstacles that have limited its research utility. First, the technique requires the costly availability of a cyclotron close at hand to generate the radioactive tracers in a timely way. Second, because PET involves the use of radioactive material, women of childbearing age are usually not enrolled in these studies. Third, while the radioactive half-life of the tracers is short, subjects can only receive a finite lifetime amount, restricting within-subject testing across multiple different conditions and/or days. Finally, due to the low signal-to-noise ratio of the PET signal, data for a given condition need to be collected over a 1–2 minute period (or longer). This poor temporal resolution is not compatible with many experimental designs involving rapid event-based stimulus presentations.

13.4.2 Initial Olfactory fMRI Studies

By the mid-1990s, cognitive neuroscience researchers were increasingly turning towards fMRI as the method of choice to elucidate human brain function. This technique takes advantage of the fact that deoxygenated (but not oxygenated) hemoglobin (Hb) is paramagnetic. Activity-dependent brain changes in the ratio of deoxygenated Hb to oxygenated Hb, arising from alterations in oxygen delivery, oxygen consumption, or cerebral blood flow per se, can be detected using specialized MRI sequences. The use of T2*-weighted fMRI to measure this hemodynamic effect, referred to as blood-oxygen-level dependent (BOLD) contrast, has made it possible to study a wide range of perceptual and cognitive tasks. Formally speaking, the increased BOLD signal following cerebral activation is due to the delivery of oxygenated Hb that effectively displaces (washes out) the deoxygenated Hb, leading to reduced T2* shortening and a brighter MR signal (Huettel et al., 2008). Based on combined neurophysiological and fMRI studies in both monkeys (Logothetis et al., 2001; Goense and Logothetis, 2008) and humans (Huettel et al., 2004; Mukamel et al., 2005), the BOLD signal appears to correlate best with the local field potential, reflecting integrated population activity and postsynaptic inputs, rather than single-unit or multi-unit spike firing reflecting the neuronal output of action potentials.

Interestingly, olfactory fMRI only lagged a few years behind fMRI of the visual, auditory, and somatosensory systems, hitting its stride in 1997 with publications from four different labs (Levy et al., 1997; Sobel et al., 1997; Yang et al., 1997; Yousem et al., 1997). Despite differences in imaging protocol, subject number, odors, stimulus delivery method, and experimental paradigm, these studies variously demonstrated fMRI activations in olfactory-related areas such as the medial temporal lobes and OFC, as well as in non-olfactory areas such as the cingulate cortex and the superior temporal gyrus. In an important early experiment, Sobel et al. showed that regions of the human olfactory brain responded not only to odor, but also to sniffing in the absence of odor (Sobel et al., 1998). The ingenious use of a topical anesthetic applied to the inside of the nose was associated with a reduction of sniff-evoked fMRI activity, consistent with the idea that it is the physical sensation of airflow across the nasal mucosa that underlies the sniff-related imaging effect (Figure 13.5b). These findings were in close accord with animal models and suggested that the act of sniffing may serve to prime the olfactory system in advance of odor receipt.

One problematic issue in much of the early olfactory fMRI work was the inconsistent and weak activation of primary olfactory cortex (i.e., piriform cortex). This was attributed to two main causes. First, the inferior frontal lobes and medial temporal lobes happen to be situated near air-tissue interfaces, which are highly susceptible to fMRI signal artifact and dropout. This anatomical circumstance made it difficult to obtain reliable activity from piriform cortex and OFC, and in fact was cited by many researchers as a good reason to use PET techniques instead (which are not subject to this problem). Fortunately, the subsequent development of new imaging protocols (Deichmann et al., 2003; De Panfilis and Schwarzbauer, 2005; Hsu and Glover, 2005; Weiskopf et al., 2006) has led to substantial improvements in signal recovery from both piriform cortex and the OFC. As a simple first step, even just tilting the axial acquisition plane at a 30° angle to the intercommissural border (rostral edge higher than caudal edge) offers a fairly dramatic improvement over standard horizontal acquisition planes, though this can come at the expense of signal artifact in lateral temporal regions.

A second reason for the inconstancy of fMRI activity in the piriform cortex was based on the fact that many olfactory studies used "blocked" task designs, in which odor and no-odor periods were each presented in alternating blocks of 30–60 s. To be fair, these blocked designs were used in virtually all of the early fMRI studies, partly because the PET functional studies had established a precedent for condition blocks, and partly because the statistical algorithms for analyzing "event-related" designs (in which different conditions, or events, rapidly presented within the same block could be distinguished) were not yet available (e.g., Dale and Buckner, 1997; Josephs et al., 1997). All other things being equal, another relative advantage was that blocked designs, with long stimulus presentations,

13.4 Functional Imaging

conferred greater statistical power to identify main effects. However, in the case of odor stimuli, a longer presentation meant greater response habituation in piriform cortex, which had been shown to exhibit marked decline in animal single-unit studies (Wilson, 1998, 2000). Sobel and colleagues highlighted this issue by showing that with 40-s blocks of odor (vs. no odor), very little fMRI activity in piriform cortex could be visualized when averaging across the whole block, but by modeling the odor block with an exponential decay function, they were able to demonstrate a robust piriform response (Sobel et al., 2000); also see Poellinger et al., 2001 for related findings). This study brought new awareness to the importance of tailoring experimental designs to the unique constraints of the olfactory system.

13.4.3 The Event-Related Era of Olfactory fMRI

Recognizing the value of short stimulus presentations, researchers began utilizing event-related olfactory task designs. The use of brief presentations (~1–3 s) of multiple different odorants, with sufficient trial-to-trial spacing (~10–20 s), minimized the problem of sensory habituation, and also enhanced the flexibility and range of experimental designs. These technical innovations helped enhance the types of studies that could be conducted. An important element in this formula was the technology to deliver these odorants with a rapid on-off time, and MRI-compatible olfactometers (smell delivery machines) were introduced by Sobel et al. (1997) and by Lorig et al. (1999) to achieve fast computer-controlled presentation of odor in the absence of auditory, tactile, or thermal confounds. In addition, given that fMRI activity in the piriform cortex could be elicited by sniffing even in the absence of odor (Sobel et al., 1998), it became imperative to collect online measures of respiration during scanning, in order to ensure that condition-specific differences in sniffing did not confound the imaging data, or if so, to covary out this information from the analysis.

In a pair of studies, Gottfried and colleagues used event-related fMRI techniques to explore odor processing and olfactory Pavlovian conditioning in the human brain, with a focus on pleasant and unpleasant odors (Gottfried et al., 2002a; Gottfried et al., 2002b). During scanning, subjects participated in an olfactory version of classical conditioning in which a neutral item (the conditioned stimulus, or CS+) acquired behavioral significance by predicting the occurrence of an emotionally salient odor (the unconditioned stimulus, or UCS) after repeated pairings. In this study, three different neutral faces represented the CS+ stimuli, and three different odors varying in pleasantness were used as the UCS stimuli. Under a 50% partial reinforcement schedule, only one-half of all CS+

trials was paired with odor, providing a way to isolate olfactory sensory processing (by comparing paired vs. unpaired trials) as well as learning-related effects without interference from the UCS (Figure 13.6a).

A key finding in this work was that discrete piriform sub-regions showed differential sensitivity to odor: valence-independent responses were observed in the posterior piriform cortex (PPC), whereas valence-dependent responses were observed in the anterior piriform cortex (APC) (Gottfried et al., 2002a) (Figure 13.6b). These results provided evidence for the functional heterogeneity of human piriform cortex, in line with rodent models demonstrating anatomical and functional differences along the rostral-caudal axis of piriform cortex (Haberly, 1998). Interestingly, the amygdala and posterior OFC also responded to all three valence classes of odor, though fMRI representations of pleasant and unpleasant odor were identified in dissociable areas of OFC. Concurrent analysis of the learning data revealed that the OFC was activated in response to the conditioned cue (in the absence of odor), with selective involvement of the medial and lateral OFC in processing of the appetitive or aversive CS+ cue, respectively (Gottfried et al., 2002b). This and related fMRI studies (Gottfried et al., 2003; Gottfried and Dolan, 2004) have highlighted the capacity of emotionally salient odors to promote cross-modal associative learning, and support the idea that both piriform cortex and its projection sites in OFC should be considered as active participants in more complex aspects of odor information processing, rather than as mere sensory relays.

The following sections review some of the key functions and mechanisms of the human olfactory system that have been established using both fMRI and PET methods, with an emphasis on studies in which the neural data are closely linked to specific olfactory perceptual or cognitive states. Since the last version of this chapter appeared ten years ago, there has been tremendous growth in olfactory functional imaging. However, due to space limitations, it is not possible to include all of these studies. In addition, topics covered in detail in other chapters of the *Handbook*, including odor memory, multisensory integration, trigeminal processing, social behavior, and clinical investigations, will be discussed only briefly.

13.4.4 Encoding Odor Valence

Psychophysical research dating back to the 1970s indicates that valence (pleasantness) is a primary dimension by which humans categorize odors, particularly for smells that are unfamiliar or hard to name (Schiffman, 1974; Khan et al., 2007). Given the intimate anatomical overlap between the olfactory system and limbic brain regions mediating emotion and aversive learning, a plausible hypothesis was that the hedonic features of an odor might

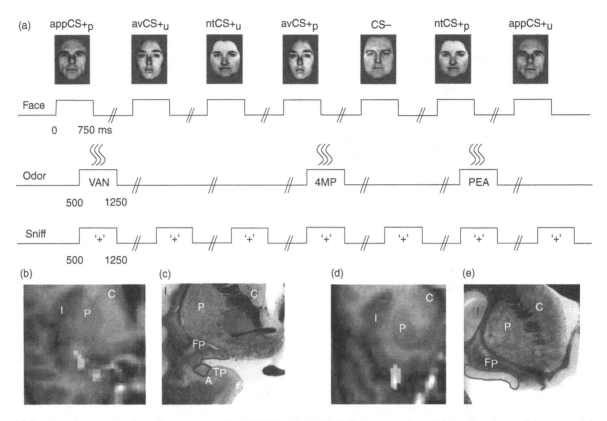

Figure 13.6 An early example of an olfactory event-related fMRI study. (a) Task. During scanning, subjects viewed one of three neutral faces (the conditioned stimuli, or CS+). Each face was paired with a distinct odorant on 50% of trials (i.e., CS+p) and was unpaired on 50% of trials (i.e., CS+u). Another face was never paired with odorant (the control stimulus, CS−). Odorants served as the unconditioned stimuli (UCS), including vanillin (VAN) as an appetitive (app) reinforcer, 4-methyl-pentanoic acid (4MP) as an aversive (av) reinforcer, and phenethyl alcohol (PEA) as a neutral (nt) reinforcer. On each trial subjects were cued to sniff upon viewing a '+' crosshair on a computer monitor. Trials recurred every 7.5 sec, with each trial type evenly interspersed over the course of the 25-min experiment. (b-c) Valence-*independent* fMRI responses. The comparison of CS+p trials (odor present) vs. CS+u trials (odor absent) elicited odor-evoked activity in posterior piriform cortex (B), spanning frontal (Fp) and temporal (Tp) areas. (c) Activations are shown beside a high-resolution anatomical atlas image. (D-E) Valence-*dependent* fMRI responses. The direct contrast of aversive and appetitive odorants ([avCS+p—avCS+u]—[appCS+p—appCS+u]) revealed fMRI activity in anterior piriform cortex (d), shown next to a high-resolution image (e). Key: C, caudate; Fp, frontal portion of piriform cortex; I, insula; P, putamen; Tp, temporal portion of piriform cortex. Data from Gottfried et al., 2002a. Atlas images in C and E are modified from Mai et al., 1997, and used with permission of Academic Press and the author. (*See plate section for color version.*)

be encoded in the amygdala. To explore how the human brain processes unpleasant odor, Zald and Pardo (1997) used H_2O^{15} PET scanning to measure rCBF while subjects smelled highly aversive sulfides (rotten egg odor). Comparison of these unpleasant smells to less aversive odorants revealed activity in the amygdala bilaterally and in the left OFC (Figure 13.7a). Although these results were highly suggestive of a selective aversive response in these brain regions, differences in intensity between the odorants complicated the interpretation of the effects. In separate PET studies, increased rCBF in the amygdala, OFC, and hypothalamus were also observed when subjects made hedonic judgments of odors (Royet et al., 2000; Zald and Pardo, 2000; Royet et al., 2001).

Subsequent fMRI investigations arrived at somewhat different conclusions. As noted above, pleasant, unpleasant, and neutral smells were each found to evoke similar magnitudes of fMRI activity in the amygdala (Gottfried et al., 2002a), arguing for a valence-independent coding scheme. In another study, one pleasant odor and one unpleasant odor were each presented at low and high intensity, providing an elegant way to dissociate these two perceptual features within the same experimental paradigm (Anderson et al., 2003). Interestingly, a main effect of intensity was found in the amygdala, with greater activity for pleasant and unpleasant odors at high (vs. low) intensity (Figure 13.7b,c). Valence-specific effects were only identified in the OFC. Finally, an fMRI study using a larger panel of odorants further complicated the story, showing that pleasantness ratings covaried with OFC activity, whereas intensity ratings covaried with activity in entorhinal cortex (Rolls et al., 2003). Together these

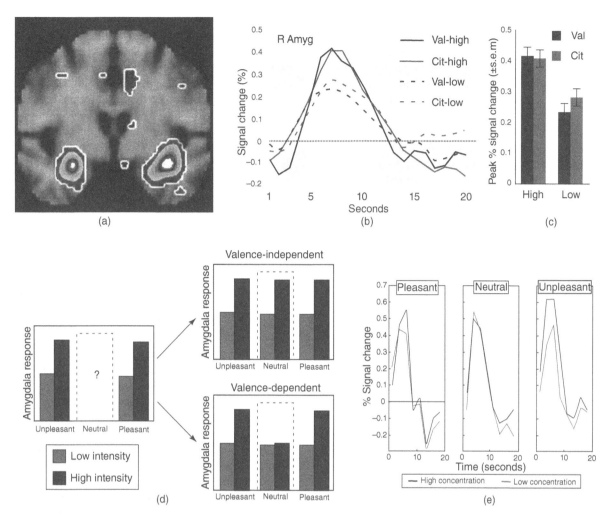

Figure 13.7 Odor valence and intensity processing in the human amygdala. (a) Aversive vs. less aversive odorants enhance PET rCBF bilaterally in the amygdala. Activations superimposed on a T1-weighted MRI scan. From Zald and Pardo, 1997. (b, c) The fMRI time-series in right amygdala (B) for high- and low-intensity versions of pleasant (Citral, Cit) and unpleasant (Valeric acid, Val) odorants demonstrates an intensity effect in the absence of a valence effect. (c) Plots represent peak odor-evoked changes in fMRI signal. From Anderson et al., 2003. (D,E) Testing the hypothesis that odor-related amygdala activity reflects an integrated response to both valence and intensity. (d) The valence-independent model predicts greater activity for high-intensity odorants at all valence levels, whereas the valence-dependent model predicts greater activity for high-intensity odorants only at the extremes of valence. (e) Odor-evoked fMRI time-series are consistent with valence-dependent coding in the amygdala. Modified from Winston et al., 2005. (*See plate section for color version.*)

findings raised questions about the precise role of the amygdala in olfactory hedonic processing.

In an effort to clarify these discrepant results, Winston et al. (2005) built upon the Anderson et al. study by including high- and low-intensity versions of a neutral odorant, in addition to pleasant and unpleasant odorants. With this design, the dimensions of valence and intensity could be dissociated across a broader hedonic spectrum, enabling a test of two competing models of amygdala function: a valence-independent hypothesis in which the amygdala response to intensity is similar at all levels of valence; and a valence-dependent hypothesis in which the amygdala is sensitive to odor intensity only at the outer bounds of valence (Figure 13.7d). Data were consistent with the latter prediction: intensity-based fMRI activity in the amygdala was observed only for the pleasant and unpleasant odors, but not for the neutral odor (Figure 13.7e). That the amygdala encodes an integrated representation of intensity and valence underscores the added importance of odor strength in appraising the behavioral saliency of valenced odors. Put differently, the more intense an emotionally salient odor, the more likely it is to be relevant for survival, which is not the case for emotionally neutral odors. These findings help bring some resolution to the mixed conclusions regarding the olfactory response profile of the human amygdala.

Across numerous functional imaging studies, odor valence-specific responses have been consistently observed in the OFC (Zald and Pardo, 1997; Gottfried et al., 2002a; Anderson et al., 2003; Rolls et al., 2003). Similar effects have been shown in an fMRI paradigm of olfactory sensory-specific satiety (O'Doherty et al., 2000). In this experiment, subjects were scanned while smelling two different food smells. Subsequently, the pleasantness of one of the two odors was reduced by having subjects eat a lunch of the corresponding food until they were sated. In parallel, the magnitude of OFC activity was reduced for the devalued odor but not for the control odor, suggesting that a representation of olfactory reward value is encoded in this region. Based on monkey single-unit studies showing that the OFC may encode the relative value of different stimulus options, Grabenhorst and Rolls (2009) showed that a given odorant was rated more pleasant when preceded by a less pleasant odorant, and was rated less pleasant when preceded by a more pleasant odorant. The degree of relative, or subjective, pleasantness between the first and second odorants on each trial was found to correlate with fMRI activity in the anterior lateral OFC (and relative unpleasantness correlated with activity in the insula). Interestingly, the absolute values of the pleasantness ratings were not associated with activity in these areas, though other regions of OFC were responsive to this measure. These findings suggest that the human brain has access to both relative and absolute representations of odor valence in order to guide olfactory-based decisions.

13.4.5 Encoding Odor Objects

The term "odor object" refers to an olfactory source (such as a rose bush) or an olfactory event (such as the aroma of rose) that is presented to the senses. This is no different from conventional notions of objects in the visual domain, such as visual sources (rose bush) and visual events (the light reflecting off a rose bush). It is the distinctive physical and ecological form of sensory objects that shapes nervous system function (Gottfried, 2009, 2010). The fact that most smells naturally encountered in the environment are complex mixtures of dozens or even hundreds of different odorous molecules means that the olfactory system needs mechanisms to weave these odor elements into perceptual wholes. For many vertebrate and invertebrate species, efficient encoding and retrieval of behaviorally salient odor objects is critical for adaptive behavior, including but not limited to food search, mating choice, and territorial defense.

Anatomical, physiological, and computational studies in animal models suggest that odor object information may take the form of distributed ensemble patterns of activity in the piriform cortex (Haberly, 1985; Hasselmo et al., 1990; Illig and Haberly, 2003), but how odor objects are represented in the human brain is unclear. PET imaging was initially used to address these questions by evaluating which brain regions were recruited during odor identification and discrimination tasks (Royet et al., 1999; Savic et al., 2000). To the extent that identifying an odor requires retrieving semantic object-based information about the stimulus, these studies showed widespread task-related activity in the inferior frontal gyrus, insula, caudate, and cerebellum. Although the use of PET techniques precluded the ability to assess odor object coding of specific stimuli, this is not a trivial goal even with fMRI. Assuming that different odor qualities (i.e., the perceptual character of a smell, such as its mintiness) evoke distributed but overlapping responses in piriform cortex, the use of standard fMRI analyses would be unable to discriminate unique representations of odor quality, because activity is averaged over all of the voxels within a region of interest, obscuring information that might be contained in individual voxels.

Recent methodological innovations have provided new ways to overcome these limitations. In one approach (Gottfried et al., 2006), an olfactory version of fMRI "cross-adaptation" was devised to dissociate the neural correlates of odor perceptual quality and molecular functional group (as a key attribute of odorant chemical structure). Cross-adaptation is based on the idea that sequential repetition of stimuli sharing a particular feature (such as object shape) causes adaptation of neural populations specifically sensitive to that feature, leading to local response decreases (Buckner et al., 1998; Kourtzi and Kanwisher, 2001; Winston et al., 2004). In this study, subjects made two successive sniffs to pairs of odors differing either in quality or functional group. The fMRI results indicated a double dissociation in piriform cortex, whereby *PPC encoded quality* (but not structure) and *APC encoded structure* (but not quality) (Figure 13.8). The identification of dual olfactory representations within discrete piriform areas accords with the known physiological organization of piriform cortex in animal models, suggesting a functional preservation across species. The presence of structure-based codes in APC would help ensure stimulus fidelity of the original sensory input, while quality-based codes in PPC suggests that more integrative mechanisms may underlie olfactory coding of odor objects in the human brain.

Taking these findings one step further, Howard et al. (2009) combined high-resolution fMRI with olfactory multivariate analysis techniques to investigate spatial ensemble coding of odor qualities and categories in human PPC. As noted above, multivariate fMRI methods differ from conventional (univariate) fMRI analyses, in which data are averaged over space (voxels), time (scans) and subjects, obscuring potentially important information that may be contained at the level of individual voxels, scans, and subjects. Pattern-based approaches provide a

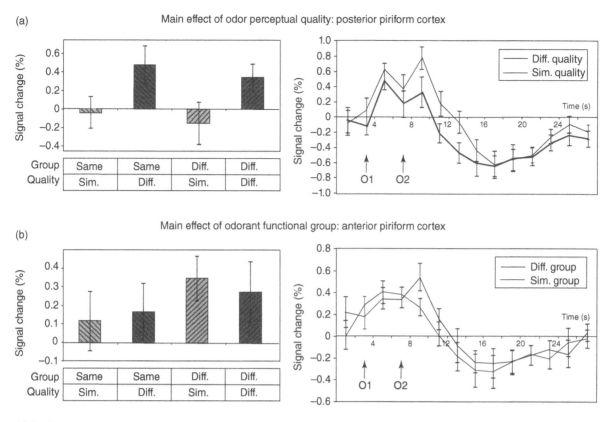

Figure 13.8 Functional dissociation of odor quality and odorant molecular functional group in human piriform cortex. (a) Condition-specific plots of fMRI activity in posterior piriform cortex (PPC) show greater responses for odorants with different (diff.) perceptual qualities than similar (sim.) qualities, in the absence of an influence of functional group. (b) The fMRI time-course in PPC shows selective decrement (adaptation) when the second odorant (O2) is similar in quality to the first odorant (O1). (c) Plots of fMRI activity in anterior piriform cortex (APC) showed an opposite effect, with greater responses to odorants differing in functional group, with no influence of perceptual quality. (d) The fMRI time-course in APC shows response adaptation when O2 shares the same functional group as O1. Modified from Gottfried et al., 2006.

robust way to characterize how (rather than just where) perceptual information is represented in the human brain (Kriegeskorte and Bandettini, 2007). During the experiment, subjects smelled sets of odorants differing in odor quality categories (e.g., minty, floral, woody). Different odor qualities evoked overlapping but unique activity patterns in PPC, and these ensemble patterns coincided with perceptual ratings of odor quality, such that odorants with more (or less) similar fMRI patterns were perceived as more (or less) alike (Figure 13.9). Such effects were not observed in APC, amygdala or OFC, indicating that ensemble coding of odor categorical perception is regionally specific for PPC. These results demonstrate that the spatial arrangement of fMRI activity in PPC is distributed and overlapping for each odorant, without any obvious local clustering, consistent with the known anatomical organization of this region. Overall, these findings suggest that distributed olfactory codes of odor object categories are arranged in much the same way that visual object categories (for example, houses, cows, and chairs) are organized in the inferotemporal cortex, highlighting the critical sensory-associative nature of PPC.

13.4.6 Odor Mixture Discrimination and Disambiguation

Several studies have begun examining odor mixture processing in the human brain. Because most naturalistic smells are complex mixtures of different odorous molecules, characterizing how the brain integrates odorant elements into perceptual wholes will be important for gaining a comprehensive understanding of olfactory perception. In mixtures containing both pleasant and unpleasant odor components, fMRI responses in discrete regions of the OFC and anterior cingulate cortex appeared to encode independent representations of positive and negative hedonic value (Grabenhorst et al., 2007). In another study, 100% citral (pleasant odor) and 100% pyridine (unpleasant odor) were mixed in proportions from 10/90 to 90/10. PET activity in the lateral OFC was associated with increasing odor impurity, whereas activity in the anterior

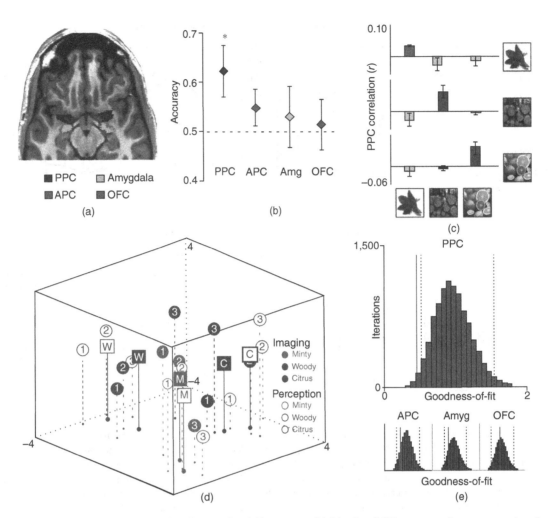

Figure 13.9 Distributed odor object representations in posterior piriform cortex. Multivariate fMRI pattern analyses were used to characterize how, rather than simply where, odor object information is encoded in the human brain. (a) Regions of interest included PPC, APC, amygdala (amg), and OFC. (B,C) Subjects were presented with nine different odorants, including three stimuli within each of three perceptual categories (minty, woody, and citrus). (b) Classification accuracy, as indexed by the proportion of within-category correlations exceeding between-category correlations, was significant (*, $p<0.05$) only in PPC. (c) This effect was observed for each odor category. (d) Perceptual ratings of odor quality similarity between all odorant pairs, and fMRI measures of odor pattern similarity between all odorant pairs, were projected onto a common 3-dimensional space. Perceptual maps and imaging maps closely overlapped, both for individual odorants and for odor categories. Squares labeled M (minty), W (woody), and C (citrus) represent centroids of each category for perceptual (empty squares) and imaging (solid squares) maps. (e) The 'goodness-of-fit', based on a comparison of the actual data to a distribution of 1,500 random permuted samples, was significant only in PPC ($p<0.05$). Modified from Howard et al., 2009. (*See plate section for color version.*)

OFC was higher for the binary mixtures than for the pure single odorants (Boyle et al., 2009). Finally, a more recent investigation combined fMRI, psychophysical techniques, and computational modeling to establish whether noisy sensory evidence is accumulated over time in the human brain to optimize olfactory perceptual decision-making (Bowman et al., 2012). In a two-alternative forced-choice task, subjects were presented with a series of odor mixture morphs ranging between 100% eugenol (clove odor) and 100% citral (lemon odor), and made as many sniffs as needed to determine whether the mixture smelled more like clove or lemon. A ramp-like increase of fMRI activity was observed in OFC, peaking at the time of decision, whereas activity in PPC reached an early plateau after odor onset and was sustained up until the decision point. These data suggest that PPC provides an ongoing sensory report of odor input, while the OFC integrates this information over time to help resolve sensory noise and ambiguity.

13.4.7 Attending to Odors

Attending to an odor in advance of its receipt is important for optimizing olfactory perception. In general, the process by which the brain transforms sensory inputs into perceptual events often begins before physical contact with the sensory stimulus. In the case of olfaction, the motor act of sniffing is an essential component of directing sensory attentional resources toward the olfactory milieu. Combining a clever psychophysical paradigm with fMRI, Zelano and colleagues (Zelano et al., 2005) asked subjects either to sniff and try to detect an odorant (which was absent on 50% of trials), or to sniff passively, knowing that no odorant was to be delivered (a less attentionally demanding task). Greater fMRI activity in frontal piriform cortex was observed in response to the no-odorant condition on the detection task, compared to the passive sniff task, suggesting a direct modulatory influence of attention on olfactory processing in the human brain. In a complementary experiment, subjects were simultaneously presented with an auditory tone and an odor stimulus, and informed either to attend to the odor and rate its intensity (ignoring the tone), or to attend to the tone and rate its pitch (ignoring the odor). Olfactory (vs. auditory) attention again revealed significant piriform activation. These findings demonstrated that the act of sniffing engages the olfactory system only when an odor is expected.

The modulatory role of attention on the human olfactory system has also been explored using fMRI effective connectivity. This analytical technique tests the influence that one neural population exerts over another, with the goal of defining how differences in brain states modulate the interactions between regions (Friston et al., 2003). In a study by Plailly and colleagues (Plailly et al., 2008), effective connectivity techniques were used to test the hypothesis that attention to smell augments functional interactions within thalamocortical networks. Prior animal studies suggest that odor information has direct access from piriform cortex to OFC without a thalamic intermediary, distinguishing olfaction from other sensory modalities. However, an indirect pathway from piriform cortex to OFC via the mediodorsal thalamus (MDT) has also been described (Price and Slotnick, 1983; Russchen et al., 1987; Ray and Price, 1993). Results showed that odor attention significantly modulated fMRI coupling within the indirect pathway, strengthening connectivity between MDT and OFC (Plailly et al., 2008). Critically, these effects were modality specific (odor > tone attention), directionally sensitive (forward > backward connections), and selective to route (indirect > direct pathway). These findings suggest that the human transthalamic pathway is functionally active in humans and is a modulatory target of olfactory attentional processing.

The idea that attentional states can set up expectations for what is likely—but not yet—to be encountered is consistent with the concept of predictive coding (Rao and Ballard, 1999; Friston, 2005; Summerfield and Egner, 2009). By setting up predictive templates or "search images," the brain can optimize perceptual processing at the time of stimulus delivery. Early work in animal models (Freeman and Schneider, 1982; Freeman, 1983) provided theoretical and physiological support for predictive coding in the olfactory bulb, though whether such mechanisms are utilized in the humans has remained unclear. Recent work by Zelano, Mohanty, and Gottfried (Zelano et al., 2011) combined an olfactory attentional search task with multivariate pattern-based fMRI analyses to examine olfactory predictive coding in the human brain. In separate task blocks, subjects needed to decide whether a predetermined target smell (either "A" or "B") was present on each trial, where odor stimuli consisted of A, B, or the mixture AB. This design enabled us to model the pre-stimulus and post-stimulus periods separately, and provided a way to examine target-related and stimulus-related activity patterns as they evolved from pre- to post-stimulus.

In APC and OFC, odor-specific ensemble patterns were formed prior to odor onset (Zelano et al., 2011). Notably, these patterns persisted even after odor stimulation, suggesting that APC and OFC encode what is being sought rather than what is being delivered. In contrast, fMRI patterns in PPC evolved from pre-stimulus target representations to post-stimulus odor representations, whereby activity patterns more closely resembled what was delivered rather than what was expected (Figure 13.10). Critically, the robustness of target-related patterns in PPC predicted subsequent identification accuracy on the odor search task. These findings were among the first to document feature-specific predictive codes in the human brain, and reinforce the idea that stimulus templates in PPC, with physical correspondence to odor-specific ensemble representations, play an instrumental role in guiding olfactory perceptual decisions. In identifying a distributed neural signature of odor templates in PPC, the results help validate longstanding models of piriform function (Freeman and Schneider, 1982; Haberly, 1985; Hasselmo et al., 1990; Wilson and Stevenson, 2003).

Interestingly, mean fMRI activity in MDT was significantly higher in response to unexpected trials (e.g., search for A, receive B) than expected trials (e.g., search for A, receive A). This profile is consistent with the idea that MDT may compare bottom-up odor input with top-down predictions to generate an error signal (Summerfield and Egner, 2009), and would be compatible with its known anatomical connectivity (Price and Slotnick, 1983; Russchen et al., 1987; Ray and Price, 1993). Additional support for this hypothesis comes from a recent patient-based MRI volumetric study (Olofsson et al., 2013), which showed that the cortical volume of MDT correlated with error rates on matching tasks between odors and pictures (and also between odors and words), with smaller thalamic volumes relating to a higher proportion of between-category errors. Together these findings highlight the putative role of MDT in coding of olfactory prediction error, and may help

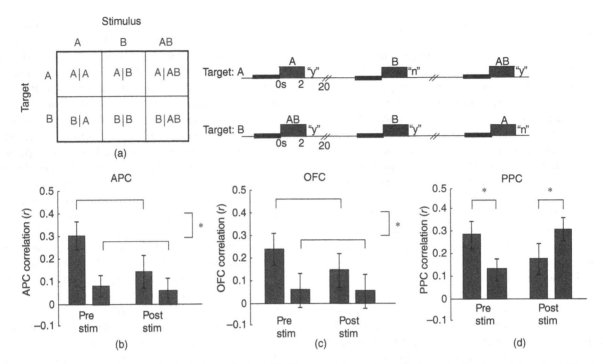

Figure 13.10 Olfactory predictive coding in the human brain. (a) Subjects participated in an olfactory attentional search task, in which stimuli consisted of odorants A, B, or AB, and the odor search target (A or B) varied across blocks. Subjects were informed at the beginning of a block whether the target odor would be A or B. On each trial, odor delivery was preceded by a 3-s pre-stimulus countdown cue. Subjects pressed a button indicating whether the target odor was present in the actual stimulus ("y" or "n"). (b-d) Target- and stimulus-related fMRI effects were computed by testing whether ensemble pattern correlations between same target/different stimulus conditions (green bars; e.g., A|A vs. A|B) differed from different target/same stimulus conditions (blue bars; e.g., A|A vs. B|A). In both APC (b) and OFC (c), target-related patterns emerged in the pre-stimulus period and persisted into the post-stimulus period. *, $p<0.05$. In contrast, in PPC (d), target-related patterns in the pre-stimulus period gave way to stimulus-specific patterns in the post-stimulus period. In PPC the pre-stimulus target-specific patterns resembled the post-stimulus odor patterns, consistent with the formation of predictive odor templates in this region (data not shown). Modified from Zelano et al., 2011.

unify disparate observations of MDT olfactory function in both animal and human studies (Eichenbaum et al., 1980; Slotnick and Kaneko, 1981; Staubli et al., 1987; Plailly et al., 2008; Sela et al., 2009; Tham et al., 2009).

13.4.8 Odor Imagery

That stimulus-specific odor templates are encoded in the human brain raises an interesting implication: internal access to these perceptual representations should favor odor imagery. Indeed, behavioral studies have shown that odor imagery can enhance odor detection in a stimulus-specific way (Djordjevic et al., 2004), and that imagery of pleasant or unpleasant odors causes valence-appropriate modulation of sniff volume (Bensafi et al., 2003). Complementary imaging work demonstrates that odor imagery elicits activations in the same brain areas that are engaged during actual stimulus encounter (Djordjevic et al., 2005; Bensafi et al., 2007). In professional perfumers, the duration of olfactory training was found to correlate inversely with activity in PPC, OFC, and hippocampus (Plailly et al., 2012), perhaps reflecting greater processing efficiency in odor experts.

13.4.9 Odor Memory

With Marcel Proust's madeleine biscuit triggering nine volumes of turgid (for some, transcendent) prose, olfactory researchers have endeavored to find the Proust spot of the brain, that is, to understand at a neural level why odor stimuli seem to have such potency to evoke intense autobiographical memories. The short answer to this conundrum is that it awaits further study. However, in an interesting fMRI experiment, Herz et al. (2004) identified subjects for whom specific perfumes triggered remote personal memories. Presentation of these same perfume smells in the scanner evoked increased activation in the amygdala, compared to the activity levels evoked by pictures of the perfume bottles or by control odors. Although this study was based on relatively few subjects, it provided interesting insights into the involvement of the amygdala in reactivating olfactory memory.

PET and fMRI studies of olfactory memory have run the full gamut, including working memory (Dade et al., 2001; Zelano et al., 2009), recognition memory (Savic et al., 2000; Dade et al., 2002; Gottfried et al., 2004; Cerf-Ducastel and Murphy, 2006; Royet et al., 2011; Lehn et al., 2012), and free recall (Kjelvik et al., 2012). In most instances, successful odor retrieval is associated with neural activity in the same brain areas that are recruited in non-olfactory retrieval tasks (such as hippocampus and prefrontal cortex), suggesting that a dedicated memory network is engaged independently of sensory modality. In some of these studies, a retrieval cue in one modality (e.g., visual) elicited activity in olfactory cortical areas, either following encoding of new cross-modal odor-object associations (Gottfried et al., 2004), or as a result of previously learned associations (Cerf-Ducastel and Murphy, 2006; Lehn et al., 2012). These findings concur with the idea that odor-specific features of a multisensory memory may reside in the olfactory system and can be reactivated by non-olfactory components of the original event. In another cross-modal memory study, subjects learned to associate the same visual object with one, and then another, set of pleasant and unpleasant odors and tones (Yeshurun et al., 2009). At recall one week later, the first associations were remembered better than the second associations, for both unpleasant odors and tones, whereas fMRI scanning revealed preferential hippocampal activation only for first odor associations, regardless of valence. Finally, in a study examining sleep-related effects of odor context on episodic memory (Rasch et al., 2007), subjects were able to recall the locations of visual objects more effectively when an odor present at initial learning was re-presented during sleep, with odor-related fMRI activity in the hippocampus observed during the sleep period.

13.4.10 Olfactory Learning, Plasticity, and Experience

Odor perception is profoundly shaped by learning and experience. Olfactory psychophysical studies have robustly demonstrated the perceptual pliability of the human sense of smell (Rabin, 1988; Wysocki et al., 1989; Jehl et al., 1995; Herz and von Clef, 2001; Dalton et al., 2002; Mainland et al., 2002), and functional imaging studies have brought new insights into the mechanisms underlying this plasticity. Sensory context can alter odor processing at the behavioral and neural levels. Across several studies, it is apparent that presentation of an odor alongside a semantically congruent flavor, picture, or color facilitates odor perception and induces greater responses in the OFC, amygdala, and hippocampus, when compared to semantically incongruent pairings (Small et al., 1997; De Araujo et al., 2003; Gottfried and Dolan, 2003; Small et al., 2004; Osterbauer et al., 2005). Hedonic shifts in verbal context

are sufficient to modulate odor-evoked brain activity: an olfactory mixture containing isovaleric acid and cheddar cheese flavor evoked greater fMRI signal in medial OFC when it was paired with the label "cheddar cheese" than when it was paired with the label "body odor" (de Araujo et al., 2005).

The mere exposure to an odor can enhance perceptual processing. In one fMRI study, a single 3.5-minute presentation of one odorant enhanced perceptual differentiation for odorants belonging to the same odor quality category, an effect that persisted for up to 24 hours (Li et al., 2006). Thus, for example, subjects exposed to a minty odorant became mint "experts," such that they were better able to tell apart different minty smells. This perceptual enhancement was paralleled by learning-induced response increases in PPC and OFC. Interestingly, the magnitude of change in OFC activation predicted subsequent improvement in perceptual differentiation on a subject-by-subject basis. These findings suggest that neural representations of odor quality can be rapidly updated through mere perceptual experience. This process of odor feature differentiation, via sensory exposure, may underlie much of the way that humans naturally learn to build up their "vocabulary" to recognize and discriminate odors.

Odor perceptual representations can also be updated through associative learning. Li et al. (2008) combined functional magnetic resonance imaging (fMRI) with multivariate analysis techniques to explore the impact of aversive Pavlovian conditioning on discrimination of predictive odor cues. The use of perceptually identical odor enantiomers (mirror-image molecules) provided a way to determine whether humans can acquire the ability to distinguish between odorous stimuli that initially smell the same. Subjects were presented with four odor enantiomers (two different pairs), one of which (the target CS+, "tgCS+") was repetitively paired with an electric shock (US) during a conditioning phase, whereas its mirrored counterpart ("chCS+") was not accompanied by shock. The second pair of odor enantiomers served as non-conditioned control stimuli. After conditioning, behavioral accuracy for distinguishing between tgCS+ and chCS+ was significantly enhanced, without any improvement in distinguishing between the control enantiomer pair. In parallel to these behavioral effects, ensemble activity patterns in PPC for the CS+ pair (tgCS+ and chCS+) became decorrelated after conditioning, particularly in comparison to the non-conditioned pair. These findings underscore the capacity of fear conditioning to update perceptual representation of predictive cues. That completely indiscriminable sensations can be transformed into discriminable percepts further accentuates the potency of associative learning to enhance sensory cue perception and support adaptive behavior.

Notably, the *lack* of experience can induce an opposite effect on odor discrimination. Following from animal models of odor deprivation to provoke changes in the olfactory system (Brunjes and Borror, 1983; Maruniak et al., 1989; Guthrie et al., 1990), Wu and colleagues developed a method to induce long-term odor deprivation in humans (Wu et al., 2012). Healthy subjects were admitted to the clinical research wing of a hospital for one week, during which time their nostrils were occluded with foam tape during waking odors. Perceptual tests and fMRI scanning were performed at baseline, immediately after deprivation, and one week later at recovery. Interestingly, olfactory perceptual performance was maintained despite a substantial reduction in odor afferent input. At the same time, the deprivation procedure elicited reversible fMRI changes in piriform and orbitofrontal cortices that may be instrumental in sustaining odor perception in the wake of disrupted input. The fact that odor perception was preserved across subjects suggests that the olfactory system possesses compensatory mechanisms that render it resistant to transient perturbations of afferent sensory input, as commonly occur during extended instances of rhinosinusitis or upper respiratory infections (Holbrook and Leopold, 2006).

13.4.11 Odors as Putative Chemosignals

The role of airborne odorants as putative human pheromones, or "chemosignals" (a less polarizing term), is hotly debated and the subject of entire chapters and books (Wysocki and Preti, 2004; Doty, 2010). Whether humans secrete odors that can be detected by the olfactory systems of other members of the same species in order to influence emotion, behavior, autonomic and hormonal states, and social judgments is an open question. Regardless of one's scientific stance, it is clear that chemosignals have served as potent triggers of functional imaging investigations. Delivery of estra-1,3,5(10),16-tetraen-3-ol (EST), an estrogen derivative found in female urine, at sub-threshold concentration elicited fMRI responses in the anterior medial thalamus and inferior frontal gyrus (Sobel et al., 1999). Seminal PET work by Savic and colleagues showed that androstadienone (AND), a testosterone derivative found in male sweat, evoked activity in the hypothalamus in women, whereas EST evoked activity in a different hypothalamic region in men (Savic et al., 2001). Follow-up work by this same group confirmed and extended these sex-specific imaging effects in revealed similar effects in line with gender and sexual orientation (Savic et al., 2005; Berglund et al., 2006).

Other studies have utilized more naturalistic stimuli collected directly from human bodily secretions, including axillary sweat and tears. Smelling a friend's body odor activates a different network of brain regions than when smelling a stranger's body odor (Lundstrom et al., 2008), including the precuneus, insula, and inferior frontal gyrus. Similarly, if women are asked to smell the body odor of their sister, compared to the body odor of a female friend, their PET rCBF responses show differential changes in the prefrontal cortex and intraparietal lobule (Lundstrom et al., 2009). These findings suggest that olfactory social-based cues can exert an impact on human brain function, lending credence to the idea of pheromonal control of kin recognition. The use of fear- or anxiety-inducing manipulations to induce production of emotionally laden chemosignals has also revealed interesting effects. Chemosensory "anxiety" signals, obtained from axillary secretions in students awaiting an important academic exam, were shown to evoke fMRI responses in numerous areas such as the precuneus, fusiform cortex, insula, and cingulate cortex (Prehn-Kristensen et al., 2009). "Fear" sweat obtained from individuals about to undergo their first skydive out of an airplane was associated with greater fMRI activity in the amygdala, cerebellum, inferior temporal gyrus, and precuneus, compared to a control exercise condition (Mujica-Parodi et al., 2009). "Sexual" sweat from the armpits of men watching erotic videos was presented in airborne form to female recipients, resulting in enhanced fMRI activation in the OFC, hypothalamus, and fusiform cortex (Zhou and Chen, 2008). Finally, sniffing "sad" tears from female donors reduced the sexual attraction of men to pictures of women's faces, and also reduced fMRI activity in the hypothalamus and fusiform cortex while male subjects viewed erotic pictures and films, in comparison to control saline (Gelstein et al., 2011).

Whether the above chemosignaling effects are mediated by the olfactory system, the vomeronasal system (unlikely based on current evidence: Trotier et al., 2000; Meredith, 2001; Witt et al., 2002), or through direct stimulus absorption into the bloodstream, has not been clearly established. Interestingly, in men with smell loss arising from nasal polyposis, the EST stimulus failed to activate the hypothalamus (Savic et al., 2009), suggesting that the ability of these putative chemosignals to modulate hypothalamic brain activity is mediated by the main olfactory pathway. In an ambitious effort to test whether the human vomeronasal organ (VNO) might be involved in behavioral or neural processing of putative pheromones, Frasnelli et al. (2011) blocked the VNO using a latex patch that was placed under endoscopic guidance. This maneuver (compared to a non-occluded patch control) had no effect on perceptual responses to AND, nor did it alter AND-evoked PET activations in the hypothalamus and other areas. These findings provide additional negative evidence for the unlikely role of a separate VNO pathway in human olfactory function.

13.5 FINAL OBSERVATIONS

The introduction of noninvasive imaging techniques ushered in a highly tractable method to elucidate the structure and function of the human olfactory system. The relative rarity of patient lesion cases and postmortem material, and the coarse spatial resolution of CT scans and scalp EEG recordings, meant that prior to MRI, our understanding of human olfaction was mostly limited to behavioral observations. Although psychophysical studies provided (and continue to provide) keen insights into human olfactory perception, a neuroscientific perspective had been wanting. Combined with a wide range of experimental designs, psychological tasks, and imaging analyses, fMRI has offered unique insights into the mechanisms by which the human brain encodes information about odor objects and odor valence, and how learning, attention, context, and experience modulate this information.

That said, it is important to bear in mind that functional imaging, and also structural imaging of network connectivity (such as DTI), represent highly indirect measures of brain activity, and the physiological principles underlying neurovascular coupling—that is, the transformation of neuronal signals into hemodynamic responses—are vague at best. Moreover, fMRI activation within a single 1-mm³ voxel represents the pooled activity of $\sim 10^5$–10^6 neurons, with a time-to-peak of 4–6 seconds, long after fast synaptic transmission has concluded. My reason for highlighting these limitations is not to denigrate the field of human functional imaging, but rather to underscore the care one must take in designing and interpreting olfactory fMRI experiments. A few general guidelines are listed below.

1. Minimize MR signal dropout. A variety of methods is available to minimize MR signal dropout and distortion in basal frontal and medial temporal lobes. Using "out-of-the-box" MR acquisition protocols is unlikely to provide robust functional data from piriform cortex, amygdala, or OFC.

2. Limit odor exposure. The olfactory system is highly prone to sensory habituation, especially in the piriform cortex, comprising fMRI signal recovery. Unless a specific experimental design calls for prolonged odor exposure, it is best to limit stimulation to 2–3 seconds, including several different odorants as well as control (no odor) trials in the design. The start of one trial to the start of the next trial should be spaced 15–20 seconds ideally, but can be shorter depending on task needs.

3. Record respirations online. Because sniffing can induce activity irrespective of odor, it is critical to obtain a measure of sniffing during the imaging experiment. Sniff size, volume, and latency may vary by odorant, condition, or task, warranting acquisition of event-related respiratory parameters. Breathing belts (respiratory effort bands), spirometers, and thermistors have all been used to collect these data.

4. Beware the lure of reverse inference. This problem is applicable to all imaging studies. The lure goes as follows: it is previously known that task A evokes activity in region X; in your new experiment, activity in region X is identified; therefore, you infer that region X is involved in task A.

As an example, one might design an odor detection paradigm in which odor-evoked activity is detected in the precuneus, an area adjacent to visual cortex that has been associated with visuospatial imagery. Therefore, it is concluded that the precuneus activity represents visual imagery that has taken place during the odor detection task. While theoretically plausible, the fact is that the original experiment was not designed to assess imagery. This is an example of reverse inference. The important point is that an imaging experiment should be designed to test specific hypotheses about specific cognitive/perceptual functions. In many instances it is also advisable to identify specific *a priori* brain regions of interest.

In conclusion, this chapter has surveyed human olfactory imaging from its historical beginnings up through 2012. The last decade has seen remarkable advances in both MRI technology and fMRI analytical approaches. Whether similar progress will occur over the next decade is less clear. The use of higher MRI field strengths will enhance signal-to-noise ratios as well as spatial resolution, but at the expense of greater susceptibility artifact that would further compromise fMRI signal in olfactory brain areas. This issue is particularly problematic for functional imaging of the human olfactory bulb, which to date has remained elusive. The development of new fMRI protocols not confounded by susceptibility artifacts may hold promise for imaging of this region. Future efforts are likely to utilize olfactory fMRI as one component of a multi-disciplinary enterprise, in which imaging is combined with other research modalities including EEG, "reversible lesion" techniques such as transcranial magnetic stimulation, and molecular genetics to generate a truly comprehensive scientific framework of human olfaction.

REFERENCES

Abolmaali, N. D., Hietschold, V., Vogl, T. J., et al. (2002). MR evaluation in patients with isolated anosmia since birth or early childhood. *AJNR. Am. J. Neuroradiol.* 23: 157–164.

Allison, A. C. (1954). The secondary olfactory areas in the human brain. *J. Anat.* 88: 481–488.

Anderson, A. K., Christoff, K., Stappen, I., et al. (2003). Dissociated neural representations of intensity and valence in human olfaction. *Nat. Neurosci.* 6: 196–202.

Assaf, Y., and Pasternak, O. (2008). Diffusion tensor imaging (DTI)-based white matter mapping in brain research: a review. *J. Mol. Neurosci.* 34: 51–61.

Bensafi, M., Porter, J., Pouliot, S., Mainland, J., Johnson, B., Zelano, C., Young, N., Bremner, E., Aframian, D., Khan, R., and Sobel, N (2003). Olfactomotor activity during imagery mimics that during perception. *Nat. Neurosci.* 6: 1142–1144.

Bensafi, M., Sobel, N., and Khan, R. M. (2007). Hedonic-specific activity in piriform cortex during odor imagery mimics that during odor perception. *J. Neurophysiol.* 98: 3254–3262.

Berglund, H., Lindstrom, P., and Savic, I. (2006). Brain response to putative pheromones in lesbian women. *Proc. Natl. Acad. Sci. U. S. A.* 103: 8269–8274.

Bowman, N. E., Kording, K. P., and Gottfried, J. A. (2012). Temporal integration of olfactory perceptual evidence in human orbitofrontal cortex. *Neuron* 75: 916–927.

Boyle, J. A., Djordjevic, J., Olsson, M. J., et al. (2009). The human brain distinguishes between single odorants and binary mixtures. *Cereb. Cortex* 19: 66–71.

Brunjes, P. C., and Borror, M. J. (1983). Unilateral odor deprivation: differential effects due to time of treatment. *Brain Res. Bull.* 11: 501–503.

Buckner, R. L., Goodman, J., Burock, M., et al. (1998). Functional-anatomic correlates of object priming in humans revealed by rapid presentation event-related fMRI. *Neuron* 20: 285–296.

Burmeister, H. P., Baltzer, P. A., Moslein, C., et al. (2011). Reproducibility and repeatability of volumetric measurements for olfactory bulb volumetry: which method is appropriate? An update using 3 Tesla MRI. *Acad. Radiol.* 18: 842–849.

Burmeister, H. P., Bitter, T., Heiler, P. M., et al. (2012). Imaging of lamination patterns of the adult human olfactory bulb and tract: in vitro comparison of standard- and high-resolution 3T MRI, and MR microscopy at 9.4 T. *Neuroimage* 60: 1662–1670.

Buschhuter, D., Smitka, M., Puschmann, S., et al. (2008). Correlation between olfactory bulb volume and olfactory function. *Neuroimage* 42: 498–502.

Buxton, R. B. (2009). *Introduction to Functional Magnetic Resonance Imaging: Principles and Techniques*, 2nd Ed. (Cambridge, UK: Cambridge University Press).

Cerf-Ducastel, B., and Murphy, C. (2006). Neural substrates of cross-modal olfactory recognition memory: an fMRI study. *Neuroimage* 31: 386–396.

Dade, L. A., Zatorre, R. J., Evans, A. C., and Jones-Gotman, M. (2001). Working memory in another dimension: functional imaging of human olfactory working memory. *Neuroimage* 14: 650–660.

Dade, L. A., Zatorre, R. J., and Jones-Gotman, M. (2002). Olfactory learning: convergent findings from lesion and brain imaging studies in humans. *Brain* 125: 86–101.

Dale, A. M., and Buckner, R. L. (1997). Selective averaging of rapidly presented individual trials using fMRI. *Hum. Brain Mapp.* 5: 329–340.

Dalton, P., Doolittle, N., and Breslin, P. A. (2002). Gender-specific induction of enhanced sensitivity to odors. *Nat. Neurosci.* 5: 199–200.

de Araujo, I. E., Rolls, E. T., Kringelbach, M. L., et al. (2003). Taste-olfactory convergence, and the representation of the pleasantness of flavour, in the human brain. *Eur. J. Neurosci.* 18: 2059–2068.

de Araujo, I. E., Rolls, E. T., Velazco, M. I., et al. (2005). Cognitive modulation of olfactory processing. *Neuron* 46: 671–679.

De Panfilis, C., and Schwarzbauer, C. (2005). Positive or negative blips? The effect of phase encoding scheme on susceptibility-induced signal losses in EPI. *Neuroimage* 25: 112–121.

Deichmann, R., Gottfried, J. A., Hutton, C., and Turner, R. (2003). Optimized EPI for fMRI studies of the orbitofrontal cortex. *Neuroimage* 19: 430–441.

Djordjevic, J., Zatorre, R. J., Petrides, M., and Jones-Gotman, M. (2004). The mind's nose: Effects of odor and visual imagery on odor detection. *Psychol. Sci.* 15: 143–148.

Djordjevic, J., Zatorre, R. J., Petrides, M., et al. (2005). Functional neuroimaging of odor imagery. *Neuroimage* 24: 791–801.

Doty, R. L., Reyes, P. F., and Gregor, T. (1987). Presence of both odour identification and detection deficits in Alzheimer's disease. *Brain Res. Bull.* 18: 597–600.

Doty, R. L. (2010). *The Great Pheromone Myth*. Baltimore: The Johns Hopkins University Press.

Duprez, T. P., and Rombaux, P. (2010). Imaging the olfactory tract (cranial nerve #1). *Eur. J. Radiol.* 74: 288–298.

Eichenbaum, H., Shedlack, K. J., and Eckmann, K. W. (1980). Thalamocortical mechanisms in odor-guided behavior. I. Effects of lesions of the mediodorsal thalamic nucleus and frontal cortex on olfactory discrimination in the rat. *Brain Behav. Evol.* 17: 255–275.

Engstrom, H., and Bloom, G. (1953). The structure of the olfactory region in man. *Acta Otolaryngol.* 43: 11–21.

Eskenazi, B., Cain, W. S., Novelly, R. A., and Friend, K. B. (1983). Olfactory functioning in temporal lobectomy patients. *Neuropsychologia* 21: 365–374.

Frasnelli, J., Lundstrom, J. N., Boyle, J. A., et al. (2011). The vomeronasal organ is not involved in the perception of endogenous odors. *Hum. Brain Mapp.* 32: 450–460.

Freeman, W. J. (1983). The physiological basis of mental images. *Biol. Psychiatry* 18: 1107–1125.

Freeman, W. J., and Schneider, W. (1982). Changes in spatial patterns of rabbit olfactory EEG with conditioning to odors. *Psychophysiology* 19: 44–56.

Friston, K. (2005). A theory of cortical responses. *Philos. Trans. R. Soc. Lond. B. Biol. Sci.* 360: 815–836.

Friston, K. J., Harrison, L., and Penny, W. (2003). Dynamic causal modelling. *Neuroimage* 19: 1273–1302.

Gelstein, S., Yeshurun, Y., Rozenkrantz, L., et al. (2011). Human tears contain a chemosignal. *Science* 331: 226–230.

Goense, J. B., and Logothetis, N. K. (2008). Neurophysiology of the BOLD fMRI signal in awake monkeys. *Curr. Biol.* 18: 631–640.

Gottfried, J. A., Deichmann, R., Winston, J. S., and Dolan, R. J. (2002a). Functional heterogeneity in human olfactory cortex: an event-related functional magnetic resonance imaging study. *J. Neurosci.* 22: 10819–10828.

Gottfried, J. A., O'Doherty, J., and Dolan, R. J. (2002b). Appetitive and aversive olfactory learning in humans studied using event-related functional magnetic resonance imaging. *J. Neurosci.* 22: 10829–10837.

Gottfried, J. A., and Dolan, R. J. (2003). The nose smells what the eye sees: crossmodal visual facilitation of human olfactory perception. *Neuron* 39: 375–386.

Gottfried, J. A., O'Doherty, J., and Dolan, R. J. (2003). Encoding predictive reward value in human amygdala and orbitofrontal cortex. *Science* 301: 1104–1107.

Gottfried, J. A., and Dolan, R. J. (2004). Human orbitofrontal cortex mediates extinction learning while accessing conditioned representations of value. *Nat. Neurosci.* 7: 1144–1152.

Gottfried, J. A., Smith, A. P., Rugg, M. D., and Dolan, R. J. (2004). Remembrance of odors past: human olfactory cortex in cross-modal recognition memory. *Neuron* 42: 687–695.

Gottfried, J. A., Winston, J. S., and Dolan, R. J. (2006). Dissociable codes of odor quality and odorant structure in human piriform cortex. *Neuron* 49: 467–479.

Gottfried, J. A. (2009). Function follows form: ecological constraints on odor codes and olfactory percepts. *Curr. Opin. Neurobiol.* 19: 422–429.

References

Gottfried, J. A. (2010). Central mechanisms of odour object perception. *Nat. Rev. Neurosci.* 11: 628–641.

Grabenhorst, F., Rolls, E. T., Margot, C., et al. (2007). How pleasant and unpleasant stimuli combine in different brain regions: odor mixtures. *J. Neurosci.* 27: 13532–13540.

Grabenhorst, F., and Rolls, E. T. (2009). Different representations of relative and absolute subjective value in the human brain. *Neuroimage* 48: 258–268.

Gudziol, V., Buschhuter, D., Abolmaali, N., et al. (2009). Increasing olfactory bulb volume due to treatment of chronic rhinosinusitis--a longitudinal study. *Brain* 132: 3096–3101.

Guthrie, K. M., Wilson, D. A., and Leon, M. (1990). Early unilateral deprivation modifies olfactory bulb function. *J. Neurosci.* 10: 3402–3412.

Haberly, L. B. (1985). Neuronal circuitry in olfactory cortex: anatomy and functional implications. *Chem. Senses* 10: 219–238.

Haberly, L. B. (1998). Olfactory cortex. In *The Synaptic Organization of the Brain*, Shepherd, G. M., (Ed). New York: Oxford University Press, pp. 377–416.

Haehner, A., Rodewald, A., Gerber, J. C., and Hummel, T. (2008). Correlation of olfactory function with changes in the volume of the human olfactory bulb. *Arch. Otolaryngol. Head Neck Surg.* 134: 621–624.

Hagmann, P., Kurant, M., Gigandet, X., et al. (2007). Mapping human whole-brain structural networks with diffusion MRI. *PLoS ONE* 2: e597.

Hasselmo, M. E., Wilson, M. A., Anderson, B. P., and Bower, J. M. (1990). Associative memory function in piriform (olfactory) cortex: computational modeling and neuropharmacology. *Cold Spring Harb. Symp. Quant. Biol.* 55: 599–610.

Herz, R. S., and von Clef, J. (2001). The influence of verbal labeling on the perception of odors: evidence for olfactory illusions? *Perception* 30: 381–391.

Herz, R. S., Eliassen, J., Beland, S., and Souza, T. (2004). Neuroimaging evidence for the emotional potency of odor-evoked memory. *Neuropsychologia* 42: 371–378.

Holbrook, E. H., and Leopold, D. A. (2006). An updated review of clinical olfaction. *Curr. Opin. Otolaryngol. Head Neck Surg.* 14: 23–28.

Howard, J. D., Plailly, J., Grueschow, M., et al. (2009). Odor quality coding and categorization in human posterior piriform cortex. *Nat. Neurosci.* 12: 932–938.

Hsu, J. J., and Glover, G. H. (2005). Mitigation of susceptibility-induced signal loss in neuroimaging using localized shim coils. *Magn. Reson. Med.* 53: 243–248.

Huettel, S. A., McKeown, M. J., Song, A. W., et al. (2004). Linking hemodynamic and electrophysiological measures of brain activity: evidence from functional MRI and intracranial field potentials. *Cereb. Cortex* 14: 165–173.

Huettel, S. A., Song, A. W., and McCarthy, G. (2008). *Functional Magnetic Resonance Imaging*, Second Edition, Sunderland, MA: Sinauer Associates.

Hummel, T., Damm, M., Vent, J., et al. (2003). Depth of olfactory sulcus and olfactory function. *Brain Res.* 975: 85–89.

Ibarretxe-Bilbao, N., Junque, C., Marti, M. J., et al. (2010). Olfactory impairment in Parkinson's disease and white matter abnormalities in central olfactory areas: A voxel-based diffusion tensor imaging study. *Mov. Disord.* 25: 1888–1894.

Illig, K. R., and Haberly, L. B. (2003). Odor-evoked activity is spatially distributed in piriform cortex. *J. Comp. Neurol.* 457: 361–373.

Jehl, C., Royet, J. P., and Holley, A. (1995). Odor discrimination and recognition memory as a function of familiarization. *Percept. Psychophys.* 57: 1002–1011.

Jezzard, P., Matthews, P. M., and Smith, S. M. (2002). *Functional MRI: An Introduction to Methods*, New York: Oxford University Press.

Jones-Gotman, M., and Zatorre, R. J. (1988). Olfactory identification deficits in patients with focal cerebral excision. *Neuropsychologia* 26: 387–400.

Josephs, O., Turner, R., and Friston, K. (1997). Event-related fMRI. *Hum. Brain Mapp.* 5: 243–248.

Khan, R. M., Luk, C. H., Flinker, A., et al. (2007). Predicting odor pleasantness from odorant structure: pleasantness as a reflection of the physical world. *J. Neurosci.* 27: 10015–10023.

Kjelvik, G., Evensmoen, H. R., Brezova, V., and Haberg, A. K. (2012). The human brain representation of odor identification. *J. Neurophysiol.* 108: 645–657.

Koss, E., Weiffenbach, J. M., Haxby, J. V., and Friedland, R. P. (1988). Olfactory detection and identification performance are dissociated in early Alzheimer's disease. *Neurology* 38: 1228–1232.

Kourtzi, Z., and Kanwisher, N. (2001). Representation of perceived object shape by the human lateral occipital complex. *Science* 293: 1506–1509.

Krause, K. F. T., and Krause, W. (1876). *Handbuch der menschlichen Anatomie* (Erster Band), 3rd Ed. Hannover, Germany: Hahn'sche Hofbuchhandlung.

Kriegeskorte, N., and Bandettini, P. (2007). Analyzing for information, not activation, to exploit high-resolution fMRI. *Neuroimage* 38: 649–662.

Lehn, H., Kjonigsen, L. J., Kjelvik, G., and Haberg, A. K. (2012). Hippocampal involvement in retrieval of odor vs. object memories. *Hippocampus.* 45: 13–23.

Levy, L. M., Henkin, R. I., Hutter, A., et al. (1997). Functional MRI of human olfaction. *J. Comput. Assist. Tomogr.* 21: 849–856.

Li, W., Luxenberg, E., Parrish, T., and Gottfried, J. A. (2006). Learning to smell the roses: experience-dependent neural plasticity in human piriform and orbitofrontal cortices. *Neuron* 52: 1097–1108.

Li, W., Howard, J. D., Parrish, T. B., and Gottfried, J. A. (2008). Aversive learning enhances perceptual and cortical discrimination of indiscriminable odor cues. *Science* 319: 1842–1845.

Logothetis, N. K., Pauls, J., Augath, M., et al. (2001). Neurophysiological investigation of the basis of the fMRI signal. *Nature* 412: 150–157.

Lorig, T. S., Elmes, D. G., Zald, D. H., and Pardo, J. V. (1999). A computer-controlled olfactometer for fMRI and electrophysiological studies of olfaction. *Behav. Res. Methods Instrum. Comput.* 31: 370–375.

Lundstrom, J. N., Boyle, J. A., Zatorre, R. J., and Jones-Gotman, M. (2008). Functional neuronal processing of body odors differs from that of similar common odors. *Cereb. Cortex* 18: 1466–1474.

Lundstrom, J. N., Boyle, J. A., Zatorre, R. J., and Jones-Gotman, M. (2009). The neuronal substrates of human olfactory based kin recognition. *Hum. Brain Mapp.* 30: 2571–2580.

Mai, J. K., Assheuer, J., and Paxinos, G. (1997). *Atlas of the Human Brain.* San Diego: Academic Press.

Mainland, J. D., Bremner, E. A., Young, N., et al. (2002). Olfactory plasticity: one nostril knows what the other learns. *Nature* 419: 802.

Maruniak, J. A., Taylor, J. A., Henegar, J. R., and Williams, M. B. (1989). Unilateral naris closure in adult mice: atrophy of the deprived-side olfactory bulbs. *Brain Res. Dev. Brain Res.* 47: 27–33.

Meredith, M. (2001). Human vomeronasal organ function: a critical review of best and worst cases. *Chem. Senses* 26: 433–445.

Meynert, T. (1872). The brain of mammals. In *Manual of Human and Comparative Histology, Vol. II,* Transl. H. Power, Stricker, S., (Ed). London: The New Sydenham Society, pp. 365–537.

Mori, S., and Zhang, J. (2006). Principles of diffusion tensor imaging and its applications to basic neuroscience research. *Neuron* 51: 527–539.

Morrison, E. E., and Costanzo, R. M. (1990). Morphology of the human olfactory epithelium. *J. Comp. Neurol.* 297: 1–13.

Morrison, E. E., and Costanzo, R. M. (1992). Morphology of olfactory epithelium in humans and other vertebrates. *Microsc. Res. Tech.* 23: 49–61.

Mueller, A., Rodewald, A., Reden, J., et al. (2005). Reduced olfactory bulb volume in post-traumatic and post-infectious olfactory dysfunction. *Neuroreport* 16: 475–478.

Mujica-Parodi, L. R., Strey, H. H., Frederick, B., et al. (2009). Chemosensory cues to conspecific emotional stress activate amygdala in humans. *PLoS ONE* 4: e6415.

Mukamel, R., Gelbard, H., Arieli, A., et al. (2005). Coupling between neuronal firing, field potentials, and FMRI in human auditory cortex. *Science* 309: 951–954.

Murphy, C., Gilmore, M. M., Seery, C. S., et al. (1990). Olfactory thresholds are associated with degree of dementia in Alzheimer's disease. *Neurobiol. Aging* 11: 465–469.

Murphy, C., Jernigan, T. L., and Fennema-Notestine, C. (2003). Left hippocampal volume loss in Alzheimer's disease is reflected in performance on odor identification: a structural MRI study. *J. Int. Neuropsychol. Soc.* 9: 459–471.

Naessen, R. (1971). The "receptor surface" of the olfactory organ (epithelium) of man and guinea pig. A descriptive and experimental study. *Acta Otolaryngol.* 71: 335–348.

Nakashima, T., Kimmelman, C. P., and Snow, J. B., Jr., (1984). Structure of human fetal and adult olfactory neuroepithelium. *Arch. Otolaryngol.* 110: 641–646.

Negoias, S., Croy, I., Gerber, J., et al. (2010). Reduced olfactory bulb volume and olfactory sensitivity in patients with acute major depression. *Neuroscience* 169: 415–421.

O'Doherty, J., Rolls, E. T., Francis, S., et al. (2000). Sensory-specific satiety-related olfactory activation of the human orbitofrontal cortex. *Neuroreport* 11: 893–897.

Olofsson, J. K., Rogalski, E., Harrison, T., et al. (2013). A cortical pathway to olfactory naming: evidence from primary progressive aphasia. *Brain* 136: 1245–1259.

Osterbauer, R. A., Matthews, P. M., Jenkinson, M., et al. (2005). Color of scents: chromatic stimuli modulate odor responses in the human brain. *J. Neurophysiol.* 93: 3434–3441.

Plailly, J., Howard, J. D., Gitelman, D. R., and Gottfried, J. A. (2008). Attention to odor modulates thalamocortical connectivity in the human brain. *J. Neurosci.* 28: 5257–5267.

Plailly, J., Delon-Martin, C., and Royet, J. P. (2012). Experience induces functional reorganization in brain regions involved in odor imagery in perfumers. *Hum. Brain Mapp.* 33: 224–234.

Poellinger, A., Thomas, R., Lio, P., et al. (2001). Activation and habituation in olfaction--an fMRI study. *Neuroimage* 13: 547–560.

Poldrack, R. A., Mumford, J. A., and Nichols, T. E. (2011). *Handbook of Functional MRI Data Analysis*, 2nd Ed. Cambridge, UK: Cambridge University Press.

Potter, H., and Butters, N. (1980). An assessment of olfactory deficits in patients with damage to prefrontal cortex. *Neuropsychologia* 18: 621–628.

Prehn-Kristensen, A., Wiesner, C., Bergmann, T. O., et al. (2009). Induction of empathy by the smell of anxiety. *PLoS ONE* 4: e5987.

Price, J. L., and Slotnick, B. M. (1983). Dual olfactory representation in the rat thalamus: an anatomical and electrophysiological study. *J. Comp. Neurol.* 215: 63–77.

Rabin, M. D. (1988). Experience facilitates olfactory quality discrimination. *Percept. Psychophys.* 44: 532–540.

Rao, R. P., and Ballard, D. H. (1999). Predictive coding in the visual cortex: a functional interpretation of some extra-classical receptive-field effects. *Nat. Neurosci.* 2: 79–87.

Rasch, B., Büchel, C., Gais, S., and Born, J. (2007). Odor cues during slow-wave sleep prompt declarative memory consolidation. *Science* 315: 1426–1429.

Rausch, R., Serafetinides, E. A., and Crandall, P. H. (1977). Olfactory memory in patients with anterior temporal lobectomy. *Cortex* 13: 445–452.

Ray, J. P., and Price, J. L. (1993). The organization of projections from the mediodorsal nucleus of the thalamus to orbital and medial prefrontal cortex in macaque monkeys. *J. Comp. Neurol.* 337: 1–31.

Rolls, E. T., Kringelbach, M. L., and De Araujo, I. E. (2003). Different representations of pleasant and unpleasant odours in the human brain. *Eur. J. Neurosci.* 18: 695–703.

Rombaux, P., Weitz, H., Mouraux, A., et al. (2006). Olfactory function assessed with orthonasal and retronasal testing, olfactory bulb volume, and chemosensory event-related potentials. *Arch. Otolaryngol. Head Neck Surg.* 132: 1346–1351.

Royet, J. P., Koenig, O., Gregoire, M. C., et al. (1999). Functional anatomy of perceptual and semantic processing for odors. *J. Cogn. Neurosci.* 11: 94–109.

Royet, J. P., Zald, D., Versace, R., et al. (2000). Emotional responses to pleasant and unpleasant olfactory, visual, and auditory stimuli: a positron emission tomography study. *J. Neurosci.* 20: 7752–7759.

Royet, J. P., Hudry, J., Zald, D. H., et al. (2001). Functional neuroanatomy of different olfactory judgments. *Neuroimage* 13: 506–519.

Royet, J. P., Morin-Audebrand, L., Cerf-Ducastel, B., et al. (2011). True and false recognition memories of odors induce distinct neural signatures. *Front. Human Neurosci.* 5: 65.

Rupp, C. I., Fleischhacker, W. W., Kemmler, G., et al. (2005). Olfactory functions and volumetric measures of orbitofrontal and limbic regions in schizophrenia. *Schizophr. Res.* 74: 149–161.

Russchen, F. T., Amaral, D. G., and Price, J. L. (1987). The afferent input to the magnocellular division of the mediodorsal thalamic nucleus in the monkey, Macaca fascicularis. *J. Comp. Neurol.* 256: 175–210.

Savic, I., Gulyas, B., Larsson, M., and Roland, P. (2000). Olfactory functions are mediated by parallel and hierarchical processing. *Neuron* 26: 735–745.

Savic, I., Berglund, H., Gulyas, B., and Roland, P. (2001). Smelling of odorous sex hormone-like compounds causes sex- differentiated hypothalamic activations in humans. *Neuron* 31: 661–668.

Savic, I., Berglund, H., and Lindstrom, P. (2005). Brain response to putative pheromones in homosexual men. *Proc. Natl. Acad. Sci. U. S. A.* 102: 7356–7361.

Savic, I., Heden-Blomqvist, E., and Berglund, H. (2009). Pheromone signal transduction in humans: what can be learned from olfactory loss. *Hum. Brain Mapp.* 30: 3057–3065.

Schellinger, D., Henkin, R. T., and Smirniotopoulos, J. G. (1983). CT of the brain in taste and smell dysfunction. *AJNR. Am. J. Neuroradiol.* 4: 752–754.

Scherfler, C., Schocke, M. F., Seppi, K., et al. (2006). Voxel-wise analysis of diffusion weighted imaging reveals disruption of the olfactory tract in Parkinson's disease. *Brain* 129: 538–542.

Schiffman, S. S. (1974). Physicochemical correlates of olfactory quality. *Science* 185: 112–117.

Schultze, M. (Nov. 1856). Über die Endigungsweise des Geruchsnerven und die Epithelialgebilde der Nasenschleimhaut. *Monatsberichte der Königlichen Preussische Akademie des Wissenschaften zu Berlin*, 504–514.

Sela, L., Sacher, Y., Serfaty, C., et al. (2009). Spared and impaired olfactory abilities after thalamic lesions. *J. Neurosci.* 29: 12059–12069.

Slotnick, B. M., and Kaneko, N. (1981). Role of mediodorsal thalamic nucleus in olfactory discrimination learning in rats. *Science* 214: 91–92.

Small, D. M., Jones-Gotman, M., Zatorre, R. J., et al. (1997). Flavor processing: more than the sum of its parts. *Neuroreport* 8: 3913–3917.

Small, D. M., Voss, J., Mak, Y. E., et al. (2004). Experience-dependent neural integration of taste and smell in the human brain. *J. Neurophysiol.* 92: 1892–1903.

Sobel, N., Prabhakaran, V., Desmond, J. E., et al. (1997). A method for functional magnetic resonance imaging of olfaction. *J. Neurosci. Methods* 78: 115–123.

Sobel, N., Prabhakaran, V., Desmond, J. E., et al. (1998). Sniffing and smelling: separate subsystems in the human olfactory cortex. *Nature* 392: 282–286.

Sobel, N., Prabhakaran, V., Hartley, C. A., et al. (1999). Blind smell: brain activation induced by an undetected air-borne chemical. *Brain* 122: 209–217.

Sobel, N., Prabhakaran, V., Zhao, Z., et al. (2000). Time course of odorant-induced activation in the human primary olfactory cortex. *J. Neurophysiol.* 83: 537–551.

Sömmerring, S. T. (1809). *Abbildungen der menschlichen Organe des Geruches,* Vol Digitized by Google, downloaded October 18, 2012 (Frankfurt am Main: Varrentrapp and Wenner).

Staubli, U., Schottler, F., and Nejat-Bina, D. (1987). Role of dorsomedial thalamic nucleus and piriform cortex in processing olfactory information. *Behav. Brain Res.* 25: 117–129.

Summerfield, C., and Egner, T. (2009). Expectation (and attention) in visual cognition. *Trends Cogn. Sci.* 13: 403–409.

Suzuki, M., Takashima, T., Kadoya, M., et al. (1989). MR imaging of olfactory bulbs and tracts. *AJNR. Am. J. Neuroradiol.* 10: 955–957.

Tham, W. W., Stevenson, R. J., and Miller, L.A. (2009). The functional role of the medio dorsal thalamic nucleus in olfaction. *Brain Res. Rev.* 62: 109–126.

Thomann, P. A., Dos Santos, V., Seidl, U., et al. (2009a). MRI-derived atrophy of the olfactory bulb and tract in mild cognitive impairment and Alzheimer's disease. *J. Alzheimers Dis.* 17: 213–221.

Thomann, P. A., Dos Santos, V., Toro, P., et al. (2009b). Reduced olfactory bulb and tract volume in early Alzheimer's disease--a MRI study. *Neurobiol. Aging* 30: 838–841.

Toga, A. W., and Mazziotta, J. C. (1996). *Brain Mapping: The Methods*, San Diego: Academic Press.

Trotier, D., Eloit, C., Wassef, M., et al. (2000). The vomeronasal cavity in adult humans. *Chem. Senses* 25: 369–380.

Turetsky, B. I., Moberg, P. J., Yousem, D. M., et al. (2000). Reduced olfactory bulb volume in patients with schizophrenia. *Am. J. Psychiatry* 157: 828–830.

Turetsky, B. I., Moberg, P. J., Arnold, S. E., et al. (2003a). Low olfactory bulb volume in first-degree relatives of patients with schizophrenia. *Am. J. Psychiatry* 160: 703–708.

Turetsky, B. I., Moberg, P. J., Roalf, D. R., et al. (2003b). Decrements in volume of anterior ventromedial temporal lobe and olfactory dysfunction in schizophrenia. *Arch. Gen. Psychiatry* 60: 1193–1200.

Wattendorf, E., Welge-Lussen, A., Fiedler, K., et al. (2009). Olfactory impairment predicts brain atrophy in Parkinson's disease. *J. Neurosci.* 29: 15410–15413.

Weiskopf, N., Hutton, C., Josephs, O., and Deichmann, R. (2006). Optimal EPI parameters for reduction of susceptibility-induced BOLD sensitivity losses: a whole-brain analysis at 3 T and 1.5 T. *Neuroimage* 33: 493–504.

Wilson, D. A. (1998). Habituation of odor responses in the rat anterior piriform cortex. *J. Neurophysiol.* 79: 1425–1440.

Wilson, D. A. (2000). Odor specificity of habituation in the rat anterior piriform cortex. *J. Neurophysiol.* 83: 139–145.

Wilson, D. A., and Stevenson, R. J. (2003). The fundamental role of memory in olfactory perception. *Trends Neurosci.* 26: 243–247.

Winston, J. S., Henson, R. N., Fine-Goulden, M. R., and Dolan, R. J. (2004). fMRI-adaptation reveals dissociable neural representations of identity and expression in face perception. *J. Neurophysiol.* 92: 1830–1839.

Winston, J. S., Gottfried, J. A., Kilner, J. M., and Dolan, R. J. (2005). Integrated neural representations of odor intensity and affective valence in human amygdala. *J. Neurosci.* 25: 8903–8907.

Witt, M., Georgiewa, B., Knecht, M., and Hummel, T. (2002). On the chemosensory nature of the vomeronasal epithelium in adult humans. *Histochem. Cell Biol.* 117: 493–509.

Wu, K. N., Tan, B. K., Howard, J. D., et al. (2012). Olfactory input is critical for sustaining odor quality codes in human orbitofrontal cortex. *Nat. Neurosci.* 15: 1313–1319.

Wysocki, C. J., Dorries, K. M., and Beauchamp, G. K. (1989). Ability to perceive androstenone can be acquired by ostensibly anosmic people. *Proc. Natl. Acad. Sci. U. S. A.* 86: 7976–7978.

Wysocki, C. J., and Preti, G. (2004). Facts, fallacies, fears, and frustrations with human pheromones. *The Anatomical Record* 281: 1201–1211.

Yang, Q. X., Dardzinski, B. J., Li, S., et al. (1997). Multi-gradient echo with susceptibility inhomogeneity compensation (MGESIC): demonstration of fMRI in the olfactory cortex at 3.0 T. *Magn. Reson. Med.* 37: 331–335.

Yeshurun, Y., Lapid, H., Dudai, Y., and Sobel, N. (2009). The privileged brain representation of first olfactory associations. *Curr. Biol.* 19: 1869–1874.

Yousem, D. M., Geckle, R. J., Bilker, W., et al. (1996a). MR evaluation of patients with congenital hyposmia or anosmia. *AJR. Am. J. Roentgenol.* 166: 439–443.

Yousem, D. M., Geckle, R. J., Bilker, W. B., et al. (1996b). Posttraumatic olfactory dysfunction: MR and clinical evaluation. *AJNR. Am. J. Neuroradiol.* 17: 1171–1179.

Yousem, D. M., Williams, S. C., Howard, R. O., et al. (1997). Functional MR imaging during odor stimulation: preliminary data. *Radiology* 204: 833–838.

Yousem, D. M., Geckle, R. J., Bilker, W. B., and Doty, R. L. (1998). Olfactory bulb and tract and temporal lobe volumes: normative data across decades. *Ann. NY Acad. Sci.* 855: 546–555.

Zald, D. H., and Pardo, J. V. (1997). Emotion, olfaction, and the human amygdala: amygdala activation during aversive olfactory stimulation. *Proc. Natl. Acad. Sci. U. S. A.* 94: 4119–4124.

Zald, D. H., and Pardo, J. V. (2000). Functional neuroimaging of the olfactory system in humans. *Int. J. Psychophysiol.* 36: 165–181.

Zatorre, R. J., Jones-Gotman, M., Evans, A. C., and Meyer, E. (1992). Functional localization and lateralization of human olfactory cortex. *Nature* 360: 339–340.

Zelano, C., Bensafi, M., Porter, J., et al. (2005). Attentional modulation in human primary olfactory cortex. *Nat. Neurosci.* 8: 114–120.

Zelano, C., Montag, J., Khan, R., and Sobel, N. (2009). A specialized odor memory buffer in primary olfactory cortex. *PLoS ONE* 4: e4965.

Zelano, C., Mohanty, A., and Gottfried, J. A. (2011). Olfactory predictive codes and stimulus templates in piriform cortex. *Neuron* 72: 178–187.

Zhou, W., and Chen, D. (2008). Encoding human sexual chemosensory cues in the orbitofrontal and fusiform cortices. *J. Neurosci.* 28: 14416–14421.

Chapter 14

Prenatal and Postnatal Human Olfactory Development: Influences on Cognition and Behavior

BENOIST SCHAAL

14.1 INTRODUCTION

Our current understanding of the functions of human olfaction largely derives from research carried out in adults. Accordingly, knowledge of this sense during the first decade and a half of life remains largely fragmentary and contradictory, reflecting, in part, limited access to infant populations and difficulties in testing the perception of preverbal subjects (for reviews, see Crook, 1987; Beauchamp et al., 1991; Doty, 1992; Schaal, 1988, 1999, 2005; Schmidt and Beauchamp, 1992). An example of contradictory evidence is the paradox that children under the age of 10 are often credited with mediocre olfactory skills, even though memories for odors present during this early period may be vividly recalled in adulthood (Chu and Downes, 2000, 2002; Willander and Larsson, 2006).

This chapter focuses on the importance of ontogeny and early experience in understanding human olfaction. The first section examines the emergence of the anatomy and function of nasal chemosensation, including its development in young organisms as they gain sophistication in sensori-motor and cognitive skills. The second summarizes adaptive responses to odor exhibited by infants and young children, including newborns, in individual and social contexts. The final section assesses the long-term consequences of early exposure to odorants. Animal studies are mentioned throughout the chapter, taking advantage of the fact that they may provide insight into the specific psycho-biological mechanisms involved in the early

development of human olfactory abilities (e.g., Cheal, 1975; Alberts, 1976, 1987; Rosenblatt, 1983; Blass, 1986; Hepper, 1991a; Leon, 1992; Schaal, 2010).

14.2 DEVELOPMENT OF OLFACTORY STRUCTURES AND PERFORMANCE

14.2.1 Nasal Chemoreceptive Structures

As shown in other chapters of the *Handbook*, chemoreception is based on multiple and anatomically distinct sensory structures associated with the assessment, intake, and utilization of volatile chemicals. Many mammals possess several distinct intranasal chemosensory systems, including the main olfactory and accessory (vomeronasal) olfactory subsystems, the trigeminal nerve, the terminal nerve (CN 0), and, in rodents, the septal organ of Masera and the Grueneberg ganglion (see Chapters 50–52). In humans, the only chemosensory systems that are functional in adulthood are those of the main olfactory (CN I) and trigeminal nerve (CN V) systems. Although a human vomeronasal system develops in utero, only non-functional remnants exist in adults. Nonetheless, it is theoretically possible that this system functions in utero in humans.

Milestones of the development of the olfactory, trigeminal, and vomeronasal systems of humans are depicted in Figure 14.1. The primordium of the *olfactory system*

Handbook of Olfaction and Gustation, Third Edition. Edited by Richard L. Doty.
© 2015 Richard L. Doty. Published 2015 by John Wiley & Sons, Inc.

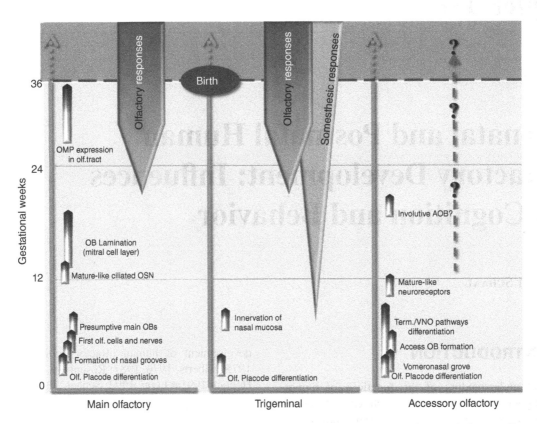

Figure 14.1 Timing of known steps of the prenatal development of nasal chemoreceptive structures and functions in humans. Data are presented for the main olfactory subsystem, the trigeminal subsystem, and the accessory (or vomeronasal) subsystem. Gestational age is given in post-ovulatory weeks. Based on Humphrey, 1978; Bossy, 1980; Chuah and Zheng, 1987; Schaal, 1988; Müller and O'Rahilly, 2004). Abbreviations: AOB: accessory olfactory bulb, Olf.: olfactory, OB: olfactory bulb, OMP: olfactory marker protein, OSN: olfactory sensory neuron, VNO: vomeronasal organ, Term.: terminal nerve.

can be seen during the first trimester, when the olfactory placodes differentiate during post-ovulatory weeks 3.5–5 and when the olfactory bulbs (OB) become visible during weeks 6–8 (Bossy, 1980). By weeks 4.5–7, the axons of the olfactory sensory neurons (OSN) develop first and define the olfactory nerves. During weeks 10 and 11, dendritic knobs of the OSN burgeon profusely into ciliary processes that express G-proteins involved in sensory transduction. Morphologically mature OSN can be seen from week 11, connecting with the OBs which are then visualized by MRI to be large relative to the entire brain (Azoulay et al. 2006). Histological analyses indicate that the OBs acquire their typical laminar appearance by week 19 (Bossy, 1980). Sensory like neurons are detected in the VNO between 5 and 13 weeks (Bossy, 1980; Ortmann, 1989; Boehm and Gasser, 1993), although the accessory olfactory bulbs to which they project undergo regression during the second trimester (Humphrey, 1940). Free endings of the *trigeminal nerve* become visible in the respiratory and olfactory mucosae in the 4-week-old embryo (Brown, 1974). They participate in tactile stimulus–response loops between weeks 7.5 and 10, when they already mediate tactile reactivity (Humphrey, 1978).

The aforementioned studies accord well with the concept that some chemoreceptive structures are functional before birth, at least during the last trimester. Although early anatomists reported that epithelial plugs obstruct external nares until 16–24 gestational weeks (Schaffer, 1910), recent ultrasonographic studies show fluid movements through upper aerial passages during fetal pseudo-respiratory activity. Studies of premature neonates suggest chemosensory function may be restricted to the last two gestational months (Schaal et al., 2004). Given their stages of embryological development, information from several sensory systems, including taste, touch, and hearing, may be woven together during the later stages of pregnancy (Lecanuet and Schaal, 1996; Schaal and Durand, 2012).

14.2.2 Nasal Chemoreceptive Functioning

The functional onset of nasochemoreception has been studied in both animal and human infants (Smotherman

14.2 Development of Olfactory Structures and Performance

and Robinson, 1987, 1995; Schaal et al., 1995B). In the rat, olfactory sensory neurons (OSN) exhibit neural activity during the last five days of gestation with further maturation occurring postnatally (Gesteland et al., 1982). Although no such electrophysiological verification is feasible in human fetuses, indirect molecular markers of activity have been used, namely the 'olfactory marker protein' (OMP). The expression of OMP coincides with selective odor receptor cell reactivity and connectivity with mitral cells in the olfactory bulb of rats (Lee et al., 2011). Although the human olfactory system exhibits OSN of adult-like appearance by gestational week 11, OMP is not expressed before week 28. Between weeks 28 and 32, it is expressed in OSN cell bodies and dendrites, and, by week 32, in the entire OSN. Finally, by week 35, OMP is detected in the OBs (Chuah and Zheng, 1987) suggesting connections have been made with secondary neurons. The pattern of neural maturation indexed by OMP appears to parallel emerging odor reactiveness in premature infants. Before week 28, they are rarely responsive to mint odor, whereas they react more and more consistently to it after 29 weeks. By 33–36 gestational weeks, premature infants respond as frequently to mint odor as do term newborns (Sarnat, 1978). But mint odor has a distinct somesthesic effect and trigeminal nerve endings may accordingly also contribute to early responses.

How central structures involved in adult odor processing develop in fetuses, infants, and children is poorly understood. Anatomical studies suggest that cerebral structures receiving primary afferents (i.e., anterior olfactory nucleus, olfactory tubercle, piriform cortex, and superficial parts of the amygdala, such as anterior cortical nucleus, nucleus of the lateral olfactory tract, and periamygdaloid cortex) and secondary or tertiary afferents (i.e., entorhinal cortex, hippocampus, orbitofrontal cortex, insula) reorganize during fetal development (Macchi, 1951; Humphrey, 1963, 1967, 1968). Several lines of evidence, albeit limited, suggest that the OBs connect with the limbic forebrain, olfactory paleocortex, and frontal neocortex during this early period of development. First, the OBs undergo anatomical changes discernible by MRI during the first decade, underlain by neuronal maturation (i.e., synaptic pruning, apoptosis, myelination) and internal reorganizations (Schneider and Floemer, 2009). Second, fetal and neonatal cortical processing competence of odor inputs is suggested by odor-induced variations in hemoglobin oxygenation (measured by Near Infrared Spectroscopy, NIRS) detected over the orbito-frontal regions in premature infants (Bartocci et al., 2000, 2001). Finally, the perceptual competence of fetuses, neonates, and infants (described in the following section) indicates that olfaction likely relies on processes that mobilize all levels of the olfactory system.

14.2.3 Olfactory Performance: Sensitivity, Reactivity, Discrimination, Recognition, Memory

14.2.3.1 Fetuses. Strong evidence for integrative processing of olfactory stimuli stems from studies on rat, mouse, and sheep fetuses tested *in utero* (Smotherman and Robinson, 1987; Schaal and Orgeur, 1992). Considerable progress was accomplished when extra-uterine live preparations of rodent fetuses were achieved by W. P. Smotherman and S. R. Robinson (e.g., Smotherman, 1982; Smotherman and Robinson, 1987, 1995). Subsequently, rat fetuses could be directly stimulated by infusing odors close to their snouts or into their mouths while their responses could be observed through the transparent amnion. These studies disclosed nasal detection of volatile compounds despite their being dissolved in fluid. In all species tested to date, fetuses exhibited acute autonomic and motor reactions when exposed to odorants. As these fetal responses differed with qualitatively and/or quantitatively distinct stimuli, discriminative abilities were demonstrated to emerge prenatally (e.g., Smotherman and Robinson, 1987; Ronca and Alberts, 1990; Schaal et al., 1991, Robinson et al., 1995).

In the human fetus, direct olfactory testing could only be run ex utero with infants born before gestational term (Peterson and Rayney, 1910-1911; Stirnimann, 1936; Sarnat, 1978; Pihet et al., 1996, 1997). Although the conditions of ligand-receptor interactions in such conditions do not match those prevailing *in utero*, it is assumed that, when tested as closely as possible to birth, preterm infants' reactions reflect the operational properties of olfaction in same-age fetuses. Thus, prematurely born infants were shown to detect (Sarnat, 1978) and discriminate odorants. For example, Pihet et al. (1996) found that the odors of nonanoic acid or eucalyptol, selected for their respective low and high trigeminal potency (but matched in subjective odor intensity), released temporally different responses in infants born within 30–37 gestational weeks. A later study using vanilla and butyric acid odors at very weak concentrations showed that prematures aged on average 31.2 weeks discriminate both stimuli in terms of respiratory response, vanilla eliciting a fast accelerative response and butyric acid a slower, more persistent decelerative response (Marlier et al., 2001). The habituation paradigm further established discriminative abilities in prematures: after habituating infants (mean age: 31.8 weeks) to odorant A, the delivery of a novel odorant B prompted an increase in facial responses (Goubet et al., 2002). Finally, in preterm infants (mean age: 33.7 weeks) odorants with trigeminal side-effects decreased oxygenated hemoglobin over the parietal region (Bartocci et al., 2001). The fetal brain can

308   Chapter 14   Prenatal and Postnatal Human Olfactory Development: Influences on Cognition and Behavior

thus bear olfaction- and trigeminal-mediated sensation as early as the last months of gestation.

14.2.3.2 Neonates[1].

Since the first experimental attempts in the 1800s (e.g., Kussmaul, 1859; Preyer, 1884) up to the present time, infants have been nasally exposed to more than 80 different odorants and irritants. Over the first 3–4 days postbirth, their chemo-reactivity increases as measured by changes in motor/respiratory responses to constant stimuli (Self et al., 1972) or to dilution series intended to appraise detection thresholds. For example, Lipsitt et al. (1963) deduced rising odor sensitivity over the first four postnatal days by noting a decrease in the concentration of *Asafoetida*[2] necessary to elicit the response criterion. Alternative interpretations may, however, reside in neonatal response amelioration or in induced sensitivity by repeated exposure to the experimental odorant (as noted in adults; Rabin, 1988; Wysocki et al., 1989).

Several lines of evidence suggest that newborns can also discern odor qualities. First, different odor stimuli have been shown to induce different autonomic (heart and respiratory rates) and general motor reactions that vary in latency, duration, rate, and amplitude (Stirnimann, 1936; Engen et al., 1963; Self et al., 1972; Soussignan et al., 1997; Lipsitt and Rovee-Collier, 2012). Second, EEG responses to odorants (e.g., citral, coffee, lavender, pyridine, rosemary, and vanillin) have been demonstrated in neonates, although not before the age of six days (Fusari and Pardelli, 1962; Fernandez et al., 2003). Third, differential cortical reactivity has been inferred for vanillin, human milk, or axillary odors as early as the first hours after birth from variations of brain hemoglobin oxygenation measured using NIRS (Bartocci et al., 2000). Fourth, using habituation paradigms, infants 3 days to 1 month of age have been found to discriminate between pure odorants (Engen et al., 1963; Guillory et al., 1980; Goubet et al., 2002) or between mixtures and their components (Engen and Lipsitt, 1965). Fifth, "mere odor familiarization" can

---

[1]Developmental psychologists are in consensus in subdividing human development into several periods. First, the *neonatal period* lasts from the day of birth to one month of age. Then, *infancy* is defined from ages 1 to 24 months, *toddlerhood* being situated between 1 and 2 years. *Early childhood* lasts from the end of the 2nd to 5th year, *mid-childhood* from the end of the 5th to the 8th year, *late childhood* (prepubertal) from 9 to 11 years, and *adolescence* from 11 to 18 years, with prepubertal, pubertal and postpubertal stages being defined individually on the variable basis of developing primary and secondary sexual characteristics (e.g., Berk, 2007, 2008).

[2]Asafoetida was often used in early studies on infant olfaction. It derives from a powder obtained from the dried sap of the plant *Ferula assa-foetida* (*Apiaceae*). Its smell is very unpleasant due to sulfur compounds (hence its name *devils dung*), but in small amounts it is used as a spice (becoming garlicky during cooking) and as a remedy (hence, its other name *food of gods*) (Teuscher et al., 2005).

establish discrimination from a novel odorant diffused in the air around infants (Balogh and Porter, 1986; Davis and Porter, 1991). Finally, associations have been conditioned between the presence of an odorant and behavioral responses made to that odorant. Such learning procedures have been carried out using a range of rewards, including breast or bottle-fed milk (Irzhanskaia and Felberbaum, 1954; Schleidt and Genzel, 1990; Delaunay-El Allam et al., 2006) or tactile stimulation (Sullivan et al., 1991). Using such procedures, very early (i.e., first postnatal day), quick (i.e., minute-wise), and more or less stable (i.e., > 1, 5 year) discrimination abilities have been demonstrated (see below).

14.2.3.3 Infancy and Early Childhood.

Olfactory development is poorly understood beyond the neonatal period, largely reflecting difficulties in devising age-adapted tests and in gaining preverbal children's compliance. Nevertheless, several lines of evidence indicate that toddlers and young children react to, and discriminate among, odorants. First, as in neonates, behavioral measures indicate that preverbal children are affected by odorants, whether presented on objects, that is, as *features of objects*, or as contextual cues. For example, Schmidt and Beauchamp (1990) videotaped 9-month-olds sequentially given three objects (rattles) scented either pleasantly, unpleasantly, or neutrally (to adults); at first sight, the responses failed to yield conclusive data in terms of hedonic reactions, but the systematic coding of the videotapes revealed better-than-chance level responses to the hedonic quality of the odor of the object being handled. The same authors showed that female infants explored the odorized object more than a non-odorized object, while boys behaved indiscriminately (Schmidt and Beauchamp, 1989). Later, Mennella and Beauchamp (1998b) and Delaunay-El Allam et al. (2010) coded infants' behavior (aged 6–13 and 5–23 months, respectively) while handling identical objects bearing different scents, one scent having been associated with the mother. The infants exhibited more positive reactions of the face/body toward the objects that smelled congruently with early odor experience. Finally, Durand et al. (2008) introduced olfactory novelty to assess odor detection in 7–15-month-olds who were videotaped while playing with scented and unscented versions of similar objects in consecutive sessions. The infants manipulated and mouthed less, and for shorter duration, the scented objects. But these responses, interpreted as indexing withdrawal, appeared only after a 2-minute delay, suggesting that it may take some time for infants to react to odor novelty and to separate odor cues from other attractive features of objects.

Infants and young children also react to the *odor quality of the background environment*. For example, Bloom

14.2 Development of Olfactory Structures and Performance

(1975; cited in Engen, 1982) diffused odorants (lavender, amyl acetate, butyric acid, dimethyl disulfide) unknowingly to 1–2-year-olds who played at a table while their facial reactions were evaluated by observers. Although 48 to 63% of the odor presentations did not elicit any detectable facial reactions, presumably due to the infant's attentional focus on playing, the remaining responses indicated nonrandom responsiveness to the odors in a way predictable from adult perception. More recently, 3-month-olds were subjected to an operant conditioning task in which they had to realize that their leg movements activate a colorful mobile, and this association was established in presence of an odor. When tested for retention of the contingency between kicking and the moving mobile 1, 3, and 5 days later, those infants re-exposed to the same odor remembered the association well. In contrast, those exposed to no odor exhibited only partial recall, whereas those exposed to a novel odor displayed no signs of remembrance (Rubin et al., 1998; Schroers et al., 2007; Suss et al, 2012). Thus, as early as 3 months of age, and probably earlier, contextual odors can facilitate subsequent memory retrieval.

Research assessing odor-induced brain activation in 3–4-month-old infants remains inconsistent. While some studies have reported undifferentiated EEG responses to food odors in waking infants (Kendal-Reed and Van Toller, 1992), others have noted differential cerebral responses to odorants in sleeping infants. For example, maternal milk odor reduced, bilaterally, β and θ band waves recorded from frontal and central electrodes, whereas orange odor increased δ wave activity from the central and right parietal positions (Yasumatsu et al., 1994), a difference attributable to qualitative and/or intensity (and trigeminal) stimulus differences. Another study recorded olfactory evoked response potentials (OERPs) to odorants administered intranasally through an olfactometer in 3.5–5-year-old children. The late positive peak of the OERP increased in latency with age, which was interpreted as reflecting deeper perceptual and semantic processing in older children (Hummel et al., 2007).

14.2.3.4 Mid-childhood to Adolescence.

Psychophysical methods with children were initiated in the late 19th century, when Toulouse and Vaschide (1899a) studied detection and recognition thresholds for camphor in 3–12-year-olds. Older children (6 and 12 years) were reported to be more sensitive than 3–5-year-olds. An additional 10-odor recognition task indicated that accurate answers improved with age. In a related study that compared children to adults, 3–5-year-olds had detection thresholds for camphor equivalent to those of adults, but older children were more sensitive and adults were better in suprathreshold odor recognition (Toulouse and Vaschide, 1899a,b). At all ages, females outperformed males in terms

of sensitivity, discrimination, and identification. Although they bear the methodological limitations of their time, these early studies nonetheless set the stage for issues on age and gender differences in human olfactory sensitivity, discrimination, recognition, naming, and memory that still remain mostly unresolved.

Since these pioneering studies, much effort has been devoted to apply psychophysical approaches to children, mostly over 4–5 years of age. Although designed for adult participants, several tests have been used to quantitatively assess children's olfaction, usually for clinical purposes [e.g., University of Pennsylvania Smell Identification Test (UPSIT; Doty et al., 1984), a match-to-sample odor identification test (Richman et al., 1995), the Alcohol Sniff test (Davidson et al., 1998), the Sniff Magnitude test (Frank et al., 2004), the Biolfa test (Chalouhi et al., 2005), the San Diego Odor Identification test (Sandford et al., 2006), the Sydney Children's Hospital Odor Identification Test (Armstrong et al., 2008), the 'Sniffin Sticks' test (Hummel et al., 2007), the Lyon Clinical Olfactory Test (Patris et al., 2009), the NIH Toolbox Pediatric Odor Identification Test (Dalton et al., 2011), and the Wheel Smell Identification Test (Cameron and Doty, 2013)]. Each of these tests has advantages and limitations in terms of validation with substantial sample sizes, practicality, and duration of administration, cultural ubiquity of odorants, dependence on memory and verbal abilities (see review in Laing and Schaal, 2012). Some of them have been used to characterize the development of olfaction over wide age-spans. For example, the UPSIT, which comprises 40 suprathreshold odorants each cued by four response alternatives as words or icons to lessen age/individual variations in mnesic proficiency, was applied cross-sectionally over the life span. The smell identification performance followed an n-shaped curve, the average recognition score of 5–9-year-olds matching those of adults in the 7–8th decades of life, when olfactory abilities sharply decline. But by age 10–19, children's UPSIT scores reached as high a level as those observed in non-elderly adults (Doty et al., 1984). Across the 6–21-year age-range, females had better UPSIT recognition scores than males (Doty et al., 1985), suggesting female advantages in either sensitivity, cognitive abilities relevant to language, or familiarization with odorants.

More basic research in child samples has been applied in various tasks, involving detection threshold, suprathreshold identification and naming ability, and/or memory. But their focus was more on child-adult comparisons (clumping early ages indiscriminately into a "children" category) than on phenomenological and mechanistic analyses of *olfactory development* (e.g., Cain et al., 1995; De Wijk and Cain, 1994; Lehrner et al. 1999a, b). First, they reported that children (>8 years) were at least as sensitive as young adults for certain odorants [e.g., 1-butanol, but also pyridine

310   Chapter 14   Prenatal and Postnatal Human Olfactory Development: Influences on Cognition and Behavior

(Dorries et al., 1989), amyl acetate (Koelega and Köster, 1974), trimethylamine (Solbu et al., 1990), R-(+)-carvone (Patris et al 2009)], although other odorants resulted in lower thresholds in younger than in older children or adults [e.g., isovaleric acid (Koelega, 1994), tetrahydrothiophene (Patris et al., 2009)]. Second, when simple discriminative tasks were required, children nearly equaled or were only slightly below adults (5–8 years, similarity judgments; Thomas and Murray, 1980; 6 years, oddity test; Stevenson et al., 2007). Further, children (from 6 years) appeared as accurate as adults in expressing quantitative judgments when required to display simple motor actions commensurate with subjective intensity ratings (Rovee-Collier et al., 1975). Thirdly, when recognition was requested, children were clearly disadvantaged relative to adults, but the magnitude of this disadvantage depended on the task, the child's age, and familiarity with the odor stimuli. Tasks dependent on lexical, semantic, or symbolic processes show increments in performance as a function of age and training. Therefore, numerous studies have sought to assess odor identification and naming in children of various ages (Larjola and Wright, 1976: 5/10/15 years; Doty et al., 1984: 5–9/10–19 years; de Wijk and Cain, 1994: 8–14 years; Richman et al., 1992: 3.5–5 years; Cain et al., 1995; Rouby et al., 1997: 5–6 years; Jehl and Murphy, 1998: 5–7/8–11/12–16 years; Stevenson et al., 2007: 6/11 years; Patris et al., 2009: 5/6/8/10.5 years; Frank et al., 2011: 4–7/8–11 years). Odor identification tasks were variable across this set of studies, generally based on free remembrance of an accurate name or on a choice among four or five words or images. In these conditions, younger children had generally lower identification performances than older children and adults. Differences with adults tended to fade when children were semantically primed by a name or picture to the identity of the odor (e.g., Cain et al., 1995), highlighting that early limitations in odor recognition may depend on producing or remembering verbal associations. The general effect of age further interacted with gender, girls being generally better than boys in odor identification and naming. This age by gender interaction is often interpreted as resulting from superior verbal fluency or broader odor-related semantic knowledge in females, even from early ages (Ferdenzi et al., 2008a; Patris et al., 2009). Developmental differences in odor distinction are certainly causally multiple, presumably encompassing sensory factors, such as changes in odor sampling ability (changing nasal aerodynamics, poor sniffing before 2.5–3 years; e.g., Schaal, 1988; Mennella and Beauchamp, 1992) or in odor detection due to changing properties of the olfactory mucosa and range of expressed olfactory receptors (e.g., Apfelbach et al., 1991). It may also depend on general or olfaction-specific cognitive factors, such as more limited experience with odors leading to reduced related semantic stores, lower ability to verbally label odor stimuli (Engen

and Engen, 1997; Frank et al., 2011; Lumeng et al., 2005; Rinck et al., 2011. Bensafi et al., 2007), shorter spans in working memory (Larjola and Wright, 1976), or poorer recognition memory (Hvastja and Zanuttini, 1989; Frank et al., 2011). The latter psychological factors are also often evoked to explain early gender differences in olfaction. A challenge for future research will be to integrate longitudinally in a single multidisciplinary study biological, psychological and environmental determinants of chemosensory performance during human ontogeny.

Younger children are clearly disadvantaged in odor identification tests requiring naming of the veridical source. But identification of higher-order categories reveals their already elaborate odor cognition, even in cultural conditions that do not favor training and teaching of olfaction. For example, when 8–14-year-old American children had to tell the identity of 17 odorants and whether they belong to edible/non-edible categories, they were less accurate than young adults in naming of odor sources, but were equal to them by the edibility criterion (de Wijk and Cain, 1994). A related study asked French 5–6-year-olds to categorize 12 common odorants along different criteria such as origin (in/outdoor, natural/artificial, vegetable/animal) and functionality (edibility, wearability) (Rouby et al., 1997). While these children categorized the odorants accurately for edibility, and partly accurately for wearability, they could not classify them using the other, more abstract, criteria. Another study required 4–11-year-old French children and adults to rate six fruity and six floral odors in terms of typicality, familiarity, liking, and edibility, and to order them as members of binary fruit or flower categories (Valentin and Chanquoy, 2012). Results indicated a relatively high classificatory agreement from age 5 for some odorants in the series, but interestingly without a necessary overlap with hedonic categorization, suggesting that young children can separate emotional from more cognitive operations while processing odors. But from age 8, most odorants were classified into the expected category, as well as given a name in the correct semantic category. The ability to assign an odor an egocentric functional meaning may thus prevail on the ability to accurately name it, although naming certainly crystallizes conceptualization. Thus, children's odor categorization abilities depend strongly on the breadth and depth of prior experience with odors, and hence can only improve with exposure (Rouby et al., 1997; Prescott et al., 2008). Exposure creates familiarity with object features, increasing their perceptual salience and differentiation, and facilitating the learning of category types and borders and the articulation of semantic relationships (e.g., Folstein et al., 2010).

The effect of unsupervised training in the acquisition of the meaning of odors is best supported by studies on olfactory recognition of items that prevail in the environment. For example, children aged 2.5–6 (Noll et al.,

1990) or 5.5–10.5 years (Fossey, 1993) recognize and identify different types of alcoholic beverages, and these differences increase with age and rate of exposure. Thus, even if (Euro-American) children do not spontaneously talk about their odor experience in the early stages of language development (Engen and Engen, 1997), they are nevertheless aware of odors in their environment and behave accordingly. A survey study conducted on Finnish and French 6–12-year-olds confirmed early attention to the odor attributes of objects, persons, and domestic and outdoor settings (Ferdenzi et al., 2008a,b). Although these children ranked olfaction after sight, touch, and hearing in importance in their everyday experience, they also highlighted the unique significance of olfaction in their attention to the environment, especially in food selection and appreciation, self-grooming, interpersonal relationships, and environmental enjoyment. In line with the psychophysical data mentioned above, this survey confirmed that age obviously influences the ability of children to scrutinize the odor facets of their environment, although the most affectively-laden items or situations went unaffected by age. Lastly, girls (between 6 and 12 years) reported higher attention and reactivity to odors in daily settings than did boys. Beyond gender differences, this study also pointed to individual differences in the power of olfaction to control attitudes and overt behavior of children. When the participants' sample surveyed above was split as a function of their odor attention score, children who scored high on 'olfactory attention in everyday settings' were more aware of implicitly-presented real odors and more extreme in hedonic judgments of real odors than subjects who scored low on olfactory attention (Ferdenzi, 2007). The children's reported attention to environmental odor cues may directly influence the way they sense objects or contexts. For example, 4–7-year-olds sniffed at odor containers more distally and cautiously when they were classified as food neophobic (Bunce and Gibson, 2012), indicating either greater sensitivity and/or higher emotional reactivity to odors. This whole area of individual differences in olfaction is especially promising to probe how sensory phenotypes emerge and fluctuate during typical and atypical development, and which early life conditions do stabilize or alter the processing of odors from the molecular to the behavioral levels (e.g., Lane et al., 2010; Schecklmann et al., 2013).

A turning point in the development of odor perception is adolescence. A puberty-related shift in olfaction was first reported by Le Magnen (1952) who noted prepubertal children and adult men found the steroid exaltolide to be less intense than did women. This age by sex interaction in the detection of certain odorants was subsequently confirmed in cross-sectional studies for other steroidal compounds, including artificial musk, androstenone, and androstadienone (Koelega and Köster, 1974; Koelega,

1994; Dorries et al., 1989; Hummel et al., 2005; Chopra et al., 2008). Typically, male sensitivity to these compounds decreased across the pubertal years, while female sensitivity remained constant. The rate of anosmia to the odor of androstenone between pubertal (9–14 years) and postpubertal (15–20 years) groups tripled (to reach 30%) in boys, while it remained low (around 10%) in girls (Dorries et al., 1989). Some studies indicated that this age by sex interaction concerned musky odorants (Koelega and Köster, 1974; Koelega, 1994; Dorries et al., 1989), but a wider range of odorants was affected by puberty-related changes in olfaction: strongly unpleasant odorants (sulfur compounds; Chopra et al., 2008) and even fruity or floral notes of strawberry or lavender (Moncrieff, 1965) were also involved, although food odors appeared generally excluded from such effects (at least in boys; Laing and Clark, 1983). The proximal causes of these peripubertal variations in olfactory processing remain little investigated and empirical attention from biological, psychological, and biosocial perspectives is needed. Related fluctuations in reproductive physiology are certainly involved in altered peripheral reception and perceptual processes, as are changes in brain maturation (e.g., Brewer et al., 2006) and related psychological processes (e.g., increased disgust sensitivity, novelty seeking and risk taking, and changing social roles and status, as suggested by variations in the valuation of certain foods, perfumes, or addictive agents).

14.2.4 Developmental Odor Hedonics

Research on adults show that nasal chemoreception can mediate a wealth of informative "dimensions" inherent either to the stimulus (i.e. identity, quality, perceived intensity, complexity, position in space, variety) or to properties derived from an individual's prior interactions with it [i.e., familiarity/novelty, hedonic valence (pleasantness/unpleasantness), utility knowledge or affordances (e.g., eatability, toxicity, stress buffering value)]. These properties are not independent, however, as quality is often linked with intensity and hedonic valence with intensity or familiarity (Engen, 1974; Cain and Johnson, 1978; Delplanque et al., 2008). The extent to which these "dimensions" are relevant in the development of the ability to segregate odors into categories is poorly understood. Most studies have sought to understand when and how the affective impact of odorants develops, often confounded in intensity or familiarity/novelty "dimensions" (Delplanque et al., 2008). Some of these studies suggested that humans of all ages and genetic/ethnic backgrounds use similar hedonic criteria in dividing up the odor space. From the first hours after birth, odors considered pleasant for adults (e.g., orange, geranium) were reported to evoke positive oral activation, whilst unpleasant-to-adult odors

312 **Chapter 14 Prenatal and Postnatal Human Olfactory Development: Influences on Cognition and Behavior**

(e.g., asafoetida) caused facial/oral expressions of disgust and negative head turning in infants aged 15 minutes to 6 hours (Peterson and Rayney, 1910–1911). Also, Stirnimann (1936) established orofacial expressions emitted by newborns, including a premature infant born at 6 G months, in response to odorants that are pleasant or unpleasant to adults. These pioneering studies were confirmed and expanded by Steiner (1977, 1979) who showed that newborns under 12 hours of age, before any postnatal ingestion, display differential responses to various intense odorants chosen to represent pleasant and unpleasant extremes for adults (banana, vanilla, milky vs. fishy and rotten egg, respectively). Steiner presented these stimuli orthonasally while neonatal faces were photographed. These photos were then seen by a panel of "blind" adult judges who had to tell whether the facial response was caused by a pleasant, unpleasant, or neutral stimulus. Generally, odorants considered pleasant by adults induced facial responses rated as expressing contentment and acceptance (relaxed facial muscles, raising of mouth corners, licking and sucking), whilst unpleasant odorants elicited expressions interpreted as reflecting dislike and disgust (lowering of mouth corners, lip and tongue protrusion, gaping). Assuming that a commonality of facial expressions between neonates and adults implies analogous emotional states, Steiner postulated that certain odors were more acceptable than others for the newborn and that the perceptual attributes infants use to differentiate odors is hedonic valence. Accordingly, Steiner (1979) postulated that neonatal facial reactivity to odors depends on a hard-wired 'hedonic monitor' that activates an olfacto-facial reflex. After testing an anencephalic infant who responded to fishy, rotten egg, and banana odors by facial expressions similar to those of normal infants, he concluded that this early stimulus–response link was mediated by brainstem structures (Steiner, 1979).

Using a more quantitative and dynamic means to code infant responses, Soussignan et al. (1997) presented 3-day-old newborns (in different arousal stages) with 12 odorants [including four unfamiliar biological odorants (amniotic fluid, human milk, and two formula milks having different qualities), vanillin and butyric acid diluted to match the perceived intensity of previous odorants, and odorless controls]. Facial responses to odors were videotaped and then systematically decoded using Ekman and Friesen's (1982) Facial Action Coding System.[3] In such conditions, neonatal facial actions to the different odorants were highly variable between neonates and could accordingly not be considered to be 'automatic,' as predicted by Steiner's olfacto-facial reflex model. However, facial responses were differentiable according to the hedonic value of the odorants, but more clearly so for odorants that were unpleasant to adults. Thus, while unpleasant butyric acid odor elicited more negatively-valenced responses than vanilla odor, the pleasant vanilla did not elicit any more positive facial response than did butyric acid. As both odorants were intensity-matched, the negative response could only be based on the neonates' aptitude to differentiate odors in qualitative and/or hedonic terms, thus confirming only partially Steiner's results. However, the odors of unfamiliar formula milks, although clearly unpleasant to adults, did not elicit negative facial responses in newborns. Therefore, the hedonic facets that newborns and adults detect in odorants are not completely superimposable.

Beyond the neonatal period, differences between infants, children, and adults in hedonic reactivity to odors have been, and still are, a source of long-standing debate (e.g., Schaal, 2012). Early studies reported young children's relative tolerance to odorants that are unanimously rejected by adults. Based on behavioral or declarative responses of (hospitalized) Hungarian participants aged 1 month to 16 years to the offensive odors of oil of *Chenopodium*, trimethylamine or asafoetida, Peto (1936) pinpointed age 5 years as a turning point in the hedonic appraisal of odor: before that age, negative responses occurred only in 3% of the subjects, whereas beyond that age 72% expressed dissatisfaction. A convergent pattern was also noted among American children sampled between 3 and 14 years (Stein et al., 1958) with fruity (amyl acetate), sweaty and fecal odorants: 3–4-year-olds found all stimuli equally likeable, whereas older children discerned fruity from sweaty and fecal odorants, and clearly rejected the latter. Engen and Katz (1968) later found that the liking response for the unpleasant odor of butyric acid drops sharply between ages 4 and 5–6 years while those for pleasant odorants remain steady. But, in an effort to explain this apparent 5-year shift in the hedonic appraisal of odors, Engen (1974) suggested the involvement of a non-specific age-related response bias in a test situation where an infant is questioned by an unfamiliar experimenter. At age 4, children exposed to butyric acid odor were more inclined to answer "yes" than "no" to the queries "Tell me if it smells pretty?" as well as "Tell me if it smells ugly?". This tendency fell sharply in 5–6-year-olds who were increasingly negative to this odor, answering "no" and "yes" to the former questions, respectively. Thus, a variation in the 'yes man bias' (Guilford, 1954) may best explain why younger children report positive responses to odorants regardless of their hedonic valence. Studies asking American or British children over the age of 6 years to rank-order sets of odorants in terms of their pleasantness resulted in responses roughly similar to those of adults (Kniep et al., 1931; Moncrieff, 1966). However, children ranked some odorants (fecal notes)

---

[3]Ekman and Friesen's (1982) Facial Action Coding System is an anatomically-based system that itemizes the movements of each of the 44 facial muscles. It is a valid and reliable tool which has been used extensively for coding adult facial movements, but has been adapted for coding neonate and infant faces.

14.2 Development of Olfactory Structures and Performance

lower in unpleasantness than adults did (Moncrieff, 1966), suggesting either that rating scales are used in different ways at different ages or that preferences assessed through relative judgments (rank ordering or pair comparisons) are prone to age-related variations (Guinard, 2001). Finally, comparing 6–12-year-old children from different cultural backgrounds (i.e., Québec, Syria, Indonesia) for their hedonic rating of a set of 14 odorants, responses were consensual across cultural groups for the stimuli classified as the most negative, while the most preferred odors were variable between cultural groups, reflecting odor sources that are locally appreciated (Schaal et al., 1996). Thus, verbal responses to pleasant and unpleasant odorants appear asymmetrical in infants and children, the latter generally eliciting more consensual verbal responses than the former (see below).

Do measurements based exclusively on behavioral variables (i.e. independent of verbal response) provide more reliable results on the developmental trends of odor preferences? In the study by Bloom (1975; cited in Engen, 1982) mentioned above, pleasant (lavender, amyl acetate) and unpleasant smells (butyric acid, dimethyl disulfide) were unobtrusively sprayed in front of 1–2-year-olds who were playing at a table. Their facial reactions to the odor stimuli, as observed from behind a one-way mirror, were rather rare as most odor puffs were inefficient to cause a visually-detectable reaction. But, when only the stimuli effective to elicit a facial response were analyzed, pleasant odorants triggered more positive than negative facial actions. The reverse was true for unpleasant odorants. The hedonic hierarchy derived from these data in 1- and 2-year-olds was consistent with that predicted by adult ratings, *viz.* lavender > amyl acetate > butyric acid > dimethydisulfide. When these same odorants were presented to 3-, 4- and 5-year-olds, their facial reactions led to the same hedonic ordering of them. Thus, at least for these four odorants, orderly preferences begin to develop in 1- and 2-year-olds and appear stable through age 5 years (Bloom, 1975; see also Barnham and Broughan, 2002). But even earlier, by 9 months, American infants act differentially on identical objects scented pleasantly (methyl salicylate), unpleasantly (butyric acid) or not at all (mineral oil) (Schmidt and Beauchamp, 1990), some expressing reliable hedonic facial actions, while others responded by seemingly disregarding the object. A similar test situation in 7- and 21-months-olds came out with more positive and less negative responses (as indexed by object mouthing and face actions, respectively) for chamomile odor to which they were familiarized in the context of breastfeeding (Delaunay-El Allam et al., 2010). But again, the facial readouts provoked by the affective odor appraisal appeared asymmetrical, negative responses being generally clearer in communicative value to observers than positive responses (e.g., Steiner, 1979; Soussignan et al., 1997;

Delaunay-El Allam et al., 2010; Mennella et al., 2001; Zeinstra et al., 2009; Booth et al., 2010).

In older children, the accuracy of facial cues in unveiling an individual's hedonic experience can be strongly mitigated by the social context (Soussignan and Schaal, 1996a, 1996b), indicating incorporation of odor-elicited facial responses into the communicative repertoire (exaggeration, inhibition). When odors were presented by an adult experimenter (as compared to when they were self-administered in social isolation), 5- to 12-year-olds' responses were more positive for pleasant odors and less negative for unpleasant odors (and girls were more sensitive to such experimenter's presence than boys). An alternative to measuring facial responses in vivo or from videotapes is electromyography (EMG), the recording of electrical activity of facial muscles even before visual detection of any actual response. EMG has been used successfully in 7-year-old children exposed to pleasant vs. unpleasant odorants (Armstrong et al., 2007). However, as with facial expressions, EMG responses are not immune of social constraints (Jäncke and Kaufmann, 1994).

Another approach to assess the development of hedonic processes in olfaction relies on the subject's ability to translate hedonic judgments from odors onto attributes of visual stimuli having a pronounced social or symbolic meaning. For example, Strickland et al. (1988) required 3–5 year-old children to inhale hedonically-contrasted odorants and make the dichotomous decision to place them on 'smiley faces.' In these conditions, the pleasant (to adults) benzaldehyde and the unpleasant dimethyl disulfide were assigned to the smiling and frowning face, respectively, with 3-year-olds' attribution decisions being less clear than those of 4–5-year-olds. In a further study, 3-year-olds and adults were asked to give nine odorants to one of two television characters famous to American children, Big Bird and Oskar the Grouch, considered positive and negative figures, respectively (Schmidt and Beauchamp, 1988). Both age groups showed overall agreement in their preference pattern, although they differed in their hedonic assignments of some odorants to the positive or negative figure. While these studies have been criticized on the basis of early semantic and linguistic skills (Engen and Engen, 1997), they corroborate the idea that young children at the fringe of language and schooling can consciously distinguish odor stimuli in hedonic terms and that they do so in relative consensus with adult judgments.

In conclusion, research summarized in Section 14.2 clearly establishes the capacities of newborns, infants, and children to detect, discriminate, and to a certain extent, categorize olfactory stimuli. On top of this is their aptitude to subdivide the odor stimuli into attractive, non-attractive,

314   Chapter 14   Prenatal and Postnatal Human Olfactory Development: Influences on Cognition and Behavior

and repulsive. The sensory and psychological dimensions at the root of these early preferences remain poorly explored in humans, but one can safely assert that any weak odor to which the infant has previously been exposed will subsequently have a higher reinforcing value than any novel odor (although sensory specific satiety or boredom effects may induce occasional preferences for olfactory novelty; e.g., Mennella and Beauchamp, 1999; Hausner et al., 2010). This hypothesis attributes a major role to experience in the acquisition of olfactory preferences, even prenatally, although chemoperceptual specializations may emerge uninfluenced (or minimally influenced) by experience, as we will see below. The next section will consider the functional implications of these early discriminative abilities.

14.3   EARLY BEHAVIORAL FUNCTIONS OF OLFACTION

Here, we will consider the adaptive roles of olfaction in the succession of transitions that are typical of mammals, emphasizing the human case. We will first describe the fetal period during which offspring are anticipatorily primed to their future environment and the birth transition when newborns need to mobilize their sensory abilities to contact the mother and successfully ingest milk. Next, we will outline neonates' and infants' responses to social odors in the nursing niche, and survey odor-based processes in self-regulation abilities and in coping with social diversification. Finally, the involvement of olfaction in infants' and children's behavior will be considered in affiliative networks within and outside the family.

14.3.1   Fetal Chemoreception

Mammalian, including human, fetuses dwell in amniotic fluid that is transplacentally permeated by odorous (and sapid) compounds. Such compounds are dependent upon the genetic, immunogenetic, and physiological constitution of the mother, as well as upon her life style, including diet, stress, activity and, in the case of humans, smoking behavior, medicaments, cosmetics, and use of addictive drugs (Schaal, 2005; Schaal et al., 1995d). This has been shown in numerous mammals, including humans. For example, pregnant ewes' or women's ingestion of garlic alters the sensory properties of amniotic fluid for adult noses (Nolte et al., 1992; Mennella et al., 1995), and odorous metabolites can be chemically traced into amniotic fluid after ovine or human mothers ate cumin in late gestation (Schaal, 2005). The mother-to-fetus transfer of dietary aromatic compounds can sometimes be so effective that the newborn's initial body odor is "spiced" accordingly (Hauser et al., 1985).

The odorous properties of the prenatal environment organize the fetuses' olfactory neural networks and induce the first 'odor images.' Integrative odor responses in fetuses were first assessed *in utero* in animal models. For example, fetal rats conditioned aversively to mint odor on embryonic day 17 retain the aversion on embryonic day 19 (Smotherman and Robinson, 1987). Thus, the neural chain for encoding, retaining, and retrieving chemosensory cues are present in the fetus in the later days of gestation. The *in utero* odor experience can carry over to the postnatal period (Smotherman, 1982; Stickrod et al., 1982; Hepper, 1988; Schaal et al., 1995d). Thus, mammalian fetuses from many species can respond to odors extracted from the womb and can retain memory for them for days to months postpartum (e.g., rats: Smotherman and Robinson, 1995; Hepper, 1988; mice: Youngentob and Glendinning, 2009; Todrank et al., 2011; rabbits: Bilko et al., 1994; cats: Becques et al., 2010; Hepper et al., 2012a; sheep: Schaal et al., 1995b; Simitzis et al., 2008; pigs: Oostindjier et al., 2009, 2010; *see* Schaal, 2005, for a review). Similar abilities have been shown in humans. For example, infants born to mothers who had eaten garlic during pregnancy show less aversion to the odor of allyl sulphide on the first postnatal day (Hepper, 1995).Infants of women that ingest anise in the last 10 days of gestation display greater appetitive responses and less aversive responses to a diluted anise odor relative to infants whose mothers did not consume anise (Schaal et al., 2000; see Figure 14.2). Finally, infants born to non-alcoholic mothers categorized as frequent drinkers during pregnancy (≥ 4 times/month) are more reactive to ethanol odor than those born to non-frequent drinkers (<4 times/month) in terms of cephalic and facial movements on postnatal day 2 (Faas et al., 2000).

The most parsimonious explanation of prenatal odor acquisition is that of simple familiarization with odorants present in the amniotic sac. Thus, stimuli merely experienced *in utero* generally cause less aversion or more attraction than any novel stimulus. But associative learning may also function *in utero* to launch appetitive or aversive processes. This has been empirically explored in the fetal rat, but remains hypothetical in the human fetus. For example, a rat fetus readily associates a negative meaning to an odor experienced when an illness state is inflicted to itself (Stickrod et al., 1982; Smotherman, 1982) or to its mother (Gruest et al., 2004). Rat fetuses can also associate odors with endogenous physiological variations. For example, they can link an odorant co-occurring with decreasing brain oxygenation or re-oxygenation by cord clamping or unclamping and this odorant later becomes laden with aversive or attractive value, respectively (Hepper, 1991b, 1993).

Otherwise, each maternal meal, especially in late gestation, provides optimal conditions for the fetus to associate

14.3 Early Behavioral Functions of Olfaction

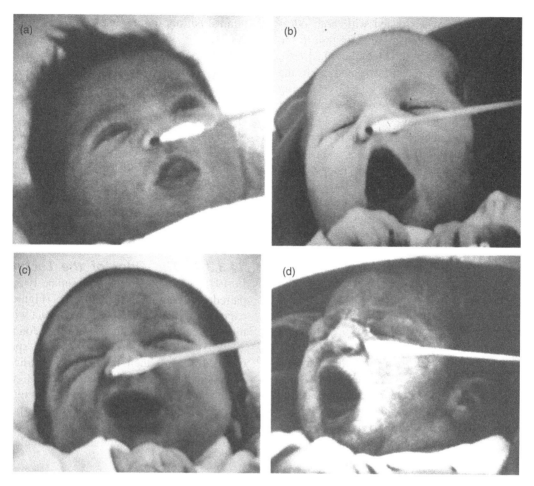

Figure 14.2 Examples of facial responses elicited by the presentation of odor stimulations to awake newborns. These infants were exposed to 10-second stimulations of a diluted anise odor. Infant a (age at test: 3 hours at the test) was born to a mother who consumed anise-flavored foodstuffs or beverages during pregnancy, and accordingly displays facial actions reflecting appetence to anise odor. Infants b, c and d (aged 1.5, 8, and 0.5 hours, respectively) were born to non-anise consuming pregnant mothers and expressed facial actions reflecting rejection (i.e., brow lowering, cheek raising, nose wrinkling, gaping, negative head turning). From B. Schaal, L. Marlier, and R. Soussignan, Human fetuses learn odours from their pregnant mother's diet. *Chemical Senses*, 25: 729–737.

inputs from chemoreception, motor activation, and brain metabolic sensing and reward processes. Exposure to an odorant *in utero* may induce epigenetic changes in olfactory receptor expression that may be actualized in biased sensitivity in the newborn (Semke et al., 1995; Youngentob et al., 2007). Prenatal olfactory experience may further facilitate neurogenesis and the synaptic organization in the OBs (Leon, 1992; Wilson and Sullivan, 1994; Todrank et al., 2011), as well as later positive investment in the newborn, for example, by strengthening an odorant's reinforcing value through postnatal reconsolidation. Such a fetal preparation process was shown in rat pups: when sensitized to citral *in utero*, they engaged a preference for citral only when re-exposed to it postnatally in particular arousal conditions (Pedersen and Blass, 1982; Alberts and Ronca, 2012). Ordinarily it is the amniotic fluid that plays the role of an olfactory "initiator" that provides a link between the prenatal and postnatal environments (see below).

14.3.2 Birth and the Rapid Learning of Odors

The birth process itself impacts the perceptual development of the 'perinate', viz. the fetus as it is becoming a newborn. Labor-related physical compression and internal (blood gases) and thermal upheavals alter the perinate's physiology with rising brain catecholamines and correspondingly high arousal levels, and facilitate learning of the current odor environment. During labor, the fetal brain seems to update the latest chemosensory profile of its amniotic environment. This is suggested in fetal rats exposed sequentially to two odorants 10 or 40 minutes before cesarean delivery. When tested with both odorants as neonates, the pups exhibit more activation to the odorant experienced closer to birth (Molina et al., 1995). Physiological and sensory effects associated with the natal transition itself further promote olfactory learning (Alberts and Ronca, 2012). This was shown by having rat fetuses externalized from the uterus and exposing them to artificial compressions mimicking labor in the presence or absence of an odor. Those fetuses exposed to odor-contingent compressions acquired the odorant as a postnatal cue to nipple seizing. In contrast, those fetuses exposed to compressions without odor or to odor without compressions did not acquire the odor as a cue to nipple grasping.

Labor-related experience has also been suggested in human infants exposed to an odor for 30 min after a cesarean delivery that was preceded or not by contractions. When these two groups of newborns were tested 1–5 days later for their attraction to that odorant relative to a novel one, only those having been exposed to labor oriented more to the familiar odor (Varendi et al., 2002). The experience of labor seems to mediate high arousal states during the first hours postbirth and to facilitate learning. Thus, exposing human neonates to an odorant for 30 min either in the first hour or after 12 hours' postpartum led to a differential response for the exposure odor two to three days later in the early-exposed infants, but not in those exposed later (Romantshik et al., 2007).

14.3.3 Odor Communication in the Nursing Niche

As structures that provide milk to offspring, nipples or teats are the vital interface between lactating females and sucklings. These structures evolved to be sensorily conspicuous to offspring, mostly through their tactile and odor properties (Blass, 1990; Schaal, 2010). Empirical validations of this, however, are of recent vintage, despite the fact that this was recognized much earlier by such notables as Galen (AD 130–200) (Daremberg, 1856) and Darwin (1877).

14.3.3.1 The Odor of the Lactating Breast.
In 1975, Macfarlane exposed supine-lying newborns to paired odor pads hung by their faces (Figure 14.3c). When exposed to a pad that had been in contact with the mother's nipple-areolar region and to a clean control pad, 17 of 20 breast-fed infants (aged 2–7 days) spent more time oriented towards their mother's pad, indicating not only detection but attraction. Such attraction was confirmed by others using variations of this test (Figure 14.3d). For example, 2-week-old bottle-fed infants spend more time oriented to an unfamiliar lactating mother's breast odor than to the odor of their familiar formula (Porter et al.,

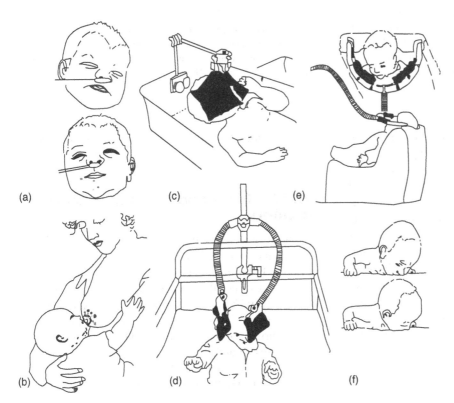

Figure 14.3 Contexts and devices used to investigate olfaction in human newborns. (a) Odorants are presented sequentially on cotton buds to assess differential oral-facial responses or reaction of the autonomous nervous system (e.g., Steiner, 1979; Soussignan et al., 1997; Mizuno et al., 2004; Doucet et al., 2009); (b) Infants are directly presented to their mother's breast after different treatments applied on the breast surface (e.g., Doucet et al., 2007). (c-d) Paired-choice devices to assess the general motor responses and relative head orientation of supine-laying infants (c: Macfarlane, 1975; d: e.g., Schaal et al., 1980). (e) Paired-odor choice test to assess differential head-turning and oral activation toward either stimulus in sitting awake infants (e.g., Schaal et al., 1995c; Delaunay-El Allam et al., 2006). (f) Differential rooting or crawling movements of the infant toward an odor source (e.g., Varendi and Porter, 2001) (Drawings: a-e: b Schaal; f: redrawn after Prechtl, 1958).

1991). One study presented 15-day-old bottle-fed infants with the breast odor of an unfamiliar lactating woman and the odors of either (A) the breast of a non-parturient woman or (B) the axilla of the same woman (Makin and Porter, 1989). Although they never directly contacted their mother's breast and were fed formula in highly reinforcing conditions, the infants oriented more to odor of the lactating breast, confirming that lactating women emit an attractive odor from the breast. Interestingly, newborns can crawl to an odor source 10–15 cm away (Figure 14.3f). When laid prone on their mother's chest, they can reach a breast within an hour from birth (Righard and Alade, 1990; Widström et al., 1987, 2011) presumably via the smell of the breast (Varendi et al., 1994). Similarly, when left bare in a warm incubator, infants creep more rapidly towards a pad odorized with their mother's breast odor than a scentless control pad (Varendi and Porter, 2001).

What are the sources of the attractive odor cues released by the breast? The human areolar-nipple area is complex in terms of multiplicity of secretions, making this difficult to determine. The breast is supplied with skin glands of eccrine, apocrine, and sebaceous types (Perkins and Miller, 1926; Montagna and MacPherson, 1974). The areolae are additionally dotted with small eminences that mark Montgomery's glands (MG) (Montgomery, 1837; Schaal et al., 2006), composed of sebaceous glands coalesced with miniature mammary acini (Montagna and Yun, 1972; Smith et al., 1982), and which can give off a latescent fluid (Schaal et al., 2006; Doucet et al., 2012). A recent study attempted to evaluate whether morphologically differentiable areas of the breast could induce distinct behavioral effects in newborns (Doucet et al., 2007). The 3-day-old infants responded (by rooting, mouth opening, or tongue protruding) similarly to whole breast vs. areola vs. nipple vs. milk odors, suggesting overlapping or equivalent attractive potencies in corresponding exocrine substrates. However, when presented separately from the other breast secretions, the Montgomerian secretion was more active in eliciting mouthing and in stimulating respiration than milk and sebum, as well as several other reference stimuli (e.g., cow's milk-based formula, fresh cow's milk, vanilla, water) (Doucet et al., 2009; Figure 14.4). Thus, Montgomerian secretions may play a special role in the infant's attraction to, and coordinated action on, the lactating breast (Schaal et al., 2008).

14.3.3.2 The Odor of Milk. Colostrum and milk add their intrinsic olfactory qualities to the nipple-areola. In both preterm and term newborns (Bingham et al., 2003a, 2007; Russell, 1976; Soussignan et al., 1997) , these fluids elicit positive head-turning (Marlier et al., 1998; Marlier and Schaal, 2005) and activate facial/oral responses indicative of their appetitive value. For example, the odor

of mother's milk, as compared to the odor of a familiar formula, increases the efficiency of sucking during regular formulae feeding (Mizuno and Ueda, 2004) as well as non-nutritive sucking in premature infants (Meza et al., 1998; Bingham et al., 2003a, 2007). This milk odor-based stimulation of oral activity has been used to train the oral musculature of prematurely born infants and to accelerate the strengthening of their oral competence (Raimbault et al., 2006; Yildiz et al., 2011). Further, the odors of colostrum or milk elicit cortical activation (as assessed by EEG or NIRS; Yasumatsu et al., 1994; Bartocci et al., 2000), and the pattern of cortical activation induced by human milk in the orbito-frontal region differs from that of the odor of formula milk in infants consuming either milk (Aoyama et al., 2010).

The chemical nature of the active odorant(s) emitted in human milk or in any mammal's milk is largely unknown. Several analytic studies have found a variety of volatile compounds in human milk (Shimoda et al., 2000; Bingham et al., 2003b; Büttner, 2007), but these were so methodologically diverse (pooled vs. individual milk samples; fresh vs. frozen; solvent extracted vs. headspace) that no clear conclusion can be drawn. A similar approach in the European rabbit was more successful, certainly because of rabbit pups' non-ambiguous responses to odorants. Rabbit pups do indeed instantaneously orally grasp a glass rod dipped into fresh milk obtained from any female (Keil et al., 1990). In one study, gas-chromatography coupled with mass spectrometry and direct sniffing led to the identification in rabbit milk of a compound, 2-methyl-but-2-enal (2MB2), which was as efficient as whole rabbit milk in eliciting oral grasping motions in pups (Schaal et al., 2003). 2MB2 presumably is added to the milk in the final portion of the nipple (Moncomble et al., 2005). 2MB2 was detected in milk samples from females of different genetic and dietary backgrounds. Since its behavioral activity was highly selective and species-specific, prenatal or postnatal exposure was apparently not necessary for its function, Schaal et al. (2003) conclude that this agent functions as a pheromone and named it "mammary pheromone." The generalization of this mode of chemocommunication to other parenting and nursing strategies than the European rabbit awaits further investigation (Schaal, 2010).

14.3.3.3 Transnatal Chemosensory Continuity. By which mechanisms do secretions from the lactating breast or milk become attractive to neonates? Apart from mechanisms which apparently do not depend on sensory induction by prior direct exposure, such as the case with 2MB2, it seems that exposure effects and learning prevail in explaining this early attractiveness. As outlined previously, the perinatal brain is an efficient learning device, which is in age-adapted optimal operating

318 Chapter 14 Prenatal and Postnatal Human Olfactory Development: Influences on Cognition and Behavior

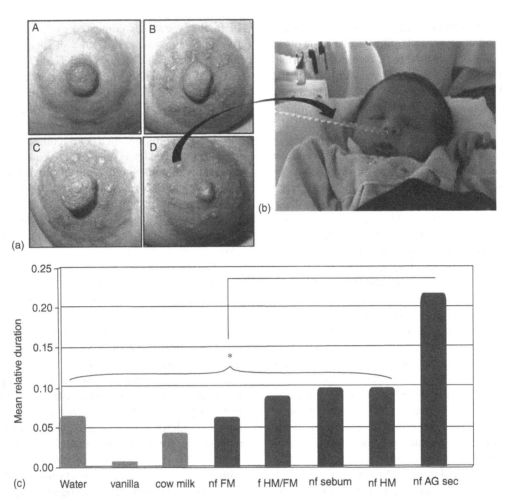

Figure 14.4 The areolar glands of Montgomery and the behavioral effects of their secretion's odor in neonates. (a) Human areolae of lactating women (3 days postpartum) endowed with different amounts of areolar glands (a: absence of glands; b, c, d: variable number of glands. Photographs: S. Doucet). (b). Presentation of Montgomerian secretions to a newborn (age: 3 days) showing lip pursing and tongue protrusion. (c). Mean relative durations of oro-cephalic responses of 3-day-old newborns (n=19) exposed (for a 10-second stimulus period) to nonhuman control stimuli [water, vanilla, non-pasteurized cow's milk, non-familiar formula milk (nfFM)] and to human odor substrates [familiar (F) milk (either human milk (HM) or formula milk (FM); non-familiar sebum (nf sebum); non-familiar human milk (nfHM); and non-familiar areolar gland secretion (nfAG sec)]. Areolar gland secretion releases more oro-cephalic responses than any other stimulus (From Doucet et al., 2009, The secretion of areolar (Montgomery's) glands from lactating women elicits selective, unconditional responses in neonates. *PLoS ONE*, 10: e7579).

conditions right before and after birth. Mammalian perinates do then face a sequence of odor substrates in highly reinforcing circumstances: amniotic fluid—secretions of the parturient mother's skin/breast—colostrum—milk. In human newborns, amniotic and lacteal fluids are attractive separately, but up to age 3 days they bear equivalent attractivity when presented concurrently as odors. However, beyond day 3, breast-fed infants turn more to the odor of their mother's milk than to the odor of their own amniotic fluid (Figure 14.5; Marlier et al., 1998). Undifferentiated responses between colostrum and amniotic fluid are also noted in other mammalian newborns that have been investigated (Figure 14.6; Schaal, 2005). This suggests a generalized chemosensory strategy among mammals that raises the hypothesis of a *transnatal chemosensory continuity* between both vital fluids.

Such chemosensory continuity can be achieved by various means. First, the physiological transfer of odorant compounds into both fetal and lacteal compartments may create a *chemical matching* between amniotic fluid and colostrum. As already mentioned above, amniotic fluid is infiltrated with odorants from the mother's diet. Such odorants are also transferred into the colostrum and milk as shown by chemical analysis (Desage et al., 1996) or flavor preferences in weanling animals (e.g., Le Magnen and Tallon, 1968; Galef and Henderson, 1972; Capretta and Rawls, 1974; Wuensch, 1975; Mainardi et al., 1989; Bilko et al., 1994; Schaal et al., 2005; Langendijk et al., 2007). In humans, aromas from

14.3 Early Behavioral Functions of Olfaction

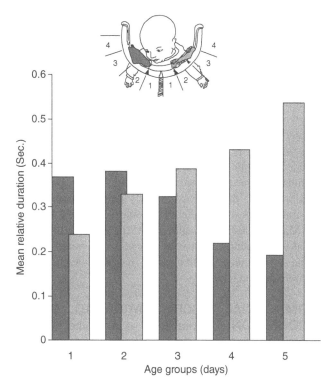

Figure 14.5 Mean relative duration of head (nose) orientation response of breastfeeding newborns within their first 5 days after birth when concurrently exposed to the odor of their prenatal environment (amniotic fluid; blue) and to the odor of the mother's milk of the day (yellow). From Marlier et al., 1997. (*See plate section for color version.*)

various foods (e.g., garlic, vanilla, alcohol, carrot) also easily pass into milk (Mennella and Beauchamp, 1991a; Hausner et al., 2008) and subsequent odor changes are detected by nurslings (Mennella and Beauchamp, 1991a, b, 1996, 1999). Maternally-inhaled odorants

(Dougherty et al., 1962), perfumes (Jirowetz et al., 1991), and burning tobacco odor (Lambers and Clark, 1996; Svensson, 1997; Karmowski et al., 1998; Mennella and Beauchamp, 1998a) also readily pass into the amniotic and mammary compartments. Finally, overlapping odor-active compounds deriving from normal metabolism were found in amniotic fluid, colostrum, and milk (e.g., androstenone: Doucet et al., 2010; fatty acids: Contreras et al., 2013; Hartmann et al., 2012).

A second mechanism of *perinatal chemosensory matching* is based on behavior. Females and their offspring deposit prenatal fluids on each other's bodies and, as already noted, these fluids are attractive to newborns of many mammals (i.e., rat: Teicher and Blass, 1977; Hepper, 1987; mouse: Al Aïn et al., 2013b; Logan et al., 2012; rabbit: Coureaud et al., 2002; sheep: Schaal et al., 1995b; and pig: Parfet and Gonyou, 1991), including human infants (Schaal et al., 1995a, 1998; Soussignan et al., 1997; Varendi et al., 1996). Finally, a process of *motivational matching* between amniotic fluid and colostrum may operate in two ways. First, while nursing, females afford redundant rewards that boost the offspring's learning of any odor cue present on the nipple. Thus, for the newborn, the mere exercise of searching, rooting, and sucking in the presence of amniotic and colostrum odors assign them similar reinforcing value. Second, common bioactive compounds may similarly impinge on the brain of the fetus and of the neonate. Both fluids convey compounds, for example, opioids, that may promote learning of any contingent stimulus (Schaal, 2005; Robinson and Méndez-Gallardo, 2010).

14.3.3.4 Olfaction and Energy Conservation.

After efficient suckling, it is essential that newborns conserve their energy. Early causes of energy loss are multiple, including separation-induced drops in body temperature,

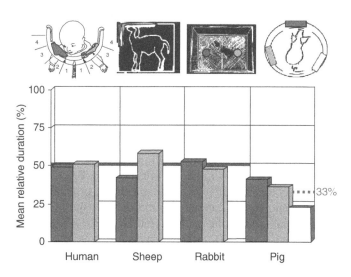

Figure 14.6 Mean relative duration of nose orientation of various neonate mammals when exposed to paired-choice tests opposing the odors of amniotic fluid and of colostrum or milk. (a) Cephalic orientation of 2-day-old breast-fed infants exposed to their own amniotic odor and to their mother's colostrum (Marlier et al., 1998). (b) search response of newborn lambs aged 2 hours post-birth to their amniotic odor and the odor of their dam's colostrum (Schaal and Orgeur, unpublished data). (c) Orientation response of newborn rabbits to the odors of placenta from their litter and of colostrum from their doe (Coureaud et al., 2002). (d) Nosing response of pig neonates in a 3-choice test opposing the odors of their own birth fluids, of the sow's milk, and of a scentless control (Parfet and Gonyou, 1991).

320   Chapter 14   Prenatal and Postnatal Human Olfactory Development: Influences on Cognition and Behavior

restlessness, crying, and jeopardizing illnesses and pain. In human neonates, crying increases oxygen consumption by 10% (Blass, 1994). Energetic losses are attenuated or arrested by sensory features of the mother, including her warmth, movements, odors, and sounds (heartbeat and voice). As mentioned above, breast odor alone can prevent neonatal agitation (Schaal et al., 1980; Sullivan and Toubas, 1998). When distress and crying are caused by acute pain induced by heel lance or venipuncture, odor (or taste) cues from human milk reduce pain responses more than water or formula milk (e.g., Mellier et al., 1997; Örs et al., 1999; Upadhyay et al., 2004; Rattaz et al., 2005; Nishitani et al., 2009). Familiar odors can have similar effects (Field et al., 2008; Sadathosseini et al., 2011; but see Marom et al., 2011). Physiological responses (salivary cortisol level) to painful procedures can be attenuated by non-biological odors such as lavender or γ-dodecalactone thought to mimic milk odor (Field et al., 2008; Kawakami et al., 1997). In addition to its analgesic action mediated through opioidergic processes (Blass and Miller, 2001), milk also conveys bioactive compounds known to induce quiet states and sleep (Sànchez et al., 2009; Graf and Kastin, 1986). However, the efficacy of odors to induce and sustain sleep has been little investigated. Preliminary assessments on whether a t-shirt worn by the mother could prevent night awakening in 3, 6, 9, and 12 month-olds obtained mixed results due to poor control of maternal odor (i.e., one monthly sampling) and of infants' sleep parameters (Goodlin-Jones et al., 1997; Burnham et al., 2004).

14.3.4 Olfaction in the Development of Social Cognition

Odor-based adaptive behavior persists well beyond the context of nursing and interactions with the mother. In fact, it participates in virtually all aspects of infant mammals' everyday lives, and the behavior of many mammalian infants cannot be fully understood without an in-depth understanding of olfaction (Alberts, 1981; Rosenblatt, 1983). Although most studies have employed altricial newborns whose perceptions are dominated by chemosensation, olfaction is also important in explaining the responses of precocial newborns of other species that also rely on vision and hearing (Olson, 1981).

14.3.4.1 Social Diversification. In most mammals, the nursing relationship favors more or less exclusive bonds between infants and their mothers. Postparturient females typically restrict their caregiving to their own offspring and reject unrelated young. Infants as well tend to respond preferentially to their own mothers and siblings (Hepper, 1991a). Such selective interactions

indicate that animals identify individuals or classes of conspecifics. However, when this occurs differs among species, being influenced by such factors as sensori-motor development at birth and the nature of the mother-infant bond, including the length of maternal dependence. For example, precocial neonates born in groups composed of several females (ungulates, pinnipeds) usually begin responding selectively to their own mother within several days after birth (Vince, 1993; Nowak, 1989; Charrier et al., 2001). In contrast, altricial neonates that remain confined in a nest may not develop individual recognition until weaning, the time when they become mobile and constrained to social selectivity (most rodents; Holmes and Mateo, 2007). The emergence of such discriminative social relationships in altricial neonates is generally based on the senses that develop earliest, that is, somesthesis and chemoreception.

FROM SOCIAL DISCRIMINATION TO SOCIAL RECOGNITION Discrimination of the mother is rooted in the newborns' first experience with her as a sensory mosaic, composed of somesthesic, olfactory, auditory and visual cues, and their temporal arrangement. As noted above, odor cues carried in amniotic fluid, milk, and mammary secretions can convey several levels of meanings such as species, class of conspecifics, and specific individuals. An example of the first level is the rabbit pups' active nuzzling directed towards the belly or milk odor of any rabbit female, but not to such odors from rats or cats (Hudson, 1985; Schaal et al., 2003). Likewise, human infants orient more insistently to the odor of the breast/milk from an unfamiliar mother rather than to cow's milk or cow-based formula milk odors (Russell, 1976; Porter et al., 1991; Marlier and Schaal, 2005; Doucet et al., 2009). A second level, as alluded to earlier in this chapter, subdivides conspecifics into classes of individuals. Thus, mouse and rabbit newborns olfactorily differentiate males from females, and non-lactating from lactating females (Al Aïn et al., 2013a; Coureaud and Schaal, 2000). Similarly, human infants discriminate body odors from unfamiliar lactating women as opposed to those of non-parturient women (Porter et al., 1991), indicating that lactation defines a class of especially attractive conspecifics. With regard to the third level, newborns appear to react discriminatively to odor cues that appear idiosyncratic to the mother. For example, while mammalian infants orient to any sample of amniotic fluid from their own species (Teicher and Blass, 1977, Schaal et al., 1995a, 1998), they selectively orient to their own amniotic fluid compared to the amniotic fluid from an unfamiliar conspecific neonate (rat: Hepper, 1987; rabbit: Coureaud et al., 2002; sheep: Vince, 1993; Schaal et al., 1995c; human: Schaal et al., 1998). Similarly, breast-fed infants aged 4 days display longer head-turning to their mother's milk odor when paired with milk odor from

14.3 Early Behavioral Functions of Olfaction

another woman in the same lactation stage (Marlier and Schaal, 1997) and 4-day-old infants mouth more to their mother's milk odor than to another mother's odor when they were in contact with the mother right after birth (Mizuno et al., 2004).

Studies using cotton pads containing the *entire secretion* from the nipple-areola region provide convergent results. By timing the head-turning of 2, 6, or 8–10-day-old infants in tests pairing the mother's breast odor with that of another mother's breast, Macfarlane (1975) showed the gradual differentiation of mother's odor: 2-day-olds responded randomly, 6-day-olds turned more to their mother's breast odor, and 8–10-day-olds even more so. A subsequent experiment found that 3-day-old breast-fed infants can already discriminate their mother's breast odor, reducing their movements when oriented to their mother's breast pad relative to a pad that had been worn by an unfamiliar lactating mother or to a clean control pad (Schaal et al., 1980). Russell (1976) likewise observed differences in the mouthing responses of 6-week-olds when their nursing mother's breast pad was held under their nostrils, rather than a strange mother's pad. Even younger newborns, at about 24 hours of age, increased their rate of mouthing when exposed to the scent of their mother's hospital gown (especially the part that had been in contact with the mammary-axillary region) as compared with another mother's gown (Sullivan and Toubas, 1998). Finally, olfactory cues that enable infants to recognize their mother can be delivered redundantly by several other sources than the breast or milk, including those of the neck (Schaal et al., 1980) and axillae (Cernoch and Porter, 1985).

The above results indicate that volatile compounds from body secretions convey various levels of cues that newborns can use in their selective decisions, including identification of species, classes of conspecifics, and individual conspecifics. However, much remains to be learned about the nature and development of such processes. For example, it is not clear whether or to what extent a newborn orients to its mother's mammary scent because it is part of a recurring experience learned as safe and rewarding, because it carries attractive cues that are common to any lactating female, or because it evokes the mother as a unique individual. Indeed, social discrimination "is not an inevitable result of social recognition" (Mateo, 2004: 730). Contingency-based processes of recognition, that is, that an odor stimulus reliably recurs in a given context, may be a prelude to the elaboration of familiarity-based processes of recognition which precede the recognition of individuals as truly unique, which in human infants probably does not occur before several months of age (Olson, 1981). The need for several days of experience to establish maternal breast odor as attractive (Macfarlane, 1975; Schaal et al., 1980) and the failure of human

neonates to discriminate axillary odors from individuals with whom they have had little direct contact (i.e., fathers; Cernoch and Porter, 1985) underscore the prominence of repetitive exposure, familiarization, and reinforcement in the establishment of early social olfactory recognition. In general, concluding that human newborns can olfactorily recognize individuals seems equivocal as the experiments conducted so far involve odor stimuli stemming from individuals differing in familiarity and/or in genetic relatedness (Schaal, 1988). Thus, the critical experiment for *natural odor*-based recognition of individuals as unique (i.e. opposing individuals that are genetically and experientially equivalent for the tested subject) waits to be carried out in neonates.

The involvement of olfaction in early social discrimination is also attested by evidence that artificial odorants presented in the context of breastfeeding rapidly gain the ability to release approach and appetitive behaviors. Thus, painting an initially neutral odorant on the mother's breast in the first days of life converts it into an attractive one (Schleidt and Genzel, 1990; Delaunay-El Allam et al., 2010). In only three days of such association with nursing can chamomile odor become as (but not more) attractive as the natural odor of mother's milk (Delaunay-El Allam et al., 2006), and only a 5-minute exposure conjointly with tactile stimulation suffices to change a neutral odorant into an attractive one within the first day of life (Sullivan et al., 1991). If one assumes that the processes converting an artificial odorant into a meaningful cue are similar to the natural processes that assign significance to biologically-produced odorants, it is a suitable paradigm to use controllable artificial odorants in social contexts with human infants and children. Such experimental odor "probes" have greatly contributed to highlight how and when odor cues are behaviorally important in nonhuman infants (e.g., Alberts, 1976; Smotherman et al., 1987; Rosenblatt, 1983; see Section 14.4).

While individual recognition based on *natural odor* cues alone remains to be ascertained in human newborns, olfaction may participate in social learning via influences on other sensory modalities. For example, odors can influence the development of visual activity. When exposed to their mother's breast odor, 3-day-old infants (especially boys) display longer periods of eye opening as compared to a similar situation where the nipple-areolar odor is masked (Doucet et al., 2007). Such early intersensory influences reflect that, from the very onset of aerial life, olfaction can be engaged into wider perceptual activity. Thus, breast odor can be instrumental in an ordered searching sequence mobilizing touch and vision to boost nipple grasping; alternatively, breast odor may become rapidly linked with expectancies for postural, tactile, visual, and auditory reinforcements, hence instigating the search for corresponding cues (cf. Schaal and Durand, 2012).

322   Chapter 14   Prenatal and Postnatal Human Olfactory Development: Influences on Cognition and Behavior

INDIVIDUAL RECOGNITION? The importance of odors in early social cognition has been assessed in several ways. One, described above, is to demonstrate discriminative abilities among different social agents. Another is to measure the effects of changing previously encoded odor features of a given conspecific. In altricial rat pups, puppies, or kittens, perturbing maternal odor can induce distress as if the mother was not present or recognized (Rosenblatt, 1983; Blass, 1990). Similarly, applying strong artificial odorants on their mother's nipple releases aversion and crying in human neonates (Kroner, 1882; Preyer, 1884); masking the natural breast odor reduces their willingness to orally grasp the nipple (Doucet et al., 2007). But beyond the breast, altering the whole mother's body odor has been used to show how critical odor cues are shaping the global "representation" of the mother. When squirrel monkey mothers (genus *Saimiri*) were sprayed with an artificial odorant, their 1–5-month-old infants did not show the typical visual preference for their mother relative to a control female (Redican and Kaplan, 1978). A previous experiment had shown that such infants rely more on olfactory than visual properties of the (anaesthetized) mother (Kaplan et al., 1977). Thus, sensorily precocial *Saimiri* infants concurrently monitor the olfactory and visual features of conspecifics (or their substitutes).

Human infants also monitor odor cues in their multisensory environment (Schaal and Durand, 2012). For example, 4-month-old infants rapidly relate to an object with a distinctive odor (Fernandez and Bahrick, 1994). Such abilities may shape the recognition of social partners. A recent study in 4-month-olds showed that the mother's body odor influenced the infant's visual exploration of faces: in presence of their mother's odor, infants' preference for faces is enhanced and the eye region is fixated longer than other facial regions (Durand et al., 2013). Thus, a faint odor can make a difference in the way infants look at a face, presumably easing the encoding of individual-specific facial cues.

At later ages, when children can reliably point out or name their choices, evidence that body odors can be recognized as belonging to individuals should become clearer, although current data remain inconsistent. For example, using a two-choice procedure between t-shirts worn by their mother or an unfamiliar woman, 45 to 58-month-old children were each given 15 to 37 trials (Schaal et al., 1980). Among the 26 children tested, 18 chose the mother's t-shirt in >60% of the trials, but statistically significant differences were observed in only eight cases. It is not known whether such variability in toddlers' recognition of mothers' body odor pertains to maternal factors (quality or intensity of body odor), child factors (odor sensitivity, attention) or both (maternal proximity, quality of attachment). However, 3–8-year-olds of either gender are able to distinguish the t-shirts of their own

full siblings from those of unrelated age mates (Porter and Moore, 1981), indicating that they can determine familiarity and/or relatedness using odor cues produced by adults or prepubertal age mates. In a novel study, Weisfeld et al. (2003) tested the ability of 4–11-year-old children in remarried families to recognize the odor of siblings differing in coefficient of consanguinity [i.e. full (0.5), half- (0.25), and stepsiblings (0)]. The children were able to identify the odors of their full siblings, but not of their half- and stepsiblings, a result interpreted as reflecting a strong genetic influence on the production and perception of body odors. However, this differential response could reflect the generally less positive affective ties between half siblings and step siblings than between full siblings (Weisfeld et al., 2003), suggesting that the level of relatedness could be confounded by the frequency of exposure to body odors and, hence, with familiarity. A further test in that same study assessed whether 6–15-year-old children from stable families comprising at least two children could olfactorily recognize a t-shirt worn by their mother, father, or sibling(s) (relative to a t-shirt worn by an unrelated participant matched in gender and age). On average, (a) children chose their mother's odor at random [but in splitting the group in younger (6–8 years) and older (>9 years) participants, the older expressed correct mother recognition, without gender difference], (b) the father's odor was unambiguously recognized by daughters and sons alike, and (c) children identified their opposite-sex sibling, but not their same-sex sibling. To add complexity to the issue, a recent study investigating odor recognition between all members of stable 2-children families (opposing t-shirts of a person from the target family with t-shirts from matched person from an unfamiliar family) found that the children were inaccurate in identifying their mother, father, or siblings (Ferdenzi et al., 2010). Thus, evidence for true individual recognition does not appear fully conclusive. In two-choice paradigms, it is difficult to establish if an odor characterizes an individual rather than higher level meanings (e.g., familiarity, kin, gender, age, shared environment, etc.) (Weisfeld et al., 2003).

Another source of data germane to odor-based social recognition comes from studies of children who are genetically unrelated and who share only limited environmental circumstances (as in temporarily stable groups that form in schools or youth camps). For example, Verron and Gaultier (1976) asked 4–5-year-old blindfolded children to identify three to five classmates by smelling their neck. While boys were accurate in their recognition responses in 33% of the tests (21/62), girls succeeded in 69% of them (36/52). A similar study on same-age preschoolers showed that 8 out of 18 participants were able to name classmates whose neck they sniffed (Marlier and Schaal, 1989). Girls were more reliable in their odor-based recognition than boys, but in addition the girls *induced* better recognition than boys

14.3 Early Behavioral Functions of Olfaction

from the odor cues of their necks. These studies highlight the fact that children's recognition performance of familiar peers is interindividually variable and doubly biased by the gender of the smeller and by the gender of the donor. In an additional study, 9-year-old children from a same school class were required to recognize the t-shirts of six target odor donors (self and five types of classmates delineated after ranking of socio-affective closeness within the class: same-sex and opposite-sex preferred peers, same-sex least liked peers, and mere acquaintances of same and opposite sex) (Mallet and Schaal, 1998). The tested personal odors reflected the habitual mix of natural and extraneous scents that impregnated t-shirts worn for three nights and one day, and to which classmates were normally exposed while interacting. Each of the six target odor donors was to be recognized among four distractor stimuli. On average, children were unable to accurately categorize their peers' body odors by gender, but the odor donors were individually recognized at a better-than-chance level. Same-sex peers were better identified than opposite-sex peers, with girls scoring higher for same-sex preferred and opposite-sex acquainted peers, and boys scoring lower for opposite-sex acquaintances. This study was confounded by the participants' cultural ecology, as their usual hygienic practices were not altered during the impregnation of the t-shirts. Personal odors are indeed the emergent result from mingling endogenous and extraneous compounds which create "culturalized" odor signatures that potentiate the communicative value of the natural odor signature (Milinski and Wedekind, 2001; Lenochova et al., 2012). Both natural and artificial "fractions" of olfactory signatures are accurately recognizable to children, as shown by the recognition of a t-shirt of an unrelated friend presented among four distractor t-shirts by adolescents (Olsson et al., 2006), gender assignation dependent on the "perfumedness" of body odor (Mallet and Schaal, 1998), and the ability of young children to identify their mother's perfume (Durand and Schaal, 2011).

SOCIAL PREFERENCES During development, odor-based preferences fluctuate until a time when they come to more or less overlap with those of adults of their community. This socialization of the hedonics of body odors may be achieved through various experiential processes involving (a) recurrent exposure to the natural odor signatures of conspecifics, as well as to added artificial odorants (perfumes), (b) passive reception of hygienic practices and learning to conform with local norms, (c) acquisition of contamination sensitivity (Rozin et al., 1986) and education of disgust for uncleanliness (Peto, 1936; Stein et al., 1958; Oaten et al., 2009), and (d) neophobia, which is best illustrated in the increased wariness for novel foodstuffs between 2 and 4 years (Cashdan, 1994), but could as well concern social odors (e.g., Stevenson and

Repacholi, 2005). Also, it is a general phenomenon that infants and children unknowingly perceive odors during emotionally-arousing events and label them as negative or positive. Such odors can subsequently re-evoke the affective states and emotional reactions in which they were acquired (Epple and Herz, 1999; Chu, 2008) and can subtly differ in relation to events concerning significant others. For example, in families where parents consume alcohol to cope with dysphoria, 5–10-year-old children dislike alcohol-related odors more than children whose parents usually drink for fun (Mennella and Garcia, 2000; Mennella and Forestell, 2008). Similarly, children whose mothers smoke to alleviate stress dislike the odor of tobacco smoke more than children of mothers who report smoking for other reasons (Forestell and Mennella, 2005). In sum, children can rely on odors not only to recognize individuals, but to differentially label the emotional context in which the individual odors are perceived.

Clearly, the developmental niche frames a child's perceptual abilities as well as normative expectations about self and others' olfactory appearance. When such expectations about personal odors are violated, faulty odor sources or odor-emitting individuals become rapidly consensual as targets of exclusion. For example, in a single case follow-up of a 13-year-old boy, violent stigmatization was reported to be induced by "bad body odors" (Todd, 1979); subsequently, this child was diagnosed defective in trimethylamine metabolism and treated with a choline-poor diet which alleviated the olfactory signature and resulted in satisfactory social integration a year later.

The most obvious way to develop selective responses to body odors would be through positive imprinting (as opposed to "negative imprinting" processes outlined below). That is, adolescents and young adults would be more attracted to odor profiles associated with the mother (caregiver) in early development or to individuals emitting elements of the parental odor phenotype. Numerous studies in nonhuman mammals have shown that infantile learning of odors in association with the mother can have long-term consequences in that they can positively influence mate choice and sexual performance. Although some studies indicate assortative mate preference and selection based on visual appearance of facial traits in humans (reviewed in Rantala and Marcinkowska, 2010), to our knowledge no such studies have been undertaken in the domain of olfaction (see below Section 14.4).

Such a framing of social odor preferences by the practices and values of local cultures was also suggested to be underlain by several kinds of more general mechanisms. First, parent-offspring interactions were conjectured to be influenced between two and five years by the oedipal process, which, according to Freud (1905-1962), is actualized by children's attraction to the opposite-sex parent and relative avoidance from the same-sex parent (Buss,

324   Chapter 14   Prenatal and Postnatal Human Olfactory Development: Influences on Cognition and Behavior

2008). Based on clinical material involving olfaction, Bieber (1959; Bieber et al., 1992) predicted an odor-based mechanism of avoidance and attraction in the family triad, between young children and their same-sex and opposite-sex parents, respectively. This point was followed up in a single case who displayed contrasting responses to each parent (Kalogerakis, 1972). Between 2.5 and 4 years of age, this boy reacted acutely to the body odors of both parents, with increasing differentiation in attraction to the mother and rejection of the father; when the mother induced occasional rejection by the child, it could be traced to the fact that her body or garments occasionally conveyed paternal odor cues. Although such casual observations have the rare merit of indicating that children as young as 2.5 years can detect and react to deviations from baseline body odor in adults, they all stem from Freudian proponents. When similar work was repeated in keeping on alternative theoretical options, it was concluded that the oedipal family situation has no empirical support (e.g., Goldman and Goldman, 1982; Daly and Wilson, 1990). A more parsimonious interpretation would be that children react more negatively to male body odors, just as adult raters who report male axillary or oral odors as clearly more intense and unpleasant than those of females (Doty, 1981; Schleidt and Hold, 1982).

Evolved odor-based processes have been proposed in a second domain of parent-offspring and between-sibling relations: the circumvention of incest. Mechanisms to prevent sexual interaction between offspring and parents operate in nonhuman primates, despite enduring affinitive bonds (Silverman and Bevc, 2005). In our own species, the primary way of incest avoidance appears to be "behavioral avoidance toward those with whom the individual has shared continuous early association" (Silverman and Bevc, 2005: 300), a mechanism of sexual uninterest triggered by physical proximity during the first five years of life and the so-called "Westermarck effect" (Westermarck, 1891). It has been repeatedly suggested that some olfactory mechanism might produce aversions during an early sensitive period. Some have predicted that such aversions, directed towards individuals rather their scent alone, will arise when adult-like sexual interests emerge and will be more potent among females (Schneider and Hendrix, 2000). Laboratory tests have sought to assess whether odor cues could mediate aversion for the opposite-sex parent and siblings in human children. Weisfeld et al. (2003) assessed preferences of each of 6–15-year-old children for the odors of t-shirts worn by their mother, father, and siblings. Each was paired with reference t-shirts from unrelated matched individuals. Overall, children expressed less preference for parents' odor than for the reference odors (with mutual aversion between father and daughter) and sisters tended to avoid their brother's odor. Mothers also expressed less preference for their children's odor than for the reference

children's odor. Interestingly, these responses raise the issue of whether or not the odor donor was recognized. Ferdenzi et al. (2010), adopting a procedure where the father's and mother's body odors were directly compared, showed that 7–18-year-old girls judged their father's odor (without explicit recognition of him) as more intense, more masculine, and more unpleasant, than did boys. In contrast, no differential hedonic rating occurred for mothers and siblings. In addition, the younger, prepubertal children of either gender (aged 7–10 years) rated their father's and sibling's odor as less pleasant than did the pubertal and post-pubertal children. Thus, an indication of an odor-mediated aversion towards the opposite-sex parent is suggested in girls, corroborating one of Schneider and Hendrix's predictions and Weisfeld et al.'s results above, but no clear pattern emerged in boys toward mothers' odor.

These initial assessments of potential odor processes underlying the Westermarck effect need further documentation both in humans and nonhuman primates. The processes involved may be secondary to puberty-related perceptual changes co-occurring along with changes in chemo-emission. During the pubertal transition, children's negative reactivity to androstenes in axillary sweat does indeed increase sharply, especially in girls (Dorries et al., 1989). At the same time, adolescents and adults express stronger aversion to the odor of t-shirts worn by unfamiliar young adults than do prepubertal (8-year-old) children (Stevenson and Repacholi, 2003). Otherwise, the negative valuation of male axillary odor by females fluctuates with the menstrual cycle, so that during the fertile phase they express less rejection (Moshkin et al., 2011). Thus, before and during adolescence, body odors from adult persons can come to evoke intense dislike, but intra-individual perceptual variations occurring after puberty may temporarily attenuate such repulsion. Clearly, the issue of olfactory influences on affectionate behavior and attitudes within families and among kin awaits more investigation.

14.4 EARLY ODOR ACQUISITIONS PERSIST LATER IN THE CHILD AND ADULT

Infantile olfactory memory has been examined in relation to (a) social and sexual preferences, (b) food preferences, and (c) fearful situations or stimuli. In dogs, for example, puppies learn the individual odor of the bitch, and after two years of total separation remain capable of recognizing it (Hepper, 1994). In Asian elephants, male calves that are separated from their mother for husbandry reasons can retain chemosensory information carried in maternal urine for 2 to 27 years (Rasmussen, 1995). Thus, early odor learning favors long-term recognition of individuals and kin.

Such an odor-based imprinting-like process has been analyzed in more detail in rodents. A pioneering

14.4 Early Odor Acquisitions Persist Later in the Child and Adult

experiment consisted in treating rat dams and pups with cologne during the first postnatal month. In adult tests, the cologne-exposed rats were less sexually competent than the controls when exposed with a normal-odor partner (Marr and Gardner, 1965). A similar result was reached by Mainardi et al. (1965) who tested female mice whose nursing mothers were scented with violet odor. When sexually mature, these young females preferred males carrying this odor, while females nursed by olfactorily-intact dams preferred normal-odor males (an effect not found in male mice). However, in rats, experiencing a given artificial odor in association with suckling modified the adult males' reactivity towards a receptive female carrying the same odor (in reducing ejaculation latency; Fillion and Blass, 1986). This effect was specific to the odor-nursing association, as an odor applied to the back of the nursing female did not alter adult sexual behavior. Further, when rodent pups were fostered to females of a different species, they showed more attraction to the odors of the foster species than do control animals, and such preferences continued into adulthood (Huck and Banks, 1980; Quadagno and Banks, 1970). Nursing-related odor learning also affects the behavior of females. For example, rat pups reared by a scented mother later, as adults, displayed increased body contact with, and licking of, pups carrying the same odor (although their latency to express maternal care was longer with the scented pups as compared with the control pups)(Shah et al., 2002). This phenomenon is not restricted to rodent species, as it occurs in ungulates as well (Müller-Schwarze and Müller-Schwarze 1971).

Clearly, early odor experience mediates adult recognition of individuality, familiarity, or kinship in various mammals, but the source and nature of the involved odorants and how information about kinship is olfactorily encoded are poorly understood. In rodents, relatedness assessment between individuals may depend on odor cues that are correlated with the major histocompatibility complex (MHC). Related individuals share similar MHC and similar odor properties, so that MHC-derived odor cues could function as an index of kinship. Mice and rats do indeed discriminate by odors among conspecifics whose genotypes only differ within MHC (Brown et al., 1987). Male mice of at least some inbred strains prefer females whose MHC differs from their own (Yamazaki et al., 1976). Such preferences for non-self MHC appear to result from early familial "imprinting." Males that had been raised until weaning by foster parents with a different MHC mated preferentially with females that shared their own MHC rather than that of their parents (Yamazaki et al., 1988, 2000; Penn and Potts, 1998). These males therefore avoided mating with females whose olfactory phenotype resembled the odor of the foster parent. In subsequent experiments with other inbred strains of mice, the effect of the rearing environment on MHC-based preferences was

greater for females than for males (Arcaro and Eklund, 1999). These results were generalized to semi-natural conditions in which early familial odor imprinting also appeared to account for the avoidance of males with wild-type females carrying the foster parent MHC genotype (Penn and Potts, 1998). Thus, in these rodents, an early olfactory imprint with MHC-based odorants can have long-term effects on choice of potential mates, either by avoiding inbreeding with mates sharing familial MHC chemosignals or by selecting mates bearing slightly different MHC genotypes.

Whether similar natural imprinting mechanisms operate in other species, including primates and humans, has begun to be addressed (Ruff et al., 2012). Humans also appear to be olfactorily sensitive to HLA[4]-correlated odor cues (Wedekind and Füri, 1997; Wedekind et al., 1995; reviewed in Havlicek and Roberts, 2009), but virtually nothing is known about the developmental course of this perceptual ability. In one experiment, Jacob et al. (2002) asked young women to rate the t-shirts of six male donors. The participating smellers, the donors and the smellers' parents were previously HLA-typed and characterized for the number of allelic matches. The participants expressed a preference for the odor of men bearing an intermediate level of dissimilarity (not those who had no or all alleles in common) and they preferred men who shared HLA alleles with their own fathers (but not with their mothers). A logical deduction would be, if fathers' HLA-type correlates with body odor cues, that adult females are more attracted to a male odor when it conveys some cues in common with their own father. But this possibility, which would run against the Westermarck effect outlined above, needs to be investigated properly before any conclusion can be made.

In both humans and nonhuman mammals long-term effects of early odor experience can further alter ingestive behaviors. Early breast or bottle feeding can affect subsequent odor or flavor preferences. Infants who have been nursed at a breast scented with chamomile during the first postnatal weeks are more attracted by a chamomile scented toy at 7 months of age, and, when re-tested 21 months later, choose to suck from a chamomile-scented bottle over a violet-scented bottle (Delaunay-El Allam et al., 2010; cf. Figure 14.7). Another element in favor of persistent effects of early experience in milk comes from a study showing that 6–13-month-old infants reacted differently to toys scented with ethanol or vanilla as compared to an unscented toy (Mennella and Beauchamp, 1998b). These authors hypothesized that the infants may have become familiar with these odors or flavors in the home, conceivably through mother's milk or breath odors. Such exposure effects appear heterogeneous during development, suggesting that early infancy constitutes a particularly sensitive

---

[4]The human MHC is named Human Leucocyte Antigen (HLA).

326 Chapter 14 Prenatal and Postnatal Human Olfactory Development: Influences on Cognition and Behavior

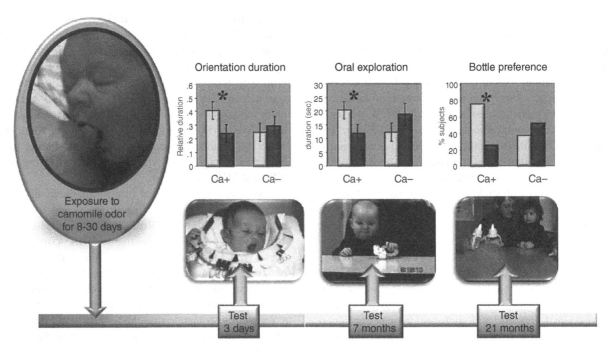

Figure 14.7 Odor experience during breastfeeding and subsequent preferences for the exposed odorant in different contexts. Infants were first exposed to chamomile odor applied directly on the areola and nipple for 8 to 30 days after delivery. They were then tested for differential responses to the chamomile odor when they were aged 3 days, and 7 and 21 months. The chamomile-exposed participants (Ca+) were compared to same-age participants who were never exposed to chamomile (Ca−). The **3-day test** consisted in a double-choice test opposing chamomile odor and a scented control stimulus; it showed that infants of the Ca+ group, but not of the Ca- group, were attracted to the chamomile odor (Delaunay-El Allam et al., 2006). The **7-month test** consisted in presenting sequentially, in counterbalanced order, same objects odorized with chamomile or violet. The Ca+ group displayed longer mouthing of the object scented chamomile than of the object scented violet, while the Ca- infants did not express a differential response (Delaunay-El Allam et al., 2010). The **21-month test** consisted in presenting the infants two bottles scented chamomile or violet, in counterbalanced order, and asking them to choose the one they would like to suck. While the Ca+ group prevalently chose the chamomile-scented over the violet-scented bottle, the Ca- group did not respond differentially (Delaunay-El Allam et al., 2010). (*p<0.05, **p<0.01).

period for chemosensory learning in humans (Beauchamp and Mennella, 1998; Mennella and Castor, 2012).

Odors acquired *in utero* can be more, or at least equally, preferred postpartum than artificial odors subsequently associated with nursing. This effect has been demonstrated in the rat during the preweaning period (e.g., Hepper, 1988; Chotro and Molina, 1990) and even as adults (Smotherman, 1982). In the rabbit, a prenatally acquired odor can release as many appetitive and consummatory responses at weaning as odors present during suckling or first solid food ingestion (e.g., Bilko et al., 1994; Hudson et al., 1999). In humans, infants of mothers who consumed carrot flavor during the last weeks of gestation more easily accept carrot-flavored cereals five months later (Mennella et al., 2001). Further, 8–9-year-old offspring of mothers who ingested garlic during pregnancy ate more of a garlic-flavored meal than did children born to mothers who did not do so (Hepper et al., 2012b).

Odor aversions or preferences encoded during infancy and childhood can last until adulthood. For example, childhood food aversions can persist for over 50 years (Garb and Stunkart, 1974). Among respondents aged 13–20, 12% traced their aversion to the 0–5 years' period, 58% to the 6–12 years' period, and 30% to the 13–20 years' period (see also Batsell et al., 2002). One study took advantage of a legal change in the flavoring of infant milk formula, which was prohibited in 1992 because of rising food allergies. Infants who were either breast- or bottle-fed before and after 1992 were considered massively or minimally exposed to vanilla, respectively. A determination was made of the relative preferences of subjects (mean age of 28, range 12–59 years) for a standard ketchup and a ketchup with slight vanilla flavor (Haller et al., 1999). While the early breast-fed participants preferred the normal ketchup to the vanilla-flavored ketchup (70.9 vs. 29.1%), those who were exposed to vanilla through formula-feeding responded the other way round (40 vs. 60%). Although this approach did not account for exposure to vanilla since weaning, it does underscore the fact that infantile memory for odors and flavors remains accessible and is sufficient to orient subsequent ingestive decisions in both juveniles and adults. A more recent study compared perceptual and brain responses to the odors of mint and rose in adults who grew up either with parents who emigrated from Algeria to France (Algerian-French, AF) or with parents from French, German, British, or Spanish origins (European-French, EF) (Poncelet et al., 2010). On average, the AF group was exposed earlier and more often to mint odor/flavor in childhood than the EF group, expressed stronger preference for mint odor over rose odor (while the EF group showed no preference), and recollected more vivid childhood events after inhaling mint. These differences were linked with chemosensory evoked response potentials indicating differences in the brain processing speed presumably due to disparities in attention-evoking and emotional memory retrieval as a function early experience.

REFERENCES

Al Aïn, S., Chraïti, A., Schaal, B., and Patris, B. (2013a). Orientation of newborn mice to lactating females: biological substrates of semiochemical interest. *Dev. Psychobiol.* 55: 113–124

Al Aïn, S., Belin, L., Schaal, B., and Patris, B. (2013b). How does a newly born mouse get to the nipple? *Odor substrates eliciting first nipple grasping and sucking responses. Dev. Psychobiol.* 55: 888–901

Alberts, J. R. (1976). Olfactory contributions to behavioral development in rodents. In R. L. Doty (Ed.), *Mammalian Olfaction, Reproductive Processes and Behavior*, pp. 67–94. Academic Press: New York.

Alberts, J. R. (1981). Ontogeny of olfaction: Reciprocal roles of sensation and behavior in the development of perception. In R. N. Aslin, J. R. Albert, and M. R. Petersen (eds), *Development of Perception: Psychological Perspectives*, Volume 1, pp. 319–357. New York: Academic Press.

Alberts, J. R. (1987). Early learning and ontogenetic adaptation. In N. A. Krasnegor, E. M. Blass, M. A. Hofer, and W. P. Smotherman (Eds.). *Perinatal Development. A Psychobiological Perspective*, pp. 11–37. Orlando: Academic Press.

Alberts, J. R., and Ronca, A. E. (2012). The experience of being born: A natural context for learning to suckle. *Int. J. Pediatr. doi:* 10.1155/2012/129328.

Aoyama, S., Toshima, T., Saito, Y. et al. (2010). Maternal breast milk odour induces frontal lobe activation in neonates. *A NIRS study. Early Hum. Dev.* 86: 541–545.

Apfelbach, R., Russ, D., and Slotnick, B. (1991). Ontogenetic changes in odor sensitivity, olfactory receptor area and olfactory receptor density in the rat. *Chem. Senses* 16: 209–218.

Arcaro, K. F. and Eklund, A. (1999). A review of MHC-based mate preferences and fostering experiments in two congenic strains of mice. *Genetica* 104: 241–244.

Armstrong, J. E., Hutchinson, I., Laing, D. G., and Jinks, A. L. (2007). Facial electromyography: responses of children to odors and taste stimuli. *Chem. Senses* 32: 611–621

Armstrong, J. E., Laing, D. G., Wilkes, F. J., and Laing D. G. (2008). Olfactory function in Australian Aboriginal children and chronic otitis media. *Chem Senses.* 33: 503–507.

Azoulay, R., Fallet-Bianco, C., Garel, C., et al. (2006). MRI of the olfactory bulbs and sulci in human fetuses. *Pediatr. Radiol.* 36: 97–107.

Balogh, R. D., and Porter, R. H. (1986). Olfactory preferences resulting from mere exposure in human neonates. *Infant Behav. Dev.* 9: 395–401.

Barnham, A. L., and Broughan, C; (2002). Sugar and spice and all things nice: the effects of odours on task performance and emotions in children. *Int. J. Aromather.* 12: 127–130.

Bartocci, M., Winberg, J., Ruggiero, C. et al. (2000). Activation of olfactory cortex in newborn infants after odor stimulation: a functional near-infrared spectroscopy study. *Pediatr. Res.* 48: 18–23.

Bartocci, M., Winberg, J., Papendieck, G., et al. (2001). Cerebral hemodynamic response to unpleasant odors in the preterm newborn measured by near-infrared spectroscopy. *Pediatr. Res.* 50: 324–330.

Batsell, W. R., Brown, A. S., Ansfield, M., and Paschall, G. (2002). "You will eat all of that!". *A retrospective analysis of forced consumption episodes. Appetite* 38: 211–219.

Beauchamp, G. K., Cowart, B. J., and Schmidt, H. J. (1991). Development of chemosensory sensitivity and reference. In T. V. Getchell, R. L. Doty, L. M. Bartoshuk, and J. B. Snow (Eds.), *Smell and Taste in Health and Disease*, pp. 405–416. Raven Press: New York.

Beauchamp, G. K. and Mennella, J. A. (1998). Sensitive periods in the development of human flavor perception and preference. *Ann. Nestlé*, 56: 19–31.

Becques, A., Larose, C., Gouat, P., and Serra, J. (2010). Effects of pre-and postnatal olfacto-gustatory experience on early preferences at birth and dietary selection at weaning in kittens. *Chem. Senses* 35: 41–45.

Bensafi, M., Rinck, F., Schaal, B., and Rouby, C. (2007). Verbal cues modulate hedonic perception of odors in 5-year-old children as well as in adults. *Chem. Senses* 32: 855–862.

Berk, L. E. (2007). *Development Through the Lifespan (4th Edition)*. Pearson, Allyn and Bacon: Boston MA.

Berk, L. E. (2008). *Infants, Children, and Adolescents (6th Edition)*. Pearson, Allyn and Bacon: Boston MA.

Bieber, I. B. (1959). Olfaction in sexual development and adult sexual organization. *Am. J. Psychother.* 13: 851–859.

Bieber, I., Bieber T. B., and Friedman, R. C. (1992). Olfaction and human sexuality: a psychoanalytic approach. In M. J. Serby and K. L. Chobor (Eds.), *Science of Olfaction*, pp. 396–409. Springer: New York.

Bilko, A., Altbäcker, V., and Hudson, R. (1994). Transmission of food preference in the rabbit: the means of information transfer. *Physiol. Behav.* 56: 907–912.

Bingham, P. M., Abassi, S. and Sivieri, E. (2003a). A pilot study of milk odor effects on nonnutritive sucking by premature infants. *Arch. Pediatr. Adol. Med.* 157: 72– 75.

Bingham, P. M., Sreven-Tuttle, D., Lavin, E., and Acree, T. (2003b). Odorants in breast milk. *Arch. Pediatr. Adolesc. Med.* 157: 1031.

Bingham, P., Churchill, D., and Ashikaga, T. (2007). Breast milk odor via olfactometer for tube-fed, premature infants. *Behav. Res. Methods* 39: 630–634.

Blass, E. M. (Ed.) (1986). *Handbook of Behavioral Neurobiology, Vol. 9, Developmental Psychobiology and Behavioral Ecology*. Plenum Press: New York.

Blass, E. M. (1990). Suckling: determinants, changes, mechanisms, and lasting impressions. *Dev. Psychol.* 26: 520–533.

Blass, E. M. (1994). Behavioral and physiological consequences of suckling in rat and human newborns. *Acta Paediatr., Suppl.* 397: 71–76.

Blass, E. M., and Miller, L. W. (2001). Effects of colostrum in newborn humans: dissociation between analgesic and cardiac effects. *Dev. Behav. Pediatr.* 22: 385–390.

Bloom, S. J. (1975). Olfaction in Children One to Five Years of Age. *Unpublished report*, Brown University: Providence, RI.

Boehm, N. and Gasser, B. (1993). Sensory-receptor like cells in the human fetal vomeronasal organ. *Neuroreport* 4: 867–870.

Booth, D. A., Higgs, S., Schneider, J., and Klinkenberg, I. (2010). Learned liking versus inborn delight. Can sweetness give sensual pleasure or is it just motivating? *Psychol. Sci.* 21: 1656–1663.

Bossy, J. (1980). Development of olfactory and related structures in staged human embryos. *Anat. Embryol.* 161: 225–236.

Brewer, W. J., Pantelis, C., De Luca, C., and Wood, S. J. (2006). Olfactory processing and brain maturation. In W. J. Brewer, Castle, D. and C. Pantelis (Eds.), *Olfaction and the Brain*, pp. 103–118. Cambridge University Press: Cambridge, UK.

Brown, J. W. (1974). Prenatal development of the human chief trigeminal nucleus. *J. Comp. Neurol.* 156: 307–336.

Brown, R. E., Singh, P. B., and Roser, B. (1987). The major histocompatibility complex and the chemosensory recognition of individuality in rats. *Physiol. Behav.* 40: 65–73.

Bunce, C., and Gibson, E. (2012). Sniffing, eating and disgust in food neophobic children. *Appetite* 59: 622.

Burnham, M., Goodlin-Jones, B. L., Gaylor, E., and Anders, T. F. (2004). Use of sleep aids during the first year of life. *Pediatrics* 109: 594–601.

Buss, D. M. (2008). *Evolutionary psychology*, 3rd Edition. Pearson: Boston, MA.

Büttner, A. (2007). A selective and sensitive approach to characterize odour-active and volatile constituents in small-scale human milk samples. *Flavour Fragr. J.* 22: 465–473.

Cain, W. S., and Johnson, F. (1978). Lability of odor pleasantness: influence of mere exposure. *Perception* 7: 459–465.

Cain, W. S., Stevens, J. C., Nickou, C., et al. (1995). Life-span development of odor identification, learning, and olfactory sensitivity. *Perception* 24: 1457–1472.

Cameron, E. L., and Doty, R. L. (2013). Odor identification testing in children and young adults using the smell wheel. *Int. J. Pediatr. Otolaryngol.* 77: 346–350.

Capretta, P. J. and Rawls, A. (1974). Establishment of a flavor preference in rats: Importance of nursing and weaning experience. *J. Comp. Physiol. Psychol.* 86: 670–673.

Cashdan, E. (1994). A sensitive period for learning about food. *Hum. Nature*, 5: 279–291.

Cernoch, J. M. and Porter, R. H. (1985). Recognition of maternal axillary odors by infants. *Child Dev.* 56: 1593–1598.

Chalouhi, C., Faulcon, P., Le Bihan, C., et al. (2005). Olfactory evaluation in children : Application to the CHARGE syndrome. *Pediatrics*, 116: e81–e88.

Charrier, I., Mathevon, N., and Jouventin, P. 2001. Mother's voice recognition by seal pups. *Nature* 412: 873.

Cheal, M. (1975). Social olfaction: a review of the ontogeny of olfactory influences on vertebrate behavior. *Behav. Biol.*, 15: 1–25.

Chopra, A., Baur, A., and Hummel, T. (2008). Thresholds and chemosensory event-related potentials to malodors before, during, and after puberty: differences related to sex and age. *NeuroImage* 40: 1257–1263.

Chotro G., and Molina, J. C. (1990). Acute ethanol contamination of the amniotic fluid during gestation day 21: Postnatal changes in alcohol responsiveness in rats. *Dev. Psychobiol.* 23: 535–547.

Chu, S. (2008). Olfactory conditioning of positive performance in humans. *Chem. Senses* 33: 65–71.

Chu, S. and Downes, J. J. (2000). Long live Proust: the odour-cued autobiographical memory bump. *Cognition* 75: 41–50.

Chu, S. and Downes, J. J. (2002). Proust nose better: odors are better cues of autobiographical memories. *Mem. Cognition* 30: 511–518.

Chuah, M. I. and Zheng, D. R. (1987). Olfactory marker protein is present in olfactory receptor cells of human fetuses. *Neurosci.* 23: 363–370.

Contreras, C. M., Guttiérrez-Garcia, A. G., Mendoza-Lopez, R., et al. (2013). Amniotic fluid elicits appetitive responses in human newborns: fatty acids and appetitive responses. *Dev. Psychobiol.* 55: 221–231.

Coureaud, G., and Schaal, B. (2000). Attraction response of newborn rabbits to body odours of adults differing in sex and physiological state. *Dev. Psychobiol.* 36: 271–281.

Coureaud G., Schaal B., Hudson R., et al. (2002). Transnatal olfactory continuity in the rabbit: behavioral evidence and short-term consequence of its disruption. *Dev. Psychobiol.* 40: 372–390.

Crook, C. (1987). Taste and olfaction. In Salapatek, P., and Cohen, L. (Eds.), *Handbook of Infant Perception, Volume 1, From Sensation to Perception*, pp. 237–264. Academic Press: New York.

Dalton, P., Mennella, J., Maute, C., et al. (2011). Development of a test to evaluate olfactory function in a pediatric population. *Laryngoscope.* 121: 1843–1850.

Daly, M. and Wilson, M. (1990). Is parental offspring conflict sex-linked? Freudian and Darwinian models. *J. Pers.* 58: 163–189.

Daremberg, C. (1856). *Oeuvres anatomiques, physiologiques et médicales de Galien, Vol, 2*. Paris.

Darwin, C. (1877). A biographical sketch of an infant. *Mind* 7: 285–294.

Davidson T. M., Freed C., Healy M. P., and Murphy C. (1998). Rapid clinical evaluation of anosmia in children: The alcohol sniff test. *Annal NY Acad Sci.* 855: 787–792.

Davis, L. B. and Porter, R. H. (1991). Persistent effects of early odour exposure on human neonates. *Chem. Senses* 16: 169–174.

de Wijk, R. A. and Cain, W. S. (1994). Odor identification by name and by edibility: life-span development and safety. *Hum. Factors* 36: 182–187.

Delaunay-El Allam, M., Marlier, L., and Schaal, B. (2006). Learning at the breast: Preference formation for an artificial scent and its attraction against to odor of maternal milk. *Infant Behav. Dev.* 29: 308–321.

Delaunay-El Allam, M., Soussignan, R., Patris, B., et al. (2010). Long lasting memory for an odor acquired at mother's breast. *Dev. Sci.* 13: 849–863.

Delplanque, S., Grandjean, D., Chrea, C., et al. (2008). Emotional processing of odors: evidence for nonlinear relation between pleasantness and familiarity evaluations. *Chem. Senses* 33: 469–479.

Desage, M., Schaal, B., Soubeyran, J., et al. (1996). Transfer of dietary odorous compounds into plasma and milk: GC-MS Methodology. *J. Chromatogr. B* 678: 205–210.

Dorries, K. M., Schmidt, H., Beauchamp, G. K., and Wysocki, C. J. (1989). Changes in sensitivity to the odor of androstenone during adolescence. *Dev. Psychobiol.* 22: 423–435.

Doty, R. L. (1981). Olfactory communication in humans. *Chem. Senses* 6: 351–376.

Doty, R. L. (1992). Ontogeny of Human Olfactory Function. In W. Breipohl (Ed.), *Ontogeny of Olfaction*, pp. 3–17. Springer: Berlin.

Doty, R. L., Shaman, P., Applebaum, S. L., et al. (1984). Smell identification ability: changes with age. *Science* 226: 1441–1443.

Doty, R. L., Newhouse, M. G., and Azzalina, J. D. (1985). Internal consistency and short-term test-retest reliability of the University of Pennsylvania Smell Identification Test. *Chem. Senses* 10: 297–300.

Doucet, S., Soussignan, R., Sagot, P., and Schaal, B. (2007). The "smellscape" of the human mother's breast: effects of odour masking and selective unmasking on neonatal arousal, oral and visual responses. *Dev. Psychobiol.* 49: 129–138.

Doucet, S., Soussignan, R., Sagot, P., and Schaal, B. (2009). The secretion of areolar (Montgomery's) glands from lactating women elicits selective, unconditional responses in neonates. *PLoS ONE* 4: e7579.

References

Doucet, S., Soussignan, R., Sagot, P., and Schaal, B. (2012). An overlooked aspect of the human breast: Areolar glands in relation with breastfeeding pattern, neonatal weight gain, and dynamics of lactation. *Early Hum. Dev.* 8: 119–128.

Doucet, S., Hartmann, C., Schaal, B., and Buettner, A. (2010). Do invariant odour constituents of amniotic fluid and human milk explain undifferentiated attraction in newborns? In T. Hofmann, W. Meyerhof and P. Schieberle (Eds.), *Advances and Challenges in Flavor Chemistry and Biology*. Deutsche Forschungsanstalt für Lebensmittelchemie, Berlin.

Dougherty, R. W., Shipe, W. F., Gudnason, G. V., et al. (1962). Physiological mechanisms involved in the transmission of flavors and odors to milk I. Contribution of eructed gases to milk flavor. *J. Dairy Sci.* 45: 472–476.

Durand, K., Baudon, G., Freydefont, L., and Schaal, B. (2008). Odorization of a novel object can influence infant's exploratory behavior in unexpected ways. *Infant Behav. Dev.* 31: 629–636.

Durand, K. and Schaal, B. (2011). *Are Young Children Knowledgeable About Perfumes? Implications for Social Odour Learning and Preferences*. 12th Meeting on Chemical signals in Vertebrates (Berlin, Germany).

Durand, K., Monnot, J., Martin, S., et al. (2013). Eye-catching odors: Familiar odors promote attention and sustained gazing to faces and eyes in 4 month-old infants *PLoS ONE* 8: 70677.

Ekman, P. and Friesen, W., 1982. *Facial Action Coding System: A Technique for Measurement of Facial Movement*. Consulting Psychologist Press: Palo Alto, CA.

Engen, T. (1974). Method and theory in the study of odor preferences. In I. W. Johnson, D. G. Moulton, and A. Türk (Eds.), *Human Responses to Environmental Odors* (pp. 121–141). London: Academic Press.

Engen, T. (1982). *The Perception of Odors*. Academic Press: New York.

Engen, T. and Lipsitt, L. P. (1965). Decrement and recovery of responses to olfactory stimuli in the human neonate, *J. Comp. Physiol. Psychol.* 59: 312–316.

Engen, T. and Katz H. I. (1968). *Odor responses and response bias in young children*. Unpublished manuscript, Brown University, Providence, RI.

Engen, T., Lipsitt, L. P., and Kaie, H. (1963). Olfactory response and adaptation in the human neonate. *J. Comp. Physiol. Psychol.* 56: 73–77.

Engen T. and Engen, E. A. (1997). Relationship between development of odor perception and language. In B. Schaal (Ed.), *L'Odorat chez l'Enfant, Perspectives croisées*, Paris: Presses Universitaires de France (*Enfance*, 1/1997, 125–140).

Epple, G. and Herz, R. (1999). Ambient odors associated to failure influence cognitive performance in children. *Dev. Psychobiol.* 35: 103–107.

Faas, A. E., Sponton, E. D., Moya, P. R., and Molina, J. C. (2000). Differential responsiveness to alcohol odor in human neonates. Effects of maternal consumption during gestation. *Alcohol* 22: 7–17.

Ferdenzi, C. (2007). *Variations interindividuelles des comportements olfactifs chez les enfants de 6–12 ans*. Unpublished Doctoral Thesis, Université de Bourgogne, Dijon, France.

Ferdenzi, C., Coureaud, G., Camos, V., and Schaal, B. (2008a). Human awareness and uses of odor cues in everyday life: results from a questionnaire study in children. *Int. J. Behav. Dev.* 32: 422–431.

Ferdenzi, C., Mustonen, S., Tuorila, H., and Schaal, B. (2008b). Children's Awareness and Uses of Odor Cues in Everyday Life: A Finland–France Comparison. *Chem. Percept.* 1: 190–198.

Ferdenzi, C., Schaal, B., and Roberts, C. S. (2010). Family scents: developmental changes in body odor perception of kin? *J. Chem. Ecol.* 36: 847–854.

Fernandez, M. and Bahrick, L. (1994). Infants' sensitivity to arbitrary object-odour pairings. *Infant Behav. Dev.* 17: 471–474.

Fernandez, M., Hernandez-Reif, M., Field, T., et al. (2003). EEG during lavender and rosemary exposure in infants of depressed and non-depressed mothers. *Inf. Behav. Dev.* 27: 91–100.

Field, T. M., Field, T., Cullen, C., et al. (2008). Lavender bath oil reduces stress and crying and enhances sleep in very young infants. *Early Hum. Dev.* 84: 399–401.

Fillion, T. J. and Blass, E. M., 1986. Infantile experience with suckling odors determines adult sexual behavior in male rats. *Science* 231: 729–731.

Folstein, J. R., Gautier, I., and Palmeri, T. J. (2010). Mere exposure alters category learning of novel objects. *Front. Psychol.* 1: doi: 10.3389/fpsyg.2010.00040.

Forestell, C. A. and Mennella, J. A. (2005). Children's hedonic judgment of cigarette smoke odor: effect of parental smoking and maternal mood. *Psychol. Addiction Behav.* 19: 423–432.

Fossey, E. (1993). Identification of alcohol by smell among young children: an objective measure of early learning in the home. *Drug Alcohol Dep.* 34: 29–35.

Frank, R. A., Dulay, M. F., Niergarth, K. A., and Gesteland, R. C. (2004). A comparison of the sniff magnitude test and the University of Pennsylvania Smell Identification Test in children and nonnative English speakers. *Physiol Behav.* 81: 475–480.

Frank, R. A., Brearton, M., Rybalsky, K., et al. (2011). Consistent flavor naming predicts recognition memory in children and young adults. *Food Qual. Pref.* 22: 173–178.

Freud, S. (1905-1962). *Three Essays on the Theory of Sexuality*. Basic Books: New York

Fusari, C. and Pardelli, C. (1962). L'olfattometria elettroencefalografica nel lattante, *Boll Mal Orechio Gola Naso* 80: 719–734.

Galef, B. G. and Henderson, P. W. (1972). Mother's milk: A determinant of the feeding preferences of weaning rat pups. *J. Comp. Physiol. Psychol.* 78: 213–219.

Garb, J. L. and Stunkart, A. J. (1974). Taste aversion in man. *Am. J. Psychiatr.* 131: 1204–1207.

Gesteland, R. C., Yancey, R. A., and Farbman, A. I. (1982). Development of olfactory receptor neuron selectivity in the rat fetus. *Neurosci.* 7: 3127–3136.

Goldman, R. and Goldman, J. (1982). *Children's Sexual Thinking*. Routledge and Kegan: London.

Goodlin-Jones, B. L., Eiben, L. A., and Anders, T. F. (1997). Maternal well-being and sleep-wake behaviors in infants: An intervention using maternal odors. *Inf. Ment. Health J.* 18: 378–393.

Goubet, N., Rattaz, C., Pierrat, V., et al. (2002). Olfactory familiarisation and discrimination in preterm and full-term newborns. *Infancy* 3: 53–75.

Graf, M. V., and Kastin, A. J. (1986). Delta-sleep-inducing peptide (DSIP): An update. *Peptides* 6: 1165–1187.

Gruest, N., Richer, P., and Hars, B. (2004). Emergence of long-term memory for conditioned aversion in the rat fetus. *Dev. Psychobiol.* 44: 189–198.

Guilford, J. P. (1954). *Psychometric Methods, 2nd edition*. McGraw-Hill, New York.

Guillory, A. W., Self, P. A., Francis, P., and Paden, L. Y. (1980). *Habituation in Studies of Neonatal Olfaction*. Biennal Meeting of the South-western Society for Research in Human Development, Lawrence, KS.

Guinard, J. X. (2001). Sensory and consumer testing with children. *Trends Food Sci. Technol.* 11: 273–283.

Haller, R., Rummel, C., Henneberg, S., et al. (1999). The effect of early experience with vanillin on food preference later in life. *Chem. Senses* 24: 465–467.

Hartmann, C., Doucet, S., Niclass, Y., et al. (2012). Human sweat odour conjugates in human milk, colostrum and amniotic fluid. *Food Chem.* 135: 228–233.

Hauser, G. J., Chitayat, D., Berns, L., et al. (1985). Peculiar odors in newborns and maternal prenatal ingestion of spicy foods. *Eur. J. Pediatr.* 144: 403.

Hausner, H., Bredie, W., Molgaard, C., et al. (2008). Differential transfer of dietary flavour compounds into human breast milk. *Physiol. Behav.* 95: 118–124.

Hausner, H., Nicklaus, S., Issanchou, S., et al. (2010). Breastfeeding facilitates acceptance of a novel dietary flavour compound. *Clin. Nutr.* 29: 141–148.

Havlicek, J., and Roberts, S. C. (2009). MHC-correlated mate choice in humans: a review. *Psychoneuroendocrinology* 34: 245–249.

Hepper, P. G. (1987). The amniotic fluid: an important priming role in kin recognition. *Anim. Behav.* 35: 1343–1346.

Hepper, P. G. (1988). Adaptive fetal learning: prenatal exposure to garlic affects postnatal preferences. *Anim. Behav.* 36: 935–936.

Hepper, P. G. (Ed.) (1991a). *Kin Recognition*, Cambridge University Press: Cambridge, UK.

Hepper, P. G. (1991b). Transient hypoxic episodes: a mechanism to support associative fetal learning, *Anim. Behav.* 41: 477–480.

Hepper, P. G. (1993). *In utero* release from a single transient hypoxic episode: a positive reinforcer? *Physiol. Behav.* 53: 309–311.

Hepper, P. G. (1994). Long-term retention of kinship recognition established during infancy in the domestic dog. *Behav. Processes* 33: 3–14.

Hepper, P. G. (1995). Human fetal "olfactory" learning. *Int. J. Prenatal, Perinatal Psychol. Med.* 7: 147–151.

Hepper, P. G., Wells, D. L., Millsopp, S., et al. (2012a). Prenatal and early sucking influences on dietary preferences in newborn, weaning and young adult cats. *Chem. Senses* 37: 755–766.

Hepper, P. G., Wells, D. L., Dornan, J. C., and Lynch, C. (2012b). Long-term flavor recognition in humans with prenatal garlic experience. *Dev. Psychobiol.* 55: 568–574.

Holmes, W. G., and Mateo, J. M. (2007). Kin recognition in rodents: issues and evidence. In J. O. Wolff and P. W. Sherman (Eds.), *Rodent Societies. An Ecological and Evolutionary Perspective*, pp. 216–228. Chicago University Press: Chicago.

Huck, U. W., and Banks, E. M. (1980). The effect of cross-fostering on the behaviour of two species of North American lemmings. I. Olfactory preferences. *Anim. Behav.* 28: 1046–1052.

Hudson, R. (1985). Do newborn rabbits learn the odour stimuli releasing nipple-search behaviour? *Dev. Psychobiol.* 18: 575–585.

Hudson, R. Schaal, B., and Bilko, A. (1999). Transmission of olfactory information from mother to young in the European rabbit. In H. O. Box and K. R. Gibson (Eds.), *Mammalian Social Learning. Comparative and Ecological Perspectives>*, pp. 141–157. Cambridge University Press: Cambridge, UK.

Hummel, T., Krone, F., Lundstrom, J., and Bartsch, O. (2005). Androstadienone odor thresholds in adolescents. *Horm. Behav.* 47: 306–310.

Hummel, T., Bensafi, M., Nikolaus, J., et al. (2007). Olfactory function in children assessed with psychophysical and electrophysiological techniques. *Behav. Brain Res.* 180: 133–138.

Humphrey, T. (1940). The development of the olfactory and the accessory formations in human embryos and fetuses, *J. Com. Neurol.* 73: 431–478.

Humphrey, T. (1963). The development of the anterior olfactory nucleus of human fetuses. *Prog. Brain Res.* 3: 170–190.

Humphrey, T. (1967). The development of the human tuberculum olfactorium during the first three months of embryonic life. *J. Hirnforsch*, 9: 437.

Humphrey, T. (1968). The development of the human amygdala during early embryonic life. *J. Comp. Neurol.* 132: 135.

Humphrey, T. (1978). *Functions of the nervous system during prenatal life in U.* Stave (Ed) *Perinatal Physiology*, pp. 651–683. Plenum, NewYork.

Hvastja, L., and Zanuttini, L. (1989). Odour memory and odour hedonics in children. *Perception* 18: 391–396.

Irzhanskaia, K. N. and Felberbaum, R. A. (1954). Conditioned reflex activity in premature children. *Fisiol. Zh. SSSR.* 40: 668–672.

Jacob, S., McClintock, M. K., Zelano, B., and Ober, C. (2002). Paternally inherited HLA alleles are associated with women's choice of male odor. *Nature Gen.* 30: 175–179.

Jäncke, L. and Kaufmann, N. (1994). Facial EMG responses to odors in solitude and with an audience. *Chem. Senses* 19: 99–111.

Jehl, C., and Murphy, C. (1998). Developmental effects on odor learning and memory in children. *Ann. NY Acad. Sci*, 855: 632–634.

Jirowetz, L., Jäger, W., Buchbauer, G., et al. (1991). Investigation of animal blood samples after fragrance drug inhalation by GC/MS with chemical ionisation and selected ion monitoring. *Biol. Mass Spectr.* 20: 801–803.

Kalogerakis, M. G. (1972). The role of olfaction in sexual development. *Psychosom. Med.* 25: 420–432.

Kaplan, J. N., Cubicciotti, D. D., and Redican, W. K. (1977). Olfactory and visual differentiation of synthetically scented surrogates by infant squirrel monkeys. *Dev. Psychobiol.* 12: 1–10.

Karmowski, A., Sobiech, K. A., Dobek, D., et al. (1998). The concentration of cotidine in urine, colostrum and amniotic fluid within the system mother-baby. *Ginekol. Polska* 69: 115–122.

Kawakami, K., Takai-Kawakami, K., Okazaki, Y., et al. (1997). The effect of odor on human newborn infant under stress. *Infant Behav. Dev.* 20: 531–535.

Keil, W., von Stralendorff, F., and Hudson, R. (1990). A behavioral bioassay for analysis of rabbit nipple-seach pheromone. *Physiol. Behav.* 47: 525–529.

Kendal-Reed, M. and Van Toller. S. (1992). Brain electrical activity mapping: an exploratory study of infant response to odours. *Chem Senses* 17: 765–777.

Kniep, H. H., Morgan, W. L., and Young, P. T. (1931). Studies in affective psychology. XI. Individual differences in affective reaction to odors. *Am. J. Psychol.* 43: 406–421.

Koelega, H. (1994). Prepubescent children may have specific deficits in olfactory sensitivity. *Percept. Mot. Skills* 78: 191–199.

Koelega, H., and Köster, E. P. (1974). Some experiments on sex differences in odor perception. *Ann. N.Y. Acad. Sci.* 237: 234–246.

Kroner, T. (1882). Uber die Sinnesempfindungen des Neugeborenen. *Breslauer ärzt. Zeitsch.* 4.

Kussmaul, A. (1859). *Untersuchungen über das Seelenleben des Neugeborenen Menschen*. Morer, Tübingen.

Laing, D. G. and Clark, P. G. (1983). Puberty and olfactory preferences of males. *Physiol Behav.* 30: 591–597.

Laing, D. G. and Schaal B. (2012). Chemosensory function in infants and children. In Welge-Lüssen, A., and Hummel, T. (Eds.), *Management of Smell and Taste Disorders: A practical guide for clinicians*, Thieme, Stuttgart.

Lane, A., Young, R., Baker, A., and Angley, M. (2010). Sensory processing subtypes in autism: Association with adaptive behavior. *J. Autism Dev. Disord.* 40: 112–122.

Lambers, D. S., and Clark, K. E. (1996). The maternal and fetal physiologic effects of nicotine. *Sem. Perinatol.* 20: 115–126.

Langendijk, P., Bolhuis, J. E., and Laurenssen, B. (2007). Effects of pre- and postnatal exposure to garlic and aniseed flavour on pre- and post-weaning feed intake in pigs. *Livest. Prod. Sci.* 108: 284–287.

Larjola, K. and Wright, J. von (1976). Memory of odors: Developmental data. *Percept. Motor Skills* 42: 1138.

Le Magnen, J. (1952). Les phénomènes olfacto-sexuels chez l'homme. *Arch. Sci. Physiol.* 6: 125–160.

Le Magnen, J. and Tallon, S. (1968). Préférence alimentaire du jeune rat induite par l'allaitement maternel. *C. R. Séances Soc. Biol.* 162: 387–390.

Lecanuet, J. P. and Schaal, B. (1996). Fetal sensory competences. *Eur. J. Obstet. Gynaecol. Reprod. Biol.* 68: 1–23.

Lee, A. C., He, J., and Ma, M. (2011). Olfactory marker protein is critical for functional maturation of olfactory sensory neurons and development of mother preference. *J. Neurosci.* 31: 2974–2982.

Lehrner, J. P., Walla, P., Laska, M., and Deecke, L. (1999a). Different forms of human memory: A developmental study. *Neurosci Lett.* 272: 17–20.

Lehrner, J. P., Gluck, J., and Laska, M. (1999b). Odor identification, consistency of label use, olfactory threshold and their relationships to odor memory over the human lifespan. *Chem. Senses* 24: 337–346.

Lenochova, P., Vohnoutova, P., Roberts, S. C., et al. (2012). Psychology of fragrance use: perception of individual odor and perfume blends reveals a mechanism of idiosyncratic fragrance choice. *PLoS One* 7: e33810.

Leon, M. (1992). Neuroethology of olfactory preference development. *J. Neurobiol.* 23: 1557–1573.

Lipsitt, L. P., Engen, T., and Kaie, H. 1963, Developmental changes in the olfactory threshold of the neonate, *Child Dev.* 34: 371–376.

Lipsitt, L. P., and Rovee-Collier, C. K. (2012). The psychophysics of olfaction in the human neonate: habituation and cross adaptation. In G. Zucco, R. Herz, and B. Schaal (Eds.), *Olfactory Cognition*, pp. 221–235. John Benjamins: Amsterdam.

Logan, D. W., Brunet, L., Webb, W., et al. (2012). Learned recognition of maternal signature odors mediates the first suckling episode in mice. *Curr. Biol.* 22: 1998–2007.

Lumeng, J., Zuckerman, M., Cardinal, T., and Kaciroti, N. (2005). The association between flavor labeling and flavor recall ability in children. *Chem. Senses* 30: 565–574.

Macchi, G. (1951). The ontogenetic development of the olfactory telencephalon in man. *J. Comp. Neurol.* 95: 245–305.

Macfarlane, A. (1975). Olfaction in the development of social preferences in the human neonate. *Ciba Found. Symp.* 33: 103–113.

Mainardi, D., Marsan, M., and Pasquali, A. (1965). Causation of sexual preferences of house mouse. The behaviour of mice reared by parents whose odour was artificially altered. *Atti Soc. It. Sci. Nat.* 104: 325–338.

Mainardi, M., Poli, M., and Valsecchi, P. (1989). Ontogeny of dietary selection in weanling mice: effects of early experience and mother's milk. *Biol. Behav.* 14: 185–194.

Makin, J. W., and Porter, R. H. (1989). Attractiveness of lactating females' breast odors to neonates. *Child Dev.* 60: 803–810.

Mallet, P. and Schaal, B. (1998). Rating and recognition of peer's personal odours in nine-year-old children: An exploratory study. *J. General Psychol.* 125: 47–64.

Marlier, L., and Schaal, B. (1989). *Olfactory, tactile and auditory recognition of individuality in children.* 21st International Ethological Conference (Utrecht, Netherlands).

Marlier, L. and Schaal, B. (1997). Familiarité et discrimination olfactive chez le nouveau-né: influence différentielle du mode d'alimentation? In B. Schaal (Ed.), *L'Odorat chez l'Enfant, Perspectives croisées*, Presses Universitaires de France, Paris (*Enfance*, 1/1997, 47–61).

Marlier, L., Schaal, B., and Soussignan, R. (1997). Orientation responses to biological odors in the human newborn. Initial pattern and postnatal plasticity. *C. R. Acad. Sci. Paris, Life Sciences* 320: 999–1005.

Marlier, L., Schaal, B., and Soussignan, R. (1998). Neonatal responsiveness to the odor of amniotic and lacteal fluids: A test of perinatal chemosensory continuity. *Child Dev.* 64: 611–623.

Marlier, L., Schaal, B., Gaugler, C., and Messer, J. (2001). Olfaction in premature newborns: Detection and discrimination abilities two months before gestational term. In Marchlewska-Koj et al. (Ed.), *Chemical Signals in Vertebrates* 9. New York: Kluwer Academic/Plenum Publishers, pp. 205–209.

Marlier, L. and Schaal, B. (2005). Human newborns prefer human milk: Conspecific milk odor is attractive without postnatal exposure. *Child Dev.* 76: 155–168.

Marom, R., Kening, T. S., Mimouni, F. B. et al. (2011). The effect of olfactory stimulation on energy expenditure in growing preterm infants. *Acta Paediatr.* 101: e11–e14.

Marr, J. N., and Gardner, L. E. Jr., (1965). Early olfactory experience and later social behavior in the rat: preference, sexual responsiveness, and care of young. *J Genetic Psychol.* 107: 167–174.

Mateo, J. M. (2004). Recognition systems and biological organization: the perception component of social recognition. *Ann. Zool. Fenn.* 41: 729–745.

Mellier, D., Bézard, S., and Caston, J. (1997). Etudes exploratoires des relations intersensorielles olfaction-douleur. In B. Schaal (Ed.), *L'Odorat chez l'Enfant, Perspectives croisées*. Presses Universitaires de France, Paris (*Enfance*, 1, 98–111).

Mennella, J. A., and Beauchamp, G. K. (1991a). Maternal diet alters the sensory qualities of human milk and the nursling's behavior, *Pediatrics* 88: 737–744.

Mennella, J. A., and Beauchamp, G. K. (1991b). The transfer of alcohol to human milk: effects on flavor and the infant's behavior. *New Eng. J. Med.* 325: 981–985.

Mennella, J. A., and Beauchamp, G. K. (1992). Developmental changes in nasal airflow patterns. *Acta Otolaryngol.* 112: 1025–1031.

Mennella, J. A., Johnson, A., and Beauchamp, G. K. (1995). Garlic ingestion by pregnant women alters the odor of amniotic fluid. *Chem. Senses* 20: 207–209.

Mennella, J. A., and Beauchamp, G. K. (1996). The human infants' responses to vanilla flavors in human milk and formula. *Infant Behav. Dev.* 19: 13–19.

Mennella, J. A., and Beauchamp, G. K. (1998a). Smoking and the flavor of milk. *New Eng. J. Med.* 339: 1559–1560.

Mennella, J. A., and Beauchamp, G. K. (1998b). Infants' exploration of scented toys: effects of prior experience. *Chem. Senses* 23: 11–17.

Mennella, J. A., and Beauchamp, G. K. (1999). Experience with a flavor in mother's milk modifies the infant's acceptance of flavored cereal. *Dev. Psychobiol.* 35: 197–203.

Mennella, J. A., and Garcia, P. L. (2000). Children's hedonic response to the smell of alcohol: effects of parental drinking habits. *Alcohol: Clin. Exp. Res.* 24: 1167–1171.

Mennella, J. A., Jagnow, C. P., and Beauchamp, G. K. (2001). Prenatal and postnatal flavor learning by human infants. *Pediatrics* 107: 1–6.

Mennella, J. A. and Forestell, C. A. (2008). Children's hedonic responses to the odors of alcoholic beverages: A window to emotions. *Alcohol*, 42: 249–260.

Mennella, J. A., and Castor, S. M. (2012). Sensitive period in flavor learning: Effects of duration of exposure to formula flavors on food likes during infancy. *Clin. Nutr.* 3: 1022–1025.

Meza, C. V., Powell, N. J., and Covington, C. (1998). The influence of olfactory intervention on non-nutritive sucking skills in a premature infant. *Occup. Ther. J. Res.* 18: 71–83.

Milinski, M., and Wedekind, C. (2001). Evidence for MHC-correlated perfume preferences in humans. *Behav. Ecol.* 12: 140–149.

Mizuno, K., and Ueda, A. (2004). Antenatal olfactory learning influences infant feeding. *Early Hum. Dev.* 76: 83–90.

Mizuno, K., Mizuno, N., Shinohara, T., and Noda, M. (2004). Mother-infant skin-to-skin contact after delivery results in early recognition of own mother's milk odour. *Acta Paediatr.* 93: 1640–1645.

Molina, J. C., Chotro, M. G., and Dominguez, H. D. (1995). Fetal alcohol learning resulting from alcohol contamination of the prenatal environment. In J. P. Lecanuet, W. P. Fifer, N. A. Krasnegor, and W. P. Smotherman (eds.), *Fetal development, A psychobiological perspective*, pp. 419–438. Lawrence Erlbaum: Hillsdale, NJ.

Moncomble, A. S., Coureaud, G., Quennedey, B., et al. (2005). The mammary pheromone of the rabbit: From where does it come? *Anim. Behav.* 69: 29–38.

Moncrieff, R. W. (1965). Changes in olfactory preferences with age. *Rev. Laryngol (Bordeaux)*, 86 (Suppl.), 895–904.

Moncrieff, R. W. (1966). *Odour Preferences.* Wiley: New York.

Montagna, W. and Yun, J. S. (1972). The glands of Montgomery. *Br. J. Dermatol.* 86: 126–133.

Montagna, W. and MacPherson, E. E. (1974). Some neglected aspects of the anatomy of human breasts. *J. Invest. Dermatol.* 63: 10–16.

Montgomery, W. F. (1987). *An Exposition of the Signs and Symptoms of Pregnancy, the Period of Human Gestation, and Signs of Delivery.* Sherwood, Gilber, and Piper: London.

Moshkin, M. P., ,Litvinova, N. A., Bedareva, A. V., et al. (2011). Odor as an element of subjective assessment of attractiveness of young males and females. *J. Evol. Biochem. Physiol.* 47: 69–82.

Müller, F. and O'Rahilly, R. (2004). Olfactory structures in staged embryos. *Cells Tissues Organs* 178: 93–116.

Müller-Schwarze, D., and Müller-Schwarze, C. (1971). Olfactory imprinting in a precocial mammal. *Nature* 229: 55–56.

Nishitani, S., Miyamura, T., Tagawa, M., et al. (2009). The calming effect of a maternal breast milk odor on the human newborn infant. *Neurosci. Res.* 63: 66–71.

Noll, R. B., Zucker, R. A., and Greenberg, G. S. (1990). Identification of alcohol smell among preschoolers: evidence for early socialization about drugs occurring in the home. *Child Dev.* 61: 1520–1527.

Nolte, D. L., Provenza, F. D., Callan, R., and Panter, K. E. (1992). Garlic in the ovine fetal environment. *Physiol. Behav.* 52: 1091–1093.

Nowak, R. (1989). Mother and sibling discrimination at a distance by three- to seven-day-old lambs. *Dev. Psychobiol.* 23: 285–295.

Oaten, M., Stevenson, R. J., and Case, T. I. (2009). Disgust as a disease avoidance mechanism. *Psychol. Bull.* 135: 303–321.

Olson, G. M. (1981). The recognition of specific persons. In M. E. Lamb and L. R. Sherrod (Eds.), *Infant Social Cognition*, pp. 37–59. Erlbaum: Hillsdale, NJ.

Olsson, S. B., Barnard, J. and Turri, L. (2006). Olfaction and identification of unrelated individuals. *J. Chem. Ecol.* 32: 1635–1645.

Oostindjier, M., Bolhuis, J. E., van den Brand, H., and Kemp, B. (2009). Prenatal flavor exposure affects flavor recognition and stress-related behavior of piglets. *Chem. Senses* 34: 775–787.

Oostindjier, M., Bolhuis, J. E., van den Brand, H., et al. (2010). Prenatal flavor exposure affects growth, health and behavior of newly weaned piglets. *Physiol. Behav.* 99: 579–586.

Örs, R., Özek, E., Baysoy, G., et al. (1999). Comparison of sucrose and human milk on pain response in newborns. *Eur. J. Pediatr.* 158: 63–66.

Ortmann, R. (1989). Über Sinneszellen am fetalen vomeronasalen Organ des Menschen. *Hals Nase Ohren* 37: 191–197.

Parfet, K. A. R. and Gonyou, H. W. (1991). Attraction of newborn piglets to auditory, visual, olfactory and tactile stimuli. *J. Anim. Sci.* 69: 125–133.

Patris, S. M., Rouby, C., Nicklaus, S., and Issanchou, S. (2009). Development of olfactory ability in children : Sensitivity and identification. *Dev. Psychobiol.* 51: 268–276.

Pedersen P. A. and Blass E. M. (1982). Prenatal and postnatal determinants of the 1st suckling episode in albino rats. *Dev. Psychobiol.* 15: 349–355.

Penn, D. and Potts, W. (1998). How do major histocompatibility genes influence odor and mating preferences? *Adv. Immunol.* 69: 411–436.

Perkins, O. M. and Miller, A. M. (1926). Sebaceous glands in the human nipple. *Am. J. Obstet.* 11: 789–794.

Peterson, F. and Rayney, L. H. (1910-1911). The beginnings of mind in the newborn. *Bull. Lying-In Hosp. NY City* 7: 99–122.

Peto, E. (1936). Contribution to the development of smell feeling. *Br. J. Med. Psychol.* 15: 314–320.

Pihet, S., Schaal, B., Bullinger, A., and Mellier, D. (1996). An investigation of olfactory responsiveness in premature newborns. *Infant Behav. Dev., ICIS Issue*: 676.

Pihet, S., Mellier, D., Bullinger, A., and Schaal, B. (1997). La compétence olfactive du nouveau-né prématuré: une étude préliminaire. In B. Schaal (Ed.), *L'Odorat chez l'Enfant, Perspectives croisées.* Presses Universitaires de France: Paris (*Enfance*, 1: 33–46).

Poncelet, J., Rinck, F., Bourgeat, F., et al. (2010). The effect of early experience on odor perception in humans: psychological and physiological correlates. *Behav. Brain Res.* 208: 458–465.

Porter, R. H. and Moore, J. D. (1981). Human kin recognition by olfactory cues. *Physiol. Behav.* 27: 493–495.

Porter, R., Makin, J. W., Davis, L. B., and Christensen, K. M. (1991). An assessment of the salient olfactory environment of formula-fed infants. *Physiol. Behav.* 50: 907–911.

Prechtl, H. F. R. (1958). The directed head-turning response and allied movements of the human baby. *Behaviour.* 13: 212–242.

Prescott, J., Kim, H., and Kim, K. O. (2008). Cognitive mediation of hedonic changes to odors following exposure. *Chem. Percept.* 1: 2–8.

Preyer, W. (1884). *Die Seele des Kindes.* Grieben: Leipzig.

Quadagno, D. M. and Banks, E. M. (1970). The effect of reciprocal cross-fosterinf on the behaviour of two species of rodents, Mus musculus and Baiomys taylori ater. *Anim. Behav.* 18: 379–390.

Rabin, M. D. (1988). Experience facilitates olfactory quality discrimination. *Percept. Psychophys.* 44: 532–540.

Raimbault, C., Saliba, E., and Porter, R. H. (2006). The effect of the odour of mother's milk on breastfeeding behaviour of premature neonates. *Acta Paediatr.* 96: 368–371.

Rantala, M. J. and Marcinkowska, U. (2010). The role of sexual imprinting and the Westermarck effect in mate choice in humans. *Behav. Ecol. Sociobiol.* 65: 859–873.

Rasmussen, L. E. L. (1995). Evidence for long-term chemical memory in Elephants. *Chem. Senses* 20: 762.

Rattaz, C., Goubet, N., and Bullinger, A. (2005). The calming effect of a familiar odor on full-term newborns. *Dev. Behav. Pediatr.* 26: 86–92.

Redican, W. K. and Kaplan, J. N. (1978). Effects of synthetic odors on filial attachement in infant squirrel monkeys. *Physiol. Behav.* 20: 79–85.

Richman, R. A., Wallace, K., and Sheehe, P. R. (1995). Assessment of an abbreviated odorant identification task for children: a rapid screening device for schools and clinics. *Acta Paediatr.* 84: 434–437.

Richman, R. A., Post, E. M., Sheehe, P. R., and Wright, H. N. (1992). Olfactory performance during childhood: I. Development of an odorant identification test for children. *J. Pediatr.* 121: 908–911.

Righard, L. and Alade, M. O. (1990). Effect of delivery routines on success of first breast-feed. *Lancet* 336: 1105–1107.

Rinck, F., Barkat-Defradas, M., Chakirian, A., et al. (2011). Ontogeny of odor liking during childhood and its relation to language development. *Chem. Senses* 36: 83–91.

References

Robinson, S. R., Wong, C., Robertson, S. S., et al. (1995). Behavioral responses of the chronically instrumented sheep fetus to chemosensory stimuli presented *in utero*. *Behav. Neurosci.* 109: 551–562.

Robinson, S. R. and Méndez-Gallardo, V. (2010). Amniotic fluid as an extended milieu intérieur. In K. E. Hood, C. T. Halpern, G. Greenberg, and R. M. Lerner (Eds.), *Handbook of Developmental Science, Behavior, and Genetics*, pp. 234–284. Blackwell: New York.

Romantshik, O., Porter, R. H., Tillmann, V. and Varendi, H. (2007). Preliminary evidence of a sensitive period for olfactory learning by human newborn. *Acta Paediatr.* 96: 372–376.

Ronca, A. E. and Alberts, J. R. (1990). Heart rate development and sensory-evoked cardiac responses in perinatal rats. *Physiol. Behav.* 47: 1075–1082.

Rosenblatt, J. S. (1983). Olfaction mediates developmental transitions in the altricial newborn of selected species of mammals. *Dev. Psychobiol.* 16: 347–375.

Rouby, C., Chevalier, G., Gautier, B., and Dubois, D. (1997). Connaissance et reconnaissance d'une série olfative chez l'enfant préscolaire. In B. Schaal (Ed.), *L'Odorat chez l'Enfant, Perspectives croisées*. Presses Universitaires de France, Paris (*Enfance*, 1: 152–171).

Rovee-Collier, C. K., Cohen, R. Y., and Shlapack, W. (1975). Life-span stability in olfactory sensitivity. *Dev. Psychol.* 11: 311–318.

Rozin, P., Hammer L., Oster, H., et al. (1986). The child's conception of food: Differentiation of categories of rejected substances in the 16 months to 5 year age range. *Appetite* 7: 141–151.

Rubin, G. B., Fagen, J. W., and Carroll, M. H. (1998). Olfactory context and memory retrieval in 3-month-old infants. *Infant Behav. Dev.* 21: 641–658.

Ruff, J. S., Nelson, A. C., Kubinak, J. L., and Potts, W. K. (2012). MHC signaling during social communication. In C. Lopez-Larrea (Ed.), *Self and Nonself*, pp. 290–313. Landes Bioscience-Springer: Berlin.

Russell, M. J. (1976). Human olfactory communication. *Nature* 260: 520–522.

Sadathosseini, A. S., Negarandeh, R., and Movahedi, Z. (2011). The effect of a familiar scent on the behavioral and physiological pain responses in neonates. *Pain Manag. Nursing* 14: e196–203.

Sànchez, C. L., Cubero, J., Sanchez, J., et al. (2009). The possible role of human milk nucleotides as sleep inducers. *Nutr. Neurosci.* 12: 2–8.

Sandford, A. A., Davidson, T., Herrera, N., et al. (2006). Olfactory dysfunction: A sequela of pediatric blunt head trauma. *Int. J. Pediatr. Otolaryngol.* 70: 1015–1025.

Sarnat, H. B. (1978). Olfactory reflexes in the newborn infant. *J Pediatr.* 92: 624–626.

Schaal, B. (1988). Olfaction in infants and children: developmental and functional perspectives. *Chem. Senses* 13: 145–190.

Schaal, B. (1999). Le développement de la sensibilité olfactive: de la période foetale à la puberté. In Y. Christen, L. Collet, and M. T. Droy-Lefaix (Eds.), *Rencontres IPSEN en Oto-Rhino-Laryngologie, Volume 3*, pp. 115–135. Irvinn Editions, Neuilly sur Seine.

Schaal, B. (2005). From amnion to colostrum to milk: Odour bridging in early developmental transitions. In B. Hopkins and S. Johnson (Eds.), *Prenatal Development of Postnatal Functions*, pp. 52–102. Praeger: Westport, CT.

Schaal, B. (2010). Mammary odor cues and pheromones: mammalian infant-directed communication about maternal state, mammae, and milk. *Vit. Horm.* 83: 81–134.

Schaal, B. (2012). Emerging chemosensory preferences: another playground for the innate-acquired dichotomy in human cognition. In G. M. Zucco, R. Herz, and B. Schaal (Eds.), *Olfactory Cognition*, pp. 237–268. John Benjamins: Amsterdam.

Schaal, B., Montagner, H., Hertling, E., et al. (1980). Les stimulations olfactives dans les relations entre l'enfant et la mère. *Reprod. Nutr. Dév.* 20: 843–858.

Schaal, B., Orgeur, P., Lecanuet, J. P., et al. (1991). *In utero* nasal chemoreception: preliminary experiments in the fetal sheep. *C.R. Acad. Sci. Paris, (Série III),* 313: 319–325.

Schaal, B. and Orgeur, P. (1992). Olfaction *in utero*: can the rodent model be generalized? *Quart. J. Exp. Psychol. B. Comp. Physiol. Psychol.* 44B: 245–278.

Schaal, B., Marlier, L., and Soussignan, R. (1995a). Neonatal responsiveness to the odour of amniotic fluid. *Biol. Neonate* 67: 397–406.

Schaal, B., Orgeur, P., and Arnould, C. (1995b). Chemosensory preferences in newborn lambs: prenatal and perinatal determinants. *Behaviour* 132: 352–365.

Schaal, B., Orgeur, P., and Marlier, L. (1995c). Amniotic fluid odor in neonatal adaptation: A summary of recent research in mammals. *Adv. Biosci.* 93: 239–245.

Schaal, B., Orgeur, P., and Rognon, R. (1995d). Odor sensing in the human fetus: anatomical, functional and chemo-ecological bases. In Lecanuet, J. P., Krasnegor, N. A., Fifer, W. A. and Smotherman, W. (Eds.), *Prenatal Development, A Psychobiological Perspective*, pp. 205–237. Lawrence Erlbaum: Hillsdale, NJ.

Schaal, B., Soussignan, R., Marlier. L., et al. (1996). Variability and invariants in early odor preferences: Comparative data from children belonging to three cultures. *Chem. Senses* 22: 212.

Schaal, B., Marlier, L., and Soussignan, R. (1998). Olfactory function in the human fetus: evidence from selective neonatal responsiveness to the odor of amniotic fluid. *Behav. Neurosci.* 112: 1438–1449.

Schaal, B., Marlier, L., and Soussignan, R. (2000). Human foetuses learn odours from their pregnant mother's diet. *Chem. Senses* 25: 729–737.

Schaal, B., Coureaud, G., Langlois, D., et al. (2003). Chemical and behavioural characterization of the mammary pheromone of the rabbit. *Nature* 424: 68–72.

Schaal, B., Hummel, T., and Soussignan, R. (2004). Olfaction in the fetal and premature infant: functional status and clinical implications. *Clin. Perinatol.* 31: 261–285.

Schaal, B., Doucet, S., Sagot, P., et al. (2006). Human breast areolae as scent organs: morphological data and possible involvement in maternal-neonatal co-adaptation. *Dev. Psychobiol.* 48: 100–110.

Schaal, B., Doucet, S., Soussignan, R., et al. (2008). The human breast as a scent organ: exocrine structures, secretions, volatile components, and possible function in breastfeeding interactions. In J. L. Hurst, R. J. Beynon, S. C. Roberts, and T. D. Wyatt (Eds.), *Chemical Signals in Vertebrates 11*, pp. 325–335. New York: Springer.

Schaal, B., and Durand, K. (2012). The role of olfaction in human multisensory development. In A. Bremner, D. Lewkowicz, and C. Spence (Eds.), *Multisensory Development*, pp. 29–62. Oxford University Press: Oxford, UK.

Schaffer, J. P. (1910). The lateral wall of the cavum nasi in man with special reference to the various developmental stages. *J. Morphol.* 21: 613–707.

Schecklmann, M., Schwenk, C., Taurines, R., et al. (2013). A systyematic review on olfaction in child and adolescent psychiatric disorders. *J. Neural Transm.* 120: 121–130.

Schleidt, M. and Genzel, C. (1990). The significance of mother's perfume fir infants in the first weeks of their life. *Ethol. Sociobiol.* 11: 145–154.

Schleidt, M.and Hold, B. (1982). Human odour and identity. In W. Breipohl (ed.), *Olfaction and endocrine regulation*, pp. 181–194. IRL Press: London.

Schmidt, H. J. and Beauchamp, G. K. (1988). Adult-like odor preferences and aversions in 3-year-old children. *Child Dev.* 59: 1136–1143.

334   Chapter 14   Prenatal and Postnatal Human Olfactory Development: Influences on Cognition and Behavior

Schmidt, H. J., and Beauchamp, G. K. (1989). Sex differences in responsiveness to odors in 9-month-old infants. *Chem. Senses* 14: 744.

Schmidt, H. J., and Beauchamp, G. K. (1990) Adult-like hedonic responses to odors in 9-month-old infants. *Chem. Senses* 15: 634.

Schmidt, H. J., and Beauchamp, G. K. (1992). Human olfaction in infancy and early childhood. In M. J. Serby and K. L. Chobor (Eds.), *Science of Olfaction*, pp. 378–395. Springer: New York.

Schneider, J. F. and Floemer, F. (2009). Maturation of olfactory bulbs: MR imaging findings. *Am. J. Neuroradiol.* 30: 1149–1152.

Schneider, M. A. and Hendrix, L. (2000). Olfactory sexual inhibition and the Westeremarck effect. *Hum. Nature* 11: 65–92.

Schroers, M., Prigot, J., and Fagen, J. (2007). The effect of a salient odor context on memory retrieval in young infants. *Infant Behav. Dev.* 30: 685–689.

Self, P. A., Horowitz, F. D., and Paden, L. Y. (1972), Olfaction in newborn infants, *Dev. Psychol.* 7: 349–363.

Semke, E., Distel, H., and Hudson, R. (1995). Specific enhancement of olfactory receptor sensitivity associated with fetal learning of food odors in the rabbit. *Naturwiss.* 82: 148–149.

Shah, A., Oxley, G., Lovic, V., and Fleming, A. S. (2002). Effects of preweaning exposure to novel maternal odors on maternal responsiveness and selectivity in adulthood. *Dev. Psychobiol.* 41: 187–196.

Shimoda, M., Yoshimura, T., Ishikawa, H., et al. (2000). Volatile compounds of human milk. *J. Fac. Agricult. Kyushu Univ.* 45: 199–206.

Silverman, I., and Bevc, I. (2005). Evolutionary origins and ontogenetic development of incest avoidance. In B. J. Ellis and D. F. Bjorklund (Eds.), *Origins of the Social Mind: Evolutionary Psychology and Child Development*, pp. 292–313. Guildford: New York.

Simitzis, P. E., Deligeorgis, S. G., Bizelis, J., and Fegeros, K. (2008). Feeding preferences in lambs influenced by prenatal flavor exposure. *Physiol. Behav.* 93: 529–536.

Smith, D. M., Peters, T. G., and Donegan, W. L. (1982). Montgomery's areolar tubercle. *A light microscopic study. Arch. Pathol. Lab. Med.* 106: 60–63.

Smotherman, W. P. (1982). Odor aversion learning by the rat fetus. *Physiol. Behav.* 29: 769–771.

Smotherman, W. P. and Robinson, S. R. (1987). Psychobiology of fetal experience in the rat. In N. E. Krasnegor, E. M. Blass, M. A. Hofer, and W. P. Smotherman (Eds.), *Perinatal Development. A Psychobiological Perspective*, pp. 39–60. Orlando: Academic Press.

Smotherman, W. P., Robinson, S. R., La Vallée, P. A., and Hennessy, M. B. (1987). Influences of the early olfactory environment on the survival, behavior and pituitary-adrenal activity of caesarean delivered preterm rat pups. *Dev. Psychobiol.* 20: 415–423.

Smotherman, W. P., and Robinson, S. R. (1995). Tracing Developmental trajectories into the prenatal period. In J. P. Lecanuet, W. P. Fifer, N. A. Krasnegor, and W. P. Smotherman, Eds), *Fetal Development: A Psychobiological Perspective*, pp. 15–32. Erlbaum: Hillsdale, N.J.

Solbu, E. H., Jellestadt, F. K., and Straetkvern, K. O. (1990). Children's sensitivity to odor of trimethylamine. *J. Chem. Ecol.* 16: 1829–1840.

Soussignan, R. and Schaal, B. (1996a). Children's facial responsiveness to odors: Influences of hedonic valence of odor, Gender, age and social presence. *Dev. Psychol.* 32: 367–379.

Soussignan, R. and Schaal, B. (1996b). Forms and social value of smiles associated to pleasant and unpleasant sensory experience. *Ethology* 102: 1020–1040.

Soussignan, R., Schaal, B., Marlier, L., and Jiang T. (1997). Facial and autonomic responses to biological and artificial olfactory stimuli in human neonates: Re-examining early hedonic discrimination of odors. *Physiol. Behav.* 62: 745–758.

Stein, M., Ottenberg, P., and Roulet, N. (1958). A study of the development of olfactory preferences. *Arch. Neurol. Psychiatr.* 80: 264–266.

Steiner, J. E. (1977). Facial expressions of the neonate infant indicating the hedonics of food-related stimuli. In Weiffenbach, J. M. (Ed.), *Taste and Development, The Genesis of Sweet Preference*, pp. 173–189. NIH-DHEW, Bethesda, MD.

Steiner, J. E. (1979). Human facial expressions in response to taste and smell stimulations. *Adv. Child Dev.* 13: 257–295.

Stevenson, R. J. and Repacholi, B. (2003). Age-related changes in children's hedonic response to male body odor. *Dev. Psychol.* 39: 670–679.

Stevenson, R. J. and Repacholi, B. (2005). Does the source of an interpersonal odour affect disgust? A disease-risk model and its alternatives. *Eur. J. Soc. Psychol.* 35: 375–401.

Stevenson, R. J., Mahmut, M., and Sundqvist, N. (2007). Age-related changes in odor discrimination. *Dev. Psychol.* 43: 253–260.

Stickrod, G., Kimble, D. P., and Smotherman, W. P. (1982). *In utero* taste-odor aversion conditioning of the rat. *Physiol. Behav.* 28: 5–7.

Stirnimann, F. (1936). Versuche über Geschmack und Geruch am ersten Lebenstag. *Jhb. Kinderhlk.* 146: 211–227.

Strickland, M., Jessee, P., and Filsinger, E. E. (1988). A procedure for obtaining young children's reports of olfactory stimuli. *Percept. Psychophys.* 44: 379–372.

Sullivan, R. M., and Toubas, P. (1998). Clinical usefulness of maternal odors in newborns: soothing and feeding preparatory movements. *Biol. Neonate* 74: 402–408.

Sullivan, R. M., Taborsky, S. B., Mendoza, R., et al. (1991). Olfactory classical conditioning in neonates. *Pediatrics* 87: 511–517.

Suss, C., Gaylord, S., and Fagen, J. (2012). Odor as a contextual cue in memory reactivation in young infants. *Inf. Behav. Dev.* 35: 580–583.

Svensson, C. K. (1997). Clinical pharmacokinetics of nicotine. *Clin. Pharmacokin.* 12: 30–40.

Teicher, M. H., and Blass, E. M. (1977). First suckling response in the newborn albino rat: the roles of olfaction and amniotic fluid. *Science* 198: 635–636.

Teuscher, E., Anton, R., and Lobstein, A. (2005). Plantes aromatiques. *Epices, aromates, condiments et huiles essentielles.* Lavoisier: Paris.

Thomas, A. M. and Murray, F. S. (1980). Taste perception in young children. *Food Technol.* 2: 38–41.

Todd, W. A. (1979). Psychosocial problems as a major complication of an adolescent with trimethylaminuria. *J. Pediatr.* 94: 936–937.

Todrank, J., Heth, G., and Restrepo, D. (2011). Effects of *in utero* odorant exposure on neuroanatomical development of the olfactory bulb and odor preferences. *Proc. Biol. Soc.* 278: 1949–1955.

Toulouse, N. and Vaschide, S. (1899a). Mesure de l'odorat chez les enfants. *Séances Mém. Soc. Biol.* 51: 487–489.

Toulouse, N. and Vaschide, S. (1899b). Mesure de l'odorat chez l'homme et la femme. *Séances Mém. Soc. Biol.* 51: 381–383.

Upadhyay, A., Aggarwal, R., Narayan, S., et al. (2004). Analgesic effect of expressed breast milk in procedural pain in term neonates: a randomized, placebo-controlled, double-blind trial. *Acta Paediatr.* 93: 518–522.

Valentin, D. and Chanquoy, L. (2012). Olfactory categorization: A developmental study. *J. Exp. Child Psychol.* 113: 337–352.

Varendi, H., Porter, R. H., and Winberg, J. (1994). Does the newborn baby find the nipple by smell? *Lancet* 344: 989–990.

Varendi, H., Porter, R. H., and Winberg, J. (1996). Attractiveness of amniotic odor: evidence for prenatal olfactory learning? *Acta Paediatr.* 85: 1223–1227.

Varendi, H. and Porter, R. H. (2001). Breast odor as the only maternal stimulus elicits crawling towards the odour source. *Acta Paediatr.* 90: 372–375.

References

Varendi, H., Porter, R. H. and Winberg, J. (2002). The effect of labor on olfactory exposure learning within the first postnatal hour. *Behav. Neurosci.* 116: 206–211.

Verron, H. and Gaultier, C. (1976). Processus olfactifs et structures relationnelles. *Psychol. Franç.* 21: 205–209.

Vince, M. A. (1993). Newborn lambs and their dams: the interaction that leads to sucking. *Adv. Study Behav.* 22: 239–268.

Wedekind, C. and Füri, S. (1997). Body odour preferences in men and women: do they aim for specific MHC combinations or simple heterozygosity? *Proc. R. Soc. Lond. B*, 264: 1471–1479.

Wedekind, C., Seebeck, T., Bettens, F., and Paepke, A. J., (1995). MHC-dependent mate preferences in humans. *Proc. R. Soc. Lond. B* 260: 245–249.

Weisfeld, G. E., Czilly, T., Phillips, K. A., et al. (2003). Possible olfaction-based mechanisms in human kin recognition and inbreeding avoidance. *J. Exp. Child Psychol.* 85: 279–295.

Westermarck, E. (1891). *The History of Human Marriage*. Macmillan: London.

Widström, A. M., Arvidson, A. B., Christensson, K., et al. (1987). Gastric suction in healthy newborn infants: effects on circulation and developing feeding behaviour. *Acta Paediatr. Scand.* 76: 566–562.

Widström, A. M., Lilja, G., Michalias, P. A., et al. (2011). Newborn behaviour to locate the breast when skin-to-skin: a possible method for enabling early self-regulation. *Acta Paediatr.* 100: 79–85.

Willander, J. and Larsson, M. (2007). Olfaction and emotion: the case of autobiographical memory. *Mem. Cognition* 35: 1659–1663.

Wilson, D. A. and Sullivan, R. M. (1994). Neurobiology of associative learning in the neonate: early olfactory learning. *Behav. Neural Biol.* 61: 1–18.

Wuensch, K. L. (1975). Exposure to onion taste in mother's milk leads to enhanced preference for onion diet among weanling rats. *J. General Psychol.* 99: 163–167.

Wysocki, C. J., Dorries, K. M., and Beauchamp, G. K., (1989). Ability to perceive androstenone can be acquired by ostensibly anosmic people. *Proc Nat. Acad. Sci. USA* 81: 7976–7978.

Yamazaki, K., Boyse, E. A., Mike, V., et al. (1976). Control of mating preferences in mice by genes in the major histocompatibility complex. *J. Exp. Med.* 144: 1324–1335.

Yamazaki, K., Beauchamp, G. K., Kupniewski, D., et al. (1988). Familial imprinting determines H-2 selective mating preferences. *Science* 240: 1331–1332.

Yamazaki, K., Beauchamp, G. K., Curran, M., et al. (2000). Parent-progeny recognition as a function of MHC odortype identity. *Proc. Nat. Acad. Sci. USA* 97: 10500–10502.

Yasumatsu, K., Uchida, S., Sugano, H., and Suzuki, T. (1994). The effect of the odour of mother's milk and orange on the spectral power of EEG in infants. *J. UOEH* 16: 71–83.

Yildiz, A., Arikan, D., Gözüm, S., et al. (2011). The effect of the odor of breast milk on the time needed for transition from gavage to total oral feeding in preterm infants. *J. Nurs. Scholarsh.* 43: 265–273.

Youngentob, S. L., and Glendinning, J. I. (2009). Fetal ethanol exposure increases ethanol intake by making it smell and taste better. *Proc Nat. Acad. Sci. USA* 106: 5359–5364.

Youngentob, S. L., Kent, P. F., Sheehe, P. R., et al. (2007). Experience-induced fetal plasticity: The effect of gestational ethanol exposure on the behavioral and neurophysiologic olfactory response to ethanol odor in early postnatal and adult rays. *Behav. Neurosci.* 121: 1293–1305.

Zeinstra, G., Koelen, M., Colindres, D., et al. (2009). Facial expressions of school-aged children are good indicators of "dislikes", but not of "likes". *Food Qual. Pref.* 20: 620–624.

Chapter 15

Olfactory Memory

THERESA L. WHITE, PER MØLLER, E. P. KÖSTER, HOWARD EICHENBAUM, and CHRISTIANE LINSTER

15.1 INTRODUCTION

Olfactory memory plays an important role in the everyday lives of both animals and humans, even if people generally attend much less than animals to incoming olfactory information. Each of the three main functions of olfaction outlined by Stevenson (2010) would be impossible without odor memory: ingestive behaviors (e.g., food detection, flavor determination, breast finding), avoiding environmental hazards (e.g., fear related, disgust related), and social communication (e.g., reproductive behaviors, emotional contagion, territorial demarcation). Whether trying to remember whether the fruit of a tree is edible or toxic, or trying to determine friend from foe, the past experiences accessed through olfactory memory are critical in making the decision.

The phrase "odor memory" generally has two meanings. The first meaning reflects "odor-evoked memory". Most of us are familiar with "Proustian effects" (Proust, 1928), or the ability of odor memory to evoke rich recollections of times and events gone by (Chu and Downes, 2000; 2002; Willander and Larsson, 2006; 2007; 2008; Herz and Schooler, 2002). Although these effects emphasize the episodic nature of olfactory memory, the memories in question are multi-sensorial, making them odor-evoked memories (Herz, 2012; Herz and Cupchik, 1992; 1995). This meaning of odor memory also illustrates the way that odors are associated with a variety of other stimuli through experiences and learning (Wilson and Stevenson, 2006). The second meaning of "odor memory" concerns the way that the odors themselves are remembered and recognized at a later time. Historically, the study of this type of odor memory has been dominated by tasks that have their counterparts in the study of vision, such as discrimination, recognition and identification (e.g., Engen and Ross, 1973; Mozell, 1972; Goldman and Seamon, 1992; Jehl et al., 1994). These studies have led to some theories related to this meaning of odor memory that place odor semantics into prominence by invoking "odor objects" (Wilson and Stevenson, 2006), a form of internal representation that can be changed by such processes as odor-odor and odor-taste learning. Thus, from this theoretical viewpoint, recognition and identification of the "odor object" seem to be the most important tasks of olfactory learning and memory. Other theoretical views concerning the way that odors are remembered have emphasized the implicit and episodic nature of olfactory cognition (e.g., Zucco, 2003), and question the role of odor percept recollection in everyday memory use (Köster, 2005).

Rather than attempting to dissociate theoretical viewpoints, this chapter focuses on recognition memory and learning paradigms useful in characterizing the olfactory memory system. Since the basic anatomy and physiology of the olfactory system is well known and is described in great detail elsewhere in this volume (Chapters 1–10), only brief mention of anatomical structures critical for odor memory is made in the sections that follow. Among the brain regions most critical for odor memory are the hippocampus and the piriform, entorhinal, and orbitofrontal cortices, each of which seems to contribute to different memory-related functional properties (Eichenbaum, 1997; Petrulis and Eichenbaum, 2003a; Gottfried, 2010).

Handbook of Olfaction and Gustation, Third Edition. Edited by Richard L. Doty.
© 2015 Richard L. Doty. Published 2015 by John Wiley & Sons, Inc.

15.2 RECOGNITION MEMORY

Adaptive behavior is characterized by the ability to remember information over both short and long time intervals. Although some memory systems (such as the verbal system) allow humans to reproduce a stimulus in order to indicate memory via recall (i.e., to repeat the to-be-remembered word out loud), demonstration of olfactory memory is generally restricted to recognition paradigms.

15.2.1 Short-term Memory

At intensities that are above threshold, a normal sense of smell enables many animals to easily distinguish the scent of two different odorants from one another, such as the odor of a strawberry from that of a lemon, even when the odorants are sniffed sequentially. This simple discrimination task requires the activation of a temporary internal representation of each odorant, and shifting attention toward different parts of each of the representations so that they can be compared, all the while keeping readily available the parts of the representations outside of the current attentional focus. Essentially this is a task that requires a short-term or working memory, a type of memory that is critical to the performance of any sort of olfactory comparison (White, 2012). Working memory is generally regarded as a type of memory that is responsible for the temporary maintenance and manipulation of information (e.g., Baddeley and Hitch, 1974). Such memory is possible because olfactory information, like visual or verbal information, can be maintained in a temporary memory state (White, 1998; White et al., 1998; Andrade and Donaldson, 2007; Zelano et al., 2009; Jönsson et al., 2011). Brain activation continues after the termination of the presentation of an odorant (Rolls et al., 2008), and the location of that activity seems to differ, depending on whether the odorant can be easily named. Odorants that are not easily named are most strongly associated with activity in primary olfactory (piriform) cortex, while odorants that have been verbally re-coded are more strongly associated with activity in other areas (Zelano et al., 2009).

15.2.1.1 Habituation/Discrimination. Habituation is one of the simplest forms of memory, seen in most animals and sensory systems (Castellucci et al., 1970; Ezzeddine and Glanzman, 2003; Kandel, 2001), including olfaction. Olfactory stimuli can elicit a variety of responses which decrease as a result of repeated stimulation over time. Such responses range from autonomic reflexes to behavioral arousal and investigation. Two primary behavioral paradigms have been used to study odor habituation in mammals: odor-evoked heart-rate reflex and active investigation of odorants.

The simplest, both in terms of underlying neural circuitry and interpretation, is the odor-evoked heart-rate orienting reflex. Novel sensory stimuli evoke a bradycardia reflex, which habituates with repeated stimulation (McDonald et al., 1964). The second major behavioral paradigm for investigating odor habituation in mammals is monitoring investigation of scented objects (Cleland et al., 2002; Hunter and Murray, 1989). An object scented with a novel odor is presented, and the duration of investigation is monitored and compared with investigation times on subsequent presentations of the same scent. Over repeated trials, investigation times decrease.

Habituation paradigms can be used to measure the formation of a simple memory, its duration, specificity, and neural correlates (Wilson and Linster, 2008). For example, after habituation to a specific stimulus, responses to novel stimuli can indicate how well an animal discriminates between the two stimuli (Cleland et al., 2002). Presentation of a habituated stimulus after varying time intervals also provides a means for testing the duration of memory. In rodents, habituation memory and specificity have been shown to be altered by such task demands as retention intervals (McNamara et al., 2008), as well as by such neuromodulators as acetylcholine and norepinephrine (Mandairon et al., 2006; Mandairon et al., 2008). Habituation paradigms are also commonly used to study species-specific communication such as recognition among males or females, discrimination between individuals, and scent marking (Petrulis, 2009; Petrulis et al., 2005).

Although olfactory sensory neurons can adapt to repeated or prolonged odor stimulation, behavioral odor habituation is a central phenomenon. As is the case in other sensory systems, central olfactory neurons show greater response decrement than peripheral receptors, with piriform cortical neurons showing rapid, nearly complete response adaptation within seconds or minutes of stimulation in both rodents (Wilson, 1998a; b) and humans (Sobel et al., 2000). Simultaneous recordings of second-order mitral cells and their target piriform cortical pyramidal cells have shown that the cortical neurons adapt to repeated or prolonged odor stimulation more rapidly and completely than their afferent mitral cells (Wilson, 1998a). Mitral cells do adapt to odors (Chaudhury et al., 2010); however, adaptation of their downstream targets is frequently fast and relatively complete. In vivo intracellular recordings have shown that the cortical adaptation is associated with depression of the glutamatergic mitral-pyramidal cell synapse and that this synaptic depression recovers with the same time course as the short-term adaptation of odor-evoked postsynaptic potentials (Wilson, 1998a). This short-term depression is mediated by group III metabotropic glutamate receptors. In addition to the short-term habituation described above, extended and/or spaced odor exposure can lead to long-term habituation. In humans exposed daily to odors in the home or workplace, habituation to the exposed odor (elevated detection threshold) can last weeks (Dalton and Wysocki, 1996). Electrophysiological recordings in rats using the same stimulus paradigm used in behavioral studies showed that olfactory bulb mitral cells decrease their response to odor stimulation applied with 5-minute intertrial intervals and that this response decrease is dependent on functioning NMDA receptors in the OB (Chaudhury et al., 2010).

While the neural correlates of olfactory habituation in mice and rats are thought to be in olfactory bulb and cortex, habituation paradigms studying species-specific interactions in hamsters have located cellular correlates to higher structures such as enthorinal cortex, frontal cortex, and amygdala (Maras and Petrulis, 2008; Petrulis et al., 2005; Petrulis et al., 1998; Petrulis and Eichenbaum, 2003b). It is most likely that, as in other memory functions, the formation of habituation memory is widely distributed across brain areas and depends on the types of olfactory stimuli used as well as on the behavioral relevance of these stimuli.

15.2.1.2 Serial Memory..
As with many types of non-olfactory stimuli, memory for a series of odors is typically characterized in both humans and animals by a tendency to remember the beginning (primacy) and ending (recency) of the series better than the odorants in the middle (see Figure 15.1). Serial position effects offer an additional opportunity to examine olfactory working memory, as the recency portion of the curve is often associated with short-term memory (Jonides et al., 2008). However, in stark contrast to verbal and visual memory, the effects of serial position are not always observed in olfactory memory (Lawless and Cain, 1975; Gabassi and Zanuttini, 1983;

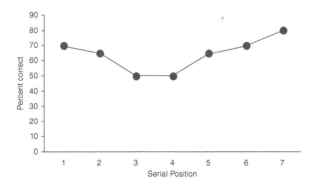

Figure 15.1 Typical serial position curves, illustrating primacy and recency effects. Memory for the first few items constitutes the primacy portion of the curve, while memory for the last few items indicates effects of recency. The shape of the curve varies considerably for studies that require remembering odors in a series.

Johnson and Miles, 2009). When serial position effects are observed, reports of the shape of the curve vary considerably across experiments (e.g., White and Treisman, 1997; Reed, 2000; Miles and Hodder, 2005). For example, although primacy is occasionally observed (e.g., Reed, 2000), the more common finding is a curve characterized by recency only (Deacon and Rawlins, 1995; Miles and Hodder, 2005; Johnson and Miles, 2009; Fortin et al., 2002; Miles and Jenkins, 2000; White and Treisman, 1997). The memory for odorants in a series appears to be dependent on the hippocampal and medial prefrontal brain regions (Fortin et al., 2002; DeVito and Eichenbaum, 2011), athough the ability to remember individual odors seems to be dependent on the parahippocampal region (Otto and Eichenbaum, 1992; Petrulis and Eichenbaum, 2000).

15.2.1.3 Interference.
In other sensory systems, short-term memory stores are conceptualized as having limited capacity. If such were the case for olfaction, incoming information should interfere with information currently residing in the temporary storage center (thus producing the characteristic "recency" portion of the serial position curve). Such retroactive interference has been observed, in which the presentation of a new odor interferes with memory for a previously presented odorant (Walk and Johns, 1984; Miles and Jenkins, 2000; Thor and Holloway, 1982; Rochefort et al., 2002). Additionally, attempts by people to perform a concurrent olfactory task impair olfactory memory performance (Andrade and Donaldson, 2007), thus demonstrating retroactive interference. Although both proactive and retroactive interference have been demonstrated with olfactory stimuli, interference in olfactory memory is not a consistently observed phenomenon. Several studies have presented results that did not show any decrements in performance due to interference (Zucco, 2003; Schab and Crowder, 1995) and some show only proactive interference (Lawless and Engen, 1977) or interference only when explicit memory is tested (Stevenson et al., 2005).

15.2.1.4 Delayed Match and Non-match to Sample Tasks.
The delayed non-match- to-sample task (DNMS) is a common and effective method of non-verbally evaluating olfactory recognition performance (Gaffan, 1974; Mishkin, 1978). This task consists of three phases: a sample phase where subjects are rewarded for investigating a novel object, a delay phase where the memory is retained, and a test phase, where the subject is rewarded for selecting a novel (non-matching) object over a familiar one (e.g., Eichenbaum et al., 2000). A variant of this task is the delayed match-to-sample which involves rewarding the animal for selecting the familiar object in both the sample and test phases (DMS; e.g., Yeshurun et al.,

2008). In both cases, the subject must remember the odor that has been previously presented over a delay interval that is sometimes brief (Otto and Eichenbaum, 1992) and sometimes extended (Ramus et al., 2007), thereby demonstrating declarative and episodic memory for that odorant (Suzuki and Eichenbaum, 2000).

The DNMS or DMS technique has been used to evaluate olfactory memory performance in a variety of animal models, including rats (e.g., Otto and Eichenbaum, 1992), monkeys (Gaffan, 1974), and humans (Kesslak et al., 1991; Yeshurun et al., 2008). Most species show increased forgetting when the size of the delay phase is increased (Koger and Mair, 1994; Otto and Eichenbaum, 1992), and when the same stimuli appear more frequently (Koger and Mair, 1994; Otto and Eichenbaum, 1992). Acquisition of this task is dependent upon an intact orbitofrontal cortex (Otto and Eichenbaum, 1992), but performance seems related to medial temporal lobe activity (Suzuki and Eichenbaum, 2000), particularly activity within the perirhinal and lateral entorhinal areas (Eichenbaum et al., 2007).

15.2.2 Long-term Memory

Long-term memory refers to representations of the past that have been stored for an extended period of time, and, in many models of memory, is theorized to be separate and distinct from working memory (e.g., Baddeley and Hitch, 1974). Some alternative theories of memory suggest a unitary memory store (e.g., Ranganath and Blumenfeld, 2005; Stevenson and Boakes, 2003) that is based entirely on activation of long-term processes. In these models, reliance on long-term processes suggests that the ability to perceive odors is dependent upon olfactory experience or knowledge (Rabin, 1988; Wilson and Stevenson, 2006).

15.2.2.1 Explicit Memory. Explicit or declarative memory refers to consciously available memories. Explicit memory is a complex function, which relies on many areas of the brain. Imaging studies in humans, as well as lesion studies in animals, show activation in the prefrontal cortex and the amygdala, hippocampus, and rhinal cortices in the temporal lobe (Zola-Morgan and Squire, 1993; Kolb and Whishaw, 2003).

FORMAL MEMORY TASKS Formal long-term memory tasks differ across species, but each seems to test memory for the "odor object" over time. Humans are often tested using an item recognition paradigm, while animals are tested with a variety of techniques (Wood et al., 2000; Otto et al., 1991; Ramus et al., 2007), including novel-odor-recognition (White and Youngentob, 2004) and juvenile recognition (which is largely odor dependent) (Thor and Holloway, 1982; Petrulis and Eichenbaum,

2003a), both of which (like the human technique below) require the subject's recognition of novelty. In both of these latter tasks, odors (or a juvenile) are presented for several minutes to a rodent, followed by a delay of varying length. The rodent is then presented with either the same odor (or juvenile) or a different one, and the amount of inspection time for each is measured. In both of these paradigms, mice are able to remember olfactory information over delays of a day or more (Kogan et al., 2000; White and Youngentob, 2004), though forgetting in rats is found at much shorter delays (on the order of 30 minutes) when juvenile conspecifics serve as the stimuli (Bluthe and Dantzer, 1990).

In the typical experiment on item recognition, human subjects smell a series of odorants and are subsequently tested with another series of odors that includes both the originally presented odors and an equal number of new ones. Participants must distinguish each odor as "old" or "new". The number of odors in the initial series, as well as the length of delay prior to the recognition test, varies considerably between studies (e.g., Engen and Ross, 1973; Lawless, 1978), yet a reasonably similar overall picture emerges from such studies. Despite poor initial encoding, humans are able to remember a large number of smells over a long period of time. For example, in one study the subjects were able to accurately recognize an average of 70% of a 48 odor set when tested after a period of 30 days (Engen and Ross, 1973). Similarly, an array of 22 odors was recognized to 75% accuracy after 28 days (Lawless and Cain, 1975), and 24 odors were recognized at the 75% level after a 4-month delay (Lawless, 1978). The large number of odors in each set that is retained over long time periods is seemingly independent of set size, indicating a large capacity for olfactory long-term memory.

FAMILIARITY VS. RECOLLECTION Recent studies on odor recognition memory have exploited findings from cognitive science and neuroscience that employ signal detection analysis. This approach is based on the widely held view that recognition is supported by two processes: episodic recollection of previously presented stimuli and/or a sense of familiarity for recently experienced stimuli (Yonelinas, 2001). A typical human signal detection investigation of odor memory uses a recognition paradigm, in which people designate odors as "old" or "new". Receiver operating characteristic (ROC) curves then relate the proportion of "hits" (correct identifications of old items) to that of false alarms (incorrect identifications of new items as "old") across a range of response criteria, as described below. Performance is then plotted as two dimensional [P(hits) vs P(false alarms)] data points. Memory is reflected where P(hits) > P(false alarms), i.e., in data points that lie above the diagonal of the ROC curve,

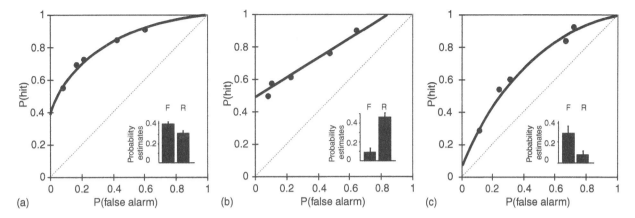

Figure 15.2 ROC analyses of odor recognition memory. Each panel shows ROC function indicating the strength of recollection as the asymmetry of the function and familiarity as the degree of curvilinearity. In addition, normalized probability estimates of recollection (R) and familiarity (F) are compared in insets. (a) item recognition. (b) associative recognition. (c) item recognition with response deadline.

which indicates chance accuracy at different response criteria levels (Figure 15.2).

An example study using an ROC approach to characterize odor recognition memory in rats is provided by Fortin et al. (2004). In this study, the memory cues were composed of a large pool of ordinary household odors (e.g., lemon, thyme) mixed in sand within small plastic cups. A series of 10 stimuli were presented during the study phase, with a bit of sweetened cereal buried in the sand of each cup (Fortin et al., 2004). On each successive stimulus presentation, animals were allowed to dig for that reward. After a 30 minute delay, a series of 20 "target cups" was presented, consisting of a random ordering of 10 old odors (those presented in the sample phase) and 10 new odors taken from the pool. The test phase involved a non-match contingency, such that rats could obtain rewards by digging only in target cups containing new odors. In addition, they could also obtain a reward in an alternate cup in the back of the cage if they refrained from digging in target cups that contained old odors. To manipulate the animal's bias for responding (or not responding) to target cups, both the height of the target cup and the ratio of reward magnitude in the target cup versus that in the alternate cup were varied. Under these conditions, the rats were more likely to refrain from digging (i.e., identify this item as "old") in a target cup containing a new odor if they could obtain only a small reward or had to apply more effort (corresponding to a liberal threshold for "old" responses in humans). Conversely, rats were more inclined to dig in a target cup (i.e., identify an odor as "new") which contained a greater reward or required less effort to acquire (corresponding to a conservative threshold in humans). The possibility that rats can smell the rewards buried in target cups was controlled for by the use of probe trials in which the reward was not present in a cup with a new odor and instead was given only after digging commenced.

Using this protocol, the ROC curve of intact rats contained both an asymmetrical component (above-zero Y-intercept) and a strong curvilinear component (Figure 15.2a). This pattern is remarkably similar to the ROC of humans in verbal recognition performance, and consistent with a combination of recollection-like and familiarity-based components of recognition in animals (see Yonelinas, 2001). To examine whether the recollection (asymmetry) component of the olfactory ROC could be dissociated from the familiarity (curvilinearity) component, an associative recognition paradigm for rats was developed, using stimulus pairs composed of combinations of an odor mixed into one of several digging media (e.g., wood chips, beads, sand) contained in a cup (Sauvage et al., 2008). Each day the animals would initially sample a series of 10 odor-medium pairings, then following a 30 minute delay, had to distinguish re-presentations of the 10 original (old) pairs from 10 rearranged (new) pairings of the same odors and media, using the same non-matching rule and manipulations of bias as in our study on item recognition described above. The resulting ROC function was highly asymmetric, indicating the presence of a strong recollection component (Figure 15.2b). Furthermore, the shape of the ROC was linear, indicating the absence of a significant familiarity component. This pattern is similar to the ROC function of human subjects when they rely selectively on recollection in associative recognition and source memory studies (Yonelinas, 2001).

To examine whether the familiarity (curvilinearity) component of the olfactory ROC can be dissociated from the recollection (symmetry) component, another variant of the protocol was developed to exploit a distinction between the speed of recollection and familiarity as observed in humans (Sauvage et al., 2010). Familiarity was characterized as a perceptually driven, pattern matching process that is completed rapidly, whereas recollection was characterized as a conceptually driven, organizational process that requires more time. A prediction from dual process theory was tested, that the addition of an early response deadline at the memory test phase would reduce the contribution of the slower recollection process while sparing that of the more rapid familiarity process. Animals were initially trained and tested using the same item recognition task without a response deadline that produces an ROC function that is both asymmetrical and curvilinear, replicating earlier results (Fortin et al., 2004). In this condition, the critical measure was the time to dig in the test cups containing new odors, or to turn away from test cups containing old odors, in order to determine normal response latencies. In subsequent testing, response deadlines that were half of the natural response latencies were invoked, using a cover to prevent digging in the test cup when the deadline arrived. Analysis of the ROC functions showed that, under the response deadline condition, performance was supported by familiarity as reflected by a curvilinear ROC function (Figure 15.2c). However, in striking contrast to the ROC function in the no-deadline condition, recollection did not contribute significantly to recognition performance such that the ROC function became fully symmetrical. Collectively, these findings support the view that familiarity-based responses can be accomplished rapidly whereas recollection based responses are slow, consistent with the dual process model.

ROLE OF THE HIPPOCAMPUS AND AMYGDALA

The ROC studies described above provide a strong foundation for examining the role of the hippocampus in episodic recollection. In the following studies, the hypothesis was that the hippocampus is selectively involved in recollection and not familiarity, and is also not involved in learning the basic rules of the non-matching task or in sensitivity to manipulations that affect response biases. ROC analyses then should show that rats with isolated hippocampal damage are able to perform the non-matching task and their response biases should not be affected. Furthermore, animals with hippocampal damage should have an intact familiarity (curvilinearity) component of the ROC function but a loss of the recollective (asymmetry) component. These expectations were fully supported in a study showing that rats with localized hippocampal damage have a fully symmetrical and curvilinear ROC function, consistent with the loss of recollection and sparing of familiarity, respectively (Fortin et al., 2004). The conclusion that the hippocampus is selectively involved in episodic recollection but not in familiarity was further confirmed by observations that, in the associative recognition variant of the task, familiarity is enhanced in the absence of normal hippocampal function (Sauvage et al., 2008). As described above, the ROC function of normal rats in associative recognition was strongly asymmetrical

and linear, consistent with strong recollection and absence of familiarity, respectively. Animals with hippocampal damage suffered a significant decrease in the asymmetry of the ROC, indicating a deficit in recollection. Furthermore, after damage to the hippocampus, the shape of the ROC function became curvilinear, consistent with enhanced and compensatory use of familiarity. The double-dissociation between reduced recollection and enhanced familiarity is entirely consistent with the hypothesis that the hippocampus plays a selective role in recollection and in its absence performance is supported by familiarity.

An additional study examined the role of the amygdala in odor recognition memory using the signal detection approach. The amygdala is connected to the perirhinal cortex and hippocampus, so it might influence the role of either area in recognition memory. However, the connection to perirhinal cortex is exceptionally robust, suggesting that the amygdala might differentially contribute to perirhinal familiarity representations by supporting an increased attractiveness of the study stimuli associated with recent "mere exposure" and rewards. Consistent with these expectations, bilateral lesions of the amygdala resulted in a decrease in the curvilinearity of the ROC function, consistent with a loss of odor familiarity, without a significant change in the ROC symmetry, consistent with no effect on recollection (Farovik et al., 2011). Combined with the findings on hippocampal damage, these studies reveal a double-dissociation between critical contributions of the hippocampus in odor recollection and of the amygdala in odor familiarity.

15.2.2.2 Implicit Memory.
Implicit memory is a form of memory in which previous experiences influence later behavior without awareness of these previous experiences. In everyday life, it may be the most important form of odor memory. We encounter smells of our surroundings on a daily basis. Each room, each part of the garden, and even each person, has a different smell. Yet we seldom notice smells consciously and, if we do so, it is either because we are surprised by a smell or because we intentionally sniff to verify the safety and/or the pleasantness of our surroundings. In everyday life, conscious olfactory perception seems to be a form of "perception by exception" that occurs spontaneously only when the inhaled air or odor does not fit expectations. This means that odor memory plays a very special role in olfactory perception: it helps people <u>not to notice</u> the well-known odors in the surroundings, thus increasing the likelihood that novel stimuli will be detected.

NOVELTY AND CHANGE DETECTION RATHER THAN RECOLLECTION
One technique for studying implicit memory for odors in people is repetition priming

(Olsson and Cain, 1995; Olsson, 1999; Olsson and Friden, 2001), which compares the reaction time for the identification of previously presented odors to that of new odors. Response latencies of the previously presented odors that later were identified (or named) by the participant were longer than the latencies of previously presented odors by controls, while the reverse was true for the unidentified odors (Olsson, 1999). Thus, in this technique, semantic knowledge about the odor, such as its name, seemed to hamper implicit memory rather than help it, suggesting that olfactory memory may be more designed for change detection than labeling. This finding has also been demonstrated in studies that used a different technique for examining implicit memory in which incidentally learned associations were formed between odors and the rooms in which they were unconsciously perceived (Degel and Köster, 1999; Degel et al., 2001).

Additional support for the idea that, at least in olfaction, novelty or change detection is more important than recollection has also been obtained in experiments on olfactory same-different detection (Møller et al., 2012). In two experiments, it was shown that in olfaction "different" decisions were made more quickly than "same" decisions (Møller et al., 2012), while in vision, the reverse is true (Posner, 1986; Luce, 1986). In those experiments, people who claimed to have the ability to imagine odors were much faster in their same-different judgments than those who could not imagine odors, though their "different" judgments were still faster than "same" judgments. This suggests that the capability for an internal representation of the odor influences decision speed, but that novelty and change detection prevail over recollection even in those who seem to be able to recollect their odor memories willfully.

SOCIAL RECOGNITION OF BODY ODORS Long-term recognition memory for olfactory stimuli can be demonstrated via social recognition of body odors, such as kin recognition and pair-bonding. Each of these social activities requires the individual to have retained memory for the odor of another individual over an extended period of time following initial encoding of that individual's scent.

Kin recognition is the ability of parents to recognize the scent of their offspring (and vice versa), and can be observed in a variety of species, including rodents (Todrank et al., 1998) and humans (Porter and Moore, 1981). This recognition of the odor of the offspring could be an evolutionary phenomenon possibly evolved to prevent inbreeding and appears to be achieved with very little conscious recognition (Lundström et al., 2009).

Shortly after mating, pair-bonding species that do not typically form long-term attachments (such as hamsters or rats) prefer the odor of a novel potential mate over the odor of the recent partner (e.g., Carr et al., 1970). This variant of the "Coolidge effect", in which sexual arousal is enhanced by exposure to new mating partners, is dependent on olfactory (as opposed to vomeronasal) perception (Johnson and Rasmussen, 1984), and does not seem to be hippocampally dependent (Petrulis and Eichenbaum, 2000). An intact perirhinal–entorhinal cortex is required for the effect, however, suggesting a strong role for that structure in the natural formation of social olfactory recognition (Petrulis and Eichenbaum, 2003b) and that the general familiarity/novelty detection process that is widely used for visual object and odor recognition tasks is involved in the effect (Petrulis, 2009). Olfactory-based recognition of mates also has been readily observed in a number of species that form long-term attachments, including prairie voles (Newman and Halpin, 1988) and humans (Hold and Schleidt, 1977).

SEMANTICS-FREE ODOR MEMORY In most explicit odor recognition experiments, the stimuli have been common odorants disconnected from the way that they are normally perceived. As a result, it is almost impossible to separate the roles of pure odor memory and verbal or semantic memory in people. In implicit memory tasks, semantic memory is avoided through the use of uncommon and unidentifiable odors presented at concentration levels that do not catch the conscious attention of the subject. Another technique is to use experimental paradigms in which the subject believes that they are participating for a different purpose than just odor exposure (Degel et al., 2001). Another way of securing semantics-free memory has been used in food-related experiments in which the subjects were exposed to target foods during a meal that was offered to them between other experiments without any reference to a possible later memory test (Mojet and Köster, 2002; 2005; Morin-Audebrand et al., 2011). When later unexpectedly tested they had to recognize the target foods they had eaten between distracters that were only slight variations of the target (differing only at just noticeable difference level in one aspect) and therefore had the same overall taste, appearance and, as a result, the same name as the target. In these experiments, the target odors were not recognized any better than chance guessing, but the distracters were significantly rejected and the subjects were more sure of these negative decisions.

DIFFERENCES IN IMPLICIT AND EXPLICIT MEMORY In one set of studies, elderly people with a still intact sense of smell performed at least as well, and in one case even slightly better, than young people in implicit memory tasks while this pattern was reversed when intentional learning was involved (Møller et al., 2004; Møller et al., 2007). This performance pattern was primarily due to a better rejection of the distracters in the intentional test phase by the young people, as their hit rate was at roughly at chance level (Møller et al., 2004; Møller et al., 2007) and could not be attributed to self-developed verbal descriptors for the odorants (Møller et al., 2004). Whether this difference in ages reflects the fact that incidents of intentional learning become less frequent later in life, despite continued incidental learning, or whether there are real impediments of intentional olfactory encoding involved in the elderly, is not yet clear.

15.3 LEARNING

15.3.1 Appetitive Conditioning of Discrimination of Odor Cues

Novel odors can be paired with positive consequences (such as food) or negative consequences (such as illness or shock) in order to gain appetitive meaning (Galef and Giraldeau, 2001; Laska et al., 2003; Li et al., 2008). For both animals and humans, previously acquired meaning for odorants can be modified through subsequent experiences, such as sensory specific satiety (Rolls et al., 1981; Rolls et al., 1986).

15.3.1.1 Discrimination Learning. In discrimination learning, animals learn to differentiate between two stimuli, usually associated with different behavioral outcomes (Nigrosh et al., 1975; Slotnick and Katz, 1974). Most often these stimuli can readily be discriminated and what is learned is the expected response to each stimulus (Slotnick, 2001; Slotnick and Bodyak, 2002; Smith and Cobey, 1994). In some cases, stimuli that cannot be easily discriminated, such as the enantiomers of limonene, alcohols with similar numbers of carbons, highly overlapping odor mixtures, or two concentrations of a same odorant, can be learned to be discriminated in a situation in which an animal is motivated to do so (Cleland and Narla, 2003; Linster et al., 2002; Linster et al., 2001; Pelz et al., 1997). Because it is difficult to differentiate between the acquisition of a rule in discrimination learning and odor discrimination, it is not easy to know how well animals can discriminate between a pair of odorants. It has been shown that rodents can acquire a rule after which they can discriminate between further odor sets much faster and one may assume that these odors can readily be discriminated (Barkai and Saar, 2001; Slotnick, 2001; Slotnick and Katz, 1974). Instead of associating different behavioral outcomes with odorants, one can probe odor discrimination by associating a reward repeatedly with a single odorant and subsequently probing the animal's response to novel odorants (Cleland et al., 2002; Daly et al., 2001; Linster and Smith, 1997; Smith and Cobey, 1994). This paradigm, often referred to as generalization, makes it possible to determine how closely animals associate odors which each other and how this changes as a function of learning parameters, stimulus parameters, and previous experience (Cleland et al., 2009). Previous experience with odorants strongly affects odor discrimination learning, as tested through latent inhibition (Meyer and Louilot, 2011; Sevelinges et al., 2009) or blocking type paradigms (Gerber and Smith, 1998; Pelz et al., 1997; Wiltrout et al., 2003).

Although odor discrimination learning can be strongly affected by manipulations in the olfactory bulb (reviewed by Mandairon and Linster in 2009), and bulbar neurons significantly modulate their activity in response to odor stimuli during and after discrimination training (Doucette and Restrepo, 2008; Kay and Laurent, 1999), neural correlates of odor discrimination can be found in many other structures. Neurons in olfactory bulb, piriform cortex, orbitofrontal cortex, and hippocampus change their response profiles to odorants as a result of discrimination learning (Calu et al., 2007; Roesch et al., 2007; Schoenbaum et al., 1998; 1999; Schoenbaum and Eichenbaum, 1995). In many cases, responses reflect the learning of odor discriminations, especially in the olfactory bulb. In addition to single cell responses, overall dynamics in olfactory structures are also strongly modulated by discrimination learning (Beshel et al., 2007; Kay and Beshel, 2010; Kay and Lazzara, 2010; Martin, et al., 2007). In invertebrates, neural correlates of olfactory discrimination learning have been reported in both the antennal lobe and mushroom body structures (Silbering and Galizia, 2007; Silbering et al., 2008; Szyszka et al., 2005). In summary, discrimination learning is a distributed process across many brain structures and is modulated by stimulus and task parameters, previous experience with odorants, and manipulations of specific neurotransmitter and neuromodulator systems.

15.3.1.2 Odor Memory in Food Preference and Food Habit Formation. Flavor is the integrated experiences of taste, smell, and texture (tactile) sensations, the latter of which are largely mediated via the trigeminal nerve. Since olfactory sensations are crucial in forming the flavor, odor memory plays a substantial role in food acceptance and in food preference formation through a number of mechanisms. For example, in mere exposure (Zajonc, 2001; 1968) the participant is exposed to the same stimulus (e.g., foods) a number of times, which often leads to a reduction of neophobia (Pliner, 1982) and an increase in liking for the stimulus, possibly due to 'the absence of ill effects' (Zajonc, 2001). Most flavor learning in food preference formation is incidental and the kind of memory involved in the changed preference responses to the stimulus is unconscious. This type of learning has been observed in fetuses (Schaal et al., 2000; Mennella et al., 2001; Mennella and Beauchamp, 2002) and pre-verbal

15.3.2 Stimulus-Stimulus Associations

In everyday life, odor memory not only helps us to feel at home in our world (McBurney et al., 2012), but plays an important role in the development of links to personal history via the strong associations between odors, other stimuli, and past places. Apart from the reactions to predator odors (e.g., Wallace and Rosen, 2000) and possibly "pheromonal" activity found in some animals (Keil et al., 1990; but see Doty, 2010), associations to most odors are not innate (Engen and Engen, 1997). Instead, these associations are learned through experience across the life-span (e.g., Schaal et al., 2000; Beauchamp and Mennella, 2009) and enhanced by emotional responses to the experiences (Engen, 1991; Herz, 2001; 1997). Hippocampal neurons across species respond generally to novelty or familiarity (Wood et al., 2000; Cahusac et al., 1993; Ekstrom et al., 2003), suggesting that hippocampal neurons encode associations between a specific odorant and its location or context (Eichenbaum et al., 2007). Thus coding in the hippocampus reflects the unique episodic combination of a stimulus, its importance to the perceiver, and the behavioral response it elicits, as well as the places and contexts in which the stimulus occurs (Eichenbaum, 2004; Goodrich-Hunsaker et al., 2009).

15.3.2.1 Associations Between Odors and Visual Stimuli, Especially Places.

Cross-modal associations between visual and olfactory stimuli have been demonstrated in a variety of species, including honey bees (Srinivasan et al., 1998), Drosophila (Guo and Guo, 2005), primates (Rolls et al., 1996), and humans (Gottfried and Dolan, 2003; Lawless and Engen, 1977). These associations have indicated that odorants can influence preference for visual stimuli, such as faces or abstract paintings (Todrank et al., 1995; Van Reekum et al., 1999), and conversely that visual cues can affect aspects of olfactory perception, such as perceived intensity (Zellner and Kautz, 1990; Dematté et al., 2009). Combining information from both senses improves task performance (Steck et al., 2011), underscoring the episodic nature of olfactory memory.

Studies with animals have evaluated the association between odors and places using a variety of experimental techniques. Open field tests indicate that olfactory cues can be salient cues for rats in identifying location, by showing that the number of correct food site identifications is increased (Lavenex and Schenk, 1995; 1998). In another technique involving a multi-armed maze, rats can be trained to sample an odor at a central location and respond as to the odor's identity by entering a location associated with the smell (Youngentob et al., 1990; 1991). Each of these techniques suggests that rodents can easily associate a place with a particular odorant.

Humans can also associate places and odors, and the odorants that people incidentally encounter in specific places and situations become a part of the memory for the entire episode (Degel and Köster, 1999; Degel et al., 2001). When implicitly tested later, the unidentified odor seems to "fit" the place where it was previously encountered (Johns et al., 2012). An odorant that is highly specific to a particular location can strongly evoke memory for the location. For example, people who had been exposed to an unusual "viking" smell in the Jorvik Viking museum an average of six years prior, remembered many more details of the museum when they were given the specific odor again (Aggleton and Wasket, 1999). It is of interest to note that most odor names reflect the source of the odor, which in one sense is the odor's location (e.g., lemon, lime, coffee, fish, flower, etc.).

15.3.2.2 Odor-Tactile Associations.

Cross-modal associations between olfactory and tactile stimuli can be observed in a variety of species, including rats and humans (Tomie and Whishaw, 1990; Dematté et al., 2006). The associations have been taught explicitly in rats, most notably in a paradigm that associated strings of different textures and thicknesses with odorants, and then required particular combinations of strings and odors in order to obtain a food reward (Tomie and Whishaw, 1990). The acquisition and retention of this task was impaired by lesions of the orbitofrontal cortex (Whishaw et al., 1992), suggesting that an alternative pathway to the hippocampus for acquiring associations may exist.

During the course of everyday life, experiences of the combination of tactile sensations and odors (such as soft clothes and floral odor) occur often enough to form an implicit association between them, through associative learning, which is possibly facilitated through shared hedonic qualities of the stimuli (Dematté et al., 2007). In people, ratings of the texture of a cloth can be altered by the presence of an odorant, and the qualities of the specific odorant can determine whether the cloth is perceived as more rough or more soft (Dematté et al., 2006). This change in the human perception of texture due to fragrance also has been observed in the differences in the perception of hair washed with shampoos that differ only in fragrance (Churchill et al., 2009). These learned associations are stable enough to affect the latencies in an Implicit Association Test (Dematté et al., 2007), which indirectly measures the association between olfaction and touch.

15.3.2.3 Odor-Odor Associations.

Memory for relationships between two odors has been used as a method of investigating declarative memory in animals and humans (Bunsey and Eichenbaum, 1996; Alvarez et al., 2001), as well as memory flexibility (Bunsey and Eichenbaum, 1993). In a typical task, subjects learn to associate two odors with each other, and then later learn additional transitive inferences (e.g., A and B are associated, B and C are associated, so A and C are associated). Acquisition of this task is disrupted by lesions to the perirhinal and entorhinal cortices (Bunsey and Eichenbaum, 1993) and prefrontal cortex (DeVito et al., 2010). In contrast, lesions of hippocampus do not disrupt formation of the initial association in this type of learning, but the associations are not transitive after that point (Bunsey and Eichenbaum, 1996; Li et al., 1999). Odor guided transitive interference tasks, which are hierarchical in nature, are also disrupted by disconnection of the hippocampus (Davis, 1992), as is a transverse patterning (A>B, B>C, C>A) task (Dusek and Eichenbaum, 1998). Thus relational memory between odors seems to be mediated by a network of connections between the hippocampus, surrounding cortical areas, and the prefrontal cortex (DeVito et al., 2010).

15.3.2.4 Odor-Taste and Odor Toxicosis Association.

Both animals and humans will avoid a flavorant that has been associated with gastric illness (Garcia and Koelling, 1966; Bernstein, 1978). This special case of single-trial classical conditioning is called "taste potentiated odor aversion," and is often mediated through the sense of smell (Bartoshuk, 1989), which can warn the individual of the potential of danger. Although both taste and smell are "near" senses, orthonasal olfactory information arrives prior to the ingestion of a flavor, and can give rise to expectations of a particular taste that accompanies the odor (White and Prescott, 2007). Aversion to ingestion of a substance that produced an illness is robust and can continue for more than 30 days after the initial gastric distress (e.g., Chapuis et al., 2007). The basolateral nucleus of the amygdala is involved in the acquisition, consolidation, and the retrieval of this type of memory (Shionoya and Datiche, 2009), but other structures, such as orbitofrontal cortex (Schoenbaum et al., 1999), hippocampus (Ferry et al., 1999), and olfactory bulb (Kay and Laurent, 1999) are also involved.

15.3.2.5 Odor-Verbal Associations.

Humans are notoriously bad at associating a name with an odorant (e.g., Engen and Ross, 1973). Many studies have reported un-cued odor identification accuracy levels that are quite low, generally falling between 44% and 55% (Cain and Potts, 1996; Engen, 1987). Despite the difficulty in attaching a name to an odor, it is possible that verbal encoding assists in olfactory memory accuracy (Paivio, 1991). Reports as to whether or not labeling assists in memory vary, with some studies observing no benefit from labeling (Engen and Ross, 1973; Møller et al., 2004) and others reporting memory improvement (Rabin and Cain, 1984; Lyman and McDaniel, 1986). Some of the differences in these reports seem related to the nature of the label, as being able to apply a label consistently (as opposed to accurately) seems to improve odor recognition memory accuracy considerably (Cain and Potts, 1996; Frank et al., 2011). The consistently applied label may seem to be incorrect, yet may simply reflect idiosyncratic odor knowledge that reflects the variability of the olfactory stimulus (Cain, 1979; Sulmont-Rosse et al., 2005).

15.4 CONCLUSIONS

Olfactory learning and memory result from brain activity in multiple memory pathways that contribute to different strategies of memory performance (McKenzie and Eichenbaum, 2012). The hippocampus is intimately involved in these pathways (DeVito et al., 2010), supporting the process of odor recollection (Fortin et al., 2004), as opposed to familiarity, which is mediated by the amygdala. The hippocampus also acts to integrate aspects of associative memory (Gottfried and Dolan, 2003; Eichenbaum et al., 2007) and is important in transitive inference. The medio-dorsal thalamus and the oribitofrontal cortex contribute to memory formation, particularly the acquisition of "rules" of learning associations with odors (Petrulis and Eichenbaum, 2003a), including the value changes that occur with reinforcers (Rolls, 2000). The olfactory bulb and piriform cortex are also integral to the olfactory memory process (Gottfried and Dolan, 2003; Zelano et al., 2009).

A well-known scent such as coffee evokes in people not only the memory for the characteristic smell of coffee, but also recall of the verbal label for "coffee" and a myriad of other associations with the hot beverage. In studies of human odor memory, this complicates the question of whether humans are endowed with memory systems for the odor per se or whether 'odor memory' is mostly based on verbal and other semantic labels, particularly when well-known odors and explicit memory paradigms are used. Implicit learning paradigms (e.g., Degel and Köster, 1998) seem to more closely model the types of episodic interactions with odors that are experienced in everyday life.

Because of the critical nature of odor memory in performing daily olfactory functions, continued examination of the parameters surrounding the learning and memory of odors is vital. Evaluation of the theoretical, behavioral, anatomical, and neurochemical substrates will continue to be important in elucidating the mechanisms of memory that are integral to olfactory perception. A major issue to be resolved in that evaluation is the level to which odor memory processes are unique or independent of the memory processes associated with other sensory systems.

REFERENCES

Aggleton, J. P. and Waskett, L. (1999). The ability of odors to serve as state-dependent cues for real-world memories: can Viking smells aid the recall of Viking experiences? *Br. J. Psychol.* 90: 1–7.

Alvarez, P., Lipton, P. A., Melrose, R., and Eichenbaum, H. (2001). Differential effects of damage within the hippocampal region on memory for a natural, nonspatial odor-odor association. *Learn. Mem.* 8: 79–86.

Andrade, J. and Donaldson, L. (2007). Evidence for an olfactory store in working memory? *Psychologia.* 50(2): 76–89.

Baddeley, A. D. and Hitch, G. J. (1974). Working memory. In *Recent Advances in Learning and Motivation Vol. VIII*, G. Bower (Ed.). Academic Press: New York, NY, pp. 47–90.

Barkai, E. and Saar, D. (2001). Cellular correlates of olfactory learning in the rat piriform cortex. *Rev. Neurosci.* 12(2): 111–120.

Bartoshuk, L. M. (1989). The functions of taste and olfaction. *Ann. NY Acad. Sci.* 575: 353–362.

Beauchamp, G. K. and Mennella, J. A. (2009). Early flavor learning and its impact on later feeding behavior. *J. Pediatr. Gastr. Nutr.* 48(Suppl. 1): S25–S30.

Bernstein, I. L. (1978). Learned taste aversions in children receiving chemotherapy. *Science* 200: 1302–1303.

Beshel, J., Kopell, N., and Kay, L. M. (2007). Olfactory bulb gamma oscillations are enhanced with task demands. *J. Neurosci.* 27(31): 8358–8365.

Bluthe, R. M. and Dantzer, R. (1990). Social recognition does not involve vasopressinergic neurotransmission in female rats. *Brain Res.* 535: 301–304.

Bunsey, M. and Eichenbaum, H. (1993). Critical role of the parahippocampal region for paired-associate learning in rats. *Behav. Neurosci.* 107: 740–747.

Bunsey, M. and Eichenbaum, H. (1996). Conservation of hippocampal memory function in rats and humans. *Nature* 379: 255–257.

Cahusac, P. M. B., Rolls, E. T., Miyashita, Y., and Niki, H. (1993). Modification of the responses of hippocampal neurons in the monkey during the learning of a conditional spatial response task. *Hippocampus.* 3: 29–42.

Cain, W. S. (1979). To know with the nose: Keys to odor identification. *Science* 203(4379): 467–470.

Cain, W. S. and Potts, B. C. (1996). Switch and bait: Probing the discriminative basis of odor identification via recognition memory. *Chem. Senses* 21(1): 35–44.

Calu, D. J., Roesch, M. R., Stalnaker, T. A., and Schoenbaum, G. (2007). Associative encoding in posterior piriform cortex during odor discrimination and reversal learning. *Cereb. Cortex*, 17(6): 1342–1349.

Carr, W. J., Krames, L., and Costanzo, D. J. (1970). Previous sexual experience and olfactory preference for novel versus original sex partners in rats. *J. Comp. Physiol. Psychol.* 71: 216–222.

Castellucci, V., Pinsker, H., Kupfermann, I., and Kandel, E. R. (1970). Neuronal mechanisms of habituation and dishabituation of the gill-withdrawal reflex in Aplysia. *Science* 167(3926): 1745–1748.

Chaudhury, D., Manella, L., Arellanos, A., et al.(2010). Olfactory bulb habituation to odor stimuli. *Behav. Neurosci*, 124(4): 490–499.

Chapuis, J., Messaoudi, B., Ferreira, G., and Ravel, N. (2007). Importance of retronasal and orthonasal olfaction for odoraversion memory in rats. *Behav. Neurosci.* 121: 1383–1392.

Chu, S. and Downes, J. J. (2000). Long live Proust: the odor-cued autobiographical memory bump. *Cognition* 75: B41–B50.

Chu, S. and Downes, J. J. (2002). Proust nose best: Odors are better cues of autobiographical memory. *Mem. Cogn.* 30: 511–518.

Churchill, A., Meyners, M., Griffiths, L., and Bailey, P. (2009). The cross-modal effect of fragrance in shampoo: Modifying the perceived feel of both product and hair during and after washing. *Food Qual. Prefer.* 20: 320–328.

Cleland, T. A., Morse, A., Yue, E. L., and Linster, C. (2002). Behavioral models of odor similarity. *Behav. Neurosci*, 116(2): 222–231.

Cleland, T. A. and Narla, V. A. (2003). Intensity modulation of olfactory acuity. *Behav. Neurosci.* 117(6): 1434–1440.

Cleland, T. A., Narla, V. A., and Boudadi, K. (2009). Multiple learning parameters differentially regulate olfactory generalization. *Behav. Neurosci.* 123(1): 26–35.

Dalton, P. and Wysocki, C. J. (1996). The nature and duration of adaptation following long-term odor exposure. *Percept. Psychophys.* 58(5): 781–792.

Daly, K. C., Chandra, S., Durtschi, M. L., and Smith, B. H. (2001). The generalization of an olfactory-based conditioned response reveals unique but overlapping odour representations in the moth Manduca sexta. *J. Exp. Biol.* 204(Pt 17): 3085–3095.

Davis, H. (1992). Transitive inference in rats (*Rattus norvegicus*). *J. Comp. Psychol.* 106: 342–349.

Deacon, R. M. J. and Rawlins, J. N. P. (1995). Serial position effects and duration of memory for nonspatial stimuli in rats. *J. Exp. Psychol. Anim. Behav.* 21(4): 285–292.

Degel, J. and Köster, E. P. (1998). Implicit memory for odors: A possible method for observation. *Percept. Motor Skills.* 86: 943–952.

Degel, J. and Köster, E. P. (1999). Odors: Implicit memory and performance effects, *Chem. Senses.* 24: 317–325.

Degel, J., Piper, D., and Köster, E. P. (2001). Implicit learning and implicit memory for odors: The influence of odor identification and retention time. *Chem. Senses.* 26: 267–280.

Demattè, M. L., Sanabria, D., and Spence, C. (2007). Olfactory–tactile compatibility effects demonstrated using a variation of the Implicit Association Test. *Acta Psychol.* 124(3): 332–343.

Demattè, M. L., Sanabria, D., Sugarman, R., and Spence, C. (2006). Cross-modal interactions between olfaction and touch. *Chem. Senses.* 31(4): 291–300.

DeVito, L. M. and Eichenbaum, H. (2011). Memory for the order of events in specific sequences: Contributions of the hippocampus and medial prefrontal cortex. *J. Neurosci.* 31(9): 3169–3175.

DeVito, L. M., Lykken, C., Kanter, B. R., and Eichenbaum, H. (2010). Prefrontal cortex: Role in acquisition of overlapping associations and transitive inference. *Learn. Mem.* 17: 161–167.

Doty, R. L. (2010). *The Great Pheromone Myth*. Baltimore: Johns Hopkins University Press.

Doucette, W. and Restrepo, D. (2008). Profound context-dependent plasticity of mitral cell responses in olfactory bulb. *PLoS Biol.* 6(10): e258.

Dusek, J. A. and Eichenbaum, H. (1998). The hippocampus and transverse patterning guided by olfactory cues. *Behav. Neurosci.* 112: 762–771.

Eichenbaum, H. (1997). Declarative memory: Insights from cognitive neurobiology. *Ann. Rev. Psychol.* 48: 547–572.

Eichenbaum, H. (2004). Hippocampus: Cognitive processes and neural representations that underlie declarative memory. *Neuron* 44: 109–20.

Eichenbaum, H., Alvarez, P., and Ramus, S. J. (2000). Animal models of amnesia. In *Handbook of Neuropsychology: Memory Disorders*, L. Cermak, (Ed.). Elsevier: Amsterdam, pp. 175–198.

Eichenbaum, H., Yonelinas, A. R., and Ranganath, C. (2007). The medial temporal lobe and recognition memory. *Annu. Rev. Neurosci.* 30: 123–152.

Ekstrom, A. D., Kahana, M.J., Caplan, J.B., et al. (2003). Cellular networks underlying human spatial navigation. *Nature.* 425: 184–187.

Engen, T. (1987). Remembering odors and their names. *Am. Sci.* 75: 497–503.

Engen, T. (1991). *Odor Sensation and Memory*. Praeger: New York.

Engen, T. and Engen, E. (1997). Relationship between development of odor perception and language. *Enfance* 50(1): 125–140.

Engen, T. and Ross, B. M. (1973). Long-term memory of odors with and without verbal descriptions. *J. Exper. Psychol.* 100: 221–227.

Ezzeddine, Y. and Glanzman, D. L. (2003). Prolonged habituation of the gill-withdrawal reflex in Aplysia depends on protein synthesis, protein phosphatase activity, and postsynaptic glutamate receptors. *J. Neurosci.*23(29): 9585–9594.

Farovik, A., Place, R., Miller, D., and Eichenbaum, H. (2011). Amygdala lesions selectively impair familiarity in recognition memory. *Nat. Neurosci.* 14: 1416–1417.

Ferry, B., Wirth, S., and DiScala, G. (1999). Functional interaction between entorhinal cortex and basolateral amygdale during trace conditioning of odor aversion in the rat. *Behav. Neurosci.* 113: 118–125.

Fortin, N. J., Agster, K. L., and Eichenbaum, H. B. (2002). Critical role of the hippocampus in memory for sequences of events. *Nat. Neurosci.* 5(5): 458–462. doi: 10.1038/nn834.

Fortin, N. J., Wright, S. P., and Eichenbaum, H. (2004). Recollection-like memory retrieval in rats is dependent on the hippocampus. *Nature* 431(7005): 188–191.

Frank, R. A., Brearton, M., Rybalsky, K., et al.(2011). Consistent flavor naming predicts recognition memory in children and young adults. *Food Qual. Prefer.* 22(1): 173–178.

Gabassi, P. G. and Zanuttini, L. (1983). Ricinosimento di stimol: olfattivi nella memoria a brevtermine. *G. Ital. Psicol.* 10: 51–60.

Gaffan, D. (1974). Recognition impaired and association intact in the memory of monkeys after transection of the fornix. *J. Comp. Physiol. Psychol.* 86: 1100–1109.

Galef, B. G. and Giraldeau, L. A. (2001) Social influences on foraging in vertebrates: Causal mechanisms and adaptive functions. *Anim. Behav.* 61: 3–15.

Garcia, J. and Koelling, R. A. (1966). Relation of cue to consequence in avoidance learning. *Psychonom. Sci.* 4: 123–124.

Gerber, B. and Smith, B. H. (1998). Visual modulation of olfactory learning in honeybees. *J. Exp. Biol.* 201(Pt 14): 2213–2217.

Goldman, W. P. and Seamon, J. G. (1992). Very long-term memory for odors: Retention of odor-name associations. *Am. J. Psychol.* 105: 549–563.

Goodrich-Hunsaker, N. J., Gilbert, P. E., and Hopkins, R. O. (2009). The role of the human hippocampus in odor-place associative memory. *Chem. Senses* 34: 513–521.

Gottfried, J. A. (2010). Central mechanisms of odour object perception. *Nat. Rev. Neurosci.* 11: 628–641.

Gottfried, J. A. and Dolan, R. J. (2003). The nose smells what the eye sees: Crossmodal visual facilitation of human olfactory perception. *Neuron* 39: 375–386.

Guo, J. and Guo, A. (2005). Crossmodal interactions between olfactory and visual learning in *Drosophila*. *Science* 309: 307–310.

Hausner, H., Nicklaus, S., Issanchou, S., et al.(2010). Breastfeeding facilitates acceptance of a novel dietary flavour compound. *Clin. Nutr.* 29: 141–148.

Herz, R. S. (1997). Emotion experienced during encoding enhances odor retrieval cue effectiveness. *Am. J. Psychol.* 110: 489–505.

Herz, R. S. (2001). Ah, sweet skunk: Why we like or dislike what we smell. *Cerebrum* 3(4): 31–47.

Herz, R. S. (2012). Odor memory. In *Olfactory Cognition*, G. M. Zucco, R. S. Herz, and B. Schaal (Eds.). John Benjamins: Amsterdam.

Herz, R. S. and Cupchik, G. C. (1992). An experimental characterization of odor evoked memories in humans. *Chem. Senses* 17: 519–528.

Herz, R. S. and Cupchik, G. C. (1995). The emotional distinctiveness of odor-evoked memories. *Chem. Senses* 20: 517–528.

Herz, R. S. and Schooler, J. W. (2002). A naturalistic study of autobiographical memories evoked by olfactory and visual cues: Testing the Proustian hypothesis. *Am. J. Psychol.* 115: 21–32.

Hold, B. and Schleidt, M. (1977). The importance of human odour in non-verbal communication. *Z. Tierpsychol.* 43(3): 225–238.

Hunter, A. J. and Murray, T. K. (1989). Cholinergic mechanisms in a simple test of olfactory learning in the rat. *Psychopharmacology* 99(2): 270–275.

Jehl, C., Royet, J. P., and Holley, A. (1994). Very short term recognition memory for odors. *Percept. Psychophys.* 56(6): 658–668.

Johns, A., Homewood, J., Stevenson, R., and Taylor, A. (2012). Implicit and explicit olfactory memory in people with and without Down syndrome. *Res. Dev. Disabil.* 33(2): 583–593.

Johnson, A. J. and Miles, C. (2009). Single-probe serial position recall: Evidence of modularity for olfactory, visual, and auditory short-term memory. *Q. J. Exp. Psych.* 62(2): 267–275.

Johnson, R. E. and Rasmussen, K. (1984). Individual recognition of female hamsters by males: role of chemical cues and of the olfactory and vomeronasal systems. *Physiol. Behav.* 33: 95–104.

Jonides, J. et al. (2008). The mind and brain of short-term memory. *Ann. Rev. Psychol.* 59: 193–224.

Jönsson, F., Møller, P., and Olsson, M. (2011). Olfactory working memory: Effects of verbalization on the 2-back task. *Mem. Cogn.* 39: 1023–1032.

Kandel, E. R. (2001). The molecular biology of memory storage: a dialogue between genes and synapses. *Science* 294(5544): 1030–1038.

Kay, L. M. and Beshel, J. (2010). A beta oscillation network in the rat olfactory system during a 2-alternative choice odor discrimination task. *J. Neurophysiol.* 104(2): 829–839.

Kay, L. M. and Laurent, G. (1999). Odor- and context-dependent modulation of mitral cell activity in behaving rats. *Nat. Neurosci.* 2(11): 1003–1009.

Kay, L. M. and Lazzara, P. (2010). How global are olfactory bulb oscillations? *J. Neurophysiol.* 104(3): 1768–1773.

Keil, W., von Stralendorff, F., and Hudson, R. (1990). A behavioral bioassay for analysis of rabbit nipple-search pheromone. *Physiol. Behav.* 47(3): 525–529.

Kesslak, J. P., Profitt, B. F., and Criswell, P. (1991). Olfactory function in chronic alcoholics. *Percept. Motor Skills.* 73: 551–554.

Kogan, J. H., Frankland, P. W., and Silva, A. J. (2000). Long-term memory underlying hippocampus-dependent social recognition in mice. *Hippocampus.* 10: 47–56.

Koger, S. M. and Mair, R. G. (1994). Comparison of the effects of frontal cortical and thalamic lesions on measures of olfactory learning and memory in the rat. *Behav. Neurosci.* 108: 1088–1100.

Kolb, B. and Whishaw, I. Q. (2003). *Fundamentals of Human Neuropsychology*, 5th ed. Worth Publishers: New York.

References

Köster, E. P. (2005). Does olfactory memory depend on remembering odors? *Chem. Senses* 30, (suppl.1), i236–i237.

Laska, M., Salazar, L. T. H., and Luna, E. R. (2003). Successful acquisition of an olfactory discrimination paradigm by spider monkeys, *Ateles geoffroyi. Physiol. Behav.* 78(2): 321–329.

Lavenex, P. and Schenk, F. (1995). Influence of local environmental olfactory cues on place learning in rats. *Physiol. Behav.* 58: 1059–1066.

Lavenex, P. and Schenk, F. (1998). Olfactory traces and spatial learning in rats. *Anim. Behav.* 56: 1129–1136.

Lawless, H. T. (1978). Recognition of common odors, pictures, and simple shapes. *Percept. Psychophys.* 24: 493–495.

Lawless, H. T. and Cain, W. S. (1975). Recognition memory for odors. *Chem. Senses Flav.* 1: 331–337.

Lawless, H. T. and Engen, T. (1977). Associations to odors: Interference, mnemonics, and verbal labeling. *J. Exp. Psychol.* 3: 52–59.

Li, H., Matsumoto, K., and Watanabe, H. (1999). Different effects of unilateral and bilateral hippocampal lesions in rats on the performance of radial maze and odor-paired associate tasks. *Brain Res. Bull.* 48: 113–119.

Li, W., Howard, J. D., Parrish, T. B., and Gottfried, J. A. (2008). Aversive learning enhances perceptual and cortical discrimination of indiscriminable odor cues. *Science* 319(5871): 1842–1845.

Linster, C., Johnson, B. A., Morse, A., et al.(2002). Spontaneous versus reinforced olfactory discriminations. *J. Neurosci.* 22(16): 6842–6845.

Linster, C., Johnson, B. A., Yue, E., et al.(2001). Perceptual correlates of neural representations evoked by odorant enantiomers. *J. Neurosci.* 21(24): 9837–9843.

Linster, C. and Smith, B. H. (1997). A computational model of the response of honey bee antennal lobe circuitry to odor mixtures: overshadowing, blocking and unblocking can arise from lateral inhibition. *Behav. Brain Res.* 87(1): 1–14.

Luce, R. D. (1986). *Response Times.* Oxford University Press: New York.

Lundström, J. N., Boyle, J. A., Zatorre, R.J., and Jones-Gotman, M. (2009). The neuronal substrates of human olfactory based kin recognition. *Hum. Brain Mapp.* 30(8): 2571–2580.

Lyman, B. J. and McDaniel, M. A. (1986). Effects of encoding strategy on longterm memory for odours. *J. Exp. Psychol. Hum. Learn. Mem.* 38: 753–765.

Mandairon, N. and Linster, C. (2009). Odor perception and olfactory bulb plasticity in adult mammals. *J. Neurophysiol.* 101(5): 2204–2209.

Mandairon, N., Ferretti, C. J., Stack, C. M., et al. (2006). Cholinergic modulation in the olfactory bulb influences spontaneous olfactory discrimination in adult rats. *Eur. J. Neurosci.* 24(11): 3234–3244.

Mandairon, N., Peace, S., Karnow, A., et al.(2008). Noradrenergic modulation in the olfactory bulb influences spontaneous and reward-motivated discrimination, but not the formation of habituation memory. *Eur. J. Neurosci.* 27(5): 1210–1219.

Maras, P. M. and Petrulis, A. (2008). The posteromedial cortical amygdala regulates copulatory behavior, but not sexual odor preference, in the male Syrian hamster (Mesocricetus auratus). *Neuroscience.* 156(3): 425–435.

Martin, C., Beshel, J., and Kay, L. M. (2007). An olfacto-hippocampal network is dynamically involved in odor-discrimination learning. *J. Neurophysiol.* 98(4): 2196–2205.

Meyer, F. F. and Louilot, A. (2011). Latent inhibition-related dopaminergic responses in the nucleus accumbens are disrupted following neonatal transient inactivation of the ventral subiculum. *Neuropsychopharmacology* 36(7): 1421–1432.

McBurney, D. H., Streeter, S., and Euler, H. A. (2012). Olfactory comfort in close relationships: You aren't the only one who does it. In *Olfactory Cognition.* G. M. Zucco, R. S. Herz, and B. Schaal (Eds.). John Benjamins: Amsterdam.

McDonald, D. G., Johnson, L. C., and Hord, D. J. (1964). Habituation of the orienting response in alert and drowsy subjects. *Psychophysiology*, 1, 163–173.

McKenzie, S. and Eichenbaum, H. (2012). New approach illuminates how memory systems switch. *Trend. Cog. Sci.* 16(2): 102–103.

McNamara, A. M., Magidson, P. D., Linster, C., et al.(2008). Distinct neural mechanisms mediate olfactory memory formation at different timescales. *Learn. Mem.* 15(3): 117–125.

Mennella, J. A. and Beauchamp, G. K. (2002). Flavor experiences during formula feeding are related to preferences during childhood. *Early Hum. Dev.* 68: 71–82.

Mennella, J. A., Jagnow, C. J., and Beauchamp, G. K. (2001). Pre- and postnatal flavor learning by human infants. *Pediatrics.* 107: e88.

Miles, C. and Hodder, K. (2005). Serial position effects in recognition memory for odors: A reexamination. *Mem. Cogn.* 33: 1303–1314.

Miles, C. and Jenkins, R. (2000). Recency and suffix effects with immediate recall of olfactory stimuli. *Memory.* 8(3): 195–205.

Mishkin, M. (1978). Memory in monkeys severely impaired by combined but not by separate removal of amygdala and hippocampus. *Nature.* 273: 297–298.

Mojet, J., and Köster, E. P. (2002). Texture and flavour memory in foods: an incidental learning experiment. *Appetite.* 38: 110–117.

Mojet, J., and Köster, E. P. (2005). Sensory memory and food texture. *Food Qual. Prefer.* 16: 251–266.

Møller, P., Köster, E. P., Dijkman, N., et al. (2012). Same-different reaction times to odors: some unexpected findings. *Chemosens. Percept.* 5(2): 158–171.

Møller, P., Mojet, J. and Köster, E. P. (2007). Incidental and intentional flavour memory in young and older subjects. *Chem. Senses* 32(6): 557–567.

Møller, P., Wulff, C., and Köster, E. P. (2004). Do age differences in odor memory depend on differences in verbal memory? *NeuroReport* 15: 915–917.

Morin-Audebrand, L., Mojet, J., Chabanet, C., et al.(2011). The role of novelty detection in food memory. *Acta Psychol.* 139: 233–238.

Mozell, M. M. (1972). The chemical senses II: Olfaction. In *Experimental Psychology*, King, J. W. and Riggs, L. A. (Eds.). Holt, Rinehart, and Winston, New York, pp. 195–222.

Newman, K. S., and Halpin, Z. T. (1988). Individual odours and mate recognition in the prairie vole, *Microtus ochrogaster. Anim. Behav.* 36: 1779–1787.

Nigrosh, B. J., Slotnick, B. M., and Nevin, J. A. (1975). Olfactory discrimination, reversal learning, and stimulus control in rats. *J. Comp. Physiol. Psychol.* 89(4): 285–294.

Olsson, M. J. (1999). Implicit testing of odor memory: Instances of positive and negative repetition priming. *Chem. Senses.* 24(3): 347–350.

Olsson, M. J. and Cain, W. S. (1995). Early temporal events in odor identification. *Chem. Senses* 20: 753.

Olsson, M. J. and Friden, M. (2001). Evidence of odor priming: Edibility judgements are primed differently between the hemispheres. *Chem. Senses* 26(2): 117–123.

Otto, T. and Eichenbaum, H. (1992). Complementary roles of the orbital prefrontal cortex and the perirhinal-entorhinal cortices in an odor-guided delayed-nonmatching-to-sample task. *Behav. Neurosci.* 106: 762–775.

Otto, T., Schottler, F., Staubli, U., et al.(1991). Hippocampus and olfactory discrimination learning: Effects of entorhinal cortex lesions on olfactory learning and memory in a successive-cue, go–no-go task. *Behav. Neurosci.* 105(1): 111–119.

Paivio, A. (1991). Dual coding theory: Retrospect and current status. *Can. J. Psychol.* 45(3): 255–287.

Pelz, C., Gerber, B., and Menzel, R. (1997). Odorant intensity as a determinant for olfactory conditioning in honeybees: roles in discrimination, overshadowing and memory consolidation. *J. Exp. Biol.* 200(Pt 4): 837–847.

Petrulis, A. (2009). Neural mechanisms of individual and sexual recognition in Syrian hamsters (*Mesocricetus auratus*). *Behav. Brain Res.* 200(2): 260–267.

Petrulis, A., Alvarez, P., and Eichenbaum, H. (2005). Neural correlates of social odor recognition and the representation of individual distinctive social odors within entorhinal cortex and ventral subiculum. *Neuroscience* 130(1): 259–274.

Petrulis, A., DeSouza, I., Schiller, M., and Johnston, R. E. (1998). Role of frontal cortex in social odor discrimination and scent-marking in female golden hamsters (*Mesocricetus auratus*). *Behav. Neurosci.* 112(1): 199–212.

Petrulis, A. and Eichenbaum, H. (2000). The hippocampal system, individual odor discrimination and the Collidge effect in golden hamsters (*Mesocricetus auratus*). *Soc. Neurosci. Abs.* 26: 468.

Petrulis, A. and Eichenbaum, H. (2003a). Olfactory memory. In *Handbook of Olfaction and Gustation*, R. L. Doty (Ed.). 2nd ed. Marcel Dekker: New York, pp. 409–438.

Petrulis, A. and Eichenbaum, H. (2003b). The perirhinal-entorhinal cortex, but not the hippocampus, is critical for expression of individual recognition in the context of the Coolidge effect. *Neuroscience* 122: 599–607.

Pliner, P. (1982). The effects of mere exposure on liking for edible substance. *Appetite.* 3: 283–290.

Porter, R. H. and Moore, J. D. (1981). Human kin recognition by olfactory cues. *Physiol. Behav.* 27: 493–495.

Posner, M. I. (1986). *Chronometric Explorations of Mind, 2nd* revised ed. Oxford University Press: Oxford.

Proust, M. (1928). *Swann's Way.* Modern Library: New York.

Rabin, M. D. (1988). Experience facilitates olfactory quality discrimination. *Percept. Psychophys.* 44: 532–540.

Rabin, M. D. and Cain, W. S. (1984). Odor recognition: Familiarity, identifiability, and encoding consistency. *J. Exper. Psychol. Learn. Mem.Cogn.* 10: 316–325.

Ramus, S. J., Davis, J. B., Donahue, R. J., et al. (2007). Interactions between the orbitofrontal cortex and the hippocampal memory system during the storage of long-term memory. *Ann. N.Y. Acad. Sci.* 1121: 216–231.

Ranganath, C. and Blumenfeld, R. S. (2005). Doubts about double dissociations between short- and long-term memory. *Trends Cogn. Sci.* 9: 374–380.

Reed, P. (2000). Serial position effects in recognition memory for odors. *J. Exp. Psychol. Learn.* 26: 411–422.

Rochefort, C., Gheusi, G., Vincent, J. D., and Lledo, P. M. (2002). Enriched odor exposure increases the number of newborn neurons in the adult olfactory bulb and improves odor memory. *J. Neurosci.* 22(7): 2679–2689.

Roesch, M. R., Stalnaker, T. A., and Schoenbaum, G. (2007). Associative encoding in anterior piriform cortex versus orbitofrontal cortex during odor discrimination and reversal learning. *Cereb Cortex.* 17(3): 643–652.

Rolls, B. J., Rolls, E. T., Rowe, E. A., and Sweeney, K. (1981). Sensory specific satiety in man. *Physiol. Behav.* 27: 137–142.

Rolls, E. T. (2000). The orbitofrontal cortex and reward. *Cereb. Cortex* 10: 284–294.

Rolls, E. T., Critchley, H. D., Mason, R., and Wakeman, E. (1996). Orbitofrontal cortex neurons: Role in olfactory and visual association learning. *J. Neurophys.* 75(5): 1970–1981.

Rolls, E. T., Grabenhorst, F., Margot, C., et al. (2008). Selective attention to affective value alters how the brain processes olfactory stimuli. *J. Cogn. Neurosci.* 20(10): 1815–1826.

Rolls, E. T., Murzi, E., Yaxley, S., et al. (1986). Sensory-specific satiety: food-specific reduction in responsiveness of ventral forebrain neurons after feeding in the monkey. *Brain Res.* 368: 79–86.

Sauvage, M. M., Beer, Z., and Eichenbaum, H. (2010). Recognition memory: Adding a response deadline eliminates recollection but spares familiarity. *Learn. Memory* 17: 104–108.

Sauvage, M. M., Fortin, N. J., Owens, C. B., et al. (2008). Recognition memory: opposite effects of hippocampal damage on recollection and familiarity. *Nat. Neurosci.* 11(1): 16–18.

Schaal, B. (2012). Emerging chemosensory preferences: Another playground for the innate-acquired. In *Olfactory Cognition*, G. M. Zucco, R. S. Herz, and B. Schaal (Eds.). John Benjamins Publishing Company: Philadelphia, PA, pp. 237–253.

Schaal, B., Marlier, L., and Soussignan, R. (2000). Human foetuses learn odors from their pregnant mother's diet. *Chem. Senses.* 25: 729–37.

Schab, F. R., and Crowder, R. G. (1995). Odor recognition memory. In *Memory for Odors*, F. R. Schab and R. G. Crowder (Eds.). Lawrence Erlbaum Associates: Mahwah, New Jersey.

Schoenbaum, G., Chiba, A. A., and Gallagher, M. (1998). Orbitofrontal cortex and basolateral amygdala encode expected outcomes during learning. *Nat. Neurosci.* 1(2): 155–159.

Schoenbaum, G., Chiba, A. A., and Gallagher, M. (1999). Neural encoding in orbitofrontal cortex and basolateral amygdale during olfactory discrimination learning. *J. Neurosci.* 19: 1876–1884.

Schoenbaum, G., and Eichenbaum, H. (1995). Information coding in the rodent prefrontal cortex. I. Single-neuron activity in orbitofrontal cortex compared with that in pyriform cortex. *J. Neurophysiol.* 74(2): 733–750.

Sevelinges, Y., Desgranges, B., and Ferreira, G. (2009). The basolateral amygdala is necessary for the encoding and the expression of odor memory. *Learn. Mem.* 16(4): 235–242.

Shionoya, K. and Datiche, F. (2009). Inactivation of the basolateral amygdala impairs the retrieval of recent and remote taste-potentiated odor aversion memory. *Neurobiol. Learn. Mem.* 92(4): 590–596.

Silbering, A. F. and Galizia, C. G. (2007). Processing of odor mixtures in the Drosophila antennal lobe reveals both global inhibition and glomerulus-specific interactions. *J. Neurosci.* 27(44): 11966–11977.

Silbering, A. F., Okada, R., Ito, K., and Galizia, C. G. (2008). Olfactory information processing in the Drosophila antennal lobe: anything goes? *J. Neurosci.* 28(49): 13075–13087.

Slotnick, B. (2001). Animal cognition and the rat olfactory system. *Trends Cogn. Sci.* 5(5): 216–222.

Slotnick, B. and Bodyak, N. (2002). Odor discrimination and odor quality perception in rats with disruption of connections between the olfactory epithelium and olfactory bulbs. *J. Neurosci.* 22(10): 4205–4216.

Slotnick, B. M. and Katz, H. M. (1974). Olfactory learning-set formation in rats. *Science* 185(4153): 796–798.

Smith, B. H. and Cobey, S. (1994). The olfactory memory of the honeybee Apis mellifera. II. Blocking between odorants in binary mixtures. *J. Exp. Biol.*195: 91–108.

Sobel, N., Prabhakaran, V., Zhao, Z., et al. (2000). Time course of odorant-induced activation in the human primary olfactory cortex. *J. Neurophysiol.* 83(1): 537–551.

Srinivasan, M. V., Zhang, S. W., and Zhu, H. (1998). Honeybees link sights to smells. *Nature.* 396: 637–638.

Steck, K., Hansson, B. S., and Knaden, M. (2011). Desert ants benefit from combining visual and olfactory landmarks. *J. Exp. Biol.* 214: 1307–1312.

Stevenson, R. J. (2010). An initial evaluation of the functions of human olfaction. *Chem. Senses.* 35(1): 3–20. doi: 10.1093/chemse/bjp083.

References

Stevenson, R. J. and Boakes, R. A. (2003). A mnemonic theory of odor perception. *Psychol. Rev.* 110(2): 340–364.

Stevenson, R. J., Case, T. I., and Boakes, R. A. (2005). Implicit and explicit tests of odor memory reveal different outcomes following interference. *Learn. Motiv.* 36(4): 353–373.

Sulmont-Rosse, C., Issanchou, S., and Koster, E. P. (2005). Odor naming methodology: Correct identification with multiple-choice versus repeatable identification in a free task. *Chem. Sens.* 30(1): 23–27.

Suzuki, W. and Eichenbaum, H. (2000). The neurophysiology of memory. *Ann. N. Y. Acad. Sci.* 911: 175–191.

Szyszka, P., Ditzen, M., Galkin, A., et al. (2005). Sparsening and temporal sharpening of olfactory representations in the honeybee mushroom bodies. *J. Neurophysiol.* 94(5): 3303–3313.

Thor, D. H. and Holloway, W. R. (1982). Social memory of the male laboratory rat. *J. Comp. Physiol. Psychol.* 96: 1000–1006.

Todrank, J., Byrnes, D., Wrzesniewski, A., and Rozin, P. (1995). Odor can change preferences for people in photographs: A cross-modal evaluative conditioning study with olfactory USs and visual CSs. *Learn. Motiv.* 26: 116–140.

Todrank, J., Heth, G., and Johnston, R. E. (1998). Kin recognition in golden hamsters: Evidence for kinship odours. *Anim. Behav.* 55(2): 377–86.

Tomie, J. A. and Whishaw, I. Q. (1990). New paradigms for tactile discrimination studies with the rat: Methods for simple, conditional, and configural discriminations. *Physiol. Behav.* 48: 225–231.

Van Reekum, C. M., Van den Berg, H., and Frijda, N. H. (1999). Cross-modal preferences acquisition: Evaluative conditioning of pictures by affective olfactory and auditory cues. *Cog. Emot.* 13: 831–836.

Walk, H. A. and Johns, E. E. (1984). Interference and facilitation in short term memory for odors. *Percept. Psychophys.* 36: 508–514.

Wallace, K. J. and Rosen, J. B. (2000). Predator odor as an unconditioned fear stimulus in rats: Elicitation of freezing by trimethylthiazoline, a component of fox feces. *Behav. Neurosci.* 114(5): 912–922.

Whishaw, I. Q., Tomie, J. A., and Kolb, B. (1992). Ventrolateral prefrontal cortex lesions in rats impair the acquisition and retention of a tactile-olfactory configural task. *Behav. Neurosci.* 106: 597–603.

White, T. L. (1998). Olfactory memory: The long and short of it. *Chem. Senses.* 23: 433–441.

White, T. L. (2012). Attending to olfactory short-term memory. In *Olfactory Cognition*, G. M. Zucco, R. S. Herz, and B. Schaal (Eds.). John Benjamins Publishing Company: Philadelphia, PA, pp. 137–152.

White, T. L., Hornung, D. E., Kurtz, D. B., et al. (1998). Phonological and perceptual components of short-term memory for odors. *Am. J. Psych.* 111: 411–441.

White, T. L. and Prescott, J. (2007). Chemosensory cross-modal Stroop effects: Congruent odors facilitate taste identification. *Chem. Sens.* 32(6): 337–341.

White, T. L. and Treisman, M. (1997). A comparison of the encoding of content and order in olfactory memory and in memory for visually presented verbal materials. *Brit. J. Psychol.* 88: 459–472.

White, T. L. and Youngentob, S. L. (2004). The effect of over-expression of NR2B receptor subunit on olfactory memory performance in the mouse. *Brain Res.* 1021(1): 1–7.

Willander J. and Larsson, M. (2006). Smell your way back to childhood: Autobiographical odor memory. *Psychon. Bull. Rev.* 13: 240–244.

Willander J. and Larsson, M. (2007). Olfaction and emotion: the case of autobiographical memory. *Mem. Cogn.* 35: 1659–1663.

Willander, J. and Larsson, M. (2008). The mind's nose and autobiographical odor memory. *Chem. Percept.* 1: 210–215.

Wilson, D. A. (1998a). Habituation of odor responses in the rat anterior piriform cortex. *J. Neurophysiol.* 79(3): 1425–1440.

Wilson, D. A. (1998b). Synaptic correlates of odor habituation in the rat anterior piriform cortex. *J. Neurophysiol.* 80(2): 998–1001.

Wilson, D. A. and Linster, C. (2008). Neurobiology of a simple memory. *J. Neurophysiol.* 100(1): 2–7.

Wilson, D. A. and Stevenson, R. J. (2006). Learning to Smell: Olfactory Perception from Neurobiology to Behavior. Baltimore: The Johns Hopkins University Press.

Wiltrout, C., Dogra, S., and Linster, C. (2003). Configurational and non-configurational interactions between odorants in binary mixtures. *Behav Neurosci.* 117(2): 236–245.

Wood, E. R., Dudchenko, P. A., Robitsek, R. J., and Eichenbaum, H. (2000). Hippocampal neurons encode information about different types of memory episodes occurring in the same location. *Neuron* 27: 623–633.

Yeshurun, Y., Dudai, Y., and Sobel, N. (2008). Working memory across nostrils. *Behav. Neurosci.* 122(5): 1031–1037.

Yonelinas, A. P. (2001). Components of episodic memory: the contribution of recollection and familiarity. *Philos. Trans. R. Soc. Lon. B. Biol. Sci.* 356(1413): 1363–1374.

Youngentob, S. L. and Glendinning, J. (2009). Fetal ethanol exposure increases ethanol intake by making it smell and taste better. *PNAS.* 106: 5359–5364.

Youngentob, S. L., Markert, L. M., Hill, T. W., et al. (1991). Odorant identification in rats: An update. *Physiol. Behav.* 49: 1293–1296.

Youngentob, S. L., Markert, L. M., Mozell, M. M., and Hornung, D. E. (1990). A method for establishing a five odorant identification confusion matrix task in rats. *Physiol. Behav.* 47: 1053–1059.

Zajonc, R. B. (1968). Attitudinal effects of mere exposures. *J. Pers. Soc. Psychol.* 9(2, Pt. 2): 1–27.

Zajonc, R. B. (2001). Mere exposure: A gateway to the subliminal. *Curr. Dir. Psychol. Sci.* 10: 224–228.

Zelano, C., Montag, J., Khan, R., and Sobel, N. (2009). A specialized odor memory buffer in primary olfactory cortex. *PLoS ONE.* 4(3): 1–11.

Zellner, D.A. and Kautz, M. A. (1990). Color affects perceived odor intensity. *J. Exp. Psychol. Hum. Percept. Perform.* 16: 391–397.

Zola-Morgan, S. and Squire, L. (1993). Neuroanatomy of memory. *Annu. Rev. Neurosci.* 16: 547–563.

Zucco, G. M. (2003). Anomalies in cognition: Olfactory memory. *Eur. Psychol.* 8(2): 77–86.

Part 4

Clinical Applications and Perspectives

Part 4

Clinical Applications
and Perspectives

Chapter 16

Nasal Patency and the Aerodynamics of Nasal Airflow in Relation to Olfactory Function

KAI ZHAO and RICHARD E. FRYE

16.1 INTRODUCTION

16.1.1 Recognition of the Nose in Respiratory Function: A Brief History

The Ebers Papyrus, the only complete Egyptian papyrus, mentions the nose as a respiratory organ: "As to the breath which enters into the nose, it enters the heart and lungs; these give to the whole belly" (Ebbell, 1937). The Egyptians knew that the nose secretes mucus, contains arteries and veins, and is responsible for olfaction, but Galen was the only Greek physician or philosopher on record to recognize the importance of the nose in respiration (Kimmelman, 1989). In 1844, Piorry espoused the importance of the nose in respiratory function (Williams et al., 1970), but it was not until Kayser (1895) that nasal airflow was quantitatively measured with the first anterior rhinomanometer.

16.1.2 Functional Physiology of the Nasal Airway: A Brief Review

The nasal airway serves to optimize gas exchanged by conditioning the inspired air; incoming air is heated, humidified, and filtered of airborne pathogens and environmental pollutants. Even in extreme environmental conditions, inspired air can be heated and humidified to nearly 90% of the intra-alveolar air before reaching the nasopharynx (Rouadi et al., 1999). Thomson and Dudley-Buxton (1923) highlighted the functional

significance of nasal morphology by demonstrating that the anthropological cephalometric nasal index varies across races in relation to indigenous climate. Approximately two-thirds of the resistance of the entire airway tract occurs in the nasal airway (Ferris et al., 1964). Nasal airway patency is functionally coordinated with pulmonary function. By matching lower airway impedance, the nose assists in control of breathing frequency and expiration length and provides positive end-expiratory pressure. Turbinate swelling is increased by pulmonary stretch, lower airway irritation, normocapnic progressive hypoxia, and respiratory drive (Lung and Wang, 1991; Maltais et al., 1991; Nishihira and McCaffrey, 1987; Series et al., 1989).

Three sets of protrusions from the lateral wall form the inferior, middle, and superior turbinates in the human nose (see Chapter 2). These structures assist in efficient transfer of heat and moisture and diverge the inspiratory and expiratory nasal airflow into complex parallel channels. The resulting airflow can exhibit dramatic intra- and interindividual differences due to congenital anatomical features, inflammation arising from acute or chronic conditions, or the presence of polyps. These anatomical deviations may have additional functional consequences. Excessive high- or low-speed flows may also induce abnormal structural changes in the nasal cavity. For example, Ogawa (1986) reports that polyps and turbinate hypotrophy occur more often on the concave side of unilateral septal deviated noses, possibly because of higher flow rates.

Handbook of Olfaction and Gustation, Third Edition. Edited by Richard L. Doty.
© 2015 Richard L. Doty. Published 2015 by John Wiley & Sons, Inc.

16.1.3 Influence of Nasal Airflow on Olfactory and Trigeminal Senses

The nasal cavity is the initial site where airborne chemicals contact olfactory receptors and trigeminal nerve endings. The olfactory neuroepithelium is a small ($\sim$2–5 cm^2) region of the nasal mucosa located below the cribriform plate, on the upper nasal septum, dorsal superior turbinate, and regions of the middle turbinate. The irregular contour of the nasal cavity, combined with high-velocity airflow, produces nonlinear aerodynamics that promote odorant mixing while producing a complicated odorant distribution. Only 15% of the incoming airstream passes near the olfactory epithelium. Subtle alternations in nasal geometry may deflect the airstream away from the olfactory epithelium; in many cases nasal resistance and patency perception are unaffected. Trigeminal nerve endings are distributed throughout the nasal mucosa; however, the anterior regions (vestibule and nasal valve) have been reported to be more sensitive to irritants and nasal airflow (Clarke and Jones, 1994; Eccles and Jones 1983; Meusel et al., 2010; Scheibe et al., 2006; Zhao et al., 2011).

16.2 BASIC CONCEPTS OF NASAL FLUID DYNAMICS

Nasal aerodynamics are influenced by nasal anatomy and serve as a basis for much of its physiological function.

16.2.1 Basic Equations

The Hagen–Poiseuille equation, which is derived from fully developed, steady, laminar flow in a circular, straight, and rigid duct, is often used to describe nasal airway resistance (NAR). This equation relates the driving pressure (p) across a conduit to the volume airflow rate (q) through the conduit. The equation can be rewritten in the form of Poiseuille resistance, R, an analog to electrical ohmic-type resistance, where l is conduit length, η is dynamic viscosity, and r_o is the radius of the tube:

$$\frac{p}{q} = R = \frac{8 \cdot \eta \cdot l}{\pi \cdot r_o^4}.$$

Since the resistance of a conduit is inversely proportional to the fourth power of the radius, small changes in the diameter of the nasal cavity can greatly influence resistance. However, the Poiseuille equation assumptions are certainly violated by the complex shape of the nasal cavity; thus, the relationship between nasal pressure and flow is often nonlinearly related even in healthy airways. Although most studies attribute this nonlinearity to the airflow turbulence, slow flow development due to the nasal cavity's irregular geometry also produces a nonlinear relationship.

16.2.2 Incompressible, Laminar, Turbulent, and Quasi-Steady Flow—What Do They Mean?

The discussion of laminar vs turbulent flow starts with the concept of viscosity, which describes the internal resistance of fluid to flow. Since air is only considered compressible at flow velocities near Mach 1 (340.3 m/s in air), nasal airflow, which is considerably below Mach 1, is considered "incompressible." As flowing air comes into contact with a wall (or sinonasal mucosa), the velocity of the air in immediate contact with the mucosa is zero, owing to viscosity (the term "no-slip" is used to describe this situation), with flow velocity gradually increasing as distance from the wall increases. When this velocity gradient is low, laminar flow is maintained, characterized by concentric layers of well-organized flow, with the outermost layer close to the wall traveling at almost zero velocity and the centermost layer traveling at maximum velocity, resulting in a parabolic flow profile. If this velocity gradient continues to rise, by either reducing viscosity or increasing bulk stream flow, random vortexes or eddies occur between different layers of flow—hence, turbulence. When turbulence occurs, greater energy (pressure) is needed to push the flow. Turbulent flow results in a flow profile that is flat, with a quick velocity drop near the wall of the conduit. The ratio of a fluid's momentum (velocity) force to inertial (viscosity) force is often described by a dimensionless number, called the Reynolds number (Re):

$$\mathrm{Re} = \frac{\rho V D}{\mu}$$

where ρ is fluid density, V is fluid velocity, D is distance or diameter of the fluid containing cavity, and μ is dynamic fluid viscosity. In general, Re < 1500 usually indicates laminar flow, Re > 2000 indicates turbulence, and 1500 < Re <2000 is transition flow—a mixture of turbulent and laminar flow. Theoretical estimates of the Re of nasal airflow range from 600 (resting breathing) to 2000 (strong sniffing) (Keyhani et al., 1995). However, the occurrence of turbulence also depends on geometry. Flow through a smooth tube can be carefully maintained in a laminar state at Re values as high 10,000, whereas turbulence may occur at Re values as low as 1000 when a conduit has irregular walls, bifurcations, bends, or abrupt changes in a cross-sectional area (CSA). The accurate characterization of nasal airflow turbulence has to come from empirical measurement. It is generally agreed that nasal airflow at a restful breathing rate <200 mL/sec is likely predominantly laminar, while strong sniffing would result in transitional or turbulent flow.

Once turbulence develops, the previously orderly forward fluid motion disperses in random eddies, thus increasing air mixing and odorant diffusivity. The turbulent diffusivity, which may facilitate the delivery of odorants to the olfactory epithelium, is often hundreds of times higher than that of laminar flow. One computational study surprisingly suggested that turbulence mixing may have a negligible effect on odorant sorption in the olfactory mucosa (Zhao et al., 2006a). Their results suggested that nasal airflow, during sniffing, is far from fully turbulent, with a ratio of turbulent viscosity to molecular viscosity <5 (in full turbulence this ratio is >100). Thus, the increase of eddy diffusivity due to turbulence in the main airstream is limited. Nevertheless, even without the effect of turbulence, the mere increase in flow rate to the olfactory region during sniffing can have a strong functional effect and can potentially differentiate odorants based on their sorptive properties (Sobel et al., 1999, 2000).

Another rather confusing term depicts the flow phase, that is, steady or quasi-steady flow. During both inspiration and expiration, incoming nasal airflow starts with acceleration, which is followed by stagnation, deceleration, and then flow reversing. Quasi-steady flow assumes that these transitional phases are relatively insignificant and assumes that the transient cycle at a given time point can be captured by the steady state of the same total flow rate. This assumption is adopted by most nasal airflow studies since it greatly eases both computational and experimental setups. Strouhal and Womersley number analyses are frequently the theoretical basis for this assumption; however, studies have increasingly suggested that even if the nasal airflow is quasi steady, the underlying transport events can be unsteady (Shi et al., 2008, Jiang and Zhao, 2010). Nevertheless, steady-state analysis remains important because one can estimate the unsteady effect averaging over a breathing cycle (Shi et al., 2008) which seems to provide sufficient functional information. However, recent advances in animal studies that focus on temporal coding of the olfactory system within or between sniffing cycles (see Section 16.5) may move future nasal airflow study in humans toward full-blown transient-state analysis, once the findings in animal models are translated to humans.

16.2.3 The Collapsible Nasal Valve

The pressure drop across a tube due to airflow is determined by the smallest cross-sectional area (CSA), also known as the minimal cross-sectional area (MCA) (Williams et al., 1970). In the normal nasal airway, this occurs at the nasal valve which accounts for 90% of the pressure drop (Jones et al., 1988), and is located approximately 2 cm posterior to the entrance of the naris. Part of the nasal valve is composed of compliant tissue, allowing it to partly collapse when the differential pressure reaches a critical value. Collapsible tubes are known as flow regulators, which prevent flow

from exceeding an upper limit (Conrad, 1969; Holt, 1969). Nasal alar muscle activity reduces nasal valve elasticity, thereby increasing the critical pressure (Cole et al., 1985). Many factors influence the onset and magnitude of alar muscle activity, including breathing rate, maximum flow rate, acceleration of airflow, sleep state, CO_2 concentration, resistive loading, and negative airway pressure (Mezzanotte et al., 1992; Strohl et al., 1980, 1982). Sagging or depression of the upper lateral nasal cartilage, anterior nasal septum buckling, or inferior turbinate inflammation can narrow the nasal valve area, leading to a lower critical pressure (Adamson, 1987; Goode, 1985; Kasperbauer and Kern, 1987). Critical pressure also depends on both upstream and downstream resistance; thus, the dynamics of the nasal valve may depend on nasal airflow resistance changes in other regions. Except in rare conditions, the nasal valve normally does not collapse during expiration.

The collapsible nasal valve and the elastic or dynamic properties of the mucosa are not considered in most experimental and computational models when studying nasal airflow. Their implications for olfactory or trigeminal function are presently unclear.

16.3 MEASURING NASAL AIRFLOW AND AIRWAY PROPERTIES

16.3.1 Rhinometry

Rhinomanometry is a technique that measures differential pressure across, and volume flow through, the nasal cavity. It was first described by Kayser (1895) and further improved upon by Spoor (1965), who incorporated electronic pressure transducers (Nakano, 1967; Randall, 1962). Rhinomanometry measures NAR, which is best measured by anterior rhinomanometry. In this procedure, nasal airflow is measured on one side of the nose while pressure is measured on the other side of the nose. However, this method is technically difficult. Thus, NAR varies significantly among subjects; within subjects, alterations in NAR can result from the nasal cycle, body position, and nasal decongestion (Eccles, 2000).

Acoustic rhinometry introduces sound pulses into the nose, and analyzes the reflections to determine nasal cross-sectional area as a function of distance from the naris. Primary reportable measurements include unilateral or total nose MCA and nasal cavity volume (NCV). This technique requires minimal cooperation from the subject and does not require breathing effort, making it particularly useful in evaluating children, even infants, or patients with severe nasal obstruction.

Both rhinomanometry and acoustic rhinometry are reliable over several weeks if performed by an experienced operator under controlled circumstances (Silkoff, 1999). Both methods are also easy to perform on large samples,

358   **Chapter 16   Nasal Patency and the Aerodynamics of Nasal Airflow in Relation to Olfactory Function**

objective, and can reflect changes in nasal cavity anatomy resulting from various clinical abnormalities (e.g., septal deviations and surgery induced alterations) and pharmacological treatments [e.g., decongestants (Hummel et al., 1998b; Masieri et al., 1997); steroids (Lund et al., 1998)]. Acoustic rhinometry and anterior rhinomanometry are most sensitive in revealing severe deviations in the anterior nasal cavity and are less sensitive in demonstrating middle and posterior deviations (Szucs and Clement, 1998). The accuracy of acoustic rhinometric measurements is greatly reduced beyond the nasal valve, due to sound wave scattering by the turbinates and sinus. Thus, NCV measured by coronal high-resolution CT correlates well with acoustic rhinometry measurements in the anterior but not posterior nasal cavity (Dastidar et al., 1999a,b). Alternatively, information from both rhinomanometry and acoustic rhinometry can be combined into a unique index. However, both methods do not provide details on airflow patterns, potential turbulence, or the dynamics of the nasal valve, and may be less sensitive in populations with less severe symptoms (Andre et al., 2009).

16.3.2   Experimental Models

16.3.2.1   Early Nasal Models. Early in vitro models were constructed using plastic casts from cadaver noses in which the septum was replaced with a flat glass to aid visualization. Airflow measurements of these models were generally crude or descriptive, accomplished by visualizing dye (Masing, 1967) or aluminum tracer (Stuiver, 1958) in water flow, or smoke in airflow (Proetz, 1951), or by using miniature Pitot tubes (Swift and Proctor, 1977), laser Doppler velocimetry (Girardin et al., 1983), or radioactive tracers (Hornung et al., 1987a). However, many of the insights that these early studies provided are still relevant today. For example, airflow to the olfactory region increased as total nasal flow rate increased (Stuiver, 1958). Various abnormalities in the nasal cavity may have differential impact on nasal airflow (Proetz, 1951). Visual comparison of nasal airflow between F344 rat and rhesus monkey revealed that a common distribution mechanism of nasal airflow does not exist across species (Morgan et al., 1991). However, the anterior portion of the nose is important for directing the incoming airstream, regardless of the species. Some early reports differed on nasal airflow being laminar (Masing, 1967; Hornung et al., 1987a) or "turbulent-like" (Swift and Proctor, 1977), which historically created some confusion. It is now clear that these different flow regimens can be attributed to individual differences. Nevertheless, how to generalize the results across potential individual differences has been a consistent problem facing nasal airflow studies, even today.

16.3.2.2   Quantitative Nasal Models. In attempts to increase the spatial resolution and accuracy of measures of nasal airflow patterns, enlarged anatomically accurate models of the nasal cavity were constructed based on coronal MRI or CT scans [Schreck et al., 1993 (3 times larger); Hahn et al., 1993 (20 times larger)]. Using a hot wire anemometer, Scherer et al. (1989) and Hahn (1993) quantified nasal air currents at physiological inspiratory flow rates ranging from restful breathing (15 L/min) to strong sniffing (120 L/min). Air velocity was found to be relatively high through the inferior and middle meatuses and lower in the olfactory cleft. At least 50% of the airflow passed through the inferior and middle meatuses, while 15% passed through the olfactory region. Turbulence intensity was also first measured in these studies. For restful breathing with Reynolds numbers roughly 400, laminar-like flow was present throughout most of the nasal cavity, with some regional transitional flow. However, at higher flow rates, moderate turbulence (turbulence intensity <5%) occured, although full turbulence was never reached even at the highest nasal flow rate.

Particle image velocimetry (PIV) is a recently developed technique in experimental fluid dynamics research. It uses suspended particles in the flow as tracers, which is in principle similar to earlier smoke or dye methods. However, PIV uses shutters and sheets of laser light to shine in one or more planes within the flow field and digitally track the movement of particles within the planes. This provides far more quantitative data and higher spatial resolution than visually tracking particles (Adrian, 1991). Park et al. (1997) and Kelly et al. (2000) were among the first to apply PIV to investigate nasal airflow, with the latter employing a model with better anatomical accuracy.

In earlier PIV investigations, the light sheet was pulsed twice, one plane at time; thus, only those particles that moved within the plane were recorded, and velocity across the plane was lost. Since most of the nasal airflow movement is in one direction (from nostril to pharynx during inspiration, and the reverse during expiration), this limitation is acceptable with careful selection of the study plane. Recent advances (e.g., stereoscopic PIV, defocusing PIV, tomographic PIV, and holographic PIV) allow three-dimensional spatial measurement of all three velocity components, although as yet there has been no application of three-dimensional PIV in nasal airflow studies. Creating an accurate transparent three-dimensional nasal model is another essential element in the PIV approach. Here, rapid prototyping and three-dimensional printing techniques, first used by Kelley (2000), represent another technological advancement. The approach first creates a computational nasal model from CT scans, which is then printed out in solid three-dimension with water-soluble starch and molded in transparent silicone. Once cured and the starch dissolved, what remains is a transparent replica of the nasal

16.3 Measuring Nasal Airflow and Airway Properties

cavity. PIV also uses much more sophisticated particles as tracers (e.g., hollow glass spheres) that are a better match for the density of the fluid, which is often a mixture of glycerin and water in order to match the light refractive index of silicon. A tidal water pump (Chung et al., 2006) can be used to simulate breathing. Kim and Chung (2004) have also shown that rapid prototyping can be used to study the effect of various turbinectomies. The results from these PIV studies showed similar levels of intersubject (model) variabilities. For example, the main flow was found in the inferior airway in Kelly's (2000) experiments but in the middle airway in those of Chung et al. (2006).

Despite the advances in in vitro models, the limitations of the approaches include greater cost, time, human effort, and in some cases specialized equipment. In addition, suspended particles may not simulate molecular dynamics accurately, and insertion of Pitot tubes or anemometers into a model may alter aerodynamics. Given that PIV and computational models have shown generally good agreement (Hörschler et al., 2006, Weinhold and Mlynski, 2004), the latter has become more common in the study of nasal airflow due to its lower cost and ease of performance. Nevertheless, validation with experimental results is still needed.

16.3.3 With Computational Fluid Dynamics (CFD) Models

Although the set of equations that govern all fluid motions—the Navier–Stokes equation—are unsolvable analytically, advances in computer technology have made it possible to simulate nasal airflow numerically. The methods work in principle by dividing the air space geometry, obtained through imaging or casting, into many small and simple volumes (mesh), for example, tetrahedrals, polyhedrons, prisms. The first attempt was made by Elad et al. (1993), who used a crude nasal cavity represented by a trapezoid outline and two curved plates as the inferior and middle turbinates. Later, Keyhani et al. (1995) and Subramaniam et al. (1999) developed anatomically accurate models constructed from CT and MRI scans, respectively. A typical CFD model for the nasal cavity contains a few hundred thousand to a few million elements (mesh). Within each element volume, the nonlinear and higher-order Navier–Stokes equations can then be approximated linearly, which leads to approximation of the solution through iterations. Obviously, the more numerous and finer the elements, the closer the approximation is to the real solution. Thus, a mesh convergence check is often necessary by comparing simulation results in models with increasing mesh size. The quality of the elements, often reported as aspect ratios (between the shortest and the longest edge), also affects the accuracy of the simulation.

Turbulence presents another mathematical challenge, and various additional numerical schemes must be used, often involving the averaging of the randomness of turbulent flow (Reynolds averaging) (Zhao et al., 2006a) or the filtering out of small eddies (Large Eddies Simulations). For these common CFD schemes, once the mesh is created, the computation afterward is rather straightforward, although incorrect settings to the software can produce distorted findings.

A few other CFD schemes, such as direct numerical solutions and Lattice Boltzmann simulations (Eitel et al., 2010), may also be applied to nasal airflow simulation. While Boltzmann simulation is a nonorthodox CFD approach and has some advantage in grid/mesh generation, direct numerical simulation is often considered the gold standard in CFD. However, both methods require significantly greater computational cost and often require in-house coding. Validation between CFD simulation and experimental models is sparse; however, good agreement has been shown when tested (Keyhani et al., 1995; Elad et al., 1993).

With the advances in computational power, the bottleneck in the CFD approach lies in the meshing process, which, to date, still cannot be fully automated. Early CFD nasal models took a few months to a few years to generate manually, element by element. Major advances occurred around the turn of twenty-first century. For example, Zhao et al. (2004) rapidly semiautomatically generated eight versions of models (>1 million meshes for each requiring only days for the generation of each model) (see Figure 16.1). However, the technique has not advanced much since that time. Without a fully automated scheme, considerable effort is required to correct and oversee the meshing process. Most studies, even the most recent ones, still involve models of one or just a few subjects.

Earlier CFD simulation results indicated that most air flows along the nasal cavity floor, while the turbinate structures confine flow into parallel channels in an anterior-posterior direction, with about 10% of the airflow passing into the olfactory cleft (Keyhani et al., 1995; Keyhani et al., 1997). However, later CFD analysis showed, in accord with other methods, that nasal airflow distribution may have significant individual variations. For example, Keyhani et al. (1995) and Ishikawa et al. (2006) both reported the peak velocity in the nasal valve, but the specific location of the valve differed. Many studies differ regarding where most of the flow occurs. [e.g., along the middle meatus (Simmen, 1999); lateral to the nasal septum at the level of the either middle or inferior turbinate (Grant et al., 2004; Hahn et al., 1993; Lee et al., 2009)]. Zhao et al. (2004) demonstrated this anatomical effect using a three-dimensional anatomically accurate nasal cavity model based on a normal subject's CT scan. These investigators varied the nasal anatomy in two critical regions (the

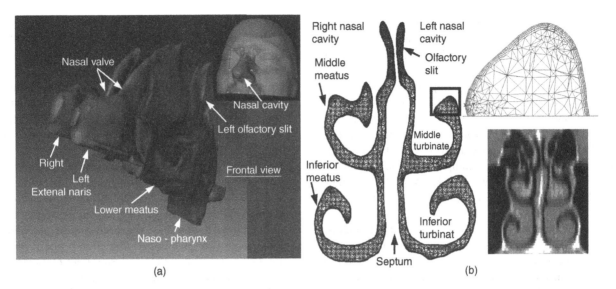

Figure 16.1 (a) an anatomically accurate computational nasal model constructed from nasal CT scans (b, lower right) of a normal adult human female. (b) A coronal cross section of the same model, which is filled with 1.7 million tetrahydral elements suitable for numerical airflow simulation. In a close-up view (b, top right), layers of small and fine elements can be seen along the wall that can capture the rapid near wall changes of air velocity and chemical concentration. Figure is adapted from Zhao et al., 2004. (*See plate section for color version.*)

nasal valve and the olfactory cleft) and showed that the overall nasal airflow pattern was highly sensitive to local nasal geometric configurations (Figure 16.2). Fractional airflow to the olfactory region was found to increase up to 50 times. This explains previously reported airflow pattern discrepancies based on different cast models. Such variant airflow patterns, independent of total nasal resistance or total nasal airflow rate, are likely to have enormous implications for the sensory function of the nose—pungency, patency, and olfaction.

In addition to inter-subject differences, CFD offers a potential tool to predict the impact of various pathologic conditions and surgical maneuvers on nasal airflow [e.g., the effects of septal deviation (Ozlugedik et al., 2008; Chen et al., 2009; Garcia et al., 2010), septal perforation (Pless et al., 2004, Grant et al., 2004), endoscopic sinus surgery (Zhao et al., 2006b), inferior turbinate reduction (Wexler, 2005), turbinectomy (Garcia et al., 2007), and turbinate hypertrophy (Lee et al., 2009)]. However, few studies to date have included odor absorption and olfactory function (Zhao et al., 2006b; see Section 16.4). Even though these CFD studies provide new insights into nasal airflow patterns in both normal and pathologic conditions, most of them are case reports and many more questions remain. Of particular interest is the identification of anatomic and flow pattern parameters that can reliably predict which patients will complain of nasal sinus problems. Such information may be useful in preoperative planning and postoperative analysis.

In summary, for the purposes of measuring aerodynamics of the nasal cavity, rhinomanometry, which indexes overall nasal resistance, and acoustic rhinometry, which indexes the anterior nasal airway cross-sectional area, remain valuable tools applicable to large population-based clinical studies, despite the fact they do not capture the full picture of nasal airflow details. In vitro models and CFD approaches remain investigational tools, with single or a small number of individuals examined in each study. They both can provide detailed quantitative information about nasal airflow patterns and properties, but the challenges of applying to the general population and determining the functional relevant parameter remain. Nasal airflow and patterns seem to have significant individual differences and are prone to various perturbations. However, it is unclear how to draw a line between the variability among healthy individuals and those with pathological conditions, and whether and which changes under pathological conditions have functional relevance or are the causal factors for symptoms.

16.4 NASAL AIRFLOW: IMPLICATIONS FOR OLFACTORY FUNCTION IN HUMANS

Nasal airflow is essential for olfactory perception and is dictated by several factors. First, the location of olfactory epithelium in human nasal anatomy is rather small and confined to a narrow region in a remote portion of the nasal cavity; thus, disturbances of airflow and odor distribution to this regional may have functional consequences. Second, the effects of nasal cavity anatomy on nasal airflow are

16.4 Nasal Airflow: Implications for Olfactory Function in Humans

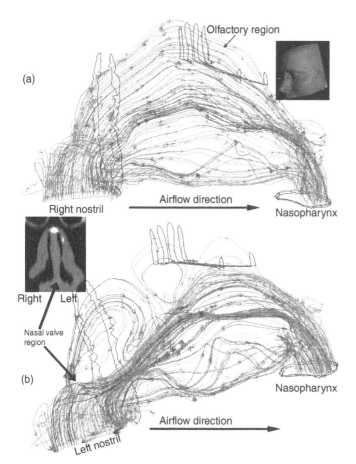

Figure 16.2 Nasal airflow pattern can be highly sensitive to local airway geometry. This figure shows the streamline patterns for inspiratory steady laminar airflow in the left and right sides of the nasal cavity depicted in Figure 16.1. Note the smooth laminar flow on the right side of the nose (a) and the eddies in the anterior part of the left side of the nose (b) induced by a small airway constriction in the nasal valve region. Note also the secondary eddy in the superior and posterior part of the nasal cavity of the left side. When the constriction was artificially removed, the streamline pattern became smooth again (figure not shown) just as in the right side. Adapted from Zhao et al., 2004. (*See plate section for color version.*)

complex. For example, since airflow travels along the path of least resistance, it is easily redistributed if one path is blocked. Thus, airstream distribution can be profoundly influenced without a substantial change in nasal anatomy, while nasal resistance, measured by rhinomanometry, is typically elevated only by severe nasal anatomic abnormalities. Finally, the receptors do not have direct access to airborne odorants, so sorption into the mucosa is necessary prior to receptor activation, while the sorption further depends on many factors such as odorant diffusivity and solubility. This section reviews evidence for the influence of nasal airflow and nasal resistance on olfactory function in humans. Studies examining this relationship either have used clinical populations of individuals with nasal disease or have studied the relationship of olfactory function and intersubject or intrasubject variations in nasal airflow and nasal resistance. Each of these approaches has advantages and limitations. The emphasis on nasal sinus disease has important health relevance, because a significant portion (~15–25%) of persons with olfactory loss have nasal sinus disease. The etiology and subsequent treatment of olfactory loss accompanied by nasal sinus disease are poorly understood. The overall findings from these studies are summarized at the end of this section.

16.4.1 Clinical Studies: Olfactory Function in Nasal Pathology and After Clinical Treatment

A suggestion that nasal airflow and nasal anatomy influence olfactory function has been based historically on the fact that olfactory dysfunction is associated with the same clinical abnormalities that affect nasal airflow, such as allergic rhinitis, polyposis, nasal sinus disease, or adenoid hypertrophy. Further support for this notion is provided by uncontrolled studies that have documented improvement in olfactory function following procedures known to improve nasal airflow and nasal resistance, such as endoscopic nasal sinus surgery (Seiden and Smith, 1988), adenoidectomy (Ghorbanian et al., 1983), septoplasty (Stevens and Stevens, 1985), and inferior turbinectomy (Ophir et al., 1986; Elwany and Harrison, 1990). Below we review the findings of more recent, higher-quality, clinical studies, including clinical trials that have examined improvement in olfaction and nasal airflow as a consequence of surgical and pharmaceutical interventions. Some of these clinical findings are inconsistent, likely reflecting the fact that the clinical effects of nasal anatomy and airflow abnormalities on olfactory function are complex.

16.4.1.1 Rhinosinusitis and Olfactory Loss.

Overall, anosmia or hyposmia occurs in 15–25% patients with underlying rhinosinusitis disease (Cullen and Leopold, 1999; Seiden and Duncan, 2001; Cowart et al., 1997), and this percentage increases up to 80% when sinus disease co-occurs with nasal polyps (Wolfensberger and Hummel, 2002). In addition, nasal polyps can differentially impair orthonasal vs retronasal olfactory acuity (Landis et al., 2003). It has been widely believed that a conductive mechanism is responsible for olfactory dysfunction in sinus disease, especially in those with polyposis. However, empirical testing of this notion has provided mixed results. In one study, NAR measured by active anterior rhinomanometry was found to correlate with both odor identification and olfactory threshold among rhinosinusitis patients (Damm et al., 2000), but other studies have not been able to document a correlation between NAR and olfactory function in rhinitis patients. For example, Simola and Malmberg (1998), in a rather large controlled study of patients with both allergic and nonallergic rhinitis, demonstrated an olfactory deficit in the rhinitis group compared with the control group, but were unable to document a relationship between olfactory thresholds and NAR as measured by active anterior rhinomanometry.

Although one potential reason for the mixed results in the above studies is that rhinomanometry may not necessarily reflect flow specifically to the olfactory region, other evidence suggests that there may be other inflammatory confounding factors. In two small prospective studies on allergic rhinitis patients, Klimek (1998) demonstrated that abnormalities in olfactory function were much better correlated with a marker of nasal inflammation than with NAR measured by rhinomanometry. Guilemany et al. (2009) demonstrated that olfactory function was correlated with nasal nitric oxide, a marker of nasal inflammation, but not with nasal inspiratory flow or acoustic rhinometry measurements in 49 patients with persistent allergic rhinitis.

These data may indicate that inflammatory processes, which are significantly involved in the pathophysiology of sinus disease, may play a larger role in olfactory dysfunction. Indeed, patients with chronic sinusitis who have inflammatory cell influx into the olfactory mucosa are much more likely to have an olfactory deficit (Kern, 2000). Chronic rhinosinusitis has been associated with pathological changes in the olfactory neural epithelium (Yee et al., 2009) and inflammatory mediators such as tumor necrosis factor alpha have been shown to cause olfactory neuron dysfunction (Sultan et al., 2011). Additional evidence for the salient role of inflammation includes observations that (a) olfactory dysfunction occurs in inflammatory disorders that are not associated with airflow obstruction of the upper nasal cavity (Laudien et al., 2009) and (b) that, in acute rhinitis, smell function, as measured by event-related potentials, recovers long after resolution of the obstructive component of the illness (Hummel et al., 1998a). Still, one has to recognize the tight interrelation between inflammation and sinus disease severity (and thus conductive airway restriction). For example, Soler et al. (2009) demonstrated that, in patients with chronic rhinosinusitis, smell identification test scores correlated with both mucosal eosinophilia and disease severity, as measured by computer tomography.

16.4.1.2 Clinical Interventions.

One approach to delineate the effect of airflow on olfactory function is to objectively demonstrate the correlation between improvements in airflow and olfaction with interventions in clinical populations.

LARYNGECTOMY One clinical condition associated with a close correspondence between rhinomanometry measurements and olfactory function is laryngectomy. Manestar et al. (2011) demonstrated a significant positive correlation between olfactory function and nasal airflow in laryngectomy patients using the polite yawning technique. Further studies by Manestar et al. (2012) demonstrated that the minimum nasal airflow required for olfactory stimulation in successfully rehabilitated patients was approximately 60 cm^3/s.

ENDOSCOPIC SINUS SURGERY Early studies with a limited number of subjects failed to document improvements in NAR or olfactory function with function endoscopic sinus surgery (Lund et al., 1991), while early follow-up studies on larger samples did document improvements in both subjective and objective symptoms of nasal airflow and olfactory function (Lund and Scadding, 1994). More recent studies have confirmed such findings. For example, Wang et al. (2002) demonstrated marked improvements in both NAR, as measured by anterior rhinomanometry, and olfactory function following endoscopic sinus surgery for chronic sinusitis and/or nasal polyps; however, they did not attempt to establish the association between the two. Other researchers have demonstrated that the relationship between nasal airflow and olfaction is more complex and may occur over different time periods. For example, Rózańska-Kudelska et al. (2010) demonstrated that improvement in olfaction lagged several months after NAR improvements following endoscopic surgery for nasal polyps. Other studies have suggested that a clear olfactory cleft is most important. Using CT staging of the sinuses, Shin et al. (1999) demonstrated that the blockage of the olfactory cleft was significantly related to the butanol olfactory threshold values in 50 patients with bilateral sinusitis. A more sophisticated study that used CFD modeling was able to predict that airflow delivery to the neuroepithelium increased by 1000 times as a result of endoscopic polyp removal (Zhao et al., 2006b). Thus, the contribution of NAR in general to improvement in olfactory function endoscopic sinus surgey is unclear at this time. Specific changes in airflow to the olfactory cleft may be more important than general changes in NAR, but are not easily measured.

SEPTOPLASTY AND TURBINOPLASTY The olfactory region is located superior in the human nasal cavity, which is not directly touched upon during septo- and turbinoplasty. It is thus worth looking at these anterior nasal surgeries specifically, as they represent an opportunity to isolate the effect of nasal airflow on olfactory function. Several turbinoplasty studies have demonstrated improvement in both olfactory function and NAR. Rhee et al. (2001) reported objective improvement in total nasal volume and NAR, as measured by acoustic rhinometry and rhinomanometry, respectively, as well as improvement in olfactory thresholds in previously hyposmic patients after temperature-monitored radiofrequency or laser vaporizing turbinoplasty. Damm et al. (2003) demonstrated differential nasal airflow and olfactory function improvement following septoplasty and partial inferior turbinectomy. Nasal airflow, as measured by anterior rhinomanometry, improved in 87% of the patients, while odor identification improved in 80%, odor discrimination improved in 70%, and olfactory threshold improved in 54%. In a longitudal study, radiofrequency inferior turbinate reduction resulted in a significant improvement in NAR, as measured by anterior rhinomanometry, as well as odor identification and threshold, two months (Garzaro et al., 2010) and two years (Garzaro et al., 2012) following the procedure.

However, not all studies have been able to document the correlation between improvement in NAR and olfactory function following turbinoplasty procedures. Three months following CO_2 laser turbinoplasty, Olszewska et al. (2010) found significant improvements in total NAR as measured by anterior rhinomanometry, but improvement in olfactory function in only a minority of patients. Modified vidian neurectomy combined with inferior turbinoplasty resulted in significant improvement in NAR, but subjective resolution of anosmia in only half of the patients (Ikeda et al., 2006).

PHARMACOLOGICAL TREATMENT Corticosteroids, antihistamines, and nasal decongestants are the primary medications in sinus disease management; however, their effects on olfactory recovery are mixed. In a double-blind, placebo-controlled trial of 17 patients with seasonal allergic rhinitis and symptoms of impaired olfaction, Sivam et al. (2010) found that the intranasal steroid mometasone improved nasal symptoms, nasal peak inspiratory flow, chemosensory quality of life, nasal

inflammation, and the number of eosinophils in the olfactory epithelium, but not olfactory function. Vaidyanathan et al. (2011) demonstrated that oral steroid prednisolone treatment of 60 adults with chronic rhinosinusitis and nasal polyps significantly decreased polyp grade and the perceived magnitude of the smell loss, but not perceived nasal patency. In 27 patients with persistent allergic rhinitis, Guilemany et al. (2012) demonstrated that levocetirizine, a third-generation antihistamine, resulted in improvement in both perceived smell loss and olfactory function that corresponded more to a reduction of nasal inflammation, as measured by nasal nitric oxide levels, than to the improvement in nasal airflow.

The lack of correspondence between improvements in nasal airflow and olfactory function is echoed by studies investigating nasal decongestants—medications that temporarily constrict blood flow, reduce nasal congestion, increase airflow, and effectively eliminate the nasal cycle (see Section 16.4.2.3), but do not have a direct impact on the inflammation. At least three studies (Hummel et al., 1998b; Temmel et al., 1999, Jung et al., 2011) have investigated the effect of two commonly used nasal decongestants, ephedrine and oxymetazoline, on olfactory function. All three studies employed double-blind random placebo-group designs; one enrolled rhinitis patients (Hummel et al., 1998b) while the other two focused on healthy populations. While all three found a significant reduction of NAR as measured by acoustic rhinometry or rhinomanometry after nasal decongestant treatment, none found any positive or negative effects on olfactory function, as measured by intensity ratings, odor identification, or thresholds. One of the studies also examined trigeminal function and found no effect (Hummel et al., 1998b), while another examined topical application of a commonly used nasal anesthetic agent lidocaine (Jung et al., 2011) and found no effect on odor identification.

Together, these studies highlight the complicated nature of the relationship between nasal airflow and olfaction. The vascular effect of intranasal applied treatments may be mostly localized in erectile tissue along the inferior turbinates and septal body, far away from the olfactory region, a potential explanation for its lack of effect on olfactory function. Another complication is that there is very little control of how much aerosol penetrates into the nasal cavity or into the olfactory region during the nasal spray. It has been reported that direct application of lidocaine onto the olfactory cleft can result in temporary anosmia (Welge-Lussen et al., 2004).

NASAL DILATOR As noted earlier, the nasal valve is a dynamic and collapsible tube that contributes significantly to nasal resistance (see Section 16.2.3). A narrower resting nasal valve or an anterior septal deviation may create high baseline valve resistance, which would make one prone to nasal obstruction. External nasal dilator strips, such as the Breathe Right strip (GSK, Inc.), can effectively increase nasal valve CSA in the pre- and postdecongested nose (Ng et al., 1998) and improve MCA and NAR in normal subjects and in patients with septal deviation that affect the nasal valve area (Roithmann et al., 1998). Nasal dilators have been shown to decrease heart rate, ventilation, and VO_2 in athletes during submaximal exercise (Griffin et al., 1997) and to improve the respiratory disturbance index in a subpopulation of patients with obstructive sleep apnea and snoring (Gosepath et al., 1999). Hornung et al. (1997 and 2001) reported improved odor intensity perception, odor identification, and thresholds after the application of Breathe Right nasal dilators in healthy subjects. They attributed the effect to increased anterior nasal airway CSA (measured by MRI scans) and increased sniffing volume and duration. Surprisingly, there have been no replicating studies with larger sample sizes or clinical populations to validate these results. Raudenbush and Meyer (2001) reported that nasal dilators increased the intensity and pleasantness of food flavor placed in the oral cavity in healthy subjects.

16.4.2 Non-clinical Human Studies

Several studies have used natural variations in nasal anatomy and nasal resistance to study their correlation with olfactory function.

16.4.2.1 Nasal Cavity Volume. To identify the relationship between nasal anatomy and olfactory dysfunction, Leopold (1988) obtained nasal cavity CT scans of 34 patients with conductive or idiopathic hyposmia. The nasal cavity superior to the middle turbinate was divided into nine sections. Interestingly, a larger volume in the region 10–15 mm below the cribriform plate and a smaller volume 1–5 mm inferior and anterior to the cribriform plate was associated with higher olfactory ability, as measured by olfactory confusion matrix. The study was confirmed by Hornung and Leopold (1999) with unilateral, as opposed to bilateral, olfactory function in another 19 patients with static conductive hyposmia due to polyposis or mucosal edema. A larger volume 5–10 mm below and anterior to the cribriform plate and a smaller nasal vestibule volume correlated with better olfactory function. Damm et al. (2002) replicated the study in 50 healthy males using MRI. Olfactory thresholds were found to correlate with the anterior segment of intranasal volume in the upper meatus below the cribriform plate, as well as the anterior segment of the inferior meatus. Other studies have also indicated that a clear olfactory cleft is important for its function (Shin et al., 1999).These studies suggest that critical areas of the nasal cavity, both near and remote from the olfactory receptors,

influence olfactory function, presumably by directing air toward the olfactory cleft. However, the exact extent of how these regions affect olfactory airflow remains unknown.

16.4.2.2 Artificial Blocking of the Nasal Cavity. While it is difficult to derive a convincing conductive mechanism to olfactory losses in clinical studies due to the complex etiology of nasal sinus disease, at least one study has attempted to artificially create blockages inside the nasal cavity and confirmed that the resulting airflow changes can induce olfactory loss. In this double-blind study, sponges with a high saline content were placed in either the anterior olfactory cleft or the lower respiratory region. The results indicated that orthonasal, but not retronasal, olfactory identification ability was lower when the anterior olfactory cleft was blocked (Pfaar et al., 2006). This confirms that the differences between orthonasal and retronasal olfactory function observed in nasal polyps patients (Landis et al., 2003) may, to some degree, be due to mechanical obstruction of the anterior portion of the olfactory cleft.

16.4.2.3 The Nasal Cycle. The nasal cycle is a periodic ultradian rhythm that results in a change in nasal cavity volume that may occur in up to 80% of the adult population. The periodicity has been reported to range from 40 minutes to 4 hours (Bojsen-Muller and Fahrenkrug, 1971; Eccles, 1978; Principato and Ozenberger, 1970). Studies using acoustic rhinometry demonstrate that the nasal cycle is present, in some form, in most adults and in children as young as three years of age, even those with structural abnormalities such as septal deviation (Gungor et al., 1999; Lund, 1996; Sung et al., 2000). Classically, the change in volume is in the opposite direction across the two nasal cavities. More recent studies have found that many people do not exhibit this classic cycle—some have fluctuations on only one side of the nose and others have left:right airflow fluctuations that are poorly correlated with one another (Gilbert and Rosenwasser, 1987; Mirza, Doty and Kruger, 1997).

To determine if a relationship exists between the nasal cycle and olfactory sensitivity, Frye and Doty (1992) measured unilateral 2-butanone olfactory thresholds and NAR for 33 men and 44 women in two sessions separated by 4 hours. In approximately half of the subjects, the nostril not sampling the odorant was occluded. If the nasal cycle had not spontaneously changed its phase at the beginning of the second session, an attempt was made to change the phase by applying pressure under the armpit or in the palm of the hand or by auditory stimulation on the side of the more patent airway. In subjects whose contralateral naris was blocked, lower right NAR was associated with decreased olfactory thresholds on both sides of the nose, whereas lower left NAR was associated with comparatively increased olfactory thresholds on both sides of the nose.

These data suggest that NAR per se was not responsible for alterations in olfactory sensitivity; rather, other factors associated with the nasal cycle such as changes in cerebral EEG or central arousal mechanisms likely accounted for these findings. Thus, augmentation of olfactory sensitivity was associated with increased arousal and greater left hemispheric integrated EEG activity, and olfactory sensitivity attenuation was associated with decreased arousal and greater right hemispheric integrated EEG activity.

To better define this relationship, NAR was measured in eight right-handed male subjects every 15 min over 6 hours (Frye, Valle, and Doty, unpublished data). Olfactory sensitivity to phenyl ethyl alcohol and subject response bias were measured using a signal detection paradigm. The perithreshold odor concentration was determined by a unilateral olfactory threshold test preceding the study; the contralateral nostril was blocked during odorant sampling. The nasal cycle and unilateral NAR were also measured. The left dominant phase of the nasal cycle was associated with increased and decreased subject response bias for odors presented to the right and left sides of the nose, respectively. Since asymmetrical changes in integrated hemispheric EEG activity are correlated with the nasal cycle, and most of the olfactory bulb's efferents project unilaterally, interhemispheric alteration in decision processing associated with olfactory recognition may explain this finding.

16.4.3 Modeling Approach to Understanding Odorant Deposition and Uptake

The absorption of airborne chemicals into the nasal mucus layer, which marks the initial step in the olfactory process, depends on the interplay of several factors, including airflow, diffusion, and solubility of the chemicals in mucus (Mozell and Jagodowicz, 1973). Very few attempts have been made to directly measure the sorption of odorants in mucosa (Amoore and Buttery, 1978; Hornung et al., 1980, 1987b); rather, most efforts have been theoretical, computational, or inferred from particle deposition study.

16.4.3.1 Theoretical Model. Hahn et al. (1994) developed an odorant transport model that included aerodynamic transport of the odorant molecule through bulk and lateral flow mechanisms and local odorant sorption onto olfactory mucosa, as well as olfactory receptor interactions. The model predicted that increasing the flow rate would increase the perceived odor intensity for highly soluble odorants but decrease odor intensity for insoluble odorants. These predictions match several experimental studies. However, if given a substantial decrease in mucus surface area for sorption, the model would predict that perceived odor intensity would decrease for all odorants regardless of the solubility, which matches electrophysiology recordings in bullfrogs (Mozell et al. 1991).

16.4.3.2 Particle Deposition and Uptake. As noted in Chapter 20, particle deposition in the olfactory region may have additional health concerns because some particles can bypass the blood–brain barrier through the olfactory nerve tract (Oberdorster et al., 2004). However, significant differences exist in the mechanism of particle versus odorant deposition. Particle deposition often assumes perfect absorption (i.e., that any particle contacting the mucosa is permanently trapped), whereas odorants may be released into the air phase from the musoca (Bocca et al., 1956). Particle transport is a function of particle size; for large, micrometer-size particles, inertial force and impaction play significant roles. In vivo experimental data suggested that an impact factor defined as $\rho d^2 Q$ (where ρ is the particle density, d is the particle diameter, and Q is the volume airflow rate) correlates well with nasal deposition (see review Yu et al., 1998). This impact factor is equivalent to the Stokes force, which suggests that inertial impaction is indeed dominant over the diffusional or gravitational effects for the deposition of micrometer sized particles.

However, most odorants are in the nanometer range, and for nanoparticles diffusion dominates the deposition mechanism. Cheng et al. (1996b), using nanoparticles ranging from 4 to 150 nm in 10 subjects at flow rates of 167 ml/s and 333 ml/s, showed that the deposition is higher for subjects with smaller mean CSA, larger total surface area, and/or larger mean perimeter. In vitro studies of nanoparticle deposition in replicas from CT or MRI scans or from casts of cadavers have also been conducted by many researchers (Cheng et al., 1988, 1993, 1996a; Cheng, 2003; Gradon and Yu, 1989; Guilmette et al., 1994; Yamada et al., 1988; Kelly et al., 2004), the results of which are in general agreement with in vivo results. Based on these data, Cheng (2003) further developed an empirical equation for the nanoparticle deposition in the nasal cavity based on the Reynolds number, Schmidt number (Sc = ν/D, where D is the diffusivity of the particles and ν, air viscocity), and the ratio of average CSA to the surface area of the nasal passage (see Cheng, 2003, eq. 12). Despite abundant experimental data on this topic, no data were available for deposition specifically in the olfactory region for either micro or nanometer particles, which necessitates computational approaches.

16.4.3.3 Computational Studies on Odorant Sorption. Computational approaches offer a way to theoretically study airflow and odorant transport to the olfactory region. Computing nanoparticle transport often involves the same equations as odorant transport—in

other words, nanoparticles are treated as a gas without inertial impaction. Thus, previous validation of computational nanoparticle transport can serve as validation of the methodology.

Keyhani et al. (1997) were among the first to investigate inspired odorant molecule transport and uptake onto the nasal and olfactory mucusa based on an anatomically accurate computational nasal model. Assuming quasi-steady state and laminar flow, total odorant flux was calculated as a function of several transport parameters, including odorant solubility and diffusivity in the air and in the mucosal lining and the thickness of the mucus layer. Odorant flux increased with inlet flow rate in a nonlinear fashion and depended on odorant solubility. For example, fractional flux decreased for poorly soluble odorants and increased for highly soluble odorants. Odorant flux decreased along the olfactory slit from anterior to posterior and from inferior to superior, with this gradient being dependent on odorant mucus solubility. Thus, different odorants generated discernibly different flux patterns across the olfactory mucosa, confirming previous predictions (Mozell et al., 1987).

Given that the human olfactory cleft receives a relatively small distribution of total airflow, it is often assumed that it may be sensitive to anatomical changes. Zhao et al. (2004) computationally demonstrated that small pertubations in olfactory cleft volume significantly changed airflow to the olfactory region but did not affect NAR. Change in regional airflow rates in the olfactory region also effected odorant absorption, especially for odorants with higher mucosal solubility (-carvone) and higher air diffusivity (methanol) as compared to odorants with low-solubility (*d*-limonene).

Applying CFD to the clinical setting, Zhao et al. (2006b) compared pre- and postpolypectomy CT scans in a patient with nasal polyps. Preoperative endoscopy revealed polyps within the olfactory cleft and along the middle turbinate in a patient complaining of nasal obstruction and anosmia. After bilateral endoscopic sinus surgery and polypectomy, CFD analysis revealed decreased overall airflow resistance and an increase in airflow to the middle and superior nasal cavities in the olfactory cleft region, with improved delivery of a highly soluble odorant (phenylethyl alcohol) at a rate greater than 1000 times what was seen preoperatively. These findings correlated with the patient's postoperative recovery of odor detection threshold to that same odorant. However, beyond this case report, few studies have directly related CFD findings to olfactory function either in healthy subjects or in clinical populations.

No direct measurement of odorant sorption in human nasal cavity has been reported. However, in one interesting attempt to combine CFD with an experimental approach (Kurtz et al., 2004), odorants of various solubilities were pumped into one nostril and out of the other, while healthy subjects performed velopharyngeal closure. The study measured the fraction of odor retained by the nasal passage at different flow rates, which depends on odorant solubility, and demonstrated good agreement with Mozell's early olfactory nerve recordings in animals. The experiments were then replicated computationally with a published CFD model, and calibrated several unknown parameters (odorant diffusivity and mucosa permeability), which were then applied to other odorants of unknown mucosa solubility. While not able to direct measure odor sorption, this study serves as initial validation and confirmation of the functional relevance of CFD approaches.

16.4.4 Summary

In summary, the clinical and empirical studies reviewed in this section provide insight to the relationship between olfaction and nasal airflow and NAR in humans, particularly in clinical populations. Studies from the laryngectomy population demonstrate that a certain amount of nasal airflow is necessary for sufficient quantities of odorants to be delivered to the olfactory cleft, whereas clinical, experimental, and computational studies demonstrate that airflow specifically to the olfactory cleft is important for odorant perception. Both experimental studies examining the relationship between nasal volume and olfaction and computational studies artificially modifying regional nasal volume also suggest that areas of the nose remote from the olfactory cleft can be important for directing airflow upward toward the olfactory cleft.

However, the application of information does not easily translate to clinic relevance. Even though studies suggest that surgical or medical intervention can improve olfaction, success is uncertain. It is difficult to predict which patients will or will not experience restoration of olfactory function following intervention. Nasal sinus disease associated with swelling of the superior nasal cavity likely obstructs airflow to the olfactory region, while disorders associated with the inferior turbinate or nasal septum presumably redistribute the airstream within the nasal cavity. However, airflow obstruction is only one mechanism of olfactory dysfunction, and changes in the olfactory epithelium due to inflammatory processes, may also have significant effects. Clearly, disorders such as rhinitis can influence olfactory function through inflammatory processes and treatment for such disorders may influence olfactory function through factors other than nasal airflow. Currently, most clinical studies perform only simple statistical analyses and use convenient rather than well-stratified samples. Clearly, more sophisticated studies are needed to account for these multiple factors, and more sophisticated analyses will help determine the contribution of each of the pathological factors to olfactory dysfunction.

Several inconsistencies and differences within the clinical studies reviewed above are important to note. Many studies suggest that, following surgery, olfactory thresholds, but not measures of odor identification, improve (Shin et al., 1999), whereas other studies claim that odor identification improves more than olfactory thresholds (Damm et al., 2003). Studies have also pointed out that quantitative and qualitative olfactory measures do not correlate (Shin et al., 1999). Computational (Keyhani et al., 1997) and animal studies (Mozell and Jagodowicz, 1973) have suggested that not all odorants follow the same airflow distribution and that changes in airflow likely affect high- more than low-mucosal-solubility odorants. However, few clinical/human studies have carefully employed odorant sorptive properties in their test designs, which may contribute to the different outcomes due to the different odorants employed.

Many studies have used anterior rhinomanometry and acoustic rhinometry to study nasal airflow and NAR. These techniques have their own limitations. For example, NAR and/or nasal airflow and subjective symptoms of nasal patency correlate well in the presence, but not the absence, of symptoms (Andre et al., 2009). Both techniques index global nasal airflow/airway properties, which may not reflect regional flow or delivery of odorants to olfactory epithelium, as shown by several CFD studies. Thus, they may be unable to mechanistically separate the confounding factors due to either nasal airflow or inflammatory processes among clinical populations. Finally, it may be advantageous to introduce techniques that are very specific to the olfactory cleft into clinical studies, such as the use of the CFD technique (Zhao et al., 2006b). However, such an approach at this time remains investigational, with applications mostly in case studies, due to the technical challenges.

16.5 ANIMAL STUDIES: IMPLICATIONS FOR HUMANS

Although this review focuses on humans, data from animal experiments, especially from rodents, are far more abundant and greatly advance our understanding of olfactory system in general. However, the differences in nasal anatomy across species are substantial (Negus, 1958) and most terrestrial mammals have much more complicated nasal anatomy than do humans. The turbinates in animals often consist of more intricate and convoluting scrolls and plates of thin bones, which diverge odorant-carrying airflow into complicated channels and circulations with varying speeds, turbulence, and directions. In this regard, Adrian (1950) postulated the existence of odorant-dependent spatiotemporal patterns at the level of the olfactory mucosa

that corresponded to each odorant's value of some physicochemical property. This notion was confirmed by Mozell et al. (1973, 1987), who showed that different odorants flowing across the mucosa do indeed produce different spatiotemporal patterns of neural activity, with this effect strongly influenced by airflow rate, depending on the solubility of the odorants (Mozell et al., 1991). This differential absorption process was even considered analogous to gas chromatography (Mozell and Jagodowicz, 1973). While early studies by Mozell's group were mostly in bullfrogs and salamanders, amphibians with simple nasal structures, later electroolfactogram (EOG) recordings in rodents confirmed the existence of such patterns (Scott et al., 1997, 2000; Scott and Brierley, 1999).

On the other hand, studies (Kauer and Moulton, 1974; Mackay-Sim et al., 1982) have demonstrated the existence of intrinsic spatial neural activity patterns that were independent of airflow constraints. Moulton (1976) coined the terms "imposed" and "inherent" patterns to differentiate the two mechanisms for odorant-induced mucosal spatial patterns. With the identification of the olfactory receptor gene family (Buck and Axel, 1991), several investigators (Ressler et al., 1993; Strotmann et al., 1994; Vassar et al., 1993) have identified different aggregated receptors zones spatially extending anteroposteriorly throughout the rat olfactory mucosa.

Experimentally quantifying the absorption process in species with more complex nasal structures is difficult. A few successful efforts have relied on the computational approach. Rodent CFD models have shown that the airflow through the nose is laminar, with a central recess of the olfactory region where airflow is high and larger lateral region where the airflow is substantially lower (Kimbell et al., 1997; Yang et al., 2007; Jiang and Zhao, 2010). Consequently, the high-solubility odorants are readily absorbed in the central zone, whereas low-solubility odorants are absorbed more laterally, due to lower flow speed and much larger surface area in the lateral region (Yang, 1999; Zhao et al., 2006a). With increased sniffing flow rate, the absorption of high-solubility odorants increases, whereas the sorption of low-solubility odorants remains unchanged or even decreases as a fraction of total inhaled odorant molecular number, which matches what Mozell described earlier (Mozell et al., 1991). Furthermore, it is interesting to note that the contours of the flow and absorption patterns in these models correspond roughly to the longitudinal mapping zones of receptor genes (Strotmann et al., 1994; Yang et al., 2007). This further implies a structure-function optimization, extending Mozell and colleague's chromatography analogy, that at least in rodents, both imposed and inherent (neuron distribution) patterning may be utilized to optimize odor perception: the turbinate structure distributes the incoming odorants to different mucosal regions based on their physiochemical properties

368   Chapter 16   Nasal Patency and the Aerodynamics of Nasal Airflow in Relation to Olfactory Function

and aerodynamics where types of receptors that are more sensitive to or better discriminative of those odorants might be located (Yang et al., 2007); reviewed by Schoenfeld and Cleland, 2005, 2006). Computational studies also showed that, proportionally, the inspiration airstream penetrates more into the olfactory region than does the expiration airstream (Zhao et al., 2006a; Yang et al., 2007). This lack of reversibility may trap inhaled air and odorants in the rat olfactory region, which is not immediately washed out on exhalation; this is likely a very important aspect of the airflow in the rat sniffing strategy and allows odorants to be retained longer in the complex ethmoid air spaces, where they have longer contact with olfactory receptors on the epithelium. It has long been noticed that rodents employ complex and precise motor regulation of sniffing behaviors during odor tasks (Youngertob et al., 1987; Wesson et al., 2008). The latest study in dogs (Lawson et al., 2012) has shown a similar, although more complex, lateral turbinate structure as that of rodents, however, the central (dorsal medial) channel and different inspiration vs expiration flow are conserved, which further implicate its potential functional role.

The spatial and temporal activation patterns at the olfactory epithelium are important because they reflect the first event where odor information is represented in the neural system. Such patterns may directly affect the activation pattern in the secondary neural circuits, as epithelial neurons converge to the olfactory bulb in an organized fashion. A number of studies have used various techniques, such as 2-deoxyglucose mapping (Johnson and Leon, 1996), calcium signals (Verhagen et al., 2007), functional MRI (Xu et al 2003), and electrophysiology (Shusterman et al., 2011), to investigate the importance of temporal and spatial patterns in the olfactory bulb during active sniffing. However, the spatial temporal pattern at the epithelial level as a function of nasal structure, aerodynamics, and sniffing parameters remain relatively understudied.

The human olfactory structure appears to diverge evolutionary compared with many other land mammals: the human turbinate structure is much simpler, and the olfactory epithelium is confined to a relatively small patch with fewer functional receptor genes. Sniffing behaviors in humans are much different than those of rodents or dogs. Whether there is a zonal distribution of human receptor genes and whether there are stable organized absorption patterns is unknown. For example, Zhao et al. (2004) has shown completely different absorption patterns between the two sides of the nasal cavity of the same healthy subject. It may be debatable whether our olfactory acuity is poorer than that of other species (Shepherd, 2004); however, these features are likely the reason that our olfactory abilities, relative to other species, may be far more prone to disruption by even small nasal anatomical changes (whether disease, developmental, or surgical,

etc.) that may restrict odorant access to the limited patch of olfactory epithelium. The differences in airflow and odor sorption patterns need to be seriously considered when translating results from animal model to human and addressing the clinical problems that may be unique to the human nose. Nevertheless, this should not deter our efforts to protect our sense of smell, which often offers us priceless enjoyment from food, flavor, and natural and man-made fragrances and, most essentially, provides early detection of environmental hazards.

16.6 NASAL PATENCY: HOW WE PERCEIVE NASAL OBSTRUCTION

"Nasal patency" can be either objective or subjective–objective nasal patency often refers to measured NAR, which can be effectively measured with rhinometry. Subjective nasal patency is the self-reported perception of nasal congestion or nasal obstruction, which is often more important to patients with nasal sinus disease than the objective one. In general, anterior defects are more symptomatic, both subjectively and objectively, although posterior nasal airway obstruction does cause airflow and symptomatic changes: Adenoid hypertrophy decreases nasopharyngeal CSA (Cho et al., 1999; Mostafa, 1997).

Nevertheless, it is the perception of a lack of nasal patency that drives these patients to seek medical treatment. More surprisingly, this subjective perception often bears little relationship to the actual physical resistance to airflow in the nose (NAR). Objective evaluation tools, such as acoustic rhinometry, rhinomanometry, CT staging scores, and endoscopic examination, provide inconsistent correlation with subjective patency (Lam et al., 2006; André et al., 2009), even though these objective tools often correlate well with each other. In general, the correlation seems better in studies examining unilateral measurements, and in patients with higher baseline NAR and/or symptoms, and following large modifications in nasal lumen size (reviewed by Andre et al., 2009). The lack of a reliable objective tool to index perceived nasal patency creates a significant clinical problem in the diagnosis and treatment of nasal obstructive symptoms, which are currently based mostly on patient subjective feedback and clinician's experience.

16.6.1 Potential Mechanisms of Nasal Flow Perception

The perception of nasal patency may involve nasal trigeminal activation by the cool inspiratory airflow rather than by NAR or effort in breathing alone. Mucosal temperature, as measured by noncontact infrared thermometry, is correlated with subjective nasal patency (Willatt, 1993; Willatt and Jones, 1996). The thermal perception is possibly mediated by TRPM8 channels expressed in trigeminal C-fiber or A-δ fiber endings such as those that innervate nasal epithelium (Lumpkin and Caterina, 2007). Pharmacological modulation of trigeminal feedback has been shown to alter patency perception. For example, topical or oral application of L-menthol produces the illusion of decongestion and improved nasal airflow without actually altering NAR (Eccles and Jones, 1983). In contrast, topical application of local anesthetics on the nasal epithelium results in an artificial sensation of nasal obstruction, presumably due to blockage of the trigeminal feedback (Jones et al., 1987). L-Menthol was widely used as an ingredient in common cold medications, nasal spray, candy, chewing gum, and cigarettes long before its target receptor, the nonselective voltage-dependent cation channel TRPM8, was identified (McKemy et al., 2002; Peier et al., 2002). It is now known that menthol and air temperature both shift TRPM8 voltage-activation dependency toward the physiological range, and their effects are additive (Voets et al., 2007). Thus, when combined, menthol and cool air can greatly enhance TRPM8 activation.

In the human, the palatine nerve (a branch of V2) innervates the turbinates and nasal septum, while the ethmoid nerve (a branch of V1) innervates the vestibule. However, the sensitivity of the nose to an air jet pulse and to temperature changes is greatest at the nasal vestibule (Clarke and Jones, 1994; Clarke and Jones, 1992), which suggests that regional aerodynamics and thermo-events might be more functionally important than overall NAR or thermo-exchange.

To further delineate the physical factors contributing to the perception of patency, a recent study asked 44 healthy subjects to rate their nasal patency while sampling untreated room air, cold air, and dry air (Zhao et al., 2011). The results indicate that both dry and cold air induce less perceived nasal obstruction compared to room air. Since the dry and room air did not differ statistically in temperature, the perceptual difference most likely resulted from differential mucosal cooling—the combination of both conductive heat loss (driven by temperature gradient) and evaporative heat loss (driven by water vapor pressure gradient), rather than from static air temperature alone.

Although there has been no direct documentation on alteration of trigeminal afferents during nasal sinus disease and/or inflammation processes, the literature does support the notion that trigeminal senses in general may undergo different adaptive processes during acute or prolonged inflammation. Chronic occupational exposure to irritants such as acetic acid or acetone specifically decreases trigeminal sensitivity to those chemicals (Dalton et al., 2007; Wysocki et al., 1997), whereas conditions such as allergic rhinitis, leading to inflammation in the nasal mucosa, are associated with greater sensitivity to nociceptive stimuli such as CO_2 and acetic acid (Shusterman et al., 2003;

Shusterman and Murphy, 2007). Benoliel et al. (2006) have shown that warm and electrical current perception thresholds were hypersensitized during the acute phase of rhinosinusitis but hyposensitized during chronic (>3 months) rhinosinusitis.

Based on these studies, it is clear that we may need to drastically change how we view the perception of nasal obstruction. A constricted airway with an insufficient airstream, as we often think about with physical nasal obstruction, produces less cooling. However, wide nasal passages with the bulk of the airstream having little contact with the mucosal wall may also produce less mucosal cooling. Thus, nasal airway abnormalities may have diverse clinical manifestations in a patient's subjective symptoms, depending on the mucosal/airflow interaction and the patient's thermosensory sensitivities, which may also undergo diverse changes during disease processes. Clinical evidence include the "empty nose" or "atrophic rhinitis" syndrome, in which patients with complete turbinate removed may still complain of the perception of nasal congestion, despite exhibiting very little objective NAR (Moore et al., 1985). Still, other, potential factors may also contribute to perceived patency. For example, the nonspecific cation channel TRPA1, which is activated at much colder (noxious) temperatures than the mechanosensors, could contribute to the sensation of patency. In addition, aging does affect nasal mechanosensitivity (Wrobel et al., 2006). TRMP8, TRPA1, and mechanoreceptors can be activated chemically, which offers a unique dual investigatory tool that has the potential to enhance our understanding of chemosensory functions in nasal/inflammatory disease.

16.7 CONCLUSIONS: AIRFLOW DYNAMICS IN RELATION TO CHEMOPERCEPTION

Nasal airflow in the human is complex and subject to both intersubject and intrasubject variability. Turbulent flow is likely present at higher sniffing rates and contributes to mixing of odorant molecules and the proper humidification and filtering of the incoming air. Although Reynolds numbers are not indicative of turbulent flow, such flow may develop in the expanded post nasal valve region during inspiration. Disorders within this region may alter vortices and turbulent flow by redirecting airstreams and augmenting nasal resistance. However, turbulence prior to reaching the olfactory cleft may not be particularly important since secondary air currents and diffusion across the boundary layer are probably dominant in the olfactory cleft during inspiration. Instead, inspiratory airflow velocity and volume flow may have a major influence on odorant deposition and the proportion of odorized air transported to the olfactory region. Under some circumstances, humans may

370 Chapter 16 Nasal Patency and the Aerodynamics of Nasal Airflow in Relation to Olfactory Function

adjust their sniff durations to optimize odorant perception (Sobel et al., 2000). Airflow rate is positively correlated with odor intensity (Rehn, 1978), the number of odorants identified in a confusion matrix (Schwartz et al., 1987), and the magnitude of the olfactory evoked potential (Kobal and Hummel, 1991). The expiratory airstream exhibits higher velocities and a more direct path to the olfactory regions than the inspiratory airstream in humans, which is contrary to rodents. While animal studies provide important insight into the human olfactory system, nasal structural and airflow differences must be taken into account.

In the clinical setting, it is widely believed that conductive airway abnormalities are responsible for some olfactory losses; however, conclusive evidence is lacking. Endoscopic operative procedures, septoplasty, turbinoplasty or adenoidectomy can improve nasal airflow and olfactory function. Nonetheless, many patients do not experience restoration of olfactory function following such interventions. Since only a small percentage of air flows through the olfactory region, abnormalities and subsequent surgery, while profoundly affecting NAR, may or may not have the same effect on delivery of odorant to the olfactory cleft. It is also likely that airflow obstruction only accounts for a subset of alterations in olfactory perception, while inflammatory processes may affect the olfactory neuroepethelium. The lack of correlation of olfactory function with changes in nasal airflow and the many factors besides nasal airflow that affect olfactory function make management of olfactory dysfunction complicated.

The clinical implications of changes in nasal airflow are clearly important but differences in the perceived vs measured nasal airflow create complexity when treating symptoms. Hopefully the studies reviewed in this chapter will be of future help, in advancing our understanding of the relationship between nasal airflow and olfactory function.

REFERENCES

Adamson, J. E. (1987). Constriction of the internal nasal valve in rhinoplasty: treatment and precention. *Ann. Plast. Surg.* 18: 114–121.

Adrian, R. J. (1991). Particle-imaging techniques for experimental fluid mechanics. *Ann. Rev. Fluid. Mech.* 23: 261–304.

Adrian, E. D. (1950). Sensory discrimination with some recent evidence from the olfactory organ. *Br Med Bull.* 6: 330–331.

Amoore, J. E., and Buttery, R. G. (1978). Partition coefficients and comparative olfactometry. *Chem. Senses Flav.* 3: 57–71.

Andre, R. F., Vuyk, H. D., Ahmed, A., et al. (2009). Correlation between subjective and objective evaluation of the nasal airway. A systematic review of the highest level of evidence. *Clin. Otolaryngol.* 34: 518–525.

Benoliel, R., Biron, A., Quek, S. Y., et al. (2006). Trigeminal neurosensory changes following acute and chronic paranasal sinusitis. *Quintessence Int.* 37: 437–443.

Bocca, E., Antonelli, A. R., and Mosciaro, O. (1956) Mechanical co-factors in olfactory stimulation. *Acta. Otolaryngol.* 59: 243–248.

Bojsen-Muller, F. and Fahrenkrug, J. (1971). Nasal swell-bodies and cyclic changes in the air passages of the rat and rabbit nose. *J. Anat.* 110: 25–27.

Buck, L. and Axel, R. (1991). A novel multigene family may encode odorant receptors: a molecular basis for odor recognition. *Cell* 65: 175–187.

Chen, X. B., Lee, H. P., Chong, F. H., and Wang, D. Y. (2009). Assessment of septal deviation effects on nasal air flow: a computational fluid dynamics model. *Laryngoscope.* 119: 1730–1736.

Cheng, Y. S., Yamada, Y., Yeh, H. C., and Swift, D. L. (1988). Diffusional deposition of ultrafine aerosols in a human nasal cast. *J. Aerosol. Sci.* 19: 741–751.

Cheng, Y. S., Su, Y. F., Yeh, H. C., and Swift, D. L. (1993). Deposition of thoron progeny in human head airways. *Aerosol. Sci. Tech.* 18: 359–375.

Cheng, K. H., Cheng, Y. S., Yeh, H. C., et al. (1996a). In vivo measurements of nasal airway dimensions and ultrafine aerosol deposition in the human nasal and oral airways. *J. Aerosol. Sci.* 27: 785–801.

Cheng, Y. S., Yeh, H. C., Guilmette, R. A., et al. (1996b). Nasal deposition of ultrafine particles in human volunteers and its relationship to airway geometry. *Aerosol. Sci. Technol.* 25: 274–291.

Cheng, Y. S. (2003). Aerosol deposition in the extrathoracic region. *Aerosol. Sci. Technol.* 37: 659–671.

Cho, J. H., Lee, D. H., Lee, N. S., et al. (1999). Size assessment of adenoid and nasopharyngeal airway by acoustic rhinometry in children. *J. Laryngol. Otol.* 113: 899–905.

Chung, S. K., Son, Y. R., Shin, S. J., and Kim, S. K. (2006). Nasal airflow during respiratory cycle. *Am. J. Rhinol.* 20: 379–384.

Clarke, R. W. and Jones, A. S. (1992). Nasal airflow receptors: the relative importance of temperature and tactile stimulation. *Clin. Otolaryngol.* 17: 388–392.

Clarke, R. W. and Jones, A. S. (1994). The distribution of nasal airflow sensitivity in normal subjects. *J. Laryngol. Otol.* 108: 1045–1047.

Cole, P., Haight, J. S. J., Love, L., and Oprysk, D. (1985). Dynamic components of nasal resistance. *Am. Rev. Respir. Dis.* 132: 1229–1232.

Conrad, W. A. (1969). Pressure-flow relations in collapsible tubes. *IEEE Trans. Bio. Med. Eng.* 16: 284–295.

Cowart, B. J., Young, I. M., Feldman, R. S., and Lowry, L. D. (1997). Clinical disorders of smell and taste. *Occupational Medicine: State of the Art Reviews. Philadelphia: Hanley & Belfus, Inc.*, 465–483.

Cullen, M. M. and Leopold, D. A. (1999). Disorders of smell and taste. *Med. Clin. North Am.* 83: 57–74.

Damm, M., Eckel, H. E., Streppel, M., et al. (2000). Dependence of uni- and bilateral olfactory capacity on nasal airflow in patients with chronic rhinosinusitis. *HNO.* 48: 436–43.

Damm, M., Vent, J., Schmidt, M., et al. (2002) Intranasal volume and olfactory function. *Chem. Senses.* 27:831–839.

Damm, M., Eckel, H. E., Jungeh Ising, M. and Hummel, T. (2003). Olfactory changes at threshold and suprathreshold levels following septoplasty with partial inferior turbinectomy. *Ann. Otol. Rhinol. Laryngol.* 112: 91–97.

Dalton, P., Lees, P. S., Gould, M., et al. (2007). Evaluation of long-term occupational exposure to styrene vapor on olfactory function. *Chem. Senses* 32: 739–747.

Dastidar, P., Heinonen, T., Numminen, J., et al. (1999a). Semi-automatic segmentation of computed tomographic images in volumetric estimation of nasal airway. *Eur. Arch. Otorhinolaryngol.* 256: 192–198.

Dastidar, P., Numminen, J., Heinonen, T., et al. (1999b). Nasal airway volumetric measurement using segmented HRCT images and acoustic rhinometry. *Am. J. Rhinol.* 13: 97–103

Ebbell, B. (1937). *The Papyrus Ebers.* Oxford University Press: London.

References

Eccles, R. (1978). The central rhythm of the nasal cycle. *Acta. Otolaryngol.* 86: 464–468.

Eccles, R., and Jones, A. S. (1983). The effect of menthol on nasal resistance to air flow. *J. Laryngol. Otol.* 97: 705–709.

Eccles, R. (2000). Nasal airflow in health and disease. *Acta. Otolaryngol.* 120: 580–595.

Eitel, G., Freitas, R. K., Lintermann, A., et al. (2010) Numerical simulation of nasal cavity flow based on a Lattice-Boltzmann method. *Springer-Verlag Berlin Heidelberg VII, NNFM.* 112: 513–520.

Elad, D., Liebenthal, R., Wenig, B. L., and Einav, S. (1993). Analysis of airflow patterns in the human nose. *Med. Biol. Eng. Comput.* 31: 585–592.

Elwany, S. and Harrison, R. (1990). Inferior turbinectomy: comparison of four techniques. *J. Laryngol. Otol.* 104: 206–209.

Ferris, B. G., Jr.,, Mead, J., and Opie, L. H. (1964). Partitioning of respiratory flow resistance in man. *J. Appl. Physiol.* 19: 653–658.

Frye, R. E. and Doty, R. L. (1992). The influence of ultradian autonomic rhythms as indexed by the nasal cycle on unilateral olfactory thresholds. In *Chemical Signals in Vertebrates VI*, R. L. Doty and D. Muller-Schwarze (Eds.) Plenum Press: New York, pp. 595–598.

Garcia, G. J., Bailie, N., Martins, D. A., and Kimbell, J. S. (2007). Atrophic rhinitis: a CFD study of air conditioning in the nasal cavity. *J. Appl. Physiol.* 103: 1082–1092.

Garcia, G. J., Rhee, J. S., Senior, B. A., et al. (2010). Septal deviation and nasal resistance: An investigation using virtual surgery and computational fluid dynamics. *Am. J. Rhinol. Allergy* 24: e46–e53.

Garzaro, M., Pezzoli, M., PecorarI, G., et al. (2010) Radiofrequency inferior turbinate reduction: an evaluation of olfactory and respiratory function. *Otolaryngol. Head Neck Surg.* 143: 348–352.

Garzaro, M., Pezzoli, M., Landolfo, V., et al. (2012) Radiofrequency inferior turbinate reduction: long-term olfactory and functional outcomes. *Otolaryngol. Head Neck Surg.* 146: 146–150.

Ghorbanian, S. N., Paradise, J. L., and Doty, R. L. (1983). Odor perception in children in relation to nasal obstruction. *Pediatrics* 72: 510–516.

Gilbert, A. N, and Rosenwasser, A. M. (1987). Biological rhythmicity of nasal airway patency: a re-examination of the 'nasal cycle'. *Acta Otolaryngol.* 104: 180–186.

Girardin, M., Bilgen, E., and Arbour, P. (1983). Experimental study of velocity fields in a human nasal fossa by laser anemometry. *Ann. Otol. Rhinol. Laryngol.* 92: 231–236.

Goode, R. L. (1985). Surgery of the incompetent nasal valve. *Laryngoscope* 95: 546–555.

Gosepath, J., Amedee, R. G., Romantschuck, S., and Mann, W. J. (1999). Breathe Right nasal strips and the respiratory disturbance index in sleep related breathing disorders. *Am. J. Rhinol.* 13: 385–389.

Gradon, L. and Yu, C. P. (1989). Diffusional particle deposition in the human nose and mouth. *Aerosol. Sci. and Technol.* 11: 213–220.

Grant, O., Bailie, N., Watterson, J., et al. (2004). Numerical model of a nasal septal perforation. *Medinfo.* 11: 1352–1356.

Griffin, J. W., Hunter, G., Ferguson, D., and Sillers, M. J. (1997). Physiologic effects of an external nasal dilator. *Laryngoscope* 107: 1235–1238.

Guilemany, J. M., García-Piñero, A., Alobid, I., et al. (2009) Persistent allergic rhinitis has a moderate impact on the sense of smell,depending on both nasal congestion and inflammation. *Laryngoscope.* 119: 233–238.

Guilemany, J. M., García-Piñero, A., Alobid, I., et al. (2012) The loss of smell in persistent allergic rhinitis is improved by levocetirizine due to reduction of nasal inflammation but not nasal congestion (the CIRANO study). *Int. Arch. Allergy Immunol.* 158: 184–190.

Guilmette, R. A., Cheng, Y.S., Yeh, H. C., and Swift, D. L. (1994). Deposition of 0.005–12 Mm monodisperse particles in a computer-milled, MRI-based nasal airway replica. *Inhalat. Toxicol.* 6: 395–399.

Gungor, A., Moinuddin, R., Nelson, R. H., and Corey, J. P. (1999). Detection of the nasal cycle with acoustic rhinometry: techniques and applications. *Otolaryngol. Head Neck Surg.* 120: 238–247.

Hahn, I., Scherer, P. W., and Mozell, M. M. (1993). Velocity profiles measured for airflow through a large-scale model of the human nasal cavity. *J. Appl. Physiol.* 75: 2273–2287.

Hahn, I., Scherer, P. W., and Mozell, M. M. (1994). A mass transport model of olfaction. *J. Theor. Biol.* 167: 115–128.

Holt, J. P. (1969). Flow through collapsible tubes and through in situ veins. *IEEE Trans. Bio. Med. Eng.* 16: 274–283.

Hornung, D. E., Mozell, M. M., and Serio, J. A. (1980). Olfactory mucus/air partitioning of odorant. In: *Olfaction and Taste VII*. London: *IRL Press*, 167–170.

Hornung, D. E., Leopold, D. A., Youngentob, S. L., et al. (1987a). Airflow patterns in a human nasal model. *Arch. Otolaryngol. Head Neck Surg.* 113: 169–172.

Hornung, D. E., Youngentob, S. L., and Mozell, M. M. (1987b). Olfactory mucosa/air partitioning of odorants. *Brain Res.* 413: 147–154.

Hornung, D. E., Chin, C., Kurtz, D. B., et al. (1997). Effect of nasal dilators on perceived odor intensity. *Chem. Senses.* 22: 177–180.

Hornung, D. E. and Leopold, D. A. (1999). Relationship between uninasal anatomy and uninasal olfactory ability. *Arch. Otolaryngol. Head Neck Surg.* 125: 53–58.

Hornung, D. E., Smith, D. J., Kurtz, D. B., et al. (2001). Effect of nasal dilators on nasal structures, sniffing strategies, and olfactory ability. *Rhinology.* 39: 84–87.

Hörschler, I., Brucker, C., Schröder, W., and Meinke, M., (2006). Investigation of the impact of the geometry on the nose flow. *Eur. J. Mech. B Fluids* 25: 471–490.

Hummel, T., Rothbauer, C., Barz, S., et al. (1998a). Olfactory function in acute rhinitis. *Ann. N. Y. Acad. Sci.* 855: 616–624.

Hummel, T., Rothbauer, C., Pauli, E., and Kobal, G. (1998b). Effects of the nasal decongestant oxymetazoline on human olfactory and intranasal trigeminal function in acute rhinitis. *Eur. J. Clin. Pharmacol.* 54: 521–528.

Ikeda, K., Oshima, T., Suzuki, M., et al. (2006) Functional inferior turbinosurgery (FITS) for the treatment of resistant chronic rhinitis. *Acta Otolaryngol.* 126: 739–745.

Ishikawa, S., Nakayama T., Watanabe, M., and Matsuzawa, T. (2006) Visualization of flow resistance in physiological - nasal respiration analysis of velocity and vorticities using numerical simulation. *Arch. Otolaryngol. Head Neck Surg.* 132: 1203–1209.

Jiang, J. B. and Zhao, K. (2010). Airflow and nanoparticle deposition in rat nose under various breathing and sniffing conditions - A computational evaluation of the unsteady and turbulent effect. *J. Aerosol Sci.* 41: 1030–1043.

Johnson, B. A. and Leon, M. (1996). Spatial distribution of [14C]2-deoxyglucose uptake in the glomerular layer of the rat olfactory bulb following early odor preference learning. *J. Comp. Neurol.* 376: 557–566.

Jones, A. S., Crosher, R., Wight, R. G., et al. (1987). The effect of local anesthesia of the nasal vestibule on nasal sensation to airflow and nasal resistance. *Clin. Otolaryngol.* 12: 461–464.

Jones, A. S., Wight, R. G., Stevens, J. C., et al. (1988). The nasal valve: a physiological and clinical study. *J. Laryngol. Otol.* 102:1089–1094.

Jung, Y. G., Ha, S. Y., Eun, Y. G., and Kim, M. G. (2011). Influence of intranasal epinephrine and lidocaine spray on olfactory function tests in healthy human subjects. *Otolaryngol. Head Neck Surg.* 145: 946–950.

Kamami, Y. V. (1997). Laser-assisted outpatient septoplasty results on 120 patients. *J. Clin. Laser Med. Surg.* 15: 123–129.

Kamami, Y. V., Pandraud, L., and Bougara, A. (2000). Laser-assisted outpatient septoplasty: results in 703 patients. *Otolaryngol. Head Neck Surg.* 122: 445–449.

Kasperbauer, J. L. and Kern, E. B. (1987). Nasal valve physiology. *Otolaryngol. Clin. North Am.* 20: 699–719.

Kauer, J. S., Moulton, D. G. (1974) Responses of olfactory bulb neurones to odour stimulation of small nasal areas in the salamander. *J Physiol* 243: 717–737.

Kayser, R. (1895). Die exacte Messung der Luftdurchgangegkeit der Nase. *Arch. Laryngol.* 3: 101–120.

Kelly, J. T., Prasad, A. K., and Wexler, A. S. (2000). Detailed flow patterns in the nasal cavity. *J. Appl. Physiol.* 89: 323–337.

Kelly, J. T., Asgharian, B., Kimbell, J. S. and Wong, B. A. (2004). Particle deposition in human nasal airway replicas manufactured by different methods. Part I: inertial regime particles. *Aerosol Sci. Technol.* 38: 1063–1071.

Kern, R. C. (2000). Chronic sinusitis and anosmia: pathologic changes in the olfactory mucosa. *Laryngoscope* 110: 1071–1077.

Keyhani, K., Scherer, P. W., and Mozell, M. M. (1995). Numerical simulation of airflow in the human nasal cavity. *J. Biomech. Eng.* 117: 429–441.

Keyhani, K., Scherer, P. W., and Mozell, M. M. (1997). A numerical model of nasal odorant transport for the analysis of human olfaction. *J. Theor. Biol.* 186: 279–301.

Kim, S. K. and Chung, S. K. (2004). An investigation on airflow in disordered nasal cavity and its corrected models by tomographic PIV. *Meas. Sci. Technol.* 15: 1090–1096.

Kimbell, J. S., Godo, M. N., Gross, E. A., et al. (1997). Computer simulation of inspiratory airflow in all regions of the F344 rat nasal passages. *Toxicol. Appl. Pharmacol.* 145: 388–398.

Kimmelman, C. P. (1989). The problem of nasal obstruction. *Otolaryngol. Clin. North Am.* 22: 253–264.

Klimek, L. (1998) Sense of smell in allergic rhinitis. *Pneumologie.* 52: 196–202.

Kobal, G., and Hummel, T. (1991). Human electro-olfactograms and brain responses to olfactory stimulation. In *The Human Sense of Smell*, D. G. Laing, R. L. Doty, and W. Breipohl (Eds.). Springer-Verlag: Berlin, pp. 135–150.

Lam, D. J., James, K. T., and Weaver, E. M. (2006). Comparison of anatomic, physiological, and subjective measures of the nasal airway. *Am. J. Rhinol.* 20: 463–470.

Landis, B. N., Giger, R., Ricchetti, A., et al. (2003). Retronasal olfactory function in nasal polyposis. *Laryngoscope.* 113: 1993–1997.

Laudien, M., Lamprecht, P., Hedderich, J., et al. (2009) Olfactory dysfunction in Wegener's granulomatosis. *Rhinology.* 47: 254–259.

Lawson, M. J., Craven, B. A., Paterson E. G. and Settles, G. S (2012) A computational study of odorant transport and deposition in the canine nasal cavity: implications for olfaction, *Chem. Senses* 37: 553–566.

Lee, H. P., Poh, H. J., Chong, F. H. and Wang, D. Y. (2009) Changes of airflow pattern in inferior turbinatehypertrophy: A computational fluid dynamics model. *Am. J. Rhinol. Allergy* 23: 153–158.

Leopold, D. A. (1988). The relationship between nasal anatomy and human olfaction. *Laryngoscope* 98: 1232–1238.

Levine, S. C., Levine, H., Jacobs, G., and Kasick, J. (1986). A technique to model the nasal airway for aerodynamic study. *Otolaryngol. Head Neck Surg.* 95: 442–449.

Lumpkin, E. A., and Caterina, M. J. (2007). Mechanisms of sensory transduction in the skin. *Nature* 445: 858–865.

Lund, V..J., Holmstrom, M., and Scadding, G. K. (1991). Functional endoscopic sinus surgery in the management of chronic rhinosinusitis. An objective assessment. *J. Laryngol. Otol.* 105: 832–835.

Lund, V. J., and Scadding, G. K. (1994). Objective assessment of endoscopic sinus surgery in the management of chronic rhinosinusitis: an update. *J. Laryngol. Otol.* 108: 749–753.

Lund, V. J. (1996). Nasal physiology: neurochemical receptors, nasal cycle, and ciliary action. *Allergy Asthma Proc.* 17: 179–184.

Lund, V. J., Flood, J., Sykes, A. P., and Richards, D. H. (1998). Effect of fluticasone in severe polyposis. *Arch. Otolaryngol. Head Neck Surg.* 124: 513–518.

Lung, M. A. and Wang, J. C. (1991). Mechanical stimulation of canine respiratory tract and nasal vascular and airway resistance. *Respir. Med.* 85 (Suppl A): 67–68.

Mackay-Sim, A., Shaman, P., Moulton, D. G. (1982) Topographic coding of olfactory quality: odorant-specific patterns of epithelial responsivity in the salamander. *J Neurophysiol* 48: 584–596.

Maltais, F., Dinh, L., Cormier, Y., and Series, F. (1991). Changes in upper airway resistance furing progressive normocapnic hypoxia in normal men. *J. Appl. Physiol.* 70: 548–553.

Manestar, D., Ticac, R., Velepic, M., et al. (2011) The significance of rhinomanometry in evaluation of postlaryngectomy olfactory rehabilitation by polite yawning technique. *Rhinology.* 49: 238–242.

Manestar, D., Tićac, R., Maričić, S., et al. (2012) Amount of airflow required for olfactory perception in laryngectomees: a prospective interventional study. *Clin. Otolaryngol.* 37: 28–34.

Masieri, S., Cavaliere, F., and Filiaci, F. (1997). Nasal obstruction improvement induced by topical furosemide in subjects affected by perennial nonallergic rhinitis. *Am. J. Rhinol.* 11: 443–447.

Masing, H. (1967). Investigations about the course of flow in the nose model. *Arch. Klin. Exp. Ohren. Nasen. Kehlkopfheilkd* 189: 371–381.

McKemy, D. D., Neuhausser,W. M., and Julius,D. (2002). Identification of a cold receptor reveals general role for TRP channels in thermosensation. *Nature* 416: 52–58.

Meusel, T., Negoias, S., Scheibe, M., and Hummel T. (2010) Topographical differences in distribution and responsiveness of trigeminal sensitivity within the human nasal mucosa. *Pain* 151: 516–521.

Mezzanotte, W. S., Tangel, D. J., and White, D. P. (1992). Mechanisms of control of alae muscle activity. *J. Appl. Physiol.* 72: 925–933.

Mirza, N., Doty, R. L., and Kruger, H. (1997). Age-related influences on the nasal cycle.. *Laryngoscope* 107: 62–66.

Moore, G. F., Freeman, T. J., Ogren, F. P., and Yonkers, A. J. (1985). Extended follow-up of total inferior turbinate resection for relief of chronic nasal obstruction. *Laryngoscope* 95: 1095–1099.

Morgan, K. T., Kimbell, J. S., Monticello, T. M., et al. (1991). Studies of inspiratory airflow patterns in the nasal passages of the F344 rat and rhesus monkey using nasal molds: relevance to formaldehyde toxicity. *Toxicol. Appl. Pharmacol.* 110: 223–240.

Mostafa, B. E. (1997). Detection of adenoidal hypertrophy using acoustic rhinomanometry. *Eur. Arch. Otorhinolaryngol. Suppl.* 1: S27–29.

Moulton, D. G. (1976) Spatial patterning of response to odors in the peripheral olfactory system. *Physiol Rev* 56: 578–593.

Mozell, M. M. and Jagodowicz, M. (1973). Chromatographic separation of odorants by the nose: retention times measured across in vivo olfactory mucosa. *Science* 181: 1247–1249.

Mozell, M. M, Sheehe, P. R., Hornung, D. E., et al. (1987). Imposed and inherent mucosal activity patterns. Their composite representation of olfactory stimuli. *J. Gen. Physiol.* 90: 625–650.

Mozell, M. M., Kent, P. F. and Murphy, S. J. (1991). The effect of flow rate upon the magnitude of the olfactory response differs for different odorants. *Chem. Senses* 16: 631–649.

Nakano, T. (1967). Influence of nozzles on pressure and flow measurement studies by means of the artificial nose and conductivity meter. *Rhinol. Int.* 5: 183–196.

References

Negus, V. E. (1958) The comparative anatomy and physiology of the nose and paranasal sinuses. London, UK: Livingstone.

Ng, B. A., Mamikoglu, B., Ahmed, M.S., and Corey, J. P. (1998) The effect of external nasal dilators as measured by acoustic rhinometry. *Ear Nose Throat J.* 77: 840–844.

Nishihira, S., and McCaffrey, T. V. (1987). Reflex control of nasal blood vessels. *Otolaryngol. Head Neck Surg.* 96: 273–277.

Oberdorster, G., Sharp, Z., Atudorei, V. et al. (2004). Translocation of inhaled ultrafine particles to the brain. *Inhalation Toxicol.* 16: 437–445.

Ogawa, H. (1986). A possible role of aerodynamic factors in nasal polyp formation. *Acta. Orolaryngol. (Stockh.)* S430: 18–20.

Olszewska, E., Sieskiewicz, A., Kasacka, I., et al. (2010). Cytology of nasal mucosa, olfactometry and rhinomanometry in patients after CO_2 laser mucotomy in inferior turbinate hypertrophy. *Folia Histochem. Cytobiol.* 48: 217–221.

Ophir, D., Gross-Isseroff, R., Lancet, D., and Marshak, G. (1986). Changes in olfactory acuity induced by total inferior turbinectomy. *Arch. Otolaryngol. Head Neck Surg.* 112: 195–197.

Ozlugedik, S., Nakiboglu, G., Sert, C., et al. (2008). Numerical study of the aerodynamic effects of septoplasty and partial lateral turbinectomy. *Laryngoscope*, 118: 330–334.

Park, K. I., BrÜcker, C., Limberg, W. (1997). Experimental study of velocity fields in a model of human nasal cavity by DPIV. Laser anemometry advances and applications. In: Ruck, B., Leder, A., Dopheide, D. (Eds.), Proceedings of the 7th International Conference. University of Karlsruhe, Karlsruhe, Germany, pp. 617–626.

Passali, D., Lauriello, M., Anselmi, M., and Bellussi, L. (1999). Treatment of hypertrophy of the inferior turbinate: long-term results in 382 patients randomly assigned to therapy. *Ann. Otol. Rhinol. Laryngol.* 108: 569–575.

Peier, A. M., Moqrich, A., Hergarden, A. C., et al. (2002). A TRP channel that senses cold stimuli and menthol. *Cell* 108: 705–715.

Pfaar, O., Landis, B. N., Frasnelli, J., et al. (2006). Mechanical obstruction of the olfactory cleft reveals differences between orthonasal and retronasal olfactory functions. *Chem. Senses.* 31: 27–31.

Pless, D., Keck, T., Wiesmiller, K. M., et al. (2004). Numerical simulation of airflow patterns and air temperature distribution during inspiration in a nose model with septal perforation. *Am. J. Rhinol.* 18: 357–362.

Principato, J. J., and Ozenberger, J. M. (1970). Cyclical changes in nasal resistance. *Acta Otolaryngol.* 91: 71–77.

Proetz, A. W. (1951). Air currents in the upper respiratory tract and their clinical importance. *Ann. Otol. Rhinol. Laryngol.* 60: 439–467.

Randall, J. E. (1962). *Elements of Biophysics*, 2nd Ed. Year Book Publishers: Chicago.

Raudenbush, B. and Meyer, B. (2001). Effect of nasal dilators on pleasantness, intensity and sampling behaviors of foods in the oral cavity. *Rhinology.* 39: 80–83.

Rehn, T. (1978). Perceived odor intensity as a function of airflow through the nose. *Sen. Proc.* 2: 198–205.

Ressler, K. J., Sullivan, S. L., Buck L. B. (1993) A zonal organization of odorant receptor gene expression in the olfactory epithelium. *Cell* 73: 597–609.

Rhee, C. S., Kim, D. Y., Won, T. B., et al. (2001) Changes of nasal function after temperature-controlled radiofrequency tissue volume reduction for the turbinate. *Laryngoscope.* 111: 153–158.

Roithmann, R., Chapnik, J., Cole, P., et al. (1998). Role of the external nasal dilator in the management of nasal obstruction. *Laryngoscope.* 108: 712–715.

Rouadi, P., Fuad, M. B., David, A., et al. (1999). A technique to measure the ability of the human nose to warm and humidify air. *J. Appl. Physiol.* 87: 400–406.

Rózańska-Kudelska, M., Sieśkiewicz, A., Rogowski, M., and Godlewska-Zołądkowska, K. (2010). Assessment of olfactory disturbances in the patients with rhinosinusitis and polypi nasi treated by endoscopic sinus surgery]. *Pol. Merkur Lekarski.* 28(166): 273–276.

Scheibe, M., Zahnert, T., and Hummel, T. (2006) Topographical differences in the trigeminal sensitivity of the human nasal mucosa. *Neuroreport.* 17: 1417–1420.

Scherer, P. W., Hahn, I. I., and Mozell, M. M. (1989). The biophysics of nasal airflow. *Otolaryngol. Clin. Am.* 22: 265–278.

Schreck, S., Sullivan, K. J., Ho, C. M. & Chang, H. K. (1993) Correlations between flow resistance and geometry in a model of the human nose. *J. Appl. Physiol.* 75(4), 1767–1775.

Schoenfeld, T. A., Cleland, T. A. (2005) The anatomical logic of smell. *Trends Neurosci* 28: 620–627.

Schoenfeld, T. A., Cleland, T. A. (2006) Anatomical contributions to odorant sampling and representation in rodents: zoning in on sniffing behavior. *Chem Senses* 31: 131–144.

Schwartz, D. N., Mozell, M. M., Youngentob, S. L., et al. (1987). Improvement of olfaction in laryngectomized patients with the larynx bypass. *Laryngoscope.* 97: 1280–1286.

Scott, J. W., Shannon, D. E., Charpentier, J., et al. (1997). Spatially organized response zones in rat olfactory epithelium. *J. Neurophysiol.* 77: 1950–1962.

Scott, J. W. and Brierley, T. (1999) A functional map in rat olfactory epithelium. *Chem. Senses* 24: 679–690.

Scott, J. W., Brierley, T., and Schmidt, F. H. (2000) Chemical determinants of the rat electro-olfactogram. *J. Neurosci.* 20: 4721–4731.

Seiden, A. M. and Smith, D. V. (1988). Endoscopic intranasal surgery on olfaction. *Chem. Senses* 13: 736.

Seiden, A. M., and Duncan, H. J. (2001). The diagnosis of a conductive olfactory loss. *Laryngoscope* 111: 9–14.

Series, F., Cormier, Y., Desmeules, M., and LaForge, J. (1989). Influence of respiratory drive on upper airway resistance in normal men. *J. Appl. Physiol.* 66: 1242–1249.

Shepherd, G. M. (2004) The human sense of smell: are we better than we think? *PLoS Biol.* 2(5): E146.

Shi, H., Kleinstreuer, C., and Zhang, Z. (2008). Dilute suspension flow with nanoparticle deposition in a representative nasal airway model. *Physics of Fluids* 20: 013301.

Shin, S. H., Park, J. Y., and Sohn, J. H. (1999) Clinical value of olfactory function tests after endoscopic sinus surgery: a short-term result. *Am. J. Rhinol.* 13: 63–66.

Shusterman, D., and Murphy, M. A. (2007). Nasal hyperreactivity in allergic and non-allergic rhinitis: a potential risk factor for non-specific building-related illness. *Indoor Air* 17: 328–333.

Shusterman, D., Murphy, M. A., and Balmes, J. (2003). Differences in nasal irritant sensitivity by age, gender, and allergic rhinitis status. *Int. Arch. Occup. Environ. Health* 76: 577–583.

Shusterman, R., Smear, M. C., Koulakov, A. A., and Rinberg, D. (2011). Precise olfactory responses tile the sniff cycle. *Nat. Neurosci.* 14: 1039–1044.

Silkoff, P. E., Chakravorty, S., Chapnik, J., et al. (1999). Reproducibility of acoustic rhinometry and rhinomanometry in normal subjects. *Am. J. Rhinol.* 13: 131–135.

Simmen, D., Scherrer, J. L., Moe, K., and Heinz, B. (1999). A dynamic and direct visualization model for the study of nasal airflow. *Arch. Otolaryngol. Head Neck Surg.* 125: 1015–1021.

Simola, M. and Malmberg, H. (1998). Sense of smell in allergic and non-allergic rhinitis. *Allergy* 53: 190–194.

Sivam, A., Jeswani, S., Reder, L., et al. (2010) Olfactory cleft inflammation is present in seasonal allergic rhinitis and is reduced with intranasal steroids. *Am. J. Rhinol. Allergy.* 24: 286–290.

Sobel, N., Khan, R., Saltman, A., et al. (1999). The world smells differently to each nostril. *Nature* 402: 35.

Sobel, N., Khan, R. M., Hartley, C. A., et al. (2000) Sniffing longer rather than stronger to maintain olfactory detection threshold. *Chem. Senses* 25: 1–8.

Soler, Z. M., Sauer, D. A., Mace, J., and Smith, T. L. (2009) Relationship between clinical measures and histopathologic findings in chronic rhinosinusitis. *Otolaryngol. Head Neck Surg.* 141: 454–461.

Spoor, A. (1965). A new method for measuring nasal conductivity. *Rhinol. Int.* 3: 27–35.

Stevens, C. N. and Stevens, M. H. (1985). Quantitative effects of nasal surgery on olfaction. *Am. J. Otolaryngol.* 6: 264–267.

Strohl, K. P., Hensley, M. J., Hallett, M., et al. (1980). Activation of upper airway muscles before onset of inspiration in normal humans. *J. Appl. Physiol.* 49: 638–642.

Strohl, K. P., O'Cain, C. F., and Slutsky, A. S. (1982). Alae nasi activation and nasal resistance in healthy subjects. *J. Appl. Physiol.* 52: 1432–1437.

Strotmann, J., Wanner, I., Helfrich, T., Beck, A., Breer, H. (1994). Rostro-caudal patterning of receptor-expressing olfactory neurones in the rat nasal cavity. *Cell Tissue Res.* 278: 11–20.

Stuiver, M. (1958). *Biophysics of the sense of smell.* Master's thesis, Groningen, The Netherlands.

Subramaniam, R. P., Richardson, R. B., Morgan, K. T., and Kimbell, J. S. (1999). Computational fluid dynamics simulations of inspiratory airflow in the human nose and nasopharynx. *Inhal. Toxicol.* 10: 91–120.

Sultan, B., May, L. A., Lane, A. P. (2011) The role of TNFalpha in inflammatory olfactory loss. *Laryngoscope.* 121: 2481–2486.

Sung, Y. W., Lee, M. H., Kim, I. J., et al. (2000). Nasal cycle in patients with septal deviation: evaluation by acoustic rhinometry. *Am. J. Rhinol.* 14: 171–174.

Swift, D. L., and Proctor, D. F. (1977). Access of air to the respiratory tract. In: Brain, J. D., Proctor, D. F. and Reid, L. M., editors, *Respiratory defense mechanism.* New York: Marcel Dekker Inc., pp. 63–91.

Szucs, E. and Clement, P. A. (1998). Acoustic rhinometry and rhinomanometry in the evaluation of nasal patency of patients with nasal septal deviation. *Am. J. Rhinol.* 12: 345–352.

Temmel, A. F., Quint, C., Toth, J., et al. (1999). Topical ephedrine administration and nasal chemosensory function in healthy human subjects. *Arch. Otolaryngol. Head Neck Surg.* 125: 1012–1014.

Thomson, A., Buxton L. H. D. (1923) Man's nasal index in relation to certain climatic conditions. *J R Anthropol Inst.* 53: 92–122.

Vaidyanathan, S., Barnes, M., Williamson, .P, et al. (2011) Treatment of chronic rhinosinusitis with nasal polyposis with oral steroids followed by topical steroids: a randomized trial. *Ann. Intern. Med.* 154: 293–302.

Vassar, R., Ngai, J., Axel, R. (1993) Spatial segregation of odorant receptor expression in the mammalian olfactory epithelium. *Cell* 74: 309–318.

Verhagen, J. V., Wesson, D. W., Netoff, T. I., et al. (2007). Sniffing controls an adaptive filter of sensory input to the olfactory bulb. *Nat. Neurosci.* 10: 631–639.

Voets, T., Owsianik, G., and Nilius, B. (2007). TRPM8. *HEP* 179: 329–344.

Wang, H., Zhang, W., Han, D., and Zhou, B. (2002). Measurement of nasal airway resistance and olfactory function before and after endoscopic sinus surgery. *Zhonghua Er Bi Yan Hou Ke Za Zhi.* 37: 177–179.

Weinhold, I. and Mlynski, G. (2004). Numerical simulation of airflow in the human nose. *Eur. Arch. Otorhinolaryngol.* 261: 452–455.

Welge-Lussen, A., Wille, C., Renner, B., and Kobal, G. (2004) Anesthesia affects olfaction and chemosensory event-related potentials. *Clin. Neurophysiol.* 115: 1384–1391.

Wesson, D. W., Donahou, T. N., Johnson, M. O., and Wachowiak, M. (2008). Sniffing behavior of mice during performance in odor-guided tasks. *Chem. Senses* 33: 581–596.

Wexler, D., Segal, R., and Kimbell, J. (2005). Aerodynamic effects of inferior turbinate reduction: computational fluid dynamics simulation. *Arch. Otolaryngol. Head Neck Surg.* 131: 1102–1107.

Willatt, D. J. (1993). Continuous infrared thermometry of the nasal mucosa. *Rhinology* 31: 63–67.

Willatt, D. J., and Jones, A. S. (1996). The role of the temperature of the nasal lining in the sensation of nasal patency. *Clin. Otolaryngol.* 21: 519–523.

Williams, H. L., Banvoetz, J. D., Brewer, D. W., et al. (1970). Report of Committee on Standardization of Definitions, Terms, Symbols in Rhinometry of the American Academy of Ophthalmology and Otolaryngology: A Handbook and Glossary, *American Academy of Ophthalmology and Otolaryngology*, Rochester, MN.

Wolfensberger, M. and Hummel, T. (2002) Anti-inflammatory and surgical therapy of olfactory isorders related to sino-nasal disease. *Chem. Senses.* 27: 617–622.

Wrobel, B. B., Bien, A. G., Holbrook, E. H., et al. (2006). Decreased nasal mucosal sensitivity in older subjects. *Am. J. Rhinol.* 20: 364–368.

Wysocki, C. J., Dalton, P., Brody, M. J., and Lawley, H. J. (1997). Acetone odor and irritation thresholds obtained from acetone-exposed factory workers and from control (occupationally non-exposed) subjects. *Am. Ind. Hyg. Assoc. J.* 58: 704–712.

Xu, F., Liu, N., Kida, I., et al. (2003) Odor maps of aldehydes and esters revealed by functional MRI in the glomerular layer of the mouse olfactory bulb. *Proc. Natl. Acad. Sci. U S A.* 2003Sep 16; 100(19): 11029–11034.

Yang, C. C. (1999). *Numerical modeling of odorant uptake in the rat and bullfrog nasal cavities. University of Pennsylvania.* Ref Type: Thesis/Dissertation.

Yang, G. C., Scherer, P. W., Zhao, K., and Mozell, M. M. (2007). Numerical modeling of odorant uptake in the rat nasal cavity. *Chem. Senses* 32: 273–284.

Yee, K. K., Pribitkin, E. A., Cowart, B. J., et al. (2009). Analysis of the olfactory mucosa in chronic rhinosinusitis. *Ann. N. Y. Acad. Sci.* 1170: 590–595.

Youngentob, S. L., Mozell, M. M., Sheehe, P. R., Hornung, D. E. (1987) A quantitative analysis of sniffing strategies in rats performing odor detection tasks. *Physiol Behav* 41: 59–69.

Yu, G., Zhang, Z., and Lessmann, R. (1998). Fluid flow and particle diffusion in the human upper respiratory system. *Aerosol Sci. Technol.* 28: 146–158.

Zhao, K., Scherer, P. W., Hajiloo, S. A., and Dalton, P. (2004). Effect of anatomy on human nasal air flow and odorant transport patterns: implications for olfaction. *Chem. Senses* 29: 365–379.

Zhao, K., Dalton, P., Yang, G. C., and Scherer, P. W. (2006a). Numerical modeling of turbulent and laminar airflow and odorant transport during sniffing in the human and rat nose. *Chem. Senses* 31: 107–118.

Zhao, K., Pribitkin, E. A., Cowart, B. J., Rosen, D., et al. (2006b). Numerical modeling of nasal obstruction and endoscopic surgical intervention: outcome to airflow and olfaction. *Am. J. Rhinol.* 20: 308–316.

Zhao, K., Blacker, K., Luo, Y., et al. (2011). Perceiving nasal patency through mucosal cooling rather than air temperature or nasal resistance. *PLoS ONE* 6(10): e24618. doi:10.1371/journal.pone.0024618

Chapter 17

Clinical Disorders of Olfaction

RICHARD L. DOTY

17.1 INTRODUCTION

Chronic olfactory dysfunction is not uncommon, being present in 1–2% of the population under the age of 65 years and in well over half the population older than 65 years (Doty et al., 1984a; Hoffman et al., 1998; Murphy et al., 2002). Such dysfunction can profoundly influence a patient's safety and quality of life. In a study of 750 consecutive patients presenting to our center with chemosensory complaints, 68% reported altered quality of life, 46% changes in appetite or body weight, and 56% adverse influences on daily living or psychological well-being (Deems et al., 1991). In another study of 445 such patients, at least one hazardous event, such as food poisoning or failure to detect fire or leaking natural gas, was reported by 45.2% of those with anosmia, 34.1% of those with severe hyposmia, 32.8% of those with moderate hyposmia, 24.2% of those with mild hyposmia, and 19.0% of those with normal olfactory function (Santos et al., 2004). In a longitudinal study of 1,162 non-demented older persons, mortality risk was 36% higher in those with low than with high scores on a 12-item odor identification test after adjusting for such variables as sex, age, and education (Wilson et al., 2010).

In this chapter I describe the olfactory deficits associated with a range of diseases and disorders, how they are classified, and how they are evaluated and treated. Since other chapters focus on olfactory dysfunction due to toxic chemical exposure (Chapters 19 and 20), nutritional disturbances (Chapter 37), and neurodegenerative diseases and schizophrenia (Chapter 18), these conditions are not specifically addressed in this chapter. The reader is referred to Chapter 39 for disorders of taste function, per se, that is, those that influence taste-bud mediated sensations.

17.2 CLASSIFICATION OF OLFACTORY DISORDERS

Olfactory disorders can be classified as follows: **Complete anosmia:** absence of smell function; **Partial anosmia:** ability to perceive some, but not all, odorants; **Hyposmia or microsmia:** decreased sensitivity to odorants; **Dysosmia** (also termed parosmia): distorted or perverted smell perception to odorant stimulation; **Phantosmia:** a dysosmia sensation perceived in the absence of an odor stimulus (a.k.a. olfactory hallucination); **Olfactory agnosia:** inability to recognize an odor sensation, even though olfactory processing, language, and general intellectual functions are essentially intact, as in some stroke patients; **Hyperosmia:** abnormally acute smell function. **Presbyosmia** is sometimes used to describe smell loss due to aging, but this term is less specific than those noted above (e.g., it does not distinguish between anosmia and hyposmia) and is laden, by definition, with the notion that it is age, per se, that is causing the age-related deficit.

It should be noted that hyperosmia is very rare and most persons claiming to be hypersensitive to odorants, such as those with "multiple chemical hypersensitivity," in fact have normal olfactory thresholds (Caccappolo et al., 2000; Doty et al.. 1988; Nordin et al., 2005). A number of such persons may suffer from environmental chemosensory responsivity (CR), a condition correlated with neuroticism (Karnekull et al., 2011). While complaints of dysosmia and phantosmia are often correlated with psychophysical test results, that is, lower test scores, the results of such tests cannot validate their presence. Nevertheless, these conditions are accompanied by enhanced brain activity in the anterior frontal cortex and temporal lobes, as determined

---

Handbook of Olfaction and Gustation, Third Edition. Edited by Richard L. Doty.
© 2015 Richard L. Doty. Published 2015 by John Wiley & Sons, Inc.

by functional imaging (fMRI) (Henkin et al., 2000; Levy and Henkin, 2000).

When possible, it is useful to classify olfactory impairment into three general classes: (1) conductive or transport impairments from obstruction of the nasal passages (e.g., by chronic nasal inflammation, polyposis); (2) sensorineural impairment from damage to the olfactory neuroepitheliumm (e.g., by viruses, airborne toxins); and (3) central olfactory neural impairment from central nervous system (CNS) damage (e.g., tumors, masses impacting on olfactory tract, neurodegenerative disease). However, such definitive classification is often not feasible, and these categories are not mutually exclusive. For example, both damage and blockage of airflow to the receptors can occur from chronic rhinosinusitis, and some viruses and nanoparticles that damage the olfactory neuroepithelium also can be transported into the CNS via the olfactory nerves, subsequently altering central elements of the system as well (see Chapters 19 and 20).

17.3 CLINICAL EVALUATION OF OLFACTORY DISORDERS

17.3.1 Medical History

The etiology of most cases of olfactory dysfunction can be ascertained from carefully questioning the patient about the nature, timing, onset, duration, and pattern of their symptoms, as well as a historical determination of antecedent events (e g., head trauma, upper respiratory infections, toxic exposures, nasal surgeries). Fluctuations in function usually reflect obstructive, rather than neural, factors. Subtle symptoms of central tumors, dementia, tremor, and seizure activity (e.g., automatisms, occurrence of blackouts, auras, and déjà vu) should be sought, given the frequent association between smell dysfunction and not only brain tumors, but a range of neurological disorders, including neurodegenerative ones (see Chapter 18). Delayed puberty in association with some degree of smell loss, with or without midline craniofacial abnormalities, deafness, and renal anomalies, suggest the possibility of Kallmann syndrome or one of its variants (Lewkowitz-Shpuntoff et al., 2012). Medications being used prior to or at the time of the symptom onset should be determined, as some may influence olfaction, although in most cases such effects are specific to taste, per se [e.g., antifungal agents, angiotensin-converting enzyme (ACE) inhibitors; see Chapter 38]. Medical conditions potentially associated with smell impairments should also be identified (e.g., liver disease, hypothyroidism, diabetes). A history of epistasis, discharge (clear, purulent, or bloody), nasal obstruction, and somatic symptoms, including headache or irritation, as well as a report as to whether the problem

seems to be more prevalent on one side of the nose or the other, may be of localizing value. Idiopathic cases that present during winter months, which is not uncommon (Konstantinidis et al., 2006), suggest the possibility of a bacterial or viral origin, even if other elements of an upper respiratory infection are not present or recognized.

It is critical for the clinician to be aware that while patients often present with the complaint of taste toss, quantitative testing usually reveals only an olfactory problem (Deems et al., 1991), reflecting decreased retronasal stimulation of the olfactory receptors during deglutition (Burdach and Doty, 1987). Importantly, the clinician should be cognizant of the fact that combinations of causal factors may be present that need to be considered. For example, older persons and persons with allergies may be more susceptible than others to smell loss from viral and other causes because of previous cumulative damage to the olfactory epithelium.

Nasal endoscopy can assess the health of the nasal cavity, although complete visualization of the olfactory epithelium is not generally possible even with the use of rhinoscopes. Modern imaging techniques can detect inflammatory processes within the nose and sinuses, as well as brain lesions and the integrity of the olfactory bulbs, tracts, and cortical parenchyma (Chapter 13). As described later in this chapter, patients complaining of never having a sense of smell, that is, congenital anosmia, typically lack normal olfactory bulbs or tracts upon appropriate magnetic resonance imaging (MRI) (Yousem et al., 1996). Some laboratory tests (e.g., blood serum tests) are helpful in detecting underlying medical conditions suggested by history and physical examination, such as infection, nutritional deficiencies (e.g., vitamins B6, B12), allergy, diabetes mellitus, and thyroid, liver, and kidney disease. Visual acuity, visual field, and optic disc examinations can aid in the detection of possible intracranial mass lesions that, in addition to producing visual deficits, impinge upon the olfactory tract.

17.3.2 Quantitative Olfactory Testing and MR Imaging

A common error made on the part of clinicians is to accept a patient's self-report of sensory dysfunction and to fail to verify the presence or magnitude of the problem via tests, despite the commercial availability of such tests (Chapter 11). Many persons, particularly the elderly and those with dementia, are unaware of their dysfunction or are inaccurate in assessing its magnitude (Figure 17.1). Standardized quantitative olfactory testing allows, in most instances, for (1) the characterization of the nature and degree of the chemosensory problem, (2) establishing the validity of the patient's complaint, including the detection of malingering, (3) monitoring changes in function over

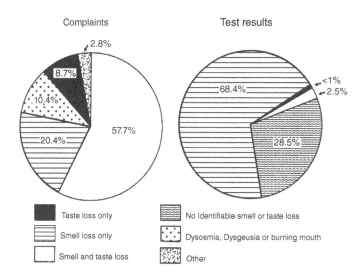

Figure 17.1 Distribution of primary chemosensory complaints (left diagram) and test results (right diagram) in 750 consecutive patients who presented to the University of Pennslyvannia Smell and Taste Center during the period extending from 1980 through 1986. Note that the complaint categories of smell loss, taste loss, and smell and taste loss include some patients with secondary complaints of chemosensory distortion and sensations of burning mouth. From Deems et al. (1991) with permission. Copyright © 1991 American Medical Association, all rights reserved.

time, and (4) providing objective data for establishing disability compensation.

Although smell can be tested using psychophysical, electrophysiological, and psychophysiological procedures, only psychophysical tests are widely used, reflecting issues of cost, practicality, and utility (Chapter 11). Despite the fact that electrical activity can be directly measured from the olfactory epithelium (electro-olfactogram or EOG), this measure cannot be obtained from all patients, is variable, and is influenced by sampling issues. Moreover, many patients have difficulty tolerating electrodes placed in their non-anesthetized noses. Since a normal EOG is present in some anosmic patients and can be recorded even after death in animals, it cannot be used alone to define dysfunction. Unfortunately, the odor event-related potential (OERP), unlike auditory brainstem evoked potentials, sheds little light on localizing anomalies within the neural pathway. As noted in Chapter 12, however, recent advances in analysis and measurement of EEG waveforms may lead to broader application of these techniques within the clinic.

Some physicians, as well as lawyers seeking to denigrate psychophysical test results, divide sensory tests into "subjective" and "objective" categories. In this context, "subjective" usually refers to tests that require a conscious response, whereas "objective" refers to electrophysiological tests. Unfortunately, this dichotomy is misleading and laden with a value judgment, since objective always trumps subjective. As pointed out for audition by the Nobel laureate Georg von Békésy 45 years ago, putative "subjective" tests can be more sensitive and reliable than putative "objective" tests (von Békésy, 1968). Although it is presumed that "subjective" tests are easier to malinger than "objective tests," forced-choice psychophysical tests can detect most malingerers on the basis of improbable responses (Doty and Crastnopol, 2010). Moreover, so-called "objective" olfactory tests are not immune to malingering, since, for example, reliable measurement of electrophysiological responses requires considerable subject cooperation, such as sitting very still during recording sessions.

Modern imaging procedures can now quantify, in vivo, the volume of the olfactory bulbs and other olfactory structures. Based on MRI measures, decreases in olfactory bulb volumes have been noted in older persons (Yousem et al., 1998) and in patients with Alzheimer's disease (Thomann et al., 2009), head trauma (Collet et al., 2009; Doty et al., 1997c; Rombaux et al., 2006a; Yousem et al., 1995), idiopathic congenital smell loss (Croy et al., 2010; Yousem et al., 1996), Kallmann syndrome (Bajaj et al., 1993; Koenigkam-Santos et al., 2011; Yousem et al., 1993), multiple sclerosis (Schmidt et al., 2011), Parkinson's disease (Brodoehl et al., 2012; Hummel et al., 2010; Kim et al., 2007), total laryngectomy (Veyseller et al., 2011), and viral-related smell dysfunction (Rombaux et al., 2006b). Patients with schizophrenia, as well as their first-order relatives, exhibit decrements in olfactory bulb size (Nguyen et al., 2011; Turetsky et al., 2000, 2003). Interestingly, one study reports that early blind persons have larger olfactory bulbs than sighted persons (Rombaux et al., 2010).

The clinical utility of volumetric measures is yet to be determined since, in most cases, correlations between the structure sizes and psychophysical test measures are moderate and the volume measures are dynamic and not disease specific. Currently standardization is lacking in defining morphological parameters to be assessed and considerable age-related variability in bulbar measures can be present in normal individuals (Yousem et al., 1998). That being said, a recent study has provided evidence that olfactory bulb volumes may have considerable prognostic value (Rombaux et al., 2012): no future improvement in olfactory function is likely in head trauma patients and patients with upper respiratory infection (URI)-related

378   **Chapter 17   Clinical Disorders of Olfaction**

smell loss whose olfactory bulb volumes are less than 40 mm^3.

17.4   CAUSES OF OLFACTORY DYSFUNCTION

As can be seen in Table 17.1, there are many reported etiologies for olfactory disturbance. Approximately two thirds of cases of chronic anosmia or hyposmia (i.e., those which are presumably permanent) that present to a clinic are due

Table 17.1   Agents, Diseases, Drugs, Interventions, and other Etiological Categories Associated in the Medical or Toxicological Literature with Olfactory Dysfunction

Drugs
Adrenal steroids (chronic use)
Amino acids (excess)
   Cysteine
   Histidine
Analgesics
   Antipyrine
Anesthetics, local
   Cocaine HCl
   Procaine HCl
   Tetracaine HCl
Anticancer agents (e.g., methotrexate)
Antihistamines (e.g., chlorpheniramine malate)
Antimicrobials
   Griseofulvin
   Lincomycin
   Macro!ides
   Neomycin
   Pencillins
   Streptomycin
   Tetracyclines
   Tyrothricin
Antirheumatics
   Mercury/gold salts
   D-Penicillamine
Antithyroids
   Methimazole
   Propylthiouracil
   Thiouracil
Antivirals
Cardiovascular/hypertensives
Gastric medications
   Cimetidine
Hyperlipoproteinemia medications
   Artovastatin calcium (Lipitor)
   Cholestyramine
   Clofibrate
Intranasal saline solutions with:
   Acetylcholine
   Acetyl, β-methylcholine

Table 17.1   (*Continued*)

Intranasal saline solutions with: (Cont.)
   Menthol
   Strychnine
   Zinc sulfate, gluconate
Local vasoconstrictors
Opiates
   Codeine
   Hydromophone HCl
   Morphine
Psychopharmaceuticals (e.g., LSD, psilocybin)
Sympathomimetics
   Amphetamine sulfate
   Fenbutrazate HCl
   Phenmetrazine theoclate

Endocrine/Metabolic
Addison's disease
Congenital adrenal hyperplasia
Cushing's syndrome
Diabetes mellitus
Froelich's syndrome
Gigantism
Hypergonadotropic hypogonadism
Hypothyroidism
Kallmann's syndrome
Pregnancy
Panhypopituitarism
Pseudohypoparathyroidism
Sjogren's syndrome
Turner's syndrome

Industrial Dusts, Metals, Volatiles
Acetone
Acids (e.g., sulfuric)
Ashes
Benzene
Benzol
Butyl acetate
Cadmium
Carbon disulfide
Cement
Chalk Chlorine
Chromium
Coke/coal
Cotton
Cresol
Ethyl acetate
Ethyl and methyl acrylate
Flour
Formaldehyde
Grain
Hydrazine
Hydrogen selenide
Hydrogen sulfide
Iron carboxyl
Lead
Manganese

17.4 Causes of Olfactory Dysfunction

Table 17.1 (*Continued*)

Industrial Dusts, Metals, Volatiles (Cont.)
Mercury
Nickel
Nitrous gases
Paint solvents
Pepper
Peppermint oil
Phosphorus oxychloride
Potash
Silicone dioxide
Spices
Trichloroethylene

Infections
Acquired immunodeficiency (AIDS)
Acute viral rhinitis
Bacterial rhinosinusitis
Bronchiectasis
Fungal
Influenza
Rickettsia
Microfilaria

Lesions of the nose/Airway blockage
Adenoid hypertrophy
Allergic rhinitis
    Perennial
    Seasonal
Atrophic rhinitis
Chronic inflammatory rhinitis
Hypertrophic rhinitis
Nasal polyposis
Rhinitis medicamentosa
Structural abnormality
    Deviated septum
    Weakness of alae nasi
Vasomotor rhinitis

Medical Interventions
Adrenalectomy
Anesthesia
Anterior craniotomy
Arteriography
Chemotherapy
Frontal lobe resection
Gastrectomy
Hemodialysis
Hypophysectomy
Influenza vaccination
Laryngectomy
Oophorectomy
Paranasal sinus exenteration
Radiation therapy
Rhinoplasty
Temporal lobe resection
Thyroidectomy

Table 17.1 (*Continued*)

Neoplasms- Intracranial
Frontal lobe gliomas and other tumors
Midline cranial tumors
    Parasagital meningiomas
    Tumors of the corpus callosum
Olfactory groove/cribriform plate meningiomas
Osteomas
Paraoptic chiasma tumors
    Aneurysms
    Craniopharyngioma
    Pituitary tumors (esp. adenomas)
    Suprasellar cholesteatoma
    Suprasellar meningioma
Temporal lobe tumors

Neoplasms-Intranasal
Neuro-olfactory tumors
    Esthesioepithelioma
    Esthesioneuroblastoma
    Esthesioneurocytoma
    Esthesioneuroepithelioma
Other benign or malignant nasal tumors
    Adenocarcinoma
    Leukemic infiltration
Nasopharyngeal tumors with extension
    Neurofibroma
    Paranasal tumors with extension
    Schwannoma

Neoplasms-Extranasal and Extracranial
Breast
Gastrointestinal tract
Laryngeal
Lung
Ovary
Testicular

Neurological
Amyotrophic lateral sclerosis
Alzheimer's disease
Cerebral abscess (esp. frontal or ethmoidal regions)
Down syndrome
Familial dysautonomia
Guam ALS/PD/dementia
Head trauma
Huntington's disease
Hydrocephalus
Korsakoff's psychosis
Migraine
Meningitis
Multiple sclerosis
Myesthenia gravis
Paget's disease
Parkinson's disease
Refsum's syndrome

Table 17.1 (*Continued*)

Restless leg syndrome
Syphilis
Syringomyelia
Temporal lobe epilepsy
Hamartomas
Mesial temporal sclerosis
Scars/previous infarcts
Vascular insufficiency/anoxia
Small multiple cerebrovascular accidents
Subclavian steal syndrome
Transient ischemic attacks

Nutritional/metabolic
Abetalipoproteinemia
Chronic alcoholism
Chronic renal failure
Cirrhosis of liver
Gout
Protein-calorie malnutrition
Total parenteral nutrition w/o adequate replacement
Trace metal deficiencies
    Copper
    Zinc
Whipple's disease
Vitamin deficiency
    Vitamins A, B_6, B_{12}

Psychiatric
Anorexia nervosa (severe stage)
Attention deficit disorder
Depressive disorders (rare)
Hysteria (rare)
Malingering
Olfactory reference syndrome
Schizophrenia
Schizotypy
Seasonal affective disorder

Pumonary
Chronic obstructive pulmonary disease

to prior upper respiratory infections (URIs), head trauma, and nasal and paranasal sinus disease (Deems et al., 1991), and most can be expected to reflect significant damage to the olfactory neuroepithelium (Holbrook et al., 2005). Major causes of smell loss, save those due to neurodegenerative diseases and schizophrenia (Chapter 18), are described below.

17.4.1 Upper Respiratory Infections

Although rarely appreciated, the most frequent cause of *chronic* and often permanent loss of smell in the adult is an URI. URIs that produce olfactory loss include those as ones associated with the common cold, influenza, pneumonia, or human immunodeficiency virus (HIV) (Akerlund et al., 1995; Deems et al., 1991; Hummel et al., 1998a). Often the respiratory illness is described as being more severe than usual and, in many cases, dysosmia or phantosmia is present for some period. Usually the dysosmia, which is generally associated with at least some smell function, subsides over time, often leaving the patient with a noticeable olfactory deficit, most commonly anosmia. It should be noted that most viral infections are either entirely asymptomatic or are so mildly symptomatic that they go unrecognized (Stroop, 1995). Thus, during seasonal epidemics, the quantity of serologically documented infections of influenza or arboviral encephalitis exceeds the number of acute cases by several hundredfold, raising the possibility that many unexplained cases of smell dysfunction are due to unrecognized viral infections.

Exactly what predisposes someone to viral or bacterial-induced smell dysfunction or the mechanisms underlying it remains unclear. Most such losses become manifest in middle or older age, suggesting the potential importance of cumulative insult and the challenge of regeneration of the neuroepithelium in advancing age when proliferation of basal cells and immature neurons is significantly reduced (Loo et al., 1996). Direct insult to the olfactory neuroepithelium is presumably the primary basis of the problem in URIs, as biopsy studies of olfactory epithelia from patients with post-URI anosmia evidence extensive cicatrisation, decreases in receptor cell number, absent or decreased numbers of cilia on remaining receptor cells, and the replacement of sensory epithelium with respiratory epithelium (Douek et al., 1975; Jafek et al., 1990b; Yamagishi 1988; Yamagishi et al., 1994). As noted in Chapters 19 and 20, many viruses have the capability of invading the CNS via the olfactory neuroepithelium, and some may damage the olfactory bulb and other central structures in addition to, or independent of, peripheral damage to the olfactory epithelium (Doty, 2008).

Recent data suggest that susceptibility to URIs is influenced by the presence of T2R38 receptors within the respiratory mucosa (Lee et al., 2012). While T2R receptors were first discovered in the taste system and are known to transduce bitter tastes via taste buds, they are also found in both the upper and lower respiratory epithelia of humans (Shah et al., 2009) where they regulate nitric oxide (NO) excretion and increase intracellar Ca^{2+} and ciliary beat frequency (Jain et al., 1993). NO, which is released at high levels into nasal cavity and the paranasal sinuses, has antimicrobial properties and plays a critical role in defending against respiratory infections. In patients with chronic rhinosinusitis, nasal levels of NO are positively correlated with olfactory threshold sensitivity (Elsherif et al., 2007).

17.4.2 Nasosinus Disease and Disorders Associated with Nasal Obstruction

Olfactory impairment that accompanies most nasosinus diseases has been traditionally viewed as being solely conductive, altering the normal airflow into the olfactory meatus, a very small cleft located in the superior sector of the nose. This narrow cleft is evident in a computerized tomography (CT) scan which also shows the general anatomy of the nose and paranasal sinuses (Figure 17.2). Although marked airflow blockage, particularly secondary to polyps, diminishes the olfactory function of some patients, surgical interventions or the administration of systemic or topical steroids often fails to return function to normal, as measured by quantitative tests. In many cases, airway blockage alone cannot completely explain the olfactory loss, and inflammatory infiltrates may alter olfaction in the absence of mucosal hypertrophy (for reviews, see Dalton, 2004; Doty and Mishra, 2001). While, in general, olfactory dysfunction is related to the severity of the rhinosinusitis [e.g., in one study, the mean scores on the University of Pennsylvania Smell Identification Test (UPSIT; see Chapter 11) were 35, 31, 26 and 23 for Kennedy Stages I to IV of the disease, respectively (Downey et al., 1996)], the defining factor may be, in fact, the severity of histopathological changes within the olfactory mucosa (Jafek et al., 1987). Kern (2000) found such severity to be related to olfactory test scores in patients with chronic rhinosinusitis. Moreover, biopsies from the olfactory mucosal region of patients with nasal disease are less likely to yield olfactory-related tissue than such biopsies from controls (Feron et al., 1998). The same is true for anosmic vs. nonanosmic rhinosinusitis patients, the former of whom exhibit a more pathological epithelium (Lee et al., 2000; Yee et al., 2010). This hypothesis is further supported by findings of weak or no associations between olfactory test scores and measures of nasal airway patency (save severe blockage), whether measured by rhinoscopy, rhinomanometry, or acoustic rhinometry (Apter et al., 1999; Cowart et al., 1993; Nordin et al., 1998; Scott et al., 1988).

17.4.2.1 Hypertrophied Adenoids.

Hypertrophied adenoid tissue can significantly block the nasal airflow of children whose airways are otherwise patent. In general, nasal resistance decreases by 20–40% following adenoidectomy (Crysdale et al., 1985; Fielder, 1985). Olfactory thresholds to phenyl ethyl alcohol measured in children with varying degrees of nasal obstruction are directly related to clinical ratings of obstruction (Figure 17.3) and improve after adenoidectomy (Figure 17.4) (Delank, 1992; Ghorbanian et al.,

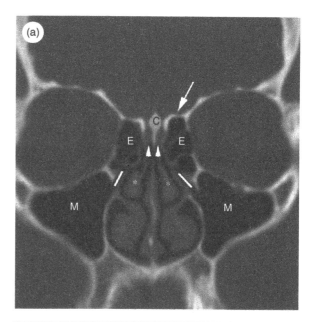

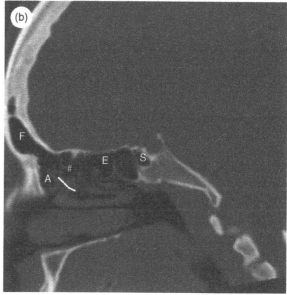

Figure 17.2 **a.** Normal sinus anatomy. Coronal computed tomographic (CT) image of the paranasal sinuses (a). The paired maxillary sinuses (M) drain through the maxillary infundibulum (white lines) into the middle meatus. Along the medial margin of the infundibulum is the uncinate process. Ethmoid air cells (E) are present on both sides, between the lamina papyracea laterally and the vertical lamella medially. The middle turbinates (*) are attached to the anterior skull base by the vertical lamella. The crista galli (C) lies in the midline between the two sides of the cribriform plate (arrowheads). The vertical lamella (arrow) is slightly elevated relative to the crista galli, and is separated by the intervening lateral lamella. **b.** Sagittal CT of the paranasal sinuses shows the frontal sinus (F), which is drained by the frontal recess (white line). Anterior and inferior to the frontal recess is the agger nasi air cell (A), and along its posterior superior margin is the ethmoid bulla (#). The sphenoid sinus (S) is located posterior to the ethmoid air cells (E). Reprinted with permission from Mossa-Basha and Blitz (2013). Copyright © 2013 Elsevier, Inc.

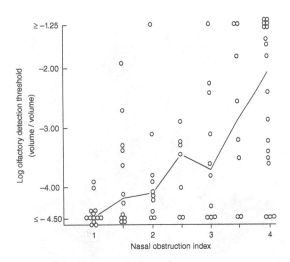

Figure 17.3 Relation between nasal obstruction index, a validated pediatric assessment from nasal sounds and other signs, and the olfactory detection threshold to phenyl ethyl alcohol in 78 children. Joined points signify median values. Reprinted with permission from Ghorbanian et al. (1983). Copyright © 1983 American Academy of Pediatrics.

1983). Such improvement is particularly evident for retronasal olfactory function (Konstantinidis et al., 2005).

17.4.2.2 Acute Viral-Related Rhinosinusitis.
Several studies have measured olfaction longitudinally following the onset of the common cold. Akerlund et al. (1995) assessed 1-butanol odor-detection thresholds in student volunteers before and four days after nasal inoculation with the coronavirus 229E. The nine students who developed a cold had elevated olfactory thresholds on the post-innoculation test relative to the controls—impairment that correlated with nasal congestion but not nasal discharge. Hummel et al. (1998b) evaluated smell function and nasal patency in 18 women and 18 men at the time of onset of a natural cold as well as 2, 4, 6, and 35 days thereafter. The cold produced a decrease in the volume of the anterior nasal cavity and increased mucus secretion, heightened olfactory thresholds, lowered odor intensity ratings, and decreased the amplitude of the odor event-related potential to both olfactory and trigeminal stimuli. Even when airway patency was normal and mucus secretion was minimal, evoked potential amplitudes to olfactory stimuli were still depressed, supporting the concept that URIs may influence olfactory function independently of nasal congestion.

17.4.2.3 Acute and Chronic Rhinosinusitis.
The first large-scale empirical study of the influences of allergic rhinitis on olfactory function was that of Cowart et al. (1993). Detection thresholds to phenyl ethyl alcohol were obtained from 91 patients with symptoms of allergic rhinitis and from 80 non-atopic controls. The allergy patients exhibited greater dysfunction than the controls, with 23% having a threshold at or above the 2.5th percentile of the controls. Clinical or radiographic evidence of rhinosinusitis or nasal polyps or both was associated with hyposmia in the allergy patients: 14% with no associated rhinosinusitis exhibited hyposmia, whereas 43% with associated rhinosinusitis did so. Similar general observations have been noted by others (Apter et al., 1992; Golding-Wood et al., 1996a; Guss et al., 2009; Kondo et al., 1998; Litvack et al., 2008; Mott et al., 1997; Paparo et al., 1996; Rydzewski et al., 2000; Simola and Malmberg, 1998). Interestingly, there are cases in which the

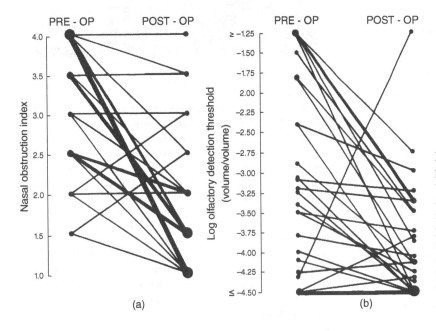

Figure 17.4 Nasal obstruction index (a) and phenyl ethyl alcohol olfactory detection threshold values in 28 children before and after adenoidectomy. Larger negative threshold values indicate better odor detection. Each line joins preoperative and postoperative values for each subject. Heavy lines denote confluence of several subjects who had same preoperative detection threshold score. Reprinted with permission from Ghorbanian et al. (1983). Copyright © 1983 American Academy of Pediatrics.

inflammatory process seems to be confined to the olfactory cleft (Trotier et al., 2007). In many instances, improvement of function occurs after topical nasal administration of corticosteroids (Sivam et al., 2010). An example of extreme acute sinusitis in which there has been invasion of the disease into the right orbit and cranium is shown in Figure 17.5. Note that the invasion has occurred through the right cribriform plate, undoubtly compromising the patient's ability to smell.

Overall, there appears to be general correspondence between a patient's self-report of olfactory loss and empirical test results. Golding-Wood et al. (1996), for example, evaluated odor identification ability before and after six weeks of betamethasone treatment in 25 well-documented perennial rhinitis patients. The patient group was initially divided into two groups: those who affirmatively answered the question "is your sense of smell impaired?" (n = 15) and those who did not (n = 10). The UPSIT scores of each of the 15 members of the former group were higher after the betamethasone treatment [respective group means (SD) = 18.93 (9.4) and 33.4 (4.01). This was not the case for those who initially felt that they had no problems smelling [respective pre-/post-treatment means (SD) = 33.40 (4.01) and 32.8 (4.94)]. Moderate correlations between the UPSIT scores and the self-ratings of olfactory function were found both before (r = −0.52) and after (r = −0.58) treatment. As has been noted by others, however, the average post-treatment UPSIT score was still indicative of a mild hyposmic condition. The UPSIT scores retained, among the patients, a similar rank order before and after treatment (Spearman r = 0.75).

Several studies have examined olfactory function before and after allergen challenges (Hilberg, 1995; Hinriksdottir et al., 1997; Klimek and Eggers, 1997; Lane et al., 1996; Moll et al., 1998). In all cases, smell function decreased as a result of the challenges, although in the few cases that examined measures of airway patency, no association was noted between the degree of smell dysfunction and the patency measure. Hilberg (1995) assessed the comparative effects of the oral antihistamine terfenadine (an H1-blocker) and the topical steroid, budesonide, on an allergen challenge in subjects with nasal allergy uncomplicated by polyposis on olfactory dysfunction and various other hay fever symptoms. Although both drugs had an effect on the non-olfactory hay fever symptoms during the nasal pollen challenge, only the budesonide improved the challenge-related decrement in olfactory sensitivity. This steroid was also more effective in increasing nasal volume. Unfortunately, the improvement in olfactory function occurred in less than half of the patients (7/17; 41%).

17.4.3 Nasal Surgery

Gross-Isseroff et al. (1989) obtained detection threshold and UPSIT measures in children and adults with choanal

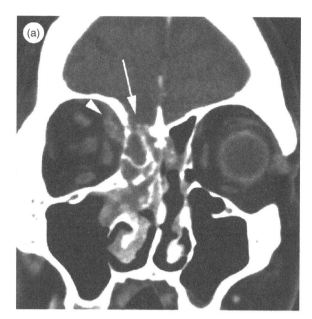

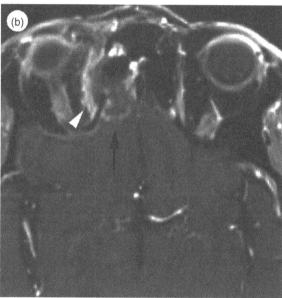

Figure 17.5 a. Coronal CT image of the orbits and paranasal sinuses of a 68-year-old man presenting with headaches, right orbital pain, and blurry vision. Acute sinusitis is apparent with right orbital and intracranial invasion. Note opacification of the right ethmoid air cells and maxillary infundibulum, with soft-tissue attenuation extending into the right maxillary ostium. There is orbital involvement with peripherally enhancing soft tissue extending into the medial superior right orbit, with involvement of the right superior oblique muscle (arrowhead). Erosion of the right cribriform plate (arrow) is present, undoubtedly influencing smell function and suggesting the possibility of intracranial involvement. b. Axial T1 postcontrast image at the level of the orbits showing a rim-enhancing collection in the medial right orbit (arrowhead), representing subperiosteal abscess and peripherally enhancing soft tissue extending intracranially (black arrow) consistent with accumulation of pus and fluid from infected tissue (empyema). Reprinted with permission from Mossa-Basha and Blitz (2013). Copyright © 2013 Elsevier, Inc.

atresia before and after surgical repair, which occurred at relatively advanced ages (8–31 years). The three patients who had suffered from bilateral atresia had permanent olfactory deficits, whereas the one patient who had suffered from unilateral atresia had normal function. These findings suggested to the authors the question as to whether early sensory exposure is needed for normal development of olfactory function.

The limited empirical data suggest that septoplasty and rhinoplasty can modestly improve olfactory function in cases where the airway is significantly constricted (Damm et al., 2003; Kimmelman, 1994; Stevens and Stevens, 1985). However, several studies suggest that such operations can damage the olfactory system (e.g., Damm et al., 2003; Pfaar et al., 2004). Olfactory function has been evaluated before and after common operative procedures for chronic rhinosinusitis and/or polyposis unresponsive to more conservative treatments (e.g., allergen avoidance, nasal corticosteroids). Among such operations are middle turbinate medialization, polypectomy, uncinectomy, anterior ethmoidectomy, posterior ethmoidectomy, and sphenoidectomy, alone or in combination. With rare exception, olfactory function improves to some degree following such surgeries (Delank and Stoll, 1998; Downey et al., 1996; Federspil et al., 2008; Gupta et al., 2014; Hoseman et al., 2000; Hu et al., 2010; Johansson et al., 2004; Katotomichelakis et al., 2010; Kim et al., 2011; Kimmelman, 1994; Klossek et al., 1997; Konstantinidis et al., 2007; Leonard et al., 1988; Litvack et al., 2009; Minovi et al., 2008; Olsson et al., 2010; Olsson and Stjarne, 2010), even though the proportion of patients regaining normal olfactory function is typically less than 40% (e.g., Delank and Stoll, 1998; Leonard et al., 1988; Min et al., 1995; Pade and Hummel, 2008) and in many, but not all, cases the dysfunction regresses within a year to preoperative levels (Briner et al., 2012; Jiang et al., 2008; Klimek et al., 1997). The degree of improvement is well exemplified by examining the median pre- and postoperative UPSIT scores calculated across six studies in which five or more patients were tested (Eichel, 1994; el Naggar et al., 1995; Friedman et al., 1999; Kimmelman, 1994; Lund and Scadding, 1994; Seiden and Smith, 1998). These values, 17.0 and 25.5, respectively, are indicative of an average change in function from total anosmia to a borderline moderate/severe microsmia, a degree of improvement also observed in subsequent studies (Kim et al., 2011; Soler et al., 2010). For a recent review of this literature, see Rudmik and Smith (2012).

17.4.4 Tumors

Olfactory dysfunction can result from a variety of intranasal and intracranial tumors. McCormack and Harris (1955) reported on five cases in which neurogenic tumors arising in the lateral nasal wall were accompanied by anosmia and noted that over 100 such cases had been cited in the literature up to that time. In one study, Ho et al. (2002) assessed the olfactory function of 48 patients undergoing radiotherapy for nasopharyngeal carcinoma before and up to 12 months after radiotherapy. Olfactory function returned to pre-radiotherapy levels by 12 months. Around 20% of the tumors of the temporal lobe or lesions of the uncinate convolution are said to produce some form of olfactory disturbance, most typically the hallucination of a bad smell (Furstenberg et al., 1943).

The olfactory bulbs and tracts are very sensitive to pressure from meningiomas from the dura of the cribriform plate and surrounding regions (e.g., olfactory groove meningiomas, suprasellar ridge meningiomas), pituitary growths that extend above the diaphragm of the sella turcica, and tumors inside or on the floor of the third ventricle. Thus, both unilateral and bilateral hyposmia and anosmia have been reported in cases of frontal lobe glioma, suprasellar meningioma, and sphenoidal ridge meningioma, as well as in cases involving non-neoplastic space-filling lesions, such as large internal carotid aneurysms extending over the pituitary fossa, aneurysms of the anterior communicating bifurcation, and hydrocephalus that pushes the floor of the third ventricle downward. Although signs other than olfactory ones are typically present in such cases, olfactory dysfunction can, in fact, be the sole sign (McCormack and Harris, 1955). In some cases, the olfactory dysfunction is relieved after tumor removal and decompression on olfactory eloquent structures (Ishimaru et al., 1999).

According to Finelli and Mair (1991), the single most egregious error of neurologists is failure to recognize the symptom of anosmia as the principal or sole feature of an olfactory groove neoplasm. However, olfactory dysfunction is often only evident after such tumors have achieved a considerable size and, even then, separate testing of each side of the nose may be needed to identify a deficit. Many such patients are not aware of an olfactory deficit and most present with headache, visual disturbances, or cognitive problems such as memory deficits (Welge-Luessen et al., 2001).

Unfortunately, the classical surgical approach to anterior skull base tumors—bifrontal craniotomy—is almost always associated with postoperative anosmia and other complications. Moreover, subcranial endoscopic approaches to such tumors can also produce olfactory deficits (Kinnunen and Aitasalo, 2006; Tam et al., 2013). Fortunately, a number of surgical approaches that better spare the olfactory system have been devised (e.g., unilateral frontal craniotomy with orbital osteotomy; the transglabellar/subcranial approach) (Babu et al., 1995; Jung et al., 1997). Nevertheless, with special care and adherence to certain procedures, the olfactory bulbs and tracts can be preserved using the bilateral sub-frontal approach to anterior skull base tumors (Cömert et al., 2011; Sepehrnia and Knopp, 1999). Moreover, different approaches to palliative whole brain radiotherapies have been found that minimize complaints of smell problems (Akhtar et al., 2012).

It is of interest that olfactory tests were employed over 80 years ago to help diagnose cribriform plate meningiomas and to localize other tumors impinging upon the olfactory nerve (Elsberg, 1935a,b). Despite the limitations of the olfactory test procedure used (i.e., the so-called blast injection technique; see Chapter 11), the olfactory recognition threshold was generally elevated on the side where a neoplasm exerted pressure on one olfactory nerve. When both nerves were involved, bilateral threshold elevation was noted, with greater elevation on the most affected side. Such changes were observed for expanding lesions on the ventral surfaces of the frontal lobes, as well as for suprasellar meningiomas and aneurysms of the internal carotid artery or the anterior part of the circle of Willis. Threshold values were not influenced by pituitary adenomas confined to the region of the sella turcica, but were heightened if growth occurred above the sella. Although most intracranial tumors were not associated with altered recognition thresholds, they were associated with prolonged duration of olfactory adaptation or fatigue. Thus, tumors in or near the midline of the cranial cavity produced long-lasting adaptation (e.g., parasagital meningliomas, tumors of the corpus callosum, and infiltrating growths extending to the medial surface of a cerebral hemisphere). When generalized intracranial pressure was observed, decreased recognition threshold sensitivity was sometimes present. Although modern imaging techniques make such testing somewhat archaic, olfactory testing may still be of value today in the early detection of some brain tumors.

Daniels et al. (2001) examined odor discrimination performance and OERPs in 20 subjects with unilateral frontal or temporal lobe brain tumors. Patients with right-side lesions exhibited deficits in odor discrimination in both the left and right sides of the nose. Patients with left-side lesions only exhibited attenuated function when the odorant was presented to the left side. The amplitude of the OERPs was decreased after left-side stimulation, but not after right-side stimulation. Interestingly, a correlation was present between olfactory and acoustic event-related potentials in patients with right-side lesions after right-side stimulation.

17.4.5 Exposures to Airborne Pollutants

Decrements in smell function are common in highly polluted urban areas (Calderon-Garciduenas et al., 2010; Guarneros et al., 2009; Hudson et al., 2006), as well as in some occupational settings where workers are exposed on a relatively continuous basis to airborne particulates, metals, and chemicals (Adams and Crabtree, 1961; Ahlstrom et al., 1986; Antunes et al., 2007; Dalton, 2003; Hisamitsu et al., 2011; Lucchini et al., 1999; Schwartz et al., 1989) (see Chapter 20). Olfactory dysfunction has been demonstrated in residents, firefighters, and workers exposed to the air pollution from the demolition and clean-up of the World Trade Center (WTC) following September 11, 2001 (Altman et al., 2011; Dalton et al., 2010). In a number of cases, damage to the olfactory epithelium, as well as other sectors of the olfactory pathway, is evident, and inflammation of both the upper and lower respiratory tracts is common (Doty and Hastings, 2001; Holt, 1996; Trevino, 1996). At autopsy, ultrafine particulate matter (UFPM <100 nm) has been found in the olfactory bulbs of dogs, children, and young adults who have been chronically exposed to extreme air pollution, reflecting the transit of materials from the nasal cavity into the brain via the olfactory fila (Calderon-Garciduenas et al., 2008a,b; 2010).

A classic study of how chemical exposure in a modern workplace can influence olfaction was performed by Schwartz et al. (1989). These investigators administered the UPSIT to 731 workers exposed to acrylates and methacrylate vapors in a chemical manufacturing plant and for whom quantitative work exposure histories were well documented. A logistic regression analysis which adjusted for multiple confounders revealed exposure odds ratios of 2.8 (1.1, 7.0) for all workers and 13.5 (2.1, 87.6) for workers who never smoked cigarettes.

17.4.6 Head Trauma

Olfactory dysfunction secondary to head trauma (HT) has been reported in the medical literature since the 19th century. In apparently the first description of this phenomenon, Jackson (1864) described a 50-year-old man who complained of loss of smell after falling off of a horse. Three years later Ogle (1870) provided case descriptions of three HT patients with smell loss and made the important point that blows to the back of the head can induce smell loss by shearing the olfactory fila at the base of the brain, a point subsequently reiterated by Althaus (1881) and others. That same year, Notta reported that anosmia could exist in patients who do not lose consciousness and mentioned that, in his experience, recovery may occur in 8 weeks to 7 months after the injury (Notta, 1870). The first description of dysosmia induced by HT was provided by Legg (1873): a man fell off a cart and thereafter reported that food had the flavor of "gas or paraffin." Although this individual reported having lost both his senses of smell and taste, upon testing his ability to detect sweet, sour, bitter, and salty tasting substances was intact. Numerous other reports subsequently appeared in the 19th century on post-traumatic loss of smell function (e.g., (Ferrier, 1876; Gowers,1893; Jacob, 1882; Mollière, 1871; Rotch, 1878).

The first large scale study of HT-related smell loss was published during World War II (Leigh, 1943). In this study of 1000 consecutive cases admitted to a military hospital for head injuries, Leigh found 72 with impaired smell function (7.2%), of which 41 had complete anosmia (4.1%). Blows to the front and the back of the head were most common (18 occipital and 30 frontal), and nearly all cases in which smell loss was present represented severe trauma. Leigh found that only six of his 72 HT cases regained smell function (8%) during this study, half of whom went through a period of dysosmia. Time to recovery of function varied from 20 days to 12 months (median = 3 months).

In a large study of 1167 unselected HT cases performed two decades later, Sumner (1964) observed a remarkably similar percentage of cases with smell loss (7.1%). Sumner examined the site of the blows in anosmics whose HT was relatively minor, i.e., who had no amnesia or amnesia lasting less than an hour. Independent of a smell problem, many more patients had HT due to frontal than to vertical or occipital blows. However, occipital blows were five times more likely than frontal blows to produce anosmia. Interestingly, an abrupt change occurred in the number of cases that improved over time at ~10 weeks (Figure 17.6). This suggested to Sumner that two different processes may be involved in recovery, one that reflects the disappearance of compression, edema, and blood clot formation, and the other presumably a neural and longer-lasting process. No meaningful association was found between the severity of the head trauma and the time required for the anosmia to resolve. Sumner notes that his study differed from earlier ones in that (a) anosmia was sought out in patients soon after their trauma, (b) his sample contained many patients with minor HT, and (c) patients were followed for a relatively long time after injury (up to five years).

Since 1980, numerous studies have appeared on this topic, many from specialized smell and taste centers (Atighechi et al., 2013; Charland-Verville et al., 2012; Collet et al., 2009; Costanzo et al., 1987; Deems et al., 1991; Doty et al., 1997c; Duncan and Seiden, 1995; Fujii et al., 2002; Haxel et al., 2008; Ikeda et al., 1995; Jafek et al., 1989; Kern et al., 2000; Kim et al., 2014; Levin et al., 1985; London et al., 2008; Mann and Vento, 2006; Reden et al., 2006; Reiter et al., 2004; Rombaux et al., 2012, 2012a; Schechter and Henkin, 1974; Vent et al., 2010a; Welge-Lüssen et al., 2012; Wise et al., 2006; Yousem et al., 1999). Unlike the earlier studies, quantitative olfactory testing was performed. Because most studies examined HT in patients who presented with complaints of smell loss, a much higher percentage of smell and taste center patient populations have olfactory dysfunction than noted above. For example, in one large study of 268 such persons, 87.3% had demonstrable smell loss, with 66.8% having anosmia

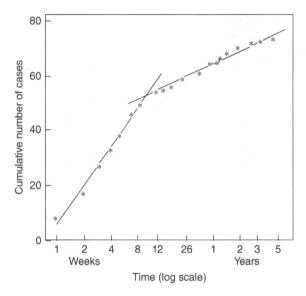

Figure 17.6 The number of patients whose smell function recovered over time since injury onset. Note abrupt change in slope of function. This may reflect two different underlying factors, i.e., the shorter-term disappearance of compression, edema, and blood clot formation, and longer-term neuronal recovery. Reprinted with permission from Sumner (1964). Copyright © 1964 Oxford University Press.

and 20.5% having microsmia (Doty et al., 1997c). In one of the more recent studies that administered olfactory tests to patients who had presented with head trauma rather than olfaction secondary to head trauma, an estimated 12.7% incidence of olfactory dysfunction was noted (Haxel et al., 2008). There is now evidence that head trauma impacts olfactory function even in young children (Bakker et al., 2014).

Several important observations accrued from the Doty et al. (1997c) study. First, of 168 patients for whom explicit information regarding a single impact focus was available, 36.3% received the primary impact to the front, 39.3% to the back, 10.1% to the right, 11.3% to the left, and 3% to the top of the head. Blows to the back or side of the head resulted in significantly ($p < 0.03$) larger deficits in function than did blows to the front of the head [mean UPSIT scores = 15.38 for the back, 16.31 for the side, and 19.90 for the front]. Second, the olfactory test scores were not meaningfully related to patient sex or age at the time of trauma or at the time of testing. Third, patients who had some smell function (UPSIT scores > 18) and who had never experienced parosmia had significantly higher UPSIT scores than their counterparts who currently or previously experienced parosmia ($p < 0.001$). This suggests that parosmia is associated with greater damage to the olfactory system. Fourth, as shown in Figure 17.7, the prevalence of parosmia decreased over time. Fifth, no association was evident between the smell test scores and the duration of loss of consciousness (LOC; a measure of HT

17.4 Causes of Olfactory Dysfunction

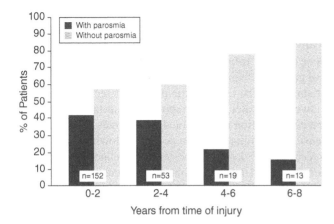

Figure 17.7 Relationship between proportion of patients reporting current parosmia (intermittent and chronic combined) and time since head injury. Years falling exactly on an even value were placed in the earlier category (i.e., in the case of 2, data are in the 0–2 year category rather than the 2–4 year category). Reprinted with permission from Doty et al. (1997c). Copyright © 1997 American Medical Association, all rights reserved.

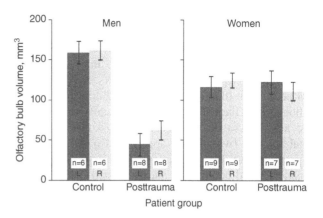

Figure 17.8 Olfactory bulb volumes (left and right) in cubic millimeters for male and female post-trauma and control subjects determined from MRI images. Vertical bars represent SEM. Reprinted with permission from Doty et al. (1997c). Copyright © 1997 American Medical Association, all rights reserved.

severity) when all of the data were included in the analysis. However, when the data from the anosmic patients were excluded, the mean UPSIT scores of patients with LOC > 24 hours were significantly lower than those of patients who had never lost consciousness or who had LOC < 24 hours. Indeed, a −0.56 correlation was present between the duration of LOC and UPSIT score ($p < 0.05$). Sixth, in the nonanosmic HT group, patients who had not been hospitalized had better olfactory function than those who had been hospitalized; no meaningful correlation was present between the UPSIT scores and the length of hospitalization ($r = -0.08$, $p = 0.76$). Seventh, no associations between UPSIT scores and psychological depression, as measured by the Beck Depression Inventory, were observed. Eighth, of 66 patients who were retested at intervals ranging from 0.5 to 13 years, nearly half showed no change in their function (respective pre/post mean UPSIT scores = 19.10 and 19.73). Test scores declined in 1/5th of the group, moving from severe microsmia to total anosmia (respective mean UPSIT scores = 20.08 and 13.08). Slightly more than a third of patients improved somewhat on retesting, moving, on average, from anosmia (17.83) to borderline moderate microsmia (25.46), although only three subjects improved into the absolute normal range (4.5%). The time between the two test sessions was statistically unrelated to whether the UPSIT scores improved, worsened, or stayed the same on the second test occasion. Ninth, in a subgroup of 30 patients for whom olfactory bulb volume measures were obtained, smaller olfactory bulbs were present in male, but not female, HT patients relative to sex- and age-matched controls (Figure 17.8).

Two studies have suggested that anosmia is a reliable predictor of vocational outcome in patients with closed head injury (CHI). In these studies, 80% to 93% of patients with *mild CHI* and anosmia were said to be vocationally disabled, that is, being employed less than 25% of the time after receiving medical clearance to return to work (Martzke et al., 1991; Muliol et al., 2008). Correia et al. (2001) sought to verify these findings, since (a) such rates are as high or higher than those observed in severe head trauma, (b) no quantitative olfactory testing was performed, and (c) a sampling bias may have been present (patients are often referred for neuropsychological testing because of concerns about adaptive functioning). Instead of focusing on patients referred to a neuropsychological clinic, they interviewed patients who had been previously referred to the University of Pennsylvania Smell and Taste Center for head trauma-related smell loss. In addition to quantitative olfactory test scores, medical histories and hospital reports were available to confirm the smell loss and the severity of the CHI. These investigators found that only 7% of mild CHI patients with severe smell loss were vocationally disabled.

17.4.7 Congenital Anosmia

A number of patients report never having ever experienced smell function.[1] In the vast majority of such cases,

---

[1] Most persons are said to be constitutionally unable to detect a few specific odorants, a phenomenon termed specific anosmia. First described by Blakeslee for the fragrance of verbena flowers (Blakeslee, 1918), dozens of specific anosmias have been reported in the literature, including ones for musks, trimethylamine, hydrogen cyanide, n-butyl mercaptan, and isovaleric acid (for review, see Takagi, 1989). However, like the blind spot of the retina, specific anosmias are not typically recognized by a patient

appropriate MRI evaluation of the cribriform plate and gyrus rectus regions reveals agenesis or dysgenesis of the olfactory bulbs and tracks (Yousem et al., 1996). While some of these cases may represent long-term degeneration from viral or traumatic insults to the olfactory epithelium or fila (reflecting the withdrawal of trophic influences on the bulb from intact olfactory neuron), most are assumed to be hereditary. Lygonis (1969) reported total anosmia (without any apparent associated disorders) in members of four generations of a family living in an isolated island community and suggested that the pattern of inheritance was likely autosomal dominant. Singh et al. (1970) described cases of familial anosmia loosely associated with premature baldness and vascular headaches, which appeared to be inherited as dominant with varying penetrance. Several congenital olfactory syndromes are accompanied by endocrine dysfunction and are discussed in the next section. Indeed, there has been suggestion that all individuals with congenital anosmia should undergo genetic testing to determine whether they harbor any of the 19 or so Kallmann-related genotypes.

17.4.8 Endocrine Disorders

Although a number of endocrine disorders have been associated with smell dysfunction, many more have not been assessed for this problem. Of those that have been examined, the mechanisms responsible for the smell loss—aside from obvious anatomical alterations in the primary olfactory pathways—are poorly understood. Listed below are disorders for which at least some quantitative olfactory test data are available.

17.4.8.1 Adrenocortical Insufficiency (Addison's Disease). Henkin and Bartter (1966) reported that patients with untreated adrenocortical insufficiency exhibited greater odor detection threshold sensitivity, relative to normal controls, not only for the odorants pyridine,

and are rarely a reason for seeking clinical help. Moreover, some specific anosmics can, in fact, detect high concentrations of odorants to which they are said to be anosmic, implying hyposmia rather than anosmia. Moreover, not all specific anosmias are really that specific. Amoore (1971), for example, tested the thresholds of 10 subjects who exhibited specific anosmia to isobutyric acid to 17 other compounds, finding heightened thresholds for straight chain fatty acids with four to seven carbon atoms. Importantly, exposure to, or repeated testing with, some odorants, including ones associated with specific anosmias, can result in an increase in their sensitivity in both humans and rats (Doty et al., 1981; Doty and Ferguson-Segall, 1989). For example, Wysocki et al. (1989) found that the ability to perceive androstenone could be induced in 10 of 20 initially insensitive subjects by systematically exposing them to this odorant, an observation confirmed by others (Möller et al., 1999). This finding led Wysocki et al. to suggest that three categories of human subjects may exist in regards to responsiveness to this substance; namely, the truly anosmic, the inducible, and those who are either constitutionally sensitive or have experienced incidental past exposure.

thiophene, and nitrobenzene, but for the vapors above aqueous solutions of six tastants ($NaCl$, KCl, HCl, $NaHCO_3$, sucrose, and urea). These effects were remarkable, with none of the distributions of scores from the patient and control groups showing any overlap. In the case of the volatiles arising from the tastants, the patients detected differences between the water and the vapors of the tastants at 1/10,000 the concentration of that noted for normals. The increased sensitivity did not return to normal following daily 20 mg injections of deoxycorticosterone acetate (DOCA) for periods up to 10 days. In contrast, treatment of the subjects with 20 mg of the carbohydrate-active steroid prednisolone returned sensitivity to normal within 24 hours.

Attempts to replicate or extend these findings have not been forthcoming. Empirically measured hypersensitivity is rare in the chemical senses literature, and numerous tests of auditory, gustatory, and olfactory function in rats before and after adrenalectomy have found no evidence of hypersensitivity of any sort. Indeed, the available data suggest that adrenalectomy may, in fact, produce not only altered odor preferences, but *decrements* in general sensory function (Brosvic et al., 1989; Conn and Mast, 1973; Doty et al., 1991; Kosten and Contreras, 1985; Weigel et al., 1989).

17.4.8.2 Chromatin Negative Gonadal Dysgenesis (Turner's Syndrome). Turner's syndrome (TS) is a form of gonadal dygenesis resulting from a 45X karyotype (X-chromosomal monosomy). It is characterized by a female phenotype, a shield-like chest, short stature, short and sometimes webbed neck, low-set ears, small mandible, a high-arched palate, and sexual infantilism. Other problems often include congenital lymphedema, skeletal anomalies, abnormalities of the nails, and cardiac and renal deficits. A pioneering study reported that nine TS patients had elevated detection and recognition thresholds to three odorants that were assessed (pyridine, thiophene, nitrobenzene) (Henkin, 1967). Gonadal hormone therapy reversed the chemosensory deficits. Interestingly, the mothers of the patients exhibited similar olfactory abnormalities, which the author suggests is in accord with a potential genetic basis for the chemosensory disorder. More recent work has suggested that TS influences odor memory and odor identification (Ros et al., 2012).

17.4.8.3 Cushing's Syndrome. Cushing's syndrome (CS) is a disease characterized by chronic excessive secretion of adrenal corticosteroids (i.e., hypercortisolism). Approximately 80% of cases reflect excessive secretion of adrenocorticotropic hormone (ACTH) (e.g., from ACTH-producing pituitary tumors), whereas the remainder reflect ACTH-independent etiologies (e.g., adrenal gland adenomas and carcinomas, cortisol-secreting tumors) (Ferrante, 1999). According to Henkin (1975),

17.4 Causes of Olfactory Dysfunction

patients with CS exhibit a decrease in sensory detection acuity for olfaction, gustation, hearing, and proprioception. Dogs given chronic dexamethasone (an animal model of Cushing's syndrome) appear to have increased detection thresholds for benzaldehyde and eugenol, in accord with this notion (Ezeh et al., 1992).

17.4.8.4 Hypothyroidism.

A number of hypothyroid patients complain of chemosensory problems (Deems et al., 1991). Lewitt et al. (1989) reported that 11 of 16 (69%) of the hypothyroid patients they evaluated complained of altered smell and taste function, and McConnell et al. (1975) found that 7 of 18 patients with untreated hypothyroidism (39%) were cognizant of some alteration in their sense of smell, with 3 (17%) experiencing dysosmia.

Intuitively, one might expect hypothyroidism to be associated with measurable chemosensory deficits, since 36–83% of hypothyroid patients reportedly experience somesthetic disturbances, and 25–45% auditory or visual disturbances, with night blindness reportedly being a common feature of the disorder (Mattes et al., 1986). However, the few available studies are not in agreement on this point. McConnell et al. (1975) reported that odor detection thresholds for pyridine and nitrobenzene, as well as various taste threshold measures, were strikingly elevated in patients with untreated hypothyroidism—a problem that resolved following thyroxine treatment. In contrast, Lewitt et al. (1989) found no detection threshold differences between 16 patients and 17 controls for the odorant phenyl ethyl alcohol and only a slight, but statistically significant, difference on a suprathreshold test of odor identification. Various measures of suprathreshold taste function, as well as auditory and visual evoked potentials, did not differ between the patients and the controls, and no pre-/post-thyroxin differences were observed for any measure. Negative findings have been reported by others for various taste measures (Pittman and Beschi, 1967).

The basis of these discrepancies is not clear, although the various studies have used different odorants and procedures, and it is not apparent whether the patients were comparable in terms of the nature, duration, and severity of their hypothyroid problems. It is noteworthy that well-controlled psychophysical studies employing rats have found no influences of hypothyroidism on their ability to detect low concentrations of either tastes or smells (Brosvic et al., 1992; Brosvic et al., 1996). However, taste preference differences were found for NaCl, HCl, and quinine, but not sucrose (Brosvic et al., 1996). While an earlier mouse study linked hypothyroidism to anosmia, the paradigm employed was, in fact, a two-bottle preference test, leaving open the possibility to dysosmia, not anosmia, as the basis for the behavioral changes (Beard and Mackay-Sim, 1987).

17.4.8.5 Pseudohypoparathyroidism.

Pseudohypoparathyroidism (PHP) is an endocrine disorder characterized by deficiencies in responsiveness to, but not in the production of, parathyroid hormone. Individuals with Type Ia PHP exhibit a generalized hormone resistance, a deficiency of the alpha chain of the stimulatory guanine nucleotide-binding protein (Gsα) of adenylyl cyclase, and an unusual constellation of skeletal and developmental abnormalities termed Albright hereditary osteodystrophy (AHO). Among the abnormalities expressed in AHO are obesity, short stature, brachydactyly, round facies, and subcutaneous ossifications. Type Ib PHP patients, in contrast, exhibit a specific hormone resistance to parathyroid hormone, do not express a deficiency in Gsα protein activity, and do not have AHO.

PHP was first associated with olfactory dysfunction in 1968 (Henkin, 1968). Subsequent studies reported that Type Ia, but not Type Ib, PHP is accompanied by decreased olfactory function (Ikeda et al., 1988; Weinstock et al., 1986). The olfactory problem was believed to reflect the Gsα protein deficiency of the Type Ia patients, since Gs proteins are involved in the first stages of olfactory transduction (see Chapters 5 and 6). However, a more recent and comprehensive study, while confirming that Type Ia PHP patients exhibit altered olfactory function, also found that Type Ib patients have olfactory dysfunction, making it unlikely that the olfactory dysfunction is related to the proposed Gsα protein deficiency (Doty et al., 1997a). In this same study, relatively normal function was documented in patients with pseudopseudohypoparathyroidism (PPHP), a disorder found in some relatives of patients with Type Ia PHP. Like PHP Type Ia, PPHP is accompanied by a Gsα protein deficiency and by AHO, but not by a generalized end organ hormone insensitivity.

17.4.8.6 Kallmann Syndrome.

Kallmann syndrome (KS) is a congenital form of idiopathic hypogonadotropic hypogonadism (IHH) and, by definition, is the form of IHH associated with anosmia (Kallmann et al., 1944). IHH is a heterogenous group of disorders characterized by frank hypogonadism secondary to deficient release of gonadotropin releasing hormone (GnRH) from the hypothalamus and, as has been shown in large-scale studies employing the UPSIT, a spectrum of olfactory deficits ranging from total anosmia (KS) to normosmia (nIHH) (Lewkowitz-Shpuntoff et al., 2012). KS is generally believed to be caused by an autosomal dominant mode of inheritance with incomplete expressivity (Quinton et al., 2001). It is much more prevalent in men than in women, and is sometimes associated with midline craniofacial abnormalities, tooth agenesis, deafness, and renal anomalies, among others. Family members related to KS individuals may be normal or have both hypogonadism and anosmia, anosmia with no hypogonadism, or hypogonadism without anosmia. Although MRI studies generally reveal bilateral agenesis of the olfactory bulbs and tracts (Bajaj et al., 1993; Klingmuller et al., 1987; Yousem et al., 1993), there is one report of a KS patient with aplasia of the left, but not the right, olfactory tract and bulb (Wustenberg et al., 2001).

The origin of IHH and the various degrees of olfactory loss stem from anomalies in the development of the olfactory and GnRH pathways (Schwanzel-Fukuda et al., 1989). During embryogenesis, olfactory and GnRH neurons germinate from the same general stem cell population within the olfactory placode. Following fate specification, the GnRH neurons migrate on structural scaffolding provided by the olfactory axons before entering the CNS via the cribriform plate. While the olfactory axons terminate within the olfactory bulb, the GnRH neurons do so within the mediobasal hypothalamus, where they eventually form a functional network that initiates pulsatile GnRH secretion (Lewkowitz-Shpuntoff et al., 2012) . To date, mutations in 19 different genes have been associated with KS and its variants. Among the most important are the Kallmann syndrome 1 sequence gene (*KAL1*), which encodes a protein important for the cell adhesion and migration in the developing brain (anosmin-1) (Franco et al., 1991), fibroblast growth factor (*FGF*) signaling pathway genes (*FGF8* and *FGFR1*), important for regulating embryonic cell proliferation, differentiation, and migration (Hu et al., 2012), *PROK2* signaling pathway genes (*PROK2* and *PROKR2*), which influence olfactory bulb morphogenesis (Pitteloud et al., 2007), and *CHD7*, a calcium dependent cell-cell adhesion glycoprotein (Kim and Layman, 2011). *CHD7* is closely associated with CHARGE syndrome, a disorder with **C**oloboma, **H**eart defects, **A**tresia choanae (also known as choanal atresia), **R**etarded growth and development, **G**enital abnormalities, and **E**ar abnormalities, as well as olfactory dysfunction (Bergman et al., 2011) and narrowed or completely blocked nasal passages (choanal stenosis or choanal atresia) (Janssen et al., 2012).

The olfactory epithelia of patients with KS have pathological changes analogous to those of mammals whose connections between the olfactory bulb and the epithelium have been severed; namely, the presence of morphologically immature receptor neurons lacking cilia, a decrease in the number of olfactory receptor cells, and the formation of intraepithelial neuromas (Schwob et al., 1993). Nevertheless, based upon calcium imaging, some mature olfactory receptor cells have been identified in nasal biopsies from KS patients (Rawson et al., 1995). In one nasal biopsy study, the total lack of an olfactory epithelium in a KS subject was reported (Jafek et al., 1990a), although this observation must be viewed conservatively in light of sampling problems endemic to such biopsy studies (Paik et al., 1992).

Unfortunately, the anosmia noted in Kallmann syndrome is not reversed by either gonadotropin or gonadal steroid therapy. However, because the GnRH deficiency can be successfully treated when administered at the appropriate time, it is important that attempts be made to detect the endocrine problem as soon as possible. Sparkes et al. (1968) stress that anosmia serves as an important diagnostic marker in the early detection of such problems, and Mroueh and Kase (1968) suggest that perhaps all pediatric patients presenting with anosmia should undergo a specific gonadal evaluation to prevent the possible development of irreversible atrophic changes. Unfortunately, many individuals lacking smell ability (particularly youngsters who do not recognize the deficit or who are too shy to mention it to their parents) do not seek help for this difficulty, and routine pediatric examinations rarely test olfactory function. Most importantly, delay in diagnosis can have significant psychological and physiological consequences, as eloquently described in a detailed description of what a young man went through before receiving a proper diagnosis of KS at the age of 22 years (Smith and Quinton, 2012).

17.4.8.7 Diabetes. Although some studies have found no alterations in olfactory function in patients with diabetes relative to controls (Naka et al., 2010; Patterso et al., 1966), others have reported decreases in such function (Le Floch et al., 1993; Weinstock et al., 1993). In one study of 111 patients, the odor identification deficit was unrelated to disease duration, type of diabetes, presence of neuropathy, retinopathy, nephropathy, hypertension or impotence (Weinstock et al., 1993). Test scores were associated with the presence of macrovascular disease, implying that the olfactory deficit may be enhanced, if not dependent upon, the development of this condition. It is noteworthy that insulin receptors are particularly prevalent throughout olfactory and limbic regions of the brain (Hill et al., 1986).

17.4.9 Psychiatric Disorders

A number of psychiatric disorders are accompanied by altered smell function, including schizophrenia and chronic hallucinatory psychoses (Chapter 18), seasonal affective disorder (Postolache et al., 1998; Postolache et al., 1999; Postolache et al., 2002), and severe stage anorexia nervosa (Aschenbrenner et al., 2008; Fedoroff et al., 1995; Rapps et al., 2010; Roessner et al., 2005). Among the more interesting of the psychiatric disorders is "olfactory reference syndrome," a condition generally reviewed as distinct from schizophrenia or affective disorders (Begum and McKenna, 2011; Bishop, Jr., 1980; Pryse-Phillips, 1971). In this disorder, the patient believes that smells emanate either from his or her person (intrinsic hallucinations) or from elsewhere (extrinsic hallucinations). The hallucinations can be "minimal," where the odor is complained about but the patient does not take steps to remove it, or the "reasonable" reaction, in which the patient takes steps to eliminate the odor (e.g., complaining to authorities, visiting a dermatologist, or plugging up the chimney with newspapers). In intrinsic hallucinatory cases, compulsive washing behavior, changing clothes, and restriction of social activity is common. In many cases, this syndrome is closely linked with an obsessive-compulsive disorder and is amenable to treatment with serotonin-uptake inhibitors (Dominguez and Puig, 1997; O'Sullivan et al., 2000; Stein et al., 1998).

17.4.10 Autoimmune Diseases

It is now apparent that a number of classic autoimmune diseases are associated with smell loss. Those with the best documented losses are as follows.

17.4.10.1 Multiple Sclerosis (see also Chapter 18).

Decreased olfactory function has been reported to be present in 20 to 70% of patients with multiple sclerosis (MS) (Doty et al., 1997b; Fleiner et al., 2010; Pinching, 1977), although some authors have not observed such dysfunction (Ansari, 1976). The frequency of impairment is higher in patients with secondary progressive MS than in those with relapsing-remitting MS (Silva et al., 2012). The dysfunction, which in rare instances can be a presenting symptom (Constantinescu et al., 1994), is correlated with the lesion load and number of demyelinating plaques within the inferior temporal and frontal lobes (Doty et al., 1997b; Zorzon et al., 2000). Interestingly, odor identification test scores wax and wane over time relative to the fluctuations in plaque activity that occur within these brain regions (Doty et al., 1999) and are significantly correlated with symptoms of depression, anxiety, and severity of neurological impairment (Zivadinov et al., 1999). MS-related smell dysfunction is observed using both psychophysical and event-related potential electrophysiological measures (Dahlslett et al., 2012; Hawkes et al., 1997).

17.4.10.2 Systemic Lupus Erythematosus.

Systemic lupus erythematosus (SLE), better known as simply lupus, is an autoimmune disorder that can damage the nervous system, blood vessels, joints, skin, and internal organs. It is more prevalent in women than in men. Like MS, periods of exacerbation (flares) often alternate with periods of remission. In a study of 50 SLE patients and 50 controls, Shoenfeld et al. (2009) found that decreased smell function was present in 46% of patients as compared to 25% of controls. Anosmia was present in 10% of the patients and in none of the controls. Among the auto-antibodies found in this disease are anti-ribosomal P

antibodies. These investigators found, in a mouse study, that injection of such antibodies into the brain preferentially penetrated the limbic system and induced decreased ability to smell menthol (Katzav et al., 2008).

17.4.10.3 Sjögren's Syndrome.

Sjögren's syndrome is an autoimmune exocrinopathy in which the salivary and lacriminal glands are destroyed. Smell loss is prevalent in such patients and is evident in tests of olfactory thresholds and odor identification (Henkin et al., 1972; Kamel et al., 2009; Weiffenbach and Fox, 1993). For example, Weiffenbach and Fox (1993) administered the UPSIT to 30 patients who met strict criteria for Sjögren's syndrome and to 60 matched normal controls. 30% of the patients and 10% of the controls evidenced some olfactory impairment. Although the median score of the patients fell within normal limits (36.5), considerable variation was present and the median was significantly lower (p < 0.02) than that of the controls (38.0). Only one of the 30 patients (3%) with Sjögren's syndrome scored perfectly on the UPSIT, in contrast to 19 (32%) of the controls (p < 0.001). The only anosmic that was identified had Sjögren's.

17.4.10.4 Myasthenia Gravis.

Myasthenia gravis (MG), an autoimmune disease identified with deficient acetylcholine receptor transmission at the post-synaptic neuromuscular junction, is accompanied by a profound loss of olfactory function. In the sole study on this topic, Leon-Sarmiento et al. (2012) administered the UPSIT to 27 MG patients, 27 matched healthy controls, and 11 patients with polymiositis (PM). PM is a disease with peripheral neuromuscular symptoms analogous to MG with no known central nervous system involvement. The UPSIT scores of the MG patients [mean (SD) = 20.15 (6.40)] were at the same level typically seen in early stage Alzheimer's disease and Parkinson's disease (see Chapter 18) and were markedly lower (ps < 0.0001) than those of the PM patients [mean (SD) = 33.30 (1.42)] and age- and sex-matched normal controls [mean (SD) = 35.67 (4.95)}. No correlations between the olfactory test scores and thymectomy, time since diagnosis, type of treatment regimen, or the presence or absence of serum anti-nicotinic or muscarinic antibodies were present.

17.4.11 Cystic Fibrosis

Cystic fibrosis (CF; mucoviscidosis) is an autosomal recessive disease due to a mutation in a protein critical for regulating exocrine gland secretions (e.g., mucus, sweat, saliva, digestive fluids). CF is accompanied by thick and viscous secretions due to abnormalities in sodium and chloride transport across epithelial membranes. This disease affects the lungs, liver, pancreas, intestine, and sweat glands, and is associated with difficult breathing and both upper and lower respiratory system infections. CF receives its name from the cysts and fibrosis (scarring) that were first noticed within the pancreas.

Although one study reported that smell sensitivity was heightened in CF (Henkin and Powell, 1962), all other studies have found evidence for the opposite, i.e., lower sensitivity (Aitken et al., 1997; Hertz et al., 1975; Lindig et al., 2013; Mueller et al., 2007; Sherman et al., 1979; Weiffenbach and McCarthy, 1984). The abnormalities may reflect the influences of nasosinus infections and decreased numbers of receptor cells, the latter of which has been observed in CF mouse models with the CF mutation (Grubb et al., 2007). One reason for the decrease in receptor cell number may be due to sodium hyperabsorption by supporting (sustentacular cells) within the olfactory epithelium.

17.5 TREATMENT OF OLFACTORY DYSFUNCTION

Spontaneous recovery of olfactory function can occur over time in a considerable number of patients without treatment. In one longitudinal study, 542 patients (234 men, 308 women) presenting with complaints of smell dysfunction were tested twice at individual intervals ranging from one month to 24 years [median (IQR) follow-up = 2.9 yrs (1.4–4.2)] (London et al., 2008). The most common etiologies were severe upper respiratory infections (n = 208), head trauma (n = 106), and nasosinus disease (n = 98). Eighty-one had no identifiable cause, and the remainder had problems due to a variety of etiologies. On average, statistically significant improvement in smell function occurred over the two test occasions—improvement that, surprisingly, did not differ among the major etiology categories. The percentages of initially anosmic and hyposmic patients who, in an absolute sense, regained normal function were 7.24% and 22.28%, respectively. Corresponding figures for those who regained normal *age-related* function was 11.76% and 42.49%. Severity of initial olfactory loss, patient age, and the time between the onset of the dysfunction and the patient's initial presentation to the clinic were significant predictors of the amount of change. Sex, time between the two test administrations, and initial smoking behavior were not meaningful predictors. Clearly, spontaneous recovery can be expected in many patients who experience smell loss from a range of causes.

As noted earlier in the chapter, there is little doubt that meaningful treatments are available for some patients whose olfactory dysfunction results from nasal inflammation or blockage of airflow to the olfactory neuroepithelium (e.g., those with allergic rhinitis, polyposis, rhinitis, or chronic sinusitis) (Doty and Mishra, 2001; Vaidyanathan et al., 2011). Effective therapies for olfactory loss

secondary to allergic rhinitis include allergy management, topical cromolyn, topical and systemic corticosteroid therapies, and surgical procedures to reduce inflammation or obstruction, including ones that allow for better mucus drainage from the sinuses to the nasal cavity. The effects of such surgeries are described earlier in the chapter. A brief course of systemic steroid therapy can help in distinguishing between conductive and sensorineural olfactory loss, as patients with the former will often respond positively to some extent to the treatment. However, longer-term systemic steroid therapy is not advised (Scott, 1989).

Once inflammation is quelled with a systemic steroid, maintaining normal olfactory function and the associated decrease in inflammation is challenging. In most cases, nasal sprays do not reach the higher recesses of the nose, although increased efficacy may be obtained by spraying the agents in an upside-down head position, such as the "head-low" or "lateral head low" position (Canciani and Mastella, 1988; Moffett, 1941). In cases where the nasal spray mechanism does not work well upside down, one can first spray the nose and then avert the head for a few minutes to allow the material to enter the olfactory meatus. Some otolaryngologists use steroid drops in place of sprays to increase steroid penetrance through the olfactory meatus, although such treatments must be used with caution to minimize high systemic steroid uptake. Recent research has found that applying topical corticosteroids with a squirt system is more effective than using a nasal spray in maintaining olfactory function, although improvement often subsides unless the therapy is combined with periodic intervals of short-term oral corticosteroids (Shu et al., 2012).

Numerous drugs and vitamin supplements have been claimed to successfully treat anosmia or hyposmia, including alpha-lipoic acid (Hummel et al., 2002), theophylline (Henkin et al., 2009; Levy et al., 1998), caroverine (Quint et al., 2002), zinc sulfate (Henkin et al., 1975), Vitamin A (Duncan and Briggs, 1962), and vitamin B complex (Heilmann et al., 2004). Unfortunately, none of the studies supporting these claims were double blinded and all lacked a control group. In most cases the agent was presented after a control baseline test, confounding test order with the outcome measure. It is well established that olfactory test scores improve upon repeated testing (Doty et al., 1984b) and, as noted above, spontaneous improvement occurs over time in many patients (London et al., 2008). In the case of Vitamin A, the rationale for efficacy was based upon the now disproved concept that pigment is essential for olfactory transduction (Briggs and Duncan 1962; Moulton 1971). Where double-blind studies have been employed, no evidence for efficacy of either zinc sulfate or Vitamin A has been found (Henkin et al., 1976; Lyckholm et al., 2012; Reden et al., 2012).

One large study has purported to show positive and dose-related effects of theophylline on smell test scores

(Henkin et al., 2009). Unfortunately, the dose-related effects were confounded by the time since the onset of the problem, subject discontinuation bias, and the frequency of testing. Moreover, no statistical correction was made for inflated alpha due to the calculation of multiple individual t-tests. When such a correction is made, no significant effects emerge, as is the case with a later study claiming even more positive effects from intranasally administered theophylline (Henkin et al., 2012).

Several studies have suggested that practicing smelling odorants at night and upon awakening improves olfactory function (Hummel et al., 2002; Damm et al., 2013; Haehner et al., 2013). Unfortunately, these studies were also not double blinded and the magnitude of the effects are not large. One recent study found no meaningful alterations of such olfactory training in older subjects (Schriever et al., 2014). Nonetheless, a strong rationale for such a concept comes from both rodent and human studies in which repeated testing to odorants increased sensitivity to them, even in cases where they were initially imperceptible (Dalton et al., 2002; Doty et al., 1981; Doty and Ferguson-Segall, 1989; Wysocki et al., 1989). Rodent studies show that 'mere exposure' to odorants increases neural activity within odor-specific regions of the olfactory bulb (Coopersmith et al., 1986). In light of evidence that experience can influence a wide range of olfactory measures, including brain processes (see Royet et al., 2013, for review), double blind studies addressing this issue are sorely needed.

Two studies have reported that persons with smell loss may benefit from acupuncture. The first is a case report of a healthy individual with no apparent nasal disease who, on the basis of self report and the presentation of a few odorants, was anosmic before treatment (Wevitavidanalage, 2003). After treatment she was reported to have accurately identified the odors of lime, apple, banana, strawberry, and perfumes upon 'crude' testing. The second study assessed olfactory function in 15 patients with post-viral smell loss before and after 10 weekly acupuncture treatments (Vent et al., 2010b). Testing was similarly made of a comparison group of post-viral patients who were being treated over the same time period with vitamin B complex (thiamine, pyridoxine, and cobalamine). The authors concluded that "Eight patients treated with TCA (traditional Chinese acupuncture) had improved olfactory function compared with two treated with vitamin B complex." The data of this study were reanalyzed by Silas and Doty (2010) using a mixed analysis of covariance with the between subject factors of sex and treatment group (TCA, vitamin B complex) and the within subject factor of treatment condition (before and after treatment). Age served as a covariate. No significant effects were observed for any of the variables ($p > 0.15$).

Since Vent et al., had originally matched the two study groups on the basis of sex and age, an ANOVA was performed without these measures in the analysis. A main effect of test occasion was found irrespective of treatment group ($p < 0.01$), but no significant interaction was present. Since a non-treatment group was not tested in this non-blind study, it is impossible to determine whether both acupuncture and the vitamin B complex both improved function in an equivalent manner or whether the putative improvement simply reflected the effects of spontaneous recovery, repeated testing, or both. A definitive double blind study (e.g., where meridian points not expected to alter nasal function are employed in a control group) is clearly needed before one can conclude that acupuncture is useful for improving olfaction. In support of possible efficacy are the results of a double blind study that found acupuncture enhances olfactory function in *normal* subjects, at least for some short period of time after treatment (Anzinger et al., 2009). This phenomenon may reflect the well-known influence of autonomic responses on nasal tissue engorgement (Cole et al., 1985) and on other processes, including general arousal (Chen et al., 2011).

One study has claimed that transcranial magnetic stimulation (TMS) is effective for treating chronic phantosmia and "global oral phantogeusia" (Henkin et al., 2011). Seventeen patients were treated on two occasions separated by 4 to 32 weeks. On each of these occasions, two "sham" TMS procedures were then followed by a "real" TMS procedure. The first sham consisted of 20 stimuli presented at 1- to 5-second intervals sequentially over the right shoulder, then the left shoulder, and then to the back of the neck, at 25% and 40% of maximum output (1.5 T). The second sham was a series of stimulations (10–15% of maximum output) to the front, back, and sides of the skull. The "real" stimulation was directed to the same skull sights in the same order with the stimulator at 40–55% of maximum output. Smell and taste tests were administered at the beginning and end of each of the two 3-stimulation sessions. Brief assessments of the magnitude of the phantosmias "and/or" phantogeusias "and/or" responses to a single odorant presentation were assessed after each sham or real stimulation. If improvement was noted at one site, stimulation was continued until no further changes were noted.

As pointed out by Silas et al. (2011), this pioneering study suffers from multiple methodological problems, including lack of double blinding and analytical rigor. Only anecdotal information was presented on potential effects of the sites that were stimulated, and multiple variables and covariates were assessed without consideration of interaction effects or correlations. Efforts to lessen the likelihood of placebo effects are absent from the study (e.g., randomization, control sites over the cortex, and questioning of the patient as to which condition "felt" most effective) (see Loo et al., 2000). Both sham procedures were perceivably distinct from the "real" procedure, being either at a different location or employing a smaller magnitude stimulus. Importantly, no correction for inflated alpha was applied for 24 t-test multiple comparisons which, when Bonferroni corrected, result in no significant effects. Clearly, as with many of the purported treatments noted in this section, well-controlled double-blind studies are needed to determine whether TMS may be of value in treating some patients with olfactory dysfunction. It is not unreasonable to assume that global magnetic stimulation might induce neurochemical or cortical electrical changes that could influence olfactory function in some patients, although this specific study fails to establish such an influence.

A recent epidemiological study suggests that exercise can reduce the risk of olfactory impairment in later life. Schubert et al. (2013) reported, in a longitudinal study of 1611 patients initially without olfactory impairment, that those who exercised twice a week and achieved a sweat had a decreased risk of developing olfactory impairment 10 years later (age- and sex-adjusted hazard ratio = 0.76 (95% CI, 0.60–0.97). These findings accord with other studies suggesting that regular exercise has significant cardiovascular and cognitive benefits.

Despite the fact that there are few bona fide treatments for the vast majority of cases of smell loss, it is important to reiterate that prognosis is better for patients with less severe microsmia than for those with anosmia or severe microsmia. In some etiologies, this reflects the less extensive damage into the basal cell layer of the epithelium and possibly less fibrosis around the foramina of the cribriform plate through which the olfactory nerve axons must pass. An often overlooked element of therapy for many patients is the quantitative determination of the true degree of olfactory loss. This helps to put into perspective the degree of dysfunction for many patients. For example, even if demonstrable smell loss is present, the test scores of a number of patients will fall above the median (50th percentile) for normal individuals of their same sex and age. It can be very therapeutic for an older person to learn that, while his or her smell function has indeed decreased and is not what it used to be, it still falls above the average of his or her peer group.

REFERENCES

Adams, R. G. and Crabtree, N. (1961) Anosmia in alkaline battery workers. *Brit. J. Indus. Med.*, 18: 216–221.

Ahlstrom, R., Berglund, B., Berglund, U., et al. (1986) Impaired odor perception in tank cleaners. *Scand. J. Work, Environ. Health*, 12: 574–581.

Aitken, M. L., Martinez, S., McDonald, G. J., et al. (1997) Sensation of smell does not determine nutritional status in patients with cystic fibrosis. *Ped. Pulmonol.*, 24: 52–56.

Akerlund, A., Bende, M. and Murphy, C. (1995) Olfactory threshold and nasal mucosal changes in experimentally induced common cold. *Acta Oto-Laryngol.*, 115: 88–92.

Akhtar, M. S., Kousar, F., Fatmi, S., et al. (2012) Quality of life and symptoms control in brain metastasis after palliative whole brain radiotherapy using two different protocols. *J. Coll. Phys. Surg. Pak.*, 22: 311–316.

Althaus, J. (1881) Lecture on the physiology and pathology of the olfactory nerve. *Lancet*, 771–773: 813–815.

Altman, K. W., Desai, S. C., Moline, J., et al. (2011) Odor identification ability and self-reported upper respiratory symptoms in workers at the post-9/11 World Trade Center site. *Int. Arch. Occup. Environ. Health*, 84: 131–137.

Amoore, J. E. (1971) Olfactory genetics and anosmia. In: Beidler, L. M., *Handbook of Sensory Physiology. Vol. IV. Chemical Senses. Part I.* Springer-Verlag: Berlin, 145–156.

Ansari, K. A. (1976) Olfaction in multiple sclerosis. With a note on the discrepancy between optic and olfactory involvement. *Eur. Neurol.*, 14: 138–145.

Antunes, M. B., Bowler, R. and Doty, R. L. (2007) San Francisco/Oakland Bay Bridge Welder Study: olfactory function. *Neurology*, 69: 1278–1284.

Anzinger, A., Albrecht, J., Kopietz, R., et al. (2009) Effects of laserneedle acupuncture on olfactory sensitivity of healthy human subjects: a placebo-controlled, double-blinded, randomized trial. *Rhinology*, 47: 153–159.

Apter, A. J., Gent, J. F. and Frank, M. E. (1999) Fluctuating olfactory sensitivity and distorted odor perception in allergic rhinitis. *Arch. Otolaryngol. Head Neck Surg.*, 125: 1005–1010.

Apter, A. J., Mott, A. E., Cain, W. S., et al. (1992) Olfactory loss and allergic rhinitis. *J. Allergy Clin. Immunol.* Oct; 90: 670–680.

Aschenbrenner, K., Scholze, N., Joraschky, P. and Hummel, T. (2008) Gustatory and olfactory sensitivity in patients with anorexia and bulimia in the course of treatment. *J. Psychiatr. Res.*, 43: 129–137.

Atighechi, S., Zolfaghari, A., Baradaranfar, M., and Dadgarnia, M. (2013) Estimation of sensitivity and specificity of brain magnetic resonance imaging and single photon emission computed tomography in the diagnosis of olfactory dysfunction after head traumas. *Am. J. Rhinol. Allergy.* 27: 403–406.

Babu, R., Barton, A. and Kasoff, S. S. (1995) Resection of olfactory groove meningiomas: technical note revisited. *Surg. Neurol. Dec*; 44: 567–572.

Bajaj, S., Ammini, A. C., Marwaha, R., et al. (1993) Magnetic resonance imaging of the brain in idiopathic hypogonadotropic hypogonadism. *Clin. Radiol.* 48: 122–124.

Bakker, K., Catroppa, C., Anderson, V. (2014) Olfactory dysfunction in pediatric traumatic brain injury: a systematic review. *J Neurotrauma*. 31: 308–14.

Beard, M. D. and Mackay-Sim, A. (1987) Loss of sense of smell in adult, hypothyroid mice. *Brain Res.*, 433: 181–189.

Begum, M. and McKenna, P. J. (2011) Olfactory reference syndrome: a systematic review of the world literature. *Psychol. Med.*, 41: 453–461.

Bergman, J. E., Bocca, G., Hoefsloot, L. H., Meiners, et al. (2011) Anosmia predicts hypogonadotropic hypogonadism in CHARGE syndrome. *J. Pediatr.*, 158: 474–479.

Bishop, E. R., Jr., (1980) An olfactory reference syndrome—monosymptomatic hypochondriasis. *J. Clin. Psychiat.*, 41: 57–59.

Blakeslee, A. F. (1918) Unlike reaction of different individuals to fragrance in verbena flowers. *Science*, 48: 298–299.

Briggs, M. H. and Duncan, R. B. (1962) Pigment and the olfactory mechanism. *Nature*, 195: 1313–1314.

Briner, H. R., Simmen, D. and Jones, N. (2003) Impaired sense of smell in patients with nasal surgery. *Clin. Otolaryngol. Allied Sci.* 28: 417–419.

Briner, H. R., Jones, N. and Simmen, D. (2012) Olfaction after endoscopic sinus surgery: long-term results. *Rhinology*, 50: 178–184.

Brodoehl, S., Klingner, C., Volk, G. F., et al. (2012) Decreased olfactory bulb volume in idiopathic Parkinson's disease detected by 3.0-Tesla magnetic resonance imaging. *Mov. Disord.*, 27: 1019–1025.

Brosvic, G. M., Doty, R. L., Rowe, M. M., et al. (1992) Influences of hypothyroidism on the taste detection performance of rats: a signal detection analysis. *Behav. Neurosci.*, 106: 992–998.

Brosvic, G. M., Risser, J. M. and Doty, R. L. (1989) No influence of adrenalectomy on measures of taste sensitivity in the rat. *Physiol. Behav.*, 46: 699–705.

Brosvic, G. M., Risser, J. M., Mackay-Sim, A. and Doty, R. L. (1996) Odor detection performance in hypothyroid and euthyroid rats. *Physiol. Behav.*, 59: 117–121.

Burdach, K. J. and Doty, R. L. (1987) The effects of mouth movements, swallowing, and spitting on retronasal odor perception. *Physiol. Behav.*, 41: 353–356.

Caccappolo, E., Kipen, H., Kelly-McNeil, K., et al. (2000) Odor perception: multiple chemical sensitivities, chronic fatigue, and asthma. *J. Occup. Environ. Med.*, 42: 629–638.

Calderon-Garciduenas, L., Franco-Lira, M., et al. (2010) Urban air pollution: Influences on olfactory function and pathology in exposed children and young adults. *Exper. Tox. Path.* 62: 91–102.

Calderon-Garciduenas, L., Mora-Tiscareno, A., Ontiveros, E., et al. (2008a) Air pollution, cognitive deficits and brain abnormalities: a pilot study with children and dogs. *Brain Cogn.*, 68: 117–127.

Calderon-Garciduenas, L., Solt, A. C., Henriquez-Roldan, C., et al. (2008b) Long-term air pollution exposure is associated with neuroinflammation, an altered innate immune response, disruption of the blood–brain barrier, ultrafine particulate deposition, and accumulation of amyloid beta-42 and alpha-synuclein in children and young adults. *Toxicol. Pathol.*, 36: 289–310.

Canciani, M. and Mastella, G. (1988) Efficacy of beclomethasone nasal drops, administered in the Moffat's position for nasal polyposis. *Acta Paediatr. Scand.*, 77: 612–613.

Charland-Verville, V., Lassonde, M., and Frasnelli, J. (2012) Olfaction in athletes with concussion. *Am. J. Rhinol. Allergy.* 26: 222–226.

Chen, M. J., Thompson, T., Kropotov, J. and Gruzelier, J. H. (2011) Beneficial effects of electrostimulation contingencies on sustained attention and electrocortical activity. *CNS Neurosci. Ther.*, 17: 311–326.

Cole, P., Haight, J. S., Love, L. and Oprysk, D. (1985) Dynamic components of nasal resistance. *Am. Rev. Respir. Dis.*, 132: 1229–1232.

Collet, S., Grulois, V., Bertrand, B. and Rombaux, P. (2009) Post-traumatic olfactory dysfunction: a cohort study and update. *B-ENT*, 5 Suppl 13, 97–107.

Cömert, A., Uğur, H. C., Kahiloğullar, G. et al. (2011) Microsurgical anatomy for intraoperative preservation of the olfactory bulb and tract. *J. Craniofac. Surg.* 22:1080–1082.

Conn, F. W. and Mast, T. E. (1973) Adrenal insufficiency and electrophysiological measures of auditory sensitivity. *Am. J. Physiol.*, 225: 1430–1436.

Constantinescu, C. S., Raps, E. C., Cohen, J. A., et al. (1994) Olfactory disturbances as the initial or most prominent symptom of multiple sclerosis. *J. Neurol. Neurosurg. Psychiat.*, 57: 1011–1012.

Coopersmith, R., Henderson, S. R. and Leon, M. (1986) Odor specificity of the enhanced neural response following early odor experience in rats. *Brain Res.*, 392: 191–197.

Correia, S., Faust, D. and Doty, R. L. (2001) A re-examination of the rate of vocational dysfunction among patients with anosmia and mild to moderate closed head injury. *Arch. Clin. Neuropsychol.*, 16: 477–488.

Costanzo, R. M., Heywood, P. G., Ward, J. D. and Young, H. F. (1987) Neurosurgical applications of clinical olfactory assessment. *Ann. N.Y. Acad. Sci.*, 510: 242–244.

Cowart, B. J., Flynn-Rodden, K., McGeady, S. J. and Lowry, L. D. (1993) Hyposmia in allergic rhinitis. *J. Allergy Clin. Immunol.*, 91: 747–751.

Croy, I., Hummel, T., Pade, A. and Pade, J. (2010) Quality of life following nasal surgery. *Laryngoscope*, 120: 826–831.

Crysdale, W. S., Cole, P. and Emery, P. (1985) Cephalometric radiographs, nasal airway resistance, and the effect of adenoidectomy. *J. Otolaryngol.*, 14: 92.

Dahlslett, S. B., Goektas, O., Schmidt, F., et al. (2012) Psychophysiological and electrophysiological testing of olfactory and gustatory function in patients with multiple sclerosis. *Eur. Arch. ORL*, 269: 1163–1169.

Dalton, P. (2003) Upper airway irritation, odor perception and health risk due to airborne chemicals.. *Toxicol. Lett.*, 140–141: 239–248.

Dalton, P. (2004) Olfaction and anosmia in rhinosinusitis. *Curr. Allergy Asthma Rep.*, 4: 230–236.

Dalton, P., Doolittle, N. and Breslin, P. A. (2002) Gender-specific induction of enhanced sensitivity to odors. *Nat. Neurosci.*, 5: 199–200.

Dalton, P. H., Opiekun, R. E., Gould, M., et al. (2010) Chemosensory loss: functional consequences of the world trade center disaster. *Environ. Health Perspect.*, 118: 1251–1256 .

Damm, M., Eckel, H. E., Jungehulsing, M. and Hummel, T. (2003) Olfactory changes at threshold and suprathreshold levels following septoplasty with partial inferior turbinectomy. *Ann. Otol. Rhinol. Laryngol.* 112: 91–97.

Damm, M., Pikart, L. K., Reimann, H., et al. (2013) Olfactory training is helpful in postinfectious olfactory loss: A randomized, controlled, multicenter study. *Laryngoscope* 124: 826–831.

Daniels, C., Gottwald, B., Pause, B. M., et al. (2001) Olfactory event-related potentials in patients with brain tumors. *Clin. Neurophysiol.*, 112: 1523–1530.

Deems, D. A., Doty, R. L., Settle, R. G., et al. (1991) Smell and taste disorders, a study of 750 patients from the University of Pennsylvania Smell and Taste Center. *Arch. Otolaryngol. Head Neck Surg.*, 117: 519–528.

Delank, K. W. (1992) Olfactory sensitivity in children with enlarged adenoids. *Laryngo-Rhino-Otol.*, 71: 293–297.

Delank, K. W. and Stoll, W. (1998) Olfactory function after functional endoscopic sinus surgery for chronic sinusitis. *Rhinology*, 36: 15–19.

Dominguez, R. A. and Puig, A. (1997) Olfactory reference syndrome responds to clomipramine but not fluoxetine: a case report. *J Clin. Psychiat.*, 58: 497–498.

Doty, R. L. (2008) The olfactory vector hypothesis of neurodegenerative disease: is it viable? *Ann. Neurol.*, 63: 7–15.

Doty, R. L. and Crastnopol, B. (2010) Correlates of chemosensory malingering. *Laryngoscope*, 120: 707–711.

Doty, R. L., Deems, D. A., Frye, R. E., et al. (1988) Olfactory sensitivity, nasal resistance, and autonomic function in patients with multiple chemical sensitivities. *Arch. Otolaryngol. Head Neck Surg.*, 114: 1422–1427.

Doty, R. L. and Ferguson-Segall, M. (1989) Influence of adult castration on the olfactory sensitivity of the male rat: a signal detection analysis. *Behav. Neurosci.*, 103: 691–694.

Doty, R. L., Fernandez, A. D., Levine, M. A., et al. (1997a) Olfactory dysfunction in type I pseudohypoparathyroidism: dissociation from Gs alpha protein deficiency. *J. Clin. Endocrinol. Metabol.*, 82: 247–250.

Doty, R. L. and Hastings, L. (2001) Neurotoxic exposure and olfactory impairment. *Clin. Occup. Environ. Med.*, 1: 547–575.

Doty, R. L., Li, C., Mannon, L. J. and Yousem, D. M. (1997b) Olfactory dysfunction in multiple sclerosis. *New Engl. J. Med.*, 336: 1918–1919.

Doty, R. L., Li, C., Mannon, L. J. and Yousem, D. M. (1999) Olfactory dysfunction in multiple sclerosis: relation to longitudinal changes in plaque numbers in central olfactory structures. *Neurology*, 53: 880–882.

Doty, R. L. and Mishra, A. (2001) Olfaction and its alteration by nasal obstruction, rhinitis, and rhinosinusitis. *Laryngoscope*, 111: 409–423.

Doty, R. L., Risser, J. M. and Brosvic, G. M. (1991) Influence of adrenalectomy on the odor detection performance of rats. *Physiol. Behav.*, 49: 1273–1277.

Doty, R. L., Shaman, P., Applebaum, S. L., et al. (1984a) Smell identification ability: changes with age. *Science*, 226: 1441–1443.

Doty, R. L., Shaman, P. and Dann, M. (1984b) Development of the University of Pennsylvania Smell Identification Test: a standardized microencapsulated test of olfactory function. *Physiol. Behav.*, 32: 489–502.

Doty, R. L., Snyder, P. J., Huggins, G. R. and Lowry, L. D. (1981) Endocrine, cardiovascular, and psychological correlates of olfactory sensitivity changes during the human menstrual cycle. *J. Comp. Physiol. Psychol.*, 95: 45–60.

Doty, R. L., Yousem, D. M., Pham, L. T., et al. (1997c) Olfactory dysfunction in patients with head trauma. *Arch. Neurol.*, 54: 1131–1140.

Douek, E., Bannister, L. H. and Dodson, H. C. (1975) Recent advances in the pathology of olfaction. *Proc. Roy. Soc. Med.*, 68: 467–470.

Downey, L. L., Jacobs, J. B. and Lebowitz, R. A. (1996) Anosmia and chronic sinus disease. *Otolaryngol. Head Neck Surg.*, 115: 24–28.

Duncan, H. J. and Seiden, A. M. (1995) Long-term follow-up of olfactory loss secondary to head trauma and upper respiratory tract infection. *Arch. Otolaryngol. Head Neck Surg.* 121: 1183–1187.

Duncan, R. B. and Briggs, M. (1962) Treatment of uncomplicated anosmia by vitamin A. *Arch. Otolaryngol.*, 75: 116–124.

Eichel, B. S. (1994) Improvement of olfaction following pansinus surgery. *Ear Nose Throat J.*, 73: 248–250.

el Naggar M., Kale, S., Aldren, C. and Martin, F. (1995) Effect of Beconase nasal spray on olfactory function in post-nasal polypectomy patients: a prospective controlled trial. *J. Laryngol. Otol.*, 109: 941–944.

Elsberg, C. A. (1935a) The sense of smell. XII. The localization of tumors of the frontal lobe of the brain by quantitative olfactory tests. *Bull. Neurol. Inst. N. Y.*, 4: 535–543.

Elsberg, C. A. (1935b) XI. The value of quantitative olfactory tests for the localization of supratentorial tumors of the brain. A preliminary report. *Bull. Neurol. Inst. N. Y.*, 4: 511–522.

Elsherif, H. S., Landis, B. N., Hamad, M. H., et al. (2007) Olfactory function and nasal nitric oxide. *Clin. Otolaryngol*, 32: 356–360.

Ezeh, P. I., Myers, L. J., Hanrahan, L. A., et al. (1992) Effects of steroids on the olfactory function of the dog. *Physiol. Behav.*, 51: 1183–1187.

Federspil, P. A., Wilhelm-Schwenk, R. and Constantinidis, J. (2008) Kinetics of olfactory function following endonasal sinus surgery for nasal polyposis. *Rhinology*, 46: 184–187.

Fedoroff, I. C., Stoner, S. A., Andersen, A. E., et al. (1995) Olfactory dysfunction in anorexia and bulimia nervosa. *Int. J. Eating Disord.*, 18: 71–77.

Feron, F., Perry, C., McGrath, J. J. and Mackay, S. (1998) New techniques for biopsy and culture of human olfactory epithelial neurons. *Arch. Otolaryngol. Head Neck Surg.*, 124: 861–866.

Ferrante, M. A. (1999) Endogenous metabolic disorders. In: Goetz, C.G. & Pappert, E. J., *Textbook of Clinical Neurology*. W.B. Saunders Company: Philadelphia, 731–767.

Ferrier, D. (1876) *The Functions of the Brain*. Smith and Elder: London.

Fielder, C. P. (1985) The effect of adenoidectomy on nasal resistance to airflow. *Acta Otolaryngol.*, 100: 444–449.

Finelli, P. F. and Mair, R. G. (1991) Disturbances of taste and smell. In: Fenichel, G.M. & Marsden, C.D., *Neurology in Clinical Practice*. Butterworth-Heinemann: Boston, 209–216.

Fleiner, F., Dahlslett, S. B., Schmidt, F., et al. (2010) Olfactory and gustatory function in patients with multiple sclerosis. *Am. J. Rhinol. Allergy*, 24: e93–e97.

Franco, B., Guioli, S., Pragliola, A., et al. (1991) A gene deleted in Kallmann's syndrome shares homology with neural cell adhesion and axonal path-finding molecules. *Nature*, 353: 529–536.

Friedman, M., Tanyeri, H., Landsberg, R. and Caldarelli, D. (1999) Effects of middle turbinate medialization on olfaction. *Laryngoscope*, 109: 1442–1445.

Fujii, M., Fukazawa, K., Takayasu, S. and Sakagami, M. (2002) Olfactory dysfunction in patients with head trauma. *Auris Nasus Larynx*, 29: 35–40.

Furstenberg, A. C., Crosby, E. and Farrior, B. (1943) Neurologic lesions which influence the sense of smell. *Arch. Otolaryngol.*, 48: 529–530.

Ghorbanian, S. N., Paradise, J. L. and Doty, R. L. (1983) Odor perception in children in relation to nasal obstruction. *Pediatrics*, 72: 510–516.

Golding-Wood, D. G., Holmstrom, M., Darby, Y., et al. (1996) The treatment of hyposmia with intranasal steroids. *J. Laryngol. Otol.* 110: 132–135.

Gowers, W. R. (1893) *A manual of diseases of the nervous system*. Churchill: London.

Gross-Isseroff, R., Ophir, D., Marshak, G., et al. (1989) Olfactory function following late repair of choanal atresia. *Laryngoscope*, 99: 1165–1166.

Grubb, B. R., Rogers, T. D., Kulaga, H. M., et al. (2007) Olfactory epithelia exhibit progressive functional and morphological defects in CF mice. *Am. J. Physiol. Cell Physiol*, 293: C574–C583.

Guarneros, M., Hummel, T., Martinez-Gomez, M. and Hudson, R. (2009) Mexico City air pollution adversely affects olfactory function and intranasal trigeminal sensitivity. *Chem. Senses*, 34: 819–826.

Gupta, D., Gulati, A., Singh, I. and Tekur, U. (2014) Endoscopic, radiological, and symptom correlation of olfactory dysfunction in pre- and postsurgical patients of chronic rhinosinusitis. *Chem. Senses*, 39: 705–710.

Guss, J., Doghramji, L., Reger, C. and Chiu, A. G. (2009) Olfactory dysfunction in allergic rhinitis. *ORL J. Otorhinolaryngol. Relat. Spec.*, 71: 268–272.

Haehner, A., Tosch, C., and Wolz, M., et al. (2013) Olfactory training in patients with Parkinson's disease. *PLoS ONE*, 8(4): e61680. doi:10.1371/journal.pone.0061680.

Hawkes, C. H., Shephard, B. C. and Kobal, G. (1997) Assessment of olfaction in multiple sclerosis: evidence of dysfunction by olfactory evoked response and identification tests. *J. Neurol. Neurosurg. Psychiat.*, 63: 145–151.

Haxel, B. R., Grant, L. and Mackay-Sim, A. (2008) Olfactory dysfunction after head injury. *J. Head Trauma Rehabil.*, 23: 407–413.

Heilmann, S., Just, T., Goktas, O., et al. (2004) [Effects of systemic or topical administration of corticosteroids and vitamin B in patients with olfactory loss]. [German]. *Laryngol. Rhinol. Otol.*, 83: 729–734.

Henkin, R. I. (1967) Abnormalities of taste and olfaction in patients with chromatin negative gonadal dysgenesis. *J. Clin. Endocrinol. Metab.*, 27: 1436–1440.

Henkin, R. I. (1968) Impairment of olfaction and of the tastes of sour and bitter in pseudohypoparathyroidism. *J. Clin. Endocrinol. Metab.*, 28: 624–628.

Henkin, R. I. (1975) The role of adrenal corticosteroids in sensory processes. In: Blaschko, H., Smith, A. D. & Sayers, G., *Handbook of Physiology*. American Physiological Society: Washington, DC, 209–230.

Henkin, R. I. and Bartter, F. C. (1966) Studies on olfactory thresholds in normal man and in patients with adrenal cortical insufficiency: the role of adrenal cortical steroids and of serum sodium concentration. *J. Clin. Invest.*, 45: 1631–1639.

Henkin, R. I., Levy, L. M. and Lin, C. S. (2000) Taste and smell phantoms revealed by brain functional MRI (fMRI). *J. Comput. Assist. Tomogr.*, 24: 106–123.

Henkin, R. I., Patten, B. M., Re, P. K. and Bronzert, D. A. (1975) A syndrome of acute zinc loss. Cerebellar dysfunction, mental changes, anorexia, and taste and smell dysfunction. *Arch. Neurol.*, 32: 745–751.

Henkin, R. I., Potolicchio, S. J., Jr., and Levy, L. M. (2011) Improvement in smell and taste dysfunction after repetitive transcranial magnetic stimulation. *Am. J. Otolaryngol.*, 32: 38–46.

Henkin, R. I. and Powell, G. F. (1962) Increased sensitivity of taste and smell in cystic fibrosis. *Science*, 138: 1107–1108.

Henkin, R. I., Schecter, P. J., Friedewald, W. T., et al. (1976) A double blind study of the effects of zinc sulfate on taste and smell dysfunction. *Am. J. Med. Sci.*, 272: 285–299.

Henkin, R. I., Schultz, M. and Minnick-Poppe, L. (2012) Intranasal theophylline treatment of hyposmia and hypogeusia: a pilot study. *Arch. Otolaryngol. Head Neck Surg.*, 138: 1064–1070.

Henkin, R. I., Talal, N., Larson, A. L. and Mattern, C. F. (1972) Abnormalities of taste and smell in Sjogren's syndrome. *Ann. Intern. Med.*, 76: 375–383.

Henkin, R. I., Velicu, I. and Schmidt, L. (2009) An open-label controlled trial of theophylline for treatment of patients with hyposmia. *Am. J. Med. Sci.*, 337: 396–406.

Hertz, J., Cain, W. S., Bartoshuk, L. M. and Dolan, T. F. Jr., (1975) Olfactory and taste sensitivity in children with cystic fibrosis. *Physiol. Behav.*, 14: 89–94.

Hilberg, O. (1995) Effect of terfenadine and budesonide on nasal symptoms, olfction, and nasal airway patency following allergen challenge. *Allergy*, 50: 683–688.

Hill, J. M., Lesniak, M. A., Pert, C. B. and Roth, J. (1986) Autoradiographic localization of insulin-receptors in rat brain - p ominence in olfactory and limbic areas. *Neuroscience*, 17: 1127–1138.

Hinriksdottir, I., Murphy, C. and Bende, M. (1997) Olfactory threshold after nasal allergen challenge. *J. Otorhinolaryngol. Relat. Spec.*, 59: 36–38.

Hisamitsu, M., Okamoto, Y., Chazono, H., et al. (2011) The influence of environmental exposure to formaldehyde in nasal mucosa of medical students during cadaver dissection. *Allergol. Int.*, 60: 373–379.

Ho, W. K., Kwong, D. L., Wei, W. I. and Sham, J. S. (2002) Change in olfaction after radiotherapy for nasopharyngeal cancer–a prospective study. *Am. J. Otolaryngol.*, 23: 209–214.

Hoffman, H. J., Ishii, E. K. and Macturk, R. H. (1998) Age-related changes in the prevalence of smell/taste problems among the United States adult population. Results of the 1994 disability supplement to the National Health Interview Survey (NHIS). *Ann. N. Y. Acad. Sci.*, 855: 716–722.

Holbrook, E. H., Leopold, D. A. and Schwob, J. E. (2005) Abnormalities of axon growth in human olfactory mucosa. *Laryngoscope*, 115: 2144–2154.

Holt, G. R. (1996) Effects of air pollution on the upper aerodigestive tract. *Otolaryngol. Head Neck Surg.*, 114: 201–204.

Hoseman, W., Goertzen, W., Wohlleben, R., et al. (2000) Olfaction after endoscopic endonasal ethmoidectomy. *Am. J. Rhinol.*, 7: 11–15.

Hu, B., Han, D., Zhang, L., et al. (2010) Olfactory event-related potential in patients with rhinosinusitis-induced olfactory dysfunction. *Am. J. Rhinol. Allergy*, 24: 330–335.

Hu, Y., Poopalasundaram, S., Graham, A. and Bouloux, P. M. (2012) GnRH neuronal migration and olfactory bulb neurite outgrowth are dependent on FGF Receptor 1 signaling, specifically via the PI3K p110alpha isoform in chick embryo. *Endocrinology*. 14: 23–27.

Hudson, R., Arriola, A., Martinez-Gomez, M. and Distel, H. (2006) Effect of air pollution on olfactory function in residents of Mexico City. *Chem. Senses*, 31: 79–85.

Hummel, T., Barz, S., Pauli, E. and Kobal, G. (1998a) Chemosensory event-related potentials change with age. *EEG Clin. Neurophysiol*, 108: 208–217.

Hummel, T., Heilmann, S. and Huttenbriuk, K. B. (2002) Lipoic acid in the treatment of smell dysfunction following viral infection of the upper respiratory tract. *Laryngoscope.*, 112: 2076–2080.

Hummel, T., Rothbauer, C., Barz, S., et al. (1998b) Olfactory function in acute rhinitis. *Ann. N. Y. Acad. Sci.*, 855: 616–624.

Hummel, T., Witt, M., Reichmann, H., et al. (2010) Immunohistochemical, volumetric, and functional neuroimaging studies in patients with idiopathic Parkinson's disease. *J. Neurol. Sci.*, 289: 119–122.

Ikeda, K., Sakurada, T., Sasaki, Y., et al. (1988) Clinical investigation of olfactory and auditory function in type I pseudohypoparathyroidism: participation of adenylate cyclase system. *J. Laryngol. Otol.*, 102: 1111–1114.

Ikeda, K., Sakurada, T., Takasaka, T., et al. (1995) Anosmia following head trauma: preliminary study of steroid treatment. *Tohoku J. Experi. Med.*, 177: 343–351.

Ishimaru, T., Miwa, T., Nomura, M., et al. (1999) Reversible hyposmia caused by intracranial tumour. *J. Laryngol. Otol.*, 113: 750–753.

Jackson, J. H. (1864) Illustrations of diseases of the nervous system. *London Hosp. Rep.*, 1: 470–471.

Jacob, E. G. (1882) Report on a case of anosmia. *Lancet*, 779.

Jafek, B. W., Eller, P. M., Esses, B. A. and Moran, D. T. (1989) Post-traumatic anosmia. Ultrastructural correlates. *Arch. Neurol.*, 46: 300–304.

Jafek, B. W., Gordon, A. S., Moran, D. T. and Eller, P. M. (1990a) Congenital anosmia. *Ear Nose Throat J*, 69: 331–337.

Jafek, B. W., Hartman, D., Eller, P. M., et al. (1990b) Postviral olfactory dysfunction. *Am. J. Rhinol.*, 4: 91–100.

Jafek, B. W., Moran, D. T., Eller, P. M., et al. (1987) Steroid-dependent anosmia. *Arch. Otolaryngol. Head Neck Surg.*, 113: 547–549.

Jain, B., Rubinstein, I., Robbins, R. A., et al. (1993) Modulation of airway epithelial cell ciliary beat frequency by nitric oxide. *Biochem. Biophys. Res. Commun.*, 191: 83–88.

Janssen, N., Bergman, J. E., Swertz, M. A., et al. (2012) Mutation update on the CHD7 gene involved in CHARGE syndrome. *Hum. Mutat.*, 33: 1149–1160.

Jiang, R. S., Lu, F. J., Liang, K. L., et al. (2008) Olfactory function in patients with chronic rhinosinusitis before and after functional endoscopic sinus surgery. *Am. J. Rhinol.*, 22: 445–448.

Johansson, L., Bramerson, A., Holmberg, K., et al. (2004) Clinical relevance of nasal polyps in individuals recruited from a general population-based study. *Acta Otolaryngol.*, 124: 77–81.

Jung, T. M., TerKonda, R. P., Haines, S. J., et al. (1997) Outcome analysis of the transglabellar/subcranial approach for lesions of the anterior cranial fossa: a comparison with the classic craniotomy approach. *Otolaryngol. Head Neck Surg.*, 116: 642–646.

Kallmann, F. J., Schoenfeld, W. A. and Barrera, S. E. (1944) The genetic aspects of primary eunuchoidism. *Am. J. Ment. Defic.*, 48: 203–236.

Kamel, U. F., Maddison, P. and Whitaker, R. (2009) Impact of primary Sjogren's syndrome on smell and taste: effect on quality of life. *Rheumatology (Oxford).*, 48: 1512–1514.

Karnekull, S. C., Jonsson, F. U., Larsson, M. and Olofsson, J. K. (2011) Affected by smells? Environmental chemical responsivity predicts odor perception. *Chem. Senses*, 36: 641–648.

Katotomichelakis, M., Gouveris, H., Tripsianis, G., et al. (2010) Biometric predictive models for the evaluation of olfactory recovery after endoscopic sinus surgery in patients with nasal polyposis. *Am. J. Rhinol. Allergy*, 24: 276–280.

Katzav, A., Ben-Ziv, T., Chapman, J., et al. (2008) Anti-P ribosomal antibodies induce defect in smell capability in a model of CNS -SLE (depression). *J. Autoimmun.*, 31: 393–398.

Kern, R. C. (2000) Chronic sinusitis and anosmia: pathologic changes in the olfactory mucosa. *Laryngoscope*, 110: 1071–1077.

Kern, R. C., Quinn, B., Rosseau, G. and Farbman, A. I. (2000) Post-traumatic olfactory dysfunction. *Laryngoscope*, 110: 2106–2109.

Kim, D. W., Kim, J. Y. and Jeon, S. Y. (2011) The status of the olfactory cleft may predict postoperative olfactory function in chronic rhinosinusitis with nasal polyposis. *Am. J. Rhinol. Allergy*, 25: e90–e94.

Kim, H. G. and Layman, L. C. (2011) The role of CHD7 and the newly identified WDR11 gene in patients with idiopathic hypogonadotropic hypogonadism and Kallmann syndrome. *Mol. Cell Endocrinol.*, 346: 74–83.

Kim, J. Y., Lee, W. Y., Chung, E. J. and Dhong, H. J. (2007) Analysis of olfactory function and the depth of olfactory sulcus in patients with Parkinson's disease. *Mov Disord.*, 22: 1563–1566.

Kim, S. W., Kim, D. W., Yim, Y. J., et al. (2014) Cortical magnetic resonance imaging findings in patients with posttraumatic olfactory dysfunction: comparison according to the interval between trauma and evaluation. *Clin. Exp. Otorhinolaryngol.* 7: 188–192.

Kinnunen, I. and Aitasalo, K. (2006) A review of 59 consecutive patients with lesions of the anterior cranial base operated on using the subcranial approach. *J. Craniomaxillofacial Surg.* 34: 405–411.

Kimmelman, C. P. (1994) The risk to olfaction from nasal surgery. *Laryngoscope*, 104: 981–988.

Klimek, L. and Eggers, G. (1997) Olfactory dysfunction in allergic rhinitis is related to nasal eosinophilic inflammation. *J. Allergy Clin. Immunol.*, 100: 158–164.

Klimek, L., Moll, B., Amedee, R. G. and Mann, W. J. (1997) Olfactory function after microscopic endonasal surgery in patients with nasal polyps. *Am. J. Rhinol.*, 11: 251–255.

Klingmuller, D., Dewes, W., Krahe, T., (1987) Magnetic resonance imaging of the brain in patients with anosmia and hypothalamic hypogonadism (Kallmann's syndrome). *J. Clin. Endocrinol. Metab.*, 65: 581–584.

Klossek, J. M., Peloquin, L., Friedman, W. H., et al. (1997) Diffuse nasal polyposis: postoperative long-term results after endoscopic sinus surgery and frontal irrigation. *Otolaryngol. Head Neck Surg.*, 117: 355–361.

Koenigkam-Santos, M., Santos, A. C., Versiani, B. R., et al. (2011) Quantitative magnetic resonance imaging evaluation of the olfactory system in Kallmann syndrome: correlation with a clinical smell test. *Neuroendocrinol.*, 94: 209–217.

Kondo, H., Matsuda, T., Hashiba, M. and Baba, S. (1998) A study of the relationship between the T and T olfactometer and the University of Pennsylvania Smell Identification Test in a Japanese population. *Am. J. Rhinol.*, 12: 353–358.

Konstantinidis, I., Haehner, A., Frasnelli, J., et al. (2006) Post-infectious olfactory dysfunction exhibits a seasonal pattern. *Rhinology*, 44: 135–139.

Konstantinidis, I., Triaridis, S., Printza, A., et al. (2007) Olfactory dysfunction in nasal polyposis: correlation with computed tomography findings. *ORL J. Otorhinolaryngol. Relat. Spec.*, 69: 226–232.

Konstantinidis, I., Triaridis, S., Triaridis, A., et al. (2005) How do children with adenoid hypertrophy smell and taste? Clinical assessment of olfactory function pre- and post-adenoidectomy. *Int. J. Pediat. Otorhinolaryngol.*, 69: 1343–1349.

Kosten, T. and Contreras, R. J. (1985) Adrenalectomy reduces peripheral neural responses to gustatory stimuli in the rat. *Behav. Neurosci.*, 1985 Aug;99: 734–741.

Lane, A. P., Zweiman, B., Lanza, D. C., et al. (1996) Acoustic rhinometry in the study of the acute nasal allergic response. *Ann. Otol. Rhinol. Laryngol.*, 105: 811–818.

Le Floch, J. P., Le Lievre, G., Labroue, M., et al. (1993) Smell dysfunction and related factors in diabetic patients. *Diabetes Care*, 16: 934–937.

Lee, R. J., Xiong, G., Kofonow, J. M., et al. (2012) T2R38 taste receptor polymorphisms underlie susceptibility to upper respiratory infection. *J. Clin. Invest.*, 122: 4145–4159.

Lee, S. H., Lim, H. H., Lee, H. M., et al. (2000) Olfactory mucosal findings in patients with persistent anosmia after endoscopic sinus surgery. *Ann. Otol. Rhinol. Laryngol.*, 109: 720–725.

Legg, J. W. (1873) A case of anosmia following a blow. *Lancet*, 2: 659–660.

Leigh, A. D. (1943) Defects of smell after head injury. *Lancet*, 1: 38–40.

Leon-Sarmiento, F. E., Bayona, E. A., Bayona-Prieto, J., et al. (2012) Profound olfactory dysfunction in myasthenia gravis. *PLoS One*, 7: e45544.

Leonard, G., Cain, W. S. and Clavet, G. (1988) Surgical correction of olfactory disorders. *Chem. Senses*, 13: 708.

Levin, H. S., High, W. M. and Elsenberg, H. M. (1985) Impairment of olfactory recognition after closed head injury. *Brain*, 108: 579–583.

Levy, L. M. and Henkin, R. I. (2000) Physiologically initiated and inhibited phantosmia: cyclic unirhinal, episodic, recurrent phantosmia revealed by brain fMRI. *J. Computer Assisted Tomog*, 24: 501–520.

Levy, L. M., Henkin, R. I., Lin, C. S., et al. (1998) Increased brain activation in response to odors in patients with hyposmia after theophylline treatment demonstrated by fMRI. *J. Comput. Assist. Tomo.*, 22: 760–770.

Lewitt, M. S., Laing, D. G., Panhuber, H., et al. (1989) Sensory perception and hypothyroidism. *Chem. Senses*, 14: 537–546.

Lewkowitz-Shpuntoff, H. M., Hughes, V. A., Plummer, L., et al. (2012) Olfactory phenotypic spectrum in idiopathic hypogonadotropic hypogonadism: pathophysiological and genetic implications. *J. Clin. Endocrinol. Metab*, 97: E136–E144.

Lindig, J., Steger, C., Beiersdorf, N., et al. (2013) Smell in cystic fibrosis. *Eu. Arch. Oto-Rhino-Laryngol.*, 270: 915–921.

Litvack, J. R., Fong, K., Mace, J., et al. (2008) Predictors of olfactory dysfunction in patients with chronic rhinosinusitis. *Laryngoscope*, 118: 2225–2230.

Litvack, J. R., Mace, J. and Smith, T. L. (2009) Does olfactory function improve after endoscopic sinus surgery? *Otolaryngol. Head Neck Surg.*, 140: 312–319.

London, B., Nabet, B., Fisher, A. R., et al. (2008) Predictors of prognosis in patients with olfactory disturbance. *Ann. Neurol.*, 63: 159–166.

Loo, A. T., Youngentob, S. L., Kent, P. F. and Schwob, J. E. (1996) The aging olfactory epithelium: neurogenesis, response to damage, and odorant-induced activity. *Int. J. Dev. Neurosci.*, 14: 881–900.

Loo, C. K., Taylor, J. L., Gandevia, S. C., et al. (2000) Transcranial magnetic stimulation (TMS) in controlled treatment studies: are some "sham" forms active? *Biol. Psychiat.*, 47: 325–331.

Lucchini, R., Apostoli, P., Perrone, C., et al. (1999) Long-term exposure to "low levels" of manganese oxides and neurofunctional changes in ferroalloy workers. *Neurotoxicology*, 20: 287–297.

Lund, V. J. and Scadding, G. K. (1994) Objective assessment of endoscopic sinus surgery in the management of chronic rhinosinusitis: an update [see comments]. *J. Laryngol. Otol.*, 108: 749–753.

Lyckholm, L., Heddinger, S. P., Parker, G., et al. (2012) A randomized, placebo controlled trial of oral zinc for chemotherapy-related taste and smell disorders. *J. Pain Palliat. Care Pharmacother.*, 26: 111–114.

Lygonis, C. S. (1969) Familial absence of olfaction. *Heredity*, 61: 413–416.

Mann, N. M. and Vento, J. A. (2006) A study comparing SPECT and MRI in patients with anosmia after traumatic brain injury. *Clin. Nucl. Med.*, 31: 458–462.

Martzke, J. S., Swan, C. S. and Varney, N. R. (1991) Posttraumatic anosmia and orbital frontal damage: neuropsychological and neuropsychiatric correlates. *Neuropsychology*, 5: 213–225.

Mattes, R. D., Heller, A. D. and Rivlin, R. S. (1986) Abnormalities in suprathreshold taste function in early hypothyroidism in humans. In: Meiselman, H. L. and Rivlin, R. S., *Clinical Measurement of Taste and Smell*. Macmillan Publishing Company: New York, 467–486.

McConnell, R. J., Menendez, C. E., Smith, F. R., et al. (1975) Defects of taste and smell in patients with hypothyroidism. *Am. J. Med.*, 59: 354–364.

McCormack, L. J. and Harris, H. E. (1955) Neurogenic tumors of the nasal fossa. *JAMA*, 157: 318–321.

Min, Y.-G., Yun, K.-S., Song, B. H., et al. (1995) Recovery of nasal physiology after functional endoscopic sinus surgery: Olfaction and mucociliary transport. *ORL*, 57: 264–268.

Minovi, A., Hummel, T., Ural, A., et al. (2008) Predictors of the outcome of nasal surgery in terms of olfactory function. *Eur. Arch. Otorhinolaryngol.*, 265: 57–61.

Moffett, A. J. (1941) Postural instillation: a method of inducing local anesthesia in the nose. *J. Laryngol. Otol.*, 56: 429–436.

Moll, B., Klimek, L., Eggers, G. and Mann, W. (1998) Comparison of olfactory function in patients with seasonal and perennial allergic rhinitis. *Allergy*, 53: 297–301.

Moller, R., Pause, B. M. and Ferstl, R. (1999) [Inducibility of olfactory sensitivity by odor exposure of persons with specific anosmia]. *Z. Exp. Psychol.*, 1999; 46: 53–59.

Mollière, D. (1871) Note pour servir à l'histoire de la pathologie du nerf olfactif. *Lyon méd.*, 8: 385.

Mossa-Basha, M. and Blitz, A. M. (2013) Imaging of the Paranasal Sinuses. *Sem. Roentgenol.* 48: 14–34.

Mott, A. E., Cain, W. S., Lafreniere, D., et al. (1997) Topical corticosteroid treatment of anosmia associated with nasal and sinus disease. *Arch. Otolaryngol. Head Neck Surg.*, 123: 367–372.

Moulton, D. G. (1971) The olfactory pigment. In: Beidler, L.M., *Handbook of Sensory Physiology*. Springer-Verlag: New York, 59–74.

Mroueh, A. and Kase, N. (1968) Olfactory-genital dysplasia. *Am. J. Obstet. Gynecol.*, 1968 Feb 15; 100: 525–527.

Mueller, C. A., Quint, C., Gulesserian, T., et al. (2007) Olfactory function in children with cystic fibrosis. *Acta Paediatrica*, 96: 148–149.

Muliol, J., Maurer, M. and Bousquet, J. (2008) Sleep and allergic rhinitis. *J. Investig. Allergol. Clin. Immunol.*, 18: 415–419.

Murphy, C., Schubert, C. R., Cruickshanks, K. J., et al. (2002) Prevalence of olfactory impairment in older adults. *JAMA*, 288: 2307–2312.

Naka, A., Riedl, M., Luger, A., et al. (2010) Clinical significance of smell and taste disorders in patients with diabetes mellitus. *Eur. Arch. Otorhinolaryngol.*, 267: 547–550.

Nguyen, A. D., Pelavin, P. E., Shenton, M. E., et al. (2011) Olfactory sulcal depth and olfactory bulb volume in patients with schizophrenia: an MRI study. *Brain Imaging Behav.*, 5: 252–261.

Nordin, S., Lotsch, J., Kobal, G. and Murphy, C. (1998) Effects of nasal-airway volume and body temperature on intranasal chemosensitivity. *Physiol. Behav.*, 63: 463–466.

Nordin, S., Martinkauppi, M., Olofsson, J., et al. (2005) Chemosensory perception and event-related potentials in self-reported chemical hypersensitivity. *Int. J. Psychophysiol.*, 55: 243–255.

Notta, A. (1870) Recherches sur la perte de l'odorat. *Arch. Gen. Med.*, 15: 385–407.

O'Sullivan, R. L., Mansueto, C. S., Lerner, E. A. and Miguel, E. C. (2000) Characterization of trichotillomania. A phenomenological model with clinical relevance to obsessive-compulsive spectrum disorders. [Review] [97 refs]. *Psychiat. Clin. N. Amer.*, 23: 587–604.

Ogle, W. (1870) Anosmia of cases illustrating the physiology and pathology of the sense of smell. *Med. Chir. Trans.*, 53: 263–290.

Olsson, P., Ehnhage, A., Nordin, S. and Stjarne, P. (2010) Quality of life is improved by endoscopic surgery and fluticasone in nasal polyposis with asthma. *Rhinology*, 48: 325–330.

Olsson, P. and Stjarne, P. (2010) Endoscopic sinus surgery improves olfaction in nasal polyposis, a multi-center study. *Rhinology*, 48: 150–155.

Pade, J. and Hummel, T. (2008) Olfactory function following nasal surgery. *Laryngoscope*, 118: 1260–1264.

Paik, S. I., Lehman, M. N., Seiden, A. M., et al. (1992) Human olfactory biopsy. The influence of age and receptor distribution. *Arch. Otolaryngol. Head Neck Surg.*, 118: 731–738.

Paparo, B. S., Leri, O., Andreoli, P., et al. (1996) Allergic rhinitis, olfactory disorders and secretory IgA. *Riv. Eur. Sci. Med. Farmacol.*, 18: 157–161.

Patterso, D. S., Turner, P. and Smart, J. V. (1966) Smell threshold in diabetes mellitus. *Nature*, 209: 625–634.

Pinching, A. J. (1977) Clinical testing of olfaction reassessed. *Brain*, 100: 377–388.

Pitteloud, N., Zhang, C., Pignatelli, D., et al. (2007) Loss-of-function mutation in the prokineticin 2 gene causes Kallmann syndrome and normosmic idiopathic hypogonadotropic hypogonadism. *Proc. Natl. Acad. Sci. U.S.A*, 104: 17447–17452.

Pfaar, O., Hüttenbrink, K. B. and Hummel T. (2004) Assessment of olfactory function after septoplasty: A longitudinal study. *Rhinology* 42: 195–199.

Pittman, J. A. and Beschi, R. J. (1967) Taste thresholds in hyper- and hypothyroidism. *J. Clin. Endocrinol. Metabol.*, 27: 895–896.

Postolache, T. T., Doty, R. L., Wehr, T. A., et al. (1998) Unirhinal olfactory identification in seasonal affective disorder. *Biol. Psychiat.*, 43: 338.

Postolache, T. T., Doty, R. L., Wehr, T. A., et al. (1999) Monorhinal odor identification and depression scores in patients with seasonal affective disorder. *J. Affecti. Disord.*, 56: 27–35.

Postolache, T. T., Wehr, T. A., Doty, R. L., et al. (2002) Patients with seasonal affective disorder have lower odor detection thresholds than control subjects. *Arch. Gen. Psychiat.*, 59: 1119–1122.

Pryse-Phillips, W. (1971) An olfactory reference syndrome. *Acta Psychiat. Scand.*, 47: 484–509.

Quint, C., Temmel, A. F., Hummel, T. and Ehrenberger, K. (2002) The quinoxaline derivative caroverine in the treatment of sensorineural smell disorders: a proof-of-concept study. *Acta Otolaryngol.*, 122: 877–881.

Quinton, R., Duke, V. M., Robertson, A., et al. (2001) Idiopathic gonadotrophin deficiency: genetic questions addressed through phenotypic characterization. [see comments]. *Clin. Endocrinol.*, 55: 163–174.

Rapps, N., Giel, K. E., Sohngen, E., et al. (2010) Olfactory deficits in patients with anorexia nervosa. *Eur. Eat. Disord. Rev.*, 18: 385–389.

Rawson, N. E., Brand, J. G., Cowart, B. J., et al. (1995) Functionally mature olfactory neurons from two anosmic patients with Kallmann syndrome. *Brain Res.*, 681: 58–64.

Reden, J., Lill, K., Zahnert, T., et al. (2012) Olfactory function in patients with postinfectious and posttraumatic smell disorders before and after treatment with vitamin A: a double-blind, placebo-controlled, randomized clinical trial. *Laryngoscope*, 122: 1906–1909.

Reden, J., Mueller, A., Mueller, C., et al. (2006) Recovery of olfactory function following closed head injury or infections of the upper respiratory tract. *Arch. Otolaryngol. Head Neck Surg.*, 132: 265–269.

Reiter, E. R., DiNardo, L. J. and Costanzo, R. M. (2004) Effects of head injury on olfaction and taste. *Otolaryngol. Clin. North Am.*, 37: 1167–1184.

Roessner, V., Bleich, S., Banaschewski, T. and Rothenberger, A. (2005) Olfactory deficits in anorexia nervosa. *Eur. Arch. Psych. Clin. Neurosci.*, 255: 6–9.

Rombaux, P., Huart, C., De Volder, A. G., et al. (2010) Increased olfactory bulb volume and olfactory function in early blind subjects. *Neuroreport*, 21: 1069–1073.

Rombaux, P., Huart, C., Deggouj, N., et al. (2012) Prognostic value of olfactory bulb volume measurement for recovery in postinfectious and posttraumatic olfactory loss. *Otolaryngol. Head Neck Surg.*, 147: 1136–1141.

Rombaux, P., Mouraux, A., Bertrand, B., et al. (2006a) Retronasal and orthonasal olfactory function in relation to olfactory bulb volume in patients with posttraumatic loss of smell. *Laryngoscope*, 116: 901–905.

Rombaux, P., Mouraux, A., Bertrand, B., et al. (2006b) Olfactory function and olfactory bulb volume in patients with postinfectious olfactory loss. *Laryngoscope*, 116: 436–439.

Ros, C., Alobid, I., Centellas, S., et al. (2012) Loss of smell but not taste in adult women with Turner's syndrome and other congenital hypogonadisms. *Maturitas.* 12: 43–50.

Rotch, T. M. (1878) A case of traumatic anosmia and ageusia, with partial loss of hearing and sight. *Bost. Med. Surg. J.*, 99: 130–132.

Royet, J-P., Plailly, J., Saive, A-L., et al. (2013) The impact of expertise in olfaction. *Front. Psychol.* 4:928 article 928. doi:10.3389/fpsyc.2013.00928.

Rudmik, L. and Smith, T. L. (2012) Olfactory improvement after endoscopic sinus surgery. *Curr. Opin. Otolaryngol. Head Neck Surg.*, 20: 29–32.

Rydzewski, B., Pruszewicz, A. and Sulkowski, W. J. (2000) Assessment of smell and taste in patients with allergic rhinitis. *Acta Otolaryngolog.*, 120: 323–326.

Santos, D. V., Reiter, E. R., DiNardo, L. J. and Costanzo, R. M. (2004) Hazardous events associated with impaired olfactory function. *Arch. Otolaryngol. Head Neck Surg.*, 130: 317–319.

Schechter, P. J. and Henkin, R. I. (1974) Abnormalities of taste and smell after head trauma. *J. Neurol. Neurosurg. Psychiat.*, 37: 802–810.

Schmidt, F. A., Goktas, O., Harms, L., et al. (2011) Structural correlates of taste and smell loss in encephalitis disseminata. *PLoS.One.*, 6, e19702.

Schriever, V. A., Lehmann, S., Prange, J., and Hummel, T. Preventing olfactory deterioration: Olfactory training may be of help in older people. *J. Am. Geriatr. Soc.*, 62: 384–386.

Schubert, C. R., Cruickshanks, K. J., Nondahl, D. M. et al. (2013) Association of exercise with lower long-term risk of olfactory impairment in older adults. *JAMA Otolaryngol. Head Neck Surg.* 139(10): 1061–1066.

Schwanzel-Fukuda, M., Bick, D. and Pfaff, D. W. (1989) Luteinizing hormone-releasing hormone (LHRH)-expressing cells do not migrate normally in an inherited hypogonadal (Kallmann) syndrome. *Brain Res.*, 12: 311–326.

Schwartz, B. S., Doty, R. L., Monroe, C., et al. (1989) Olfactory function in chemical workers exposed to acrylate and methacrylate vapors. *Am. J. Pub. Health*, 79: 613–618.

Schwob, J. E., Szumowski, K. E., Leopold, D. A. and Emko, P. (1993) Histopathology of olfactory mucosa in Kallmann's syndrome. *Ann. Otol. Rhinol. Laryngol.*, 102: 117–122.

Scott, A. E. (1989) Caution urged in treating "steroid dependent anosmia.". *Arch. Otolaryngol. Head Neck Surg.*, 115: 109–110.

Scott, A. E., Cain, W. S. and Clavet, G. (1988) Topical corticosteroids can alleviate olfactory dysfunction. *Chem. Senses*, 13: 735.

Seiden, A. M. and Smith, D. V. (1998) Endoscopic intranasal surgery as an approach to restoring olfactory function. *Chem. Senses*, 13: 736.

Sepehrnia, A. and Knopp, U. (1999) Preservation of the olfactory tract in bifrontal craniotomy for various lesions of the anterior cranial fossa. *Neurosurgry*, 44: 113–117.

Shah, A. S., Ben-Shahar, Y., Moninger, T. O., et al. (2009) Motile cilia of human airway epithelia are chemosensory. *Science*, 325: 1131–1134.

Sherman, A. H., Amoore, J. E. and Weigel, V. (1979) The pyridine scale for clinical measurement of olfactory threshold: a quantitative reevaluation. *Otolaryngol. Head Neck Surg.*, 87: 717–733.

Shoenfeld, N., gmon-Levin, N., Flitman-Katzevman, I., et al. (2009) The sense of smell in systemic lupus erythematosus. *Arthritis Rheum.*, 60: 1484–1487.

Shu, C. H., Lee, P. L., Shiao, A. S., et al. (2012) Topical corticosteroids applied with a squirt system are more effective than a nasal spray for steroid-dependent olfactory impairment. *Laryngoscope*, 122: 747–750.

Silas, J. and Doty, R. L. (2010) No evidence for specific benefit of acupuncture over vitamin B complex in treating persons with olfactory dysfunction. *Otorhinolaryngol. Head Neck Surg.*, 6: 143–603.

Silas, J., Atif, M. A. and Doty, R. L. (2011) Transcranial magnetic stimulation: a treatment for smell and taste dysfunction. *Am. J. Otolaryngol.*, 32: 177–180.

Silva, A. M., Santos, E., Moreira, I., et al. (2012) Olfactory dysfunction in multiple sclerosis: association with secondary progression. *Mult. Scler.*, 18: 616–621.

Simola, M. and Malmberg, H. (1998) Sense of smell in allergic and nonallergic rhinitis. *Allergy*, 53: 190–194.

Singh, N., Grewal, M. S. and Austin, J. H. (1970) Familial anosmia. *Arch. Neurol.*, 22: 40–44.

Sivam, A., Jeswani, S., Reder, L., et al. (2010) Olfactory cleft inflammation is present in seasonal allergic rhinitis and is reduced with intranasal steroids. *Am. J. Rhinol. Allergy*, 24: 286–290.

Smith, N. and Quinton, R. (2012) Kallmann syndrome. *BMJ*, 345: e6971.

Soler, Z. M., Sauer, D. A., Mace, J. C. and Smith, T. L. (2010) Ethmoid histopathology does not predict olfactory outcomes after endoscopic sinus surgery. *Am. J. Rhinol. Allergy*, 24: 281–285.

Sparkes, R. S., Simpson, R. W. and Paulsen, C. A. (1968) Familial hypogonadotropic hypogonadism with anosmia. *Arch. Int. Med.*, 1968 121: 534–538.

Stein, D. J., Le Roux, L., Bouwer, C. and Van Heerden, B. (1998) Is olfactory reference syndrome an obsessive-compulsive spectrum disorder?: two cases and a discussion. *J. Neuropsychiat. Clin. Neurosci.*, 10: 96–99.

Stevens, C. N. and Stevens, M. H. (1985) Quantitative effects of nasal surgery on olfaction. *Am. J. Otolaryngol.*, 6: 264–267.

Stroop, W. G. (1995) Viruses and the olfactory system. In: Doty, R.L., *Handbook of Olfaction and Gustation*. Marcel Dekker: New York, 367–393.

Sumner, D. (1964) Post-traumatic anosmia. *Brain*, 87: 107–120.

Tam, S., Duggal, N, and Rotenberg, B.W. (2013) Olfactory outcomes following endoscopic pituitary surgery with or without septal flap reconstruction: a randomized controlled trial. *Int. Forum Allergy Rhinol.* 3:62–65.

Thomann, P. A., Dos, S. V., Seidl, U., et al. (2009) MRI-derived atrophy of the olfactory bulb and tract in mild cognitive impairment and Alzheimer's disease. *J. Alzheimers Dis.*, 17: 213–221.

Trevino, R. J. (1996) Air pollution and its effect on the upper respiratory tract and on allergic rhinosinusitis. *Otolaryngol. Head Neck Surg.*, 114: 239–241.

References

Trotier, D., Bensimon, J.L., Herman, P., et al. (2007) Inflammatory obstruction of the olfactory clefts and olfactory loss in humans: a new syndrome? *Chem. Senses*, 32: 285–292.

Turetsky, B. I., Moberg, P. J., Arnold, S. E., et al. (2003) Low olfactory bulb volume in first-degree relatives of patients with schizophrenia. *Am. J. Psychiat.*, 160: 703–708.

Turetsky, B. I., Moberg, P. J., Yousem, D. M., et al. (2000) Reduced olfactory bulb volume in patients with schizophrenia. *Am. J. Psychiat.*, 157: 828–830.

Vaidyanathan, S., Barnes, M., Williamson, P., et al. (2011) Treatment of chronic rhinosinusitis with nasal polyposis with oral steroids followed by topical steroids: a randomized trial. *Ann. Intern. Med.*, 154: 293–302.

Vent, J., Koenig, J., Hellmich, M., et al. (2010a) Impact of recurrent head trauma on olfactory function in boxers: a matched pairs analysis. *Brain Res.*, 1320: 1–6.

Vent, J., Wang, D. W. and Damm, M. (2010b) Effects of traditional Chinese acupuncture in post-viral olfactory dysfunction. *Otolaryngol. Head Neck Surg.*, 142: 505–509.

Veyseller, B., Aksoy, F., Yildirim, Y. S., et al. (2011) Reduced olfactory bulb volume in total laryngectomy patients: a magnetic resonance imaging study. *Rhinology*, 49: 112–116.

von Bekésy, G. (1968) Problems relating psychological and electrophysiological observations in sensory perception. *Perspect. Biol. Med.*, 11: 179–194.

Weiffenbach, J. M. and Fox, P. C. (1993) Odor identification ability among patients with Sjogren's syndrome. *Arthritis Rheum.*, 36: 1752–1754.

Weiffenbach, J. M. and McCarthy, V. P. (1984) Olfactory deficits in cystic fibrosis: distribution and severity. *Chem. Senses*, 9: 193–199.

Weigel, M. T., Prazma, G. and Pillsbury, H. C. (1989) Auditory acuity in adrenocorticoid insufficiency. *Am. J. Otol.*, 10: 267–271.

Weinstock, R. S., Wright, H. N. and Smith, D. U. (1993) Olfactory dysfunction in diabetes mellitus. *Physiol. Behav.*, 53: 17–21.

Weinstock, R. S., Wright, H. N., Spiegel, A. M., et al. (1986) Olfactory dysfunction in humans with deficient guanine nucleotide-binding protein. *Nature*, 322: 635–636.

Welge-Lüssen, A., Hilgenfeld, A., Meusel, T. and Hummel, T. (2012) Long-term follow-up of posttraumatic olfactory disorders. *Rhinology*. 50: 67–72.

Welge-Luessen, A., Temmel, A., Quint, C., et al. (2001) Olfactory function in patients with olfactory groove meningioma. *J. Neurol. Neurosurg. Psychiat.*, 70: 218–221.

Wevitavidanalage, M. (2003) Anosmia treated with acupuncture. *Acupunct. Med.*, 21: 153–154.

Wilson, R. S., Yu, L. and Bennett, D. A. (2010) Odor identification and mortality in old age. *Chem. Senses*, 57:12–24.

Wise, J. B., Moonis, G. and Mirza, N. (2006) Magnetic resonance imaging findings in the evaluation of traumatic anosmia. *Ann. Otol. Rhinol. Laryngol.*, 115: 124–127.

Wustenberg, E. G., Fleischer, A., Gerbert, B., et al. (2001) Lateralized normosmia in a patient with Kallmann's syndrome. *Laryngo-Rhino-Otologie*, 80: 85–89.

Wysocki, C. J., Dorries, K. M. and Beauchamp, G. K. (1989) Ability to perceive androstenone can be acquired by ostensibly anosmic people. *Proc. Nat. Acad. Sci. U.S.A.*, 86: 7976–7978.

Yamagishi, M. (1988) [Immunohistochemical study of the olfactory epithelium in the process of regeneration]. [Japanese]. *Nippon Jibiinkoka Gakkai Kaiho [Journal of the Oto-Rhino-Laryngological Society of Japan]*, 91: 730–738.

Yamagishi, M., Fujiwara, M. and Nakamura, H. (1994) Olfactory mucosal findings and clinical course in patients with olfactory disorders following upper respiratory viral infection. *Rhinology*, 32: 113–118.

Yee, K. K., Pribitkin, E. A., Cowart, B. J., et al. (2010) Neuropathology of the olfactory mucosa in chronic rhinosinusitis. *Am. J. Rhinol. Allergy*, 24: 110–120.

Yousem, D. M., Geckle, R. and Doty, R. L. (1995) MR of patients with post-traumatic olfactory deficits. *Chem. Senses*, 20: 338.

Yousem, D. M., Geckle, R. J., Bilker, W., et al. (1996) MR evaluation of patients with congenital hyposmia or anosmia. *Am. J. Roentgenol.*, 166: 439–443.

Yousem, D. M., Geckle, R. J., Bilker, W. B. and Doty, R. L. (1998) Olfactory bulb and tract and temporal lobe volumes. Normative data across decades. *Ann. N.Y. Acad. Sci.*, 855: 546–555.

Yousem, D. M., Geckle, R. J., Bilker, W. B., et al. (1999) Posttraumatic smell loss: relationship of psychophysical tests and volumes of the olfactory bulbs and tracts and the temporal lobes. *Acad. Radiol.*, 6: 264–272.

Yousem, D. M., Turner, W. J. D., Cheng, L., et al. (1993) Kallmann Syndrome - Mr Evaluation of Olfactory System. *Am. J. Neuroradiol.*, 14: 839–843.

Zivadinov, R., Zorzon, M., Monti, B. L., et al. (1999) Olfactory loss in multiple sclerosis. *J. Neurol. Sci.*, 168: 127–130.

Zorzon, M., Ukmar, M., Bragadin, L. M., et al. (2000) Olfactory dysfunction and extent of white matter abnormalities in multiple sclerosis: a clinical and MR study. *Mult. Scler.*, 6: 386–390.

Chapter 18

Odor Perception and Neuropathology in Neurodegenerative Diseases and Schizophrenia

RICHARD L. DOTY, CHRISTOPHER H. HAWKES, KIMBERLEY P. GOOD, and JOHN E. DUDA

18.1 INTRODUCTION

It is now well established that the ability to smell is compromised in a number of neurodegenerative and neurodevelopmental diseases, including Alzheimer's disease (AD), Down's syndrome (DS), Huntington's disease (HD), idiopathic Parkinson's disease (PD), multiple sclerosis (MS; see Chapter 17), schizophrenia (SZ), and the Parkinsonism-dementia complex of Guam (PDG). While reports of alterations in the ability to smell in patients with AD and PD appeared in the 1970s (e.g., Waldton, 1974; Ansari and Johnson, 1975), the breadth of smell losses in neurodegenerative and neurodevelopmental diseases only became apparent after the development and widespread use of standardized and commercially available odor identification tests in the early 1980s (see Chapter 11).

This chapter is a review of what is known about the olfactory function of a range of neurodegenerative diseases and schizophrenia. The physiological bases for the losses observed in such disorders are explored, including, when available, their associations with known genetic mutations and other factors. The diseases are presented in alphabetical order to make it easy for the reader to locate a particular entity.

18.2 ALZHEIMER'S DISEASE

The diagnosis of AD is pathologically based, possible only at autopsy. Probable AD, a diagnosis based upon a set of well-defined clinical criteria (e.g., idiopathic slowly developing memory loss), is typically referred to as AD in living persons (McKhann et al., 1984). Early diagnosis is important, not only for long-term care and planning, but for the development and application of therapeutics that may eliminate or delay disease progression. Hence, markers for early AD, such as smell dysfunction, are of significant value to the physician and patient, and are focus of active research throughout the world.

Much is known about the olfactory dysfunction of AD, as a large number of studies have been performed on this topic (Table 18.1). A number of generalizations can be made from these studies. **First**, the loss – which is usually not total – is present in both nasal chambers and in the earliest stages of the disease, including in many cases of so-called mild cognitive impairment (MCI). **Second**, the deficit is robust, present in a large proportion of even early stage patients (85–90%) and detected by a wide range of nominally distinct olfactory tests, including tests of odor identification and detection threshold sensitivity (Table 18.1). Interestingly, suprathreshold

---

Handbook of Olfaction and Gustation, Third Edition. Edited by Richard L. Doty.
© 2015 Richard L. Doty. Published 2015 by John Wiley & Sons, Inc.

Table 18.1 Procedural details of Alzheimer's disease olfactory function studies. P values reflect lower AD-related performances. See footnote for abbreviations

Author and Year	No. AD	No. Control	AD Mean Age (SD or Range)	Control Mean Age (SD or Range)	Olfactory Test	Odorant Types	P Value
Waldton 1974	66 F	50 F	72.2 (66–93)	77.3 (65–93)	Ability to "apprehend"	6 types	NST
Richard and Bizzini. 1981	2 M, 6 F	1 M, 5 F	74.3 (9.5)	70.33 (3.37)	D-threshold	n-propanol	NST
Corwin and Serby, 1985. Same as: Serby et al., 1985a	11	20 elderly	NR	NR	Odor ID	10 pairs	0.0001
Peabody and Tinklenberg, 1985	14 M, 4 F	16 M, 10 F	66 (53–79)	M: 65 (59–77); F: 65 (59–72)	Odor ID	5 types	NST
Knupfer and Spiegel 1986	6 M, 12 F	7 M, 12 F	81.6 (5.8)	72.2 (3.2)	D-threshold D-threshold D-threshold Odor memory, Odor ID, Naming	Eucalyptol Citral Prunolide Multiple Multiple	0.001 0.001 0.001 0.001 0.001
Warner et al., 1986	12 M, 5 F	17 M, 5 F	66.7 (55–82)	67.9 (60–80)	UPSIT	40 types	0.001
Doty et al., 1987	14 M, 11 F	14 M, 11 F	M: 69.0 (8.5) F: 66.4 (9.5)	Matched	UPSIT D-threshold	40 types PEA	0.001 0.001
Koss et al., 1987	10 (sexes NR)	8	NR	NR	UPSIT D-threshold	40 types Pyridine	0.01 NS
Moberg et al., 1987b	42 (sexes NR)	42 (sexes NR)	72.5 (7.8)	71.4 (5.8)	Odor Memory	20 types	0.001
Rezek 1987	18 (60–80)	26 (sexes NR)	70.0 (4.8)	70.4 (3.7)	Noncued ID Cued ID D-threshold D-threshold	5 types 5 types Pentanol Cinnamon oil	0.002 0.01 NS 0.001
Kesslak et al., 1988	8 M, 10 F	8 M, 10 F	64.2 (1.7)	63.4 (1.8)	UPSIT Odor matching	40 types 15 sets	0.0001 0.0001
Koss et al., 1988	10 M	10 M	61.7 (8.7)	66.3 (7.8)	UPSIT D-threshold	40 types Pyridine	0.001 NS
Green et al., 1989	5 M, 7 F	4 M, 8 F	68.4 (6.3)	64.3 (7.1)	Intensity ratings Odor matching	PEA, Eugenol 10 types	Normal 0.05
Murphy et al., 1990	10 M, 11 F	10 M, 11 F	72.8 (5.4)	72.3 (5.4)	D-threshold	n-butanol	0.001
Schiffman et al., 1990	30 (sexes NR)	12 (sexes NR)	69.7 (6.9)	68.9 (4.1)	Detection of Odorants	14 types	NST

Study	N (sex)	Age	Comparison N (sex)	Comparison Age	Test	Stimuli	p
Buchsbaum et al., 1991	2 M, 4 F	72.1 (3.5)	3 M, 3 F	67.8 (5.2)	Odor matching	30 trials	0.001
Doty et al., 1991	10 M, 14 F	68.7 (8.5)	Compared w/ Other Neurol. Diseases	Compared w/ Other Neurol. Diseases	UPSIT	40 types	0.001
Kesslak et al., 1991	5 M, 3 F	72.3 (8.4)	4 M, 3 F	68.5 (7.1)	UPSIT Odor matching	40 types 15 sets	0.0001 0.0001
Serby et al., 1991	55 (sexes NR)	60–79	57 (sexes NR)	60–79	UPSIT	40 types	0.004
Almkvist et al., 1992	3 M, 11 F	72.6 (7.0)	7 M, 9 F	80.4 (2.6)	D-threshold	Pyridine	0.01
Perl et al., 1992	2 M, 18 F	67–91	4 M, 16 F	66–91	Facial Reaction	6 types	0.04 to 0.002
Solomon, 1994	5 M, 5 F	76.1 (4.9)	None	None	PST	3 types	9 of 10 impaired
Morgan et al., 1995	15 M, 3 F	73.5 (9.5)	15 M, 3 F	76.4 (6.0)	UPSIT SDOID D-threshold	40 types 8 types n-butanol	0.001 0.001 0.01
Nordin et al., 1995a	42 M, 38 F	74.0 (6.7)	51 M, 29 F	73.1 (7.0)	D-threshold	n-butanol	0.01
Nordin and Murphy, 1996	4 M, 12 F	76.4 (10.4)	4 M, 12 F	76.9 (7.8)	Odor Memory	10–15	0.05
Lehrner et al., 1997	2 M, 20 F	77.4 (8.8)	4 M, 15 F	67.8 (15.1)	D-threshold Identification D-threshold Memory	n-butanol 20 types n-butanol 20 types	0.05 0.001 0.05 0.05
Moberg et al., 1997a	4 M, 16 F	73.9 (9.4)	6 M, 14 F	72.5 (6.4)	UPSIT	40 types	0.001
Nordin et al., 1997	4 M, 14 F	70.6 (8.7)	7 M, 9 F	80.4 (2.6)	D-threshold	pyridine	0.007
Bacon et al., 1998	8 (sexes NR)	78.5 (8.5)	62 (sexes NR)	73.6 (7.5)	D-threshold	n-butanol	0.03
Hawkes and Shephard, 1998a	8 (sexes NR)	NR	vs norms	vs norms	UPSIT	40 types	0.001
Solomon et al., 1998	8 (sexes NR) 8 M, 12 F	(sexes NR) 74.5 (7.8)	7 M, 13 F (depressed)	69.4 (7.7)	OERP PST	2 types 3 types	0.001
Bacon-Moore et al., 1999	22 M, 18 F	72.8 (7.6)	28 M, 27 F	73.2 (9.7)	D-threshold Odor Fluency	n-butanol 10 types	0.001 0.05
Larsson et al., 1999	11 F	69.7 (8.2)	11 F	73.7 (5.8)	Identification D-threshold	20 types 1-butanol	0.001 NS
Lehrner and Deecke, 2000	20 NR	63–94	38 NR	60–92	WOTB	20 types	0.001

(continued)

Table 18.1 (*Continued*)

Author and Year	No. AD	No. Control	AD Mean Age (SD or Range)	Control Mean Age (SD or Range)	Olfactory Test	Odorant Types	P Value
Niccoli-Waller et al., 1999	15 M, 17 F	15 M, 17 F	76.0 (9.1)	75.9 (7.3)	D-threshold	n-butanol	0.001
McCaffrey et al., 2000	7 M, 13 F	9 M, 11 F	74.2 (7.9)	67.55 (7.29)	PST	3 types	0.001
Broggio et al., 2001	8 M, 12 F	5 M, 11 F	75 (66–80)	71 (60–79)	Discrim and ID of Oral Stimuli	multiple	0.001
Gray et al., 2001	4 M, 9 F	4 M, 9 F	75.4 (71.6–79.2)	75.6 (72.8–79.2)	UPSIT	40 types	0.001
Kareken et al., 2001	4 M, 3 F	2 M, 6 F	73.1 (8.3)	71.8 (6.8)	UPSIT D-threshold	40 types PEA	0.001 NS
McShane et al., 2001	39 M, 53 F	39 M, 55 W	75.6 (8.0)	75.3 (5.9)	Ability to "perceive"	Lavender water	NS
Royet et al., 2001	3 M, 12 F	3 M, 12 F	68.4 (8.4)	71.7 (11.6)	ID: Intensity Rating Hedonic Rating Familiarity Rating Edibility Rating	12 types 12 types 12 types 12 types 12 types	0.005 NS 0.025 0.05 NS
Chan et al., 2002	2 M,10 F	2 M, 10 F	75.9 (5.4)	74.3 (5.80)	SDOIT AST	6 types Isopropyl alcohol	0.01 0.05
Duff et al., 2002	6 M, 14 F	5 M, 15 F VD 8 M, 12 F MDD	73.9 (8.9)	74.4 (6.5) 71.4 (5.4)	PST	3 types	0.001
Lange et al., 2002	17 M, 31 F	1 M, 72 F	NR	NR	UPSIT	40 types	0.001
Morgan and Murphy, 2002	8 M, 4 F	8 M, 4 F	72.8 (7.3)	73.9 (6.4)	Threshold UPSIT SDOIT	n-butanol 40 types 6 types	NS 0.001 0.001
Wang et al., 2002	13 M, 15 F MCI	14 M, 16 F	71.9 (7.8)	73.8 (5.9)	B-SIT	12 types	0.01
Murphy et al., 2003	8 M, 5 F	10 M, 12 F	73.1 (2.2)	72.5 (1.8)	D-threshold SDOIT	n-butanol 6 types	NS 0.0001
Getchell et al., 2003	16 M, 2 F	3 M, 3 F	75.2 (9.2)	78.0 (4.2)	AST	1 type	0.0001
Peters et al., 2003	5 M, 9 F 8 M, MCI	5 M, 3 F	72.2 (5.7)	72.5 (5.0) 73.9 (9.4)	SS	12 types	0.001
Westervelt et al., 2003	3 M, 6 F	vs norms	77.3 (3.8)	vs norms	B-SIT	12 types	23% to 83% correct

Study			Age		Test	Number	p-value
Gilbert et al., 2004	8 M, 4 F AD / 8 M, 4 F (LBV)	8 M, 4 F	74.7 (1.8) / 74.5 (1.9)	75.4 (1.8)	Odor Mem, Threshold		0.001 / LBV<AD <NC 0.05
Suzuki et al., 2004	17 M, 68 F	12 M, 18 F	76.3 (7.2)	74.8 (8.5)	B-SIT and PST	12 types	0.0001
Eibenstein et al., 2005	10 M, 19 F MCI	11 M, 18 F MCI	71.6 (5.3)	68.8 (5.4)	SS	16 types	0.0001
Sparks et al., 2005	9 (sexes NR)	40 (NC) / 21 (MCI)	NR	NR	UPSIT	40 types	0.001
Tabert et al., 2005	37 M, 63 F	66 M, 81 F (MCI) / 29 M, 34 F (NC)	71.7 (9.5)	67.3 (9.9) / 65.7 (9.4)	UPSIT / B-SIT	40 types / 12 types	0.001 / 0.001
Motomura and Tomota, 2006	12 (NR)	30 (sexes NR)	69 (3.8)	70 (4.5)	B-SIT	12 types	0.01
Kjelvik et al., 2007	14 M, 25 FM (all MCI)	15 M, 37 F	75 (54–89)	78 (55–91)	B-SIT	12 types	0.0005
Luzzi et al., 2007	7 M, 7 F	10 M, 10 F	71 (8)	65 (7)	Discrim. / Naming / O-P Matching	16 pairs / 16 types / 16 types	0.0001 / 0.0001 / 0.001
Pentzek et al., 2007	5 M, 15 F	5 M, 15 F MDD / 6 M, 24 F (NC)	76.0 (9.1)	74.5 (5.6) / 77.1 (6.8)	SS	16 types	AD<MDD= NC / 0.001
Wilson et al., 2007a	51 M, 126 F (incident MCI)	84 M, 328 F No cognitive impairment	79.2 (6.8)	81.7 (6.7)	B-SIT	12 types	0.001
Devanand et al., 2008	17 M, 22 F (converters)	49 M, 60 F (non-converters) / 29 M, 34 F (NC)	73.2 (7.1)	64.9 (9.9) / 65.7 (9.3)	UPSIT	40 types	0.001
Djordjevic et al., 2008	14 M, 13 F	17 M, 16 F	77.0 (55–88)	73.7 (63–87)	UPSIT / D-threshold / Discrim	40 types / 1 type	0.001 / 0.001 / 0.001
McLaughlin and Westervelt, 2008	6 M, 8 F	6 M, 8 F (FTD) / 6 M, 8 F NC	68.8 (8.9)	64.9 (10.0) / 65.9 (8.2)	B-SIT	12 types	AD<MDD= NC / 0.001

(continued)

Table 18.1 (Continued)

Author and Year	No. AD	No. Control	AD Mean Age (SD or Range)	Control Mean Age (SD or Range)	Olfactory Test	Odorant Types	P Value
Westervelt et al., 2008	12 M, 32 F / 9 M, 12 F NC	43 M, 45 F MCI	76.3 (5.4)	74.6 (8.2) / 71.3 (5.6)	B-SIT	12 types	AD < MCI, 0.001 / MCI < NC 0.01
Jungwirth et al., 2009	38 M, 52 F	150 M, 238 F	78.3 (0.4)	78.3 (0.5)	PST	3 types	0.002
Lehrner et al., 2009	4 M, 7 F A-MCI-SD / 8 M, 11 F A-MCI-MD / 5 M, 16 F NA-MCI-SD / 4 M, 9 F NA-MCI-MD	19 M, 21 F NC	69.6 (7.5) / 68.4 (9.0) / 64.6 (9.3) / 64.1 (10.2)	66.5 (8.2)	UPSIT	40 types	A-MCI-MD v. NC 0.05
Devanand et al., 2010	12 M, 19 F (converters)	43 M, 53 F (non-converters) / 27 M, 32 F NC	72.8 (7.0)	64.5 (9.7) / 65.7 (9.5)	UPSIT	40 types	0.0001
Forster et al., 2010	12 M, 12 F	14 M, 14 F	71.4 (7.9)	68.2 (3.9)	SS	TDI 16	0.001
Steinbach et al., 2010	13 M, 17 F	MCI 17 M, 12 F / NC 12 M, 17 F	73.3 (7.8)	71.7 (7.7) / 68.2 (3.9)	SS	TDI 16	0.05
Wang et al., 2010	5 M, 7 F	8 M, 5 F	74.5 (7.5)	67.8 (9.8)	UPSIT	40 types	0.001
Bahar-Fuchs et al., 2011	12 M, 13 F	MCI: 16 M, 9 F / NC:10 M, 12 F	73.0 (8.5)	74.4 (9.3) / 71.7 (7.1)	UPSIT	6 types Unirhinal	AD=MCI
Hidalgo et al., 2011	9 M, 6 F	9 M, 6 F	78 (NR)	78 (NR)	n-butanol JND	1 odor	0.02
Jimbo et al. 2011	31 M, 69 F	6 M, 11 F	79.3 (5.8)	75.7 (8.0)	OSIT-J	12 types	<0.001
Makowska et al., 2011	15 M, 15 F	15 M, 15 F	72.3 (6.1)	72.3 (6.3)	PST	3 types	0.001
Schofield et al., 2012	10 M, 4 F	NC 9 M, 20 F / MCI 6 M, 7 F	73.5 (4.6)	74.0 (6.6) / 77.1 (5.6)	UPSIT	20 types unirhinal	NC= MCI >AD, 0.001
Conti et al., 2013	39 M, 28 F (MCI)	20 M, 26 F	74.0 (6.4)	73.7 (7.3)	UPSIT	40 types	< 0.001
Stamps et al., 2013	3 M, 15 F	11 M, 17 F	75.5 (9.7)	69.1 (9.6)	PBT	1 type	<0.0001
Velayvalhan et al., 2013	22 M, 35 F	10 M, 14 F	81.4 (5.4)	77.3 (6.6)	UPSIT	40 types	0.001

Abbreviations – AD, Alzheimer's disease; **A – MCI – MD:** Amnesic MCI multiple domain; **A – MCI – SD:** Amnesic MCI single domain; **AST,** Alcohol Sniff Test; **B-SIT,** Brief Smell Identification Test; **Conv,** Converted from MCI to AD; **D-threshold,** detection threshold; **Discrim,** Discrimination Test; **F,** female; **FTD,** frontotemporal dementia; **ID,** identification; **JND,** just noticeable difference; **LBV,** Lewy body variant; **M,** male; **MCI,** mild cognitive impairment; **MDD,** major depressive disorder; **NA,** not applicable; **NA_MCI – MD:** Non – amnestic MCI multiple domain; **NA-MCI-SD:** Non-amnestic MCI single domain; **NC,** normal controls; **NR,** not reported; **NS,** not significant; **NST,** no statistical test applied; **O-P,** odor-picture; **OSIT-J,** Japanese Odor Smell Identification Test; **PBT,** Peanut butter test of nostril asymmetry; **P-SIT,** Picture Smell Identification Test; **PST,** Pocket Smell Test; **R-threshold,** recognition threshold; **SD,** Standard deviation; **SDOID,** San Diego Odor Identification Test; **SS,** Sniffin' Sticks; **TDI,** addition of values from the threshold, discrimination, and intensity subtests of the SS; **UPSIT,** University of Pennsylvania Smell Identification Test; **VD,** vascular dementia; **WOTB,** Wiener Olfaktorischen Testbatterie.

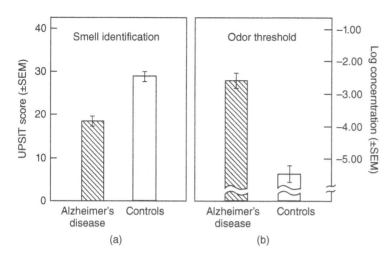

Figure 18.1 An example showing that both odor identification (a) and detection (b) are clearly influenced by Alzheimer's disease. Identification is measured by the University of Pennsylvania Smell Identification Test (UPSIT) and detection by the phenyl ethyl alcohol single-staircase odor detection threshold test. Controls are matched on the basis of age, gender, and ethnicity. Data from Doty et al. (1987). Copyright © 2013 Richard L. Doty.

intensity ratings appear to be little influenced by the deficit (Green et al., 1989; Royet et al., 2001), conceivably reflecting divergence of such ratings from other types of olfactory measurements (Doty et al., 1994). In one meta-analysis of olfactory/AD studies, effect sizes as large as 12.15 (median = 2.17) were noted (effect sizes > 0.80 are considered "enormous") (Mesholam et al., 1998). **Third**, the olfactory dysfunction is accompanied by decreased odor-related activation of central structures (e.g., subfrontal temporal lobe), as measured by functional imaging procedures such as positron emission tomography (PET) (Buchsbaum et al., 1991; Kareken et al., 2001). **Fourth**, many AD patients are unaware of their olfactory deficit until formal testing (Doty et al., 1987; Fusetti et al., 2010; Nordin et al., 1995a). For example, Doty et al. (1987) reported that only 2 of 34 early stage AD patients (6%) responded affirmatively to the question, posed before olfactory testing, "Do you suffer from smell and/or taste problems?", despite the fact that over 90% of the patients exhibited abnormal scores on the 40-item University of Pennsylvania Smell Identification Test (UPSIT). **Fifth**, the olfactory loss is, on average, equivalent to that observed in PD, PDG, Down's syndrome, and even myasthenia gravis, begging the question as to whether the olfactory loss reflects a common pathological substrate of a number of diseases (Doty et al., 1991; Doty, 2012; Leon-Sarmiento et al., 2012; Moberg et al., 1997a). That being said, inverse associations have been noted between odor identification scores and tau-related pathology within the hippocampus and entorhinal cortex (Wilson et al., 2007). **Sixth**, the olfactory dysfunction appears to progress as AD progresses over time (Corwin and Serby, 1985; Jimbo et al., 2011; Knupfer and Spiegel, 1986; Murphy et al., 1990; Nordin et al., 1997; Serby et al., 1985a; see, however, Tissingh et al., 2001), although controls for the confounding of dementia, per se, on non-olfactory elements of psychophysical tests are often lacking and the reliability of findings from significantly demented patients must be questioned. **Seventh**, olfactory testing is useful in the differential diagnosis of AD from disorders commonly misdiagnosed as AD, such as depression (Pentzek et al., 2007). Indeed, even a 3-item odor identification test accurately discriminates patients with AD from patients with depression (McCaffrey et al., 2000; Solomon et al., 1998). **Eighth**, there is evidence that smell loss is present in some inherited forms of AD (Nee and Lippa, 2001) and, moreover, that the loss in idiopathic AD correlates with a family history of dementia (Schiffman et al., 1990), implying genetic linkage. Indeed, persons with the ApoE ε4 allele, a risk factor for AD, are more likely to exhibit olfactory dysfunction than those not having this allele (Bacon et al., 1998; Gilbert and Murphy, 2004; Handley et al., 2006; Hozumi et al., 2003; Salerno-Kennedy et al., 2005; Wang et al., 2002; Wetter and Murphy, 2001).[1]

Olfactory dysfunction – particularly in conjunction with other risk factors – may aid, to some degree, in the prediction of subsequent development of cognitive decline and AD in older persons (Bacon et al., 1998; Fusetti et al., 2010; Schubert et al., 2008; Sohrabi et al., 2012; Wilson et al., 2007b). For example, Devanand et al. (2000) evaluated both the cognitive and olfactory function of 90 outpatients with MCI at 6-month intervals for both cognitive and olfactory function. Scores on the UPSIT were lower in patients with MCI than in healthy controls. Of 77 patients that were followed over a two-year period, those with mild, moderate, or severe smell loss, as well as those with low UPSIT scores who reported no subjective problems smelling, were more likely to develop AD than other individuals. Although low UPSIT scores alone did not forecast the time until development of AD, low UPSIT scores accompanied by lack of awareness of olfactory

---

[1] Apolipoprotein E is a widely distributed cholesterol transport protein that circulates in the plasma after synthesis by the liver, spleen and kidneys. In the CNS, it is synthesized by macrophages, neurons, and glia. In the PNS, it is synthesized by macrophages, nonmyelinated Schwann cells, and ganglionic satellite cells. Common isoforms include ε2, ε3, and ε4, with ε2 seemingly having a neuroprotectant action, and E4 being associated with a higher risk of not only developing some neurodegenerative diseases, such as AD, but in having a more malignant course of degeneration (Bedlack et al., 2000).

problems did so. Even in patients with relatively high Mini-Mental State Exam scores ($\geq$27), low UPSIT scores with lack of deficit awareness remained a significant, albeit imperfect, predictor of AD. UPSIT scores of 30–35 showed moderate to strong sensitivity and specificity for diagnosis of AD at follow-up.

More recently, this group administered the UPSIT to 1092 non-demented older persons (average age 80 years) from a multi-ethnic community in New York (Devanand et al., 2010). UPSIT scores varied as a function of the degree of memory impairment [respective mean (SD) scores for no mild cognitive impairment (MCI), non-amnestic MCI, and amnestic MCI = 26.6 (6.8), 24.4 (7.2), and 23.5 (6.7)]. The scores correlated weakly with hippocampal volume (r = 0.16, p < 0.001), but not with entorhinal cortex volume or total number of white matter hyperintensities. They also correlated negatively with age (r = −0.24, p < 0.001) and positively with several cognitive measures, including Selective Reminding Test immediate recall (r = 0.33, p < 0.001), delayed recall (r = 0.28, p < 0.001), category fluency (r = 0.28, p < 0.001), and the Boston Naming Test (r = 0.23, p < 0.001).

In a similarly large study, Graves et al. (1999) administered the 12-item version of the UPSIT (the Brief Smell Identification Test™ or B-SIT; see Chapter 11) to 1,604 non-demented community-dwelling senior citizens 65 years of age or older. Over a subsequent 2-year time period, the B-SIT scores were found to be a better predictor of cognitive decline than scores on a global cognitive test. Persons who were anosmic and possessed at least one ApoE ε4 allele had 4.9 times the risk of having cognitive decline than normosmic persons not possessing this allele (i.e., an odds ratio of 4.9). This is in contrast to the 1.23 times greater risk for cognitive decline in normosmic individuals possessing at least one such APOE allele – a risk rate that may be greater in older persons (Devanand et al., 2005). When the data were stratified by sex, women who were anosmic and possessed at least one ApoE ε4 allele had an odds ratio of 9.71, compared to an odds ratio of 1.90 for women who were normosmic and possessed at least one allele. The corresponding odds ratios for men were 3.18 and 0.67, respectively.

In light of such findings, it is of interest that some close relatives of AD patients exhibit less smell function than non-relatives of the same age and sex. Serby et al. (1996) administered the UPSIT to 28 first-degree relatives of AD patients and to 28 healthy controls. Despite similar global cognitive scores (MMSE), the relatives exhibited lower UPSIT scores than those of the controls (p < 0.01). Although olfactory testing was suggested by Nee and Lippa (2001) as not being useful in predicting the symptom onset of one relatively rare genetically-determined form of AD (presenilin-1 AD), such data, along with evidence of (a) poorer odor detection or identification performance in individuals having one or more ApoE ε4 alleles (Bacon et al., 1998; Bahar-Fuchs et al., 2010) and (b) delayed latencies to olfactory event-related potentials (OERPs) in individuals with the ApoE ε4 allele (Corby et al., 2012; Wetter and Murphy, 2001), are in agreement with the concept of genetic vulnerability to olfactory dysfunction in this disease. The finding that olfactory event-related potentials from an odor/visual congruency paradigm exhibit high sensitivity and specificity in differentiating non-demented ApoE ε4 carriers from non-carriers also accords well with this concept (Kowalewski and Murphy, 2012). Interestingly, one study found an association between younger age of onset of AD and the presence of high copy number of submicroscopic DNA segments in the olfactory receptor region of 14q11.2 (Shaw et al., 2011). This association was independent of and in addition to that observed for the ApoE ε4 allele.

There is evidence that older persons with olfactory loss also are more likely to report that they experience more memory problems (Sohrabi et al., 2009). Hence, olfactory and cognitive impairment may be interrelated in older persons. Some investigators have, in fact, increased the sensitivity and specificity of predicting conversion of at-risk older patients to AD by combining olfactory test scores with psychological measures, as well as with MRI-determined volumes of limbic brain structures. Devanand et al. (2008) found, for example, that at a specificity point of 90%, the UPSIT alone had a sensitivity of 49% in making this prediction in 148 outpatients with MCI. When combined with age, hippocampal volume, entorhinal cortex volume, and scores from the MMSE, the Pfeiffer Functional Activities Questionnaire (FAQ), and the Selective Reminding Test (SRT) immediate recall (verbal memory), this value increased to 81.3%. More recently, Lojkowska et al. (2011) reported that, in a 24-month follow up study of 49 MCI patients, combining the results of odor identification tests with those of neuropsychological measures markedly increased the sensitivity of prediction. The odor identification test alone had a sensitivity and specificity of 57% and 88%, respectively, in predicting conversion. By combining the olfactory test results with neuropsychological measures, sensitivity and specificity increased to 100% and 84%, respectively.

The specific physiological basis or bases for the olfactory loss of AD is unknown and probably multifactorial. AD-related pathology is seen throughout the olfactory pathways, as reviewed in more detail elsewhere (Hawkes and Doty, 2009; Smutzer et al., 2003). Thus, dystrophic neurites and neurofibrillary tangles have been found in the olfactory epithelia of AD patients, although their density is not marked and these entities are also found in the epithelia of healthy older persons, as well as in that of persons with other neurodegenerative diseases, including progressive supranuclear palsy (PSP), a disorder with relatively

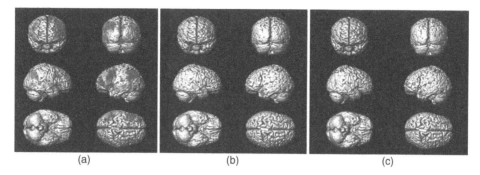

Figure 18.2 Three-dimensional representation of average brain activation maps in response to lavender odor. (a) Healthy control group (one-sample *t*-test, voxel-wise threshold $p < 0.005$, uncorrected with extent threshold = 6) at lowest and first exposed odorant concentration (0.10%). (b) Early AD group at same statistical threshold and at the highest odorant concentration exposure (1.0%). (c) AD group with more relaxed statistical threshold (one-sample *t*-test, voxel-wise threshold $p < 0.01$, uncorrected with extent threshold = 0) and at the highest odorant concentration exposure (1.0%). From Wang et al. (2010). (*See plate section for color version.*)

little olfactory dysfunction (Trojanowski et al., 1991). In AD, neurofibrillary tangles and amyloid plaques are evident throughout the olfactory bulb, the anterior olfactory nucleus, and in higher brain regions, most notably those that receive projections from the olfactory bulb (e.g., the piriform cortex and the corticomedial nucleus of the amygdala). Functional imaging studies clearly demonstrate decreased odor-induced activity in a range of brain regions (Figure 18.2).

Bahar-Fuchs et al., (2010) sought to determine whether olfactory identification test scores were associated, in patients with amnestic MCI, with in vivo measures of β-amyloid burden, as measured by PET imaging using Pittsburgh Compound B (PiB). The olfactory test scores did not differ between the patients who were PiB-positive and those who were PiB-negative, leading the authors to conclude that AD-related olfactory identification deficits are probably not directly related to levels of β-amyloid. This is in agreement with the conclusions of earlier studies of both non-demented and demented older subjects who received olfactory testing and postmortem brain analyses (Pentzek et al., 2007; Wilson et al., 2007). No meaningful associations were evident between olfactory test scores, obtained prior to death, and levels of β-amyloid in olfactory eloquent brain regions. However, an inverse correlation was found between odor identification test scores and tau-related NFT accumulation within the entorhinal cortex and the CA1/subiculum area of the hippocampus.

Although there are proponents of the concept that AD-related pathology begins within the olfactory bulb and spreads to other brain regions, this is a matter of controversy. Pathological studies by Braak and Braak (1991) suggested that the earliest pathology is found in the transentorhinal cortex, a brain region between the entorhinal cortex and the hippocampus. On the other hand, Kovacs et al. (2001) have noted cases of AD in which neurofibrillary tangles (NFTs) were most evident in the olfactory bulb, but not in other brain structures. Moreover, they found cases where the primary olfactory cortex was less severely affected than the so-called secondary olfactory cortex (i.e., the medial orbitofrontal cortex), implying the NFT formation developed independently of synaptic connections. In other words, it need not be a process that advanced from the periphery along established neural pathways.

As noted in Chapter 20, there is evidence that environmental factors, such as severe air pollution, can induce AD-like pathology in human brains, most notably the olfactory bulbs. Moreover, there is evidence that some, even young, residents in metropolitan areas with very high air pollution exhibit smell dysfunction (Calderon-Garciduenas et al., 2010; Guarneros et al., 2009). While it is unknown whether the AD-like pathology is secondary to olfactory dysfunction induced by the pollutants, it is of interest that rats whose olfactory bulbs are removed and transgenic mice lacking intact olfactory CNG channels exhibit increased neurofilament hyperphosphorylation in both the hippocampus and cortex (Hu et al., 2012). Such observations are in general accord with elements of the olfactory vector hypothesis (Doty, 2008).

It is important to note that neurotransmitter and neuromodulator anomalies may be present within the olfactory system of early stage AD patients before meaningful degeneration occurs. Thus, cholinergic dysfunction is present in some early stage AD patients and in patients with mild cognitive impairment (MCI) unaccompanied by significant cell loss (Schliebs and Arendt, 2011). Associated with this dysfunction are imbalances in the expression of nerve growth factor (NGF), its precursor proNGF, and in NGF receptors (trkA and p75NTR, respectively), as well as changes in acetylcholine release, high-affinity choline uptake, and alterations in muscarinic and nicotinic acetylcholine expression. Massive loss of cholinergic centrifugal inputs to all layers of the olfactory bulb have

412   Chapter 18   Odor Perception and Neuropathology in Neurodegenerative Diseases and Schizophrenia

been demonstrated in patients with AD, as well as PD and dementia with Lewy bodies (Mundinano et al., 2013).

18.3   AMYOTROPHIC LATERAL SCLEROSIS

Amyotrophic lateral sclerosis (ALS), popularly known as Lou Gehrig's disease, is the most common form of motor neuron disease. It is typified by degeneration of neurons within the anterior horn of the spinal cord and cortical neurons with which they interact. Although most forms are considered sporadic, the A4V mutation of the superoxide dismutase SOD1 gene along with FUS positive basophilic inclusions have been described in early onset forms of ALS (Aksoy et al., 2003; Baumer et al., 2010; Juneja et al., 1997). Accumulation of lipofuscin in cells has been noted within the anterior olfactory nucleus (AON) and the mitral and tufted cells of the olfactory bulb of ALS patients (Hawkes et al., 1998).

It is now apparent that ALS is more than just a motor neuron disease. In a pioneering study, Elian (1991) found nine male and six female ALS patients of varying disease severity to have UPSIT scores significantly lower than those of age- and sex-matched controls. However, the magnitude of the deficit was modest, with the scores averaging around 30. Subsequently, Sajjadian et al. (1994) administered the UPSIT bilaterally to 17 female and 20 male ALS patients, and unilaterally (i.e., each nostril separately) to 7 male and 7 female ALS patients. Age-, gender-, smoking habit-, and race-matched controls were also evaluated. While the UPSIT scores on the ALS patients were significantly lower than those of the controls, no left:right differences were observed and the degree of dysfunction was similar to that seen by Elian. Only 11% of the ALS patients had UPSIT scores indicative of total or near total anosmia and nearly half had scores falling within the normal range. Nevertheless, three quarters of the patients scored below their individually matched controls. Significant correlations were found between UPSIT scores and neurophysiological measures of peripheral nerve conductance, conceivably reflecting an association with cholinergic dysfunction.

Evidence for a modest loss of smell function in ALS was also found in a study of 58 ALS patients and 135 controls to whom the UPSIT was administered (Hawkes et al., 1998). Olfactory event-related potentials (OERPs) in response to H_2S were recorded in 15 patients. Only the bulbar patients exhibited significant UPSIT decrements, although, overall, the UPSIT scores were slightly worse in the ALS patients. OERPs were normal in nine patients and delayed in one; those of the remaining five subjects were not able to be recorded. Histological analysis of olfactory bulb tissue from eight deceased ALS patients revealed, in all cases, excessive lipofuscin deposition.

18.4   DOWN SYNDROME

Down syndrome (DS), a trisomy 21 disorder, is the most common form of mental disability, accounting for ~17% of the intellectually disabled population. It was suggested as early as 1928 that individuals with DS have difficulty smelling (Brousseau and Brainerd, 1928). More modern studies have found adult DS patients to have olfactory deficits, as measured by tests of odor identification, detection, memory, and EEG responses to odorants (Hemdal et al., 1993; Murphy and Jinich, 1996; Warner et al., 1988; Wetter and Murphy, 1999; Zucco and Negrin, 1994). Such observations are of significance, given that the average smell loss observed in DS is very close to that observed in AD (i.e., UPSIT scores ~20; McKeown et al. 1996; Warner et al., 1988) and DS patients who live into early adulthood inevitably develop the neuropathological features of AD, such as senile plaques and neurofibrillary tangles (Oliver and Holland, 1986). Importantly, deposition of amyloid in the form of senile plaques or diffuse amyloid deposits occurs in cortical brain regions associated with olfactory processing (e.g., entorhinal cortex) (Hof et al., 1995).

That being said, classic AD pathology – neuritic plaques and neurofibrillary tangles – is unlikely to be the basis for the smell loss of DS, since olfactory dysfunction appears in children with DS who would not yet be expected to exhibit such pathology. McKeown et al. (1996) administered the UPSIT and a 16-item odor discrimination test to (a) 20 children with DS [mean age (SD) = 13.89 yrs (1.98)], (b) 20 non-DS intellectually disabled children matched on the basis of mental age (Peabody Picture Vocabulary Test-Revised (PPVT-R; Dunn, 1981), and (c) 20 non-intellectually disabled children also matched on mental age. Although no meaningful differences in olfactory function were found among the three study groups, the test scores of both the DS and non-DS intellectually disabled subjects were markedly lower than non-intellectually disabled children matched on chronologic, rather than mental, age. Moreover, the UPSIT scores of these young people were similar in magnitude to those of adult DS subjects (~20).

The basis for the olfactory loss of Down syndrome is not known, although it conceivably reflects abnormal development of the cholinergic basal forebrain which is known to be involved in both cognition and olfaction (Berger-Sweeney, 1998). There is a growing body of evidence that suggests dysregulation of cholinergic circuits could explain the olfactory dysfunction observed in a number of diseases, including AD, PD, and DS (Doty, 2012; Schliebs and Arendt, 2011). Although the Down syndrome critical gene region 1 (DSCR1) interacts with the Fragile X mental retardation protein (FMRP) to regulate local protein synthesis and dendritic spine morphogenesis in numerous brain regions (Wang et al., 2012), the olfactory dysfunction seen in FMRP and other trinucleotide repeat disorders, such as spinocerebellar ataxias (see Section 18.14), is much less than that observed in DS (Juncos et al., 2012).

18.5 ESSENTIAL TREMOR

Unlike PD, essential tremor (ET) is usually associated with an action tremor, not a resting tremor, and is unaccompanied by stooped posture, slow movement, or a shuffling gait. Unlike the tremors of PD, which predominate in the hands, those of ET are more likely to involve the legs, hands, voice, and head. Nonetheless, persons with ET are more at risk than those in the general population for future development of AD and PD (Laroia and Louis, 2011; Louis, 2009). No studies have uniformly found neuropathology of any sort within the olfactory bulbs or higher olfactory structures of ET patients.

Patients with ET have no meaningful olfactory dysfunction, unlike the case of PD. In 1992, Busenbark reported that 15 ET patients scored normally on the UPSIT (Busenbark et al., 1992). Although a later study claimed that a number of ET patients had mild impairment on the UPSIT (Louis et al., 2002), all subsequent studies have found normal UPSIT scores in ET patients (Shah et al., 2008; Ward, 2003). Shah et al. (2008), for example, compared UPSIT scores of 59 ET patients to those of 64 tremor-dominant PD patients. Nearly complete separation of the two groups was made on the basis of UPSIT scores and to a lesser degree on measures from OERPs. Surprisingly, when ET subjects were separated by family history of tremor in a first degree relative, this group scored significantly *better* than age- and gender-matched controls, a finding that begs replication.

18.6 HUNTINGTON'S DISEASE

Huntington's disease (HD) is an autosomal genetic disorder of dysfunctional movement, cognitive deterioration, and altered behavior that becomes phenotypically expressed progressively relatively late in life. Among its primary motor symptoms are hyperkinesias that initially take the form of chorea, being characterized by fleeting movements which, in some persons, appear semi-purposively within the context of overall heightened activity and motor restlessness (Shoulson, 1986). Like a number of other neurodegenerative diseases, HD is associated with marked deficits in odor identification, discrimination, detection, and odor memory (Bylsma et al., 1997; Hamilton et al., 1999; Lazic et al., 2007; Moberg et al., 1987a; Moberg and Doty, 1997; Nordin et al., 1995b). The P3 component of the olfactory event-related potential (OERP), like that of the auditory event-related potential, is significantly delayed in HD, with an amplitude inversely correlated with the number of CAG repeats and UHDRS movement scores (Wetter et al., 2005). A study employing diffusion tensor MRI found negative correlations between UPSIT scores and mean diffusivity in multiple subcortical and cortical brain regions, including the insula, anterior caudate, and medial temporal cortex (Delmaire et al., 2013), supporting the concept of involvement of central olfactory processing neural circuits.

The question arises as to whether asymptomatic individuals who carry the HD mutation have decreased olfactory function. In one study of 25 probands with HD, 12 at-risk offspring, and 37 unrelated controls, significant UPSIT and PEA threshold deficits were apparent only in the HD group (Moberg and Doty, 1997). Since HD is an autosomal dominant genetic disorder with 100% penetrance, on average half of the at-risk offspring would have been expected to have the mutation and to evidence smell dysfunction if it was a very early marker of HD. A subsequent study administered the UPSIT to 20 HD patients who had the disease for a mean of 8.0 years (range: 4 to 14 years), 20 normal subjects with the genetic mutation that causes HD, and 20 mutation-negative adults (Bylsma et al., 1997). Again, only the patients with clinical signs of HD exhibited depressed olfaction.

While others have similarly seen no odor identification or detection deficits in non-symptomatic HD gene carriers (Larsson et al., 2006; Pirogovsky et al., 2007), subtle deficits in some other olfaction-related measures have been reported. For example, Larsson et al. (2006) found 10 gene carriers to perform more poorly than 10 controls on a match-to-sample odor quality discrimination test in which 12 different combinations of four fruit-like odors were presented (p < 0.03). Pirogovsky et al. (2007) found that HD patients performed more poorly than controls on visual and olfactory measures of "source memory" (i.e., whether a male or female had presented a given stimulus on earlier trials), but not for the memory of the smell or pictured item itself ("item memory").

18.7 IDIOPATHIC RAPID EYE MOVEMENT SLEEP BEHAVIOR DISORDER

Idiopathic rapid eye movement sleep behavior disorder (RBD) is a unique parasomnia associated with degeneration of pontine and medullary brain regions, such as the sublaterodorsal nucleus, and their glutaminergic projections to the medulary or magnocellular reticular formation (Postuma et al., 2012). The majority of patients with this disorder eventually develop some type of neurodegenerative disease, including PD, MSA, or LBD. Importantly, many exhibit olfactory dysfunction, as measured by a variety of olfactory tests (Miyamoto et al., 2010; Postuma

414 **Chapter 18 Odor Perception and Neuropathology in Neurodegenerative Diseases and Schizophrenia**

et al., 2011). In a five-year prospective follow-up study of 62 patients with idiopathic RBD, those with impaired olfaction at baseline had a 65% 5-year risk of developing a defined neurodegenerative disease, compared to a 14% risk in those with normal olfaction (Postuma et al., 2011). As with PD, olfactory dysfunction was present in a number of the patients up to five years before diagnosis.

18.8 LEWY BODY DISEASE

Lewy body disease (LBD), commonly termed "diffuse Lewy body disease," is the second most common dementing illness, being surpassed only by AD. In this disorder, Lewy bodies (LBs) – spherical intracellular cytoplasmic inclusions made up of aberrant α-synuclein – are present throughout the cortex, in contrast to the neuritic plaques and neurofibrillary tangles of AD. Related structures – Lewy neurites – extend for considerable distances within axons. LBD is associated with dementia. If the dementia occurs before, during, or within a year of the motor symptoms, the syndrome is termed *Dementia with Lewy bodies* (Lippa et al., 2007).

Smell dysfunction is common in LBD and appears unrelated to disease duration or severity (Liberini et al., 2000; McShane et al., 2001; Olichney et al., 2005; Williams et al., 2009; Wilson et al., 2011). McShane et al. (2001) reported that anosmia was present in 9 of 22 autopsy confirmed LBD patients (41%), 7 of 43 autopsy confirmed AD patients (16%), and 6 of 94 age-matched controls (6%). Unfortunately, the olfactory test consisted of simply sniffing a bottle of lavender water. Westervelt et al. (2003) administered the B-SIT to 9 LBD patients and 9 AD patients matched on the basis of age, gender, and dementia level. Five of the 9 LBD patients were anosmic (56%), as compared to only one AD patient (11%), and the B-SIT scores were markedly lower in the LBD than in the AD group. More recently, Wilson et al. (2011) identified 26 persons with LBD from a sample of 201 autopsied brains of older persons. None had PD prior to death and all had been tested with the B-SIT. Those without Lewy bodies performed at the level expected for someone of their age [mean (SD) age at time of death = 88.0 (6.54) yrs], whereas those with Lewy bodies exhibited test scores similar to those expected from older patients with PD (Aden et al., 2011). Those with Lewy bodies located within limbic or cortical regions exhibited decreased smell function, whereas those with Lewy bodies confined to the substantia nigra showed no meaningful deficit. Although the olfactory bulbs were not examined for Lewy pathology, such pathology is known to exist, along with neurofibrillary tangles and tau pathology, in the bulbs of patients with LBD and patients with PD (Mundinano et al., 2011; Tsuboi et al., 2003).

It is noteworthy that Ross et al. (2006), in the Honolulu-Asian aging study, found an association between the presence of post-mortem incidental Lewy bodies and pre-mortem B-SIT scores in persons without Parkinsonism or dementia during life. The age-adjusted relative odds of ILB in the lowest v. the highest B-SIT tertile was 11.0 (p = 0.02).

18.9 PARKINSON'S DISEASE AND PARKINSON-LIKE DISORDERS

18.9.1 Sporadic Parkinson's Disease

Until the 1980s, Parkinson's disease (PD) was widely considered to be solely a disease of the extrapyramidal motor system. It is now apparent that PD is also associated with a range of autonomic, cognitive, and sensory changes. Some occur during its so-called preclinical or premotor stage, when motor symptoms are not evident. Among the most salient of these changes are alterations in the ability to smell.

At the time of this writing, nearly 100 peer-reviewed studies have been published in which olfactory test scores of sporadic or idiopathic PD patients were compared to those of controls. In essentially every study, test scores were lower in PD patients than in controls (Table 18.2). From this literature, a number of generalizations can be made. **First**, the smell dysfunction is bilateral and occurs in 90% or more of such cases (Doty et al., 1984; Hawkes and Doty, 2009). **Second**, it is robust and detectable with both ortho- and retro-nasal testing (Landis et al., 2009). In one meta-analysis, Cohen effect sizes > 3 were found for tests of odor identification, discrimination, detection, and memory (Mesholam et al., 1998). One study examining the sensitivity and specificity of the UPSIT in differentiating between clinically-diagnosed PD patients and normal controls found these values to be quite high (0.91 and 0.88, respectively, in males ≤ 60 years of age) (Doty et al., 1995a). **Third**, women with PD generally outperform their male counterparts on olfactory tests (Stern et al., 1994), a sex difference seen not only in the general population (Doty and Cameron, 2009), but in diseases ranging from AD (Doty et al., 1987) to schizophrenia (Good et al., 2007; Seidman et al., 1997). **Fourth**, anosmia is not the norm. In one study, only 38% of 81 PD patients had UPSIT scores suggestive of anosmia and 87% of 38 patients could reliably detect the highest stimulus concentration presented in an odor detection threshold test (Doty et al., 1988). In another study, all but one of 41 PD patients reported that 35 or more of the 40 UPSIT items had some type of odor, even though the perceived sensation did not correspond to any of the response alternatives (Doty et al., 1992b).

Table 18.2 Procedural details of Parkinson's disease olfactory function studies

Author and Year	No. PD	No. Control	PD Mean Age (SD or Range)	Control Mean Age (SD or Range)	Olfactory Test	Odorant Types	P Value
Constantinidis and de Ajuriaguerra 1970	1 M, 1 F	None	47 and 39	None	NR	Unknown	NA
Ansari and Johnson 1975	22 M	37 M	58 (41–67)	53 (43–68)	D-threshold	P-acetate	0.05
Kissel and Andre. 1976	2 twins	None	37	None	None	None	smell problem
Korten and Meulstee, 1980	39 M, 41 F	80 (sexes NR)	61.5 (sexes NR)	"Matched"	Questionnaire	Unknown	0.001
Ward et al., 1983	45 M, 27 F	18 M, 35 F	60 (21–85)	58 (21–85)	D-threshold D-threshold Discrimination	PEMEC p- acetate 4 types	0.01 0.03 0.01
Corwin and Serby, 1985; Serby et al., 1985b	5 (sexes NR)	54 (sexes NR)	NR	NR	Odor ID	10 pairs	0.0001
Quinn et al., 1987b	50 M, 28 F	27 F, 13 F	61.5 (10.3)	61.5 (10.3)	D-threshold	P-acetate	0.001
Doty et al., 1988	46 M, 35 F	NR; Matched	M: 65.7; (10.1)	NR; Matched	UPSIT	40 types	0.0001
			F: 67.2 (9.2)		D-threshold	PEA	0.0001
Kesslak et al., 1988	8 M, 10 F	10 M, 4 F	65.4 (3.2)	63.4 (1.8)	UPSIT Odor matching task	40 types 15 sets	0.0001 0.028
Doty et al., 1989	33 M, 25 F	Relative to Norms	M: 65.8 (9.0) F: 69.2 (8.7)	From Norms	UPSIT	40 types	0.0001
Bostantjopoulou et al., 1991	22 M, 22 F	17 M, 13 F	61.9 (9.4)	62.5 (11.7)	D-threshold Odor Naming	P-acetate 4 types	0.001 0.01
Murofushi et al., 1991	11 M, 7 F	5 M, 5 F	59.6 (41–76)	56.7 (31–81)	D- & R-thresholds	5 types	0.01 and .05
Zucco et al., 1991	5M, 3 F	7M, 9F	61.5 (9.3)	80.0 (10.2)	Odor Memory & ID	10 types	0.02 (ID only)
Doty et al., 1992a	9 M, 4 F	4 M, 6 F	39.1 (4.8)	38.2 (7.3)	UPSIT	40 types	0.005
Doty et al., 1992b	30 M, 10 F	15 M. 5 F	61.9 (10.0)	65.4 (9.7)	UPSIT D-threshold	40 types PEA	0.001 0.001
Doty et al., 1993	12 M, 9 F	12 M, 9 F	M: 68.3 (8.2); F: (66.8 (6.0)	M: 67.4 (6.8); F: 67.0 (7.0)	UPSIT	40 types	0.001

(*continued*)

Table 18.2 (*Continued*)

Author and Year	No. PD	No. Control	PD Mean Age (SD or Range)	Control Mean Age (SD or Range)	Olfactory Test	Odorant Types	P Value
Hawkes and Shephard 1993	49 M, 47 F	39 M, 57 F	57 (27–81)	43 (18–78)	UPSIT	40 types	0.0001
Wenning et al. 1993	15 M, 8 F	15 M, 8 F	60.9 (7.1)	60.3 (7.0)	UPSIT	40 types	0.001
Stern et al., 1994	68 M, 50 F	vs norms	64.3 (9.34)	vs norms	UPSIT	40 types	0.001
Doty et al., 1995a	78 M, 78 F	vs norms	64: ≤ 60 yr 55: 61–71 yr 47: ≥ 71 yr	vs norms	UPSIT	40 types	0.0001
Lehrner et al., 1995	13 (sexes NR)	NR	64.7 (11.4)	NR	D-threshold ID Odor Memory	n-butanol 20 types 20 types	NR NR NR
Wenning et al., 1995	116 M 2 F	82 M 41 F	59.4 (11.6)	46.4 (17.3)	UPSIT	40 types	0.001
Barz et al., 1997	No drugs: 8 M, 10 F; Drugs: 5 M, 8 F	No drugs: 12 M, 11 F; Drugs: 5 M, 10 F	No drugs: 63 (44–80); Drugs: 72 (56–81)	No drugs: 63 (50–85); Drugs:67 (52–82)	Discrim. ID OERP	8 pairs 8 types vanillin, H_2S	NS 0.001 0.05
Hawkes et al., 1997	49 M, 47 F 66 (sex nr)	18 M, 29 F	57 (27–81)	45.6 (17–74)	UPSIT OERP	40 types H_2S	0.0001 0.001
Ahlskog et al., 1998	7 M, 2 F	15 M, 38 F	66.1 (6.4)	62.1 (10.8)	UPSIT – Modified	20 types	0.01
Hawkes and Shephard 1998a	78 M, 77 F 19 M, 18 F	57 M, 99 F	61 (34–84) 62 (27–77)	44 (17–90)	UPSIT OERP	40 types vanillin, H_2S	0.0001 NR
Daum et al., 2000	28 M, 12 F	28 M	63.6 (8.7)	64.1 (8.7)	SS	16 types	0.001
Lehrner and Deecke 2000	NR	59	(45–85)	(45–92)	Viennese olfactory test battery	20 types	0.001
		12 F			D-threshold	n-butanol	0.001
Montgomery, Jr. et al., 2000a	9 M, 9 F	9 M, 10 F	64 (45–79)	64 (43–77)	UPSIT	40 types	0.001
Montgomery et al., 2000b	32 M, 26 F, 1 Unknown	19 M, 21 F	69 (38–83)	72.7 (61–87)	UPSIT	40 types	0.001
Sobel et al., 2001	10 M, 10 F	10 M, 10 F	67.1 (9.9)	68.0 (7.6)	UPSIT D-threshold	40 types	0.0001 0.007
Tissingh et al., 2001	25 M, 16 F	8 M, 10 F	56.4 (8.9)	55.3 (9.1)	B-SIT Discrimination D-threshold	12 types 32 trials PEA	0.001 0.001 0.001
Zucco et al., 2001	5 M, 3 F	7 M, 9 F	61.5 (9.3)	80.0 (10.2)	Odor naming and matching	10 types	0.01 0.13

Muller et al., 2002	37 (sexes NR)	NA	NA	vs norms	SS	Comb thresh and ID	37/37 deficient; compared to MSA p = 0.001
Double et al., 2003	26 M, 23 F	22 M, 30 F	68 (8)	71 (10)	B-SIT	12 types	0.001
Hudry et al., 2003	12 M, 12 F	12 M, 12 F	M: 63.2 (8.3); F: 67.5 (7.3)	M: 63.9 (6.4) F: 66.9 (8.7)	ID: Intensity R Hedonics R Familiarity R Edibility R	12 types	All<0.0001 except for Edibility R
Katzenschlager et al., 2004	8 M, 5 F	27 (sexes NR)	70.6 (64–85)	72.6 (63–85)	UPSIT	40 types	0.0001
Khan et al., 2004	11 M, 7 F 10 M, 1 F	23 M, 5 F	68.1 (5.1) 55.1 (11.7)	50.5 (13.0)	UPSIT UPSIT	40 types 40 types	0.0001 0.0001
Hummel et al., 2005	6 M, 5 F (On, Off, DBS)	NA	57.3 (42–64)	NA	SS	D-threshold Discrim. (on vs. off)	NS 0.03
Ondo and Lai 2005	13 M, 7 F TD FH+ 13 M, 2 F TD FH− PD 21 M, 4 F	NA	68.7 (7.8) 73.2 (4.4) 69.0 (8.2)	NA	UPSIT	40 types	TD FH− > TD FH+ 0.007; PD < TD FH+ 0.035
Marras et al., 2005	26 M	26 M (unaffected twin)	72.6 (2.5)	72.7 (2.5)	UPSIT	40 types	0.001
Siderowf et al., 2005	17 M, 8 F	vs norms	58.7 (12.6)	vs norms	UPSIT	40 types	6 anosmic, 10 microsmic, 8 normal
Ross et al., 2006b	17 M with incidental Lewy bodies but not PD	147 M	83.5 (6.1)	81.5 (5.1)	B-SIT	12 types	0.011
Deeb et al., 2006	34 M, 16 F PD 14 M, 9 F Non-PD	311 (sex NR)	63 (33–80) 63 (40–78)	NR	UPSIT	40 types	PD < NC 0.001 PD < NonPD 0.01 NonPD < NC 0.01
Lee et al., 2006	13 M, 13 F PD 12 M, 8 F MSA	5 M, 10 F	64 (10) 54 (6)	59 (11)	B-SIT	12 types	PD < MSA, NC 0.001
Bohnen et al., 2007	20 M, 7 F	20 M, 7 F	60.0 (11.1)	60.9 (12.3)	UPSIT	40 types	0.0001

(continued)

Table 18.2 (*Continued*)

Author and Year	No. PD	No. Control	PD Mean Age (SD or Range)	Control Mean Age (SD or Range)	Olfactory Test	Odorant Types	P Value
Ferreira et al., 2007	5 M, 6 F	vs norms	66.1 (9.6)	NA	UPSIT	40 types	9 of 11 impaired
Kim et al., 2007	37 M, 22 F	10 M, 15 F	63.4 (12.1)	59.9 (11.6)	B-SIT	12 types	0.001
Lee et al., 2007	12 M, 12 F PD 3 M, 12 F DIP	7 M, 8 F	64.4 (10.7) 68.7 (6.9)	63.7 (10.1)	B-SIT	12 types	PD < DIP (0.001) DIP = NC (ns)
Quagliato et al., 2007	25 M, 25 F	26 M, 30 F	64.8 (8.8)	61.0 (10.3)	B-SIT	12 types	0.047 to 0.011
Boesveldt et al., 2008a	253 M, 151 F	87 M, 63 F	61.5 (40–90)	59.2 (45–78)	SS	16 types	0.001
Goldstein et al., 2008	57 M, 21 F	MSA 36 M, 21 F	65 (1)	61 (1)	UPSIT	40 types	0.0002
Guo et al., 2008	9 M, 6 F	8 M, 7 F	61.1 (7.8)	62.9 (5.4)	D-, 1-thresholds	5 types	0.01
Herting et al., 2008	22 M, 5 F	NA	49 (27–64)	NA	SS	12 odorants	All deficient
Iijima et al., 2008	33 M, 21 F	30 M, 20 F	69.7 (8.1)	69.3 (7.3)	OSIT-J	12 types	0.0001
Lotsch et al., 2008	81 M, 21 F	916 M, 1160 F	60.8 (10.1)	35.2 (16.2)	SS D-threshold Discrim.	TDI n-butanol 16 types	0.001 0.001
Louis et al., 2008	NA	MPS 26 M 115 F Non MPS 275 M, 626 F	NA	82.8 (5.7) 80.1 (5.6)	UPSIT	40 types	MPS < no MPS 0.001
Ross et al., 2008	NA	1st quartile 549 (sex NR) 2nd quartile 515 (sex NR) 3rd quartile 622 (sex NR) 4th quartile 581 (sex NR)	NA	81.2 (4.5) 80.2 (4.2) 79.2 (3.7) 78.4 (3.3)	B-SIT	12 types	1st quartile higher incidence of PD 0.001 Adjusted OR 5.2 (95%CI: 1.5-25.6) during the first 4 years
Shah et al., 2008	44 M, 20 F	84 M, 161 F	67.2 (39–82)	49.4 (17–90)	UPSIT	40 types	0.0001
Verbaan et al., 2008	192 M, 103 F	150 (sexes NR)	60.2 (10.6)	NR	SS	16 types	0.001
Wilson et al., 2008	188 M, 554 F	NA	80.9 (7.0)	NA	B-SIT	12 types	Score correlated with change in parkinsonism 0.002
Boesveldt et al., 2009a	29 M, 23 F	27 M, 23 F	61.8 (7.0)	59.5 (7.6)	SS D-threshold Discrimin.	16 types n-butanol 16 & 32 types	0.001 0.013 0.001

Study	Patients	Controls	Patient age	Control age	Test	Types	Significance
Boesveldt et al., 2009b	12 M, 8 F	9 M, 12 F	61.5 (50–73)	56.3 (49–73)	SS D-threshold	TDI n-butanol	0.001 0.001
Chou and Bohnen 2009	31 M, 13 F	31 M, 13 F	59.3 (10.1)	59.4 (10.8)	UPSIT PD-3 types AD-10 types	40 types 3 types 10 types	0.0001 0.0001 0.0001
Haehner et al., 2009	263 M, 137 F	vs norms	64.3 (33–85)	NA	SS	TDI	97% microsmic; 74% w/age adjustment
Landis et al., 2009	23 M, 22 F	vs norms	60.7 (12.0)	NA	SS	TDI	6% normosmia 65% hyposmia 29% anosmia
Miyamoto et al., 2009	12 M, 9 F	28 M, 6 F	63.9 (9.4)	62.2 (6.7)	OSIT-J	12 types	0.001
Postuma et al., 2009	PD-RBD: 13 M, 8 F	RBD: 53 M, 15 F	69.8 (49–83)	68.0 (44–93)	B-SIT	12 types	RBD < NC 0.001
	PD+RBD: 28 M, 6 F	NC 28 M, 8 F	68.8 (49–94)	65.8 (46–87)	UPSIT	40 types	PD-RBD < NC (0.001) PD+RBD < NC (0.001) PD+RBD<RB (0.001) PD-RBD<RB 0.05
Shah et al., 2009	75 (sex NR)	74 (sex NR)	NR	NR	UPSIT	40 types	0.0001
Silveira-Moriyama et al., 2009a	113 M, 78 F	NC 77 M, 68 F PAF: 6 M, 10 F MSA: 8 M, 6 F	65.4 (11.1)	63.8 (9.5) 66.1 (8.6) 61.3 (8.6)	UPSIT	40 types	PAF < NC 0.001 PAF < MSA 0.001 PD < PAF 0.005 PD < MSA 0.006 MSA = PAF
Silveira-Moriyama et al., 2009b	81 M, 64 F 8 M, 11 F PARK8	71 M, 84 F	68.4(6.2) 68.6(9.0)	62.3 (12.3)	UPSIT,	40 types	NC >PARK8 0.004 PARK8=PD (NS)
Silveira-Moriyama et al., 2009c	60 M, 39 F	52 M, 48 F	60.4(8.8)	61.3 (6.9)	SS	16 types	0.001
Wattendorf et al., 2009	11 M, 4 F early PD 6 M, 6 F Moderate PD	9 M, 8 F	60.3 (44–71) 61.7 (46–69)	63.1 (52–72)	SS	12 types	0.00001

(continued)

Table 18.2 (*Continued*)

Author and Year	No. PD	No. Control	PD Mean Age (SD or Range)	Control Mean Age (SD or Range)	Olfactory Test	Odorant Types	P Value
Bohnen et al., 2010	49 M, 9 F	17 M, 9 F	69.0 (7.6)	67.2 (10.5)	UPSIT	40 types	0.0001
Bovi et al., 2010	PD 4 M, 9 F DIP 4 M, 12 F	5 M, 14 F	67.8 (7.7) 66 (NR)	66.7 (NR)	SS	TDI	0.001
Cramer et al., 2010	50 M, 20 F	NA	70.1(11.5)	NA	B-SIT	12 types	0.03 apathetic PD <non-apathetic PD
Deeb et al., 2010	34 M, 16 F PD	14 M, 9 F	63 (33–80)	63 (40–78)	UPSIT	40 types	0.001
Hummel et al. 2010a	8 (sex NR)	12 (sex NR)	60 (10.9)	58 (10.9)	SS	12 types	NT
Kertelge et al., 2010	61 M, 39 F	62 M, 42 F	63.7 (10.3)	58.7 (10.9)	UPSIT	40 types	0.004
McKinnon et al., 2010	12 M, 7 F	59 M, 148 F	71.7 (9.0)	77.0 (8.3)	UPSIT	40 types	0.001
Meusel et al., 2010	14 M, 5 F	None	65 (29–76)	None	SS	12 types Threshold Discrim.	0.008 NS 0.06
Oka et al., 2010	32 M, 34 F	11 M, 15 F	70.5 (9.3)	70.3 (9.8)	OSIT-J	12 types	0.001
Ramjit et al., 2010	58 (sex NR)	51(sex NR)	69.3 (6.9)	66.5 (9.2)	UPSIT	40 types	0.001
Santin et al., 2010	38 M, 32 F	38 M, 32 F	61.61	63.8 (10.5)	SS	12 types	0.001
Sedig et al., 2010	61 (sex nr)	51 (sex nr)	69 (7)	66 (9)	B-SIT	12 types	0.08
Silveira-Moriyama et al., 2010a	PD: 83 M, 57 F PSP: 20 M, 16 F	63 M, 63 F	65.6 (11.4) 69.2 (6.3)	59.6 (12.9)	UPSIT	40 types	PD < PSP 0.001 PSP < NC 0.001
Silveira-Moriyama et al., 2010b	iPD: 71 M, 35 F; LRRK2: 2 M, 11 F	103 M, 15 F	61.3 (11.0); 66 (14.6)	63 (9.8)	SS	12 types	LRKK2 >PD 0.001 LRKK2 < C 0.007
Aden et al., 2011	47 M, 40 F	14 M, 14 F	69.3(8.7)	69.5 (5.3)	B-SIT	12 types	0.001
Alcalay et al., 2011	13 M, 12 F PARK2 non-carriers; 8 M, 2 F heterozygotes 3 M, 6 F compound heterozygotes	NA	56 (36–66) 49 (36–59) 48 (30–59)	NA	UPSIT	40 types	Compound heterozygotes > heterozygotes & non-carriers 0.001

Berendse et al., 2011	59 M, 37 F	64.9 (9.7)	vs norms	NA	UPSIT	40 types	0.0001
Damholdt et al., 2011	42 M, 21 F	69.4 (6.3)	13 M, 15 F	68.4 (6.79)	B-SIT	12 types	0.0001
Iijima et al., 2011	53 M, 37 F	68.3 (8.4)	53 M, 37 F	68.4 (8.9)	OSIT-J	12 types	0.0001
Kim et al., 2011	39 M, 25 F	63.4 (12.1)	10 M. 15 F	59.9 (11.6)	B-SIT	12 types	0.001
Moessnang et al., 2011	8 M, 8 F	58.4 (9.5)	8 M, 8 F	57.4 (7.4)	SS	D-threshold Discrimin. ID	0.003 0.012 0.01
Muller et al., 2011	122 M, 85 F	67.9 (42.3–88.1)	104 M, 71 F	67.5 (39.4–85)	Questionnaire	Questionnaire	0.001
Rodriguez-Violante et al., 2011	41 M, 29 F	66.2 (8.8)	41 M, 29 F	66.2(8.8)	B-SIT	12 types	0.0001
Rolheiser et al., 2011	8 M, 6 F	55.9 (4.8)	8 M, 6 F	55.2(6.2)	UPSIT	40 types	0.0001
Ruiz-Martinez et al., 2011	81 M, 65 F LRRK2 – 23 M, 21 F LRRK2 +	70.3 (9.8) 70.0(9.6)	vs norms	NA	B-SIT	12 types	0.001
Saunders-Pullman et al., 2011	17 M, 13 F LRRK2 – PD 15 M, 16 F LRRK2 + PD 13 M, 15 F NMC	63.4(7.8) 64.7(9.8) 58.0(22.4)	22 M, 24 F	57.9(16.7)	UPSIT	40 types	0.0001 0.0001 0.006
Suzuki et al., 2011	47 M, 47 F	68.6 (5.1)	9 M, 20 F	66.1 (NR)	OSIT-J	12 types	0.0001
Valldeoriola et al., 2011	8 M, 6 F IPD 8 M, 6 F LRRK2	62.1 (11.6) 61.9 (12.6)	7 M, 6 F	63.5 (11.9)	UPSIT	40 types	0.01 0.01
Wang et al., 2011b	15 M, 15 F	61.7 (43–78)	15 M, 15 F	62.6 (42–81)	T&T thresholds	5 types	0.001
Wu et al., 2011	12 M, 14 F	59.0 (10.0)	12 M, 14 F	58.9 (9.5)	D & I threshold	5 types	0.0001 & 0.023
Yoritaka et al., 2011	15 (NR)	69.6 (6.6)	10 (nr)	46.0 (15.3)	SS	12 types	0.0001

(continued)

Table 18.2 (*Continued*)

Author and Year	No. PD	No. Control	PD Mean Age (SD or Range)	Control Mean Age (SD or Range)	Olfactory Test	Odorant Types	P Value
Zhang et al., 2011	11M, 14F	11M, 14 F	58.4 (9.8)	58.4 (9.3)	D & I threshold	5 types	0.0001
Busse et al., 2012.	257 M, 128 F	67 M, 65 F[20]	65.5 (8.9)	64.9 (11.3)	SS	12 types	0.001
Chen et al., 2012	66 M, 44 F	66 M, 44 F	64.6 (7.1)	64.4 (7.8)	SS	16 types	0.01
Kang et al., 2012	11 M, 4 F	9 M, 9 F	65.7 (12.3)	60.3 (13.5)	UPSIT	40 types	0.01
Maremmani et al., 2012	83 M, 50 F	254 M, 257 F	67.2 (7.0)	56.8 (13.6)	Italian Odor ID Test	33 types	0.0001
Parrao et al., 2012	44 (NR)	17	63.5 (10.2)	63.1 (9.5)	SS D-thresholds	12 types 2 types	0.0001 0.0001
Siderowf et al., 2012	NA	Normosmic: 2220 M, 2130 F; Microsmic: 355 M, 414 F	NA	63.9 (9.6) 64.5 (9.3)	UPSIT	40 types	Microsmics: prodromal features
Hakyemez et al., 2013	5 M, 23 F	8 M, 11 F	65.0 (6.2)	62.8 (6.0)	UPSIT	40 types	0.001
Sierra et al., 2013	29 M, 21 F	17 M, 33 F	68.9 (9.9)	60.11 (12.1)	UPSIT	40 types	0.001
Antsov et al., 2014	21 M, 29 F	21 M, 29 F	73.5 (58–87)	71.9 (55–89)	SS	12 types	0.001
Driver-Dunckley, et al. 2014	5 M, 5 F	45 M, 24 F	79.7 (8.0)	84.2 (5.9)	UPSIT	40 types	0.001
Gaig et al., 2014	30 M, 36 F (half LRRK2)	15 M, 18 F	65.0 (9.9)	64.8 (10.7)	UPSIT	40 types	0.001
Navarro-Otano et al., 2014	10 M, 5 F	6 M, 3 F	66.2 (9.5)	68.1 (8.2)	UPSIT	40 types	0.001
Picillo et al., 2014	40 M, 28 F	27 M, 34 F	61.8 (8.5)	59.5 (8.5)	UPSIT	40 types	0.001
Johansen et al, 2014	189 M, 103 F	28 M, 23 F	62.5 (8.7)	51.2 (13.3)	B-SIT	12 types	0.0009

Abbreviations: DIP, drug induced parkinsonism; **D-threshold**, detection threshold; **F**, female; **I**, increased performance relative to control; **ID**, identification; **I-threshold**, Identification threshold; **iPD**, sporadic or idiopathic PD; **M**, male; **MSA**, multi system atrophy; **MPS**, mild Parkinson symptoms; **NA**, not applicable; **NC**, normal controls; **NMC**, non-manifesting (LRRK2) carrier; **NST**, no statistical test applied; **NR**, not reported; **NS**, not statistically significant; **OERP**, Odor Event Related Potential; **OSIT-J**, Japanese Odor Smell Identification Test; **P-Acetate**, Pentyl (Amyl) acetate; **PAF**, pure autonomic failure; **PD**, Parkinson's disease; **PEA**, phenyl ethyl alcohol; **PEMEC**, Phenyl ethyl methyl carbinol; **R**, Rating; **R-Threshold**, recognition threshold; **RBD**, REM behavior sleep disorder; **SS**, Sniffin' Sticks Test; **TD FH+**, Parkinson, tremor dominant, positive family history of tremor; **TD FH−**, Parkinson, tremor dominant, negative family history of tremor; **TDI**, combined threshold, discrimination & identification; **T&T**, **T UPSIT**, University of Pennsylvania Smell Identification Test

Interestingly, unlike what appears to be the case in AD and schizophrenia, PD patients evidence a decrement in the build-up of perceived intensity across increasing suprathreshold concentrations of an odor (Doty et al., 2014). **Fifth**, the olfactory deficit is not confined to specific odors. Although there are reports that some odorants differentiate PD from controls better than others, little consistency has been found among studies and confounding from such variables as culture, odorant intensity, odorant type, and the choice of response alternatives is likely present (Boesveldt et al., 2008b; Bohnen et al., 2007; Daum et al., 2000; Double et al., 2003; Hawkes et al., 1997; Silveira-Moriyama et al., 2005). Since most stimuli are comprised of a melody of chemicals, and different sets of chemicals can sometimes produce the same odor quality, the focus on nominal odor qualities appears questionable. **Sixth**, the same degree of dysfunction found in PD occurs in some other neurodegenerative diseases. For example, the UPSIT scores of patients with early stage AD, PD, and the Parkinson-Dementia Complex of Guam (PDG) are essentially equivalent (Doty et al., 1991; Mesholam et al., 1998). However, olfactory testing can aid in differential diagnosis among selected movement disorders, since some have comparatively little or no olfactory dysfunction, including progressive supranuclear palsy (PSP), MPTP-induced parkinsonism (MPTP-P), multiple system atrophy (MSA), and essential tremor (ET) (Busenbark et al., 1992; Doty et al., 1993; Doty et al., 1995a; Ondo and Lai, 2005; Shah et al., 2008; Wenning et al., 1995). **Seventh**, PD-related olfactory dysfunction is not meaningfully influenced by medications used to mitigate the motor dysfunction, i.e., L-DOPA, dopamine agonists, and anticholinergic compounds (Doty et al., 1992b; Quinn et al., 1987a; Roth et al., 1998). **Eighth**, in well-established PD the dysfunction is surprisingly stable over time and typically unrelated to disease stage or duration (Barz et al., 1997; Doty et al., 1988; Doty et al., 1992b; Hawkes et al., 1997; Oka et al., 2010; Verbaan et al., 2008), although exceptions occur (Sharma and Turton, 2012) and associations have been noted between olfactory test scores and the degree of damage to dopaminergic brain regions, as inferred from functional imaging of the dopamine transporter within the caudate and putamen (Berendse et al., 2011; Herting et al., 2008; Siderowf et al., 2005). **Ninth**, PD patients with the most olfactory dysfunction tend to be those with the lowest body weight (Sharma and Turton, 2012). **Tenth**, suboptimal sniffing may contribute to the olfactory problem, particularly in later stages of the disease (Sobel et al., 2001). However, the degree of this contribution is small and in most investigations is null so long as adequate sniffing has occurred. **Eleventh**, the olfactory dysfunction is associated with some other PD-related changes, such as those related to cardiac sympathetic function. For example, strong correlations have been noted between UPSIT scores and cardiac ^{123}I-metaiodobenzylguanidine (MIBG) uptake

(Goldstein et al., 2008). Such associations appear to be independent of disease duration and clinical ratings of motor function (Lee et al., 2006). **Twelfth**, olfactory test scores are weakly correlated with some cognitive measures, most notably ones involving verbal memory and executive function (Bohnen et al., 2010; Morley et al., 2011; Parrao et al., 2012; Devanand et al., 2014). Although strong relationships among olfactory and cognitive test scores are generally lacking in most PD cohorts (Doty et al., 1989), longitudinal studies suggest that olfactory dysfunction is associated with a much higher risk of subsequent development of dementia (Baba et al., 2012). **Thirteenth**, some studies report that olfactory bulb volume is decreased in patients with PD (Brodoehl et al., 2012; Wang et al., 2011b), although others have not found this association (Hummel et al., 2010b; Mueller et al., 2005; Mundinano et al., 2011). If this does occur, it should be pointed out that decreased olfactory bulb volumes have also been observed in older persons (Yousem et al., 1998) and individuals with AD (Thomann et al., 2009), head trauma (Collet et al., 2009; Doty et al., 1997b; Rombaux et al., 2006a; Yousem et al., 1995), multiple sclerosis (Schmidt et al., 2011), and viral-related smell dysfunction (Rombaux et al., 2006b), implying lack of specificity (see Chapter 17). **Fourteenth**, the olfactory deficit often precedes the classical clinical PD motor signs by a considerable period, serving as a 'pre-clinical' or 'pre-motor' marker (Haehner et al., 2007; Ponsen et al., 2004; Ross et al., 2005). **Fifteenth**, olfactory testing is sensitive to a number of risk factors associated with future development of PD, including age (Doty et al., 1984) and head trauma (Doty et al., 1997b). **Sixteenth**, such factors as cigarette smoking and lifetime intake of caffeinated beverages may mitigate to some degree the olfactory dysfunction observed in PD (Lucassen et al., 2014; Sharer et al., 2015; Siderowf et al., 2007). **Seventeenth**, some asymptomatic first-degree relatives of patients with sporadic PD exhibit olfactory dysfunction that predicts, to some degree, future development of PD (Berendse et al., 2001; Montgomery, Jr. et al., 1999; Montgomery, Jr. et al., 2000b; Ponsen et al., 2004). In one study, Ponsen et al. (2004) tested the olfactory function of 361 asymptomatic relatives of PD patients. Those who performed in the top 10% and bottom 10% were then scanned for dopamine transporter function within the striatum using the [123]I-ß-CIT labelled DA transporter. A number of those in the bottom 10% exhibited substantial reduction in DA transporter uptake, whereas none of those in the top 10% did so. When assessed two years later, none of the 38 relatives who scored in the top 10% had developed clinically defined PD or exhibited DA transporter dysfunction. All of those in the bottom 10% exhibited DA transporter dysfunction and four had developed clinically defined PD.

As with AD, the physiological basis for the olfactory loss of PD is obscure, although most likely multifactorial.

Figure 18.3 Lewy pathology of the olfactory bulb demonstrated with alpha-synuclein immunohistochemistry. Lewy body and Lewy neurites are seen throughout the bulb, but particularly in the internal plexiform layer (IPL) and the intrabulbar anterior olfactory nucleus (iAON). Scale bar = 100 μm. Abbreviations: ONL: olfactory nerve layer; GLOM: glomerular layer; EPL: external plexiform layer; MCL: mitral cell layer; IPL: internal plexiform layer; GRAN: granular cell layer; iAON: intrabulbar anterior olfactory nucleus. From Duda (2010) (*See plate section for color version.*)

Among suggested causes are decrements in cholinergic function and the presence of Lewy bodies and neurites within the olfactory system (Figure 18.3) (Doty, 2011; Duda, 2010).

18.9.2 Familial Parkinson's Disease

Twin studies generally fail to find meaningful heritability of olfactory loss in PD patients (Tanner, 2003; Ward et al., 1983; Ward et al., 1984). Nevertheless, specific mutations have been identified in a number of familial forms of PD associated with olfactory system compromise. Such monogenetic forms of PD are rare in most, but not all, populations. For example, in PD cases within the Arab-Berbers of North Africa, over 40% have a mutation in the leucine-rich repeat kinase 2 gene (LRRK2) (Lesage et al., 2005). In North American, only around 1% of PD cases have this mutation (Correia et al., 2010). Identified genes associated with familial PD are listed in Table 18.3.

As with PD itself, genetic forms of PD exhibit heterogeneity in olfactory dysfunction. As discussed below for each of the known PD-related mutations, in many instances the olfactory test scores are very similar to those observed in sporadic PD. In others, no meaningful smell dysfunction is evident.

18.9.2.1 PARK1/PARK4 Locus Mutations (α-synuclein).
In PARK1/PARK4, an autosomal-dominant form of PD, the motor phenotype is typically exhibited early in life, usually in the 30's and 40's, although later onset may occur in some with the A30P mutation (Kruger et al., 1998). α-synuclein is the encoded protein. Several studies have demonstrated olfactory dysfunction in subjects with the PARK1 mutation in small numbers of subjects (Bostantjopoulou et al., 2001; Kertelge et al., 2010; Nishioka et al., 2009). For example, Bostantjopoulou et al. (2001) demonstrated that two out of the eight Greek PD patients carrying the G209A PARK1 mutation were anosmic, whereas the others were normal. No evidence of smell loss was observed by Tijero et al. (2010) in an *asymptomatic* carrier of the E46K substitution in the α-synuclein gene.

18.9.2.2 PARK2 Locus Mutations (Parkin gene).
The most common form of *early-onset* parkinsonism carries the autosomal recessive Parkin gene (PARK2). This gene is closely associated with the function of actin filaments within the cytoplasm and neuronal processes. Parkin mutations appear to have little influence on olfaction. Thus, Khan et al. (2004) found normal UPSIT scores in all 27 PARK2 patients they tested. Deficits were observed in a PARK2 negative PD group and a sporadic PD patient group. Normal UPSIT scores were also reported by Alcalay et al. for compound PARK2 heterozygotes – that is, individuals who had PARK2 mutations on both the paternal and material alleles (Alcalay et al., 2011). PD patients who did not carry the PARK2 gene and non-compound PARK2 heterozygotes had UPSIT scores analogous to those observed in sporadic PD. Yoritaka et al. (2011) reported that all but one of three men and three women with PARK2 mutations, apparently a mixture of homozygous and heterozygous mutations or deletions, had elevated thresholds relative to controls, although the patients were older than the controls which may explain their positive finding [respective mean (range) ages = 69.6 (60–89) and 46.0 (39–79)].

18.9.2.3 PARK6 Locus Mutations (PINK1 gene).
This autosomal-recessive disorder, whose clinical features are similar to those of PARK2, exhibits mutations in the PTEN-induced putative kinase 1 (PINK1) gene. After *Parkin*, *PINK1* mutations are the most frequent cause of autosomal recessive parkinsonism. This gene appears to inhibit the formation of reactive oxygen species

18.9 Parkinson's Disease and Parkinson-Like Disorders

Table 18.3 Genetic forms of PD or parkinsonism. Those forms for which olfactory tests have been administered are indicated in the light gray rows. Other genetic forms, namely PARK5 (4p14), PARK11 (2q36-q37) and PARK13 (2p12), are not indicated since they are now considered to have questionable associations with the expression of parkinsonism.

Locus (Map Position)	Protein	Inheritance Pattern	Mutations	Clinical Features
PARK1/PARK4 (4q21-q23)	α-synuclein	Autosomal Dominant	A30P, A53T, E46K; duplications and triplications	Early onset; dementia; autonomic dysfunction. Rare and not yet observed in sporadic PD cases.
PARK2 (6q25.2-q27)	Parkin	Autosomal Recessive	A wide variety of mutations, exonic deletions, duplications and triplications	Juvenile and early onset,; slow progression; good L-DOPA response
PARK3 (2p13)	Unknown	Autosomal Dominant	Not Identified	Late onset; typical PD features
PARK6 (1p35-p36)	Pink-1	Autosomal Recessive	G309D, exonic deletions	Early onset; slow progression; good L-DOPA response
PARK7 (1p36)	DJ-1	Autosomal Recessive	Homozigotic exonic deletion, L166P	Early onset, slow progression; very rare
PARK8 (12q12)	LRRK2	Autosomal Dominant (incomplete penetrance)	G2019S (most common), R1441C/G/H, Y1699C, I2020T, G2385R, others	Late onset, Tremor dominant PD symptoms
PARK9 (1p36)	ATP13A2	Autosomal Recessive	Loss-of-function mutations	Atypical PD features; Kufor-Rakeb syndrome; juvenile and early onset (11–16 yrs); dementia; pyramidal degeneration; spasticity; paralysis of gaze
PARK10 (1p32)	Unknown	Not clear	Not identified	Late onset, typical PD features
PARK12 (Xq21-25)	Unknown	Not clear	Not identified	Late onset PD
PARK14 (22q13.1)	PLA2G6	Autosomal Recessive	Two missense mutations	Early cerebellar signs and visual disturbances; late-onset dystonia with PD features; good L-DOPA response
PARK15 (22q12–q13)	FBXO7	Autosomal Recessive	Three point mutations	Early onset, progressive pallidio-pyramidal syndrome
PARK16 (1q32)	RAB7	Risk	RAB7L1, SLC41A1	Late onset PD
PARK17 (4P16)	GAK	Risk	Not identified	Late onset PD
PARK18 (6p21.3)	HLA-DRA	Risk	Not identified	Late onset PD

426   Chapter 18   Odor Perception and Neuropathology in Neurodegenerative Diseases and Schizophrenia

(Wang et al., 2011a). Ferraris et al. (2009) administered olfactory tests to seven patients with homozygous PARK6 PD, six patients with heterozygous PARK6 PD, and 12 healthy individuals heterozygous for the PINK1 gene. All 13 PARK6 cases, and 8 of the 12 healthy heterozygotes, scored below normal relative to non-PD persons of their age. In contrast, Eggers et al. (2010) reported that none of four homozygous PARK6 PD patients they tested had UPSIT scores falling outside of the 10th percentile of age- and sex-matched normal controls. Although 4 of 10 heterozygotes did so, only one was at or below the 5th percentile, suggesting to these authors that PARK6 mutations have only modest influences on smell function.

18.9.2.4 PARK 8 Locus Mutations (LRRK2 gene). This is the most common form of dominantly inherited PD. Mutations are found in the leucine-rich repeat kinase 2 gene (LRRK2/Dardarin). The absence of a gene dose effect is suggested by the fact that both homozygotes and heterozygotes have the same clinical features (Ishihara et al., 2006). Penetrance of the most common LRRK2 mutation (G2019S on exon 41) is incomplete and age-dependent (Lesage et al., 2007).

The olfactory dysfunction of PARK8 cases is variable but generally similar to that of sporadic PD. However, despite having abnormal smell scores, several studies have noted that G2019S PD mutation carriers have olfactory test scores somewhat higher than those of non-mutation carrying PD patients (Berg et al., 2005; Kertelge et al., 2010; Khan et al., 2005; Lin et al., 2008; Saunders-Pullman et al., 2011; Silveira-Moriyama et al., 2010b). *Asymptomatic* LRRK2 mutation carriers generally exhibit no smell dysfunction (Johansen et al., 2011; Saunders-Pullman et al., 2011; Silveira-Moriyama et al., 2008).

18.9.3 Other Forms of Parkinsonism

18.9.3.1 Corticobasal Degeneration. In corticobasal degeneration (CBD), parkinsonian features are evident along with limb dystonia, ideomotor apraxia, myoclonus, and ultimately, cognitive decline. This disease is noted for the accumulation of tau protein largely in the basal ganglia and fronto-parietal cortex. The olfactory bulbs and temporal cortex appear to be spared from meaningful tau pathology (Tsuboi et al., 2003). In one study of seven patients with clinically suspected CBD, UPSIT scores did not differ significantly from those of age-matched controls (Wenning et al., 1995). A similar finding was noted in a more recent study of seven patients with clinically defined CBD for odor discrimination, although mild deficits in odor naming and odor picture matching were reported (Luzzi et al., 2007).

18.9.3.2 Drug-induced Parkinsonism. Drug-induced parkinsonism, more commonly termed drug-induced Parkinson's disease (DPD), is usually caused by neuroleptic drugs. The prevalence of DPD is lower today than 25 years ago, reflecting the introduction of selective D_2 dopamine receptor and 5-hydroxytryptamine (5HT) blocking agents for treating psychotic disorders. The influences of DPD on smell function are not entirely clear, as the range of involved drugs is large and only a few studies with relatively small sample sizes have addressed this topic in detail.

In a study by Lee et al. (2007a), normal B-SIT scores were observed in 15 DPD patients that were tested. The DPD had been induced by levosulpiride, haloperidone, flunarizine, perphenazine, metoclopramide, or risperidone. In contrast, Hensiek et al. (2000) tested 10 patients whose DPD had been induced by phenothiazine. Five had abnormal UPSIT scores and did not completely recover even after medication cessation or change. The UPSIT scores in all but one of the other five patients returned to normal after such adjustment. More recently, Bovi et al. (2010) administered olfactory tests to 16 DPD patients (seven haloperidol, five amisulpride, two perphenazine, one fluphenazine, one clomipramine), 13 PD patients, and 19 age- and sex-matched normal controls. The DPD patients were divided into those with normal (n = 9) or abnormal (n = 7) putamen DA transporter binding as determined from ^{123}I-FP-CIT SPECT imaging. Only those patients with a pathological putamen uptake had abnormal olfactory test scores – scores which, like those for PD, correlated with putamen uptake values, suggesting that the olfactory losses were more closely associated with central DA damage than with drug-induced DA receptor blockade.

The most publicized case of drug-induced parkinsonism is that which appeared in young drug addicts in northern California in the early 1980s. These cases resulted in an error in the synthesis of a heroin-like substance which produced 1-methyl-4-phenyl-1,2,3,6-tetrahydropyridine (MPTP). MPTP is metabolically converted within the brain, mainly in glial cells, to the toxic ion 1-methyl-4-phenyl-piperidinium (MPP$^+$) by monoamine oxidase B. MPP$^+$ selectively damages striatal DA neurons by interfering with complex 1 of the electron transport chain. MPP$^+$ has a high binding affinity to the dopamine transporter and readily enters dopaminergic terminals. Mice that lack this transporter are protected from MPTP toxicity (Bezard et al., 1999).

In the only human study to assess olfactory function in patients with MPTP-induced parkinsonism (Doty et al., 1992a), the UPSIT and a phenyl ethyl alcohol detection threshold test were administered to six of the original young MPTP-induced parkinsonism patients. Scores were compared to those of 13 young PD patients and

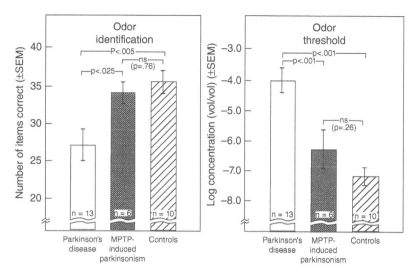

Figure 18.4 University of Pennsylvania Smell Identification Test (UPSIT) and phenyl ethyl alcohol odor detection threshold test scores for patients with MPTP-induced parkinsonism (MPTP-P), young patients with idiopathic Parkinson's disease (PD), and matched normal controls. From Doty et al. (1992a).

ten normal subjects. The UPSIT scores of the MPTP-P patients did not differ significantly from those of normal controls, which was not the case for the young PD patients (Figure 18.4). It is conceivable that olfactory dysfunction would have been detected in the MPTP patients if testing had been performed sooner after their MPTP exposure or if greater MPTP exposure had occurred, either via higher doses or frequency of exposures. The olfactory bulbs of MPTP-treated mice exhibit inflammatory microgliosis and increased expression of cytokines associated with apoptosis such as interleukin-1α (IL-1α) and IL-1β (Vroon et al., 2007). The olfactory dysfunction of mice who receive intranasal MPTP recovers over time (Prediger et al., 2006), as does at least one non-motor measure of MPTP-treated monkeys which correlates with olfactory dysfunction in humans (i.e., cardiac denervation) (Elsworth et al., 2000; Goldstein et al., 2003).

18.9.3.3 Glucocerebrosidase-Related Parkinsonism.

Glucocerebrosidase (GBA) is a gene implicated in parkinsonism that causes Gaucher's disease (GD), the most prevalent autosomal recessive lysosomal disorder. Although parkinsonism can be a presenting feature of GD, the most prominent features are bone, hematologic, and pulmonary abnormalities (Saunders-Pullman et al., 2010). In one study of six GBA mutation carriers, three were anosmic, two severely microsmic, and one moderately microsmic, with the mean UPSIT score falling within the range expected for patients with sporadic PD (Goker-Alpan et al., 2008). In a second study, 20 PD patients heterozygous for one of the two GBA mutations (N370S, L444P) exhibited lower scores on a 12-item odor identification test, although the effects were not significant at the 0.05 α level (p = 0.08) (Brockmann et al., 2011). In a third study, one of two GBA cases, both of whom were anosmic, was a 54-year-old man who reported that the smell dysfunction appeared when he was a teenager (Saunders-Pullman et al., 2010). The initial PD sign, a right-hand rest tremor, appeared at 48 years of age. Such symptoms as anxiety, depression, orthostasis, urinary urgency, and medication sensitivity followed soon thereafter.

18.9.3.4 Multiple System Atrophy.

Multiple system atrophy (MSA) is a rapidly progressing type of parkinsonism which is often misdiagnosed as PD in its earliest stages, but evolves to include prominent autonomic and cerebellar dysfunction. Clinical features typically include dysarthria, stridor, contractures, dystonia, altered orthostatic blood pressure, digestive problems, bladder control issues, sexual dysfunction, and rapid eye movement (REM) sleep behavior disorder (Kaufmann and Biaggioni, 2003). MSA's pathology is found within the basal ganglia, cortex and spinal cord, but peripheral autonomic neurons are spared. Pathological changes have been noted in the olfactory bulbs of MSA (Kovacs et al., 2003). MSA-Parkinsonism (MSA-P), the most common form of MSA (~80% of cases), is defined by akinesia and rigidity. MSA-Cerebellar (MSA-C) is a form in which cerebellar ataxia is dominant. The discovery in 1989 of glial cytoplasmic inclusions (GCIs) in MSA brains confirmed beliefs that striatonigral degeneration, sporadic olivopontocerebellary atrophy, and the Shy-Drager syndrome are, in fact, different clinical expressions of MSA (Papp et al., 1989).

Wenning et al. (1995) administered the UPSIT to 29 patients with MSA and 123 controls. Moderate loss of function was observed in the MSA patients relative to the controls (respective means = 26.7 and 33.5). No differences were apparent between the MSA-P and MSA-C types. Similar findings have subsequently been reported by others (Abele et al., 2003; Garland et al., 2011; Goldstein et al., 2008; Muller et al., 2002; Nee et al., 1993). No meaningful relationship has been found between odor identification test scores and measures of cardiac ^{123}I-metaiodobenzylguanidine (MIBG) uptake, unlike with PD (Lee et al., 2006).

18.9.3.5 Parkinson Dementia Complex of Guam.

The Parkinson Dementia Complex of Guam (PDG), also known as the ALS/parkinsonism-dementia Complex of Guam (ALS/PDG), is a progressive neurodegenerative disease that has the features and pathology, in variable combinations, of atypical parkinsonism, dementia, and ALS. It is largely confined to the residents of Guam, the Mariana islands, the Kii peninsula of Japan, and the coastal plain of West New Guinea (McGeer and Steele, 2011). Although this disease accounted for at least 15% of adult deaths in the Chamorro population between 1957 and 1965 (Reed et al., 1966; Reed and Brody, 1975), its prevalence has since dropped markedly and by 1999 the ALS component was no longer evident (Plato et al., 2003). Significant olfactory bulb pathology is present in PDG, with nearly complete loss of the cells within the AON (Perl and Doty, unpublished). A retinopathy with an appearance similar to a larval migration has been noted in a significant number of PDG patients (Cox et al., 1989; Kato et al., 1992; Kokubo et al., 2006). This unique retinopathy has not been found in any other part of the world or in any other neurodegenerative disease.

Doty et al. (1991) administered the UPSIT to 24 PDG patients. Their test scores were significantly depressed, being equivalent to those from 24 AD and 24 PD North American patients matched on smoking behavior, gender, and age. More recently, Ahlskog et al. (1998) administered an abbreviated version of the UPSIT to 9 Guamanians with symptoms of ALS, 9 with symptoms of pure parkinsonism, 11 patients with pure dementia, and 31 patients with PDG, as well as to 53 neurologically normal Guamanians and 25 neurologically normal North American controls. The UPSIT scores were markedly and equivalently depressed in the four disease groups.

18.9.3.6 Vascular Parkinsonism.

Vascular parkinsonism (VP) occurs in patients with extensive cerebrovascular disease involving the basal ganglia. It differs from PD in that it rarely has resting tremor, is variably responsive to L-DOPA, and exhibits variable DA transporter deficits on SPECT or PET imaging. Acute onset cases have disproportionately more lesions in the subcortical gray nuclei (striatum, globus pallidus, and thalamus), whereas insidious onset cases tend to have more diffusely distributed lesions (Zijlmans et al., 1995).

Katzenschlager et al. (2004) administered the UPSIT to 14 VP patients, 18 PD patients, and 27 normal controls of similar age. The UPSIT scores of the VP patients did not differ significantly from those of the controls (respective means = 26.1 and 27.6), unlike the scores of the PD patients (mean = 17.1; ps < 0.0001). These findings, along with more recent research (Navarro-Otano et al., 2014), suggest that olfactory testing may be useful in differentiating VP from PD.

18.9.3.7 X-linked Recessive Dystonia-parkinsonism.

X-linked Recessive Dystonia-parkinsonism, also termed 'Lubag', is found mostly among adult male Filipinos with maternal roots from the Philippine Island of Panay. Dystonia, mainly in the jaw, neck, trunk, and eyes, but sometimes in the limbs, tongue, pharynx and larynx, becomes salient as the disease progresses. The disease, whose average age of onset is approximately 40 years, can rapidly progress and often results in death from pneumonia or other infections within a few years of onset. In the sole olfactory study on this topic, slight dysfunction on a culturally modified 25-item version of the UPSIT was observed in 20 affected men relative to 20 control men (respective means adjusted to 40-item UPSIT scale = 29 and 33, p < 0.001) (Evidente et al., 2004). Smell loss was present in the earliest disease stages and was unrelated to disease duration, severity, and the degree of dystonia.

18.10 PROGRESSIVE SUPRANUCLEAR PALSY

Progressive supranuclear palsy (PSP; also termed the Steele, Richardson, and Olszewski syndrome) accounts for approximately 4% of patients who exhibit parkinsonian symptoms. Based upon a large clinicopathological study of 103 cases, Williams et al. (2005) divided PSP into two classes which they termed the 'Richardson Syndrome' (RS) and 'PSP-P'. RS accounted for 54% of their cases and is defined by an early onset of postural instability and falls, supranuclear vertical gaze palsy, and cognitive dysfunction. PSP-P accounted for 32% of their patients and is characterized by asymmetrical onset tremor and moderate initial therapeutic response to levodopa. According to these investigators, PSP-P patients are frequently misdiagnosed as classical PD. The remaining 14% of their patients were difficult to assign to these two categories.

Like PD, PSP is associated with widespread accumulation of tau protein in degenerating nerves, although the olfactory bulbs appear to be spared of both tau and α-synuclein pathology (Tsuboi et al., 2003). Unlike PD, however, the parkinsonian features are less responsive to anti-PD medications (Jackson et al., 1983) and PSP is typically characterized by comparatively more frontal lobe dysfunction, more neuronal degeneration within the basal ganglia and upper brain stem, and less involvement of mesolimbic and mesocortical dopamine systems than PD (Cambier et al., 1985; Jankovic, 1989).

Most patients with PSP have a relatively normal sense of smell, although moderate losses appear to be present in some individuals, conceivably reflecting the PSP-P cohort. In the first study on this topic, the UPSIT and a phenyl ethyl alcohol odor detection threshold test were administered to 21 and 17 PSP patients, respectively. All scored well on a test analogous to the UPSIT in which pictures of the UPSIT odor sources, rather than smells, were presented (Doty et al., 1993). The UPSIT scores were compared to those from equivalent numbers of PD patients and neurologically normal age-, gender-, and race-matched controls. The performance of the PSP patients was markedly superior to that of the PD patients [respective UPSIT means (SD): 31.59 (7.18) and 18.82 (6.94)], with approximately half of the PSP patients scoring within the normal range. While the PSP threshold values tended to be higher than those of controls, the difference was not significant at the $0.05\ \alpha$ level (p = 0.086). Similar results have been noted by others (Wenning et al., 1995), with the general conclusion being that PSP-related smell loss, while present on average, is not as great as that seen in PD (Silveira-Moriyama et al., 2010a). Some investigators report not being able to detect any PSP-related olfactory deficits on tests of odor identification, detection or discrimination (Lang et al., 2011). Baker and Montgomery (2001) found 23 relatives of PSP patients to have lower UPSIT scores than 23 matched controls, although both sets of scores were within the normal range.

18.11 PURE AUTONOMIC FAILURE

Pure autonomic failure (PAF) is a slowly progressing degenerative disease in which failure of the autonomic nervous system is the sole clinical finding (Hague et al., 1997). Lewy bodies are present in the brain stem and pre- and postganglionic autonomic neurons, as well as in the peripheral sympathetic and parasympathetic nerves (Hague et al., 1997). It is conceivable that this disorder is a precursor to PD, but given its slow progression most patients die before significant PD-related CNS involvement becomes evident (Kaufmann and Biaggioni, 2003). Patients with PAF exhibit UPSIT scores essentially equivalent or only slightly above those of patients with PD (Garland et al., 2011; Goldstein and Sewell, 2009; Silveira-Moriyama et al., 2009a).

18.12 SCHIZOPHRENIA

Schizophrenia (SZ) is a complex and debilitating psychiatric disease typically first appearing in late adolescence and early adulthood. Like a number of other diseases discussed in this chapter, SZ is a heterogeneous disorder that is difficult to completely characterize by current classification systems. Between 1–3% of the population suffers from SZ and related disorders, including those with positive symptoms (e.g., hallucinations, delusions, unusual thought content) and negative symptoms (e.g., alogia, anhedonia, social withdrawal). Many aspects of cognitive functioning can be altered in SZ, including attention, memory, language ability, visuospatial integration, executive function, and perceptual ability (Harvey and Sharma, 2002). While SZ has been typically conceptualized as a neurodevelopmental disorder (Weinberger, 1987), neuropathological changes occur in some cases even after disease onset (Church et al., 2002; Hulshoff Pol and Kahn, 2008). Active psychotic episodes potentially damage the brain, including brain regions associated with olfactory processing, a process most salient early in the disease (de Haan and Bakker, 2004).

Since the time of Kraepelin (1913), olfactory hallucinations, most typically of an unpleasant nature, have been reported in SZ. They are common (11–34%) (Kopala et al., 1995b; Meats, 1988; Mueser et al., 1990; Stevenson et al., 2011), likely depend upon a relatively intact olfactory system (Kopala et al., 1994), and exceed the prevalence of auras of temporal lobe epilepsy (Gupta et al., 1983), a disease that shares with SZ medial temporal lobe pathological substrates (Hayashi, 2004; McAbee et al., 2000). Olfactory hallucinations may be predictive of a poorer prognosis (Kwapil et al., 1996) and earlier age of SZ onset (Lewandowski et al., 2009), underscoring their clinical significance.

Unlike AD, PD, and some other disorders described in this chapter, there is less uniformity of findings regarding the nature of the olfactory deficits observed in SZ when such deficits are measured by tests of odor detection, discrimination, memory, and hedonics (Table 18.4). This may reflect the more modest olfactory dysfunction of SZ and non-olfactory related task demands that differ among such tests (e.g., the ability to attend, discriminate, remember, or identify independent of the ability to smell) (Doty, 2007).

In an early study, *increased* threshold sensitivity to a "sexually relevant" odor was described (Bradley, 1984). However, such hypersensitivity has yet to be replicated, and most investigators find either normal (Geddes et al., 1991; Kopala et al., 1989; Moberg et al., 1995) or elevated (Gross-Isseroff et al., 1987; Serby et al., 1990; Turetsky and Moberg, 2009) thresholds in this disease. Similar inconsistencies have been noted in studies in which nominal measures of odor discrimination have been assessed (e.g., Dunn and Weller, 1989; Rupp et al., 2005b), as well as in studies of hedonic ratings, where there have been reports of impaired (Crespo-Facorro et al., 2001; Hudry et al., 2002; Moberg et al., 2003a; Plailly et al., 2006), enhanced (Becker et al., 1993; Rupp et al., 2005b), or unaltered (Rupp et al., 2005a) hedonic sensations. In one

Table 18.4 Procedural details of studies of olfactory function in patients with schizophrenia (SZ)

Author and Year	No. SZ (sex)	No. Control	SZ Mean Age (SD or Range)	Control Mean Age (SD or Range)	Olfactory Test	Odorant Types	P Value
Campbell and Gregson 1972	20 (sex NR)	10 (sex NR)	17+	17+	Memory for Odor Positions	12 types	NST
Bradley 1984	5 M, 6 F	4 F	(20–50)	NR	D-threshold	Androstenone	0.05 M Enhanced NS F
Gross-Isseroff et al., 1987	22 M, 20 F	20 M, 20 F	30.1 (6.4)	30.0 (6.4)	D-threshold	Androsteone	NS
Sreenivasan et al., 1987	32 (sex NR)	17 w/ neurosis	NR	NR	Odor Matching Task D-threshold	4 types Pentyl acetate	NST 0.01
Hurwitz et al. 1988	15 M, 3 F	7 M, 3 F	23.9 (17–41)	33.6 (21–43)	UPSIT	40 types	0.03
Dunn and Weller 1989	13 M, 2 F	13 M, 2 F	54.2 (28–71)	52.9 (25–65)	Discrimination test	4 odor sets	NS
Kopala et al., 1989	26 M 15 F	23 M 20 F	26.3 25.3 (18–54)	29.6 29.6 (18–54)	UPSIT	40 types	0.005
Serby et al., 1990	14 M	14 M	(40–49)	(40–49)	UPSIT D- threshold	40 types Geraniol	0.001 0.001
Warner et al., 1990	12 M	14 M	34 (20–42)	34 (20–50)	UPSIT	40 types	0.03
Clark et al. 1991	20 M, 6 F	8 M*	26.6 (7.9)	24.3 (6.2)*	UPSIT	40 types	NST
Geddes et al. 1991	16 M, 8 F	17 M 6 F	38.5 (20–66)	37.5 (21–62)	D-threshold	Musk Ketone	NS, 0.03 (between - and + syndromes)
Seidman et al., 1992	15 M, 1 F	16 M, 1 F	36.5 (8.1)	31.7(8.9)	UPSIT	40 types	0.001
Kopala et al., 1993	30 M 10 F	28 M 30 F	25.9 (7.1) 27.3 (8.3)	28.0 (7.5) 29.0 (8.1)	UPSIT D-threshold D-threshold	40 types n-butanol PEA	0.002 I (NS) NS
Wu et al., 1993	19 M, 1 F	23M, 1F	32.1 (9.3)	27.7 (7.2)	UPSIT Odor memory	40 types 10 sets	0.01 0.001
Houlihan et al., 1994	25 M 22 F	21 M, 15 F	32.7 (5.8) 33.8 (7.9)	30.3 (6.3) 32.9 (8.7)	UPSIT	40 types	0.004
Kopala et al., 1994	131 (NR)	77 (NR)	27.3 (7.8)	32.5 (11.1)	UPSIT	40 types	0.001

Malaspina et al., 1994	15 M, 5 F	15 M, 5 F	33.3 (5.9)	33.3 (5.9)	UPSIT	40 types	0.01
Kopala et al., 1995a	49 M 16 F	14 M 16 F	26.9 (8.0) 32.7 (8.4)	31.5 (7.7) 37.0 (8.6)	UPSIT	40 types	0.02
Kopala et al., 1995b	15 F pre 12 F post	17 F pre 8 F post	33.7 (8.6) 54.3 (5.1)	36.8 (10.1) 56.6 (6.4)	UPSIT	40 types	0.001
Seidman et al., 1995	14 M, 4 F	14 M, 4 F	39.8 (8.3)		UPSIT	40 types	0.006
Brewer et al., 1996	27 M	19 M	31.8 (8.5)	34.8 (12.5)	UPSIT	40 types	0.001
Moberg et al., 1997a	4 M, 12 F	6 M, 14 F	77.9 (6.5)	72.5 (6.4)	UPSIT	40 types	0.001
Moberg et al., 1997c	18 M, 20 F	18 M, 22 F	50.6 (25.5)	49.6 (24.6)	UPSIT	40 types	0.001
Seidman et al., 1997	24 M, 16 F	15 M, 17 F	38.5 (6.5)	36.4 (9.3)	UPSIT	40 types	0.02
Good et al., 1998	65 M	59 M	28.1 (8.4)	32.2 (7.8)	UPSIT (unilateral) D-threshold (unilateral)	40 types PEA	0.05 NS
Kopala et al., 1998a	9 M (polydipsic) 9 M (non-polydipsic)	9 M	38.2 (6.7) 41.0 (4.2)	39.2 (8.5)	UPSIT D-Threshold	40 types n-butanol	0.001 0.05
Kopala et al., 1998b	12 (sex NR)	12 (sex NR)	36.8 (4.9)	37.5 (4.6)	UPSIT	40 types	0.02
Malaspina et al., 1998	5 M, 1 F	6 M, 1 F	34.6 (14.8)	33.3 (7.0)	UPSIT	40 types	0.05
Purdon 1998	18 M, 3 F	vs norms	37.1 (9.6)	NA	UPSIT	40 types	(in microsmic range of norms)
Saoud et al., 1998	24 M	21 M	31.3 (8.9)	30.2 (9.2)	'Perfume Boxes'	12 odorants	0.013
Stedman and Clair, 1998	37 M, 9 F	vs norms	36.4 (8.1)	NA	UPSIT	40 types	0.001
Sirota et al. 1999b	19 M 12 M	10 M 20 M	36.5 (11.9) 34.2 (11.9)	36.7 (15.3) 36.0 (12.2)	D-threshold D-threshold	Pentyl acetate Androstenone Pentyl acetate Androstenone	0.0001 0.02 E (NS) E (0.02)
Purdon and Flor-Henry, 2000	11 M, 6 F	11 M, 6 F	27.7 (7.8)	27.4 (6.1)	UPSIT D-threshold	40 types n-butanol	0.05 0.05

(continued)

Table 18.4 (*Continued*)

Author and Year	No. SZ (sex)	No. Control	SZ Mean Age (SD or Range)	Control Mean Age (SD or Range)	Olfactory Test	Odorant Types	P Value
Brewer et al., 2001	55 M, 19 F	27M, 11F	22.3 (3.7)	21.0 (4.2)	UPSIT	40 types	0.001
Kohler et al., 2001	22 M, 18 F	15M, 10F	31.8 (9.3)	27.7 (8.4)	UPSIT D-threshold	40 types PEA	0.008 (NS)
Kopala et al., 2001	11 M, 8F	18M,25F	43.4 (9.9)	42.5 (10.9)	UPSIT	40 types	0.001
Hudry et al., 2002	20 M 20 F	20 M 20 F	33.4 (8.6) 38.4 (10.2)	33.8 (9.4) 37.6 (9.7)	Hedonic judgments Identification	12 odorants	0.05 0.0005
Goudsmit et al. 2003	52 M, 31 F	69 (sex NR)	33.9 (10.7)	40.3 (14.9)	UPSIT B-SIT	40 types 12 types	0.001 0.001
Moberg et al. 2003a	15 M, 15 F	11 M, 19 F	33.4 (8.8)	35.8 (15.5)	Odor hedonics Odor intensity	Amyl acetate	0.02 M E (NST)
Turetsky et al. 2003b	14 M, 7 F	10 M, 10 F	30.9 (8.4)	30.5 (10.9)	UPSIT (unilateral) D-threshold	40 types PEA	0.08 0.029
Szeszko et al., 2004	11 M, 4 F	13 M, 4 F	24.1 (5.2)	26.2 (5.1)	UPSIT (unilateral)	40 types	0.002
Corcoran et al. 2005a	16 M, 10 F	Relative to norms	15.0 (1.8)	NA	UPSIT	40 types	23rd percentile
Rupp et al., 2005a	33 M	40 M	25.4 (4.7)	26.2 (4.7)	SS Odor Judgments	16 odorants 16 odorants	0.001 0.002
Rupp et al., 2005b	30 M	30 M	31.5 (6.1)	32.5 (6.9)	Munich Olf Test (unilateral) Discrimination Identification Familiarity Edibility	PEA DMDS 8 odorants 8 odorants 8 odorants	NS 0.017 0.001 0.26 0.035
Ugur et al. 2005	10 (sex NR)[11]	10 (sex NR)	34.5 (11.4)	35.3 (11.1)	SS Discrimination ID D-threshold	16 odorants n-butanol	0.01 NS 0.01
Doop and Park 2006	17 (sex NR)	14 (sex NR)	37.0 (9.8)	33.8 (10.7)	UPSIT	40 types	0.05
Roalf et al., 2006	10 M, 12 F	21 M, 24 F	34.9 (12.5)	34.2 (12.7)	UPSIT D-threshold	40 types PEA	0.008 0.029
Moberg et al., 1995	13 M, 8 F	10 M, 10 F	24.6 (5.4)	28.6 (8.3)	UPSIT D-threshold	40 types PEA	0.026 NR

Study	Patients (n)	Controls (n)	Patient age	Control age	Test	Odorant	E (NR)
Compton et al., 2006	41 (sex NR)	38 (sex NR)	33.8 (10.0)	51.7 (10.3)	UPSIT	40 types	NR
Brewer et al., 2007	23 M, 8 F FEP 28 M, 4 F CSZ	19 M, 5 F	20.2 (3.9) 33.1 (8.4)	20.2 (4.19)	Yes/No D-threshold UPSIT	Steroids Butanol 40 types MHA	NR NR 0.001 CSZ < NC 0.002
Good et al. 2007	258 M 95 F	45 M 44 F	23.6 (5.0) 23.8 (5.5)	23.6 (5.5) 23.3 (6.6)	UPSIT	40 types	0.004
Ishizuka et al. 2010	9 M, 6 F	14 M, 5 F	37.5 (10.7)	36.0 (10.9)	UPSIT	40 types	0.01
Strauss et al., 2010	28 M, 13 F	4 M, 18 F	41.9 (11.8)	40.0 (12.1)	B-SIT Odor hedonics	12 odorants Pleasant Unpleasant	0.001 0.04 0.04
Cumming et al. 2011	27 (sex NR)	22 (sex NR)	37.1 (8.4)	35.5 (9.8)	UPSIT	40 types	0.0005
Kamath et al., 2011	19 M, 4 F	13 M, 8 F	42.8 (10.7)	43.9 (12.8)	B-SIT	12 odorants	0.01
Kamath et al. 2012	23 M, 19 F	18 M, 12 F	35.4 (10.8)	30.83 -12.1	UPSIT (unilateral)	40 types	0.00001
Malaspina et al., 2012	31 M 27 F	18 M 24 F	32.3 (10.0) 33.0 (11.0)	29.5 (8.3) 34.5 (13.2)	UPSIT D-threshold	40 types PEA	NS NS
Kamath et al. 2013	36 M, 28 F	30 M, 24 F	37.0(10.8)	33.2(10.9)	SS	16 types	0.0001
Kästner et al., 2013	881 (sexes NR)	102 (sexes NR)	NR	NR	Odor Naming & Interpretation	10 types	0.000001
Kamath et al., 2014	36 M, 29 F	34 M, 32 F	37.3 (10.4)	33.9 (11.4)	SS	16 types	0.001

Abbreviations –**B-SIT**, Brief Smell Identification Test; **CSZ**, chronic schizophrenia; **D-threshold**, detection threshold; **Discrim**, Discrimination Test; **DMDS**, dimethyl disulfide; **E**, essentially equivalent (no statistical test performed); **F**, female; **FEP**, first episode psychosis; **I**, increased performance relative to control; **ID**, identification; **M**, male; **MHA**, *trans*-3-methyl-2-hexenoic acid; **NA**, not applicable/available; **NC**, normal controls; **NR**, not reported; **NS**, not statistically significant; **NST**, no statistical test applied; **PEA**, Phenylethanol; **SD**, Standard deviation; **SS**, Sniffin' Sticks; **SZ**, Schizophrenia; **UPSIT**, University of Pennsylvania Smell Identification Test.

*These controls did not receive olfactory testing, only imaging.

434   Chapter 18   Odor Perception and Neuropathology in Neurodegenerative Diseases and Schizophrenia

case, lower SZ-related hedonic ratings were seen in men, but not in women, adding complexity to this issue (Moberg et al., 2003a). Only one study has assessed odor memory in SZ, reporting impairment in a predominantly male group of SZ patients for whom neuroleptic medications had been withdrawn (Wu et al., 1993).

Unlike the aforementioned test measures, consensus is present for tests of odor identification, conceivably reflecting their higher reliability (see Chapter 11). Thus, impairments in odor identification have been reliably reported in patients with psychotic disorders (Brewer et al., 1996; Compton et al., 2006; Corcoran et al., 2005a; Good et al., 2010; Hurwitz et al., 1988; Malaspina et al., 2002; Moberg et al., 1999). Odor identification in SZ has been shown to correlate with measures of verbal memory (Compton et al., 2006), quality of smooth pursuit eye movements in patients with deficit syndrome SZ (Malaspina et al., 1994; Malaspina et al., 2002), disease duration (Moberg et al., 1997b; Moberg et al., 2006), negative disease symptoms (Brewer et al., 2001; Corcoran et al., 2005a; Good et al., 2010; Strauss et al., 2010), and measures of social drive and intelligence (Corcoran et al., 2005b; Malaspina and Coleman, 2003; Moberg et al., 2006). A more recent study found SZ patients may be impaired in their accuracy for identifying UPSIT odorants that are perceived as pleasant, but not for ones that are perceived as unpleasant, although differences in stimulus intensity could be a mitigating factor (Kamath et al., 2011). Most studies have found the SZ-related deficits to be unrelated to smoking history (Brewer et al., 1996; Houlihan et al., 1994; Kopala et al., 1993), antipsychotic medication treatment (Kopala et al., 1989; Szeszko et al., 2004; Wu et al., 1993), cannabis use (Brewer et al., 1996), or age of symptom onset (Corcoran et al., 2005a).

The data that have emerged regarding specific subgroups of patients with psychotic disorders shed light on the fact that not all patients are impaired on olfactory testing. Depending on the sample composition, rates of olfactory impairment differ. For example, in a sample of antipsychotic drug naïve, first-episode psychosis patients, Kopala et al. (1993) demonstrated that only 30% met criteria for 'microsmia' according to standardization data (Doty, 1995). In contrast, in patients with longstanding illness and comorbid polydipsia (water intoxication), 100% of the sample would have been considered impaired (Kopala et al., 1998a) and in those patients who meet criteria for the deficit syndrome, 72% of patients would also be considered impaired (Malaspina et al., 2002).

It is noteworthy that earlier research had noted sex differences in the odor identification test scores of SZ patients and that exposure to estrogen may 'protect' the olfactory system from damage (Kopala et al., 1989; Kopala and Clark, 1990). However, it has since become clear that both sexes demonstrate these deficits (for a review, see Moberg et al. 1999). Nonetheless, women in the general population, as well as women with AD or PD, outperform their male counterparts on numerous olfactory tests (Doty and Cameron, 2009). This sex difference extends to patients with psychotic disorders in general, albeit not beyond the generalized male–female difference (Good et al., 2007; Malaspina et al., 2012). Interestingly, in female patients with psychotic disorders, estrogen deficiency, such as occurs in the menopause, eliminates the female:male disparity in olfactory performance seen in later life (Kopala et al., 1995b). Postmenopausal women, who are otherwise healthy and using estrogen replacement therapy (ERT), reportedly have better olfactory memory skills than postmenopausal women who have never used ERT (Sundermann et al., 2006; Doty et al., 2015a). Such therapy has also been noted to improve psychotic symptoms in both male and female patients with psychotic disorders (Kulkarni et al., 2008; Le et al., 2000).

The aforementioned studies have largely assessed olfaction bilaterally, that is, with both sides of the nose being evaluated at the same time. However, testing each nasal chamber separately, as can be done by occluding the contralateral nostril, makes it possible to examine, to some degree, lateralized pathology since the output neurons of the the olfactory bulb project mainly to ipsilateral brain structures (Doty et al., 1997a; Moberg et al., 2003b). There is a suggestion that the right nostril and right hemisphere may be dominant for olfactory processing, in particular for olfactory discrimination (Zatorre and Jones-Gotman, 1990). Nonetheless, a functional lateralization of odor identification ability has not been consistently observed (Hornung et al., 1990) and unilateral testing has not generally found left:right differences in the olfactory function of SZ patients (Good et al., 2002; Roalf et al., 2006; Szeszko et al., 2004). That being said, when Good et al. (2002) divided their male patients into those who were normosmic (normal sense of smell) and those who were microsmic (impaired sense of smell), the microsmic patients were more likely to have relatively poorer left nostril identification, whereas the opposite was true of the normosmics. This finding, which suggests that a specific subgroup of male patients may have significant left hemisphere abnormalities, is in accord with the observation of Szeszko et al. (2004) that patients with left nostril impairment exhibit more problems with personal hygiene.

Reduced olfactory identification performance has been noted in unaffected first-degree relatives of patients with psychotic disorders (Keshavan et al., 2009; Kopala et al., 2001) and in non-psychotic twins of monozygotic twin pairs discordant for SZ (Kopala et al., 1998b; Ugur et al., 2005). These gene-based findings suggest that olfactory processing is a trait marker of psychosis, rather than simply being state-related. Individuals who score highly on schizotypy measures are thought to be genetically similar to patients with psychosis and share genetic vulnerability. Indeed, olfactory test scores are similar in schizotypy patients when compared with those who are psychotic and are impaired relative to control subjects (Kamath and Bedwell, 2008; Mohr et al., 2001; Park and Schoppe, 1997). Further support for a gene-based olfactory deficit in SZ comes from evidence that children with the 22q11 deletion syndrome, who often develop SZ in adulthood, uniformly exhibit olfactory dysfunction (Sobin et al., 2006).

Like olfactory hallucinations, impaired olfactory identification ability may be a marker for poorer outcomes in patients with SZ. Good et al. (2006) have shown that those SZ patients who had higher negative scores on the Positive and Negative Syndrome Scale (PANSS; Kay and Sevy, 1990) after one year of treatment had lower initial (drug naïve) UPSIT scores at baseline assessment. In a subsequent study, these researchers also found that patients with poorer olfactory test scores at first episode had poorer functional outcome than those with normal olfactory function after an approximate four year follow up (Good et al., 2010). Testing olfactory function in individuals who are either at genetic high risk or who are at clinically high risk may allow for the identification of those who are likely to convert to frank psychosis (Brewer et al., 2003; see, however, Gill et al., 2014).

A number of studies have documented structural, electrophysiological, and metabolic alterations, as measured by functional imaging, within the olfactory system of patients with SZ. SZ has been associated with an overall 23% reduction in olfactory bulb volume (Turetsky et al., 2000), smaller nasal cavity volumes (Moberg et al., 2004), and less deep olfactory sulci (Turetsky et al., 2009a). Right olfactory bulb volumes have been reported to correlate positively with UPSIT scores (Nguyen et al., 2011). The 23% reduction of olfactory bulb volume noted by Turetsky et al. is substantially greater in magnitude than what has been demonstrated for generalized brain volumetric changes in schizophrenia. Cell loss, such as what may occur as a result of the interruption of the rostral migratory stream regeneration, has been proposed (Chazal et al., 2000). In an animal model of schizophrenia, secretion of a glycoprotein assumed to be involved in the regulation of neuronal migration (reelin) appears to be compromised in the olfactory bulb (Pappas et al., 2003). Reelin expression has been noted to be significantly reduced in patients with psychotic disorders (Pantazopoulos et al., 2010) and thus may have a bearing on olfactory processing in this patient population.

Olfactory evoked potentials (ERPs) were examined in a group of patients with psychotic disorders by Turetsky et al. (2003a). Increasing concentrations of hydrogen sulfide (an unpleasant odorant) were passively presented while ERPs were measured. Patients demonstrated N1 and P2 latency prolongation and P2 amplitude reduction, implicating the primary olfactory cortex (N1) and secondary integration regions (P2), respectively. N1 prolongation was associated with olfactory threshold deficits while P2 amplitude reduction was correlated with decreased olfactory identification accuracy. However, these data cannot rule out more peripheral abnormalities that may be at the heart of these findings. Indeed, *larger* summated surface potentials have been recorded from the olfactory epithelia of patients with SZ (i.e., the electro-olfactogram) (Turetsky et al., 2009b), suggesting deficits at either the epithelial or bulbar level, the latter possibly reflecting release of inhibition on the olfactory receptor cells.

The earliest neuroimaging examination of patients with SZ used positron emission tomography (PET) to compare a group of patients with an olfactory agnosia (which the authors defined as normal detection threshold but impaired olfactory identification test scores) with a group of normosmic patients and a group of healthy control subjects (Clark et al., 1991). The combined normosmic and agnosic patient groups had lower rates of frontal metabolism than control subjects. Relative to those with normal olfactory test scores, the patients with olfactory agnosia had hypometabolism in the right thalamus and basal ganglia. A subsequent study that used single photon emission computed tomography (SPECT) found that patients with psychotic disorders exhibited hypometabolism in the right frontal, superior temporal, supramarginal, and angular gyri during an odor matching task (Malaspina et al., 1998). Bertollo et al. (1996) also noted, in patients with psychotic disorders, non-dominant abnormalities in the lateral posterior orbitofrontal cortex, a key odor processing area of the brain.

Wu et al. (1993) examined olfactory function and performed a PET scan on a group of patients with SZ. Significant correlations between UPSIT scores and left hemisphere regional brain metabolism were noted for the middle frontal, inferior frontal, and inferior gyri. Significant correlations between scores on an olfactory match to sample test and metabolism in left temporal regions, including the amygdala and hippocampus, were also observed. Healthy controls, however, did not undergo PET scanning, and, as such, the normal pattern of metabolic correlations is unknown. The regional cerebral blood flow (rCBF) was measured while patients were engaged in a test of attention and monitoring, confusing the interpretation of these findings.

Using odors they deemed pleasant and unpleasant, Crespo-Facorro et al. (2001) found that, for pleasant odors, SZ patients and control subjects demonstrated similar patterns of neural activation. Conversely, for unpleasant odors, patients invoked a compensatory pattern of brain activation that included distributed regions of the frontal cortex instead of the predicted limbic/paralimbic regions.

436   Chapter 18   Odor Perception and Neuropathology in Neurodegenerative Diseases and Schizophrenia

A PET study performed by these investigators employed numerous odorants to examine the neural substrates of olfactory detection, hedonics, and familiarity. The patients with SZ evidenced reduced odor-induced activation in frontal and temporal regions, including the pyriform and orbitofrontal cortices and insula. Consistent with the findings of Wu et al. (1993), more abnormalities were apparent in the left hemisphere.

Only one fMRI study has been published examining olfactory activation in schizophrenia (Schneider et al., 2007). In this study, male patients with schizophrenia (along with unaffected brothers and healthy controls) were presented with an unpleasant and a pleasant odorant in a block design. Blood oxygen level dependent (BOLD) effects were examined. In patients presented with an unpleasant odor, reduced activation was noted in frontal (middle frontal gyrus) and temporal (middle temporal gyrus) regions, but there was increased activation (compared to controls) in the anterior cingulate. For a pleasant odor, patients demonstrated reduced activation of the thalamus compared with control subjects. A major limitation of this and most extant SZ imaging data is that only males (Clark et al., 1991; Plailly et al., 2006; Schneider et al., 2007) or a disproportionate number of males (Crespo-Facorro et al., 2001; Wu et al., 1993) were tested and only those with a diagnosis of pure schizophrenia.

18.13 SPINOCEREBELLAR ATAXIAS

Spinocerebellar ataxias (SCAs) are a heterogenous group of diseases of the cerebellum and its afferent and efferent connections. Typical symptoms include progressive deterioration of limb movements, gait, and speech. However, a more diffuse neurodegenerative process is present in most instances, as evidenced by autonomic dysfunctions, extra pyramidal signs, sleep disturbances, visual attentional deficits, and impairments of saccades and smooth pursuit movements (Rub et al., 2008). Most forms are inherited, with over 30 genes identified to date. The most common SCA is Machado-Joseph disease (SCA3).

Since the pioneering functional imaging study of Yousem et al. (1997), it has become apparent that the cerebellum is somehow associated with olfactory processing, both in terms of motor and sensory function (Sobel et al., 1998). However, the first evidence for an olfactory deficit in SCAs came from a threshold study by Satya-Murti and Crisostomo (1988). On average, recognition thresholds for seven patients with Friedreich's ataxia (FRDA), an SCA associated with degeneration of the afferent cerebellar pathways, were elevated relative to those of matched controls.

Three studies appeared in 2003 noting olfactory impairment in patients with hereditary ataxias. In the first of these studies, Connelly et al. (2003) administered the UPSIT to 23 patients with Friedreich's ataxia (FRDA), a disorder associated with degeneration of the afferent cerebellar pathways, and 12 patients with other ataxias associated with cerebellar degeneration. The latter were designated as CNS ataxias (two with Spinocerebellar Ataxia type 2 or SCA2; five with SCA3; one with SCA7; and four whose genetic makeup was unknown). Both of these patient groups, whose scores did not differ significantly from one another, exhibited significant (p < 0.001) olfactory deficits relative to matched controls, although the effects were modest (Friedreich ataxia and control UPSIT medians = 36 and 38; CNS ataxias and control UPSIT medians 34.5 and 37). Seventy percent of the 23 FRDA patients scored below their matched controls, whereas 83% of the 12 CNS ataxia patients did so. In FRDA, UPSIT scores did not correlate with the length of the shorter pathological expanded GAA triplet repeat in the FRDA gene, disease duration, or ambulatory ability. In the CNS ataxias, no meaningful correlations were present between the UPSIT scores and disease duration or ambulatory ability.

In the second of these studies, the UPSIT was administered to 12 patients with SCA2, 5 patients with Machado-Joseph disease (SCA3/MJD), 5 patients with autosomal recessive ataxia, and 5 patients with sporadic ataxia (Fernandez-Ruiz et al., 2003). Relative to matched controls, subjects from all groups except the Machado-Joseph disease group exhibited significantly lower UPSIT scores than controls (median scores of all ataxia patients and controls = 26.0 and 34, respectively).

In the third study, Abele et al. (2003) administered tests of odor identification, detection (n-butanol thresholds), and odor discrimination to 19 patients with cerebellar ataxia (8 with multiple system atrophy of the cerebellar type and 11 with late onset sporadic ataxia of unknown etiology. While significant impairment was noted for the odor identification and discrimination tests (74% and 44% of the patients, respectively), this was not the case for the odor threshold test, where only 16% of the patients scored in the abnormal range. No differences in test scores were apparent between the patients with the two classes of ataxia.

Subsequent studies have essentially confirmed the findings of the aforementioned studies, save those for detection thresholds. In a study of 53 patients with spinocerebellar ataxia type 2, Velazquez-Perez et al. (2006) found significant decrements in odor identification, discrimination, recognition, and detection, although like the earlier studies, the magnitude of the deficits were much less than those seen in AD or PD (e.g., the mean UPSIT score for the patients was 27.3 compared to 31.7 for the controls). Weak, but statistically significant, correlations were noted between UPSIT scores and ataxia scores, age, and smoking history (rs < 0.30). However, only the correlation with age remained after omitting scores of patients with MMSE scores below 25. Braga-Neto et al. (2011) tested 41 patients with SCA3 and 46 controls for odor identification and found the expected reduction in function which, as was the case with the Velazquez-Perez et al. (2006) study, was related to the MMSE scores. Moscovich et al. (2012) identified olfactory deficits in a study of 37 genetically determined autosomal dominant ataxia and 31 patients with familial ataxia of unknown genetic basis. When statistically adjusted for cognitive function, however, no significant difference was observed between the ataxia and control groups, suggesting to these authors that previous findings were likely due to general cognitive deficits rather than to specific olfactory problems. This seems unlikely, however, since at least some of the previous studies took cognitive factors into account (e.g., Connelly et al., 2003).

18.14 TEMPORAL LOBE EPILEPSY

Although early threshold studies reported heightened overall bilateral sensitivity in patients with temporal lobe epilepsy (TLD), particularly prior to an ictal event (e.g. Campanella et al., 1978; De Michele G. et al., 1976; Dimov, 1973; Santorelli and Marotta, 1964), more recent studies have not observed this phenomenon. In fact, most have failed to observe threshold changes (Eskenazi et al., 1986; Martinez et al., 1993). However, relatively few subjects have been tested and most studies have used threshold tests of questionable reliability, such as the simple ascending method of limits (see Chapter 11). An exception is a recent study of 141 epilepsy patients which employed a single staircase detection threshold test for the rose-like odorant phenyl ethyl alcohol (Doty et al., 2015b). As shown in Table 18.5, a significant threshold *deficit* was found on both the right and left sides of the nose, with a slightly larger deficit on the left. This finding is an example of why it is wrong to infer that olfactory thresholds are a measure of "peripheral" dysfunction, since in epilepsy the lesions are clearly within the CNS.

Suprathreshold deficits in epileptic patients have also been observed. For example, Abraham and Mathai (1983) noted decreased performance on an odor-matching task in 28 epileptic patients with right-sided foci. Interestingly,

patients with left-sided foci did not exhibit the same difficulty in performing the bilaterally-administered olfactory matching task. Eskenazi et al. (1986) found a small bilateral odor identification deficit in 18 epileptic patients relative to 17 normal controls, which was apparently independent of the size of the seizure focus. Carroll et al. (1993) observed a decrease in immediate odor memory to common odorants (e.g., vinegar, coconut, coffee, nail varnish, and garlic) in 10 patients with right-sided, but not left-sided, TLE, seemingly in accord with the findings of Abraham and Mathai. In contrast, Martinez et al. (1993) found no left:right differences in measures odor discrimination, short- and long-term odor memory, and odor naming in 21 patients with epilepsy, although, relative to controls, deficits were observed on all of these measures.

More recently, Haehner et al. (2012) reported that while 13 patients with unilateral epileptic foci (number of left and right side foci not mentioned) tended to have more impaired odor identification deficits on the side of the lesion ($p = 0.057$), this was not the case for odor discrimination and detection measures. However, in the aforementioned Doty et al. (2015b) study, in which 141 epilepsy patients were evaluated along with matched controls, no evidence for lower UPSIT scores on the side of the lesion was evident, despite the fact that, overall, these scores were clearly depressed in the epilepsy patients (Table 18.6). Temporal lobe resection, however, was associated with ipsilateral deficits.

18.15 VASCULAR DEMENTIA

Vascular dementia (VD), the second most common form of dementia, is associated with vascular disease and reduced blood flow to the brain. Patients with VD have considerable loss of smell function. In the pioneering study on this topic, Knupfer and Spiegel (1986) found 18 VD patients, relative to 19 normal controls, exhibited elevated thresholds to eucalyptol, citral, and prunolide, and to perform more poorly on tests of odor identification, naming, and recognition. For all tests, however, the degree of dysfunction was less than that observed in 18 patients with probable AD, a finding supported by another study using a three-odor

Table 18.5 PEA odor detection threshold values with temporal lobe epilepsy and matched normal controls. Threshold values log vol/vol in USP grade light mineral oil. Negative signs are omitted from threshold values, so larger numbers signify greater sensitivity. From Doty et al. (2015b)

Epilepsy Focus	Nostril	Sample Size	Epilepsy Mean (SEM)	Control Mean (SEM)	P-value	Age and Education Adjusted P-value
Left	Left	35	4.51 (0.25)	5.39 (0.19)	0.02	0.06
	Right	35	4.60 (0.30)	5.58 (0.23)	0.04	0.04
Right	Left	36	4.76 (0.24)	5.30 (0.24)	0.13	0.26
	Right	36	4.95 (0.17)	5.72 (0.24)	0.007	0.01

438   **Chapter 18 Odor Perception and Neuropathology in Neurodegenerative Diseases and Schizophrenia**

Table 18.6 UPSIT scores of patients with temporal lobe epilepsy and matched controls. Data from Doty et al. (2015b)

Epilepsy Focus	Nostril	Sample Size	Epilepsy Mean (SEM)	Control Mean (SEM)	P-value	Age and Education Adjusted P-value
Left	Left	35	30.0 (1.35)	35.5 (0.49)	0.0006	0.005
	Right	35	29.0 (1.46)	36.3 (0.49)	0.0002	0.001
Right	Left	36	32.5 (0.87)	35.5 (0.43)	0.006	0.01
	Right	36	30.2 (0.95)	36.1 (0.55)	<0.0001	0.0002

identification test (Duff et al., 2002). Although Gray et al. (2001) found that odor identification performance, as measured by the UPSIT, was lower in 13 VD patients relative to 13 controls, their test scores did not differ significantly from those of the 13 AD patients they tested. Recently, Lee et al. (2010) demonstrated that, on average, 27 patients with a 'pure' form of vascular dementia, namely cerebral autosomal dominant arteriopathy with subcortical infarcts and leukoencephalopathy (CADASIL), scored below matched normal controls on an odor identification test. The olfactory test scores correlated significantly with scores on the Mini-Mental Status Examination (MMSE; r = 0.43) and the Controlled Oral Word Association Test (COWAT; r = 0.49).

18.16 CONCLUSIONS AND CONSIDERATIONS

In this chapter, we have reviewed a number of studies demonstrating olfactory dysfunction in a wide range of neurological diseases. While the focus has been on those for which the most olfactory data are available, it should be noted that olfactory dysfunction has been observed in a number of disorders not discussed in detail in this chapter. Thus, smell loss is present in frontotemporal dementia (Heyanka et al., 2014; Pardini et al., 2009), hypogonadopropic hypogonadism (Lewkowitz-Shpuntoff et al., 2012), restless leg syndrome (Adler et al., 1998), Korsakoff's psychosis (Mair et al., 1986), alcoholism (Shear et al., 1992), seasonal affective disorder (Faulcon et al., 1999; Postolache et al., 2002), head trauma (Doty et al., 1997b), attention deficit/hyper-reactivity disorder (Gansler et al., 1998), stroke (Rousseaux et al., 1996), and human immunodeficiency or acquired immune deficiency syndrome (Brody et al., 1991), to name a few (also see Chapter 17). It is noteworthy that a number of neurological disorders are accompanied by no, or only modest, alterations in smell function. In addition to those mentioned earlier in this chapter (e.g., MPTP-P), such disorders include depression (Amsterdam et al., 1987) and panic disorder (Kopala and Good, 1996). However, lack of involvement in neurological diseases seems to be the exception rather than the rule, as most neurological – particularly degenerative – diseases appear to be associated with at least some degree of olfactory dysfunction.

It is important to note that several studies of AD and SZ patients have reported lack of significant effects on odor detection threshold tests, in contrast to highly significant effects on such tests as the UPSIT. In some such cases, the lack of a threshold effect has been assumed to signify that olfactory system is, in effect, functioning normally and that the deficit on identification reflects a central olfactory agnosia. A case in point for AD is work by Koss et al. These investigators reported, in two publications largely based upon the same set of patients, significant UPSIT, but not pyridine threshold, deficits in AD patients relative to controls (Koss et al., 1987; Koss et al., 1988). They concluded that the olfactory deficit of AD is associated with a "central" rather than a "peripheral" anomaly. While, ultimately, their overall conclusion may be correct (i.e., that the cause of olfactory deficits in AD reflects damage in structures more central than the olfactory neuroepithelium), the basis upon which they arrive at this conclusion is suspect for a number of reasons. First, the vast majority of studies find threshold deficits in AD patients, and even their own data point to this conclusion. Thus, three of their subjects (30%) were either anosmic or hyposmic by their own criteria and, in fact, the average threshold value for the AD patients was higher (i.e., indicative of less sensitivity) than that of the controls, even though statistical significance at the 0.05 level was not achieved. Second, compared to most studies on this topic (see Table 18.1), their sample size was very small, compromising statistical power. In general, single ascending series detection threshold tests, such as those administered in their studies, have test-retest reliability values below 0.40, compared to those of the UPSIT, which are uniformly above 0.90 (Doty et al., 1995b). Hence, the power of such a threshold test to detect a deficit is much less than that of a 40-item odor identification test, and is particularly compromised with small subject samples. Third, the general assumption that threshold tests reflect peripheral damage and identification tests reflect central damage is untenable. Thus, central lesions, in fact, have been associated with threshold deficits (Doty et al., 2015b) and peripheral (i.e., neuroepithelial) lesions are known to influence both tests of odor detection threshold and odor identification (Deems et al., 1991).

Such issues are important for interpreting divergent findings in the literature (see Chapter 11). Thus, while the test scores of most nominally distinct olfactory tests are positively correlated, olfactory tests differ in terms of their relative reliability and sensitivity, making direct comparisons among them problematic. Moreover, they also differ in terms of the types and number of odorants they employ and in their non-olfactory task demands. Differences in psychophysical paradigms, including instructions to the subject, the time between stimulus presentations, the use or lack of use of forced-choice trials, the differences in odorants used, and the criteria employed for establishing the final test measure (e.g., the number of reversals in a staircase threshold procedure), all can influence the final outcome. This is particularly true in diseases, such as SZ, where the olfactory deficits are comparatively modest (Turetsky and Moberg, 2009). In such disorders, a variety of additional factors can influence the observed effects, including disparities in illness severity from sample to sample (Kopala et al., 1998a) and the involved medications (Sirota et al., 1999a).

REFERENCES

Abele, M., Riet, A., Hummel, T., et al. and (2003) Olfactory dysfunction in cerebellar ataxia and multiple system atrophy. *J. Neurol.*, 250: 1453–1455.

Abraham, A. and Mathai, K. V. (1983) The effect of right temporal lobe lesions on matching of smells. *Neuropsychologia*, 21: 277–281.

Aden, E., Carlsson, M., Poortvliet, E., et al. (2011) Dietary intake and olfactory function in patients with newly diagnosed Parkinson's disease: a case–control study. *Nutr. Neurosci.*, 14: 25–31.

Adler, C. H., Gwinn, K. A. and Newman, S. (1998) Olfactory function in restless legs syndrome. *Mov. Disord.*, 13: 563–565.

Ahlskog, J. E., Waring, S. C., Petersen, R. C., et al. (1998) Olfactory dysfunction in Guamanian ALS, parkinsonism, and dementia. *Neurology*, 51: 1672–1677.

Aksoy, H., Dean, G., Elian, M., et al. (2003) A4T mutation in the SOD1 gene causing familial amyotrophic lateral sclerosis. *Neuroepidemiology*, 22: 235–238.

Alcalay, R. N., Siderowf, A., Ottman, R., et al. (2011) Olfaction in Parkin heterozygotes and compound heterozygotes: the CORE-PD study. *Neurology*, 76: 319–326.

Almkvist, O., Berglund, B., Nordin, S. and Wahlund, L. O. (1992) Is pyridine odor deficit related to progression of dementia in Alzheimer's disease? *Reports from the Department of Psychology, Stockholm University*, 748: 1–15.

Amsterdam, J. D., Settle, R. G., Doty, R. L., et al. (1987) Taste and smell perception in depression. *Bio. Psychiat.*, 22: 1481–1485.

Ansari, K. A. and Johnson, A. (1975) Olfactory function in patients with Parkinson's disease. *J. Chronic. Dis.*, 28: 493–497.

Antsov, E., Silveira-Moriyama, L. Kilk, S. et al. (2014) Adapting the Sniffin' Sticks olfactory test to diagnose Parkinson's disease in Estonia. *Parkinsonism Relat. Disord* 20: 1419–1422.

Baba, T., Kikuchi, A., Hirayama, K., et al. 2012) Severe olfactory dysfunction is a prodromal symptom of dementia associated with Parkinson's disease: a 3 year longitudinal study. *Brain*, 135: 161–169.

Bacon, A. W., Bondi, M. W., Salmon, D. P. and Murphy, C. (1998) Very early changes in olfactory functioning due to Alzheimer's disease and the role of apolipoprotein E in olfaction. *Ann. N. Y. Acad. Sci.*, 855: 723–731.

Bacon-Moore, A. S., Paulsen, J. S. and Murphy, C. (1999) A test of odor fluency in patients with Alzheimer's and Huntington's disease. *J. Clin. Exp. Neuropsychol.*, 21: 341–351.

Bahar-Fuchs, A., Chetelat, G., Villemagne, V. L., et al. (2010) Olfactory deficits and amyloid-beta burden in Alzheimer's disease, mild cognitive impairment, and healthy aging: a PiB PET study. *J. Alz. Dis.*, 22: 1081–1087.

Bahar-Fuchs, A., Moss, S., Rowe, C. and Savage, G. (2011) Awareness of olfactory deficits in healthy aging, amnestic mild cognitive impairment and Alzheimer's disease. *Intl. Psychogeratrics*, 23: 1097–1106.

Baker, K. B. and Montgomery, E. B., Jr., (2001) Performance on the PD test battery by relatives of patients with progressive supranuclear palsy. *Neurology*, 56: 25–30.

Barz, S., Hummel, T., Pauli, E., et al. (1997) Chemosensory event-related potentials in response to trigeminal and olfactory stimulation in Parkinson's disease. *Neurology*, 49: 1424–1431.

Baumer, D., Hilton, D., Paine, S. M., et al. (2010) Juvenile ALS with basophilic inclusions is a FUS proteinopathy with FUS mutations. *Neurology*, 75: 611–618.

Becker, E., Hummel, T., Piel, E., et al. (1993) Olfactory event-related potentials in psychosis-prone subjects. *Int. J. Psychophysiol.*, 15: 51–58.

Bedlack, R. S., Strittmatter, W. J. and Morgenlander, J. C. (2000) Apolipoprotein E and neuromuscular disease: a critical review of the literature. *Arch. Neurol.*, 57: 1561–1565.

Berendse, H. W., Booij, J., Francot, C. M., et al. (2001) Subclinical dopaminergic dysfunction in asymptomatic Parkinson's disease patients' relatives with a decreased sense of smell. *Ann. Neurol.*, 50: 34–41.

Berendse, H. W., Roos, D. S., Raijmakers, P. and Doty, R. L. (2011) Motor and non-motor correlates of olfactory dysfunction in Parkinson's disease. *J Neurol. Sci.* 310: 21–24.

Berg, D., Schweitzer, K., Leitner, P., et al. (2005) Type and frequency of mutations in the LRRK2 gene in familial and sporadic Parkinson's disease. *Brain*, 128: 3000–3011.

Berger-Sweeney, J. (1998) The effects of neonatal basal forebrain lesions on cognition: towards understanding the developmental role of the cholinergic basal forebrain. *Int. J. Dev. Neurosci.*, 16: 603–612.

Bertollo, D. N., Cowen, M. A. and Levy, A. V. (1996) Hypometabolism in olfactory cortical projection areas of male patients with schizophrenia: an initial positron emission tomography study. *Psychiat. Res.*, 60: 113–116.

Bezard, E., Gross, C. E., Fournier, M.C., et al. (1999) Absence of MPTP-induced neuronal death in mice lacking the dopamine transporter. *Exp. Neurol.*, 155: 268–273.

Boesveldt, S., Verbaan, D., Knol, D. L., et al. (2008a). A comparative study of odor identification and odor discrimination deficits in Parkinson's disease. *Mov. Disord.* 23: 1984–1990.

Boesveldt, S., de Muinck Keizer, R. J., Knol, D. L., et al. (2009a) Extended testing across, not within, tasks raises diagnostic accuracy of smell testing in Parkinson's disease. *Mov. Disord.*, 24: 85–90.

Boesveldt, S., de Muinck Keizer, R. J. O., Wolters, E. Ch. et al. (2009) Odor recognition memory is not independently impaired in Parkinson's disease. *J. Neural Transm.* 116: 575–578.

Boesveldt, S., Stam, C. J., Knol, D. L., et al. (2009b) Advanced time-series analysis of MEG data as a method to explore olfactory function in healthy controls and Parkinson's disease patients. *Hum. Brain Mapp.*, 30: 3020–3030.

Boesveldt, S., Verbaan, D., Knol, D. L., et al. (2008b) Odour identification and discrimination in Dutch adults over 45 years. *Rhinology* 46: 131–136.

Bohnen, N. I., Gedela, S., Kuwabara, H., et al. (2007) Selective hyposmia and nigrostriatal dopaminergic denervation in Parkinson's disease. *J. Neurol.*, 254: 84–90.

Bohnen, N. I., Muller, M. L., Kotagal, V., et al. (2010) Olfactory dysfunction, central cholinergic integrity and cognitive impairment in Parkinson's disease. *Brain*, 133: 1747–1754.

Bostantjopoulou, S., Katsarou, Z., Mentenopoulos, G. and Logothetis, J. (1991) Olfactory disturbances in patients with Parkinson's disease. *Neurol. Psychiatr.*, 12: 13–15.

Bostantjopoulou, S., Katsarou, Z., Papadimitriou, A., et al. (2001) Clinical features of parkinsonian patients with the alpha-synuclein (G209A) mutation. *Mov. Disord.*, 16: 1007–1013.

Bovi, T., Antonini, A., Ottaviani, S., et al. (2010) The status of olfactory function and the striatal dopaminergic system in drug-induced parkinsonism. *J. Neurol.* 257: 1882–1889.

Braak, H. and Braak, E. (1991) Neuropathological stageing of Alzheimer-related changes. *Acta Neuropathol.*, 82: 239–259.

Bradley, E. A. (1984) Olfactory acuity to a pheromonal substance and psychotic illness. *Biol. Psychiat.*, 1984 Jun;19: 899–905.

Braga-Neto, P., Felicio, A. C., Pedroso, J. L., et al. (2011) Clinical correlates of olfactory dysfunction in spinocerebellar ataxia type 3. *Parkinsonism Relat. Disord.*, 17: 353–356.

Brewer, W. J., Edwards, J., Anderson, V., et al. (1996) Neuropsychological, olfactory, and hygiene deficits in men with negative symptom schizophrenia. *Biol. Psychiat.*, 40: 1021–1031.

Brewer, W. J., Pantelis, C., Anderson, V., et al. (2001) Stability of olfactory identification deficits in neuroleptic-naive patients with first-episode psychosis. *Am. J. Psychiatry.*, 158: 107–115.

Brewer, W. J., Wood, S. J., McGorry, P. D., et al. (2003) Impairment of olfactory identification ability in individuals at ultra-high risk for psychosis who later develop schizophrenia. *Am. J. Psychiat.*, 160: 1790–1794.

Brewer, W. J., Wood, S. J., Pantelis, C., et al. (2007) Olfactory sensitivity through the course of psychosis: Relationships to olfactory identification, symptomatology and the schizophrenia odour. *Psychiat. Res.*, 149: 97–104.

Brockmann, K., Srulijes, K., Hauser, A. K., et al. (2011) GBA-associated PD presents with nonmotor characteristics. *Neurology*, 77: 276–280.

Brodoehl, S., Klingner, C., Volk, G. F., et al. (2012) Decreased olfactory bulb volume in idiopathic Parkinson's disease detected by 3.0-Tesla magnetic resonance imaging. *Mov. Disord.*, 27: 1019–1025.

Brody, D., Serby, M., Etienne, N. and Kalkstein, D. S. (1991) Olfactory identification deficits in HIV infection [see comments]. *Am. J. Psychiat.*, 148: 248–250.

Broggio, E., Pluchon, C., Ingrand, P. and Gil, R. (2001) [Taste impairment in Alzheimer's disease]. [French]. *Revue Neurologique*, 157: 409–413.

Brousseau, K. and Brainerd, H. G. (1928) *Mongolism: a study of the physical and mental characteristics of mongolian imbeciles.* Williams and Wilkins: Baltimore.

Buchsbaum, M. S., Kesslak, J. P., Lynch, G., et al. (1991) Temporal and hippocampal metabolic rate during an olfactory memory task assessed by positron emission tomography in patients with dementia of the Alzheimer type and controls. Preliminary studies. *Arch. Gen. Psychiat.*, 1991 Sep; 48: 840–847.

Busenbark, K. L., Huber, S. I., Greer, G., et al. (1992) Olfactory function in essential tremor. *Neurology*, 42: 1631–1632.

Busse, K., Heilmann, R., Kleinschmidt, S., et al. (2012) Value of combined midbrain sonography, olfactory and motor function assessment in the differential diagnosis of early Parkinson's disease. *J. Neurol. Neurosurg. Psychiat.*, 83: 441–447.

Bylsma, F. W., Moberg, P. J., Doty, R. L. and Brandt, J. (1997) Odor identification in Huntington's disease patients and asymptomatic gene carriers. *J. Neuropsychiat. Clin. Neurosci.*, 9: 598–600.

Calderon-Garciduenas, L., Franco-Lira, M., Henriquez-Roldan, C., et al. (2010) Urban air pollution: Influences on olfactory function and pathology in exposed children and young adults. *Exp. Toxicol. Pathol.*, 62: 91–102.

Cambier, J., Masson, M., Viader, F. et al. (1985). [Frontal syndrome of progressive supranuclear palsy]. *Rev. Neurol.* (Paris). 141: 528–536.

Campanella, G., Filla, A. and De, M. G. (1978) Smell and taste acuity in epileptic syndromes. *Eur. Neurol.*, 17: 136–141.

Campbell, I. M. and Gregson, R. A. M. (1972) Olfactory short term memory in normal, schizophrenic and brain-damaged cases. *Aust. J. Psychol.*, 24: 179–185.

Carroll, B., Richardson, J. T. and Thompson, P. (1993) Olfactory information processing and temporal lobe epilepsy. *Brain Cogn.*, 22: 230–243.

Chan, A., Tam, J., Murphy, C., et al. (2002) Utility of olfactory identification test for diagnosing Chinese patients with Alzheimer's disease. *J. Clin. Exp. Neuropsychol.*, 24: 251–259.

Chazal, G., Durbec, P., Jankovski, A., et al. (2000) Consequences of neural cell adhesion molecule deficiency on cell migration in the rostral migratory stream of the mouse. *J. Neurosci.*, 20: 1446–1457.

Chen, W., Chen, S., Kang, W. Y., et al. (2012) Application of odor identification test in Parkinson's disease in China: a matched case–control study. *J. Neurol. Sci.*, 316: 47–50.

Chou, K. L. and Bohnen, N. I. (2009) Performance on an Alzheimerselective odor identification test in patients with Parkinson's disease and its relationship with cerebral dopamine transporter activity. *Parkinsonism Relat Disord.* 15: 640–643.

Church, S. M., Cotter, D., Bramon, E. and Murray, R. M. (2002) Does schizophrenia result from developmental or degenerative processes? *J Neural. Transm. Suppl*, 129–147.

Clark, C., Kopala, L., Hurwitz, T. and Li, D. (1991) Regional metabolism in microsmic patients with schizophrenia. *Can. Psychiatry-Revue Canadienne de Psychiatrie*, 36: 645–650.

Collet, S., Grulois, V., Bertrand, B. and Rombaux, P. (2009) Post-traumatic olfactory dysfunction: a cohort study and update. *B-ENT*, 5 Suppl 13: 97–107.

Compton, M. T., McKenzie, M. L., Esterberg, M. L., et al. (2006) Associations between olfactory identification and verbal memory in patients with schizophrenia, first-degree relatives, and non-psychiatric controls. *Schizoph. Res.*, 86: 154–166.

Connelly, T., Farmer, J. M., Lynch, D. R. and Doty, R. L. (2003) Olfactory dysfunction in degenerative ataxias. *J. Neurol. Neurosurg. Psychiat.*, 74: 1435–1437.

Constantinidis, J. and de Ajuriaguerra, J. (1970) [Familial syndrome with parkinsonian tremor and anosmia and its therapy with L-dopa associated with a decarboxylase inhibitor]. [French]. *Therapeutique*, 46: 263–269.

Conti, M. Z., Vicini-Chilovi, B., Riva, M., et al. (2013) Odor identification deficit predicts clinical conversion from mild cognitive impairment to dementia due to Alzheimer's disease. *Arch. Clin. Neuropsychol.*, 28: 391–399.

Corby, K., Morgan, C. D. and Murphy, C. (2012) Abnormal event-related potentials in young and middle-aged adults with the ApoE epsilon4 allele. *Int. J. Psychophysiol.*, 83: 276–281.

Corcoran, C., Whitaker, A., Coleman, E., et al. (2005a) Olfactory deficits, cognition and negative symptoms in early onset psychosis. *Schizophr. Res.*, 80: 283–293.

Corcoran, C., Whitaker, A., Coleman, E., et al. (2005b) Olfactory deficits, cognition and negative symptoms in early onset psychosis. *Schizophr. Res.*, 80: 283–293.

Correia, G. L., Ferreira, J. J., Rosa, M. M., et al. (2010) Worldwide frequency of G2019S LRRK2 mutation in Parkinson's disease: a systematic review. *Parkinsonism. Relat. Disord.*, 16: 237–242.

Corwin, J. and Serby, M. (1985) Olfactory recognition deficit in Alzheimer's and Parkinsonian dementias. *IRCS Med Sci*, 13: 260.

Cox, T. A., McDarby, J. V., Lavine, L., et al. (1989) A retinopathy on Guam with high prevalence in Lytico-Bodig. *Ophthalmology*, 96: 1731–1735.

Cramer, C. K., Friedman, J. H. and Amick, M. M. (2010) Olfaction and apathy in Parkinson's disease. *Parkinsonism. Relat. Disord.*, 16: 124–126.

Crespo-Facorro, B., Paradiso, S., Andreasen, N. C., et al. (2001) Neural mechanisms of anhedonia in schizophrenia: a PET study of response to unpleasant and pleasant odors. *JAMA*, 286: 427–435.

Cumming, A. G., Matthews, N. L. and Park, S. (2011) Olfactory identification and preference in bipolar disorder and schizophrenia. *Eur. Arch. Psychiat. Clin. Neurosci.*, 261: 251–259.

Damholdt, M. F., Borghammer, P., Larsen, L. and Ostergaard, K. (2011) Odor identification deficits identify Parkinson's disease patients with poor cognitive performance. *Mov. Disord.*, 26: 2045–2050.

Daum, R. F., Sekinger, B., Kobal, G. and Lang, C. J. (2000) Riechprufung mit "sniffin' sticks" zur klinischen Diagnostik des Morbus Parkinson. *Nervenarzt*, 71: 643–650.

de Haan, L. and Bakker, J. M. (2004) Overview of neuropathological theories of schizophrenia: from degeneration to progressive developmental disorder. *Psychopathology*, 37: 1–7.

De Michele G., Filla, A. and Campanella, G. (1976) Ulteriori dati sull'acuitá olfattiva negli epilettici. *Acta Neurol.*, 31: 250–256.

Deeb, J., Findley, L. J., Shah, M., et al. (2006) Smell tests compared to dopamine transporter imaging in diagnosis of idiopathic Parkinson's disease: A pilot study. *J. Neurol. Neurosurg. Psychiat.*, 77: 128.

Deeb, J., Shah, M., Muhammed, N., et al. (2010) A basic smell test is as sensitive as a dopamine transporter scan: comparison of olfaction, taste and DaTSCAN in the diagnosis of Parkinson's disease. *Q. J. Med.*, 103: 941–952.

Deems, D. A., Doty, R. L., Settle, R. G., et al. (1991) Smell and taste disorders, a study of 750 patients from the University of Pennsylvania Smell and Taste Center. *Arch. Otolaryngol. Head Neck Surg.*, 117: 519–528.

Delmaire, C., Dumas, E. M., Sharman, M. A., et al. (2013) The structural correlates of functional deficits in early huntington's disease. *Hum. Brain Mapp.* 34: 2141–2153.

Devanand, D. P., Liu, X., Tabert, M. H., et al. (2008) Combining early markers strongly predicts conversion from mild cognitive impairment to Alzheimer's disease. *Biol. Psychiat.*, 64: 871–879.

Devanand, D. P., Michaels-Marston, K. S., Liu, X., et al. (2000) Olfactory deficits in patients with mild cognitive impairment predict Alzheimer's disease at follow-up. *Am. Psychiat.*, 157: 1399–1405.

Devanand, D. P., Pelton, G. H., Zamora, D., et al. (2005) Predictive utility of apolipoprotein E genotype for Alzheimer disease in outpatients with mild cognitive impairment. *Arch. Neurol.*, 62: 975–980.

Devanand, D. P.,Van Heertum, R. L., Kegeles, L. S. et al. (2010) ^{99m}Tc hexamethyl-propylene-aminoxine single-photon emission computed tomography prediction of conversion from mild cognitive impairment to Alzheimer disease. *Am. J. Geriat. Psychiat.* 18: 959–972.

Devanand, D. P., Lee, S., Manly, J. et al. Olfactory deficits predict cognitive decline and Alzheimer's dementia in an urban community. *Neurology*, 2014, in press.

Dimov, D. (1973) The condition of olfaction in epilepsy. *Vestnik Otorinolaringol.*, 35: 22–23.

Djordjevic, J., Jones-Gotman, M., De, S. K. and Chertkow, H. (2008) Olfaction in patients with mild cognitive impairment and Alzheimer's disease. *Neurobiol. Aging.*, 29: 693–706.

Doop, M. L. and Park, S. (2006) On knowing and judging smells: identification and hedonic judgment of odors in schizophrenia. *Schizoph. Res.*, 81: 317–319.

Doty, R. L. (1995) *The Smell Identification Test^{TM} Administration Manual–3rd Edition*. Sensonics, Inc.: Haddon Hts., NJ.

Doty, R. L. (2007) Office procedures for quantitative assessment of olfactory function. *Am. J. Rhinol.*, 21: 460–473.

Doty, R. L. (2008) The olfactory vector hypothesis of neurodegenerative disease: is it viable? *Ann. Neurol.*, 63: 7–15.

Doty, R. L. (2011) Olfaction in Parkinson's disease and related disorders. *Neurobiol. Dis.* 46: 527–552, 2012.

Doty, R. L. (2012) Olfactory dysfunction in Parkinson disease. *Nat. Rev. Neurol.*, 8: 329–339.

Doty, R. L., Beals, E., Osman, A., et al. (2014) Suprathreshold odor intensity perception in early-stage Parkinson's disease. *Mov. Disord.* 29: 1208–1212.

Doty, R. L., Bromley, S. M., Moberg, P. J. and Hummel, T. (1997a) Laterality in human nasal chemoreception. *In:* Christman, S., *Cerebral Asymmetries in Sensory and Perceptual Processing*. North Holland Publishing Co.: Amsterdam, 497–542.

Doty, R. L., Bromley, S. M. and Stern, M. B. (1995a) Olfactory testing as an aid in the diagnosis of Parkinson's disease: development of optimal discrimination criteria. *Neurodegeneration*, 4: 93–97.

Doty, R. L. and Cameron, E. L. (2009) Sex differences and reproductive hormone influences on human odor perception. *Physiol. Behav.*, 97: 213–228.

Doty, R. L., Deems, D. A. and Stellar, S. (1988) Olfactory dysfunction in parkinsonism: a general deficit unrelated to neurologic signs, disease stage, or disease duration. *Neurology*, 38: 1237–1244.

Doty, R. L., Golbe, L. I., McKeown, D. A., et al. (1993) Olfactory testing differentiates between progressive supranuclear palsy and idiopathic Parkinson's disease. *Neurology*, 43: 962–965.

Doty, R. L., McKeown, D. A., Lee, W. W. and Shaman, P. (1995b) A study of the test-retest reliability of ten olfactory tests. *Chem. Senses.*, 20: 645–656.

Doty, R. L., Perl, D. P., Steele, J. C., et al. (1991) Odor identification deficit of the parkinsonism-dementia complex of Guam: Equivalence to that of Alzheimer's and idiopathic Parkinson's disease. *Neurology*, 41: 77–80.

Doty, R. L., Reyes, P. F. and Gregor, T. (1987) Presence of both odor identification and detection deficits in Alzheimer's disease. *Brain Res. Bull.*, 18: 597–600.

Doty, R. L., Riklan, M., Deems, D. A., et al. (1989) The olfactory and cognitive deficits of Parkinson's disease: evidence for independence. *Ann. Neurol.*, 25: 166–171.

Doty, R. L., Shaman, P., Applebaum, S. L., et al. (1984) Smell identification ability: changes with age. *Science*, 226: 1441–1443.

Doty, R. L., Singh, A., Tetrud, J. and Langston, J. W. (1992a) Lack of major olfactory dysfunction in MPTP-induced parkinsonism. *Ann. Neurol.*, 32: 97–100.

Doty, R. L., Smith, R., McKeown, D. A. and Raj, J. (1994) Tests of human olfactory function: principal components analysis suggests that most measure a common source of variance. *Percept.Psychophys.*, 56: 701–707.

Doty, R. L., Stern, M. B., Pfeiffer, C., et al. (1992b) Bilateral olfactory dysfunction in early stage treated and untreated idiopathic Parkinson's disease. *J. Neurol. Neurosurg. Psychiat.*, 55: 138–142.

Doty, R. L., Tourbier, I., Ng, V., et al. (2015a) Influences of hormone replacement therapy on olfactory and cognitive function in the menopause. *Neurobiol. Aging*, in press.

Doty, R. L., Tourbier, I., Silas, J., et al. (2015b) Influences of temporal lobe epilepsy and temporal lobe resection on olfactory function. Submitted.

Doty, R. L., Yousem, D. M., Pham, L.T., et al. (1997b) Olfactory dysfunction in patients with head trauma. *Arch. Neurol.*, 54: 1131–1140.

Double, K. L., Rowe, D. B., Hayes, M., et al. (2003) Identifying the pattern of olfactory deficits in Parkinson disease using the brief smell identification test. *Arch. Neurol.*, 60: 545–549.

Duda, J. E. (2010) Olfactory system pathology as a model of Lewy neurodegenerative disease. *J. Neurol. Sci.*, 289: 49–54.

Duff, K., McCaffrey, R. J. and Solomon, G. S. (2002) The Pocket Smell Test: successfully discriminating probable Alzheimer's dementia from vascular dementia and major depression. *J. Neuropsychiat. Clin. Neurosci.*, 14: 197–201.

Dunn, L. M. (1981) *Peabody Picture Vocabulary Test-Revised Manual for Forms L and M*. American Guidance Service: Circle Pines, MN.

Dunn, T. P. and Weller, M. P. (1989) Olfaction in schizophrenia. *Percept. Mot. Skills*, 69: 833–834.

Eggers, C., Schmidt, A., Hagenah, J., et al. (2010) Progression of subtle motor signs in PINK1 mutation carriers with mild dopaminergic deficit. *Neurology*, 74: 1798–1805.

Eibenstein, A., Fioretti, A. B., Simaskou, M. N., et al. (2005) Olfactory screening test in mild cognitive impairment. *Neurol. Sci.*, 26: 156–160.

Elian, M. (1991) Olfactory impairment in motor neuron disease: a pilot study. *J. Neurol. Neurosurg. Psychiat.* 54: 927–928.

Elsworth, J. D., Taylor, J. R., Sladek, J. R., Jr.,, et al. (2000) Striatal dopaminergic correlates of stable parkinsonism and degree of recovery in old-world primates one year after MPTP treatment. *Neuroscience*, 95: 399–408.

Eskenazi, B., Cain, W. S., Novelly, R. A. and Mattson, R. (1986) Odor perception in temporal lobe epilepsy patients with and without temporal lobectomy. *Neuropsychologia*, 24: 553–562.

Evidente, V. G., Esteban, R. P., Hernandez, J. L., et al. (2004) Smell testing is abnormal in 'lubag' or X-linked dystonia-parkinsonism: a pilot study. *Parkinsonism Relat. Disord.*, 10: 407–410.

Faulcon, P., Portier, F., Biacabe, B. and Bonfils, P. (1999) [Anosmia secondary to acute rhinitis: clinical signs and course in a series of 118 patients]. *Ann. Oto. Laryngol. Chir. Cervico Fac.*, 116: 351–357.

Ferraris, A., Ialongo, T., Passali, G. C., et al. (2009) Olfactory dysfunction in Parkinsonism caused by PINK1 mutations. *Mov. Disord.*, 24: 2350–2357.

Ferreira, J. J., Guedes, L. C., Rosa, M. M., et al. (2007) High prevalence of LRRK2 mutations in familial and sporadic Parkinson's disease in Portugal. *Mov. Disord.*, 22: 1194–1201.

Fernandez-Ruiz, J., Díaz, R., Hall-Haro, C. et al. (2003). Olfactory dysfunction in hereditary ataxia and basal ganglia disorders. *Neuroreport.* 14: 1339–1341.

Forster, S., Vaitl, A., Teipel, S. J., et al. (2010) Functional representation of olfactory impairment in early Alzheimer's disease. *J. Alz.Dis.*, 22: 581–591.

Fusetti, M., Fioretti, A. B., Silvagni, F., et al. (2010) Smell and preclinical Alzheimer disease: study of 29 patients with amnesic mild cognitive impairment. *J. Otolaryngol. Head Neck Surg.*, 39: 175–181.

Gaig, C., Vilas, D., Infante, J. et al. (2014) Nonmotor symptoms in LRRK2 G2019S associated Parkinson's disease. PLoS ONE 9(10): e108982.

Gansler, D. A., Fucetola, R., Krengel, M., et al. (1998) Are there cognitive subtypes in adult attention deficit/hyperactivity disorder?. *J. Nerv. Ment. Dis.*, 186: 776–781.

Garland, E. M., Raj, S. R., Peltier, A. C., et al. (2011) A cross-sectional study contrasting olfactory function in autonomic disorders. *Neurology*, 76: 456–460.

Geddes, J., Huws, R. and Pratt, P. (1991) Olfactory acuity in the positive and negative syndromes of schizophrenia. *Biol. Psychiat.*, 29: 774–778.

Getchell, M. L., Shah, D. S., Buch, S. K., et al. (2003) 3-Nitrotyrosine immunoreactivity in olfactory receptor neurons of patients with Alzheimer's disease: implications for impaired odor sensitivity. *Neurobiol. Aging.*, 2003 Sep; 24: 663–673.

Gilbert, P. E., Barr, P. J. and Murphy, C. (2004) Differences in olfactory and visual memory in patients with pathologically confirmed Alzheimer's disease and the Lewy body variant of Alzheimer's disease. *J. Int. Neuropsychol. Soc.*, 10: 835–842.

Gilbert, P. E. and Murphy, C. (2004) The effect of the ApoE epsilon4 allele on recognition memory for olfactory and visual stimuli in patients with pathologically confirmed Alzheimer's disease, probable Alzheimer's disease, and healthy elderly controls. *J. Clin. Exper. Neuropsychol.*, 26: 779–794.

Gill, K. E., Evans, E., Kayser, J. et al. (2014) Smell identification in individuals at clinical high risk for schizophrenia. *Psychiatry Res.* 220: 201–204.

Goker-Alpan, O., Lopez, G., Vithayathil, J., et al. (2008) The spectrum of parkinsonian manifestations associated with glucocerebrosidase mutations. *Arch. Neurol.*, 65: 1353–1357.

Goldstein, D. S., Holmes, C., Bentho, O., et al. (2008) Biomarkers to detect central dopamine deficiency and distinguish Parkinson disease from multiple system atrophy. *Parkinsonism. Relat Disord.*, 14: 600–607.

Goldstein, D. S., Li, S. T., Holmes, C. and Bankiewicz, K. (2003) Sympathetic innervation in the 1-methyl-4-phenyl-1,2,3,6-tetrahydropyridine primate model of Parkinson's disease. *J Pharmacol. Exp. Ther.*, 306: 855–860.

Goldstein, D. S. and Sewell, L. (2009) Olfactory dysfunction in pure autonomic failure: Implications for the pathogenesis of Lewy body diseases. *Parkinsonism Relat. Disord.*, 15: 516–520.

Good, K. P., Leslie, R. A., McGlone, J., et al. (2007) Sex differences in olfactory function in young patients with psychotic disorders. *Schizoph. Res.*, 97: 97–102.

Good, K. P., Martzke, J. S., Honer, W. G. and Kopala, L. C. (1998) Left nostril olfactory identification impairment in a subgroup of male patients with schizophrenia. *Schizophr. Res.*, 33: 35–43.

Good, K. P., Martzke, J. S., Milliken, H. I., et al. (2002) Unirhinal olfactory identification deficits in young male patients with schizophrenia and related disorders: association with impaired memory function. *Schizophr. Res.*, 56: 211–223.

Good, K. P., Tibbo, P., Milliken, H., et al. (2010) An investigation of a possible relationship between olfactory identification deficits at first episode and four-year outcomes in patients with psychosis. *Schizophr. Res.* 124: 60–65.

Good, K. P., Whitehorn, D., Rui, Q., et al. (2006) Olfactory identification deficits in first-episode psychosis may predict patients at risk for persistent negative and disorganized or cognitive symptoms. *Am. J. Psychiat.*, 163: 932–933.

Goudsmit, N., Coleman, E., Seckinger, R. A., et al. (2003) A brief smell identification test discriminates between deficit and non-deficit schizophrenia. *Psychiatry Res*, 120: 155–164.

Graves, A. B., Bowen, J. D., Rajaram, L., et al. (1999) Impaired olfaction as a marker for cognitive decline: interaction with apolipoprotein E epsilon4 status. *Neurology*, 53: 1480–1487.

Gray, A. J., Staples, V., Murren, K., et al. (2001) Olfactory identification is impaired in clinic-based patients with vascular dementia and senile dementia of Alzheimer type. *Int. Geriatr. Psychiatry*, 16: 513–517.

Green, J. E., Songsanand, P., Peretz, S., et al. (1989) Dissociation between basic and high order olfactory cacpities in Alzheimer's disease. *In:* Wurtman, R. J., Corkin, S. H., Growden, J. H. et al., *Proceedings of the Fifth Meeting of the International Study Group on the Pharmacology of Memory Disorders Associated with Aging.* Center for Brain Sciences and Metabolism Cahritable Trust: Cambridge, MA, 449–455.

Gross-Isseroff, R. G., Stoler, M., Ophir, D., et al. (1987) Olfactory sensitivity to androstenone in schizophrenic patients. *Biol. Psychiatry,* 22: 922–925.

Guarneros, M., Hummel, T., Martinez-Gomez, M. and Hudson, R. (2009) Mexico City air pollution adversely affects olfactory function and intranasal trigeminal sensitivity. *Chem. Senses.,* 34: 819–826.

Guo, X., Gao, G., Wang, X., et al. (2008) Effects of bilateral deep brain stimulation of the subthalamic nucleus on olfactory function in Parkinson's disease patients. *Stereotact. Funct. Neurosurg.,* 86: 237–244.

Gupta, A. K., Jeavons, P. M., Hughes, R. C. and Covanis, A. (1983) Aura in temporal lobe epilepsy: clinical and electroencephalographic correlation. *J. Neurol. Neurosurg. Psychiat.,* 46: 1079–1083.

Haehner, A., Boesveldt, S., Berendse, H. W., et al. (2009) Prevalence of smell loss in Parkinson's disease–a multicenter study. *Parkinsonism. Relat D.,* 15: 490–494.

Haehner, A., Henkel, S., Hopp, P., et al. (2012) Olfactory function in patients with and without temporal lobe resection. *Epilepsy Behav.,* 25: 477–480.

Haehner, A., Hummel, T., Hummel, C., et al. (2007) Olfactory loss may be a first sign of idiopathic Parkinson's disease. *Mov. Disord.,* 22: 839–842.

Hague, K., Lento, P., Morgello, S., et al. (1997) The distribution of Lewy bodies in pure autonomic failure: autopsy findings and review of the literature. *Acta Neuropathol.,* 94: 192–196.

Hamilton, J. M., Murphy, C. and Paulsen, J. S. (1999) Odor detection, learning, and memory in Huntington's disease. *J. Int. Neuropsychol. Soc,* 5: 609–615.

Handley, O. J., Morrison, C. M., Miles, C. and Bayer, A. J. (2006) ApoE gene and familial risk of Alzheimer's disease as predictors of odour identification in older adults. *Neurobiol. Aging,* 27: 1425–1430.

Harvey, P. D. and Sharma, T. (2002) *Understanding and Treating Cognition in Schizophrenia.* Martin Dunitz, Ltd.: London.

Hakyemez, H. A., Veyseller, B., Ozer, F. et al. (2013) Relationship of olfactory function with olfactory bulbus volume, disease duration and Unified Parkinson's disease rating scale scores in patients with early stage idiopathic Parkinson's disease. *J. Clin. Neurosci.* 20: 1469–1470.

Hawkes, C. H. and Doty, R. L. (2009) *The Neurology of Olfaction.* Cambridge University Press: Cambridge.

Hawkes, C. H. and Shephard, B.C. (1993) Selective anosmia in Parkinson's disease?. *Lancet,* 341: 435–436.

Hawkes, C. H. and Shephard, B. C. (1998a) Olfactory evoked responses and identification tests in neurological disease. *Ann. N. Y. Acad. Sci.,* 855: 608–615.

Hawkes, C. H., Shephard, B. C. and Daniel, S. E. (1997) Olfactory dysfunction in Parkinson's disease. *J. Neurol. Neurosurg. Psychiat.,* 62: 436–446.

Hawkes, C. H., Shephard, B. C., Geddes, J. F., et al. (1998) Olfactory disorder in motor neuron disease. *Exp. Neurol.,* 150: 248–253.

Hayashi, R. (2004) Olfactory illusions and hallucinations after right temporal hemorrhage. *Eur. Neurol.,* 51: 240–241.

Hemdal, P., Corwin, J. and Oster, H. (1993) Olfactory identification deficits in Down's syndrome and idiopathic mental retardation. *Neuropsychologia,* 31: 977–984.

Hensiek, A. E., Bhatia, K. and Hawkes, C. H. (2000) Olfactory function in drug induced parkinsonism. *J. Neurol.* 82: 247. Suppl 3 (2000): 303.

Herting, B., Schulze, S., Reichmann, H., et al. (2008) A longitudinal study of olfactory function in patients with idiopathic Parkinson's disease. *J. Neurol.,* 255: 367–370.

Heyanka, D. J., Golden, C. J., McCue, R. B. et al. (2014) Olfactory deficits in frontotemporal dementia as measured by the Alberta Smell Test. *Appl. Neuropsychol. Adult* 21: 176–182.

Hidalgo, J., Chopard, G., Galmiche, J., et al. (2011) Just noticeable difference in olfaction: a discriminative tool between healthy elderly and patients with cognitive disorders associated with dementia. *Rhinology,* 49: 513–518.

Hof, P. R., Bouras, C., Perl, D. P., et al. (1995) Age-related distribution of neuropathologic changes in the cerebral cortex of patients with Down's syndrome. Quantitative regional analysis and comparison with Alzheimer's disease. *Arch. Neurol.,* 52: 379–391.

Hornung, D. E., Leopold, D. A., Mozell, M. M., et al. (1990) Impact of left and right nostril abilities on binasal olfactory performance. *Chem. Senses,* 15: 233–237.

Houlihan, D. J., Flaum, M., Arnold, S. E., et al. (1994) Further evidence for olfactory identification deficits in schizophrenia. *Schizophr. Res.,* 12: 179–182.

Hozumi, S., Nakagawasai, O., Tan-No, K., et al. (2003) Characteristics of changes in cholinergic function and impairment of learning and memory-related behavior induced by olfactory bulbectomy. *Behav. Brain Res.,* 138: 9–15.

Hu, J., Wang, X., Liu, D., et al. (2012) Olfactory deficits induce neurofilament hyperphosphorylation. *Neurosci. Lett.,* 506: 180–183.

Hudry, J., Saoud, M., d'Amato, T., Dalery, J. and Royet, J. P. (2002) Ratings of different olfactory judgements in schizophrenia. *Chem. Senses,* 27: 407–416.

Hudry, J., Thobois, S., Broussolle, E., et al. (2003) Evidence for deficiencies in perceptual and semantic olfactory processes in Parkinson's disease. *Chem. Senses.,* 28: 537–543.

Hulshoff Pol, H. E. and Kahn, R. S. (2008) What happens after the first episode? A review of progressive brain changes in chronically ill patients with schizophrenia. *Schizophr. Bull.,* 34: 354–366.

Hummel, T., Fliessbach, K., Abele, M., et al. (2010a) Olfactory FMRI in patients with Parkinson's disease. *Front. Integr. Neurosci.,* 4: 125.

Hummel, T., Jahnke, U., Sommer, U., et al. (2005) Olfactory function in patients with idiopathic Parkinson's disease: effects of deep brain stimulation in the subthalamic nucleus. *J. Neural Transm.,* 112: 669–676.

Hummel, T., Witt, M., Reichmann, H., et al. (2010b) Immunohistochemical, volumetric, and functional neuroimaging studies in patients with idiopathic Parkinson's disease. *J. Neurol. Sci.,* 289: 119–122.

Hurwitz, T., Kopala, L., Clark, C. and Jones, B. (1988) Olfactory deficits in schizophrenia. *Biol. Psychiatry,* 23: 123–128.

Iijima, M., Kobayakawa, T., Saito, S., et al. (2008) Smell identification in Japanese Parkinson's disease patients: using the odor stick identification test for Japanese subjects. *Internal Med.,* 47: 1887–1892.

Iijima, M., Kobayakawa, T., Saito, S., et al. (2011) Differences in odor identification among clinical subtypes of Parkinson's disease. *Eur. J. Neurol.,* 18: 425–429.

Ishihara, L., Warren, L., Gibson, R., et al. (2006) Clinical features of Parkinson disease patients with homozygous leucine-rich repeat kinase 2 G2019S mutations. *Arch. Neurol.,* 63: 1250–1254.

Ishizuka, K., Tajinda, K., Colantuoni, C., et al. (2010) Negative symptoms of schizophrenia correlate with impairment on the University of Pennsylvania smell identification test. *Neurosci Res.,* 66: 106–110.

Jackson, J.A., Jankovic, J., Ford, J. (1983). Progressive supranuclear palsy: clinical features and response to treatment in 16 patients. *Ann. Neurol.* 13: 273–278.

Jankovic, J. (1989). J. Parkinsonism-plus syndromes. *Mov. Disord.* 4 Suppl 1:S95–119.

Jimbo, D., Inoue, M., Taniguchi, M. and Urakami, K. (2011) Specific feature of olfactory dysfunction with Alzheimer's disease inspected by the Odor Stick Identification Test. *Psychogeriatrics.*, 11: 196–204.

Johansen, K. K., Waro, B. J. and Aasly, J. O. (2014) Olfactory dysfunction in sporadic Parkinson's Disease and LRRK2 carriers. *Acta Neurol. Scand.* 129: 300–306.

Johansen, K. K., White, L. R., Farrer, M. J. and Aasly, J. O. (2011) Subclinical signs in LRRK2 mutation carriers. *Parkinsonism Relat. Disord.*, 17: 528–532.

Juncos, J. L., Lazarus, J. T., Rohr, J., et al. (2012) Olfactory dysfunction in fragile X tremor ataxia syndrome. *Mov. Disord.*, 27: 1556–1559.

Juneja, T., Pericak-Vance, M.A., Laing, et al. (1997) Prognosis in familial amyotrophic lateral sclerosis: progression and survival in patients with glu100gly and ala4val mutations in Cu,Zn superoxide dismutase. *Neurology*, 48: 55–57.

Jungwirth, S., Zehetmayer, S., Bauer, P., et al. (2009) Screening for Alzheimer's dementia at age 78 with short psychometric instruments. *Int Psychogeriatrics*, 21: 548–559.

Kamath, V. and Bedwell, J. S. (2008) Olfactory identification performance in individuals with psychometrically-defined schizotypy. *Schizophr. Res.*, 100: 212–215.

Kamath, V., Bedwell, J. S. and Compton, M. T. (2011) Is the odour identification deficit in schizophrenia influenced by odour hedonics? *Cogn. Neuropsychiatry*, 16: 448–460.

Kamath, V., Moberg, P. J., Gur, R. E., et al. (2012) Effects of the val(158)met catechol-O-methyltransferase gene polymorphism on olfactory processing in schizophrenia. *Behav. Neurosci.*, 126: 209–215.

Kamath, V., Turetsky, B. I., Calkins, M. E., et al. (2013) The effect of odor valence on olfactory performance in schizophrenia patients, unaffected relatives and at-risk youth. *J. Psychiatr. Res.* 47: 1636–1641.

Kamath, V., Turetsky, B. I., Calkins, M. E. et al. (2014) Olfactory processing in schizophrenia, non-ill first-degree family members, and young people at-risk for psychosis. *World J. Biol. Psychiat.* 15: 209–218.

Kang, P., Kloke, J. and Jain, S. (2012) Olfactory dysfunction and parasympathetic dysautonomia in Parkinson's disease. *Clin. Auton. Res.* 22: 161–166.

Kareken, D. A., Doty, R. L., Moberg, P. J., et al. (2001) Olfactory-evoked regional cerebral blood flow in Alzheimer's disease. *Neuropsychology*, 15: 18–29.

Kästner, A., Malzahn, D., Begemann, M. et al. (2013) Odor naming and interpretation performance in 881 schizophrenia subjects: association with clinical parameters. *BMC Psychiatry* 13:218

Kato, S., Hirano, A., Llena, J. F., et al. (1992) Ultrastructural identification of neurofibrillary tangles in the spinal cords in Guamanian amyotrophic lateral sclerosis and parkinsonism-dementia complex on Guam. *Acta Neuropathol.*, 83: 277–282.

Katzenschlager, R., Zijlmans, J., Evans, A., et al. (2004) Olfactory function distinguishes vascular parkinsonism from Parkinson's disease. *J Neurol. Neurosurg. Psychiat.*, 75: 1749–1752.

Kaufmann, H. and Biaggioni, I. (2003) Autonomic failure in neurodegenerative disorders. *Semin. Neurol.*, 23: 351–363.

Kay, S. R. and Sevy, S. (1990) Pyramidical model of schizophrenia. *Schizophr. Bull.*, 16: 537–545.

Kertelge, L., Bruggemann, N., Schmidt, A., et al. (2010) Impaired sense of smell and color discrimination in monogenic and idiopathic Parkinson's disease. *Mov. Disord.*, 25: 2665–2669.

Keshavan, M. S., Vora, A., Montrose, D., et al. (2009) Olfactory identification in young relatives at risk for schizophrenia. *Acta Neuropsychiatr.*, 21: 121–124.

Kesslak, J. P., Cotman, C. W., Chui, H. C., et al. (1988) Olfactory tests as possible probes for detecting and monitoring Alzheimer's disease. *Neurobiol. Aging*, 9: 399–403.

Kesslak, J. P., Nalcioglu, O. and Cotman, C. W. (1991) Quantification of magnetic resonance scans for hippocampal and parahippocampal atrophy in Alzheimer's disease [see comments]. *Neurology*, 41: 51–54.

Khan, N. L., Jain, S., Lynch, J. M., et al. (2005) Mutations in the gene LRRK2 encoding dardarin (PARK8) cause familial Parkinson's disease: clinical, pathological, olfactory and functional imaging and genetic data. *Brain*, 128: 2786–2796.

Khan, N. L., Katzenschlager, R., Watt, H., et al. (2004) Olfaction differentiates parkin disease from early-onset parkinsonism and Parkinson disease. *Neurology*, 62: 1224–1226.

Kim, J. Y., Lee, W. Y., Chung, E. J. and Dhong, H. J. (2007) Analysis of olfactory function and the depth of olfactory sulcus in patients with Parkinson's disease. *Mov. Disord.* 22: 1563–1566.

Kim, Y. H., Lussier, S., Rane, A., et al. (2011) Inducible dopaminergic glutathione depletion in an alpha-synuclein transgenic mouse model results in age-related olfactory dysfunction. *Neuroscience*, 172: 379–386.

Kissel, P. and Andre, J. M. (1976) [Parkinson's disease and anosmia in monozygotic twin sisters (author's transl)]. *J. Genet. Hum.*, 24: 113–117.

Kjelvik, G., Sando, S. B., Aasly, J., et al. (2007) Use of the Brief Smell Identification Test for olfactory deficit in a Norwegian population with Alzheimer's disease. *Int. J. Geriatr. Psychiat.*, 22: 1020–1024.

Knupfer, L. and Spiegel, R. (1986) Differences in olfactory test performance between normal aged, Alzheimer and vascular type dementia individuals. *Int. J. Geriat. Psychiat*, 1: 3–14.

Kohler, C. G., Moberg, P. J., Gur, R. E., et al. (2001) Olfactory dysfunction in schizophrenia and temporal lobe epilepsy. *Neuropsychiat. Neuropsychol. Behav. Neurol.*, 14: 83–88.

Kokubo, Y., Ito, K., Fukunaga, T., et al. (2006) Pigmentary retinopathy of ALS/PDC in Kii. *Ophthalmology*, 113: 2111–2112.

Kopala, L. and Clark, C. (1990) Implications of olfactory agnosia for understanding sex differences in schizophrenia. *Schizophr. Bull*, 16: 255–261.

Kopala, L., Clark, C. and Hurwitz, T. A. (1989) Sex differences in olfactory function in schizophrenia. *Am. J. Psychiat.*, 146: 1320–1322.

Kopala, L., Good, K., Martzke, J. and Hurwitz, T. (1995a) Olfactory deficits in schizophrenia are not a function of task complexity. *Schizophr. Res.*, 17: 195–199.

Kopala, L. C., Clark, C. and Hurwitz, T. (1993) Olfactory deficits in neuroleptic naive patients with schizophrenia. *Schizophr. Res.*, 8: 245–250.

Kopala, L. C., Good, K. and Honer, W. G. (1995b) Olfactory identification ability in pre- and postmenopausal women with schizophrenia. *Biol. Psychiatry*, 38: 57–63.

Kopala, L. C. and Good, K. P. (1996) Olfactory identification ability in patients with panic disorder. *J. Psychiat. Neurosci.*, 21: 340–342.

Kopala, L. C., Good, K. P. and Honer, W. G. (1994) Olfactory hallucinations and olfactory identification ability in patients with schizophrenia and other psychiatric disorders. *Schizophr. Res.*, 12: 205–211.

Kopala, L. C., Good, K.P., Koczapski, A. B. and Honer, W. G. (1998a) Olfactory deficits in patients with schizophrenia and severe polydipsia. *Biol. Psychiat.*, 43: 497–502.

Kopala, L. C., Good, K. P., Morrison, K., et al. (2001) Impaired olfactory identification in relatives of patients with familial schizophrenia. *Am. J. Psychiat.*, 158: 1286–1290.

Kopala, L. C., Good, K. P., Torrey, E. F. and Honer, W. G. (1998b) Olfactory function in monozygotic twins discordant for schizophrenia. *Am. J. Psychiat.*, 155: 134–136.

Korten, J. J. and Meulstee, J. (1980) Olfactory disturbances in Parkinsonism. *Clin. Neurol. Neurosurg.*, 82: 113–118.

Koss, E., Weiffenbach, J. M., Haxby, J. V. and Friedland, R. P. (1987) Olfactory detection and recognition in Alzheimer's disease [letter]. *Lancet*, 1: 622.

Koss, E., Weiffenbach, J. M., Haxby, J. V. and Friedland, R. P. (1988) Olfactory detection and identification performance are dissociated in early Alzheimer's disease. *Neurology*, 38: 1228–1232.

Kovacs, T., Cairns, N. J. and Lantos, P. L. (2001) Olfactory centres in Alzheimer's disease: olfactory bulb is involved in early Braak's stages. *Neuroreport*, 12: 285–288.

Kovacs, T., Papp, M. I., Cairns, N. J., et al. (2003) Olfactory bulb in multiple system atrophy. *Mov. Disord.*, 18: 938–942.

Kowalewski, J. and Murphy, C. (2012) Olfactory ERPs in an odor/visual congruency task differentiate ApoE epsilon4 carriers from non-carriers. *Brain Res.*, 1442: 55–65.

Kraepelin, E. (1913) *Dementia Praecox and Paraphrenia*. E&S Livingstone: Edinburgh.

Kruger, R., Kuhn, W., Muller, T., et al. (1998) Ala30Pro mutation in the gene encoding alpha-synuclein in Parkinson's disease. *Nat. Genet.*, 1998 18: 106–108.

Kulkarni, J., de, C. A., Fitzgerald, P. B., et al. (2008) Estrogen in severe mental illness: a potential new treatment approach. *Arch. Gen. Psychiatry*, 65: 955–960.

Kwapil, T. R., Chapman, J. P., Chapman, L. J. and Miller, M. B. (1996) Deviant olfactory experiences as indicators of risk for psychosis. *Schizophr. Bull.*, 22: 371–382.

Landis, B. N., Cao, V. H., Guinand, N., et al. (2009) Retronasal olfactory function in Parkinson's disease. *Laryngoscope*, 119: 2280–2283.

Lang, C. J., Schwandner, K. and Hecht, M. (2011) Do patients with motor neuron disease suffer from disorders of taste or smell? *Amyotroph. Lateral. Scler.*, 12: 368–371.

Lange, R., Donathan, C. L. and Hughes, L. F. (2002) Assessing olfactory abilities with the University of Pennsylvania smell identification test: a Rasch scaling approach. *J. Alzheimer Dis.*, 4: 77–91.

Laroia, H. and Louis, E. D. (2011) Association between essential tremor and other neurodegenerative diseases: what is the epidemiological evidence? *Neuroepidemiology*, 37: 1–10.

Larsson, M., Lundin, A. and Robins Wahlin, T. B. (2006) Olfactory functions in asymptomatic carriers of the Huntington disease mutation. *J. Clin. Exper. Neuropsychol.*, 28: 1373–1380.

Larsson, M., Semb, H., Winblad, B., et al. (1999) Odor identification in normal aging and early Alzheimer's disease: effects of retrieval support. *Neuropsychology.*, 13: 47–53.

Lazic, S. E., Goodman, A. O., Grote, H. E., et al. (2007) Olfactory abnormalities in Huntington's disease: decreased plasticity in the primary olfactory cortex of R6/1 transgenic mice and reduced olfactory discrimination in patients. *Brain Res.*, 1151: 219–226.

Le, G. G., Argenti, A. M., Duyme, M., (2000) Statistical sulcal shape comparisons: application to the detection of genetic encoding of the central sulcus shape. *Neuroimage.*, 11: 564–574.

Lee, J. S., Choi, J. C., Kang, S. Y., et al. (2010) Olfactory identification deficits in cerebral autosomal dominant arteriopathy with subcortical infarcts and leukoencephalopathy. *Eur. Neurol.*, 64: 280–285.

Lee, P. H., Yeo, S. H., Kim, H. J. and Youm, H. Y. (2006) Correlation between cardiac 123I-MIBG and odor identification in patients with Parkinson's disease and multiple system atrophy. *Mov. Disord.*, 21: 1975–1977.

Lee, P. H., Yeo, S. H., Yong, S. W. and Kim, Y. J. (2007) Odour identification test and its relation to cardiac 123I-metaiodobenzylguanidine in patients with drug induced parkinsonism. *J Neurol. Neurosurg. Psychiat.*, 78: 1250–1252.

Lehrner, J., Brucke, T., Kryspin-Exner, I., et al. (1995) Impaired olfactory function in Parkinson's disease. *Lancet*, 345: 1054–1055.

Lehrner, J. and Deecke, L. (2000) The Viennese olfactory test battery - A new method for assessing human olfactory functions. *Aktuelle Neurologie*, 27: 170–177.

Lehrner, J., Pusswald, G., Gleiss, A., et al. (2009) Odor identification and self-reported olfactory functioning in patients with subtypes of mild cognitive impairment. *Clin. Neuropsychol.*, 23: 818–830.

Lehrner, J. P., Brucke, T., Dal-Bianco, P., et al. (1997) Olfactory functions in Parkinson's disease and Alzheimer's disease. *Chem Senses.*, 22: 105–110.

Leon-Sarmiento, F. E., Bayona, E. A., Bayona-Prieto, J., et al. (2012) Profound olfactory dysfunction in myasthenia gravis. *PLoS One*, 7, e45544.

Lesage, S., Durr, A. and Brice, A. (2007) LRRK2: a link between familial and sporadic Parkinson's disease? *Pathol. Biol. (Paris)*, 55: 107–110.

Lesage, S., Ibanez, P., Lohmann, E., et al. (2005) G2019S LRRK2 mutation in French and North African families with Parkinson's disease. *Ann. Neurol.*, 58: 784–787.

Lewandowski, K. E., DePaola, J., Camsari, G. B., et al. (2009) Tactile, olfactory, and gustatory hallucinations in psychotic disorders: a descriptive study. *Ann. Acad. Med. Singapore*, 38: 383–385.

Lewkowitz-Shpuntoff, H. M., Hughes, V. A., Plummer, L., et al. (2012) Olfactory phenotypic spectrum in idiopathic hypogonadotropic hypogonadism: pathophysiological and genetic implications. *J. Clin. Endocrinol. Metab*, 97: E136–E144.

Liberini, P., Parola, S., Spano, P. F. and Antonini, L. (2000) Olfaction in Parkinson's disease: methods of assessment and clinical relevance. *J. Neurol.*, 247: 88–96.

Lin, C. H., Tzen, K. Y., Yu, C. Y., et al. (2008) LRRK2 mutation in familial Parkinson's disease in a Taiwanese population: clinical, PET, and functional studies. *J Biomed. Sci.*, 15: 661–667.

Lippa, C. F., Duda, J. E., Grossman, M., et al. (2007) DLB and PDD boundary issues - Diagnosis, treatment, molecular pathology, and biomarkers. *Neurology*, 68: 812–819.

Lojkowska, W., Sawicka, B., Gugala, M., et al. (2011) Follow-up study of olfactory deficits, cognitive functions, and volume loss of medial temporal lobe structures in patients with mild cognitive impairment. *Curr. Alz. Res.*, 8: 689–698.

Lotsch, J., Reichmann, H. and Hummel, T. (2008) Different odor tests contribute differently to the evaluation of olfactory loss. *Chem. Senses*, 33: 17–21.

Louis, E. D. (2009) Essential tremors: a family of neurodegenerative disorders? *Arch. Neurol.*, 66: 1202–1208.

Louis, E. D., Bromley, S. M., Jurewicz, E. C. and Watner, D. (2002) Olfactory dysfunction in essential tremor: a deficit unrelated to disease duration or severity. *Neurology.*, 59: 1631–1633.

Louis, E. D., Marder, K., Tabert, M. H. and Devanand, D. P. (2008) Mild parkinsonian signs are associated with lower olfactory test scores in the community-dwelling elderly. *Mov. Disord.*, 23: 524–530.

Lucassen, E. B., Sterling, N. W., Lee, E. Y., et al. (2014) History of smoking and olfaction in Parkinson's disease. *Mov. Disord.* 29:1069–74.

Luzzi, S., Snowden, J. S., Neary, D., Coccia, et al. (2007) Distinct patterns of olfactory impairment in Alzheimer's disease, semantic dementia, frontotemporal dementia, and corticobasal degeneration. *Neuropsychologia*, 45: 1823–1831.

Mair, R. G., Doty, R. L., Kelly, K. M., et al. (1986) Multimodal sensory discrimination deficits in Korsakoff's psychosis. *Neuropsychologia*, 24: 831–839.

Makowska, I., Kloszewska, I., Grabowska, A., et al. (2011) Olfactory deficits in normal aging and Alzheimer's disease in the polish elderly population. *Arch. Clin. Neuropsychol.*, 26: 270–279.

Malaspina, D. and Coleman, E. (2003) Olfaction and social drive in schizophrenia. *Arch. Gen. Psychiat.*, 60: 578–584.

Malaspina, D., Coleman, E., Goetz, R. R., et al. (2002) Odor identification, eye tracking and deficit syndrome schizophrenia. *Biol. Psychiat.*, 51: 809–815.

Malaspina, D., Goetz, R., Keller, A., et al. (2012) Olfactory processing, sex effects and heterogeneity in schizophrenia. *Schizophr. Res.*, 135: 144–151.

Malaspina, D., Perera, G. M., Lignelli, A., et al. (1998) SPECT imaging of odor identification in schizophrenia. *Psychiat. Res.*, 82: 53–61.

Malaspina, D., Wray, A. D., Friedman, J. H., et al. (1994) Odor discrimination deficits in schizophrenia: association with eye movement dysfunction. *J. Neuropsychiat. Clin. Neurosci.*, 6: 273–278.

Maremmani, C., Rossi, G., Tambasco, N., et al. (2012) The validity and reliability of the Italian Olfactory Identification Test (IOIT) in healthy subjects and in Parkinson's disease patients. *Parkinsonism. Relat. Disord.*, 18: 788–793.

Marras, C., McDermott, M. P., Rochon, P. A., et al. (2005) Survival in Parkinson disease: thirteen-year follow-up of the DATATOP cohort. *Neurology*, 64: 87–93.

Martinez, B. A., Cain, W. S., de Wijk, R. A., et al. (1993) Olfactory functioning before and after temporal lobe resection for intractable seizures. *Neuropsychology*, 7: 351–363.

McAbee, G., Sagan, A. and Winter, L. (2000) Olfactory hallucinations during migraine in an adolescent with an MRI temporal lobe lesion. *Headache*, 40: 592–594.

McCaffrey, R. J., Duff, K. and Solomon, G. S. (2000) Olfactory dysfunction discriminates probable Alzheimer's dementia from major depression: a cross-validation and extension. *J. Neuropsychiat. Clin. Neurosci.*, 12: 29–33.

McGeer, P. L. and Steele, J. C. (2011) The ALS/PDC syndrome of Guam: Potential biomarkers for an enigmatic disorder. *Prog. Neurobiol.* 95: 663–669.

McKeown, D. A., Doty, R. L., Perl, D. P., et al. (1996) Olfactory function in young adolescents with Down's syndrome. *J. Neurol. Neurosurg. Psychiat.*, 61: 412–414.

McKhann, G. D., Drachman, D., Folstein, M., et al. (1984) Clinical diagnosis of Alzheimer's disease: Report of the NINCDS-ADRDA work group under the auspices of Department of Health and Human Services Task Force on Alzheimer's disease. *Neurology*, 34: 939–944.

McKinnon, J., Evidente, V., Driver-Dunckley, E., et al. (2010) Olfaction in the elderly: a cross-sectional analysis comparing Parkinson's disease with controls and other disorders. *Int. J Neurosci*, 120: 36–39.

McLaughlin, N. C. and Westervelt, H. J. (2008) Odor identification deficits in frontotemporal dementia: a preliminary study. *Arch. Clin. Neuropsychol.*, 23: 119–123.

McShane, R. H., Nagy, Z., Esiri, M. M., et al. (2001) Anosmia in dementia is associated with Lewy bodies rather than Alzheimer's pathology. *J. Neurol. Neurosurg. Psychiat.*, 70; 739–743.

Meats, P. (1988) Olfactory hallucinations. *Br. Med. J.(Clin. Res. Ed)*, 296; 645.

Mesholam, R. I., Moberg, P. J., Mahr, R. N. and Doty, R. L. (1998) Olfaction in neurodegenerative disease: a meta-analysis of olfactory functioning in Alzheimer's and Parkinson's diseases. *Arch. Neurol.*, 55: 84–90.

Meusel, T., Westermann, B., Fuhr, P., et al. (2010) The course of olfactory deficits in patients with Parkinson's disease--a study based on psychophysical and electrophysiological measures. *Neurosci. Lett.*, 486: 166–170.

Miyamoto, T., Miyamoto, M., Iwanami, M., et al. (2010) Olfactory dysfunction in idiopathic REM sleep behavior disorder. *Sleep Med.*, 11: 458–461.

Miyamoto, T., Miyamoto, M., Iwanami, M., et al. (2009) Odor identification test as an indicator of idiopathic REM sleep behavior disorder. *Mov. Disord.*, 24: 268–273.

Moberg, P. J., Agrin, R., Gur, R. E., et al. (1999) Olfactory dysfunction in schizophrenia: a qualitative and quantitative review. *Neuropsychopharmacology*, 21: 325–340.

Moberg, P. J., Arnold, S. E., Doty, R. L., et al. (2006) Olfactory functioning in schizophrenia: relationship to clinical, neuropsychological, and volumetric MRI measures. *J. Clin. Exp. Neuropsychol.*, 28: 1444–1461.

Moberg, P. J., Arnold, S. E., Doty, R. L., et al. (2003a) Impairment of odor hedonics in men with schizophrenia. *Am. J. Psychiat.*, 160: 1784–1789.

Moberg, P. J. and Doty, R. L. (1997) Olfactory function in Huntington's disease patients and at-risk offspring. *Int. J. Neurosci.*, 89: 133–139.

Moberg, P. J., Doty, R. L., Mahr, R. N., et al. (1997a) Olfactory identification in elderly schizophrenia and Alzheimer's disease. *Neurobiol. Aging*, 18: 163–167.

Moberg, P. J., Doty, R. L., McKeown, D. A., et al. (1995) Olfactory function in schizophrenia: Relationship to clinical neuropsychological and MRI volumetric measures. *Chem. Senses*, 20: 199.

Moberg, P. J., Doty, R. L., Turetsky, B. I., et al. (1997b) Olfactory identification deficits in schizophrenia: correlation with duration of illness. *Am. J. Psychiatry*, 154: 1016–1018.

Moberg, P. J., Pearlson, G. D., Speedie, L. J., et al. (1987a) Olfactory recognition: differential impairments in early and late Huntington's and Alzheimer's diseases. *J. Clin. Exp. Neuropsychol.*, 9: 650–664.

Moberg, P. J., Roalf, D. R., Gur, R. E. and Turetsky, B. I. (2004) Smaller nasal volumes as stigmata of aberrant neurodevelopment in schizophrenia. *Am. J. Psychiat.*, 161: 2314–2316.

Moberg, P. J., Turetsky, B. I., Johnson, S., et al. (2003b) Unirhinal olfactory performance in schizophrenia: Laterality and relationship to clinical, neuropsychological and MRI volumetric measures. *Schizophr. Res.*, 60: 149.

Moessnang, C., Frank, G., Bogdahn, U., et al. (2011) Altered activation patterns within the olfactory network in Parkinson's disease. *Cereb. Cortex*, 21: 1246–1253.

Mohr, C., Rohrenbach, C. M., Laska, M. and Brugger, P. (2001) Unilateral olfactory perception and magical ideation. *Schizophr. Res.*, 47: 255–264.

Montgomery, E. B., Jr.,, Baker, K. B., Lyons, K. and Koller, W. C. (1999) Abnormal performance on the PD test battery by asymptomatic first-degree relatives. *Neurology*, 52: 757–762.

Montgomery, E. B., Jr.,, Koller, W. C., LaMantia, T. J., et al. (2000a) Early detection of probable idiopathic Parkinson's disease: I. Development of a diagnostic test battery. *Mov. Disord.*, 15: 467–473.

Montgomery, E. B., Jr.,, Lyons, K. and Koller, W. C. (2000b) Early detection of probable idiopathic Parkinson's disease: II. A prospective application of a diagnostic test battery. *Mov. Disord.*, 15: 474–478.

Morgan, C. D. and Murphy, C. (2002) Olfactory event-related potentials in Alzheimer's disease. *J. Int. Neuropsychol. Soc.*, 2002 Sep; 8: 753–763.

Morgan, C. D., Nordin, S. and Murphy, C. (1995) Odor identification as an early marker for Alzheimer's disease: impact of lexical functioning and detection sensitivity. *J. Clin. Exp. Neuropsychol.*, 17: 793–803.

Morley, J. F., Weintraub, D., Mamikonyan, E., et al. (2011) Olfactory dysfunction is associated with neuropsychiatric manifestations in Parkinson's disease. *Mov. Disord.* 26:2051–2057

Moscovich, M., Munhoz, R. P., Teive, H. A., et al. (2012) Olfactory impairment in familial ataxias. *J. Neurol. Neurosurg. Psychiatry*, 83: 970–974.

Motomura, N. and Tomota, Y. (2006) Olfactory dysfuntion in dementia of Alzheimer's type and vascualr dementia. *Psychogeriatrics.*, 6: 19–20.

Mueller, A., Abolmaali, N. D., Hakimi, A. R., et al. (2005) Olfactory bulb volumes in patients with idiopathic Parkinson's disease a pilot study. *J. Neural. Trans.*, 112: 1363–1370.

Mueser, K. T., Bellack, A. S. and Brady, E. U. (1990) Hallucinations in schizophrenia. *Acta Psychiatr. Scand.*, 82: 26–29.

Muller, A., Mungersdorf, M., Reichmann, H., et al. (2002) Olfactory function in Parkinsonian syndromes. *J. Clin. Neurosci.*, 9: 521–524.

Muller, B., Larsen, J. P., Wentzel-Larsen, T., et al. (2011) Autonomic and sensory symptoms and signs in incident, untreated Parkinson's disease: frequent but mild. *Mov. Disord.*, 26: 65–72.

Mundinano, I. C., Caballero, M. C., Ordonez, C., et al. (2011) Increased dopaminergic cells and protein aggregates in the olfactory bulb of patients with neurodegenerative disorders. *Acta Neuropathol.*, 122: 61–74.

Mundinano, I. C., Hernandez, M., Dicaudo, C., et al. (2013) Reduced cholinergic olfactory centrifugal inputs in patients with neurodegenerative disorders and MPTP-treated monkeys. *Acta Neuropathol.* 126: 411–425.

Murofushi, T., Mizuno, M., Osanai, R. and Hayashida, T. (1991) Olfactory dysfunction in Parkinson's disease. *ORL J. Otorhinolaryngol. Relat. Spec.*, 53: 143–146.

Murphy, C., Gilmore, M. M., Seery, C. S., (1990) Olfactory thresholds are associated with degree of dementia in Alzheimer's disease. *Neurobiol. Aging*, 11: 465–469.

Murphy, C., Jernigan, T. L. and Fennema-Notestine, C. (2003) Left hippocampal volume loss in Alzheimer's disease is reflected in performance on odor identification: a structural MRI study. *J. Int. Neuropsychol. Soc.*, 2003 Mar; 9: 459–471.

Murphy, C. and Jinich, S. (1996) Olfactory dysfunction in Down's Syndrome. *Neurobiol. Aging*, 17: 631–637.

Nee, L. E. and Lippa, C. F. (2001) Inherited Alzheimer's disease PS-1 olfactory function: a 10-year follow-up study. *Am. J. Alzheimers Dis. Other Demen.*, 16: 83–84.

Navarro-Otano, J., Gaig, C., Muxi, A., et al. (2014) ^{123}I-MIBG cardiac uptake, smell identification and ^{123}I-FP-CIT SPECT in the differential diagnosis between vascular parkinsonism and Parkinson's disease. *Parkinsonism Rel. Disord.* 20: 192–197.

Nee, L. E., Scott, J. and Polinsky, R. J. (1993) Olfactory dysfunction in the Shy-Drager syndrome. *Clin. Auton. Res.*, 3: 281–282.

Nguyen, A. D., Pelavin, P. E., Shenton, M. E., et al. (2011) Olfactory sulcal depth and olfactory bulb volume in patients with schizophrenia: an MRI study. *Brain Imaging Behav.*, 5: 252–261.

Niccoli-Waller, C. A., Harvey, J., Nordin, S. and Murphy, C. (1999) Remote odor memory in Alzheimer's disease: deficits as measured by familiarity. *J. Adult Dev.*, 6: 131–136.

Nishioka, K., Ross, O. A., Ishii, K., et al. (2009) Expanding the clinical phenotype of SNCA duplication carriers. *Mov. Disord.*, 24: 1811–1819.

Nordin, S., Almkvist, O., Berglund, B. and Wahlund, L. O. (1997) Olfactory dysfunction for pyridine and dementia progression in Alzheimer disease. *Arch. Neurology*, 54: 993–998.

Nordin, S., Monsch, A. U. and Murphy, C. (1995a) Unawareness of smell loss in normal aging and Alzheimer's disease: discrepancy between self-reported and diagnosed smell sensitivity. *J. Gerontol.*, 50: 187–192.

Nordin, S. and Murphy, C. (1996) Impaired sensory and cognitive olfactory function in questionable Alzheimer's disease. *Neuropsychology*, 10: 113–119.

Nordin, S., Paulsen, J. S. and Murphy, C. (1995b) Sensory- and memory-mediated olfactory dysfunction in Huntington's disease. *J. Int. Neuropsychol. Soc.*, 1: 281–290.

Oka, H., Toyoda, C., Yogo, M. and Mochio, S. (2010) Olfactory dysfunction and cardiovascular dysautonomia in Parkinson's disease. *J. Neurol.*, 257: 969–976.

Olichney, J. M., Murphy, C., Hofstetter, C. R., et al. (2005) Anosmia is very common in the Lewy body variant of Alzheimer's disease. *J. Neurol. Neurosurg. Psychiat.*, 76: 1342–1347.

Oliver, C. and Holland, A. J. (1986) Down's syndrome and Alzheimer's disease: a review. *Psychol. Med.*, 16: 307–322.

Ondo, W. G. and Lai, D. (2005) Olfaction testing in patients with tremor-dominant Parkinson's disease: is this a distinct condition? *Mov. Disord.*, 20: 471–475.

Pantazopoulos, H., Woo, T. U., Lim, M. P., et al. (2010) Extracellular matrix-glial abnormalities in the amygdala and entorhinal cortex of subjects diagnosed with schizophrenia. *Arch. Gen. Psychiatry*, 67: 155–166.

Papp, M. I., Kahn, J. E. and Lantos, P. L. (1989) Glial cytoplasmic inclusions in the CNS of patients with multiple system atrophy (striatonigral degeneration, olivopontocerebellar atrophy and Shy-Drager syndrome). *J Neurol. Sci*, 94: 79–100.

Pappas, G. D., Kriho, V., Liu, W. S., et al. (2003) Immunocytochemical localization of reelin in the olfactory bulb of the heterozygous reeler mouse: an animal model for schizophrenia. *Neurol. Res.*, 25: 819–830.

Pardini, M., Huey, E.D., Cavanagh, A. L. et al. (2009) Olfactory function in corticobasal syndrome and frontotemporal dementia. *Arch Neurol.* 66: 92–96.

Park, S. and Schoppe, S. (1997) Olfactory identification deficit in relation to schizotypy. *Schizophr. Res.*, 26: 191–197.

Parrao, T., Chana, P., Venegas, P., et al. (2012) Olfactory Deficits and Cognitive Dysfunction in Parkinson's Disease. *Neurodegener. Dis.*, 10: 179–182.

Peabody, C. A. and Tinklenberg, J. R. (1985) Olfactory deficits and primary degenerative dementia. *Am. J. Psychiat.*, 142: 524–525.

Pentzek, M., Grass-Kapanke, B. and Ihl, R. (2007) Odor identification in Alzheimer's disease and depression. *Aging Clin. Exp. Res.*, 19: 255–258.

Perl, E., Shay, U., Hamburger, R. and Steiner, J. E. (1992) Taste- and odor-reactivity in elderly demented patients. *Chem. Senses*, 17: 779–794.

Peters, J. M., Hummel, T., Kratzsch, T., et al. (2003) Olfactory function in mild cognitive impairment and Alzheimer's disease: an investigation using psychophysical and electrophysiological techniques. *Am. J. Psychiat.*, 160: 1995–2002.

Picillo, M., Pellecchia, M. T., Erro, R., et al. (2014) The use of University of Pennsylvania Smell Identification Test in the diagnosis of Parkinson's disease in Italy. *Neurol. Sci.* 35: 379–383.

Pirogovsky, E., Gilbert, P. E., Jacobson, M., et al. (2007) Impairments in source memory for olfactory and visual stimuli in preclinical and clinical stages of Huntington's disease. *J. Clin. Exp. Neuropsychol.*, 29: 395–404.

Plailly, J., d'Amato, T., Saoud, M. and Royet, J. P. (2006) Left temporo-limbic and orbital dysfunction in schizophrenia during odor familiarity and hedonicity judgments. *Neuroimage*, 29: 302–313.

Plato, C. C., Garruto, R. M., Galasko, D., et al. (2003) Amyotrophic lateral sclerosis and parkinsonism-dementia complex of Guam: changing incidence rates during the past 60 years. *Am. J. Epidemiol*, 157: 149–157.

Ponsen, M. M., Stoffers, D., Booij, J., et al. (2004) Idiopathic hyposmia as a preclinical sign of Parkinson's disease. *Ann. Neurol.*, 56: 173–181.

Postolache, T. T., Wehr, T. A., Doty, R. L., et al. (2002) Patients with seasonal affective disorder have lower odor detection thresholds than control subjects. *Arch. Gen. Psychiat.*, 59: 1119–1122.

Postuma, R. B., Gagnon, J. F. and Montplaisir, J. Y. (2012) REM sleep behavior disorder: From dreams to neurodegeneration. *Neurobiol. Dis.*, 46: 553–558.

Postuma, R. B., Gagnon, J. F., Vendette, M., et al. (2011) Olfaction and color vision identify impending neurodegeneration in rapid eye movement sleep behavior disorder. *Ann. Neurol.*, 69: 811–818.

Postuma, R. B., Gagnon, J. F., Vendette, M. and Montplaisir, J. Y. (2009) Markers of neurodegeneration in idiopathic rapid eye movement sleep behaviour disorder and Parkinson's disease. *Brain.*, 132: 3298–3307.

Prediger, R.D., Batista, L.C., Medeiros, R. et al. (2006). The risk is in the air: Intranasal administration of MPTP to rats reproducing clinical features of Parkinson's disease. *Exp. Neurol.* 202: 391–403.

Purdon, S. E. (1998) Olfactory identification and Stroop interference converge in schizophrenia. *J. Psychiat. Neurosci.* 23(3):163–171.

Purdon, S. E. and Flor-Henry, P. (2000) Asymmetrical olfactory acuity and neuroleptic treatment in schizophrenia. *Schizophr. Res.*, 44: 221–232.

Quagliato, L. B., Viana, M. A., Quagliato, E. M. A. B. and Simis, S. (2007) Olfactory dysfunction in Parkinson's disease. *Arq. Neuro Psiquiatr*, 65: 647–652.

Quinn, N. P., Rossor, M. N. and Marsden, C. D. (1987a) Olfactory threshold in Parkinson's disease. *J. Neurol. Neurosurg. Psychiat.*, 50: 88–89.

Quinn, N. P., Rossor, M. N. and Marsden, C. D. (1987b) Olfactory threshold in Parkinson's disease. *J. Neurol. Neurosurg. Psychiat*, 50: 88–89.

Ramjit, A. L., Sedig, L., Leibner, J., et al. (2010) The relationship between anosmia, constipation, and orthostasis and Parkinson's disease duration: results of a pilot study. *Int. J. Neurosci.*, 120: 67–70.

Reed, D., Plato, C., Elizan, T. and Kurland, L. T. (1966) The amyotrophic lateral sclerosis/parkinsonism-dementia complex—A ten-year follow-up on Guam. I. *Epidemiological studies. Am. J. Epidemiol.*, 83: 54–73.

Reed, D. M. and Brody, J. A. (1975) Amyotrophic lateral sclerosis and parkinsonism-dementia on Guam 1945–1972. I. Descriptive epidemiology. *Amer J Epidemiol*, 101: 287–301.

Rezek, D. L. (1987) Olfactory deficits as a neurologic sign in dementia of the Alzheimer type. *Arch. Neurol.*, 44: 1030–1032.

Richard, J. and Bizzini, L. (1981) [Olfaction and dementia. Preliminary results of a clinical and experimental study with N-propanol]. [French]. *Acta Neurol Belgica*, 81: 333–351.

Roalf, D. R., Turetsky, B. I., Owzar, K., et al. (2006) Unirhinal olfactory function in schizophrenia patients and first-degree relatives. *J. Neuropsychiat. Clin. Neurosci.*, 18: 389–396.

Rodriguez-Violante, M., Lees, A. J., Cervantes-Arriaga, A., et al. (2011) Use of smell test identification in Parkinson's disease in Mexico: a matched case–control study. *Mov. Disord.*, 26: 173–176.

Rolheiser, T. M., Fulton, H. G., Good, K. P., et al. (2011) Diffusion tensor imaging and olfactory identification testing in early-stage Parkinson's disease. *J. Neurol.*, 258: 1254–1260.

Rombaux, P., Mouraux, A., Bertrand, B., et al. (2006a) Retronasal and orthonasal olfactory function in relation to olfactory bulb volume in patients with posttraumatic loss of smell. *Laryngoscope*, 116: 901–905.

Rombaux, P., Mouraux, A., Bertrand, B., et al. (2006b) Olfactory function and olfactory bulb volume in patients with postinfectious olfactory loss. *Laryngoscope*, 116: 436–439.

Ross, G. W., Abbott, R. D., Petrovitch, H., et al. (2006a) Association of olfactory dysfunction with incidental Lewy bodies. *Mov. Disord.*, 21: 2062–2067.

Ross, G. W., Petrovitch, H., Abbott, R. D., et al. (2008) Association of olfactory dysfunction with risk for future Parkinson's disease. *Ann. Neurol.*, 63: 167–173.

Ross, W., Petrovitch, H., Abbott, R. D., et al. (2005) Association of olfactory dysfunction with risk of future Parkinson's disease. *Mov. Disord.*, 20: S129–S130.

Roth, J., Radil, T., Ruzicka, E., et al. (1998) Apomorphine does not influence olfactory thresholds in Parkinson's disease. *Funct. Neurol.*, 13: 99–103.

Rousseaux, M., Muller, P., Gahide, I., et al. (1996) Disorders of smell, taste, and food intake in a patient with a dorsomedial thalamic infarct. *Stroke*, 27: 2328–2330.

Royet, J. P., Croisile, B., Williamson-Vasta, R., et al. (2001) Rating of different olfactory judgements in Alzheimer's disease. *Chem. Senses*, 26: 409–417.

Rub, U., Brunt, E. R. and Deller, T. (2008) New insights into the pathoanatomy of spinocerebellar ataxia type 3 (Machado-Joseph disease). *Curr. Opin. Neurol.*, 21: 111–116.

Ruiz-Martinez, J., Gorostidi, A., Goyenechea, E., et al. (2011) *Olfactory deficits and cardiac (123)* I-MIBG in Parkinson's disease related to the LRRK2 R1441G and G2019S mutations. *Mov. Disord* 26: 2026–2031.

Rupp, C. I., Fleischhacker, W. W., Kemmler, G., et al. (2005a) Olfactory functions and volumetric measures of orbitofrontal and limbic regions in schizophrenia. *Schizophr. Res.*, 74: 149–161.

Rupp, C. I., Fleischhacker, W. W., Kemmler, G., et al. (2005b) Various bilateral olfactory deficits in male patients with schizophrenia. *Schizophr. Bull.*, 31: 155–165.

Sajjadian, A., Doty, R. L., Gutnick, D. N., Chirurgi, R. J., et al. (1994) Olfactory dysfunction in amyotrophic lateral sclerosis. *Neurodegeneration*, 3: 153–157.

Salerno-Kennedy, R., Cusack, S. and Cashman, K. D. (2005) Olfactory function in people with genetic risk of dementia. *Ir. J. Med. Sci.*, 174: 46–50.

Santin, R., Fonseca, V. F., Bleil, C. B., et al. (2010) Olfactory function and Parkinson's disease in Southern Brazil. *Arq. Neuro. Psiquiatr.*, 68: 252–257.

Santorelli, G. and Marotta, A. (1964) La sogli olfattometrica dell'epilettico in condizioni di base e dopo crisi. *Osp. Psichiatr.*, 32: 185–190.

Saoud, M., Hueber, T., Mandran, H., et al. (1998) Olfactory identification deficiency and WCST performance in men with schizophrenia. *Psychiat. Res.*, 81: 251–257.

Satya-Murti, S. and Crisostomo, E. A. (1988) Olfactory threshold in Friedreich's ataxia. *Muscle Nerve*, 11: 406–407.

Saunders-Pullman, R., Hagenah, J., Dhawan, V., et al. (2010) Gaucher disease ascertained through a Parkinson's center: imaging and clinical characterization. *Mov. Disord.*, 25: 1364–1372.

Saunders-Pullman, R., Stanley, K., Wang, C., et al. (2011) Olfactory dysfunction in LRRK2 G2019S mutation carriers. *Neurology*, 77: 319–324.

Schiffman, S. S., Clark, C. M. and Warwick, Z. S. (1990) Gustatory and olfactory dysfunction in dementia: not specific to Alzheimer's disease. *Neurobiol. Aging*, 11: 597–600.

Schliebs, R. and Arendt, T. (2011) The cholinergic system in aging and neuronal degeneration. *Behav. Brain Res.*, 221: 555–563.

Schmidt, F. A., Goktas, O., Harms, L., et al. (2011) Structural correlates of taste and smell loss in encephalitis disseminata. *PLoS. One.*, 6, e19702.

Schneider, F., Habel, U., Reske, M., et al. (2007) Neural substrates of olfactory processing in schizophrenia patients and their healthy relatives. *Psychiat. Res.*, 155: 103–112.

Schofield, P. W., Ebrahimi, H., Jones, A. L., et al. (2012) An olfactory 'stress test' may detect preclinical Alzheimer's disease. *BMC Neurol.*, 12: 24.

Schubert, C. R., Carmichael, L. L., Murphy, C., et al. (2008) Olfaction and the 5-year incidence of cognitive impairment in an epidemiological study of older adults. *J. Am. Geriatr. Soc.*, 56: 1517–1521.

Sedig, L., Leibner, J., Ramjit, A. L., et al. (2010) Is rhinorrhea an under-recognized intrinsic symptom of Parkinson disease? A prospective pilot study. *Int. J. Neurosci.*, 120: 258–260.

Seidman, L. J., Goldstein, J. M., Goodman, J. M., et al. (1997) Sex differences in olfactory identification and Wisconsin Card Sorting performance in schizophrenia: relationship to attention and verbal ability. *Biol. Psychiat.*, 42: 104–115.

Seidman, L. J., Oscar-Berman, M., Kalinowski, A. G., et al. (1995) Experimental and clinical neuropsychological measures of prefrontal dysfunction in schizophrenia. *Neuropsychology*, 9: 481–490.

Seidman, L. J., Talbot, N. L., Kalinowski, A. G., et al. (1992) Neuropsychological probes of fronto-limbic system dysfunction in schizophrenia. Olfactory identification and Wisconsin Card Sorting performance. *Schiz. Res.*, 6: 55–65.

Serby, M., Corwin, J., Conrad, P. and Rotrosen, J. (1985a) Olfactory dysfunction in Alzheimer's disease and Parkinson's disease. *Am. J. Psychiat.*, 142: 781–782.

Serby, M., Corwin, J., Novatt, A., et al. (1985b) Olfaction in dementia. *J. Neurol. Neurosurg. Psychiatry*, 48: 848–849.

Serby, M., Larson, P. and Kalkstein, D. (1990) Olfactory sense in psychoses. *Biol. Psychiat*, 28: 829–830.

Serby, M., Larson, P. and Kalkstein, D. (1991) The nature and course of olfactory deficits in Alzheimer's disease. *Am. J. Psychiat.*, 148: 357–360.

Serby, M., Mohan, C., Aryan, M., et al. (1996) Olfactory identification deficits in relatives of Alzheimer's disease patients. *Biol. Psychiat.*, 39: 375–377.

Shah, M., Deeb, J., Fernando, M., et al. (2009) Abnormality of taste and smell in Parkinson's disease. *Parkinsonism Relat. Disord.*, 15: 232–237.

Shah, M., Muhammed, N., Findley, L. J. and Hawkes, C. H. (2008) Olfactory tests in the diagnosis of essential tremor. *Parkinsonism Relat. Disord.*, 14: 563–568.

Sharer, J. D., Leon-Sarmiento, F. E., Morley, J. F. et al. (2015) Olfactory dysfunction in Parkinson's disease: Positive effect of cigarette smoking. *Mov. Disord.*, in press.

Sharma, J. C. and Turton, J. (2012) Olfaction, dyskinesia and profile of weight change in Parkinson's disease: identifying neurodegenerative phenotypes. *Parkinsonism Relat. Disord.*, 18: 964–970.

Shaw, C. A., Li, Y., Wiszniewska, J., et al. (2011) Olfactory copy number association with age at onset of Alzheimer disease. *Neurology*, 76: 1302–1309.

Shear, P. K., Butters, N., Jernigan, T. L., et al. (1992) Olfactory loss in alcoholics: correlations with cortical and subcortical MRI indices. *Alcohol*, 9: 247–255.

Shoulson, I. (1986) Huntington's disease. *In:* Asbury, A. K., McKhann, G. M. and McDonald, W. I., *Diseases of the nervous system*. W. B. Saunders: Philadelphia, 1258–1267.

Siderowf, A., Jennings, D., Connolly, J., et al. (2007) Risk factors for Parkinson's disease and impaired olfaction in relatives of patients with Parkinson's disease. *Mov. Disord.*, 22: 2249–2255.

Siderowf, A., Jennings, D., Eberly, S., et al. (2012) Impaired olfaction and other prodromal features in the Parkinson At-Risk Syndrome study. *Mov. Disord.*, 27: 406–412.

Siderowf, A., Newberg, A., Chou, K. L., et al. (2005) [99mTc]TRODAT-1 SPECT imaging correlates with odor identification in early Parkinson disease. *Neurology*, 64: 1716–1720.

Sierra, M. et al. (2013) Olfaction and imaging biomarkers in premotor LRRK2 G2019S-associated Parkinson's disease. *Neurology* 80: 621–625.

Silveira-Moriyama, L., Guedes, L. C., Kingsbury, A., et al. (2008) Hyposmia in G2019S LRRK2-related parkinsonism: clinical and pathologic data. *Neurology*, 71: 1021–1026.

Silveira-Moriyama, L., Hughes, G., Church, A., et al. (2010a) Hyposmia in progressive supranuclear palsy. *Mov. Disord.*, 25: 570–577.

Silveira-Moriyama, L., Mathias, C., Mason, L., et al. (2009a) Hyposmia in pure autonomic failure. *Neurology*, 72: 1677–1681.

Silveira-Moriyama, L., Munhoz, R. P., de, J.C., Raskin, S., et al. (2010b) Olfactory heterogeneity in LRRK2 related Parkinsonism. *Mov. Disord.*, 25: 2879–2883.

Silveira-Moriyama, L., Schwingenschuh, P., O'Donnell, A., et al.(2009b) Olfaction in patients with suspected parkinsonism and scans without evidence of dopaminergic deficit (SWEDDs). *J. Neurol. Neurosurg. Psychiat.*, 80: 744–748.

Silveira-Moriyama, L., Sirisena, D., Gamage, P., et al. (2009c) Adapting the Sniffin' Sticks to diagnose Parkinson's disease in Sri Lanka. *Mov. Disord.*, 24: 1229–1233.

Silveira-Moriyama, L., Williams, D., Katzenschlager, R. and Lees, A. J. (2005) Pizza, mint, and licorice: Smell testing in Parkinson's disease in a UK population. *Mov. Disord.*, 20: S139.

Sirota, P., Davidson, B., Mosheva, T., et al. (1999b) Increased olfactory sensitivity in first episode psychosis and the effect of neuroleptic treatment on olfactory sensitivity in schizophrenia. *Psychiat. Res.*, 86: 143–153.

Smutzer, G. S., Doty, R. L., Arnold, S. E. and Trojanowski, J. Q. (2003) Olfactory system neuropathology in Alzheimer's disease, Parkinson's disease, and schizophrenia. *In:* Doty, R.L., *Handbook of Olfaction and Gustation*. Marcel Dekker: New York, 503–523.

Sobel, N., Prabhakaran, V., Hartley, C. A., et al. (1998) Odorant-induced and sniff-induced activation in the cerebellum of the human. *J. Neurosci.*, 18: 8990–9001.

Sobel, N., Thomason, M. E., Stappen, I., et al. (2001) An impairment in sniffing contributes to the olfactory impairment in Parkinson's disease. *PNAS USA*, 98: 4154–4159.

Sobin, C., Kiley-Brabeck, K., Dale, K., et al. (2006) Olfactory disorder in children with 22q11 deletion syndrome. *Pediatrics*, 118: e697–e703.

Sohrabi, H. R., Bates, K. A., Rodrigues, M., et al. (2009) Olfactory dysfunction is associated with subjective memory complaints in community-dwelling elderly individuals. *J. Alz. Dis.*, 17: 135–142.

Sohrabi, H. R., Bates, K. A., Weinborn, M. G., et al. (2012) Olfactory discrimination predicts cognitive decline among community-dwelling older adults. *Transl. Psychiat.*, 2: e118.

Solomon, G. S. (1994) Anosmia in Alzheimer disease. *Percept. Mot. Skills.*, 79: 1249–1250.

Solomon, G. S., Petrie, W. M., Hart, J. R. and Brackin, H. B., Jr., (1998) Olfactory dysfunction discriminates Alzheimer's dementia from major depression. *J. Neuropsychiat. Clin. Neurosci.*, 10: 64–67.

Sparks, D. L., Petanceska, S., Sabbagh, M., et al. (2005) Cholesterol, copper and Aβ in controls, MCI, AD and the AD cholesterol-lowering treatment trial (ADCLT). *Cur. Alz. Res.*, 2: 527–539.

Sreenivasan, K. V., Abraham, A. and Verghese, A. (1987) Right temporal lobe functions in psychiatric disorders. *Ind. J. Clin. Psychol.*, 14: 40–42.

Stamps, J. J., Bartoshuk, L. M. and Heilman, K. M. (2013) A brief test for Alzheimer's disease. *J. Neurol. Sci.* 333: 19–24.

Stedman, T. J. and Clair, A. L. (1998) Neuropsychological, neurological and symptom correlates of impaired olfactory identification in schizophrenia. *Schizophr. Res.*, 32: 23–30.

Steinbach, S., Hundt, W., Vaitl, A., et al. (2010) Taste in mild cognitive impairment and Alzheimer's disease. *J. Neurol.*, 257: 238–246.

Stern, M. B., Doty, R. L., Dotti, M., et al. (1994) Olfactory function in Parkinson's disease subtypes. *Neurology*, 44: 266–268.

Stevenson, R. J., Langdon, R. and McGuire, J. (2011) Olfactory hallucinations in schizophrenia and schizoaffective disorder: a phenomenological survey. *Psychiat. Res.*, 185: 321–327.

Strauss, G. P., Allen, D. N., Ross, S. A., et al. (2010) Olfactory hedonic judgment in patients with deficit syndrome schizophrenia. *Schizophr. Bull.*, 36: 860–868.

Sundermann, E., Gilbert, P. E. and Murphy, C. (2006) Estrogen and performance in recognition memory for olfactory and visual stimuli in females diagnosed with Alzheimer's disease. *J. Int. Neuropsychol. Soc.*, 12: 400–404.

Suzuki, M., Hashimoto, M., Yoshioka, M., et al. (2011) The odor stick identification test for Japanese differentiates Parkinson's disease from multiple system atrophy and progressive supranuclear palsy. *BMC Neurol.*, 11: 157.

Suzuki, Y., Yamamoto, S., Umegaki, H., et al. (2004) Smell identification test as an indicator for cognitive impairment in Alzheimer's disease. *Int. J. Geriatr. Psychiat.*, 19: 727–733.

Szeszko, P. R., Bates, J., Robinson, D., et al. (2004) Investigation of unirhinal olfactory identification in antipsychotic-free patients experiencing a first-episode schizophrenia. *Schizophr. Res.*, 67: 219–225.

Tabert, M. H., Liu, X., Doty, R. L., et al. (2005) A 10-item smell identification scale related to risk for Alzheimer's disease. *Ann Neurol.*, 58: 155–160.

Tanner, C. M. (2003) Is the cause of Parkinson's disease environmental or hereditary? Evidence from twin studies. *Adv. Neurol.*, 91: 133–142.

Thomann, P. A., Dos, S., V, Seidl, U., et al. (2009) MRI-derived atrophy of the olfactory bulb and tract in mild cognitive impairment and Alzheimer's disease. *J. Alz. Dis.*, 17: 213–221.

Tijero, B., Gomez-Esteban, J. C., Llorens, V., et al. (2010) Cardiac sympathetic denervation precedes nigrostriatal loss in the E46K mutation of the alpha-synuclein gene (SNCA). *Clin. Auton. Res*, 20: 267–269.

Tissingh, G., Berendse, H. W., Bergmans, P., et al. (2001) Loss of olfaction in de novo and treated Parkinson's disease: possible implications for early diagnosis. *Mov. Disord.*, 16: 41–46.

Trojanowski, J. Q., Newman, P. D., Hill, W. D. and Lee, V. M. (1991) Human olfactory epithelium in normal aging, Alzheimer's disease, and other neurodegenerative disorders. *J. Comp. Neurol.*, 1991 Aug 15; 310: 365–376.

Tsuboi, Y., Wszolek, Z. K., Graff-Radford, N. R., et al. (2003) Tau pathology in the olfactory bulb correlates with Braak stage, Lewy body pathology and apolipoprotein epsilon4. *Neuropathol. Appl. Neurobiol.*, 29: 503–510.

Turetsky, B. I., Crutchley, P., Walker, J., et al. (2009a) Depth of the olfactory sulcus: a marker of early embryonic disruption in schizophrenia? *Schizophr. Res.*, 115: 8–11.

Turetsky, B. I., Hahn, C. G., Arnold, S. E. and Moberg, P. J. (2009b) Olfactory receptor neuron dysfunction in schizophrenia. *Neuropsychopharmacology*, 34: 767–774.

Turetsky, B. I. and Moberg, P. J. (2009) An odor-specific threshold deficit implicates abnormal intracellular cyclic AMP signaling in schizophrenia. *Am. J. Psychiatry*, 166: 226–233.

Turetsky, B. I., Moberg, P. J., Owzar, K., et al. (2003b) Physiologic impairment of olfactory stimulus processing in schizophrenia. *Biol. Psychiat.*, 53: 403–411.

Turetsky, B. I., Moberg, P. J., Yousem, D. M., et al. (2000) Reduced olfactory bulb volume in patients with schizophrenia. *Am. J. Psychiat.*, 157: 828–830.

Ugur, T., Weisbrod, M., Franzek, E., et al. (2005) Olfactory impairment in monozygotic twins discordant for schizophrenia. *Eur. Arch. Psychiat. Clin. Neurosci.*, 255: 94–98.

Velayudhan, L., Pritchard, M. Powell, J. F. et al. (2013) Smell identification function as a severity and progression marker in Alzheimer's disease. *Int. Psychogeriat.* 25: 1157–1166.

Valldeoriola, F., Gaig, C., Muxi, A., et al. (2011) (123)I-MIBG cardiac uptake and smell identification in parkinsonian patients with LRRK2 mutations. *J. Neurol.*, 258: 1126–1132.

Velazquez-Perez, L., Fernandez-Ruiz, J., Diaz, R., et al. (2006) Spinocerebellar ataxia type 2 olfactory impairment shows a pattern similar to other major neurodegenerative diseases. *J. Neurol.*, 253: 1165–1169.

Verbaan, D., Boesveldt, S., van Rooden, S. M. et al. (2008) Is olfactory impairment in Parkinson disease related to phenotypic or genotypic characteristics? *Neurology* 71: 1877–1882.

Vroon, A., Drukarch, B., Bol, J. G. J. M., et al. (2007) Neuroinflammation in Parkinson's patients and MPTP-treated mice is not restricted to the nigrostriatal system: Microgliosis and differential expression of interleukin-1 receptors in the olfactory bulb. *Exp. Gerontol.*, 42: 762–771.

Waldton, S. (1974) Clinical observations of impaired cranial nerve function in senile dementia. *Acta Psychiatr. Scand.*, 50: 539–547.

Wang, H. L., Chou, A. H., Wu, A. S., et al. (2011a) PARK6 PINK1 mutants are defective in maintaining mitochondrial membrane potential and inhibiting ROS formation of substantia nigra dopaminergic neurons. *Biochim. Biophy. Acta*, 1812: 674–684.

Wang, J., Eslinger, P. J., Doty, R. L., et al. (2010) Olfactory deficit detected by fMRI in early Alzheimer's disease. *Brain Res.*, 1357: 184–194.

Wang, J., You, H., Liu, J. F., et al. (2011b) Association of olfactory bulb volume and olfactory sulcus depth with olfactory function in patients with Parkinson disease. *AJNR Am. J. Neuroradiol.*, 32: 677–681.

Wang, Q. S., Tian, L., Huang, Y. L., et al. (2002) Olfactory identification and apolipoprotein E epsilon 4 allele in mild cognitive impairment. *Brain Res.*, 2002 Sep 27; 951: 77–81.

Wang, W., Zhu, J. Z., Chang, K. T. and Min, K. T. (2012) DSCR1 interacts with FMRP and is required for spine morphogenesis and local protein synthesis. *EMBO J.*, 31: 3655–3666.

Ward, C. d. (2003) Neuropsychiatric interpretations of postencephalitic movement disorders. *Mov. Disord.*, 18: 623–630.

Ward, C. d., Duvoisin, R. C., Ince, S. E., et al. (1984) Parkinson's disease in twins. *Adv. Neurol.*, 1984;40: 341–344.

Ward, C. d., Hess, W. A. and Calne, D. B. (1983) Olfactory impairment in Parkinson's disease. *Neurology*, 33: 943–946.

Warner, M. D., Peabody, C. A. and Berger, P. A. (1988) Olfactory deficits in Down's syndrome. *Biol. Psychiat.*, 23: 836–839.

Warner, M. D., Peabody, C. A. and Csernansky, J. G. (1990) Olfactory functioning in schizophrenia and depression. [letter; comment]. [see comments]. *Biol. Psychiat.*, 27: 457–458.

Warner, M. D., Peabody, C. A., Flattery, J. J. and Tinklenberg, J. R. (1986) Olfactory deficits and Alzheimer's disease. *Biol. Psychiat.*, 21: 116–118.

Wattendorf, E., Welge-Lussen, A., Fiedler, K., et al. (2009) Olfactory impairment predicts brain atrophy in Parkinson's disease. *J. Neurosci.*, 29: 15410–15413.

Weinberger, D. R. (1987) Implications of normal brain development for the pathogenesis of schizophrenia. *Arch. Gen. Psychiatry*, 44: 660–669.

Wenning, G. K., Shephard, B., Hawkes, C., et al. (1995) Olfactory function in atypical parkinsonian syndromes. *Acta Neurol. Scand.*, 91: 247–250.

Westervelt, H. J., Bruce, J. M., Coon, W. G. and Tremont, G. (2008) Odor identification in mild cognitive impairment subtypes. *J. Clin. Exp. Neuropsychol.*, 30: 151–156.

Westervelt, H. J., Stern, R. A. and Tremont, G. (2003) Odor identification deficits in diffuse lewy body disease. *Cogn. Behav. Neurol.*, 16: 93–99.

References

Wetter, S. and Murphy, C. (1999) Individuals with Down's syndrome demonstrate abnormal olfactory event-related potentials. *Clin. Neurophysiol.*, 110: 1563–1569.

Wetter, S. and Murphy, C. (2001) Apolipoprotein E epsilon4 positive individuals demonstrate delayed olfactory event-related potentials. *Neurobiol, Aging*, 22: 439–447.

Wetter, S., Peavy, G., Jacobson, M., et al. (2005) Olfactory and auditory event-related potentials in Huntington's disease. *Neuropsychology*, 19: 428–436.

Williams, D. R., de Silva, R., Paviour, D. C., et al. 2005) Characteristics of two distinct clinical phenotypes in pathologically proven progressive supranuclear palsy: Richardson's syndrome and PSP-parkinsonism. *Brain*, 128: 1247–1258.

Williams, S. S., Williams, J., Combrinck, M., et al. (2009) Olfactory impairment is more marked in patients with mild dementia with Lewy bodies than those with mild Alzheimer disease. *J. Neurol. Neurosurg. Psychiat.*, 80: 667–670.

Wilson, R. S., Arnold, S. E., Buchman, A. S., et al. (2008) Odor identification and progression of parkinsonian signs in older persons. *Exp. Aging Res.*, 34: 173–187.

Wilson, R. S., Arnold, S. E., Schneider, J. A., et al. (2007) The relationship between cerebral Alzheimer's disease pathology and odour identification in old age. *J. Neurol. Neurosurg. Psychiat.*, 78: 30–35.

Wilson, R. S., Schneider, J. A., Arnold, S. E. et al. (2007a) Olfactory identification and incidence of mild cognitive impairment in older age. *Arch. Gen. Psychiat.* 64: 802–808.

Wilson, R. S., Yu, L., Schneider, J. A., et al. (2011) Lewy bodies and olfactory dysfunction in old age. *Chem. Senses.*, 36: 367–373.

Wu, J., Buchsbaum, M. S., Moy, K., et al. (1993) Olfactory memory in unmedicated schizophrenics. *Schizophr. Res.*, 9: 41–47.

Wu, X., Yu, C., Fan, F., et al. (2011) Correlation between progressive changes in piriform cortex and olfactory performance in early Parkinson's disease. *Eur. Neurol.*, 66: 98–105.

Yoritaka, A., Shimo, Y., Shimo, Y., et al. (2011) Nonmotor Symptoms in Patients with PARK2 Mutations. *Parkinsons Dis.*, 2011, 473640.

Yousem, D. M., Geckle, R. and Doty, R. L. (1995) MR of patients with post-traumatic olfactory deficits. *Chem. Senses*, 20:, 338.

Yousem, D. M., Geckle, R. J., Bilker, W. B. and Doty, R. L. (1998) Olfactory bulb and tract and temporal lobe volumes. Normative data across decades. *Ann. N. Y. Acad. Sci.*, 855: 546–555.

Yousem, D. M., Williams, S. C., Howard, R. O., et al. (1997) Functional MR imaging during odor stimulation: preliminary data. *Radiology*, 204: 833–838.

Zatorre, R. J. and Jones-Gotman, M. (1990) Right-nostril advantage for discrimination of odors. *Percept. Psychophys.*, 47: 526–531.

Zhang, K., Yu, C., Zhang, Y., et al. (2011) Voxel-based analysis of diffusion tensor indices in the brain in patients with Parkinson's disease. *Eur. J. Radiol.*, 77: 269–273.

Zijlmans, J. C. M., Thijssen, H. O. M., Vogels, O. J. M., et al. (1995) MRI in patients with suspected vascular parkinsonism. *Neurology*, 45: 2183–2188.

Zucco, G., Zeni, M. T., Perrone, A. and Piccolo, I. (2001) Olfactory sensitivity in early-stage Parkinson patients affected by more marked unilateral disorder. *Percept. Mot. Skills*, 2001 92: 894–898.

Zucco, G. M. and Negrin, N. S. (1994) Olfactory deficits in Down subjects: a link with Alzheimer disease. *Percept. Mot. Skills*, 78: 627–631.

Zucco, G. M., Zaglis, D. and Wambsganss, C. S. (1991) Olfactory deficits in elderly subjects and Parkinson patients. *Percept. Mot. Skills*, 73: 895–898.

Chapter 19

The Olfactory System as a Route of Delivery for Agents to the Brain and Circulation

Mary Beth Genter, Mansi Krishan, and Rui Daniel Prediger

19.1 INTRODUCTION

The important role the olfactory system plays in food acquisition and mating behaviors is reflected in the substantial development of this sensory modality in the majority of vertebrate species with, perhaps, the exception of primates. In view of these life-sustaining roles, the common dogma posits that a system as important to survival as olfaction must have mechanisms to protect the sensory cells within the olfactory receptor epithelium from the materials in inspired air, both toxicants and odorants. Also, a means should exist to prevent access of extrinsic substances to the central nervous system (CNS). As discussed below (see also Chapter 3), several mechanisms have been proposed that would protect the epithelium from environmental toxic exposure. Included among these mechanisms are intracellular detoxification of molecules, removal of substances by ligand-specific binding proteins contained in mucosal secretions, surveillance by immune system cells and degeneration and replacement of damaged receptor neurons with new cells derived from basal stem cells. Recent data indicate, however, that these protective mechanisms can be circumvented, resulting in access of xenobiotics from the nares to the CNS and to the systemic circulation.

Figure 19.1 indicates three major routes of entry for agents to the CNS and systemic circulation from the nasal cavity. The first route occurs via internalization of xenobiotics into the receptor neurons of the olfactory epithelium.

Substances are transported into the brain through the axons of olfactory receptor neurons in a process also called the *olfactory nerve pathway* by Mathison et al. (1998). In an older publication, McMartin et al. (1987) referred to both *transcytosis* (uptake of the agent into vesicles with subsequent discharge into interstitial space) and *transcellular transport* (passage of agents via pores or carriers across the cell membrane, diffusion across the cytoplasm, and transport out of the cell to the glomerular region of the olfactory bulb). Within the glomerulus, axodendritic contacts between receptor cells and mitral cells serve as one transport pathway. For some xenobiotics, transneuronal transport has been observed in widespread areas of the brain. Both anterograde and retrograde transport have been reported, the latter perhaps a result of the large centrifugal afferent innervation of the olfactory glomerular region where receptor afferent fibers terminate. Anterograde labeling can then be observed in mitral cell terminal fields such as the piriform cortex. Retrograde transport is also observed in brain regions with projections to the olfactory bulb including the horizontal limb of the diagonal band, substantia nigra, locus coeruleus, and dorsal raphe nucleus.

Xenobiotics also may gain access to the CNS by two other routes as indicated in Figure 19.1. A number of drugs (see below) and metals are hypothesized to reach the subarachnoid space by movement through perineural spaces, a process also called the *olfactory epithelial pathway* or *paracellular transport* (Jackson et al., 1979; Mathison et al., 1998; McMartin et al., 1987). In fact, many drugs

---

Handbook of Olfaction and Gustation, Third Edition. Edited by Richard L. Doty.
© 2015 Richard L. Doty. Published 2015 by John Wiley & Sons, Inc.

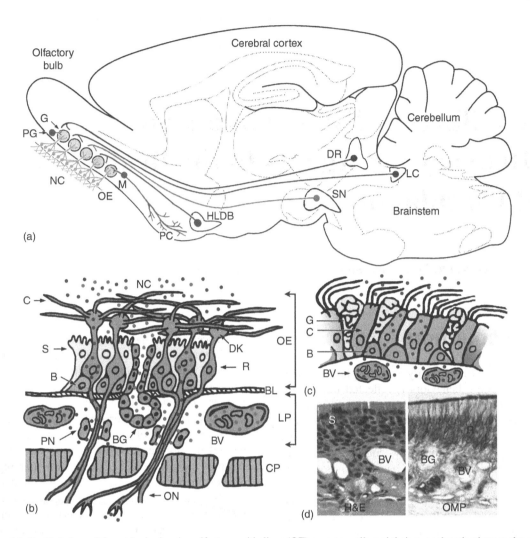

Figure 19.1 (a). Sagittal view of the rat brain showing olfactory epithelium (OE) receptor cells and their axonal projections to the olfactory glomeruli (G). In the glomerulus, receptor cell axons contact the dendrites of periglomerular (PG) and mitral (M) cells. Mitral cells project to the piriform cortex (PC). Centrifugal afferent innervation comes from the horizontal limb of the diagonal band (HLDB), the substantia nigra (SN), the dorsal raphe (DR) and the locus ceruleus (LC). (b). The olfactory mucosa includes an epithelial cell layer (OE) and the lamina propria (LP) separated by the basal lamina (BL). The OE contains sustentacular (S), basal (B) and receptor (R) cells. Receptor cells have a dendritic knob (DN) from which cilia (c) project into the nasal cavity (NC). Receptor cell axons fasciculate to form the olfactory nerve (ON) that crosses the cribriform plate (CP) to enter the CNS. The axon and nerve are surrounded by a perineural sheath that forms the perineural space (PN). The lamina propria contains mucus secreting Bowman's glands (BG), axons of receptor cells and numerous blood vessels (BV). Red and green dots depict possible entry pathways through neurons, glands and blood vessels. (c). Respiratory epithelium consists of columnar ciliated (C), goblet (G) and basal (B) cells and is highly vascular (BV). (d). H & E stained section illustrating the layers of the OE and LP showing the numerous blood vessels in the lamina propria. Olfactory marker protein (OMP) immunostained section showing that only mature receptor neurons and their axons, not basal or sustentacular cells, contain OMP (B and C were adapted from Lewis and Dahl, 1995). (*See plate section for color version.*)

that otherwise would not have access to the brain can be applied intranasally, including antiviral drugs for the treatment of HIV infections that have spread to the CNS. The third route is a consequence of the highly vascular nature of the respiratory and olfactory epithelia. This pathway provides access of intranasally applied agents to the systemic circulation and thus to the brain. Another consequence of the extensive vascularity of the olfactory epithelium is that some viruses (Charles et al., 1995; Oliver and Fazakerley, 1998) and a variety of organic compounds (Brandt et al., 1990; Brittebo, 1988; Feng et al., 1990) are concentrated in the olfactory epithelium following systemic application. Subsequently, there may be either transport to the brain or destruction of the epithelium (Table 19.1).

Substances that have been reported to be internalized, transported, and accumulated through the olfactory system

19.1 Introduction

Table 19.1 Substances analyzed for transport from the nasal cavity to the brain

Substance	Species Studied	Application Method	Transneuronal Transport	Reference
Metals				
Aluminum lactate	Rabbit	IN	Indirect pathological evidence	Perl and Good, 1987
Aluminum silicate	Rabbit	Bedding	No	Hayek and Waite, 1991
Aluminum acetyl-acetonate	Rat	IN	Yes	Zatta et al., 1993
Cadmium	Pike	IN	No	Gottofrey and Tjälve, 1991
Cadmium chloride	Rat	IN	No	Tjälve et al., 1996; Tjälve and Henriksson, 1999; Evans and Hastings, 1992
Cadmium oxide	Rat	Aerosol	No	Hastings, 1990; Sun et al., 1992
Cobalt	Salmon	IN	Possible	Sun et al., 1996; Bazer et al, 1987
	Rats	IN	Yes	Persson et al., 2003
Gold	Squirrel monkey	IN	Yes	DeLorenzo, 1970
	Rabbit	Mucosal	No	Czerniawska, 1970
Manganese	Rat	IN	Yes	Gianutsos et al, 1997; Tjälve et al., 1996; Tjälve and Henriksson, 1999; Henriksson et al., 1999.
	Rat	Inhalation	Yes	Brenneman et al., 2000; Dorman et al., 2001
	Rats	IN	ND	Henriksson and Tjälve 2000
Mercury	Pike	IN	No	Borg-Neczak and Tjälve, 1996.
Mercury	Rat	IN	No	Henriksson and Tjälve, 1998
Thallium	Mice	IN	Yes	Kinoshita et al., 2008
Nanoparticles				
Copper (23.5 nm)	Mice	IN	Yes	Liu et al., 2009
Diesel exhaust (21.5 nm)	Rats	IN	Yes	Yokota et al., 2011
Ferric oxide	Mice	IN	Yes	Wang Y et al., 2011
Manganese oxide (30 nm)	Rats	Inhalation	Yes	Elder et al., 2006
Silica	Rats	IN	Yes	Wu et al., 2011;
	Mice	IN, IV	ND	Higashisaka et al., 2011
Silver (25 nm)	Mice	IN	Yes	Genter et al., 2012
Titanium dioxide	Mice	IN	Yes	Zhang et al., 2011
Lectins				
WGA-HRP	Rat	IN	Yes	Baker and Spencer, 1986; Broadwell and Balin, 1985; Balin et al., 1986; Itaya, 1987; Shipley, 1985; Stewart, 1985; Thorne et al., 1995
WGA-HRP	Mouse	IN	Yes	Baker, 1995
Barley	Mouse	Genetic	Yes	Horowitz et al., 1999
Concanavilin A	Rat	IN	No	Wiley et al., 1984
Viruses				
Adeno (Recominant)	Rat	IN	Yes	Draghia et al., 1995; Zhao et al., 1996
Borna disease	Rat	IN	Yes	Morales et al., 1988
Herpes simplex	Rat	IN	Yes	Esiri and Tomlinson, 1984; McLean et al., 1989.
Herpes simplex	Mouse	Corneal	Yes	Stroop et al., 1984
Herpes simplex	Mouse	Facial skin	Limited	Stroop et al., 1984
Hepatitis	Mouse	IN	Yes	Perlman et al., 1990, Barnett et al., 1993; Lavi et al., 1988.
Polio	Chimpanzee	IN	Yes	Bodian and Howe, 1941a,b
Rabies	Mouse	IN	Yes	Astic et al., 1993. Lafay et al., 1991

(continued)

456   **Chapter 19   The Olfactory System as a Route of Delivery for Agents to the Brain and Circulation**

Table 19.1   (*Continued*)

Substance	Species Studied	Application Method	Transneuronal Transport	Reference
St. Louis encephalitis	Hamster	IP	Yes	Monath et al., 1983
Sendai	Mice	IN	Limited	Mori et al., 1995
Semiliki Forest	Mice	IN	Age-Dependent	Oliver and Fazakerley, 1998
Venezuelan Equine	Mouse	Subcutaneous	Yes	Charles et al., 1995; Ryzhikov et al., 1995
Dyes				
Procion yellow	Catfish	IN	No	Holl, 1980
	Eel	IN	No	Holl, 1981
Lucifer yellow	Lamprey	IN	No	Suzuki, 1984
	Frog	IN	No	Suzuki, 1984
Evans blue	Mouse	IN	No	Kristensson and Olsson, 1971
Prussian Blue	mouse	IN	No	Rake, 1937
Trypan blue	Mouse	IN	No	Seki, 1941
	Bullhead	IN	No	Holl, 1965
Amino acids				
Leucine	Toad	IN	No	Weiss and Holland, 1967
	Garfish	IN	No	Gross and Beidler, 1973
	Rabbit	IN	No	Land and Shepherd, 1974
	Pike	IN	No	Weiss and Buchner, 1988
Alanine	Mouse	IN	No	Margolis and Grillo, 1977
	Hamster	IN	No	Burd et al., 1982
Taurine	Mouse	IN	No	Brittebo and Ericksson, 1995; Lindquist et al., 1983
Drugs				
Cocaine	Mouse	IN	No	Brittebo, 1988
Valproic acid	Rat	IN	No	Hoeppner, 1990
Levodopa	Rat	IN	Yes	de Souza Silva et al., 1997a,b; Chao et al., 2012;
Growth Differentiation factor 5	Rat	IN	Yes	Hanson et al., 2012
Insulin-like growth factor-I (IGF-I)	Rat	IN	Yes	Liu et al., 2001; 2004
Miscellaneous				
Aflatoxin B$_1$	rat	IN	Limited	Larsson and Tjälve, 2000
Ameba	mouse	IN	Yes	Jarolim et al., 2000
Horseradish Peroxidase	Rat	IN	No	Balin et al., 1986; Kristensson and Olsson, 1971;
	Mouse	IN	No	Stewart, 1985; Meredith and O'Connell, 1988
Immunoglobulins	Mouse	IN	No	Baker and Maruniak, 1990
Polychlorinated biphenyls (PCBs)	Ferret	Ambient air	ND	Apfelbach et al., 1998
Solvents	Mouse	Inhalation	No	Ghantous et al., 1990

IN, intranasal; IP, intraperitoneal; IV, intravenous; ND, not determined.

are quite varied, including lectins, dyes, solvents, metals, viruses, amino acids, and drugs. Table 19.1, a compilation of the range of transported substances, also indicates those for which the evidence suggests transneuronal transport. The consequences of such transport have only begun to be assessed (see below). One hypothesis suggests that the olfactory system may be a route by which some,

as yet undefined, pathogen or toxin reaches the brain to produce degenerative disorders such as Alzheimer's disease (Baker and Margolis, 1986; Crapper McLachlan, 1986) and Parkinson's disease (Prediger et al., 2006, 2010, 2011, 2012; Tjälve et al., 1996), as well as learning and behavioral abnormalities (Becker, 1995). This internalization may have effects reaching through the food chain

19.2 Nature of Substances Transported 457

Table 19.2 Systemically-administered agents associated with olfactory mucosal damage in rodents

Compound	Route	Endpoint	References
Acetaminophen	i.p., oral	OMD[1]	Jeffery et al., 1988; Genter et al., 1998
Alachlor	oral	tumors	U.S. Environmental Protection Agency, 1985; Genter et al., 2000
Bromobenzene	i.v., i.p.	OMD	Brittebo et al., 1990
Caffeine-derived N-nitroso cmpds.	Oral	tumors	Ivankovic et al., 1998
Carbimazole	i.p.	OMD	Genter, 1998
Coumarin	i.p., oral	OMD	Gu et al., 1997
p-Cresidine	oral	tumors	National Cancer Institute, 1979
2,6-dichlorobenzamide	i.p.	OMD	Brittebo et al., 1991
2,6-dichlorobenzonitrile	i.p., i.v., dermal	OMD	Brandt et al., 1990 Deamer et al., 1994
2,6-dichlorothiobenzamide	i.p.	OMD	Brittebo et al., 1991
Dihydropyridines	i.p.	protoporphyria	Reed et al., 1989
Diethyldithiocarbamate	i.p.	OMD	Deamer and Genter, 1995
Disulfiram	i.p.	OMD	Deamer and Genter, 1995
1,4-dithiane	gavage	OMD	Schieferstein et al., 1988
N-bis(2-hydroxypropyl)nitrosamine	s.c.	tumors	Koujitani et al., 1999
N-hydroxy-IDPN	i.p.	OMD	Crofton et al., 1996
3,3′iminodipropionitrile	i.p.	OMD	Genter et al., 1992
Methimazole	i.p., oral	OMD	Genter et al., 1995
3-methylindole	i.p.	OMD	Turk et al., 1986
Methylsulfonyl-2,6-dichlorobenzene	i.p.	OMD, metaplasia	Bahrami et al., 2000
Naphthalene	i.p.	OMD	Plopper et al., 1992
N-nitrosodiethylamine	i.p.	OMD	Jensen and Sleight, 1987
NNK[2]	s.c.	OMD, tumors	Belinsky et al., 1987
2-Pentenenitrile	i.p.	OMD	Genter and Crofton, 2000
Phenacetin	oral	OMD, tumors	Bogdanffy et al., 1989 Isaka et al., 1979

Routes of administration: i.p.=intraperitoneal injection; i.v.=intravenous injection; s.c.=subcutaneous

[1]OMD=olfactory mucosal degeneration

[2]NNK=4-(N-methyl-N-nitrosamino)-1-(3-pyridyl)-1-butanone

as some fish apparently can accumulate materials such as metals found in water by this route (Bazer et al., 1987). Table 19.2 lists compounds that when administered systemically either destroy or produce tumors in the olfactory epithelium. There are, however, benefits to the nasal route of application since in the past two decades, a number of encouraging outcomes have been reported in the treatment of diseases of the brain or CNS through nasal administration of drugs (see Table 19.3). Intranasal drug administration as a non-invasive brain delivery method has developed quickly in terms of safety, efficiency, and convenience. These advantages have made intranasal delivery a hot topic in the exploration of the brain drug delivery.

First to be discussed will be the specific substances and pathogens transported through olfactory receptor cells. Second, the proposed mechanisms of internalization and transport will be reviewed with special reference as to why transneuronal transport occurs only in limited instances. Intranasal application of drugs will be discussed as a means of administering a number of agents for direct entry to the brain. Next, the consequences of xenobiotic access to the brain will be assessed from the disease perspective. Finally, efforts to enhance the delivery of intranasally-adminstered drugs to the CNS and systemic circulation will be summarized.

19.2 NATURE OF SUBSTANCES TRANSPORTED

19.2.1 Metals

Aluminum, cadmium, cobalt, gold, and manganese are among the metals that are internalized in the olfactory receptor epithelium and transported either directly to the brain via olfactory axons or perhaps, as reviewed by Jackson et al. (1979), through either the subarachnoid space or perineurally. The evidence, both direct and circumstantial, for their transneuronal transport to the brain has supported the controversial hypothesis that some of these metals, for example, aluminum and manganese, may be etiologically significant in pathological syndromes such as Alzheimer's and Parkinson's diseases (Crapper McLachlan, 1986; Roberts, 1986; Tjälve et al., 1996). Described below is the method of application and evidence for uptake and transport of each of the metals.

458 Chapter 19 The Olfactory System as a Route of Delivery for Agents to the Brain and Circulation

Table 19.3 Drugs/toxicants administered via the nasal cavity for CNS and systemic effects

Purpose	Drug	Model	Reference
Drugs administered via the nasal cavity for CNS effects			
Migraine headaches	Sumatriptan (Imitrex®)	Dog, Human	Barrow et al., 1997; Dahlof et al., 1998
	BMS-181885	monkey	Srinivas et al. 1998
	ergotamine derivatives	human	Gallagher 1996; Lipton 1997; Ziegler et al., 1994
		rabbit	Marttin et al., 1997
	butorphanol	human	Melanson et al., 1997
	lidocaine	human	Lane, 1996; Mills and Scoggin 1997; Sachs, 1996
	capsaicin	human	Levy 1995
Sedation	Midazolam	human	Björkman et al., 1997; Louon and Reddy, 1994
	Diazepam	rabbit	Bechgaard et al., 1997
Seizure/epilepsy control	Midazolam	human	Jeannet et al., 1999; Wallace, 1997
Analgesia	oxycodone	human	Takala et al., 1997
	Oxymorphone	dog	Hussain and Aungst, 1997
	Acetylsalicylic acid	rat	Hussain et al., 1992
	Tizanidine	Mice	Patel et al., 2012
	Neurotoxin II	Mice	Ruan et al., 2012
	Fentanyl	Human	Mudd, 2011, Mystakidou et al., 2011
Psychiatric diseases	Olanzapine	Mice	Seju U et al., 2011
	Rispiradone	Mice	Patel S et al., 2011
Cancer chemotherapy	5-fluorouracil	rat	Sakane et al., 1999
	Monoterpene perillyl alcohol	human	da Fonseca et al., 2011
Treatment/prevention of neurodegenerative diseases	Lipophilic analogue of vasoactive intestinal peptide ([St-Nle^{17}]VIP)	rat	Gozes et al., 1996, 1997
	Dextromethorphan	Rat	Char et al., 1992
	Neostigmine	human	Di Costanzo et al., 1993
	Nerve growth factor	Rat	Frey et al., 1997
	Synthetic acetylcholine receptor epitopes	Mouse	Karachunski et al., 1998
	Urocortin	Rat	Wen et al., 2011
	Insulin	Human	Craft et al., 2012
	Levodopa	Rat, Human	De Souza Silva et al., 1997a,b; Kao et al., 2000; Chao et al., 2012
Treatment/prevention of CNS infections	zidovudine (AZT)	Rat	Seki et al. 1994
	D4T	Rat	Yajima et al. 1998
	cephalexin	Rat	Sakane et al., 1991b
Increased brain activity/ memory improvement	arginine-vasopressin	Human	Pietrowsky et al., 1996 Perras et al., 1997
Appetite control/weight loss	cholecystokinin (CCK) agonists	Dog	Pierson et al., 1997
	α-melanocyte-stimulating hormone	Humans	Hallschmid et al., 2004
	Insulin	Humans	Hallschmid et al., 2004
Cerebral hypoxia/stroke	erythropoietin	Gerbils	Gao et al., 2011
	E-selectin	Mice	Li et al., 2011
	Novel caspase 9 inhibitor	Rat	Akpan et al., 2011
Respiratory infection	siRNA against respiratory Syncicial virus (RSV)	Humans	Barik, 2011

(continued)

19.2 Nature of Substances Transported

Table 19.3 (*Continued*)

Purpose	Drug	Model	Reference
Drugs administered via the nasal cavity for systemic effects			
Diabetes	Insulin	human, rabbit, rat	Fernandez-Urrusuno et al., 1999; Hilsted et al., 1995 Valensi et al., 1996; Jintapattanakit et al., 2010
Smoking cessation	Nicotine nasal sprays	human	Gourlay and Benowitz, 1997; Jones et al. 1998; Schneider et al., 1995.
Vaccines	Outer membrane vesicles (*N. meningitides*)	human	Haneberg et al., 1998
	Diptheria-tetanus boosters	human	Aggerbeck et al., 1997
	Bordatella pertussis BPZE1	Human	Li et al., 2011
	Shigella flexneri	Human	Riddle et al., 2011
Normalization of vitamin B_{12}	hydroxocobalamin	human	Slot et al., 1997; Swain, 1995
Motion sickness	scopolamine	human	Putcha et al., 1996
	promethazine	dog	Ramanathan et al., 1998
Nocturia	desmopressin	human	Butler et al., 1998; Kallas et al., 1999
Contraception	gonadotrophin-releasing hormone agonist	human	Bergquist et al., 1979
Endometriosis, ovarian estrogen receptor stimulation	nafarelin, buserelin	human	Dantas et al., 1994; Jacobson et al., 1994; Lemay et al., 1984
Prostate cancer, Oligoasthenozoospermia	buserelin	human	Matsumiya et al., 1998; Tolis et al., 1983
Prevention of bone resorption	calcitonin	human	Silverman 1997
Enhanced lactation	oxytocin	human	Ruis et al., 1981
Anemia	erythropoietin	rat	Shimoda et al., 1995
Erectile dysfunction	udenafil	Rat, human nasal epithelial cells in vitro	Cho et al., 2012

19.2.1.1 Aluminum. Several studies have shown that aluminum can reach the CNS after intranasal application (Hayek and Waite, 1991; Perl and Good, 1987; Zatta et al., 1993). Aluminum silicate, presented as a powder in the bedding, was enriched specifically in the rabbit olfactory epithelium and olfactory bulb, but not in cerebellum (Hayek and Waite, 1991). Perl and Good (1987) demonstrated that, following intranasal application of aluminum lactate, granulomas were observed in the olfactory bulb and cerebral cortex. The pathological changes observed in the brain suggested that aluminum reaches these brain regions by transport through olfactory neurons. Inhalation of aluminum acetylacetonate resulted in widespread dissemination of the metal complex in diverse brain regions (Zatta et al., 1993). Aluminum was also observed in the neurofibrillary tangles associated with Alzheimer's disease (for review see Crapper McLachlan, 1986; Good et al., 1992), and on intraventricular administration was shown to produce neurofibrillary degeneration in rabbit and rat brain (Kowall et al., 1989; Shigematsu and McGeer, 1992). Additionally, the demonstration that brain regions having direct connections to the olfactory system are primarily affected in Alzheimer's disease has suggested the provocative hypothesis that transport of aluminum via the olfactory system may play a role in this degenerative syndrome (Roberts, 1986).

Brains from patients with Parkinson's disease have, relative to control brains, elevated levels of aluminum in the substantia nigra, caudate nucleus, and globus pallidus (Yasui et al., 1992). Moreover, increased levels of aluminum have been noted in Lewy bodies of patients with Parkinson's disease. However, whether aluminum plays a causative role is presently enigmatic (Hirsch et al., 1991).

19.2.1.2 Iron. Interest in iron as a potential risk factor for Parkinson's disease initially stemmed from reports of heightened accumulation of iron within the substantia nigra in postmortem Parkinson's disease brains (Dexter et al., 1987). In general, evidence for the hypothesis that iron is involved in the etiology of Parkinson's disease is much stronger than that for aluminum, possibly because iron, unlike aluminum, is needed for cellular metabolism. For example, epidemiological studies suggest that too much dietary iron is a risk factor for Parkinson's disease (Lan and Jiang, 1997; Johnson et al., 1999; Powers et al., 2003, 2009). In animals, iron significantly contributes to oxygen free radical formation during degeneration of nigrostriatal pathways. The neurotoxins 1-methyl-4-phenyl-1,2,3,6-tetrahydropyridine (MPTP) and 6-hydroxydopamine (6-OHDA), widely used in experimental models of Parkinson's disease, potentiate an increase in iron within the substantia nigra and striatum in mice, rats, and monkeys (Gal et al., 2010). Iron injected directly into the dorsal striatum of rats selectively damages dopaminergic neurons, inducing behavioral and biochemical changes akin to parkinsonism (Youdim et al., 1993). Moreover, mice fed diets low in iron (4 ppm) for 6 weeks, unlike mice on a normal (48 ppm) or high (400 ppm) iron diets, have increased serum total iron binding capacity and exhibit protection against motor impairments induced by MPTP (Levenson et al., 2004). Iron chelation therapies have restored nigrostriatal dopaminergic neurons in MPTP animal models of PD (Gal et al., 2010). Interestingly, a mutation in the divalent metal transporter-1 (DMT1), which ferries iron across cell membranes, protects rodents from the effects of MPTP and 6-OHDA within the substantia nigra, implying a critical role for DMT1 in the degenerative process in this brain region (Salazar et al., 2008).

19.2.1.3 Cadmium.

Cadmium transport from the olfactory epithelium to the olfactory bulb has been demonstrated in both fish and rodents. A number of reports suggest an association between cadmium exposure and anosmia in humans, and also accumulation and transport of the metal (Hastings, 1990). In the pike *(Esox lucius)*, cadmium appeared to move at a velocity consistent with processing by the fast axonal transport system (Gottofrey and Tjälve, 1991). Cadmium accumulated in anterior regions of the olfactory bulb, but did not appear to move transsynaptically to mitral cell dendrites (Tjälve and Henriksson, 1999; Tjälve et al., 1996). In rats, exposure to both cadmium oxide and chloride, in aerosol form, resulted in accumulation of cadmium in the olfactory bulb (Evans and Hastings, 1992; Hastings, 1990; Hastings and Evans, 1988). Cadmium oxide did not produce anosmia (Hastings, 1990; Hastings and Evans, 1988). Hastings (1990) suggested that the lack of a functional deficit in rodents following cadmium exposure, in view of the purported effects in humans, might reflect a requirement for protracted exposure to produce anosmia.

19.2.1.4 Cobalt.

Intranasal irrigation with cobalt lysine in king salmon fry *(Oncorhynchus tshawytscha, Walbaum)* resulted in transport of the conjugate to the olfactory bulb as well as to both ventral and lateral regions of the ventral telencephalon (Bazer et al., 1987). The latter projections may represent transneuronal transport of the metal conjugate (Bazer et al., 1987) as no direct epithelial connections to this brain region have previously been reported. There is evidence that cobalt is able to cross neuronal membranes (Fredman and Jahan-Pawar, 1980).

19.2.1.5 Gold.

Electron microscopic studies demonstrated that colloidal gold was internalized by olfactory receptor cells within 15 min of intranasal application in squirrel monkeys (DeLorenzo, 1970). Gold was thought to be incorporated into receptor cells by the pinocytotic vacuoles normally observed in the olfactory rod region, more commonly referred to as the dendritic knob (DeLorenzo, 1970). Subsequently, gold particles were observed in the cytoplasm of olfactory axons and within 1 hr in the olfactory glomerulus. Evidence was also presented for transneuronal transport of the gold particles from receptor cell axon terminals within the glomeruli to the dendrites of mitral cells (DeLorenzo, 1970). Within the olfactory bulb the particles were preferentially associated with mitrochondria. The rate of appearance in the glomeruli was consistent with the movement of the gold by fast axonal transport (estimated in these experiments at about 2.5 mm/hr). Czerniawska, (1970) also demonstrated uptake of gold into the cerebrospinal fluid, especially in the region of the cribriform plate and the olfactory bulbs, following either mucous membrane or intravenous administration of the metal.

19.2.1.6 Manganese.

A recent series of papers documents both possible routes of entry of manganese into the CNS and the neurological consequences of exposure. Definitive evidence now exists in rats for transport of manganese from the olfactory epithelium into the brain (Tjälve and Henriksson, 1999; Tjälve et al., 1996). The transport of intranasal administered [54]Mn to the olfactory bulb was significantly reduced by the transection of olfactory nerve fibers (Kinoshita et al., 2008). In contrast to other metals applied by intranasal irrigation (e.g., nickel and cadmium), manganese was transported to all regions of the CNS, including the spinal cord, and could still be detected 12 weeks after initial exposure. Manganese may also reach the CNS when administered by parenteral routes, but in much lower concentrations. Manganism, or manganese toxicity, is associated with a neurological syndrome characterized by features of Parkinson's disease. Interestingly, Lucchini et al. (2009) suggested that acute exposure to high levels of manganese may damage the globus pallidus and induce manganism, whereas chronic low levels of manganese exposure may result in cumulative damage to the substantia nigra and to the expression of Parkinson's disease. Occupational exposure occurs in the mining industry, in steel manufacturing and in welding and is primarily through inhalation of metal-containing dusts and fumes (Gorell et al., 1999; Tjälve and Henriksson, 1999). Taken together with the widespread CNS distribution following nasal exposure, the association with Parkinson's disease is consistent with specific toxicity towards dopamine neurons. One proposed mechanism (Tjälve and Henriksson, 1999) suggests that the metal may be involved in the formation of free radicals in the presence of catecholamines.

19.2.1.7 Other Metals.

Several studies investigated the transport of nickel and mercury presented intranasally. Inorganic mercury exhibited limited transneuronal transport restricted to the olfactory bulb following nasal application in rats (Henriksson and Tjälve, 1998). Long term occupational exposure was not associated with increased risk for Parkinson's disease (Gorell et al., 1999). Transneuronal transport of nickel was studied in rats and pike. The transport rate for this metal was 20 times slower than for cadmium and manganese. Nickel could be demonstrated in several forebrain regions suggesting that the metal is transported slowly trans-synaptically (Tallkvist et al., 1998; Tjälve and Henriksson, 1999). Nickel exposure has been associated with cell loss in the olfactory epithelium but did not alter olfactory function as indicated by measurement of either threshold or odor discrimination (Evans et al., 1995). Uptake of thallium along the olfactory nerve is strongly suggested by the observation that the olfactory nerve transection greatly reduces the transport of intranasally-administered thallium into the brain in mice (Kinoshita et al., 2008). Reports that monkeys could be protected against intranasal inoculation with poliomyelitis virus by nasal lavage with solutions of alum, picric acid or zinc sulfate (Armstrong and Harrison, 1935; Schultz and Gebhardt, 1936) led to prophylactic treatment of children with nasal zinc sulfate during poliomyelitis epidemics (Peet et al., 1937; Schultz and Gebhardt, 1937; Tisdall et al., 1937). An undesirable consequence of these treatments was long-lasting anosmia in many individuals (Tisdall et al., 1938) presumably produced by irreversible destruction of the olfactory sensory epithelium; unfortunately, the incidence of polio in the treated children was no different than that in untreated children. More recently, Lim et al. (2009) demonstrated that intranasal administration of Zicam (zinc gluconate; Matrixx Initiatives, Inc), a homeopathic product marketed to alleviate cold symptoms, showed significant cytotoxicity in both mouse and human nasal tissue. Specifically, Zicam-treated mice had disrupted sensitivity of olfactory sensory neurons to odorant stimulation and were unable to detect novel odorants in behavioral testing. These findings were long-term, as no recovery of function was observed after two months (Lim et al., 2009). A causality analysis supported the Bradford Hill criteria and demonstrated that zinc gluconate therapy caused hyposmia and anosmia in humans (Davidson and Smith, 2010).

19.2.2 Lectins

Although native and derivatized lectins have been utilized by a number of investigators to delineate neuronal connectivity, especially in the visual system (Itaya and Van Hausen, 1982; Spencer et al., 1982), transport studies of these oligosaccharide-specific proteins in the olfactory system have been limited primarily to concanavalin A (Wiley et al., 1984) and wheatgerm agglutinin (WGA). A number of investigators (Baker and Spencer, 1986; Balin et al., 1986; Broadwell and Balin, 1985; Itaya, 1987; Shipley, 1985; Stewart, 1985; Thorne et al., 1995) have shown that WGA conjugated to horseradish peroxidase (WGA-HRP) is internalized by olfactory receptor cells following binding to surface receptors and transported to the olfactory bulb. Uptake and transport occurred primarily ipsilaterally following intranasal application (Baker and Spencer, 1986; Shipley, 1985). Both anterograde and retrograde transneuronal transport have been observed from the olfactory bulb. Labeled primary olfactory axons containing WGA-HRP reaction product were observed in the nerve and glomerular layers of the main olfactory bulb within 6 hr of intranasal irrigation (Broadwell and Balin, 1985; Itaya, 1987; Shipley, 1985). At longer survival times (four to seven days), labeled neurons and terminals were reported in a number of brain regions known to either receive input from or send axons to the olfactory bulb. Anterograde and retrograde label were observed in the anterior olfactory nucleus (Itaya, 1987; Shipley, 1985). The piriform and entorhinal cortices, as well as the olfactory tubercle, also contained significant label (Baker and Spencer, 1986; Itaya, 1987; Shipley, 1985). Brain regions that send centrifugal afferent innervation primarily to the glomerular region of the olfactory bulb also contained retrogradely labeled cells (Baker and Spencer, 1986; Shipley, 1985). These data indicate the extent of transport that occurs from the olfactory epithelium to those brain regions with connections to the olfactory bulb. More recently, the smallest known lectin, odorranalectin, was shown to enhance the delivery of a potential Parkinson's disease treatment, urocortin peptide, from the nasal cavity into the brain (Wen et al., 2011).

19.2.3 Viruses

From a historical perspective, a neural route of viral invasion of the CNS was postulated as a mechanism for the development of rabies as early as the 18th century and demonstrated in the late 19th century (see Stroop, 1995 for review and references). The nasal route as a site of viral entry to the CNS was established during the 1920s and 1930s (Clark, 1929; Hurst, 1936). Poliomyelitis virus was

462   Chapter 19   The Olfactory System as a Route of Delivery for Agents to the Brain and Circulation

subsequently shown to reach the olfactory bulbs through the olfactory neurons following intranasal application in primates (Bodian and Howe, 1941a,b).

Subsequently, transport to the CNS has been reported for several types of viruses following intranasal inoculation, including herpes simplex (Esiri and Tomlinson, 1984; McLean et al., 1989; Stroop et al., 1984), murine coronavirus (Barnett et al., 1993; Perlman et al., 1989; Perlman et al., 1988), Borna disease virus (Morales et al., 1988), Nile virus (Nir et al., 1965), Venezuelan equine encephalitis virus (Charles et al., 1995), vesicular stomatitis virus (Huneycutt et al., 1994), Semiliki virus (Oliver and Fazakerley, 1998), sendai virus (Mori et al., 1995) and rabies virus (Astic et al., 1993; Lafay et al., 1991). Olfactory bulb ablation prevented the spread of the neurotropic coronavirus mouse hepatitis virus into the brain (Perlman et al., 1990). Two neurotropic viruses, herpes simplex virus type 1 and mouse hepatitis virus, spread along different neural pathways following intranasal inoculation, suggesting that uptake in specific neurotransmitter systems influences viral distribution in the CNS (Barnett et al., 1993). Selective lesions of neural pathways also contributed to spatial learning deficits (McLean et al., 1993). Learning and behavioral deficits have recently been associated with HSV-1 brain infection, presumably by the olfactory nerve route (Becker, 1995). Olfactory labeling was observed in several instances where the viruses were applied at non-olfactory sites, including intraperitoneal (Monath et al., 1983) and subcutaneous routes (Charles et al., 1995). Virus titers were observed first in the olfactory epithelium (four days) and then in the olfactory bulb (five days) followed by the rest of the brain. Olfactory labeling was observed following corneal inoculation and intradermal application in the mouse (Stroop et al., 1984). In the latter instances the virus may have reached the olfactory epithelium through tears entering the naris or nasopharynx and then entering the nasal receptor epithelium. Applying sensitive immunohistochemical techniques, herpes simplex virus produced Golgi-like, transneuronal labeling in the CNS (Blessing et al., 1994; McLean et al., 1989) and the authors suggested that virus injection may be useful for tracing neuronal pathways. Many recent reports demonstrate that pseudorabies virus also produced significant transneuronal transport and can be used to label CNS pathways (Sams et al., 1995; Strack and Loewy, 1990; Ugolini, 1995). These and many other studies (Table 19.1) demonstrate that the olfactory system may serve as an efficient route of entry of viruses into the brain.

19.2.4 Dyes

19.2.4.1 Procion and Lucifer Dyes. Several dyes, including procion, lucifer, and Evans blue, were internalized into olfactory epithelial cells. In those species tested, procion and lucifer dyes specifically labeled olfactory receptor neurons. Two procion dyes, brilliant yellow M4RAN and yellow M4R, labeled the receptor neurons of the catfish, *Ictalurus nebulosus* (Holl, 1980). Similar intravital staining was observed in olfactory receptor neurons of the eel, *Anguilla anguilla* (Holl, 1981). In both the lamprey, *Entosphenus*, and the grass frog, *Rana*, procion yellow MX4R and lucifer yellow VS exhibited internalization specifically by olfactory receptor neurons in contrast to either supporting cells or the glandular components of the epithelium (Suzuki, 1984). Since the stained receptor neurons retained their dye through dissociation procedures, Suzuki (1984) suggested that these fluorescent dyes may be useful cell markers for identifying viable receptor cells in dissociated epithelial preparations. Interestingly, not all dyes showed specific uptake. Lucifer yellow CH stained the receptor epithelium nonspecifically, and procion red H3B at high concentrations resulted in dye aggregations on the surface of the olfactory epithelium. The basis for the difference in labeling between the dyes was thought to reflect binding to specific membranous molecules.

19.2.4.2 Evans Blue. Inhalation of a complex of Evans blue dye with albumin resulted in strong red fluorescence within numerous cells of the olfactory epithelium of mice (Kristensson and Olsson, 1971). After 24 hr, the filia olfactoria of the mice exhibited fluorescent label. Although labeling was found in the olfactory bulbs, especially in young mice, there was no evidence of transneuronal transport to secondary neurons such as mitral cells. Evidence for label within the submucosal area and leptomeninges surrounding the olfactory bulb suggested that some of the dye could reach the CNS by other routes in addition to transport through the olfactory nerves (Kristensson and Olsson, 1971).

19.2.4.3 Trypan Blue. Trypan blue dye was studied in both the brown bullhead, *Ictalurus natalis* (Holl, 1965) and the mouse (Seki, 1941). The investigations were undertaken to demonstrate the time course and conditions for specific uptake of the dye into olfactory receptor cells and not into other cell types in the olfactory epithelium. Transneuronal transport was not investigated.

19.2.5 Amino Acids

19.2.5.1 Leucine. The transport of leucine in the olfactory system has been studied by a number of investigators (for review see Weiss and Buchner, 1988). For the most part, uptake, metabolism, and transport of the amino acid were used to characterize the kinetics of protein movement in normal and regenerating nerve fibers. The olfactory system was an ideal model since in many fish species the nerve is long and regenerates

19.2 Nature of Substances Transported · 463

readily. One of the pioneering studies used histochemical techniques to demonstrate axoplasmic transport of labeled proteins following intranasal application of tritiated leucine in the toad, *Bufo americanus* (Weiss and Holland, 1967). Gross and Beidler (1973) studied fast axonal transport in C-fibers of the garfish, *Lepisosteus osseus.* Experiments performed in the pike, *Esox lucius,* demonstrated similar dynamics of fast axonal transport (Gross and Kreutzberg, 1978; Weiss et al., 1978). The long olfactory nerve of this species was used to study temperature dependence of axonal transport, especially slow axonal transport in normal and regenerating olfactory nerve (Cancalon, 1979a,b). The relative uniformity of transport in both the antero- and retrograde direction was shown in the pike olfactory nerve (Weiss and Buchner, 1988). Transport of tritiated leucine was also utilized to analyze the projections of olfactory receptor neurons in the adult rabbit (Land and Shepherd, 1974). In all studies the majority of the leucine was incorporated into amino acids following a short delay. Transport was monitored either autoradiographically or biochemically within dissected nerve segments. Although both slow (Cancalon, 1979a,b) and fast axonal transport (Gross and Beidler, 1973; 1975) were demonstrable using these techniques, transneuronal transport was not observed under the experimental conditions utilized (Land and Shepherd, 1974; Weiss and Holland, 1967).

19.2.5.2 *Alanine.*

In the olfactory system the dipeptide carnosine (β-alanyl-L-histidine) may be either a neurotransmitter or neuromodulator of the primary olfactory neurons (Burd et al., 1982). Therefore, β-alanine presented at the mouse external naris was incorporated specifically into carnosine (Margolis and Grillo, 1977). Subsequent experiments in hamsters demonstrated that the radioactivity associated with the β-alanine was converted to carnosine (>82% of the radioactivity was in the carnosine fraction at 24 hr) and transported to the olfactory bulb (Burd et al., 1982). The rate of appearance in the olfactory bulb was consistent with movement of the β-alanine by the fast axonal transport system.

Tritiated α-alanine exhibited a pattern of transport consistent with a less specific internalization in the olfactory epithelium and incorporation into a number of proteins after intranasal administration. Both sensory and nonsensory cells and the lamina propria were labeled. Peak labeling in the olfactory bulb was less intense and delayed as compared to β-alanine, suggesting that the protein was carried by slow axonal transport. Transneuronal transport was not observed for either α- or β-alanine (Burd et al., 1982).

19.2.5.3 *Taurine.*

Taurine accumulated specifically in the olfactory bulb after intravenous injection (Lindquist et al., 1983). The ability of the olfactory system to concentrate this amino acid was even more pronounced following intranasal irrigation. An intra-axonal transport mechanism, as opposed to a perineural one, was suggested by the restricted accumulation in the bulb ipsilateral to the application.

19.2.6 Drugs

Valproic acid (Hoeppner, 1990), cocaine (Brittebo, 1988), and a number of other drugs (Dahl and Hadley, 1991) were concentrated in the olfactory epithelium following intravenous administration. The relative enrichment of cocaine in the epithelium was maintained for at least 24 hr. The mechanism for the concentration of these drugs in the epithelium is unknown. The presence of dopamine D_2 receptors in the receptor epithelium (Guthrie et al., 1991; Shipley et al., 1991) suggests that cocaine, a dopamine uptake blocker, binds to a specific receptor site in the epithelium. The presence of numerous drug-metabolizing enzymes (Dahl and Hadley, 1991) also may contribute to the bioactivation of drugs in to reactive metabolites in the olfactory epithelium. Transport has not been studied following intranasal administration of the drugs.

Folic acid and its derivatives alone or in combination with other drugs, such as cholinesterase inhibitors and acetylcholinesterase inhibitors, can directly enter the CNS and provide a rapid effect on the prevention or treatment of Alzheimer's disease and stroke (Hussain et al., 2002). However, the transport mechanism of folic acid-mediated nose-to-brain delivery is still under investigation. Intranasal insulin administration improved memory in healthy volunteers and Alzheimer's patients (Benedict et al., 2004; Reger et al., 2008). Intranasal oxytocin administration reduced stress in monkeys (Parker et al., 2005) and intranasal insulin-like growth factor-I (IGF-I) decreased infarct volume and improved neurologic function in stroke models in rats (Liu et al., 2001, 2004).

Dopamine seems also to bypass the BBB when applied into the nose and has been shown to have behavioral and neurochemical effects in rats. For example, intranasal dopamine administration increased dopamine levels in the neostriatum and nucleus accumbens (de Souza Silva et al., 2008). Rodents showed an antidepressant-like behavior in a forced-swimming task and higher locomotor activity in the open field following intranasal dopamine administration (Buddenberg et al., 2008). Intranasal dopamine administration also reduced activity and improved attention in an animal model of attention-deficit-hyperactivity disorder (ADHD) (Ruocco et al., 2009). Interestingly, recent studies demonstrate that intranasally applied dopamine and levodopa alleviate parkinsonian symptoms in rats with unilateral nigro-striatal 6-OHDA lesions, one of the most

464 Chapter 19 The Olfactory System as a Route of Delivery for Agents to the Brain and Circulation

used animal models of Parkinson's disease (Pum et al., 2009; Chao et al., 2012).

19.2.7 Miscellaneous Substances

19.2.7.1 Horseradish Peroxidase. The transport of HRP differed from that observed for the conjugate, WGA-HRP. HRP, like WGA-HRP, was internalized by receptor cells and labeled the olfactory glomeruli (Balin et al., 1986; Kristensson and Olsson, 1971; Meredith and O'Connell, 1988; Stewart, 1985). However, transneuronal transport was not observed. Movement also occurred through the intercellular junctions to label the leptomeninges (Balin et al., 1986). HRP uptake was not observed in the accessory olfactory system unless a large dose of epinephrine was administered to activate the vomeronasal organ pumping mechanism (Meredith and O'Connell, 1988). Mechanistic considerations discussed below may underlie the differences between WGA-HRP and HRP.

19.2.7.2 Immunoglobulins. Baker and Maruniak (1990) demonstrated that mouse olfactory receptor neurons internalized IgG molecules from a number of species. Twenty-four hours after intranasal application, receptor neurons, microvillar cells, and nerve bundles in the olfactory mucosa contained IgG, but transport into the olfactory bulb was not observed.

19.2.7.3 Solvents. Uptake and metabolism were investigated for several solvents, including benzene, toluene, xylene, and styrene (Ghantous et al., 1990). All the aromatic hydrocarbons were found in the olfactory epithelium, but only toluene, xylene, and styrene were found in the olfactory bulb. Transport of the three solvents in olfactory receptor cells was thought to require conversion to either their aromatic acids or their conjugates. In support of this hypothesis, benzene does not exhibit biotransformation to aromatic acids and was not transported (Ghantous et al., 1990). The cytochrome P-450 enzymes necessary for these biotransformations are known to be present in the olfactory epithelium (Dahl, 1988; Dahl and Hadley, 1991).

19.2.8 Olfactory Toxicants Delivered via the Bloodstream

Multiple xenobiotics that are systemically administered (i.e., by non-inhalation routes of exposure) have toxic effects on the olfactory mucosa of experimental animals (Table 19.2). For example, intraperitoneal administration of the analgesic acetaminophen or the anti-hyperthyroid drugs

methimazole or carbimazole to rodents causes degeneration of the olfactory mucosa (Genter, 1998; Genter et al., 1995; Genter et al., 1998; Jeffery and Haschek, 1988). Some anti-hyperthyroid drugs have been associated with impaired olfaction in humans (Schiffman, 1983). The herbicide dichlobenil (2,6-dichlorobenzonitrile) is particularly interesting from the point of view that intravenous (i.v.), intraperitoneal (i.p.), or dermal exposure results in olfactory mucosal degeneration in mice, with dermal exposure representing an occupationally-relevant route of exposure (Brandt et al., 1990; Deamer et al., 1994). In general, there is partial to nearly full recovery of olfactory mucosa which has undergone chemically-induced degeneration (see chemicals labeled "OMD," referring to those causing olfactory mucosal degeneration in Table 19.2). However, a toxicant that destroys the olfactory basal cells and causes persistent metaplasia of the olfactory mucosa has recently been described (Bahrami et al., 2000). A number of drugs (e.g., phenacetin), industrial and agricultural chemicals, for example, (p-cresidine, alachlor) and environmental pollutants (e.g., NNK) are associated with the development of nasal cancer in rodents (Table 19.2). It is interesting to note that humans are exposed to low levels of many of these, or related, agents every day, but none has been linked definitively to human nasal cancer; the most widely-recognized risk factors for human nasal cancer are exposure to wood dust or nickel, both via the inhalation route (Barceloux, 1999; Hayes et al., 1986).

19.3 MECHANISMS OF TRANSPORT

19.3.1 Uptake

The preceding discussion demonstrates that olfactory receptor cells can internalize many types of substances. The issue is how and why these cells should internalize a wide range of xenobiotics. The answer to both these questions may be the same. Since the olfactory mucosa comes in contact with many substances in inspired air, a means of either elimination of detoxification has likely been developed as a protective mechanism. In fact, the olfactory epithelium contains high concentrations of a wide variety of xenobiotic-metabolizing enzymes, even when compared to levels in the liver (Dahl, 1988; Dahl and Hadley, 1991; Thornton-Manning and Dahl, 1997). A number of enzymes have been found in the olfactory epithelium, including numerous isozymes of P450, monooxygenases, aldehyde dehydrogenases, alcohol dehydrogenase, carboxyesterases, epoxide hydrolases, UDP-glucuronyl transferase, glutathione S-transferase and rhodanese (Dahl and Hadley, 1991) (see also Chapter 1). For some of these enzymes, isozymes have been found that

19.3 Mechanisms of Transport

are specific to the olfactory mucosa (Dahl, 1989; Jones and Reed, 1989; Lazard et al., 1991; Nef et al., 1989; Zupko et al., 1991). Thus, uptake into receptor, supporting, and glandular cells of xenobiotics, including odorants, may serve a detoxification as well as a clearance function. These enzymes also may play a role in defining the characteristic odor of a substance through the production of active metabolites, thereby creating new odorants (Dahl, 1988; Dahl and Hadley, 1983; Getchell et al., 1984).

The specificity of the uptake mechanism has not been established. For WGA-HRP, receptor-mediated processes have been suggested, since the conjugate, which is known to bind to surface membrane oligosaccharides, is found associated with vesicles (Baker and Spencer, 1986; Broadwell and Balin, 1985). Unconjugated HRP, on the other hand, is thought to be internalized via bulk endocytosis (Broadwell and Balin, 1985; Kristensson and Olsson, 1971). Surprisingly, one study reported that some preparations of WGA-HRP were not transneuronally transported, suggesting that the method of HRP conjugation may alter the ability of WGA to bind with specific membrane receptors (Russell et al., 1991). The association of gold particles with vesicles suggested that uptake of gold occurred subsequent to the formation of vesicles by pinocytosis (De Lorenzo, 1970). Demonstration of a vesicular association of several intranasally applied tracers indicated that the olfactory epithelium possesses a well-developed system of endocytic vesicles, perhaps as a result of a rapid rate of membrane turnover (Bannister and Dodson, 1992). These authors suggested that materials may be trapped during normal membrane processing. The internalization of materials into the olfactory receptor cells thus may take place by either receptor- or non-receptor-mediated processes. In the former case, internalization may occur at low xenobiotic concentrations, and in the latter, high concentrations may be necessary.

A superfamily of putative odorant receptors was cloned by Buck and Axel (1991), leading to their receiving the Nobel Prize in Physiology or Medicine. The large number of these receptors indicates the presence of cell-type-specific surface molecules that interact with odorants. The binding of odorants to the receptor proteins may result in their internalization. Specific transport of odorant receptor mRNA to the olfactory bulb was used to demonstrate that cells expressing a specific odorant receptor project to a single glomerulus (Ressler et al., 1994; Ressler et al., 1993; Sullivan et al., 1995; Vassar et al., 1994). Olfactory marker protein mRNA shows similar transport to the glomerular layer but no transneuronal transport (Wensley et al., 1995). Interestingly, transneuronal transport was demonstrated for barley lectin (a close relative of WGA) synthesized in olfactory receptor cells of a transgenic mouse expressing the lectin under control of the olfactory marker protein gene promoter

(Horowitz et al., 1999). The function of the intra-axonal mRNA is unknown, but the ability of olfactory receptor neurons to transport numerous other molecules may be a consequence of this endogenous transport capacity. Also, the dopamine D_2 receptors reported in olfactory receptor neurons may play a role in the uptake of molecules that bind to this receptor (Guthrie et al., 1991; Mansour et al., 1990; Shipley et al., 1991). Glycoproteins, as indicated by lectin binding, also have been demonstrated on olfactory receptor cells using both in vivo and in vitro techniques (Allen and Akeson, 1985; Barber, 1989; Foster et al., 1991; Key and Akeson, 1990; Key and Giorgi, 1986a; Key and Giorgi, 1986b; Lundh et al., 1989; Mori et al., 1985; Polak et al., 1989; Shirley et al., 1983). Uptake of antibodies to IgG into receptor cells has been reported (Baker and Spencer, 1986; Baker and Maruniak, 1990). However, immunoreactivity for constituents of the immune system was not found in receptor cells, but primarily in Bowman's glands, the mucociliary complex, lymphocytes clustered near the basement membrane and scattered in the lamina propria, as well as in mast cells of the human olfactory mucosa (Mellert et al., 1992). A similar distribution of the secretory immune system components was observed in salamanders and rats (Getchell and Getchell, 1991). Monoclonal antibodies also recognize different cell types within the olfactory mucosa (Hempstead and Morgan, 1985), suggesting differences in cell surface constituents that might direct differential binding to olfactory receptors as opposed to either supporting or glandular cells.

Finally, odorant-binding proteins synthesized by mucus-secreting glands of the olfactory and respiratory mucosa have been demonstrated (Baldaccini et al., 1986; Dal Monte et al., 1991; Pevsner et al., 1986). The binding proteins may be necessary for either the transport of odorants to olfactory receptors or their clearance after signal transduction. The odorant specificity of these proteins suggests that a similar binding specificity on the receptor cells may underlie ligand (either odorant of other xenobiotic) uptake into receptor, supporting, and glandular cells. Transfer between these compartments also cannot be ruled out especially since many of the degradative enzymes described above are found in glandular and supporting cells and not receptor cells (Dahl and Hadley, 1991; Thornton-Manning and Dahl, 1997).

19.3.2 Transneuronal Transport

Once internalized, transport in the olfactory receptor cells can occur by either the slow or rapid transport systems characteristic of neuronal systems (Weiss and Buchner, 1988). The distinction between molecules transported only to the olfactory bulb and those that are transneuronally transported may lie in the nature of their internalization and thus the organelles with which the molecules are

466   Chapter 19   The Olfactory System as a Route of Delivery for Agents to the Brain and Circulation

associated. For example, WGA-HRP, which is internalized by receptor-mediated endocytosis, is transported intra-axonally in tubulovesicular profiles (Baker and Spencer, 1986) and transneuronally following processing in the transmost Golgi saccule (Baker and Spencer, 1986; Broadwell and Balin, 1985). The latter compartment processes vesicles destined for axonal terminals (Broadwell and Balin, 1985). Gold is vesicle-associated and is transneuronally transported (DeLorenzo, 1970). Viruses are transported in axonal vacuoles within receptor cells and appear to move transsynaptically to invade other brain regions synaptically connected to the olfactory bulb (Lavi et al., 1988; Monath et al., 1983). The relative lack of glial labeling for WGA-HRP and some viruses (Baker and Spencer, 1986; Lavi et al., 1988) indicates limited release into the extracellular space and accounts for the restricted transfer to olfactory-related brain regions. In contrast, HRP, which does not exhibit transneuronal transport, is taken up by bulk endocytosis and is not processed through the Golgi saccule. Leucine, which also does not appear to be transported transneuronally (Land and Shepherd, 1974; Weiss and Holland, 1967), is primarily processed into cellular protein components (Weiss and Buchner, 1988). Taken together these data suggest that transneuronal transport requires receptor-mediated uptake into the olfactory sensory cells, followed by vesicular transport of unmodified materials through the axon, with subsequent release and uptake in association with synaptic specializations.

19.3.3 Specificity of CNS Transport

As mentioned above, transport from the olfactory bulb to other brain regions occurs along specific pathways. The ability of centrifugal afferents to support specific transport is suggested by data demonstrating selectivity of retrograde axonal transport of radioactive transmitters from the olfactory bulb to their nuclei of origin (Araneda et al., 1983; Bonnet-Font and Bobillier, 1990; Watanabe and Kawana, 1984). Intrabulbar application of radioactive serotonin or norepinephrine results in specific labeling respectively of the dorsal raphe nucleus and the locus coeruleus (Araneda et al., 1983; Bonnet-Font and Bobillier, 1990). Similarly, the receptor-mediated retrograde transfer of nerve growth factor from the olfactory bulb to forebrain cholinergic nuclei indicates the ability of these neurons to transport exogenously applied substances (Altar and Bakhit, 1991).

19.3.4 Transporters in Nasal Epithelia

The observation that nucleoside drugs are transported to the CNS following intranasal delivery (Seki et al., 1994; Yajima et al., 1998) prompted the question as to whether the members of the family of transporters known

to transport these drugs (i.e., the equilibrative nucleoside transporters [ENT]) were expressed in nasal epithelia. Branched DNA technology was used to compare the level of expression of nine transporters [ENT1 and ENT2; organic cation transporter (OCT)1, 2, and 3; OCTN1; organic anion-transporting polypeptide (OATP3); and multidrug resistance proteins (MRP)1 and MRP4] in nasal respiratory mucosa, olfactory mucosa, and olfactory bulb. ENT1 and ENT2 expression was relatively high in nasal epithelia and olfactory bulb, which may explain the uptake of intranasally administered nucleoside derivatives observed by other investigators. OATP3 immunoreactivity was high in olfactory epithelium and olfactory nerve bundles, which suggests that substrates transported by OATP3 may be candidates for intranasal administration (Genter et al., 2010). Metal uptake transporters, including divalent metal transporter 1 (DMT1), ZIP8, and ZIP14 are also expressed in the olfactory epithelium (Thompson et al., 2007; Ma et al., 2008; Genter et al., 2009, 2010). Zinc transporters, which either efflux or recompartmentalize cytosolic zinc, are also expressed in olfactory epithelium (Sekler et al., 2002). Amino acid, organic anion, and organic cation transporters have also been shown to be present and functional in olfactory epithelium (Kaler et al., 2006; Eraly et al., 2006; Schnabolk et al., 2006; Bröer, 2006; Chemuturi, 2007).

19.4 THE NASAL CAVITY AS A ROUTE OF ADMINISTRATION FOR THERAPEUTIC DRUGS

19.4.1 Background

There is currently considerable interest in administration of therapeutic drugs via the nasal cavity for delivery into the brain or into the systemic circulation. This route of administration can be particularly useful for drugs which are readily degraded in the gastrointestinal tract (e.g., small, readily-digestible peptides) or drugs which are extensively inactivated by liver metabolism following oral ingestion. This is not to say that the nasal respiratory and olfactory epithelia are metabolically inert; in fact, there is a vast body of literature documenting the presence of multiple Phase I and Phase II metabolic enzymes in rodent, dog, and human nasal mucosa (Dahl and Hadley, 1991; Gervasi et al., 1991; Sarkar, 1992; Thornton-Manning and Dahl, 1997).

Of the three major routes of entry of drugs from the nasal cavity and into the brain or systemic circulation, the first, "transneuronal transport", has been dealt with to this point in this chapter. Another putative mechanism for transport of agents from the nasal cavity and into the CNS has been termed the *olfactory epithelial pathway* (Mathison et al., 1998) or *paracellular transport* (McMartin et al.,

1987). Regardless of the designation, the concept is that molecules access the CNS from the nasal cavity via supporting cells, cell-to-cell junctions and/or spaces between cells. A great deal of research effort has gone into our understanding of the factors that are important in this process. For example, molecular weight, lipophilicity and ionization state/pH of the agent are all variables that appear to contribute to the ability of agents to access the CNS from the nasal cavity (McMartin et al., 1987; Sakane et al., 1991a; Sakane et al., 1994; Shimoda et al., 1995). The concentration of sulfonamides in the CSF of the rat resulting from intranasal administration was demonstrated to increase with the lipophilicity of the sulfonamide derivative (Sakane et al., 1991a). In contrast, lipophilicity did not enhance the absorption of a series of peptides from the nasal cavity (Donnelly et al., 1998; Sakane et al., 1991). Using fluorescein isothiocyanate-labeled dextran with molecular weights ranging from 4.4 to 40 kD, it was determined that drugs with a molecular weight of 20 kD or less could reasonably be expected to be transported from the nasal cavity to the CSF (Sakane et al., 1995).

19.4.2 Examples of Intranasally-administered Therapeutic Agents

Intranasally-administered drugs can be roughly divided into three main categories: (1) Drugs or other agents administered via the nasal cavity for CNS effects; (2) Drugs administered via the nasal cavity for systemic effects; and (3) Drugs administered via the nasal cavity for nasal and/or respiratory tract effects. Although the development of drugs for treatment of respiratory tract symptoms is undoubtedly the strategy most intuitively associated with intranasal drug delivery, it will not be dealt with in this chapter because of our interest here in transport. The development of drugs administered via the nasal cavity for CNS and/or systemic effects has exploded in recent years (Table 19.3).

19.4.2.1 Drugs and Other Agents Administered via the Nasal Cavity for CNS Effects.

ANTIMIGRAINE MEDICINES Therapeutic drugs formulated for intranasal administration for CNS effects include migraine headache treatments, sedatives/anti-epileptics, neuroprotective agents and drugs to suppress cravings for smoking and eating. While some efforts are still largely in the experimental stages, migraine patients using medications administered via the intranasal route have realized significant relief. Due to the high incidence of nausea, vomiting, and visual disturbances during migraine attacks, intranasal administration of migraine medicines

is proving to be superior to pills or self-injections. While ergotamine derivatives have been extensively used in the past (Gallagher, 1996; Lipton, 1997; Marttin et al., 1997), newer generation drugs which act as 5-HT_{1D} receptor agonists (e.g., sumatriptan, or Imitrex®) are also proving to be potent, highly specific anti-migraine agents when administered via the intranasal route. The bioavailability of intranasally-administered sumatriptan in dogs was equivalent to an oral dose (Barrow et al., 1997). Similarly, BMS-181885, another $5\text{-HT}_{1\text{-like}}$ receptor agonist, was demonstrated to be quantitatively absorbed from the nasal cavity of cynomolgus monkeys, with the highest observed plasma concentration attained in 30 min (Srinivas et al., 1998). There has also been smaller-scale use of intranasal lidocaine or butorphanol as abortive treatments for migraines, as well as a case study reporting the effectiveness of intranasal capsaicin in migraine relief (Levy, 1995; Melanson et al., 1997; Mills and Scoggin, 1997; Sachs, 1996).

SEDATIVES AND ANALGESICS The intranasal route is also a practical alternative to injections/intravenous administration of sedatives and medicines for chronic pain. Midazolam is administered in several medical situations, including induction of anxiolysis and sedation particularly for endoscopic and dental procedures. Midazolam absorption from the nasal mucosa was nearly complete in a study involving adult surgical patients who were administered midazolam in small intranasal doses to preclude swallowing. This study was important in that it revised previous observations that the bioavailability following intranasal administration was approximately 55%, thus reducing the risk of accidental overdoses (Bjorkman et al., 1997). Nasal midazolam also appeared to be safe and effective in treatment of epileptic seizures in children (Jeannet et al., 1999). Intranasally-administered oxycodone and oxymorphone are effectively absorbed from the nasal cavity for treatment of chronic pain (Hussain and Aungst, 1997; Takala et al., 1997). Experiments to investigate the feasibility of administering acetyl salicylic acid by the nasal route have also been conducted in rat with promising results (Hussain et al., 1992). Johnston et al. (2011) discovered that co-administration of inhaled methoxyflurane with intranasal fentanyl improved prehospital management of visceral pain in an Australian ambulance service.

ANTIVIRALS AND ANTIBIOTICS Drugs with activity toward the human immunodeficiency virus, namely zidovudine (AZT; 3'-azido-2',3'-dideoxythymidine) and D4T (2',3'-didehydro-3'-deoxythymidine) are absorbed from the rat nasal epithelia and transported into the cerebral spinal fluid (CSF; Seki et al., 1994; Yajima et al., 1998). Under normal administration regimens, these drugs cannot cross the blood brain barrier and therefore are ineffective against AIDS dementia when administered orally or by injection. The antibiotic cephalexin appeared rapidly in the CSF following intranasal administration (Sakane et al., 1991b). Favorable pharmacokinetics were observed when acyclovir was administered intranasally to rats (Chavanpatil and Vavia, 2004).

DRUGS TO MODEL PARKINSON'S DISEASE (PD) IN LABORATORY ANIMALS

The presence of smell loss and the pathological involvement of the olfactory pathways in the early stages of PD are in accord with the tenants of the olfactory vector hypothesis. This hypothesis postulates that some forms of PD may be caused or catalyzed by environmental agents that enter the brain via the olfactory mucosa. Some compounds, such as rotenone, paraquat, and 6-hydroxydopamine, have limited capacity to reach and damage the nigrostriatal dopaminergic system via the intranasal route (see Prediger et al., 2012). Others, such as MPTP, readily enter the brain via this route in some species and influence the function of the nigrostriatal pathway. Intranasal infusion of MPTP in some rodents elicits a developmental sequence of behavioral and neurochemical changes that closely mimics that seen in PD (Prediger et al., 2006, 2009, 2011). Effects noted included motor deficits that correlated with a progressive and severe depletion of striatal dopamine levels, and loss of tyrosine hydroxylase and dopamine transporter staining in substantia nigra and striatum, as well increased levels of Mn-superoxide dismutase. However, alpha-synuclein aggregation was not observed (Rojo et al., 2006, 2007; Prediger et al., 2012).

TREATMENT/PREVENTION OF NEURODEGENERATIVE DISEASES AND BRAIN INJURY

Vasoactive intestinal peptide (VIP) is widely distributed in regions of the brain associated with learning and memory. VIP antagonists are associated with learning impairments in experimental systems, and in vitro studies have demonstrated that VIP exerts neuroprotective effects, including protection of cortical neurons from the cytotoxic effects of β-amyloid peptide. Intranasal administration of a lipophilic analogue of VIP ([St-Nle^{17}]VIP) to rats has been shown to deliver unmetabolized ([St-Nle^{17}]VIP to the brain and to prevent impairments in spatial learning and memory associated with cholinergic blockade in rats (Gozes et al., 1997; Gozes et al., 1996). Similarly, dextromethorphan and neostigmine were found to be well-absorbed from the rat and human nasal cavities, with the latter more effective in relieving symptoms of myasthenia gravis as a nasal spray than when administered intravenously (Char et al., 1992; Di Costanzo et al., 1993). Administration of vasopressin to humans to increase brain activity and improve memory has resulted in reports of marginal improvement (Perras et al., 1997; Pietrowsky et al., 1996).

Recent studies have demonstrated that stem cells can migrate to the brain following intranasal instillation. This finding is the basis for great excitement about the possibility of stem cell therapies for treatment of neurodegenerative diseases. Wang et al. (2012) showed behavioral recovery in 6-hydroxydopamine-treated rats following administration of human adult olfactory epithelial-derived neural progenitor cells which had been differentiated into dopaminergic neurons. Human fibroblasts were used as controls and they did not improve parkinsonian endpoints. In addition to behavioral improvement, increased numbers of tyrosine hydroxylase positive cells were detected in the engrafted brains; these remained viable for at least six months in vivo. Higher dopamine levels were detected in the striatum of behaviorally improved rats than in equivalent regions of rats that did not recover. These results support the hypothesis that human adult olfactory epithelial-derived progenitors represent a unique autologous cell type with promising potential for future use in a cell-based therapy for patients with Parkinson's disease.

High mobility group box-1 (HMGB1) is an endogenous "danger signal molecule" which is released after ischemic injury in the brain and triggers inflammation and apoptosis. Kim et al. (2012) demonstrated that intranasal delivery of a HMGB1-binding heptamer peptide, but not a scrambled heptamer, was extremely effective in neuroprotection following ischemic brain injury, suggesting that this novel approach coupled with the intranasal route, may be an emerging stroke therapy.

Traumatic brain injury (TBI) is receiving heightened attention in light of reports of athletes developing TBI following repeated blows to the head and the prevalence of TBI in soldiers following combat (Kontos et al., 2012; Sahler and Greenwald, 2012). Therefore, therapies to treat this condition once it has developed will be of great interest. Tian et al. (2012) demonstrated in rats that intranasal administration of nerve growth factor ameliorated β-amyloid deposition after tramatic brain injury. Likewise, Won et al. (2012) demonstrated that intranasal delivery of nicotinamide adenine dinucleotide prevented TBI in rats.

SMOKING CESSATION AND APPETITE SUPPRESSION

Nicotine nasal sprays have been evaluated as agents to assist individuals in smoking cessation. It is possible that the sprays are efficacious due both to the direct delivery of nicotine to the brain, as well as maintenance of blood nicotine levels such that the majority of symptoms of withdrawal are avoided. While there have been reports that the irritation associated with nicotine nasal sprays precluded their use to attain therapeutic blood levels (Gourlay and Benowitz, 1997), another study showed that successful "quitters" who most frequently used nicotine nasal sprays as part of a smoking cessation

19.4 The Nasal Cavity as a Route of Administration for Therapeutic Drugs

program had higher blood nicotine and cotinine levels than those using the sprays occasionally or not at all (Jones et al., 1998). Experimentally, cholecystokinin (CCK) agonists show promise as appetite suppressants (Pierson et al., 1997). More recently, Hallschmid et al. (2004) studied the regulation of food intake and body weight in humans treated intranasally with several neuropeptides. Intranasal administration of neuropeptide Y was found to increase food intake (acting as an orexigenic messenger), while prolonged administration of insulin or melanocyte stimulating hormone induced weight loss in healthy humans. Interestingly, the latter treatments did not cause weight loss in overweight subjects.

BRAIN TUMOR THERAPY Monoterpene perillyl alcohol (POH), a Ras inhibitor and anti-angiogenic factor with potential capacity to arrest gliomagenesis, was investigated in a phase I/II clinical trial in adults with recurrent malignant glioma. The efficacy of intranasal administration of monoterpene POH upon survival rate of patients with recurrent glioblastoma (GBM) was demonstrated in several trials. One study included 89 adults with recurrent GBM who received daily intranasal administration of 440 mg POH. The prognosis of these patients was compared to that of 52 matched GBM patients as historical control untreated group who received only supportive treatment. Intranasal administration of POH significantly increased the overall survival of patients with recurrent GBM in comparison with historical untreated controls. The side effects of POH treatment were minimal, even in patients treated for over four years (da Fonseca et al., 2008, 2011; Fischer et al., 2008).

In another study suggesting the future importance of intranasally-administered stem cells, Reitz et al. (2012) demonstrated that intranasal administration of neural stem/progenitor cells (NSPCs) led to a rapid, targeted movement of the stem cells toward experimentally-induced gliomas in mice. While most of the NSPCs reached the tumor via the olfactory nerve, some were delivered via blood vessels, but importantly, the cells accumulated with high efficiency in the tumors but not peripheral to the tumors. The goal moving forward is to take advantage of the observed specificity to develop treatment regimens.

19.4.2.2 Drugs Administered via the Nasal Cavity for Systemic Effects.
There is a similarly long and varied list of therapeutic drugs which are designed for intranasal delivery to achieve a desired systemic effect (see reviews by Jones et al., 1997; Mathison et al., 1998).

PEPTIDES Delivery of peptides such as insulin and calcitonin via the nasal route would in theory improve

their bioavailability over oral administration. There have been mixed results with insulin administration via the nasal route (Hilsted et al., 1995), possibly because of peptidases in nasal secretions that have high activity toward insulin peptides. Methods to encapsulate insulin peptides appear promising in the effort to improve systemic delivery of insulin following intranasal administration (Fernandez-Urrusuno et al., 1999). Intranasal calcitonin is a potent agent for reducing bone resorption (Silverman, 1997).

HORMONE ANALOGUES Buserelin and nafarelin are being used as nasal sprays for treatment of endometriosis, prostate cancer, and low sperm count, and show favorable results as contraceptives (Bergquist et al., 1979). Oxytocin nasal sprays are associated with high blood levels of the drug and enhanced lactation (Landgraf, 1985; Ruis et al., 1981).

VACCINES The intranasal route is also becoming recognized as a route of immunization against agents such as influenza viruses, *N. meningitides*, diphtheria, and tetanus (Liu, 1998; Maassab and DeBorde, 1985).

OTHERS Vitamin B_{12} derivatives are well absorbed following intranasal administration, as are drugs for treatment of motion sickness and "bedwetting" (Butler et al., 1998; Kallas et al., 1999; Putcha et al., 1996; Ramanathan et al., 1998; Slot et al., 1997; Swain, 1995).

19.4.2.3 Enhancement of Delivery of Intranasally-Administered Drugs.
Considerable effort has gone into identifying agents that can enhance the absorption of drugs from the nasal cavity without causing clinically-relevant damage to the mucosal surface of the nasal cavity (Examples are included in Table 19.4). The putative absorption enhancers sodium tauro-24,25-dihydrofusidate (STDHF) and dimethyl-β-cyclodextran (DM βCD) exhibited some epithelial toxicity, whereas ethyleneamine diamine tetraacetic acid (EDTA) was not associated with enhanced absorption or epithelial toxicity (Donnelly et al., 1998). Methylated β−cyclodextrins were less toxic to nasal epithelia than sodium glycocholate, STDHF, laureth-9 (a detergent) and L-α-phosphatidylcholine (Marttin et al., 1998). Free amine chitosans and soluble chitosan salts were evaluated for their efficacy and safety as nasal enhancers of peptide absorption and were found to be comparable in their action, if not more potent, than cyclodextrins (Tengamnuay et al., 2000). Cyclodextrins are believed to enhance absorption of intranasally-administered drugs by transiently opening tight junctions between nasal epithelial cells (Marttin et al., 1998). High concentrations of certain

Table 19.4 Examples of strategies to increase delivery of intranasally-administered drugs

Molecule/Drug Treatment	Enhancer	Disease	Species	Ref.
Acyclovir	Hydroxypropyl b-cyclodextrin > sodium deoxycholate > sodium >caprate > sodium > tauroglycocholate > EDTA	Herpes simplex, varicella zoster, cytomegalovirus, and Epstein Barr virus infections	Rat	Chavanpatil and Vavia, 2004
AT1002, a six-amino acid fragment of Zonula occludens toxin	Carrageenan as bioadhesive polymer and dextrose as stabilizer	Proof of concept	Rat	Song and Eddington, 2012
Calcitonin	N- acetyl-L-cysteine	Proof of concept	Rat, rabbit, dog	Matsuyama et al., 2006b
Cannabidiol	Polyethylene glycol 400	Chronic and breakthrough pain	Rat	Paudel et al., 2010
Erythropoietin, human growth hormone, and salmon calcitonin	Lipoamino acids	Proof of concept	In vitro	Bijani et al., 2012
Exendin-4	Polyethylene glycol	Type 2 diabetes	Mouse	Kim et al., 2012
Fentanyl	Pectin, chitosan and chitosan-poloxamer 188	Proof of concept	Human	Fisher et al., 2010
Fexofenadine	Chitosan-coated liposome	Allergic rhinitis	In vitro, rat	Qiang et al., 2012
Fluorescein isothiocyanate-labeled 4 kDa dextran (FD-4)	Sodium hyaluronate	Proof of concept	In vitro, rat	Horvát et al., 2009
Fluorescent isothiocyanate-labeled dextran	N-acetyl-l-cysteine + nonionic surfactant, polyoxyethylene lauryl ether (laureth-25)	Proof of concept	Rat	Matsuyama et al., 2006a
Galanin-like peptide (^{125}I-GALP)	Cyclodextrins	obesity and related conditions	Mouse	Nonaka et al., 2008
Gemcitabine	Polybutylcyanoacrylate nanoparticles coated with polysorbate-80	Anti-tumor	Rat	Wang et al., 2009a
Gemcitabine	Papaverine	Proof of concept	Rat	Krishan et al., 2014
Granisetron	Hydroxypropyl-cyclodextrin and sodium carboxymethylcellulose	Treatment of nausea and vomiting following chemotherapy	Primary human nasal epithelial cells in vitro	Cho et al., 2010
Growth differentiation factor 5	Lipid microemulsion of olive oil and phosphatidylserine	Parkinson's disease	Rat	Hanson et al., 2012
Indomethacin	Polyvinylpyrolidone, citric acid, sodium taurocholate	Anti-inflammatory effects	In vitro, ex vivo, rat	Karasulu et al., 2008

Insulin	Oligoarginine-linked poly (N-vinylacetamide-co-acrylic acid)	Diabetes/obesity	Mouse	Sakuma et al., 2010
Insulin	Starch nanoparticles with sodium glycocholate	Streptozotocin- induced diabetic rats	Rat	Jain et al., 2008
Insulin	Aminated gelatin	Proof of concept	Rat	Seki et al., 2005
Insulin	PEG-grafted chitosan nanoparticles	Proof of concept	Rabbit	Zhang et al., 2008
Insulin	Chitosan-N-acetyl-L-cysteine nanoparticles	Proof of concept	Rat	Wang et al., 2009
Insulin	Polysaccharide nanoparticles consisting of chitosan and cyclodextrin (CD) derivatives-(sulfobutylether-β-CD or carboximethyl-β-CD)	Proof of concept	Rabbit, in vitro	Teijeiro-Osorio et al., 2009
Isosorbide dinitrate	Poloxamer 188	Angina pectoris	Rat	Na et al., 2010
Lorazepam	Vinyl polymer-coated microparticles	Maximize speed of onset and minimize carry-over sedation	Rabbit	Zhao et al., 2012
Lysozyme	Hydroxypropyl-methylcellulose, hydroxypropyl-cyclodextrin, d-alpha-tocopheryl polyethylene glycol 1000 succinate	Delivery of therapeutic proteins	Human nasal epithelium in vitro	Cho et al., 2011
Midazolam	Cyclodextrins and chitosan	Proof of concept	Human	Haschke et al., 2010
Morphine	Nanoparticles consisting of a polysaccharide core surrounded by a lipid bilayer	Analgesia	Mice	Betbeder et al., 1998
Morphine, fentanyl	Pressurized olfactory delivery device	Analgesia	Rats	Hoekman and Ho, 2011a
Nelfinavir, mannitol	Pressurized olfactory delivery device	Proof of concept	Rat	Hoekman and Ho, 2011b
Nerve growth factor	Mutation at residue R100	Neurodegenerative diseases	Mouse	Capsoni et al., 2012
	Peppermint oil		Rat	Vaka and Murthy, 2010
Olanzapine	Poly(lactic-c-glycolic acid) nanoparticles	CNS disorders	Ex vivo	Seju et al., 2011

(continued)

Table 19.4 (Continued)

Molecule/Drug Treatment	Enhancer	Disease	Species	Ref.
Ondansetron HCl	Solid lipid nanoparticles	Management of chemotherapy induced postoperative nausea and vomiting	In vitro	Joshi et al., 2012
Peptide YY 3–36 (PYY3–36)	Novel tight junction modulating peptide (PN159)	Proof of concept	Rabbit	Chen et al., 2006
Pituitary adenylate cyclase activating polypeptide	Cyclodextrins	Alzheimer's disease	Mouse	Nonaka et al., 2012
Recombinant hirudin-2 (rHV2)	Chitosan	Anticoagulant, antithrombotic activities	Rat, in vitro	Zhang et al., 2005
Recombinant human growth hormone (rhGH)	N-trimethyl chitosan chloride and a patented fatty acid-based delivery system	Proof of concept	Rat	Steyn et al., 2010
Recombinant iduronidase (IDUA; in the form of laronidase)	Adeno-associated virus vector encoding IDUA	Lysosomal storage diseases	Mouse	Wolf et al., 2012
Rivastigmine (RHT)	Chitosan nanoparticles	Alzheimer's disease	Rat, in vitro	Fazil et al., 2012
Pituitary adenylate cyclase activating polypeptide	Cyclodextrins	Alzheimer's disease	Mouse	Nonaka et al., 2012
Streptococcus equi proteins	Poly (L-lactic acid) and glycol-chitosan nanospheres	Strangles	Mouse	Florindo et al., 2009
Tacrine	Mucoadhesives (albumin, cyclodextrins)	Alzheimer's disease	In vitro, proof of concept	Luppi et al., 2011
Thymopentin (TP5)	Chitosan and bacitracin	Modulator for immunodeficiencies	Rat	Wang et al., 2006
Urocortin peptide	Odorranalectin	Parkinsonism	Rats	Wen et al., 2011

cyclodextrins are associated with ciliary damage, an end-point also associated with benzalkonium chloride, which has been used as a preservative in many nasal formulations (Bernstein, 2000; Uchenna Agu et al., 2000). Luppi et al. (2011) used mucoadhesives (albumin, cyclodextrins to improve delivery of intranasally-administered tacrine. Wen et al. (2011) took advantage of prior knowledge of lectin biology and the olfactory system to enhance the delivery of urocortin peptide, a potential Parkinson's disease therapeutic, using odorranalectin, a very small lectin molecule.

19.5 CONSEQUENCES OF TRANSPORT

One of the early changes associated with several neurodegenerative disorders, including Parkinson's and Alzheimer's diseases, is a substantial deficit in olfactory function of a discriminatory nature (Doty et al., 1988; Doty et al., 1991; Doty et al., 1987). These data in conjunction with recent findings of pathological changes in the olfactory epithelium (Tabaton et al., 1991; Talamo et al., 1989) and olfactory bulb (Esiri and Wilcock, 1984; Ohm and Braak, 1987), as well as brain areas with close synaptic relationships to the olfactory bulb (Ferreya-Moyano and Barragan, 1989; Saper et al., 1987), formed the basis for the hypothesis that the olfactory system may be a route of entry for either a toxic agent or a virus that is important to the etiology of these diseases.

For most substances, the evidence is at best circumstantial. Several reviews have outlined the data suggesting that environmental aluminum could contribute to the loss of olfactory functioning and the ensuing disease process in Alzheimer's disease. The findings of high levels of aluminum in plaques and the ability to produce lesions following inhalation of aluminum support this hypothesis. The data could reflect, however, other abnormalities in Alzheimer's disease, such as changes in the blood–brain barrier that permit accumulation of this metal in neurons (Crapper McLachlan, 1986; Roberts, 1986). Recent studies have provided anatomical demonstration of transport to the CNS of intranasally applied manganese (Tjälve and Henriksson, 1999) as well as epidemiological evidence for a Parkinson's-like syndrome following environmental exposure to the metal (Gorell et al., 1999). While there is no definitive evidence that the olfactory system serves as a route of entry for metals and other toxic agents in humans, there is strong support for an environmental role in sporadic Parkinson's disease (Langston, 1998; Tanner and Langston, 1990).

Currently, there is no evidence for viral involvement in neurodegenerative disorders such as Alzheimer's or Parkinson's disease. To postulate that the olfactory system serves as a route of entry for a virus in these diseases is only conjectural. The fact that in some degenerative disorders the affected neuronal populations appear to be restricted to interconnected pathways has been clearly established (Saper et al., 1987), and transneuronal transport of viruses does occur along some of these routes especially in the olfactory system. Trophic substances also were shown to be transported along these same pathways. In the cholinergic system, for example, which innervates the olfactory bulb, nerve growth factor is retrogradely transported (Altar and Bakhit, 1991). Thus, an abnormality in transport of a trophic factor, either inborn or toxicant (viral?) induced, could result in neuronal degeneration. Alterations in peripheral afferent innervation of the olfactory bulb, both denervation and odorant deprivation, produce profound anatomical and biochemical consequences. Both neonates and adults show reductions in bulb size (Maruniak et al., 1989a,b; Meisami, 1976), granule cell spine densities (Benson et al., 1984), number of granule cells (Frazier and Brunjes, 1988), and neurotransmitter expression, including substance P and dopamine, in juxtaglomerular cells, (Baker et al., 1983; Kream et al., 1984). The latter changes occur without apparent cell death indicating the presence of phenotypic plasticity (Baker et al., 1984; Cho et al., 1996; Stone et al., 1991; Stone et al., 1990). These data indicate the importance of afferent innervation as a trophic regulator of bulb function and suggest that either exogenous agents or viral internalization may alter this trophic balance and result in the loss of olfactory function with eventual consequences in synaptically interconnected brain regions.

On a positive note, recent successes in the treatment of human CNS disorders using intranasally-administered agents continues to justify expanding the use of this route of exposure, as well as further studies into transiently and safely increasing the permeability of the route between the nasal cavity and the brain.

19.6 CONCLUSIONS

The intensive study of the olfactory mucosa over the years has done much to expand our knowledge in many seemingly-unrelated areas. The ill-advised intranasal zinc treatments of the 1920s and 1930s, intended as a polio prophylactic treatment, helped us refine our knowledge of the natural course of polio infection and ultimately lead to the development of an effective vaccination. The original identification of the putative structure of olfactory receptors by Buck and Axel culminated in their wining the Nobel Prize in Physiology or Medicine in 2004. Studies by Brunet et al. (1996) in a knockout mouse model determined that cyclic nucleotide gated channels subserve excitatory olfactory signal transduction and suggested that cAMP is the critical second messenger in olfaction. Still unsolved is the mystery underlying the 'plasticity' of the olfactory epithelium. While many investigators have identified patterns of gene expression that make the olfactory mucosa unique (e.g., Genter et al., 2003), the definitive answer as to why these neurons not only have a basal level of turnover, but can regenerate after toxicant- (and some injury-) mediated damage is still unknown; yet this answer is still highly sought after, with the hope that this understanding could improve our ability to treat injuries in other regions of the nervous system. Important observations that many neurodegenerative diseases are associated with functional and structural alterations of the olfactory system (e.g., Doty et al., 1988; Talamo et al., 1989) have led to testable hypothyses about the possibility that causative agents for neurodegenerative diseases, particularly Parkinson's disease, may attack the brain from the nasal cavity. Convincing evidence has been provided earlier in this chapter with MPTP, an important model of toxicant-induced Parkinson's disease in rodents (e.g., Prediger et al., 2006, 2009, 2010, 2011, 2012). With regard to the topic of this chapter, the nose as a route of exposure/delivery, our knowledge continues to expand. Intranasal vaccinations are becoming more routine. Abundant recent experimental and clinical data show that the nasal route may provide an ideal route for delivery of drugs and even stem cells to the CNS for treatment of brain injuries, neurodegenerative diseases, and brain tumors. The story of the intranasal, mechanism-based treatment for patients with recurrent, inoperable glioblastomas (Fischer et al., 2008; da Fonseca, 2008, 2011) should serve as an inspiration to all of us as we attempt to take our bench research to the bedside. The nose knows a lot—it is up to us to ask the right questions!

ACKNOWLEDGMENTS

Dr. Harriet Baker was a major contributor to the previous version of this chapter, and given that much of her content has been kept in this version, we acknowledge these contributions.

REFERENCES

Aggerbeck, H., Gizurarson, S., Wantzin, J., and Heron, I. (1997). Intranasal booster vaccination against diphtheria and tetanus in man. *Vaccine.* 15: 307–316.

Akpan, N., Serrano-Saiz, E., Zacharia, B. E., et al. (2011). Intranasal delivery of caspase-9 inhibitor reduces caspase-6-dependent axon/neuron loss and improves neurological function after stroke. *J. Neurosci.* 31: 8894–8904.

Allen, W. K. and Akeson, R. (1985). Identification of a cell surface glycoprotein family of olfactory receptor neurons with a monoclonal antibody. *J. Neurosci.* 5: 284–296.

Altar, C. A. and Bakhit, C. (1991). Receptor-mediated transport of human recombinant nerve growth factor from olfactory bulb to forebrain cholinergic nuclei. *Brain Res.* 541: 82–88.

Apfelbach, R., Engelhart, A., Behnisch, P., and Hagenmaier, H. (1998). The olfactory system as a portal of entry for airborne polychlorinated biphenyls (PCBs) to the brain? [letter]. *Arch. Toxicol.* 72: 314–317.

Araneda, S., Font, C., Pujol, J. F., and Bobillier, P. (1983). Retrograde axonal transport after radioactive hydroxyindole injections into the olfactory bulb-an autoradiographic study. *Neurochem. Int.* 5: 741–750.

Armstrong, C. and Harrison, W. T. (1935). Prevention of intranasally inoculated poliomyelitis of monkeys by instillation of alum into nostrils. *Public Health Rep.* 50: 725–730.

Astic, L., Saucier, D., Coulon, P., et al. (1993). The CVS strain of rabies virus as transneuronal tracer in the olfactory system of mice. *Brain Res.* 619: 146–156.

Bahrami, F., Bergman, U., Brittebo, E. B., and Brandt, I. (2000). Persistent olfactory mucosal metaplasia and increased olfactory bulb glial fibrillary acidic protein levels following a single dose of methylsulfonyl-dichlorobenzene in mice: comparison of the 2,5- and 2, 6- dichlorinated isomers. *Toxicol. Appl. Pharmacol.* 162: 49–59.

Baker, H. (1995)Transport phenomena within the olfactory system. In: Doty, R., ed. *Handbook of Clinical Olfaction and Gustation.* New York: Marcel Dekker,: 173–190.

Baker, H., Kawano, T., Albert, V. R., et al. (1984). Olfactory bulb dopamine neurons survive deafferentiation induced loss of tyrosine hydroxylase. *Neuroscience.* 11: 605–615.

Baker, H., Kawano, T., Margolis, F. L., and Joh, T. H. (1983). Transneuronal regulation of tyrosine hydroxylase expression in olfactory bulb of mouse and rat. *J. Neurosci.* 3: 69–78.

Baker, H. and Margolis, F. L. (1986). Deafferentation induced alterations in olfactory bulb as a model for the etiology of Alzheimer's disease. *Neurobiol. Aging.* 7: 568–569.

Baker, H. and Spencer, R. F. (1986). Transneuronal transport of peroxidase-conjugated wheat germ agglutinin (WGA-HRP) from the olfactory epithelium to the brain of the adult rat. *Exp. Brain Res.* 63: 461–473.

Baker, T. A. and Maruniak, J. (1990). Uptake of immunoglobulins by olfactory receptor neurons. *Chem. Senses.* 15: 549.

Baldaccini, N. E., Gagliardo, A., Pelosi, P., and Topazzini, A. (1986). Occurrence of a pyrazine binding protein in the nasal mucosa of some vertebrates. *Comp. Biochem. Physiol. [B].* 84: 249–253.

Balin, B. J., Broadwell, R. D., Salcman, M., and el-Kalliny, M. (1986). Avenues for entry of peripherally administered protein to the central nervous system in mouse, rat, and squirrel monkey. *J. Comp. Neurol.* 251: 260–280.

Bannister, L. H. and Dodson, H. C. (1992). Endocytic pathways in the olfactory and vomeronasal epithelia of the mouse: ultrastructure and uptake of tracers. *Microsc. Res. Tech.* 23: 128–141.

Barber, P. C. (1989). *Ulex europus* agglutinin I binds exclusively to primary olfactory neurons in the rat nervous system. *Neuroscience.* 30: 1–9.

Barceloux, D. G. (1999). Nickel. *J. Toxicol. Clin. Toxicol.* 37: 239–258.

Barik, S. (2011). Intranasal delivery of antiviral siRNA. *Methods Mol. Biol.* 721: 333–338.

Barnett, E. M., Cassell, M. D., and Perlman, S. (1993). Two neurotropic viruses, herpes simplex virus type 1 and mouse hepatitis virus, spread along different neural pathways from the main olfactory bulb. *Neuroscience.* 57: 1007–1025.

Barrow, A., Dixon, C. M., Saynor, D. A., et al. (1997). The absorption, pharmacodynamics, metabolism and excretion of ^{14}C- sumatriptan following intranasal administration to the beagle dog. *Biopharm. Drug Dispos.* 18: 443–458.

Bazer, G. T., Ebbesson, S. O. E., Reynolds, J. B., and Bailey, R. P. (1987). A cobalt-lysine study of primary olfactory projections in king salmon fry (Oncorhynchus tshawytscha Walbaum). *Cell. Tissue Res.* 248: 499–503.

Bechgaard, E., Gizurarson, S., and Hjortkjaer, R. K. (1997). Pharmacokinetic and pharmacodynamic response after intranasal administration of diazepam to rabbits. *J. Pharm. Pharmacol.* 49: 747–750.

Becker, Y. (1995). HSV-1 brain infection by the olfactory nerve route and virus latency and reactivation may cause learning and behavioral deficiencies and violence in children and adults: a point of view. *Virus Genes.* 10: 217–226.

Belinsky, S., Walker, V., Maronpot, R., et al. (1987). Molecular dosimetry of DNA adduct formation and cell toxicity in rat nasal mucosa following exposure to the tobacco specific nitrosamine 4-(N-methyl-N-nitrosamino)-1-(3-pyridyl)-1-butanone and their relationship to induction of neoplasia. *Cancer Res.* 47: 6058–6065.

Benedict, C., Hallschmid, M., Hatke, A., et al. (2004). Intranasal insulin improves memory in humans. *Psychoneuroendocrinology.* 29: 1326–1334.

Benson, T. E., Ryugo, D. K., and Hinds, J. W. (1984). Effects of sensory deprivation on the developing mouse olfactory system: a light and electron microscopic, morphometric analysis. *J. Neurosci.* 4: 238–253.

Bergquist, C., Nillius, S. J., and Wide, L. (1979). Intranasal gonadotropin-releasing hormone agonist as a contraceptive agent. *Lancet.* 2: 215–217.

Bernstein, I. L. (2000). Is the use of benzalkonium chloride as a preservative for nasal formulations a safety concern? A cautionary note based on compromised mucociliary transport. *J. Allergy Clin. Immunol.* 2000; 105: 39–44.

Betbeder, D., Sperandio, S., deNadai, J., et al. (1998). The analgesic activity of nasal morphine in mice increases with biovector™ nanoparticles. *Fundam. Clin. Pharmacol.* 3: 314.

Bijani, C., Arnarez, C., Brasselet, S., et al. (2012). Stability and structure of protein–lipoamino acid colloidal particles: Toward nasal delivery of pharmaceutically active proteins. *Langmuir.* 28: 5783–5794.

Bjorkman, S., Rigemar, G., and Idvall, J. (1997). Pharmacokinetics of midazolam given as an intranasal spray to adult surgical patients [see comments]. *Br. J. Anaesth.* 79: 575–580.

Blessing, W. W., Ding, Z. Q., Li, Y. W., et al. (1994). Transneuronal labelling of CNS neurons with herpes simplex virus. *Prog. Neurobiol.* 44: 37–53.

Bodian, D. and Howe, H. A. (1941a). Experimental studies on intraneural spread of poliomyelitis virus. *Bull. Johns Hopkins Hosp.* 68: 248–267.

Bodian, D. and Howe, H. A. (1941b). The rate of progression of poliomyelitis virus in nerves. *Bull. Johns Hopkins Hosp.* 69: 79–85.

Bogdanffy, M. S., Mazaika, T. J., and Fasano, W. J. (1989). Early cell proliferative and cytotoxic effects of phenacetin on rat nasal mucosa. *Toxicol. Appl. Pharmacol.* 98: 100–112.

Bonnet-Font, C. and Bobillier, P. (1990). Retrograde axonal transport specificity in the locus coeruleus neurons after [^{3}H] Noradrenaline injection into the rat olfactory bulb. *Neurochem. Int.* 16: 523–532.

Borg-Neczak, K., and Tjälve, H. (1996). Uptake of ^{203}Hg2+ in the olfactory system in pike. *Toxicol. Lett.* 84: 107–112.

Brandt, I., Brittebo, E. B., Feil, V. J., and Bakke, J. E. (1990). Irreversible binding and toxicity of the herbicide dichlobenil (2,6- dichlorobenzonitrile) in the olfactory mucosa of mice. *Toxicol. Appl. Pharmacol.* 103: 491–501.

Brenneman, K. A., Wong, B. A., Buccellato, M. A., et al. (2000). Direct olfactory transport of inhaled manganese (^{54}MnCl$_2$) to the rat brain: toxicokinetic investigations in a unilateral nasal occlusion model. *Toxicol. Appl. Pharmacol.* 169: 238–248.

Brittebo, E. B. (1988). Binding of cocaine in the liver, olfactory mucosa, eye, and fur of pigmented mice. *Toxicol. Appl. Pharmacol.* 96: 315–323.

Brittebo, E. B. and Eriksson, C. (1995). Taurine in the olfactory system: effects of the olfactory toxicant dichlobenil. *Neurotoxicology.* 16: 271–280.

Brittebo, E. B., Eriksson, C., Feil, V., et al. (1991). Toxicity of 2,6-dichlorothiobenzamide (chlorthiamid) and 2,6- dichlorobenzamide in the olfactory nasal mucosa of mice. *Fundam. Appl. Toxicol.* 17: 92–102.

Brittebo, E. V., Eriksson, C., and Brandt, I. (1990). Activation and toxicity of bromobenzene in nasal tissue in mice. *Arch. Toxicol.* 64: 54–60.

Broadwell, R. D. and Balin, B. J. (1985). Endocytic and exocytic pathways of the neuronal secretory process and trans-synaptic transfer of wheat germ agglutinin-horseradish peroxidase in vivo. *J. Comp. Neurol.* 242: 632–250.

Bröer, S. (2006). The SLC6 orphans are forming a family of amino acid transporters. *Neurochem. Int.* 48: 559–567.

Brunet, L. J., Gold, G. H., and Ngai, J. (1996). General anosmia caused by a targeted disruption of the mouse olfactory cyclic nucleotide-gated cation channel. *Neuron.* 17: 681–693.

Buck, L. and Axel, R. (1991). A novel multigene family may encode odorant receptors: A molecular basis for odor recognition. *Cell.* 65: 175–187.

Buddenberg, T. E., Topic, B., Mahlberg, E. D., et al. (2008). Behavioral actions of intranasal application of dopamine: effects on forced swimming, elevated plus-maze and open field parameters. *Neuropsychobiology.* 57: 70–79.

Burd, G. D., Davis, B. J., Macrides, F., et al. (1982). Carnosine in primary afferents of the olfactory system: an autoradiographic and biochemical study. *J. Neurosci.* 2: 244–255.

Butler, R., Holland, P., Devitt, H., et al. (1998). The effectiveness of desmopressin in the treatment of childhood nocturnal enuresis: predicting response using pretreatment variables. *Br. J. Urology.* 81: 29–36.

Cancalon, P. (1979a). Influence of temperature on the velocity and on the isotope profile of slowly transported labeled proteins. *J. Neurochem.* 1979a; 32: 997–1007.

Cancalon, P. (1979b). Subcellular and polypeptide distributions of slowly transported proteins in the garfish olfactory nerve. *Brain Res.* 161: 115–130.

Cancalon, P. F. (1988b). Axonal transport in the garfish optic nerve: comparison with the olfactory system. *J. Neurochem.* 51: 266–276.

Capsoni, S., Marinelli, S., Ceci, M., et al. (2012). Intranasal "painless" human nerve growth factors slows amyloid neurodegeneration and prevents memory deficits in App X PS1 mice. *PLoS ONE.* 7: e37555.

Chao, O. Y., Mattern, C., Silva, A. M., et al. (2012). Intranasally applied L-DOPA alleviates parkinsonian symptoms in rats with unilateral nigro-striatal 6-OHDA lesions. *Brain Res. Bull.* 87: 340–345.

Cho, H. J., Balakrishnan, P., Shim, W. S., et al. (2010). Characterization and in vitro evaluation of freeze-dried microparticles composed of granisetron–cyclodextrin complex and carboxymethylcellulose for intranasal delivery. *Int. J. Pharm.* 400: 59–65.

Cho, H. J., Ku, W. S., Termsarasab, U., et al. (2012). Development of udenafil-loaded microemulsions for intranasal delivery: in vitro and in vivo evaluations. *Int. J. Pharm.* 423(2): 153–160.

Char, H., Kumar, S., Patel, S., et al. (1992). Nasal delivery of [14C] dextromethorphan hydrochloride in rats: levels in plasma and brain. *J. Pharm. Sci.* 81: 750–752.

Charles, P. C., Walters, E., Margolis, F., and Johnston, R. E. (1995). Mechanism of neuroinvasion of Venezuelan equine encephalitis virus in the mouse. *Virology.* 208: 662–671.

Chavanpatil, M. D. and Vavia, P. R. (2004). The influence of absorption enhancers on nasal absorption of acyclovir. *Eur. J. Pharm. Biopharm.* 57: 483–487.

Chemuturi, N. V., Haraldsson, J. E., Prisinzano, T., and Donovan, M. (2006). Role of dopamine transporter (DAT) in dopamine transport across the nasal mucosa. *Life Sci.* 79: 1391–1398.

Chen, S. C., Eiting, K., Cui, K., et al. (2006). Therapeutic utility of a novel tight junction modulating peptide for enhancing intranasal drug delivery. *J. Pharm. Sci.* 95: 1364–1371.

Cho, H. J., Balakrishnan, P., Chung, S. J., et al. (2011). Evaluation of protein stability and in vitro permeation of lyophilized polysaccharides-based microparticles for intranasal protein delivery. *Int. J. Pharm.* 416: 77–84.

Cho, J. Y., Min, N., Franzen, L., and Baker, H. (1996). Rapid down-regulation of tyrosine hydroxylase expression in the olfactory bulb of naris-occluded adult rats. *J. Comp. Neurol.* 369: 264–276.

Clark, W. E. L. (1929). Anatomical investigation into the routes by which infections may pass from the nasal cavities into the brain. *Rep. Public Health Med. Subjects No. 54. London* 1–27.

Craft, S., Baker, L. D., Montine, T. J., et al. (2012). Intranasal insulin therapy for Alzheimer disease and amnestic mild cognitive impairment: a pilot clinical trial. *Arch. Neurol.* 69: 29–33.

Crapper McLachlan, D. R. (1986). Aluminum and alzheimer's disease. *Neurobiol. Aging.* 7: 525–532.

Crofton, K. M., Zhao, X., Sayre, L. M., and Genter, M. B. (1996). Characterization of the effects of N-hydroxy-IDPN on the auditory, vestibular, and olfactory systems in rats. *Neurotoxicol. Teratol.* 18: 297–303.

Czerniawska, A. (1970). Experimental investigations on the penetration of ^{198}Au from nasal mucous membrane into cerebrospinal fluid. *Acta Otolaryngol.* 70: 58–61.

da Fonseca, C. O., Linden, R., Futuro, D., et al. (2008). Ras pathway activation in gliomas: a strategic target for intranasal administration of perillyl alcohol. *Arch. Immunol. Ther. Exp. (Warsz).* 56: 267–276.

da Fonseca, C. O., Simão, M., Lins, I. R., et al. (2011). Efficacy of monoterpene perillyl alcohol upon survival rate of patients with recurrent glioblastoma. *J. Cancer Res. Clin. Oncol.* 137: 287–293.

Dahl, A. R. (1989). The cyanide-metabolizing enzyme rhodanese in rat nasal respiratory and olfactory mucosa. *Toxicol. Lett.* 45: 199–205.

Dahl, A. R. (1988). The effect of cytochrome P-450-dependent metabolism and other enzyme activities on olfaction. In: Margolis, F.L. and Getchell, T.V., eds. *Molecular Neurobiology of the Olfactory System.* New York: Plenum Press, 51–70.

Dahl, A. R. and Hadley, W. M. (1983). Formaldehyde production promoted by rat nasal cytochrome P-450-dependent monooxygenases with nasal decongestants, essences, solvents, air pollutants, nicotine, and cocaine as substrates. *Toxicol. Appl. Pharmacol.* 67: 200–205.

Dahl, A. R. and Hadley, W. M. (1991). Nasal cavity enzymes involved in xenobiotic metabolism: effects on the toxicity of inhalants. *Toxicology.* 21: 345–372.

Dahlof, C. G., Boes-Hansen, S., Cederberg, C. G., et al. (1998). How does sumatriptan nasal spray perform in clinical practice? *Cephalalgia.* 18: 278–282.

Dal Monte, M., Andrieni, I., Revoltella, R., and Pelosi P. (1991). Purification and characterization of two odorant-binding proteins from nasal tissue of rabbit and pig. *Comp. Biochem. Physiol.* 2: 445–451.

Dantas, Z. N., Vicino, M., Balmaceda, J. P., et al. (1994). Comparison between nafarelin and leuprolide acetate for in vitro fertilization: preliminary clinical study. *Fertil. Steril.* 61: 705–708.

Davidson, T. M. and Smith, W. M. (2010). The Bradford Hill criteria and zinc-induced anosmia. *Arch. Otolaryngol Head Neck Surg.* 2010; 136: 673–676.

Deamer, N. J. and Genter, M. B. (1995). Olfactory toxicity of diethyldithiocarbamate (DDTC) and disulfiram and the protective effect of DDTC against the olfactory toxicity of dichlobenil. *Chem. Biol. Interact.* 95: 215–226.

Deamer, N. J., O'Callaghan, J. P., and Genter, M. B. (1994). Olfactory toxicity resulting from dermal application of 2,6- dichlorobenzonitrile (dichlobenil) in the C57Bl mouse. *Neurotoxicology.* 15: 287–293.

de Souza Silva, M. A., Mattern, C., Hacker, R., et al. (1997a). Intranasal administration of the dopaminergic agonists l-DOPA, amphetamine, and cocaine increases dopamine activity in the neostriatum: a microdialysis study in the rat. *J. Neurochem.* 68: 233–239.

de Souza Silva, M. A., Mattern, C., Hacker, R., et al. (1997b). Increased neostriatal dopamine activity after intraperitoneal or intranasal administration of l-DOPA: on the role of benserazide pretreatment. *Synapse.* 27: 294–302.

de Souza Silva, M. A., Topic, B., Huston, J. P., and Mattern, C. (2008). Intranasal dopamine application increases dopaminergic activity in the neostriatum and nucleus accumbens and enhances motor activity in the open field. *Synapse.* 62: 176–184.

DeLorenzo, A. J. D. (1970). The olfactory neuron and the blood brain barrier. In: Wolstenholme, G.E.W. and Knight, J., eds. *Taste and Smell in Vertebrates.* London: Churchill Livingstone, 151–175.

Dexter, D. T., Wells, F. R., Agid, F., et al. (1987). Increased nigral iron content in postmortem parkinsonian brain. *Lancet.* 2: 1219–1220.

Di Costanzo, A., Toriello, A., Mannara, C., et al. (1993). Intranasal versus intravenous neostigmine in myasthenia gravis: assessment by computer analysis of saccadic eye movements. *Clin. Neuropharmacol.* 16: 511–517.

Donnelly, A., Kellaway, I. W., Taylor G, and Gibson M. (1998). Absorption enhancers as tools to determine the route of nasal absorption of peptides. *J. Drug Targeting.* 5: 121–127.

Dorman, D. C., Struve, M. F., James, R. A., et al. (2001). Influence of particle solubility on the delivery of inhaled manganese to the rat brain: manganese sulfate and manganese tetroxide pharmacokinetics following repeated (14-day) exposure. *Toxicol. Appl. Pharmacol.* 170: 79–87.

Doty, R. L., Deems, D. A., and Stellar, S. (1988). Olfactory dysfunction in parkinsonism: a general deficit unrelated to neurologic signs, disease stage, or disease duration. *Neurology.* 38: 1237–1244.

Doty, R. L., Perl, D. P., Steele, J. C., et al., (1991). Olfactory dysfunction in three neurodegenerative diseases. *Geriatrics.* 1: 47–51.

Doty, R. L., Reyes, P., and Gregor, T. (1987). Presence of both odor identification and detection deficits in Alzheimer's disease. *Brain Res. Bull.* 18: 597–600.

Draghia, R., Caillaud, C., Manicom, R., et al. (1995). Gene delivery into the central nervous system by nasal instillation in rats. *Gene Ther.* 2: 418–423.

Elder, A., Gelein, R., Silva, V., et al. (2006). Translocation of inhaled ultrafine manganese oxide particles to the central nervous system. *Environ. Health Perspect.* 114: 1172–1178.

Eraly, S. A., Vallon, V., Vaughn, D. A., et al. (2006). Decreased renal organic anion secretion and plasma accumulation of endogenous organic anions in OAT1 knock-out mice. *J. Biol. Chem.* 281: 5072–5083.

Esiri, M. and Wilcock, G. (1984). The olfactory bulbs in Alzheimer's disease. *J. Neurol. Neurosurg. Psychiatry.* 47: 56–60.

Esiri, M. M. and Tomlinson, A. H. (1984). Immunohistological demonstration of spread of virus via olfactory and trigeminal pathways after infection of facial skin in mice. *J. Neurol. Sci.* 64: 213–217.

Evans, J. and Hastings, L. (1992). Accumulation of Cd (II) in the CNS depending on the route of administration: intraperitoneal, intratracheal, or intranasal. *Fundam. Appl. Toxicol.* 19: 275–278.

Evans, J. E., Miller, M. L., Andringa, A., and Hastings, L. (1995). Behavioral, histological, and neurochemical effects of nickel (II) on the rat olfactory system. *Toxicol. Appl. Pharmacol.* 130: 209–220.

Fazil, M., Md, S., Haque, S., et al. (2012). Development and evaluation of rivastigmine loaded chitosan nanoparticles for brain targeting. *Eur. J. Pharm. Sci.* 47: 6–15.

Feng, P. C., Wilson, A. G., McClanahan, R. H., et al. (1990). Metabolism of alachlor by rat and mouse liver and nasal turbinate tissues. *Drug Metab. Dispos.* 18: 373–377.

Fernandez-Urrusuno, R., Calvo, P., Remunan-Lopez, C., et al. (1999). Enhancement of nasal absorption of insulin using chitosan nanoparticles. *Pharm. Res.* 16: 1576–1581.

Ferreya-Moyano, H. and Barragan, E. (1989). The olfactory system and Alzheimer's disease. *Int. J. Neurosci.* 49: 157–197.

Fischer, Jde S., Carvalho, P. C., Neves-Ferreira, A. G., et al. (2008). Anti-thrombin as a prognostic biomarker candidate for patients with recurrent glioblastoma multiform under treatment with perillyl alcohol. *J. Exp. Ther. Oncol.* 7: 285–290.

Fisher, A., Watling, M., Smith, A., and Knight, A. (2010). Pharmacokinetic comparisons of three nasal fentanyl formulations; pectin, chitosan and chitosan–poloxamer 188. *Int. J. Clin. Pharmacol. Ther.* 48: 138–145.

Florindo, H. F., Pandit, S., Goncalves L. M. D., et al. (2009). New approach on the development of a mucosal vaccine against strangles: Systemic and mucosal immune responses in a mouse model. *Vaccine.* 27: 1230–1241.

Foster, J. D., Getchell, M. L., and Getchell, T. V. (1991). Identification of sugar residues in secretory glycoconjugates of olfactory mucosae using lectin histochemistry. *Anat. Rec.* 229: 525–544.

Frazier, L. I. and Brunjes, P. C. (1988). Unilateral odor deprivation: early postnatal changes in olfactory bulb cell density and number. *J. Comp. Neurol.* 269: 355–370.

Fredman, S. M. and Jahan-Pawar, B. (1980). Cobalt mapping of the nervous system: evidence that cobalt can cross a neuronal membrane. *J. Neurobiol.* 11: 209–214.

Frey, W. H., Liu, J., Chen, X., et al. (1997). Delivery of 125I-NGF to the brain via the olfactory route. *Drug Deliv.* 4: 87–92.

Gal, S., Zheng, H., Fridkin, M., and Youdim, M. B. (2010). Restoration of nigrostriatal dopamine neurons in post-MPTP treatment by the novel multifunctional brain-permeable iron chelator-monoamine oxidase inhibitor drug, M30. *Neurotox. Res.* 17: 15–27.

Gallagher, R. M. (1996). Acute treatment of migraine with dihydroergotamine nasal spray. *Dihydroergotamine Working Group. Arch. Neurol.* 53: 1285–1291.

Gao, Y., Mengana, Y., Cruz, Y. R., et al. (2011). Different expression patterns of Ngb and EPOR in the cerebral cortex and hippocampus revealed distinctive therapeutic effects of intranasal delivery of Neuro-EPO for ischemic insults to the gerbil brain. *J. Histochem. Cytochem.* 59: 214–227.

Genter, M. B., Llorens, J., O'Callaghan, J. P., et al. (1992). Olfactory toxicity of beta,beta'-iminodipropionitrile in the rat. *J. Pharmacol. Exp. Ther.* 263: 1432–1439.

Genter, M. B., Deamer, N. J., Blake, B. L., et al. (1995). Olfactory toxicity of methimazole: dose–response and structure-activity studies and characterization of flavin-containing monooxygenase activity in the Long-Evans rat olfactory mucosa. *Toxicol. Pathol.* 23: 477–486.

Genter, M. B. (1998). Evaluation of olfactory and auditory system effects of the antihyperthyroid drug carbimazole in the Long-Evans rat. *J. Biochem. Mol. Toxicol.* 12: 305–314.

Genter, M. B., Liang, H. C., Gu, J., et al. (1998). Role of CYP2A5 and 2G1 in acetaminophen metabolism and toxicity in the olfactory mucosa of the *Cyp1a2(−/−)* mouse. *Biochem. Pharmacol.* 55: 1819–1826.

Genter, M. B., Burman, D. M., Dingeldein, M. W., et al. (2000). Evolution of alachlor-induced nasal neoplasms in the Long-Evans rat. *Toxicol. Pathol.* 28: 770–781.

Genter, M. B. and Crofton, K. M. (2000). 2-Pentenenitrile. In: *Experimental and Clinical Neurotoxicology*, 2nd edition. P. S. Spencer, H. H. Schaumberg, and A. C. Ludolph, eds. New York: Oxford University Press, 968–969.

Genter, M. B., Van Veldhoven, P. P., Jegga, A. G., et al. (2003). Microarray-based discovery of highly expressed olfactory mucosal genes: potential roles in the various functions of the olfactory system. *Physiol. Genomics.* 16: 67–81.

Genter, M. B., Kendig, E. L., and Knutson, M. D. (2009). Uptake of materials from the nasal cavity into the blood and brain: are we finally beginning to understand these processes at the molecular level? *Ann. N.Y. Acad. Sci.* 1170: 623–628.

Genter, M. B., Krishan, M., Augustine, L. M., and Cherrington, N. J. (2010). Drug transporter expression and localization in rat nasal respiratory and olfactory mucosa and olfactory bulb. *Drug Metab. Dispos.* 38: 1644–1647.

Genter, M. B., Newman, N. C., Shertzer, H. G., et al. (2012). Distribution and systemic effects of intranasally administered 25 nm silver nanoparticles in adult mice. *Toxicol. Pathol.* 40: 1004–1013.

Gervasi, P. G., Longo, V., Naldi, F., et al. (1991). Xenobiotic-metabolizing enzymes in human respiratory nasal mucosa. *Biochem. Pharmacol.* 41: 177–184.

Getchell, M. L. and Getchell T. V. (1991). Immunohistochemical localization of components of the immune barrier in the olfactory mucosae of salamanders and rats. *Anat. Rec.* 231: 358–374.

Getchell, T. V., Margolis, F. L, and Getchell, M. L. (1984). Perireceptor and receptor events in vertebrate olfaction. *Prog. Neurobiol.* 23: 317–345.

Ghantous, H., Dencker, L., Gabrielsson, J., et al. (1990). Accumulation and turnover of metabolites of toluene and xylene in nasal mucosa and olfactory bulb in the mouse. *Pharmacol. Toxicol.* 66: 87–92.

Gianutsos, G., Morrow, G. R., and Morris, J. B. (1997). Accumulation of manganese in rat brain following intranasal administration. *Fundam. Appl. Toxicol.* 37: 102–105.

Good, P. F., Olanow, C. W., and Perl, D. P. (1992). Neuromelanin-containing neurons of the substantia nigra accumulate iron and aluminum in Parkinson's disease: a LAMMA study. *Brain Res.* 593: 343–346.

Gorell, J. M., Johnson, C. C., Rybicki, B. A., et al. (1999). Occupational exposure to manganese, copper, lead, iron, mercury and zinc and the risk of Parkinson's disease. *Neurotoxicology.* 20: 239–247.

Gottofrey, J. and Tjälve, H. (1991). Axonal transport of cadmium in the olfactory nerve of the pike. *Pharmacol. Toxicol.* 69: 242–252.

Gourlay, S. G. and Benowitz, N. L. (1997). Arteriovenous differences in plasma concentration of nicotine and catecholamines and related cardiovascular effects after smoking, nicotine nasal spray, and intravenous nicotine. *Clin. Pharmacol. Ther.* 62: 453–463.

Gozes, I., Bardea, A., Bechar, M., et al. (1997). Neuropeptides and neuronal survival: neuroprotective strategy for Alzheimer's disease. *Ann. N.Y. Acad. Sci.* 814: 161–166.

Gozes, I., Bardea, A., Reshef, A., et al. (1996). Neuroprotective strategy for Alzheimer disease: intranasal administration of a fatty neuropeptide. *Proc. Natl. Acad. Sci. USA.* 93: 427–432.

Gross, G. W. and Beidler, L. M. (1973). Fast axonal transport in the c-fibers of the garfish olfactory nerve. *J. Neurobiol.* 4: 413–428.

Gross, G. W. and Beidler, L. M. (1975). A quantitative analysis of isotope concentration profiles and rapid transport velocities in the C-fibers of the garfish olfactory nerve. *J. Neurobiol.* 6: 213–232.

Gross, G. W., and Kreutzberg, G. W. (1978). Rapid axoplasmic transport in the olfactory nerve of the pike: I. *Basic transport parameters for proteins and amino acids. Brain Res.* 139: 65–76.

Gu, J., Walker, V. E., Lipinskas, T. W., et al. (1997). Intraperitoneal administration of coumarin causes tissue-selective depletion of cytochromes P450 and cytotoxicity in the olfactory mucosa. *Toxicol. Appl. Pharmacol.* 146: 134–143.

Guthrie, K. M., Pullara, J. M., Marshall, J. F., and Leon M. (1991). Olfactory deprivation increases dopamine D2 receptor density in the rat olfactory bulb. *Synapse.* 8: 61–70.

Hallschmid, M., Benedict, C., Born, J., et al. (2004). Manipulating central nervous mechanisms of food intake and body weight regulation by intranasal administration of neuropeptides in man. *Physiol. Behav.* 83: 55–64.

Haneberg, B., Dalseg, R., Oftung, F., et al. (1998). Towards a nasal vaccine against meningococcal disease, and prospects for its use as a mucosal adjuvant. *Dev. Biol. Stand.* 92: 127–133.

Hanson, L. R., Fine, J. M., Hoekman, J. D., et al. (2012). Intranasal delivery of growth differentiation factor 5 to the central nervous system. *Drug Deliv.* 19: 149–154.

Haschke, M., Suter, K., Hofmann, S., et al. (2010). Pharmacokinetics and pharmacodynamics of nasally delivered midazolam. *Br. J. Clin. Pharmacol.* 69: 607–616.

Hastings, L. (1990). Sensory neurotoxicology: use of the olfactory system in the assessment of toxicity. *Neurotoxicol. Teratol.* 12: 455–459.

Hastings, L. and Evans, J. E. (1992). Olfactory primary neurons as a route of entry for toxic agents into the CNS. *Neurotoxicology.* 12: 707–714.

Hayek, R and Waite, P. M. E. (1991). The olfactory pathway as a possible route for aluminum entry to the brain. *J. Neurochem.* 57: S113.

Hayes, R. B., Gerin, M., Raatgever, J. W., and de Bruyn, A. (1986). Wood-related occupations, wood dust exposure, and sinonasal cancer. *Am. J. Epidemiol.* 124: 569–577.

Hempstead, J. L. and Morgan, J. I. (1985). A panel of monoclonal antibodies to the rat olfactory epithelium. *J. Neurosci.* 5: 438–449.

Henriksson, J. and Tjälve, H. (1998). Uptake of inorganic mercury in the olfactory bulbs via olfactory pathways in rats. *Environ. Res.* 77: 130–140.

Henriksson, J., Tallkvist, J., and Tjälve, H. (1999). Transport of manganese via the olfactory pathway in rats: dosage dependency of the uptake and subcellular distribution of the metal in the olfactory epithelium and the brain. *Toxicol. Appl. Pharmacol.* 156: 119–128.

Henriksson, J. and Tjälve, H. (2000). Manganese taken up into the CNS via the olfactory pathway in rats affects astrocytes. *Toxicol. Sci.* 55: 392–398.

Higashisaka, K., Yoshioka, Y., Yamashita, K., et al. (2011). Acute phase proteins as biomarkers for predicting the exposure and toxicity of nanomaterials. *Biomaterials.* 32: 3–9.

Hilsted, J., Madsbad, S., Hvidberg, A., et al. (1995). Intranasal insulin therapy: the clinical realities. *Diabetologia.* 38: 680–684.

Hirsch, E. C., Brandel, J. P., Galle, P., et al. (1991). Iron and aluminum increase in the substantia nigra of patients with Parkinson's disease: an X-ray microanalysis. *J. Neurochem.* 56: 446–451.

Hoekman, J. D. and Ho, R. J. (2011a). Enhanced analgesic responses after preferential delivery of morphine and fentanyl to the olfactory epithelium in rats. *Anesth. Analg.* 113: 641–651.

Hoekman, J. D. and Ho, R. J. (2011b). Effects of localized hydrophilic mannitol and hydrophobic nelfinavir administration targeted to olfactory epithelium on brain distribution. *AAPS PharmSciTech.* 12: 534–543.

Hoeppner, T. J. (1990). The anticonvulsant valproic acid concentrates in the olfactory bulb: selective laminar localization. *Brain Res.* 532: 326–328.

Holl A. (1981). Marking of olfactory axons of fishes by intravital staining with procion brilliant yellow. *Stain Technol.* 56: 67–70.

Holl, A. (1980). [Selective staining by procion dyes of olfactory sensory neurons in the catfish Ictalurus nebulosus (author's transl)]. *Z. Naturforsch [C].* 35: 526–528.

Holl, A. (1965). Vital staining by trypan blue; its selectivity for olfactory receptor cells of the brown bullhead, *Ictalurus natalis. Stain Technol.* 40: 269–273.

Horowitz, L. F., Montmayeur, J. P., Echelard, Y., and Buck, L. B. (1999). A genetic approach to trace neural circuits. *Proc. Natl. Acad. Sci. USA.* 96: 3194–3199.

Horvát, S., Fehér, A., Wolburg, H., et al. (2009). Sodium hyaluronate as a mucoadhesive component in nasal formulation enhances delivery of molecules to brain tissue. *Eur. J. Pharm. Biopharm.* 72: 252–259.

Huneycutt, B. S., Plakhov, I. V., Shusterman, Z., et al. (1994). Distribution of vesicular stomatitis virus proteins in the brains of BALB/c mice following intranasal inoculation: an immunohistochemical analysis. *Brain Res.* 635: 81–95.

Hurst, E. W. (1936). New knowledge of virus diseases of nervous system: review and interpretation. *Brain.* 59: 1–34.

Hussain, A. A., Iseki, K., Kagoshima, M., and Dittert, L. W. (1992). Absorption of acetylsalicylic acid from the rat nasal cavity. *J. Pharm. Sci.* 81: 348–349.

Hussain, A. A., Dittert, L. W., and Traboulsi, A. (2002). Brain delivery of folic acid for the prevention of Alzheimer's disease and stroke. US Patent 6369058.

Hussain, M. A. and Aungst, B. J. (1997). Intranasal absorption of oxymorphone. *J. Pharm. Sci.* 86: 975–976.

Isaka, H., Yoshii, H., Otsuji, A., et al. (1979). Tumors of Sprague–Dawley rats induced by long-term feeding of phenacetin. *Gann.* 70: 29–36.

Itaya, S. K. (1987). Anterograde transsynaptic transport of WGA-HRP in rat olfactory pathways. *Brain Res.* 409: 205–214.

Itaya, S. K. and Van Hausen, G. W. (1982). WGA-HRP as a transneuronal marker in the visual pathways of monkey and rat. *Brain Res.* 236: 199–204.

Ivankovic, S., Seibel, J., Komitowski, D., et al. (1998). Caffeine-derived N-nitroso compounds. V. *Carcinogenicity of mononitrosocaffeidine and dinitrosocaffeidine in bd-ix rats. Carcinogenesis.* 19: 933–937.

Jackson, R. T., Tigges, J., and Arnold, W. (1979). Subarachnoid space of the CNS, nasal mucosa, and lymphatic system. *Arch. Otolaryngol.* 105: 180–184.

Jacobson, J., Harris, S. R., and Bullingham, R. E. (1994). Low dose intranasal nafarelin for the treatment of endometriosis. *Acta Obstet. Gynecol. Scand.* 73: 144–150.

Jain, A. K., Khar, R. K., Ahmed, F. J., and Diwan, P. V. (2008). Effective insulin delivery using starch nanoparticles as a potential trans-nasal mucoadhesive carrier. *Eur. J. Pharm. Biopharm.* 69: 426–435.

Jarolim, K. L., McCosh, J. K., Howard, M. J., and John, D. T. (2000). A light microscopy study of the migration of Naegleria fowleri from the nasal submucosa to the central nervous system during the early stage of primary amebic meningoencephalitis in mice. *J. Parasitol.* 86: 50–55.

Jeannet, P. Y., Roulet. E., Maeder-Ingvar, M., et al. (1999). Home and hospital treatment of acute seizures in children with nasal midazolam. *Europ. J. Paediatr. Neurol.* 3: 73–77.

Jeffery, E. H. and Haschek W. M. (1988). Protection by dimethylsulfoxide against acetaminophen-induced hepatic, but not respiratory toxicity in the mouse. *Toxicol. Appl. Pharmacol.* 93: 452–461.

Jensen, R. K. and Sleight, S. D. (1987). Toxic effects of N-nitrosodiethylamine on nasal tissues of Sprague- Dawley rats and golden Syrian hamsters. *Fundam. Appl. Toxicol.* 8: 217–229.

Jintapattanakit, A., Peungvicha, P., Sailasuta, A., et al. (2010). Nasal absorption and local tissue reaction of insulin nanocomplexes of trimethyl chitosan derivatives in rats. *J. Pharm. Pharmacol.* 62: 583–591.

Johnson, C. C., Gorell, J. M., Rybicki, B. A., et al. (1999). Adult nutrient intake as a risk factor for Parkinson's disease. *Int. J. Epidemiol.* 28: 1102–1109.

Johnston, S., Wilkes, G. J., Thompson, J. A., et al. (2011). Inhaled methoxyflurane and intranasal fentanyl for prehospital management of visceral pain in an Australian ambulance service. *Emerg. Med. J.* 28: 57–63.

Jones, D. T. and Reed, R. R. (1989). G_{olf}: An olfactory neuron specific-G protein involved in odorant signal transduction. *Science.* 244: 790–795.

Jones, N. S., Quraishi, S., and Mason, J. D. (1997). The nasal delivery of systemic drugs. *Int. J. Clin. Pract.* 51: 308–311.

Jones, R. L., Nguyen, A., and Man, S. F. (1998). Nicotine and cotinine replacement when nicotine nasal spray is used to quit smoking. *Psychopharmacology (Berl).* 137: 345–350.

Joshi, A. S., Patel, H. S., Belgamwar, V. S., et al. (2012), Solid lipid nanoparticles of ondansetron HCl for intranasal delivery: development, optimization and evaluation. *J. Mater. Sci: Mater. Med.* DOI 10.1007/s10856-012-4702-7.

Kaler, G., Truong, D. M., Sweeney, D. E., et al. (2006). Olfactory mucosa-expressed organic anion transporter, Oat6, manifests high affinity interactions with odorant organic anions. *Biochem. Biophys. Res. Commun.* 351: 872–876.

Kallas, H. E., Chintanadilok, J., Maruenda, J., et al. (1999). Treatment of nocturia in the elderly. *Drugs Aging.* 15: 429–437.

Kao, H. D., Traboulsi, A., Itoh, S., et al. (2000). Enhancement of the systemic and CNS specific delivery of L-dopa by the nasal administration of its water soluble prodrugs *Pharm. Res.* 17: 978–984.

Karachunski, P. I., Ostlie, N. S., Okita, D. K., and Conti-Fine, B. M. (1998). Nasal administration of synthetic acetylcholine receptor T epitopes affects the immune response to the acetylcholine receptor and prevents experimental myasthenia gravis. *Ann. N.Y. Acad. Sci. USA.* 841: 560–564.

Karasulu, H. Y., Şanal, Z. E., Sözer, S., et al. (2008). Permeation studies of indomethacin from different emulsions for nasal delivery and their possible anti-inflammatory effects. *AAPS PharmSciTech.* 9: 342–348.

Key. B. and Akeson. R. A. (1990). Olfactory neurons express a unique glycosylated form of the neural cell adhesion molecule (N-CAM). *J. Cell. Biol.* 110: 1729–1743.

Key, B. and Giorgi, P. P. (1986a). Selective binding of soybean agglutinin to the olfactory system of *Xenopus. Neuroscience.* 18: 507–515.

Key, B. and Giorgi, P. P. (1986b). Soybean agglutinin binding to the olfactory systems of the rat and mouse. *Neurosci. Lett.* 69: 131–136.

Kim, I. D., Shin, J. H., Lee, H. K., et al. (2012). Intranasal delivery of HMGB1-binding heptamer peptide confers a robust neuroprotection in the postischemic brain. *Neurosci. Lett.* 525: 179–183.

Kim, T. H., Park, C. W., Kim, H. Y., et al. (2012). Low molecular weight (1 kDa) Polyethylene glycol conjugation markedly enhances the hypoglycemic effects of intranasally administered exendin-4 in type 2 diabetic *db/db* mice. *Biol. Pharm. Bull.* 35: 1076–1083.

Kinoshita, Y., Shiga, H., Washiyama, K., et al. (2008). Thallium transport and the evaluation of olfactory nerve connectivity between the nasal cavity and olfactory bulb. *Chem. Senses.* 33: 737–738.

Kontos, A. P., Kotwal, R. S., Elbin, R., et al. (2012). Residual effects of combat-related mild traumatic brain injury. *J. Neurotrauma.* 2012 Oct 2. [Epub ahead of print] PubMed PMID: 23031200.

Koujitani, T., Yasuhara, K., Kobayashi, H., et al. (1999). Tumor-promoting activity of 2,6-dimethylaniline in a two-stage nasal carcinogenesis model in N-bis(2-hydroxypropyl)nitrosamine-treated rats. *Cancer Lett.* 142: 161–171.

Kowall, N. W., Pendlebury, W. W., Kessler, J. B., et al. (1989). Aluminum-induced neurofibrillary degeneration affects a subset of neurons in rabbit cerebral cortex, basal forebrain and upper brainstem. *Neuroscience.* 29: 320–337.

Kream, R. M., Davis, B. J., Kawano, T., et al. (1984). Substance P and catecholaminergic expression in neurons of the hamster main olfactory bulb. *J. Comp. Neurol.* 222: 140–154.

Krishan, M., Gudelsky, G., Desai, P., and Genter, M. B. (2014). Manipulation of olfactory tight junctions using papaverine to enhance intarnasal delivery of gemcitabine to the brain. *Drug Delivery* 21:8–16.

Kristensson, K. and Olsson, Y. (1971). Uptake of exogenous proteins in mouse olfactory cells. *Acta Neuropathol.* 19: 145–154.

Lafay, F., Coulon, P., Astic, L., et al. (1991). Spread of the CVS strain of rabies virus and of the antivirulent mutant AvO1 along the olfactory pathways of the mouse after intranasal inoculation. *Virology.* 183: 320–330.

Lan, J. and Jiang, D. H. (1997). Excessive iron accumulation in the brain: a possible potential risk of neurodegeneration in Parkinson's disease. *J. Neural. Transm.* 104: 649–660.

Land, L. J. and Shepherd, G. M. (1974). Autoradiographic analysis of olfactory receptor projections in the rabbit. *Brain Res.* 70: 506–510.

Landgraf, R. (1985). Plasma oxytocin concentrations in man after different routes of administration of synthetic oxytocin. *Exp. Clin. Endocrinol.* 85: 245–248.

Lane, R. J. (1996). Intranasal lidocaine for treatment of migraine [letter]. *JAMA* 276: 1553; discussion 1554.

Langston, J. W. (1998). Epidemiology versus genetics in Parkinson's disease: Progress in resolving an age-old debate. *Ann. Neurol.* 44 (Suppl. 1): S45–S52.

Larsson, P. and Tjälve, H. (2000). Intranasal instillation of aflatoxin B(1) in rats: Bioactivation in the nasal mucosa and neuronal transport to the olfactory bulb. *Toxicol. Sci.* 55: 383–391.

Lavi, E., Fishman, P. S., Highkin, M. K., and Weiss, S. R. (1988). Limbic encephalitis after inhalation of a murine coronavirus. *Lab. Invest.* 58: 31–36.

Lazard, D., Zupko, K., Poria, Y., et al. (1991). Odorant signal termination by olfactory UDP glucuronosyl transferase. *Nature.* 349: 790–793.

Lemay, A., Maheux, R., Faure, N., et al. (1984). Reversible hypogonadism induced by a luteinizing hormone-releasing hormone (LH-RH) agonist (Buserelin) as a new therapeutic approach for endometriosis. *Fertil. Steril.* 41: 863–871.

Levenson, C. W., Cutler, R. G., Ladenheim, B., et al. (2004). Role of dietary iron restriction in a mouse model of Parkinson's disease. *Exp. Neurol.* 190: 506–514.

Levy, R. L. (1995). Intranasal capsaicin for acute abortive treatment of migraine without aura [letter]. *Headache.* 35: 277.

Lewis, J. L. and Dahl, A. R. (1995). Olfactory mucosa: composition, enzymatic localization and metabolism. In: Doty, R. L., ed. *Handbook of Olfaction and Gustation.* New York: Marcel Dekker. 33–52.

Li, X., Johnson, K. R., Bryant, M., et al. (2011). Intranasal delivery of E-selectin reduces atherosclerosis in ApoE−/− mice. *PLoS One.* 6(6): e20620.

Lim, J. H., Davis, G. E., Wang, Z., et al. (2009). Zicam-induced damage to mouse and human nasal tissue. *PLos One.* 4: e7647.

Lindquist, N. G., Lyden, A., Narfstrom, K., and Samaan, H. (1983). Accumulation of taurine in the nasal mucosa and the olfactory bulb. *Experientia.* 39: 797–799.

Lipton, R. B. (1997). Ergotamine tartrate and dihydroergotamine mesylate: safety profiles. *Headache.* 37: S33–41.

Liu, M. A. (1998).*Vaccine developments. Nat. Med.* 4(5 Suppl): 515–519.

480   Chapter 19   The Olfactory System as a Route of Delivery for Agents to the Brain and Circulation

Liu, X. F., Fawcett, J. R., Hanson, L. R., and Frey, II. W. H. (2004). The window of opportunity for treatment of focal cerebral ischemic damage with noninvasive intranasal insulin-like growth factor-I in rats. *J. Stroke Cerebrovasc. Dis.* 13: 16–23.

Liu, X. F., Fawcett, J. R., Thorne, R. G., and Frey, II. W..H. (2001). Non-invasive intranasal insulin-like growth factor-I reduces infarct volume and improves neurologic function in rats following middle cerebral artery occlusion. *Neurosci. Lett.* 308: 91–94.

Liu, Y., Gao, Y., Zhang, L., et al. (2009). Potential health impact on mice after nasal instillation of nano-sized copper particles and their translocation in mice. *J. Nanosci. Nanotechnol.* 9: 6335–6343.

Louon, A. and Reddy, V. G. (1994). Nasal midazolam and ketamine for paediatric sedation during computerised tomography. *Acta Anaesthesiol. Scand.* 38: 259–261.

Lucchini, R. G., Martin, C. J., and Doney, B. C. (2009). From manganism to manganese-induced parkinsonism: a conceptual model based on the evolution of exposure. *Neuromolecular Med.* 11: 311–332.

Lundh, B., Brockstedt, U., and Kristensson, K. (1989). Lectin-binding pattern of neuroepithelial and respiratory epithelial cells in the mouse nasal cavity. *Histochem. J.* 21: 33–43.

Luppi, B., Bigucci, F., Corace, G., et al. (2011). Albumin nanoparticles carrying cyclodextrins for nasal delivery of the anti-Alzheimer drug tacrine. *Eur. J. Pharm. Sci.* 44: 559–565.

Ma, C., Schneider, S. N., Miller, M., et al. (2008). Manganese accumulation in the mouse ear following systemic exposure. *J. Biochem. Mol. Toxicol.* 22: 305–310.

Maassab, H. F. and DeBorde, D. C. (1985). Development and characterization of cold-adapted viruses for use as live virus vaccines. *Vaccine.* 3: 355–369.

Mansour, A., Meador-Woodruff, J. H., Bunzow, J. R., et al. (1990). Localization of dopamine D_2 receptor mRNA and D_1 and D_2 receptor binding in the rat brain and pituitary: an in situ hybridization-receptor autoradiographic analysis. *J. Neurosci.* 10: 2587–2600.

Margolis, F. L. and Grillo, M. (1977). Axoplasmic transport of carnosine (β-alanyl-L-histidine) in the mouse olfactory pathway. *Neurochem. Res.* 2: 507–519.

Maruniak, J. A., Lin, P. J., and Henegar, J. R. (1989). Effects of unilateral naris closure on the olfactory epithelia of adult mice. *Brain Res.* 490: 212–218.

Maruniak, J. A., Taylor, J. A., Henegar, J. R., and Williams, M. B. (1989). Unilateral naris closure in adult mice: atrophy of the deprived-side olfactory bulbs. *Dev. Brain Res.* 47: 27–33.

Marttin, E., Romeijn, S. G., Verhoef, J. C., and Merkus, F. W. (1997). Nasal absorption of dihydroergotamine from liquid and powder formulations in rabbits. *J. Pharm. Sci.* 86: 802–807.

Marttin, E., Verhoef, J. C., and Merkus, F. W. (1998). Efficacy, safety and mechanism of cyclodextrins as absorption enhancers in nasal delivery of peptide and protein drugs. *J. Drug Target.* 6: 17–36.

Matsuyama, T., Morita, T., Horikiri, Y., et al. (2006a). Enhancement of nasal absorption of large molecular weight compounds by combinationof mucolytic agent and nonionic surfactant. *J. Control. Release.* 110: 347–352.

Matsuyama, T., Morita, T., Horikiri, Y., et al. (2006b). Improved nasal absorption of salmon calcitonin by powdery formulation with N-acetyl-L-cysteine as a mucolytic agent. *J. Control. Release.* 115: 183–188.

Mathison, S., Nagilla, R., and Kompella, U. B. (1998). Nasal route for direct delivery of solutes to the central nervous system: fact or fiction? *J. Drug Target.* 5: 415–441.

Matsumiya, K., Kitamura, M., Kishikawa, H., et al. (1998). A prospective comparative trial of a gonadotropin-releasing hormone analogue with clomiphene citrate for the treatment of oligoasthenozoospermia. *Int. J. Urol.* 5: 361–363.

McLean, J. H., Shipley, M. T., and Bernstein, D. I. (1989). Golgi-like, transneuronal retrograde labeling with CNS injections of Herpes simplex virus type 1. *Brain Res. Bull.* 22: 867–881.

McLean, J. H., Shipley, M. T., Bernstein, D. I., and Corbett, D. (1993). Selective lesions of neural pathways following viral inoculation of the olfactory bulb. *Exp. Neurol.* 122: 209–222.

McMartin, C., Hutchinson, L. E., Hyde, R., and Peters, G. E. (1987). Analysis of structural requirements for the absorption of drugs and macromolecules from the nasal cavity. *J. Pharm. Sci.* 76: 535–540.

Meisami, E. (1976). Effects of olfactory deprivation on postnatal growth of the rat olfactory bulb utilizing a new method for production of neonatal anosmia. *Brain Res.* 107: 437–444.

Melanson, S. W., Morse, J. W., Pronchik, D. J., and Heller, M. B. (1997). Transnasal butorphanol in the emergency department management of migraine headache. *Am. J. Emerg. Med.* 15: 57–61.

Mellert, T. K., Getchell, M. L., Sparks. L., and Getchell, T. V. (1992). Characterization of the immune barrier in human olfactory mucosa. *Otolaryngol. Head Neck Surg.* 106: 181–188.

Meredith, M. and O'Connell, R. J. (1988). HRP uptake by olfactory and vomeronasal receptor neurons: use as an indicator of incomplete lesions and relevance for non-volatile chemoreception. *Chem. Senses.* 13: 487–515.

Mills, T. M. and Scoggin, J. A. (1997). Intranasal lidocaine for migraine and cluster headaches. *Ann. Pharmacother.* 31: 914–915.

Monath, T. P., Cropp, C. B., and Harrison, A. K. (1983). Mode of entry of a neurotropic arbovirus into the central nervous system. *Lab. Invest.* 48: 399–410.

Morales, J. A., Herzog, S., Kompter, C., et al. (1988). Axonal transport of Borna disease virus along olfactory pathways spontaneously and experimentally infected rats. *Med. Microbiol. Immunol.* 177: 51–68.

Mori, I., Komatsu, T., Takeuchi, K., et al. (1995). Parainfluenza virus type 1 infects olfactory neurons and establishes long-term persistence in the nerve tissue. *J. Gen. Virol.* 76: 1251–1254.

Mori, K., Fujita, S. C., Imamura, K., and Obata, K. (1985). Immunohistochemical study of subclasses of olfactory nerve fibers and their projections to the olfactory bulb in the rabbit. *J. Comp. Neurol.* 242: 214–229.

Mudd, S. (2011). Intranasal fentanyl for pain management in children: a systematic review of the literature. *J. Pediatr. Health Care.* 25: 316–322.

Mystakidou, K., Panagiotou, I., and Gouliamos, A. (2011). Fentanyl nasal spray for the treatment of cancer pain. *Expert Opin. Pharmacother.* 12: 1653–1659.

Na, L., Mao, S., Wang, J., and Sun, W. (2010). Comparison of different absorption enhancers on the intranasal absorption of isosorbide dinitrate in rats. *Int. J. Pharm.* 397: 59–66.

National Cancer Institute. (1979). Bioassay of p-cresidine for possible carcinogenicity. Technical Report No. 142. Bethesda, MD: Carcinogenesis Testing Program, NCI, National Institutes of Health.

Nef, P., Heldman, H., Lazard, D., et al. (1989). Olfactory-specific cytochrome P-450. *J. Biol. Chem.* 264: 6780–6785.

Nir, Y., Beemer, A., and Goldwasser, R. A. (1965). West Nile virus infection in mice following exposure to a viral aerosol. *Br. J. Exp. Pathol.* 46: 443–448.

Nonaka, N., Farr, S. A., Nakamachi, T., et al. (2012). Intranasal administration of PACAP: Uptake by brain and regional brain targeting with cyclodextrins. *Peptides.* 36: 168–175.

References

Nonaka, N., Farr, S. A., Kageyama, H., et al. (2008). Delivery of galanin-like peptide to the brain: Targeting with intranasal delivery and cyclodextrins. *J. Pharmacol. Exp. Therap.* 325: 513–519.

Ohm, T. and Braak, H. (1987). Olfactory bulb changes in Alzheimer's disease. *Acta Neuropathol.* 73: 365–369.

Oliver, K. R. and Fazakerley, J. K. (1998). Transneuronal spread of Semliki Forest virus in the developing mouse olfactory system is determined by neuronal maturity. *Neuroscience.* 82: 867–877.

Parker, K. J., Buckmaster, C. L., Schatzberg, A. F., and Lyons, D. M. (2005). Intranasal oxytocin administration attenuates the ACTH stress response in monkeys. *Psychoneuroendocrinology.* 30: 924–929.

Patel, D., Naik, S., and Misra, A. (2012). Improved transnasal transport and brain uptake of tizanidine HCl-loaded thiolated chitosan nanoparticles for alleviation of pain. *J. Pharm. Sci.* 101: 690–706.

Patel, S., Chavhan, S., Soni, H., et al. (2012). Brain targeting of risperidone-loaded solid lipid nanoparticles by intranasal route. *J. Drug Target.* 19: 468–474.

Paudel, K. S., Hammell, D. C., Agu, R. U., et al. (2010). Cannabidiol bioavailability after nasal and transdermal application: effect of permeation enhancers. *Drug Dev. Ind. Pharm.* 36: 1088–1097.

Peet, M. M., Echols, D. H., and Richter, H. J. (1937). The chemical prophylaxis for poliomyelitus: the technique of applying zinc sulfate intranasally. *JAMA* 108: 2184–2187.

Perl, P. T. and Good, P. F. (1987). Uptake of aluminum into central nervous system along nasal-olfactory pathways. *Lancet.* 1028.

Perlman, S., Evans, G., and Affifi, A. (1990). Effect of olfactory bulb ablation on spread of a neurotropic coronavirus into the mouse brain. *J. Exp. Med.* 172: 1127–1132.

Perlman, S., Jacobsen, G., and Afifi, A. (1989). Spread of a neurotropic murine coronavirus into the CNS via the trigenial and olfactory nerves. *Virology.* 170: 556–560.

Perlman, S., Jacobsen, G., and Moore, S. (1988). Regional localization of virus in the central nervous system of mice persistently infected with murine coronavirus JHm. *Virology.* 166: 328–338.

Perras, B., Droste, C., Born, J., et al. (1997). Verbal memory after three months of intranasal vasopressin in healthy old humans. *Psychoneuroendocrinology.* 22: 387–396.

Persson, E., Henriksson, J., and Tjälve, H. (2003). Uptake of cobalt from the nasal mucosa into the brain via olfactory pathways in rats. *Toxicol. Lett.* 145: 19–27.

Pevsner, J., Sklar, P. B., and Snyder, S. H. (1986). Odorant-binding protein: Localization to nasal glands and secretions. *Proc. Natl. Acad. Sci. USA.* 83: 4942–4946.

Pierson, M. E., Comstock, J. M., Simmons, R. D., et al. (1997). Synthesis and biological evaluation of potent, selective, hexapeptide CCK-A agonist anorectic agents. *J. Med. Chem.* 40: 4302–4307.

Pietrowsky, R., Struben, C., Molle, M., et al. (1996). Brain potential changes after intranasal vs. intravenous administration of vasopressin: evidence for a direct nose-brain pathway for peptide effects in humans. *Biol. Psychiatry.* 39: 332–340.

Plopper, C. G., Suverkropp, C., Morin, D., et al. (1992). Relationship of cytochrome P-450 activity to Clara cell cytotoxicity. I. Histopathologic comparison of the respiratory tract of mice, rats and hamsters after parenteral administration of naphthalene. *J. Pharmacol. Exp. Therap.* 261: 353–363.

Polak, E. H., Shirley, S. G., and Dodd, G. H. (1989). Concanavalin A reveals olfactory receptors which discrimitate between alkane odorants on the basis of size. *Biochem. J.* 262: 475–478.

Powers, K. M., Smith-Weller, T., Franklin, G. M., et al. (2003). Parkinson's disease risks associated with dietary iron, manganese, and other nutrient intakes. *Neurology.* 60: 1761–1766.

Powers, K. M., Smith-Weller, T., Franklin, G. M., et al. (2009). Dietary fats, cholesterol and iron as risk factors for Parkinson's disease. *Parkinsonism Relat. Disord.* 15: 47–52.

Prediger, R. D., Aguiar, A. S. Jr, Matheus, F. C., et al. (2012). Intranasal administration of neurotoxicants in animals: support for the olfactory vector hypothesis of Parkinson's disease. *Neurotox. Res.* 21: 90–116.

Prediger, R. D., Aguiar, A. S. Jr, Moreira, E. L., et al. (2011). The intranasal administration of 1-methyl-4-phenyl-1,2,3,6-tetrahydropyridine (MPTP): a new rodent model to test palliative and neuroprotective agents for Parkinson's disease. *Curr. Pharm. Des.* 17: 489–507.

Prediger, R. D., Aguiar, A. S. Jr, Rojas-Mayorquin, A. E., et al. (2010) Single intranasal administration of 1-methyl-4-phenyl-1,2,3,6-tetrahydropyridine in C57BL/6 mice models early preclinical phase of Parkinson's disease. *Neurotox. Res.* 17: 114–129.

Prediger, R. D., Batista, L. C., Medeiros, R., et al. (2006). The risk is in the air: Intranasal administration of MPTP to rats reproducing clinical features of Parkinson's disease. *Exp. Neurol.* 202: 391–403.

Prediger, R. D., Rial, D., Medeiros, R., et al. (2009). Risk is in the air: an intranasal MPTP (1-methyl-4-phenyl-1,2,3,6-tetrahydropyridine) rat model of Parkinson's disease. *Ann. N.Y. Acad. Sci.* 1170: 629–636.

Pum, M. E., Schable, S., Harooni, H. E., et al. (2009). Effects of intranasally applied dopamine on behavioral asymmetries in rats with unilateral 6-hydroxydopamine lesions of the nigrostriatal tract. *Neuroscience.* 162: 174–183.

Putcha, L., Tietze, K. J., Bourne, D. W., (1996). Bioavailability of intranasal scopolamine in normal subjects. *J. Pharm. Sci.* 85: 899–902.

Qiang, F., Shin, H. J., Lee, B. J., and Han, H. K. (2012). Enhanced systemic exposure of fexofenadine via the intranasal administration of chitosan-coated liposome. *Int. J. Pharm.* 430: 161–166.

Rake, G. (1937). The rapid invasion of the body through the olfactory mucosa. *J. Exp. Med.* 65: 303–315.

Ramanathan, R., Geary, R. S., Bourne, D. W., and Putcha, L. (1998). Bioavailability of intranasal promethazine dosage forms in dogs. *Pharmacol. Res.* 38: 35–39.

Reed, C. J., van den Broeke, L. T., and De Matteis, F. (1989). Drug-induced protoporphyria in the olfactory mucosa of the hamster. *J. Biochem. Toxicol.* 4: 161–164.

Reger, M. A., Watson, G. S., Green, P. S., et al. (2008). Intranasal insulin improves cognition and modulates beta-amyloid in early AD. *Neurology.* 70: 440–448.

Reitz, M., Demestre, M., Sedlacik, J., Meissner, H., et al. (2012). Intranasal delivery of neural stem/progenitor cells: A noninvasive passage to target intracerebral glioma. *Stem Cells Transl. Med.* Dec. 2012; 1(12): 866–873.

Ressler, K. J., Sullivan, S. L., and Buck, L. B. (1994). Information coding in the olfactory system: evidence for a stereotyped and highly organized epitope map in the olfactory bulb. *Cell.* 79: 1245–1255.

Ressler, K. J., Sullivan, S. L., and Buck, L. B. (1993). A zonal organization of odorant receptor gene expression in the olfactory epithelium. *Cell.* 73: 597–609.

Riddle, M. S., Kaminski, R. W., Williams, C., et al. (2011). Safety and immunogenicity of an intranasal Shigella flexneri 2a Invaplex 50 vaccine. *Vaccine.* 29: 7009–7019.

Roberts, E. (1986). Alzheimer's disease may begin in the nose and may be caused by aluminosilicates. *Neurobiol. Aging.* 7: 561–567.

Rojo, A. I., Cavada, C., de Sagarra, M. R., and Cuadrado, A. (2007). Chronic inhalation of rotenone or paraquat does not induce Parkinson's disease symptoms in mice or rats. *Exp. Neurol.* 208: 120–126.

Rojo, A. I., Montero, C., Salazar, M., et al. (2006). Persistent penetration of MPTP through the nasal route induces Parkinson's disease in mice. *Eur. J. Neurosci.* 24: 1874–1884.

Ruan, Y., Yao, L., Zhang, B., et al. (2012). Nanoparticle-mediated delivery of neurotoxin-II to the brain with intranasal administration: an effective strategy to improve antinociceptive activity of neurotoxin. *Drug Dev. Ind. Pharm.* 38: 123–128.

Ruis, H., Rolland, R., Doesburg, W., et al. (1981). Oxytocin enhances onset of lactation among mothers delivering prematurely. *Br. Med. J. (Clin. Res. Ed.)* 283: 340–342.

Ruocco, L. A., de Souza Silva, M. A., Topic, B., et al. (2009). Intranasal application of dopamine reduces activity and improves attention in Naples high excitability rats that feature the mesocortical variant of ADHD. *Eur. Neuropsychopharmacol.* 19: 693–701.

Russell, M. J., Liu, H., Nunes, R. J., and Vijayan, V. (1991). Transsynaptic and non-transsynaptic forms of WGA-HRP. *Soc. Neurosci. Abstr.* 17: 60.

Ryzhikov, A.B., Ryabchikova, E. I., Sergeev, A. N., and Tkacheva, N. V. (1995). Spread of Venezuelan equine encephalitis virus in mice olfactory tract. *Arch. Virol.* 140: 2243–2254.

Sachs, C. J. (1996). Intranasal lidocaine for treatment of migraine [letter; comment]. *JAMA.* 276: 1553–1554; discussion 1554.

Sahler, C. S. and Greenwald, B. D. (2012). Traumatic brain injury in sports: a review. *Rehabil. Res. Pract.* 2012: 659652. Epub 2012 Jul 9.

Sakane, T., Akizuki, M., Taki, Y., et al. (1995). Direct drug transport from the rat nasal cavity to the cerebrospinal fluid: the relation to the molecular weight of drugs. *J. Pharm. Pharmacol.* 47: 379–381.

Sakane, T., Akizuki, M., Yamashita, S., et al. (1991a). The transport of a drug to the cerebrospinal fluid directly from the nasal cavity: the relation to the lipophilicity of the drug. *Chem. Pharm. Bull. (Tokyo).* 39: 2456–2458.

Sakane, T., Akizuki, M., Yamashita, S., et al. (1994). Direct drug transport from the rat nasal cavity to the cerebrospinal fluid: the relation to the dissociation of the drug. *J. Pharm. Pharmacol.* 46: 378–379.

Sakane, T., Akizuki, M., Yoshida, M., et al. (1991b). Transport of cephalexin to the cerebrospinal fluid directly from the nasal cavity. *J. Pharm. Pharmacol.* 43: 449–451.

Sakane, T., Yamashita, S., Yata, N., and Sezaki, H. (1999). Transnasal delivery of 5-fluorouracil to the brain in the rat. *J. Drug Target.* 7: 233–240.

Sakuma, S., Suita, M., Masaoka, Y., et al. (2010). Oligoarginine-linked polymers as a new class of penetration enhancers. *J. Control. Release.* 148: 187–196.

Salazar, J., Mena, N., Hunot, S., et al. (2008). Divalent metal transporter 1 (DMT1) contributes to neurodegeneration in animal models of Parkinson's disease. *Proc. Nat. Acad. Sci. USA.* 105: 18578–18583.

Sams, J. M., Jansen, A. S., Mettenleiter, T. C., and Loewy, A. D. (1995). Pseudorabies virus mutants as transneuronal markers. *Brain Res.* 687: 182–190.

Saper, C. B., Wainer, B. H., and German, D. C. (1987). Axonal and transneuronal transport in the transmission of neurological disease: potential role in system degenerations, including Alzheimer's disease. *Neuroscience.* 23: 389–398.

Sarkar, M. A. (1992). Drug metabolism in the nasal mucosa. *Pharm. Res.* 9: 1–9.

Schieferstein, G. J., Sheldon, W. G., Cantrell, S. A., and Reddy, G. (1988). Subchronic toxicity study of 1,4-dithiane in the rat. *Fundam. Appl. Toxicol.* 11: 703–714.

Schiffman, S. S. (1983). Taste and smell in disease (first of two parts). *N. Engl. J. Med.* 308: 1275–1279.

Schnabolk, G. W.,Youngblood, G. L., and Sweet, D. H. (2006). Transport of estrone sulfate by the novel organic anion transporter Oat6 (Slc22a20). *Am. J. Physiol. Renal Physiol.* 291: F314–321.

Schneider, N. G., Olmstead, R., Mody, F. V., et al. (1995). Efficacy of a nicotine nasal spray in smoking cessation: a placebo- controlled, double-blind trial. *Addiction.* 90: 1671–1682.

Schultz, E. W. and Gebhardt, L. P. (1936). Prevention of intranasally inoculated poliomyelitis in monkeys by previous intranasal irrigation with chemical agents. *Proc. Soc. Exp. Biol. Med.* 34: 133–135.

Schultz, W. W. and Gebhardt, L. P. (1937). Zinc sulfate prophylaxis in poliomyelitis. *JAMA.* 108: 2182–2184.

Seju, U., Kumar, A., and Sawant, K. K. (2011). Development and evaluation of olanzapine-loaded PLGA nanoparticles for nose-to-brain delivery: in vitro and in vivo studies. *Acta Biomater.* 7: 4169–4176.

Seki, M. (1941). Vitalfarburng des riechepithels der maus mit trypanblau. *Z. Zellforsch.* 31: 218–223.

Seki, T., Kanbayashi, H., Nagao, T., et al. (2005). Effect of aminated gelatin on the nasal absorption of insulin in rats. *Biol. Pharm. Bull.* 28: 510–514.

Seki, T., Sato, N., Hasegawa, T., et al. (1994). Nasal absorption of zidovudine and its transport to cerebrospinal fluid in rats. *Biol. Pharm. Bull.* 17: 1135–1137.

Sekler, I., Moran, I., Hershfinkel, M., et al. (2002). Distribution of the zinc transporter ZnT-1 in comparison with chelatable zinc in the mouse brain. *J. Comp. Neurol.* 447: 201–209.

Shigematsu, K. and McGeer, P. L. (1992). Accumulation of amyloid precursor protein in damaged neuronal processes and microglia following intracerebral administration of aluminum salts. *Brain Res.* 593: 117–123.

Shimoda, N., Maitani, Y., Machida, Y., and Nagai, T. (1995). Effects of dose, pH and osmolarity on intranasal absorption of recombinant human erythropoietin in rats. *Biol. Pharm. Bull.* 18: 734–739.

Shipley, M. T. (1985). Transport of molecules from nose to brain: Transneuronal anterograde and retrograde labeling in the rat olfactory system by wheat germ agglutinin-horseradish peroxidase applied to the nasal epithelium. *Brain Res. Bull.* 15: 120–142.

Shipley, M. T., Nickell, W. T., Norman, A. B., and Gerfen, C. (1991). Localization of D2 DA receptor mRNA in primary olfactory neurons. *Soc. Neurosci. Abst.* 17: 1091.

Shirley, S., Polak, E., and Dodd, G. (1983). Selective inhibition of rat olfactory receptors by Concanavalin A. *Biochem. Soc. Trans.* 11: 780–781.

Silverman, S. L. (1997). Nasal calcitonin. *Endocrine.* 6: 199–202.

Slot, W. B., Merkus, F. W., Van Deventer, S. J., and Tytgat, G. N. (1997). Normalization of plasma vitamin B12 concentration by intranasal hydroxocobalamin in vitamin B12-deficient patients. *Gastroenterology.* 113: 430–433.

Song, K. H. and Eddington, N. D. (2012). The influence of stabilizer and bioadhesive polymer on the permeation- enhancing effect of AT1002 in the nasal delivery of a paracellular marker. *Arch. Pharm. Res.* 2: 359–366.

Spencer, R. F., Baker, H., and Baker, R. (1982). Evaluation of wheat germ agglutinin immunohistochemistry as a neuroanatomical method for retrograde, anterograde, and anterograde transsynaptic labeling in the cat visual oculomotor systems. *Soc. Neurosci. Abst.* 8: 785.

Srinivas, N. R., Shyu, W. C., Soong, C. W., and Greene, D. (1998). Absolute bioavailability and dose proportionality of BMS-181885, an antimigraine agent, following the administration of single intranasal doses to cynomolgus monkeys. *J. Pharm. Sci.* 87: 1170–1172.

Stewart, W. B. (1985). Labelling of olfactory bulb glomeruli following horseradish peroxidase lavage of the nasal cavity. *Brain Res.* 347: 200–203.

Steyn, D., Plessis, L., and Kotze, A. (2010). Nasal delivery of recombinant human growth hormone: In vivo evaluation with pheroid™ technology and N-trimethyl chitosan chloride. *J. Pharm. Pharmaceut. Sci.* 13: 263–273.

Stone, D. M., Grillo, M., Margolis, F. L., et al. (1991). Differential effect of functional olfactory bulb deafferentation on tyrosine hydroxylase and glutamic acid decarboxylase messenger RNA levels in rodent juxtaglomerular neurons. *J. Comp. Neurol.* 311: 223–233.

References

Stone, D. M., Wessel, T., Joh, T. H., and Baker, H. (1990). Decrease in tyrosine hydroxylase, but not aromatic L-amino acid decarboxylase, messenger RNA in rat olfactory bulb following neonatal, unilateral odor deprivation. *Mol. Brain Res.* 8: 291–300.

Strack, A. M. and Loewy, A. D. (1990). Pseudorabies virus: a highly specific transneuronal cell body marker in the sympathetic nervous system. *J. Neurosci.* 10: 2139–2147.

Stroop, W. G. (1995). Viruses and the olfactory system. In: Doty, R.L., ed. *Handbook of Olfaction and Gustation*. New York: Marcel Dekker. 367–393.

Stroop, W. G., Rock, D. L., and Fraser, N. W. (1984). Localization of herpes simplex virus in the trigeminal and olfactory systems of the mouse central nervous system during acute and latent infections by *in situ* hybridization. *Lab. Invest.* 51: 27–38.

Sullivan, S. L., Ressler, K. J., and Buck, L. B. (1995). Spatial patterning and information coding in the olfactory system. *Curr. Opin. Genet. Dev.* 5: 516–523.

Sun, T.-S., Miller, M. L., and Hastings, L. (1996). Effects of inhalation of cadmium on the rat olfactory system; Behavior and morphology, Neurotoxicol. *Teratol* 18: 89–98.

Suzuki, N. (1984). Anterograde fluorescent labeling of olfactory receptor neurons by Procion and Lucifer dyes. *Brain Res.* 311: 181–185.

Swain, R. (1995). An update of vitamin B12 metabolism and deficiency states. *J. Fam. Pract.* 41: 595–600.

Tabaton, M., Cammarata, S., Mancardi, G. L., et al. (1991). Abnormal tau-reactive filaments in olfactory mucosa in biopsy specimens of patients with probable Alzheimer's disease. *Neurology.* 41: 391–394.

Takala, A., Kaasalainen, V., Seppala, T., et al. (1997). Pharmacokinetic comparison of intravenous and intranasal administration of oxycodone. *Acta Anaesthesiol. Scand.* 41: 309–312.

Talamo, B. R., Rudel, R. A., Kosik, K., et al. (1989). Pathology of olfactory neurons in patients with Alzheimer's disease. *Nature.* 37: 736–739.

Tallkvist, J., Henriksson, J., d'Argy, R., and Tjälve, H. (1998). Transport and subcellular distribution of nickel in the olfactory system of pikes and rats. *Toxicol. Sci.* 43: 196–203.

Tanner, C. M. and Langston, J. W. (1990). Do environmental toxins cause Parkinson's disease? A critical review. *Neurology.* 40: 17–30.

Teijeiro-Osorio, D., Remunan-Lopez, C., and Alonso, M. J. (2009). New generation of hybrid poly/oligosaccharide nanoparticles as carriers for the nasal delivery of macro- molecules. *Biomacromolecules.* 10: 243–249.

Tengamnuay, P., Sahamethapat, A., Sailasuta, A., and Mitra, A. K. (2000). Chitosans as nasal absorption enhancers of peptides: comparison between free amine chitosans and soluble salts. *Int. J. Pharm.* 197: 53–67.

Thompson, K., Molina, R. M., Donaghey, T., et al. (2007). Olfactory uptake of manganese requires DMT1 and is enhanced by anemia. *FASEB J.* 21: 223–230.

Thorne, R. G., Emory, C. R., Ala, T. A., and Frey, W. H., 2nd. (1995). Quantitative analysis of the olfactory pathway for drug delivery to the brain. *Brain Res.* 692: 278–282.

Thornton-Manning, J. R. and Dahl, A. R. (1997). Metabolic capacity of nasal tissue interspecies comparisons of xenobiotic-metabolizing enzymes. *Mutat. Res.* 380: 43–59.

Tian, L., Guo, R., Yue, X., et al. (2012). Intranasal administration of nerve growth factor ameliorate β-amyloid deposition after traumatic brain injury in rats. *Brain Res.* 1440: 47–55.

Tisdall, F. F., Brown, A., and Defries, R. D. (1938). Persistent anosmia following zinc sulfate nasal spraying. *J. Pediatr.* 13: 60–62.

Tisdall, F. F., Brown, A., Defries, R. D., et al. (1937). Nasal spraying as preventitive of poliomylitis. *Can. Public Health J.* 28: 431–434.

Tjälve, H. and Henriksson, J. (1999). Uptake of metals in the brain via olfactory pathways. *Neurotoxicology.* 20: 181–195.

Tjälve, H., Henriksson, J., Tallkvist, J., et al. (1996). Uptake of manganese and cadmium from the nasal mucosa into the central nervous system via olfactory pathways in rats. *Pharmacol. Toxicol.* 79: 347–356.

Tolis, G., Faure, N., Koutsilieris, et al. (1983). Suppression of testicular steroidogenesis by the GnRH agonistic analogue Buserelin (HOE-766) in patients with prostatic cancer: studies in relation to dose and route of administration. *J Steroid Biochem.* 19: 995–998.

Turk, M. A., Flory, W., and Henk, W. G. (1986). Chemical modulation of 3-methylindole toxicosis in mice: effect on bronchiolar and olfactory mucosal injury. *Vet. Pathol.* 23: 563–570.

Uchenna A. R., Jorissen, M., Willems, T., et al. (2000). Safety assessment of selected cyclodextrins - effect on ciliary activity using a human cell suspension culture model exhibiting in vitro ciliogenesis. *Int. J. Pharm.* 193: 219–226.

Ugolini, G. (1995). Specificity of rabies virus as a transneuronal tracer of motor networks: transfer from hypoglossal motoneurons to connected second- order and higher order central nervous system cell groups. *J. Comp. Neurol.* 356: 457–480.

U S Environmental Protection Agency. (1985). *Special Review of certain pesticide products. Alachlor: Position document 1.* Springfield, VA: Office of Pesticide Programs.

Vaka, S. R. and Murthy, S. N. (2010). Enhancement of nose-brain delivery of therapeutic agents for treating neurodegenerative diseases using peppermint oil. *Pharmazie,* 65: 690–692.

Valensi, P., Zirinis, P., Nicolas, P., et al. (1996). Effect of insulin concentration on bioavailability during nasal spray administration. *Pathol. Biol. (Paris).* 44: 235–240.

Vassar, R., Chao, S. K., Sitcheran, R., et al. (1994). Topographic organization of sensory projections to the olfactory bulb. *Cell.* 79: 981–991.

Wallace, S. J. (1997). Nasal benzodiazepines for management of acute childhood seizures? *Lancet.* 349: 222.

Wang, C. X., Huang, L. S., Hou, L. B., et al. (2009). Antitumor effects of polysorbate-80 coated gemcitabine polybutylcyanoacrylate nanoparticles in vitro and its pharmacodynamics in vivo on C6 glioma cells of a brain tumor model. *Brain Res.* 1261: 91–99.

Wang, J., Lu, W. L., Liang, G. W., et al. (2006). Pharmacokinetics, toxicity of nasal cilia and immunomodulating effects in Sprague–Dawley rats following intranasal delivery of thymopentin with or without absorption enhancers. *Peptides.* 27: 826–835.

Wang, M., Lu, C., and Roisen, F. (2012). Adult human olfactory epithelial-derived progenitors: a potential autologous source for cell-based treatment for Parkinson's disease. *Stem Cells Transl. Med.* 1: 492–502. doi: 10.5966/sctm.2012-0012.

Wang, X., Zheng, C., Wu, Z. M., et al. (2009b). Chitosan–NAC nanoparticles as a vehicle for nasal absorption enhancement of insulin. *J. Biomed. Mater. Res. B, Appl. Biomater.* 88B: 150–161.

Wang, Y., Wang, B., Zhu, M. T., et al. (2011). Microglial activation, recruitment and phagocytosis as linked phenomena in ferric oxide nanoparticle exposure. *Toxicol. Lett.* 205: 26–37.

Watanabe, K. and Kawana, E. (1984). Selective retrograde transport of tritiated D-aspartate from the olfactory bulb to the anterior olfactory nucleus, pyriform cortex and nucleus of the lateral olfactory tract in the rat. *Brain Res.* 296: 148–151.

Weiss, D. G, and Buchner, K. (1988). Axoplasmic transport in olfactory receptor neurons. In: Margolis, FL and Getchell, TV, eds. *Molecular Neurobiology of the Olfactory System*. New York: Plenum Press. 217–236.

Weiss, D. G., Krygier-Brevart, V., Gross, G. W., and Kreutzberg, G. W. (1978). Rapid axoplasmic transport in the olfactory nerve of the pike: II. Analysis of transported proteins by SDS gel electrophoresis. *Brain Res.* 139: 77–87.

Weiss, P. and Holland, Y. (1967). Neuronal dynamics and axonal flow in the olfactory nerve as model test object. *Proc. Natl. Acad. Sci. USA.* 57: 258–264.

Wen, Z., Yan, Z., Hu, K., et al. (2011). Odorranalectin-conjugated nanoparticles: preparation, brain delivery and pharmacodynamic study on Parkinson's disease following intranasal administration. *J. Control. Release.* 151: 131–138.

Wensley, C. H., Stone, D. M., Baker, H., et al. (1995). Olfactory marker protein mRNA is found in axons of olfactory receptor neurons. *J. Neurosci.* 15: 4827–4837.

Wiley, R., Baker, H., and Baker, R. (1984). Concavalin A-horseradish peroxidase (CON A-HRP) conjugate is a useful neuroanatomical tracer. *Soc. Neurosci. Abst.* 10: 420.

Wolf, D. A., Hanson, L. R., Aronovich, E. L., et al. (2012). Lysosomal enzyme can bypass the blood–brain barrier and reach the CNS following intranasal administration. *Mol. Genet. Metab.* 106: 131–134.

Won, S. J., Choi, B. Y., Yoo, B. H., et al. (2012). Prevention of traumatic brain injury-induced neuron death by intranasal delivery of nicotinamide adenine dinucleotide. *J. Neurotrauma.* 29: 1401–1409.

Wu, J., Wang, C., Sun, J., and Xue, Y. (2011). Neurotoxicity of silica nanoparticles: brain localization and dopaminergic neurons damage pathways. *ACS Nano.* 5: 4476–4489.

Yajima, T., Juni, K,. Saneyoshi. M., et al. (1998). Direct transport of 2′,3′-didehydro-3′-deoxythymidine (D4T) and its ester derivatives to the cerebrospinal fluid via the nasal mucous membrane in rats. *Biol. Pharm. Bull.* 21: 272–277.

Yasui, M., Kihira, T., and Ota, K. (1992). Calcium, magnesium and aluminum concentrations in Parkinson's disease. *Neurotoxicology.* 13: 593–600.

Yokota, S., Takashima, H., Ohta, R., et al. (2011). Nasal instillation of nanoparticle-rich diesel exhaust particles slightly affects emotional behavior and learning capability in rats. *J. Toxicol. Sci.* 36: 267–276.

Youdim, M. B., Ben-Shachar, D., and Riederer, P. (1993). The possible role of iron in the etiopathology of Parkinson's disease. *Mov. Disord.* 8: 1–12.

Zatta, P., Favarato, M., and Nicolini, M. (1993). Deposition of aluminum in brain tissues of rats exposed to inhalation of aluminum acetylacetonate. *Neuroreport.* 4: 1119–1122.

Zhang, L., Bai, R., Li, B., et al. (2011). Rutile TiO particles exert size and surface coating dependent retention and lesions on the murine brain. *Toxicol. Lett.* 207: 73–81.

Zhang, X. G., Zhang, H. J., Wu, Z. M., (2008). Nasal absorption enhancement of insulin using PEG-grafted chitosan nanoparticles. *Eur. J. Pharm. Biopharm.* 2008; 68: 526–534.

Zhang, Y. J., Ma, C. H., Lu, W. L., et al. (2005). Permeation-enhancing effects of chitosan formulations on recombinant hirudin-2 by nasal delivery *in vitro* and *in vivo*. *Acta Pharmacologica Sinica.* 26: 1402–1408.

Zhao, H., Otaki, J. M., and Firestein S. (1996). Adenovirus-mediated gene transfer in olfactory neurons in vivo. *J. Neurobiol.* 30: 521–530.

Zhao, Y., Brown, M. B., Khengar, R. H., et al. (2012). Pharmacokinetic evaluation of intranasally administered vinyl polymer-coated lorazepam microparticles in rabbits. *The AAPS Journal.* 14: 218–224.

Ziegler, D., Ford, R., and Kriegler, J., et al. (1994). Dihydroergotamine nasal spray for the acute treatment of migraine. *Neurology.* 44: 447–453.

Zupko, K., Poria, Y., and Lancet, D. (1991). Immunolocalization of cytochromes P-450olf1 and P-450olf2 in rat olfactory mucosa. *Eur. J. Biochem.* 196: 51–58.

Chapter 20

Influence of Toxins on Olfactory Function and their Potential Association with Neurodegenerative Disease

LILIAN CALDERÓN-GARCIDUEÑAS

20.1 INTRODUCTION

Olfactory impairment significantly affects many facets of well-being, including safety, quality of life, and social behavior. As noted throughout this volume, smell dysfunction is common in a significant proportion of the population and can reflect numerous factors, including epithelial damage or airway blockage from local nasal disease, neurotransmitter alterations, metabolic derangements, exposures to environmental hazards and toxins, brain lesions, trauma, tumors, neurodegeneration, and occlusion of the foramina of the cribriform plate through which the olfactory receptor cell axons pass from the nose into the brain. In practice, the three most common causes of smell disorders are upper respiratory infections, sinonasal diseases, and head trauma (Deems et al., 1991). Although up to half of patients presenting with chemosensory dysfunction exhibit some return of smell function over time, less than 12% of those with anosmia return to normal age-related function. Prognosis is better for those with some initial function, with over 20% regaining normal age-related function (London et al., 2008).

Of importance to this chapter is the fact that large segments of the population are exposed to a myriad of toxic substances on a daily basis that have the potential for harming the olfactory system and, in some cases,

penetrating the brain via the olfactory epithelium (OE) (Tjälve et al., 1996, 1999; Dorman et al., 2006; Wang et al., 2011). Extreme instances of such exposures include the massive dust cloud following the September 11th, 2001, terrorist attack in New York City, smoke and debris from wildfires, dust and particulate matter encountered in desert storms, exposures to airborne herbicides and pesticides in farming communities, and pollutants from vehicle exhaust and manufacturing enterprises in major metropolitan areas. Nearly all people in industrialized societies are exposed, on a cumulative basis, to such airborne agents.

The amount, that is, dose, of xenobiotics that reaches the nasal passages depends upon their concentration in air, the volume of the inhaled air, the chemical nature of the xenobiotic (e.g., diffusivity, solubility), airflow characteristics of the nose, and the absorption and metabolic properties of the nasal lining (Kimbell et al., 2006). The inhaled air traverses the complex and highly variable nasal passages, that is, the folds and protrusions of the nasal walls and the inferior, middle, and superior turbinates, where its contents are largely deposited (Chapter 2). The volume of inhaled air is variable and depends on such subject factors as age, gender, weight, height, metabolic activity, and health (Becquemin et al., 1991; Vearrier and Greenberg, 2011). The integrity of the nasal and olfactory epithelia can further predispose them to damage and, as discussed later in this chapter, a

---

Handbook of Olfaction and Gustation, Third Edition. Edited by Richard L. Doty.
© 2015 Richard L. Doty. Published 2015 by John Wiley & Sons, Inc.

disrupted or "sick" epithelium may increase the likelihood of xenobiotics penetrating the brain.

20.2 THE OLFACTORY EPITHELIUM: A KEY TARGET FOR DAMAGE FROM TOXIC XENOBIOTICS

The olfactory epithelium (OE), which is described in detail in Chapter 4, is particularly susceptible to damage from environmental agents given its rather direct exposure to the inhaled air and associated outside environment. Thus, cellular damage to this neuroepithelium, including its stem cells near the basement membrane, has been repeatedly documented following exposure to various environmental pollutants, toxic gases, nanoparticles, and even chemotherapeutic agents (Féron et al., 2001; Steinbach et al., 2009; Faure et al., 2010; Chen et al., 2012; de Gabory et al., 2011; Inthavong et al., 2011; Holbrook et al., 2011; Altman et al., 2011; González-Pérez et al., 2011; Girard et al., 2011; Prediger et al., 2012; Nazarenko et al., 2012).

Computer models of airflow during normal respiration predict that approximately 3% of the inhaled air reaches the OE of most adults (Kimbell et al., 2006; see Chapter 16). Sniffing dramatically increases this percentage. While children inhale less air than adults, their nasal surface is smaller, resulting in essentially the same degree of epithelial exposure. However, their noses are not as efficient filters as those of adults, even though both groups exhibit similar patterns of nasal vapor uptake (Becquemin et al., 1991; García et al., 2009). For some toxic agents (e.g., formaldehyde), most nasal absorption occurs in the mid-septal area. For others (e.g., ozone), the major impact occurs elsewhere. Importantly, the toxic effects do not affect the same anatomical regions of each individual.

Nanoparticles are very effective in reaching the OE (Oberdorster et al., 2004). Deposition of particulate matter (PM) depends upon a number of factors, including access to the nasal region, speed of particle diffusion from the air stream, and particle size (see Chapter 16). Even slight geometric differences in the nasal passages can have significant effects on nasal deposition (Schroeter et al., 2011a), a factor to take into account when comparing nasal exposures among individuals and when attempting to extrapolate animal data to humans. Human nasal computational fluid dynamic models using size particles in the range of 5–50 μm and volumetric flow rates of 7.5, 15, and 30 L/min have found a 3% maximum PM deposition efficiency in the olfactory region (Schroeter et al., 2006). For micrometer-sized PM, inertia is a key factor influencing deposition, while for nano size PM (<100 nm) Brownian diffusion is a critical factor (García and Kimbell, 2009). Estimates of nanoparticle deposition

in the rat olfactory region based on anatomically accurate fluid dynamic models of the nasal passages have shown maximal olfactory deposition of 6–9% with a size particle from 3 to 4 nm (García and Kimbell, 2009). PM <3 nm deposits predominantly in the anterior nose, while PM >30 nm distributes more uniformly throughout the nasal passages.

To protect against cellular damage, as well as the entrance of some xenobiotics into the brain via the nose, the OE is equipped with Phase I and Phase II metabolizing enzymes and Phase III transporters (see Chapter 3). The Phase I mixed-function oxidases include members of the cytochrome P450 family, aldehyde dehydrogenases (ALDHs), and epoxide hydrolases (EPHXs). Among the Phase II enzymes found in the OE are UDP-glucoronosyl-transferases (UGTs), sulfotransferases (SULTs), and glutathione-S-transferases (GSTs) (Su et al., 1996; Ding and Dahl, 2003; Su and Ding, 2004; Genter et al., 2009, 2010; Heydel et al., 2010; Thiebaud et al., 2011). Transport proteins that facilitate Phase III xenobiotic metabolism, thus allowing conjugates and their metabolites to be excreted from cells, play a critical role in the OE and olfactory bulb (OB). Metal transporters, multidrug resistance P-glycoproteins (MDR), multidrug resistance-related proteins (MRP), and organic anion and cation transporters are the most transcriptionally expressed transporters in the OE (Genter et al., 2010; Thiebaud et al., 2011). Many of the detoxification proteins are located at the apical part of the olfactory mucosa, olfactory cilia, and in sustentacular and Bowman cells. Given that excreted metabolites of odorants and other xenobiotics are disposed of within the nasal mucus (a further elimination step), a significant change in the nasal mucus amount and alterations under pathological circumstances (e.g., squamous metaplastic changes) can serve to decrease the mucus concentration and, hence, maximize the impact of the deposited xenobiotics.

20.3 MOVEMENT OF XENOBIOTICS INTO THE BRAIN AND GENERAL CIRCULATION FROM THE NOSE

It has been known for nearly a hundred years that viruses and microorganisms can be transported via olfactory neurons to the OB glomeruli, where movement can occur across synapses to dendrites of second order neurons (Doty, 2008). Among the documented mechanisms by which xenobiotics enter the brain via the OE are paracellular transport, axonal movement, and uptake through specific transporters (Thiebaud et al., 2011). Metal transporters, such as those with divalent cation transport function, play a critical role in cases where exposures to metals are prevalent, as in polluted urban environments and in certain occupational settings (Lewis et al., 2005; Genter et al., 2010; Prediger et al., 2012).

Additional uptake of xenobiotics into the brain can occur through the nasal vasculature. The nasal cavity is lined with capillaries from the ethmoid branches of the internal carotid arteries and the facial and internal maxillary divisions of the external carotid arteries. Such capillaries serve a primary role in the transport of substances from the nose to the systemic circulation. From there, such transport can occur to all sectors of the body (Turker et al., 2004). In fact, nasal drug delivery of hormones and nano-size particles have been found to be an effective means for accessing both the circulatory system and the brain (Ali et al., 2010; Craft et al., 2012). The long-term effects of such delivery on nasal (e.g., ciliary transport) and olfactory (e.g., sensitivity) function, however, has not been thoroughly assessed.

It has only recently been appreciated that damage to the OE in a prion host can result in a substantial increase in the release of prion infectivity into the nasal mucosa. Such prion shedding may, in fact, further infect intranasal neural pathways (Bessen et al., 2012). In a hamster prion model, infected brain homogenates are transported paracellularly across the respiratory, olfactory and follicle associated epithelia of the nasal cavity. Prions cross the nasal mucosa via multiple routes and quickly enter lymphatics where they can spread systemically via lymph draining the nasal cavity (Kincaid et al., 2012). Axonal pathways from the nasal region to the brain include the trigeminal nasal branches (V1 ophthalmic and V2 maxillary branches) and their transganglionic projections to the trigeminal brainstem nuclear complex (Anton and Peppel, 1991; Allavena et al., 2011). The drainage of lymphatic nasal turbinate channels to the CSF through the perineural olfactory nerve space (Zakharov et al., 2004; Walter et al., 2006; Koh et al., 2006) also provides a further connection between the nose and the brain.

The second order olfactory system neurons, namely the mitral and tufted cells, project to allocortical regions, most notably the anterior olfactory nucleus and the piriform and entorhinal cortices, as well as to sectors of the amygdala (Kalinke et al., 2011). The observation that the piriform and entorhinal cortices share with the OE and the OB the presence of specific metabolizing enzymes, that is, UGTs (Heydel et al., 2010), is particularly important and may be associated, in part, with the olfactory deficits observed in some neurodegenerative diseases. Subjects with mutations in the P450 cytochrome CYP2D6-debrisoquine hydroxylase gene have an increased risk of developing Parkinson's disease (PD), in accord with hypotheses that impairment in local metabolism can aid xenobiotic toxicity, facilitate penetration of toxins into the brain via the nasal/olfactory pathway, and produce neuronal damage in the brain beyond

the initial nasal route (Iscan et al., 1990; Ding et al., 1992; Xie et al, 2010).

20.4 CRITICAL NEURONAL RESERVE AREA IN THE SUBVENTRICULAR ZONE AND ITS IMPACT ON OLFACTION

The neuronal reserve area of the subventricular zone (SVZ) is critical when discussing the influence of toxins and neurodegenerative diseases on olfactory function and pathology. Some toxins may damage this reserve early in life and alter its capacity to generate neuroblasts which, in turn, are responsible for development of new OB neurons, such as periglomerular and granule cells (Conover and Shook, 2011). Although the human SVZ is small, recent studies have shown there are clusters of heterogeneous cells along the anterior horn and body of the ventricular surface. Critical to this chapter, a ventral extension of the anterior horn has the potential to generate neuroblasts capable of reaching the OB along the rostral migratory stream (Curtis et al., 2007; Guerrero-Cázares et al., 2011). The rostral migratory stream itself plays a key role in intranasal delivery of drugs into the CNS and is a potential pathway for the access of hazardous materials and organisms into the brain (Scranton et al., 2011).

20.5 IMMUNOLOGICAL RESPONSES TO XENOBIOTICS WITHIN THE OLFACTORY SYSTEM AND THEIR LIKELY ROLE IN DAMAGING THE SYSTEM

The immune system plays a critical role in thwarting the penetration of unwelcome organisms or xenobiotics into the brain via the OE. An excellent model of immunological responses that occur in the olfactory system is the system's responses to neurotropic viruses reaching the nose (Huneycutt et al., 1993; Reiss et al., 1998; Detje et al., 2009; Kalinke et al., 2011; Harberts et al., 2011). Intranasal vesicular stomatitis virus (VSV) rapidly spreads throughout the olfactory system of mice whose functional interferon (IFN) system has been genetically compromised. In contrast, in wild-type mice of the same strain the virus stops at the interface between glomerular structures and periglomerular cells (Detje et al., 2009; Kalinke et al., 2011). These observations are relevant to humans and suggest the OB offers an effective barrier to restrain the passage of microorganisms through a Type I IFN pathway (Kalinke et al., 2011).

The fact that viruses cross into and stay within the olfactory bulb for long periods of time has been shown in human autopsy specimens. For example, human herpes virus 6 (HHV-6) exhibits a high prevalence in OBs and the nasal cavity, as assessed in human cadavers (Harberts et al., 2011). HHV-6 is capable of replicating in olfactory-ensheathing cells *in vitro*. Of utmost importance is that olfactory-ensheathing cells guide the connections between the OE and the OB and have relevant neuroregenerative properties (Chapter 4). In consequence, any pathology affecting olfactory-ensheathing cells could potentially alter their critical functions.

The bridge between the OE and the immunologic system is shown by the dual presence of the TAC4 gene that encodes a tachykinin related to substance P in both the OE and in B cells (Kurtz et al., 2002; Tran et al., 2011). The inhibitory role on developing B cells and the role of TLR4 in inducing differentiation to mature B cells would be relevant in human pathological processes involving upregulation of TAC4 in the olfactory region. Moreover, inflammatory mediators and infiltration of inflammatory cells can contribute to olfactory loss. In the setting of inflammation, a significant number of cytokines and chemokines participate in cell proliferation, survival, migration, and differentiation of neural stem cells, including ones that repopulate the olfactory bulbs. Thus, an imbalance of inflammatory mediators can have an impact on neurogenic niches (González-Pérez et al., 2010; Sultán et al., 2011).

20.6 NEUROGENIC INFLAMMATION AND THE IMPACT OF AUTONOMIC MODULATION ON THE OLFACTORY EPITHELIUM

The sensory, sympathetic, and parasympathetic nerves within the olfactory neuroepithelium are highly responsive to environmental and endogenous stimuli (Sarin et al., 2006). Inflammatory processes trigger neural hyperresponsiveness with the release of neuropeptides (i.e., tachykinins) by sensory nerve endings and their resultant vasodilatation, leukocyte recruitment, and immune cell activation in a process known as neurogenic inflammation (Baraniuk and Kaliner, 1990; Sanico et al., 1997; Sarin et al., 2006; Jornot et al., 2008). Nerve activity is transmitted to the CNS and can further travel to secondary synapses to activate efferent motor and autonomic neurons (Figure 20.1) (Sarin et al., 2006).

The aforementioned scenario is important in pathological states such as allergic rhinitis and is potentially relevant to the nasal (including olfactory) responses to xenobiotics. Vascular dilation – an intrinsic part of neurogenic inflammation – may aid in facilitating the transport of toxic substances into the vascular network. The sensory nerves can act as carriers for the transport of nanoparticles to the sympathetic, parasympathetic, and trigeminal ganglia. Further, inflammatory nasal responses, including

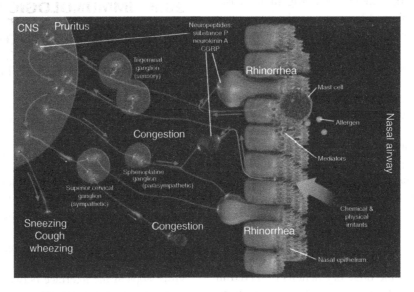

Figure 20.1 Neural pathways and nasal symptoms. Sensory nerves can be stimulated by products of allergic reactions and by external physical and chemical irritants. Action potentials traveling through parasympathetic efferent nerves can lead to glandular activation and rhinorrhea, as well as to vasodilation. Suppression of sympathetic neural output results in vasodilation and nasal congestion. Antidromic stimulation of sensory nerves with release of tachykinins and other neuropeptides at the nasal mucosa contributes to symptom development with glandular activation, vasodilation, and plasma extravasation. Neuropeptide release can also lead to leukocyte recruitment and activation. Collectively, events generated by the antidromic stimulation of sensory nerves constitute the phenomenon of "neurogenic inflammation". From Sarin et al. (2006). Reprinted with permission. (*See plate section for color version.*)

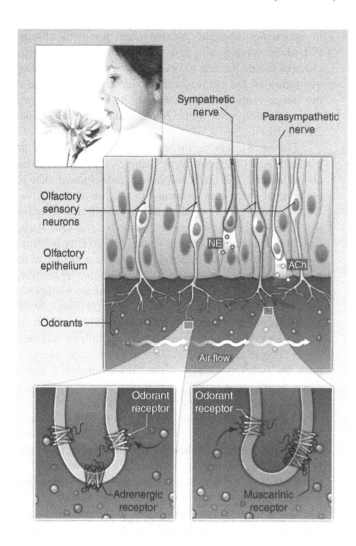

Figure 20.2 Autonomic nervous system modulation of the earliest steps of olfactory signaling. Norepinephrine (NE) released by sympathetic nerves and acetylcholine (ACh) released by parasympathetic nerves both can modulate the responses of olfactory sensory neurons to odorants. Stimulation of adrenergic receptors, in either the cilia or cell body, results in downstream changes in ion channel activity and olfactory sensory neuron output, whereas muscarinic acetylcholine receptors can form complexes with odorant receptors in the olfactory cilia to directly control odorant receptor activity. From Hall (2011). Reprinted with permission. (*See plate section for color version.*)

the production of inflammatory cytokines, can affect the OE with an enhanced sensory neuropeptide release, a phenomenon that has been described following exposure to asphalt fumes in a study by Sikora et al. (2003). These investigators found that the activation of sensory C-fibers in the nasal cavity was followed by increases in the levels of substance P and calcitonin-gene-related peptide in trigeminal ganglia neurons innervating the nasal mucosa.

The importance of the autonomic system is evident in the modulation of the responses of olfactory sensory neurons to odorants and in diseases such as Parkinson's (PD), where autonomic imbalance and olfactory deficits precede the motor symptoms by years (Bohnen et al., 2008, 2010; Dickson et al., 2009a, b; Hall, 2011; Doty, 1988a, b, 2012a, b). A recent review describes the extensive innervation of the OE by sympathetic and parasympathetic nerve endings that release norepinephrine and acetylcholine, respectively, which can subsequently modulate the responses of the olfactory sensory neurons to odorants (Figure 20.2) (Hall,

2011). It is conceivable that the olfactory deficits observed in the early pre-motor stages of PD (Braak stages I and II) reflect disturbances in the autonomic nervous system's innervation of the olfactory system. From the toxicological point of view, any drug, chemical agent, or pathological process which produces a dysbalance and/or directly damages the autonomic system could result in olfactory dysregulation (Anderson et al., 2012; Balali-Mood et al., 2011; Hall, 2011; Liu et al., 2011; Doty, 2008, 2012b).

20.7 AIR POLLUTION, OLFACTION, AND NEURODEGENERATION

20.7.1 Urban Air Pollution and Olfaction

Air pollution is a complex mixture of particulate matter (PM), gases, and both organic and inorganic compounds that are present in outdoor and indoor air. Urban outdoor

490   **Chapter 20   Influence of Toxins on Olfactory Function and Neurodegeneration**

pollution is a well-documented global health problem (Molina et al., 2007). The role of indoor air pollution (e.g., cooking with biomass) is widely recognized to adversely affect the health of millions of people, particularly in underdeveloped countries. Both short and long-term exposure to air pollutants produce respiratory and cardiovascular morbidity and mortality (Brook et al., 2010; Coneus and Spiess, 2011; Rückerl et al., 2011; Anderson et al., 2012). It is now evident that the CNS is also a target of air pollutants (Calderón-Garcidueñas et al., 2004, 2008a, b; 2011a, b; Block and Calderón-Garcidueñas, 2009; Levesque et al., 2011; Guxens and Sunyer, 2012).

PM is broadly defined by the diameter of the aerodynamic particles, being classified into coarse particles (<10 μm; PM_{10}), fine particles (<2.5 μm; $PM_{2.5}$), and ultrafine particles (<100 nm; UFPM). Fine and ultrafine PM are of particular interest given their capability to reach the brain. Their major sources in the environment are fires, industrial processes, metals, biological materials, and the combustion of gas, oil, and coal (Brook et al., 2010). The smaller the particle, the greater its penetration, diffusion, and deposition into the respiratory tract and its direct translocation into the brain. Man-made nanoparticles <100 nm are a significant threat given their increased availability. It is not widely appreciated that nano-sized materials are present in many consumer products to which large segments of the population are exposed (e.g., toothpastes, cosmetics, sunscreens, food additives, and laser printer emissions) (Tang et al., 2011; Fröhlich and Roblegg, 2012).

Chronic inflammation is the link between air pollution and brain damage, including damage to the olfactory pathway. Such inflammation can involve the upper and lower respiratory tracts and can induce the production of inflammatory mediators capable of reaching the brain. Continuous expression of inflammatory mediators in the CNS promotes the formation of reactive oxygen species (ROS). The uptake of ultrafine particulate matter associated with lipopolysaccharides (PM-LPS) and metals can take place through olfactory neurons, cranial nerves such as the trigeminal and vagus, the systemic circulation, and macrophages loaded with PM from the lungs (Calderón-Garcidueñas, 2002, 2003a, b, 2004, 2008a, b, 2010, 2011a, b, 2012a, b). Activation of innate immune responses within the brain may follow the interactions between circulating cytokines and the constitutively expressed cytokine receptors of brain endothelial cells. Such responses may, in turn, be followed by activation of cells involved in adaptive immunity (Simart and Rivest, 2006; Calderón-Garcidueñas, 2012a). Monocytes are the main innate immune response mediator cells, producing and secreting TNF-α, interleukin-6 (IL-6), and IL-1β, which in turn recruit and increase the activity of other immune cells. Thus, in the sustained respiratory tract chronic inflammatory process elicited by air pollutants

in megacities such as Mexico City, fine and ultrafine PM likely serve as the first falling dominos, initiating a chain of events leading to brain endothelial cell activation, disruption of the neurovascular unit, altered response of the innate immune system, neuroinflammation, and neurodegeneration (Neuwelt, 2004; Calderón-Garcidueñas, 2012 a, b).

The nasal epithelia, including the olfactory region, are prime targets of exposures to complex mixtures of urban air pollutants. The nasal respiratory pathological changes are critical since they prevent the respiratory epithelium from executing its normal protective functions, including the formation of a protective barrier for the OE. Pathological nasal changes seen in some Mexico City urbanites only a few months after arrival include severe loss of the normal ciliated pseudostratified respiratory epithelium, prominent basal cell hyperplasia, squamous metaplasia, and submucosal vascular proliferation (Figure 20.3a). Such effects are evident even in children (Figure 20.3b and 20.3d) and dogs, where progressive pathology has been noted (Figure 20.4). The end result is the easy access of particles and toxins through the broken epithelial barrier and a compromised OE.

The severe pathological changes in the nasal respiratory epithelium go hand and hand with a marked decrease in olfactory neurons, significant changes in Bowman's glands, and pathologic changes within the olfactory bulbs (OBs). In one study, the OBs of 35 young Mexico City residents and 9 controls (average ages ~20 years) from a minimally polluted control city were examined. The Mexico City residents exhibited ill-defined and fragmented organization of the olfactory bulb layers, including small acellular glomeruli (Figure 20.5b, c). In contrast, the basic laminar OB organization of the glomerular, external plexiform, mitral cell, internal plexiform, and granular cell layers of the controls was generally intact (Figure 20.5a). Mexico City residents, but not controls, exhibited neuronal periglomerular accumulation of particulate material (Figure 20.6a) and immunoreactivity to alpha-synuclein, amyloid beta $A\beta_{42}$, and hyperphosphorilated tau (Figure 20.6b–d). Interestingly, the disruption of the olfactory bulb architecture was more severe in the Mexico City APOE 4 allele carriers than in the non-carriers (Figures 20.5d and e).

The finding of PM in the glomerular OB region of the Mexico City residents suggests that air pollution-related particles are readily transported from the nasal cavity into the brain via the olfactory nerves. It seems plausible that PM is largely responsible for the accumulation of Aβ and α synuclein within their olfactory bulbs and other olfaction-related brain structures (Cedervall et al., 2007; Dell'Orco et al., 2010). Interactions between PM and normally soluble Aβ, tau, and α synuclein, conceivably changes their conformation and, thus, their function.

20.7 Air Pollution, Olfaction, and Neurodegeneration

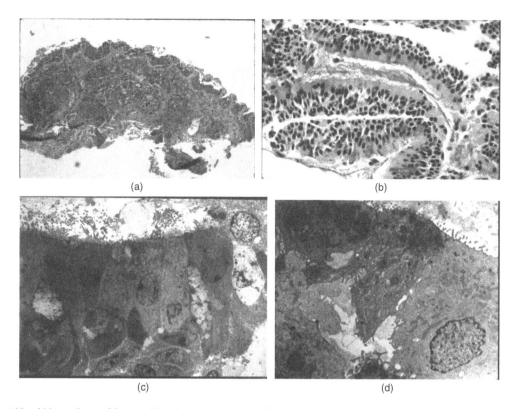

Figure 20.3 (a) Nasal biopsy from a 30-year-old male who had been residing in MC for 3 months. The normal pseudostratified ciliated columnar epithelium has been replaced by a squamous metaplastic epithelium sitting on an abnormally thick basement membrane and accompanied by an extensive submucosal chronic inflammatory infiltrate. This subject has lost all normal nasal mucosa properties, i.e., secretion of mucus, cilia movement, the capacity to warm or cool the passing air within one degree of his body temperature, the inhaled air is no longer humidified, and a strong mucosal barrier is no longer present. (b) This biopsy comes from an 11-year-old Mexico City boy. A moderate basal cell hyperplasia is present, along with a reduction in the number of cilia. (c) An electron micrograph of a child resident in a low polluted city shows an intact nasal epithelium with abundant intact cilia. (d) In sharp contrast, this Mexico City boy shows an abnormal epithelium with wide intercellular spaces (that are synonymous with a severe breakdown of the epithelial barrier) and a severe reduction in the number of cilia. (*See plate section for color version.*)

Alpha-synuclein – an abundant brain 140 residue protein – is the main culprit in PD, while hyperphosphorilated tau is a key feature of both PD and Alzheimer's disease (AD) (Jellinger, 2012; Braak et al., 2011a, b). The scenario of neuron-to-neuron transmission and trans-synaptic transport of tau protein aggregates, Aβ and other proteins, has been contemplated as a major pathogenic mechanism in AD and PD. These mechanisms could be playing a critical role in the OB pathology of exposed urbanites (Jellinger, 2012; Doty, 2012b; Prediger et al., 2012).

It is important to note that, in Mexico City residents, we have observed elevation of indices of neuroinflammation and oxidative stress, AD and PD-associated pathology, and down-regulation of the PrP(C) not only in their olfactory bulbs, but also in their frontal cortices. The inducible regulation of critical genes suggests they are evolving different mechanisms in an attempt to cope with the constant state of inflammation and oxidative stress related to their environmental exposures. The down-regulation of the PrP(C) is key given its important roles for neuroprotection, neurodegeneration, and mood disorder states.

Our data and those of others suggest that exposure to PM can activate pathogen sensors, and that signaling by ROS can drive inflammatory processes (e.g., Martinon, 2010). Asbestos and silica activate the NALP 3 inflammasome and NALP3 deficient mice have a significant reduction of their lung inflammatory responses (Dostert et al., 2008). The innate immune system rapidly detects invading pathogenic microbes and eliminates them. It is biologically plausible that PM with lipopolysaccharides reaching the brain initiates an inflammatory response, resulting in detrimental effects. Toll-like receptors sense "extracellular microbes" (e.g., PM-LPS) and trigger anti-pathogen signaling cascades (Martinon, 2010). Both LPS responses and systemic inflammation are important for the understanding of how the sensing of "microbial invaders" could translate into signaling pathways that culminate in the transcriptional regulation of immune responsive genes and how the activation of inflammasomes could be a contributing factor

492 Chapter 20 Influence of Toxins on Olfactory Function and Neurodegeneration

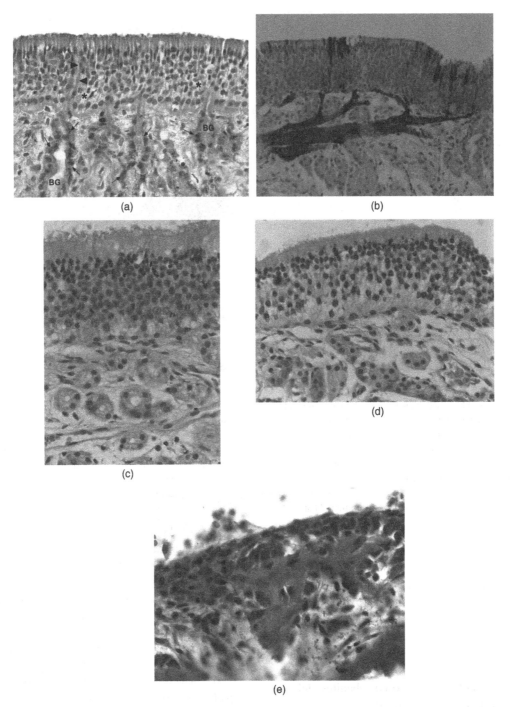

Figure 20.4 (a) The olfactory epithelium of a 6-month-old Mexico City dog shows the presence of particulate material in the cytoplasm of numerous Bowman's glands (arrows), while particles are also seen in the glands' ducts (arrow heads). In addition, there are scattered areas where cellular elements are missing (*), including neurons and sustentacular cells. (b) The olfactory epithelium of a 5-month-old Mexico City dog exhibits a marked expression of metallothionein I-II in sensory neurons and sustentacular cells. The olfactory axons in the lamina propia are also strongly stained. Sequestration of heavy metals by metallothioneins is an important olfactory defense mechanism employed to minimize adverse effects caused by heavy metal contaminants in polluted air (Tallkvist et al., 2002). (c) The olfactory epithelium from a 3-year-old dog living in Mexico City. Note the moderate decrease in thickness and several areas devoid of viable cells, both neurons and sustentacular cells. The Bowman glands are smaller than those of the 6-month-old dog denoted in Figure 4a. (d) The olfactory epithelium of a 12-year-old Mexico City dog showing a severe decrease in thickness and extensive areas devoid of viable cells, both neurons and sustentacular cells. This dog likely had significant olfaction deficits. (e) Same dog as in D, showing an abnormal nasal respiratory epithelium with an area of metaplastic bone replacing the normal ciliated pseudostratified epithelium. (*See plate section for color version.*)

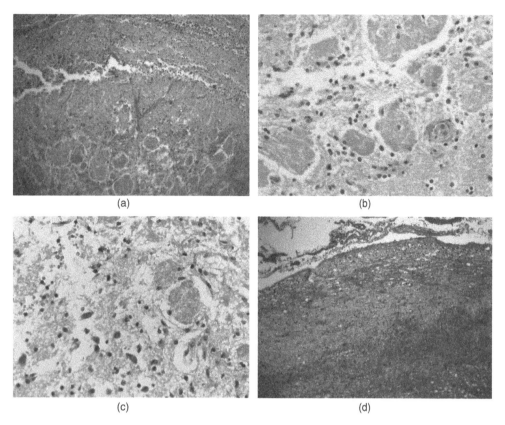

Figure 20.5 (a) Olfactory bulb from a 21-year-old control male. The laminar organization is intact and there are abundant glomeruli. (b) Olfactory bulb from a 19-year-old Mexico City woman showing small glomeruli and arteriolar vessels with thick walls. (c) Olfactory bulb from a 26-year-old Mexico City woman with few remaining acellular glomeruli, scarce periglomerular cells, and an extensive fragmentation of periglomerular neuropil. (d) The olfactory bulb from a 23-year-old woman with an APOE 3/4 haplotype. There is premature accumulation of abundant corporae amylacea (lower right corner). Glycoproteinaceous inclusions in astrocytic processes play an important role in the sequestration of toxic cellular metabolites. (*See plate section for color version.*)

for CNS inflammatory responses. The inflammasome activation results in caspase 1 activation, leading to processing and secretion of pro-inflammatory cytokines like IL1β to engage innate immune defenses (Latz, 2010).

In keeping with the neuropathological findings, olfactory dysfunction was recently documented in 62 young persons (average age 21 years) who were living in a highly polluted area of Mexico City (Calderón-Garcidueñas et al., 2010). Abnormal UPSIT scores were present in 35.5% of the highly exposed subjects and in only 12% of controls of similar age who lived in a less polluted control city. Moreover, the influence of the APOE epsilon 4 genotype was related to the test scores of the Mexico City residents. Thus, highly exposed APOE epsilon 4 carriers failed, on average, 2.4 of the 10 UPSIT items identified in one study as being most strongly related to AD (Tabert et al., 2005), while APOE 2/3 and 3/3 subjects failed only 1.36 such items ($p = 0.01$).

In our studies, the early olfactory deficits appear to be associated with the aforementioned presence of beta amyloid, alpha synuclein, and hyperphosphorilated tau in the olfactory bulbs and the frontal cortices. The extensive olfactory bulb pathology likely affects proteins with critical functions. Recently, Fernández-Irigoyen et al. (2012) have characterized 1466 human olfactory bulb proteins. Proteins with catalytic and nucleotide binding activities were over-represented and the identification of a subset of proteins mainly involved in axon guidance, opioid signaling, neurotransmitter receptor binding, and synaptic plasticity were in evidence. In a study of the frontal cortices (Figure 20.7), 35 high pollution-exposed and 8 low-pollution exposed children and young adults, the exposed cohort displayed differential (>2-fold) regulation of 134 genes. Up-regulated gene network clusters included IL1, NFκB, TNF, IFN, and TLRs. A 15-fold frontal down-regulation of the prion-related protein [PrP(C)] was also present. Forty percent of the exposed individuals exhibited tau hyperphosphorylation with pre-tangle material and 51% had Aβ diffuse plaques compared with 0% in controls. In keeping with the OB findings, APOE4

494 Chapter 20 Influence of Toxins on Olfactory Function and Neurodegeneration

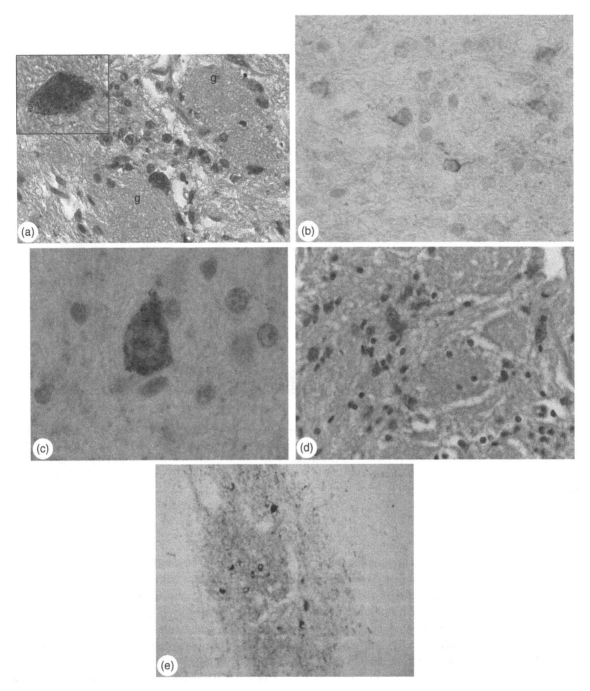

Figure 20.6 (a) Olfactory bulb from a 14-year-old boy living in Mexico City. Neurons in the glomerular region exhibit abundant intracytoplasmic particulate matter. The inset shows a close-up of a PM-loaded neuron with a positive immunoreactivity to beta amyloid$_{42}$ (red product). (b) Olfactory bulb from an 11-year-old living in Mexico City with a APOE 3/3 haplotype. Note neuronal granular positive immunoreactivity for α synuclein (red product). (c) A close-up of an olfactory bulb neuron with abundant granular α-synuclein, a marker of Parkinson's disease (red product). (d) Olfactory bulb from a 23-year-old woman with an APOE 3/4 haplotype. There are abundant neurons, glial cells and endothelial cells positive for β amyloid (4G8 monoclonal antibody reactive to amino acid residues 17–24 of beta amyloid). (e) Same 23-year-old woman as in d. Olfactory neurons with immunoreactivity to tau 8 PHF-Tau-8 (Innogenetics, Belgium, AT-8). AT8 reacts with tau only when multiple sites around Ser202, including Ser199, Scr202, and Thr205, are phosphorylated. (*See plate section for color version.*)

20.7 Air Pollution, Olfaction, and Neurodegeneration

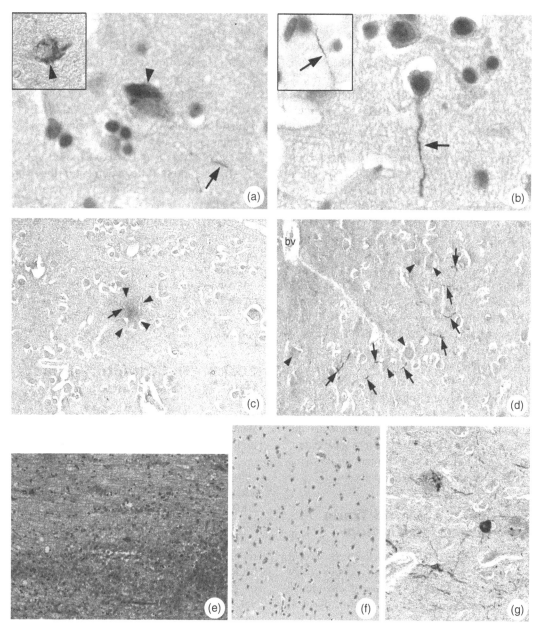

Figure 20.7 (a) Sample from frontal cortex of a 15-year-old Mexico City APOE 3/3 boy. Note that the cortex exhibits abnormal tau protein immunoreactivity in the neuronal body (insert and head arrow) as well as in neurites (arrow). Immunohistochemistry for AT8 counterstained with hematoxilin x 100. The insert has no counterstaining. (b) Sample from frontal cortex of a 15-year-old Mexico City APOE 3/4 boy. Small varicosities are visible in the IR abnormal neuritic tau (arrow). IHC for AT8 counterstained with hematoxilin. (c) Same boy as 3C. HP τ + neurites (arrow) were associated with diffuse amyloid plaques (surrounded by head arrows) in dual stained HP τ and Aβ sections. (d) Sample from frontal cortex of an 18-year-old Mexico City APOE 3/4 male. Clusters of HP τ + neurites up to 12 per 20x field (arrows) are seen in a dual HP τ and Aβ stained sections. IR for Aβ is seen in neuronal cytoplasm (head arrows). IHC for AT8 (brown product) and Aβ 6E10 antibody (red product). (e) Frontal one micron toluidine blue section from the white matter in a 14-year-old Mexico City woman. Arterioles exhibited numerous macrophage-like perivascular cells (arrow) and prominent expansion of the Virchow-Robin spaces (*), along with lipid deposits. The insert shows an arteriolar white matter vessel in a control 17-year-old boy. WM (white matter), (a) axons, and bv is the lumen of the blood vessel. Stain: Toluidine blue. (f) Frontal cortex from a control 17-year-old man stained with Aβ 6E10 antibody and counterstained with hematoxylin. The frontal cortex is unremarkable and no diffuse plaques were identified. (g) Frontal cortex from a 75-year-old Mexican woman with a diagnosis of Alzheimer's disease. HP τ + neurites and tangles surrounding Aβ positive mature senile plaques are seen in a dual HP τ and Aβ stained sections. IHC for AT8 (brown product) and Aβ 6E10 (red product) x 40. (*See plate section for color version.*)

carriers had greater hyperphosphorylated tau and diffuse Aβ plaques versus E3 carriers (p = 0.005).

The observations noted above are in potential accord with the tenants of the olfactory vector hypothesis (Doty, 2008; 2009b; Prediger et al., 2012). This hypothesis posits that intranasal exposure via the olfactory pathways to environmental agents, including viruses, toxins, agricultural chemicals, dietary nutrients, and metals, may lead to the development of some neurodegenerative diseases. Whether the olfactory loss is due to the transit of such agents or is secondary to the disease processes themselves is unknown. As discussed in detail in Chapter 18, olfactory dysfunction is among the earliest 'pre-clinical' features of AD and PD, occurring in ~90% of early onset cases (Doty, 2012a, b). Importantly, equivalent smell loss is evident in Down syndrome, non-Down mental retarded individuals, the PD-Dementia complex of Guam, and, most recently, myasthenia gravis (Leon-Sarmiento et al., 2012). Such observations have led Doty to suggest that a common pathological substrate may be involved (Doty, 2012a, b). There are, in fact, significant correlations of olfactory dysfunction with decreased numbers of neurons in structures such as the locus coeruleus, the raphe nuclei, and the nucleus basalis of Meynart (Doty, 2012a). In the case of PD, the involvement of the autonomic nervous system in numerous PD-related symptoms suggests that deficits in cholinergic, noradrenergic, and serotonergic function may contribute to the olfactory loss (Doty, 2012b). It is not known to what extent damage to such neurotransmitter/neuromodular systems contributes to neurodegenerative disease pathology.

20.7.2 The Effect of World Trade Center Exposures on Olfaction

The September 11th, 2001, terrorist attack on the World Trade Center exposed more than 50,000 people to an intense dust cloud with massive amounts of chemicals, xenobiotics (including dioxins, furans), asbestos, plastics, and construction debris (Betts, 2010; Wisniveskyn et al., 2011). These exposed subjects are currently at higher risk for heart disease, systemic inflammation, airflow obstruction, mental health symptoms, asthma, cancer, and chronic lung disease including "sarcoid-like" granulomatous pulmonary disease (Wisniveskyn et al., 2011; CDC, 2012). Airborne dust contaminants recovered from the WTC surrounding area 24–48h after the act demonstrated direct cytotoxicity to airway epithelial and smooth muscle cells, airway cell types most directly exposed to dust (Xu et al., 2011).

Two studies have evaluated olfactory function in rescue, demolition, and cleanup WTC workers. Many of these workers were not properly protected with gas masks or ventilators. In the first study, the prevalence of decreased olfactory and trigeminal sensitivity was found to be higher in

102 workers than in a similar number of non-exposed controls (Dalton et al., 2010). Testing took place, on average, two years after the exposures. WTC responders exposed directly to the dust cloud from the collapsed buildings exhibited the most serious trigeminal loss. Chronic nasal inflammation was a key finding in exposed subjects. In the second study, UPSIT scores of 99 WTC subjects were compared to controls obtained from a large normative database (Altman et al., 2011). The mean UPSIT scores were significantly lower in the WTC exposed cohort (30.05 vs. 35.94; p = 0.003). Fifteen percent of the exposed subjects had severe microsomia, but only 3% anosmia. The authors concluded that exposure to severe massive air pollution impacts the capacity to identify odors in a manner similar to that found in Mexico City residents (Calderón-Garcidueñas et al., 2010). Based on the clinical and neuropathological findings in such residents, the most highly exposed WTC subjects with the most olfactory deficits and genetic risk factors (e.g., APOE 4 allele) may be those at higher risk for the future development of AD- and/or PD-related pathology (Calderón-Garcidueñas et al., 2012a).

20.7.3 The Olfactory Effects of the Gulf War Syndrome

There is a significant amount of information in the Gulf War Syndrome (GWS) literature to suggest that physiological, psychological, and environmental factors play a role in the presentation and severity of the illness (Iannacchione et al., 2011; Li B et al., 2011; Li X et al., 2011; Chao et al., 2010; Fiedler et al., 2004; Ferguson et al., 2001/2002, 2004). Given that impaired metabolism of toxic chemicals is a postulated mechanism underlying multiple chemical sensitivity (MCS), it is important to record patients' war exposure history (Fiedler et al., 2004; McKeown-Eyssen et al., 2004). Interestingly, one study of 72 GW US veterans and 33 military personnel activated during the GW, but not deployed to the war zone, showed no evidence that performances on olfactory or neurocognitive measures were related to war-zone duty or to self-reported exposure to GW toxicants (Vasterling et al., 2003). In relation to such findings, it should be pointed out that empirical studies have found no evidence that patients complaining of MCS evidence altered smell function (Doty et al., 1988; Hummel et al., 1996; Dalton and Hummel, 2000).

Also relevant to this subject are reports of hippocampal neurophysiological and volume changes in Gulf War veterans (Li X et al., 2011). It remains to be seen if the available quantitative olfaction tests exhibit differences between the three target GWS groups: impaired cognition, confusion-ataxia, and central neuropathic pain. The hippocampal dysfunction could be a part of an immunoexcitotoxicity setting to explain the evolving pathological changes in patients with GWS (Blaylock and Maroon, 2011). The interaction between immune receptors within the CNS and excitatory glutamate receptors trigger a series of events, including ROS/RNS generation, accumulation of lipid peroxidation products, and prostaglandin activation, which then leads to dendritic retraction, synaptic injury, damage to microtubules, and mitochondrial suppression.

20.7.4 Metals and Olfaction

Metal ions are readily transported from the OE to the brain (Tjälve et al., 1996; Tjälve and Henriksson, 1999; Divine et al., 1999; Sundermann, 2001; Persson et al., 2003a, b; Bondier et al., 2008; Jia et al., 2010). The issue of metals and toxicity to the olfactory system and neurodegeneration has been discussed in epidemiological and occupational studies, in the olfactory vector hypothesis, and in fish models where emerging evidence shows very low sub-lethal levels of metal ions can adversely affect the sense of smell (Doty, 2008; 2009a, b; Flynn and Susi, 2009; Sriram et al., 2010; Tierney et al., 2010; Kiaune and Singhasemanon, 2011; Tilton et al., 2011; Prediger et al., 2012). The issue of metals and olfaction is particularly relevant to the field of occupational medicine. Despite the fact that OSHA works to assure safe working conditions by setting and enforcing standards and by providing training, and both outreach education and assistance, occupational exposures still remain harmful for millions of US workers (Rose et al., 1992; Bowler et al., 2011). Listed below are the major metals to which exposures are common in the workplace and their realized or potential influences on the olfactory system.

20.7.4.1 *Cadmium (Cd)*. Cadmium is a heavy
metal widely used in industry and is a widespread environmental pollutant. Exposures to Cd include the ingestion of contaminated cereals and vegetables, inhalation of cigarette smoke, and airborne exposures within occupational and contaminated environmental settings. Cd has been associated with fatalities and adverse health effects, including cancer, kidney damage, low bone mineral density and fractures, hearing loss, and impaired olfaction (Tallkvist et al., 2002; Mascagni et al., 2003; Gobba, 2006; Bondier et al., 2008; Shargorodsky et al., 2011; Hossain et al., 2012; Chang et al., 2012). The association between Cd and impaired olfaction was well established in the 1950s (reviewed in Gobba, 2006). Workers in alkaline battery factories, welders, smelters and brazing workers were found to be at high risk for olfaction dysfunction (Adams and Crabtree, 1961; Ahlström et al., 1986; Rose et al., 1992; Sulkowski et al., 2000; Mascagni et al., 2003).

20.7.4.2 *Chromium (Cr)*. Chromium is used in
the manufacture of steel and non ferrous alloys, dyes and pigments, leather tanning, photographic processing, wood fabrication, and in various medical applications, including prostheses in surgery (hip, knee). Hexavalent chromium is the form with the greatest occupational and environmental health concern (Pellerin and Booker, 2000). Chronic exposures can result in nasal septum perforations, nasal ulcers, and olfaction deficits (Lin et al., 1994; Kuo et al., 1997; Kitamura et al., 2003; Goullé et al., 2012). In the United States, industrial exposure to hexavalent Cr via inhalation of fume dusts is the most likely basis for reported nasal and olfaction deficits. Cr is the primary contaminant of over 50% of Superfund hazardous waste sites. Numerous waste sites are found near chromate factories. Although OSHA and NIOSH have established permissible exposure limits (PELs) and recommended exposure limits (RELs) for Cr in the workplace, thousands of workers are likely exposed to higher levels of Cr in welding and other workshop operations in places lacking adequate ventilation, while their family members are exposed through contact with contaminated clothes, shoes, and other items.

20.7.4.3 *Manganese (Mn)*. Manganese is an
essential trace element. Distinct pathways are known to contribute to Mn influx into the brain and to Mn olfactory neurotoxicity (Tjälve and Henriksson, 1999; Aschner et al., 2007; Leavens et al., 2007; Dorman et al., 2006, 2008; Erikson et al., 2008; Bouchard et al., 2007, 2008; Burton and Guilarte, 2009; Guilarte, 2010; Kim et al., 2011; Bowler et al., 2011; Schroeter et al., 2011b; Racette et al., 2012). Mn occupational neurotoxicity is very common worldwide (Antonini, 2003; Criswell et al., 2011). It was first described in the 19th century in two workers with high exposures to Mn oxide (Couper, 1837).

There are presently ~466,000 full-time welding and soldering workers in the US. More than a million workers perform welding as part of their job requirements (Antonini, 2003; Criswell et al., 2011). Although current human Mn exposures are usually below the current NIOSH recommended exposure limit (REL) of 1 mg/m^3 and the American Conference of Govermental Industrial Hygienists (ACGIH) Threshold Limit Value (TLV) of 0.2 mg/m^3, there are common examples of high exposures in the occupational setting (Bowler et al., 2011; Racettte et al., 2012).

Mn homeostasis is critical in light of the fragile relationship between its essentiality and toxicity. Movement abnormalities are found in non-human primates exposed to moderate levels of Mn – abnormalities associated with the inability of dopaminergic neurons to release dopamine (Guilarte, 2010). In addition to subtle deficits in motor function, deficits in working memory and neurodegenerative changes are seen in the prefrontal cortex (Guilarte et al., 2008, Guilarte, 2010; Racette et al., 2012; Nelson et al., 2012). Manganism – a disorder characterized by bradykinesia, masked facies, and gait impairment – is the most common syndrome associated with Mn neurotoxicity, although PD, per se, may also result from long-term low-level exposure to this metal.

Olfactory deficits are a critical early finding in idiopathic PD and in Mn-associated toxicity (Ansari and Johnson, 1975; Doty, 1988a, 2008, 2009b, 2012a, b; Lucchini et al., 1997, 2012; Antunes et al., 2007; Bowler et al., 2007, 2011; Bohnen et al., 2008, 2010; Dickson et al., 2009b; Baba et al., 2011). Bowler et al. (2007) studied 43 welders working at the San Francisco-Oakland Bay bridge under confined spaces with little protection equipment (see also Antunes et al., 2007). Workers, who were on average 43.8 years of age, had spent on average of 16.5 months working on the bridge. The mean time weighted average of Mn in air ranged from 0.11 to 0.46 mg/m^3, while Mn in blood was 9.6± 2.5 µg/L. Significant cognitive impairment and mood disturbances and a high number of days with poor health were found, with 80% of the welders with lower scores than their matched controls p<0.001. In their 2011 follow-up study and after 3.5 years of cessation of welding in a confined space, there was some improvement of their cognitive function, but olfactory, extrapyramidal, and mood disturbances remained unchanged (Bowler et al., 2011). Thus, in spite of an overall decrease of Mn exposure, some workers continued to have higher concentrations of Mn in blood and exhibited higher than normal tremor intensity on the CATSYS (quantifies coordination ability, reaction time, tremor and postural stability). The lack of improvement in neurological functions involving the olfactory pathway and the limbic system (olfaction and mood), basal ganglia (neuromotor and movement alterations), in conjunction with improvement of frontal cortex and the hippocampus (short term memory), strongly suggest the presence of intrinsic brain Mn vulnerabilities and the potential problem that after a certain level of exposure, the Mn neurotoxicity reaches a point of no return (Bowler et al., 2011).

Non-primate studies have provided invaluable insights into the CNS effects of Mn (Eriksson et al., 2008; Burton and Guilarte, 2009; Schneider et al., 2009; Guilarte, 2010). A spectrum of cognitive deficits have also been described in children exposed to Mn in drinking water and living in the proximity of a ferro-manganese alloy plant, as well as in occupationally exposed workers (Bouchard et al., 2007, 2008; Menezes-Filho et al., 2011). Children are a susceptible population, as shown by Lucchini in their 11–14-year-old cohort in Valcamonica, Italy, a region that had high ferro-alloy plant emissions for a century before 2001 (Lucchini et al., 2012). Impairment of motor coordination, hand dexterity and odor identification were associated with the measures of soil manganese. Tremor intensity was positively associated with blood and hair Mn. This study illustrates a very important aspect of Mn toxicity in children with high exposures: olfactory deficits are readily detected along with motor dysfunction.

20.7.4.4 Nickel (Ni).
Nickel (Ni) is an environmental pollutant with an impact in the occupational setting mainly through inhalation and in dietary exposures secondary to water and food chain-induced bioaccumulation (Liapi et al., 2011). Ni is actively transferred across the blood-placenta-barrier into the fetus. The placenta has a high affinity for nickel and the placental barrier does not confer the fetus protection from Ni exposure (Hou et al., 2011). Children are at risk as well, especially those exposed to polluting steel mills in industrial areas (Kasper-Sonnenberg et al., 2011). Also critical is the Ni ion bioavailability as a key factor determining toxicity (Schaumlöffel, 2012; Henderson et al., 2012).

The altered olfaction and anosmia resulting from Ni exposures are well known in occupational medicine and in studies of experimental animals (Tjälve and Henriksson, 1999; Evans et al., 1995). Jia et al. (2010) exposed adult mice to intranasally instilled NiSO$_4$ or saline followed by ATP, purinergic receptor antagonists, or saline. Location-dependent atrophy of the OE was found to be associated with sustentacular cell loss and apoptosis of OSNs. This study illustrates the selective susceptibility of sustentacular cells to local damage from NiSO$_4$. The location of the sustentacular cell somas in the apical portion of the epithelium forms a protective barrier for the rest of the OE – a barrier crucial for mitigating the effects of toxins. Menco (1980) had observed a zone of tight junctions close to the apical surfaces of sustentacular cells that prevents paracellular diffusion of lipophobic compounds. Ni^{2+} induces the formation of hydroxyl radicals which can produce single DNA strand breaks and DNA-protein cross-links consistent with Jia et al.'s (2010) observations of DNA fragmentation in sustentacular cells one day following the instillation of NiSO$_4$ (Torrielles and Guerin, 1990; Patierno and Costa,1985). Evans also suggested the mechanism of NiSO$_4$-induced sustentacular cell loss could be attributed to the expression of metabolic enzymes given that biotransformation enzymes are preferentially expressed in sustentacular cells (Evans et al., 1995; Banger et al., 1994).

Other olfactory neuronal damage mechanisms related to Ni toxicity include apoptosis via activation of death receptor 3 and caspase-8, leading to activation of caspase-3 (Zhao et al., 2009). Ni^{2+} reduces odorant-induced calcium transients in the dendritic knob, the dendrite, and the soma of OSNs via inhibition of T-type Ca^{2+} channels (Gautam et al., 2007). Ni's inhibition of the T-type channels may result in inhibition of odorant signal transduction that ultimately leads to hyposmia and anosmia (Jia et al., 2010).

20.7.4.5 Zinc (Zn). There is an extensive literature on the loss of smell function resulting from intranasal exposure to zinc ions (Tisdall et al., 1938; McBride et al., 2003; Seidman, 2006; Alexander and Davidson, 2006; Slotnick et al., 2007; Lim et al., 2009; Davidson and Smith, 2010; Duncan-Lewis et al., 2011). Recently the potential association of zinc with olfactory dysfunction has come to the fore as a result of claims that a popular zinc-based intranasal agent, Zicam (Matrixx Initiatives, Inc), causes hyposmia or anosmia (e.g., Alexander and Davidson, 2006). Considerable contention has arisen around this issue, in part because (a) it is not entirely clear how much of the zinc preparation, particularly from gels, actually reaches the OE, (b) the most frequent cause of smell loss is the common cold for which Zicam is often employed, and (c) many of the clinical reports specific to this agent have been driven by litigation. Moreover, some animal studies have found that olfactory deficits secondary to Zn exposure of their olfactory epithelia are largely reversible (Slotnick et al., 2007) and that low doses of zinc nanoparticles actually enhance physiological measures of olfactory function (Viswaprakash et al., 2009; Moore et al., 2012). Systemic zinc has been used for years to treat human olfactory dysfunction. Except in cases of frank zinc deficiencies, however, such treatments are generally viewed as ineffectual (Price, 1986). In light of the numerous reports that intranasal application of zinc ions is toxic to olfactory epithelia, the use of nasal preparations containing zinc is not generally recommended.

20.7.4.6 Aluminum (Al). The neurotoxicity of aluminum has been well known for years (Dollken, 1897; Klatzo et al., 1965). Exposure to Al is accompanied by oxidative stress, apoptosis, and the perikaryal accumulation of tangles of phosphorylated neurofilaments (NFs). A prolonged interaction induced among phospho-NFs by the trivalent Al impairs axonal transport and promotes perikaryal aggregation (Bharathi et al., 2008; Kushkuley et al., 2010). Olfactory deficits have been observed in rabbits who received intraventricular injections of aluminum chloride (Polonskaia et al., 1990). In a *Drosophila* model with chronic dietary Al overloading, Wu et al. (2012) observed general neurodegeneration and a number of behavioral changes. These authors indicated that the Al toxicity is likely mediated through ROS production and iron accumulation and suggested a remedial route to reduce Al toxicity. Al neurotoxicity is important to consider in this chapter because the reports dealing with olfaction deficits associated with glue sniffing (glues and their containers often have high Al content) in teenagers who inadvertently expose themselves to such metals and other neurotoxic chemicals with the intent of feeling high (Akay et al., 2008).

20.8 INFLUENCES OF ALCOHOL, TOBACCO, CHEMOTHERAPY, AND DIALYSIS ON OLFACTION

20.8.1 Alcohol

Alcohol, particularly during development, has multiple negative effects on the brain, including the olfactory system (Vent et al., 2004; Bragulat et al., 2008; Youngentob and Glendinning, 2009; Middleton et al., 2009; Eade and Youngentob, 2010; Akers et al., 2011; Maurage et al., 2011; Huang et al., 2012; Balaraman et al., 2012; Munner et al., 2012; Valenzuela et al., 2012). Prenatal alcohol exposure can result in abnormal olfactory bulb development, as reflected by smaller bulbs, fewer bulbar cells, and fewer neural precursory cells within the sub-ependimal zone. In later life, prenatal exposure to alcohol is associated with impaired odor discrimination and altered olfactory bulb expression of neurotransmission genes, in addition to general alterations in learning, memory, motor coordination, social behavior, and stress responses (Youngentob and Glendinning, 2009; Middleton et al., 2009; Akers et al., 2011; Valenzuela et al., 2012). This is an issue of relevance to healthcare providers since approximately 1% of children in the US exhibit symptoms of Fetal Alcohol Spectrum Disorder (FASD) (Sampson et al., 1997).

It is of interest that the effects of alcohol on the fetal brain can result in life-long increased vulnerability to substance abuse and impairments in neurotransmitter systems, neuromodulators, and/or synaptic plasticity in numerous brain regions (Valenzuela et al., 2012). Although acute exposure to alcohol in healthy subjects has no effect on odor identification, alcohol ingestion to the level of legal intoxication markedly and selectively alters olfactory sensitivity to ethanol, and depresses scores on a match-to-sample odor discrimination/memory test (Patel et al., 2004). In chronic alcoholics, brain structures critical for the secondary "cognitive" processing of odors are frequently damaged (Maccioni et al., 2007; Maurage et al., 2011), and irreversible smell loss is common in the alcohol-related Wernicke-Korsakoff syndrome – a loss secondary to poor nutrition and the lack of adequate thiamine in the diet (Mair et al., 1987).

The concept of *in utero* chemosensory plasticity has been recently proposed (Youngentob and Glendinning, 2009; Middleton et al., 2009). The olfactory bulb's sensitivity to ethanol exposure depends on the dose and time of the exposure (Barron and Riley, 1992), with large effects occurring prior to when the sensory axons reach the olfactory bulb and prior to when the fetal olfactory sensory neurons begin to transduce sensory information. Alcohol causes selective effects on GABAergic and glutamatergic neurons, as well as genes involved in synaptic transmission, synaptic development, and cell fate (Middleton et al.,

2009). Specifically, *in utero* exposure to alcohol results in a significant decrease in the OB expression of the GABA B1 receptor and the mGluR2 receptor, as well as increased expression of calcium/calmodulin-dependent protein kinase IV and potassium channel. Long-term changes in the OB expression of a gene subset temporally associated with an enhanced unconditioned sniffing response to ethanol odor have been reported (Middleton et al., 2009).

It is well known that fetal ethanol exposure is highly predictive of the propensity to consume ethanol in adulthood. Indeed, the smell of alcohol in later life among early exposed individuals can lead to the development and maintenance of alcoholism, including, in some instances, to stronger alcohol craving responses and desires to drink. Presently there are 17.6 million people in the United States who abuse alcohol or who are alcohol dependent, which translates into 1 in every 12 adults. Since alcohol intake is usually associated with tobacco consumption, it is important to keep in mind the synergistic effects of both toxins to the olfactory system (Vent et al., 2004).

20.8.2 Tobacco

Tobacco is the best example of a legal substance whose addiction affects millions of people around the world (Talhout et al., 2011; Dorman et al., 2012). Cigarette smoking causes considerable morbidity and mortality by inducing cancer, chronic respiratory disease, cardiovascular disease, and nasal and sinus disorders. The effects of tobacco on children and fetuses are well known and second-hand smoke (SHS) exposure has long been correlated with disease processes (Röösli, 2011). The term third-hand smoke (THS) applies to the residual tobacco smoke-related pollutants that remain on surfaces and in dust which can be reemitted in the gas phase and can interact with other compounds. THS poses adverse health risks, particularly for children (Tillett, 2011; Matt et al., 2011). Among the mix of more than 5,000 chemicals in cigarette smoke, a number have well documented effects on the OE, including acrylic acid, acrolein, acetaldehyde, and Cr (VI) (Matt et al., 2011). Active components of tobacco smoke such as lipopolysaccharides (LPS) add significant contributions to OE damage (Yagi et al., 2007). Apoptosis within the olfactory epithelium as a result of exposure to tobacco vapors is well documented and is a plausible explanation for the olfactory deficits observed in smokers (Nageris, et al., 2001, 2002; Vent et al., 2004).

The inflammatory nasal response elicited by tobacco exposure is another likely cause of damage within the olfactory epithelium. De et al. (2011) found increased concentrations of matrix metalloproteinase (MMP) 9, a gelatinase associated with tissue remodeling, in the nasal secretions of children exposed to second-hand smoke. Passive smoking might alter the inflammatory response within the nasal mucosa in a similar way to allergy. Smoking is clearly associated with a significant dose-related decrease in olfactory function that affects both current and previous smokers. Current smokers are twice as likely as non-smokers to have olfactory deficits, reflecting, in part, the fact that cessation of smoking can reverse its detrimental effects over time (Frye et al., 1990).

20.8.3 Chemotherapy

Millions of people undergo chemotherapeutic treatments each year. Among the main clinical goals are to identify, delay, and relieve chemotherapy-induced side effects and optimize the quality of life, especially after first-line therapy (Doty and Bromley, 2004; Steinbach et al., 2009, 2010, 2012). Some chemotherapy cocktail combinations are particularly bothersome for patients, such as carboplatinum-containing chemotherapy for ovarian cancer (Steinbach et al., 2012). Among 136 chemotherapy outpatients from Kanazawa University, 55% complained of a gustatory and 19% of an olfactory disorder. Olfactory alterations were significantly greater in patients who also had gustatory disorders. Complaints of taste disorders were significantly higher among those taking docetaxel (85%) than those on other regimens (Suga et al., 2011).

In advanced terminal patients, the most common complaints include persistent bad taste in the mouth, taste distortion, and heightened sensitivity to odors. Patients with severe chemosensory complaints show substantially lower energy intakes (by 900–1,100 kcal/day), higher rates of weight loss, and lower quality of life scores compared to subjects with mild or moderate chemosensory complaints (Hutton et al., 2007). Although such symptoms are common side effects of chemo- and radiotherapy, their physiological basis remains mostly unexplained (Faure et al., 2010). Unfortunately, self-reports of taste and smell function are notoriously inaccurate (Soter et al., 2008), and a large number of patients confuse altered flavor sensations with taste disorders (Deems et al., 1991).

Some chemotherapy patients report experiencing only smell disturbances, including increased sensitivity to one or several odors, without apparent decreases in general olfactory sensitivity (Bernhardson et al., 2009, 2012). In the 2009 Bernhardson et al. survey, those patients who reported experiencing only smell problems also reported gaining weight and having fewer oral problems. Reports of increased olfactory sensitivity during chemotherapy were often unpredictable and were commonly associated with emotional sequelae. Many chemotherapy patients report having specific complaints related to eating (Bernhardson et al., 2012).

Difficulties in recognizing tastes, particularly sweet tastes, occur in about 33% of patients undergoing chemotherapy and taste thresholds, as assessed by electrogustometry, are often elevated (Berteretche et al., 2004; DeWys, 1979; Ovesen et al., 1991). Taste dysfunction can result in food avoidance and, consequently, can negatively impact the nutritional status and quality of life of patients (Bartoshuk, 1990; Berteretche et al., 2004). Many cancer patients complain that taste is "abnormal, bad, or new" during and for some months after termination of chemotherapy treatments.

It is possible that such changes in taste or flavor makes the affected foods "novel" and susceptible to conditioned taste aversions (CTAs), that is, unpleasant and nausea-associated aversions to foods that often develop in children and adults undergoing chemo- or radio-therapy (Scalera and Bavier, 2009). As noted in detail in Chapter 37, CTAs arise from the delayed pairing of the flavor of a meal with the nausea induced by the therapy (Bernstein, 1978; Bernstein and Webster, 1980). Fortunately, many CTAs can be mitigated by presenting a novel and distinctive flavored food that is rarely or never before been encountered (e.g., a root beer Lifesaver) after a meal but before the initiation of the chemo- or radio-therapy. For some unknown reason, this flavor becomes the "scapegoat" for the CTA, minimizing treatment-associated aversions and hypophagia (Broberg and Bernstein, 1987; Mattes et al., 1987). As noted above, one school of thought assumes that the cancer itself causes the flavor of foods to be distorted or "novel", making them particularly susceptible to such conditioning. By presenting an even more novel or distinct food, the CTA focuses on the more novel flavor and less on the more familiar flavor. In most cases, chemotherapy related food aversions last only a few weeks or months after cessation of treatment (Mattes et al., 1992). Perhaps their transience is due to the reversal of the taste or smell dysfunction and consequently the disappearance of the "novel" food.

20.8.4 Anesthetics

The effect of anesthetics on olfactory function remains an obscure topic and no double blind study has been performed on the potential influences of anesthetics on such function. Anesthetic drugs could adversely affect olfaction at either central or peripheral levels, including direct toxicity to the olfactory neuroepithelium, nasal vasodilatation, or mucus hypersecretion (Doty and Bromley, 2004). An already compromised olfactory epithelium could explain rare cases where anesthetics have been reported to alter chemosensory function.

Different mechanisms apply for the associations between local anesthetics and smell impairment. Salvinelli reported the case of permanent olfactory dysfunction after endonasal local anaesthesia with lidocaine 4% that was due to direct contact of the anesthetic with the OE (Salvinelli et al., 2005). Johnson et al. (2004) has shown that mitochondrial dysfunction with activation of apoptotic pathways could be the key factor associated with lidocaine neurotoxicity (Johnson et al., 2004).

It is hard to determine with certainty the causative role of specific anesthetics in the development of olfaction impairment (Konstantinidis et al., 2009). Kostopanagiotou compared the effects of general and regional anaesthesia on olfactory acuity and memory in the immediate post-operative period. In this study, olfactory memory deficits were accompanied by a significant post-operative reduction of plasma melatonin levels in the sevoflurane patients, suggesting that the misinterpretation of odors in the immediate post-operative period by sevoflurane could be mediated by the decreased levels of melatonin (Kostopanagiotou et al., 2011). It is important to note, however, that sevoflurane, as well as propofol, affect subcortical and cortical c-aminobutyric acid (GABA) receptor ligand binding (Salmi et al., 2004). Since GABA is a neurotransmitter present at many areas of the olfactory system, any alteration of the GABA dependant pathways could potentially impair olfactory transmission within the CNS (Levy and Henkin, 2004). However, it is not clear why such effects should lead to chronic dysfunction.

20.8.5 Dialysis

More than 485,000 Americans are being treated for end stage renal disease, including 341,000 dialysis patients. Each year, 90,000 new patients with kidney failure are registered (Krishnan and Kiernan, 2009). One of the main problems of dialysis patients is their low quality of life, with malnutrition and anorexia being key contributors. Uremic anorexia and olfactory dysfunction are at the core of their nutritional problems (Schiffman et al., 1978; Raff et al., 2008; Bomback and Raff, 2011).

In the setting of renal insufficiency, impaired olfactory function may serve a dual role: First, as an easily quantifiable measure of neurological dysfunction and second as an indicator of detrimental effects of uremia (Bomback and Raff, 2011). The OE, as well as higher olfactory processing pathways, are potential urea targets. The constant turnover of olfactory neurons throughout the life span presents with an easy target for systemic toxins (Schiffman et al., 1978; Conrad et al., 1987; Suzuki and Farbman, 2000; Raff et al., 2008).

Studies in hemodialysis, peritoneal dialysis, transplanted patients, and patients with varying degrees of renal insufficiency have shown a significant negative correlation between odor perception and serum concentration of urea and protein catabolic rate (Griep et al., 1997). Frasnelli et al. (2002) found reduced olfactory function in chronic renal failure patients. Raff et al. (2008) demonstrated a critical association between impaired olfactory function (UPSIT) and poor nutritional status in such patients.

20.9 SUMMARY

Olfactory dysfunction disrupts people's lives and can, in some instances, be an early sign of an ominous disease. Different components of the olfactory system may be involved in the pathological process and the list of toxics and xenobiotics is a long one. Olfactory bulb pathology relates to significant smell loss and carries a significant risk of a neurodegenerative process. The role of air pollution in the production of olfactory system pathology is enigmatic; however, current knowledge points towards a significant impact of air pollutants – especially particulate matter – in the induction of olfactory system pathology, neuroinflammation, and subsequent neurodegeneration. The issue is a serious one, since millions of people are exposed to concentrations of air pollutants above current standards and young residents of highly polluted cities are showing early olfaction deficits associated to the hallmarks of diseases such as AD and PD. Gene-environmental influences are at play in olfaction disorders, as illustrated by the significant impact of air pollutants upon the CNS in APOE 4 carriers.

20.9.1 Summary of Things to Remember on the Subject of Air Pollution and Olfaction

1. Smell deficits are seen in highly exposed young urbanites, including children. Such deficits appear to be most profound in those that spend significant amounts of time outdoors.

2. The lesions seen in children and young adults involve the nasal respiratory mucosa, the olfactory epithelium, and the olfactory bulb.

3. The olfactory bulb neuropathology associated with urban exposures is very similar to that described in early stages of AD and PD (Jellinger et al., 1991; Jellinger, 2012; Braak et al., 1991, 2003a, b, 2011a, b, c). The neuropathological hallmarks of AD (e.g., hyperphosphorilated tau and amyloid plaques) are present in the frontal cortex of children as young as 10 years, and these children already have evidence of olfactory and cognitive deficits, prefrontal white matter hyperintensities, and white matter volumetric changes (Calderón-Garcidueñas et al., 2012a, b).

4. The brunt of the inflammatory brainstem changes in Mexico City young residents involves the medulla oblongata, as well as accumulation of β amyloid and alpha synuclein in key olfactory nuclei, the dorsal motor nucleus of the vagus, the nucleus of the solitary tract, arcuate nucleus, raphe midline, and extra-raphe medial and lateral tegmentum neurons. These lesions are very similar to the PD stages I and II of Braak (Braak et al., 2003a, b).

5. Olfactory deficits in highly exposed young urbanites must be kept in mind by clinicians. Such deficits are worse in APOE 4 carriers.

6. Air pollution plays a key role in the development of CNS damage and has detrimental effects upon the developing brain. Pollution potentially plays a role in the etiology of AD, PD, and mood disorders, although direct evidence for this is currently lacking.

7. Subjects exposed to massive air pollution, such as occurred in the aftermath of the WTC tragedy, may be at higher risk for the future development of AD and/or PD. Those at greatest risk may be the most highly exposed subjects with the most olfactory deficits and genetic risk factors (e.g., APOE 4 allele carriers).

20.9.2 Summary of Things to Remember Regarding Occupational Settings where Exposure to Ionized Metals is Prevalent

1. Adverse effects on olfaction in workers exposed to metals is dose-related and the duration of exposure is critical.

2. Early toxic effects of the metal can occur at low levels, underlying the importance of olfactory tests for identifying early effects.

3. Metal toxicity mechanisms include oxidative stress, interactions with sulfhydryl groups, dysregulation of DNA methylation, mitochondrial damage and direct neuronal effect to release neurotransmitters.

4. Welding fumes contain heavy metals with aerodynamic diameters <10μm, such as chromium, manganese, and nickel. Such fumes can be toxic. Compliance of established exposure limits should be enforced.

5. A word of caution should be raised for the use of Mn nanoparticles as contrast probes for magnetic resonance imaging (Zhen and Xie, 2012) based on the occupational and non-primate Mn literature.

As is clear from the material described in this chapter, much more research is needed to elucidate the olfactory toxicity mechanisms of a variety of everyday health hazards, including exposures to nanoparticles and environmental xenobiotics. Unfortunately, until such hazards are better understood and appropriate environmental safeguards are in place, many people will continue to be exposed to xenobiotics that potentially have long-lasting effects on their health, including the brain and the olfactory pathways.

REFERENCES

Adams, R. G. and Crabtree, N. (1961). Anosmia in alkaline battery workers. *Br. J. Ind. Med.* 18: 216–221.

Ahlström, R., Berglund, B., Berglund, U., et al. (1986). Impaired odor perception in tank cleaners. *Scand. J. Work Environ. Health* 12: 574–581.

Akay, C., Kalman, S., Dündaröz, R., et al. (2008). Serum aluminium levels in glue-sniffer adolescent and in glue containers. *Basic Clin. Pharmacol. Toxicol.* 102(5): 433–436.

Akers, K. G., Kushner, S. A., Leslie, A. T., et al. (2011). Fetal alcohol exposure leads to abnormal olfactory bulb development and impaired odor discrimination in adult mice. *Mol. Brain* Jul 7; 4: 29.

Allavena, R. E., Desai, B., Goodwin, D., et al. (2011). Pathologic and virologic characterization of neuroinvasion by HSV-2 in a mouse encephalitis model. *J. Neuropath. Exp. Neurol.* 70(8): 724–734.

Alexander, T. H., and Davidson, T. M. (2006). Intranasal zinc and anosmia: the zinc-induced anosmia syndrome. *Laryngoscope* 116: 217–220.

Ali, J., Ali, M., Baboota, S., et al. (2010). Potential of nanoparticulate drug delivery systems by intranasal administration. *Curr. Pharm. Des.* 16(14): 1644–1653.

Altman, K. W., Desai, S. C., Moline, J., et al. (2011). Odor identification ability and self-reported upper respiratory symptoms in workers at the post-9/11 World Trade Center site. *Int. Arch. Occup. Environ. Health* 84: 131–137.

Anderson, J. O., Thundiyil, J. G., and Stolbach, A. (2012). Clearing the Air: A Review of the Effects of Particulate Matter Air Pollution on Human Health. *J. Med. Toxicol.* 8(2): 166–175.

Ansari, K. A., and Johnson, A. (1975). Olfactory function in patients with Parkinson's disease. *J. Chronic Dis.* 28: 493–497.

Anton, F., and Peppel, P. (1991). Central projections of trigeminal primary afferents innervating the nasal mucosa: a horseradish peroxidase study in the rat. *Neurosci.* 41(2–3): 617–628.

Antonini, J. M. (2003). Health effects of welding. *Crit. Rev. Toxicol.* 33: 61–103.

Antunes, M. B., Bowler, R., and Doty, R. L. (2007). San Francisco/Oakland Bay Bridge Welder Study: olfactory function. *Neurology* 69(12): 1278–1284.

Aschner, M., Guilarte, T. R., Schneider, J. S., and Zheng, W. (2007). Manganese: recent advances in understanding its transport and neurotoxicity. *Toxicol. Appl. Pharmacol.* 221(2): 131–147.

Baba, T., Takeda, A., Kikuchi, A., et al. (2011). Association of olfactory dysfunction and brain. Metabolism in Parkinson's disease. Mov. Disord. 26(4): 621–628.

Balali-Mood, M., Afshari, R., Zojaji, R., et al. (2011). Delayed toxic effects of sulfur mustard on respiratory tract of Iranian veterans. *Hum. Exp. Toxicol.* 30(9): 1141–1149.

Balaraman, S., Winzer-Serhan, U. H., and Miranda, R. C. (2012). Opposing Actions of Ethanol and Nicotine on MicroRNAs are Mediated by Nicotinic Acetylcholine Receptors in Fetal Cerebral Cortical-Derived Neural Progenitor Cells. *Alcohol Clin.* Exp. Mar 28.

Banger, K. K., Foster, J. R., Lock, E. A., and Reed, C. J. (1994). Immuno-histochemical localisation of six glutathione S-transferases within the nasal cavity of the rat. *Arch. Toxicol.* 69(2): 91–98.

Baraniuk., J., and Kaliner, M. (1990). Neuropeptides and nasal secretion. *Am. J. Physiol.* 261: 223–235.

Barron, S., and Riley, E. P. (1992). The effects of prenatal alcohol exposure on behavioral and neuroanatomical components of olfaction. *Neurotoxicol. Teratol.* 14(4): 291–297.

Bartoshuk, L. M. (1990). Chemosensory alterations and cancer therapies. *Natl. Cancer Inst. Monogr.* 9: 179–184.

Becquemin, M. H., Swift, D. L., Bouchikhi, A., et al. (1991). Particle deposition and resistance in the noses of adults and children. *Eur. Respir. J.* 4(6): 694–702.

Bernhardson, B. M., Tishelman, C., and Rutqvist, L. E. (2009). Olfactory changes among patients receiving chemotherapy. *Eur. J. Oncol. Nurs.* 13(1): 9–15.

Bernhardson, B. M., Olson, K., Baracos, V. E., and Wismer, W. V. (2012). Reframing eating during chemotherapy in cancer patients with chemosensory alterations. *Eur. J. Oncol. Nurs.* Jan 19.

Bernstein, I. L. (1978). Learned taste aversions in children receiving chemotherapy. *Sci.* 200: 1302–1303.

Bernstein, I. L., and Webster, M. M. (1980). Learned taste aversions in humans. *Physiol. Behav.* 25: 363–366.

Berteretche, M. V., Dalix, A. M., Cesar D'Ornano, A. M., et al. (2004). Decreased taste sensitivity in cancer patients under chemotherapy. *Supportive Care Cancer* 12: 571–576.

Bessen, R. A., Wilham, J. M., Lowe, D., et al. (2012). Accelerated shedding of prions following damage to the olfactory epithelium. *J. Virol.* 86: 1777–1788.

Betts, K. (2010). Signature of high exposure to WTC toxics identified. *Environ. Sci. Technol.* 44: 4834–4835.

Bharathi, V. P., Govindaraju, M., Palanisamy, A. P., et al. (2008). Molecular toxicity of aluminium in relation to neurodegeneration. *Indian J. Med. Res.* 128(4): 545–556.

Blaylock, R. L., and Maroon, J. (2011). Immunoexcitotoxicity as a central mechanism in chronic traumatic encephalopathy-A unifying hypothesis. *Surg. Neurol. Int.* 2: 107.

Block, M. L., and Calderón-Garcidueñas, L. (2009). Air pollution: mechanisms of neuroinflammation and CNS disease. *Trends Neurosci.* 32(9): 506–516.

Bohnen, N. I., Gedela, S., Herath, P., et al. (2008). Selective hyposmia in Parkinson disease: association with hippocampal dopamine activity. *Neurosci. Lett.* 447(1): 12–16.

Bohnen, N. I., Müller, M. L., Kotagal, V., et al. (2010). Olfactory dysfunction, central cholinergic integrity and cognitive impairment in Parkinson's disease. *Brain* 133(Pt 6): 1747–1754.

Bomback, A. S., and Raff, A. C. (2011). Olfactory function in dialysis patients: a potential key to understanding the uremic state. *Kidney Int.* 80(8): 803–805.

Bondier, J. R., Michel, G., Propper, A., and Badot, P. M. (2008). Harmful effects of cadmium on olfactory system in mice. *Inhal. Toxicol.* 20(13): 1169–1177.

Bouchard, M., Laforest, F., Vandelac, L., et al. (2007). Hair manganese and hyperactive behaviors: pilot study of school-age children exposed through tap water. *Environ. Health Perspect.* 115(1): 122–127.

Bouchard, M., Mergler, D., Baldwin, M. E., and Panisset, M. (2008). Manganese cumulative exposure and symptoms: a follow-up study of alloy workers. *Neurotoxicology* 29(4): 577–583.

Bowler, R. M., Roels, H. A., Nakagawa, S., et al. (2007). Dose-effect relationships between manganese exposure and neurological, neuropsychological and pulmonary function in confined space bridge welders. *Occup. Environ. Med.* 64(3): 167–177.

Bowler, R. M., Gocheva, V., Harris, M., et al. (2011). Prospective study on neurotoxic effects in manganese-exposed bridge construction welders. *Neurotoxicology* 32(5): 596–605.

Bragulat, V., Dzemidzic, M., Talavage, T., et al. (2008). Alcohol sensitizes cerebral responses to the odors of alcoholic drinks: an fMRI study. *Alcohol Clin. Exp. Res.* 32(7): 1124–1134.

Braak, H., and Braak, E. (1991). Neuropathological staging of Alzheimer-related changes. *Acta Neuropath.* 82: 239–259.

Braak, H., Del Tredici, K., Rüb, U., et al. (2003a). Staging of brain pathology related to sporadic Parkinson's disease. *Neurobiol. Aging* 24(2): 197–211.

Braak, H., Rüb, U., Gai, W. P., and Del Tredici, K. (2003b). Idiopathic Parkinson's disease: possible routes by which vulnerable neuronal types may be subject to neuroinvasion by an unknown pathogen. *Neural Transm.* 110(5): 517–536.

Braak, H., and Del Tredici, K. (2011a). Alzheimer's pathogenesis: is there neuron-to-neuron propagation? *Acta Neuropathol.* 121(5): 589–595.

Braak, H., Thal, D. R., Ghebremedhin, E., and Del Tredici, K. (2011b). Stages of the pathologic process in Alzheimer disease: age categories from 1 to 100 years. *J. Neuropath. Exp. Neurol.* 70(11): 960–969.

Braak, H., and Del Tredici, K. (2011c). The pathological process underlying Alzheimer's disease in individuals under thirty. *Acta Neuropathol.* 121(2): 171–181.

Broberg, D. J., and Bernstein, I. L. (1987). Candy as a scapegoat in the prevention of food aversions in children receiving chemotherapy. *Cancer* 60: 2344–2347.

Brook, R. D., Rajagopalan, S., Pope, C. A. 3rd, et al. (2010). Particulate matter air pollution and cardiovascular disease: An update to the scientific statement from the American Heart Association. *Circulation* 121(21): 2331–2378.

Burton, N. C., and Guilarte, T. R. (2009). Manganese neurotoxicity: lessons learned from longitudinal studies in nonhuman primates. *Environ. Health Perspect.* 117(3): 325–332.

Calderón-Garcidueñas, L., Azzarelli, B., Acuña, H., et al. (2002). Air pollution and brain damage. *Toxicol. Pathol.* 30: 373–389.

Calderón-Garcidueñas, L., Mora-Tiscareño, A., Fordham, L. A., et al. (2003a). Respiratory damage in children exposed to urban pollution. *Pediatric Pulmonol.* 36: 148–161.

Calderón-Garcidueñas, L., Maronpot, R. R., Torres-Jardon, R., et al. (2003b). DNA damage in nasal and brain tissues of canines exposed to air pollutants is associated with evidence of chronic brain inflammation and neurodegeneration. *Toxicol. Pathol.* 31: 524–538.

Calderón-Garcidueñas, L., Reed, W., Maronpot, R. R., et al. (2004). Brain inflammation and Alzheimer's-like pathology in individuals exposed to severe air pollution. *Toxicol. Pathol.* 32: 650–658.

Calderón-Garcidueñas, L., Solt, A. C., Henríquez-Roldán, C., et al. (2008a). Long-term air pollution exposure is associated with neuroinflammation, an altered innate immune response, disruption of the blood–brain-barrier, ultrafine particle deposition, and accumulation of amyloid beta 42 and alpha synuclein in children and young adults. *Toxicol. Pathol.* 36: 289–310.

Calderón-Garcidueñas, L., Mora-Tiscareño, A., Ontiveros, E., et al. (2008b). Air pollution, cognitive deficits and brain abnormalities: A pilot study with children and dogs. *Brain and Cognition* 68: 117–127.

Calderón-Garcidueñas, L., Franco-Lira, M., Henríquez-Roldán, C., et al. (2010). Urban air pollution: Influences on olfactory function and pathology in exposed children and young adults. *Exp. Toxicol. Path.* 62: 91–102.

Calderón-Garcidueñas, L., D'Angiulli, A., Kulesza, R. J., (2011a). Air pollution is associated with brainstem auditory nuclei pathology and delayed brainstem auditory evoked potentials. *Int. J. Dev. Neurosci.* 29: 365–375.

Calderón-Garcidueñas, L., Engle, R., Mora-Tiscareño, A., et al. (2011b). Exposure to severe urban air pollution influences cognitive outcomes, brain volume and systemic inflammation in clinically healthy children. *Brain & Cognition* 77: 345–355.

Calderón-Garcidueñas, L., Kavanaugh, M., Block, M., et al. (2012a). Neuroinflammation, hyperphosphorilated tau, diffuse amyloid plaques and down-regulation of the cellular prion protein in air pollution exposed children and young adults. *J. Alzheimer Dis.* 28: 93–107.

Calderón-Garcidueñas, L., Mora-Tiscareño, A., Styner, M., et al. (2012b). White matter hyperintensities, systemic inflammation, brain growth, and cognitive functions in children exposed to air pollution. *J. Alzheimer Dis.* 31(1): 183–191.

Cedervall, T., Lynch, I., Lindman, S., et al. (2007). Understanding the nanoparticle-protein corona using methods to quantify exchange rates and affinities of proteins for nanoparticles. *Proc. Natl. Acad. Sci. U. S. A.* 104(7): 2050–2055.

Centers for Disease Control and Prevention, HHS. (2012). World Trade Center Health Program requirements for the addition of new WTC-related health conditions. *Final rule. Fed. Regist.* 77: 24628–24632.

Chang, Y. F., Wen, J. F., Cai, J. F., et al. (2012). An investigation and pathological analysis of two fatal cases of cadmium poisoning. *Forensic Sci. Int.* Feb 1.

Chao, L. L., Rothlind, J. C., Cardenas, V. A., et al. (2010). Effects of low-level exposure to sarin and cyclosarin during the 1991 Gulf War on brain function and brain structure in US veterans. *Neurotoxicol.* 31(5): 493–501.

Chen, T. M., Malli, H., Maslove, D. M., et al. (2012). Toxic Inhalational Exposures. *J. Intensive Care Med.* Mar 22.

Coneus, K., and Spiess, C. K. (2012). Pollution exposure and child health: Evidence for infants and toddlers in Germany. *J. Health Econ.* 31(1): 180–196.

Conover, J. C., and Shook, B. A. (2011). Aging of the subventricular zone neural stem cell niche. *Aging Dis.* 2(1): 49–63.

Conrad, P., Corwin, J., Katz, L., et al. (1987). Olfaction and hemodialysis: baseline and acute treatment decrements. *Nephron* 47(2): 115–118.

Couper, J. (1837). On the effects of black oxide of manganese when inhaled into the lungs. *Br Ann. Med. Pharmacol.* 1: 41–42.

Craft, S., Baker, L. D., Montine, T. J., et al. (2012). Intranasal insulin therapy for Alzheimer disease and amnestic mild cognitive impairment: a pilot clinical trial. *Arch. Neurol.* 69(1): 29–38.

Criswell, S. R., Perlmutter, J. S., Videen, T. O., et al. (2011). Reduced uptake of $[^1F]$ FDOPA PET in asymptomatic welders with occupational manganese exposure. *Neurology* 76(15): 1296–1301.

Curtis, M. A., Kam, M., Nannmark, U., et al. (2007). Human neuroblasts migrate to the olfactory bulb via a lateral ventricular extension. *Science* 315(5816): 1243–1249.

Dalton, P., and Hummel, T. (2000). Chemosensory function and response in idiopathic environmental intolerance. *Occup. Med.* 15(3): 539–556.

Dalton, P. H., Opiekun, R. E., Gould, M., et al. (2010). Chemosensory loss: functional consequences of the world trade center disaster. *Environ. Health Perspect.* 118(9): 1251–1256.

Davidson, T. M., and Smith, W. M. (2010). The Bradford Hill criteria and zinc-induced anosmia: a causality analysis. *Arch. Otolaryngol. Head Neck Surg.* 136(7): 673–676.

Deems, D. A., Doty, R. L., Settle, R. G., et al. (1991). Smell and taste disorders, a study of 750 patients from the University of Pennsylvania Smell and Taste Center. *Arch. Otolaryngol. Head Neck Surg.* 117(5): 519–528.

Detje, C. N., Meyer, T., Schmidt, H., et al. (2009). Local type I IFN receptor signaling protects against virus spread within the central nervous system. *J. Immunol.* 182: 2297–2304.

de Gabory, L., Bareille, R., Daculsi, R., et al. (2011). Carbon nanotubes have a deleterious effect on the nose: the first in vitro data. *Rhinology* 49(4): 445–452.

Dell'Orco, D., Lundqvist, M., Oslakovic, C., Cedervall, T., and Linse, S. (2010). Modeling the time evolution of the nanoparticle-protein corona in a body fluid. *PLOS One* 5(6): e10949.

De, S., Leong, S. C., Fenton, J. E., et al. (2011). The effect of passive smoking on the levels of matrix metalloproteinase 9 in nasal secretions of children. *Am. J .Rhinol. Allergy* 25(4): 226–230.

DeWys, W. D. (1979). Anorexia as a general effect of cancer. *Cancer* 43: 2013–2019.

Dickson, D. W., Fujishiro, H., Orr, C., et al. (2009a). Neuropathology of non-motor features of Parkinson disease. *Parkinsonism Relat. Disord.* 15 Suppl 3: S1–5.

Dickson, D. W., Braak, H., Duda, J. E., et al. (2009b). Neuropathological assessment of Parkinson's disease: refining the diagnostic criteria. *Lancet Neurol.* 8(12): 1150–1157.

Ding, X., and Dahl, A. R. (2003). Olfactory mucosa: composition, enzymatic localization and metabolism, in *Handbook of Olfaction and Gustation* (Doty, R. L., editor) pp. 51–73, Marcel Dekker: New York.

Ding, X., Peng, H. M., and Coon, M. J. (1992). Cytochromes P450 NMa, NMb (2G1), and LM4 (1A2) are differentially expressed during development in rabbit olfactory mucosa and liver. *Mol. Pharmacol.* 42(6): 1027–1032.

Divine, K. K., Lewis, J. L., Grant, P. G., and Bench, G. (1999). Quantitative particle-induced X-ray emission imaging of rat olfactory epithelium applied to the permeability of rat epithelium to inhaled aluminum. *Chem. Res. Toxicol.* 12(7): 575–581.

Dollken, V. (1897). LTber die Wirkung des Aluminiums mit besonderer Berucksichtigung der durch das Aluminium verursachten Lasionen im Centralnervensystem. *Arch. Exp. Pathol. Pharmaco.* 98–120.

Dorman, D. C., Struve, M. F., Marshall, M. W., et al. (2006). Tissue manganese concentrations in young male rhesus monkeys following subchronic manganese sulfate inhalation. *Toxicol. Sci.* 92(1): 201–210.

Dorman, D. C., Struve, M. F., Norris, A., and Higgins, A. J. (2008). Metabolomic analyses of body fluids after subchronic manganese inhalation in rhesus monkeys. *Toxicol. Sci.* 106(1): 46–54.

Dorman, D. C., Mokashi, V., Wagner, D. J., et al. (2012). Biological responses in rats exposed to cigarette smoke and Middle East sand (dust). *Inhal. Toxicol.* 24(2): 109–124.

Dostert, C., Petrilli, V., van Bruggen, R., et al. (2008) Innate immune activation through Nalp3 inflammasome sensing of asbestos and silica. *Science* 320: 674–677.

Doty, R. L., Deems, D. and Stellar, S. (1988a). Olfactory dysfunction in Parkinson's disease: A general deficit unrelated to neurologic signs, disease stage, or disease duration. *Neurology* 38: 1237–1244.

Doty, R. L., Deems, D. A., Frye, R. E., et al. (1988b). Olfactory sensitivity, nasal resistance, and autonomic function in patients with multiple chemical sensitivities. *Arch .Otolaryngol. Head Neck Surg.* 114: 1422–1427.

Doty, R. L., and Bromley, S. M. (2004). Effects of drugs on olfaction and taste. *Otolaryngol. Clin. North Am.* 37: 1229–1254.

Doty, R. L. (2008). The olfactory vector hypothesis of neurodegenerative disease: is it viable? *Ann. Neurol.* 63(1): 7–115.

Doty, R. L. (2009a). The olfactory system and its disorders. *Seminars Neurology* 29: 74–81.

Doty, R. L. (2009b). Symposium overview: Do environmental agents enter the brain via the olfactory mucosa to induce neurodegenerative diseases? *Ann. N. Y. Acad. Sci.* 1170: 610–614.

Doty, R. L. (2012a). Olfactory dysfunction in Parkinson Disease. *Nat. Rev. Neurol.* 8(6): 329–339.

Doty, R. L. (2012b). Olfaction in Parkinson's disease and related disorders. *Neurobiol. Dis.* 46: 527–552.

Duncan-Lewis, C. A., Lukman, R. L., and Banks, R. K. (2011). Effects of zinc gluconate and 2 other divalent cationic compounds on olfactory function in mice. *Comp. Med.* 61: 361–365.

Eade, A. M., and Youngentob, S. L. (2010). The interaction of gestational and postnatal ethanol experience on the adolescent and adult odor-mediated responses to ethanol in observer and demonstrator rats. *Alcohol Clin. Exp. Res.* 34(10): 1705–1713.

Erikson, K. M., Dorman, D. C., Lash, L. H., and Aschner, M. (2008). Duration of airborne-manganese exposure in rhesus monkeys is associated with brain regional changes in biomarkers of neurotoxicity. *Neurotoxicol.* 29(3): 377–385.

Evans, J. E., Miller, M. L., Andringa, A., and Hastings, L. (1995). Behavioral, histological, and neurochemical effects of nickel (II) on the rat olfactory system. *Toxicol. Appl. Pharmacol.* 130(2): 209–220.

Faure, F., Da Silva, S. V., Jakob, I., et al. (2010). Peripheral olfactory sensitivity in rodents after treatment with docetaxel. *Laryngoscope* 120(4): 690–697.

Ferguson, E., and Cassaday, H. J. (2001–2002). Theoretical accounts of Gulf War Syndrome: from environmental toxins to psychoneuroimmunology and neurodegeneration. *Behav. Neurol.* 13(3–4): 133–147.

Ferguson, E., Cassaday, H. J., and Bibby, P. A. (2004). Odors and sounds as triggers for medically unexplained symptoms: a fixed-occasion diary study of Gulf War veterans. *Ann. Behav. Med.* 27(3): 205–214.

Fernández-Irigoyen, J., Corrales, F. J., and Santamaría, E. (2012). Proteomic atlas of the human olfactory bulb. *J. Proteomics* 75(13): 4005–4016.

Féron, V. J., Arts, J. H., Kuper, C. F., et al. (2001). Health risks associated with inhaled nasal toxicants. *Crit. Rev. Toxicol.* 31(3): 313–347.

Fiedler, N., Giardino, N., Natelson, B., et al. (2004). Responses to controlled diesel vapor exposure among chemically sensitive Gulf War veterans. *Psychosom. Med.* 66(4): 588–598.

Flynn, M. R., and Susi, P. (2009). Neurological risks associated with manganese exposure from welding operations-a literature review. *Int. J. Hyg. Environ. Health* 212(5): 459–469.

Frasnelli, J. A., Temmel, A. F., Quint, C., et al. (2002). Olfactory function in chronic renal failure. *Am. J. Rhinol.* 16(5): 275–279.

Fröhlich, E., and Roblegg, E. (2012). Models for oral uptake of nanoparticles in consumer products. *Toxicol.* 291(1–3): 10–17.

Frye, R. E., Schwartz, B. S., and Doty, R. L. (1990). Dose-related effects of cigarette smoking on olfactory function. *JAMA* 263: 1233–1236.

García, G. J., and Kimbell, J. S. (2009). Deposition of inhaled nanoparticles in the rat nasal passages: dose to the olfactory region. *Inhal. Toxicol.* 21: 1165–1175.

García, G. J., Schroeter, J. D., Segal, R. A., et al. (2009). Dosimetry of nasal uptake of water-soluble and reactive gases: a first study of interhuman variability. *Inhal. Toxicol.* 21(7): 607–618.

Gautam, S. H., Otsuguro, K. I., Ito, S., et al. (2007). T-type Ca^{2+} channels mediate propagation of odor-induced Ca^{2+} transients in rat olfactory receptor neurons. *Neurosci.* 144(2): 702–713.

Genter, M. B., Kendig, E. L., and Knutson, M. D. (2009). Uptake of materials from the nasal cavity into the blood and brain: are we finally beginning to understand these processes at the molecular level? *Ann. NY Acad. Sci.* 1170: 623–628.

Genter, M. B., Krishan, M., Augustine, L. M., and Cherrington, N. J. (2010). Drug transporter expression and localization in rat nasal respiratory and olfactory mucosa and olfactory bulb. *Drug Metab. Dispos.* 38(10): 1644–1647.

Girard, S. D., Devéze, A., Nivet, E., et al. (2011). Isolating nasal olfactory stem cells from rodents or humans. *J. Vis. Exp.* (54) pii: 2762. doi: 10.3791/2762.

Gobba, F. (2006). Olfactory toxicity: long-term effects of occupational exposures. *Int. Arch. Occup. Environ.* 79(4): 322–331.

González-Pérez, O., Quiñones-Hinojosa, A., and García-Verdugo, J. M. (2010). Immunological control of adult neural stem cells. *J. Stem Cells* 5(1): 23–31.

Goullé, J. P., Saussereau, E., Grosjean, J., et al. (2012). Accidental potassium dichromate poisoning. Toxicokinetics of chromium by ICP-MS-CRC in biological fluids and in hair. *Forensic Sci. Int. 10*; 217(1–3): e8-e12.

Griep, M. I., Van der Niepen, P., Sennesael, J. J., et al. (1997). Odour perception in chronic renal disease. *Nephrol. Dial. Transplant.* 12(10): 2093–2098.

Guerrero-Cázares, H., González-Pérez, O., Soriano-Navarro, M., et al. (2011). Cytoarchitecture of the lateral ganglionic eminence and rostral extension of the lateral ventricle in the human fetal brain. *J. Comp. Neurol.* 519(6): 1165–1180.

Guilarte, T. R., Burton, N. C., Verina, T., et al. (2008). Increased APLP1 expression and neurodegeneration in the frontal cortex of manganese-exposed non-human primates. *J. Neurochem.* 105: 1948–1959.

Guilarte, T. R. (2010). APLP1, Alzheimer's-like pathology and neurodegeneration in the frontal cortex of manganese-exposed non-human primates. *Neurotoxicol.* 31(5): 572–574.

Guxens, M., and Sunyer, J. (2012). A review of epidemiological studies on neuropsychological effects of air pollution. *Swiss Med. Wkly.* 141: w13322.

Hall, R. A. (2011). Autonomic modulation of olfactory signaling. *Sci.* Signal 4(155).

Harberts, E., Yao, K., Wohler, J. E., et al. (2011). Human herpesvirus-6 entry into the central nervous system through the olfactory pathway. *Proc. Natl. Acad. Sci. USA* 108(33): 13734–13739.

Henderson, R.G., Durando, J., Oller, A.R., Merkel, D.J., Marone, P.A., and Bates, H.K. (2012). Acute oral toxicity of nickel compounds. *Regul. Toxicol. Pharmacol.* 62(3): 425–432.

Heydel, J. M., Holsztynska, E. J., Legendre, A., et al. (2010). UDP-glucuronosyltransferases (UGTs) in neuro-olfactory tissues: expression, regulation, and function. *Drug Metab. Rev.* 42(1): 74–97.

Holbrook, E. H., Wu, E., Curry, W. T., et al. (2011). Immunohistochemical characterization of human olfactory tissue. *Laryngoscope* 121: 1687–1701.

Hossain, M. B., Vahter, M., Concha, G., and Broberg, K. (2012). Low-Level Environmental Cadmium Exposure is Associated with DNA Hypomethylation in Argentinean Women. *Environ. Health Perspect.* 120(6): 879–884.

Hou, Y. P., Gu, J. Y., Shao, Y. F., et al. (2011). The characteristics of placental transfer and tissue concentrations of nickel in late gestational rats and fetuses. *Placenta* 32(3): 277–282.

Huang, C., Titus, J. A., Bell, R. L., et al. (2012). A Mouse Model for Adolescent Alcohol Abuse: Stunted Growth and Effects in Brain. *Alcohol Clin. Exp. Res.* doi: 10.1111/j.1530-0277.2012.01759.

Hummel, T., Roscher, S., Jaumann, M. P., and Kobal, G. (1996). Intranasal chemoreception in patients with multiple chemical sensitivities: a double-blind investigation. *Regul. Toxicol. Pharmacol.* 24(1 Pt 2): S79–86.

Huneycutt, B. S., Bi, Z., Aoki, C. J., and Reiss, C. S. (1993). Central neuropathogenesis of vesicular stomatitis virus infection of immunodeficient mice. *J. Virol.* 67(11): 6698–6706.

Hutton, J. L., Baracos, V. E., and Wismer, W. V. (2007). Chemosensory dysfunction is a primary factor in the evolution of declining nutritional status and quality of life in patients with advanced cancer. *J. Pain Symptom. Manage.* 33(2): 156–165.

Iannacchione, V. G., Dever, J. A., Bann, C. M., (2011). Validation of a research case definition of Gulf War illness in the 1991 US military population. *Neuroepidemiology* 37(2): 129–140.

Inthavong, K., Zhang, K., and Tu, J. (2011). Numerical modelling of nanoparticle deposition in the nasal cavity and the tracheobronchial airway. *Comput. Methods Biomech. Biomed. Engin.* 14(7): 633–643.

Iscan, M., Reuhl, K., Weiss, B., and Maines, M. D. (1990). Regional and subcellular distribution of cytochrome P-450-dependent drug metabolism in monkey brain: the olfactory bulb and the mitochondrial fraction have high levels of activity. *Biochem. Biophys. Res. Commun.* 169(3): 858–863.

Jellinger, K., Braak, H., Braak, E., and Fischer, P. (1991). Alzheimer lesions in the entorhinal region and isocortex in Parkinson's and Alzheimer's diseases. *Ann. N. Y. Acad. Sci.* 640: 203–209.

Jellinger, K. A. (2012). Interaction between pathogenic proteins in neurodegenerative disorders. *J. Cell Mol. Med.* 16(6): 1166–1183.

Jia, C., Roman, C., and Hegg, C. C. (2010). Nickel sulfate induces location-dependent atrophy of mouse olfactory epithelium: protective and proliferative role of purinergic receptor activation. *Toxicol. Sci.* 115: 547–556.

Johnson, M. E., Uhl, C. B., Spittler, K. H., et al. (2004). Mitochondrial injury and caspase activation by the local anesthetic lidocaine. *Anesthesiology* 101: 1184–1194.

Jornot, L., Lacroix, J. S., and Rochat, T. (2008). Neuroendocrine cells of the nasal mucosa are a cellular source of brain-derived neurotrophic factor. *Eur. Respir. J.* 32: 769–774.

Kalinke, U., Bechmann, I., and Detje, C. N. (2011). Host strategies against virus entry via the olfactory system. *Virulence* 2: 367–370.

Kasper-Sonnenberg, M., Sugiri, D., Wurzler, S., et al. (2011). Prevalence of nickel sensitization and urinary nickel content of children are increased by nickel ambient air. *Environ. Res.*111: 266–273.

Kiaune, L., and Singhasemanon, N. (2011). Pesticidal copper (I) oxide: environmental fate and aquatic toxicity. *Rev. Environ. Contam. Toxicol.* 213: 1–26.

Kimbell, J. S. (2006). Nasal dosimetry of inhaled gases and particles: Where do inhaled agents go in the nose? *Toxicologic Pathology* 34: 270–273.

Kim, Y., Jeong, K. S., Song, H. J., et al. (2011). Altered white matter microstructural integrity revealed by voxel-wise analysis of diffusion tensor imaging in welders with manganese exposure. *Neurotox.* 32: 100–109.

Kincaid, A. E., Hudson, K. F., Richey, M. W., and Bartz, J. C. (2012). Rapid transepithelial transport of prions following inhalation. *J. Virol.* Sept 12.

Kitamura, F., Yokoyama, K., Araki, S., et al. (2003). Increase of olfactory threshold in plating factory workers exposed to chromium in Korea. *Ind. Health* 41: 279–285.

Klatzo, I., Wisniewski, H. M., and Streicher, E. (1965). Experimental production of neurofibrillary degeneration. Light microscopic observations. J. Neuropath. Exp. Neur. 24: 187–199.

Koh, L., Zakharov, A., Nagra, G., et al. (2006). Development of cerebrospinal fluid absorption sites in the pig and rat: connections between the subarachnoid space and lymphatic vessels in the olfactory turbinates. *Anat. Embryol. (Berl)* 211: 335–344.

Konstantinidis, I., Tsakiropoulou, E., Iakovou, I., et al. (2009). Anosmia after general anesthesia: a case report. *Anesthesia* 64: 1367–1370.

Kostopanagiotou, G., Kalimeris, K., Kesides, K., et al. (2011). Sevoflurane impairs post-opreative olfactory memory but preserves olfactory function. *Eur. J. Anaesthesiol.* 28: 63–68.

References

Krishnan, A. V., and Kiernan, M. C. (2009). Neurological complications of chronic kidney disease. *Nat. Rev.Neurol.* 5: 542–551.

Kuo, H. W., Lai, J. S., and Lin, T. I. (1997). Nasal septum lesions and lung function in workers exposed to chromium acid in electroplating factories. *Int. Arch. Occup. Environ. Health* 70: 272–276.

Kurtz, M. M., Wang, R., Clements, M. K., et al. (2002). Identification, localization and receptor characterization of novel mammalian substance P-like peptides. *Gene* 296: 205–212.

Kushkuley, J., Metkar, S., Chan, W. K., et al. (2010). Aluminum induces neurofilament aggregation by stabilizing cross-bridging of phosphorilated c-terminal sidearms. *Brain Res.* 1322: 118–123.

Latz, E. (2010). The inflammasomes: mechanisms of activation and function. *Curr. Opinion Immunol.* 22: 28–33.

Leavens, T. L., Rao, D., Andersen, M. E., and Dorman, D. C. (2007). Evaluating transport of manganese from olfactory mucosa to stratium by pharmacokinetic modeling. *Tox. Sci.* 97: 265–278.

Leon-Sarmiento, F. E., Bayona, E. A., Bayone-Prieto, J., et al. (2012). Profound olfactory dysfunction in myasthenia gravis. PLoS One 7(10): e45544.

Levesque, S., Surace, M. J., McDonald, J., and Block, M. L. (2011). Air pollution and the brain: Subchronic diesel exhaust exposure causes neuroinflammation and elevates early markers of neurodegenerative disease. *J. Neuroinflammation* 8: 105.

Levy, L., and Henkin, R. (2004). Brain gamma-aminobutyric acid levels are decreased in patients with phantageusia and phantosmia demonstrated by magnetic resonance spectroscopy. *J. Computer Assisted Tomography* 28: 721–727.

Lewis, J., Bench, G., Myers, O., et al. (2005). Trigeminal uptake and clearance of inhaled manganese chloride in rats and mice. *Neurotox.* 26: 113–123.

Liapi, C., Zarros, A., Theocharis, S., et al. (2011). Short-term exposure to nickel alters the adult rat brain antioxidant status and the activities of crucial membrane-bound enzymes: neuroprotection by L-cysteine. *Biol. Trace Elem. Res.*143: 1673–1681.

Li, B., Mahan, C. M., Kang, H. K., et al. (2011). Longitudinal health study of US 1991 Gulf War veterans: changes in health status at 10 year follow-up. *Am. J. Epidemiol.* 174: 761–768.

Li, X,, Spence, J. S., Buhner, D. M., et al. (2011). Hippocampal dysfunction in Gulf War veterans: investigation with ASL perfusion MR imaging and physostigmine challenge. *Radiology* 261: 218–225.

Lim, J. H., Davis, G. E., Wang, Z., et al. (2009). Zicam-induced damage to mouse and human nasal tissue. *PLOS ONE* 4: 10 e7647

Lin, S. C., Tai, C. C., Chan, C. C., and Wang, J. D. (1994). Nasal septum lesions caused by chromium exposure among chromium electroplating workers. *Am. J. Ind. Med.* 26, 221–228.

Liu, P., Aslan, S., Li X., et al. (2011). Perfusion deficit to cholinergic challenge in veterans with Gulf War illness. *Neurotoxicology* 32: 242–246.

London, B., Nabet, B., Fisher, A. R., et al. (2008). Predictors of prognosis in patients with olfactory disturbance. *Ann. Neurol.* 63: 159–166.

Lucchini, R., Bergamaschi, E., Smargiassi, A., et al. (1997). Motor function, olfactory threshold and hermatological indices in manganese-exposed ferroally workers. *Environ. Res.* 73: 175–180.

Lucchini, R., Guazzetti, S., Zoni, S., et al. (2012). Tremor, olfactory and motor changes in Italian adolescents exposed to historical ferro-manganese emission. *Neurotox.* 33(4): 687–696.

McBride, K., Slotnick, B., and Margolis, F. L. (2003). Does intranasal application of zinc sulfate produce anosmia in the mouse? An olfactometric and anatomical study. *Chem. Senses* 28: 659–670.

Maccioni, P., Orrú, A., Korkosz, A., et al. (2007). Cue-induced reinstatement of ethanol seeking in Sardinian alcohol preferring rats. *Alcohol* 41: 31–39.

Mair, R. G., Doty, R. L., Kelly, K. M., et al. (1986). Multimodal sensory discrimination deficits in Korsakoff's psychosis. *Neuropsychologia* 24: 831–839.

Martinon, F. (2010). Signaling by ROS drives inflammasome activation. *Eur. J .Immunol.* 40: 595–653.

Mascagni, P., Consonni, D., Bregante, G., et al. (2003). Olfactory function in workers exposed to moderate airborne cadmium levels. *Neurotox.* 24: 717–724.

Matt, G. E., Quintana, P. J., Destaillats, H., et al. (2011). Thirdhand tobacco smoke: emerging evidence and arguments for a multidisciplinary research agenda. *Environ. Health Perspect.* 119(9): 1218–1226.

Mattes, R. D., Arnold, C., and Boraas, M. (1987). Learned food aversions among cancer chemotherapy patients: incidence, nature and clinical implications. *Cancer* 60: 2576–2580.

Mattes, R. D., Curran, Jr.,, W. J., Alavi, J., et al. (1992). Clinical implications of learned food aversions in patients with cancer treated with chemotherapy or radiation therapy. *Cancer* 70: 192–200.

Maurage, P., Callot, C., Philippot, P., et al. (2011). Chemosensory event-related potentials in alcoholism: a specific impairment for olfactory function. *Biol. Psychol.* 1: 28–36.

McKeown-Eyssen, G., Baines, C., Cole, D. E., et al. (2004). Case–control study of genotypes in multiple chemical sensitivity: CYP2D6, NAT1, NAT2, PON1, PON2 and MTHFR. *Int. J. Epidemiol.* 33: 971–978.

Menco, B. P. (1980). Qualitative and quantitative freeze-fracture studies on olfactory and nasal respiratory epithelial surfaces of frog, ox, rat, and dog. III. Tight-junctions. Cell Tissue Res. 211: 361–373.

Menezes-Filho, J. A., Novaes, C de O., Moreira, J. C., et al. (2011). Elevated manganese and cognitive performance in school-aged children and their mothers. *Environ. Res.* 111: 156–163.

Middleton, F. A., Carrierfenster, K., Mooney, S. M., and Youngentob, S. L. (2009). Gestational ethanol exposure alters the behavioral response to ethanol odor and the expression of neurotransmission genes in the olfactory bulb of adolescent rats. *Brain Res.* 1252; 105–116.

Molina, L. T., Kolb, C. E., and de Foy, B. (2007). Air quality in North America's most populous city-overview of the MCMA-2003 campaign. *Atmos. Chem. Phys.* 7: 2447–2473.

Moore, C. H., Pustovyy, O., Dennis, J. C., et al. (2012). Olfactory responses to explosives associated odorants are enhanced by zinc nanoparticles. *Talanta* 88: 730–733.

Muneer, P. M., Alikunju, S., Szlachetka, A. M., and Haorah, J. (2012). The mechanisms of cerebral vascular dysfunction and neuroinflammation by MMP-mediated degradation of VEGFR-2 in alcohol ingestion. *Arterioscler. Thromb. Vasc. Biol.* 32(5): 1167–1177.

Nageris, B., Braverman, I., Hadar, T., et al. (2001). Effects of passive smoking on odour identification in children.*J. Otolaryngol.* 30: 263–265.

Nageris, B., Hadar, T., and Hansen, M. C. (2002). The effects of passive smoking on olfaction in children. *Re.v Laryngol. Otol. Rhinol.*123: 89–91.

Nazarenko, Y., Zhen, H., Han, T., et al. (2012). Potential for inhalation exposure to engineered nanoparticles from nanotechnology-based cosmetic powders. *Environ. Health Perspect.* 120(6): 885–892.

Nelson, G., Criswell, S. R., Zhang, J., et al. (2012). Research capacity development in South African manganese mines to bridge exposure and neuropathologic outcomes. *Neurotoxicol.* 33(4): 683–686.

Neuwelt, E. A. (2004). Mechanisms of disease: the blood–brain-barrier. *Neurosurg.* 54: 131–140.

Oberdorster, G., Sharp, Z., Atudorei, V., et al. (2004). Translocation of inhaled ultrafine particles to the brain. *Inhal. Toxicol.* 16: 437–445.

Ovesen, L., Sorensen, M., Hannibal, J., and Allingstrup, L. (1991). Electrical taste detection thresholds and chemical smell detection thresholds in patients with cancer. *Cancer* 68: 2260–2265.

Patel, S. J., Bollhoefer, A. D., and Doty, R. L. (2004). Influences of ethanol ingestion on olfactory function in humans. *Psychopharmacology* 171: 429–434.

Patierno, S. R., and Costa, M. (1985). DNA-protein cross-links induced by nickel compounds in intact cultured mammalian cells. *Chem. Biol Interact.* 55: 75–91.

Pellerin, C., and Booker, S. M. (2000). Reflections on Hexavalent Chromium: Health Hazards of an Industrial Heavyweight. *Environ. Health Perspect.* 108: 402–407.

Persson, E., Henriksson, J., and Tjälve, H. (2003a). Uptake of cobalt from the nasal mucosa into the brain via olfactory pathways in rats. *Toxicol. Lett.* 145: 19–27.

Persson, E., Henriksson, J., Tallkvist, J., et al. (2003b). Transport and subcellular distribution of intranasally administered zinc in the olfactory system of rats and pikes. *Toxicol.* 191: 97–108.

Prediger, R. D. S., Aguiar, A. S., Matheus, F. C., et al. (2012). Intranasal administration of neurotoxicants in animals: Support for the olfactory vector hypothesis of Parkinson's disease. *Neurotox. Res.* 21: 90–116.

Price, S. (1986). The role of zinc in taste and smell. H. L. Meiselman and R. S. Rivlin (Eds), *Clinical Measurement of Taste and Smell*. New York: Macmillan Publishing Company, pp. 443–445.

Polonskaia, E. L., Butikova, V. I., and Zhirov, S. V. (1990). Pathological changes in the central nervous system studied on the model of aluminum encephalopathy. *Zh. Nevropatol. Psikhiatr. Im. S. S. Korsakova* 90: 71–74.

Racette, B. A., Aschner, M., Guilarte, T. R., et al. (2012). Pathophysiology of manganese-associated neurotoxicity. *Neurotoxicol.* 33(4): 881–886.

Raff, A.C., Lieu, S., Melamed, M. L., et al. (2008). Relationship of impaired olfactory function in ESRD to malnutrition and retained uremic molecules. *Am. J. Kidney Dis.* 52: 102–110.

Reiss, C. S., Plakhov, I. V., and Komatsu, T. (1998). Viral replication in olfactory receptor neurons and entry into the olfactory bulb and brain. *Ann. N.Y. Acad. Sci.* 855: 751–761.

Rose, C. S., Heywood, P. G., and Costanzo, R. M. (1992). Olfactory impairment after chronic occupational cadmium exposure. *J. Occup. Med.* 34: 600–605.

Röösli, M. (2011). Non-cancer effects of chemical agents on children's health. *Prog. Biophys. Mol. Biol.* 107: 315–322.

Rückerl, R., Schneider, A., Breitner, S., et al. (2011). Health effects of particulate air pollution: A Review of epidemiological evidence. *Inhal. Toxicol.* 23: 555–592.

Salmi, E., Kaisti, K. K,, Metsahonkala, L., et al. (2004). Sevoflurane and propofol increase 11C-flumazenil binding to gammaaminobutyric acidA receptors in humans. *Anesthesia and Analgesia* 99: 1420–1426.

Salvinelli, F., Casale, M., Hardy, J. F., et al. (2005). Permanent anosmia after topical nasal anaesthesia with lidocaine 4%. *British Journal of Anaesthesia* 95: 838–839.

Sampson, P. D., Streissguth, A. P., Bookstein, F. L., et al. (1997). Incidence of fetal alcohol syndrome and prevalence of alcohol related neurodevelopmental disorder. *Teratology* 56: 317–326.

Sanico, A., Atsuta, S., Proud, D., and Togias, A. (1997). Dose-dependent effects of capsaicin nasal challenge: in vivo evidence of human airway neurogenic inflammation. *J. Allergy Clin. Immunol.* 100: 632–641.

Sarin, S., Undem, B., Sanico, A., and Togias, A. (2006). The role of the nervous system in rhinitis. *J. Allergy Clin. Immunol.* 118: 999–1016.

Schaumlöffel, D. (2012). Nickel species: Analysis and toxic effects. *J. Trace Elem. Med. Bio.l* Feb 24.

Schiffman, S. S., Nash, M. L., and Dackis, C. (1978). Reduced olfactory discrimination in patients on chronic hemodyalisis. *Physiol. Behav.* 21: 239–242.

Schneider, J. S., Decamp, E., Clark, K., et al. (2009). Effects of chronic manganese exposure on working memory in non-human primates. *Brain Res.* 1258: 86–95.

Schroeter, J. D., Kimbell, J. S., and Asgharian, B. (2006). Analysis of particle deposition in the turbinate and olfactory regions using a human nasal computational fluid dynamics model. *J. Aerosol Sci.* 19: 301–313.

Schroeter, J. D., García, G. J., and Kimbell, J. S. (2011a). Effects of surface smoothness on inertial particle deposition in human nasal models. *J. Aerosol Sci.* 42: 52–63.

Schroeter, J. D., Nong, A., Yoon, M., et al. (2011b). Analysis of manganese tracer kinetics and target tissue dosimetry in monkeys and humans with multi-route physiologically based pharmacokinetic models. *Toxicol. Sci.*120: 481–498.

Scranton, R. A., Fletcher, L., Sprague, S., et al. (2011). The rostral migratory stream plays a key role in intranasal delivery of drugs into the CNS. *PLOS ONE* 6 (4): el18711

Seidman, M. D. (2006). RE: Alexander TH, Davidson TM, Intranasal zinc and anosmia: the zinc-induced anosmia syndrome. *Laryngoscope* 116: 217–220. *Laryngoscope* 116: 1720–1721.

Shargorodsky, J., Curhan, S. G., Henderson, E., (2011). Heavy metals exposure and hearing loss in US adolescents. *Arch. Otolaryngol. Head Neck Surg.* 137: 1183–1189.

Sikora, E. R., Stone, S., Tomblyn, S., et al. (2003). Asphalt exposure enhances neuropeptide levels in sensory neurons projecting to the rat nasal epithelium. *J. Toxicol. Environ. Health A.* 66: 1015–1027.

Simard, A. R., and Rivest, S. (2006). Neuroprotective properties of the innate immune system and bone marrow stem cells in Alzheimer's disease. *Mol. Psychiatry* 11: 327–335.

Slotnick, B., Sanguino, A., Husband, S., et al. (2007). Olfaction and olfactory epithelium in mice treated with zinc gluconate. *Laryngoscope* 117: 743–749.

Soter, A., Kim, J., Jackman, A., et al. (2008). Accuracy of self-report in detecting taste dysfunction. *Laryngoscope*. 118:611–617

Sriram, K., Lin, G. X., Jefferson, A. M., et al. (2010). Mitochondrial dysfunction and loss of Parkinson's disease-linked proteins contribute to neurotoxicity of manganese-containing welding fumes. *FASEB J.* 24: 4989–5002.

Steinbach, S., Hummel, T., Bohner, C., et al. (2009). Qualitative and quantitative assessment of taste and smell changes in patients undergoing chemotherapy for breast cancer or gynecological malignancies. *J. Clin. Oncol.* 27: 1899–1905.

Steinbach, S., Hundt, W., Schmalfeldt, B., et al. (2012). Effect of platinum-containing chemotherapy on olfactory, gustatory, and hearing function in ovarian cancer patients. *Arch. Gynecol. Obstet.* 286(2): 473–480.

Steinbach, S., Hundt, W., Zahnert, T., et al. (2010). Gustatory and olfactory function in breast cancer patients. *Support Care Cancer* 18: 707–713.

Su, T., and Ding, X. (2004). Regulation of the cytochrome P4502A genes. *Toxicol. Appl. Pharmacol.* 199: 285–294.

Su, T., Sheng, J. J., Lipinskas, T. W., and Ding, X. (1996). Expression of CYP2A genes in rodent and human nasal mucosa. *Drug Metab. Dispos.* 24: 884–890.

Suga, Y., Kitade, H., Kawagishi, A., et al. (2011). Investigation for relation of gustatory and olfactory impairment in patients receiving cancer chemotherapy. *Gan. To Kagaku Ryoho* 38: 2617–2621.

Sultan, B., May, L. A., and Lane, A. P. (2011). The role of TNFα in inflammatory olfactory loss. *Laryngoscope* 121(11): 2481–2486.

Sundermann, F. W. Jr., (2001). Nasal toxicity, carcinogenicity and olfactory uptake of metals. *Ann. Clin. Lab. Sci.* 31: 3–24.

Sulkowski, W. J., Rydzewski, B., and Miarzynska, M. (2000). Smell impairment in workers occupationally exposed to cadmium. *Acta Otolaryngol.* 120: 316–318.

Suzuki, Y., and Farbman, A. I. (2000). Tumor necrosis factor-alpha induced apoptosis in olfactory epithelium in vitro: possible roles of caspase 1 (ICE), caspase 2 (ICH-1) and caspase 3 (CPP32). *Exp. Neurol.* 165: 35–45.

Tabert, M. H., Liu, X., Doty, R. L., et al. (2005). A 10-item smell identification scale related to risk for Alzheimer's disease. *Ann. Neurol.* 58: 155–160.

Talhout, R., Schulz, T., Florek, E., et al. (2011). Hazardous compounds in tobacco smoke. *Int. J. Environ. Res. Public Health* 8: 613–628.

Tallkvist, J., Persson, E., Henriksson, J., and Tjälve, H. (2002). Cadmium-metallothionein interactions in the olfactory pathways of rats and pikes. *Toxicol. Sci.* 67: 108–113.

Tang, T., Hurraβ, J., Gminski, R., and Mersch-Sundermann, V. (2011). Fine and ultrafine particles emitted from laser printers as indoor air contaminants in German offices. *Environ. Sci. Pollut. Res. Int. Nov* 18.

Thiebaud, N., Menetrier, F., Belloir, C., et al. (2011). Expression and differential localization of xenobiotic transporters in the rat olfactory neuro-epithelium. *Neurosci. Lett.* 505: 180–185.

Tierney, K. B., Baldwin, D. H., Hara, T. J., et al. (2010). Olfactory toxicity in fishes. *Aquat. Toxicol.* 96: 2–26.

Tillett, T. (2011). Third hand smoke in review: Research needs and recommendations. *Environ. Health Perpect.* 119(9): a399.

Tilton, F. A., Manler, T. K., and Gallagher, E. P. (2011). Swimming impairment and acetylcholinesterase inhibition in zebrafish exposed to copper or chlorpyrifos separately or as mixtures. *Comp. Biochem. Phsyiol. C. Toxicol. Pharmacol.* 153: 9–16.

Tisdall, F. F., Brown, A., and Defries, R. D. (1938). Persistent anosmia following zinc sulfate nasal spraying. *Pediatrics* 18: 60–62.

Tjälve, H., Henriksson, J., Tallkvist, J., et al. (1996). Uptake of manganese and cadmium from the nasal mucosa into the central nervous system via olfactory pathways in rats. *Pharmacol. Toxicol.* 79: 347–356.

Tjälve, H., and Henriksson, J. (1999). Uptake of metals in the brain via olfactory pathways. *Neurotoxicology* 20: 181–195.

Torreilles, J., and Guerin, M. C. (1990). Nickel (II) as a temporary catalyst for hydroxyl radical generation. *FEBS Lett.* 272: 58–60.

Tran, A. H., Berger, A., Wu, G. E., et al. (2011). Early B-cell factor regulates the expression of Hemokinin-1 in the olfactory epithelium and differentiating B lymphocytes. *J. Neuroimmunol.* 232: 41–50.

Turker, S., Onur, E., and Ozer, Y. (2004). Nasal route and drug delivery systems. *Pharm. World Sci.* 26: 137–142.

Valenzuela, C. F., Morton, R. A., Diaz, M. R., and Topper, L. (2012). Does moderate drinking harm the fetal brain? Insights from animal models. *Trends Neurosci.* 35(5): 284–292.

Vasterling, J. J, Brailey, K., Tomlin, H., et al. (2003). Olfactory functioning in Gulf War-era veterans: relationships to war-zone duty, self reported hazards exposures and psychological stress. *J .Int. Neuropsychol. Soc.* 9: 407–418.

Vearrier, D., and Greenberg, M. I. (2011). Occupational health of miners at altitude: adverse health effects, toxic exposures, pre-placement screening, acclimatization, and worker surveillance. *Clin. Toxicol. (Phila)* 49: 629–640.

Vent, J., Robinson, A. M., Gentry-Nielsen, M. J., et al. (2004). Pathology of the olfactory epithelium: smoking and ethanol exposure. *Laryngoscope* 114: 1383–1388.

Viswaprakash, N., Dennis, J. C., Globa, L., et al. (2009). Enhancement of odorant-induced responses in olfactory receptor neurons by zinc nanoparticles. *Chem. Senses* 34: 547–557.

Walter, B. A., Valera, V. A., Takahashi, S., and Ushiki, T. (2006). The olfactory route for cerebrospinal fluid drainage into the peripheral lymphatic system. *Neuropath. Appl. Neurobiol.*32: 388–396.

Wang, Y., Wang, B., Zhu, M. T., et al. (2011). Microglial activation, recruitment and phagocytosis as linked phenomena in ferric oxide nanoparticle exposure. *Toxicol. Lett.* 205: 26–37.

Wisnivesky, J. P., Teitelbaum, S. L., Todd, A. C., et al. (2011). Persistence of multiple illnesses in World Trade Center rescue and recovery workers: a cohort study. *Lancet* 378: 888–897.

Wu, Z., Du, Y., Xue, H., et al. (2012). Aluminum induces neurodegeneration and its toxicity arises from increased iron accumulation and reactive oxygen species (ROS) production. *Neurobiol. Aging* 33(1): 199. e1–12.

Xie, F., Zhou, X., Behr, M., et al. (2010). Mechanisms of olfactory toxicity of the herbicide 2, 6-dichlorobenzonitrile: essential roles of CYP2A5 and target-tissue metabolic activation. *Toxicol. Appl. Pharmacol.* 249: 101–106.

Xu, A., Prophete, C., Chen, L. C., et al. (2011). Interactive effects of cigarette smoke extract and World Trade Center dust particles on airway cell cytotoxicity. *J. Toxicol. Environ. Health A.* 74: 887–902.

Yagi, S., Tsukatani, T., Yata, T., et al. (2007). Lipopolysaccharide-induced apoptosis of olfactory receptor neurons in rats. *Acta Otolaryngol.* 127: 748–753.

Youngentob, S. L., and Glendinning, J. I. (2009). Fetal ethanol exposure increases ethanol intake by making it smell and taste better. *Proc. Natl Acad. Sci. USA* 106: 5359–5364.

Zakharov, A., Papaiconomou, C., and Johnston, M. (2004). Lymphatic vessels gain access to cerebrospinal fluid through unique association with olfactory nerves. *Lymphat. Res. Biol.*2: 139–146.

Zhao, J., Bowman, L., Zhang, X., et al. (2009). Metallic nickel nano- and fine particles induce JB6 cell apoptosis through a caspase-8/AIF mediated cytochrome c-independent pathway. *J. Nanobiotechnol* 7: 2.

Zhen, Z., and Xie, J. (2012). Development of manganese-based nanoparticles as contrast probes for magnetic resonance imaging. *Theranostics* 2: 45–54.

Part 5

Olfaction in Nonhuman Forms

Chapter 21

Microbial Chemical Sensing

JUDITH VAN HOUTEN

21.1 INTRODUCTION

Microorganisms are capable of detecting a host of chemicals in their environments and they display a very wide range of responses. If the chemicals are nutrients, the cells transport and use them for growth in size and number. However, many cues signal the quality of the environment, that is, the presence of oxygen, attractants, optimal pH, toxins or conditions that inhibit growth. The means by which microbes use their motility to accumulate in or disperse from the areas of these cues is the subject of this review. Using this latter narrow definition of microbial chemical sensing and response still leaves an enormous variety of systems to explore from bacteria to protozoa.

21.2 BACTERIAL CHEMOTAXIS

Because our current understanding of bacterial chemosensory transduction developed over more than 40 years, it is extremely detailed and deep. While bacterial chemoresponse has been observed since the 1880s (Adler, 1975; Berg, 1975), genetic and cell biological studies began in the 1960s with genetic dissections of how motile *Escherichia coli* move up gradients of attractants like aspartate or down gradients of repellents like acetate, and provided a host of very useful mutants (Armstrong and Adler, 1969; Hazelbauer et al., 1969). The field has matured to our current understanding of chemosensory transduction of several bacterial species in addition to *E. coli* (Porter et al., 2011; Rao and Ordal, 2009) and how they use chemical sensing in pathogen colonization and interaction with the host immune system, quorum sensing for parasitic life styles and virulence regulation, and biofilm formation (Wadhams and Armitage, 2004).

E. coli or *Salmonella typhimurium* swim by rotating their 5–8 flagella that arise from the sides of the cell. Unlike eukaryotic flagella, bacterial flagella are in a fixed helix. When moving counterclockwise, the flagella form a bundle at a pole and propel the cell forward. Spontaneously there will be a switch of rotation to clockwise, making the flagellar bundle fly apart and the cell to tumble for a short time. When the sense of the rotation returns to counterclockwise, the cell moves forward again but on a different trajectory (Berg, 2008). Chemical stimuli bias this process so that what is a random walk of smooth runs with random turns becomes a biased random walk (Berg, 1975). Movement up a gradient of attractant results in long runs with few tumbles until the cell eventually arrives at the attractant. Repellents bias the walk in the opposite direction, causing more frequent tumbles and net movement away from the repellent. While other bacteria move by rotating the flagella at their poles or glide with no flagella at all, their chemosensory transduction pathways that modulate motility have many components in common (Adler, 1975; Berg, 1975; Porter et al., 2011; Wadhams and Armitage, 2004). Although the field refers to these behaviors as chemotaxis, they qualify as biased random walks (Berg, 1975) or kinesis because the cells do not orient in the gradient. We will see true chemotaxis in the later example from *Dictyostelium discoideum*.

Bacteria respond to spatial gradients as they swim and also to temporal changes in chemical stimuli, making it possible to see the change in flagellar rotation as stimuli are abruptly introduced into the solution surrounding the cells. Runs of smooth swimming and tumbles can be seen in the rotation and reversal of rotation of bacterial cell bodies that are tethered to slides by antibodies against flagellar proteins (Adler, 1975). Similarly the tracking microscope allowed visualization of the cells in three dimensions (Berg, 1975) as stimuli were applied.

Handbook of Olfaction and Gustation, Third Edition. Edited by Richard L. Doty.
© 2015 Richard L. Doty. Published 2015 by John Wiley & Sons, Inc.

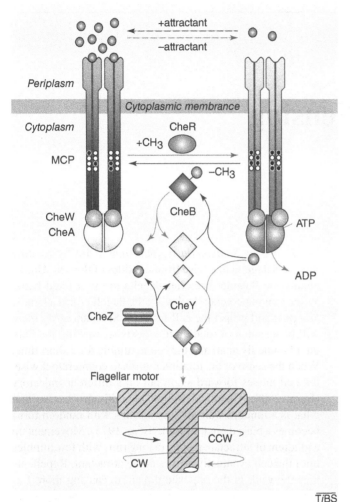

Figure 21.1 The chemoreceptor signaling pathway in E. coli. Components and reactions in red promote counter clockwise (CCW) flagellar rotation; those in green promote clockwise (CW) flagellar rotation. Components in grey represent inactive forms. Solid lines represent enzymatic reactions; broken lines indicate binding interactions. CheA-derived phosphoryl groups are shown as blue spheres. Receptor modification sites are shown as white (unmethylated) and black (methylated) circles. (*See plate section for color version.*)

In *E. coli* the five transmembrane receptors for attractants and repellents are called methyl-accepting proteins (MCPs)(Grebe and Stock, 1998) because methylation is part of their adaptation process (discussed below). The most abundant MCP is Tar that binds the ligand aspartate directly and another ligand, maltose, indirectly through a periplasmic binding protein (Figure 21.1). It also functions as the MCP for repulsion by nickel and cobalt. Other MCPs Trg, Tsr, Tap, and Aer mediate responses to ribose and galactose, serine, peptides, and oxygen respectively. The crystal structures of the receptors, their multimeric structures and ligand binding pockets have been established in great detail (Hazelbauer et al., 2008). Upon ligand binding, the transmembrane MCPs signal through the lipid bilayer to the cell interior by a piston-like molecular movement that has been described on a detailed level (Hall et al., 2011). The conformation changes in the transmembrane helices that link the binding domain to the domains involved in the intracellular signal cascade are still under investigation.

The signal transduction cascade for *E. coli* chemotaxis is outlined in Figure 21.1. At first glance, the pathway looks like that found in eukaryotic signaling with proteins interacting and becoming phosphorylated by kinases. In bacteria the phosphorylation cascade starts with ATP bound to the kinase domain in one protein of a dimer (e.g., see CheA Figure 21.1), and the γ phosphate of ATP is transferred to the other subunit of the dimer. However, the amino acid phosphorylated is histidine and the same protein goes on to transfer this same phosphate to an aspartate on a response regulator protein (e.g., see CheY in Figure 21.1). This is a two component system (sensor histidine kinase and response regulator) that is well represented in bacteria for detecting environmental signals (Krell et al., 2010) but can also be found in eukaryotes such as yeast, protozoa, and plants (Schaller et al., 2011).

In the cytoplasm CheW couples the histidine kinase CheA to the receptors that influence CheA's phosphorylation state. The autophosphorylation of CheA is critical

21.2 Bacterial Chemotaxis

in determining whether the signal to the flagellar rotor will cause clockwise or counterclockwise rotation, that is, tumbling or smooth swimming that underlie repulsion and attraction. In the presence of repellents or the removal of attractants (right side of Figure 21.1), the kinase domain of CheA that binds ATP auto-phosphorylates a histidine residue in a regulatory domain of the other member of the dimer. This phosphorylated form of CheA-P in turn transfers its phosphate to response regulator CheY. The phosphorylated form of CheY (CheY-P) interacts with the proteins of the flagellar motor and promotes clockwise movement. A phosphatase CheZ turns over the CheY-P to allow continual response to gradient sensing.

Binding of an attractant to the receptor accomplishes the converse of this transduction pathway for repellents (left side of Figure 21.1). Receptors bound with attractant do not allosterically activate the auto-phosphorylation of CheA and there is no subsequent phospho-relay to CheY. The flagellar motor consequently turns counterclockwise and the cell swims smoothly.

In addition to this phospho-relay signaling pathway, there is an adaptation process of addition and removal of methyl groups on glutamates of the receptors. Addition of the methyl groups is catalyzed by CheR and removal by CheB-P. CheA-P transfers a phosphate from its histidine to the CheB protein; CheB-P in turn removes methyl groups from the receptors. The methylation level of the receptor reflects the recent past of the cell's exposure to attractants and repellents. Methylation of the cytoplasmic domain of a receptor by CheR increases the ability of the receptor to activate CheA; the CheA-P form promotes clockwise rotation and tumbling through CheY-P. CheB-P opposes bias toward tumbling by removing methyl groups; CheB activity increases about 100-fold after phosphorylation by CheA-P. Interestingly, CheB also catalyzes the deamidation of glutamines to glutamates which can then be methylated. Thus there is a feedback system: methylation of receptors by CheR promotes tumbling through the receptor activation of CheA-P and CheA-P in turn reduces methylation by activating its phosphor-relay substrate CheB-P. Adaptation is thought to reset the pre-stimulus receptor and rotor bias very precisely back to the pre-stimulus levels. In this way, the cell continues to respond over a large dynamic range of stimulus concentrations that it encounters in a gradient.

Chemoreceptors of bacteria exhibit "exquisite sensitivity, extensive dynamic range and precise adaptation" (Hazelbauer et al., 2008). These characteristics arise in part from the physical arrangement of the receptors in large arrays primarily at a pole of the cell. A large gain is achieved in part by linking one receptor to multiple CheA molecules, allowing for large amplifications of the signaling to the kinase (Li and Hazelbauer, 2005). Additional gain and sensitivity are achieved from aggregation of the cell's 7,500 or so receptors into several large

patches of trimers of homodimers (Figure 21.2 (Bray, 2002; Kentner and Sourjik, 2006), see (Hazelbauer et al., 2008)for review). The activation of a receptor in a cluster appears to influence its neighbors and amplify the signal, explaining in part why a 1% change in receptor occupancy can result in a 50% change in motor response. There is thought to be a conformational spread of allosteric interactions and cooperativity among the receptors that account for sensitivity to small changes in concentration over several orders of magnitude of background stimulus concentration (Keymer et al., 2006). Rao and Ordal (Rao and Ordal, 2009) provide an excellent description of the role of the signaling arrays in which the receptors are highly cooperative at low ligand concentration and act as solitary receptors at high concentration thereby contributing to high sensitivity and also large dynamic range. They also discuss the coupling of mixtures of receptors in arrays that allow for integration of signaling information and tuning of receptor output to the concentration of attractant.

Networking of multiple receptor types contributes to signal integration in clusters as described above (Keymer et al., 2006) and can also be seen in the adaptation process that must reset the motor bias back precisely to pre-stimulus levels in order for the cells to respond in the presence of persistent stimuli. In the receptor clusters of trimers of dimers (Endres, 2009) there can be mixtures of homodimers of more than one receptor type in a trimer, such as Tar receptor homodimers with Tsr receptor homodimers (Li and Hazelbauer, 2005). The high abundance receptors such asTar or Tsr have the sequence NWETF that is required for efficient adaptation modification by methylation, demethylation, and deamidation to create the glutamate methylation sites. Low abundance receptors like Trg lack this sequence and are inefficiently modified, and, similar to mutants lacking this sequence, are also defective in adaptation. However, in mixtures of the high and low abundance receptors, the modifications for both receptors and adaptation are timely and efficient because the high abundance receptors provide assistance to the low abundance receptors in "assistance neighborhoods" (Endres and Wingreen, 2006; Li and Hazelbauer, 2005). The neighborhoods are five to seven near-by interacting receptors, subsets within the clusters composed of many more interacting receptors (Kentner and Sourjik, 2006; Li and Hazelbauer, 2005).

Adaptation at the level of the flagellar motor has been suspected for some time, and now has been demonstrated by Berg and co-workers (Yuan et al., 2012). Methylation and demethylation clearly are the major mechanisms for adaptation, but mutants that lack these enzymatic activities can still adapt slowly. The motor of the flagellum is extremely sensitive to levels of CheY-P, the diffusible protein that binds the receptor and the FliM proteins of the motor. Recent studies have uncovered the signal-dependent

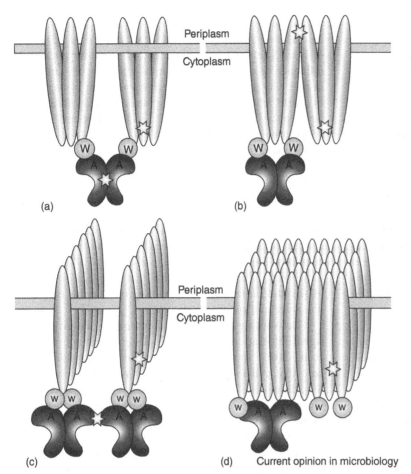

Figure 21.2 Models of cluster organization. Crucial interactions are marked with a yellow star. (a) Trimers of receptor dimers, formed by interactions at the cytoplasmic receptor tip, bind CheA through CheW and are connected by CheA dimerization. (b) Trimers-of-dimers interact at the periplasmic side. (c) Receptors form one-dimensional hedgerows through lateral interactions between their cytoplasmic domains. CheA dimers bind receptors via CheW and connect adjacent hedgerows through interactions between the P5 domains. (d) Receptors self-organize into a two-dimensional lattice through lateral interactions between their cytoplasmic domains. CheW and CheA bind with a variable stoichiometry and stabilize clustering. (*See plate section for color version.*)

turnover of the FLiM protein that remodels the target of signaling. More FliM allows more sensitivity to the decrease in CheY-P from repellents or reduction in attractants that promote smooth swimming from counterclockwise flagellar rotation. The process of adaptation by adjusting levels of FliM to adjust sensitivity is a much slower process than the adaptation by methylation levels.

While the focus above has been on *E. coli*, other bacteria have been thoroughly studied, such as *Bacillus subtilis, Rhodobacter sphaeroides*, and *Myxococcus xanthus* (see (Porter et al., 2011; Rao and Ordal, 2009; Wadhams and Armitage, 2004) for broader views beyond *E. coli*). The role of chemotaxis of *Rhizobium* for symbiosis that is beneficial to plants, likewise, has been documented (Miller et al., 2007).

Bacteria make use of a second messenger not shared with eukaryotes or Archaea, cyclic di-GMP (see (Hengge, 2009; Ryan et al., 2012; Schirmer and Jenal, 2009) for reviews). This signaling molecule is of interest to this overview of chemical sensing because it is a regulator of processes like virulence and biofilm formation that also depend upon bacterial motility and chemotaxis signaling systems (Jenal and Malone, 2006; Tamayo et al., 2007; Williams et al., 2007) Cyclic di-GMP is a ubiquitous second messenger in bacteria and controls a large and diverse range of targets that have a PilZ or one of three other domains. Cyclic di-GMP binds to these domains it acts allosterically to promote protein-protein interactions or binding to DNA elements. Remarkably, it is known to regulate at transcriptional, translational, and posttranslational levels. Motile cells generally have low levels of unbound c-di-GMP that at high levels serve as second messengers in the switch between planktonic motile and sedentary behavior through controlling flagellar synthesis (Tamayo et al., 2007). The switch to sedentary life style is critical for biofilm formation, as is promotion of biofilm matrix synthesis by cyclic di-GMP (Figure 21.3 shows an example of *V. cholera* (Tamayo et al., 2007)). This second messenger is implicated in quorum sensing, biofilm formation, exopolysaccharide production, attachment to surfaces, and attenuation of motiliy and virulence (Jenal and Malone, 2006).

Bacteria regulate the synthesis and degradation of the second messenger cyclic di-GMP with diguanylate cyclases and specific phosphodiesterases by allosteric effectors, phosphorylation, transcription control, and

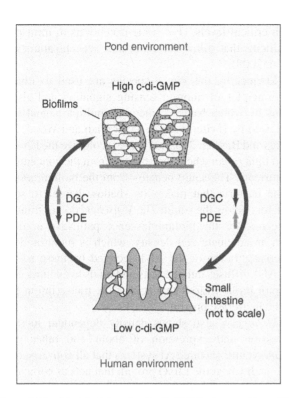

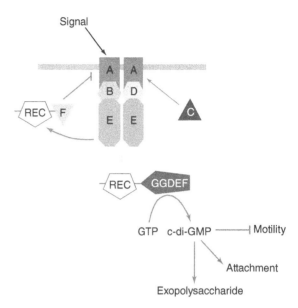

Figure 21.3 Model of the role of c-di-GMP in the transition of V. cholerae from persistence in aquatic reservoirs to survival in the human host. C-di-GMP is predicted to be high in V. cholera existing in biofilms attached to biotic and abiotic surfaces in the pond environment. Upon entry into the human host, induction of PDE and/or repression of DGC activities can lower c-di-GMP and allow dispersion from the biofilm and maximal expression of the virulence genes. Conversely, c-di-GMP must be elevated once again through activation of DGC and/or repression of PDE to resume the biofilm lifestyle.

Figure 21.4 The Wsp chemosensory pathway. The Wsp chemosensory pathway is homologus to the chemosensory system in E. coli. In this model, WspE kinase activity is activated via signal-dependent conformational changes in the WspA/WspB or WspA/WspD complex. WspE phosphorylates WspR leading to downstream signal propagation. Adaptation to external signals is mediated by the phosphorylation-dependent methylesterase activity of WspF and the methyltransferase WspC. Intracellular information transfer is indicated by arrows (25, 50, 88).

restriction to microdomains (Tamayo et al., 2007). Because cyclic di-GMP is thought to transmit and integrate information about environmental stimuli to the cell, it is presumed that cyclic di-GMP synthesis pathway is regulated by environmental signals, but the mechanisms of this regulation are far from clear. There is only a small list of environmental signals that affect the levels of cyclic di-GMP including oxygen, blue light, nutrient starvation, and mucin (Jenal and Malone, 2006). Where the cyclic di-GMP pathway intersects most clearly with bacterial chemotaxis is in *Pseudomonas fluorescens* and *P. aeruginosa* that are pathogens in lung infection in patients with cystic fibrosis. The pathway in Figure 21.4 (Jenal and Malone, 2006) shows the signal transduction pathway from the *Pseudomonas* chemoreceptor that is homologous to those in *E. coli* for the cyclic di-GMP pathway that controls motility and attachment. However, the external ligands for the receptor for *Pseudomonas* are not known.

Cyclic di-GMP promotes the adhesion of pathogenic bacteria to intestinal cells and expression of virulence factors in *Vibrio cholera* and *Clostridium difficile* by regulating riboswitches (Hengge, 2009; Ryan et al., 2012). Since cyclic di-GMP is so ubiquitous in bacteria it is not surprising that it is recognized by the mammalian innate immune sensor STING and serves as a microbial signal of infection (Burdette et al., 2011; Yin et al., 2012). There are still more variations on the cyclic di-nucleotide signaling in bacteria as cyclic di-AMP is found as a recognition signal in the mammalian innate immune system and a mixed cyclic di-AMP-GMP regulates *V. cholerae* virulence (Davies et al., 2012).

We have discussed bacteria as though they live in pure cultures, but that is clearly not the case when one examines the recently delineated microbiomes of every human surface (skin, gut, vagina, mouth, nares) (Balter, 2012; Relman, 2012). In the gut alone, our microorganisms interact with host cells in very complex and important ways. The microbes are involved in the regulation of our metabolism, nutrition, and immune-inflammation systems (Hooper et al., 2012; Nicholson et al., 2012). As we saw with cyclic di-GMP, the microbes interact with our immune system through signal transduction pathways that should be better understood in the near future if we are to comprehend our relationship with our own microbes (Kau et al., 2011; Rak and Rader, 2011).

Bacteria of our human biome and other ecosystems interact in communities with other bacteria and other microbes (Little et al., 2008; West et al., 2007). Two social activities that have already been mentioned in this review are quorum sensing and biofilm formation that are described in more depth below.

21.3 QUORUM SENSING

Bacteria monitor the density of their populations through the process of quorum sensing (Federle, 2009; Jayaraman and Wood, 2008; Little et al., 2008; Waters and Bassler, 2005). The bacteria communicate and coordinate among themselves through secreted cues called Auto Inducers (AIs) that allow the bacteria to act in unison in transcription or motility when the population density has reached the appropriate level. Quorum sensing is very specific, allowing bacteria to communicate among cells of the same species when in mixtures of microbes. Gram positive bacteria utilize a variety of homoserine lactones called AI-1s; Gram negative bacteria utilize a variety of peptides called AI-Ps. These AIs are specific for intraspecies communication, in contrast to a relatively newly discovered autoinducers (AI-2) that allow inter-species communication among Gram positive and negative bacteria. AI-2 synchronizes transcription and behavior so that the bacteria may act in unison in biofilm formation or virulence in pathogenesis. The outcomes of quorum sensing also include bioluminescence, antibiotic production, and sporulation, but key to all of these very varied outcomes is the communication among bacteria that allows them to respond only when their secreted signals and population densities reach critical levels. (For some exceptions to intraspecies specificity that allow for cross-talk see (Jayaraman and Wood, 2008).

Frequently, the *Vibrio* species are used to illustrate the principles of quorum sensing signaling and also the means by which bacteria integrate multiple signal inputs (reviewed by (Federle, 2009; Jayaraman and Wood, 2008; Waters and Bassler, 2005). *V. fischeri* colonize the Hawaiian squid light organ where they benefit from the concentration of nutrients. The squid benefits from the bioluminescence of the bacteria that makes its shadow harder to see by predators below the squid. The *V. fischeri* do not transcribe the genes for the bioluminescence pathways until they reach an adequate cell density, which is monitored from the concentration of the AI-1 secreted by these bacteria. The AI-1 diffuses into the cells and binds to an intracellular receptor protein LuxR that acts as the transcription factor for the luciferase operon.

V. harveyi also shows density dependent luciferase expression and expression of about 100 other genes. *V. harveyi* utilizes three AI systems that all converge on the same pathway at the LuxO protein that acts as coincidence detector. Three AIs that are secreted by *V. harveyi* include two different acyl homoserine lactones (HAI-1 and CAI-1), and a furanosyl borate diester autoinducer-2 (AI-2) (Jayaraman and Wood, 2008). Each binds to a surface receptor histidine kinase or a periplasmic protein that in turn binds to the receptor kinase (Figure 21.5 (Waters and Bassler, 2005)). The most detail of the signaling pathways is known for the HAI-1 and AI-2 pathways that utilize the two component histidine receptor kinases LuxN and LuxQ, respectively (Figure 21.6 (Neiditch et al., 2006)). One notable difference from the chemotaxis histidine kinases is

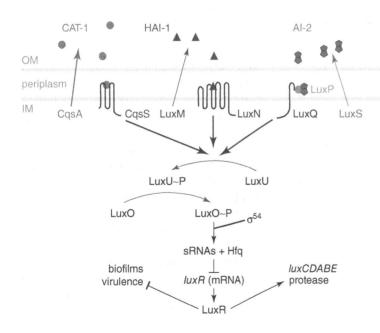

Figure 21.5 Vibrio harveyi produces and responds to three distinct autoinducers. The sensory information is fed into a shared two-component response regulatory pathway. The arrows indicate the direction of phosphate flow in the low-cell-density state. CAI-1, HAI-1, and AI-2 are respectively represented by green circles, red triangles, and blue double pentagons. OM, outer membrane; IM, inner membrane. (*See plate section for color version.*)

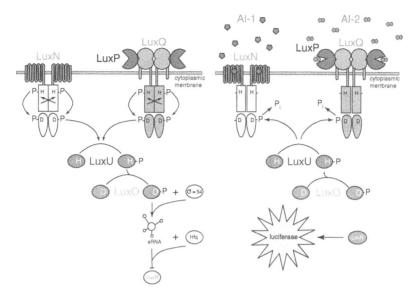

Figure 21.6 The Vibrio harveyi Quorum-Sensing Signal Transduction Circuit Under conditions of low cell density (i.e., in the absence of appreciable concentrations of autoinducers; left), LuxN and LuxPQ act as kinases that catalyze histidine (H) phosphorylation, presumably across dimer pairs (arrows). Phosphate is subsequently transferred to a conserved aspartate (D) in the receiver domains of LuxN or LuxQ and then to histidine and aspartate residues on LuxU and LuxO, respectively. LuxO-phosphate, in conjunction with the sigma factor o^{54}, promotes transcription of genes encoding small regulatory RNAs (sRNAs). The sRNAs, together with the chaperone Hfq, destabilize the mRNA encoding the transcription factor LuxR. Under high cell density conditions, (i.e., in the presence of high concentrations of autoinducers; right), LuxN and LuxPQ bind their respective autoinducer ligands (AI-1; pentagons; AI-2: double pentagons) and are converted from kinases to phosphatases. Phosphate is stripped from LuxO and LuxU and is hydrolyzed to inorganic phosphate (P_i). Because dephosphorylated LuxO is inactive, LuxR is produced and activates the expression of the luciferase operon. As a result, the bacteria produce light. (*See plate section for color version.*)

the periplasmic binding protein for AI-2 called LuxP. LuxP interacts in turn with LuxQ, the transmembrane receptor kinase. At low concentration of either AI-1, the receptors are active histidine kinases that auto-phosphorylate across dimer pairs and, through phosphor-relay, transfer the phosphate to an aspartate of the receiver domain of the same protein. The phosphates from both receiver response domains are then transferred to histidines of LuxU and LuxO where the pathways converge. Phosphorylated LuxO with a specific sigma factor for production of sRNA and a chaperone Hfq represses LuxR synthesis by destabilizing its mRNA, thus repressing transcription of the luciferase operon. At high AI concentrations, the sensor kinases switch from being kinases to phosphatases. The result is that LuxO and LuxU are no longer phoshorylated, LuxR is produced, and the genes for bioluminescence among others are expressed. The same basic principles of signaling also apply to the CAI-1 pathway.

The quorum sensing pathways of *V. harveyi* converge on the same protein, LuxO, which serves as a coincidence detector for environmental signals, but the three pathways do not have equal consequences for production of LuxR. In the absence of HAI-1, the cells continue to control light production but at 10,000 times lower light production per cell. Therefore, they appear to pay much more attention to the species specific HAI-1 signals than to the inter-species common signal AI-2.

V. cholerae is responsible for endemic diarrhea disease. This species has a no HAI-1 system, but does possess CAI-1 and AI-2 systems similar to that of *V. harveyi* (Miller et al., 2002), but, instead of controlling luciferase operons, this system promotes virulence factor production and biofilm formation at low concentration and represses them at high concentration. Virulence factor production regulated by quorum sensing is not unique to *V. cholerae* but in these other pathogenic bacteria, virulence factors are typically induced at high population densities (Waters and Bassler, 2005). Waters and Bassler (Waters and Bassler, 2005) point out that the low density induction of virulence factors and biofilm formation in *V. cholerae* fits its life style in that the diarrhea from the *V. cholerae* infection flushes the bacteria out into the environment and suppression of biofilm formation at high cell density would promote dissemination.

Quorum sensing is not restricted to microbe to microbe signaling. Inter-kingdom signaling between microbe and host is illustrated by the histidine sensory kinase QseC that binds both the bacterial signal autoinducer AI-3 and epinephrine or norepinephrine from the host (Clarke et al., 2006). Epinephrine activates the same set of genes through QseC as does AI-3 (Jayaraman and Wood, 2008).

Enterohemorrhagic *E. coli* 0157:H7 that colonize the colon use this signaling to activate its virulence genes; loss of QseC sensory kinase receptor reduces this expression and pathogenicity (Clarke et al., 2006; Jayaraman and Wood, 2008). Even more levels of interaction are among the indigenous microbiota of the gut with the host in resisting *C. difficile* colonization (Britton and Young, 2012). See (Jayaraman and Wood, 2008) for more inter-kingdom communications and (Waters and Bassler, 2005) for examples of quenching of the microbial quorum sensing by the host that can be algae, legumes, or humans.

21.4 BIOFILMS

Biofilms are fascinating in their variety and complexity. They are communities of microbes encased in a matrix of polysaccharide. They form on many kinds of surfaces and have extreme health consequences when they form on indwelling catheters or the lung epithelia of cystic fibrosis patients (Lynch and Robertson, 2008; Waters and Bassler, 2005). Microbes by living in a biofilm increase their resistance to antibiotics, extremes of pH, dessication, uv damage, and predation. There are several stages of biofilm formation: initial attachment, irreversible attachment, maturation stages 1 and 2, dispersion (Figure 21.7 (Monroe, 2007; Sauer et al., 2007)). The microbes begin as planktonic free swimming forms and disperse back into this life style at the end of the biofilm cycle. See also the NSF funded biofilm teaching tool http://www.hypertextbookshop.com/biofilmbook/v004/r003/. There are many bacteria that can form biofilms, in monocultures or in mixtures that can also include fungi (Lynch and Robertson, 2008).

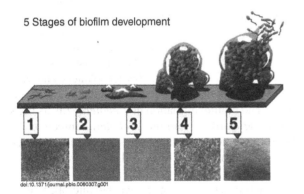

Figure 21.7 Biofilm maturation is a complex developmental process involving five stages. Stage 1, initial attachment; stage 2, irreversible attachment; stage 3, maturation I; stage 4, maturation II; stage 5, dispersion. Each stage of development in the diagram is paired with a photomicrograph of a developing P. aeruginosa biofilm. All photomicrographs are shown to same scale.

The sensor systems (quorum sensing and cyclic di-GMP) that we have already examined have roles to play in biofilm formation. Quorum sensing allows intra- and inter-species communication in communities of biofilm microbes. Cyclc di-GMP at high levels promotes biofilm formation, in both *Pseudomonas aeuriginosa* and *P. fluorescens*, but the mechanism varies with species. In *P. aeuriginosa*, cyclic di-GMP controls extracellular polysaccharide deposition and flagellar function while in *P. fluorescens* it controls secretion and surface localization of LapA that controls the commitment to a biofilm existence (Monds and O'Toole, 2009). See also Figure 21.3 above for cyclic-di-GMP in *V. cholerae*.

An illustration of the intersection of quorum sensing and cyclic di-GMP signaling is in *Pseudomonas fluorescens* biofilm formation. The adhesion protein LapA must be transported to the outer membrane for the formation of microcolonies in the evolution of the biofilm. Two regulators of LapA are cyclic di-GMP and inorganic phosphate (P_i). P_i is sensed through a two component system of PhoBR (Monds and O'Toole, 2009) that activates target genes including rapA, a phosphodiesterase that reduces cyclic di-GMP. Under conditions of low cyclic di-GMP, the secretion and transport of LapA to the outer membrane is inhibited as is biofilm formation.

21.5 EUKARYOTIC MICROBES

As we see from the varied communication and chemical sensing strategies that prokaryotic microbes live social lives (West et al., 2007). Extremely rapid and sensitive chemotaxis allows bacteria to exploit marine niches (Stocker et al., 2008). Quorum sensing has fitness benefits for microbe populations (Darch et al., 2012) and also allows bacteria to behave as though a multicellular organism. There are interesting parallels in an eukaryotic microbe *Dictyostelium discoideum* that uses cell to cell communication and quorum sensing to bring together individual amoebae cells into a multicellular structure with multiple cellular phenotypes (Parent and Devreotes, 1996).

Dictyostelium amoebae are social. They communicate among vegetative, free moving cells with secreted molecules that signify density of the population and serve quorum sensing functions (see (Golé et al., 2011) for analysis of quorum sensing kinetics). The *Dictyostelium* amoebae feed on the bacteria in soil and decaying detritus of the forest floor. As they grow they secrete multiple factors that have been harvested from growth media and shown to affect transcription, motility, and adhesion, for example (Kolbinger et al., 2005).

When bacteria become scarce, the *Dictyostelium* cells begin to starve and aggregate to form a multicellular slug that differentiates into a base and stalk topped with a

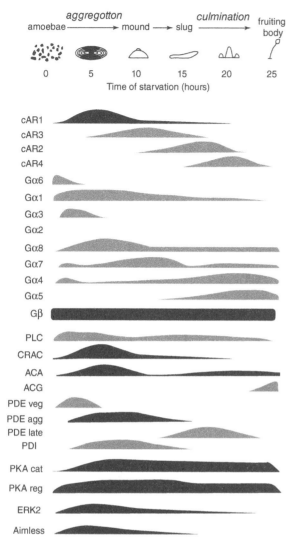

Figure 21.8 Time course of expression of major RNA transcripts. Top panel illustrates the developmental stages appearing after the onset of starvation. Receptors, G proteins, regulators, and several effectors are shown below. Components essential for early development are heavily shaded.

fruiting body with spores (Figure 21.8). The spores will disperse and form new amoebae when conditions improve. The way in which the starving amoebae find each other is to move by chemotaxis up gradients of cyclic AMP (cAMP) that is initially secreted by a founder cell, followed by secretion of cAMP by more distant cells in a relay [see for review (Parent and Devreotes, 1996)]. The cells polarize and move in a spiral pattern up the cAMP gradient, eventually aggregating in sufficient numbers to form a slug. The ligand cAMP binds to the cAR1 receptor that signals through G proteins to activate adenylyl and guanylyl cyclases, PI3Kinase, PLC and inhibit PTEN (Figure 21.9). There also are signals independent of G proteins to increase calcium uptake, for example[see for reviews (Parent and Devreotes, 1996); (Aubry and Firtel, 1999); (Kimmel and Parent, 2003; Manahan et al., 2004)].

In addition to the stages of slug and fruiting body formation, Figure 21.8 shows a time course of expression of many genes involved in signaling initiated by cAMP and development of differentiated cell types. Figures 21.9 and 21.10 (Kimmel and Parent, 2003; Manahan et al., 2004) illustrate the complex signaling that occurs upon binding of cAMP to the cAR1 receptors. These figures also show the spatial organization of the signaling pathways that lead to polarization of the cell and chemotaxis. The cells experience a gradient of cAMP with the highest concentration denoting the direction in which the amoeba should move to join other cells in aggregation. The area of the cell exposed to the highest amounts of cAMP forms pseudopods and becomes the leading edge of the cell that moves up the gradient. The back end of the cell suppresses pseudopod formation and undergoes myosin-based retraction. The steps in this process of polarization and motility are conserved in chemotaxis of neutrophils as they move up gradients of bacterially derived peptides like fMLP that signify infection (Figure 21.11) (Parent, 2004).

The aggregation of the *Dictyostelium* amoebae is not simply dependent upon the single signal, cAMP. Prior to starvation, the cells secrete a "prestarvation" factor that accumulates in proportion to the cell density, but is inhibited by the presence of bacteria (Rathi et al., 1991). When the prestarvation factor is high and bacteria are scarce, the stage is set for the aggregation of starving cells. As the cells begin to starve, prior to secreting cAMP, they secrete a glycoprotein conditioned medium factor (CMF) that integrates quorum sensing with the chemotaxis pathway (Gomer et al., 1991). When the concentration of CMF reaches a threshold indicating a minimum density of starving cells, the receptor for CMF is activated, the cAMP receptor cAR1 is expressed, and phosphodiesterase B is inhibited, preparing the cells for migration and aggregation. Neither CMF nor cAMP by themselves is sufficient for chemotaxis and aggregation, but both bound to their respective receptors are necessary for these processes (Van Haastert et al., 1996). These receptors integrate through a G-protein ensuring that cAMP cannot activate cAR1 without CMF and manipulation of cAR1 levels leads to concomitant changes in the CMF receptors. Dependence of chemotaxis upon CMF prevents an asynchronous cAMP-mediated aggregation of a small number of cells, insufficient to form a slug. In late aggregation the group size is further controlled by a secreted repellent and still more counting factors that regulate the cytoskeleton, motility, and cell adhesion so that the group size of about 10^5 cells form aggregate together to form a slug [(Gomer et al., 2011); (Tang et al., 2001)].

522   Chapter 21   Microbial Chemical Sensing

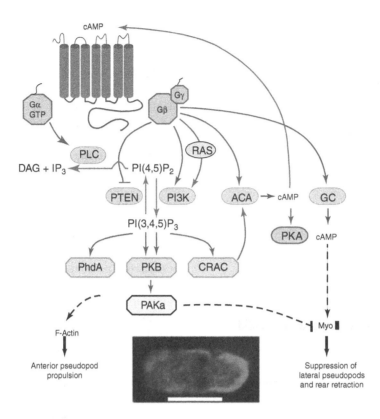

Figure 21.9 Signal transduction events leading to the spatial and temporal regulation of chemotaxis. A simplified model depicting the major effectors (in blue) activated by cAMP receptors. Receptor stimulation ultimately gives rise to actin polymerization at the front for propulsion, myosin II assembly at the sides to suppress lateral pseudopod formation, and myosin II assembly also at the rear for retraction. The molecular details leading to the regulation of PTEN, PI3K, GC, and AC by Gβγ are still being elucidated. Ras appears to be directly required for the activation of PI3K, although its role in PI3K localization is not known. Green arrows, enzyme activation; blue arrows, membrane localization; red arrows, product relation; dashed arrow, complex regulations that have yet to be fully established. (Image) Myosin heavy chain (MHC) and actin filament distribution in polarized cells. Cells expressing MHC fused with green fluorescent protein (GFP) were fully differentiated, fixed, and stained with tetramethyl rhodamine isothiocyanate (TRITC)-phalloidin. The MHC-GFP posterior signal appears in green, and the actin filaments are in red. Bar, 10 μm. (*See plate section for color version.*)

Following aggregation and slug-formation of *Dictyostelium* amoebae, the cells differentiate to form a sessile fruiting structure made of a basal disk, supporting a stalk with spherical spore mass on top (Figure 21.8). The recent identification of a morphogen for the differentiation of stalk cells brings us full circle back to di-c-GMP, the bacterial signaling molecule that plays roles in motility, quorum sensing, and biofilm formation (Chen and Schaap, 2012). Analysis of the genomes of all Dictyostelia groups has shown that they have orthologs of the enzyme required for d-c-GMP formation. However, the signature GGDEF motif of the prokaryotic di-guanylate cyclases is not found in the genomes of ciliated protozoa like *Paramecium* that, nonetheless, have chemo-sensing components in common with *Dictyostelium*, such as receptors for attractants cAMP and folic acid. *Paramecium* also have in common with bacteria a biased random walk mechanism for attraction and repulsion (Sasner and Van Houten, 1989); (Preston and Van Houten, 1987; Van Houten et al., 2010b; Van Houten et al., 2000).

21.6 CILIATED PROTOZOA

Free living ciliates use extracellular cues for location of mates, food, and optimal environments. *Paramecium tetraurelia*, for example, swim by beating its thousands of cilia (Figure 21.12) and change this beating as they respond to attractants that signify that their food source (bacteria) is in the area and to repellents that signify toxic conditions such as extremes of pH or salt (Valentine et al., 2008; Van Houten, 1998; Van Houten and Preston, 1988). Unlike *Dictyostelium* that accumulate by chemotaxis, *P. tetraurelia* accumulate or disperse by a biased random walk similar to behavior of bacteria. As in bacterial chemoresponse, chemical stimuli elicit changes in ciliates' swimming behavior with attractants causing long runs of smooth swimming and few random turns and repellents causing short runs and frequent turns (Figure 21.13). In contrast to bacteria, *P. tetraurelia* also modulate speed of swimming

21.6 Ciliated Protozoa

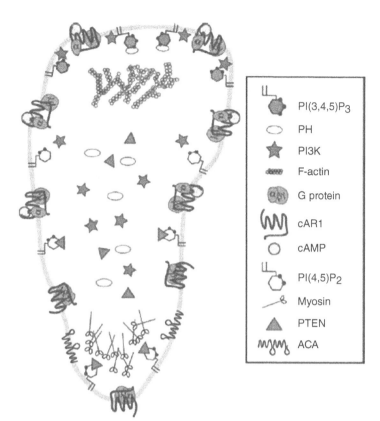

Figure 21.10 Spatial localization of key components of the signaling pathway in Dictyostelium. The cAMP receptor cAR1 is uniformly distributed and G protein dissociation is believed to mirror receptor occupancy in the presence of an external cAMP gradient. In contrast, several components are found preferentially on the front [phosphatidylinositol 3-kinase (PI3K), PI(3,4,5)P$_3$, PH domain-containing proteins, F-actin] or back [PI3-phosphatase (PTEN), adenylyl cyclase (ACA), myosin II] of migrating cells. The cell is migrating toward the top of the page.

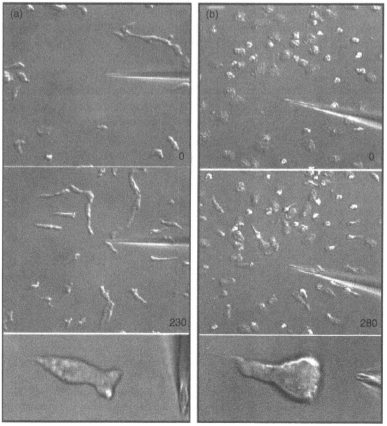

Current opinion in cell biology

Figure 21.11 D. discoideum and neutrophils migrating in response to chemoattractant gradients. (a) Differentiated wild-type D. discoideumcells were plated on a chambered cover glass, and subjected to a cAMP gradient (supplied by a micropipette containing 1 μM cAMP). The bottom panel shows a high-magnification image of a polarized cell. (b) Peripheral blood neutrophils were plated on a chambered cover glass and subjected to an fMLP gradient (supplied by a micropipette containing 1 μM fMLP). The bottom panel shows a high magnification image of a polarized cell. For both (a) and (b), the numbers on the lower right corner indicate elapsed time in seconds. The images on the upper panels were taken using 20X objectives and the high magnification bottom panel images were captured using 63X objectives.

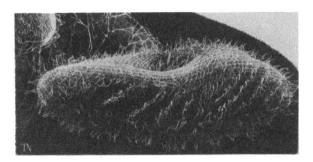

Figure 21.12 Paramecium tetraurelia showing the waves of cilia beating toward the posterior of the cell. From Grass Calendar, 1985.

that reinforces the biased random walk by making the runs longer in attractants and shorter in repellents.

An important feature of ciliates like *P. tetraurelia* is that their ciliary beat frequency and form are under electrical control earning them the nickname of swimming neurons (Kung and Saimi, 1982; Machemer, 1989). Therefore, the swimming behaviors that underlie attraction and repulsion are determined by the membrane electrical changes that the stimuli induce (Valentine et al., 2008; Van Houten, 1979; Van Houten and Preston, 1988). The cilia beat toward the posterior of the cell, driving the cell forward, but at random times the cilia will change in unison and transiently reverse the power stroke of the beat toward the anterior. This change in power stroke sends the cell backward for a short time. The beat quickly returns to normal with the power stroke toward the posterior and the cell swims forward again in a new direction. There is a connection of this swimming behavior with membrane potential: forward swimming occurs as the membrane potential V_m is at rest; hyperpolarization increases beat frequency and the cell swims faster forward; depolarization decreases beat frequency and slows the cell (Machemer, 1974; Machemer, 1989); stronger depolarization above a threshold triggers a regenerative calcium action potential that causes the cell to turn in its swimming path. The voltage gated calcium channels that are exclusively in the cilia carry the action potential current while K channels repolarize the cell to return the cell's V_m to rest and behavior to forward swimming state (Eckert, 1972; Eckert and Naito, 1972). It is the calcium entering the cilia through these channels that reverses the power stroke causing the cells to swim backward until ciliary calcium is sequestered or removed. Thus, the V_m controls the swimming behavior of the cells by controlling ciliary beat frequency for speed and frequency of action potentials for turns.

As the cells move through their environment of pond or stream in a random walk of smooth runs and random turns, chemical stimuli bias that walk. Attractants hyperpolarize the cells causing smoother swimming and fewer turns; repellents depolarize causing action potentials, slower swimming and more turns (Preston and Van Houten, 1987; Van Houten, 1978; Van Houten, 1979). These seemingly subtle changes in behavior are very effective in accumulating groups of cells, paramecia or bacteria, in areas of attractants or dispersing them away from repellents. (See (Valentine et al., 2008) and (Bell and Van Houten, 2008) for a historical perspective and foundational research that cannot be covered here).

A special feature of *P. tetraurelia* is the numerous behavioral mutants that have proven to be very helpful in dissecting the swimming components essential for chemoresponse. Pawn mutants that cannot swim backward for lack of the ciliary voltage-gated calcium conductance (Kung and Saimi, 1982; Saimi and Kung, 1987) cannot accumulate or disperse from attractants and repellents, except in rare situations where there are extremes of hyper- or de-polarization that cause attraction or repulsion by modulating speed alone (Van Houten, 1978). Other behavioral mutants made it possible to dissect the behavior

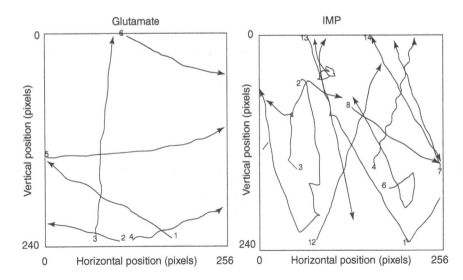

Figure 21.13 Tracks of swimming cells in 5 mM K-L-glutamate in buffer and 1 mM K2IMP in buffer. Tracks were generated by analysis of videotapes of the swimming cells that swim fast and smoothly in glutamate and more slowly with frequent turns in IMP (Van Houten, 1998).

21.6 Ciliated Protozoa

and membrane conductances involved in accumulation in the attractants biotin and acetate (Bell et al., 2007). The mutants helped to determine that the immediate onset of smooth fast swimming when cells are first exposed to acetate (on-response) is critical for successful attraction. In contrast, the off-response, when biotin is first removed and turning starts, is very pronounced and critical for accumulation of cells in this attractant. Because these mutants have been characterized by electrophysiology as well as swimming behavior (Preston et al., 1991; Preston and Kung, 1994), it was possible to use them to attribute ion conductances to the on- and off-responses in these stimuli (Bell et al., 2007).

Chemical stimuli for *P. tetraurelia* come in a variety of structures. Attractants include acetate, biotin, D- and L-glutamate, lactate, extracellular cyclic AMP, ammonium, and peptides like glutathione that are secreted or derived from bacteria. Since *P. teraurelia* feed on bacteria, these attractants likely signify that food is present. Repellents include extremes of pH, high ionic strength, organic compounds like quinine and quinidine, inosine monophosphate (IMP), GTP, and ATP, and some compounds that serve as odorants and tastants in vertebrates (Clark et al., 1993; Rodgers et al., 2008; Van Houten et al., 2010a; Van Houten, 2000). Other than signifying sub-optimal conditions for growth, there is less of a simple rationale for repellents than for attractants.

There are at least three pathways for attractant chemoresponse in *P. tetraurelia* (Figure 21.14, reviewed in (Valentine et al., 2008; Van Houten, 2000)). In common to all of the pathways is the activation of a K conductance that hyperpolarizes the cell and causes smooth fast swimming characteristic of attractants. The stimuli folate, acetate, cyclic AMP, and biotin initiate the first pathway that includes Ca/calmodulin as the second messenger and also activation of the plasma membrane calcium pumps by Ca/calmodulin to sustain the hyperpolarization.

The second pathway has one known stimulus, L-glutamate. Glutamate rapidly activates a K conductance (Preston and Usherwood, 1988) and an adenylyl cyclase resulting in elevation of intracellular cAMP levels within 30 msec (Yang et al., 1997). This increase in intracellular cAMP is not seen in cells stimulated by acetate or ammonium, for example. The second messenger cAMP activates protein kinase A (PKA), whose participation in glutamate chemoresponse was shown through kinase inhibitor and RNA interference studies (Valentine et al., 2008; Van Houten et al., 2000). In turn, the increased cAMP and activation of PKA should be sufficient to cause the changes in ciliary beat that underlie the fast smooth swimming in glutamate because it has been established long ago that ciliary beat frequency increases with cAMP and activation of PKA (Bonini et al., 1986; Hamasaki

et al., 1991; Nakaoka and Machemer, 1990; Noguchi et al., 2003; Pech, 1995; Schultz et al., 1992).

Mammals can taste amino acids, including L-glutamate, which is a stimulus for umami taste. L-glutamate is detected by mammals through at least two types of 7-trans-membrane receptors: heterodimers of taste type 1 receptors (T1R1 and T1R3) as well as metabotropic glutamate receptors 1 and 4 (Chaudhari et al., 2009). A hallmark of umami taste is the synergy of glutamate with 5'ribonucleotides, which can be explained in part by the structure of the T1R receptors (Zhang et al., 2008). While the mechanism is not yet known, *P. tetraurelia* chemoresponse, likewise, shows synergy of glutamate and GMP stimulation on attraction behavior. However, in a departure from the mammalian pattern, IMP does not synergize with glutamate and even serves as a strong repellent (Van Houten et al., 2000). If we look for parallels between protozoan and mammalian systems, perhaps the *P. tetraurelia* response to glutamate, IMP and GMP are more similar to those of taste cells in the gut (Egan and Margolskee, 2008) that respond to glutamate and IMP with different temporal and signaling patterns (Tsurugizawa et al., 2011).

The third pathway *P. tetraurelia* chemoresponse is for the stimulus ammonium and has no apparent receptor (Figure 21.14) (Davis et al., 1998). Ammonium diffuses into the cell and very rapidly increases the intracellular pH. The cells swim fast and smoothly in ammonium suggesting once again that the attractant stimulus elicits a K conductance to hyperpolarize the cells. Trimeric G proteins, although present in *P. tetraurelia*, are only indirectly implicated by pertussis toxin effects on chemoresponse to acetate and ammonium (de Ondarza et al., 2003).

The varied attractant chemical stimuli for *P. tetraurelia* have a variety of cognate chemo-receptors. Receptors associated with Pathway 1 include a folate receptor that is a GPI anchored peripheral protein and redundant cAMP receptors that are expressed from two related proteins both of which appear to code for 7-transmembrane proteins (Paquette et al., 2001; Yano et al., 2003) Czapla thesis (Weeraratne, 2007). The Pathway 2 that signals the presence of glutamate is mediated by an NMDA Receptor Associated Protein (Jacobs, 2007; Romanovitch, 2012; Valentine et al., 2008). Cells of a different species, *P. primaurelia*, respond to agonists of γ-aminobutyric acid receptors (GABA$_B$) with reduction in duration of backward swimming, implying a shortened duration of voltage gated calcium channels conductances (Ramoino et al., 2003). The source of the natural ligand may be the GABA that is released from these cells.

There is no evidence for bacteria-style quorum sensing in *Paramecium*, which was recently reinforced by experiments with *P. caudatum* that implicated physical

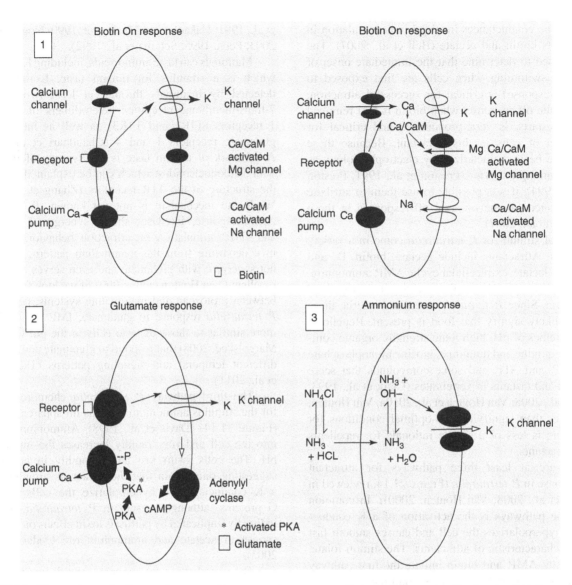

Figure 21.14 Three pathways for *Paramecium tetraurelia* attractant chemoresponse. Pathway 1) shows signaling for stimuli biotin, acetate, cyclic AMP; Pathway 2) shows signaling for stimulus L-glutamate; and Pathway 3) shows signaling for ammonium as a stimulus. (Modified from Van Houten, 2000).

interactions rather than soluble cues secreted into the medium in the dispersal of paramecia when they find themselves in dense populations (Fellous et al., 2012). An interesting contrast is the process by which swimming zoospores of the marine seaweed, *Enteromorpha*, select suitable substrates for settlement and development. The zoospores attach to bacteria of marine biofilms as choice substrates, and identify biofilms by exploiting the bacteria's own sensory system. The zoospores sense the bacterial quorum sensing acyl-homoserine lactone that plays a role in biofilm formation in the zoospores' quest for optimal substrates (Joint et al., 2002).

Tetrahymena, like *Paramecium*, move by beating the cilia that cover their surface and show attraction and dispersal from chemical stimuli. Cells of these two genuses share repellent responses to lysozyme and nuceltide triphosphates, possibly through a polycation receptor (Keedy et al., 2003). While it has been known for a long time that *Paramecium* are repelled by compounds like quinine and quinidine that are bitter to mammals (see for *P. caudatum* (Oami, 1996; Oami, 1998)) and for review of *P. tetraurelia* behavior and electrophysiological response to quinidine (Valentine et al., 2008), more recently bitter and irritating stimuli were shown to elicit repellent swimming responses from *Tetrahymena* (Rodgers et al., 2008). *Tetrahymena* also share with mammals the ability to respond to bacterial peptides and even platelet derived growth factor, β-endorphin, and insulin (Christensen et al.,

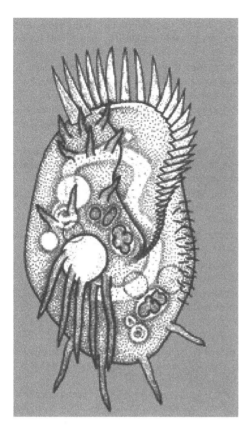

Figure 21.15 *Euplotes* (Encyclopedia of Life, http://eol.org/pages/62158/overview, D. Patterson).

2003; Kohidai et al., 2002; Koppelhus et al., 1994; Leick and Lindemose, 2007; Schneider et al., 2005).

Euplotes do not swim through their watery home as *Paramecium* cells do, but rather these ciliates creep around on the substratum using cilia fused into cirri for propulsion (Figure 21.15). While creeping in their marine or fresh water environment, the cells secrete their pheromones, which signify their mating type. Because these cells are diploid, they can secrete up to two different pheromones that are coded for in the repertoire of mating type genes in their genomes. When cells of different mating types come into the cloud of each other's pheromones, they become mating reactive and pairs form between cells of different or the same mating type. The chances that the roving *Euplotes* will come upon a cell of a different mating type are good because for *E. raikovi*, a marine ciliate, there are 12 possible alleles of the mating type locus and for *E. octocarinatus* there are 10 mating types determined by combinations expressed from four co-dominant alleles.

The structures of the pheromones and their receptors for *E. raikovi* provided an unexpected twist on pheromone function. The peptide pheromones of *E. raikovi* are 38–50 amino acids long and are secreted into the surrounding medium where they bind to surface receptors that are derived from alternative splicing of the same transcript that codes for the pheromone. A pre-pro-pheromone is processed into the secreted pheromone and a pre-pro-pheromone sequence, which is longer than the pheromone by a novel N-terminal sequence, is processed into the membrane receptor. When the cells' receptors are bound to their pheromone that they secrete, the cells are stimulated to grow vegetatively and divide by mitosis. (DiGiuseppe et al., 2002; Ortenzi et al., 2000; Vallesi et al., 1995; Zahn et al., 2001). When the peptide pheromone is from a different mating type, this pheromone binds to the heterotypic receptor and shuts down the mitogenic signaling, allowing the cell to enter into mating and nuclear divisions for meiosis. Thus the pheromone serves an autocrine function, and, much like a mammalian peptide growth factor, binds to a single pass membrane receptor. However, in mammals, the peptide hormone sequence is unrelated to the receptor sequence. Additionally in mammals, the peptide hormone does not promote mating or sexual behavior, as in the case of *E. raikovi* (Luporini et al., 1995).

E. vannus and *E. crassus* takes a different behavioral strategy for attraction to bacteria, their food. The cells move forward slowly and jerk backward called side stepping (Ricci et al., 1987), a tactic that keeps them in the area of signals from their food. As for *Paramecium*, the behavior has electrophysiological correlates. *E. vannus* cells performing this "side stepping" motion to remain near food experience long depolarizations (Schwab et al., 2008; Stock et al., 1997).

21.7 CONCLUSION

To bring this review to a conclusion, it seems appropriate to reiterate that prokaryotic microbes have complex interactomes of proteins that detect stimuli and signal to the cell about the environment's collection of other organisms, food, hosts, and conditions. The output of the signaling often is a change in motility, but the cells also use chemical signaling to temporarily become multi-cellular and make longer term adjustments to the environment through changes in gene expression. Bacteria that interact with human epithelia have very complex effects upon the host and vice versa through secreted signals. We are beginning to appreciate this complexity and the implications that chemical signals from bacteria have for human health and disease: we influence our biota and they influence us.

The eukaryotic microbes are varied in their chemoresponse systems, and, as for bacteria, the output of chemical signal detection often is modification of motility. Perhaps the most striking differences between prokaryotic and eukaryotic microbe chemical signaling are the structures of the receptors, adaptation through methylation used by bacteria only, and kinase mechanisms. Bacteria extensively utilize two-component histidine kinase signaling in which the same phosphate is transferred from one protein to another in the signaling pathway. Eukaryotic microbes chemoreponses primarily utilize kinases that derive phosphate from ATP. While histidine kinases are known in plants and animals, they do not primarily play roles in chemical signaling. However, there continue to be new commonalities uncovered, such as the use of di-cGMP as a secreted morphogen by *Dictyostelium*.

REFERENCES

Adler, J. (1975). Chemotaxis in bacteria. *Ann. Rev. Biochem.* 44: 341–356.

Armstrong, J. B., and Adler, J. (1969). Location of genes for motility and chemotaxis on the Escherichia coli genetic map. *J. Bacteriol.* 97: 156–161.

Aubry, L., and Firtel, R. (1999). Integration of signaling networks that regulate Dictyostelium differentiation. *Ann. Rev. Cell. Dev. Biol.* 15: 469–517.

Balter, M. (2012). Taking stock of the human microbiome and disease. *Science.* 336: 1246–1247.

Bell, W., and Van Houten, J. (2008). Chemical sense: Protozoa. *Encycl. Neuroscience:* 813–818.

Bell, W. E., Preston, R. R., Yano, J., and Van Houten, J.L. (2007). Genetic dissection of attractant-induced conductances in Paramecium. *J. Experi. Biol.* 210: 357–365.

Berg, H. C. (1975). Chemotaxis in bacteria. *Ann. Rev. Biophysics Bioengin.* 4: 119–136.

Berg, H. C. (2008). Bacterial flagellar motor. *Current Biology.* 18: R689–R691.

Bonini, N. M., Gustin, M. C., and Nelson, D. L. (1986). Regulation of ciliary motility by membrane potential in Paramecium: a role for cyclic AMP. *Cell. Motility Cytoskel.* 6: 256–272.

Bray, D. (2002). Bacterial chemotaxis and the question of gain. *Proc. Nat. Acad. Sciences.* 99: 7–9.

Britton, R. A., and Young, V. B. (2012). Interaction between the intestinal microbiota and host in Clostridium difficile colonization resistance. *Trends Microbiol.* 20: 313–319.

Burdette, D. L., Monroe, K. M., Sotelo-Troha, K., et al. (2011). STING is a direct innate immune sensor of cyclic di-GMP. *Nature.* advance online publication.

Chaudhari, N., Pereira, E., and Roper, S. D. (2009). Taste receptors for umami: the case for multiple receptors. *Amer. J. Clin. Nutr.* 90: 738S–742S.

Chen, Z.-h., and Schaap, P. (2012). The prokaryote messenger c-di-GMP triggers stalk cell differentiation in Dictyostelium. *Nature.* 488: 680–683.

Christensen, S., Guerra, C., Awan, A., et al. (2003). Insulin receptor-like proteins in Tetrahymena thermophila ciliary membranes. *Current Biol.* 13: r50–452.

Clark, K. D., Hennessey, T. M., and Nelson, D. L. (1993). External GTP alters the motility and elicits an oscillating membrane depolarization in Paramecium tetraurelia. *Proc. Natl. Acad. Sci. U. S. A.* 90: 3782–3786.

Clarke, M., Hughes, D., Zhu, C., et al. (2006). The OseC sensory kinase: a bacterial adrenergic receptor. *Proc. Natl. Acad. Sci. U. S. A.* 103: 10420–10425.

Darch, S. E., West, S. A., Winzer, K., and Diggle, S. P. (2012). Density-dependent fitness benefits in quorum-sensing bacterial populations. *Proc. Natl. Acad. Sci.* 109: 8259–8263.

Davies, B. W., Bogard, R. W., Young, T. S., and Mekalanos, J. J. (2012). Coordinated regulation of accessory genetic elements produces cyclic di-nucleotides for V. cholerae virulence. *Cell.* 149: 358–370.

Davis, D. P., Fiekers, J., and Van Houten, J. (1998). Intracellular pH and chemoresponse to NH4+ in Paramecium. *Cell. Motil. Cytoskel.* 40: 107–118.

de Ondarza, J., Symington, S., Van Houten, J., and Clark, J. (2003). G-protein modulators alter the swimming behavior and calcium influx of Paramecium tetraurelia. *J. Euk. Microbiol.* 50: 349–355.

DiGiuseppe, G., Miceli, C., Zahne, R., et al. (2002). A structurally deviant member of the Euplotes raikovi pheromone family: Er-23. *J. Euk. Microbiol.* 49: 86–92.

Eckert, R. (1972). Bioelectric control of ciliary activity. *Science.* 176: 473–481.

Eckert, R., and Naito, Y. (1972). Bioelectric control of locomotion in the ciliates. *J. Protozoology.* 19: 237–243.

Egan, J., and Margolskee, R. (2008). Taste cells of the gut and gastrointestinal chemosensation. *LMol. Interv.* 8: 78–81.

Endres, R. G. (2009). Polar Chemoreceptor Clustering by Coupled Trimers of Dimers. *Biophys. J.* 96: 453–463.

Endres, R. G., and Wingreen, N. S. (2006). Precise adaptation in bacterial chemotaxis through "assistance neighborhoods". *Proc. Natl. Acad. Sci. U. S. A.* 103: 13040–13044.

Federle, M. J. (2009). *Autoinducer-2-Based Chemical Communication in Bacteria: Complexities of Interspecies Signaling.* Karger, Basel.

Fellous, S., Duncan, A., Coulon, A., and Kaltz, O. (2012). Quorum Sensing and Density-Dependent Dispersal in an Aquatic Model System. *PloS one.* 7: e48436.

Grass Calendar 1985. Barry W. Ache, Ph.D. The Chemical Senses: From Microbes to Man. http://www.grasstechnologies.com/knowledgebase/calendars/1985_chemical_senses_calendar/1985_chemical_senses_calendar.html.

Golé, L., Rivière, C., Hayakawa, Y., and Rieu, J. P. (2011). A quorum-sensing factor in vegetative *dictyostelium discoideum* cells revealed by quantitative migration analysis. *PloS one.* 6: e26901.

Gomer, R. H., Jang, W., and Brazill, D. (2011). Cell density sensing and size determination. *Dev. Growth Differ.* 53: 482–494.

Gomer, R. H., Yuen, I. S., and Firtel, R. A. (1991). A secreted 80 × 10(3) Mr protein mediates sensing of cell density and the onset of development in Dictyostelium. *Development.* 112: 269–278.

Grebe, T. W., and Stock, J. (1998). Bacterial chemotaxis: the five sensors of a bacterium. *Curr. Biol.* 8: R154–157.

Hall, B. A., Armitage, J. P., and Sansom, M. S. P. (2011). Transmembrane helix dynamics of bacterial chemoreceptors supports a piston model of signalling. *PLoS Comput. Biol.* 7: e1002204.

Hamasaki, T., Barkalow, K., Richmond, J., and Satir, P. (1991). cAMP-stimulated phosphorylation of an axonemal polypeptide that copurifies with the 22S dynein arm regulates microtubule translocation velocity and swimming speed in Paramecium. *Proc. Natl. Acad. Sci. U. S. A.* 88: 7918–7922.

Hazelbauer, G. L., Falke, J. J., and Parkinson, J. S. (2008). Bacterial chemoreceptors: high-performance signaling in networked arrays. *Trends Biochem. Sci.* 33: 9–19.

Hazelbauer, G. L., Mesibov, R. E., and Adler, J. (1969). Escherichia coli mutants defective in chemotaxis toward specific chemicals. *Proc. Natl. Acad. Sci. U. S. A.* 64: 1300–1307.

Hengge, R. (2009). Principles of c-di-GMP signalling in bacteria. *Nat. Rev. Micro.* 7: 263–273.

Hooper, L., Littman, D. R., and Macpherson, A. J. (2012). Interactions between the microbiota and the immune system. *Science.* 336: 1268–1273.

Jacobs, C. (2007). NMDA receptor associate protein in Paramecium and its involvement in glutamate chemoresponse *Biology.* Vol. MS. University of Vermont.

Jayaraman, A., and Wood, T. K. (2008). Bacterial quorum sensing: signals, circuits, and implications for biofilms and disease. *Ann. Rev. Biomed. Eng.* 10: 145–167.

Jenal, U., and Malone, J. (2006). Mechanisms of cyclic-di-GMP signaling in bacteria. *Ann. Rev. Genet.* 40: 385–407.

Joint, I., Tait, K., Callow, M. E., et al. (2002). Cell-to-cell communication across the prokaryote-eukaryote boundary. *Science.* 298: 1207.

Kau, A. L., Ahern, P. P., Griffin, N. W., et al. (2011). Human nutrition, the gut microbiome and the immune system. *Nature.* 474: 327–336.

Keedy, M., Yorgey, N., Hilty, J., et al. (2003). Pharmacological evidence suggesets that the lysozyme/PACAP receptor of Tetrahymena thermophila is a polycation receptor. *Acta. Protozoologica.* 42: 11–17.

Kentner, D., and Sourjik, V. (2006). Spatial organizatin of the bacterial chemotaxis system. *Curr. Opin. Microbiol.* 9: 619–624.

Keymer, J. E., Endres, R. G., Skoge, M., et al. (2006). Chemosensing in Escherichia coli: Two regimes of two-state receptors. *Proc. Natl. Ac. Sciences of USA.* 103: 1786–1791.

Kimmel, A. R., and Parent, C. A. (2003). The signal to move: D. discoideum go orienteering. *Science.* 300: 1525–1529.

Kohidai, L., Kovacs, K., and Csaba, G. (2002). Direct chemotactic effect of bradykinin and related peptides - significance of amino and carboxyterminal character of oligopeptides in chemotaxis of Tetrahymena pyriformis. *Cell. Biol. International.* 26: 55–62.

Kolbinger, A., Gao, T., Brock, D., et al. (2005). A Cysteine-Rich Extracellular Protein Containing a PA14 Domain Mediates Quorum Sensing in Dictyostelium discoideum. *Eukaryotic Cell.* 4: 991–998.

Koppelhus, U., Hellung-Larsen, P., and Leick, V. (1994). An improved quantitative assay for chemokinesis in Tetrahymena. *Biol. Bull.* 187: 8–15.

Krell, T., Lacal, J., Busch, A., et al. (2010). Bacterial Sensory Kinases; Diversity in the Recognition of Environmental Signals. *Annu. Rev. Microbiol.* 64: 539–559.

Kung, C., and Saimi, Y. (1982). The physiological basis of taxes in Paramecium. *Ann. Rev. Physiol.* 44: 519–534.

Leick, V., and Lindemose, S. (2007). Chemokinesis by Tetrahymena in response to bacterial oligopeptides. *J. Euk. Microbiol.* 54: 271–274.

Li, M., and Hazelbauer, G. L. (2005). Adaptational assistance in clusters of bacterial chemoreceptors. *Mol. Microbiol.* 56: 1617–1626.

Little, A., Robinson, C., Peterson, S., et al. (2008). Rules of engagement: interspecies interactions that regulate microbialcommunities. *Ann. Reb. Microbiol.* 62: 375–401.

Luporini, P., Vallesi, A., Miceli, C., and Bradshaw, R. A. (1995). Chemical signaling in ciliates. *J. Euk. Microbiol.* 42: 208–212.

Lynch, A. S., and Robertson, G. T. (2008). Bacterial and fungal biofilm infections. *Ann. Rev. Med.* 59: 415–428.

Machemer, H. (1974). Frequency and directional responses of cilia to membrane potential changes in Paramecium. *J. Comp. Physiol.* 92: 293–316.

Machemer, H. (1989). Cellular behavior modulated by ions: electrophysiological implications *J. Protozool.* 36: 463–487.

Manahan, C., Iglesias, P., Long, Y., and Devreotes, P. (2004). Chemoattractant signaling in Dictyostelium discoideum. *Ann. Rev. Cell. Devel. Biology.* 20: 223–254.

Miller, L. D., Yost, C. K., Hynes, M. F., and Alexandre, G. (2007). The major chemotaxis gene cluster of Rhizobium leguminosarum bv. viciae is essential for competitive nodulation. *Mole. Microbiol.* 63: 348–362.

Miller, M. B., Skorupski, K., Lenz, D. H., et al. (2002). Parallel Quorum Sensing Systems Converge to Regulate Virulence in Vibrio cholerae. *Cell.* 110: 303–314.

Monds, R., and O'Toole, G. (2009). The developmental model of microbial biofilms: ten years of a paradigm up for review. *Trends in Microbiol.* 17: 73–87.

Monroe, D. (2007). Looking for Chinks in the Armor of Bacterial Biofilms. *PLoS Biol.* 5: e307.

Nakaoka, Y., and Machemer, H. (1990). Effects of cyclic nucleotides and intracellular Ca on voltage-activated ciliary beating in Paramecium. *J. Comp. Physiol.* 166: 401–406.

Neiditch, M. B., Federle, M. J., Pompeani, A. J., et al. (2006). Ligand-induced asymmetry in histidine sensor kinase complex regulates quorum sensing. *Cell.* 126: 1095–1108.

Nicholson, J. K., Holmes, E., Kinross, J, et al. (2012). Host-Gut Microbiota Metabolic Interactions. *Science.* 336: 1262–1267.

Noguchi, M., Sasaki, J. Y., Kamachi, H., and Inoue, H. (2003). Protein phosphatase 2C is involved in the cAMP-dependent ciliary control in Paramecium caudatum. *Cell. Motil. Cytoskel.* 54: 95–104.

Oami, K. (1996). Distribution of chemoreceptors to quinine of the cell surface of Paramecium caudatum. *J. Comp. Physiol. A* 179: 345–352.

Oami, K. (1998). Membrane potential response of Paramecium caldatum to bitter substances: existence of multiple pathways for bitter responses. *J. Exp. Biol.* 201: 13–20.

Ortenzi, C., Alimenti, C., Vallesi, A., et al. (2000). The autocrine mitogenic loop of the ciliate Euplotes raikovi: the pheromone membrane-bound forms are the cell binding sites and potential signaling receptors of soluble pheromones. *Mol. Biol. Cell.* 11: 1445–1455.

Paquette, C., Rakochy, V., Bush, A., and Van Houten, J. L. (2001). glycosylphosphaticylinositol-anchored proteins in Paramecium tetraurelia: possible role in chemoresponse. *J. Exp. Biol.* 204: 2899–2910.

Parent, C. A. (2004). Making all the right moves: chemotaxis in neutrophils and Dictyostelium. *Curr. Opin. Cell. Biol.* 16: 4–13.

Parent, C. A., and Devreotes, P. N. (1996). Molecular genetics of signal transduction in Dictyostelium. *Ann. Rev. Biochem.* 65: 411–440.

Pech, L. L. (1995). REgulation of ciliary motility in Paramecium by cAMP and cGMP. *Comp. Biochem. Physiol.* 111A: 31–37.

Porter, S. L., Wadhams, G. H., and Armitage, J. P. (2011). Signal processing in complex chemotaxis pathways. *Nat. Rev. Micro.* 9: 153–165.

Preston, R., and Van Houten, J. (1987). Chemoreception in Paramecium: acetate - and folate-induced membrane hyperpolarization. *J. Comp. Physiol.* 160: 525–536.

Preston, R. R., Kink, J. A., Hinrichsen, R. D., et al. (1991). Calmodulin mutants and Ca2(+)-dependent channels in Paramecium. *Ann. Rev. Physiol.* 53: 309–319.

Preston, R. R., and Kung, C. (1994). Isolation and characterization of paramecium mutants defective in their response to magnesium. *Genetics.* 137: 759–769.

Preston, R. R., and Usherwood, P. N. (1988). L-glutamate-induced membrane hyperpolarization and behavioural responses in Paramecium tetraurelia. *J. Comp. Physiol. A, Sens, Neur. Behav. Physiolog.* 164: 75–82.

Rak, K., and Rader, D. J. (2011). Cardiovascular disease: The diet-microbe morbid union. *Nature.* 472: 40–41.

Ramoino, P., Usai, C., Beltrame, F., et al. (2003). Swimming behavior regulation by GABAb receptors in Paramecium. *Exp. Cell. Res.* 291: 398–405.

Rao, S., and Ordal, G. (2009). *The Molecular Basis of Excitation and Adaption During Chemotactis Sensory Transduction in Bacteria.* Karger: Basel.

Rathi, A., Kayman, S., and Clarke, M. (1991). Induction of gene expression in Dictyostelium by prestarvation factor, a factor secreted by growing cells. *Dev. Genet.* 12: 82–87.

Relman, D. A. (2012). Microbiology: Learning about who we are. *Nature.* 486: 194–195.

Ricci, N. R., Gianetti,C., and C. Miceli. (1987). The ethogramm of Euplotes carssusI. The wild type. *Eur. J. Protistol.* 23: 129–140.

Rodgers, L. F., Markle, K. L., and Hennessey, T. M. (2008). Responses of the ciliates Tetrahymena and Paramecium to vertebrate odorants and tastants. *J. Euk. Microbiol.* 55: 27–33.

Romanovitch, M. (2012). *The L-glutamate Receptor in Paramecium Tetraurelia.* Vol. MS. University of Vermont.

Ryan, R. P., Tolker-Nielsen, T., and Dow, J. M. (2012). When the PilZ don t work: effectors for cyclic di-GMP action in bacteria. *Trends Microbiol.* 20: 235–242.

Saimi, Y., and Kung, C. (1987). Behavioral genetics of Paramecium. *Ann. Rev. Genetics.* 21: 47–65.

Sasner, J., and Van Houten, J. L. (1989). Evidence for a Paramecium folate chemoreceptor. *Chem. Senses.* 14: 587–595.

Sauer, K., Rickard, A., and Davies, D. (2007). Biofilms and biocomplexity. *Microbe.* 2: 347–353.

Schaller, G. E., Shiu, S.-H., and Armitage, J. P. (2011). Two-component systems and their co-option for eukaryotic signal transduction. *Curr. Biol.* 21: R320–R330.

Schirmer, T., and Jenal, U. (2009). Structural and mechanistic determinants of c-di-GMP signalling. *Nat. Rev. Micro.* 7: 724–735.

Schneider, L., Clement, C. A., Teilmann, S. C., et al. (2005). PDGFRal-phaalpha signaling is regulated through the primary cilium in fibroblasts. *Curr. Biol.* 15: 1861–1866.

Schultz, J. E., Klumpp, S., Benz, R., et al. (1992). Regulation of adenylyl cyclase from Paramecium by an intrinsic potassium conductance. *Science.* 255: 600–603.

Schwab, A., Hanley, P., Fabian, A., and Stock, C. (2008). Potassium channels keep mobile cells on the go. *Physiology.* 23: 212–220.

Stock, C., Kruppel,T., and Leuken, T. (1997). Kinesis in Euplotes vannus -ethological and electrophysiological characteristics of chemosensory beahavior. *J. Euk. Microbiol.* 44: 427–433.

Stocker, R., Seymour, J. R., Samadani, A., et al. (2008). Rapid chemotactic response enables marine bacteria to exploit ephemeral microscale nutrient patches. *Proc. Natl. Acad. Sci. U. S. A.* 105: 4209–4214.

Tamayo, R., Pratt, J., and Camilli, A. (2007). Roles of cyclic diguanylate in the regulation of bacterial pathogenesis. *Annu. Rev. Microbiol.* 61: 131–148.

Tang, L., Ammann, R., Gao, T., and Gomer, R. H. (2001). A cell number-counting factor regulates group size in Dictyostelium by differentially modulating cAMP-induced cAMP and cGMP pulse sizes. *J. Biolog. Chem.* 276: 27663–27669.

Tsurugizawa, T., Uematsu, A., Uneyama, H., and Torii, K. (2011). Different BOLD responses to intragastric load of L-glutamate and inosine monophosphate in conscious rats. *Chem. Senses.* 36: 169–176.

Valentine, M., Yano, J., and Van Houten, J. (2008). Chemosensory transduction in paramecium. *Jpn. J. Protozool.* 41: 1–7.

Vallesi, A., Giuli, G., Bradshaw, R. A., and Luporini, P. (1995). Autocrine mitogenic activity of pheromones produced by the protozoan ciliate Euplotes raikovi. *Nature.* 376: 522–524.

Van Haastert, P., Bishop, J., and Gomer, R. H. (1996). The cell density factor CMF regulates the chemotattractant receptor cAR1 in dictyostelium. *J. Cell. Biol.* 134: 1543–1549.

Van Houten, J., Valentine, M., and Yano, J. (2010a). Behavioral genetics of paramecium. *Encyclopedia of Animal Behavior:* 677–682.

Van Houten, J., Valentine, M., and Yano, J. (2010b). Chemosensory transduction in paramecium. *Jpn. J. Protozool.* 41: 1–8.

Van Houten, J., Yang, W., and Bergeron, A. (2000). Glutamate chemosensory signal transduction in Paramecium. *J. Nutr.* 130: 946S–949S.

Van Houten, J. L. (1978). Two mechanisms of chemotaxis in Paramecium. *J. Comp. Physiol. A.* 127: 167–174.

Van Houten, J. L. (1979). Membrane potential changes during chemokinesis in Paramecium. *Science.* 204: 1100–1103.

Van Houten, J. L. (1998). Chemosensory transduction in Paramecium. *Eur. J. Protistol.* 34: 301–307.

Van Houten, J. L. (2000). Chemoreception in microorganisms In *The Neurobiology of Taste and Smell*, T. Finger, W. Silver, and D. Restrepo, editors. Wiley-Liss. 11–40.

Van Houten, J. L., and Preston, R. (1988). Chemokinesis. *In Paramecium.* H.-D. Gortz, editor. Springer-Verlag: Berlin. 282–300.

Wadhams, G. H., and Armitage, J. P. (2004). Making sense of it all: bacterial chemotaxis. *Nat. Rev. Mol. Cell. Biol.* 5: 1024–1037.

Waters, C. M., and Bassler, B. L. (2005). Quorum sensing: cell-to-cell communication in bacteria. *Annu. Rev. Cell. Dev. Biol.* 21: 319–346.

Weeraratne, S. (2007). GPI-anchored chemoreceptors in folate chemosensory transduction in Paramecium tetraurelia. *In Biology.* Vol. PhD. University of Vermont.

West, S. A., Diggle, S., Buckling, A., Garrdner, A., and Griffin, A. (2007). The social lives of microbes. *Annu. Rev. Ecol. Evol. Syst.* 38: 53–77.

Williams, S. M., Chen, Y. T., Andermann, T. M., et al. (2007). Helicobacter pylori chemotaxis modulates inflammation and bacterium-gastric epithelium interactions in infected mice. *Infect. Immun.* 75: 3747–3757.

Yang, W., Braun, C., Plattner, H., et al. (1997). Cyclic nucleotides in gluamate chemosensory signal transduction of Paramecium. *J. Cell. Science.* 110: 1567–1572.

Yano, J., Rachochy, V., and Van Houten, J. L. (2003). Glycosyl phosphatidylinositol-anchored proteins in chemosensory signaling: antisense manipulation of Paramecium tetraurelia PIG-A gene expression. *Euk. Cell.* 2: 1211–1219.

Yin, Q., Tian, Y., Kabaleeswaran, V., et al. (2012). Cyclic di-GMP Sensing via the innate immune signaling protein STING. *Mol. Cell.* 46: 735–745.

Yuan, J., Branch, R. W., Hosu, B. G., and Berg, H. C. (2012). Adaptation at the output of the chemotaxis signalling pathway. *Nature.* 484: 233–236.

Zahn, R., Damberger, F., Ortenzi, C., et al. (2001). NMR structure of the Euplotes raikovi pheromone Er-23 and identification of its five disulfide bonds. *J. Mol. Biol.* 313: 923–931.

Zhang, F., Klebansky, B., Fine, R. M., et al. (2008). Molecular mechanism for the umami taste synergism. *Proc. Natl. Acad. Sciences.* 105: 20930–20934.

Chapter 22

Olfaction in Insects

Paul Szyszka and C. Giovanni Galizia

22.1 INTRODUCTION

Insects are the most species-rich group of animals on the planet. They have colonized almost every terrestrial habitat, each of which is characterized by its own unique odor profile. Although the olfactory systems of different insect species must deal with these diverse olfactory environments, their olfactory systems share remarkable similarities across species and even with mammals (Ache and Young, 2005). In this chapter we describe the challenges faced by the olfactory system of terrestrial insects and how insects cope with such challenges.

Most of what we know about insect olfaction derives from studying a few species, each with its own set of motivational influences. Honeybees are the most important pollinators and use odors to find flowers, moths are among the most devastating crop pests which cause enormous losses to the economy each year, mosquitoes are disease vectors that cause millions of deaths every year, locusts form large swarms that wipe out crops, and fruit flies have become the most important insect model species to study molecular mechanisms of olfaction. Since all of these insects use olfaction as a primary sensory modality, it is important to know how their olfactory systems work in order to control them. In this chapter we describe the basic circuitry of odor coding in the insect brain, and present example where odors are used in an insect's life.

22.2 HOW INSECTS RESPOND TO ODORS

For many insects, olfaction is *the* main sensory system – they live in an olfactory world. In this section a few examples of how odors are used by different insects are provided, taking into account the large variety of behaviors across species. Most commonly, insects use odorants for habitat localization, foraging, sexual reproduction, social communication and orientation.

Like mammals, including humans, insects need to identify odors that have innate meanings, learn new odors and their relevance, extract high-speed temporal information, and either distinguish or generalize across odors and their different concentrations. The main difference may consist in the temporal structure: mammals synchronize odor-driven activity to their breathing cycle, which is coupled to the inhalation of odorants (Macrides and Chorover, 1972). It has been proposed that insects fan their wings to induce a similar temporal structure (Loudon and Koehl, 2000), but that remains to be shown. Most importantly, the neural networks that process odor-information in the insect's brain follow the same logic as those of mammals. Thus, knowledge obtained from insects helps us in understanding our own olfactory system. However, understanding insect olfaction is also important in its own right: mosquitoes use odors to find us, with all the consequences for many mosquito transmitted diseases such as malaria and Dengue fever. Crop pests (e.g., different moth and beetle species) destroy an important portion of our food staples, both on the fields and in the warehouses, and they too use odors both to find their substrate, and to proliferate. And many pollinators – including honeybees – are essential for our harvest: they all find the flowers by smell. Thus, insect olfaction is important for agriculture, food supply, and public health. Understanding how it works will put us in a position to better protect, influence, and control our environment.

A generalist insect, such as the honeybee *Apis mellifera* (Figure 22.1a), exploits a variety of plant species.

Handbook of Olfaction and Gustation, Third Edition. Edited by Richard L. Doty.
© 2015 Richard L. Doty. Published 2015 by John Wiley & Sons, Inc.

Figure 22.1 Insect model species for studying olfaction. Insects differ greatly in their behavior and environment. Note the different shapes of their antennae, which are their olfactory organs. (a) *Apis mellifera*, (b) *Schistocerca gregaria*, (c) *Manduca sexta* larva, (d) *Bombyx mori*, (e) *Drosophila melanogaster*, (f) *Atta vollenweideri*, (g) *Manduca sexta*, (h) *Aedes albopictus* & *A. japonicus*. Images were taken by John C. Abbott (a), David Gustav (b), Krushnamegh Kunte (c), R. Alexander Steinbrecht (d), C. Giovanni Galizia (e), Christoph J. Kleineidam (f), Armin Hinterwirth (g) and Ary Farajollahi (h). (*See plate section for color version.*)

During foraging a honeybee learns the odor of a flower that provides nectar and pollen. The relationship between odors and food can change within a short period of time. Old food sources run dry and new ones appear. Thus, the meaning of odors is ephemeral and generalist insects must code for many different and unpredictable odors. A specialist insect, in contrast, interacts with just a single or few plants, and thus has to code for a few, evolutionary determined odors. Pest insects like the desert locust *Schistocerca gregaria* (Figure 22.1b) or the hawk moth *Manduca sexta* (a larva shown in Figure 22.1c, adult in 1g) are innately attracted by odors associated with their favorite foods and host plants. Intraspecific communication often makes use of pheromones, specialized odors that are directly related to behavior.[1] A sexually active male silk moth *Bombyx mori* (Figure 22.1d) detects and is attracted by single molecules of bombykol, the major sex pheromone component emitted by females (Butenandt, Groschel et al., 1959). Male fruit flies, *Drosophila melanogaster*, (Figure 22.1e) initiate courtship behavior by releasing the pheromone cis-vaccenyl acetate (Bartelt, 1985). Leafcutter ants of the genus *Atta* (Figure 22.1f) use pheromones to mark trails between their nest and food sources. These pheromones contain two major components which are sufficient to elicit trail-following. The ratio of both defines the species-specificity and allows different *Atta* species to distinguish between con- and heterospecific trails (Kleineidam et al., 2007). Flying insects, for example, moths (Figure 22.1g), possess a remarkable ability to identify and track a single odor in a highly turbulent, multi-odor background. This ability relies on analyzing the temporal structure of odor plumes, which contains information about the distance and location of the odor source (Vickers, 2000; Carde and Willis, 2008). The intermittent pulses of the sex pheromone released by a female moth provide the searching male with orientation cues (Justus et al., 2002). Mosquitoes are the most important vectors for arboviruses (arthropod-borne viruses). Mosquitoes such as *Aedes* (Figure 22.1h) find their blood meal following specific odors including ammonia and lactic acid, and fluctuating concentrations of CO_2 (Geier et al., 1999). Tritrophic interactions are particularly intriguing: As a response to feeding caterpillars, mediated by the saliva of the caterpillars, plants can attract parasitoid wasps by releasing particular semiochemicals. The wasps lay their eggs into the caterpillar, eventually killing it. A famous example of such an interaction is that of the butterfly *Pieris* and Brussels sprouts *Brassica oleracea* (Vet and Dicke, 1992).

22.2.1 The Olfactory Pathway: from Antenna to Mushroom Body

One of the important differences between humans and insect olfaction is the contact surface: while we have our odor-receptor cells in the mucosa in the nose, insects have a rigid exoskeleton, a chitinous cuticle with no mucosae.

---

[1] The term pheromone was coined in 1959 to replace the term ectohormone (external hormone) in insects. Karlson and Lüscher (1959) defined pheromones as "substances which are secreted to the outside by an individual and received by a second individual of the same species, in which they release a specific reaction, for example, a definite behavioral or a development process." See Doty (2010) for a critical discussion of the pheromone concept, especially when applied to mammals.

Their olfactory neurons are housed in so-called sensilla, which are often hair-like structures with pores (de Bruyne and Baker, 2008) (Figure 22.2). Odors enter the pores and reach the inner lumen of the sensillum, which is filled with sensillar lymph. The dendrites of olfactory sensory neurons (OSN) protrude into the sensillar lymph, so that odor molecules can reach them and elicit a neural response. Thus, in a way similar to mammals, odors have to pass a liquid phase (the sensillar lymph in insects, the mucous in mammals), before reaching the sensory neurons. Not all sensilla are hair-shaped. Indeed, there are many different sensilla shapes, short and long, flat and invaginated. Also, the number of sensory cells in each sensillum can vary from two to several dozens. The sensory neuron types are not randomly distributed across sensilla. For example, in male moths, which are highly sensitive to the female pheromone, the pheromone receptors are housed in very long sensilla trichodea (hair-like sensilla) which form a sieve-like structure that is capable of capturing every single odor molecule in the air (Figure 22.2a-c). Usually, each sensillum trichodeum has two OSNs, one for each of the two main pheromone components, thus allowing the animal to efficiently identify the species-specific pheromone blend. Olfactory sensilla are generally located on the insects' antennae, though in some species other body parts also have olfactory neurons, for example, the maxillary palps in flies.

Olfactory processing begins in OSNs. In the fruitfly *Drosophila*, each OSN expresses one or a few olfactory receptor proteins which differ in their affinity to odorants (Larsson et al., 2004; Couto et al., 2005; Fishilevich and Vosshall, 2005). Insect olfactory receptors are 7-transmembrane proteins with their N-terminus facing to the cytosol (Benton et al., 2006). This is an inverted topology as compared to G-protein coupled 7-transmembrane proteins, which is the family of molecules of mammalian olfactory receptors, implying that insect olfactory receptors and mammalian olfactory receptors have evolved independently. In fact, insect olfactory receptors belong to a family of gustatory receptors that is very ancient in evolution but which has been lost in vertebrates. In order to function, each olfactory receptor protein needs to form a heterodimer with another 7-transmembrane protein called Orco (Sato et al., 2008; Wicher et al., 2008; Vosshall and Hansson, 2011). In contrast to mammalian receptors which are metabotropic, insect olfactory receptors act as ligand gated channels, with metabotropic auto-regulation (Nakagawa and Vosshall, 2009; Silbering and Benton, 2010). Some OSNs do not express olfactory receptors, but sense odors with ligand-gated ion channels (IR) which share some homology with ionotropic glutamate receptors (Benton et al., 2009; Grosjean et al., 2011; Silbering et al., 2011).

OSNs send their axons into the primary olfactory brain area, the antennal lobe (Figure 22.3a). In structure and wiring logic the antennal lobe is very similar to the mammalian olfactory bulb. The antennal lobe is subdivided into glomeruli. OSNs with the same receptor proteins

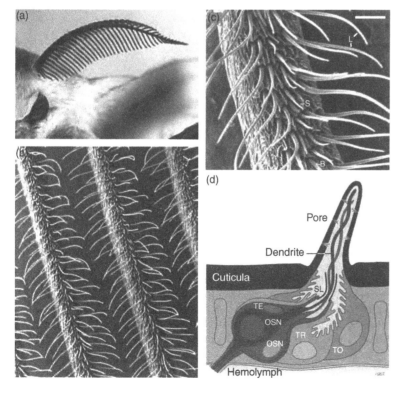

Figure 22.2 Olfactory sensory neurons are housed in sensilla. (a) Antenna of a male silk moth *Bombyx mori*. (b) Scanning electron micrograph of the sieve-like antenna. (c) Single antennal branch with different sensillum types: long sensilla trichodea (L), short sensilla trichodea (S) and sensilla basiconica (B). (d) Basic structure of a sensillum. Pores in the cuticle allow odors to enter the sensillar lymph (SL). Odor receptors are located in the dendrites of olfactory sensory neurons (OSN). Non-neuronal accessory cells, the thecogen (TE), tormogen (TO) and the trichogen (TR), produce the sensillar lymph and shield it from OSNs' somata and inner part of their dendrites and from the hemolymph (insects' circulatory fluid). (Images were kindly provided by R. Alexander Steinbrecht and modified)

534   Chapter 22   Olfaction in Insects

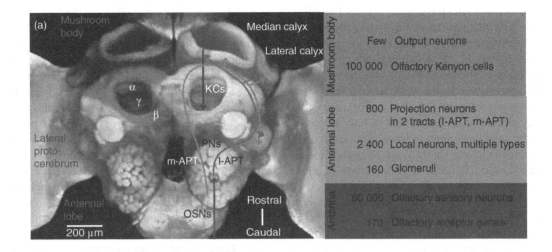

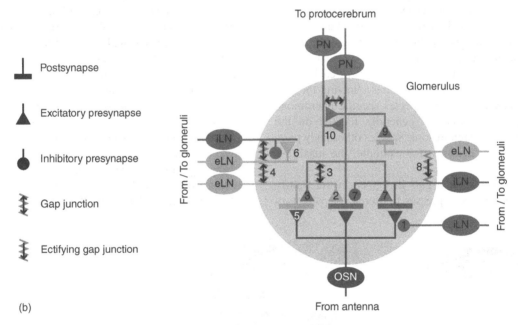

Figure 22.3 The insect olfactory pathway. (a) View of the honeybee central olfactory system (Kindly provided by Randolf Menzel and modified). Olfactory sensory neurons (ORN, blue arrow) terminate in the antennal lobe glomeruli. Uniglomerular projection neurons (PN, green) transmit odor information from the antennal lobe to the mushroom body and to the lateral protocerebrum. PN axons project along the lateral or median antenno-protocerebral tract (l-APT, m-APT) to the lateral and medial calyces. The mushroom body intrinsic Kenyon cells (KC, red) send their axons into the mushroom body lobes, where they contact mushroom body extrinsic neurons. The yellow area in the calyx represents the Kenyon cells' dendritic field. (b) Collection of synaptic connections within a glomerulus in antennal lobe of *Drosophila*. Left: types of synaptic connections. Right: known synaptic connections between olfactory sensory neurons (OSN), excitatory local neurons (eLN), inhibitory neurons (iLN) and projection neurons (PN). Numbers refer to studies which described the respective synaptic connections: 1. lateral, interglomerular presynaptic inhibition by GABAergic iLN-to-OSN synapses (Olsen and Wilson, 2008; Root, Masuyama et al., 2008); 2. lateral, interglomerular excitation by cholinergic eLN-to-PN synapses (Olsen, Bhandawat et al., 2007; Shang, Claridge-Chang et al., 2007); 3. reciprocal excitatory krasavietz eLN-PN connections through gap junctions and cholinergic synapses (Huang, Zhang et al., 2010; Yaksi and Wilson, 2010); 4. reciprocal excitatory krasavietz eLN-eLN connections through gap junctions and cholinergic synapses (Huang, Zhang et al., 2010); 5. excitatory OSN-to-krasavietz eLN synapses (Huang, Zhang et al., 2010); 6. reciprocal krasevietz eLN-iLN connections through gap junctions and cholinergic synapses (Yaksi and Wilson, 2010); 7. reciprocal PNs-iLN connections through cholinergic and GABAergic synapses (Sudhakaran, Holohan et al., 2012); 8. rectifying electrical eLN-iLN synapses (Sudhakaran, Holohan et al., 2012) 9. excitatory PN-to-eLN synapses (Sudhakaran, Holohan et al., 2012) 10. reciprocal, intraglomerular PN-PN connections through gap junctions and cholinergic synapses (Yaksi and Wilson, 2010) (*See plate section for color version.*)

converge onto the same glomerulus, and thus provide every glomerulus with a distinct response profile (Vosshall et al., 2000). The output of a glomerulus is carried via projection neurons (PN) to higher order brain centers. Within the antennal lobe a network of intra- and inter-glomerular inhibitory local neurons (iLN) and excitatory local neurons (eLN) is involved in odor processing (Sachse and Galizia, 2002; Olsen et al., 2007; Shang et al., 2007; Silbering and Galizia, 2007; Silbering et al., 2008) (Figure 22.3b). Odor representations in PNs are therefore a result of the interaction of many neurons in the antennal lobe. PNs relay odor information to various areas in the brain. The most prominent ones are the mushroom bodies, named this way because their shape is reminiscent of the stipes and caps of mushrooms, and the lateral protocerebrum ("protocerebrum" is the name for the most anterior part of the insect brain). The mushroom bodies play an important role in odor learning and memory formation (Menzel, 2001; Heisenberg, 2003; Davis, 2011), the lateral protocerebrum is thought to be the structure with the first premotoneurons – thus controlling the insect's behavior. In many species the mushroom bodies are multimodal and not only receive information from the olfactory system, but also from visual, tactile, and other modalities (Mobbs, 1982; Strausfeld, 2002; Strausfeld et al., 2003).

Mushroom bodies consist of densely packed neurons, the Kenyon cells. While Kenyon cells' somata surround the calyces, their dendrites branch into them, and their axons travel along the lobes – that is, the "stipe" of the mushroom. The population of Kenyon cells is generally very large, ranging from about 2500 in the fruitfly *Drosophila melanogaster* to about 170,000 in honeybees (with a total of 340,000 Kenyon cells, they represent every third neuron in the bee's 1,000,000 neuron large brain). The main input sides are the calyces, while the lobes are the main output sides (Figure 22.3a). The connectivity between PNs and Kenyon cells is characterized by divergence and convergence. Each PN branches over extensive portions of the calyces, and synapses onto many Kenyon cells. In the honeybee, for example, about 800 PNs diverge onto roughly 100,000 olfactory Kenyon cells. Each Kenyon cell in turn receives input from multiple PNs. In locusts, each Kenyon cell gets input from about 50% of about 800 PNs (Jortner, Farivar et al., 2007).

The glomerular map of the antennal lobe is transformed into a different map in the mushroom body. While inputs from different OSNs are spatially segregated in different glomeruli in the antennal lobe, they appear to partially overlap in the mushroom body, where information from distinct glomeruli is distributed onto many Kenyon cells across the calyces. Nevertheless, each odor stimulus creates a characteristic, combinatorial activity pattern in the antennal lobe, and also across Kenyon cells in the mushroom bodies (Perez-Orive et al., 2002; Szyszka et al., 2005; Turner et al., 2008). Just as in the antennal lobe, there are complex internal networks within mushroom bodies. There are local networks within the calyx which involve reciprocal microcircuits between the bouton-like PN terminals, G Neurons employing gamma-aminobutyric acid (GABAergic neurons) and Kenyon cells (Ganeshina and Menzel, 2001; Christiansen, Zube et al., 2011), and there are more widespread circuits with feedback from the lobes back to the calyx (Grunewald, 1999; Grunewald, 1999; Strausfeld, 2002; Papadopoulou et al., 2011).

Odor representations in the lateral protocerebrum follow a different logic. It is often assumed that the lateral protocerebrum encodes innately meaningful odors and translates them into behavioral response. However, a recent study in locusts did not find support for this view and rather suggests that it plays a role in extracting general stimulus features and integrating bilateral and multimodal sensory information (Gupta and Stopfer, 2012). In the *Drosophila* lateral protocerebrum projection neurons (PNs) target different areas dependent of whether they are activated by fruit odors or pheromones (Jefferis et al., 2007). In fact, one area of the lateral protocerebrum appears to be dedicated to appetitive behavior, one to aversive behavior, and one to sexual behavior. The detailed circuitry, however, remains to be elucidated.

22.2.2 The Coding of General Odors: Combinatorial Odor Coding

How are general, non-pheromonal odors encoded in insect brains, along the antenna-antennal lobe-higher brain centers pathway? To start with: how do odor responses look like in the antennal lobe? First, the morphology and size of antennal lobes differs widely among insects: *Drosophila melanogaster* has about 50 glomeruli (Laissue et al., 1999), moths such as *Heliothis virescens* and *Manduca sexta* about 70 (Rospars and Hildebrand, 1992; Berg et al., 2002), honeybees about 160 (Flanagan and Mercer, 1989a; Galizia et al., 1999a), and several ant species over 300 (Kelber et al., 2009). Most glomeruli are stereotypic across individuals within a species and can therefore be identified, often already on the basis of their shape, relative position and response profile (Figure 22.4a) (Galizia et al., 1999b). This stereotypical arrangement reflects the one-to-one mapping of receptor populations onto glomeruli, dictated by the number of genes in the genome for odorant-sensor proteins. But despite this variability, it appears that the basic coding mechanisms are the same, and indeed very similar to that of mammals, including humans. The simple one-to-one connectivity between OSNs and glomeruli is

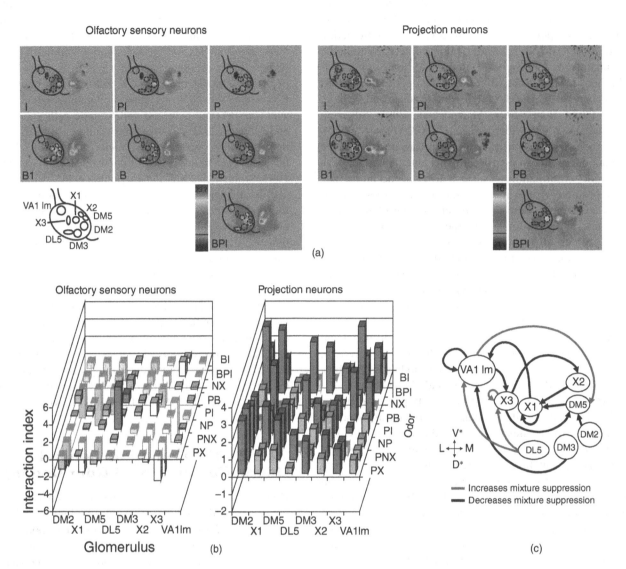

Figure 22.4 Analytic and synthetic odor representations in the antennal lobe of *Drosophila*. (a) Optically recorded glomerular calcium responses in OSNs (left) and in PNs (right) in *Drosophila*. The numbers on the color scale indicate the maximum and minimum δF/F (%). Odors were isopentyl acetate (I), propionic acid (P), 1-butanol (B) and all binary mixtures (BI, PI, PB), and the ternary mixture (BPI). The mask over the left AL shows the position of eight identified glomeruli. Note that the best odor for a given mixture in a given glomerulus was not always the same for OSNs and PNs, indicating that the antennal lobe network shapes the antennal lobe output. (b) Mixture interaction index ([response to the best component for the corresponding mixture] − [response to the mixture]) for OSNs and PNs. Positive values show inhibitory mixture interactions. OSNs represent mixtures in an *analytic* way as their responses could mostly be predicted from the components' responses, thus preserving component information. PNs show strong inhibitory mixture interactions, showing that the antennal lobe network processes mixtures in a *synthetic* way, which leads to a loss of component information. (c) Glomerulus-specific interactions shape antennal lobe output. Circles indicate identified glomeruli (X1, X2, VA1lm, etc.) and arrows indicate functional connectivity between the glomeruli. The black arrows reflect decreased mixture suppression and the gray arrows increased mixture suppression. For example, the black arrow from glomerulus X1 to glomerulus VA1lm indicates that OSN activity in glomerulus X1 positively affects the PN activity in glomerulus VA1lm, thereby decreasing mixture suppression in VA1lm. Similarly, for example, the gray arrow from glomerulus VA1lm to glomerulus DM5 indicates that OSN activity in VA1lm negatively affects the PN activity in glomerulus DM5, thereby increasing mixture suppression in DM5. (modified from Silbering and Galizia, 2007). (*See plate section for color version.*)

contrasted by a complex intra- and interglomerular network within the antennal lobe (Figure 22.3b), suggesting that the antennal lobe, like its mammalian counterpart, the olfactory bulb, is far more than a mere relay station for odor information. We will look at a few exemplary tasks of this circuit: signal-to-noise improvement, contrast enhancement, concentration coding, odor-mixture processing, and odor discrimination.

22.2.3 Signal to Noise Improvement

Among the most important properties in any sensory system is sensitivity. Our own photoreceptors are so sensitive that a single photon can elicit a neural response. Similarly, a single molecule of a pheromone can elicit a signal in the sensory neuron of a moth. However, these single events are not sufficient to recognize an odor source: as noted above, pheromones come in mixtures at a given concentration ratio, and thus many single molecule events have to occur. The glomerular structure in the antennal lobe is ideally suited to create the appropriate readout: all OSNs with equal response properties converge onto a single glomerulus. Assuming that the temporal distribution of action potentials would be a random (Poisson) event, then the inherent noise is related to the total number of action potentials as the square root, that is, with 100 action potentials, the standard deviation would be 10, and the signals-to-noise (s/n) would be 100:10, that is, 10% noise. With 10,000 action potentials, the noise would be 100 (square root), and s/n be 10,000:100, that is, 1% noise. Whether these action potentials come from a single neuron or from many does not matter, if they are all collected in the same glomerulus, and therefore, the more sensory neurons a system has, the higher the potential to increase s/n. It is no surprise, then, that sexual-pheromone sensitive glomeruli are particularly large: they collect the largest number of OSN axons, and achieve the highest s/n in the system!

22.2.4 Contrast Enhancement

Most OSNs have broad response profiles, that is, they do not respond to a single substance only, but rather to many (de Bruyne et al., 2001; Hallem et al., 2004; Galizia et al., 2010). This is an important property of olfactory systems: assume a fruit fly with 60 OSN types would only have highly selective OSNs. The olfactory environment of this animal would consist only of odors consisting of these substances, that is, 60 distinct odorants and their mixtures (still, a vast landscape of potential stimuli), and any odor-source that would not emit at least one of

these 60 substances would be odorless to the animal. This problem is solved by using generalist OSNs, which respond to many substances. Here, the requirements to the network, however, increase, because extracting odor identity becomes more difficult. For example, take a two receptor case: substance A would activate receptor 1 strongly, and receptor 2 weakly, substance B would elicit the opposite pattern. Thus, both substances elicit activity in both receptors, creating a somewhat ambiguous situation. (In fact, with such properties, it is impossible to solve ambiguity: in most cases different odor-mixtures exist that could elicit the same response pattern. The situation is similar to our color vision, where different color spectra can elicit identical color perception even under otherwise identical conditions). The antennal lobe increases the contrast in this situation, in two distinct ways. A network of interglomerular inhibitory neurons is activated by all odor-inputs, with the capacity of inhibiting all glomeruli (Figure 22.4) (Sachse and Galizia, 2002; Silbering and Galizia, 2007). Thus, weak glomeruli are reduced below threshold, while strong glomeruli remain active, creating a more distinctive activity pattern (Sachse and Galizia, 2003; Silbering et al., 2008) (this is related to "winner-take-all," but different in that the winner is not a glomerulus, but a pattern). Furthermore, each local neuron has a distinctive innervation pattern: some innervate all glomeruli, others only a subpopulation (Flanagan and Mercer, 1989b; Sun et al., 1993; Chou et al., 2010). The latter group of neurons can compute odor-specific contrasts, and therefore optimize the representation of odor combinations that are innately important. Because these connections can change during the life time, also odors that are learned can be optimized by the experience of the animal (Fernandez et al., 2009; Rath et al., 2011).

22.2.5 Concentration Coding and Gain Control

The concentration of olfactory stimuli can span a wide range. A honeybee experiences a flower odor at high concentration when she learns the odor while sitting inside the flower and drinking the nectar. On a subsequent foraging flight, however, she has to detect and identify the same odor in the air at a concentration that is several magnitudes lower in concentration. Thus insects have to generalize across a wide range of concentrations. On the other hand, honeybees and *Drosophila* can be trained to discriminate between small concentration differences of the same odorants (Ditzen et al., 2003; Yarali et al., 2009). Individual OSNs typically have a dynamic range of two to three orders of magnitude (which is larger than most OSNs described in mammals). For example, a dOR22A receptor may begin responding to ethyl hexanoate at a concentration of 10^{-9} and may already saturate at a concentration of 10^{-6}, with higher concentrations leading to the saturated response, or even to a decreased response (here, concentrations are given as dilutions of the odor in the olfactometer, not as molecules in stimulus air) (Pelz et al., 2006).

We have seen that s/n improvement (above) can lower the perceptual threshold. Adaptation of OSNs can increase the upper range: an animal in a smelly environment shifts its dose–response curve, so that odor-concentration differences can be perceived. Within the antennal lobe the network also improves concentration coding via a gain control mechanism: when no activity is present, excitatory activity "regulates" the projection neurons to be just active, creating a pattern of spontaneous activity (Sachse and Galizia, 2003; Root et al., 2007, 2008; Olsen and Wilson, 2008). Thus, even minute stimuli will already create a suprathreshold excitation: the signal is amplified. Conversely, at high concentration, the inhibitory lateral network pushes activity down (Das et al., 2011), enhancing interglomerular contrast (Silbering et al., 2008), and flattening the dose–response curves (Sachse and Galizia, 2002). This leads to a great enhancement of the dynamic range of odor concentrations that can be processed within the system. The interplay between excitatory local neurons and inhibitory local neurons is essential for gain control and concentration processing (Root et al., 2008; Silbering et al., 2008; Acebes et al., 2012). The olfactory system needs to extract both, information about the odor's concentration, and a concentration-invariant information about odor quality. Interestingly, PNs with concentration independent odor-quality responses have been reported (Stopfer et al., 2003; Yamagata et al., 2009).

22.2.6 Odor-mixture Processing

Most naturally occurring odors are mixtures. Wine odor (so attractive to both humans and fruit flies) contains hundreds of different volatiles. It is this very high number of contributing substances that allows us to enjoy the large variety of wine odors, as well as many other odors. Just as occurs in humans, it is important to realize that an animal likely experiences the mixture as the perceptual unit (or Gestalt) of an odor. Thus, the brain does not extract the single odor components, but rather creates a synthetic representation of the odor mixture. We have seen examples where this concept is realized, and examples where it is not. In many species sexual pheromones are only behaviorally active when given at the right mixture (i.e., when the corresponding glomeruli have the right relative activity pattern). In contrast, CO_2 is processed by highly selective receptors. However, whenever the corresponding glomerulus is active, the animal will react to CO_2, independently of co-occurring odors. At this

point it is important to note that even these "hard-wired" circuits are not strictly hard-wired. All olfactory processing is modulated by the internal state of the animal (sexual arousal, hunger, etc.), states that are generally mediated by large modulatory neurons that branch in many brain areas. Depending on their activity, a "stereotypic" olfactory stimulus may have a very strong behavioral effect, or even no effect at all.

Within the circuitry of the antennal lobes, inhibitory connections create mixture-unique patterns (Figure 22.4). Thus, both in behavior and in physiology, a mixture of two substances A and B does not elicit an overlay of A and B, but rather a new situation AB, with some similarity to A and to B, but also with distinct features (sometimes termed "unique cue") characteristic of the mixture itself (Deisig et al., 2006, 2010; Carlsson et al., 2007; Silbering and Galizia, 2007; Riffell et al., 2009; Kuebler et al., 2011; Su et al., 2011). The amount of inhibition, and the similarity of the resulting pattern AB to its component patterns, differs widely depending on species and odor. In fact, this is in part driven by evolution, and hard wired (as with the mixture-specific cells that have been found in the sexual pheromone system of moths), and in part driven by experience. When a honeybee is trained to discriminate between two similar odor mixtures, the representation of these mixtures gradually changes in the brain to make them more dissimilar to each other, thus aiding their discrimination (Fernandez et al., 2009).

22.2.7 Processing of Concurrent Odors from Multiple Sources

The spatial and temporal structure of airborne odorants is determined by the physics of atmospheric dispersion (Murlis et al., 1992): close to the ground odors distribute mainly via diffusion and olfactory stimuli dynamics are comparably slow, due to the lack of turbulences. However, this changes completely above the laminar boundary layer: airborne odorants distribute in plumes that fluctuate at different timescales (Riffell et al., 2008). Individual odor-plumes can last and occur at intervals ranging from milliseconds to minutes. Since odor-plumes of different odor sources intermingle, the olfactory system needs to segregate concurrent odors from independent sources in order to recognize them as different odor-objects (Hopfield, 1991). Insects can exploit short temporal asynchronies between odor plume onsets to segregate odorants from different sources (Szyszka and Stierle, 2014). For honeybees, a 6-millisecond temporal difference in stimulus coherence is sufficient to segregate an odor-object from a background odor (Szyszka et al., 2012), and moths and beetles can distinguish plumes of attractive sex pheromone from plumes of a behavioral antagonistic odorant when they are separated by a few milliseconds (Baker et al., 1998; Andersson et al., 2011). It is not clear how odor-background segregation based on millisecond stimulus onset-asynchrony is internally represented. In locusts and honeybees, neural representations of mixtures partly match those evoked by the individual components if their onsets differ by a 6 to 100 milliseconds (Broome et al., 2006, Stierle et al., 2013).

22.2.8 Discrimination and Generalization

Generalization and discrimination are the main challenges for an animal encountering mixtures. For example, the odors of flowers of the same plant species can differ due to flowering stages, pollination status and the time of the day (Phamdelegue et al., 1989; Schiestl et al., 1997). Discrimination is important to avoid landing on a flower that has no nectar. However, restricting odor recognition to a precise stimulus would not work either: in a natural situation, no two odors are alike. A flower growing on the sunny side of the meadow will smell slightly different from the same species on the shady side, or on a more alkaline soil. This creates a problem for insects such as the honeybee, which are flower constant during their foraging trips, that is, which harvest nectar from a single species as long as the food quality is good. For some very specialized insects the system is so strict that a dependent mutualism has evolved between the flower and the insect. Consequently, the insect has to generalize from one odor to different similar odors. Generalization is the counterpart to discrimination, and it is creating this borderline between "just similar enough to be judged as the same" and "so different that it must be something else" that creates one of the major coding tasks in the olfactory system. In *Drosophila*'s olfactory lobe, the network of local interneurons modifies the responses in PNs based on their valence for the fly: odors that belong to the same behavioral meaningful category become more similar to each other in the combinatorial activity pattern, while odors that belong to another category become more dissimilar, thus aiding the brain in creating the generalization-discrimination divide (Niewalda et al., 2011; Knaden et al., 2012). Another example for the necessity to discriminate well: the leaf-cutting ants *Atta vollenweideri* and *Atta sexdens* mark their trails with pheromones that only differ in the ratio of their components, but they have no difficulties in discriminating between conspecific and heterospecific trails (Kleineidam et al., 2007).

How relevant is the similarity of odor-evoked activity patterns for the animal? The perceived odor similarity correlates well with the physiological similarity between odorant-evoked combinatorial glomerular activity patterns in the honeybee (Guerrieri et al., 2005). In *Drosophila*,

processing within the antennal lobe increases the correspondence between perceived similarity and physiological similarity in PNs as compared to OSNs (Niewalda et al., 2011). Indeed, blocking inhibitory connections within the antennal lobe using pharmacological tools changes both temporal and spatial properties of odor-response patterns and reduces the capacity of animals to differentiate among similar odors (Stopfer et al., 1997; Sachse and Galizia, 2002). Thus, both innate circuitry and learned components modify the "olfactory landscape" of odor representations to create a landscape where odors that belong together are more closely associated with each other.

22.2.9 The Coding of Special Odors: Sexual Pheromones

Among the best studied cases in insect olfaction is the capacity of male moths to find a female by following the odor trail of a sexual pheromone over very long distances (up to several km). Several aspects are important to understand this system. Most pheromones are mixtures of more than one substance (Kaissling, 1996). The hypothesis is that a mixture allows for better coding of odor-quality over a large range of concentrations. Assume a pheromone would consist of only a single substance, and that an animal had a single, highly sensitive neuron for this substance. Even the most selective neuron might still have some cross-reaction to other substances, when given at high concentration. Consequently, a single neuron type does not give sufficient information to differentiate between a very low concentration of the best ligand (maybe a female on a tree far away), and a very high concentration of a weak ligand (maybe the female of a closely related species with a similar, but slightly different pheromone, sitting on the next branch). A combination of two neurons is much more distinctive. Furthermore, this robustness transports to the sending signal: some closely related species do not differ in their pheromone components, but only in the relative ratio, that is, whether substance A or substance B is the major component. Thus, again, a mixture gives a safer and more robust communication signal for pheromones. Moths have even evolved the capacity of detecting time-onset differences down to a millisecond (Baker et al., 1998): only the conspecific female will release all components at exactly the same time. Onset asynchrony therefore indicates that the odor plume must come from different sources, and not from a conspecific female – following this signal would be a waste of resources.

Generally, for each pheromone component, there is a dedicated OSN population. Each population innervates a glomerulus. Therefore, activity in a glomerulus is a clear signal that a particular chemical substance has been smelled (this is sometimes called a "labeled line," though some scholars use "labeled line" in a more restrictive way, that is, only for signals that have a dedicated processing stream all the way to eliciting a stereotype behavior). Since the components of the pheromone are detected by different OSNs, which innervate different glomeruli, the conspecific pheromone signal is uniquely identified by a pattern of active glomeruli: in the case of a pheromone composed of two substances, the signal consists of the two corresponding glomeruli lighting up. We do not yet know exactly how this combinatorial pattern is read out and then converted into behavior (i.e., an upwind flight towards the pheromone source), but we do know that already within the antennal lobe there are mixture-specific neurons (Riffell et al., 2009), and that synchrony of neural activity in the glomeruli involved is crucial for a reliable recognition of the pheromone (Heinbockel et al., 2004). Interestingly, the necessary combinatorial pattern is not limited to the presence of some glomeruli – the absence of others is also required. In the moth *Heliothis*, pheromone components of closely related species act as direct inhibitors (Vickers et al., 1998). The glomeruli involved in sexual pheromone processing in this species are four: two codes for the two main components of the conspecific pheromone, and two for components of other, closely related species. The relevant combinatorial pattern thus consists of the first ones being active, and the other two being silent – activity in one of the latter is sufficient to block upwind flight, that is, search for a female.

In moths, the sexual pheromone-sensitive glomeruli are in the antennal lobe close to the entrance of the antennal nerve. Since they are innervated by a very large number of OSNs (to increase sensitivity), these glomeruli are generally large, and the complex of all sexual pheromone-sensitive glomeruli is called the macroglomerular complex. The number and shape of glomeruli in the macroglomerular complex differs widely across species, in most species studied it is in the range of three to four.

22.2.10 The Coding of Special Odors: CO$_2$

Carbon dioxide (CO$_2$) is a special odor. Humans cannot smell it, even though we have internal sensors for CO$_2$: the concentration of CO$_2$ is tightly controlled in our blood, and sensed as dissolved bicarbonate.[2] Controlling CO$_2$ in our blood is important for homeostasis, that is, keeping bodily conditions constant. In some insect species, CO$_2$ plays a similar role: ants and bees live in large colonies, some of which are located in burrows dug several meters

deep into the ground. Controlling oxygen and CO$_2$ in these conditions is vital, and when CO$_2$ levels rise above permissive values, the animals open ventilation paths to aerate the hive (Kleineidam and Tautz, 1996; Weidenmuller et al., 2002). The CO$_2$ receptors have properties that match this necessity to measure absolute concentrations: their responses are mostly tonic, that is, when in a certain CO$_2$ concentration, the neurons fire with a given frequency that is constant, so that the brain can directly translate firing frequency into CO$_2$ concentration (Lacher, 1964; Kleineidam et al., 2000). The sensilla are also adapted to slow rather than fast measurements: the receptor neurons are located within sensilla consisting of a long, invaginated tube, so that fast fluctuations of external CO$_2$ are filtered out. The gene of the CO$_2$ receptor is not yet known.

The situation is totally different in mosquitoes: they use CO$_2$ to find hosts. Humans and other mammals exhale CO$_2$ in a rhythmic pattern, and some mosquito species have evolved to sense such fluctuations in CO$_2$ (Geier et al., 1999). Their receptors are mostly phasic, that is, their action potentials increase when the CO$_2$ concentration changes but are reduced to a small tonic component when the concentration remains constant (Grant et al., 1995). The receptor is known: it is not directly related to other odorant receptors, but rather evolved from a gustatory receptor, suggesting that it is not sensing CO$_2$ but a reaction product, for example, bicarbonate. The same CO$_2$ receptor is expressed in the fruit fly *Drosophila*, a species for which many details of its brain circuitry are known (de Bruyne et al., 2001; Jones et al., 2007; Kwon et al., 2007). It appears that CO$_2$ is processed using a labeled-line strategy: all receptors innervate a single glomerulus (Suh et al., 2004). Deleting these receptors makes the fly anosmic to CO$_2$. CO$_2$ is a repellent odor for flies, leading to a rapid escape response. Interestingly, highly aversive odors activate adjacent brain areas, and it appears that odors that are innately negative and/or repulsive have their distinct pre-motor area here (Jefferis et al., 2007), an observation that is paralleled in the antennal lobe, where odors of positive and negative valence have their preferred coding regions (Knaden et al., 2012).

22.2.11 Odor Representation is Transformed Between Antennal Lobe and Mushroom Bodies

The logic of odor coding is transformed between antennal lobes and mushroom bodies. The most apparent change is a sparsening of activity: odors usually activate a high percentage of PNs in the antennal lobe, but only activate a small percentage of Kenyon cells (Perez-Orive et al., 2002; Wang et al., 2003; Szyszka et al., 2005; Honegger et al., 2011). This population sparseness (i.e., few Kenyon cells

---

[2] In humans, CO$_2$ at high concentrations can be detected intranasally by the trigeminal (CN V), but not by the olfactory (CN I), nerve. The intranasal trigeminal system detects somatosensory sensations elicited by volatiles, such as cooling, burning, and stinging (see Chapter 50).

active for each stimulus) goes hand-in-hand with lifetime sparseness (i.e., each Kenyon cell responds only to a small number of odors). Thus, Kenyon cells are narrowly tuned, and some are even selective for particular stimulus concentrations (Stopfer et al., 2003). To analyze the transformation of odor representations in the mushroom body and to evaluate the contribution of pre- and postsynaptic processing in honeybees, we characterized three consecutive processing stages: first in PNs of the antennal lobe, second in PNs' synaptic terminals in the mushroom body calyces, and third in the postsynaptic Kenyon cells (Szyszka et al., 2005). The percentage of activated units that responded to one odor only (out of four tested odors) increased across the three processing stages from 55, to 70, to 92%, while the units that responded to two odors or more dropped from 45, to 30, to 8%. The sparsening of the Kenyon cell code is a progressive process and involves a series of transformations. These are accomplished by the divergent–convergent connectivity between PNs and Kenyon cells, by the weak synaptic connections between PNs and Kenyon cells, and by the odor-driven inhibition of Kenyon cells (Perez-Orive et al., 2002; Jortner et al., 2007; Gupta and Stopfer, 2011; Papadopoulou et al., 2011). Moreover, local microcircuits in the mushroom body calyx modify the PN output (Szyszka et al., 2005; Yamagata et al., 2009). However, even though the Kenyon cells' odor representations are sparse, the odor code still remains combinatorial. In fact, the dimensionality is even increased: 50% activity chance across 160 glomeruli means 80 active glomeruli, while 1% activity chance across 100,000 Kenyon cells means 1000 active Kenyon cells. It is this expansion of possible combinatorial patterns from the antennal lobe to the mushroom bodies that increases the coding capacity, and more importantly affords the possibility to create complex odor categories.

22.2.12 Odor Coding Dynamics

Kenyon cell responses also differ from PN responses in temporal aspects: PNs respond to odors with long lasting, often temporally complex responses, which are superimposed onto a high level of background activity (Abel et al., 2001; Perez-Orive et al., 2002; Sachse and Galizia, 2002; Galan et al., 2006; Szyszka et al., 2011; Nawrot, 2012). Kenyon cells, in contrast, show brief odor responses of only a few spikes, generally at odor onset and at odor offset only, and background activity is nearly absent (Perez-Orive et al., 2002; Stopfer et al., 2003; Szyszka et al., 2005; Ito, Ong et al., 2008; Turner et al., 2008). The combinatorial pattern across Kenyon cells is not equal at odor-onset and at odor-offset. A honeybee flying through an odor plume will get information both when it enters the plume and when it exits the plume. The reduced level of activity during the stimulus raises a conundrum: the temporal complexity created in the odor response of PNs by the antennal lobe network appears to be ignored by the readout in the mushroom bodies. Indeed, behavioral responses suggest that animals do not need a long time to recognize an odor (Ditzen et al., 2003), suggesting that slow activity patterns across PNs cannot be relevant for odor recognition under normal circumstances. Associative learning experiments in honeybees and moths confirm that only the onset of odor-elicited activity carries behaviorally relevant odor information: In bees the first bout of PN activity is relevant, whereas ongoing post-odor activities are not (Szyszka et al., 2011), and in moths only a few Kenyon cell spikes at odor-stimulus onset are relevant (Ito et al., 2008). It appears as if ongoing PN activity would be wasted. However, this activity may be important for other brain areas, for example the lateral protocerebrum, or it may be used for establishing the contiguity between an odor and a meaningful stimulus during associative learning (Szyszka et al., 2008) – a hypothesis that remains to be investigated.

Spiking activity in individual neurons is always fast, and creates a situation for the brain where synchrony can be used for information coding. In the antennal lobe, synchrony of spikes across glomeruli is important for binding information across glomerular patterns (MacLeod and Laurent, 1996; Christensen et al., 2000; Tanaka et al., 2009). Odor-evoked oscillations have been found across the animal kingdom, and may be important to create time-windows to read-out spike synchrony across PNs. The sharp onset and the temporal precision of action potentials across Kenyon cells in the mushroom bodies may also be related to the necessity of reading out across-neuron activity patterns based on synchrony. A detailed review of these fast components in odor-evoked activity is beyond the scope of this review, and can be found elsewhere (Laurent, 2002).

22.3 CONCLUSION: FROM ODOR TO BEHAVIOR

In this review, we have followed odor representation along the olfactory system from olfactory sensory neurons, through the antennal lobe, into the mushroom bodies. This is not yet the behavioral readout, and more research is needed to close the circuit. We have investigated two coding systems: innate-meaning odor-coding (CO_2 and sexual pheromones) and general odor coding. The latter are situations that involve many odor sources with different smells, while the more specialized systems use few, and innately predetermined processing circuits. While innate systems often use labeled line like odor-coding schemes (see discussion of CO_2 coding, above), the combinatorial logic of general odor coding prevails: odor stimulus, pattern of activity across glomeruli in the antennal lobes, a transformed and different pattern of activated cells in the mushroom bodies, and a resulting behavioral output Galizia CG 2014. It is the latter that now enters into the focus of olfactory research in insects: while in few specialized cases the lateral protocerebrum may already be the place where an olfactory input elicits the premotor command, the link from odor representation to behavior remains to be elucidated in insects. It will become apparent that this link is not simple: First, natural odors occur in a temporally fluctuating, multi-odor background, yet insects possess a remarkable ability to extract intelligible odors under such complex conditions. The neural basis of odor-background segregations remains currently unknown. Second, even a plain odor stimulus does not elicit a stereotype response. Rather, the state of the animal is important: circadian rhythms, sexual state, thirst or hunger, and previous experience all influence in a decisive manner how an odor is responded to by the insect. All of these modulatory systems are represented in the brain by dedicated neural circuits, and these circuits interact with the olfactory pathway along all its stages: at the antenna, in the antennal lobe, in the mushroom bodies and in the lateral protocerebrum. It is in this interaction that we see a strong need for further research. Understanding these circuits will help us understand our own olfactory system, but most importantly it will help us understand the olfactory world of insects: how do they find the right flower to visit, the right leaf to pierce, or the chosen victim to attack.

ACKNOWLEDGMENTS

We thank Jacob S. Stierle for valuable comments.

REFERENCES

Abel, R., J. Rybak, et al. (2001). Structure and response patterns of olfactory interneurons in the honeybee, *Apis mellifera. J. Comp. Neurol.* 437(3): 363–383.

Acebes, A., J. M. Devaud, et al. (2012). Central adaptation to odorants depends on PI3K levels in local interneurons of the antennal lobe. *J. Neurosci.* 32(2): 417–422.

Ache, B. W. and J. M. Young (2005). Olfaction: Diverse species, conserved principles. *Neuron* 48(3): 417–430.

Andersson, M. N., M. Binyameen, et al. (2011). Attraction modulated by spacing of pheromone components and anti-attractants in a bark beetle and a moth. *J. Chem. Ecol.* 37(8): 899–911.

Baker, T. C., H. Y. Fadamiro, et al. (1998). Moth uses fine tuning for odour resolution. *Nature* 393(6685): 530–530.

Bartelt, R. J., Schaner, A.M, and Jackson, L. L. (1985). cis-Vaccenyl acetate as an aggregation pheromone in *Drosophila melanogaster J. Chem. Ecol.* 11(12): 1747–1756.

Benton, R., S. Sachse, et al. (2006). Atypical membrane topology and heteromeric function of Drosophila odorant receptors in vivo. *PLoS. Biol.* 4(2): e20.

Benton, R., K. S. Vannice, et al. (2009). Variant ionotropic glutamate receptors as chemosensory receptors in Drosophila. *Cell* 136(1): 149–162.

Berg, B. G., C. G. Galizia, et al. (2002). Digital atlases of the antennal lobe in two species of tobacco budworm moths, the Oriental *Helicoverpa assulta* (male) and the American *Heliothis virescens* (male and female). *J. Comp. Neurol.* 446(2): 123–134.

Broome, B. M., V. Jayaraman, et al. (2006). Encoding and decoding of overlapping odor sequences. *Neuron* 51(4): 467–482.

Butenandt, A., U. Groschel, et al. (1959). [N-acetyl tyramine, its isolation from Bombyx cocoons & its chemical & biological properties]. *Arch Biochem. Biophys.* 83(1): 76–83.

Carde, R. T. and M. A. Willis (2008). Navigational strategies used by insects to find distant, wind-borne sources of odor. *J. Chem. Ecol.* 34(7): 854–866.

Carlsson, M. A., K. Y. Chong, et al. (2007). Component information is preserved in glomerular responses to binary odor mixtures in the moth *Spodoptera littoralis. Chem. Senses* 32(5): 433–443.

Chou, Y. H., M. L. Spletter, et al. (2010). Diversity and wiring variability of olfactory local interneurons in the Drosophila antennal lobe. *Nat. Neurosci.* 13(4): 439–449.

Christensen, T. A., V. M. Pawlowski, et al. (2000). Multi-unit recordings reveal context-dependent modulation of synchrony in odor-specific neural ensembles. *Nat. Neurosci.* 3(9): 927–931.

Christiansen, F., C. Zube, et al. (2011). Presynapses in kenyon cell dendrites in the mushroom body calyx of Drosophila. *J. Neurosci.* 31(26): 9696–9707.

Couto, A., M. Alenius, et al. (2005). Molecular, anatomical, and functional organization of the Drosophila olfactory system. *Curr. Biol.* 15(17): 1535–1547.

Das, S., M. K. Sadanandappa, et al. (2011). Plasticity of local GABAergic interneurons drives olfactory habituation. *Proc. Natl. Acad. Sci. USA* 108(36): E646–E654.

Davis, R. L. (2011). Traces of Drosophila memory. *Neuron* 70(1): 8–19.

de Bruyne, M. and T. C. Baker (2008). Odor detection in insects: volatile codes. *J. Chem. Ecol.* 34(7): 882–897.

de Bruyne, M., K. Foster, et al. (2001). Odor coding in the Drosophila antenna. *Neuron* 30(2): 537–552.

Deisig, N., M. Giurfa, et al. (2006). Neural representation of olfactory mixtures in the honeybee antennal lobe. *Eur. J. Neurosci.* 24(4): 1161–1174.

Deisig, N., M. Giurfa, et al. (2010). Antennal lobe processing increases separability of odor mixture representations in the honeybee. *J. Neurophysiol.* 103(4): 2185–2194.

Ditzen, M., J. F. Evers, et al. (2003). Odor similarity does not influence the time needed for odor processing. *Chem. Senses* 28(9): 781–789.

Doty, R. L. (2010). The Great Pheromone Myth. Baltimore: Johns Hopkins University Press.

Fernandez, P. C., F. F. Locatelli, et al. (2009). Associative conditioning tunes transient dynamics of early olfactory processing. *J. Neurosci.* 29(33): 10191–10202.

Fishilevich, E. and L. B. Vosshall (2005). Genetic and functional subdivision of the Drosophila antennal lobe. *Curr. Biol.* 15(17): 1548–1553.

Flanagan, D. and A. R. Mercer (1989). An atlas and 3-D reconstruction of the antennal lobes in the worker honey bee, *Apis mellifera* L (Hymenoptera, Apidae). *Internat J. Insect Morphol. Embryol.* 18(2-3): 145–159.

Flanagan, D. and A. R. Mercer (1989). Morphology and response characteristics of neurons in the deutocerebrum of the brain in the honeybee *Apis mellifera. J. Comp. Physiol. - Sensory Neural. Behav. Physiol.* 164(4): 483–494.

References

Galan, R. F., M. Weidert, et al. (2006). Sensory memory for odors is encoded in spontaneous correlated activity between olfactory glomeruli. *Neural Comput.* 18(1): 10–25.

Galizia, C. G., S. L. McIlwrath, et al. (1999a). A digital three-dimensional atlas of the honeybee antennal lobe based on optical sections acquired by confocal microscopy. *Cell Tissue Res.* 295(3): 383–394.

Galizia, C. G., D. Munch, et al. (2010). Integrating heterogeneous odor response data into a common response model: A DoOR to the complete olfactome. *Chem. Senses* 35(7): 551–563.

Galizia, C. G., S. Sachse, et al. (1999b). The glomerular code for odor representation is species specific in the honeybee *Apis mellifera*. *Nat. Neurosci.* 2(5): 473–478.

Galizia, C. G. (2014) Olfactory coding in the insect brain: data and conjectures. *Eur. J. Neurosci.* 39(11): 1784–1795.

Ganeshina, O. and R. Menzel (2001). GABA-immunoreactive neurons in the mushroom bodies of the honeybee: An electron microscopic study. *J. Comp. Neurol.* 437(3): 335–349.

Geier, M., O. J. Bosch, et al. (1999). Influence of odour plume structure on upwind flight of mosquitoes towards hosts. *J. Exp. Biol.* 202 (Pt 12): 1639–1648.

Grant, A. J., B. E. Wigton, et al. (1995). Electrophysiological responses of receptor neurons in mosquito maxillary palp sensilla to carbon-dioxide. *J. Comp. Physiol. A.* 177(4): 389–396.

Grosjean, Y., R. Rytz, et al. (2011). An olfactory receptor for food-derived odours promotes male courtship in Drosophila. *Nature* 478(7368): 236–240.

Grunewald, B. (1999a). Morphology of feedback neurons in the mushroom body of the honeybee, Apis mellifera. *J. Comp. Neurol.* 404(1): 114–126.

Grunewald, B. (1999b). Physiological properties and response modulations of mushroom body feedback neurons during olfactory learning in the honeybee, *Apis mellifera. J. Comp. Physiol. A.* 185(6): 565–576.

Guerrieri, F., M. Schubert, et al. (2005). Perceptual and neural olfactory similarity in honeybees. *PLoS. Biol.* 3(4): e60.

Gupta, N. and M. Stopfer (2011). Olfactory coding: giant inhibitory neuron governs sparse odor codes. *Curr. Biol.* 21(13): R504–R506.

Gupta, N. and M. Stopfer (2012). Functional analysis of a higher olfactory center, the lateral horn. *J. Neurosci.* 32(24): 8138–8148.

Hallem, E. A., M. G. Ho, et al. (2004). The molecular basis of odor coding in the drosophila antenna. *Cell* 117(7): 965–979.

Heinbockel, T., T. A. Christensen, et al. (2004). Representation of binary pheromone blends by glomerulus-specific olfactory projection neurons. *J. Comp. Physiol. A.* 190(12): 1023–1037.

Heisenberg, M. (2003). Mushroom body memoir: from maps to models. *Nat. Rev. Neurosci.* 4(4): 266–275.

Honegger, K. S., R. A. A. Campbell, et al. (2011). Cellular-resolution population imaging reveals robust sparse coding in the Drosophila mushroom body. *J. Neurosci.* 31(33): 11772–11785.

Hopfield, J. J. (1991). Olfactory computation and object perception. *Proc. Natl. Acad. Scie. USA* 88(15): 6462–6466.

Huang, J., W. Zhang, et al. (2010). Functional connectivity and selective odor responses of excitatory local interneurons in Drosophila antennal lobe. *Neuron* 67(6): 1021–1033.

Ito, I., R. C. Y. Ong, et al. (2008). Sparse odor representation and olfactory learning. *Nat. Neurosci.* 11(10): 1177–1184.

Jefferis, G. S., C. J. Potter, et al. (2007). Comprehensive maps of Drosophila higher olfactory centers: spatially segregated fruit and pheromone representation. *Cell* 128(6): 1187–1203.

Jones, W. D., P. Cayirlioglu, et al. (2007). Two chemosensory receptors together mediate carbon dioxide detection in Drosophila. *Nature* 445(7123): 86–90.

Jortner, R. A., S. S. Farivar, et al. (2007). A simple connectivity scheme for sparse coding in an olfactory system. *J. Neurosci.* 27(7): 1659–1669.

Justus, K. A., S. W. Schofield, et al. (2002). Flight behaviour of *Cadra cautella* males in rapidly pulsed pheromone plumes. *Physiol. Entomol.* 27(1): 58–66.

Kaissling, K. E. (1996). Peripheral mechanisms of pheromone reception in moths. *Chem. Senses* 21(2): 257–268.

Karlson, P. and Lüscher, M. (1959). "Pheromones": a new term for a class of biologically active substances. *Nature* 183:55–56.

Kelber, C., W. Rossler, et al. (2009). The antennal lobes of rungus-growing ants (Attini): neuroanatomical traits and Evolutionary trends. *Brain Beh. Evol.* 73(4): 273–284.

Kleineidam, C., R. Romani, et al. (2000). Ultrastructure and physiology of the CO2 sensitive sensillum ampullaceum in the leaf-cutting ant *Atta sexdens. Arthropod Struct. Dev.* 29(1): 43–55.

Kleineidam, C. and J. Tautz (1996). Perception of carbon dioxide and other "air-condition" parameters in the leaf cutting ant *Atta cephalotes. Naturwissenschaften* 83(12): 566–568.

Kleineidam, C. J., W. Rossler, et al. (2007). Perceptual differences in trail-following leaf-cutting ants relate to body size. *J. Insect Physiol.* 53(12): 1233–1241.

Knaden, M., A. Strutz, et al. (2012). Spatial representation of odorant valence in an insect brain. *Cell Rep.* 1(4): 392–399.

Kuebler, L. S., S. B. Olsson, et al. (2011). Neuronal processing of complex mixtures establishes a unique odor representation in the moth antennal lobe. *Front Neural Circuits* 5:7.

Kwon, J. Y., A. Dahanukar, et al. (2007). The molecular basis of CO_2 reception in Drosophila. *Proc. Nat. Acad. Sci. USA* 104(9): 3574–3578.

Lacher, V. (1964). Elektrophysiologische Untersuchungen an einzelnen Rezeptoren fur Geruch, Kohlendioxyd, Luftfeuchtigkeit und Temperatur auf den Antennen der Arbeitsbiene und der Drohne (*Apis mellifica* L). *Z. Vergleich Physiol.* 48(6): 587–623.

Laissue, P. P., C. Reiter, et al. (1999). Three-dimensional reconstruction of the antennal lobe in *Drosophila melanogaster. J. Comp. Neurol.* 405(4): 543–552.

Larsson, M. C., A. I. Domingos, et al. (2004). Or83b encodes a broadly expressed odorant receptor essential for Drosophila olfaction. *Neuron* 43(5): 703–714.

Laurent, G. (2002). Olfactory network dynamics and the coding of multidimensional signals. *Nat. Rev. Neurosci.* 3(11): 884–895.

Loudon, C. and M. A. Koehl (2000). Sniffing by a silkworm moth: wing fanning enhances air penetration through and pheromone interception by antennae. *J. Exp. Biol.* 203(Pt 19): 2977–2990.

MacLeod, K. and G. Laurent (1996). Distinct mechanisms for synchronization and temporal patterning of odor-encoding neural assemblies. *Science* 274(5289): 976–979.

Macrides, F. and S. L. Chorover (1972). Olfactory bulb units: activity correlated with inhalation cycles and odor quality. *Science* 175(4017): 84–87.

Menzel, R. (2001). Searching for the memory trace in a mini-brain, the honeybee. *Learn Mem.* 8(2): 53–62.

Mobbs, P. G. (1982). The Brain of the honeybee *Apis mellifera* .1. The connections and spatial-organization of the mushroom bodies. *Phil. T Roy. Soc. B* 298(1091): 309–354.

Murlis, J., J. S. Elkinton, et al. (1992). Odor plumes and how insects use them. *Ann. Rev. Entomol.* 37: 505–532.

Nakagawa, T. and L. B. Vosshall (2009). Controversy and consensus: noncanonical signaling mechanisms in the insect olfactory system. *Curr. Opin. Neurobiol.* 19(3): 284–292.

Nawrot, M. P. (2012). Dynamics of sensory processing in the dual olfactory pathway of the honeybee. *Apidologie* 43(3):269–291.

Niewalda, T., T. Voller, et al. (2011). A combined perceptual, physico-chemical, and imaging approach to 'odour-distances' suggests a categorizing function of the Drosophila antennal lobe. *PLoS. One* 6(9) e24300.

Olsen, S. R., V. Bhandawat, et al. (2007). Excitatory interactions between olfactory processing channels in the Drosophila antennal lobe. *Neuron* 54(1): 89–103.

Olsen, S. R. and R. I. Wilson (2008). Lateral presynaptic inhibition mediates gain control in an olfactory circuit. *Nature* 452(7190): 956–960.

Papadopoulou, M., S. Cassenaer, et al. (2011). Normalization for sparse encoding of odors by a wide-field interneuron. *Science* 332(6030): 721–725.

Pelz, D., T. Roeske, et al. (2006). The molecular receptive range of an olfactory receptor in vivo (*Drosophila melanogaster* Or22a). *J. Neurobiol.* 66(14): 1544–1563.

Perez-Orive, J., O. Mazor, et al. (2002). Oscillations and sparsening of odor representations in the mushroom body. *Science* 297(5580): 359–365.

Phamdelegue, M. H., P. Etievant, et al. (1989). Sunflower volatiles involved in honeybee discrimination among genotypes and flowering stages. *J. Chem. Ecol.* 15(1): 329–343.

Rath, L., C. Giovanni Galizia, et al. (2011). Multiple memory traces after associative learning in the honey bee antennal lobe. *Eur. J. Neurosci.* 34(2): 352–360.

Riffell, J. A., L. Abrell, et al. (2008). Physical processes and real-time chemical measurement of the insect olfactory environment. *J. Chem. Ecol.* 34(7): 837–853.

Riffell, J. A., H. Lei, et al. (2009). Characterization and coding of behaviorally significant odor mixtures. *Curr. Biol.* 19(4): 335–340.

Root, C. M., K. Masuyama, et al. (2008). A presynaptic gain control mechanism fine-tunes olfactory behavior. *Neuron* 59(2): 311–321.

Root, C. M., J. L. Semmelhack, et al. (2007). Propagation of olfactory information in Drosophila. *Proc. Natl. Acad. Sci. USA* 104(28): 11826–11831.

Rospars, J. P. and J. G. Hildebrand (1992). Anatomical identification of glomeruli in the antennal lobes of the male sphinx moth *Manduca sexta*. *Cell Tissue Res.* 270(2): 205–227.

Sachse, S. and C. G. Galizia (2002). Role of inhibition for temporal and spatial odor representation in olfactory output neurons: a calcium imaging study. *J. Neurophysiol.* 87(2): 1106–1117.

Sachse, S. and C. G. Galizia (2003). The coding of odour-intensity in the honeybee antennal lobe: local computation optimizes odour representation. *Eur. J. Neurosci.* 18(8): 2119–2132.

Sato, K., M. Pellegrino, et al. (2008). Insect olfactory receptors are heteromeric ligand-gated ion channels. *Nature* 452(7190): 1002–1006.

Schiestl, F. P., M. Ayasse, et al. (1997). Variation of floral scent emission and postpollination changes in individual flowers of *Ophrys sphegodes* subsp. sphegodes. *J. Chem. Ecol.* 23(12): 2881–2895.

Shang, Y., A. Claridge-Chang, et al. (2007). Excitatory local circuits and their implications for olfactory processing in the fly antennal lobe. *Cell* 128(3): 601–612.

Silbering, A. F. and C. G. Galizia (2007). Processing of odor mixtures in the Drosophila antennal lobe reveals both global inhibition and glomerulus-specific interactions. *J. Neurosci.* 27(44): 11966–11977.

Silbering, A. F., R. Okada, et al. (2008). Olfactory information processing in the Drosophila antennal lobe: anything goes? *J. Neurosci.* 28(49): 13075–13087.

Silbering, A. F., R. Rytz, et al. (2011). Complementary function and integrated wiring of the evolutionarily distinct Drosophila olfactory subsystems. *J. Neurosci.* 31(38): 13357–13375.

Silbering A. F. and Benton, R., (2010) Ionotropic and metabotropic mechanisms in chemoreception: 'chance or design'? *EMBO reports* 11(3): 173–179.

Stierle JS, Galizia CG, & Szyszka P (2013) Millisecond stimulus onset-asynchrony enhances information about components in an odor mixture. *J. Neurosci.* 33(14): 6060–6069.

Stopfer, M., S. Bhagavan, et al. (1997). Impaired odour discrimination on desynchronization of odour-encoding neural assemblies. *Nature* 390(6655): 70–74.

Stopfer, M., V. Jayaraman, et al. (2003). Intensity versus identity coding in an olfactory system. *Neuron* 39(6): 991–1004.

Strausfeld, N. J. (2002). Organization of the honey bee mushroom body: representation of the calyx within the vertical and gamma lobes. *J. Comp. Neurol.* 450(1): 4–33.

Strausfeld, N. J., I. Sinakevitch, et al. (2003). The mushroom bodies of *Drosophila melanogaster*: an immunocytological and golgi study of Kenyon cell organization in the calyces and lobes. *Microsc. Res. Tech.* 62(2): 151–169.

Su, C. Y., C. Martelli, et al. (2011). Temporal coding of odor mixtures in an olfactory receptor neuron. *Proc. Natl. Acad. Sci. USA* 108(12): 5075–5080.

Sudhakaran, I. P., E. E. Holohan, et al. (2012). Plasticity of recurrent inhibition in the Drosophila antennal lobe. *J. Neurosci.* 32(21): 7225–7231.

Suh, G. S. B., A. M. Wong, et al. (2004). A single population of olfactory sensory neurons mediates an innate avoidance behaviour in Drosophila. *Nature* 431(7010): 854–859.

Sun, X. J., C. Fonta, et al. (1993). Odor quality processing by bee antennal lobe interneurons. *Chem. Senses* 18(4): 355–377.

Szyszka, P., C. Demmler, et al. (2011). Mind the gap: olfactory trace conditioning in honeybees. *J. Neurosci.* 31(20): 7229–7239.

Szyszka, P., M. Ditzen, et al. (2005). Sparsening and temporal sharpening of olfactory representations in the honeybee mushroom bodies. *J. Neurophysiol.* 94(5): 3303–3313.

Szyszka, P., A. Galkin, et al. (2008). Associative and non-associative plasticity in Kenyon cells of the honeybee mushroom body. *Front Syst. Neurosci.* 2: 3.

Szyszka, P., J. S. Stierle, et al. (2012). The speed of smell: odor-object segregation within milliseconds. *PLoS. One* 7(4): e36096.

Szyszka P, Stierle JS (2014). Mixture processing and odor-object segregation in insects. *Prog. Brain Res.* 208 63-85. doi: 10.1016/B978-0-444-63350-7.00003-6

Tanaka, N. K., K. Ito, et al. (2009). Odor-evoked neural oscillations in Drosophila are mediated by widely branching interneurons. *J. Neurosci.* 29(26): 8595–8603.

Turner, G. C., M. Bazhenov, et al. (2008). Olfactory representations by Drosophila mushroom body neurons. *J. Neurophysiol.* 99(2): 734–746.

Vet, L. E. M. and M. Dicke (1992). Ecology of infochemical use by natural enemies in a tritrophic context. *Ann. Rev. Entomol.* 37: 141–172.

Vickers, N. J. (2000). Mechanisms of animal navigation in odor plumes. *Biol. Bull.* 198(2): 203–212.

Vickers, N. J., T. A. Christensen, et al. (1998). Combinatorial odor discrimination in the brain: attractive and antagonist odor blends are represented in distinct combinations of uniquely identifiable glomeruli. *J. Comp. Neurol.* 400(1): 35–56.

References

Vosshall, L. B. and B. S. Hansson (2011). A unified nomenclature system for the insect olfactory coreceptor. *Chem. Senses* 36(6): 497–498.

Vosshall, L. B., A. M. Wong, et al. (2000). An olfactory sensory map in the fly brain. *Cell* 102(2): 147–159.

Wang, J. W., A. M. Wong, et al. (2003). Two-photon calcium imaging reveals an odor-evoked map of activity in the fly brain. *Cell* 112(2): 271–282.

Weidenmuller, A., C. Kleineidam, et al. (2002). Collective control of nest climate parameters in bumblebee colonies. *Anim. Behav.* 63: 1065–1071.

Wicher, D., R. Schafer, et al. (2008). Drosophila odorant receptors are both ligand-gated and cyclic-nucleotide-activated cation channels. *Nature* 452(7190): 1007–1011.

Yaksi, E. and R. I. Wilson (2010). Electrical coupling between olfactory glomeruli. *Neuron* 67(6): 1034–1047.

Yamagata, N., M. Schmuker, et al. (2009). Differential odor processing in two olfactory pathways in the honeybee. *Front Syst. Neurosci.* 3: 16.

Yarali, A., S. Ehser, et al. (2009). Odour intensity learning in fruit flies. *Proc. R. Soc. B* 276(1672): 3413–3420.

Chapter 23

Olfaction in Aquatic Vertebrates

KEITH B. TIERNEY

23.1 INTRODUCTION

Vertebrate olfaction evolved underwater and it has remained underwater ever since, regardless of whether it has come to be used in terrestrial environments. The reason for this is simple: wherever vertebrates are, their olfactory epithelium is bathed in mucous. This is not to say that aquatic and terrestrial vertebrates can detect the same bouquet of odorants – quite the opposite. Terrestrial vertebrates were given the "evolutionary option" to sense odorants that were volatile and not soluble in water. For terrestrial vertebrates, "smelling air" brought certain adaptations such as olfactory binding proteins (OBPs) to chaperone volatile and frequently hydrophobic odorants to a diversity of olfactory receptor proteins located on olfactory receptor cells (Firestein, 2004; Millery et al., 2005). For aquatic vertebrates, "smelling water" brought developments in its own right, not only in terms of odorant receptor proteins (Eisthen, 1997; Hansen and Zielinski, 2005), but also in terms of anatomical adaptations that deal with moving water (Cox, 2008).

For vertebrates that live either in or around water, a division can be made between those who have noses that interface with air and those who have noses that interface with water. A major consideration is the nature of the medium that conveys the odorants. The far greater relative density and viscosity of water vs. air has necessitated numerous adaptations in aquatic species to direct water flow over the olfactory epithelium, as described later in this chapter. Importantly, in light such factors, evolutionary specializations have defined the classes of odorants to which the noses of water- and air-living species can detect, with dissolved vs. volatile categories often being quite distinct. In fish, where volatility is not a major issue, odorants are commonly classified in relation to structure,

such as amine-containing odorants (Hara, 2005; Miklavc and Valentinčič, 2012; Rolen et al., 2003), nucleobase odorants (Brown et al., 2001; Parra et al., 2009; Rolen et al., 2003), prostaglandins (Belanger et al., 2010; Cole and Stacey, 1984), steroid (cholesterol)-based odorants (Cole and Stacey, 2006; Stacey et al., 2003), sugar-based odorants (Mathuru et al., 2012), and ions (Hubbard et al., 2002). Although many airborne odorants detected by air-living species are also named on the basis of structure (e.g., mercaptan containing), other classification schemes are nascent, such as ones based upon hedonics (Djordjevic et al., 2012). Alternative means of detecting chemicals have evolved in some terrestrial vertebrates, such as the vomeronasal organ (VNO), the Grueneberg ganglion, and other specialized systems (see Chapters 51 and 52).

Although an argument can be made that the behavioral and physiological responses of vertebrates that smell in water meaningfully differ from those that smell in air, it is noteworthy that olfaction enables a similar diversity of responses in both groups. In water and air, olfactory-based activities include searching, be it for food (Hirsch, 2010; Miklavc and Valentinčič, 2012), a home locale (Benvenuti et al., 1993; Yamamoto et al., 2010), or the right environmental conditions (Hubbard et al., 2002; Nevitt, 2011), providing information of importance to relatedness (Leinders-Zufall et al., 2004; Olsén et al., 2002), a predator's presence (Amo et al., 2008; Idler et al., 1956), and the timing and preparation of reproduction (Stacey et al., 2003; Veyrac et al., 2011).

A primary way that air and water smellers differ is in the number and types of olfactory receptor proteins they possess. Bony fishes, which gave rise to terrestrial vertebrates, have ~150 distinct olfactory receptor proteins (Firestein, 2004). However, terrestrial vertebrates generally have many more such proteins. For example, there

---

Handbook of Olfaction and Gustation, Third Edition. Edited by Richard L. Doty.
© 2015 Richard L. Doty. Published 2015 by John Wiley & Sons, Inc.

appear to be ~1200 distinct receptor proteins in mice (Firestein, 2004; Zhang et al., 2004). Although this suggests that water-smelling vertebrates may be comparably disadvantaged over land-dwelling vertebrates, this is likely incorrect: fishes have increased variability in their receptor repertoire which confer the ability to detect a large array of odors (Hashiguchi and Nishida, 2005, 2006). This chapter begins with a historical account of landmark scientific studies of olfaction in fish. The chapter's primary goal is to succinctly describe the olfactory systems and abilities of all aquatic vertebrates, that is, vertebrates that are active underwater. This includes ones that (1) develop and live underwater, (2) develop underwater but leave the aquatic environment, and (3) develop on land yet spend appreciable time living in and around aquatic environments. In the first two groups, there are fishes and amphibians that spend either the majority or portions of their lives underwater yet may leave for periods from the brief to the extended (e.g., mudskippers, frogs, and salamanders). In the last group, numerous examples exist of animals that have returned to an aquatic existence, including birds, reptiles, and marine mammals. Across all vertebrates, while much of the cellular machinery has homology, there is a diversity of odorant detection abilities and incredible morphological-structural differences that warrant description.

23.2 HISTORY OF THE SCIENTIFIC STUDY OF OLFACTION IN FISH

In 1909, Burne authored "The anatomy of the olfactory organ of teleostean fishes" (Burne, 1909). This lengthy paper (~54 pages) is remarkable in that it was arguably the first to showcase the diversity of anatomical structures associated with fish olfaction. The paper contains several detailed illustrations with cutaways showing not only the olfactory structures but also those underneath. It also contains a brief review of other sources known at the time (all of ten references). It still gathers citations today, but owing to its age, it is sometimes not cited appropriately by databases, and so the actual number of citations is not known (current Google Scholar citations number 94).

The ability of migrating salmon to home to their natal streams is now widely accepted to be driven by olfaction. The first paper to experimentally test this theory was authored by E. Horne Craigie (Craigie, 1926). In his study, a sample of 513 returning Pacific sockeye salmon were captured, tagged, and released several kilometers before they entered the riverine portion of their migration. However, before being released, half had their olfactory nerve tracts surgically severed. The results of the study were not conclusive, but loss of olfaction clearly reduced the number of salmon who returned to their native stream. The suggestion to disable olfaction (and thus the likelihood that it was

the sense) was credited to George H. Parker, a zoologist at Harvard. It wasn't until 25 years later that the olfactory imprinting hypothesis was formalized by Arthur D. Hasler and Warren Wisby (Hasler and Wisby, 1951). This area of research remains active, with recent studies identifying the nature of the critical odorants (Ueda, 2012; Yamamoto et al., 2010). In 1956, a game-changing methods paper was published by Ottoson entitled "Analysis of the electrical activity of the olfactory epithelium" (Ottoson, 1956). While research had already been published of recordings of the electrical properties of the olfactory bulb, this was the first paper to describe the activity of the peripheral olfactory epithelium. Also noteworthy is the species tested, a facultative aquatic vertebrate – the common frog (*Rana temporaria*). This paper was invaluable to the study of olfaction as it gave rise to what became known as the electro-olfactogram (EOG). This paper also has citation issues (and can be incorrectly listed as 1955) and but still is widely cited (Google Scholar citations: 322). Years later, Evans and Hara (1985) authored a paper on how to use this technique in fish with gel-filled electrodes. This paper remains the standard operating procedure for assessing olfaction in fish.

The 1970s saw a revolution in terms of the significance of olfaction to physiological and behavioral responses. One of the first papers to demonstrate the role of odorants to mating processes tested the ability of male goldfish (*Carassius auratus*) to discern between ovulated and unovulated females based on compounds within ovarian fluid (Partridge et al., 1976). This work arguably established the foundation for a legacy of studies exploring sexual signaling molecules in fishes. It also pioneered the use of a Y-maze to determine odorant preferences of fish, a procedure which became widely adopted. Even recently, a large in-field Y-maze was used to demonstrate the ability of species-specific molecules to control the movement of an invasive fish species, the sea lamprey (*Petromyzon marinus*) (Johnson et al., 2009). With olfaction being indisputably invaluable to fishes, it is easy to make a case that its disruption has far reaching influences on fish populations. The first data to show that human-produced compounds could impair olfaction came from a technical report from the Fisheries Research Board of Canada (Sutterlin et al., 1971). This paper showed that surfactants, such as those found in soaps that are often released into the aquatic environment, could reduce the olfactory responses of fish. The paper was also unique in several other ways – it measured numerous compounds (150), it endeavored to link the toxic effects to chemical structure, and it explored whether the compounds could themselves evoke olfactory responses (e.g., can fish smell novel compounds that would not be encountered in their evolutionary history?). This paper arguably was the first of many to look at "olfactory toxicity" in fish (reviewed in Tierney et al., 2010).

A theme in many present olfactory studies is the linking of odorants to their cellular machinery and to the neurological circuits of the brain. While this sort of information is intriguing on its own, much has been accomplished under the auspices of fish as models for human research. In this, no fish is more popular than the zebrafish. A recent study used gene trapping in mutant zebrafish lines to link a specific olfactory sensory neuron (OSN) class (microvillus) to locations in the olfactory bulb, and to the odorants and behaviors they evoke (Koide et al., 2009). With a high degree of homology between zebrafish and mammals, future studies may be useful in helping to elucidate the neurologic processes responsible for such neurodegenerative diseases as Alzheimer's and Parkinson's (see Chapter 18).

23.3 ANATOMY

Olfaction depends on the transport of odors to the receptors located in epithelia; this involves bulk flow of the medium (convection) and diffusion of the odors. Aquatic organisms face a challenge their terrestrial counterparts do not: water is far denser (800×) and more viscous (50×); hence, molecules diffuse more slowly than in air (e.g., 10,000× for CO_2). However, water has at least one advantage: it facilitates the dissolution of molecules that air cannot. As a result of the challenges water presents, fishes have developed several adaptations to deal with its motion. These include specialized nostrils, structures to guide water flow, and cilia and accessory sacs to pump it. With the diversity of fish species, the hydrodynamics of underwater olfaction are largely underexplored (Cox, 2008). Unfortunately, space limitations of this chapter allow for only for a cursory examination of such processes.

23.3.1 Nares

Aside from the monorhynic (one nostril) adaptation of ancient jawless fishes, most fishes have bilateral nostrils or "nares" located rostral (anterior) to the eyes, as occurs in air smelling vertebrates. However, unlike air smellers, the nares of most fish consist of two openings, an anterior incurrent opening and a posterior excurrent opening (Zielinski and Hara, 2007) (Figure 23.1d). These openings are often structured to direct flow into a chamber that contains the olfactory epithelium. Thus, the incurrent naris may not be just a hole, but a tube (sometimes referred to as a vestibule) (Døving et al., 1977), such as occurs in the tube nose goby (*Proterorhinus marmoratus*). The excurrent naris may be both a tube and a slit. An example of such a tube/slit combination can be found in the striped eel catfish (*Plotosus lineatus*) (Theisen et al., 1991). Some species

do not have a covering over their olfactory epithelium at all, such as garfish (*Belone belone*) (Theisen et al., 1980) (Figure 23.1a) and flying fishes of suborder Exocoetoidei (Burne, 1909). Other species may have adaptations associated with their nares, such as hammerhead sharks (family Sphyrnidae) that possess prenarial grooves that presumably act to direct flow (Kajiura et al., 2004) (Figure 23.1e).

By virtue of the location and number of their nares, water smellers can be uniquely adapted for tropotaxis or klinotaxis. Tropotaxis is evident in fairly linear movement up a concentration gradient and is used in organisms able to detect a gradient between two or more anatomically separated receptors. Klinotaxis is typified by relatively less linear movement up a concentration gradient and is used by organisms that must establish a gradient by moving one receptor through it. The hammerhead sharks mentioned above, for example, have nares spaced far apart which will facilitate the detection of an odor gradient over their cephalofoil head (Kajiura et al., 2004). Conceivably this "stereo-olfaction" provides a means for more direct movements towards an odor source (e.g., prey). Conversely, the jawless fishes that have a single naris serving one olfactory chamber (Zielinski and Hara, 2007) may need to physically move their sensory epithelia through the gradient. This type of movement will be evident in pronounced "side-to-side" searching.

23.3.2 Olfactory Chambers

As a general rule, fishes have blind olfactory chambers, that is, chambers that do not connect to any pharyngeal spaces. An exception again is in the jawless fishes, specifically the hagfishes, which have a chamber that is connected to the gill openings (Holmes et al., 2011). However, there are other oddities found in fishes. For example, in lamprey several small spherical cavities exist below, and are connected to, the main olfactory chamber via tubes. These "tubular diverticula" are referred to as the accessory olfactory organ (AOO) and are thought to be analogous to the VNO (Thornhill, 1972). These cavities contain ciliated cells and are believed to function in the communication of reproductive or migratory information (Ren et al., 2009). Additionally, while they do have a single chamber, there is a central tissue fold that may functionally separate this chamber (Theisen, 1973). A truly unique anatomy exists in the puffer (*Spheroides maculatus*): the chamber resides at the end of a 4 mm papilla (Copeland, 1912) (Figure 23.1b). This has the obvious advantage that the olfactory tissue is placed directly in the flow about the fish, but it also places the tissue in harm's way. Perhaps for this slow swimming fish that has its own unique method of protection, placing the sensitive olfactory epithelium in a refuge is unnecessary.

550　Chapter 23　Olfaction in Aquatic Vertebrates

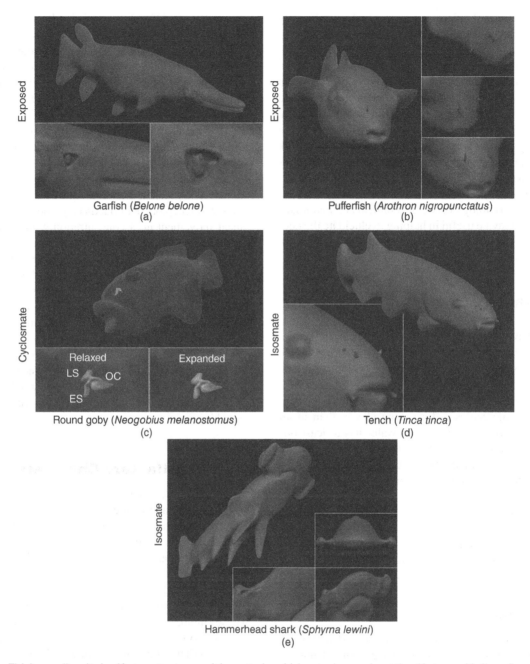

Figure 23.1 Fish have a diversity in olfactory structures and the routes by which water is moved past the olfactory epithelium. Some species have their olfactory epithelium exposed to the environment, either in pits (a) or on stalks (b). Other species, referred to as cyclosmates may have blind sacs connected to the olfactory chamber and associated with bones of the jaw, such that the action of opening and closing the mouth expands the sacs to pull water through the chamber (c). Many active swimming species referred to as isomates do not have sacs, but rather use the pressure associated with swimming to create nasal water flow. Some of these fish may use tubes (d) or prenarial grooves (e) to direct nasal flow. Olfactory chamber, OC; accessory nasal sacs: ethmoidal sac, ES; lachrymal sac, LS. The olfactory sac was modelled from Belanger et al. (2003). Illustration by Kofi Garbrah.

23.3.3　Nasal Flow

For many active fishes, swimming and the action of cilia serve are the primary ways to move water through the nose. For more sedentary fishes, water must be circulated through the chamber by the pumping action of accessory sacs (Figure 23.1c). These strategies are typically associated with pelagic (water-column) and demersal (bottom-dwelling) species; however, this is not always the case. Pelagic species may have sacs and demersal species may not (Burne, 1909; Zielinski and Hara, 2007). Døving et al. (1977) proposed a classification system for the two ventilatory strategies, categorizing fish as either isosmates or cylcosmates, respectively. In terms of efficiency with isosmates, swimming action may increase nasal ventilation fivefold (Døving et al., 1977). For isosmates, the energy associated with swimming can be utilized for "olfactory ventilation", thereby eliminating the cost associated with building specialized olfactory structures. For cyclosmates that rely on pumping water over the olfactory epithelium, they are likewise coupling another energetic activity with another – in their case, respiratory ventilation. For these fishes, the expanding and collapsing of the olfactory sacs is coupled to the ventilatory cycle (Nevitt, 1991). In this way, their sense of smell has a rhythmic component, much like air smellers. Additionally, opening and closing of the mouth may effectively allow cyclosmates to "sneeze" (Døving et al., 1977; Kapoor and Ojha, 1972; Nevitt, 1991).

There is diversity in the sacs that cyclosmates have to generate flow. Some species possess two sacs: the ethmoidal and lachrymal sacs (Figure 23.1c), while others have only one sac (Zielinski and Hara, 2007). The lachrymal sac is so named from its association with the lachrymal bone of the jaw and may be more ventral; the ethmoidal sac is named from its association with the ethmoid bone that separates the nasal cavity from the brain and may be more dorsal (Burne, 1909). Both sacs typically flank the rear of the olfactory chamber and are proximal to the openings of the caudal nares (i.e., are distal from the openings of the rostral nares). In terms of size, the lachrymal sac is typically larger (Nevitt, 1991). It has been hypothesized that the two-sac system, perhaps in conjunction with a collapsible caudal nares flap that acts as a valve, allows for unidirectional flow in the olfactory chamber (Døving et al., 1977; Pfeiffer, 1964). Some fish achieve pumping using a one sac/valve system (Kux et al., 1988). In either case, what likely occurs is that with inhalation (jaw opening) water is pulled in through the openings of the anterior nares, past the olfactory epithelia, and into the rear sac(s). With exhalation (jaw relaxation), water is forced out of the sac(s) and through the caudal nares. What is unknown is whether the complete filling and emptying of the nasal chamber occurs with every ventilatory cycle (i.e., a "one cycle" system). With fishes with a lengthy olfactory chamber (Kux et al., 1988), perhaps multiple ventilatory cycles are needed, but this remains to be determined.

A curiosity is that some of the most ancestral fish have an olfactory ventilatory strategy that is most comparable to that of air smellers. Thus, hagfish have an array of olfactory receptors within an olfactory epithelium that is in line with their gills (Holmes et al., 2011). With this system these fishes have an oscillating, albeit unidirectional, water flow

in the olfactory chamber directly related to the respiratory cycle.

A question arises as to whether fish can sniff? In fact, they can increase the flow of water to their olfactory organs which, in some cases, occurs in a manner similar to sniffing in various land dwelling mammals. There are arguably two routes by which this process occurs. First as mentioned above, they can increase the pumping of the accessory olfactory sacs by modulating their jaw movement, as if they were coughing (Nevitt, 1991). Second, they can repeat sample an area of high odorant concentration by physically moving into and out of such an area. This has been observed in zebrafish (*Danio rerio*) when an odor is presented into a tank using a point source (Tierney et al., 2011). Such behavior may extend over periods of minutes. This form of sniffing is likely utilized to a large extent in actively swimming species who dwell within water columns.

23.3.4 Olfactory Epithelium

The olfactory epithelium of fishes can be described as ranging from simple to elaborate, from having no folds to having complex two-tiered lamellae arranged in rosettes. This variation may reflect the relative importance of olfaction (the larger the surface area, the more important olfaction), but it may also have to do with habitat niche and a trade-off with rigidity or perhaps even drag. At the extreme end of surface area are reef-associated fishes such as the barred snapper (*Hoplopagrus guentherii*), which has been reported to have more lamellae than any other fish, with up to 230 per nare (est. surface area of 2450 mm^2) (Pfeiffer, 1964). These numbers outshine migratory salmon, species that indisputably depend on their noses to guide them over tremendous distances. In chum salmon (*Oncorhynchus keta*), for example, there may be 18 lamellae with as many as 14 million olfactory cells per naris (Kudo et al., 2009). These numbers appear to be dynamic and increase allometrically as fish mature. An example of a fish at the other extreme of surface area is the invasive round goby (*Neogobius melanostomus*), which have a single lamella (Belanger et al., 2003). This species provides an argument against the surface area-olfactory importance hypothesis, since the success of the round goby in colonizing new areas has been at least partially attributed to its use of semiochemicals (Katare et al., 2011). However, this species is a cyclosmate with two olfactory sacs. Cyclosmates may not need a large surface area since they can increase their water sampling rate (Zielinski and Hara, 2007).

The olfactory epithelia of some fishes are exposed to the water on rigid papillae or stalks. For example, in fishes of the suborder Exocoetoidei, which include flying fishes and garfishes, the olfactory epithelia are present on papilla

552   **Chapter 23   Olfaction in Aquatic Vertebrates**

within recessed, but exposed, cavities (Theisen et al., 1980) (Figure 23.1a). Other fish, such as the puffer mentioned above, lack a cavity altogether and have their tissue on stalks in a manner similar to that seen in some arthropods (Copeland, 1912) (Figure 23.1b). For the Exocoetoidei, it is possible that the relatively simple and open structure may be an adaptation to reduce drag or facilitate olfactory perfusion. An argument for the perfusion theory can be made based on the ontogeny and life history of a species of garfish (*Belone belone*): non-sensory ciliated cells exist in the less active juveniles and not the active adults. In fishes with asymmetry, such as flatfishes, the olfactory epithelium may also be asymmetrical. For example, in the sole (*Solea senegalensis*), which has one naris in contact with the substrate (and likely interstitial fluid) and one in contact with the water column, olfactory detection of social cues appears to be greater on the water column side (Velez et al., 2007).

The cellular structure of the olfactory epithelium of fish is as complex as that seen in other vertebrates. In species with rosettes, the bases and tips of lamellae are typically dominated by non-sensory cells. The sensory cells on the "tongues" of the lamellae are typically evenly distributed across the epithelium (Hara and Zhang, 1996), but may also have patchy distribution, such as in the striped eel catfish (*Plotosus lineatus*) (Theisen et al., 1991). A way to group cells of the olfactory epithelium is whether or not they serve a sensory function. Non-sensory cells include at least four types: supporting cells (sustentacular or nonciliated supporting cells), stratified epithelial cells, ciliated cells, and goblet cells, all of which are morphologically distinguishable (Datta and Bandyopadhyay, 1997; Theisen et al., 1980). Sustentacular cells are apparent as columnar cells lacking projections, but may include microvillus supporting cells (Byrd and Brunjes, 1995). Ciliated, non-sensory cells may be present and likely generate flow (mucociliary clearance), but they have also been hypothesized to serve structural or hydrodynamic roles (Theisen et al., 1980). Goblet cells are the mucus secreting cells of fish. Unlike mammals, fish lack Bowman's glands. Goblet cells and thus mucus production is distributed across the olfactory epithelium (Byrd and Brunjes, 1995).

With sensory cells, there are also three morphologically distinct types: ciliated, microvillus, and crypt cells (Hansen and Zielinski, 2005; Zielinski and Hara, 2007). Unlike the non-sensory cells, all of these are bipolar, first order neurons. Of the three types of OSNs, the ciliated cell is a major component of the olfactory epithelium and resembles ciliated cells of other vertebrates (Eisthen, 1997). This cell type has a perikaryon that is sunken into the olfactory epithelium and a dendrite that extends to the surface which terminates in several apical cilia. The other major cell type – the microvillus cell – differs principally in two ways: it has a more superficial body

and apically has a few microvilli (short protuberances). While microvillus cells were historically thought to be absent in mammals (Eisthen, 1997), data suggest otherwise (Moran et al., 1982; Elsaesser and Paysan, 2007). The third cell type – the crypt cell – occurs in lower numbers, differs markedly in appearance and appears to be unique to fishes (Bettini et al., 2012; Ferrando et al., 2006; Hansen and Zielinski, 2005; Yoshihara, 2009). Crypt cells are an ovoid cell that interfaces directly with the olfactory epithelium and that has an apical cavern with abbreviated cilia (Hansen and Zeiske, 1998). Not all three sensory cell types are present in all fishes: cartilaginous fishes do not have ciliated OSNs (Eisthen, 1997) and crypt cells are absent in lobe-finned (sarcopterygian) fishes (Hansen and Finger, 2000). Curiously, all three cell types or analogs are present in the lamprey (Laframboise et al., 2007; Theisen, 1973). A final category of cell present in the olfactory epithelium but not included in the above is the basal, stem cells which give rise to all.

23.3.5 Olfactory Tracts and Connections

The OSNs send their axons along three distinct bundles in the bilateral olfactory tracts to innervate the mitral cells of the bulb (Hamdani and Døving, 2007). Each bundle appears to be made up of axons of a distinct OSN type. The lateral olfactory tract (LOT) contains microvillus axons, the lateral-medial olfactory tract (LMOT) contains crypt axons, and the medial-medial olfactory tract (MMOT) contains ciliated axons (Hamdani and Døving, 2007). These bundles project to different areas of the olfactory bulb (OB). Ciliated OSNs project to the medial area of the OB (Sato et al., 2007), microvillus OSNs to the anterior dorsal OB (Morita and Finger, 1998), and crypt OSNs to two ventral or possibly dorsomedial areas of the OB (Gayoso et al., 2011, 2012; Hansen et al., 2003). For the lamprey and its unique ciliated cells of the accessory olfactory organ, the axons terminate in the medial region of the OB (Ren et al., 2009).

The OSN types, the tracts, and the OB regions they reach can be associated with specific behaviors. The LOT may correspond to feeding responses (Koide et al., 2009), the MMOT may correspond to alarm reactions (Hamdani et al., 2000), and the LMOT may correspond to reproductive behaviors (Doving and Selset, 1980; Weltzien et al., 2003).

23.3.6 Olfactory Bulb

In most vertebrate species, the OB is comprised of two bilaterally symmetrical structures positioned at the front or base of the brain. While fishes do not appear to differ in the number of bulbs, they may differ in OB placement and morphology (Zielinski and Hara, 2007). In fishes, the

OB may be "pedunculated", that is, placed away from the brain proximal to the rosettes. This arrangement is believed to be apomorphic (derived) (Northcutt, 1986) and can be found in species such as the sea catfish (*Ariopsis felis*) (Morgan, 1975) and the coelacanth (*Latimeria chalumnae*) (Northcutt and Bemis, 1993). Alternately, the OB may be "sessile" and adjacent to the telencephalon (the plesiomorphic, ancestral condition). This arrangement can be found in fishes such as zebrafish (Byrd and Brunjes, 1995) and sea bass (*Dicentrarchus labrax*) (Cerdá-Reverter et al., 2001). The shape of the OB may vary as well – in the hammerhead shark mentioned above, the OB is coupled to and runs the length of the rosette tissue in the cephalofoil (Kajiura et al., 2004). The utility of a pedunculated OB remains to be determined, but perhaps it is related to cranial space considerations.

Fishes have four main concentric OB layers that are observed in other vertebrates: the olfactory nerve (axon), glomerular, mitral (external plexiform), and granule cell layers. However, there are some important differences in fish OB neurons and their organization. The morphology of the mitral cells need not have the typical mitre shape and the glomeruli (neuropil) can be less defined (Byrd and Brunjes, 1995). Additionally, fishes have a fourth cell type in their OB that mammals do not – the ruffed cell. This cell has an unmyelinated axon that contains numerous complex protrusions (Kosaka, 1980). Ruffed cells receive excitatory input from mitral cells (not OSNs) (Zippel et al., 1999) and are closely associated with them, conceivably serving a role similar to that of a glial cell (Byrd and Brunjes, 1995; Satou, 1990). Conversely, mammals have two classes of cells which fish may not: periglomerular cells (three types) (Kosaka and Kosaka, 2005) and tufted cells (Arévalo et al., 1991). However, recently cells found in the small-spotted catshark (*Scyliorhinus canicula*) were referred to as tufted-like and periglomerular-like (Ferrando et al., 2012).

23.4 GENETICS AND PHYSIOLOGY

Vertebrate olfaction appears to have evolved in fish, as the tunicate (sea squirt *Ciona intestinalis*), a chordate sister to fishes and other vertebrates, do not have genes orthologous to vertebrate olfactory receptors (Churcher and Taylor, 2009; Niimura, 2009). Present day fishes include examples that represent the base of vertebrate evolution and those that have continued their evolution in parallel with terrestrial radiation. However, not even the most ancient groups of the jawless fishes are lacking olfactory abilities. Lampreys, for example, guide remarkable migrations using olfaction, albeit using a direct coupling between olfaction and swimming activity (Derjean et al., 2010). In another ancient, arguably basal group – the cartilaginous "elasmobranchs",

which include chimaeras, rays, and sharks – olfaction can be very well tuned, especially to amino acids that serve as food odorants (Tricas et al., 2009).

Vertebrate odorant receptors were first named by their location in terrestrial species, with those from the main olfactory epithelium (MOE), or what we consider the "nose", being labelled "OR" (for odorant receptor) (Mombaerts, 1999) and those from the vomeronasal organ (VNO) being labelled "VR" (for vomeronasal receptor) (Dulac and Axel, 1995). By the nature of these locations, these receptors were believed to serve different roles: ORs were assumed to detect volatile odorants and VRs non-volatile odorants (Keverne, 1999). While generally true, this dogma is no longer supported; in fishes ORs and VRs are known to receive soluble odorants (Sato et al., 2005; Speca et al., 1999). Furthermore, OR and VR receptors are typically found dispersed across the olfactory epithelium in fishes which makes functional distinctions difficult. It is likely that VR and OR sequestration was an adaptation for terrestrial colonization, with the VNO receptors being placed in a location such that they may receive soluble odorants via the tongue (Halpern, 1987). Therefore, the receptor nomenclature is an artifact of classifying terrestrial species irrespective of their aquatic origins (Cao et al., 1998). A final group of odorant receptors present in both fish and other vertebrates is the trace amine-associated receptor (TAAR) family, which detects compounds similar to monoamine neurotransmitters (e.g., norepinephrine or dopamine) at very low concentrations (Borowsky et al., 2001; Hussain et al., 2009).

Discussions of receptor proteins and their associated cellular machinery often include the ligands that bind them (odorants in this case). However, with olfaction, receptor-odorant combinations are often not known or can only be speculated upon (Table 23.1). An additional complication is that odorants may bind to more than one receptor. Odorants can be grouped by class and include amine-containing (amino acids, polyamines, trace amines, urea, and ammonia; Dallas et al., 2010; Michel et al., 2003; Rolen et al., 2003; Tricas et al., 2009), nucleobase (nucleic acids, nucleosides, and nucleotides; (Rolen et al., 2003; Yacoob and Browman, 2007), prostaglandins and steroid-based (steroids and bile salts) (Belanger et al., 2010; Meredith et al., 2012; Stacey et al., 2003; Zhang and Hara, 1994), as well as sugar-based (e.g., chondroitin; Huertas et al., 2007), and ions (Ca^{2+} and Mg^{2+}, possibly K^+; Hubbard et al., 2002). The roles that many of these odorants play are still unknown (Table 23.2), and vary among fishes.

23.4.1 OR Receptors

The OR genes are classified as Type 1 or 2 and are found in all vertebrates (Niimura and Nei, 2005). The Type 1 OR

554   **Chapter 23   Olfaction in Aquatic Vertebrates**

Table 23.1 Olfactory receptors, their classes and the types of odorants they respond to.

Olfactory Receptor Type		Odorant Class	Organism	Reference
Odorant Receptors (ORs)				
	Type 1			
	α	Volatile	*T	Niimura, 2009
	β	Soluble or Volatile	T, A	
	δ	Soluble	A	
	γ	Volatile	T	
	ε	Soluble	A	
	ζ	Soluble	A	
	Lamp-a	Soluble	Lamprey	Niimura and Nei, 2005
	Lamp-b	Soluble	Lamprey	
	Type 2			
	η	Soluble	T, A	Niimura, 2009
	θ1	Unknown	T, A	
	θ2	Unknown	T, A	
	κ	Unknown	T, A	
	λ	Unknown	T, A	
Trace amine-associated receptors (TAARs)		Volatile, Soluble	T, A	Liberles and Buck, 2006
Vomeronasal receptors				
	Type 1 (V1R)	Volatile, Soluble	T, A	Bazáes and Schmachtenberg, 2012; Dulac and Axel, 1995; Johansson and Banks, 2011; Pfister and Rodriguez, 2005
	Type 2 (V2R)	Soluble	T, A	Hashiguchi and Nishida, 2005; Kimoto et al., 2005; Leinders-Zufall et al., 2004; Matsunami and Buck, 1997; Ryba and Tirindelli, 1997; Speca et al., 1999
	Formyl peptide receptor like (FPRL)	Likely soluble	T, unknown	Riviere et al., 2009

*T = terrestrial; A = aquatic

families appear to have evolved after jawless fishes and include six classes named alpha through to zeta (α, β, δ, γ, ε and ζ) (Niimura, 2009). There are two additional Type 1 ORs in the lamprey (lamp-a and lamp-b), whereas the other six are absent (Niimura and Nei, 2005). The Type 2 OR families appear to have evolved before jawless fishes and include five classes (η, θ1, θ2, κ and λ) (Niimura, 2009). If we consider the presence and absence of the classes in aquatic vertebrates (fish and amphibians) vs. terrestrial vertebrates (reptiles, birds, and mammals) as an indication of what type of odorants are detectable, it appears that δ, ε, ζ, and η detect soluble odorants, while α and γ detect volatile odorants. The presence of β across phyla argues for its role in detecting either odorant type. The θ1, θ2, and κ ORs may not be OR genes (Niimura, 2009). Overall, fishes can vary appreciably in the ORs that they have, from two in the elephant shark (*Callorhinchus milii*) to 176 in zebrafish (Niimura, 2009).

23.4.2 VR Receptors

As mentioned previously, fishes lack a VNO. However, they do not lack VRs as they can be found in microvillus and crypt cells (Oka et al., 2011; Yoshihara, 2009). There are two classes of VRs – V1R and V2R – and they are structurally dissimilar and have been hypothesized to receive very different odorants (Shi and Zhang, 2007). In mice, the V1R family is known to be associated with volatile odorants (e.g., 2-heptanone) (Boschat et al., 2002), whereas in fishes, V1R ligands remain unknown. However, research suggests that this receptor family recognizes a small number of odorants that likely do not serve species-specific roles, since these receptors are highly conserved across species (Johansson and Banks, 2011). The presence of V1Rs in sea lamprey (Grus and Zhang, 2009) supports this and also indicates an early evolution of this receptor class. Their occurrence in fishes also suggests that either V1R receptors respond to soluble odorants, or that volatile

Table 23.2 Odorant classes and the behavioral responses they are associated with in fishes.

Behavior	Odorant Type								
	Amino Acids	Polyamines	Trace Amines	Nucleobase-	Saccharide-	Prostaglandins	Steroids	Bile Salts	Ions
Food searching	Yes[1]	Yes[2]	Likely[3]	Unknown	Unknown	Unknown	Unknown	Unknown	Unknown
Migration	Yes[3]	Unknown	Unknown	Unknown	Unknown	Possibly[4]	Yes[5]	Possibly[6]	Possibly[7]
Environmental sensing	Unknown	Unknown	Unknown	Unknown	Unknown	Unknown	Unknown	Unknown	Yes[8]
Predator avoidance	Yes[9]	Unknown	Unknown	Unknown	Unknown	Unknown	Unknown	Unknown	Unknown
Alarm response	Unknown	Unknown	Unknown	Yes[10]	Yes[11]	Unknown	Unknown	Unknown	Unknown
Social cues	Unknown	Unknown	Unknown	Unknown	Likely[12]	Unknown	Unknown	Likely[13]	Unknown
Reproduction	Unknown	Unknown	Unknown	Unknown	Unknown	Yes[14]	Yes[15]	Unknown	Unknown

References or rationale: [1]Jones and Hara, 1985; Miklavc and Valentinčič, 2012; Shamushaki et al., 2011; [2]Rolen et al. 2003; [3]Yamamoto et al., 2010; [4]Foster, 1985; Nordeng, 1977; [5]Sorensen et al., 2005; [6]Li et al., 1995; [7]Calcium sensing neurons may allow fish to move in or out of estuaries; [8]Hubbard et al., 2002; [9]Idler et al., 1956; [10]Brown et al., 2001; [11]Huertas et al., 2007; [12]mucus is comprised of polysaccharides and it would be reasonable to hypothesize it would make an economical social cue; [13]Quinn and Hara, 1986; [14]Cole and Stacey, 1984; [15]Bhatt and Sajwan, 2001.

odorants are present underwater. Regardless, an expectation for volatile odorant receptors would be that their diversity is greater in tetrapods, and this appears to be the case. For example, the mouse has 187 V1Rs and zebrafish has just two (Shi and Zhang, 2007). V1Rs may present a curiosity in fishes: a recent study found that a single V1R gene may be expressed only in crypt OSNs, which for the first time suggests that a cell type can be specific to a receptor (Oka et al., 2011), akin to the rod/cone paradigm.

Unlike V1Rs, V2Rs appear specialized for soluble odorants, regardless of organism. In the mouse V2Rs detect water soluble peptides (Kimoto et al., 2005; Leinders-Zufall et al., 2004) and in fishes they detect amino acids (Speca et al., 1999). Since V2Rs receive soluble odorants, it might be reasonable to expect similar numbers in fishes and tetrapods. Indeed, the numbers are not dissimilar (e.g., 44 in zebrafish vs. 70 in mouse) (Shi and Zhang, 2007). However, in fishes, V2Rs represent a unique receptor group, the diversity of which expanded after tetrapods evolved (Hashiguchi and Nishida, 2006). This evolution appears to be the result of gene duplication within and across chromosomes (Hashiguchi and Nishida, 2005).

23.4.3 TAAR Receptors

This receptor class exists across all vertebrates, but fishes appear to be especially well endowed with >100 TAARs present in some species (Gloriam et al., 2005). In fishes, there are three classes of TAARs (I, II, and II) (Hussain et al., 2009). Classes I and II appear to have developed in jawed fishes and are not present in jawless fishes (Hussain et al., 2009).[1] Classes I and II carried forward to other vertebrates (amphibian, birds, mammals) (Hussain et al., 2009), while Class III appears to be a teleost-specific young class of olfactory receptors (Hussain et al., 2009). Considering fishes have an entire class of TAARs that terrestrial vertebrates do not, the olfactory abilities of fishes might be far better than implied by the relatively lower overall number of receptors.

The odorants that classes I and II TAAR receive include food cues, since trace amines are found in arthropods, bacteria, and plants (Axelrod and Saavedra, 1977; Branchek and Blackburn, 2003). Alternately, since trace amines may be excreted in urine (Reese et al., 2007), they may serve a role in social responses. What is interesting is that class III is missing an aminergic ligand binding motif found in the other two TAAR classes and so the ligands are unknown (Hussain et al., 2009). Overall, much remains to be discovered concerning the role of TAARs in fishes; however, their diversity suggests the importance of their ligands (Niimura, 2009).

[1] Sea lampreys do, in fact, have 21 functional TAAR genes. However, they are in their own class (Hashiguchi and Nishida, 2007).

23.4.4 Signal Transduction

All odorant receptor proteins are coupled to guanosine-nucleotide-binding proteins and so are G-protein coupled receptors (GPCRs). However, the G_α subunits that activate (or inhibit) the second messenger systems and evoke generator potentials differ in the OSN classes. Specifically, ciliated OSNs have $G_{\alpha olf}$, microvillus have $G_{\alpha q}$, and crypt have two forms of $G_{\alpha o}$ (Hansen et al., 2003; Oka and Korsching, 2011). The first two G_α subunits are stimulatory, working to activate adenylate cyclase (AC) and phospholipase C (PLC), respectively, while the second messenger cascade of the last ($G_{\alpha o}$) is not clear. The $G_{\alpha o}$ subunit is known to have an inhibitory function, but it is uncertain if it acts through AC or PLC (Dellacorte et al., 1996; Hansen et al., 2003). Conceivably $G_{\alpha o}$ may be in a sub-population of spontaneously active amino acid-sensitive neurons (Kang and Caprio, 1995). These OSNs showed increased, decreased or no change in firing in response to amino acids (Kang and Caprio, 1995). Regardless of G_α subunit type, some amino acid detecting neurons are narrowly "tuned", responding to a single compound, while others are more broadly "tuned", responding to more than one compound (Kang and Caprio, 1995; Nikonov and Caprio, 2007). A major difference with fish OSNs is that some break the one neuron-one odorant receptor rule that was proposed for vertebrates (Serizawa et al., 2005; Yoshihara, 2009). Specifically, a study on zebrafish found some ciliated OSNs express two or three types of ORs: zOR103-1 and one or both of zOR103-2 and zOR103-5 (Sato et al., 2007).

23.4.5 Biotransformation

Odorants and other compounds are deactivated through phase I (PI) and II (PII) biotransformation enzymes. Several PI enzymes, that is, cytochrome P450 (CYP) isoforms (Matsuo et al., 2008), and at least three PII enzymes, for example, glutathione S-transferase (GST) isoforms, can exist in olfactory tissues (π – Starcevic and Zielinski, 1995; Yanagi et al., 2004; μ – Pérez-Lopez et al., 2000; θ – Pérez-Lopez et al., 2000). In fishes, CYP1A1 has been localized to lamellar tips (the non-sensory epithelium), and, with exposure to a CYP inducer (β-naphthoflavone; BNF), in the sensory epithelium as well (Monod et al., 1994; Saucier et al., 1999). In mice, PI and PII enzymes colocalize to non-neuronal cells (Weech et al., 2003). The ability of compounds to induce PI enzymes (or markers thereof) has also been used as a toxicological endpoint. For example, ethoxyresorufin O-deethylase (EROD; a CYP 1A1 substrate) activity increased in the olfactory epithelium by >5-fold following exposure to 3,3′, 4,4′-tetrachlorobiphenyl (TCB; a PCB) (Cravedi et al., 1998). In the olfactory bulb, BNF exposure was used to characterize the CYP induction time course, as CYP1A mRNA was increased within 4 hr of exposure (Chung-Davidson et al., 2004). Studies characterizing olfactory PII responses are more rare. One study noted that inducing cell death (through severing the bulbar innervation) caused a rapid decrease in GST activity that returned after two months (Starcevic and Zielinski, 1997). Another study found that insecticide exposure increased GST activity (Tierney et al., 2008).

23.5 FUNCTIONAL ASSESSMENT

With such a complex tissue and such a diversity of structures and cells, it is unsurprising that a diversity of methods have been used to explore underwater olfaction. Studies have measured the ability of odorants to: (1) excite the olfactory epithelium (Evans and Hara, 1985), (2) excite specific OSNs (Lastein et al., 2008; Valentincic and Koce, 2000), (3) activate brain neurons and form topographical maps (Bandoh et al., 2011; Nikonov and Caprio, 2005; Nikonov et al., 2005; Zou et al., 2005), (4) evoke downstream motor neurons (Derjean et al., 2010), (5) evoke physiological responses (Olsén et al., 2006), and (6) evoke behavioral responses (Bhinder and Tierney, 2012; Stacey, 2003). In fishes, the method of choice has been a technique pioneered in frogs (Ottoson, 1956) – the electro-olfactogram (EOG) – which assesses changes in potential across the olfactory epithelium, that is, generator potentials (Baldwin and Scholz, 2005). The EOG has been used to quantify the detection threshold for a diversity of odorants in a vast array of fishes. Examples include $\sim 10^{-5}$ M for L-alanine and L-glutamine in hagfish (*Myxine glutinosa*) (Døving and Holmberg, 1974), $\sim 10^{-10}$ M (alanine and cysteine) in scalloped hammerhead shark (*Sphyrna lewini*) (Tricas et al., 2009), and 10^{-14} M (3kPZS) lamprey (*Petromyzon marinus*) (Johnson et al., 2009).

A second area of functional assessment that has received considerable attention in aquatic olfaction is olfactory-evoked behaviors. Behavioral responses have been measured in numerous ways: Y-mazes have been used to characterize attraction to streams (Scholz et al., 2000), siblings (Olsén and Winberg, 1996), foods (Rehnberg et al., 1985), and mates (Yambe and Yamazaki, 2001); avoidance troughs have helped determine whether fishes avoid contaminants, such as metals (Giattina et al., 1982; Hansen et al., 1999) and pesticides (Saglio et al., 2001; Tierney et al., 2007b), or predators (Rehnberg et al., 1985); open arenas have been used to characterize alarm responses (Brown, 2003; Mirza et al., 2009; Pollock et al., 2003; Speedie and Gerlai, 2008). In recent years, tracking software has helped with measuring behaviors. For example, changes in larval zebrafish swimming can be measured in 96 well plates (Baraban et al., 2005) and fish movements can be quantified in three dimensions (Cachat et al., 2011). A challenge with behaviors is that individual variation can be considerable: some fishes may be active and grow inactive when presented with an odorant while a sibling may do the exact opposite (Shamchuk and Tierney, 2012).

23.6 TERRESTRIAL OLFACTORY ADAPTATIONS

Amphibians are ideal organisms to showcase the transition from aquatic to terrestrial existence as they contain life phases adapted to both. Furthermore, some amphibians remain adapted to both environments into adulthood, possessing unique elements of olfactory anatomy and physiology. In the African clawed frog *Xenopus*, for example, the principle olfactory chamber is subdivided into the lateral and medial diverticula (a VNO is also present) (Altner, 1962; Hofmann and Meyer, 1991). These two chambers can be selectively opened to the environment using a flap of skin. While the head is underwater, the more ventral lateral diverticulum is open; while above water, the more dorsal medial diverticulum is open. A study indicated that the olfactory receptors of these subdivisions were specialized for their environments, with fish-like receptors expressed ventrally and mammalian-like receptors expressed dorsally (Freitag et al., 1995). This divide therefore potentially represents a structural means to selectively receive soluble vs. volatile odorants (Freitag et al., 1995). Larval *Xenopus* lack this structure and express both types of receptors (and cells later expressed in the VNO) in the principal chamber (Hansen et al., 1998), and so are ontogenetically akin to their origins. As the receptor locations are remodelled during development (Hansen et al., 1998), conceivably this is a limitation of developmental control.

An expectation for amphibians would be that they switch from detecting soluble compounds to volatile compounds between juvenile aquatic and adult terrestrial phases. However, this does not appear to be the case (Shibuya and Takagi, 1963). A secondary expectation might be that the activity of the more aquatic receptors would be down regulated between these phases. This also does not appear to be the case (Inoue and Nakatani, 2010). In fact, the opposite may be true: the detection of an amino acid mixture (alanine, arginine, proline, and glutamic acid) increased up to 115 hr after land transfer in the Japanese newt (*Cynops pyrrhogaster*) (Inoue and Nakatani, 2010). The authors noted that olfactory cilia grow in length over this time period (Shibuya and Takagi, 1963) and so postulated that there had been an increase in receptor number. However, the method of olfactory assessment they used (EOG) is dependent on many factors, and the duration of the experiment may have been insufficient to observe longer-term changes. If amino acid reception remains throughout terrestrial life, what would be its purpose?

Several other terrestrial adaptations first emerge with amphibians. For example, ORs used for detection of volatile odorants are first seen and they can have several (*Xenopus* has 1638; Niimura, 2009), olfactory binding proteins can be found, but only as tadpoles transition to land (Millery et al., 2005), similarly a Bowman's gland (Jungblut et al., 2011).

23.7 APPLICATIONS AND PERSPECTIVES

Olfaction has found relevance as an endpoint in many fish toxicology studies as it is of such importance and is exquisitely sensitive to waterborne contaminants (Tierney et al., 2010). This thinking was presciently arrived at, as one of the first functional measurements of fishes olfactory physiology explored the effects of surfactants (from detergents) on the olfactory responses of Atlantic salmon (Sutterlin et al., 1971). This study set the tone for decades of research demonstrating the effects of dissolved metals and natural and synthetic organic compounds on fish olfaction. A limitation with many of these studies is that they determine short term effects and not the ability of the tissue to cope (physiologically adapt) with the agent. Nevertheless, these studies can determine two types of OSN impairment: general impairment of all OSNs and impairment of specific subtypes (Tierney et al., 2007a). As to what these two inhibitory routes mean to olfaction, each can be framed as changes in vision: a general OSN impairment may be akin to myopia, that is, only being able to sense things "close up" (loss of ability to detect low odorant concentrations), while specific impairment is akin to colour blindness, that is, perceiving odorant information incorrectly. What these changes might mean to fishes such as migratory salmon remain unknown. A caveat is that it can be difficult to relate changes in olfaction to OSN damage. Many synthetic contaminants are structurally similar to natural compounds that act as odorants, for example, the herbicide glyphosate and the amino acid glycine (Tierney et al., 2006). Additionally, natural organic compounds can also interfere with olfaction, such as humic acid (Fabian et al., 2007).

23.8 SUMMARY OF KEY DIFFERENCES IN FISH VS. MAMMALIAN OLFACTION

Unlike mammals, in fish:

- There are typically two openings per nostril, one for incurrent flow and the other for excurrent flow.

- The olfactory chambers are generally blind sacs.
- Bowman's glands are absent but mucus producing goblet cells are distributed across the olfactory epithelium.
- Nasal flow is typically achieved in one of two ways:
  - By the pumping of two sacs whose filling is coupled to ventilation or "coughing"
  - By the action of cilia and/or the pressure associated with swimming
- Unique cell types exist in the olfactory epithelium and the olfactory bulb. Examples include crypt and ruffed cells, respectively.
- There is a diversity of vomeronasal and trace-amine associated receptors not present in tetrapods.
- The one-to-one, olfactory sensory neuron-to-odorant receptor protein rule does not hold for all fishes.
- A morphologically distinct cell type (crypt) may express just one receptor gene, and so is akin to photoreceptor cells.

REFERENCES

Altner, H. (1962). Untersuchungen über leistungen und bau der nase des südafrikanischen krallenfrosches *Xenopus laevis* (Daudin, 1803). *Z. Vergl. Physiol.* 45: 272–306.

Amo, L., Galvan, I., Tomas, G., and Sanz, J. J. (2008). Predator odour recognition and avoidance in a songbird. *Funct. Ecol.* 22: 289–293.

Arévalo, R., Alonso, J. R., Lara, J., et al. (1991). Ruffed cells in the olfactory bulb of freshwater teleosts. II. A Golgi/EM study of the ruff. *J. Hirnforsch.* 32: 477–484.

Axelrod, J., and Saavedra, J. M. (1977). Octopamine. *Nature* 265: 501–504.

Baldwin, D. H., and Scholz, N. L. (2005). The electro-olfactogram: an *in vivo* measure of peripheral olfactory function and sublethal neurotoxicity in fish. In Techniques in Aquatic Toxicology, Ostrander, G.K. (Ed.), Boca Raton: CRC Press, pp. 257–276.

Bandoh, H., Kida, I., and Ueda, H. (2011). Olfactory responses to natal stream water in sockeye salmon by BOLD fMRI. *PLoS ONE* 6.

Baraban, S. C., Taylor, M. R., Castro, P. A., and Baier, H. (2005). Pentylenetetrazole induced changes in zebrafish behavior, neural activity and c-fos expression. *Neuroscience* 131: 759–768.

Bazáes, A., and Schmachtenberg, O. (2012). Odorant tuning of olfactory crypt cells from juvenile and adult rainbow trout. *J. Exp. Biol.* 215: 1740–1748.

Belanger, R., Smith, C. M., Corkum, L. D., and Zielinski, B. S. (2003). Morphology and histochemistry of the peripheral olfactory organ in the round goby, *Neogobius melanostomus* (Teleostei: Gobiidae). *J. Morphol.* 257: 62–71.

Belanger, R. M., Pachkowski, M. D., and Stacey, N. E. (2010). Methyltestosterone-induced changes in electro-olfactogram responses and courtship behaviors of cyprinids. *Chem. Senses* 35: 65–74.

Benvenuti, S., Ioale, P., and Massa, B. (1993). Olfactory experiments on cory shearwater (*Calonectris diomedea*) - the effect of intranasal zinc-sulfate treatment on short-range homing behavior. *Boll. Zool.* 60: 207–210.

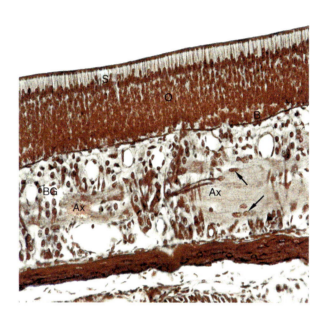

Figure 4.1

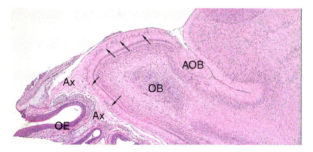

Figure 4.13

Handbook of Olfaction and Gustation, Third Edition. Edited by Richard L. Doty.
© 2015 Richard L. Doty. Published 2015 by John Wiley & Sons, Inc.

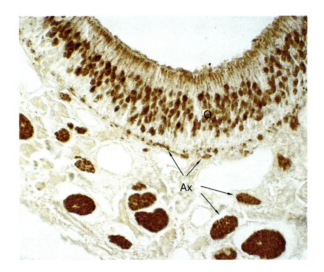

Figure 4.14

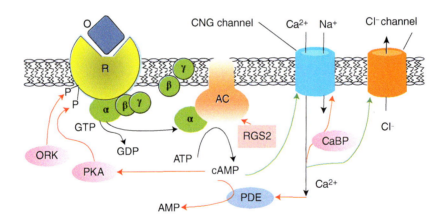

Figure 4.22

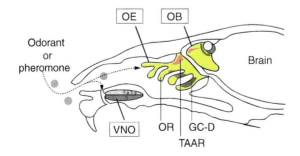

Figure 5.1

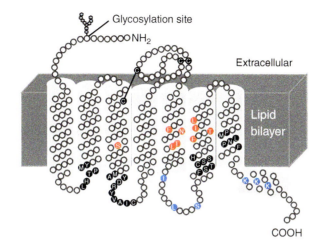

Figure 5.2

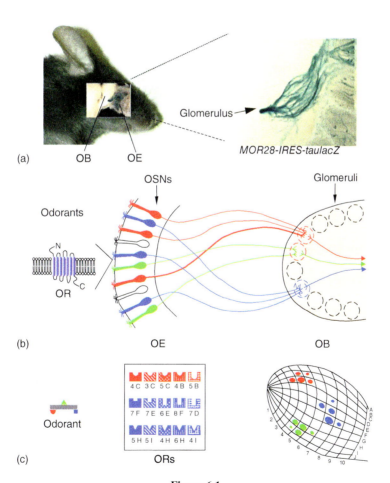

Figure 6.1

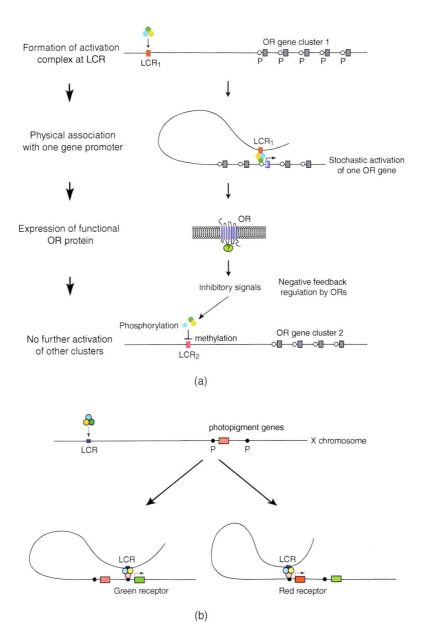

Figure 6.2

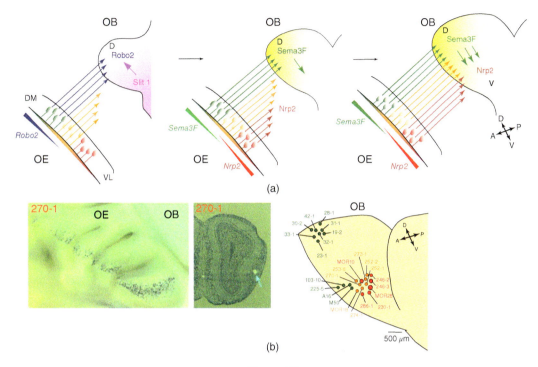

Figure 6.3

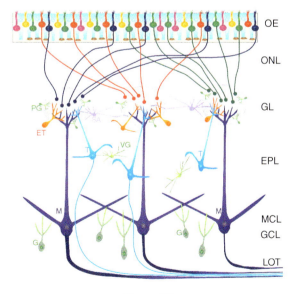

Figure 8.2

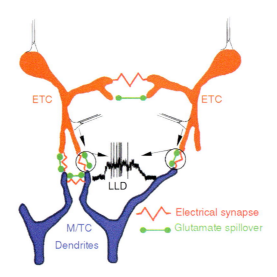

Figure 8.5

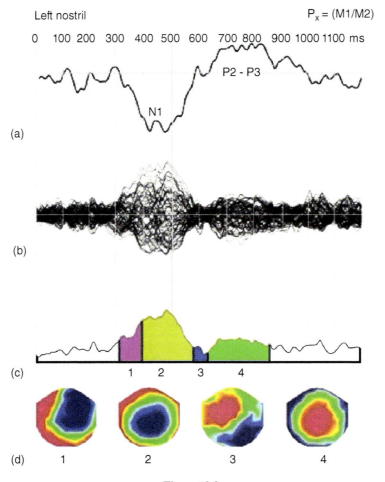

Figure 12.2

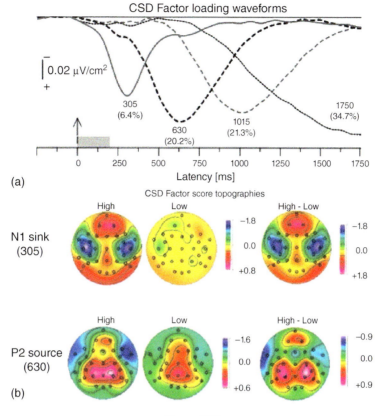

Figure 12.3

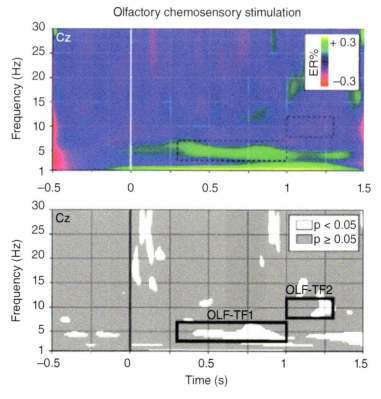

Figure 12.6

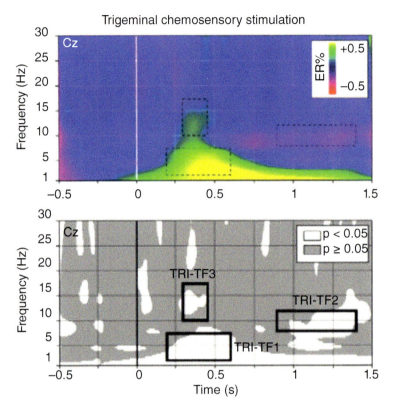

Figure 12.7

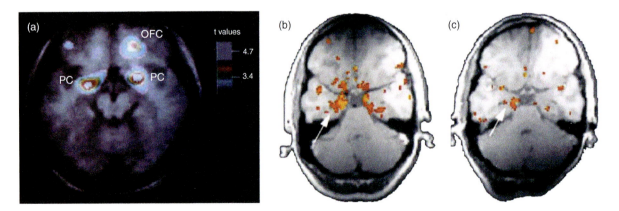

Figure 13.5

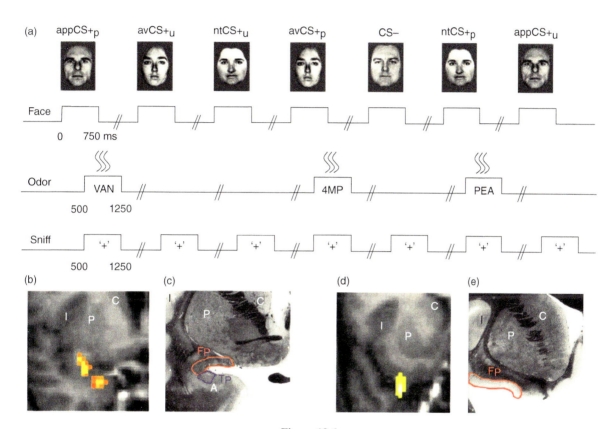

Figure 13.6

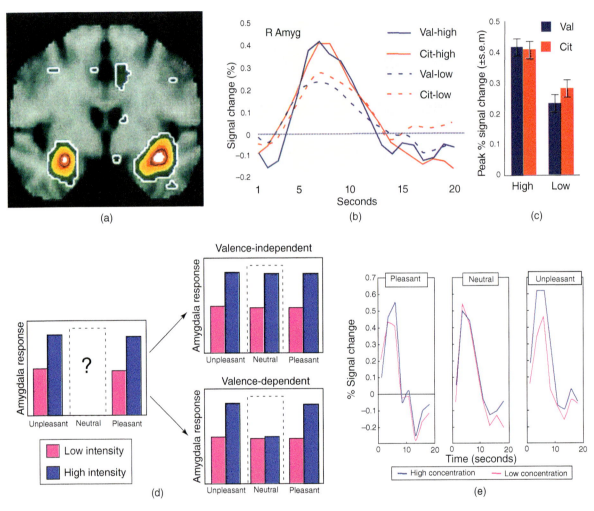

Figure 13.7

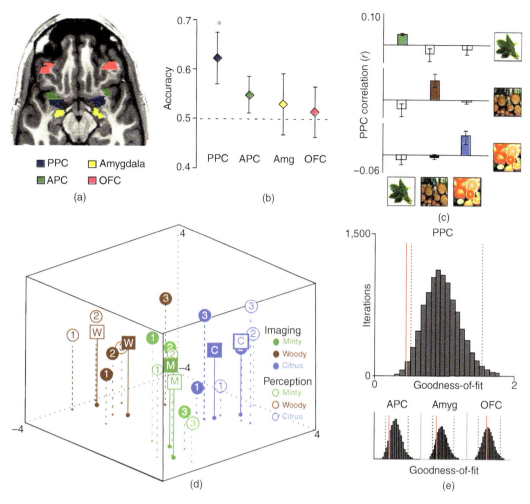

Figure 13.9

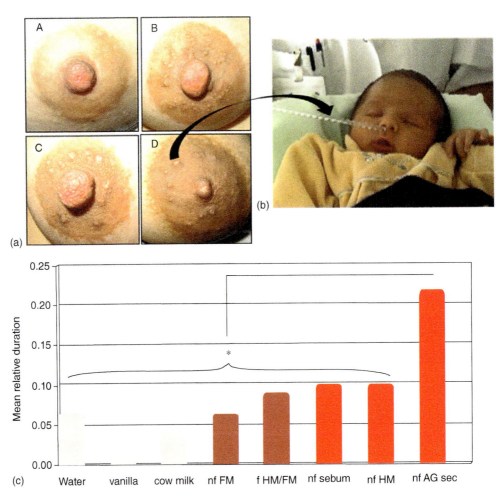

Figure 14.4

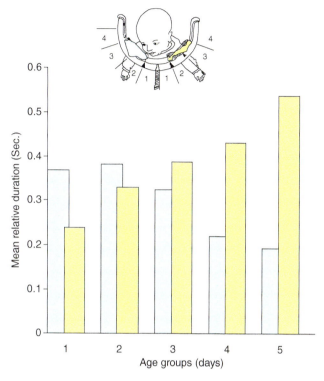

Figure 14.5

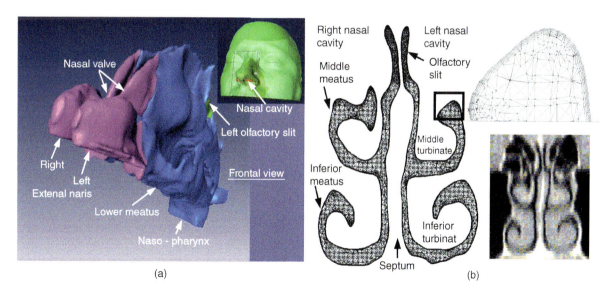

Figure 16.1

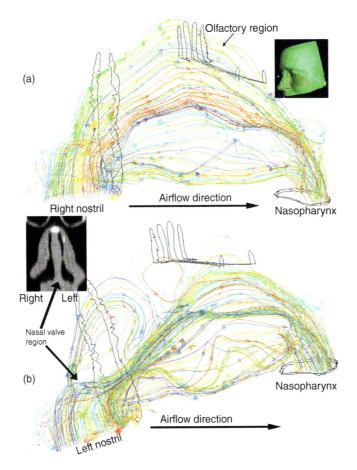

Figure 16.2

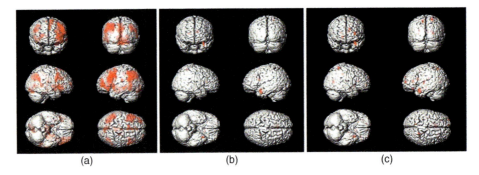

Figure 18.2

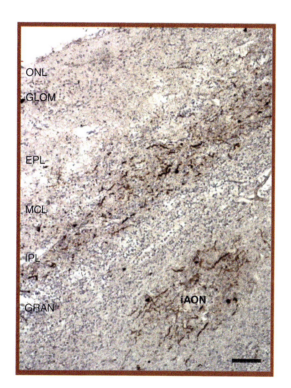

Figure 18.3

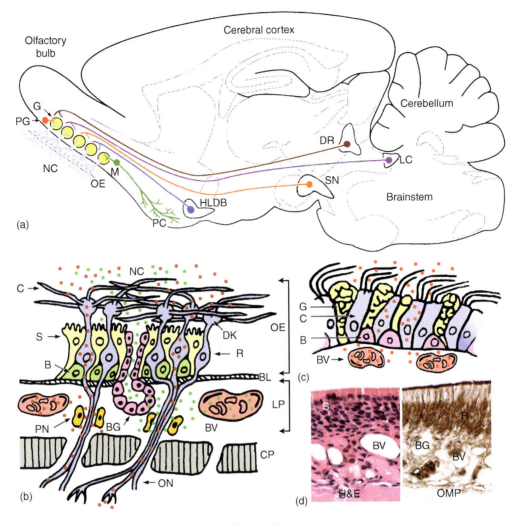

Figure 19.1

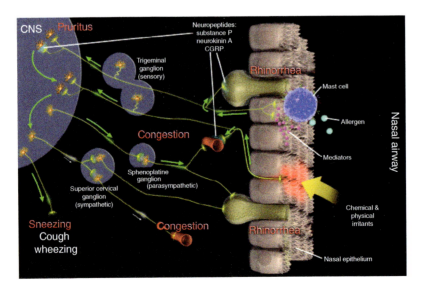

Figure 20.1

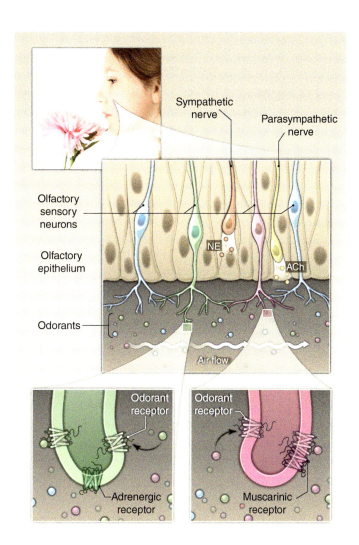

Figure 20.2

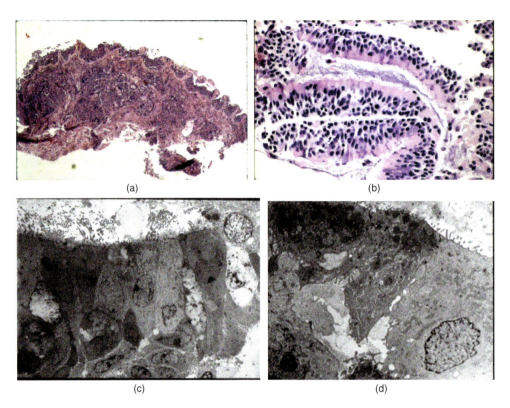

Figure 20.3

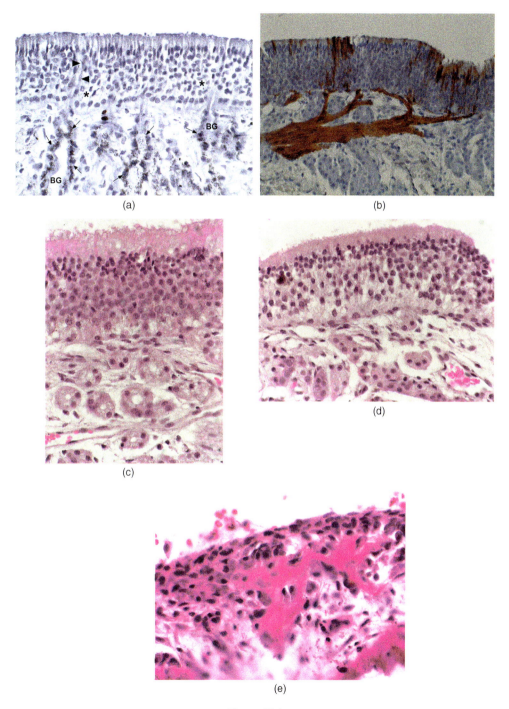

Figure 20.4

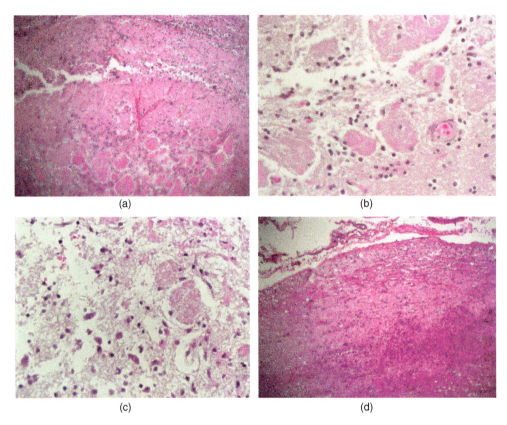

Figure 20.5

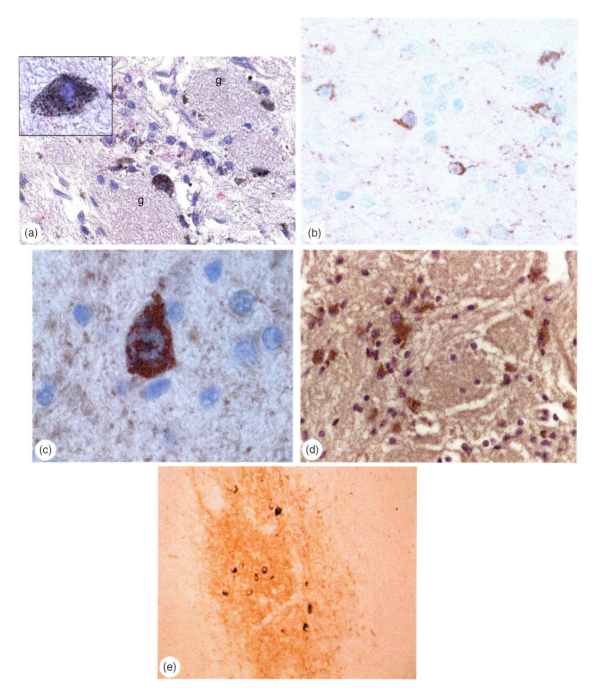

Figure 20.6

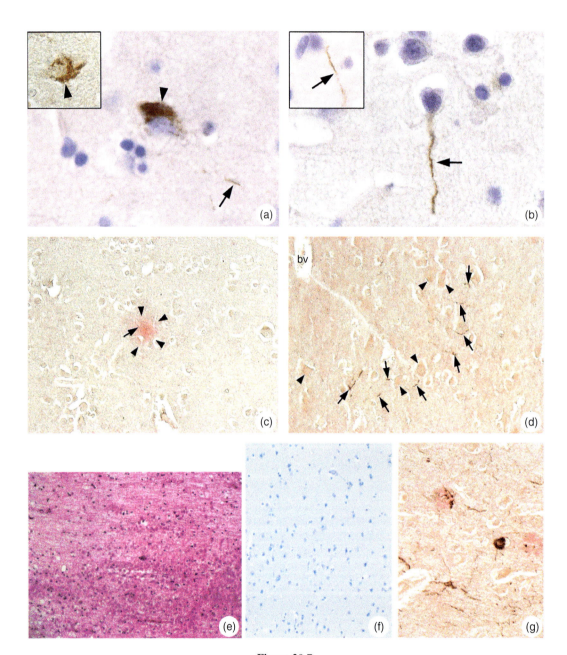

Figure 20.7

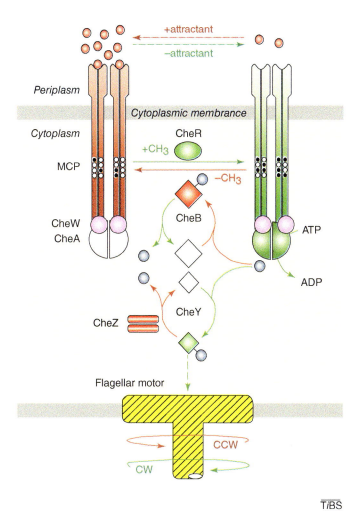

Figure 21.1

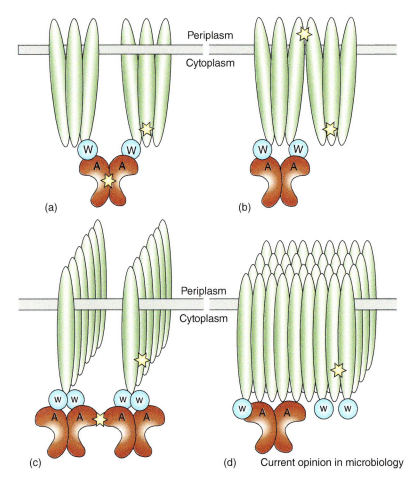

Figure 21.2

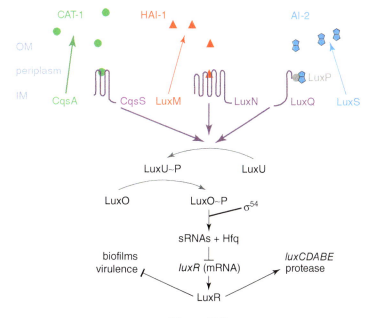

Figure 21.5

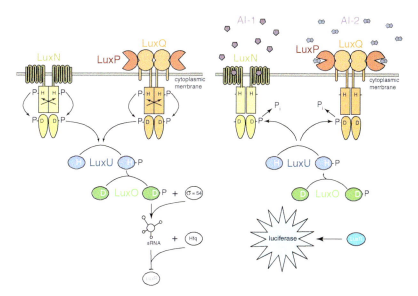

Figure 21.6

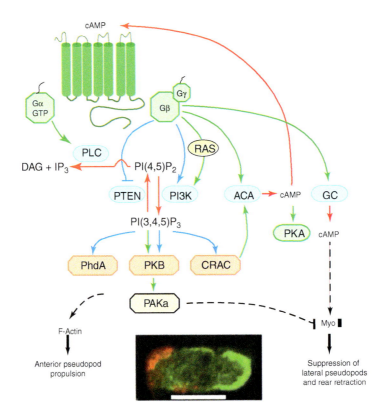

Figure 21.9

Figure 22.1

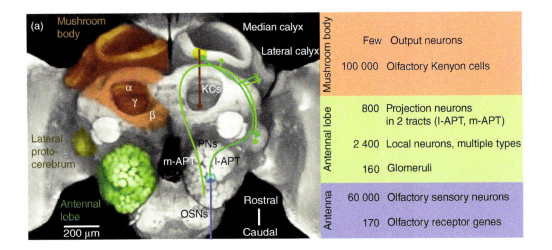

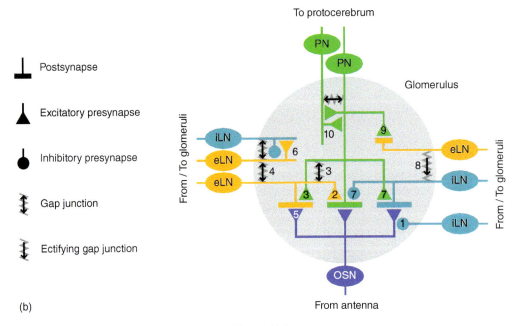

Figure 22.3

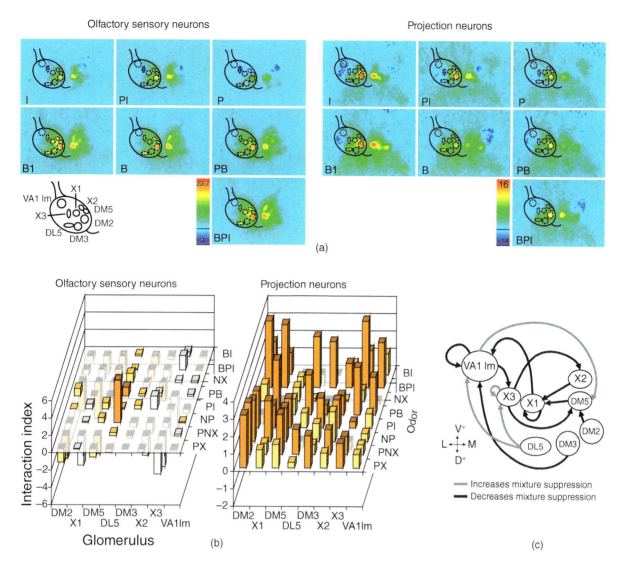

Figure 22.4

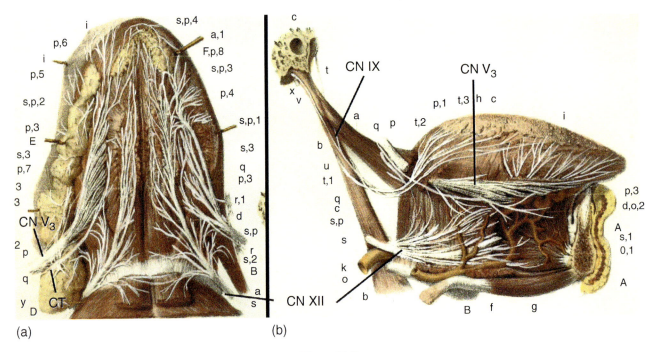

Figure 29.3

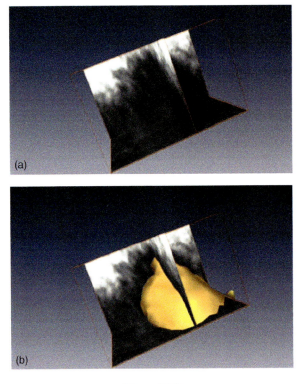

Figure 29.8

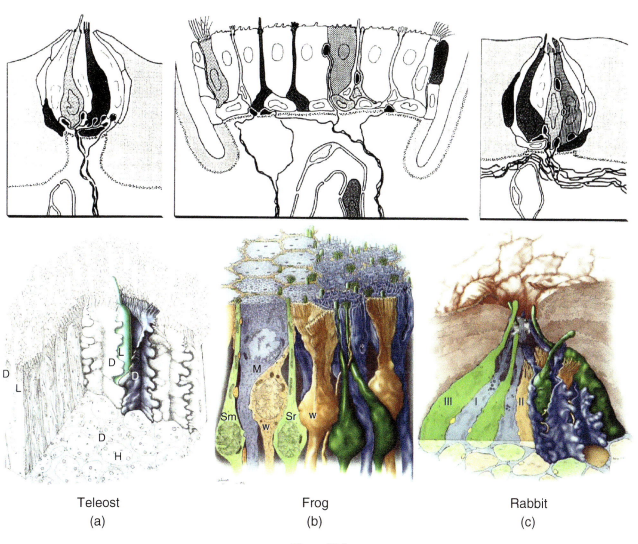

Teleost
(a)

Frog
(b)

Rabbit
(c)

Figure 29.9

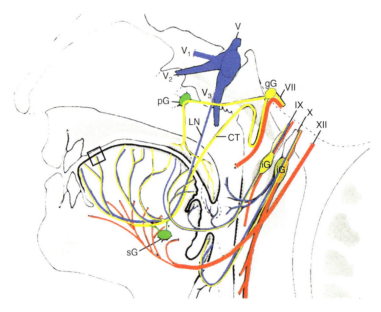

Figure 29.15

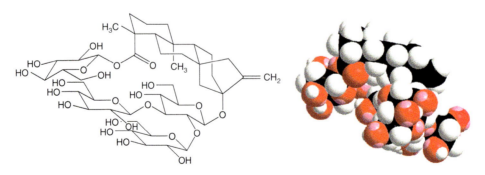

Figure 30.2

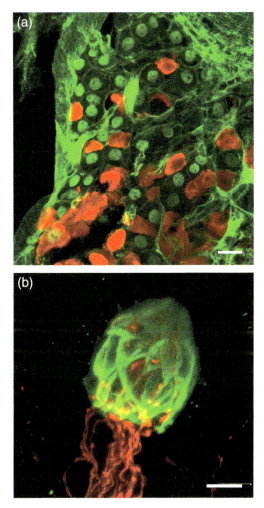

Figure 33.2

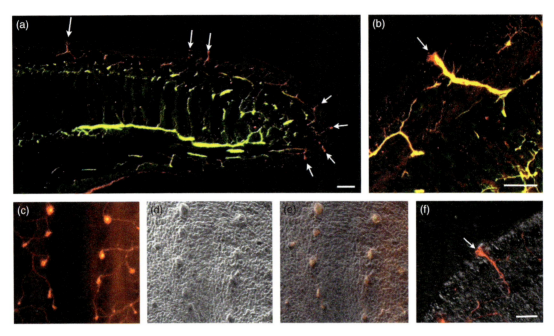

Figure 33.3

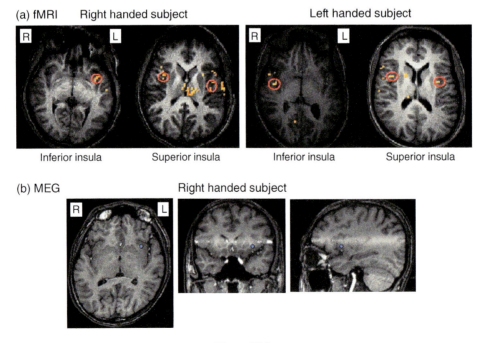

Figure 35.1

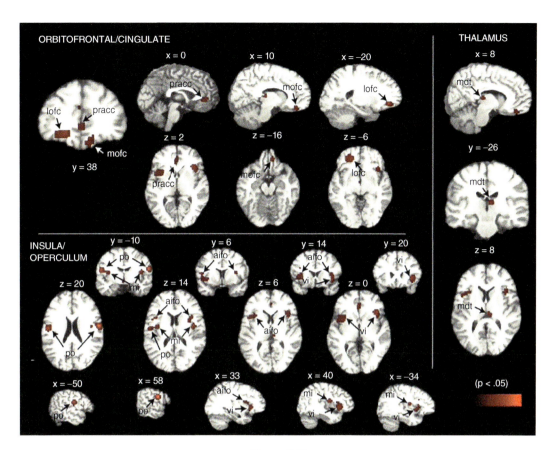

Figure 35.2

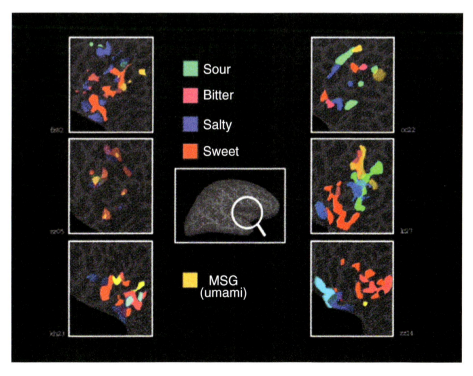

Figure 35.3

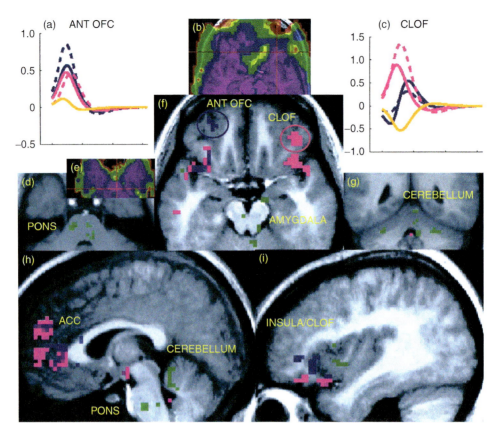

Figure 35.4

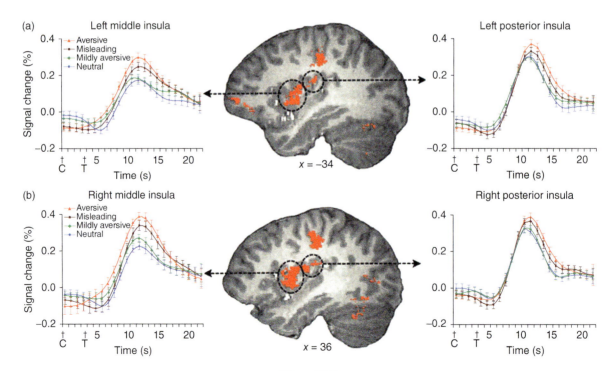

Figure 35.5

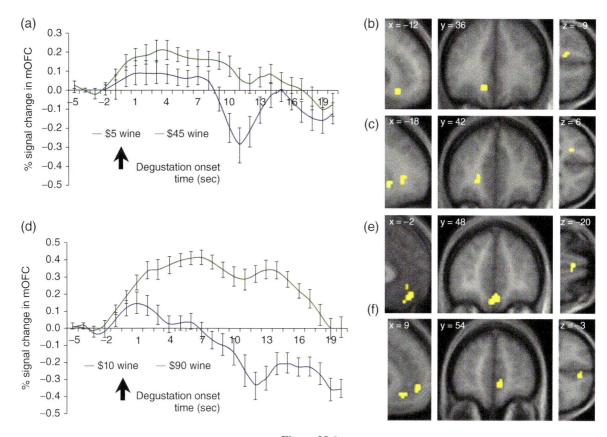

Figure 35.6

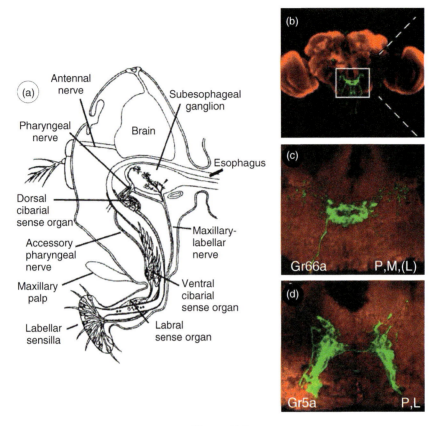

Figure 41.3

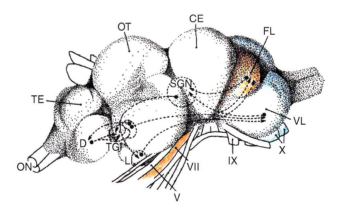

Figure 42.2

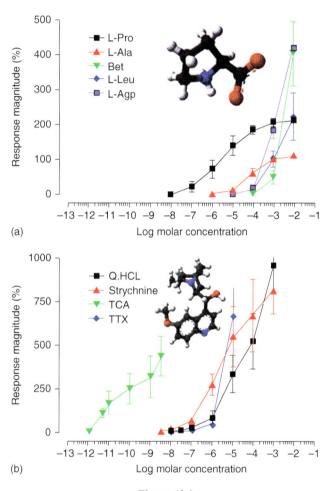

Figure 42.3

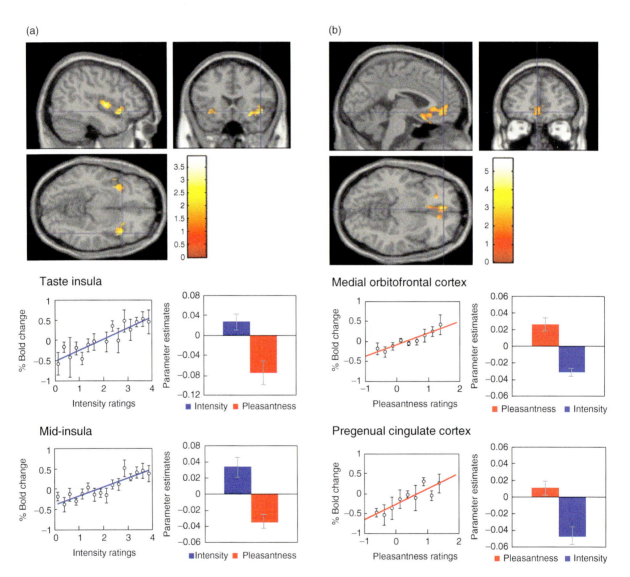

Figure 46.9

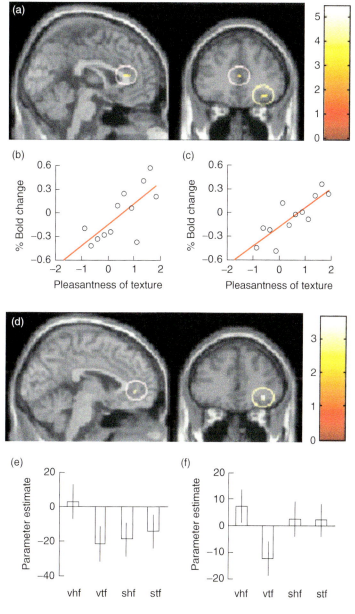

Figure 46.10

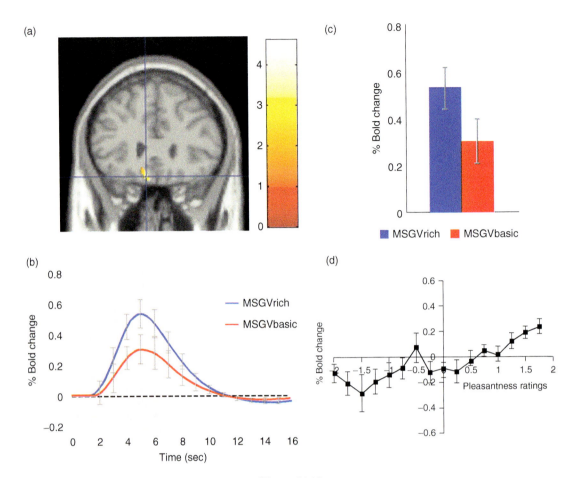

Figure 46.11

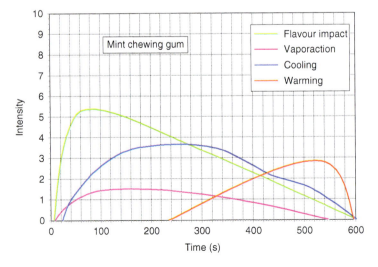

Figure 48.3

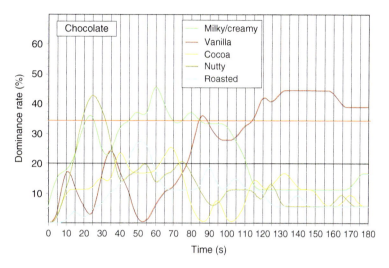

Figure 48.4

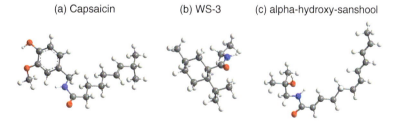

Figure 50.6

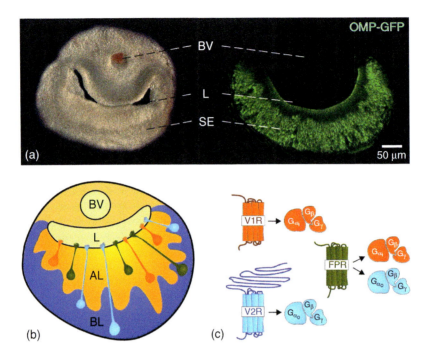

Figure 51.1

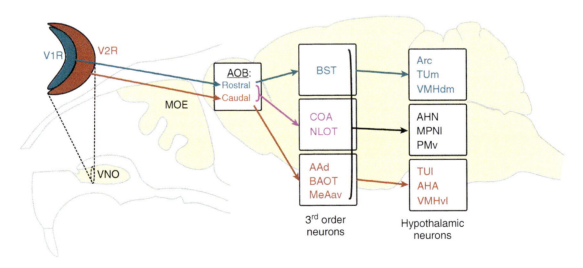

Figure 51.2

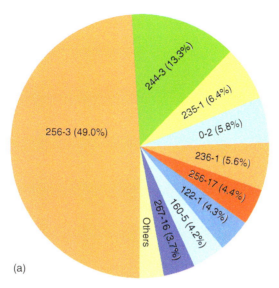

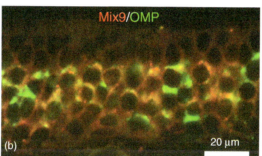

Figure 52.2

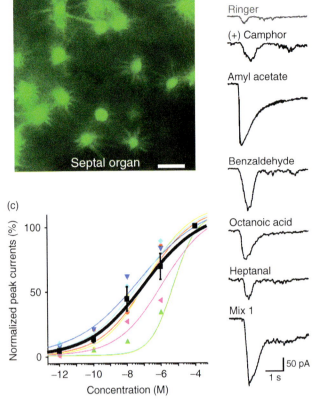

Figure 52.3

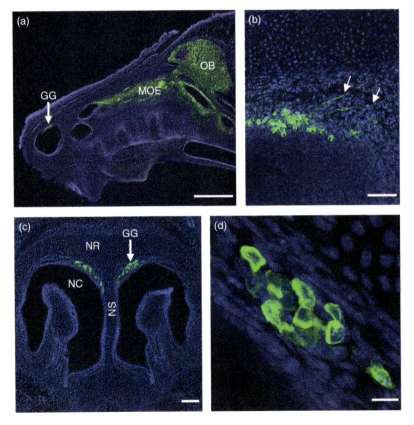

Figure 52.4

Bettini, S., Lazzari, M., and Franceschini, V. (2012). Quantitative analysis of crypt cell population during postnatal development of the olfactory organ of the guppy, *Poecilia reticulata* (Teleostei, Poecilidae), from birth to sexual maturity. *J. Exp. Biol.* 215: 2711–2715.

Bhatt, J. P., and Sajwan, M. S. (2001). Ovarian steroid sulphate functions as priming pheromone in male *Barilius bendelisis* (Ham.). *J. Biosci.* 26: 253–263.

Bhinder G., and Tierney, K. B. (2012) Olfactory evoked activity assay for larval zebrafish. In *Zebrafish Protocols for Neurobehavioral Research. Series: Neuromethods.* Kalueff A.V., and Stewart, A.M. (Eds). New York: Springer/Humana Press, 66: 71–84.

Borowsky, B., Adham, N., Jones, K. A., et al. (2001). Trace amines: Identification of a family of mammalian G protein-coupled receptors. *PNAS* 98: 8966–8971.

Boschat, C., Pelofi, C., Randin, O., et al. (2002). Pheromone detection mediated by a V1r vomeronasal receptor. *Nat. Neurosci.* 5: 1261–1262.

Branchek, T. A., and Blackburn, T. P. (2003). Trace amine receptors as targets for novel therapeutics: legend, myth and fact. *Curr. Opin. Pharm.* 3: 90–97.

Brown, G. E. (2003). Learning about danger: chemical alarm cues and local risk assessment in prey fishes. *Fish Fish.* 4: 227–234.

Brown, G. E., Adrian, J. C., and Shih, M. L. (2001). Behavioural responses of fathead minnows to hypoxanthine-3-N-oxide at varying concentrations. *J. Fish Biol.* 58: 1465–1470.

Burne, R. H. (1909). The anatomy of the olfactory organ of Teleostean fishes. *Proc.Zool. Soc. Lond.* 2: 610–663.

Byrd, C. A., and Brunjes, P. C. (1995). Organization of the olfactory system in the adult zebrafish - histological, immunohistochemical, and quantitative analysis. *J. Comp. Neurol.* 358: 247–259.

Cachat, J., Stewart, A., Utterback, E., et al. (2011). Three-dimensional neurophenotyping of adult zebrafish behavior. *PLoS ONE* 6: e17597.

Cao, Y., Oh, B. C., and Stryer, L. (1998). Cloning and localization of two multigene receptor families in goldfish olfactory epithelium. *PNAS* 95: 11987–11992.

Cerdá-Reverter, J. M., Zanuy, S., and Muñoz-Cueto, J. A. (2001). Cytoarchitectonic study of the brain of a perciform species, the sea bass (*Dicentrarchus labrax*). I. The telencephalon. *J. Morphol.* 247: 217–228.

Chung-Davidson, Y.-W., Rees, C. B., Wu, H., et al. (2004). {beta}-naphthoflavone induction of CYP1A in brain of juvenile lake trout (*Salvelinus namaycush* Walbaum). *J. Exp. Biol.* 207: 1533–1542.

Churcher, A. M., and Taylor, J. S. (2009). Amphioxus (*Branchiostoma floridae*) has orthologs of vertebrate odorant receptors. *BMC Evol. Biol.* 9: 242.

Cole, K. S., and Stacey, N. E. (1984). Prostaglandin induction of spawning behavior in *Cichlasoma bimaculatum* (Pisces Cichlidae). *Horm. Behav.* 18: 235–248.

Cole, T. B., and Stacey, N. E. (2006). Olfactory responses to steroids in an African mouth-brooding cichlid, *Haplochromis burtoni* (Günther) *J. Fish Biol.* 68: 661–680.

Copeland, M. (1912). The olfactory reactions of the puffer or swellfish, *Spheroides maculatus* (Bloch and Schneider). *J. Exp. Zool.* 12: 363–368.

Cox, J. P. L. (2008). Hydrodynamic aspects of fish olfaction. *J. Royal Soc. Inter.* 5: 575–593.

Craigie, E. H. (1926). A preliminary experiment upon the relation of the olfactory sense to the migration of the sockeye salmon (*Oncorhynchus nerka*, Walbaum). *Trans. Royal Soc. Can.* 3rd Ser.: 215–224.

Cravedi, J. P., Perdu-Durand, E., and Poupin, E. (1998). Effects of a polycyclic aromatic hydrocarbon and a PCB on xenobiotic metabolizing activities in rainbow trout (*O. mykiss*). *Bull. Fr. Peche Piscic.* 350–51: 563–570.

Dallas, L. J., Shultz, A. D., Moody, A. J., et al. (2010). Chemical excretions of angled bonefish *Albula vulpes* and their potential use as predation cues by juvenile lemon sharks *Negaprion brevirostris*. *J. Fish Biol.* 77: 947–962.

Datta, N. C., and Bandyopadhyay, S. K. (1997). Ultrastructure of cell types of the olfactory epithelium in a catfish, *Heteropneustes fossilis* (Bloch). *J. Biosci. (Bangalore)* 22: 233–245.

Dellacorte, C., Restrepo, D., Menco, B. P. M., et al. (1996). G alpha(q)/G alpha(11): Immunolocalization in the olfactory epithelium of the rat (*Rattus rattus*) and the channel catfish (*Ictalurus punctatus*). *Neuroscience* 74: 261–273.

Derjean, D., Moussaddy, A., Atallah, E., et al. (2010). A novel neural substrate for the transformation of olfactory inputs into motor output. *PLoS Biol.* 8.

Djordjevic, J., Boyle, J. A., and Jones-Gotman, M. (2012). Pleasant or unpleasant: attentional modulation of odor perception. *Chem.Percept.* 5: 11–21.

Doving, K., and Selset, R. (1980). Behavior patterns in cod released by electrical stimulation of olfactory tract bundlets. *Science* 207: 559–560.

Døving, K. B., Dubois-Dauphin, M., Holley, A., and Jourdan, F. (1977). Functional anatomy of the olfactory organ of fish and the ciliary mechanism of water transport. *Acta Zool.* 58: 245–255.

Døving, K. B., and Holmberg, K. (1974). A note on the function of the olfactory organ of the hagfish *Myxine glutinosa*. *Acta Physiol. Scand.* 91: 430–432.

Dulac, C., and Axel, R. (1995). A novel family of genes encoding putative pheromone receptors in mammals. *Cell* 83: 195–206.

Eisthen, H. L. (1997). Evolution of vertebrate olfactory systems. *Brain Behav. Evol.* 50: 222–233.

Elsaesser, R., and Paysan, J. (2007). The sense of smell, its signalling pathways, and the dichotomy of cilia and microvilli in olfactory sensory cells. *BMC Neurosci* 8 Suppl 3: S1.

Evans, R. E., and Hara, T. J. (1985). The characteristics of the electro-olfactogram (EOG): its loss and recovery following olfactory nerve section in rainbow trout (*Salmo gairdneri*). *Brain Res.* 330: 65–75.

Fabian, N., Albright, L., Gerlach, G., et al. (2007). Humic acid interferes with species recognition in zebrafish (*Danio rerio*). *J. Chem. Ecol.* 33: 2090–2096.

Ferrando, S., Bottaro, M., Gallus, L., et al. (2006). Observations of crypt neuron-like cells in the olfactory epithelium of a cartilaginous fish. *Neurosci. Lett.* 403: 280–282.

Ferrando, S., Gallus, L., Gambardella, C., et al. (2012). Neuronal nitric oxide synthase (nNOS) immunoreactivity in the olfactory system of a cartilaginous fish. *J. Chem. Neuroanat.* 43: 133–140.

Firestein, S. (2004). A code in the nose. *Sci. STKE* 2004: 15.

Foster, N. R. (1985). Lake trout reproductive behavior: influence of chemosensory cues from young-of-the-year by-products. *Trans. Am. Fish. Soc.* 114: 794–803.

Freitag, J., Krieger, J., Strotmann, J., and Breer, H. (1995). Two classes of olfactory receptors in *Xenopus laevis*. *Neuron* 15: 1383–1392.

Gayoso, J., Castro, A., Anadón, R., and Manso, M. J. (2012). Crypt cells of the zebrafish *Danio rerio* mainly project to the dorsomedial glomerular field of the olfactory bulb. *Chem. Senses* 37: 357–369.

Gayoso, J. Á., Castro, A., Anadón, R., and Manso, M. J. (2011). Differential bulbar and extrabulbar projections of diverse olfactory receptor neuron populations in the adult zebrafish (*Danio rerio*). *J. Comp. Neurol.* 519: 247–276.

Giattina, J. D., Garton, R. R., and Stevens, D. G. (1982). Avoidance of copper and nickel by rainbow trout *Salmo gairdneri* as monitored by a computer based data acquisition system. *Trans. Am. Fish. Soc.* 111: 491–504.

Gloriam, D. E. I., Bjarnadottir, T. K., Yan, Y. L., et al. (2005). The repertoire of trace amine G-protein-coupled receptors: large expansion in zebrafish. *Mol. Phylogen. Evol.* 35: 470–482.

Grus, W. E., and Zhang, J. (2009). Origin of the genetic components of the vomeronasal system in the common ancestor of all extant vertebrates. *Mol. Biol. Evol.* 26: 407–419.

Halpern, M. (1987). The organization and function of the vomeronasal system. *Annu. Rev. Neurosci.* 10: 325–362.

Hamdani, E.-H., Stabell, O. B., Alexander, G., and Døving, K. B. (2000). Alarm reaction in the crucian carp is mediated by the medial bundle of the medial olfactory tract. *Chem. Senses* 25: 103–109.

Hamdani, E. H., and Døving, K. B. (2007). The functional organization of the fish olfactory system. *Prog. Neurobiol.* 82: 80–86.

Hansen, A., and Finger, T. E. (2000). Phyletic distribution of crypt-type olfactory receptor neurons in fishes. *Brain Behav. Evol.* 55: 100–110.

Hansen, A., Reiss, J. O., Gentry, C. L., and Burd, G. D. (1998). Ultrastructure of the olfactory organ in the clawed frog, *Xenopus laevis*, during larval development and metamorphosis. *J. Comp. Neurol.* 398: 273–288.

Hansen, A., Rolen, S. H., Anderson, K., et al. (2003). Correlation between olfactory receptor cell type and function in the channel catfish. *J. Neurosci.* 23: 9328–9339.

Hansen, A., and Zeiske, E. (1998). The peripheral olfactory organ of the zebrafish, *Danio rerio*: an ultrastructural study. *Chem. Senses* 23: 39–48.

Hansen, A., and Zielinski, B. (2005). Diversity in the olfactory epithelium of bony fishes: Development, lamellar arrangement, sensory neuron cell types and transduction components. *J. Neurocytol.* 34: 183–208.

Hansen, J. A., Marr, J. C. A., Lipton, J., et al. (1999). Differences in neurobehavioral responses of chinook salmon (*Oncorhynchus tshawytscha*) and rainbow trout (*Oncorhynchus mykiss*) exposed to copper and cobalt: behavioral avoidance. *Environ. Toxicol. Chem.* 18: 1972–1978.

Hara, T. J. (2005). Olfactory responses to amino acids in rainbow trout: revisited, In *Fish Chemosenses*, Reutter, K., and Kapoor, B. G. (Eds.), Enfield: Science Publishers, pp. 31–64.

Hara, T. J., and Zhang, C. (1996). Spatial projections to the olfactory bulb of functionally distinct and randomly distributed primary neurons in salmonid fishes. *Neurosci. Res.* 26: 65–74.

Hashiguchi, Y., and Nishida, M. (2005). Evolution of vomeronasal-type odorant receptor genes in the zebrafish genome. *Gene* 362: 19–28.

Hashiguchi, Y., and Nishida, M. (2006). Evolution and origin of vomeronasal-type odorant receptor gene repertoire in fishes. *BMC Evol. Biol.* 6: 76.

Hashiguchi, Y., and Nishida, M. (2007). Evolution of trace amine–associated receptor (TAAR) gene family in vertebrates: lineage-specific expansions and degradations of a second class of vertebrate chemosensory receptors expressed in the olfactory epithelium. *Mol. Biol. Evol.* 24: 2099–2107.

Hasler, A. D., and Wisby, W. J. (1951). Discrimination of stream odors by fishes and its relation to parent stream behavior. *Am. Nat.* 85: 223–238.

Hirsch, B. T. (2010). Tradeoff between travel speed and olfactory food detection in ring-tailed coatis (*Nasua nasua*). *Ethology* 116: 671–679.

Hofmann, M. H., and Meyer, D. L. (1991). Functional subdivisions of the olfactory system correlate with lectin-binding properties in *Xenopus*. *Brain Res.* 564: 344–347.

Holmes, W. M., Cotton, R., Xuan, V. B., et al. (2011). Three-dimensional structure of the nasal passageway of a hagfish and its implications for olfaction. *Anat. Rec.: Advances Integrat. Anat. Evolut. Biol.* 294: 1045–1056.

Hubbard, P. C., Ingleton, P. M., Bendell, L. A., et al. (2002). Olfactory sensitivity to changes in environmental [Ca^{2+}] in the freshwater teleost *Carassius auratus*: an olfactory role for the Ca2+−sensing receptor? *J. Exp. Biol.* 205: 2755–2764.

Huertas, M., Hubbard, P. C., Canario, A. V. M., and Cerda, J. (2007). Olfactory sensitivity to conspecific bile fluid and skin mucus in the European eel *Anguilla anguilla* (L.). *J. Fish Biol.* 70: 1907–1920.

Hussain, A., Saraiva, L. R., and Korsching, S. I. (2009). Positive Darwinian selection and the birth of an olfactory receptor clade in teleosts. *PNAS* 106: 4313–4318.

Idler, D. R., Fagerlund, U. H. M., and Mayoh, H. (1956). Olfactory perception in migrating salmon I. L-serine, a salmon repellent in mammalian skin. *J. Gen. Physiol.* 39: 889–892.

Inoue, R., and Nakatani, K. (2010). Changes in the olfactory response to amino acids in Japanese newts after transfer from an aquatic to a terrestrial habitat. *Zool. Sci.* 27: 369–373.

Johansson, M. L., and Banks, M. A. (2011). Olfactory receptor related to class A, type 2 (V1r-Like Ora2) genes are conserved between distantly related rockfishes (genus *Sebastes*). *J. Hered.* 102: 113–117.

Johnson, N. S., Yun, S.-S., Thompson, H. T., et al. (2009). A synthesized pheromone induces upstream movement in female sea lamprey and summons them into traps. *PNAS* 106: 1021–1026.

Jones, K. A., and Hara, T. J. (1985). Behavioral responses of fishes to chemical cues results from a new bioassay. *J. Fish Biol.* 27: 495–504.

Jungblut, L. D., Pozzi, A. G., and Paz, D. A. (2011). Larval development and metamorphosis of the olfactory and vomeronasal organs in the toad *Rhinella* (Bufo) *arenarum* (Hensel, 1867). *Acta Zool.* 92: 305–315.

Kajiura, S. M., Forni, J. B., and Summers, A. P. (2004). Olfactory morphology of carcharhinid and sphyrnid sharks: Does the cephalofoil confer a sensory advantage? *J. Morphol.* 264: 253–263.

Kang, J., and Caprio, J. (1995). In vivo responses of single olfactory receptor neurons in the channel catfish, *Ictalurus punctatus*. *J. Neurophysiol.* 73: 172–177.

Kapoor, A. S., and Ojha, P. P. (1972). Studies on ventilation of the olfactory chambers of fishes with a critical reevaluation of the role of accessory nasal sacs. *Arch. Biol. (Liege)* 83: 167–178.

Katare, Y. K., Scott, A. P., Laframboise, A. J., et al. (2011). Release of free and conjugated forms of the putative pheromonal steroid 11-oxo-etiocholanolone by reproductively mature male round goby (*Neogobius melanostomus* Pallas, 1814). *Biol. Reprod.* 84: 288–298.

Keverne, E. B. (1999). The vomeronasal organ. *Science* 286: 716–720.

Kimoto, H., Haga, S., Sato, K., and Touhara, K. (2005). Sex-specific peptides from exocrine glands stimulate mouse vomeronasal sensory neurons. *Nature* 437: 898–901.

Koide, T., Miyasaka, N., Morimoto, K., et al. (2009). Olfactory neural circuitry for attraction to amino acids revealed by transposon-mediated gene trap approach in zebrafish. *PNAS* 106: 9884–9889.

Kosaka, K., and Kosaka, T. (2005). Synaptic organization of the glomerulus in the main olfactory bulb: Compartments of the glomerulus and heterogeneity of the periglomerular cells. *Anatom. Sci. Internat.* 80: 80–90.

Kosaka, T. (1980). Ruffed cell: A new type of neuron with a distinctive initial unmyelinated portion of the axon in the olfactory bulb of the goldfish (*Carassius auratus*): II. Fine structure of the ruffed cell. *J. Comp. Neurol.* 193: 119–145.

Kudo, H., Shinto, M., Sakurai, Y., and Kaeriyama, M. (2009). Morphometry of olfactory lamellae and olfactory receptor neurons during the life history of chum salmon (*Oncorhynchus keta*). *Chem. Senses* 34: 617–624.

Kux, J., Zeiske, E., and Osawa, Y. (1988). Laser doppler velocimetry measurement in the model flow of a fish olfactory organ. *Chem. Senses* 13: 257–265.

Laframboise, A. J., Ren, X., Chang, S., et al. (2007). Olfactory sensory neurons in the sea lamprey display polymorphisms. *Neurosci. Lett.* 414: 277–281.

References

Lastein, S., Hamdani, E. H., and Doving, K. B. (2008). Single unit responses to skin odorants from conspecifics and heterospecifics in the olfactory bulb of crucian carp *Carassius carassius*. *J. Exp. Biol.* 211: 3529–3535.

Leinders-Zufall, T., Brennan, P., Widmayer, P., S., et al. (2004). MHC class I peptides as chemosensory signals in the vomeronasal organ. *Science* 306: 1033–1037.

Li, W., Sorensen, P. W., and Gallaher, D. D. (1995). The olfactory system of migratory adult sea lamprey (*Petromyzon marinus*) is specifically and acutely sensitive to unique bile acids released by conspecific larvae. *J. Gen. Physiol.* 105: 569–587.

Liberles, S. D., and Buck, L. B. (2006). A second class of chemosensory receptors in the olfactory epithelium. *Nature* 442: 645–650.

Mathuru, A. S., Kibat, C., Cheong, W..F., et al. (2012). Chondroitin fragments are odorants that trigger fear behavior in fish. *Curr. Biol.* 22.

Matsunami, H., and Buck, L. B. (1997). A multigene family encoding a diverse array of putative pheromone receptors in mammals. *Cell* 90: 775–784.

Matsuo, A. Y. O., Gallagher, E. P., Trute, M., et al. (2008). Characterization of phase I biotransformation enzymes in coho salmon (*Oncorhynchus kisutch*). *Comp. Biochem. Physiol.* 147C:78–84.

Meredith, T. L., Caprio, J., and Kajiura, S. M. (2012). Sensitivity and specificity of the olfactory epithelia of two elasmobranch species to bile salts. *J. Exp. Biol.* 215: 2660–2667.

Michel, W. C., Sanderson, M. J., Olson, J. K., and Lipschitz, D. L. (2003). Evidence of a novel transduction pathway mediating detection of polyamines by the zebrafish olfactory system. *J. Exp. Biol.* 206: 1697–1706.

Miklavc, P., and Valentinčič, T. (2012). Chemotopy of amino acids on the olfactory bulb predicts olfactory discrimination capabilities of zebrafish *Danio rerio*. *Chem. Senses* 37: 65–75.

Millery, J., Briand, L., Bézirard, V., et al. (2005). Specific expression of olfactory binding protein in the aerial olfactory cavity of adult and developing Xenopus. *Eur. J. Neurosci.* 22: 1389–1399.

Mirza, R. S., Green, W. W., Connor, S., et al. (2009). Do you smell what I smell? Olfactory impairment in wild yellow perch from metal-contaminated waters. *Ecotoxicol. Environ. Saf.* 72: 677–683.

Mombaerts, P. (1999). Seven-transmembrane proteins as odorant and chemosensory receptors. *Science* 286: 707–711.

Monod, G., Saucier, D., Perdu-Durand, E., et al. (1994). Biotransformation enzyme activities in the olfactory organ of rainbow trout (*Oncorhynchus mykiss*). Immunocytochemical localization of cytochrome P4501A1 and its induction by β-naphthoflavone. *Fish Physiol. Biochem.* 13: 433–444.

Moran, D. T., Rowley, J. C., Jafek, B. W., and Lovell, M. A. (1982). The fine structure of the olfactory mucosa in man. *J. Neurocytol.* 11: 721–746.

Morgan, G. C. (1975). The telencephalon of the sea catfish *Galeichthys felis*. *J. Hirnforsch.* 16: 131–150.

Morita, Y., and Finger, T. E. (1998). Differential projections of ciliated and microvillous olfactory receptor cells in the catfish, *Ictalurus punctatus*. *J. Comp. Neurol.* 398: 539–550.

Nevitt, G. A. (1991). Do fish sniff? A new mechanism of olfactory sampling in pleuronectid flounders. *J. Exp. Biol.* 157: 1–18.

Nevitt, G. A. (2011). The neuroecology of dimethyl sulfide: a global-climate regulator turned marine infochemical. *Integr. Comp. Biol.* 51: 819–825.

Niimura, Y. (2009). On the origin and evolution of vertebrate olfactory receptor genes: comparative genome analysis among 23 chordate species. *Gen. Biol. Evol.* 1: 34–44.

Niimura, Y., and Nei, M. (2005). Evolutionary dynamics of olfactory receptor genes in fishes and tetrapods. *PNAS* 102: 6039–6044.

Nikonov, A. A., and Caprio, J. (2005). Processing of odor information in the olfactory bulb and cerebral lobes. *Chem. Senses* 30: i317–i318.

Nikonov, A. A., and Caprio, J. (2007). Highly specific olfactory receptor neurons for types of amino acids in the channel catfish. *J. Neurophysiol.* 98: 1909–1918.

Nikonov, A. A., Finger, T. E., and Caprio, J. (2005). Beyond the olfactory bulb: An odotopic map in the forebrain. *PNAS* 102: 18688–18693.

Nordeng, H. (1977). A pheromone hypothesis for homeward migration in anadromous salmonids. *Oikos* 28: 155–159.

Northcutt, R. G. (1986). Lungfish neural characters and their bearing on sarcopterygian phylogeny. *J. Morphol.*: 277–297.

Northcutt, R. G., and Bemis, W. E. (1993). Cranial nerves of the coelacanth, *Latimeria chalumnae* [Osteichthyes: Sarcopterygii: Actinistia], and comparisons with other craniata. *Brain Behav. Evol.* 42 (Suppl 1): 1–76.

Oka, Y., and Korsching, S. I. (2011). Shared and unique G alpha proteins in the zebrafish versus mammalian senses of taste and smell. *Chem. Senses* 36: 357–365.

Oka, Y., Saraiva, L. R., and Korsching, S. I. (2011). Crypt neurons express a single V1R-related ora gene. *Chem. Senses.* DOI: 10.1093/chemse/bjr095

Olsén, K. H., Grahn, M., and Lohm, J. (2002). Influence of MHC on sibling discrimination in Arctic char, *Salvelinus alpinus* (L.). *J. Chem. Ecol.* 28: 783–796.

Olsén, K. H., Sawisky, G. R., and Stacey, N. E. (2006). Endocrine and milt responses of male crucian carp (*Carassius carassius* L.) to periovulatory females under field conditions. *Gen. Comp. Endocrinol.* 149: 294–302.

Olsén, K. H., and Winberg, S. (1996). Learning and sibling odor preference in juvenile Arctic char, *Salvelinus alpinus* (L.). *J. Chem. Ecol.* 22: 773–786.

Ottoson, D. (1956). Analysis of the electrical activity of the olfactory epithelium. *Acta Physiol. Scand.* 35: 1–83.

Parra, K. V., Adrian Jr, J. C., and Gerlai, R. (2009). The synthetic substance hypoxanthine 3-N-oxide elicits alarm reactions in zebrafish (*Danio rerio*). *Behav. Brain Res.* 205: 336–341.

Partridge, B. L., Liley, N. R., and Stacey, N. E. (1976). Role of pheromones in sexual-behavior of goldfish. *Anim. Behav.* 24: 291–299.

Pérez-Lopez, M., Anglade, P., Bec-Ferté, M. P., et al. (2000). Characterization of hepatic and extrahepatic glutathione *S*-transferases in rainbow trout (*Oncorhynchus mykiss*) and their induction by 3,3′,4,4′-tetrachlorobiphenyl. *Fish Physiol. Biochem.* 22: 21–32.

Pfeiffer, W. (1964). The morphology of the olfactory organ of *Hoplopagrus guesntheri* Gill 1862. *Can. J. Zool.* 42: 235–237.

Pfister, P., and Rodriguez, I. (2005). Olfactory expression of a single and highly variable V1r pheromone receptor-like gene in fish species. *PNAS* 102: 5489–5494.

Pollock, M. S., Chivers, D. P., Mirza, R. S., and Wisenden, B. D. (2003). Fathead minnows, *Pimephales promelas*, learn to recognize chemical alarm cues of introduced brook stickleback, *Culaea inconstans*. *Environ. Biol. Fishes* 66: 313–319.

Quinn, T. P., and Hara, T. J. (1986). Sibling recognition and olfactory sensitivity in juvenile coho salmon (*Oncorhynchus kisutch*). *Can. J. Zool.* 64: 921–925.

Reese, E. A., Bunzow, J. R., Arttamangkul, S., et al. (2007). Trace amine-associated receptor 1 displays species-dependent stereoselectivity for isomers of methamphetamine, amphetamine, and para-hydroxyamphetamine. *J. Pharmacol. Exp. Ther.* 321: 178–186.

Rehnberg, B. G., Jonasson, B., and Schreck, C. B. (1985). Olfactory sensitivity during parr and smolt developmental stages of coho salmon. *Trans. Am. Fish. Soc.* 114: 732–736.

Ren, X., Chang, S., Laframboise, A., et al. (2009). Projections from the accessory olfactory organ into the medial region of the olfactory bulb in the sea lamprey (*Petromyzon marinus*): a novel vertebrate sensory structure? *J. Comp. Neurol.* 516: 105–116.

Riviere, S., Challet, L., Fluegge, D., et al. (2009). Formyl peptide receptor-like proteins are a novel family of vomeronasal chemosensors. *Nature* 459: 574–577.

Rolen, S. H., Sorensen, P. W., Mattson, D., and Caprio, J. (2003). Polyamines as olfactory stimuli in the goldfish *Carassius auratus*. *J. Exp. Biol.* 206: 1683–1696.

Ryba, N. J. P., and Tirindelli, R. (1997). A new multigene family of putative pheromone receptors. *Neuron* 19: 371–379.

Saglio, P., Olsén, K. H., and Bretaud, S. (2001). Behavioral and olfactory responses to prochloraz, bentazone, and nicosulfuron-contaminated flows in goldfish. *Arch. Environ. Contam. Toxicol.* 41: 192–200.

Sato, Y., Miyasaka, N., and Yoshihara, Y. (2005). Mutually exclusive glomerular innervation by two distinct types of olfactory sensory neurons revealed in transgenic zebrafish. *J. Neurosci.* 25: 4889–4897.

Sato, Y., Miyasaka, N., and Yoshihara, Y. (2007). Hierarchical regulation of odorant receptor gene choice and subsequent axonal projection of olfactory sensory neurons in zebrafish. *J. Neurosci.* 27: 1606–1615.

Satou, M. (1990). Synaptic organization, local neuronal circuitry, and functional segregation of the teleost olfactory bulb. *Prog. Neurobiol.* 34: 115–142.

Saucier, D., Julliard, A. K., Monod, G., et al. (1999). CYP1A1 immunolocalization in the olfactory organ of rainbow trout and its possible induction by β-naphthoflavone: analysis in adults and embryos around hatching. *Fish Physiol. Biochem.* 21: 179–192.

Scholz, N. L., Truelove, N. K., French, B. L., et al. (2000). Diazinon disrupts antipredator and homing behaviors in chinook salmon (*Oncorhynchus tshawytscha*). *Can. J. Fish. Aquat. Sci.* 57: 1911–1918.

Serizawa, S., Miyamichi, K., and Sakano, H. (2005). Negative Feedback Regulation Ensures the One Neuron–One Receptor Rule in the Mouse Olfactory System. *Chem. Senses*. 30: i99–i100.

Shamchuk, A. L., and Tierney, K. B. (2012). Phenotyping stimulus evoked responses in larval zebrafish. *Behav.* 149: 1177–1203.

Shamushaki, V. A. J., Abtahi, B., and Kasumyan, A. O. (2011). Olfactory and taste attractiveness of free amino acids for Persian sturgeon juveniles, *Acipenser persicus*: a comparison with other acipenserids. *J. Appl. Ichthyol.* 27: 241–245.

Shi, P., and Zhang, J. Z. (2007). Comparative genomic analysis identifies an evolutionary shift of vomeronasal receptor gene repertoires in the vertebrate transition from water to land. *Genome Res.* 17: 166–174.

Shibuya, T., and Takagi, S. F. (1963). Electrical response and growth of olfactory cilia of the olfactory epithelium of the newt in water and on land. *J. Gen. Physiol.* 47: 71–82.

Sorensen, P. W., Fine, J. M., Dvornikovs, V., et al. (2005). Mixture of new sulfated steroids functions as a migratory pheromone in the sea lamprey. *Nat. Chem. Biol.* 1: 324–328.

Speca, D. J., Lin, D. M., Sorensen, P. W., et al. (1999). Functional identification of a goldfish odorant receptor. *Neuron* 23: 487–498.

Speedie, N., and Gerlai, R. (2008). Alarm substance induced behavioral responses in zebrafish (*Danio rerio*). *Behav. Brain Res.* 188: 168–177.

Stacey, N. (2003). Hormones, pheromones and reproductive behavior. *Fish Physiol. Biochem.* 28: 229–235.

Stacey, N., Chojnacki, A., Narayanan, A., et al. (2003). Hormonally derived sex pheromones in fish: exogenous cues and signals from gonad to brain. *Can. J. Physiol. Pharmacol.* 81: 329–341.

Starcevic, S. L., and Zielinski, B. S. (1995). Immunohistochemical localization of glutathione S-transferase pi in rainbow trout olfactory receptor neurons. *Neurosci. Lett.* 183: 175–178.

Starcevic, S. L., and Zielinski, B. S. (1997). Glutathione and glutathione S-transferase in the rainbow trout olfactory mucosa during retrograde degeneration and regeneration of the olfactory nerve. *Exp. Neurol.* 146: 331–340.

Sutterlin, A., Sutterlin, N., and Rand, S. (1971). The influence of synthetic surfactants on the functional properties of the olfactory epithelium of Atlantic salmon. *Fish. Res. Board Can. Tech. Rep.* 287.

Theisen, B. (1973). The olfactory system in the hagfish *Myxine glutinosa*. *Acta Zool.* 54: 271–284.

Theisen, B., Breucker, H., Zeiske, E., and Melinkat, R. (1980). Structure and development of the olfactory organ in the garfish *Belone belone* (L.) (Teleostei, Atheriniformes). *Acta Zool.* 61: 161–170.

Theisen, B., Zeiske, E., Silver, W. L., et al. (1991). Morphological and physiological studies on the olfactory organ of the striped eel catfish, *Plotosus lineatus. Mar. Biol.* 110: 127–135.

Thornhill, R. A. (1972). The ultrastructure of the accessory olfactory organ in the river lamprey (*Lampetra jiuviatilis*). *Acta Zool.* 53: 49–56.

Tierney, K. B., Baldwin, D. H., Hara, T. J., et al. (2010). Review: Olfactory toxicity in fishes. *Aquat. Toxicol.* 96: 2–26.

Tierney, K. B., Ross, P. S., Jarrard, H. E., et al. (2006). Changes in juvenile coho salmon electro-olfactogram during and after short-term exposure to current-use pesticides. *Environ. Toxicol. Chem.* 25: 2809–2817.

Tierney, K. B., Ross, P. S., and Kennedy, C. J. (2007a). Linuron and carbaryl differentially impair baseline amino acid and bile salt olfactory responses in three salmonids. *Toxicol.* 231: 175–187.

Tierney, K. B., Sampson, J. L., Ross, P. S., et al. (2008). Salmon olfaction is impaired by an environmentally realistic pesticide mixture. *Environ. Sci. Technol.* 42: 4996–5001.

Tierney, K. B., Sekela, M. A., Cobbler, C. E., et al. 2011). Evidence for behavioral preference towards environmental concentrations of urban-use herbicides in a model adult fish. *Environ. Toxicol. Chem.* 30: 2046–2054.

Tierney, K. B., Singh, C. R., Ross, P. S., and Kennedy, C. J. (2007b). Relating olfactory neurotoxicity to altered olfactory-mediated behaviors in rainbow trout exposed to three currently-used pesticides. *Aquat. Toxicol.* 81: 55–64.

Tricas, T., Kajiura, S., and Summers, A. (2009). Response of the hammerhead shark olfactory epithelium to amino acid stimuli. *J. Comp. Physiol.* 195A: 947–954.

Ueda, H. (2012). Physiological mechanisms of imprinting and homing migration in Pacific salmon *Oncorhynchus* spp. *J. Fish Biol.* 81: 543–558.

Valentincic, T., and Koce, A. (2000). Coding principles in fish olfaction as revealed by single unit, EOG and behavioral studies. *Eur. J. Physiol.* 439: R193–R195.

Velez, Z., Hubbard, P. C., Barata, E. N., and Canario, A. V. M. (2007). Differential detection of conspecific-derived odorants by the two olfactory epithelia of the Senegalese sole (Solea senegalensis). *Gen. Comp. Endocrinol.* 153: 418–425.

Veyrac, A., Wang, G., Baum, M. J., and Bakker, J. (2011). The main and accessory olfactory systems of female mice are activated differentially by dominant versus subordinate male urinary odors. *Brain Res.* 1402: 20–29.

Weech, M., Quash, M., and Walters, E. (2003). Characterization of the mouse olfactory glutathione S-transferases during the acute phase response. *J. Neurosci. Res.* 73: 679–685.

Weltzien, F.-A., Höglund, E., Hamdani, E. H., and Døving, K. B. (2003). Does the lateral bundle of the medial olfactory tract mediate reproductive behavior in male crucian carp? *Chem. Senses* 28: 293–300.

References

Yacoob, S. Y., and Browman, H. I. (2007). Olfactory and gustatory sensitivity to some feed-related chemicals in the Atlantic halibut (*Hippoglossus hippoglossus*). *Aquacult.* 263: 303–309.

Yamamoto, Y., Hino, H., and Ueda, H. (2010). Olfactory imprinting of amino acids in lacustrine sockeye salmon. *PLoS ONE* 5: e8633.

Yambe, H., and Yamazaki, F. (2001). Species-specific releaser effect of urine from ovulated female masu salmon and rainbow trout. *J. Fish Biol.* 59: 1455–1464.

Yanagi, S., Kudo, H., Doi, Y., et al. (2004). Immunohistochemical demonstration of salmon olfactory glutathione S-transferase class pi (N24) in the olfactory system of lacustrine sockeye salmon during ontogenesis and cell proliferation. *Anat. Embryol.* 208: 231–238.

Yoshihara, Y. (2009). Molecular genetic dissection of the zebrafish olfactory system. In *Chemosensory systems in mammals, fishes, and insects,* Korsching, S., and Meyerhof, W. (Eds). Berlin / Heidelberg: Springer, pp. 1–24.

Zhang, C., and Hara, T. J. (1994). Multiplicity of salmonid olfactory receptors for bile acids as evidenced by cross-adaptation and ligand binding assay. *Chem. Senses* 19: 579.

Zhang, X., Rodriguez, I., Mombaerts, P., and Firestein, S. (2004). Odorant and vomeronasal receptor genes in two mouse genome assemblies. *Genomics* 83: 802–811.

Zielinski, B. S., and Hara, T. J. (2007). Olfaction. In *Fish Physiology,* Zielinski, B.S., and Hara T.J. (Eds.). New York: Elsevier/Academic Press, pp. 1–43.

Zippel, H. P., Reschke, C., and Korff, V. (1999). Simultaneous recordings from two physiologically different types of relay neurons, mital cells and ruffer cells, in the olfactory bulb of goldfish. *Cell. Mol. Biol.* 45: 327–337.

Zou, Z., Li, F., and Buck, L. B. (2005). Odor maps in the olfactory cortex. *PNAS* 102: 7724–7729.

Chapter 24

The Chemistry of Avian Odors: An Introduction to Best Practices

GABRIELLE A. NEVITT and PAOLA A. PRADA

24.1 INTRODUCTION

Over the last century, birds have been important models in sexual selection and evolutionary theory. Consequently, multiple aspects of the visual ornaments and songs of birds have been catalogued and studied for literally hundreds of species, while any mention of scent is routinely either ridiculed or denied in most textbooks. This was not the case, however, for ornithologists and natural historians living in the early 1800s. At this time, a controversy concerning the Turkey vulture's *(Cathartes aura)* sense of smell was raging mostly because John James Audubon defied the then commonly accepted observation that Turkey vultures scavenged using their sense of smell.

But why was Audubon so set on disproving something so obvious? Some might say he was biased, but we hold that it was his ignorance of chemistry that really got him into trouble. According to biographer William Souder (Souder, 2005), Audubon held to the theory that "nature is parsimonious—that creatures may, for example, rely on keen eyesight or a sensitive ability to smell, but not both." A 'definitive' experiment involved using a painting of a dead sheep as bait: "A coarse painting on canvass was made, representing a sheep skinned and cut-open." Upon discovering it, vultures "commenced tugging at the painting. They seemed much disappointed and surprised, and after having satisfied their curiosity, flew away." Among other problems with experimental design beyond the scope of this review, we now know that Turkey vultures *(Cathartes aura)* are sensitive to sulfides emitted by freshly decaying proteinaceous flesh, similar to those sulfides found in pigments of oil based paints (Luxan and Dorrego, 1999; White et al., 2006). These compounds are

so attractive to vultures that they have been purposefully used to scent natural gas, so that pipeline leaks in remote areas can be located by the kettles of vultures overhead (Stager, 1964; for review, see DeBose and Nevitt, 2008).

From the vantage point of 2012, Souder's descriptions seem almost comical, but the ideas put forth in Audubon's publications, and the popular attention that followed, set the stage for both academic and public denial that any bird could smell, with a lasting impact on the development of behavioral ecology as a discipline, and the potential influence of chemical senses in such behaviors, from foraging to personal odor recognition. If Audubon could 'prove' that even Turkey vultures lacked a sense of smell, then how could olfaction be important to the behaviors of other birds? This simple Type 1 error – rejection of the null hypothesis even though it is true – changed the course of the study of chemical senses. The result was that birds, a class of vertebrates that arguably presents some of the most tractable experimental models for the study of chemical senses, have been passed over in favor of rodents, salamanders, and flies. Just as importantly, with renewed interest in avian olfaction, researchers getting into this field should be cautious of Type II errors, or false positive error, particularly in an era where manuscripts are published rapidly, and analytical tools and technologies are just as rapidly developed beyond the expertise of even conscientious reviewers.

With avian chemical senses experiencing a hey-day, there have been several recent reviews of avian olfaction, and the interested reader is referred to these (Campagna et al., 2012; Hagelin and Jones, 2007; Hagelin, 2007; Nevitt and Bonadonna, 2005; Nevitt and Hagelin, 2009;

Handbook of Olfaction and Gustation, Third Edition. Edited by Richard L. Doty.
© 2015 Richard L. Doty. Published 2015 by John Wiley & Sons, Inc.

566   Chapter 24   The Chemistry of Avian Odors: An Introduction to Best Practices

Nevitt, 2000, 2008; Rajchard, 2007, 2008; Roper, 1999; Wallraff, 2004; Wenzel, 2007). Our goal is not to repeat these reviews, but rather provide interested readers with background information about potential sources of avian odor, and to educate behavioral ecologists about the relative strengths and weaknesses of chemical approaches in a user-friendly manner. We are focusing on this problem because one of the most rapidly growing areas of avian chemical ecology is the study of personal odor and how it may or may not contribute to chemical signaling in birds. Much of this work is being carried out by laboratories that do not focus on the chemical senses, and by young researchers in behavioral ecology who need to be able to build bridges with analytical chemists to design effective and realistic experiments to test hypotheses. Many studies are being published even though they have not used calibrated methods. Just as Audubon's study did, they may make headlines, but they do not necessarily authentically progress scientific understanding.

In this review, we first introduce readers to potential sources of avian scent, and provide an overview of functional chemistry necessary to understand the literature. We then review analytical techniques focusing on understanding best practices, and finally highlight representative studies that we feel have used these methods appropriately. In non-prejudicial fashion to achieve quality, we have purposely avoided studies that have been highlighted in recent reviews, but have included them in table form for the interested reader (Table 24.1).

24.2 SOURCES OF AVIAN ODOR CHEMICALS

Several different physiological sources likely contribute to the scent of any given bird. These sources include: the skin and feathers, glandular and skin secretions, dietary products, and digestive oils and feces.

24.2.1 Skin and Feathers

Avian skin is distinguished from that of other terrestrial vertebrates by its lipophilic properties (reviewed by Menon, 1984) and that it is characteristically much thinner in order to conserve weight as an adaptation for flight. Although it tends to be only a few cell layers thick, it is composed of a dermis and an epidermis, and functions primarily to provide a permeable barrier to regulate water loss and to prevent dehydration (reviewed by Pass, 1995). In birds, invaginations of the skin also form feather follicles, but these follicles do not contain glands as hair follicles do in mammals (reviewed by Stettenheim, 2000). Instead, due to its lipogenic nature, the entire skin can be regarded as a fat-producing holocrine gland (Lucas, 1970, 1980).

Menon and Menon (2000) provide an excellent overview of the gross anatomy and typical pattern of epidermal lipogenesis, secretion and sequestration of lipid within the stratum corneum, and the following discussion should be regarded as only a cursory review of a truly fascinating and involved subject.

Given the diversity of birds, avian skin can be highly variable depending on the region of the body, the reproductive state of the animal and the species (Wrench et al., 1980; Menon et al., 1981), but there are common features. Among other functions, the dermal layer stores fat, contains larger blood vessels and nerves associated with the feather follicles, and otherwise supports the epidermal layers and integument. Following Menon and Menon (2000), the epidermis is composed of multiple (typically four to five) layers of cells called sebokeratinocytes due to their tendency to produce lipid (sebum). As cells move upward towards the epidermal surface, they change their form and function. Thus, cells progress from basal (germinitive) and sub-basal layers, to layers such as the stratum transitivum, characterized by free lipid droplets and multigranular bodies analogous to epidermal lamellar bodies found in mammals. The outer-most layer is the stratum corneum or keratin layer which gives rise to specialized structures such as feathers, scales, claws, spurs, beaks, and horny sheaths.

Feathers are an obvious feature of avian skin. Among numerous functions, feathers provide insulation, aerodynamic flight power, water repellency, cleanliness, body support, and sex appeal (reviewed by Gill, 1994). The basic feather structure consists of a shaft with parallel branches on either side. This shaft has a short, tubular portion implanted in the follicle and a longer medullar portion above the skin.

Feathers are periodically replaced, but molt patterns are complex and vary by life history of the species in question (Stettenheim, 2000). Although this, to our knowledge, has not yet been convincingly shown, the constraint to molt suggests that developmental changes might be reflected in the chemical profile of the feather, skin, and preen gland secretions, or of products from bacterial, fungal, or other microbial interactions that adhere to the feather surface (reviewed by Rajchard, 2010). Specific feathers such as the powder downs also have the potential to contribute to the generation of plumage odor by the production of a talcum-like substance that complements preen gland secretions in preening and waterproofing functions. This powder is composed of keratinized cells with lipid-type characteristics (Menon and Menon, 2000).

24.2.2 The Preen Gland

Unlike mammals, most species of birds do not have skin glands aside from the uropygial or preen gland, glands of the ear canal (Menon and Salinukul, 1989), and glands in

24.2 Sources of Avian Odor Chemicals

Table 24.1 Summary of Avian Odor Chemicals Classified by Extraction Technique.

Extraction Technique	Biological Specimen		
Solid Phase Microextraction (SPME)	Feather	Uropygial Glandular Secretion	Avian Species
Douglas H. D. et al. 2001	x		Crested Auklet (*Aethia cristatella*)
Hagelin J. C. et al. 2003	x		Crested Auklet (Aethia cristatella)
Williams C. R. et al. 2003	x		Chicken (Gallus gallus domesticus)
Burger B. V. et al. 2004		x	Green Woodhoopoe (*Phoeniculus purpureus*)
Whelan R. J. et al. 2010		x	Gray Catbird (*Dumetella carolinensis*)
Shaw C. L. et al. 2011		x	Gray Catbird (*Dumetella carolinensis*)
Stir-bar Sorptive Extraction (SBSE)			
Soini H. A. et al. 2007	x	x	Dark-eyed Junco (*Junco hyemalis*)
Whittaker D. J. et al. 2010		x	Dark-eyed Junco (*Junco hyemalis*)
Solvent Extraction			
Kolattukudy PE and Sawaya WN 1974		x	Chicken (*Gallus gallus domesticus*)
Dekker et al. 1999		x	Red Knot (*Calidris canutus*)
Piersma et al. 1999	x	x	Red Knot (*Calidris canutus*)
Douglas H. D. et al. 2001	x		Crested Auklet (*Aethia cristatella*)
Reneerkens et al. 2002		x	Sandpipers (*Scolopacidae*) 19 species
Sandilands V. et al. 2004	x	x	Chicken (*Gallus gallus domesticus*)
Sweeney R. J. et al. 2004	x		Range of 91 species of passerines
Bonadonna F. et al. 2007	x		Antarctic Prion (*Pachyptila desolata*)
Zhang J. X. et al. 2008	x	x	Budgerigar (Melopsittacus undulatus)
Karlsson A. C. et al. 2010		x	Red Junglefowl (*Gallus gallus*)
Mardon J. et al. 2010		x	Antarctic Prion (*Pachyptila desolata*) Blue Petrel (*Halobaena caerulea*)
Thomas R. H. et al. 2010	x		White-throated Sparrow (*Zonotrichia albicollis*)
Zhang J. X. et al. 2010	x	x	Budgerigar (*Melopsittacus undulatus*)
Leclaire S. et al. 2011	x	x	Black-legged kittiwake (*Rissa tridactyla*)
Mardon J. et al. 2011	x	x	Blue Petrel (*Halobaena caerulea*)
Amo L. et al. 2012		x	Starling (*Sturnus unicolor*)
Leclaire S. et al. 2012		x	Black-legged kittiwake (*Rissa tridactyla*)

the vent region (Quay, 1967; King, 1981; Menon et al., 1987). However, due mostly to its size, accessibility, and morphological diversity (Jacob and Ziswiler, 1982; Johnston, 1988), the preen gland has been frequently investigated as a potential source of personal odor in birds. The preen gland is an integumentary, bi-lobate gland found in most species of birds (with some exceptions, including ostriches, emus, cassowaries, bustards, and some species of parrots). The preen gland produces a waxy oil that has been historically studied for its various roles in feather maintenance (Salibian and Montalti, 2009). Although the detailed anatomy varies between species, in general, preen oil is secreted from the gland and passes to the surface via ducts that lead to a papilla resembling a small nipple situated dorsally near the base of the tail (for a more detailed description, see Salibian and Montalti, 2009; Jacob

and Ziswiler, 1982). Oil from the preen gland collects on the skin and feathers; birds collect this oil by rubbing the back of the head against the opening of the gland, and then distribute it throughout the feathers and skin through the action of preening (Elder, 1954). In a variety of species, the preen gland is fringed by a feather tuft, and it has been suggested that the tuft functions as collection spot for oils (Jacob and Ziswiler, 1982).

Preen gland secretions tend to be a complex and highly variable mixture of natural esters, fatty acids, and long chain alcohols (Jacob, 1992). Extensive evidence from a wide range of species suggests that glandular secretions help to maintain plumage by conditioning the feathers and, in some cases, by controlling skin fungi, bacteria, and parasites through the actions of specific compounds (see discussion in Moyer et al., 2003; Shawkey et al., 2003).

568   **Chapter 24   The Chemistry of Avian Odors: An Introduction to Best Practices**

Preen glands tend to be well developed in aquatic birds, so extensive research has also focused on the potential role that preen gland secretions play in waterproofing, thermoregulation, and in creating a protective barrier on the feather surface against various forms of environmental perturbation (Salibian and Montalti, 2009). Waxy secretions chemically repel water, but even more importantly they maintain the structural integrity of the feather, which, in turn, preserves loft and the associated airspaces to provide an insulating barrier (Kostina et al., 1996). While this has not, to our knowledge, been considered in the avian literature, preserving loft would also assist in presenting scented volatiles via the feather matrix from odor sources other than the preen gland itself.

24.2.3 Sources of Avian Scent that are Not Usually Considered

While the preen gland has received considerably more attention than avian skin as a source for personal scent, we suggest that, as a holocrine secretion unit (Lucas, 1970), the skin has great potential for odor production, storage, and rapid presentation through the secretion of lipids. This hypothesis predicts that lipid secretion might be highest in tissues associated with sexual displays or other aspects of reproductive behavior where scent might be adaptive for advertising quality. There is already correlative evidence for this idea in the literature, in that lipid secretion is higher in featherless, glabrous regions of skin. At least some of these skin regions tend to be sexually dimorphic, hormonally induced ornaments that are often used in courtship displays (for example, the combs and waddles of galliformes). Lipid production is also higher in brood patches, and, in some groups (storks and ibisis) where feathers are permanently lost during maturation. This suggests a possible avenue whereby lipid production, and any associated changes in scent, could conceivably be influenced by reproductive or metabolic hormones, and even transferred to the egg or chick (Sneddon et al., 1998; Burne and Rogers, 1999; O'Dwyer and Nevitt, 2009; Cunningham and Nevitt, 2011). Such possibilities need to be explored, but interestingly, outside the field of avian chemical senses, it has already been suggested that avian skin may be pre-adapted for sequestering lipophilic dietary compounds due to its high turnover rate. This concept has perhaps been most convincingly illustrated by detailed studies of the Pitohuis birds of New Guinea. Members of this passerine group sequester dietary-derived neurotoxic steroidal alkaloids (primarily homobatrachotoxin) in their skin and feathers as an antipredator behavior (see Dumbacher et al., 1992; Dumbacher et al., 2009). According to these reports, these batrachotoxins interact with sodium channels of potential predators, and also function to chemically defend the bird from parasites. Pitohuis birds advertise their toxicity through brightly colored plumage and by emitting a "sour" scent, detectable to humans. Although the origins of this scent have not been well worked out, the dietary origins of the neurotoxins alone suggest that consumed products can influence compounds presented in avian skin, and suggests to us a mechanism by which volatile compounds from diet or digestion might also be contributing to feather odor.

The ability of many seabirds, specifically the order Procellariiforms, to regurgitate stomach oil to their young or as a defense mechanism against predators, provides yet another basis for chemical odor generation (Clarke and Prince, 1976). It was thought at one point that these stomach oils could be another type of secretion; however, the high variation in chemical composition between individuals of the same species suggested instead that stomach oil is derived from dietary sources (Lewis, 1969). For example, pristane (2, 6, 10, 14-Tetramethylpentadecane), which is common in stomach oil, is a metabolic derivative of phytol found in different forms of plankton that these birds consume. Stomach oils reflect diet, and we routinely find these compounds on feathers of Leach's storm-petrels that haven't obviously regurgitated (Prada and Nevitt, unpublished observation), suggesting that at least some of these scented products are used to perfume feathers. Whether scents from stomach oil serve as communication signals about prey resources has not been demonstrated.

Despite being one of the most obvious attributes of birds to fieldworkers, feces are commonly (and somewhat ironically) overlooked as a source of personal odor in birds. It is recognized that feces of stressed chickens differ in both consistency and smell from that of unstressed birds (Jones and Roper, 1997). Jones and Roper (1997) suggested that this scent might function as a 'non-vocal' alarm call, though this has not, to our knowledge, been demonstrated. Avian fecal products have not been studied with respect to individual odor, but should be considered in light of studies in mammals (e.g., mouse urinary proteins or MUPS; reviewed by Beynon and Hurst, 2004).

In summary, although we provide only a brief introduction to the topic, identifying the origin of any individual scent is critical to understanding its biological function in behavioral contexts, yet this is one of the greatest challenges of this field. Equally important is that once odors are produced, they are generally trapped in some form of biological matrix that needs to be considered in context. That matrix can be a physical structure, like a feather, or a secretion, such as preen wax. In birds, both of these matrices are likely to contain endogenous or intrinsic compounds produced from the individual itself, or exogenous compounds that are environmentally absorbed or derived through a behavioral or developmental process. For example, although it is produced endogenously, preen wax will inherently interact with chemicals in the environment, which together may contribute to the birds' personal scent. Microbial activity may further interact with one or both intrinsic and exogenous compounds and these interactions have yet to be worked out in any species of bird. Since identifying the chemical composition of the scent is usually an inherent goal, understanding the chemical properties that render compounds volatile should be a fundamental concern. The analytical presentation of the scent will also impact what compounds are accessible in high enough concentrations to be measured, and care must be taken to design experimental protocols and instrumental analyses that incorporate this understanding. We focus on these aspects in the next section.

24.3 WHAT MAKES SCENTS AND WHAT (USUALLY) DOES NOT?

The scent of a bird can be made up of hundreds of detectable compounds (Campagna et al., 2012), which can be confounding to analytical analysis because detection threshold by instrumental methods varies from compound to compound. The instrumental response is directly affected by the intensity (concentration) of the odor in the matrix being sampled. So, for example, a sample might yield different amounts and types of compounds when sampled from a secretion versus a feather simply because the physical matrix has different properties. Although scented compounds may be presented in a non-volatile matrix such as a secretion or on an organic substrate, whether or not a particular molecule contributes to an odor signature in air is likely to depend on the relative volatility of that compound. Volatile organic compounds (VOCs) that make up scent in air are characterized by their high vapor pressure at room temperature. These molecules have low boiling points, which is the physical property that allows the molecule to evaporate from its liquid/solid matrix and enter into the surrounding air. This measure should be reported in any serious investigation of odor chemistry in air.

A general classification of vapor pressures and volatility is presented in Table 24.2. Based on numerous studies in other terrestrial vertebrates, one might expect that, for a bird, molecules of interest should be the ones with the highest volatility because they may be the most accessible in air. Interestingly, a common feature of avian chemical odor analysis is that the detection and subsequent identification of compounds in many cases focuses on non-volatiles which may be useful in other contexts, such as working out metabolic pathways for production, but their relevance to scent is difficult to determine using standard analytical methods. In nearly every published analysis of bird odor, volatility is not reported, and rarely acknowledged (a notable exception is Reneerkens et al., 2007a, b).

Compounds of potential biological interest in birds include a range of functional groups, which impact odor. Appreciating the differences between the chemical functional groups becomes important for evaluating odor profiles, and this is a common obstacle for the non-chemist. With the exception of hydrocarbons, most of these functional groups (acids, alcohols, esters, amides, aldehydes, ketones) have a carbonyl structure that is defined by a double bond between carbon and oxygen (Table 24.3). Although the metabolic or behavioral role of many of these compounds is only beginning to be explored in most cases, we include here a brief tutorial on these common functional groups (Figure 24.1).

1. Carboxylic acids are defined by the carbonyl group and are attached to a hydroxyl group ($-OH$). Carboxylic acids are common in both feather and preen gland secretions, and it has been suggested that at least two carboxylic acids (propanoic and 2-methylpropanoic acids) are modulated by sex steroids in grey catbirds (Whelan et al., 2010).

2. Alcohols are characterized by a hydroxyl group ($-OH$) bound to a carbon atom. Alcohols tend to be derived from fatty acid reduction processes (Solomons and Fryhle, 2002), thus explaining their high occurrence in preen gland secretions and feathers.

3. When a carboxylic acid reacts with an alcohol, an ester is formed. Esters are made up of a mixture of fatty acids and alcohols. They are thought to be important to the

Table 24.2 Volatility and Vapor Pressure Classification.

	Pressure (Torr)
Volatile	1.00E+02
	1.00E+01
	1.00E+00
	1.00E−01
Semi-volatile	1.00E−02
	1.00E−03
	1.00E−04
	1.00E−05
	1.00E−06
	1.00E−07
Non-volatile	1.00E−08
	1.00E−09
	1.00E−10
	1.00E−11
	1.00E−12
	1.00E−13

570   **Chapter 24   The Chemistry of Avian Odors: An Introduction to Best Practices**

Table 24.3 Structure of Simple Functional Groups in Odorous Chemical Compounds.

Functional Group	Structure
Acid	
Alcohol	
Aldehyde	
Ketone	
Ester	
Amide	
Alkane	
Alkene	
Alkyne	
Aromatic	

biosynthesis of the complex molecules of preen secretions (e.g. Dekker et al., 1999; Reneerkens et al., 2002).

4. Amides have a carbonyl group attached to a nitrogen atom. Amides are typically absent or occur in trace amounts in studies of secretions or feathers, but this may be due to limitations of the extraction methods typically used (Campagna et al., 2012).

5. Aldehydes are defined by a carbonyl group bonded to a hydrogen atom and an alkyl chain. They tend to be characterized by strong aromas, and have been shown to be a dominant component of feather odor in some species. For example, aldehyde content in feathers from Crested auklets (*Aethia psittacula*) constituted 73.7% of the identified molecules, with aldehydes such as n-octanal and n-hexanal contributing to their unique citrus scent (Douglas et al., 2001). Possible roles include chemical defense (Douglas et al., 2001) and social signaling (Hagelin et al., 2003, 2007).

6. Ketones are defined by a carbonyl group bound to two carbon atoms. For example, ketones have

been detected in preen gland secretion, some linear methyl ketones (tridecan-2-one, tetradecan-2-one and pentadecan-2-one) have been linked to seasonal changes in dark-eyed juncos (*Junco hyemalis*; e.g., Soini et al., 2007).

7. Hydrocarbons consist entirely of hydrogen and carbon and can be further characterized into subcategories such as aromatic, cyclic alkane, alkene, alkyne, and terpene hydrocarbons. Hydrocarbons are one of the most abundant functional groups in feathers and glandular secretion samples (Campagna et al., 2012), and may be derived from bacterial or other environmental sources (Yamashita et al., 2007; Jaspers et al., 2008).

24.3.1 Analytical Techniques for the Isolation of Avian VOCs

Many researchers come into this field with strong backgrounds in behavioral ecology but without the necessary background in analytical chemistry to evaluate and compare studies, and this has led to difficulties in interpreting and comparing results. Here we provide a basic framework of the methods used in the analytical laboratory that should help researchers to obtain the most accurate results, and hopefully to better understand the analytical techniques involved. Because the heart of a good analysis is sample preparation, the three major types of extractions currently being used for both volatile and non-volatile analysis of feather and preen gland are also described in detail in this section. However, these methods could be applied to other types of biological matrices, including skin and feces.

Regardless of the sampling method used, analytical analysis requires that certain conditions be met. First and foremost, to ensure optimal accuracy in any analysis, calculating the sensitivity of the instrument is critical. The sensitivity of an instrument is the measure of its ability to detect small differences in sample concentration. Without this step, compounds cannot be identified and concentrations cannot be measured. Despite its importance, several prominent papers have been published recently without this step.

To achieve this step, the response of the analytical instrument needs to be calibrated to a known sample analyte and concentration (Skoog et al., 1998). Two of the most common ways to do this are to: (1) produce a calibration curve using a variety of known compounds, or 'standards', or (2) to use internal standards as a means of calibrating the instrumental response. A calibration curve is derived from measuring several chemical standards (i.e., chemical compounds that have been putatively identified in the sample) at different known concentrations. Collection methods used must be identical to those used for the sample. A plot of response versus concentration can then

24.3 What Makes Scents and what (Usually) Does Not?

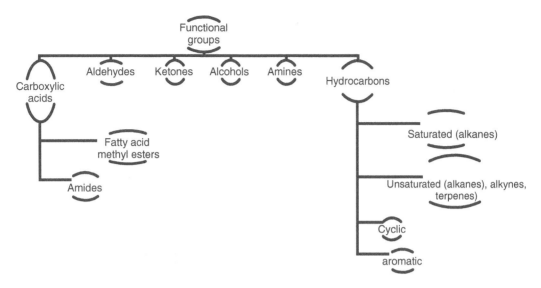

Figure 24.1 Classification of Chemical Functional Groups.

be used to determine the concentrations of those same compounds, if they are present in the sample being studied.

The second method is generally applied in cases where it is difficult to match the standards with unknown compounds. In such cases, in addition, the relative concentration of compounds can be established using an internal standard. This approach involves adding a known concentration of a chemical compound to all samples, blanks, and calibration standards. This internal standard should not be present in the sample, but should have similar chemical properties to at least some of the unknown compounds. Accurate calibration is achieved by plotting the ratio of the sample signal to the internal standard signal as a function of the concentration of the standard. This method helps to compensate for both random and systematic errors, because the sample and the internal standards will respond proportionally to any variation, and serve as a robust control (Skoog et al., 1998). Precise external or internal calibration not only provides the analyst with the capability to accurately identify the analytes in an unknown sample, but also monitors whether the analytical instrument is functioning properly.

Finally, when performing volatile identification in any sample matrix, it is standard practice to report retention indices that can be compared to known retention values. This is commonly done using Kovat's retention indices, which allows the retention times to be converted into system-independent constants. The retention index of a chemical compound can be defined as its reported retention time normalized to the retention times of adjacently eluting n-alkanes. While retention times can vary with specific chromatographic parameters such as column type, diameter, thickness and/or length, the calculated retention indices are independent of these parameters and allow compounds to be compared and their identity to be further confirmed.

In terms of optimal practices for handling and processing samples, when working in odor analysis, it is critical to wear nitrile gloves to collect or handle samples to prevent cross-contamination of any kind. This also requires extreme care in the sterilization and cleaning of all glassware, including the vials used to collect the samples, to make sure volatiles being detected originate from the sample source and not from background contamination. In the search for individual volatile odor markers from a specific population, species, gender, or individual, researchers in this field also tend to misunderstand the importance of sampling in replicate. It is well established from studies in humans, for example, that establishing a baseline for an individual is critical before individuals within a population can be compared (Gallagher et al., 2008; Curran et al., 2010; Prada et al., 2011). Thus, in the context of avian odor signatures, multiple samples from the same bird should be collected in replicate to capture the most accurate chemical odorprint from the individual being studied. Without replicate samples, it is difficult to determine if the detected compounds are characteristic of the individual, or due to contamination.

Although different extraction procedures have been used to study odor profiles of feather and preen gland secretions, detailed techniques are rarely reported. This suggests that future reviews comparing results among studies should be more explicit about techniques. Table 24.1 provides a list of recent studies that have used various extraction techniques to study avian VOCs from matrices such as feathers and glandular secretions. To familiarize non-chemists interested in avian olfaction with the advantages and disadvantages of methodologies, we present an overview of the three most commonly used extraction techniques.

24.3.2 Solvent Extraction

Solvent extraction uses organic solvents at room temperature, and is the most common and conventional separation method. This method is used to extract potentially scented compounds from biological matrices such as glandular secretions or feathers (Table 24.1). This method has been frequently used to study bird odors (Amo et al., 2012; Bonadonna et al., 2007; Dekker et al., 1999; Douglas et al., 2001; Karlsson et al., 2010; Kolattukudy and Sawaya, 1974; Leclaire et al., 2011, 2012; Mardon et al., 2010, 2011; Piersma et al., 1999; Reneerkens et al., 2002, Sandilands et al., 2004; Sweeney et al., 2004; Thomas et al., 2010; Zhang et al., 2008, 2010) and offers the advantage of being an efficient and robust method to extract complex, high molecular weight molecules. Samples can easily be preserved for analyses because the extraction does not need to be done immediately. The major variable to consider when performing this type of extraction is the polarity and type of solvent used, and this will depend on the types of compounds to be extracted and identified. However, this extraction method has several drawbacks for studying avian scent. Perhaps most importantly, if the goal of a study is to identify VOCs, it may not be the most appropriate method for the question because solvent extraction does not specifically target VOCs. Consequently some of the extracted compounds will not contribute to the odor. Instead, solvent extraction targets more complex, sometimes viscous, high molecular weight compounds that are not volatile at room temperature. The non-volatile constituents may also potentially interfere with the recovery of semi-volatile compounds, which, together with solvent impurities, can compromise the analysis of the sample. Finally, organic solvents can pose health hazards when used under field conditions, and transport and disposal can be costly.

24.3.3 Solid Phase Microextraction

Solid Phase Microextraction (SPME) is a useful technique for the rapid and accurate identification of volatile compounds in the air space ("headspace") surrounding either a solid or liquid matrix. In headspace extraction, the headspace should be as small as possible as this increases the concentration of analyte and thus reduces extraction time. SPME has been used effectively to measure VOCs from mammals (Cablk et al., 2012), including human skin (Curran et al., 2010; Prada et al., 2011). In the context of birds, SPME has been used to extract odors from feathers (Douglas et al., 2001; Hagelin et al., 2003; Williams et al., 2003) and preen gland secretion (Burger et al., 2004;

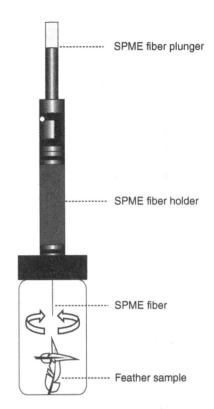

Figure 24.2 Solid phase microextraction schematic.

Shaw et al., 2011; Whelan et al., 2010). SPME is typically used for the collection and subsequent identification of VOCs by chromatographic techniques, and has several advantages in that it is easy to use, fast, portable, and inexpensive (Penalver et al., 1999). SPME employs a fiber, which is a short, thin, solid rod of fused silica. These fibers are typically 1 cm in length and have approximately 0.11 mm as an outer diameter. SPME fibers are coated with different types of sorbents that vary by polarity and thickness; however fiber selection should be optimized for the types of compounds of interest (Table 24.4). A general consideration is that "like dissolves like" when selecting a fiber for an application. Polydimethylsiloxane (PDMS) and polyacrylate (PA) were the first fiber coatings utilized for SPME fibers. PDMS is apolar and has a high affinity for non-polar compounds. The PA fiber coatings are more polar and thus extract polar compounds. Fiber coatings containing more porous and adsorbent materials include blends of divinylbenzene (DVB) or Carbowax (CW) with PDMS. Fibers used for SPME technique can also be categorized as either absorption or adsorption fibers. Absorption (film) fibers extract analytes whereas adsorbent (particle) fibers physically trap or chemically react with the analytes. The latter are made from a porous material and analytes compete for sites on the fiber coating. Adsorbent fibers have a limited capacity (Wercinski, 1999). Table 24.4 provides a concise description of the most recommended

24.4 Examples of Appropriate Practices 573

Table 24.4 Summary of Fiber Coatings available for SPME Fibers.

Fiber Coating	Fiber Type	Application
Polydimethylsiloxane (PDMS)	Absorption	Used for non-polar organic compounds; analyte group volatiles MW 60–275
Polyacrylate (PA)	Absorption	Used for polar organic compounds; MW 80–300
Polydimethylsiloxane/ divinylbenzene (PDMS/DVB)	Adsorption	Used for aromatic hydrocarbons and small volatile analytes such as solvents; MW 50–300
Carboxen/ Polydimethylsiloxane (CAR/PDMS)	Adsorption	Used for gases and low MW compounds MW 30–225
DVB/CAR/PDMS	Adsorption	Used for volatiles and semi volatiles C3–C20; MW 40–275

fiber type depending on the volatility and molecular weight of the target analyte.

The fiber is attached to a metal rod, which is protected by a metal sheath that allows the fiber to be covered when not in use. For sampling purposes, the SPME fiber is withdrawn from the syringe needle and, by pressing down on the plunger, the fiber is then exposed to the headspace above the sample of interest for a predetermined time to absorb analytes. After sampling has concluded, the fiber is retracted and directly injected into a gas chromatograph (GC) for instrumental analysis (King et al., 2003). A schematic diagram of the SPME fiber assembly is shown in Figure 24.2 depicting the extraction of a feather sample. To our knowledge, Douglas et al. (2001) were the first investigators to use the SPME technique to extract odor volatiles from a feather matrix; they used fibers coated with 65 μm polydimethylsiloxane/Divinylbenzene coatings to measure VOCs from Crested auklet *(Aethia psittacula)* feathers. SPME fibers with 75 μm Carboxen-PDMS coatings have more recently been used for the collection and analysis of VOCs from uropygial glandular secretions of both *in situ* and dissected glands of euthanized Gray catbirds *(Dumetella carolinensis*; Shaw et al., 2011; Whelan et al., 2010).

24.3.4 Stir-bar Sorptive Extraction

A second, newer form of solvent-less extraction is the stir bar sorptive extraction (SBSE) technique, commercially known as Twister®. In SBSE, the extraction is performed using a magnetic stir bar coated with sorbent material. Currently, only polydimethylsiloxane and modified ethylene glycol phases are available. The design of the stir bar allows the device to contain more sorbent material on its surface, thus providing space (capacity) for analyte

extraction up to one hundred times greater than in SPME (Baltussen et al., 2002). SBSE can be used to extract VOCs from either solid or liquid matrices. After extraction procedures, the bar is removed from the sample vial and placed in a thermal desorption inlet of a GC for analyte desorption. The extracted analytes are cryofocused at the front of the column for GC analysis. One major disadvantage when using SBSE is that only the thermal desorption step is currently automated, thus rendering this technique cumbersome for large sample sizes (Kloskowski et al., 2007). SBSE has been successfully used to analyze VOCs from uropygial gland and the surface of wing feathers (via direct wiping of the SBSE on the wing) of dark eyed juncos *(Junco hyemalis;* Soini et al., 2007; Whittaker et al., 2010).

24.4 EXAMPLES OF APPROPRIATE PRACTICES

Chemical ecology of birds is in its infancy as a discipline and researchers are grappling with the problem of how to select appropriate techniques. Reviewers and editors rarely have the expertise to critically evaluate the validity of the analytical chemistry used, and literature reviews commonly fail to recognize fundamental differences or limitations of the techniques used. Here we review examples of chemical studies chosen to highlight the techniques we have described. In our opinion, these particular studies exemplify best practices in action.

24.4.1 A Study of Anti-predator Behavior in Red Knots

A study focusing on Red knot *(Calidris canutus)* anti-predator behavior by Reneerkens and co-workers

574   Chapter 24   The Chemistry of Avian Odors: An Introduction to Best Practices

exemplifies what we consider to be the optimal convergence of biology, chemistry, and a solid experimental design rooted in natural history and a simple observation by the authors. Sandpipers (family: Scolopacidae) are ground nesting birds, which make them vulnerable to mammalian predators, and yet hunting dogs were often observed to overlook vulnerable females on eggs. The authors wanted to know whether this was because the nesting Red knots' scent was somehow masked. This hypothesis predicted that the odor chemistry of a bird changed predictably during the breeding season.

Reneerkens and coworkers went on to test whether the composition of preen oil changed with breeding phenology. They first evaluated the preen oil composition of 19 different sandpiper species using a novel analytical approach for this application – gas chromatography and tandem mass spectrometry (GC- MS/MS; Dekker et al., 2000). Gas chromatography/mass spectrometry (GC/MS) is more commonly used to analyze VOCs, but GC/MS targets fatty acids and alcohols that are released after the sample is hydrolyzed. GC- MS/MS allowed the investigators to elucidate the chemical structure of the waxy esters in the preen gland matrix, and how it contributed to scent. They demonstrated a shift from monoesters (lower molecular weight/more volatility) to diesters (higher molecular weight/less volatility) in preen gland secretions during the breeding season, and suspected that one function of this switch might be olfactory camouflage for nesting birds. The gas chromatograms were characteristic for the wax type, and offered a robust demonstration of seasonal shift in the chemistry of preen gland secretion with a clear functional advantage to the bird. Reneerkens et al. (2007a, b) note, however, that the wax composition is not volatile and thus the monoesters themselves may not be present in the odor. This result does, however, highlight that simple modifications in the chemical structure of these compounds can contribute to avian scent in an evolutionarily adaptive context, and this was a remarkable result. As a further test of their hypothesis, they used a sniffer dog to confirm the low volatile character of the diester secretions (Reneerkens et al., 2007a, b). The canine had a greater difficulty detecting this heavy weight diester mixture thus providing experimental support for olfactory camouflage using an appropriate biological detector for the question at hand – the canine nose.

24.4.2 Dark-eyed Juncos and Crested Auklets

Reneerken's studies are currently state-of-the-art in terms of marrying a relevant hypothesis-driven question to solid chemical and field investigations. Investigations into other avian systems are much less developed, but offer great promise. As mentioned above, Douglas et al. (2001) were

the first to report the use of solvent-less SPME extraction technique for headspace analysis of feathers from Crested auklets *(Aethia cristatella)*. In this study, the fiber was positioned near the specimen's neck and extracted for a period of 4.5 hours. Interestingly, they also performed a second extraction of feather samples using traditional solvent extraction (either methylene chloride or methanol) for comparative purposes. Although identification of specific compounds (e.g., hexanal) differed between methods, in both cases aldehydes such as n-octanal, n-decanal, and Z-4-decenal were detected, thereby confirming SPME as a novel chemical approach for studying feather odor.

More recently, Soini et al. (2006; 2007) were the first to apply the Twister (stir bar) method to study feather and preen gland odors. They used the stir bar technique to examine VOCs from biological samples, including human skin, fruits, fingerprints, and bird feathers (Soini et al., 2006). They used embedded internal standards to achieve highly reproducible and quantitative results for a wide variety of trace compounds. A year later, a more complete investigation focused on preen gland secretions from dark-eyed junco *(Junco hyemalis)* using a quantitative stir bar approach (Soini et al., 2007). The study provided insights into the effects of seasonal changes on chemical composition. Authentic standards were used for the verification of detected compounds, as well as an internal standard (7-tridecanone). Although studies were done on captive birds, replicate sampling was conducted for 20 individuals for both breeding and non-breeding study groups, thereby taking into consideration intra-individual variation over time. Multiple sampling yields a greater level of certainty in the analysis and identification of the detected VOCs. The GC-MS VOC profiles obtained from the secretions included linear alcohols, as well as a number of aldehydes, acids, and methyl ketones, thereby demonstrating the potential of the technique for volatile analysis on an important model species in field endocrinology. Figure 24.3 shows a brief comparison of chromatographic results obtained through the various extraction techniques.

24.5 CONCLUSIONS

The scent of a bird is generally a complex fragrance made up of a suite of volatile organic compounds (VOCs). Identifying these VOCS and their relative contributions to a species' or individual's scent is central to understanding their chemical nature, their potential biological source, and, eventually, whether or how they influence behavior. Not surprisingly, the instrumental analysis of VOCs is a rapidly expanding field in the race to find compounds of potential biological significance among birds, but robust methodological standards for odor collection and analysis are still being established. In this review we provided a

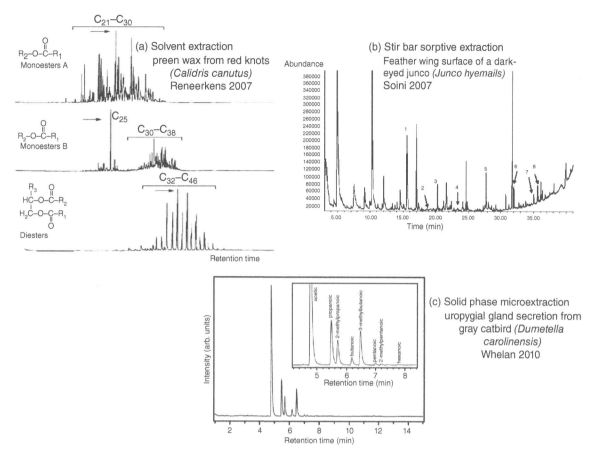

Figure 24.3 Representative chromatographic results via different extraction techniques.

brief overview of the general sources of odors in birds, a description of the functional groups of interest for the 'non-chemist,' and a critique of the instrumental methods currently being used for this purpose.

The science of avian scent is growing in popularity, and this offers great promise for the future of this discipline. Still, much of the published chemical literature is preliminary due to inconsistencies or misapplication of analytical chemical techniques. We wrote this chapter to provide information to ornithologists who, like Audubon, are passionate about avian biology but may not have the foundation in the chemical classifications, techniques, and best practices necessary to most effectively examine hypotheses about avian scent. The wide array of avian species, chemical analytical tools, methods of extractions, and statistical methods present obstacles to biologists seeking to work in this field. This is a brief attempt to introduce foundational concepts that may be useful in getting around some of them.

To summarize, it is critical to select an appropriate species and extraction technique for the hypothesis being tested, and to analyze replicate samples. Despite the considerable extra time and effort required, chemical calibrations are necessary to translate chromatograms into credible data. Finally, characterizing or differentiating one chemical profile from another does not, in itself, imply a function or a role for behavior. Differences in odor profiles, and any potential role in communication, needs to be demonstrated by an appropriate biological detector in an ecological context, in addition to using analytical detectors, before a scent can be called a cue. We hope this attempt to highlight best practices will help the field to progress to the point where avian scent can be examined in a comparative context, from production and regulation, to adaptation and function in the natural world.

REFERENCES

Amo, L., Aviles, J. M., Parejo, D., et al. (2012). Sex recognition by odour and variation in the uropygial gland secretion in starlings. *J. Anim. Ecol.* 81: 605–613.

Baltussen, E., Cramers, C. A., Sandra, P. J. F. (2002). Sorptive sample preparation- a review. *Anal. Bioanalyt. Chem.* 373: 3–22.

Beynon, R. J. and Hurst, J. L. (2004). Review: Urinary proteins and the modulation of chemical scents in mice and rats. *Peptides.* 25: 1553–1563.

Bonadonna, F., Miguel, E., Grosbois, V., et al. (2007). Individual odor recognition in birds: an endogenous olfactory signature on petrel's feathers? *J. Chem. Ecol.* 33: 1819–1829.

Burger, B. V., Reiter, B. Borzyk, O., DuPlessis, M. A. (2004). Avian exocrine secretions I. Chemical characterization of the volatile fraction of the uropygial secretion of the Green woodhoopoe, *Phoeniculus purpureus. J. Chem. Ecol.* 30: 1603–1611.

Burne, T. H. J. and Rogers, L. J. (1999). Changes in olfactory responsiveness by the domestic chick after early exposure to odorants. *Anim. Behav.* 58: 329–336.

Cablk, M. E., Szelagowski, E. E., and Sagebiel, J. C. (2012). Characterization of the volatile organic compounds present in the headspace of decomposing animal remains, and compared with human remains. *Forensic Sci. Int.* 220: 118–125.

Campagna, S., Mardon, J., Celerier, A., and Bonadonna, F. (2012). Potential semiochemical molecules from birds: a practical and comprehensive compilation of the past 20 years studies. *Chem. Senses.* 37: 3–25.

Clark, A. and Prince, P. A. (1976). The origin of stomach oil in marine birds: analyses of the stomach oil from six species of subantarctic procellariiform birds. *J. Exper. Marine Biol. Ecol.* 23: 15–30.

Cunningham, G. B. and Nevitt, G. A. (2011). Evidence for olfactory learning in procellariiform seabird chicks. *J. Avian Biol.* 42: 85–88.

Curran, A. M., Prada, P. A., and Furton, K. G. (2010). The differentiation of the volatile organic signatures of individuals through SPME-GC/MS of characteristic human scent compounds. *J. Forensic Sci.* 55: 50–57.

DeBose, J. L. and Nevitt, G. A. (2008). The use of odors at different spatial scales: comparing birds with fish. *J. Chem. Ecol.* 34: 867–881.

Dekker, M. H. A., Piersma, T., and Damster, J. S. S. (2000). Molecular analysis of intact preen waxes of *Calidris canutus* (Aves: Scolopacidae) by gas chromatography/mass spectrometry. *Lipids.* 35: 533–541.

Dekker, M. H. A., Piersma, T., and Sinninghe Damste J. S. (1999). Molecular analysis of intact preen waxes of *Calidris canutus* (Aves: Scolopacidae) by Gas Chromatography/Mass Spectrometry. *Lipids.* 35: 533–541.

Douglas, H. D., Co, J. E., Jones, T. H., and Conner, W. E. (2001). Heteropteran chemical repellents identified in the citrus odor of a seabird (Crested auklet: *Aethia cristatella*): Evolutionary convergence in chemical ecology. *Naturwissenschaften.* 88: 330–332.

Dumbacher, J. P., Beehler, B. M., Spande, T. E et al. (1992). Homobatrachotoxin in the genus *Pitohui*: chemical defense in birds? *Science.* 258: 799–801.

Dumbacher, J. P., Menon, G. K., and Daly, J. W. (2009). Skin as toxin storage organ in the endemic New Guinea genus *Pitohui*. *Auk.* 126(3): 520–530.

Elder, W. H. (1954). The oil gland of birds. *Wilson Bulletin.* 66: 6–31.

Gallagher, M., Wysocki, C. J., Leyden, J. J. et al. (2008). Analysis of volatile organic compounds from human skin. *British J. Dermatol.* 159: 780–791.

Gill, F. B. (1994). *Ornithology.* New York: W. H. Freeman and Company.

Hagelin, J. C. (2007). Odors and chemical signaling. In *Reproductive Behavior and Phylogeny of Aves*. Vol. 6B, B. G. M. Jamieson (Ed.) Enfield, NH: Science Publishers, pp. 76–119.

Hagelin, J. C. and Jones, I. L. (2007). Bird odors and other chemical substances: a defense mechanism or overlooked mode of intraspecific communication? *The Auk.* 124: 741–761.

Hagelin, J. C., Jones, I. L., and Rasmussen, L. E. L. (2003). A tangerine-scented social odour in a monogamous seabird. *Proc. R. Soc. B.* 270: 1323–1329.

Jacob, J. (1992). Systematics and the analysis of integumental lipids. *Bull. Br. Orn. Club.* 112A: 159–167.

Jacob, J. and Ziswiler, V. (1982). The uropygial gland. In *Avian Biology*, Vol. VI, Farner, D. S., King, J. R., and Parkes, K. C. (Eds.), New York: Academic Press, pp. 199–324.

Jaspers, V. L. B., Covaci, A., Deleu, P. et al. (2008). Preen oil as the main source of external contamination with organic pollutants onto feathers of the common magpie (*Pica pica*). *Environ. Int.* 34: 741–748.

Johnston, D. W. (1988). A morphological atlas of the avian uropygial gland. *Bull. Br. Mus. (Nat. Hist.) Zool.* 54: 199–259.

Jones, R. B. and Roper, T. J. (1997). Olfaction in the domestic fowl: A critical review. *Physiol. Behav.* 62: 1009–1018.

Karlsson, A. C., Jensen, P., Elgland, M. et al. (2010). Red junglefowl have individual body odors. *J. Exp. Biol.* 213: 1619–1624.

King, A. J., Readman, J. W., and Zhou, J. L. (2003). The application of solid-phase micro-extraction (SPME) to the analysis of polycyclic aromatic hydrocarbons (PAHs). *Environ. Geochem. Health*, 25: 69–75.

King, A. S. (1981). Cloaca. In *Form and Function in Birds*, Vol. 2A. King, S. and Mc-Lelland, J. (Eds.). London: Academic Press, pp. 63–105.

Kloskowski, A., Chrzanowski, W., and Pilarczyk, M. (2007). Modern techniques of sample preparation for determination of organic analytes by gas chromatography. *Crit. Rev. Analyt. Chem.* 37: 15–38.

Kolattukudy, P. E. and Sawaya, W. N. (1974). Age dependent structural changes in the diol esters of uropygial glands of chicken. *Lipids.* 9: 290–292.

Kostina, G. N., Sokolov, V. E., Romanenko, E. V. et al. (1996). Hydrophobicity of penguin feather structures (*Aves, Sphenisciformes*). *Zool. Zhur.*75: 237–248.

Leclaire, S., Merkling, T., Raynaud, C. et al. (2011). An individual and a sex odor signature in kittiwakes? Study of the semiochemical composition of preen secretion and preen down feathers *Naturwissenschaften.* 98: 615–624.

Leclaire, S., Merkling, T., Raynaud, C. et al. (2012). Semiochemical compounds of preen secretion reflect genetic make-up in a seabird species. *Proc. R. Soc. B.* 279: 1185–1193.

Lewis, R. W. (1969). Studies on the stomach oils of marine animals II. Oils of some procellariiform birds. *Comp. Biochem. Physiol.* 31: 725–731.

Lucas, A. M. (1970). Avian functional anatomical problems. *Fed. Proc.* 29: 1641–1648.

Lucas, A. M. (1980). Lipoid secretion by the body epidermis in avian skin. In *The Skin of Vertebrates*, Spearman, R. I. C. and Riley P. A. (Eds.), London: Symp. Linn. Soc. Lond., No. 9. Academic Press, pp. 33–45.

Luxan, M. P. and Dorrego, F. (1999). Reactivity of earth and synthetic pigments with linseed oil. *Surface Coatings Intl.* 82: 390–402.

Mardon, J., Saunders, S. M., Anderson, M. J. et al. (2010). Species, gender, and identity: cracking petrel's sociochemical code. *Chem. Senses.* 35: 309–321.

Mardon, J., Saunders, S. M., and Bonadonna, F. (2011). From preen secretion's to plumage: the chemical trajectory of Blue petrel's *Halobaena caerulea* social scent. *J. Avian. Biol.* 42: 29–38.

Menon, G. K. (1984). Glandular functions of Avian Integument: An overview. *Yamashina Choruigaku Zasshi.*16: 1–12.

Menon, G. K., Aggarwal, S. K., and Lucas A. M. (1981). Evidence for the holocrine nature of lipoid secretion by avian epidermal cells: A histochemical and ultrastructural study of rictus and the uropygial gland. *J. Morphol.* 167: 185–199.

Menon, G. K. and Menon, J. (2000). Avian epidermal lipids: Functional considerations and relationship to feathering. *Am. Zool.* 40: 540–552.

Menon, G. K. and Salinukul, N. (1989). Ceruminous glands in ear canal of Domestic Fowl: Morphology, histochemistry and ultrastructure. *Zoolog. Anz.* 222: 110–121.

References

Menon, G. K., Salinukul, N., and Jani, M. B. (1987). Epidermal glands of the vent during breeding season in the House Sparrow: Histology, glycosaminoglycan content and ultrastructure. *Zoolog. Anz.* 218: 93–103.

Moyer, B. R., Rock, A. N., and Clayton, D. H. (2003). An experimental test of the importance of preen oil in rock doves (*Columba livia*). *Auk.* 120: 490–496.

Nevitt, G. A. (2000). Olfactory foraging by Antarctic Procellariiform seabirds: life at high Reynolds numbers. *Biol. Bull.* 198: 245–253.

Nevitt, G. A. (2008). Sensory ecology on the high seas: the odor world of Procellariiform seabirds. *J. Exper. Biol.* 211: 1706–1713.

Nevitt, G. A. and Bonadonna, F. (2005). Seeing the world through the nose of a bird: new developments in the sensory ecology of Procellariiform seabirds. *Mar. Ecol. Prog. Ser.* 287: 263–307.

Nevitt, G. A. and Hagelin, J. C. (2009). Olfaction in birds: a dedication to the pioneering spirit of Bernice Wenzel and Betsy Bang. *Intl. Symp. Olfaction and Taste: Ann. N.Y. Acad. Sci.* 1170: 424–427.

O'Dwyer, T. W. and Nevitt, G. A. (2009). Individual odor recognition in Procellariiform chicks. *Int. Symp. Olfaction and Taste: Ann. N.Y. Acad. Sci.* 1170: 442–446.

Pass, D. A. (1995). Normal anatomy of the avian skin and feathers. *Semin. Avian Exot. Pet Med.* 4(4): 152–160.

Penalver, A., Pocurull, E., Borrull, F., and Marce, R. M. (1999). Trends in solid phase Microextraction for determining organic pollutants in environmental samples. *TrAC.* 18: 557–568.

Piersma, T., Dekker, M., and Sinninghe Damste J. S. (1999). An avian equivalent of make-up? *Ecol. Letts.* 2: 201–203.

Prada, P. A., Curran, A. M., and Furton, K. G. (2011). The evaluation of human hand odor volatiles on various textiles: a comparison between contact and noncontact sampling methods. *J. Forens. Sci.* 56: 866–881.

Quay, W. B. (1967). Comparative survey of the anal glands of birds. *Auk* 84: 379–389.

Rajchard, J. (2007). Intraspecific and interspecific chemosignals in birds: a review. *Vet. Med-Czech.* 52: 385–391.

Rajchard, J. (2008). Exogeneous chemical substances in bird perception: a review. *Vet. Med-Czech.* 53: 412–419.

Rajchard, J. (2010). Biologically active substances of bird skin: a review. *Vet. Med-Czech.* 55: 413–421.

Reneerkens, J., Almeida, J. B., Lank, D. B., et al. (2007a). Parental role division predicts avian preen wax cycles. *Ibis.* 149: 721–729.

Reneerkens, J., Piersma, T., and Damste, J. S. S. (2007b). Expression of annual cycles in preen wax composition in Red knots: constraints on the changing phenotype. *J. Exp. Zool.* 307A: 127–139.

Reneerkens, J., Piersma, T., and Sinninghe Damste, J. S. (2002). Sandpipers (Scolopacidae) switch from monoester to diester preen waxes during courtship and incubation, but why? *Proc. R. Soc. B.* 269: 2135–2139.

Roper, T. J. (1999). Olfaction in birds. In *Advances in the Study of Behavior*, Slater, P. J. B., Rosenblat, J. S., Snowden, C. T., and Roper, T. J. (Eds). London: Academic Press, pp. 247–332.

Salibian, A. and Montalti, D. (2009). Physiological and biochemical aspects of the avian uropygial gland. *Braz. J. Biol.* 69(2): 437–446.

Sandilands, V., Powell, K., Keeling, L., and Savory, C. J. (2004). Preen gland function in layer fowls: Factors affecting preen oil fatty acid composition. *Br. Poult. Sci.* 45(1): 109–115.

Shaw, C. L., Rutter, J. E., Austin, A. L., et al. (2011). Volatile and semivolatile compounds in Gray catbird uropygial secretions vary with age and between breeding and wintering grounds. *J. Chem. Ecol.* 37: 329–339.

Shawkey, M. D., Pillai, S. R., and Hill, G. E. (2003). Chemical warfare? Effect of uropygial oil on feather-degrading bacteria. *J. Avian Biol.* 34: 345–349.

Skoog, D. A., Holler, F. J., and Nieman, T. A. (1998). *Principles of Instrumental Analysis*. USA: Thomson Learning, Inc.

Sneddon, H., Hadden, R., and Hepper, P. G. (1998). Chemosensory learning in the chicken embryo. *Physiol. Behav.* 64: 133–139.

Soini, H. A., Bruce, K. E., Klouckova, I. et al. (2006). In situ surface sampling of biological objects and preconcentration of their volatiles for chromatographic analysis. *Analyt. Chem.* 78: 7161–7168.

Soini, H. A., Schrock, S. E., Bruce, K. E. et al. (2007). Seasonal variation in volatile compound profiles of preen gland secretions of the Dark-eyed Junco (*Junco hyemalis*). *J. Chem. Ecology.* 33: 183–198.

Solomons, T. W. G. and Fryhle, C. B. (2002). *Organic Chemistry*. New York: John Wiley & Sons, Inc.

Souder, W. (2005). *Under a Wild Sky: John James Audubon and Making of the Birds of America*. New York: North Point Press.

Stager, K. E. (1964). The role of olfaction in food location by the turkey vulture (*Cathartes aura*). *Los Angeles City. Museum Contribution Science.* 81: 3–63.

Stettenheim, P. R. (2000). The integumentary morphology of modern birds – an overview. *Amer. Zool.* 40: 461–477.

Sweeney, R. J., Lovette, I. J., and Harvey, E. L. (2004). Evolutionary variation in feather waxes of passerine birds. *The Auk.* 121: 435–445.

Thomas, R. H., Price, E. R., Seewagen, C. L. et al. (2010). Use of TLC-FID and GC-MS/FID to examine the effects of migratory state, diet, and captivity on preen wax composition in White-throated sparrows *Zonotrichia albicollis. Ibis.* 152: 782–792.

Wallraff, H. G. (2004). Avian olfactory navigation: its empirical foundation and conceptual state. *Anim. Behav.* 67: 189–204.

Wenzel, B. M. (2007). Avian olfaction: then and now. *J. Ornithol.* 148: S191–194.

Wercinski, S. A. (1999). *Solid Phase Microextraction A Practical Guide*. New York: Marcel Dekker, Inc.

Whelan, R. J., Levin, T. C., Owen, J. C., and Garvin, M. C. (2010). Short-chain carboxylic acids from Gray catbird (*Dumetella carolinensis*) uropygial secretions vary with testosterone levels and photoperiod. *Comp. Biochem. Physiol.* 156: 183–188.

White, R., Phillips, M. R., Thomas, P., and Wuhrer, R. (2006). *In-situ* investigation of discolouration processes between historic oil paint pigments. *Microchim. Acta.* 155: 319–322.

Whittaker, D. J., Soini, H. A., Atwell, J. W. et al. (2010). Songbird chemosignals: volatile compounds in preen gland secretions vary among individuals, sexes, and populations. *Behav. Ecol.* 21: 608–614.

Williams, C. R., Kokkinn, M. J., and Smith, B. P. (2003). Intraspecific variation in odor-mediated host preference of the mosquito *Culex annulirostris. J. Chem. Ecol.* 29: 1889–1903.

Wrench, R., Hardy, J. A., and Spearman, R. I. C. (1980). Sebokeratocytes of avian epidermis with mammalian comparisons. In *The Skin of Vertebrates*, Spearman, R. I. C. and Riley, P. A. (Eds.), London: Academic Press.

Yamashita, R., Hideshige, T., Murakami, M. et al. (2007). Evaluation of non-invasive approach for monitoring PCB pollution of seabirds using preen gland oil. *Environ. Sci. Technol.* 41: 4901–4906.

Zhang, J. X., Wei, W., and Zhang, J. H. (2008). Preen gland-secreted alkanols enhance male attractiveness in parrots. *Nature.* http://hdl.handle.net/10101/npre.2008.2305.1

Zhang, J. X., Wei, W., Zhang, J. H., and Yang, W. H. (2010). Uropygial gland-secreted alkanols contribute to olfactory sex signals in budgerigars. *Chem. Senses.* 35: 375–382.

Chapter 25

Olfactory Communication in Rodents in Natural and Semi-Natural Habitats

DANIEL W. WESSON

25.1 INTRODUCTION

Olfactory communication in rodents involves the passive or active release of a chemical signal into the environment for the purpose of influencing behavior. This signal must have the capacity to elicit neural responses in one of the several systems for olfaction in the brain, including most prominently the main and accessory olfactory systems (see Chapters 8, 10, 51, and 52). Both active and passive odor cue emission are governed by sex, hormonal status, age, behavioral state, past experience, and other factors (nutrient intake, illness) which together provide a highly dynamic source of signals not only in their release but also in the behaviors they elicit. Indeed, in "intraspecific" communication, the dynamics of these signals translate into numerous possible "messages" being conveyed onto conspecifics wherein they guide proper behavioral responses.

Traditionally, research into odor-guided behaviors in rodents has explored the occurrence of discrete stereotyped behaviors in artificially reduced laboratory environments. This most commonly occurs within the context of a small cage or an enclosed testing arena. This has been a highly successful framework for studies into odor-behavior relationships, resulting in elucidation of fundamental aspects of odor learning, odor detection, and odor discrimination. The laboratory environment results in an optimal context to collect behavioral and physiological data and provides control over environmental variations (temperature, light cycle, nutrients, other animals) which compounds the experimenter's dependent variables. This is especially useful for studies of olfaction since not only can "real-world"

environmental factors affect the emission of odors and the motivation to gather odor information, but it can also have a direct impact on the odor itself. For instance, if one were to measure the impact of female ground squirrel scent marking on male ground squirrel behavior, a finding that the male was not attracted to the female scent mark could be interpreted as meaning it is not a socially-relevant cue. However, if it was a cold and windy day, now perhaps an alternative hypothesis could be approached: the scent mark dissipated in the wind and failed to evoke behavior in the male because he never received the signal. Compounding matters further, perhaps the male recently mated and thus is not in an appetitive motivational state. These and other issues have resulted in the commonplace occurrence of olfactory research in rodents to occur within the controllable confines of a laboratory.

However, rodents did not evolve to perform complex behaviors in the context of limited cage environments. More particularly, rodents did not evolve remarkable mechanisms of olfactory communication for them to occur within the confines of a simple laboratory cage. Indeed, even basic aspects of social behaviors are altered dramatically, depending upon the physical context (Wills et al., 1983). Arguably, some forms of odor communication, such as the scent marking for signaling social status, are for most purposes useless in a laboratory cage where social status of a conspecific is worked out within a short period of group housing and, unless a new conspecific is introduced, not reassessed by cage-mates. Thus, it is difficult to imagine conclusions on odor-behavior relationships are as ethologically-relevant as possible when derived from experiments wherein animals are not only housed,

---

Handbook of Olfaction and Gustation, Third Edition. Edited by Richard L. Doty.
© 2015 Richard L. Doty. Published 2015 by John Wiley & Sons, Inc.

but also tested, in a highly artificial environment. In this context, not only are we as researchers likely missing highly significant information about how the environment impacts behavior, but also how animals have evolved to most optimally perform olfactory communication.

The purpose of this chapter, therefore, is to introduce main concepts of rodent olfactory neuroethology and draw upon some particularly elegant examples of studies in rodents within natural and semi-natural environments to explore principles of ethologically-relevant rodent olfactory communication. Examples from laboratory-based studies will also be used throughout to exemplify cases of olfactory communication not yet reported in natural contexts or those methodologically difficult to study in natural contexts. Together, these studies will be used to highlight means by which animals use odors to communicate with each other and other animals, as well as to validate (with some exceptions) laboratory-based studies. This chapter is not meant to be an exhaustive review of all studies on ethologically-relevant olfactory communication, and sincere apologies are made to those researchers whose studies I have not included, yet have also contributed to a greater understanding of the olfactory basis of rodent behavior. For those interested in further reading on complimentary aspects of odor-guided behavior in rodents and other mammals, see (Doty, 1986, 2010; Brennan et al., 1990; Leon, 1992; McClintock, 1998; Meredith, 1998; Slotnick, 2001; Wilson and Linster, 2008; Keller et al., 2009; Wachowiak, 2011).

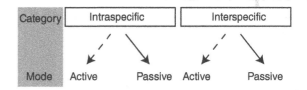

Figure 25.1 Framework for the understanding rodent olfactory communication. Rodents emit chemical signals which may be received by either members of their own species (intraspecific), or members of other species (interspecific). The emission of odors in both of these categories may be active (volitional) or passive. Rodents have evolved to display and also functionally utilize these different forms of olfactory communication adaptively.

the probability of a desired behavior to occur, whether to mate or to flee.

While the emission of olfactory signals may have initially evolved for intraspecific purposes, the release of these cues can have unwanted consequences. Indeed, olfactory cues enhance predation. In other cases, olfactory cues may signal to a predator unsuitability as a food source, and thereby prolong survival. Alternatively, in a more complex sequence of events, perception of a predator odor signal or perception of an odor from a stressed conspecific may result in an "alarm" signal in rodents to aid conspecifics in survival. These forms of olfactory communication, wherein chemical signals are emitted by a member of one species to be received by a member of another species, are referred to as interspecific and will be discussed in detail in Section 25.4 of this chapter.

25.2 INTRA- VS INTER-SPECIES OLFACTORY COMMUNICATION

As an initial framework for the understanding of rodent olfactory communication (Figure 25.1), it is useful to consider that rodents emit chemical signals which may be received by either members of their own species, or members of other species. Transfer of olfactory information between members of the same species can be referred to as intraspecific olfactory communication. Intraspecific olfactory communication is by far the most commonly studied form and entails commonly observed sociosexual behaviors including scent marking, face-to-face sniffing, anogenital investigation, and maternal behaviors of pup-retrieval and suckling. The role of olfactory signals in the orchestration of these behaviors will be discussed in detail later within this chapter (see Section 25.3). In intraspecific communication, rodents may utilize odors to influence behavior of a recipient conspecific, or may investigate received odors to explore the nature of a conspecific. In this context, the odor itself must have either an innate or a learned meaning to the recipient animal, which increases

25.3 ETHOLOGICALLY-RELEVANT INTRASPECIFIC OLFACTORY COMMUNICATION

25.3.1 Active Intraspecific Olfactory Communication

Active intraspecific olfactory communication behavior involves an animal engaging in an olfactory behavior with the intent to modulate conspecific behavioral and/or physiological responses. This is the equivalent of a human talking to another human or a dog barking at a dog. One conspecific will "say" a particular message with, in many cases, a desired response latent in their mind. In consequence to this, assuming the message is received, the receiver conspecific will engage in a divergent behavior from their existing behavioral state.

25.3.1.1 Scent Marking.
Perhaps the most common form of olfactory communication in rodents within their natural environment is active intraspecific communication. Here I draw upon the wealth of studies regarding

25.3 Ethologically-Relevant Intraspecific Olfactory Communication

scent marking behavior in rodents to exemplify common principles of active intraspecific odor communication. Indeed, this behavior is by far the clearest example of active olfactory communication in rodents and serves a variety of critical roles, mainly in terms of sociosexual behavior.

Scent marking is generally considered to be employed for the signaling of social status (reviewed in detail by (Hurst and Beynon, 2004; Roberts, 2007; Arakawa et al., 2008)). However, multiple views, perhaps not mutually exclusive, exist regarding the function of scent marks and why rodents so frequently lay down scent. Marking behavior in monogamous rodents (e.g., prairie voles, *Microtus ochrogaster*) and other monogamous animals has been interpreted to aid in maintaining the pair bond (Peters and Mech, 1975). More frequent dispersal of one's scent not only reinforces the bond but also might discourage unwanted intruder males from attempting to intervene. Scent marks may also aid animals in orienting and navigating within their territory. Thus, individual marks lay out a road map to navigate burrows and localize previous sources of food (Kleiman, 1966). Scent marking may also function by signaling fitness of the emitter to potential intruders (Gosling, 1982). Whether an invader would flee or stay upon detection of a scent mark may be determined by a mixture of possible costs of staying (possible bodily harm), the incentive to stay (how attractive the territory is), and the certainty of the animal that the detected signal was interpreted correctly (Gosling, 1982). For instance, detection of the scent mark of an identifiably fit male would deter intruders from invading territory whereas scent of a weak male (as described below) would be less of a deterrent. However, if the intruder was uncertain about the mark's meaning, and the territory was prime real estate, the likelihood that the intruder would stay and fight for the territory would increase. While the above examples are raised in terms of male competition, this similar theory extends into the function of scent marking in terms of mating and the female reception of scent.

An extensive amount of research is available on the regulation of scent marking, making this a strong model system for ethologically-relevant olfactory communication. Scent marks are composed of odorants originating from a variety of species-specific sources, including urine and glandular secretions (McIntosh et al., 1979). Accordingly, male rodents may scent mark by rubbing their preputial area along the ground or along objects they climb over. This allows chemical signals to be deposited from the preputial glands around the genitals onto the environmental substrate. The remnant scent mark is thus made available for immediate reception (by a proximate conspecific) or for future reception by a distant conspecific. The acquisition of odor information from a scent mark may occur from the emitted volatile odorants, although in many cases the scent

mark's efficacy is greatest when the animal's nose actually contacts the mark. Indeed, male mice engage in more frequent scent mark displays themselves when allowed to come into direct contact with a conspecifics scent mark, versus when the scent mark is covered with a wire mesh (Nevison et al., 2003).

The nature and function of scent marking displayed by a male is closely linked to his social status. As shown in Figure 25.2, male mice that are defeated or subjugated in social interactions will inhibit future scent marking displays (Lumley et al., 1999).

Dominant males in a semi-natural open environment will mark more frequently and in a more dispersed pattern compared to subordinate males (McIntosh et al., 1979). Indeed, in some animals (e.g., the bank vole *Clethrionomys glareolus*) the size of the preputial gland is positively correlated with aggressive dominance (Gustafsson et al., 1980). Across strains of mice, the less aggressive the strain, the smaller the preputial gland (Yamashita et al., 1992), reflecting a drive to increase marking capacity in aggressive/dominant males. It is thus not surprising that male deer mice have larger midventral sebaceous glands than and that their size and secretions are influenced directly by male testosterone levels (Doty and Kart, 1972). Additionally, in semi-natural environments, scent marks by dominant males are most likely found in regions associated with intruder origin – areas with greatest need to deter threats (Gosling, 1982). The frequency and pattern of scent marking are both taken as evidence of the males' motivation to establish territory and also to attract potential mates. Indeed, the prevalence of scent marking among juvenile mice (*Mus musculus*) is a strong predictor of dominance in adulthood (Collins et al., 1997).

Laboratory studies in reduced environments have shown that both scent marking behavior and the behavior of investigating scent marks are gonadal hormone dependent. Investigation of an open area is inhibited by the presence of scent marks from gonadally intact males, however, castration of these marking males results in no difference in behavior of the investigator males (Jones and Nowell, 1973), likely, as discussed below, due to changes in the chemical composition of the marking male's urine or preputial secretions. Indeed, gonadectomy even results in decreased size of androgen dependent preputial tissue (Lumia et al., 1994; Djeridane, 2002). Males of many rodent species treated with testosterone and other testosterone analogs mark more frequently than castrated or physiologically intact (untreated) males (McIntosh et al., 1979; Lumia et al., 1994; Wesson and McGinnis, 2006). Further, treating male rats with androgen receptor inhibitors reduces scent marking behavior along with other sociosexual behaviors (Vagell and McGinnis, 1998). This androgen dependence of scent marking behavior is in

582 Chapter 25 Olfactory Communication in Rodents in Natural and Semi-Natural Habitats

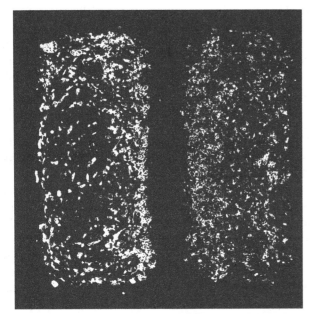

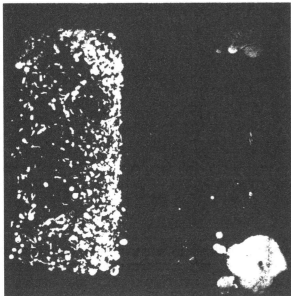

Figure 25.2 *Upper photo:* Ultraviolet photographs of overnight urine marking patterns of socially naïve, male, wild house mice, one in each compartment of a 12 x 12 inch cage, separated from each other by a wire mesh partition. Bottom photo: Urine marking patterns following establishment of dominance in an agonistic encounter, with marking of dominant animal shown on left and subordinate shown on right. Note large urine deposit from the subordinate. From Bronson (1976); used with permission, Copyright Elsevier, 1976.

accordance with the escalation of scent marking behavior in pubertal males (Arakawa et al., 2008).

Relatedly, scent marks play a large role in regulating female choice of mates. Female deer mice (*Peromyscus maniculatus*) display differential preference for male urine odor depending upon the state of their estrus cycle (Doty, 1972). This finding is highly intuitive as one would expect a sexually-motivated female to seek a dominant male who would pass on viable genes. Female meadow voles (*Microtus pennsylvanicus*) spend more time near the scent marks of a well-nourished male versus a malnourished one (Ferkin et al., 1997), in a manner evidently due to differential protein content of restricted diets. This is thought to be due to the females' motivation to select viable and successful mates with whom to reproduce (Roberts, 2007). Females of many rodent species, including mice, prefer to mate with males with whom they have had prior experience via their scent mark (Hurst, 2009). This familiarity effect is something of a paradox when considered in relation to the utility of odor cues to aid in dispersing genes and selecting mates with slightly differing body odor composition (Isles et al., 2001). However, the preference of females to mate with males whose scent mark is familiar might provide a mechanism for generalization in simply seeking more available mates.

While not addressed in detail herein, the chemical composition of scent marks which animals emit is becoming increasingly understood. To explore the chemical nature of the scent marks, researchers have utilized both liquid and head-space gas chromatography. In head-space chromatography, one can explore the types and amounts of volatiles emitted from a liquid odor source. Chromatographic evidence for differences in urinary compounds between dominant and subordinate mice exists (Harvey et al., 1989). In one study, Harvey and colleagues (1989) found 16 compounds which differ in concentration based upon social status of male mice, including dihydrofurans, acetates, and ketones which decrease in the urine of submissive mice and α- and β-farnesene which increase in urine from dominant mice. Some of this research has focused on the presence of major urinary proteins (MUPs) which are heavy weight large molecules that possess considerable stability and resist volatility over small urinary molecules and additionally appear to regulate the factors involved in social signaling within the urine (Sharrow et al., 2002).

Complementing their overt behavioral actions, male scent marks have the capacity to exert physiological modulations to the female reproductive system, which enhances the probability of successful copulation. These include the ability for dominant adult male scents to accelerate puberty in juvenile females (i.e., Vandenbergh effect) (Lombardi and Vandenbergh, 1977) and to induce synchronous estrus (i.e., Whitten effect) (Whitten, 1956). Interestingly, in the Vandenbergh effect, the physical presence of a dominant male can apparently suppress levels of a chemical or chemicals found in the urine of subordinate conspecifics. The authors found that it was this suppressed substance or substances which was responsible for accelerating puberty in females (Lombardi and Vandenbergh, 1977) when upregulated in other males. Similar findings have been observed in other animals in addition to rodents

(Vandenbergh, 1989). Thus, physiological modulations of odor emission can occur by social factors which can in turn modulate endocrine functions in other conspecifics. Some of these have been found mostly or if not exclusively to be the result of a single urinary constituent. For instance, 6-hydroxy-6-methyl-3-heptanone has a strong capacity to accelerate puberty in female mice (Novotny et al., 1999). Thus, both male urine and scent marks can profoundly affect female behavior and physiology through the olfactory system. Notably, alternative hypotheses exist regarding some of these "pheromone-mediated" findings, including the Vandenbergh effect (for further details, see Doty, 2010).

Counter-marks can occur when a dominant male encounters a scent mark from a subordinate conspecific or when a potentially-dominant male enters an environment where scent markers from other males have been deposited (Rich and Hurst, 1999; Hurst and Beynon, 2004). In these cases, the dominant or new male will place a scent mark on top of the existing scent mark in a competitive behavior. While this "over-marking" would seem to create a complex chemical signal that may be confusing to future recipients, counter-marking is a successful way for the dominant male to actually *replace* a subordinates scent within his own (Hurst and Beynon, 2004). Indeed, female and even male mice show remarkable preferences for males whose scent mark was on top of the existing mark (Hurst, 2009). This is likely an important behavioral preference since it would aid females in finding the closest males. While counter-marking is commonly observed in mice and rats, it is not universal for all rodents, and voles in particular do not seem to actively counter-mark (Mech et al., 2003).

Given their critical role in behavior, it is perhaps not surprising that the function of scent marks varies with experience. Olfactory experience with scent marks has been studied extensively in voles, wherein voles of both sexes spend more time investigating novel scent marks (those from unfamiliar animals) versus those from familiar conspecifics (Ferkin et al., 2010). Novel scent mark investigation and investigation of flank gland odors are also subject to habituation (Beauchamp and Wellington, 1984). These basic principles of experience-dependent scent mark investigation behavior are also apparent in semi-natural contexts. In one very powerful example of this, ground squirrels (*Spermophilus beldingi*) clearly discriminate between familiar and novel conspecific odors even after hibernating for the winter (Mateo and Johnston, 2000).

While scent marking is most widely studied in males (likely due to the incorrect assumption that male rodents play the greatest role in initiating and controlling mating; Doty, 1974), female rodents also scent mark as a form of olfactory communication. However, the functions of female scent marks seem to differ greatly from that of males, and their occurrence is rarer, making studies into the modulation and function of female scent marking behaviors difficult. Similar to that discussed above for male mice and rats, hormones influence the scent marking of females, with the most scent marking of females occurs during proestrus or early estrus (Hurst, 1990; Matochik et al., 1992). Even in laboratory conditions female scent marks are dictated by social status, with virgin females marking less frequently than their mothers (Heise and Rozenfeld 1999). In those species that have been evaluated, female scent marks appear to play a salient role in female to female interactions. Indeed, in female hamsters more scent marking behavior occurs upon reception of female scent than male scent (Johnston, 1977). In mice, female scent marking is more prevalent when in the proximity of a gonadally intact male than in the presence of a castrated one (Maruniak et al., 1975). Female urinary scent is highly attractive to sexually experienced males (Lydell and Doty 1972). Scent marking behavior appears, for both sexes, as a means for coordinating reproductive success.

Overall, the aforementioned studies on scent marking behavior highlight the fact that this form of olfactory communication is critical for rodent survival within nature. Scent marking patterns and strategies are highly plastic, being modulated by age, hormonal status, and social experience. Further, recognition of conspecifics based upon their scent marks is mediated by highly robust memories, which likely translate into better adaptive behavioral responses to both familiar and unfamiliar conspecifics.

25.3.2 Passive Intraspecific Olfactory Communication

While evidently emitted with no perceivable effort, passive odor cues robustly modulate a great number of conspecific behaviors, including conspecific sexual, aggressive, maternal, and food seeking/preference behaviors. While rodents spend considerable risk and energy depositing scent marks, in the following passive odor communication examples, information is exchanged without major energy loss.

25.3.2.1 Sociosexual Behaviors. Rodent body odors emitted from a variety of glandular sources modulate the occurrence of sociosexual behaviors. Each animal has an individualized odor-type, which is based upon numerous factors (Beauchamp, 1988; Yamazaki et al., 1999). Far more studies have been performed highlighting the role of body odors on mediating conspecific preference and sociosexual behaviors than can be nearly addressed within this chapter (but see Keller et al., 2009). Indeed, it has been known for many years that odors modulate sociosexual preferences of a wide variety of animals (e.g., Edwards and Burge, 1973; Beach, 1974). Further, appetitive and consummatory sociosexual behaviors are strongly

584   Chapter 25   Olfactory Communication in Rodents in Natural and Semi-Natural Habitats

modulated by olfaction in a variety of rodents. Olfactory bulbectomy or nasal epithelium lesions, for instance, markedly influence copulatory behaviors of hamsters, rats and mice of both sexes (e.g., Doty et al., 1971; Edwards and Burge, 1973; Meisel et al., 1980; Wood and Newman, 1995; Keller et al., 2006).

In one study, Drickamer (1989) tested the role of male or female urinary odor, soiled bedding odor, and whole body odor on female mouse preferences. Different groups of females were tested at different ages, thereby allowing a test of the influence of female reproductive status on conspecific passive odor preference. Before puberty, females avoided the odors of males and showed a preference for grouped female odors. This shifted at the time of puberty, resulting in a strong attraction for male odors and an avoidance of odor cues from grouped females.

This same group of researchers later went on to use a semi-natural enclosure paradigm to explore a similar question of impact of reproductive state on odor preferences (Drickamer and Brown, 1998). In this study, relatively large populations of male and female mice were housed in a large wire mesh arena (25 × 40 meters) with numerous "mouse cities" that provided shelter and burrowing opportunities for the mice. These mouse colonies were allowed to flourish under no experimental intervention for 30 weeks. Throughout the study, live traps containing cotton nesting materials were placed within each enclosure, making it possible to capture random mice from the entire population during the course of the study. Cleverly, the authors utilized live traps that allowed for measuring odor preferences. They found that mice are attracted to traps previously holding another mouse. Using this approach, they inferred, for example, that a mouse that was caught in trap that previously held a female mouse preferred the smell of that female mouse. It was found that, as females mature, they change their odor preferences. Young female and male mice both avoided traps previously housing adult males. However, upon entering adulthood, female mice shifted their preference for traps that previously held adult males. Similarly, adult male mice preferred traps previously holding adult females. Thus in this semi-natural design, Drickamer and Brown (1998) were able to demonstrate that several basic principles of odor-guided social preference in mice seen in the laboratory were also present in a more natural setting.

Just as an attractive odor may result in social preference, an "unattractive" body odor may result in avoidance. For instance, a female may avoid mating with a male based upon his body odor composition and instead prefer to mate with a male with a different odor. This has been modeled in several ways, including treatment of rodents with *E. coli* lipopolysaccharide (LPS) (Arakawa et al., 2009) to induce a sickness odor phenotype. In one study, Arakawa and colleagues (2009) treated male rats with LPS. LPS-treated male rat odor was avoided by both male and female conspecifics. Interestingly, the authors found that this avoidance was dependent upon the status of the stimulus animal, with the odor of prepubertal males treated with LPS failing to elicit avoidance behavior (Arakawa et al., 2009). Thus, perhaps conspecific avoidance of "sick" conspecifics only occurs when the conspecific is of age to be a threat or mate. Alternatively, this finding may be an experimental artifact of an influence of endogenous hormones on the immune-mediated expression of the sickness odor (Arakawa et al., 2009). This finding reflects that sickness odors transmitted through passive bodily means are potent modulators of sociosexual behaviors and highlight the utility of such odors in sociosexual behaviors.

25.3.2.2 Aggression.

Aggressive behavior in rodents is tightly linked to body odor composition. Again, this behavior is largely dependent upon olfaction in a variety of rodents and has been documented from studies that differentially perturbed olfactory sensory input (Hull et al., 1974; Lumia et al., 1975; Edwards et al., 1993; Kolunie and Stern, 1995; Stowers et al., 2002). The odor-dependence of aggressive behavior, taking into account differential body odors of animals (Beauchamp, 1988), provides an efficient mechanism for defending territory. In an early and elegant example of this phenomenon, the urine of a novel male was rubbed on a familiar male who was part of a dyad (Mackintosh and Grant, 1966). The conspecific in the dyad failed to recognize the familiar male whose odor had been masked with that the unfamiliar male, resulting in escalated agonistic behaviors direct towards the previously familiar male. This is a powerful example of how body odors dictate social encounters, since the resident male is seemingly attending exclusively to bodily odors. The submissive males, however, are emitting little to no other communicative cues, such as ultrasonic vocalizations (Thomas et al., 1983). In another study, resident males displayed less aggression to novel males rubbed with odors of previously familiar males (Gosling and McKay, 1990). These studies suggest that rodents match the scents of animals (Gosling and McKay, 1990). This appears to be adaptive, since it allows rodents to generalize between animals efficiently as long as their body odor matches those of previously familiar animals. Such a mechanism would be highly useful and relevant within rodent burrow systems, where light is lacking and non-visual communication is required for communicating threat signals.

25.3.2.3 Maternal Behavior.

In many species of newborn rodents, the young are born without functional sensory systems beyond that of olfaction. Thus, the recognition and care of newborns by dams is dependent upon emission of olfactory cues by pups (Sullivan, 2003).

Similarly, pups identify dams and localize nipples for nourishment based also upon olfactory cues (Teicher and Blass, 1976; Blass and Teicher, 1980; Logan et al., 2012). While the majority of research on this subject requires a laboratory environment for experimental manipulation and close observations, these olfactory dependent behaviors position odor cues as highly critical for survival in natural contexts and thus are discussed in more detail here (also see (Leon, 1992)).

Perhaps one of the most powerful models is the finding that the pup sucking response in rodents is largely odor-dependent (Stewart et al., 1983). Pups have trouble localizing and attaching to maternal nipples when the nipples have their natural maternal odors removed (Teicher and Blass, 1976). But do pups learn that maternal odors are ethologically relevant? Or instead are pups born seeking maternal odors? In one early investigation, Pederson and Blass explored this question by addressing when an odor must be experienced to determine the pup's first suckling response in albino rats (Pedersen and Blass, 1982). In their first experiment, the authors exposed pups prenatally and immediately postnatal to the odorant citral. These pups readily suckled nipples of an anesthetized dam which were cleaned of natural odors, and instead scented with citral. In a different experiment, Pedersen and Blass (1982) exposed pups to the smell of citral in utero, immediately following birth, or both in utero and immediately following birth. Only pups exposed to the citral odor in utero and immediately upon birth suckled. These results are highly informative of the mechanisms of early olfactory learning in an ultimately ethologically-relevant task. Interestingly, while not relevant to the current chapter, these results also lent evidence for prenatal events in olfaction (Pedersen et al., 1983), wherein memories for odors may even occur in utero. These experiments have been supported more recently in mice (Logan et al., 2012) and together demonstrate that the maternal odor cues serve as experience-dependent learnt odors (Leon et al., 1987) to promote pup survival.

Maternal behaviors (nursing, licking, and grooming) are required for thermoregulation and returning the pup to the nest for nursing (Fleming et al., 2002). Thus it is adaptive that pup body odor facilitates maternal behavior by aiding in pup localization/identification and retrieval. In one study, Mayer and Rosenblatt (1993) sought to test the critical nature of pup odors on maternal behavior. The authors abolished olfactory sensory input to the olfactory bulbs of lactating females with a zinc sulfate lesion of the nasal epithelium and later explored their preference for pup odors (Mayer and Rosenblatt, 1993). Bedding from cages containing the lactating female's pups was placed in two out of four holes of a hole board apparatus, and the remaining two holes were filled with clean bedding. Females with sham epithelial lesions (intact sense of smell) displayed a strong preference for the holes containing bedding with pup odor, however, anosmic females did not. Thus, attraction of females to their pups is largely mediated by pup bodily odor (Benuck and Rowe, 1975), in a manner similar to that seen in other mammals (Lévy and Keller, 2009).

25.3.2.4 Food Foraging and Preference.

Semi-natural experimental designs lend themselves quite well to understanding the modulation of food foraging by odors. In one such study, Verplancke and colleagues used an artificially enclosed arena which was situated within a wooded area free of predators and human intervention (Verplancke et al., 2010). Within the enclosure, the authors placed feeders which allowed for both video recordings of the number of visits by bank voles and also the amount of food how much food was consumed. Relevant to this discussion, the authors scented the food with conspecific odors (male or female) to determine how conspecific odor modulates food intake. It was found that conspecific odor increases the number of visits to the feeders and the amount of foraging in this semi-natural context (Verplancke et al., 2010). Although, interestingly, in this paradigm the authors noted a several fold difference in non-scented food consumption compared to earlier laboratory studies. While only in one rodent species, the finding in the above study that animals enhance food consumption and proximity/activity around sites which smell like conspecifics is highly intuitive and it is reasonable to assume that rodents, just like humans, do this as a behavior of familiarity and comfort.

Within conspecific feeding conditions, odor of consumed foods originating from the facial region or breath of one conspecific are well known to modulate future food preferences and consumption behaviors in other conspecifics. While the majority of studies into this social transmission of food preference have occurred in reduced laboratory environments (Galef, 1985), this paradigm is highly relevant for the ethology of rodents and thus discussed here. In the classic design, a hungry rat (the "demonstrator") is allowed to eat food laced with an odor, for instance, cinnamon, for several days. Immediately after one of the food consumption days, a hungry conspecific is placed with the demonstrator and they are allowed to interact for a period of time. It is thought that during this time, the conspecific samples the odors emanating from the facial region and breath of the demonstrator and forms a learned association with that odor. Then the hungry conspecific is placed into a context where it has the choice to consume food laced (for example) with cinnamon odor versus cocoa. Significantly more consumption of cinnamon laced food indicates the social transmission of food preference. This paradigm is conserved across a considerable number of rodents (Galef, 1985; Bunsey and Eichenbaum, 1995;

586   **Chapter 25   Olfactory Communication in Rodents in Natural and Semi-Natural Habitats**

Faulkes, 1999) and can be used to study odor preferences and aversions alike. However, some animals, like Norway rats, do not learn odor-based aversions within the natural social context, but instead readily form preferences for the breath of a sick/poisoned conspecific (Noble et al., 2001).

Intraspecific passive odor cues can also modulate the localization of food. This has been elegantly studied in eusocial naked mole rats (*Heterocephalus glaber*) allowed to live among their colony in a semi-natural burrow system constructed of clear plastic tubes (Faulkes, 1999). This labyrinth of clear plastic tubes when kept in a dark environment mimics the subterranean system of tunnels and burrows the naked mole rats naturally develop to live in. In this paradigm, colony mates prefer to spend time at a site within the tunnel system whereabouts a conspecific previously found and consumed food. They would even bypass other sites with the same food, to go to the site where the conspecific previously found their food. Conspecifics favored tunnels which were traveled in by the initial foraging mole, versus tunnels other colony members traversed while carrying the same food, thus suggesting a major role for passive olfactory cues in this foraging behavior. This finding reflects that the voles are not honing in on the smell of the food itself, nor even a generalized conspecific, but the initial conspecific who had consumed food. Indeed, this preference for the initial consumed tunnel site was abolished if that tunnel was replaced with a clean tunnel (Faulkes, 1999). This finding suggests that naked mole rats follow individualized odor trails (viz., those from 1 preferential conspecific) to localize food.

25.3.2.5 Warning Signals. Olfactory warning signals are sometimes referred to as "alarm pheromones," and are thought to be emitted by rodents as a physiological response to stress, fear, or generalized anxiety. The method for odor emission in this context would likely be vastly different than active communication signals, like scent marking and thus, for the purpose of this chapter, I classify olfactory warning signals or alarm pheromones as a form of passive intraspecific olfactory communication. Very little is understood about this variation of olfactory communication, although evidence exists for the physiological transduction of these warning signals via the Gruenberg ganglion (Brechbühl et al., 2008) (see Chapter 52). This method of olfactory communication notably does not lend itself well to the natural environment and instead, many rodents have developed more visible or audible warning signals, most of which are actively communicated, to signal to conspecifics the presence of a predator or other threat. While these signals are risky in drawing the attention of predators, they would allow an animal to escape better than a scent trail which may be advantageous for predation.

This said, several examples in the literature have been interpreted to support the existence of alarm pheromones.

In one particular example, mice were subjected to different stressful paradigms and later the behavior of conspecifics was measured when they were placed within the chamber which previously housed the "stressed" mouse (Zalaquett and Thiessen, 1991). The authors of this study reported that mice avoided the side of an experimental tube which previously a stressed mouse had entered through and similarly that being in the environment of the previously stressed mouse resulted in generalized arousal and vigilance. Ultimately, perception of warning odors by design should aid in promoting survival of conspecifics by warning of predation or other threats.

25.4 ETHOLOGICALLY-RELEVANT INTERSPECIFIC OLFACTORY COMMUNICATION

While rodents have evolved to emit odor cues to influence conspecific behavior, a downside of this occurs when predators learn (or have developed instinct) to utilize the same cues for predation. Indeed, basically all of the examples above can be utilized in predation scenarios to the benefit of other animals. This highlights how while there are benefits involved in olfactory communication, there is also considerable risk involved. Animals therefore must adjust their patterns of olfactory communication to adapt in the presence of conspecifics versus when predation is likely. Therefore, within this section, I discuss a few particularly powerful examples of these forms of interspecific olfactory communication.

25.4.1 Active Interspecific Olfactory Communication

The idea that an animal might actively release an odor to communicate with a member of a different species seems counterintuitive. This is especially true in the case of rodents who may find predation among a large variety of animals in their ethological niche. Scent marks for instance, as discussed in detail above in terms of their ability to modulate an array of conspecific behaviors, are also readily used by predators to localize or "eavesdrop" on prey (Roberts, 2007; Hughes and Banks, 2010). Indeed, there is a strong connection between scent marking and an increased risk of predation. Unlike many other signaling systems in animals, olfactory signaling such as scent marking can be "turned off" at will. For example, in the field, experimentally increasing the amount of vole scent marking leads to increased predation of voles (Klemola et al., 2000; Kiovula and Korpimaki, 2001) due to the ultraviolet reflection given off by proteins within the scent marks and detected by some birds of prey. So, how could a rodent volitionally modulate scent marking in a manner to increase survival?

While rare, there are examples of cases wherein a rodent emits an odor, or changes the deposition of an odor, with apparent insight into modulating the behavior of a potential predator. As a basic example, mice within laboratory environments decrease scent marking intensity when in the suspected presence of predators, although these findings do not uphold as strongly in semi-natural tests (for review see Roberts, 2007). For example, when the smell of cat (*Felis catus*) was placed in a large arena, neither scent marking rates nor the patterns of scent marking were impacted in mice (Hughes and Banks, 2010).

25.4.2 Passive Interspecific Olfactory Communication

In what manners might passively emitted rodent odors perform interspecific communication? This is a far less studied realm of olfactory communication. While a considerably vast amount of work has been performed demonstrating the influence of other animal odors on rodent behavior and physiological responses (e.g., Dielenberg et al., 2001; Wallace and Rosen, 2001; Campeau et al., 2008 and see below), beyond simply being used to locate rodents as prey, much less is known about the details of passive rodent odors in terms of interspecific communication.

25.4.2.1 *Food Foraging and Preference.* In the study of Verplancke and colleagues discussed above (Verplancke et al., 2010), in addition to using conspecific odors, the authors sprayed predator odors and "neutral" odors on food to explore the influence of these odors on bank vole food intake. Again, in this study the authors used an artificially enclosed arena which was situated within a wooded area free of predators and human intervention (Verplancke et al., 2010). Within the enclosure, the authors placed feeders which allowed video recordings of the number of visits by bank voles and also the amount of food consumed. The authors found that predator odor decreases the number of visits to the feeders and the amount of foraging in this semi-natural context (Verplancke et al., 2010). This study nicely illustrates how odor from another animal may modulate rodent behavior.

25.4.2.2 *Other Examples: Nesting and Interspecific Preference/Avoidance.* The nest site is a likely major source of olfactory signals which may enhance predation risks among rodents. Some animals have developed strategies for reducing this unwanted consequence. For instance, female deer mice (*Peromyscus maniculatus*) carry newborn pups from the birthing nest to novel nest sites to reduce unwanted nest odors (Sharpe and Millar, 1980). Defecation (which has no apparent active or volitional component) can also enhance predation risk. Seemingly to reduce this risk, lemmings (*Dicrostonyx groenlandicus*) defecate in underground sites in the summer, but not during the winter when the snow offers coverage of odor signals (Boonstra et al., 1996).

Many rodent species are sympatric and while some do not have evident predator–prey relationships, they must deal with each other's presence in their ecological niche. Examining the olfactory modulation of this, Doty (1973) explored the role of odors in interspecific preference/aversion behavior in white footed mice (*P. leucopus noveboracensis*) and deer mice (*P. maniculatus bairdi*). In this study, virgin female deer mice were found to prefer conspecific male odor more than heterospecific male odor, as reflected by an increased time spent near the odor of the conspecific mouse. In contrast, white-footed female mice failed to show a preference or aversion for either stimulus (Doty, 1973). Thus, while with differences between types of rodents, rodents can readily discriminate interspecific from intraspecific odors, and use these distinctions to guide behaviors.

25.5 SUMMARY

Throughout this chapter, examples of how rodents utilize olfactory communication methods have been outlined. This by drawing upon work which stems from research performed in semi-natural or natural rodent "field" habitats and in some cases laboratory-based experiments which compliment or extend more ethologically-relevant paradigms. Both active and passive odor cue emission are governed by sex, hormonal status, age, behavioral state, and other factors (nutrient intake, illness) which together provide a highly dynamic source of signals not only in their release but also in the behaviors they elicit.

Future work addressing and extending major questions of olfaction within more ethologically-relevant and ideally natural habitats will be critical in bringing about a greater understanding of how rodents and other animals utilize smells (by releasing them or acquiring them) to adaptively survive in their environments. Little has been elucidated regarding how natural rearing, and the use of odors in nature, might entail considerably different neural responses to odors in a manner adaptive to survival in the wild. Further, major questions exist in terms of how rodents acquire (through sniffing for instance) odors in their natural habitat or even in ethologically-relevant social contexts. Answers to these and other questions will bring about greater understanding and appreciation for the complex olfactory world rodents naturally negotiate.

ACKNOWLEDGEMENTS

This work was supported by NSF grant ISO-1121471.

REFERENCES

Arakawa, H., Arakawa, K., and Deak, T. (2009) Acute illness induces the release of aversive odor cues from adult, but not prepubertal, male rats and suppresses social investigation by conspecifics. *Behav. Neurosci.* 123(5): 964–978.

Arakawa, H., Blanchard, D. C., Arakawa, K., et al. (2008) Scent marking behavior as an odorant communication in mice. *Neurosci. Biobehav. Rev.* 32(7): 1236–1248.

Beach, F. A. (1974) Effects of gonadal hormones on urinary behavior in dogs. *Physiol. Behav.* 12(6): 1005–1013.

Beauchamp, G. K., and Wellington, J. L. (1984) Habituation to individual odors occurs following brief, widely-spaced presentations. *Physiol. Behav.* 32(3): 511–514.

Beauchamp, G. K., Yamazaki, K. and Boyse, E. A. (1988) The chemosensory recognition of genetic individuality. *Sci. Am.* 253: 86–92.

Benuck, I., and Rowe, F. A. (1975) Centrally and peripherally induced anosmia: influences on maternal behavior in lactating female rats. *Physiol. Behav.* 14(4): 439–447.

Blass, E. M., and Teicher, M. H. (1980) Suckling. *Science* 210(4465): 15–22.

Boonstra, R., Krebs, C.J., and Kenney, A. (1996) Why lemmings have indoor plumbing in the winter. *Canadian J. Zool.* 74: 1947–1949.

Brechbühl, J., Klaey, M., and Broillet, M.-C. (2008) Grueneberg ganglion cells mediate alarm pheromone detection in mice. *Science* 321(5892): 1092–1095.

Brennan, P., Kaba, H., and Keverne, E. B. (1990) Olfactory recognition: a simple memory system. *Science* 250(4985): 1223–1226.

Bronson, F. H. (1976) Urine marking in mice: causes and effects. In R. L. Doty (Ed.), *Mammalian Olfaction, Reproductive Processes, and Behavior.* New York: Academic Press, pp. 119–143.

Bunsey, M., and Eichenbaum, H. (1995) Selective damage to the hippocampal region blocks long-term retention of a natural and nonspatial stimulus-stimulus association. *Hippocampus* 5(6): 546–556.

Campeau, S., Nyhuis, T. J., Sasse, S. K., et al. (2008) Acute and chronic effects of ferret odor exposure in Sprague–Dawley rats. *Neurosci. Biobehav. Rev.* 32(7): 1277–1286.

Collins, S. A., Gosling, L. M., Hudson, J., and Cowan, D. (1997) Does behavior after weaning affect the dominance status of adult male mice (*Mus musculus*)? *Behaviour* 134: 989–1002.

Dielenberg, R. A., Hunt, G. E., and McGregor, I. S. (2001) "When a rat smells a cat": the distribution of Fos immunoreactivity in rat brain following exposure to a predatory odor. *Neuroscience* 104(4): 1085–1097.

Djeridane, Y. (2002) Integumentary odoriferous glands in *Meriones-libycus*: a histological study and related behavior. *Folia Zoologica.* 51(37–58): 12–34.

Doty, R. L. (1972) Odor preferences for female Peromyscus maniculatus bairdii for male mouse odors of *P. m. bairdii* and *P. leucopus noveboracensis* as a function of estrous state. *J. Comp. Physiol. Psych.* 81: 191–197.

Doty, R. L. (1973) Reactions of deermice (*Peromyscus maniculatus*) and white-footed mice (*Peromyscus leucopus*) to homospecific and heterospecific urine odors. *J. Comp. Physiol. Psych.* 84: 296–303.

Doty, R. L. (1974) A cry for the liberation of the female rodent: Courtship and copulation in rodentia. *Psychol. Bull.* 81(3): 159–172.

Doty, R. L. (1986) Odor-guided behavior in mammals. *Experientia.* 42: 257–271.

Doty, R. L. (2010) *The Great Pheromone Myth.* Johns Hopkins University Press: Baltimore.

Doty, R. L., Carter, C. S., and Clemens, L. G. (1971) Olfactory control of sexual behavior in the male and early-androgenized female hamster. *Horm. Behav.* 2(4): 325–335.

Doty, R. L., and Kart, R. (1972) A comparative and developmental analysis of the midventral sebaceous glands in 18 taxa of *Peromyscus*, with an examination of gonadal steroid influences in *Peromyscus maniculatus bairdii. J. Mammalogy* 53(1): 83–99.

Drickamer, L. C. (1989) Odor preferences of wild stock female house mice (*Mus domesticus*) tested at three ages using urine and other cues from conspecific males and females. *J. Chem. Ecol.* 15(7): 1971–1987.

Drickamer, L. C., and Brown, P. L. (1998) Age-related changes in odor preferences by house mice living in semi-natural enclosures. *J. Chem. Ecol.* 24(11): 1745–1756.

Edwards, D. A., and Burge, K. G. (1973) Olfactory control of the sexual behavior of male and female mice. *Physiol. Behav.* 11(6): 867–872.

Edwards, D. A., Nahai, F. R., and Wright, P. (1993) Pathways linking the olfactory bulbs with the medial preoptic anterior hypothalamus are important for intermale aggression in mice. *Physiol. Behav.* 53(3): 611–615.

Faulkes, C. G. (1999) Social transmission of information in a eusocial rodent, the named-mole rat. In *Mammalian Social Learning: Comparative and Ecological Perspectives.* (Eds. H. O. Box and K. R. Gibson) pp. 205–20. Cambridge University Press: Cambridge.

Ferkin, B. D., Sorokin, E. S., Johnston, R. E., and Lee, C. J. (1997) Attractiveness of scents varies with protein contentof the diet in meadow voles, *Microtus pennsylvanicus. Animal Behav.* 53: 133–141.

Ferkin, M. H., Ferkin, D. A., Ferkin, B. D., and Vlautin, C. T. (2010) Olfactory experience affects the response of meadow voles to the opposite-sex scent donor of mixed-sex over-marks. *Ethology* 116(9): 821–831.

Fleming, A. S., Kraemer, G. W., Gonzalez, A., et al. (2002) Mothering begets mothering: the transmission of behavior and its neurobiology across generations. *Pharm. Biochem. Behav.* 73(1): 61–75.

Galef, B. G. (1985) Direct and indirect behavioral pathways to the social transmission of food avoidancea. *Ann. N.Y. Acad. Sci.* 443(1): 203–215.

Gosling, L. M. (1982) A reassessment of the function of scent marking in territories. *Z. Tierpsychol.* 60: 89–118.

Gosling, L. M., and McKay, H. V. (1990) Competitor assessment by scent matching: An experimental test. *Behav. Ecol. Sociobiol.* 26: 415–420.

Gustafsson, T., Andersson, B., and Meurling, P. (1980) Effect of social rank on the growth of the preputial glands in male bank voles, *Clethrionomys glareolus. Physiol. Behav.* 24(4): 689–692.

Harvey, S., Jemilio, B., and Novotny, M. V. (1989) Pattern of volatile compounds in dominant and subordinate male mouse urine. *J. Chem. Ecol.* 15: 2061–2072.

Heise, S. R., and Rozenfeld, F. E. (1999) Reproduction and urine marking in laboratory groups of female common voles, *Microtus arvalis. J. Chem. Ecol.* 25: 1671–1685.

Hughes, N. K., and Banks, P. B. (2010) Interacting effects of predation risk and signal patchiness on activity and communication in house mice. *J. Animal Ecol.* 79(1): 88–97.

Hull, E. M., Hamilton, K. L., Engwall, D. B., and Rosselli, L. (1974) Effects of olfactory bulbectomy and peripheral deafferentation on reactions to crowding in gerbils (*Meriones unguiculatus*). *J. Comp. Physiol. Psychol.* 86(2): 247–254.

Hurst, J. L. (1990) Urine marking in populations of wild house mice *Mus domesticus* Rutty. II. Communication between females. *Animal Behav.* 40: 223–232.

Hurst, J. L. (2009) Female recognition and assessment of males through scent. *Behav. Brain Res.* 200(2): 295–303.

Hurst, J. L., and Beynon, R. J. (2004) Scent wars: the chemobiology of competitive signalling in mice. *BioEssays* 26(12): 1288–1298.

Isles, A. R., Baum, M. J., Ma, D., et al. (2001) Urinary odour preferences in mice. *Nature* 409: 783–784.

Johnston, R. E. (1977) The causation of two scent marking behavior patterns in female hampsters (*Mesocricetus auratus*). *Animal Behav.* 25: 317–327.

Jones, R. B., and Nowell, N. W. (1973) Aversive effects of the urine of a male mouse upon the investigatory behavior of its defeated opponent. *Animal Behav.* 21(4): 707–710.

Keller, M., Baum, M., Brock, O., et al. (2009) The main and the accessory olfactory systems interact in the control of mate recognition and sexual behavior. *Behav. Brain Res.* 200(2): 268–276.

Keller, M., Douhard, Q., Baum, M. J., and Bakker, J. (2006) Destruction of the main olfactory epithelium reduces female sexual behavior and olfactory investigation in female mice. *Chem. Senses* 31(4): 315–323.

Kiovula, M., and Korpimaki, E. (2001) Do scent marks increase predation risk of microtine rodents? *Oikos* 95: 275–281.

Kleiman, D. (1966) Scent marking in the Canidae. *Symp. Zool. Soc. London* 18: 167–177.

Klemola, T., Kiovula, M., Korpimaki, E., and Norrdahl, K. (2000) Experimental tests of predation and food hypotheses for population cycles of voles. *Proc. Biol. Sci.* 267(1441): 351–356.

Kolunie, J. M., and Stern, J. M. (1995) Maternal aggression in rats: effects of olfactory bulbectomy, ZnSO$_4$-induced anosmia, and vomeronasal organ removal. *Horm. Behav.* 29(4): 492–518.

Leon, M. (1992) Neuroethology of olfactory preference development. *J. Neurobiol.* 23(10): 1557–1573.

Leon, M., Coopersmith, R., Lee, S., et al. (1987) Neural and behavioral plasticity induced by early olfactory learning. In *Perinatal Development: A Psychobiological Perspective*. (Eds. N. A. Krasnegor, E. M. Blass, M. Hofer and W. P. Smotherman). (Academic Press, Inc.)

Lévy, F., and Keller, M. (2009) Olfactory mediation of maternal behavior in selected mammalian species. *Behav. Brain Res.* 200(2): 336–345.

Logan, D. W., Brunet, L. J., Webb, W. R., et al. (2012) Learned recognition of maternal signature odors mediates the first suckling episode in mice. *Curr. Bio.: CB* 22(21): 1998–2007.

Lombardi, J. R., and Vandenbergh, J. G. (1977) Pheromonally induced sexual maturation in females: regulation by the social environment of the male. *Science* 196(4289): 545–546.

Lumia, A. R., Thorner, K. M., and McGinnis, M. Y. (1994) Effects of chronically high doses of the anabolic androgenic steroid, testosterone, on intermale aggression and sexual behavior in male rats. *Physiol. Behav.* 55(2): 331–335.

Lumia, A. R., Westervelt, M. O., and Rieder, C. A. (1975) Effects of olfactory bulb ablation and androgen on marking and agonistic behavior in male Mongolian gerbils (*Meriones unguiculatus*). *J. Comp. Physiol. Psychol.* 89(9): 1091–1099.

Lumley, L. A., Sipos, M. L., Charles, R. C., et al. (1999) Social stress effects on territorial marking and ultrasonic vocalizations in mice. *Physiol. Behav.* 67(5): 769–775.

Lydell, K., and Doty, R. L. (1972) Male rat of odor preferences for female urine as a function of sexual experience, urine age, and urine source. *Horm. Behav.* 3(3): 205–212.

Mackintosh, J. H., and Grant, E. C. (1966) The effect of olfactory stimuli on the agonistic behaviour of laboratory mice. *Z. Tierpsychol.* 23: 584–587.

Maruniak, J. A., Owen, K., Bronson, F. H., and Desjardins, C. (1975) Urinary marking in female house mice: effects of ovarian steroids, sex experience, and type of stimulus. *Behav. Biol.* 13(2): 211–217.

Mateo, J. M., and Johnston, R. E. (2000) Retention of social recognition after hibernation in Belding's ground squirrels. *Animal Behav.* 59(3): 491–499.

Matochik, J. A., White, N. R., and Barfield, R. J. (1992) Variations in scent marking and ultrasonic vocalizations by Long-Evans rats across the estrous cycle. *Physiol. Behav.* 51(4): 783–786.

Mayer, A. D., and Rosenblatt, J. S. (1993) Peripheral olfactory deafferentation of the primary olfactory system in rats using ZnSO$_4$ nasal spray with special reference to maternal behavior. *Physiol. Behav.* 53(3): 587–592.

McClintock, M. K. (1998) On the nature of mammalian and human pheromones. *Ann. N.Y. Acad. Sci.* 855(1): 390–392.

McIntosh, T. K., Davis, P. G., and Barfield, R. J. (1979) Urine marking and sexual behavior in the rat (*Rattus norvegicus*). *Behav. Neural. Bio.* 26(2): 161–168.

Mech, S. G., Dunlap, A. S., and Wolff, J. O. (2003) Female prairie voles do not choose males based on their frequency of scent marking. *Behav. Proc.* 61(3): 101–108.

Meisel, R. L., Lumia, A. R., and Sachs, B. D. (1980) Effects of olfactory bulb removal and flank shock on copulation in male rats. *Physiol. Behav.* 25(3): 383–387.

Meredith, M. (1998) Vomeronasal, olfactory, hormonal convergence in the brain: cooperation or coincidence? *Ann. N.Y. Acad. Sci.* 855(1): 349–361.

Nevison, C. M., Armstrong, S., Beynon, R. J., et al. (2003) The ownership signature in mouse scent marks is involatile. *Proc. Biol. Sci.* 270(1527): 1957–1963.

Noble, J., Tobb, P. M., and Tuci, E. (2001) Explaining social learning of food preferences without aversions: an evolutionary simulation model of Norway rats. *Proc. Biol. Sci.* 268(1463): 141–149.

Novotny, M. V., Jemiolo, B., Wiesler, D., et al. (1999) A unique urinary constituent, 6-hydroxy-6-methyl-3-heptanone, is a pheromone that accelerates puberty in female mice. *Chem. Bio.* 6(6): 377–383.

Pedersen, P. E., and Blass, E. M. (1982) Prenatal and postnatal determinants of the 1st suckling episode in albino rats. *Dev. Psychobiol.* 15(4): 349–355.

Pedersen, P. E., Stewart, W. B., Greer, C. A., and Shepherd, G. M. (1983) Evidence for olfactory function in utero. *Science* 221(4609): 478–480.

Peters, R. P., and Mech, L. D. (1975) Scent-marking in wolves. *Am. Sci.* 63(6): 628–637.

Rich, T. J., and Hurst, J. L. (1999) The competing countermarks hypothesis: reliable assessment of competitive ability by potential mates. *Anim. Behav.* 58(5): 1027–1037.

Roberts, S. C. (2007) Scent marking. In *Rodent Societies: An Ecological & Evolutionary Perspective*. (Eds. J. O. Wolff and P. W. Sherman), pp. 255–266. University of Chicago Press: Chicago.

Sharpe, S. T., and Millar, J. S. (1980) Relocation of nest sites by female deer mice, *Peromyscuc maniculatus borealis. Can. J. Zool.* 68: 2364–2367.

Sharrow, S. D., Vaughn, J. L., Zidek, L., et al. (2002) Pheromone binding by polymorphic mouse major urinary proteins. *Protein Sci.* 11(9): 2247–2256.

Slotnick, B. (2001) Animal cognition and the rat olfactory system. *Trends Cogn. Sci.* 5(5): 216–222.

Stewart, W. B., Greer, C. A., and Teicher, M. H. (1983) The effect of intranasal zinc sulfate treatment on odor-mediated behavior and on odor-induced metabolic activity in the olfactory bulbs of neonatal rats. *Brain Res.* 284(2–3): 247–259.

Stowers, L., Holy, T. E., Meister, M., et al. (2002) Loss of sex discrimination and male-male aggression in mice deficient for TRP2. *Science* 295(5559): 1493–1500.

Sullivan, R. M. (2003) Developing a sense of safety: the neurobiology of neonatal attachment. *Ann. N.Y. Acad. Sci.* 1008: 122–131.

Teicher, M. H., and Blass, E. M. (1976) Suckling in newborn rats: eliminated by nipple lavage, reinstated by pup saliva. *Science* 193(4251): 422–425.

Thomas, D. A., Takahashi, L. K., and Barfield, R. J. (1983) Analysis of ultrasonic vocalizations emitted by intruders during aggressive encounters among rats (*Rattus norvegicus*). *J. Comp. Psychol.* 97(3): 201–206.

Vagell, M. E., and McGinnis, M. Y. (1998) The role of gonadal steroid receptor activation in the restoration of sociosexual behavior in adult male rats. *Horm. Behav.* 33(3): 163–179.

Vandenbergh, J. G. (1989) Coordination of social signals and ovarian function during sexual development. *J. Anim. Sci.* 67(7): 1841–1847.

Verplancke, G., Le Boulengé, É., and Diederich, C. (2010) Differential foraging in presence of predator and conspecific odors in bank voles: a field enclosure study. *Ecol. Res.* 25(5): 973–981.

Wachowiak, M. (2011) All in a sniff: olfaction as a model for active sensing. *Neuron* 71(6): 962–973.

Wallace, K. J., and Rosen, J. B. (2001) Neurotoxic lesions of the lateral nucleus of the amygdala decrease conditioned fear but not unconditioned fear of a predator odor: comparison with electrolytic lesions. *J. Neurosci.* 21(10): 3619–3627.

Wesson, D. W., and McGinnis, M. Y. (2006) Stacking anabolic androgenic steroids (AAS) during puberty in rats: A neuroendocrine and behavioral assessment. *Pharm. Biochem. Behav.* 83(3): 410–419.

Whitten, W. K. (1956) Modification of the oestrous cycle of the mouse by external stimuli associated with the male. *J. Endocrinology* 13(4): 399–404.

Wills, G. D., Wesley, A. L., Moore, F. R., and Sisemore, D. A. (1983) Social interactions among rodent conspecifics: a review of experimental paradigms. *Neurosci. Biobehav. Rev.* 7(3): 315–323.

Wilson, D. A., and Linster, C. (2008) Neurobiology of a simple memory. *J. Neurophys.* 100(1): 2–7.

Wood, R., and Newman, S. (1995) Integration of chemosensory and hormonal cues is essential for mating in the male Syrian hamster. *J. Neurosci.* 15(11): 7261–7269.

Yamashita, J., Hirata, Y., and Hayashi, S. (1992) Low intermale aggression associated with small submandibular and preputial glands in goldthioglucose-obese mice. *Physiol. Behav.* 52(1): 91–95.

Yamazaki, K., Beauchamp, G. K., Singer, A., et al. (1999) Odortypes: their origin and composition. *Proc. Natl. Acad. Sci. U. S. A.* 96(4): 1522–1525.

Zalaquett, C., and Thiessen, D. (1991) The effects of odors from stressed mice on conspecific behavior. *Physiol. Behav.* 50(1): 221–227.

Chapter 26

Olfaction in the Order Carnivora: Family Canidae

PETER HEPPER and DEBORAH WELLS

26.1 INTRODUCTION

This chapter reviews olfactory behavior in the Carnivora, concentrating on the family Canidae. Although there have been "pockets" of study on various aspects of olfactory behavior in other Carnivora (e.g., Mellon, 1993; Rozhnov and Rozhnov, 1998; Sliwa and Richardson, 1998; Soini et al., 2012), most research has focused on canids and it is these studies that form the focus of this review. Canids are also of interest because of their unique status in 'serving' humans using their sense of smell. Indeed, the olfactory abilities of wolves were recognized by early humans and used to aid hunting and this contributed to the development of cooperation between man and wolves and the process of the domestication of this species (Ruusila and Pesonen, 2004).

The chapter is divided into three main sections. First, aspects of the olfactory system that may contribute to the canids' acute sense of smell are considered. Second, the olfactory behavior of canids is discussed, and, finally, the applications of dogs' olfactory abilities are reviewed. We will use selective examples to illustrate olfactory behavior in canids and highlight issues which require further study in order to fully understand the role of odors for these species. With regard to terminology, we use the term "canid(s)" to refer to the family Canidae and "dog" when referring to the domestic dog, *Canis familiaris*, whether this be a pet, kennelled, or experimental animal. When referring to domestic dogs that are wild or feral, we use the term "free-ranging dog."

26.2 CONSIDERATIONS OF OLFACTION IN THE CARNIVORA

We commence with two observations regarding olfaction in the Carnivora as a whole. First, olfactory behavior generally has been subjected to little study in this Order. As a result, caution must be exercised in making assumptions about the role, or lack of role, of odors in Carnivoran behavior. For example, studies of olfactory structures in Pinnipeds (walrus, seals, and sea-lions) indicate they are under-developed in comparison to other, land-based, Carnivora. This has led to a view that they have poor olfactory abilities and that odor plays little role in guiding their behavior. Recent research has challenged this notion. Psychophysical studies have demonstrated that South African fur seals and harbour seals have good olfactory discrimination abilities and sensitivity to various odors (Kowalewsky et al., 2006; Laska et al., 2008). Further studies have demonstrated the use of olfaction in the natural behavior of pinnipeds. Female Australian sea-lions distinguish their own pups from others by their odor (Pitcher et al., 2011) and the South American sea-lion locates its pups by smell (Trimble and Insley, 2010).

Second, the absence of study has led to some "facts" regarding the abilities of certain Carnivora becoming widely accepted. Perhaps the best example of this is the widely held view that bears have excellent olfactory abilities, for example, polar bears can smell a seal from over 1 km (Seaworld, 2012) or 30 km (Wiki, 2012) away. Whilst published papers report observations of bears in the wild using odors to detect prey (e.g., Blanco et al., 2011),

---

Handbook of Olfaction and Gustation, Third Edition. Edited by Richard L. Doty.
© 2015 Richard L. Doty. Published 2015 by John Wiley & Sons, Inc.

there are no reports demonstrating that the sense of smell in bears is more or less acute than that of other species. Whether the bear's olfactory abilities are exceptional, or better than other Carnivora, therefore requires experimental study.

Available evidence, albeit sparse, suggests that all members of the Order Carnivora have a rich olfactory-guided repertoire of behavior. However, much further scientific study is needed to evaluate the olfactory abilities of, and role played by odor in, various Carnivora species.

26.2.1 Olfactory Sensitivity

Early interactions between humans and wolves are speculated to have resulted from encounters when hunting the same prey. Observations of the hunting behavior of canids may have alerted man to the animals' use of odors and contributed to an association that lasts today. Interest in the olfactory abilities of the canids has arisen because of their reputed excellent sensitivity to smells, far superior to that of man. Evidence for this has been derived from observations of the dog being able to detect the presence of odors, such as explosives and drugs (see Section 26.4), at extremely low levels. Other studies have attempted to quantify the sensitivity of the dog's olfactory system through more conventional psychophysical tests. As a general conclusion, such experiments have found that dogs detect odors at a much lower concentration than do humans (e.g., Neuhaus, 1953; Marshall et al., 1981; Marshall and Moulton, 1981; Krestel et al., 1984; Walker et al., 2006).

A major issue in evaluating the threshold levels of olfactory detection reported by different studies is that varied methodologies have been used and give rise to different threshold estimates for the same odor. For example, when assessing thresholds for n-amyl acetate using a naturalistic "find the target task," Walker et al. (2006) reported threshold detection levels of 1–2 parts per trillion, over 30–20,000 times lower than those reported by Krestel et al. (1984) of 52–326,000 parts per trillion using a conditioned suppression paradigm. The performance of dogs may vary according to the paradigm used and also the training provided. A further significant factor is that different studies have used different breeds, e.g., standard schnauzer and Rottweilers (Walker et al., 2006), beagles (Krestel et al., 1984), and German shepherds and fox terriers (Marshall et al., 1981). Not only may there be breed differences in olfactory abilities, there may also be differences between individuals of the same breed (Marshall and Moulton, 1981).

Despite these problems, studies are consistent in indicating that dogs respond to odors at much lower levels (1,000–1,000,000 times lower) than humans. Much further study is needed, controlling for among other variables, test paradigm and dog breed, to document the sensitivity of the dog's olfactory system.

26.2.2 A Canid is not just a Canid

One note of caution must be added here. The family Canidae comprises some 38 different species. The amount of study examining olfactory behavior in each of these species varies considerably. For example, the experimental studies described above examining olfactory sensitivity have focused exclusively on the dog. It is assumed that similar levels of sensitivity are displayed by other canids. Whether this is true, however, is unknown. The extent that findings found for one species of canid generalize to other canid species is an area requiring further investigation for all aspects of olfaction.

26.3 CANID OLFACTORY SYSTEM(S)

The excellent olfactory abilities of canids have long been established, but the question of what the underlying structural and/or neural mediation of these abilities are remains unknown. Although neural structures of the dogs' olfactory system were being uncovered at the start of the 20th century, more recent developments in genetic research may contribute to our understanding of the factors mediating its olfactory abilities. Here, we briefly review the olfactory system of the canids, focusing on those elements that may account for their acute olfactory sense. Having an excellent ability to detect odors suggests that there are scents present in the animal's environment that impart important information for the individual and its behavior. This section thus also discusses the odors produced by canids.

26.3.1 Olfactory Sensory Structures

The olfactory system of the canids, and indeed the Carnivora, is basically similar to that found in all vertebrates (Stoddart, 1980). Canids possess both main and accessory olfactory bulbs and a vomeronasal organ. Olfactory bulb volume, when compared to total brain volume, is higher in dogs (1.95%), than in herbivores (goat 0.77%) and humans (0.03%) (Kavoi and Jameela, 2011). Compared to rodents, the accessory olfactory bulb is less well developed and appears as a much simpler structure with less lamination (e.g., Salazar et al., 1994). Whilst dogs possess a vomeronasal organ that comprises both sensory and non-sensory epithelium and expresses markers of neuronal activity (Dennis et al., 2003), it is unclear whether the vomeronasal organ serves any function in canids. The characteristic facial expression associated with processing chemosensory stimuli by the vomeronasal organ, flehmen, is not observed in the dog but is present in the coyote, jackal, and bushdog.

The area covered by the olfactory epithelium is much larger in dogs than in humans (e.g., German shepherd 150 cm^2 cf. humans 3–5 cm^2), moreover the receptor cells appear to be more densely packed in dogs (Dodd and Squirrel, 1980). As a consequence, dogs possess a significantly greater number of olfactory receptor cells than do humans, 220 m–2 billion cf. to 12–14 million (Miklosi, 2007). Other structures implicated in transducing chemosensory stimuli, for example, the trigeminal nerve, the septal organ of Masera, Grueneberg ganglion, are all present in the canids (McCotter, 1913; Pettigrew et al., 2009). However, whether these have any functional significance for olfactory behavior in these animals is unknown.

26.3.2 Olfactory Receptor Genes

Since the discovery of olfactory receptor genes (Buck and Axel, 1991), there has been considerable interest in exploring the olfactory receptor gene repertoire of different animals, including the dog; such information may contribute to our understanding of their olfactory abilities. Current analysis has identified around 1100 olfactory receptor (OR) genes in the dog, of which 20–25% are thought to be pseudogenes (Quignon et al., 2012). The overall number of OR genes is higher than identified in humans (600–900), but lower than that found in the mouse (1200–1400) or rat (1600). A similar percentage of pseudogenes (20–25%) are observed in dogs as in rats and mice; humans, by contrast, have more (~50%).

OR gene proteins may be classified into families and sub-families according to their amino acid characteristics. Dogs exhibit a greater number of sub-families compared to the rat and mouse (300 cf. 280 and 246, respectively, Quignon et al., 2006). Two hundred of these genes are Class 1, a high percentage compared to other species. It has been suggested that this may be significant in mediating the olfactory abilities of the dog (Olender et al., 2004).

In contrast to OR genes, only nine intact vomeronasal V1R genes have been observed in the dog (Young et al., 2005). This is a similar number to that identified in humans, but significantly fewer than the 100 plus found in rodents. It is suggested that this smaller number is linked to the reduced structures of the accessory olfactory bulb and vomeronasal organ found in the dog (compared to rodents) and the associated decrease in functionality of these organs (Quignon et al., 2006). The same V1R pseudogenes are found in both the wolf and dog, suggesting that the reduction in the number of V1R genes is not a result of domestication (Young et al., 2010) and may be a feature of canids more generally.

Exactly how the olfactory system of canids accounts for their acute olfactory abilities has yet to be determined. The greater number and more densely packed olfactory receptors may have a role in detection thresholds. The greater repertoire of OR gene sub-families may enable a greater range of odors to be transduced. Although more research is required, some preliminary studies do support the links between structures and function suggested above. For example, the olfactory ability of dogs differs widely between breeds. Boxers, which have a less acute sense of smell than poodles, possess a higher percentage of OR pseudogenes (Quignon et al., 2006). Breeds generally noted for good olfactory abilities (e.g., German Shepherd, Belgian Malinois) exhibit greater OR gene polymorphism compared to breeds with poorer olfactory abilities (e.g., greyhound and Pekingese) (Robin et al., 2009). Trained detection dogs that possess a glycine/arginine substitution at one specific locus on the cOR52N9 gene exhibit poorer performance on explosive detection tasks (Lesniak et al., 2008), indicating the potential importance of specific genes for the detection of specific odors. However, much more work is needed before the link(s) between structure/genes and olfactory function is fully elucidated. It should also be noted that the emphasis to date has been on peripheral sensory receptor mechanisms. Little attention has been paid to how more central processing of olfactory information may contribute to the dog's olfactory abilities.

26.3.3 Sniffing

Whilst the effective transduction of odors to neural impulses is a vital process, the odors have to be received by the receptors. It appears that the mechanism of delivery has been optimized through the form of the dog's nasal structures and its sniffing behavior.

The olfactory epithelium, situated in a recess at the rear of the canine nose, is separated by the *lamina transversa* from the main airways through which air traverses during breathing. This position enables air to travel across the epithelium in one direction during inspiration (allowing the possibility of spatial coding) and no movement of air during expiration, thus providing longer access for the odor molecules to the receptors (Craven et al., 2010). Sniffing is independent from respiration. Dogs sniff at a much higher rate than they respire, and sniffing rate may increase upon encountering novel or weak odors. Whilst breed size affects respiration rate, the rate of sniffing (3–12 Hz) is similar across all breeds of dog. It would thus appear that the odors are delivered to the receptor cells in a manner that optimizes their transduction, facilitating both detection and discrimination (Lawson et al., 2012).

26.3.4 Glands

As well as possessing structures to detect odors, canids also produce odors that may be important for guiding their behavior. Odors are produced by waste products: urine and faeces, and also from a variety of glands distributed over the animal's body. Canids possess sebaceous, apocrine, and eccrine glands (Harrington and Asa, 2003) which can be found singly or in aggregations and emit odor-producing secretions which can be detected by others. In many cases, little is known about the role these odors play or the functional chemicals emitted, and much more work is needed to understand their significance for canid behavior. Here we concentrate on describing the sources of odors and will deal with their possible functions in more detail later.

Sweat (eccrine) glands are found on the footpads and apocrine glands between the toes (Nielsen, 1953). Thus, the animals may leave an odor cue every time the paw comes into contact with a surface or through the specific behavior of ground scratching which may accompany urine or faecal scent marking. Apocrine glands are also spread over the hair-covered skin in most carnivores and may impart a distinctive whole body odor, for example, as noted in African wild dogs (van Heerden, 1981).

The anal region of some canids has three distinct odor producing areas: the supracaudal gland, situated on the top of the tail; the circumanal glands, which surround the anus; and the anal sac, situated internally with ducts to the anal canal. These odor sources are described in the following three paragraphs.

The supracaudal (tail) gland, or "violet" gland in the fox, is comprised of both sebaceous and apocrine cells, although it is the former that is regarded as the main odor-producing component (Albone and Flood, 1976). Most canids possess this gland; although it may be present in the dog, it appears not to function and is absent in the African wild dog. In the red fox, it produces a sweet, pleasant 'violet'-scented odor. The odor is weaker in wolves and coyotes, but smells similar to that produced by the red fox. In the gray fox, the odor is described as "musky" (Fox, 1971). The odor can be easily distinguished from that produced by the fox's anal gland (see below), which is described as "rancid," in contrast to the "pleasant" smell produced by the supracaudal gland. Although this gland is present in wolves, individuals do not appear to investigate this area in conspecifics (Harrington and Asa, 2003).

Anal sacs are paired reservoirs with ducts that lead to the anus and are found in most carnivores. The sacs are surrounded by smooth muscle and it is believed that, to a certain extent, secretions from the anal sac are under voluntary control (Harrington and Asa, 2003). Secretions can accompany some defecations, but not all, and may occur on their own in the absence of defecation, especially when the animal is acutely stressed. In wolves, it is the dominant males that leave the majority of anal sac scent on their faeces (Harrington and Asa, 2003). In skunks and pole cats, anal gland secretions are sprayed on others as a means of defense. In beagles, consistent individual differences in the odors of anal sac secretions are present. Thus, a sharp acrid odor is associated with light colored, less viscous, and more rapidly secreted anal sac secretions, whereas a dull, somewhat neural or slightly pleasant, dog-like odor is associated with dark, pasty, and less rapidly secreted anal sac secretions (Doty and Dunbar, 1974a).

Circumanal glands are located around the anus in canids and are made up of single apocrine glands, sebaceous glands, and hepatoid glands (Shabadash and Zelikina, 2002). With the concentration of odor sources distributed around the anal region, it is not surprising that this area receives significant investigation when conspecifics meet.

In addition to the anal region, the head provides a concentrated source of odors, and the head and muzzle region are sites of interest and investigation between conspecifics. Wolves appear to pay more attention to the head region than do dogs. Apocrine and sebaceous glands are found in high concentrations in the ears, where they combine to produce cerumen (ear wax). Eccrine (sweat) glands are found on the skinless end of the nose and saliva may be distributed around the mouth and jaw area.

Fox (1971) suggested that there is an inverse relationship between the intensity of body odor, especially that produced by the supracaudal gland, and the level of sociality exhibited by the animal. More intense odors are produced by the least social species. Thus foxes, which are less social than wolves and dogs, produce a more powerful odor. One exception to this is the African wild dog, which is highly social and produces a strong odor, but this may result from apocrine secretions on the body as opposed to anal region odors. A more intense odor may be required in more solitary species, as the opportunity for exposure to conspecific odors is poorer and therefore odors need to persist for longer to serve their communication function.

26.3.5 Summary

Whilst the role of some aspects of the canid's olfactory system that contribute to its sensitivity are understood, much still remains to be elucidated. The role of the animal's odors in shaping its olfactory system has yet to be explored. It may be that the information produced by various secretions requires a greater olfactory sensitivity to fully detect the information and this has, in part, driven the development of the canid's olfactory system.

26.4 OLFACTORY-GUIDED BEHAVIOR

The fact that canids have excellent olfactory abilities and possess a variety of sources for producing odors suggests that scent plays a significant role in canid behavior. The following section reviews olfactory-guided behavior in the canids. We consider the development of olfactory behaviors, the use of odors in social recognition, and scent-marking behavior. Again, the review will highlight areas where further research is needed.

26.4.1 Early Development of Olfactory Behavior

Dogs are able to detect chemosensory substances before birth and respond to these odors postnatally (Wells and Hepper, 2006). Dogs exposed to aniseed via their mother's diet prenatally exhibit a preference for this stimulus immediately after birth (Wells and Hepper, 2006). This preference is specific to the "odor" experienced before birth. Such learning may be important for the acceptance of milk (Wells and Hepper, 2006). At birth, pups exhibit rooting behavior, but whether this is controlled by olfactory stimuli is unknown. Pups detect and respond to olfactory stimuli at birth (e.g., aniseed, fish scent), even though the olfactory nerves are poorly myelinized (Scott and Marston, 1950; Jones, 2007). Day-old pups are able to acquire conditioned responses to olfactory stimuli (e.g., Volokhov, 1959). The odor of amniotic fluid may be important for mothers to accept pups, since mothers reject offspring which have been washed, but accept those bathed in amniotic fluid (Abitol and Inglis, 1997). Odors thus play an important role in canid behavior, beginning before birth and this continues throughout the individual's life.

26.4.2 Social Recognition

The ability of conspecifics to recognize others by individual, group or kinship cues is widespread amongst animals. There is no reason to expect these abilities are not present in the canids; however, little attention has been devoted to this issue.

It is generally believed that canids are able to recognize individual conspecifics by odors. It has been speculated that individually identifiable odors may originate from the faeces, urine, feet, anal sacs/gland, supracaudal gland, and skin. However, very little study has explored odor sources of individuality other than that presented by the animal's urine. Interestingly, dogs spend more time investigating the urine from a colony mate than they do their own urine, and longer again when presented with urine from a dog outside their colony (Dunbar and Carmichael, 1981). They also over-mark their own urine less than urine from another dog (Bekoff, 2001). Wolves are able to discriminate individuals by differences in their urine (Brown and Johnston, 1983) and also increase urine over-marking upon encountering urine marks from conspecifics (Peters and Mech, 1975). Red foxes investigate both the urine and anal sac secretions of unfamiliar foxes more than familiar foxes (Blizard and Perry, 1979), although whether this reflects individual recognition, or recognition of familiarity, is unknown. There is additional evidence that dogs produce individually identifying odors. Wells and Hepper (2000) found that dog owners were able to identify their own dogs from their odors deposited on blankets. Detection dogs are able to identify individual maned wolves by their faeces (Wasser et al., 2009).

Dogs are able to recognize their siblings and mother at 5 weeks of age (Hepper, 1986, 1994), most likely using odor cues. Further, dogs recognize their mother by olfactory cues, even after two years of separation, although they only recognize unfamiliar siblings from whom they have been separated for two years if they have been living with another sibling during that time (Hepper, 1994). Contradictory evidence was obtained from a study of dogs examining the mother's ability to recognize her pups using a retrieval task: one female tested retrieved only her own pups, whereas another female retrieved both her own and alien pups (Dunbar et al., 1981). In a simultaneous choice task, mothers investigated their own 5-week-old pups more than unrelated offspring (Hepper, 1994).

Bitches in heat elicit much more investigation from sexually experienced male conspecifics than when not in heat, indicating their odor, especially urine, reflects their reproductive status (Beach and Gilmore, 1949; Doty and Dunbar, 1974b). Moreover, females urinate more when in heat (Beach and Gilmore, 1949), perhaps advertising their receptive status.

It is apparent that much more research is needed to determine the social recognition abilities of the canids. In particular, it will be of interest to establish if there are any differences between species that live in social groups (e.g., hunting dogs, wolves) and those that are more solitary (e.g., foxes).

26.4.3 Scent Marking

One behavior that has elicited investigation in the canids is scent marking. Scent marking, that is, the deposition of an odor through defecation, urination, and/or glandular deposits, is common to all canids. It has been argued that scent marks may signify individual or group identity and/or some aspect of current status, for example, dominance, reproductive state, age. These in turn may serve to determine ownership of territory, mark food caches, maintain group cohesion, and signal status. Whilst the behavior of scent marking has been documented amongst canids, the information presented by marks and how this "acts" on the recipient has been less well elucidated.

Scent marking is not performed equally by all individuals. In most canid species, the dominant individual or pair is responsible for more acts of scent marking than lower ranked individuals, for example, coyotes (Gese and Ruff, 1997), gray wolves (Peters and Mech, 1975), dogs (Lisberg and Snowdon, 2011) and Ethiopian wolves (Sillero-Zubiri and Macdonald, 1998). It is unknown whether the marks contain information of dominance directly or by association, that is, others learn the individual identity of the dominant male/female and associate their individual signature left in the mark with their dominance status.

Individuals in groups or packs mark more than lone individuals, for example, coyotes (Gese and Ruff, 1997) and wolves (Mech and Boitani, 2003). Lone male wolves have been observed defecating away from paths used by packs, whereas packs tend to deposit marks on trails, especially at junctions (Peters and Mech, 1975), presumably places where others can find them. Males urine mark more than females, for example, dogs and free-ranging dogs (Pal, 2003), dholes (Paulraj et al., 1992) and wolves (Peters and Mech, 1975).

Marking may also be influenced by context. A higher rate of marking is observed during the mating season when females come into estrous, for example, dholes (Paulraj et al., 1992) and free-ranging dogs (Pal, 2003), suggesting that this behavior has a role in the maintenance of the pair bond (Rothman and Mech,1979). Indeed, newly formed pairs exhibit an increase in marking. An increased frequency of urine marking is observed when the scent of other conspecifics is present, for example, as in dogs (Hart, 1974) and wolves (Peters and Mech, 1975). Some canids mark food caches that have been emptied with urine, for example, red fox (Henry, 1977), Arctic wolf (Mech, 2006), wolf (Harrington, 1981), and coyote (Allen et al., 1999).

Whilst wolf packs mark frequently in their own territory, they do not mark in the territories of their neighbors (Mech and Boitani, 2003). Increased marking is observed around the periphery of the animals'/group's territory, for example, wolves (Peters and Mech, 1975), free-ranging dogs (Pal, 2003), coyotes (Allen et al., 1999), and foxes (Arnold et al., 2011), suggesting a role in territory demarcation. Faeces may be deposited in latrines throughout the territory, for example, golden jackals (Macdonald, 1979a), dholes (Davidar, 1975), and Ethiopian wolves (Sillero-Zubiri and Macdonald, 1998).

Scent marking by urination in male canids is accompanied by a specific raised leg urination posture and often only small amounts of urine are expelled (see Figure 26.1). This posture may be important as the animal sometimes displays this pose without urinating. The raised leg position may also serve to raise the position of the mark on the vertical object, perhaps reflecting something about the scent marker and

Figure 26.1 The characteristic 'raised leg display' posture adopted by male canids when depositing urine scent marks, here exemplified by the coyote (by permission J.W. Wall, jwallphoto.blogspot.com).

also to position the mark at nose height, thereby increasing the chances of detection. However, given the acute olfactory sense of the canid, this latter explanation seems unlikely and the function of raised leg urination is probably related more to its spatial position than odor. Canids, for example, red fox (Macdonald, 1979b), maned wolves (Macdonald, 1980), and black-backed jackals (Haywood and Haywood, 2010) deposit faecal marks on conspicuous, often raised, locations. Although the function of this is unknown, it is believed to increase the visual conspicuousness of the mark.

One specific pattern of urine marking that is observed is "double marking," in which a scent mark deposited by one individual of a dominant pair is quickly over-marked by the other, for example, Arctic wolves (Harrington, 2006), gray wolves (Rothman and Mech, 1979), and coyotes (Gese and Ruff, 1997). It is speculated that this aids in courtship, the formation and maintenance of pair bonds, mate guarding, and territory formation (Macdonald, 1985). These hypotheses require further examination.

Observations of scent marking by canids are abundant and some of the parameters that affect marking, for example, dominance, and its context, edge of territories, are well established. As more studies emerge, increased variability in scent-marking behavior is becoming apparent. For example, the ground scratching rates of wolves in Poland (Zub et al., 2003) are much higher than those reported for wolves in North America (Peters and Mech, 1975). Individual differences are observed in marking behavior (Wirant and McGuire, 2004) and differences are seen between the behavior of captive and wild animals. The information provided by scent marks remains somewhat of a mystery. It is clear that these odor marks have a function and present information to other conspecifics. The nature of this information has yet to be elucidated. Such information is essential if the functions of scent marking are to be fully understood. Scent-marking behavior is ubiquitous in canids yet much remains to be discovered about how scent marks exert their function.

26.4.4 Scent Rolling

A final behavior worth noting is scent rolling. Many canids roll in particularly pungent substances, such as rotten food or faeces (Reiger, 1979). Such rolling is often undertaken in a ritualistic manner. A number of possible functions have been suggested: to learn about the new odors; to camouflage and hide their own odors; to enhance their social standing; an evolutionary remnant of behavior that now serves no function; and, a comforting/reinforcing behavior. Much further work is needed to explore this. In African wild dogs, females roll in the urine of the males of packs they attempt to join (Frame et al., 1979). Presumably this enhances their acceptability, but exactly how remains to be determined.

26.4.5 Summary

Odors are important in guiding the behavior of canids, from birth and throughout their lives. However, our understanding of the full range of behaviors influenced by odors and how they function in different canid species remains sparse and much further work is needed before the extent of the role of odor in canids is fully understood.

26.5 USES OF THE DOG'S NOSE

The olfactory abilities of dogs, combined with their trainability, have led to their use to detect and discriminate odors in a wide variety of settings and for a variety of purposes. The following section will review these uses. First, however, a number of issues relating to the performance of dogs on these tasks are considered.

26.5.1 A Dog is Not Just a Dog

The use of dogs in applied settings is widespread, however factors that may affect their performance have yet to be fully explored. The performance of the dog will be influenced by two main factors: the individual animal and its training. These factors may singly, or in combination, influence how well the dog does on any particular task.

Different breeds of dog differ in their olfactory abilities, moreover, even within a particular breed, there are differences between individuals in their sensitivity and ability to discriminate odors. As yet, there has been no systematic study of breed or individual differences upon olfactory abilities within dogs. Related to this, is the dog's motivation to undertake the required task. Scott and Fuller (1965) assessed the interest of different breeds in locating

a live mouse in a 1 acre field. Beagles found the mouse in 1 minute, whilst fox terriers took 15 minutes. Scottish terriers never found the mouse and indeed it is reported that one even stood on the rodent but failed to notice it. Dogs will differ in their motivation and this also has to be factored in to assessments of performance.

The vast majority of detection dogs undergo training to ensure they are able to perform the required task. That said, one of the first reported instances of a dog detecting skin cancer was by an untrained animal (Williams and Pembroke, 1989). There are a number of issues surrounding training that should be considered. First, as with olfactory abilities, there are individual differences in the trainability of dogs (e.g., Maejima et al., 2007). Second, there is no standardized method for training dogs (e.g., Jezierski et al., 2010). Thus, the most appropriate procedure for obtaining optimum performance in dogs is unknown. Further, there is no standardized method for assessing performance (e.g., Rooney et al., 2007). Often organizations using dogs will have their own criteria, but these may vary between organizations and between applications.

Due to the potential variation introduced by these factors it is difficult to evaluate the meaning of a specific performance achievement (e.g., x% success rate) for any particular task. For example, if a dog is reported as being 80% successful, this does not necessarily reflect the absolute capability of *this* dog (e.g., training may have hindered its performance) or *all* dogs (e.g., this dog could have superior/inferior olfactory abilities). In the absence of standardized procedures for evaluating performance, the reported figures of success are little more than indications that, under some circumstances, a particular dog can complete, to some extent, the task required. Hence we do not report absolute success rates here. Comparisons within studies are reported where relevant as the same techniques (for good or bad) have been used and enable some conclusions to be drawn about relative performance.

The use of detection dogs is considered below under four main headings: forensic; economic, conservation, and health.

26.5.2 Forensic Applications

It has been widely assumed that one of the earliest uses of dogs by man was for tracking prey, thus assisting hunting. A variety of canids are reported to use odors (amongst other senses) in their hunting (Fox, 1975). Indeed dogs have been specifically bred for their ability to follow an odor trail. Romans found the Celtic tribes of the UK using the Agassian, a small dog renowned for its ability to air scent. In 2012, The Federation Cynologique Internationale recognized over 70 breeds of scent hounds.

Dogs have exceptional abilities to follow a route travelled by an individual on foot (e.g., Wells and Hepper,

2003). Dogs may use one, or a combination, of three methods to follow an individual (Hepper and Wells, 2005). They may: *track*, where they follow the route of the individual with their head down and nose very close to the ground; *trail*, when they follow the route with their head down when moving with the wind, but with their head up when moving into the wind; or *air scent*, where they follow the route with their heads up sampling the air.

The odor cues used by dogs may originate either from the individual directly or from ground disturbance (Hepper and Wells, 2005). The fact that individuals can follow the trail of one individual when crossed by many others suggests that there are individual odor cues that are available to the dog to use (Kalmus, 1955). The literature, however, is conflicting. Some studies suggest dogs cannot track in the absence of odor from ground disturbance (e.g., Budgett, 1933), others that dogs cannot follow a trail in the absence of the individual's odor (e.g., Pearsall and Verbruggen, 1982).

Whilst dogs can follow a track, exactly how they do so remains poorly understood. Some studies have attempted to explore how dogs use odor information to detect the direction of a track. For example, Hepper and Wells (2005) found that there was sufficient information contained in five consecutive footprints to determine the direction of a track. An evaluation of the time to deposit the five footsteps indicated that it took approximately 1.9 seconds from the 1st to the 5th step. Assuming that the dog makes some comparison between the footsteps (e.g., based on the quality of odor), then there is a change in the olfactory information presented by a time difference of under 2 seconds that is detectable by dogs. What also seems vital for dogs to track is discontinuity in the olfactory signal. Whilst dogs can easily determine directionality from a track formed of discrete footsteps, they find it much more difficult from a continuous track (Steen and Wilson, 1990). This is similar to insect behavior. Moths, for example, are unable to locate the source of odors when presented with a continuous plume, but can do so when the odor plume presents intermittent information (Justus and Cardé, 2002).

The ability of dogs to discriminate individuals (e.g., Hepper, 1988) has led to proposals that such animals may be used to detect individual odor "fingerprints" from crime scenes and use this information to match to the owner of the scent and possibly the perpetrator of the crime (Schoon, 1996). Scent identification line-ups, in which the dog is required to match the odor obtained from a crime scene to one of a number of odors presented in a row, is now accepted as part of proceedings in a number of European Courts (Schoon and Haak, 2002).

The ability of the dog to detect human odor is also used by search and rescue dogs, ranging from locating lost hikers to individuals buried under avalanches or earthquake debris. Others have been trained to locate cadavers and

parts of bodies (e.g., Komar, 1999). Of all the detections that the dog can perform, we still consider one of the most impressive to be the location of bodies submerged underwater (e.g., Osterkamp, 2011).

It is worth mentioning here another issue to be addressed in assessing detection dogs' performance. In most cases, dogs work as part of a team, the other team-member being their handler. It is known that handlers influence the reliability of their dog's performance (e.g., Lasseter et al., 2003). Thus, a further complication in assessing how well dogs perform is to assess the role of the handler. It may be appropriate to consider that in tasks in which the dog performs "off-lead," the animal is considered the unit of assessment, however when the dog performs the task "on-lead," it is the dog and handler that is considered the unit of assessment.

The ability of the dog to be trained and recognize specific odors has led to it being used in a number of forensic settings to detect the presence of a particular trained substance. Such detections include explosives (Gazit et al., 2005) and even the scent of the trip wires that trigger mines (Hayter, 2003). Following the detonation of explosives, for example, IEDs, dogs are able to match human odors obtained from the blast debris to the humans depositing scent on the device pre-explosion (e.g., Curran et al., 2010). Dogs are able to discriminate accelerants that have been used in arson, despite the presence of additional burnt material at the scene (e.g., Gialamas, 1996). Dogs are used to detect drugs, counterfeit and smuggled goods, and mobile phones in the prison population, amongst others. Indeed, with appropriate training, dogs appear able to detect virtually anything, provided it emits an odor at a level that can be detected by the dog.

26.5.3 Economic Applications

Detector dogs have been used in a variety of ways to reduce economic costs arising from damage to buildings or crops and to improve productivity. For example, dogs have been trained to detect the presence of dry rot fungi and decay caused by fungi (Kauhanen et al., 2002) and termite infestations (Brooks et al., 2003). Both causes of damage may be difficult to detect early and by visual inspection and result in considerable economic cost if not treated. When the performance of dogs to detect termites was compared against an electronic odor-sensing device, dogs performed better (81% vs 48–62% successful detection), but only when there was more than 50 termites present (Lewis et al., 1997). Below this number, the dog's performance became poorer, possibly due to the lack of sufficient odor.

Dogs have also been used in agricultural settings to detect the presence of insects that may damage crops, for example, red palm weevil (Nakash et al., 2000) and parasites which cause damage to livestock, for example, screwworm and its pupae (Welch, 1990) and nematodes (Richards et al., 2008). Dogs can also, more effectively than other methods, determine estrous in dairy cows, thereby enhancing breeding success (Fischer-Tenhagen et al., 2011). Finally, dogs have been trained to detect bed bugs (Pfiester et al., 2008), a cause of concern for travellers and individuals prone to allergic reactions.

26.5.4 Conservation Applications

Detection dogs have found extensive use in conservation applications to detect the presence of other animals, a feat they perform more effectively than other methods. For example, Long et al. (2007) found that dogs were able to detect black bears, fishers, and bobcats from their faeces. Dogs detected more of these species than either hair snares or cameras; however they were more expensive to employ. Overall, however, the increased cost was more than made up for by increased detection rates and dogs were found to be the most cost effective method.

Dogs have been used to detect the presence of a wide variety of animals to enable assessment of their number in a particular area. This includes, for example, species large, Amur tiger (Kerley and Salkina, 2007), small, rats and mice (Gsell et al., 2010), and minute, fire ants (Lin et al., 2011). As well as detecting feces to provide estimates of numbers of animals, dogs have also been used to help humans detect and collect a sufficient quantity of fecal material to enable detailed testing to be undertaken to provide information on the health of the individuals, for example, Northern right whales (Rolland et al., 2006).

Dogs have been used to detect the presence of invasive species which may cause devastation to their new environment. For example, brown tree-snakes in Guam (Savidge et al., 2011), Quagga and zebra mussels in California (California Dept Fish and Game, 2009), and knapweed in Montana (Goodwin, 2010).

Finally, dogs may also be used to provide information about the effects of human changes to the environment on animal species. One area of concern has been the effect of wind farms on bird life. Dogs have been found to be far better at locating carcasses of birds around wind farms than humans, thus improving estimates of the impact of wind farms on the bird population (Paula et al., 2011).

26.5.5 Health and Medical Applications

One area of considerable interest in recent years has been the use of the dog's nose as a diagnostic instrument for evaluating human health (Wells, 2007). An initial paper (Williams and Pembroke, 1989) reported the case of a border collie/Doberman cross-breed persistently sniffing at a mole on the patient's skin. Following investigation,

the mole was found to be a malignant melanoma. The dog had undergone no training and appeared to be attracted to something about the odor of the mole. Since then, other studies have reported the ability of dogs to detect melanoma (Wells, 2012).

Based on the assumption that the presence of cancer changes the odor signature of the individual in a manner that is consistent across patients with the same condition, studies have explored the ability of dogs to discriminate samples from patients with cancer from those without. Dogs are able to detect the presence of bladder cancer from urine samples (Willis et al., 2004), although it has been questioned as to whether this is due to detecting the odor arising from the cancer or other factors that increase the risk of cancer, for example, smoking (Leahy, 2007). Dogs have been reported to be able to detect prostate cancer from urine (Cornu et al., 2011), breast and lung cancer from breath samples (McCulloch et al., 2006), colorectal cancer from breath or stool sample (Sonoda et al., 2011), and ovarian cancer from tissue (Horvath et al., 2008).

A second line of research has explored whether dogs can predict the onset of serious medical conditions. Based on owners' accounts of their dogs' untrained behavior, Wells et al. (2008) report that pet dogs displayed behavioral changes when their diabetic owners became hypoglycaemic, in many cases before the patients knew they were becoming low in sugar. Similar reports of dogs' behavior changing prior to the onset of migraine attacks have also been documented (Marcus, 2012). Studies indicate dogs can be trained to detect the onset of an epileptic seizure (Strong et al., 1999). Whether these alerts are based on the dog detecting changes in the individual's odor, as opposed to some other cue (e.g., visual), is unknown.

26.5.6 Summary

In summary, if it produces an odor the likelihood is that the dog can be trained to detect it and discriminate it from other odors. The key task is to correctly identify the odor for the dog. Despite many studies demonstrating that dogs undertake remarkable detections, the lack of standardization in, and standards of, training and appropriate evaluations of dog olfactory abilities preclude accurate assessment of the effectiveness of dogs, in general, performing these tasks. Further, great coverage is given when search and rescue dogs locate an individual alive following an appalling natural disaster. However, the fact that it is a natural disaster and a scene of devastation and confusion means it is virtually impossible to determine if dogs missed other individuals. Thus, assessing performance in the field may be difficult and this further emphasizes the need for standardized assessment and evaluation during training.

26.6 IMPROVING WELFARE THROUGH ODORS

Lately, some attention has been directed towards the use of so-called pheromones as a tool for enhancing animal welfare (see Doty, 2010, for a critique of the pheromone concept in mammals). Synthetic analogues of domestic species' facial secretions have been successfully produced to this end, with studies reporting efficacy in producing more normal patterns of behavior in animals who are exposed to them (e.g., Spielman, 2000). A "dog appeasing pheromone" (DAP, Ceva Sante Animale, France) has been reported to be useful in treating some behavior problems (e.g., firework phobia, travel-related excitement, separation anxiety) in pet dogs, although its effectiveness has been questioned (Frank et al., 2010). DAP has also been reported as useful in enriching the environment of captive canids. Thus, Tod et al. (2005) found a significantly reduced frequency of barking in sheltered dogs subjected to DAP exposure for 7 days, although maximum barking amplitude was not significantly altered.

Other odors have been used to provide enrichment for animals in captivity (e.g., cats and gorillas, Wells, 2009), but only one study has evaluated the use of odors for enriching the welfare of dogs (Graham et al., 2005). Odors were diffused into dogs' cages in a rescue shelter to assess whether they influenced the animals' behavior in ways suggestive of enhanced welfare. Lavender and chamomile encouraged more behaviors suggestive of increased relaxation (more resting and less barking), while peppermint and rosemary increased locomotion and barking. Given the significance of odors for canids, manipulation of their odor environment undoubtedly has the potential to improve their welfare. This is an area worthy of significant further study.

26.7 CANINE OLFACTION: A STUDY IN CONTRADICTION

Ever since man encountered canids, the acute olfactory sense of the wolf, and then the dog, has been recognized, selected for, and put to use. Recent years have seen the development of applications for the dog's nose, expanding from the earliest uses of tracking prey to the more recent detection of cancerous tumors. However, although the behavioral evidence is clear, our understanding of the factors underlying the canid's sense of smell and its functions is poor. Even basic questions, such as how good is the sense of smell across canids have yet to be answered. The challenge for future research is to explain what factors, both proximate and ultimate, have led to the acute sense of smell in the canid. We need to identify what information is provided by the olfactory cues left by canids, often in a stylised and ritualistic manner. This will not only enhance our understanding of the olfactory world of the canid, but may also improve our opportunities to benefit from this sensory skill, thereby extending the cooperation that has been evident since *Homo* and *Canis* first encountered each other.

REFERENCES

Abitol, M. L., and Inglis, S. R. (1997). Role of amniotic fluid in newborn acceptance and bonding in canines. *J. Mat-Fetal. Med.* 6: 49–52.

Albone, E. S., and Flood, P. F. (1976). The supracaudal scent gland of the red fox *Vulpes vulpes*. *J. Chem. Ecol.* 2: 167–175.

Allen, J. J., Bekoff, M., and Crabtree, R. L. (1999). An observational study of coyote (*Canis latrans*) scent-marking and territoriality in Yellowstone National Park. *Ethol.* 105: 289–302.

Arnold, J., Soulsbury C. D., and Harris, S. (2011). Spatial and behavioral changes by red foxes (*Vulpes vulpes*) in response to artificial territory intrusion. *Can. J. Zool.* 89: 808–815.

Beach, F. A. and Gilmore, R. W. (1949). Response of male dogs to urine from females in heat. *J. Mammal.* 30: 391–392.

Bekoff, M. (2001). Observations of scent-marking and discriminating self from others by a domestic dog (*Canis familiaris*): tales of displaced yellow snow. *Behav. Proc.* 55: 75–79.

Blanco, J. C., Ballesteros, F., Garcia-Serrano, A., et al. (2011). Behaviour of brown bears killing wild ungulates in the Cantabrian Mountains, Southwestern Europe. *Eur. J. Wildlf. Res.* 57: 669–673.

Blizard, R. A. and Perry, G. C. (1979). Response of captive male red foxes (*Vulpes vulpes* L.) to some conspecific odors. *J. Chem. Ecol.* 5: 869–880.

Brooks, S. E., Oi, F. M., and Koehler, P. G. (2003). Ability of canine termite detectors to locate live termites and discriminate them from non-termite material. *J. Econ. Entomol.* 96: 1259–1266.

Brown, D. S., and Johnston, R. E. (1983) Individual discrimination on the basis of urine in dogs and wolves. In Chemical Signals in Vertebrates 3, Müller-Schwarze, D., and Silverstein, R. M. (Eds.). New York, Plenum, pp. 343–346.

Buck, L., and Axel, R. (1991). A novel multigene family may encode odorant receptors: a molecular basis for odor recognition. *Cell* 65: 175–187.

Budgett, H. M. (1933) Hunting by Scent. New York, Charles Scribner.

California Department of Fish and Game (2009). K-9 Programme. www.dfa.ca.gov/enforcement/K9/.

Cornu, J-N., Cancel-Tassin, G., Ondet, V., et al. (2011). Olfactory detection of prostate cancer by dogs sniffing urine: a step forward in early diagnosis. *Eur. Urol.* 59: 197–201.

Craven, B. A., Paterson, E. G., and Settles, G. S. (2010). The fluid dynamics of canine olfaction: unique nasal airflow patterns as an explanation of macrosmia. *J.R.Soc. Interface* 7: 933–943.

Curran, A. M., Prada, P. A., and Furton, K. G. (2010). Canine human scent identifications with post-blast debris collected from improvised explosive devices. *Forensic Sci. Int.* 199: 103–108.

Davidar, E. R. C. (1975). Ecology and behavior of the Dhole or Indian wild dog *Cuon alpinus* (Pallus). In The Wild Canids. Their systematics, behavioral ecology and evolution, Fox, M. W. (Ed.). New York, Van Nostrand Reinhold, pp. 109–119.

Dennis, J. C., Allgier, J. G., Desouza, L. S., et al. (2003). Immunohistochemistry of the canine vomeronasal organ. *J. Anat.* 203: 329–338.

Dodd, G. H., and Squirrel, D. J. (1980). Structure and mechanism in the mammalian olfactory system. *Symp. Zool. Soc. Lond.* 45: 35–36.

Doty, R. L. (2010) The Great Pheromone Myth. Baltimore: Johns Hopkins University.

References

Doty, R. L., and Dunbar, I. A. (1974a). Color, odor, consistency and secretory rate of anal sac secretions from male, female and early-androgenized female Beagles. *Am. J. Vet. Res.* 12: 825–833.

Doty, R. L., and Dunbar, I. (1974b). Attraction of beagles to conspecific urine, vaginal and anal sac secretions. *Physiol. Behav.* 12: 825–833.

Dunbar, I., and Carmichael, M. (1981). The response of male dogs to urine from other males. *Behav. Neur. Biol.* 31: 465–470.

Dunbar, I., Ranson, E., and Buehler, M. (1981). Pup retrieval and maternal attraction to canine amniotic fluids. *Behav. Proc.* 6: 249–260.

Fischer-Tenhage, C., Wetterholm, L., Tenhagen, B-A., and Heuwieser, W. (2011). Training dogs on a scent platform for oestrous detection in cows. *Appl. Anim. Behav. Sci.* 131: 63–70.

Fox, M. W. (1971). Behaviour of Wolves, Dogs, and Related Canids. London, Jonathan Cape Ltd.

Fox, M. W. (1975). The Wild Canids. *Their systematics, Behavioral ecology and evolution.* New York, Van Nostrand Reinhold.

Frame, L. H., Malcolm, J. R., Frame, G. W., and van Lawick, H. (1979). Social organization of African wild dogs (*Lycaon pictus*) on the Serengeti Plains. *Z. Tierpsychol.* 50: 225–249.

Frank, D., Beauchamp, G., and Palestrini, C. (2010). Systematic review of the use of pheromones for treatment of undesirable behaviour in cats and dogs. *JAVMA* 236: 1308–1316.

Gazit, I., Goldblatt, A., and Terkel, J. (2005). Formation of an olfactory search image for explosives odours in sniffer dogs. *Ethol.* 111: 669–680.

Gese, E. M., and Ruff, R. L. (1997). Scent-marking by coyotes, *Canis latrans*: the influence of social and ecological factors. *Anim. Behav.* 54: 1155–1166.

Gialamas, D. M. (1996). Enhancement of fire scene investigations using accelerant detection canines. *Sci. Justice* 36: 51–54.

Goodwin, K. M. (2010). *Using canines to detect spotted knapweed: field surveys and characterization of plant volatiles.* MSc Thesis, Montana State University.

Graham, L., Wells, D. L., and Hepper, P. G. (2005). The influence of olfactory stimulation on the behaviour of dogs housed in a rescue shelter. *Appl. Anim. Behav. Sci.* 91: 143–153.

Gsell, A., Innes, J., de Monchy, P., and Brunton, D. (2010). The success of using trained dogs to locate sparse rodents in pest-free structures. *Wildlf. Res.* 37: 39–46.

Harrington F. H. (1981). Urine-marking and food caching behaviour in the wolf. *Behav.* 76: 280–288.

Harrington, F. H. (2006). Double marking in Arctic wolves, *Canis lupus arctos*: Influence of order on posture. *Can. Field. Natural.* 120: 417–473.

Harrington, F., and Asa, C. S. (2003). Wolf communication. In Wolves. Behavior, Ecology, and Conservation, Mech, L. D. and Boitani, L. (Eds.). London, University of Chicago Press, pp. 66–103.

Hart, B. L. (1974). Environmental and hormonal influences on urine marking behaviour in the adult male dog. *Behav. Biol.* 11: 167–176.

Hayter, D. (2003). Training dogs to detect tripwires. In Mine Detection Dogs: Training, Operations and Odour Detection, McLean, I. G. (Ed.). Geneva, GICHD, pp. 109–138.

Haywood, M. W., and Haywood, G. J. (2010). Potential amplification of territorial advertisement markings by black-backed jackals (*Canis mesomelas*). *Behav.* 147: 979–992.

Hepper, P. G. (1986). Sibling recognition in the domestic dog. *Anim. Behav.* 34: 288–289.

Hepper, P. G. (1988). The discrimination of human odour by the dog. *Perception* 17: 549–554.

Hepper, P. G. (1994). Long-term retention of kinship recognition established during infancy in the domestic dog. *Behav. Proc.* 33: 3–14.

Hepper, P. G., and Wells, D. L. (2005). How many footsteps do dogs need to determine the direction of an odour trail? *Chem. Sens.* 30: 291–298.

Henry, J. D. (1977). The use of urine marking in the scavenging behaviour of the red fox (*Vulpes vulpes*). *Behav.* 61: 82–105.

Horvath, G., Jarverud, G. A., Jarverud, S., and Horvath, I. (2008). Human ovarian carcinomas detected by specific odor. *Integr. Cancer Therap.* 7: 76–80.

Jezierski, T., Górecka-Bruzda, A., Walczak, M., et al. (2010). Operant conditioning of dogs (*Canis familiaris*) for identification of humans using scent lineup. *Anim. Sci. Pap. Rep.* 28: 81–93.

Jones, A. C. (2007). Sensory development in puppies (*Canis lupus f. familiaris*): implications for improving canine welfare. *Anim. Welf.* 16: 319–329.

Justus, K. A., and Cardé, R. T. (2002). Flight behaviour of males in two moths, *Cadra cautella* and *Pectinophora gossypiella*, in homogenous clouds of pheromone. *Physiol. Entomol.* 27: 67–75.

Kalmus, H. (1955). The discrimination by the nose of the dog of individual human odours and in particular of the odours of twins. *Brit. J. Anim. Behav.* 3: 25–31.

Kauhanen, E., Harri, M., Nevalainen, A., and Nevalainen, T. (2002). Validity of detection of microbial growth in buildings by trained dogs. *Environ. Intl.* 28: 153–157.

Kavoi, B. M., and Jameela, H. (2011). Comparative morphometry of the olfactory bulb, tract and stria in the human, dog and goat. *Int. J. Morph.* 29: 939–946.

Kerley, L. L., and Salkina, G. P. (2007). Using scent-matching dogs to identify individual Amur tigers from scats. *J. Wildlf. Manag.* 71: 1349–1356.

Krestel, D., Passe, D., Smith, J. C., and Jonsson, L. (1984). Behavioral determination of olfactory thresholds to amyl acetate in dogs. *Neurosci. Biobehav. Rev.* 8: 169–174.

Komar, D. (1999). The use of cadaver dogs in locating scattered scavenged human remains: preliminary field test results. *J. For. Sci.* 44: 405–408.

Kowalewsky, S., Dambach, M., Mauck, B., and Denhardt, G. (2006). High olfactory sensitivity for dimethyl sulphide in harbour seals. *Biol. Lett.* 2: 106–109.

Laska, M., Svelander, M., and Amundin, M. (2008). Successful acquisition of an olfactory discrimination paradigm by South African fur seals, *Arctocephalus pusillus*. *Physiol. Behav.* 93: 1033–1038.

Lasseter, A. E., Jacobi, K. P., Farley, R., and Hensel, L. (2003). Cadaver dog and handler team capabilities in the recovery of buried human remains in the southeastern United States. *J. For. Sci.* 48: 617–621.

Lawson, M. J., Craven, B. A., Paterson, E. G., and Settles, G. S. (2012). A computational study of odorant transport and deposition in the canine nasal cavity: implications for olfaction. *Chem. Sens.* 37: 553–566.

Leahy, M. (2007). Olfactory detection of human bladder cancer by dogs: cause or association? *BMJ* 329: 1286.

Lesniak, A., Walczak, M., Jezierski, T., et al. (2008). Canine olfactory receptor gene polymorphism and its relation to odor detection performance by sniffer dogs. *J. Heredit.* 99: 518–527.

Lewis, V. R., Fouche, C. F., and Lemaster, R. L. (1997). Evaluation of dog-assisted searches and electronic odor devices for detecting the Western subterranean termite. *Forest Prod. J.* 47: 79–84.

Lisberg, A. E., and Snowdon, C. T. (2011). Effects of sex, social status and gonadectomy on countermarking by domestic dogs, *Canis familiaris*. *Anim. Behav.* 81: 757–764.

Lin, H-M., Chi, W-L., Lin, C-C., et al. (2011). Fire ant-detecting canines: A complementary method in detecting red imported fire ants. *J. Econ. Entomol.* 104: 225–231.

Long, R. A., Donovan, T. M., Mackay, P., et al. (2007). Comparing scat detection dogs, cameras, and hair snares for surveying carnivores. *J. Wildlf. Man.* 71: 2018–2025.

Macdonald, D. W. (1979a). The flexible social system of the golden jackal, *Canis aureus*. *Behav. Ecol. Sociobiol.* 5: 17–38.

Macdonald, D. W. (1979b). Some observations and field experiments on the urine marking behaviour of the red fox, *Vulpes vulpes*. *Z. Tierpsychol.* 51: 1–22.

Macdonald, D. W. (1980). Patterns of scent marking with urine and faeces amongst carnivore communities. *Symp. Zool. Soc. Lond.* 45: 107–139.

Macdonald, D. W. (1985). The carnivores: order Carnivora. In Social Odours in Mammals, Brown, R. E. and Macdonald, D. W. (Eds.). Oxford: Clarendon Press, pp. 619–722.

Maejima, M., Inoue-Murayama M., Tonosaki, K., et al. (2007). Traits and genotype may predict the successful training of drug detection dogs. *Appl. Anim. Behav. Sci.* 107: 287–298.

Marcus, D. A. (2012). Canine responses to impending migranes. *J. Alt. Comp. Med.* 18: 106–108.

Marshall, D. A., Blumer, L., and Moulton, D. G. (1981). Odor detection curves for *n*-pentanoic acid in dogs and humans. *Chem. Sens.* 6: 445–453.

Marshall, D. A., and Moulton, D. G. (1981). Olfactory sensitivity to α–ionone in humans and dogs. *Chem. Sens.* 6: 53–61.

McCotter, R. E. (1913) The nervus terminalis in the adult dog and cat. *J. Comp. Neurol.* 23: 145–152.

McCulloch, M., Jezierski, T., Broffman, M., et al. (2006). Diagnostic accuracy of canine scent detection in early- and late-stage lung and breast cancers. *Int. Can. Therap.* 5: 30–39.

Mech, L. D., and Boitani, L. (2003). Wolf social ecology. In The Wild Canids. Their systematics, Behavioral ecology and evolution, Fox, M. W. (Ed.). New York, Van Nostrand Reinhold, pp. 1–34.

Mech, L. D. (2006). Urine-marking and ground scratching by free-ranging Arctic wolves, *Canis lupus arctos*, in summer. *Can. Field. Natural.* 120: 466–470.

Mellon, J. D. (1993). A comparative-analysis of scent-marking, social and reproductive-behavior in 20 species of small cats (Felis). *Am. Zool.* 33: 151–166.

Miklosi, A. (2007). Dog Behaviour, Evolution and Cognition. Oxford, Oxford University Press.

Nakash, J., Osem., Y., and Kehat, M. (2000). A suggestion to use dogs for detecting red palm weevil (*Rhynchophorus ferrugineus*) infestation in date palms in Israel. *Phytoparasitica* 28: 153–155.

Neuhaus, W. (1953). Uber die Riechscharfe des Hundes fur Fettsauren. *Z. vergl. Physiol.* 35: 527–552.

Nielsen, S. W. (1953). Glands of the canine skin morphology and distribution. *Am. J. Vet. Res. Assoc.* 14: 448–454.

Olender, T., Fuchs, T., Linhart, C., et al. (2004). The canine olfactory subgenome. *Genomics* 83: 361–372.

Osterkamp, T. (2011). K9 water searches: scent and scent transport considerations. *J. For. Sci.* 56: 907–912.

Pal, S. K. (2003). Urine marking by free-ranging dogs (*Canis familiaris*) in relation to sex, season, place and posture. *Appl. Anim. Behav. Sci.* 80: 45–59.

Paula, J., Leal, M. C., Silva, M. J., et al. (2011). Dogs as a tool to improve bird-strike mortality estimates at wind farms. *J. Nat. Conserv.* 19: 202–208.

Paulraj, S., Sundararajan, N., Manimozhi, A., and Walker, S. (1992). Reproduction of the Indian Wild dog (*Cuon alpinus*) in captivity. *Zoo Biol.* 11: 235–241.

Pearsall, M. D., and Verbruggen, H. (1982). Scent - Training to Track, Search and Rescue. Alpine Publications, Colorado.

Peters, R. P., and Mech, L. D. (1975). Scent-marking in wolves. *Amer. Sci.* 63: 628–637.

Pettigrew, R., Rylander, H., and Schwarz, T. (2009). Magnetic resonance imaging contrast enhancement of the trigeminal nerve in dogs without evidence of trigeminal neuropathy. *Vet. Radiol. Ultrasound* 50: 276–278.

Pfiester, M., Koehler, P. G., and Pereira, R. M. (2008). Ability of bed bug-detecting canines to locate live bed bugs and viable bed bug eggs. *J. Econ. Entomol.* 101: 1389–1396.

Pitcher, B. J., Harcourt, R. G., Schaal., B., and Charrier, I. (2011). Social olfaction in marine mammals: wild female Australian sea lions can identify their pup's scent. *Biol. Lett.* 7: 60–62.

Quignon, P., Rimbault, M., Robin, S., and Galibert, F. (2012). Genetics of canine olfaction and receptor diversity. *Mamm. Genome* 23: 132–143.

Quignon, P., Tacher, S., Rimbault, M., and Galibert, F. (2006). The dog olfactory and vomeronasal receptor repertoires. In The Dog and its Genome, Ostrander, E. A., Giger, U., and Lindblad-Toh, K. (Eds.). Cold Spring Harbor, Cold Spring Harbor Laboratory Press, pp. 221–231.

Reiger, I. (1979). Scent rubbing in carnivores. *Carnivores* 2: 7–25.

Richards, K. M., Cotton, S. J., and Sandeman, R. M. (2008). The use of detector dogs in the diagnosis of nematode infections in sheep feces. *J. Vet. Behav.* 3: 25–31.

Robin, S., Tacher, S., Rimbault, M., et al. (2009). Genetic diversity of canine olfactory receptors. *BMC Genomics* 10: 21.

Rolland, R. M., Hamilton, P. K., Kraus, S. D., et al. (2006). Faecal sampling using detetion dogs to study reproduction and health in North Atlantic right whales (*Eubalaena glacialis*). *J. Cetacean Res. Manage.* 8: 121–125.

Rooney, N. J., Gaines, S. A., Bradshaw, J. W. S., and Penman, S. (2007). Validation of a method for assessing the ability of trainee specialist search dogs. *Appl. Anim. Behav. Sci.* 103: 90–104.

Rothman, R. J., and Mech, L. D. (1979). Scent-marking in lone wolves and newly formed pairs. *Anim. Behav.* 27: 750–760.

Rozhnov, V. V., and Rozhnov, Y. V. (1998). Materials on olfactive communication of common palm civet, *Paradoxurus hermaphrodites*. *Zool. Zh.* 77: 1032–1040.

Ruusila, V., and Pesonen, M. (2004). Interspecific cooperation in human (*Homo sapiens*) hunting: the benefits of a barking dog (*Canis familiaris*). *Ann. Zool. Fennici* 41: 545–549.

Salazar, I., Cifuentes, J. M., Quinteiro, P. S., and Caballero, T. M. (1994). Structural, morphometric, and immunohistological study of the accessory olfactory bulb in the dog. *Anat. Rec.* 240: 277–285.

Savidge, J. A., Stanford, J. W., Reed, R. N., et al. (2011). Canine detection of free-ranging brown treesnakes in Guam. *N.Z. J. Ecol.* 35: 174–181.

Schoon, A., and Haak, R. (2002). K9 Suspect Discrimination, Canada, Detselig Enterprises Ltd.

Schoon, G. A. A. (1996). Scent identification line ups by dogs (*Canis familiaris*): experimental design and forensic application. *Appl. Anim. Behav. Sci.* 49: 257–267.

Scott, J. P., and Fuller, J. L. (1965). Genetics and the Social Behavior of the Dog. Chicago, Chicago University Press.

Scott, J. P., and Marston, M. (1950). Critical periods affecting the development of normal and mal-adjustive social behaviour in puppies. *J. Gen. Psychol.* 77: 25–60.

Seaworld. (2012). http://www.seaworld.org/animal-info/info-books/polar-bear/senses.htm

Shabadash, S. A., and Zelikina, T. I. (2002). Once more about the hepatoid circumanal glands of dogs. History of their discovery and reasons for revision of the structural and functional data. *Izv. Akad. Nauk. Ser. Biol.* Mar–Apr: 176–185.

Sillero-Zubiri, C., and Macdonald, D. W. (1998). Scent-marking and territorial behaviour of Ethiopian wolves *Canis simensis*. *J. Zool. Lond.* 245: 351–361.

References

Sliwa, A., and Richardson, P. R. K. (1998). Responses of aardwolves, *Proteles cristatus*, Sparrman 1783, to translocated scent marks. *Anim. Behav.* 56: 137–146.

Soini, H. A., Linville, S. U., Wiesler, D., et al. (2012). Investigation of scents on cheeks and foreheads of large felines in connection to the facial marking behaviour. *J. Chem. Ecol.* 38: 145–156.

Sonoda, H., Kohnoe, S., Yamazato, T., et al. (2011). Colorectal cancer screening with odour material by canine scent detection. *Gut* 60: 814–819.

Spielman, J. S. (2000). *Olfactory enrichment for captive tigers (*Panthera tigris*) and lions (*Panthera leo*), using a synthetic analogue of feline facial pheromone.* MSc Thesis. University of Edinburgh, UK.

Steen, J. B., and Wilson, E. (1990). How do dogs determine the direction of tracks? *Acta Physiol. Scand.* 139: 531–534.

Stoddart, D. M. (1980). The Ecology of Vertebrate Olfaction. London, Chapman Hall.

Strong, V., Brown, S.W., and Walker, R. (1999). Seizure-alert-dogs – fact or fiction. *Seizure* 8: 62–65.

Tod, E., Brander, D., and Warran, N. (2005). Efficacy of dog appeasing pheromone in reducing stress and fear related behaviour in shelter dogs. *Appl. Anim. Behav. Sci.* 93: 295–308.

Trimble, M., and Insley, S. J. (2010). Mother-offspring reunion in the South American sea lion *Otaria flavescens* at Isla de Lobos (Uruguay): use of spatial, acoustic and olfactory cues. *Ethol. Ecol. Evol.* 22: 233–246.

van Heerden, J. (1981). The role of the integumental glands in the social and mating behaviour of the hunting dog, *Lycaon pictus* (Temmick, 1820). *Onders. J. Vet. Res.* 48: 19–21.

Volokhov, A. A. (1959). Comparative-physiological investigation of conditioned and unconditioned reflexes during ontogeny. *J. Higher Ner. Activ.* 9: 49–60.

Walker, D. B., Walker, J. C., Cavnar, P. J., et al. (2006). Naturalistic quantification of canine olfactory sensitivity. *Appl. Anim. Behav. Sci.* 97: 241–254.

Wasser, S. K., Smith, H., Madden, L., et al. (2009). Scent-matching dogs determine number of unique individuals from scat. *J. Wildlf. Man.* 73: 1233–1240.

Welch, J. (1990). A dog detector for screwworms (*Diptera: Calliphoridae*). *J. Econ. Entomol.* 85: 1932–1934.

Wells, D. L. (2007). Domestic dogs and human health: an overview. *Brit. J. Health Psych.* 12: 145–156.

Wells, D. L. (2009). Sensory stimulation as environmental enrichment for captive animals: A review. *Appl. Anim. Behav. Sci.* 118: 1–11.

Wells, D. L. (2012). Dogs as a diagnostic tool for ill-health in humans. *Alt. Ther. Health Med.* 18: 12–17.

Wells, D. L., and Hepper, P. G. (2000). The discrimination of dog odours by humans. *Perception* 29: 111–115.

Wells, D. L., and Hepper, P. G. (2003). Directional tracking in the domestic dog, *Canis familiaris*. *Appl. Anim. Behav. Sci.* 84: 297–305.

Wells, D. L., and Hepper, P. G. (2006). Prenatal olfactory learning in the domestic dog. *Anim. Behav.* 72: 681–686.

Wells, D. L., Lawson, S. W., and Siriwardena, A. N. (2008). Canine responses to hypoglycaemia in patients with Type 1 diabetes. *J. Alt. Comp. Med.* 14: 1235–1241.

Wiki. (2012). http://wiki.answers.com/Q/How_sensitive_is_a_bear's_sense_of_smell.

Williams, H., and Pembroke, A. (1989). Sniffer dogs in the melanoma clinic? *Lancet* 1: 734.

Willis, C. M., Church, S. M., Guest, C. M., et al. (2004). Olfactory detection of human bladder cancer by dogs: proof of principle study. *BMJ* 329: 712–718.

Wirant, S. C. and McGuire, B. (2004). Urinary behaviour of female domestic dogs (*Canis familiaris*): influence of reproductive status, location, and age. *Appl. Anim. Behav. Sci.* 85: 335–348.

Young, J. M., Kambere, M., Trask, B. J., and Lane R. P. (2005). Divergent V1R repertoires in five species: Amplification in rodents, decimation in primates, and a surprisingly small repertoire in dogs. *Genome Res.* 15: 231–240.

Young, J. M., Massa, H. F., Hsu, L., and Trask, B. J. (2010). Extreme variability among mammalian V1R gene families. *Genome Res.* 20: 10–18.

Zub, K., Theuerkauf, J., Jędrzejewski, W., et al. (2003). Wolf pack territory marking in the Białowieża primeval forest (Poland). *Behav.* 140: 635–648.

Chapter 27

Olfaction in Nonhuman Primates

MATTHIAS LASKA and LAURA TERESA HERNANDEZ SALAZAR

27.1 INTRODUCTION

Primates are traditionally considered as primarily "visual" animals with a poorly developed sense of smell. This view, however, is mainly, if not exclusively, based on an interpretation of neuroanatomical and recent genetic findings and not on behavioral or physiological evidence. In fact, an increasing number of studies now suggest that olfaction may play a significant role in regulating a wide variety of primate behaviors and that the olfactory capabilities of both human and nonhuman primates are not generally inferior to that of nonprimate species believed to have a keen sense of smell. This chapter therefore aims at summarizing the current knowledge about the anatomy, physiology, genetics, and behavior concerning the sense of smell in nonhuman primates.

27.2 ANATOMY

The anatomy of the nose has been studied in a variety of primate species including members of all major groups within this order of mammals (Smith et al., 2007; see Chapter 2). The anatomy of the external nose of primates allows for a classification into two subgroups based on the shape of the nostrils: the strepsirrhines, meaning "curved nose," which include the lemurs, lorises, and galagos, and the haplorhines, meaning "simple nose," which include the tarsiers, New World primates, Old World primates, and hominoids. Strepsirrhines are characterized by the presence of a rhinarium, the moist and naked surface around the nostrils which is also present in a number of nonprimate mammals such as cats and dogs. The rhinarium is believed to play a role in olfaction, as its presence may increase the probability of volatiles being absorbed by the thin layer of mucus that covers the rhinarium, thereby increasing the number of molecules that can be detected.

Another classification of primates also refers to a feature of their nostrils: the catarrhines, meaning "downward-nosed," which include the Old World primates and hominoids have downward-pointing nostrils and a narrow septum whereas the platyrrhines, meaning "flat-nosed," which include the New World primates have outward-pointing nostrils and a wide septum (Figure 27.1). Both of these nose-based classifications are widely used in primate taxonomy.

The anatomy of the internal nose of primates shows all the features that are considered as typical for mammalian noses in general: the nasal fossae are paired passageways located on either side of a midline septum that connect the nostrils (that is: the external nasal openings) and the choanae (the internal nasal openings) leading to the pharynx. From the lateral nasal walls, turbinals invade and subdivide the nasal fossae, thus increasing their internal surface. Depending on the species, these projections may vary from simple ridges to highly elaborated scrolls. Comparative studies suggest that the turbinals of strepsirrhines have a more complex morphology than those of haplorhines (Smith and Rossie, 2006). Whereas most strepsirrhines have four ethmoturbinals in addition to a well-developed maxilloturbinal and nasoturbinal, the haplorhines have only between one and three ethmoturbinals and less-developed maxilloturbinals and nasoturbinals. Within the haplorhines there seems to be no further systematic trend towards a reduction in the complexity of the turbinals from New World primates over Old World primates to hominoids (Figure 27.2).

Handbook of Olfaction and Gustation, Third Edition. Edited by Richard L. Doty.
© 2015 Richard L. Doty. Published 2015 by John Wiley & Sons, Inc.

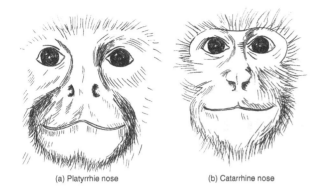

Figure 27.1 (a) Platyrrhine nostrils pointing outward and separated by a wide nasal septum. (b) Catarrhine nostrils pointing downward and separated by a narrow nasal septum. (Redrawn from Ankel-Simons, 2007).

The mucosa that lines the nasal fossae includes respiratory, stratified, and olfactory epithelia. The olfactory epithelium in primates is composed of the same elements (olfactory sensory neurons, supporting cells, basal cells) as that of nonprimate mammals (Loo, 1977). The proportion of olfactory epithelium relative to total nasal fossa area appears to be reduced in haplorhines compared to strepsirrhines. In the latter group, olfactory epithelium has been found to cover all ethmoturbinals whereas in the former group it is usually limited to the posterosuperior-most ethmoturbinals. However, recent detailed histological studies showed marked inter- and intraspecific variations in the extent of olfactory epithelium inside the nasal fossae. In some anthropoid primates, for example, olfactory epithelium has been found more anteriorly than previously described. Similar to the turbinals, there seems to be no further systematic trend within the haplorhines for a further reduction in the distribution of olfactory epithelium from New World primates over Old World primates to hominoids (Smith et al. 2004).

27.3 NEUROANATOMY

The comparative neuroanatomical studies of Stephan and co-workers on mammalian brains and brain structures provide data on the absolute and the relative size of the main olfactory bulbs in 45 species of primates representing all major groups within this order (Table 27.1). Several evolutionary trends can be deduced when comparing main olfactory bulb sizes between species:

1. There is a clear trend for a decrease in the relative size (but not in the absolute size) of the main olfactory bulbs from prosimians over New World primates and Old World primates to hominoids. However, this

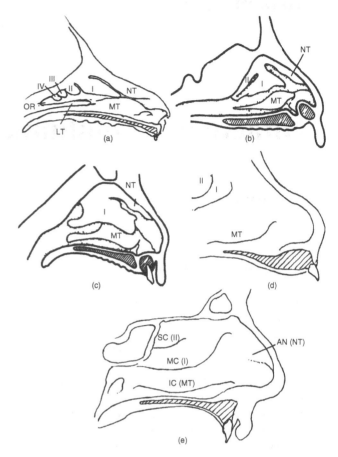

Figure 27.2 Variations in the lateral nasal wall of different primates. (a) Lemurs and most other strepsirrhines have four ethmoturbinals (indicated from anterior to posterior as I–IV). The ethmoturbinals II–IV are found entirely within the "olfactory recess" (OR) and are lined with olfactory epithelium. Note that the ethmoturbinals, nasoturbinal (NT) and maxilloturbinal (MT) are oriented in a more vertical arrangement in anthropoids (B.-E.), with some variation. Note a reduction in the number of ethmoturbinals and proportionally reduced NT in New World primates (B. Saimiri) and Old World primates (C. Macaca). In the apes, the ethmoturbinals are more vertically oriented, as shown in the gibbon (D. Hylobates) and especially in humans (E.). In some anthropoids, the NT may be absent or extremely reduced, as in the homologous aggar nasi (E. AN). IC, inferior nasal concha (MT homologue); MC, middle nasal concha; SC, superior nasal concha; MT, maxilloturbinal; hard palate (hatched lines). Specimens are drawn to similar sizes, not to scale. (Redrawn from Smith and Rossie 2006).

can mainly be attributed to a progressive increase in neocortex size across these primate groups.

2. Primates tend to have smaller main olfactory bulbs relative to total brain volume than basal insectivores such as hedgehogs and shrews which are considered to resemble the common ancestor of today's mammals. However, this too, can mainly be attributed to the massively increased neocortex in primates relative to basal insectivores.

27.3 Neuroanatomy

Table 27.1 Absolute and relative size of the main olfactory bulbs (MOBs) in primates.

Common Name	Scientific Name	MOB Volume (mm³)	Brain Volume (mm³)	MOB ‰ of Brain
Gray mouse lemur	*Microcebus murinus*	43.0	1654	26
Greater dwarf lemur	*Cheirogaleus major*	158	6320	25
Fat-tailed dwarf lemur	*Cheirogaleus medius*	102	3000	34
White-fronted brown lemur	*Lemur albifrons*	207	22258	9.3
Ruffed lemur	*Varecia variegata*	369	30750	12
Red-tailed sportive lemur	*Lepilemur ruficaudatus*	113	7062	16
Eastern woolly lemur	*Avahi laniger laniger*	89.2	9802	9.1
Western woolly lemur	*Avahi l. occidentalis*	70.1	9104	7.7
Verreaux's sifaka	*Propithecus verreauxi*	147	24915	5.9
Indri	*Indri indri*	168	36522	4.6
Aye-aye	*Daubentonia madagascar.*	685	42813	16
Slender loris	*Loris tardigradus*	85.8	6129	14
Slow loris	*Nycticebus coucang*	159	11357	14
Potto	*Perodicticus potto*	310	13478	23
Senegal galago	*Galago senegalensis*	79.2	4400	18
Greater galago	*Otolemur crassicaudatus*	166	9764	17
Demidoff's galago	*Galagoides demidoff*	83.1	3196	26
Philippine tarsier	*Tarsius syrichta*	18.8	3418	5.5
Common marmoset	*Callithrix jacchus*	22.6	7290	3.1
Pygmy marmoset	*Cebuella pygmaea*	12.3	4241	2.9
Red-handed tamarin	*Saguinus midas*	17.3	9611	1.8
Cotton-top tamarin	*Saguinus oedipus*	19.1	9550	2.0
Goeldi's marmoset	*Callimico goeldii*	27.0	10385	2.6
Capuchin monkey	*Cebus albifrons*	39.9	66500	0.6
Owl monkey	*Aotus trivirgatus*	61.0	16053	3.8
Dusky tit monkey	*Callicebus moloch*	19.2	17455	1.1
Squirrel monkey	*Saimiri sciureus*	26.8	22333	1.2
Monk saki	*Pithecia monachus*	34.8	31636	1.1
Red howler monkey	*Alouatta seniculus*	41.4	51750	0.8
Spider monkey	*Ateles geoffroyi*	90.4	100444	0.9
Woolly monkey	*Lagothrix lagotricha*	73.4	91750	0.8
Rhesus monkey	*Macaca mulatta*	84.3	84300	1.0
Gray-cheeked mangabey	*Cercocebus albigena*	121	100833	1.2
Olive baboon	*Papio anubis*	287	191333	1.5
Red-tail monkey	*Cercopithecus ascanius*	99.5	62188	1.6
Blue monkey	*Cercopithecus mitis*	117	68823	1.7
Talapoin monkey	*Miopithecus talapoin*	27.8	39714	0.7
Patas monkey	*Erythrocebus patas*	51.8	103600	0.5
Red colobus monkey	*Colobus badius*	51.3	73286	0.7
Douc langur	*Pygathrix nemaeus*	10.7	71333	0.15
Proboscis monkey	*Nasalis larvatus*	30.3	101000	0.3
White-handed gibbon	*Hylobates lar*	43.9	97556	0.45
Chimpanzee	*Pan troglodytes*	257	367143	0.7
Gorilla	*Gorilla gorilla*	316	451429	0.7
Humans	*Homo sapiens*	114	1266665	0.09

Data from Stephan et al. (1988)

3. Folivorous primates tend to have smaller main olfactory bulbs relative to total brain volume than frugivorous primates. Exactly the same trend can be found in other mammalian orders suggesting that dietary specialization is an important determinant of the size of the main olfactory bulbs.

A re-analysis of Stephan et al.'s data by Barton (2006) generally confirmed the above-mentioned evolutionary trends and additionally found activity period (whether a primate species is diurnal or nocturnal or cathemeral) to correlate with relative size of the main olfactory bulb. Further, the social/mating system (whether a primate species is monogamous or polygynous) was found to correlate with the relative size of both the main and accessory olfactory bulbs. Some authors have argued that the absolute size of the olfactory bulbs should be indicative of a species' olfactory capabilities and have taken the available data as an argument for a "greatly reduced sense of smell in primates" relative to other mammals (Rouquier et al., 2000). However, if the simple reasoning according to which a larger olfactory bulb would necessarily be accompanied by a better sense of smell was true, then mice and rats should have a markedly poorer sense of smell than humans and most of the primate species listed in Table 27.1 as the absolute size of the olfactory bulbs is 8 and 24 mm^3, respectively, in the two rodent species and 114 mm^3 in humans – a conclusion which is not exactly intuitive.

Other authors have argued that the relative size rather than the absolute size of the olfactory bulbs would allow for conclusions as to the olfactory capabilities of different species. Here, too, the simple reasoning that an increase in the size of non-olfactory brain structures would necessarily lead to a decrease in the efficiency of olfactory brain structures and thus to a decrease in olfactory capabilities is neither logical from a theoretical point of view nor supported by physiological data. In this context it is also interesting to note that nobody ever claimed that primates would have a "greatly reduced sense of hearing" although comparative studies also showed that the relative size of their subcortical auditory brain structures is clearly smaller than that of nonprimate mammals (Glendenning and Masterton, 1998).

The size of the olfactory epithelium is another neuroanatomical feature that has been suggested by some authors to be indicative of a species' olfactory capabilities. Unfortunately, no systematic data on olfactory epithelium sizes in nonhuman primate species are at hand. However, as a caveat concerning the plausibility of the idea that a larger olfactory epithelium would necessarily be accompanied by a better sense of smell it should be mentioned that the size of the human olfactory epithelium has been estimated to be around 5 cm^2 whereas that of the rat is only 4 cm^2 (Güntherschulze, 1979), leading to the interesting conclusion that based on this neuroanatomical feature humans should have a better sense of smell than rats. Comparisons of olfactory detection thresholds between species with known sizes of their olfactory epithelium do not generally support a correlation between olfactory sensitivity and epithelium size (Smith and Bhatnagar, 2004).

Using cranial endocasts of 150 species of mammals including 16 species of primates, Pihlström et al. (2005) measured total skull area as well as the area of the cribriform plate of the ethmoid bone, through which the axons of the olfactory sensory neurons reach the olfactory bulbs (Table 27.2). Based on a subsample of 15 nonprimate species, the authors found that the size of the cribriform plate is directly proportional to the size of the olfactory

Table 27.2 Size of the cribriform plate in primates.

Common Name	Scientific Name	mm^2
Potto	*Perodicticus potto*	41.5
Greater dwarf lemur	*Cheirogaleus major*	17.6
Fork-marked lemur	*Phaner furcifer*	23.7
Ruffed lemur	*Varecia variegata*	59
Black lemur	*Eulemur macaco*	78
Wooly lemur	*Avahi laniger*	36
Southern muriqui	*Brachyteles arachnoides*	36
Tufted capuchin	*Cebus apella*	24.3
Red-headed tamarin	*Saguinus midas*	1.94
Blue monkey	*Cercopithecus mitis*	16.1
Diana monkey	*Cercopithecus diana*	20.6
Barbary macaque	*Macaca sylvanus*	27.2
Mantled guereza	*Colobus guereza*	11
Orangutan	*Pongo pygmaeus*	75
Gorilla	*Gorilla gorilla*	135
Human	*Homo sapiens*	132

Data from Pihlström et al. (2005)

epithelium. As the authors believe that the size of the olfactory epithelium would be a reliable indicator of a species' olfactory performance (see previous paragraph on the trustworthiness of this notion), they conclude that the relative size of the cribriform plate would be a reliable indicator of olfactory capabilities, too. With regard to primates, they mention that humans, apes, and monkeys would have small olfactory organs in relation to skull size, while prosimians would have olfactory organs of a size that was typical for mammals of their body size.

27.4 NEUROPHYSIOLOGY

Using a combination of electrophysiological recordings, lesioning of circumscribed brain areas, and behavioral assays, Takagi and co-workers systematically studied the central olfactory projections in members of an Old World primate genus, the macaques (Takagi, 1986). They were able to establish that the basic architecture and central connections of the olfactory pathway in these nonhuman primates are very similar to those reported in rodents such as mice and rats. At least three additional important findings resulted from these studies:

1. The authors demonstrated the presence of an olfactory area in the neocortex of the macaques (Tanabe et al., 1975a). More specifically, they found that the lateral and posterior areas of the orbitofrontal cortex are involved in the processing of olfactory information.

2. The authors demonstrated the presence of a transthalamic olfactory pathway in macaques (Yarita et al., 1980). Whereas a neuronal connection between paleocortical areas involved in olfactory processing and

neocortical areas bypassing the thalamus had been well-established in a variety of animal models, the existence of a thalamic connection between paleo- and neocortex as part of the olfactory pathway had been a matter of debate for many years.

3. The authors demonstrated that the orbitofrontal cortex is involved in odor discrimination (Tanabe et al., 1975b). Among the neocortical neurons they recorded from, 50% responded to only one of eight odorants presented to the animals and none of the neurons responded to more than four of the eight odorants. Lesioning of these neocortical areas led to an impairment in olfactory discrimination performance at the behavioral level.

The high degree of homology between the central olfactory projections in rat and macaque monkey were also confirmed histologically using anterograde and retrograde neuronal tracers (Carmichael et al., 1994) as well as by functional PET imaging in alert animals (Kobayashi et al., 2002).

Rolls and co-workers continued to explore the specific functions of olfactory cortical areas in macaques (Rolls, 2004). They demonstrated that subpopulations of neurons in the primate orbitofrontal cortex that respond to odor stimuli are also involved in associative learning (Rolls et al., 1996) and in the perception of flavor, that is, the combined sensations of smell, taste, and touch in the context of food ingestion (Rolls, 2006). Further, they showed that physiological states such as hunger and satiety may affect the responsiveness of olfactory neurons in the orbitofrontal cortex of macaques (Critchley and Rolls, 1996).

Studies on central olfactory projections in nonhuman primates other than macaques are sparse. Using neuronal tracers, Liebetanz et al. (2002) found the architecture of the olfactory pathway in the common marmoset, a New World primate, to be similar to that in the macaque. Using functional magnetic resonance imaging, Boyett-Anderson et al. (2003) reported odorant-induced activation in awake squirrel monkeys, also a New World primate, in the same brain areas previously implicated in conscious human olfactory processing. Together, these studies suggest that the olfactory pathway and the brain areas involved in processing of olfactory information do not differ fundamentally between New and Old World primates, or between primates and non-primate mammals.

27.5 GENETICS

Olfactory receptor (OR) genes comprise the largest multigene family in mammalian genomes. Bioinformatic analyses of the draft genome sequences of various mammals revealed that both the total number of functional OR genes and the fraction of OR pseudogenes may differ considerably among species. Gilad et al. (2004) sequenced 100 randomly chosen OR genes in 19 primate species and found that the fraction of OR pseudogenes in Old World primates and the howler monkey was significantly higher than in New World primates (Table 27.3). From this observation they hypothesized that OR genes in primates lost their functionality due to the acquisition of full trichromatic vision, which is only found in Old World primates and the howler monkey.

However, using high-coverage whole-genome sequences, Matsui et al. (2010) analyzed the complete OR gene repertoires of seven primate species (Table 27.4) and found no significant differences in the number of functional OR genes and in the fraction of OR pseudogenes between New World primates and Old World primates and hominids. From this the authors concluded that the degeneration of OR genes in primates cannot simply be explained by the acquisition of trichromatic vision.

Analyzing the evolutionary processes underlying the between-species differences in OR gene repertoires, Dong et al. (2009) reported dramatic variations and dynamic turnovers in the evolution of OR genes in primates with evidence for both relaxed selective pressure on some OR

Table 27.3 Fraction of OR pseudogenes in primates, based on 100 randomly chosen OR genes.

Common Name	Scientific Name	%
Human	*Homo sapiens*	51
Chimpanzee	*Pan troglodytes*	34
Gorilla	*Gorilla gorilla*	33
Orangutan	*Pongo pygmaeus*	32
Siamang gibbon	*Hylobates syndactylus*	33
Guinea Baboon	*Papio papio*	29
Rhesus macaque	*Macaca mulatta*	28
Silver langur	*Trachypithecus auratus*	33
Mona monkey	*Cercopithecus mona*	27
Agile Mangabey	*Cercocebus agilis*	29
Mantled gueraza	*Colobus guereza*	31
Tufted capuchin	*Cebus apella*	17
Owl monkey	*Aotus azarae*	15
Spider monkey	*Ateles fusciceps*	18
Black howler monkey	*Alouatta caraya*	31
Squirrel monkey	*Saimiri sciureus*	17
Wooly monkey	*Lagothrix lagotricha*	19
Common marmoset	*Callithrix jacchus*	17
Mongoose lemur	*Eulemur mongoz*	18
Mouse	*Mus musculus*	18
Great apes	mean ± SD	33.0 ± 0.8
Old World Monkeys	mean ± SD	29.3 ± 2.4
New World Monkeys	mean ± SD	18.4 ± 5.6

Data from Gilad et al. (2004)

610   **Chapter 27   Olfaction in Nonhuman Primates**

Table 27.4 Number of functional OR genes and OR pseudogenes from high-coverage whole-genome sequences.

Common Name	Scientific Name	OR Genes	
		Functional	Pseudogenized
Human	*Homo sapiens*	396	425
Chimpanzee	*Pan troglodytes*	380	414
Orangutan	*Pongo pygmaeus*	296	488
Rhesus macaque	*Macaca mulatta*	309	280
Common marmoset	*Callithrix jacchus*	366	231
Bush baby	*Otolemur garnettii*	356	370
Mouse lemur	*Microcenus murinus*	361	339

Data from Matsui et al. (2010)

genes and positive selection for other OR genes in the same species. From this the authors concluded that OR gene repertoires in primates may have evolved in such a way to adapt to their respective chemical environments. This idea is supported by findings from comparative studies that assessed olfactory capabilities such as threshold sensitivity or discrimination performance and found that the frequency of occurrence of odorants in a species' chemical environment or the behavioral relevance of odorants may affect their detectability and discriminability (Laska et al., 2005a,2005b). Additional evidence for unusually rapid processes in the evolution of primate OR genes is provided by a comparison of the repertoires of functional OR genes between humans and chimpanzees. Within only 6 million years since their last common ancestor about 25% of their functional OR genes are now species-specific (Go and Niimura, 2008).

Some authors have argued that the number of functional OR genes and the fraction of OR pseudogenes should be indicative of a species' olfactory capabilities and have taken the limited data available so far as an argument for a "greatly reduced sense of smell in primates" relative to other species (Rouquier et al., 2000). However, if the simple reasoning according to which a high number of functional OR genes would necessarily be accompanied by a keen sense of smell and a low number of such genes by a poor sense of smell was true, then dogs would have to be considered having markedly poorer olfactory capabilities than rats as the former has only 822 functional OR genes whereas the latter has 1259 (Nei et al., 2008) – a conclusion which is not exactly intuitive. Thus, it should be emphasized here that there is no physiological evidence whatsoever supporting the hypothesis of a correlation between the size of the olfactory receptor repertoire and olfactory performance. Rather, comparative studies of olfactory performance in species with known numbers of functional OR genes generally fail to support the idea of "more functional OR genes equal a better sense of smell" (Kjeldmand et al., 2011).

27.6   BEHAVIOR

The traditional belief that primates would be primarily "visual" animals with a poorly developed sense of smell prevented behavioral studies specifically addressing the role of olfaction in regulating primate behavior for a long time. However, there is now accumulating evidence for olfactory involvement in a variety of behavioral contexts including foraging and food selection, predator avoidance, social communication, and reproduction in nonhuman primates.

With regard to the role of olfaction in foraging and food selection, numerous anecdotal reports described free-ranging primates of all major groups within this order of mammals to smell at potential food prior to deciding for or against its consumption. Experimental studies demonstrated that nocturnal mouse lemurs are able to find hidden fruit using olfactory cues alone (Siemers et al., 2007). Similarly, nocturnal owl monkeys, but not diurnal capuchin monkeys, succeeded in finding hidden fruit based on olfactory cues alone (Bolen and Green, 1997). However, Ueno (1994a) showed that capuchin monkeys are at least able to discriminate between food odors. Diademed sifakas have been shown to use olfactory cues to forage for the inflorescences of subterranean parasitic plants (Irwin et al., 2007). Laska et al. (2007a) studied the food selection behavior in captive squirrel monkeys and spider monkeys. They found that both species use olfactory, gustatory, and tactile cues in addition to visual information to evaluate novel food, whereas familiar food items were mainly assessed visually prior to consumption. The importance of olfactory cues for food selection was further indicated by the finding that familiar food which was experimentally modified by adding a non-matching odorant was treated, that is, sniffed at as if it was unfamiliar (Laska et al., 2007a). Interestingly, dichromatic and trichromatic wild spider monkeys were found not to differ in their use of olfactory cues when assessing potential food items (Hiramatsu et al., 2009). The odor of potential food has also been demonstrated to serve as an effective cue in food avoidance learning by nonhuman primates. Squirrel monkeys and common marmosets learned to reliably avoid unpalatable food based on olfactory cues alone after only one or a few trials and showed retention of the significance of these olfactory cues for up to four weeks (Laska and Metzker, 1998).

Information about the edibility of food is also transferred between individuals by using olfactory cues: in three species of Old World primates, mandrills, drills, and olive baboons, animals were found to consume the same type of food as a conspecific after they had sniffed at the conspecific's mouth. This information transfer by actively sniffing at the mouth of an eating conspecific occurred most often between young and older animals suggesting olfactory learning about edibility of food (Laidre, 2009).

With regard to the role of olfactory cues for predator avoidance, Buchanan-Smith et al. (1993) demonstrated that cotton-top tamarins respond differentially to the fecal odors of sympatric predators and non-predators. They concluded that the discrimination of predator and non-predator odors was innate in this primate species as their animals were captive-born and the observed behavioral responses were unaffected by whether or not the parent animals were wild-caught or captive-born. Similarly, olfactory predator recognition has been demonstrated in both wild (Kappel et al., 2011) and captive-born (Sündermann et al., 2008) mouse lemurs. Interestingly, the odor of owl feces, that is, of an aerial predator, was not avoided by the mouse lemurs, whereas this was not the case for the odor of fossa feces, that is, of a terrestrial predator. This difference may be explained either by a lack of predator-typical volatiles in owl feces or by a lower behavioral significance for olfactory recognition of aerial predators compared to terrestrial predators.

Gisela Epple and co-workers were the first to systematically study olfactory social communication in nonhuman primates. Using a combination of behavioral, chemo-analytical, and endocrinological approaches, they demonstrated that marmosets and tamarins, a group of New World primates, strongly rely on olfactory cues in the regulation of their social behavior. They described a variety of scent-marking behaviors (Epple, 1986), analyzed the chemical composition of scent gland secretions (Belcher et al., 1986), unraveled the informational content and the behavioral functions of scent-marks (Epple et al., 1993), and thus contributed significantly to our understanding of the role of chemical signals in the control of social behavior in nonhuman primates. Further, they demonstrated that, similar to nonprimate mammals, the scent-marks of marmosets and tamarins contain information about species, social group, gender, age, reproductive status, social rank, health status, individual identity, and genetic relatedness of the odor donor (Epple et al., 1989).

Whereas specialized scent glands are a common feature of prosimians and New World primates, their presence in Old World primates and hominoids appears to be rare. However, it should be noted that several species of Old World primates, such as mandrills (Setchell et al., 2010), guenons (Loireau and Gautier-Hion, 1988), vervet monkeys (Freeman et al., 2012), and even hominoids such as gibbons (Geissmann 1987), have been reported to possess fully functional specialized scent glands and to perform scent-marking behaviors. Recent studies have shown that the secretion of the sternal gland of the mandrill contains information about its producer's major histocompatibility complex and thus about its individual genetic quality which is thought to play a crucial role in mate choice and kin selection (Setchell et al., 2011). Primate species possessing specialized scent glands often display conspicuous scent-marking behaviors which may include, but are not restricted to, nasal rubbing, throat rubbing, chest rubbing, back rubbing, suprapubic rubbing, and anogenital rubbing, depending on the localization of the scent glands on the body (Figure 27.3).

These behaviors may be performed to deposit scent gland secretions either on a substrate such as branches, or on a conspecific, or on parts of the odor donor's own body. An example for this latter case are the ring-tailed lemurs which impregnate their bushy tails with the secretions of their antebrachial (that is: forearm) glands and then engage in ritualized "stink-fights" which serve to establish a stable hierarchy within a social group and to limit the degree of physical aggression (Kappeler, 1998).

Other primate species that are lacking specialized scent glands may nevertheless use olfactory cues for social communication, for example by employing urine as a signal. A wide variety of primate species including galagos, slender loris, mouse lemurs, capuchins, squirrel monkeys, owl monkeys (Andrew and Klopman, 1974), woolly monkeys (Milton, 1985), howler monkeys (Jones, 2003), and moustached tamarins (Heymann, 1995) engage in urine washing. This conspicuous behavioral pattern consists of urination onto the palm of the hand with subsequent distribution of the urine either onto the soles of the feet, or onto the fur, or onto the substrate, usually a branch. Interestingly, some of the species mentioned above do have specialized scent glands, but nevertheless display urine washing in certain behavioral contexts.

The variety of information that has been identified in the scent marks of nonhuman primates has also been demonstrated to serve a variety of functions including the advertisement of presence, the establishment and maintenance of social rank (e.g., Kappeler, 1998), territorial defense (e.g., Palagi and Dapporto, 2007), recognition of group members (e.g., Braune et al., 2005), advertisement of sexual receptivity (e.g., Clarke et al., 2009), synchronization or monopolization of reproductive activity (e.g.,

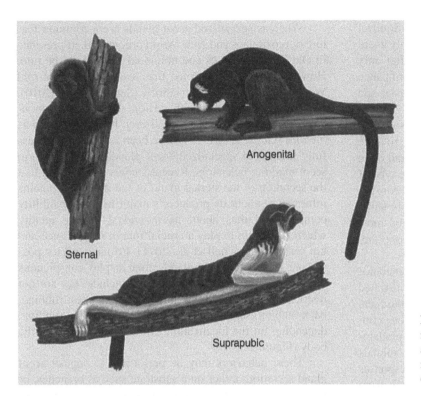

Figure 27.3 Scent-marking behaviors in callithrichid monkeys. Sternal marking (Goeldi's monkey), anogenital marking with simultaneous sniffing of marking site (moustached tamarin), suprapubic marking (Geoffroyi's tamarin). Redrawn from Heymann (2003).

Barrett et al., 1993), establishment and maintenance of an inbreeding barrier (e.g., Charpentier et al., 2010), directed altruism (e.g., Boulet et al., 2009), and social bonding (e.g., Kaplan et al., 1977), to name but a few.

27.7 PHYSIOLOGY

The systematic experimental assessment of olfactory performance in nonhuman primates has begun only two decades ago. Measurements of olfactory sensitivity and discrimination ability, as well as of olfactory learning and memory performance under controlled conditions, are clearly necessary to address the apparent discrepancy between (a) anatomical and genetic findings that suggest primates have a poorly developed sense of smell and (b) behavioral findings that suggest the sense of smell plays a crucial role in regulating primate behavior.

Laska and co-workers developed behavioral tests based on operant conditioning paradigms and using established psychophysical procedures which made it possible to assess key aspects of olfactory performance in squirrel monkeys and spider monkeys, two New World primate species, and in pigtail macaques, an Old World primate species. Figures 27.4 and 27.5 illustrate the behavioral tests.

By the time this chapter was printed, Laska and co-workers have published olfactory detection thresholds for 74 odorants in spider monkeys, for 61 odorants in squirrel monkeys, and for 60 odorants in pigtail macaques (Table 27.5). These numbers are considerably higher than the numbers of odorants that have been tested so far with any other nonhuman species.

Two tentative conclusions can be drawn from these olfactory detection threshold data:

1. Within-species comparisons suggest that olfactory sensitivity may correlate with the behavioral relevance of the odor stimuli. All three primate species tested were found to be particularly sensitive for odorants occurring in the contexts of food selection, predator avoidance, social communication, and reproduction. This is in line with findings from non-primate species which also suggest that the behavioral relevance of odor stimuli may affect their detectability.

2. Between-species comparisons suggest that the olfactory sensitivity of all three primate species is not generally inferior to that reported in species presumed to have a keen sense of smell, but in several cases matches or even surpasses the olfactory sensitivity of mice, rats, or dogs. This, in turn, suggests that neither anatomical features such as the absolute or the relative size of olfactory brain structures nor genetic features such as the number of functional olfactory receptor genes are reliable predictors of a species' olfactory sensitivity.

For the sake of completeness it should be mentioned that Glaser and co-workers (1994) determined the olfactory detection threshold for one odorant in the pygmy

27.7 Physiology 613

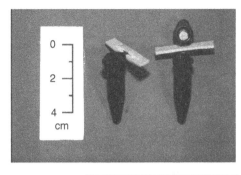

Figure 27.4 Olfactory conditioning method used with squirrel monkeys. *Upper left hand:* picture of a squirrel monkey *(Saimiri sciureus)*. *Upper right hand:* Eppendorf® cups equipped with absorbent paper strips. The Eppendorf® cups serve as manipulation objects ("artificial nuts") that are either baited with a piece of peanut or not, depending on the odorant applied on the absorbent paper strip. *Lower right hand:* experimental set-up. An "artificial nut tree" is used to present numerous "artificial nuts", half of them baited with a piece of peanut and bearing an odorant used as rewarded stimulus, and half of them empty and bearing an odorant used as unrewarded stimulus. *Lower left hand:* a squirrel monkey inspecting an "artificial nut" on a branch of the "artificial nut tree". (A detailed description of the method is given in Laska and Hudson, 1993b).

Figure 27.5 Olfactory conditioning method used with spider monkeys. *Upper left hand:* picture of a spider monkey *(Ateles geoffroyi)*. *Upper right hand:* the two-choice apparatus used, viewed from the animal's side. It consists of two manipulation boxes of which one is baited with a Kellogg's® honeyloop while the other is empty, depending on the odorant applied on the absorbent paper strip attached to the box. *Lower right hand:* a spider monkey smelling at one of the absorbent paper strips bearing an odorant used either as a rewarded stimulus or as an unrewarded stimulus. *Lower left hand:* a spider monkey indicating his decision for one of the two simultaneously presented odorants by opening the corresponding manipulation box. (A detailed description of the method is given in Laska et al., 2003a).

614 **Chapter 27 Olfaction in Nonhuman Primates**

Table 27.5 Olfactory detection thresholds (log ppm) in three species of nonhuman primates.

Chemical Class Odorant	Spider Monkey	Squirrel Monkey	Pigtail Macaque
aliphatic alcohols			
ethanol		+2.04 / +2.57	+2.04 / +2.57
1-propanol	−0.29 / +0.71	+0.25 / +1.25	+0.73 / +1.73
1-butanol	−0.59 / −0.06	−0.06 / +0.94	−0.06 / +0.94
1-pentanol	−3.39 / −1.39	−0.90 / −0.38	−0.90 / +1.10
1-hexanol	−2.22 / −1.22	−2.70 / −1.70	−2.22 / −0.70
1-heptanol	−3.50 / −2.50	−1.50 / +0.02	−2.50 / −0.98
1-octanol	−2.32 / −2.32	−1.32 / −0.85	−2.32 / −1.32
2-propanol		+1.50 / +2.50	+1.98 / +1.98
2-butanol		+0.22 / +0.70	+0.70 / +1.70
2-pentanol		−1.01 / −1.01	−0.53 / −0.01
3-pentanol		−0.03 / −0.03	−0.03 / +0.97
aliphatic aldehydes			
n-butanal	−1.40 / −0.40	−2.40 / −0.40	−3.40 / −1.40
n-pentanal	−2.83 / −1.83	−1.83 / +0.17	−0.83 / −0.83
n-hexanal	−2.29 / −1.29	−1.29 / −0.29	−3.29 / −1.29
n-heptanal	−2.63 / −0.63	−2.63 / −2.63	−2.63 / −1.63
n-octanal	−3.79 / −0.79	−0.79 / −0.32	−2.79 / −2.32
n-nonanal	−2.23 / −1.23	−0.23 / +0.78	−2.23 / −1.75
aliphatic esters			
ethyl acetate	−1.43 / −0.96	−1.43 / +0.04	+1.56 / +2.04
n-propyl acetate	−1.35 / −0.34	−0.82 / +0.18	−0.34 / +1.18
n-butyl acetate	−4.22 / −3.22	−4.22 / −1.22	−3.22 / −1.22
n-pentyl acetate	−4.57 / −3.57	−4.57 / −2.09	−2.09 / −0.57
n-hexyl acetate	−3.88 / −2.40	−4.88 / −0.88	−1.40 / −1.40
n-heptyl acetate	−3.75 / −2.23	−7.23 / −2.75	−2.75 / −1.75
n-octyl acetate	−2.48 / −2.01	−3.48 / −0.01	−1.01 / −0.48
iso-propyl acetate	−2.63 / −2.15	−2.62 / −2.15	+0.07 / +1.37
iso-butyl acetate	−0.58 / −0.07	+0.94 / +0.94	−2.58 / −0.58
iso-pentyl acetate	−3.96 / −3.44	−3.96 / −0.44	−3.44 / −0.44
aliphatic carboxylic acids			
n-propanoic acid	−1.44 / −1.44	−0.44 / +0.04	−0.44 / +0.03
n-butanoic acid	−3.83 / −2.83	−1.31 / −0.79	−4.83 / −3.35
n-pentanoic acid	−2.12 / −2.12	−1.66 / −1.14	−3.66 / −2.12
n-hexanoic acid	−3.48 / −3.48	−3.42 / −1.94	−2.00 / −1.00
n-heptanoic acid	−4.73 / −3.73	−3.74 / −3.26	−0.73 / +0.27
aliphatic ketones			
2-butanone		+1.54 / +3.02	+2.02 / +2.54
2-pentanone		+0.65 / +2.17	−0.83 / +0.17
2-hexanone		−0.23 / +1.26	−1.23 / −0.23
2-heptanone		−0.61 / +0.39	−0.61 / −0.61
2-octanone		−0.91 / +0.56	−2.91 / −0.91
2-nonanone		−0.15 / +0.32	−4.15 / −3.15
3-pentanone		+1.17 / +1.63	+0.17 / +1.17
3-heptanone		−0.09 / +0.91	−0.57 / −0.57
4-heptanone		−1.03 / +0.97	−0.51 / −0.51
5-nonanone		−3.68 / −2.68	−0.68 / −0.68
6-undecanone		−1.32 / −0.32	−1.80 / −1.32

27.7 Physiology 615

Table 27.5 (*Continued*)

Chemical Class Odorant	Spider Monkey	Squirrel Monkey	Pigtail Macaque
thiols			
ethanethiol	−6.02 / −4.02	−2.02 / −0.02	−4.02 / −1.02
1-propanethiol	−2.29 / −1.29	−2.29 / −0.29	−2.29 / −2.29
1-butanethiol	−3.80 / −1.80	−3.80 / −2.32	−3.80 / −3.80
1-pentanethiol	−3.20 / −1.20	−3.20 / −2.72	−3.72 / −2.72
indols			
indol	−3.52 / −2.52	−7.52 / −5.03	−7.52 / −7.03
3-methyl indol	−5.43 / −4.43	−4.95 / −3.95	−4.43 / −3.43
monoterpene alcohols			
geraniol	−2.39 / −1.39	−4.86 / −1.86	−4.39 / −2.39
nerol	−1.35 / −0.83	−4.83 / −2.83	−3.83 / −1.83
linalool	−0.84 / +0.16	−3.32 / −1.32	−2.32 / −0.84
citronellol	−1.73 / −1.73	−4.73 / −3.73	−3.73 / −1.26
myrcenol	−2.32 / −1.80	−4.32 / −2.80	−1.32 / −0.32
lavandulol	−2.43 / −1.94	−4.43 / −3.94	−0.94 / +0.06
alkylpyrazines			
pyrazine	+1.44 / +1.44		
2-methylpyrazine	−1.35 / +0.12		
2-ethylpyrazine	−1.15 / −0.15		
2,5-dimethylpyrazine	−4.18 / −3.18		
2,6-dimethylpyrazine	−2.63 / −1.63		
2,3,5,6-tetramethylpyrazine	−3.20 / −2.72		
aromatic aldehydes			
bourgeonal	−4.61 / −3.61		
lilial	−2.13 / −1.13		
3-phenylpropanal	−3.29 / −2.76		
cyclamal	−1.49 / −1.02		
canthoxal	−0.88 / −0.35		
helional	−2.23 / −0.23		
lyral	−1.61 / −1.13		
monoterpenes			
(+)-carvone	−2.83 / −1.83		
(−)-carvone	−1.83 / −1.83		
(±)-carvone	−4.83 / −3.35		
(+)-limonene	−2.53 / −1.53		
(−)-limonene	−2.02 / −0.02		
(±)-limonene	−5.02 / −3.02		
1,8-cineole		−2.98 / −1.98	
"green" odors			
cis-3-hexenol	−1.23 / −0.75		
cis-3-hexenal	−0.26 / −0.26		
trans-3-hexenol	−1.75 / −0.75		
trans-3-hexenal	−0.74 / −0.26		
trans-2-hexenol	−1.26 / −0.78		
trans-2-hexenal	−1.09 / −0.09		
thiazoles			
2,4,5-trimethylthiazoline	−3.86 / −2.86	−2.39 / −1.86	−3.39 / −2.39

(*continued*)

Table 27.5 (*Continued*)

Chemical Class Odorant	Spider Monkey	Squirrel Monkey	Pigtail Macaque
amino acids			
L-cysteine	−2.88 / −2.35		
D-cysteine	−2.88 / −2.88		
L-methionine	−2.95 / −1.95		
D-methionine	−2.95 / −2.43		
L-proline	−1.69 / −0.69		
D-proline	−2.69 / −2.69		
steroids			
5-α-androst-16-en-3-one	7.3 /25 μM	25 / 250 μM	25 / 73 μM
5-α-androst-16-en-3-ol	6.2 / 6.2 mM	0.62 / 6.2 mM	6.2 / 6.2 mM
androsta-4,16-dien-3-one	18.5 / 63 μM	0.63 / 18.5 μM	6.3 / 6.3 μM
estra-1,3,5(10),16-tetraen-3-ol		0.63 / 63.3 mM	6.3 / 63.3 mM

Data from Hernandez Salazar et al. 2003; Joshi et al. 2006; Laska and Seibt 2002a,2002b; Laska et al. 2000, 2003b, 2004, 2005a, 2005b, 2005c, 2005d, 2006a, 2006b, 2006c, 2007b, 2009; Løtvedt et al. 2012; Wallén et al. (2012)

The first value in each data pair refers to the lowest individual threshold, and the second value to the highest individual threshold determined with a given odorant.

Threshold values for the steroids are given in liquid concentrations (μM and mM) rather than in gas phase concentrations (log ppm) as no experimentally determined vapor pressures are available for these odorants.

marmoset. Using an operant conditioning procedure and an automated olfactometer, they found that this smallest of all primate species is able to detect vanillin at concentrations as low as −3.68 log ppm. Unfortunately, no further studies on olfactory capabilities in this primate species and using this method were published.

Studies on olfactory discrimination capabilities in squirrel monkeys using structurally related monomolecular odorants showed that this New World primate species possesses a well-developed ability to distinguish between members of homologous series of aliphatic acetic esters (Laska and Freyer, 1997), n-carboxylic acids (Laska and Teubner, 1998), 1-alcohols, n-aldehydes, and 2-ketones (Laska et al., 1999b). However, the squirrel monkeys failed to discriminate between some odorants that differed from each other by only one carbon atom, whereas mice and Asian elephants tested on the same odor pairs did not.

Squirrel monkeys were also found capable of discriminating between binary odor mixtures that vary only slightly in the concentration ratios of their components (Laska and Grimm, 2003), and between complex odor mixtures from which only one or a few components have been removed (Laska and Hudson, 1993a). Further, squirrel monkeys are able to discriminate between urine odors of individual conspecifics, irrespective of the gender of and the test animal's familiarity with the odor donor (Laska and Hudson, 1995). Ueno (1994b) reported that tufted capuchin monkeys are able to discriminate between the urine odors of five different species. The ability of both squirrel monkeys and pigtail macaques to discriminate between the odors of the (+)- and (−)-forms of enantiomers, that is, of optical isomers, was

found to be comparable to that of human subjects but inferior to that reported in mice and Asian elephants tested on the same set of stimuli (Laska et al., 1999a, 2005b).

With regard to olfactory long-term memory, Laska et al. (1996) demonstrated that squirrel monkeys show excellent retention of the reward value of previously learned odor pairs up to 15 weeks and above-chance level performance up to 30 weeks, the longest interval tested with this species. Similarly, pigtail macaques and spider monkeys showed no signs of forgetting of previously learned odor pairs after three and four weeks, respectively, the longest intervals tested with these two species (Hübener and Laska, 1998; Laska et al., 2003a).

Measurements of olfactory learning capabilities yielded mixed results in nonhuman primates: with regard to the speed of initial task acquisition, that is, mastering of the very first odor pair presented to the animals, squirrel monkeys needed 450–750 stimulus contacts for reaching a pre-set learning criterion, spider monkeys needed 660–720, and pigtail macaques 960–1800 stimulus contacts using the operant conditioning paradigms mentioned above (Hübener and Laska, 2001; Laska and Hudson, 1993b; Laska et al., 2003a). These numbers are comparable to those reported in South African fur seals (480–880), but are considerably higher than those reported in mice, rats, dogs, and Asian elephants which all needed <150 stimulus contacts to learn to criterion their very first odor pair. This suggests that nonhuman primates may not be as prepared as non-primate mammals to acquire the basic concept underlying olfactory discrimination paradigms. It should be mentioned, however, that gray mouse lemurs, a prosimian primate, needed only 160–560 stimulus contacts to acquire a two-odorant discrimination paradigm using an operant conditioning procedure and apparatus similar to one employed with mice and rats (Joly et al., 2004).

The number of stimulus contacts needed to master subsequent odor pairs rapidly decreased in all four primate species to a level comparable with that reported in mice, rats, dogs, and Asian elephants. Thus, the speed of learning intramodal transfer tasks using olfactory stimuli displayed by nonhuman primates is not inferior to that of non-primate species believed to strongly rely on their sense of smell.

27.8 THE VOMERONASAL SYSTEM

The vomeronasal organ (VNO) has been demonstrated to play a role in sociosexual chemical communication in a wide variety of vertebrates (see Chapter 51, this volume). In mammals, the organ is situated at the anterior base of the nasal septum and consists of two elongated, liquid-filled and blindly-ending tubes whose inner surface is lined with a neuroepithelium which receives chemical stimuli via openings in the nasal cavity, or the oral cavity, or in both. The VNO has been found to be anatomically present and fully developed in all prosimians and New World primates studied so far (Smith et al., 2011). In contrast, Old World primates and hominoids seem to lack a fully developed VNO, at least postnatally (Smith et al., 2001). Recent reports of a VNO being anatomically present in postnatal humans (Meredith, 2001) and chimpanzees (Smith et al., 2002) suggest this structure to be vestigial and non-functional in these hominoids as a neuroepithelium as well as a neuronal connection to the brain seem to be absent. It should be emphasized, however, that in the majority of primate species known to have a fully developed VNO, physiological evidence of its functionality is lacking (but see Schilling et al., 1990). Nevertheless, behavioral studies suggest the VNO to be functional in at least some of these species (Evans, 2006). Similarly, studies reporting behavioral and/or hormonal changes as a consequence of VNO removal suggest the organ to be functional in primate species such as the grey mouse lemur (Aujard, 1997) or the common marmoset (Barrett et al., 1993).

The axons of the vomeronasal chemosensory neurons project to the accessory olfactory bulbs (AOBs), the first relay station of the vomeronasal pathway. The systematic neuroanatomical studies of Stephan and co-workers on the comparative size of mammalian brains and brain structures provide data on the absolute and the relative size of the AOBs in 45 species of primates representing all major groups within this order of mammals (Table 27.6). All 31 species of prosimians and New World primates studied were found to possess AOBs whereas this structure

was absent in all 14 species of Old World primates and hominoids studied.

Several evolutionary trends can be described when comparing AOB sizes between species:

1. Prosimians tend to have larger accessory olfactory bulbs relative to total brain volume than New World primates.
2. Within the prosimians, nocturnal species tend to have larger accessory olfactory bulbs relative to total brain volume than diurnal species. This is consistent with the idea that nocturnal species might rely to a higher degree on olfactory signals in the context of reproduction compared to diurnal species.
3. No interdependency between the size of the main olfactory bulbs and the size of the accessory olfactory bulbs was found. This is consistent with the idea that the main and the accessory olfactory systems serve different functions, that is, process olfactory information in different behavioral contexts.
4. No sex differences in the size of the accessory olfactory bulbs were found in any of the primate species studied. This is remarkable given that the vomeronasal organ is thought to be involved in the behavioral context of reproduction and given that sex differences in the production and deposition of scent marks as well as in flehmen-like behaviors have been reported in several species of nonhuman primates.
5. No correlation between the size of the accessory olfactory bulbs and dietary specialization was found. This is consistent with the idea that the vomeronasal organ is involved in the behavioral context of reproduction but not in the context of foraging and food selection.

Flehmen, that is, the curling of the upper lip in response to the detection of certain odorants has been reported in a variety of mammals that possess a vomeronasal system and is thought to facilitate the access of signaling chemicals secreted by conspecifics to the vomeronasal organ. The behavior is thought to play a role in the social and reproductive context. Flehmen-like facial expressions have been described in some prosimian species such as the ring-tailed lemur (Bailey, 1978) and in New World primates such as marmosets (Epple et al., 1993), but not in Old World primates and hominoids. When sniffing the urine of female conspecifics, some New World primate species such as howler monkeys (Van Belle et al., 2009) and owl monkeys (Wolovich and Evans, 2007) have been reported to show rhythmic tongue protrusions reminiscent of those in lizards and snakes. Since this behavior in reptiles has been linked with vomeronasal organ function it has been suggested that these rapid oscillating protrusions of the tongue are also involved in the perception of odorants via the vomeronasal organ in New World primates. Consistent with this idea,

Chapter 27 Olfaction in Nonhuman Primates

Table 27.6 Absolute and relative size of the accessory olfactory bulbs (AOBs) in primates.

Common Name	Scientific Name	AOB Volume (mm^3)	Brain Volume (mm^3)	AOB ‰ of Brain
Gray mouse lemur	*Microcebus murinus*	1.6	1654	0.970
Greater dwarf lemur	*Cheirogaleus major*	4.0	6320	0.630
Fat-tailed dwarf lemur	*Cheirogaleus medius*	2.7	3000	0.900
White-fronted brown lemur	*Lemur albifrons*	3.2	22258	0.015
Ruffed lemur	*Varecia variegata*	5.3	30750	0.018
Red-tailed sportive lemur	*Lepilemur ruficaudatus*	2.7	7062	0.038
Eastern woolly lemur	*Avahi laniger laniger*	3.0	9802	0.031
Western woolly lemur	*Avahi l. occidentalis*	2.6	9104	0.028
Verreaux's sifaka	*Propithecus verreauxi*	3.4	24915	0.013
Indri	*Indri indri*	2.3	36522	0.006
Aye-aye	*Daubentonia madagascar.*	8.3	42813	0.020
Slender loris	*Loris tardigradus*	2.3	6129	0.037
Slow loris	*Nycticebus coucang*	4.3	11357	0.036
Potto	*Perodicticus potto*	2.7	13478	0.021
Senegal galago	*Galago senegalensis*	2.6	4400	0.057
Greater galago	*Otolemur crassicaudatus*	2.5	9764	0.026
Demidoff's galago	*Galagoides demidoff*	1.3	3196	0.039
Philippine tarsier	*Tarsius syrichta*	0.8	3418	0.024
Common marmoset	*Callithrix jacchus*	0.3	7290	0.005
Pygmy marmoset	*Cebuella pygmaea*	0.4	4241	0.009
Red-handed tamarin	*Saguinus midas*	0.3	9611	0.003
Cotton-top tamarin	*Saguinus oedipus*	0.4	9550	0.004
Goeldi's marmoset	*Callimico goeldii*	0.5	10385	0.005
Capuchin monkey	*Cebus albifrons*	0.7	66500	0.001
Owl monkey	*Aotus trivirgatus*	0.2	16053	0.001
Dusky tit monkey	*Callicebus moloch*	0.3	17455	0.002
Squirrel monkey	*Saimiri sciureus*	1.0	22333	0.004
Monk saki	*Pithecia monachus*	0.8	31636	0.003
Red howler monkey	*Alouatta seniculus*	0.6	51750	0.001
Spider monkey	*Ateles geoffroyi*	2.2	100444	0.002
Woolly monkey	*Lagothrix lagotricha*	1.4	91750	0.002

Data from Stephan et al. (1982)

there are no reports on similar facial motor patterns in Old World primates and hominoids.

REFERENCES

Andrew, R. J., and Klopman, R. B. (1974) Urine-washing: comparative notes. In: Martin, R. D., Doyle, G. A., Walker, A. C. (Eds.) *Prosimian Biology*. Duckworth, Gloucester Crescent, pp. 303–312.

Ankel-Simons, F. (2007) *Primate Anatomy – An Introduction*, 3rd edition. New York: Academic Press.

Aujard, F. (1997) Effect of vomeronasal organ removal on male socio-sexual responses to female in a prosimian primate *(Microcebus murinus)*. *Physiol. Behav.* 62: 1003–1008.

Bailey, K. (1978) Flehmen in the ring-tailed lemur *(Lemur catta)*. *Behaviour* 65: 309–319.

Barrett, J., Abbott, D. H., and George, L. M. (1993) Sensory cues and the suppression of reproduction in subordinate female marmoset monkeys, *Callithrix jacchus*. *J. Reprod. Fertil.* 97: 301–310.

Barton, R. A. (2006) Olfactory evolution and behavioral ecology in primates. *Am. J. Primatol.* 68: 545–558.

Belcher, A. M., Smith, A. B., Jurs, P. C., et al. (1986) Analysis of chemical signals in a primate species *(Saguinus fuscicollis)*: use of behavioral, chemical, and pattern recognition methods. *J. Chem. Ecol.* 12: 513–531.

Bolen, R. H., and Green, S. M. (1997) Use of olfactory cues in foraging by owl monkeys *(Aotus nancymai)* and capuchin monkeys *(Cebus apella)*. *J. Comp. Psychol.* 111: 152–158.

Boulet, M., Charpentier, M. J. E., and Drea, C. M. (2009) Decoding an olfactory mechanism of kin recognition and inbreeding avoidance in a primate. *BMC Evol. Biol.* 9: 281.

Boyett-Anderson, J. M., Lyons, D. M., Reiss, A. L., et al. (2003) Functional brain imaging of olfactory processing in monkeys. *Neuroimage* 20: 257–264.

Braune, P., Schmidt, S., and Zimmermann, E. (2005) Spacing and group coordination in a nocturnal primate, the golden brown mouse lemur *(Microcebus ravelobensis)*: the role of olfactory and acoustic signals. *Behav. Ecol. Sociobiol.* 58: 587–596.

Buchanan-Smith, H. M., Anderson, D. A., and Ryan, C. W. (1993) Responses of cotton-top tamarins *(Saguinus oedipus)* to faecal scents of predators and non-predators. *Animal Welfare* 2: 17–32.

Carmichael, S. T., Clugnet, M. C., and Price, J. L. (1994) Central olfactory connections in the macaque monkey. *J. Comp. Neurol.* 346: 403–434.

References

Charpentier, M. J. E., Crawford, J. C., Boulet, M., and Drea, C. M. (2010) Message "scent": lemurs detect the genetic relatedness and quality of conspecifics via olfactory cues. *Anim. Behav.* 80: 101–108.

Clarke, P. M. R., Barrett, L., and Heinzi, S. P. (2009) What role do olfactory cues play in chacma baboon mating? *Am. J. Primatol.* 71: 493–502.

Critchley, H. D., and Rolls, E. T. (1996) Hunger and satiety modify the responses of olfactory and visual neurons in the primate orbitofrontal cortex. *J. Neurophysiol.* 75: 1673–1686.

Dong, D., He, G., Zhang, S., and Zhang, Z. (2009) Evolution of olfactory receptor genes in primates dominated by birth-and-death process. *Genome Biol. Evol.* 2009: 258–264.

Epple, G. (1986) Communication by chemical signals. In: Mitchell, G., Erwin, J. (Eds.) *Comparative Primate Biology*, Vol. 2A. Alan R. Liss, New York, pp. 531–580.

Epple, G., Belcher, A. M., Greenfield, K. L., et al. (1989) Scent mixtures used as social signals in two primate species: *Saguinus fuscicollis* and *Saguinus o. oedipus*. In: Laing, D. G., Cain, W. S., McBride, R. L., Ache, B. W. (Eds.) *Perception of Complex Smells and Tastes*. Academic Press: New York, pp. 1–25.

Epple, G., Belcher, A. M., Küderling, I., et al. (1993) Making sense out of scents: species differences in scent glands, scent-marking behaviour, and scent-mark composition in the *Callitrichidae*. In: Rylands, A. B. (Ed.) *Marmosets and Tamarins: Systematics, behavior, and ecology*. Oxford University Press: Oxford, pp. 123–151.

Evans, C. S. (2006) Accessory chemosignaling mechanisms in primates. *Am. J. Primatol.* 68: 525–544.

Freeman, N. J., Pasternak, G. M., Rubi, T. L., et al. (2012) Evidence for scent marking in vervet monkeys? *Primates*, 53: 311–315.

Geissmann, T. (1987) A sternal gland in the siamang gibbon *(Hylobates syndactylus)*. *Int. J. Primatol.* 8: 1–15.

Gilad, Y., Wiebel, V., Przeworski, M. et al. (2004) Loss of olfactory receptor genes coinincides with the acquisition of full trichromatic vision in primates. *PLoS Biol.* 2: 120–125.

Glaser, D., Etzweiler, F., Graf, R., et al. (1994) The first odor threshold measurement in a non-human primate (*Cebuella pygmaea*; Callitrichidae) with a computerized olfactometer. *Adv. Biosci.* 93: 445–455.

Glendenning, K. K., and Masterton, R. B. (1998) Comparative morphometry of mammalian central auditory systems: variation in nuclei and form of the ascending system. *Brain Behav. Evol.* 51: 59–89.

Go, Y., and Niimura, Y. (2008) Similar numbers but different repertoires of olfactory receptor genes in humans and chimpanzees. *Mol. Biol. Evol.* 25: 1897–1907.

Güntherschulze, J. (1979) Studies about the region olfactoria of the wild boar *(Sus scrofa)* and of the domestic pig *(Sus scrofa domestica)*. *Zool. Anz.* 202: 256–279.

Hernandez Salazar, L. T., Laska, M., and Rodriguez Luna, E. (2003) Olfactory sensitivity for aliphatic esters in spider monkeys, *Ateles geoffroyi*. *Behav. Neurosci.* 117: 1142–1149.

Heymann, E. W. (1995) Urine washing and related behaviour in wild moustached tamarins, *Saguinus mystax* (Callitrichidae). *Primates* 36: 259–264.

Heymann, E. W. (2003) New World monkeys. II: Marmosets, tamarins, and Goeldi's monkey (Callitrichidae). In: Kleiman, D. G., Geist, V., Hutchins, M., McDade, M. (Eds.). *Grzimek's Animal Life Encyclopedia*, Vol. 14, pp. 115–133.

Hiramatsu, C., Melin, A. D., Aureli, F., et al. (2009) Interplay of olfaction and vision in fruit foraging of spider monkeys. *Anim. Behav.* 77: 1421–1426.

Hübener, F., and Laska, M. (1998) Assessing olfactory performance in an Old World primate, *Macaca nemestrina*. *Physiol. Behav.* 64: 521–527.

Hübener, F., Laska, M. (2001) A two-choice discrimination method to assess olfactory performance in pigtail macaques, *Macaca nemestrina*. *Physiol. Behav.* 72: 511–519.

Irwin, M. T., Raharison, F. J., Rakotoarimanana, H., et al. (2007) Diademed sifakas *(Propithecus diadema)* use olfaction to forage for the inflorescences of subterranean parasitic plants. *Am. J. Primatol.* 69: 471–476.

Joly, M., Michel, B., Deputte, B., and Verdier, J. M. (2004) Odor discrimination assessment with an automated olfactometric method in a prosimian primate, *Microcebus murinus*. *Physiol. Behav.* 82: 325–329.

Jones, C. B. (2003) Urine-washing behaviors as condition-dependent signals of quality by adult mantled howler monkeys *(Alouatta palliata)*. *Lab. Primate Newsl.* 42: 12–14.

Joshi, D., Völkl, M., Shepherd, G. M., and Laska, M. (2006) Olfactory sensitivity for enantiomers and their racemic mixtures – a comparative study in CD-1 mice and spider monkeys. *Chem. Senses* 31: 655–664.

Kaplan, J. N., Cubicciotti, D., and Redican, W. K. (1977) Olfactory discrimination of squirrel monkey mothers by their infants. *Dev. Psychobiol.* 10: 447–453.

Kappel, P., Hohenbrink, S., and Radespiel, U. (2011) Experimental evidence for olfactory predator recognition in wild mouse lemurs. *Am. J. Primatol.* 73: 928–938.

Kappeler, P. (1998) To whom it may concern: the transmission and function of chemical signals in *Lemur catta*. *Behav. Ecol. Sociobiol.* 42: 411–421.

Kjeldmand L., Hernandez Salazar L. T., and Laska, M. (2011) Olfactory sensitivity for sperm-attractant aromatic aldehydes: a comparative study in human subjects and spider monkeys *J. Comp. Physiol. A* 197: 15–23.

Kobayashi, M., Sasabe, T., Takeda, M., et al. (2002) Functional anatomy of chemical senses in the alert monkey revealed by positron emission tomography. *Eur. J. Neurosci.* 16: 975–980.

Laidre, M. E. (2009) Informative breath: olfactory cues sought during social foraging among Old World monkeys *(Mandrillus sphinx, M. leucophaeus, and Papio anubis)*. *J. Comp. Psychol.* 123: 34–44.

Laska, M., Alicke, T., and Hudson, R. (1996) A study of long-term odor memory in squirrel monkeys, *Saimiri sciureus*. *J. Comp. Psychol.* 110: 125–130.

Laska, M., Fendt, M., Wieser, A., et al. (2005a) Detecting danger – or just another odorant ? Olfactory sensitivity for the fox odor component 2,4,5-trimethylthiazoline in four species of mammals. *Physiol. Behav.* 84: 211–215.

Laska, M., Freist, P., and Krause, S. (2007a) Which senses play a role in nonhuman primate food selection? A comparison between squirrel monkeys and spider monkeys. *Am. J. Primatol.* 69: 282–294.

Laska, M., and Freyer, D. (1997) Olfactory discrimination ability for aliphatic esters in squirrel monkeys and humans. *Chem. Senses* 22: 457–465.

Laska, M., Genzel, D., and Wieser, A. (2005b) The number of functional olfactory receptor genes and the relative size of olfactory brain structures are poor predictors of olfactory discrimination with enantiomers. *Chem. Senses* 30: 171–175.

Laska, M., and Grimm, N. (2003) SURE, why not? The SUbstitution-REciprocity method for measurement of odor quality discrimination thresholds: replication and extension to nonhuman primates. *Chem. Senses* 28: 105–111.

Laska, M., Hernandez Salazar, L. T., and Rodriguez Luna, E. (2003a) Successful acquisition of an olfactory discrimination paradigm by spider monkeys *(Ateles geoffroyi)*. *Physiol. Behav.* 78: 321–329.

Laska, M., Höfelmann, D., Huber, D., and Schumacher, M. (2006a) The frequency of occurrence of acyclic monoterpene alcohols in the chemical environment does not determine olfactory sensitivity in nonhuman primates. *J. Chem. Ecol.* 32: 1317–1331.

Laska, M., Hofmann, M. and Simon, Y. (2003b) Olfactory sensitivity for aliphatic aldehydes in squirrel monkeys and pigtail macaques. *J. Comp. Physiol. A* 189: 263–271.

Laska, M., and Hudson, R. (1993a) Discriminating parts from the whole: Determinants of odor mixture perception in squirrel monkeys, *Saimiri sciureus. J. Comp. Physiol. A.* 173: 249–256.

Laska, M., and Hudson, R. (1993b) Assessing olfactory performance in a New World primate, *Saimiri sciureus. Physiol. Behav.* 53: 89–96.

Laska, M., and Hudson, R. (1995) Ability of female squirrel monkeys (*Saimiri sciureus*) to discriminate between conspecific urine odours. *Ethology* 99: 39–52.

Laska, M., Liesen, A., and Teubner, P. (1999a) Enantioselectivity of odor perception in squirrel monkeys and humans. *Am. J. Physiol.* 277: R1098–R1103

Laska, M., and Metzker, K. (1998) Food avoidance learning in squirrel monkeys and common marmosets. *Learning & Memory* 5: 193–203.

Laska, M., Miethe, V., Rieck, C., and Weindl, K. (2005c) Olfactory sensitivity for aliphatic ketones in squirrel monkeys and pigtail macaques. *Exp. Brain Res.* 160: 302–311.

Laska, M., Persson, O., and Hernandez Salazar, L. T. (2009) Olfactory sensitivity for alkylpyrazines – a comparative study in CD-1 mice and spider monkeys. *J. Exp. Zool. A* 311: 278–288.

Laska, M., Rivas Bautista, R. M., and Hernandez Salazar, L. T. (2006b). Olfactory sensitivity for aliphatic alcohols and aldehydes in spider monkeys, *Ateles geoffroyi. Am. J. Phys. Anthropol.* 129: 112–120.

Laska, M., Rivas Bautista, R. M., Höfelmann, D., et al. (2007b) Olfactory sensitivity for putrefaction-associated thiols and indols in three species of non-human primate. *J. Exp. Biol.* 210: 4169–4178.

Laska, M., and Seibt, A. (2002a) Olfactory sensitivity for aliphatic esters in squirrel monkeys and pigtail macaques. *Behav. Brain Res.* 134: 165–174.

Laska, M., and Seibt, A. (2002b) Olfactory sensitivity for aliphatic alcohols in squirrel monkeys and pigtail macaques. *J. Exp. Biol.* 205: 1633–1643.

Laska, M., Seibt, A., and Weber, A. (2000) "Microsmatic" primates revisited—Olfactory sensitivity in the squirrel monkey. *Chem. Senses* 25: 47–53.

Laska, M., and Teubner, P. (1998) Odor structure-activity relationships of carboxylic acids correspond between squirrel monkeys and humans. *Am. J. Physiol.* 274: R1639–R1645.

Laska, M., Trolp, S., and Teubner, P. (1999b) Odor structure-activity relationships compared in human and non-human primates. *Behav. Neurosci.* 113: 98–1007.

Laska, M., Wieser, A., and Hernandez Salazar, L. T. (2005d) Olfactory responsiveness to two odorous steroids in three species of nonhuman primates. *Chem. Senses* 30: 505–511.

Laska, M., Wieser, A., and Hernandez Salazar, L. T. (2006c) Sex-specific differences in olfactory sensitivity for putative human pheromones in nonhuman primates. *J. Comp. Psychol.* 120: 106–112.

Laska, M., Wieser, A., Rivas Bautista, R. M., and Hernandez Salazar, L. T. (2004) Olfactory sensitivity for carboxylic acids in spider monkeys and pigtail macaques. *Chem. Senses* 29: 101–109.

Liebetanz, D., Nitsche, M. A., Fromm, C., and Reyher, C. K. H. (2002) Central olfactory connections in the microsmatic marmoset monkey (*Callithrix jacchus*). *Cells Tissues Organs* 172: 53–69.

Loireau, J. N., and Gautier-Hion, A. (1988) Olfactory marking behaviour in guenons and its implications. In: Gautier-Hion, A., Bourliere, F., Gautier, J. P. (Eds.) *A Primate Radiation: Evolutionary biology of the African guenons.* Cambridge University Press, Cambridge, pp. 246–254.

Loo, S. K. (1977) Fine structure of the olfactory epithelium in some primates. *J. Anat.* 123: 135–145.

Løtvedt, P. K., Murali, S. K., Hernandez Salazar, L. T., and Laska, M. (2012) Olfactory sensitivity for "green odors" (aliphatic C_6 alcohols and C_6 aldehydes) – A comparative study in male CD-1 mice (*Mus musculus*) and female spider monkeys (*Ateles geoffroyi*). *Pharmacol. Biochem. Behav.* 101: 450–457.

Matsui, A., Go, Y., and Niimura, Y. (2010) Degeneration of olfactory receptor gene repertoires in primates: no direct link to full trichromatic vision. *Molec. Biol. Evol.* 27: 1192–1200.

Meredith, M. (2001) Human vomeronasal organ function: a critical review of best and worst cases. *Chem. Senses* 26: 433–445.

Milton, K. (1985) Urine washing behavior in the woolly spider monkey (*Brachyteles arachnoides*). *Z. Tierpsychol.* 67: 154–160.

Nei, M., Niimura, Y., and Nozawa, M. (2008) The evolution of animal chemosensory receptor gene repertoires: roles of chance and necessity. *Nat. Rev. Genet.* 9: 951–963.

Palagi, E., and Dapporto, L. (2007) Females do it better. Individual recognition experiments reveal sexual dimorphism in *Lemur catta* olfactory motivation and territorial defence. *J. Exp. Biol.* 210: 2700–2705.

Pihlström, H., Fortelius, M., Hemilä, S., et al. (2005) Scaling of mammalian ethmoid bones can predict olfactory organ size and performance. *Proc. Roy. Soc. B* 272: 957–962.

Rolls, E. T. (2004) The functions of the orbitofrontal cortex. *Brain Cognit.* 55: 11–29.

Rolls, E. T. (2006) Brain mechanisms underlying flavour and appetite. *Phil. Trans. R. Soc. B* 361: 1123–1136.

Rolls, E. T., Critchley, H. D., Mason, R., and Wakeman, E. A. (1996) Orbitofrontal cortex neurons: role in olfactory and visual association learning. *J. Neurophysiol.* 75: 1970–1981.

Rouquier, S., Blancher, A., and Giorgi, D. (2000) The olfactory receptor gene repertoire in primates and mouse: evidence for reduction of the functional fraction in primates. *Proc. Natl. Acad. Sci. USA* 97: 2870–2874.

Schilling, A., Serviere, J., Gendrot, G., and Perret, M. (1990) Vomeronasal activation by urine in the primate *Microcebus murinus*: a 2 DG study. *Exp. Brain Res.* 81: 609–618.

Setchell, J. M., Vaglio, S., Abbott, K. M., et al. (2011) Odour signals major histocompatibility complex genotype in an Old World monkey. *Proc. Roy. Soc. Lond. B* 278: 274–280.

Setchell, J. M., Vaglio, S., Moggi-Cecchi, J., et al. (2010) Chemical composition of scent-gland secretions in an Old World monkey (*Mandrillus sphinx*): influence of sex, male status, and individual identity. *Chem. Senses* 35: 205–220.

Siemers, B. M., Goerlitz, H. R., Robsomanitrandrasana, E., et al. (2007) Sensory basis of food detection in wild *Microcebus murinus. Int. J. Primatol.* 28: 291–304.

Smith, T. D., and Bhatnagar, K. P. (2004) Microsmatic primates: reconsidering how and when size matters. *Anat. Rec. B* 279: 24–31.

Smith, T. D., Bhatnagar, K. P., Shimp, K. L., et al. (2002) Histological definition of the vomeronasal organ in humans and chimpanzees, with a comparison to other primates. *Anat. Rec.* 267: 166–176.

Smith, T. D., Bhatnagar, K. P., Tuladhar, P., and Burrows, A. M. (2004) Distribution of olfactory epithelium in the primate nasal cavity: are microsmia and macrosmia valid morphological concepts? *Anat. Rec. A* 281: 1173–1181.

Smith, T. D., Garrett, E. C., Bhatnagar, K. P., et al. (2011) The vomeronasal organ of New World monkeys (Platyrrhini). *Anat. Rec.* 294: 2158–2178.

Smith, T. D., and Rossie J. (2006) Primate olfaction: anatomy and evolution. In: Brewer, W. J., Castle, D., Pantelis, C. (Eds.) *Olfaction and the Brain.* Cambridge University Press: Cambridge, pp. 135–166.

References

Smith, T. D., Rossie, J., and Bhatnagar, K. P. (2007) Evolution of the nose and nasal skeleton in primates. *Evol. Anthropol.* 16: 132–146.

Smith, T. D., Siegel, M. I., and Bhatnagar, K. P. (2001) Reappraisal of the vomeronasal system of Catarrhine primates: ontogeny, morphology, functionality, and persisting questions. *Anat. Rec.* 265: 176–192.

Stephan, H., Baron, G., and Frahm, H. D. (1982) Comparison of brain structure volumes in insectivora and primates. II. Accessory olfactory bulb (AOB). *J. Hirnforsch.* 23: 575–591.

Stephan, H., Baron, G., and Frahm, H. D. (1988) Comparative size of brains and brain components. In: Steklis, H., Erwin, J. (Eds.) *Comparative Primate Biology*, Vol. 4. Alan R. Liss: New York, pp. 1–38.

Sündermann, D., Scheumann, M., and Zimmermann, E. (2008) Olfactory predator recognition in predator-naive gray mouse lemurs *(Microcebus murinus)*. *J. Comp. Psychol.* 122: 146–155.

Takagi, S. F. (1986) Studies on the olfactory nervous system of the Old World monkey. *Prog. Neurobiol.* 27: 195–250.

Tanabe, T., Iino, M., Ooshima, Y., and Takagi, S. F. (1975a) An olfactory projection area in orbitofrontal cortex of the monkey. *J. Neurophysiol.* 38: 1269–1283.

Tanabe, T., Iino, M., and Takagi, S. F. (1975b) Discrimination of odors in olfactory bulb, pyriform-amygdaloid areas, and orbitofrontal cortex of the monkey. *J. Neurophysiol.* 38: 1284–1296.

Ueno, Y. (1994a) Olfactory discrimination of eight food flavors in the capuchin monkey *(Cebus apella)*: comparison between fruity and fishy odors. *Primates* 35: 301–310.

Ueno, Y. (1994b) Olfactory discrimination of urine odors from five species by tufted capuchin *(Cebus apella)*. *Primates* 35: 311–323.

Van Belle, S., Estrada, A., Ziegler, T. E., and Strier, K. B. (2009) Sexual behavior across ovarian cycles in wild black howler monkeys *(Alouatta pigra)*: male mate guarding and female mate choice. *Am. J. Primatol.* 71: 153–164.

Wallén, H., Engström, I., Hernandez Salazar, L. T., and Laska, M. (2012) Olfactory sensitivity for six amino acids – A comparative study in CD-1 mice and spider monkeys. *Amino Acids* 42: 1475–1485.

Wolovich, C. K., and Evans, S. (2007) Sociosexual behavior and chemical communication of *Aotus nancymaae. Int. J. Primatol.* 28: 1299–1313.

Yarita, H., Iino, M., Tanabe, T., et al. (1980) A transthalamic olfactory pathway to orbitofrontal cortex in the monkey. *J. Neurophysiol.* 43: 69–85.

Part 6

Gustation
Taste Anatomy and Neurobiology

Part 6

Chapter 28

The Role of Saliva in Taste Transduction

RYUJI MATSUO and GUY H. CARPENTER

28.1 INTRODUCTION

Saliva is the first digestive fluid in the alimentary canal, and is produced by three pairs of major salivary glands (parotid, submandibular, and sublingual) and by hundreds of minor salivary glands (labial, lingual, buccal, and palatal) spread over most parts of the oral mucosa. This fluid usually coats the surface of teeth and oral mucosa and has multiple functions, including the maintenance of taste and somatosensory functions of the oral cavity.

The role of saliva in taste was first documented by McBurney and Pfaffmann (1963), who showed that salivary constituents (sodium and chloride) could affect taste sensitivity to NaCl by human psychophysical experiment. Gurkan and Bradley (1988) recorded the taste nerve (the glossopharyngeal nerve) response in rats and have suggested that saliva not only acts as a solvent of taste substances, but also interacts in some way with the receptor mechanism. On the basis of these studies, many human psychophysical and animal behavioral and electrophysiological studies have been employed and provided evidence that saliva can widely affect the oral sense including the five basic tastes and the taste of fat, starch, astringency, and tactile sensation.

Since Ludwig (1850) and Bernard (1856) discovered the secretory role of nerves to salivary glands, it has been known that salivary secretion is dependent on reflex activity. The salivary reflex is regulated by the autonomic nerves innervating salivary glands. Whilst we are awake there is a resting flow of saliva by all glands into the mouth. The most copious flow of saliva is produced during eating with inputs from taste, chewing, and smell, and to a lesser degree the texture of the food. Activation of the autonomic nerves is initiated by multiple sensory afferents, including taste bud activation, mechanoreceptors in the gingivae and bare nerve endings in the mucosa as well as the olfactory bulbs in the nasal cavity. These stimuli for salivation are conveniently called gustatory, masticatory, olfactory, and oesophageal salivary reflexes (Hector and Linden, 1999). This reflex saliva, or stimulated saliva, particularly assists in the breakup and bolus formation of food ready for swallowing. Resting saliva, in contrast, is more suited to the tasting of foods by having a different composition to reflex-stimulated saliva.

Unstimulated, or resting, saliva is secreted in the absence of apparent sensory stimuli related to eating, but is actually the greatest contributor to total salivary secretion during the diurnal cycle (Sreebny, 1989). Although this saliva is produced at rest it is still an active secretion process and is rich in glycoproteins, including mucins and IgA, mainly excreted from the submandibular and sublingual salivary glands (approximately 60–70%). Although the parotid glands are the largest salivary glands in humans they have a very low resting flow but contribute the greatest amount to whole mouth saliva (the sum of all salivary secretions into the mouth) during reflex activation. Whereas the submandibular and sublingual glands have large quantities of mucins, the parotid has none. Mucins have the properties of high viscosity and strong adhesiveness to oral tissues and are often perceived as molecules that create the thickness and elasticity of whole mouth saliva. The viscosity of resting saliva is two to three times

---

Handbook of Olfaction and Gustation, Third Edition. Edited by Richard L. Doty.
© 2015 Richard L. Doty. Published 2015 by John Wiley & Sons, Inc.

that of stimulated saliva induced by chewing paraffin wax in healthy adults (Rantonen and Meurman, 1998). These findings imply that resting saliva contributes to maintaining the health of the oral cavity by lubricating and protecting oral tissues, and acting as a barrier against chemical and mechanical irritants. In patients with decreased salivary flow (xerostomia), whether caused by prescribed drugs, disease or following head and neck irradiation, the result is clinically significant oral discomfort that may manifest as increased caries, susceptibility to infection of the oral mucosa, or altered oral sensation (Hershkovich and Nagler, 2004; Mese and Matsuo, 2007).

In normal healthy subjects, saliva constantly coats the oral mucosa, protecting taste receptors from their external environment, but most also mediate the initial events in taste transduction (Spielman, 1990; Bradley, 1991; Matsuo, 2000; Bradley and Beidler, 2003). To reach and stimulate taste receptors, taste stimuli initially have to dissolve in saliva and then pass through the salivary layer towards the receptor. The fluid layer is composed of various organic and inorganic substances, some of which can stimulate taste receptors and/or chemically interact with taste substances. These reactions eventually affect perceived taste intensity and/or quality.

Food contains various substances. Oral sensory receptors distinguish food chemicals into the five fundamental taste qualities of sweet, salty, sour, bitter, and umami. Other tastes, such as astringency, metallic and fatty sensations, are probably detected by oral mechanoceptors and bare nerve endings within the mucosa.

28.2 MILIEU OF SENSORY RECEPTORS IN THE ORAL CAVITY

28.2.1 Composition of Saliva

Saliva is a mixture of saliva from three pairs of major glands and hundreds of minor glands, all of which have a different profile of proteins. Some proteins are universal to all glands, such as the secretory component, the transporter of IgA (the main antibody in saliva). Mucins (Muc 5b and Muc 7 gene products) are common to the submandibular, sublingual, and most minor glands, but are not expressed by the serous glands, parotid, and von Ebner's glands. Basic PRPs (proline-rich proteins) appear to be exclusive to the parotid glands, whilst acidic PRPs appear in submandibular and parotid glands. Furthermore, the relative contribution of each gland to saliva varies between resting and stimulated states. Hence, the composition of saliva can vary greatly within a person and vary even more among individuals due to genetic polymorphisms. For example, the proline-rich proteins, which form nearly 70% of parotid total protein, are encoded by only six or so genes but can be rearranged and have so many post-transcriptional modifications that a family of nearly 30 separate proteins appear in saliva (Maeda, 1985; Azen and Maeda, 1988).

Compared to reflex-secreted stimulated saliva, resting saliva is ideally modified for the initial tasting of foods. Although salivary secretion involves the active secretion of chloride ions into the ducts, to be followed by sodium ions and water, it has a low salt content as the low flow rate allows the striated ducts to reabsorb most of the sodium and chloride ions. This energy-demanding process changes the primary saliva, which is initially isotonic with serum, into a hypotonic solution (Table 28.1). The benefit to taste is that, because taste buds become adapted to their ionic environment, the taste threshold for salt is much lower than it would be if the ion reabsorption process did not occur. In addition, resting saliva has low buffering ability since the secretion of both bicarbonate ions and carbonic anhydrase, the main buffering agents in saliva, is related to flow rate. Thus, in unstimulated resting saliva, the low buffering capacity (although proteins do provide some buffering capacity) allows the taste buds to detect low concentrations of protons that create the sour taste sensation.

Table 28.1 Electrolytes, pH, and osmolality of human saliva.

	Resting Saliva	Stimulated Saliva	Source of Saliva
Na^+	2.7 mEq/l	63.3 mEq/l	parotid
	3.3 mEq/l	45.5 mM/l	submandibular
K^+	46.3 mEq/l	18.7 mEq/l	parotid
	13.9 mEq/l	17.8 mM/l	submandibular
Cl^-	31.5 mEq/l	35.9 mEq/l	parotid
	12.0 mEq/l	23.4 mM/l	submandibular
HCO_3^-	0.60 mEq/l	29.7 mEq/l	parotid
pH	5.47	6.48	parotid
	6.73	–	submandibular
Osmolality	85.7 mOsm/kg	136.9 mOsm/kg	parotid

Data for parotid gland are cited from Shannon et al., 1974, and those for the submandibular gland are from Ferguson and Fort, 1974, and Ferguson and Botchway, 1979.

Table 28.2 Salivary proteins related to oral senses and their main functions.

Proteins	Main Function	Effect on Oral Sense
Mucins	lubrication and pellicle formation	emulsifying of fats/oils, and tactile
Amylase	starch digestion	taste of carbohydrate and glucose, and tactile
Lipase	fat digestion	taste of fatty acids
Carbonic anhydrase	buffering action	acid taste
Proline-rich proteins	pellicle formation	astringency
Histatin	anti-bacterial action	astringency
Statherin	remineralization of teeth	lowering of surface tension

A further modification of saliva that facilitates the initial tasting of foods is the low surface tension. This is a feature of resting and stimulated saliva which allows the rapid interaction of the tastant with taste buds. The low surface tension of saliva is generated by the proteins within saliva and, in particular, surface active proteins such as statherin (Proctor et al., 2005a) (Table 28.2).

For many tastants, for example, salt and acids, saliva will be the solvent used to convey the ions to the taste buds. It is surprising then that an aqueous solution, such as saliva, is readily able to detect tastes locked up in an oil/fat liquid. The interaction of oil in water emulsions stabilized by different emulsifiers is a subject of considerable interest. Oil is well known to become emulsified by saliva, probably by interaction with mucins (Silletti et al., 2007). The deposition of oil onto the tongue surface, studied by *in vivo* confocal microscopy (Adams et al., 2007) and simpler paper sampling methods (Pivk et al., 2008) again relates to the emulsifying behavior of saliva. Further interactions of emulsions stabilized with different proteins reveals that the nature of the interaction with the oral mucosa governs taste and the mouthfeel of the food (Vingerhoeds et al., 2008; Vingerhoeds et al., 2009). Whey protein-stabilized emulsions did not readily interact with the tongue and formed creamy-type sensations, whereas lysozyme-stabilized emulsions formed almost astringent-like sensations and remained on the tongue.

28.2.2 Salivary Layer on the Mucosal Membrane

Saliva exists in the mouth as a heterogeneous (Sas and Dawes, 1997) thin film (Pramanik et al., 2010) that coats and interacts with the mucosa to form a highly lubricating layer (Bongaerts et al., 2007). The structure of the oral mucosa varies from non-keratinized squamous cells in the buccal region to keratinized gingivae and papillae on the dorsum of the tongue. The thickness of the salivary film varies from 5 μm on the hard palate to 25 μm on the tongue, as calculated by measuring the amount of saliva per unit area (Pramanik et al., 2010). The thickness of saliva in different areas of the mouth probably reflects a number of variables, including local flow rates, salivary viscoelastic properties, and the surface roughness of the mucosa. The dorsal surface of the tongue, for example, has the thickest film, probably because of the papillae forming pronounced ridges. On the hard surfaces in the mouth, salivary proteins readily and specifically interact with the mineral components of teeth. The acquired enamel pellicle (Dawes et al., 1963) is a subset of salivary proteins that rapidly bind to calcium phosphate salts within teeth, mostly *via* post-translational modification of the protein, that is, a phosphate group. The most abundant salivary proteins in acquired enamel pellicles include statherin, acidic proline-rich proteins, and histatins (Jensen et al., 1992; Li et al., 2004). Modelling teeth with hydroxyapatite, mucins were also found to form strongly lubricating layers (Cardenas et al., 2007; Hahn Berg et al., 2003; Halthur et al., 2010). The presence of statherin, a calcium carrier, enhances the local concentration of calcium around teeth and thus prevents the dissolution of teeth (Hay and Bowen, 1996). The other proteins presumably play a role in managing the microflora that also bind to teeth.

As well as the enamel pellicle there is also evidence of a mucosal pellicle, that is, a select group of salivary proteins that are bound, to varying degrees, to oral mucosal cells. Initial observations (Bradway et al., 1989; Bradway et al., 1992) suggested that mucins, amylase, cystatin S, and acidic PRPs were present. The absorption of mucins to any surface in the mouth would be important for lubrication (Stokes and Davies, 2007). Once absorbed, the film can be modified by the ionic composition of saliva. In a study of absorbed mucins to control surfaces mimicking the oral mucosa, the absorbed mucin layer was found to vary in thickness between a low and high ionic strength solution (e.g., resting vs. stimulated saliva). This mucin film collapsed in a completely deionised environment, suggesting that static repulsion between molecules can influence film hydration/thickness (Macakova et al., 2010). It would be interesting to know whether the salivary film ever collapses *in vivo*. A study of dry-mouth patients indicated that the salivary film was still present but thinner (Pramanik et al., 2010), perhaps suggesting a collapse of the mucosal pellicle, leading to the loss of lubrication and feelings of dryness. Over these pellicles (enamel and mucosal) flows the fluid phase of saliva. The flow starts from the opening of the ducts leading from the major glands (sublingual and buccal areas) and travels over all surfaces, at varying speeds, towards the throat for swallowing (or spitting out). This flow of saliva is important for the clearance of food and bacteria from the mouth and so maintains a normal healthy mouth.

28.2.3 Saliva and Taste Pore

28.2.3.1 Taste Buds and Flow of Salivas are.
Taste buds, consisting of taste receptor cells, are unevenly distributed in the oral cavity. Many taste buds are located in the epithelial folds of the foliate and circumvallate papillae, and a smaller number are situated in the mucosa of the fungiform papillae and soft palate (Figure 28.1). It is likely that taste buds in various areas are exposed to different fluid environments, since secretions produced by each salivary gland differ in composition and, moreover, they are not always well-mixed or well-distributed around the mouth.

Lingual minor salivary glands on the posterior tongue, von Ebner's glands, drain saliva into the grooves of the circumvallate and foliate papillae. Saliva from von Ebner's glands may exclusively fill the cleft and can effectively rinse away taste stimuli, as demonstrated by electrophysiological recordings of taste responses from circumvallate taste buds (Gurkan and Bradley, 1988).

Parotid saliva is secreted into the mouth from the oral opening of the main excretory duct at the buccal mucosa near the upper molar teeth. The main duct of the submandibular gland and several small ducts of the sublingual glands open onto the sublingual mucosa of the floor of the mouth. Due to this anatomical feature, parotid and submandibular/sublingual saliva is not evenly distributed throughout the mouth (Dawes and Macpherson, 1993). Mixed saliva collected from posterior regions of the mouth contains the greatest proportion of parotid saliva, whereas saliva from anterior regions contains the least under both stimulated and unstimulated conditions in humans (Sas and Dawes, 1997); a similar result was observed in rats (Matsuo et al., 1994). When a dye was infused into the mouth through the duct of the parotid or submandibular gland, the anterior and posterior parts of the tongue were mainly stained by infusion from the submandibular and parotid ducts, respectively. These findings imply that submandibular/sublingual saliva contributes to maintaining the external milieu of taste buds in the fungiform papillae more than parotid saliva.

28.2.3.2 Milieu in the Taste Pores.
Taste receptor sites are located on microvilli or club-shaped apical

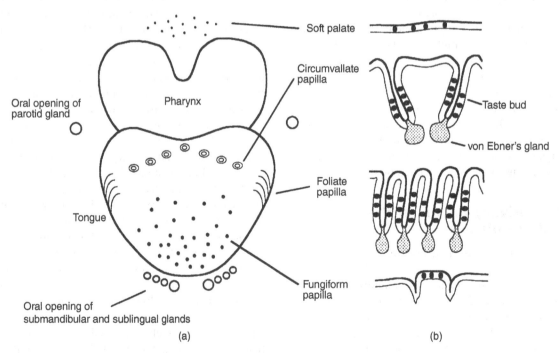

Figure 28.1 Locations of oral openings of the major salivary glands and taste papillae (a), and taste buds in the soft palate and taste papillae (b). (From Matsuo, 2000).

processes that emerge from taste cells. The apical membrane is exposed to the oral cavity in the taste pore (about 20 μm in diameter) at the top of the taste bud (Segovia et al., 2002). Taste pores are usually filled with mucous fluid that is derived from saliva and perhaps secretions from taste bud cells (Brouwer and Wiersma, 1980). Histochemical analyses of rabbits revealed that the taste pore content has a different carbohydrate composition from that of mucous substances overlying the adjacent epithelium, and that the composition of taste pore materials differs between fungiform and circumvallate/foliate taste buds (Witt and Miller, 1992). These findings suggest that secretions from taste buds immediately contact with the taste receptor. One such substance may be amylase, secreted from taste buds of the circumvallate papillae, as described later.

28.3 ROLE OF SALIVA IN TASTE

Taste substances should be dissolved in the salivary fluid layer to reach and stimulate taste receptors. During the process, saliva may affect taste substances by solubilization, diffusion, dilution, and chemical interaction. Some salivary components also affect the taste receptor and can alter its sensitivity by acting as stimulants and by affecting the affinity of the receptor to taste substances. Moreover, salivary digestive enzymes interact with certain components of food and produce tastants probably other than the five basic taste modalities. However, it is known that these salivary actions vary among animal species due to differences in taste transduction mechanisms and salivary compositions.

Based on recent molecular genetic studies on G protein-coupled receptors in various animals, it is suggested that species difference in taste receptor function intimately relates to taste behavior and feeding ecology (Li et al., 2005, 2009; Jiang et al., 2012). For example, cats and lions (obligate carnivores) lack the sweet taste receptor and are indifferent to sweeteners. Dogs (omnivore) and sheep (herbivore) do not show preferences to some of natural sugars including maltose (Ferrell, 1984; Glaser, 2002; Ginane et al., 2011), whilst rats and mice (omnivores) recognize and prefer most natural sugars including maltose (Li et al., 2002; Damak et al., 2003). On the other hand, it is well known that species have differences in salivary glands (e.g., Young and van Lennep, 1978) and salivary constituents (e.g., Erickson, 1964; Young and Schneyer, 1981). For example, in cats and dogs the zygomatic gland is prominent enough to be considered as the fourth major salivary gland, and in sheep the parotid gland is exceptionally large perhaps reflecting habitually ingesting dry food. Interestingly amylase is in low concentration or absent from the saliva of animals lacking the sweet taste receptors (e.g., cats) and those showing indifference to maltose (e.g., dogs and sheep). These findings suggest that, in evolution

of dietary habits, the function of taste and saliva might be designed to be suitable for detection, mastication, and digestion of food. Laboratory animals such as rats, mice, and hamsters secrete saliva of different concentration of ionic salts, amylase activity, and pH. There is a possibility that such variations may reflect species difference in taste sensitivity or taste transduction.

28.3.1 Saliva as a Solvent

The process of the dissolution and diffusion of taste substances in the pre-receptor fluid layer can influence the onset times of taste responses after the intake of food and beverages. Once taste substances dissolve into ions and molecules in the salivary layer, their movement to taste receptors by diffusion is relatively rapid, as shown by studies on taste response latencies (Beidler, 1961; Kelling and Halpern, 1993). Thus, the process of dissolution rather than diffusion may govern most of the pre-receptor time. The typical effect of solubilization can be observed after the intake of dried solid food and aqueous taste solutions (Matsuo et al., 1994). As seen in taste responses of the chorda tympani innervating the taste buds of the fungiform papillae in rats, a far longer time to reach the peak response was induced by eating food pellets than that induced by licking a taste solution (Figure 28.2). This suggests that chewing reduces the size of food particles and mixes them with saliva, gradually increasing the concentration of taste stimulants in the salivary layer.

28.3.2 Salty Taste

Of the salivary constituents of human mixed saliva, the principal ions of Na^+, K^+, Cl^-, and HCO_3^- can reach concentrations above their taste-detection threshold (see Table 28.1). These ions constantly stimulate the apical membranes of taste receptor cells. Eventually, taste receptor cells are adapted to such salivary components, and we are normally unaware of this background taste. In particular, Na^+ and Cl^- may largely contribute to the adaptation or the basal activity of taste receptor cells, since Na^+ and Cl^- (NaCl) induce greater taste responses than other salivary ions (KCl, $KHCO_3$, or $NaHCO_3$). Sodium can enter taste receptor cells through epithelial sodium channels, the putative Na^+ receptor, built into their apical membranes. Moreover, the tight junctions between the apical tips of taste bud cells and/or the epithelium are permeable to Na^+ and Cl^-; the permeated Cl^- may contribute to evoking stronger saltiness (Ye et al., 1993; Roper, 2007).

It is well documented that salivary Na^+ and Cl^- can influence the detection of sodium chloride, the principal salty tastant, contained in food and beverages by the process of self-adaptation. Psychophysical studies in human (McBurney and Pfaffmann, 1963; Delwiche and

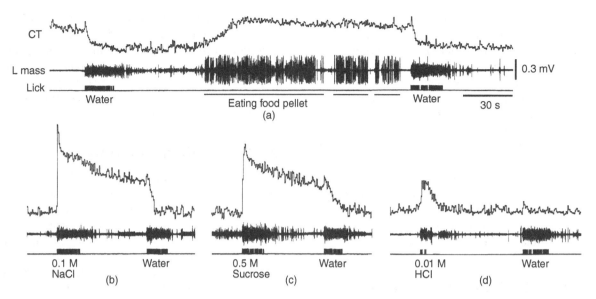

Figure 28.2 Taste responses of the rat chorda tympani while eating a food pellet (A), licking 0.1 M NaCl (B), 0.5 M sucrose (C), and 0.01 M HCl (D). Each set of recordings is digitally summated chorda tympani discharge (CT), electromyographic activity of the left masseter muscle (L Mass), and lick signal (Lick). (From Matsuo et al., 1994).

O'Mahony, 1996; Bartoshuk, 1978) and animal experiments (Matsuo and Yamamoto, 1992; Rehnberg et al., 1992; Matsuo et al., 1994) have shown that salivary ions elevate the taste threshold and decrease the supra-threshold intensity of NaCl.

28.3.3 Sour Taste

Sour taste is elicited by acid, of which the proton and/or protonated molecular forms of organic acids in aqueous solutions are the ligand for sourness. Acids can stimulate both taste and trigeminal sensory receptors. In particular, non-dissociated acid molecules may enter the lingual epithelium and liberate protons. These protons could reach and stimulate the trigeminal free nerve endings innervating the epithelium, including the taste bud (Yamasaki et al., 1984; Bryant and Moore, 1995). In humans, acids are a powerful simulator evoking reflex salivation. A copious flow of saliva will dilute the taste stimulant. Moreover, saliva acts as a buffer by neutralizing acids with its components of bicarbonate, phosphate, and amphoteric proteins and enzymes. Bicarbonate is the most important buffering system of saliva. Since the concentration of bicarbonate and pH of saliva increase as the flow rate of saliva increases, the buffering action works more efficiently during stimulated high flow rates than during unstimulated low flow rates.

In animal experiments using rats, electrophysiological recordings of the taste nerve (chorda tympani) showed lower responses to HCl solution in the presence of saliva or bicarbonate salts on the tongue surface than in the absence of saliva (Matsuo and Yamamoto, 1992; Matsuo et al., 1994). In humans, the pH of HCl solution increased after subjects sipped and expectorated the solution (Christensen et al., 1987). Human subjects who have greater sensitivity to acids tend to show a higher salivary flow rate, and eventually their perceived taste intensity will be decreased by dilution and buffering of the acid stimuli (Christensen et al., 1987; Lugaz et al., 2005).

28.3.4 Sweet Taste

Salivary effects on sweet taste differ among animal species, reflecting differences in the salivary composition and taste receptor mechanism. In rats, sweet taste responses were recorded from the chorda tympani under anesthetized and unanesthetized conditions, and the results showed that the magnitude of sweet responses in awake animals, or when the tongue was adapted to saliva in anesthetized animals, was about twice that obtained when the tongue was adapted to distilled water (Matsuo and Yamamoto, 1992; Matsuo et al., 1994). The response to licking 0.5 M sucrose solution is comparable to that of 0.1 M NaCl solution (Figure 28.2). The second-order taste relay neurons, the rostral nucleus of the solitary tract, also showed a large response magnitude when awake rats licked sucrose solution, partly because of the effect of saliva (Nakamura and Norgren, 1991). This enhancement is common to sweet compounds having various chemical structures, including sugars such as mono- and di-saccharides, amino acids, and artificial sweeteners such as saccharin and cyclamate (Matsuo and Yamamoto, 1990). This salivary effect can be mimicked by artificial saliva containing HCO_3^- and solutions with the pH adjusted to 8–10; meanwhile, sweet taste responses were suppressed when the tongue was adapted to solutions of pH 3–4 (Matsuo and Yamamoto, 1992). A possible explanation for these findings is that salivary pH can change the charge distribution on the putative receptor, T1R2-T1R3 heterodimers of G protein-coupled receptors, and the allosteric H^+ binding of sweeteners with the receptor molecule is optimal at about pH 9.

In hamsters, no enhanced sucrose responses in the chorda tympani occur in the presence of saliva on the tongue (Rehnberg et al., 1992), but suppression of sucrose responses under acidic conditions of the tongue surface is suggested (Formaker et al., 2009). The lack of sweet taste enhancement in hamsters may be because the pH values of whole saliva are higher in hamsters (8.5–9.0) than rats (7.5–8.4) (Erickson, 1964), and chorda tympani responses to sweet taste stimuli are naturally much stronger in hamsters than rats (Ogawa et al., 1968).

In dogs (Kumazawa and Kurihara, 1990) and rats (Matsuo and Yamamoto, 1992), sweet taste responses can be slightly enhanced by the presence of Na^+ on the tongue at a low concentration equivalent to salivary levels, although the molecular mechanism of this effect is not known. In humans, psychophysical experiments reveal that the epithelial Na^+ channel blocker, amiloride, suppresses sweet taste-perceived intensity, suggesting the enhancement of sweet taste by Na^+ (e.g., Schiffman et al., 1983). Amiloride, however, can bind to the human sweet taste receptor and acts as its inhibitor (Imada et al., 2010). The effect of salivary Na^+ or pH on sweet taste intensity is not yet known in humans.

28.3.5 Umami Taste

Representative umami taste substances are 5'-mononucleotides, such as inosine 5'-monophosphate (IMP) and guanosine 5'-monophosphate (GMP), and some amino acids including glutamate whose sodium salt was the first umami substance identified. One candidate umami receptor is a heterodimer of G-protein-coupled receptors, T1R1-T1R3. This receptor may relate to the synergy reported for glutamate and nucleotide monophosphates in behavioural and nerve-recording studies (Roper, 2007; Zhan et al., 2008); for example, the taste intensity of a mixture of glutamate and IMP is larger than the sum of the individual taste intensities. This synergy is a unique characteristic of umami taste and also appears with mixtures of IMP and various amino acids in addition to glutamate. Among such umami amino aids, glutamate, glicyne, proline, and serine are the main amino acids found in human saliva. A psychophysical study indicates that the low concentration of salivary glutamate is enough to generate a synergistic taste-enhancing effect in the presence of IMP (Yamaguchi, 1991). It is possible that IMP and/or GMP in foods can also boost taste intensities of other salivary amino acids,

since human saliva contains more glicyne, proline, and serine than glutamate (Pobozy et al., 2006).

28.3.6 Bitter Taste

Saliva can act as a depressor of bitter taste, as suggested in animal experiments, but has not been tested in human psychophysical experiments. In rats, the neural activity from the chorda tympani was almost negligibly small when the rat licked a bitter solution, quinine-HCl, with normal levels of saliva present on the tongue (Matsuo et al., 1994). In anesthetized rats, the quinine response was smaller when the tongue was adapted to collected saliva rather than to deionized water. This suppressive effect was mimicked by adaptation of the tongue with solutions containing the predominant electrolytes of rat saliva, but not with dialyzed rat saliva (Matsuo and Yamamoto, 1992). A similar result has been reported in recordings from the chorda tympani of hamsters, using artificial saliva containing various ions (Rehnberg et al., 1992). Hence, salivary electrolytes may suppress the bitter taste transduction at the receptor level; however there is no explanation of this effect because of the wide diversity of bitter compounds and little knowledge of the molecular mechanism of bitter stimulation.

As for the bitter taste from the foliate and circumvallate papillae, a protein secreted from von Ebner's gland was a candidate carrier of bitter taste substances (Schmale et al., 1990). This salivary protein belongs to the lipocalin superfamily including odorant-binding proteins. Since these proteins can bind a variety of hydrophobic molecules, it was initially postulated that the von Ebner protein might bind lipophilic compounds, most likely of bitter substances, and carry them to or concentrate them at, taste receptor sites; however, the protein did not bind any of the various lipophilic bitter stimuli (Schmale et al., 1993). Thus, the von Ebner protein may be involved in non-taste-related activities such as protection against harmful lipophilic dietary chemicals, facilitation of lingual lipase activity (Kock et al., 1994), or general protection of epithelia such as fluids from tears, the nasal mucosa and tracheal mucosa; an identical protein has been found in these fluids (Redl, 2000). To define the role of von Ebner protein, it may be necessary to elucidate the ligand-binding specificity or the three-dimensional structure.

28.3.7 Taste Produced by Amylase

Amylase is the major digestive enzyme secreted by the parotid and von Ebner's glands, which breaks down starch and glycogen into di- and mono-saccharides, that is, maltose and glucose. Recently, another source of lingual amylase has been found in taste buds of the circumvallate papillae by a histochemical study of rats (Merigo et al., 2009); that is, amylase immunoreactivity is observed in the epithelium lining the top of the taste buds and in the taste pore and, moreover, in some taste bud cells. This finding suggests that the taste bud cells may secrete amylase and collaborate in carbohydrate digestion with the concerted action of salivary amylase, and sugars generated by digestion, especially in the taste pore, may stimulate taste receptors. This potential role of amylase probably contributes to the detection of the 'taste' of starch, which is commonly assumed to be tasteless. In animal experiments, it has been postulated that the starch taste is detected by two types of carbohydrate taste receptors: one for the sweet taste of sugars and another for the taste of polysaccharides (e.g., Sclafani and Mann, 1987). Unfortunately, the latter receptor has not been discovered.

In addition to taste, the digestion of starch by amylase may result in the modification of texture sensations, that is, increased melting and decreased thickness sensations of starch-based food. An *in vitro* experiment has demonstrated that human whole saliva markedly reduces the viscosity of custard desserts within a few seconds (de Wijk et al., 2004). Such action of salivary amylase will effectively progress while amylase is mixed properly into the food bolus by mastication.

28.3.8 Taste Produced by Lipase

Salivary lipase secreted only by von Ebner's glands breaks down emulsified dietary fat, triacylglycerides, in the salivary layer to free fatty acids and monoacylglycerol. The free fatty acids liberated in the oral cavity likely act as signalling molecules for the detection of fat taste, since triacylglycerides are tasteless, but recent human psychophysical evidence has suggested that various free fatty acids have a taste whose quality is probably different from the five basic tastes (Chalé-Rush et al., 2007; Stewart et al., 2010). Some candidate receptors for free fatty acids have now been isolated from taste cells of foliate and circumvallate papillae in rodents, and one in humans (Mattes, 2011).

The activity of lipase in stimulated whole saliva is sufficient to break down fat and produce free fatty acids at concentrations over their taste detection thresholds in humans (Stewart et al., 2010). This is supported by an experiment in rats indicating that fat digestion occurs within 1–5 s in the circumvallate papillae, and the amount of the generated free fatty acids is enough to be detected in the mouth. Moreover, a lipase inhibitor diminished the preference for triacylglycerides but not for free fatty acids (Kawai and Fushiki, 2003).

28.4 ROLE OF SALIVA IN ASTRINGENCY

Tannins are naturally occurring secondary metabolites of some plants and are found in the human diet, in the greatest amounts in (green or black) tea and red wine. Although adding an interesting astringent sensation when consumed in larger quantities, tannins are harmful since they bind the enzymes and proteins necessary for the normal uptake of nutrients (Manach et al., 2005). Salivary proteins, histatins, and proline-rich proteins (PRPs), which make up nearly 70% of all parotid salivary protein (Kauffman and Keller, 1979), bind very well to dietary polyphenols/tannins (Baxter et al., 1997; Bennick, 2002; Wroblewski et al., 2001). The high affinity of these proteins to tannins are thought to have a protective effect since animals that lack proline-rich proteins do not survive on high tannin diets (McArthur et al., 1995). Although dietary polyphenols can cause tooth staining (Carpenter et al., 2005; Proctor et al., 2005b), their role in causing astringency generates most interest. Astringency, at least at low levels, adds refreshment to a drink. Although tannins are known to interact with salivary proline-rich proteins and histatins, it is not understood how this leads to a dry puckering sensation of the entire mucosa. A common view is that astringency is due to the loss of lubrication of the salivary film. Recent studies have suggested that whilst there is a loss of interfacial rheology of saliva following interactions with polyphenols (Rossetti et al., 2008) it could not account for the astringency of all astringent compounds tested. The possible interaction of proline-rich proteins already bound to buccal cells, as part of the mucosal pellicle, was explored in a recent report by the author (Nayak and Carpenter, 2008). We speculated that if proline-rich proteins and histatins are involved in astringency, then washing the mouth out with water (to reduce the levels of these proteins) prior to the ingestion of a tannin solution should reduce the perception of astringency; in fact, astringency increased. This reinforces the view that saliva acts principally as a barrier to astringency. Analysis of the proteins precipitated by the tannin solution after several oral rinses revealed proline-rich proteins and mucins, presumably adhering to the mucosa. This would suggest that any food/chemical that disrupted this adherent layer of protein would cause an astringent feeling. Detergents, for example, sodium/ lauryl sulphate, or metal ions such as, copper (Hong et al., 2009), or whey proteins (Vardhanabhuti et al., 2010) can all cause astringent sensations without obvious interactions with or precipitation of salivary proline-rich proteins; hence, the continued research for mechanisms underlying the physiology of astringency.

28.5 LONG-TERM RELATIONSHIP BETWEEN SALIVA AND ORAL SENSES

Besides the initial process of sensory stimulation, saliva has roles in the long-term protection of the oral mucosa and maintenance of the health and integrity of the taste function. Taste cells are continuously replaced with new cells about every nine days (Zeng and Oakley, 1999). Saliva contains candidates for trophic factors of regeneration. One candidate is zinc and/or a zinc-binding protein, carbonic anhydrase (gustin), which has been postulated as a trophic factor that promotes the growth and development of taste buds (Henkin et al., 1999, 2010). This hypothesis is based on the treatment of patients with taste disorders by zinc supplementation; however, other clinical studies (Matson et al., 2003; Osaki et al., 1996) and reviews (Bradley and Beidler, 2003; Matsuo, 2000) cast doubts on zinc therapy and such hypotheses. No physiological basic study has been performed on the effects of salivary zinc and/or carbonic anhydrase on the renewal of taste buds. The other candidate is epidermal growth factor (EGF), which has been widely studied as one of the major accelerators of wound healing (e.g., Jahovic et al., 2004; Noguchi et al., 1991). As for taste buds in rats, supplemental EGF in drinking water can prevent the loss of taste buds in the fungiform papillae after removal of the submandibular and sublingual glands (Morris-Wiman et al., 2000). This suggests that salivary EGF may always act as a growth factor for maintaining the normal renewal of taste buds. Taste bud cells, like epithelial cells, arise from stem cells near the basal area of the local epithelium (Stone et al., 1995) and therefore, the role of EGF in taste buds involves the oral epithelium.

Sjögren's syndrome patients having chronic failure of salivary gland function display only decreased taste threshold sensitivity (Weiffenbach et al., 1995); however, animal models of severe hypo-salivation (removal of the major salivary gland) exhibited more measurable taste impairments, such as poor performance in taste discrimination tests, hyperkeratosis of the tongue epithelium, shrunken and disorganized taste buds (Cano and Rodriguez-Echandia, 1980; Nanda and Catalanotto, 1981), loss of taste buds (Morris-Wiman et al., 2000), and decreased responses of the taste nerve (Matsuo et al., 1997). Moreover, relating to the renewal of taste buds, taste sensitivity can be modified by one-week manipulation of the external milieu of the taste buds (Matsuo et al., 1997). In contrast with such a role of saliva in taste, long-term alteration of oral sensitivity can affect saliva and salivation; for example, increased salivary PRP by feeding a tannin-containing diet (Mehansho et al., 1992; Jasman et al., 1994), the effect of food texture (Kurahashi and Inomata, 1999) and starchy food (Perry et al., 2007) on amylase synthesis, and the effect of food texture on the size of the salivary gland (atrophy or hypertrophy) (Proctor, 1998); hence, saliva and oral sensitivity have the characteristic of affecting each other. We should be aware of this reciprocal relationship in studies on the taste acuity of older persons or individual differences in taste sensitivity.

28.6 CONCLUDING REMARKS

More than 99% of saliva is water, which acts as a solvent used to convey taste substances to the receptor. Only less than 1% of saliva is inorganic ions and organic molecules. Nevertheless, some inorganic salts can stimulate taste receptors and affect the taste threshold and supra-threshold intensity of taste. The organic constituents (proteins) themselves cannot stimulate taste receptors; however, some of them may affect the initial process of oral sensory transduction *via* the digestion of starches and fats, emulsification of fats/oils, lowering of surface tension, and chemical interaction with food, for example, tannins. Saliva also forms the milieu of taste receptors and can influence the sensitivity of renewed taste receptors. To maintain these roles of saliva, a sufficient amount of resting saliva is necessary. The flow rate of resting saliva positively correlates with that of stimulated saliva (Rantonen and Meurman, 1998), bite force (Yeh et al., 2000), and the sizes of parotid and submandibular salivary glands (Ono et al., 2006). These findings suggest that, for the secretion of sufficient resting saliva and maintenance of oral senses, it is necessary to maintain the ability to produce sufficient stimulated saliva, which requires sound salivary glands, neural circuits, and chewing capability.

REFERENCES

Adams, S., Singleton, S., Juskaitis, R., and Wilson, T. (2007). In-vivo visualization of mouth-material interactions by video rate endoscopy. *Food Hydrocolloids* 21: 986–995.

Azen, E. A., and Maeda, N. (1988). Molecular genetics of human salivary proteins and their polymorphisms. *Advan. Human Genet.* 17: 141–199.

Bartoshuk, L. M. (1978). The psychophysics of taste. *Am. J. Clin. Nutr.* 31: 1068–1077.

Baxter, N. J., Lilley, T. H., Haslam, E., and Williamson, M. P. (1997). Multiple interactions between polyphenols and salivary proline-rich proteins repeat result in complexation and precipitation. *Biochem.* 36: 5566–5577.

Beidler, L. M. (1961). Taste Receptor Stimulation. In *Progress in Biophysics Biophysical Chemistry, XII*, J.A.V. Butler, H.E. Huxley, and R.E. Zirkle (Eds.). Pergamon Press: New York, pp. 107–151.

Bennick, A. (2002). Interaction of plant polyphenols with salivary proteins. *Crit. Rev. Oral Biol. Med.* 13: 184–196.

Bernard, C. (1856). *Leçons de physiologie expérimentale. Appliwuée à la Médecine.* Vol II. Bailliére: Paris.

Bongaerts, J. H. H., Fourtouni, K., and Stokes, J. R. (2007). Soft-tribology: Lubrication in a compliant PDMS-PDMS contact. *Tribol. Int.* 40: 1531–1542.

Bradley, R. M. (1991). Salivary secretion. In *Smell and taste in health and disease*, T.V. Getchell (Ed.). Raven Press: New York, pp. 127–144.

Bradley, R. M., and Beidler, L. M. (2003). Salivary: Its role in taste function. In *The Handbook of Olfaction and Gustation*, 2nd edition, R. L. Doty (Ed.). Marcel Dekker, New York, pp. 639–650.

Bradway, S. D., Bergey, E. J., Jones, P. C., and Levine, M. J. (1989). Oral mucosal pellicle. Adsorption and transpeptidation of salivary components to buccal epithelial cells. *Biochem. J.* 261: 887–896.

Bradway, S. D., Bergey, E. J., Scannapieco, F. A., et al. (1992). Formation of salivary-mucosal pellicle: the role of transglutaminase. *Biochem. J.* 284: 557–564.

Brouwer, J. N., and Wiersma, A. (1980). Stimulus-induced appearance of proteinaceous material in the taste pore. In *Olfaction and Taste, VII*, H. van der Starre (Ed.). IRL Press: London, pp. 179–182.

Bryant, B. P., and Moore, P. A. (1995). Factors affecting the sensitivity of the lingual trigeminal nerve to acids. *Am. J. Physiol.* 268: R58–R65.

Cano, J., and Rodriguez-Echandia, E. L. (1980). Degenerating taste buds in sialectomized rats. *Acta Anat.* 106: 487–492.

Cardenas, M., Elofsson, U., and Lindh, L. (2007). Salivary mucin MUC5B could be an important component of in vitro pellicles of human saliva: an *in situ* ellipsometry and atomic force microscopy study. *Biomacromol.* 8: 1149–1156.

Carpenter, G. H., Pramanik, R., and Proctor, G. B. (2005). An *in vitro* model of chlorhexidine-induced tooth staining. *J. Periodontl. Res.* 40: 225–230.

Chalé-Rush, A., Burgess, J. R., and Mattes, R. D. (2007). Evidence for human orosensory (taste?) sensitivity to free fatty acids. *Chem. Senses* 32: 426–431.

Christensen, C. M., Brand, J. G., and Malamud, D. (1987). Salivary changes in solution pH: a source of individual differences in sour taste perception. *Physiol. Behav.* 40: 221–227.

Damak, S., Rong, M., Yasumatsu, K., et al. (2003). Detection of sweet and umami taste in the absence of taste receptor T1r3. *Science* 301: 850–853.

Dawes, C., Jenkins, G. N., and Tonge, C. H. (1963). The nomenclature of the integuments of the enamel surface of teeth. *Br. Dent. J.* 115: 65–68.

Dawes, C., and Macpherson, L. M. D. (1993). The distribution of saliva and sucrose around the mouth during the use of chewing gum and the implications for the site-specificity of caries and calculus deposition. *J. Dent. Res.* 72: 852–857.

Delwiche, J., and O'Mahony, M. (1996). Changes in secreted salivary sodium are sufficient to alter salt taste sensitivity: use of signal detection measures with continuous monitoring of the oral environment. *Physiol. Behav.* 59: 605–611.

de Wijk, R. A., Prinz, J. F., Engelen, L., and Weenen, H. (2004). The role of alpha-amylase in the perception of oral texture and flavour in custards. *Physiol. Behav.* 83: 81–91.

Erickson, Y. (1964). The role of some salivary constituents in oral pathology, with special regard to caries experiments with rodents. In *Salivary glands and their secretions*, L.M. Sreebny, and J. Meyer (Eds.). Pergamon Press: New York, pp. 281–297.

Ferguson, D. B., and Botchway, C. A. (1979). Circadian variations in flow rate and composition of human stimulated submandibular saliva. *Arch. Oral Biol.* 24: 433–437.

Ferguson, D. B., and Fort, A. (1974). Circadian variations in human resting submandibular saliva flow rate and composition. *Arch. Oral Biol.* 19: 47–55.

Ferrell, F. (1984). Preference for sugars and nonnutritive sweeteners in young beagles. *Neurosci. Biobehav. Rev.* 8: 199–203.

Formaker, B. K., Lin, H., Hettinger, T. P., and Frank, M. E. (2009). Responses of the hamster chorda tympani nerve to sucrose+acids and sucrose+citrate taste mixtures. *Chem. Senses* 34: 607–616.

Ginane, C., Baumont, R., and Favreau-Peigné, A. (2011). Perception and hedonic value of basic tastes in domestic ruminants. *Physiol. Behav.* 104: 666–674.

Glaser, D. (2002). Specialization and phyletic trends of sweetness reception in animals. *Pure Appl. Chem.* 74: 1153–1158.

Gurkan, S., and Bradley, R. M. (1988). Secretions of von Ebner's glands influence responses from taste buds in rat circumvallate papilla. *Chem. Senses* 13: 655–661.

Hahn Berg, I. C., Rutland, M. W., and Arnebrant, T. (2003). Lubricating properties of the initial salivary pellicle – an AFM Study. *Biofouling* 19: 365–369.

Halthur, T. J., Arnebrant T., Macakova, L., and Feiler, A. (2010). Sequential adsorption of bovine mucin and lactoperoxidase to various substrates studied with quartz crystal microbalance with dissipation. *Langmuir* 26: 4901–4908.

Hay, D. I., and Bowen, W. H. (1996). The functions of salivary proteins. In *Saliva and Oral Health*, W. Edgar, and D. O'Mullane, pp. 105–122. Thanet Press: Margate, UK.

Hector, M. P. and Linden, R. W. A. (1999). Reflexes of salivary secretion. In *Neural Mechanisms of Salivary Gland Secretion*, J. R. Garrett, J. Ekström, and L. C. Anderson (Eds.). Karger: Basel, pp. 196–217.

Henkin, R. I., Martin, B. M., and Agarwal, R. P. (1999). Efficacy of exogenous oral zinc in treatment of patients with carbonic anhydrase VI deficiency. *Am. J. Med. Sci.* 318: 392–405.

Henkin, R. I., Potolicchio, S. J., Levy, L. M., et al. (2010). Carbonic anhydrase I, II, and VI, blood plasma, erythrocyte and saliva zinc and copper increase after repetitive transcranial magnetic stimulation. *Am. J. Med. Sci.* 339: 249–257.

Hershkovich, O., and Nagler, R. M. (2004). Biochemical analysis of saliva and taste acuity evaluation in patients with burning mouth symdrome, xerostomia and/or gustatory disturbances. *Arch. Oral Biol.* 49: 515–522.

Hong, J. H., Duncan, S. E., Dietrich, A. M., et al. (2009). Interaction of copper and human salivary proteins. *J. Agric. Food Chem.* 57: 6967–6975.

Imada, T., Misaka, T., Fujiwara, S., et al. (2010). Amiloride reduces the sweet taste intensity by inhibiting the human sweet taste receptor. *Biochem. Biophys. Res. Commun.* 397: 220–225.

Jahovic, N., Güzel, E., Arbak, S., and Yegen, B. C. (2004). The healing-promoting effect of saliva on skin burn is mediated by epidermal growth factor (EGF): role of the neurophils. *Burns* 30: 531–538.

Jasman, A. J. M., Frohlich, A. A., and Marquardt, R. R. (1994). Production of proline-rich proteins by the parotid glands of rats is enhanced by feeding diets containing tannins from faba beans (*Vicia faba L.*). *J. Nutr.* 124: 249–258.

Jensen, J. L., Lamkin, M. S., and Oppenheim, F. G. (1992). Adsorption of human salivary proteins to hydroxyapatite: a comparison between whole saliva and glandular salivary secretions. *J. Dent. Res.* 71: 1569–1576.

Jiang, P., Josue, J., Li, X., et al. (2012). Major taste loss in carnivorous mammals. *Proc. Nat. Acad. Sci. USA* 109: 4956–4961.

Kauffman, D. L., and Keller, P. J. (1979). The basic proline-rich proteins in human parotid saliva from a single subject. *Arch. Oral Biol.* 24: 249–256.

Kawai, T., and Fushiki, T. (2003). Importance of lipolysis in oral cavity for orosensory detection of fat. *Am. J. Regul. Integr. Comp. Physiol.* 285: R447–R454.

Kelling, S. T., and Halpern, B. P. (1993). Gustatory neural response latencies in the frog. *Chem. Senses* 18: 169–187.

Kock, K., Morley, S. D., Mullins, J. J., and Schmale, H. (1994). Denatonium bitter tasting among transgenic mice expressing rat von Ebner's gland protein. *Physiol. Behav.* 56: 1173–1177.

Kumazawa, T., and Kurihara, K. (1990). Large enhancement of canine taste responses to sugars by salts. *J. Gen. Physiol.* 95: 1007–1018.

Kurahashi, M., and Inomata, K. (1999). Effects of dietary consistency and water content on parotid amylase secretion and gastric starch digestion in rats. *Arch. Oral Biol.* 11: 1013–1019.

Li, J., Heknergirst, E.J., Yao, Y., et al. (2004). Statherin is *in vivo* pellicle constituent: identification and immuno-quantification. *Arch. Oral Biol.* 49: 379–385.

References

Li, X., Glaser, D., Li, W., et al. (2009). Analyses of sweet receptor gene (*Tas1r2*) and preference for sweet stimuli in species of Carnivora. *J. Hered.* 100(Suppl. 1): S90–S100.

Li, X., Li, W., Wang, H., et al. (2005). Pseudogenization of a sweet-receptor gene accounts for cat's indifference toward sugar. *PLoS Genet.* 1: 27–35.

Li, X., Staszewski, L., Xu, H., et al. (2002). Human receptors for sweet and umami taste. *Proc. Nat. Acad. Sci. USA* 99: 4692–4696.

Ludwig, C. (1850). Neue Versuche über die Beihilfe der Nerven zu der Speichelsekretion. *Naturforsh. Ges. Zürich* 53/54: 210–239.

Lugaz, O., Pillias, A. M., Boireau-Ducept, N., and Faurion, A. (2005). Time-intensity evaluation of acid taste in subjects with saliva high flow and low flow rates for acids of various chemical properties. *Chem. Senses* 30: 89–103.

Macakova, L., Yakubov, G. E., Plunkett, M. A., and Stokes, J. R. (2010). Influence of ionic strength changes on the structure of pre-adsorbed salivary films. A response of a natural multi-component layer. *Colloids Surf. B: Biointerfaces* 77: 31–39.

Maeda, N. (1985). Inheritance of the human salivary proline-rich proteins: a reinterpretation in terms of six loci forming two subfamilies. *Biochem. Genet.* 23: 455–464.

Manach, C., Williamson, G., Morand, C., et al. (2005). Bioavailability and bioefficacy of polyphenols in humans. I. Review of 97 bioavailability studies. *Ame. J. Clin. Nutri.* 81: 230S–242S.

Matson, A., Wright, M., Oliver, A., et al. (2003). Zinc supplementation at conventional doses not improve the disturbance of taste perception in hemodialysis patients. *J. Renal Nutri.* 13: 224–228.

Matsuo, R. (2000). Role of saliva in the maintenance of taste sensitivity. *Crit. Rev. Oral Biol.* 11: 216–229.

Matsuo, R., and Yamamoto, T. (1990). Taste nerve responses during licking behavior in rats: importance of saliva in responses to sweeteners. *Neurosci. Lett.* 108: 121–126.

Matsuo, R., and Yamamoto, T. (1992). Effects of inorganic constituents of saliva on taste responses of the rat chorda tympani nerve. *Brain Res.* 583: 71–80.

Matsuo, R., Yamamoto, T., Ikehara, A., and Nakamura, O. (1994). Effect of salivation on neural taste responses in freely moving rats: analyses of salivary secretion and taste responses of the chorda tympani nerve. *Brain Res.* 649: 136–146.

Matsuo, R., Yamauchi, Y., and Morimoto, T. (1997). Role of submandibular and sublingual saliva in maintenance of taste sensitivity recorded in the chorda tympani of rats. *J. Physiol.* 498: 797–807.

Mattes, R. D. (2011). Accumulating evidence supports a taste component for free fatty acids in humans. *Physiol. Behav.* 104: 624–631.

McArthur, C., Sanson, G. D., and Beal, A. D. (1995). Salivary proline-rich proteins in mammals: roles in oral homeostasis and counteracting dietary tannin. *J. Chem. Ecol.* 21: 663–691.

McBurney, D. H., and Pfaffmann, C. (1963). Gustatory adaptation to saliva and sodium chloride. *J. Exp. Psychol.* 65: 523–529.

Mehansho, H., Asquith, T. N., Butler, L. G., et al. (1992). Tannin-mediated induction of proline-rich protein synthesis. *J. Agric. Food Chem.* 40: 93–97.

Merigo, F., Benati, D., Cecchini, M. P., et al. (2009). Amylase expression in taste receptor cells of rat circumvallate papillae. *Cell Tissue Res.* 336: 411–421.

Mese, H., and Matsuo, R. (2007). Salivary secretion, taste and hyposalivation. *J. Oral Rehabil.* 34: 711–723.

Morris-Wiman, J., Sego, R., Brinkley, L., and Dolce, C. (2000). The effects of sialoadenectomy and exogenous EGF on taste bud morphology and maintenance. *Chem. Senses* 25: 9–19.

Nakamura, K., and Norgren, R. (1991). Gustatory responses of neurons in the nucleus of the solitary tract of behaving rats. *J. Neurophysiol.* 66: 1232–1248.

Nanda, R., and Catalanotto, F. A. (1981). Long-term effects of surgical desalivation upon taste acuity, fluid intake, and taste buds in the rat. *J. Dent. Res.* 60: 69–76.

Nayak, A., and Carpenter, G. H. (2008). A physiological model of tea-induced astringency. *Physiol. Behav.* 95: 290–294.

Noguchi, S., Ohba, Y., and Oka, T. (1991). Effect of salivary epidermal growth factor on wound healing of tongue in mice. *Am. J. Physiol.* 260: E620–E625.

Ogawa, H., Sato, M., and Yamashita, S. (1968). Multiple sensitivity of chorda tympani fibers of the rat and hamster to gustatory and thermal stimuli. *J. Physiol.* 199: 223–240.

Ono, K., Morimoto, Y., Inoue, H., et al. (2006). Relationship of the unstimulated whole saliva flow rate and salivary gland size estimated by magnetic resonance image in health young humans. *Arch. Oral Biol.* 51: 345–349.

Osaki, T., Tomita, M., Matsugi, N., and Nomura, Y. (1996). Clinical and physiological investigations in patients with taste abnormality. *J. Oral Pathol. Med.* 25: 38–43.

Perry, G. H., Dominy, N. J., Claw, K. G., et al. (2007). Diet and the evolution of human amylase gene copy number variation. *Nat. Genet.* 39: 1256–1260.

Pivk, U., Ulrih, N. P., Juillerat, M. A., and Raspor, P. (2008). Assessing lipid coating of the human oral cavity after ingestion of fatty foods. *J. Agric. Food Chem.* 56: 507–511.

Pobozy, E., Czarkowska, W., and Trojanowicz, M. (2006). Determination of amino acids in saliva using capillary electrophoresis with fluorimetric detection. *J. Biochem. Biophys. Methods* 67: 37–47.

Pramanik, R., Osailan, S. M., Challacombe, S. J., et al. (2010). Protein and mucin retention on oral mucosal surfaces in dry mouth patients. *Eur. J. Oral Sci.* 118: 245–253.

Proctor, G. B. (1998). Secretory protein synthesis and constitutive (vesicular) secretion by salivary glands. In *Glandular Mechanisms of Salivary Secretion*, J. R. Garrett, J. Ekström, and L. C. Anderson (Eds.). Karger, Basel, pp. 73–88.

Proctor, G. B., Hamdan, S., Carpenter, G. H., and Wilde, P. (2005a). A statherin and calcium enriched layer at the air interface of human parotid saliva. *Biochem. J.* 389: 111–116.

Proctor, G. B., Pramanik, R., Carpenter, G. H., and Rees, G. D. (2005b). Salivary proteins interact with dietary constituents to modulate tooth staining. *J. Dent. Res.* 84: 73–78.

Rantonen, P. J. F., and Meurman, J. H. (1998). Viscosity of whole saliva. *Acta. Odont. Scand.* 56: 210–214.

Redl, B. (2000). Human tear lipocalin. *Biochem. Biophys. Acta.* 1482: 241–248.

Rehnberg, B. G., Hettinger, T. P., and Frank, M. E. (1992). Salivary ions and neural taste responses in the hamster. *Chem. Senses* 17: 179–190.

Roper, S. D. (2007). Signal transduction and information processing in mammalian taste buds. *Pflugers Arch. – Eur. J. Physiol.* 454: 759–776.

Rossetti, D., Yakubov, G. E., Stokes, J. R., et al. (2008). The interaction of human whole saliva (hws) and astringent dietary compounds investigated by interfacial shear rheology. *Food Hydrocolloids* 22: 1068–1078.

Sas, R., and Dawes, C. (1997). The intra-oral distribution of unstimulated and chewing-gum-stimulated parotid saliva. *Arch. Oral Biol.* 42: 469–474.

Schiffman, S. S., Lockhead, E., and Maes, F. W. (1983). Amiloride reduce the taste intensity of Na+ and Li+ salts and sweeteners. *Proc. Nat. Acad. Sci. USA* 80: 6136–6140.

Schmale, H., Holtgreve-Grez, H., and Christiansen, H. (1990). Possible role for salivary gland protein in taste reception indicated by homology to lipophilic-ligand carrier proteins. *Nature* 343: 366–369.

Schmale, H., Ahlers, C., Bläker, M., et al. (1993). Perireceptor events in taste. In *The Molecular Basis of Smell and Taste Discoders*, D. Chadwick, J. Marsh, and J. Goode (Eds.). John Wiley: Chichester, pp. 167–185.

Sclafani, A., and Mann, S. (1987). Carbohydrate taste preferences in rats: glucose, sucrose, maltose, fructose, and Polycose compared. *Physiol. Behav.* 40: 563–568.

Segovia, C., Hutchinson, I., Laing, D. G., and Jinks, A. L. 2002. A quantitative study of fungiform papillae and taste pore density in adults and children. *Dev. Brain Res.* 138: 135–146.

Shannon, I. L., Suddick, R. P., and Dowd, F. J. (1974). Saliva: composition and secretion. In *Monographs in Oral Science*, Vol. 2. Karger: Basel, pp. 3–42.

Silletti, E., Vingerhoeds, M. H., Norde, W., and Van Aken, G. A. (2007). The role of electrostatics in saliva-induced emulsion flocculation. *Food Hydrocolloids* 21: 596–606.

Spielman, A. I. (1990). Interaction of saliva and taste. *J. Dent. Res.* 69: 838–843.

Sreebny, L. M. (1989). Recognition and treatment of salivary induced conditions. *Int. J. Dent. Res.* 39: 197–294.

Stewart, J. E., Feinle-Bisset, C., Golding, M., et al. (2010). Oral sensitivity to fatty acids, food consumption and BMI in human subjects. *Brit. J. Nutr.* 104: 145–152.

Stokes, J. R., and Davies, G. A. (2007). Viscoelasticity of human whole saliva collected after acid and mechanical stimulation. *Biorheol.* 44: 141–160.

Stone, L. M., Finger, T. E., Tam, P. P. L., and Tan, S-S. (1995). Taste receptor cells arise from local epithelium, not neurogenic ectoderm. *Proc. Natl. Acad. Sci. USA* 92: 1916–1920.

Vardhanabhuti, B., Kelly, M. A., Luck, P. J., et al. (2010). Roles of charge interactions on astringency of whey proteins at low pH. *J. Dairy Sci.* 93: 1890–1899.

Vingerhoeds, M. H., de Wijk, R. A., Zoet, F. D., et al. (2008). How emulsion composition and structure affect sensory perception of low-viscosity model emulsins. *Food Hydrocolloids* 22: 631–646.

Vingerhoeds, M. H., Silletti, E., de Groot, J., et al. (2009). Relating the effect of saliva-induced emulsion flocculation on rheological properties and retention on the tongue surface with sensory perception. *Food Hydrocolloids* 23: 773–785.

Weiffenbach, J. M., Schwartz, L. K., Atkinson, J. C., and Fox, P. C. (1995). Taste performance in Sjögren's syndrome. *Physiol. Behav.* 57: 89–96.

Witt, M., and Miller, I. J. (1992). Comparative lectin histochemistry on taste buds in foliate, circumvallate and fungiform papillae of the rabbit tongue. *Histochemistry* 98: 173–182.

Wroblewski, K., Muhandiram, R., Chakrabartty, A., and Bennick, A. (2001). The molecular interaction of human salivary histatins with polyphenolic compounds. *Eur. J. Biochem.* 268: 4384–4397.

Yamaguchi, S. (1991). Basic properties of umami and effects on humans. *Physiol. Behav.* 49: 833–841.

Yamasaki, H., Kubota, Y., Tagaki, H., and Toyama, M. (1984). Immunoelectron-microscopic study on the fine structure of substance-P-containing fibers in the taste buds of the rat. *J. Comp. Neurol.* 227: 380–392.

Ye, Q., Heck, G. L., and DeSimone, J. A. (1993). Voltage dependence of the rat chorda tympani response to Na+ salts: implications for the functional organization of taste receptor cells. *J. Neurophysiol.* 70: 167–178.

Yeh, C. K., Johnson, D. A., Dodds, M. W. J., et al. (2000). Association of salivary flow rates with maximal bite force. *J. Dent. Res.* 79: 1560–1565.

Young, J. A., and Schneyer, C. A. (1981). Composition of saliva in mammalia. *Aust. J. Exp. Biol. Med. Sci.* 59: 1–53.

Young, J. A, and van Lennep, E. W. (1978). *The Morphology of Salivary Glands*. Academic Press: London.

Zeng, Q., and Oakley, B. (1999). p53 and Bax: putative death factors in taste cell turnover. *J. Comp. Neurol.* 413: 168–180.

Zhan, F., Klebansky, B., Fine, R. M., et al. (2008). Molecular mechanism for the umami taste synergism. *Proc. Nat. Acad. Sci. USA* 105: 20930–20934.

Chapter 29

Anatomy of the Tongue and Taste Buds

MARTIN WITT and KLAUS REUTTER

29.1 INTRODUCTION

"... Many papillae are evident, I might say, innumerable, and the appearance is so elegant that they catch the view and thoughts of the observer, and control him for a long time and not without enjoyment ... All this delights the curious mind, when observing with an engyscope; and if anyone asks to what they are similar, I am unsure whether I should compare this huge number of papillae first with grapes or fruits of the bay, or innumerable mushrooms emerging between fine, densely standing blades of grass ..."

These are the first detailed and enthusiastic words on papillae and their "membranes" in human tongues by Lorenzo Bellini (1665) (Figure 29.1), who knew that Marcello Malpighi (1628–94) had reported on lingual papillae the year before. Malpighi discovered mucosal elevations associated with nerve fibers, naming these elevations "papillae," as the morphological substrates for gustatory sensation (Figure 29.2).

The peripheral taste apparatus includes the gustatory sensory organs or taste buds and their innervation, and the specific papillae within which taste buds are assembled. What one needs to know about taste buds in relation to how one perceives tastants depends on the approach to taste perception. One aim of this chapter is to provide a sufficient knowledge on peripheral gustatory anatomy as a basis for understanding other chapters of this book. Some structural details about the human peripheral taste system are well known, but it is also worthwhile to provide comparative anatomical information to fill in the gaps or understand and establish basic principles. The fundamental question of how tastants are perceived has been addressed for more than two millenia, and the majority of concepts, theories, and experimental "proofs" that have been proposed have since given

way to present-day concepts. This chapter incorporates this rich history which sculpted our contemporary views of gustatory anatomy and physiology.

Aristotle (384–322 BC), applying Platonic concepts, argued that taste sensation was carried from the tongue via the blood, to the liver or heart, which was the common seat of the soul and all sense perception [De sensu 438b 26; On youth and old age, 469a 5–13; cited after Siegel (1970)]. Galen's (Claudius Galenus, 129–201 AD) anatomical analyses challenged this notion: his detailed studies on the innervation of the tongue describe correctly the different functions of the three principal nerves supplying the tongue (lingual, glossopharyngeal, and hypoglossal nerves), and demonstrate their origin at the base of the brain. Galen posited that the lingual nerve communicated gustatory sensations, a concept yet resonant in contemporary neurobiology. One strand of nerve fibers corresponding to cranial nerve IX (CN IX or glossopharyngeal nerve) of present terminology (rediscovered in humans by Panizza, 1834 was already known to Galen as the principal gustatory nerve of the tongue. CN IX also carries some motor fibers to the pharynx. Galen also noted excretory ducts of the lingual and sub-mandibular glands (in ox). The particular structure of the tongue surface was initially described by Casserius (1609) and, later by Malpighi (1686) and Bellini (1665) (reviewed by Jurisch, 1922. Further evidence for the significance of the papillary epithelium and its cells came from observations in taste organs of the frog (Fixsen, 1857, Waller, 1847, 1849). Taste buds were identified initially on the barbels and skin of fishes by Leydig (1851) and described as *"becherförmige Organe"* (goblet-shaped organs), whose function he associated with tactile sensitivity. Schulze (1863) subsequently suggested they were chemosensory structures. Similar organs in mammals were described as *Schmeckbecher* (taste goblets)

Handbook of Olfaction and Gustation, Third Edition. Edited by Richard L. Doty.
© 2015 Richard L. Doty. Published 2015 by John Wiley & Sons, Inc.

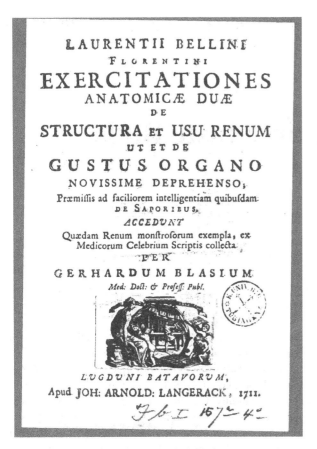

Figure 29.1 The cover sheet of Lorenzo Bellini's book that summarizes examinations on the anatomy of the kidney and the taste organ. (Langerack, Leiden, 1711).

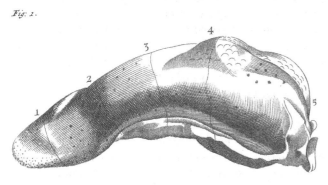

Figure 29.2 Depiction of a bovine tongue by Marcello Malpighi (1686) showing "patches" where papillae were observed. Note the concentration of fungiform papillae (dots) on the tip of the tongue. The drawing of vallate and foliate papillae is still rather vague.

(Lovén, 1868) and *Geschmacksknospen* (taste buds) or *Geschmackszwiebeln* ("taste onions") (Schwalbe, 1868). Herrick (1904) translated "Geschmackskospen" as "taste buds." Their location within lingual papillae, the latter already associated with the loci of taste perception, lent credence to their identity as taste sensor organs.

Nineteenth-century studies focused on cytological features, nerve supply (Figures 29.3 and 29.5), and the development of taste buds. While the nerve-dependent nature of taste sensation was known since Galen's time, its significance for sensory organ physiology blossomed in the mid-1800s. With development of the neuron doctrine (see Koelliker, 1844), previously described "ganglion globules" (*Ganglienkugeln*: Ehrenberg, 1833) could now be acknowledged as part of a specialized cellular system (cell theory of Schleiden and Schwann) (Schwann, 1839). Two major prerequisites favored the expansion of scientific knowledge in the nineteenth century: (1) replacement of the often speculative natural philosophy by the experimentally based natural sciences, mainly represented by Francois Magendie (1783–1855) in France and Johannes Müller (1801–58) in Germany, and (2) technical advances which included improved microscopes (K. Zeiss, 1816–88, together with Ernst Abbé, 1840–1905, and E. Leitz, 1843–1920) as well as use of histological techniques, for example, introduction of chromic acid for histological examination of neural tissue (Hannover, 1840), allowing for the distinction between cells and fibers. Further, gold chloride staining facilitated discriminating finely ramifying nerve fibers (Gerlach, 1858). Helmholtz (1842) observed a direct continuity between nerve cells (globuli gangliosi) and their fibers in evertebrates. The demand for the neuron doctrine in vertebrates was established by Koelliker (1844). Introduction of the methylene blue staining method facilitated observations that *fine nerve* fibers in the frog taste disc widened with varicosities, penetrated the gustatory epithelium and approached the sensory cells "with extremely sharp small knobs, [which] … connect taste cells not continuously, but *per contiguitatem*" (Ehrlich, 1886). This pioneering observation of secondary sensory cells in a taste organ that was clearly different from primary sensory cells, as had already been described in the olfactory epithelium, albeit without the understanding of the "contact" nature of taste bud cells and their innervating afferents, as later described ultrastructurally, was opposed by Retzius (1892). Using the Golgi silver impregnation method (Golgi, 1873), Retzius attributed a sensory function only to free nerve endings (Figure 29.5b). Later, Krause (1911) clearly demonstrated that the finest nerve fibers enter the taste bud and ascend to the taste pore, but do not merge with taste bud cells whose basal processes end near the basal cells.

29.2 LINGUAL PAPILLAE AND TASTE BUD DISTRIBUTION

29.2.1 Nongustatory Papillae

Approximately 60 years before taste buds were identified as gustatory organs, an illustration of the human tongue

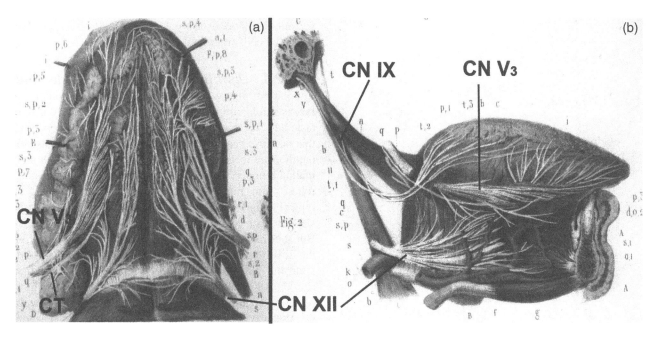

Figure 29.3 These illustrations of Bourgery and Jacob (1839) show the human tongue including nerve and blood supply, as well as the lingual muscle system. In spite of precise macroscopical observations on the innervation of glands and muscles, it still represents indirectly the (nowadays revised) morphological concept of gustation in the early nineteenth century, according to which the principal taste nerves are the lingual (*p* in (a) and (b); CN V$_3$) and glossopharyngeal (*t* in (b); CN IX) nerves. The chorda tympani, although depicted near the submandibular ganglion (*q* in (a) and (b); CT), does not reach the lingual dorsum. Ebner's glands are not known to the authors. (a): *D*, submandibular gland with *q*, chorda tympani and submandibular ganglion and *p*, lingual nerve; *s*, hypoglossal nerve (CN XII); *E*, sublingual gland; *F*, Nuhn's gland. (b): *C* styloid process; *b*, stylopharyngeus muscle, the leading muscle for *t*, glossopharyngeal nerve (CN IX), which overlaps partly with the innervation area of the lingual nerve; *k*, lingual artery; *p*, lingual nerve, interrupted to show the glossopharyngeal nerve; *q*, part of the chorda tympani nerve; *s*, hypoglossal nerve (CN XII); *X*, trunk of the facial nerve. (*See plate section for color version.*)

by Sömmering (1806) (Figure 29.4) accurately showed the regional distribution of lingual papillae. A line called the sulcus terminalis (an ontogenetic remnant, see below), which is located posterior to the vallate papillae, separates the body of the tongue from the lingual root. It may be seen that the sulcus terminalis extends laterally to the pharyngeal wall from the foramen caecum (also an ontogenetic remnant) near the midline (see Figures 29.4 and 29.11). The root of the tongue is covered by a papilla- free smooth epithelium, and beneath this epithelium lie mucous glands and a reticular connective tissue filled with lymphatic follicles which lead to the designation of "lingual tonsil." Ducts of intralingual salivary glands (Ebner, 1873) empty into the troughs of vallate and foliate papillae (see below).

The dorsal surface of the tongue is covered with filiform and conical papillae from the sulcus terminalis to the tongue tip. Filiform papillae are the most prevalent type, while the number of conical papillae may vary. Both types of papillae are sparse along the lingual margin and abundant in the middle regions. Conical papillae have a cylindrical base, and taper to a sharp point at their apex. Filiform papillae (L. filum = thread) have a pyramidal shape, and a narrow tail of cornified cells extending from their apical tips as a pennant. The fila are part of the fibrous mat on the tongue's surface in the hypertrophic condition called "hairy tongue."

29.2.2 Gustatory Papillae

Taste buds occur in distinct papillae of the tongue, the epithelium of the palate, oropharynx, larynx (epiglottis), and the upper esophagus. Taste buds of most vertebrates are bulb-shaped structures which are composed of about 50–120 bipolar cells (see Figures 29.9 and 29.10). With the exception of basal cells, the slender taste bud cells arise from an interrupted basal membrane and converge with their apical protrusions, the microvilli, into the mucus-filled taste pit. Together, these cells form the organ's sensory epithelium. The nuclei of the cells are located in the lower third of the taste bud, which is approximately the region where most afferent nerve fiber terminals are distributed. Sensory cells possess transmembrane receptors and/or ion channels for specific taste stimuli at apical and lateral portions of the cell membrane. Taste buds are demarcated from

640 Chapter 29 Anatomy of the Tongue and Taste Buds

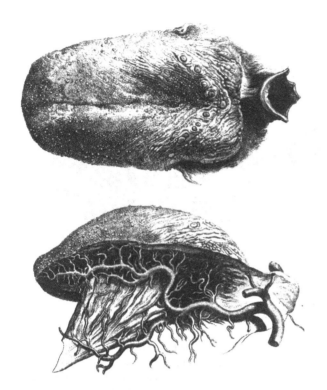

Figure 29.4 The first precise depiction of a human tongue by Samuel Sömmering (1806). (Above) The anterior part (left side) of the tongue shows numerous fungiform papillae. Behind the V-like arrangement of vallate papillae (right side), the lingual tonsils and the intrance to larynx with the epiglottis are visible. (Below) Lateral view of the tongue shows the left lingual artery and its ramifications into gustatory papillae, which appear as red dots after injection of the artery with a red dye (observation of Sömmering).

surrounding nongustatory epithelial cells by specialized epithelial cells (marginal cells).

29.2.2.1 Distribution of Lingual Taste Buds.

The pattern of taste bud distribution over the tongue surface is similar among humans and other mammals. Lingual taste buds are found exclusively within gustatory papillae, that is, those bearing taste buds. Similar types of gustatory papillae are located on homologous regions of adult mammalian tongues. The gustatory papillae include the vallate, foliate, and fungiform papillae. As the term suggests, typical fungiform papillae are mushroom-shaped, with a slender neck and an enlarged head (Figures 29.6–29.8). But the majority of fungiform papillae vary in form and the filiform papillae are intermingled among them. Shortly after the published discovery of taste buds in humans (Lovén, 1868; Schwalbe, 1868), the first systematic investigations on the distribution of human taste buds within the oral cavity were carried out by the medical student Hoffmann (1875). He emphasized that taste buds are more sparse within foliate papillae and the soft palate including the uvula. Hoffmann concluded that the development of taste perception is dependent on the number of taste buds on a particular location.

Of the approximately 4,600 total taste buds in all three lingual fields in humans, vallate buds comprise about 48% (2200), foliates about 28% (1280) and fungiforms 24% (1120). However, taste bud numbers vary greatly among individuals (Miller and Reedy, 1990a), with some adults possessing a total of only 500 taste buds

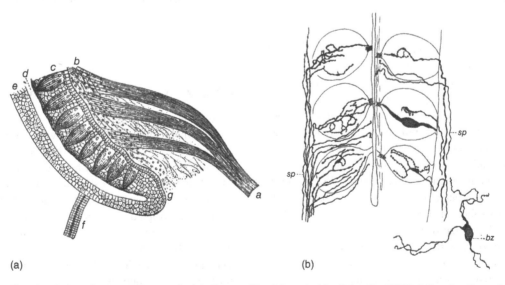

Figure 29.5 (a) First description of a mammalian taste bud (vallate papilla of the swine) by Schwalbe (1868). Minor bundles and fibrils of nerves are lost in the *"interior of taste goblets"*. Note the incorrect arrangement of basal cells. According to Schwalbe, two cell types, pin cells (*Stiftchenzellen*), and rod cells (*Stabzellen*), might mediate different taste sensations. (b) Distinguished observation of intra- and intergemmal nerve terminals in the rabbit foliate papillae by Retzius (1892) using the Golgi silver impregnation method. Taste buds are directed to the trench of the papilla. Retzius also depicts a slender bipolar "sensory" intragemmal cell and a multipolar cell (bz) (right side). Direct contacts between nerve fibers and taste cells were not noticed. Retzius believed that exclusively nerve fibers rather than taste cells were responsible for gustatory sensations.

29.2 Lingual Papillae and Taste BUD Distribution

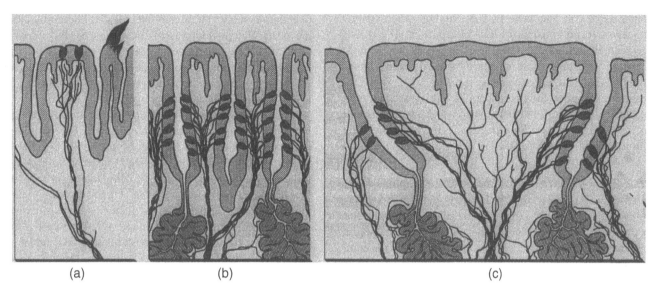

Figure 29.6 (a)–(c). Schematic drawings of taste bud-bearing, mammalian lingual papillae (longitudinal sections). (a) Fungiform papilla (papilla fungiformis) with two apically situated taste buds and their innervation. On the right side, a (taste bud- free) filiform papilla. (b) Foliate papillae (pp. foliatae). (c) Vallate papilla (p. vallata). In B and C, the taste buds are directed to the lateral trenches of the papillae. Serous Ebner glands drain to the trenches. Dermal connective tissue is rich in nerve fibers; below the taste buds they form subgemmal plexus.

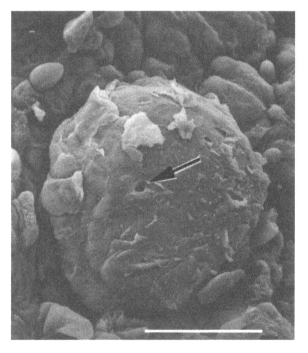

Figure 29.7 Human fungiform papilla during development, week 15. Only one taste pore is visible (arrow). Scale bar = 50 μm.

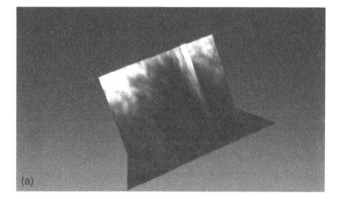

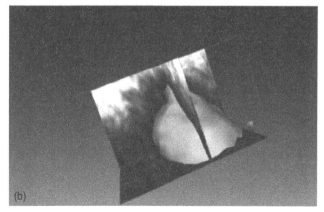

Figure 29.8 *In vivo* confocal laser scanning microscopy of a human fungiform taste bud. (a) 3D-reconstruction of a single taste bud. (b) The taste bud contours were delineated that allows the calculation of the taste bud volume. From Srur et al. (2010). (*See plate section for color version.*)

(Linden, 1993). The taste bud density of *foliate* papillae seems to be constant in life (Hou-Jensen, 1933), but age-related differences have been reported for *vallate* papillae, which are more numerous and containing more taste buds in younger individuals (Jurisch, 1922). There are also more marginal fungiform papillae during the late fetal and newborn period, but they usually lack taste buds and are referred to as "sucking papillae" (Habermehl, 1952; Yamasaki and Takahashi, 1982).

29.2.2.2 Vallate Papillae.
Vallate papillae, first comprehensively described by Haller (1766) and Sömmering (1806), lie directly anterior to the sulcus terminalis and extend in a V-shaped line across the root of the tongue (Figures 29.4 and 29.12). They are round and measure between 2 and 8 mm in diameter. The pores of taste buds open into the trenches around the bases of each vallate papilla (Figure 29.6). The papillae are innervated by an enormously large nerve fiber plexus originating from CN IX (see below) compared to foliate or fungiform papillae. The number of vallate papillae per human tongue varies between 4 and 18 ($n = 2,264$ tongues), with an average of 9.2 ± 1.8 papillae (Münch, 1896). Ninety-eight percent of all tongues have a central median papilla (Figure 29.4). The presence of three or four lateral papillae on each tongue half was observed most often (20%). Atrophic changes were observed in papillae of some men > 40 years old and some women > 55–60 years old, though Jurisch (1922) reported that the number of vallate papillae did not appear to change systematically as a function of age. The average numbers of taste buds per papilla are summarized in Table 29.1.

29.2.2.3 Foliate Papillae.
Foliate papillae in humans were first reported by Albinus (1754) and histologically described by Hönigschmied (1873), but did not become the focus of scientific attention in humans until the twentieth century. These papillae, located bilaterally along the posterolateral margins of the tongue surface, consist of parallel rows of ridges (folia) and valleys which lie adjacent to the lower molar teeth. Ducts located between the folia transmit secretions from mostly serous lingual glands within the root of the tongue.

Scanning electron microscopy of human foliate papillae and transmission electron microscopy of their resident taste buds was reported by Svejda and Janota (1974) and Azzali et al. (1996). The number of taste buds in human foliate papillae was reported by Hou-Jensen (1933) and Mochizuki (1939) (see Table 29.1). Confusion in finding taste buds within the folds of human foliate papillae reflects papilla structure. As many as 20 parallel ridges and furrows are found on the posterolateral margin of the human tongue. The rostralmost furrows (lateral rugae) contain no glandular ducts or taste buds, and their epithelia are more cornified than that between foliate papillae (Hou-Jensen, 1933). Fungiform papillae can be found on the tops of these lateral rugae. Although human foliate papillae were thought to be "rudiments" (in comparison to the well developed rabbit foliate papillae), Mochizuki (1939) calculated an average of 1300 taste buds per tongue, which exceeds the number of fungiform taste buds [800 buds: Braus, 1940; <1000 buds: Miller and Bartoshuk, 1991]. Indeed, contiguous taste buds within the same cleft may form a more functional unit, since they share access to a common taste stimulus pool. Foliate papillae are innervated by branches of the glossopharyngeal nerve (CN IX), but the more anterior portion also receives nerve fibers from the chorda tympani (Oakley, 1970; Pritchard, 1991).

29.2.2.4 Fungiform Papillae.
Due to their morphological heterogeneity, fungiform papillae have been variously described as papillae clavatae, capitatae, lenticulares, obtusae, majores, and mediae (Haller, 1766; Sömmering, 1806). These papillae can be easily identified

Table 29.1 Distribution of taste bud numbers within the oral cavity of adult humans.

Vallate Papillae	Fungiform Papillae	Foliate Papillae	Palate	Larynx	Author
–	20–30	?	15–20		Hoffmann, 1875
400					Wyss, 1870
252±151*					Arey, Tremaine, and Monzingo, 1935
234±114*		1279			Mochizuki, 1939
240±125*	<1000 (total)	–			Miller and Bartoshuk, 1991
	800 (total)				Braus, 1940
		708–1328			Hou-Jensen, 1933
			>2500 (neonate)		Lalonde and Eglitis, 1961
				<25 (senile)	Jowett and Shrestha, 1998
	585				Cheng and Robinson, 1991
	1120				Miller and Reedy, 1990b

*per papilla.

as pink elevations about 0.5 mm in diameter on the anterior portion of the living human tongue. Notwithstanding its convenient location, the fungiform taste bud population has been difficult to quantify since fungiform papillae vary in appearance and are distributed over a large area of tongue surface. The anterior portion of the tongue extends from the line of vallate papillae to the tongue tip (Figures 29.4 and 29.12). This region contains about 30 cm^2 of surface area, depending on the size of the person, and the fungiform papillae are spread unevenly over it. The number of taste buds differs among fungiform papillae, and there are large variations among human subjects in fungiform taste bud distribution. Most papillae on the 5 mm margin of the tongue tip are shorter than those on more posterior regions. Following the surface in a posterior direction from the midline of the tip toward the back of the tongue, fungiform papillae become progressively larger in size, being largest in the more posterior lingual regions.

Small, rounded papillae are present on the margin. Some of them contain taste buds, while others are comparable in size to filiform papillae but lack fila. Some papillae on the margin of the tip are elongated like conical papillae, and these, generally, lack taste buds. Fungiform papillae vary in size and shape: some are short and cuboidal, and others are tall with expanded heads like mushrooms. Among papillae on the margin of shorter height and smaller diameter (<0.5 mm), the distinction between filiform and fungiform papillae becomes obscure. Fungiform papillae occasionally have projections on their apices like small fila on filiform papillae.

Taste buds have been quantified in terms of tongue surface areas, referred to as "taste bud density," or number of taste buds per cm^2 of tongue surface. There are about 145 gustatory (fungiform) papillae per lateral half of the tongue, with about 30 papillae per cm^2 on the tip, but only about 3 papillae per cm^2 on the posterolateral area. There are about 30 large fungiform papillae in the posteromedial region, for an estimated total of about 320 fungiform papillae per tongue (Miller and Reedy, 1990b; Shahbake et al., 2005). On 320 fungiform (gustatory) papillae, an average of about 3.5 taste buds per papilla has been estimated, for a total of 1120 fungiform taste buds (Miller and Reedy, 1990a, 1990b) (Table 29.1).

Most investigators who study human fungiform papillae report the existence of papillae without taste buds, which rarely occurs in other mammals (Mistretta and Baum, 1984). Studies in humans show, depending on the methods used, that fungiform papillae lacking taste pores comprise 1–67% of the fungiform population per subject (Arvidson and Friberg, 1980; Cheng and Robinson, 1991; Miller and Reedy, 1990a; Segovia et al., 2002). This wide range may reflect how investigators decide which papillae are "fungiform." Children possess a higher density of fungiform papillae, which correlates well with

increased sensitivity for sucrose (Segovia et al., 2002). Modern non-invasive imaging techniques, for example, confocal laser microscopy, allow the *in vivo* visualization of fungiform taste buds and their dynamics over a certain period of time, even buds that have not yet presented a taste pore (approx. 10%; Just et al., 2005; Srur et al., 2010) (Figure 29.8).

29.3   EXTRALINGUAL TASTE BUDS

There are "extralingual" taste buds in regions of the oral, pharyngeal, and laryngeal cavities. Interestingly, Magendie (1820) and Carus (1849) erroneously associated the teeth with taste perception. Verson (1868) described the first "goblet-like organs" within the dorsal epithelium of the epiglottis, and Davis (1877) and Wilson (1905) observed taste buds in and the elicitation of taste perception from the human larynx. Lalonde and Eglitis (1961) counted more than 2,500 taste buds on the epiglottis, soft palate, laryngeal pharynx, and oral pharynx of one human neonate. Taste buds are evident in the epiglottis of one neonatal specimen (Rabl, 1895), and esophagus in human fetuses (Ponzo, 1907) and adults (Burkl, 1954; Schinkele, 1942). Taste buds are also found near the openings of sublingual salivary gland ducts in some other primates (Hofer et al., 1979), and near the ducts of the molar glands in rodents (Iida et al., 1983). In chickens and quail, taste buds in nonlingual parts of the oral cavity are almost always associated with salivary gland ducts (Ganchrow and Ganchrow, 1987). Miller and Smith (1984) estimate that about 25% of the hamster's total taste buds are extralingual, and Mistretta and Baum (1984) accounted for a similar proportion of extralingual taste buds in the rat. It is not known whether extralingual taste buds are functionally different from those on the tongue. Taste buds of the epiglottis and/or uvula could be involved in initiation of upper airway reflexes (Bradley et al., 1983) and in the pharyngolaryngeal water response, possibly mediated by receptors signaling the absence of chloride ions (reviewed by Lindemann, 1996). Similarly, taste buds of the larynx seem not to play a role in gustation but detect chemicals that are not saline-like in composition, for example, CO_2 (Bradley, 2000; Nishijima and Atoji, 2004). Apparently, structure and immunohistochemical properties of taste buds are remarkably conserved despite their different locations and innervation patterns (Kinnamon, 2011).

29.4   SALIVARY GLANDS OF THE TONGUE

Lingua sicca non gustat (*A dry tongue does not taste*). This declaration of Haller (1766) refers to the dependence of taste ability on solutions, within which tastants are dissolved and transported to the taste bud. Extralingual saliva is secreted by small, mostly mucous glands embedded in the epithelium of the cheek and palate. More saliva is produced by the serous parotid gland and the muco-serous sublingual and submandibular glands (Figure 29.3), whose secretory ducts open at the tongue frenulum just underneath its tip. Intralingual saliva originates from the mucoserous anterior lingual glands [glands of Blandin and Nuhn, (Tandler et al., 1994), Figure 29.3], and deep posterior serous salivary glands (Ebner) located in the submucous connective tissue below the foliate and vallate papillae of the tongue (Ebner, 1873; Riva et al., 1999). Their excretory ducts lead to the deepest sites of the papillar furrows (Figure 29.6). The gland lobules lie deeply in large patches of connective tissue which, in turn, are separated from each other by muscle fiber bundles. In addition, adjacent to Ebner glands in vallate papillae lie mucous (Weber's) glands (Nagato et al., 1997), which in humans open into the crypts of the lingual tonsils (Zimmermann, 1927). Neither Weber's nor Blandin-Nuhn glands lie in close proximity to taste buds and their particular significance for taste perception is unknown because of the difficulties in collecting saliva from these glands (Tandler et al., 1994).

There is biochemical and histochemical evidence that the saliva of Ebner's glands, as well as that of other non-lingual salivary glands, has more functions than that of a serious "washing solution." Binding proteins such as Ebnerin (Li and Snyder, 1995) are supposed to modulate sensations. Schmale et al. (1990) isolated a protein from rat Ebner's glands that is structurally similar to odorant binding proteins in Bowman's glands of the olfactory mucosa. The gland is under autonomic control (Fukami and Bradley, 2005; Gurkan and Bradley, 1987). Interestingly, a recent report seems to point at a dependency of the hypoglossal nerve on sympathetic innervation of posterior lingual glands in hamsters (Cheng et al., 2009). For reports on specific ligand-receptor interaction with taste qualities, see Azen et al. (1990), Schmale et al. (1993), Spielman, (1990), Toto et al. (1993), and Chapter 31 of this volume.

29.5 BLOOD SUPPLY TO GUSTATORY PAPILLAE

The mammalian tongue receives its blood supply from the lingual artery, which is usually a branch of the external carotid artery (Figures 29.3 and 29.4). Study of the tongue's vascular system historically parallels that of the lingual papillae. For example, Albinus (1754), Sömmering (1806) (Figure 29.4) and Arnold (1839) performed intravascular injections in order to visualize the papillary surface. More recently, distribution of the blood supply to different regions of the tongue and different types of lingual papillae has been described by Hellekant (1976). Each type of gustatory papillae is supplied by a characteristic capillary configuration (rat: Ohshima et al., 1990; cat, rabbit: Ojima et al., 1997a-c) and fine capillary networks are found adjacent to taste buds. The capillary loops of larger papillae in rats and dogs often show a constriction, maybe sphincter-like structures, but rarely arteriovenous anastomoses (Hu et al., 1996; Selliseth and Selvig, 1993). Taste stimuli injected systemically elicit responses in gustatory nerves as the bolus passes through the tongue (Bradley, 1973).

29.6 SOLITARY CHEMOSENSORY CELLS

In addition to taste buds and free nerve endings, the solitary chemosensory cells (SCC) comprise another chemosensory system in vertebrates. They are not assembled in clusters, but are dispersed across the surface of the animal. SCC are related to taste bud cells in the sense that the former are secondary sensory cells with a slender, bipolar phenotype (Finger, 1997). "Classical" SCC have been studied first in teleosts (Whitear, 1992). The evolutionary benefits of these cells are still in question: In sea robins (*Trigon*), they are, beside taste buds, involved in finding food; in rocklings (*Gaidropsarus*) they are assumed to be important for predator avoidance (Kotrschal, 1996). Since the arginine-like receptor in catfish taste buds also occurs in SCC, Finger (1997) suggests that taste buds might include SCC within them. During development of fish, SCC seem to precede the development of taste buds. In mammals, however, SCC-like cells have been observed only transiently, during development. In newborn rats, single gustducin-immunopositive cells are seen in locations where later-developing vallate papillae will appear (Sbarbati et al., 1999). Individual slender cells, immunopositive for cytokeratin 20 (Witt and Kasper, 1999), an intermediate filament protein that is exclusively present in taste bud and epidermal Merkel cells (Moll, 1993; Zhang and Oakley, 1996; Zhang et al., 1995), are seen occasionally during early ontogenesis of the human tongue. Gustatory epithelia *sensu strictu* of adult mammals have not yet been reported to possess SCC. However, alpha-gustducin-immunoreactive SCC occur in the nasal mucosa of mice and have been interpreted as sentinels in the anterior nasal air passages (Finger et al., 2003). Taste receptors (Tas1R, Tas2R) in SCC along with their synaptic connectivity to CGRP-positive polymodal pain fibers of the trigeminal nerve indicate a role in detection of irritants and foreign substances by triggering trigeminally-mediated reflexes (Tizzano et al., 2011).

29.7 CELL TYPES OF VERTEBRATE TASTE BUDS

Peripheral taste organs differ in number, size, and shape in different vertebrate taxa, according to their importance for the particular species. For the sake of brevity, the following overview is restricted to some functionally well-characterized vertebrate species. Details are available in reviews by Reutter and Witt (1993) or Chaudhari and Roper (2010). Figure 29.9 shows a scheme representing the organization of cells in fish, frog, and mammalian taste buds.

29.7.1 Taste Buds of Lower Vertebrates – Cell Types

29.7.1.1 Fish.
In fish, and especially in some teleosts that are well-adapted to the dark, the taste organ is significantly more important for food intake than in amphibians and mammals. Thus, these fish, like the *Siluridae*, possess many more taste buds than representatives of the latter classes (Atema, 1971; Finger et al., 1996; Miller and Bartoshuk, 1991). The fish taste bud is generally pear-shaped and similar to that of mammals (Figure 29.9). Electron microscopic studies of teleost taste buds have

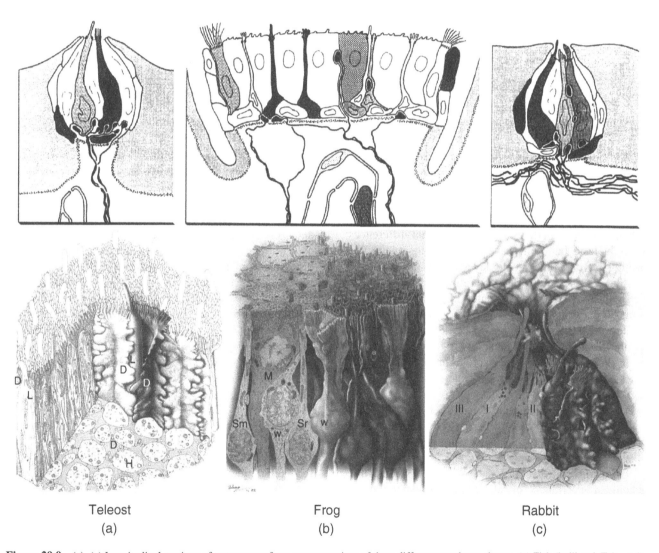

Figure 29.9 (a)–(c) Longitudinal sections of taste organs from representatives of three different vertebrate classes. (a) Fish (bullhead, Teleostei), (b) Amphibian (frog, Anura), (c) Mammal (rabbit). In the schematic drawing, each sensory epithelial cell type is represented once with a distinct grey-step. The organs lie in squamous epithelium of different height, on top of dermal papillae which are also of different height. Each dermal papilla contains nerve fibers and a capillary vessel. (Below) 3D reconstructions of the respective apical taste bud portions. In all species, glia-like cells (blue in teleost and rabbit; sepia in frog "wing" cells) enwrap sensory cells. Sensory cells (green; sepia in rabbit as type II receptor cell). Nerve fibers entangle all cells, but synapses are seen only with type III cells in rabbit and microvillar (Sm) and rod-like receptor cells (Sr) in frog. (*See plate section for color version.*)

attempted, somewhat incompletely, to relate the ultrastructure and functions of taste bud cell types (for reviews see Jakubowski and Whitear, 1986; Reutter, 1986; Reutter and Witt, 1993). This is also apparent in the nonuniform usage of nomenclatures. Most authors refer to elongated "light" and "dark" taste bud cells, as well as "intermediate" and "degenerative" cells (Connes et al., 1988; Desgranges, 1965; Reutter, 1971, 1978; Welsch and Storch, 1969; Whitear, 1970). Generally, light cells are supposed to be "sensory" (receptor) cells, while the dark cells are regarded as "supporting" or "sustentacular" cells (e.g., Desgranges, 1965; Hirata, 1966; Whitear, 1970). However, Reutter (1992) considers the dark cells also as sensory because they exhibit synaptic contacts with nerve profiles or with basal cells. These observations, however, are not supported by the work of Grover Johnson and Farbman (1976) and Jakubowski and Whitear (1986). According to these authors the differentiation "light" and "dark" in combination with functional terms such as "supporting cells" and "sensory (receptor) cells" are misleading and should be avoided. According to Hirata (1966), Merkel (1880) and Reutter (1973) the taste bud may also have mechanoreceptive functions, particularly in view of the morphology of basal cells (Reutter, 1971, 1986) or serotonergic Merkel-like cells (Zachar and Jonz, 2011). While the existence of synaptic contacts of light *and* dark taste bud cells has not yet been finally proven, the different lengths of their microvilli perhaps suggest functional differences: The light taste bud cells have a single long microvillus reaching far into the mucous layer of the taste bud surface, the receptor area. By penetrating this layer, longer microvilli may be exposed to quite different "perireceptor events" (Getchell et al., 1984) than the small microvilli of dark taste bud cells that do not penetrate the mucous layer (Reutter, 1980; Witt and Reutter, 1990).

In larvae of lampreys, representatives of fish-like jawless vertebrates (Petromyzontidae), taste bud cells were characterized by Retzius (1893) using the Golgi silver impregnation technique. Recently, Barreiro-Iglesias et al. (2010) reported three different cell types including serotonergic bi-ciliated cells, basal cells, and sustentacular cells.

Ultrastructural (Reutter and Witt, 1993) and neurochemical (Ferrando et al., 2012) investigations in non-teleostean fish reveal clearly that taste buds differ within the main vertebrate taxa, which is partly represented in the differential expression of G protein-coupled receptors (Oka and Korsching, 2011). Further, in different systematic groups of fish the taste buds do not follow only one structural design. Thus, taxon-specific taste buds or cell types do not exist. Similar differences among mammalian taste buds point to a similar inevitable conclusion regarding particular phenotypes: There are only species-specific taste bud types, and a general "model" seems difficult to find among vertebrates. This underlines the thesis that morphological phenotypes and the structural organization of taste bud cells do not necessarily reveal a general bauplan, but rather, reflect specific environmental conditions and/or feeding behaviors (Reutter and Witt, 1999).

The influence of environmental differences has been studied in two closely related teleosts, one of which is sighted (*Astynax mexicanus*) and the other of which is a blind cave fish (*Astynax jordani*). Whereas taste bud morphology is rather similar, the cave fish compensates for blindness by significantly more gustatory axon profiles (Boudriot and Reutter, 2001) and an expanded expression of Prox 1 gene in developing taste buds (Jeffery et al., 2000; Varatharasan et al., 2009).

29.7.1.2 Amphibians. The taste organs of post-metamorphotic Salientia (=Anura), unlike piscine and mammalian taste bud bulb-like formations, are relatively large disk-like epithelial differentiations of the dorsal lingual and palatal mucosa (Figure 29.9). Waller (1847, 1849) and Engelmann (1872) called these structures "Geschmacksscheibe" or "taste disc." In contrast to Salientia, the taste buds of the urodeles [= Caudata, e.g., mudpuppy (*Necturus*), newt (*Triturus*) or Axolotl (*Ambystoma*)] have a bulb-like shape (Cummings et al., 1987; Delay and Roper, 1988; Fährmann, 1967; Farbman and Yonkers, 1971; Toyoshima et al., 1987; Toyoshima and Shimamura, 1987). The cellular elements of the taste disc, or "Endscheibe" (Merkel, 1880), of the frog were subject to numerous investigations and received various designations: After Waller (1847, 1849) had first distinguished between *papillae conicae* (=*P. filiformes*) and *papillae fungiformes*, Fixsen (1857) described two different cell types, the so-called "*cellulae cylindricae*" and "*cellulae fusiformes*," the processes of which pass through the whole sensory epithelium to reach the connective tissue core of the papilla. Engelmann (1872) and Merkel (1880) further developed the terminology: Merkel distinguished between "*cylindrical cells*" (Cylinderzellen) situated on the epithelial surface and surrounding "*wing cells*" (Flügelzellen), the nuclei of which lie deeper in the epithelium. After Graziadei and DeHan (1971) had described only two cell types ("associate cells" and "sensory cells") in electron microscopy, the close relationship between "rod cells" (Stäbchenzellen) and cylindrical cells (Merkel, 1880) was recently re-introduced by von Düring and Andres (1976). In addition, the latter authors first described basal cells and Merkel cells of the frog taste disc, which are the only cells that do not contact the epithelial surface. The taste disc in adult frogs contains up to eight cell types (Reutter and Witt, 1993): mucus cells, wing cells, two types of sensory cells (cylindrical and rod-like type), two types of basal cells [stem and

Merkel cell-like basal cells (Zancanaro et al., 1995)], and marginal cells and ciliated cells (Toyoshima et al., 1999) (Figure 29.9). The cell types and the history of the nomenclature are described in detail by Jaeger and Hillman (1976), Reutter and Witt (1993), Witt (1993), Osculati and Sbarbati (1995), and Li and Lindemann (2003).

Tadpoles possess so-called premetamorphic papillae which bear bud-like taste organs at their tops. During metamorphosis, these structures wholly disappear and are replaced by fungiform papillae with large taste disks (Nomura et al., 1979; Żuwała, 1997; Żuwała and Jakubowski, 1991). Tadpole taste discs consist of sensory and supporting cells, and basal cells are lacking.

Taste buds of the mudpuppy (*Necturus*, Urodela) are similar to those of fishes. They are composed of dark and light cells and possess serotonergic Merkel cell-like basal cells which are synaptically connected with either nerve fibers or dark and light cells (Delay et al., 1994, 1993; Delay and Roper, 1988).

29.7.1.3 *Reptiles and Birds.*

Lingual taste buds in reptiles have been described in turtles (Iwasaki et al., 1996; Korte, 1979), tortoises (Pevzner and Tikhonova, 1980), and some lizards (Uchida, 1980). Lizards (Gekkonidae or Anguidae), as well as snakes, have virtually no taste buds on the tongue, but rather on the buccal floor and oral epithelia of the mandible and maxilla (Toubeau et al., 1994). In shape, reptilian taste buds resemble those of mammals. There are up to five different types of taste bud cells, classified into light and dark cells and as types 1,2,3,A,B, or as types I, II, II, and basal cells (Reutter and Witt, 1993).

Reptiles and birds belong to the same superclass, and one might expect a similar organization of avian taste buds. However, the few examples of bird taste buds show great variability among species. Buds in grain-eating birds appear in the posterior part of the tongue, near the pharynx, and in the distal palatal mucosa (Ganchrow et al., 1991; Ganchrow and Ganchrow, 1987; Saito, 1966; Sprissler, 1994). Unlike mammals, avian taste buds do not reside within lingual bud-bearing papillae. In addition, taste buds contain tubular-like channels circumscribed by elongated cells grouped in a rosette configuration, with the channel lumen continuous apically with the taste pore (Berkhoudt, 1985; Ganchrow et al., 1993). Taste buds in chicken are richly innervated (Ganchrow et al., 1986), contain gustducin in a subset of taste cells (Kudo et al., 2010), and synapses are seen between all cell types (light cells and dark cells) and nerve fibers (Reutter and Witt, 1993; Sprissler, 1994). Stornelli et al. (2000) observed four cell types in duck taste buds: light cells, dark cells, intermediate cells, and basal cells.

29.7.2 Mammalian Taste Buds – Cell Types

Early histological investigators of mammalian taste buds described two types of elongated, fusiform cells in taste buds of human vallate papillae (Schwalbe, 1868): "supporting" and "taste cells," the latter divided into Stiftchenzellen (*pin cells*) and Stabzellen (*rod cells*) with differences in contrast and brightness (Figure 29.5).

Classification into "light" and "dark" taste bud cells was also used in early electron microscopic analyses (Engström and Rytzner, 1956). Farbman (1965) considered dark, fusiform taste bud cells of human fungiform papillae as sensory cells (type I), whereas other investigators such as Paran et al. (1975) described a type II-cell that contains numerous vacuoles and mitochondria, especially in apical areas. This latter cell is not believed to be sensory. Cottler Fox et al. (1987) speculated that the different electron densities are due to irreproducible fixation artifacts. In general, ultrastructural and immunohistochemical criteria are considered more important for the classification of cells than the evaluation of the electron density of the cytoplasm. Moreover, present understanding in mammalian taste bud cytology leads to using a rather heterogeneous nomenclature, based on morphological and functional differences across species.

The basis for the current nomenclature was established by the Murrays and colleagues (Murray, 1986; Murray and Murray, 1967, 1971; Murray et al., 1969) in taste bud cells of the rabbit foliate papillae. However, these cell types differ in some respects from rodent taste buds (Kinnamon et al., 1985, 1993, 1994; Pumplin and Getschman, 2000; Pumplin et al., 1997, 1999). Thus, Royer and Kinnamon (1988) observed considerable deviations in the cytoarchitecture of murine foliate taste buds compared to that of other mammals. For example, they did not find type III cells, and all bud cells had synaptic connections with nerve fibers. Most current information on chemoreception of taste stimuli and information transfer to primary afferent axons have been established in mice (Dando and Roper, 2009; Huang et al., 2009; Roper, 2006), but there are significant species differences (Ma et al., 2007), and the precise organization of the taste bud in humans remains to be determined (Azzali, 1996, 1997; Paran et al., 1975; Witt and Reutter, 1996). An example for a more generalized taste bud is depicted in Figure 29.10. Work of the last two decades has revealed that taste stimuli are received by type II receptor cells (Tomchik et al., 2007), while type III (presynaptic) taste cells are the only ones that contain synaptic proteins and form synapses with nerve fibers (Kinnamon et al., 1988). The information transfer from type II (receptor) cells to type III (presynaptic) cells is mediated through ATP and pannexin hemichannels (reviewed by Chaudhari and Roper, 2010) or directly

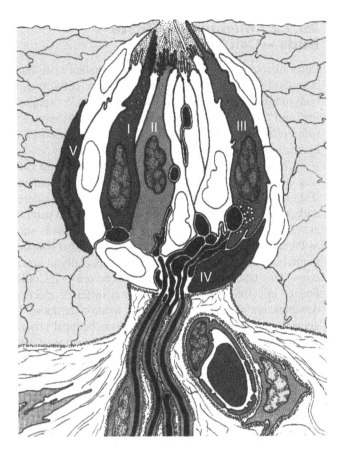

Figure 29.10 Mammalian taste bud in longitudinal section, idealized schematic drawing according to electron microscopical findings. Each cell type is depicted once. Following Murray's nomenclature, cells of type I, II and III are elongated and form the buds sensory epithelium proper. Apically, these cells end with different types of microvilli within the taste pit and may reach the taste pore. Type I cells are glia-like, type II cells comprise receptor cells for bitter, sweet, and umami stimuli. Synapses are only found at the bases of type III (presynaptic) cells. Type IV cells are basal cells, type V marginal cells. Nerve fibers within the short dermal papilla are slightly myelinated, and within the taste bud they form an unmyelinated plexus. Note the basal lamina between dermis (which contains a capillary) and the epithelium.

from type II cells to nerve fibers, as gustatory afferents also contain purinergic receptors P2X2 and P2X3 for the interaction with ATP (Finger et al., 2005). A survey representing the most important substrates associated with taste cell types in vertebrates is shown in Table 29.2.

29.7.2.1 Type I Cells (Glia-Like Cells).
These cells are the most frequent. They are spindle-shaped and have a basal process that envelops the axons in a Schwann cell-like manner. Type I cells protrude with brush-like, long microvilli (1–2 μm) into the taste pit. These cells ensheath type II and type III cells with cellular protrusions and may insulate them (Figures 29.9 and 29.10). Apically, they contain large granules, 100–400 nm in diameter (Murray, 1986). The nuclei of type I cells are irregularly shaped. According to Murray and Murray (1971) and Murray (1986) these cells have a secretory (supporting cells) and possibly phagocytotic function, and probably produce the amorphous material of the taste pit (e.g., (Farbman, 1965; Menco, 1989; Ohmura et al., 1989; Witt, 1996). Type I cells are involved in neurotransmitter clearance and ion redistribution, similar to the function of glia cells in the central nervous system. They contain GLAST, a glutamate transporter (Lawton et al., 2000), as well as enzymes degrading extracellular ATP (Bartel et al., 2006). Some type I cells possess the K+ channel ROMK that may eliminate accumulated extracellular potassium ions to retain the excitability of type II and type III cells (Dvoryanchikov et al., 2009).

29.7.2.2 Type II Cells (Receptor Cells).
Most of these cells are located in the periphery of the taste bud. They are fusiform, but do not possess enveloping processes, or granules. Their cytoplasm is moderately electron-dense, and nuclei are round to oval. Synapses are not observed (Farbman et al., 1985, 1987). Toyoshima and Tandler (1987) describe a modified endoplasmic reticulum with specialized sub-surface cisterns adjacent nerve profiles. In mice, vallate and foliate "light" (type II-) and "dark" (type I- and type II-) taste bud cells exhibit synaptic contacts with nerve fibers, thus suggesting a gustatory function. However, type I- and type II- cells do not form synapses with the same nerve fibers (Kinnamon et al., 1985, 1988; Royer and Kinnamon, 1988). These cells are equipped with G-protein coupled receptors and tuned to sweet, bitter, and umami, but not to sour and salty taste (Matsunami et al., 2000; Tomchik et al., 2007).

29.7.2.3 Type III Cells (Presynaptic Cells).
Taste buds contain only 5–7% of type III cells. They have unbranched basal and apical processes. Their apical portion protrudes with a single large microvillus into the taste pit and may reach the taste pore. Type III cells are the only cells that have synaptic contacts with intragemmal nerve fibers. Near the cell nucleus there are numerous dense-cored vesicles (80–140 nm in diameter) which are involved in neurotransmitter synthesis (GABA, serotonin, norepinephrine) (Fujimoto et al., 1987; Hokfelt et al., 1980; Huang et al., 2005; Nada and Hirata, 1977). Type III cells contain several synaptic proteins and are therefore termed "presynaptic cells" (DeFazio et al., 2006). Their apical process apparently possesses channels for sour taste reception (Huang et al., 2008).

29.7.2.4 Type IV Cells (Basal Cells).
Type IV cells (Murray, 1973; Murray et al., 1969) include basal cells (Nemetschek-Ganssler and Ferner, 1964) or "pregustatory cells" (Scalzi, 1967). These are relatively small

29.7 Cell Types of Vertebrate Taste Buds

Table 29.2 Survey of selected substances (transmitters, channel or transporter proteins) associated with certain types of taste bud cells in various species.

Taste Bud Cell Type	Substance/Transmitter	Species	Reference
Basal Cells	5HT (serotonin)	catfish	Reutter, 1971
basal cells	5Ht, Glu, GABA	necturus	Kim and Roper, 1995; Nagai et al., 1996
basal cells	5HT	frog	Jain and Roper, 1991
I	Glutamate/aspartate transporter (GLAST)	mouse	Lawton et al., 2000
I	Ecto-ATPases	mouse	Bartel et al., 2006
I	Epithelial Na+ channel (ENAC)	mouse	Lin and Kinnamon, 1999; Vandenbeuch, Clapp, and Kinnamon, 2008
I	K+ channel ROMK	mouse	Dvoryanchikov et al., 2009
II	5HT, Serotonin transporter	rat	Ren et al., 1999
II	Acetylcholine	mouse	Dando and Roper, 2009
II	Acetylcholine transporter, Choline-acetyltransferase (ChAT),	mouse	Ogura et al., 2007
II	Vasoactive intestinal peptide (VIP)	rat, carp	Shen et al., 2005; Witt, 1995
II	Cholecystokinin	rat	Herness et al., 2002
II	Neuropeptide Y	rat	Zhao et al., 2005
II	Type III IP3 receptor	rat, mouse	Clapp et al., 2001
n.d.	Gustducin	chicken	Kudo et al., 2010
II	Gustducin	mouse, rat, hamster	Boughter Jr. et al., 1997; Hoon et al., 1995; McLaughlin, McKinnon, and Margolskee, 1992; Ruiz-Avila et al., 1995
II	TAS2R38 (bitter taste receptor)	human	Behrens et al., 2012
II	NGF, trkA	mouse	Yee, Bartel, and Finger, 2005
II	PLCbeta2	Mouse	Clapp et al., 2004
II	PLCbeta2	Human	Behrens et al., 2012
II	TRPM5	mouse	Perez et al., 2002
II	T2R		Matsunami, Montmayeur, and Buck, 2000
II	T1R		Hoon et al., 1999
II	GPCRs signal molecules for bitter, sweet and umami	mouse	DeFazio et al., 2006
II (III?)	ATP release	mouse	Bartel et al., 2006; Finger et al., 2005
II (Gustducin-neg.)	PGP 9.5	rat	Yee et al., 2001
(II)	Snap 25	rat	Oike, Matsumoto, and Abe, 2006; Pumplin and Getschman, 2000; Ueda et al., 2006
(III)	Snap 25	rat	Yang et al., 2000
III	SNAP25, synapsin II, NCAM,AADC.	mouse	DeFazio et al., 2006
II	GABA, glutamate decarboxylase (GAD)	rat	Cao et al., 2009
(II)	GABA	rat	Obata et al., 1997
II (Gustducin neg.)	GABAB2	mouse	Starostik et al., 2010
III	GAD, GABA-B Receptor	mouse	Starostik et al., 2010
III	GAD	mouse	DeFazio et al., 2006
III	5HT	mouse, rat, rabbit, monkey	Fujimoto, Ueda, and Kagawa, 1987; Kaya et al., 2004; Nada and Hirata, 1975; Nada and Hirata, 1977
III	5HT	mouse	Huang et al., 2005
III	5HT	monkey, rb	Fujimoto, Ueda, and Kagawa, 1987
III	NCAM	rat	Nelson and Finger, 1993
III	BDNF, trkB	mouse	Yee, Bartel, and Finger, 2005
III (5HT-neg.)	Protein gene product 9.5 (PGP 9.5)	rat	Yee et al., 2001

650	**Chapter 29 Anatomy of the Tongue and Taste Buds**

undifferentiated cells that lie at the taste bud's base, which do not form processes that reach the pore. They contain numerous bundles of intermediate filaments (Royer and Kinnamon, 1991), and differ from Merkel cell-like basal cells of fishes and amphibians. Type IV cells are considered to be undifferentiated stem cells of their bud cell progeny (Chaudhari and Roper, 2010; Murray, 1973; Murray, 1986; Roper, 1989). In embryonic and adult vallate taste buds of mice, basal cells also express the transcription factors Hes6 and Mash-1, the latter possibly being involved in specification of a type III lineage (Seta et al., 2003, 2011). Using genetic lineage tracing in murine taste buds, Miura et al. (2014) found sonic hedgehog (shh) in most basal cells, which indicates a fate as postmitotic, immediate precursors of all three cell types rather than a stem cell.

Some authors report on "intermediate cells" (e.g., Kinnamon et al., 1985). It has been pointed out (Farbman et al., 1985; Roper, 1989) that differences in the electron density of the cytoplasm could also reflect different stages in the maturation of the same cell type which indicate different states of function.

29.7.2.5 Type V- Cells (Marginal Cells).

"Marginal cells" (also "perigemmal cells," and, in extension of Murray's nomenclature, "type V cells") have been described (Beidler and Smallman, 1965; Farbman, 1980; Gurkan and Bradley, 1987; Reutter and Witt, 1993). However, they have nothing in common with the secretory marginal cells of taste organs in fish and frog, and may possibly be taste bud stem cells (Beidler and Smallman, 1965; Farbman, 1980) which express particular non-taste receptor proteins, for example, CD44 isoforms (Witt and Kasper, 1998) during human taste bud ontogenesis.

29.7.3 Molecular Markers of Taste Bud Cells

One of the most intriguing challenges for suggesting possible functional properties of taste bud cells is to identify subsets of these cells by morphological features as well as molecular properties, many of which can be traced even in enriched primary taste bud cell cultures (Kishi et al., 2001, 2002; Ozdener et al., 2006). Histochemical evidence on the neurochemical nature of taste cells have identified the panneuronal markers, neuron-specific enolase (NSE) and protein gene product 9.5 (PGP 9.5) (Astbäck et al., 1997; Montavon et al., 1996; Yee et al., 2001; Yoshie et al., 1988), carbohydrate- binding proteins, the lectins (Witt and Miller, 1992; Witt and Reutter, 1988). Immunoelectron microscopic studies have tried to match functional parameters with those of conventional electron density. For example, cell adhesion molecules (Nolte and Martini, 1992; Smith et al., 1993, 1994) and several blood-group antigens (Pumplin et al., 1997, 1999; Smith et al., 1999) characterize subsets of type II taste bud cells. A subset of the (light) type II cell contains partly the G protein gustducin (Menco et al., 1997; Ruiz-Avila et al., 1995), which is involved in the perception of sweet and bitter taste (Tomonari et al., 2012; Wong et al., 1996).

Choline acetyl transferase, an enzyme involved in the synthesis of the neurotransmitter acetylcholine, has been identified in rat type II cells (Menco et al., 1997). Whereas the putative neurotransmitter serotonin is confined to basal cells of fish taste buds (Reutter, 1971) and Merkel cell-like basal cells of amphibian taste organs (Delay et al., 1997; Hamasaki et al., 1998; Toyoshima and Shimamura, 1987), serotonin in mammals has been described in type III cells in the rabbit (Fujimoto et al., 1987; Kim and Roper, 1995) and human taste buds (Azzali, 1997). This led to the hypothesis that these cell types were equivalent in both taxa (Kim and Roper, 1995). Lindemann (1996) suggests the term "serotonergic cells" instead of type III cells (<10% of all cells).

Generally, neuropeptides are located in intragemmal nerve fibers rather than in particular bud cells. An exception is vasoactive intestinal peptide (VIP) that has been detected in a subset of rat type II cells (Herness, 1989; Shen et al., 2005), and in light taste bud cells in the carp (Witt, 1995) by electron microscopy. Though most of these markers are expressed only in differentiated cells and are not evident after nerve dissection (Smith et al., 1993; Whitehead et al., 1998), their functional correlation with taste perception data is mostly unknown.

To avoid the present Babylonian confusion of tongues with regard to taste bud cell nomenclature, present research directions try to associate electrophysiologically-characterized, isolated taste bud cells with a particular cell type based on its specific substrate expression. Modern cell biological approaches, for example, introduction of green fluorescent protein chimeras in vitro (Landin et al., 2005) or calcium imaging after application of specific stimuli (Caicedo et al., 2000), have contributed to a solution of this problem.

Several authors report on morphological and immunohistochemical differences between vallate/foliate and fungiform taste buds within the same species. For example, mouse taste bud cells of fungiform papillae contain more synapses and presynaptic vesicles than those of vallate papillae (Kinnamon et al., 1993), and the number of taste bud cells containing group H blood antigen and gustducin is three times higher in vallate than in fungiform papillae (Smith et al., 1993). The coexpression pattern of taste receptors (T2R and T3R) and gustducin differs between fungiform and vallate taste buds in mice (Kim et al., 2003). In rabbit, lectin carbohydrate profiles of both taste bud populations differ as well (Witt and Miller, 1992). The reasons and significance of these differences between fungiform and vallate/foliate taste cells are not clear, but

29.8 Development of the Human Peripheral Taste System

factors determining their varying phenotype could include a different local saliva composition (Schmale and Bamberger, 1997; Schmale et al., 1990; Shatzman and Henkin, 1981) or morphogenetic conditions of local epithelium (Smith et al., 1999).

Evidence for communication between taste cells, apart from purinergic transmission (Finger et al., 2005), includes the presence of cell adhesion molecules (Nolte and Martini, 1992; Smith et al., 1993), heparin-binding proteins (Wakisaka et al., 1998), and membrane receptors that influence the intracellular signal transduction cascades. For example, the hyaluronan receptor, CD44, was identified in a subset of human fetal taste bud cells (type V, marginal cells) and most of adult human taste bud cells (Witt and Kasper, 1998). This transmembrane protein is linked to a series of actin-associated microfilaments, for example, ezrin and ankyrin, which are located in microvilli of type I cells and might influence the function of ion-translocating membrane proteins (Höfer and Drenckhahn, 1999). In light of efferent neural control, taste bud cell communication may be mediated via local axon reflexes between sensory cells (Caicedo et al., 2000, Reutter and Witt, 2004).

29.8 DEVELOPMENT OF THE HUMAN PERIPHERAL TASTE SYSTEM

Morphogenesis of the mouth cavity is characterized by the development of the tongue anlage which appears prior to, and is a prerequisite of, the formation of gustatory papillae.

At the embryonic age of 4 weeks, the first structure of the tongue anlage to appear is the tuberculum impar which is situated between the first (mandibular) and second (hyoid) branchial arches (Figure 29.11). Then, anterolateral to the tuberculum impar, the paired lingual swellings (which derive from the medial parts of the mandibular arches) fuse with the tuberculum impar. The tongue's base is formed by the hypobranchial eminence (copula of *His)* forming within the third and fourth branchial arches. The border between the caudal part and the body of the tongue is demarcated by a V-shaped rim, the sulcus terminalis (Bradley, 1972; Witt and Reutter, 1997). The innervation pattern of cranial nerves, which later supplies particular lingual regions, reflects the early innervation of branchial arches (see Figure 29.11): The pretrematic nerve of the first branchial arch is the lingual nerve (from CN V_3); that of the second arch, the chorda tympani (from the intermedio-facial nerve); and that of the third arch constitutes later lingual rami of the glossopharyngeal nerve (for details see textbooks on embryology, for example, Hinrichsen, 1990; Williams et al., 1989).

The first detailed developmental studies on the surface appearance of the tongue were carried out by Froriep (1828), and continued histologically by Tuckerman (1889), Gråberg (1898), and Hellman (1922). Hermann (1885) described the stages of karyokinesis in developing taste buds. It was unclear to Hermann if supporting, or neuroepithelial, cells were being replaced. Vallate papillae start to develop earlier than fungiform papillae, and begin with the appearance of a central midline papilla just behind the foramen caecum around the sixth postovulatory week. From week 7 on, there develop many hillock-like epithelial elevations on the tongue's dorsum, as seen with scanning electron microscopy (Figure 29.12). Some of these elevations are precursors of fungiform papillae and are especially densely distributed near the midline and the lateral ridges of the tongue (Habermehl, 1952; Hersch and Ganchrow, 1980; Witt and Reutter, 1997). Analysis of serial sections of the tongue encompassing this critical developmental age (weeks 6–8) demonstrates that not every dermal elevation will be the target of nerve fibers. First, around week 7–8, nerve fibers, migrating towards the periphery, form a large intragemmal plexus. Our own studies show that there are no taste bud anlagen without approaching nerve fibers (Witt and Reutter, 1996; Witt and Kasper, 1998), but recent studies revealed the nerve-independent development of taste buds and their preforming papillae (Barlow, 2003; Ito and Nosrat, 2009; Nosrat et al., 2012; Stone et al., 1995; Thirumangalathu et al., 2009). Also, taste bud primordia without dermal papillae are evident, as well as individual bipolar epithelial cells resembling solitary chemosensory cells. These individual cells are immunopositive for cytokeratin 20, a marker for lingual taste bud cells (Witt and Kasper, 1999; Zhang and Oakley, 1996). Temporal correlation, which would suggest dependence of taste bud development on nerve ingrowth, has not as yet been seen.

Lingual taste bud primordia first occur around the 7th and 8th postovulatory weeks (Bradley, 1972; Bradley and Stern, 1967). Taste pores, commonly acknowledged as a sign of taste bud maturity, appear between the 10th and 14th week. The presence of a taste pore is not always associated with a fully mature taste bud because the bottoms of early taste pits may be covered by flat epithelial cells (Witt and Reutter, 1997). However, transmission electron microscopical studies show that early taste bud primordia (week 8) synaptically contact nerve fibers, suggesting the potential for neurotransmission precedes the exposure of sapid molecules to the apical surface of the taste bud cell. At its base, the developing human taste bud (weeks 12–15) contains processes of dark and light cells, as well as processes resembling type III cells (exhibiting synapses with nerve fibers). At their apical ends, taste bud cells cannot be distinguished by their electron density (week 15, Figure 29.13). There are cells with long, slender microvilli, but, in contrast to adult taste buds (Azzali, 1997), there are no type I cells

Chapter 29 Anatomy of the Tongue and Taste Buds

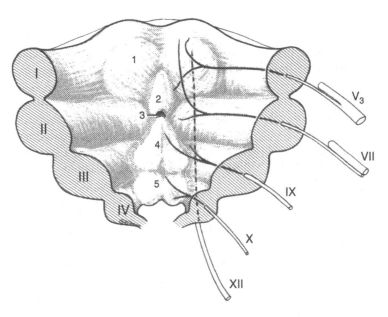

Figure 29.11 Development of the human tongue, 5th postovulatory week. The schematic drawing was done by compiling Hinrichsen's (1990) and our own data. – By horizontal section the floor of the forecoming buccal cavity of a human embryo is removed and viewed from dorsally. The floor relief is derived from the branchial arches I (mandibular arch), II (hyoidal arch), III (3rd pharyngeal arch) and IV (4th pharyngeal arch), and by their derivatives which are 1- the paired lingual swellings, 2- the impar tubercle and 4- the hypobranchial eminence. These three structures form the tongue anlage. Between 2 and 4, the anlage of the thyroid gland invaginates, and as its remnant the 3- foramen caecum is left. 5- is the anlage of the epiglottis. The branchial arches, as well as the tongue anlage, are innervated by the cranial nerves V$_3$ (mandibular nerve), VII (facial nerve), IX (glossopharyngeal nerve) and X (vagal nerve). (CN XII -hypoglossal nerve- invades the tongue anlage as well, and innervates its muscular system). Later, the lingual nerve (from V$_3$) and the chorda tympani (running with VII) join each other and to supply the anterior two- thirds of the tongue with somatosensory and gustatory nerve fibers, whereas IX and X carry taste fibers for the posterior third of the tongue, the epiglottis and the pharynx.

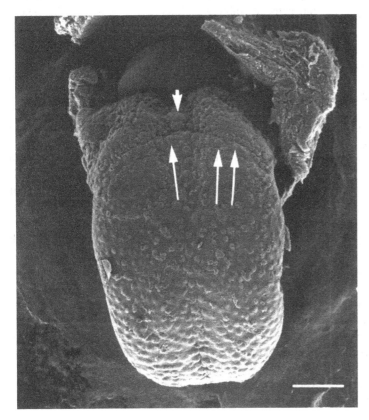

Figure 29.12 Scanning electron micrograph of a human embryonic tongue, 7th postovulatory week. Fine dots in the dorsal surface demarcate later fungiform papillae. Anlagen of vallate papillae (arrows) lie in front of the sulcus terminalis. Short arrow indicates the median vallate papillae, which originates first. Scale bar: 0.5 mm.

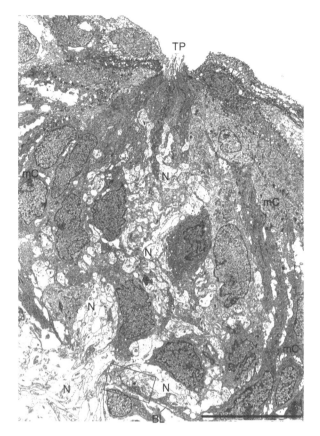

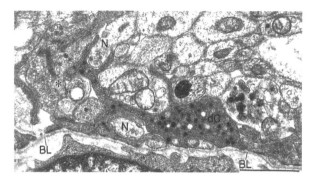

Figure 29.14 Basal portion of a developing human taste bud, week 15 (detail from Figure 29.13). Numerous processes of an electron-dense cell (dC) rich in dense cored vesicles surround nerve fibers (N), but synapses and signs of release of the vesicular contents are not observed. One cell process is filled with dense-cored and clear vesicles (asterisk). BL = basal lamina. Bar = 1 μm. From Witt and Reutter (1996), with permission of John Wiley & Sons.

Figure 29.13 Transmission electron micrograph of a human fungiform taste bud during development (week 15). The taste pore (TP) is already open, and some elongated cells stick into the taste pit. However, differences between cell types in the apical portion of the taste bud cannot be made yet. There is no mucus in the taste pit. Approximately two-thirds of the taste bud are filled with ramifications of nerve fibers (N). A part of the basal portion is outlined with a rectangle (see Figure 29.14). mC = marginal cell, BL = basal lamina; scale bar: 10 μm. From Witt and Reutter (1996), with permission of Wiley & Sons.

with typical dark granules believed to secrete the mucous material that fills the taste pit (Witt and Reutter, 1996).

29.9 INNERVATION OF THE HUMAN TONGUE AND TASTE BUDS

The tongue is innervated by (1) motor nerve fibers constituting the hypoglossal nerve (CN XII), which supplies the inner (intrinsic) and the hyoidal tongue muscles; (2) somatosensory nerve fibers, composed of divisions of the trigeminal (CN V_3) and glossopharyngeal (CN IX) nerves; (3) autonomic nerve fibers which stem from the intermedio-facial nerve (CN VII), the glossopharyngeal nerve, and the vagus nerve (CN X); and (4) sensory (gustatory) nerve fibers that transmit taste information centrally from taste buds, namely (a) the chorda tympani, a branch of the intermediate nerve (generally considered a part of the facial nerve, CN VII), (b) the greater (=superficial) petrosal nerve, also part of the intermedio-facial complex, (c) glossopharyngeal nerve (CN IX), and (d) vagus nerve (CN X) (summarized in Figure 29.15).

Classical research papers of the nineteenth century explored the clinical consequences in taste perception associated with diagnostic features of the cranial nerves. These papers have formed the bases for the neurological examination. Two prominent themes were explored: (1) Which cranial nerves are associated with functional attributes of the taste system? (2) How is taste perception affected by neurological diseases? In physiological experiments, Magendie (1820) dissected the lingual nerve of living, unanaesthetized animals and observed loss of taste, but intact movement and sensation of palate, gingival, and buccal mucosa. Panizza (1834) performed dissection of the glossopharyngeal nerve resulting in loss of taste. Alcock (1836) described the role of the chorda tympani and the sphenopalatine (pterygopalatine) ganglion. Lussana (1869, 1872) traced the target tissue of the chorda tympani nerve to the anterior two-thirds of the tongue. The dependence of taste buds on nerve supply was experimentally shown by von Vintschgau and Hönigschmied (1877): 40 days after dissection of the glossopharyngeal nerve, the number of taste buds dramatically decreased. In similar experiments, Ranvier (1888) observed that taste bud sensory cells degenerate, and supporting cells pushed through the pore to the tongue surface. During this process, he observed *cellules migratrices* [phagocytic fibroblasts? (Suzuki et al., 1996)] loaded with fed particles, which were "*probably responsible for removal of old material.*" Soon it became evident that experiments based on vivisection were not

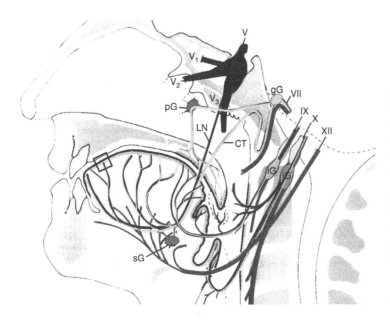

Figure 29.15 Innervation of the human tongue and the taste bud-bearing epithelia (hatched regions), compiled from (Feneis, 1985) and (Sobotta, 1993). The cranial nerves VII (which includes the intermediate nerve with its branches, greater petrosal nerve and chorda tympani), IX and X contain sensory gustatory fibers (yellow). V, trigeminal nerve (blue) with its divisions V_1, ophthalmic, V_2, maxillary, and V_3, mandibular nerves. XII, hypoglossal nerve (red), is the motor nerve of the intrinsic tongue muscles. gG, geniculate ganglion; pG, pterygopalatine (sphenopalatine) ganglion; iG inferior ganglion of the glossopharyngeal (IX) and vagal (X) nerves, respectively; sG, submandibular ganglion with postganglionic autonomic nerve fibers of the chorda tympani to supply the submandibular and sublingual glands. (*See plate section for color version.*)

only painful for the animal (mostly dogs or cats), but also unreliable in their results (Alcock, 1836; Wagner, 1837). Nevertheless, the overall conclusions derived from these nineteenth century nerve dissection experiments (reviewed by Jägel, 1991; Parker, 1922) cannot deny their import for current knowledge of cranial nerve supply and taste sensitivity.

An elegant review of the background of peripheral taste pathways in humans was written by Lewis and Dandy (1930). They examined both the neurological and neuroanatomical literature on gustatory pathways. The sensory distribution of the facial nerve and its clinical importance was described by Hunt (1915). He disentangled the overlapping sensory fields of the facial nerve (including the chorda tympani) from the trigeminal nerve by documenting the distribution of herpes zoster inflammation. The herpetic eruptions outlined the sensory fields of the geniculate ganglion on the tongue, soft palate, and ear. Another basis for evaluating the involvement of the chorda tympani nerve with lingual taste buds came from patients who had undergone middle ear surgery (Borg et al., 1967; Bull, 1965). Contemporary reviews of human (Norgren, 1990; Prichard, 2011) and primate (Pritchard, 1991) taste pathways have incorporated observations from the second half of the twentieth century, particularly those derived from electrophysiological studies.

1. *Chorda tympani and greater petrosal nerve*: Distal to the intermediate nerve branch of the facial nerve, peripheral axons of some geniculate ganglion somata (the chorda tympani nerve) take a recurrent course within the facial canal in the petrosal part of the temporal bone, pass through the middle ear, and exit the skull via the petrotympanic fissure to join the lingual division of the trigeminal nerve, the lingual nerve. Both intermedio-facial (gustatory) and trigeminal (somatosensory) fibers run in the lingual nerve and distribute to the fungiform papillae on the anterior two thirds of the tongue and may reach also the anterior portion of the foliate papillae. Taste buds on the soft palate are innervated by the greater petrosal branch of the intermedio-facial nerve, whose somata also lie within the geniculate ganglion (Harris, 1952; Miller and Spangler, 1982). Some chorda tympani fibers are reported to anastomose with the greater petrosal nerve via the otic ganglion (Pritchard, 1991; Schwartz and Weddell, 1938). Both the greater petrosal and chorda tympani nerves also carry parasympathetic fibers to their associated salivary glands: The greater petrosal nerve serves the palatine and the lacrimal glands, while the chorda tympani innervates the submandibular and sublingual glands via the submandibular ganglion.

2. *Glossopharyngeal nerve*: Axons of the glossopharyngeal nerve originate from ganglion cells mainly in the inferior (petrosal) glossopharyngeal ganglion. These peripheral axons supply both taste buds and general sensory innervation to the vallate and foliate papillae. Salivary glands (Ebner) are supplied by parasympathetic fibers via an intrinsic ganglion (Remak, 1852). Probably, the glossopharyngeal nerve also supplies taste buds in the pharynx. Bourgery and Jacob (1839) and Doty et al. (2009) observed that some CN IX fibers extend into the middle part of the tongue to overlap with innervation areas of the chorda tympani. However, it is unclear if they innervate taste buds.

3. *Vagus nerve*: Taste buds on the laryngeal surface of the epiglottis, larynx, and proximal portion of the esophagus are innervated by the superior laryngeal

branch of the vagus nerve, which has the perikarya of its chemosensory neurons in the inferior (nodose) vagal ganglion.

4. *Trigeminal nerve*: The possible role of trigeminal nerve fibers in taste perception has been discussed for two millenia, and there is no end in sight. Via the lingual nerve, this nerve conveys somatosensory and viscerosensory information from the tongue to trigeminal brain stem nuclei. In fact, most of the nerve fibers entering the fungiform papillae are trigeminal, while a few fibers originate from the chorda tympani (25% in rat: Farbman and Hellekant, 1978).

Trigeminal fibers may respond to sapid stimuli, as revealed by trigeminal transection experiments (Berridge and Fentress, 1985), electrophysiology (Harada and Smith, 1992), and trigeminal ganglion cell response (Liu and Simon, 1998). Taste receptors also occur in a variety of nongustatory cells proper, outside the oral cavity (Tizzano et al., 2011). Finally, taste qualities may be influenced by nonsapid stimuli, for example, temperature: approximately one-half of the nerve fibers involved in taste transduction respond to temperature (Cruz and Green, 2000). The interaction of both gustatory and somatosensory qualities may be as tight-knit as their anatomical proximity. Katz et al. (2000) suggest that gustation should be thought of as an integral part of a distributed, interacting multimodal system.

In contrast to most other sensory systems, gustatory function is distributed among three different cranial nerves, which makes taste difficult to eliminate and, secondly, also difficult to distinguish from trigeminal perception (Prichard, 2011).

The observation that taste buds degenerate after dissection of their sensory innervation and, subsequently, reappear after regeneration of their peripheral nerves has been a major focus of research in the peripheral taste system. Nineteenth- and early twentieth-century literature on taste bud degeneration, regeneration, and development was reviewed comprehensively by Parker (1922). Olmsted (1920) proposed that trophic maintenance of fish taste buds depended on the transmission of a putative trophic material from nerve to epithelium. Cross reinnervation of the glossopharyngeal nerve to fungiform taste buds (which are normally supplied by the chorda tympani) had no effect on the usual immunohistochemical properties of fungiform versus vallate taste buds (Smith et al., 1999). As a consequence, these authors believe that the protein expression in and subsequent function of taste buds depend on the epithelium from which the cells arise, and not the buds' specific nerve supply. Meanwhile, Nosrat and Olson (1995) and Nosrat et al. (2000) detected mRNA of brain-derived neurotrophic factor (BDNF) and neurotrophins in developing anterior tongue epithelium, before nerve fibers were observed. In BDNF-overexpressing mice, larger taste buds and more taste cells have been observed (Nosrat et al., 2012). This argues for the hypothesis that trophic factors act as target-derived chemoattractants for the early nerve fibers. These, in turn, initiate the formation of taste buds. BDNF-null mutant mice fail to develop taste buds (Oakley et al., 1998). Sensory ganglia involved in taste bud innervation (see above) are reduced by 40% in volume compared to about 20% of trigeminal ganglion under the same condition (Mistretta et al., 1999). Taste buds do not develop after injection of ß-bungarotoxin into the amniotic fluid in fetal mice. This neurotoxin abolishes motor and sensory nerve development (Morris-Wiman et al., 1999).

Although nerve fibers are required to maintain taste buds once the latter are formed and start to function (e.g., Hosley et al., 1987; Yee et al., 2005), nerves seem not to be necessary to initiate taste bud development. Initial taste bud development is nerve-independent, suggested by a series of studies in salamanders (Stone, 1940), axolotl (Barlow et al., 1996; Barlow and Northcutt, 1998a,b) and mouse (Mbiene and Roberts, 2003). Taste buds seem to develop from local epithelium and not from neurogenic ectoderm [axolotl: Barlow and Northcutt (1997), mouse: Stone et al. (1995)]. It may be that mechanisms of differentiation of the same receptor organ vary among vertebrate taxa (Barlow, 2003). Growth factors other than BDNF may contribute to the maintenance of gustatory papillae, for example, epidermal growth factor (EGF) supplied by salivary glands (Liu et al., 2008; Morris-Wiman et al., 2000). More detailed studies on developmental aspects of the peripheral gustatory system including whether taste buds may develop without the stimulation of nerves are described later in this book (Chapter 33).

ACKNOWLEDGEMENTS

The authors are indebted to Dr. Inglis Miller, Jr. who wrote a previous version of this chapter in the first edition, and Mihnea Nicolescu, who provided the schematic drawings. Drs. Judith and Donald Ganchrow helped with critical reading of an earlier version of the manuscript.

REFERENCES

Albinus (1754). *Academicarum annotationum libri I-VIII*. liber I, Tab.I, Lugdunum Bataviensis (Leiden).

Alcock, B. (1836). Determination of the question, which are the nerves of taste. *Dublin. J. Med. Chem. Sci.* 10: 256–279.

Arey, L., Tremaine, M., and Monzingo, F. (1935). The numerical and topographical relation of taste buds to human circumvallate papillae throughout the life span. *Anat. Rec.* 64: 9–25.

Arnold, F. (1839). *Tabulae Anatomicae. Icones organum sensuum. Organon gustus*. Fasc. secundus. Orellii, Fuesslini, Zürich.

Arvidson, K., and Friberg, U. (1980). Human taste: response and taste bud number in fungiform papillae. *Science* 209: 807–808.

Astbäck, J., Arvidson, K., and Johansson, O. (1997). An immunohistochemical screening of neurochemical markers in fungiform papillae and taste buds of the anterior rat tongue. *Arch. Oral. Biol.* 42(2): 137–147.

Atema, J. (1971). Structures and functions of the sense of taste in the catfish (Ictalurus natalis). *Brain. Behav. Evol.* 4(4): 273–294.

Azen, E. A., Hellekant, G., Sabatini, L. M., and Warner, T. F. (1990). mRNAs for PRPs, statherin, and histatins in von Ebner's gland tissues. *J. Dent. Res.* 69: 1724–1730.

Azzali, G., Gennari, P. U., Maffei, G., and Ferri, T. (1996). Vallate, foliate and fungiform human papillae gustatory cells. An immunocytochemical and ultrastructural study. *Minerva Stomatol.* 45(9): 363–379.

Azzali, G. (1997). Ultrastructure and immunocytochemistry of gustatory cells in man. *Anat. Anz.* 179(1): 37–44.

Barlow, L. A. (2003). Toward a unified model of vertebrate taste bud development. *J. Comp. Neurol.* 457(2): 107–110.

Barlow, L. A., Chien, C. B., and Northcutt, R. G. (1996). Embryonic taste buds develop in the absence of innervation. *Development* 122: 1103–1111.

Barlow, L. A., and Northcutt, R. G. (1997). Taste buds develop autonomously from endoderm without induction by cephalic neural crest or paraxial mesoderm. *Development* 124(5): 949–957.

Barlow, L. A., and Northcutt, R. G. (1998a). The role of innervation in the development of taste buds: insights from studies of amphibian embryos. *Ann. N. Y. Acad. Sci.* 855: 58–69.

Barlow, L. A., and Northcutt, R. G. (1998b). Vertebrate taste-bud development: are salamanders the model? Reply. *Trends Neurosci.* 21(8): 338–338.

Barreiro-Iglesias, A., Anadon, R., and Rodicio, M. C. (2010). The gustatory system of lampreys. *Brain Behav. Evol.* 75(4): 241–250.

Bartel, D. L., Sullivan, S. L., Lavoie, E. G., et al.(2006). Nucleoside triphosphate diphosphohydrolase-2 is the ecto-ATPase of type I cells in taste buds. *J. Comp. Neurol.* 497(1): 1–12.

Behrens, M., Born, S., Redel, U., et al. (2012). Immunohistochemical Detection of TAS2R38 Protein in Human Taste Cells. *PLoS One* 7(7): e40304.

Beidler, L. M., and Smallman, R. L. (1965). Renewal of cells within taste buds. *J. Cell Biol.* 27: 263–272.

Bellini, L. (1665). *Gustus organum novissime deprehensum praemissis ad faciliorem intelligentiam quibusdam de saporibus (Taste organs newly observed; with certain premises about the senses given for easier understanding (translation from Latin into German by Jurisch (1922)).* Mangetus Bibliotheca anat.2, Bologna.

Berkhoudt, H. (1985). Special sense organs: Structure and function of avian taste receptors. In *Form and function in birds*, King, A. S. and McIelland, J. (Eds). Vol. 3, Academic Press, New York, pp. 463–496.

Berridge, K. C., and Fentress, J. C. (1985). Trigeminal-taste interaction in palatability processing. *Science* 228(4700): 747–750.

Borg, G., Diamant, H., Oakley, B., et al. (1967). A comparative study of neural and psychophysical responses to gustatory stimuli. In *Olfaction and Taste 2*, Hayashi, T. (Ed). Pergamon Press, Oxford, pp. 253–265.

Boudriot, F., and Reutter, K. (2001). Ultrastructure of the taste buds in the blind cave fish Astyanax jordani ("Anoptichthys") and the sighted river fish Astyanax mexicanus (Teleostei, Characidae). *J. Comp. Neurol.* 434(4): 428–444.

Boughter Jr.,, J. D., Pumplin, D. W., Yu, C., et al. (1997). Differential Expression of alpha—Gustducin in Taste Bud Populations of the Rat and Hamster. *J. Neurosci.* 17(8): 2852–2858.

Bourgery, J. M., and Jacob, N. H. (1839). Organes de la digestion, de la dépuration urinaire et de la génération. Embryotomie. In *Traité complet de l'anatomie de l'homme*, Delaunay, C. A. (Ed). Vol. 5, Jules Didot L'Ainé, Paris, pp.

Bradley, R. M. (1972). Development of the taste bud and gustatory papillae in human fetuses. In *Third symposium on oral sensation and perception*, Bosma, J. F. (Ed). Charles C Thomas, Springfield, Ill., USA, pp. 137–162.

Bradley, R. M. (1973). Electrophysiological investigations of intravascular taste using perfused rat tongue. *Am. J. Physiol.* 224(2): 300–304.

Bradley, R. M. (2000). Sensory receptors of the larynx. *Am. J. Med.* 108 Suppl 4a: 47S–50S.

Bradley, R. M., Stedman, H. M., and Mistretta, C. M. (1983). Superior laryngeal nerve response patterns to chemical stimulation of sheep epiglottis. *Brain Res.* 276(1): 81–93.

Bradley, R. M., and Stern, I. B. (1967). The development of the human taste bud during the foetal period. *J. Anat.* 101: 743–752.

Braus, H. (1940). *Handbuch der Mikroskopischen Anatomie des Menschen.* Springer, Berlin.

Bull, T. R. (1965). Taste and the chorda tympani. *J. Laryngol. Otol.* 79: 479–493.

Burkl, W. (1954). Über das Vorkommen von Geschmacksknospen im mittleren Drittel des Oesophagus. *Anat. Anz.* 100: 320–321.

Caicedo, A., Kim, K. N., and Roper, S. D. (2000). Glutamate-induced cobalt uptake reveals non-NMDA receptors in rat taste cells. *J. Comp. Neurol.* 417(3): 315–324.

Cao, Y., Zhao, F. L., Kolli, T., et al. (2009). GABA expression in the mammalian taste bud functions as a route of inhibitory cell-to-cell communication. *Proc. Natl. Acad. Sci. U.S.A.* 106(10): 4006–4011.

Carus, C. G. (1849). *System der Physiologie.* 2nd ed. Brockhaus, Leipzig.

Casserius (1609). *Penthaesteseion (cited after Jurisch, 1922).*

Chaudhari, N., and Roper, S. D. (2010). The cell biology of taste. *J. Cell Biol.* 190(3): 285–296.

Cheng, L. H., and Robinson, P. P. (1991). The distribution of fungiform papillae and taste buds on the human tongue. *Arch. Oral Biol.* 36: 583–589.

Cheng, S. J., Huang, C. F., Chen, Y. C., et al. (2009). Ultrastructural changes of posterior lingual glands after hypoglossal denervation in hamsters. *J. Anat.* 214(1): 163–170.

Clapp, T. R., Stone, L. M., Margolskee, R. F., and Kinnamon, S. C. (2001). Immunocytochemical evidence for co-expression of Type III IP3 receptor with signaling components of bitter taste transduction. *BMC Neurosci.* 2: 6.

Clapp, T. R., Yang, R., Stoick, C. L., et al. (2004). Morphologic characterization of rat taste receptor cells that express components of the phospholipase C signaling pathway. *J. Comp. Neurol.* 468(3): 311–321.

Connes, R., Granie-Prie, M., Diaz, J. P., and Paris, J. (1988). Ultrastructure des bourgeons du gout du téléostéen marin Dicentrarchus labrax L. *Can. J. Zool.* 66: 2133–2142.

Cottler Fox, M., Arvidson, K., Hammarlund, E., and Friberg, U. (1987). Fixation and occurrence of dark and light cells in taste buds of fungiform papillae. *Scand. J. Dent. Res.* 95: 417–427.

Cruz, A., and Green, B. G. (2000). Thermal stimulation of taste. *Nature* 403(6772): 889–892.

Cummings, T. A., Delay, R. J., and Roper, S. D. (1987). Ultrastructure of apical specializations of taste cells in the mudpuppy, Necturus maculosus. *J. Comp. Neurol.* 261(4): 604–615.

Dando, R., and Roper, S. D. (2009). Cell-to-cell communication in intact taste buds through ATP signalling from pannexin 1 gap junction hemichannels. *J. Physiol.* 587(Pt 24): 5899–5906.

Davis, C. (1877). Die becherförmigen Organe des Kehlkopfs. *Arch. mikr. Anat.* 14: 158–167.

DeFazio, R. A., Dvoryanchikov, G., Maruyama, Y., et al. (2006). Separate populations of receptor cells and presynaptic cells in mouse taste buds. *J. Neurosci.* 26(15): 3971–3980.

Delay, R. J., Kinnamon, S. C., and Roper, S. D. (1997). Serotonin modulates voltage-dependent calcium current in Necturus taste cells. *J. Neurophysiol.* 77(5): 2515–2524.

Delay, R. J., Mackay Sim, A., and Roper, S. D. (1994). Membrane properties of two types of basal cells in Necturus taste buds. *J. Neurosci.* 14: 6132–6143.

Delay, R. J., and Roper, S. D. (1988). Ultrastructure of taste cells and synapses in the mudpuppy Necturus maculosus. *J. Comp. Neurol.* 277: 268–280.

Delay, R. J., Taylor, R., and Roper, S. D. (1993). Merkel-like basal cells in Necturus taste buds contain serotonin. *J. Comp. Neurol.* 335(4): 606–613.

Desgranges, J. C. (1965). Sur l'existence de plusieurs types de cellules sensorielles dans les bourgeons du gout des barbillons du Poisson-chat. *C.R. Acad. Sc.(Paris)* 261: 1095–1098.

Doty, R. L., Cummins, D. M., Shibanova, A., et al. (2009). Lingual distribution of the human glossopharyngeal nerve. *Acta Otolaryngol.* 129(1): 52–56.

Dvoryanchikov, G., Sinclair, M. S., Perea-Martinez, I., et al. (2009). Inward rectifier channel, ROMK, is localized to the apical tips of glial-like cells in mouse taste buds. *J. Comp. Neurol.* 517(1): 1–14.

Ebner, V. (1873). *Die acinösen Drüsen der Zunge und ihre Beziehungen zu den Geschmacksorganen.* Leuschner & Lubensky, Graz.

Ehrenberg, C. G. (1833). Notwendigkeit einer feineren mechanischen Zerlegung des Gehirns und der Nerven vor der chemischen, dargestellt an Beobachtungen von C.G.Ehrenberg. *Poggendorfs Annl. Physik. Chemie.* XXVIII: 450.

Ehrlich, P. (1886). Über die Methylenblaureaktion der lebenden Nervensubstanz. *Deutsche Med. Wochenschr.* 12: 49–52.

Engelmann, T. W. (1872). Die Geschmacksorgane. In *Handbuch der Lehre von den Geweben des Menschen und der Thiere*, Stricker, S. (Ed). Vol. 2, Engelmann, Leipzig, pp. 822–838.

Engström, H., and Rytzner, C. (1956). *The fine structure of taste buds and taste fibers.* 361–375 65.

Fährmann, W. (1967). [Light and electron microscopical studies on the taste bud of neotene axolotls (Siredon mexicanum Shaw)]. *Z. Mikrosk. Anat. Forsch.* 77: 117–152.

Farbman, A. I. (1965). Electron microscope study of the developing taste bud in rat fungiform papilla. *Dev. Biol.* 11: 110–135.

Farbman, A. I. (1980). Renewal of taste bud cells in rat circumvallate papillae. *Cell Tissue Kinet.* 13: 349–357.

Farbman, A. I., and Hellekant, G. (1978). Quantitative analyses of the fiber population in rat chorda tympani nerves and fungiform papillae. *Am. J. Anat.* 153: 509–521.

Farbman, A. I., Hellekant, G., and Nelson, A. (1985). Structure of taste buds in foliate papillae of the rhesus monkey, Macaca mulatta. *Am. J. Anat.* 172(1): 41–56.

Farbman, A. I., Ogden Ogle, C. K., Hellekant, G., et al. (1987). Labeling of sweet taste binding sites using a colloidal gold- labeled sweet protein, thaumatin. *Scanning. Microsc.* 1: 351–357.

Farbman, A. I., and Yonkers, J. D. (1971). Fine structure of the taste bud in the mud puppy, Necturus maculosus. *Am. J. Anat.* 131: 353–369.

Feneis, H. (1985). *Pocket Atlas of human anatomy.* Thieme, New York.

Ferrando, S., Gallus, L., Gambardella, C., et al. (2012). First detection of taste buds in a chimaeroid fish (Chondrichthyes: Holocephali) and their Galphai-like immunoreactivity. *Neurosci. Lett.* 517(2): 98–101.

Finger, T. E. (1997). Evolution of taste and solitary chemoreceptor cell systems. *Brain Behav. Evol.* 50(4): 234–243.

Finger, T. E., Bottger, B., Hansen, A., et al. (2003). Solitary chemoreceptor cells in the nasal cavity serve as sentinels of respiration. *Proc. Natl. Acad. Sci. U.S.A.* 100(15): 8981–8986.

Finger, T. E., Bryant, B. P., Kalinoski, D. L., et al. (1996). Differential localization of putative amino acid receptors in taste buds of the channel catfish, Ictalurus punctatus. *J. Comp. Neurol.* 373(1): 129–138.

Finger, T. E., Danilova, V., Barrows, J., et al. (2005). ATP signaling is crucial for communication from taste buds to gustatory nerves. *Science* 310(5753): 1495–1499.

Fixsen, C. (1857). *De linguae raninae textura disquisitiones microscopicae. Diss. inaug.* Dorpati Livonor.

Froriep, R. (1828). De lingua anatomica quaedam et semiotica.

Fujimoto, S., Ueda, H., and Kagawa, H. (1987). Immunocytochemistry on the localization of 5-hydroxytryptamine in monkey and rabbit taste buds. *Acta Anat.* 128: 80–83.

Fukami, H., and Bradley, R. M. (2005). Biophysical and morphological properties of parasympathetic neurons controlling the parotid and von Ebner salivary glands in rats. *J. Neurophysiol.* 93(2): 678–686.

Ganchrow, D., Ganchrow, J. R., and Goldstein, R. S. (1991). Ultrastructure of palatal taste buds in the perihatching chick. *Am. J. Anat.* 192(1): 69–78.

Ganchrow, J. R., and Ganchrow, D. (1987). Taste bud development in chickens (Gallus gallus domesticus). *Anat. Rec.* 218: 88–93.

Ganchrow, J. R., Ganchrow, D., and Oppenheimer, M. (1986). Chorda tympani innervation of anterior mandibular taste buds in the chicken (Gallus gallus domesticus). *Anat. Rec.* 216(3): 434–439.

Ganchrow, J. R., Ganchrow, D., Royer, S. M., and Kinnamon, J. C. (1993). Aspects of vertebrate gustatory phylogeny: morphology and turnover of chick taste bud cells. *Microsc. Res. Tech.* 26: 106–119.

Gerlach, J. (1858). *Mikroskopische Studien aus dem Gebiet der menschlichen Morphologie.* F. Enke, Erlangen.

Getchell, T. V., Margolis, F. L., and Getchell, M. L. (1984). Perireceptor and receptor events in vertebrate olfaction. *Prog. Neurobiol.* 23: 317–345.

Golgi, C. (1873). Sulla struttura della sostanza grigia del cervello (comunicazione preventiva). *Gazzetta Medica Italiana-Lombardia* 33: 244 –246.

Gråberg, J. (1898). Beiträge zur Genese des Geschmacksorgans des Menschen. *Morphol. Arb.* 8: 117–134.

Graziadei, P. P., and DeHan, R. S. (1971). The ultrastructure of frogs' taste organs. *Acta Anat. (Basel)* 80(4): 563–603.

Grover Johnson, N., and Farbman, A. I. (1976). Fine structure of taste buds in the barbel of the catfish, Ictalurus punctatus. *Cell Tissue Res.* 169: 395–403.

Gurkan, S., and Bradley, R. M. (1987). Autonomic control of von Ebner's lingual salivary glands and implications for taste sensation. *Brain Res.* 419: 287–293.

Habermehl, K. H. (1952). Über besondere Randpapillen an der Zunge neugeborener Säugetiere. *Z. Anat. Entwickl. Gesch.* 116: 355–372.

Haller, A. v. (1766). Gustus. In *Elementa physiologiae* Vol. IV, V, F.Grasset, Lausanne, pp. 99–124.

Hamasaki, K., Seta, Y., Yamada, K., and Toyoshima, K. (1998). Possible role of serotonin in Merkel-like basal cells of the taste buds of the frog, Rana nigromaculata. *J. Anat.* 193(4): 599–610.

Hannover, A. (1840). Die Chromsäure, ein vorzügliches Mittel bei mikroskopischen Untersuchungen. *Müllers Archiv:* 549–558.

Harada, S., and Smith, D. V. (1992). Gustatory sensitivities of the hamster's soft palate. *Chem. Senses* 17: 37–51.

Harris, W. (1952). The fifth and seventh nerves in relation to the nervous mechanism of taste sensation: a new approach. *Brit. Med. J.* 1: 831–836.

Hellekant, G. (1976). The blood circulation of the tongue. In *Frontiers of Oral Physiology*, Kawamura, Y. (Ed) S. Karger, Basel, pp. 130–145.

Hellman, T. J. (1922). Die Genese der Zungenpapillen beim Menschen. *Ups. Läkaref. Förh.* 26,5–6: 1–69.

Helmholtz (1842). *De fabrica systematis nervosi evertebratorum.* Dissertation, University of Berlin.

Hermann, F. (1885). Beittrag zur Entwicklungsgeschichte des Geschmacksorgans beim Kaninchen. *Arch. f. Mikroskop. Anat. Entwicklungsmech.* 24: 216–229.

Herness, M. S. (1989). Vasoactive intestinal peptide-like immunoreactivity in rodent taste cells. *Neuroscience* 33(2): 411–419.

Herness, S., Zhao, F. L., Lu, S. G., et al. (2002). Expression and physiological actions of cholecystokinin in rat taste receptor cells. *J. Neurosci.* 22(22): 10018–10029.

Herrick, C. J. (1904). The organ and sense of taste in fishes. *U.S. Fish Comm. Bull. 1902 (abstracted in J. Comp. Neurol. Psych. 277–278, 1904)*: 237–272.

Hersch, M., and Ganchrow, D. (1980). Scanning electron microscopy on human embryonic and fetal tongue. *Chem. Senses* 5: 331–341.

Hinrichsen, K. V. (1990). *Humanembryologie. Lehrbuch und Atlas der vorgeburtlichen Entwicklung des Menschen.* (Hinrichesen, K. V., ed.) Berlin, Heidelberg, New York, Springer Verlag.

Hirata, Y. (1966). Fine structure of the terminal buds on the barbels of some fishes. *Arch. Histol. Jpn.* 26(5): 507–523.

Höfer, D., and Drenckhahn, D. (1999). Localisation of actin, villin, fimbrin, ezrin and ankyrin in rat taste receptor cells. *Histochem. Cell Biol.* 112(1): 79–86.

Hofer, H., Meinel, W., and Rommel, C. (1979). Taste buds in the epithelium of the plica sublingualis of New World monkeys. *Anat. Anz.* 145(1): 17–31.

Hoffmann, A. (1875). Ueber die Verbreitung der Geschmacksknospen beim Menschen. *Arch. pathol. Anat. Physiol. klin. Med.* 62: 516–530.

Hokfelt, T., Lundberg, J. M., Schultzberg, M., et al. (1980). Cellular localization of peptides in neural structures. *Proc. R. Soc. Lond B. Biol. Sci.* 210(1178): 63–77.

Hönigschmied, J. (1873). Beiträge zur mikroskopischen Anatomie über die Geschmacksorgane der Säugethiere. *Zeitschr. Wiss. Zoologie* 23: 414–434.

Hoon, M. A., Adler, E., Lindemeier, J., et al. (1999). Putative mammalian taste receptors: a class of taste-specific GPCRs with distinct topographic selectivity. *Cell* 96(4): 541–551.

Hoon, M. A., Northup, J. K., Margolskee, R. F., and Ryba, N. J. (1995). Functional expression of the taste specific G-protein, alpha-gustducin. *Biochem. J.* 309(Pt 2): 629–636.

Hosley, M. A., Hughes, S. E., and Oakley, B. (1987). Neural induction of taste buds. *J. Comp. Neurol.* 260: 224–232.

Hou-Jensen, H. (1933). Die Papillae foliatae des Menschen. *Z. Anat. Entwicklgesch.* 102: 348–388.

Hu, Z. L., Masuko, S., and Katsuki, T. (1996). Distribution and origins of nitric oxide-producing nerve fibers in the dog tongue: correlated NADPH-diaphorase histochemistry and immunohistochemistry for calcitonin gene-related peptide using light and electron microscopy. *Arch. Histol. Cytol.* 59(5): 491–503.

Huang, Y.-J., Maruyama, Y., Lu, K.-S., et al. (2005). Mouse Taste Buds Use Serotonin as a Neurotransmitter. *J. Neurosci.* 25(4): 843–847.

Huang, Y. A., Dando, R., and Roper, S. D. (2009). Autocrine and paracrine roles for ATP and serotonin in mouse taste buds. *J. Neurosci.* 29(44): 13909–13918.

Huang, Y. A., Maruyama, Y., Stimac, R., and Roper, S. D. (2008). Presynaptic (Type III) cells in mouse taste buds sense sour (acid) taste. *J. Physiol.* 586(12): 2903–2912.

Hunt, J. R. (1915). The sensory field of the facial nerve: a further contribution to the symptomatology of the geniculate ganglion. *Brain* 38: 418–446.

Iida, M., Yoshioka, I., and Muto, H. (1983). Taste bud papillae on the retromolar mucosa of the rat, mouse and golden hamster. *Acta Anat. Basel.* 117: 374–381.

Ito, A., and Nosrat, C. A. (2009). Gustatory papillae and taste bud development and maintenance in the absence of TrkB ligands BDNF and NT-4. *Cell Tissue Res.* 337(3): 349–359.

Iwasaki, S., Yoshizawa, H., and Kawahara, I. (1996). Three-dimensional ultrastructure of the surface of the tongue of the rat snake, Elaphe climacophora. *Anat. Rec.* 245(1): 9–12.

Jägel. (1991). Zur Geschichte der Anatomie und Physiologie des Geschmackssinnes. *Diss. Univ. Kiel*: 1–118.

Jaeger, C.B., Hillmann, D.E. (1976). Gustatory system. Morphology of gustatory organs. In Frog neurobiology A Handbook, R. Llinás, and W. Precht, eds. (Berlin, Heidelberg, New York: Springer), pp. 588–606.

Jain, S., and Roper, S. D. (1991). Immunocytochemistry of gamma-aminobutyric acid, glutamate, serotonin, and histamine in Necturus taste buds. *J. Comp. Neurol.* 307(4): 675–682.

Jakubowski, and Whitear, M. (1986). Ultrastructure of taste buds in fishes. *Folia Histochem. Cytobiol.* 24: 310–311.

Jowett, A., and Shrestha, R. (1998). Mucosa and taste buds of the human epiglottis. *J. Anat.* 193(4): 617–618.

Jeffery, W. R., Strickler, A. G., Guiney, S., et al. (2000). Prox 1 in eye degeneration and sensory organ compensation during development and evolution of the cavefish Astyanax. *Dev. Genes. Evol.* 210(5): 223–230.

Jurisch, A. (1922). Studien über die Papillae vallatae beim Menschen. *Z. Anat. Entwicklungsgesch.* 66: 1–149.

Just, T., Pau, H. W., Bombor, I., et al. (2005). Confocal microscopy of the Peripheral Gustatory System: Comparison between Healthy Subjects and Patients Suffering from Taste Disorders during Radiochemotherapy. *Laryngoscope* 115(12): 2178–2182.

Kaya, N., Shen, T., Lu, S.-g., et al. (2004). A paracrine signaling role for serotonin in rat taste buds: expression and localization of serotonin receptor subtypes. *American Journal of Physiology-Regulatory, Integrative and Comparative Physiology* 286(4): R649–R658.

Katz, D. B., Nicolelis, M. A., and Simon, S. A. (2000). IV. There is more to taste than meets the tongue. *Am. J. Physiol. Gastrointest Liver Physiol.* 278(1): G6–G9.

Kim, D. J., and Roper, S. D. (1995). Localization of serotonin in taste buds: a comparative study in four vertebrates. *J. Comp. Neurol.* 353: 364–370.

Kim, M. R., Kusakabe, Y., Miura, H., et al. (2003). Regional expression patterns of taste receptors and gustducin in the mouse tongue. *Biochem. Biophys. Res. Commun.* 312(2): 500–506.

Kinnamon, J. C., Henzler, D. M., and Royer, S. M. (1993). HVEM ultrastructural analysis of mouse fungiform taste buds, cell types, and associated synapses. *Microsc. Res. Tech.* 26: 142–156.

Kinnamon, J. C., McPheeters, M. M., and Kinnamon, S. C. (1994). Structure/Function correlates in taste buds. *In: Olfaction and Taste (K.Kurihara, N Suzuki, H Ogawa, eds), Springer-Verlag*, Tokyo, Berlin, Heidelberg, New York, London, Paris XI: 9–12.

Kinnamon, J. C., Sherman, T. A., and Roper, S. D. (1988). Ultrastructure of mouse vallate taste buds: III. Patterns of synaptic connectivity. *J. Comp. Neurol.* 270: 1–10, 56.

Kinnamon, J. C., Taylor, B. J., Delay, R. J., and Roper, S. D. (1985). Ultrastructure of mouse vallate taste buds. I. Taste cells and their associated synapses. *J. Comp. Neurol.* 235: 48–60.

Kinnamon, S. C. (2011). Taste receptor signalling—from tongues to lungs. *Acta Physiol. (Oxf)* 204(2): 158–168.

References

Kishi, M., Emori, Y., Tsukamoto, Y., and Abe, K. (2001). Primary culture of rat taste bud cells that retain molecular markers for taste buds and permit functional expression of foreign genes. *Neuroscience* 106(1): 217–225.

Kishi, M., Emori, Y., Tsukamoto, Y., and Abe, K. (2002). Changes in cell morphology and cell-to-cell adhesion induced by extracellular Ca2+ in cultured taste bud cells. *Biosci. Biotechnol. Biochem.* 66(2): 484–487.

Koelliker, R. A. (1844). *Die Selbständigkeit und Abhängigkeit des sympathischen Nervensystems durch anatomische Beobachtungen bewiesen. Ein akademisches Programm.* U.Zeller, Zürich.

Korte, G. E. (1979). Unusual association of 'chloride cells' with another cell type in the skin of the glass catfish, Kryptopterus bicirrhis. *Tissue Cell* 11(1): 63–68.

Kotrschal, K. (1996). Solitary chemosensory cells: Why do primary aquatic vertebrates need another taste system. *Trends Ecol.Evol.* 11: 110–114.

Krause, R. (1911). *Kursus der normalen Histologie. (American edition: Rebman Co., N.Y.) ed.* Urban und Schwarzenberg, Berlin and Wien.

Kudo, K., Wakamatsu, K., Nishimura, S., and Tabata, S. (2010). Gustducin is expressed in the taste buds of the chicken. *Anim. Sci. J.* 81(6): 666–672.

Lalonde, E., and Eglitis, J. (1961). Number and distribution of taste buds on the epiglottis, pharynx, larynx, soft palate and uvula in a human newborn. *Anat.Rec.* 140: 91–95.

Landin, A. M., Kim, J. W., and Chaudhari, N. (2005). Liposome-mediated transfection of mature taste cells. *J. Neurobiol.* 65(1): 12–21.

Lawton, D. M., Furness, D. N., Lindemann, B., and Hackney, C. M. (2000). Localization of the glutamate-aspartate transporter, GLAST, in rat taste buds. *Eur. J. Neurosci.* 12(9): 3163–3171.

Lewis, D., and Dandy, W. E. (1930). The course of the nerve fibers transmitting sensation of taste. *Arch.Surg.* 21: 249–288.

Leydig, F. (1851). Über die Haut einiger Süßwasserfische. *Z. Wiss. Zool.* 3: 1–12.

Li, J. H., and Lindemann, B. (2003). Multi-photon microscopy of cell types in the viable taste disk of the frog. *Cell Tissue Res.* 313(1): 11–27.

Li, X. J., and Snyder, S. H. (1995). Molecular cloning of Ebnerin, a von Ebner's gland protein associated with taste buds. *J. Biol. Chem.* 270(30): 17674–17679.

Lin, W., and Kinnamon, S. C. (1999). Co-localization of epithelial sodium channels and glutamate receptors in single taste cells. *Biol. Signals Recept.* 8(6): 360–365.

Lindemann, B. (1996). Taste reception. *Physiol. Rev.* 76: 719–766.

Linden, R. W. (1993). *Taste. Br. Dent J.* 175(7): 243–253.

Liu, H. X., Henson, B. S., Zhou, Y., et al. (2008). Fungiform papilla pattern: EGF regulates inter-papilla lingual epithelium and decreases papilla number by means of PI3K/Akt, MEK/ERK, and p38 MAPK signaling. *Dev. Dyn.* 237(9): 2378–2393.

Liu, L., and Simon, S. A. (1998). Responses of cultured rat trigeminal ganglion neurons to bitter tastants. *Chem. Senses* 23(2): 125–130.

Lovén, C. (1868). Beiträge zur Kenntnis vom Bau der Geschmackswärzchen der Zunge. *Arch. Mikrosk. Anat. IV:* 96–110.

Lussana, F. (1869). Recherches expérimentales et observations pathologiques sur les nerfs du gout. *Arch. phys. (Paris)* 2: 197–210.

Lussana, F. (1872). Sur les nerfs du gout. Observations et expériences nouvelles. *Arch. phys. (Paris)* 4: 334–350.

Ma, H., Yang, R., Thomas, S. M., and Kinnamon, J. C. (2007). Qualitative and quantitative differences between taste buds of the rat and mouse. *BMC Neurosci.* 8: 5.

Magendie, F. (1820). *Grundriss der Physiologie (translated from French by C.F. Heusinger).* Baercke, Eisenach.

Malpighi, M. (1686). Exercitatio epistolica de lingua. (1664; Jo. Alphonso Borellio). In *Opera omnia*, Malpighi, M. (Ed). R.Scott & G. Wells, Londini, pp. 13–20.

Matsunami, H., Montmayeur, J. P., and Buck, L. B. (2000). A family of candidate taste receptors in human and mouse. *Nature* 404(6778): 601–604.

Mbiene, J. P., and Roberts, J. D. (2003). Distribution of keratin 8-containing cell clusters in mouse embryonic tongue: evidence for a prepattern for taste bud development. *J. Comp. Neurol.* 457(2): 111–122.

McLaughlin, S. K., McKinnon, P. J., and Margolskee, R. F. (1992). Gustducin is a taste-cell-specific G protein closely related to the transducins. *Nature* 357(6379): 563–569.

Menco, B. P. (1989). Olfactory and nasal respiratory epithelia, and foliate taste buds visualized with rapid-freeze freeze-substitution and Lowicryl K11M embedding. Ultrastructural and initial cytochemical studies. *Scanning. Microsc.* 3: 257–272.

Menco, B. P. M., Yankova, M. P., and Simon, S. A. (1997). Freeze- substitution and postembedding immunocytochemistry on rat taste buds: G-proteins, calcitonin gene-related peptide, and choline acetyl transferase. *Microsc. Microanal.* 3: 53–69.

Merkel, F. (1880). *Ueber die Endigungen der sensiblen Nerven in der Haut der Wirbelthiere.* Stiller, Rostock.

Miller, I. J., Jr., and Bartoshuk, L. M. (1991). Taste perception, taste bud distribution, and spatial relationships. 1 ed. In *Smell and taste in health and disease*, Getchell, T. V., Bartoshuk, L. M., Doty, R. L. and Snow, J. B. J., Jr., (Eds). New York, Raven Press, pp. 205–233.

Miller, I. J., Jr., and Reedy, F. E., Jr., (1990a). Variations in human taste bud density and taste intensity perception. *Physiol. Behav.* 47(6): 1213–1219.

Miller, I. J., Jr., and Reedy, F. E. J. (1990b). Quantification of fungiform papillae and taste pores in living human subjects. *Chem. Senses* 15: 281–294.

Miller, I. J., Jr., and Smith, D. V. (1984). Quantitative taste bud distribution in the hamster. *Physiol.Behav.* 32: 275–285.

Miller, I. J., Jr., and Spangler, K. M. (1982). Taste bud distribution and innervation on the palate of the rat. *Chem. Senses* 7: 99–108.

Mistretta, C. M., and Baum, B. J. (1984). Quantitative study of taste buds in fungiform and circumvallate papillae of young and aged rats. *J. Anat.* 138: 323–332.

Mistretta, C. M., Goosens, K. A., Farinas, I., and Reichardt, L. F. (1999). Alterations in size, number, and morphology of gustatory papillae and taste buds in BDNF null mutant mice demonstrate neural dependence of developing taste organs. *J. Comp. Neurol.* 409(1): 13–24.

Miura, H., Scott, J. K., Harada, S., Barlow, L. A. (2014). Sonic hedgehog-expressing basal cells are general post-mitotic precursors of functional taste receptor cells. *Dev. Dyn.* 243: 1286–1297.

Mochizuki, Y. (1939). Studies of the papilla foliata of Japanese. *Okajimas Folia Anat. Jpn.* 18: 334–369.

Moll, R. (1993). Cytokeratins as markers of differentiation: Expression profiles in epithelia and epithelial tumors. In *Progress in Pathology*, Seifert, G. (Ed). Vol. 142, G. Fischer, Stuttgart, Jena, New York, pp. 1–197.

Montavon, P., Hellekant, G., and Farbman, A. (1996). Immunohistochemical, electrophysiological, and electron microscopical study of rat fungiform taste buds after regeneration of chorda tympani through the non-gustatory lingual nerve. *J. Comp. Neurol.* 367: 491–502.

Morris-Wiman, J., Basco, E., and Du, Y. (1999). The effects of beta-bungarotoxin on the morphogenesis of taste papillae and taste buds in the mouse. *Chem. Senses* 24(1): 7–17.

Morris-Wiman, J., Sego, R., Brinkley, L., and Dolce, C. (2000). The effects of sialoadenectomy and exogenous EGF on taste bud morphology and maintenance [In Process Citation]. *Chem. Senses* 25(1): 9–19.

Münch, F. (1896). Die Topographie der Papillen Zunge des Menschen und der Säugethiere. *Morphol. Arb. Schwalbe* 6: 605.

Murray, R. G. (1973). The ultrastructure of taste buds. In *The Ultrastructure of Sensory Organs*, Friedmann, I. (Ed). North Holland publishing company, Amsterdam, London, pp. 1–81.

Murray, R. G. (1986). The mammalian taste bud type III cell: a critical analysis. *J. Ultrastruct. Mol. Struct. Res.* 95: 175–188.

Murray, R. G., and Murray, A. (1967). Fine structure of taste buds of rabbit foliate papillae. *J. Ultrastruct. Res.* 19: 327–353.

Murray, R. G., and Murray, A. (1971). Relations and possible significance of taste bud cells. *Contrib. Sens. Physiol.* 5: 47–95.

Murray, R. G., Murray, A., and Fujimoto, S. (1969). Fine structure of gustatory cells in rabbit taste buds. *J.Ultrastruct.Res.* 27: 444–461.

Nada, O., and Hirata, K. (1977). The monoamine-containing cell in the gustatory epithelium of some vertebrates. *Arch. Histol. Jpn.* 40 Suppl: 197–206.

Nada, O., and Hirata, K. (1975). The occurrence of the cell type containing a specific monoamine in the taste bud of the rabbit's foliate papila. *Histochemistry* 43(3): 237–240.

Nagai, T., Kim, D. J., Delay, R. J., and Roper, S. D. (1996). Neuromodulation of transduction and signal processing in the end organs of taste. *Chem. Senses* 21(3): 353–365.

Nagato, T., Ren, X. Z., Toh, H., and Tandler, B. (1997). Ultrastructure of Weber's salivary glands of the root of the tongue in the rat. *Anat. Rec.* 249(4): 435–440.

Nelson, G. M., and Finger, T. E. (1993). Immunolocalization of different forms of neural cell adhesion molecule (NCAM) in rat taste buds. *J. Comp. Neurol.* 336:507–516.

Nemetschek-Ganssler, H., and Ferner, H. (1964). Über die Ultrastruktur der Geschmacksknospen. *Z. Zellforsch.* 63: 155–178.

Nishijima, K., and Atoji, Y. (2004). Taste buds and nerve fibers in the rat larynx: an ultrastructural and immunohistochemical study. *Arch. Histol. Cytol.* 67(3): 195–209.

Nolte, C., and Martini, R. (1992). Immunocytochemical localization of the L1 and N-CAM cell adhesion molecules and their shared carbohydrate epitope L2/HNK-1 in the developing and differentiated gustatory papillae of the mouse tongue. *J.Neurocytol.* 21: 19–33.

Nomura, S., Shiba, Y., Muneoka, Y., and Kanno, Y. (1979). A scanning and transmission electron microscope study of the premetamorphic papillae: possible chemoreceptive organs in the oral cavity of an anuran tadpole (Rana japonica). *Arch. Histol. Jpn.* 42(5): 507–516.

Norgren, R. (1990). Gustatory system. In *The Human Nervous System*, Paxinos, G. (Ed). Academic Press, San Diego, pp. 845–861.

Nosrat, C. A., and Olson, L. (1995). Brain-derived neurotrophic factor mRNA is expressed in the developing taste bud-bearing tongue papillae of rat. *J. Comp. Neurol.* 360(4): 698–704.

Nosrat, I. V., Lindskog, S., Seiger, A., and Nosrat, C. A. (2000). Lingual BDNF and NT-3 mRNA expression patterns and their relation to innervation in the human tongue: similarities and differences compared with rodents. *J. Comp. Neurol.* 417(2): 133–152.

Nosrat, I. V., Margolskee, R. F., and Nosrat, C. A. (2012). Targeted taste cell-specific overexpression of brain-derived neurotrophic factor in adult taste buds elevates phosphorylated TrkB protein levels in taste cells, increases taste bud size, and promotes gustatory innervation. *J. Biol. Chem.* 287(20): 16791–16800.

Oakley, B. (1970). Reformation of taste buds by crossed sensory nerves in the rat's tongue. *Acta. Physiol. Scand* 79(1): 88–94.

Oakley, B., Brandemihl, A., Cooper, D., et al. (1998). The morphogenesis of mouse vallate gustatory epithelium and taste buds requires BDNF-dependent taste neurons. *Dev. Brain Res.* 105(1): 85–96.

Obata, H., Shimada, K., Sakai, N., and Saito, N. (1997). GABAergic neurotransmission in rat taste buds: immunocytochemical study for GABA and GABA transporter subtypes. *Brain Res. Mol. Brain Res.* 49(1–2): 29–36.

Ogura, T., Margolskee, R. F., Tallini, Y. N., et al. (2007). Immuno- localization of vesicular acetylcholine transporter in mouse taste cells and adjacent nerve fibers: indication of acetylcholine release. *Cell Tissue Res.* 330(1): 17–28.

Ohmura, S., Horimoto, S., and Fujita, K. (1989). Lectin cytochemistry of the dark granules in the type 1 cells of Syrian hamster circumvallate taste buds. *Arch. Oral Biol.* 34: 161–166.

Ohshima, H., Yoshida, S., and Kobayashi, S. (1990). Blood vascular architecture of the rat lingual papillae with special reference to their relations to the connective tissue papillae and surface structures: a light and scanning electron microscope study. *Acta. Anat. (Basel)* 137(3): 213–221.

Oike, H., Matsumoto, I., and Abe, K. (2006). Group IIA phospholipase A(2) is coexpressed with SNAP-25 in mature taste receptor cells of rat circumvallate papillae. *J. Comp. Neurol.* 494(6): 876–886.

Ojima, K., Matsumoto, S., Takeda, M., et al. (1997a). Numerical variation and distributive pattern on microvascular cast specimens of vallate papillae in the crossbred Japanese cat tongue. *Ann. Anat.* 179(2): 117–126.

Ojima, K., Takahashi, T., Matsumoto, S., et al. (1997b). Angioarchitectural structure of the fungiform papillae on rabbit tongue anterodorsal surface. *Ann. Anat.* 179(4): 329–333.

Ojima, K., Takeda, M., Matsumoto, S., et al. (1997c). Functional role of V form distribution seen in microvascular cast specimens of the filiform and fungiform papillae on the posterior central dorsal surface of the cat tongue. *Ann. Anat.* 179(4): 321–327.

Oka, Y., and Korsching, S. I. (2011). Shared and unique G alpha proteins in the zebrafish versus mammalian senses of taste and smell. *Chem. Senses* 36(4): 357–365.

Olmsted, J. M. D. (1920). The results of cutting the seventh cranial nerve in *Amiurus nebulosus* (Leseur). *J.Exp.Zool.* 31: 369–401.

Osculati, F., and Sbarbati, A. (1995). The frog taste disc: a prototype of the vertebrate gustatory organ. Prog Neurobiol *46*, 351–399.

Ozdener, H., Yee, K. K., Cao, J., et al. (2006). Characterization and Long-Term Maintenance of Rat Taste Cells in Culture. *Chem. Senses* 31(3): 279–290.

Panizza, B. (1834). *Ricerche sperimentali sopra i nervi.* Bizzoni, Pavia.

Paran, N., Mattern, C. F., and Henkin, R. I. (1975). Ultrastructure of the taste bud of the human fungiform papilla. *Cell Tissue Res.* 161: 1–10.

Parker, G. H. (1922). *Smell, Taste, and Allied Senses in the Vertebrates.* Lippincott, London.

Perez, C. A., Huang, L., Rong, M., et al. (2002). A transient receptor potential channel expressed in taste receptor cells. *Nat. Neurosci.* 5(11): 1169–1176.

Pevzner, R. A., and Tikhonova, N. A. (1980). [Cytochemical demonstration of cyclic nucleotide phosphodiesterases in the taste buds of Testudo horsefieldi turtles upon exposure to flavored substances]. *Zh. Evol. Biokhim Fiziol.* 16(2): 133–136.

Ponzo, M. (1907). Sulla presenza di organi del gusto nella parte laringea della faringe, nel tratto cervicale dell'ésofago e nel palato duro del feto umano. *Anat.Anz.* 31: 570–575.

Prichard, T. C. (2011). Gustatory system. 3rd ed. In *The Human Nervous System*, Mai, J. K. and Paxinos, G. (Eds). Elsevier, Amsterdam, Boston, Heidelberg, pp. 1187–1218.

Pritchard, T. C. (1991). The Primate Gustatory System. In *Smell and Taste in Health and Disease*, Getchell, T. V., Bartoshuk, L. M., Doty, R. L. and Snow, J. B. J., Jr., (Eds). Raven Press, New York, pp. 109–125.

Pumplin, D. W., and Getschman, E. (2000). Synaptic proteins in rat taste bud cells: appearance in the Golgi apparatus and relationship to alpha-gustducin and the Lewis(b) and A antigens. *J. Comp. Neurol.* 427(2): 171–184.

Pumplin, D. W., Getschman, E., Boughter, J. D., Jr., et al. (1999). Differential expression of carbohydrate blood-group antigens on rat taste-bud cells: relation to the functional marker alpha-gustducin. *J. Comp. Neurol.* 415(2): 230–239.

Pumplin, D. W., Yu, C., and Smith, D. V. (1997). Light and dark cells of rat vallate taste buds are morphologically distinct cell types. *J. Comp. Neurol.* 378(3): 389–410.

Rabl, H. (1895). Notiz zur Morphologie der Geschmacksknospen auf der Epiglottis. *Anat.Anz.* 11: 153–156.

Ranvier, L. (1888). *Technisches Lehrbuch der Histologie (German translation by W.Nicati and H.v.Wyss).* F.C.W.Vogel, Leipzig.

Remak, R. (1852). Ueber die Ganglien der Zunge bei Säugethieren und beim Menschen. In *Archiv für Anatomie, Physiologie und Wissenschaftliche Medicin*, Müller, J. (Ed). Veit et Comp., Berlin, pp. 58–62.

Ren, Y., Shimada, K., Shirai, Y., et al. (1999). Immunocytochemical localization of serotonin and serotonin transporter (SET) in taste buds of rat. *Brain Res. Mol. Brain Res.* 74(1–2): 221–224.

Retzius, G. (1892). Die Nervenendigungen in dem Geschmacksorgan der Säugethiere und Amphibien. In *Biologische Untersuchungen Neue Folge IV*, Samson & Wallin, Stockholm, pp. 26–32.

Retzius, G. (1893). Ueber Geschmacksknospen bei Petromyzon. In *Biologische Untersuchungen. Neue Folge* Vol. V, Samson & Wallin, Stockholm, pp. 69–70.

Reutter, K. (1971). The taste-buds of Amiurus nebulosus (Lesueur). Morphological and histochemical investigations. *Z. Zellforsch. Mikrosk. Anat.* 120(2): 280–308.

Reutter, K. (1973). [The types of taste buds in fishes. I. Morphological and neurohistochemical investigations on Xiphophorus helleri Heckel (Poeciliidae, Cyprinodontiformes, Teleostei) (author's transl)]. *Z. Zellforsch Mikrosk Anat.* 143: 409–423.

Reutter, K. (1978). Taste organ in the bullhead (Teleostei). *Adv. Anat. Embryol. Cell Biol.* 55(1): 3–94.

Reutter, K. (1980). SEM- study of the mucus layer on the receptor-field of fish taste buds. In *Olfaction and Taste VII*, van der Starre, H. (Ed). IRL Press, London, Washington DC, pp. 107.

Reutter, K. (1986). Chemoreceptors. In *Biology of the Integument*, Bereiter-Hahn, J., Matoltsy, A. G. and Richards, K. S. (Eds). Vol. 2, Springer, Berlin, Heidelberg, pp. 586–604.

Reutter, K. (1992). Structure of the peripheral gustatory organ, represented by the siluroid fish Plotosus lineatus (Thunberg). In *Fish chemoreception*, Hara, T. J. (Ed). Chapman & Hall, London, pp. 60–78.

Reutter, K., and Witt, M. (1993). Morphology of vertebrate taste organs and their nerve supply. In *Mechanisms of taste transduction*, Simon, S. A. and Roper, S. D. (Eds). CRC Press, Boca Raton, Ann Arbor, London, Tokyo, pp. 29–82.

Reutter, K., and Witt, M. (1999). Comparative aspects of fish taste bud ultrastructure. In *Advances in chemical signals in vertebrates*, Johnston, E., Müller-Schwarze, D. and Sorensen, P. W. (Eds). Kluwer Academic/Plenum Publishers, New York, Boston, Dordrecht, London, Moscow, pp. 573–581.

Reutter, K., and Witt, M. (2004). Are there efferent synapses in fish taste buds? *J. Neurocytol.* 33(6): 647–656.

Riva, A., Loffredo, F., Puxeddu, R., and Testa Riva, F. (1999). A scanning and transmission electron microscope study of the human minor salivary glands. *Arch. Oral Biol.* 44 Suppl 1: S27–31.

Roper, S. D. (1989). The cell biology of vertebrate taste receptors. *Annu. Rev. Neurosci.* 12: 329–353.

Roper, S. D. (2006). Cell communication in taste buds. *Cell Mol. Life Sci.* 63(13): 1494–1500.

Royer, S. M., and Kinnamon, J. C. (1988). Ultrastructure of mouse foliate taste buds: synaptic and nonsynaptic interactions between taste cells and nerve fibers. *J. Comp. Neurol.* 270: 11–24, 58.

Royer, S. M., and Kinnamon, J. C. (1991). HVEM serial-section analysis of rabbit foliate taste buds: I. Type III cells and their synapses. *J. Comp. Neurol.* 306: 49–72.

Ruiz-Avila, L., McLaughlin, S. K., Wildman, D., et al. (1995). Coupling of bitter receptor to phosphodiesterase through transducin in taste receptor cells. *Nature* 376: 80–85.

Saito, I. (1966). Comparative anatomical studies of the oral organs of the poultry. V. Structures and distribution of taste buds of the fowl. (In Japanese). *Bull Fac. Agric. Miyazahi. Univ.* 13: 95–102.

Sbarbati, A., Crescimanno, C., Bernardi, P., and Osculati, F. (1999). Alpha-gustducin-immunoreactive solitary chemosensory cells in the developing chemoreceptorial epithelium of the rat vallate papilla. *Chem. Senses* 24(5): 469–472.

Scalzi, H. A. (1967). The cytoarchitecture of gustatory receptors from the rabbit foliate papillae. *Z. Zellforsch Mikrosk Anat.* 80(3): 413–435.

Schinkele, O. (1942). Über das Vorkommen von Geschmacksknospen im kranialen Drittel des Oesophagus. *Z. Mikrosk.-anat. Forsch.* 51: 498–501.

Schmale, H., Ahlers, C., Blaker, M., et al. (1993). Perireceptor events in taste. *Ciba Found Symp.* 179: 167–80; discussion 180–185.

Schmale, H., and Bamberger, C. (1997). A novel protein with strong homology to the tumor suppressor p53 [In Process Citation]. *Oncogene* 15(11): 1363–1367.

Schmale, H., Holtgreve-Grez, H., and Christiansen, H. (1990). Possible role for salivary gland protein in taste reception indicated by homology to lipophilic-ligand carrier proteins. *Nature* 343(6256): 366–369.

Schulze, F. E. (1863). Über die becherförmigen Organe der Fische. *Z. Wiss. Zool.* 12: 218–222.

Schwalbe, G. A. (1868). Über die Geschmacksorgane der Säugethiere und des Menschen. *Arch. Mikr. Anat.* 4: 154–187.

Schwann, T. (1839). *Mikroskopische Untersuchungen über die Übereinstimmung in der Struktur und dem Wachstum der Thiere und Pflanzen.* Sander'sche Buchhandlung (Reimer), Berlin.

Schwartz, H. G., and Weddell, G. (1938). Observations on the pathways transmitting the sensation of taste. *Brain* 61: 99–115.

Segovia, C., Hutchinson, I., Laing, D. G., and Jinks, A. L. (2002). A quantitative study of fungiform papillae and taste pore density in adults and children. *Brain Res. Dev. Brain Res.* 138(2): 135–146.

Selliseth, N. J., and Selvig, K. A. (1993). Microvasculature of the dorsum of the rat tongue: a scanning electron microscopic study using corrosion casts. *Scand J. Dent. Res.* 101(6): 391–397.

Seta, Y., Oda, M., Kataoka, S., et al. (2011). Mash1 is required for the differentiation of AADC-positive type III cells in mouse taste buds. *Dev. Dyn.* 240(4): 775–784.

Seta, Y., Seta, C., and Barlow, L. A. (2003). Notch-associated gene expression in embryonic and adult taste papillae and taste buds suggests a role in taste cell lineage decisions. *J. Comp. Neurol.* 464(1): 49–61.

Shahbake, M., Hutchinson, I., Laing, D. G., and Jinks, A. L. (2005). Rapid quantitative assessment of fungiform papillae density in the human tongue. *Brain Res.* 1052(2): 196–201.

Shatzman, A. R., and Henkin, R. I. (1981). Gustin concentration changes relative to salivary zinc and taste in humans. *Proc. Natl. Acad Sci. U. S. A.* 78(6): 3867–3871.

Shen, T., Kaya, N., Zhao, F.-L., et al. (2005). Co-expression patterns of the neuropeptides vasoactive intestinal peptide and cholecystokinin with the transduction molecules [alpha]-gustducin and T1R2 in rat taste receptor cells. *Neuroscience* 130(1): 229–238.

Siegel, R. E. (1970). *Galen on sense perception. His doctrines, observations and experiments on vision, hearing, smell, taste, touch and pain, and their historical sources*. S.Karger, Basel, New York.

Smith, D. V., Akeson, R. A., and Shipley, M. T. (1993). NCAM expression by subsets of taste cells is dependent upon innervation. *J. Comp. Neurol.* 336: 493–506.

Smith, D. V., Klevitsky, R., Akeson, R. A., and Shipley, M. T. (1994). Expression of the neural cell adhesion molecule (NCAM) and polysialic acid during taste bud degeneration and regeneration. *J. Comp. Neurol.* 347: 187–196.

Smith, D. V., Som, J., Boughter, J. D., Jr,., et al. (1999). Cellular expression of alpha-gustducin and the A blood group antigen in rat fungiform taste buds cross-reinnervated by the IXth nerve. *J. Comp. Neurol.* 409(1): 118–130.

Sobotta, J. (1993). *Atlas der Anatomie des Menschen*. Vol.1. Urban und Schwarzenberg, München.

Sömmering, S. T. (1806). *Abbildungen der menschlichen Organe des Geschmackes und der Stimme*. Varrentrapp und Wenner, Frankfurt.

Spielman, A. I. (1990). Interaction of saliva and taste. *J. Dent. Res.* 69(3): 838–843.

Sprissler, C. (1994). *Ultrastruktur der Geschmacksknospe der japanischen Wachtel (Coturnix coturnix japonica)*. University of Tübingen.

Srur, E., Stachs, O., Guthoff, R., et al. (2010). Change of the human taste bud volume over time. *Auris. Nasus. Larynx.* 37(4): 449–455.

Starostik, M. R., Rebello, M. R., Cotter, K. A., et al. (2010). Expression of GABAergic receptors in mouse taste receptor cells. *PLoS One* 5(10): e13639.

Stone, L. M., Finger, T. E., Tam, P. P., and Tan, S. S. (1995). Taste receptor cells arise from local epithelium, not neurogenic ectoderm. *Proc. Natl. Acad. Sci. U.S.A.* 92: 1916–1920.

Stone, L. S. (1940). The origin and development of taste organs in salamanders observed in the living condition. *J. Exp. Zool.* 83: 481–506.

Stornelli, M. R., Lossi, L., and Giannessi, E. (2000). Localization, morphology and ultrastructure of taste buds in the domestic duck (Cairina moschata domestica L.) oral cavity. *Ital. J. Anat. Embryol.* 105(3): 179–188.

Suzuki, Y., Takeda, M., Obara, N., and Nagai, Y. (1996). Phagocytic cells in the taste buds of rat circumvallate papillae after denervation. *Chem. Senses* 21(4): 467–476.

Svejda, J., and Janota, M. (1974). Scanning electron microscopy of the papillae foliatae of the human tongue. *Oral Surg. Oral Med. Oral Pathol.* 37(2): 208–216.

Tandler, B., Pinkstaff, C. A., and Riva, A. (1994). Ultrastructure and histochemistry of human anterior lingual salivary glands (glands of Blandin and Nuhn). *Anat. Rec.* 240: 167–177.

Thirumangalathu, S., Harlow, D. E., Driskell, A. L., et al. (2009). Fate mapping of mammalian embryonic taste bud progenitors. *Development* 136(9): 1519–1528.

Tizzano, M., Cristofoletti, M., Sbarbati, A., and Finger, T. E. (2011). Expression of taste receptors in solitary chemosensory cells of rodent airways. *BMC Pulm. Med.* 11: 3.

Tomchik, S. M., Berg, S., Kim, J. W., et al. (2007). Breadth of Tuning and Taste Coding in Mammalian Taste Buds. *J. Neurosci.* 27(40): 10840–10848.

Tomonari, H., Miura, H., Nakayama, A., et al. (2012). GÎ±−gustducin Is Extensively Coexpressed with Sweet and Bitter Taste Receptors in both the Soft Palate and Fungiform Papillae but Has a Different Functional Significance. *Chemical Senses* 37(3): 241–251.

Toto, P. D., Nadimi, H., and Martinez, R. (1993). von Ebner's gland. an immunohistochemical study. *Ann. N. Y. Acad. Sci.* 694: 322–324.

Toubeau, G., Cotman, C., and Bels, V. (1994). Morphological and kinematic study of the tongue and buccal cavity in the lizard *Anguis fragilis* (Reptilia:Anguidae). *Anat. Rec.* 240: 423–433.

Toyoshima, K., Miyamoto, K., and Shimamura, A. (1987). Fine structure of taste buds in the tongue, palatal mucosa and gill arch of the axolotl, Ambystoma mexicanum. *Okajimas Folia Anat. Jpn.* 64(2–3): 99–109.

Toyoshima, K., Seta, Y., Toyono, T., and Takeda, S. (1999). Merkel cells are responsible for the initiation of taste organ morphogenesis in the frog. *J. Comp. Neurol.* 406(1): 129–140.

Toyoshima, K., and Shimamura, A. (1987). Monoamine-containing basal cells in the taste buds of the newt Triturus pyrrhogaster. *Arch. Oral Biol.* 32: 619–621.

Toyoshima, K., and Tandler, B. (1987). Modified smooth endoplasmic reticulum in type II cells of rabbit taste buds. *J. Submicrosc. Cytol.* 19: 85–92.

Tuckerman, F. (1889). On the development of the taste-organs of man. *J. Anat. Physiol.* 23: 559–582.

Uchida, T. (1980). Ultrastructural and histochemical studies on the taste buds in some reptiles. *Arch. Histol. Jpn.* 43: 459–478.

Ueda, K., Ichimori, Y., Okada, H., et al. (2006). Immunolocalization of SNARE proteins in both type II and type III cells of rat taste buds. *Arch. Histol. Cytol.* 69(4): 289–296.

Vandenbeuch, A., Clapp, T. R., and Kinnamon, S. C. (2008). Amiloride-sensitive channels in type I fungiform taste cells in mouse. *BMC Neurosci.* 9: 1.

Varatharasan, N., Croll, R. P., and Franz-Odendaal, T. (2009). Taste bud development and patterning in sighted and blind morphs of Astyanax mexicanus. *Dev. Dyn.* 238(12): 3056–3064.

Verson, E. (1868). Beiträge zur Kenntnis des Kehlkopfes und der Trachea. *Sitzungsber. Wiener Acad. Wissenschaft, Math.-naturwiss.Klasse* 57: 1093–1102.

von Düring, M. V., and Andres, K. H. (1976). The ultrastructure of taste and touch receptors of the frog's taste organ. *Cell Tissue Res.* 165(2): 185–198.

von Vintschgau, M., and Hönigschmied, J. (1877). Nervus glossopharyngeus und Schmeckbecher. *Arch. Physiol.* 14: 443–448.

Wagner, R. (1837). Bestätigung des Panizzaschen Lehrsatzes, dass das 9te Nervenpaar (n.glossopharyngeus) der Geschmacksnerv ist. *Frorieps Notizen* 4: 129–131.

Wakisaka, S., Tabata, M. J., Maeda, T., et al. (1998). Immunohistochemical localization of pleiotrophin and midkine in the lingual epithelium of the adult rat. *Arch. Histol. Cytol.* 61(5): 475–480.

Waller, A. (1847). Microscopic examination of the papillae and nerves of the tongue of the frog, with observations on the mechanism of taste. *London, Edinburgh, and Dublin Philosoph. Magazine J. Sci.* Vol XXX: 277–289.

Waller, A. (1849). Minute structure of the papillae and nerves of the tongue of the frog and the toad. Communicated by R. Owen. *Philosoph.Transact.Royal Soc.London* Pt.I: 139–149.

Welsch, U., and Storch, V. (1969). [Fine structure of the taste buds of catfish (Clarias batrachus (L) and Kryptopterus bicirrhis (Cuvier and Valenciennes)]. *Z. Zellforsch Mikrosk. Anat.* 100(4): 552–559.

Whitear, M. (1970). The skin surface of bony fishes. *J.Zool.London* 160: 437–454.

Whitear, M. (1992). Solitary chemosensory cells. In *Fish chemoreception*, Hara, T. (Ed). Chapman & Hall, London, pp. 103–125.

Whitehead, M. C., Ganchrow, J. R., Ganchrow, D., and Yao, B. (1998). Neural cell adhesion molecule, neuron-specific enolase and calcitonin gene-related peptide immunoreactivity in hamster taste buds after chorda tympani lingual nerve denervation. *Neuroscience* 83(3): 843–856.

Williams, P. L., Warwick, R., Dyson, M., and Bannister, L. H. (1989). *Gray's Anatomy*. Curchill Livingstone, Edinburgh London.

Wilson, J. G. (1905). The structure and function of the taste-buds of the larynx. *Brain* 28: 339–351.

Witt, M. (1993). Ultrastructure of the taste disc in the red-bellied toad Bombina orientalis (Discoglossidae, Salientia). *Cell Tissue Res.* 272: 59–70.

Witt, M. (1995). Distribution of vasoactive intestinal peptide-like immunoreactivity in the taste organs of teleost fish and frog. *Histochem. J.* 27: 161–165.

Witt, M. (1996). Carbohydrate histochemistry of vertebrate taste organs. *Progr. Histochem. Cytochem.* 30/4: 1–172.

Witt, M., and Kasper, M. (1998). Immunohistochemical distribution of CD44 and some of its isoforms during human taste bud development. *Histochem. Cell Biol.* 110: 95–113.

Witt, M., and Kasper, M. (1999). Distribution of cytokeratin filaments and vimentin in developing human taste buds. *Anat. Embryol.* 199: 291–299.

Witt, M., and Miller, I. J., Jr., (1992). Comparative lectin histochemistry on taste buds in foliate, circumvallate and fungiform papillae of the rabbit tongue. *Histochemistry* 98(3): 173–182.

Witt, M., and Reutter, K. (1988). Lectin histochemistry on mucous substances of the taste buds and adjacent epithelia of different vertebrates. *Histochemistry* 88: 453–461.

Witt, M., and Reutter, K. (1990). Electron microscopic demonstration of lectin binding sites in the taste buds of the European catfish Silurus glanis (Teleostei). *Histochemistry* 94: 617–628.

Witt, M., and Reutter, K. (1996). Embryonic and early fetal development of human taste buds: a transmission electron microscopical study. *Anat. Rec.* 246(4): 507–523.

Witt, M., and Reutter, K. (1997). Scanning electron microscopical studies of developing gustatory papillae in humans. *Chem. Senses* 22: 601–612.

Wong, G. T., Ruiz-Avila, L., Ming, D., et al. (1996). Biochemical and transgenic analysis of gustducin's role in bitter and sweet transduction. *Cold Spring Harb. Symp. Quant. Biol.* 61: 173–184.

Wyss, H. v. (1870). Die becherförmigen Organe der Zunge. *Arch. mikrosk. Anat.* 6: 237–260.

Yamasaki, F., and Takahashi, K. (1982). A description of the times of appearance and regression of marginal lingual papillae in human fetuses and newborns. *Anat.Rec.* 204: 171–173.

Yang, R., Crowley, H. H., Rock, M. E., and Kinnamon, J. C. (2000). Taste cells with synapses in rat circumvallate papillae display SNAP-25-like immunoreactivity. *J. Comp. Neurol.* 424(2): 205–215.

Yee, C., Bartel, D. L., and Finger, T. E. (2005). Effects of glossopharyngeal nerve section on the expression of neurotrophins and their receptors in lingual taste buds of adult mice. *J. Comp. Neurol.* 490(4): 371–390.

Yee, C. L., Yang, R., Böttger, B., et al. (2001). "Type III" cells of rat taste buds: immunohistochemical and ultrastructural studies of neuron-specific enolase, protein gene product 9.5, and serotonin. *J. Comp. Neurol.* 440(1): 97–108.

Yoshie, S., Wakasugi, C., Teraki, Y., et al. (1988). Immunocytochemical localizations of neuron-specific proteins in the taste bud of the guinea pig. *Arch. Histol. Cytol.* 51: 379–384.

Zachar, P. C., and Jonz, M. G. (2011). Confocal imaging of Merkel–like basal cells in the taste buds of zebrafish. *Acta Histochem.* 114(2): 101–115.

Zancanaro, C., Sbarbati, A., Bolner, A., et al. (1995). Biogenic amines in the taste organ. *Chem.Senses* 20: 329–335.

Zhang, C., and Oakley, B. (1996). The distribution and origin of keratin 20-containing taste buds in rat and human. *Differentiation* 61(2): 121–127.

Zhang, C. X., Cotter, M., Lawton, A., et al. (1995). Keratin 18 is associated with a subset of older taste cells in the rat. *Differentiation* 59: 155–162.

Zhao, F. L., Shen, T., Kaya, N., et al. (2005). Expression, physiological action, and coexpression patterns of neuropeptide Y in rat taste-bud cells. *Proc. Natl. Acad. Sci. U.S.A.* 102(31): 11100–11105.

Zimmermann, K. W. (1927). Die Speicheldrüsen der Mundhöhle und die Bauchspeicheldrüse. In *Handbuch der Mikroskopischen Anatomie des Menschen*, Möllendorf, W. v. (Ed). Vol. V, Part 1, Springer, Berlin, pp. 61–244.

Żuwała, K. (1997). Ultrastructure of premetamorphic taste organs of the Bombina variegata. *Rocz. Akad. Med. Bialymst.* 42 Suppl 2: 204–207.

Żuwała, K., and Jakubowski, M. (1991). Development of taste organs in Rana temporaria. Transmission and scanning electron microscopic study. *Anat. Embryol. (Berl)* 184(4): 363–369.

Chapter 30

Chemical Modulators of Taste

JOHN A. DESIMONE, GRANT E. DUBOIS, and VIJAY LYALL

30.1 INTRODUCTION

In the consumer's lexicon, taste is a word that captures the sum total of perceptions following taking a substance into the mouth. Of course, this includes olfactory and tactile as well as gustatory sensations. A more limited definition of taste focuses on sensations emanating from the oral cavity which certainly would include sweet, bitter, salty, sour, and umami tastes, but also other sensations (e.g., astringent, cooling, hot, prickle, licorice-like, fatty, and mouthfeel). However, consensus among scientists limits taste to sweet, bitter, salty, sour, and umami and, accordingly, we limit our discussion to modulators of these tastes.

Now that the cellular basis of taste transduction has been described for each of the taste qualities, we have a structural foundation upon which to develop quality-specific taste modulators. Before transduction was understood, taste modulators were discovered largely by chance. Apart from being interesting curiosities they sometimes allowed some insight into taste mechanism. A good example is gymnemic acid, which blocks sweet taste in humans and closely related primates, but not in lower primates and other mammals (Hellekant et al., 1996). This suggested a mutation in the sweet taste receptor structure among species, long before this could be demonstrated in molecular terms. The epithelial sodium transport inhibitor, amiloride, led to a major paradigm shift in our thinking about the sodium taste receptor. The earlier view was that the sodium taste receptor was an equilibrium sodium-binding protein (Beidler, 1954). However, the lingual epithelium was later shown to actively transport sodium and that this sodium current could be blocked by ouabain and the sodium channel blocker amiloride (DeSimone et al., 1981; Stewart et al., 1997). When

amiloride was shown to block the chorda tympani taste response to NaCl in rats and other mammals, the cumulative data indicated that the sodium taste receptor was most likely the epithelial sodium channel, also found in lung and kidney (Stewart et al., 1997). Sodium taste reception is, therefore, a nonequilibrium process that results directly in a depolarizing sodium current that excites the release of the afferent neurotransmitter from receptor cells. Because of the high biochemical reactivity of hydrogen ions at the cellular level, it was perhaps not unexpected that at least for a time there might be a confusing plethora of proposed sour taste receptor mechanisms (DeSimone et al., 2001). Notwithstanding, human psychophysical studies have for more than a century shown that sourness does not in general correlate with stimulus pH, that is, sour taste receptors are not extracellular pH sensors (DeSimone et al., 2001). Consistent with the psychophysics, work over the last decade shows that sour taste receptors are rather intracellular pH sensors and that their responsiveness is especially sensitive to modulators of intracellular pH, as will be explained in detail below.

As we begin, it is essential that we reach consensus as to the meaning of "chemical modulator of taste." We define them to be chemical compounds which, at the level employed, are tasteless but which either enhance or inhibit one of the five tastes. Sweetness enhancers, bitterness inhibitors, saltiness enhancers, sourness inhibitors, and umami enhancers have obvious food and beverage applications. However, uses for sweetness inhibitors, bitterness enhancers, saltiness inhibitors, sourness enhancers, and umami inhibitors are less clear. Being as it may, applications for sweetness and saltiness inhibitors have been identified, whereas those for bitterness enhancers, sourness enhancers, and umami inhibitors remain to be identified. That being said, the future may lead to uses for such mod-

---

Handbook of Olfaction and Gustation, Third Edition. Edited by Richard L. Doty.
© 2015 Richard L. Doty. Published 2015 by John Wiley & Sons, Inc.

666 Chapter 30 Chemical Modulators of Taste

ulators and therefore we include discussion on enhancers and inhibitors for all of the five primary tastes, where such information is available.

30.2 MODULATORS OF SWEET TASTE

In a rational program targeted at sweet taste modulators, it is essential to understand the biochemical processes which mediate activation of the taste bud cells (tbcs) which signal sweet taste to the CNS. Enabled by the human genome sequence, tremendous progress was made in elucidation of the biochemistry and physiology of sweet taste as well as in understanding the Structure-Activity-Relationships (SAR) of sweeteners (DuBois et al., 2008a). To summarize, it is generally accepted that sweet taste is initiated by the single Class C heterodimeric G Protein-Coupled Receptor (GPCR) T1R2/T1R3, where the native agonists (e.g., sucrose, fructose, etc.) bind in large extracellular Venus Fly Trap (VFT) domains. Transduction of receptor activation is mediated by: (1) the G protein gustducin, (2) the enzyme phospholipase $C\beta_2$ ($PLC\beta_2$), (3) inositol trisphosphate (IP_3), (4) the ribosomal inositol trisphosphate receptor (IP_3R), (5) Ca^{2+}, (6) a transient receptor potential ion channel $TRPM_5$, as well as other factors. Sweet-sensitive tbcs are Type II tbcs characterized by apical microvillar projections which are the receptor loci. Efforts to identify synapses with afferent nerve fibers have been unsuccessful leading to the proposal (Huang et al., 2007) that excited sweet-sensitive tbcs secrete ATP through gap junction channels thus exciting Type III tbcs which subsequently communicate with afferent nerve fibers. However, as an experiment resulting in deletion of Type III cells (Huang et al., 2006) eliminated sour taste while not affecting sweet taste, the full details of the sweet-sensitive tbc signaling to afferent nerve fibers remain to be clarified.

As noted, all sweeteners activate T1R2/T1R3. However, to explain multiple observations not explained by this pathway, the Margolskee and Gilbertson laboratories provided evidence that carbohydrate (CHO) sweeteners have an additional activation mechanism involving glucose transporters and other glucose sensing proteins (Yee et al., 2011). However, these results are preliminary and much more remains to be done to validate this pathway.

Interest in modulators of sweet taste is primarily related to technologies which will enable improved replication of sucrose taste. In one approach to improved sucrose taste simulation, focus has been placed on enhancers for CHO sweeteners (e.g., sucrose, fructose and glucose) where the expectation is *retention of CHO sweetener taste*. Senomyx (San Diego, CA) was the first to develop a cell-based assay based on the human sweetener receptor (Li et al., 2002) and this assay enabled

high-throughput-screening (HTS) of large libraries of chemical compounds and discovery of the first sweetener enhancers, which really are more properly termed sweetener receptor Positive Allosteric Modulators (PAMs). Since then, other organizations have initiated PAM discovery programs including Chromocell of North Brunswick, NJ (Gravitz, 2012). Chromocell HTS technology is based on cells which naturally express the sweetener receptor, thereby more accurately recapitulating sweet-sensitive tbc physiology and improving assay fidelity for PAM discovery.

In a second approach to improved sucrose taste simulation, efforts have been focused on modulators of the negative taste attributes of high-potency (HP) sweeteners. Due to the non-sugar-like tastes of HP sweeteners like aspartame, saccharin, and stevioside (DuBois, 2008a), there is interest in modulation so as to enable more sugar-like tastes. HP sweeteners are often disadvantaged by: 1) low maximal sweetness (i.e., maximal response, R_m), 2) bitter off taste, 3) sweetness which develops slowly and then lingers (i.e., non-sugar-like temporal profile), and 4) sweetness which causes desensitization of the sensory system. The discussion here is focused on modulation of the temporal profiles of HP sweeteners to enable more sugar-like tastes. At the present time, there is limited understanding of the biochemical rationale for low R_m and preferential desensitization of the responses of HP sweeteners relative to CHO sweeteners. Therefore, these topics are not further discussed in this chapter.

Research has also been carried out to identify modulators which inhibit sweetness. Such ingredients may have application for reduction of the sweetness of food and beverage products which are too sweet due to the natural levels of CHO sweeteners present (e.g., lactase-treated milk) or because of a need for elevated levels of CHO sweeteners to provide some physical property to a food (e.g., crispness in crackers enabled by sucrose).

30.2.1 Positive Allosteric Modulators (PAMs) as Modulators of Sweet Taste

The principal challenge in sweetener research is accurate replication of sucrose taste. In principle, a very simple approach to this objective would be enhancement of the sweetness potency of sucrose and other CHO sweeteners. It is reasonable to expect that this could be accomplished by modulators acting at any step in the sweet-sensitive tbc excitation cascade. Many reports of modulators of sweet-sensitive tbc activity have appeared with the following being exemplary: (1) inhibitory activity of cholecystokinin and neuropeptide Y (Herness and Zhao, 2009), (2) inhibitory activity of leptin and enhancement activity of endocannabinoids (Niki et al., 2010),

30.2 Modulators of Sweet Taste

(3) enhancement activity by TRPm$_5$ conductance potentiators (Bryant et al., 2008), and (4) enhancement activity by adenosine (Dando et al., 2012). However, these modulators are inactive in human sensory assays, presumably because they cannot readily access their molecular targets. And so, it appears that the only practical approach to enhancement of the activity of sweet-sensitive tbcs may be PAMs acting at extracellular sites on T1R2/T1R3.

The foregoing rationale provided justification for scientists at The Coca-Cola Company (TCCC) to initiate a PAM discovery program (DuBois, 2011). At the outset of this work, there were literature reports of CHO sweetener enhancers. However, none of the claims could be replicated. Nonetheless, in continuation of the program, 2, 4-dihydroxybenzoic acid (**1** in Figure 30.1) was evaluated, based on a report by Holland Sweetener Company (HSC) scientists of aspartame (APM) sweetness enhancement (Britton et al., 1999). HSC's interest in **1** was improvement of APM taste, but not enhancement. Thus it was found that formulation of 6% sucrose with **1** showed an unambiguous >1.3-fold enhancement and, importantly, *with retention of sugar taste*. This proof of concept for a sweetener receptor PAM was an early breakthrough which justified an intensive discovery program to follow. The CHO sweetener enhancement activity of **1** was strongest with sucrose, weaker with fructose and negligible with glucose, selectivity consistent with action as a PAM.

The Senomyx T1R2/T1R3 cell-based assay (Li et al., 2002) enabled a marked acceleration of sweetener receptor PAM discovery. T1R2/T1R3 is one of 12 Class C GPCRs and, beginning in the late 1990s, reports on discoveries of Class C GPCR PAMs as well as negative allosteric modulators (NAMs) began to appear, such that today PAMs and/or NAMs are known for all of them (Urweyler, 2011). Thus, in 2002, a PAM discovery program was initiated at Senomyx, in collaboration with TCCC. The program led to the sucralose-selective PAM **2** (Servant et al., 2010), the first high-potency/high-efficacy T1R2/T1R3 PAM. PAM **2** @ 9.7 mg/L enhances the sweetness of 100 mg/L sucralose to that of 600 mg/L sucralose (i.e., 6-fold). PAM **2** was quickly followed by sucrose-selective PAM **3** (Tachdjian et al., 2008). PAM **3** @ 8.8 mg/L enhances 6% sucrose sweetness to that of 10.6% sucrose (i.e., 1.8-fold) and importantly, *with retention of sugar-like taste*. Both PAMs

2 and **3** now have regulatory approvals. In recent work (Zhang et al., 2010), evidence was presented that PAM **2** acts in the T1R2 VFT with concurrent binding to the receptor and sucralose. Analogous binding for PAM **3** and sucrose seems likely.

Subsequent to the pioneering work by Senomyx, other companies reported the discovery of sweetness enhancers as follows: 1) Cadbury Adams USA (Bingley et al., 2007), 2) Givaudan (Slack et al., 2007, Jia et al., 2008, Jia and Yang, 2009, Bouter et al., 2009, Wang and Daniher, 2009, Flamme et al., 2011, Flamme and Bom, 2011), 3) Nutrinova (Krohn and Zinke, 2009a; 2009b), 4) Redpoint Bio Corporation (Salemme et al., 2010) and 5) Symrise (Ley et al., 2006; 2007; Krammer et al., 2007, Ley et al., 2008; Wessjohann, 2010). However, the enhancements reported are very modest and, in all cases, the enhancers are sweet themselves. Compounds exhibiting both agonist (e.g., sweetness) and PAM activities are termed "agoenhancers" (Schwartz and Holst, 2006) and so, since the claimed enhancements are small, it is not certain that they are truly enhancements.

It is noteworthy that many sweeteners are agoenhancers. However, in the sweetener literature, they are referred to as synergistic sweeteners. In effort to better understand "sweetness synergy," Schiffman and coworkers carried out a systematic study on binary blends (Schiffman et al., 1995). Many synergistic pairs were identified and the sweeteners in these blends may be thought of as PAMs which are also agonists (i.e., sweeteners) and which are selective in their enhancement action. The structural similarities between PAMs **2** and **3** and well known sweeteners saccharin and acesulfame-K is striking. And thus it is logical to expect saccharin and/or acesulfame-K synergism with sucrose and sucralose. However, such synergy was not observed in the Schiffman study. It is interesting to note, however, that cyclamate is synergistic with 11 of the 14 sweeteners evaluated by Schiffman including sucrose, fructose, and glucose. And since agoenhancers are PAMs, cyclamate analogues may be a promising structural class of compounds for PAM discovery.

In summary, Senomyx has demonstrated tasteless T1R2/T1R3 PAMs (i.e., pure PAMs rather than agoenhancers) and it is possible that PAM ingredients could be a "breakthrough technology" (Servant et al., 2011).

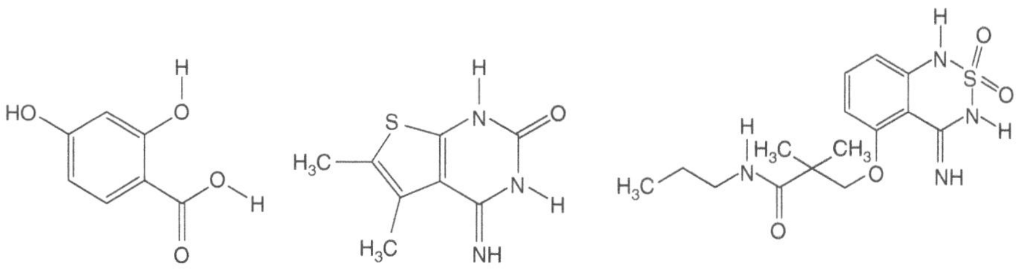

Figure 30.1 Human sweetener receptor Positive Allosteric Modulators.

Yet much remains unknown. Questions include the following: (1) Can PAMs with >/=20-fold enhancement effects be found as required to enable 0-calorie CHO ingredient labeling? (2) Can PAMs active with all CHO sweeteners be found? (3) Can natural product PAMs be found so as to permit all-natural ingredient labeling? (4) Can PAMs for natural HP sweeteners (e.g., rebaudioside A, mogroside V, etc.) be found and, if so, will they enable improved taste qualities for natural HP sweetener systems?

30.2.2 Osmolytes as Modulators of Sweet Taste

The singularly most challenging problem in replication of CHO sweetener taste with HP sweeteners is delivering on CHO sweetener temporal profile. HP sweeteners exhibit delays in sweetness onset and, especially, sweetness which lingers. Methods have been developed to quantify the temporal aspects of sweetener taste (DuBois and Lee, 1983).

In 2003, a program was initiated at TCCC for the development and commercialization of a natural product sweetener rebaudioside A (Figure 30.2) derived from the Paraguayan shrub *Stevia rebaudianna* (Bertoni) and often referred to as REBA (Prakash et al., 2008). A significant challenge in this program was modulation of the temporal profile of REBA so as to enable acceptable taste quality. This challenge was addressed by consideration of the most likely mechanistic rationale for REBA's slow sweetness onset and sweetness linger (DuBois, 2011). Clearly, given their higher potencies, HP sweeteners bind with higher affinity to the receptor than do CHO sweeteners. However, REBA's sweetness potency of ca. 200X sucrose is insufficient to explain a sweetness which persists for minutes. On consideration of the apparent equilibrium constant for REBA/receptor dissociation (K_d) of 210 μM (DuBois et al., 1991) and given that 1) K_d is the quotient of the rate constants for REBA/receptor dissociation (k_d) and association (k_a) and 2) rate constants for binding of small molecules to receptors reflect diffusion-controlled kinetics (i.e., $k_a \sim 10^{10}$ $M^{-1}sec^{-1}$), a k_d for REBA/receptor dissociation is estimated to be ca. 2×10^6 sec^{-1}. Thus, if we assume first order kinetics for REBA/receptor dissociation, at a time 60 sec after the time of maximal sweetness, the percentage of REBA still bound is ≪ 1%. Thus, it is clear that the potency of REBA cannot explain sweetness which lingers for many minutes. Michael Naim and colleagues (Hebrew University) proposed that sweetness linger may be a consequence of sweetener inhibition of G Protein Receptor Kinases (GRKs) involved in termination of sweetener receptor signaling (Zubare-Samuelov et al., 2005). However, this rationale does not explain slow sweetness onset, which always co-occurs with sweetness linger, and therefore is unlikely. Other explanations were considered as well. However, it was concluded that the most probable rationale for sweetness linger, which also accounts for slow sweetness onset, is that HP sweeteners, particularly large saponin-like molecules like REBA (Figure 30.2), engage in extensive non-specific binding to cell membrane sites throughout the oral cavity. If this is the case, it would be expected that much of the sweetener in a solution entering the mouth would bind weakly and rapidly to non-receptor sites and the concentration on the receptor would only reach its maximum after a delay. If this mechanistic rationale is operative, then, upon HP sweetener dissociation from the receptor, it would likely bind non-specifically to nearby non-receptor sites and thus may re-activate the receptor over and over, thus leading to lingering sweetness. This rationale for HP sweetener atypical temporal profiles was referred to as the Non-Specific Binding (NSB) hypothesis (DuBois, 2011).

If the NSB hypothesis is accurate, then the challenge in replication of the temporal profile of CHO sweeteners becomes one of formulation to attenuate non-specific binding. The tonicity of media is well understood to dramatically affect biological cells and, for this reason, they are normally incubated in isotonic buffered saline systems. If media tonicity is reduced, cells swell and may even lyse if sufficiently hypotonic. And conversely, if tonicity is increased to become hypertonic, cells shrink. Recognizing that hypertonic formulations of HP sweeteners may have such an effect on oral epithelia, it was considered that hypertonic formulation may be an effective approach for attenuation of HP sweetener non-specific binding, thereby

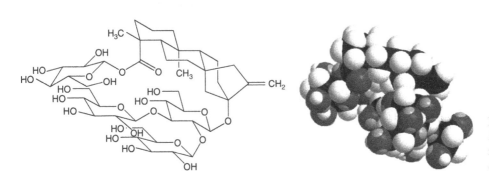

Figure 30.2 Rebaudioside A illustrated in 3D structural representations. (*See plate section for color version.*)

accelerating sweetness onset and attenuating sweetness linger.

Support for the NSB hypothesis was found in observations of Maruzen Pharmaceutical Company (MPC) scientists (Crammer and Ikan, 1987). They reported that hypertonic NaCl enhanced the sweetness potency of a steviol glycoside (SG) sweetener composition by ca. 3-fold and of a SG/glycyrrhizic acid (GA) blend by 4- to 5-fold. NaCl and other salts have been reported to function as PAMs for a number of GPCRs (Christopoulos and Kennakin, 2002) and therefore this mechanism was considered as an explanation for the observed potency enhancements. However, no precedent exists for such a strong enhancement by salts (i.e., 3–5 fold) and so PAM action by NaCl was considered unlikely. SG and GA sweeteners normally exhibit notable delays in sweetness onsets and pronounced sweetness linger effects. However, in hypertonic NaCl, SG and GA temporal properties were observed to be similar to sucrose. In effort to explain the apparent enhancement effects, earlier findings on the sensory properties of SG sweeteners and GA were considered. Stevioside (STV) and REBA, both SG sweeteners, were determined to exhibit maximal sweetness responses of 9.9 and 10.0% SE, respectively, when evaluated in water (DuBois et al., 1991). And GA, under the same conditions was determined to exhibit a maximal response of 7.3% SE. Also, STV and GA in water were determined to require 22 and 69 seconds, respectively, for sweetness decay to a 2% sucrose equivalency (SE), while sucrose sweetness decay requires only 13 seconds (DuBois and Lee, 1983). However, as noted above, in hypertonic NaCl, SGs and GA show rapid sweetness onset and do not show sweetness linger. Thus, in hypertonic NaCl, the entire sweetness signals for SGs and GA are observed over a short window of time as is the case for sucrose, and this modulatory effect of NaCl leads to an increase in SG and GA R_ms, an observation interpreted as enhancement by the MPC scientists. In subsequent work on REBA temporal profile modulation (Prakash et al., 2007), it was demonstrated that NaCl was not unique in its effects on the temporal profiles of REBA and other HP sweeteners. It was demonstrated that osmolytes, in general, cause this effect. To illustrate, 500 mOsM NaCl, 500 mOsM KCl, and 500 mOsM erythritol were found to be equally effective in accelerating the sweetness onset of 500 mg/L REBA and in attenuating its sweetness linger.

Information presented above generally supports the NSB hypothesis as the mechanism for non-sugarlike temporal profiles of HP sweeteners. Direct support for the interactions of HP sweeteners with biological membranes was recently provided (Miyano et al., 2010) by a surface plasmon resonance study of CHO sweeteners and HP sweeteners (i.e., APM, STV, thaumatin, and single-chain monellin). The levels of interaction of these sweeteners with model membranes correlate with observed levels of sweetness linger.

30.2.3 Antagonists as Modulators of Sweet Taste

In pharmacology, inhibitors historically have been very important for identification of receptor subtypes. Early sweetness inhibitors discovered were triterpenoid natural products including gymnemic acid, ziziphins, and hodulcin (Suttisri et al., 1995). None show sweetener selectivity and, generally, their inhibitory effects develop slowly and are long lasting. In the 1980s, Lindley of Tate and Lyle (Reading, England) initiated a focused program on sweetness inhibitors and identified multiple inhibitors (Lindley, 1991). One of these is phenoxyalkanoic acid salt **5** (Figure 30.3), now commonly known as lactisole. Interestingly, Lindley found that lactisole exhibited equivalent inhibitory effects for all sweeteners tested. At the time of this work, the general consensus was that multiple receptors mediate human sweetness. However, based on the absence of selectivity for lactisole, Lindley argued for a single receptor and now it appears that he was substantially correct. Later, in a cell-based assay, Senomyx scientists substantiated Lindley's conclusion based on sensory experiments and demonstrated lactisole to be a T1R2/T1R3 NAM (Jiang et al., 2005).

In experimentation with lactisole, DuBois and coworkers (D'Angelo et al., 1999) observed lactisole to exhibit a curious "sweet water aftertaste" (SWA). To be clear, water used to rinse the mouth after tasting lactisole @ 200 mg/L tastes strikingly sweet. Such a SWA effect had earlier been reported (Bartoshuk and Lee, 1972) as a taste observed on eating artichokes while drinking water. However, the biochemical mechanism for artichoke SWA has not been explained.

In the course of a sweetener discovery program at TCCC, on occasion new sweetener target compounds were found to be non-sweet, but, surprisingly, to exhibit SWA. And, based on the observation that the sweetness inhibitor lactisole exhibits SWA, other SWA compounds were evaluated for sweetness inhibition and, conversely, other known inhibitors were evaluated for SWA. Included in this study was lactisole (**5**) and indole acetic acid (**6**) which was also claimed a sweetness inhibitor by Lindley, the known phenylurea sweetness inhibitor **7** (Muller et al., 1992) and two non-sweet SWA compounds **8** and **9**. In this study, sweetness inhibitors **6** (@ 1000 mg/L) and **7** (@ 2000 mg/L) exhibited SWA. And both SWA compounds **8** (@ 5000 mg/L) and **9** (@ 500 mg/L) inhibited the sweetness of seven sweeteners tested.

In summary, the correlation between sweetness inhibition and SWA is clear. And, of course, this finding suggests a simplified approach to identification of novel

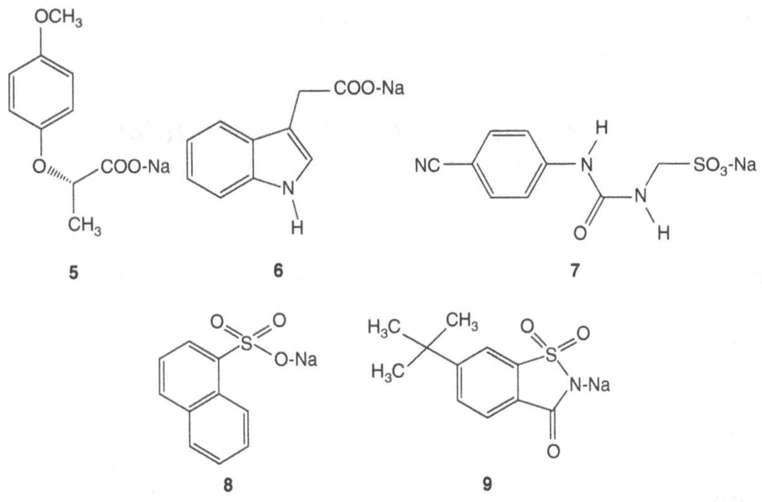

Figure 30.3 Sweetness inhibitors which exhibit Sweet Water Aftertaste.

sweetness inhibitors. However, the question remains as to the mechanism of the SWA effect. A hypothesis was proposed by DuBois and coworkers (D'Angelo et al., 1999) based on three considerations: (1) biological receptors, independent of agonist presence, exist in both active (R_A) and inactive (R_I) states in rapid equilibrium (Monod et al., 1965), (2) most inhibitors (i.e., antagonists) are "inverse agonists," compounds which bind to R_I, thereby inhibiting conformational change to R_A, and (3) some receptors are "constitutively active," whereby, in the absence of agonist, they are substantially in the R_A state. Now, if the human sweetener receptor is constitutively active, then antagonists, acting as inverse agonists, will shift the equilibrium to R_I.

An adaptation of the cell to a reduced population of R_A may then prime the cell to signal following a water rinse causing a restoration of the R_A/R_I equilibrium with a concomitant increase in intracellular affector levels. Of course such signaling could only occur with a constitutively active receptor. Support for this hypothesis was provided by the Breslin group based on observation of SWA and sweetness inhibition for saccharin at high concentrations (Galindo-Cuspinera and Breslin, 2007). Ca^{2+} levels were demonstrated to drop in a T1R2/T1R3 cell system upon exposure to lactisole, thus demonstrating constitutive activity for T1R2/T1R3.

Many salts (e.g., NaCl, KCl, LiCl, LiSO$_4$, and K$_2$SO$_4$), at or below their taste thresholds, exhibit weak sweet taste (Shallenberger, 1993) and, when formulated with CHO sweeteners, give small sweetness enhancements. Interestingly, however, Breslin and coworkers reported ZnSO$_4$ to be an effective sweetness inhibitor (Keast et al., 2005). ZnSO$_4$ @ 25 mM inhibited the sweetness of 10 sweeteners while not affecting the sweetness of sodium cyclamate, a selectivity effect not presently understood.

30.3 BITTER TASTE MODULATORS

Modulators of bitter taste are of interest for the elimination of the bitter off tastes of HP sweeteners (e.g., saccharin, REBA, etc.) as well as some foods (e.g., cruciferous vegetables), drugs and nutritional products. Early work on bitter taste inhibition has been reviewed elsewhere (Beauchamp, 1994; Roy, 1997). Here we summarize more recent progress.

As for other tastes, great progress has recently been made in elucidation of the biochemistry and physiology of bitter taste, as well as in understanding the breadth of bitterant SAR (DuBois et al., 2008a). To summarize, we now understand that bitter taste emanates from a subset of bitter-specific Type II tbcs expressing 25 Class A GPCRs known as T2Rs, thought to be dimeric. Demonstrating the potential complexity of bitterant receptors, Meyerhof and coworkers provided evidence that T2Rs may hetero- as well as homo-dimerize such that the total possible number of receptors is 325 (Kuhn et al., 2010). As with sweet taste, T2R receptors, upon activation, transduce their signals through gustducin, PLCβ$_2$, IP$_3$, IP$_3$R, Ca^{2+}, TRPM$_5$, etc. en route to activation of afferent nerve fibers and communication to the CNS via labeled lines. As for sweet taste, it is unclear if activated bitterant-sensitive tbcs communicate directly with afferent nerve fibers or if Type III tbcs are intermediates (Huang et al., 2007). In any case, since bitter taste modulators must work on the same timescale as the bitterants themselves, research is focused on compounds which act at the receptor. In bitter taste research, substantial effort has been expended on T2R deorphanization. Exemplary of such work are the findings the groups of: (1) Meyerhof: (a) T2R16 – bitter glucopyranosides (e.g., salicin) and T2R10 – strychnine (Bufe et al., 2002), (b) T2R46 – sesquiterpenoid lactones, labdane diterpenoids, strychnine and denatonium salts (Brockhoff et al., 2007),

30.3 Bitter Taste Modulators

and (c) T2R50 – terpenoids andrographolide and amarogentin (Behrens et al., 2009), (2) Gruppen: T2R14 and T2R39 – soy isoflavones and other isoflavonoids (Roland et al., 2011), (3) Pronin: T2R44 – denatonium chloride and 6-nitrosaccharin and T2R61 – 6-nitrosaccharin (Pronin et al., 2004) and (4) Meyerhof and Hofmann: T2R4 and T2R14 – steviol glycoside (SG) sweeteners (Hellfritsch et al., 2012). The finding that SG sweeteners with bitter off taste activate >1 T2R is noteworthy, as activation of multiple T2Rs by individual bitterants is likely to be commonplace. Further, illustrating the complexity of the bitterant receptor situation, recently evidence was provided that the nicotinic acetylcholine receptor also activates bitterant-sensitive tbcs (Oliveira-Maia et al., 2009).

In summary, the number of receptors mediating the activities of bitterants is likely to be large. Nonetheless, enabled by a basic understanding of these receptors, rational cell-based HTS discovery programs for bitterant inhibitors are possible. Examples of progress on this approach are provided below. Also summarized are alternative approaches to bitterness inhibition where the mechanistic details are not understood or which may operate without receptor involvement.

30.3.1 T2R Antagonists

Saccharin was the first HP sweetener to be commercially developed, although usage was limited due to bitter "off" taste. Recently, advantaged by knowledge of T2Rs responsive to saccharin, Givaudan scientists conducted a cell-based HTS program and identified carboxylic acid **10** (Figure 30.4) as a saccharin bitterness antagonist (Slack et al., 2010). Disappointingly, however, while **10** is very effective in the single T2R cell-based assay, it is not as effective in sensory panel testing. It was interesting to find that **10** also inhibits the bitterness of REBA (Slack and Evans-Pennimpede, 2009). Other interesting reports include that related compound **11** inhibits (Brune et al.,

2008) sucralose bitterness and that coumarin **12** inhibits the bitterness of REBA as well as of aspartame and sucralose (Ungureanu and Van Omerren, 2010). And finally, gout drug probenecid (**13**) was found effective in antagonism of multiple bitterant receptors including T2R16, T2R31, T2R38, and T2R43, thus enabling inhibition of the bitter tastes of salicin, saccharin, phenylthiocarbamide (PTC)/6-propyl-2-thiouracil (PROP) and aloin, respectively (Greene et al., 2011). In general, loci of action for antagonists on their receptors are unknown. However, studies on probenecid suggest allosteric action. Senomyx scientists have also carried out cell-based discovery programs on specific T2Rs and their work has led to the two bitterness inhibitors **14** and **15**, both now with regulatory approvals for use in foods and beverages (Li et al., 2011). They are claimed effective in inhibition of the bitter off taste of REBA as well as bitterants in coffee, cocoa, soy protein, and whey protein.

It is important to reflect on the possible rationale for the poor effectiveness of saccharin bitterness antagonist **10** in sensory panel assays while very effective in a cell-based assay. The most plausible rationale may be that, *in vivo*, bitterant-sensitive tbc activation may be mediated by a complex array of receptors, while *in vitro*, the response to a single homodimeric GPCR is observed. Thus, while discovery of potent antagonists for individual T2Rs may be routine, it seems probable that the discovery of antagonists that are fully effective in human taste will remain a challenge.

Today work is in progress to find bitterness antagonists for natural HP sweeteners (e.g., steviol glycosides, mogrol glycosides, etc.). However, while searching for broad-activity antagonists may appear rational, the practical utility of such antagonists may be limited. REBA and many HP sweeteners exhibit low R_ms and the major application for HP sweeteners is in beverages where high sweetness intensity is required. And since REBA exhibits an R_m of 10.0% SE (DuBois et al., 1991), it may be

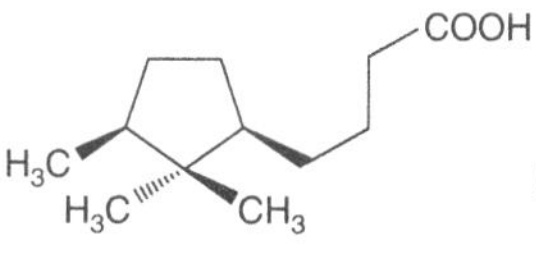

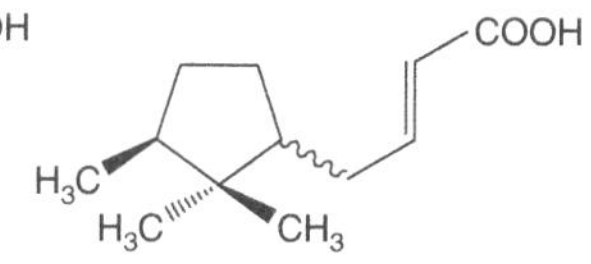

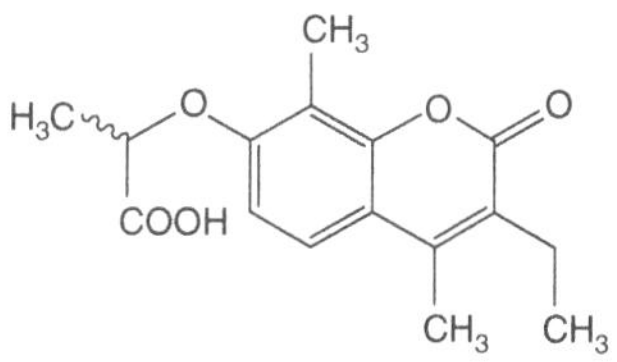

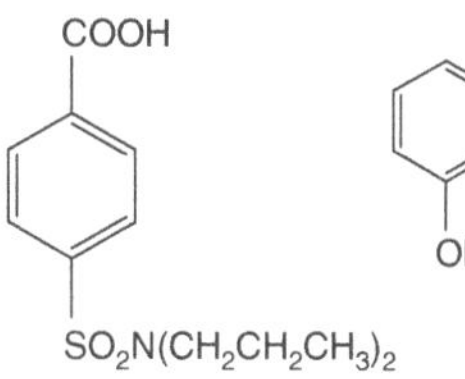

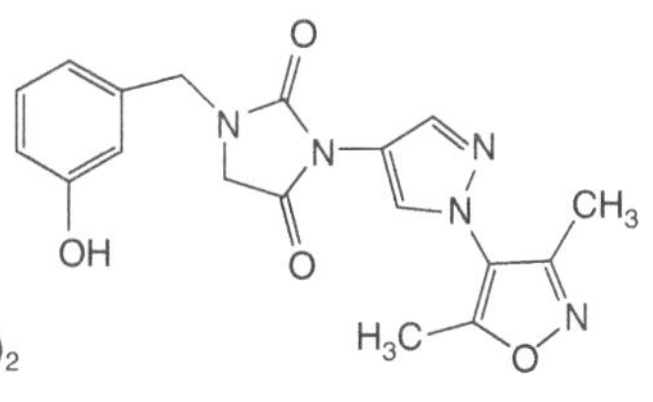

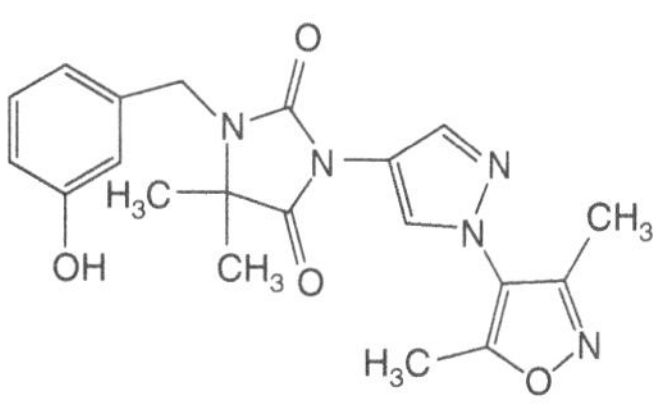

Figure 30.4 Inhibitors of the bitter off tastes of HP sweeteners: Saccharin (**10** and **13**), Aspartame (**12**), Sucralose (**11** and **12**) and Rebaudioside A (**10, 12, 14** and **15**).

impractical to formulate a beverage with REBA as the sole sweetener. Even if a 9.0% SE level of sweetness is adequate, it can be calculated that 1800 mg/L would be required, while the maximal allowable level for beverages (US) is 600 mg/L. A practical solution to this challenge with a low R_m sweetener like REBA is blending with a second sweetener. Thus, if in a binary blend, REBA is used to provide a 5% SE level of sweetness intensity, only 200 mg/L is required, where bitter off taste is negligible.

30.3.2 No-Known-Mechanism Inhibitors as Modulators of Bitter Taste

A very old approach to bitterness inhibition is based on a phenomenon known as *mixture suppression* (Lawless and Heymann, 1998). Thus compositions with bitter off taste are often formulated with sweeteners with the result that they taste less bitter. The mechanism of this effect is not known and much debate has centered on the question as to whether the effect occurs at the periphery (i.e., the taste bud) or in the CNS. Most scientists believe that both peripheral and central processes are operative.

Some salts attenuate bitterness. Thus NaCl reduces caffeine bitterness (Pangborn, 1960) and also quinine bitterness (Bartoshuk, 1975; Frijters and Schifferstein, 1994). Later, LiCl as well as NaCl were shown effective in suppression of urea bitterness, but KCl was not (Breslin and Beauchamp, 1995). Interestingly, the bitterness suppressive effect for sodium salts was found independent of saltiness since the less salty sodium acetate and sodium gluconate are equally potent in urea bitterness suppression. In continuing work (Keast et al., 2004), it was demonstrated that the sodium salt effect on bitterant taste suppression is not universal, not all bitterants being affected. In these studies, it was clearly demonstrated that sodium salts are effective at concentrations below their salty taste thresholds, and therefore, it is clear that their effect on bitter taste perception is not mixture suppression. The mechanism mediating bitterness suppression by sodium salts remains to be determined.

Zinc salts are also effective in inhibition of bitter taste (Keast, 2004). However, a problem in their use is that they also inhibit the sweetness of most sweeteners, thus limiting application to non-sweet products. However, a formulation of a zinc salt with sodium cyclamate is a possible solution to this problem, since cyclamate sweeteners are unaffected by zinc salts (Keast and Breslin, 2005). The mechanism for zinc salt inhibition of bitter taste is not known at this time.

Phosphatidic acid (PA) and PA/protein complexes (e.g., β-lactoglobulin) are active in inhibition of the bitter tastes of many compounds including quinine, caffeine, propranolol, promethazine, denatonium salts, etc. (Katsuragi and Kurihara, 1995). The mechanism is thought to be via sorption of bitterants into PA lipid vesicles rendering bitterants unavailable to bitterant receptors. Application of this approach to HP sweeteners (e.g., saccharin, REBA, etc.), however, is impractical since sweetness would also be suppressed.

30.4 MODULATORS OF UMAMI TASTE

In humans, umami taste is essentially the taste of glutamic acid salts, where monosodium glutamate (MSG) is the most common umami stimulus. Umami is a Japanese word meaning "delicious taste" and, in English, the word "savory" best describes the sensation. Interest in umami taste modulators is focused on two types of compounds: (1) MSG enhancers and (2) MSG enhancers which are also umami agonists. As is the case for the other taste modalities, great progress has been made over the last decade in elucidation of the biochemistry and physiology of umami taste, as well as in understanding the SAR of umami tastants (DuBois et al., 2008a). To summarize current knowledge, we now understand that umami taste emanates from a subset of umami-specific Type II tbcs expressing the heterodimeric Class C GPCR T1R1/T1R3. It is generally accepted that T1R1/T1R3 binds glutamate, its native agonist, in the hinge region of the extracellular VFT domain of T1R1. T1R1/T1R3, upon activation, transduces its signal through the same affecters as involved in sweet and bitter taste (i.e., gustducin, PLCβ$_2$, IP$_3$, IP$_3$R, Ca^{2+}, TRPM$_5$, etc.) en route to activation of afferent nerve fibers which communicate to the CNS via a labeled line process. As for sweet and bitter tastes, it is unclear at the present time if activated umami-sensitive tbcs communicate directly with afferent nerve fibers or if Type III tbcs are intermediate (Huang et al., 2007). Clearly umami taste modulator research could be focused at many points in the cellular activation cascade. However, since they must work on the same timescale as MSG, research on umami receptor PAMs and PAM/agonists has been focused on compounds acting at the receptor.

30.4.1 Pure PAMs as Modulators of the Umami Receptor

The purine nucleotides inosine monophosphate (IMP (**16**)), guanosine monophosphate (GMP (**17**)), and adenosine monophosphate (AMP (**18**)), as well as many structural analogues, have been reported to exhibit umami taste, as has been reviewed (Yamaguchi, 1979). In addition to exhibiting umami tastes themselves, these compounds are claimed to be strongly synergistic with MSG, with relative enhancement potencies of 1.0, 2.3, and 0.18, respectively. However, *in vitro* studies on the rat (Nelson et al., 2002)

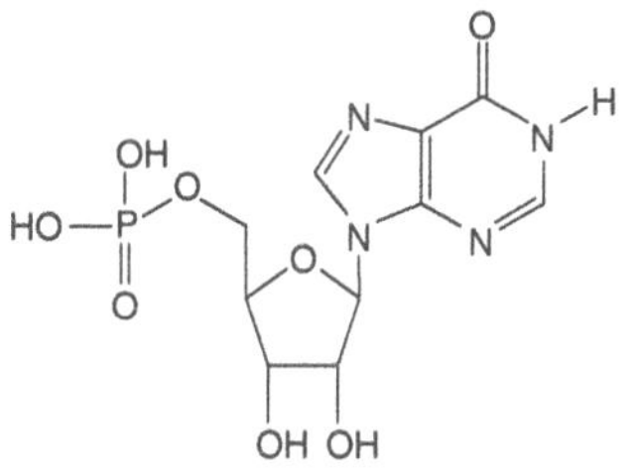

30.4 Modulators of Umami Taste

Figure 30.5 Pure PAMs as modulators of the human umami receptor.

16

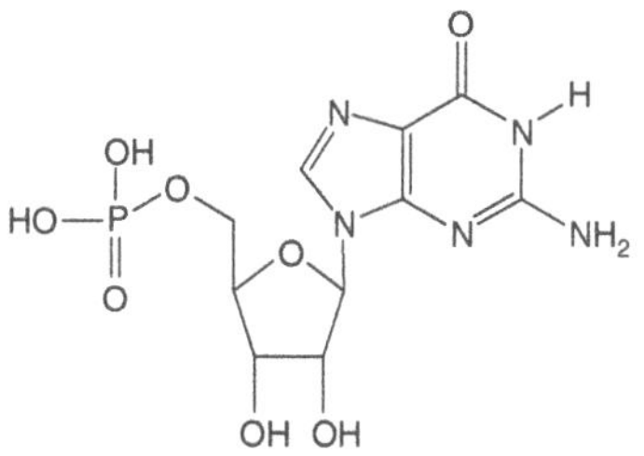

17

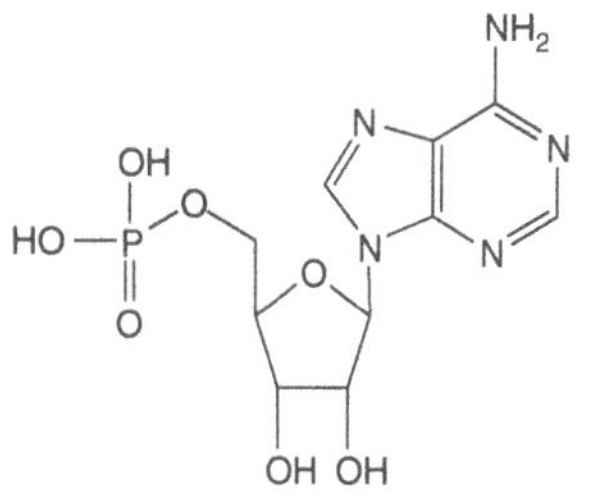

18

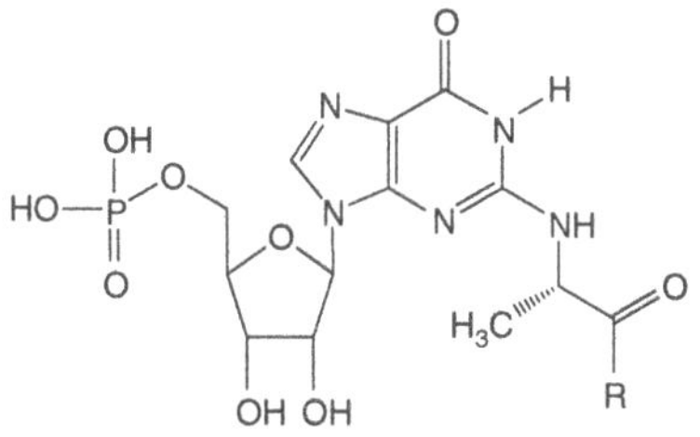

19: R = OH
20: R = NHCH$_2$CH(CH$_3$)$_2$

and human receptors (Li et al., 2002), showed that these nucleotides are inactive if evaluated in the absence of glutamate or other amino acids with umami taste. Normally, in the *in vitro* receptor assays carried out in heterologous cell systems, the extracellular fluid contains low levels of amino acids. And, under these conditions, purine nucleotides were observed to exhibit activity. However, to ensure that the activity observed was due to the purine nucleotides rather than an enhancement effect of endogenous amino acids, the cell preparations were washed with saline immediately before assay. Under these conditions, the purine nucleotides give no responses, while MSG responds in the normal manner. In summary, purine nucleotides **16–18** are not umami receptor agonists but are umami receptor PAMs for MSG. Supporting this conclusion is that, while 0.1 mM IMP exhibits a weak umami taste, no increase in umami taste intensity is observed for 1.0 and 10.0 mM solutions (DuBois and San Miguel, 2005). It appears that 0.1 mM IMP is sufficient to maximally enhance the umami taste of endogenous glutamate in saliva and thus increased IMP levels are without effect. It is noteworthy that IMP, GMP, and AMP exhibit quite variable efficacies as PAMs, as consistent with a receptor-mediated effect. Studies involving chimeric receptors, site-directed mutagenesis and molecular modeling aimed at elucidation of the mechanism for purine nucleotide enhancement of glutamate activity at the umami receptor have been reported (Zhang et al., 2008). And these studies suggest that purine nucleotides bind within the T1R1 VFT domain concurrently with the receptor and the glutamate agonist. Apparently, purine nucleotides are unable to affect closure of the VFT domain

by themselves, but, in the presence of glutamate, are able to increase the stability of the closed VFT domain conformation in comparison to the situation in the absence of the nucleotide.

In recent work in the laboratories of Meyerhof and Hofmann, a series of 25 structural analogues of GMP were prepared by reaction of GMP with dihydroxyacetone and glyceraldehyde under Maillard Reaction type conditions (Festring et al., 2011). Of the 25 compounds prepared, 12 were found to be more efficacious as MSG-enhancing PAMs than GMP itself. Exemplary of the GMP analogues studied are carboxyalkyl derivative **19** and carboxamidoalkyl derivative **20**, where in a T1R1/T1R3 cell-based assay, the shifts in the MSG C/R function observed were 2.9- and 3.2-fold larger, respectively, than that of GMP.

30.4.2 Agoenhancers as Modulators of the Umami Receptor

Concurrent with the discovery of the human sweetener receptor, the human umami receptor was discovered by Senomyx scientists (Li et al., 2002) thus enabling a marked acceleration of umami receptor agonist and PAM discovery. This program led to the discoveries of many MSG enhancers including compounds **21–24** shown in Figure 30.6 (Tachdjian et al., 2012), which now have regulatory approvals. Benzodioxole **21** was reported to exhibit an EC$_{50}$ of 200 nM in a T1R1/T1R3 cell-based assay and, when employed at 30 nM in the same assay, caused a 6.92 shift in the MSG C/R function. Oxalamides **22**, **23**, and

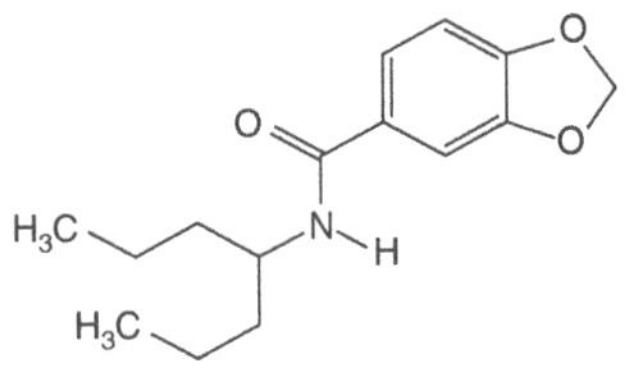

674   **Chapter 30   Chemical Modulators of Taste**

Figure 30.6   Agoenhancer modulators of the human umami receptor.

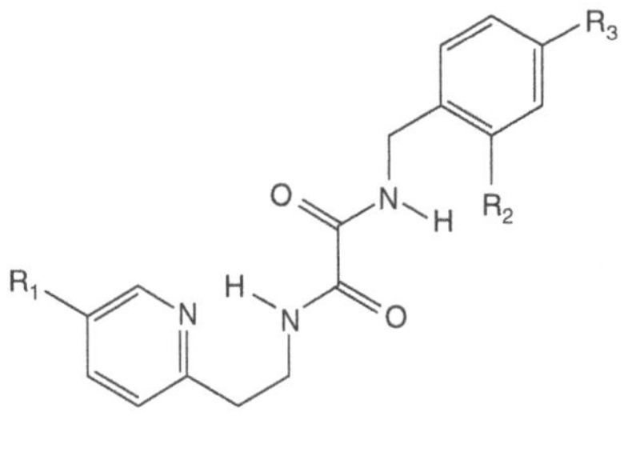

22 : R_1 = H, R_2 = OCH_3 and R_3 = CH_3

23 : R_1 = H and R_2 = R_3 = OCH_3

24 : R_1 = CH_3, R_2 = OCH_3 and R_3 = CH_3

21

24 were reported to exhibit EC_{50}s of 40, 90 and 30 nM, respectively. And **23** @ 300 nM, caused a 6.51 shift in the MSG C/R function. In a sensory panel assay, benzodioxole **21** @ 3 µM was reported to provide an enhancement of 12 mM MSG taste equivalent to that enabled by 100 µM IMP. And oxalamides **23** @ 1 µM and **24** @ 0.3 µM were reported to provide enhancements of 12 mM MSG taste equivalent to that enabled by 100 µM IMP. Also, in sensory panel assays (Shigemura et al., 2012), oxalamides **23** at 1 µM and **24** at 0.3 µM were reported to be equal in umami taste intensity to 12 mM MSG.

30.4.3   Antagonists as Modulators of the Umami Receptor

It is not clear that there is a practical utility for a umami taste antagonist. Nonetheless, it is noteworthy to point out that lactisole, a known antagonist for sweet taste that binds in the TMD region of T1R3, is also an antagonist for umami taste (Xu et al., 2004). Based on studies in a T1R1/T1R3 cell-based assay and a human sensory study on MSG taste thresholds, lactisole was demonstrated to be a umami antagonist. Lactisole antagonist activity for MSG was also demonstrated at suprathreshold levels (Galindo-Cuspinera and Breslin, 2006).

30.4.4   Glycoconjugated Peptides as Modulators of the Umami Receptor

During the aging and/or cooking process, a reaction called the "Maillard reaction" occurs between reducing carbohydrate and amino acids. Maillard peptides (1000–5000 Da) generated during the aging of soy-pesto and soy-sauce affected not only basic taste qualities but also enhanced "Kokumi," a sensation of mouthfulness and continuity. Naturally occurring Maillard peptides (1000–5000 Da) were purified from Korean soy sauce matured for four years. Alternately, Maillard peptides were synthesized by conjugating a peptide fraction (1000–5000 Da) purified from soy protein hydrolysate with galacturonic acid, glucosamine, xylose, fructose, or glucose. In an *in vivo* assay using chorda tympani (CT) taste nerve response in rats, wildtype (WT) mice, and TRPV1 knockout (KO) mice, Maillard peptides were found to modulate salty taste and umami taste. At low concentrations (0.05 to 1%) Maillard peptides conjugated with galacturonic acid, glucosamine, and xylose, modulated the benzamil (Bz)-insensitive NaCl CT responses in a biphasic manner (Katsumata et al., 2008; see below). However, at concentrations above those which modulate the neural responses to salt taste, Maillard peptides conjugated with xylose (at 2.5%) enhanced the CT response to mono sodium glutamate (MSG). In these studies the CT response to MSG was monitored in the presence of Bz (a blocker of the epithelial Na^+ channel; ENaC) and SB-366791 (a blocker of TRPV1/TRPV1t non-specific cation channel) to eliminate the contribution of Na^+ to the CT response. As expected, in these studies IMP enhanced the CT response to glutamate, however, in the presence of IMP+Maillard peptides conjugated with xylose, the magnitude of the CT response to glutamate was less than additive relative to the CT response to IMP alone or Maillard peptides conjugated with xylose. In contrast, Maillard peptides conjugated with galacturonic acid or glucosamine did not modulate the CT response to glutamate (Katsumata et al., 2008). These results suggest that IMP and Maillard peptides conjugated with xylose produce their effects on the CT response to glutamate by acting on the umami receptor. In human sensory studies naturally occurring Maillard peptides have recently been shown to enhance umami taste (Rhyu et al., unpublished observations).

30.4.5   NGCC as a Modulator of the Umami Receptor

Surprisingly, N-geranyl cyclopropylcarboxamide (NGCC), a compound synthesized by International Flavors and

Fragrances, also has similar effects on salty and umami taste as Maillard peptides. At low concentration range (1–10 μM) it modulates NaCl+Bz CT responses in rats and WT mice in a biphasic manner (see below). Above 20 μM NGCC enhanced the CT response to glutamate (Dewis et al., 2013). In human sensory evaluation, NGCC enhanced the saltiness of 80 mM NaCl at 5 and 10 μM and enhanced the umami taste perception of chicken broth containing 60 mM Na^+. However, in the CT studies and human sensory studies several structurally related compounds N-cyclopropyl E2, Z6-nonadienamide, N-geranyl isobutanamide, N-geranyl 2-methylbutanamide, and allyl N-geranyl carbamate did not mimic the effects observed with NGCC (Dewis et al., 2013). These studies raise the possibility that the umami receptor can interact with a very diverse and broad range of compounds and peptides.

30.5 MODULATORS OF SOUR TASTE

30.5.1 Taste Cell Subset in Sour Transduction

Sour taste reception is evoked by both organic and mineral acids and is mediated by a subset of taste receptor cells that contain characteristic transient receptor potential polycystic (TRPP) marker proteins, PKD2L1 and PKD1L3 (DuBois et al., 2008). PKD2L1 is found in Type III taste cells in both the fungiform papillae of the anterior tongue and the circumvallate (CV) papillae of the posterior tongue while PKD1L3 is confined to CV taste cells (Ishimaru et al., 2006; Huang et al., 2006). Type III cells make conventional synapses with afferent taste nerve fibers, so encoding for sour taste probably involves this cell type exclusively (Vandenbeuch and Kinnamon, 2009). Eliminating the subset of taste cells containing PKD2L1 in mice eliminates neural responses specifically to sour stimuli (Huang et al., 2006). Accordingly, it was hypothesized that PKD2L1 and PKD1L3 might be sour taste receptors. However, in mice lacking PKD2L1 expression, CT responses to acids from the anterior tongue were only partially inhibited while PKD1L3 suppression had no effect (Horio et al., 2011). Single and double knock-out mice gave glossopharyngeal nerve responses to acids from the posterior tongue that were not different from wild type mice (Horio et al., 2011). Overall these results indicate that PKD2L1 and PKD1L3 are reliable markers of sour-sensitive taste cells, PKD2L1 may have some role in sour transduction in fungiform taste cells, but neither PKD2L1 nor PKD1L3 plays a role in sour transduction in CV taste cells. Sour ageusic human subjects lack transcripts for PKD2L1 and PKD1L3 in their fungiform papillae cDNA suggesting that the taste cells expressing these proteins are also sour taste transducing cells in humans (Huque et al., 2009).

30.5.2 Stimulus Flow Rate Modulates Phasic and Tonic Sour Taste Responses

A CO_2/HCO_3^- buffer at pH 7.4 is a potent sour taste stimulus, as determined by its CT response in rats, in spite of its alkaline pH (Lyall et al., 2001). This itself is proof that stimulus extracellular pH is not directly related to sourness, a conclusion supported by diverse physiological and psychophysical studies for over a century (DeSimone et al., 2001; DeSimone and Lyall, 2006). When applied at a flow rate of 1 ml/s the response consists of a rapid phasic increase in neural activity followed by a slower decrease in response that asymptotically approaches a steady-state, the so-called tonic response level. This phasic-tonic character and its relation to stimulus flow rate have been previously noted for other taste qualities (Pfaffmann, 1955; Smith and Bealer, 1975; Lyall et al., 2001; Lyall et al., 2002a). At the slower flow rate of 1 ml/min the phasic response is notably absent (Lyall et al., 2001). Rather the initial response is a decaying exponential rise that nonetheless attains the same tonic steady state (DeSimone and Heck, 1980). The results suggest the phasic and tonic parts of the response occur by independent processes, perhaps in separate receptor cell types and mediated by different transduction mechanisms (Lyall et al., 2006; McCaughey, 2007).

30.5.3 Calcium Modulates the Tonic Response to Acids but the not Phasic

Acidification of a subset of isolated taste cells is accompanied by an increase in intracellular Ca^{2+} (Liu and Simon, 2001; Richter et al., 2003). The role of Ca^{2+} was found to be confined to the tonic phase of the response (Lyall et al., 2006). Only the tonic part of the CT response in rat to HCl (Lyall et al., 2006), to a CO_2/HCO_3^- buffer (pH 7.4) and acetic acid (DeSimone et al., 2012) was blocked after treatment of rat tongue with BAPTA-AM, precursor to the Ca^{2+} chelator, BAPTA. The tonic response to each acid could be restored by adding additional intracellular Ca^{2+} (using the Ca^{2+} ionophore, ionomycin) to exceed the Ca^{2+}-buffer capacity of the sequestered BAPTA (DeSimone et al., 2012). These results suggest that an increase in intracellular Ca^{2+} is necessary for transduction during the tonic response to all acids and that the phasic response occurs by a Ca^{2+}-independent mechanism.

30.5.4 Cell Shrinkage Modulates the Phasic Response to Acids

Intracellular acidification causes rapid taste cell shrinkage independently of stimulus osmolarity and involves the conversion of taste cell cytoskeleton F-actin to G-actin which is driven by a decrease in intracellular pH (pH_i) (Lyall et al., 2006). This then activates flufenamic acid-sensitive shrinkage-activated nonselective cation channels in the basolateral membrane of the taste cells that permit a Na^+ current to depolarize the cells yielding the phasic response (Lyall et al., 2006).

30.5.5 Mannitol and Cytochalasin B Modulate Cell Shrinkage/Phasic Response

pH_i-decrease-induced cell shrinkage can be modulated by preshrinking the cells by presenting the acid stimulus in a hypertonic medium (e.g., 0.5 M mannitol) or by disrupting the F-actin-G-actin equilibrium by pre-treating the tongue with cytochalasin B (Lyall et al., 2006). Following cytochalasin B pre-treatment stimulating the tongue with CO_2 in hypertonic mannitol, the phasic response was virtually eliminated, while the tonic response was unaffected. Mannitol suppression of the phasic response was completely reversible. The suppressive effect of cytochalasin B could be reversed by subsequent treatment of the tongue with phalloidin (Lyall et al., 2006). Phalloidin stabilizes F-actin and, therefore, opposes the action of cytochalasin B leading to recovery of the phasic response. These results suggest that the phasic and tonic responses are pharmacologically separable and arise from distinct transduction mechanisms that very likely occur in separate receptor cell types (McCaughey, 2007).

30.5.6 Tonic Response Modulated by Cell NHE1: A Balance Between Acid Entry and Exit

The sour tonic response level is proportional to the time-invariant decrease in pH_i achieved when the rate of acid equivalents entering the taste cells across their apical membranes is just balanced by acid efflux across their basolateral membranes. The manner in which acids enter the sour taste cells differs for weak and strong acids (see next section), but the mechanism by which taste cells remove acid from their cytosol is the same for all types of acid stimulation.

Protons are removed from the sour-sensing cells by the sodium-hydrogen exchanger isoform 1 (NHE1) (Lyall et al., 2004a; Vinnikova et al., 2004). Measurement of taste cell pH_i changes in isolated rat polarized fungiform taste buds shows that NHE1 activity increases with increasing intracellular Ca^{2+} (Lyall et al., 2004a). Thus increasing intracellular Ca^{2+} has two effects: it sustains the tonic response and also stimulates NHE1 to remove H^+ ions from the cytosol. The final intracellular Ca^{2+} concentration achieved is, therefore, critical in determining the final tonic response level or the sour adaptation level. For each of three

acids: HCl, CO_2, and acetic acid treatment of rat tongue with ionomycin, a Ca^{2+} ionophore, allowed increased Ca^{2+} entry which in each case significantly suppressed the tonic response (Lyall et al., 2004a). That this suppression arises from Ca^{2+}-stimulated increased H^+ efflux through basolateral membrane NHE1 was demonstrated by treating the tongue with zoniporide, a specific inhibitor of NHE1 activity (Lyall et al., 2004a). In each case the response was fully restored. These results suggest that irrespective of the acid, it is the *decrease in intracellular pH* that is the proximate driver of the sour CT response, and that the steady state level of intracellular pH is, in turn, determined by the balance of acid influx across sour taste cell apical membranes and acid efflux through Ca^{2+}-regulated NHE1 across taste cell basolateral membranes. Increasing proton influx across the apical membranes of taste cells using the K^+-H^+ ion exchanger, nigericin shifted the pH threshold of observing a CT response by 2 pH-units in the alkaline direction, reinforcing the idea that a decrease in intracellular pH of sour-sensing cells is the proximate signal for sour taste transduction (Sturz et al., 2011).

30.5.7 NADPH Oxidase-and cAMP-linked Membrane Proton Channels Mediate Sour Transduction of Strong Acids

Strong acids (e.g., HCl) are fully dissociated, so protons must be able to cross the apical membranes of taste cells through ion channels or other ion transporters. Recent evidence indicates that at least two types of proton channel mediate H^+ entry into taste cells. One of these is an NADPH oxidase-dependent proton channel and another is a channel regulated by cAMP (DeSimone et al., 2011). An apical membrane proton conductance unique to PKD2L1-expressing taste cells has been demonstrated (Chang et al., 2010). That one or more proton channels actually do mediate strong acid sour transduction was confirmed by monitoring CT nerve responses. First, the response to HCl can be completely blocked by Zn^{2+}, a specific blocker of proton conductance, (DeCoursey, 2003; DeSimone et al., 2011) but not the response to the nearly undissociated acetic acid (DeSimone et al., 2011). Second, the response to HCl is voltage sensitive, and that sensitivity is also blocked by Zn^{2+} (DeSimone et al., 2011). The rat CT response to HCl (but not weak acids) is enhanced by the NADPH oxidase activators H_2O_2, PMA, and nitrazepam (DeSimone et al., 2011). Knock-out mice lacking the gp^{91phox} subunit of NADPH oxidase had CT responses to HCl that were diminished by 64% relative to wild type mice and were unaffected by H_2O_2 (DeSimone et al., 2011). However, the remaining response was still inhibited by Zn^{2+} and significantly enhanced by treating the KO mouse tongue with a membrane permeable cAMP analogue (DeSimone et al., 2011).

30.5.8 Modulation of CT Response by Zn^{2+}: Distinguishing Weak and Strong Acid Transduction Mechanisms

The CT response to HCl can be blocked completely by Zn^{2+} indicating that all of transduction depends on proton channels (DeSimone et al., 2011). None of the response to acetic acid or CO_2 is blocked by Zn^{2+}, suggesting that proton channels play no role in transduction of acids that are primarily presented as lipid permeable neutral molecules (DeSimone et al., 2011). Phosphoric acid is an acid that has both strong and weak acid properties. At 20 mM, phosphoric acid at pH 2.3 is about 60% dissociated, that is, 60% of the acid behaves as a strong acid and 40% as a weak acid. Consistent with this about 65% of the response to 20 mM phosphoric acid was blocked by Zn^{2+} (DeSimone et al., 2011). Therefore, phosphoric acid is both a strong and weak acid stimulus, a fact that suggests that its sour taste quality may vary strongly over a narrow pH range as transduction for it engages strong as well as weak acid mechanisms.

30.5.9 Carbonic Anhydrase (CA) Mediates CO_2 Sour Taste Transduction

Carbon dioxide mobilization is essential in regulating respiration and acid–base balance and not surprisingly various isoforms of carbonic anhydrase are found in cells throughout the body, which catalyze the conversion of CO_2 to bicarbonate or the reverse. Carbon dioxide rapidly diffuses across lipid bilayers, so the acid that stimulates sour taste receptors (carbonic acid), is formed inside the cells by carbonic anhydrase catalyzed hydration of CO_2. This will result in a decrease in intracellular pH (Lyall et al., 2001) provided the bicarbonate, which is also formed, can be transported out of the cell. Consistent with this, CT responses in rat were inhibited when the rat tongue was treated with a membrane permeable carbonic anhydrase inhibitor (Lyall et al., 2001). The CT response to CO_2 was shown to be voltage-sensitive (Lyall et al., 2001) and both the CT response to CO_2 and its voltage sensitivity could be blocked by the bicarbonate transport inhibiter, 4,4′diisothiocyanostilbene-2,2′-disulfonic acid (DIDS) (unpublished observation). This indicates that bicarbonate is removed by an electrogenic transporter in the taste cell apical membrane. Membrane bicarbonate transporters often form a metabolon by linking to both intracellular and extracellular forms of carbonic anhydrase (Sterling et al., 2002). This appears to be the case for taste cells which

have the extracellular carbonic anhydrase-IV (CAIV) in addition to intracellular forms of CA (Chandrashekar et al., 2009).

It has been suggested that CAIV by itself acts as a CO_2 taste receptor by catalyzing the formation of H^+ in the apical extracellular solution and detecting it by an unspecified mechanism (Chandrashekar et al., 2009). However, were that the mechanism CAIV knockout mice would not give a CT response to CO_2. They give a reduced response, but only after treating the mice with a membrane permeable carbonic anhydrase inhibitor is the response completely inhibited. This supports the metabalon concept of CO_2 detection where the functional unit that is responsible for lowering intracellular pH by CO_2 is a bicarbonate transporter linking CAIV with an intracellular CA. CO_2 responses result from an intracellular pH decrease regulated by NHE1 just like other acids (Lyall et al., 2004a). Were H^+ formation and detection occurring in the apical extracellular space, NHE1 regulation would not be expected. Finally, responses to a buffer system in which CO_2, HCO_3^-, and H^+ are in chemical equilibrium are routinely recorded (Lyall et al., 2004a). A response is observed because CO_2 rapidly enters the taste cell interior where CO_2, HCO_3^-, and H^+ are not in equilibrium. This is the necessary condition that enables intracellular CA to accelerate the hydration of CO_2. If CAIV were to make contact with and act exclusively on the stimulus solution as hypothesized (Chandrashekar et al., 2009), then CAIV action cannot, by itself, account for the observed response to CO_2 buffers in equilibrium. Catalysts of any sort do not drive reactions when the component species are in chemical equilibrium. Thus the mechanism of Chandrashekar et al. has the serious problem of violating the second law of thermodynamics. CAIV is important in CO_2 taste transduction, but it must function in concert with HCO_3^- transport out of the taste cells and intracellular CA action.

30.6 MODULATORS OF SALT TASTE

Maintaining proper Na^+ balance is a survival challenge for many terrestrial mammals especially herbivores. The discovery of edible salts in the environment by Na^+ seeking mammals is made possible by salt taste receptors including one that is Na^+ specific. For humans salt taste is generally appetitive and only in recent times has dietary salt been available in large quantities. For these reasons people in developed countries often ingest far more salt than is required to maintain a normal Na^+ balance (Weinberger, 1996). In the USA the clinical and public health recommendations by the Institute of Medicine of the National Academies in 2010 are that Na^+ intake be no more than 2,300 mg/day for persons of two or more years of age.

678   **Chapter 30   Chemical Modulators of Taste**

Because of a lack in food flavor low Na^+ diets are difficult to maintain and, therefore, do not work to lower Na^+ intake over an extended period of time. For that reason the identification of the salt taste transduction mechanisms leading to possible discovery of natural and synthetic agonists of salt taste, could result in ways of making low Na^+ diets more palatable by maintaining an acceptable sensation of salty at reduced dietary Na^+ levels (Wenner, 2008).

In mammals at least two salt taste receptors have been characterized so far: one that is Na^+ specific, and a second that does not discriminate among Na^+, K^+, NH_4^+, and Ca^{2+}. The Na^+-specific receptor in fungiform Type I taste cells is the amiloride- and benzamil(Bz)-sensitive epithelial Na^+ channel (ENaC) and is the predominant salt taste transducer in rats, many mice species and hamsters (DeSimone and Lyall, 2008; Chandrashekar et al., 2010). The non-specific salt taste receptor is amiloride- and Bz-insensitive, and is the predominant transducer of salt taste in some mammalian species, including humans (Ossebaard and Smith, 1995). The best evidence to date suggests that it is a constitutively active, non-selective cation channel that shares many similarities with TRPV1 (TRPV1t) (Lyall et al., 2004b). However, additional amiloride-insensitive salt taste receptors may be present in taste receptive fields apart from the anterior tongue (Treesukosol et al., 2007). Thus, modulators of ENaC and the putative TRPV1t in taste cells are, therefore, potential targets for modifying the taste of Na^+.

30.6.1 Modulators of ENaC-Dependent Salt Taste

30.6.1.1 Osmolarity.
Rat chorda tympani (CT) responses to NaCl are reversibly enhanced in mixtures containing NaCl + hypertonic mannitol or cellobiose. Both the NaCl response and the subsequent enhancement in the presence of hypertonic NaCl stimulating solutions are blocked by amiloride (Lyall et al., 1999). This effect is most likely due to a sustained decrease in taste cell volume in hypertonic mannitol or cellobiose and the shrinkage-induced activation of ENaC.

30.6.1.2 Intracellular pH.
Similar to other tissues, in taste cells ENaC is regulated by changes in intracellular pH (pH_i). A decrease in taste cell pH inhibited, and an increase cell pH increased the Bz-sensitive NaCl CT response. Diethylpyrocarbonate and Zn^{2+}, modification reagents for histidine residues in proteins, attenuated the pH_i-induced inhibition of the Bz-sensitive NaCl CT responses (Lyall et al., 2002b).

30.6.1.3 Intracellular Calcium.
In taste cells ENaC is regulated by changes in intracellular Ca^{2+} ($[Ca^{2+}]_i$). A decrease in taste cell $[Ca^{2+}]_i$ enhanced and an increase $[Ca^{2+}]_i$ inhibited the Bz-sensitive NaCl CT response. Since changes in pH_i show an inverse relationship with $[Ca^{2+}]_i$ in a subset of taste cells, an increase in pH_i and the related decrease in $[Ca^{2+}]_i$ may produce additive effects in increasing the Bz-sensitive NaCl CT response (Lyall et al., 2003).

30.6.1.4 3'-5'-cyclic Adenosine Monophosphate (cAMP).
Increasing cAMP in taste cells directly by the topical lingual application of membrane permeable form of cAMP (8-(4-chlorophenylthio)-adenosine 3', 5'-cyclic monophosphate) enhanced the Bz-sensitive NaCl CT response (Lyall et al., 2011). Arginine vasopressin and cAMP also enhance currents through amiloride-sensitive ENaC in isolated hamster taste cells (Gilbertson et al., 1993). These results suggest that in taste cells ENaC and the amiloride-sensitive NaCl CT responses are regulated by the same hormonal intracellular signaling mechanisms found in other tight epithelia.

30.6.1.5 Na^+-Self Inhibition.
In taste cells topical lingual application of Zn^{2+} enhanced the Bz-sensitive NaCl CT response at low NaCl concentrations (DeSimone and Lyall, 2008). Zn^{2+} most likely enhances NaCl CT response by interfering with the Na^+-self inhibition of the ENaC in taste cells (Gilbertson and Zhang, 1998). Ouabain, a blocker of the Na^+-K^+-ATPase, inhibited the NaCl CT response (Lyall et al., unpublished observations). This suggests that the Na^+-self inhibition of the ENaC in taste cells is most likely involved in adaptation of the NaCl CT response.

30.6.1.6 Channel Activating Proteases and Small Molecule Activators.
Treating the lingual surface with trypsin for 20 min enhanced the CT response to NaCl, suggesting that similar to other tissues, in taste cells ENaC activity is regulated by proteases (DeSimone and Lyall, 2008). A small molecule activator, N-(2-hydroxyethyl)-4-methyl-2-((4-methyl-1H-indol-3-yl)thio)pentanamide, that activates human ENaC (Lu et al. 2008) enhanced the rat Bz-sensitive NaCl CT response (Mummalaneni et al., 2014). This suggests that ENaC modulators can be useful in modulating salt taste in humans.

30.6.1.7 Temperature.
Increasing the temperature from 23°C to 30°C enhanced the Bz-sensitive NaCl CT response. However, progressively increasing the temperature from 30°C to 36.8°C and above inhibited the Bz-sensitive NaCl CT response in a temperature-dependent manner (DeSimone and Lyall, 2008). This suggests that at temperatures above 37°C ENaC does not contribute significantly to salt taste perception.

30.6.2 Modulators of Putative TRPV1t-dependent Salt Taste

30.6.2.1 TRPV1 Agonists and Antagonists.

The Bz-insensitive NaCl CT response is modulated by several agonists that also modulated TRPV1. To date all of the agonists tested have one property in common, that is, they modulate the Bz-insensitive NaCl CT response in a biphasic manner. At low concentrations they enhance and at high concentrations they inhibit the Bz-insensitive NaCl CT response. These agonists include resiniferatoxin, capsaicin, cetylpyridinium chloride, nicotine, ethanol, N-geranylcyclopropylcarboximide, and both synthetic and naturally occurring glycoconjugated peptides (Katsumata et al., 2008). Both the constitutive Bz-insensitive NaCl CT response and its modulation by the above agonists are inhibited by TRPV1 blockers, capsazepine, and N-(3-methoxyphenyl)-4-chlorocinnamid and by high concentration of the above agonists (Lyall et al., 2004b). In contrast to the wildtype mice, the TRPV1 knockout mice (KO) demonstrate no constitutive Bz-insensitive NaCl CT response and the response remains at the rinse baseline level in the presence of the above agonists. Unlike TRPV1, the constitutive Bz-insensitive NaCl CT response is observed at room temperature, in the absence of an agonist, is pH-insensitive and is modulated by cetylpyridinium chloride, which is most likely not a TRPV1 agonist. This suggests that the Bz-insensitive NaCl CT response may be dependent upon a putative variant of TRPV1, TRPV1t (Lyall et al., 2004b). Although, the Bz-insensitive constitutive NaCl CT response is pH-independent, the relationship between pH and the agonist-induced increase in the neural response is bell shaped, with the maximum increase in the response being observed around pH 6.1.

Although in behavioral studies TRPV1 KO mice did not show significant differences in response to NaCl+amiloride relative to the wildtype mice (Treesukosol et al., 2007), both glycoconugated peptides (Katsumata et al., 2008) and N-geranylcyclopropylcarboximide (Dewis et al., 2013) modulate human salt taste perception. In alcohol-preferring (P) rats the Bz-insensitive NaCl CT response is spontaneously upregulated and further enhancing its magnitude by glycoconjugated peptide decreased consumption of 150 mM NaCl+Bz and NaCl preference relative to the control alcohol non-preferring (NP) rats (Coleman et al., 2011).

30.6.2.2 Intracellular Calcium.

Similar to the ENaC-dependent NaCl CT response, the Bz-insensitive NaCl response is also modulated by changes in taste cell $[Ca^{2+}]_i$. A decrease in $[Ca^{2+}]_i$ enhanced the constitutive Bz-insensitive NaCl CT response and shifted the relationship between the agonist concentration and the magnitude of the CT response upwards relative to control. Under these conditions, no inhibition of the Bz-insensitive NaCl CT response was observed at high agonist concentrations. In contrast, increasing taste cell $[Ca^{2+}]_i$ did not alter the constitutive Bz-insensitive NaCl CT response but attenuated the magnitude of the neural response at all agonist concentrations (Lyall et al., 2009). Altering the phosphorylation-dephosphorylation state of the putative TRPV1t in taste cells produced similar changes in the Bz-insensitive NaCl CT response in the absence and presence of resiniferatoxin. This suggests that changes in $[Ca^{2+}]_i$ modulate the Bz-insensitive NaCl CT responses partly by altering the phosphorylation-dephosphorylation state of the putative TRPV1t channel(Lyall et al., 2009).

30.6.2.3 Phosphatidylinositol 4, 5-bisphosphate (PIP₂).

Increasing cell membrane PIP_2 levels by topical lingual application of U73122 (a non-specific blocker of phospholipase C) or direct application of DiC8-PIP_2 (a synthetic PIP_2) not only inhibited the constitutive Bz-insensitive NaCl CT response but also attenuated the agonist-induced enhancement of the neural response. In contrast, decreasing cell membrane PIP_2 levels by the topical application of phenylarsine oxide (a blocker of PI_3 kinase) enhanced both the constitutive Bz-insensitive NaCl CT response and the agonist-induced enhancement in the neural response (Lyall et al., 2010). Under these conditions no inhibition of the neural response was observed at high agonist concentrations.

30.6.2.4 Temperature.

Increasing the temperature of the NaCl stimulating solutions progressively from 36°C to 43°C, enhanced the Bz-insensitive NaCl CT response in a dose-dependent manner. However, above 43°C the magnitude of the neural response was attenuated relative to 43°C (Lyall et al., 2004b; DeSimone and Lyall, 2008). In mixtures containing NaCl+Bz and varying resiniferatoxin concentrations, the relationship between temperature and magnitude of the neural response was higher and shifted to lower temperatures. This suggests that both temperature and resiniferatoxin produce additive effects on the Bz-insensitive NaCl CT response.

In summary, many intracellular signaling effectors and agonists of ENaC and putative TRPV1t channel in the anterior taste receptive field that modulate the Bz-sensitive and Bz-insensitive neural responses are potential targets for developing safe and effective salt taste enhancers that can help reduce salt intake in humans.

30.7 CONCLUSIONS

In conclusion, molecular biochemical methods have revolutionized the chemosensory sciences providing answers to long standing questions regarding the transduction mechanisms of the five basic qualities. These techniques, supplemented by electrophysiology and pharmacology, will play an important role in the further investigation of sensory properties such as taste adaptation and taste modulation. Subtle, but important, integrative effects such as the modulation of the time necessary to extinguish a primary taste response by properties of the rinse solution, such as ionic strength, can now be understood as a coupling of physical and primary biochemical events at the taste bud level. It is through such integrative effects that primary tastes may be modulated in the taste periphery, combining to create temporal effects that influence overall flavor perception.

ABBREVIATIONS

AMP	adenosine monophosphate
APM	aspartame
Bz	benzamil
CHO	carbohydrate
CT	chorda tympani
GPCR	G Protein-Coupled Receptor
DIDS	4,4'diisothiocyanostilbene-2,2'-disulfonic acid
ENaC	epithelial Na+ channel
CAIV	extracellular carbonic anhydrase-IV
GA	glycyrrhizic acid
GRKs	G Protein Receptor Kinases
GMP	guanosine monophosphate
HTS	high-throughput-screening
HP	high-potency
IMP	inosine monophosphate
IP3	inositol trisphosphate
IP3R	inositol trisphosphate receptor
pHi	intracellular pH
Rm	maximal response
MSG	monosodium glutamate
NAMs	negative allosteric modulators
NGCC	N-geranyl cyclopropylcarboxamide
NSB	non-specific binding
REBA	Rebaudioside A
PTC	phenylthiocarbamide
PA	Phosphatidic acid
PIP2	Phosphatidylinositol 4, 5-bisphosphate
PROP	6-propyl-2-thiouracil
PAMs	Positive Allosteric Modulators
NHE1	sodium-hydrogen exchanger isoform 1
SG	steviol glycoside
STV	Stevioside
SAR	Structure-Activity-Relationships

SWA	sweet water aftertaste
tbcs	taste bud cells
TRPP	transient receptor potential polycystic marker proteins
VFT	Venus Fly Trap

REFERENCES

Bartoshuk, L. M. (1975). Taste mixtures: Is mixture suppression related to compression? *Physiol. Behav.* 14: 643–649.

Bartoshuk, L. M., and Lee, C.-H. (1972). Sweet taste of water induced by artichoke (Cynara scolymus). *Science* 178: 988–989.

Beauchamp, G. K., Kurihara, K., Ninomiya, Y., et al. (1994). Kirin International Symposium on Bitter Taste, Tarrytown, NY: Elsevier Science, Special Issue: *Physiol. Behav.* 56: 1121–1266.

Behrens, M., Brockhoff, A., Batram, C., et al. (2009). The Human Bitter Taste Receptor hTAS2R50 is Activated by the Two Natural Bitter Terpenoids Andrographolide and Amarogentin. *J. Agric. Fd. Chem.* 57: 9860–9866.

Beidler, L. M. (1954). A theory of taste stimulation. *J. Gen. Physiol.* 38: 133–139.

Bingley, C. A., Olcese, G., and Darnell, K. C. (2007). Taste potentiator compositions and beverages containing same, *U.S. Patent Appln. No. US 2007/0054023 A1* (March 8, 2007).

Bouter, N., Koenig, T., van Ommeren, E., et al. (2009). Sweet flavour modulating carboxyalkyl-substituted phenyl derivatives. *Eur. Patent Appn. EP 2 055 196 A1* (June 6, 2009).

Breslin, P. A. S., and Beauchamp, G. K. (1995). Suppression of Bitterness by Sodium: Variation Among Bitter Taste Stimuli. *Chem. Senses* 20: 609–623.

Britton, S. J., Fry, J. C., Lindley, M. G., and Marshall, S. (1999). Sweetening composition comprising aspartame and 2, 4-dihydroxybenzoic acid. *PCT Patent Appln. No. WO 99/15032* (April 1, 1999).

Brockhoff, A., Behrens, M., Massarotti, A., et al. (2007). Broad tuning of the human bitter taste receptor hTAS2R46 to various sesquiterpene lactones, clerodane and labdane diterpenoids, strychnine, and denatonium. *J. Agric. Fd. Chem.* 55(15), 6236–6243.

Brune, N. E. I., Slack, J. P., Ungureanu, I. M., et al. (2008). Methods to identify modulators. *PCT Patent Appln. WO 2008/119195 A1* (October 9, 2008).

Bryant, R. W., Atwal, K. S., Bakaj, I., et al. (2008). *Development of transient receptor potential melanostatin 5 modulators for sweetness enhancement, In* Sweetness and Sweeteners: Biology, Chemistry and Psychophysics, *edited by Deepthi Weerasinghe and Grant DuBois, ACS Symposium Series 979*, Washington, DC: American Chemical Society, pp. 386–399.

Bufe, B., Hofmann, T., Krautwurst, D., et al. (2002). The human TAS2R16 receptor mediates bitter taste in response to β-glucopyanosides. *Nature Genet.* 32, 397–401.

Chandrashekar, J., Yarmolinsky, D. A., von Buchholtz, L., et al. (2009). The taste of carbonation. *Science* 326: 443–445.

Chandrashekar, J., Kuhn, C., Oka, Y., et al. (2010). The cells and peripheral representation of sodium taste in mice. *Nature* 464: 297–301.

Chang, R. B., Waters, H., and Liman, E. R. (2010). A proton current drives action potentials in genetically identified sour taste cells. *PNAS* 107: 22320–22325.

References

Christopoulos, A., and Kennakin, T. (2002). G-Protein coupled receptor allosterism and complexation. *Pharmacol. Rev.* 54: 323–374.

Coleman, J., Williams, A., Phan, T. H. T., et al. (2011). Strain differences in the neural, behavioral and molecular correlates of sweet and salty taste in naïve, ethanol- and sucrose-exposed P and NP rats. *J. Neurophysiol.* 106: 2606–2621.

Crammer, B., and Ikan, R. (1987). Progress in the chemistry and properties of rebaudiosides. In Developments in Sweeteners, Vol. 3, Chapman and Hall: London, pp. 45–64.

D'Angelo, L. L., King, G. A., and DuBois, G. E. (1999). Selective sweetness inhibitors and a biological mechanism for sweet water aftertaste. Poster presentation at the XXI Association of Chemoreception Sciences Meeting, Sarasota, FL, April 16, 1999, Abstract No. 307.

Dando, R., Dvoryanchikov, G., Pereira, E., et al. (2012). Adenosine enhances sweet taste through A2B receptors in the taste. *J. Neurosci.* 32: 322–330.

DeCoursey, T. E. (2003). Voltage-gated proton channels and other proton transfer pathways. *Physiol. Rev.* 83: 475–579.

DeSimone, J. A., and Heck, G. L. (1980). An analysis of the effects of stimulus transport and membrane charge on salt, acid, and water-response of mammals. *Chem. Senses* 5: 295–316.

DeSimone, J. A., Heck, G. L., and DeSimone, S. K. 1981. Active ion transport in dog tongue: a possible role in taste. *Science* 214: 1039–1041.

DeSimone, J. A., Lyall, V., Heck, G. L., and Feldman, G. M. (2001). Acid detection by taste receptor cells. *Respir. Physiol.* 129: 231–245.

DeSimone, J. A., and Lyall, V. (2006). Taste receptors in the gastrointestinal tract III. Salty and sour taste: sensing of sodium and protons by the tongue. *Am. J. Physiol. Gastrointest. Liver Physiol.* 291: 1005–1010.

DeSimone, J. A., Lyal, I. V. (2008) Amiloride-sensitive Ion Channels. In: Basbaum, A. I., Kaneko, A., Shepherd, G. M. and Westheimer, G. (Eds) The Senses: A Comprehensive Reference. *Vol. 4*, Olfaction & Taste, Eds. S. Firestein and G. Beauchamp. San Diego: Academic Press;. pp. 281–288.

DeSimone, J. A., Phan, T-H. T., Heck, G. L., et al. (2011). Involvement of NADPH-dependent and cAMP-PKA sensitive H^+ channels in the chorda tympani nerve responses to strong acids. *Chem. Senses* 36: 389–403.

DeSimone, J. A., Ren, Z., Phan. T. H., et al. (2012). Changes in taste receptor cell [Ca2+]i modulate chorda tympani responses to salty and sour taste stimuli. *J. Neurophysiol.* 108: 3206–3220.

Dewis, M. L., DeSimone, J. A., Phan, T-H. T., Heck, G. L., Lyall, V. (2006) Effect of N-geranyl cyclopropylcarboxamide (NGCC) on TRPV1 variant salt taste receptor (TRPV1t). *Chem. Senses* 31: A105 (Abstr.).

Dewis M. L., Phan T. H., Ren Z., et al. (2013). N-geranyl cyclopropyl-carboximide modulates salty and umami taste in humans and animal models. *J. Neurophysiol.* 109: 1078–1090.

DuBois, G. E., DeSimone, J. A., and Lyall, V. (2008a). Chemistry of gustatory stimuli. In The Senses: A Comprehensive Reference, Vol 4, Olfaction & Taste, Basbaum, A. I., Keneko, A., Shepherd, G. M., and Westheimer, G. (Eds). San Diego: Academic Press pp. 27–74.

DuBois, G. E. (2008b). *Sweeteners and sweetness modulators: Requirements for commercial viability, In* Sweetness and Sweeteners: Biology, Chemistry and Psychophysics, *edited by Deepthi Weerasinghe and Grant DuBois, ACS Symposium Series 979*, Washington, DC: American Chemical Society, pp. 444–462.

DuBois, G. E. (2011). Validity of early indirect models of taste active sites and advances in new taste technologies enabled by improved models. *Flavour Frag. J.* 26: 239–253.

DuBois, G. E., and Lee, J. (1983). A simple technique for the evaluation of temporal taste properties. *Chem. Senses* 7: 237–247.

DuBois, G. E., and San Miguel, R. (2005). *The Coca-Cola Company*, Atlanta, GA, Unpublished Results

DuBois, G. E., Walters, D. E., Schiffman, S. S., et al. (1991). Concentration-response relationships of sweeteners: A systematic study. In *Sweeteners: Discovery, Molecular Design and Chemoreception*, edited by D. Eric Walters, Frank T. Orthoefer, Grant E. DuBois, ACS Symposium Series 450, Chapter 20, American Chemical Society: Washington, DC, pp. 261–276.

Festring, D., Brockhoff, A., Meyerhof, W. and Hofmann, T. (2011). Stereoselective synthesis of amides sharing the guanosine 5'-monophosphate scaffold and umami enhancement studies using human sensory and hT1R1/rT1R3 receptor assays. *J. Agric. Fd. Chem.* 59: 8875–8885.

Flamme, E, Ungureanu, I. M., and Kohlen, E. (2011). Organic compounds. *PCT Patent Application WO 2011/032967 A1* (March 24, 2011).

Flamme, E., and Bom, D. C. (2011). 2-Methoxy-5-(phenoxymethyl)-phenol. *PCT Patent Application WO 2011/067313 A1* (June 9, 2011).

Frijters, J. E. R., and Schifferstein, H. N. J. (1994). Perceptual interactions in mixtures containing bitter tasting substances. *Physiol. Behav.* 56: 1243–1249.

Galindo-Cuspinera, V., and Breslin, P. A. S. (2007). Taste after-images: the science of "water tastes". *Cell Mol. Life Sci.* 64: 2049–2052.

Galindo-Cuspinera, V., and Breslin, P. A. S. (2006). The liaison of sweet and savory. *Chem. Senses* 31: 221–225.

Galindo-Cuspinera, V., and Breslin, P. A. S. (2008). *What can psychophysical studies with sweetness inhibitors teach us about taste? In* Sweetness and Sweeteners: Biology, Chemistry and Psychophysics, *edited by Deepthi Weerasinghe and Grant DuBois, ACS Symposium Series 979*, Washington, DC: American Chemical Society, pp. 170–184.

Gilbertson, T. A., Roper, S. D., and Kinnamon, S. C. (1993). Proton currents through amiloride-sensitive Na channels in isolated hamster taste cells: Enhancement by vasopressin and cAMP. *Neuron* 10: 931–942.

Gilbertson, T. A., and Zhang, H. (1998). Self-inhibition in amiloride-sensitive sodium channels in taste receptor cells. *J. Gen. Physiol.* 111: 667–677.

Gravitz, L. (2012). Taste bud hackers: Scientists and psychologists are trying to trick our mouths and minds into enjoying foods that are better for us. *Nature* 486: 14–15.

Greene, T. A., Alarcon, S., Thomas, A., et al. (2011). Probenecid inhibits the human bitter taste receptor TAS2R16 and suppresses bitter taste perception of salicin. *PLoS ONE* 6(5): e20123.

Hellekant, G., Ninomiya, Y., DuBois, G. E., and Roberts, T. W. (1996). Taste in chimpanzee: I. The summated response to sweeteners and the effect of gymnemic acid. *Physiol. and Behav.* 60: 469–479.

Hellfritsch, C., Brockhoff, A., Stahler, F., et al. (2012). Human psychometric and taste receptor responses to steviol glycosides. *J. Agric. Fd. Chem.* 60: 6782–6793.

Herness, S., and Zhao, Fang-li. (2009). The neuropeptides CCK and NPY and the changing view of cell-to-cell communication in the taste bud, *Physiol. Behav.* 97: 581–591.

Horio, N., Yoshida, R., Yasumatsu, K., et al. (2011). Sour taste responses in mice lacking PKD channels. *PLoS ONE* 6: 1–10.

Huang, A. L., Chen, X., Hoon, M. A., Chandrashekar, J., et al. (2006). The cells and logic for mammalian sour taste detection. *Nature* 242: 934–938.

Huang, Y.-J., Maruyama, Y., Dvoryanchikov, G., et al.(2007). The role of pannexin 1 hemichannels in ATP release and cell-cell communication in mouse taste buds. *PNAS* 104: 6436–6441.

Huque, T., Cowart, B. J., Dankulich-Nagrudny, L., et al.(2009). Sour ageusia in two individuals implicates ion channels of the ASIC and PKD families in human sour taste perception at the anterior tongue. *PLoS ONE* 4: 1–12.

Ishimaru, Y., Inada, H., Kubota, M., et al. (2006). Transient receptor potential family members PKD1L3 and PKD2L1 form a candidate sour taste receptor. *PNAS* 103: 12569–12574.

Jia, Z., and Yang, X. (2009). Compositions and their use. *PCT Patent Appln. WO 2009/023975 A2* (February 26, 2009).

Jia, Z., Yang, X., Hansen, C. A., et al. (2008). Consumables. *PCT Patent Appln. WO 2008/148239 A1* (December 11, 2008).

Jiang, P., Cui, M., Zhao, B., et al. (2005). Lactisole interacts with the transmembrane domains of human T1R3 to inhibit sweet taste. *J. Biol. Chem.* 280: 15238–15246.

Katsuragi, Y., and Kurihara, K. (1997). Specific inhibitor for bitter taste. In Modifying Bitterness: Mechanism, Ingredients and Applications, Roy, G. (Ed). Lancaster, PA: Technomic Publishing Co., pp. 255–281.

Katsumata, T., Nakakuki, H., Tokunaga, C., et al. (2008). Effect of Maillard reacted peptides on human salt taste and the amiloride-insensitive salt taste receptor (TRPV1t). *Chem. Senses* 33: 665–680.

Keast, R. S. J. (2004). The effect of zinc on human taste perception. *J. Food Sci.* 68: 1871–1877.

Keast, R. S. J., and Breslin, P. A. S. (2005). Bitterness suppression with zinc sulfate and na-cyclamate: a model of combined peripheral and central neural approaches to flavor modification. *Pharmaceut. Res.* 22: 1970–1977.

Krammer, G., Ley, J., Riess, T., et al. (2007). Use of 4-hydroxydihydrochalcones and their salts for enhancing an impression of sweetness. *PCT Patent Application WO 2007/107596 A1* (September 27, 2007).

Krohn, M., and Zinke, H. (2009a). Oligopeptides for use as taste modulators. *PCT Patent Application WO 2009/098016 A1* (August 13, 2009).

Krohn, M., and Zinke, H. (2009b). Cyclic lipopeptides for use as taste modulators. *PCT Patent Application WO 2009/156112 A1* (December 30, 2009).

Kuhn, C., Bufe, B., Batram, C., and Meyerhof, W. (2010). Oligomerization of TAS2R bitter taste receptors. *Chem. Senses* 35: 395–406.

Lawless, H. T., and Heymann, H. (1998). Sensory Evaluation of Food: Principles and Practices. New York: Springer Science, pp. 39–49.

Ley, J. P., Blings, M., Paetz, S., et al. (2008). Enhancers for sweet taste from the world of non-volatiles: polyphenols as taste modifiers. In *Sweetness and Sweeteners: Biology, Chemistry, and Psychophysics, ACS Symposium Series 979*, edited by Deepthi Weerasinghe and Grant DuBois), Washington, DC: American Chemical Society, pp. 400–409.

Ley, J., Kindel, G., Paetz, S., and Krammer, G. (2006). Hydroxydeoxybenzoins and their use to mask a bitter taste. *PCT Patent Application WO 2006/106023 A1* (October 12, 2006).

Ley, J., Kindel, G., Paetz, S., et al. (2007). Use of hesperetin for enhancing sweet taste. *PCT Patent Application WO 2007/014879 A1* (February 8, 2007).

Li, X., Patron, A., Tachdjian, C., et al. (2011). *"Compounds that inhibit (block) bitter taste in composition and use thereof", US Patent 7,939,671 (May 10, 2011).*

Li, X., Staszewski, L., Xu, H., et al. (2002). Human receptors for sweet and umami taste. *PNAS* 99: 4692–4696.

Lindley, M. G. (1991). Phenoxyalkanoic acid sweetness inhibitors. In *Sweeteners: Discovery, Molecular Design and Chemoreception*, edited by D. Eric Walters, Frank T. Orthoefer, Grant E. DuBois, ACS Symposium Series 450, Chapter 19, American Chemical Society: Washington, DC, pp. 251–260.

Liu, L., and Simon, S. A. (2001). Acidic stimuli activates (sic) two distinct pathways in taste receptor cells from rat fungiform papillae. *Brain Res.* 923: 58–70.

Lu, M., Echeverri, F., Kalabat, D., et al. (2008). Small molecule activator of the human epithelial sodium channel. *J. Biol. Chem.* 283: 11981–11994.

Lyall, V., Alam, R. I., Phan, D. Q., et al. (2001). Decrease in rat taste receptor cell intracellular pH is the proximate stimulus in sour taste transduction. *Am. J. Physiol. Cell Physiol.* 281: C1005–C1013.

Lyall, V., Alam, R. I., Phan, T-H. T., et al. (2002a). Excitation and adaptation in the detection of hydrogen ions by taste receptor cells: a role for cAMP and Ca^{2+}. *J. Neurophysiol.* 87: 399–408.

Lyall, V., Alam, R. I., Phan, T. H. T., et al. (2002b). Modulation of rat chorda tympani NaCl responses and intracellular Na^+ activity in polarized taste receptor cells by pH. *J. Gen. Physiol.* 120: 793–815.

Lyall V., Malik S. A., Alam, R.I., et al. (2003) Relationship between intracellular pH and Ca^{2+} in fungiform rat taste receptor cells. *Chem. Senses* 28: A83, (Abstr).

Lyall, V., Alam, R. I., Malik, S. A., et al. (2004a). Basolateral Na^+-H^+ exchanger-1 in rat taste receptor cells is involved in neural adaptation to acidic stimuli. *J. Physiol.* 556: 159–173.

Lyall, V., Heck, G. L., DeSimone, J. A., Feldman, G. M. (1999). Effects of osmolarity on taste receptor cell size and function. *Am. J. Physiol. Cell Physiol.* 277:C800–C813.

Lyall, V., Heck, G. L.,Vinnikova, A. K., et al. (2004b). The mammalian amiloride-insensitive non-specific salt taste receptor is a vanilloid receptor variant. *J. Physiol.* 558: 147–159.

Lyall, V., Pasley, H., Phan, T-H. T., et al.(2006). Intracellular pH modulates taste receptor cell volume and the phasic part of the chorda tympani response to acids. *J. Gen. Physiol.* 127: 15–34.

Lyall, V., Phan, T. H. T., Mummalaneni, S., et al. (2009). Regulation of amiloride-insensitive NaCl chorda tympani responses by intracellular Ca^{2+}, protein Kinase C and calcineurin. *J. Neurophysiol.* 102: 1591–1605.

Lyall, V., Phan, T. H. T., Ren, Z. J., et al. (2010). Regulation of the amiloride-insensitive NaCl chorda tympani responses by Phosphatidylinositol 4, 5-bisphosphate. *J. Neurophysiol.* 103: 1337–1349.

Lyall, V., Phan, T. H. T., Mummalaneni, S., and DeSimone, J. A. (2011). Effect of cAMP on the NaCl Chorda Tympani (CT) taste nerve response profile of young rats. *Chem. Senses* 31: A29 (Abstr. #96).

McCaughey, S. A. (2007). Taste-evoked responses to sweeteners in the nucleus of the solitary tract differ between C57BL/6ByJ and 129P3/J mice. *J. Neurosci.* 27: 35–45.

Miyano, M., Yamashita, H., Sakurai, T., et al. 2010). Surface plasmon resonance analysis on interactions of food components with a taste epithelial cell model. *J. Agric. Food Chem.* 58: 11870–11875.

Monod J., Wyman J., and Changeux J.-P. (1965). On the nature of allosteric transitions: a plausible model. *J. Mol. Biol.* 12: 88–118.

Muller, G. W., Culberson, J. C., Roy, G., et al. (1992). Carboxylic acid replacement structure-activity relationship in suosan type sweeteners. a sweet taste antagonist. *J. Med. Chem.* 35: 1747–1751.

Mummalaneni, S., Qian, J., Phan, TH., et al. (2014). Effect of ENaC modulators on rat neural responses to NaCl. *PLoS One* 9(5): e98049.

Nelson, G., Chandrashekar, J., Hoon, M. A., et al. (2002). An amino-acid taste receptor. *Nature* 416: 199–202.

Niki, M., Jyotaki, M., Yoshida, R., and Ninomiya, Y. (2010). Reciprocal modulation of sweet taste by leptin and endocannabinoids. *Results Probl. Cell Differ.*, 52: 101–114.

Oliveira-Maia, A. J., Stapleton-Kotloski, J. R., Lyall, V., et al. 2009). *PNAS* 106: 1596–1601.

Ossebaard, C. A., and Smith, D. V. (1995). Effect of amiloride on the taste of NaCl, Na-gluconate and KCl in humans: implications for Na^+ receptor mechanisms. *Chem. Senses* 20: 37–46.

Pangborn, R. M. (1960). Taste interrelationships. *Food Res.* 25: 245–256.

Pfaffmann, C. (1955). Gustatory nerve impulses in rat, cat and rabbit. *J. Neurophysiol.* 18: 429–440.

Prakash, I., DuBois, G. E., Clos, J. F., et al. (2008). Development of rebiana, a natural non-caloric sweetener. *Food Chem. Toxicol.* 46 (Supplement 7S): 75–82.

References

Prakash, I., DuBois, G. E., Jella, P., et al. (2007). Natural high-potency sweetener compositions with improved temporal profile, method for their formulation, and uses. *U.S. Patent Appln. 20070128311* (June 7, 2007).

Pronin, A. N., Tang, H., Connor, J., and Keung, W. (2004). Identification of ligands for two bitter T2R receptors. *Chem. Senses* 29: 583–593.

Richter, T. A., Caicedo, A., and Roper, S. D. (2003). Sour taste stimuli evoke Ca^{2+} and pH responses in mouse taste cells. *J. Physiol.* 547: 475–483.

Roland, W. S. U., Vincken, J.-P., Goukas, R. J., et al. (2011). Soy isoflavones and other isoflavonoids activate the human bitter taste receptors hTAS2R14 and hTAS2R39. *J. Agric. Fd. Chem.* 59: 11764–11771.

Roy, G. (1997). Modifying Bitterness: Mechanism, Ingredients and Applications. Lancaster, PA: Technomic Publishing Co.

Salemme, F. R., Long, D., Palmer, R. K., et al. (2010). Rebaudioside C and its stereoisomers as natural product sweetness enhancers. *PCT Patent Application WO 2010/135378 A1* (November 25, 2010).

Schiffman, S. S., Booth, B. J., Carr, B. T., et al. (1995). Investigation of synergism in binary blends of sweeteners, *Brain Res. Bull.* 38: 105–120.

Schwartz, T. W., and Holst, B. (2006). Ago-allosteric modulation and other types of allostery in dimeric 7TM receptors. *J. Recept. Signal Transduct.* 26: 107–128.

Servant, G., Tachdjian, C., Li, X., and Karanewsky, D. S. (2011). The sweet taste of true synergy: positive allosteric modulation of the human sweet taste receptor. *Trends Pharmacolog. Sci.* 32: 631–636.

Servant, G., Tachdjian, C., Tang, X.-Q., et al. (2010). Positive allosteric modulators of the human sweet taste receptor enhance sweet taste. *PNAS* 107: 4746–4751.

Shallenberger, R. S. (1993). Taste Chemistry. Glasgow: Blackie Academic and Professional, pp. 132–137.

Shigemura, R., Chen, Q., Darmohusodo, V., and Dean, A. (2012). Comestible compositions comprising high potency savory flavorants, and processes for producing them. US Patent 8,148,536B2 (April 3, 2012).

Slack, J. P., Brockhoff, A., Batram, C., et al. (2010). Modulation of bitter taste perception by a small molecule hTAS2R antagonist. *Current Biology* 7: 1104–1109.

Slack, J. P.,and Evans-Pennimpede, J. E. (2009). Methods of identifying modulators of the bitter taste receptor TAS2R44. PCT Patent Appln. WO 2009/149577 A1 (December 17, 2009).

Slack, J. P., Simons, C. T., and Hansen, C. A. (2007). Method relating to sweetness enhancement. *PCT Patent Application WO 2007/121604 A2* (November 1, 2007).

Smith, D. V., and Bealer, S. L. (1975). Sensitivity of the rat gustatory system to the rate of stimulus onset. *Physiol. Behav.* 15: 303–314.

Sterling, D., Alvarez, B. V., and Casey J. R. (2002). The extracellular component of a transport metabolon. Extracellular loop 4 of the human AE1 Cl^-/HCO_3^- exchanger binds carbonic anhydrase IV. *J. Biol. Chem.* 277: 25239–25246.

Stewart, R. E., DeSimone, J. A., and Hill, D. L. 1997. New perspectives in gustatory physiology: transduction, development, and plasticity. *Am. J. Physiol. Cell Physiol.* 272: C1–C26.

Sturz, G. R., Phan, T-H. T., Mummalaneni, S., et al. (2011). The K^+-H^+ exchanger, nigericin, modulates taste cell pH and chorda tympani taste nerve responses to acidic stimuli. *Chem. Senses* 36: 375–388.

Suttisri, R., Lee, I.-S. and Kinghorn, A. D. (1995). Plant-derived triterpenoid sweetness inhibitors. *J. Ethnopharmacol.* 47: 9–26.

Tachdjian, C., Karanewsky, D. S., Tang, X., et al. (2008). *Modulation of chemosensory receptors and ligands associated therewith.* WO/2008/154221 (2008).

Tachdjian, C., Patron, A. P., Adamski-Werner, S. L., et al. (2012). Flavors, flavor modifiers, tastants, taste enhancers, umami or sweet tastants, and/or enhancers and use thereof. US Patent 8,124,121B2 (February 28, 2012).

Treesukosol, Y., Lyall, V., Heck, G. L., et al. (2007). A psychophysical and electrophysiological analysis of salt taste in Trpv1 null mice. *Am. J. Physiol. Regul. Integr. Comp. Physiol.* 292: R1799–R1809.

Ungureanu, I. M., and Van Ommeren, E. (2010). Off-taste masking. *PCT Patent Appln. WO 2010/100158 A1* (September 10, 2010).

Urweyler, S. (2011). Allosteric modulation of family C G-protein-coupled receptors: from molecular insights to therapeutic perspectives. *Pharmacol. Rev.* 63: 59–126.

Vandenbeuch, A., and Kinnamon, S. C. (2009). Why do taste cells generate action potentials? *J. Biol.* 8: 1–5.

Vinnikova, A. K., Alam, R. I., Malik, S. A., et al. (2004). Na^+-H^+ exchange activity in taste receptor cells. *J. Neurophysiol.* 91: 1297–1313.

Wang, Y., and Daniher, A. (2009). Flavor molecules. *PCT Patent Application WO 2009/105906 A1* (September 3, 2009).

Weinberger, M. H. (1996). Salt sensitivity of blood pressure in humans. *Hypertension* 27: 481–490.

Wenner M. (2008). Magnifying taste. *Sci. Am.* 299: 96–99.

Wessjohann, L., Backes, M, Ley, J. P., et al. (2010). Use of hydroxyflavan derivatives for taste modification. *U.S. Patent Application US 2010/0292175 A1* (November 18, 2010).

Xu, H., Staszewski, L., Tang, H., et al. (2004). Different functional roles of T1R subunits in the heteromeric taste receptors. *PNAS* 101: 14258–14263.

Yamaguchi, S. (1979). The umami taste. (ed. J. C. Boudreau), pp. 33–51, *ACS Symposium Series 115*, American Chemical Society, Washington, D.C.

Yee, K. K., Sukumaran, S. K., Kotha, R., et al. (2011). Glucose transporters and ATP-gated K^+ (K_{ATP}) metabolic sensors are present in type 1 taste receptor 3 (T1r3)-expressing taste cells. *PNAS* 108: 5431–5436.

Zhang, F., Klebansky, B., Fine, R. M., et al. (2010). Molecular mechanism of the sweet taste enhancers, *PNAS* 107: 4752–4757.

Zhang, F., Klebansky, B., Fine, R. M., et al. (2008). Molecular mechanism for the umami taste synergism. *PNAS* 105: 20930–20934.

Zubare-Samuelov, M., Shaul, M. E., Peri, I., et al. (2005). Inhibition of signal termination-related kinases by membrane-permeant bitter and sweet tastants: Potential role in taste signal termination. *Am. J. Physiol.* 289: C483–C492.

Chapter 31

The Molecular Basis of Gustatory Transduction

STEVEN D. MUNGER and WOLFGANG MEYERHOF

31.1 INTRODUCTION

The initial stage of taste perception is the detection and transduction of taste stimuli within the oral cavity. These stimuli are generally nonvolatile and are components of foods. They engage specific detection mechanisms present in taste cells of the gustatory epithelium to trigger cellular changes that result in neural signals being sent to the brain. Over the last two decades, key components of the taste transduction apparatus have been identified, including many of the receptors, ion channels, and intracellular signaling molecules that couple stimulus detection to an electrochemical response (Yarmolinsky et al., 2009). In this chapter, we will discuss a number of the important findings that have illuminated the molecular basis of gustatory transduction in mammals as well as the implication of these findings for understanding how we make sense of what we choose to ingest or reject.

Taste stimuli have long been grouped according to the perceptual qualities they elicit in humans. The existence of sweet, bitter, sour, and salty tastes has been universally accepted for years. More recently, umami taste (the savory taste of glutamate) has gained broad acceptance as a fifth taste quality (Beauchamp, 2009). Fat, calcium, starch, and even water have been suggested to elicit distinct taste perceptions, but each remains controversial to different degrees (Bachmanov and Beauchamp, 2007). Prior to the identification of many of the molecular components of taste transduction mechanisms, a clear understanding of the strategies used by mammals to detect and encode different types of taste stimuli remained elusive.

Organizational principles of taste processing in the gustatory epithelium are now being defined, at least in part as a result of our growing understanding of the molecular basis of taste transduction (Yarmolinsky et al., 2009). Although they will be discussed in greater detail below, these principles include: (1) stimuli that elicit the same perceptual quality are detected by the same receptor or by a family of closely related receptors; (2) taste receptors that mediate different taste qualities do so because they are segregated into distinct taste cell subpopulations that are dedicated ("tuned") to a single taste quality. This latter characteristic of the taste system was most clearly demonstrated in mice engineered to express a specific human bitter receptor, called T2R16, in either of two populations of taste cells (Mueller et al., 2005): one that normally express sweet taste receptors, and another that normally express bitter taste receptors. Wild-type mice are indifferent to the cognate agonists of T2R16, β-glucopyranosides, although humans find these compounds to be extremely bitter (Bufe et al., 2002). However, mice in which T2R16 was expressed in bitter taste cells acquired strong avoidance of β-glucopyranosides, whereas mice in which T2R16 was expressed in sweet taste cells found this normally innocuous stimulus to be highly appetitive (Mueller et al., 2005). Thus, while the receptor type dictates the stimulus selectivity of a particular taste cell, it is the cell itself (presumably because of its particular connections to the nervous system) that dictates the *meaning* of the stimulus (i.e., sweet, bitter, umami, salty, or sour). Data such as these also illustrate that the rapid advances in molecular biology, genetics, and genomics has not only allowed

---

Handbook of Olfaction and Gustation, Third Edition. Edited by Richard L. Doty.
© 2015 Richard L. Doty. Published 2015 by John Wiley & Sons, Inc.

686   **Chapter 31   The Molecular Basis of Gustatory Transduction**

researchers to identify the molecular components of the taste transduction apparatus, but has revealed basic principles of taste coding in the periphery that can help us to understand why both sugars and artificial sweeteners taste sweet, why we have trouble discriminating most bitter stimuli, and even why umami should be considered a distinct taste quality.

31.2   TASTE RECEPTORS AND TRANSDUCTION MECHANISMS

The molecular mechanisms employed by mammals to detect and transduce taste stimuli vary with the taste quality they elicit. Sweet, bitter, and umami stimuli bind to different G protein-coupled receptors (GPCRs; each expressed in different subpopulations of taste cells) (see Table 31.1), but engage similar intracellular signaling cascades (Bachmanov and Beauchamp, 2007; Yarmolinsky et al., 2009). These GPCR-mediated transduction mechanisms will be discussed later in this section. Salty and sour stimuli, by contrast, seem to require ion channels as their primary means of detection (Bachmanov and Beauchamp, 2007; Yarmolinsky et al., 2009).

31.2.1   Ion Channels as Taste Receptors

31.2.1.1 Salt Taste. Sodium chloride is the prototypical salt stimulus, although many other salts have a salty taste (e.g., sodium gluconate, KCl, LiCl) (DeSimone and Lyall, 2006). NaCl is appetitive to mammals at concentrations below ~150 mM, but becomes aversive at higher concentrations (Duncan, 1962). This complex response likely reflects the fact that while sodium is an important nutrient that we must obtain from the environment, the ingestion of high concentrations of sodium and other salts (such as in seawater) can lead to severe dehydration, kidney damage and death. The concentration-dependent differences in the hedonic quality of sodium salts as well as the chemically diverse array of salty-tasting compounds suggest multiple mechanisms for the detection of salty stimuli. This indeed appears to be the case.

EPITHELIAL SODIUM CHANNELS (ENaCS) AND SODIUM TASTE   The recognition that mammalian taste responses to NaCl were sensitive to the diuretic amiloride offered a pharmacological clue to the identity of the sodium salt sensor (DeSimone et al., 1981; Schiffman et al., 1983; Heck et al., 1984; Brand et al., 1985). Amiloride was known to inhibit sodium transport across membranes, and was later found to act on a broadly expressed type of Na^+-selective channel know as the epithelial sodium channel (ENaC; a.k.a., sodium channel non-neuronal 1,

Scnn1) (Canessa et al., 1993, 1994). The ENaCs are multimeric channel complexes typically composed of α, β, and γ subunits (although the α subunit is replaced by a related δ subunit in some human tissues); all three subunits are required for full channel function (Kashlan and Kleyman, 2011). But while the physiological properties of sodium taste, including the sensitivity to amiloride, are consistent with a role for ENaCs, the evidence was not definitive. For example, amiloride has effects on molecular targets other than ENaCs including acid-sensing ion channels (ASICs) and Na^+-H^+-exchangers. Furthermore, ENaCs are found in many tissues (Kashlan and Kleyman, 2011), begging the question of whether the expression of ENaC subunits in taste buds is indicative of a specialized taste role. A more direct test of the role of ENaCs in sodium taste would be to delete one of the subunits in mice and assess the impact on taste function. However, because ENaC gene deletions result in embryonic lethality, such assessments awaited a conditional knockout strategy that could largely restrict the deletion to taste buds. This was recently achieved by deleting the *Scnn1a* gene (which encodes ENaCα) in cells expressing a marker for mature taste cells, cytokeratin 19. Mice lacking ENaCα in taste cells exhibited no amiloride-sensitive sodium taste responses in behavioral assays or in chorda tympani nerve recordings (Chandrashekar et al., 2010). Remarkably, these mice retained an amiloride-insensitive NaCl response as well as normal responses to other salts and to sweet, umami, bitter, and sour stimuli (Chandrashekar et al., 2010). Furthermore, NaCl-sensitive, ENaCα-expressing taste cells comprise a distinct subpopulation of taste bud cells, indicating that they are dedicated sodium sensors (Chandrashekar et al., 2010). Although it has taken nearly 30 years and a multitude of approaches, a primary role for ENaCs in the detection and transduction of sodium salts is now firmly established.

MEDIATORS OF AVERSIVE SALT TASTE   Both the taste of non-sodium salts and the aversive taste of high concentrations of sodium salts appear to rely on one or more ENaC-independent mechanisms. ENaCs are highly selective for Na^+ and Li^+, but not K^+. The ENaC blocker amiloride inhibits chorda tympani responses and raises the detection threshold for NaCl but not KCl (Heck et al., 1984; Spector et al., 1996; Geran et al., 1999; Geran and Spector, 2000; Eylam and Spector, 2003). Furthermore, mice lacking a functional ENaC channel in taste cells retain behavioral and chorda tympani nerve responses to NaCl (albeit only for higher concentrations) and KCl (Chandrashekar et al., 2010). The exact molecular identity of the ENaC-independent salt taste mechanism(s) is unknown. One candidate, a splice variant of the capsaicin-sensitive transient receptor potential V1 (TRPV1) ion channel, has been suggested based on the observation that this channel can be amplified from taste

31.2 Taste Receptors and Transduction Mechanisms

Table 31.1 Symbols for human and mouse bitter taste receptor genes approved by Hugo Gene Nomenclature Committee and International Committee on Standardized Genetic Nomenclature for Mice, along with any previous aliases (adopted from Behrens and Meyerhof, 2011).

Human		Mouse	
Approved Gene Symbols	*Previous Aliases*	*Approved Gene Symbols*	*Previous Aliases*
TAS2R1	T2R1, TRB7	Tas2r102	T2R102, mT2R51, STC9-7
TAS2R3	T2R3	Tas2r103	T2R103, T2R10, TRB2
TAS2R4	T2R4	Tas2r104	T2R104, mT2R45
TAS2R5	T2R5	Tas2r105	T2R105, T2R5, T2R9
TAS2R7	T2R7, TRB4	Tas2r106	T2R106, mT2R44
TAS2R8	T2R8, TRB5	Tas2r107	T2R107, mT2R43, T2R4, STC5-1
TAS2R9	T2R9, TRB6	Tas2r108	T2R4, T2R8
TAS2R10	T2R10, TRB2	Tas2r109	T2R109, mT2R62
TAS2R13	T2R13, TRB3	Tas2r110	T2R110, mT2r57, STC 9-1
TAS2R14	T2R14, TRB1	Tas2r113	T2R113, mT2R58
TAS2R16	T2R16	Tas2r114	T2R114, mT2R46
TAS2R19	TAS2R48, TAS2R23, T2R19, T2R23	Tas2r115	
TAS2R20	TAS2R49, T2R20, T2R56	Tas2r116	T2R116, mT2R56, TRB1, TRB4
TAS2R30	TAS2R47, T2R30	Tas2r117	T2R117, mT2R54
TAS2R31	TAS2R44, T2R31, T2R53	Tas2r118	T2R16, T2R18, mt2r40
TAS2R38	PTC, T2R61	Tas2r119	T2R119, T2R19
TAS2R39		Tas2r120	T2R120, mT2R47
TAS2R40	GPR60	Tas2r121	T2R13, T2R121, mT2R48
TAS2R41	T2R59	Tas2r122	
TAS2R42	T2R24, T2R55, hT2R55, TAS2R55	Tas2r123	T2R123, T2R23, mT2R55, STC9-2
TAS2R43	T2R52	Tas2r124	T2R124, mT2R50
TAS2R45	ZG24P, GPR59	Tas2r125	T2R125, mT2R59
TAS2R46	T2R54	Tas2r126	T2R41, T2R12, T2R26
TAS2R50	T2R51	Tas2r129	T2R129, mT2R60
TAS2R60	T2R60	Tas2r130	T2R7, T2R6, T2R30, STC7-4, mT2R42
		Tas2r131	
		Tas2r134	T2R134, T2R34
		Tas2r135	T2R135, T2R35, mT2r38
		Tas2r136	T2R136, T2R36, mT2r52
		Tas2r137	T2R3, T2R137, T2R37, mT2r41
		Tas2r138	T2R38, T2R138, mT2R31
		Tas2r139	T2R39, mT2R34
		Tas2r140	T2R140, T2R40, T2R8, T2R13, mT2r64, mTRB3, mTRB5
		Tas2r143	T2R143, T2R43, mT2R36
		Tas3r144	T2R40, mT2R33

tissue and amiloride-insensitive salt taste electrophysiological responses are reduced or eliminated in TRPV1 knockout mice (Lyall et al., 2004; Treesukosol et al., 2007). However, these same mice show normal behavioral responses to NaCl and KCl in the presence or absence of amiloride (Ruiz et al., 2006; Treesukosol et al., 2007), casting doubt on the necessity of the TRPV1 channel for amiloride-insensitive salt taste responses. More recent studies suggest that both non-sodium salts and high concentrations of sodium salts activate sour and bitter-sensitive taste cells in an amiloride-independent manner (Oka et al., 2013). Even so, the molecular basis of aversive salt taste transduction remains unknown.

31.2.1.2 *Sour Taste.*

The information conveyed by sourness is not clear, but may indicate spoilage or a lack of ripeness. In any case, sour is predominantly an aversive taste. Most sour-tasting stimuli are acids. However, the proximate stimulus for sourness remains controversial. One model holds that a proton conductance contributes directly to cellular depolarization of sour-sensing taste cells. Support for this model includes the identification of a proton conductance specifically expressed in sour-sensing taste cells that can lead to cellular depolarization and the generation of action potentials (Chang et al., 2010). By contrast, other studies favor a role for intracellular acidification of signaling proteins in sour-sensing taste cells (Lyall et al., 2001; Huang et al., 2008). The observation that weak acids (e.g., acetic acid) are perceived as more sour than strong acids (e.g., HCl) at the same concentration (Taylor, 1928) suggests that sourness is not strictly dependent on H^+ concentration. In fact, both mechanisms may contribute to sour taste: strong acids (which are fully dissociated in solution) may depolarize sour-sensing taste cells via the conductance of extracellular protons, while weak acids may diffuse through taste cell membranes before acidifying proteins that influence cellular physiology.

Although a number of candidates for sour taste sensors have been suggested, the molecular basis of sour taste remains unclear (Bachmanov and Beauchamp, 2007). One of the most intriguing candidates has been a TRP-family channel composed of the subunits polycystic kidney disease 2L1 (PKD2L1) and PKD1L3 (Huang et al., 2006; Ishimaru et al., 2006; LopezJimenez et al., 2006). The PKD2L1 subunit is certainly expressed in most, if not all, sour-sensitive cells, as genetic ablation of the PKD2L1-expressing taste cell population in mice disrupts sour taste behavioral and electrophysiological responses (Huang et al., 2006). However, while PKD2L1 is a marker of sour taste cells, it appears unlikely that this channel is the sour taste sensor itself. For example, while PKD2L1 requires coexpression with PKD1L3 to confer acid sensitivity in heterologous cells (Ishimaru et al., 2006), PKD1L3 is not expressed in PKD2L1-expressing fungiform or palatal taste buds (Huang et al., 2006; Ishii et al., 2009). Furthermore, electrophysiological studies in heterologous cells indicate that the PKD2L1-PKD1L3 channel is activated by the removal, not the application, of acids (Inada et al., 2008). Finally, sour taste responses are unaffected by deletion of the *Pkd1l3* gene and only partially reduced in mice lacking the *Pkd2l1* gene (Nelson et al., 2010; Horio et al., 2011).

While the search continues for the primary detector of sour stimuli, a specialized mechanism appears to mediate the sour taste of CO_2 (Lyall et al., 2001). Sour-sensitive cells express the membrane-associated, extracellular carbonic anhydrase, Car4. CO_2-dependent activation of sour taste cells (identified by their expression of PKD2L1) requires Car4 (Chandrashekar et al., 2009), which likely mediates sour taste by catalyzing the conversion of CO_2 and water to bicarbonate and H^+, thus increasing the local proton concentration.

31.2.2 G Protein-Coupled Taste Receptors

31.2.2.1 *Sweet and Umami Taste.* The primary value of food is the nutrients it provides. The gustatory system plays a critical role in sampling the nutrient content of foods before we ingest them. As described above,

amiloride-sensitive salt taste is critical for assessing the sodium content of foods. Similarly, sweet and umami taste are essential for the detection of certain carbohydrates and amino acids. Sweet-tasting stimuli include a number of mono- and disaccharides (e.g., glucose, fructose, galactose, sucrose, maltose, lactose) commonly found in many fruits, vegetables, and dairy products. Certain amino acids, including d-phenylalanine and glycine, also have a sweet taste. For humans, umami stimuli include l-glutamate, and the 5'-ribonucleotides inosine monophosphate and guanosine monophosphate. These substances are commonly found in fish, meat, fermented foods (e.g., cheeses) and certain fruits and vegetables (e.g., tomatoes, mushrooms). Although sweet and umami stimuli exhibit great chemical diversity, it is now established that these two taste qualities both rely on Class C GPCRs for recognition by the gustatory system.

TYPE 1 TASTE RECEPTORS (T1RS) Driven by the nutritional value of sugars and the structural diversity of the various sweet tasting substances—including mono- and disaccharides, amino acids, peptides, proteins, plant secondary metabolites as well as various types of synthetic substances—there have been numerous attempts to identify and characterize the receptors responsible for sweet taste. Based on structure activity relationships, Shallenberger and Acree developed the acid-hydrogen-base model in which appropriately spaced hydrogen donor and acceptor were thought to form the receptors' active site (Shallenberger and Acree, 1967). Although refined several times, the model failed to explain sweetness of large molecules or of molecules with flexible structures, leading to the assumption that different types of sweet receptors exist. This assumption was reinforced by structure-activity investigations, human psychophysical experiments and nerve recordings using various sweeteners (Faurion, 1987). Only recently, sophisticated molecular biological studies and mining of genome databases allowed for the identification of the sweet taste receptor as well as the structurally and functionally related umami receptor.

Vertebrate genomes contain a small family of genes (*TAS1Rs* in humans, *Tas1rs* in rodents) that encode three GPCRs known as type 1 taste receptors (T1Rs): T1R1, T1R2, and T1R3 (Hoon et al., 1999; Vigues et al., 2009; Yarmolinsky et al., 2009). The *Tas1R3* gene is expressed in a subset of taste bud cells in all three types of gustatory papillae as well as in the palate (Kitagawa et al., 2001; Max et al., 2001; Montmayeur et al., 2001; Nelson et al., 2001; Sainz et al., 2001). Rodent T1R3-expressing cells often coexpress either T1R1 (in fungiform and palatal taste buds) or T1R2 (in foliate, vallate, and palatal taste buds), suggesting that T1Rs, like some other class C GPCRs, function as heterodimers. Indeed, heterologous expression studies indicate that T1R3 must be coexpressed with either

T1R1 or T1R2 to produce a fully functional umami or sweet receptor, respectively (Nelson et al., 2001, 2002; Li et al., 2002). Thus, sweet and umami receptors contain one common subunit (T1R3) and one specific subunit (T1R1 or T1R2). Both the sweet and umami receptors are activated by quite high agonist concentrations (>10 mM) as compared to typical GPCRs, with their dose-response function matching the nutritionally relevant concentration range of carbohydrate and amino acid stimuli. This relatively low efficacy is consistent with the proposed roles of these receptors as nutrient detectors that prevent animals from wasting time on foods with insufficient caloric or nutritional content.

The sweet taste receptor The sweet taste receptor, T1R2–T1R3, responds in functional receptor assays to mono- and disaccharides but also to all other sweet tasting compounds tested (Vigues et al., 2009; Behrens and Meyerhof, 2011). The role of T1R2 and/or T1R3 in sensing sweeteners is strongly supported by various independent lines of evidence (Chandrashekar et al., 2006; Vigues et al., 2009). First, T1R3 corresponds to *Sac*, a locus in the mouse genome that determines differences in sweet taste sensitivity across strains of mice (Bachmanov et al., 2001). Second, a "non-taster" strain of mouse has been rescued to respond with exquisite sensitivity to sweet substances by transgenic expression of the "taster" T1R3 variant (Nelson et al., 2001). Third, knockout mice with eliminated *Tas1r2* and/or *Tas1r3* genes show largely diminished or abolished gustatory responses to a number of tested sweet stimuli (Damak et al., 2003; Zhao et al., 2003). Fourth, differences in T1R2–T1R3 agonist profiles across species establish interspecies sweet taste differences (Li et al., 2002; Xu et al., 2004). This is impressively demonstrated in cats, which are obligatory carnivores. Cats do not sense sweet stimuli and do not make a functional sweet taste receptor due to pseudogenization of the *Tas1r2* gene (Li et al., 2005). A similar correlation of defective *Tas1r* genes with diet is seen in other carnivores including the giant panda, vampire bat, otter, and sea lion (Zhao et al., 2010; Jiang et al., 2012). Moreover, the chicken genome lacks *Tas1r2* while frogs possess no *Tas1r* genes at all, thus predicting that these species will have vastly different responses to sweeteners and amino acids as compared with humans and rodents (Shi and Zhang, 2009). Depending on species, bony fish do have 1–4 functional *Tas1r* genes. In cell-based receptor assays, both fish T1R1–T1R3 and fish T1R2–T1R3 responded to amino acids but not sugars (Oike et al., 2007), suggesting that ancestral T1Rs were tuned to detect amino acids and evolution only later brought about T1R2–type receptors for sugars. Intriguingly, human and rodent T1R2–T1R3 is sensitive to the amino acids glycine and l-alanine, highlighting its evolutionary origin (Behrens et al., 2011). In addition, interspecific sequence differences in functional *TAS1R* genes are associated with both differences in the agonist profiles of the sweet receptors and phenotypic variation in sweet taste perception. For example, sweet proteins and aspartame, both of which taste sweet to humans but to which rodents are "taste-blind," are effective agonists for human but not rodent T1R2–T1R3 receptors (Li et al., 2002).

Although T1Rs have not been crystallized and diffracted, combinations of various experimental strategies have provided some insights into their structure-function relationship. These approaches include computational structure prediction and ligand-docking methods, analysis of mutated receptor subunits in heterologous expression systems, and biochemical/biophysical characterization of recombinant amino-terminal domains (Li, 2009; Vigues et al., 2009; Behrens et al., 2011; Servant et al., 2011). As typical class C GPCRs, T1Rs possess a heptahelical transmembrane domain that is connected by a cysteine-rich domain (CRD) to a large amino-terminal venus flytrap domain (VFTD). This VFTD consists of two lobes separated by a hinge region. The transmembrane domain, CRD and VFTD are each able form ligand binding sites. Orthosteric binding of agonists can occur in the VFTD of all T1R protomers, thereby switching the VFTD from inactive 'open' to active 'closed' configurations. Biophysical measurements of the purified recombinant VFTDs indicates that sucrose, sucralose, and glucose bind to the VFTD of both the T1R2 and T1R3 protomers (Nie et al., 2005). By contrast, analyses of chimeric receptors indicates that the amino acid derivatives aspartame and neotame bind only to T1R2 (Xu et al., 2004), while neoculin, a pH-sensitive, sweet-tasting protein, interacts primarily with the VFTD of T1R3 (Koizumi et al., 2007).

The T1R2–T1R3 receptor is also a target of allosteric agonists and modulators. The sweet tasting proteins monellin and brazzein interact with the CRD of T1R3 (Jiang et al., 2004), while the sweet agonists neohesperidin dihydrochalcone and cyclamate and the sweet taste inhibitor lactisole bind to an overlapping set of amino acids that form an allosteric binding pocket in the TMD of T1R3. This latter site of agonist/inhibitor binding is consistent with the observations that cyclamate and lactisole modulate the activity of both T1R1–T1R3 and T1R1–T1R3 (Xu et al., 2004). Finally, researchers have recently synthesized several small molecule positive allosteric modulators that bind, in an agonist-specific fashion, close to orthosteric sugar binding site in the VFTD of T1R2 (Servant et al., 2010; Zhang et al., 2010a,b). The presence of these modulators synergistically enhances activity of the sweet receptor dimers (Servant et al., 2011).

The umami taste receptor The T1R1–T1R3 receptor shares a number of functional properties with human and rodent umami/amino acid taste, making it the top candidate to mediate this taste quality. In heterologous expression assays, the human variant of T1R1–T1R3 is narrowly tuned to l-glutamate; by contrast, several l-amino acids stimulate the rodent receptor (Li et al., 2002; Nelson et al., 2002). The activity of the glutamate-bound T1R1–T1R3 is further enhanced by the simultaneous presence of the 5'-ribonucleotides IMP, AMP or GMP (Li et al., 2002; Nelson et al., 2002; Li, 2009). The importance of T1R1–T1R3 for umami taste is further supported by the observation that mice in which the *Tas1r1* and/or *Tas1r3* genes have been deleted show little to no response to umami stimuli (the extent to which umami responses are lost in *Tas1r* knockouts continues to be the subject of debate; see below). Analogous to the orthosteric binding site for sugars on the sweet taste receptor, glutamate binds to the hinge region of the VFTD in T1R1. This binding induces the 'closed' configuration and subsequent receptor activation. The VFTD of T1R1 also contains the allosteric binding site for 5'ribonucleotides: IMP binds within the VFTD, adjacent to the binding site for glutamate, where it acts to stabilize the closed conformation (Zhang et al., 2008).

While a dominant role for T1R1–T1R3 in umami taste is undisputed, other receptors have been suggested to contribute (Chaudhari et al., 2009; Yasumatsu et al., 2009). For example, while one group of studies finds that umami taste responses are completely abolished in T1R3 knockout mice (Zhao et al., 2003), others find residual taste responses to glutamate (Yasumatsu et al., 2012). These residual glutamate responses are unresponsive to ribonucleotide synergism and predominantly associated with taste nerves innervating the posterior tongue. Leading candidates for T1R-independent umami taste receptors include isoforms of the metabotropic glutamate receptors (mGluRs), which, like T1Rs, are Class C GPCRs. Evidence supporting a role for mGluRs in umami taste includes the expression of two isoforms, mGluR1 and mGluR4, in taste cells and the ability of mGluR-active pharmacological agents on glutamate-induced taste responses. The fact that umami stimuli are weakly appetitive in mice has probably contributed to the difficulty in clearly defining the full repertoire of glutamate-responsive receptors involved in the detection of umami tastants. This continues to be an area of active investigation.

31.2.2.2 Bitter Taste.

Bitterness is believed to indicate the presence of potential toxins in food. In fact, numerous toxins elicit a bitter taste (Meyerhof, 2005). Strong bitter taste is highly aversive and causes food rejection. Compared with sweet and umami taste stimuli, which only elicit taste responses in the millimolar range, bitter stimuli are perceived over a broad range reaching from millimolar to upper nanomolar concentrations (Meyerhof et al., 2010). This high sensitivity means that food contaminated with very low concentrations of some bitter compounds, including the highly toxic strychnine and aristolochic acid, will be rejected. By contrast, food containing less potent bitter substances that are less toxic or nontoxic, such as glycosides, can still be consumed. In this way, vital calories can be consumed even though the foods in which they are found may be tolerably spoiled with harmful substances. Despite their potential toxicity, humans and other animals adapt to bitter compounds and tolerate or even prefer some bitterness in their diets (Drewnowski, 2001). Thus, adaptation can be considered another way to enable organisms to ingest calories despite the presence of acceptable amounts of harmful substances or of repulsive but otherwise harmless compounds. Bitter compounds are often of plant origin but are also produced during food processing or from animal sources (Drewnowski, 2001). Bitter substances are numerous and ubiquitous and of amazing structural diversity, observations that led to the early proposal that multiple receptors mediate bitter detection (Belitz and Wieser, 1985; DuBois et al., 2008). This proposal was supported by psychophysical experiments showing cross-adaptation for some bitter compounds but not for others (McBurney et al., 1972). Cross-adapting bitter substances were thought to activate the same receptor, whereas the absence of cross-adaptation suggested that the test substances used different bitter receptors. The observation that some genetic loci determine bitter sensitivity to specific bitter compounds in mice was also in agreement with the existence of several bitter receptors (Lush and Holland, 1988). Finally, this assumption was fully supported by molecular biology approaches involving data base mining, mouse genetics, expression analysis and demonstration of receptor function *in vitro* that identified a family of GPCRs expressed in taste cells and responsive to bitter stimuli (Adler et al., 2000; Chandrashekar et al., 2000; Matsunami et al., 2000).

TYPE 2 TASTE RECEPTORS (T2RS) Bitter taste receptors are encoded by a small GPCR gene family called type 2 taste receptors (*TAS2Rs* in human or *Tas2rs* in rodents, encoding the T2Rs). The *TAS2R/Tas2r* nomenclatures are confusing because these receptors have been cloned independently and simultaneously in several laboratories and have been subjected to nomenclature changes several times (Behrens and Meyerhof, 2011). Humans possess 25, and mice 35, functional bitter taste receptor genes, which are present in clusters at only a few chromosomal loci. Segregating pseudogenes and copy number variations create interindividual variability in the number of *TAS2R* genes per human genome (Kim et al., 2005; Pronin et al., 2007; Roudnitzky et al., 2011); similar events may have occurred in other species. Moreover, numerous single nucleotide polymorphisms in the *TAS2R* genes predict considerable functional diversity for the encoded receptors,

31.2 Taste Receptors and Transduction Mechanisms

thus accounting for population differences in bitterness perception. The clearest example of such differences is provided by the *TAS2R38* gene, which encodes two major receptor variants: a fully functional one, T2R38-PAV (with proline, alanine and valine at amino acid positions 49, 262 and 296, respectively), and a non-functional variant, T2R38-AVI (with alanine, valine and isoleucine at those positions). PAV carriers are sensitive to the bitter tasting compounds phenylthiocarbamide, propylthiouracil and similar thiourea compounds, whereas homozygotes for the AVI variant display virtually no perception of these bitter chemicals (Kim et al., 2003b; Bufe et al., 2005). Similar observations have also been made for the genes encoding T2R9, T2R16, T2R31, and T2R43, suggesting that genetically determined differences in bitter taste perception are widespread (Soranzo et al., 2005; Pronin et al., 2007; Dotson et al., 2008; Roudnitzky et al., 2011). These genetic differences in T2Rs may shape preferences for foods and beverages or affect other phenotypical properties including alcohol consumption, glucose homeostasis, colorectal cancer, longevity, eating behaviors, or even responsiveness to bitter-tasting pharmaceuticals (Hinrichs et al., 2006; Dotson et al., 2008, 2009, 2012; Carrai et al., 2011; Clark et al., 2012).

While T2Rs are GPCRs, they differ considerably in sequence from, and lack the signatures of, other GPCR subfamilies (Meyerhof et al., 2011). T2Rs are quite small (only 290–330 amino acids), but are encoded by genes containing a single open reading frame. They are characterized by very short amino-termini and are glycosylated at a highly conserved asparagine-linked glycosylation site in the extracellular loop 2 (GPCRs are more typically glycosylated within the amino terminus). Glycosylation of T2Rs is required for proper folding and maturation. T2Rs generally require the addition of amino terminal plasma membrane targeting sequences or coexpression with various auxiliary proteins to facilitate their functional expression in heterologous cells (Chandrashekar et al., 2000; Bufe et al., 2002; Behrens et al., 2006; Ilegems et al., 2010). This suggests that such factors are active in native taste receptor cells and required for proper T2R expression.

Expression studies in heterologous cells have revealed key functional properties of T2Rs. Humans possess at least three very broadly tuned T2Rs, each sensitive to about one-third, and together recognizing ~50%, of ~100 tested bitter substances (Meyerhof et al., 2010). The cognate agonists for these receptors exhibit amazing structural diversity, making it impossible to identify common structural features required for T2R activation. Eight T2Rs exhibit intermediately broad and moderately overlapping agonist spectra (Meyerhof et al., 2010). Again, no common structural motifs responsible for receptor activation have been identified. By contrast, seven T2Rs are narrowly tuned to a very few (1–3) compounds. However, two T2Rs

do show predictable ligand selectivity: T2R38 responds to with a –N=C=S moiety, while T2R16 responds to β-glucopyranosides (Bufe et al., 2002; Bufe et al., 2005; Meyerhof et al., 2010). Together, these response properties explain how countless and structurally diverse bitter tasting chemicals can be recognized.

Despite their generally broad tuning, T2Rs do respond quite specifically to their cognate agonists. For example T2R16 is sensitive to β-gluco- and mannopyranosides, whereas α-glucopyranosides and galactosides fail to activate this receptor (Bufe et al., 2002). Similarly, whereas many sesquiterpene lactones activate T2R46, others do not even though they differ only slightly in structure from the agonists (Brockhoff et al., 2007). These properties likely account for pronounced differences in bitterness between structurally similar chemicals, such as between the bitter tasting phenyl-β-D-glucopyranoside and the non-bitter phenyl-β-D-galactopyranoside.

Each T2R recognizes its cognate compounds with different potencies and efficacies (Meyerhof et al., 2010). Potency and efficacy ranking can be moderate to extreme. For example, various glucopyranosides differ only about 20-fold in potency and little in efficacy for T2R16 activation. By contrast, aristolochic acid ($EC_{50} \approx 80$ nM) and acesulfame K ($EC_{50} \approx 10$ mM) demonstrate a $\sim 10^6$-fold difference in potency and a substantial difference in efficacy at T2R43, making the former a full agonist and the latter only a weak partial agonist. Finally, potencies for of a single agonist can vary across receptors. Denatonium benzoate, for example, activates T2R47, T2R10, and T2R46 with EC_{50} values of 0.27, 120, and 240 μM, respectively. Thus, a given receptor could make a major contribution to the bitterness of one compound and only a minor contribution to the bitterness of another.

Genes encoding T2Rs are found in numerous species (Shi and Zhang, 2009). Surprisingly, gene numbers vary largely across species, ranging from only three in chicken to 49 in clawed frog. This broad range in *TAS2R* number could reflect differences in the poisonous load of food or the efficacy of detoxification mechanisms across species. In other words, some species might require only restricted protection against food intoxication and thus relatively few bitter taste receptors. Alternatively, T2Rs in species with few genes could all be broadly tuned to numerous substances whereas those in species with many genes could include more narrowly tuned receptors. In this latter case, different species could have comparable bitter perception despite having different numbers of *TAS2R* genes. Finally, it must be considered that, although the numbers of genes are known, the numbers of functional receptors remains unclear. Human T2Rs can oligomerize (Kuhn et al., 2010), suggesting that the 25 *TAS2R* gene products could results in 325 different heterodimeric receptors (or even more if higher oligomeric structures are adopted). However,

692   **Chapter 31   The Molecular Basis of Gustatory Transduction**

the functional significance of the oligomers has not been determined.

Bitter taste inhibitors are now being reported. Based on sensory experiments in humans, bitter taste masking substances have been identified (Ley, 2008). Certainly, such compounds could optimize the taste of some foods and beverages. However, the T2Rs targeted by these substances have never been identified. The first T2R-specific antagonist, 4-(2,2,3-trimethylcyclopentyl) butanoic acid, or GIV3727, was isolated by screening a synthetic compound library for molecules suppressing saccharin-activation of T2R31 in a cell-based assay (Slack et al., 2010). This molecule acts in the micromolar range as a competitive antagonist. The observation that GIV3727 prevents the activation of T2R31 by all of its agonists reinforces the notion that T2Rs possess only a single agonist binding pocket. GIV3727 also blocks the activation of five other T2Rs by their cognate agonists. The drug probenicid, which is used in certain cell assays, has also been implicated as an inhibitor of certain T2Rs, including T2R16 and T2R38 (Greene et al., 2011). Interestingly, several natural compounds from edible plants also have been identified as T2R antagonists with properties similar to those of GIV3727 (Brockhoff et al., 2011). As animals and humans usually have numerous compounds present in the mouth during chewing, it is likely that T2R antagonists are often present. Thus, the normal presence of bitter blocking agents could regularly impact the bitterness of foods. Moreover, the presence of bitter blockers in food likely shaped the molecular evolution T2Rs. The existence of multiple T2Rs with partially overlapping ligand spectra ensures protection against a complete block of bitter perception and potentially fatal outcomes (Brockhoff et al., 2011).

The amazing responsiveness of some T2Rs to many structurally diverse compounds has prompted a number of studies addressing the question of how T2Rs accommodate their agonists. Traditional receptor binding studies are problematic because of low predicted affinities and relatively poor expression in heterologous cells. Instead, studies have relied on computational structure prediction and agonist docking and on functional expression of wild type and mutated receptors. One such study focused on the closely related receptors T2R31, T2R43, and T2R46, which, despite high sequence similarities, demonstrate unique agonist spectra (Brockhoff et al., 2010). It was reasoned that amino acids that differ in the receptors could mediate interactions with the respective cognate ligands. Indeed, such residues have been identified in T2R46 by mutation and functional expression. Transfer of these few amino acid residues that impaired agonist responsiveness of T2R46 to the same positions in T2R43 or T2R31 switched the agonist spectra of the recipient receptors to that of the donor receptor, providing strong evidence for

the existence of a single binding pocket for all agonists. Essentially the same conclusions were drawn from similar studies on T2R16 and T2R38 (Biarnes et al., 2010; Sakurai et al., 2010). Computational studies predict that the T2R binding pockets are formed by residues within the transmembrane helices, even though differences are apparent regarding the specific transmembrane (TM) regions that contribute. Whereas residues of TM III and TM VI form the binding site of T2R38, residues of TM III, TM V, and TM VI are responsible in T2R16 and residues of TM II, TM III, TM V, and TM VII assemble the binding pocket in T2R46. Not only do the TMs that form the binding pockets differ across receptors but the amino acid residues that interact with agonists vary. Only a single residue, $N^{3.36}$, appears to be involved in agonist interaction in all three of these T2Rs. In marked contrast to these studies, another mutational study on T2R31 and T2R43 found evidence for agonist specific interactions of amino acid residues in the extracellular loop I (Pronin et al., 2004). At present this apparent disparity cannot be resolved but could be due to differences in experimental techniques or agonists. Alternatively, extracellular loop residues could regulate the access of some agonists to the binding pocket or, when mutated, result in overall conformational changes that impact agonist-dependent activation.

Whereas the variable residues in the upper, extracellularly-directed parts of the TM bundle appear to interact with ligands, conserved residues in the lower, intracellularly-directed regions have been suggested to participate in T2R activation. Experimental support for this idea has been obtained in studies of T2R1 and T2R4, results from which suggested a network of interhelical hydrogen bonds that stabilize the receptors in the inactive conformation (Singh et al., 2010; Pydi et al., 2012). Disruption of this network through mutation or agonist binding would thus be part of a receptor activation mechanism.

31.2.2.3 *Other GPCRS Expressed in Taste Cells.* Other GPCRs, including GPR40, GPR120 and the calcium-sensing receptor (CaSR), have been described in the gustatory system (San Gabriel et al., 2009; Cartoni et al., 2010; Galindo et al., 2011). GPR40 and GPR120 are sensitive to stimulation with medium and long-chain fatty acids and have been implicated to mediate a chemosensory component of fat detection and in preference for fatty acids (Cartoni et al., 2010; Galindo et al., 2011). Indeed, GPR40 knockout mice and GPR120 knockout mice show diminished preferences for linoleic acid and oleic acid and attenuated nerve responses to several fatty acids. The CaSR is not only activated by calcium ions but also by amino acids, glutathione and so-called kokumi peptides (Ohsu et al., 2009; Maruyama et al., 2012). Kokumi, also referred to as mouthfeel or richness, appears to be a sensation

that enhances the classical sweet, salty and umami taste perception. Although under debate and requiring further research, these receptors may play modulatory roles in gustation rather than mediating separate taste qualities.

31.3 TRANSDUCTION OF SWEET, UMAMI AND BITTER TASTE STIMULI

Although they are segregated in different subsets of taste cells, sweet, umami and bitter taste receptors employ very similar signal transduction pathways (Zhang et al., 2003). Agonist activation of both T1Rs and T2Rs leads to stimulation of heterotrimeric G proteins, activation of effector enzymes, release of Ca^{2+} from intracellular stores, gating of ion channels in the plasma membrane, cellular depolarization and neurotransmitter release.

31.3.1 G Proteins

Several G protein α-subunits have been implicated in mammalian taste transduction. The most compelling evidence supports a role for the $G\alpha_{i/o}$ family protein α-gustducin in bitter, sweet and umami taste. Alpha-gustducin is closely related to α-transducin, the G protein subunit that is essential for phototransduction in the vertebrate retina. Originally isolated from bovine taste epithelia (McLaughlin, 1992), α-gustducin is expressed in almost all T2R-expressing taste cells (Adler et al., 2000) and in a subset of T1R-expressing cells. It is clear that α-gustducin plays a critical role in bitter taste transduction: T2Rs functionally couple to α-gustducin (Chandrashekar et al., 2000) and α-gustducin knockout mice have vastly diminished bitter responses (Wong et al., 1996). Residual bitter taste responses in these mice may be due to the ability of other G protein α-subunits to partially compensate for α-gustducin. In particular, α-transducin, which is also expressed in taste cells, may contribute to bitter taste responses (He et al., 2002; Ozeck et al., 2004).

Alpha-gustducin knockout mice are also compromised in their sweet and umami responses (Glendinning et al., 2005), albeit to a lesser extent than for bitter responses. These differences, and the observation that sweet taste is less affected than umami taste in α-gustducin knockouts, may reflect a more variable and regionally restricted coexpression with T1Rs (Kim et al., 2003a). Other G proteins likely also play important roles in these appetitive responses. One canddiate is $G\alpha_{14}$, which is coexpressed with T1R3 on the posterior tongue (Shindo et al., 2008; Tizzano et al., 2008). However, verification of a role for $G\alpha_{14}$ in sweet and umami responses must await studies in mice lacking this subunit.

Specific G protein β and γ subunits have also been implicated in taste transduction. Specifically, both $\beta3/\gamma13$ or $\beta1/\gamma13$ complexes have been identified in taste cells and implicated in taste transduction (Huang et al, 1999). As will be discussed below, it is these β/γ subunits that act to stimulate effector enzyme activity that is critical to taste transduction.

31.3.2 Enzymes and Second Messengers

Sweet, bitter and umami stimuli are transduced via a phosphoinositide-dependent signaling cascade (Yarmolinsky et al., 2009). Upon activation of T1R and T2R receptors by taste ligands, the associated heterotrimeric G proteins will dissociate and the β/γ subunits can act to stimulate the plasma membrane-associated enzyme phospholipase C $\beta2$ (PLC$\beta2$) (Rossler et al., 1998; Huang et al., 1999). PLC$\beta2$ can then cleave the membrane phospholipid phosphatidylinositol 4,5-bisphosphate (PIP$_2$) into two second messengers: diacylglycerol and inositol 1,4,5-trisphosphate (IP$_3$). The actions of diacylglycerol in taste cells are not clear. However, IP$_3$ is thought to open type III IP$_3$ receptors (IP$_3$R3) on the endoplasmic reticulum (Clapp et al., 2001), thus releasing Ca^{2+} from intracellular stores and elevating cytoplasmic Ca^{2+} concentrations (Akabas et al., 1988; Hwang et al., 1990). The obligatory role of PLC$\beta2$-mediated signaling in sweet, bitter and umami taste was confirmed by studies in PLC$\beta2$ knockout mice (Zhang et al, 2005). In the absence of PLC$\beta2$, mice do not behaviorally differentiate sweeteners, amino acids or bitter tastants from water, nor do they exhibit electrophysiological responses in chorda tympani nerve recordings. However, overexpression of PLC$\beta2$ under the control of a T2R gene promoter can rescue bitter, but not sweet or umami, responses.

Even so, sequence similarities between α-gustducin and α-transducin had suggested that taste stimuli might also influence cyclic nucleotide signaling in taste cells (McLaughlin et al., 1992). Indeed, biochemical measurements of membrane preparations obtained from taste tissues indicated that both sweet and bitter stimuli regulate cAMP levels *in vitro* (Yan et al. 2001; Striem et al., 1989). However, the functional relevance of this second messenger remains unclear. An intriguing possibility is that cAMP modulates the PLC$\beta2$-mediated transduction cascade, perhaps by regulating protein kinase A-dependent phosphorylation of IP$_3$R3 or PLC$\beta2$ (Chaudhari and Kinnamon, 2008).

31.3.3 Effector Channels

The IP$_3$-mediated elevation of intracellular Ca^{2+} is not sufficient to elicit neurotransmitter release. Rather, this Ca^{2+} response must be coupled to membrane depolarization. In taste cells, this occurs through the Ca^{2+}-gating of the transient receptor potential channel isoform TRPM5, which is expressed in sweet, bitter and umami-sensitive taste cells (Perez et al, 2002; Zhang et al, 2003). This channel is selective for monovalent cations and activated by micromolar concentrations of intracellular Ca^{2+} (Hofmann et al., 2003; Liu and Liman, 2003). Separate analyses from two different lines of TRPM5 knockout mice have arrived at slightly different conclusions about the role of TRPM5 in sweet, bitter and umami taste transduction. One group found that taste electrophysiological responses (recordings from the chorda tympani and glosspharyngeal nerves) and behavioral responses to sugars, amino acids and bitter tastants were completely abolished in mice lacking TRPM5 (Zhang et al., 2003), indicating that TRPM5 is required for sweet, bitter and umami taste transduction. By contrast, others have reported that while TRPM5 deletion causes significant functional impairment, some taste responses (primarily at the level of nerve activity) remain (Damak et al., 2006). Simultaneous experiments in both mouse lines are needed to resolve these discrepancies. Even so, it is clear that TRPM5 is required for normal taste responses to sweet, bitter and umami stimuli.

31.4 NEUROTRANSMITTERS AND NEUROPEPTIDES

While taste receptor cells are excitable, they are not neurons and do not send afferent projections to the central nervous system. Rather, these taste cells must communicate with cranial nerve fibers innervating the taste bud, and it is these nerves that will carry taste information from the taste bud to the gustatory areas of the brainstem. Unfortunately, the neurochemical mechanisms of intercellular communication within the taste bud are anything but simple (Yarmolinsky et al., 2009; Chaudhari and Roper, 2010). First, there are at least two different types of synaptic communication between taste cells and afferent nerve fibers. One cell population (largely comprised of sour-sensitive cells) exhibits typical presynaptic endings containing synaptic vesicles, while a second population (the sweet, bitter and umami-sensitive cells) does not contain obvious presynaptic modifications. Second, there is extensive evidence for paracrine signaling between cells within individual taste buds. Finally, taste buds express a large number of classical, non-classical and peptide neurotransmitters. Because of this neurochemical complexity, making sense of neurotransmission in the taste bud has been a challenge.

A major breakthrough came with the identification of adenosine triphosphate (ATP) as a major neurotransmitter in taste buds. Based on the observation that the purinergic receptors $P2X_2$ and $P2X_3$ are localized to taste nerves in rat (Bo et al., 1999), taste responses were assessed in $P2X_2/P2X_3$ double knockout mice (Finger et al., 2005).

Chorda tympani and glossopharyngeal nerve responses to stimuli of all taste qualities were completely abolished in these mice. However, behavioral responses were not as clear: while sweet, umami and salty stimuli were ineffective, sour and some bitter (i.e., quinine hydrochloride) showed residual responses. Subsequent studies demonstrated that mice use additional orosensory systems to recognize acids and quinine HCl: these stimuli also activate the superior laryngeal nerve in the $P2X_2/P2X_3$ knockout mice (Ohkuri et al., 2012). ATP release is not vesicular, but rather occurs through hemichannels in response to an elevation of intracellular Ca^{2+} and membrane depolarization (Huang and Roper, 2010). Several membrane channels have been implicated at the ATP release channel in taste cells. The leading candidate is the voltage-sensitive ion channel calcium modulated homeostasis 1 (CALHM1); deletion of this channel in mice results in the loss of sweet, bitter and umami taste (Taruno et al., 2013). However, there is also experimental support for both connexin and pannexin hemichannels in these functions (Huang et al., 2007; Romanov et al., 2007; Huang and Roper, 2010; Romanov et al., 2012). The mechanistic details of purinergic neurotransmission in the taste bud are yet to be fully understood.

Although ATP plays the primary role in the communicating taste information from taste cells to afferent nerves, both it and other small molecule neurotransmitters are found in taste buds and have been implicated in intercellular signaling. For example, serotonin, norepinephrine and γ-aminobutyric acid (GABA) are released by sour-sensitive cells (Obata et al., 1997; Huang et al., 2008a,b, 2009, 2011; Cao et al., 2009; Dvoryanchikov et al., 2011). Acetylcholine, on the other hand, is released from cells that respond to sweet, bitter or umami stimuli (Ogura, 2002; Dando and Roper, 2012). The primary targets of all of these neurotransmitters are taste cells themselves. Autocrine and/or paracrine signaling through these neurotransmitters (and perhaps others) provides the opportunity to modulate taste cell function and thus shape the output of the taste bud before signals are transmitted to afferent nerves.

Taste buds also express a large number of peptides that may serve as autocrine, paracrine or endocrine signals. These include glucagon (Elson et al., 2010), glucagon-like peptide-1 (GLP-1) (Shin et al., 2008), neuropeptide Y (NPY) (Zhao et al., 2005), vasoactive intestinal peptide (VIP) (Shen et al., 2005), cholecystokinin (CCK) (Shen et al., 2005) and ghrelin (Shin et al., 2010). When produced in the stomach or intestine, these peptides serve as satiety signals or regulators of metabolism. While the precise roles of these peptides in taste coding remain unclear, immunohistochemical and behavioral studies have contributed some intriguing insights. For example, GLP-1 and glucagon are produced in partially-overlapping subpopulations of taste cells that also express the sweet and umami receptor subunit T1R3 (GLP-1 is also found in a subset of serotonergic taste cells), and genetic and/or pharmacological disruption of GLP-1 or glucagon signaling in mice results in reduced sweet taste responsiveness in behavioral assays (Shin et al., 2008; Elson et al., 2010). Cognate receptors for GLP-1 are found on cranial nerve fibers closely apposed to taste cells, while glucagon receptors are expressed on the same cells that produce the glucagon peptide (Shin et al., 2008; Elson et al., 2010). Thus, these peptides may function as local signals to modulate the processing of sweet taste signals within the taste bud and to communicate this information to afferent nerve fibers. However, it cannot be ruled out that peptides produced in the taste bud serve as endocrine signals, entering the bloodstream to act at distant sites. Similarly, the peptide receptors present in taste buds may be targeted by hormones released from the gut or other organs, thus allowing peripheral taste function to be regulated in the context of the body's nutritional needs or metabolic state. Indeed, both the anorectic peptide hormone leptin, which is produced by fat cells, and the appetite-stimulating endocannaboanoids, can act on their cognate receptors expressed on taste cells to reciprocally modulate sweet taste responses (Niki et al., 2010).

31.5 TASTE TRANSDUCTION MOLECULES OUTSIDE THE TASTE SYSTEM

Although α-gustducin, TRPM5, T1Rs and T2Rs were discovered in the taste system, it took little time to find that these molecules are not taste specific. Indeed, there is growing evidence for the expression of "taste transduction" molecules throughout the gastrointestinal (GI) and respiratory systems (and perhaps other tissues, as well). The first extraoral cells found to express α-gustducin were brush cells of the stomach and intestine (Hofer et al., 1996). Subsequently, evidence of α-gustducin expression (at the message and/or protein level) was found in populations of gut enteroendocrine cells, pancreatic islets, solitary chemoreceptors of the nasal cavity and upper airways, ciliated respiratory epithelia and airway smooth muscle (Behrens and Meyerhof, 2011). Similarly, T1R1, T1R2 and/or T1R3, as well as numerous T2Rs, are reported expressed in many of these same cell types (Behrens and Meyerhof, 2011). TRPM5 expression has particularly been described in brush cells of the GI tract and solitary chemosensory cells of the respiratory epithelium (Bezencon et al., 2008).

In most cases, direct evidence supporting an extragustatory function of these proteins is circumstantial at best. Exceptions include enteroendocrine L cells, pancreatic β cells, and solitary chemosensory cells, where experimental support for the roles of T1Rs, α-gustducin and/or

TRPM5 is more robust. The use of knockout mice lacking α-gustducin, T1R2 or T1R3 have shown that both this G protein and the sweet taste receptor play critical roles in coupling the detection of luminal glucose (and other sweet-tasting compounds) to the secretion of the incretin hormone glucagon-like peptide-1 (GLP-1) (Jang et al., 2007; Geraedts et al., 2012), the regulation of glucose absorption (Margolskee et al., 2007), and the maintenance of glucose homeostasis (Jang et al., 2007; Geraedts et al., 2012). The sweet taste receptor is also implicated the modulation of insulin secretion from pancreatic β cells: T1R2 knockout mice are deficient in fructose-potentiation of glucose-stimulated insulin secretion (Kyriazis et al., 2012), while islets from T1R3 knockout mice display slowed kinetics of insulin granule fusion (Geraedts et al., 2012). Finally, α-gustducin and TRPM5 knockout mice lack the acyl-homoserine lactone-induced respiratory changes seen in wildtype animals, presumably due to a functional loss in the solitary chemosensory cells of the respiratory epithelium that express these proteins (Tizzano et al., 2010).

31.6 SUMMARY/CONCLUSIONS/IMPLICATIONS

The great advances made over the last two decades in dissecting the molecular basis of taste transduction have already led to surprising insights into the way we detect and respond to tastes and nutrients. For example, the identification of the T1R and T2R families of taste receptors has allowed researchers to employ high throughput systems to screen thousands of natural and synthetic compounds for their ability to activate or inhibit receptor function. Indeed, candidate sweet taste enhancers and bitter taste blockers have already been identified based on their actions on taste receptors expressed in heterologous systems (Servant et al., 2010; Slack et al., 2010; Zhang et al., 2010). These compounds, or ones like them, could be important tools in combating obesity (by reducing the amount of sugar needed to achieve sweetness in processed foods) or increasing tolerance for pharmaceuticals (by blocking the bitter taste of many common drugs). The identification of taste receptors and taste transduction molecules has also provided researchers with important tools needed to begin to tease apart the neural strategies for stimulus, quality and hedonic coding by the peripheral and central gustatory system and to better assess the contribution of specific taste functions to ingestive behaviors and post-ingestive responses (Sclafani, 2006; Simon et al., 2006; Fernstrom et al., 2012). And of course, the observations that receptors and other transduction molecules first identified in the gustatory epithelium may also play significant physiological roles in other tissues could have a far-reaching impact on how we understand diverse physiological and

696   Chapter 31   The Molecular Basis of Gustatory Transduction

pathophysiological responses to diverse chemostimuli such as ingested nutrients, inhaled pathogens and the metabolites of symbiotic microorganisms found throughout the body (Yarmolinsky et al., 2009; Behrens and Meyerhof, 2011; Clark et al., 2012). In these ways, and many others, investigations into the molecular basis of taste transduction promises to stimulate important research for years to come.

REFERENCES

Adler, E., Hoon, M. A., Mueller, K. L., et al. (2000). A novel family of mammalian taste receptors. *Cell.* 100: 693–702.

Akabas, M. H., Dodd, J., and Al-Awqati, Q. (1988). A bitter substance induces a rise in intracellular calcium in a subpopulation of rat taste cells. *Science.* 242: 1047–1050.

Bachmanov, A. A., Li, X., Reed, D. R., et al. (2001). Positional cloning of the mouse saccharin preference (Sac) locus. *Chem. Senses.* 26: 925–933.

Bachmanov, A. A. and Beauchamp, G. K. (2007). Taste receptor genes. *Ann. Rev. Nutr.* 27: 389–414.

Beauchamp, G. K. (2009). Sensory and receptor responses to umami: an overview of pioneering work. *Am. J. Clin. Nutr.* 90: 723S–727S.

Behrens, M., Bartelt, J., Reichling, C., et al. (2006). Members of RTP and REEP gene families influence functional bitter taste receptor expression. *J. Biol. Chem.* 281: 20650–20659.

Behrens, M. and Meyerhof, W. (2011). Gustatory and extragustatory functions of mammalian taste receptors. *Physiol. Behav.* 105: 4–13.

Behrens, M., Meyerhof, W., Hellfritsch, C., and Hofmann, T. (2011). Sweet and umami taste: natural products, their chemosensory targets, and beyond. *Angew. Chem. Int. Ed. Engl.* 50: 2220–2242.

Belitz, H.-D. and Wieser, H. (1985). Bitter compounds: occurrence and structure-activity relationship. *Food Reviews International.* 1: 271–354.

Bezencon, C., Furholz, A., Raymond, F., et al. (2008). Murine intestinal cells expressing Trpm5 are mostly brush cells and express markers of neuronal and inflammatory cells. *J. Comp. Neurol.* 509: 514–525.

Biarnes, X., Marchiori, A., Giorgetti, A., et al. (2010). Insights into the Binding of Phenyltiocarbamide (PTC) Agonist to Its Target Human TAS2R38 Bitter Receptor. *PLoS One.* 5: e12394.

Bo, X., Alavi, A., Xiang, Z., et al. (1999). Localization of ATP-gated P2X2 and P2X3 receptor immunoreactive nerves in rat taste buds. *Neuroreport.* 10: 1107–1111.

Brand, J. G., Teeter, J. H., and Silver, W. L. (1985). Inhibition by amiloride of chorda tympani responses evoked by monovalent salts. *Brain Res.* 334: 207–214.

Brockhoff, A., Behrens, M., Massarotti, A., et al. (2007). Broad tuning of the human bitter taste receptor hTAS2R46 to various sesquiterpene lactones, clerodane and labdane diterpenoids, strychnine, and denatonium. *J. Agric. Food Chem.* 55: 6236–6243.

Brockhoff, A., Behrens, M., Niv, M. Y., and Meyerhof, W. (2010). Structural requirements of bitter taste receptor activation. *Proc. Natl. Acad. Sci. U.S.A.* 107: 11110–11115.

Brockhoff, A., Behrens, M., Roudnitzky, N., et al. (2011). Receptor agonism and antagonism of dietary bitter compounds. *J. Neurosci.* 31: 14775–14782.

Bufe, B., Hofmann, T., Krautwurst, D., et al. (2002). The human TAS2R16 receptor mediates bitter taste in response to β-glucopyranosides. *Nat. Genet.* 32: 397–401.

Bufe, B., Breslin, P. A., Kuhn, et al. (2005). The molecular basis of individual differences in phenylthiocarbamide and propylthiouracil bitterness perception. *Curr. Biol.* 15: 322–327.

Canessa, C. M., Horisberger, J. D., and Rossier, B. C. (1993). Epithelial sodium channel related to proteins involved in neurodegeneration. *Nature.* 361: 467–470.

Canessa, C. M., Schild, L., Buell, G., et al. (1994). Amiloride-sensitive epithelial Na+ channel is made of three homologous subunits. *Nature.* 367: 463–467.

Cao, Y., Zhao, F. L., Kolli, T., et al. (2009). GABA expression in the mammalian taste bud functions as a route of inhibitory cell-to-cell communication. *Proc. Natl. Acad. Sci. U.S.A.* 106: 4006–4011.

Carrai, M., Steinke, V., Vodicka, P., et al. (2011). Association Between TAS2R38 Gene Polymorphisms and Colorectal Cancer Risk: A Case-control Study in Two Independent Populations of Caucasian Origin. *PLoS One.* 6: e20464.

Cartoni, C., Yasumatsu, K., Ohkuri, T., et al. (2010). Taste preference for fatty acids is mediated by GPR40 and GPR120. *J. Neurosci.* 30: 8376–8382.

Chandrashekar, J., Mueller, K. L., Hoon, M. A., et al. (2000). T2Rs function as bitter taste receptors. *Cell.* 100: 703–711.

Chandrashekar, J., Hoon, M. A., Ryba, N. J., and Zuker, C. S. (2006). The receptors and cells for mammalian taste. *Nature.* 444: 288–294.

Chandrashekar, J., Yarmolinsky, D., von Buchholtz, et al. (2009). The taste of carbonation. *Science.* 326: 443–445.

Chandrashekar, J., Kuhn, C., Oka, Y., et al. (2010). The cells and peripheral representation of sodium taste in mice. *Nature.* 464: 297–301.

Chang, R. B., Waters, H., and Liman, E. R. (2010). A proton current drives action potentials in genetically identified sour taste cells. *Proc. Natl. Acad. Sci. U.S.A.* 107: 22320–22325.

Chaudhari, N. and Kinnamon, S. C. (2008). cAMP: a role in sweet taste adaptation. *ACS Symp. Series.* 979: 220–229.

Chaudhari, N., Pereira, E., and Roper, S. D. (2009). Taste receptors for umami: the case for multiple receptors. *Am. J. Clin. Nutr.* 90: 738S–742S.

Chaudhari, N. and Roper, S. D. (2010). The cell biology of taste. *J Cell Biol.* 190: 285–296.

Clapp, T. R., Stone, L. M., Margolskee, R. F., and Kinnamon, S. C. (2001). Immunocytochemical evidence for co-expression of Type III IP3 receptor with signaling components of bitter taste transduction. *BMC Neurosci.* 2: 6.

Clark, A. A., Liggett, S. B., and Munger, S. D. (2012). Extraoral bitter taste receptors as mediators of off-target drug effects. *FASEB J.* doi: 10.1096/fj.12-215087.

Damak, S., Rong, M., Yasumatsu, K., et al. (2003). Detection of sweet and umami taste in the absence of taste receptor T1r3. *Science.* 301: 850–853.

Damak, S., Rong, M., Yasumatsu, K., et al. (2006). Trpm5 null mice respond to bitter, sweet, and umami compounds. *Chem. Senses.* 31: 253–264.

Dando, R. and Roper, S. D. (2012). Acetylcholine is released from taste cells, enhancing taste signalling. *J. Physiol.* 590: 3009–3017.

DeSimone, J. A., Heck, G. L., and DeSimone, S. K. (1981). Active ion transport in dog tongue: a possible role in taste. *Science.* 214: 1039–1041.

DeSimone, J. A. and Lyall, V. (2006). Taste receptors in the gastrointestinal tract III. Salty and sour taste: sensing of sodium and protons by the tongue. *Am. J. Physiol. Gastrointest. Liver Physiol.* 291: G1005–1010.

Dotson, C. D., Zhang, L., Xu, H., et al. (2008). Bitter taste receptors influence glucose homeostasis. *PLoS One.* 3: e3974.

Dotson, C. D., Shaw, H. L., Mitchell, B. D., et al. (2009). Variation in the gene TAS2R38 is associated with the eating behavior disinhibition in Old Order Amish women. *Appetite.* 54: 93–99.

Dotson, C. D., Wallace, M. R., Bartoshuk, L. M., and Logan, H. L. (2012). Variation in the Gene TAS2R13 is Associated with Differences in Alcohol Consumption in Patients with Head and Neck Cancer. *Chem. Senses.* 37: 737–744.

Drewnowski, A. (2001). The science and complexity of bitter taste. *Nutr Rev.* 59: 163–169.

DuBois, G. E., DeSimone, J. A., and Lyall, V. (2008). Chemistry of Gustatory Stimuli. In *Olfaction and taste*, Firestein, S. and Beauchamp, G.K.(Eds). Amsterdam: Elsevier, pp. 27–74.

Duncan, C. J. (1962). Salt preferences of birds and mammals. *Physiol. Zool.* 35: 120–132.

Dvoryanchikov, G., Huang, Y. A., Barro-Soria, R., et al. (2011). GABA, its receptors, and GABAergic inhibition in mouse taste buds. *J. Neurosci.* 31: 5782–5791.

Elson, A. E., Dotson, C. D., Egan, J. M., and Munger, S. D. (2010). Glucagon signaling modulates sweet taste responsiveness. *FASEB J.* 24: 3960–3969.

Eylam, S. and Spector, A. C. (2003). Oral amiloride treatment decreases taste sensitivity to sodium salts in C57BL/6J and DBA/2J mice. *Chem. Senses.* 28: 447–458.

Faurion, A. (1987). Physiology of the Sweet Taste. In *Progress in Sensory Physiology*, Ottoson, D. (Ed Heidelberg: Springer-Verlag, pp. 132–201.

Fernstrom, J. D., Munger, S. D., Sclafani, A., et al. (2012). Mechanisms for sweetness. *J. Nutr.* 142: 1134S–1141S.

Finger, T. E., Danilova, V., Barrows, J., et al. (2005). ATP signaling is crucial for communication from taste buds to gustatory nerves. *Science.* 310: 1495–1499.

Galindo, M. M., Voigt, N., Stein, J., et al. (2011). G protein-coupled receptors in human fat taste perception. *Chem. Senses.* 37: 123–139.

Geraedts, M. C., Takahashi, T., Vigues, S., et al. (2012). Transformation of postingestive glucose responses after deletion of sweet taste receptor subunits or gastric bypass surgery. *Am. J. Physiol. Endocrinol. Metab.* 303: E464–474.

Geran, L. C., Guagliardo, N. A., and Spector, A. C. (1999). Chorda tympani nerve transection, but not amiloride, increases the KCl taste detection threshold in rats. *Behav. Neurosci.* 113: 185–195.

Geran, L. C. and Spector, A. C. (2000). Amiloride increases sodium chloride taste detection threshold in rats. *Behav. Neurosci.* 114: 623–634.

Glendinning, J. I., Bloom, L. D., Onishi, M., et al. (2005). Contribution of {alpha}-Gustducin to Taste-guided Licking Responses of Mice. *Chem. Senses.* 30: 299–316.

Greene, T. A., Alarcon, S., Thomas, A., et al. (2011). Probenecid inhibits the human bitter taste receptor TAS2R16 and suppresses bitter perception of salicin. *PLoS One.* 6: e20123.

He, W., Danilova, V., Zou, S., et al. (2002). Partial rescue of taste responses of alpha-gustducin null mice by transgenic expression of alpha-transducin. *Chem. Senses.* 27: 719–727.

Heck, G. L., Mierson, S., and DeSimone, J. A. (1984). Salt taste transduction occurs through an amiloride-sensitive sodium transport pathway. *Science.* 223: 403–405.

Hinrichs, A. L., Wang, J. C., Bufe, B., et al. (2006). Functional variant in a bitter-taste receptor (hTAS2R16) influences risk of alcohol dependence. *Am. J. Hum. Genet.* 78: 103–111.

Hofer, D., Puschel, B., and Drenckhahn, D. (1996). Taste receptor-like cells in the rat gut identified by expression of α-gustducin. *Proc. Natl. Acad. Sci. U.S.A.* 93: 6631–6634.

Hofmann, T., Chubanov, V., Gudermann, T., and Montell, C. (2003). TRPM5 is a voltage-modulated and Ca(2+)-activated monovalent selective cation channel. *Curr. Biol.* 13: 1153–1158.

Hoon, M. A., Adler, E., Lindemeier, J., et al. (1999). Putative mammalian taste receptors: a class of taste-specific GPCRs with distinct topographic selectivity. *Cell.* 96: 541–551.

Horio, N., Yoshida, R., Yasumatsu, K., et al.2011). Sour taste responses in mice lacking PKD channels. *PLoS One.* 6: e20007.

Huang, A. L., Chen, X., Hoon, M. A., et al. (2006). The cells and logic for mammalian sour taste detection. *Nature.* 442: 934–938.

Huang, L., Shanker, Y. G., Dubauskaite, J., et al. (1999). Ggamma13 colocalizes with gustducin in taste receptor cells and mediates IP3 responses to bitter denatonium. *Nat. Neurosci.* 2: 1055–1062.

Huang, Y. A., Maruyama, Y., and Roper, S. D. (2008). Norepinephrine is coreleased with serotonin in mouse taste buds. *J. Neurosci.* 28: 13088–13093.

Huang, Y. A., Maruyama, Y., Stimac, R., and Roper, S. D. (2008b). Presynaptic (Type III) cells in mouse taste buds sense sour (acid) taste. *J. Physiol.* 586: 2903–2912.

Huang, Y. A., Dando, R., and Roper, S. D. (2009). Autocrine and paracrine roles for ATP and serotonin in mouse taste buds. *J. Neurosci.* 29: 13909–13918.

Huang, Y. A. and Roper, S. D. (2010). Intracellular Ca(2+) and TRPM5-mediated membrane depolarization produce ATP secretion from taste receptor cells. *J. Physiol.* 588: 2343–2350.

Huang, Y. A., Pereira, E., and Roper, S. D. (2011). Acid stimulation (sour taste) elicits GABA and serotonin release from mouse taste cells. *PLoS One.* 6: e25471.

Huang, Y. J., Maruyama, Y., Dvoryanchikov, G., et al. (2007). The role of pannexin 1 hemichannels in ATP release and cell-cell communication in mouse taste buds. *Proc. Natl. Acad. Sci. U.S.A.* 104: 6436–6441.

Hwang, P. M., Verma, A., Bredt, D. S., and Snyder, S. H. (1990). Localization of phosphatidylinositol signaling components in rat taste cells: role in bitter taste transduction. *Proc. Natl. Acad. Sci. U.S.A.* 87: 7395–7399.

Ilegems, E., Iwatsuki, K., Kokrashvili, Z., et al. (2010). REEP2 enhances sweet receptor function by recruitment to lipid rafts. *J. Neurosci.* 30: 13774–13783.

Inada, H., Kawabata, F., Ishimaru, Y., et al. (2008). Off-response property of an acid-activated cation channel complex PKD1L3-PKD2L1. *EMBO Rep.* 9: 690–697.

Ishii, S., Misaka, T., Kishi, M., et al. (2009). Acetic acid activates PKD1L3-PKD2L1 channel-a candidate sour taste receptor. *Biochem. Biophys. Res. Commun.* 385: 346–350.

Ishimaru, Y., Inada, H., Kubota, M., et al. (2006). Transient receptor potential family members PKD1L3 and PKD2L1 form a candidate sour taste receptor. *Proc. Natl. Acad. Sci. U.S.A.* 103: 12569–12574.

Jang, H. J., Kokrashvili, Z., Theodorakis, M. J., et al. (2007). Gut-expressed gustducin and taste receptors regulate secretion of glucagon-like peptide-1. *Proc. Natl. Acad. Sci. U.S.A.* 104: 15069–15074.

Jiang, P., Ji, Q., Liu, Z., et al. (2004). The cysteine-rich region of T1R3 determines responses to intensely sweet proteins. *J. Biol. Chem.* 279: 45068–45075.

Jiang, P., Josue, J., Li, X., et al. (2012). Major taste loss in carnivorous mammals. *Proc. Natl. Acad. Sci. U.S.A.* 109: 4956–4961.

Kashlan, O. B. and Kleyman, T. R. (2011). ENaC structure and function in the wake of a resolved structure of a family member. *Am. J. Physiol. Renal. Physiol.* 301: F684–696.

Kim, M. R., Kusakabe, Y., Miura, H., et al. (2003). Regional expression patterns of taste receptors and gustducin in the mouse tongue. *Biochem. Biophys. Res. Commun.* 312: 500–506.

Kim, U., Wooding, S., Ricci, D., et al. (2005). Worldwide haplotype diversity and coding sequence variation at human bitter taste receptor loci. *Hum. Mutat.* 26: 199–204.

Kim, U. K., Jorgenson, E., Coon, H., et al. (2003). Positional cloning of the human quantitative trait locus underlying taste sensitivity to phenylthiocarbamide. *Science.* 299: 1221–1225.

Kitagawa, M., Kusakabe, Y., Miura, H., et al. (2001). Molecular genetic identification of a candidate receptor gene for sweet taste. *Biochem. Biophys. Res. Comm.* 283: 236–242.

Koizumi, A., Nakajima, K., Asakura, T., et al. (2007). Taste-modifying sweet protein, neoculin, is received at human T1R3 amino terminal domain. *Biochem Biophys Res. Comm.* 358: 585–589.

Kuhn, C., Bufe, B., Batram, C., and Meyerhof, W. (2010). Oligomerization of TAS2R bitter taste receptors. *Chem. Senses.* 35: 395–406.

Kyriazis, G. A., Soundarapandian, M. M., and Tyrberg, B. (2012). Sweet taste receptor signaling in beta cells mediates fructose-induced potentiation of glucose-stimulated insulin secretion. *Proc. Natl. Acad. Sci. U.S.A.* 109: E524–532.

Ley, J. P. (2008). Masking Bitter Taste by Molecules. *Chem. Percept.* 1: 58–77.

Li, X., Staszewski, L., Xu, H., et al. (2002). Human receptors for sweet and umami taste. *Proc. Natl. Acad. Sci. U.S.A.* 99: 4692–4696.

Li, X., Li, W., Wang, H., et al. (2005). Pseudogenization of a sweet-receptor gene accounts for cats' indifference toward sugar. *PLoS Genet.* 1: 27–35.

Li, X. (2009). T1R receptors mediate mammalian sweet and umami taste. *Am. J. Clin. Nutr.* 90: 733S–737S.

Liu, D. and Liman, E. R. (2003). Intracellular Ca2+ and the phospholipid PIP2 regulate the taste transduction ion channel TRPM5. *Proc. Natl. Acad. Sci. U.S.A.* 100: 15160–15165.

LopezJimenez, N. D., Cavenagh, M. M., Sainz, E., et al. (2006). Two members of the TRPP family of ion channels, Pkd1l3 and Pkd2l1, are co-expressed in a subset of taste receptor cells. *J. Neurochem.* 98: 68–77.

Lush, I. E. and Holland, G. (1988). The genetics of tasting in mice. V. Glycine and cycloheximide. *Genet. Res.* 52: 207–212.

Lyall, V., Alam, R. I., Phan, D. Q., et al. (2001). Decrease in rat taste receptor cell intracellular pH is the proximate stimulus in sour taste transduction. *Am. J. Physiol. Cell Physiol.* 281: C1005–1013.

Lyall, V., Heck, G. L., Vinnikova, A. K., et al. (2004). The mammalian amiloride-insensitive non-specific salt taste receptor is a vanilloid receptor-1 variant. *J. Physiol.* 558: 147–159.

Margolskee, R. F., Dyer, J., Kokrashvili, Z., et al. (2007). T1R3 and gustducin in gut sense sugars to regulate expression of Na+−glucose cotransporter 1. *Proc. Natl. Acad. Sci. U.S.A.* 104: 15075–15080.

Maruyama, Y., Yasuda, R., Kuroda, M., and Eto, Y. (2012). Kokumi substances, enhancers of basic tastes, induce responses in calcium-sensing receptor expressing taste cells. *PLoS One.* 7: e34489.

Matsunami, H., Montmayeur, J. P., and Buck, L. B. (2000). A family of candidate taste receptors in human and mouse. *Nature.* 404: 601–604.

Max, M., Shanker, Y. G., Huang, L., et al. (2001). Tas1r3, encoding a new candidate taste receptor, is allelic to the sweet responsiveness locus Sac. *Nat. Genet.* 28: 58–63.

McBurney, D. H., Smith, D. V., and Shick, T. R. (1972). Gustatory cross adaptation: sourness and bitterness. *Percept. Psychophys.* 11: 228–232.

McLaughlin, S. K., McKinnon, P. J., and Margolskee, R. F. (1992). Gustducin is a taste-cell-specific G protein closely related to the transducins. *Nature.* 357: 563–569.

Meyerhof, W. (2005). Elucidation of mammalian bitter taste. In *Rev. Physiol. Biochem. Pharmacol*, pp. 37–72.

Meyerhof, W., Batram, C., Kuhn, C., et al. (2010). The molecular receptive ranges of human TAS2R bitter taste receptors. *Chem. Senses.* 35: 157–170.

Meyerhof, W., Born, S., Brockhoff, A., and Behrens, M. (2011). Molecular biology of mammalian bitter taste receptors. A review. *Flavour Fragr. J.* 26: 260–268.

Montmayeur, J. P., Liberles, S. D., Matsunami, H., and Buck, L. B. (2001). A candidate taste receptor gene near a sweet taste locus. *Nat. Neurosci.* 4: 492–498.

Mueller, K. L., Hoon, M. A., Erlenbach, I., et al. (2005). The receptors and coding logic for bitter taste. *Nature.* 434: 225–229.

Nelson, G., Hoon, M. A., Chandrashekar, J., et al. (2001). Mammalian sweet taste receptors. *Cell.* 106: 381–390.

Nelson, G., Chandrashekar, J., Hoon, M. A., et al. (2002). An amino-acid taste receptor. *Nature.* 416: 199–202.

Nelson, T. M., Lopezjimenez, N. D., Tessarollo, L., et al. (2010). Taste function in mice with a targeted mutation of the pkd1l3 gene. *Chem. Senses.* 35: 565–577.

Nie, Y., Vigues, S., Hobbs, J. R., et al. (2005). Distinct contributions of T1R2 and T1R3 taste receptor subunits to the detection of sweet stimuli. *Curr. Biol.* 15: 1948–1952.

Niki, M., Jyotaki, M., Yoshida, R., and Ninomiya, Y. (2010). Reciprocal modulation of sweet taste by leptin and endocannabinoids. *Res. Probl. Cell Differ.* 52: 101–114.

Obata, H., Shimada, K., Sakai, N., and Saito, N. (1997). GABAergic neurotransmission in rat taste buds: immunocytochemical study for GABA and GABA transporter subtypes. *Brain Res. Mol. Brain Res.* 49: 29–36.

Ogura, T. (2002). Acetylcholine increases intracellular Ca2+ in taste cells via activation of muscarinic receptors. *J. Neurophysiol.* 87: 2643–2649.

Ohkuri, T., Horio, N., Stratford, J. M., et al. (2012). Residual chemoresponsiveness to acids in the superior laryngeal nerve in "taste-blind" (P2X2/P2X3 double-KO) mice. *Chem. Senses.* 37: 523–532.

Ohsu, T., Amino, Y., Nagasaki, H., et al. (2009). Involvement of the calcium-sensing receptor in human taste perception. *J. Biol. Chem.* 285: 1016–1022.

Oike, H., Nagai, T., Furuyama, A., et al. (2007). Characterization of ligands for fish taste receptors. *J. Neurosci.* 27: 5584–5592.

Oka Y, Butnaru M, von Buchholtz L, Ryba NJ, Zuker CS. (2013). High salt recruits aversive taste pathways. *Nature.* 494, 472–475.

Ozeck, M., Brust, P., Xu, H., and Servant, G. (2004). Receptors for bitter, sweet and umami taste couple to inhibitory G protein signaling pathways. *Eur. J. Pharmacol.* 489: 139–149.

Pronin, A. N., Tang, H., Connor, J., and Keung, W. (2004). Identification of Ligands for Two Human Bitter T2R Receptors. *Chem. Senses.* 29: 583–593.

Pronin, A. N., Xu, H., Tang, H., et al. (2007). Specific alleles of bitter receptor genes influence human sensitivity to the bitterness of aloin and saccharin. *Curr. Biol.* 17: 1403–1408.

Pydi, S. P., Bhullar, R. P., and Chelikani, P. (2012). Constitutively active mutant gives novel insights into the mechanism of bitter taste receptor activation. *J. Neurochem.* 122: 537–544.

Romanov, R. A., Rogachevskaja, O. A., Bystrova, M. F., et al. (2007). Afferent neurotransmission mediated by hemichannels in mammalian taste cells. *EMBO J.* 26: 657–667.

Romanov, R. A., Bystrova, M. F., Rogachevskaya, O. A., et al. (2012). Dispensable ATP permeability of Pannexin 1 channels in a heterologous system and in mammalian taste cells. *J. Cell Sci. doi:* 10.1242/jcs.111062.

Rossler, P., Kroner, C., Freitag, J., et al. (1998). Identification of a phospholipase C beta subtype in rat taste cells. *Eur. J. Cell Biol.* 77: 253–261.

Roudnitzky, N., Bufe, B., Thalmann, S., et al. (2011). Genomic, genetic and functional dissection of bitter taste responses to artificial sweeteners. *Hum. Mol. Genet.* 20: 3437–3449.

Ruiz, C., Gutknecht, S., Delay, E., and Kinnamon, S. (2006). Detection of NaCl and KCl in TRPV1 knockout mice. *Chem. Senses.* 31: 813–820.

Sainz, E., Korley, J. N., Battey, J. F., and Sullivan, S. L. (2001). Identification of a novel member of the T1R family of putative taste receptors. *J. Neurochem.* 77: 896–903.

Sakurai, T., Misaka, T., Ishiguro, M., et al. (2010). Characterization of the beta-d-glucopyranoside binding site of the human bitter taste receptor hTAS2R16. *J. Biol. Chem.* 285: 28373–28378.

References

San Gabriel, A., Uneyama, H., Maekawa, T., and Torii, K. (2009). The calcium-sensing receptor in taste tissue. *Biochem. Biophys. Res. Commun.* 378: 414–418.

Schiffman, S. S., Lockhead, E., and Maes, F. W. (1983). Amiloride reduces the taste intensity of Na+ and Li+ salts and sweeteners. *Proc Natl Acad Sci U.S.A.* 80: 6136–6140.

Sclafani, A. (2006). Oral, post-oral and genetic interactions in sweet appetite. *Physiol. Behav.* 89: 525–530.

Servant, G., Tachdjian, C., Tang, X. Q., et al. (2010). Positive allosteric modulators of the human sweet taste receptor enhance sweet taste. *Proc Natl Acad Sci U.S.A.* 107: 4746–4751.

Servant, G., Tachdjian, C., Li, X., and Karanewsky, D. S. (2011). The sweet taste of true synergy: positive allosteric modulation of the human sweet taste receptor. *Trends Pharmacol. Sci.* 32: 631–636.

Shallenberger, R. S. and Acree, T. E. (1967). Molecular theory of sweet taste. *Nature.* 216: 480–482.

Shen, T., Kaya, N., Zhao, F. L., et al. (2005). Co-expression patterns of the neuropeptides vasoactive intestinal peptide and cholecystokinin with the transduction molecules alpha-gustducin and T1R2 in rat taste receptor cells. *Neuroscience.* 130: 229–238.

Shi, P. and Zhang, J. (2009). Extraordinary diversity of chemosensory receptor gene repertoires among vertebrates. *Res. Probl. Cell Differ.* 47: 1–23.

Shin, Y. K., Martin, B., Golden, E., et al. (2008). Modulation of taste sensitivity by GLP-1 signaling. *J. Neurochem.* 106: 455–463.

Shin, Y. K., Martin, B., Kim, W., et al. (2010). Ghrelin is produced in taste cells and ghrelin receptor null mice show reduced taste responsivity to salty (NaCl) and sour (citric acid) tastants. *PLoS One.* 5: e12729.

Shindo, Y., Miura, H., Carninci, P., et al. (2008). G alpha14 is a candidate mediator of sweet/umami signal transduction in the posterior region of the mouse tongue. *Biochem. Biophys. Res. Commun.* 376: 504–508.

Simon, S. A., de Araujo, I. E., Gutierrez, R., and Nicolelis, M. A. (2006). The neural mechanisms of gustation: a distributed processing code. *Nat. Rev. Neurosci.* 7: 890–901.

Singh, N., Pydi, S. P., Upadhyaya, J., and Chelikani, P. (2010). Structural basis of activation of bitter taste receptor T2R1 and comparison with Class A G-protein-coupled receptors (GPCRs). *J. Biol. Chem.* 286: 36032–36041.

Slack, J. P., Brockhoff, A., Batram, C., et al. (2010). Modulation of Bitter Taste Perception by a Small Molecule hTAS2R Antagonist. *Curr. Biol.* 20: 1104–1109.

Soranzo, N., Bufe, B., Sabeti, P. C., et al. (2005). Positive selection on a high-sensitivity allele of the human bitter-taste receptor TAS2R16. *Curr. Biol.* 15: 1257–1265.

Spector, A. C., Guagliardo, N. A., and St John, S. J. (1996). Amiloride disrupts NaCl versus KCl discrimination performance: implications for salt taste coding in rats. *J. Neurosci.* 16: 8115–8122.

Striem, B. J., Pace, U., Zehavi, U., Naim, M., and Lancet, D. (1989). Sweet tastants stimulate adenylate cyclase coupled to GTP-binding protein in rat tongue membranes. *Biochem. J.* 260: 121–126.

Taruno A., Vingtdeux V., Ohmoto M., et al. (2013). CALHM1 ion channel mediates purinergic neurotransmission of sweet, bitter and umami tastes. *Nature.* 495, 223–226.

Taylor, N. W. (1928). Acid Penetration into Living Tissues. *J. Gen. Physiol.* 11: 207–219.

Tizzano, M., Dvoryanchikov, G., Barrows, J. K., et al. (2008). Expression of Galpha14 in sweet-transducing taste cells of the posterior tongue. *BMC Neurosci.* 9: 110.

Tizzano, M., Gulbransen, B. D., Vandenbeuch, A., et al. (2010). Nasal chemosensory cells use bitter taste signaling to detect irritants and bacterial signals. *Proc Natl Acad Sci U.S.A.* 107: 3210–3215.

Treesukosol, Y., Lyall, V., Heck, G. L., et al. (2007). A psychophysical and electrophysiological analysis of salt taste in Trpv1 null mice. *Am. J. Physiol. Regul. Integr. Comp. Physiol.* 292: R1799–1809.

Vigues, S., Dotson, C. D., and Munger, S. D. (2009). The receptor basis of sweet taste in mammals. *Res. Probl. Cell Differ.* 47: 187–202.

Wong, G. T., Gannon, K. S., and Margolskee, R. F. (1996). Transduction of bitter and sweet taste by gustducin. *Nature.* 381: 796–800.

Xu, H., Staszewski, L., Tang, H., et al. (2004). Different functional roles of T1R subunits in the heteromeric taste receptors. *Proc Natl Acad Sci U.S.A.* 101: 14258–14263.

Yarmolinsky, D. A., Zuker, C. S., and Ryba, N. J. (2009). Common sense about taste: from mammals to insects. *Cell.* 139: 234–244.

Yasumatsu, K., Horio, N., Murata, Y., et al. (2009). Multiple receptors underlie glutamate taste responses in mice. *Am. J. Clin. Nutr.* 90: 747S–752S.

Yasumatsu, K., Ogiwara, Y., Takai, S., et al. (2012). Umami taste in mice uses multiple receptors and transduction pathways. *J. Physiol.* 590: 1155–1170.

Zhang, F., Klebansky, B., Fine, R. M., et al. (2008). Molecular mechanism for the umami taste synergism. *Proc Natl Acad Sci U.S.A.* 105: 20930–20934.

Zhang, F., Klebansky, B., Fine, R. M., et al. (2010). Molecular mechanism of the sweet taste enhancers. *Proc Natl Acad Sci U.S.A.* 107: 4752–4757.

Zhang, Y., Hoon, M. A., Chandrashekar, J., et al. (2003). Coding of sweet, bitter, and umami tastes. Different receptor cells sharing similar signaling pathways. *Cell.* 112: 293–301.

Zhao, F. L., Shen, T., Kaya, N., et al. (2005). Expression, physiological action, and coexpression patterns of neuropeptide Y in rat taste-bud cells. *Proc Natl Acad Sci U.S.A.* 102: 11100–11105.

Zhao, G. Q., Zhang, Y., Hoon, M. A., et al. (2003). The receptors for Mammalian sweet and umami taste. *Cell.* 115: 255–266.

Zhao, H., Yang, J. R., Xu, H., and Zhang, J. (2010). Pseudogenization of the umami taste receptor gene Tas1r1 in the giant panda coincided with its dietary switch to bamboo. *Mol. Biol. Evol.* 27: 2669–2673.

Zhao, H., Zhou, Y., Pinto, C. M., et al. (2010). Evolution of the sweet taste receptor gene Tas1r2 in bats. *Mol. Biol. Evol.* 27: 2642–2650.

Chapter 32

Central Taste Anatomy and Physiology of Rodents and Primates

THOMAS C. PRITCHARD and PATRICIA M. DI LORENZO

32.1 INTRODUCTION

Our understanding of the mammalian gustatory system, though incomplete, has grown tremendously over the last 15–20 years. Rodents and nonhuman primates have been among the many species that have served as animal models over this period, but it is interesting that they have played almost complementary roles in this common effort. Most of what is known about the hindbrain has been obtained from rodents; in the forebrain, primate research has prevailed, but the balance is rapidly shifting to rodents. Many reviews have been written about the anatomy and physiology of the rodent and primate gustatory systems; few have attempted to present this information in a side-by-side fashion. Juxtaposing the data from rodents and primates, level by level, from the periphery through frontal cortex, is revealing. For all of the differences between the two species, the similarities are more striking.

32.2 GUSTATORY NEURAL CODING: THE BASICS

A prerequisite to any discussion of gustatory neural coding is an understanding of how taste responses are measured. These methods are common to electrophysiological studies directed at single fibers of peripheral nerves as well as evoked activity in individual neurons in the central nervous system. The basic data provided by these analytical techniques forms the foundation for all theories of gustatory neural coding as well as our understanding of how the sense of taste supports ingestive behavior.

32.2.1 Response Intervals and Sensory Confounds

The chemical senses of gustation and olfaction analyze peripheral stimulation that usually lasts several seconds. Excitatory responses are most common, but the time course is quite variable. Some gustatory evoked responses are coincident with the application of the sapid stimulus; others stop before or after taste stimulation has ceased. Gustatory responses usually begin with a phasic burst of action potentials followed by a sustained period of activity that exceeds the cell's or fiber's average spontaneous rate. The fluctuating nature of taste responses makes the determination of the response interval – the time period during which the response is analyzed – a critical decision. Most investigators count the number of action potentials during a fixed time interval (~3–5 sec.), beginning at the moment of stimulus onset and including the period when the sapid stimulus bathes the tongue. After subtracting the cell's spontaneous rate from the response evoked by the tastant, the resultant net increase or decrease in activity reflects the response attributable solely to the taste stimulus. Because many if not most taste-responsive cells also respond to tactile or thermal stimuli (Ogawa et al., 1988; Halsell et al., 1993), a more conservative approach is to apply distilled water or artificial saliva to control for potential tactile or thermal stimulation. Subtracting the firing rate during the water (or saliva) presentation from the response evoked by sapid stimulus yields the net or corrected taste response. The decision to correct for spontaneous rate or water depends on the response characteristics of the neurons under investigation. When neurons respond to the topical flow of fluids, per se, it is problematic to use water or

---

Handbook of Olfaction and Gustation, Third Edition. Edited by Richard L. Doty.
© 2015 Richard L. Doty. Published 2015 by John Wiley & Sons, Inc.

702   **Chapter 32   Central Taste Anatomy and Physiology of Rodents and Primates**

saliva to correct for thermal and tactile sensitivity (Rosen et al., 2010a; McDonald et al., 2012; de Araujo et al., 2003a).

32.2.2 Temporal and Ensemble Coding

Calculating the average firing rate over the predetermined response interval assumes that the nervous system also averages neural activity across time. Use of the average response rate across time ignores the temporal fluctuations of action potentials that carry as much or more information than the average firing rate (brainstem: Chen et al., 2011; Di Lorenzo and Victor, 2003, 2007; Di Lorenzo et al., 2009a; Roussin et al., 2008, 2012; gustatory cortex (GC): Katz et al., 2001; Stapleton et al., 2006). The recognition that spike tempo can convey taste information has spawned a subspecialty of sensorineural analysis known as temporal coding. Sole reliance on spike count also precludes analysis of the potential contribution of synchronized or coordinated neural activity to the coding of taste. This is important. Katz et al. (2002) and Stapleton et al. (2007) have reported that taste quality information in the GC of rats is coded by neuron ensembles.

32.2.3 Stimulus Selection

Electrophysiologists design experiments knowing that they have only a limited time to test the chemical sensitivity of the neurons they isolate. When large stimulus arrays are used, it is unlikely that all of the chemicals can be tested before electrical isolation is lost. Use of a small stimulus set makes it likely that testing can be completed, but it limits the scope of the experiment. These completing issues make choice of the stimulus array a vitally important decision, one that will not only affect data interpretation, but also whether the goals of the experiment can be achieved.

Depending on the questions that are asked, some investigators have tested as few as four tastants, while others have examined more than a dozen. Investigators who opt for just a few stimuli always include representatives of the four "basic" taste qualities: salty (NaCl or N), sweet (sucrose/glucose or S/G), sour (HCl or H), and bitter (QHCl or Q). With increasing evidence that monosodium glutamate (MSG) represents a fifth primary taste quality (umami or savory; Yarmolinsky et al., 2009), many investigators now include MSG in their stimulus battery. Larger stimulus arrays may include amino acids (e.g., alanine, glycine), organic and inorganic salts, sugars, artificial sweeteners, and even chemicals such as oils and lipids whose status as gustatory stimuli is being actively debated (Gilbertson et al., 1998). Today, the search and even the meaning of what constitutes a primary taste quality is being assisted by molecular biological tools that have provided detailed descriptions of taste receptors and the enzymatic cascades that transduce taste stimuli. Knowledge about the receptor proteins to which tastants bind in order for gustatory transduction to take place will inevitably expand and refine the criteria for defining primary taste qualities.

After the chemical array has been selected, the concentration of each stimulus must be chosen. Again, because of the limited time for testing, many electrophysiological studies have tested each chemical at only one concentration. The chosen concentration usually lies in the mid-range of the concentration-response function obtained from either a multi-neuron or multi-fiber recording; not surprisingly, individual neurons may vary considerably from the population average. The response demographics of individual neurons, as well as between-neuron analyses (e.g., breadth of tuning, interneuronal response correlations, multidimensional analyses), depend on the choice of tastants and their concentration(s). In short, stimulus choice has a profound influence on how we interpret the organization of the nervous system and its strategy for coding taste quality.

32.2.4 Quality Coding – Within or Across Neurons?

The collective taste literature over the last half century shows that the majority of gustatory neurons at every level of the neuraxis respond to exemplars of more than one taste quality. Despite sharing this common trait, the breadth of tuning across the four or five basic taste qualities, as well as the relative sensitivity to one taste quality over others, varies widely. This diversity is seen in the chorda tympani (CT) and glossopharyngeal (GP) nerves which convey gustatory information to the caudal brainstem. It is telling that the theories of taste coding based on peripheral nerve recordings have been applied without modification to taste responsive neurons in the central nervous system.

In early work on single axons of the CT nerve, it was shown that the taste stimulus which evokes the most vigorous response was a good predictor of the relative response magnitudes of the stimuli representing the other taste qualities (Frank, 1973). The heuristic value of the best stimulus analysis is reflected in the fact that it forms the basis for the labeled line theory, which is really a working hypothesis that axons and neurons convey qualitative taste information about only the most effective (or best) stimulus. Thus, NaCl-best cells encode saltiness, sucrose-best cells encode sweetness, and so on, while "sideband" responses to the second or third most effective stimuli are irrelevant and represent neural noise. A corollary to the labeled line theory is that each group of cells identified by their best stimulus represents a separate neuron type. This argument has been buttressed by observations that manipulations such as sodium deprivation (Contreras, 1977; Contreras and Frank, 1979; Scott and Giza, 1990), conditioned taste aversion learning (Chang and Scott, 1984), taste adaptation (Di Lorenzo and Lemon, 2000; Smith et al., 1996), and sex hormones (Di Lorenzo and Monroe, 1989, 1990) affect one class of best stimulus neurons, but not others, in predictable ways. These studies demonstrated that the best stimulus classification system is relevant to taste-guided behavior, and by extension, give credence to the assertion that members of each best stimulus group are the neural messengers for their signature taste quality.

In contrast to the labeled line theory, investigators have proposed that the responses of the entire population to all stimuli, including the sideband responses, contribute to the neural code for taste quality (Erickson, 2008). Because the majority of taste-responsive cells respond to more than one stimulus, each tastant generates a unique pattern or envelope of spike activity across the neural population. This inclusive view of quality coding defines the across neuron (or fiber) pattern (ANP or AFP) theory of taste quality coding. The ANP theory (actually, another hypothesis) gathers support from studies showing that similar tasting stimuli evoke comparable (i.e., positively correlated) patterns of activity across the population of fibers (or neurons) under study (Doetsch and Erickson, 1970). More recent variants of the ANP theory propose that taste quality is coded by the pattern of responsiveness across a subset of cells (Lemon and Smith, 2006).

Recent studies have provided data that support a partial merging of the labeled line and the AFP/ANP theories. These studies blocked transduction of NaCl with the Na$^+$ channel blocker amiloride and then recorded the electrophysiological response evoked by topical application of NaCl to the tongue. Although some reports concluded that amiloride blockade only affects NaCl responses in NaCl-best neurons in the nucleus tractus solitarius (NTS) of the rat (Giza and Scott, 1991; Scott and Giza, 1990), more recent studies in the hamster concluded that amiloride affects NaCl responses in both NaCl-best and sucrose-best NTS neurons (Boughter and Smith, 1998; Smith et al., 1996). Smith et al. (2000) concluded that the code for NaCl is represented by the neural activity in both salt-best and sucrose-best neurons. Thus, the idea of neuron types is retained, but the pattern of activity across both sets of neurons (instead of the entire population) conveys the information. More recently, Lemon and Smith (2006) demonstrated that NTS cells could be classified as NaCl-, HCl-, or sucrose-sensitive based on their best responses across taste qualities. However, they also showed that the total spike count of the best stimulus neurons was an unreliable measure of taste identity unless these data were supplemented by information about the pattern of firing across cells.

Both the labeled line and ANP theories are based on the sum or average number of action potentials evoked by taste stimuli over a defined period; both ignore the information conveyed by the temporal pattern of neural activity. Of course, temporal coding can mean several things (see Hallock and Di Lorenzo, 2006, for a review). Spike count, for example, reflects an average firing rate across a response interval, so in the strictest sense of the word, it can be considered a global measure of the temporal characteristics of a response. It is more common, however, for temporal coding to be defined as the dynamic changes in firing during the response interval. The rate envelope, or "time course" of the response, reflects the phasic-tonic firing rate changes during a sustained response. The ratio of the phasic to tonic components of taste responses conveys information about taste quality and/or hedonics in several taste-related brain areas (Scott and Mark, 1987; Travers and Norgren, 1989; Verhagen et al., a).

Recently, it has been shown that spike timing in individual cells and the cooperative activity within neural ensembles contribute to quality coding in the brainstem and cerebral cortex. In a series of studies, the information conveyed by spike timing was compared to that carried by spike count for stimuli that varied in taste quality (Di Lorenzo and Victor, 2003; Roussin et al., 2008, 2012), flow rate (Di Lorenzo and Victor, 2007) or concentration (Chen et al., 2011) in the NTS. In about half of all neurons, spike timing was significantly more informative than spike count alone. Spike timing was also sufficient to enable individual neurons to discriminate the components of taste mixtures from one another (Di Lorenzo et al., 2009a). In the pontine parabrachial nucleus (PBN), spike timing was also more effective than spike count alone in conveying information about taste quality (Rosen et al., 2011). In the GC, ensemble coding, a process that depends heavily on the temporal activity of networked cells, contributes to taste quality coding (Katz et al., 2002; Stapleton et al. 2007).

To demonstrate that the temporal characteristics of neural responses contribute to sensory neural coding and behavior, Di Lorenzo and colleagues tested awake, behaving rats during lick-contingent patterned electrical pulse trains that mimic the responses of NTS neurons to the taste of sucrose (Di Lorenzo et al., 2003a; Di Lorenzo and Hecht, 2003) or quinine (Di Lorenzo et al., 2009b). In these experiments, conditioned aversions learned to the patterned "sucrose simulation" or the "quinine simulation" generalized specifically to real sucrose or quinine, respectively. Randomized electrical pulse trains with the same number of pulses and the same compliment of interpulse intervals were ineffective. These data provide strong direct evidence that the temporal pattern of brainstem activity conveys information about taste quality and underscore the functional significance of temporal coding.

32.3 ANATOMY AND PHYSIOLOGY OF THE GUSTATORY SYSTEM

32.3.1 Taste Bud Innervation

The most striking fact about the features of taste buds, their distribution throughout the oral cavity, and their peripheral innervation in rodents and primates, is their similarity. The facial, GP, and vagus nerves innervate taste buds located throughout the oral cavity (see Figure 32.1). The facial nerve contains two branches: the CT nerve and the greater superficial petrosal (GSP) nerve. Innervation of the taste buds on the rostral 2/3 of the tongue and the anterior two or three foliate papillae located along the lateral margin of the tongue is provided by the CT nerve (McManus et al., 2011; Taillibert et al., 1998). Taste buds in the nasoincisor duct of the rat and along part of the soft palate of rodents and primates are innervated by the GSP nerve (Hamilton and Norgren, 1984). Taste buds on the caudal 1/3 of the tongue, within the posterior two or three foliate papillae, and on the nasopharynx are innervated by the lingual branch of the GP nerve. The GP nerve and the superior laryngeal branch of the vagus nerve innervate taste buds located on the soft palate. Cell bodies of the facial, GP, and vagus nerves are located in the geniculate, petrosal, and nodose ganglia, respectively; the central processes of these nerves terminate in the NTS in a roughly topographical order, with considerable overlap (Hamilton and Norgren, 1984; Halsell et al., 1993; May and Hill, 2006; McPheeters et al., 1990; Travers and Norgren, 1995).

32.3.2 Caudal Brainstem

32.3.2.1 Intrinsic Anatomy of the NTS.

RODENT Taste-responsive neurons are located in the rostral third of the NTS, which, based upon functional and morphological criteria, consists of four distinct subnuclei: rostral central, rostral medial, rostral lateral, and rostral ventral (Halsell, et al. 1996; Whitehead, 1988; see Figure 32.2). In both rat and hamster, the rostral central subnucleus of the NTS receives most of its projections from the three nerves that innervate the taste buds in the oral cavity (Whitehead, 1988; Lundy and Norgren, 2013, 2014) and contains the majority of NTS taste neurons (Whitehead and Frank, 1983). Primary afferent glutamatergic neurons terminate within glomeruli located along the distal dendrites

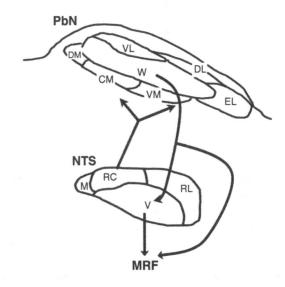

Figure 32.2 Subnuclear organization and reciprocal projections of the NTS (bottom) and PbN (top) in the rat. Abbreviations in PbN: VL = ventrolateral nucleus; DL = dorsolateral nucleus; EL = external lateral nucleus; DM = dorsomedial nucleus; W = waist area; CM = central medial nucleus.; VM = ventromedial nucleus; in NTS: M = medial nucleus; RC = rostrocentral nucleus; RL = rostrolateral nucleus, V = ventral nucleus.

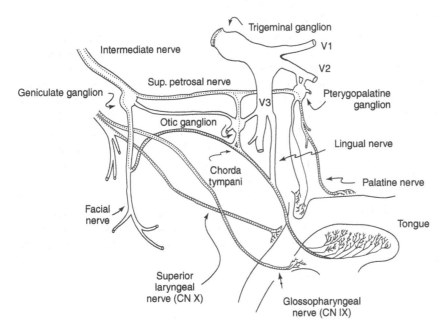

Figure 32.1 Gustatory innervation of the oral cavity in primates. (Diagram adapted with permission from Oxford University Press).

32.3 Anatomy and Physiology of the Gustatory System

and spines in the rostral central and rostral lateral NTS (Li and Smith, 1997; Wang and Bradley, 1995, 2010b; May and Hill, 2006; Whitehead, 1986; Whitehead and Frank, 1983). The rostral central subnucleus consists of neurons responsive to convergent taste and tactile stimulation; these are projection neurons that relay taste information centrally to the PBN (Halsell et al., 1993; Norgren and Pfaffmann, 1975). Cells in the NTS that project rostrally are separate from those with local brainstem projections (Halsell et al., 1993). This anatomical segregation of function has also been observed in the geniculate ganglion where different groups of cells project to areas of the NTS that are part of the ascending taste pathway related to sensory discrimination and homoeostatic functions, or elements of the local or descending circuits associated with orofacial reactivity (Zaidi et al., 2008). Neurons in the rostral ventral NTS project to the intermediate reticular formation and other brainstem areas involved with taste-related reflexes (Beckman and Whitehead, 1991; Travers, 1988).

Morphological studies of neurons in the rostral NTS have identified three general cell types: fusiform, stellate, and ovoid. The fusiform and stellate cells are the projection neurons of the rostral central NTS (Whitehead, 1993). Fusiform cells have at least two primary dendrites which can extend for several hundred micra in the mediolateral plane, perpendicular to the incoming fibers from the solitary tract (Whitehead, 1986; Whitehead and Frank, 1983). This arrangement maximizes synaptic opportunities with incoming taste fibers (Davis and Jang, 1986; Whitehead, 1986; Whitehead and Frank, 1983). There is conflicting evidence using 3-D reconstruction of neurobiotin-labeled cells, however, suggesting that the NTS actually contains very few bipolar neurons (Renehan et al., 1994). This group proposed a more intricate cell classification based on cluster analyses of six quantified morphological features. One of these features, cell size, can be associated with immunohistochemical characteristics. For example, large neurons show immunoreactivity to tyrosine hydroxylase (Davis, 1998), suggesting that these are dopaminergic cells. Small cells are believed to be GABAergic, inhibitory interneurons (Davis, 1993; Lasiter and Kachele, 1988b; Whitehead, 1986). Based upon vesicle distribution, high density (GABA-HD) and low density (GABA-LD) GABAergic terminals have been identified within the rostral NTS (Leonard et al., 1999; Whetherton et al., 1998). Proximal dendrites and soma contain both LD and HD variants of GABA terminals; only GABA-HD terminals are found at distal dendrites (Leonard et al., 1999). In general, GABAergic synapses are concentrated on the proximal dendrites and soma of NTS cells in the rostral central NTS; afferent glutamatergic input from peripheral nerves synapse on distal dendrites (Leonard et al., 1999; Whitehead, 1993).

Wang and Bradley (2010a), using a transgenic mouse strain where the expression of a green fluorescent protein is under the control of a glutamic acid decarboxylase promoter (GAD67–GFP), identified a subset of inhibitory neurons that express somatostatin. Based upon their morphological and physiological characteristics, this subset of GABAergic neurons consists of two groups. The most common type of GAD67-GFP cells (73%) are small ovoid or fusiform that fire in phasic bursts; larger fusiform or multipolar GAD67-GFP cells fire tonically when depolarized (Wang and Bradley, 2010a). Because these GAD-67-GFP neurons are concentrated in the ventral subdivision of the NTS, they have little direct interaction with cells in the CT or GP terminal fields; they may, however, influence taste-responsive cells through polysynaptic connections.

The widespread distribution of GABAergic terminals in the NTS underscores the importance of inhibitory circuits in medullary taste processing. Indeed, over half of the neurons in the rostral central and rostral lateral subdivisions of the NTS respond to exogenously applied GABA (Davis, 1993; Liu et al., 1993; Smith and Li, 1998, 2000). Although both $GABA_A$ and $GABA_B$ receptors are present in the NTS (Liu et al., 1993; Smith and Li, 1998), most (84%) NTS GABAergic cells are affected by the $GABA_A$ antagonist bicuculline; only 14% of GABA-responsive NTS neurons respond to baclofen, a $GABA_B$ agonist.

At the physiological level, inhibition provides a backdrop for incoming excitatory signals. In fact, Smith and Li (1998) have shown that NTS cells are tonically inhibited. Electrical stimulation of the solitary tract evokes both excitatory (mean 4.8 ms latency) and inhibitory (mean 8.8 ms latency) postsynaptic potentials (PSPs) in NTS cells *in vitro* (Grabauskas and Bradley, 1996; Bradley, 2007). Bradley (2007) has suggested that the time course of each of these PSPs indicates monosynaptic excitation followed by polysynaptic recurrent inhibition.

The idea that excitatory input to the NTS initiates an inhibitory process was supported by a series of experiments using electrical stimulation of the CT nerve. After Grabasukas and Bradley (1998, 1999) demonstrated that tetanic stimulation of NTS cells can potentiate inhibitory PSPs *in vitro*, Lemon and Di Lorenzo (2002) showed that tetanic stimulation of the CT nerve *in vivo* can suppress or enhance subsequent taste responses in NTS neurons. These effects are stimulus-specific, with QHCl responses most frequently affected; NaCl responses are typically unaffected. However, the relationship of this effect to natural taste stimulation is unclear. To study this issue, Di Lorenzo et al. (2003b) used brief (100 ms) pulses of taste stimuli to provide a natural counterpart to tetanic electrical stimulation, followed at 1 or 5 s by the same or a different taste stimulus. Results showed that these "pre-pulses" of taste stimuli have a similar effect on taste responses in the NTS as tetanic stimulation of the CT nerve. That is, pre-pulses of some taste stimuli, like tetanic stimulation of the CT nerve, suppress or enhance the responses to subsequently presented taste stimuli in a stimulus-specific

706   **Chapter 32   Central Taste Anatomy and Physiology of Rodents and Primates**

manner. This effect is so pronounced that it changed the cell's most effective taste stimulus in 39% of the sample. Pre-pulses are effective when the pre-pulse/taste stimulus interval is 1 s, but not when the interval is increased to 5 s, suggesting that the pre-pulse potentiation effect dissipates over time.

Experiments using paired pulse stimulation have provided further insight into the function of NTS inhibition. For example, Grabauskas and Bradley (2003), using paired pulse electrical stimulation of the solitary tract *in vitro* in newborn rats, showed that the inhibitory postsynaptic current (IPSC) evoked by a conditioning pulse summates with the IPSC evoked by a subsequent test pulse when the pulses are <200 ms apart. In adult rats, field potential recordings from the NTS following paired-pulse CT stimulation showed that the response to the test pulse is suppressed by the conditioning pulse for about 2 s (Lemon and Di Lorenzo, 2002). Paired-pulse stimulation of the CT nerve attenuates the extracellular CT-stimulation evoked response of single NTS cells (Rosen and Di Lorenzo, 2009, 2012; Rosen et al., 2010b). These experiments described two types of paired-pulse attenuation: 1) a short latency (~20 ms) response that degrades quickly (~100 ms); and 2) a long latency (~50 ms) response that degrades slowly (~500 ms). During sapid stimulation of the tongue, the short latency, fast decay NTS neurons are generally more narrowly tuned to the four basic taste stimuli and exhibit larger responses than the other group of neurons that respond to paired-pulse stimulation with long latencies and a prolonged decay. These observations prompted Rosen et al. (2010b) to propose a model of NTS circuitry that incorporates a series of taste-stimulus specific cell assemblies in which the activation of one assembly inhibits others, perhaps through lateral inhibition. Research using simultaneous recordings from multiple NTS cells provides further support for this type of organization (Rosen and Di Lorenzo, 2012).

PRIMATE Although the general organization of the NTS in nonhuman primates is very similar to the pattern described above for rodents, less is known about the subnuclear organization and chemoarchitecture in primates. The most common parcellation of the primate NTS uses the solitary tract as a landmark that divides the nucleus into medial and lateral subdivisions (Beckstead and Norgren, 1979). Besides ease of use, dividing the NTS in this fashion creates a lateral, primarily gustatory, subdivision that dominates the rostral half of the nucleus and a medial, mostly viscerosensory, region in the caudal and commissural regions. Fibers of the facial nerve terminate almost exclusively in the lateral division, while vagal terminals dominate the medial division; axons of the GP nerve terminate in both medial and lateral divisions (Beckstead and Norgren, 1979; Satoda et al., 1996; Rhoton, 1968). Thus, the rostrocaudal terminal pattern of the CT, GP, and vagus nerves within the NTS of primates follows the same sequence as the rat, with the addendum that the lateral gustatory division of primates extends rostrally beyond the entry point of the facial nerve (Beckstead and Norgren, 1979; Rhoton, 1968). The anatomical organization of primates, including humans, is described in more detail in Pritchard (2012).

32.3.2.2 *Efferent Projections of the NTS.*

RODENT The efferent projections from the NTS are complex, involving multiple subnuclei which project to areas that control orofacial and ingestive behaviors, affect autonomic function, and contribute to neural coding of taste quality and intensity. This review focuses on the ascending projections that lead to the forebrain.

In the rat, about a third of the cells in the NTS have direct projections to the PBN (Monroe and Di Lorenzo, 1995; Ogawa et al., 1984; Ogawa et al., 1982); in the hamster, about 85% of NTS cells project to the PBN (Cho et al. (2002). In addition to this species difference, NTS projects to the contralateral solitary nucleus in the hamster (Li et al., 2008), but not in the rat (Lundy and Norgren, 2014).

PRIMATE Axons of gustatory neurons in the rostral NTS enter the central tegmental tract, ascend ipsilaterally through the pons and midbrain, and finally terminate in the parvicellular division of the ventroposteromedial nucleus of the thalamus (VPMpc; Beckstead et al., 1980; see Figure 32.3). The gustatory portion of VPMpc occupies the medial aspect of the ventrobasal thalamus, immediately medial to the oral somatosensory representation. The projection from the NTS to VPMpc in nonhuman primates bypasses the PBN, which is an obligate relay in rodents for gustatory information ascending from the medulla to the thalamus (Loewy and Burton, 1978; Norgren, 1978; Norgren and Leonard, 1973). In primates, the PBN relays general visceral and nociceptive information from the caudal NTS (Beckstead et al., 1980) and the spinal cord (Westlund and Craig, 1996) to the forebrain. There is no evidence of a gustatory presence in the PBN of primates (Pritchard et al., 2000).

32.3.2.3 *Subnuclear Organization of the PBN in the Rat.* Most of the neurons in the rostral central subnucleus of the NTS project to the "waist" region of the PBN, so named because this cellular area occupies the somewhat constricted part near the mid-section of the brachium conjunctivum (BC; see Figure 32.2). The

32.3 Anatomy and Physiology of the Gustatory System

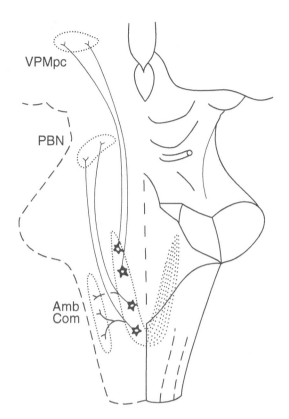

Figure 32.3 Efferent projections of the NST to the caudal brainstem and thalamus of the monkey. Whereas, the caudal visceroceptive NST projects to the PBN, the rostral gustatory NST bypasses the PBN and terminates in VPMpc. (Used with permission from John Wiley and Sons, Inc.).

waist region contains several PBN subnuclei, including the central medial and ventral lateral nuclei; other gustatory neurons are scattered among the fascicles of the BC itself (Davis, 1991; Lasiter and Kachele, 1988a; Norgren, 1978; Travers, 1988). Gustatory neurons in the waist region are classified as fusiform and multipolar (Davis, 1991; Lasiter and Kachele, 1988a); compared with cells in the NTS, they have more elaborate, but less expansive, dendritic arborizations (Davis, 1991).

32.3.2.4 4 Neural Coding in the Caudal Brainstem.

RODENT Neural activity, both spontaneous and evoked, is generally sluggish in the gustatory system compared to other sensory systems. In the CT nerve of the rat, the spontaneous rate averages only 1.09 spikes/s (Pritchard and Scott, 1982), but it increases to 14.2 spikes/s in the NTS (Giza and Scott, 1996). Using the published reports of Erickson and Doetsch (1965) and Doetsch and Erickson (1970) on taste responses in the CT nerve and NTS, respectively, Scott and Erickson (1971) concluded that NTS response magnitudes exceed those of the CT nerve by a factor of four.

Salt-and acid-best neurons account for approximately 60% and 30% of NTS taste neurons, respectively (Di Lorenzo and Monroe, 1997; Giza and Scott, 1991). Neither of these studies, nor Di Lorenzo and Monroe (1997), found any neurons that gave their largest response to QHCl. Despite this, all three studies reported that medullary taste neurons are broadly tuned with an average entropy of approximately 0.82.[1]

Gustatory response profiles of PBN neurons are similar to those in the NTS, although whether PBN responses are slightly smaller (Verhagen et al., 2003) or slightly larger (Di Lorenzo and Monroe, 1997; Di Lorenzo et al., 2009c) depends on the detail of analysis used to address this issue. When the responsiveness of PBN neurons is calculated separately for NTS relay and intrinsic neurons, the response magnitude of the former group is significantly higher (Di Lorenzo and Monroe, 1997; Monroe and Di Lorenzo 1995; Cho et al., 2002; Ogawa et al., 1984). Similarly, taste responses in PBN-thalamic relay neurons are larger than non-relay cells (Ogawa et al., 1987).

Data on the functional specialization of the waist area and external subnuclei of the PBN are conflicting. Based on results from c-*fos* studies, Yamamoto et al. (1994) proposed that the PBN waist area processes hedonically positive taste stimuli (NaCl, sucrose), while unpleasant gustatory stimuli (QHCl) are routed through the external lateral subnucleus. King et al. (2003), however, showed that intraoral infusions of QHCl produce *fos*-like immunoreactivity in both areas; Travers et al. (1999) reported the same results in decerebrate rats.

Just as in the NTS, the PBN of anesthetized rats is dominated by NaCl- and HCl-best neurons (Di Lorenzo and Monroe, 1997); in awake rats, NaCl-best neurons predominate (Nishijo and Norgren, 1997). Regardless of anesthetic state, PBN neurons are broadly responsive to the four basic taste stimuli (H = 0.58 to 0.87; Halsell and Travers, 1997; Di Lorenzo and Monroe, 1997; Nishijo and Norgren, 1991, 1997). These similarities aside, PBN gustatory neurons in the awake rat have a higher spontaneous rate and greater sucrose sensitivity than their anesthetized counterparts (Nishijo and Norgren, 1990, 1997). Both effects could be related to relief from anesthetic depres-

---

[1] The entropy coefficient (H) reflects the breadth-of-tuning of a neuron, that is, the degree to which an individual neuron responds to each of the four prototypical taste stimuli: NaCl, sucrose, HCl, and QHCl (Smith and Travers, 1979). The entropy equation is: $H = -k \Sigma p_i \log p_i$ Where H = breadth of tuning; k = the scaling constant; p_i = the proportional response to each of the four stimuli; i = 1-4 stimuli. When k, the scaling constant, is set to 1.661, the entropy coefficient will equal 1.0 when the cell's response to all four stimuli is identical.

708   **Chapter 32   Central Taste Anatomy and Physiology of Rodents and Primates**

sion, either locally or in forebrain areas that modulate PBN taste activity (Lundy and Norgren, 2014). An alternate explanation for the prevalence of sucrose responsivity is that a wider area of the oral cavity, including the taste buds around the nasoincisor duct, is stimulated when the awake rats drink the sapid stimuli. Taste buds located around the nasoincisior duct are preferentially sensitive to sucrose (Travers et al., 1986) and are seldom effectively stimulated when sapid stimuli are delivered by the investigator.

Data from simultaneous recordings of NTS and PBN cells in the anesthetized rat reveal that PBN cells with a given response profile, for example, NaCl-best or sucrose-best, receive direct projections from NTS neurons that have a variety of response profiles, and sometimes different best stimuli (Di Lorenzo and Monroe, 1997). However, NTS-PBN cell pairs that have the same best stimulus communicate more effectively than those with dissimilar best stimuli (Di Lorenzo and Monroe, 1997). That is, action potentials of NTS neurons are more likely to be followed by a spike in a PBN neuron if these functionally connected neurons share a best stimulus.

PRIMATE Recording from the NTS in nonhuman primates is such a daunting task that few investigators have attempted it. The NTS is a long, narrow target, buried deep in the posterior cranial fossa and protected dorsally by the fibrous tentorium. When acute rodent preparations are used, some investigators remove the overlying cerebellum and resect the tentorium; this option is not available to primate researchers who, for a variety of reasons, employ recording systems that permit repeated (chronic) experiments on the same animal. The primary advantage to using chronic recording techniques is that the monkeys can be tested while awake, which enables the animals to actively engage the investigator and the sapid stimuli.

For these reasons, only two papers have described the neural coding properties of taste neurons in the primate NTS. Scott et al. (1986a) reported that taste neurons in the NTS have concentration-response functions and threshold sensitivity that match human psychophysical data (Collings, 1974). Gustatory neurons in the NTS are broadly tuned (H = 0.87) and respond more often to glucose (40%) and NaCl (35%) than to QHCl (17%) or HCl (8%).

In the second report, Yaxley et al. (1985) reported that allowing a monkey to drink glucose to the point of satiety does not affect the glucose-evoked response by individual NTS taste neurons. This paradigm, and the significance of these negative data, will be described in more detail in the section on the orbitofrontal cortex (OFC) where the responses of taste neurons are modulated by internal factors such as satiety.

32.3.2.5   Projections of the PBN in the Rat.

DESCENDING PROJECTIONS Neurons in the waist area of the PBN send a reciprocal projection to the ventral subnucleus of the NTS (see Figure 32.2), which in turn, projects to the adjacent medullary reticular formation (MRF), another recipient of direct projections from the PBN. The MRF contains premotor circuits that contribute to taste-evoked orofacial reflexes (Halsell et al., 1996; Karimnamazi and Travers, 1998; Travers and Norgren, 1983). The external lateral and external medial subnuclei of the PBN also receive taste-related input from the NTS, but they project to the MRF rather than the solitary nucleus (Herbert et al., 1990; Halsell and Travers, 1997; Karimnamazi et al., 2002; Yamamoto et al., 1994).

ASCENDING PROJECTIONS Axons of gustatory neurons in the PBN join the central tegmental tract and terminate bilaterally in VPMpc (Bester et al., 1999; Karimnamazi and Travers, 1998; Krout and Loewy, 2000; Norgren, 1974). Electrophysiological experiments have confirmed that VPMpc gustatory receptive fields are located bilaterally on the tongue as well as on the hard and soft palates (Nomura and Ogawa, 1985; Ogawa and Nomura, 1988). Other PBN efferents terminate in the GC (Saper, 1982), lateral hypothalamus (LH: Bester et al., 1997), the central nucleus of the amygdala (CNA: Bernard et al., 1993; Karimnamazi and Travers, 1998), substantia innominata (Karimnamazi and Travers, 1998) and the bed nucleus of the stria terminalis (BNST: Alden et al., 1994). See Lundy and Norgren (2014) for a review of the central taste pathways of the rat.

32.3.2.6   Centrifugal Modulation of Gustatory Responses in the Brainstem.

Top down modulation is a feature of every sensory system, including gustation. In rodents, projections originating in the GC terminate in VPMpc (Norgren and Grill, 1976; Shi and Cassell, 1998; Wolf, 1968), the CNA (Norgren and Grill, 1976; Shi and Cassell, 1998), the PBN (Lasiter et al., 1982; Norgren and Grill, 1976; Wolf, 1968), and the NTS (Norgren and Grill, 1976; Whitehead et al., 2000). The CNA, in turn, projects to the NTS (Whitehead et al., 2000) and the PBN (Kang and Lundy, 2009; Tokita et al., 2009). The PBN and NTS also receive direct projections from the LH (Bereiter et al., 1980; Hosoya and Matsushita, 1981; Whitehead et al., 2000), and the BNST (Kang and Lundy 2009; Tokita et al., 2009). In hamster, projections from the contralateral NTS have also been reported (Whitehead et al., 2000); electrical stimulation of these fibers modulates taste responses in the NTS (Li et al., 2008).

Although the anatomy of centrifugal input to taste-related structures has been well described, less is known about the physiology of these projections and their

32.3 Anatomy and Physiology of the Gustatory System

functional significance. Both inhibitory and facilitatory responses have been recorded in VPMpc following electrical stimulation of the GC (Ganchrow and Erickson, 1972; Yamamoto et al., 1980). Both long and short duration effects were noted: long duration (~60 ms) responses that are either inhibitory or excitatory and short duration (~10 ms) responses which are strictly inhibitory (Yamamoto et al., 1980). Subsequent research in the somatosensory system suggests that these effects might be mediated by a disynaptic cortico-reticular pathway (Lam and Sherman, 2011; see also Guillery et al. 1998). Yamamoto et al. (1984) also studied the effects of GC electrical stimulation on CNA neurons. In that study, 13 of 16 CNA neurons were facilitated and/or inhibited by GC stimulation with a mean latency of 16 ms. Such a long latency suggests that the GC feedback to the CNA is mediated by a polysynaptic pathway. Similarly, evidence for a polysynaptic influence of the GC on the LH was reported by Kita and Oomura (1981), who observed excitation followed by inhibition, or inhibition alone.

There is considerable evidence that descending projections from the forebrain affect taste-evoked activity in the brainstem. For example, Di Lorenzo and Monroe (1995) showed that infusion of procaine, a short acting local anesthetic, into either side of the GC in rats decreases the number of basic taste stimuli that evoke responses in the NTS. The largest effects of GC inactivation are seen in the NTS intrinsic neurons, that is, the neurons that do not project to the PBN. Di Lorenzo and Monroe (1995) showed that ipsilateral GC inactivation preferentially affects NTS responses to sucrose and saccharin, both sweet tastes, while contralateral infusions enhance differences between palatable and unpalatable taste stimuli. Smith and Li (2000) reported that chemical or electrical stimulation of the GC produces excitation and inhibition in separate groups of cells in the NTS of hamster. The excitatory effects are distributed across all best stimulus types, but the inhibition targets NTS neurons that are more responsive to NaCl. Electrical stimulation of the contralateral GC is more effective than stimulation of the ipsilateral side; this relationship is reversed in the rat (Di Lorenzo and Monroe, 1995).

Taste responses in the NTS are also modulated by input from the BNST, the CNA, and the LH (reviewed in Lundy, 2008). In hamster, inhibitory influences originate predominantly in the BNST; excitatory effects most often originate in the LH and the CNA; these data have been reviewed by Smith et al. (2005a). Stimulation of the BNST, which inhibits 36% of NTS cells, mostly targets citric acid-best neurons (Smith et al., 2005b). Cho et al. (2002) reported that both ipsilateral and contralateral LH stimulation affects taste-responsive NTS cells, although contralateral stimulation is more effective. Ipsilateral stimulation can have either an excitatory or inhibitory effect on NTS cells, but contralateral LH stimulation

is always excitatory. In general, stimulation of the LH during sapid stimulation of the tongue more than doubles the magnitude of NTS taste responses. Li et al., (2002) report that only a third of NTS neurons are affected by CNA stimulation; this influence targets NTS cells that are only weakly taste-responsive. Subthreshold stimulation of the CNA, however, enhances taste responses in NTS neurons that receive CNA input (Li et al., 2002). Cho et al. (2003) reported that 24% of NTS cells are affected by stimulation of both the LH and the CNA; in rat, however, the influence of the CNA on NTS taste responses is more widespread, with more than half of NTS neurons affected (Kang and Lundy, 2010). Collectively, these studies show that descending forebrain projections, by differentially targeting best stimulus cell types, influence taste-evoked activity in the NTS in complex ways.

In both rat (Lundy and Norgren, 2001; 2014) and hamster (Li et al., 2005; Li and Cho, 2006), the PBN receives substantial centrifugal projections from the GC, the BNST, the LH, and the CNA (reviewed in Lundy, 2008). Interestingly, for each of these forebrain sites, the centrifugal projections to the PBN and NTS originate from different groups of neurons (Kang and Lundy, 2009). In hamster, nearly all PBN cells project to the CNA and/or the LH, but only a third receive reciprocal projections from these nuclei (Li et al., 2005). Similarly, in rat, about a third of taste-responsive PBN cells receive descending projections from the GC (Di Lorenzo and Monroe, 1992). Electrical stimulation of the LH and CNA of the hamster inhibits or facilitates PBN taste-evoked responses (Li et al., 2005). Inhibition narrows tuning of NaCl-best cells and broadens tuning of citric acid-best neurons (Lundy and Norgren, 2001). Electrical stimulation of the GC and CNA inhibits responses in approximately two-thirds of PBN taste neurons; about half of the PBN taste cells are either inhibited or facilitated by LH stimulation (Lundy and Norgren, 2014). Overall, centrifugal input sharpens PBN response profiles.

To summarize, descending projections to the PBN and NTS is both complex and varied, often depending on the best stimulus classification of the target neuron (Lundy, 2008). By changing the across neuron pattern of activity, descending projections often accentuate differences between taste qualities. The ability of centrifugal input to modulate taste-response profiles and, at times, the best stimulus classification of individual neurons, strongly suggests that neural coding of taste quality is a network dependent, fluid process and not a stable, immutable characteristic of individual neurons.

32.3.3 Neural Coding in the Forebrain

The thalamic sensory nuclei have, historically, been considered passive relays to the cerebral cortex. This is certainly true for the gustatory thalamus which has received much less attention than either the gustatory peripheral nerves, the NTS, or the PBN. Even though this view is changing, the role of the thalamus in perception is still not well-understood (see Saalmann and Kastner, 2009; Sherman, 2007; Bruno, 2011). Aside from this issue, from a strictly technical standpoint, the greatest challenge to studies on neural coding in the forebrain is the impact of general anesthesia. General anesthetics depress and distort neural activity throughout the visual, auditory, and somatosensory systems (Land et al., 2012; Detsch et al. 1999; Chapin and Lin, 1984), but the effect is particularly pronounced in the gustatory forebrain (Scott and Erickson, 1971). Even ketamine, which is generally considered to have a lesser impact on neural activity than barbiturates, virtually eliminates spontaneous and evoked taste responses in the thalamus of macaques (Pritchard, unpublished observations, 1987). The challenge presented by general anesthesia is best summed up by Scott and Erickson (1971) who after reporting that thalamic gustatory neurons were "quite unresponsive," switched to the paralytic gallamine triethiodide (Flaxedil). These recordings in paralyzed rats were better, but still sluggish. After the use of paralyzing agents for invasive procedures fell into disfavor in the research community, there was little incentive to study taste coding in the forebrain of mammals. Finally, chronic recording techniques, which had been used in primate sensory and motor research for years, began to see widespread use in gustatory research (e.g., Pritchard et al., 1989; Scott et al., 1986a).

32.3.4 Thalamus

32.3.4.1 Anatomy.

RODENT In the rat, axons originating in the PBN terminate in VPMpc, the medial small-celled extension of the oral somatosensory relay, VPM. Cell-size changes so gradually along the mediolateral dimension of the ventrobasal thalamus that the VPMpc/VPM boundary cannot be identified solely on the basis of neuron morphology. Electrophysiological studies have shown that the taste area is bounded laterally by lingual, thermosensitive neurons, which lie medial to oral tactile neurons, but determination of the boundaries between these areas awaits a study that combines anatomical reconstruction with detailed electrophysiological mapping of the region with taste, thermal, and tactile stimuli. Size estimates of the taste-responsive region of VPMpc in the mediolateral dimension approach 2.0 mm, but a more realistic width would be 0.5–1.0 mm (Lundy and Norgren, 2014).

PRIMATE In primates, gustatory projections to VPMpc originate in the rostral NTS (Beckstead et al., 1980). Just as in rodents, the boundaries between the gustatory, thermal, and tactile regions of the ventrobasal thalamus are, at best, rough estimates, but based on electrophysiological recordings, the taste area is almost cube-shaped, measuring at least 0.75 mm on each side (Pritchard et al. 1989).[2]

32.3.4.2 Physiology.

In both rat (Verhagen et al., 2003) and monkey (Benjamin, 1963; Pritchard et al., 1989), several different sensory modalities are represented within the limited confines of VPMpc. In the anesthetized rat, VPMpc gustatory neurons (9%) co-mingle with cells that respond to thermal (8%) or tactile (6%) stimulation. Multimodal neurons (41%) are common and 36% of the neurons in the thalamic taste area have no identifiable adequate stimulus. In awake, behaving macaques, gustatory neurons account for 36% of the population, with another 11% responsive to tactile stimulation (Pritchard et al., 1989). Thermosensitive neurons are scarce (<1%) and unresponsive neurons (18%) still account for a significant fraction of the population. Similar to the observations in the rat, about half of the thalamic taste neurons also respond to either thermal or tactile stimulation. In 35% of VPMpc neurons, spontaneous activity is inhibited immediately prior to fluid delivery. These neurons, which may provide the link between taste perception and ingestive behavior, are unlikely to be responsive in anesthetized animals.

RODENT The spontaneous rate of taste neurons in the thalamus is slower than in either the CT, the NTS, or the PBN (Verhagen et al., 2003, 2005). It is widely believed that anesthetics and paralytics suppress neural activity more in the forebrain than in the hindbrain, but it is uncertain to what degree the low spontaneous activity in the thalamus reflects inherent properties of these neurons, or is a byproduct of drug restraint. In the thalamus, spontaneous activity in paralyzed and anesthetized rats ranges from 3–6 spikes/s (Scott and Yalowitz, 1978; Nomura and Ogawa, 1985; Ogawa and Nomura, 1988; Verhagen et al., 2003, 2005). In awake monkeys, the spontaneous rate of thalamic taste neurons is 9.4 spikes/s, which is significantly higher than in the NTS (1.2 spikes/s; Scott et al., 1986a).

Verhagen et al. (2003) reported that gustatory stimuli activate 42% of the neurons in the thalamus of rats anesthetized with a urethan/-chloralose mixture. Only 9% of the sample respond exclusively to sapid stimulation of the oral cavity; the remaining 33% respond to thermal stimulation. Interestingly, only 6% of the sample responds to tactile stimulation and only one of these was a gustatory

---

[2] This estimate is based upon sapid stimulation limited to the anterior tip of the tongue.

neuron. Nomura and Ogawa (1985), using paralyzed and artificially ventilated rats, reported that 88% of a small sample (N = 16) of thalamic taste neurons respond to "strong mechanical stimulation, that is pinching with non-serrated forceps." In most cases, the tactile and taste receptive fields did not overlap. Most gustatory neurons respond to bilateral stimulation of the tongue (43%), with the receptive fields of the remaining 57% restricted to either the ipsilateral (28.5%) or contralateral (2.5%) side (Nomura and Ogawa (1985; see Ogawa and Nomura, 1988).

Based upon the entropy statistic, thalamic gustatory neurons are broadly sensitive (H = 0.79), but chemosensitivity is not evenly distributed across the four prototypical taste stimuli (Verhagen et al., 2003). Approximately half (51%) of the gustatory cells in the thalamus respond best to 0.1 M NaCl, with 0.01 M HCl evoking the largest responses in only a quarter (26%) of the sample; sensitivity to 0.5 M sucrose (16%) and 0.01 M QHCl (7%) was low. This is strikingly similar to the pattern observed in the NTS (Giza and Scott, 1991; Giza et al., 1991).

PRIMATE Pritchard et al. (1989) reported that the gustatory receptive fields in the VPMpc of macaques are located on the ipsilateral side of the tongue, thereby confirming the anatomical observations of Beckstead et al. (1980) and Pritchard et al. (2000). A subsequent electrophysiological report by Lenz et al. (1997) also shows that taste localizes to the ipsilateral side of the tongue of human neurosurgical patients. The laterality of the ascending gustatory pathway is an important issue because of its clinical implications, yet because the data from human stroke patients are so contradictory, this question remains unsettled (Pritchard, 2012).

Gustatory-evoked responses are typically more sluggish in VPMpc than in the NTS of anesthetized rats (Scott and Erickson, 1971; Scott and Yalowitz, 1978), but when this issue is examined in awake, behaving monkeys, rates are higher in the thalamus (Pritchard et al., 1989) than in the medulla (Scott et al., 1986a). Despite having a moderate spontaneous rate of 9.4 spikes/s, inhibition is uncommon (9.1% of all responses). Approximately 60% of all taste neurons in VPMpc respond to 1.0 M sucrose, with 1.0 M NaCl a distant second (26%); approximately 7% of the taste-responsive neurons respond best to either 0.01 M HCl or 0.001 M QHCl. This pattern contrasts sharply with the thalamic data collected from the rodent thalamus, which is dominated by salt- and acid-best neurons (see above), but these differences are confounded by both species and anesthesia issues. When anesthesia effects are eliminated through the use of awake, behaving primates, a valid comparison can be made between medullary and thalamic taste neurons. Unlike the thalamus where sucrose-best neurons dominate, 1.0M NaCl is more effective than 1.0M glucose (43% vs 31%) in the NTS. Thalamic taste cells are broadly tuned in both the NTS (H = 0.87) and VPMpc (H = 0.73), but given the nonlinearity of the entropy statistic (see Pritchard et al., (1989), medullary neurons are considerably more broadly tuned those in the thalamus.

32.3.5 Primary Taste Cortex

32.3.5.1 Anatomy.

RODENT In rats, fourth-order gustatory neurons in VPMpc project to a narrow band of cortex dorsal to the rhinal fissure that overlies the claustrum (Kosar et al., 1986b; Norgren and Wolf, 1975; see Figure 32.4). This area contains so few granule cells that it has been described as "agranular" by some investigators (e.g., Kosar et al., 1986a) and "dysgranular" by others (e.g., Cechetto and Saper, 1987). The gustatory strip lies ventrally to a dysgranular band of thermosensitive neurons, which is ventral to a larger granular cortical zone that contains the lingual representation of primary and secondary somatosensory cortices (Kosar et al., 1986a; Guldin and Markowitsch, 1983; Shi and Cassell, 1998). The CT field is rostral to the GP field, which partially overlaps regions containing visceroceptive and nociceptive neurons (Hanamori et al., 1997, 1998a, b). As described above, neurons in GC project to each of the subcortical taste relays including VPMpc, the PBN, the NTS, and affiliated gustatory nuclei, including the LH and the CNA (Kang and Lundy, 2009; Di Lorenzo and Monroe, 1995; Lundy and Norgren, 2014; Yamamoto et al., 1980).

PRIMATE In primates, taste neurons in VPMpc project to a small crescent of granular neocortex that straddles the inner operculum and dorsal insula (Benjamin and Burton, 1968; Benjamin et al., 1968; Pritchard et al., 1986; see

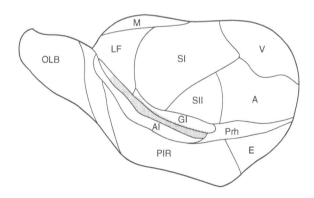

Figure 32.4 Diagram showing the location of the gustatory area in dysgranular cortex (shaded area) of the rat. Abbreviations: AI-agranular insular cortex; E = entorhinal cortex; GI = granular insular cortex; LF-lateral frontal cortex; M = motor cortex; OLB = olfactory bulb; PIR = piriform cortex; Prh = perirhinal cortex; SI = primary somatosensory cortex; SII = secondary somatosensory cortex. (Used with permission from John Wiley and Sons, Inc).

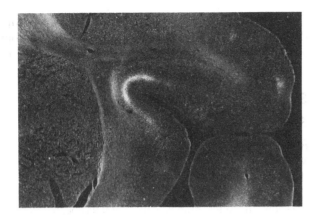

Figure 32.5 Autoradiographic label in monkey cortex following an injection of tritiated amino acids into the thalamic gustatory relay, VPMpc. (Used with permission from John Wiley and Sons, Inc.).

Figure 32.5). Because this area receives projections from the thalamic taste relay, by definition, it wears the mantle of primary taste cortex. From this robust terminal field at the fundus of the superior limiting sulcus, the gustatory projection area tapers into the inner operculum and the body of the rostral insula.

Taste evoked activity has been recorded from a number of regions beyond the boundaries of primary taste cortex: Brodmann areas (BAs) 1 and 2 on the inner operculum, the precentral extension of BA 3b on the lateral convexity, and the middle and posterior insulae (imaging studies); all of these areas, as well as the dysgranular and agranular regions of the insula and the OFC, are populated predominantly, if not exclusively, by second-, third-, and even higher-order cortical taste neurons. Based upon a series of electrophysiological experiments in the insular and opercular cortices, Scott and Plata-Salamán (1999) estimate that the taste-responsive area in the rostrodorsal insula encompasses 50 mm^3, which is slightly larger than the area defined by less sensitive, anatomical methods. See Pritchard (2012) for a more detailed description of cortical taste anatomy in primates.

Every investigator who has described single neuron responses in primary taste cortex dating back to Sudakov et al. (1971) has commented on the paucity of taste neurons located there. Based upon their extensive recordings, Scott and Plata-Salamán (1999) estimate that only 6.5% of the cells in the rostral insular-opercular cortices respond to sapid situation of the tongue. It is worth noting that these experiments sampled neurons within primary taste cortex (anatomically defined) as well as within the dysgranular body of the insula, where many if not most cells are second, or even third-order cortical taste neurons. As in subcortical taste areas, gustatory neurons are scattered among cells that respond to oral tactile stimulation.

32.3.5.2 Physiology.

RODENT There has been unprecedented interest in the response properties of cortical taste neurons during the last decade. Just as in the thalamus, previous studies of cortical taste activity in anesthetized rats struggled with obstinate neural spindling and lethargic responses. Because many of the early studies on anesthetized rats (see Yamamoto (1984) for an early review) and one study on awake rats (Yamamoto et al., 1989) were done in the same laboratory, evaluating the impact of general anesthesia on cortical taste activity is relatively straightforward.

Spontaneous activity of cortical taste neurons in awake rats is almost twice as high as in anesthetized and paralyzed rats (4.4 vs 2.4 spikes/s, respectively; Yamamoto et al., 1989). Likewise, the responses evoked by the four basic taste stimuli are twice as robust in awake rats (8.0 spikes/s). Cortical taste neurons in rats are broadly tuned (H = 0.54), a characteristic that is not affected by anesthesia. Yamamoto et al. (1989), using awake rats, divided his sample of 35 cortical taste neurons into two groups: type I cells that they hypothesized code taste quality, and type 2 neurons that attend to hedonic value. The type I neurons are almost equally divided among the four best stimulus categories: sweet (N = 6), salty (N = 7), sour (N = 7), and bitter (N = 9). Such uniform chemosensitivity had not been reported previously, regardless of anesthetic state or species, at any level of the gustatory neuraxis. The responses of type 2 neurons vary according to the palatability of the stimulus rather than the chemical moiety.

Unlike thalamic taste cells in both rats and monkeys, cortical gustatory neurons do not respond to tactile stimulation in either anesthetized (Yamamoto et al., 1989) or awake rats (Yamamoto et al., 1981); thermal responses are observed under both conditions.

Descriptive information about taste neurons in awake animals is important because it not only provides baseline information about the system and its fundamentals, but it also provides a basis for cross-synaptic and across species comparisons. The real advantage to using awake, behaving animals in electrophysiological experiments, however, is the ability to examine the brain-behavioral interface.

Beginning about a decade ago, there has been renewed interest in temporal coding in the gustatory system by individual neurons as well as by neuron ensembles. Collectively, these studies have shown that a substantial amount of information is embedded within the temporal firing pattern of individual neurons, local neuron pools, and even functionally-related but anatomically distant sites (Katz et al., 2002; Gutierrez et al., 2010). Although it remains to be demonstrated to what degree the brain uses this information, it is clear that summing the number of action potentials across several seconds provides an

32.3 Anatomy and Physiology of the Gustatory System

incomplete view of neural processing (Katz et al., 2001, 2002).

Katz et al. (2001) demonstrated that the temporal discharge pattern of individual neurons in the GC of awake rats contains important data that are lost when the spike record is collapsed across several seconds. For example, when the summed activity during a 2.5 sec. period is analyzed, Katz et al. (2001) reported that 14% of GC neurons respond to gustatory stimulation. When the same neural records were divided into 500 ms periods and reanalyzed, 28% of the cells met the criterion for a taste neuron. This doubling in the number of taste-responsive neurons was enough to more than offset the loss of 38% of the neurons that originally met the taste criterion across the 2.5 s period, but were now reclassified as non-responsive. When this temporal analysis switched to a running average of 500 ms bins, the percentage of gustatory neurons increased further to 41%. Stapleton et al. (2006), comparing the lick-by-lick responses in a fixed ratio 5 experiment, reported that 34% of the neurons in the GC respond to sapid stimuli.

All experiments that use awake subjects include some provision for discarding trials in which part or all of the evoked response is caused by concurrent somatosensory stimulation. The fine-grained temporal analysis used by Katz et al. (2001) makes it possible to dissociate the contributions of the gustatory and somatosensory system during each 500 ms epoch. Using this approach, Katz et al. (2001) demonstrated that activity during the first 200 ms following stimulus onset is driven by somatosensory rather than gustatory stimulation. The intensity of the sapid stimulus dominates the next 600 ms, but after 800 ms, the neural record reflects stimulus palatability (Katz et al., 2001; Sadacca et al., 2012).

This new appreciation for the importance of temporal coding combined with advances in multi-electrode recording technology have provided a unique opportunity to investigate how taste information is processed by neural ensembles. Although the picture is far from complete, a decade of research has shown that coincident activity among taste neurons is commonplace, and we believe, networked processing will eventually be seen as the rule rather than the exception in sensory coding.

Early studies that examined the responses of cortical taste neuron pairs in anesthetized/paralyzed rats reported that the responses of 55% of these dyads are correlated with one another (Nakamura and Ogawa, 1997). Although more than 60% of correlated neuron pairs respond best to the same basic taste stimulus, coincident activity is also evident in 14% of the taste/non-taste neuron pairs (see also Katz et al., 2002). Because this synchronous activity is marked by tight cross-correlations and was derived from single electrodes to neuron pairs separated by less than 50 m (Nakamura and Ogawa, 1997), these dyads were most likely elements of local neuron ensembles.

Halpern (1985) demonstrated that taste perception and taste-guided behavior may take place as quickly as a few hundred milliseconds in the case of a rat licking a fluid spout. Many other experiments have shown that gustatory information also operates across many seconds for more complex behaviors such eating a meal. There is evidence that neural ensembles function within this broad, but behaviorally relevant, time scale. For example, Stapleton et al. (2007), using electrode microarrays implanted in the GC of awake rats, showed that small neural ensembles respond differentially within 150 ms of a single lick on a fluid spout. McDonald et al. (2012) reported that rats can identify the quality and intensity of a NaCl stimulus on the basis of two licks. It is noteworthy that 18% of the neurons they sampled responded to water, but not salt. Insular activation by water also has been reported in fMRI studies of human subjects (de Araujo et al., 2003a).

Katz et al. (2002), using similar techniques, demonstrated that neuron pairs show coincident activity during the 75–500 ms following the onset of sapid stimulation. Dyads of taste neurons show a mean cross-correlation half-width of 300 ms, which indicates that these cells are separated from one another by multiple synapses and perhaps considerable distances. Indeed, a number of these coupled neurons were located in opposite hemispheres, which is essential for a complex process like ingestive behavior that involves multiple brain areas. Simultaneous recordings in the GC and the gustatory OFC during a taste-guided motor task show that ensemble responses are not only coordinated, but also plastic. Whereas activity in the GC is related to properties of the taste cue that guides response selection, OFC activity reflects the spatiotemporal properties of taste-guided the motor behavior. A more expansive set of experiments by Gutierrez et al. (2010) used simultaneous recordings in the GC, OFC, amygdala, and nucleus accumbens to demonstrate that synchronous activity in this multi-site ensemble is entrained by the rat's licking, increases as a function of learning, and improves taste discrimination in a GO/NOGO task (see also Pennartz et al., 2012).

Ensemble recording, by exploiting the temporal information within the neural record, portrays taste coding – and by extension taste-guided behavior and ingestive behavior – as a dynamic process that employs multiple cortical and subcortical brain areas. It remains to be determined whether the information extracted in these studies is utilized by the nervous system or if it is simply a derivative analysis that merely benefits the experimenter/observer (Katz et al., 2001). However, the utility of this information for the rat has been shown. Time-shifting the individual spikes in a neural record lessens the ability of an ensemble to discriminate between different tastants (Katz et al., 2002). It is hard to imagine that networked temporal information is not being used by the nervous system, especially since ensembles have better predictive value of taste quality than either the rate or temporal components of individual neurons (Stapleton et al., 2007). It is also worth noting that predictive value increases directly with ensemble size (Stapleton et al., 2007).

The ensemble coding literature has important implications for the labeled line and across neuron pattern theories of taste quality coding. Both theories use a static measure of taste coding: action potentials over an extended time period. The dynamic and labile nature of ensemble activity creates a shifting landscape that undermines the neural basis of best-stimulus categories and, by extension, the labeled line theory of quality coding. By definition, the concept of ensemble or network processing supports and builds upon the across neuron theory of quality coding (Erickson, 1967) by adding temporal processing.

PRIMATE Numerous electrophysiological studies have described the properties of taste-responsive neurons in the insula and operculum of alert nonhuman primates (Scott et al., 1986b; Ogawa et al., 1989; Yaxley et al., 1988; Scott et al., 1991; Ito and Ogawa, 1994; Rolls and Baylis, 1994). Because most of these studies were done by Scott and colleagues, it is relatively easy to make study-to-study comparisons of their data. These experiments examined thresholds, concentration-response functions (Scott et al., 1991) and quality coding (Smith-Swintosky et al., 1991), as well as responses to arrays of sweet (Plata-Salamán et al., 1993), salty (Scott et al., 1994), sour (Plata-Salamán et al., 1995), and bitter (Scott et al., 1999) stimuli. Other studies examined complex gustatory stimuli such as amino acids (Plata-Salamán et al., 1992) and taste mixtures (Plata-Salamán et al., 1996).

The spontaneous rate of gustatory neurons in the insula is a languid 3.2 spikes/s, which from a statistical standpoint, provides few opportunities for inhibition and, predictably, few (1.9%) are found. This spontaneous rate places the insula between the NTS (1.2 spikes/s) and VPMpc (9.4 spikes/s). Response rates are also slow with the most effective stimulus, 1.0M glucose, only evoking a net 5.1 spikes/s. Like the thalamus (Pritchard et al., 1989, which tested 1.0M sucrose), most neurons in the insula respond best to a sweet stimulus (1.0M glucose = 38%), with 0.3M NaCl-best neurons (34%) a close second; the remainder respond best to 0.001M QHCl (22%) or 0.01M HCl (5%). Insular taste neurons, like those in the medulla and thalamus, are broadly tuned (H = 0.70). One of the most important findings from this prodigious literature is that the electrophysiological data collected from the monkey for this array of salty, sweet, sour, and bitter stimuli are highly correlated with the human psychophysical data (r = +0.91; see Figure 32.6; Scott and Plata-Salamán, 1999). The close correspondence between the primate neural data and the human psychophysical ratings suggests that the nervous system completes taste quality coding within the insula, if not earlier.

Just as in the NTS (Yaxley et al., 1985), the shift in hedonic value from positive to negative that occurs over the course of a meal has no effect on the response magnitude of gustatory neurons in the insular-opercular cortex (Rolls et al., 1988; Yaxley et al., 1988; Ifuku et al., 2003; Ogawa, et al., 2005). Humans also report that the hedonic value of food decreases over the course of a meal (Rolls et al., 1981), but noninvasive imaging studies have obtained contradictory data (Small et al., 2001; Tataranni et al., 1999; Gautier, 2001; Del Parigi et al., 2002); thus, the neural basis of this effect may not be the same in humans and monkeys.

As in the thalamus, the insula also contains many non-gustatory neurons (Scott and Plata-Salamán, 1999). Cells that respond during mouth movements are common (23.8%) and interspersed with neurons that respond to tactile stimulation of the oral cavity (3.7%), approach of

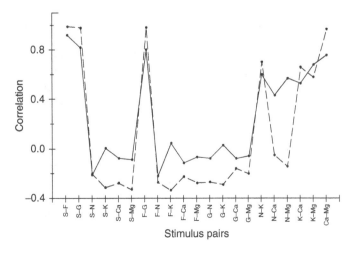

Figure 32.6 Correlations between taste quality between electrophysiological data from the monkey insula (solid lines; Smith-Swintosky et al., 1991) and human psychophysical reports (dotted lines; Kuznicki and Ashbaugh, 1979). (Used with permission from the American Physiological Society).

the stimulus delivery syringe (1.2%), tongue extension (0.3%) and olfactory stimulation. Other cells respond to stimulus viscosity (53%), temperature (35%), fat content (8%), grittiness (8%), and astringency (8%) of sapid stimuli and a subset of neurons respond to capsaicin (Critchley and Rolls, 1996; Rolls, 2005). Half of all insular neurons respond to at least one of these characteristics (Verhagen et al., 2004[3]). The proximity of gustatory, olfactory, and somesthetic (touch, temperature, nociception) neurons in the insula raises the possibility that this may be one locus involved with flavor perception.

32.3.6 Higher-order Gustatory Cortices

32.3.6.1 Anatomy.
The OFC in rodents and primates, which receives projections from most if not all sensory modalities, including taste, has been implicated in executive functions related to emotion, reward evaluation, social behavior, and decision-making (Schoenbaum et al., 2006; Frederique and Cattarelli, 1996; Ongur and Price, 2000).

RODENT In rats and hamsters, gustatory afferent projections to the orbital cortex originate in the insula (Jasmin et al., 2004; Reep and Winans, 1982; Reep et al., 1996).

PRIMATE In the macaque, the higher-order taste cortices form an elaborate matrix that stretches from the body of the insula, through the orbital surface, and across much of the posterior OFC (Carmichael and Price, 1995; Mesulam and Mufson, 1982; Morecraft et al., 1992; Barbas, 1993; see Figure 32.7). Most studies have focused on the cortex between the medial orbital sulcus and the precentral operculum (area PrCO), which contains the medial orbitofrontal (mOFC) and the caudolateral orbitofrontal (clOFC) taste areas.

Gustatory afferent information reaches the OFC through a multisynaptic pathway that originates in primary GC. The initial segment of this insulo-orbital circuit terminates in the dysgranular and agranular regions of the insula. Neurons in the dysgranular insula also project to the agranular insula, and then both areas project to parts of the posterior orbital cortex (Carmichael and Price, 1995). Tracking the ascent of gustatory information to the OFC is complicated by repeated convergence, divergence, and overlap of the individual projections. Although the entire circuit between primary taste cortex and the OFC is not fully understood, it is clear that the final leg of the pathway to the OFC originates within the dysgranular and agranular insulae, and the posterior orbital cortex rather than primary taste cortex (Carmichael and Price, 1995, 1996; Mufson and Mesulam, 1982; Mesulam and Mufson, 1985; Barbas,

[3] All of the stimulus modalities listed above were tested, except astringency.

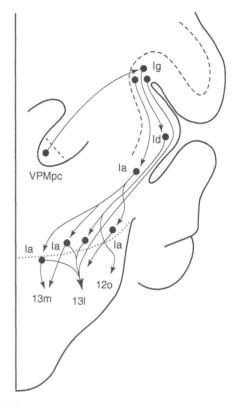

Figure 32.7 Diagram of the gustatory thalamocortical and cortico-cortical circuitry of the monkey. The thalamus and insula, depicted in the coronal plane, have been blended with the ventral surface of the OFC, shown in the horizontal plane. The dotted line marks the insulo-orbitofrontal boundary. As described in the text, most of the gustatory projections to the OFC originate in the agranular insula rather than primary taste cortex or the dysgranular insula. Abbreviations: Brodmann areas 12o, 13l, 13m; Ia = agranular insula; Id = dysgranular insula; Ig = granular insula.

1993; Baylis et al., 1995; Pritchard, unpublished observations, 2005). Thus, gustation resembles other sensory systems whose projections to the OFC originate in sensory association cortices rather than in the primary sensory cortex (Carmichael and Price, 1995; Barbas, 1993, 2007). These studies also suggest that the route from primary taste cortex to the mOFC is more direct, that is, it involves fewer synapses than the pathway that eventually leads to the clOFC taste area.

The mOFC is located between the medial and lateral orbital sulci and consists primarily of BAs 13m and 13a (Pritchard et al., 2005). Unlike the rest of the OFC where only 3.8% (range = 1.6 to 7.3%; Thorpe et al., 1983; Rolls et al., 1990, 1996; Rolls and Baylis, 1994) of the cells respond to sapid stimulation of the tongue, taste stimuli activate 20% of the neurons in this part of the mOFC. In the surrounding cortex, primarily BAs 13b, 13l, 12m, 11l, and 11m, and the medial agranular insula, 8% of the neurons respond to taste stimulation[4] (see Figure 32.8). Lara et al. (2009) reported that taste neurons account for 11% of the cells in the mOFC. A second taste area, the clOFC, is located in BA47/12o, but only 2% of the neurons there respond to taste stimulation (Rolls et al., 1990).

32.3.6.2 Physiology.

RODENT The OFC is next "frontier" for ensemble-based research. Complicated surgical and experimental paradigms using free-moving rats that include ensemble recordings from the OFC and GC and probably the amygdala and the nucleus accumbens as well, will be needed to determine how these anatomically intertwined areas integrate and differentiate sensory, motivational, and experiential information as rodents transition sensation into action (see Gutierrez et al., 2010, above).

PRIMATE The electrophysiological data accumulated by Scott and colleagues in the insula and its good correspondence with the human psychophysical data of Kuznicki and Ashbaugh (1979) suggest that the brain has deciphered the code for taste quality either within the insula, or at the subcortical level. Regardless of whether some residual quality coding takes place in the dysgranular and agranular gustatory association cortices, it appears that their primary function is to mobilize taste information that guides feeding behavior and nourishment. The progression from sensory analysis to sensory support of behavior that began in the insula continues in the OFC where it rises to new levels.

The widespread but sparse distribution of taste-responsive neurons in the posterior OFC has been a challenge for systematic, in-depth analysis of gustatory neurons in this area of the brain. To date, the neural coding properties of OFC taste neurons have only been examined in two areas: the mOFC (Pritchard et al., 2005; Lara et al., 2009) and the clOFC (Rolls et al., 1990). The mOFC is one of the earliest OFC sites to receive projections from the gustatory association areas in the insula (Carmichael and Price, 1995).

Gustatory neurons in the mOFC, like those in the insula, are broadly tuned to representative stimuli of the four basic taste qualities (mOFC $H = 0.79$; insula $H = 0.70$), with each stimulus activating 50–60% of the neurons sampled. The broad tuning in the insula and the mOFC contrasts sharply with neurons in the clOFC where the entropy coefficient is extremely low ($H = 0.39$) and 82% of the population responds to glucose; salty, sour, and bitter stimuli each activate less than 12% of the population. It is not clear how meaningful the entropy statistic is for clOFC neurons, given that they are more intimately involved with reward, emotion, and other executive functions than sensory processing, per se (see below). To point, despite such narrow tuning to gustatory stimuli, clOFC neurons showed significantly more sideband sensitivity to visual, olfactory, auditory, and somatosensory information than neurons in the mOFC (Pritchard et al., 2005) or the insula (Rolls and Baylis, 1994; Critchley and Rolls, 1996; Rolls, 2004; Thorpe et al., 1983; Scott and Plata-Salamán, 1999). The multimodal character of clOFC neurons has

---

[4] It is possible the percentage of taste neurons in BA 13l has been underestimated due to limited sampling.

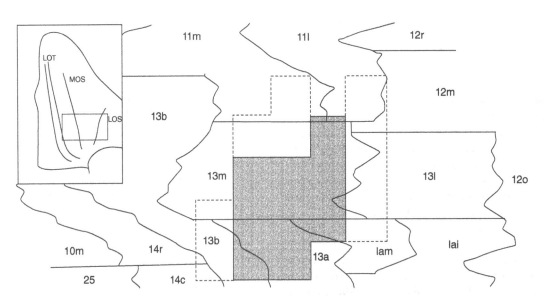

Figure 32.8 Diagram of the mOFC (unfolded) showing the core and perimeter gustatory areas. The inset indicates the location of this map relative to the ventral surface of frontal cortex. (Adapted from Pritchard et al., 2005).

fueled speculation that the area plays a role in flavor perception (de Araujo et al., 2003b).

Recent insights into the function of the OFC have used its longstanding role in emotion as a point of departure for a related role: evaluation of reward and risk related to decision-making (Tremblay and Schultz, 1999; O'Neill and Schultz, 2010; Rempel-Clower, 2007; Summerfield and Tsetsos, 2012). This view of the OFC is supported by studies in both the animal and human literatures.

The presence of so many non-responsive neurons and low to middling percentages of gustatory neurons in the forebrain "taste" areas strongly suggests that the optimal stimuli or paradigms are not being tested. This point has been driven home by electrophysiological experiments that engage a much higher percentage of OFC neurons when the test paradigm examines cue-guided behavior, reward evaluation, and reward contingencies instead of passive consumption of sapid stimuli (e.g., Pritchard et al., 2005). By using a delayed color matching test, Hosokawa et al. (2007) demonstrated that 31% of clOFC neurons code the relative preference of positive and negative reinforcement. Other mOFC neurons show proportional responses when presented with cues associated with two different rewards (Schultz et al., 1998; Padoa-Schioppa and Assad, 2006; Padoa-Schioppa, 2007). Changing the reward contingencies caused some neurons to increase (or decrease) their responses; these fluctuations resulted in some cells changing from non-responders to responders (Tremblay and Schultz, 2000a,b).

Baeg et al. (2009) has also reported that OFC neurons can track a monkey's behavioral response when the relationship between cue relevance and reward availability changes. The ability to adjust to fluctuating reward contingencies is critical for both man and beast. Using a manipulation known as reward devaluation, the neural correlate of this behavior has now been demonstrated. Responses of OFC neurons are modulated by postingestive factors (perhaps stomach distension or elevated blood glucose levels) that decrease the rewarding value of food which, in turn, reduces the subject's incentive to eat or drink. For example, when a monkey is allowed to drink or eat one food to the point of satiety (sensory specific satiety), the rewarding value of that item is devalued and loses its ability to act as a reinforcer. This paradigm was used by Scott et al. (1986a) and Yaxley et al. (1985) to determine if the responses of neurons in the NTS and insula, respectively, are affected by satiety; neither was. The taste-evoked responses of neurons in the clOFC, however, decrease as a function of satiety (Rolls et al., 1989). In this experiment, the responses evoked by glucose decreased as the monkey consumed it to the point of satiety; the responses evoked by NaCl were unaffected. This is an example of how reward devaluation, which reflects a decrement in the response magnitude to consumption

of the satiating fluid, impacts behavior. The relationship between satiety-induced reward devaluation and changes in neural activity is more complex and not as decisive in the mOFC (Pritchard et al., 2008). Response magnitude decreases as satiety develops for most neurons, but in a more graded fashion than in the clOFC. Response magnitude increases with satiety for a small group of neurons, but the most interesting cells are those initially classified as non-gustatory that begin responding to sapid stimulation of the tongue after the monkeys reached satiety. The presence of emergent taste neurons suggests that satiety of not strictly a passive process. The ability of OFC neurons to track changes in reward contingencies caused by internal and external factors enables it to support taste-guided reward-evaluation, risk assessment, and decision-making. Damage to the OFC should undermine these processes, and as predicted, OFC-lesioned monkeys demonstrate normal satiety, but show no evidence of satiety-induced reward devaluation (Izquierdo et al., 2004; Murray and Izquierdo, 2007).

32.4 CONCLUSION

Rodents and humans do not share the same station in life, but our gustatory systems evolved for the same purpose: identification and consumption of safe, nutritious food, while rejecting hazardous and toxic substances. A side-by-side comparison of the rodent and primate gustatory systems reveals many structural similarities, some differences, and yet a commonality of purpose. As noted above, our knowledge of the gustatory system of both the rat and primate is uneven within and between species. More has been learned about the anatomy and physiology of the gustatory hindbrain from rodents than humans; conversely, most of our knowledge about taste processing in the forebrain has been derived from primates, although research on rats is beginning to make important inroads there. If we put aside the differences that enable each species to survive within its ecological niche, what can we learn about the functional organization of taste by looking at the common features that have survived millions of years of evolution? In other words, are there enough common features in rodents and primates to support assembly of a complete, generic species blueprint of the gustatory system?

The peripheral gustatory apparatus, including the anatomy of individual taste buds and their distribution in the oral cavity, are remarkably similar, not only in rodents and humans, but also across the class Mammalia. The most striking interspecies difference, tongue shape, is driven by non-gustatory functions such as grooming, dietary choices, and mastication rather than taste issues, per se (Bradley, 1971).

In rodents and primates, taste buds are innervated by the same three cranial nerves, VII, IX, and X, which terminate in the NTS of the medulla. Although it is possible that the rodent NTS has more (or perhaps slightly different) subnuclei than the primate, functionally, it seems to matter little. In both species, gustatory neurons dominate the rostrolateral NTS with more medial and caudal areas dedicated to visceral sensation.

Many published reports in rodents have demonstrated that multiple forebrain areas, including the cerebral cortex (e.g., Di Lorenzo and Monroe, 1995; Cho et al., 2002; see also Lundy, 2008), modulate gustatory responsivity of NTS neurons; similar studies have demonstrated that the cerebral cortex modulates auditory (Winer, 2006), vestibular (Akbarian et al., 1994; Nishiike et al., 2000), and somatosensory (Wise and Jones, 1977) transmission through the brainstem. The widespread nature of this phenomenon, across both species and sensory modalities, makes it likely that it is common practice. Descending projections may improve information transmission of relevant information across multiple synapses (Nuñez and Malmierca, 2007) or act as a cognitive filter by sharpening the responses properties of individual neurons (Di Lorenzo and Monroe, 1995).

A side-by-side comparison of the rodent and primate gustatory systems is complicated by the PBN, a visceral and nociceptive relay in both rodents and primates that does not play a role in taste perception in primates (Pritchard et al., 2000; Scott and Small, 2009). This significant interspecies difference raises an obvious question: what does the PBN contribute to taste-guided behavior in rodents? Lesion-behavioral studies in rats are addressing this issue (Grigson et al., 1994, 1998), but for the present, it appears that either the PBN is essential for one or more taste-guided behaviors unique to rodents, or that neural control of these behaviors in primates resides in other brain areas. Ironically, this anatomical issue is most likely to be resolved with behavioral experiments.

Electrophysiological studies directed at the thalamic gustatory relay have revealed very little about its function in either rodents or primates, but from what is known about other sensory systems, the gustatory thalamus in both species may filter information destined for primary taste cortex (Saalmann and Kastner, 2009). Lesion-behavioral experiments may provide additional insight into how the thalamus contributes to taste-guided behavior (Grigson et al., 2000; Mungarndee et al., 2008; Reilly and Pritchard, 1996a,b,c), but the most revealing experiments are likely to involve simultaneous stimulation and recording from the thalamus and primary gustatory cortex.

For many years, knowledge about cortical processing of gustatory information lagged behind what was known about other sensory systems. That began to change after chronic recording techniques developed in other sensory systems were introduced to the field of taste. Electrophysiological investigations on awake, behaving monkeys have shown that primates are an excellent surrogate for human taste perception. Insights provided by primate research are guiding studies of human taste perception and decision-making being conducted with state of the art non-invasive imaging techniques. Recent electrophysiological experiments in awake rats have provided compelling evidence that the gustatory system is a very dynamic, labile, and distributed system that shapes ingestive behavior.

In summary, rats have different needs and in some ways their neural repertoire is more limited than humans, but after comparing the two species side-by-side, their functional and even the anatomical similarities are more striking than their differences.

REFERENCES

Akbarian., S., Grüsser, O. J., and Guldin, W. O. (1994). Corticofugal connections between the cerebral cortex and brainstem vestibular nuclei in the macaque monkey. *J. Comp. Neurol.* 339: 421–437.

Alden, M., Besson, J. M., and Bernard, J. F. (1994). Organization of the efferent projections from the pontine parabrachial area to the bed nucleus of the stria terminalis and neighboring regions: a PHA-L study in the rat. *J. Comp. Neurol.* 341: 289–314.

Baeg, E. H., Jackson, M. E., Jedema, H. P., and Bradberry, C. W. (2009). Orbitofrontal and anterior cingulate cortex neurons selectively process cocaine-associated environmental cues in the rhesus monkey. *J. Neurosci.* 29: 11619–11627.

Barbas, H. (1993). Organization of cortical afferent input to orbitofrontal areas in the rhesus monkey. *Neurosci.* 56: 841–864.

Barbas, H. (2007). Specialized elements of orbitofrontal cortex in primates. *Ann. N.Y. Acad. Sci.* 1121: 10–32.

Baylis, L. L., Rolls, E. T., and Baylis, G. C. (1995). Afferent connections of the caudolateral orbitofrontal cortex taste area of the primate. *Neurosci.* 64: 801–812.

Beckman, M. E., and Whitehead, M. C. (1991). Intramedullary connections of the rostral nucleus of the solitary tract in the hamster. *Brain Res.* 557(1–2): 265–279.

Beckstead, R. M., and Norgren, R. (1979). Central distribution of the trigeminal, facial glossopharyngeal and vagus nerves in the monkey. *J. Comp. Neurol.* 184: 455–472.

Beckstead, R. M., Morse, J. R., and Norgren, R. (1980). The nucleus of the solitary tract in the monkey: projections to the thalamus and brain stem nuclei. *J. Comp. Neurol.* 190: 259–282.

Benjamin, R. M. (1963). Some thalamic and cortical mechanisms of taste. In *Olfaction and Taste*, Zotterman Y. (Ed.) New York: Pergamum. vol. 1: pp 309–329.

Benjamin, R. M., and Burton, H. (1968). Projection of taste nerve afferents to anterior opercular-insular cortex in squirrel monkey (*Saimiri sciureus*). *Brain Res.* 7: 221–231.

Benjamin, R. M., Emmers, R., and Blomquist, A. J. (1968). Projection of tongue nerve afferents to somatic sensory area I in squirrel monkey (*Saimiri sciureus*). *Brain Res.* 7: 208–220.

Bereiter, D. A., Berthoud, H. R., and Jeanrenaud, B. (1980). Hypothalamic input to brain stem neurons responsive to oropharyngeal stimulation. *Exp. Brain Res.* 39: 33–39.

References

Bernard, J. F., Alden, M., and Besson, J. M. (1993). The organization of the efferent projections from the pontine parabrachial area to the amygdaloid complex: a Phaseolus vulgaris leucoagglutinin (PHA-L) study in the rat. *J. Comp. Neurol.* 329: 201–229.

Bester, H., Besson, J. M., and Bernard, J. F. (1997). Organization of efferent projections from the parabrachial area to the hypothalamus: a *Phaseolus vulgaris-leucoagglutinin* study in the rat. *J. Comp. Neurol.* 383: 245–281.

Bester, H., Bourgeais, L., Villanueva, L., et al. (1999). Differential projections to the intralaminar and gustatory thalamus from the parabrachial area: a PHA-L study in the rat. *J. Comp. Neurol.* 405: 421–449.

Boughter, J. D., and Smith, D. V. (1998). Amiloride blocks acid responses in NaCl-best gustatory neurons of the hamster solitary nucleus. *J. Neurophysiol.* 80: 1362–1372.

Bradley, R. M. (1971). Tongue topography. In *Taste*, Beidler, L.M. (Ed.) In *Handbook of Sensory Physiology*, New York: Springer-Verlag. *vol. IV, Chemical Senses*, pt. 2: pp 1–30.

Bradley R. M. (1996). Neurotransmitter and neuromodulator activity in the gustatory zone of the nucleus tractus solitarius. *Chem. Senses* 21: 377–385.

Bradley R. M. (2007). rNST Circuits. In: *The Role of the Nucleus of the Solitary Tract in Gustatory Processing*. Bradley R.M. (Ed.) Boca Raton (FL): CRC Press, pp. 137–151.

Bruno, R. M. (2011). Synchrony in sensation. *Curr. Opin. Neurobiol.* 21: 701–708.

Carleton, A., Accolla, R,, and Simon, S. A. (2010). Coding in the mammalian gustatory system. *Trends Neurosci.* 33: 326–334.

Carmichael, S. T., and Price, J. L. (1995). Sensory and premotor connections of the orbital and medial prefrontal cortex in macaque monkeys. *J. Comp. Neurol.* 363: 642–664.

Carmichael, S. T., and Price, J. L. (1996). Connectional networks within the orbital and medial prefrontal cortex of macaque monkeys. *J. Comp. Neurol.* 371: 179–207.

Cechetto, D. F., and Saper, C. B. (1987). Evidence for a viscerotopic sensory representtaion in the cortex and thalamus in the rat. *J. Comp. Neurol.* 262: 27–45.

Chang, F.-C. T. and Scott, T. R. (1984). Conditioned taste aversions modify neural responses in the rat nucleus tractus solitarius. *J. Neurosci.* 4: 1850–1862.

Chapin, J. K., and Lin, C. S. (1984). Mapping the body representation in the SI cortex of anesthetized and awake rats. *J. Comp. Neurol.* 229: 199–213.

Chen, J. Y., Victor, J. D, and Di Lorenzo, P. M. (2011). Temporal coding of intensity of NaCl and HCl in the nucleus of the solitary tract of the rat. *J. Neurophysiol.* 105: 697–711.

Cho, Y. K., Li, C.-S., and Smith, D. V. (2002). Gustatory projections from the nucleus of the solitary tract to the parabrachial nuclei in the hamster. *Chem. Senses* 27: 81–90.

Cho, Y. K., Li, C. S., and Smith, D. V. (2003). Descending influences from the lateral hypothalamus and amygdala converge onto medullary taste neurons. *Chem. Senses* 28: 155–171.

Collings, V. B. (1974). Human taste response as a function of locus of stimulation on the tongue and soft palate. *Percept. Psychophys.* 16: 169–174.

Contreras, R. J. (1977). Changes in gustatory nerve discharges with sodium deficiency: A single unit analysis. *Brain Res.* 121: 373–378.

Contreras, R. J. and Frank, M. E. (1979). Sodium deprivation alters neural responses to gustatory stimuli. *J. Gen. Physiol.* 73: 569–594.

Critchley, H. D., and Rolls, E. T. (1996). Hunger and satiety modify the responses of olfactory and visual neurons in the primate orbitofrontal cortex. *J. Neurophysiol.* 75: 1673–1686.

Davis, B. J. (1991). The ascending gustatory pathway: a Golgi analysis of the medial and lateral parabrachial complex in the adult hamster. *Brain Res. Bull.* 27: 63–73.

Davis, B. J. (1993). GABA-like immunoreactivity in the gustatory zone of the nucleus of the solitary tract in the hamster: light and electron microscopic studies. *Brain Res. Bull.* 30: 69–77.

Davis, B. J. (1998). Synaptic relationships between the chorda tympani and tyrosine hydroxylase-immunoreactive dendritic processes in the gustatory zone of the nucleus of the solitary tract in the hamster. *J. Comp Neurol.* 392: 78–91.

Davis, B. J., and Jang, T. (1986). The gustatory zone of the solitary tract in the hamster. *Light microscopic morphometric studies. Chem. Senses* 11: 213–228.

de Araujo, I. E., Kringelbach, M. L., Rolls, E. T., and McGlone, F. (2003a). Human cortical responses to water in the mouth, and the effects of thirst. *J. Neurophysiol.* 90: 1865–1876.

de Araujo, I. E., Rolls, E. T., Kringelbach, M. L., et al. (2003b). Taste-olfactory convergence, and the representation of the pleasantness of flavour, in the human brain. *Eur. J. Neurosci.* 18: 2059–2068.

Del Parigi, A., Chen, K., Salbe, A. D., et al. (2002). Tasting a liquid meal after a prolonged fast is associated with preferential activation of the left hemisphere. *Neuroreport* 13: 1141–1145.

Detsch, O., Vahle-Hinz, C., Kochs, E., et al. (1999). Isoflurane induces dose-dependent changes of thalamic somatosensory information transfer. *Brain Res.* 829: 77–89.

Di Lorenzo, P. M. and Monroe, S. (1989). Taste responses in the parabrachial pons of male, female and pregnant rats. *Brain Res. Bull.* 23: 219–227.

Di Lorenzo, P. M., and Monroe, S. (1990). Taste responses in the parabrachial pons of ovariectomized rats. *Brain Res. Bull.* 25: 741–748.

Di Lorenzo, P. M., and Monroe, S. (1992). Corticofugal input to taste-responsive units in the parabrachial pons. *Brain Res. Bull.* 29: 925–930.

Di Lorenzo, P. M., and Hecht, G. S. (1993). Perceptual consequences of electrical stimulation in the gustatory system. *Behav. Neurosci.* 107: 130–138.

Di Lorenzo, P. M., and Monroe, S. (1995). Corticofugal influence on taste responses in the nucleus of the solitary tract in the rat. *J. Neurophysiol.* 74: 258–272.

Di Lorenzo, P. M., and Monroe, S. (1997). Transfer of information about taste from the nucleus of the solitary tract to the parabrachial nucleus of the pons. *Brain Res.* 763: 167–181.

Di Lorenzo, P. M., and Lemon, C. H. (2000). The neural code for taste in the nucleus of the solitary tract of the rat: Effects of adaptation. *Brain Res.* 852: 383–397.

Di Lorenzo, P. M., and Victor, J. D. (2003). Taste response variability and temporal coding in the nucleus of the solitary tract of the rat. *J. Neurophysiol.* 90: 1418–1431.

Di Lorenzo, P. M., Hallock, R. M., and Kennedy, D. P. (2003a). Temporal coding of sensation: Mimicking taste quality with electrical stimulation of the brain. *Behav. Neurosci.* 117: 1423–1433.

Di Lorenzo, P. M., Lemon, C. H., and Reich, C. G. (2003b). Dynamic coding of taste stimuli in the brain stem: effects of brief pulses of taste stimuli on subsequent taste responses. *J. Neurosci.* 23: 8893–8902.

Di Lorenzo, P. M., Chen, J. Y., and Victor, J. D. (2009a). Quality time: Representation of a multidimensional sensory domain through temporal coding. *J Neurosci.* 29: 9227–9238.

Di Lorenzo, P. M., Leshchinskiy, S., Moroney, D. N., and Ozdoba, J. M. (2009b). Making time count: Functional evidence for temporal coding of taste sensation. *Behav. Neurosci.* 123: 14–25.

720   Chapter 32   Central Taste Anatomy and Physiology of Rodents and Primates

Di Lorenzo, P. M., Platt, D., and Victor, J. D. (2009c). Information processing in the parabrachial nucleus of the pons: Temporal relationships of input and output. *Ann. N.Y. Acad. Sci.* 1170: 365–371.

Doetsch, G. S., and Erickson, R. P. (1970). Synaptic processing of taste-quality information in the nucleus tractus solitarius of the rat. *J. Neurophysiol.* 33: 490–507.

Erickson, R. P. (1967). Neural coding of taste quality. In M. Kare and O. Maller (Eds.) *The Chemical Senses and Nutrition.* Baltimore: Johns Hopkins Press, pp. 313–327.

Erickson, R. P. (2008). A study of the science of taste: on the origins and influence of the core ideas. *Behav. Brain Sci.* 31: 59–75.

Erickson, R. P. and Doetsch, G.S . (1965). The gustatory neural response function. *J. Gen. Physiol.* 49: 247–268.

Frank, M. (1973). An analysis of hamster afferent taste nerve response functions. *J. Gen. Physiol.* 61: 588–618.

Frederique, F., and Cattarelli, M. (1996). Reciprocal and topographic connections between the piriform and prefrontal cortices in the rat: a tracing study using the B subunit of the cholera toxin. *Brain Res. Bull.* 41: 391–398.

Ganchrow, D., and Erickson, R. P. (1972). Thalamocortical relations in gustation. *Brain Res.* 36: 289–305.

Gautier, J.-F., Del Parigi, A., Chen, K., et al. (2001). Effect of satiation on brain activity in obese and lean women. *Obesity Res.* 9: 676–684.

Gilbertson, T. A. (1998). Gustatory mechanisms for the detection of fat. *Curr. Opin. Neurobiol.* 8: 447–452.

Giza, B. K., and Scott, T. R. (1991). The effect of amiloride on taste-evoked activity in the nucleus tractus solitarius of the rat. *Brain Res.* 550: 247–256.

Giza, B. K., Scott, T. R., Sclafani, A., and Antonucci, R. F. (1991). Polysaccharides as tastes stimuli: their effect in the nucleus tractus solitarius of the rat. *Brain Res.* 555: 1–9.

Giza, B. K., and Scott, T. R. (1996). Taste responses in the nucleus of the solitary tract in saccharin-preferring and saccharin-averse rats. *Chem. Senses* 21: 147–157.

Grabauskas, G., and Bradley, R. M. (1996). Synaptic interactions due to convergent input from gustatory afferent fibers in the rostral nucleus of the solitary tract. *J. Neurophysiol.* 76: 2919–2927.

Grabauskas, G., and Bradley, R. M. (1998). Tetanic stimulation induces short-term potentiation of inhibitory synaptic activity in the rostral nucleus of the solitary tract. *J. Neurophysiol.* 79: 595–604.

Grabauskas, G., and Bradley, R. M. (1999). Potentiation of GABAergic synaptic transmission in the rostral nucleus of the solitary tract. *Neurosci.* 94: 1173–1182.

Grabauskas, G., and Bradley, R. M. (2003). Frequency-dependent properties of inhibitory synapses in the rostral nucleus of the solitary tract. *J. Neurophysiol.* 89: 199–211.

Grigson, P. S., Spector, A. C., and Norgren, R. (1994). Lesions of the pontine parabrachial nuclei eliminate successive negative contrast effects in rats. *Behav. Neurosci.* 108: 714–723.

Grigson, P. S., Reilly, S., Scalera, G., and Norgren, R. (1998). The parabrachial nucleus is essential for acquisition of a conditioned odor aversion in rats. *Behav. Neurosci.* 112: 1104–1113.

Grigson, P. S., Lyuboslavsky, P., and Tanase, D. (2000). Bilateral lesions of the gustatory thalamus disrupt morphine- but not LiCl-induced intake suppression in rats: evidence against the conditioned taste aversion hypothesis. *Brain Res.* 858: 327–337.

Guillery, R. W., Feig, S. L., and Lozsádi, D. A. (1998). Paying attention to the thalamic reticular nucleus. *Trends Neurosci.* 21: 28–32.

Guldin, W. O., and Markowitsch, H. J. (1983). Cortical and thalamic afferent connections of the insular and adjacent cortex of the rat. *J. Comp. Neurol.* 215: 135–153.

Gutierrez, R., Simon, S. A., and Nicolelis, M. A. L. (2010). Licking-induced synchrony in the taste-reward circuit improves cue discrimination during learning. *J. Neurosci.* 30: 287–303.

Hallock, R. M., and Di Lorenzo, P. M. (2006). Temporal coding in the gustatory system. *Neurosci. Biobehav. Rev.* 30: 1145–1160.

Halpern, B. P. (1985). Time as a factor in gustation: temporal patterns of taste stimulation and response. In: *Taste, Olfaction, and the Central Nervous System.* Pfaff, D.W. (Ed,) New York: Rockefeller Univ. Press, pp 181–209.

Halsell, C. B., Travers, J. B., and Travers, S. P. (1993). Gustatory and tactile stimulation of the posterior tongue activate overlapping but distinctive regions within the nucleus of the solitary tract. *Brain Res.* 632: 161–173.

Halsell, C. B., Travers, S. P., and Travers, J. B. (1996). Ascending and descending projections from the rostral nucleus of the solitary tract originate from separate neuronal populations. *Neurosci.* 72: 185–197.

Halsell, C. B. and Travers, S. P. (1997). Anterior and posterior oral cavity responsive neurons are differentially distributed among parabrachial subnuclei in rat. *J. Neurophysiol.* 78: 920–938.

Hamilton, R., and Norgren, R. (1984). Central projections of gustatory nerves in the rat. *J. Comp. Neurol.* 222: 560–577.

Hanamori, T., Kunitake, T., Kazuo, K., and Kannan, H. (1997). Convergence of afferent inputs from the chorda tympani, lingual-tonsillar and pharyngeal branches of the glossopharyngeal and superior laryngeal nerve on the neurons in the insular cortex in rats. *Brain Res.* 763: 267–270.

Hanamori, T., Kunitake, T., Kato, K., and Kannan, H. (1998a). Responses of neurons in the insular cortex to gustatory, visceral, and nociceptive stimuli in rats. *J. Neurophysiol.* 79: 2535–2545.

Hanamori, T., Kunitake, T., Kazuo, K., and Kannan, H. (1998b). Neurons in the posterior insular cortex are responsive to gustatory stimulation of the pharyngeolarynx, baroreceptor and chemoreceptor stimulation, and tail pinch in rats. *Brain Res.* 785: 97–106.

Herbert, H., Moga, M. M., and Saper, C. B. (1990). Connections of the parabrachial nucleus with the nucleus of the solitary tract and the medullary reticular formation in the rat. *J. Comp. Neurol.* 293: 540–580.

Hosokawa, T., Kato, K., Inoue, M., and Mikami, A. (2007). Neurons in the macaque orbitofrontal cortex code relative preference of both rewarding and aversive outcomes. *Neurosci. Res.* 57: 434–445.

Hosoya, Y., and Matsushita, M. (1981). Brainstem projections from the lateral hypothalamic area in the rat, as studied with autoradiography. *Neurosci. Lett.* 24: 111–116.

Ifuku, H., Hirata, S.I., Nakamura, T., and Ogawa, H. (2003). Neuronal activities in the monkey primary and higher-order gustatory cortices during a taste discrimination delayed GO/NOGO task and after reversal. *Neurosci. Res.* 2: 161–175.

Ito, S., and Ogawa, H. (1994). Neural activities in the frontal-opercular cortex of macaque monkeys during tasting and mastication. *Jap. J. Physiol.* 44: 141–156.

Izquierdo, A., Suda, R. K., and Murray, E. A. (2004). Bilateral orbital prefrontal cortex lesions in rhesus monkeys disrupt choices guided by both reward value and reward contingency. *J. Neurosci.* 24: 7540–7548.

Jasmin, L., Granato, A, and Ohara, P. T. (2004). Rostral agranular insular cortex and pain areas of the central nervous system: a tract-tracing study in the rat. *J. Comp. Neurol.* 468: 425–440.

Kang, Y., and Lundy, R. F. (2009). Terminal field specificity of forebrain efferent axons to brainstem gustatory nuclei. *Brain Res.* 1248: 76–85.

Kang, Y., and Lundy, R. F. (2010). Amygdalofugal influence on processing of taste information in the nucleus of the solitary tract of the rat. *J. Neurophysiol.* 104: 726–741.

Karimnamazi, H., and Travers, J. B. (1998). Differential projections from gustatory responsive regions of the parabrachial nucleus to the medulla and forebrain. *Brain Res.* 813: 283–302.

Karimnamazi, H., Travers, S. P., and Travers, J. B. (2002). Oral and gastric input to the parabrachial nucleus of the rat. *Brain Res.* 957: 193–206.

Katz, D. B., Simon, S. A., and Nicolelis, M. A. (2001). Dynamic and multimodal responses of gustatory cortical neurons in awake rats. *J. Neurosci.* 21: 4478–4489.

Katz, D. B., Simon, S. A., and Nicolelis, M. A. (2002). Taste-specific neuronal ensembles in the gustatory cortex of awake rats. *J. Neurosci.* 22: 1850–1857.

King, C. T., Deyrup, L. D., Dodson, S. E., et al. (2003). Effects of gustatory nerve transection and regeneration on quinine-stimulated Fos-like immunoreactivity in the parabrachial nucleus of the rat. *J. Comp. Neurol.* 465: 296–308.

Kita, H., and Oomura, Y. (1981). Functional synaptic interconnections between the lateral hypothalamus and frontal and gustatory cortices in the rat. In *Brain Mechanisms of Sensation*, Katsuki, Y., Norgren, R., and Sato, M. (Eds.) New York: John Wiley and Sons, pp 307–322.

Kosar, E. M., Grill, H. J., and Norgren, R. (1986a). Gustatory cortex in the rat. *I. Physiological properties and cytoarchitecture. Brain Res.* 379: 329–341.

Kosar, E. M., Grill, H. J., and Norgren, R. (1986b). Gustatory cortex in the rat. *II. thalamocortical projections. Brain Res.* 379: 342–352.

Krout, K. E., and Loewy, A. D. (2000). Parabrachial nucleus projections to midline and intralaminar thalamic nuclei of the rat. *J. Comp. Neurol.* 428: 475–494.

Kuznicki, J., and Ashbaught, N. (1979). Taste quality differences within the sweet and salty taste categories. *Sens. Process.* 3: 157–182.

Lam, Y.-W., and Sherman, S. M. (2011). Functional organization of the thalamic input to the thalamic reticular nucleus. *J. Neurosci.* 31: 6791–6799.

Land, R., Engler, G., Kral, A., and Engel, A. K. (2012). Auditory evoked bursts in mouse visual cortex during isoflurane anesthesia. Public Library of Science (PLoS One) 2012;7(11):e49855. doi: 10.1371/journal.pone.0049855. Epub 2012 Nov 21.

Lara, A. H., Kennerley, S. W., and Wallis, J. D. (2009). Encoding of gustatory memory by orbitofrontal neurons. *J. Neurosci.,* 29: 774–765.

Lasiter, P. S., Glanzman, D. L., and Mensah, P. A. (1982). Direct connectivity between pontine taste areas and gustatory neocortex in rat. *Brain Res.* 234: 111–121.

Lasiter, P. S., and Kachele, D. L. (1988a). Postnatal development of the parabrachial gustatory zone in rat: dendritic morphology and mitochondrial enzyme activity. *Brain Res. Bull.* 21: 79–94.

Lasiter, P. S., and Kachele, D. L. (1988b). Organization of GABA and GABA-transaminase-containing neurons in the gustatory zone of the nucleus of the solitary tract. *Brain Res. Bull.* 21: 623–636.

Lemon, C. H. and Di Lorenzo, P. M. (2002). Effects of electrical stimulation of the chorda tympani nerve on taste responses in the nucleus of the solitary tract of the rat. *J. Neurophysiol.* 88: 2477–2489.

Lemon, C. H., and Smith, D. V. (2006). Influence of response variability on the coding performance of central gustatory neurons. *J. Neurosci.* 26: 7433–7443.

Lenz, F. A., Gracely, R. H., Zirh, T. A., et al. (1997). Human thalamic nucleus mediating taste and multiple other sensations related to ingestive behavior. *J. Neurophysiol.* 77: 3406–3409.

Leonard, N. L., Renehan, W. E., and Schweitzer, L. (1999). Structure and function of gustatory neurons in the nucleus of the solitary tract. IV. The morphology and synaptology of GABA-immunoreactive terminals. *Neurosci.* 92: 151–162.

Li, C.-S., and Smith, D. V. (1997). Glutamate receptor antagonists block gustatory afferent input to the nucleus of the solitary tract. *J. Neurophysiol.* 77: 1514–1525.

Li, C.-S., Cho, Y. K., and Smith, D. V. (2002). Taste responses of neurons in the hamster solitary nucleus are modulated by the central nucleus of the amygdala. *J. Neurophysiol.* 88: 2979–2992.

Li, C.-S., Cho, Y. K., and Smith, D. V. (2005). Modulation of parabrachial taste neurons by electrical and chemical stimulation of the lateral hypothalamus and amygdala. *J. Neurophysiol.* 93: 1183–1196.

Li, C.-S., and Cho, Y. K. (2006). Efferent projection from the bed nucleus of the stria terminalis suppresses activity of taste-responsive neurons in the hamster parabrachial nuclei. *Am. J. Physiol. Regul., Integr. Comp. Physiol.* 291: R914–926.

Li, C.-S., Mao, L., and Cho, Y. K. (2008). Taste-responsive neurons in the nucleus of the solitary tract receive gustatory information from both sides of the tongue in the hamster. *Am. J. Physiol. Regul. Integr. Comp. Physiol.* 294: R372–381.

Liu, H., Behbehani, M. M., and Smith, D. V. (1993). The influence of GABA on cells in the gustatory region of hamster solitary nucleus. *Chem. Senses* 18: 285–305.

Loewy, A. D., and Burton, H. (1978). Nuclei of the solitary tract: Efferent projections to the lower brain stem and spinal cord of the cat. *J. Comp. Neurol.* 181: 421–450.

Lundy, R. F. (2008). Gustatory hedonic value: potential function for forebrain control of brainstem taste processing. *Neurosci. Biobehav. Rev.* 32: 1601–1606.

Lundy, R. F., and Norgren, R. (2001). Pontine gustatory activity is altered by electrical stimulation in the central nucleus of the amygdala. *J. Neurophysiol.* 85: 770–783.

Lundy, R. F., and Norgren, R. (2014). Gustatory system. In: *The Rat Nervous System*, Paxinos, G. (Ed.) Amsterdam: Elsevier Academic, pp 731–750..

May, O. L., and Hill, D. L. (2006). Gustatory terminal field organization and developmental plasticity in the nucleus of the solitary tract revealed through triple-fluorescence labeling. *J. Comp. Neurol.* 497: 658–669.

MacDonald, C. J., Meck, W. H., and Simon, S. A. (2012). Distinct neural ensembles in the rat gustatory cortex encode salt and water tastes. *J. Physiol.* 590: 3169–3184.

McManus, L. J., Dawes, P. J. D., and Stringer, M. D. (2011). Clinical anatomy of the chorda tympani: a systematic review. *J. Larygol. Otol.* 125: 1101–1108.

McPheeters, M., Hettinger, T., Nuding, S.C., et al. (1990). Taste-responsive neurons and their locations in the solitary nucleus of the hamster. *Neurosci.* 34: 745–758.

Mesulam, M.-M., and Mufson, E. J. (1982). Insula of the old world monkey. III: Efferent cortical output and comments on function. *J. Comp. Neurol.* 212: 38–52.

Mesulam, M.-M., and Mufson, E. J. (1985). The insula of Reil in man and monkey. *In Cerebral Cortex*, Peters, A. and Jones, E.G. (Eds.) 4, pp. 179–226.

Monroe, S., and Di Lorenzo, P. M. (1995). Taste responses in neurons in the nucleus of the solitary tract that do and do not project to the parabrachial pons. *J. Neurophysiol.* 74: 249–257.

Morecraft, R. J., Geula, C., and Mesulam, M.-M. (1992). Cytoarchitecture and neural afferents of orbitofrontal cortex in the brain of the monkey. *J. Comp. Neurol.* 323: 341–358.

Mufson, E. J., and Mesulam, M.-M. (1982). Insula of the Old World monkey. II: Afferent cortical input and comments on the claustrum. *J. Comp. Neurol.* 212: 23–37.

Mungarndee, S. S., Lundy, R. F. Jr., Norgren, R. (2008). Expression of Fos during sham sucrose intake in rats with central gustatory lesions. *Am. J. Physiol. Regul. Integr. Comp. Physiol.* 295: R751–763.

Murray, E. A., and Izquierdo, A. (2007). Orbitofrontal cortex and amygdala contributions to affect and action in primates. *Ann. N.Y. Acad. Sci.* 1121: 273–296.

Nakamura, T., and Ogawa, H. (1997). Neural interaction between cortical taste neurons in rats: a cross-correlation analysis. *Chem. Senses* 22: 517–528.

Nishijo, H., and Norgren, R. (1990). Responses from parabrachial gustatory neurons in behaving rats. *J. Neurophysiol.* 63: 707–724.

Nishijo, H., and Norgren, R. (1991). Parabrachial gustatory neural activity during licking by rats. *J. Neurophysiol.* 66: 974–985.

Nishijo, H., and Norgren, R. (1997). Parabrachial neural coding of taste stimuli in awake rats. *J. Neurophysiol.* 78: 2254–2268.

Nishiike, S., Guldin, W. O., and Bäurle, J. (2000). Corticofugal connections between the cerebral cortex and the vestibular nuclei in the rat. *J. Comp. Neurol.* 420: 363–372.

Nomura, T., and Ogawa, H. (1985). The taste and mechanical response properties of neurons in the parvicellular part of the thalamic posteromedial ventral nucleus of the rat. *Neurosci. Res.* 3: 91–105.

Norgren, R. (1974). Gustatory afferents to ventral forebrain. *Brain Res.* 81: 285–295.

Norgren, R. (1978). Projections from the nucleus of the solitary tract in the rat. *Neurosci.* 3: 207–218.

Norgren, R., and Leonard, C. M. (1973). Ascending central gustatory pathways. *J. Comp. Neurol.* 150: 217–237.

Norgren, R., and Pfaffmann, C. (1975). The pontine taste area in the rat. *Brain Res.* 91: 99–117.

Norgren, R., and Wolf, G. (1975). Projections of thalamic gustatory and lingual areas in the rat. *Brain Res.* 92: 123–129.

Norgren, R., and Grill, H. J. (1976). Efferent distribution from the cortical gustatory area in rats. *Neurosci. Abstr.* 2: 124.

Nuñez, A, and Malmierca, E. (2007). Corticofugal modulation of sensory information. *Adv. Anat. Embryol. Cell. Biol.* 187: 1–74.

Ogawa, H., Imoto, T., Hayama, T., and Kaisaka, J. (1982). Afferent connections to the pontine taste area: Physiologic and anatomic studies. In *Brain Mechanisms of Sensation*, Katsuki, Y., Norgren, R. and Sato, M. (Eds.), New York: John Wiley and Sons, pp. 161–176.

Ogawa, H., Imoto, T., and Hayama, T. (1984). Responsiveness of solitario-parabrachial relay neurons to taste and mechanical stimulation applied to the oral cavity in rats. *Exp. Brain Res.* 54: 349–358.

Ogawa, H., Hayama, T., and Ito, S. (1987). Response properties of the parabrachio-thalamic taste and mechanoreceptive neurons in rats. *Exp. Brain Res.* 68: 449–457.

Ogawa, H., Hayama, T., and Yamashita, Y. (1988). Thermal sensitivity of neurons in a rostral part of the rat solitary tract nucleus. *Brain Res.* 454: 321–331.

Ogawa, H., and Nomura, T. (1988). Receptive field properties of thalamo-cortical taste relay neurons in the parvicellular part of the posteromedial ventral nucleus in rats. *Exp. Brain Res.* 73: 364–370.

Ogawa, H., Ito, S.I., and Nomura, T. (1989). Oral cavity representation at the frontal operculum of macaque monkeys. *Neurosci Res.* 6: 283–298.

Ogawa, H., Ifuku, H., Nakamura, T., and Hirata, S. (2005). Possible changes in information from the primary to higher-order gustatory cortices studied by recording neural activities during a taste discrimination GO/NOGO task in monkeys. *Chem. Senses* 30: i78–i79

O'Neill, M. and Schultz, W. (2010). Coding of reward risk distinct from reward value by orbitofrontal neurons. *Neuron* 68: 789–800.

Öngür, D., and Price, J. L. (2000). The organization of networks within the orbital and medial prefrontal cortex of rats, monkeys and humans. *Trends Neurosci.* 29: 116–124.

Padoa-Schioppa, C., and Assad, J. A. (2006). Neurons in orbitofrontal cortex encode economic value. *Nature* 441: 223–226.

Padoa-Schioppa, C. (2007). Orbitofrontal cortex and the computation of economic value. *Ann. N.Y. Acad. Sci.* 1121: 232–253.

Pennartz, C. M. A., van Wingerden, M., and Vinck, M. (2011). Population coding and neural rhythmicity in the orbitofrontal cortex. *Ann. N.Y. Acad. Sci.* 1239: 149–161.

Plata-Salamán, C. R., Scott, T. R. and Smith, V. L. (1992). Gustatory neural coding in the monkey cortex: L-amino acids. *J. Neurophysiol.* 67: 1552–1561.

Plata-Salamán, C. R., Scott, T. R., and Smith-Swintosky, V. L. (1993). Gustatory neural coding in the monkey cortex: The quality of sweetness. *J. Neurophysiol.* 60: 482–493.

Plata-Salamán, C. R., Scott, T. R., and Smith-Swintosky, V. L. (1995). Gustatory neural coding in the monkey cortex: Acid stimuli. *J. Neurophysiol.* 74: 556–564.

Plata-Salamán, C. R., Smith-Swintosky, V. L., and Scott, T. R. (1996). Gustatory neural coding in the monkey cortex: Mixtures. *J. Neurophysiol.* 75: 2369–2379.

Pritchard, T. C., and Scott, T. R. (1982). Amino acids as taste stimuli. *II. Taste quality. Brain Res.* 253: 93–104.

Pritchard, T. C., Hamilton, R., Morse, J., and Norgren, R. (1986). Projections from thalamic gustatory and lingual areas in the monkey, *Macaca fascicularis. J. Comp. Neurol.* 244: 213–228.

Pritchard, T. C., Hamilton, R. B., and Norgren, R. (1989). Neural coding of gustatory information in the thalamus of *Macaca mulatta. J. Neurophysiol.* 61: 1–14.

Pritchard, T. C., Hamilton, R. B., and Norgren, R. (2000). Projections of the parabrachial nucleus in the Old World monkey. *Exp. Neurol.* 165: 101–117.

Pritchard, T. C., Edwards, E. M., Smith, C. A., et al. (2005). Gustatory neural responses in the medial orbitofrontal cortex of the Old World monkey. *J. Neurosci.* 25: 6047–6056.

Pritchard, T. C., Nedderman, E. N., Edwards, E. M., and Petticoffer, A. C. (2008). Satiety-responsive neurons in medial orbitofrontal cortex of the macaque. *Behav. Neurosci.* 122: 174–182.

Pritchard, T. C. (2012). Gustatory system. In Paxinos, G. and Mai, J. (Eds.) *The Human Nervous System.* 3rd edition, New York: Elsevier, pp. 1187–1218.

Reep, R. L., and Winans, S. S. (1982). Efferent connections of dorsal and ventral agranular insular cortex in the hamster, *Mesocricetus auratus. Neurosci.* 7: 2609–2635.

Reep, R. L., Corwin, J. V., and King, V. (1996). Neuronal connections of orbital cortex in rats: topography of cortical and thalamic afferents. *Exp. Brain Res.* 111: 215–232.

Reilly, S., and Pritchard, T. C. (1996a). Gustatory thalamus lesions in the rat: I. Innate taste preferences and aversions. *Behav. Neurosci.* 110: 737–745.

Reilly, S., and Pritchard, T. C. (1996b). Gustatory thalamus lesions in the rat: II. Aversive and appetitive taste conditioning. *Behav. Neurosci.* 110: 746–759.

Reilly, S., and Pritchard, T. C. (1996c). Gustatory thalamus lesions in the rat: III. Simultaneous contrast and autoshaping. *Physiol. Behav.* 62: 1355–1363.

References

Rempel-Clower, N. L. (2007). Role of orbitofrontal cortex connections in emotion. *Ann. N.Y. Acad Sci.* 1121: 72–86.

Renehan, W. E., Jin, Z., Zhang, X., and Schweitzer, L. (1994). The structure and function of gustatory neurons in the nucleus of the solitary tract. I. A classification of neurons based on morphological features. *J. Comp. Neurol.* 347: 531–544.

Rhoton, A. L. Jr. (1968). Afferent connections of the facial nerve. *J. Comp. Neurol.* 133: 89–100.

Rolls, B. J., Rolls, E. T., Rowe, E. A., and Sweeney, K. I. (1981). Sensory specific satiety in man. *Physiol. Behav.* 27: 137–142.

Rolls, E. T. (2004). Convergence of sensory systems in the orbitofrontal cortex of primates and brain design for emotion. *Anat. Rec. A Discov. Mol. Cell Evol. Biol.* 281: 1212–1225.

Rolls, E. T. (2005). Taste and related systems in primates including humans. *Chem. Senses* 30: i76–i77.

Rolls, E. T., Scott, T. R., Sienkiewicz, Z. J., and Yaxley, S. (1988). The responsiveness of neurons in the frontal opercular gustatory cortex of the macaque monkey is independent of hunger. *J. Physiol.* 397: 1–12.

Rolls, E. T., Sienkiewicz, Z. J., and Yaxley, S. (1989). Hunger modulates the responses to gustatory stimuli of single neurons in the caudolateral orbitofrontal cortex of the macaque monkey. *Eur. J. Neurosci.* 1: 53–60.

Rolls, E. T., Yaxley, S., and Sienkiewicz, Z. J. (1990). Gustatory responses of single neurons in the caudolateral orbitofrontal cortex of the macaque monkey. *J. Neurophysiol.* 64: 1055–1066.

Rolls, E. T., and Baylis, L. L. (1994). Gustatory, olfactory, and visual convergence within the primate orbitofrontal cortex. *J. Neurosci.* 14: 5437–5452.

Rolls, E. T., Critchley, H. D., Wakeman, E. A., and Mason, R. (1996). Responses of neurons in the primate taste cortex to the glutamate ion and to inosine 5'-monophosphate. *Physiol. Behav.* 59: 991–1000.

Rosen, A. M., and Di Lorenzo, P. M. (2009). Two types of inhibitory influences target different groups of taste-responsive cells in the nucleus of the solitary tract of the rat. *Brain Res.* 1275: 24–32.

Rosen, A. M., Roussin, A. R., and Di Lorenzo, P. M. (2010a). Water as an independent taste modality. *J. Enteric Neurosci.* 4: 175–185.

Rosen, A. M., Sichtig, H., Schaffer, J. D., and Di Lorenzo, P. M. (2010b). Taste-specific cell assemblies in a biologically informed model of the nucleus of the solitary tract. *J. Neurophysiol.* 104: 4–17.

Rosen, A. M., Victor, J. D., and Di Lorenzo, P. M. (2011). Temporal coding of taste in the parabrachial nucleus of the pons of the rat. *J. Neurophysiol.* 105: 1889–1896.

Rosen, A. M., and Di Lorenzo, P. M. (2012). Neural coding of taste by simultaneously recorded cells in the nucleus of the solitary tract of the rat. *J. Neurophysiol.* 108: 3301–3312.

Roussin, A. T., Victor, J. D., Chen, J. Y., and Di Lorenzo, P. M. (2008). Variability in responses and temporal coding of tastants of similar quality in the nucleus of the solitary tract of the rat. *J. Neurophysiol.* 99: 644–655.

Roussin, A. T., D'Agostino, A. E., Fooden, A. M., et al. (2012). Taste coding in the nucleus of the solitary tract of the awake, freely licking rat. *J. Neurosci.* 32: 10494–10506.

Saalmann, Y. B., and Kastner, S. (2009). Gain control in the visual thalamus during perception and cognition. *Curr. Opin. Neurobiol.* 19: 408–414.

Sadacca, B. P., Rothwax, J. T., and Katz, D. B. (2012). Sodium concentration coding gives way to evaluative coding in cortex and amygdala. *J. Neurosci.* 32: 9999–10011.

Saper, C. B. (1982). Reciprocal parabrachial-cortical connections in the rat. *Brain Res.* 242: 33–40.

Satoda, T., Takahashi, O., Murakami, C., et al. (1996). The sites of origin and termination of afferent and efferent components in the lingual and pharyngeal branches of the glossopharyngeal nerve in the Japanese monkey (*Macaca fuscata*). *Neurosci. Res.* 24: 385–392.

Schoenbaum, G., Roesch, M. R., and Stalnaker, T. A. (2006). Orbitofrontal cortex, decision-making and drug addiction. *Trends Neurosci.* 29: 116–124.

Schultz, W., Tremblay, L., and Hollerman, J. R. (1998). Reward prediction in primate basal ganglia and frontal cortex. *Neuropharmacol.* 37: 421–429.

Scott, T. R., and Erickson, R. P. (1971). Synaptic processing of taste quality information in thalamus of the rat. *J. Neurophysiol.* 34: 868–884.

Scott, T. R., and Yalowitz, M. S. (1978). Thalamic taste responses to changing stimulus concentration. *Chem. Senses Flavor* 3: 167–175.

Scott, T. R., Yaxley, S., Sienkiewicz, Z. J., and Rolls, E. T. (1986a). Gustatory responses in the nucleus tractus solitarius of the alert cynomolgus monkey. *J. Neurophysiol.* 55: 182–200.

Scott, T. R., Yaxley, S., Sienkiewicz, Z. J., and Rolls, E. T. (1986b). Gustatory responses in the frontal operculum of the alert cynomolgus monkey. *J. Neurophysiol.* 56: 876–890.

Scott, T. R., and Mark, G. P. (1987). The taste system encodes stimulus toxicity. *Brain Res.* 414: 197–203.

Scott, T. R., and Giza, B. (1990). Coding channels in the rat taste system. *Science* 249: 1585–1587.

Scott, T. R., Plata-Salamán, C.R ., Smith, V. L., and Giza, B. K. (1991). Gustatory neural coding in the monkey cortex: Stimulus intensity. *J. Neurophysiol.* 65: 76–86.

Scott, T. R., Plata-Salamán, C. R., and Smith-Swintosky, V. L. (1994). Gustatory coding in the monkey cortex: The quality of saltiness. *J. Neurophysiol.* 71: 1692–1701.

Scott, T. R., Giza, B. K., and Yan, J (1999). Gustatory neural coding in the cortex of the alert cynomolgus macaque: The quality of bitterness. *J. Neurophysiol.* 81: 60–71.

Scott, T. R., and Plata-Salamán, C. R. (1999). Taste in the monkey cortex. *Physiol. Behav.* 67: 489–511.

Scott, T. R., and Small, D. M. (2009). The role of the parabrachial nucleus in taste processing and feeding. *Ann. N.Y. Acad. Sci.* 1170: 372–377.

Sherman, S. M. (2007). The thalamus is more than just a relay. *Curr. Opin. Neurobiol.* 17: 417–422.

Shi, C. J., and Cassell, M. D. (1998). Cortical, thalamic, and amygdaloid connections of the anterior and posterior insular cortices. *J. Comp. Neurol.* 399: 440–468.

Small, D. M., Zatorre, R. J., Dagher, A., et al. (2001). Changes in brain activity related to eating chocolate: from pleasure to aversion. *Brain* 124: 1720–1733.

Smith, D. V., Liu, H., and Vogt, M. B. (1996). Responses of gustatory cells in the nucleus of the solitary tract of the hamster after NaCl or amiloride adaptation. *J. Neurophysiol.* 76: 47–58.

Smith, D. V. and Li, C.-S. (1998). Tonic GABAergic inhibition of taste-responsive neurons in the nucleus of the solitary tract. *Chem. Senses* 23: 159–169.

Smith, D. V., and Li, C.-S. (2000). GABA-mediated corticofugal inhibition of taste-responsive neurons in the nucleus of the solitary tract. *Brain Res.* 858: 408–415.

Smith, D. V., St. John, S. J., and Boughter, J. D. (2000). Neuronal cell types and taste quality coding. *Physiol. Behav.* 69: 77–85.

Smith, D. V., Li, C.-S., and Cho, Y. K. (2005a). Forebrain modulation of brainstem gustatory processing. *Chem. Senses* 30 Suppl 1: i176–i177.

Smith, D. V., Ye, M. K., and Li, C.-S. (2005b). Medullary taste responses are modulated by the bed nucleus of the stria terminalis. *Chem. Senses* 30: 421–434.

Smith-Swintosky, V. L., Plata-Salamán, C. R., and Scott, T. R. (1991). Gustatory neural coding in the monkey cortex: Stimulus quality. *J. Neurophysiol.* 66: 1156–1165.

Stapleton, J. R., Lavine, M. L., Wolpert, R. L., et al. (2006). Rapid taste responses in the gustatory cortex during licking. *J. Neurosci.* 26: 4126–4138.

Stapleton, J. R., Lavine, M. L., Nicolelis, M. A., and Simon, S. A. (2007). Ensembles of gustatory cortical neurons anticipate and discriminate between tastants in a single lick. *Front. Neurosci.* 1: 161–174.

Sudakov, K., MacLean, P. D., Reeves, A., and Marino, R. (1971). Unit study of exteroceptive inputs to claustrocortex in awake, sitting, squirrel monkey. *Brain Res.* 28: 19–34.

Summerfield, C., and Tsetsos , K. (2012). Building bridges between perceptual and economic decision-making: neural and computational mechanisms. *Front. Neurosci.* 6: 70. doi:10.3389/fnins.2012.00070.

Taillibert, S., Bazin, B., and Pierrot-Deseilligny, C. J. (1998). *Neurol. Neurosurg. Psychiat.* 64: 691–692.

Tataranni, P. A., Gautier, J.-F., Chen, K., et al. (1999). Neuroanatomical correlates of hunger and satiation in humans using positron emission tomography. *Proc. Nat. Acad. Sci.* 96: 4569–4574.

Thorpe, S. J., Rolls, E. T., and Maddison, S. (1983). The orbitofrontal cortex: Neuronal activity in the behaving monkey. *Exp. Brain Res.* 49: 93–115.

Tokita, K., Inoue, T., and Boughter, J. D. Jr. (2009). Afferent connections of the parabrachial nucleus in C57BL/6J mice. *Neurosci.* 161: 475–848.

Travers, J. B. (1988). Efferent projections from the anterior nucleus of the solitary tract of the hamster. *Brain Res.* 457: 1–11.

Travers, J. B., and Norgren, R. (1983). Afferent projections to the oral motor nuclei in the rat. *J. Comp. Neurol.* 220: 280–298.

Travers, S. P., Pfaffmann, C., and Norgren, R. (1986). Convergence of lingual and palatal gustatory neural activity in the nucleus of the solitary tract. *Brain Res.,* 365: 305–320.

Travers, S. P., and Norgren, R. (1989). The time course of solitary nucleus gustatory response: influence of stimulus and site of application. *Chem. Senses* 14: 55–74.

Travers, S. P., and Norgren, R. (1995). Organization of orosensory responses in the nucleus of the solitary tract of rat. *J. Neurophysiol.* 73: 2144–2162.

Travers, J. B., Urbanek, K., and Grill, H. J. (1999). Fos-like immunoreactivity in the brain stem following oral quinine stimulation in decerebrate rats. *Am. J. Physiol.* 277(2 Pt 2): R384–394.

Tremblay, L., and Schultz, W. (1999). Relative reward preference in primate orbitofrontal cortex. *Nature* 398: 704–708.

Tremblay, L., and Schultz, W. (2000a). Reward-related neuronal activity during go-nogo task performance in primate orbitofrontal cortex. *J. Neurophysiol.* 83: 1864–1876.

Tremblay, L., and Schultz, W. (2000b). Modifications of reward expectation-related neuronal activity during learning in primate orbitofrontal cortex. *J. Neurophysiol.* 83: 1877–1885.

Verhagen, J. V., Giza, B. K., and Scott, T. R. (2003). Responses to taste stimulation in the ventroposteromedial nucleus of the thalamus in rats. *J. Neurophysiol.* 89: 265–275.

Verhagen, J. V., Kadohisa, M., and Rolls, E. T. (2004). The primate insular/opercular taste cortex: Neuronal representation of the viscosity, fat, texture, grittiness, temperature, and taste of foods. *J. Neurophysiol.* 92: 1685–1699.

Verhagen, J. V., Giza, B. K., and Scott, T. R. (2005). Effect of amiloride on gustatory responses in the ventroposteromedial nucleus in rats. *J. Neurophysiol.* 93: 157–166.

Wang, L., and Bradley, R. M. (1995). In vitro study of afferent synaptic transmission in the rostral gustatory zone of the rat nucleus of the solitary tract. *Brain Res.* 702: 188–198.

Wang, M., and Bradley, R. M. (2010a). Synaptic characteristics of rostral nucleus of the solitary tract neurons with input from the chorda tympani and glossopharyngeal nerves. *Brain Res.* 1328: 71–78.

Wang, M., and Bradley, R. M. (2010b). Properties of GABAergic neurons in the rostral solitary tract nucleus in mice. *J. Neurophysiol.* 103: 3205–3218.

Westlund, K. N., and Craig, A. D. (1996). Association of spinal lamina I projections with brainstem catecholamine neurons in the monkey. *Exp. Brain Res.* 110: 151–162.

Wetherton, B. M., Leonard, N. L., Renehan, W. E., and Schweitzer, L. (1998). Structure and function of gustatory neurons in the nucleus of the solitary tract. III. Classification of terminals and GABAergic synapses using postembedding immunoelectronmicroscopy. *J. Biotech. Histochem.* 98: 164–173.

Whitehead, M. C. (1986). Anatomy of the gustatory system in the hamster: synaptology of facial afferent terminals in the solitary nucleus. *J. Comp. Neurol.* 244: 72–85.

Whitehead, M. C. (1988). Neuronal architecture of the nucleus of the solitary tract in the hamster. *J. Comp. Neurol.* 276: 547–572.

Whitehead, M. C. (1993). Distribution of synapses on identified cell types in a gustatory subdivision of the nucleus of the solitary tract. *J. Comp. Neurol.* 332: 326–340.

Whitehead, M. C., and Frank, M. E. (1983). Anatomy of the gustatory system in the hamster: central projections of the chorda tympani and the lingual nerve. *J. Comp. Neurol.* 220: 378–395.

Whitehead, M. C., Bergula, A., and Holliday, K. (2000). Forebrain projections to the rostral nucleus of the solitary tract in the hamster. *J. Comp. Neurol.* 422: 429–447.

Winer, J.A. (2006). Decoding the auditory corticofugal systems. *Hear. Res.* 212: 1–8.

Wise, S. P., and Jones, E. G. (1977). Cells of origin and terminal distribution of descending projections of the rat somatic sensory cortex. *J. Comp. Neurol.* 175: 129–157.

Wolf, G. (1968). Projections of thalamic and cortical gustatory areas in the rat. *J. Comp. Neurol.* 132: 519–530.

Yamamoto, T. (1984). Taste responses of cortical neurons. *Prog. Neurobiol.* 23: 273–315.

Yamamoto, T., Matsuo, R., and Kawamura, Y. (1980). Corticofugal effects on the activity of thalamic taste cells. *Brain Res.* 193: 258–262.

Yamamoto, T., Yuyama, N., and Kawamura, Y. (1981). Cortical neurons responding to tactile, thermal, and taste stimulation of the rat's tongue. *Brain Res.* 221: 202–206.

Yamamoto, T., Azuma, S., and Kawamura, Y. (1984). Functional relations between the cortical gustatory area and the amygdala: Electrophysiological and behavioral studies in rats. *Exp. Brain Res.* 56: 23–31.

Yamamoto, T., Matsuo, R., Kiyomitsu, Y., and Kitamura, R. (1989). Taste response of cortical neurons in freely moving rats. *J. Neurophysiol.* 61: 1244–1258.

References

Yamamoto, T., Shimura, T., Sakai, N., and Ozaki, N. (1994). Representation of hedonics and quality of taste stimuli in the parabrachial nucleus of the rat. *Physiol. Behav.* 56: 1197–1202.

Yarmolinsky, D. A., Zuker, C. S., and Ryba, N. J. (2009). Common sense about taste: from mammals to insects. *Cell* 139: 234–244.

Yaxley, S., Rolls, E. T., Sienkiewicz, Z. J., and Scott, T. R. (1985). Satiety does not affect gustatory activity in the nucleus of the solitary tract of the alert monkey. *Brain Res.*, 347: 85–93.

Yaxley, S., Rolls, E. T., and Sienkiewicz, Z. (1988). The responsiveness of neurones in the insular gustatory cortex of the macaque monkey is independent of hunger. *Physiol. Behav.* 42: 223 –229.

Zaidi, F. N., Todd, K., Enquist, L., and Whitehead, M. C. (2008). Types of taste circuits synaptically linked to a few geniculate ganglion neurons. *J. Comp. Neurol.* 511: 753–772.

Chapter 33

Development of the Taste System

ROBIN F. KRIMM, SHOBA THIRUMANGALATHU, and LINDA A. BARLOW

33.1 A GENERAL INTRODUCTION TO THE PERIPHERY OF THE TASTE SYSTEM

Taste or gustation is a contact mediated chemosense, where chemical stimuli are detected and transduced by clusters of specialized epithelial cells known as taste buds. In most vertebrates, taste buds reside in the epithelium of the oral cavity and the pharynx, although some fish have external taste buds embedded in epidermis of head and the trunk. In mammals, the majority of taste buds are found on the dorsal surface of the tongue, and in rodents, taste buds are present in the soft palate. Regardless of location, taste buds share a common organization in that each bud comprises a heterogeneous collection of ~60–100 fusiform cells. Taste buds receive afferent innervation from sensory neurons of the cranial nerve ganglia; those located anteriorly are innervated by VIIth cranial nerve fibers – as are external taste buds in fishes, those situated in the posterior tongue are innervated by sensory neurons of cranial nerve IX, whereas taste buds in pharynx and larynx are supplied by cranial nerve X. Taste information from all three cranial nerves is transmitted to the first taste relay in the hindbrain, the nucleus of the solitary tract (NST). Each of these elements, taste buds, sensory neurons, and hindbrain relay, develop initially independently from one another, each arising from different regions of the embryonic head. As development proceeds, these elements must establish proper connections so that a functional taste system is present at birth, and when feeding commences.

Here we review the recent data on the cellular and molecular regulation of taste system development, focusing primarily on the development of taste buds and their innervation; showcasing advances resulting from use of cutting edge molecular genetic tools. Specifically, we discuss new data on cellular, tissue-level, and molecular mechanisms of taste bud development, where the historical view of a nerve dependent model has shifted to the current one where taste buds develop via tongue-intrinsic processes. Taste buds house a collection of three cell types (I, II, and III), which are present in consistent ratios, and we discuss significant advances in our understanding of molecular regulation of taste cell fate within buds. We further highlight accumulating evidence for distinct molecular regulation of early development of taste buds in the anterior versus posterior tongue. Finally, we review recent major advances in our understanding of the development of gustatory innervation, and the molecular and cellular mechanisms that govern sensory neuron development, emphasizing advances in our understanding of how these cells connect with their proper targets, that is, taste buds peripherally, and the NST centrally.

33.2 ANATOMY OF THE TASTE EPITHELIUM

The majority of studies of taste epithelium development have focused on lingual taste buds of rodents. In mice and rats, taste buds reside in specialized epithelial structures, termed gustatory or taste papillae, on the tongue surface. Anteriorly, a single taste bud is situated in each of a dispersed array of fungiform papillae, while posteriorly, several hundred taste buds are housed in a single midline circumvallate papilla, as well as in bilateral foliate papillae. Additionally, rodents possess a collection of taste buds embedded directly in the epithelium of the soft palate, in the posterior palatine field, as well as in a region known as the Geschmackstreifen (literally "taste stripe"); these palatal buds are not housed in papillae. Fungiform papillae

Handbook of Olfaction and Gustation, Third Edition. Edited by Richard L. Doty.
© 2015 Richard L. Doty. Published 2015 by John Wiley & Sons, Inc.

728 Chapter 33 Development of the Taste System

are relatively simple structures; each is mushroom-shaped with the epithelium surrounding a mesenchymal core comprising neural and vascular components, as well as connective tissue. The posterior taste papillae, by contrast, are significantly more complex. The circumvallate papilla possesses a large connective tissue core, which is surrounded bilaterally by deep epithelial invaginations or trenches. The trenches, replete with taste buds, are continuous with the epithelial ducts of paired sublingual salivary glands, which empty their contents into the taste trenches. Likewise, foliate papillae possess hundreds of taste buds embedded in invaginated epithelial folds that are contiguous with the ducts of lingual serous glands. As posterior papillae are fed by lingual salivary glands, the development of the glands and posterior papillae is linked, and thus more complex. Here, we first address development of the more simply organized fungiform field, and then review posterior taste papillae development, emphasizing similarities and important differences between the two taste domains.

33.3 ESTABLISHING THE EMBRYONIC ORIGINS OF TASTE BUDS AND PAPILLAE

In mice, development of fungiform papillae is evident at mid-gestation (embryonic day (E) 12.5), when taste placodes (epithelial thickenings) appear in bilateral rows along the lingual surface. In addition to their columnar morphology, taste placode cells express a number of gene products that distinguish them from non-placodal cells, including expression of a secreted protein, Sonic hedgehog (Shh) (Hall et al., 1999). Subsequently, taste placodes undergo morphogenesis beginning at E14.5, to form papillae, each with an invaginated epithelium surrounding a mesenchymal core. However, fungiform taste buds do not differentiate until the first postnatal week, when each of the three taste bud cell types is identifiable via expression of marker proteins (El-Sharaby et al., 2001a,b; Ohtubo et al., 2012; Zhang et al., 2006; Glover, Nguyen and Barlow, pers obs). Thus, taste bud development is intricately entangled with that of the taste papillae. For this reason, taste placodes have been considered to be taste papillae precursors within which a subset of cells is eventually established as taste bud cells. Alternatively, however, taste placodes could be taste bud precursor cells, which instruct local epithelium and mesenchyme to form the surrounding papilla during embryogenesis, and then differentiate into functional taste bud cells after birth. To distinguish between these two hypotheses, we employed an inducible genetic lineage tracing approach to follow the fate of taste placode cells and their progeny (Thirumangalathu et al., 2009). In fact, we showed that taste placodes which express Shh (Shh+)

are precursors for taste receptor cells within buds, and do not contribute to the papilla structure housing buds.

Taste bud cells possess numerous characteristics consistent with their epithelial location, including epithelial morphology as well as regular renewal of new taste cells throughout adult life (Miura and Barlow, 2010). However, taste cells also have significant similarities with neurons; a subset is electrically excitable, fire action potentials and form synapses on afferent nerve fibers (Chaudhari and Roper, 2010). Nonetheless, historically taste buds were considered to arise from the local epithelium (Cook and Neal, 1921; Johnston, 1910; Landacre, 1907; Stone, 1940). These descriptive results were confirmed experimentally some time ago using complementary approaches in amphibian and mouse models (Barlow and Northcutt, 1995; Stone et al., 1995). However, recently, a report has emerged ascribing neural crest cells as the embryonic source of taste buds (Liu et al., 2012b).

The neural crest is a fascinating embryonic population of cells, which arises from the dorsal region of the developing brain and spinal cord early in development. While these cells initially reside in the early neural ectoderm, they undergo an epithelial-to-mesenchymal transition, emerge from the neural tube and migrate in stereotypic streams to a variety of locations in the head and trunk. Once they arrive at their ultimate locations, neural crest cells contribute to a diverse array of cell types including pigment cells, as well as sensory and autonomic neurons and glia in both the head and trunk. However, the neural crest repertoire is greatly expanded in the head; cranial neural crest gives rise to the cartilage and bones, as well as all the connective tissues of the face, supplies the dentine-producing cells of teeth, is the source of all smooth muscle wrapping the arterial vasculature of the head, and further contributes to numerous specialized glands of the head and neck (Le Douarin et al., 2007; Le Douarin and Kalcheim, 1999). Because the neural crest contributes to such a wide array of tissues in the adult, it had also been suggested that it supplies cells to taste buds (Reutter, 1978). This hypothesis was particularly attractive, given the neural characteristics of taste cells, and the fact that *bona fide* sensory neurons in the cranial ganglia derive from the neural crest (D'Amico-Martel, 1982; Narayanan and Narayanan, 1980) and see below).

In previous studies, however, a neural crest contribution to taste buds had been ruled out. In both mouse and amphibian models, when neural crest cells were genetically labeled prior to their emergence from the neural tube and then followed through development, labeled cells were never detected in taste buds (mouse: (Thirumangalathu et al., 2009); amphibian: (Barlow and Northcutt, 1995)). Moreover, in oral epithelial explants from amphibian embryos, taste buds developed despite the complete absence of neural crest cells (Barlow, 2000b,2001; Barlow and Northcutt, 1997; Parker et al.,

2004). However, using molecular genetic labeling in mice, Mistretta and colleagues now report large numbers of labeled cells in postnatal taste buds (Liu et al., 2012b).

The difference between results from this new study compared with those from the previous ones are likely attributable to differences in the genetic models employed. In both studies (Liu et al., 2012b, Thirumangalathu et al., 2009), Cre lox systems were used where a tissue-specific promoter, in this case Wnt1, was used to drive Cre recombinase in the neural crest prior to its migration from the neural tube. In embryos carrying Wnt1Cre and a Cre reporter allele, Rosa26-LacZ, most pre-migratory neural crest cells are indelibly labeled, such that all neural crest derivatives are likewise labeled (Chai et al., 2000, Danielian et al., 1998, Jiang et al., 2002). Using this approach, we found no neural crest-descendent cells in the lingual epithelium of embryos nor in the taste buds, taste papillae or tongue epithelium in postnatal mice (Thirumangalathu et al., 2009). Mistretta and colleagues, by contrast, report that although exceedingly rare, occasional single neural crest cells were encountered associated with taste placodes in embryos. Predominantly, however, they employed a less well-characterized Cre allele, P0-Cre, to map the fate of neural crest cells in postnatal mice. Specifically, in this transgenic allele, a cassette encoding Cre driven by a fragment of the promoter for P0, an important myelin protein, has been randomly incorporated in the mouse genome (Yamauchi et al., 1999). While native P0 is not expressed in migrating neural crest (Hagedorn et al., 1999), P0-Cre is (Kawakami et al., 2011; Yamauchi et al., 1999), resulting in labeling of early migrating crest in embryos that possess a reporter allele. Using this Cre-lox approach, Mistretta and coworkers found that lingual epithelium and multiple taste placodes were labeled in embryos, and that the extent of labeled cells in the tongue surface was tremendously expanded in postnatal mice. Importantly, however, P0-Cre also labels other embryonic domains in addition to neural crest, including the early oral endoderm and ectoderm, as well as the notochord, indicating that not all genetically labeled cells within P0-Cre embryos derive from neural crest precursors (Yamauchi et al., 1999). This ectopic expression pattern further raises the question of whether P0-Cre is expressed in lingual epithelium postnatally, which would account for the dramatic differences in the results obtained with Wnt1Cre, where epithelial labeling is extraordinarily infrequent, versus those obtained with P0-Cre, where over 80% of taste buds and large swaths of tongue epithelium are labeled. These questions remain to be addressed through a full characterization of the P0-Cre transgenic allele. See Buckingham and Meilhac (2011) for a general discussion of the use of genetic tools for lineage tracing.

33.4 CELL AND TISSUE-LEVEL INTERACTIONS GOVERN EARLY TASTE BUD DEVELOPMENT

33.4.1 The Role of Gustatory Nerves

Historically, the primary model for taste bud development had been one of neural induction, where gustatory nerve fibers invade the embryonic lingual epithelium and induce taste bud formation (Mistretta and Liu, 2006; Oakley and Witt, 2004; Olmsted, 1920; Torrey, 1940). This hypothesis is supported by descriptive data, where in all cases, nerve fibers penetrate taste papilla epithelium long before taste buds differentiate, as well as by denervation studies in postnatal rats, where interruption of the nerve supply within the first postnatal week permanently reduces taste bud number despite recovery of the gustatory innervation (see Krimm and Barlow (2007) for review). However, as reviewed previously (Barlow, 2000a; Krimm and Barlow, 2007), early taste bud development has now been shown to be nerve independent. Specifically, in amphibian embryos, taste buds complete their entire developmental program, including taste cell differentiation, in the complete absence of nerves (Barlow, 2001; Barlow et al., 1996; Barlow and Northcutt, 1997). Similarly, in rodents, embryonic development of taste bud precursor cells occurs in cultured tongue explants, including specification and patterning of taste precursors, as well as initial aspects of papilla morphogenesis (Hall et al., 2003; Ito and Nosrat, 2009; Ito et al., 2010; Mistretta et al., 2003; Nosrat et al., 2001). Moreover, these same features of taste bud development occur independently of innervation in vivo; specifically, fungiform taste buds are induced and patterned, and papillae undergo morphogenesis in mouse embryos null for the neurotrophin, BDNF or its receptor TrkB, which are required for development of gustatory innervation (Fritzsch et al., 1997; Ito and Nosrat, 2009; Ito et al., 2010; Thirumangalathu et al., 2009) and see below). In contrast to amphibians, however, taste cell differentiation in postnatal rodents likely depends upon an intact innervation (Krimm et al., 2001; Liebl et al., 1999; Mistretta et al., 1999; Nosrat et al., 1997; Oakley et al., 1998).

33.4.2 Signaling Between the Embryonic Epithelium and Mesenchyme

Taste buds and their papillae are often categorized as ectodermal appendages, similar to teeth, feathers, and hair follicles (Mistretta and Liu, 2006; Pispa and Thesleff, 2003). Development of these later ectodermal specializations requires extensive reciprocal signaling between epithelial and mesenchymal compartments, interruption of which results in failure to develop teeth, feathers, or hair (Morgan et al., 1998; Thesleff et al., 1991; Tucker and Sharpe, 1999). The tongue possesses an extensive subepithelial mesenchyme, the lamina propria, which is derived from the cranial neural crest (Chai et al., 2000). In the anterior third of the tongue, fungiform papillae, as well adjacent non-gustatory filiform papillae (the vast majority of papillae covering the tongue surface) all have a similar structure with epithelium bounding mesenchyme in the core. More posteriorly, in the middle third of the rodent tongue, fungiform taste buds are mostly absent, and papillae are non-gustatory (Mistretta, 1972). It appears that signals from the mesenchyme, TGFß pathway inhibitors in particular and see below, restrict fungiform taste papillae to the anterior domain, as genetic loss of this inhibition results in development of taste bud precursors in the tongue's middle third (Beites et al., 2009). Similarly, culture studies suggest a role for lingual mesenchyme in the pattern of taste buds that form in the anterior third of the tongue (Kim et al., 2003). Thus, signals from lingual mesenchyme appear to regulate the pattern of developing taste buds in rodents. Interestingly, in amphibian embryos, taste buds differentiate in the absence of cranial mesenchyme (Barlow and Northcutt, 1997; Parker and Barlow, 2000), suggesting that in this vertebrate group taste bud specification is independent of mesenchymal signals. In mammals, it has not been determined whether initial formation of taste buds is likewise mesenchyme-independent, nor has a patterning role for mesenchyme been ruled out for amphibian taste buds. Resolution of these questions will likely reveal which aspects of epithelial-mesenchymal interactions are common to all vertebrates, and which may regulate specialized features of taste epithelium in each vertebrate sub-class.

33.4.3 Molecular Regulation of Anterior Taste Bud Development

As mentioned earlier, taste buds have both epithelial and neural characteristics, and thus molecular regulation of the development of these receptor organs may resemble that of skin or neural epithelium, or may comprise an amalgam of these mechanisms that is unique to taste. An additional consideration is that the developmental profile for taste epithelium may be more comparable to that of gut. Specifically, the oropharyngeal cavity is the gateway to the GI tract, and as such is contiguous with the gut axis. The idea that taste receptors are more "gut-like" than "skin appendage-like" is reinforced by the fact that taste receptor and transduction proteins are expressed by chemosensory cells in the stomach and intestine (Höfer et al., 1996; Kokrashvili et al., 2009; Sternini et al., 2008). However, even within a broad tissue type, such as the gut, the mechanisms governing development are organ specific (Udager et al., 2010; Zorn and Wells, 2009), likely

reflecting the divergence in the function of each region during evolution, for example, stomach – digestion, small intestine – absorption. Thus the tissue level and molecular mechanisms for taste bud development are unlikely to mirror those driving formation of only one these specialized tissues, but rather will represent a combination of mechanisms entirely unique to taste.

A host of important signaling pathways are expressed in the developing tongue epithelium, commencing as taste placodes form, and we have significant functional data for several of these. Of prime importance to date, are the Sonic hedgehog (Shh) and Wnt/ß-catenin pathways.

33.4.3.1 Sonic Hedgehog (Shh). Sonic hedgehog is a key signaling molecule that regulates multiple aspects of vertebrate organogenesis (Ingham and McMahon, 2001), and in particular, functions as a morphogen in the development of epithelial appendages, that is, hair, feathers, and teeth (Chuong et al., 2000). In the developing tongue, Shh is expressed by taste placode cells, while expression of its receptor (and target gene) Ptch1, and downstream target Gli1, are initially broad, and then become restricted to the epithelium and mesenchyme immediately surrounding placodes; these expression patterns suggest that Shh+ taste placodes signal to the local epithelium and mesenchyme during embryonic development. As Shh null mutants have major oral malformations including absence of the tongue (Incardona and Roelink, 2000), functional studies of Shh's role in taste epithelial development have been limited to the manipulation of embryonic tongue explants. Pharmacological inhibition of Shh, with the steroidal alkaloid cyclopamine or a function-blocking antibody, results in larger and more fungiform placodes, and triggers taste placode formation ectopically, in the middle third of the tongue normally devoid of taste buds (Hall et al., 2003; Liu et al., 2004; Mistretta et al., 2003). Treatment of cultured tongues with excess Shh ligand, by contrast, blocks placode development (Iwatsuki et al., 2007). In sum, these data indicate that Shh acts to repress taste fate in the anterior and middle zones of the tongue, serving an important function in taste organ patterning. Interestingly, inhibition of Shh results in increased Shh expression in these excess and enlarged taste papillae, suggesting that Shh is an indirect negative regulator of its own expression, and implicating additional molecular players in taste patterning.

33.4.3.2 Wnt/ß-Catenin. Wnt signaling molecules are a family of evolutionarily conserved secreted glycoproteins, which function in the specification, patterning, differentiation, and regeneration of numerous epithelial appendages (Gordon and Nusse, 2006). Wnt ligands bind specifically to members of the Frizzled family of receptors

33.4 Molecular Regulation of Anterior Taste bud Development

and to LRP5/6 co-receptors to activate several signaling cascades, of which the β-catenin or "canonical" pathway is best studied and understood (Angers and Moon, 2009; Bejsovec, 2005). In brief, Wnt ligand binding to its receptors results in the dissociation of an intracellular destruction complex, which normally acts to maintain low levels of cytosolic ß-catenin. When this complex is inactivated, ß-catenin accumulates and moves to the nucleus, binding with TCF/LEF transcription factors to turn on Wnt target genes. Several reporter constructs have been built to identify cells experiencing active Wnt signaling, and these have been employed in both in vitro and in vivo models. Specifically, the TOPGAL mouse line carries an engineered allele where TCF/LEF response elements turn on expression of the LacZ gene when Wnt signaling drives ß-catenin into the nucleus (DasGupta and Fuchs, 1999).

Use of the TOPGAL reporter mouse has revealed that Wnt signaling is active in lingual epithelium of embryos, including in taste placodes as they first form at E12.5, while the lingual subepithelial mesenchyme is devoid of Wnt/ß-catenin-responsive cells (Iwatsuki et al., 2007; Liu et al., 2007). Tests of Wnt function in vivo therefore have entailed either activating or deleting the canonical Wnt effector gene, ß-catenin, exclusively within the lingual epithelium (Liu et al., 2007). Mice carrying transgenes with promoters for human cytokeratins 5 or 14 (K5 (Diamond et al., 2000); K14 (Andl et al., 2004) were used to drive expression of Cre recombinase in developing tongue epithelium, coincident with the development of taste placodes (Farbman and Mbiene, 1991; Mbiene and Roberts, 2003). When the Wnt pathway is triggered by Cre-mediated activation of a non-degradable allele of ß-catenin (Harada et al., 1999), the entire tongue surface is transformed into enlarged taste papillae, each housing large clusters of Shh+ taste bud precursor cells. By contrast, in embryos with epithelial deletion of ß-catenin, Shh+ taste placodes fail to form; likewise, Shh+ taste bud precursor cells are absent in embryos in which canonical Wnt signaling is broadly inhibited via epithelial over-expression of a secreted Wnt antagonist, Dkk-1 (Liu et al., 2007). Thus, Wnt signaling is both required and sufficient for taste bud precursor and papilla development, and functions as an initiator of taste placode formation, genetically upstream of Shh. However, Shh is also a negative regulator of Wnt signaling, as cultured tongues treated with Shh inhibitors exhibit expanded placodes with commensurate increased TOPGAL signaling (Iwatsuki et al., 2007).

To date, the Wnt gene(s) involved in taste bud initiation has not been identified. One obvious candidate is *Wnt10b*, a ligand in the canonical pathway, whose expression is restricted to taste placodes (Iwatsuki et al., 2007). However loss of Wnt10b does not phenocopy the loss of ß-catenin or overexpression of Dkk-1, as taste placodes still form in Wnt10b KO tongues, inferring that other Wnts must

be involved (Iwatsuki et al., 2007). Likewise, deletion of Lef1, one of several downstream transcriptional activators in the canonical pathway, results in reduced taste papillae with reduced Shh expression, yet taste papillae are not completely absent implicating additional Wnt pathway transcription factors (Iwatsuki et al., 2007). Additionally, *Wnt5a*, a ligand for the planar cell polarity pathway (Wansleeben and Meijlink, 2011), is expressed in deep lingual mesenchyme including muscle layers (Cervantes et al., 2009; Paiva et al., 2010), but taste placodes are unaffected in Wnt5a KO embryos; although the tongue is dramatically shortened likely due to failure of development of the lingual musculature (Liu et al., 2009). Overall, of the 19 Wnt ligands in mouse, a total of 5 have been shown to be expressed in the developing tongue, including both canonical ligands, that is, signal via ß-catenin (Wnt4, Wnt10a, Wnt10b) (Iwatsuki et al., 2007; Liu et al., 2007), and non-canonical ligands, that is, signal via alternative pathways (Wnt5a, Wnt11) (Kim et al., 2009; Lin et al., 2011). This extensive ligand expression repertoire suggests that Wnts function in numerous aspects of tongue and taste bud development, hinting that a complex balance among canonical and non-canonical Wnt pathways is required (Liu and Millar, 2010).

33.4.3.3 Bone Morphogenetic Proteins (Bmps).

BMPs belong to the large superfamily of TGFß secreted proteins, comprising ~20 genes, of which BMP2, 4, and 7 are expressed in the developing taste epithelium (Beites et al., 2009; Hall et al., 2003; Jung et al., 1999). BMPs bind specific heterodimeric transmembrane receptors, resulting in serial phosphorylation events, including that of SMAD transcription factors. Upon phosphorylation, these SMADs translocate to the nucleus to activate target gene expression (Guo and Wu, 2012). Most tests of BMP function in embryonic taste bud development to date have utilized cultured tongue explants. Specifically, BMP 2, 4, and 7 have shifting roles in taste bud development (Zhou et al., 2006). When tongue rudiments are treated with any of the three BMP ligands commencing prior to formation of taste placodes, the number and size of placodes are expanded. However, if tongue explants are exposed to BMP after taste placodes have been specified, then the number of placodes declines. Intriguingly, the secreted BMP antagonist, Noggin, has the same effect on taste placodes regardless of the timing of treatment; more taste placodes form than in untreated control explants. By contrast, follistatin, another secreted blocker of BMP function, has no impact on taste placode development in vitro (Zhou et al., 2006).

In contrast to the culture studies, Calof and colleagues have shown that repression of BMP4 signaling by follistatin restricts taste placode formation to the anterior tongue. In follistatin knockout embryos, taste placodes form in the middle third of the tongue, a domain typically devoid of taste organs (Beites et al., 2009). The difference between the in vivo and in vitro results may be due to the likelihood that follistatin manipulation in vitro occurred well after follistatitin's first function in vivo. Finally, expression of BMP4 has been placed genetically downstream of Wnt/ß-catenin (Liu et al., 2007), and is impacted by Shh signaling in vitro; inhibition of Shh in lingual explants results in expanded BMP4 expression in expanded papillae (Hall et al., 2003). Likewise, BMP ligand treatment represses Shh expression in tongue cultures (Zhou et al., 2006). Thus, the Shh, BMP and Wnt/ß-catenin pathways interact extensively, and likely repeatedly, as an organized taste epithelium emerges through the stages of specification, patterning, morphogenesis and cell differentiation.

33.4.3.4 *Fibroblast Growth Factors (Fgfs).*
The Fgf family of secreted factors comprises over 20 ligands, which bind 4 Fgf receptor proteins, each with multiple alternative splice variants resulting in distinct proteins and functions (Itoh and Ornitz, 2011; Thisse and Thisse, 2005). Among these, Fgf8 is expressed transiently in the early lingual epithelium prior to placode formation (Jung et al., 1999), whereas subsequent Fgf2, Fgf7, and Fgf10 expression is restricted to the tongue mesenchyme and muscle layers (Nie, 2005). Likewise, distinct Fgf receptors are expressed in a temporally complex pattern in both epithelium and mesenchyme (Nie, 2005).

Although Fgf pathway members are expressed throughout the anterior and posterior tongue, the most extensive examinations of Fgf function to date have focused on development of posterior taste buds (Kapsimali et al., 2011; Petersen et al., 2011), and thus will be discussed below in the section dedicated to the posterior taste papillae. Nonetheless, in the tongues of mouse embryos null for Fgf10, we found qualitatively that anterior fungiform taste papillae are expanded, whereas in tongues in which a pair of negative regulators of Fgf signaling, Spry1 and Spry2, have been deleted, fungiform taste papillae are reduced (Petersen et al., 2011). These data indicate that Fgf signaling, at least in the case of Fgf10, negatively regulates development of taste buds in the anterior tongue, but does so by mechanisms that have yet to be defined.

33.4.3.5 *Epidermal Growth Factor (Egf).*
EGF is the founding member of the EGF family of secreted factors that include the neuregulins, and TGF-alpha, among others (Normanno et al., 2005). EGF ligands bind to EGF and/or ErbB receptors to selectively activate specific intracellular pathways, including PI3 kinase/AKT, MEK/ERK, and p38 MAPK. Both EGF ligand and the

Egf receptor are initially expressed broadly in the lingual epithelium, but only the Egf receptor is then progressively excluded from taste placodes, suggesting a role for this pathway in non-taste epithelium (Liu et al., 2008). EGF function was implicated in taste bud development over a decade ago, where EGF KO embryos were found to have fewer fungiform papillae (Threadgill et al., 1995), a finding reinforced by examination of Egfr KO tongues of postnatal mice (Sun and Oakley, 2002). Most recently, EGF function in early taste bud development has been assessed pharmacologically in cultured embryonic tongues (Liu et al., 2008). Excess EGF represses fungiform papilla formation in explants, whereas inhibition of Egfr resulted in expanded and fused papillae. Each of the three pathways downstream of Egfr activation appear to contribute to EGF's role in taste placode development; EGF repression of taste placodes could only be rescued when all three downstream pathways, PI3K/AKT, MEK/ERK, and p38 MAPK, were simultaneously inhibited. In sum, EGF likely functions to promote non-taste epithelial fate, thus the interaction between EGF and Wnt/ß-catenin signaling, which promotes taste fate, should prove to be an exciting area to investigate.

33.4.4 Development of the Circumvallate Taste Papilla and Von Ebner's Gland Complex (CVP/VEG)

In contrast to the anterior fungiform papillae, taste buds in the posterior tongue reside in large and more complex structures, the circumvallate and foliate taste papillae. Each of these papillae houses several hundred taste buds, and is supplied by lingual salivary glands. The majority of studies of development of posterior taste buds has focused on the single circumvallate papilla (CVP) of rodents, which is associated with specific tubuloacinar serous glands, the Von Ebner glands – VEG; the ducts of these glands are continuous with the trench lumens of the CVP such that together they form the CVP/VEG complex. Biochemical (Hamosh and Burns, 1977) and immunocytochemical studies (Roberts and Jaffe, 1986) have demonstrated that the CVP/VEG complex initially produces the digestive enzyme lingual lipase. As nutrition for neonatal mammals comprises primarily milk triglycerides, these are initially digested with the help of lingual lipase during the period when the pancreas is still immature and pancreatic lipase is scarce. Whether this enzymatic profile shifts with weaning, as is the case for the enteroendocrine cells in the intestine (Henning, 1981), remains to be determined.

Like fungiform papillae, the CVP is first evident as a large placode in the posterior midline of the embryonic tongue. In mice, the CV placode forms by E12.0, followed by morphogenesis of bilateral epithelial trenches and a central mesenchymal core emerges replete with vascular and neural components (AhPin et al., 1989; Graziadei and Graziadei, 1978 Jitpukdeebodintra et al., 2002; Petersen et al., 2011; Sbarbati et al., 2000; Wakisaka et al., 1996). Precursors of the VEGs are thought to be located at the tips of the invaginating trenches as early as E15.5, based upon keratin gene expression patterns (Lee et al., 2006), while branching morphogenesis of the glands begins around birth (Jitpukdeebodintra et al., 2002; Lee et al., 2006; Sbarbati et al., 2000,2001; Sohn et al., 2011). As is the case for fungiform papillae, taste buds first differentiate within the CV trench epithelia during the first postnatal week, and over several weeks, the number and cell complement of CV taste buds in rodents increases to the adult levels (El-Sharaby et al., 2001b; Hosley and Oakley, 1987; Miller and Smith, 1988; Oakley et al., 1991; Sbarbati et al., 1999,2001; Zhang et al., 2008).

33.4.5 A Comparison of the Molecular Regulation of Early Development of the Circumvallate Versus Fungiform Taste Papillae

As discussed above, the Wnt/ß-catenin, Shh, and BMP pathways regulate taste bud development in the anterior tongue, and these signaling pathways are also involved in development of the CVP. However, in a number of instances, these pathways appear to have quite divergent functions in the specification and patterning of the anterior versus posterior taste fields.

33.4.5.1 Wnt/ß-catenin.
The first example is the role of Wnt/ß-catenin signaling. As discussed earlier, Wnt/ß-catenin is both required and sufficient for initiation of fungiform taste bud development (Iwatsuki et al., 2007; Liu et al., 2007). Consistent with this requirement, Wnt signaling is evident in the earliest stages of fungiform placode development (Iwatsuki et al., 2007; Liu et al., 2007). By contrast, we find that Wnt/ß-catenin activity is not present in the circumvallate placode when it first forms (data not shown), nor is it evident as morphogenesis begins at E13.5 (Figure 33.1A) (Iwatsuki et al., 2007). Instead, Wnt activity appears in the CVP at E14.5 (Figure 33.1B), thus implicating the Wnt/ß-catenin pathway in later aspects of CVP development.

Interestingly, *Wnt10b*, a ligand for the canonical Wnt pathway, is expressed by both CV and fungiform placodes early on, however, loss of *Wnt10b* had no effect on the CVP, while fungiform papillae were reduced in number and size (Iwatsuki et al., 2007). Similarly, loss of Lef1, an effector of Wnt/β-catenin signaling resulted in fewer and smaller fungiform papillae at birth, while the CVP appeared normal at E14.5 (Iwatsuki et al., 2007). These results suggest that compared with fungiform development where Wnt initiates placode formation, Wnt signaling is not required for

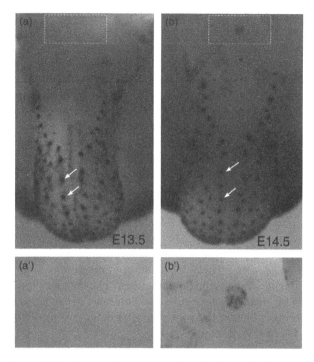

Figure 33.1 Tongues from mouse embryos harboring the Wnt reporter allele, BATGAL, when reacted for X-gal histochemistry, reveal extensive Wnt/ß-catenin signaling in the anterior fungiform taste placodes at both E13.5 (a) and E14.5 (b) (examples indicated with white arrows). However, BATGAL reporting is not evident in the posterior CVP until E14.5. Compare E13.5 where no X-gal product is evident (a, a'), with E14.5 where the CVP has extensive staining (b, b'). Areas inside dashed boxes in a and b are shown in a' and b' respectively.

CV placode specification, and instead may function in CVP morphogenesis and/or differentiation of taste buds within the CVP trench epithelium.

The Wnt planar cell polarity (PCP) pathway has also been investigated in taste epithelial development. While loss of Wnt5a, a ligand for the PCP pathway, has no impact on fungiform taste papillae, the CVP is smaller in Wnt5a mutants (Liu et al., 2012a), suggesting that the Wnt/PCP may play a role in morphogenesis of the CVP but not of fungiform papillae. Consistent with this observation, Wnt 11, another PCP pathway ligand, appears to direct cell shape changes in the apical CVP epithelium through its interaction with the Rho kinases (Kim et al., 2009). These observations are intriguing, and indicate that the role(s) of non-canonical pathway in the development of the CVP needs further exploration.

33.4.5.2 Sonic Hedgehog (Shh).
Shh is an important negative regulator of specification and patterning of fungiform taste buds. In the anterior tongue, Shh is expressed in taste placode cells as these structures

734   Chapter 33 Development of the Taste System

first form. Shh is also expressed in the CV placode at E12.0 (in mice), and Shh expression is evident in the epithelial trenches of the CVP by late gestation (Hall et al., 1997; Lee et al., 2006; Petersen et al., 2011). While inhibition of Shh results in excess and enlarged fungiform placodes and papillae, Shh inhibition during CVP development results in stunted CVP trenches with reduced proliferation and differentiation, suggesting that Shh promotes CVP and VEG development (Kim et al., 2009), as is the case for Shh function in submandibular salivary gland morphogenesis (Jaskoll et al., 2004). Examination of Ectodysplasin (Eda) signaling also sheds light on Shh's role in CVP/VEG development. The Eda pathway comprises the tumor necrosis factor-like ligand, EDA A1, and its receptor EDAR. Eda pathway mutants have an abnormal CVP with diminished Shh expression, suggesting that Shh functions downstream of Eda signaling. However, addition of exogenous Shh to the tongue explants did not fully rescue gland development in these mutants suggesting that Shh must act in concert with other factors to support normal development of the CVP/VEG (Wells et al., 2010).

33.4.5.3 Bone Morphogenetic Proteins (Bmps).
In cultured tongue explants BMP2, 4, and 7 and their secreted antagonist Noggin have opposing inhibitory or activating roles thereby balancing their functions in setting up a stereotypical array of fungiform taste field (Zhou et al., 2006). Follistatin (Fst), a BMP antagonist, is expressed in lingual mesenchymal and regulates the size, pattern, and differentiation of fungiform taste papillae through its interaction with epithelial BMP7 (Beites et al., 2009); genetic deletion of follistatin results in development of ectopic fungiform taste papillae. However, the CVP and VEG appear completely normal in $Fst^{-/-}$ mutant tongues, revealing yet another signaling pathway that differentially regulates the posterior and anterior taste fields.

33.4.5.4 Fibroblast Growth Factors (Fgfs).
As discussed earlier, several FGFs and their receptors are expressed in the developing tongue, suggestive of a role in taste development (Jung et al., 1999; Nie 2005). In fact, loss of Fgf10 results in enlarged and more numerous fungiform papillae, whereas enhanced Fgf signaling, via deletion of negative regulators of receptor tyrosine kinase signaling (Spry2 and Spry1), results in diminished fungiform papillae (Petersen et al., 2011). By contrast, the CVP is lost in Fgf10 null embryos, whereas multiple CVPs develop in Spry mutants. In fact, loss of Spry2 results in a duplicated CVP, replete with differentiated taste buds, and in Spry1/2 double knockout tongues, multiple CVP form. As in mouse CVP development, Fgf signaling is also required for development of the posterior taste buds in the pharyngeal cavity of zebrafish (Kapsimali and Barlow,

2012; Kapsimali et al., 2011). The timing and pattern of expression of Fgfs and their receptors are consistent with a role in taste bud development. In fact, loss of Fgf signaling (genetically or pharmacologically) blocks pharyngeal taste bud development. Thus, as pharyngeal taste buds in zebrafish are not housed in papillae, results of this study support the results in mouse, that there is an evolutionarily conserved requirement of Fgf signaling for posterior taste epithelium development.

In sum, the results of the above studies provide strong evidences for differential regulation of taste bud specification and patterning in the anterior versus posterior tongue. In fact, in some cases, pathway function in the anterior tongue may be opposite to that in the posterior tongue, for example, Shh, Fgf. What could account for these differences? One likely explanation is that the anterior tongue and resident fungiform taste buds are derived from ectoderm, whereas the posterior lingual epithelium, housing the CVP and foliate papillae, arises from the endoderm. Certainly this is the case in amphibians, where lineage tracing in embryos of an aquatic salamander revealed that most taste buds derive from endoderm (Barlow and Northcutt, 1995), yet those located most anteriorly are ectodermal in origin (Barlow, 2000b). However, in mammals, endoderm versus ectoderm assignations have been based solely on descriptive differences in lingual epithelial morphology. Only recently has the germ layer of origin for the disparate taste papillae been defined in mice, using molecular genetic lineage tracing. Abigail Tucker and colleagues employed a Sox17-2A-iCre allele that drives Cre recombinase expression in early endoderm crossed with a reporter line to demonstrate an endodermal origin for the posterior foliate and CV papillae, as well as for their associated minor lingual glands. The anterior fungiform taste papillae were not derived from the labeled endoderm, and thus were assigned an ectodermal origin (Rothova et al., 2012). Thus, differences in early patterning of taste epithelium may reflect the different embryonic histories of the two germ layers; an idea that is now readily testable.

33.5 MOLECULAR REGULATION OF TASTE BUD CELL LINEAGE

While initial patterning of anterior and posterior taste papillae are differentially regulated, mechanisms governing cell fate among specified taste precursors may be more similar with respect to taste bud location. However, no studies have addressed establishment of taste bud cell fate in a regionally comprehensive way, so this question remains an open one. Nonetheless, the Notch pathway, which acts widely in cell fate specification in a variety of embryonic tissues (Gridley, 1997; Muskavitch, 1994), clearly functions in taste cell fate choice during development.

33.5.1 Notch Signaling

In contrast to the previous sections where we discussed the roles of secreted factors in taste development, the Notch pathway is contact-mediated signaling system (Bray, 2006). Notch transmembrane receptors, of which there are four in mammals (N1, N2, N3, and N4), bind membrane-spanning ligands on adjacent cells. These ligands comprise four Delta-like members (Dll-1, Dll-2, Dll-3 and Dll-4), and two Jagged genes (Jag1, Jag2). Notch receptor binding to ligand triggers a succession of proteolytic cleavages of the N receptor, resulting in a free intracellular element (NICD), which translocates to the nucleus to affect gene expression in conjunction with CBF1/RBPkappaJ. In particular, expression of several bHLH transcription factors, including Ascl1 (previously named Mash1) and Hes1, are regulated by Notch signaling; activation of the pathway promotes Hes1 expression which in turn represses Ascl1 transcription. Notch pathway genes are expressed in the developing taste epithelium in mouse, although these expression data have been generated primarily for the developing CVP (Seta et al., 2003). In particular, expression of N pathway genes commences once the CV placode begins morphogenesis, suggestive of a role in cell fate decisions in developing taste epithelium, rather than in specification or patterning of taste versus non-taste fields. Similarly, Ascl1 is expressed in the developing taste placodes of the anterior tongue and of the soft palate but commencing after placode-specific expression of Shh has peaked, again suggestive of cell fate decisions within already specified taste placodes (Nakayama et al., 2008). Notch function in the anterior tongue has not been explored, however, the Notch pathway has been shown to play a role in selection of cell fate in taste buds located in the posterior oral cavity in zebrafish (Kapsimali et al., 2011). Specifically, a reduction in Notch signaling results in an overproduction of serotonin+ taste receptor cells (likely homologs of Type III cells in mammals), with a commensurate absence of calbindin 2b+ (Calb2b) taste bud cells (potential homologs of mammalian Type II taste cells), while enhanced Notch signaling favors a Calb2b+ over a serotonin+ taste cell fate. See Kapsimali and Barlow (2013) for a recent discussion of fish versus mammalian taste cell types.

33.5.2 Transcription Factors and Taste Bud Development

Cell signaling via diffusible and cell contact-mediated pathways are non-cell autonomous regulators of taste epithelial development, while the response of cells to developmental signaling is mediated by cell intrinsic regulation of gene expression. Thus, a wide array of transcription factors is expressed at progressive stages in taste placode and papillae, the timing of which are suggestive of functions in discrete aspects of taste epithelial development. Specifically, several transcription factors are expressed in taste precursor cells coincident with the first emergence of taste placodes, suggesting roles in specification and patterning; these include Six1, Six4, Sox2, and Prox1. An additional set of transcription factors, comprising Mash1/Ascl1 and Hes1, which are typically regulated by Notch signaling, are expressed later on, once papilla morphogenesis is underway, and likely function in cell fate choice within specified taste precursors. These two groups are discussed in turn below.

Six1 is a homeobox transcription factor, which has crucial functions in vertebrate organogenesis and in specification of sensory fields as well as in cell fate decisions in sensory organs such as the inner ear (Ahmed et al., 2012) and olfactory epithelium (Chen et al., 2009). Six1 is initially expressed broadly in the epithelium and mesenchyme of the developing tongue (Suzuki et al., 2010a), focalizes to taste placodes by mid-gestation, and persists in fungiform epithelial cells through embryogenesis. Fungiform taste papillae are affected in Six1 knockout embryos, as both their number and size dramatically increase (Suzuki et al., 2010b). Importantly, the distribution of fungiform papillae is distorted compared with that of wild type, and this may in part be due the impact of loss of Six1 on tongue development in general, as tongue morphology is also disturbed. Six4, another homeobox transcription factor, has a pattern of expression similar to that of Six1 in the developing tongue and fungiform papillae, but loss of Six4 alone does not affect the tongue nor the taste papillae. However, embryos null for both Six1 and Six4 have a more severe taste phenotype than the Six1 knockouts; not only are taste papillae enlarged and more numerous, these papillae appear to undergo precocious morphogenesis (Suzuki et al., 2011). However, it is not clear if Six1 and Six4 loss of function impacts taste placodes cell-autonomously, or if early Six1/Six4 dependent events in the tongue epithelium and mesenchyme indirectly impact patterning and morphogenesis of taste papillae in the anterior tongue.

Sox2 is an HMG-box transcription factor, most well known for its function in maintenance of stem cell fate in a variety of adult tissues. However, Sox2 is also required for many aspects of embryonic development, including differentiation of taste bud cells (Okubo et al., 2006). Sox2 is expressed broadly throughout the lingual epithelium of the developing tongue, and is most highly expressed in taste placodes and papillae. While taste placodes form in the tongues of Sox2 hypomorphic embryos, that is, those possessing an allele of Sox2 with reduced expression, these placodes fail to give rise to taste bud precursor cells, as evidenced by the lack of placode-specific Shh expression. Intriguingly, while Sox2 is required for taste bud development, it is not sufficient; overexpression of

Sox2 within lingual epithelium represses differentiation of non-gustatory (filiform) papillae, but does not trigger taste bud development. Importantly, Sox2 is genetically downstream of canonical Wnt signaling, and thus the Wnt pathway likely requires downstream factors in addition to Sox2 in order to initiate the full developmental program of taste placodes.

A third transcription factor, Prox1, is also expressed in early taste placodes, and in particular, in Shh+ taste precursor cells (Nakayama et al., 2008). While Prox1 function has not yet been assessed in taste development, it is known to control cell fate decisions in both the retina (Dyer et al., 2003; Dyer, 2003) and the inner ear (Dabdoub et al., 2008) of vertebrates, and its Drosophila homolog, *prospero*, was originally identified as a determinant of neural fate in the central nervous system (CNS) of flies (Doe et al., 1991). Thus, one potential role for Prox1 would be to regulate cell fate assignment in the taste cell precursor population, an hypothesis that remains to be tested.

Ascl1 (formerly Mash1) is a bHLH transcription factor that is regulated by Notch pathway activity, and functions in neural fate selection in a wide variety of tissues, including the olfactory epithelium (Beites et al., 2005; Calof et al., 2002; Cau et al., 2002, Murray et al., 2003; Shou et al., 1999). Ascl1 expression and function have not been defined in the fungiform papillae, but have been assessed in developing palatal taste buds and those of the CVP. *Ascl1* expression initiates after that of Shh and *Prox1* in palatal taste placodes (Nakayama et al., 2008), suggesting a role in cell fate determination, rather than in the initial specification of taste bud precursors. Consistent with this hypothesis, cells expressing AADC, a marker of Type III taste cells, are absent in Ascl1 knockout embryos examined immediately prior to birth, whereas gustducin–immunoreactive Type II taste cells are present, and appear in normal numbers in mutant soft palate taste buds (Seta et al., 2011). Thus, Ascl1 appears to function in the selection of Type III taste cell fate in rodents, a result consistent with those from studies of taste bud cell fate decisions in zebrafish embryos (Kapsimali et al., 2011). Specifically in zebrafish, *Ascl1-/-* mutants fail to develop serotonin+ taste cells, which are likely homologous to the Type III cells of mammals, but Calb2b+ taste cells (likely Type II cells) do form, and may even be increased in number in Ascl1 mutants.

Hes1 is another bHLH transcription factor which is transcriptional target of the Notch pathway. Hes1 expression has only been reported for the posterior CVP where it localizes to the epithelium of the papilla late in embryonic development (Ota et al., 2009; Seta et al., 2003), again consistent with a function in taste bud cell lineage decisions. In fact, embryos that lack Hes1 have an overabundance of Type II taste cells at birth, compared to wild type or heterozygous null littermates, suggesting that Hes1 acts

in a dose-dependent manner to repress selection of Type II fate (Ota et al., 2009). This hypothesis is particularly attractive as these authors further showed via chromatin immunoprecipitation analysis, that Hes1 binds to regulatory sites upstream of well known Type II taste cell markers, IP3R3 and PLCß2 (Ota et al., 2009).

Most recently, the homeodomain transcription factor, Skn-1a (Pou2f3), has been shown to be required cell autonomously for Type II taste cell fate (Matsumoto et al., 2011). Specifically, Skn-1a is expressed in taste cells responsive to sweet, bitter and umami, and Skn-1a$^{-/-}$ mice fail to form type II taste cells. Instead, in the absence of Skn-1a, the number of type III cells is expanded.

33.6 FORMATION OF THE GUSTATORY SENSORY GANGLIA

Gustatory neurons are contained within the geniculate, petrosal, and nodose cranial ganglia. These neurons project into the tongue and palate to innervate taste buds and into the hindbrain to innervate specific neurons within the nucleus of the solitary tract (NTS). Cranial ganglia tend to have mixed embryonic origins, with some neurons arising from neural crest and others from epibranchial placodes, which are transient ectodermal thickenings. All cranial placodes, including the epibranchial placodes, arise from a common panplacodal primordium (Schlosser, 2006; Schlosser, 2010). Multiple transcription and signaling factors regulate panplacodal primordium formation and specify the epibranchial placodes that will form the geniculate, petrosal, and nodose ganglion (recent reviews include Ladher et al., 2010; McCabe and Bronner-Fraser, 2009; Schlosser, 2010). Until recently, it was unclear whether the gustatory ganglia also contained neural crest-derived neurons in addition to placodal-derived neurons. However, recent fate mapping experiments in salamanders revealed that gustatory neurons are derived entirely from epibranchial placodes and that neural crest-derived neurons within the gustatory ganglia are strictly somatosensory (Harlow and Barlow, 2007). Furthermore, the bulk of somatosensory neurons within the geniculate ganglion of mice are placodal in origin (Harlow et al., 2011; Quina et al., 2012; Yang et al., 2008). Genetic labeling of ectoderm specifically demonstrated that geniculate neurons are of placodal origin (Harlow and Barlow, 2007), while genetically labeled neural crest cells (using Wnt1-Cre) differentiate primarily into glial cells (Harlow et al., 2011). There are a few neural crest-derived neurons within the geniculate, and while they are likely somatosensory neurons as in the axolotl, they are insufficient to account for most of the geniculate somatosensory neurons. Interestingly, the reason neural crest cells do not form neurons in the geniculate ganglion is because neuronal differentiation is suppressed by the transcription factor Homeobox A2 (Hoxa2), which is specifically in the cells derived from neural crest (Yang et al., 2008). Removal of Hoxa2 results in a substantial increase in neural crest-derived neurons in the geniculate. Furthermore, these neurons express other markers typical of trigeminal somatosensory neurons (i.e., Neurogenin-1 and TrkA). Together, these findings indicate that gustatory neurons are derived solely from epibranchial placodes and that Hoxa2 expression suppresses neural crest cells from becoming somatosensory neurons in gustatory ganglia.

To form the gustatory ganglia, epibranchial placode cells differentiate into neuroblasts and subsequently delaminate, migrate, and coalesce (Baker and Bronner-Fraser, 2001). In chick, placodally derived neurons differentiate en route and arrive at their final location as differentiated neurons (Blentic et al., 2011). This means that all gustatory neurons start out in placodes and migrate to their ultimate destination. This is very different than neural crest-derived dorsal root ganglia (DRG), where migrating progenitors populate the ganglia after their arrival (Carr and Simpson, 1978; Krispin et al., 2010). This difference requires that placodally derived neurons to migrate to the ganglion for a much longer time than is required for neural crest migration. The path that gustatory neurons take during their inward migration is determined by neural crest cells (Begbie and Graham, 2001). Specifically, neural crest cells from rhombomere 2 normally facilitate the inward migration of the geniculate ganglion toward the hindbrain (Begbie and Graham, 2001). However, the migration path of these neurons appears to be plastic; genetic deletion of the rhombomere 2 neural crest population in double mutant $Hoxa1^{-/-}/Hoxb1^{-/-}$ embryos does not prevent geniculate ganglion cell migration. Instead, geniculate neurons follow a slightly different path and merge with the trigeminal ganglion. This indicates that in the absence of rhombomere 2 neural crest cells, geniculate ganglia follow rhombomere 4 neural crest cells, which is the same path used by the trigeminal ganglion. When placodally derived neurons aggregate, adherens junctions form between the cells (Shiau and Bronner-Fraser, 2009). N-cadherin is required for normal placodal neuron aggregation, and the location and distribution of N-cadherin appear to be controlled by Slit1-Robo2 signaling. Based on these findings, it appears that gustatory ganglia are formed from neurons that migrate from epibranchial placodes following a path delineated by neural crest cells and then coalesce under N-cadherin control.

Neuronal differentiation within the gustatory ganglia is controlled by transcription factors, including neurogenin 2 (Neurg2). In the absence of Neurog2, placodal neuroblasts that give rise to the geniculate and petrosal ganglia fail to delaminate, migrate, and express neural differentiation markers (Fode et al., 1998; Takano-Maruyama et al., 2012). Several transcription factors, including Islet 1 and Tlx3, are expressed in most sensory neurons and may be important for specifying sensory neuronal fate in developing gustatory neurons (D'Autreaux et al., 2011; Logan et al., 1998; Sun et al., 2008). The geniculate ganglion contains both a visceral sensory subpopulation that innervates taste buds and a somatic subpopulation, which innervates the pinna of the ear. Recently, it has been demonstrated that the transcription factor Phox2b is expressed in roughly half of neurons in the geniculate ganglion, and the transcription factor Brn3a (Pou4f1) is expressed in the other half (D'Autreaux et al., 2011; Quina et al., 2012); Figure 33.2A). Based on the expression of

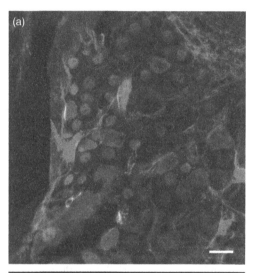

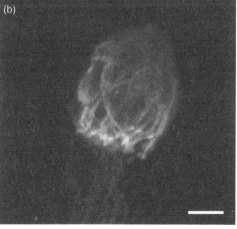

Figure 33.2 (a) An adult geniculate ganglion labeled with anti-Brn3a (green) and Phox2b-Cre; tdtomato (red). Although the populations are somewhat mixed the Brn3a positive nuclei appear to be primarily on one side of the ganglion, while the Phox2b positive neurons are concentrated on the other side of the geniculate ganglion. (b) An adult taste bud showing Phox2b positive (red) nerve fibers innervating an anti-troma 1 labeled taste bud. Scale bar in A = 20 μm, scale bar B = 10 μm. (*See plate section for color version.*)

these two factors in other ganglia and in the CNS, it is likely that Phox2b defines and regulates gustatory neuron formation, whereas the somatic transcription factor Brn3a is expressed in the somatic subpopulation. Consistent with this hypothesis, neurons that express Phox2b during development innervate taste buds in adulthood (Figure 33.2B). It was originally reported that in the absence of Phox2b, the geniculate and/or petrosal ganglia degenerate (Dauger et al., 2003). However, more recent studies report that these ganglia are not lost; instead, Brn3a is up-regulated in the visceral neurons of the geniculate, nodose, and petrosal ganglia, which stimulates differentiation of somatosensory neuron characteristics at the expense of gustatory characteristics (D'Autreaux et al., 2011). Therefore, it could be that Phox2b specifies visceral neurons by suppressing Brn3a. It is unclear whether these neurons also change their projection path and innervate somatosensory targets instead of taste buds in the absence of Phox2b. There also appears to be some overlap in the Phox2b and Brn3a neuron subpopulations, and it is unclear whether these neurons are taste or somatosensory. Another variant homeodomain factor (Hmx1) is expressed in many but not all of Brn3a-positive neurons (Quina et al., 2012). Hmx1 appears to be important for the development of the somatosensory neurons that project through the posterior auricular nerve to innervate the ear pinna (Quina et al., 2012). Therefore, it is possible that the Brn3a-positive, Hmx1-negative neurons are also Phox2b-positive and gustatory in nature. In conclusion, Phox2b influences the differentiation of epibranchial placodal sensory neurons into visceral neurons, and co-expression of Brn3a and Hmx1 promote somatosensory neuron development in the same ganglia.

33.7 REGULATION OF GUSTATORY NEURON NUMBER

Waves of apoptosis and proliferation delineate epibranchial placodes and facilitate their formation (Washausen et al., 2005). As mentioned earlier, recent evidence in chick indicates that unlike the DRG, all gustatory neurons are produced in the epibranchial placodes and migrate into the developing geniculate and petrosal ganglia as differentiated neurons. If this is also true in rodents, then dividing cells in the geniculate (peak division is E12 in rat and approximately E10-E11 in mice) would primarily be glia (Altman and Bayer, 1982). Consistent with this idea is the finding that no ganglion cells double-labeled with BrdU and anti-neurofilament 150, which marks neuronal precursors, can be found in the geniculate ganglion at E11.5 or E13.5 (Patel et al., 2010; Patel and Krimm, 2012). However, cell death occurs over a very long period in the developing rodent without any corresponding decrease in total neuron

number (Carr et al., 2005). This finding would indicate that neurons are continually being added to the ganglia, even in late embryonic development. Perhaps neurons are still migrating from placodes at this time, but neuronal proliferation seems more likely. Therefore, whether the rodent gustatory ganglia acquire additional neurons by proliferation following migration from placodes within the ganglia is still an issue to be resolved.

The final number of neurons within the gustatory ganglion that are available to innervate taste buds is determined by proliferation and migration of neural precursors, as well as cell death within the ganglia. Cell death in the geniculate ganglion appears to peak at two time points: before target innervation (E12.5 in rats and E10-E11 in mice) and during target innervation (E16.5 in rats and E14-E15 in mice) (Carr et al., 2005). Neuron survival within the gustatory ganglia is regulated by neurotrophins. Mice lacking brain-derived neurotrophin (BDNF) and neurotrophin 4 (NT4) lose approximately half of their geniculate ganglion and nodose/petrosal ganglion complex during development (Conover et al., 1995; Erickson et al., 1996; Liebl et al., 1997; Liu et al., 1995; Patel and Krimm, 2010). Hybrid $Bdnf^{-/-}/Ntf4^{-/-}$ animals lose 90–94% of their geniculate and nodose/petrosal neurons and also more taste buds than either group alone (Ito and Nosrat, 2009; Patel and Krimm, 2010). Mice lacking the primary receptor for both BDNF and NT4, TrkB, also lose most of the geniculate ganglion early in development (Fritzsch et al., 1997). In addition to BDNF and NT4, neurotrophin 3 (NT3) also regulates neuronal survival in gustatory ganglia. Mice lacking NT3 lose approximately 47% of neurons in the geniculate ganglion and 44% in the nodose/petrosal ganglion complex (Liebl et al., 1997), and mice lacking both NT3 and BDNF lose more taste buds than BDNF mutants (Ito et al., 2010). Thus, at least three different neurotrophins regulate the number of neurons within the gustatory ganglia.

The timing of neurotrophin dependence differs among the neurotrophins and may be indicative of their individual mechanisms (Farinas et al., 1998; Farinas et al., 1996; Liebl et al., 1997). In the mouse geniculate ganglia, neurons become NT4-dependent on E11.5 and BDNF-dependent at E13.5 (Patel and Krimm, 2010; Patel and Krimm, 2012). Because gustatory axons have reached their targets by E13.5, but not by E11.5, BDNF is a target-derived neurotrophin, while NT4 is not. However, BDNF is also expressed in the geniculate ganglion and may support another aspect of gustatory neuron development in this location (Harlow et al., 2011; Huang and Krimm, 2010). While neuron survival is one possibility, ganglionic BDNF seems more likely to influence neuron differentiation, because the loss of BDNF specifically in the geniculate ganglion appears to slow gustatory neuron development (Harlow et al., 2011). BDNF and NT4 function at different but overlapping times to support gustatory neuron survival. Specifically, NT4 dependence continues until E16.5, and BDNF dependence continues through E18.5 (Patel and Krimm, 2010; Patel and Krimm, 2012). Therefore, neurons do not switch from being NT4-dependent to being BDNF-dependent. It is also likely that the neuronal subpopulations supported by NT4 and BDNF are overlapping, because each neurotrophin supports neurons innervating the tongue and the palate to the same degree (Patel et al., 2010). These two neurotrophins bind to the same receptors: TrkB and p75. It may be that instead of separate NT4- and BDNF-dependent subpopulations, some gustatory neurons are TrkB-ligand sensitive and die if either neurotrophin is removed, while others are TrkB-ligand insensitive and only die when both TrkB ligands are removed. Interestingly, although much of the functions of NT4 and BDNF appear to be overlapping, they support neuron survival through different mechanisms. BDNF prevents cell death by inhibiting activated caspase 3 activation, whereas NT4 does not prevent activated caspase 3 activation and instead prevents cell death through an unknown mechanism (Patel and Krimm, 2010; Patel and Krimm, 2012). Therefore, it appears that gustatory neurons do not simply switch dependence from one neurotrophin to another; rather, they have overlapping dependence on these factors, which regulate cell survival via different mechanisms.

33.8 DEVELOPMENT OF THE GUSTATORY INNERVATION OF THE TONGUE

Neurite extension occurs as neuronal precursors become post-mitotic and begin to differentiate. *In vitro*, geniculate axon outgrowth requires the addition of an exogenous neurotrophin. BDNF, NT4/5, and GDNF are all capable of supporting neurite outgrowth, but NT3 and NGF are not (Rochlin and Farbman, 1998; Runge et al., 2012). However, axons *in vivo* still reach the tongue in the absence of both BDNF and NT4 (Ma et al., 2009), indicating that neurotrophins redundantly support axon growth *in vivo*. To innervate the tongue, gustatory axons must navigate from the geniculate ganglion to the lingual epithelium of the dorsal tongue and the petrosal ganglion to the CV placode. Gustatory axons grow into the tongue as it develops (Mbiene, 2004; Rochlin and Farbman, 1998). Axons of the chorda tympani reach the epithelial surface as early as E11.5 (in mouse), and axonal endings are both lateral and caudal to the fungiform papillae they will eventually innervate (Patel and Krimm, 2012; Rochlin and Farbman, 1998; Rochlin et al., 2000). From E11.5 to E13.5, axons grow from lateral/caudal to rostral/medial as the tongue expands. Gustatory axon pathways are precisely restricted from the time of tongue formation through papillae morphogenesis,

resulting in distinctive lingual territories for each nerve (Mbiene and Mistretta, 1997). By E13.5, chorda tympani axons have reached the fungiform placodes but have not innervated the epithelial surface. From E14 to E15, fibers defasciculate and penetrate the epithelial surface, forming a distinctive ending called a neural bud (Lopez and Krimm, 2006b). The progression of neural bud formation occurs from caudal to rostral on the tongue surface, and epithelial surface innervation is complete by E15.5 (Figure 33.3). Innervation of the CVP placode occurs at roughly the same time as fungiform placode innervation.

Because chorda tympani axons follow precise, spatially restricted pathways to the tongue (Mbiene and Mistretta, 1997), and because initial targeting is so accurate (Lopez and Krimm, 2006b), a series of molecular cues from the environment must guide these axons to the lingual epithelium. Two families of factors that mediate target innervation in the taste system are the semaphorins and the neurotrophins, which are repulsive and attractive, respectively. The semaphorin, Sema3A, is expressed in the developing tongue (Giger et al., 1996) in a gradient that decreases from medial to lateral across the tongue surface (Rochlin and Farbman, 1998, Rochlin et al., 2000). Because chorda tympani innervation progresses from lateral to medial, Sema3A is in the appropriate location to prevent premature and aberrant growth of trigeminal and gustatory fibers into the mid-region of the tongue. Consistent with this idea, sensory axons in $Sema3a^{-/-}$ mutants prematurely reach the epithelium (Dillon et al., 2004). In addition, as geniculate axons near the epithelial surface, Sema3A prevents premature penetration of the epithelium (Dillon et al., 2004; Vilbig et al., 2004). Another member of this family, Sema3F, is also expressed by lingual epithelium, although its function remains unclear (Vilbig et al., 2004).

The neurotrophin, BDNF, is the only specific chemoattractant that has been identified for gustatory neuron development. It is expressed in developing fungiform placodes and circumvallate before they are innervated (Nosrat et al., 1996,2001; Nosrat and Olson, 1995: Figure 33.4A). Gustatory axons will grow toward a BDNF source (Hoshino et al., 2010). The neurotrophin, NT4, which shares the same receptors as BDNF, is also chemoattractive (Runge et al., 2012), but is not specifically expressed in fungiform palcodes during development (Huang and Krimm, 2010; Nosrat et al., 1996; Nosrat and Olson, 1995). Overexpression of BDNF in inappropriate locations disrupts the normal gustatory innervation pattern to the tongue surface (Krimm et al., 2001; Lopez and Krimm, 2006a; Ringstedt et al., 1999). Specifically, when BDNF is overexpressed in the entire tongue epithelium, axons preferentially innervate the non-taste filiform papillae-containing regions of the tongue, which are ectopically expressing BDNF, instead of gustatory regions (Lopez and Krimm, 2006a). Removal of functional BDNF also disrupts innervation to the tongue

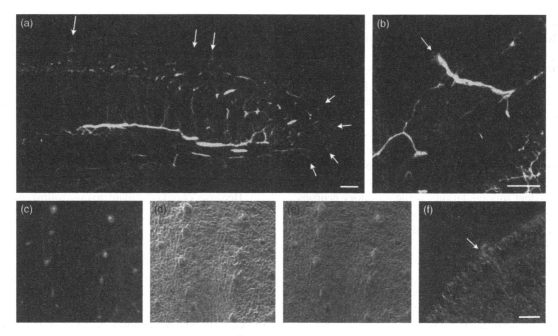

Figure 33.3 (a) A section though an E16.5 mouse tongue in which innervation is labeled with anti-beta tubulin (green) and P2X3 (red). Branches from the chorda tympani can be seen branching toward the dorsal tongue surface. The epithelium is innervated in specific regions by P2X3 positive nerve bundles (arrows), which can be seen at higher power in (b). A region of the E16.5 tongue can be viewed from the dorsal surface in C, D, and E. DiI-labeled fiber bundles innervate the tongue surface and form a "neural bud" in (c), a Scanning Electron Microscope (SEM) image of the same region (d) can be aligned with the SEM (e) to determine which papillae are innervated and which are not innervated. A section through the DiI-labeled tongue illustrates that "neural buds" for where fibers innervate the epithelial surface. The scale bar in A = 200 μm, the scale bar in b and f = 50 μm. The scale bar in f also applies to c, d, and e. (*See plate section for color version.*)

surface (Figure 33.4B, 33.4C). Mice lacking BDNF show aberrant innervation patterns with increased branching (Ma et al., 2009). Furthermore, there is a critical period for this effect. When BDNF is conditionally removed from the lingual epithelium under the control of the Keratin 14 promoter, innervation to the tongue tip is disrupted, similar to what is observed in *Bdnf* null mutants. However, innervation to the mid-region and back of the tongue is normal. This is because Keratin 14 expression develops in a rostral to caudal fashion across the tongue surface, which prevents *Bdnf* expression in the tongue tip by E13.5 but not in the more caudal region of the tongue until E14.5. Meanwhile, innervation develops first in the caudal portion of the tongue and proceeds rostrally and medially as the tongue develops. Therefore, BDNF is removed from the tongue tip before fungiform placodes are innervated, whereas the more caudal part of the tongue is innervated before BDNF is removed by K14-Cre mediated gene recombination. These findings demonstrate that although BDNF is expressed in multiple tissues, it is specifically BDNF in the lingual epithelium in developing fungiform placodes that is important for target innervation. This study also demonstrates that BDNF removal must occur before chorda tympani fibers innervate the fungiform placodes for targeting to be disrupted. Taken together, these studies show that BDNF is both necessary and sufficient for inducing chorda tympani axons to innervate their appropriate targets during development.

Although the neurotrophin BDNF is well established as an important chemoattractant required for gustatory axons to locate and innervate the fungiform placodes, other unknown chemoattractive factors must play roles. Gustatory axons have no difficulty reaching the tongue when both BDNF and NT4 are removed *in vivo* (Ma et al., 2009), and the *in vitro* chemoattractive properties of the mammalian tongue are not affected by the presence of BDNF or NT4 (Vilbig et al., 2004). These studies demonstrate that neither of these TrkB-ligands are required to guide axons to the tongue. Therefore, it is likely that another chemoattractant present in tongue is involved. Both the mammalian tongue (Vilbig et al., 2004) and the axolotl oropharyngeal endoderm (Gross et al., 2003) are chemoattractive. In addition to the tongue, developing gustatory regions probably also produce an additional chemoattractant that is not a neurotrophin. In the absence of BDNF a few gustatory axons successfully innervate fungiform placodes (Ma et al., 2009). However, this does not occur until approximately E18.5, which is four days after these neurons would normally innervate the correct target. It is clear that navigating axons from the ganglion to their

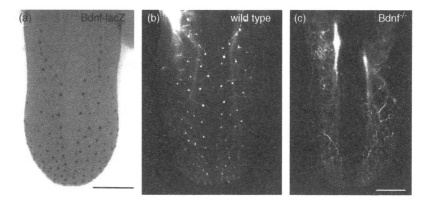

Figure 33.4 (a) Beta-galactosidase-staining in of a E14.5 Bdnf-lacZ mouse tongue illustrates the pattern of Bdnf expression in the tongue during embryonic development. Bdnf is specifically expressed in the fungiform placodes, but not the surrounding epithelium. This pattern of expression is important for chorda tympani axons to correctly innervate the tongue surface. The normal pattern of innervation to the tongue surface is illustrated in (b) at E16.5. In mice lacking Bdnf, axons fail to innervate specific regions of the tongue and form a neural bud (c), instead there is a great deal of branching to the surface of the tongue. Scale bar = 500 μm, the scale bar in c also applies to c.

proper gustatory targets is a complicated developmental task, which likely involves many different guidance cues.

Once the gustatory epithelium is innervated, the taste system continues to develop. Taste buds differentiate and continue to grow in size, and these processes are influenced in part by geniculate neurons. During postnatal development, taste buds grow to a size that can be predicted by the number of neurons that innervated them at an earlier age (Krimm and Hill, 2000). Because gustatory nerve fibers are required to maintain a full complement of taste buds, even in adulthood (Guth, 1957; Nagato et al., 1995; Oakley et al., 1990), it seems likely that factors released from nerve fibers influence the postnatal growth of taste buds. The peripheral taste system is also plastic, in that while the number of circumvallate taste buds increases through a long period of postnatal development (Hosley and Oakley, 1987), the number of fungiform taste buds concomitantly decreases (Liebl et al., 1999; Patel and Krimm, 2012). It is not clear what causes these changes in taste bud number, but one possibility is a change in innervation. Petrosal neurons innervating the CVP may show increased branching, while geniculate neurons innervating fungiform taste buds may lose branches postnatally. These alterations could influence the number of taste buds. Consistent with this hypothesis, sheep geniculate neurons have a postnatal reduction in branching (Mistretta et al., 1988; Nagai et al., 1988). The mechanisms for this phenomenon are unclear, but it could be regulated at least in part by BDNF. At birth, BDNF is present in all taste bud cells (Huang and Krimm, 2010), but it becomes restricted to a subset of taste cells during postnatal development (Yee et al., 2003). It is possible that this normal reduction in BDNF results in reduced taste bud innervation. Consistent with this scenario, BDNF overexpression in the taste cells that would normally lose BDNF expression postnatally, results in more innervation and slightly larger taste buds in the adulthood (Nosrat et al., 2012). Interestingly, the increase in taste bud size was primarily due to increases in one subpopulation of taste cells: an NCAM-positive subpopulation that also normally expresses BDNF (Yee et al., 2003), rather than the gustducin-positive taste cells in which BDNF was overexpressed (Nosrat et al., 2012). This indicates that the postnatal role of BDNF is taste cell specific. Whether this is a direct effect of BDNF on taste cells or whether it is due to the increased taste bud innervation observed in BDNF overexpressors is still unclear.

33.9 DEVELOPMENT OF GUSTATORY AXON PROJECTIONS TO THE CNS

In addition to innervating taste buds during development, primary gustatory neurons project to specific locations in the CNS. Central branches of gustatory axons from the geniculate, petrosal, and nodose ganglia terminate in the rostral portion of the nucleus of the solitary tract (NST). The terminal field for petrosal neurons is caudal to that of the geniculate neurons, although some overlap exists. The terminal field begins to form by E15, when geniculate ganglion axons invade the rat NST (Zhang and Ashwell, 2001). The first synaptic thickenings in the rostral NST are detectable at E17, and synaptic vesicles are observed by E19 (Zhang and Ashwell, 2001). Interestingly by E14.5, before synapses are structurally mature, the postsynaptic cell in the NTS develops voltage gated channels and has functional neurotransmitter receptors for both Glutamate and GABA (Suwabe et al., 2012). These receptors may initially respond to paracrine stimulation, but their presence early in development indicates that functional connections can be made as soon as afferent fibers enter the rostral NTS and before synapses show a mature morphology. As a result, in addition to extrinsic growth factors and intrinsic transcription factors, activity may shape the initial formation of these circuits.

Virtually nothing is known about the factors that coordinate the formation of central connections in the taste solitary nucleus. However, the transcription factor Phox2b may play a role. The central projections of Phox2b-expressing gustatory neurons presumably contact Phox2b-expressing NST neurons (D'Autreaux et al., 2011; Dauger et al., 2003). As with peripheral neurons, NST development is also disrupted in the absence of Phox2b (D'Autreaux et al., 2011; Dauger et al., 2003), which suggests that Phox2b may coordinate the initial development of peripheral and central gustatory connections.

The terminal fields of all three gustatory nerves continue to grow through postnatal day 15 (PN15) and then decrease in size (Mangold and Hill, 2008; Sollars et al., 2006). In fact, there is almost total overlap of the chorda tympani (CT), greater superficial petrosal (GSP), and glossopharyngeal terminal (IX) fields in rat on PN15. Each field decreases in size sometime after PN15 to occupy a distinctive area in the NST, and the zone of overlap is considerably reduced (Mangold and Hill, 2008). The age at which the terminal fields reach their mature size varies for the different primary afferents. Both the GSP and IX nerve fields reach maturity at PN25, but the chorda tympani is not fully mature until PN35 (Mangold and Hill, 2008). This postnatal refinement of terminal field size is likely due to competition between terminal fields for the same synaptic space (Corson and Hill, 2011). The chorda tympani terminal field is refined by the elimination branches in one NST subdivision (the rostrolateral region), while branches in another subdivision (the rostrocentral regions) are stable (Wang et al., 2012). In addition to terminal field refinement, the number of chorda tympani synapses onto GABAergic NST targets is also reduced between postnatal days 15 and 25 of development (Wang et al., 2012). On the postsynaptic cell side, dendrites acquire spines during early postnatal development (Suwabe et al., 2011). There are also postnatal changes in NST neuron dendritic fields (Lasiter et al., 1989; Liu et al., 2000; Suwabe et al., 2011). It is unclear whether changes in the terminal fields influence dendritic arbor development or if changes in dendritic arbors reduce available synaptic space for developing axons during postnatal development.

Postnatal development of the gustatory system is heavily influenced by environmental factors. For example, maternal sodium restriction during development specifically blocks the development of chorda tympani nerve responses to sodium (Hill and Przekop, 1988) and results in enlarged chorda tympani and glossopharyngeal nerve terminal fields (King and Hill, 1991; Krimm and Hill,

1997; May and Hill, 2006; Sollars et al., 2006). Furthermore, the density of chorda tympani arbors and synapses is also greater in sodium-restricted animals (May et al., 2008). Because sodium restriction also results in a loss of sodium-specific activity in developing nerve fibers (Hill and Przekop, 1988), it is likely that a reduction in nerve activity is responsible for preventing gustatory terminal field refinement under the conditions of developmental sodium restriction.

Unfortunately, understanding the mechanism of sodium deprivation on terminal field development has been complicated by the fact that sodium deprivation has multiple effects during development. A long period of sodium deprivation results in a loss of sodium driven activity in the chorda tympani nerve. However, a brief period of embryonic deprivation (E3-E12) results in an enlarged chorda tympani field later in development (Krimm and Hill, 1997; Mangold and Hill, 2008). This occurs even though the CT shows normal levels of activity. Furthermore, this effect is not specific for sodium deprivation; protein deprivation during development can also lead to enlarged gustatory terminal fields (Thomas and Hill, 2008). Embryonic dietary restriction (E3-E12), which does not disrupt CT activity, does not impact chorda tympani field development in the same way as dietary sodium restriction throughout development (E3-PN28), which does disrupt CT activity (Mangold and Hill, 2007). Surprisingly, early dietary restriction has a much greater effect on gustatory terminal field size than chronic dietary restriction (Mangold and Hill, 2007). It is possible that the timing of sodium repletion at E12 causes the development of an enlarged terminal field during the postnatal period. Sodium repletion occurring during a critical developmental phase could influence the expression of specific growth factors, which in turn could impact gustatory terminal field development. Manipulations that separate the roles of activity versus growth factors on the postnatal development of the rostral NST terminal field, would contribute considerably to our understanding of how dietary manipulations and/or taste sensory function influence development of the rostral NST.

REFERENCES

Ahmed, M., Xu, J., and Xu, P.X. (2012). EYA1 and SIX1 drive the neuronal developmental program in cooperation with the SWI/SNF chromatin-remodeling complex and SOX2 in the mammalian inner ear. *Development* 139: 1965–1977.

AhPin, P., Ellis, S., Arnott, C., and Kaufman, M. H. (1989). Prenatal development and innervation of the circumvallate papilla in the mouse. *J. Anat.* 162: 33–42.

Altman, J., and Bayer, S. (1982). Development of the cranial nerve ganglia and related nuclei in the rat. *Adv. Anat. Embryol. Cell Biol.* 74: 1–90.

Andl, T., Ahn, K., Kairo, A., et al. (2004). Epithelial Bmpr1a regulates differentiation and proliferation in postnatal hair follicles and is essential for tooth development. *Development* 131: 2257–2268.

References

Angers, S., and Moon, R. T. (2009). Proximal events in Wnt signal transduction. *Nat. Rev. Mol. Cell Biol.* 10: 468–477.

Baker, C. V., and Bronner-Fraser, M. (2001). Vertebrate cranial placodes I. Embryonic induction. *Dev. Biol.* 232: 1–61.

Barlow, L. A. (2000a). Gustatory Development. In *Neurobiology of Taste and Smell*, ed. T. E. Finger, W. Silver, D. Restrepo, pp. 393–422: Wiley & Sons.

Barlow, L. A. (2000b). Taste buds in ectoderm are induced by endoderm: Implications for mechanisms governing taste bud development. In *Regulatory Processes in Development: The Legacy of Sven Hörstadius. Proceedings of the Wenner-Gren International Symposium.*, ed. L. Olsson, C-O. Jacobson, pp. 185–190: Portland Press.

Barlow, L. A. (2001). Specification of pharyngeal endoderm is dependent on early signals from axial mesoderm. *Development* 128: 4573–4583.

Barlow, L. A., Chien, C.-B., and Northcutt, R. G. (1996). Embryonic taste buds develop in the absence of innervation. *Development* 122: 1103–1111.

Barlow, L. A., and Northcutt, R. G. (1995). Embryonic origin of amphibian taste buds. *Dev. Biol.* 169: 273–285.

Barlow, L. A., and Northcutt, R. G. (1997). Taste buds develop autonomously from endoderm without induction by cephalic neural crest or paraxial mesoderm. *Development* 124: 949–957.

Begbie, J., and Graham, A. (2001). Integration between the epibranchial placodes and the hindbrain. *Science* 294: 595–598.

Beites, C. L., Hollenbeck, P. L., and Kim, J., et al. (2009). Follistatin modulates a BMP autoregulatory loop to control the size and patterning of sensory domains in the developing tongue. *Development* 136: 2187–2197.

Beites, C. L., Kawauchi, S., Crocker, C. E., and Calof, A. L. (2005). Identification and molecular regulation of neural stem cells in the olfactory epithelium. *Exp. Cell Res.* 306: 309–316.

Bejsovec, A. (2005). Wnt pathway activation: new relations and locations. *Cell* 120: 11–14.

Blentic, A., Chambers, D., Skinner, A., et al. (2011). The formation of the cranial ganglia by placodally-derived sensory neuronal precursors. *Mol. Cell Neurosci.* 46: 452–459.

Bray, S. J. (2006). Notch signalling: a simple pathway becomes complex. *Nat. Rev. Mol. Cell Biol.* 7: 678–689.

Buckingham, M. E., and Meilhac, S. M. (2011). Tracing cells for tracking cell lineage and clonal behavior. *Dev. Cell* 21: 394–409.

Calof, A. L., Bonnin, A., Crocker, C., et al. (2002). Progenitor cells of the olfactory receptor neuron lineage. *Microsc. Res. Tech.* 58: 176–188.

Carr, V. M., and Simpson, S. B., Jr., (1978). Proliferative and degenerative events in the early development of chick dorsal root ganglia. I. Normal development. *J. Comp. Neurol.* 182: 727–739.

Carr, V. M., Sollars, S. I., and Farbman, A. I. (2005). Neuronal cell death and population dynamics in the developing rat geniculate ganglion. *Neuroscience* 134: 1301–1308.

Cau, E., Casarosa, S., and Guillemot, F. (2002). Mash1 and Ngn1 control distinct steps of determination and differentiation in the olfactory sensory neuron lineage. *Development* 129: 1871–1880.

Cervantes, S., Yamaguchi, T. P., and Hebrok, M. (2009). Wnt5a is essential for intestinal elongation in mice. *Dev. Biol.* 326: 285–294.

Chai, Y., Jiang, X., Ito, Y., et al. (2000). Fate of the mammalian cranial neural crest during tooth and mandibular morphogenesis. *Development* 127: 1671–1679.

Chaudhari, N., and Roper, S. D. (2010). The cell biology of taste. *J. Cell Biol.* 190: 285–296.

Chen, B., Kim, E. H., and Xu, P. X. (2009). Initiation of olfactory placode development and neurogenesis is blocked in mice lacking both Six1 and Six4. *Dev. Biol.* 326: 75–85.

Chuong, C. M., Patel, N., Lin, J., et al. . (2000). Sonic hedgehog signaling pathway in vertebrate epithelial appendage morphogenesis: perspectives in development and evolution. *Cell Mol. Life Sci.* 57: 1672–1681.

Conover, J. C., Erickson, J. T., Katz, D. M., et al. (1995). Neuronal deficits, not involving motor neurons, in mice lacking BDNF and/or NT4. *Nature* 375: 235–238.

Cook, M. H., and Neal, H. V. (1921). Are the taste buds of elasmobranchs endodermal in origin? *J. Comp. Neurol.* 33: 45–63.

Corson, S. L., and Hill, D. L. (2011). Chorda tympani nerve terminal field maturation and maintenance is severely altered following changes to gustatory nerve input to the nucleus of the solitary tract. *J. Neurosci.* 31: 7591–7603.

D'Amico-Martel, A. (1982). Temporal patterns of neurogenesis in avian cranial sensory and autonomic ganglia. *Am. J. Anat.* 163: 351–372.

D'Autreaux, F., Coppola, E., Hirsch, M. R.,et al.. (2011). Homeoprotein Phox2b commands a somatic-to-visceral switch in cranial sensory pathways. *Proc. Natl. Acad. Sci. U. S. A.* 108: 20018–20023.

Dabdoub, A., Puligilla, C., Jones, J. M., et al. (2008). Sox2 signaling in prosensory domain specification and subsequent hair cell differentiation in the developing cochlea. *Proc. Natl. Acad. Sci. U. S. A.* 105: 18396–18401.

Danielian, P. S., Muccino, D., Rowitch, D. H.,et al. (1998). Modification of gene activity in mouse embryos in utero by a tamoxifen-inducible form of Cre recombinase. *Curr. Biol.* 8: 1323–1326.

DasGupta, R., and Fuchs, E. (1999). Multiple roles for activated LEF/TCF transcription complexes during hair follicle development and differentiation. *Development* 126: 4557–4568.

Dauger, S., Pattyn, A., Lofaso, F., et al. (2003). Phox2b controls the development of peripheral chemoreceptors and afferent visceral pathways. *Development* 130: 6635–6642.

Diamond, I., Owolabi, T., Marco, M., et al. (2000). Conditional gene expression in the epidermis of transgenic mice using the tetracycline-regulated transactivators tTA and rTA linked to the keratin 5 promoter. *J. Invest. Dermatol.* 115: 788–794.

Dillon, T. E., Saldanha, J., Giger, R., et al. (2004). Sema3A regulates the timing of target contact by cranial sensory axons. *J. Comp. Neurol.* 470: 13–24.

Doe, C. Q., Chu-LaGraff, Q., Wright, D. M., and Scott, M. P. (1991). The prospero gene specifies cell fates in the Drosophila central nervous system. *Cell* 65: 451–464.

Dyer, M. A. (2003). Regulation of proliferation, cell fate specification and differentiation by the homeodomain proteins Prox1, Six3, and Chx10 in the developing retina. *Cell Cycle* 2: 350–357.

Dyer, M. A., Livesey, F. J., Cepko, C. L., and Oliver, G. (2003). Prox1 function controls progenitor cell proliferation and horizontal cell genesis in the mammalian retina. *Nat. Genet.* 34: 53–58.

El-Sharaby, A., Ueda, K., Kurisu, K., and Wakisaka, S. (2001a). Development and maturation of taste buds of the palatal epithelium of the rat: histological and immunohistochemical study. *Anat. Rec.* 263: 260–268.

El-Sharaby, A., Ueda, K., and Wakisaka, S. (2001b). Differentiation of the lingual and palatal gustatory epithelium of the rat as revealed by immunohistochemistry of alpha-gustducin. *Arch. Histol. Cytol.* 64: 401–409.

Erickson, J. T., Conover, J. C., Borday, V., et al. (1996). Mice lacking brain-derived neurotrophic factor exhibit visceral sensory neuron losses distinct from mice lacking NT4 and display a severe developmental deficit in control of breathing. *J. Neurosci.* 16: 5361–5371.

Farbman, A. I., and Mbiene, J.-P. (1991). Early development and innervation of taste bud-bearing papillae on the rat tongue. *J. Comp. Neurol.* 304: 172–186.

Farinas, I., Wilkinson, G. A., Backus, C., et al. (1998). Characterization of neurotrophin and Trk receptor functions in developing sensory ganglia: direct NT-3 activation of TrkB neurons in vivo. *Neuron.* 21: 325–334.

Farinas, I., Yoshida, C. K., Backus, C., and Reichardt, L. F. (1996). Lack of neurotrophin-3 results in death of spinal sensory neurons and premature differentiation of their precursors. *Neuron.* 17: 1065–1078.

Fode, C., Gradwohl, G., Morin, X., et al. (1998). The bHLH protein NEU-ROGENIN 2 is a determination factor for epibranchial placode-derived sensory neurons. *Neuron.* 20: 483–494.

Fritzsch, B., Sarai, P.A., Barbacid, M., and Silos-Santiago, I. (1997). Mice lacking the neurotrophin receptor trkB lose their specific afferent innervation but do develop taste buds. *Intl. J. Dev. Neurosci.* 15: 563–576.

Giger, R., Wolfer, D. P., De Wit, G. M., and Verhaagen, J. (1996). Anatomy of rat semaphorin III/collapsin-1 mRNA expression and relationship to developing nerve tracts during neuroembryogenesis. *J. Comp. Neurol.* 375: 378–392.

Gordon, M. D., and Nusse, R. (2006). Wnt signaling: multiple pathways, multiple receptors, and multiple transcription factors. *J. Biol. Chem.* 281: 22429–22433.

Graziadei, P. P., and Graziadei, G. A. (1978). Observations on the ultrastructure of ganglion cells in the circumvallate papilla of rat and mouse. *Acta. Anat. (Basel).* 100: 289–305.

Gridley, T. (1997). Notch signaling in vertebrate development and disease. *Molec. Cell. Neurosci.* 9: 103–108.

Gross, J. B., Gottlieb, A. A., and Barlow, L. A. (2003). Gustatory neurons derived from epibranchial placodes are attracted to, and trophically supported by, taste bud-bearing endoderm in vitro. *Dev. Biol.* 264: 467–481.

Guo, J., and Wu, G. (2012). The signaling and functions of heterodimeric bone morphogenetic proteins. *Cytokine Growth Factor Rev.* 23: 61–67.

Guth, L. (1957). The effects of glossopharyngeal nerve transection on the circumvallate papilla of the rat. *Anat. Rec.* 128: 715–731.

Hagedorn, L., Suter, U., and Sommer, L. (1999). P0 and PMP22 mark a multipotent neural crest-derived cell type that displays community effects in response to TGF-beta family factors. *Development* 126: 3781–3794.

Hall, J., Anderson, K., Hooper, J., and Finger, T. E. (1997). Expression of *sonic hedgehog* (*shh*) and *patched* (*ptc*) in developing taste papillae of the mouse. *Chem. Senses* 22: 692–693.

Hall, J. M., Bell, M. L., and Finger, T. E. (2003). Disruption of sonic hedgehog signaling alters growth and patterning of lingual taste papillae. *Dev. Biol.* 255: 263–277.

Hall, J. M., Hooper, J. E., and Finger, T. E. (1999). Expression of *Sonic hedgehog, Patched* and *Gli1* in developing taste papillae of the mouse. *J. Comp. Neurol.* 406: 143–155.

Hamosh, M., and Burns, W. A. (1977). Lipolytic activity of human lingual glands (Ebner). *Laboratory investigation; a journal of technical methods and pathology* 37: 603–608.

Harada, N., Tamai, Y., Ishikawa, T. et al. (1999). Intestinal polyposis in mice with a dominant stable mutation of the beta-catenin gene. *Embo. J.* 18: 5931–5942.

Harlow, D. E., and Barlow, L. A. (2007). Embryonic origin of gustatory cranial sensory neurons. *Dev. Biol.* 310: 317–328.

Harlow, D. E., Yang, H., Williams, T., and Barlow, L. A. (2011). Epibranchial placode-derived neurons produce BDNF required for early sensory neuron development. *Dev. Dyn.* 240: 309–323.

Henning, S. J. (1981). Postnatal development: coordination of feeding, digestion, and metabolism. *Am. J. Physiol.* 241: G199–214.

Hill, D. L., and Przekop, P. R. (1988). Influences of dietary sodium on functional taste receptor development: a sensitive period. *Science* 241: 1826–1827.

Höfer, D., Püschel, B., and Drenckhahn, D. (1996). Taste receptor-like cells in the rat gut identified by expression of alpha-gustducin. *Proc. Natl. Acad. Sci. USA* 93: 6631–6634.

Hoshino, N., Vatterott, P., Egwiekhor, A., and Rochlin, M. W. (2010). Brain-derived neurotrophic factor attracts geniculate ganglion neurites during embryonic targeting. *Dev. Neurosci.* 32: 184–196.

Hosley, M. A., and Oakley, B. (1987). Postnatal development of the vallate papilla and taste buds in rats. *Anat. Rec.* 218: 216–222.

Huang, T., and Krimm, R. F. (2010). Developmental expression of Bdnf, Ntf4/5, and TrkB in the mouse peripheral taste system. *Dev. Dyn.* 239: 2637–2646.

Incardona, J. P., and Roelink, H. (2000). The role of cholesterol in Shh signaling and teratogen-induced holoprosencephaly. *Cell Mol. Life Sci.* 57: 1709–1719.

Ingham, P. W., and McMahon, A. P. (2001). Hedgehog signaling in animal development: paradigms and principles. *Genes Dev.* 15: 3059–3087.

Ito, A., and Nosrat, C. A. (2009). Gustatory papillae and taste bud development and maintenance in the absence of TrkB ligands BDNF and NT-4. *Cell Tissue Res.* 337: 349–359.

Ito, A., Nosrat I. V., and Nosrat, C. A. (2010). Taste cell formation does not require gustatory and somatosensory innervation. *Neurosci. Lett.* 471: 189–194.

Itoh, N., and Ornitz, D. M. (2011). Fibroblast growth factors: from molecular evolution to roles in development, metabolism and disease. *J. Biochem.* 149: 121–130.

Iwatsuki, K., Liu, H.X., Gronder, A., et al. (2007). Wnt signaling interacts with Shh to regulate taste papilla development. *Proc. Natl. Acad. Sci. U. S. A.* 104: 2253–2258.

Jaskoll, T., Leo, T., Witcher, D., et al. (2004). Sonic hedgehog signaling plays an essential role during embryonic salivary gland epithelial branching morphogenesis. *Dev. Dyn.* 229: 722–732.

Jiang, X., Iseki, S., Maxson, R. E., et al. (2002). Tissue origins and interactions in the mammalian skull vault. *Dev. Biol.* 241: 106–116.

Jitpukdeebodintra, S., Chai, Y., and Snead, M. L. (2002). Developmental patterning of the circumvallate papilla. *Int. J. Dev. Biol.* 46: 755–763.

Johnston, J. B. (1910). The limit between the ectoderm and entoderm in the mouth, and the origin of taste buds. *I. Amphibians. Am. J. Anat.* 10: 41–67.

Jung, H. S., Oropeza, V., and Thesleff, I. (1999). Shh, Bmp-2, Bmp-4 and Fgf-8 are associated with initiation and patterning of mouse tongue papillae. *Mech. Dev.* 81: 179–182.

Kapsimali, M., and Barlow, L. A. (2012). *Developing a sense of taste. Semin Cell Dev Biol*

Kapsimali, M., Kaushik A.L., Gibon G., et al. (2011). Fgf signaling controls pharyngeal taste bud formation through miR-200 and Delta-Notch activity. *Development* 138: 3473–3484.

Kawakami, M., Umeda, M., Nakagata, N., et al. (2011). Novel migrating mouse neural crest cell assay system utilizing P0-Cre/EGFP fluorescent time-lapse imaging. *BMC Dev. Biol.* 11: 68.

Kim, J. Y., Lee, M. J., Cho, K. W., et al. (2009). Shh and ROCK1 modulate the dynamic epithelial morphogenesis in circumvallate papilla development. *Dev. Biol.* 325: 273–280.

Kim, J. Y., Mochizuki, T., Akita, K., and Jung, H. S. (2003). Morphological evidence of the importance of epithelial tissue during mouse tongue development. *Exp. Cell Res.* 290: 217–226.

King, C. T., and Hill, D. L. (1991). Dietary sodium chloride deprivation throughout development selectively influences the terminal field organization of gustatory afferent fibers projecting to the rat nucleus of the solitary tract. *J. Comp. Neurol.* 303: 159–169.

Kokrashvili, Z., Mosinger, B., and Margolskee, R. F. (2009). Taste signaling elements expressed in gut enteroendocrine cells regulate nutrient-responsive secretion of gut hormones. *Am. J. Clin. Nutr.* 90(3): 822S–825S.

Krimm, R. F., and Barlow, L. A. (2007). Development of the taste system. In *Olfaction and Taste*, ed. DV Smith, S Firestein, GK Beauchamp. San Diego: Academic Press.

Krimm, R. F., and Hill, D. L. (1997). Early prenatal critical period for chorda tympani nerve terminal field development. *J. Comp. Neurol.* 378: 254–264.

Krimm, R. F., and Hill, D. L. (2000). Neuron/target matching between chorda tympani neurons and taste buds during postnatal rat development. *J. Neurobiol.* 43: 98–106.

Krimm, R. F., Miller, K. K., Kitzman, P. H., et al. (2001). Epithelial overexpression of BDNF or NT4 disrupts targeting of taste neurons that innervate the anterior tongue. *Dev. Biol.* 232: 508–521.

Krispin, S., Nitzan, E., Kassem, Y., and Kalcheim, C. (2010). Evidence for a dynamic spatiotemporal fate map and early fate restrictions of premigratory avian neural crest. *Development* 137: 585–595.

Ladher, R. K., O'Neill, P., and Begbie, J. (2010). From shared lineage to distinct functions: the development of the inner ear and epibranchial placodes. *Development* 137: 1777–1785.

Landacre, F. L. (1907). On the place of origin and method of distribution of taste buds in Ameirus melas. *J. Comp. Neurol.* 17: 1–66.

Lasiter, P. S., Wong, D. M., and Kachele, D. L. (1989). Postnatal development of the rostral solitary nucleus in rat: dendritic morphology and mitochondrial enzyme activity. *Brain Res. Bull.* 22: 313–321.

Le Douarin, N. M., Brito, J. M., and Creuzet, S. (2007). Role of the neural crest in face and brain development. *Brain Res. Rev.* 55: 237–247.

Le Douarin, N. M., and Kalcheim, C. (1999). *The Neural Crest*. Cambridge: Cambridge University Press. 445 pp.

Lee, M. J., Kim, J. Y., Lee, S. I., et al. (2006). Association of Shh and Ptc with keratin localization in the initiation of the formation of circumvallate papilla and von Ebner's gland. *Cell Tissue Res.* 325(2): 253–261.

Liebl, D. J., Mbiene, J. P., and Parada, L. F. (1999). NT4/5 mutant mice have deficiency in gustatory papillae and taste bud formation. *Dev. Biol.* 213: 378–389.

Liebl, D. J., Tessarollo, L., Palko, M. E., and Parada, L. F. (1997). Absence of sensory neurons before target innervation in brain-derived neurotrophic factor-, neurotrophin 3-, and trkC-deficient embryonic mice. *J. Neurosci.* 17: 9113–9121.

Lin, C., Fisher, A. V., Yin, Y., et al. (2011). The inductive role of Wnt-beta-Catenin signaling in the formation of oral apparatus. *Dev. Biol.* 356: 40–50.

Liu, F., and Millar, S. E. (2010). Wnt/beta-catenin signaling in oral tissue development and disease. *J. Dent. Res.* 89: 318–330.

Liu, F., Thirumangalathu, S., Gallant, N. M., et al. (2007). Wnt-beta-catenin signaling initiates taste papilla development. *Nat. Genet.* 39: 106–112.

Liu, H. X., Grosse, A. M., Walton, K. D., et al. (2009). WNT5a in tongue and fungiform Papilla development. *Ann. N.Y. Acad. Sci.* 1170: 11–17.

Liu, H. X., Grosse, A. S., Iwatsuki, K., et al. (2012a)a. Separate and distinctive roles for Wnt5a in tongue, lingual tissue and taste papilla development. *Dev. Biol.* 361: 39–56.

Liu, H. X., Henson, B. S., Zhou, Y., et al. (2008). Fungiform papilla pattern: EGF regulates inter-papilla lingual epithelium and decreases papilla number by means of PI3K/Akt, MEK/ERK, and p38 MAPK signaling. *Dev. Dyn.* 237: 2378–2393.

Liu, H. X., Komatsu, Y., Mishina, Y., and Mistretta, C. M. (2012b). Neural crest contribution to lingual mesenchyme, epithelium and developing taste papillae and taste buds. *Dev. Biol.* 368(2): 294–303.

Liu, H. X., Maccallum, D. K., Edwards, C., et al. (2004). Sonic hedgehog exerts distinct, stage-specific effects on tongue and taste papilla development. *Dev. Biol.* 276: 280–300.

Liu, X., Ernfors, P., Wu, H., and Jaenisch, R. (1995). Sensory but not motor neuron deficits in mice lacking NT4 and BDNF. *Nature* 375: 238–241.

Liu, Y. S., Schweitzer, L., and Renehan, W. E. (2000). Development of salt-responsive neurons in the nucleus of the solitary tract. *J. Comp. Neurol.* 425: 219–232.

Logan, C., Wingate, R. J., McKay, I. J., and Lumsden, A. (1998). Tlx-1 and Tlx-3 homeobox gene expression in cranial sensory ganglia and hindbrain of the chick embryo: markers of patterned connectivity. *J. Neurosci.* 18: 5389–5402.

Lopez, G. F., and Krimm, R. F. (2006a). Epithelial overexpression of BDNF and NT4 produces distinct gustatory axon morphologies that disrupt initial targeting. *Dev. Biol.* 292: 457–468.

Lopez, G. F., and Krimm, R. F. (2006b). Refinement of innervation accuracy following initial targeting of peripheral gustatory fibers. *J. Neurobiol.* 66: 1033–1043.

Mangold, J.E., Hill, D.L. (2007). Extensive reorganization of primary afferent projections into the gustatory brainstem induced by feeding a sodium-restricted diet during development: less is more. *J. Neurosci.* 27: 4650–4662.

Ma, L., Lopez G. F., and Krimm, R. F. (2009). Epithelial-derived brain-derived neurotrophic factor is required for gustatory neuron targeting during a critical developmental period. *J. Neurosci.* 29: 3354–3364.

Mangold, J. E., and Hill, D. L. (2008). Postnatal reorganization of primary afferent terminal fields in the rat gustatory brainstem is determined by prenatal dietary history. *J. Comp. Neurol.* 509: 594–607.

Matsumoto, I., Ohmoto, M., Narukawa, M., et al. (2011). Skn-1a (Pou2f3) specifies taste receptor cell lineage. *Nat. Neurosci.* 14(6): 685–687.

May, O. L., Erisir, A., and Hill, D. L. (2008). Modifications of gustatory nerve synapses onto nucleus of the solitary tract neurons induced by dietary sodium-restriction during development. *J. Comp. Neurol.* 508: 529–541.

May, O. L., and Hill, D. L. (2006). Gustatory terminal field organization and developmental plasticity in the nucleus of the solitary tract revealed through triple-fluorescence labeling. *J. Comp. Neurol.* 497: 658–669.

Mbiene, J.-P., and Mistretta, C. M. (1997). Initial innervation of embryonic rat tongue and developing taste papillae: nerves follow distinctive and spatially restricted pathways. *Acta. Anat.* 160: 139–158.

Mbiene, J. P. (2004). Taste placodes are primary targets of geniculate but not trigeminal sensory axons in mouse developing tongue. *J. Neurocytol.* 33: 617–629.

Mbiene, J. P., and Roberts, J. D. (2003). Distribution of keratin 8-containing cell clusters in mouse embryonic tongue: evidence for a prepattern for taste bud development. *J. Comp. Neurol.* 457: 111–122.

McCabe, K. L., and Bronner-Fraser, M. (2009). Molecular and tissue interactions governing induction of cranial ectodermal placodes. *Dev. Biol.* 332: 189–195.

Miller, I. J., and Smith, D. V. (1988). Proliferation of taste buds in the foliate and vallate papillae of postnatal hamsters. *Growth, Dev. Aging* 52: 123–131.

Mistretta, C. M. (1972). Topographical and histological study of the developing rat tongue, palate and taste buds. In *Third Symposium on Oral Sensation and Perception. The Mouth of the Infant.*, ed. J. F. Bosma, pp. 163–187. Springfield, IL: Charles C. Thomas.

Mistretta, C. M., Goosens, K. A., Farinas, I., and Reichardt, L. F. (1999). Alterations in size, number, and morphology of gustatory papillae and taste buds in BDNF null mutant mice demonstrate neural dependence of developing taste organs. *J. Comp. Neurol.* 409: 13–24.

Mistretta, C. M., Gurkan, S., and Bradley, R. M. (1988). Morphology of chorda tympani receptive fields and proposed neural rearrangements during development. *J. Neurosci.* 8: 73–78.

Mistretta, C. M., and Liu, H. X. (2006). Development of fungiform papillae: patterned lingual gustatory organs. *Arch. Histol. Cytol.* 69: 199–208.

Mistretta, C. M., Liu, H. X., Gaffield, W., and MacCallum, D. K. (2003). Cyclopamine and jervine in embryonic rat tongue cultures demonstrate a role for Shh signaling in taste papilla development and patterning: fungiform papillae double in number and form in novel locations in dorsal lingual epithelium. *Dev. Biol.* 254: 1–18.

Miura, H., and Barlow, L. A. (2010). Taste bud regeneration and the search for taste progenitor cells. *Arch. Ital. Biol.* 148: 107–118.

Morgan, B. A., Orkin, R. W., Noramly, S., and Perez, A. (1998). Stage-specific effects of sonic hedgehog expression in the epidermis. *Dev. Biol.* 201: 1–12.

Murray, R. C., Navi, D., Fesenko, J., et al. (2003). Widespread defects in the primary olfactory pathway caused by loss of Mash1 function. *J. Neurosci.* 23: 1769–1780.

Muskavitch, M. A. T. (1994). Delta-Notch signaling and *Drosophila* cell fate choice. *Dev. Biol.* 166: 415–430.

Nagai, T., Mistretta, C. M., and Bradley, R. M. (1988). Developmental decrease in size of peripheral receptive fields of single chorda tympani nerve fibers and relation to increasing NaCl taste sensitivity. *J. Neurosci.* 8: 64–72.

Nagato, T., Matsumoto, K., Tanioka, H., et al. (1995). Effect of denervation on morphogenesis of the rat fungiform papilla. *Acta. Anat.* 153: 301–309.

Nakayama, A., Miura, H., Shindo, Y., et al. (2008). Expression of the basal cell markers of taste buds in the anterior tongue and soft palate of the mouse embryo. *J. Comp. Neurol.* 509: 211–224.

Narayanan, C. H., and Narayanan, Y. (1980). Neural crest and placodal contributions in the development of the glossopharyngeal-vagal complex in the chick. *Anat. Rec.* 196: 71–82.

Nie, X. (2005). Apoptosis, proliferation and gene expression patterns in mouse developing tongue. *Anat. Embryol. (Berl.)* 210: 125–132.

Normanno, N., Bianco, C., Strizzi, L., et al. (2005). The ErbB receptors and their ligands in cancer: an overview. *Current Drug Targets* 6: 243–257.

Nosrat, C. A., Blomlöf, J., ElShamy, W. M., et al. (1997). Lingual deficits in BDNF and NT3 mutant mice leading to gustatory and somatosensory disturbances, respectively. *Development* 124: 1333–1342.

Nosrat, C. A., Ebendal, T., and Olson, L. (1996). Differential expression of brain-derived neurotrophic factor and neurotrophin 3 mRNA in lingual papillae and taste buds indicates roles in gustatory and somatosensory innervation. *J. Comp. Neurol.* 376: 587–602.

Nosrat, C. A., MacCallum, D. K., and Mistretta, C. M. (2001). Distinctive spatiotemporal expression patterns for neurotrophins develop in gustatory papillae and lingual tissues in embryonic tongue organ cultures. *Cell Tissue Res.* 303: 35–45.

Nosrat, C. A., and Olson, L. (1995). Brain-derived neurotrophic factor mRNA is expressed in the developing taste bud-bearing tongue papillae of rat. *J. Comp. Neurol.* 360: 698–704.

Nosrat, I. V., Margolskee, R. F., and Nosrat, C. A. (2012). Targeted taste cell-specific overexpression of brain-derived neurotrophic factor in adult taste buds elevates phosphorylated TrkB protein levels in taste cells, increases taste bud size, and promotes gustatory innervation. *J. Biol. Chem.* 287: 16791–16800.

Oakley, B., Brandemihl, A., Cooper, D., et al. (1998). The morphogenesis of mouse vallate gustatory epithelium and taste buds requires BDNF-dependent taste neurons. *Dev. Brain Res.* 105: 85–96.

Oakley, B., LaBelle, D. E., Riley, R. A., et al. (1991). The rate and locus of development of rat vallate taste buds. *Dev. Brain Res.* 58: 215–221.

Oakley, B., and Witt, M. (2004). Building sensory receptors on the tongue. *J. Neurocytol.* 33: 631–646.

Oakley, B., Wu, L. H., Lawton, A., and deSibour, C. (1990). Neural control of ectopic filiform spines in adult tongue. *Neuroscience* 36: 831–838.

Ohtubo, Y., Iwamoto, M., and Yoshii, K. (2012). Subtype-dependent postnatal development of taste receptor cells in mouse fungiform taste buds. *Eur. J. Neurosci.* 35(11): 1661–1671.

Okubo, T., Pevny, L. H., and Hogan, B. L. (2006). Sox2 is required for development of taste bud sensory cells. *Genes Dev.* 20: 2654–2659.

Olmsted, J. M. D. (1920). The nerve as a formative influence in the development of taste-buds. *J. Comp. Neurol.* 31: 465–468.

Ota, M. S., Kaneko, Y., Kondo, K., et al. (2009). Combined in silico and in vivo analyses reveal role of Hes1 in taste cell differentiation. *PLoS. Genet.* 5: e1000443

Paiva, K. B., Silva-Valenzuela, M. G., Massironi, S. M., et al. (2010). Differential Shh, Bmp and Wnt gene expressions during craniofacial development in mice. *Acta. Histochem.* 112: 508–517.

Parker, M. A., and Barlow, L. A. (2000). The role of cell contacts in the development of amphibian taste buds. *Chem. Senses* In press (abstract).

Parker, M. A., Bell, M., and Barlow, L. A. (2004). Cell contact-dependent mechanisms specify taste bud number and size during a critical period early in embryonic development. *Dev. Dyn.* 230: 630–642.

Patel, A. V., Huang, T., and Krimm, R. F. (2010). Lingual and palatal gustatory afferents each depend on both BDNF and NT-4, but the dependence is greater for lingual than palatal afferents. *J. Comp. Neurol.* 518: 3290–3301.

Patel, A. V., and Krimm, R. F. (2010). BDNF is required for the survival of differentiated geniculate ganglion neurons. *Dev. Biol.* 340: 419–429.

Patel, A. V., and Krimm, R. F. (2012). Neurotrophin-4 regulates the survival of gustatory neurons earlier in development using a different mechanism than brain-derived neurotrophic factor. *Dev. Biol.* 365: 50–60.

Petersen, C. I., Jheon, A. H., Mostowfi, P., et al. (2011). FGF signaling regulates the number of posterior taste papillae by controlling progenitor field size. *PLoS. Genet.* 7: e1002098.

Pispa, J., and Thesleff, I. (2003). Mechanisms of ectodermal organogenesis. *Dev. Biol.* 262: 195–362.

Quina, L. A., Tempest, L., Hsu, Y. W., et al. (2012). Hmx1 is required for the normal development of somatosensory neurons in the geniculate ganglion. *Dev. Biol.* 365: 152–163.

Reutter, K. (1978). Taste organ in the bullhead (Teleostei). *Adv. Anat. Embryol. Cell Biol.* 55: 1–98.

Ringstedt, T., Ibanez, C. F., and Nosrat, C. A. (1999). Role of brain-derived neurotrophic factor in target invasion in the gustatory system. *J. Neurosci.* 19: 3507–3518.

Roberts, I. M., and Jaffe, R. (1986). Lingual lipase: immunocytochemical localization in the rat von Ebner gland. *Gastroenterology* 90: 1170–1175.

Rochlin, M. W., and Farbman, A. I. (1998). Trigeminal ganglion axons are repelled by their presumptive targets. *J. Neurosci.* 18: 6840–6852.

Rochlin, M. W., O'Connor, R., Giger, R. J., et al. (2000). Comparison of neurotrophin and repellent sensitivities of early embryonic geniculate and trigeminal axons. *J. Comp. Neurol.* 422: 579–593.

Rothova, M., Thompson, H., Lickert, H., and Tucker, A. S. (2012). Lineage tracing of the endoderm during oral development. *Dev. Dyn.* 241: 1183–1191.

Runge, E. M., Hoshino, N., Biehl, M. J., et al. (2012). Neurotrophin-4 Is more potent than brain-derived neurotrophic factor in promoting, attracting and suppressing geniculate ganglion neurite outgrowth. *Dev. Neurosci.* 34: 389–401.

Sbarbati, A., Crescimanno, C., Bernardi, P., et al. (2000). Postnatal development of the intrinsic nervous system in the circumvallate papilla-vonEbner gland complex. *Histochem. J.* 32: 483–488.

Sbarbati, A., Crescimanno, C., Bernardi, P., and Osculati, F. (1999). Alpha-gustducin-immunoreactive solitary chemosensory cells in the developing chemoreceptorial epithelium of the rat vallate papilla. *Chem. Senses* 24: 469–472.

References

Sbarbati, A., Crescimanno, C., Merigo, F., et al. (2001). A brief survey of the modifications in sensory-secretory organs of the neonatal rat tongue. *Biol. Neonate* 80: 1–6.

Schlosser, G. (2006). Induction and specification of cranial placodes. *Dev. Biol.* 294: 303–351.

Schlosser, G. (2010). Making senses development of vertebrate cranial placodes. *International review of cell and molecular biology* 283: 129–234.

Seta, Y., Oda, M., Kataoka, S., et al. (2011). Mash1 is required for the differentiation of AADC-positive type III cells in mouse taste buds. *Dev. Dyn.* 240: 775–784.

Seta, Y., Seta, C., and Barlow, L. A. (2003). Notch-associated gene expression in embryonic and adult taste papillae and taste buds suggests a role in taste cell lineage decisions. *J. Comp. Neurol.* 464: 49–61.

Shiau, C. E., and Bronner-Fraser, M. (2009). N-cadherin acts in concert with Slit1-Robo2 signaling in regulating aggregation of placode-derived cranial sensory neurons. *Development* 136: 4155–4164.

Shou, J., Rim, P. C., and Calof, A. L. (1999). BMPs inhibit neurogenesis by a mechanism involving degradation of a transcription factor. *Nat. Neurosci.* 2: 339–345.

Sohn, W. J., Gwon, G. J., An, C. H., et al. (2011). Morphological evidences in circumvallate papilla and von Ebners' gland development in mice. *Anat. Cell Biol.* 44: 274–283.

Sollars, S. I., Walker, B. R., Thaw, A. K., and Hill, D. L. (2006). Age-related decrease of the chorda tympani nerve terminal field in the nucleus of the solitary tract is prevented by dietary sodium restriction during development. *Neuroscience* 137: 1229–1236.

Sternini, C., Anselmi, L., and Rozengurt, E. (2008). Enteroendocrine cells: a site of 'taste' in gastrointestinal chemosensing. *Curr. Opin. Endocrinol. Diabetes Obes.* 15: 73–78.

Stone, L. M., Finger, T. E., Tam, P. P. L., and Tan, S.-S. (1995). Taste receptor cells arise from local epithelium, not neurogenic ectoderm. *Proc. Natl. Acad. Sci. USA* 92: 1916–1920.

Stone, L. S. (1940). The origin and development of taste organs salamanders observed in the living condition. *J. Exp. Zool.* 83: 481–506.

Sun, H., and Oakley, B. (2002). Development of anterior gustatory epithelia in the palate and tongue requires epidermal growth factor receptor. *Dev. Biol.* 242: 31–43.

Sun, Y., Dykes, I. M., Liang, X., et al. (2008). A central role for Islet1 in sensory neuron development linking sensory and spinal gene regulatory programs. *Nat. Neurosci.* 11: 1283–1293.

Suwabe, T., Mistretta, C. M., and Bradley, R. M. (2012). Excitatory and inhibitory synaptic function in the rostral nucleus of the solitary tract in embryonic rat. *Brain Res.* 1490: 117–127.

Suwabe, T., Mistretta, C. M., Krull, C., and Bradley, R. M. (2011). Pre- and postnatal differences in membrane, action potential, and ion channel properties of rostral nucleus of the solitary tract neurons. *J. Neurophysiol.* 106: 2709–2719.

Suzuki, Y., Ikeda, K., and Kawakami, K. (2010a). Expression of Six1 and Six4 in mouse taste buds. *J. Mol. Histol.* 41: 205–214.

Suzuki, Y., Ikeda, K., and Kawakami, K. (2010b). Regulatory role of Six1 in the development of taste papillae. *Cell Tissue Res.* 339: 513–525.

Suzuki, Y., Ikeda, K., and Kawakami, K. (2011). Development of gustatory papillae in the absence of Six1 and Six4. *J. Anat.* 219: 710–721.

Takano-Maruyama, M., Chen, Y., and Gaufo, G. O. (2012). Differential contribution of Neurog1 and Neurog2 on the formation of cranial ganglia along the anterior-posterior axis. *Dev. Dyn.* 241: 229–241.

Thesleff, I., Partanen, A. M., and Vainio, S. (1991). Epithelial-mesenchymal interactions in tooth morphogenesis: the roles of extracellular matrix, growth factors, and cell surface receptors. *J. Craniofac. Genet. Dev. Biol.* 11: 229–237.

Thirumangalathu, S., Harlow, D. E., Driskell, A. L., et al. (2009). Fate mapping of mammalian embryonic taste bud progenitors. *Development* 136: 1519–1528.

Thomas, J.E., Hill, D.L. (2008). The effects of dietary protein restriction on chorda tympani nerve taste responses and terminal field organization. *Neuroscience* 157, 329–339.

Thisse, B., and Thisse, C. (2005). Functions and regulations of fibroblast growth factor signaling during embryonic development. *Dev. Biol.* 287: 390–402.

Threadgill, D. W., Dlugosz, A. A., Hansen, L. A., et al. (1995). Targeted disruption of mouse EGF receptor: effect of genetic background on mutant phenotype. *Science* 269: 230–234.

Torrey, T. W. (1940). The influence of nerve fibers upon taste buds during embryonic development. *Proc. Natl. Acad. Sci. USA* 26: 627–634.

Tucker, A. S., and Sharpe, P. T. (1999). Molecular genetics of tooth morphogenesis and patterning: the right shape in the right place. *J. Dent. Res.* 78: 826–834.

Udager, A., Prakash, A., and Gumucio, D. L. (2010). Dividing the tubular gut: generation of organ boundaries at the pylorus. *Prog. Mol. Biol. Transl. Sci.* 96: 35–62.

Vilbig, R., Cosmano, J., Giger, R., and Rochlin, M. W. (2004). Distinct roles for Sema3A, Sema3F, and an unidentified trophic factor in controlling the advance of geniculate axons to gustatory lingual epithelium. *J. Neurocytol.* 33: 591–606.

Wakisaka, S., Miyawaki, Y., Youn, S. H., et al. (1996). Protein gene-product 9.5 in developing mouse circumvallate papilla: comparison with neuron-specific enolase and calcitonin gene-related peptide. *Anat. Embryol. (Berl).* 194: 365–372.

Wang, S., Corson, J., Hill, D., and Erisir, A. (2012). Postnatal development of chorda tympani axons in the rat nucleus of the solitary tract. *J. Comp. Neurol.* 520: 3217–3235.

Wansleeben, C., and Meijlink, F. (2011). The planar cell polarity pathway in vertebrate development. *Dev. Dyn.* 240: 616–626.

Washausen, S., Obermayer, B., Brunnett, G., et al. (2005). Apoptosis and proliferation in developing, mature, and regressing epibranchial placodes. *Dev. Biol.* 278: 86–102.

Wells, K. L., Mou, C., Headon, D. J., and Tucker, A. S. (2010). Defects and rescue of the minor salivary glands in Eda pathway mutants. *Dev. Biol.* 349: 137–146.

Yamauchi, Y., Abe, K., Mantani, A., et al. (1999). A novel transgenic technique that allows specific marking of the neural crest cell lineage in mice. *Dev. Biol.* 212: 191–203.

Yang, X., Zhou, Y., Barcarse, E. A., and O'Gorman, S. (2008). Altered neuronal lineages in the facial ganglia of Hoxa2 mutant mice. *Dev. Biol.* 314: 171–188.

Yee, C. L., Jones, K. R., and Finger, T. E. (2003). Brain-derived neurotrophic factor is present in adult mouse taste cells with synapses. *J. Comp. Neurol.* 459: 15–24.

Zhang, G. H., Deng, S. P., Li, L. L., and Li, H. T. (2006). Developmental change of alpha-gustducin expression in the mouse fungiform papilla. *Anat. Embryol. (Berl).* 211(6): 625–630

Zhang, G. H., Zhang, H. Y., Deng, S. P., et al. (2008). Quantitative study of taste bud distribution within the oral cavity of the postnatal mouse. *Arch. Oral Biol.* 53: 583–589.

Zhang, L. L., and Ashwell, K. W. (2001). The development of cranial nerve and visceral afferents to the nucleus of the solitary tract in the rat. *Anat. Embryol. (Berl)* 204: 135–151.

Zhou, Y., Liu, H. X., and Mistretta, C. M. (2006). Bone morphogenetic proteins and noggin: Inhibiting and inducing fungiform taste papilla development. *Dev. Biol.* 297(1): 198–213.

Zorn, A. M., and Wells, J. M. (2009). Vertebrate endoderm development and organ formation. *Annu. Rev. Cell Dev. Biol.* 25: 221–251.

Part 7

Human Taste Measurement, Physiology, and Development

Chapter 34

Psychophysical Measures of Human Oral Sensation

DEREK J. SNYDER, CHARLES A. SIMS, and LINDA M. BARTOSHUK

34.1 INTRODUCTION

The measurement of conscious experience defines psychophysics, a field of study that has significantly advanced our understanding of sensory and hedonic processes. In the chemical senses, psychophysical tests are used extensively to guide product development and quality control efforts in food science (e.g., Lawless and Heymann, 2010; Chapters 11, 47 and 48 this volume), and their use in basic research has revealed the broad influence of oral sensation and dysfunction on health-related behaviors and overall quality of life (e.g., Bartoshuk et al., 2004a). From a medical perspective, oral sensory disturbances may be relatively benign, but sometimes they are profoundly life-altering (e.g., Bromley and Doty, 2003; Chapter 39 this volume). As such, chemosensory experience and its consequences represent a vital clinical and consumer issue requiring careful evaluation.

Measuring this experience, however, is an extremely challenging task. By definition, individual experience is subjective: We can *describe* our experiences and track them over time, but we cannot directly *share* the experiences of another person. This observation presents a serious dilemma, as it is meaningless to *compare* descriptions that mean different things to different people. Because we use comparisons of real-world experience throughout life to convey what is acceptable (e.g., pleasure) and what is not (e.g., pain, disgust), it is important to measure these experiences well: We need to know how the sensory properties of food change with illness and age, how the sweet taste of sucrose differs between women and men, and so on. Early scholars in psychophysics relied on threshold measures to address these questions, and some researchers still believe

they can be answered only in vague terms — but recent advances in suprathreshold scaling capture sensory and affective differences with improved reliability and precision, supporting the notion that perceptual experiences can be measured and compared.

This chapter evaluates contemporary methods of oral sensory assessment, noting their origins and evolution. Threshold psychophysical tests remain popular and useful for many types of oral sensory evaluation, even as suprathreshold intensity scales provide an increasingly comprehensive and realistic picture of overall function. Efforts to identify suitable suprathreshold tools include an examination of labeled scales, which are used (often inappropriately) to compare experiences between individuals and groups. When performed correctly, psychophysical oral sensory measurement is especially informative in concert with parallel methods (e.g., multiple standards, genetic testing, oral anatomy), an approach that permits confirmation of results and (where necessary) correction for unanticipated scaling problems. Finally, the chapter describes tests of suprathreshold oral sensory function, including their potential for clinical and industrial use.

34.2 THRESHOLDS VS. INTENSITY: HOW SHOULD ORAL SENSATION BE MEASURED?

When we enjoy a meal, we can tell easily if the soup is too salty or the cocktails watered down. These judgments demonstrate that intensity is a continuous concept (i.e., intervals of strength) rather than a binary one (i.e., present or absent), yet debate continues regarding the ability of various measurement strategies to capture sensory

---

Handbook of Olfaction and Gustation, Third Edition. Edited by Richard L. Doty.
© 2015 Richard L. Doty. Published 2015 by John Wiley & Sons, Inc.

experience. Some psychophysicists avoid suprathreshold methods because they believe that degrees of intensity are immeasurable or at best ordinal (e.g., Brindley, 1960; Laming, 1997), while others contend that thresholds offer insufficient or inaccurate information about the range of sensory function (e.g., Moskowitz, 1977a). In the case of oral sensation, neither of these views is entirely correct: Taste threshold differences may carry important clinical and research implications when interpreted carefully, and suprathreshold oral sensory measures possess unique diagnostic and predictive capabilities when used correctly.

34.2.1 Threshold Procedures

Thresholds have been used for sensory evaluation ever since Fechner described them in his *Elemente der Psychophysik* (1860/1966), one of the first published works of experimental psychology. An expansion of work by Weber, Fechner's elaboration of the three threshold methods — the method of limits, the method of adjustment, and the method of constant stimuli — established psychophysics as a science. Although thresholds present both conceptual and practical challenges (e.g., Engen, 1972; Lawless and Heymann, 2010), their basic definition is straightforward: The *absolute* or *detection threshold* for a stimulus is the lowest concentration at which its presence can be detected as something, whether or not it is qualitatively discernible. The *recognition threshold*, which is often slightly higher than absolute threshold, is the lowest concentration at which the primary quality of a stimulus (e.g., sweet, painful) can be identified. Finally, the *difference threshold* is the smallest increase in suprathreshold stimulus concentration that can be detected (i.e., the "just noticeable difference" or jnd). Clinical assessments of gustatory function have historically focused on the absolute threshold, while food scientists often use difference thresholds to evaluate flavor changes or off-tastes.

Of the classical threshold methods, various iterations of the method of limits are used for taste evaluation most often. In the most basic version of this procedure, participants report whether or not they perceive a taste sensation from a given stimulus (i.e., "yes" or "no"). Concentration is increased from an undetectable level until participants report a sensation, and then it is decreased from a clearly detectable level until participants report an absence of sensation. (Taste stimuli are generally presented in logarithmic steps, as this spacing approximates equal intervals of perceived intensity.) Over several ascending and descending runs, the average concentration of transition points serves as an estimate of threshold (McBurney and Collings, 1977). One popular variant of this method involves the presentation of ascending series only, introduced to address concerns about adaptation to suprathreshold stimuli during descending series (e.g., Pangborn et al., 1963).

While elegant, this procedure raises concerns that warrant serious consideration. Test efficiency is extremely important; too few trials make the procedure unreliable, but too many trials can lead to stimulus adaptation and participant fatigue. In addition, signal detection theory (Green and Swets, 1966) shows that individuals set different criteria when deciding how much sensory change warrants a response: One person may anticipate a difference and respond to the slightest hint of change, while another may want to be highly confident that change has occurred before responding. As a result, participants tend to show bias toward responses already made (i.e., habituation errors), yet they may also change their responses prematurely when they believe an actual change is forthcoming (i.e., anticipation errors). Shifts in these criteria can occur over many trials, which further highlights the need for concise testing protocols.

Based on these concerns, efforts to balance task difficulty, bias, and test duration have yielded several refinements in the reliability of threshold measurement. In the "up-down" or "staircase" method, the concentration of the target stimulus depends on the outcome of the previous trial; a positive response proceeds to a lower target concentration, while a negative response proceeds to a higher one (Cornsweet, 1962). This arrangement improves test efficiency because it focuses on observations around threshold. Meanwhile, forced-choice discrimination paradigms address response bias by requiring participants to identify a target stimulus on each trial. When up-down and forced-choice elements are combined (e.g., Jesteadt, 1980), an increase in the number of correct vs. incorrect trials required to change concentration reduces the likelihood of chance (i.e., false-positive) performance at that level (Wetherill and Levitt, 1965), and a reduction in the ratio of target vs. background stimuli in each trial lowers chance performance and may curb adaptation effects (e.g., Lawless et al., 1995; Murphy et al., 1995). These modifications have proven especially useful in the chemical senses, which adapt and fatigue easily compared to other sensory systems (e.g., Linschoten et al., 1996). On the other hand, the fact that thresholds are so susceptible to change reflects a fundamental idea that is often overlooked: Thresholds are a statistical construct rather than a fixed biological constant. As such, they must be considered in the context of the specific tasks used to derive them, and they can be compared only with values measured under similar conditions.

Two general approaches dominate modern taste threshold measurement: Chemical taste thresholds typically involve whole-mouth sampling of dilute solutions (although regional testing is also feasible), while electro-gustometry involves localized taste sensations elicited by weak electrical current applied to specific regions of the tongue.

34.2 Thresholds vs. Intensity: How Should Oral Sensation be Measured?

34.2.1.1 Chemical Taste Thresholds. Building on the advances in methodology described above, one of the most widely accepted and rigorous methods for determining taste detection thresholds is an up-down, two-alternative forced-choice (2-AFC) staircase (McBurney and Collings, 1977; see also Wetherill and Levitt, 1965). This test comprises multiple trials of two stimuli in which participants distinguish tastant from water. Testing begins at a stimulus concentration that can be perceived, proceeding to a lower concentration (in logarithmic steps) following two correct trials and a higher concentration following one incorrect trial. Such an arrangement produces a series of "reversals" over time, and the geometric mean of the second through seventh reversal concentrations is defined as threshold. Subsequent testing has revealed potential confounds that may be avoided with careful test administration. For example, saliva contains sufficient NaCl to act as an adapting solution (McBurney and Pfaffmann, 1963), which can raise NaCl thresholds and induce a water taste that may be confused with solute taste (e.g., Bartoshuk, 1974; Hertz et al., 1975). To avoid this possibility, participants should rinse thoroughly before sampling any test solution.

In the clinic, streamlined methods for determining taste detection thresholds include the 3-drop and 8-cup techniques — which, like the up-down 2-AFC staircase, differ from Fechner's original methods mainly in terms of stimulus delivery and criterion performance. In the 3-drop technique, one drop of tastant and two drops of water are presented in each trial, and threshold is defined as the lowest concentration at which the participant chooses the target correctly in at least two of three trials (Henkin et al., 1963). The 8-cup technique is a sorting task in which four cups of tastant and four cups of water are presented; threshold is defined as the lowest concentration at which the participant separates the cups into tastant and water groups without error (Harris and Kalmus, 1949b). When directly compared, the 3-drop technique produces significantly higher detection thresholds for NaCl and sucrose than does the 8-cup technique (Weiffenbach, 1983), a difference attributed mainly to adaptation state (i.e., water rinses occur in the 8-cup task, but not in the 3-drop task) and the volume of liquid sampled (Lawless, 1987). Nevertheless, both are far less stringent than the up-down 2-AFC staircase (Frank et al., 2003).

Recognition thresholds for taste are distinct from detection thresholds, as a concentration range exists in which dilute solutions can be discriminated from water but cannot be identified in terms of taste quality. However, the challenge in determining taste detection and recognition thresholds for a given person is that the two test procedures must be equivalent in difficulty to permit valid comparison. Signal detection theory has been used successfully to assess the influences of age and other factors on the ability to taste (e.g., Matsuda and Doty, 1993). While it has also been used to develop up-down, forced-choice recognition threshold measures that match the task demands of the detection threshold (Collings, 1974), this application remains biased because individuals do not choose randomly among taste qualities when guessing (Weiffenbach, 1983). Moreover, the fact that taste solutions can produce tactile sensations (e.g., tingle, burn) suggests that individuals with severe taste loss retain some ability to discriminate tastant vs. water, which clouds the interpretation of taste threshold results. Accordingly, instructions for participants should be explicit in terms of the types of sensation to be reported and judged.

Chemical taste thresholds enjoy widespread use in research and clinical settings, where they have been used extensively to document regional and whole-mouth losses associated with aging, health status, and medication use (e.g., Cowart, 1989; Doty et al., 2001; Heath et al., 2006; Ileri-Gurel et al., 2013; Mojet et al., 2001; Schiffman, 1997), as well as individual differences associated with genetic taste status (e.g., Desai et al., 2011; Reed et al., 1995). One reason for the apparent popularity of thresholds is that their values are expressed in terms of a physical unit of intensity (i.e., molarity), which gives the impression of objectivity compared to suprathreshold ratings. In reality, thresholds are neither more reliable nor more accurate than other sensory measures, since they are statistical approximations of ever-changing sensitivity (Lawless and Heymann, 2010). Reflecting this fact, test-retest reliability for chemical taste thresholds is low (McMahon et al., 2001; Stevens et al., 1995); as with thresholds in other sensory systems, many trials are often required to achieve a stable result. Thus, even with abbreviated methods, the time required to measure taste thresholds remains a discouraging feature, particularly considering that they yield only the lower boundary of the psychophysical function. By comparison, suprathreshold procedures approximate the entire taste function in much less time.

A broader concern with taste thresholds regards their validity as a measure of overall taste function, particularly at suprathreshold levels approximating real-world sensation. In general, the decision to use thresholds in lieu of suprathreshold measures is reasonable provided there is strong concordance between threshold and suprathreshold experiences. While such a relationship appears to occur for olfactory sensation (e.g., Cain and Stevens, 1989), taste thresholds and suprathreshold taste intensity can dissociate in substantial and unexpected ways, compromising predictions of suprathreshold sensation from threshold values alone. As Figure 34.1 shows, some manipulations (e.g., exposure to sodium lauryl sulfate, a detergent found in many toothpastes) suppress taste sensation uniformly across the perceptible range (Panel a; DeSimone et al., 1980), but others affect specific regions of

754 Chapter 34 Psychophysical Measures of Human Oral Sensation

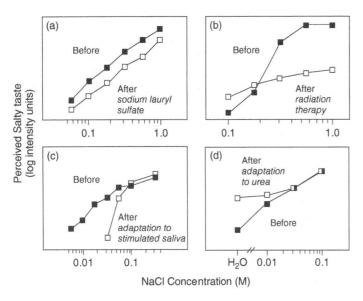

Figure 34.1 Examples of alterations in the psychophysical functions for taste stimuli. All panels show magnitude estimate ratings for several concentrations of NaCl. Panel (a): NaCl intensity following exposure to 0.05% sodium lauryl sulfate; similar effects occur for sucrose and quinine (DeSimone et al., 1980). Panel (b): NaCl intensity in a patient during (day 5) and after (day 94) radiation treatment for head/neck cancer; similar effects occur for other taste qualities (Bartoshuk, 1978). Panel (c): NaCl intensity following adaptation to stimulated saliva (0.051 M sodium); average of five replicates (Bartoshuk, 1978). Panel (d): NaCl intensity following adaptation to 0.82 M urea (McBurney and Bartoshuk, 1973).

the dose-response function: Radiation therapy can impair taste sensation at high concentrations while leaving it intact near threshold (Panel b; Bartoshuk, 1978), and taste thresholds can shift without affecting suprathreshold taste intensity during both aging (Bartoshuk et al., 1986) and adaptation (Panels c-d; Bartoshuk, 1978; McBurney and Bartoshuk, 1973; see also Bartoshuk, 1974; McBurney, 1966). This disagreement alone does not preclude the use of threshold testing, as it remains useful to identify taste anomalies regardless of their effect on real-world experience: Compensatory oral sensory interactions may sustain whole-mouth sensation following nerve damage, rendering individuals unaware of underlying sensory disturbance (e.g., Bartoshuk et al., 1987; Lehman et al., 1995), so taste threshold dysfunction signals a need for further examination. On the other hand, a normal taste threshold should not be interpreted by itself as evidence of stable function, but should be confirmed with suprathreshold and/or spatial testing.

34.2.1.2 Electrogustometry.
An alternative to chemical measures of taste sensitivity, electrogustometry involves the application of weak anodal electric currents to specific regions of the mouth (Krarup, 1958; Mackenzie, 1955). Several mechanisms for electric taste have been proposed (Bujas, 1971; DeSimone et al., 1981; DeSimone et al., 1984; Kashiwayanagi et al., 1981), the most likely of which involves "ion accumulation": Electric current indirectly stimulates taste by transporting positively charged ions in saliva toward taste receptors, concentrating them to levels where they can be detected (Herness, 1985). Because saliva is mildly acidic and contains salts, electrogustometry sometimes evokes sour or salty taste sensations (e.g., Bujas, 1971; Grant et al., 1987; Murphy et al., 1995).

The chief advantage of electrogustometry is its convenience; it is portable, avoids the use of chemical solutions, permits regional stimulation of taste bud fields, and provides values that can be compared across individuals, time points, locations within the mouth, or treatment conditions (e.g., Frank and Smith, 1991). In fact, electrogustometry shows high test-retest reliability and bilateral correspondence compared to chemical taste thresholds (Murphy et al., 1995), suggesting that it may be superior for clinical use. Accordingly, normative data have been described for some groups (e.g., Tomita et al., 1986), and electric taste thresholds have been used to identify substantial taste losses associated with aging, denervation, and disease (Grant et al., 1987; Groves and Gibson, 1974; Le Floch et al., 1990; Murphy et al., 1995; Nakazato et al., 2002; Ovesen et al., 1991; Pavlidis et al., 2013).

That said, electric taste threshold data should be interpreted with care, especially regarding their ability to capture the full breadth of taste function (e.g., Stillman et al., 2003). First, taste loss varies across quality, but electrogustometry does not stimulate all taste qualities, particularly non-ionic modalities like sweetness and bitterness (Frank and Smith, 1991). Mounting data indicate that oral sensory alterations are often quality-specific, especially for bitter taste (Bartoshuk et al., 2002a; Grushka et al., 1986; Lehman et al., 1995; Yanagisawa et al., 1998), so methods that fail to assess bitter taste may fail to identify clinically relevant damage. Second, electrogustometric thresholds correlate well with regional (Krarup, 1958; Tomita et al., 1986), but not whole-mouth chemical taste thresholds (Murphy et al., 1995), and suprathreshold functions for electrical and chemical taste also show poor agreement (Salata et al., 1991). Again, as with chemical taste thresholds, electrogustometry should not be discounted simply because it is discontinuous with real-world taste experience, but its limitations should be considered as part of a comprehensive taste evaluation.

34.2.2 Direct Scaling of Suprathreshold Intensity

While threshold methods provide only the lower limit of physical energy that can be perceived (e.g., decibels of sound, molar concentration), suprathreshold or "direct" scaling methods measure perceived intensity across the full dynamic range of sensation (e.g., Stevens, 1946).

34.2.2.1 Magnitude Estimation. Fechner assumed that the jnd was the basic unit of psychological intensity, so he created a suprathreshold scale in which stimulus intensity was described in terms of the number of jnds above absolute threshold. Though not universally accepted, this view prevailed for nearly a century until S. S. Stevens observed that the jnd does not behave like a proper unit: An 8 jnd stimulus is not twice as intense as a 4 jnd stimulus; it is actually more intense (Stevens, 1961). This finding shows that suprathreshold scales based on threshold values provide a distorted view of intensity experiences.

Stevens introduced direct scaling methods with ratio properties, the most popular of which is magnitude estimation (Stevens, 1956). In the most popular version of this procedure, subjects provide a number reflecting the perceived intensity of a stimulus; they then give a number twice as large to a stimulus that is twice as intense, a number half as large to a stimulus half as intense, and so on. The size of the numbers is irrelevant; only the ratios among numbers carry meaning. Accordingly, when magnitude estimate data are pooled, they are typically "normalized" to bring ratings into a common register while preserving the ratios among them. This practice ensures that one individual's data are not unduly weighted just because they used larger numbers (Marks, 1974).

Over several years, Stevens used magnitude estimation to compare the growth rates of psychophysical functions across sensory modalities (Stevens, 1959; Stevens, 1962; Stevens and Galanter, 1957), but he was not particularly interested in studying how functions for a specific stimulus vary among people. This distinction is important because magnitude estimates describe only how perceived intensity varies with stimulus intensity *within* an individual; they cannot reflect meaningful differences of absolute perceived intensity *between* individuals or groups (Marks, 1974). Because group comparisons are such a basic element of scientific inquiry, this limitation has not been fully appreciated, but its consequences are severe. As the following section illustrates, the identification of robust variation in oral sensation has figured prominently in efforts to measure group differences accurately.

34.2.2.2 Measuring Oral Sensory Differences: Magnitude Matching.

INDIVIDUAL DIFFERENCES IN ORAL SENSATION
Discovered by the chemist A. L. Fox (1931), individuals differ significantly in their ability to taste thiourea compounds like phenylthiocarbamide (PTC) and 6-n-propylthiouracil (PROP) (e.g., Barnicot et al., 1951; Fox, 1932); most individuals perceive bitterness (i.e., tasters), but others are "taste blind" and perceive nothing (i.e., nontasters). Early reports suggested that nontasting is a recessive trait with a single genetic locus (Blakeslee, 1932; Snyder, 1931), while other studies measured the proportion of tasters by race, sex, and disease (e.g., Kalmus and Farnsworth, 1959; Parr, 1934; Sunderland and Cartwright, 1968; Whissell-Buechy and Wills, 1989). In response to the variety of methods used to measure PTC/PROP sensitivity, Harris and Kalmus (1949b) introduced a threshold technique that dominated research on taste blindness for over 25 years, largely because it produces a bimodal distribution of PTC/PROP detection thresholds that easily distinguishes nontasters and tasters (Olson et al., 1989; Whissell-Buechy, 1990). In the 1960s, Fischer used this technique to explore the behavioral implications of taste blindness, revealing associations between PTC/PROP sensitivity and food preferences, alcohol and tobacco use, and body weight (Fischer, 1967; Fischer et al., 1961, 1963). These studies were among the first to use PROP as a marker of taste blindness; unlike PTC, PROP is odorless and has well-defined safety limits as a thyroid medication (Lawless, 1980; Wheatcroft and Thornburn, 1972).

Encouraged by the potential benefits of direct scaling, Bartoshuk sought to compare the suprathreshold bitterness of PTC between nontasters and tasters. However, this comparison presents a problem: Magnitude estimates have *relative* meaning when subjects are used as their own controls (e.g., McBurney and Bartoshuk, 1973), but how can *absolute* bitterness be compared across groups? The answer to this question involves measuring PTC bitterness relative to an unrelated standard. As described previously, magnitude estimates are often normalized to obtain group functions, but this procedure addresses a different issue: Bringing data into common range invokes an arbitrary standard that conveys nothing about absolute perceived intensity, but group comparisons reference a standard that is assumed to be equally intense (on average) to the groups being compared. If this assumption holds, normalization yields valid across-group differences in absolute perceived intensity for stimuli of interest: If "10" denotes the intensity of a standard to nontasters and tasters, PTC ratings of

"40" for tasters and "20" for nontasters reflect a twofold intensity difference.

To quantify oral sensory variation accurately, appropriate standards for comparison must be identified. This pursuit has been strongly informed by research on "cross-modality matching," the ability to match qualitatively different sensations (Stevens and Marks, 1965). Because thioureas share the N-C=S chemical group, Bartoshuk reasoned that the taste intensity of a compound lacking the N-C=S group should be equal, on average, to nontasters and tasters. If so, group averages of PTC bitterness can be compared by rating it relative to, for example, NaCl saltiness. With this procedure, tasters find PTC and PROP more bitter than do nontasters (Hall et al., 1975; Lawless, 1980), and subsequent reports extensively document oral sensory variation extending far beyond the N-C=S group: Tasters perceive more intense taste and oral tactile sensations overall (e.g., Bartoshuk, 2000), and a subset of tasters known as "supertasters" consistently gives the highest ratings to taste stimuli, oral irritants (e.g., capsaicin, alcohol), fats, and retronasal odors (e.g., Bartoshuk et al., 2004b; Prescott et al., 2004).

MAGNITUDE MATCHING While the discovery of supertasters reinforced the breadth of individual differences in oral sensation, it also presented a problem: If supertasters of PROP perceive NaCl more intensely than do others, then NaCl is a poor standard for PROP-related comparisons, rendering observed differences based on PROP status inaccurate (albeit conservatively so). This problem was resolved by introducing stimuli from non-taste modalities as standards. For example, if taste and hearing are assumed to be unrelated, taste stimuli can be rated relative to auditory intensity. This procedure, formalized as "magnitude matching" (Marks and Stevens, 1980; Marks et al., 1988; Stevens and Marks, 1980) confirmed suspicion that the saltiness of NaCl varies with taster status (Bartoshuk et al., 1998). Magnitude matching addresses the problem of group comparisons by changing the task: Oral sensations cannot be compared directly across PROP taster groups, so subjects rate stimuli of interest relative to a non-oral sensory standard. As long as variability in the standard remains unrelated to variability in PROP bitterness, oral sensory experiences are comparable across taster groups.

The ability to observe accurate sensory differences in oral sensation has enabled identification of associations between sensory experience, dietary behavior, and disease risk (e.g., Duffy, 2007; Snyder et al., 2008a; Tepper et al., 2009). For example, PROP bitterness associates with decreased vegetable preference and intake (Dinehart et al., 2006; Drewnowski et al., 2000; Duffy et al., 2001), a known risk factor for colon cancer; it also associates with an increased number of colon polyps (Basson et al., 2005). PROP intensity also predicts avoidance of high-fat foods,

and some studies have shown that supertasters have lower body mass indices and more favorable cardiovascular profiles (Duffy, 2004; Tepper and Ullrich, 2002).

GENETIC FACTORS IN ORAL SENSORY VARIATION

Based on family studies showing that nontaster parents produce nontaster children, taste blindness was thought for many years to arise from a single gene (Blakeslee, 1932; Snyder, 1931; Whissell-Buechy, 1990). However, modern genetic analysis suggests that the genetic contribution to PTC/PROP bitterness is much more complex (e.g., Bufe et al., 2005). Linkage analysis has identified candidate genes for taste blindness on chromosomes 5p15, 7, and 16p (Drayna et al., 2003; Reed et al., 1999), and subsequent mapping of chromosome 7q has identified sequence polymorphisms in a taste receptor gene (T2R38) that account for up to 85% of observed differences in PTC threshold sensitivity (Kim et al., 2003; Wooding et al., 2004). These relationships distinguish between nontasters and tasters, but their ability to identify supertasters is limited: Medium tasters and supertasters have similar PTC/PROP thresholds (e.g., Bartoshuk, 2000), and T2R38 variation does not fully predict the robust differences in suprathreshold PROP intensity observed between these groups (Hayes et al., 2008). In other words, PROP supertasting cannot be explained completely by threshold sensitivity or T2R38 genetics.

Taking a broader view, rising PROP intensity is associated with greater intensity for virtually all oral sensory stimuli, as described previously (e.g., Bartoshuk et al., 2004b; Prescott et al., 2004). Moreover, supertasters express the highest density of fungiform papillae, structures on the anterior tongue that contain taste buds and oral somatosensory end organs (Bartoshuk and Duffy, 1994). These differences suggest that PTC/PROP taster status is a function of two independent conditions: T2R38 expression and fungiform papilla density (Hayes et al., 2008). In this model, PROP supertasters carry the taster variant of the T2R38 gene (which confers the ability to taste PTC/PROP), but they also express a high density of fungiform papillae (which maximizes oral sensory input of all kinds) (Bartoshuk et al., 2001). Several gene products govern the development of fungiform papillae (e.g., sonic hedgehog, bone morphogenic proteins, epidermal growth factor, Wnt/ß-catenin signaling), but their combinatorial impact on the quantity of papillae produced has yet to be fully determined (e.g., Liu et al., 2009). Nevertheless, it now appears that oral sensory variation arises from the expression of multiple genes mediating both receptor activity and underlying anatomical structure.

While the full breadth of oral sensory variation has been characterized by association with PTC/PROP bitterness, it is much more likely that PTC/PROP bitterness is but a single, highly robust example of broader oral sensory

34.2 Thresholds vs. Intensity: How Should Oral Sensation be Measured?

variation. In other words, individual differences in sensation for any oral stimulus probably result from genetic variation in a specific receptor or transduction element, which is then amplified by differences in fungiform papilla density. Recent identification of polymorphisms associated with differential sweet, umami, and fat perception supports this view (Fushan et al., 2010; Keller et al., 2012; Raliou et al., 2009; Shigemura et al., 2009), as does an emerging interaction between PROP bitterness, T2R38 genetics, and sequence variation in the taste bud trophic factor gustin (Calò et al., 2011). As new oral sensory gene variants and expression patterns are discovered, it is highly likely that taster status will be defined more broadly, based on genotype-phenotype relationships across multiple aspects of oral sensation and anatomy.

TASTER STATUS CLASSIFICATION Overall, valid consensus values for the classification of PROP status are lacking, mainly because steady improvements in genetic and psychophysical testing have eclipsed previous estimates. Consequently, existing criteria are both idiosyncratic and variable, resulting in vigorous debate over which classification scheme best reflects differences in oral sensation (e.g., Drewnowski, 2003; Prutkin et al., 2000; Rankin et al., 2004). At the center of this issue, the validity of any boundary value depends on the instrument used to measure it, so when suprathreshold psychophysical tools produce distorted comparisons among subjects, the sorting criteria derived from those tools are also distorted. (Thresholds have remained a popular clinical measure for precisely this reason, even though they also have features that distort real-world sensory experience.) In other words, measurement scales may be numerous, but not all are created equal: For oral sensory evaluation, magnitude matching remains the "gold standard" technique.

Broadly speaking, the most effective assessment strategies integrate multiple correlates of function. As advances in anatomy and genetics permit more nuanced study, the best methods for oral sensory evaluation will encompass an array of techniques that complement and enrich sophisticated psychophysical measurement. Based on laboratory and questionnaire data collected over many years, this multivariate approach has yielded working guidelines for the determination of PROP status (Snyder et al., 2006):

- Nontasters and tasters of PROP are easily distinguished by genetic analysis of T2R38 (i.e., nontasters are recessive, tasters are dominant) (Kim et al., 2003), which reliably predicts PROP threshold differences (i.e., > 0.2 mM for nontasters, < 0.1 mM for tasters; Bartoshuk, 1979; Bartoshuk et al., 1994a). In a sample of 1400 healthy lecture participants, this difference corresponds roughly to a boundary value on the general Labeled Magnitude Scale (gLMS; see below) of

"weak" (i.e., 17 out of 100) for filter papers impregnated with saturated PROP ($\sim$0.058 M). Consistent with previous estimates (Bartoshuk et al., 1994a; Harris and Kalmus, 1949a), this cutoff yields $\sim$25% nontasters in the sample.

- Supertasters of PROP are distinguished from medium tasters by psychophysical criteria. Existing population estimates of PROP taster status are based on a single-locus model, so this boundary value will remain arbitrary until the genetic basis of PROP bitterness is fully understood. If nontasters represent the lowest 25% of PROP paper ratings, a working definition of supertasting might include the top 25% of ratings; this logic suggests a gLMS boundary value of 80.

- Individuals with taster genotypes and nontaster PROP ratings probably reflect oral sensory pathology (see below). In these cases, oral anatomy can often be used to identify supertasters (e.g., Bartoshuk et al., 2004a), who show high fungiform papilla density (i.e., over 100 papillae/cm^2; Bartoshuk et al., 1994a) and low PROP responses.

In recent years, as the parameters of oral sensory variation have expanded, the use of thiourea bitterness as a proxy for overall oral sensory ability has been questioned at considerable length, mainly because correlations with it are imperfect and sometimes quite low. Proposed alternatives to PROP intensity include the ability to perceive specific taste cues from thermal or irritant stimuli (Green and George, 2004; Green and Hayes, 2004); these methods show promise, but their broader predictive ability remains unclear because, like thiourea bitterness, they emphasize associations with a particular taste modality. Meanwhile, a more generalized concept of taster status has emerged, one that frames supertasting as an elevated response across multiple stimuli rather than a single class of compounds (Hayes and Keast, 2011; Lim et al., 2008; Reed, 2008). New measures like these may prove more veridical than PROP intensity, but they will require extensive study to determine boundary values, interactions with genetics and anatomy, and clinical significance. Thus, despite its flaws, PROP bitterness remains a useful screening tool, as it is currently the best characterized index of individual differences in oral sensation.

Whether taster group classifications are based on PROP bitterness or more general criteria, some researchers persistently claim that oral sensation has little effect on sensation, food behavior, or health (e.g., Drewnowski, 2003; Kranzler et al., 1998). Upon closer examination, many of these dissenting reports fail to show effects of interest because they incorporate methods known to distort or eliminate main effects, including inappropriate group comparisons, poor scale instructions, and the use of threshold rather than suprathreshold taste tests (e.g., Bartoshuk, 2000; Bartoshuk et al., 2005). These problems are often reinforced by the inappropriate use of labeled intensity scales.

34.2.2.3 Measuring Oral Sensory Differences: Labeled Scales.

Measurement scales labeled with intensity descriptors (e.g., weak, strong, very strong) are used widely throughout the medical, scientific, and consumer disciplines. Although many of these category scales have been "validated," the fact that a scale measures what it was intended to measure does not guarantee its ability to produce valid group comparisons. In fact, several reports have determined that category scales distort results when their labels fail to denote equal perceived intensities to everyone.

Category scales are much older than the formal psychophysics of Fechner, dating back at least to the astronomer Hipparchus (190–120 BC). In the modern era, category scales in common use include the Likert scale (1932), which was originally developed to measure attitudes, and the Natick 9-point scale, which was created by the military to quantify food preferences (Jones et al., 1955; Peryam and Girardot, 1952). When adapted for sensory use, these scales are typically anchored by descriptors spaced equally along a line (e.g., 1 = none, 3 = slight, 5 = moderate, 7 = strong, 9 = extreme), as shown in Figure 34.2 (Kamen et al., 1961). The visual analog scale (VAS), a line labeled at its endpoints with the minimum and maximum intensity of a particular experience (e.g., Aitken et al., 1963; Hetherington and Rolls, 1987), came into widespread use during the 1960s and remains the favored sensory scale for medical assessments of pain; it is essentially a category scale without categories. Multiple versions of these scales have been used in oral sensory research, differing in terms of gradation, labeling, and stimulus specificity (Lawless and Heymann, 2010).

PROPERTIES OF INTENSITY LABELS We commonly use intensity descriptors to compare our experiences with the experiences of those around us (e.g., "This solution tastes *strong* to me. Does it taste *strong* to you?"). Because we use these words so frequently, they were incorporated as anchors on intensity scales.

Ratings from category scales generally have ordinal but not ratio properties (Stevens and Galanter, 1957), primarily because the equidistant spacing of labels does not reflect their actual perceived intensity (e.g., Berry and Huskisson, 1972; Lasagna, 1960): A sensation rated "8" on the Natick 9-point scale is more intense than a sensation rated "4," but it is not necessarily twice as intense. Consequently, the 9-point scale is especially prone to ceiling effects (Bartoshuk, 2000; Lucchina et al., 1998). Borg (1970, 1982) addressed this problem by deriving the first category scale with ratio properties, and subsequent efforts

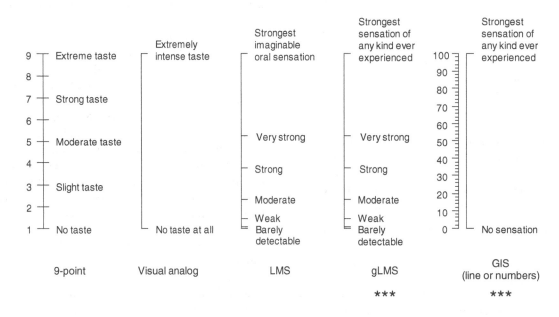

Figure 34.2 Examples of labeled scales used for oral sensory evaluation. Asterisks indicate scales for group comparisons of oral sensation; scales without asterisks require an independent standard for such comparisons. Note the change in context from imaginable to experienced events in recent versions of the gLMS/GIS. LMS = Labeled Magnitude Scale; gLMS = general Labeled Magnitude Scale; GIS = Global Intensity Scale.

to define scale labels empirically have generated similar "quasi-logarithmic" spacing across multiple sensory and hedonic attributes (Borg, 1990; Gracely et al., 1978; Green et al., 1993; Moskowitz, 1977a; Schutz and Cardello, 2001), indicating that sensory and hedonic experiences possess similar intensity properties.

Another important feature of intensity labels is that they convey relative and variable magnitude. Adjectives modify nouns, so their absolute meaning is flexible, particularly in terms of size. S. S. Stevens (1958, p. 633) illustrated this concept, noting: "Mice may be called large or small, and so may elephants, and it is quite understandable when someone says it was a large mouse that ran up the trunk of the small elephant." As this observation shows, "large" and "small" carry only relative meaning until their nouns are specified. Only when the context is defined can these words be placed in their proper perspective: A large mouse will always be smaller than a small elephant.

A similar point applies to intensity descriptors: The word "strong" varies in magnitude depending on whether it refers to a strong odor, a strong gust of wind, or a strong pain. Nevertheless, many comparisons made with labeled scales implicitly assume that scale descriptors denote the same absolute intensity regardless of the object described. Range theory, proposed by Borg (1982) and Teghtsoonian (1973), extended this idea by proposing that sensory ranges are essentially the same for all modalities and all people. While this idea would make across-subject comparisons much easier, it is incorrect (Bartoshuk et al., 2002b).

Indeed, intensity descriptor meanings vary among groups of people just as they do among different sensory modalities. This concept is illustrated by a study assessing the magnitudes denoted by scale descriptors for taste perception (Bartoshuk et al., 2004b). Participants used magnitude estimation to rate a variety of taste stimuli, a series of intensity descriptors, and a series of tones; taste intensity descriptors were then matched to sounds for nontasters and supertasters of PROP. As Figure 34.3 shows, the spacing among descriptors appears proportional for both groups, but the supertaster range is expanded: The descriptor "very strong" is almost twice as intense for supertasters as for nontasters, consistent with the idea that supertasters have a broader taste intensity range than do nontasters.

In short, labeled scales maintain their relative spacing, but they are elastic in terms of the domain to be measured and the individual's experience with that domain. Because conventional labeled scales fail to account for this elasticity, they can be used for within-subject comparisons or for across-group comparisons in which groups are assigned randomly. However, across-group comparisons are invalid whenever subject classification (e.g., sex, age, weight, clinical status) produces groups for which scale labels denote different absolute intensities.

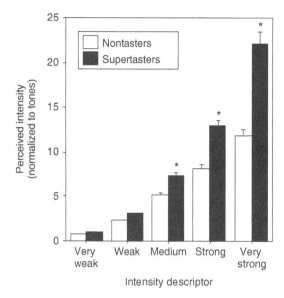

Figure 34.3 Taste intensity ratings associated with adjective descriptors for nontasters and supertasters of PROP. Data were collected using magnitude estimation and normalized to ratings for a 1000 Hz auditory stimulus. (Modified from Bartoshuk et al., 2004b).

CONSEQUENCES OF INVALID COMPARISONS

Figure 34.4 shows errors resulting from the false assumption that intensity descriptors denote the same absolute intensity to everyone. (This figure is idealized, but effects have been verified using taste and food stimuli; Bartoshuk et al., 2004a). The left side of the figure shows stimuli that produce equal perceived intensities to PROP nontasters. The diverging lines connecting nontaster and supertaster ratings indicate PROP effects of differing sizes; the intensity difference between groups for the label "very strong taste" is the same difference shown in Figure 34.3. When the label "very strong taste" is treated as if it denotes the same average intensity to nontasters and supertasters, supertaster data are compressed relative to nontaster data, as shown on the right side of Figure 34.4:

- Stimulus A appears more intense to supertasters than to nontasters, but the magnitude of the effect is blunted.
- The difference between nontasters and supertasters for stimulus C is equal to the difference between the labels, so it disappears.
- For stimulus D, the actual difference between nontasters and tasters is smaller than the difference in meaning for "very strong taste," so group differences appear to go in the opposite direction. This phenomenon is known as a *reversal artifact* (e.g., Bartoshuk and Snyder, 2004).

Although the limitations of labeled scales for group comparison have been discussed at length (e.g., Bartoshuk et al., 2002b, 2005; Biernat and Manis, 1994; Birnbaum,

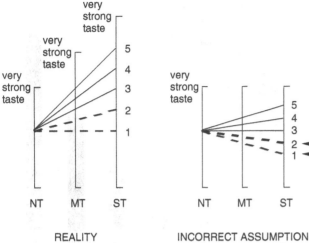

Figure 34.4 Consequences of invalid comparisons. The left panel shows taste functions reflecting real differences between nontasters (NT), medium tasters (MT), and supertasters (ST) measured with magnitude matching. The right panel shows the consequences of incorrectly assuming that "very strong taste" indicates the same absolute perceived intensity to NT, MT, and ST: Valid effects appear truncated and may reverse direction inaccurately. (Modified from Bartoshuk et al., 2004b).

1999; Manis et al., 1991; Narens and Luce, 1983), some critics have argued that group effects with significant biological impact should be sufficiently robust to be detected with any and all methods (Drewnowski, 2003). Claims like these are distortions themselves: Biological effects exist whether they are measured or not, but measurement tools are useless if they cannot detect those effects realistically. Moreover, the popularity of a scale does not necessarily make it the right tool for the task at hand. Although improved labeled scales show promise, contrary reports arising from invalid scaling methods remain significant obstacles to research efforts in the chemical senses.

GENERAL LABELED MAGNITUDE SCALE (gLMS)

Category scales assume ratio properties when real-world experience defines the spacing among labels. Given that intensity categories maintain their relative differences across individuals and modalities, might stretching this common intensity scale to its maximum produce a labeled scale allowing valid comparisons of oral sensory intensity? By this point, the Labeled Magnitude Scale (LMS) had been developed specifically to measure oral sensations (Green et al., 1993). As shown in Figure 34.2, the LMS is a ratio scale anchored by empirically-spaced intensity descriptors, including the top anchor "strongest imaginable [oral] sensation." To generalize the LMS for experiences beyond oral sensation, Bartoshuk and colleagues replaced its top anchor with the label "strongest imaginable sensation of any kind."

This scale, now known as the general LMS (gLMS), shares similar logic to magnitude matching: The top anchor of the gLMS is meant to function as a standard, so it must remain unrelated to oral sensation to ensure valid comparisons of chemosensory function. This requirement appears to be satisfied for studies of oral sensory experience: Participants rarely describe oral sensations as their strongest imaginable sensory experience (Bartoshuk et al., 2002b), and the gLMS and magnitude matching produce similar oral sensory differences among PROP taster groups (Bartoshuk et al., 2004c). Data collected with the gLMS also show strong test-retest reliability (Galindo-Cuspinera et al., 2009).

Although it was developed for sensory use, the gLMS also shows promise as a hedonic scale. Building on the empirical finding that intensity labels for sensation and affect are similarly spaced (e.g., Moskowitz, 1977b; Schutz and Cardello, 2001), a "hedonic gLMS" was created by extending two gLMSs in opposite directions from a common midpoint; "neutral" is in the center, "strongest imaginable disliking" anchors one end, and "strongest imaginable liking" anchors the other (Bartoshuk et al., 2002b). This scale, shown in Figure 34.5, has proven quite useful in recent studies of food affect: Overall food acceptance increases with body mass index, suggesting that the obese experience greater palatability from foods than do the non-obese (Bartoshuk et al., 2006); supertasters of PROP experience greater overall food liking and disliking than do nontasters, which may explain their selective food preferences (Bartoshuk et al., 2010). As with its sensory counterpart, the hedonic gLMS is appropriate for group comparisons because oral sensory cues rarely elicit affective ratings at the endpoints of the scale (Bartoshuk et al., 2010). Other hedonic scales may perform similarly to the gLMS in terms of group comparisons, provided their

34.2 Thresholds vs. Intensity: How Should Oral Sensation be Measured?

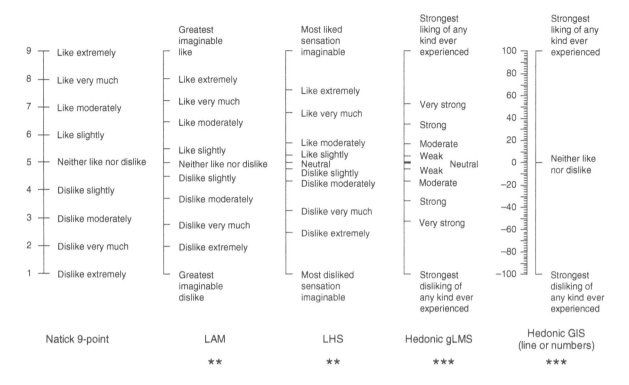

Figure 34.5 Examples of labeled scales used for hedonic evaluation. Asterisks indicate scales suitable for group comparisons of food-related affect; scales without asterisks require an independent standard for such comparisons. Note the change in context from imaginable events (**) to experienced events (***) in recent versions of the gLMS/GIS. LAM = Labeled Affective Magnitude scale; LHS = Labeled Hedonic Scale; gLMS = general Labeled Magnitude Scale; GIS = Global Intensity Scale.

boundary labels are explicitly framed in terms of all affective experience; promising candidates shown in Figure 34.5 include the Labeled Affective Magnitude Scale (Schutz and Cardello, 2001) and the Labeled Hedonic Scale (Lim et al., 2009).

Because the gLMS contains an internal standard, data collected with it should not require normalization. However, when the top of the gLMS is related to an experience of interest, comparisons using raw gLMS ratings are invalid. With regard to chemosensation, this problem may be especially relevant in studies of trigeminal function; the gLMS fails as a standard for measures of pain intensity because pain is often associated with the top of the scale. Accordingly, some researchers embed several candidate standards in their experiments; if one standard fails, others may succeed. For example, "brightest light ever seen" has emerged as a suitable standard for pain psychophysics (Bartoshuk et al., 2004a).

The standards used in the laboratory often require cumbersome and expensive equipment. Because scale labels rely on memories of perceived intensity, remembered sensations have been proposed as standards. Although the precise relationship between real and remembered intensity is unclear (e.g., Algom, 1992), remembered oral sensations appear to reflect effects observed with actual stimuli (e.g., Fast et al., 2001; Stevenson and Prescott, 1997). As indicated above, the incorporation of real and remembered stimuli as standards brings flexibility to data management; raw gLMS scores may be used (which invokes an internal standard), or they may be normalized to other variables (which converts the gLMS to a magnitude matching task), allowing confirmation of effects across a variety of assumptions. As a whole, these multimodal ratings coalesce into a snapshot of one's sensory and affective world, enabling individual differences of interest to resolve on a diffuse yet stable background (Bartoshuk et al., 2004b).

34.2.3 Global Intensity Scales: Beyond the gLMS

The similarity of intensity descriptor usage across widespread modalities (sensory and hedonic, real and remembered) suggests a cognitive model for intensity perception. We learn the relative spacing among intensity descriptors early in life, and these ratios remain constant. However, the absolute sensory range covered by intensity descriptors is variable, so we stretch or compress our descriptive vocabulary to fit the extremes of whatever we

762   Chapter 34   Psychophysical Measures of Human Oral Sensation

wish to describe. As a result, we may use the same words to describe just about anything, but our experience with the object at hand is critical in guiding our frame of reference. This concept raises several interesting psychophysical questions: How many intensity labels does a scale require? Are some labels more important than others? Which ones?

For example, the term "imaginable" is included in the top anchors of many intensity scales (e.g., "worst imaginable pain"), but is it really useful? Do all people imagine the same strongest sensation regardless of what they actually experience? When subjects rated both the most intense sensation of any kind ever experienced and the strongest sensation imaginable, the two were highly correlated when expressed relative to the brightest light ever seen (Fast et al., 2002). As such, "imaginable" sensations may be inappropriate standards for comparisons of experience, and the term has been removed from recent iterations of the gLMS, as shown in Figure 34.5; boundary labels now refer to the strongest intensity ever experienced.

With respect to other labels, the utility of remembered sensations suggests a more radical approach: Perhaps all labels should be abandoned except for those at the ends of the scale. The result (for sensation) would be a line denoting the distance from "no sensation" to the "strongest sensation of any kind ever experienced." Alternatively, one could simply use numerical ratings (e.g., 0 to 100 for sensation, −100 to +100 for affect) provided the appropriate endpoint meanings were conveyed. Recent data indicate that these minimalist "global intensity scales" (GIS), shown in Figure 34.5, perform similarly to the gLMS (Snyder et al., 2008b). As with the gLMS, GIS values could be used to make group comparisons unless the strongest sensation ever experienced showed a group difference, in which case an alternative standard would be used.

34.3   REGIONAL VS. WHOLE-MOUTH SENSATION: WHERE SHOULD ORAL SENSATION BE MEASURED?

Given its broad behavioral and health implications, oral sensory variation shows considerable promise as a diagnostic aid in epidemiology, sensory and consumer evaluation, and the health professions. To date, comprehensive assessments of oral sensory function have been used mainly in basic and clinical research, but there is growing appreciation in food science that information regarding baseline oral sensory viability aids a full understanding of chemosensory response.

Several afferent nerves carry sensory information from the mouth, each carrying a particular array of information

from a particular area (e.g., Pritchard and Norgren, 2004). The chorda tympani (CT), a branch of the facial nerve (cranial nerve VII), carries taste information from the anterior, mobile tongue; the lingual branch of the trigeminal nerve (V) carries pain, tactile, and temperature information primarily from the same region (Lewis and Dandy, 1930; Zahm and Munger, 1985). Multimodal information (i.e., taste, touch, pain, temperature) is carried from the posterior tongue by the glossopharyngeal nerve (IX), from the palate by the greater superficial petrosal nerve (another branch of VII), and from the throat by the vagus (X) (Fay, 1927; Kanagasuntheram et al., 1969; Norgren, 1984; Reichert, 1934). Taste and oral somatosensory cues combine centrally with retronasal olfaction to produce the composite experience of flavor (e.g., McBurney, 1986, Rozin, 1982). This spatial distribution of input has led researchers to consider the impact of localized oral sensory damage, so modern protocols for oral sensory evaluation typically include judgments of intensity and quality for both regional and whole-mouth stimuli.

34.3.1   Whole-Mouth Oral Sensation

In whole-mouth gustatory testing, chemical stimuli are sampled and moved throughout the mouth, stimulating all oral taste bud fields simultaneously; subjects rinse with water prior to each stimulus. Laboratory tests of oral sensation involve the presentation of chemical solutions at multiple concentrations spanning the functional range of perception (e.g., Bartoshuk et al., 1998; Tepper et al., 2001), but most clinical tests have streamlined this process to a single concentration for each of the common taste qualities (i.e., sucrose, NaCl, citric acid, quinine hydrochloride) (e.g., Frank et al., 2003). In addition, multiple concentrations may be used to derive suprathreshold taste functions, and other oral stimuli may be included to evaluate oral tactile sensation (e.g., capsaicin, alcohol) or individual differences (e.g., PROP). Having discussed the disadvantages of thresholds, we favor suprathreshold measurements involving magnitude matching or the gLMS/GIS with appropriate standards, so stimuli unrelated to oral sensation (e.g., sound, remembered sensations) should be incorporated.

Aqueous solutions are inconvenient for clinical, field, or large-scale use, so alternative methods of stimulus delivery have been explored, including paper strips, tablets, and edible films (Adler, 1972; Hummel et al., 1997; Mueller et al., 2003; Smutzer et al., 2008). In early studies, PTC crystals were placed directly on the tongue (Fox, 1932) or delivered on saturated filter papers (Blakeslee and Fox, 1932). As threshold techniques became dominant, the filter paper method faded from use until it was revived as a rough screening tool for lecture attendees (Bartoshuk et al., 1996). Today, PROP papers are made by soaking

34.3 Regional vs. Whole-Mouth Sensation: Where Should Oral Sensation be Measured?

laboratory-grade filter papers in a supersaturated solution of PROP heated to just below boiling. When dry, each paper contains ~1.6 mg (Bartoshuk et al., 1996); by comparison, patients with hyperthyroidism are prescribed 100–300 mg PROP daily (Cooper, 2005). Variants of this method have been described (e.g., Tepper et al., 2001; Zhao et al., 2003), but all share the common goal of introducing a small amount of crystalline PROP to the tongue surface.

Although intended as a rough estimate of PROP taster status, the filter paper technique shows promise because of its simplicity. Nevertheless, the technical disadvantages of this technique warrant careful consideration. To produce a taste, the paper must be completely moistened with saliva, which requires both healthy salivary function and a sufficient period of contact with the tongue. Some studies have reported excessive false-positive and false-negative responses to PTC/PROP filter papers, but these response rates may reflect minor variations in testing that confer bias (Azevedo et al., 1965; Lawless, 1980). Other critics have observed that filter paper testing shows only moderate concordance with threshold sensitivity (Hartmann, 1939), but threshold and suprathreshold measures of PROP bitterness almost always dissociate when proper scaling is used (e.g., Bartoshuk, 1989).

Despite these concerns, filter paper ratings and laboratory assessments of PROP bitterness show significant agreement (Bartoshuk et al., 1996; Kaminski et al., 2000; Zhao et al., 2003) and high test-retest reliability (Ly and Drewnowski, 2001; Zhao et al., 2003). One of the reasons that filter paper delivery functions so well may be that the exact concentration of PROP on each paper is trivial provided it is high. As filter papers soaked in saturated PROP dry, PROP crystals adhere to the paper; when a dried PROP paper is placed in the mouth, the crystals on the paper dissolve in saliva. Comparisons of PROP bitterness produced by filter papers vs. solution suggest that the concentration of PROP dissolved from the paper into saliva approaches the solubility limit of PROP. Because functions of PROP bitterness for nontasters, medium tasters, and supertasters diverge (Bartoshuk, 2000; Bartoshuk et al., 1994a), the most efficient way to sort subjects into these groups is to use the highest concentration possible. Based on this logic, papers made from saturated PROP (Bartoshuk et al., 1996) may be preferable to those made from lower concentrations (Zhao et al., 2003).

34.3.2 Oral Sensory Anatomy: Videomicroscopy of the Tongue

Several lines of evidence indicate that taste sensation varies with the number of taste buds stimulated. The most direct example of this relationship is that threshold and suprathreshold taste function increase with the size of the area stimulated on the tongue, presumably because a larger stimulus area captures more taste buds (Doty et al., 2001; Hara, 1955; Linschoten and Kroeze, 1991; Smith, 1971). Clinical data paint a complementary picture, as patients with lingual nerve damage show parallel losses of fungiform papillae and taste perception (Bull, 1965; Cowan, 1990; Ogden, 1989) that recover partially with nerve regeneration (Robinson et al., 2000; Zuniga et al., 1997). Finally, inbred strains of mice expressing differences in bitter avoidance (i.e., SWR/J tasters of sucrose octaacetate vs. C57BL/6J nontasters; Harder et al., 1984) show corresponding differences in taste bud density (Miller and Whitney, 1989) and gustatory nerve responses (Shingai and Beidler, 1985).

To explore variation in human oral anatomy, Miller and Reedy (1990a, 1990b) developed a method for visualizing the tongue in vivo (see also Shahbake et al., 2005). The anterior surfaces of human tongues are covered with two distinct types of papillae (i.e., raised structures); the larger fungiform papillae hold taste buds, but the smaller filiform papillae do not (e.g., Miller and Bartoshuk, 1991). When blue food coloring is applied to the tongue surface, it fails to stain fungiform papillae, which appear as pink circles against a blue background of filiform papillae. Fungiform papillae can then be counted with a magnifying glass and a flashlight. Videomicroscopy at higher magnification allows resolution of small blue dots on the surfaces of fungiform papillae; these dots are pores that function as conduits to the apical tips of taste buds.

Measured in this fashion, lingual anatomy has revealed robust positive associations between PROP intensity, fungiform papillae density, and taste bud density (Bartoshuk et al., 1994a; Delwiche et al., 2001; Miller and Reedy, 1990b; Zuniga et al., 1993). Fungiform papillae are dually innervated by CT and V (Gairns, 1953), which accounts for the elevated taste and oral tactile sensations experienced by supertasters (e.g., Essick et al., 2003; Prutkin et al., 2000). Videomicroscopy of the tongue has also enabled identification of some forms of oral sensory pathology. Human fungiform papillae and taste buds degenerate with lingual nerve damage (Zuniga et al., 1994 1997), but CT damage alone sacrifices taste buds while leaving fungiform papillae intact (Schwartz, 1998). Consequently, in individuals with taster genotypes and nontaster PROP ratings, high fungiform papilla counts signal anterior taste damage (e.g., Bartoshuk et al., 2004a).

Finally, the association between taste intensity and oral anatomy among healthy subjects can be used to evaluate the ability of various scales to provide valid across-group comparisons. To illustrate, Figure 34.6 compares two experiments in which the same stimuli were tasted. Subjects in one study rated perceived intensity with the gLMS, while subjects in the other study used a category scale. Note that the gLMS produces robust correlations between taste intensity and fungiform papillae density, while the category scale

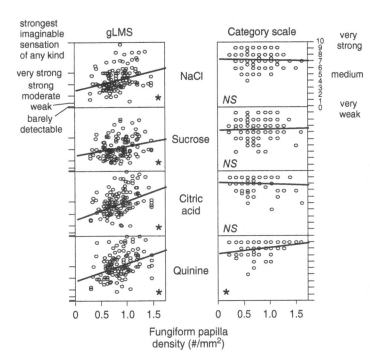

Figure 34.6 Perceived taste intensity as a function of fungiform papillae density. Graphs on left were obtained with the gLMS; graphs on right were obtained with a 9-point category scale. Asterisks indicate statistically significant effects ($p < 0.05$); NS = not significant. All correlations on the left are significant, but only the correlation for quinine is significant on the right. (Modified from Snyder et al., 2004).

does so only for bitter taste (Snyder et al., 2004). By showing that conventional labeled scales can distort a valid and meaningful difference, this result further confirms that the gLMS (and, by extension, magnitude matching) is superior for comparisons of group experience.

34.3.3 Spatial Taste Testing

Because different nerves innervate different regions of the oral cavity, oral sensation can be absent in one area but remain intact in others. Remarkably, individuals with extensive taste damage are often unaware of it unless it is accompanied by tactile loss (Bull, 1965; House, 1963; Moon and Pullen, 1963; Pfaffmann and Bartoshuk, 1989, 1990; Rice, 1963), presumably because taste cues are referred perceptually to sites in the mouth that are touched, whether or not taste receptors at those sites are present or functional (Delwiche et al., 2000; Green, 2002; Todrank and Bartoshuk, 1991). As a result of this "tactile referral," regional taste loss rarely produces whole-mouth taste loss, yet it remains clinically significant as a precursor to altered, heightened, and phantom oral sensations. Measures of regional taste function are an important tool for identifying the source of these complaints.

The integrity of specific taste nerves is assessed clinically via spatial testing (Bartoshuk, 1989), in which suprathreshold solutions of sweet, sour, salty, and bitter stimuli are applied with cotton swabs onto the anterior tongue tip, foliate papillae (i.e., posterolateral edges of tongue), circumvallate papillae (i.e., raised circular structures on posterior tongue), and soft palate. (A spatial taste test involving filter paper strips impregnated with taste stimuli has also been described; Mueller et al., 2003.) Stimuli are presented on the right and left sides at each locus, and subjects make quality and intensity judgments using magnitude matching or the gLMS/GIS. Special care must be taken to avoid stimulating both sides of the mouth simultaneously (which impedes localization), triggering a gag reflex during circumvallate stimulation, or allowing palate stimuli to reach the tongue surface (which leads to inflated palate ratings). Following regional testing, subjects swallow a small volume of each solution and rate its intensity, thereby enabling comparisons of regional and whole-mouth sensation. Comparisons of psychophysical functions across oral loci indicate that taste cues are perceived at similar intensity on all tongue areas holding taste buds, but less so on the palate (Bartoshuk, 1988). Thus, oral sensory losses can be identified as significant local variations from otherwise stable perception across the tongue surface.

For many years, virtually the only feature of taste perception mentioned in textbooks was a map showing the areas on the tongue sensitive to each of the four basic tastes: sweet on the tip, salt and sour on the edges, and bitter at the rear. Spatial taste testing demonstrates that this "tongue map" is inaccurate, and further study reveals that it was based on faulty evidence all along. As a student, Hänig (1901) showed that thresholds for the basic tastes show slight variation across various tongue loci, but he never proposed that individual taste modalities are restricted to specific regions of the tongue surface. The mistake occurred years later when Boring (1942) plotted

34.3 Regional vs. Whole-Mouth Sensation: Where Should Oral Sensation be Measured?

the reciprocal of Hänig's threshold values as a measure of the sensitivity of each tongue area. Boring did not include numbers on his graph, so subsequent readers failed to realize that his data actually represented very small threshold differences, and a myth was born.

34.3.3.1 Clinical Correlates of Localized Taste

Loss. Disorders of oral sensation are both widespread and variable, yet useful resources and appropriate medical treatment are frustratingly sparse (e.g., Deems et al., 1991; Hoffman et al., 1998). Because taste cues influence nutritional health, metabolism, and affect, their loss can be traumatic (e.g., Giduck et al., 1987; Mattes and Cowart, 1994; Tennen et al., 1991), yet in many cases the absence is hardly noticed (e.g., Bull, 1965; Goto et al., 1983; Tomita et al., 1986). To add another layer of complexity, gustatory disturbances are often associated with identifiable disorders and treatment interventions, but just as often they are of unknown origin and unpredictable onset (e.g., Bromley and Doty, 2003). Thus, a full oral sensory evaluation requires thorough examination of physical (e.g., oral anatomy, oral and salivary pathology, non-gustatory neurological damage), sensory (e.g., taste, oral somatosensation, retronasal olfaction), and emotional aspects of chemosensation (e.g., psychopathology, quality of life).

Spatial taste testing is most powerful when used in combination with genetic and anatomical data, as it reveals discrepancies between heredity and experience that arise via pathology (e.g., Snyder et al., 2006). Careful psychophysical study has revealed a great deal about the consequences of localized oral sensory damage; with the use of magnitude matching and the gLMS/GIS for valid across-group comparisons, oral sensory alterations have been implicated in an ever-growing array of health conditions. These relationships do not necessarily validate the methods used to discover them, but their concurrent relationship to measures of oral anatomy strongly suggests that improved psychophysical methods reflect meaningful differences in human experience. The examples that follow illustrate conditions in which modern oral sensory testing may facilitate diagnosis or intervention.

ORAL DISINHIBITION The term *dysgeusia* refers to a chronic taste that occurs in the absence of obvious stimulation (Snow et al., 1991). Many clinical complaints of dysgeusia result from taste stimuli that are not readily apparent to the patient (e.g., medications tasted in saliva, crevicular fluid, or blood; Alfano, 1974; Bradley, 1973; Fetting et al., 1985; Stephen et al., 1980), but some chronic taste sensations, known as phantoms, appear to arise centrally.

Neurological disorders can lead to taste phantoms (El-Deiry and McCabe, 1990; Hausser-Hauw and Bancaud, 1987), but CT damage appears to be a primary factor

in clinical accounts (Bull, 1965; Jones and Fry, 1984; Yamada and Tomita, 1989). Moreover, electrophysiological recordings from rodents and dogs show that blocking CT input produces elevated activity in brain regions receiving input from IX (Halpern and Nelson, 1965; Ninomiya and Funakoshi, 1982; Norgren and Pfaffmann, 1975; Ogawa and Hayama, 1984; Sweazey and Smith, 1987). These data indicate that CT inhibits IX normally, so CT loss should disinhibit IX. Human psychophysical data support this model: In both patient cohorts (e.g., head injury, craniofacial tumors, ear infections) and healthy subjects under anesthesia, unilateral CT loss leads to increased whole-mouth perceived bitterness via increased contralateral taste sensation at IX (Catalanotto et al., 1993; Kveton and Bartoshuk, 1994; Lehman et al., 1995; Yanagisawa et al., 1998), an effect that may occur preferentially for PROP supertasters (Snyder, 2010). The majority of oral sensory input rises ipsilaterally into the CNS (Norgren, 1990), so contralateral effects strongly implicate central modulation.

While under CT anesthesia, about 40% of healthy subjects experience taste phantoms contralaterally at IX; these sensations vary in quality and intensity, fading with the anesthetic (Yanagisawa et al., 1998). Topical anesthesia at the site of sensation abolishes these *release-of-inhibition* phantoms (Yanagisawa et al., 1998), presumably by suppressing spontaneous neural activity at their source (Norgren and Pfaffmann, 1975). Also, in one case report (Bartoshuk et al., 1994b), a bitter phantom arose bilaterally at IX following tonsillectomy. Spatial testing indicated complete IX loss, yet the phantom became more intense with whole-mouth topical anesthesia. This *nerve-stimulation phantom* was probably caused by surgical damage to IX and further disinhibited by CT anesthesia.

Similar to its interaction with IX, CT taste input also appears to inhibit trigeminal cues. This interaction may suppress oral pain during intake, and it may facilitate tactile referral of taste information following localized taste damage. Nevertheless, because supertasters have the most taste and trigeminal input, CT damage may lead to adverse sensory consequences due to extreme trigeminal disinhibition: Following unilateral CT anesthesia, PROP supertasters show increased ratings for the burn of capsaicin on the contralateral anterior tongue (Tie et al., 1999). In addition, severe childhood ear infections may compromise CT (e.g., Bartoshuk et al., 1996; Gedikli et al., 2001), leading to elevated tactile sensations from dietary fat that may promote sweet-fat food acceptance and long-term weight gain (e.g., Bartoshuk et al., 2012; Catalanotto et al., 2009; Kim et al., 2007; Snyder et al., 2003; Ventura et al., 2009).

Oral pain phantoms are another serious consequence of V disinhibition. The identification of burning mouth syndrome (BMS) as such a phantom vividly illustrates the clinical relevance of modern oral sensory testing. BMS, a condition most often found in postmenopausal women, is characterized by severe oral pain in the absence of visible pathology (Grushka and Sessle, 1987). BMS is often described as psychogenic, but systematic psychophysical testing tells a different story. Most BMS patients show significantly reduced bitterness for quinine on the anterior tongue, consistent with CT damage (Bartoshuk et al., 1999; Grushka and Bartoshuk, 2000). Nearly 50% of BMS patients experience taste phantoms at IX (Grushka et al., 1986), and topical anesthesia usually intensifies BMS-related taste and oral pain (Ship et al., 1995). Finally, the peak intensity of BMS pain correlates with fungiform papillae density, indicating that BMS is most prevalent among supertasters. Taken together, these data strongly suggest that BMS is an oral pain phantom generated by CT damage. Grushka has shown that agonists to the inhibitory neurotransmitter γ-aminobutyric acid (GABA) suppress BMS pain (Grushka et al., 1998), presumably by restoring lost inhibition from absent taste cues.

ORAL ANESTHESIA IN CLINICAL EVALUATION
Laboratory and clinical data support the use of topical anesthesia in the mouth to determine the locus of oral sensory dysfunction. However, interpretations of topical anesthesia must be made carefully, as incomplete anesthesia will hamper differential diagnosis. In topical anesthesia, patients hold ~5 mL of 0.5% dyclone in the mouth for 60 seconds, rest for 60 seconds, rinse with water, and describe any oral sensations experienced for the duration of the sensory block (Bartoshuk et al., 1994b). Anesthesia should always be accompanied by a thorough medical history, physical examination, and careful evaluation of spatial and whole-mouth oral sensation.

If a taste or oral pain complaint becomes *more intense* as a result of oral anesthesia, then it does not arise from normal stimulation of oral sensory receptors. Venous taste sensations and other dysgeusias may be caused by intake of certain therapeutic agents (Bradley, 1973; Fetting et al., 1985; Stephen et al., 1980), so the patient's use of medications and supplements should be reviewed. Another possibility is that the nerve innervating the region of sensory disturbance has sustained physical damage. If damage is peripheral to ganglion cell bodies, the resulting neuroma may produce a nerve-stimulation phantom; topical anesthesia exacerbates nerve-stimulation phantoms via central disinhibition (Bartoshuk et al., 1994b). Conclusions involving nerve damage should be confirmed by further neurological examination.

When local anesthesia *abolishes* a taste or oral pain complaint, an actual stimulus may be present in the mouth. To test for the presence of such a stimulus, the patient should attempt to rinse it from the mouth; if the offending sensation subsides at all, an actual stimulus should be considered. Alternatively, nerve damage unrelated to the altered sensation may have disinhibited input related to it, resulting in a central release-of-inhibition phantom; topical anesthesia further intensifies these phantoms (Bartoshuk et al., 1994b). Spatial testing should reveal localized taste loss at a site distant from the phantom.

34.3.4 Retronasal Olfaction

Odorants reach the nasal mucosa by two different routes: Orthonasal olfaction refers to odorants that enter the nostrils during sniffing, while retronasal olfaction refers to odors that are emitted by foods and forced behind the palate into the nasal cavity during mastication and swallowing (Rozin, 1982). This distinction is not merely conceptual, as orthonasal and retronasal olfaction are processed in different brain regions (e.g., Small et al., 2005; Small et al., 1997). Perceptually, retronasal odor cues fuse with taste and oral tactile sensations, culminating in a unitary sensation of flavor that is localized to the mouth. This process was long believed to involve oral touch alone (Hollingworth and Poffenberger, 1917), but growing data indicate that taste input plays a significant role: Retronasal and taste cues enhance one another when mixed (e.g., Frank et al., 1989; Kuo et al., 1993; Murphy et al., 1977), PROP supertasters experience stronger retronasal sensations than do nontasters (Bartoshuk et al., 2002a; Snyder et al., 2007), and regional taste loss compromises retronasal olfaction and whole-mouth taste in parallel (Bartoshuk et al., 2012; Snyder, 2010). Clinical accounts of chemosensory disturbance often commence with complaints of flavor loss, and industrial applications of oral sensory testing typically involve flavor effects of altered food composition or manufacturing, so retronasal olfactory testing has emerged as a useful element of oral sensory evaluation.

Experimental vehicles for the presentation of retronasal olfactory stimuli vary considerably, including odors infused into water or tastant solution, vapor-phase odorants released into the mouth, and sampled food items. All of these models are confounded to varying degrees by enhancement or "dumping" effects arising from odor-taste integration (e.g., Frank and Byram, 1988; Stevenson et al., 1999), compromising the validity of retronasal intensity ratings. As such, consistent procedures for retronasal olfactory evaluation have not been established, although there is general agreement that an oral stimulus must be present to induce sufficient levels of retronasal sensation. One promising approach used recently involves the spatial and temporal dissection of flavor components: Foods are sniffed (orthonasal olfaction) and then sampled with the nose plugged (taste and oral somatosensation), and the nose is released to elicit a distinct retronasal cue (Snyder, 2010). Testing in this fashion permits evaluation of flavor loss induced by oral sensory insult vs. other factors (e.g., olfactory deficits, nasal blockade).

The ability to evaluate distinct components of flavor experience holds promise for industrial efforts to enhance food palatability. For economic reasons, the food industry has long added taste stimuli to food to enhance its flavor, as commodities like sugar and salt are typically less expensive than flavor extracts (e.g., Noble, 1996). However, rising consumer awareness of health risks associated with added sugar and salt has raised interest in alternative strategies for flavor enhancement. Recent data from horticulture offer an intriguing possibility: Analysis of chemical and sensory measurements from tomatoes reveals that some flavor volatiles enhance sweetness independently of sugar (Tieman et al., 2012). In other words, some of the sweetness perceived in fruits and vegetables arises from sugars, and some of it arises from retronasal odor cues associated with sweetness. Thus, the addition of volatiles to food may represent a novel way to enhance sweetness, flavor, and palatability without adding sugar.

34.4 ORAL SENSORY EVALUATION IN FOOD SCIENCE

In research and clinical settings, the goal of oral sensory evaluation is to measure the perceptual ability of chemosensory systems. Although within-subject comparisons are an important tool in this arena — as when sensation is compared before and after a manipulation, such as surgery or adaptation — group comparisons are especially prominent as a means of linking oral sensation to broader health-related themes. As such, much of the present chapter has focused on the use of psychophysics to make sensory or affective comparisons among people whose chemosensory experiences vary in some way: Does the taste of sucrose differ between males and females? Do PROP supertasters experience greater pleasure from sucrose than do others?

A different purpose guides industrial applications of chemosensation, as observed by Lawless and Heymann (2010: 5): "The primary concern of any sensory evaluation specialist is to insure that the test method is appropriate to answer the questions being asked about the product in the test." Here, the focus is on the product rather than the individual, and healthy chemosensory function is either assumed or confirmed prior to testing. Consequently, differences among people are considered secondary to within-subject shifts following stimulus change: "From a practical point of food testing, the issue is not to uncover group differences, but rather to find differences among food products that will help workers in the food industry, food science or nutrition to provide food products with high acceptability to consumers" (Cardello et al., 2008:

474). In the future, this view may change with increasing use of sensory profiling in market segmentation to develop food products aimed at specific consumer groups.

Despite this fundamental difference, comparisons across foods and comparisons across consumers of food both require measurement of oral sensory experiences — for example, healthy chemosensory function must be confirmed in panelists, and manufacturers must identify optimal ingredient levels in their products. Relevant measures include threshold sensitivity, suprathreshold intensity across the dynamic range of a stimulus, and hedonic value. While the present chapter has outlined these methods, their specific use in food science is discussed at length in contemporary texts of sensory evaluation (e.g., Lawless and Heymann, 2010; Meilgaard et al., 2007).

34.5 CONCLUSION

Human psychophysics is a powerful and necessary aspect of clinical, basic, and consumer science, offering a window onto neurobehavioral processes that are often inaccessible by other means. In the chemical senses, significant progress has been made in crafting measurement tools that render individual differences accurately and allow adaptive use in clinical, research, industrial, and field settings. Conservative approaches to this task emphasize threshold measures, but suprathreshold scaling has evolved into a sophisticated means of measuring and comparing biologically relevant experiences at real-world levels. When performed correctly, these methods have enabled the study of oral sensory variation in health and disease, and the systematic use of techniques from psychophysics, anatomy, neurology, and genetics has revealed complex relationships between oral sensation, affect, behavior, and pathology. In short, ongoing refinements in oral chemosensory assessment have strong clinical, food science, and basic research implications, as they yield highly predictive and highly comparable measures of sensory and hedonic experience.

REFERENCES

Adler, J. (1972). Chemoreception in bacteria. In *Olfaction and Taste IV*, D. Schneider (Ed.). Stuttgart: Wissenschaftliche Verlagsgesellschaft MBH, pp. 70–80.

Aitken, R. C. B., Ferres, H. M., and Gedye, J. L. (1963). Distraction from flashing lights. *Aerosp. Med.*, 34: 302–306.

Alfano, M. (1974). The origin of gingival fluid. *J. Theor. Biol.*, 47: 127–136.

Algom, D. (1992). Memory psychophysics: An examination of its perceptual and cognitive prospects. In *Psychophysical Approaches to Cognition*, D. Algom (Ed.). New York: North-Holland, pp. 441–513.

Azevedo, E., Krieger, H., Mi, M. P., and Morton, N. E. (1965). PTC taste sensitivity and endemic goiter in Brazil. *Am. J. Hum. Genet.*, 17: 87–90.

Barnicot, N. A., Harris, H., and Kalmus, H. (1951). Taste thresholds of further eighteen compounds and their correlation with P.T.C. thresholds. *Ann. Eugen.*, 16: 119–128.

Bartoshuk, L. M. (1974). NaCl thresholds in man: Thresholds for water taste or NaCl taste? *J. Comp. Physiol. Psychol.*, 87: 310–325.

Bartoshuk, L. M. (1978). The psychophysics of taste. *Am. J. Clin. Nutr.*, 31: 1068–1077.

Bartoshuk, L. M. (1979). Bitter taste of saccharin: Related to the genetic ability to taste the bitter substance 6-*n*-propylthiouracil (PROP). *Science*, 205: 34–935.

Bartoshuk, L. M. (1988). Clinical psychophysics of taste. *Gerodontics*, 4: 249–255.

Bartoshuk, L. M. (1989). Clinical evaluation of the sense of taste. *Ear Nose Throat J.*, 68: 331–337.

Bartoshuk, L. M. (2000). Comparing sensory experiences across individuals: Recent psychophysical advances illuminate genetic variation in taste perception. *Chem. Senses*, 25: 447–460.

Bartoshuk, L. M., Catalanotto, F. A., Hoffman, H., et al. (2012). Taste damage (otitis media, tonsillectomy, and head and neck cancer), oral sensations, and BMI. *Physiol. Behav.*, 107: 516–526.

Bartoshuk, L. M., Chapo, A. K., Duffy, V. B., et al. (2002a). Oral phantoms: Evidence for central inhibition produced by taste. *Chem. Senses*, 27: A52.

Bartoshuk, L. M., Desnoyers, S., Hudson, C. A., et al. (1987). Tasting on localized areas. *Ann. N. Y. Acad. Sci.*, 510: 166–168.

Bartoshuk, L. M., and Duffy, V. B. (1994). Supertasting and earaches: Genetics and pathology alter our taste worlds. *Appetite*, 23: 292–293.

Bartoshuk, L. M., Duffy, V. B., Chapo, A. K., et al. (2004a). From psychophysics to the clinic: Missteps and advances. *Food Qual. Pref.*, 15: 617–632.

Bartoshuk, L. M., Duffy, V. B., Fast, K., et al. (2002b). Labeled scales (e.g., category, Likert, VAS) and invalid across-group comparisons: What we have learned from genetic variation in taste. *Food Qual. Pref.*, 14: 125–138.

Bartoshuk, L. M., Duffy, V. B., Fast, K., et al. (2001). What makes a supertaster? *Chem. Senses*, 26: 1074.

Bartoshuk, L. M., Duffy, V. B., Fast, K., and Snyder, D. J. (2004b). Genetic differences in human oral perception: Advanced methods reveal basic problems in intensity scaling. In *Genetic Variation in Taste Sensitivity: Measurement, Significance, and Implications*, J. Prescott and B. J. Tepper (Eds.). New York: Marcel Dekker, pp. 1–42.

Bartoshuk, L. M., Duffy, V. B., Green, B. G., et al. (2004c). Valid across-group comparisons with labeled scales: The gLMS vs. magnitude matching. *Physiol. Behav.*, 82: 109–114.

Bartoshuk, L. M., Duffy, V. B., Hayes, J. E., et al. (2006). Psychophysics of sweet and fat perception in obesity: Problems, solutions, and new perspectives. *Philos. Trans. R. Soc., B*, 361: 1137–1148.

Bartoshuk, L. M., Duffy, V. B., Lucchina, L. A., et al. (1998). PROP (6-*n*-propylthiouracil) supertasters and the saltiness of NaCl. *Ann. N. Y. Acad. Sci.*, 855: 793–796.

Bartoshuk, L. M., Duffy, V. B., and Miller, I. J. (1994a). PTC/PROP tasting: Anatomy, psychophysics, and sex effects. *Physiol. Behav.*, 56: 1165–1171.

Bartoshuk, L. M., Duffy, V. B., Reed, D. R., and Williams, A. L. (1996). Supertasting, earaches, and head injury: Genetics and pathology alter our taste worlds. *Neurosci. Biobehav. Rev.*, 20: 79–87.

Bartoshuk, L. M., Fast, K., and Snyder, D. J. (2005). Differences in our sensory worlds: Invalid comparisons with labeled scales. *Curr. Dir. Psychol. Sci.*, 14: 122–125.

Bartoshuk, L. M., Grushka, M., Duffy, V. B., et al. (1999). Burning mouth syndrome: Damage to CN VII and pain phantoms in CN V. *Chem. Senses*, 24: 609.

Bartoshuk, L. M., Kalva, J. J., Puentes, L. A., et al. (2010). Valid comparisons of food preferences. *Chem. Senses*, 35: A20.

Bartoshuk, L. M., Kveton, J. F., Yanagisawa, K., and Catalanotto, F. A. (1994b). Taste loss and taste phantoms: A role of inhibition in taste. In *Olfaction and Taste XI*, K. Kurihara, N. Suzuki, and H. Ogawa (Eds.). New York: Springer-Verlag, pp. 557–560.

Bartoshuk, L. M., Rifkin, B., Marks, L. E., and Bars, P. (1986). Taste and aging. *J. Gerontol.*, 41: 51–57.

Bartoshuk, L. M., and Snyder, D. J. (2004). Psychophysical measurement of human taste experience. In *Handbook of Behavioral Neurobiology (Vol. 14: Neurobiology of Food and Fluid Intake)*, 2nd ed., E. M. Stricker and S. C. Woods (Eds.). New York: Plenum Press, pp. 89–107.

Basson, M. D., Bartoshuk, L. M., Dichello, S. Z., et al. (2005). Association between 6-*n*-propylthiouracil (PROP) bitterness and colonic neoplasms. *Dig. Dis. Sci.*, 50: 483–489.

Berry, H., and Huskisson, E. C. (1972). Treatment of rheumatoid arthritis. *Clin. Trials J.*, 9: 13–14.

Biernat, M., and Manis, M. (1994). Shifting standards and stereotype-based judgements. *J. Pers. Soc. Psychol.*, 66: 5–20.

Birnbaum, M. H. (1999). How to show that 9 > 221: Collect judgements in a between-subjects design. *Psychol. Methods*, 4, 243–249.

Blakeslee, A. F. (1932). Genetics of sensory thresholds: Taste for phenyl thio carbamide. *Proc. Natl. Acad. Sci.*, 18: 120–130.

Blakeslee, A. F., and Fox, A. L. (1932). Our different taste worlds. *J. Hered.*, 23: 97–107.

Borg, G. (1970). Perceived exertion as an indicator of somatic stress. *Scand. J. Rehab. Med.*, 2: 92–98.

Borg, G. (1982). A category scale with ratio properties for intermodal and interindividual comparisons. In *Psychophysical Judgment and the Process of Perception*, H.G. Geissler & P. Petzold (Eds.). Berlin: VEB Deutscher Verlag der Wissenschaften, pp. 25–34.

Borg, G. (1990). Psychophysical scaling with applications in physical work and the perception of exertion. *Scand. J. Work Environ. Health*, 16 (Suppl 1): 55–58.

Boring, E. G. (1942). *Sensation and perception in the history of experimental psychology*. New York: Appleton.

Bradley, R. M. (1973). Electrophysiological investigations of intravascular taste using perfused rat tongue. *Am. J. Psychol.*, 224: 300–304.

Brindley, G. S. (1960). *Physiology of the Retina and the Visual Pathway*. London: Edward Arnold Ltd.

Bromley, S. M., and Doty, R. L. (2003). Clinical disorders affecting taste: Evaluation and management. In *Handbook of Olfaction and Gustation*, R. L. Doty (Ed.). New York: Marcel Dekker, pp. 935–957.

Bufe, B., Breslin, P. A. S., Kuhn, C., et al. (2005). The molecular basis of individual differences in phenylthiocarbamide and propylthiouracil bitterness perception. *Curr. Biol.*, 15: 322–327.

Bujas, Z. (1971). Electrical taste. In *Handbook of Sensory Physiology (Vol. 4, Part 2: Chemical Senses — Taste)*, L. M. Beidler (Ed.). Berlin: Springer, pp. 180–199.

Bull, T. R. (1965). Taste and the chorda tympani. *J. Laryngol. Otol.*, 79: 479–493.

Cain, W. S., and Stevens, J. C. (1989). Uniformity of olfactory loss in aging. *Ann. N. Y. Acad. Sci.*, 561: 29–38.

Calò, C., Padiglia, A., Zonza, A., et al. (2011). Polymorphisms in TAS2R38 and the taste bud trophic factor gustin gene co-operate in modulating PROP taste phenotype. *Physiol. Behav.*, 104: 1065–1071.

Cardello, A. V., Lawless, H. T., and Schutz, H. G. (2008). Effects of extreme anchors and interior label spacing on labeled affective magnitude scales. *Food Qual. Pref.*, 19: 473–480.

Catalanotto, F. A., Bartoshuk, L. M., Östrom, K. M., et al. (1993). Effects of anesthesia of the facial nerve on taste. *Chem. Senses*, 18: 461–470.

Catalanotto, F. A., Broe, E. T., Bartoshuk, L. M., et al. (2009). Otitis media and intensification of non-taste oral sensations. *Chem. Senses*, 34: A120.

Collings, V. B. (1974). Human taste response as a function of locus of stimulation on the tongue and soft palate. *Percept. Psychophys.*, 16: 169–174.

Cooper, D. S. (2005). Antithyroid drugs. *New England Journal of Medicine*, 352: 905.

Cornsweet, T. N. (1962). The staircase method in psychophysics. *Am. J. Psychol.*, 75: 485–491.

Cowan, P. W. (1990). "Atrophy of fungiform papillae following lingual nerve damage" — A suggested mechanism. *Br. Dent. J.*, 168: 95.

Cowart, B. J. (1989). Relationships between taste and smell across the adult life span. *Ann. N. Y. Acad. Sci.*, 561: 39–55.

Deems, R. O., Friedman, M. I., Friedman, L. S., and Maddrey, W. C. (1991). Clinical manifestations of olfactory and gustatory disorders associated with hepatic and renal disease. In *Smell and Taste in Health and Disease*, T. V. Getchell, R. L. Doty, L. M. Bartoshuk, & J. B. Snow, Jr., (Eds.). New York: Raven Press, pp. 805–816.

Delwiche, J. F., Buletic, Z., and Breslin, P. A. S. (2001). Relationship of papillae number to bitter intensity of quinine and PROP within and between individuals. *Physiol. Behav.*, 74: 329–337.

Delwiche, J. F., Lera, M. F., and Breslin, P. A. S. (2000). Selective removal of a target stimulus localized by taste in humans. *Chem. Senses*, 25: 181–187.

Desai, H., Smutzer, G., Coldwell, S. E., and Griffith, J. W. (2011). Validation of edible taste strips for identifying PROP taste recognition thresholds. *Laryngoscope*, 121: 1177–1183.

DeSimone, J. A., Heck, G. L., and Bartoshuk, L. M. (1980). Surface active taste modifiers: A comparison of the physical and psychophysical properties of gymnemic acid and sodium lauryl sulfate. *Chem. Senses*, 5: 317–330.

DeSimone, J. A., Heck, G. L., and DeSimone, S. K. (1981). Active ion transport in dog tongue: A possible role in taste. *Science*, 214: 1039–1041.

DeSimone, J. A., Heck, G. L., Mierson, S., and DeSimone, S. K. (1984). The active ion transport properties of canine lingual epithelia in vitro: Implications for gustatory transduction. *J. Gen. Physiol.*, 83: 633–656.

Dinehart, M. E., Hayes, J. E., Bartoshuk, L. M., et al. (2006). Bitter taste markers explain variability in vegetable sweetness, bitterness, and intake. *Physiol. Behav.*, 87: 304–313.

Doty, R. L., Bagla, R., Morgenson, M., and Mirza, N. (2001). NaCl thresholds: Relationship to anterior tongue locus, area of stimulation, and number of fungiform papillae. *Physiol. Behav.*, 72: 373–378.

Drayna, D., Coon, H., Kim, U.-K., et al. (2003). Genetic analysis of a complex trait in the Utah Genetic Reference Project: A major locus for PTC taste ability on chromosome 7q and a secondary locus on chromosome 16p. *Hum. Genet.*, 112: 567–572.

Drewnowski, A. (2003). Genetics of human taste perception. In *Handbook of Olfaction and Gustation*, 2nd ed., R. L. Doty (Ed.). New York: Marcel Dekker, pp. 847–860.

Drewnowski, A., Henderson, S. A., Hann, C. S., et al. (2000). Genetic taste markers and preferences for vegetables and fruit of female breast care patients. *J. Am. Diet. Assoc.*, 100: 191–197.

Duffy, V. B. (2004). Associations between oral sensation, dietary behaviors, and risk of cardiovascular disease (CVD). *Appetite*, 43: 5–9.

Duffy, V. B. (2007). Variation in oral sensation: Implications for diet and health. *Curr. Opin. Gastroenterol.*, 23: 171–177.

Duffy, V. B., Phillips, M. N., Peterson, J. M., and Bartoshuk, L. M. (2001). Bitterness of 6-*n*-propylthiouracil (PROP) associates with bitter sensations and intake of vegetables. *Appetite*, 37: 137–138.

El-Deiry, A., and McCabe, B. F. (1990). Temporal lobe tumor manifested by localized dysgeusia. *Ann. Otol. Rhinol. Laryngol.*, 99: 586–587.

Engen, T. (1972). Psychophysics. I. Discrimination and Detection. In *Woodworth and Schlosberg's Experimental Psychology (Vol. 1: Sensation and Perception)*, J. W. Kling & L. A. Riggs (Eds.). New York: Holt, Rinehart and Winston, Inc, pp. 11–46.

Essick, G. K., Chopra, A., Guest, S., and McGlone, F. (2003). Lingual tactile acuity, taste perception, and the density and diameter of fungiform papillae in female subjects. *Physiol. Behav.*, 80: 289–302.

Fast, K., Green, B. G., and Bartoshuk, L. M. (2002). Developing a scale to measure just about anything: Comparisons across groups and individuals. *Appetite*, 39: 75.

Fast, K., Green, B. G., Snyder, D. J., and Bartoshuk, L. M. (2001). Remembered intensities of taste and oral burn correlate with PROP bitterness. *Chem. Senses*, 26: 1069.

Fay, T. (1927). Observations and results from intracranial section of the glossopharyngeus and vagus nerves in man. *J. Neurol. Psychopathol.*, 8, 110–123.

Fechner, G. (1860/1966). *Elements of Psychophysics* (H.E. Adler, Trans.). New York: Holt, Rinehart and Winston.

Fetting, J. H., Wilcox, P. M., Sheidler, V. R., et al. (1985). Tastes associated with parenteral chemotherapy for breast cancer. *Cancer Treat. Rep.*, 69: 1249–1251.

Fischer, R. (1967). Genetics and gustatory chemoreception in man and other primates. In *The Chemical Senses and Nutrition*, M. R. Kare and O. Maller (Eds.). Baltimore: John Hopkins Press, pp. 621–681.

Fischer, R., Griffin, F., England, S., and Garn, S. (1961). Taste thresholds and food dislikes. *Nature*, 191: 1328.

Fischer, R., Griffin, F., and Kaplan, A. R. (1963). Taste thresholds, cigarette smoking, and food dislikes. *Med. Exp.*, 9: 151–167.

Fox, A. L. (1931). Six in ten "tasteblind" to bitter chemical. *Sci. News Lett.*, 9, 249.

Fox, A. L. (1932). The relationship between chemical constitution and taste. *Proc. Natl. Acad. Sci.*, 18: 115–120.

Frank, M. E., Hettinger, T. P., and Barry, M. A. (2003). Contemporary measurement of human gustatory function. In *Handbook of Olfaction and Gustation*, 2nd ed., R. L. Doty (Ed.). New York: Marcel Dekker, pp. 783–804.

Frank, M. E., and Smith, D. V. (1991). Electrogustometry: A simple way to test taste. In *Smell and Taste in Health and Disease*, T. V. Getchell, R. L. Doty, L. M. Bartoshuk, & J. B. Snow, Jr., (Eds.). New York: Raven Press, pp. 503–514.

Frank, R. A., and Byram, J. (1988). Taste-smell interactions are tastant and odorant dependent. *Chem. Senses*, 13, 445–455.

Frank, R. A., Ducheny, K., and Mize, S. J. S. (1989). Strawberry odor, but not red color, enhances the sweetness of sucrose solutions. *Chem. Senses*, 14: 371–377.

Fushan, A. A., Simons, C. T., Slack, J. P., and Drayna, D. (2010). Association between common variation in genes encoding sweet taste signaling components and human sucrose perception. *Chem. Senses*, 35: 579–592.

Gairns, F. W. (1953). Sensory endings other than taste buds in the human tongue. *J. Physiol.*, 121: 33P–34P.

Galindo-Cuspinera, V., Waeber, T., Antille, N., et al. (2009). Reliability of threshold and suprathreshold methods for taste phenotyping: Characterization with PROP and sodium chloride. *Chemosens. Percept.*, 2: 214–228.

Gedikli, O., Doğru, H., Aydin, G., et al. (2001). Histopathological changes of chorda tympani in chronic otitis media. *Laryngoscope*, 111: 724–727.

Giachetti, I., and MacLeod, P. (1975). Cortical neuron responses to odors in th rat. In D.A. Denton and J. P. Coghlan (Eds.), Olfaction and Taste V. New York: Academic Press, pp. 303–307.

Giachetti, I., and MacLeod, P. (1977). Olfactory input to the thalamus: evidence for a ventroposteromedial projection. *Brain Res.* 125: 166–169.

Giduck, S. A., Threatte, R. M., and Kare, M. R. (1987). Cephalic reflexes: Their role in digestion and possible roles in absorption and metabolism. *J. Nutr.*, 117: 1191–1196.

Goto, N., Yamamoto, T., Kaneko, M., and Tomita, H. (1983). Primary pontine hemorrhage and gustatory disturbance: Clinicoanatomic study. *Stroke*, 14: 507–511.

Gracely, R. H., McGrath, P., and Dubner, R. (1978). Validity and sensitivity of ratio scales of sensory and affective verbal pain descriptors: Manipulation of affect by diazepam. *Pain*, 5: 19–29.

Grant, R., Ferguson, M. M., Strang, R., et al. (1987). Evoked taste thresholds in a normal population and the application of electrogustometry to trigeminal nerve disease. *J. Neurol. Neurosurg. Psychiatry*, 50: 12–21.

Green, B. G. (2002). Studying taste as a cutaneous sense. *Food Qual. Pref.*, 14, 99–109.

Green, B. G., and George, P. (2004). "Thermal taste" predicts higher responsiveness to chemical taste and flavor. *Chem. Senses*, 29: 617–628.

Green, B. G., and Hayes, J. E. (2004). Individual differences in perception of bitterness from capsaicin, piperine, and zingerone. *Chem. Senses*, 29: 53–60.

Green, B. G., Shaffer, G. S., and Gilmore, M. M. (1993). A semantically-labeled magnitude scale of oral sensation with apparent ratio properties. *Chem. Senses*, 18: 683–702.

Green, D. M., and Swets, J. A., (1966). *Signal Detection Theory and Psychophysics*. New York: Wiley.

Groves, J., and Gibson, W. P. (1974). Significance of taste and electrogustometry in assessing the prognosis of Bell's (idiopathic) facial palsy. *J. Laryngol. Otol.*, 88: 855–861.

Grushka, M., and Bartoshuk, L. M. (2000). Burning mouth syndrome and oral dysesthesias. *Can. J. Diagn.*, 17: 99–109.

Grushka, M., Epstein, J., and Mott, A. (1998). An open-label, dose escalation pilot study of the effect of clonazepam in burning mouth syndrome. *Oral Surg. Oral Med. Oral Pathol. Oral Radiol. Endod.*, 86: 557–561.

Grushka, M., and Sessle, B. J. (1987). Burning mouth syndrome: A historical review. *Clin. J. Pain*, 2: 245–252.

Grushka, M., Sessle, B. J., and Howley, T. P. (1986). Psychophysical evidence of taste dysfunction in burning mouth syndrome. *Chem. Senses*, 11: 485–498.

Hall, M. J., Bartoshuk, L. M., Cain, W. S., and Stevens, J. C. (1975). PTC taste blindness and the taste of caffeine. *Nature*, 253: 442–443.

Halpern, B. P., and Nelson, L. M. (1965). Bulbar gustatory responses to anterior and to posterior tongue stimulation in the rat. *Am. J. Physiol.*, 209: 105–110.

Hänig, D. P. (1901). Zur psychophysik des geschmackssinnes. *Philos. Stud.*, 17: 576–623.

Hara, S. (1955). Interrelationship among stimulus intensity, stimulated area, and reaction time in the human gustatory sensation. *Bull. Tokyo Med. Dent. Univ.*, 2: 147–158.

Harder, D. B., Whitney, G., Frye, P., et al. (1984). Strain differences among mice in taste psychophysics of sucrose octaacetate. *Chem. Senses*, 9: 311–323.

Harris, H., and Kalmus, H. (1949a). Chemical sensitivity in genetical differences of taste sensitivity. *Ann. Eugen.*, 15: 32–45.

Harris, H., and Kalmus, H. (1949b). The measurement of taste sensitivity to phenylthiourea (P.T.C.). *Ann. Eugen.*, 15: 24–31.

Hartmann, G. (1939). Application of individual taste difference towards phenyl-thio-carbamide in genetic investigations. *Ann. Eugen.*, 9: 123–135.

Hausser-Hauw, C., and Bancaud, J. (1987). Gustatory hallucinations in epileptic seizures. *Brain*, 110: 339–359.

Hayes, J. E., Bartoshuk, L. M., Kidd, J. R., and Duffy, V. B. (2008). Supertasting and PROP bitterness depends on more than the TAS2R38 gene. *Chem. Senses*, 33: 255–265.

Hayes, J. E., and Keast, R. S. J. (2011). Two decades of supertasting: Where do we stand? *Physiol. Behav.*, 104: 1072–1074.

Heath, T. P., Melichar, J. K., Nutt, D. J., and Donaldson, L. F. (2006). Human taste thresholds are modulated by serotonin and noradrenaline. *J. Neurosci.*, 26: 12664–12671.

Henkin, R. I., Gill, J. R., and Bartter, F. C. (1963). Studies on taste thresholds in normal man and in patients with adrenal cortical insufficiency: the role of adrenal cortical steroids and of serum sodium concentration. *J. Clin. Invest.*, 42, 727–735.

Herness, M. S. (1985). Neurophysiological and biophysical evidence on the mechanism of electric taste. *J. Gen. Physiol.*, 86: 59–87.

Hertz, J., Cain, W. S., Bartoshuk, L. M., and Dolan, T. F. (1975). Olfactory and taste sensitivity in children with cystic fibrosis. *Physiol. Behav.*, 14: 89–94.

Hetherington, M. M., and Rolls, B. J. (1987). Methods of investigating human eating behavior. In *Feeding and Drinking*, F. M. Toates & N. E. Rowland (Eds.). New York: Elsevier, pp. 77–109.

Hoffman, H. J., Ishii, E. K., and Macturk, R. H. (1998). Age-related changes in the prevalence of smell/taste problems among the United States adult population: Results of the 1994 Disability Supplement to the National Health Interview Survey (NHIS). *Ann. N. Y. Acad. Sci.*, 855: 716–722.

Hollingworth, H. L., and Poffenberger, A. T. (1917). *The Sense of Taste*. New York: Moffat, Yard and Company.

House, H. P. (1963). Early and late complications of stapes surgery. *Arch. Otolaryngol.*, 78: 606–613.

Hummel, T., Erras, A., and Kobal, G. (1997). A test for screening of taste function. *Rhinology*, 35: 146–148.

Ileri-Gurel, E., Pehlivanoglu, B., and Dogan, M. (2013). Effect of acute stress on taste perception: In relation with baseline anxiety level and body weight. *Chem. Senses*, 38: 27–34.

Jesteadt, W. (1980). An adaptive procedure for subjective judgments. *Percept. Psychophys.*, 28(1): 85–88.

Jones, L. V., Peryam, D. R., and Thurstone, L. L. (1955). Development of a scale for measuring soldier's food preferences. *Food Res.*, 20: 512–520.

Jones, R. O., and Fry, T. L. (1984). A new complication of prosthetic ossicular reconstruction. *Arch. Otolaryngol.*, 110: 757–758.

Kalmus, H., and Farnsworth, D. (1959). Impairment and recovery of taste following irradiation of the oropharynx. *J. Laryngol. Otol.*, 73: 180–182.

Kamen, J. M., Pilgrim, F. J., Gutman, N. J., and Kroll, B. J. (1961). Interactions of suprathreshold taste stimuli. *J. Exp. Psychol.*, 62: 348–356.

Kaminski, L. C., Henderson, S. A., and Drewnowski, A. (2000). Young women's food preferences and taste responsiveness to 6-*n*-propylthiouracil (PROP). *Physiol. Behav.*, 68: 691–697.

Kanagasuntheram, R., Wong, W. C., and Chan, H. K. (1969). Some observations of the innervation of the human nasopharynx. *J. Anat.*, 104: 361–376.

Kashiwayanagi, M., Yoshii, K., Kobatake, Y., and Kurihara, K. (1981). Taste transduction mechanism: Similar effects of various modifications of gustatory receptors on neural responses to chemical and electrical stimulation in the frog. *J. Gen. Physiol.*, 78: 259–265.

References

Keller, K. L., Liang, L. C., Sakimura, J., et al. (2012). Common variants in the CD36 gene are associated with oral fat perception, fat preferences, and obesity in African Americans. *Obesity*, 20: 1066–1073.

Kim, J. B., Park, D. C., Cha, C. I., and Yeo, S. G. (2007). Relationship between pediatric obesity and otitis media with effusion. *Arch. Otolaryngol. Head Neck Surg.*, 133: 379–382.

Kim, U.-K., Jorgenson, E., Coon, H., et al. (2003). Positional cloning of the human quantitative trait locus underlying taste sensitivity to phenylthiocarbamide. *Science*, 299: 1221–1225.

Kranzler, H. R., Skipsey, K., and Modesto-Lowe, V. (1998). PROP taster status and parental history of alcohol dependence. *Drug Alcohol Depend.*, 52, 109–113.

Krarup, B. (1958). Taste reactions of patients with Bell's palsy. *Acta Oto-Laryngol.*, 49: 389–399.

Kuo, Y.-L., Pangborn, R. M., and Noble, A. C. (1993). Temporal patterns of nasal, oral, and retronasal perception of citral and vanillin and interaction of these odorants with selected tastants. *Int. J. Food Sci. Tech.*, 28: 127–137.

Kveton, J. F., and Bartoshuk, L. M. (1994). The effect of unilateral chorda tympani damage on taste. *Laryngoscope*, 104: 25–29.

Laming, D. (1997). *The Measurement of Sensation*. Oxford, New York: Oxford University Press.

Lasagna, L. (1960). The clinical measurement of pain. *Ann. N. Y. Acad. Sci.*, 86: 28–37.

Lawless, H. T. (1987). Gustatory psychophysics. In *Neurobiology of Taste and Smell*, T.E. Finger & W.L. Silver (Eds.). New York: Wiley, pp. 401–420.

Lawless, H. T. (1980). A comparison of different methods used to assess sensitivity to the taste of phenylthiocarbamide (PTC). *Chem. Senses*, 5: 247–256.

Lawless, H. T., and Heymann, H. (2010). *Sensory Evaluation of Food: Principles and Practices*, 2nd ed. New York: Springer.

Lawless, H. T., Thomas, C. J. C., and Johnston, M. (1995). Variation in odor thresholds for l-carvone and cineole and correlations with suprathreshold intensity ratings. *Chem. Senses*, 20: 9–17.

Le Floch, J. P., Le Lievre, G., Verroust, J., et al. (1990). Factors related to the electric taste threshold in type 1 diabetic patients. *Diabet. Med.*, 7: 526–531.

Lehman, C. D., Bartoshuk, L. M., Catalanotto, F. A., et al. (1995). The effect of anesthesia of the chorda tympani nerve on taste perception in humans. *Physiol. Behav.*, 57: 943–951.

Lewis, D., and Dandy, W. E. (1930). The course of the nerve fibers transmitting sensation of taste. *Arch. Surg.*, 21: 249–288.

Likert, R. (1932). A technique for the measurement of attitudes. *Arch. Psychol.*, 140: 5–55.

Lim, J., Urban, L., and Green, B. G. (2008). Measures of individual differences in taste and creaminess perception. *Chem. Senses*, 33: 493–501.

Lim, J., Wood, A., and Green, B. G. (2009). Derivation and evaluation of a labeled hedonic scale. *Chem. Senses*, 34: 739–751.

Linschoten, M. R., Harvey, L. O., Eller, P. A., and Jafek, B. W. (1996). Rapid and accurate measurement of taste and smell thresholds using an adaptive maximum-likelihood staircase procedure. *Chem. Senses*, 21: 633–634.

Linschoten, M. R., and Kroeze, J. H. A. (1991). Spatial summation in taste: NaCl thresholds and stimulated area on the anterior tongue. *Chem. Senses*, 16: 219–224.

Liu, H.-X., Staubach-Grosse, A. M., Walton, K. D., et al. (2009). WNT5a in tongue and fungiform papilla development. *Ann. N. Y. Acad. Sci.*, 1170: 11–17.

Lucchina, L. A., Curtis, O. F., Putnam, P., et al. (1998). Psychophysical measurement of 6-*n*-propylthiouracil (PROP) taste perception. *Ann. N. Y. Acad. Sci.*, 855: 817–822.

Ly, A., and Drewnowski, A. (2001). PROP (6-*n*-Propylthiouracil) tasting and sensory responses to caffeine, sucrose, neohesperidin dihydrochalcone, and chocolate. *Chem. Senses*, 26: 41–47.

Mackenzie, I. C. K. (1955). A simple method of testing taste. *Lancet*, 265: 377–378.

Manis, M., Biernat, M., and Nelson, T. F. (1991). Comparison and expectancy processes in human judgment. *J. Pers. Soc. Psychol.*, 61: 203–211.

Marks, L. E. (1974). *Sensory Processes: The New Psychophysics*. New York: Academic Press.

Marks, L. E., and Stevens, J. C. (1980). Measuring sensation in the aged. In *Aging in the 1980s: Psychological Issues*, L.W. Poon (Ed.). Washington, D.C.: American Psychological Association, pp. 592–598.

Marks, L. E., Stevens, J. C., Bartoshuk, L. M., et al. (1988). Magnitude matching: The measurement of taste and smell. *Chem. Senses*, 13: 63–87.

Mattes, R. D., and Cowart, B. J. (1994). Dietary assessment of patients with chemosensory disorders. *J. Am. Diet. Assoc.*, 94: 50–56.

Matsuda, T., and Doty, R. L. (1995). Regional taste sensitivity to NaCl: Relationship to subject age, tongue locus, and area of stimulation. *Chem. Senses*, 20: 283–290.

McBurney, D. H. (1966). Magnitude estimation of the taste of sodium chloride after adaptation to sodium chloride. *J. Exp. Psychol.*, 72: 869–873.

McBurney, D. H. (1986). Taste, smell, and flavor terminology: Taking the confusion out of fusion. In *Clinical Measurement of Taste and Smell*, H.L. Meiselman and R.S. Rivlin (Eds.). New York: Macmillan, pp. 117–125.

McBurney, D.H., and Bartoshuk, L. M. (1973). Interactions between stimuli with different taste qualities. *Physiol. Behav.*, 10: 1101–1106.

McBurney, D. H., and Collings, V. B. (1977). *Introduction to Sensation/Perception*. Englewood Cliffs, NJ: Prentice-Hall.

McBurney, D. H., and Pfaffmann, C. (1963). Gustatory adaptation to saliva and sodium chloride. *J. Exp. Psychol.*, 65: 523–529.

McMahon, D. B. T., Shikata, H., and Breslin, P. A. S. (2001). Are human taste thresholds similar on the right and left sides of the tongue? *Chem. Senses*, 26: 875–883.

Meilgaard, M. C., Civille, G. V., and Carr, B. T. (2007). *Sensory Evaluation Techniques*, 4th ed. Boca Raton, FL: CRC Press.

Miller, I. J., and Bartoshuk, L. M. (1991). Taste perception, taste bud distribution, and spatial relationships. In *Smell and Taste in Health and Disease*, T.V. Getchell, R.L. Doty, L.M. Bartoshuk, & J.B. Snow (Eds.). New York: Raven Press, pp. 205–233.

Miller, I. J., and Reedy, F. E. (1990a). Quantification of fungiform papillae and taste pores in living human subjects. *Chem. Senses*, 15: 281–294.

Miller, I. J., and Reedy, F. E. (1990b). Variations in human taste bud density and taste intensity perception. *Physiol. Behav.*, 47: 1213–1219.

Miller, I. J. and Whitney, G. (1989). Sucrose octaacetate-taster mice have more vallate taste buds than non-tasters. *Neurosci. Lett.*, 100: 271–275.

Mojet, J., Christ-Hazelhof, E., and Heidema, J. (2001). Taste perception with age: Generic or specific losses in threshold sensitivity to the five basic tastes? *Chem. Senses*, 26: 845–860.

Moon, C. N., and Pullen, E. W. (1963). Effects of chorda tympani section during middle ear surgery. *Laryngoscope*, 73: 392–405.

Moskowitz, H. R. (1977a). Magnitude estimation: Notes on what, how, when, and why to use it. *J. Food Qual.*, 1: 195–228.

Moskowitz, H. R. (1977b). Psychophysical and psychometric approaches to sensory evaluation. *Crit. Rev. Food Sci. Nutr.*, 9: 41–79.

Mueller, C., Kallert, S., Renner, B., et al. (2003). Quantitative assessment of gustatory function in a clinical context using impregnated "taste strips". *Rhinology*, 41: 2–6.

Murphy, C., Quinonez, C., and Nordin, S. (1995). Reliability and validity of electrogustometry and its application to young and elderly persons. *Chem. Senses*, 20: 499–503.

Murphy, C., Cain, W.S., and Bartoshuk, L.M. (1977). Mutual action of taste and olfaction. *Sens. Proc.*, 1: 204–211.

Nakazato, M., Endo. S., Yoshimura, I., and Tomita, H. (2002). Influence of aging on electrogustometry thresholds. *Acta Otolaryngol.*, 546: 16–26.

Narens, L., and Luce, R. D. (1983). How we may have been misled into believing in the interpersonal comparability of utility. *Theory Decis.*, 15: 247–260.

Ninomiya, Y., and Funakoshi, M. (1982). Responsiveness of dog thalamic neurons to taste stimulation of various tongue regions. *Physiol. Behav.*, 29: 741–745.

Noble, A. C. (1996). Taste-aroma interactions. *Trends Food Sci. Tech.*, 7: 439–444.

Norgren, R. (1984). Central neural mechanisms of taste. In *Handbook of Physiology (Section 1, Vol. III: The Nervous System — Sensory Processes)*, J. M. Brookhart and V. B. Mountcastle (Eds.). Washington, DC: American Physiological Society, pp. 1087–1128.

Norgren, R. (1990). Gustatory system. In *The Human Nervous System*, G. Paxinos (Ed.). New York: Academic Press, pp. 845–861.

Norgren, R., and Pfaffmann, C. (1975). The pontine taste area in the rat. *Brain Res.*, 91: 99–117.

Ogawa, H., and Hayama, T. (1984). Receptive fields of solitario-parabrachial relay neurons responsive to natural stimulation of the oral cavity in rats. *Exp. Brain Res.*, 54: 359–366.

Ogden, G. R. (1989). Atrophy of fungiform papillae following lingual nerve damage — A poor prognosis? *Br. Dent. J.*, 167: 332.

Olson, J. M., Boehnke, M., Neiswanger, K., et al. (1989). Alternative genetic models for the inheritance of the phenylthiocarbamide taste deficiency. *Genet. Epidemiol.*, 6: 423–434.

Ovesen, L., Sorensen, M., Hannibal, J., and Allingstrup, L. (1991). Electrical taste detection thresholds and chemical smell detection thresholds in patients with cancer. *Cancer*, 10: 2260–2265.

Pangborn, R. M., Berg, H. W., Roessler, E. B., and Webb, A. D. (1964). Influence of methodology on olfactory response. *Percept. Mot. Skills*, 18: 91–103.

Pavlidis, P., Gouveris, H., Anogeianaki, A., et al. (2013). Age-related changes in electrogustometry thresholds, tongue tip vascularization, density, and form of the fungiform papillae in humans. *Chem. Senses*, 38: 35–43.

Parr, L. W. (1934). Taste blindness and race. *J. Hered.*, 25: 187–190.

Peryam, D. R., and Girardot, N. F. (1952). Advanced taste test method. *Food Eng.*, 24, 58–61: 194.

Pfaffmann, C., and Bartoshuk, L. M. (1989). Psychophysical mapping of a human case of left unilateral ageusia. *Chem. Senses*, 14: 738.

Pfaffmann, C., and Bartoshuk, L. M. (1990). Taste loss due to herpes zoster oticus: An update after 19 months. *Chem. Senses*, 15: 657–658.

Prescott, J., Bartoshuk, L. M., and Prutkin, J. M. (2004). 6-*n*-propylthiouracil tasting and the perception of non-taste oral sensations. In *Genetic Variation in Taste Sensitivity: Measurement, Significance, and Implications*, J. Prescott & B. J. Tepper (Eds.). New York: Marcel Dekker, pp. 89–104.

Pritchard, T. C., and Norgren, R. (2004). Gustatory system. In *The Human Nervous System*, 2nd ed., G. Paxinos & J.K. Mai (Eds.). San Diego: Elsevier Academic Press, pp. 1171–1196.

Prutkin, J. M., Duffy, V. B., Etter, L., et al. (2000). Genetic variation and inferences about perceived taste intensity in mice and men. *Physiol. Behav.*, 69: 161–173.

Raliou, M., Wiencis, A., Pillias, A. M., et al. (2009). Nonsynonomous single nucleotide polymorphisms in human tas1r1, tas1r3, and mGluR1 and individual taste sensitivity to glutamate. *Am. J. Clin. Nutr.*, 90: 789S–799S.

Rankin, K. M., Godinot, N., Christensen, C. M., et al. (2004). Assessment of different methods for 6-*n*-propylthiouracil status classification. In *Genetic Variation in Taste Sensitivity: Measurement, Significance, and Implications*, J. Prescott & B.J. Tepper (Eds.). New York: Marcel Dekker, pp. 63–88.

Reed, D. R. (2008). Birth of a new breed of supertaster. *Chem. Senses*, 33: 489–491.

Reed, D. R., Bartoshuk, L. M., Duffy, V. B., et al. (1995). Propylthiouracil tasting: Determination of underlying threshold distributions using maximum likelihood. *Chem. Senses*, 20: 529–533.

Reed, D. R., Nanthakumar, E., North, M., et al. (1999). Localization of a gene for bitter taste perception to human chromosome 5p15. *Am. J. Hum. Genet.*, 64: 1478–1480.

Reichert, F. L. (1934). Neuralgias of the glossopharyngeal nerve: With particular reference to the sensory, gustatory, and secretory functions of the nerve. *Arch. Neurol. Psychiatry*, 32: 1030–1037.

Rice, J. C. (1963). The chorda tympani in stapedectomy. *J. Laryngol. Otol.*, 77: 943–944.

Robinson, P. P., Loescher, A. R., and Smith, K. G. (2000). A prospective, quantitative study on the clinical outcome of lingual nerve repair. *Br. J. Oral Maxillofac. Surg.*, 38: 255–263.

Rozin, P. (1982). "Taste-smell confusions" and the duality of the olfactory sense. *Percept. Psychophys.*, 31: 397–401.

Salata, J. A., Raj, J. M., and Doty, R. L. (1991). Differential sensitivity of tongue areas and palate to chemical stimulation: A suprathreshold cross-modal matching study. *Chem. Senses*, 16: 483–489.

Schiffman, S. S. (1997). Taste and smell losses in normal aging and disease. *JAMA*, 278: 1357–1362.

Schutz, H.G., and Cardello, A.V. (2001). A labeled affective magnitude (LAM) scale for assessing food liking/disliking. *J. Sens. Stud.*, 16: 117–159.

Schwartz, S. R. (1998). *The effects of chorda tympani nerve transection on the human tongue: Anatomic and somatosensory alterations*. M.D. thesis, Yale University School of Medicine, New Haven, CT.

Shahbake, M., Hutchinson, I., Laing, D. G., and Jinks, A. L. (2005). Rapid quantitative assessment of fungiform papillae density in the human tongue. *Brain Res.*, 1052: 196–201.

Shigemura, N., Shirosaki, S., Sanematsu, K., et al. (2009). Genetic and molecular basis of individual differences in human umami taste perception. *PLoS One*, 4: e6717.

Shingai, T., and Beidler, L. M. (1985). Inter-strain differences in bitter taste responses in mice. *Chem. Senses*, 10: 51–55.

Ship, J. A., Grushka, M., Lipton, J. A., et al. (1995). Burning mouth syndrome: An update. *J. Am. Dent. Assoc.*, 126: 842–853.

Small, D. M., Gerber, J. C., Mak, Y. E., and Hummel, T. (2005). Differential neural responses evoked by orthonasal versus retronasal odorant perception in humans. *Neuron*, 47: 593–605.

Small, D. M., Jones-Gotman, M., Zatorre, R. J., et al. (1997). Flavor processing: More than the sum of its parts. *Neuroreport*, 8: 3913–3917.

Smith, D. V. (1971). Taste intensity as a function of area and concentration: Differentiation between compounds. *J. Exp. Psychol.*, 87: 163–171.

Smutzer, G., Lam, S., Hastings, L., et al. (2008). A test for measuring gustatory function. *Laryngoscope*, 118: 1411–1416.

Snow, J. B., Doty, R. L., Bartoshuk, L. M., and Getchell, T. V. (1991). Categorization of chemosensory disorders. In *Smell and Taste in Health and Disease*, T. V. Getchell, R. L. Doty, L. M. Bartoshuk, & J. B. Snow (Eds.). New York: Raven Press, pp. 445–447.

Snyder, D. J. (2010). *Multimodal interactions supporting oral sensory capture and referral*. PhD thesis, Yale University, New Haven, CT.

Snyder, D. J., Clark, C. J., Catalanotto, F. A., and Bartoshuk, L. M. (2007). Oral anesthesia specifically impairs retronasal olfaction. *Chem. Senses*, 32: A15.

Snyder, D. J., Duffy, V. B., Chapo, A. K., et al. (2003). Food preferences mediate relationships between otitis media and body mass index. *Appetite*, 40: 360.

Snyder, D. J., Duffy, V. B., Marino, S. E., and Bartoshuk, L. M. (2008a). We are what we eat, but why? Relationships between oral sensation, genetics, pathology, and diet. In *Sweetness and Sweeteners: Biology, Chemistry, and Psychophysics*, D. K. Weerasinghe & G. E. DuBois (Eds.). Washington, DC: American Chemical Society, pp. 258–284.

Snyder, D. J., Fast, K., and Bartoshuk, L. M. (2004). Valid comparisons of suprathreshold stimuli. *J. Conscious. Stud.*, 11: 40–57.

Snyder, D. J., Prescott, J., and Bartoshuk, L. M. (2006). Modern psychophysics and the assessment of human oral sensation. *Adv. Otorhinolaryngol.*, 63: 221–241.

Snyder, D. J., Puentes, L. A., Sims, C. A., and Bartoshuk, L. M. (2008b). Building a better intensity scale: Which labels are essential? *Chem. Senses*, 33: S142.

Snyder, L. H. (1931). Inherited taste deficiency. *Science*, 74: 151–152.

Stephen, K. W., McCrossan, J., Mackenzie, D., et al. C.F. (1980). Factors determining the passage of drugs from blood into saliva. *Br. J. Clin. Pharmacol.*, 9: 51–55.

Stevens, J. C. (1959). Cross-modality validation of subjective scales for loudness, vibration, and electric shock. *J. Exp. Psychol.*, 57: 201–209.

Stevens, J. C., Cruz, L. A., Hoffman, J. M., and Patterson, M. Q. (1995). Taste sensitivity and aging: High incidence of decline revealed by repeated threshold measures. *Chem. Senses*, 20: 451–459.

Stevens, J. C., and Marks, L. E. (1965). Cross-modality matching of brightness and loudness. *Proc. Natl. Acad. Sci.*, 54: 407–411.

Stevens, J. C., and Marks, L. E. (1980). Cross-modality matching functions generated by magnitude estimation. *Percept. Psychophys.*, 27: 379–389.

Stevens, S. S. (1946). On the theory of scales of measurement. *Science*, 103: 677–680.

Stevens, S. S. (1956). The direct estimation of sensory magnitudes-loudness. *Am. J. Psychol.*, 69: 1–25.

Stevens, S. S. (1958). Adaptation-level vs the relativity of judgment. *Am. J. Psychol.*, 71: 633–646.

Stevens, S. S. (1961). To honor Fechner and repeal his law. *Science*, 133: 80–86.

Stevens, S. S. (1962). The surprising simplicity of sensory metrics. *Am. Psychol.*, 17: 29–39.

Stevens, S. S., and Galanter, E. H. (1957). Ratio scales and category scales for a dozen perceptual continua. *J. Exp. Psychol.*, 54: 377–411.

Stevenson, R. J., and Prescott, J. (1997). Judgments of chemosensory mixtures in memory. *Acta Psychol.*, 95: 195–214.

Stevenson, R. J., Prescott, J., and Boakes, R. A. (1999). Confusing tastes and smells: How odours can influence the perception of sweet and sour tastes. *Chem. Senses*, 24: 627–635.

Stillman, J. A., Morton, R. P., Hay, K. D., et al. (2003). Electrogustometry: Strengths, weaknesses, and clinical evidence of stimulus boundaries. *Clin. Otolaryngol. Allied. Sci.*, 28: 406–410.

Sunderland, E., and Cartwright, R. A. (1968). Iodine estimations, endemic goitre, and phenylthiocarbamide (PTC) tasting ability. *Acta Genet. Stat. Med.*, 18: 593–598.

Sweazey, R. D., and Smith, D. V. (1987). Convergence onto hamster medullary taste neurons. *Brain Res.*, 408: 173–184.

Teghtsoonian, R. (1973). Range effects in psychophysical scaling and a revision of Stevens' law. *Am. J. Psychol.*, 86: 3–27.

Tennen, H., Affleck, G., and Mendola, R. (1991). Coping with smell and taste disorders. In *Smell and Taste in Health and Disease*, T. V. Getchell, R. L. Doty, L. M. Bartoshuk, & J. B. Snow (Eds.). New York: Raven Press, pp. 787–802.

Tepper, B. J., Christensen, C. M., and Cao, J. (2001). Development of brief methods to classify individuals by PROP taster status. *Physiol. Behav.*, 73: 571–577.

Tepper, B. J., and Ullrich, N. V. (2002). Influence of genetic taste sensitivity to 6-*n*-propylthiouracil (PROP), dietary restraint, and disinhibition on body mass index in middle-aged women. *Physiol. Behav.*, 75: 305–312.

Tepper, B. J., White, E. A., Koelliker, Y., et al. (2009). Genetic variation in taste sensitivity to 6-*n*-propylthiouracil and its relationship to taste perception and food selection. *Ann. N. Y. Acad. Sci.*, 1170: 126–139.

Tie, K., Fast, K., Kveton, J. F., et al. (1999). Anesthesia of chorda tympani nerve and effect on oral pain. *Chem. Senses*, 24: 609.

Tieman, D., Bliss, P., McIntyre, L. M., et al. (2012). The chemical interactions underlying tomato flavor preferences. *Curr. Biol.*, 22: 1035–1039.

Todrank, J., and Bartoshuk, L. M. (1991). A taste illusion: Taste sensation localized by touch. *Physiol. Behav.*, 50: 1027–1031.

Tomita, H., Ikeda, M., and Okuda, Y. (1986). Basis and practice of clinical taste examinations. *Auris Nasus Larynx*, 13: S1–S15.

Ventura, A. K., Reed, D. R., and Mennella, J. A. (2009). Chronic otitis media is associated with a marker of taste damage and higher weight status in children. *Chem. Senses*, 34: A37.

Weiffenbach, J. M. (1983). Taste quality recognition and forced-choice response. *Percept. Psychophys.*, 33: 251–254.

Wetherill, G. B., and Levitt, H. (1965). Sequential estimation of points on a psychometric function. *Br. J. Math. Stat. Psychol.*, 18: 1–10.

Wheatcroft, P. E. J., and Thornburn, C. C. (1972). Toxicity of the taste testing compound phenylthiocarbamide. *Nature New Biol.*, 235: 93–94.

Whissell-Buechy, D. (1990). Genetic basis of the phenylthiocarbamide polymorphism. *Chem. Senses*, 15: 27–37.

Whissell-Buechy, D., and Wills, C. (1989). Male and female correlations for taster (P.T.C.) phenotypes and rate of adolescent development. *Ann. Hum. Biol.*, 16: 131–146.

Wooding, S., Kim, U.-K., Bamshad, M. J., et al. (2004). Natural selection and molecular evolution in PTC, a bitter-taste receptor gene. *Am. J. Hum. Genet.*, 74: 637–646.

Yamada, Y., and Tomita, H. (1989). Influences on taste in the area of chorda tympani nerve after transtympanic injection of local anesthetic (4% lidocaine). *Auris Nasus Larynx*, 16: S41–S46.

Yanagisawa, K., Bartoshuk, L. M., Catalanotto, F. A., et al. (1998). Anesthesia of the chorda tympani nerve and taste phantoms. *Physiol. Behav.*, 63: 329–335.

Zahm, D. S., and Munger, B. L. (1985). The innervation of the primate fungiform papilla — Development, distribution, and changes following selective ablation. *Brain Res. Rev.*, 9: 147–186.

Zhao, L., Kirkmeyer, S. V., and Tepper, B. J. (2003). A paper screening test to assess genetic taste sensitivity to 6-*n*-propylthiouracil. *Physiol. Behav.*, 78: 625–633.

Zuniga, J. R., Chen, N., and Miller, I. J. (1994). Effects of chorda-lingual nerve injury and repair on human taste. *Chem. Senses*, 19: 657–665.

Zuniga, J. R., Chen, N., and Phillips, C. L. (1997). Chemosensory and somatosensory regeneration after lingual nerve repair in humans. *J. Oral Maxillofac. Surg.*, 55: 2–13.

Zuniga, J. R., Davis, S. H., Englehardt, R. A., et al. (1993). Taste performance on the anterior tongue varies with fungiform taste bud density. *Chem. Senses*, 18: 449–460.

Chapter 35

Mapping Brain Activity in Response to Taste Stimulation

DANA M. SMALL and ANNICK FAURION

35.1 INTRODUCTION

Before *functional* cerebral imaging was made accessible for human experimentation, the knowledge of neurophysiological pathways and the determination of the functions of the brain depended on animal electrophysiology. However, evolution across species modifies neural functions and anatomies. Clinical observations in patients constituted another source of information but lesions often affect multiple brain regions making definitive conclusions problematic. Positron Emission Tomography (PET), functional Magnetic Resonance Imaging (fMRI), and Magnetoencephalography (MEG) afforded, for the first time, the ability to non-invasively study the neurophysiological mechanisms of taste perception in humans. In this chapter, we review what we have learned about gustatory representation of the human brain from neuroimaging experiments. We also review relevant results from lesion studies in humans where appropriate and briefly describe important points of convergence and divergence with gustatory neurophysiology in rodents and primates.

35.2 LOCALIZING THE HUMAN GUSTATORY CORTEX

In non-human primates, single-cell recording (Ogawa et al., 1989; Yaxley et al., 1990; Scott and Plata-Salaman, 1999) and anatomical tract tracing (Pritchard et al., 1986, 2000) have identified the dorsal anterior insula and adjacent frontal operculum as the "primary" gustatory cortex (the cortical area where the taste message arrives first).

Gustatory information is subsequently conveyed to the ventral insula and to the caudomedial and caudolateral orbitofrontal cortex (OFC) (Baylis et al., 1995; Pritchard et al., 2005), which are generally considered secondary and higher-order gustatory cortical regions (Rolls et al., 1990; Rolls and Baylis, 1994). An obvious first step for human neuroimaging investigations was to determine whether there are homologous representations of gustatory information in the human brain.

Kinomura was the first to show brain activation in response to taste using PET, observing activation within the thalamus, anterior insular cortex, anterior cingulate gyrus, parahippocampal gyrus, lingual gyrus, caudate nucleus, and temporal gyrus (Kinomura et al., 1994). Subsequently, Cerf and collaborators developed a stimulating method using 100 μl bolus repeatedly sent to the subject's mouth every 3 sec which allowed them to exclude somatosensory stimulation and adaptation as potential confounds while minimizing head movements consecutive to swallowing. A linear potentiometer enabled the subject, in the magnet, to record his/her "Time-Intensity perception profile" (finger-span); this profile of perceived intensity, continuously recorded over time, was used as a template and correlated to the magnetic resonance signal of each voxel to retrieve activation (Cerf et al., 1996, 1998; Van de Moortele et al., 1997; Faurion et al., 1998, 1999). The waveform of the recorded intensity profiles confirmed a slow rate of rise of the taste perception in these conditions. Such quantitative psychophysical evaluation is an important factor for retrieving activated areas because the Time Intensity profile allows gustatory perception to be dissociated from the tactile stimulation inherent in the act

---

Handbook of Olfaction and Gustation, Third Edition. Edited by Richard L. Doty.
© 2015 Richard L. Doty. Published 2015 by John Wiley & Sons, Inc.

of tasting. At 3 Teslas, fMRI offers a spatial resolution of $3 \times 3 \times 3$ mm projected on an anatomy resolution of 1 mm^3. The slow rate of rise of the signal in the Blood Oxygenation Level Dependent (BOLD) method is of the same order as the psychophysical signal. Hence, the stimulation technique and the perception recording were coherent with the BOLD method. Using single subject analysis, taste areas were localized in various parts of the insula and in frontal, temporal, Rolandic opercula, including pre- and post-central gyri (Cerf et al., 1996; Cerf et al., 1998; Cerf-Ducastel et al., 2001), as was expected from electrophysiological recordings in the monkey (Ruch and Patton, 1946; Bagshaw and Pribram, 1953; Benjamin and Burton, 1968; Ogawa et al., 1985; Pritchard et al., 1986; Yaxley et al., 1990; Scott and Plata-Salaman, 1991; Kadohisa et al., 2005; Ogawa et al., 2005b) and clinical observations in humans (Bornstein, 1940; Lewey, 1943; Penfield and Faulk, 1955; Motta, 1959; Hausser-Hauw and Bancaud, 1987). Moreover, an unexpected activation was observed in the most inferior and anterior part of the insula in the dominant hemisphere of subjects which was in the direction of the orbitofrontal cortex (Faurion et al., 1999).

Simultaneously or soon after, other authors confirmed and extended these data with PET and MEG. Several studies reported activation of multiple insular, opercular, and caudal medial and lateral orbitofrontal regions during gustatory stimulation (Kobayakawa et al., 1996, 1999; Zald et al., 1998; Frey and Petrides, 1999). Small and colleagues (1997) elucidated a role for the right anterior temporal lobe in taste quality recognition and taste intensity perception coding (Small et al., 2001b; c). Zald and Pardo (2002) showed that water activated a similar region of the insula and operculm as taste. O'Doherty et al. (2001) reported activation in the insula and opercula, without chemotopical organization. O'Doherty et al. (2002) and De Araujo et al. (2003b) showed unimodal neuronal activity for taste in the anterior superior insula (Y = 3 to 14) and frontal operculum, indicating that although most of the chemoreceptive neurons in these areas are polymodal neurons (taste and somatic, taste and olfactory etc.), a given neuronal activity can signal pure taste. These authors also confirmed activation of the caudolateral orbitofrontal cortex as well as a part of the rostral anterior cingulate cortex (De Araujo et al., 2003b).

Barry et al. (2001) with fMRI, and Yamamoto et al. (2003) with MEG, both used electrogustatory stimulation – or iontophoretic stimulation – to study cortical representation of taste. Small currents produced by a generator induce a movement of the cations in the subject's own saliva towards his/her receptors. With intensities below those eliciting tingling, these authors showed insular and opercular activation. For higher intensities, activation of the somatosensory cortex was also obtained, confirming

trigeminal stimulation. The electrogustatory stimulation (the same which is used in electrogustometry, EGM) furthermore confirmed that actual taste and not somatosensory responses were found in the postcentral gyrus at a more ventral location of the postcentral gyrus than where Pardo et al. (1997) found both taste and somatosensory senses.

Collectively, then, the early gustatory neuroimaging literature demonstrates that taste is represented, not only in the thalamus, anterior insula/frontal operculum, and caudal orbitofrontal cortex, but also in several other regions of the insula and overlying operculum, as well as in the anterior cingulate cortex and anterior medial temporal lobe.

35.2.1 Time Resolution: Magnetoncephaloraphy (MEG) Contributions to Describing Human Taste Pathways

BOLD fMRI suffers from the long latency of the hemodynamic response (several seconds). Magnetoencephalography (MEG) records, in a series of SQUID (Superconducting QUantum Interference Device) sensors positioned on the scalp, the magnetic fields produced by the minute electrical currents elicited locally in the brain by any taste stimulation. Results are positioned on a MRI map of the subject's brain with a high spatial resolution. As MEG records the actual electrophysiological brain activity, its time resolution can easily be in the millisecond range. However, MEG suffers from a spatial drawback as the result depends on the solution of the inverse problem to localize the spatial origin of these local activities using first a "guess," then minimization algorithms. The strategy, which consists of associating both MEG and fMRI, guarantees, when finding the same areas activated, a good spatial resolution (1–3mm) without an *a priori* guess (with fMRI) and with an excellent time resolution (millisecond) in the same areas (with MEG).

To use MEG, Kobayakawa et al. (1996; 1999) developed a novel method so that stimulus delivery could be precisely time-locked to brain response. Tastants were delivered as boluses embedded in a stream of water that flowed through a Teflon tube. A small (3 x 9 mm) opening in the tube was created and placed across the tip of the subject's tongue so that the tastant could stimulate taste receptors on this region of the tongue. Importantly, the bolus was injected with a colored dye that could be detected by an optical sensor enabling stimulus onset to be time-locked to brain response. In addition, since the gustatory stimulus is embedded in a constant tactile stimulus (flow of liquid over the tongue) taste and tactile stimulation are dissociated. This offers the advantage of being able to detect brain response to pure taste signals. However, it is an artificial way to experience taste, which could also be taken as a disadvantage because during everyday tasting the gustatory and tactile signals co-occur. As such, similar flow

35.2 Localizing the Human Gustatory Cortex

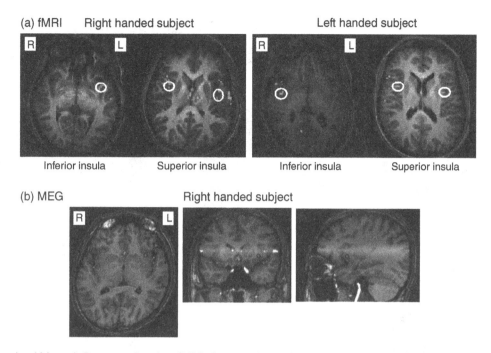

Figure 35.1 Functional Magnetic Resonance Imaging (fMRI) showed activation in the left inferior anterior insula in right-handers (Oldenburg score 0) and in the right inferior anterior insula in the left-handers (Oldenburg score > 16) for taste stimulations. MagnetoEncephalography (MEG) showed activation in the left inferior anterior insula of right-handers at late latencies (800–1300ms). At this time, subjects begin to perceive the quality of the stimulus. This left inferior anterior insular area was also demonstrated, with fMRI, to be co-activated with the left angular gyrus in right-handers, a structure known to be involved in semantic treatments. NB: the right hemisphere appears in the left hand side of all images. (fMRI images from Faurion A, Cerf B et al. *Neuroscience Letters*, 1999, in collaboration with D. Le Bihan, SHFJ, CEA, DRM, Orsay, France, with permission from Elsevier; MEG images from Kobayakawa T and collaborators, Human Technology Research Institute, AIST, Tsukuba, Japan, by courtesy). (*See plate section for color version.*)

methods have been associated with reduced gustatory sensitivity in psychophysical assessments (Meiselman, 1971).

Using this method Kobayakawa and colleagues found that the area first activated by taste was located at the transition between the parietal operculum and the posterior superior insular cortex which they named area G for gustatory area, and in the Rolandic operculum, including the buried part of the central sulcus. The average onset latency was 93 ms for NaCl and 172 ms for saccharin (Kobayakawa et al., 1996), whereas no response was obtained for pure water used as a stimulus. The longer latency for the activation elicited by saccharin compared to NaCl was suggested to reflect the difference of events at the peripheral level, including the time for diffusion to receptors and transduction events for organic molecules compared to mere channel opening for Na^+ cations (Kobayakawa et al., 1999). Kobayakawa and colleagues suggested "area G" represented the primary taste cortex since it was the first region to be activated. They also described other regions that were activated subsequently to area G in the frontal operculum and the anterior part of the insula, the hippocampus, the parahippocampal gyrus and the superior temporal sulcus.

Compared to electrophysiological results localizing the "primary" taste area in the anterior insula in monkeys or other neuroimaging results in human, this result from MEG and EEG recordings (Mizoguchi et al., 2002) seems contradictory at first. However, many fMRI studies do report activity in this region. For example, Ogawa et al. (2005a) reported activation in both the superior posterior insula (area G) and the buried part of the parietal operculum at right coordinates x;y;z: 46;−6;12 and left −38;−8;12, in the left frontal operculum at: −52;26;4 and Rolandic operculum at −60;−18;24 using fMRI. As such, the three meta-analyses of taste studies that have been conducted all conclude that there are multiple taste-responsive regions in the insula and overlying operculum (Small et al., 1999; Verhagen and Engelen, 2006; Veldhuizen et al., 2011a). For example, Veldhuizen and colleagues reviewed 15 studies in a meta analysis and observed the involvement of multiple cortical areas in gustatory processing constituting a taste evoked probability map including the bilateral anterior insula and frontal operculum, the bilateral mid dorsal insula and Rolandic operculum, the bilateral posterior insula/parietal operculum/postcentral gyrus, the left lateral orbitofrontal cortex (OFC), the right medial

778   Chapter 35 Mapping Brain Activity in Response to Taste Stimulation

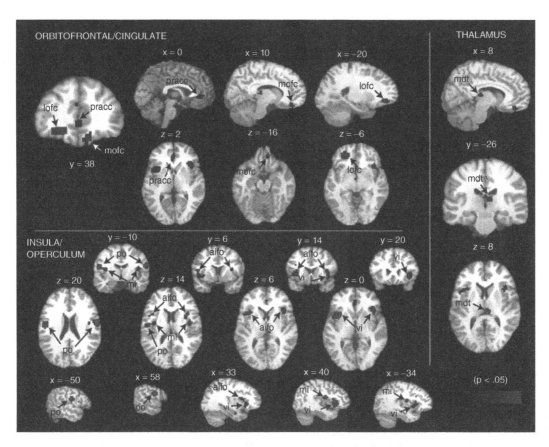

Figure 35.2 Localization of significant ALE values ($p < 0.05$) of gustatory stimulation projected onto a standard template (colin1.1.nii) in Talairach space. Location of selected slice is denoted with stereotactic coordinates above each of them. Only clusters exceeding 100 cubic mm in size are labeled. Abbreviations of anatomical areas: mdt = mediodorsal thalamus, pracc = pregenual anterior cingulate cortex, mofc = medial orbitofrontal cortex, lofc = lateral orbitofrontal cortex, aifo = anterior insula and overlying frontal operculum, po = postcentral gyrus, posterior insula and overlying parietal operculum, mi = mid insula and overlying Rolandic operculum, and vi = ventral insula. Reprinted from Veldhuizen et al. (2011a) with permission from John Wiley & Sons. (*See plate section for color version.*)

OFC, the pregenual anterior cingulate cortex (prACC) and the right mediodorsal thalamus (Figure 35.2). Thus it is clear that the human neuroimaging literature supports the possibility that there is taste representation at the transition between the parietal operculum and the insular cortex. Also supporting this possibility is a clinical report of a patient with a gustatory disorder following a left posterior insular infarct (Cereda et al., 2002). However, Small has argued that designating this region as primary gustatory cortex in humans remains problematic because of the lack of anatomical evidence for a projection from the gustatory thalamus to this region in humans (Small, 2010). Neither histology studies on the primate gustatory organization (Pritchard et al., 1986), nor the studies of central connectivity (Mesulam and Mufson, 1982a; b; 1985) in the monkey or humans, have found connections of area G with other taste regions. As such, Small and colleagues have suggested that responses in this region of the posterior insula and parietal operculum likely reflect attentional orienting to the mouth during tasting (Veldhuizen et al., 2007). It is also important to note that more recently, Ohla and colleagues using multiple EEG recordings noted pronounced activations of the bilateral anterior insula and adjacent frontal operculum as early as 70–160 ms (Ohla et al., 2010). This time frame is in line with the early activations noted by Kobayakawa and colleagues. There are possibilities that MEG, fMRI or multiple EEG alone may not detect all activation for technical reasons. MEG will detect activation in the sulci, which is the case for area G thanks to the specially efficient detection of orthogonal dipoles, but there are known limitations in dipole modeling that result in failure to detect two simultaneous signals. However, this early double information both in the anterior insula and frontal operculum (Ohla et al., 2010) and at the transition with the parietal operculum and the far posterior insula including the central sulcus (Kobayakawa et al., 1996) reminds one of the discovery of the bifurcating taste neurons projecting from the thalamus VPMpc to both opercular insular area and the somatosensory area I with a 2–3 ms delay (Benjamin and Burton, 1968).

In any case, caution is called for in discounting area G, as interspecies differences clearly exist in the gustatory system. The most prominent example is the pontine parabrachial taste nucleus, which is of critical importance for rodent taste and ingestive behavior (Di Lorenzo, 1988; Di Lorenzo and Monroe, 1989; Di Lorenzo, 1990; Di Lorenzo and Monroe, 1992), and which is lacking in human (Topolovec et al., 2004) as in non-human primates (Pritchard et al., 1986; Scott and Small, 2009; Small and Scott, 2009). In addition, Craig (2009) has suggested the far anterior insular cortex may be a newly evolved area unique to humans. If true, this would mean that the anterior insula/opercular gustatory region is further caudal within the insular region in humans compared to monkeys. Consistent with this possibility, activations observed in human neuroimaging studies generally exclude the far anterior insular region (Small, 2010).

In summary, there has been, and continues to be debate over the precise location of the "primary" gustatory cortex in humans. This controversy exists for a number of reasons. First, tasting is inherently multi-modal. The primary taste cortex is not taste specific, most single neurons responding to taste and olfaction or to taste and somatic sensation. When we taste we also touch the food or drink in our mouths and sense its odor via retronasal olfaction (Giachetti and Mac Leod, 1975, 1977). Taste and touch are particularly intimately related with evidence for overlapping representation at virtually every level of the neuroaxis from the nucleus tractus solitarius (Boucher et al., 2003; review in (Small and Green, 2011). As such, the signals can be difficult to disentangle, or their successful disentanglement leads to issues with interpretation, since sensation is not ecological (cf. Gibson, 1979). Second, interspecies differences likely exist. In non-human primates, single-cell recording (Ogawa et al., 1989; Yaxley et al., 1990; Scott and Plata-Salaman, 1999), and anatomical tract tracing (Pritchard et al., 1986, 2000), have identified the dorsal anterior insula and adjacent frontal operculum as the primary gustatory cortex. The human neuroimaging studies reviewed in this chapter clearly demonstrate that taste is represented not only in this region, but also in several other regions of the insula and overlying operculum. However, there are limited anatomical studies in humans so it is difficult to rely on convergence between anatomy and electrophysiology to determine which of these regions reflect the first cortical representation. In addition, activations generally exclude the far anterior insular region, suggesting that gustatory information may be represented further caudally in the human compared to the monkey insular cortex. This is consistent with the evolutionary hypotheses developed by Craig (2009), which suggest that the taste area may appear to be moved further caudally compared

to monkeys because the far anterior insular cortex may be a newly evolved area unique to humans. Third, not only do conclusions about the primary gustatory cortex vary as a function of interpretation of data generated from different techniques (e.g., Kobayakawa et al., 1999 versus Veldhuizen et al., 2007), but the same gustatory stimuli can produce responses in the insula and overlying operculum at different locations depending upon the task a subject is asked to perform during stimulation (Bender et al., 2009).

35.2.2 Beyond the Primary Gustatory Cortex

One of the advantages afforded by the MEG sensitivity to temporal events is that it allows for viewing the spatio-temporal pattern of activation across brain areas. Each dipole can be followed during a few tens of milliseconds and new dipoles appear in various regions. As such, it is possible to gain insight into the flow of gustatory information from primary to higher-order regions (Kobayakawa et al., 1996, 1999, 2012). Four periods can be distinguished.

1. The first dipoles are observed in the superior posterior insula (area G), Rolandic operculum and pre-central gyrus. In a given example with NaCl, the earliest dipoles start bilaterally in the left post-central gyrus (−50; 15; 21) at 76 ms for example and in the right post-central gyrus (44;−24;27) at 80 ms, traveling respectively for 8 mm during 32 ms and 8.6 mm during 28 ms to reach both superior posterior insulae at 108 ms.

2. During a period between 200 ms and 500 ms, many areas are activated including various substructures in the insula, the frontal operculum, parietal operculum, middle temporal gyrus, cingulate gyrus, but also hippocampus and amygdala.

3. From 500 ms to 750 ms the activation decreases somehow even sooner for NaCl than for saccharin: signals disappear progressively in the post-central gyrus, the temporal transverse gyrus, the supramarginal gyrus, the superior temporal gyrus, and the inferior insula. During these periods, dipoles move within a given area, stop, and start flashing elsewhere and eventually come back.

4. Between 750 ms and 1500 ms, the nearly silent brain resumes a very late activation in the inferior far anterior insula (Figure 35.1) in the dominant left hemisphere of right-handers (Faurion et al., 1999), which was found to be co-activated with the left angular gyrus (Cerf-Ducastel et al., 2001) which is involved in semantic tasks. For comparison, in many behavioral experiments, a one second period is sufficient to consciously identify the taste stimulus and name it.

Hence, not only a spatial pattern is depicted for taste stimulation but also a dynamic spatio-temporal pattern is observed. Tracking dipoles from their appearance to their extinction exhibits numerous inter-talks between brain areas and within areas during the first second and a half period studied in MEG, with a climax between 200 and 500 ms (Faurion et al., 2008). Conspicuously absent from the spatio-temporal map produced by the MEG results is the orbitofrontal cortex. This is because the signals originating from this region lie in the radial direction relative to the SQUID sensors and thus escape detection (Kobayakawa et al., 1999).

35.2.3 Laterality of Taste Projections

For a long time, taste was thought to project contralaterally because researchers who used electrical stimulation of the tongue in rodents to activate taste nerves and localize their cortical projection elicited trigeminal evoked potentials instead of taste evoked potentials. Benjamin and Burton (1968) stimulated the chorda tympani and the lingual-tonsillar branch of the glossopharyngeal nerve in the squirrel monkey, producing exclusively ipsilateral evoked potentials in the anterior opercular-insular cortex. However, Norgren (1978) showed with histology tracing techniques that a few fibers cross from the NTS to the opposite parabrachial nucleus. But in cats and monkeys contrary to the rat, receptive fields for thalamic neurons are exclusively ipsilateral on the tongue (Norgren, 1995).

In the rat, the pontine taste or parabrachial nucleus (PBN) is a mandatory relay. In contrast, in primates neurons from the rostral NTS ascend directly to the thalamus (Beckstead et al., 1980), leading to the conclusion that primates lack the taste PBN relay. Consistent with this possibility, Norgen and colleagues were unable to identify monkey PBN neurons responding to the sapid stimulation of the anterior tongue (Norgren, 1981). However, he could obtain evoked potentials in the PBN with electrical stimulation of the glossopharyngeal nerve innervating circumvallate and foliate papillae, raising the possibility of a primate PBN taste relay for the posterior part of the tongue. In humans, Topolovec et al. (2004) used fMRI to study brain stem functions in healthy humans. They reported activation in the NTS to whole mouth stimulation with sucrose in all eight subjects (two bilateral, five right, and one left), but failed to isolate response in the PBN. Although taste deficits are occasionally reported following pontine lesions (Goto et al., 1983; Nakajima et al., 1983; Uesaka et al., 1998; Onoda and Ikeda, 1999), Small (Small, 2006) suggested that these disturbances likely reflected damage to fibers of passage rather than to cell bodies, which would imply the presence of a relay.

Information about laterality in humans either comes from observing unilateral neurological lesions with MRI and their functional consequences or from a few fMRI or MEG experiments in healthy subjects with hemi-lingual stimulation. So far, as described below, a wealth of clinical observations supports the hypothesis of an ipsilateral taste pathway.

Weidemann and Sparing (2002) reported the rare case of isolated hemiageusia due to a spontaneous hemorrhage in the ipsilateral solitary tract resulting from a cavernous haemangioma. Landis et al. (2006) also saw an ipsilateral transient hemiageusia after a thrombosis of the anterior inferior cerebellar artery. Uesaka et al. (1998) examined a patient with an ipsilateral ageusia and a small lesion localized with MRI around the brachium conjunctivum, involving a small area of the pontine tegmentum clearly dorsal to the right medial lemniscus (the medial lemniscus is the pathway for trigeminal sensitivity). Sato and Nitta (2000) reported a woman with ageusia on the right half of the tongue and a small lesion revealed by MRI in the right dorsolateral tegmentum of the middle pons, lateral to the medial lemniscus. These cases suggest that the central gustatory pathway (central tegmental pathway) projecting from the nucleus of the solitary tract to the thalamus is ipsilateral. Ipsilateral loss of taste is also observed following pontine (e.g., (Nakajima et al., 1983; Takeuchi et al., 2006) lesions?. Such clinical observations are not rare but, as indicated above, it is difficult to know whether they affect fibers rather than neuronal cells. Hence, they definitely argue for an ipsilateral projection of taste from the periphery to the PBN but they cannot be taken as evidence of a taste relay in the human pons on its way to the cortex.

As far as higher brain pathways are concerned, Berlucchi et al. (2004), studying patients with right brain damage compared to patients with left brain damage, suggested a predominantly ipsilateral cortical projection for taste and a predominantly contralateral projection for lingual tactile information. At the thalamic level, the laterality of the gustatory representation remains unresolved for Sánchez-Juan and Combarros (2001) because taste is preserved in patients with a corpus callosum section and strokes or tumors involving the insula. The authors collected arguments in support of the hypothesis of a gustatory representation of both hemi-tongues in the left cerebral hemisphere, and of only the right hemi-tongue in the right hemisphere. Similarly, Cereda et al. (2002) show taste disorders after insular infarct in the left side only. In Sánchez-Juan and Combarros (2001), the loss of taste after surgery for trigeminal neuralgia suggests an accessory gustatory pathway might exist through the trigeminal sensory root. But, Boucher et al. (2003) have shown the convergence of trigeminal afferents on second order taste responding neurons in the NTS. The mere suppression of these afferents drastically reduces taste sensitivity, as shown in subjects with dental de-afferentation (Boucher et al., 2006). Both trigeminal sensitivity and taste are in neurophysiological intimate interaction (Cerf-Ducastel et al., 2001; Lugaz et al., 2005; Faurion, 2006).

Among studies supporting a partially bilateral taste projection, Barry et al. (2001), using fMRI and electro-gustometric stimulation at low current intensities which excludes tingling sensation, in right-handed subjects, could not show a simple relationship between the side of the applied stimulus and the side where the activation was found in insular and opercular areas at the level of the individual. However, in the group analysis, they found a predominant right activation in the frontal opercular area. Finally, in this study taste appears to project more to the right hemisphere than lingual touch. They also divided the insula into nine parts divided the insula into nine parts (i.e. anterior, middle, posterior x superior, central, inferior) and concluded, in right handers, that taste-evoked activation in the superior part seems to predominate in the right hand side, activation in the central part appears bilateral, and activation in the inferior part is in the dominant left hemisphere. An alternated right–left prevalence, depending on anteriority, was observed with equal left and right projection in the anterior insula/opercular area at $20<y<30$ (Talairach coordinates) and a more pronounced right insular activation at $0<y<10$ (less anterior) which may also look subject-dependent (Cerf-Ducastel et al., 2001). Faurion et al. (1999) showed bilateral activation in middle insula and unilateral activation in the far inferior anterior insula. These results do not prove, however, the existence of initial contralateral projection.

Onoda et al. (2005) observed more frequent bilateral than ipsilateral or contralateral responses at the transitional cortex between the insula and the parietal operculum (area G) in patients whose chorda tympani nerve had been severed unilaterally. Averaged activation latencies were not significantly different in both hemispheres. These results suggest that unilateral gustatory stimulation activates the transitional cortex between the insula and the parietal operculum bilaterally in humans. A temporary conclusion of this development was brought by Onoda et al. (2012) who observed more ipsilateral cases for lesions located from the medulla to the pons versus ipsilateral, contralateral, and bilateral cases for lesions located above the midbrain. Bilateral cases were more frequently detected, suggesting the taste pathway branches and ascends bilaterally from midbrain to cortex in humans.

Conversely, in more recent fMRI studies by Iannilli et al. (2012), the unilateral stimulation with supra-threshold concentrations showed a high predominance of ipsilateral activation at the thalamus and insular levels, and also in the frontal operculum and the lateral prefrontal cortex. Convergingly, Stephani et al. (2011) presented patients with implanted electrodes in the insula, which produced often unpleasant phenomena ("bad," "nasty" taste ($n = 7$) or "metallic" or "like aluminium," with all effects of lateralized electrodes found ipsilateral to the side of stimulation.

What can be concluded about the laterality of the taste system given these contradictory findings? We argue that one critical consideration is that many of the tests used for evaluating deficits in cases of brain lesions are not conclusive because they cannot rule out higher-order contributions that confound the findings. For example, a bilateral deficit evaluated by the failure to recognize a tastant in response to a stimulus applied on either side of the tongue, associated to a unilateral lesion in the insula, is no proof of bilateral primary projection because taste "recognition" involves integrative functions such as learning, semantics, memorization and retrieval, processes far beyond the core of the primary gustatory projections. Moreover, methods of evaluation of taste disorders could be improved at the clinical level, considering quantitative evaluations are closer to peripheral chemoreceptive events and practically devoid of cognitive integration, hence being better adapted to track the "primary taste projection". Secondly, there is no doubt that the gustatory neuroaxis ascends ipsilaterally, but the converging–diverging transfer of information to a series of subcortical regions may prevent a clear-cut picture with the experimental tools and observations so far in use.

35.3 FUNCTION

The sense of taste contributes to ingestive behavior by enabling organisms to detect, quantify, and identify critical nutrients or toxins in foods and drinks. In this section, we focus on reviewing the neurophysiology of taste intensity, quality, familiarity, and pleasantness coding. We also consider the role of top-down influences, such as expectation and beliefs, on brain representation of the perceptual dimensions of taste. However, taste sensations rarely occur in isolation. Rather, taste co-occurs with other oral sensations to produce unitary flavor perceptions (Prescott, 1999). In addition, sensory information must be integrated with the nutritional state of the organism to appropriately guide behavior. As such, neural coding of taste is not simply a reflection of peripheral input but also a multimodal integration where gustation is processed among sensory and visceral inputs from other modalities to inform feeding behavior relevant to survival (de Araujo et al., 2012). We refer the reader to the following recent reviews for further details the integration of taste with flavor and homeostatic processing (Stevenson and Tomiczek, 2007; de Araujo et al., 2012; Small, 2012).

35.3.1 Intensity Coding of Taste

Responses to tastants increase with stimulus concentration at all levels of the neuroaxis (Zotterman, 1935; Scott et al., 1986a,b; Spector, 1995; Smith and Scott, 2003). However, the *perception* of taste intensity likely emerges from neural circuits within the insula and amygdala. Response magnitude functions generated from taste-responsive cells in the anterior insula and frontal operculum in monkeys conform best to the slopes reported in human psychophysical experiments of perceived intensity (Smith-Swintosky et al., 1991). Changes in suprathreshold taste intensity perception have also been observed following lesions to the insula in humans. Pritchard and colleagues (1999) asked subjects with insular lesions to rate the intensity of tastes applied to either the right or the left side of the tongue. Patients with right insular damage showed decreased sensitivity to tastes applied to the right side of the tongue and patients with left insular damage showed decreased sensitivity to tastes applied to the left side of the tongue. Mak and colleagues also reported decreased taste intensity perception on the side of the tongue ipsilateral to an insular lesion (Mak et al., 2005). However, when they compared their data to matched controls they noted that the change was not due to an ipsilateral decrease but rather to a contralateral increase in taste intensity perception. One possible explanation for this result is that a release from interhemispheric reciprocal inhibition occurred. Similar observations have been reported following damage of the CT in humans (Kveton and Bartoshuk, 1994), suggesting that inhibition may be a general property of the gustatory system. For example, Yanagisawa reported that anesthetizing the CT nerve in humans resulted in increased intensity perception of bitter following application of quinine to the regions of the oral cavity innervated by IX (Yanagisawa et al., 1998), providing evidence for inhibition between these cranial nerves. Also Berteretche et al. (2008) showed patients whose chorda tympani had been slightly iinjured during stapedectomy surgery complained of painful sensations on the tongue, suggesting a desinhibition of the lingual (trigeminal) nerve.

Clinical studies also implicate the amygdala in taste intensity coding. Small and colleagues reported that patients with resection of the right anterior temporal lobe showed decreased taste intensity ratings (Small et al., 2001b, c). Patients tested with solutions applied to discrete loci of the tongue gave more intense ratings than did the control group, with the exception of sucrose ratings. The authors argued that their pattern of findings might reflect differential intensity coding of appetitive vs. aversive tastes, as suggested by the dual affective architecture model of affect proposed by Cacioppo and Bernston (1999). In this model, the appetitive system has a weak positive output, whereas the activation function for the aversive system has a lower offset and rises more quickly. They referred to this as a "negativity bias." Such a model conforms well to early gustatory psychophysical experiments. For example, Wundt (1897) noted, "that for equal intensities, sour, and more especially bitter, produce much stronger feelings than sweet." With this in mind Small and colleagues developed an fMRI study in which the intensity of a pleasant sweet taste and an unpleasant bitter taste were carefully matched for perceived intensity (Small et al., 2003). This matching procedure is important because it ensures that the difference in intensity perception between the strong and weak solutions is similar for the appetitive and the aversive taste. Under these carefully controlled perceptual conditions, amygdala activation was observed following the comparison of the strong sweet vs. the weak sweet, as well as the comparison of the strong bitter vs. the weak bitter. Similar effects can also be observed with olfactory stimuli (Anderson et al., 2003) that are either positive or negative, but not neutral in valence (Winston et al., 2005). Thus we concluded that the processing in the amygdala reflects an interaction between valence and intensity, and as such, that encoding here may underlie the negativity bias for unpleasant tastes.

Neuroimaging studies of taste intensity perception also implicate the insula and overlying operculum (Small et al., 2003; Nitschke et al., 2006; Grabenhorst et al., 2008; Spetter et al., 2010; Cerf-Ducastel et al., 2012; Kobayakawa et al., 2012). In addition to amygdala responses to intensity, Small and colleagues reported preferential responses in the cerebellum, pons, middle insula, operculum, and posterior insula to bitter and sweet tastes that were rated strong compared to weak (Small et al., 2003). In all cases these responses reflected intensity perception irrespective of hedonic valence. In contrast, valence-specific responses were observed in the anterior inferior insula (xyz: 36;12;−3 and −27;12;−15. MNI) which extends to the OFC. Notably, the OFC responses were independent of intensity, suggesting that coding in the OFC is concerned with other dimensions of taste than intensity. Grabenhorst and Rolls (2008) asked subjects to rate intensity or pleasantness of 0.1M monosodium glutamate with 0.005 M inosine monophosphate on separate trials. Individual differences in rated perceived pleasantness were associated with responses in the medial OFC but not in the insula, while individual differences in perceived intensity were associated with responses in the middle and anterior insula but not in the medial OFC. Corresponding differential responses were also observed in these regions as a function of attention to pleasantness vs. intensity. When subjects attended to the intensity of the stimulus, greater activation was produced in the insula. When they attended to the pleasantness of the stimulus, greater activation was produced in the medial OFC. More recently,

Spetter and colleagues reported increased activation in the middle insula associated with increasing concentrations for both NaCl and sucrose which were independent of the tastant (Spetter et al., 2010). The right putamen and the precentral gyrus also represented intensity, but the right amygdala was modulated only by salt, which may again reflect the role of the amygdala in maintaining a negativity bias. Finally, Ge et al. (2012), have also reported intensity-dependent responses in the right anterior insula (42;18;-14) and right middle insula (40;-2;4) where the activation can be top-down modulated from anterior lateral prefrontal areas.

Responses to taste stimulus concentration (Kobayakawa et al., 2008) and intensity perception (Kobayakawa et al., 2012) have also been investigated with MEG. In these studies, concentration or intensity parametrically modulates responses in area G rather than in more anterior insular regions. However, in keeping with the fMRI reports, intensity perception modulated responses in the hippocampus, anterior cingulate cortex, and prefrontal sulcus. Amygdala responses were not observed, likely because this nucleus does not have layered structure and therefore does not generate signals that can be detected with MEG.

Thus, neuroimaging investigations of taste intensity perception in humans clearly implicate the insula, but the exact region again appears dependent upon the methodology employed. Studies also indicate that there is a larger network of regions involved in intensity variations, including the amygdala and hippocampus of the anterior medial temporal lobe, the cerebellum, the anterior cingulate cortex, and the prefrontal sulcus.

35.3.2 Quality Coding of Taste

As in rodents (Katz et al., 2002a; Katz et al., 2002b) and primates (Smith-Swintosky et al., 1991), the coding of taste quality perceptions in humans appears dependent upon insular circuits (Prescott, 1999; Schoenfeld et al., 2004; Adolphs et al., 2005). However, whether chemotopy exists, and the extent to which other regions contribute to the quality code, remain uncertain. Several clinical studies found that resection of the anterior medial temporal lobe resulted in elevated recognition thresholds for a sour taste (Henkin et al., 1977; Small et al., 1997) and that a patient with bilateral damage to this region displayed gustatory agnosia (Small et al., 2005), which is the complete inability to identify taste quality despite intact ability to discriminate stimuli based on intensity or affective differences. These reports are reminiscent of the "Kluver-Bucy syndrome" in which monkeys with bilateral amygdala lesions have been observed to indiscriminately place objects (edible and inedible) in their mouths (Kluver and Bucy, 1938). Patients with insular lesions similarly show deficits in taste quality

identification (Pritchard et al., 1999), with one case of agnosia reported where the individual was able to indicate preferences for taste stimuli without being able to report the taste quality (Adolphs et al., 2005). Thus, the clinical literature highlights involvement of the anterior medial temporal lobe and insula in taste quality coding.

Whereas many studies report tastant-specific activations in the insula, overlying operculum, and anterior medial temporal lobe, there is a lack of obvious chemotopy, and most studies have attributed segregated responses to differences in stimulus valence or physiological significance rather than quality (Zald et al., 1998, 2002; O'Doherty et al., 2001; Small et al., 2003; Haase et al., 2009). For example, O'Doherty et al. (2001) looked for a topographic representation of sweet and salty sensations and concluded that the patterns of activation showed considerable overlap. Nevertheless, they were able to map distinct localizations for glucose and salt in accordance with valence in the OFC. Small and colleagues reported distinct responses for quinine and sucrose in the insula, operculum, and OFC, but also attributed these differential responses to valence rather than quality. Also congruent with these findings are studies reporting overlapping responses to different taste qualities in the insula, OFC, amygdala, and anterior cingulate cortex (de Araujo et al., 2003a; Small et al., 2003; Rudenga et al., 2010).

After de Araujo et al. (2003) had shown that monosodium L-glutamate (MSG), which elicits the so-called "umami" taste, activates the insular/opercular cortex, researchers showed that isomolar solutions of MSG and sodium chloride (NaCl), two compounds having the sodium cation in common, induce fMRI activation in various ROIs in the insula, some of them being common to both stimuli (Nakamura et al., 2011; Iannilli et al., 2012). In the middle insular cortex, peaks of activation for MSG were very close to the activation produced by NaCl. Moreover, using a personalized mouth piece to ensure stimulating a wide tongue area, Nakamura et al. (2011) showed that MSG and NaCl induced different but reproducible coordinates of activation patterns in each individual. Hence, the "response" to a chemical seems to be constituted by several dispersed activated areas with reproducibility possibly observed at individual level.

Schoenfeld et al. (2004) have argued that individual differences in quality specific responses might account for failure to observe chemotopy (Figure 35.3). In their study, they used a cortical flattening technique to rule out uncertainties due to the high degree of cortical folding in insula/operculum taste regions and focused analyses on the patterns of responses within subjects. They found that taste specific patterns of activated ROIs were stable over time in each subject and hence, reproducible. Moreover they noted a high inter-individual topographical variability versus

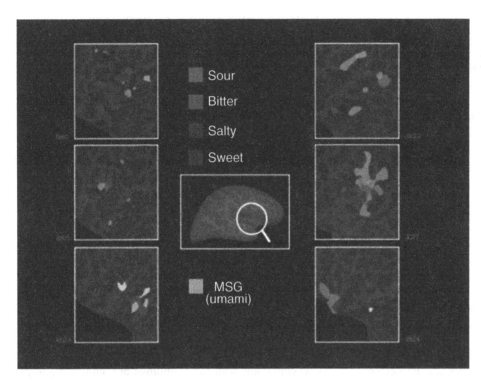

Figure 35.3 The figure shows the individual activity maps for all investigated taste stimuli on the flattened cortical surfaces of the insular/opercular cortex of the right hemisphere of six subjects. In the center of the figure the picture shows the location of the patches on an unfolded brain. Note the high inter-individual variability of the activations. Reprinted from Schoenfeld et al. (2004) with permission from Elsevier. (*See plate section for color version.*)

individual reproducibility. Two groups of researchers (James et al., 2009; Ge et al., 2012) further confirmed inter-individual differences of topographical activity for a given tastant in the brain. Thus, it is likely that individual data treatment leads to better-refined information. This result matches well with (i) the great inter-individual differences of quantitative sensitivity profiles which are reproducible in individuals (Faurion, 1980; 2008), with (ii) different neuronal sensitivity profiles also reproducible in non inbred hamster chorda tympani (Faurion and Vayssettes-Courchay, 1990; Faurion, 2008) and with (ii) differences in quality perceptions among numerous tastants (Schiffman, 1979).

In summary, activation is overlapping for different tastants; however, a pattern of such ROIs may be specific for each tastant. Although this information may differ from subject to subject, it is individually reproducible and the various ROIs can be collected as probability maps of activation for a given compound. These features are relevant to a wealth of electrophysiological data in monkeys showing that cells which respond to a given stimulus are intermingled among cells responding to other stimuli (e.g., Baylis and Rolls, 1991), or even other modalities. Neurons also respond to several stimuli and the parameter evaluating the breadth of tuning of single taste neurons does not show increased specificity at the insular cortex level. No chemotopic organization is observed (e.g., (Yaxley et al., 1990; Smith-Swintosky et al., 1991; Scott and Plata-Salaman, 1999). Similarly, Accolla et al. (2007), using in vivo optical imaging in the rat taste cortex, showed

taste modalities were represented by distinctive patterns but that no region was specific to a single modality.

It is also worth noting, however, that there are several complicating factors in considering the taste quality code – especially in humans. First, the spatial resolution of neuroimaging methods may not be sufficient to resolve tastant-specific neural responses. For example, Chen and colleagues have described segregated responses to taste qualities at the cellular level in rodents using biphoton microscopy and a 540 × 540 μm frame window in each animal (Chen et al., 2011). The spatial resolution of human fMRI imaging studies of taste are generally on the order of 6–10 mm, with the average voxel containing on the order of hundreds of thousands of neurons. Second, quality specific responses can be identified across large ensembles of spatially segregated neurons (Katz et al., 2002a,b) and these ensembles are very possibly modulated by the temporal patterns arising from earlier gustatory relays in the brainstem (Di Lorenzo and Victor, 2003). Thus, a quality-specific neural signature for taste may escape detection because it is represented as a distributed spatial pattern or temporal profile across this network. Finally, one important confound for all studies attempting to isolate quality-specific responses is that there is a very tight coupling between quality and physiological significance. As such, it is often impossible to designate a BOLD, electrical, optical, or psychophysical response as reflecting quality separately from physiological significance. This point is elegantly demonstrated in the classic rodent study by Chang and Scott (1984) in which they

35.3 Function 785

show that the representation of sweet quality, portrayed by multi-dimensional scaling, shifts following pairing of the sweet taste with nausea to resemble the pattern produced by bitter taste. It is also echoed in the more recent work using in vivo optical imaging in rodents to show that the spatial distinction between insular responses to saccharin and quinine fade considerably after an aversion is conditioned for the saccharin (Accolla and Carleton, 2008). Thus, caution is warranted in interpreting spatially segregated insular responses as evidence for the existence of chemotopy (Chen et al., 2011). To conclude, in our opinion the existence of chemotopy in gustation is still unresolved.

35.3.3 Familiarity Coding of Taste

Other factors than intensity coding and quality coding are clearly represented in the insula. Testing novel stimuli repeatedly induces an increase in perceived intensity in humans and an increase of chorda tympani response amplitude – both of which stabilize after these repeated exposures to the stimulus. Furthermore, this effect is stimulus–specific, suggesting peripheral events such as receptor expression (Berteretche et al., 2005). These authors looked at plasticity on repetitive imaging in subjects who underwent a protocol designed to familiarize them to novel tastes. They found that the number of activated pixels significantly changed in the insular/opercular area between the first fMRI session, in which the subject tasted the novel stimuli for the first time, and the second fMRI session, which occurred after the subjects had been repeatedly exposed to the stimuli for hundreds of trials. No change occurred between the second and third fMRI sessions, even though repeated exposures to the stimuli had again been made. The results pointed to a relationship between the evolution (increase versus decrease) of the number of activated pixels in the insular/opercular area and the orientation of the evolution of the hedonic assessment (increase versus decrease) during the process of familiarization to the novel tastants. For a group of experiments (an experiment = "one subject and one stimulus"), the hedonic assessments shifted from positive to less positive and this was interpreted as a diminution of interest for the repeated stimulus, although palatable. Simultaneously, the number of activated pixels diminished. Conversely, for the other group of experiments, the hedonic assessments shifted from negative to less negative or even to positive values and this was interpreted as extinction of neophobia. Simultaneously, the number of activated pixels increased. What was impressive, albeit unexpected, was the exact parallel between the evolution of hedonics and of the number of activated voxels at the individual level (Faurion et al., 1998, 2002).

35.3.4 Affective Coding of Taste

The affective value of a taste depends upon numerous factors, including its perceived intensity, quality, and familiarity, and the subject's internal state, beliefs, expectations, and prior experience (Veldhuizen et al., 2008). As such, computation of the perceived pleasantness of a taste requires sensory, homeostatic, reward, cognitive, and learning circuits. Viewed in this light, perceived pleasantness must be an emergent property of the interaction, not only of particular brain regions, but also of whole brain networks. Indeed there is clear evidence that the value of taste and flavor relies upon contributions from many different systems, including systems involved in beliefs about brand or price (McClure et al., 2004; Passamonti et al., 2009), intensity (Nitschke et al., 2006), and uncertainty (Berns et al., 2001), as well as gut hormones (Malik et al., 2008). However, studies examining these disparate factors converge in indicating that the perceived pleasantness of taste and flavor emerges from OFC circuits and that OFC responses can be a good indicator of subsequent behavior (Hare et al., 2009, 2011; Spetter et al., 2012). That said, it is important to keep in mind that reward is multifaceted. How one defines these facets is often debated (Berridge and Robinson, 2003; Veldhuizen et al., 2008; de Araujo, 2011; Salamone and Correa, 2012), but what is agreed upon is that they are distinct processes. For example, different regions may code anticipation vs. receipt of a pleasant taste (O'Doherty et al., 2002). Also relevant is the fact that there are numerous individual factors that influence responses in classic reward regions during taste and flavor stimulation such as obesity (Stice et al., 2008a; Stice et al., 2008b; Green et al., 2011), genotype (Felsted et al., 2010; Stice and Dagher, 2012), smoking status (Geha et al., 2012), dietary composition (Rudenga and Small, 2011; Burger and Stice, 2012; Green and Murphy, 2012), a history of chronic stress (Rudenga et al., 2012), or a history or presence of an eating disorder (Wagner et al., 2008; Frank et al., 2012). Here we focus on the role of the OFC as the site of convergence of the relevant signals and its role in representing taste pleasantness and unpleasantness.

Zald and Pardo were the first to investigate brain responses to aversive vs. pleasant tastes and flavors (Zald et al., 1998). A concentrated saline solution represented the aversive stimulus and a 3 g piece of Hershey's Symphony Chocolate (Hershey's, Hershey, PA., USA) represented the pleasant stimulus. Greater responses were observed to saline vs. chocolate in the pregenual cingulate cortex, motor cortex, insula, OFC, and amygdala, prompting the authors to conclude that these regions were particularly sensitive to aversive stimuli. However, a subsequent study comparing brain responses to saline vs. glucose solutions observed similar responses when subjects tasted unpleasant saline and pleasant glucose in the OFC, insula, operculum, amygdala, anterior cingulate cortex, anterior temporal lobe cortex, and inferior prefrontal cortex (O'Doherty et al., 2001). The only region showing valence-specific responses was the OFC, where overlapping as well as distinct responses to the two tastants were observed. The authors concluded that pleasant and aversive tastes were represented in the amygdala and OFC.

In support of this conclusion are the results of Small and colleagues who designed a study to dissociate brain response to intensity and pleasantness perception of taste (Small et al., 2003). Subjects first rated the perceived intensity and pleasantness of a series of sucrose and quinine solutions. Only those individuals who gave ratings in a pre-defined target range were included in the neuroimaging experiment that assessed brain response to four stimuli: weak sucrose, strong sucrose, weak quinine and strong quinine. Critically, the affective intensity of the stimuli was matched so that the sweet tastes were rated as pleasant as the bitter tastes were rated unpleasant. In addition, the strong solutions were rated as similarly strong in intensity, and the weak solutions as similarly weak. Response to the weak and strong pleasant sucrose vs. a tasteless control stimulus was observed in the putamen, claustrum, hypothalamus, thalamus OFC, subcallosal cingulate cortex, anterior cingulate cortex and anterior ventral insula. Response to the weak and strong unpleasant quinine was observed in the midbrain, hypothalamus, claustrum, insula/operculum, anterior OFC, medial prefrontal cortex, subcallosal cingulate cortex and ventral lateral prefrontal cortex. Directly contrasting the pleasant vs. unpleasant tastes resulted in valence-specific responses in the OFC, with the left anterior OFC responding preferentially to the unpleasant taste and the right caudolateral OFC responding preferentially to the pleasant taste (Figure 35.4). As mentioned in the section on intensity coding, both effects were independent of intensity. In contrast, response in the amygdala was observed to the strong stimuli compared to the weak stimuli. The authors therefore proposed that responses in the amygdala may reflect stimulus saliency, which could be determined by affective value, intensity or familiarity, whereas responses in the OFC reflected affective value. Consistent with this possibility, in the Zald and Pardo (1998) study, the saline was rated as highly intense, whereas the chocolate was rated as only moderately intense, raising the possibility that differences in intensity or saliency rather than perceived pleasantness drove differential brain responses.

The perceived pleasantness of a taste can also be modulated by changes in internal state, a phenomenon termed alliesthesia (Cabanac, 1971). Consistent with reports in primates (Critchley and Rolls, 1996), response in the OFC to the consumption of pure taste (Haase et al., 2009), and flavors (Small et al., 2001a; Kringelbach et al., 2003; Smeets et al., 2006) is attenuated by feeding to satiety and positive correlations are observed between changes in

rated pleasantness produced by feeding and responses in the OFC. These studies also report associations between pleasantness ratings and/or internal state manipulations in other regions including the insula, hypothalamus, thalamus, hippocampus, parahippocampus, amygdala and midbrain, consistent with the idea that signals from reward and homeostatic circuits contribute to orbital computation of pleasantness. Finally, it is clear that OFC attenuation by feeding is not just a generalized effect of internal state, but relates specifically to the devaluation of the sensory properties of taste, since response is modulated significantly more if subjects are sated on food related to the taste or flavor (Kringelbach et al., 2003), a phenomenon termed sensory specific satiety (Rolls et al., 1981).

There is also evidence that simply evaluating pleasantness, as opposed to other dimensions of taste, can engage orbital circuits (Bender et al., 2009; Grabenhorst and Rolls, 2010). As mentioned above, tasting MSG while attending to pleasantness produces greater medial OFC responses than does tasting MSG while attending to intensity. It may also be the case that a taste stimulus is not required for this engagement. Bender and colleagues reported that the caudolateral OFC responds to taste (sucrose, citric acid and NaCl) and tasteless solutions preferentially when subjects evaluate tastant pleasantness compared to stimulus presence or quality. However, greater connectivity between the caudolateral OFC and earlier gustatory relays (caudomedial OFC and anterior insula) was observed when subjects judged pleasantness and sampled tastants as opposed to the tasteless solution. It was therefore suggested that the lateral OFC organizes the retrieval of gustatory information from earlier relays in the service of computing perceived pleasantness or unpleasantness (Bender et al., 2009).

Also of note, is the potent influence of beliefs and expectations on the perceived pleasantness and brain encoding of taste and flavor. For example, when subjects are inaccurately lead to believe that a taste is the weaker of two bitter solutions they rate it as less unpleasant and show reduced responses in the insula and operculum compared to when they sample the same bitter stimulus and are accurately told that it is the stronger bitter taste (Nitschke et al., 2006) (Figure 35.5). Moreover, activation in the rostral anterior cingulate cortex, the OFC, and dorsal prefrontal cortex to the misleading cue predicts decreased response in the insula and amygdala to the aversive taste (Sarinopoulos et al., 2006). These data suggest that these regions can modulate insular coding of taste based on expectation to alter perceptual experiences of intensity and aversion.

In addition, a region of anterior ventral insula has been identified where overlapping responses to pleasant and unpleasant oral stimulation occur but preferential connectivity with the hypothalamus and ventral striatum occurs when the oral stimulus is potentially nutritive (sweet or salty) compared to potentially harmful (bitter or burn)

35.3 Function

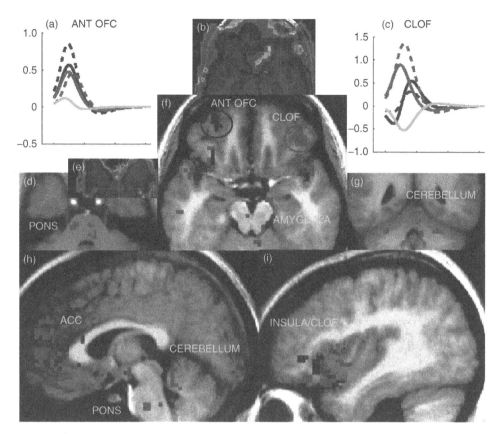

Figure 35.4 Dissociation of taste intensity and affect. Results from the three main random effects analyses color coded and superimposed on the group mean image (d, f, g, h, and i) (green, intensity [(IP + IUP) − (WP + WUP)]; magenta, pleasant [(IP + WP) − tasteless]; blue, unpleasant [(IUP + WUP) − tasteless]). The activations highlight the dissociation between areas responsive to both concentrations of pleasant (magenta) or unpleasant (blue) taste versus those areas responsive to intensity irrespective of valence (green). Note the lack of intensity activation (green) throughout the orbitofrontal cortex (f) and anterior cingulate cortex (ACC) (h). Activation related to either pleasant (pink) or unpleasant (blue) taste is seen extending from the orbitofrontal cortex into the anterior ventral insula (i). In contrast, only activation related to intensity is observed in the middle insula (i), amygdala, pons (d), and cerebellum (g). The right graph (a) displays the fitted hrf extracted from the anterior orbitofrontal cortex region (ANT OFC) corresponding to the circled blue activation in the same representative subject (y axis, response; x axis, peristimulus time with each line indicating a 5 s interval). As per Figure 35.3, blue solid line, IUP; blue dotted line, WUP; magenta solid line, IP; magenta dotted line, WP; and yellow line, tasteless. Note that the response is greatest for WUP, illustrating that valence and not intensity drives the response in this region. The right graph (c) displays the fitted hemodynamic response function (hrf) extracted from the caudolateral orbitofrontal corresponding to the circled magenta activation in a representative subject (y axis, response; x-axis, peristimulus time with each line indicating a 5 s interval). In (b), the BOLD detectability map for the orbitofrontal cortex is presented (Parrish et al., 2000), indicating that we are able to detect a 0.5%–1% signal change throughout a large portion of orbitofrontal cortex, although signal dropout is still evident within the caudomedial region. Purple indicates ability to detect a 0.5% signal change; blue, 1%; green, 2%; and yellow, 4%. The crosshatch is centered on the right caudolateral orbitofrontal cortex corresponding to the peak identified in [(IP + WP) − tasteless], which is activated to pleasant compared to unpleasant taste irrespective of valence. In (e), the BOLD detectability map is displayed for the brainstem, showing that we are able to detect greater than 0.5% signal change throughout this region. Reprinted from Small et al. (2003) with permission from Elsevier. (*See plate section for color version.*)

(Rudenga et al., 2010). This suggests that interactions between the anterior ventral insula and the hypothalamus and striatum are important for integrating sensory signals with homeostatic and affective circuits to guide feeding.

Misleading subjects about the identification of a taste can also alter responses, not only in the insula and operculum but also the thalamus, ventral striatum, OFC, anterior cingulate cortex, inferior frontal gyrus and intraparietal sulcus (Veldhuizen et al., 2011b). In all of these regions response to sucrose and tasteless solutions are greater if individuals believe they will receive a sweet solution but instead receive a tasteless solution, or if they believe they will receive a tasteless solution but instead receive a sweet solution compared to when they receive the solution they are expecting. This suggests that unexpected oral stimulation results in an up-regulation of sensory, attention and reward regions to support the orientation, identification and learning about salient stimuli. Unfortunately, perceived

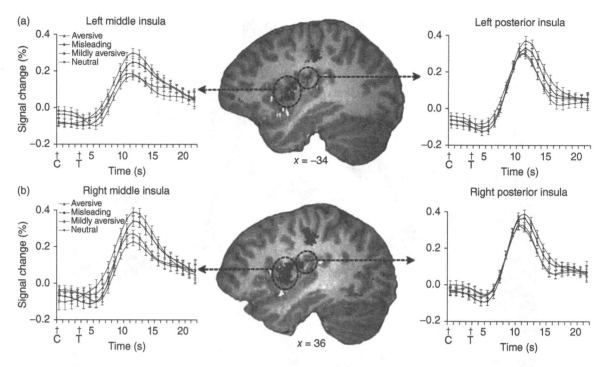

Figure 35.5 (a,b) The left and right insula and operculum clusters shown here were also more strongly activated by the highly aversive taste following the highly aversive cue than by the same taste following the misleading (mildly aversive) cue. Time courses illustrate mean percent signal change across all time points of the trials in each of the four conditions: highly aversive taste following aversive cue (red), highly aversive taste following misleading mildly aversive cue (brown), mildly aversive taste following mildly aversive cue (green) and neutral taste following neutral cue (blue). The onset of the 2-s cue (C) preceded delivery of the tastant (T) by 2–5 s. Error bars are for confidence intervals (95%) around the mean after adjusting for between-subject variance. Reprinted from Nitschke et al. (2006) with permission from the Nature Publishing Group. (*See plate section for color version.*)

pleasantness was not evaluated so it is unclear how these modulations are related to affective value.

More complex cognitive manipulations can also dramatically alter perceived pleasantness and brain responses. McClure and colleagues delivered Coke and Pepsi to subjects in behavioral taste tests and during fMRI (McClure et al., 2004). When the brand was anonymous they found a positive correlation between response in the ventromedial prefrontal cortex and subjects' behavioral preferences. In contrast, when the brand was revealed, preferences changed to reflect the preferred brand and associations with preference were observed in the hippocampus, dorsolateral prefrontal cortex, and midbrain, rather than in the ventromedial prefrontal cortex. In another study, brain responses and the perceived pleasantness of wine were manipulated by providing accurate and fictitious information about price. Subjects sampled three Cabernet Sauvignon wines and a tasteless solution. The wines were administered in random order with the appearance of a price identifier. Unbeknownst to the subjects, two of the wines were administered twice, once identified by their actual retail price and once by a 900% mark-up ($5 retail vs. 45$ fictitious) or 900% mark-down ($90 retail, $10 fictitious). Greater responses were observed to the same wine when sampled at the high vs. low price in a large network of regions that included (among others) the medial OFC, ventromedial prefrontal cortex, and dorsolateral prefrontal cortex. Strikingly, in the medial OFC the magnitude of the BOLD responses reflected the magnitude of the price differential; there were larger differences for $90 vs. $10 than for $45 vs. $5 (Figure 35.6). Moreover, price similarly influenced perceived pleasantness, which also correlated with medial OFC response (Plassmann et al., 2008).

Finally, activation in the anterior inferior insula (mostly left) and the adjacent frontal operculum was shown for disgust resulting from tasting a bitter tastant or from looking at someone tasting and experiencing disgust or by imagining being disgusted. This shared region was connected to distinct and diverging functional circuits allowing different emotional reactions to these different situations (Jabbi et al., 2008).

Collectively these studies indicate that beliefs and expectations exert powerful influences upon gustatory and flavor information processing and resulting perceptions of affective value. They also converge in highlighting a role for the OFC and insular cortex in computing perceptions of the pleasantness and unpleasantness of gustatory stimuli.

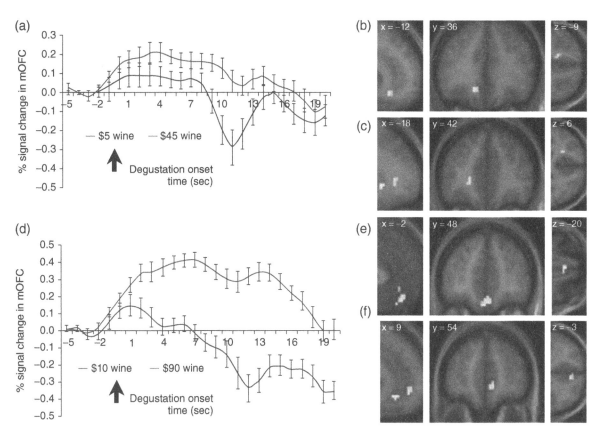

Figure 35.6 The effect of price on each wine. (a) Wine 1: averaged time courses in the medial OFC voxels shown in b (error bars denote standard errors). (b) Wine 1: activity in the mOFC was higher for the high- ($45) than the low-price condition ($5). Activation maps are shown at a threshold of P < 0.001 uncorrected and with an extend threshold of five voxels. (c) Wine 1: activity in the vmPFC was also selected by the same contrast. (d) Wine 2: averaged time courses in the medial OFC voxels shown in e. (e) Wine 2: activity in the mOFC was higher for the high- ($90) than for the low-price condition ($10). (f) Wine 2: activity in the vmPFC was higher for the same contrast. Reprinted from Plassmann et al. (2008) with permission. (*See plate section for color version.*)

35.4 CONCLUSION

In humans, taste information travels from receptors in the oral cavity to the cortex in a predominantly ipsilateral projection. The human gustatory neuroaxis appears to closely parallel the monkey pathway, with the bulk of evidence suggesting that primate evolution has favored the omission of the PBN relay. However, such a profound divergence in mammalian evolution is surprising and this question has not received the intensive interrogation it deserves given its potential importance. Although functional neuroimaging techniques are not without their weakness, one distinct advantage is the ability to examine whole brain responses to sapid stimulation. It is this advantage that has enabled human imaging studies to determine that a far larger region of insula and overlying operculum is devoted to gustatory functions than previously suspected based upon electrophysiological work. The possibility of functional specialization within the distinct sub-regions of the insula and operculum is still under investigation. Also elusive, due in part to the spatial and temporal limitations of human neuroimaging techniques, and in part to the difficulty of dissociating quality and physiological significance, are the neural correlates of taste quality in humans. However, clinical work suggests that interactions between amygdala and insula may be important. What is clear is that many of the insula/opercular taste regions are sensitive to experience, stimulus concentration, and to top-down modulation with some indication that these factors can interact to influence perceived pleasantness. Also established is the interaction between insular sensory coding and homeostatic and affective circuits. This is particularly evident in the anterior ventral insula, which is likely where flavor perceptions begin to emerge (Small, 2012). Key nodes of the gustatory network beyond the insula and operculum are the amygdala, the caudomedial and caudolateral OFC, and the subcallosal and anterior cingulate cortex. These regions play a role in establishing the value and saliency of gustatory sensations based upon stimulus features, internal state, beliefs and expectations.

REFERENCES

Accolla, R., Bathellier, B., Petersen, C. C. and Carleton, A. (2007) Differential spatial representation of taste modalities in the rat gustatory cortex. *J. Neurosci.* 27(6): 1396–404.

Accolla, R. and Carleton, A. (2008) Internal body state influences topographical plasticity of sensory representations in the rat gustatory cortex. *Proc. Natl. Acad. Sci. U. S. A.* 105: 4010–4015.

Adolphs, R., Tranel, D., Koenigs, M. and Damasio, A. R. (2005) Preferring one taste over another without recognizing either. *Nature Neurosci.* 8: 860–861.

Anderson, A. K., Christoff, K., Stappen, I., et al. (2003) Dissociated neural representations of intensity and valence in human olfaction. *Nature Neurosci.* 6: 196–202.

Bagshaw, M. H. and Pribram, K. H. (1953) Cortical organization in gustation (Macaca mulatta). *J. Neurophysiol.* 16: 499–508.

Barry, M. A., Gatenby, J. C., Zeiger, J. D. and Gore, J. C. (2001) Hemispheric Dominance of Cortical Activity Evoked by Focal Electrogustatory Stimuli. *Chem. Senses* 26: 471–482.

Baylis, L. L. and Rolls, E. T. (1991) Responses of neurons in the primate taste cortex to glutamate. *Physiol. Behav.* 49: 973–979.

Baylis, L. L., Rolls, E. T. and Baylis, G. C. (1995) Afferent connections of the caudolateral orbitofrontal cortex taste area of the primate. *Neuroscience.* 64: 801–812.

Beckstead, R. M., Morse, J. R. and Norgren, R. (1980) The nucleus of the solitary tract in the monkey: projections to the thalamus and brain stem nuclei. *J. Comp. Neurol.* 190: 259–282.

Bender, G., Veldhuizen, M. G., Meltzer, J. A., et al. (2009) Neural correlates of evaluative compared with passive tasting. *Eur. J. Neurosci.* 30: 327–338.

Benjamin, R. M. and Burton, H. (1968) Projection of taste nerve afferents to anterior opercular-insular cortex in squirrel monkey (Saimiri sciureus). *Brain Res.* 7: 221–231.

Berlucchi, G., Moro, V., Guerrini, C. and Aglioti, S. M. (2004) Dissociation between taste and tactile extinction on the tongue after right brain damage. *Neuropsychologia* 42: 1007–1016.

Berns, G. S., McClure, S. M., Pagnoni, G. and Montague, P. R. (2001) Predictability modulates human brain response to reward. *J. Neurosci.* 21: 2793–2798.

Berridge, K. C. and Robinson, T. E. (2003) Parsing reward. *Trends Neurosci.* 26: 507–513.

Berteretche, M. V., Boireau-Ducept, N., Pillias, A. M. and Faurion, A. (2005) Stimulus-induced increase of taste responses in the hamster chorda tympani by repeated exposure to 'novel' tastants. *Appetite* 45: 324–333.

Berteretche, M.-V., Eloit, C., Dumas, H., Talmain, G., Herman, P., Tran Ba Huy, P., Faurion, A. (2008) Taste deficits after middle ear surgery for otosclerosis: taste somatosensory interactions. *Eur J Oral Sci* 116: 394–404.

Bornstein, W. S. (1940) Cortical representation of taste in man and monkey: ii. the localization of the cortical taste area in man and a method of measuring impairment of taste in man. *Yale J. Biol. Med.* 13: 133–156.

Boucher, Y., Berteretche, M. V., Farhang, F., et al. (2006) Taste deficits related to dental deafferentation: an electrogustometric study in humans. *Eur. J. Oral Sci.* 114: 456–464.

Boucher, Y., Simons, C. T., Faurion, A., et al. (2003) Trigeminal modulation of gustatory neurons in the nucleus of the solitary tract. *Brain Res.* 973: 265–274.

Burger, K. S. and Stice, E. (2012) Frequent ice cream consumption is associated with reduced striatal response to receipt of an ice cream-based milkshake. *Am. J. Clin. Nutr.* 95: 810–817.

Cabanac, M. (1971) Physiological role of pleasure. *Science* 173: 1103–1107.

Cacioppo, J. T. and Berntson, G. G. (1999) The affect system: Architecture and operating characteristics. *Curr. Dir. Psychol. Sci.* 8: 133–137.

Cereda, C., Ghika, J., Maeder, P. and Bogousslavsky, J. (2002) Strokes restricted to the insular cortex. *Neurology* 59: 1950–1955.

Cerf, B., Lebihan, D., Moortele, P. F., et al. (1998) Functional lateralization of human gustatory cortex related to handedness disclosed by fMRI Study. *Ann. N. Y. Acad. Sci.* 855: 575–578.

Cerf, B., Van de Moortele, P. F., Giacomini, E., et al. (1996) *Correlation of perception to temporal variations of fMRI signal: a taste study. Proceedings ISMRM,* New York 280.

Cerf-Ducastel, B., Haase, L. and Murphy, C. (2012) Effect of magnitude estimation of pleasantness and intensity on fMRI actiation to taste. *Chemosens. Percept.* 5: 100–109.

Cerf-Ducastel, B., Van de Moortele, P. F., MacLeod, P., et al. (2001) Interaction of gustatory and lingual somatosensory perceptions at the cortical level in the human: a functional magnetic resonance imaging study. *Chem. Senses* 26: 371–383.

Chang, F. C. and Scott, T. R. (1984) Conditioned taste aversions modify neural responses in the rat nucleus tractus solitarius. *J. Neurosci.* 4: 1850–1862.

Chen, X., Gabitto, M., Peng, Y., et al. (2011) A gustotopic map of taste qualities in the mammalian brain. *Science* 333: 1262–1266.

Craig, A. D. (2009) How do you feel - now? The anterior insula and human awareness. *Nat. Rev. Neurosci.* 10: 59–70.

Critchley, H. D. and Rolls, E. T. (1996) Hunger and satiety modify the responses of olfactory and visual neurons in the primate orbitofrontal cortex. *J. Neurophysiol.* 75: 1673–1686.

de Araujo, I. (2011) Multiple reward layers in food reinforcement. In Gottfried, J.A. (ed) *Neurobiology of Sensation and Reward.* CRC Press, Boca Raton, pp. 263–286.

de Araujo, I., Geha, P. and Small, D. (2012) Orosensory and homeostatic functions of the insular taste cortex. *Chemosens. Percept.* 5: 1–16.

de Araujo, I. E., Kringelbach, M. L., Rolls, E. T. and Hobden, P. (2003a) Representation of umami taste in the human brain. *J. Neurophysiol.* 90: 313–319.

de Araujo, I. E., Rolls, E. T., Kringelbach, M. L., et al. (2003b) Taste-olfactory convergence, and the representation of the pleasantness of flavour, in the human brain. *Eur. J. Neurosci.* 18: 2059–2068.

Di Lorenzo, P. M. (1988) Taste responses in the parabrachial pons of decerebrate rats. *J. Neurophysiol.* 59: 1871–1887.

Di Lorenzo, P. M. (1990) Corticofugal influence on taste responses in the parabrachial pons of the rat. *Brain Res.* 530: 73–84.

Di Lorenzo, P. M. and Monroe, S. (1989) Taste responses in the parabrachial pons of male, female and pregnant rats. *Brain Res. Bull.* 23: 219–227.

Di Lorenzo, P. M. and Monroe, S. (1992) Corticofugal input to taste-responsive units in the parabrachial pons. *Brain Res. Bull.* 29: 925–930.

Di Lorenzo, P. M. and Victor, J. D. (2003) Taste response variability and temporal coding in the nucleus of the solitary tract of the rat. *J. Neurophysiol.* 90: 1418–1431.

Faurion, A. (1980) Sweet taste involves several distinct receptor mechanisms. *Chem. Senses* 5: 107–121.

Faurion, A. (2006) Sensory interactions through neural pathways. *Physiol. Behav.* 89: 44–46.

Faurion, A. (2008) Interindividual differences of sensitivityin Humans and hamsters and multiple receptor sites for organic molecules: the sweet example. In Weerasinghe, D., DuBois, G.E. (eds) *Sweetness and Sweeteners.* ACS book, Washington, pp. 296–334.

References

Faurion, A., Cerf, B., Le Bihan, D. and Pillias, A. M. (1998) fMRI study of taste cortical areas in humans. *Ann. N. Y. Acad. Sci.* 855: 535–545.

Faurion, A., Cerf, B., Van De Moortele, P. F., et al. (1999) Human taste cortical areas studied with functional magnetic resonance imaging: evidence of functional lateralization related to handedness. *Neurosci. Lett.* 277: 189–192.

Faurion, A., Cerf, B., Pillias, A. M., Boireau, N. (2002). Increased taste sensitivity by familiarization to "novel" stimuli. Psychophysics, fMRI and electrophysiological techniques suggest modulations at peripheral and central levels. In: Olfaction, Taste and Cognition, Rouby, C, Schaal, B., Dubois, D., Gervais, R. Holley, A., Eds., Cambridge University Press, New York, p 350–366.

Faurion, A., Kobayakawa, T. and Cerf-Duastel, B. (2008) Functional magnetic resonance imaging study of taste. In Basbaum, A. I., Kaneko, A., Shepherd, G., Westeheimer, G. (ed) *The Senses: A Comprehensive Reference*. Academic Press, San Diego.

Faurion, A. and Vayssettes-Courchay, C. (1990) Taste as a highly discriminative system: a hamster intrapapillar singlue unit study with 18 compounds. *Brain Res.* 512: 317–332.

Felsted, J. A., Ren, X., Chouinard-Decorte, F. and Small, D. M. (2010) Genetically determined differences in brain response to a primary food reward. *J. Neurosci.* 30: 2428–2432.

Frank, G. K., Reynolds, J. R., Shott, M. E., et al. (2012) Anorexia nervosa and obesity are associated with opposite brain reward response. *Neuropsychopharmacology.* 37: 2031–2046.

Frey, S. and Petrides, M. (1999) Re-examination of the human taste region: a positron emission tomography study. *Eur. J. Neurosci.* 11: 2985–2988.

Ge, T., Feng, J., Grabenhorst, F. and Rolls, E. T. (2012) Componential Granger causality, and its application to identifying the source and mechanisms of the top-down biased activation that controls attention to affective vs sensory processing. *NeuroImage* 59: 1846–1858.

Geha, P., Aschenbrenner, K., Felsted, J., et al. (2012) Smokers have differential brain response to food in regions that perdicted weight gain in non-smokers. *Am. J. Clin. Nut.* 24: 14–18.

Giachetti, I. and Mac Leod, P. (1975) Cortical neuron responses to odours in the rat. In: Denton, D. A., Coghlan J. P., Olfaction and Taste V. Academic Press New York, 303–307.

Giachetti, I. and Mac Leod, P. (1977) Olfactory input to the thalamus: evidence from a ventroposteromedial projection. *Brain Res.* 125: 166–169.

Gibson, J. J. (1979) *The ecological approach to visual perception.* Boston, MA: Houghton Mifflin.

Goto, N., Yamamoto, T., Kaneko, M. and Tomita, H. (1983) Primary pontine hemorrhage and gustatory disturbance: clinicoanatomic study. *Stroke: J. Cereb. Circ.* 14: 507–511.

Grabenhorst, F. and Rolls, E. T. (2008) Selective attention to affective value alters how the brain processes taste stimuli. *Eur. J. Neurosci.* 27: 723–729.

Grabenhorst, F. and Rolls, E. T. (2010) Attentional modulation of affective versus sensory processing: functional connectivity and a top-down biased activation theory of selective attention. *J. Neurophysiol.* 104: 1649–1660.

Grabenhorst, F., Rolls, E. T. and Bilderbeck, A. (2008) How cognition modulates affective responses to taste and flavor: top-down influences on the orbitofrontal and pregenual cingulate cortices. *Cereb. Cortex* 18: 1549–1559.

Green, E., Jacobson, A., Haase, L. and Murphy, C. (2011) Reduced nucleus accumbens and caudate nucleus activation to a pleasant taste is associated with obesity in older adults. *Brain Res.* 1386: 109–117.

Green, E. and Murphy, C. (2012) Altered processing of sweet taste in the brain of diet soda drinkers. *Physiol. Behav.* 12:40–52.

Haase, L., Cerf-Ducastel, B. and Murphy, C. (2009) Cortical activation in response to pure taste stimuli during the physiological states of hunger and satiety. *NeuroImage* 44: 1008–1021.

Hare, T. A., Camerer, C. F. and Rangel, A. (2009) Self-control in decision-making involves modulation of the vmPFC valuation system. *Science* 324: 646–648.

Hare, T. A., Malmaud, J. and Rangel, A. (2011) Focusing attention on the health aspects of foods changes value signals in vmPFC and improves dietary choice. *J. Neurosci.* 31: 11077–11087.

Hausser-Hauw, C. and Bancaud, J. (1987) Gustatory hallucinations in epileptic seizures: electrophysiological, clinical and anatomical correlates. *Brain* 110: 339–359.

Henkin, R. I., Comiter, H., Fedio, P. and O'Doherty, D. (1977) Defects in taste and smell recognition following temporal lobectomy. *Trans. Am. Neurol. Assoc.* 102: 146–150.

Iannilli, E., Singh, P.B., Schuster, B., et al. (2012) Taste laterality studied by means of umami and salt stimuli: An fMRI study. *NeuroImage* 60: 426–435.

Jabbi, M., Bastiaansen, J. and Keysers, C. (2008) A common anterior insula representation of disugst observation, experience, and imagination shows divergent functional connectivity pathways. *PLoS One* 3: e2939.

James, G. A., Li, X., DuBois, G. E., et al. (2009) Prolonged insula activation during perception of aftertaste. *Neuroreport.* 20: 245–250.

Kadohisa, M., Rolls, E. T. and Verhagen, J. V. (2005) Neuronal representations of stimuli in the mouth: the primate insular taste cortex, orbitofrontal cortex and amygdala. *Chem. Senses* 30: 401–419.

Katz, D. B., Nicolelis, M. A. and Simon, S. A. (2002a) Gustatory processing is dynamic and distributed. *Curr. Opin. Neurobiol.* 12: 448–454.

Katz, D. B., Simon, S. A. and Nicolelis, M. A. (2002b) Taste-specific neuronal ensembles in the gustatory cortex of awake rats. *J. Neurosci.* 22: 1850–1857.

Kinomura, S., Kawashima, R., Yamada, K., et al. (1994) Functional anatomy of taste perception in the human brain studied with positron emission tomography. *Brain Res.* 659: 263–266.

Kluver, H. and Bucy, P. (1938) An analysis of certain effectsof bilateral temporal lobectomy in the rhesus monkey, with special reference to "psychic blindness". *J. Psychol.* 5: 33–54.

Kobayakawa, T., Endo, H., Ayabe-Kanamura, S., et al. (1996) The primary gustatory area in human cerebral cortex studied by magnetoencephalography. *Neurosci. Lett.* 212: 155–158.

Kobayakawa, T., Ogawa, H., Kaneda, H., et al. (1999) Spatio-temporal analysis of cortical activity evoked by gustatory stimulation in humans. *Chem. Senses* 24: 201–209.

Kobayakawa, T., Saito, S. and Gotow, N. (2012) Temporal characteristics of neural activity associated with perception of gustatory stimulus intensity in humans. *Chemosens. Percept.* 5: 80–86.

Kobayakawa, T., Saito, S., Gotow, N. and Ogawa, H. (2008) Representation of salty taste stimulus concentrations in the primary gustatory area in humans. *Chemosens. Percept.* 1: 227–234.

Kringelbach, M. L., O'Doherty, J., Rolls, E. T. and Andrews, C. (2003) Activation of the human orbitofrontal cortex to a liquid food stimulus is correlated with its subjective pleasantness. *Cereb. Cortex* 13: 1064–1071.

Kveton, J. F. and Bartoshuk, L. M. (1994) The effect of unilateral chorda tympani damage on taste. *Laryngoscope* 104: 25–29.

Landis, B. N., Leuchter, I., San Millan Ruiz, D., et al. (2006) Transient hemiageusia in cerebrovascular lateral pontine lesions. *J. Neurol. Neurosurg. Psychiat.* 77: 680–683.

Lewey, F. H. (1943) Aura of taste preceding convulsions associated with a lesion of the parietal operculum: report of a case. *Arch. Neurol. Psychiat.* 50: 575–578.

Lugaz, O., Pillias, A. M., Boireau-Ducept, N. and Faurion, A. (2005) Time-intensity evaluation of acid taste in subjects with saliva high flow and low flow rates for acids of various chemical properties. *Chem. Senses* 30: 89–103.

Mak, Y. E., Simmons, K. B., Gitelman, D. R. and Small, D. M. (2005) Taste and olfactory intensity perception changes following left insular stroke. *Behav. Neurosci.* 119: 1693–1700.

Malik, S., McGlone, F., Bedrossian, D. and Dagher, A. (2008) Ghrelin modulates brain activity in areas that control appetitive behavior. *Cell Metabol.* 7: 400–409.

McClure, S. M., Li, J., Montague, L. M. and Montague, P. R. (2004) Neural correlates of behavioral preference for culturally familiar drinks. *Neuron* 44: 379–387.

Meiselman, H. (1971) Effect of presentation procedure on taste intensity functions. *Percept. Psychophys.* 10: 1–18.

Mesulam, M. M. and Mufson, E. J. (1982a) Insula of the old world monkey. I. Architectonics in the insulo-orbito-temporal component of the paralimbic brain. *J. Comp. Neurol.* 212: 1–22.

Mesulam, M. M. and Mufson, E. J. (1982b) Insula of the old world monkey. III: Efferent cortical output and comments on function. *J. Comp. Neurol.* 212: 38–52.

Mesulam, M. M. and Mufson, E. J. (1985) The Insula of Reil in man and monkey: Architectonics, connectivity and function. In Peters, A., Jones, E.G. (eds) Cerebral Cortex. Plenum Publishing, pp. 221–222.

Mizoguchi, C., Kobayakawa, T., Saito, S. and Ogawa, H. (2002) Gustatory Evoked Cortical Activity in Humans Studied by Simultaneous EEG and MEG Recording. *Chem. Senses* 27: 629–634.

Motta, G. (1959) The cortical taste centers. *Bull. Sci. Med. (Bologna)* 131, 480–493.

Nakajima, Y., Utsumi, H. and Takahashi, H. (1983) Ipsilateral disturbance of taste due to pontine hemorrhage. *J. Neurol.* 229: 133–136.

Nakamura, Y., Goto, T. K., Tokumori, K., et al. (2011) Localization of brain activation by umami taste in humans. *Brain Res.* 1406: 18–29.

Nitschke, J. B., Dixon, G. E., Sarinopoulos, I., et al. (2006) Altering expectancy dampens neural response to aversive taste in primary taste cortex. *Nat. Neurosci.* 9, 435–442.

Norgren, R. (1978) Projections from the nucleus of the solitary tract in the rat. *Neuroscience* 3: 207–218.

Norgren, R. (1981) The central organization of the gustatory and visceral afferent systems in the nucleus of the solitary tract. In Katsuki, Y., Norgren, R., Sato, M. (ed) *Brain Mechanisms of Sensation*. Wiley, New York, pp. 143–160.

Norgren, R. (1995) Gustatory System. In Paxinos, G. (ed) *The Rat Nervous System*. Academic Press, San Diego, pp. 751–771.

O'Doherty, J., Rolls, E. T., Francis, S., et al. (2001) Representation of pleasant and aversive taste in the human brain. *J. Neurophysiol.* 85: 1315–1321.

O'Doherty, J. P., Deichmann, R., Critchley, H. D. and Dolan, R. J. (2002) Neural responses during anticipation of a primary taste reward. *Neuron* 33: 815–826.

Ogawa, H., Ifuku, H., Nakamura, T. and Hirata, S. (2005a) Possible changes in information from the primary to higher-order gustatory cortices, studied by recording neural activities during a taste discrimination GO/NOGO task in monkeys. *Chem. Senses* 30 Suppl 1: i78–79.

Ogawa, H., Ito, S. and Nomura, T. (1985) Two distinct projection areas from tongue nerves in the frontal operculum of macaque monkeys as revealed with evoked potential mapping. *Neurosci. Res.* 2: 447–459.

Ogawa, H., Ito, S. and Nomura, T. (1989) Oral cavity representation at the frontal operculum of macaque monkeys. *Neurosci. Res.* 6: 283–298.

Ogawa, H., Wakita, M., Hasegawa, K., et al. (2005b) Functional MRI detection of activation in the primary gustatory cortices in humans. *Chem. Senses* 30: 583–592.

Ohla, K., Toepel, U., le Coutre, J. and Hudry, J. (2010) Electrical neuroimaging reveals intensity-dependent activation of human cortical gustatory and somatosensory areas by electric taste. *Biol. Psychol.* 85: 446–455.

Onoda, K. and Ikeda, M. (1999) Gustatory disturbance due to cerebrovascular disorder. *Laryngoscope* 109: 123–128.

Onoda, K., Ikeda, M., Sekine, H. and Ogawa, H. (2012) Clinical study of central taste disorders and discussion of the central gustatory pathway. *J. Neurol.* 259: 261–266.

Onoda, K., Kobayakawa, T., Ikeda, M., et al. (2005) Laterality of human primary gustatory cortex studied by MEG. *Chem. Senses* 30: 657–666.

Pardo, J. V., Wood, T. D., Costello, P. A., et al. (1997) PET study of the localization and laterality of lingual somatosensory processing in humans. *Neurosci. Lett.* 234: 23–26.

Parrish, T. B., Gitelman, D. R., LaBar, K. S., Mesulam, M. M. (2000) Impact of signal-to-noise on functional MRI. *Magn Reson Med.* 44(6): 925–932.

Passamonti, L., Rowe, J. B., Schwarzbauer, C., et al. (2009) Personality predicts the brain's response to viewing appetizing foods: the neural basis of a risk factor for overeating. *J. Neurosci.* 29: 43–51.

Penfield, W. and Faulk, M. E. J. (1955) The insula: further observations on its function. *Brain* 78: 445–470.

Plassmann, H., O'Doherty, J., Shiv, B. and Rangel, A. (2008) Marketing actions can modulate neural representations of experienced pleasantness. *Proc. Natl. Acad. Sci. U. S. A.* 105: 1050–1054.

Prescott, J. (1999) Flavour as a psychological construct: implications for perceiving and measuring the sensory qualities of foods. *Food Qual. Prefer.* 10: 349–356.

Pritchard, T. C., Edwards, E. M., Smith, C.A., et al. (2005) Gustatory neural responses in the medial orbitofrontal cortex of the old world monkey. *J. Neurosci.* 25: 6047–6065.

Pritchard, T. C., Hamilton, R. B., Morse, J. R. and Norgren, R. (1986) Projections of thalamic gustatory and lingual areas in the monkey, Macaca fascicularis. *J. Comp. Neurol.* 244: 213–228.

Pritchard, T. C., Hamilton, R. B. and Norgren, R. (2000) Projections of the parabrachial nucleus in the Old World monkey. *Experimental Neurology* 165: 101–117.

Pritchard, T. C., Macaluso, E. and Eslinger, P. J. (1999) Taste perception in patients with insular cortex lesions. *Behav. Neurosci.* 113: 663–671.

Rolls, B. J., Rolls, E. T., Rowe, E. A. and Sweeney, K. (1981) Sensory specific satiety in man. *Physiol. Behav.* 27: 137–142.

Rolls, E. T. and Baylis, L. L. (1994) Gustatory, olfactory, and visual convergence within the primate orbitofrontal cortex. *J. Neurosci.* 14: 5437–5452.

Rolls, E. T., Yaxley, S. and Sienkiewicz, Z. J. (1990) Gustatory responses of single neurons in the caudolateral orbitofrontal cortex of the macaque monkey. *J. Neurophysiol.* 64: 1055–1066.

Ruch, T. C. and Patton, H. D. (1946) The relation of the deep opercular cortex to taste. *Fed. Proc.* 5: 89.

Rudenga, K., Green, B. G., Nachtigal, D. and Small, D. M. (2010) Evidence for an intgrated oral sensory module in the human ventral insula. *Chem. Senses* 35: 693–703.

Rudenga, K. and Small, D. M. (2011) Amygdala response to sucrose consumption is inversely related to artificial sweetner use. *Appetite* 58 504–507.

References

Rudenga, K. J., Sinha, R. and Small, D. M. (2012) Acute stress potentiates brain response to milkshake as a function of body weight and chronic stress. *Int. J. Obesity* 10: 5–12.

Salamone, J. D. and Correa, M. (2012) Dopamine and food addiction: lexicon badly needed. *Biol. psychiat.* 24: 3–13

Sanchez-Juan, P. and Combarros, O. (2001) Gustatory nervous pathway syndromes. *Neurologia* 16: 262–271.

Sarinopoulos, I., Dixon, G. E., Short, S. J., et al. (2006) Brain mechanisms of expectation associated with insula and amygdala response to aversive taste: implications for placebo. *Brain Behav. Immun.* 20: 120–132.

Sato, K. and Nitta, E. (2000) A case of ipsilateral ageusia, sensorineural hearing loss and facial sensorimotor disturbance due to pontine lesion. *Rinsho Shinkeigaku* 40: 487–489.

Schiffman, S. S. (Year) *Preference Behaviour and Chemoreception.* In: Kroeze, J. H. A. (ed). Information Retrieval Limited London, City.

Schoenfeld, M. A., Neuer, G., Tempelmann, C., et al. (2004) Functional magnetic resonance tomography correlates of taste perception in the human primary taste cortex. *Neuroscience* 127: 347–353.

Scott, T. R. and Plata-Salaman, C. R. (1991) Coding of taste quality. In Getchell, T.V. et al. (Ed.), *Smell and Taste in Health and Disease.* Raven Press, New York. pp. 345–368.

Scott, T. R. and Plata-Salaman, C. R. (1999) Taste in the monkey cortex. *Physiol. Behav.* 67: 489–511.

Scott, T. R. and Small, D. M. (2009) The role of the parabrachial nucleus in taste processing and feeding. *Ann. N. Y. Acad. Sci.* 1179: 372–377.

Scott, T. R., Yaxley, S., Sienkiewicz, Z.J . and Rolls, E. T. (1986a) Gustatory responses in the frontal opercular cortex of the alert cynomolgus monkey. *J. Neurophysiol.* 56: 876–890.

Scott, T. R., Yaxley, S., Sienkiewicz, Z. J. and Rolls, E. T. (1986b) Gustatory responses in the nucleus tractus solitarius of the alert cynomolgus monkey. *J. Neurophysiol.* 55: 182–200.

Small, D. M. (2006) Central gustatory processing in humans. *Adv. Otorhinolaryngol.* 63: 191–220.

Small, D. M. (2010) Taste representation in the human insula. *Brain Struct. Func.* 214: 551–561.

Small, D. M. (2012) Flavor is in the Brain. *Physiol. Behav.* 107: 540–552.

Small, D. M., Bernasconi, N., Bernasconi, A., et al. (2005) Gustatory agnosia. *Neurology* 64: 311–317.

Small, D. M. and Green, B. G. (2011) A proposed model of a flavor modality. In Murray, M., Wallace, M.T. (eds) *Frontiers in the Neural Bases of Multisensory Processes.* Taylor and Francis.

Small, D. M., Gregory, M. D., Mak, Y. E., et al. (2003) Dissociation of neural representation of intensity and affective valuation in human gustation. *Neuron* 39: 701–711.

Small, D. M., Jones-Gotman, M., Zatorre, R. J., et al. (1997) A role for the right anterior temporal lobe in taste quality recognition. *J. Neurosci.* 17: 5136–5142.

Small, D. M. and Scott, T. R. (2009) What happens to the pontine processing? Repercussions of interspecies differences in pontine taste representation for tasting and feeding. *Ann. N. Y. Acad. Sci.* 1170: 343–346.

Small, D. M., Zald, D. H., Jones-Gotman, M., et al. (1999) Human cortical gustatory areas: a review of functional neuroimaging data. *Neuroreport.* 10, 7–14.

Small, D. M., Zatorre, R. J., Dagher, A., et al. (2001a) Changes in brain activity related to eating chocolate: from pleasure to aversion. *Brain* 124: 1720–1733.

Small, D. M., Zatorre, R. J. and Jones-Gotman, M. (2001b) Changes in taste intensity perception following anterior temporal lobe removal in humans. *Chem. Senses* 26: 425–432.

Small, D. M., Zatorre, R. J. and Jones-Gotman, M. (2001c) Increased intensity perception of aversive taste following right anteromedial temporal lobe removal in humans. *Brain* 124: 1566–1575.

Smeets, P. A., de Graaf, C., Stafleu, A., et al. (2006) Effect of satiety on brain activation during chocolate tasting in men and women. *Am. J. Clin. Nutrit.* 83: 1297–1305.

Smith, D. V. and Scott, T. R. (2003) Gustatory neural coding. In Doty, R.L. (Ed.), *Handbook of Olfaction and Gustation.* Marcel Dekker, Philaedelphia, pp. 731–758.

Smith-Swintosky, V. L., Plata-Salaman, C. R. and Scott, T. R. (1991) Gustatory neural coding in the monkey cortex: stimulus quality. *J. Neurophysiol.* 66: 1156–1165.

Spector, A. C. (1995) Gustatory function in the parabrachial nuclei: implications from lesion studies in rats. *Rev. Neurosci.* 6: 143–175.

Spetter, M. S., de Graaf, C., Viergever, M. A. and Smeets, P. A. (2012) Anterior cingulate taste activation predicts ad libitum intake of sweet and savory drinks in healthy, normal-weight men. *J. Nutrition* 142: 795–802.

Spetter, M. S., Smeets, P. A., de Graaf, C. and Viergever, M. A. (2010) Representation of sweet and salty taste intensity in the brain. *Chem. Senses* 35: 831–840.

Stephani, C., Fernandez-Baca Vaca, G., Maciunas, R., et al. (2011) Functional neuroanatomy of the insular lobe. *Brain Struct. Funct.* 216: 137–149.

Stevenson, R. J. and Tomiczek, C. (2007) Olfactory-induced synesthesias: A review and model. *Psych. Bull.* 133: 294–309.

Stice, E., Bohon, C., Veldhuizen, M. G. and Small, D. M. (2008a) Relation of reward from food intake and anticipated food intake to obesity: A functional magnetic resonance imaging study. *J. Abnorm. Psychol.* 117: 924–935.

Stice, E. and Dagher, A. (2012) Genetic variation in dopaminergic reward in humans. *Forum Nutr.* 63: 176–185.

Stice, E., Spoor, S., Bohon, C. and Small, D. M. (2008b) Relation between obesity and blunted striatal respone to food is moderated by *Taq*IA A1 allele. *Science* 322: 449–452.

Takeuchi, S., Takasato, Y., Masaoka, H., et al. (2006) Case of gustatory disturbance caused by pontine infarction. *No To Shinkei* 58: 1005–1007.

Topolovec, J. C., Gati, J. S., Menon, R. S., (2004) Human cardiovascular and gustatory brainstem sites observed by functional magnetic resonance imaging. *J. Comp. Neurol.* 471: 446–461.

Uesaka, Y., Nose, H., Ida, M. and Takagi, A. (1998) The pathway of gustatory fibers of the human ascends ipsilaterally in the pons. *Neurology* 50: 827–828.

Van de Moortele, P.-F. C. B, Lobel, E., Paradis, A.-L., et al. (1997) Latencies in fMRI time-series: effect of slice acquisition order and perception. *NMR Biomed.* 10: 230–236.

Veldhuizen, M. G., Albrecht, J., Zelano, C., et al. (2011a) Identification of human gustatory cortex by activation likelihood estimation. *Hum. Brain Mapp.* 32: 2256–2266.

Veldhuizen, M. G., Bender, G., Constable, R. T. and Small, D. M. (2007) Trying to detect taste in a tasteless solution: modulation of early gustatory cortex by attention to taste. *Chem. Senses* 32: 569–581.

Veldhuizen, M. G., Douglas, D., Aschenbrenner, K., et al. (2011b) The anterior insular cortex represents breaches of taste identity expectation. *J. Neurosci.* 31: 14735–14744.

Veldhuizen, M. G., Rudenga, K. and Small, D. M. (2008) The pleasure of taste, flavor and food. In Berridge, K.C., Kringelbach, M.L. (eds) *The Neurobiology of Pleasure.* Oxford University Press, Oxford.

Verhagen, J. V. and Engelen, L. (2006) The neurocognitive bases of human multimodal food perception: sensory integration. *Neurosci. Biobehav. Rev.* 30: 613–650.

Wagner, A., Aizenstein, H., Mazurkewicz, L., et al. (2008) Altered insula response to taste stimuli in individuals recovered from restricting-type anorexia nervosa. *Neuropsychopharmacology* 33: 513–523.

Weidemann, J. and Sparing, R. (2002) Hemiageusia resulting from a cavernous haemangioma in the brain stem. *J. Neurol. Neurosurg. Psychiat.* 73: 319.

Winston, J. S., Gottfried, J. A., Kilner, J. M. and Dolan, R. J. (2005) Integrated neural representations of odor intensity and affective valence in human amygdala. *J. Neurosci.* 25: 8903–8907.

Wundt, W. (1897). *Outlines of Psychology.* Translated by C. H. Judd. Leipzig: Wilhelm Engelmann (Reprinted Bristol: Thoemmes, 1999); first published in German as Wundt, W. (1896). *Grundriss der Psychologie.* Leipzig: Wilhelm Engelmann.

Yamamoto, C., Takehara, S., Morikawa, K., et al. (2003) Magnetoencephalographic study of cortical activity evoked by electrogustatory stimuli. *Chem. Senses* 28: 245–251.

Yanagisawa, K., Bartoshuk, L.M., Catalanotto, F.A., et al. (1998) Anesthesia of the chorda tympani nerve and taste phantoms. *Physiol. Behav.* 63: 329–335.

Yaxley, S., Rolls, E. T. and Sienkiewicz, Z. J. (1990) Gustatory responses of single neurons in the insula of the macaque monkey. *J. Neurophysiol.* 63: 689–700.

Zald, D. H., Hagen, M. C. and Pardo, J. V. (2002) Neural correlates of tasting concentrated quinine and sugar solutions. *J. Neurophysiol.* 87: 1068–1075.

Zald, D. H., Lee, J. T., Fluegel, K. W. and Pardo, J. V. (1998) Aversive gustatory stimulation activates limbic circuits in humans. *Brain* 121: 1143–1154.

Zotterman, Y. (1935) Action potentials in the glossopharyngeal nerve and in the chorda tympani. *Scand. Arch. Physiol.* 72: 73–77.

Chapter 36

The Ontogeny of Taste Perception and Preference Throughout Childhood

CATHERINE A. FORESTELL and JULIE A. MENNELLA

36.1 INTRODUCTION

The study of the ontogeny of human taste perception has been the focus of scientific investigations for more than a century (as summarized in Table 36.1). From this knowledge base, we now know that humans are born with a well-developed peripheral taste system—a taste system that, by the last trimester of pregnancy, is capable of conveying information to the central nervous system. This information is made available to brain structures critical for organizing a variety of consummatory (e.g., intake) and reflex-like (e.g., facial expressions) behaviors; (see Ganchrow and Mennella, 2003, for an earlier review). Clearly, taste plays an important role in determining whether a foreign substance should be accepted or rejected, and informs the gastrointestinal system about the quality and quantity of the impending rush of nutrients (Egan and Margolskee, 2008; Sclafani, 2007). Although the taste system involves innate responses, these responses change and mature throughout childhood, as will be described in this chapter (e.g., Cowart, 1981; Mennella et al., 2010b; Stein et al., 1994).

The sense of taste has taken on great interest in recent years because it is a major determinant of the food acceptance patterns of children (Birch 1998). Thus, our accumulating knowledge about taste plays an important role in understanding the basis for food choices in children. Although the modernization and industrialization of the food supply have produced many benefits, diets rich in sugars and salt and low in fruits and vegetables have become increasingly commonplace, thereby producing unanticipated consequences. For example, excessive intake of foods containing high amounts of salt or sugars (and, consequently, foods that taste salty and sweet) causes or exacerbates a number of illnesses, including hypertension, diabetes, and metabolic syndrome (Mokdad et al., 2004). Reduced intake of fruits and vegetables increases the risk of a number of chronic diseases, including cardiovascular diseases and certain cancers (Knai et al. 2006). Because eating vegetables as part of an overall healthy diet can aid in weight management for children and adults (Blanck et al., 2008; Lakkakula et al., 2008; McCrory et al., 1999), their reduced intake likely contributes to the obesity epidemic.

In this chapter, we focus on the approaches and discoveries that have contributed to our current knowledge of human development of the five "basic" tastes of sweet, sour, salty, bitter, and umami (savory) and the impact they have had on our understanding of the sensory world of children. This review is not meant to be representative of all research conducted to date; rather, it highlights the breadth of methodologies employed, the convergence of findings, and extant gaps in knowledge. Although a detailed treatment of taste anatomy is beyond the scope of this chapter, we begin by providing an overview of the mechanisms underlying taste sensations and when they emerge during development.

Handbook of Olfaction and Gustation, Third Edition. Edited by Richard L. Doty.
© 2015 Richard L. Doty. Published 2015 by John Wiley & Sons, Inc.

36.2 THE ONTOGENY OF TASTE PERCEPTION

Taste, or *gustation*, refers to the sensation that occurs when chemicals stimulate taste receptors located on a large portion of the tongue's dorsum and other parts of the oropharynx, such as the larynx, pharynx, and epiglottis. *Flavor*, in contrast, is elicited by a combination of tastes and odors experienced orthonasally and retronasally while consuming substances. Taste is initiated at taste receptors, which are localized in taste buds and are innervated by branches of three cranial nerves: the facial (VIIth), glossopharyngeal (IXth), and vagal (Xth) nerves. The taste stimuli that interact with these receptors are often separated into a small number of "primary" taste qualities: sweet, salty, bitter, sour, and umami.

Contemporary knowledge about the development of the five basic taste qualities stems from research that has examined receptor morphogenesis and functional maturity (for review, see Bradley and Mistretta, 1975; Cowart et al., 2004; Ganchrow and Mennella, 2003). Early research conducted in this area revealed that taste buds begin to emerge throughout the oral cavity, including the dorsal surface of the tongue, palatoglossal arches, palate, posterior surface of the epiglottis, and posterior wall of the oropharynx, just 8 weeks after conception. By 13 weeks, taste buds begin to resemble those of the adult and are found throughout the oral cavity (Bradley and Mistretta, 1975; Bradley and Stern, 1967; Hersch and Ganchrow, 1980; Witt and Reutter, 1996; 1998). Taste pores, which provide access for tastants to interact with taste bud receptor cells, are generally considered as markers of functional maturity (Mistretta, 1972). Such pores have been identified in fetal fungiform papillae before the end of the fourth month (Witt and Reutter, 1996; 1998). At birth, the newborn infant is endowed with a rich population of gustatory receptors associated with such pores.

It is probably not a coincidence that the timing of these developmental changes corresponds to the emergence of fetal sucking and swallowing behavior. Ultrasound studies have revealed early development of swallowing and oral sensorimotor function at approximately 15 weeks gestational age (Miller et al., 2003; Ross and Nijland, 1997). By term, infants are actively swallowing between 500 and 1000 ml/day of amniotic fluid (Pritchard, 1965; Ross and Nijland, 1997). This not only represents a major route of amniotic fluid absorption but also serves to stimulate the taste buds and influence their synaptic connections. The chemosensory composition of the amniotic fluid is in constant flux throughout pregnancy, especially once the fetus begins to urinate, exposing the newly developed taste buds to varying concentrations of a variety of chemicals, including sugars, sodium and potassium salts, various acids (Liley, 1972), and flavors from the mother's diet (Mennella et al., 1995; Mennella et al., 2001).

36.3 RESEARCH APPROACHES AND METHODOLOGIES

Table 36.1 presents an extensive summary of studies of taste sensitivity and preference during development, organized by taste quality, age group, and methodology. A variety of psychophysical methodologies have been employed to assess taste perception and preference throughout infancy, childhood, and adolescence. The method chosen depends on the objective of the study as well as the age (and, in turn, cognitive and language abilities) of the participants under study. These psychophysical studies on taste provide data relevant to two separate aspects of sensation: (1) the *sensitivity* of the system to chemical stimuli and (2) the *hedonic valence*, or pleasantness, of the sensation (Cowart, 1981; Cowart et al., 2004).

For preverbal children, the tools used often focus on reflex-like responses (e.g., salivation, sucking, heart rate, or orofacial responses) or consummatory responses. Interestingly, many of the experimental paradigms for this age group are similar to those used in animal studies (Berridge and Kringelbach, 2008; Boughter and Bachmanov, 2007; Glendinning et al., 2002; Grill and Norgren, 1978; Spector et al., 1981; Steiner et al., 2001). Because virtually all of these measures can be associated with acceptance or rejection, they presumably involve a hedonic component. As a result, sensitivity and hedonics are difficult to distinguish in animals and in nonverbal infants (Cowart et al., 2004).

For older children, a variety of methodological tools have been used, which vary according to their language and cognitive development. Measures of sensitivity include threshold detection and suprathreshold scaling procedures, while hedonic measurements include ratings or ranking of stimuli. Several of these sensory methodological tools have been published in standard guides (ASTM International, 2003) and review articles (Popper and Kroll, 2005) and often are modifications of methods used in adult populations. However, the validity and reproducibility of many of these procedures have not been assessed for children.

Findings from this expanding collection of methodological tools are remarkably consistent across studies and laboratories throughout the world. In the sections that follow, we review the methodologies used with children of different ages and highlight findings that support the hypothesis that behaviors associated with gustatory function are robust during early development and contribute significantly to adaptation and survival of the developing infant and child.

36.4 TASTE SENSITIVITY IN THE FETUS

A key question in the ontogeny of taste perception is whether function is present prior to birth. As summarized in Table 36.1, two approaches have been used to answer this important question: (1) indirect observations of the fetus and (2) direct observations of premature infants. Although the response of human fetuses to taste has never been directly investigated, indirect evidence from studies in fetuses together with findings from studies in premature infants, suggests that prenatal monitoring of some taste stimuli is possible during late gestation. Moreover, as alluded to earlier in the chapter, the fetus is exposed to a unique chemosensory environment prior to birth (for review, see (Mennella, 2007).

There are two reports of increased fetal swallowing following the injection of sweet stimuli such as saccharin into the amniotic fluid of pregnant women (DeSnoo, 1937; Liley, 1972), Decreased swallowing has been noted following analogous injection of bitter-tasting poppy seed oil (Liley, 1972). Such findings provide indirect evidence that the fetus is capable of tasting and differentially responding to bitter- and sweet-tasting substances.

Stronger evidence that the fetus can respond to taste stimuli comes from research on premature infants. To overcome methodological limitations caused by premature infants' immature suck-swallow coordination, innovative methods have been developed that avoid the risk of fluid aspiration. These include brief, intermittent presentations of small quantities of tastants (sometimes on absorbent cotton) directly into the oral cavity (Eckstein, 1927; Tatzer et al., 1985) and embedding tastants in a nipple-shaped gelatin media that release small amounts of tastants during mouthing or sucking (Maone et al., 1990). By measuring the strength, length, and latency of nonnutritive sucking to these tastants relative to a control, such as water, researchers have been able to infer sensitivity to, and preferences for, various tastants. This body of research revealed that newborns delivered preterm and tested postnatally produced stronger and more frequent sucking responses when offered a sucrose-sweetened nipple compared to an unsweetened gelatin-based nipple (Maone, Mattes et al., 1990), or to drops of sweetened solution compared to water (Tatzer, Schubert et al., 1985).

When small amounts of bitter or sour tastants were placed directly on the tongue of preterm infants, differential responding has been observed. Eckstein (1927) found, for example, that various concentrations of bitter-tasting quinine retarded sucking in most of the infants he studied, whereas a drop of pure lemon juice increased reflex salivation that was accompanied by increases in sucking vigor and, in some cases, retching. Taken together, this body of research suggests that by the last trimester, humans have a well-developed taste system capable of detecting and hedonically responding to sweet, bitter, and sour tastants.

36.5 TASTE SENSITIVITY AND PREFERENCE DURING INFANCY

36.5.1 Sucking and Consummatory Responses

Sucking and ingestive responses have been successfully used to study patterns of responses that occur as a function of individual and age-related differences in taste perception and preferences throughout infancy (summarized in Table 36.1). Methods include delivering small quantities of taste solutions (1–2 ml) applied directly to the tongue (Crook and Lipsitt, 1976; Crook, 1978), embedding tastants in a gelatin-based nipple (Maone et al., 1990), or providing brief (30–180 s) access to multiple bottles in succession which contain various taste or diluent solutions (Beauchamp et al., 1994; Beauchamp et al., 1986; Beauchamp and Moran, 1982; Beauchamp and Pearson, 1991; Kajuira et al., 1992). One can infer that the infant can detect the tastes, and to some extent that they prefer the taste solution (e.g., sweetened water) more than the diluent (e.g., water), if they consume more from the bottles that contain the taste solution than from those that contain the diluent solution (Beauchamp et al., 1986; Desor et al., 1973). Consistent with the findings for premature infants, research has repeatedly demonstrated strong acceptance of sweet-tasting sugars by the newborn. Within days after birth, they can not only detect dilute sweet solutions, but can also differentiate varying degrees of sweetness and different kinds of sugars (Desor et al., 1973). These findings converge with physiological findings that sweet tasting stimuli administered to newborn infants produced greater relative left-sided activation in both frontal and parietal regions of the brain, which is considered to be a reliable indicator of positive emotion (Fox and Davidson, 1986).

The early studies of Desor, Maller, and colleagues (Desor et al., 1975b; Maller and Desor, 1973) found that newborns did not differentially consume bitter and sour solutions relative to water. Thus, it was not clear whether newborns failed to detect the taste solution or were indifferent to it. This conundrum was overcome by adding sugar to the diluent making it slightly sweet (0.07 M), raising its baseline ingestion above that of water, which is typically aversive to newborns. When citric acid was added to this weak sweet diluent, consumption of the solution was reduced, indicating that the citric acid could be detected and was aversive (Desor, Maller et al., 1975). However, they did not differentially consume a urea solution, which has a bitter taste. According to a later study, differential

798   Chapter 36   The Ontogeny of Taste Perception and Preference Throughout Childhood

consumption of urea does not occur until infants are approximately 2 weeks of age, at which point they began to reduce their consumption of urea in a sweetened solution relative to the sweet diluent alone (Kajuira et al., 1992).

Another example of the importance of employing an appropriate diluent for taste testing in infants can be found in the research on umami taste. The early work of Steiner (1987) revealed that babies reject umami taste (e.g., amino acid L-glutamate, 5'-ribonucleotides) in water, presumably because the addition of glutamate decreases the palatability of water solutions. However, because glutamate tends to increase the palatability of food, more recent research with infants used the same method as for older children and adults, by providing the umami taste in a soup broth (Beauchamp and Pearson, 1991; Steiner, 1987; Vazquez et al., 1982) or an infant cereal base (Mennella et al., 2009). When the umami taste was presented in these food bases, infants exhibited differential preference for umami by consuming more when glutamate was added to the diluent.

A further methodological consideration for consummatory studies is that failure to control for basic stimuli (e.g., concentration, temperature, volume, and type of diluent) and methodological parameters (e.g., degree of satiation, experience, and age of the participants) may produce data that are variable and inconsistent. Many of these issues can be resolved by using a within-subject design. Mennella et al., (2009) fed, on different days, a familiar cereal that was adulterated with D-lactose (sweet), sodium chloride (salty), urea (bitter), citric acid (sour), or monosodium glutamate (umami). The presentation order of the cereal preparations was counterbalanced across days. During each test session, which occurred at approximately the same time of day and 30–60 minutes before the next scheduled feeding, mothers fed their infants while wearing a mask and refrained from talking, thereby eliminating potential influences of their facial or verbal responses on the infants' behaviors. Infants determined the pacing and duration of the feeding, and the experimenter, who was out of view of the mother–infant dyad, ended the feeding when the infant rejected the food on at least three consecutive occasions. Results indicated that infants' consumption of the different cereals was determined to some extent by their early feeding experiences: infants who had been fed protein hydrolysate formula (a formula that tastes bitter, savory and sour due to high levels of free amino acids) ate significantly more savory, bitter, sour, and plain cereals than did those who had been breastfed or formula-fed, and in formula-fed infants who were eating table foods, intake of the cereals with each of the basic tastes reflected the types of foods they had been fed. Thus, when grouping participants for data analyses, it is important to consider their early feeding history so that differences in taste preferences will not be masked by experience-related factors.

To what extent do early experiences alter or modulate the development of sweet preferences later in life? Early feeding patterns have been shown to modulate infants' innate preference for sweet tastes. Longitudinal studies have found that babies who were routinely fed sweetened water during the first months of life exhibited a greater preference for sweetened water during brief-access tests when tested at 6 months and then again at 2 years of age (Beauchamp and Moran, 1984; 1982) compared to those who had little or no experience with sweetened water. A more recent cross-sectional study of 6- to 10-year-old children supported these findings and revealed that such feeding practices may have longer term effects on the preference for sweetened water than previously realized (Pepino and Mennella, 2005b).

For salt, the story is more complex. Unlike innate responses to the other basic tastes, differential responses to salt are not exhibited by newborns. Rather, responsivity matures postnatally. While infants younger than 4 months of age consumed moderate concentrations of salt solution (0.10–0.20 M) and water in equal amounts, those between 4 and 24 months of age consumed more of the salt solutions relative to water (Beauchamp et al., 1986; Desor et al., 1975b). Beauchamp and colleagues have argued that postnatal experience with salty tastes probably does not play a major role in this shift from apparent indifference to salt at birth to acceptance at 4 months of age. This instead may reflect postnatal maturation of central or peripheral mechanisms that underlie salt taste perception (Beauchamp and Cowart, 1985; Cowart and Beauchamp, 1990). Later research has demonstrated that infants younger than 4 months of age may not be completely indifferent to salt taste. When presented with moderate (0.17 M) and strong (0.34 M) concentrations of a salted solution relative to water at 2 months of age, they consumed significantly less of the strong salt solution relative to water, and their acceptance of both concentrations of salt was negatively associated with their birth weight (Stein, Cowart and Beauchamp, 2006).

Infants' acceptance of salty taste is also affected by prenatal experiences. Infants born to women who reported suffering from moderate to severe symptoms of morning sickness had significantly higher relative intake of salt solutions at 4 months of age than did those whose mothers reported having no more than mild morning sickness (Crystal and Bernstein, 1998). That these effects last beyond infancy is suggested by the finding that adolescents and young adults preferred higher concentrations of salt in a soup base in the laboratory, and reported significantly higher daily salt use, if either they had suffered from infantile vomiting or diarrhea themselves or their mothers had suffered from morning sickness while pregnant with them (Lesham, 1998).

Like experience with sweet, experiences with salt provide the infant with opportunities to learn about the level of saltiness to be expected in foods. As has been observed with sweet taste, 6-month-old infants' consumption of moderate (0.17 M) and strong (0.34 M) salt solutions (Stein et al., 2012) or salted cereals (Harris, Thomas et al., 1990) can be modified by their dietary patterns once they start eating table foods. Relative to those who did not eat starchy foods, which are a significant source of sodium in infants' diets, those who ate these foods at home consumed significantly more of a solution or cereal when salt was added. Thus, it appears that although infants' hedonic responses to basic stimuli are innate, early experiences, some of which occur before birth, are important modulators of responses to these stimuli.

36.5.2 Orofacial Expressions

Reflexive orofacial changes of both humans and animals to tastes and flavors have also been widely studied (see Table 36.1) (Berridge and Kringelbach, 2008; Forestell and Mennella, 2012; Grill and Norgren, 1978; Mennella et al., 2009; Spector et al., 1981; Spector and Travers, 2005). Similarities between species in their facial responses to taste stimuli (as reviewed in Berridge, 1996) suggest that such behavioral responsivity has evolved in a range of mammals and is clearly adaptive. It has been hypothesized that the small number of taste qualities evolved because of the functional importance of the perception of basic taste qualities in nutrient selection. Preference for salty, sweet, and savory tastes is thought to have evolved to attract us to that which is beneficial: salty-tasting minerals, energy-producing sugars, and vitamin- and protein-rich foods; in contrast, rejection of bitter-tasting substances may have evolved to protect mammals and other organisms from poisons, many of which are bitter (Glendinning, 1994; Jacobs, 1978).

Steiner (1973) was among the first to examine facial reactivity in detail to tastants in human infants. Evidence supporting the concept that these orofacial responses were innate and unlearned came from observations that they were present in both normal and anencephalic infants whose brainstems were intact within hours of birth and before their first postnatal feeding. In these pioneering studies, a drop of solution representing one of the basic taste qualities (i.e., sweet, sour, salty, bitter) was placed on the infant's tongue. Responses of normal and anencephalic infants were strikingly similar: when tasting sweet, their faces relaxed and they began suckling and smiling; in contrast, bitter tastes elicited gaping responses, while sour elicited squinting and lip pursing (Steiner, 1973, 1977, 1979). No facial response was evident to salt before the age of 4 months, which is consistent with findings from the consumption studies discussed above e.g., (Beauchamp et al., 1986). That normal as well as anencephalic infants demonstrated similar orofacial responses suggests that

these responses are mediated in the region of the hindbrain and not in the cerebral cortex, where voluntary movement is controlled (Steiner, 1973, 1977).

Subsequently, Rosenstein and Oster (1988) employed a more precise method of describing orofacial responses with Ekman and Friesen's anatomically based Facial Action Coding System (FACS; Ekman and Friesen, 1978). With this system, virtually any facial expression can be dissected into its constituent action units (AUs). In Rosenstein and Oster's study, the percentages of infants who emitted particular reactions, or AUs, to sweet, sour, bitter, and salt solutions were measured. When initially tasting a sweet substance, infants transiently showed negative mid-face actions, such as cheek raising (AU 6) or nose wrinkling (AU 9). This was followed by more positive and sustained responses such as facial relaxation and sucking. In contrast to the observations of Steiner and colleagues (e.g., Ganchrow et al., 1983; Steiner, 1973), this study did not find marked smile-like facial responses to the presentation of sweet tasting stimuli (AU12). In response to sour and bitter tastants, infants reacted more negatively, as demonstrated mainly by their lower-face actions. For example, while both sour and bitter solutions elicited similar upper and midface AUs (which included cheek raises (AU 6)), sour solutions elicited lip pursing (AU 18), and bitter solutions elicited gaping (AU 26 and AU 27). Although no such FACS analysis has been conducted on the newborn's response to umami taste, Steiner (1987) has reported that the presentation of umami-tasting stimuli results in facial relaxation, sucking, and smiling in infants.

In more recent studies, slow-motion video analysis (Berridge, 1996) has been used to quantify the actual number of affective reactions infants express to a taste stimulus as a measure of the valence and intensity of an affective reaction (Mennella et al., 2009). A necessary requirement of this method is the need for trained individuals (preferably certified in FACS) to analyze the video images and establish reliability between their scores, which can be time-consuming and costly. In one study, the number of AUs each infant expressed in response to sweet, sour, bitter, savory, salty, and plain cereal over the first two minutes of each feeding session was recorded (Mennella et al., 2009). Because of marked individual differences in the display of faces, the analyses focused on the total number of facial expressions of distaste made for each spoonful offered, as well as the incidence of specific facial responses such as gapes and smiles. In general, infants made more facial responses related to distaste while eating the bitter cereal than while eating the other cereals. Moreover, those who had been fed protein hydrolysate formula or breast milk, both of which are higher in glutamate content than are formulas based on cow's milk, smiled more when fed the savory cereal than did those fed milk-based formulas. The effects of early exposure to flavors appear to be particularly

800 Chapter 36 The Ontogeny of Taste Perception and Preference Throughout Childhood

persistent, lasting for several years. That is, children who were fed hydrolysate formulas as infants appear to be programmed to like not only the taste of protein hydrolysates but also the taste of foods that are more savory (e.g., chicken), sour (e.g., lemon), and bitter (e.g., broccoli) compared with children who were fed milk-based formulas (Liem and Mennella, 2002; Mennella and Beauchamp, 2002; Mennella et al., 2006; Schuett et al., 1985). These findings provide further evidence of the plasticity of the taste system such that the early diet can determine later preferences and dietary habits.

In summary, studies on orofacial responsivity to taste demonstrate that, like adults, infants can control the muscles required to express primary emotions and are thus well equipped to convey a wide range of emotional states and reactivity to tastes (Ekman and Oster, 1979). Moreover, the ability to discriminate all but one of the basic tastes (salt) is evident early in life, and infants can respond with characteristic facial responses that correspond to the hedonic valence of the taste stimulus. Thus, infants' facial expressions can provide caregivers with important information about whether they like or reject a food (Forestell and Mennella, 2007, 2012). Visually striking facial displays of gaping are readily identified by caregivers and thus provide preverbal infants with an efficient means to communicate their dislike of the taste of what they are eating.

36.5.3 Analgesic Responses

Sucrose acts as an opioid-receptor-dependent calming agent; thus, its ability to act as an analgesic has been the subject of extensive investigations in preterm and term infants, beginning with the early work of Blass and Hoffmeyer (1991). As demonstrated in a recently updated systematic review (Harrison et al., 2010), studies performed over the past 20 years have shown that sweet tastants can be effective in reducing spontaneous crying and pain in preterm and term infants during an array of painful procedures such as a heal lance, venipuncture, or intramuscular injections, with minimal to no side effects. In these studies, crying, and in some cases heart rate/vagal tone, sucking, and facial grimacing, was evaluated before, during, and immediately after infants received a sweetened pacifier or drops (i.e., 0.05–0.5 ml) of nutritive (e.g., sucrose, glucose, or fructose) (for review, see Stevens et al., 2010) or nonnutritive (aspartame; Barr et al., 1999) sweeteners in solutions and compared to responses of infants given unsweetened pacifiers or drops of water, respectively.

This body of research, as summarized in Table 36.1, has revealed that, in general, when a small amount of a sweet substance is placed on the tongue of a crying infant, a rapid, calming effect ensues that persists for several minutes, whereas no such response is observed when water

or an unsweetened pacifier is administered. This reduction in crying is concordant with changes in measures of other physiological reflexes, such as heart rate and orofacial responses. Sucrose has also been shown to attenuate a negative electroencephalographic (EEG) response to a painful procedure (Fernandez et al., 2003), suggesting that sucrose blocks pain afferents, which in turn diminishes stress and cardiac changes. Because noncaloric sweet substances such as aspartame mimic the calming effects of sucrose (Barr et al., 1999), and the administration of sucrose by direct stomach loading is ineffective (Ramenghi et al., 1999), it is generally accepted that afferent signals from the mouth, rather than gastric or metabolic changes, are responsible for the analgesic properties of sweet tastes. Interestingly, such effects may be also induced by some non-sweet tasting stimuli. One study found that quinine transiently decreased crying (Graillon et al., 1997), implying that in addition to positive hedonic valence, taste salience may account for initial crying reduction. It is presently unknown whether other tastes, such as salty or sour tasting agents, would produce similar analgesic effects.

36.6 TASTE SENSITIVITY AND PREFERENCE IN CHILDREN

The ability to reliably measure changes in the taste system throughout childhood requires a thorough understanding of developmental abilities of children at various ages so that developmentally appropriate tasks may be chosen after careful consideration of the objectives of the study and the methodological issues. There is a need for more experimental investigations of children's taste sensitivity and preference; however, a major challenge in studying children's behavioral responsivity is to identify developmentally appropriate tasks that are fun for children and that produce reliable and valid results (Mennella and Beauchamp 2008). Below we discuss some of the methodological issues to consider when designing experiments to measure taste sensitivity and preference in 2- to 16-year-old children.

When testing children, care should be taken to create a friendly and inviting atmosphere and to employ an age-appropriate experimental protocol to ensure that both the child and the parent are comfortable with the setting. Upon arrival, initial interactions with the child and parent should be devoted to developing rapport and allowing them to habituate to the testing environment. To encourage the child to engage in the testing procedure, test sessions should not begin with an unpleasant taste. It is also important to remember that children are sensitive to demand characteristics and will often alter their behavior to conform to the experimenter's expectations.

For example, children will often answer questions in the affirmative. This can be avoided by using a forced-choice categorization procedure, which requires that they choose which solution they like better (or is more concentrated), rather than indicating whether they like (or can detect) the taste of a solution. In general, experimental tasks should be embedded in the context of games that are fun for the child and minimize the impact of language and cognitive development.

Relative to older children and adults, young children have difficulties with concept formation (e.g., sweetness) and classification (e.g., like/dislike), and because of their limited linguistic skills, they may have difficulty understanding directions or the abstract nature of symbols and pictures. Therefore, prior to actual testing and after a period of acclimation, the experimenter should ascertain whether or not the child comprehends the task. Even when children understand the task, they are more prone to attention lapses and limited memory spans, which may limit their performance. For example, 3-year-old children can understand a standardized sorting task, but only about half have the attention span to successfully complete the task. Any method relying on sustained attention that places demands on memory could yield spurious findings. For this reason, it is important to establish the reproducibility of measures over time. Unfortunately, few peer-reviewed research studies have systematically determined the validity of their methods among children of varying ages (Chambers, 2005). Moreover, only a handful of studies reviewed herein include methods to assess how reproducible and stable the child's response is over time.

In the following sections, we describe the methodologies used to measure 2- to 16-year-old children's reflex-like behaviors to tastants and the behavioral tasks performed to measure their taste sensitivities and preferences. A summary of this research is provided in Table 36.1.

36.6.1 Reflex-like Behaviors

36.6.1.1 Orofacial Responses.
Although studies with infants have used orofacial responses to measure differential responses to the basic tastes (Rosenstein and Oster, 1988; Steiner, 1973, 1977), fewer studies have used this approach to measure hedonic responses in older children. As children mature they learn to control and manage their facial expressions to satisfy display rules that are consistent with societal norms (Ekman and Freisen, 2003; Larochette et al., 2006). Because of such emotional masking, children's attempts to conceal or exaggerate their actual responses to particular tastes may lead to biased or inaccurate data. According to Ekman and Freisen (Ekman and Freisen, 2003), although individuals attempt to manage their facial responses, micro-expressions that reflect their true emotions often "leak" into their overall expression for a fraction of a second. Although these microexpressions are difficult to measure, subtle changes in facial responses can be detected using facial electromyography, which measures the electrical activity of facial muscles and can detect movements that are too discrete for the eye. This procedure has been used to measure hedonic, emotional, and cognitive responses to tastes (Hu, Luo et al., 2000; Armstrong, Hutchinson et al., 2007; Armstrong et al., 2007; Epstein and Paluch, 1997; Hu et al., 1999, 2000). The three most commonly used muscles for detecting changes with EMG are the levator labii, contraction of which results in lifting of the upper lip (AU10) and wrinkling of the nose (AU9), and the corrugator supercilii, which furrows the brows (AU4). Both of these muscles respond strongly to unpleasant stimuli and poorly to pleasant stimuli. The third muscle is the zygomatic major muscle, which pulls up the corners of the lips (AU12) and responds strongly to hedonically positive and negative taste stimuli (Armstrong et al., 2007). Future research that incorporates the monitoring of facial expressions will provide important insights into children's sensitivity and liking of basic tastes.

36.6.1.2 Analgesic Responses.
Only a few published studies have investigated whether the analgesic effect of sweet tastes extends from infancy into later childhood (Miller, Barr et al., 1994; Lewkowski et al., 2003; Pepino and Mennella, 2005b; Mennella, et al., 2010b). In order to induce pain in 8- to 11-year-old children in an ethical manner, these studies have used the cold pressor test in which participants submerge a forearm in a cold water bath (10 degrees Celsius) while holding equal volumes (15–20 ml) of water during one session or 24% sucrose solution in the mouth during another session.

Such studies have demonstrated that the pain threshold (the time in seconds at which discomfort was first indicated by raising the non-immersed hand) and pain tolerance (the number of seconds the hand was kept in the cold water) are both increased when children hold sucrose, but not water, in the oral cavity (Mennella et al., 2010b; Miller et al., 1994; Pepino and Mennella, 2005b). Interestingly, sucrose's efficacy in reducing pain threshold and tolerance was related to the hedonic value of sweet taste for the child: the more a child liked sucrose, the more tolerant the child was of the pain of the cold water while tasting sucrose (Mennella et al., 2010b; Pepino and Mennella, 2005b). Further work has demonstrated that children who have depressive symptomology report eating more sweet-tasting foods than do those who are not depressed; however, they had similar pain thresholds regardless of whether they held sucrose or water in the mouth during the cold pressor task, demonstrating that sucrose is an ineffective analgesic for these children (Mennella et al., 2010b). Thus, it appears that depression antagonizes the analgesic properties of sucrose.

36.6.2 Behavioral Methodologies for Measuring Taste Sensitivity and Preference

36.6.2.1 Paired-Comparison and/or Forced-Choice Procedures.
In older children, it is possible to use methods that do not confound sensitivity and hedonics. Often these methods involve forced choice, a procedure that requires respondents to compare two or more options and choose one. In taste studies with children, forced-choice procedures often involve the presentation of a series of trials that present solutions that vary in concentration of a particular tastant. Each of the solutions in the series may be presented alone, in which case the children are asked to indicate whether they can detect or like the tastant in each sample. Alternatively, each solution may be paired with another solution (e.g., the diluent, or another concentration of the tastant). In these studies children are asked to compare the solutions on each trial and to choose which they like best (preference), or which contains the tastant under study (sensitivity). In some cases, especially when the subject population consists of children younger than 5 years, this forced-choice task is embedded in the context of a game (Mennella et al., 2005; Schmidt and Beauchamp, 1988) or story (Visser et al., 2000). In general, forced-choice, paired-comparison procedures, unlike tests where the subject is required to remember sequentially presented stimuli to make relative judgments, are especially useful for young children because memory requirements are minimized. Below we highlight some examples of procedures that have utilized these methods.

Using a forced-choice procedure that was presented in the context of a game, Mennella et al. (2010a) determined children's sensitivity thresholds to 6-n-propylthiouracil (PROP). Based on the procedures of Anliker et al. (1991), children were presented, in succession, with a sample of water and then three increasing concentrations of PROP (56, 180, and 560 µM) and asked to taste the sample without swallowing. If the solution tasted like "water" or "nothing," then they were asked to give the sample to Big Bird, a popular television character. If the sample tasted "bad," "yucky," or "bitter," children were asked to give it to Oscar the Grouch so he could throw it in his trash can. Children were classified into groups based on the concentration of the first sample, if any, that was given to Oscar the Grouch. This research has shown that children who are heterozygous at the *TASR38* gene locus are more sensitive to taste of PROP than were heterozygous adults, with the thresholds of heterozygous adolescents being intermediate (Mennella et al., 2010b), whereas homozygous children and adults did not differ in threshold. This task is one of the few that has been shown to be reliable (>85%; (Mennella et al., 2005) and valid, with 76% concordance between genotype and phenotype (Keller et al., 2010),

and is effective for children as young as 3 years of age (Mennella et al., 2010a).

In other studies, children were presented with a series of pairs of solutions, with one member of the pair being water and the other containing an aqueous tastant (i.e., paired comparisons). In some cases the aqueous tastants increased in concentration with each pair presented, and the child was required to indicate which sample of the pair contained the tastant, or tasted stronger. The lowest concentration that was detected successfully in one or two consecutive trials was recorded as the detection threshold (James et al., 1997). In other cases, the concentration of the tastant in each pair was determined by the child's ability to detect the tastant in the previous pair of samples. For example, in the staircase procedure, if the child incorrectly indicated which sample contained the tastant, in the next trial the next stronger tastant solution was presented; if the child was correct, a second presentation with the same concentration was given (Anliker et al., 1991).

A similar procedure, called the forced-choice tracking procedure, has been adapted to assess suprathreshold concentrations of tastants. In these studies, participants are instructed to taste, without swallowing, each solution and then to point to which of the pair they like better (or tastes stronger). For example, Mennella et al. (2011b) presented children with pairs of sucrose solutions that differed in concentration (3–36%). They were first presented with a pair of samples chosen from the middle range (6% vs. 24%) and were asked to taste each without swallowing and then to point to which of the pair they liked better. Like the staircase procedure described above, each subsequent pair was then determined by the subject's preceding preference choice. The procedure continued until the subject either had chosen a given concentration of sucrose when it was paired with both a higher and lower concentration or had chosen the highest (36% w/v) or lowest (3% w/v) sucrose solution two consecutive times. The entire task was then repeated with stimulus pairs presented in reverse order (e.g., weaker stimulus presented first in series 1; stronger stimulus first in series 2), thus preventing children from reaching criterion response based on bias toward first or second position. The geometric mean of the two sucrose concentrations chosen in series 1 and 2 estimated the participant's preferred level of sucrose.

By comparing the solution chosen in the first and second (reverse-order) task, one can achieve reliable results within each participant. As reviewed in Table 36.1, this procedure has been shown to be sensitive enough to examine individual differences (e.g., race, age, genetic) in sweet and salt taste preferences in children as young as 5 years of age and how differences in preferences relate to a variety of factors, including analgesia, obesity, dietary habits, and depression (Cowart and Beauchamp 1990; Mennella et al.,

36.6 Taste Sensitivity and Preference in Children 803

2010b, 2011b, 2005; Pepino and Mennella 2005a; Coldwell et al., 2013).

36.6.2.2 Scaling Techniques.
Various types of scaling methods, that is, methods in which sensations to varying concentrations of suprathreshold stimuli are quantified, have been used to determine children's preferences and sensitivity to tastes. Depending on the child's age, these procedures involve presenting the child with a line or other type of scale that contains pictorial or verbal descriptors in a graded order. Although there has been no systematic determination of what scaling test is most appropriate for children at what age, some researchers have concluded that use of scales in children younger than 5 years can be problematic because they have not mastered the ability to rank things in order of magnitude (Guinard, 2001).

For children as young as 3 years of age, researchers typically use hedonic rating scales. Depending on the age of the child, these scales can consist of two to nine faces where one of the anchors is a happy face and the other is a frowning face. While the increase in the number of faces potentially allows for finer-grain analyses of children's hedonic responses, preschool children do not appear to be able to reliably use scales that contain more than five faces (Chen and Resurreccion, 1996). However, little is known about the validity and reliability of these scales and at what age children begin using the entire scale rather than the anchors only.

Due to this gap in our knowledge about the use of hedonic face scales with children, researchers typically use them in combination with other measures. For example, when de Graaf and Zandstra (1999) tested the pleasantness of sucrose (0.03–0.88 M) in water and in orangeade in 9- to 10-year-old children, adolescents, and adults with a 5-point hedonic face scale, they found ratings of pleasantness correlated moderately with intake, and the strength of these correlations increased with age. The authors suggest that the increasing reliability observed with increasing age could be related to the development of cognitive capacity, which allowed older children to more consistently process and code the sensory information.

In addition to hedonic face scales, researchers have used other scales that include verbal descriptors (ranging from "no taste" to "very strong taste" (Anliker et al., 1991), and in some cases these verbal descriptors have been used with the Labeled Magnitude Scale (Coldwell et al., 2009). Here, children are shown a visual analog scale labeled with "barely detectable," "weak," "moderate," "strong," "very strong," and "strongest imaginable" and are asked to compare the intensity of each taste to oral sensations of any kind. Typically participants are trained on this scale by asking them to rate examples, including the sweetness

of cotton candy, the bitterness of celery, or biting into a raw lemon or chili pepper. Although the use of verbal descriptors appears to be effective for measuring sensitivity in 9- to 10-year-old children (Coldwell et al., 2009; Graaf and Zandstra, 1999), younger children may require a more concrete activity to visualize these descriptors. This has been done by asking children to place a ball into buckets that are placed at different heights to exemplify increasing taste intensities (Anliker et al., 1991; Turnbull and Matisoo-Smith, 2002). Despite gaps in knowledge about the reliability and validity of scaling techniques, there is some evidence that reliable results can be obtained from children if care is taken to use developmentally appropriate scales.

36.6.2.3 Rank Order.
In rank-order methods, the goal is either to determine the child's most preferred concentration from a series of solutions, or to place all of the solutions in order of preference or stimulus intensity. In some cases children may be asked to taste all of the solutions in random order, after which they are asked to point to the one they like best (sweet preference inventory). This procedure has been used to determine whether children who chose higher concentrations of sucrose in tea have more dental caries (Jamel et al., 1997; Maciel et al., 2001a,b; Tomita et al., 1999). Their results suggest that socioeconomic status mediates the relationship between taste preference and number of dental caries in 4- to 6-year-old children (Tomita et al., 1999). However, one weakness of this methodology, which may limit the sensitivity of their results, is that children are required to remember and track a series of solutions in order to indicate their preferences, which may be difficult for preschool children.

Other approaches that are less cognitively demanding employ paired-comparison techniques in which all possible pairs of samples are presented and children are asked to choose which they prefer or which tastes stronger in each pair (e.g., Liem and Mennella, 2003). In a similar paradigm called rank by elimination, the sample that is selected from a pair in a trial is removed from the set, and children are asked to taste the remaining samples again and to point to the sample they like best or that tastes strongest. This procedure continues until a rank-order preference or intensity is established. Liem and Mennella (2003) used this procedure to determine sour preferences and sensitivity. A series of four citric acid (0.00–0.25 M) samples prepared in a gelatin base were tasted without swallowing by 5- to 9-year-old children and their mothers. They pointed to the sample that they liked best during trials of one session, and the sample that tasted most sour during trials of another session. This study demonstrated that children who showed preferences for higher sour concentrations were more likely to report that they had tried and liked sour candies compared

804 Chapter 36 The Ontogeny of Taste Perception and Preference Throughout Childhood

to those who preferred lower concentrations of citric acid in gelatin. These findings are consistent with other research that has shown that preference for sour taste is correlated with intake of fruits in infants (Blossfeld et al., 2007) and with previous consumption of protein hydrolysate formulas (Liem and Mennella, 2002) in 4- to 5-year-old children.

Other studies have used a bifurcated approach in which children are first asked to place the sample into either a "good/like" or a "bad/dislike" category. The degree to which they like the stimuli is determined using forced-choice methods, in which the children are presented with pairs of stimuli from each of the categories and asked to indicate which sample in the pair they prefer. This methodology has been used with children as young as 4 years of age to rank preferences to various foods with added sugar or salt (Sullivan and Birch, 1990), including sweetened beverages (De Graaf and Zandstra, 1999; Liem and de Graaf, 2004; Liem et al., 2004). It is important to note, however, that like the procedures previously discussed, little work has been done to validate ranking procedures with children.

36.6.2.4 Stimulus Presentation Methods.

When testing children, it is important to consider the method of presentation of the taste stimuli to the tongue. Studies in which evaluation of taste solutions were directly applied to small portions of the tongue versus the whole mouth have revealed inconsistencies in sucrose sensitivity in children relative to adults. Whereas children demonstrated high sucrose sensitivity relative to adults in local anterior regions (Temple et al., 2002b), this pattern of results was not observed when a whole-mouth procedure was used (James et al., 1997). These findings suggest that taste information may not be integrated as efficiently peripherally and/or centrally in children relative to adults. Moreover, research has shown that different regions of the tongue follow different patterns of growth throughout development: growth of the posterior region continues until approximately 15 years, whereas the fungiform-papilla-rich anterior region of the tongue attains adult size by the time children are 8–10 years of age (Temple et al., 2002a). Cessation of growth of the anterior region of the tongue by mid-childhood, however, does not mean that the anatomy and functionality of this region have reached adult levels of functioning. In a series of studies using videomicroscopy, Laing and colleagues have shown that in 8- to 9-year-old children, fungiform papillae and taste pores are smaller, papillae are rounder and more homogeneous in shape (Segovia et al., 2002), and the densities of fungiform papillae and taste buds are significantly higher compared to adults (Stein et al., 1994). As a result of the uneven pattern of growth of the tongue, care must be taken not to generalize evaluation of taste solutions directly applied to small portions of the tongue to other regions, which may be at a different stage of anatomical growth and function. This could lead to misperceptions regarding children's taste sensitivity.

36.7 IMPLICATIONS

36.7.1 Children's Different Sensory Worlds

Findings from experimental research during the past century, and especially in the past 20 years, have revealed that age is a particularly good predictor of individual differences in taste preferences. The convergence of findings from basic research reviewed herein reveals that children live in different sensory worlds than adults, preferring sweeter and saltier (and in some cases more sour) tastes than do adults and more intensely disliking bitter tastes. There is a striking paucity of research on umami taste in children; whether there are age-related changes, as observed for the other four basic tastes, remains unknown.

Further support for the concept that such hedonic responses are innate is suggested by research that has observed neonates within a couple hours of birth before their first feeding (Rosenstein and Oster 1988; Steiner, 1987). Such responses persist throughout childhood and influence dietary behaviors. For example, children generally dislike bitter-tasting foods such as leafy green vegetables, and as a result fewer than 15% of children between the ages of 4 and 8 years consume the recommended levels of vegetables (Guenther et al., 2006). Such responses are due in part to their genotype, which is related to their acceptance of a variety of foods (e.g., broccoli, brussels sprouts, cabbage) that contain bitter-tasting moieties such as glucosinolates. However, regardless of their genotype, just as children's responses to salt and sweet are influenced by early experiences, so are children's responses to bitter-tasting foods. Through repeated exposure to bitter-tasting foods, and to a variety of other foods from an early age, they can learn to like the taste of these once disliked foods. This has been demonstrated in infants who have been exposed to bitter-tasting protein hydrolysate formulas throughout infancy. Relative to those infants who had been fed milk-based formulas, these infants consumed more of and showed fewer orofacial responses of distaste to a bitter-tasting food (Mennella et al., 2009). This inherent plasticity of the chemical senses, which serves to mediate the effects of genetics through maturation and experience, ensures that we are not restricted to a narrow range of foodstuffs by virtue of a few preferences for or strong aversions to foods.

Although children are also born with an innate distaste for sour tastes, for some children this initial negative

Table 36.1 Summary of research findings on the five basic tastes during stages of development.

Development Stage/Methodology	Outcomes	References
SWEET		
Fetuses		
Consummatory responses		
Volume of amniotic fluid determined before and after injection of saccharin in amniotic sac.	↓ volume of amniotic fluid postinjection compared to pre-injection, suggesting ↑ intake of amniotic fluid after saccharin injections.	DeSnoo 1937; Liley 1972
Preterm infants (24–37 weeks gestational age at time of test; 1–29 days postnatal age)		
Reflex-like responses		
Sucking responses measured for 1 min after drops of sweetened solution (33% glucose) or water were placed on tongue of newborns (30–36 weeks gestational age) over 5 trials.	↑ strength and frequency of sucking to sweetened solutions relative to water. ↑ strength and frequency of sucking to sweetened solutions as trials progressed. No such increase was observed with water.	Tatzer et al., 1985
Sucking responses measured when sweetened (30% sucrose + corn syrup) gelatin nipples and unsweetened nipples were placed in oral cavity of newborns (25–36 weeks gestational age).	↑ strength and frequency of sucking to sweetened relative to unsweetened nipple.	Maone et al., 1990
Spontaneous crying measured relative to baseline crying when 0.1 ml/min of aqueous 10% glucose, or water was placed in oral cavity of newborns (24–36 weeks gestational age) over 5 min.	↓ crying relative to baseline in newborns that received glucose but not those who received water.	Smith and Blass 1996
Variety of pain responses measured during and after painful procedures (e.g., heel lance, subcutaneous injection, or venipuncture) when a sweetened substance (glucose- or sucrose-sweetened water and/or pacifier), or unsweetened substance (water and/or pacifier), or no substance was present in oral cavity of newborns (24–37 weeks gestational age).	↓ pain response (e.g., crying, facial grimacing, limb movements) to sweetened relative to unsweetened or mildly sweetened pacifier or solution, or no solution.	Abad et al., 1996; Acharya et al., 2004; Bucher et al., 1995; Carbajal et al., 2002; Deshmukh and Udani 2002; Gibbins et al., 2002; Okan et al., 2007; Stevens et al., 2005

(*continued*)

805

Table 36.1 (*Continued*)

Development Stage/Methodology	Outcomes	References
Infants (newborn to 2 years)		
Reflex-like responses		
Sucking responses measured when a sweetened gelatin nipple, drops of aqueous sucrose (2–20%), an unsweetened nipple, or drops of water were placed in oral cavity of newborn infants (24–72 h old).	↑ sucking burst length to 15% aqueous sucrose relative to 5% sucrose or water. ↓ sucking latency to all sucrose concentrations relative to water. ↑ strength and frequency of sucking to sweetened relative to unsweetened nipples.	Crook 1978; Crook and Lipsitt 1976
Sucking responses measured when two quantities of response-contingent aqueous sucrose (5%) were delivered in consecutive 4-min periods to newborns (48–72 h old). One group of infants received the larger amount (0.03 ml/suck) whereas another received smaller amounts (0.01 ml/suck) during both periods.	↑ sucking bursts and inter-burst pausing for infants feeding larger versus smaller amounts.	Crook 1976
EEG was recorded from the frontal and parietal scalp regions in newborns (2–3 days old) who were presented with water followed by a sucrose solution.	↑ left-side activation relative to right-side activation in frontal and parietal regions to aqueous sucrose compared with water.	Fox and Davidson 1986
A variety of pain responses, movement, EEG, and heart rate measured during heel lance, circumcision, or subcutaneous injection when a sweetened substance (nutritive or nonnutritive), unsweetened substance, or no substance was present in oral cavity in newborns ranging from < 1 day to 30 days old.	↓ crying, facial grimacing, heart rate, recovery time, vagal tone, and limb movements to sweetened substance than to unsweetened substance, or no substance. Effects of nonnutritive sweeteners were similar to those of nutritive sweeteners. ↑ left frontal and parietal EEG activation to sweetened substance (indicating positive emotion) relative to water.	Abad et al., 1996; Bellieni et al., 2002; Blass and Hoffmeyer 1991; Blass and Watt 1999; Bucher et al., 2000; Carbajal et al., 1999; Fernandez et al., 2003; Fox and Davidson 1986; Gradin et al., 2002; Greenberg 2002; Ramenghi et al., 1996
Orofacial responses measured when a sweet solution or water was placed on the tongue in normal and anencephalic newborn infants who were 2–96 h old.	↑ sucking-like and mouthing responses, licking, facial relaxation, smiling, and hand-mouth contact to sweet solutions but not to water in normal and anencephalic infants. (Rosenstein and Oster (1988) report that fewer infants demonstrated raised cheeks (AU 6), narrowed or tightly closed eyes (AU 43 or AU 7 + 43), nose wrinkles (AU 9), or brows raised (AU 1 + 2) and/or lowered (AU 4) to sweet solutions than to nonsweet solutions.) ↑ amplitude of mouthing responses as concentration of sweet solution increased.	Barr et al., 1999; Ganchrow et al., 1983; Graillon et al., 1997; Nowlis and Kessen 1976; Peterson and Rainey 1910; Rosenstein and Oster 1988; Steiner 1973, 1977, 1979; Steiner et al., 2001

(*continued*)

Table 36.1 (Continued)

Development Stage/Methodology	Outcomes	References
Spontaneous crying measured relative to baseline crying when a sweet nutritive (i.e., 6%–24% sucrose, 10% glucose, 9% fructose, 17.5% lactose) or nonnutritive (e.g., 0.12% aspartame) solution or water was placed in oral cavity of infants who ranges from 1day to 12 weeks old.	↓ crying relative to baseline for some sweet nutritive (sucrose, glucose, and fructose) or nonnutritive solutions but not for water. No change in crying relative to baseline for lactose relative to water	Barr et al., 1999; Blass and Camp 2003; Blass and Smith 1992; Smith and Blass 1996; Zeifman et al., 1996
Heart rate measured when drops of aqueous sucrose (5% or 15%) or water were placed on the tongue of newborns (48–72 h old) in calm state.	↑ heart rate to 15% sucrose solution relative to 5% or water.	Crook 1976; Crook and Lipsitt 1976
Consummatory responses		
Intake measured in newborns (1–3 days old) during brief-access (3 min) tests; bottles contained aqueous glucose, fructose, lactose, or sucrose (each ranging from 0.05 to 0.30 M) or water.	↑ intake of all sugar solutions relative to water. ↑ intake of sucrose and fructose than of lactose and glucose solutions. ↑ intake of 0.20 and 0.30 M solutions than of 0.05 and 0.10 M solutions	Desor, Maller et al., 1973
A longitudinal study measured intake in infants soon after birth (1–9 days old) and again at 6 months of age during brief-access (180 s) tests; bottles contained 0.2 or 0.6 M sucrose solution or water. Beauchamp and Moran (1984) tested a subset of these children again at 2 years of age using cups in brief-access (30 sec) tests. Cups contained 0.2 or 0.6 M sucrose or diluent (water or unsweetened Kool-Aid). Seven-day dietary records were also obtained from mothers at 6 months and 2 years.	↑ intake of sucrose relative to water in 6-month-olds and 2-year-olds who were regularly fed sugar water compared to those who were not regularly fed sugar water. Compared to intake at birth, ↓ intake of sucrose relative to water in 6-month-old infants who were not regularly fed sugar water. At 2 years of age, intake of sweetened Kool-Aid was higher than for unsweetened Kool-Aid regardless of whether they had been regularly fed sugar water.	Beauchamp and Moran 1982; Beauchamp and Moran 1984
Children (2–19 years) **Reflex-like responses**		
Pain response measured when children (5–12 years) immersed a forearm in a cold water bath while holding 24% sucrose or water in the oral cavity. Measures included threshold (latency to first report of pain) and tolerance (latency to remove arm).	↑ pain threshold to sucrose relative to water.	Mennella et al., 2010b; Miller et al., 1994; Pepino and Mennella 2005b
Facial EMG recorded in children (6–9 years) after placing drops of one of 8 concentrations of sucrose (0.05–1.34 M), 8 concentrations of quinine hydrochloride (0.08–2.31 M), NaCl (0.18 M), or citric acid (0.01 M) solutions or water on the posterior portion of the tongue.	↑ increased EMG activity in the zygomaticus and levator labii for sucrose relative to water. No difference in EMG responses in the zygomaticus to the four tastants, but for the levator labii, the difference in EMG activity was significantly lower for sucrose than for the other three tastants.	Armstrong et al., 2007

(continued)

807

Table 36.1 (Continued)

Development Stage/Methodology	Outcomes	References
"Add to taste" preference procedure Adolescents (11–19 years) were presented two cups of tea, one sweetened with 30% sucrose and one unsweetened. After tasting the solutions in each cup, they were provided with a third cup and asked to pour one-half the unsweetened tea into this cup and then to add sweetened and unsweetened tea until the mixture was "tasty." At the end of the session the cups were weighed to determine the sugar concentration of the final mixture. Mothers were queried about health history during focal pregnancy and the child's history of infantile vomiting and diarrhea.	↓ concentrations of sweetened tea preferred by adolescents who experienced infantile vomiting and diarrhea, or whose mothers reported experiencing morning sickness during their pregnancy with them.	Lesham 1998
Five-point hedonic face rating scale Children (11–15 years) rated liking for six concentrations of aqueous sucrose (0.056–1.000 M) using a *five-point hedonic face rating scale*, ranging from smiling (extreme like), to neutral (just ok), to frowning (extreme dislike). Ratings for the two lowest concentration solutions were subtracted from those of the two highest concentration solutions. Positive difference scores were designated "high preference"; negative difference scores were designated "low preference."	↑ levels of biomarkers for bone growth in the high-sugar preference group compared to low-sugar preference group.	Coldwell et al., 2009
Sweet preference inventory Children (4–15 years) tasted five sweetened aqueous solutions in random order and pointed to the solution they liked best.	Children generally preferred the higher concentrations of sucrose. Children who preferred ↑ sucrose were from lower socioeconomic backgrounds, which in turn was associated with dental caries.	Jamel et al., 1997; Maciel et al., 2001a; Maciel et al., 2001b; Tomita et al., 1999
Ranking procedure A cross-sectional comparison of children (9–15 years, a subset of whom were twins) and adults who sipped without swallowing concentrations of sucrose (0.075–0.60 M) or lactose (0.30–0.40 M) and then ranked them from most to least preferred.	Most-preferred level of sweetness ↑ in male than in female children. Most-preferred level of sweetness ↑ in black than in white children. Most preferred level of sweetness ↑ in children than in adults. No significant differences between the monozygotic and dizygotic twin groups in level of sweetness most preferred.	Desor and Beauchamp 1987; Desor et al., 1975a; Greene et al., 1975

(*continued*)

Table 36.1 (*Continued*)

Development Stage/Methodology	Outcomes	References
A longitudinal study tested participants during adolescence (11–15 years) and adulthood, who sipped without swallowing four concentrations of aqueous sucrose (0.075–0.60 M) and then ranked them from most to least preferred.	↓ sucrose concentration rated highest as age increased; 50% of the children and adolescents, but only 25% of the adults, ranked the strongest sucrose concentration (0.60 M) highest.	Desor and Beauchamp 1987
Two-step, forced-choice ranking procedure Children (4–10 years), adolescents (14–16 years), and adults were presented five to seven concentrations of sucrose (0.03–0.88M) in water or orangeade (or six sweetened, salted, and plain foods in Sullivan and Birch (1990)). Stimuli were presented in random order, and participants were asked to categorize each using a hedonic face scale ranging from a smiling face for tastes they liked to a frowning face for tastes they did not like. After all stimuli were categorized, participants were asked to rank the beverages or foods from most preferred to least preferred within each category.	Most-preferred level of sweetness ↑ in children than in adolescents and ↑ in adolescents than in adults. Most-preferred level of sweetness ↑ in children who were repeatedly exposed to sweet foods at home. Children preferred sweetened ricotta cheese to salted or plain ricotta cheese. Preference for a sweetened novel food increased after it was repeatedly exposed at home.	De Graaf and Zandstra 1999; Liem et al., 2004; Sullivan and Birch 1990
Children (8–11 years) and adults were presented with sweetened (0.42 M added sucrose) orangeade with six concentrations of citric acid (0–0.065 M citric acid) or a sweetened (0.42 M added sucrose) yogurt with six concentrations of citric acid (0–0.17 M) in random order. Participants categorized each into categories using hedonic face scales ranging from smiling (most preferred), to neutral (just ok), to frowning (least preferred). Next, participants ranked their liking of the foods within each category. They then received repeated exposure to orangeade that contained 0.42 M sucrose and 0.02 M added citric acid over 8 days. The posttest was conducted in the same manner as the pretest.	↑ ranking for the sweetened orangeade with no added citric acid in both children and adults after the 8-day exposure period. preferences for the sweetened yoghurt with no added citric acid did not reach significance.	Liem and de Graaf, 2004
Paired-comparison, forced-choice ranking procedure Children (4–7 years) and adults indicated the level of sweetness they "liked best" in pairs of five or six concentrations (0–0.93M) sucrose-sweetened beverages (apple juice or orangeade). Parents completed questionnaires that asked about their children's consumption of sweet foods. In Liem et al. (2004) participants' responses in this test were compared to responses in the two-step, forced choice ranking procedure as described above.	Most-preferred level of sweetness ↑ in children than in adults. Most-preferred level of sweetness ↑ in those who were exposed to sweet foods at home. In Liem et al., (2004) 5-year-olds' and adults' responses were consistent between procedures (76% and 87%), whereas 4-year-olds showed 61.2% consistency.	Liem et al., 2004; Liem and Mennella 2002

(*continued*)

Table 36.1 (*Continued*)

Development Stage/Methodology	Outcomes	References
Paired-comparison, forced-choice tracking procedure Children (5–12 years) and adults tasted pairs of aqueous sucrose solutions (3–36%) or pudding samples that varied in sweetness (13.4–36.2%) and chose the solution or pudding they liked better. Each subsequent pair contained the participant's preceding preferred concentration paired with an adjacent stimulus concentration, until the participant chose either the same sucrose concentration paired with a higher and lower concentration, or the highest or lowest concentration on two consecutive trials.	Most-preferred level of sweetness in water and in pudding $\uparrow$ in children and adolescents compared to adults. Most-preferred level of sweetness differed among children by genotypes at the TAS2R38 locus. Most-preferred level of sweetness $\uparrow$ in black children than in white children. Most preferred level of sweetness was associated with genetic variation within the TAS1R3 gene in adults but not children.	Mennella et al., 2005; 2010b; 2011b; 2012; Pepino and Mennella 2005b
Forced-choice sensitivity procedure Tested ability of children (8–9 years) and to detect sweet taste in binary mixtures of sucrose and salt solutions was measured. Samples were randomly ordered and consisted of sucrose concentrations (0–0.224 M) that were presented in water on eight trials and with 0.125M salt on eight trials.	Children identified both components (sweet and salty) in binary mixtures, except those containing the two lowest concentrations of sucrose. Fewer correct responses occurred with the mixtures than with sucrose alone. Similar results were found for adults.	Watson et al., 2001
Two-step, forced-choice discrimination procedure Children (4–5 years), were presented five concentrations of sucrose (0.22–0.29M) in orangeade in random order, and participants were asked to categorize each using a scale that visualized "most sugar" and "least sugar" by means of different numbers of sugar cubes. After all stimuli were categorized, participants were asked to order the stimuli within each category according to sweetness intensity.	While 5-year-old children were able to identify differences in sweetness across the five different sugar concentrations, 4-year-olds were not able to identify differences in sweetness across the five different sugar concentrations.	Liem et al., 2004
Paired-comparison, forced-choice sensitivity procedure Children (8–9 years) and adultsassessed a series of 12 concentrations of aqueous sucrose solutions (1.17 x 10^{-4} to 7.5 x 10^{-2} M) paired with water, presented in ascending order, with left-right sample position randomized. Participants were instructed to indicate which of the solutions tasted stronger.	$\uparrow$ sucrose detection threshold for boys than for girls. $\uparrow$ sucrose detection threshold for boys than for men and women.	James et al., 1997

(*continued*)

Table 36.1 (*Continued*)

Development Stage/Methodology	Outcomes	References
Labeled Magnitude Scale (LMS)		
Children (11–15 years) rated the intensity of six concentrations of aqueous sucrose (0.056–1.000 M) along a visual analog scale labeled with "barely detectable", "weak", "moderate", "strong", "very strong", and "strongest imaginable". Subjects were asked to compare the intensity of each taste to oral sensations of any kind and were given examples including the sweetness of cotton candy, biting into a raw lemon, the bitterness of celery, and biting into a chili pepper.	Children who had previously indicated that they liked lower concentrations of sucrose more than the higher concentrations (classified as low preference) using the five-point hedonic face rating scale (described above) perceived the intensity of sucrose to be higher than children who were classified as high preference. This difference was significant for three of the six sucrose solutions tested (0.056 M, 0.1 M, and 0.32 M)	Coldwell et al., 2009
Time-intensity scale		
Children (8–9 years) and adults rated the intensity and duration of sweetness over 90 s while holding sweetened water, orange drink, or sweetened custard in the oral cavity.	Children rated sweetness intensity higher and recorded shorter sweetness durations compared to adults.	Temple et al., 2002b
BITTER		
Fetuses		
Consummatory responses		
Volume of amniotic fluid determined before and after injection of Lipiodol (iodized poppy seed oil, a radiocontrast agent).	↑ volume of amniotic fluid post- versus preinjection, suggesting ↓ intake of amniotic fluid after Lipiodol injections.	DeSnoo 1937; Liley 1972
Preterm infants (**24-37 weeks of gestational age**)		
Sucking responses measured when drops of a quinine or water were placed on the tongue of newborns.	↓ sucking frequency in all but one infant.	Eckstein 1927
Infants (**newborn to 9 months**)		
Reflex-like responses		
Orofacial responses measured when drops of quinine or urea solutions, gentian tincture, or water were placed on the tongue of normal and anencephalic newborn infants who were 2–96 h old.	↑ Facial grimacing and gaping (AU 26 or AU 27) to bitter solutions relative to water in normal and anencephalic infants. Transient ↑ in mouthing to quinine in crying infants.	Ganchrow et al., 1983; Graillon et al., 1997; Peterson and Rainey 1910; Rosenstein and Oster 1988; Steiner 1973, 1977, 1979; Steiner et al., 2001
Consummatory responses		
Intake measured in newborn to 24-week-old infants during brief-access (30–180 s) tests; bottles contained aqueous solutions of urea (0.12–0.48 M) or diluent (e.g., water, or weak (0.07 M) sucrose solution).	No differential consumption of aqueous urea relative to water, or urea in sucrose relative to sucrose alone in newborns. ↓ intake of urea relative to diluent alone in 2- to 24-week-olds.	Desor et al., 1975b; Kajiura et al., 1992; Maller and Desor 1973; Schwartz et al., 2009

(*continued*)

Table 36.1 (*Continued*)

Development Stage/Methodology	Outcomes	References
Intake and facial responses measured while infants (4–9 months old) were fed urea-flavored (0.24 M) and plain cereal by their mothers in a laboratory setting.	↑ intake of bitter-tasting cereals and ↓ facial expressions of distaste in infants routinely fed hydrolyzed casein formula than among those fed breast milk or milk-based formula.	Mennella et al., 2009
Children (3–19 years) **Five-point hedonic face rating scale** Children (4–5 years) were classified as either PROP tasters or nontasters according to whether they could taste a 0.56 mM PROP solution. They then rated liking for various food items ranging in perceived bitterness using a five-point hedonic face scale: smiling (extreme like), to neutral, to frowning (extreme dislike).	Children classified as PROP tasters gave ↓ ratings to raw broccoli and American cheese than did PROP nontasters.	Keller et al., 2002
Paired-comparison, forced-choice ranking procedure After a session in which children (7–10 years) were trained to understand differences in basic tastes, they were presented three bitter agents (urea (0.5 M), caffeine (0.08 M), tetralone (1.37 x 10⁻⁴ M)), one pair at a time, in all possible pairs of four solutions (bitter alone, sodium gluconate (0.03 M), bitter + sodium gluconate, water) and asked to indicate which of the pair tasted more bitter during one test session and which tasted better during another.	Addition of a sodium gluconate significantly ↓ the perceived bitterness and ↑ the acceptance of urea and caffeine. Addition of sodium gluconate significantly ↑ the perceived bitterness and ↓ acceptance of tetralone.	Mennella et al., 2003
Forced-choice sensitivity procedure Children (3–10 years), adolescents (11–19 years), and adults (mostly mothers) were asked to taste solutions of PROP (56, 180, and 560 μM). If the solution tasted like "water" or "nothing," children were asked to give it to Big Bird. If the drink tasted "bad," "yucky," or "bitter," children were asked to give the drink to Oscar the Grouch so he could throw it in his trash can. Participants were classified into four groups based on which was the first sample, if any, given to Oscar the Grouch. The A49P alleles of the *TAS2R38* gene were also genotyped. (In Keller et al. (2010), children were classified as tasters according to whether they gave 560 μM PROP to Big Bird or Oscar the Grouch.)	Children who were heterozygous *TAS2R38*/PROP tasters perceived ↓ PROP concentrations than did heterozygous adults, and heterozygous adolescents were intermediate. ↑ BMI in boys who were homozygous *TAS2R38* AA genotype/PROP nontaster relative to homozygous females. In 76% of children, *TAS2R38* genotype predicted PROP phenotype. Mennella et al. (2010a) conducted PROP testing a second time on a random sample of children all of whom were sensitive to PROP; reliability was 85.7%.)	Keller et al., 2010; Mennella et al., 2010a, 2005

(*continued*)

Table 36.1 (*Continued*)

Development Stage/Methodology	Outcomes	References
Staircase threshold procedure Sensitivity of children (3–6 and 10–11 years old) to PROP was measured with 15 samples (0.0051–3.2 mM), offered to the child with a sample of distilled water. With the *staircase procedure*, if the child incorrectly indicated which sample contained PROP, the next stronger PROP solution was presented; if the child was correct, a second presentation with the same PROP concentration was given. If two correct choices were made at this concentration, testing continued at the next weaker PROP concentration. Four to five reversals were required to complete the test. PROP threshold was defined as the mean of the last four reversals. Children's liking of seven or eight foods was also assessed. (In Anliker et al. (1991), liking for foods in those classified as PROP tasters (i.e., those with thresholds < 0.05 mmol/L) was compared with nontasters, whereas in Turnbull and Matisoo-Smith (2002), ratings for foods were correlated with PROP thresholds.)	Children classified as PROP nontasters had ↓ thresholds for PROP than did adult nontasters. Children classified as tasters provided ↓ ratings for cheese. Children sensitive to the bitter taste of PROP were more likely to dislike the taste of raw spinach.	Anliker, Bartoshuk et al., 1991; Turnbull and Matisoo-Smith 2002
Sensitivity of children (3–6 years) to urea (3.75–3000 mM) was measured using the staircase procedure described above. The children were told a story about a goblin that makes drinks for baby goblins. A little goblin, who wanted to help make the drinks, put water in some of the bottles instead of "goblin drinks." Children were asked to help find the goblin drink in each of the pairs. A moderate concentration of urea was presented first. To shorten the procedure for younger children, concentrations were lowered by two steps until the first incorrect response, and then always increased or decreased one concentration step. The test stopped when two incorrect responses were given on the same concentration; detection threshold was defined as the first level above this concentration.	There were no age or gender differences in sensitivity to urea. The mean detection threshold for urea was 59.4 mmol/L.	Visser, Kroeze et al., 2000

(*continued*)

Table 36.1 (*Continued*)

Development Stage/Methodology	Outcomes	References
Suprathreshold rating tests Children (5–7 years) who had already been identified as PROP tasters using the staircase procedure described above were presented 15 PROP samples (56–1800 μM) one at a time in random order and asked to rate the intensity of the solutions on a scale from 1 (no taste) to 5 (very strong taste).	Children identified as PROP tasters provided ↑ intensity ratings to the PROP solutions (i.e., perceived them to be stronger) than did PROP nontasters.	Anliker et al., 1991
Children (3–6 years) tasted four PROP solutions (0.056 to 1.8 mM) and two samples of distilled water. After tasting each solution they were instructed to place a ball into one of three buckets that were placed at different heights to exemplify increasing taste intensities: mild, quite strong, and very strong. Intensity ratings for the strongest PROP solution (1.8 mM) and ratings for a variety of foods and beverages were analyzed.	Children more sensitive to PROP were more likely to dislike the taste of raw spinach.	Turnbull and Matisoo-Smith 2002
Labeled Magnitude Scaling Paper impregnated with PROP was placed on children's (10–11 years) tongues for 10 s, after which they rated its bitterness on a 10-cm scale: 0 = no sensation to 10 = most intense sensation they had ever experienced (e.g., the loudest sound they had ever heard). Those who scored > 4.2 were classified as PROP tasters.	PROP tasters were more likely to be described by their mothers as fussy eaters than were PROP nontasters.	Golding et al., 2009
SALT		
Preterm infants (**24–37 weeks of gestational age**) Orofacial responses measured when drops of salty solutions or water were placed on the tongue.	Indifference in two-thirds of the infants and rejection in one-third.	Eckstein 1927
Infants (**newborns to 24 months**) **Reflex-like responses** Orofacial responses measured when drops of salt solutions or water were placed the on tongue of newborn infants within hours of birth.	Indifference to salty solutions relative to water. Peterson & Rainey (1910) report that results were mixed; while some neonates grimaced, others exhibited contented sucking.	Peterson and Rainey, 1910; Rosenstein and Oster 1988; Steiner 1979

(*continued*)

Table 36.1 (*Continued*)

Development Stage/Methodology	Outcomes	References
Sucking responses measured when drops of NaCl (0.1–0.6 M) or water were placed in oral cavity of newborn infants (46–78 h old).	↓ burst length as the concentration of salt increased.	Crook, 1978
A cross-sectional comparison of newborns' and 4- to 8-month-olds' sucking responses during brief-access (30 s) multiple-bottle tests with aqueous salt solution (0.2 or 0.4 M) or water.	↓ number of sucks to 0.2 M salt solutions than to water in newborns. ↓ number of sucks to 0.4 M salt solution than to water in newborns and 4- to 8-month-olds.	Beauchamp, Cowart et al., 1994
A longitudinal study measured sucking responses monthly in 2- to 8-month-olds during brief-access (60 s) multiple-bottle tests of salted (0.15 M) or unsalted formula.	Indifference to salted versus unsalted formula in 2- to 3-month-olds. ↓ number of sucks to salted relative to unsalted formula in 6- to 7-month-olds (presumably from sweet suppression by salt). ↑ number of sucking bursts that were shorter in duration to salted relative to unsalted formulas in 4- to 7-month-olds.	Beauchamp, Cowart et al., 1994
Consummatory responses		
Cross-sectional comparisons of intake of varying aqueous solutions of salt (0.10–0.40 M) or diluent (water or weak (0.07 M) sucrose solutions) in infants (newborn to 8 months) during multiple-bottle brief-access (30–180 s) tests. (Mothers in Crystal and Bernstein (1998) study were also asked to report experiences of nausea or vomiting during their focal pregnancy.)	No differential consumption of 0.1 M and 0.2 M salt solutions relative to diluent (water or weak sucrose solution) in infants < 4 months. Infants between 4 and 7 months consumed ↑ 0.1 M and 0.2 M salt solutions relative to water. Infants < 8 months of age consume less 0.4 M salt than water Infants of women who reported moderate or severe vomiting during early pregnancy showed ↑ preferences for moderate concentrations of salt solutions over water compared to infants of women who reported no symptoms.	Beauchamp et al., 1986, 1994; Crystal and Bernstein 1998; Desor et al., 1975b
Longitudinal studies measured intake of salted solutions (0.17 M and 0.34 M) or water in 2- and 6-month-olds during multiple-bottle brief-access (60–120 s) tests. Mothers reported their infants' birth weight and whether they had been exposed to starchy table foods.	Negative correlation between birth weight and acceptance of 0.017 M and 0.34 M salt solutions relative to water at 2 months but not at 6 months. ↑ intake of 0.17 M and 0.34 M salt solutions relative to water in 6-month-olds whose mothers reported that they fed their infants starchy table food.	Stein, Cowart et al., 2006, 2012
Intake measured when infants (16–25 weeks) were fed salted or plain cereal by their mothers on four separate days.	↑ intake of salted relative to plain cereal at 16 weeks. ↑ consumption of salted than unsalted cereal in younger infants than in older children.	Harris et al., 1990

(*continued*)

815

Table 36.1 (*Continued*)

Development Stage/Methodology	Outcomes	References
Intake and facial responses measured while infants (4- to 9-months) were fed salted (0.1 M) or unsalted cereal by their mothers in a laboratory setting. Mothers reported whether their children had been exposed to table foods, and if so, which foods.	↑ intake of salted relative to unsalted cereals in infants who were eating cheese at home versus those who were not.	Mennella et al., 2009
***Children* (7 months to 19 years)** **Consummatory responses**		
A cross-sectional study measured intake in children (7–72 months) during brief (30 s) access to multiple cups containing 0.17 or 0.34 M salt solutions or diluent (i.e., water or soup).	↑ intake of 0.17 M salt solution relative to water, but intake of 0.34 M did not differ from that of water in 7- to 24-month-olds. ↓ intake of 0.17 M and 0.34 M salt solutions relative to water in 31- to 72-month-olds. ↑ intake of 0.34 M salted soup relative to 0.17 M salted and the unsalted soup in 36- to 72-month-olds.	Beauchamp et al., 1986; Cowart and Beauchamp 1986
"Add to taste" preference procedure		
Adolescents (11–19 years) were presented two cups of tomato soup, one salted (3.3% w/w) and one unsalted. After tasting the soups in each cup, they were provided with a third cup and asked to pour one-half the unsalted soup into this cup and then to add salted and unsalted soup until the mixture was "tasty." At the end of the session the cups were weighed to determine the salt concentration of the final mixture. Mothers were queried about health history during the focal pregnancy and the child's history of infantile vomiting and diarrhea.	↑ concentrations of salted soup preferred by adolescents who had experienced infantile vomiting and diarrhea or whose mothers reported experiencing morning sickness during their pregnancy with them. Adolescents preferred ↑ concentration of salt in soup if they reported adding more salt to food.	Lesham 1998
Ranking procedure		
Children (9–15 years, a subset of whom were twins) and adults tasted four or five concentrations of salt in water (0.05–0.4 M) or in soup (0.025–0.10 M) and either indicated which was their favorite or ranked them from most to least preferred.	Most-preferred salt concentration ↑ for children than for adults. Most-preferred salt concentration ↑ in black than in white children. No significant differences in preference for salt solutions between monozygotic and dizygotic twin groups. Most-preferred level of salt in soup in Korean adolescents was ↑ if they frequently ate at fast food restaurants.	Desor and Beauchamp 1987; Desor et al., 1975a; Greene et al., 1975; Kim and Lee 2009

(continued)

Table 36.1 (*Continued*)

Development Stage/Methodology	Outcomes	References
Two-step forced-choice ranking procedure Children (4–5 years) were presented salted, sweetened, and plain tofu and ricotta cheese in random order and asked to categorize each into hedonic categories (e.g., like, dislike) using face scales and then to rank the foods from most preferred to least preferred within each category. Children were then exposed to salted, sweetened, or plain tofu 15 times over 9 weeks. The ranking procedure was conducted two additional times throughout the exposure phase and at the end of 9 weeks.	Children initially ranked salted ricotta cheese ↑ than plain ricotta. Children exposed to salted tofu showed ↑ preference for salted tofu and ↓ preference for plain tofu. Children exposed to plain tofu showed ↑ preference for plain tofu and ↓ preference for salted tofu.	Sullivan and Birch 1990
Paired-comparison, forced-choice ranking procedure Children (3–7.5 years) ranked the level of saltiness they preferred in all possible pairs of samples chosen from three or four concentrations (0–0.56 M) of salt solutions or soup.	Most children preferred plain water more than salt water solutions. Most children preferred salted to unsalted soup. Most-preferred concentration of salt in soup was ↑ than typically observed with adults.	Beauchamp and Cowart 1990; Cowart and Beauchamp 1986
Paired-comparison, forced-choice tracking procedure Children (3–11 years) and adults tasted pairs of salted soups (0.01 M–1.00 M) and chose the sample they liked better. Each subsequent pair contained the participant's preceding preferred concentration paired with an adjacent stimulus concentration, until the participant chose either the same concentration of salt paired with a higher and lower concentration, or the highest or lowest concentration on two consecutive trials. The task was then repeated with the members of the stimulus pairs presented in reverse order. The geometric mean of the salt concentrations chosen in the two trial series provided the estimate of the child's most preferred level of salt.	Children preferred ↑ concentrations of aqueous salt than did adults. No racial differences between children were evident, nor did degree of salt exposure affect preferences.	Beauchamp and Cowart 1990
Forced-choice sensitivity procedure Children (8–9 years) and adults were presented 16 test trials in random order presenting sucrose solutions (0.056–0.224 M), and binary mixtures of NaCl (0.032–0.125 M) mixed with each of the concentrations of sucrose, or water, to determine if they could detect salt taste in aqueous binary mixtures of sucrose and salt.	Children identified both components (sweet and salty) in all binary mixtures except the two lowest concentrations of sucrose. Fewer correct responses occurred with the mixtures than with the single tastant. Similar results were found for adults.	Watson et al., 2001

(*continued*)

Table 36.1 (*Continued*)

Development Stage/Methodology	Outcomes	References
Paired-comparison, forced-choice sensitivity procedure		
Adolescents (12–13 years) were presented a three-alternative, paired-comparison, forced-choice sensitivity procedure in which each of seven aqueous salt solutions (0.86–25.6 mM) was presented in ascending order with two additional cups containing water. The participants were informed that only one cup of each triad contained a very dilute salt solution, and that the others contained just water. The lowest salt concentration that was detected successfully in one or two consecutive trials was recorded as the detection threshold.	↑ salt detection threshold in adolescents who reported liking soup.	Kim and Lee 2009
After identifying a threshold for salt using the paired-comparison, forced-choice sensitivity procedure described above, children (5–16 years) with cystic fibrosis and matched controls were presented four cups that contained the threshold concentration and four cups containing water. If the subject succeeded in accurately distinguishing the threshold solutions from the water, the procedure was repeated with the next lower salt concentration. This was continued until the participant was no longer able to distinguish the salt concentration from water, at which point the next higher concentration was deemed to be the salt concentration threshold.	Threshold concentrations of salt did not differ between children with cystic fibrosis of the pancreas and age-matched controls.	Wotman et al., 1964
Suprathreshold rating test		
Children (7–12 years) and adults rated the saltiness and preference for five bags of popcorn that differed in saltiness (0–3 M) using a seven-point Likert scale: 1 = least salty/preference, 7 = most salty/preference.	Children liked popcorn with ↑ concentrations of salt than did adults.	Verma et al., 2007

(continued)

Table 36.1 (*Continued*)

Development Stage/Methodology	Outcomes	References
SOUR		
***Preterm infants* (24–37 weeks of gestational age)**		
Sucking responses measured when drops of pure lemon juice or water were placed on the tongue of newborn infants.	↑ reflex salivation accompanied by apparent increases in sucking vigor and sometimes retching.	Eckstein 1927
***Infants* (newborns to 24 months)**		
Reflex-like responses		
Orofacial responses measured when drops of citric acid or water were placed on the tongue of normal and anencephalic newborn infants within hours after birth.	↑ facial squinting (AU 6 + 7) and pursing (AU 18) responses to sour solutions relative to water.	Ganchrow et al., 1983; Peterson and Rainey 1910; Rosenstein and Oster 1988; Steiner 1973; Steiner 1977, 1979; Steiner et al., 2001
Consummatory responses		
Intake in infants (newborn and 15–20 months) was measured during brief-access tests with bottles or cups contained aqueous solutions of citric acid (0.013–0.065 M) or diluents (water, weak 0.07 M sucrose). The older infants were grouped according to their intake of these solutions. Those who consumed more of the 0.029 M or the 0.065 M solutions relative to either water or the 0.013 M solution were grouped in the high-sour group. Infants' food intake was assessed with 3-d food diaries.	No differential consumption of aqueous citric acid relative to water, but addition of citric acid to a sucrose diluent ↓ consumption relative to aqueous sucrose alone. ↑ fruit intake in infants with high sour preferences relative to those with low sour preferences.	Blossfeld et al., 2007; Desor et al., 1975b
Intake and facial responses in infants (4–9 months) measured while their mother fed plain cereal during one session and sour-flavored cereal (0.006 M citric acid added) during another. Infants' milk feeding history was recorded.	↑ intake of sour-tasting cereals in infants fed hydrolyzed casein formula compared to those fed breast milk or milk-based formula.	Mennella et al., 2009

(continued)

Table 36.1 *(Continued)*

Development Stage/Methodology	Outcomes	References
Children (3–12 years) **Rank-by-elimination procedure** Children (5–9 years) and their mothers were presented four concentrations of citric acid (0.00–0.25 M) in a gelatin base. They tasted without swallowing each gelatin and pointed to the one that they liked best during one session or the one that tasted most sour during another; that gelatin was then removed, and children were asked to taste the remaining three gelatins again and indicate which of the three was most preferred or tasted most sour. This procedure continued until a rank-order preference and sour intensity were established.	More than one-third of the children, but virtually none of the adults, ranked the highest concentration of citric acid in gelatin as one of their most preferred. Those children who preferred the highest concentration of citric acid in gelatin were more likely to report that they had tried and liked sour candies compared to those who preferred lower concentrations of citric acid in gelatin.	Liem and Mennella 2003
Paired-comparison, forced-choice ranking procedure Children (4–7 years) were presented six concentrations of citric acid (0–0.07 M) in apple juice and ranked the concentration of citric acid most preferred in all possible pairs of solutions. Mothers were queried about the type of formula children had been fed as infants.	Children 4–5 years old who were fed protein hydrolysate formulas preferred ↑ concentrations of citric acid in juice than did 6- to 7-year-old children fed similar formulas.	Liem and Mennella 2002
Two-step forced-choice ranking procedure Children (8–11 years) and adults were presented with six concentrations of citric acid dissolved in either orangeade (0–0.065 M citric acid) or a yogurt base (0–0.17 M) in random order and asked to categorize each into hedonic categories using hedonic face scales ranging from smiling (most preferred), to neutral (just ok), to frowning (least preferred). Next, participants ranked their liking of the foods within each category. They then received repeated exposure to orangeade that contained 0.42 M sucrose and 0.02 M added citric acid over 8 days, and posttests were conducted in the same manner as the pretest.	Children's most-preferred orangeade during pretest was similar to that observed during the posttest. Similar results were obtained for yogurt.	Liem and de Graaf 2004

(continued)

Table 36.1 (*Continued*)

Development Stage/Methodology	Outcomes	References
UMAMI		
Infants (newborns to 24 months)		
Reflex-like responses		
Orofacial responses measured when drops of MSG solutions or diluent (i.e., water or soup) were placed on the tongue of normal newborn infants within hours of birth.	↑ sucking-like and mouthing responses, licking, facial relaxation, smiling, hand-mouth contact to soup with MSG relative to soup alone. MSG in water did not elicit positive facial expressions relative to water alone.	Steiner 1987
Consummatory responses		
Measured intake of MSG solutions (0.05–0.40%) or diluent (i.e., soup, water) in 2- to 24-month-old, well-nourished and/or malnourished infants during multiple-bottle brief-access (60 s) tests.	Malnourished and well-nourished infants consumed ↑ soup containing MSG relative to soup containing no MSG. Well-nourished infants consumed ↓ aqueous MSG relative to water.	Beauchamp and Pearson 1991; Vazquez et al., 1982
Intake and facial responses were measured while infants (4–9 months) were fed glutamate-flavored (0.02 M) and plain cereal or broth by their mothers in an infant-led feeding paradigm. Infants had been routinely fed hydrolyzed casein formula or cow's milk for varying amounts of time.	↑ intake of savory-tasting cereals in infants routinely fed hydrolyzed casein formula than in those fed milk-based formula or breast milk. ↑ smiling during feeding savory-tasting cereals in breastfed infants and those routinely fed hydrolyzed casein formula than in those fed milk-based formula. ↑ acceptance of savory-tasting broth relative to plain broth in infants fed hydrolyzed casein formula for 3 or 8 months but not for 1 month.	Mennella et al., 2009; Mennella and Castor, 2012

Abbreviations: AU, action units (from (Ekman and Friesen 1978) Facial Action Coding System); EEG, electroencephalogram; EMG, electromyography; PROP, 6-*n*-propylthiouracil; *TAS2R38*, taste receptor, type 2, member 38 (a bitter-receptor gene); *TAS1R3*, taste receptor, type 1, member 3 (a sweet-receptor gene); *TAS1R3*, taste receptor, type 1, member 3 (a sweet-receptor gene); MSG, monosodium glutamate; M, molarity or moles per liter of solute.

response to sour transforms into preference by 18 months of age (Blossfeld et al., 2007). By the time children have reached middle childhood, approximately one-third display a strong preference for concentrated sour tastes (e.g., 0.25 M citric acid; (Liem and Mennella 2003). These sour preferences have been shown to be related to intake of sour-tasting foods such as fruit throughout infancy (Blossfeld et al., 2007) and childhood (Liem and Mennella 2003), as well as personality characteristics, such as sensation seeking (Urbick 2000).

36.7.2 Preferences Counter-adaptive to Our Modern World

Why is it that children reject the very tastes that are associated with so many healthful foods, and prefer those that are associated with high-calorie, fatty foods? From an evolutionary perspective, it has been argued that children's responses to tastes enhance survival by attracting them to foods that contain energy, minerals, and vitamins (e.g., mother's milk, fruits) during periods of maximal growth and inhibiting ingestion of potentially toxic substances, many of which are bitter. Children's innate taste preferences, which reflect their biological needs, help to explain why they like to eat the foods and beverages they do. For example, Coldwell et al., (2009) have shown that preferences for sweet-tasting foods are correlated with periods of high growth. In our modern society, this once adaptive link between sweet preference and biological need may leave children especially vulnerable to negative long-term consequences, including obesity and dental caries.

Although children's hedonic responses to the basic tastes may have once enhanced their survival, today children's predisposition to prefer sugar and salt and to avoid bitter-tasting foods, such as green vegetables, is no longer adaptive in an obesogenic environment that contains an overabundance of processed carbohydrate-rich and salty foods. Approximately a third of children more than 2 years of age are overweight (Ogden et al., 2010); this excess weight is associated with a number of serious health complications that were previously considered rare in children, such as type 2 diabetes and high blood pressure (Lee et al., 2006; Maynard et al., 2003). Given that unhealthful dietary patterns established in childhood track into adulthood (Mikkila et al., 2004), it is necessary to develop a thorough understanding of the basic biology of taste and the mechanisms involved in the development of taste and flavor preferences throughout childhood.

The innate rejection of bitter-tasting substances in children, in addition to reducing consumption of green vegetables, also plays a central role in children's hesitance to take oral medications (for review, see Mennella and Beauchamp 2008). Although many solid oral dosage forms (e.g., pills, tablets) have the advantage of masking or encapsulating the bitter tastes of active compounds, such methods are ineffective for many children because they often cannot or will not swallow pills or tablets. Most children, at some point in their lives, are required to take medicine, which often tastes bitter, and some will refuse to take it. When children refuse to take a single dose–let alone a full course of medication–because it tastes bad to them, that bitter taste thwarts the benefits of even the most powerful drug. Thus, the main challenge of administering medicine to children is a matter of taste because drugs, by their very nature, often taste unpleasant and bitter. The unpleasant taste of a medicine is a sensory expression of its pharmacological activity–the more potent the drug, the more bitter-tasting and/or irritating it will be. And the more bitter-tasting and irritating its flavor, the more likely the drug will be rejected by children.

A recent study explored the relationship between genotype of only one of the 25 bitter receptor genes (*TAS2R38*) and medication history (Lipchock et al., 2012) and found that children younger than 10 years with at least one bitter-sensitive allele (PP or AP genotype) were more likely to have taken medicine in solid formulation than were bitter-insensitive (AA genotype) children. It was hypothesized that the resistance to taking bitter liquid formulations may be related to compliance and that bitter-sensitive children may be resistant to taking bitter liquid formulations and are more motivated to try medicine in pill form as an alternative. These findings open up another important area for taste psychophysics and genetics, suggesting that taste genetics and individual differences in bitter sensitivity could be an important factor in formulation choice and compliance in the pediatric population.

36.8 CONCLUSIONS

Although this area of research has enjoyed more than a century of basic research, we have many gaps in knowledge that have important implications for the health of our children. Although many developmentally appropriate methodologies and approaches have been created to test taste sensitivity and preference throughout childhood, more work needs to be done to validate and standardize these procedures across taste modalities. In some cases, the question may need to be asked several ways and the convergence of findings from different methodologies is required to give confidence to the conclusions.

While each measure has its limitations, the convergence of research findings supports the conclusion that the ability to detect sweets is evident very early in human development and that its hedonic tone changes throughout childhood. With respect to bitter, newborns respond with highly negative facial expressions to concentrated quinine and urea, but they do not reject moderate concentrations of urea. The reason for this difference remains unclear. Newborns also reject the sour taste of strong citric acid solutions; however, at least for some children, this initial rejection transforms into a preference before two years of age. Because only a handful of studies on sour tastes have been conducted, it is not known whether there are developmental changes in sensitivity or preference for sour-tasting foods and beverages. Conclusions regarding the response to savory and salty tastes are more problematic, and further research is needed. With regard to salt taste, studies measuring intake and facial expressions suggest that the newborn infant is indifferent to and may not detect salt. However, in tests where salt has been presented within an appropriate context (i.e., soup), children younger than two years of age have shown preferences for this taste.

The bottom line is that unhealthful behaviors, such as the overconsumption of sweet and salty foods and the rejection of bitter green vegetables or unpalatable medications, are a reflection of the child's basic biology. From an evolutionary perspective, the senses that evaluate what is put into the mouth have likely evolved to reject that which is harmful and seek out that which is beneficial. In particular, rejection of bitter-tasting and irritating substances is thought to have evolved to protect the animal from being poisoned and the plant producing these chemicals from being eaten.

That being said, our biology is not our destiny; children learn through repeated experiences with various types of foods. A case in point is sweet taste. Research has shown that children learn in which contexts sugar is typically found through experience (Beauchamp and Cowart 1985; Beauchamp and Moran 1982). That is, the sensation of sweetness may be context dependent and can acquire meaning through learning. In a recent studies, children who were exposed to a sweetened orange-flavored beverage (Liem and de Graaf 2004) or a sweetened food (Sullivan and Birch 1990) over the course of eight to fifteen days exhibited a stronger preference for the sweetened beverage or food at the end of the exposure period, but not a significant preference for the sweet taste in another context (e.g. yogurt or ricotta cheese), suggesting that this learning may be context-dependent (Liem and de Graaf 2004). That is, through familiarization, children develop a sense of what should, and what should not, taste sweet (Beauchamp and Cowart 1985; Pepino and Mennella 2005a).

Such learning is not confined to sweet taste, as evidenced by studies of salt taste perception (Beauchamp, Cowart et al., 1986; Cowart and Beauchamp 1986), as well as by studies examining more complex food media such as extensively hydrolyzed protein infant formulas (ePHF), which taste bitter, sour, and savory and have unpleasant aromas (Mennella and Beauchamp 2005). In a series of randomized clinical trials, we have shown that there is

a sensitive period before 4 months of age during which experience can shift the hedonic tone to the flavor of the formula (Mennella et al., 2004; Mennella et al., 2011a). Evidence that such flavor programming lasts beyond childhood is suggested by the finding that children who were fed ePHF as infants are programmed to like the taste of not only ePHF when they are older, but also the taste of foods that are more savory (e.g., chicken), sour (e.g., lemon), and bitter (e.g., broccoli) (Liem and Mennella 2002; Mennella 1998; Mennella and Beauchamp 2002; Mennella et al., 2009; Mennella et al., 2006).

In conclusion, it is a public health priority to develop standardized psychophysical methods that are appropriate for children of all ages. More insight into the "taste world" of children will initiate new perspectives that will aid in the development of evidence-based strategies for improving children's food preferences and eating behaviors and help ameliorate bitter taste, which is the primary culprit for children's refusal to accept liquid formulations.

ACKNOWLEDGMENTS

This project was funded in part by Awards R01DC011287 from the National Institute on Deafness and Other Communication Disorders and R01HD37119 from the Eunice Kennedy Shriver National Institute of Child Health & Human Development to J.A.M. and R15HD071486 from the Eunice Kennedy Shriver National Institute of Child Health & Human Development to C.A.F. The content is solely the responsibility of the authors and does not necessarily represent the official views of the Eunice Kennedy Shriver National Institute of Child Health & Human Development or the National Institutes of Health. We acknowledge the expert editorial assistance of Ms. Patricia Watson.

REFERENCES

Abad, F., Diaz, N. M., Domenech, E., et al. (1996). Oral sweet solution reduces pain-related behaviour in preterm infants. *Acta. Paediatr.* 85: 854–858.

Acharya, A. B., Annamali, S., Taub, N. A. and Field, D. (2004). Oral sucrose analgesia for preterm infant venepuncture. *Arch. Dis. Child Fetal. Neonatal. Ed.* 89: F17–18.

Anliker, J. A., Bartoshuk, L., Ferris, A. M. and Hooks, L. D. (1991). Children's food preferences and genetic sensitivity to the bitter taste of 6-n-propylthiouracil (PROP). *Am. J. Clin. Nutr.* 54: 316–320.

Armstrong, J. E., Hutchinson, I., Laing, D. G. and Jinks, A. L. (2007). Facial electromyography: responses of children to odor and taste stimuli. *Chem. Senses.* 32: 611–621.

ASTM International 2003. E2299-03 Standard guide for sensory evaluation of products by children. Vol. 15.08, ASTM International, West Conshohocken, PA. Available at http:www.astm.org. Accessed July 2012.

Barr, R.G., Pantel, M.S., Young, S.N., et al. (1999). The response of crying newborns to sucrose: is it a "sweetness" effect? *Physiol. Behav.* 66: 409–417.

Beauchamp, G. K. and Cowart, B. J. (1985). Congenital and experiential factors in the development of human flavor preferences. *Appetite.* 6: 357–372.

Beauchamp, G. K. and Cowart, B. J. (1990). Preference for high salt concentrations among children. *Dev. Psychol.* 26: 539–545.

Beauchamp, G. K., Cowart, B. J., Mennella, J. A. and Marsh, R. R. (1994). Infant salt taste: developmental, methodological, and contextual factors. *Dev. Psychobiol.* 27: 353–365.

Beauchamp, G. K., Cowart, B. J. and Moran, M. (1986). Developmental changes in salt acceptability in human infants. *Dev. Psychobiol.* 19: 17–25.

Beauchamp, G. K. and Moran, M. (1984). Acceptance of sweet and salty tastes in 2-year-old children. *Appetite.* 5: 291–305.

Beauchamp, G. K. and Moran, M. (1982). Dietary experience and sweet taste preference in human infants. *Appetite.* 3: 139–152.

Beauchamp, G. K. and Pearson, P. (1991). Human development and umami taste. *Physiol. Behav.* 49: 1009–1012.

Bellieni, C. V., Bagnoli, F., Perrone, S., et al. (2002). Effect of multisensory stimulation on analgesia in term neonates: a randomized controlled trial. *Pediatr. Res.* 51: 460–463.

Berridge, K. C. (1996). Food reward: brain substrates of wanting and liking. *Neurosci. Biobehav. Rev.* 20: 1–25.

Berridge, K. C. and Kringelbach, M. L. (2000). Affective neuroscience of pleasure: reward in humans and animals. *Psychopharmacology (Berl).* 199: 457–480.

Birch, L. L. (1998). Psychological influences on the childhood diet. *J. Nutr.* 128: 407S–410S.

Blanck, H. M., Gillespie, C., Kimmons, J. E., et al. (2008). Trends in fruit and vegetable consumption among U.S. men and women, 1994–2005. *Prev. Chronic. Dis.* 5: A35.

Blass, E. M. and Camp, C. A. (2003). Changing determinants of crying termination in 6 to 12week-old human infants. *Dev. Psychobiol.* 42: 312–316.

Blass, E. M. and Hoffmeyer, L. B. (1991). Sucrose as an analgesic for newborn infants. *Pediatrics.* 87: 215–218.

Blass, E. M. and Smith, B. A. (1992). Differential effects of sucrose, fructose, glucose, and lactose on crying in 1- to 3-day-old human infants: Qualitative and quantitative considerations. *Dev. Psychol.* 28: 804–810.

Blass, E. M. and Watt, L. B. (1999). Suckling- and sucrose-;induced analgesia in human newborns. *Pain.* 83: 611–623.

Blossfeld, I., Collins, A., Boland, S., et al. (2007) Relationships between acceptance of sour taste and fruit intakes in 18-month-old infants. *Br. J. Nutr.* 98: 1084–1091.

Boughter, J. D. J. and Bachmanov, A. A. (2007). Behavioral genetics and taste. *BMC Neurosci.* 8 Suppl 3: S3.

Bradley, R. M. and Mistretta, C. M. (1975). The developing sense of taste. In Denton, D.A. and Coghlan, J.P. (eds.), *Olfaction and Taste V.* Academic Press, New York, pp. 91–98.

Bradley, R. M. and Stern, I. B. (1967). The development of the human taste bud during the foetal period. *J. Anat.* 101: 743–752.

Bucher, H. U., Baumgartner, R., Bucher, N., et al. (2000). Artificial sweetener reduces nociceptive reaction in term newborn infants. *Early Hum. Dev.* 59: 51–60.

Bucher, H. U., Moser, T., von Siebenthal, K., et al. (1995). Sucrose reduces pain reaction to heel lancing in preterm infants: a placebo-controlled, randomized and masked study. *Pediatr. Res.* 38: 332–335.

Carbajal, R., Chauvet, X., Couderc, S. and Olivier-Martin, M. (1999). Randomised trial of analgesic effects of sucrose, glucose, and pacifiers in term neonates. *BMJ.* 319: 1393–1397.

Carbajal, R., Lenclen, R., Gajdos, V., et al. (2002). Crossover trial of analgesic efficacy of glucose and pacifier in very preterm neonates during subcutaneous injections. *Pediatrics.* 110: 389–393.

Chambers, E. (2005). Commentary: Conducting sensory research in children. *J. Sens. Stud.* 20: 90–92.

Chen, A. W. and Resurreccion, A. V. (1996). Age appropriate hedonic scales to measure food preferences of young children. *J. Sens. Stud.* 11: 141–163.

Coldwell, S. E., Mennella, J. A., Duffy, V. B., Pelchat, M. L., Griffith, J. W., Smutzer G., Cowart, B. J., Breslin, P. A., Bartoshuk, L. M., Hastings, L., Vistorson, D., Hoffman, H. J. (2013). Gustation assessment using the NIH Toolbox. *Neurology.* 80: S20–4.

Coldwell, S. E., Oswald, T. K. and Reed, D. R. (2009). A marker of growth differs between adolescents with high vs. low sugar preference. *Physiol. Behav.* 96: 574–580.

Cowart, B. J. (1981). Development of taste perception in humans: sensitivity and preference throughout the life span. *Psychol. Bull.* 90: 43–73.

Cowart, B. J. and Beauchamp, G. K. (1990). Early development of taste perception. In McBride, R. and MacFie, H. (eds.), *Psychological Basis of Sensory Evaluation*, London, England, pp. 1–17.

Cowart, B. J. and Beauchamp, G. K. (1986). The importance of sensory context in young children's acceptance of salty tastes. *Child Dev.* 57: 1034–1039.

Cowart, B. J., Beauchamp, G. K. and Mennella, J. A. (2004). Development of taste and smell in the neonate., *Fetal and Neonatal Physiology*, Volume 2, 3rd edition. Saunders, Philadelphia, PA, pp. 1819–1827.

Crook, C. K. (1976). Neonatal sucking: effects of quantity of the response-contingent fluid upon sucking rhythm and heart rate. *J. Exp. Child Psychol.* 21: 539–548.

Crook, C. K. (1978). Taste perception in the newborn infant. *Infant Behav. Dev.* 1: 52–69.

Crook, C. K. and Lipsitt, L. P. (1976). Neonatal nutritive sucking: effects of taste stimulation upon sucking rhythm and heart rate. *Child Dev.* 47: 518–522.

Crystal, S. R. and Bernstein, I. L. (1998). Infant salt preference and mother's morning sickness. *Appetite.* 30: 297–307.

de Graaf, C. and Zandstra, E. H. (1999). Sweetness intensity and pleasantness in children, adolescents, and adults. *Physiol. Behav.* 67: 513–520.

Deshmukh, L. S. and Udani, R. H. (2002). Analgesic effect of oral glucose in preterm infants during venipuncture--a double-blind, randomized, controlled trial. *J. Trop. Pediatr.* 48: 138–141.

DeSnoo, K. (1937). Das trinkende kind im uterus. *Monoats. Geburtsh. Gynaekol.* 105: 88–97.

Desor, J., Maller, O. and Turner, R. E. (1973). Taste in acceptance of sugars by human infants. *J. Comp. Physiol. Psychol.* 84: 496–501.

Desor, J. A. and Beauchamp, G. K. (1987). Longitudinal changes in sweet preferences in humans. *Physiol. Behav.* 39: 639–641.

Desor, J. A., Greene, L. S. and Maller, O. (1975a). Preferences for sweet and salty in 9- to 15-year-old and adult humans. *Science.* 190: 686–687.

Desor, J. A., Maller, O. and Andrews, K. (1975b). Ingestive responses of human newborns to salty, sour, and bitter stimuli. *J. Comp. Physiol. Psychol.* 89: 966–970.

Eckstein, A. (1927). Zur Physiologie der Geschmacksempfindung und des Saugreflexes bei Säuglingen. *Z. Kinderheilk.* 45: 1–18.

Egan, J. M. and Margolskee, R. F. (2008). Taste cells of the gut and gastrointestinal chemosensation. *Mol. Interv.* 8: 78–81.

Ekman, P. and Freisen, W. V. (2003). *Unmasking the face: a guide to recognizing emotions from facial expressions.* Malor Books, Cambridge, MA.

References

Ekman, P. and Friesen, W. (1978). *The facial action coding system.* Consulting Psychologists Press, Palo Alto, CA.

Ekman, P. and Oster, H. (1979). Facial expressions of emotion. *Annual Review of Psychology.* 30: 527–554.

Epstein, L. H. and Paluch, R. A. (1997). Habituation of facial muscle responses to repeated food stimuli. *Appetite.* 29: 213–224.

Fernandez, M., Blass, E. M., Hernandez-Reif, M., et al. (2003). Sucrose attenuates a negative electroencephalographic response to an aversive stimulus for newborns. *J. Dev. Behav. Pediatr.* 24: 261–266.

Forestell, C. A. and Mennella, J. A. (2007). Early determinants of fruit and vegetable acceptance. *Pediatrics.* 120: 1247–1254.

Forestell, C. A. and Mennella, J. A. (2012). More than just a pretty face. The relationship between infant's temperament, food acceptance, and mothers' perceptions of their enjoyment of food. *Appetite.* 58: 1136–1142.

Fox, N. A. and Davidson, R. J. (1986). Taste-elicited changes in facial signs of emotion and the asymmetry of brain electrical activity in human newborns. *Neuropsychologia.* 24: 417–422.

Ganchrow, J. R. and Mennella, J. A. (2003). The ontogeny of human flavor perception. In Doty, R.L. (ed.), *Handbook of Olfaction and Gustation*, 2nd edition. Marcel Dekker, Inc., New York, pp. 823–946.

Ganchrow, J. R., Steiner, J. E., and Daher, M. (1983). Neonatal facial expressions in response to different qualities and intensities of gustatory stimuli. *Infant Behav. Dev.* 6: 189–200.

Gibbins, S., Stevens, B., Hodnett, E., et al. (2002). Efficacy and safety of sucrose for procedural pain relief in preterm and term neonates. *Nurs. Res.* 51: 375–382.

Glendinning, J. I. (1994). Is the bitter rejection response always adaptive? *Physiol. Behav.* 56: 1217–1227.

Glendinning, J. I., Gresack, J. and Spector, A. C. (2002). A high-throughput screening procedure for identifying mice with aberrant taste and oromotor function. *Chem. Senses.* 27: 461–474.

Golding, J., Steer, C., Emmett, P., et al. (2009). Associations between the ability to detect a bitter taste, dietary behavior, and growth: A preliminary report. *Ann. N.Y. Acad. Sci.* 1170: 553–557.

Gradin, M., Eriksson, M., Holmqvist, G., et al. (2002). Pain reduction at venipuncture in newborns: oral glucose compared with local anesthetic cream. *Pediatrics.* 110: 1053–1057.

Graillon, A., Barr, R. G., Young, S. N., et al. (1997). Differential response to intraoral sucrose, quinine and corn oil in crying human newborns. *Physiol. Behav.* 62: 317–325.

Greenberg, C. S. (2002). A sugar-coated pacifier reduces procedural pain in newborns. *Pediatr. Nurs.* 28: 271–277.

Greene, L. S., Desor, J. A. and Maller, O. (1975). Heredity and experience: their relative importance in the development of taste preference in man. *J. Comp. Physiol. Psychol.* 89: 279–284.

Grill, H. J. and Norgren, R. (1978). The taste reactivity test. I. Mimetic responses to gustatory stimuli in neurologically normal rats. *Brain Res.* 143: 263–279.

Guenther, P. M., Dodd, K. W., Reedy, J. and Krebs-Smith, S. M. (2006). Most Americans eat much less than recommended amounts of fruits and vegetables. *J. Am. Diet Assoc.* 106: 1371–1379.

Guinard, J. X. (2001). Sensory and consumer testing with children. *Trends Food Sci. Technol.* 11: 273–283.

Harris, G., Thomas, A. and Booth, D. A. (1990). Development of salt taste in infancy. *Dev. Psychol.* 26: 534–538.

Harrison, D., Stevens, B., Bueno, M., et al. (2010). Efficacy of sweet solutions for analgesia in infants between 1 and 12 months of age: a systematic review. *Arch. Dis. Child.* 95: 406–413.

Hersch, M. and Ganchrow, D. (1980). Scanning electron microscopy of developing papillae on the tongue of human embryos and fetuses. *Chemical Senses.* 5: 331–341.

Hu, S., Luo, Y. J. and Hui, L. (2000). Preliminary study of associations between objective parameters of facial electromyography and subjective estimates of taste palatability. *Percept Mot. Skills.* 91: 741–747.

Hu, S., Player, K. A., McChesney, K. A., et al. (1999). Facial EMG as an indicator of palatability in humans. *Physiol. Behav.* 68: 31–35.

Jacobs, W. W. (1978). Taste responses in wild and domestic guinea pigs. *Physiol. Behav.* 20: 579–588.

James, C. E., Laing, D. G. and Oram, N. (1997). A comparison of the ability of 8-9-year-old children and adults to detect taste stimuli. *Physiol. Behav.* 62: 193–197.

Jamel, H. A., Sheiham, A., Watt, R. G. and Cowell, C. R. (1997). Sweet preference, consumption of sweet tea and dental caries; studies in urban and rural Iraqi populations. *Int. Dent. J.* 47: 213–217.

Kajuira, H., Cowart, B. J. and Beauchamp, G. K. (1992). Early developmental change in bitter taste responses in human infants. *Dev. Psychobiol.* 25: 375–386.

Keller, K. L., Reid, A., MacDougall, M. C., et al. (2010). Sex differences in the effects of inherited bitter thiourea sensitivity on body weight in 4-6-year-old children. *Obesity* (Silver Spring). 18: 1194–1200.

Keller, K. L., Steinmann, L., Nurse, R. J. and Tepper, B. J. (2002). Genetic taste sensitivity to 6-n-propylthiouracil influences food preference and reported intake in preschool children. *Appetite.* 38: 3–12.

Kim, G. H. and Lee, H. M. (2009). Frequent consumption of certain fast foods may be associated with an enhanced preference for salt taste. *J. Hum. Nutr. Diet.* 22: 475–480.

Knai, C., Pomerleau, J., Lock, K. and McKee, M. (2006). Getting children to eat more fruit and vegetables: a systematic review. *Prev. Med.* 42: 85–95.

Lakkakula, A. P., Zanovec, M., Silverman, L., et al. (2008). Black children with high preferences for fruits and vegetables are at less risk of being at risk of overweight or overweight. *J. Am. Diet Assoc.* 108: 1912–1915.

Larochette, A. C., Chambers, C. T. and Craig, K. D. (2006). Genuine, suppressed and faked facial expressions of pain in children. *Pain.* 126: 64–71.

Lee, J. M., Herman, W. H., McPheeters, M. L. and Gurney, JG. (2006). An epidemiologic profile of children with diabetes in the U.S. *Diabetes Care.* 29: 420–421.

Lesham, M. (1998). Salt preference in adolescence is predicted by common prenatal and infantile mineral fluid loss. *Physiol. Behav.* 63: 699–704.

Lewkowski, M. D., Barr, R. G., Sherrard, A., et al. (2003). Effects of chewing gum on responses to routine painful procedures in children. *Physiol. Behav.* 79: 257–265.

Liem, D. G. and de Graaf, C. (2004). Sweet and sour preferences in young children and adults: role of repeated exposure. *Physiol. Behav.* 83: 421–429.

Liem, D. G., Mars, M. and De Graaf, C. (2004). Sweet preferences and sugar consumption of 4- and 5-year-old children: role of parents. *Appetite.* 43: 235–245.

Liem, D. G. and Mennella, J. A. (2003). Heightened sour preferences during childhood. *Chem. Senses.* 28: 173–180.

Liem, D. G. and Mennella, J. A. (2002). Sweet and sour preferences during childhood: role of early experiences. *Dev. Psychobiol.* 41: 388–395.

Liley, A. W. (1972). Disorders of amniotic fluid. In Assali, N.S. (ed.), *Pathophysiology of gestation: fetal placental disorders.* Academic Press, New York, pp. 157–206.

Lipchock, S. V., Reed, D. R. and Mennella, J. A. (2012). Relationship between bitter-taste receptor genotype and solid medication formulation usage among young children: a retrospective analysis. *Clin. Ther.* 34: 728–733.

Maciel, S. M., Marcenes, W. and Sheiham, A. (2001a). The relationship between sweetness preference, levels of salivary mutans streptococci and caries experience in Brazilian pre-school children. *Int. J. Paediatr. Dent.* 11: 123–130.

Maciel, S. M., Marcenes, W., Watt, R. G. and Sheiham, A. (2001b). The relationship between sweetness preference and dental caries in mother/child pairs from Maringa-Pr, *Brazil. Int. Dent. J.* 51: 83–88.

Maller, O. and Desor, J. A. (1973). Effect of taste on ingestion by human newborns. *Symp. Oral Sens. Percept:* 279–291.

Maone, T. R., Mattes, R. D., Bernbaum, J. C. and Beauchamp, G. K. (1990). A new method for delivering a taste without fluids to preterm and term infants. *Dev. Psychobiol.* 13: 179–191.

Maynard, M., Gunnell, D., Emmett, P., et al. (2003). Fruit, vegetables, and antioxidants in childhood and risk of adult cancer: the Boyd Orr cohort. *J. Epidemiol. Community Health.* 57: 218–225.

McCrory, M. A., Fuss, P. J., McCallum, J. E., et al. (1999). Dietary variety within food groups: association with energy intake and body fatness in men and women. *Am. J. Clin. Nutr.* 69: 440–447.

Mennella, J. A. (2007). The chemical senses and the development of flavor preferences in humans. In Hale, T.W. and Hartmann, P.E. (eds.), *Textbook on Human Lactation.* Hale Publishing, Texas, pp. 403–414.

Mennella, J. A. (1998). Development of the chemical senses and the programming of flavor preference. Physiologic/Immunologic Responses to Dietary Nutrients: Role of elemental and Hydrolysate Formulas in Management of the Pediatric Patient. Report of the 107th Conference on Pediatric Research. Ross Products Division, Abbott Laboratories, Columbus, OH, pp. 201–208.

Mennella, J. A. and Beauchamp, G. K. (2002). Flavor experiences during formula feeding are related to preferences during childhood. *Early Hum. Dev..* 68: 71–82.

Mennella, J. A. and Beauchamp, G. K. (2008). Optimizing oral medications for children. *Clin. Ther.* 30: 2120–2132.

Mennella, J. A. and Beauchamp, G. K. (2005). Understanding the origin of flavor preferences. *Chem. Senses.* 30 Suppl 1: i242–i243.

Mennella, J. A. and Castor, S. M. (2012). Sensitive period in flavor learning: Effects of duration of exposure to formula flavors on food likes during infancy. *Clin. Nutr.* In Press doi:10.1016/j.clnu.2012.05.005

Mennella, J. A., Finkbeiner, S. and Reed, D. R. (2012). The proof is in the pudding: children prefer lower fat but higher sugar than do mothers. *Int. J. Obes. (Lond):* 1–7.

Mennella, J. A., Forestell, C. A., Morgan, L. K. and Beauchamp, G. K. (2009). Early milk feeding influences taste acceptance and liking during infancy. *Am. J. Clin. Nutr.* 90: 780–788S.

Mennella, J. A., Griffin, C. E. and Beauchamp, G. K. (2004). Flavor programming during infancy. *Pediatrics.* 113: 840–845.

Mennella, J. A., Jagnow, C. P. and Beauchamp, G. K. (2001). Prenatal and postnatal flavor learning by human infants. *Pediatrics.* 107: E88.

Mennella, J. A., Johnson, A. and Beauchamp, G. K. (1995). Garlic ingestion by pregnant women alters the odor of amniotic fluid. *Chem. Senses.* 20: 207–209.

Mennella, J. A., Kennedy, J. M. and Beauchamp, G. K. (2006). Vegetable acceptance by infants: effects of formula flavors. *Early. Hum. Dev.* 82: 463–468.

Mennella, J. A., Lukasewycz, L. D., Castor, S. M. and Beauchamp, G. K. (2011a). The timing and duration of a sensitive period in human flavor learning: a randomized trial. *Am. J. Clin. Nutr.* 93: 1019–1024.

Mennella, J. A., Lukasewycz, L. D., Griffith, J. W. and Beauchamp, G. K. (2011b). Evaluation of the Monell forced-choice, paired-comparison tracking procedure for determining sweet taste preferences across the lifespan. *Chem. Senses.* 36: 345–355.

Mennella, J. A., Pepino, M. Y. and Beauchamp, G. K. (2003). Modification of bitter taste in children. *Dev. Psychobiol.* 43: 120–127.

Mennella, J. A., Pepino, M. Y., Duke, F. F. and Reed, D. R. (2010a). Age modifies the genotype-phenotype relationship for the bitter receptor TAS2R38. *BMC Genet.* 11: 60.

Mennella, J. A., Pepino, M. Y., Lehmann-Castor, S. M. and Yourshaw, L. M. (2010b). Sweet preferences and analgesia during childhood: effects of family history of alcoholism and depression. *Addiction.* 105: 666–675.

Mennella, J. A., Pepino, M. Y. and Reed, D. R. (2005). Genetic and environmental determinants of bitter perception and sweet preferences. *Pediatrics.* 115: e216–222.

Mikkila, V., Rasanen, L., Raitakari, O. T., et al. (2004). Longitudinal changes in diet from childhood into adulthood with respect to risk of cardiovascular diseases: The Cardiovascular Risk in Young Finns Study. *Eur. J. Clin. Nutr.* 58: 1038–1045.

Miller, A., Barr, R. G. and Young, S. N. (1994). The cold pressor test in children: methodological aspects and the analgesic effect of intraoral sucrose. *Pain.* 56: 175–183.

Miller, J. L., Sonies, B. C. and Macedonia, C. (2003). Emergence of oropharyngeal, laryngeal and swallowing activity in the developing fetal upper aerodigestive tract: an ultrasound evaluation. *Early Hum. Dev.* 71: 61–87.

Mistretta, C. M. (1972). Topographical and histological study of the developing rat tongue, palate and taste buds. In Bosma, J.F. (ed.), Third Symposium on Oral Sensation and Perception: The Mouth of the Infant. C.C. Thomas, Springfield, IL, pp. 163–186.

Mokdad, A. H., Marks, J. S., Stroup, D. F. and Gerberding, J. L. (2004). Actual causes of death in the United States, 2000. *JAMA.* 291: 1238–1245.

Nowlis, G. H. and Kessen, W. (1976). Human newborns differentiate differing concentrations of sucrose and glucose. *Science.* 191: 865–866.

Ogden, C. L., Carroll, M. D., Curtin, L. R., et al. (2010). Prevalence of high body mass index in US children and adolescents, 2007–2008. *JAMA.* 303: 242–249.

Okan, F., Coban, A., Ince, Z., et al. (2007). Analgesia in preterm newborns: the comparative effects of sucrose and glucose. *Eur. J. Pediatr.* 166: 1017–1024.

Pepino, M. Y. and Mennella, J. A. (2005a). Factors contributing to individual differences in sucrose preference. *Chem. Senses.* 30 Suppl 1: i319–i320.

Pepino, M. Y. and Mennella, J. A. (2005b). Sucrose-induced analgesia is related to sweet preferences in children but not adults. *Pain.* 119: 210–218.

Peterson, F. and Rainey, L. H. (1910). The beginnings of mind in the newborn. *Bull Lying-In Hosp.* 7: 99–122.

Popper, R. and Kroll, J. J. (2005). Conducting sensory research with children. *J. Sens. Stud.* 20: 75–87.

Pritchard, J. A. (1965). Deglutition by Normal and Anencephalic Fetuses. *Obstet. Gynecol.* 25: 289–297.

Ramenghi, L. A., Evans, D. J. and Levene, M. I. (1999). "Sucrose analgesia": absorptive mechanism or taste perception? *Arch. Dis. Child Fetal. Neonatal. Ed.* 80: F146–147.

Ramenghi, L. A., Griffith, G. C., Wood, C. M. and Levene, M. I. (1996). Effect of non-sucrose sweet tasting solution on neonatal heel prick responses. *Arch. Dis. Child Fetal. Neonatal. Ed.* 74: F129–131.

Rosenstein, D. and Oster, H. (1988). Differential facial responses to four basic tastes in newborns. *Child Dev.* 59: 1555–1568.

Ross, M. G. and Nijland, M. J. (1997). Fetal swallowing: relation to amniotic fluid regulation. *Clin. Obstet. Gynecol.* 40: 352–365.

Schmidt, H. J. and Beauchamp, G. K. (1988). Adult-like odor preferences and aversions in three-year-old children. *Child Dev.* 59: 1136–1143.

References

Schuett, V. E., Brown, E. S. and Michals, K. (1985). Reinstitution of diet therapy in PKU patients from twenty-two clinics. *Am. J. Public Health.* 75: 39–42.

Schwartz, C., Issanchou, S. and Nicklaus, S. (2009). Developmental changes in the acceptance of the five basic tastes in the first year of life. *Br. J. Nutr.* 102: 1375–1385.

Sclafani, A. (2007). Sweet taste signaling in the gut. *Proc. Natl. Acad. Sci. USA.* 104: 14887–14888.

Segovia, C., Hutchinson, I., Laing, D. G. and Jinks, A. L. (2002). A quantitative study of fungiform papillae and taste pore density in adults and children. *Brain Res. Dev. Brain Res.* 138: 135–146.

Smith, B. A. and Blass, E. M. (1996). Taste-mediated calming in premature, preterm, and full-term human infants. *Dev. Psychol.* 32: 1084–1089.

Spector, A. C., Smith, J. C. and Hollander, G. R. (1981). A comparison of dependent measures used to quantify radiation-induced taste aversion. *Physiol. Behav.* 27: 887–901.

Spector, A. C. and Travers, S. P. (2005). The representation of taste quality in the mammalian nervous system. *Behav. Cogn. Neurosci. Rev.* 4: 143–191.

Stein, L. J., Cowart, B. J. and Beauchamp, G. K. (2006). Salty taste acceptance by infants and young children is related to birth weight: longitudinal analysis of infants within the normal birth weight range. *Eur J Clin Nutr.* 60: 272–279.

Stein, L. J., Cowart, B. J. and Beauchamp, G. K. (2012). The development of salty taste acceptance is related to dietary experience in human infants: a prospective study. *Am. J. Clin. Nutr.* 95: 123–129.

Stein, N., Laing, D. G. and Hutchinson, I. (1994). Topographical differences in sweetness sensitivity in the peripheral gustatory system of adults and children. *Brain Res. Dev Brain Res.* 82: 286–292.

Steiner, J. E. (1973). The gustofacial response: Observation on normal and anencephalic newborn infants. In Bosma, J.F. (ed.), Fourth Symposium On Oral Sensation and Perception. US Department of Health, Education and Welfare, Bethesda, pp. 254–278.

Steiner, J. E. (1977). Facial expressions of the neonate infant indicating the hedonics of food-related chemical stimuli. *Taste and development: the genesis of sweet preference.* U.S. Government Printing Office, Washington, DC.

Steiner, J. E. (1979). Human facial expressions in response to taste and smell stimulation. *Adv. Child Dev. Behav.* 13: 257–295.

Steiner, J. E. (1987). *What the Neonate Can Tell Us About Umami.* In Kawamura, Y. and Kare, M.R. (eds.), *Umami: A Basic Taste.* Marcel Dekker, New York, pp. 97–103.

Steiner, J. E, Glaser, D., Hawilo, M. E. and Berridge, K. C. (2001). Comparative expression of hedonic impact: affective reactions to taste by human infants and other primates. *Neurosci. Biobehav. Rev.* 25: 53–74.

Stevens, B., Yamada, J., Beyene, J., et al. (2005). Consistent management of repeated procedural pain with sucrose in preterm neonates: Is it effective and safe for repeated use over time? *Clin. J. Pain.* 21: 543–548.

Stevens, B., Yamada, J. and Ohlsson, A. (2010). Sucrose for analgesia in newborn infants undergoing painful procedures. *Cochrane Database Syst Rev:* CD001069.

Sullivan, S. A. and Birch, L. L. (1990). Pass the sugar, pass the salt: experience dictates preference. *Dev. Psychol.* 26: 546–551.

Tatzer, E., Schubert, M. T., Timischl, W. and Simbruner, G. (1985). Discrimination of taste and preference for sweet in premature babies. *Early Hum. Dev.* 12: 23–30.

Temple, E. C., Hutchinson, I., Laing, D. G. and Jinks, A. L. (2002a). Taste development: differential growth rates of tongue regions in humans. *Brain Res. Dev. Brain Res.* 135: 65–70.

Temple, E. C., Laing, D. G., Hutchinson, I. and Jinks, A. L. (2002b). Temporal perception of sweetness by adults and children using computerized time-intensity measures. *Chem. Senses.* 27: 729–737.

Tomita, N. E., Nadanovsky, P., Vieira, A. L. and Lopes, E. S. (1999). [Taste preference for sweets and caries prevalence in preschool children]. *Rev. Saude Publica.* 33: 542–546.

Turnbull, B. and Matisoo-Smith, E. (2002). Taste sensitivity to 6-n-propylthiouracil predicts acceptance of bitter-tasting spinach in 3-6-y-old children. *Am. J. Clin. Nutr.* 76: 1101–1105.

Urbick, B. (2000). Part two: what we have learned about kids. In Urbrick, B. (ed.), What about Kids: Food and Beverages. Leatherhead Publishing, Surrey, pp. 17–39.

Vazquez, M., Pearson, P. B. and Beauchamp, G. K. (1982). Flavor preferences in malnourished Mexican infants. *Physiol. Behav.* 28: 513–519.

Verma, P., Mittal, S., Ghildiyal, A., et al. (2007). Salt preference: age and sex related variability. *Indian J. Physiol. Pharmacol.* 51: 91–95.

Visser, J., Kroeze, J. H., Kamps, W. A. and Bijleveld, C. M. (2000). Testing taste sensitivity and aversion in very young children: development of a procedure. *Appetite.* 34: 169–176.

Watson, W. L., Laing, D. G., Hutchinson, I. and Jinks, A. L. (2001). Identification of the components of taste mixtures by adults and children. *Dev. Psychobiol.* 39: 137–145.

Witt, M. and Reutter, K. (1996). Embryonic and early fetal development of human taste buds: a transmission electron microscopical study. *Anat. Rec.* 246: 507–523.

Witt, M. and Reutter, K. (1998). Innervation of developing human taste buds. An immunohistochemical study. *Histochem. Cell Biol.* 109: 281–291.

Wotman, S., Mandel, I. D., Khotim, S., et al. (1964). Salt taste thresholds and cystic fibrosis. *Am. J. Dis. Child.* 108: 372–374.

Zeifman, D., Delaney, S. and Blass, E. M. (1996). Sweet taste, looking, and calm in 2- and 4-week-old infants: the eyes have it. *Dev. Psychol.* 22: 1090–1099.

Part 8

Clinical Applications and Perspectives

Chapter 37

Nutritional Implications of Taste and Smell Dysfunction

JANICE LEE, ROBIN M. TUCKER, SZE YEN TAN, CORDELIA A. RUNNING, JOSHUA B. JONES and RICHARD D. MATTES

37.1 INTRODUCTION

The chemical senses are intimately tied to nutrition. These senses have high metabolic demands due to the high turnover rates of their receptive elements and, as a consequence, require a regular supply of nutrients (Nawroth et al., 2007; Niven and Laughlin, 2008). At the same time, the chemical senses aid in the detection and ingestion of nutrients and avoidance of toxins. Eating, while essential, is a dangerous behavior in omnivores like humans as each eating event entails exposure to a different array of chemicals. Decisions must be made about when, what, how, and how much to consume. Such choices are largely guided by sensory input (Woods, 1991).

There are inherently unpleasant sensations mediated by each sensory system. Examples include bitter taste, rancid odors, and irritancy that can signal potentially dangerous food ingredients. The functionality of such aversive signals is intuitive but also limited. Many toxic compounds still pose a threat to health because they evade chemosensory detection systems (e.g., botulinum toxin). These undetected chemicals cause millions of cases of food poisoning and thousands of deaths annually (CDC, 2011). Additionally, sensory warning signals may be purposefully ignored as culturally determined safe feeding practices are learned. This can expand the available food supply, but can also lead to complications (alcohol intoxication, caffeine insomnia).

Food-borne chemicals also promote ingestive behaviors by multiple mechanisms. Flavor chemicals guide food choice. A fundamental tenet of the food industry is that in any attempt to alter food attributes such as price, nutritional value, or convenience, sensory properties must be optimized or the product will fail (i.e., not promote repeated purchases and ingestion) (IFICF, 2011). In the United States, less than 10% of discretionary income is devoted to food (USDA, 2011), and food is accessible, abundant, diverse, and convenient, so ingestion of unpalatable foods is easily avoided.

In addition to influencing food selection, the sensory properties of foods modulate their own digestion, nutrient absorptive efficiency, and peripheral metabolism. Until recently, the prevailing view held that the receptors mediating pre-ingestive chemical detection were unique and limited to the oropharyngeal region. However, the discovery of these same receptors and transduction systems in other areas of the body has prompted a more holistic perspective of an integrated, common chemical communication system playing both protective (Green, 2012) and probably nutritive (Kokrashvili et al., 2009a) roles. This raises intriguing new questions about the interaction between the chemical senses and nutrition. With the advent of food ingredients (e.g. high intensity sweeteners, fat replacers) that uncouple their sensory properties from typically associated metabolic challenges, the functionality of the system is uncertain. Animal studies (Swithers et al., 2009) and human imaging studies (Smeets et al., 2011) suggest corruption of the veracity of the signal may hold implications for food choice and nutritional status. With diet underlying the etiology, manifestation, and management of most chronic diseases, the importance of better characterization of the interaction between the chemical senses and nutrition has never been greater. The present

---

Handbook of Olfaction and Gustation, Third Edition. Edited by Richard L. Doty.
© 2015 Richard L. Doty. Published 2015 by John Wiley & Sons, Inc.

chapter summarizes the current understanding of the field though much change is anticipated in knowledge and practice in the near future.

37.2 HUNGER AND SENSORY FUNCTION

Appetitive and flavor sensations are components of multiple inter-related mechanisms that influence energy and nutrient balance. Hunger promotes the initiation of feeding, satiation influences its termination, and satiety affects the inter-meal interval whereas sensory factors guide food choice. Electrophysiological studies in primates (Scott, 1992; Rolls, 1997) and common experience indicate these four factors are interactive. Often it is assumed that a common factor (e.g., blood glucose, fatty acids, or amino acids) influences all of these sensations, but strong support for any such driver is lacking. Several hormones linked to energy metabolism have significant effects on taste responses. For example, in mouse models, plasma leptin binds to receptors on taste cells and inhibits responses to sweet stimuli (Sanematsu et al., 2009). Conversely, glucagon-like peptide-1 (GLP-1) (Jyotaki et al., 2010) and glucagon (Elson et al., 2010) increase sensitivity to sweet stimuli. How this relates to appetitive sensations and ingestive behavior is not straightforward.

37.2.1 The Chemical Senses and Appetite

Appetite is a multidimensional sensation, reflecting both metabolic hunger as well as just the desire to eat for sensory gratification (often referred to as desire to eat). While both sensations are related to the motivation to eat, they can vary independently. A good example is the desire to indulge in a palatable dessert after having eaten a satiating meal when metabolic hunger is low.

Evidence that sweeteners stimulate hunger stems from the observation that preloads of high intensity sweeteners in water produce a more rapid rebound in hunger sensations than does water alone (Blundell et al., 1988). However, this response is not sweet-specific as a palatable salty food has the same effect (Mattes, 1994a), nor is it robust as it is not observed with ingestion of energy-yielding sweet foods and beverages (Rolls et al., 1989; Drewnowski et al., 1994). A heightened appetite has been reported after chewing gum sweetened with aspartame, although the effect was not dose related and varied according to sex and time post-exposure (Tordoff and Alleva, 1990a). Further, several other experiments have failed to replicate heightened appetite (Julis and Mattes, 2007; Hetherington

et al., 2008). Neuroimaging studies suggest there are differential responses to taste stimuli during states of hunger and satiety, adding another dimension to the possible correlation between sensory function and hunger (Small et al., 2001; Spetter et al., 2012). The type of sweet stimulus can modify neural processing. Specifically, saccharin and sucrose differentially activate reward areas of the brain, and frequency of saccharin consumption affects this activation, suggesting a learning effect (Green and Murphy, 2012).

Several early studies reported a reliable direct relationship between olfactory thresholds and appetite (Goetzl and Stone, 1947; Goetzl et al., 1950). This has been expanded by more recent evidence indicating hunger can increase olfactory thresholds to neutral stimuli while satiated participants are better able to detect food stimuli (Stafford and Welbeck, 2011). Other work using similar methodology reported parallel results for food-related odors, but no impact of hunger on thresholds for non-food odors (Albrecht et al., 2009).

Textural attributes of food have also been shown to affect satiety. True (Mattes and Rothacker, 2001; Mars et al., 2009) and perceived (Cassady et al., 2012) viscosity are directly related to the expected satiety value of a food or beverage.

That being said, there are also data showing no associations between the different facets of chemosensory function and hunger (Janowitz and Grossman, 1949; Pangborn, 1959; Pasquet et al., 2006). In some work, scaling of glucose, sodium chloride, citric acid, and quinine sulfate was invariant with hunger status (Moskowitz et al., 1976; Scherr and King, 1982; Rolls et al., 1983; Looy and Weingarten, 1991; Laeng et al., 1993). These inconsistent findings are not surprising given the multitude of factors, including sensory system sensitivity and responsiveness, influencing hunger and the limited contribution of each.

37.2.2 The Chemical Senses and Food Hedonics

Food preference is driven by both inherent and acquired impressions of sensory stimuli. Though humans are drawn to certain qualities (e.g., sweet and salty) apparently independent of learning, this does not necessarily result in high levels of consumption of the palatable stimuli. Cultural practices and personal experience typically guide the use of these ingredients. This is reflected in different cuisines. One powerful determinant of the preferred concentration of a sensory stimulus in a food is the frequency of exposure to its quality. This has been documented for salt (Bertino et al., 1986), sugar (Liem and de Graaf, 2004), and fat (Mattes, 1993). However, acquired liking is interactive

37.3 SENSORY INFLUENCES ON MACRONUTRIENT DETECTION AND SELECTION

with inherent proclivities as amelioration of neophobia (i.e., fear or avoidance of novelty) is more rapid for liked compared to unpleasant qualities (Mattes, 1994).

37.2.3 The Chemical Senses and Food Intake

Sensory factors have been implicated in the promotion of positive energy balance. Associative learning linking a food's sensory properties with its post-ingestive consequences is an important mechanism to guide food choice and portion size. This need not be conscious as specific flavor preferences can be conditioned in rodents by pairing flavors with intra-gastric infusions of energy (Sclafani and Glendinning, 2005; Yiin et al., 2005).

Recent work suggests corruption of such learning through inclusion of energy-modified foods in the diet can lead to eating dysregulation. Rodents provided with chow randomly modified in energy content eat more total energy than animals provided food with predictable associations between sensory properties and metabolism (Swithers and Davidson, 2008). Additionally, neuroimaging studies with humans suggest ingestion of products that have sensory properties not consistent with their metabolic consequences blunts neural signaling (Green and Murphy, 2012). However, behavioral data from humans fail to confirm effects on energy intake (Appleton and Blundell, 2007) or suggests such effects may hold for only a subset of the population (Shaffer and Tepper, 1994).

Palatable and unpalatable foods have been implicated in feeding in excess of need. Saltiness is inherently pleasant, like sweetness, as well as nutritionally required and so would be expected to stimulate reward centers and drive intake of foods containing the nutrient. Sodium may also stimulate intake indirectly. For instance, high sodium intake may prompt increased consumption of energy-yielding beverages, and these beverages evoke weak compensatory dietary responses, potentially leading to weight gain (He et al., 2012). Inherently unpleasant sensory properties can promote increased intake by prompting the addition of energy-yielding substances (e.g., sugar, fat) to foods to mask the unpleasant properties of foods consumed for purposes such as health beliefs (spinach) or psychoactive properties (coffee or alcohol).

Sensory variety is another determinant of intake. Variety tends to promote intake (Rolls, 1986; Murphy et al., 2006) and can lead to weight gain (Meiselman et al., 2000; Zandstra et al., 2000; Remick et al., 2009). In contrast, monotony in sensory input tends to suppress consumption over time (Zandstra et al., 2000; Sorensen et al., 2003; Raynor et al., 2005).

37.3 SENSORY INFLUENCES ON MACRONUTRIENT DETECTION AND SELECTION

37.3.1 Carbohydrate

The primary sensation associated with carbohydrates is sweetness, though starch may have its own taste (at least in rats (Sclafani, 2004)). When given orally, the sweetness from sucrose intensifies odors (Green et al., 2012) and suppresses other tastes (Green et al., 2010). Sweetness is inherently pleasing, as shown by increased fetal swallowing rates of amniotic fluid after intrauterine saccharin injection (DeSnoo, 1937; Liley, 1972) contrasted with decreased swallowing after injection of a bitter compound (Liley, 1972). Preterm infants also display stronger and more frequent sucks when exposed to glucose or sucrose (Tatzer et al., 1985; Maone et al., 1990). Full-term newborns can discriminate among different sugars and concentrations of a single sugar (Desor et al., 1975a) and exhibit positive mimetic (e.g. smiling) (Bergamasco and Beraldo, 1990) and autonomic (e.g. increased heart rate) (Crook, 1979) responses following exposure to sweet taste. Additionally, sweetness of sugars enhances the reward value of food (Epstein et al., 2011).

The preferred sweetness intensity is high during childhood and diminishes with age (Desor, 1975a; Desor and Beauchamp, 1987; Ventura and Mennella, 2011). This preference, however, may also be modifiable with increased exposure to sweetness, at least in children (Liem and de Graaf, 2004). The basis for children's liking of more intensely sweet foods is unknown, though several theories have been proposed. Some evidence indicates that children, especially males, may be less sensitive to sweetness in complex stimuli than adults (James et al., 1997; James et al., 1999). This may lead to greater intake to achieve a desired level of reward. However, physiologically reduced sensitivity to sweetness in children is unlikely, as children have also been shown to rate the intensities of sweet products higher than adults (Temple et al., 2002). Another theory for childhood sweet preference is that children have a greater need for energy to support growth (Coldwell et al., 2009). Genetics may also contribute more strongly to sweet liking in children than in adults as adult preferences may be influenced more markedly by culture and cognition (Mennella et al., 2005; Pepino and Mennella, 2005; 2006).

Concern has been expressed that early exposure to sweet foods high in carbohydrates may lead to a preference for such foods and subsequent overconsumption of sugars. Several studies indicate that infants show greater preference for flavors consumed by mothers during pregnancy or breastfeeding, due to exposure to the flavors through amniotic fluid or breast milk (Mennella and Beauchamp, 1993; Mennella et al., 1995; Schaal et al., 2000; Forestell and Mennella, 2007; Beauchamp and Mennella, 2009). Less evidence exists for linking early, overall sweet exposure to adult sweet intake. Studies in rats imply that sweet foods in infancy may lead to increased consumption of such foods and weight gain in adulthood (Frazier et al., 2008; Silveira et al., 2008). Additionally, human infants exposed to sugar sweetened water during the first six months after birth drank more sweetened water at 6 months than infants not previously exposed (Beauchamp and Moran, 1982). However, at two years of age these same children maintained their reactions to sugar water but did not have significantly different responses to sugar sweetened fruit drinks (Beauchamp and Moran, 1984). Consequently, context and familiarity of the food may change acceptability and preference more than sweetness itself.

Numerous studies have attempted to elucidate the relationship among intensity ratings, liking, exposure frequency, and intake of sugary foods. With regard to sweetness exposure, low, medium, and high consumers of sweet foods assigned comparable intensity ratings for prepared samples, indicating no effect of frequency of exposure on ability to assign relative intensity ratings (Pangborn and Giovanni, 1984). Other work reveals significant positive associations between use of sweeteners and peak hedonic ratings for fruit-flavored solutions with graded levels of sucrose, fructose, or aspartame, but only in women with gestational diabetes mellitus. No significant associations were noted in healthy controls (Tepper et al., 1996). More recent work shows that in people who originally had low liking for sucrose, four weeks of sweetened soft drink supplementation increased their preferred concentration of sucrose (Sartor et al., 2011). This change was not evident in the participants who reported higher liking for sucrose initially. Overall, selection of food and preferred sweetness level depends greatly on familiarity. While a penchant for sugary foods can undoubtedly lead to high consumption of sugars, sweetness preference in specific foods is difficult to predict without knowledge of individual exposure to the food product and context.

37.3.2 Protein

Proteins encompass the whole range of taste qualities, though sourness or saltiness of peptides is likely attributable to acidic and charged residues rather than to the peptides themselves (Temussi, 2011). Sensory feedback through tastes and other properties, including odor and texture, may aid in the identification of protein sources. Infants (Vazquez et al., 1982) and elderly people (Murphy and Withee, 1987) who have inadequate or marginal protein intake show higher preferences for protein supplemented

soups, a preference not seen in adequately nourished individuals. Ingestion of an adequate quantity and balance of amino acids is central to survival. Rats demonstrate an ability to maintain this balance by adjusting their food intake and dietary choices to obtain necessary amounts of both individual amino acids and overall protein (Leung and Rogers, 1986; White et al., 2000). Proposed mechanisms to explain this dietary behavior in rats include associative learning, where sensory properties of foods that correct the amino acid deficiency become more hedonically pleasing (Gibson and Booth, 1986; Fromentin et al., 1997; DiBattista and Mercier, 1999), and also direct sensory cues of proteins or amino acids leading to a rapid modification of ingestive behavior (Vazquez et al., 1982; Deutsch et al., 1989). However, familiarity may also be involved in protein selection, even at birth, as the maternal diet of rats during pregnancy influences protein selection of pups (Leprohon and Anderson, 1980; Morris and Anderson, 1986).

Umami, characterized by the taste of monosodium glutamate (MSG), is pleasing to rats but may be either pleasing (in soup), neutral (in water for infants) (Vazquez et al., 1982; Beauchamp and Pearson, 1991; Schwartz et al., 2009) or unpleasant (in water for adults) (Beauchamp et al., 1998). The reason for MSG's augmentation of palatability when mixed with other foods is unknown. Nevertheless, this has led to this compound's widespread use, especially in savory applications. Similar to using fat or sugar, adding MSG to soup can better condition liking for a novel flavor compared to associative learning for the same flavor without MSG (Prescott, 2004).

Inverse associations between thresholds for MSG and inosine 5'-phosphate and preferences for protein have been observed (Luscombe-Marsh et al., 2008). Whether this increased MSG sensitivity is a cause, effect, or confounder of protein preference is unknown. Cultural components of perception of amino acids have also been documented. For example, Japanese, relative to Australians or Americans, have lower threshold sensitivity for and give higher hedonic ratings to MSG and 5' nucleotides (Ishii et al., 1992; Prescott et al., 1992). This raises the possibility that preferences for umami may be modifiable through exposure frequency.

37.3.3 Fat

Triacylglycerols, the predominant form of fats in foods, are perceived primarily through textural cues, but non-esterified fatty acids may also be detected through olfaction, irritation, and gustation (Chale-Rush et al., 2007). Evidence for gustatory detection, in particular, has been building over the past few years (for review, see Mattes, 2011a). Several potential mechanisms for oral fatty acid detection have been proposed, including various fatty acid receptors and passive diffusion (Gilbertson, 1998;

Khan and Besnard, 2009; Cartoni et al., 2010; Gilbertson et al., 2010; Martin et al., 2012). Mice lacking functional oral fatty acid receptors, including CD36 (Laugerette et al., 2005; Sclafani et al., 2007a; Gaillard et al., 2008), GPR120, and GPR40 (Cartoni, 2010) exhibit decreased preference for fat, though this decreased preference can be eliminated by post-oral ingestive feedback (Sclafani, 2007a).

Oral detection thresholds for a range of short to long chain non-esterified fatty acids are not correlated with somatosensory or other taste thresholds. Studies in rodents indicate that an aversion to a non-esterified fatty acid, linoleic acid, does not generalize to compounds widely accepted as eliciting the other basic taste qualities (Pittman, 2010) or to non-nutritive lipids (Smith et al., 2000). Such studies imply the sensory quality of fatty acids is unique. Mixed results have been reported on whether fatty acids can affect intensities for other taste modalities, with some rodent studies showing increased sensitivity to sweet but not salty, bitter, or sour tastes (Pittman et al., 2006), and human trials failing to replicate these findings (Mattes, 2007; Reckmeyer et al., 2010).

Humans do not appear to have an inherent preference for fat. Although fetal drinking is inhibited by injection of iodinated poppy seed oil into amniotic fluid (Liley, 1972), the fat component may or may not be responsible. If the fat is appealing, then it is not sufficiently so to override one or more negative components in this stimulus. Additionally, infants provided breast milk or formula varying in fat content do not differentially consume them (Chan et al., 1979; Woolridge et al., 1980; Nysenbaum and Smart, 1982). One study noted a stronger sucking response to a higher fat formula (Nysenbaum and Smart, 1982), but this has not been consistently replicated (Woolridge et al., 1980; Drewett, 1982). In adults, fat texture is palatable, high fat more so than low fat (Drewnowski, 1998; Grabenhorst et al., 2010), and cephalic phase responses are greater for higher fat concentrations (Crystal and Teff, 2006).

Early exposure to a high-fat diet results in a stable preference for that diet in rats and mice (Warwick and Schiffman, 1990; Teegarden et al., 2009). However, this does not necessarily translate to greater energy intake (Teegarden et al., 2009). In contrast, humans who cannot easily discern differences in fat content may consume more fat in their diet (Liang et al., 2012). Also of note in this study, these same "non-discriminators" of fat content also showed increased use of reduced- and low-fat foods, indicating that this population may be more accepting of lower fat alternatives. Findings are mixed on whether obese compared to lean individuals are more or just equally sensitive to NEFA compared to those that are lean (Stewart et al., 2011; Tucker et al., 2014). Individuals hypersensitive to oral fatty acids report reduced fat intake compared to hyposensitive subjects (Stewart et al., 2010, 2011). Preference for dietary fat may also be altered through experience.

After adherence to a reduced fat diet for 12 weeks, the preferred concentration declines (Mattes, 1993; Grieve and Vander, 2003; Ledikwe et al., 2007). Such hedonic shifts are not attributable to altered perceptual ability (Mattes, 1993). However, promotion of a broad shift may require modification of the total diet rather than selected individual foods.

37.3.4 Ethanol

Ethanol is an irritant and tastes both bitter and sweet. For rats, in utero exposure to ethanol alters responsivity to its odor postnatally (Dominguez et al., 1996; Arias and Chotro, 2005; Youngentob et al., 2009). Similarly, infants born to mothers who frequently consumed alcohol during pregnancy react more strongly to the odor of ethanol than infants born to mothers who infrequently drank (Faas et al., 2000). Breastfeeding infants are also able to detect ethanol in milk, leading to increased sucking (Mennella, 1997) and decreased consumption (Mennella, 2001). The latter may be related to the delayed time to ejection of breast milk (Chien et al., 2009) and/or reduced milk availability (Mennella, 2001) after ethanol ingestion by mothers. The increased sucking could also be due to familiarity with the stimulus from exposure in utero. Since bitter stimuli are typically rejected by newborns (Rosenstein and Oster, 1988), it is unlikely that ethanol is inherently pleasing. However, it is not known how newborns perceive the flavor of the low concentrations of ethanol found in breast milk.

While a connection between ethanol liking and sweet liking has been proposed (Kampov-Polevoy et al., 1997; Kampov-Polevoy et al., 1998; Bachmanov et al., 2003), most of the mechanistic work for this association has been conducted in mice or rats (for review, see Bachmanov, 2003). Because there are marked species differences in the perceived quality of ethanol, the suitability of animal studies for understanding human responsiveness is uncertain. Although other factors are also involved, some species (e.g., C57BL/6ByJ mice, Roman high-avoidance rats) derive a predominantly sweet note and consume high levels when available; whereas, others (e.g., 129/J mice, Roman low-avoidance rats) sense it as bitter and tend to reject it (Stewart et al., 1994; Razafimanalina et al., 1996; Coleman et al., 2011). Human data are available supporting the presence of both sweetness and bitterness of ethanol at and above threshold concentrations (Diamant et al., 1963; Martin and Pangborn, 1970; Wilson et al., 1973; Scinska et al., 2000). Approximately 60% of 50 non-alcoholic adults rated ethanol as predominantly bitter at threshold; whereas, only 6% reported sweetness (Mattes and DiMeglio, 2001). At one dilution step above threshold, a doubling of concentration, the proportions were 72% and 5%. At higher concentrations, ethanol is perceived as bitter and enhances the bitterness of other bitter compounds in foods (Noble, 1994). While some research groups have shown increased sweet liking in male alcoholics (Kampov-Polevoy, 1997; Krahn et al., 2006), others have shown no correlation between alcohol intake (Mattes and DiMeglio, 2001) or alcoholism (Bogucka-Bonikowska et al., 2001; Wronski et al., 2007) and response to sweetness.

Genetics could potentially play a role in the sensory preference for ethanol. Selective breeding studies in rats clearly highlight a genetic component to ethanol taste which results in different behavioral responses to sensory qualities (Kiefer et al., 1995; Bachmanov et al., 1996a, b; Razafimanalina, 1996; Blednov et al., 2008; Brasser et al., 2012). In mice, changes in sweet receptor genes result in reduction of ethanol intake (Blednov, 2008; Brasser et al., 2010). However, other rat research indicates that changes in sweet or bitter susceptibility do not lead to changes in ethanol preference (Wolstenholme et al., 2011).

A genetic component has also been proposed in humans. For example, the genetically determined sensitivity to phenylthiocarbamide (PTC) or 6-n-propylthiouracil (PROP) has been associated with ethanol sensitivity (Bartoshuk, 1993; Duffy et al., 2004; Duffy et al., 2004). Another study showed that in college students, sensitivity to PROP was directly related to perceived bitterness and inversely related to perceive sweetness in alcoholic beverages; decreased bitterness and increased sweetness were then correlated with alcohol intake, implying a possible indirect effect of PROP taster status on alcohol consumption (Lanier et al., 2005). Some evidence indicates there is a higher proportion of PROP non-tasters among the children of alcoholics than non-alcoholics (Pelchat and Danowski, 1992). Additionally, male alcoholics (Wronski, 2007), women (Pepino and Mennella, 2007), and depressed children (Mennella et al., 2010) with a family history of alcoholism have increased preference for high concentrations of sucrose. However, conflicting data regarding PROP status, paternal history, and ethanol preference have also been published (Noble, 1994; Kranzler et al., 1996; Scinska et al., 2001). Consequently, while taste genetics may play a role in alcohol consumption, the importance of this role compared to other factors, such as cultural and pharmacological aspects of ethanol, remains disputed.

Thus, the importance of the sensory properties of ethanol on ingestion by humans has not been established. Among non-alcoholics, dislike of the sensory properties of ethanol is a commonly cited reason for abstinence (Moore and Weiss, 1995). Not liking the taste of alcohol also may contribute to lighter drinking among non-abstaining college students (Huang et al., 2011). On the other hand, taste is also frequently expressed as a motivation for consuming alcoholic beverages (Patrick et al., 2011). Mixed findings have been reported on an association between

taste thresholds for ethanol in model solutions and ingestion of the compound (Mattes, 1994b; Kranzler, 1996). This may reflect a pharmacological or exposure effect since abstinence for three weeks results in a reduction in sensitivity (Settle, 1978). Chronic exposure to ethanol leads to increased acceptance in rats (Kiefer et al., 1994). A pharmacological basis for differences in human intake has been proposed as well (Dewit et al., 1989; Mattes, 1994) but may ultimately relate back to sensory factors if the rewarding properties serve to condition a preference to the chemosensory characteristics of ethanol (Settle, 1978). Chronic alcohol abuse is associated with olfactory impairment (i.e., elevated thresholds, reduced scaling and/or identification performance) as well as reduced gustatory intensity ratings that may persist even after detoxification (Jones et al., 1975, 1978; Ditraglia et al., 1991; Rupp et al., 2003, 2004, 2006; Maurage et al., 2011).

37.3.5 Sodium Chloride

In most studies, human neonates show similar ingestive and facial responses to saline and water, suggesting they are either insensitive or indifferent to the taste of salt (Desor et al., 1975b; Crook, 1978; Beauchamp et al., 1986; Rosenstein and Oster, 1988). However, a few studies indicate rejection of, or mixed responses to, saline within a few days of birth (Beauchamp et al., 1994; Zinner et al., 2002). Nevertheless, there is clear evidence of sensitivity and positive hedonic responsiveness to saline solutions by 4–6 months of age continuing through 2 years of age (Beauchamp, 1986; Schwartz, 2009), and toddlers have been shown to consume more carrots (Beauchamp and Moran, 1984) and green beans (Bouhlal et al., 2011) when salt is added. Despite the delay in overt acceptability, the prevailing view holds that the liking for saltiness is inherent (Beauchamp et al., 1991). In rats and sheep, the developmental lag in salt taste sensitivity coincides with maturation of the gustatory system for this quality (Hill and Mistretta, 1990). A precise understanding of the development of human salt taste receptors is still lacking. However, exposure and learning appear to play a large role in the development of salt preference among adults.

There has been considerable concern regarding the influence of early salt taste exposure on salt preferences throughout the lifecycle. In 6 month old infants, those who had been exposed to starchy table food showed a greater preference for saline (0.34M) than infants not exposed to such foods (Stein et al., 2012). All of the infants in this study were indifferent to saline at 2 months of age, prior to any table food exposure. Additionally, short-term acceptability of foods with varying salt content has been associated with salt exposure in 6 month olds, but this may simply reflect recent consumption (Harris and Booth,

1985). In the longer term, one study reported 3 month olds who ingested formula containing either 4.78 M Na/kJ or 21.5 M Na/kJ for five months did not have differing salt use or preferences at eight years of age (Whitten and Stewart, 1980). Another study reported an association between salt use at 2 and 4 years of age (Yeung et al., 1984), but whether this was indicative of familial salt use or of establishing a specific salt preference is unclear. Context and learning are also important for salt preference as 31-month-old infants reject saline solutions with questionable dietary relevance but prefer salted over unsalted foods (Beauchamp and Moran, 1984).

Marked early sodium depletion is associated with heightened preference for and intake of salt later in life. This has been demonstrated in infants experiencing dietary chloride deficiency (Stein et al., 1996), in preterm neonates treated with diuretics (Leshem et al., 1998), and in infants with vomiting and diarrhea or whose mothers experienced severe vomiting while pregnant (Crystal and Bernstein, 1998; Leshem, 1998; Kochli et al., 2005; Shirazki et al., 2007). The enhanced salt preference in offspring due to severe morning sickness during pregnancy may extend into adulthood (Crystal and Bernstein, 1995).

Dietary experience can also modify the preferred level of salt in foods. As shown in a twin study, salt preference is primarily environmentally based (Wise et al., 2007). Furthermore, salt preference can be altered through changes in sensory exposure. Chronic (e.g., 8–12 weeks) reduction of dietary sodium consumption elicits a hedonic shift toward lower salt content of foods (Bertino et al., 1982; Thaler et al., 1982; Jeffery et al., 1984; Witschi et al., 1985; Blais et al., 1986; Beauchamp et al., 1987; Stamler et al., 1987; Mattes, 1990). Increased exposure leads to preferences for higher salt levels in foods (Bertino et al., 1986). The shift is not attributable to a change of taste sensitivity or taste intensity scaling ability (Bertino et al., 1982; Teow et al., 1984; Mattes, 1990). The shift also occurs when total sodium intake is unchanged but sensory exposure is curtailed (Mattes, 1990) and does not occur under the reverse conditions (Beauchamp et al., 1987). These observations suggest exposure frequency dominates metabolic factors in eliciting hedonic shifts for salt.

37.4 CHEMOSENSORY INFLUENCES ON NUTRIENT UTILIZATION

Sensory, especially chemosensory, stimulation initiates an array of physiological processes, termed first phase, pre-absorptive, or cephalic phase responses. These responses mimic those occurring during food (nutrient) ingestion, digestion, absorption, and metabolism and are neurally mediated (primarily through the parasympathetic nervous system), occurring within seconds to minutes following exposure to foods. These cephalic responses are transient and typically small in magnitude relative to responses occurring during actual food ingestion. Although there are exceptions, there generally is a hierarchy in stimulus efficacy in eliciting these responses: swallowing > mastication > taste > smell > sight, audition, or thought. Further, concurrent stimulation of multiple sensory systems typically elicits a larger response than stimulation of a single system.

The functional role of cephalic phase responses is largely unexplored, but their importance as a signaling system is supported by evidence that their magnitude correlates with post-prandial responses. Thus, the prevailing view is that cephalic phase responses initiate and augment the efficacy of the normal digestive, absorptive, and metabolic processes associated with eating (Jordan, 1969; Nicolaidis, 1977). These responses have been hypothesized to influence food selection (Simon et al., 1986; Steffens et al., 1990) and have been implicated in varied diet-related disorders including insulin resistance and obesity (Ionescu et al., 1988; Konturek et al., 1994), eating disorders (LeGoff et al., 1988), and ulcers (Konturek et al., 1981) and have been the topic of a number of reviews (Brand et al., 1982; Powley and Berthoud, 1985; Mattes, 1997; Zafra et al., 2006; Power and Schulkin, 2008). The effective stimuli and nutritional significance of the better-documented and most recently identified cephalic phases are described in this section.

37.4.1 Cephalic Phase Salivary Response

Study of the effective stimuli for salivary flow and, as a consequence, composition, is limited but a rough ordering can be derived from the literature. Cognitive control over salivation has been documented in humans, though not consistently (Jenkins and Daws, 1966; White, 1978). Visual stimulation elicits a 1.5-fold increment over resting or cognitive stimulation (Hayashi, 1968; Wooley and Wooley, 1973; Pangborn et al., 1979). Salivation significantly increases after smelling and viewing (Aspen et al., 2012) or tasting without swallowing a stimulus (Bond et al., 2009). Relative to viewing and smelling food, mastication leads to about a 5-fold increment in flow (Richardson and Feldman, 1986). Because flow rate influences salivary composition, pH, and buffering capacity, this response can markedly influence sensory perception of stimuli, via dilution and chemical interactions, as well as digestive processes, through enzyme concentrations and activation (Pedersen et al., 2002; Engelen et al., 2003). There are some suggestions that salivation patterns are significantly different between lean and obese individuals. In humans, exposure to the sight and smell of food elicited greater salivary response in overweight adults than their normal-weight counterparts, which subsequently led to a greater desire to eat (Ferriday and Brunstrom, 2011).

Repeated olfactory and gustatory stimulation leads to reduced salivary flow (Epstein et al., 1996) due to adaptation to the sensory stimulus rather than gland fatigue (Wisniewski et al., 1992; Epstein, 1996). Habituation to a taste stimulus through repeated oral exposure is significantly delayed in morbidly obese adults when compared to lean controls (Bond, 2009). Similar phenomena have been reported in overweight and obese children (Aspen et al., 2012). A decrease in salivation after habituation to repeated stimulus exposure can promote satiation (Swithers and Hall, 1994) and result in the termination of an eating episode (Wisniewski, 1992). Failure to habituate among the obese may therefore have implications for energy intake and body weight.

Because both appealing and aversive stimuli can enhance salivary flow, palatability is not a straightforward predictor of response (Pangborn and Berggren, 1973; Klajner et al., 1981; Sahakian et al., 1981; Christensen and Navazesh, 1984). Potent sialagogues include items that are sour, bitter, physically sharp, or difficult to clear from the mouth (Guinard et al., 1998). Experience can modulate the cephalic phase salivary response (Christensen et al., 1984); however, this is not a robust predictor (Pangborn, 1979). Although salivary enzymes contribute to macronutrient digestion, protein, fat, and carbohydrate are not potent sialagogues (Christensen, 1986; Matsuo and Yamamoto, 1989; Myers and Epstein, 1997); rather, the primary influence of these macronutrients may be on salivary composition. For example, sucrose and fructose prompt the release of amylase-rich saliva (Kemmer and Malfertheiner, 1983; 1985).

The nutritional significance of the cephalic phase component of salivary flow has not been well characterized, but saliva plays a number of key roles (Mattes, 2000). It facilitates swallowing, promotes oral health and functionality of taste receptor cells (Cano and Rodriguez-Echandia, 1980; Thesleff et al., 1988; Bodner, 1991; Matsuo et al., 1997), alters the sensory properties and palatability of foods, initiates macronutrient digestion, and may alter gastric and intestinal phases of digestion.

37.4.2 Cephalic Phase Gastrointestinal Motor and Secretory Response

The thought of foods can elicit a weak cephalic phase gastric response (Moor and Motoki, 1979), but progressively higher levels of secretion are observed with olfactory and gustatory stimulation. The taste and smell of foods stimulate the release of gastrin and gastric acid (Feldman and Richardson, 1986) at levels up to a third of those occurring upon food ingestion. Food palatability increases these secretions (Janowitz et al., 1950). Sensory stimulation, including differential effects by the macronutrients, also elicits: (1) release of digestion enzymes (trypsin, lipase, and chymotrypsin are elevated by 35–66, 16–50, and 25–60%, respectively, during modified sham feeding) (Novis et al., 1971), (2) secretion of gut peptides (such as cholecystokinin, somatostatin, neurotensin (Wisen et al., 1992; Konturek, 1994), glucagon-like peptide-1 (GLP-1) (Vahl et al., 2010; Williams, 2010), and ghrelin (Monteleone et al., 2008, 2010; Bello et al., 2010; Massolt et al., 2010) that are involved with intestinal absorption and satiety, (3) stimulation of a fed-like motor pattern in the stomach (Katschinski et al., 1992), and (4) gastric emptying (Rogers et al., 1993; Nagao et al., 1998; Cecil et al., 1999). Studies also suggest oral exposure to dietary fat augments and prolongs the postprandial triacylglycerol concentration (Mattes, 1996), possibly through enhanced fat absorption (Ramirez, 1985) or decreased clearance (Mattes, 2001). These responses presumably optimize food digestion and absorption of nutrients with the potential to contribute to overall health. At the same time, contributions to adverse health outcomes have also been suggested. Given the evidence that postprandial lipemia is an independent risk factor for cardiovascular disease (Bansal et al., 2007), sensory exposure to fats may contribute to this problem. The cephalic phase gastrointestinal response has also been implicated in the etiology or severity of duodenal ulcers (Konturek et al., 1978; Thirlby and Feldman, 1984; Feldman and Richardson, 1986), irritable bowel syndrome, ulcerative colitis, and colonic diverticular disease (Rogers, 1993).

37.4.3 Cephalic Phase Pancreatic Exocrine Response

Combined visual, tactile, and olfactory stimulation elicits a pancreatic exocrine response but not as large as that following gustatory stimulation (Sarles et al., 1968). Protein concentrations are selectively increased (Novis, 1971), and under certain conditions, trypsin secretion may increase four-fold (Anagnostides et al., 1984) while lipase and amylase secretion rates can exceed those obtained with pharmacological agents (Sarles, 1968). Oral detection of dietary fat stimulates pancreatic exocrine secretion (Hiraoka et al., 2003; Laugerette et al., 2005; Gaillard et al., 2008). Studies in dogs indicate the exocrine response is enhanced by palatable (i.e., sweet) relative to less palatable (sour, bitter) stimuli (Ohara et al., 1988). Food acceptability also modulates the response in humans (Sarles, 1968). The sensory-mediated release of an enzyme-rich pancreatic secretion could enhance digestive processes, but its importance is uncertain due to the large reserve capacity for enzyme secretion by this organ (Brannon, 1990).

37.4.4 Cephalic Phase Pancreatic Endocrine Responses

Insulin, pancreatic polypeptide, and glucagon (Teff et al., 1991a; Teff and Engelman, 1996a) are all released following appropriate sensory stimulation (Schwartz et al., 1979; Katschinski, 1992; Glasbrenner et al., 1995).

The relative potency of different sensory stimuli has not been established. Human studies reveal insulin secretion is stimulated by the sight, smell, and sound of cooking food (Rodin, 1978), as well as sweet taste from both nutritive and non-nutritive sweeteners (Just et al., 2008). Recent work documents the efficacy of oral fat exposure to stimulate insulin secretion (Chavez-Jauregui et al., 2010). The role of palatability is uncertain with some work supporting a direct association (Rodin, 1981; Bellisle et al., 1985; Lucas et al., 1987; Teff and Engelman, 1996a) and other work failing to note an effect (Rodin, 1978; Bellisle, 1985; Teff and Engelman, 1996a). Although effects remain to be confirmed, individual characteristics may modulate responses as greater insulin release occurs in restrained (i.e., individuals who cognitively controlling food intake) compared to non-restrained, females (Teff and Engelman, 1996a) and African Americans relative to Caucasians (Saad et al., 1990). Across studies, 10–15% of subjects exhibit no response to sensory stimulation, which suggests there may be a subgroup of non-responders, but this has not been established. Whether and to what degree oral stimulation promotes insulin secretion through neural activation or secondarily via promotion of GLP-1 secretion is uncertain. GLP-1 is the most potent incretin hormone and oral stimulation promotes release of GLP-1 directly from enteroendocrine cells on the tongue and in the intestinal tract (Jang et al., 2007).

Cephalic phase glucagon release has long been reported in animals (Fischer et al., 1976; Nilsson and Uvnas-Wallensten, 1977), but only more recently in humans (Secchi et al., 1995). However, this response was not observed in other animal (Berthoud and Jeanrenaud, 1982) or human studies (Teff et al., 1991b; Abdallah et al., 1997). Insulin release suppresses glucagon secretion (Meier et al., 2006), and this may explain why cephalic phase glucagon release was not observed in these studies. Cephalic stimulation of both insulin and glucagon secretions may appear counter-intuitive, but glucagon release helps to maintain appropriate blood glucose concentrations (Teff, 2000).

The effective stimulus for pancreatic polypeptide release is controversial. It is often measured as an index of neural activation, and its release is viewed as independent of the nutrient composition (Katschinski, 2000; Robertson et al., 2002; Heath et al., 2004) or physical form (Teff, 2010) of a stimulus. However, recent modified sham feeding studies in humans demonstrate that cephalic phase pancreatic polypeptide secretion is stimulated only by mixed foods that are high in fat (Crystal and Teff, 2006) or with sweet and salty (Teff, 2010) taste qualities.

Accumulating evidence supports a functional role of the cephalic phase insulin response. Glucose tolerance is improved when sham feeding accompanies a glucose infusion (Lorentzen et al., 1987; Teff and Engelman, 1996b; Ahren and Holst, 2001). In contrast, when cephalic phase insulin release is inhibited by somatostatin, hyperglycemia and hyperinsulinemia result (Calles-Escandon and Robbins, 1987). Concurrent administration of insulin and meals leads to a two-fold decrease in incremental blood glucose among insulin-dependent diabetics (Kraegen et al., 1981) and a 33% lower glycemic response among non-insulin-dependent diabetics (Bruce et al., 1988) as compared to responses when sensory stimulation is delayed for 30 minutes after meal ingestion. A role for circulating glucose and/or insulin in appetitive sensations has been proposed (Smith and Campfield, 1993), but euglycemic clamp studies indicate independent manipulation of glucose and insulin is not associated with changes in hunger or fullness (Chapman et al., 1998). Insulin, pancreatic polypeptide, and glucagon also influence aspects of the gastric and intestinal phases of digestion as well as post-absorptive energy metabolism.

37.4.5 Cephalic Phase Thermogenic Response

Exposure to the sight and smell of food leads to a thermogenic response in dogs that is comparable to the level observed following food ingestion (Diamond and LeBlanc, 1985). Work in humans indicates the response is positively related to food palatability (LeBlanc and Brondel, 1985; LeBlanc and Cabanac, 1989; Westrate et al., 1990; Soucy and LeBlanc, 1999), especially in lean individuals (Hashkes et al., 1997). Sham-feeding of a high-fat meal leads to increased energy expenditure comparable to the values observed when the same meal is actually ingested (Smeets et al., 2009). Intragastric feeding is not associated with a rise in thermogenesis, and blocking the cephalic component with somatostatin or atropine infusion (Calles-Escandon and Robbins, 1987) diminishes the postprandial increment in heat production. The thermic effect of food represents about 8–10% of total energy expenditure but can be greater with imbalanced diets (Stock, 1999). Consequently, this response may contribute to energy balance and body weight.

37.4.6 Cephalic Phase Cardiovascular Response

Cardiac output and stroke volume transiently decline by 11–13% and heart rate increases by 5–10% in response

840 **Chapter 37 Nutritional Implications of Taste and Smell Dysfunction**

to the sight and smell of food (Andersen et al., 1992; Fagius and Berne, 1994). Exposure to food also increases systolic and diastolic blood pressure (Nederkoorn et al., 2000) as well as microcirculation (e.g., resting and peak functional capillary density, red blood cell velocity, flow, vasomotility; Buss et al., 2012). These responses are associated with an increase in pancreatic polypeptide secretion (Buss, 2012) suggesting they are neurally mediated. A 31% increase in mesenteric resistance has been noted in dogs viewing and smelling food (Vatner et al., 1970). While nutrient absorption from the gastrointestinal tract is very efficient, it is directly related to blood flow (Winne, 1979). Consequently, the cephalic phase hemodynamic effects may influence the kinetics of nutrient absorption. This, in turn, may alter endocrine responses (e.g., insulin and vasopressin release are largely determined by the rate of glucose and ethanol absorption, respectively).

37.4.7 Cephalic Phase Renal Response

Oral exposure to graded saline solutions alters the acute diuretic and natriuretic response in a stepwise fashion (Akaishi et al., 1991). Urine osmolality decreases or increases by 10–20% with exposure to hypotonic and hypertonic solutions, respectively. No response occurs following isotonic saline exposure. Comparable responses are observed with sodium chloride and mannitol challenges, suggesting the salient stimulus attribute is osmolality (Gebruers et al., 1985). However, palatability also alters the response as decreased diuresis is observed with ingestion of a less palatable beverage (Nagao et al., 1999). The nutritional significance of this response is not clear, but it may contribute to fluid and electrolyte balance.

37.5 TASTE RECEPTORS IN THE ORAL CAVITY AND ELSEWHERE

Taste receptors were long believed to reside solely on taste receptor cells in the orpharyngeal region. However, they have now been identified throughout the body and serve functionally diverse purposes. In the gastrointestinal (GI) tract, enteroendocrine cells express receptors and intracellular signaling proteins resembling those in taste receptor cells on the tongue, but when activated, they lead to responses relevant for the intestinal phase of digestion rather than the pre-ingestive phase. Thus, questions are now being asked about how food-related chemosensory signals in the gastrointestinal lumen influence the digestive and absorptive process, appetite, and toxin disposal. Emerging knowledge of chemosensory signaling by food constituents throughout the GI tract is highlighted below.

37.5.1 Sweet "Taste" Receptors

T1R2, T1R3, α-gustducin, and TRPM5 are co-localized in gut enteroendocrine cells in the human gastrointestinal tract spanning from the esophagus to the colon (Steinert et al., 2011b; Widmayer et al., 2011). However, TRPM5-positive cells are more characteristic of tuft cells than enteroendocrine cells; hence, it is still unclear which cell types contain taste receptors and if all components of the transduction pathway are located in the same cell (Kokrashvili et al., 2009b; Tolhurst et al., 2012). T1R3 and T1R2 are also expressed in the intrahepatic bile duct, interlobular septa, and liver lobule (Taniguchi, 2004; Toyono et al., 2007). T1R2 and T1R3 are documented in cultured human and murine pancreas, and administration of a sweet receptor antagonist, such as lactulose, abolishes fructose and glucose stimulated insulin release (Kyriazis et al., 2012).

37.5.2 Fat "Taste" Receptors

G-protein coupled receptor 120 (GPR120) and GPR40 bind fatty acids and have been localized to enteroendocrine cells (Covington et al., 2006; Vinolo et al., 2012). In the murine STC-1 enteroendocrine cell line, unsaturated long chain fatty acids stimulate GPR120 and elicit GLP-1 secretion (Hirasawa et al., 2005). Recent studies also indicate that GPR120 expression is significantly higher in obese individuals than in lean controls (Widmayer, 2011; Ichimura et al., 2012), suggesting a role of GPR120 in lipid/energy balance. In vitro, linolenic acid increases cholecystokinin (CCK) secretion through GPR40 while GPR40 knock-outs do not display this effect (Liou et al., 2011). In humans, lauric acid (12 carbon fatty acid), but not capric acid (10 carbon fatty acid), significantly increased plasma CCK, suppressed appetite, modulated antropyloroduodenal motility (stimulated basal pyloric pressure waves and suppressed antral and duodenal pressure waves), and decreased energy intake (capric: 4,109 ± 588 kJ, and lauric: 1,747 ± 632 kJ) (Feltrin et al., 2004). However, five out of eight subjects experienced nausea with lauric acid exposure but not with capric acid exposure, suggesting the differential response may be non-physiological. Fatty acid translocase cluster determinant 36 (CD36) is a long chain fatty acid scavenger that mediates fatty acid sensing in the gustatory papillae in rodents and humans (Laugerette, 2005; Simons et al., 2011). In the human gastrointestinal tract, CD36 plays a functional role in the small intestine and the liver (Lobo et al., 2001; Miquilena-Colina et al., 2011). In the murine intestine, CD36 mediates the uptake of very long chain fatty acids (Drover et al., 2008). CD36 knockout mice have reduced chylomicron formation (Nauli et al., 2006), reduced fatty acid and cholesterol uptake (Chen et al., 2001), and decreased preference for and intake of fat (Sclafani et al., 2007a), but whether these outcomes are specifically related to the absence of CD36 in the gut remains to be determined.

37.5.3 Amino Acid "Taste" Receptors

T1R1/T1R3 receptors are expressed in enteroendocrine cells (Bezencon et al., 2007) in the murine and human intestinal tracts. These receptors respond to aromatic amino acids and umami- L-glutamate (Li et al., 2002; Nelson et al., 2002). Despite these findings, the physiological relevance of T1R1/T1R3 in the gut remains uncertain. One functional response of the T1R1 receptor is regulation of peptide transporter 1 (PepT1) and glucose transporter 2 (GLUT2) in the rat small intestine, suggesting a role in regulating glucose absorption (Mace et al., 2009). The wide range of protein degradation products might contribute to numerous amino acid physiologic sensing abilities (Tolhurst, 2012). Although certain amino acids induce CCK release through GPR93 (Choi et al., 2007), more recent studies suggest that CCK release could possibly be stimulated through sodium-coupled neutral amino acid transporter 2 (SNAT2) (Young et al., 2010). Another study noted L-phenylalanine and L-tryptophan induce CCK release through activation of the calcium sensing receptor (CaSR) (Wang et al., 2010). The amino acid profile most effective for SNAT matches the amino acid specificity in STC-1 cells but does not match amino acid selectivity found earlier in T1R1/T1R3 STC-1 cells (Nelson, 2002).

37.5.4 Bitter "Taste" Receptors

T2R receptors and their transduction proteins (e.g., gustducin) are co-expressed in STC-1 cells (Wu et al., 2002; Rozengurt, 2006). Bitter compounds such as denatonium benzoate (DB), PTC, PROP, and cyclohexamide (CYX) elicit rapid increases in the intracellular Ca^{2+} concentration, suggesting T2R's role in mediating sensing of bitter compounds in these cells (Wu, 2002). T2R and its signaling transduction cascades have also been proposed to mediate CCK release in response to the bitter stimulants DB, PTC and CYX (Chen et al., 2006). However, as mentioned above, more recent studies suggest the role of CaSR and SNAT2 as potential stimulators of CCK release.

37.6 IS THERE SENSORY-DIETARY SYNERGY?

Appetite and food intake are guided by multiple environmental factors that overlay highly conserved endogenous systems. The gut is not a passive player. The frequent finding that co-localization of taste transduction signaling molecules and gut hormones occurs in human enteroendocrine cells raises the possibility that effective taste stimuli also activate processes in the gastrointestinal lumen that modify the gastric and intestinal phases of digestion (Rozengurt et al., 2006; Jang et al., 2007; Hass et al., 2010; Gerspach et al., 2011; Steinert, 2011). However, evidence for the nutritional implications of "taste" transduction signaling pathways post-deglutition are not straightforward.

37.6.1 Gut "Taste" Receptors, Flavor Preference, and Appetite

In contrast to the known function of the T1R2 and T1R3 heterodimers as sweet receptors on the tongue, rodent studies suggest that in the gut they do not mediate post-oral sweet preference conditioning. Unlike sucrose, infusion of sucralose into the gut does not condition a flavor preference. Furthermore, T1R3 knockout mice exhibit similar flavor conditioning responses to sucrose infusion as normal mice (Sclafani et al., 2010).

Modulation of appetitive sensations with nutritive and non-nutritive sweeteners also varies. Earlier human studies indicate that acute exposure to non-nutritive sweeteners in mediums of little or no energy increases hunger (Rogers et al., 1988; Tordoff and Alleva, 1990b). In rats, the ingestion of non-nutritive sweeteners may also lead to increased food intake and body weight (Swithers et al., 2010). However, human evidence fails to support modification of appetitive sensations with non-nutritive sweetener use (Steinert et al., 2011a).

37.6.2 Gut "Taste" Receptors, and Gastric Emptying

A primary role for enteroendocrine cells in the GI tract is to secrete peptides that regulate the passage of ingesta through the organ. Receptors on these cells that mimic those on taste receptor cells in the oral cavity bind bitter and sweet food chemicals that elicit the release of multiple peptides (e.g., GLP-1, CCK, PYY) that slow gastric emptying (Chen, 2006; Glendinning et al., 2008). The finding that a sweet stimulus, that reportedly signals the presence of a safe and available energy source, as well as a bitter stimulus, that theoretically signals toxicity, result in the same physiological response challenges an ecological interpretation of the phenomenon. Additionally, there may be species differences. Intragastric administration of bitter taste compounds such as PTC and DB delays gastric emptying (Janssen et al., 2011) in mice whereas similar administration of quinine and naringin does not modulate gastric emptying relative to saline in humans (Little et al., 2009).

A primary role for sensory versus nutritive signaling in gastric emptying is uncertain. Intragastric infusion of a nutritive sweetener (e.g., sucrose) in humans delays

842   **Chapter 37   Nutritional Implications of Taste and Smell Dysfunction**

emptying, but no effect is observed with a non-nutritive sweetener (e.g., sucralose) (Ma et al., 2009). Similarly, the addition of saccharin to an intragastric infusion of glucose did not alter gastric emptying compared to the glucose solution alone (Little, 2009).

37.6.3 Gut Peptides

Ghrelin, an orexigenic gut peptide, increases hunger and promotes energy intake when intravenously administered in humans (Wren et al., 2001). A greater number of ghrelin-positive cells among obese individuals has been reported, with speculation that there is a relationship between energy intake and the number of candidate chemosensory cells in the gastric mucosa (Widmayer, 2011). However, plasma ghrelin concentration peaks have also been observed in anticipation of eating a meal in people with habitual meal patterns rather than the ghrelin spikes eliciting feeding (Frecka and Mattes, 2008). Significantly greater changes in plasma ghrelin levels were also observed when food related pictures were shown to subjects with regular eating patterns (Schussler et al., 2012). Other work noted that ghrelin concentrations lagged behind hunger sensations (Lemmens et al., 2011), suggesting that concentrations of this appetitive hormone are also regulated through non-taste receptor signaling pathways.

Certain bitter taste receptor agonists such as DB and PTC increase active ghrelin (octanonyl ghrelin) secretion. However, relative mRNA expression of ghrelin remained unchanged with and without the T2R agonists (Janssen, 2011). Furthermore, α-gustducin knockout mice did not express lower concentrations of active ghrelin in response to bitter gavage relative to wild-type counterparts, questioning the relevance of T2R's role in ghrelin regulation (Janssen, 2011).

Cholecystokinin (CCK), a gut hormone that regulates appetite by reducing meal size and delaying gastric emptying, is particularly responsive to bitter taste compounds, long chain fatty acids, and certain amino acids (Dockray, 2012). CCK may also be released in response to sweet compounds. Infusion of the CCK1a antagonist, dexloxiglumide, inhibits gastric emptying, suggesting a role for CCK in carbohydrate regulation in humans (Little et al., 2010).

GLP-1 is a gut peptide that rises in anticipation of a meal and in response to nutrient intake, exerting both an anorectic and incretin effect (Baggio and Drucker, 2007). One study reported that human duodenal L-cells respond to glucose in the gut lumen by secreting GLP-1 (Jang, 2007). This study also reported dysregulation of plasma insulin and glucose in humans with compromised GLP-1 secretion. In addition, gastric gavage with glucose, sucrose, 2-deoxy-glucose (non-metabolized non-sweet sugar), and sucralose of mice elicits concentration-dependent release of GLP-1. However, conflicting data are also reported.

Activation of orosensory systems with sucralose, Stevia, Saccharin, D-tryptophan and Acesulfame K in Zucker diabetic fatty rats had no effect on GLP-1 and gastric inhibitory peptide −1 (GIP-1) concentrations (Fujita et al., 2009). Further, studies in humans have generally failed to confirm sensory influences on GLP-1 secretion or appetite (Ma, 2009; Ma et al., 2010; Wu et al., 2012). Exceptions are two trials showing oral exposure to a non-nutritive sweet beverage prior to an oral glucose tolerance test augments the GLP-1 response. However, there was no concomitant change of plasma c-peptide or insulin (Brown et al., 2009, 2012). The veracity of this finding has not been established.

37.6.4 Gut "Taste" Receptors and Glucose Homeostasis

In rodents, high intensity sweetener (i.e., sucralose, acesulfame K and, to a lesser extent, saccharin) ingestion at concentrations as low as 1 mM increases glucose transporter insertion in the apical membrane of enterocytes (Mace et al., 2007). Results from this and another study confirm that intestinal glucose transporters GLUT2, GLUT5, and SGLT1 were mostly expressed in cells with ultrastructural characteristics of taste receptor cells, and that concentrations of T1R1, T1R2, T1R3, α-gustducin and PLA-β2 changed significantly with exposure to sucrose and sucralose (Mace, 2007; Merigo et al., 2011). High levels of apical GLUT2 are a characteristic of experimental insulin resistance states (Tobin et al., 2008). The absence of apical GLUT2 in lean participants but presence in 76% of obese participants suggests that GLUT2 accumulation might be an adaptation to metabolic pathology (Ait-Omar et al., 2011). Based on the premise that sucralose increases GLUT2 in the apical membrane, ingestion of sucralose should increase blood glucose concentrations. However, infusions of sucralose into the stomach or proximal small intestine of healthy adults result in no change of plasma glucose (Ma et al., 2009, 2010). This may reflect the very high efficiency of glucose absorption, so there is little variance to no treatment effects.

Taken together, the emerging evidence on chemodetection in the GI tract suggests the potential for many nutritionally relevant sensory-dietary interactions. Presently, there is no clear picture of the interface, but this will likely change with improved testing methods.

37.7 EFFECTS OF PRIMARY CHEMOSENSORY DISORDERS ON FOOD ACCEPTABILITY, DIET, AND NUTRITIONAL STATUS

Data on the dietary and nutritional implications of specific primary chemosensory disorders are available from

37.7 Effects Of Primary Chemosensory Disorders On Food Acceptability, Diet, And Nutritional Status

several chemosensory clinical research centers. However, the number of patients in specific diagnostic groupings is often limited and evaluation and assessment measures differ among centers. In addition, patient characteristics may vary among clinics and account for apparent discrepancies in reported findings. Thus, while the following summary represents the current state of knowledge on this issue, the findings must be interpreted cautiously. It must also be noted that the numerous published reports of chemosensory abnormalities associated with specific pathologies are not reviewed here since many of the dietary and nutritional complications observed in these patients may be attributable to aspects of their pathology and not directly to their chemosensory complaint.

37.7.1 Anosmia

A total loss of the ability to smell is one of the most common complaints of patients presenting for treatment to chemosensory clinical research centers (Ferris et al., 1986; Goodspeed et al., 1987; Deems et al., 1991; Mattes and Cowart, 1994; Chapters 17 and 39). Given the importance of olfactory cues from food for judgments of food safety and palatability, anosmics would be predicted to be at nutritional risk. However, the evidence is mixed. There are reports that such patients do not markedly alter their dietary practices (Aschenbrenner et al., 2008; Schubert et al., 2012), while other reports note that between 10 and 40% of patients with non-congenital anosmia increased their use of sugar and seasonings after onset of olfactory dysfunction (Ferris et al., 1985; Mattes and Cowart, 1994; Aschenbrenner, 2008). While this dietary adjustment may enhance the appeal of foods and eating for some, it also poses a potential problem for individuals who must control their intake of sugar (e.g., diabetics) and sodium (e.g., patients with salt-sensitive hypertension). The prevalence of food complaints is inversely related to the duration of the sensory loss and is lowest in congenital anosmics whose responses resemble control subjects (Ferris et al., 1985). The food preference ratings of anosmics are only slightly lower than those of healthy controls. Noted differences varied by food classes, with the largest discrepancy occurring for homogenous items (e.g., puddings, sherbets) (Doty, 1977). Mean energy and macronutrient intake, as estimated by diet records, are similar in patients with anosmia and healthy controls (Ferris et al., 1985; Mattes and Cowart, 1994); however, a small sample of anosmics were less likely to be overweight compared to the healthy population (Robert-Koch-Institut, 2006). A subset of patients experiencing olfactory disorders, including anosmia, are at risk for either weight gain or loss. One study reported that approximately 14% of anosmic patients experience a gain of body weight exceeding 10% of their pre-disorder weight and about 6.5% lose at least this amount (Mattes

and Cowart, 1994). A weight change of this magnitude would not be expected in a healthy non-dieting individual.

37.7.2 Hyposmia

Most studies examining the effects of olfactory dysfunction on nutritional status and behaviors fail to distinguish between anosmia and hyposmia when reporting results. The few studies that do separate the various types of olfactory dysfunction estimate the prevalence of food complaints in hyposmic patients to range from 31% to almost 80% (Ferris et al., 1986; Mattes and Cowart, 1994). This wide range in prevalence estimates may be attributable to variations in sample sizes, differences in diagnostic criteria, or patient characteristics. For example, a diagnosis of hyposmia may include patients with small reductions in sensitivity to a limited number of odors as well as individuals who are nearly anosmic (Schubert et al., 2012).

Findings from several studies indicate that more than half of hyposmic patients express food complaints (Ferris et al., 1985; Mattes and Cowart, 1994; Miwa et al., 2001; Nordin et al., 2011). Logically, one might engage in dietary modification (e.g., increasing flavor by adding sweeteners or spices) in the attempt to garner more enjoyment from foods. Evidence of these practices is inconsistent. A recent trial noted no association between olfactory dysfunction and the frequency of adding salt or sugar to foods (Schubert et al., 2012), and another reported no differences in dietary modification practices between normosmic and hyposmic individuals (Aschenbrenner, 2008). In contrast, other work indicates that 20 to 50% of patients with hyposmia report altering their eating patterns and using seasonings. The prevalence of food complaints is inversely related to the duration of the sensory loss (Ferris et al., 1985). Decreased appetite is reported by 10–20% of such patients and a small proportion experience an increase. These patients apparently learn to compensate for their dysfunction by accentuating other aspects (i.e., taste, temperature, texture) of foods to enhance palatability (Davidson et al., 1987).

A decrease in appetite was reported by 56% of patients experiencing olfactory dysfunction (Temmel et al., 2002). However, how this translates into changes in intake and, ultimately, body weight is variable. In one report, over 85% of such patients maintained stable body weight subsequent to the onset of their disorder and indices of nutritional status were comparable between hyposmic patients and healthy controls. Among hyposmic patients who experienced excursions exceeding 10% of their pre-disorder weight, there was a tendency towards weight loss, with 10.6% having a loss of this magnitude or greater and 1.5% experiencing an increase. Another report found 21% of patients exhibiting olfactory dysfunction self-reported an average weight gain of 2.5 kg while 11% of patients reported weight loss averaging 5.1 kg; patients with weight gain had significantly higher odor identification scores, were more likely to engage in dietary alterations (e.g., use more or less spices), and were younger than those who lost weight (Aschenbrenner et al., 2008). One study found that individuals who experienced marked changes of body weight did not differ by age, gender, or etiology (Mattes and Cowart, 1994). The influence of duration of the disorder on weight changes is variable, with findings of no impact (Mattes and Cowart, 1994) or a positive correlation (Aschenbrenner et al., 2008). At present, there is no predictive index of nutritional risk for patients with recent onset of symptoms.

37.7.3 Dysosmia

In one study, 24% of patients diagnosed as parosmic, a form of dysosmia, reported a disorder-related decrease of appetite and 83% indicated that they enjoyed food less (Mattes and Cowart, 1994). Approximately 60% stated that they had altered their eating patterns and use of seasonings. The majority of patients with dysosmia maintain adequate nutritional status, but individual responses may warrant aggressive dietary intervention. At present there are no means to identify patients at high nutritional risk.

37.7.4 Ageusia

Complete loss of taste is rare, accounting for less than 1% of patients reporting to chemosensory clinical research centers (Ferris, 1986; Goodspeed, 1987; Deems et al., 1991; Mattes and Cowart, 1994). With available data, it is impossible to draw any general conclusions about the effect of this sensory disorder on diet and nutritional status. Loss of appetite has been observed (Ferris et al., 1986), but increased and decreased body weight has also been reported.

37.7.5 Hypogeusia

The incidence of this taste disorder, independent of other chemosensory abnormalities, is low. Some, but not all, report decreased enjoyment of food and decreased appetite (Ferris et al., 1986). When present in combination with hyposmia or anosmia, hypogeusia leads to only a slightly higher incidence of food complaints relative to patients with hyposmia or anosmia alone (Ferris et al., 1986).

37.7.6 Dysgeusia

The only data on patients with a single diagnosis of dysgeusia indicate the incidence of decreased appetite is approximately 30% and decreased enjoyment of food is about 70% (Mattes and Cowart, 1994). Roughly 60% of such patients alter their eating patterns and 40% modify their use of seasonings. Based upon reports from patients who may have had concurrent diminutions of sensory function, the foods most distorted in taste are meats, fresh fruits, coffee, eggs and carbonated beverages (Markley et al., 1983). However, this is highly idiosyncratic and affected items are not necessarily consumed in lesser amounts. A significant decline in fruit and vegetable intake has been observed, especially for individuals with the most severe dysgeusic symptoms. Mean nutrient and energy intake among dysgeusics with no additional sensory disorders were within normal ranges. These intakes were comparable to control subjects in one report (Mattes and Cowart, 1994) and reduced in another (Mattes-Kulig and Henkin, 1985). In neither of these studies was energy intake related to the duration of symptoms. Approximately 15–20% of dysgeusic patients experience a clinically significant change of body weight and/or body composition (Mattes-Kulig and Henkin, 1985; Mattes and Cowart, 1994).

In summary, a high proportion of patients with a primary complaint of a taste or smell disorder experience a loss of appetite and decreased enjoyment of food. Nutrient and energy intake is generally maintained at adequate levels, but individual patients with any of the above diagnoses may experience a marked increase or decrease of body weight. Increases may occur if patients are attempting to achieve a missed level of sensory stimulation or to mask an unpleasant sensation. A decline in intake often occurs if foods contribute to an unpleasant sensation or because of frustration with the lack of appeal of foods. Questioning patients about their compensatory dietary response may provide the best prognostic index of dietary change. Patients reporting increased or decreased intake or more dietary alterations are disproportionately represented in the sub-group of individuals experiencing shifts of body weight that could result in deleterious health consequences. In a study of 396 patients (Mattes and Cowart, 1994), 34 individuals (8.7% of the sample) reported a compensatory increase in intake and represented 35.3% of the group that gained more than 10% of their pre-disorder weight. The 47 patients (12.0% of the sample) who reported decreased intake constituted 38.3% of the sub-group losing this amount of weight.

37.8 NUTRITIONAL IMPLICATIONS OF CHEMOSENSORY ABNORMALITIES IN PATIENTS WITH SELECTED CHRONIC DISORDERS

37.8.1 Cancer

There is a high incidence of chemosensory complaints among patients with untreated cancers with 86% of cancer patients ($n=66$, excluding nasal, oral, and esophageal cancers) receiving palliative care reporting some sort of

37.8 Nutritional Implications Of Chemosensory Abnormalities In Patients With Selected Chronic Disorders   845

chemosensory abnormality (Hutton et al., 2007). This is especially true for those with more advanced cancer (DeWys, 1974). Such symptoms are not a reliable prognostic index (DeWys, 1974; Walsh et al., 1982) in patients receiving treatment. However, some work suggests tumor stage corresponds with taste alterations between 6–12 months after successful radiation treatment of head and neck cancer patients (Epstein et al., 1999). Severity of chemosensory symptoms in hospice patients has been associated with time to death (Hutton, 2007). The nature of reported complaints is highly variable, even among patients with similar pathologies, and the causes are not well understood (Hovan et al., 2010). Reports range from enhanced or diminished sensitivity to specific taste qualities or odors to generalized alterations (Trant et al., 1982; Hutton, 2007; Sanz Ortiz et al., 2008; Boer et al., 2010). The relationship between chemosensory complaints and diminished appetite, altered food preferences, or loss of body weight in untreated cancer patients is variable, with some finding no association in patients undergoing active treatment (Carson and Gormican, 1977; Trant, 1982; Bruera et al., 1984) and others suggesting a negative relationship between the presence, number, and severity of chemosensory symptoms and intake in palliative-care patients (Hutton et al., 2007; Bovio et al., 2009). Patients who reported either dysgeusia or hypogeusia consumed about 300 kcal/day less than patients without the complaint (Bovio, 2009). Patients with the most severe complaints consumed between 900–1,100 kcal less per day than patients with no complaints, were more likely to report decreased appetite, and consumed more energy from carbohydrates, less from fat and a comparable amount, of protein (Hutton et al., 2007). Weight loss in the past six months was greater among those reporting a chemosensory problem than those without a chemosensory complaint.

External radiation therapy involving the oropharyngeal region diminishes the number of taste buds (Yamashita et al., 2006b), induces changes to receptors (Goldberg et al., 2005), and damages supporting tissues (e.g., salivary glands) (Cooper, 1968; Conger and Wells, 1969; Mossman, 1986; Ophir et al., 1988). Such changes can lead to dysgeusia and/or hypogeusia. Taste detection thresholds in oral cancer patients did not differ from healthy controls prior to radiation treatment; however, taste sensitivity was reduced relative to baseline when tested one month after radiation treatment and had recovered to baseline levels at six and twelve months after treatment for all four tastes (Sandow et al., 2006). Others report reduced taste function may persist for years after radiation therapy (Mossman et al., 1982). Advances in treatment may reduce these side effects. When partial loss occurs, salt and bitter are the most severely affected qualities. Taste impairment is generally noted during the second week of treatment (Mossman et al., 1994) but can start within two to three days of treatment initiation (Sandow et al., 2006). Taste impairment is noted among patients with the greatest amount of tongue exposed to radiation (Yamashita et al., 2006a). Because the recovery rates for different aspects of gustatory function vary and individual awareness of these disorders differs, it is difficult to counsel patients regarding the expected duration of symptoms. Typically, the ability to detect or recognize low concentrations of a taste quality recovers within several months whereas suprathreshold sensations may require a year or more to normalize (Mossman et al., 1982; Epstein, 1999; Epstein et al., 2001) or, in some cases, fail to return to pre-treatment levels (Ripamonti et al., 1998; Epstein et al., 1999, 2001). Olfactory loss can occur if olfactory receptors are in the field of radiation and may be persistent (Ophir et al., 1988), but increased sensitivity to unpleasant odors is more frequently self-reported (for a review, see (Hong et al., 2009). Reduced food intake and loss of body weight are more common in patients experiencing radiotherapy-related chemosensory disturbances than in those with alterations attributable to the pathology alone (Mossman and Henkin, 1978; Bolze et al., 1982).

Chemotherapy regimens may also produce gustatory disturbances (Carson and Gormican, 1977; Mulder et al., 1983), with nearly 70% of patients reporting taste alterations during at least one point during treatment (Wickham et al., 1999; Ravasco et al., 2005; Zabernigg et al., 2010). Resolution of acute dysgeusia, typically bitter or metallic sensations, associated with chemotherapy treatment may occur between several hours of administration to three weeks (Wickham et al., 1999). Dysgeusia has also been linked to reduced energy intake and a higher incidence of food aversions, especially for meats (Markley et al., 1983). However, it cannot be assumed that this is primarily attributable to the chemosensory disturbance. Other complications of the disease or treatment (e.g., oral lesions, difficulty swallowing) may contribute to a reduction of appetite. Some efficacy of zinc sulfate administration in managing treatment-related taste disorders in patients with low zinc levels has been reported. However, changes in taste were not associated with changes of body weight (Ripamonti, 1998), and it must be emphasized benefits may be restricted to those with compromised zinc status. Current recommendations do not support the use of zinc gluconate to prevent chemotherapy- or radiation-related dysguesia (Hovan, 2010).

37.8.2   Hypertension

Various aspects of salt taste have been evaluated in selected sub-groups of hypertensive patients. Detection thresholds are similar in hypertensive and normotensive individuals. In the subset of studies noting a discrepancy between patients and controls for recognition thresholds, values

846 **Chapter 37 Nutritional Implications of Taste and Smell Dysfunction**

were higher for the former group (Mattes, 1984; Isezuo et al., 2008), with one study finding differences only in women (Michikawa et al., 2009). Hypertensive patients have normal or reduced salivary sodium levels (Niedermeier et al., 1956). The latter should enhance sensitivity to external stimulation by salt. No explanation for the conflicting evidence is apparent. Comparable intensity ratings for suprathreshold salt stimuli have been obtained from healthy controls and: (1) unselected samples of hypertensive patients (Mattes et al., 1983; Barylko-Pikielna et al., 1985; Little and Brinner, 1985), (2) patients with low normal plasma renin activity (Bernard et al., 1980), and (3) individuals classified as salt-sensitive (Mattes and Falkner, 1989; Mattes et al., 1999). There is suggestive evidence that hypertensive subjects show larger shifts in salt intensity ratings when on a sodium-restricted diet as compared to controls (Zumkley et al., 1987), but the reliability and implications of this observation have not been determined.

While a heightened preference for salt has been noted in several studies of hypertensive patients, the preponderance of evidence indicates hypertensive and normotensive individuals do not differ on this attribute (Mattes, 1984; Little and Brinner, 1985). Similar hedonic ratings have also been observed between salt-sensitive and non-salt-sensitive subjects (Mattes and Falkner, 1989; Mattes, 1999), although the latter group shows a greater increment in acceptability of reduced salt foods when on a salt restricted diet (Mattes, 1999). There is no evidence of salt cravings in untreated or treated patients with hypertension.

A large number of drugs are known to cause taste disturbances (Tomita and Yoshikawa, 2002), including antihypertensive medications that do not only involve salty taste (Clee and Burrow, 1983; Schiffman, 1983; Coulter, 1988; Doty et al., 2003). Thiazide diuretics reportedly increase salt taste sensitivity (Langford et al., 1977), although other studies have failed to confirm this effect (Mattes et al., 1988; Mattes and Engelman, 1992). Angiotensin converting enzyme (ACE) inhibitors decrease taste sensitivity, with 70% of patients reporting symptoms (Doty, 2003). Zinc deficiency may be a consequence of ACE inhibitors as well as diuretics (Zumkley et al., 1985; Abu-Hamdan et al., 1988; Golik et al., 1998; Trasobares et al., 2007; Kusaba et al., 2009), but findings vary (O'Connor et al., 1987). Different measures of zinc deficiency and dosages of ACE inhibitors may explain the conflicting results. Angiotensin-II receptor blockers decrease detection threshold sensitivity for sweet, sour, bitter, and salty taste qualities (Tsuruoka et al., 2005b) even in healthy volunteers (Tsuruoka et al., 2005a). While taste disturbances generally resolve when the medication is eliminated, this is not always an option for patients.

37.8.3 Hypothyroidism

Case histories indicate that hypothyroidism, hyperthyroidism, or pharmacologic treatment of these conditions can result in alterations of olfactory sensitivity and odor recognition (McConnell et al., 1976; Lewitt et al., 1989; MacKay-Sim, 1991). Olfactory thresholds were elevated for some (e.g., pyridine, nitrobenzene), but not all (2-phenylethanol) test stimuli, suggesting the impairment is not generalized. Thresholds improve following hormone replacement therapy (Deems et al., 1991), but patient complaints about sensation may persist. The mechanism(s) underlying olfactory changes in hypothyroidism have not been determined but may include altered secretions in the nasal cavity, changes of the olfactory epithelium, and/or modified function of the olfactory bulb.

The incidence of taste dysfunction in hypothyroid patients is not known, but estimates range up to 83% (McConnell et al., 1976; Mattes et al., 1986; MacKay-Sim, 1991). Patients with hypothyroidism express little interest in eating, but available evidence is variable in terms of hedonic responses to foods with some reporting no change (Mattes, 1986; Lewitt et al., 1989) while others find decreased hedonic value for sweetness but increased values for salty and bitter (Bhatia et al., 1991). Dysgeusia may occur in as many as 39% of hypothyroid patients (Mattes, 1986). Consistent with studies in rats showing no change in taste sensitivity with propylthiouracil treatment (Brosvic et al., 1992), taste thresholds of patients with recent onset of hypothyroidism are comparable to controls (Mattes, 1986). In contrast, longer-standing deficiency may be associated with measurable changes (McConnell, 1976). Suprathreshold function may be impaired before changes of threshold sensitivity are manifest (Mattes, 1986). The degree of taste impairment is not correlated with disease severity (Pittman and Beschi, 1967). Ambiguous results have been obtained on measures of taste preferences, as different tests yield discrepant findings. Studies in rats have also noted inconsistent effects of hypothyroidism on hedonic responses (Rivlin et al., 1977; Brosvic et al., 1992). Restoration of normal hormone status leads to a normalization of threshold sensitivity and suprathreshold function as well as taste preferences (Mattes, 1986; Deems et al., 1991). However, burning mouth syndrome and dysgeusia have been observed in 23.1% and 20.5%, respectively, of patients treated with levothyroxine (Deems, 1991). Burning mouth syndrome has been associated with decreased body mass index (Deems et al., 1991). Medications for the treatment of thyroid disease are frequently associated with gustatory complaints (for a review, see (Doty et al., 2008)). The mechanisms underlying the gustatory changes associated with hypothyroidism have not been elucidated but may include alterations of salivary secretion, central

37.8 Nutritional Implications Of Chemosensory Abnormalities In Patients With Selected Chronic Disorders **847**

or peripheral neural function, cellular metabolism and/or epithelial integrity.

37.8.4 Obesity

Studies of the sensory function of obese individuals have largely focused on responses to sweets and fats. Threshold sensitivity and suprathreshold intensity judgments of sweet stimuli are similar in the lean and obese (Grinker et al., 1972; Grinker, 1978; Malcolm et al., 1980; Rodin, 1980; Witherly et al., 1980; Frijters and Rasmussen-Conrad, 1982; Drewnowski, 1987). Reported hedonic responses to sweetness are more varied. Both heightened and depressed pleasantness ratings have been noted in obese individuals (Rodin et al., 1976; Drewnowski, 1987; Mela, 2006), but the preponderance of evidence indicates there is no association between such ratings and body weight (Spitzer and Rodin, 1981; Frijters and Rasmussen-Conrad, 1982). In contrast, hedonic ratings of sweetness reportedly rise during restriction of energy intake (Cabanac, 1971; Rodin et al., 1976). Whether or not this hedonic shift influences adherence to energy restricted diets is unknown. In one of the few prospective studies on hedonic ratings and weight gain, pleasantness of sweet solutions did not change from baseline after weight was gained (Salbe et al., 2004).

Accumulating evidence supports the claim that humans can taste non-esterified fatty acids (NEFA) (for a review, see Tucker et al., 2014). Associations between sensitivity and BMI are variable with some finding negative associations between BMI and sensitivity to either NEFA in model solutions or triacylglycerols in foods (Stewart et al., 2010; Stewart and Keast, 2012) and others finding no difference (Liang et al., 2012). A retrospective analysis of earlier studies also found no evidence of association between BMI and detection thresholds to caproic, lauric, or stearic acid (Mattes, 2011a). Habitual dietary fat intake may influence sensitivity. A low-fat diet over four weeks improved sensitivity to oleic acid in both lean and overweight subjects, but consumption of a high-fat diet decreased sensitivity only in the lean group (Stewart and Keast, 2012). How NEFA detection influences dietary fat intake remains unknown.

The obese have exhibited a greater preference for high-fat foods than for low-fat items compared to lean controls in some (Drewnowski et al., 1985), but not all (Pangborn et al., 1985; Warwick and Schiffman, 1990), studies. These apparently conflicting findings may be attributable to the types of stimuli used in testing. Hedonic responses to fat by the lean and obese may be differentially influenced by other characteristics of the stimulus such as sweetness and rheology (Drewnowski, 1985; Drewnowski et al., 1989). In addition, sensory responsiveness may vary with the timing and duration of obesity (Drewnowski et al., 1990).

While earlier studies focused on hedonics, more recent work has focused on differences in chemosensory sensitivity between the lean and obese. Olfactory deficits occur more frequently in type IV obesity (BMI>45) than leaner subjects (BMI<45) (Richardson et al., 2004). Prior to gastric bypass surgery, patients were more likely to suffer from olfactory dysfunction than controls undergoing cholecystectomy; these deficits were still present one year after surgery, and 20% of patients with normal olfactory function before gastric bypass surgery developed olfactory disorders after surgery (Richardson et al., 2012). Thus, olfactory dysfunction seems to be independent of weight loss or gain and may be present regardless of weight status. Retronasal olfaction may also influence intake, with a trend for higher sensation to discourage ad libitum intake (Ruijschop et al., 2009). However, no studies have been conducted on this issue in an obese population. The exact time course of olfactory dysfunction and obesity onset has yet to be characterized.

Following gastric bypass surgery, increased taste sensitivity and decreased hedonic responses to sweet tastes have been noted (Rodin, 1980; Burge et al., 1995; Tichansky et al., 2006) (for a review, see (Miras and le Roux, 2010)). Sensory thresholds during weight loss by dieting are variable with one report noting no changes in recognition thresholds (Burge, 1995), and another documenting improvements in recognition thresholds (Umabiki et al., 2010). Changes in liking do not typically occur after weight loss due to dieting (Rodin, 1976; Johnson et al., 1979; Drewnowski and Holden-Wiltse, 1992), but others suggest that changes depend on the amount of weight lost and the amount of time that has passed since the weight loss (Kleifield and Lowe, 1991). More recent "big" losers liked sucrose solutions more than recent "smaller" losers, and long-term "big" losers liked sucrose solutions less than long-term "smaller" losers.

37.8.5 Diabetes

Disturbances of taste and smell occur in over 60% of patients with diabetes (Settle, 1991). It should be noted that these patients frequently suffer from comorbidities including hypertension and obesity, both of which may alter chemosensory function as discussed above. The most consistent sensory change is an elevated glucose taste threshold in patients with glucose intolerance and in individuals with a family history of diabetes. Taste perception of other qualities remains largely intact (Perros et al., 1996). The mechanisms have not been identified but may involve nerve degeneration, alterations in salivary glucose concentrations, and various biochemical changes. The specificity of changes and apparent genetic influence suggest an abnormality of glucose receptors in some cases, but evidence that the severity of gustatory symptoms increases with progressing neuropathy (Abbasi, 1981) supports a neural explanation in others. Patients with uncomplicated diabetes had taste recognition and olfactory identification scores comparable to controls in one study (Naka et al., 2010), but others report taste recognition deficits in both controlled and uncontrolled diabetics compared to controls (Gondivkar et al., 2009). Differences in taste testing methodology might contribute to these contradictory findings. Aguesia to sweet taste was significantly more likely to occur in uncontrolled diabetics, and hypogeusia to sweet, sour, and salty tastes was significantly more likely to occur in diabetics compared to controls (Gondivkar, 2009). There is a direct relationship between disease severity and the risk for chemosensory complaints. There are also reports of olfactory deficits associated with degenerative complications and comorbidities (Le Floch et al., 1993; Weinstock et al., 1993; Naka, 2010). It should be noted that as the disease progresses, the likelihood of a patient taking medications that can influence chemosensory function increases. The influence of sensory changes associated with diabetes on food selection has not been determined, but there is evidence of heightened sweet preference and sweet food intake among women with gestational diabetes (Tepper and Seldner, 1999; Belzer et al., 2009).

37.8.6 Chronic Kidney Disease

Chronic kidney disease (CKD) patients often suffer from hypertension and problems with chemosensory function, especially due to medication use, likely extend to these patients as well. Elevated taste recognition thresholds may occur in patients with renal disease. However, the qualities affected have varied across studies and patient groups. Some indicate sweet and sour are the most severely affected (Burge et al., 1979; Anon, 1981) whereas others observed greater impairment for bitter and salty tasting stimuli (Fernstrom et al., 1996; Middleton and Allman-Farinelli, 1999). CKD patients reporting poor or very poor appetite were more likely to report taste and smell changes and have lower BMI scores compared to those with good appetites (Bossola et al., 2011). Dialysis leads to an improvement of chemosensory function (Anon, 1981; Burge et al., 1984), but dialysis patients may also experience dysgeusia in the form of a metallic taste (Dirschnabel et al., 2011). Renal transplantation may normalize taste sensitivity 12–18 months following surgery (Mahajan et al., 1984). Early studies suggested the impairment in taste sensitivity was related to reduced zinc status (Mahajan et al., 1980). The preponderance of data does not support this claim, although it may hold merit in the sub-group of patients with a documented marked deficiency (Anon, 1981; Burge et al., 1984). An explanation for the shift in threshold sensitivity has not been identified. In contrast, intensity ratings for suprathreshold concentrations of taste stimuli are similar in uremic patients and controls as well as within patients over the course of a dialysis treatment (Shapera et al., 1986; Shepherd et al., 1987). Patients on hemodialysis and continuous ambulatory peritoneal dialysis report avoiding red meat, with hemodialysis patients reporting more foods, including poultry and fish, that were less pleasant than peritoneal dialysis or control subjects (Dobell et al., 1993). Dialysis patients are typically counseled to consume a high-protein diet; these taste changes may make compliance difficult. A recent trial noted one week of sodium restriction improved salty recognition thresholds but not detection thresholds in a group of CKD patients (Kusaba et al., 2009).

Olfactory quality recognition performance is typically lower in renal patients than healthy controls (Schiffman et al., 1978; Conrad et al., 1987; Griep et al., 1997; Landis et al., 2011); although, one study found that renal patients without malnutrition performed as well as healthy controls (Raff et al., 2008). Those patients suffering from olfactory dysfunction may not be aware of the decline (Frasnelli et al., 2002; Landis, 2011). However, in one study (Schiffman et al., 1978) patients rated food odors as less pleasant. Poor performance in odor perception was associated with higher serum urea and higher protein catabolism but not with BMI (Griep, 1997). Yet, another study found that patients who did poorer at identifying odors were more likely to be malnourished (Raff et al., 2008). Dialysis exerts no measurable effect on performance in some studies (Schiffman et al., 1978; Griep et al., 1997), further decrement in others (Conrad et al., 1987; Corwin, 1989), and improvement in a more recent study (Landis, 2011), which could reflect advances in hemodialysis compared to the earlier trials. These observations on dialyzed patients indicate that uremic toxins either do not account for the olfactory disturbances, or there are retained uremic toxins that have not yet been identified (Raff et al., 2008). The mechanism remains unknown, but an association between higher levels of C-reactive protein and poorer olfactory performance has been reported, suggesting that a pro-inflammatory state negatively impacts olfaction (Raff et al., 2008). Olfactory function normalizes after transplant (Griep et al., 1997). The effect of olfactory changes on food habits in this population is unknown.

37.8.7 Liver diseases

Decreased taste and smell threshold sensitivity occurs in patients with cirrhosis of the liver (Burch et al., 1978; Weismann et al., 1979; Garrett-Laster et al., 1984; Pabinger et al., 2000; Temmel et al., 2005), acute viral hepatitis (Smith et al., 1976), virus-associated hepatitis (Zucco et al., 2006; Nagao et al., 2010), and liver disease, cause not specified (Landis et al., 2004). Liver transplantation results in quality-specific (salt, phenethyl alcohol) improvement of these disorders (Bloomfeld et al., 1999). While correlations between chemosensory function and plasma levels of zinc, vitamin A, and protein have not been observed, supplementation with these nutrients may result in a normalization of sensation. Whether or not alterations of chemosensory function contribute to the reduction of appetite and changes of food preferences associated with hepatic pathology is unknown (Deems et al., 1993). In one study of hepatitis C patients receiving treatment with pegylated interferon and ribavirin, recognition of sweetness and saltiness decreased and bitter tastes were rated as more unpleasant over the 12 weeks of treatment (Klimacka-Nawrot et al., 2010). Appetite decreased in these patients over the course of the treatment, but food preferences did not change. It has also been hypothesized that taste disorders accompanying liver disease have a neural origin (Bergasa, 1998).

In summary, chemosensory abnormalities have been documented in patients with a variety of health disorders. The extent to which a change of taste and/or smell alters food intake and thereby contributes to the etiology or sequelae of these diseases or hampers recovery by adversely influencing adherence to therapeutic diets or recuperative processes remains unknown.

37.9 GUIDELINES FOR THE DIETARY MANAGEMENT OF PATIENTS WITH CHEMOSENSORY DISORDERS

Because no predictable dietary response has been identified for any chemosensory abnormality, the dietary management of patients must be individualized. The approach used must be based upon sound nutritional principals (i.e., balance, variety, and moderation) because for many patients, adherence will be required for extended periods of time. To be effective, it is necessary to address the full range of environmental and physiological factors influencing ingestive behavior. To date, no specialized dietary management approach has been developed and tested, but several recommendations may be made based upon the available data. One important issue concerns the use of nutrient supplements. It is preferable to obtain all nutrition from foods as this route is most likely to provide nutrients in biologically available forms and in reasonably balanced proportions. If this is not feasible, a supplement that provides up to 100% of the recommended dietary allowances for the individual may be appropriate. Patients should be advised that ingestion of high levels of specific nutrients, even water soluble vitamins, may be harmful and exacerbate their problem. Peripheral neuropathy has been reported following excessive intake of pyridoxine

(Schaumburg et al., 1983; Gdynia et al., 2008; Scott et al., 2008).

Evidence that zinc deficiency may lead to chemosensory abnormalities has prompted recommendations to treat such disorders with high levels of this nutrient. However, this practice may only be effective in individuals where a frank zinc deficiency is the underlying basis for the sensory disorder, and symptoms may resolve as quickly as one month after initiating treatment, provided the therapy is started within six months of deficiency (Takaoka et al., 2010). Double-blind studies indicate that for the majority of patients, zinc supplements will offer no therapeutic benefit (Henkin et al., 1976; Greger and Geissler, 1978; Gibson et al., 1989), and studies of patients with chemosensory dysfunction taking or not taking zinc supplements reveal no differences (Deems et al., 1991). Indeed, at the doses often recommended, other adverse effects such as gastric distress, iron-deficiency anemia, copper deficiency, neutropenia, and impaired immune function may occur (Fosmire, 1990; Takaoka et al., 2010).

A recent case report suggested that vitamin D administration in a vitamin D deficient cancer patient undergoing chemotherapy ameliorated dysgeusia (Fink, 2011). Vitamin D's role in cell proliferation, differentiation, and death may explain its effectiveness, as taste receptor cells turn over approximately every 10 days. The prevalence of vitamin D deficiency worldwide is estimated to be 50% (Nair and Maseeh, 2012).

Fortification of foods with natural food odors and tastes has been proposed as a means for enhancing the sensory appeal of items for individuals with diminished chemosensory ability (Schiffman, 2000). However, the available evidence is not strongly supportive of the efficacy of this approach (Mattes, 2011a). If a decrement of intake is not attributable to a decline of sensory function, this approach may not be effective. This method would also be ineffective with anosmic and/or ageusic patients and those with chemosensory distortions. Flavor amplification can reduce the appeal of foods if flavor notes unaffected by the chemosensory disturbance are amplified. Determination of the optimal level of flavor amplification for individual patients may be problematic, but the concept warrants further consideration. Rather than amplifying a flavor component that is not perceived normally by a patient, it may be useful to emphasize the other sensory characteristics of foods. Thermal, textural, irritant, and visual characteristics of foods are important determinants of acceptability and can be accentuated to provide greater variety and appeal to foods that cannot be tasted or smelled. However, this approach must be implemented under appropriate guidance since increased use of certain seasonings and food constituents (e.g., salt, sugar, fiber) can have adverse nutritional effects in selected individuals.

850   **Chapter 37 Nutritional Implications of Taste and Smell Dysfunction**

There are a number of additional modifications to the diet or eating process that may help reduce the impact of chemosensory disorders that are both easy and inexpensive to implement. The use of plastic utensils may help to diminish metallic sensations in the oral cavity (Hong et al., 2009). Consuming fluids while eating may help dissolve taste compounds and aid in reaching the taste bud, and chewing slowly may help with flavor release and saliva production (Mosel et al., 2011). Artificial saliva may also help with dysgeusia resulting from xerostomia (Diaz-Arnold and Marek, 2002) while candies or sugar-free gum may be recommended to increase salivary flow (Vissink et al., 1988). Case reports suggest improvement in dysgeusia with administration of an ice cube before eating (Fujiyama et al., 2010) and improvement in parosmia with the use of nose clips (Muller et al., 2006). Serving food at room temperature or chilled can reduce offensive odors.

REFERENCES

Abbasi, A. A. (1981). Diabetes: diagnostic and therapeutic significance of taste impairment. *Geriatrics* 36(12): 73–78.

Abdallah, L., Chabert, M., Louis-Sylvestre, J. (1997). Cephalic phase responses to sweet taste. *Am. J. Clin. Nutr.* 65(3): 737–743.

Abu-Hamdan, D. K., Desai, H., Sondheimer, J., et al.(1988). Taste acuity and zinc metabolism in captopril-treated hypertensive male patients. *Am. J. Hypertens.* 1(3 Pt 3): 303S–308S.

Ahren, B. and Holst, J. J. (2001). The cephalic insulin response to meal ingestion in humans is dependent on both cholinergic and noncholinergic mechanisms and is important for postprandial glycemia. *Diabetes Care* 50: 1030–1038.

Ait-Omar, A., Monteiro-Sepulveda, M., Poitou, C., et al. (2011). GLUT2 accumulation in enterocyte apical and intracellular membranes: a study in morbidly obese human subjects and ob/ob and high fat-fed mice. *Diabetes* 60(10): 2598–2607.

Akaishi, T., Shingai, T.,and Miyaoka, Y., S., (1991). Antidiuresis immediately caused by drinking a small volume of hypertonic saline in man. *Chem. Senses* 16(3): 277–281.

Albrecht, J., Schreder, T., Kleemann, A. M., et al.(2009). Olfactory detection thresholds and pleasantness of a food-related and a non-food odour in hunger and satiety. *Rhinology* 47(2): 160–165.

Anagnostides, A., Chadwick, V. S., Selden, A. C.,and Maton, P. N. (1984). Sham feeding and pancreatic secretion. Evidence for direct vagal stimulation of enzyme output. *Gastroenterology* 87(1): 109–114.

Andersen, H. B., Jensen, E. W., Madsbad, S., et al.(1992). Sham-feeding decreases cardiac output in normal subjects. *Clin. Physiol.* 12(4): 439–442.

Anon. (1981). Decreased taste acuity in chronic renal patients. *Nutr. Rev.* 39(5): 207–210.

Appleton, K. M. and Blundell, J. E. (2007). Habitual high and low consumers of artificially-sweetened beverages: effects of sweet taste and energy on short-term appetite. *Physiol. Behav.* 92(3): 479–486.

Arias, C. and Chotro, M. G. (2005). Increased preference for ethanol in the infant rat after prenatal ethanol exposure, expressed on intake and taste reactivity tests. *Alcoholism* 29(3): 337–346.

Aschenbrenner, K., Hummel, C., Teszmer, K., et al. (2008). The influence of olfactory loss on dietary behaviors. *Laryngoscope* 118(1): 135–144.

Aspen, V. A., Stein, R. I.,and Wilfley, D. E. (2012). An exploration of salivation patterns in normal weight and obese children. *Appetite* 25: 539–542.

Bachmanov, A. A., Kiefer, S. W., Molina, J. C., et al. (2003). Chemosensory factors influencing alcohol perception, preferences, and consumption. *Alcoholism* 27(2): 220–231.

Bachmanov, A. A., Reed, D. R., Tordoff, M. G., et al. (1996a). Intake of ethanol, sodium chloride, sucrose, citric acid, and quinine hydrochloride solutions by mice: A genetic analysis. *Behav. Genet.* 26(6): 563–573.

Bachmanov, A. A., Tordoff, M. G.,and Beauchamp, G. K. (1996b). Ethanol consumption and taste preferences in C57BL/6ByJ and 129/J mice. *Alcoholism* 20(2): 201–206.

Baggio, L. L. and Drucker, D. J. (2007). Biology of incretins: GLP-1 and GIP. *Gastroenterology* 132(6): 2131–2157.

Bansal, S., Buring, J. E., Rifai, N., et al.(2007). Fasting compared with nonfasting triglycerides and risk of cardiovascular events in women. *JAMA* 298(3): 309–316.

Bartoshuk, L. M. (1993). The biological basis of food perception and acceptance. *Food. Qual. Prefer.* 4: 21–32.

Barylko-Pikielna, N., Zawadzka, L., Niegowska, J., et al. (1985). Taste perception of sodium chloride in suprathreshold concentration related to essential hypertension. *J. Hypertens Suppl.* 3(3): S449–452.

Beauchamp, G. K., Bachmanov, A., and Stein, L. J. (1998). Development and genetics of glutamate taste preference. *Olfaction and Taste XII: An International Symposium*, C. Murphy (Eds), New York: New York Academy of Sciences, pp. 412–416.

Beauchamp, G. K., Bertino, M., and Engelman, K. (1987). Failure to compensate decreased dietary-sodium with increased table salt usage. *JAMA* 258(22): 3275–3278.

Beauchamp, G. K., Bertino, M., and Engelman, K. (1991). Human salt appetite. *Chemical Senses, Volume 4: Appetite and Nutrition*, M. I. Friedman, M. G. Tordoff and M. R. Kare (Eds), New York: Marcel Dekker, pp. 85–108.

Beauchamp, G. K., Cowart, B. J., Mennella, J. A., and Marsh, R. R. (1994). Infant salt taste - Developmental, methodological, and contextual factors. *Dev. Psychobiol.* 27(6): 353–365.

Beauchamp, G. K., Cowart, B. J., and Moran, M. (1986). Developmental-Changes in Salt Acceptability in Human Infants. *Dev. Psychobiol.* 19(1): 17–25.

Beauchamp, G. K. and Mennella, J. A. (2009). Early flavor learning and its impact on later feeding behavior. *J. Pediat. Gastroenterol. Nutr.* 48: S25–S30.

Beauchamp, G. K. and Moran, M. (1982). Dietary experience and sweet taste preference in human infants. *Appetite* 3(2): 139–152.

Beauchamp, G. K. and Moran, M. (1984). Acceptance of sweet and salty tastes in 2-year-old children. *Appetite* 5(4): 291–305.

Beauchamp, G. K. and Pearson, P. (1991). Human-development and umami taste. *Physiol. Behav.* 49(5): 1009–1012.

Bellisle, F., Louis-Sylvestre, J., Demozay, F., et al. (1985). Cephalic phase of insulin secretion and food stimulation in humans: a new perspective. *Am. J. Physiol. Endocrinol. Metabol.* 249: E639–E645.

Bello, N. T., Coughlin, J. W., Redgrave, G. W., et al. (2010). Oral sensory and cephalic hormonal responses to fat and non-fat liquids in bulimia nervosa. *Physiol. Behav.* 99: 611–617.

Belzer, L. M., Smulian, J. C., Lu, S. E., and Tepper, B. J. (2009). Changes in sweet taste across pregnancy in mild gestational diabetes mellitus: relationship to endocrine factors. *Chem. Senses* 34(7): 595–605.

Bergamasco, N. H. P. and Beraldo, K. E. A. (1990). Facial expressions of neonate infants in response to gustatory stimuli. *Braz. J. Med. Biol. Res.* 23(3–4): 245–249.

Bergasa, N. V. (1998). Hypothesis: taste disorders in patients with liver disease may be mediated in the brain: potential mechanisms for a central phenomenon. *Am. J. Gastroenterol.* 93(8): 1209–1210.

Bernard, R. A., Doty, R. L., Engelman, K., and Weiss, R. A. (1980). Taste and salt intake in human hypertension. *Biological and Behavioral Aspects of Salt Intake*, M. R. Kare, M. J. Fregly and R. A. Bernard (Eds), New York: Academic Press, pp. 397–409.

Berthoud, H. R. and Jeanrenaud, B. (1982). Sham feeding-induced cephalic phase insulin release in the rat. *Am. J. Physiol. Endocrinol. Metabol.* 5: E280–E285.

Bertino, M., Beauchamp, G. K., and Engelman, K. (1982). Long-term reduction in dietary-sodium alters the taste of salt. *Am. J. Clin. Nutr.* 36(6): 1134–1144.

Bertino, M., Beauchamp, G. K., and Engelman, K. (1986). Increasing dietary salt alters salt taste preference. *Physiol. Behav.* 38(2): 203–213.

Bezencon, C., le Coutre, J., and Damak, S. (2007). Taste-signaling proteins are coexpressed in solitary intestinal epithelial cells. *Chem. Senses* 32(1): 41–49.

Bhatia, S., Sircar, S. S., and Ghorai, B. K. (1991). Taste disorder in hypo and hyperthyroidism. *Indian J. Physiol. Pharmacol.* 35(3): 152–158.

Blais, C. A., Pangborn, R. M., Borhani, N. O., et al. (1986). Effect of dietary-sodium restriction on taste responses to sodium-chloride - a longitudinal-study. *Am. J. Clin. Nutr.* 44(2): 232–243.

Blednov, Y. A., Walker, D., Martinez, M., et al. (2008). Perception of sweet taste is important for voluntary alcohol consumption in mice. *Genes Brain Behav.* 7(1): 1–13.

Bloomfeld, R. S., Graham, B. G., Schiffman, S. S., and Killenberg, P. G. (1999). Alterations of chemosensory function in end-stage liver disease. *Physiol. Behav.* 66(2): 203–207.

Blundell, J. E., Rogers, P. J., and Hill, A. J. (1988). Uncoupling sweetness and calories - methodological aspects of laboratory studies on appetite control. *Appetite* 11: 54–61.

Bodner, L. (1991). Effect of parotid submandibular and sublingual saliva on wound healing in rats. *Comp. Biochem. Physiol. A: Physiol.* 100A(4): 887–890.

Boer, C. C., Correa, M. E., Miranda, E. C., and de Souza, C. A. (2010). Taste disorders and oral evaluation in patients undergoing allogeneic hematopoietic SCT. *Bone Marrow Transplant.* 45(4): 705–711.

Bogucka-Bonikowska, A., Scinska, A., Koros, E., et al. (2001). Taste responses in alcohol-dependent men. *Alcohol Alcohol.* 36(6): 516–519.

Bolze, M. S., Fosmire, G. J., Stryker, J. A., et al. (1982). Taste acuity, plasma zinc levels, and weight loss during radiotherapy: a study of relationships. *Radiology* 144(1): 163–169.

Bond, D. S., Raynor, H. A., Vithianathan, S., et al. (2009). Differences in salivary habituation to a taste stimulus in bariatric surgery candidates and normal-weight controls. *Obesity Sur.* 19: 873–878.

Bossola, M., Luciani, G., Rosa, F., and Tazza, L. (2011). Appetite and gastrointestinal symptoms in chronic hemodialysis patients. *J. Ren. Nutr.* 21(6): 448–454.

Bouhlal, S., Issanchou, S., and Nicklaus, S. (2011). The impact of salt, fat and sugar levels on toddler food intake. *Br. J. Nutr.* 105(4): 645–653.

Bovio, G., Montagna, G., Bariani, C., and Baiardi, P. (2009). Upper gastrointestinal symptoms in patients with advanced cancer: relationship to nutritional and performance status. *Support. Care Cancer* 17(10): 1317–1324.

Brand, J. G., Cagan, R. H., and Naim, M. (1982). Chemical senses in the release of gastric and pancreatic secretions. *Ann. Rev. Nutr.* 2: 249–276.

Brannon, P. M. (1990). Adaptation of the exocrine pancreas to diet. *Ann. Rev. Nutr.* 10: 85–105.

Brasser, S. M., Norman, M. B., and Lemon, C. H. (2010). T1r3 taste receptor involvement in gustatory neural responses to ethanol and oral ethanol preference. *Physiol. Genomics* 41(3): 232–243.

Brasser, S. M., Silbaugh, B. C., Ketchum, M. J., et al. (2012). Chemosensory responsiveness to ethanol and its individual sensory components in alcohol-preferring, alcohol-nonpreferring and genetically heterogeneous rats. *Addict. Biol.* 17(2): 423–436.

Brosvic, G. M., Doty, R. L., Rowe, M. M., et al. (1992). Influences of hypothyroidism on the taste detection performance of rats: a signal detection analysis. *Behav. Neurosci.* 106(6): 992–998.

Brown, R. J., Walter, M., and Rother, K. I. (2009). Ingestion of diet soda before a glucose load augments glucagon-like Peptide-1 secretion. *Diabetes Care* 32(12): 2184–2186.

Brown, R. J., Walter, M., and Rother, K. I. (2012). Effects of diet soda on gut hormones in youths with diabetes. *Diabetes Care* 35(5): 959–964.

Bruce, D. G., Chisholm, D. J., Storlien, L. H., and Kraegen, E. W. (1988). Physiological importance of deficiency in early prandial insulin secretion in non-insulin dependent diabetes. *Diabetes* 37: 736–744.

Bruera, E., Carraro, S., Roca, E., et al. (1984). Association between malnutrition and caloric intake, emesis, psychological depression, glucose taste, and tumor mass. *Cancer Treat Rep.* 68(6): 873–876.

Burch, R. E., Sackin, D. A., Ursick, J. A., et al. (1978). Decreased taste and smell acuity in cirrhosis. *Arch. Intern. Med.* 138(5): 743–746.

Burge, J. C., Park, H. S., Whitlock, C. P., and Schemmel, R. A. (1979). Taste acuity in patients undergoing long-term hemodialysis. *Kidney Int.* 15(1): 49–53.

Burge, J. C., Schaumburg, J. Z., Choban, P. S., et al. (1995). Changes in patients' taste acuity after Roux-en-Y gastric bypass for clinically severe obesity. *J. Am. Diet. Assoc.* 95(6): 666–670.

Burge, J. C., Schemmel, R. A., Park, H. S., and Greene, J. A., 3rd., (1984). Taste acuity and zinc status in chronic renal disease. *J. Am Diet. Assoc.* 84(10): 1203–1206, 1209.

Buss, C., Kraemer-Aguiar, L. G., Maranhao, P. A., et al. (2012). Novel findings in the cephalic phase of digestion: a role for microcirculation? *Physiol. Behav.* 105: 1082–1087.

Cabanac, M. (1971). Physiological role of pleasure. *Science* 173(4002): 1103–1107.

Calles-Escandon, J. and Robbins, D. C. (1987). Loss of early phase of insulin release in humans impairs glucose intolerance and blunts thermic effect of glucose. *Diabetes* 36: 1167–1172.

Cano, J. and Rodriguez-Echandia, E. L. (1980). Degenerating taste buds in sialectomized rats. *Acta Anat.* 106: 487–492.

Carson, J. A. and Gormican, A. (1977). Taste acuity and food attitudes of selected patients with cancer. *J. Am. Diet. Assoc.* 70(4): 361–365.

Cartoni, C., Yasumatsu, K., Ohkuri, T., et al. (2010). Taste preference for fatty acids is mediated by GPR40 and GPR120. *J. Neurosci.* 30(25): 8376–8382.

Cassady, B. A., Considine, R. V., and Mattes, R. D. (2012). Beverage consumption, appetite, and energy intake: what did you expect? *Am. J. Clin. Nutr.* 95(3): 587–593.

CDC (2011). *Vital signs: incidence and trends of infection with pathogens transmitted Commonly Through Food - Foodborne Diseases Active Surveillance Network, 10 U.S. Sites, 1996–2010.* Atlanta, Centers for Disease Control.

Cecil, J. E., Francis, J., and Read, N. W. (1999). Comparison of the effects of a high-fat and high-carbohydrate soup delivered orally and intragastrically on gastric emptying, appetite, and eating behaviour. *Physiol. Behav.* 67(2): 299–306.

Chale-Rush, A., Burgess, J. R., and Mattes, R. D. (2007). Multiple routes of chemosensitivity to free fatty acids in humans. *Am. J. Physiol. Gastroint. Liver Physiol.* 292(5): G1206–G1212.

Chan, S., Pollitt, E., and Leibel, R. (1979). Effects of nutrient cues on formula intake in 5-week-old infants. *Infant Behav. Dev.* 2(3): 201–208.

Chapman, I. M., Goble, E. A., Wittert, G. A., et al. (1998). Effect of intravenous glucose and euglycemic insulin infusions on short-term appetite and food intake. *Am. J. Physiol. Reg. Integrat. Comp. Physiol.* 274(3): R596–R603.

Chavez-Jauregui, R. N., Mattes, R. D., and Parks, E. J. (2010). Dynamics of fat absorption and effect of sham feeding on postprandial lipema. *Gastroenterology* 139: 1538–1548.

Chen, M., Yang, Y., Braunstein, E., Georgeson, K. E., and Harmon, C. M. (2001). Gut expression and regulation of FAT/CD36: possible role in fatty acid transport in rat enterocytes. *Am. J. Physiol. Endocrinol. Metab.* 281(5): E916–923.

Chen, M. C., Wu, S. V., Reeve, J. R., and Rozengurt, E. (2006). Bitter stimuli induce Ca2+ signaling and CCK release in enteroendocrine STC-1 cells: role of L-type voltage-sensitive Ca2+ channels. *Am. J. Physiol. Cell Physiol.* 291(4): C726–C739.

Chien, Y. C., Huang, Y. J., Hsu, C. S., et al. (2009). Maternal lactation characteristics after consumption of an alcoholic soup during the postpartum 'doing-the-month' ritual. *Public Health Nutr.* 12(3): 382–388.

Choi, S., Lee, M., Shiu, A. L., et al. (2007). GPR93 activation by protein hydrolysate induces CCK transcription and secretion in STC-1 cells. *Am. J. Physiol. Gastrointest. Liver Physiol.* 292(5): G1366–1375.

Christensen, C. M. (1986). Importance of saliva in diet-taste relationships. *Interaction of the chemical senses with nutrition*, M. R. Kare and J. G. Brand (Eds), Orlando: Academic Press, pp. 3–24.

Christensen, C. M. and Navazesh, M. (1984). Anticipatory salivary flow to the sight of different foods. *Appetite* 5: 307–315.

Clee, M. D. and Burrow, L. (1983). Taste and smell in disease. *N. Engl. J. Med.* (309): 1062.

Coldwell, S. E., Oswald, T. K., and Reed, D. R. (2009). A marker of growth differs between adolescents with high vs. low sugar preference. *Physiol. Behav.* 96(4–5): 574–580.

Coleman, J., Williams, A., Phan, T. H. T., et al. (2011). Strain differences in the neural, behavioral, and molecular correlates of sweet and salty taste in naive, ethanol- and sucrose-exposed P and NP rats. *J. Neurophysiol.* 106(5): 2606–2621.

Conger, A. D. and Wells, M. A. (1969). Radiation and aging effect on taste structure and function. *Radiat Res.* 37(1): 31–49.

Conrad, P., Corwin, J., Katz, L., et al. (1987). Olfaction and hemodialysis: baseline and acute treatment decrements. *Nephron* 47(2): 115–118.

Cooper, G. P. (1968). Receptor origin of the olfactory bulb response to ionizing radiation. *Am. J. Physiol.* 215(4): 803–806.

Corwin, J. (1989). Olfactory identification in hemodialysis: acute and chronic effects on discrimination and response bias. *Neuropsychologia* 27(4): 513–522.

Coulter, D. M. (1988). Eye pain with nifedipine and disturbance of taste with captopril: a mutually controlled study showing a method of post-marketing surveillance. *Br. Med. J.* 296: 1084–1088.

Covington, D. K., Briscoe, C. A., Brown, A. J., and Jayawickreme, C. K. (2006). The G-protein-coupled receptor 40 family (GPR40-GPR43) and its role in nutrient sensing. *Biochem. Soc. Trans.* 34(Pt 5): 770–773.

Crook, C. K. (1978). Taste perception in the newborn infant. *Inf. Behav. Dev.* 1: 52–69.

Crook, C. K. (1979). The organization and control in infant suckling. *Advances in Child Development and Behavior*, H. Reeve and L. P. Lipsitt (Eds), New York: Academic Press, pp. 299–351.

Crystal, S. R. and Bernstein, I. L. (1995). Morning sickness - Impact on offspring salt preference. *Appetite* 25(3): 231–240.

Crystal, S. R. and Bernstein, I. L. (1998). Infant salt preference and mother's morning sickness. *Appetite* 30(3): 297–307.

Crystal, S. R. and Teff, K. L. (2006). Tasting fat: cephalic phase hormonal responses and food intake in restrained and unrestrained eaters. *Physiol. Behav.* 89: 213–220.

Davidson, T. M., Jalowayski, A., Murphy, C., and Jacobs, R. D. (1987). Evaluation and treatment of smell dysfunction. *West J. Med.* 146(4): 434–438.

Deems, D. A., Doty, R. L., Settle, R. G., et al. (1991). Smell and taste disorders, a study of 750 patients from the University of Pennsylvania Smell and Taste Center. *Arch. Otolaryngol. Head Neck Surg.* 117(5): 519–528.

Deems, R. O., Friedman, M. I., Friedman, L. S., et al. (1993). Chemosensory function, food preferences and appetite in human liver disease. *Appetite* 20(3): 209–216.

DeSnoo, K. (1937). Das trinkende Kind in Uterus. *Monatsschr fur Gerburtshilfe Gynaekol* 105.

Desor, J. A. and Beauchamp, G. K. (1987). Longitudinal changes in sweet preferences in humans. *Physiol. Behav.* 39(5): 639–641.

Desor, J. A., Greene, L. S., and Maller, O. (1975a). Preferences for sweet and salty in 9-year-old to 15-year-old and adult humans. *Science* 190(4215): 686–687.

Desor, J. A., Maller, O., and Andrews, K. (1975b). Ingestive responses of human newborns to salty, sour, and bitter stimuli. *J. Comp. Physiol. Psychol.* 89(8): 966–970.

Deutsch, J. A., Moore, B. O., and Heinrichs, S. C. (1989). Unlearned specific appetite for protein. *Physiol. Behav.* 46(4): 619–624.

Dewit, H., Pierri, J., and Johanson, C. E. (1989). Assessing individual-differences in ethanol preference using a cumulative dosing procedure. *Psychopharmacology* 98(1): 113–119.

DeWys, W. D. (1974). A spectrum of organ systems that respond to the presence of cancer. Abnormalities of taste as a remote effect of a neoplasm. *Ann. N. Y. Acad. Sci.* 230: 427–434.

Diamant, H., Funakoshi, M., Strom, L., and Zotterman, Y. (1963). Electrophysiological studies on human taste nerves. Olfaction and Taste, Y. Zotterman (Eds), New York: Macmillan Co., pp. 193–203.

Diamond, P. and LeBlanc, J. (1985). Palatability and postprandial thermogenesis in dogs. *Am. J. Physiol. Endocrinol. Metabol.* 248: E75–E79.

Diaz-Arnold, A. M. and Marek, C. A. (2002). The impact of saliva on patient care: A literature review. *J. Prosthet. Dent.* 88(3): 337–343.

DiBattista, D. and Mercier, S. (1999). Role of learning in the selection of dietary protein in the golden hamster (Mesocricetus auratus). *Behav. Neurosci.* 113(3): 574–586.

Dirschnabel, A. J., Martins Ade, S., et al. (2011). Clinical oral findings in dialysis and kidney-transplant patients. *Quintessence Int.* 42(2): 127–133.

Ditraglia, G. M., Press, D. S., Butters, N., et al. (1991). Assessment of olfactory deficits in detoxified alcoholics. *Alcohol* 8(2): 109–115.

Dobell, E., Chan, M., Williams, P., and Allman, M. (1993). Food preferences and food habits of patients with chronic renal failure undergoing dialysis. *J. Am. Diet. Assoc.* 93(10): 1129–1135.

Dockray, G. J. (2012). Cholecystokinin. *Curr. Opin. Endocrinol. Diabetes Obes.* 19(1): 8–12.

Dominguez, H. D., Lopez, M. F., Chotro, M. G., and Molina, J. C. (1996). Perinatal responsiveness to alcohol's chemosensory cues as a function of prenatal alcohol administration during gestational days 17–20 in the rat. *Neurobiol. Learn. Mem.* 65(2): 103–112.

Doty, R. L. (1977). Food preference ratings of congenitally anosmic humans. *Chemical Senses and Nutrition*, M. Kare and O. Mailer (Eds), New York: Academic Press, pp. 315–325.

Doty, R. L., Philip, S., Reddy, K., and Kerr, K. L. (2003). Influences of antihypertensive and antihyperlipidemic drugs on the senses of taste and smell: a review. *J. Hypertens.* 21(10): 1805–1813.

Doty, R. L., Shah, M., and Bromley, S. M. (2008). Drug-induced taste disorders. *Drug Saf.* 31(3): 199–215.

Drewett, R. F. (1982). Returning to the suckled breast - a further test of Halls hypothesis. *Early Hum. Dev.* 6(2): 161–163.

Drewnowski, A. (1987). Sweetness and Obesity. *Relationship between obesity, weight loss, and taste responsiveness*, J. Dobbing (Eds), New York: Springer-Verlag, pp.

Drewnowski, A. (1998). Energy density, palatability, and satiety: Implications for weight control. *Nutr. Rev.* 56(12): 347–353.

Drewnowski, A., Brunzell, J. D., Sande, K., et al. (1985). Sweet tooth reconsidered: taste responsiveness in human obesity. *Physiol. Behav.* 35(4): 617–622.

Drewnowski, A. and Holden-Wiltse, J. (1992). Taste responses and food preferences in obese women: effects of weight cycling. *Int. J Obes. Relat. Metab. Disord.* 16(9): 639–648.

Drewnowski, A., Kurth, C., and Rahaim, J. E. (1990). Human obesities and sensory preferences for sugar/fat mixtures: Effects of weight cycling. Paper read at Assoc. Chemoreception Sci. XI Annual Mtg, Sarasota, FL.

Drewnowski, A., Massien, C., Louissylvestre, J., et al. (1994). Comparing the effects of aspartame and sucrose on motivational ratings, taste preferences, and energy intakes in humans. *Am. J. Clin. Nutr.* 59(2): 338–345.

Drewnowski, A., Shrager, E. E., Lipsky, C., et al. (1989). Sugar and fat: sensory and hedonic evaluation of liquid and solid foods. *Physiol. Behav.* 45(1): 177–183.

Drover, V. A., Nguyen, D. V., Bastie, C. C., et al. (2008). CD36 mediates both cellular uptake of very long chain fatty acids and their intestinal absorption in mice. *J. Biol. Chem.* 283(19): 13108–13115.

Duffy, V. B., Davidson, A. C., Kidd, J. R., et al. (2004a). Bitter receptor gene (TAS2R38), 6-n-propylthiouracil (PROP) bitterness and alcohol intake. *Alcoholism* 28(11): 1629–1637.

Duffy, V. B., Peterson, J. M., and Bartoshuk, L. M. (2004b). Associations between taste genetics, oral sensation and alcohol intake. *Physiol. Behav.* 82(2–3): 435–445.

Elson, A. E. T., Dotson, C. D., Egan, J. M., and Munger, S. D. (2010). Glucagon signaling modulates sweet taste responsiveness. *FASEB J.* 24(10): 3960–3969.

Engelen, L., de Wijk, R. A., Prinz, J. F., et al. (2003). The relation between saliva flow after different stimulations and the perception of flavor and texture attributes in custard desserts. *Physiol. Behav.* 78(1): 165–169.

Epstein, J. B., Emerton, S., Kolbinson, D. A., et al. (1999). Quality of life and oral function following radiotherapy for head and neck cancer. *Head Neck* 21(1): 1–11.

Epstein, J. B., Robertson, M., Emerton, S., et al. (2001). Quality of life and oral function in patients treated with radiation therapy for head and neck cancer. *Head Neck* 23(5): 389–398.

Epstein, L. H., Carr, K. A., Lin, H., and Fletcher, K. D. (2011). Food reinforcement, energy intake, and macronutrient choice. *Am. J. Clin. Nutr.* 94(1): 12–18.

Epstein, L. H., Paluch, R., and Coleman, K. J. (1996). Differences in salivation to repeated food cues in obese and nonobese women. *Psychosom. Med.* 58: 160–164.

Faas, A. E., Sponton, E. D., Moya, P. R., and Molina, J. C. (2000). Differential responsiveness to alcohol odor in human neonates - Effects of maternal consumption during gestation. *Alcohol* 22(1): 7–17.

Fagius, J. and Berne, C. (1994). Increase in muscle nerve sympathetic activity in humans after food intake. *Clin. Sci.* 86: 159–167.

Feldman, M. and Richardson, C. T. (1986). Total 24-hour gastric acid secretion in patients with duodenal ulcer: Comparison with normal subjects and effects of cimetidine and parietal cell vagotomy. *Gastroenterology* (90): 540–544.

Feltrin, K. L., Little, T. J., Meyer, J. H., et al. (2004). Effects of intraduodenal fatty acids on appetite, antropyloroduodenal motility, and plasma CCK and GLP-1 in humans vary with their chain length. *Am. J. Physiol. Regul. Integr. Comp. Physiol.* 287(3): R524–533.

Fernstrom, A., Hylander, B., and Rossner, S. (1996). Taste acuity in patients with chronic renal failure. *Clin. Nephrol.* 45(3): 169–174.

Ferriday, D. and Brunstrom, J. M. (2011). 'I Just can't help myself': effects of food-cue exposure in overweight and lean individuals. *Int. J. Obes.* 35: 142–149.

Ferris, A. M., Schlitzer, J. L., and Schierberl, M. J. (1986). Nutrition and taste and smell deficits: a risk factor or an adjustment. *Clinical Measurement of Taste and Smell*, H. Meiselman and R. Rivlin (Eds), New York: MacMillan Publishing Company, pp. 264–278.

Ferris, A. M., Schlitzer, J. L., Schierberl, M. J., et al. (1985). Anosmia and nutritional status. *Nut. Res.* 5(2): 149–156.

Fink, M. (2011). Vitamin D deficiency is a cofactor of chemotherapy-induced mucocutaneous toxicity and dysgeusia. *J. Clin. Oncol.* 29(4): e81–82.

Fischer, U., Hommel, H., Gottschiling, H.-D., and Nowak, W. (1976). The effect of meal feeding and sham-feeding on insulin secretion in dogs. *Eur. J. Clin. Invest.* 6: 465–471.

Forestell, C. A. and Mennella, J. A. (2007). Early determinants of fruit and vegetable acceptance. *Pediatrics* 120(6): 1247–1254.

Fosmire, G. J. (1990). Zinc toxicity. *Am. J. Clin. Nutr.* 51(2): 225–227.

Frasnelli, J. A., Temmel, A. F., Quint, C., et al. (2002). Olfactory function in chronic renal failure. *Am. J. Rhinol.* 16(5): 275–279.

Frazier, C. R. M., Mason, P., Zhuang, X. X., and Beeler, J. A. (2008). Sucrose exposure in early life alters adult motivation and weight gain. *PLoS One* 3(9).

Frecka, J. M. and Mattes, R. D. (2008). Possible entrainment of ghrelin to habitual meal patterns in humans. *Am. J. Physiol. Gastrointest. Liver Physiol.* 294(3): G699–707.

Frijters, J. E. and Rasmussen-Conrad, E. L. (1982). Sensory discrimination, intensity perception, and affective judgment of sucrose-sweetness in the overweight. *J. Gen. Psychol.* 107(2d Half): 233–247.

Fromentin, G., Gietzen, D. W., and Nicolaidis, S. (1997). Aversion-preference patterns in amino acid- or protein-deficient rats: A comparison with previously reported responses to thiamin-deficient diets. *Br. J. Nutr.* 77(2): 299–314.

Fujita, Y., Wideman, R. D., Speck, M., et al. (2009). Incretin release from gut is acutely enhanced by sugar but not by sweeteners in vivo. *Am. J. Physiol. Endocrinol. Metabol.* 296(3): E473–E479.

Fujiyama, R., Ishitobi, S., Honda, K., et al. (2010). Ice cube stimulation helps to improve dysgeusia. *Odontology* 98(1): 82–84.

Gaillard, D., Laugerette, F., Darcel, N., et al. (2008). The gustatory pathway is involved in CD36-mediated orosensory perception of long-chain fatty acids in the mouse. *FASEB J.* 22(5): 1458–1468.

Garrett-Laster, M., Russell, R. M., and Jacques, P. F. (1984). Impairment of taste and olfaction in patients with cirrhosis: the role of vitamin A. *Hum. Nutr. Clin. Nutr.* 38(3): 203–214.

Gdynia, H. J., Muller, T., Sperfeld, A. D., et al. (2008). Severe sensorimotor neuropathy after intake of highest dosages of vitamin B6. *Neuromuscul Disord.* 18(2): 156–158.

Gebruers, E. M., Hall, W. J., O'Brien, M. H., et al. (1985). Signals from the oropharynx may contribute to the diuresis which occurs in man to drinking isotonic fluids. *J. Physiol.* 363(21–33).

Gerspach, A. C., Steinert, R. E., SchÃ¶nenberger, L., et al. (2011). The role of the gut sweet taste receptor in regulating GLP-1, PYY, and CCK release in humans. *Am. J. Physiol. Endocrinol. Metabol.* 301(2): E317–E325.

Gibson, E. L. and Booth, D. A. (1986). Acquired protein appetite in rats - Dependence on a protein-specific need state. *Experientia* 42(9): 1003–1004.

Gibson, R. S., Vanderkooy, P. D., MacDonald, A. C., et al. (1989). A growth-limiting, mild zinc-deficiency syndrome in some southern Ontario boys with low height percentiles. *Am. J. Clin. Nutr.* 49(6): 1266–1273.

Gilbertson, T. A. (1998). Gustatory mechanisms for the detection of fat. *Curr. Opin. Neurobiol.* 8(4): 447–452.

Gilbertson, T. A., Yu, T., and Shah, B. P. (2010). Gustatory mechanisms for fat detection. Fat Detection: Taste, Texture, and Post Ingestive Effects, J. P. Montmayeur and J. Le Coutre (Eds), Boca Raton: CRC Press, pp. 83–104.

Glasbrenner, B., Bruckel, J., Gritzmann, R., and Adler, G. (1995). Cephalic phase of pancreatic polypeptide release: a valid test of autonomic neuropathy in diabetics? *Diabetes Res. Clin. Pract.* 30(2): 117–123.

Glendinning, J. I., Yiin, Y. M., Ackroff, K., and Sclafani, A. (2008). Intragastric infusion of denatonium conditions flavor aversions and delays gastric emptying in rodents. *Physiol. Behav.* 93(4–5): 757–765.

Goetzl, F. R., Ahokas, A. J., and Payne, J. G. (1950). Occurrence in normal individuals of diurnal variations in acuity of the sense of taste of sucrose. *J. Appl. Physiol.* 2(11): 619–626.

Goetzl, F. R. and Stone, F. (1947). Diurnal variations in acuity of olfaction and food intake. *Gastroenterology* 9(4): 444–453.

Goldberg, A. N., Shea, J. A., Deems, D. A., and Doty, R. L. (2005). A chemosensory questionnaire for patients treated for cancer of the head and neck. *Laryngoscope* 115(12): 2077–2086.

Golik, A., Zaidenstein, R., Dishi, V., et al. (1998). Effects of captopril and enalapril on zinc metabolism in hypertensive patients. *J. Am. Coll. Nutr.* 17(1): 75–78.

Gondivkar, S. M., Indurkar, A., Degwekar, S., and Bhowate, R. (2009). Evaluation of gustatory function in patients with diabetes mellitus type 2. *Oral Surg. Oral Med. Oral Pathol. Oral Radiol. Endod.* 108(6): 876–880.

Goodspeed, R. B., Gent, J. F., and Catalanotto, F. A. (1987). Chemosensory dysfunction. Clinical evaluation results from a taste and smell clinic. *Postgrad. Med.* 81(1): 251–257, 260.

Grabenhorst, F., Rolls, E. T., Parris, B. A., and d'Souza, A. A. (2010). How the brain represents the reward value of fat in the mouth. *Cereb. Cortex* 20(5): 1082–1091.

Green, B. G. (2012). Chemesthesis and the chemical senses as components of a "chemosensor complex". *Chem. Senses* 37(3): 201–206.

Green, B. G., Lim, J., Osterhoff, F., et al. (2010). Taste mixture interactions Suppression, additivity, and the predominance of sweetness. *Physiol. Behav.* 101(5): 731–737.

Green, B. G., Nachtigal, D., Hammond, S., and Lim, J. (2012). Enhancement of retronasal odors by taste. *Chem. Senses* 37(1): 77–86.

Green, E. and Murphy, C. (2012). Altered processing of sweet taste in the brain of diet soda drinkers. *Physiol Behav.* 107: 560–567.

Greger, J. L. and Geissler, A. H. (1978). Effect of zinc supplementation on taste acuity of the aged. *Am. J. Clin. Nutr.* 31(4): 633–637.

Griep, M. I., Van der Niepen, P., Sennesael, J. J., et al. (1997). Odour perception in chronic renal disease. *Nephrol. Dial. Transplant* 12(10): 2093–2098.

Grieve, F. G. and Vander, M. W. (2003). Desire to eat high- and low-fat foods following a low-fat dietary intervention. *J. Nutr. Educ. Behav.* 35(2): 98–104.

Grinker, J. (1978). Obesity and sweet taste. *Am. J. Clin. Nutr.* 31(6): 1078–1087.

Grinker, J., Hirsch, J., and Smith, D. V. (1972). Taste sensitivity and susceptibility to external influence in obese and normal weight subjects. *J. Pers. Soc. Psychol.* 22(3): 320–325.

Guinard, J.-X., Zoumas-Morse, C., and Walchak, C. (1998). Relation between parotid saliva flow and somposition and the perception of gustatory and trigeminal stimulus in foods. *Physiol. Behav.* 63(1): 109–118.

Harris, G. and Booth, D. A. (1985). Sodium preference in food and previous dietary experience in 6-month-old infants. *IRCS Med. Sci. Biochem.* 13(12): 1177–1178.

Hashkes, P. J., Gartside, P. S., and Blondheim, S. H. (1997). Effect of food palatability on early (cephalic) phase of diet-induced thermogenesis in nonobese and obese man. *Int. J. Obes.* 21: 608–613.

Hass, N., Schwarzenbacher, K., and Breer, H. (2010). T1R3 is expressed in brush cells and ghrelin-producing cells of murine stomach. *Cell Tissue Res.* 339(3): 493–504.

Hayashi, T. (1968). Experimental evidence of the second signaling system of man. *Cond. Reflex* 3(1): 18–28.

He, F. J., Marrero, N. M., and MacGregor, G. A. (2012). Salt intake is related to soft drink consumption in children and adolescents–a link to obesity? *Hypertension* 51: 629–634.

Heath, R. B., Jones, R., Frayn, K. N., and Robertson, M. D. (2004). Vagal stimulation exaggerates the inhibitory ghrelin response to oral fat in humans. *J. Endocrinol.* 180(2): 273–281.

Henkin, R. I., Schecter, P. J., Friedewald, W. T., et al. (1976). A double blind study of the effects of zinc sulfate on taste and smell dysfunction. *Am. J. Med. Sci.* 272(3): 285–299.

Hetherington, M. M., Regan, M. F., and Boyland, E. (2008). Chewing it over: effects of chewing gum on appetite. *Appetite* 50(2–3): 560.

Hill, D. L. and Mistretta, C. M. (1990). Developmental neurobiology of salt taste sensation. *Trends Neurosci.* 13(5): 188–195.

Hiraoka, T., Fukuwatari, T., Imaizumi, M., and Fushiki, T. (2003). Effects of oral stimulation with fats on the cephalic phase of pancreatic enzyme secretion in esophagostomized rats. *Physiol. Behav.* 79: 713–717.

Hirasawa, A., Tsumaya, K., Awaji, T., et al. (2005). Free fatty acids regulate gut incretin glucagon-like peptide-1 secretion through GPR120. *Nat. Med.* 11(1): 90–94.

Hong, J. H., Omur-Ozbek, P., Stanek, B. T., et al. G. (2009). Taste and odor abnormalities in cancer patients. *J. Support. Oncol.* 7(2): 58–65.

Hovan, A. J., Williams, P. M., Stevenson-Moore, P., et al. (2010). A systematic review of dysgeusia induced by cancer therapies. *Support Care Cancer* 18(8): 1081–1087.

Huang, J. H., DeJong, W., Schneider, S. K., and Towvim, L. G. (2011). Endorsed reasons for not drinking alcohol: a comparison of college student drinkers and abstainers. *J. Behav. Med.* 34(1): 64–73.

Hutton, J. L., Baracos, V. E., and Wismer, W. V. (2007). Chemosensory dysfunction is a primary factor in the evolution of declining nutritional status and quality of life in patients with advanced cancer. *J. Pain Symptom Manag.* 33(2): 156–165.

Ichimura, A., Hirasawa, A., Poulain-Godefroy, et al. (2012). Dysfunction of lipid sensor GPR120 leads to obesity in both mouse and human. *Nature* 483(7389): 350–354.

IFICF (2011). 2011 Food and Health Survey: Consumer attitudes toward food safety, nutrition & health. International Food Information Council Foundation. Washington, DC. 2012.

Ionescu, E., Rohner-Jeanrenaud, F., Proietto, J., et al. (1988). Taste-induced changes in plasma insulin and glucose turnover in lean and genetically obese rats. *Diabetes* 37: 773–779.

Isezuo, S. A., Saidu, Y., Anas, S., et al. (2008). Salt taste perception and relationship with blood pressure in type 2 diabetics. *J. Hum. Hypertens.* 22(6): 432–434.

Ishii, R., Yamaguchi, S., and Omahony, M. (1992). Measures of taste discriminability for sweet, salty and umami stimuli - Japanese verses Americans. *Chem. Senses* 17(4): 365–380.

James, C. E., Laing, D. G., and Hutchinson, I. (1999). Perception of sweetness in simple and complex taste stimuli by adults and children. *Chem. Senses* 24(3): 281–287.

James, C. E., Laing, D. G., and Oram, N. (1997). A comparison of the ability of 8-9-year-old children and adults to detect taste stimuli. *Physiol. Behav.* 62(1): 193–197.

Jang, H.-J., Kokrashvili, Z., Theodorakis, et al. (2007). Gut-expressed gustducin and taste receptors regulate secretion of glucagon-like peptide-1. *PNAS* 104(38): 15069–15074.

Janowitz, H. D. and Grossman, M. I. (1949). Gusto-olfactory thresholds in relation to appetite and hunger sensations. *J. Appl. Physiol.* 2(4): 217–222.

Janowitz, H. D., Hollander, F., Orringer, D., et al. (1950). A quantitative study of the gastric secretory response to sham feeding in a human subject. *Gastroenterology* 16: 104–116.

Janssen, S., Laermans, J., Verhulst, P. J., et al. (2011). Bitter taste receptors and alpha-gustducin regulate the secretion of ghrelin with functional effects on food intake and gastric emptying. *Proc. Natl. Acad. Sci. U. S. A.* 108(5): 2094–2099.

Jeffery, R. W., Pirie, P. L., Elmer, P. J., et al. (1984). Low-sodium, high-potassium diet - feasibility and acceptability in a normotensive population. *Am. J. Pub. Health* 74(5): 492–494.

Jenkins, G. N. and Daws, C. (1966). The psychic flow of saliva in man. *Arch. Oral Biol.* 11: 1203–1204.

Johnson, W. G., Keane, T. M., Bonar, J. R., and Downey, C. (1979). Hedonic ratings of sucrose solutions: effects of body weight, weight loss and dietary restriction. *Addict. Behav.* 4(3): 231–236.

Jones, B. P., Butters, N., Moskowitz, H. R., and Montgomery, K. (1978). Olfactory and gustatory capacities of alcoholic Korsakoff patients. *Neuropsychologia* 16(3): 323–337.

Jones, B. P., Moskowitz, H. R., Butters, N., and Glosser, G. (1975). Psychophysical scaling of olfactory, visual, and auditory-stimuli by alcoholic Korsakoff patients. *Neuropsychologia* 13(4): 387–393.

Jordan, H. (1969). Voluntary intragastric feeding: oral and gastric contributions to food intake and hunger in man. *J. Comp. Physiol. Psychol.* 68(4): 498–506.

Julis, R. A. and Mattes, R. D. (2007). Influence of sweetened chewing gum on appetite, meal patterning and energy intake. *Appetite* 48(2): 167–175.

Just, T., Pau, H. W., Engel, U., and Hummel, T. (2008). Cephalic phase insulin release in healthy humans after taste stimulation? *Appetite* 51(3): 622–627.

Jyotaki, M., Shigemura, N., and Ninomiya, Y. (2010). Modulation of sweet taste sensitivity by orexigenic and anorexigenic factors. *Endocrine J.* 57(6): 467–475.

Kampov-Polevoy, A. B., Garbutt, J. C., Davis, C. E., and Janowsky, D. S. (1998). Preference for higher sugar concentrations and tridimensional personality questionnaire scores in alcoholic and nonalcoholic men. *Alcoholism* 22(3): 610–614.

Kampov-Polevoy, A. B., Garbutt, J. C., and Janowsky, D. (1997). Evidence of preference for a high-concentration sucrose solution in alcoholic men. *Am. J. Psychiat.* 154(2): 269–270.

Katschinski, M. (2000). Nutritional implications of cephalic phase gastrointestinal responses. *Appetite* 34(2): 189–196.

Katschinski, M., Dahmen, G., Reinshagen, M., et al. (1992). Cephalic stimulation of gastrointestinal secretory and motor responses in humans. *Gastroenterology* 103(2): 383–391.

Kemmer, T. and Malfertheiner, P. (1983). Der differenzierte einflurB der beschmacks-qualitaten "suB" and "sauer" auf die parotissekretion. *Res. Exp. Med.* 183: 35–46.

Kemmer, T. and Malfertheiner, P. (1985). Influence of atropine on taste-stimulated parotid secretion. *Res. Exp. Med.* 185: 495–502.

Khan, N. A. and Besnard, P. (2009). Oro-sensory perception of dietary lipids: New insights into the fat taste transduction. *Biochim. Biophys. Acta Mol. Cell Biol. Lipids* 1791(3): 149–155.

Kiefer, S. W., Badiaelder, N., and Bice, P. J. (1995). Taste reactivity in high alcohol-drinking and low alcohol-drinking rats. *Alcoholism* 19(2): 279–284.

Kiefer, S. W., Bice, P. J., and Badiaeder, N. (1994). Alterations in taste reactivity to alcohol in rats given continuous alcohol access followed by abstinence. *Alcoholism* 18(3): 555–559.

Klajner, F., Herman, C. P., Polivy, J., and Chhabra, R. (1981). Human obesity, dieting, and anticipatory salivation to food. *Physiol. Behav.* 27(2): 195–198.

Kleifield, E. I. and Lowe, M. R. (1991). Weight loss and sweetness preferences: the effects of recent versus past weight loss. *Physiol. Behav.* 49(6): 1037–1042.

Klimacka-Nawrot, E., Musialik, J., Suchecka, W., et al. (2010). Taste disturbances during therapy with pegylated interferon-alpha 2b and ribavirin in patients with chronic hepatitis C. *Wiad Lek* 63(4): 289–299.

Kochli, A., Tenenbaum-Rakover, Y., and Leshem, M. (2005). Increased salt appetite in patients with congenital adrenal hyperplasia 21 hydroxylase deficiency. *Am. J. Physiol.-Regul. Integr. Comp. Physiol.* 288(6): R1673–R1681.

Kokrashvili, Z., Mosinger, B., and Margolskee, R. F. (2009a). Taste signaling elements expressed in gut enteroendocrine cells regulate nutrient-responsive secretion of gut hormones. *Am. J. Clin. Nutr.* 90(3): 822S–825S.

Kokrashvili, Z., Rodriguez, D., Yevshayeva, V., et al. (2009b). Release of endogenous opioids from duodenal enteroendocrine cells requires Trpm5. *Gastroenterology* 137(2): 598–606, 606 e591–592.

Konturek, J. W., Thor, P., Maczka, M., et al. (1994). Role of cholecystokinin in the control of gastric emptying and secretory response to a fatty meal in normal subjects and duodenal ulcer patients. *Scand. J. Gastroenterol.* 29: 583–590.

Konturek, S. J., Kwiecien, N., Obtulowicz, W., et al. (1978). Cephalic phase of gastric secretion in healthy subjects and duodenal ulcer patients: role of vagal innervation. *Gut.* 20: 875–881.

Konturek, S. J., Swierczek, J., Kwiecien, N., et al. (1981). Gastric secretory and plasma hormonal responses to sham-feeding of varying duration in patients with duodenal ulcer. *Gut.* 22: 1003–1010.

Kraegen, E. W., Chisholm, D. J., and McNamara, M. E. (1981). Timing of insulin delivery with meals. *Hormone and Metabolic Research* 13(7): 365–367.

Krahn, D., Grossman, J., Henk, H., et al. (2006). Sweet intake, sweet-liking, urges to eat, and weight change: Relationship to alcohol dependence and abstinence. *Addict. Behav.* 31(4): 622–631.

Kranzler, H. R., Moore, P. J., and Hesselbrock, V. M. (1996). No association of PROP taster status and paternal history of alcohol dependence. *Alcoholism* 20(8): 1496–1500.

Kusaba, T., Mori, Y., Masami, O., et al. (2009). Sodium restriction improves the gustatory threshold for salty taste in patients with chronic kidney disease. *Kidney Int.* 76(6): 638–643.

Kyriazis, G. A., Soundarapandian, M. M., and Tyrberg, B. (2012). Sweet taste receptor signaling in beta cells mediates fructose-induced potentiation of glucose-stimulated insulin secretion. *Proc. Natl. Acad. Sci. U. S. A.* 109(8): E524–532.

Laeng, B., Berridge, K. C., and Butter, C. M. (1993). Pleasantness of a sweet taste during hunger and satiety - effects of gender and sweet tooth. *Appetite* 21(3): 247–254.

Landis, B. N., Konnerth, C. G., and Hummel, T. (2004). A study on the frequency of olfactory dysfunction. *Laryngoscope* 114(10): 1764–1769.

Landis, B. N., Marangon, N., Saudan, P., et al. (2011). Olfactory function improves following hemodialysis. *Kidney Int.* 80(8): 886–893.

Langford, H. G., Watson, R. L., and Thomas, J. G. (1977). Salt intake, diuretics, and the treatment of hypertension. *Trans. Am. Clin. Climatol. Assoc.* 88: 32–37.

Lanier, S. A., Hayes, J. E., and Duffy, V. B. (2005). Sweet and bitter tastes of alcoholic beverages mediate alcohol intake in of-age undergraduates. *Physiol. Behav.* 83(5): 821–831.

Laugerette, F., Passilly-Degrace, P., Patris, B., et al. (2005). CD36 involvement in orosensory detection of dietary lipids, spontaneous fat preference, and digestive secretions. *J. Clin. Invest.* 115(11): 3177–3184.

Le Floch, J. P., Le Lievre, G., Labroue, M., et al. (1993). Smell dysfunction and related factors in diabetic patients. *Diabetes Care* 16(6): 934–937.

LeBlanc, J. and Brondel, L. (1985). Role of palatability on meal-induced thermogenesis in human subjects. *Am. J. Physiol. Endocrinol. Metabol.* 248(3): E333–E336.

LeBlanc, J. and Cabanac, M. (1989). Cephalic postprandial thermogenesis in human subjects. *Physiol. Behav.* 46(3): 479–482.

Ledikwe, J. H., Ello-Martin, J., Pelkman, C. L., et al. (2007). A reliable, valid questionnaire indicates that preference for dietary fat declines when following a reduced-fat diet. *Appetite* 49(1): 74–83.

LeGoff, D. B., Leichner, P., and Spigelman, M. N. (1988). Salivary response to olfactory food stimuli in anorexics and bulimics. *Appetite* 11: 15–25.

Lemmens, S. G., Martens, E. A., Kester, A. D., and Westerterp-Plantenga, M. S. (2011). Changes in gut hormone and glucose concentrations in relation to hunger and fullness. *Am. J. Clin. Nutr.* 94(3): 717–725.

Leprohon, C. E. and Anderson, G. H. (1980). Maternal diet affects feeding-behavior of self-selecting weanling rats. *Physiol. Behav.* 24(3): 553–559.

Leshem, M., Maroun, M., and Weintraub, Z. (1998). Neonatal diuretic therapy may not alter children's preference for salt taste. *Appetite* 30(1): 53–64.

Leung, P. M. B. and Rogers, Q. R. (1986). Effect of amino-acid imbalance and deficiency on dietary choice patterns of rats. *Physiol. Behav.* 37(5): 747–758.

Lewitt, M. S., Liang, D. G., Panhuber, H., et al. (1989). Sensory perception and hypothyroidism. *Chem. Senses* 14(4): 537–546.

Li, X., Staszewski, L., Xu, H., Durick, K., et al. (2002). Human receptors for sweet and umami taste. *Proc. Natl. Acad. Sci. U. S. A.* 99(7): 4692–4696.

Liang, L. C., Sakimura, J., May, D., et al. (2012). Fat discrimination: a phenotype with potential implications for studying fat intake behaviors and obesity. *Physiol. Behav.* 105(2): 470–475.

Liem, D. G. and de Graaf, C. (2004). Sweet and sour preferences in young children and adults: role of repeated exposure. *Physiol. Behav.* 83(3): 421–429.

Liley, A. W. (1972). Disorders of amniotic fluid. *Pathophysiology of Gestation*, N. S. Assali (Eds), New York: Academic Press, pp. 157–206.

Liou, A. P., Lu, X., Sei, Y., et al. (2011). The G-protein-coupled receptor GPR40 directly mediates long-chain fatty acid-induced secretion of cholecystokinin. *Gastroenterology* 140(3): 903–912.

Little, A. C. and Brinner, L. J. (1985). Taste responses to saltiness among hypertensive subjects under different therapeutic regimens. *J. Am. Diet. Assoc.* 85(5): 557–563.

Little, T. J., Gopinath, A., Patel, E., et al. (2010). Gastric emptying of hexose sugars: role of osmolality, molecular structure and the CCK(1) receptor. *Neurogastroenterol. Motil.* 22(11): 1183–1190, e1314.

Little, T. J., Gupta, N., Case, R. M., et al. (2009). Sweetness and bitterness taste of meals per se does not mediate gastric emptying in humans. *Am. J. Physiol. Regul. Integr. Comp. Physiol.* 297(3): R632–639.

Lobo, M. V., Huerta, L., Ruiz-Velasco, N., et al. (2001). Localization of the lipid receptors CD36 and CLA-1/SR-BI in the human gastrointestinal tract: towards the identification of receptors mediating the intestinal absorption of dietary lipids. *J. Histochem. Cytochem.* 49(10): 1253–1260.

Looy, H. and Weingarten, H. P. (1991). Effects of metabolic state on sweet taste reactivity in humans depend on underlying hedonic response profile. *Chem. Senses* 16(2): 123–130.

Lorentzen, M., Madsbad, S., Kehlet, H., and Tronier, B. (1987). Effect of sham-feeding on glucose tolerance and insulin secretion. *Acta Endocrinologica* 115: 84–86.

Lucas, F., Bellisle, F., and Di Maio, A. (1987). Spontaneous insulin fluctuations and the preabsorptive insulin response to food ingestion in humans. *Physiol. Behav.* 40(5): 631–636.

Luscombe-Marsh, N. D., Smeets, A., and Westerterp-Plantenga, M. S. (2008). Taste sensitivity for monosodium glutamate and an increased liking of dietary protein. *Br. J. Nutr.* 99(4): 904–908.

Ma, J., Bellon, M., Wishart, J. M., et al. (2009). Effect of the artificial sweetener, sucralose, on gastric emptying and incretin hormone release in healthy subjects. *Am. J. Physiol. Gastrointest. Liver Physiol.* 296(4): G735–G739.

Ma, J., Chang, J., Checklin, H. L., et al. (2010). Effect of the artificial sweetener, sucralose, on small intestinal glucose absorption in healthy human subjects. *Br. J. Nutr.* 104(6): 803–806.

Mace, O. J., Affleck, J., Patel, N., and Kellett, G. L. (2007). Sweet taste receptors in rat small intestine stimulate glucose absorption through apical GLUT2. *J. Physiol.* 582(1): 379–392.

Mace, O. J., Lister, N., Morgan, E., et al. (2009). An energy supply network of nutrient absorption coordinated by calcium and T1R taste receptors in rat small intestine. *J. Physiol.* 587(Pt 1): 195–210.

MacKay-Sim, A. (1991). Changes in smell and taste function in thyroid, parathyroid, and adrenal diseases. *Smell and Taste in Health and Disease*, T. Getchell (Eds), New York: Raven, pp. 817–827.

Mahajan, S. K., Abraham, J., Migdal, S. D., et al. (1984). Effect of renal transplantation on zinc metabolism and taste acuity in uremia. A prospective study. *Transplantation* 38(6): 599–602.

Mahajan, S. K., Prasad, A. S., Lambujon, J., et al. (1980). Improvement of uremic hypogeusia by zinc: a double-blind study. *Am. J. Clin. Nutr.* 33(7): 1517–1521.

Malcolm, R., O'Neil, P. M., Hirsch, A. A., et al. (1980). Taste hedonics and thresholds in obesity. *Int. J. Obes.* 4(3): 203–212.

Maone, T. R., Mattes, R. D., Bernbaum, J. C., and Beauchamp, G. K. (1990). A new method for delivering a taste without fluids to preterm and term infants. *Dev. Psychobiol.* 23(2): 179–191.

Markley, E. J., Mattes-Kulig, D. A., and Henkin, R. I. (1983). A classification of dysgeusia. *J. Am. Diet. Assoc.* 83(5): 578–580.

Mars, M., Hogenkamp, P. S., Gosses, A. M., et al. (2009). Effect of viscosity on learned satiation. *Physiol. Behav.* 98(1–2): 60–66.

Martin, C., Chevrot, M., Poirier, H., et al. (2012). CD36 as a lipid sensor. *Physiol. Behav.* 105(1): 36–42.

Martin, S. and Pangborn, R. M. (1970). A note on responses to ethyl alcohol before and after smoking. *Percept. Psychophys.* 8(3): 169–173.

Massolt, E. T., van Haard, P. M., Rehfeld, J. F., et al. (2010). Appetite suppression through smelling of dark chocolate correlates with changes in ghrelin in young women. *Reg. Peptides* 161(1–3): 81–86.

Matsuo, R. and Yamamoto, T. (1989). Salivary secretion elicited by taste stimulation with umami substances in human adults. *Chem. Senses* 14(1): 47–54.

Matsuo, R., Yamauchi, Y., and Morimoto, T. (1997). Role of submandibular and sublingual saliva in maintenance of taste sensitivity recorded in the chorda tympani of rats. *J. Physiol.* 498: 797–807.

Mattes-Kulig, D. A. and Henkin, R. I. (1985). Energy and nutrient consumption of patients with dysgeusia. *J. Am. Diet. Assoc.* 85(7): 822–826.

Mattes, R. D. (1984). Salt taste and hypertension: a critical review of the literature. *J. Chronic. Dis.* 37(3): 195–208.

Mattes, R. D. (1990). Discretionary salt and compliance with reduced sodium diet. *Nutrition Res.* 10(12): 1337–1352.

Mattes, R. D. (1993). Fat preference and adherence to a reduced-fat diet. *Am. J. Clin. Nutr.* 57(3): 373–381.

Mattes, R. D. (1994a). Influences on acceptance of bitter foods and beverages. *Physiol. Behav.* 56(6): 1229–1236.

Mattes, R. D. (1994b). Interaction between the energy content and sensory properties of foods. *Synergy:* 39–51.

Mattes, R. D. (1996). Oral fat exposure alters postprandial lipid metabolism in humans. *Am. J. Clin. Nutr.* 63(6): 911–917.

Mattes, R. D. (1997). Physiologic responses to sensory stimulation by food: nutritional implications. *J. Am. Diet. Assoc.* 97(4): 406–413.

Mattes, R. D. (2000). Nutritional implications of the cephalic-phase salivary response. *Appetite* 34(2): 177–183.

Mattes, R. D. (2001). Oral exposure to butter, but not fat replacers elevates postprandial triacylglycerol concentration in humans. *J. Nutr.* 131(5): 1491–1496.

Mattes, R. D. (2007). Effects of linoleic acid on sweet, sour, salty, and bitter taste thresholds and intensity ratings of adults. *Am. J. Physiol.-Gastroint. Liver Physiol.* 292(5): G1243–G1248.

Mattes, R. D. (2011a). Accumulating evidence supports a taste component for free fatty acids in humans. *Physiol. Behav.* 104(4): 624–631.

Mattes, R. D. (2011b). Oral fatty acid signaling and intestinal lipid processing: Support and supposition. *Physiol. Behav.* 105: 27–35.

Mattes, R. D. (2011c). Spices and energy balance. *Physiol. Behav.* 107: 584–590.

Mattes, R. D., Christensen, C. M., Engelman, K. (1988). Effects of therapeutic doses of amiloride and hydrochlorothiazide on taste, saliva and salt intake in normotensive adults. *Chem. Senses* 13: 33–44.

Mattes, R. D. and Cowart, B. J. (1994). Dietary assessment of patients with chemosensory disorders. *J. Am. Diet. Assoc.* 94(1): 50–56.

Mattes, R. D. and DiMeglio, D. (2001). Ethanol perception and ingestion. *Physiol. Behav.* 72(1–2): 217–229.

Mattes, R. D. and Engelman, K. (1992). Effects of combined hydrochlorothiazide and amiloride versus single drug on changes in salt taste and intake. *Am. J. Cardiol.* 70: 91–95.

Mattes, R. D. and Falkner, B. (1989). Salt taste and salt sensitivity in black adolescents. *Chem. Senses* 14: 673–679.

Mattes, R. D., Heller, A. D., and Rivlin, R. S. (1986). Abnormalities in suprathreshold taste function in early hypothyroidism in humans. Clinical Measurement of Taste and Smell, H. L. Mieselman and R. S. Rivlin (Eds), New York: MacMillan, pp. 467–486.

Mattes, R. D., Kumanyika, S. K., and Halpern, B. P. (1983). Salt taste responsiveness and preference among normotensive, prehypertensive and hypertensive adults. *Chem. Senses* 8(1): 27–40.

Mattes, R. D. and Rothacker, D. (2001). Beverage viscosity is inversely related to postprandial hunger in humans. *Physiol. Behav.* 74(4–5): 551–557.

Mattes, R. D., Westby, E., De Cabo, R., and Falkner, B. (1999). Dietary compliance among salt-sensitive and salt-insensitive normotensive adults. *Am. J. Med. Sci.* 317(5): 287–294.

Maurage, P., Callot, C., Chang, B., et al. (2011). Olfactory impairment is correlated with confabulation in alcoholism: towards a multimodal testing of orbitofrontal cortex. *PLoS One* 6(8).

McConnell, R. J., Menendez, C. E., Smith, F. R., et al. (1976). Defects of taste and smell in patients wtih hypothyroidism. *Am. J. Med.* 59: 354–364.

Meier, J. J., Kjems, L. L., Veldhuis, J. D., et al. (2006). Postprandial suppression of glucagon secretion depends on intact pulsatile insulin secretion: further evidence for the intraislet insulin hypothesis. *Diabetes* 55(4): 1051–1056.

Meiselman, H. L., deGraaf, C., and Lesher, L. L. (2000). The effects of variety and monotony on food acceptance and intake at a midday meal. *Physiol. Behav.* 70(1–2): 119–125.

Mela, D. J. (2006). Eating for pleasure or just wanting to eat? Reconsidering sensory hedonic responses as a driver of obesity. *Appetite* 47(1): 10–17.

Mennella, J. A. (1997). Infants' suckling responses to the flavor of alcohol in mothers' milk. *Alcoholism* 21(4): 581–585.

Mennella, J. A. (2001). Regulation of milk intake after exposure to alcohol in mothers' milk. *Alcoholism* 25(4): 590–593.

Mennella, J. A. and Beauchamp, G. K. (1993). The effects of repeated exposure to garlic-flavored milk on the nurslings behavior. *Pediat. Res.* 34(6): 805–808.

Mennella, J. A., Johnson, A., Staley, C., and Beauchamp, G. K. (1995). Garlic ingestion by pregnant women alters the odor of amniotic fluid. *Chem. Senses* 20(6): 192–192.

Mennella, J. A., Pepino, M. Y., Lehmann-Castor, S. M., and Yourshaw, L. M. (2010). Sweet preferences and analgesia during childhood: effects of family history of alcoholism and depression. *Addiction* 105(4): 666–675.

Mennella, J. A., Pepino, Y., and Reed, D. R. (2005). Genetic and environmental determinants of bitter perception and sweet preferences. *Pediatrics* 115(2): E216–E222.

Merigo, F., Benati, D., Cristofoletti, M., et al. (2011). Glucose transporters are expressed in taste receptor cells. *J. Anat.* 219(2): 243–252.

Michikawa, T., Nishiwaki, Y., Okamura, T., et al. (2009). The taste of salt measured by a simple test and blood pressure in Japanese women and men. *Hypertens. Res.* 32(5): 399–403.

Middleton, R. A. and Allman-Farinelli, M. A. (1999). Taste sensitivity is altered in patients with chronic renal failure receiving continuous ambulatory peritoneal dialysis. *J. Nutr.* 129(1): 122–125.

Miquilena-Colina, M. E., Lima-Cabello, E., Sánchez-Campos, S., et al. (2011). Hepatic fatty acid translocase CD36 upregulation is associated with insulin resistance, hyperinsulinaemia and increased steatosis in non-alcoholic steatohepatitis and chronic hepatitis C. *Gut.* 60(10): 1394–1402.

Miras, A. D. and le Roux, C. W. (2010). Bariatric surgery and taste: novel mechanisms of weight loss. *Curr. Opin. Gastroenterol.* 26(2): 140–145.

Miwa, T., Furukawa, M., Tsukatani, T., et al. (2001). Impact of olfactory impairment on quality of life and disability. *Arch. Otolaryngol. Head Neck Surg.* 127(5): 497–503.

Monteleone, P., Serritella, C., Martiadis, V., and Maj, M. (2008). Deranged secretion of ghrelin and obestatin in the cephalic phase of vagal stimulation in women with anorexia nervosa. *Biol. Psychiat.* 64: 1005–1008.

Monteleone, P., Serritella, C., Scognamiglio, P., and Maj, M. (2010). Enahnced ghreline secretion in the cephalic phase of food ingestion in women with bulimia nervosa. *Psychoneuroendocrinology* 35: 284–288.

Moor, J. G. and Motoki, D. (1979). Gastric secretory and humoral responses to anticipated feeding in five men. *Gastroenterology* 76: 71–75.

Moore, M. and Weiss, S. (1995). Reasons for non-drinking among Israeli adolescents of 4 religions. *Drug Alcohol Depend.* 38(1): 45–50.

Morris, P. and Anderson, G. H. (1986). The effects of early dietary experience on subsequent protein selection in the rat. *Physiol. Behav.* 36(2): 271–276.

Mosel, D. D., Bauer, R. L., Lynch, D. P., and Hwang, S. T. (2011). Oral complications in the treatment of cancer patients. *Oral Diseases* 17(6): 550–559.

Moskowitz, H. R., Kumraiah, V., Sharma, K. N., et al. (1976). Effects of hunger, satiety and glucose-load upon taste intensity and taste hedonics. *Physiol. Behav.* 16(4): 471–475.

Mossman, K., Shatzman, A., and Chencharick, J. (1982). Long-term effects of radiotherapy on taste and salivary function in man. *Int. J. Radiat. Oncol. Biol. Phys.* 8(6): 991–997.

Mossman, K. L. (1986). Gustatory tissue injury in man: radiation dose response relationships and mechanisms of taste loss. *Br. J. Cancer. Suppl.* 7: 9–11.

Mossman, K. L. (1994). Frequent short-term oral complications of head and neck radiotherapy. *Ear Nose Throat J.* 73(5): 316–320.

Mossman, K. L. and Henkin, R. I. (1978). Radiation-induced changes in taste acuity in cancer patients. *Int. J. Radiat Oncol. Biol. Phys.* 4(7–8): 663–670.

Mulder, N. H., Smit, J. M., Kreumer, W. M., et al. (1983). Effect of chemotherapy on taste sensation in patients with disseminated malignant melanoma. *Oncology* 40(1): 36–38.

Muller, A., Landis, B. N., Platzbecker, U., et al. (2006). Severe chemotherapy-induced parosmia. *Am. J. Rhinol.* 20(4): 485–486.

Murphy, C. and Withee, J. (1987). Age and biochemical status predict preference for casein hydrolysate. *J. Gerontol.* 42(1): 73–77.

Murphy, S. P., Foote, J. A., Wilkens, L. R., et al. (2006). Simple measures of dietary variety are associated with improved dietary quality. *J. Am. Diet. Assoc.* 106(3): 425–429.

Myers, M. D. and Epstein, L. H. (1997). The effect of dietary fat on salivary habituation and satiation. *Physiol. Behav.* 62(1): 155–161.

Nagao, Y., Kodama, H., Yamaguchi, T., et al. (1999). Reduced urination rate while drinking beer with and unpleasant taste and off-flavor. *Biosci. Biotechnol. Biochem.* 63(3): 468–473.

Nagao, Y., Kodama, H., Yonezawa, T., et al. (1998). Correlation between the drinkability of beer and gastric emptying. *Biosci. Biotechnol. Biochem.* 62(5): 846–851.

Nagao, Y., Matsuoka, H., Kawaguchi, T., and Sata, M. (2010). Aminofeel improves the sensitivity to taste in patients with HCV-infected liver disease. *Med. Sci. Monit.* 16(4): PI7–12.

Nair, R. and Maseeh, A. (2012). Vitamin D: The "sunshine" vitamin. *J. Pharmacol. Pharmacother.* 3(2): 118–126.

Naka, A., Riedl, M., Luger, A., et al. (2010). Clinical significance of smell and taste disorders in patients with diabetes mellitus. *Eur. Arch. Otorhinolaryngol.* 267(4): 547–550.

Nauli, A. M., Nassir, F., Zheng, S., et al. (2006). CD36 is important for chylomicron formation and secretion and may mediate cholesterol uptake in the proximal intestine. *Gastroenterology* 131(4): 1197–1207.

Nawroth, J. C., Greer, C. A., Chen, W. R., et al. (2007). An energy budget for the olfactory glomerulus. *J. Neurosci.* 27(36): 9790–9800.

Nederkoorn, C., Smulders, F. T., and Jansen, A. (2000). Cephalic phase responses, craving and food intake in normal subjects. *Appetite* 35(1): 45–55.

Nelson, G., Chandrashekar, J., Hoon, M. A., et al. (2002). An amino-acid taste receptor. *Nature* 416(6877): 199–202.

Nicolaidis, S. (1977). Sensory-neuroendocrine reflexes and their anticipatory optimizing role on metabolism. The chemical senses and nutrition, M. R. Kare and O. Maller (Eds), New York: Acedemic Press, pp. 123–140.

Niedermeier, W., Dreizen, S., Stone, R. E., and Spies, T. D. (1956). Sodium and potassium concentrations in the saliva of normotensive and hypertensive subjects. *Oral. Surg. Oral. Med. Oral. Pathol.* 9(4): 426–431.

Nilsson, G. and Uvnas-Wallensten, K. (1977). Effect of teasing and sham feeding on plasma glucagon concentration in dogs. *Acta. Physiol. Scand.* 100(3): 298–302.

Niven, J. E. and Laughlin, S. B. (2008). Energy limitation as a selective pressure on the evolution of sensory systems. *J. Exp. Biol.* 211(11): 1792–1804.

Noble, A. C. (1994). Bitterness in wine. *Physiol. Behav.* 56(6): 1251–1255.

Nordin, S., Heden Blomqvist, E., Olsson, P., et al. (2011). Effects of smell loss on daily life and adopted coping strategies in patients with nasal polyposis with asthma. *Acta. Oto-laryngologica* 131(8): 826–832.

Novis, B. H., Banks, S., and Marks, I. N. (1971). The cephalic phase of pancreatic secretion in man. *Scand. J. Gastroenterol.* 6: 417–422.

Nysenbaum, A. N. and Smart, J. L. (1982). Sucking behavior and milk intake of neonates in relation to milk-fat content. *Early Hum. Dev.* 6(2): 205–213.

O'Connor, D. T., Strause, L., Saltman, P., et al. (1987). Serum zinc is unaffected by effective captopril treatment of hypertension. *J. Clin. Hypertens.* 3(4): 405–408.

Ohara, I., Otsuka, S. I., and Yugar, Y. (1988). Cephalic phase response of pancreatic exocrine secretion in conscious dogs. *Am. J. Phys. Gastrointest. Liver Physiol.* 254(17): G424–G428.

Ophir, D., Guterman, A., and Gross-Isseroff, R. (1988). Changes in smell acuity induced by radiation exposure of the olfactory mucosa. *Arch. Otolaryngol. Head Neck Surg.* 114(8): 853–855.

Pabinger, S., Temmel, A. F. P., Quint, C., et al. (2000). Olfactory function and cirrhosis of the liver. *Chem. Senses* 25(5): 657.

Pangborn, R. M. (1959). Influence of hunger on sweetness preferences and taste thresholds. *Am. J. Clin. Nutr.* 7(3): 280–287.

Pangborn, R. M. and Berggren, B. (1973). Human parotid secretion in reponse to pleasant and unpleasant odorants. *Psychophysiology* 10(231–237).

Pangborn, R. M., Bos, K. E., and Stern, J. S. (1985). Dietary fat intake and taste responses to fat in milk by under-, normal, and overweight women. *Appetite* 6(1): 25–40.

Pangborn, R. M. and Giovanni, M. E. (1984). Dietary-intake of sweet foods and of dairy fats and resultant gustatory responses to sugar in lemonade and to fat in milk. *Appetite* 5(4): 317–327.

Pangborn, R. M., Witherly, S. A., and Jones, F. (1979). Parotid and whole-mouth secretion in response to viewing, handling, and sniffing food. *Perception* 9: 339–346.

Pasquet, P., Monneuse, M. O., Simmen, B., et al. (2006). Relationship between taste thresholds and hunger under debate. *Appetite* 46(1): 63–66.

Patrick, M. E., Schulenberg, J. E., O'Malley, P. M., et al. (2011). Age-related changes in reasons for using alcohol and marijuana from ages 18 to 30 in a national sample. *Psychol. Addict. Behav.* 25(2): 330–339.

Pedersen, A. M., Bardow, A., Beier Jensen, S., and Nauntofte, B. (2002). Saliva and gastrointestinal functions of taste, mastication, swallowing and digestion. *Oral Dis.* 8(3): 117–129.

Pelchat, M. L. and Danowski, S. (1992). A possible genetic association between prop-tasting and alcoholism. *Physiol. Behav.* 51(6): 1261–1266.

Pepino, M. Y. and Mennella, J. A. (2005). Factors contributing to individual differences in sucrose preference. *Chem. Senses* 30: I319–I320.

Pepino, M. Y. and Mennella, J. A. (2006). Children's liking of sweet tastes and its biological basis. Optimising Sweet Taste in Foods, W. J. Spillane (Eds), Cambridge: Woodhead Publ Ltd, pp. 54–65.

Pepino, M. Y. and Mennella, J. A. (2007). Effects of cigarette smoking and family history of alcoholism on sweet taste perception and food cravings in women. *Alcoholism* 31(11): 1891–1899.

Perros, P., MacFarlane, T. W., Counsell, C., and Frier, B. M. (1996). Altered taste sensation in newly-diagnosed NIDDM. *Diab. Care* 19(7): 768–770.

Pittman, D. W. (2010). Role of the gustatory system in fatty acid detection in rats. Fat Detection: Taste, Texture, and Post Ingestive Effects, J. P. Montmayeur and J. Le Coutre (Eds), Boca Raton, FL: CRC Press, pp.

Pittman, D. W., Labban, C. E., Anderson, A. A., and O'Connor, H. E. (2006). Linoleic and oleic acids alter the licking responses to sweet, salt, sour, and bitter tastants in rats. *Chem. Senses* 31(9): 835–843.

Pittman, J. A. and Beschi, R. J. (1967). Taste thresholds in hyper- and hypothyroidism. *J. Clin. Endocrinol. Metab.* 27(6): 895–896.

Power, M. L. and Schulkin, J. (2008). Anticipatory physiological regulation in feeding biology: cephalic phase responses. *Appetite* 50: 194–206.

Powley, T. L. and Berthoud, H. R. (1985). Diet and cephalic phase insulin responses. *Am. J. Clin. Nutr.* 42(5): 991–1002.

Prescott, J. (2004). Effects of added glutamate on liking for novel food flavors. *Appetite* 42(2): 143–150.

Prescott, J., Laing, D., Bell, G., et al. (1992). Hedonic responses to taste solutions: A cross-cultural study of Japanese and Australians. *Chem. Senses* 17(6): 801–809.

Raff, A. C., Lieu, S., Melamed, M. L., et al. (2008). Relationship of impaired olfactory function in ESRD to malnutrition and retained uremic molecules. *Am. J. Kidney Dis.* 52(1): 102–110.

Ramirez, I. (1985). Oral stimulation alters digestion of intragastric oil meals in rats. *Am. J. Physiol. Reg. Integrat. Comp. Physiol.* 248(4 Pt 2): R459–463.

Ravasco, P., Monteiro-Grillo, I., Marques Vidal, P., and Camilo, M. E. (2005). Impact of nutrition on outcome: A prospective randomized controlled trial in patients with head and neck cancer undergoing radiotherapy. *Head Neck* 27(8): 659–668.

Raynor, H. A., Jeffery, R. W., Phelan, S., et al. (2005). Amount of food group variety consumed in the diet and long-term weight loss maintenance. *Obes. Res.* 13(5): 883–890.

Razafimanalina, R., Mormede, P., and Velley, L. (1996). Gustatory preference aversion profiles for saccharin, quinine and alcohol in Roman high- and low-avoidance lines. *Behav. Pharmacol.* 7(1): 78–84.

Reckmeyer, N. M., Vickers, Z. M., and Csallany, A. S. (2010). Effect of free fatty acids on sweet, salty, sour and umami tastes. *J. Sens. Stud.* 25(5): 751–760.

Remick, A. K., Polivy, J., and Pliner, P. (2009). Internal and external moderators of the effect of variety on food intake. *Psychol. Bull.* 135(3): 434–451.

Richardson, B. E., Vander Woude, E. A., Sudan, R., et al. (2004). Altered olfactory acuity in the morbidly obese. *Obes. Surg.* 14(7): 967–969.

Richardson, B. E., Vanderwoude, E. A., Sudan, R., et al. (2012). Gastric bypass does not influence olfactory function in obese patients. *Obes. Surg.* 22(2): 283–286.

Richardson, C. T. and Feldman, M. (1986). Salivary response to food in humans and its effect on gastric acid secretion. *Amer. J. Physiol. Gastrointest. Liver Physiol.* 250: G85–G91.

Ripamonti, C., Zecca, E., Brunelli, C., et al. (1998). A randomized, controlled clinical trial to evaluate the effects of zinc sulfate on cancer patients with taste alterations caused by head and neck irradiation. *Cancer* 82(10): 1938–1945.

Rivlin, R. S., Osnos, M., Rosenthal, S., and Henkin, R. I. (1977). Abnormalities in taste preference in hypothyroid rats. *Am. J. Physiol.* 232(1): E80–84.

Robert-Koch-Institut (2006). *Gesundheitsberichterstattung des Bundes: Gesundheit in Deutschland*. Berlin, Staistisches Bundesamt: 112–115.

Robertson, M. D., Mason, A. O., and Frayn, K. N. (2002). Timing of vagal stimulation affects postprandial lipid metabolism in humans. *Am. J. Clin. Nutr.* 76: 71–77.

Rodin, J. (1978). Has the distinction between internal versus external control of feeding outlived its usefulness? *Recent Advances in Obesity Research*, G. A. Bray (Eds), London: Newman, pp. 75–85.

Rodin, J. (1980). Changes in perceptual responsiveness following jejunoileostomy: their potential role in reducing food intake. *Am. J. Clin. Nutr.* 33(2 Suppl): 457–464.

Rodin, J. (1981). Current status of the internal-external hypothesis for obesity: what went wrong? *The American Psychologist* 36(4): 361–372.

Rodin, J., Moskowitz, H. R., and Bray, G. A. (1976). Relationship between obesity, weight loss, and taste responsiveness. *Physiol. Behav.* 17(4): 591–597.

Rogers, J., Raimundo, A. H., and Misiewicz, J. J. (1993). Cephalic phase of colonic pressure response to food. *Gut.* 34: 537–543.

Rogers, P. J., Carlyle, J. A., Hill, A. J., and Blundell, J. E. (1988). Uncoupling sweet taste and calories: comparison of the effects of glucose and three intense sweeteners on hunger and food intake. *Physiol. Behav.* 43(5): 547–552.

Rolls, B. J. (1986). Sensory-specific satiety. *Nutr. Rev.* 44(3): 93–101.

Rolls, B. J., Laster, L. J., and Summerfelt, A. (1989). Hunger and food-intake following consumption of low-calorie foods. *Appetite* 13(2): 115–127.

Rolls, E. T. (1997). Taste and olfactory processing in the brain and its relation to the control of eating. *Crit. Rev. Neurobiol.* 11(4): 263–287.

Rolls, E. T., Rolls, B. J., and Rowe, E. A. (1983). Sensory-specific and motivation-specific satiety for the sight and taste of food and water in man. *Physiol. Behav.* 30(2): 185–192.

Rosenstein, D. and Oster, H. (1988). Differential facial responses to 4 basic tastes in newborns. *Child Dev.* 59(6): 1555–1568.

Rozengurt, E. (2006). Taste receptors in the gastrointestinal tract. I. Bitter taste receptors and α−gustducin in the mammalian gut. *Am. J. Physiol. Gastrointest. Liver Physiol.* 291(2): G171–G177.

Rozengurt, N., Wu, S. V., Chen, M. C., et al. (2006). Colocalization of the alpha-subunit of gustducin with PYY and GLP-1 in L cells of human colon. *Am. J. Physiol. Gastrointest. Liver Physiol.* 291(5): G792–802.

Ruijschop, R. M., Burgering, M. J., Jacobs, M. A., and Boelrijk, A. E. (2009). Retro-nasal aroma release depends on both subject and product differences: a link to food intake regulation? *Chem. Senses* 34(5): 395–403.

Rupp, C. I., Fleischhacker, W. W., Drexler, A., et al. (2006). Executive function and memory in relation to olfactory deficits in alcohol-dependent patients. *Alcohol. Clin. Exp. Res.* 30(8): 1355–1362.

Rupp, C. I., Fleischhacker, W. W., Hausmann, A., et al. (2004). Olfactory functioning in patients with alcohol dependence: Impairments in odor judgements. *Alcohol Alcohol.* 39(6): 514–519.

Rupp, C. I., Kurz, M., Kemmler, G., et al. (2003). Reduced olfactory sensitivity, discrimination, and identification in patients with alcohol dependence. *Alcohol. Clin. Exp. Res.* 27(3): 432–439.

Saad, M. J. A., Monte-Alegre, S., and Saad, S. T. O. (1990). Differences in first-phase insulin release between normal blacks and whites. *Braz. J. Med. Biol. Res.* 23: 655–657.

Sahakian, B. J., Lean, M. E. J., Robbins, T. W., and James, W. P. T. (1981). Salivation and insulin release in response to food in nonobese men and women. *Appetite* 2: 209–216.

Salbe, A. D., DelParigi, A., Pratley, R. E., et al. (2004). Taste preferences and body weight changes in an obesity-prone population. *Am. J. Clin. Nutr.* 79(3): 372–378.

Sandow, P. L., Hejrat-Yazdi, M., and Heft, M. W. (2006). Taste loss and recovery following radiation therapy. *J. Dent. Res.* 85(7): 608–611.

Sanematsu, K., Horio, N., Murata, Y., et al. (2009). Modulation and transmission of sweet taste information for energy homeostasis. *International Symposium on Olfaction and Taste* 1170: 102–106.

Sanz Ortiz, J., Moreno Nogueira, J. A., and Garcia de Lorenzo y Mateos, A. (2008). Protein energy malnutrition (PEM) in cancer patients. *Clin. Transl. Oncol.* 10(9): 579–582.

Sarles, H., Dani, R., Prezelin, G., et al. (1968). Cephalic phase of pancreatic secretion in man. *Gut.* 9(2): 214–221.

Sartor, F., Donaldson, L. F., Markland, D. A., et al. (2011). Taste perception and implicit attitude toward sweet index and soft drink supplementation. *Appetite* 57(1): 237–246.

Schaal, B., Marlier, L., and Soussignan, R. (2000). Human foetuses learn odours from their pregnant mother's diet. *Chem. Senses* 25(6): 729–737.

Schaumburg, H., Kaplan, J., Windebank, A., et al. (1983). Sensory neuropathy from pyridoxine abuse. *N. Engl. J. of Med.* 309(8): 445–448.

Scherr, S. and King, K. R. (1982). Sensory and metabolic feedback in the modulation of taste hedonics. *Physiol. Behav.* 29(5): 827–832.

Schiffman, S. S. (1983). Taste and smell in disease (Part I). *N. Engl. J. Med.* 308: 1275–1279.

Schiffman, S. S. (2000). Taste quality and neural coding: Implications from psychophysics and neurophysiology. *Physiol. Behav.* 69(1–2): 147–159.

Schiffman, S. S., Nash, M. L., and Dackis, C. (1978). Reduced olfactory discrimination in patients on chronic hemodialysis. *Physiol. Behav.* 21(2): 239–242.

Schubert, C. R., Cruickshanks, K. J., Fischer, M. E., et al. (2012). Olfactory impairment in an adult population: the Beaver Dam Offspring Study. *Chem. Senses* 37(4): 325–334.

Schussler, P., Kluge, M., Yassouridis, A., et al. (2012). Ghrelin levels increase after pictures showing food. *Obesity (Silver Spring)* 20(6): 1212–1217.

Schwartz, C., Issanchou, S., and Nicklaus, S. (2009). Developmental changes in the acceptance of the five basic tastes in the first year of life. *Br. J. Nutr.* 102(9): 1375–1385.

Schwartz, T. W., Stenquist, B., and Olbe, L. (1979). Cephalic phase of pancreatic-polypeptide secretion studied by sham feeding in man. *Scand. J. Gastroenterol.* 14: 313–320.

Scinska, A., Bogucka-Bonikowska, A., Koros, E., et al. (2001). Taste responses in sons of male alcoholics. *Alcohol Alcohol.* 36(1): 79–84.

Scinska, A., Koros, E., Habrat, B., et al. (2000). Bitter and sweet components of ethanol taste in humans. *Drug and Alcohol Dependence* 60(2): 199–206.

Sclafani, A. (2004). The sixth taste? *Appetite* 43(1): 1–3.

Sclafani, A., Ackroff, K., Abumrad, N. A. (2007a). CD36 gene deletion reduces fat preference and intake but not post-oral fat conditioning in mice. *Am. J. Physiol. Regul. Integr. Comp. Physiol.* 293(5): R1823–R1832.

Sclafani, A., Glass, D. S., Margolskee, R. F., and Glendinning, J. I. (2010). Gut T1R3 sweet taste receptors do not mediate sucrose-conditioned flavor preferences in mice. *Am. J. Physiol. Regul. Integr. Comp. Physiol.* 299(6): R1643–1650.

Sclafani, A. and Glendinning, J. I. (2005). Sugar and fat conditioned flavor preferences in C57BL/6J and 129 mice: oral and postoral interactions. *Am. J. Physiol. Regul. Integr. Comp. Physiol.* 289(3): R712–R720.

Scott, K., Zeris, S., and Kothari, M. J. (2008). Elevated B6 levels and peripheral neuropathies. *Electromyogr. Clin. Neurophysiol.* 48(5): 219–223.

Scott, T. R. (1992). Taste: the neural basis of body wisdom. Nutritional Triggers for Health and in Disease. *World. Rev. Nutr. Dietet.* 67: 1–39.

Secchi, A., Caldara, R., Caumo, A., et al. (1995). Cephalic-phase insulin and glucagon release in normal subjects and in patients receiving pancreas transplantation. *Metabolism* 44(9): 1153–1158.

Settle, R. G. (1978). The alcoholic's taste perception of alcohol: preliminary findings. *Currents in Alcoholism*, M. Galanter (Eds), New York: Grune and STratton, pp. 257–267.

Settle, R. G. (1991). The chemical senses in diabetes mellitus. *Smell and Taste and Disease* , T. V. Getchell, L. M. Bartoshuk, R. L. Doty and J. B. Snow (Eds), New York: Raven Press, pp. 829–843.

Shaffer, S. E. and Tepper, B. J. (1994). Effects of learned flavor cues on single meal and daily food-intake in humans. *Physiol. Behav.* 55(6): 979–986.

Shapera, M. R., Moel, D. I., Kamath, S. K., et al. (1986). Taste perception of children with chronic renal failure. *J. Am. Diet. Assoc.* 86(10): 1359–1362, 1365.

Shepherd, R., Farleigh, C. A., Atkinson, C., and Pryor, J. S. (1987). Effects of haemodialysis on taste and thirst. *Appetite* 9(2): 79–88.

Shirazki, A., Weintraub, Z., Reich, D., et al. (2007). Lowest neonatal serum sodium predicts sodium intake in low birth weight children. *Am. J. Physiol. Regul. Integr. Comp. Physiol.* 292(4): R1683–R1689.

Silveira, P. P., Portella, A. K., Crema, L., et al. (2008). Both infantile stimulation and exposure to sweet food lead to an increased sweet food ingestion in adult life. *Physiol. Behav.* 93(4–5): 877–882.

Simon, C., Schlienger, J. L., Sapain, R., and Imler, M. (1986). Cephalic phase insulin secretion in relation to food presentation in normal and overweight subjects *Physiol. Behav.* 36: 465–469.

Simons, P. J., Kummer, J. A., Luiken, J. J., and Boon, L. (2011). Apical CD36 immunolocalization in human and porcine taste buds from circumvallate and foliate papillae. *Acta. Histochem.* 113(8): 839–843.

Small, D. M., Zatorre, R. J., Dagher, A., et al. (2001). Changes in brain activity related to eating chocolate - From pleasure to aversion. *Brain* 124: 1720–1733.

Smeets, A. J., Lejeune, M. P., and Westerterp-Plantenga, M. S. (2009). Effects of oral fat perception by modified sham feeding on energy expenditure, hormones and appetite profile in the postprandial state. *Br. J. Nutr.* 101(9): 1360–1368.

Smeets, P. A. M., Weijzen, P., de Graaf, C., and Viergever, M. A. (2011). Consumption of caloric and non-caloric versions of a soft drink differentially affects brain activation during tasting. *Neuroimage* 54(2): 1367–1374.

Smith, F. J. and Campfield, L. A. (1993). Meal initiation occurs after experimental induction of transient declines in blood glucose. *Am. J. Physiol. Reg. Integrat. Comp. Physiol.* 34: R1423–R1429.

Smith, F. R., Henkin, R. I., and Dell, R. B. (1976). Disordered gustatory acuity in liver disease. *Gastroenterology* 70(4): 568–571.

Smith, J. C., Fisher, E. M., Maleszewski, V., and McClain, B. (2000). Orosensory factors in the ingestion of corn oil/sucrose mixtures by the rat. *Physiol. Behav.* 69(1–2): 135–146.

Sorensen, L. B., Moller, P., Flint, A., et al. (2003). Effect of sensory perception of foods on appetite and food intake: a review of studies on humans. *Int. J. Obes.* 27(10): 1152–1166.

Soucy, J. and LeBlanc, J. (1999). Protein meals and postprandial thermogenesis. *Physiol. Behav.* 65(4–5): 705–709.

Spetter, M. S., de Graaf, C., Viergever, M. A., and Smeets, P. A. M. (2012). Anterior cingulate taste activation predicts ad libitum intake of sweet and savory drinks in healthy, normal-weight men. *J. Nutr.* 142(4): 795–802.

Spitzer, L. and Rodin, J. (1981). Human eating behavior: A critical review of studies in normal weight and overweight individuals. *Appetite* 2: 293–329.

Stafford, L. D. and Welbeck, K. (2011). High hunger state increases olfactory sensitivity to neutral but not food odors. *Chem. Senses* 36(2): 189–198.

Stamler, R., Stamler, J., Grimm, R., et al. (1987). Nutritional therapy for high blood-pressure - final report of a 4-year randomized controlled trial - the Hypertension Control Program. *JAMA* 257(11): 1484–1491.

Steffens, A. B., Strubbe, J. H., Balkan, B., and Scheurink, A. J. W. (1990). Neuroendocrine mechanisms involved in regulation of body weight, food intake and metabolism. *Neurosci. Biobehav. Rev.* 14: 305–313.

Stein, L. J., Cowart, B. J., and Beauchamp, G. K. (2012). The development of salty taste acceptance is related to dietary experience in human infants: a prospective study. *Am. J. Clin. Nutr.* 95(1): 123–129.

Stein, L. J., Cowart, B. J., Epstein, A. N., et al. (1996). Increased liking for salty foods in adolescents exposed during infancy to a chloride-deficient feeding formula. *Appetite* 27(1): 65–77.

Steinert, R. E., Frey, F., Topfer, A., et al. (2011a). Effects of carbohydrate sugars and artificial sweeteners on appetite and the secretion of gastrointestinal satiety peptides. *Br. J. Nutr.* 105(9): 1320–1328.

Steinert, R. E., Gerspach, A. C., Gutmann, H., et al. (2011b). The functional involvement of gut-expressed sweet taste receptors in glucose-stimulated secretion of glucagon-like peptide-1 (GLP-1) and peptide YY (PYY). *Clin. Nutr.* 30(4): 524–532.

Stewart, J. E., Feinle-Bisset, C., Golding, M., et al. (2010). Oral sensitivity to fatty acids, food consumption and BMI in human subjects. *Br. J. Nutr.* 104(1): 145–152.

Stewart, J. E. and Keast, R. S. J. (2012). Recent fat intake modulates fat taste sensitivity in lean and overweight subjects. *Int. J. Obes.* 36(6): 834–842.

Stewart, J. E., Newman, L. R., and Keast, R. S. J. (2011a). Oral sensitivity to oleic acid is associated with fat intake and body mass index. *Clin. Nutr.* 30(6): 838–844.

Stewart, J. E., Seimon, R. V., Otto, B., et al. (2011b). Marked differences in gustatory and gastrointestinal sensitivity to oleic acid between lean and obese men. *Am. J. Clin. Nutr.* 93(4): 703–711.

Stewart, R. B., Russell, R. N., Lumeng, L., et al. (1994). Consumption of sweet, salty, sour, and bitter solutions by selectively bred alcohol-preferring and alcohol-nonpreferring lines of rats. *Alcoholism* 18(2): 375–381.

Stock, M. J. (1999). Gluttony and thermogenesis revisited. *Int. J. Obes.* 23(11): 1105–1117.

Swithers, S. E., Baker, C. R., and Davidson, T. L. (2009). General and persistent effects of high-intensity sweeteners on body weight gain and caloric compensation in rats. *Behav. Neurosci.* 123(4): 772–780.

Swithers, S. E. and Davidson, T. L. (2008). A role for sweet taste: Calorie predictive relations in energy regulation by rats. *Behav. Neurosci.* 122(1): 161–173.

Swithers, S. E. and Hall, W. G. (1994). Does oral experience terminate ingestion? *Appetite* 23: 113–138.

Swithers, S. E., Martin, A. A., and Davidson, T. L. (2010). High-intensity sweeteners and energy balance. *Physiol. Behav.* 100(1): 55–62.

Takaoka, T., Sarukura, N., Ueda, C., et al. (2010). Effects of zinc supplementation on serum zinc concentration and ratio of apo/holo-activities of angiotensin converting enzyme in patients with taste impairment. *Auris Nasus Larynx* 37(2): 190–194.

Taniguchi, K. (2004). Expression of the sweet receptor protein, T1R3, in the Human liver and pancreas. *J. Vet. Med. Sci.* 66(11): 1311–1314.

Tatzer, E., Schubert, M. T., Timischl, W., and Simbruner, G. (1985). Discrimination of taste and preference for sweet in premature babies. *Early Hum. Dev.* 12(1): 23–30.

Teegarden, S. L., Scott, A. N., and Bale, T. L. (2009). Early life exposure to a high fat diet promotes long-term changes in dietary preferences and central reward signaling. *Neuroscience* 162(4): 924–932.

Teff, K. (2000). Nutritional implications of the cephalic-phase reflexes: endocrine responses. *Appetite* 34(2): 206–213.

Teff, K. L. (2010). Cephalic phase pancreatic polypeptide responses to liquid and solid stimuli in humans. *Physiol. Behav.* 99: 317–323.

Teff, K. L. and Engelman, K. (1996a). Oral sensory stimulation improves glucose tolerance in humans: effects on insulin, C-peptide, and glucagon. *Am. J. Physiol. Reg. Integrat. Comp. Physiol.* 27: R1371–R1379.

Teff, K. L. and Engelman, K. (1996b). Palatability and dietary restraint: effect on cephalic phase insulin release in women. *Physiol. Behav.* 60(2): 567–573.

Teff, K. L., Mattes, R. D., and Engelman, K. (1991a). Cephalic phase insulin release in normal weight males: verification and reliability. *Am. J. Physiol. Endocrinol. Metabol.* 261(4): E430–E436.

Teff, K. L., Mattes, R. D., and Engelman, K. (1991b). Oral sensory stimulation improves glucose tolerance in humans: effects on insulin, C-peptide and glucagon. *Am. J. Physiol. Endocrinol. Metabol.* 261: E430–E436.

Temmel, A. F., Pabinger, S., Quint, C., et al. (2005). Dysfunction of the liver affects the sense of smell. *Wien Klin Wochenschr.* 117(1–2): 26–30.

Temmel, A. F., Quint, C., Schickinger-Fischer, B., et al. (2002). Characteristics of olfactory disorders in relation to major causes of olfactory loss. *Arch. Otolaryngol. Head Neck Surg.* 128(6): 635–641.

Temple, E. C., Laing, D. G., Hutchinson, I., and Jinks, A. L. (2002). Temporal perception of sweetness by adults and children using computerized time-intensity measures. *Chem. Senses* 27(8): 729–737.

Temussi, P. A. (2011). The good taste of peptides. *J. Pept. Sci.* 18(2): 73–82.

Teow, B. H., Dinicolantonio, R., and Morgan, T. O. (1984). Sodium detection threshold and preference for sodium-salts in humans on high and low salt diet. *Chem. Senses* 8(3): 267–267.

Tepper, B. J., Hartfiel, L. M., and Schneider, S. H. (1996). Sweet taste and diet in type II diabetes. *Physiol. Behav.* 60(1): 13–18.

Tepper, B. J. and Seldner, A. C. (1999). Sweet taste and intake of sweet foods in normal pregnancy and pregnancy complicated by gestational diabetes mellitus. *Am. J. Clin. Nutr.* 70(2): 277–284.

Thaler, B. I., Paulin, J. M., Phelan, E. L., and Simpson, F. O. (1982). A pilot-study to test the feasibility of salt restriction in a community. *NZ Med. J.* 95(721): 839–842.

Thesleff, I., Viinikka, L., Saxen, L., and Perheentupa, J. (1988). The parotid gland is the main source of human salivary epidermal growth factor. *Life Sci.* 43: 13–18.

Thirlby, R. C. and Feldman, M. (1984). Effect of chronic sham-feeding on maximal gastric acid secretion in the dog. *J. Clin. Invest.* 73: 566–569.

Tichansky, D. S., Boughter, J. D., Jr., and Madan, A. K. (2006). Taste change after laparoscopic Roux-en-Y gastric bypass and laparoscopic adjustable gastric banding. *Surg. Obes. Relat. Dis.* 2(4): 440–444.

Tobin, V., Le Gall, M., Fioramonti, X., et al. (2008). Insulin internalizes GLUT2 in the enterocytes of healthy but not insulin-resistant mice. *Diabetes* 57(3): 555–562.

Tolhurst, G., Reimann, F., and Gribble, F. M. (2012). Intestinal sensing of nutrients. *Handb. Exp. Pharmacol.* (209): 309–335.

Tomita, H. and Yoshikawa, T. (2002). Drug-related taste disturbances. *Acta Otolaryngol.* 122(4): 116–121.

Tordoff, M. G. and Alleva, A. M. (1990a). Effect of drinking soda sweetened with aspartame or high-fructose corn syrup on food intake and body weight. *Am. J. Clin. Nutr.* 51(6): 963–969.

Tordoff, M. G. and Alleva, A. M. (1990b). Oral-stimulation with aspartame increases hunger. *Physiol. Behav.* 47(3): 555–559.

Toyono, T., Seta, Y., Kataoka, S., and Toyoshima, K. (2007). CCAAT/ Enhancer-binding protein β regulates expression of human T1R3 taste receptor gene in the bile duct carcinoma cell line, HuCCT1. *Biochim. Biophys. Acta.* 1769: 641–648.

Trant, A. S., Serin, J., and Douglass, H. O. (1982). Is taste related to anorexia in cancer patients? *Am. J. Clin. Nutr.* 36(1): 45–58.

Trasobares, E., Corbaton, A., Gonzalez-Estecha, et al. (2007). Effects of angiotensin-converting enzyme inhibitors (ACE i) on zinc metabolism in patients with heart failure. *J. Trace. Elem. Med. Biol.* 21 Suppl 1: 53–55.

862 Chapter 37 Nutritional Implications of Taste and Smell Dysfunction

Tsuruoka, S., Wakaumi, M., Araki, N., et al. (2005a). Comparative study of taste disturbance by losartan and perindopril in healthy volunteers. *J. Clin. Pharmacol.* 45(11): 1319–1323.

Tsuruoka, S., Wakaumi, M., Ioka, T., et al. (2005b). Angiotensin II receptor blocker-induces blunted taste sensitivity: comparison of candesartan and valsartan. *Br. J. Clin. Pharmacol.* 60(2): 204–207.

Tucker, R. M., Edlinger, C., Craig, B. A., and Mattes, R. D. (2014). Associations between BMI and fat taste sensitivity in humans. *Chem. Senses.* 39(4): 349–3571.

Umabiki, M., Tsuzaki, K., Kotani, K., et al. (2010). The improvement of sweet taste sensitivity with decrease in serum leptin levels during weight loss in obese females. *Tohoku. J. Exp. Med.* 220(4): 267–271.

USDA (2011). ERS briefing room on food CPI, prices, and expenditures. United States Department of Agriculture Economic Research Service. Washington, DC. http://www.ers.usda.gov/briefing/cpifoodandexpenditures, Accessed June 19, 2012.

Vahl, T. P., Drazen, D. L., Seeley, R. J., et al. (2010). Meal-anticipatory glucagon-like peptide-1 secretion in rats. *Endocrinology* 151(2): 569–575.

Vatner, S. F., Franklin, D., and Van Critters, R. L. (1970). Mesenteric vasoactivity associated with eating and digestion in the conscious dog. *Am. J. Physiol.* 219: 170–174.

Vazquez, M., Pearson, P. B., and Beauchamp, G. K. (1982). Flavor preferences in malnourished Mexican infants. *Physiol. Behav.* 28(3): 513–519.

Ventura, A. K. and Mennella, J. A. (2011). Innate and learned preferences for sweet taste during childhood. *Curr. Opin. Clin. Nutr. Metab. Care* 14(4): 379–384.

Vinolo, M. A., Hirabara, S. M., and Curi, R. (2012). G-protein-coupled receptors as fat sensors. *Curr. Opin. Clin. Nutr. Metab. Care.* 15(2): 112–116.

Vissink, A., Johannes's-Gravenmade, E., Panders, A. K., and Vermey, A. (1988). Treatment of hyposalivation. *Ear Nose Throat J.* 67(3): 179–185.

Walsh, T. D., Bowman, K., and Jackson, G. P. (1982). Taste changes in advanced cancer. *Proc. Nutr. Soc.* 41(3): 108A.

Wang, Y., Chandra, R., Samsa, L. A., et al. (2010). Amino acids stimulate cholecystokinin release through the Ca2+-sensing receptor. *Am. J. Physiol. Gastrointest. Liver Physiol.* 300(4): G528–537.

Warwick, Z. S. and Schiffman, S. S. (1990). Sensory evaluations of fat-sucrose and fat-salt mixtures: relationship to age and weight status. *Physiol. Behav.* 48(5): 633–636.

Weinstock, R. S., Wright, H. N., and Smith, D. U. (1993). Olfactory dysfunction in diabetes mellitus. *Physiol. Behav.* 53: 17–21.

Weismann, K., Christensen, E., and Dreyer, V. (1979). Zinc supplementation in alcoholic cirrhosis. A double-blind clinical trial. *Acta. Med. Scand.* 205(5): 361–366.

Westrate, J. A., Dopheide, T., Robroch, L., et al. (1990). Does variation in palatability affect the postprandial response in energy expenditure? *Appetite* 15: 209–219.

White, B. D., Porter, M. H., and Martin, P. J. (2000). Protein selection, food intake, and body composition in response to the amount of dietary protein. *Physiol. Behav.* 69(4–5): 383–389.

White, K. D. (1978). Salivation: The significance of imagery in its voluntary control. *Psychophysiology* 15(3): 196–203.

Whitten, C. F. and Stewart, R. A. (1980). The effect of dietary-sodium in infancy on blood-pressure and related factors - studies of infants fed salted and unsalted diets for 5 months at 8 months and 8 years of age. *Acta Paediatr. Scand.:* 1–17.

Wickham, R. S., Rehwaldt, M., Kefer, C., et al. (1999). Taste changes experienced by patients receiving chemotherapy. *Oncol. Nurs. Forum.* 26(4): 697–706.

Widmayer, P., Kuper, M., Kramer, M., et al. (2011). Altered expression of gustatory-signaling elements in gastric tissue of morbidly obese patients. *Int. J. Obes. (Lond).*

Williams, D. L. (2010). Expecting to eat: glucagon-like peptide-1 and the anticipation of meals. *Endocrinology* 151(2): 445–447.

Wilson, C. W. M., Obrien, C., and Macairt, J. G. (1973). Effect of metronidazole on human taste threshold to alcohol. *Br. J. Addict.* 68(2): 99–110.

Winne, G. I. (1979). Influence of blood flow on intestinal absorption of drugs and nutrients. *J. Pharmacol. Exp. Ther.* 6: 333–393.

Wise, P. M., Hansen, J. L., Reed, D. R., and Breslin, P. A. S. (2007). Twin study of the heritability of recognition thresholds for sour and salty taste. *Chem. Senses.* 32(8): 749–754.

Wisen, O., Bjorvell, H., Cantor, P., et al. (1992). Plasma concentrations of regulatory peptides in obesity following modified sham feeding (MSF) and a liquid test meal. *Regulatory Peptides* 39(1): 43–54.

Wisniewski, L., Epstein, L. H., and Caggiula, A. R. (1992). Effect of food change on consumption, hedonics, and salivation. *Physiol. Behav.* 52: 21–26.

Witherly, S. A., Pangborn, R. M., and Stern, J. S. (1980). Gustatory responses and eating duration of obese and lean adults. *Appetite* 1: 53–63.

Witschi, J. C., Ellison, R. C., Doane, D. D., et al. (1985). Dietary-sodium reduction among students - feasibility and acceptance. *J. Am. Diet. Assoc.* 85(7): 816–821.

Wolstenholme, J. T., Warner, J. A., Capparuccini, M. I., et al. (2011). Genomic analysis of individual differences in ethanol drinking: evidence for non-genetic factors in C57BL/6 mice. *PLoS One* 6(6): 13.

Woods, S. C. (1991). The eating paradox: how we tolerate food. *Psycho. rev.* 98(4): 488–505.

Wooley, S. C. and Wooley, O. W. (1973). Salivation to the sight and thought of food: a new measure of appetite. *Psychosom. Med.* 35: 136–142.

Woolridge, M. W., Baum, J. D., and Drewett, R. F. (1980). Does a change in the composition of human-milk affect sucking patterns and milk intake. *Lancet* 2(8207): 1292–1294.

Wren, A. M., Seal, L. J., Cohen, M. A., et al. (2001). Ghrelin enhances appetite and increases food intake in humans. *J. Clin. Endocrinol. Metab.* 86(12): 5992.

Wronski, M., Skrok-Wolska, D., Samochowiec, J., et al. (2007). Perceived intensity and pleasantness of sucrose taste in male alcoholics. *Alcohol* 42(2): 75–79.

Wu, S. V., Rozengurt, N., Yang, M., et al. (2002). Expression of bitter taste receptors of the T2R family in the gastrointestinal tract and enteroendocrine STC-1 cells. *Proc. Natl. Acad. Sci. U. S. A.* 99(4): 2392–2397.

Wu, T., Zhao, B. R., Bound, M. J., et al. (2012). Effects of different sweet preloads on incretin hormone secretion, gastric emptying, and postprandial glycemia in healthy humans. *Am. J. Clin. Nutr.* 95(1): 78–83.

Yamashita, H., Nakagawa, K., Nakamura, N., et al. (2006a). Relation between acute and late irradiation impairment of four basic tastes and irradiated tongue volume in patients with head-and-neck cancer. *Int. J. Radiat. Oncol. Biol. Phys.* 66(5): 1422–1429.

Yamashita, H., Nakagawa, K., Tago, M., et al. (2006b). Taste dysfunction in patients receiving radiotherapy. *Head Neck* 28(6): 508–516.

Yeung, D. L., Leung, M., and Pennell, M. D. (1984). Relationship between sodium-intake in infancy and at 4 years of age. *Nutr. Res.* 4(4): 553–560.

Yiin, Y. M., Ackroff, K., and Sclafani, A. (2005). Flavor preferences conditioned by intragastric nutrient infusions in food restricted and free-feeding rats. *Physiol. Behav.* 84(2): 217–231.

Young, S. H., Rey, O., Sternini, C., and Rozengurt, E. (2010). Amino acid sensing by enteroendocrine STC-1 cells: role of the Na+−coupled neutral amino acid transporter 2. *Amer. J. Physiol. Cell Physiol.* 298(6): C1401–C1413.

References

Youngentob, S. L., Glendinning, J. I., and Bartoshuk, L. M. (2009). Fetal ethanol exposure increases ethanol intake by making it smell and taste better. *Proc. Natl. Acad. Sci. U.S.A.* 106(13): 5359–5364.

Zabernigg, A., Gamper, E.-M., Giesinger, J. M., et al. (2010). Taste alterations in cancer patients receiving chemotherapy: a neglected side effect? *The Oncologist* 15(8): 913–920.

Zafra, M. A., Molina, F., and Puerto, A. (2006). The neural/cephalic phase reflexes in the physiology of nutrition. *Neurosci. Biobehav. Rev.* 30(7): 1032–1044.

Zandstra, E. H., De Graaf, C., and van Trijp, H. C. M. (2000). Effects of variety and repeated in-home consumption on product acceptance. *Appetite* 35: 113–119.

Zinner, S. H., McGarvey, S. T., Lipsitt, L. P., and Rosner, B. (2002). Neonatal blood pressure and salt taste responsiveness. *Hypertension* 40(3): 280–285.

Zucco, G. M., Amodio, P., and Gatta, A. (2006). Olfactory deficits in patients affected by minimal hepatic encephalopathy: a pilot study. *Chem. Senses* 31(3): 273–278.

Zumkley, H., Bertram, H. P., Vetter, H., et al. (1985). Zinc metabolism during captopril treatment. *Horm. Metab. Res.* 17(5): 256–258.

Zumkley, H., Vetter, H., Mandelkow, T., and Spieker, C. (1987). Taste sensitivity for sodium chloride in hypotensive, normotensive and hypertensive subjects. *Nephron* 47 Suppl 1: 132–134.

Chapter 38

Conditioned Taste Aversions

KATHLEEN C. CHAMBERS

38.1 INTRODUCTION

38.1.1 Description of Conditioned Taste Aversion

Conditioned taste aversions (CTAs) are those that develop on the basis of experience with a food or drink. The experience on which such aversions are based is often the consumption of a food or drink prior to a bout of illness. The illness can be a consequence of ingestion or it can be coincidentally associated with consumption of a food or drink, such as happens when one acquires aversions during a bout of flu. There is a strong propensity for humans and other animals to "blame" any illness on some ingested food (Garb and Stunkard, 1974). Learned aversions tend to be idiosyncratic, that is, individuals differ from one another with respect to the targeted food or drink that becomes associated with the illness. Most meals include a number of different food items and thus identifying the culprit can be problematic. Blame is more likely to be directed towards less preferred and less familiar foods. There also are certain types of foods that are more likely to be targets; close to half of all learned taste aversions are directed towards major protein sources (red meats, poultry, fish, and eggs) while infrequent targets include sweet (cakes and pies) and nonsweet carbohydrates (bread, crackers, rice, and potatoes; Logue, 1985; Midkiff and Bernstein, 1985; Mooney and Walbourn, 2001). Some humans have reported aversions persisting since childhood or more than 50 years (Garb and Stunkard, 1974). However, it has been shown in animals that learned aversion can be extinguished and palatability restored if the food is experienced without subsequent illness (Dwyer, 2009; Spector et al., 1983).

In the laboratory, CTA is most commonly induced by administering a chemical agent (usually the agent lithium chloride, LiCl) that produces some type of illness

(unconditioned stimulus, US), after consumption of a food or drink (usually a highly palatable solution such as sucrose or saccharin flavored water) that has a novel taste (conditioned stimulus, CS). After such a pairing, individuals learn an association between the sensory properties of the novel food or drink and the sensory properties of the toxic agent (CS-US association). As a result of this learned association, there is a change in the behaviors expressed when individuals again encounter the food or drink. These conditioned responses (CR) can be divided into three categories (see Figure 38.1).

First, individuals will exhibit responses that are similar to the unconditioned responses (URs) elicited by the illness-inducing agent. In rats, LiCl elicits lying-on-belly, hypothermia, and decreased heart rate and after a novel taste has been paired with LiCl, the conditioned taste will elicit these same responses (Bull et al., 1991; Kosten and Contreras, 1989; Meachum and Bernstein, 1990). In humans, simply hearing or thinking about a conditioned taste elicits reports of nausea (Garcia, 1989). This type of learned response to the CS is referred to as a conditioned illness response ($CR_{CS-LLNESS}$).

Second, individuals will show reduction in the consumption of the conditioned taste. Measurements of this reduction include comparisons before and after CS-US pairing in the amount consumed, the number of ingestive orofacial responses (a series of rhythmical mouth movements and alternations between tongue protrusions and tongue retractions that result in swallowing a liquid), the number of licks from a drinking spout, and the preference for the conditioned taste when it is made available along with another taste substance (Bernstein and Goehler, 1983; Grill and Berridge, 1985; Kent et al., 2002). This type of learned response to the CS is referred to as a conditioned avoidance response ($CR_{CS-AVOIDANCE}$).

Handbook of Olfaction and Gustation, Third Edition. Edited by Richard L. Doty.
© 2015 Richard L. Doty. Published 2015 by John Wiley & Sons, Inc.

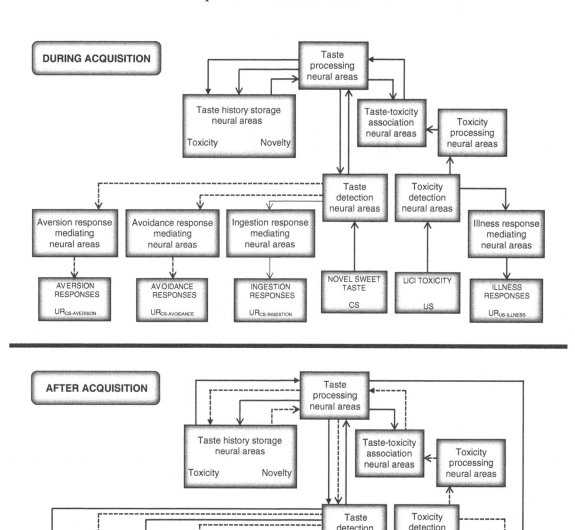

Figure 38.1 A simplified schematic showing the hypothesized basic neural components that mediate conditioned taste aversion during and after acquisition. During acquisition, a taste conditioned stimulus (CS) is recognized by taste detection areas, which send the information to taste processing areas, which access storage areas for taste history, determine that the taste is novel, and then send this information to taste-toxicity association areas as well as back to the taste detection areas. The taste detection areas send a weak (neophobia) activation (innate sweet preference) signal to the ingestion unconditioned response (UR) areas but not the avoidance/aversion (allesthesia) UR areas. A toxic unconditioned stimulus (US) in the food is recognized by the toxicity detection areas, which send the information to toxicity processing areas, which send processed toxicity information to the taste-toxicity association areas and this converged information, is sent to taste storage areas via taste processing areas. Upon encountering the taste again, the forging of new connections during acquisition based on taste-toxicity integration are revealed: taste toxicity history conveyed to CS response mediating areas, which activate aversion/avoidance responses and to illness response mediating areas, which activate illness responses. Lines and Arrows: solid lines represent a connection that is activated, dotted lines represent a connection that is not activated, the open arrows represent the direction of the flow of taste or toxicity information, and the closed arrows represent the direction of the flow of integrated taste-toxicity information. Abbreviations: $UR_{CS\text{-}INGESTION}$, ingestion UR in response to the CS; $UR_{US\text{-}ILLNESS}$, an illness UR in response to the US; $CR_{CS\text{-}AVOIDANCE}$, an avoidance CR in response to the CS; $CR_{CS\text{-}AVERSION}$, an aversion CR in response to the CS; $CR_{CS\text{-}ILLNESS}$, an illness CR in response to the CS.

Third, individuals will display responses that are indicative of a negative hedonic shift or an aversion. When rats are poisoned after consuming a sweet taste such as sucrose, which is liked at birth, their subsequent behavioral responses to sucrose resemble those exhibited after consumption of a bitter taste such as quinine, which is disliked at birth. These behavioral responses include increases in aversive orofacial/somatic responses, spillage of food from eating containers, and defensive burying of drinking spouts (Grill and Berridge, 1985; Parker, 1988; Rozin, 1967). In a series of studies, Parker and her associates have presented compelling evidence that the combination of the orofacial/somatic responses paw treading, chin rubbing, and mouth gaping (tongue retraction followed by long-duration tongue protrusion and mouth closure) best represent conditioned aversion in rats and that gaping is the predominant aversive response (see Parker et al., 2009). The muscular movements of the gaping response shown by rats who do not vomit are similar to those made by emetic species and orofacial topography of gaping is strikingly similar to the retch response that precedes vomiting in the insectivore shrew, *Suncus murinus*, suggesting that for the rat, gaping is an incipient vomiting response[1] (Parker et al., 2009; Travers and Norgren, 1986). Other species have their own species typical aversive responses. For example, coyotes respond to beef previously paired with poison by urinating on and rolling over it (Garcia et al., 1985). Anecdotal reports in humans indicate that thoughts of learned food aversions elicit facial expressions of loathing similar to those expressed in response to innately disliked foods and those who avoid food that previously had been followed by emesis report that the avoided food is distasteful (Garcia, 1989; Steiner, 1979). This type of learned response to the CS is referred to as a conditioned aversion response ($CR_{CS-AVERSION}$).

The majority of investigations of CTA have assessed aversion learning by only measuring $CR_{CS-AVOIDANCE}$. However, there is consensus that true CTA involves a hedonic shift and thus some evidence of aversion is required before one can call the reduced consumption a CTA (Garcia, 1989; Hunt and Amit, 1987; Parker, 1995). The importance of this verification is demonstrated by two types of studies, one type showing that some agents induce avoidance but not aversion and the other type showing that conditioned reductions in consumption can occur for reasons other than learned aversion. With respect to the first type of study, there are drugs such as apomorphine that are regarded as rewarding because they are effective

positive reinforcers in a drug self-administration paradigm (Hunt and Amit, 1987). The fact that they do not produce $CR_{CS-AVERSION}$ but induce $CR_{CS-AVOIDANCE}$ at the same doses that are rewarding suggests that the learned reductions in consumption are based on something other than taste aversion conditioning (Parker, 1995; Parker and Brosseau, 1990). It has been suggested it is based on reward comparison (Grigson 1997). In a typical example of this type of situation, animals are given access to a saccharin solution, which is less preferred, before they are given access to a sucrose solution, which is more preferred. They learn that the presence of the saccharin solution predicts the future availability of the sucrose solution, and they consequently reduce their consumption of the saccharin solution (Flaherty and Checke, 1982; Lucas et al., 1990). When applied to rewarding drugs, animals learn that the saccharin CS, which is given first, predicts the future availability of the rewarding drug US, which neurally activates similar hedonic areas as highly preferred sucrose.

With respect to the second type of study, reduced consumption can be learned on the basis of satiety. In both rats and humans, pairing a food with satiating properties can produce learned reduction in the consumption of that food (Booth, 1985). In one study, human subjects were given either high-calorie or low-calorie soups along with a plate of small sandwiches on each of two separate occasions, for a total of four exposures (Booth et al., 1972). The high- and low-calorie soups had different and distinctive flavorings and the caloric value of the high-calorie soup was obtained by adding a starch that could not be detected by taste. During the first exposure to each of the soups, the subjects ate the same number of sandwiches. However, during the second exposure to each of the soups, the subjects ate fewer sandwiches with the high-calorie soup and more sandwiches with the low-calorie soup. Thus, in one trial, the subjects were able to learn to associate the taste of the soup with the satiating properties of the food and to adjust their consumption levels accordingly. The implication of these studies is that the neural mechanisms mediating learned reduction in consumption based on illness, reward comparison, or satiety are different.

38.1.2 What Kind of Learning?

Although CTAs have been of scientific interest since 1887 (Poulton, 1887; Rzoska, 1953; Steiniger, 1950), it was not until Garcia and his associates presented it as a challenge to existing theories of learning that the study of this topic mushroomed (Garcia et al., 1966; Garcia and Koelling, 1966). Countless discussions of whether CTA constituted a new kind of learning or whether it could be considered a

---

[1] Nausea has been placed in the illness response category for humans because LiCl can induce nausea and vomiting in humans and as such it is a UR. In contrast, there are no reports in rats that the retch-like gaping response is a UR to LiCl; therefore it has been placed in the aversion response category.

form, albeit anomalous, of instrumental or classical conditioning ensued. Most investigators placed CTA within the Pavlovian or classical conditioning framework. In classical conditioning paradigms, a US that elicits a UR is paired with a CS that does not elicit the UR_{US} and learning is inferred when presentation of the CS produces a CR similar to the UR_{US}. The critical problem when thinking about CTA within the framework of classical conditioning is that behaviors that are generally regarded as the CR, avoidance and aversion responses, are entirely part of the repertoire of unconditioned elicited responses for the taste CS, that is, the UR_{CS} (Chambers, 1990). Under most circumstances before conditioning, a sweet taste will elicit consumption and positive reactions, such as ingestive orofacial/somatic responses, in rats, but when they are satiated, this taste will elicit negative reactions, such as avoidance and aversive orofacial/somatic responses (Grill and Berridge, 1985). Furthermore, although the reduction in consumption of the CS after conditioning appears similar to the unconditioned anorexia elicited by the illness US, it is specific to the CS rather than generally applied to most foods as is the case during illness[2]. This information led some investigators to suggest that CTA is different from many other forms of classical conditioning. However, there is no need to invoke a new form of learning for CTA (Chambers and Bernstein, 2003).

In studies of traditional classical conditioning situations, the CR_{CS} often closely resembles the UR_{US}, although it generally does not become identical to the UR_{US} (Holland, 1984; Rescorla, 1988). It can lack the intensity and some of the repertoire of the UR_{US}. There are cases, however, when the CR_{CS} produced by a given CS can include some responses that are not part of the UR_{US} and can even include a CR_{CS} that is opposite that of the UR_{US}, e.g., increases in heart rate elicited by a shock US and decreases in heart rate elicited by a tone CS (de Toledo and Black, 1966; Rescorla, 1988). These results led Holland (1984) to suggest that the CR_{CS} is composed of two behavioral elements: one that is similar to the UR_{US} or at least in some way appropriate to the US and one that is similar to the responses elicited by the CS prior to conditioning (UR_{CS}). Thus, comparable to traditional classical situations, the CR in CTA is composed of behavioral elements similar to the UR elicited by the US ($CR_{US-ILLNESS}$) and behavioral elements similar to responses that can be elicited by the CS prior to conditioning ($CR_{CS-AVOIDANCE}$ and $CR_{CS-AVERSION}$).

[2]In this chapter, the term anorexia will refer to reductions in eating that are found during illness, dysfunctional states, or chronic disease states while hypophagia will refer to normal reductions in eating that are the result of satiety signals in normal functioning individuals.

38.2 CONDITIONED TASTE AVERSION IN THE CONTEXT OF THE FEEDING SYSTEM

38.2.1 Plasticity of Preferences and Aversions

Food and some drinks provide nutrients and are a source of calories necessary for survival. For omnivores such as humans, a wide range of substances potentially can serve as food. Some of these substances have nutritive and caloric value, some have no food value, and some are harmful. Particular characteristics associated with the substance, such as taste, odor, texture, and appearance, serve as identifiers and become signals of the consequences of ingestion. For most mammals, taste or flavor (a combination of taste, smell, and texture) appears to be the primary cue used as an identifier. Preferences for sweet tastes and for the taste of salt, starch, and fat and aversions to bitter tastes and the flavor of chili peppers are thought to be innate at the time of birth in that individuals show the preferences or aversions without having had prior postnatal experience with the tastes (Ackroff et al., 1990; Ganchrow et al., 1986; Mattes, 1997; Moe, 1986; Rozin and Schiller, 1980; Sclafani, 1987; Steiner, 1973, 1979). This use of the word innate does not preclude the possibility that fetal learning processes contributed to the preference or aversion shown at birth. In rats, the ability to acquire taste avoidance is present during near-term fetal life (Gruest et al., 2004), so if fetal learning is involved, it would have to have occurred prior to that period.

The neural pathways mediating these innate preferences and aversions are plastic, as postnatal experience can strengthen, weaken, or reverse them and new preferences and aversions can be established through experience with different foods. Four processes have been suggested to contribute to increases in existing innate preferences and establishment of new preferences: (1) repeated exposure without negative consequences (Booth, 1985; Sullivan and Birch, 1990); (2) learned associations that occur when consumption of a food is followed by calorie or nutrient repletion (Birch et al., 1990; Gietzen et al., 1992); (3) learned associations that occur when a food is paired with another food that has high positive hedonic value (Bolles, 1985); and (4) learned association that occur when a food inhibits a negative hedonic process, as exemplified by sweet tastes which elicit reductions in emotional distress and physical pain in rat pups and in infants and prepubertal children, but not adults (Kehoe and Sakurai, 1991; Papino and Mennella, 2005; Segato et al., 1997).

Innate preferences for highly preferred tastes can be reversed and changed to aversions if consumption of these tastes is followed by illness (Garcia et al., 1955). In contrast, reversals of innate aversions are rare in animals, although they are frequently found in humans (Rozin et al., 1979). For example, prolonged exposure to chili flavoring in rats does not alter the aversion and repeated pairings of a chili-flavored food with recovery from thiamine deficiency illness attenuates, but does not reverse the innate aversion to the taste of chili. However, millions of people in many parts of the world have come to prefer the taste of chili peppers. The mechanism that accounts for this change remains unknown, although it is likely that at least one of the processes described above is involved.

Although animals come into life with innate predispositions and learning mechanisms that aid them in choosing amongst a vast array of potentially edible substances, this is not the end of story. Socially interactive animals have at their disposal help from conspecifics. Verbalizations and behavioral modeling by significant adults and peers influence food preferences in children (Birch, 1980; Capretta et al., 1973). In rats, the mother and adult conspecifics convey information to their young about what constitutes acceptable foods through gustatory cues in the rat mother's milk that reflect the taste of her diet and the physical presence of adult rats at feeding sites (Galef, 1977; Galef and Clark, 1971; Galef and Henderson, 1972). Social cues continue to play a role throughout adulthood. For example, adult rats can detect illness in a conspecific and will use that information to avoid consumption of a diet. Mother rats will decrease consumption of a novel taste when ingestion of that substance is followed by the illness of her pups (Gemberling, 1984) and similar effects are found with adult cage mates (the "poisoned-partner" effect; Lavin et al., 1980).

Apart from social guidance, as young mammals gain more independence, they begin to sample new and different foods that are present in their environment on their own. They rely on strategic mechanisms to guide their selection of foods. For example, opportunistic omnivores such as rats exhibit neophobia to new foods, even to those that have a sweet taste (Barnett, 1958; Domjan, 1977). Neophobia also is found in nonhuman primates (Itani, 1958; Weiskrantz and Cowey, 1963) and 2-year-old children (Birch and Marlin, 1982). The advantage of neophobia is that if the food is toxic, cautious sampling might lead to illness but not death.

Food selection, then, involves a complex interaction of innate predispositions, personal and social experiences, and strategic mechanisms that eventually lead to a relatively stable pattern of food preferences and aversions. Although stable, the hedonic value of these preferences and aversions does not remain constant. Preferred foods can become temporarily less positive and aversive foods can become temporarily less negative as the internal state of the individual changes (Cabanc, 1971, 1979; Fantino, 1984). The shift in hedonic value is temporary because when the body-state is restored, the hedonic value also is restored. As mentioned above, reactions to calorie-rich sweet solutions change from positive to negative as rats and humans go from food depletion to food repletion (see *What Kind of Learning*; Cabanac, 1971; Grill and Berridge, 1985). In addition, deprivation of specific substances such as sodium increases preferences for foods or solutions that contain these substances, even when the solution contains concentrations that are normally aversive (Berridge et al., 1984). These temporary shifts in hedonic value are referred to as allesthesia.

38.2.2 Estradiol, Satiety, CTA, and Anorexia Nervosa

Estradiol is a member of a group of steroid hormones called estrogens. It is present in both males and females and its primary source in the circulatory system of rodents is the granulosa cells in the ovaries of females and the aromatization of testosterone in the testes of males (de Jong et al., 1973; Shoham and Schachter, 1996). Secondary sources include neural sites such as the lateral parabrachial, hypothalamic preoptic area, and amygdala (Balthazart et al., 1990; Chambers et al., 1991; Saleh et al., 2002).

38.2.2.1 *Estradiol and Food Intake.* Substantial evidence unequivocally establishes estradiol as a hypophagic hormone. In a number of species, food intake decreases during the follicular phase of the cycle when estradiol levels are elevated and increases during the luteal phase when estradiol levels are lower (see Chambers and Bernstein, 2003). Ovariectomized rats and rhesus monkeys exhibit hyperphagia with increased body weight and estradiol treatment that mimics physiological levels of the hormone is sufficient to normalize both eating and body weight (Czaja and Goy, 1975; Geary and Asarian, 1999). Estradiol acts as a satiety hormone in that it reduces food intake by decreasing the size of meals rather than the number of meals (Blaustein and Wade, 1976). In addition, it is regarded as a long-term satiety hormone because it induces hypophagia as a means of regulating body weight, causing only a transient decrease in eating when administered to ovariectomized females, but a lasting decrease in body weight (Wade, 1972).

The area postrema (AP) and paraventricular hypothalamus (PVH) have been implicated in mediation of estradiol hypophagia. Production of a marker for neuronal activity, c-fos like immunoreactivity (c-FLI), is activated in the AP of ovariectomized female rats 24 hours after injection of estradiol benzoate, which is the time period when estradiol-induced hypophagia is expressed (Chambers and Hintiryan, 2009) and lesions of the AP eliminate the hypophagia expressed by male rats placed under chronic estradiol treatment (Bernstein et al., 1986). Eating increases expression of c-FLI in the PVH of food-deprived ovariectomized rats and activation of c-FLI is found within this same neural area when ovariectomized females are given pretreatment with a dose of estradiol that mimics follicular phase blood levels (Eckel and Geary, 2001). Furthermore, lesions of the PVH reduce the responsiveness of ovariectomized rats to estradiol, intracranial application of estradiol into the PVH, but not other hypothalamic regions, induces hypophagia in ovariectomized rats, and implanting the protein inhibitor anisomycin into the PVH disrupts the hypophagic effects of estradiol (Butera and Beikirch, 1989; Butera et al., 1990, 1992, 1993).

38.2.2.2 Estradiol and CTA.

Estradiol strengthens acquisition of LiCl-induced taste avoidance when it is present during acquisition, accelerates extinction when it is present before acquisition or before extinction, and prolongs extinction when it is present during extinction (Chambers and Hayes, 2002; Yuan and Chambers, 1999). We have suggested that all of these data can be accounted for by estradiol acting as an illness inducing agent (Chambers and Hintiryan, 2009). First, supraphysiological doses of estradiol can induce CTA in rats as evidenced by expression of both $CR_{CS-AVOIDANCE}$ and $CR_{CS-AVERSION}$ (Ossenkopp et al., 1996; Yuan and Chambers, 1999). Second, evidence demonstrates that when two USs are administered after CS consumption, they induce stronger taste avoidance than when either is administered alone (Hayes and Chambers, 2005b). This suggests that estradiol strengthened acquisition when present during acquisition because of the additive effect of its US properties with those of LiCl. Third, evidence indicates that acquisition can be strengthened by giving multiple CS-US pairings (Garcia et al., 1956). This suggests that estradiol prolonged extinction because its presence during extinction constituted multiple pairings of the sucrose CS with a US, LiCl in the first pairing, and estradiol in subsequent pairing. Fourth, studies have established that substances capable of inducing $CR_{CS-AVOIDANCE}$ can serve as preexposure agents to attenuate acquisition and accelerate extinction when they are given prior to acquisition in the absence of the taste substance targeted for future conditioning or prior to extinction in the absence of the conditioned taste stimulus (Cannon et al., 1977; Mikulka et al., 1977; Rabin et al., 1989). This occurs when the preexposure agent is the same as or different than the agent used to induce the taste avoidance (Cannon et al., 1977). This suggests that estradiol accelerated extinction when it was present before acquisition or extinction of the LiCl-induced taste avoidance because it was acting as a preexposure agent.

Due to the timing of the unconditioned effects exerted on eating by estradiol, the role of this hormone as a true conditioning agent has been challenged (Flanagan-Cato et al., 2001). In studies using estradiol as the US, taste avoidance has been observed when the first test for acquisition was given 24, 48, or 72 hours after CS-US pairing (De Beun et al., 1991, 1993; Hintiryan et al., 2005; Merwin and Doty, 1994; Ossenkopp et al., 1996). Unconditioned hypophagia also has been shown to occur within the first 24 hours following administration of estradiol in ovariectomized rats and to persist for at least 72 hours (Asarian and Geary, 2002; Chambers and Hintiryan, 2009; Rivera and Eckel, 2005; Tarttellin and Gorski, 1973). Consequently, it is possible that the taste avoidance observed after pairing a novel sweet taste with estradiol may have been due to the unconditioned hypophagia produced by estradiol and not to conditioning capabilities of the hormone. However, a series of studies conducted in our lab strongly indicates that the reduced consumption of sucrose after sucrose-estradiol pairing is a conditioned response (Hintiryan et al., 2009). First, when the initial post-pairing test was given 9 days after sucrose-estradiol pairing, reduced sucrose consumption was expressed even though unconditioned lab chow hypophagia was not evident. Second, when sucrose was paired with a physiological dose of estradiol, reduced sucrose consumption was not expressed at the time of the post-pairing test even though unconditioned lab chow hypophagia was evident. Third, when sucrose-estradiol pairing was non-contingent, reduced sucrose consumption was not expressed even though the post-pairing test was held at a time when unconditioned lab chow hypophagia is evident.

Gustavson and his colleagues have suggested that estradiol-induced nausea parallels the nausea induced by LiCl (Gustavson and Gustavson, 1987; Gustavson et al., 1989b). Consistent with this hypothesis is the similarity between LiCl and supraphysiological estradiol in their aversiveness. Both agents can induce nausea and vomiting (Gustavson et al., 1989b; Schou, 1968), elicit $CR_{CS-AVERSION}$ after it is paired with a novel taste stimulus (Breslin et al., 1992; Ossenkopp et al., 1996), and produce conditioned place avoidance (Cunningham and Niehus, 1993; De Beun et al., 1991). Also consistent are the results of US preexposure studies. Because attenuation of taste avoidance when preexposed to a US is more easily obtained if the same agent is used as both the preexposure agent and the conditioning agent, Rabin and colleagues (1989) proposed that the closer the neural pattern activated by the preexposure agent is to the conditioning agent the greater the likelihood that attenuation of the taste avoidance would occur. It has been demonstrated that preexposure to estradiol weakens acquisition and accelerates extinction of taste avoidance induced by LiCl as well as estradiol and preexposure to LiCl attenuates acquisition of taste avoidance induced by estradiol as well as LiCl (Chambers and Hayes, 2002; De Beun et al., 1993). This symmetrical preexposure attenuation of taste avoidance suggests that the aversive effects of estradiol and LiCl are mediated by the same neural pathway.

38.2.2.3 *Estradiol and Anorexia Nervosa.*

Anorexia nervosa is an eating disorder with unknown physical or emotional causes that occurs most often in young women and is characterized by a failure to maintain body weight within the normal range for age and height, an intense fear of gaining weight, and a disturbance of body image (DSM-IV-TR, 2000). A number of theories, ranging from psychosocial to physiological, have been proposed to account for anorexia nervosa, but most theories remain unsubstantiated or inadequate. In general, psychological therapy has had limited success as evidenced by follow-up studies twenty years after treatment showing that 71% of patients had not recovered and of these, 21% had died of suicide or complications of the disorder (Ratnasuriya et al., 1991). Many different kinds of chemical interventions also have been employed with little success (Mitchell, 1989). In recent analyses of psychotropic medication use, it was concluded that although a high number of individuals with anorexia nervosa use these medications, there is no conclusive evidence supporting their efficacy (Fazeli et al., 2012; Kishi et al., 2012).

There is an estrogen based hypothesis that does have supporting evidence (Chambers and Bernstein, 2003; Gustavson and Gustavson, 1986; Gustavson et al., 1989b; Young, 1991). At its core is the proposition that the brain of some females is altered during fetal and early postnatal development such that estrogen responsive tissues become hypersensitive and consequently some of the normal physiological and behavioral effects of this hormone become exaggerated. Four effects that could contribute to anorexia are: (1) triggering unconditioned satiety at lower levels of estradiol such as found during the luteal phase of the reproductive cycle (see Estradiol and Food Intake); (2) resetting existing conditioned satieties and acquiring new ones (see Introduction); (3) inducing unconditioned illness anorexia at lower levels of estradiol such as found during the reproductive cycle (see Anorexia and Illness); and (4) inducing CTAs (see Estradiol and CTA). These four factors could interact at puberty in individuals with estrogen hypersensitivity to induce anorexia and precipitate substantial weight loss.

The evidence supporting the estrogen hypersensitivity hypothesis is as follows. First, anorexia nervosa is more prevalent in women than men, with more than 90% of cases of anorexia nervosa occurring in females (DSM-IV-TR, 2000). Second, early satiety has been reported in patients with anorexia nervosa, suggesting an involvement of unconditioned and/or conditioned satiety in this disorder (Holt et al., 1981). Third, diethylstilbesterol (DES), a synthetic non-steroidal estrogen, alters the brains of female rats when administered during fetal/neonatal development such that when they are adults, they acquire taste avoidance with lower doses of estradiol, thus indicating a heightened sensitivity to the conditioning properties of estradiol (Gustavson et al., 1989a). Fourth, there is an association between DES usage and increased incidence of anorexia nervosa. The agricultural usage of DES began around 1950 (Nash et al., 1983; Rico, 1983; Rumsey, 1983) and the medical usage began in 1947 when physicians began to prescribe it to pregnant women in hopes of avoiding complications associated with pregnancy (Hammes and Laitman, 2003). This usage peaked in 1964 and continued through 1971 when the FDA encouraged the discontinuation of its use because of its association with vaginal and cervical cancer. The usage corresponds with the increase in the incidence of anorexia nervosa in 14–20-year-old females, but not other age groups, in the US and Switzerland beginning about 1965 and continuing through 1976 (Jones et al., 1980; Willi and Grossmann, 1983). Fifth, women who were exposed to DES fetally are five times more likely to show inexplicable weight loss and to be diagnosed with an eating disorder such as anorexia nervosa or bulimia nervosa, compared to women who were not exposed to the drug (Gustavson et al., 1989a, 1991). Although one cannot draw a causal relationship between these two findings, they do allow the possibility that there is one and that this relationship was forged by chemically induced alteration of the developing fetal/neonatal brain. Sixth, an association among estrogen, illness, anorexia nervosa, and CTA has been reported in humans. In patients with anorexia nervosa, episodes of vomiting are associated with estrogenic vaginal smears and estrogen therapy worsens the vomiting episodes (Moulton, 1942). Consistent with this report is a case study of a 15-year-old girl who suffered incidences of nausea and repeated vomiting subsequent to being placed on birth control pills containing ethinyl estradiol (Gustavson et al., 1988, 1989b). Over the course of months, she eventually met the criteria for anorexia nervosa; the range of foods that she would eat became limited and her body weight decreased from 26% more than expected for her age and height to more than 27% below her expected weight. For a few of the foods she indicated a negative shift in palatability, which also has been reported in other patients with anorexia nervosa (Drewnowski, 1989), and identified food as the source of her illness. When she was taken off the birth control pills, the vomiting subsided. She then participated in a CTA experiment one month after she stopped taking the birth control pills (Gustavson et al., 1988, 1989b). She showed suppression in consumption of flavored water that had been paired with an ethinyl estradiol tablet but not to flavored water that had been paired with a sucrose tablet. The estradiol tablet had triggered nausea and vomiting responses. There is enough support for an estrogen-based hypothesis to warrant serious consideration and further study, especially in light of the fact that anorexia nervosa remains a serious and intractable eating disorder. This hypothesis does not preclude the contribution of other factors. Certainly, social and cultural factors are powerful forces that could help maintain and exacerbate the anorexia that estrogen triggered.

38.3 CONDITIONED TASTE AVERSION IN THE CONTEXT OF ILLNESS SYSTEMS

38.3.1 Anorexia and Illness

38.3.1.1 Systemic Infections. Animals exhibit a constellation of physiological and behavioral changes during the acute phase response to infection, including activation of the hypothalamic-pituitary-adrenal (HPA) axis, fever, lethargy and reduction in motor activity, depression and reduction in general environmental responsiveness, decreases in social exploration, increases in sleepiness and slow-wave sleep, and anorexia as well as reductions in food motivated behaviors (see Hart, 1988 and Kent et al., 1992). These responses are thought to be part of a homeostatic behavioral strategy that has evolved to facilitate survival of individuals suffering from systemic illnesses caused by a variety of microorganisms, including bacteria and viruses (Asarian and Langhans, 2010; Hart, 1988; Murray and Murray, 1981). The importance of anorexia in fighting against infection is demonstrated by the following: force-feeding mice during bacterial infections reduces survival time and increases mortality (Murray and Murray, 1979), while depriving mice of food two to three days before bacterial infection increases survival rates (Wing and Young, 1982). Three cytokines that are part of the cytokine cascade triggered by endotoxin infection have been implicated in the anorexia accompanying infection. Lipopolysaccharides (LPS) from gram-negative bacteria cell walls trigger release of interleukin-1 (IL-1), tumor necrosis factor-α (TNF-α), and high mobility group-1 (HMG-1) protein and reduce food intake in rodents (Langhans et al., 1993b; Nathan, 1987; Wang et al., 1999). Evidence suggests that this reduction is mediated by the released cytokines as administration of each of these cytokines produces anorexia (Agnello et al., 2002; Langhans et al., 1993a; McCarthy et al., 1984). The specific role of each of these cytokines and the nature of their interaction in producing anorexia is yet to be fully delineated.

38.3.1.2 The Chronic Disease State Cancer. It is well established that anorexia is a significant problem for individuals suffering from various kinds of cancer. It has been suggested that early satiety and symptoms of abdominal fullness as well as altered taste quality, which are commonly found in cancer patients, promote anorexia (DeWys, 1985). However, these factors may be secondary to the contribution of cytokines. A series of experimental and clinical studies have demonstrated that advanced-stage cancer patients with anorexia have high serum levels of Il-1 and TNF-α and treatment with megestrol acetate, which downregulates the synthesis and release of cytokines, increases appetite (Mantovani, et al., 1998).

38.3.1.3 Toxicity. LiCl triggers nausea and vomiting and as such is regarded as an illness-inducing agent. However, it also triggers many of the same physiological and behavioral responses as infectious illnesses, including activation of the HPA axis, decreases in motor activity, general environmental responsiveness, and social exploration, increases in sleepiness and slow-wave sleep, and anorexia (see Chambers and Bernstein, 2003; Eckel and Ossenkopp, 1993). This aftermath of exposure to LiCl and infectious agents is so strikingly similar that it is tempting to view illness as a unitary phenomenon with one mediating mechanism that controls its multifaceted expression. However, differences have been found in the effects of LiCl and LPS exposure. For example, both of these agents induce a fall in body temperature within a few minutes after injection, but while fever then ensues within the next 6 minutes for LPS, it is not a consequence of LiCl injection (Bull et al., 1991). Furthermore, LiCl promotes anorexia through a mechanism that does not involve cytokines; it acts via vasopressin V_1 receptor-mediated activation of an α-adrenergic mechanism, in support of this hypothesis (Langhans et al., 1991). Vasopressin also mediates anorexia provoked by cytotoxic chemotherapy, which compounds the anorexia associated with cancer itself (see Scalera and Bavier, 2009). Although it has been proposed that cytokines mediate chemotherapy-induced anorexia, animal research implicates vasopressin (Sinno et al., 2010). During methotrexate treatment, food intake is reduced and c-FLI is expressed in vasopressin neurons located in the hypothalamic supraoptic and paraventricular nuclei. In contrast, there is no evidence of IL-1 activation. These data indicate that different illness promoting agents employ different mediating mechanisms to activate similar response mechanisms. Illness, then, appears to be a multifaceted phenomenon, each illness defined by the specific sensory, mediating, and response mechanisms involved.

38.3.2 CTA and Illness

A number of investigators have suggested that the formation of CTAs could contribute to the anorexia observed during systemic infections, cancer, and chronic toxicity exposure. Evidence supports this hypothesis in that taste avoidance can be induced by pairing a novel diet or solution with LPS, IL-1, HMG-1, implantation of experimental sarcoma, and chronic infusion of LiCl (Agnello et al., 2002; Bauer et al., 1995; Bernstein and Goehler, 1983; Bernstein et al., 1985; Weingarten et al., 1993). However the ability of these agents to induce $CR_{CS\text{-}AVOIDANCE}$ does not extend to familiar diets, which limits the contribution that conditioning makes to anorexia and suggests that CTA may not play a role under normal feeding conditions. Still, there is a caveat to this conclusion. In all of the above studies, the animals had limited experience with different diets before they had been given the novel diet. Nothing is known about the impact of CTAs when animals are maintained from an early age on varied diets, a situation that is more typical of the eating patterns of humans and of rodents living in the wild or kept as pets. This issue is important since it is possible that for individuals on varied diets, the rules of conditioning change. Although conditioned avoidance of substances consumed since weaning is difficult to obtain in animals with limited diet variation (Elkins, 1974; Kalat and Rozin, 1973), for animals exposed to a variety of flavored foods and drinks, familiarity may not so easily exclude a particular food/drink from conditioning.

It also has been concluded that learned food aversion plays an inconsequential role in the anorexia of cancer patients undergoing chemotherapy (Mattes et al., 1992). This conclusion was based on the failure to find associations between the experiences of nausea/vomiting and the incidence of food aversions, the degree of loss of appetite and the likelihood of developing food aversions, and the presence of food aversions and the presence of anorexia. However, it has been demonstrated that aversions to specific foods consumed before treatment are learned in patients receiving gastrointestinally toxic chemotherapy (Bernstein, 1978; Bernstein and Webster, 1980). Furthermore, when a distinctive flavored food (such as coconut or root beer Lifesavers) is used as a target food by giving it to children and adults after consumption of a meal but before administration of chemotherapy, this "scapegoat" interferes with the development of aversions to items in the meal (Broberg and Bernstein, 1987; Mattes et al., 1987). Thus, the impact of chemotherapy on preferences for normal meal items can be reduced. Considering all of the data, it is likely that food aversions learned as a consequence of chemotherapy do play a role in the anorexia of cancer patients. This role may be more important in young patients since aversions develop more easily before the age of 20 (Garb and Stunkard, 1974; Logue et al., 1981).

38.4 CONDITIONED TASTE AVERSION IN THE CONTEXT OF STRESS SYSTEMS

38.4.1 The Hormonal Stress Response

Most stress researchers agree that stress is composed of three components (see Levine, 2005):

1. a stimulus or stressor, which is an environmental event that can be sensed by an individual, has aversive properties, is potentially harmful to the individual, and elicits an acute and/or chronic response;

2. an evaluation, which is the ability of an individual to process the stimulus, detect its hedonic value, and translate this information into a response; and

3. the stress response, which includes a complex interplay of physiological, behavioral, emotional, and cognitive changes in the individual.

Hormones play a vital role in activating the physiological changes and neural mechanisms necessary for responding effectively to stressors. Increased release of corticotropin releasing factor (CRF) and arginine vasopressin (VP) occur in less than 1 min, followed by increased secretion of adrenocorticotropin hormone (ACTH), renin, catecholamines, prolactin, glucagon, and growth hormone in less than 10 min and glucocorticoids in less than 1 hr (see Sapolsky et al., 2000). Four of these hormones have been found to play a role in CTA: CRF, VP, ACTH, and glucocorticoids.

Stressors, acting via neural systems dedicated to sensory detection and processing, induce an increase in the release of CRF and VP from the parvocellular portion of the PVH into the portal vessels. Vasopressin has little ACTH-releasing capability on its own but acts synergistically with CRF on corticotrophs in the anterior pituitary to increase release of ACTH into the general circulation. ACTH, in turn, triggers increased synthesis and release of glucocorticoids (predominantly cortisol or corticosterone) from the adrenal cortex (see Fulford and Harbuz, 2005). Besides its synergistic role as part of the HPA system, VP has diffuse effects on peripheral tissues via increased release into the general circulatory system from magnocellular neurons in the PVH and supraoptic hypothalamus whose axons terminate near capillaries in the pars nervosa of the posterior pituitary (Swab et al., 1975; Swanson and Sawchenko, 1980). In addition to their endocrine roles, both CRF and VP have been localized in and are synthesized by a number of neural sites outside the PVH, which indicates their ability to function as neurotransmitters or neuromodulators (see Steckler, 2005 and Douglas, 2005).

38.4.2 Stress Hormones and Acquisition of CTA

Exposure to LiCl triggers increases in blood levels of corticosterone and VP in rats and it induces c-FLI expression in VP neurons in the PVH and supraoptic nucleus (Ader, 1976; Hennessy et al., 1976a; Olsewski et al., 2001; Verbalis et al., 1986). However, an elevation in corticosterone after CS-US pairing is not necessary for acquisition of CTA because neither adrenalectomy nor hypophysectomy eliminate the ability of animals to acquire taste avoidance (Ader et al., 1978; Rabin et al., 1983). Whether the same is true of VP is unknown, although it is likely to be the case.

Evidence suggests that increased secretion of both hormones serve the function of reducing the impact of LiCl. Studies of glucocorticoids have shown the following: (1) glucocorticoids (cortisol, prednisone, methylprednisone, and dexamethasone phosphate) attenuate LiCl-induced taste avoidance when administered before or between CS-US pairing (Hennessy et al., 1976a; Revusky and Martin, 1988); (2) dexamethasone phosphate alleviates emetic distress in humans when administered before cancer chemotherapy (Cassileth et al., 1983; Markman et al., 1984); and (3) corticosterone fails to induce taste avoidance even after 12 pairings with a sweet solution (Smotherman and Levine, 1978). Similar to glucocorticoids, studies have shown that low doses of VP: (1) attenuate LiCl-induced taste avoidance (as measured by accelerated extinction) when administered 50 min after sucrose-LiCl pairing (Chambers and Hayes, 2005; Hayes and Chambers, 2002, 2005a); (2) reduce the effect of the illness-inducing cytokine IL-1 by partially reversing its suppression of social exploration of conspecific juveniles by adult animals (Dantzer et al., 1991); and (3) fail to induce taste avoidance even after multiple pairings. Furthermore, VP accelerates the completion of extinction, not the initiation, which is the same change that occurs when the dose of LiCl is reduced from 1.5 to 0.3 mEq/kg (Chambers and Wang, 2004). It has been suggested that one of the ways VP reduces exposure to the toxic effects of LiCl is by reducing intestinal blood flow through constriction of splanchnic blood vessels (Stricker and Verbalis, 1991).

CRF may play an opposite role during acquisition as there is a positive association between the neural CRF system and measures of aversive states. Intracerebroventricular administration of CRF induces conditioned place avoidance (Cador et al., 1992) and unlike corticosterone and VP, it can serve as a US to induce taste avoidance (Benoit et al., 2000; Gosnell et al., 1983). Further evidence for the aversive role of CRF is demonstrated by peripheral administration of urocortin 1 (Ucn1, a member of the CRF-peptide family that is a non-selective CRF receptor agonist) which increases corticosterone levels, produces diarrhea, reduces meal frequency and size, and induces taste avoidance (Fekete et al., 2011). These results are noteworthy as it has been hypothesized that the CRF neural system plays a role in mediating the aversive state during stress and activating negatively valenced avoidance tendencies (Cador et al., 1992; Heinrichs and Koob, 2004). This could mean that the CRF neural system plays a role in defining whether a particular agent is aversive and that the ability of CRF to induce taste avoidance is related to its ability to directly activate those circuits involved in evaluating and concluding that certain properties of an agent are harmful.

38.4.3 Stress Hormones and Extinction of CTA

Extinction is delayed when ACTH and VP are given before each extinction trial (Kendler et al., 1976; Rigter and Popping, 1976; Vawter and Green, 1980). These hormones have an immediate suppressive effect on CS consumption during extinction. When ACTH was administered prior to the second extinction test under forced drinking conditions (given one bottle containing the CS after 48 hours of water deprivation), consumption of the CS was reduced on that test and when VP was given before the fifth extinction test, CS consumption on that test was suppressed despite the fact that the animals had begun extinguishing their avoidance (Cooper et al., 1980; Hennessy et al., 1976b). Given that endogenous ACTH and VP are released after LiCl injection, it may be that the exogenous presence of these hormones before encountering the CS during extinction served to as a reminder cue which evoked taste avoidance memory. It has been suggested that ACTH facilitates retrieval of the avoidance memory (Hennessy et al., 1976b). Support for this hypothesis is provided by a study with infant rats that were given three different CS-US intervals during acquisition (0-hr, 1-hr, and 24-hr) and injected with $ACTH_{4-10}$ or saline prior to the retention test (Dray and Taylor, 1982). The $ACTH_{4-10}$ and saline animals acquired equivalent taste avoidances with the 0-hr interval and no taste avoidances with the 24-hr interval. With the 1-hr interval, $ACTH_{4-10}$ infants acquired taste avoidance while the saline rats did not. Thus, the avoidance memory was present in the 1-hr interval infants, but hard to retrieve without $ACTH_{4-10}$ augmentation. This effect is not dependent on the glucocorticoid response that accompanies ACTH treatment because $ACTH_{4-10}$ is devoid of steroidogenic activity. The same is true for the delayed extinction effect in adults as $ACTH_{4-10}$ also suppresses CS consumption during extinction (Rigter and Popping, 1976).

Retention of other types of inhibitory avoidance behavior, such as step-through shock avoidance, also is augmented by $ACTH_{4-10}$ (van Wimersma Greidanus et al., 1979; Wan et al., 1992). Evidence suggests that this peptide produces its effect via modulatory effects on the release of neurotransmitters in the limbic system. Infusion of $ACTH_{4-10}$ directly into the hippocampus increases concentrations of serotonin (van Wimersma Greidanus et al., 1983). Furthermore, there is a positive association between increased hippocampal serotonin and the ability to retrieve and express retention of step-through avoidance (Ramakekers et al., 1978). In untreated rats, levels of serotonin are elevated 24 hr after acquisition and retention is expressed. If rats are subjected to amnesic carbon dioxide treatment immediately after acquisition, levels of serotonin fail to rise and avoidance is not expressed. However, if amnesic rats are administered $ACTH_{4-10}$ 1 hour before the retention test, serotonin levels rise and retention of the avoidance is expressed. It is unclear how ACTH achieves its effect on extinction of taste avoidance but a similar type of mechanism is possible, although not necessarily the same neurotransmitter or neural structure. For CTAs, a functional amygdala appears to be necessary as bilateral lesions eliminate the ability of ACTH to delay extinction (Burt and Smotherman, 1980).

38.5 NEURAL MECHANISMS FOR CONDITIONED TASTE AVERSIONS

38.5.1 Acquisition of the CS-US Association

If rats are force-fed with saccharin and then poisoned with LiCl while under pentobarbital anesthesia or bilateral reversible decortication via cortical spreading depression (CSD), they do not acquire taste avoidance but they do exhibit unconditioned signs of illness (Bures and Buresova, 1989; Buresova and Bures, 1973). Similarly, taste avoidance cannot be learned when rats are decerebrated at the supracollicular level, which leaves a functional nucleus of the solitary tract (NST) and parabrachial nucleus (PBN) that are cut off from forebrain intercommunication (Grill and Norgren, 1978). However, when either anesthesia or CSD is given after exposure to saccharin but before LiCl administration, taste avoidance learning is not prevented (Bures and Buresova, 1989; Buresova and Bures, 1973). These results suggest that acquisition of taste avoidance requires cortical involvement in taste processing, but subcortical areas are sufficient for illness detection and processing.

38.5.1.1 *Taste Stimulus Detection.* Taste information is transmitted from taste receptor cells located primarily in the tongue to the chorda tympani and superficial petrosal branches of the facial nerve, the lingual branch of the glossopharyngeal nerve, and the vagus nerve (Hamilton and Norgren, 1984). The gustatory afferent fibers from these nerves terminate in the ipsilateral rostral NST. Second order gustatory neurons in the rostral NST project ipsilaterally to neurons in the medial PBN (mPBN), ventral subnucleus of the lateral PBN (lPBN), and waist region in rodents and to neurons in the parvocellular ventroposteromedial thalamic nucleus in primates (VPMpc; see Pritchard and Norgren, 2004; Herbert et al., 1990; Travers et al., 1987). Neural areas rostral to the PBN are not critical for innate hedonic reactions to taste in rats as intraoral taste stimulation elicits the same ingestive and aversive patterns of taste responsiveness at the same concentrations in intact and supracollicular decerebrated rats (Grill and Berridge, 1985).

38.5.1.2 *Toxicity Stimulus Detection.* The stimuli that activate other sensory systems such as vision, audition, gustation, and olfaction, are well known and the sensations produced by activation of these systems are clearly definable. This is not the case for the illness sensory system; we do not have a clear picture of what stimuli constitute illness. Within the context of the studies reviewed in this chapter, identification of stimuli that can activate the illness sensory system has been done almost entirely on the basis of whether a given stimulus can induce $CR_{CS-AVOIDANCE}$. Given this criterion, the list of stimuli that can activate the illness sensory system and thus are considered illness-inducing agents seems limitless, at least for the rat.

The sensation of illness is a conscious awareness of negative changes in the body. One sensation that is commonly reported during illness, and specifically reported after lithium administration, is nausea (Karniol et al., 1978; Schou, 1968). Garcia (1989) has held that nausea is the unifying symptom that ties together all the agents that can induce CTA, as defined by both reduced consumption and negative hedonic shift. Of course investigations of nausea in animals are hampered by the inability to assess what constitutes feelings of nausea. This issue has been circumvented by making two assumptions: (1) any drug that causes nausea in humans will induce nausea in animals, and (2) retching and vomiting in emetic species and aversive orofacial/somatic responses in the nonemetic rat comprise the behavioral manifestation of nausea. In this vein, it has been shown that both lactose, which does not produce nausea in humans, and LiCl, which does, induce comparable decreases in consumption in rats, but only LiCl produces aversive orofacial/somatic responses (Pelchat et al., 1983). This suggests that nausea mediates $CR_{CS-AVERSION}$ but not $CR_{CS-AVOIDANCE}$. Support for this hypothesis comes from studies examining the effects of antinausea drugs on LiCl-induced $CR_{CS-AVOIDANCE}$ and $CR_{CS-AVERSION}$ in rats (Parker, 2003). Acquisition of taste aversion, but not taste avoidance, is disrupted by administration of antinausea drugs that reduce serotonin availability (Limebeer and Parker, 2000, 2003). Furthermore, 5,7-DHT lesions of the dorsal and median raphe nuclei, which effectively reduce serotonin availability, also prevent acquisition of taste aversion, but not taste avoidance (Limebeer et al., 2004). Parker has suggested that LiCl-induced nausea may be mediated, at least in part, by serotonergic pathways (Parker et al., 2009).

If nausea mediates $CR_{CS-AVERSION}$ in rats than what mediates $CR_{CS-AVOIDANCE}$? Studies using VP as an illness-inducing agent have suggested its vascular pressor property (vasoconstriction) as a possible stimulus mediating learned taste avoidance. This hypothesis is supported by three pieces of evidence. First, doses of VP that do not alter systemic blood pressure also do not induce $CR_{CS-AVOIDANCE}$ when VP is infused intracerebroventricularly (De Wied et al., 1984). Second, VP analogs with weak or no pressor-agonist activity do not induce $CR_{CS-AVOIDANCE}$ (Bluthe et al., 1985; Ettenberg, 1984). Third, VP-induced $CR_{CS-AVOIDANCE}$ can be blocked by subcutaneous or intracerebroventricular administration of the V_1 pressor antagonist dPTyr(Me)AVP and the amount required to block avoidance also blocks peripheral changes in blood pressure (Bluthe et al., 1985; Lebrun et al., 1985). That the pressor property of VP also mediates LiCl-induced $CR_{CS-AVOIDANCE}$ is supported by the fact that LiCl triggers increases in both VP and blood pressure (O'Connor et al., 1987; Verbalis et al., 1986).

Because LiCl is the putative illness-inducing agent, knowledge of the neural illness sensory system involved in CTA has been based primarily on studies using this agent as the US. The investigation has led to the AP (area postrema) as a necessary and sufficient site of action for illness detection. The evidence is as follows:

1. permanent lesions of the AP prevent expression of many of the UR effects of LiCl, including lying-on-belly, decreased motor activity, delayed stomach emptying, and hypothermia (Bernstein et al., 1992; Ladowsky and Ossenkopp, 1986);

2. c-FLI is induced in the AP after CS-US pairing (Swank et al., 1995);

3. permanent lesions of the AP prevent acquisition of taste avoidance and aversion (Eckel and Ossenkopp, 1996; Ritter et al., 1980);

4. acquisition of taste avoidance is blocked when reversible cooling lesions start immediately after sucrose consumption and continue through the peak effectiveness of LiCl but are not administered during post-acquisition testing (Wang et al., 1997a,b); and

5. electrical stimulation of the AP can serve as a US in a taste avoidance paradigm (Gallo et al., 1988).

Although Li is concentrated throughout the brain after administration of LiCl (Muckherjee et al., 1976), the fact that this agent is unable to induce CTA when the AP is lesioned suggests that the only CTA-associated neurons directly activated by LiCl are in the AP. It is thought that LiCl information is transmitted to other neural areas involved in CTA via general visceral projections from the AP. It should be noted that the AP is not a necessary and sufficient site of action for all agents. For example, nicotine also induces CTA, but lesions of the AP do not reduce $CR_{CS-AVOIDANCE}$ (Ossenkopp and Giugno, 1990; Parker, 1995).

It has been suggested that the pressor property of VP induces $CR_{CS-AVOIDANCE}$ by acting via baroreceptor vagal connections to the brainstem (Bluthe et al., 1985; Hreash, et al., 1990). However, bilateral subdiaphragmatic vagotomy does not prevent acquisition of LiCl-induced taste avoidance (Martin et al., 1978). It is more likely that LiCl-triggered VP acts directly on pressor regulating neurons in the AP. This circumventricular structure has neurons that contain V_1 receptors, is directly activated by VP, and has been implicated in pressor regulation (Borison, 1989; Carpenter et al., 1988; Phillips et al., 1988; Skoog et al., 1990; Skoog and Mangiapane, 1988; Tribollet et al., 1998).

38.5.1.3 Integration of Taste and Toxicity.
Convergence of gustatory and LiCl information at a particular neural site or sites and ultimately on single neurons in the site(s) is essential for association to take place. Evidence of such a site should include strong evidence that it is involved in CTA acquisition and that it receives both gustatory and toxicity information. Furthermore, based on the anesthesia and CSD data discussed in the introduction to this section, the site should be a subcortical area and the gustatory information it receives should come either directly or indirectly from gustatory processing areas in the cortex.

A substantial amount of evidence has established the involvement of three neural areas in acquisition of taste avoidance, the PBN (lateral and medial), basolateral amygdala (BLA, an anatomical unit composed of lateral and basal nuclei), and the central insular cortex (cIC). After pairing saccharin with LiCl, c-FLI is induced in all three areas, including both lPBN and mPBN (Koh and Bernstein, 2005). Furthermore, bilateral electrolytic and ibotenic lesions abolish expression of LiCl-induced $CR_{CS-AVOIDANCE}$ in both the lPBN and mPBN (Agüero et al., 1993; Grigson et al., 1998; Reilly and Trifunovic, 2000; Sakai and Yamamoto, 1998; Spector et al., 1992) but only attenuate expression in the BLA and cIC (see Braun et al., 1982 and Reilly, 2009; Dunn and Everitt, 1988; Morris et al., 1999; Nachman and Ashe, 1974; Nerad et al., 1996; Sakai and Yamamoto, 1999; Simbayi et al., 1986; St. Andre and Reilly, 2007).

THE CENTRAL INSULAR CORTEX The cIC is an area that receives both ascending gustatory and visceral input (Hanamori, 1998). However, because neither CSD nor reversible tetrodotoxin lesions of the cIC applied during the CS-US interval impairs acquisition of taste avoidance, the cIC plays a gustatory role (Burešova and Bureš, 1973; Gallo et al., 1992; Roldan and Bures, 1994). This role is not simple sensory detection as lesioned rats are able to display a significant preference for saccharin when given a choice between water and a saccharin solution (Dunn and Everitt, 1988). The predominant hypothesis at present is that the role is associated with determination of CS novelty

38.5 Neural Mechanisms For Conditioned Taste Aversions

and that cIC lesions reduce the degree to which the CS is perceived as novel (see Bernstein et al., 2009 and Reilly, 2009). This hypothesis is supported by the following: (1) animals with cIC lesions show normal acquisition when the CS is familiar but attenuated taste avoidance when it is novel (Kiefer and Braun, 1977; Roman and Reilly, 2007); (2) stronger c-FLI induction is found in the cIC after novel saccharin-LiCl pairing than after familiar saccharin-LiCl pairing (Koh and Bernstein, 2005); and (3) both electrolytic and ibotenic acid lesions of the cIC reduce the neophobic response of animals to the CS (Dunn and Everitt, 1988). Koh and Bernstein (2005) have suggested that the cIC performs two functions during taste conditioning: processing taste to be stored in memory as it progresses from novel to familiar and providing online comparisons between incoming tastes and stored taste memories.

Two studies have demonstrated the importance of connections between the IC and the amygdala and PBN for taste avoidance acquisition. Contralateral lesions of the IC (by tetrodotoxin before CS presentation) and amygdala (by tetrodotoxin before LiCl injection) disrupt acquisition of taste avoidance but ipsilateral lesions have no effect (Bielavska and Roldan, 1996). Likewise, contralateral lesions of the IC (by reversible CSD before CS presentation) and PBN (by reversible tetrodotoxin before LiCl injection) disrupt avoidance learning while ipsilateral lesions have no effect (Gallo and Bures, 1991). If the function of the cIC is to inform the amygdala and PBN of the novelty status of the CS, then the contralateral data indicate that for one side of the brain, the CS consumed before illness was not novel (IC lesioned but amygdala and PBN not lesioned) and for the other side of the brain, illness did not follow consumption of a novel CS (amygdala and PBN lesioned but IC not lesioned). In either case, acquisition of taste avoidance would be compromised. The question to be answered is whether the disrupted acquisition is based on CS-US convergent neurons in the amygdala and PBN some other factor.

THE BASOLATERAL AMYGDALA Looking first at the amygdala, there are gustatory projections from the IC to the lateral and basal amygdala (Ottersen, 1982; Nakashima et al., 2000), the involvement of the BLA is limited to novel tastes as lesions attenuate learned avoidance to novel but not familiar tastes (Morris et al., 1999; St. Andre and Reilly, 2007), BLA lesions disrupt taste neophobia (Reilly and Bornovalova, 2005), and stronger c-FLI induction is found in the BLA after pairing LiCl with novel saccharin than with familiar saccharin (Koh and Bernstein, 2005). Furthermore, lesions of the cIC reduce the amount of c-FLI induction in the BLA in response to novel saccharin-LiCl pairing to the level found after familiar saccharin-LiCl pairing and reversible lesions of the cIC with muscimol during familiarization (CS preexposure) of saccharin increases c-FLI induction in the BLA

in response to familiar saccharin-LiCl to the level found after novel saccharin-LiCl pairing (Koh and Bernstein, 2005). Because the BLA does not respond differentially to novelty in the absence of the cIC, the BLA is informed by the cIC as to novelty status. It has been suggested that BLA lesions reduce the degree to which the CS is perceived as novel thus producing a CS preexposure-like (latent inhibition-like) effect (see Reilly, 2009).

The BLA also receives LiCl information; c-FLI is induced in the BLA after LiCl (Koh and Bernstein, 2005) and evidence suggests there are two indirect general visceral pathways from the lPBN to the BLA (Groenewegen and Berendse, 1994; Sakai and Yamamoto, 1999). These data taken together with the contralateral/ipsilateral data described above (Bielavska and Roldan, 1996) suggest that receipt of LiCl information by the BLA from the lPBN is important for successful acquisition. Although the data are consistent with an integration hypothesis, it also is possible that the BLA only conveys CS novelty and LiCl information to the site(s) of integration.

THE PARABRACHIAL NUCLEUS Receipt of both taste and LiCl information by the PBN is important for successful taste avoidance acquisition (Bielavska and Rolday, 1996). The most compelling evidence for integration of that information comes from electrophysiological and anatomical studies that have identified convergence of taste and general visceral sensory projections within the PBN. Visceral neurons in the NST project to neurons within the gustatory zone (Herbert et al., 1990) and the threshold of the gustatory evoked responses of some neurons in the PBN can be modified by visceral (vagal) afferent stimulation (Hermann and Rogers, 1985). Although not directly applicable to LiCl-induced CTA, vagal rather than postremal stimulation, it does identify the PBN as an area of neuronal convergence. Direct support for the integration hypothesis awaits studies showing actual convergence of gustatory input from the cIC (either directly or indirectly) on neurons receptive to LiCl.

Most investigators are in agreement that the PBN is a site of CS-US integration, but they differ in their view as to whether the specific site is in the lPBN or mPBN (Chambers and Wang, 2004; Reilly, 1999; Spector et al., 1992; Yamamoto et al., 1994). The AP projects directly and indirectly (via the caudal NST) to the lPBN (Herbert et al., 1990; Shapiro and Miselis, 1985) and although stimulation of the AP is an effective US in intact rats, it is ineffective in inducing $CR_{CS-AVOIDANCE}$ in animals with electrolytic lesions of the lPBN (Agüero et al., 1993). However, induction of c-FLI is found in both the lPBN and mPBN after LiCl administration, indicating the LiCl information does reach the mPBN (Koh and Bernstein, 2005). Because expression of $CR_{CS-AVOIDANCE}$ can be blocked when temporary and reversible cooling lesions of the lPBN are applied only during LiCl processing, the lPBN is not necessary for CS detection and processing or CR retrieval. This leaves the possibility that the lPBN is involved in making and storing the taste-illness association as well as in processing LiCl (Wang and Chambers, 2002). Three specific subnuclei in the lPBN have been implicated in acquisition of taste avoidance. After sucrose-LiCl pairing, c-FLI is induced in the central, external, and crescent lPBN and there is strong correspondence between amount of c-FLI induction and strength of $CR_{CS-AVOIDANCE}$ in each of these three subnuclei (Chambers and Wang, 2004). With respect to the mPBN, failure to acquire taste avoidance after mPBN lesions cannot be accounted for by deficits in taste or illness detection and processing because animals with mPBN lesions are capable of learning a conditioned taste preference for a noncaloric flavor paired with calorie-rich sucrose and a learned place avoidance based on LiCl administration (Reilly et al., 1993). This leaves the possibility that mPBN lesions produce deficits in the ability to make, store, retrieve, or express the taste-illness association. At the present time, there is no conclusive data that eliminates either one of these two PBN areas from consideration as an integrative site.

AN EMERGING PICTURE The gustatory information the PBN receives is sent to the cIC to be assessed for novelty status. That information is sent back to convergent neurons in the PBN where it can wait for hours for the arrival of toxicity information. If toxicity information arrives (from the AP in the case of LiCl), an association is formed and when the taste is encountered again, taste avoidance ensues. These three neural components are essential for acquisition of LiCl-induced CTA. The BLA also receives gustatory information from the cIC and it receives toxicity information from the lPBN. Thus, integration may take place in this structure as well as the PBN. If so, the integration in the BLA is not necessary for acquisition of CTA, as lesions only attenuate learned taste avoidance. Instead it serves an additive function and as a result, a stronger CTA is achieved.

38.5.2 Expression of the CS-US Association

When an animal encounters a taste stimulus after it has been paired with a toxin, expression of illness, avoidance, and aversion responses is a reflection of successful association, storage of the association, and retrieval of the association from storage. Two approaches have been taken in the study of neural pathways mediating conditioned responses. One is the examination of changes in electrophysiological responses and c-FLI induction in response to the CS after acquisition and the other is the determination of whether lesioning a discrete neural area after acquisition abolishes expression of a conditioned response.

It is unlikely that the AP is involved in conditioned responses to the CS. Lesions of the AP after acquisition do not prevent expression of $CR_{CS-AVOIDANCE}$ or the decrease in heart rate that occurs in response to a CS (Kosten and Contreras, 1989; Rabin et al., 1984). Also, after pairing a sweet solution with LiCl, c-fos-like immunoreactivity is not induced in the AP when rats are exposed to the CS (Houpt et al., 1994; Swank and Bernstein, 1994).

There is evidence of the involvement of the NST in the mediation of conditioned responses. The pattern of activity of sugar responsive NST neurons changes after conditioning such that exposure to a saccharin CS triggers a brief burst of activity that is absent in unconditioned rats (Chang and Scott, 1984). Likewise, intraoral infusion of sweet tastes before conditioning does not induce c-FLI in the medial intermediate NST, but c-FLI is induced in this area when exposed to sucrose after it has been paired with LiCl (Houpt et al., 1994). A change in induction of c-FLI in response to saccharin also has been found in the PBN after conditioning. Saccharin induces c-FLI in the dorsal and central lPBN in naive rats, but in rats with acquired taste avoidance to saccharin, c-FLI is induced predominantly in the external lPBN (Yamamoto, 1993, 1994; Yamamoto et al., 1994). Lesions of the rostral gustatory NST and mPBN do not affect expression of $CR_{CS-AVOIDANCE}$ acquired prior to lesioning, which indicates that these areas of the NST and mPBN are not essential for expression of this conditioned response (Grigson et al., 1997). On the other hand, rats decerebrated at the supracollicular level do not express an avoidance acquired prior to transection (Grill and Norgren, 1978). Taken together, these data suggest that regions within the NST and/or PBN but outside the gustatory zones lesioned in the Grigson study are necessary for expression of $CR_{CS-AVOIDANCE}$.

Both the BLA and cIC also have been implicated in expression of conditioned responses. Electrolytic/ablation and cytotoxic lesions of both of these structures disrupt expression of taste avoidance learned prior to lesioning (Braun et al., 1981; Nachman and Ashe, 1974; Ormsby et al., 1998; Simbayi et al., 1986; Yamamoto, 1994; Yamamoto et al., 1980). The cIC probably is involved in retrieval rather than storage since fetal transplants can restore expression of an avoidance learned prior to lesioning (Ormsby et al., 1998). This suggests that the cIC plays a role in accessing an experiential memory of the CS that includes not only a novel-familiar dimension but a safe-toxic dimension.

38.5.3 Extinction of the CS-US Association

Extinction is a reflection of the integration of two different processes. In a CTA situation, it is a reflection of the strength of the taste-illness association, which impacts the vigor with which the association is retained in storage and the facility with which the association is retrieved from storage. The course of extinction also reflects the ability to forge a new memory, as extinction is a learning process (Hull, 1943). The prevailing view is that animals learn a new association while retaining memory of the CTA, rather than unlearning or forgetting the original aversive learning experience and consequently replacing that memory. Evidence for this assertion comes from other learning paradigms in which the rate of reacquisition after extinction is more rapid than the original acquisition, spontaneous recovery of the conditioned behavior occurs after exposure to the US in the absence of the CS (Bouton, 1993; Leung et al., 2007; Pavlov 1927; Rescorla, 2001). What constitutes the US in the new association formed during extinction of CTA is a subject of debate, but generally, the new association is referred to as learned safety. Whether no-illness is an identifiable neural response is at the heart of the debate.

After extinction, a conditioned sweet taste no longer induces c-FLI in the medial intermediate NST, which means that there is a reversal back to the c-FLI response to the sweet taste prior to conditioning (see Expression of the CS-US Association; Houpt et al., 1994). Furthermore, when c-Fos antisense is infused into the NST prior to the first post-acquisition test, it prevents initiation of extinction (Swank et al., 1996). Taken together, these results suggest that the NST plays an important role in extinction. It is noteworthy that some neural responses in the NST remain after extinction. Examination of the subpopulation of sugar responsive neurons showing brief bursts of activity in response to saccharin after conditioning has revealed that after expression of the $CR_{CS-AVOIDANCE}$ is fully extinguished, these neurons continue to respond to saccharin with a well-defined burst of activity that has reduced amplitude and slightly longer latency than in fully conditioned rats (see Expression of the CS-US Association; McCaughey et al., 1997). The authors have suggested that this neural response may be a permanent marker for saccharin as a predictor of illness that could influence any subsequent reacquisition of the taste avoidance.

Amygdalar and cortical areas also have been implicated in extinction. Initiation of extinction is prevented when the protein synthesis inhibitor anisomycin is infused into the BLA and cIC before and into the infralimbic-prelimbic areas of the VPMpc after the first post-acquisition test (Akirav et al., 2012; Bahar et al., 2003; Berman and Dudai, 2001). In addition, if a second infusion into the cIC is given prior to the second post-acquisition test, extinction remains blocked, but if no infusion is given, initiation of extinction proceeds (Berman and Dudai, 2001). Increases in c-FLI induction also have been found in these three areas during extinction (Mickley et al., 2002, 2005). The cortex, however, may not be critical for extinction. Animals are able to extinguish a previously learned avoidance when they are anesthetized or under CSD (Bures and Buresova, 1979).

38.6 CONCLUSION

Conditioned taste aversion is a member of the classical conditioning category of learning paradigms and as such the responses elicited by the taste change from ingestive responses to illness, avoidance, and aversion responses after the taste-illness association has been learned. These learned responses play a critical role throughout our lifetime in protecting us from eating harmful substances. Although CTA typically is learned from firsthand experience, we also can learn to avoid foods that have made others ill. In addition to its role in the learning and feeding systems, CTA intersects with a number of other neural systems, including illness and stress systems. When infected by bacteria or viruses, a constellation of physiological and behaviors responses are elicited that play a critical role in defending against the invasion and increasing our ability to survive. The toxins that induce CTA elicit some of the same physiological and behavioral responses triggered by infectious microorganisms and it is possible that some of the response components of this illness system are employed to effect the taste-illness association. In turn, the illness system uses CTA to promote its anorectic defense against infection. Stress hormones also are triggered by CTA-inducing toxins and although they may not play a critical role in acquiring the taste-illness association, they produce physiological changes that reduce exposure to the toxin and potentially save us from lethal exposure. Because of the intersection of CTA with these neural systems, the study of this learned behavior and the neural mechanisms that mediate it could provide insight into how the different systems function and interact with one another to promote our well-being.

REFERENCES

Ackerman, B. H. and Kasbekar, N. (1997). Disturbances of taste and smell induced by drugs. *Pharmacotherapy*. 17: 482–496.

Ackroff, K., Vigorito, M., and Sclafani, A. (1990). Fat appetite in rats: the response of infant and adult rats to nutritive and non-nutritive oil emulsions. *Appetite*. 15: 171–188.

Ader, R. (1976). Conditioned adrenocortical steroid elevations in the rat. *J. Comp. Physiol. Psychol*. 90: 1156–1163.

Ader, R., Grota, L. J., and Buckland, R. (1978). Effects of adrenalectomy on taste aversion learning. *Physiol. Psychol.* 6: 359–361.

Agnello, D., Wang, H., Yang, H., et al. (2002). HMGB-1, a DNA-binding protein with cytokine activity, induces brain TNF and IL-6 production, and mediates anorexia and taste aversion. *Cytokine.* 18: 231–236.

Agüero, A., Arnedo, M., Gallo, M., and Puerto, A. (1993). The functional relevance of the lateral parabrachial nucleus in lithium chloride-induced aversion learning. *Pharmacol. Biochem. Behav.* 45: 973–978.

Akirav, I., Khatsrinov, V., Vouimba, R.-M., et al. (2012). Extinction of conditioned taste aversion depends on functional protein synthesis but not on NMDA receptor activation in the ventromedial prefrontal cortex. *Lrn. Mem.* 13: 254–258.

Asarian, L., and Geary, N. (2002). Cyclic estradiol treatment normalizes body weight and restores physiological patterns of spontaneous feeding and sexual receptivity in ovariectomized rats. *Horm. Behav.* 42: 461–471.

Asarian L., and Langhans, W. (2010). A new look on brain mechanisms of acute illness anorexia. *Physiol. Behav.* 100: 464–471.

Bahar, A., Samuel, A., Hazvi, S., and Dudai, Y. (2003). The amygdalar circuit that acquires taste aversion memory differs from the circuit that extinguishes it. *Eur. J. Neurosci.* 17: 1527–1530.

Balthazart, J., Foidart, A., and Hendrick, J. C. (1990). The induction by testosterone of aromatase activity in the preoptic area and activation of copulatory behavior. *Physiol. Behav.* 47: 83–94.

Barnett, S. A. (1958). Experiments on "neophobia" in wild and laboratory rats. *Br. J. Psychol.* 49: 195–201.

Bauer, C., Weingarten, S., Senn, M., and Langhans, W. (1995). Limited importance of a learned aversion in the hypophagic effect of interleukin-1β. *Physiol. Behav.* 57: 1145–1153.

Benoit, S. C., Thiele, T. E., Heinrichs, S. C., et al. (2000). Comparison of central administration of corticotropin-releasing hormone and urocortin on food intake, conditioned taste aversion, and *c-fos* expression. *Peptides.* 21: 345–351.

Berman, D. E., and Dudai, Y. (2001). Memory extinction, learning anew, and learning the new: dissociations in the molecular machinery of learning in cortex. *Science* 291: 2417–2419.

Bernstein, I. L. (1978). Learned taste aversions in children receiving chemotherapy. *Science.* 200: 1302–1303.

Bernstein, I. L., and Goehler, L. E. (1983). Chronic LiCl infusions: conditioned suppression of food intake and preference. *Behav. Neurosci.* 97: 290–298.

Bernstein, I. L., and Webster, M. M. (1980). Learned taste aversions in humans. *Physiol. Behav.* 25: 363–366.

Bernstein, I. L., Treneer, C. M., Goehler, L. E., and Murowchick, E. (1985). Tumor growth in rats: conditioned suppression of food intake and preference. *Behav. Neurosci.* 99: 818–830.

Bernstein, I. L., Courtney, L., and Braget, D. J. (1986). Estrogens and the Leydig LTW(m) tumor syndrome: anorexia and diet aversions attenuated by area postrema lesions. *Physiol. Behav.* 38: 159–163.

Bernstein, I. L., Chavez, M., Allen, D., and Taylor, E. M. (1992). Area postrema mediation of physiological and behavioral effects of lithium chloride in the rat. *Brain Res.* 575: 132–137.

Bernstein, I. L., Wilkins, E. E., and Barot, S. K. (2009). Mapping conditioned taste aversion associations through patterns of c-Fos expression. In *Conditioned Taste Aversion: Behavioral and Neural Processes*, S. Reilly and T. R. Schachtman (Eds.). Oxford University Press, pp. 328–340.

Berridge, K. C., Flynn, F. W., Schulkin, J., and Grill, H. J. (1984). Sodium depletion enhances salt palatability in rats. *Behav. Neurosci.* 98: 652–660.

Bielavska, E., and Roldan, G. (1996). Ipsilateral connections between the gustatory cortex, amygdala and parabrachial nucleus are necessary for acquisition and retrieval of conditioned taste aversion in rats. *Behav. Brain Res.* 81: 25–31.

Birch, L. L. (1980). Effects of peer models' food choice and eating behavior on preschoolers' food preferences. *Child Dev.* 51: 489–496.

Birch, L. L., and Marlin, D. W. (1982). I don't like it; I never tried it: effects of exposure on two-year old children's food preferences. *Appetite* 3: 353–360.

Birch, L. L., McPhee, L., Steinberg, L., and Sullivan, S. (1990). Conditioned flavor preferences in young children. *Physiol. Behav.* 47: 501–505.

Blaustein, J., and Wade, G. (1976). Ovarian influences on the meal pattern of female rats. *Physiol. Behav.* 17: 201–208.

Bluthe, R. M., Dantzer, R., Mormede, P. and LeMoal, M. (1985). Specificity of aversive stimulus properties of vasopressin. *Psychopharmacol.* 87: 238–241.

Bolles, R. C. (1985). Introduction: associative processes in the formation of conditioned food aversions-an emerging functionalism? *Ann. N. Y. Acad. Sci.* 443: 1–7.

Booth, D. A. (1985). Food-conditioned eating preferences and aversions with interoceptive elements: Conditioned appetites and satieties. *Ann. N. Y. Acad. Sci.* 443: 22–41.

Booth, D. A., Lovett, D., and McSherry, G. M. (1972). Postingestive modulation of the sweetness preference gradient in the rat. *J. Comp. Physiol. Psychol.* 78: 485–512.

Borison, H. L. (1989). Area postrema: Chemoreceptor circumventricular organ of the medulla oblongata. *Prog. Neurobiol.* 32: 351–390.

Bouton, M. (1993). Context, time, and memory retrieval in the interference paradigms of Pavlovian learning. *Psychol. Bull.* 114: 80–99.

Braun, J. J., Kiefer, S. W., and Ouellet, J. V. (1981). Psychic ageusia in rats lacking gustatory neocortex. *Exp. Neurol.* 72: 711–716.

Braun, J. J., Lasiter, P. S., and Kiefer, S. W. (1982). The gustatory neocortex of the rat. *Physiol. Psychol.* 10: 13–45.

Breslin, P. A. S., Spector, A. C., and Grill, H. J. (1992). A quantitative comparison of taste reactivity behaviors to sucrose before and after lithium chloride pairings: A unidimensional account of palatability. *Behav. Neurosci.* 106: 820–836.

Broberg, D. J., and Bernstein, I. L. (1987). Candy as a scapegoat in the prevention of food aversions in children receiving chemotherapy. *Cancer* 60: 2344–2347.

Bull, D. F., Brown, R., King, M. G., and Husband, A. J. (1991). Modulation of body temperature through taste aversion conditioning. *Physiol. Behav.* 49: 1229–1233.

Bures, J., and Buresova, O. (1979). Neurophysiological analysis of conditioned taste aversion. In *Brain Mechanisms in Memory and Learning: From the Single Neuron to Man*, M. A. Brazier (Ed.). Raven Press, New York, pp. 127–138.

Bures, J., and Buresova, O. (1989). Conditioned taste aversion to injected flavor: differential effect of anesthesia on the formation of the gustatory trace and on its association with poisoning in rats. *Neurosci. Ltr.* 98: 305–309.

Buresova, O., and Bures, J. (1973). Cortical and subcortical components of the conditioned saccharin aversion. *Physiol. Behav.* 11: 435–439.

Burt, G., and Smotherman, W. P. (1980) Amygdalectomy induced deficits in conditioned taste aversion: possible pituitary-adrenal involvement. *Physiol. Behav.* 24: 645–649.

Butera, P. C., and Beikirch, R. J. (1989). Central implants of diluted estradiol: independent effects on ingestive and reproductive behaviors of ovariectomized rats. *Brain Res.* 491: 266–273.

References

Butera, P. C., Beikirch, R. J., and Willard, D. M. (1990). Changes in ingestive behaviors and body weight following intracranial application of 17α-estradiol. *Physiol. Behav.* 47: 1291–1293.

Butera, P. C., Willard, D. M., and Raymond, S. A. (1992). Effects of PVN lesions on the responsiveness of female rats to estradiol. *Brain Res.* 576: 304–310.

Butera, P. C., Campbell, R. B., and Bradway, D. M. (1993). Antagonism of estrogenic effects on feeding behavior by central implants of anisomycin. *Brain Res.* 624: 354–356.

Cabanac, M. (1971). Physiological role of pleasure. *Science.* 173: 1103–1107.

Cabanac, M. (1979). Sensory pleasure. *Q. Rev. Biol.* 54: 1–29.

Cador, M., Ahmed, S. H., Koob, G.F., et al. (1992). Corticotropin-releasing factor induces a place aversion independent of its neuroendocrine role. *Brain Res.* 597: 304–309.

Cannon, D. S., Baker, T. B., and Berman, R. F. (1977). Taste aversion disruption by drug pretreatment: dissociative and drug-specific effects. *Pharm. Biochem. Behav.* 6: 93–100.

Capretta, P. J., Moore, M. J., and Rossiter, T. R. (1973). Establishment and modification of food and taste preferences: effects of experience. *J. Gen. Psychol.* 89: 27–46.

Carpenter, D. O., Briggs, D. B., Knox, A. P., and Strominger, N. (1988) Excitation of area postrema neurons by transmitters, peptides and cyclic nucleotides. *J. Neurophysiol.* 59: 358–369.

Cassileth, P. A., Lusk, E. J., Torri, S., et al. (1983). Antiemetic efficacy of dexamethasone therapy in patients receiving chemotherapy. *Arch. Int. Med.* 143: 1347–1349.

Chambers, K. C. (1990). A neural model for conditioned taste aversions. *Ann. Rev. Neurosci.* 13: 373–385.

Chambers, K. C., & Bernstein, I. L. (2003). Conditioned flavor aversions. In *Handbook of Olfaction and Gestation, 2nd Edition*, R.L. Doty (Ed.). New York: Marcel Dekker, pp. 905–933.

Chambers, K. C., and Hayes, U. (2002). Exposure to estradiol before but not during acquisition of LiCl-induced conditioned taste avoidance accelerates extinction. *Horm. Behav.* 41: 297–305.

Chambers, K. C., and Hayes, U. L. (2005). The role of vasopressin in behaviors associated with aversive stimuli. In *Handbook of Stress and the Brain. Vol. 15, Part 1*, T. Steckler, N. H. Kalin, and J. M. H. M. Reul (Eds.). Elsevier, Amsterdam, pp. 231–262.

Chambers, K. C., and Hintiryan, H. (2009). Hormonal modulation of taste avoidance: the role of estradiol. In *Conditioned Taste Aversion: behavioral and neural processes*, S. Reilly and T.R. Schachtman (Eds.), Oxford University Press, Oxford, pp. 364–386

Chambers, K. C. and Wang, Y. (2004). Role of the lateral parabrachial nucleus in apomorphine-induced conditioned consumption reduction: Cooling lesions and relationship of c-Fos-like-immunoreactivity to strength of conditioning. *Behav. Neurosci.* 118: 199–213.

Chambers, K. C., Thornton, J. E., and Roselli, C. E. (1991). Age-related deficits in brain androgen binding and metabolism, testosterone, and sexual behavior of male rats. *Neurobiol. Aging* 12: 123–130.

Chang, F-C. T., and Scott, T. R. (1984). Conditioned taste aversions modify neural responses in the rat nucleus tractus solitarius. *J. Neurosci.* 4: 1850–1862.

Cooper, R. L., McNamara, M. C., and Thompson, W. G. (1980). Vasopressin and conditioned flavor aversion in aged rats. *Neurobiol. Aging.* 1: 53–57.

Cunningham, C. L., and Niehus, J. S. (1993). Drug-induced hypothermia and conditioned place aversion. *Behav. Neurosci.* 107:468–479.

Czaja, J. A., and Goy, R. W. (1975). Ovarian hormones and food intake in female guinea pigs and rhesus monkeys. *Horm. Behav.* 6: 329–349.

Dantzer, R., Bluthe, R. M., and Kelly, K. W. (1991). Androgen-dependent vasopressinergic neurotransmission attenuates interleukin-1-induced sickness behavior. *Brain Res.* 557: 115–120.

De Beun, R., Jansen, E., Smeets, M. A. M., et al. (1991). Estradiol-induced conditioned taste aversion and place aversion in rats: sex- and dose-dependent effects. *Physiol. Behav.* 50: 995–1000.

De Beun, R., Peeters, B. W. M. M., and Broekkamp, C. L. E. (1993). Stimulus characterization of estradiol applying a crossfamiliarization taste aversion procedure in female mice. *Physiol. Behav.* 53: 715–719.

De Jong, F. H., Hey, A. H., and van der Molen, H. J. (1973). Effect of gonadotrophins on the secretion of oestradiol- and testosterone by the rat testis. *J. Endocrinol.* 57, 277–284.

de Toledo, L., and Black, A. H. (1966). Heart rate: changes during conditioned suppression in rats. *Science* 152: 1404–1406.

De Wied, D., Gaffori, O., van Ree, J. M., and De Jong, W. (1984). Central target for the behavioural effects of vasopressin neuropeptides. *Nature* 308: 276–278.

DeWys, W. D. (1985). Nutritional problems in cancer patients: overview and perspective. In *Cancer, Nutrition, and Eating Behavior: A Biobehavioral Perspective*, T. G. Burish, S. M. Levy, and B. E. Meyerowitz (Eds.). Lawrence Erlbaum, Hillsdale, NJ, pp. 135–148.

Domjan, M. (1977). Attenuation and enhancement of neophobia for edible substances. In *Learning Mechanisms in Food Selection*, L. M. Barker, M. R. Best, and M. Domjan (Eds.). Baylor University Press, Waco, TX, pp. 151–179.

Douglas, A. J. (2005). Vasopressin and oxytocin. In *Handbook of Stress and the Brain. Vol. 15, Part 1*, T. Steckler, N. H. Kalin, and J. M. H. M. Reul (Eds.). Elsevier, Amsterdam, pp. 205–229.

Dray, S. M., and Taylor, A. N. (1982). $ACTH_{4-10}$ enhances retention of conditioned taste aversion learning in infant rats. *Behav. Neural Biol.* 35: 147–158.

Drewnowski, A. (1989). Taste responsiveness in eating disorders. *Ann. NY Acad. Sci.* 42: 399–409.

DSM-IV-TR. (2000). *American Psychiatric Association*, Washington, DC, pp. 583–589.

Dunn, L. T., and Everitt, B. J. (1988). Double dissociations of the effects of amygdala and insular cortex lesions on conditioned taste aversion, passive avoidance, and neophobia in the rat using the excitotoxin ibotenic acid. *Behav. Neurosci.* 102: 3–23.

Dwyer, D. M. (2009). Microstructural analysis of ingestive behavior reveals no contribution of palatability to the incomplete extinction of a conditioned taste aversion. *Quart. J. Exp. Psychol.* 62: 9–17.

Eckel, L. A., and Geary, N. (2001). Estradiol treatment increases feeding-induced cfos expression in the brains of ovariectomized rats. *Amer. J. Physiol.* 281: R738–R746.

Eckel, L. A., and Ossenkopp, K-P. (1993). Novel diet consumption and body weight gain are reduced in rats chronically infused with lithium chloride: mediation by the chemosensitive area postrema. *Brain Res. Bull.* 31: 613–619.

Eckel, L. A., and Ossenkopp, K-P. (1996). Area postrema mediates the formation of rapid, conditioned palatability shifts in lithium-treated rats. *Behav. Neurosci.* 110: 202–212.

Elkins, R. L. (1974). Conditioned flavor aversions to familiar tap water in rats: an adjustment with implications for aversion therapy treatment of alcoholism and obesity. *J. Abnorm. Psychol.* 83: 411–417.

Ettenberg, A. (1984). ICV application of a vasopressin antagonist peptide prevents the behavioral actions of vasopressin. *Behav. Brain Res.* 14: 201–211

Fantino, M. (1984). Role of sensory input in the control of food intake. *J. Autonom. Nerv. Sys.* 10: 347–358.

Fazeli, P. K., Clader, G. L., Miller, K. K., et al. (2012). Psychotropic medication use in anorexia nervosa between 1997 and 2009. *Int. J. Eat. Disord.* 45: 970–976.

Fekete, É. M., Zhao, Y., Szücs, A., et al. (2011). Systemic urocortin 2, but not urocortin 1 or stressin1-A, suppresses feeding via CRF2 receptors without malaise and stress. *Brit. J. Pharmacol.* 164: 1959–1975.

Flaherty, C. F., and Checke, S. (1982). Anticipation of incentive gain. *Anim. Learn. Behav.* 10: 177–182.

Flanagan-Cato, L. M., Grigson, P. S., and King, J. L. (2001). Estrogen-induced suppression of intake is not mediated by taste aversion in female rats. *Physiol. Behav.* 72: 549–558.

Fulford, A. J., and Harbuz, M. S. (2005). An introduction to the HPA axis. In *Handbook of Stress and the Brain. Vol. 15, Part 1*, T. Steckler, N. H. Kalin, and J. M. H. M. Reul (Eds.). Elsevier, Amsterdam, pp. 43–65.

Galef, B. G., Jr., (1977). Mechanisms for the social transmission of acquired food preferences from adult to weanling rats. In *Learning Mechanisms in Food Selection*, L. M. Barker, M. R. Best, and M. Domjan (Eds.). Baylor University Press, Waco, TX, pp. 123–148.

Galef, B. G., Jr.,, and Clark, M. M. (1971). Social factors in the poison avoidance and feeding behavior of wild and domesticated rat pups. *J. Comp. Physiol. Psychol.* 75: 341–357.

Galef, B. G., Jr.,, and Henderson, P. W. (1972). Mother's milk: a determinant of the feeding preferences of weaning rat pups. *J. Comp. Physiol. Psychol.* 78: 213–219.

Gallo, M., and Bures, J. (1991). Acquisition of conditioned taste aversion in rats is mediated by ipsilateral interaction of cortical and mesencephalic mechanisms. *Neurosci. Lett.* 133: 187–190.

Gallo, M., Arnedo, M. Agüero, A., and Puerto, A. (1988). Electrical intracerebral stimulation of the area postrema on taste aversion learning. *Behav. Brain Res.* 30: 289–296.

Gallo, M., Roldan, G., and Bures, J. (1992). Differential involvement of gustatory insular cortex and amygdala in the acquisition and retrieval of conditioned taste aversion in rats. *Behav. Brain Res.* 52: 91–97.

Ganchrow, J. R., Steiner, J. E., and Canetto, S. (1986). Behavioral displays to gustatory stimuli in newborn rat pups. *Dev. Psychobiol.* 19: 163–174.

Garb, J. L., and Stunkard, A. J. (1974). Taste aversions in man. *Amer. J. Psychiatry* 131: 1204–1207.

Garcia, J. (1989). Food for Tolman: cognition and cathexis in concert. In *Aversion, Avoidance and Anxiety*, T. Archer and L-G. Nilson (Eds.). Lawrence Erlbaum, Hillsdale, NJ, pp. 45–85.

Garcia, J., and Koelling, R. A. (1966). Relation of cue to consequence in avoidance learning. *Psychon. Sci.* 4: 123–124.

Garcia, J., Kimeldorf, D. J., and Koelling, R. A. (1955). Conditioned aversion to saccharin resulting from exposure to gamma radiation. *Science* 122: 157–158.

Garcia, J., Kimeldorf, J., and Hunt, E. L. (1956). Conditioned responses to manipulative procedures resulting from exposure to gamma radiation. *Rad. Res.* 5: 79–87.

Garcia, J., Ervin, R. R., and Koelling, R. A. (1966). Learning with prolonged delay of reinforcement. *Psychon. Sci.* 5: 121–122.

Garcia, J., Lasiter, P. S., Bermudez-Rattoni, F., and Deems, D. A. (1985). A general theory of avoidance learning. *Ann. N.Y. Acad. Sci.* 443: 8–21.

Geary, N., and Asarian, L. (1999). Cyclic estradiol treatment normalizes body weight and test meal size in ovariectomized rats. *Physiol. Behav.* 67: 141–147.

Gemberling, G. A. (1984). Ingestion of a novel flavor before exposure to pups injected with lithium chloride produces a taste aversion in mother rats *(Rattus norvegicus). J. Comp. Psychol.* 98: 285–301.

Gietzen, D. W., McArthur, L. H., Theisen, J. C., and Rogers, Q. R. (1992). Learned preference for the limiting amino acid in rats fed a threonine-deficient diet. *Physiol. Behav.* 51: 909–914.

Gosnell, B. A., Morley J. E., and Levine, A. S. (1983). A comparison of the effects of corticotropin releasing factor and sauvagine on food intake. *Pharm. Biochem. Behav.* 19: 771–775.

Grigson, P. S. (1997). Conditioned taste aversions and drugs of abuse: a reinterpretation. *Behav. Neurosci.* 111: 129–136.

Grigson, P. S., Shimura, T., and Norgren, R. (1997). Brainstem lesions and gustatory function: III. The role of the nucleus of the solitary tract and the parabrachial nucleus in retention of a conditioned taste aversion in rats. *Behav. Neurosci.* 111: 180–187.

Grigson, P. S., Reilly, S., Shimura, T., and Norgren, R. (1998). Ibotenic acid lesions of the parabrachial nucleus and conditioned taste aversion: further evidence for an associative deficit in rats. *Behav. Neurosci.* 112: 160–171.

Grill, H. J., and Berridge, K. C. (1985). Taste reactivity as a measure of the neural control of palatability. *Prog. Psychobiol. Physiol. Psychol.* 11: 1–61.

Grill, H. J., and Norgren, R. (1978). Chronically decerebrate rats demonstrate satiation but not bait shyness. *Science* 201: 267–269.

Groenewegen, H. J., and Berendse, H. W. (1994). The specificity of the 'nonspecific' midline and intralaminar thalamic nuclei. *Trends Neurosci.* 17: 52–57.

Gruest, N., Richer, P., and Benard, H. (2004). Emergence of long-term memory for conditioned aversion in the rat fetus. *Dev. Psychobiol.* 43: 189–198.

Gustavson, C. R., and Gustavson, J. C. (1986). Estrogen based conditioned taste aversion: a neuroethological model of anorexia nervosa. *Soc. Neurosci. Abstr.* 12: 1451.

Gustavson, C. R., and Gustavson, J. C. (1987). Estrogen induced anorexia in the rat: an LiCl comparison suggests it is estrogen induced nausea. *Soc. Neurosci. Abstr.* 13: 555.

Gustavson, C. R., Reinarz, D. E., Gustavson, C. R., and Pumariega, A. J. (1988). *Estrogen based conditioned taste aversion in a single human subject.* West. Psychol. Assoc. Mtg. San Francisco, CA.

Gustavson, C. R., Gustavson, J. C., Noller, K. L., et al. (1989a). Fetal diethylstilbestrol exposure: a possible risk factor in the development of anorexia nervosa. *Soc. Neurosci. Abstr.* 15: 1131.

Gustavson, C. R., Gustavson, J. C., Young, J. K., et al. (1989b). Estrogen induced malaise. In *Neural Control of Reproductive Function*, J. M. Lakoski, J. R. Perez-Polo, and D. K. Rassin (Eds.). Alan R. Liss, New York, pp. 501–523.

Gustavson, C. R., Gustavson, J. C., Noller, K. L., et al. (1991). Increased risk of profound weight loss among women exposed to diethylstilbestrol in utero. *Behav. Neural Biol.* 55: 307–312.

Hamilton, R. B., and Norgren, R. (1984). Central projections of gustatory nerves in the rat. *J. Comp. Neurol.* 222: 560–577.

Hammes, B., and Laitman, C. J. (2003). Diethylstilbesterol (DES) update: recommendations for the identification and management of DES-exposed individuals. *J. Midwif. Wom. Health* 48: 19–29.

Hanamori, T., Kunitake, T., Kato, K., and Kannan, H. (1998). Responses of neurons in the insular cortex to gustatory, visceral and nociceptive stimuli in rats. *J. Neurophysiol.* 79: 2535–2545.

Hart, B. L. (1988). Biological basis of the behavior of sick animals. *Neurosci. Biobehav. Rev.* 12: 123–137.

Hayes, U. L., and Chambers, K. C. (2002). Central infusion of vasopressin in male rats accelerates extinction of conditioned taste avoidance induced by LiCl. *Brain Res. Bull.* 57: 727–733.

Hayes, U. L., and Chambers, K. C. (2005a). Peripheral vasopressin accelerates extinction of conditioned taste avoidance. *Physiol. Behav.* 84: 147–156.

References

Hayes, U. L., and Chambers, K. C. (2005b). High doses of vasopressin delay the onset of extinction and strengthen acquisition of LiCl-induced conditioned taste avoidance. *Physiol. Behav.* 84: 625–633.

Heinrichs, C., and Koob, G. F. (2004). Corticotropin-releasing factor in brain: a role in activation, arousal, and affect regulation. *J. Pharmacol. Exp. Ther.* 311: 427–440.

Hennessy, J. W., Smotherman, W. P., and Levine, S. (1976a). Conditioned taste aversion and the pituitary-adrenal system. *Behav. Biol.* 16:413–424.

Hennessy, J. W., Smotherman, W. P., and Levine, S. (1976b). Investigations into the nature of the dexamethasone and ACTH effects upon learned taste aversion. *Physiol. Behav.* 24:645–649.

Herbert, H., Moga, M. M., and Saper, C.B. (1990). Connections of the parabrachial nucleus with the nucleus of the solitary tract and the medullary reticular formation in the rat. *J. Comp. Neurol.* 293: 540–580.

Hermann, G. E., and Rogers, R. C. (1985). Convergence of vagal and gustatory afferent input within the parabrachial nucleus of the rat. *J.Auton. Nerv. Sys.* 13: 1–17.

Hintiryan, H., Hayes, U. L., and Chambers, K. C. (2005). The role of histamine in estradiol-induced conditioned consumption reductions. *Physiol. Behav.* 84, 117–128.

Hintiryan, H., Foster, N. N., and Chambers, K. C. (2009). Dissociating the conditioning and the anorectic effects of estradiol in female rats. *Behav. Neurosci.* 123: 1226–1237

Holland, P. C. (1984). Origins of behavior in Pavlovian conditioning. In *The Psychology of Learning and Motivation, Vol. 18*, G. Bower (Ed.). Academic Press, Orlando, FL, pp. 129–174.

Holt, S., Ford, M. J., Grant, S., and Heading, R. C. (1981). Abnormal gastric emptying in primary anorexia nervosa. *Br. J. Psychiatr.* 139: 550–552.

Houpt, T. A., Philopena, J. M, Wessel, T. C., et al. (1994). Increased c-*fos* expression in nucleus of the solitary tract correlated with conditioned taste aversion to sucrose in rats. *Neurosci. Ltr.* 172: 1–5.

Hreash, R., Keil, L. C., Chou, L., and Reid, I. A. (1990). Effects of carotid occlusion and angiotensin II on vasopressin secretion in intact and vagotomized conscious rabbits. *Endocrinology.* 127: 1160–1166.

Hull, C. L. (1943). *Principles of Behavior*. Appleton: New York.

Hunt, T., and Amit, Z. (1987). Conditioned taste aversion induced by self-administered drugs: paradox revisited. *Neurosci. Biobehav. Rev.* 1: 107–130.

Itani, J. (1958). On the acquisition and propagation of a new food habit in the natural group of Japanese monkeys at Takasaki Yama. *J. Primatol.* 1: 84–98.

Jones, D. J., Fox, M. M., Babigian, H. M., and Hutton, H. E. (1980). Epidemiology of anorexia nervosa in Monroe County, New York: 1960–1976. *Psychosom. Med.* 42: 551–558.

Kalat, J. W., and Rozin, P. (1973). "Learned safety" as a mechanism in long-delay taste-aversion learning in rats. *J. Comp. Physiol. Psychol.* 83: 198–207.

Karniol, I. G., Dalton, J., and Lader, M. H. (1978). Acute and chronic effects of lithium chloride on physiological and psychological measures in normals. *Psychopharmacol.* 57: 289–294

Kehoe, P., and Sakurai, S. (1991). Preferred tastes and opioid-modulated behaviors in neonatal rats. *Develop. Psychobiol.* 24: 135–148.

Kendler, K., Hennessey, J. W., Smotherman, W. P., and Levine, S. (1976). An ACTH effect on recovery from conditioned taste aversion. *Behav. Biol.* 17: 225–229.

Kent, S., Bluthe, R-M., Kelley, K. W., and Dantzer, R. (1992). Sickness behavior as a new target for drug development. *Trends Pharmacol. Sci.* 13: 24–28.

Kent, W. D. T., Cross-Mellor, S. K., Kavaliers, M., Ossenkopp, K-P. (2002). Acute effects of corticosterone on LiCl-induced rapid gustatory conditioning in rats: a microstructural analysis of licking patterns. *Behav. Brain Res.* 136: 143–150.

Kiefer, S. W., and Braun, J. J. (1977). Absence of differential associative responses to novel and familiar taste stimuli in rats lacking gustatory neocortex. *J. Comp. Physiol. Psychol.* 91: 498–507.

Kishi T., Kafantaris V., Sunday S., Sheridan E. M., and Correll, C. U. (2012). Are antipsychotics effective for the treatment of anorexia nervosa? results from a systematic review and meta-analysis. *J. Clin. Psychiatry.* 73:e757–e766.

Koh, M. T., and Bernstein, I. L. (2005). Mapping conditioned taste aversion associations using c-Fos reveals a dynamic role for insular cortex. *Behav. Neurosci.* 119: 388–398.

Kosten, T., and Contreras, R. J. (1989). Deficits in conditioned heart rate and taste aversion in area postrema-lesioned rats. *Behav. Brain Res.* 35: 9–21.

Ladowsky, R. L., and Ossenkopp, K-P. (1986). Conditioned taste aversions and changes in motor activity in lithium-treated rats: mediating role of the area postrema. *Neuropharmacology.* 25: 71–77.

Langhans, W., Delprete, E., and Scharrer, E. (1991). Mechanisms of vasopressin's anorectic effect. *Physiol. Behav.* 49: 169–176.

Langhans, W., Balkowski, G., and Savoldelli, D. (1993a). Further characterization of the feeding responses to interleukin-1 and tumor necrosis factor. In *Endocrine and Nutritional Control of Basic Biological Functions*, J. Lehnert, R. Murison, and H. Weiner, (Eds.). Hogrefe & Huber, Seattle, WA, pp. 135–142.

Langhans, W., Savoldelli, D., and Weingarten, S. (1993b). Comparison of the feeding responses to bacterial lipopolysaccharide and interleukin-1β. *Physiol. Behav.* 53: 643–649.

Lavin, M. J., Freise, B., and Coombes, S. (1980). Transferred flavor aversions in adult rats. *Behav. Neural Biol.* 28: 15–33.

Lebrun, C., Le Moal, M., Koob, G. F. and Bloom, F. E. (1985). Vasopressin pressor antagonist injected centrally reverses behavioral effects of peripheral injection of vasopressin, but only at doses that reverse increase in blood pressure. *Reg. Peptides.* 11: 173–181.

Leung, H. T., Bailey, G. K., Laurent, V., and Westbrook, R. F. (2007). Rapid reacquisition of fear to a completely extinguished context is replaced by transient impairment with additional extinction training. *J. Exp. Psychol.: Anim. Behav. Proc.* 33: 299–313

Levine, S. (2005). Stress: an historical perspective. In *Handbook of Stress and the Brain. Vol. 15*, T. Steckler, N. H. Kalin, and J. M. H. M. Reul (Eds.). Elsevier, Amsterdam, pp. 3–23.

Limebeer, C. L., and Parker, L. A. (2000). Ondansetron interferes with lithium-induced conditioned rejection reactions, but not lithium-induced taste avoidance. *J. Exp. Psychol.: Anim. Behav. Proc.* 26: 371–384.

Limebeer, C. L., and Parker, L. A. (2003). The 5-HT$_{1A}$ agonist 8-OH-DPAT dose-dependently interferes with the establishment and expression of lithium-induced conditioned rejection reactions in rats. *Psychopharmacol.* 166: 120–126.

Limebeer, C. L., Parker, L. A., and Fletcher, P. J. (2004). 5,7-Dihydroxytryptamine lesions of the dorsal and median raphe nuclei interfere with lithium-induced conditioned gaping, but not conditioned taste avoidance, in rats. *Behav. Neurosci.* 118: 1391–1399.

Logue, A. W. (1985). Conditioned food aversion learning in humans. *Ann. N.Y. Acad. Sci.* 443: 316–329.

Logue, A. W., Ophir, I., and Strauss, K. E. (1981). The acquisition of taste aversions in humans. *Behav. Res. Therapy*, 19: 319–333.

Lucas, G. A., Timberlake, W., Gawley, D. J., and Drew, J. (1990). Anticipation of future food: suppression and facilitation of saccharin intake depending on the delay and type of future food. *J. Exp. Psychol.: Anim. Behav. Proc.* 16: 169–177.

Mantovani, G., Macciò, A., Lai, P., et al. (1998). Cytokine involvement in cancer anorexia/cachexia: role of megestrol acetate and medroxyprogesterone acetate on cytokine downregulation and improvement of clinical symptoms. *Crit. Rev. Oncog.* 9: 99–106.

Markman, M., Sheidler, V., Ettinger, D. S., et al. (1984). Antiemetic efficacy of dexamethasone. *N. Engl. J. Med.* 311:549–552.

Martin, J. R., Cheng, F. Y., and Novin, D. (1978). Acquisition of learned taste aversion following bilateral subdiaphragmatic vagotomy in rats. *Physiol. Behav.* 21: 13–17.

Mattes, R. D. (1997). The taste for salt in humans. *Amer. J. Clin. Nutr.* 65: 692S–697S.

Mattes, R. D., Arnold, C., and Boraas, M. (1987). Learned food aversions among cancer chemotherapy patients: incidence, nature and clinical implications. *Cancer* 60: 2576–2580.

Mattes, R. D., Curran, Jr., ,W. J., Alavi, J., et al. (1992). Clinical implications of learned food aversions in patients with cancer treated with chemotherapy or radiation therapy. *Cancer* 70: 192–200.

McCarthy, D. O., Kluger, M. J., and Vander, A. J. (1984). Suppression of food intake during infection: is interleukin-1 involved? *Am. J. Clin. Nutr.* 42: 1179–1182.

McCaughey, S. A., Giza, B. K., Nolan, L. J., and Scott, T. R. (1997). Extinction of a conditioned taste aversion in rats: II. Neural effects in the nucleus of the solitary tract. *Physiol. Behav.* 61: 373–379.

Meachum, C. L., and Bernstein, I. L. (1990). Conditioned responses to a taste CS paired with LiCl administration. *Behav. Neurosci.* 104: 711–715.

Merwin, A. A., and Doty, R. L. (1994). Early exposure to low levels of estradiol (E2) mitigates E,-induced conditioned taste aversions in prepubertally ovariectomized female rats. *Physiol. Behav.* 55: 185–187.

Mickley, G. A., Kenmuir, C. L., McMullen, C. A., et al. (2002). Dynamic processing of taste aversion extinction in the brain. *Brain Res.* 1016: 79–89.

Mickley, G. A., Kenmuir, C. L., Yocom, A. M., et al. (2005). A role for prefrontal cortex in the extinction of a conditioned taste aversion. *Brain Res.* 1051: 176–182.

Midkiff, E. E., and Bernstein, I. L. (1985). Targets of learned food aversions in humans. *Physiol. Behav.* 34: 839–841.

Mikulka, P. J., Leard, B., and Klein, S. B. (1977). Illness-alone exposure as a source of interference with the acquisition and retention of a taste aversion. *J. Exp. Psychol. Anim. Behav. Proc.* 3: 189–201.

Mitchell, J. E. (1989). Psychopharmacology of eating disorders. *Ann. N.Y. Acad. Sci.* 575: 41–49.

Moe, K. E. (1986). The ontogeny of salt preference in rats. *Dev. Psychobiol.* 19: 185–196.

Mooney, K. M., and Walbourn, L. (2001). When college students reject food: not just a matter of taste. *Appetite* 36: 41–50.

Morris, R., Frey, S., Kasambir, T., and Petrides, M. (1999). Ibotenic acid lesions of the basolateral, but not the central, amygdala interfere with conditioned taste aversion: evidence from a combined behavioral and anatomical tract-tracing investigation. *Behav. Neurosci.* 113: 291–302.

Moulton, R. (1942). A psychosomatic study of anorexia nervosa, including the use of vaginal smears. *Psychosom. Med.* 4: 62–74.

Mukherjee, B. P., Bailey, P. T., and Pradhan, S. N. (1976). Temporal and regional differences in brain concentrations of lithium in rats. *Psychopharmacology.* 48: 119–121.

Murray, M. J., and Murray, A. B. (1979). Anorexia of infection as a mechanism of host defense. *Am. J. Clin. Nutr.* 32: 593–596.

Murray, J., and Murray, A. (1981). Toward a nutritional concept of host resistance to malignancy and intracellular infection. *Perspect. Biol. Med.* 25: 290–301.

Nachman, M., and Ashe, J. H. (1974). Effects of basolateral amygdala lesions on neophobia, learned taste aversions and sodium appetite in rats. *J. Comp. Physiol. Psychol.* 87: 622–643.

Nakashima, M., Uemura, M., Yasui, K., et al. (2000). An anterograde and retrograde tract-tracing study on the projections from the thalamic gustatory area in the rat: distribution of neurons projecting to the insular cortex and amygdaloid complex. *Neurosci. Res.* 36: 297–309.

Nash, S., Tilley, B. C., Kurland, L. T., et al. (1983). Identity and tracing a population at risk: The DESAD project experience. *Amer. J. Publ. Health*, 73: 253–259.

Nathan, C. F. (1987). Secretory products of macrophages. *J.Clin. Invest.* 79: 319–326.

Nerad, L., Ramirez-Amaya, V., Ormsby, C. E., and Bermudez-Rattoni, F. (1996). Differential effects of anterior and posterior insular cortex lesions on the acquisition of conditioned taste aversion and spatial learning. *Neurobiol. Lrn. Mem.* 66: 44–50.

O'Connor, E. F., Cheng, S. W. T., and North, W. G. (1987). Effects of intraperitoneal injection of lithium chloride on neurohypophyseal activity: implication for behavioral studies. *Physiol. Behav.* 40: 91–95.

Olsewski, P. K., Wirth, M. M., Shaw, T. J., et al. (2001). Role of alpha-MSH in the regulation of consummatory behavior: immunohistochemical evidence. *Amer. J. Physiol.- Reg. Integr. Comp. Physiol.* 281: R673–680.

Ormsby, C. E., Ramirez-Amaya, V, and Bermudez-Rattoni, F. (1998). Long-term memory retrieval deficits of learned taste aversions are ameliorated by cortical fetal brain implants. *Behav. Neurosci.* 112: 172–182.

Ossenkopp, K-P., and Giugno, L. (1990). Nicotine-induced conditioned taste aversions are enhanced in rats with lesions of the area postrema. *Pharmacol. Biochem. Behav.* 36: 625–630.

Ossenkopp, K-P., Rabi, Y. J., and Eckel, L. A. (1996). Oestradiol-induced taste avoidance is the result of a conditioned palatability shift. *NeuroRep.* 7: 2777–2780.

Ottersen, O. P. (1982). Connections of the amygdala of the rat. IV: Corticoamygdaloid and intraamygdaloid connections as studied with axonal transport of horseradish peroxidase. *J. Comp. Neurol.* 205: 30–48.

Papino, M. Y., and Mennella, J. A. (2005). Sucrose-induced analgesia is related to sweet preferences in children but not adults. *Pain* 119: 210–218.

Parker, L. A. (1988). Defensive burying of flavors paired with lithium but not amphetamine. *Psychopharmacology.* 96: 250–252.

Parker, L. A. (1995). Rewarding drugs produce taste avoidance, but not taste aversion. *Neurosci. Biobehav. Rev.* 19: 143–151.

Parker, L. A. (2003). Taste avoidance and taste aversion: evidence for two different processes. *Lrn. Behav.* 31: 165–172.

Parker, L. A. and Brosseau, L. (1990). Apomorphine-induced flavor-drug associations: A dose-;response analysis by the taste reactivity test and the conditioned taste avoidance test. *Pharm. Biochem. Behav.* 35: 583–587.

Parker, L. A., Limebeer, C. L., S.A. (2009). Conditioned disgust, but not conditioned taste avoidance, may reflect conditioned nausea in rats. In *Conditioned Taste Aversion: Behavioral and Neural Processes*, S. Reilly and T.R. Schachtman (Eds.). Oxford University Press, pp. 92–113.

Pavlov, I. P. (1927). *Conditioned Reflexes.* Oxford University Press: Oxford, England.

Pelchat, M., Grill, H. J., Rozin, P., and Jacobs, J. (1983). Quality of acquired responses to tastes by *Rattus norvegicus* depends on type of associated discomfort. *J. Comp. Physiol. Psychol.* 97: 140–153.

Phillips, P. A., Abrahams, J. M., Kelly, J., et al. (1988). Localization of vasopressin binding sites in rat brain by *in vitro* autoradiography using a radioiodinated V_1 receptor antagonist. *Neurosci.* 27: 749–761.

Poulton, E. B. (1887). The experimental proof of the protective value of color and marking in insects in reference to their vertebrate enemies. *Proc. Zool. Soc. Lond.* 197–274.

Pritchard, T. C., and Norgren, R. (2004). Gustatory system. In *The Human Nervous System*, 2nd Edition, G. Paxinos and J.K. Mai (Eds.). Academic Press, San Diego, CA, pp. 1171–1196.

Rabin, B. M., Hunt, W. A., and Lee, J. (1983). Acquisition of lithium chloride- and radiation-induced taste aversions in hypophysecomized rats. *Pharmacol. Biochem. Behav.* 18: 463–465.

Rabin, B. M., Hunt, W. A., and Lee, J. (1984). Recall of a previously acquired conditioned taste aversion in rats following lesions of the area postrema. *Physiol. Behav.* 32: 503–506.

Rabin, B. M., Hunt, W. A., and Lee, J. (1989). Attenuation and cross-attenuation in taste aversion learning in the rat: studies with ionizing radiation, lithium chloride and ethanol. *Pharm. Biochem. Behav.* 31: 909–918.

Ramakekers, F., Rigter, H., and Leonard, B. E. (1978). Parallel changes in behavior and hippocampal monoamine metabolism in rats after administration of ACTH-analogues. *Pharm. Biochem. Behav.* 8: 547–551.

Ratnasuriya, R. H., Eisler, I., Szmukler, G. I., and Russell, G. F. M. (1991). Anorexia nervosa: Outcome and prognostic factors after 20 years. *Brit. J. Psychiat.* 158: 495–502.

Reilly, S. (1999). The parabrachial nucleus and conditioned taste aversion. *Brain Res. Bull.* 48: 239–254.

Reilly, S. (2009). Central gustatory system lesions and conditioned taste aversion. In *Conditioned Taste Aversion: Behavioral and Neural Processes*, S. Reilly and T.R. Schachtman (Eds.). Oxford University Press, pp. 309–327.

Reilly, S., and Bornovalova, M. (2005). Conditioned taste aversion and amygdala lesions in the rat: a critical review. *Neurosci. Biobehav. Rev.* 29: 1067–1088.

Reilly, S. and Trifunovic, R. (2000). Lateral parabrachial nucleus lesions in the rat: aversive and appetitive gustatory conditioning. *Brain Res. Bull.* 52: 269–278.

Reilly, S., Grigson, P. S., and Norgren, R. (1993). Parabrachial nucleus lesions and conditioned taste aversion: evidence supporting an associative deficit. *Behav. Neurosci.* 107: 1005–1017.

Rescorla, R. A. (1988). Pavlovian conditioning: it's not what you think it is. *Amer. Psychol.* 43: 151–160.

Rescorla, R. A. (2001). Retraining of extinguished Pavlovian stimuli. *J. Exp. Psychol. Anim. Behav. Proc.* 27:115–124.

Revusky, S, and Martin, G. M. (1988). Glucocorticoids attenuate taste aversions produced by toxins in rats. *Psychopharmacol.* 96:400–407.

Rico, A. G. (1983). Metabolism of endogenous and exogenous anabolic agents in cattle. *J. Anim. Sci.* 57: 226–232.

Rigter, H. and Popping, A. (1976). Hormonal influences on the extinction of conditioned taste aversion. *Psychopharmacologia (Berl.).* 46: 255–261.

Ritter, S., McGlone, J. J., and Kelley, K. W. (1980). Absence of lithium-induced taste aversion after area postrema lesion. *Brain Res.* 201: 501–506.

Rivera, H. M., and Eckel, L. A. (2005). The anorectic effect of fenfluramine is increased by estradiol treatment in ovariectomized rats. *Physiol. Behav.* 86: 331–337.

Roldan, G., and Bures, J. (1994). Tetrodotoxin blockade of amygdala overlapping with poisoning impairs acquisition of conditioned taste aversion in rats. *Behav. Brain Res.* 65: 213–219.

Roman, C., and Reilly, S. (2007). Effects of insular cortex lesions on conditioned taste aversion and latent inhibition in the rat. *Eur. J. Neurosci.* 26: 2627–2632.

Rozin, P. (1967). Specific aversions as a component of specific hungers. *J. Comp. Physiol. Psychol.* 64: 237–242.

Rozin, P., and Schiller, D. (1980). The nature and acquisition of a preference for chili pepper by humans. *Motiv. Emot.* 4: 77–101.

Rozin, P., Gruss, L., and Berk, G. (1979). Reversal of innate aversions: attempts to induce a preference for chili peppers in rats. *J. Comp. Physiol. Psychol.* 93: 1001–1014.

Rumsey, T. S. (1983). Experimental approaches to studying metabolic fate of xenobiotics in food animals. *J. Anim. Sci.* 56: 222–234.

Rzoska, J. (1953). Bait shyness, a study in rat behavior. *Br. J. Anim. Behav.* 1: 128–135.

Sakai, N., and Yamamoto, T. (1998). Role of the medial and lateral parabrachial nucleus in acquisition and retention of conditioned taste aversion in rats. *Behav. Brain Res.* 93: 63–70.

Sakai, N., and Yamamoto, T. (1999). Possible routes of visceral information in the rat brain in formation of conditioned taste aversion. *Neurosci. Res.* 35: 53–61.

Saleh, T. M., Saleh, M. C., Deacon, C. L., and Chisholm, A. (2002). 17β-estradiol release in the parabrachial nucleus of the rat evoked by visceral afferent activation. *Mol. Cell. Endocrinol.* 186: 101–110.

Sapolsky, R. M., Romero, L. M., and Munck, A. U. (2000). How do glucocorticoids influence stress responses: Integrating permissive, suppressive, stimulatory, and preparative actions. *Endoc. Rev.* 21:55–89.

Scalera, G., and Bavier, M. (2009). Role of conditioned taste aversion on the side effects of chemotherapy in cancer patients. In *Conditioned Taste Aversion: behavioral and neural processes*, S. Reilly and T.R. Schachtman (Eds.), Oxford University Press, Oxford, pp. 513–541.

Schou, M. (1968). Lithium in psychiatry - a review. In *Pharmacology: a review of progress 1957–1967*, E.H. Efron (Ed). Public Health Service Publication No 1836, pp.701–718.

Sclafani, A. (1987). Carbohydrate taste, appetite, and obesity: an overview. *Neurosci. Biobehav. Rev.* 11: 131–153.

Segato, F. N., Castro-Souza, C., Segato, E. N., et al. (1997). Sucrose ingestion causes opioid analgesia. *Braz. J. Med. Biol. Res.* 30: 981–984.

Shapiro, R. E., and Miselis, R. R. (1985). The central neural connections of the area postrema of the rat. *J. Comp. Neurol.* 234: 344–364.

Shoham, Z., and Schachter, M. (1996). Estrogen biosynthesis-regulation, action, remote effects, and value of monitoring in ovarian stimulation cycles. *Fert. Steril.* 65: 687–701

Simbayi, L. C., Boakes, R. A., and Burton, M. J. (1986). Effects of basolateral amygdala lesions on taste aversions produced by lactose and lithium chloride in the rat. *Behav. Neurosci.* 100: 455–465.

Sinno, M. H., Coquerel, Q., Boukhettala, N., et al. (2010). Chemotheraphy-induced anorexia is accompanied by activation of brain pathways signaling dehydration. *Physiol. Behav.* 101: 639–648.

Skoog, K. M., and Mangiapane, M. L. (1988). Area postrema and cardiovascular regulation in rats. *Am. J. Physiol.* 254: H963–H969.

Skoog, K. M., Blair, M. L., Sladek, C. D., et al. (1990). Area postrema: Essential for support of arterial pressure after hemorrhage in rats. *Am. J. Physiol.* 258: R1472–R1478.

Smotherman, W. P., and Levine, S. (1978). ACTH and ACTH$_{4-10}$ modification of neophobia and taste aversion responses in the rat. *J. Comp. Physiol. Psychol.* 92: 22–33.

Spector, A. C., Smith, J. C., and Hollander, G. R. (1983). The effect of postconditioning CS experience on recovery from radiation-induced taste aversion. *Physiol. Behav.* 30: 647–649.

Spector, A. C., Norgren, R., and Grill, H. J. (1992). Parabrachial gustatory lesions impair taste aversion learning in rats. *Behav. Neurosci.* 106: 147–161.

St. Andre, J., and Reilly, S. (2007). Effects of central and basolateral amygdala lesions on conditioned taste aversion and latent inhibition. *Behav. Neurosci.* 121: 90–99.

Steckler, T. (2005). CRF antagonists as novel treatment strategies for stress-related disorders. In *Handbook of Stress and the Brain. Vol. 15, Part 2*, T. Steckler, N. H. Kalin, and J. M. H. M. Reul (Eds.). Elsevier, Amsterdam, pp. 43–65.

Steiner, J. E. (1973). The gustofacial response: observation on normal and anencephalic newborn infants. In *Development in the Fetus and Infant. Vol. 4. 4th Symposium on Oral Sensation and Perception*, J.F. Bosma (Ed.). NIH DHEW, Bethesda, MD, pp. 254–278.

Steiner, J. E. (1979). Human facial expressions in response to taste and smell stimulation. In *Advances in Child Development and Behavior*, H. W. Reese and L. P. Lipsitt (Eds.). Academic Press, New York, pp. 257–295.

Steiniger, von, F. (1950). Beitrage zur Soziologie und sonstigen Biologie der Wanderratte. *Zeitschr. Tierpsychol.* 7: 356–379. Cited in Galef (1977).

Stricker, E. M. and Verbalis, J. G. (1991). Caloric and noncaloric controls of food intake. *Brain Res. Bull.* 27: 299–303.

Sullivan, S. A., and Birch, L. L. (1990). Pass the sugar, pass the salt: experience dictates preference. *Dev. Psychol.* 26: 546–551.

Swaab, D. F., Nijveldt, F., and Pool, C. W. (1975). Distribution of oxytocin and vasopressin in the rat supraoptic and paraventricular nucleus. *J. Endocrinol.* 67: 461–462.

Swank, M. W., and Bernstein, I. L. (1994). c-Fos induction in response to a conditioned stimulus after single trial taste aversion learning. *Brain Res.* 636: 202–208.

Swank, M. W., Schafe, G. E., and Bernstein, I. L. (1995). c-Fos induction in response to taste stimuli previously paired with amphetamine or LiCl during taste aversion learning. *Brain Res.* 673: 251–261.

Swank, M. W., Ellis, A. E., and Cochran, B. N. (1996). c-Fos antisense blocks acquisition and extinction of conditioned taste aversion in mice. *NeuroRep.* 7: 1866–1870.

Swanson, L. W. and Sawchenko, P. E. (1980). Paraventricular nucleus: A site for integration of neuroendocrine and autonomic mechanism. *Neuroendocrinol.* 31: 410–417.

Tarttelin, M. F., and Gorski, R. A. (1973). The effects of ovarian steroids on food and water intake and body weight in the female rat. *Acta Endocrinol.* 72: 551–568.

Travers, J. B., and Norgren, R. (1986). Electromyographic analysis of the ingestion and rejection of sapid stimuli in the rat. *Behav. Neurosci.* 100: 544–555.

Travers, J. B., Travers, S. P., and Norgren, R. (1987). Gustatory neural processing in the hindbrain. *Ann. Rev. Neurosci.* 10: 595–632.

Tribollet, E., Arsenijevic, Y., and Barberis, C. (1998). Vasopressin binding sites in the central nervous system: distribution and regulation. *Prog. Brain Res.* 119: 45–55.

van Wimersma Greidanus, Tj.B., Croiset, G., Bakker, E., and Bouman, H. (1979). Amygdaloid lesions block the effect of neuropeptides (vasopressin, $ACTH_{4-10}$) on avoidance behavior. *Physiol. Behav.* 22: 291–295.

van Wimersma Greidanus, Tj.B., Bohus, B., Kovacs, G. L., et al. (1983). Sites of behavioral and neurochemical action of ACTH-like peptides and neurohypophyseal hormones. *Neurosci. Biobehav. Rev.* 7: 453–463.

Vawter, M. P. and Green, K. F. (1980). Effects of desglycinamide-lysine vasopressin on a conditioned taste aversion in rats. *Physiol. Behav.* 25: 851–854.

Verbalis, J. G., McHale, C. M., Gardiner, T. W., and Stricker, E. M. (1986). Oxytocin and vasopressin secretion in response to stimuli producing learned taste aversions in rats. *Behav. Neurosci.* 100: 466–475.

Wade, G. N. (1972). Gonadal hormones and behavioral regulation of body weight. *Physiol. Behav.* 8: 523–534.

Wan, R., Diamant, M, De Jong, W., and De Wied, D. (1992). Differential effects of $ACTH_{4-10}$, DG-AVP, and DG-OXT on heart rate and passive avoidance behavior in rats. *Physiol. Behav.* 51: 507–513.

Wang, H., Bloom, O., Zhang, M., et al. (1999). HMG-1 as a late mediator of endotoxin lethality in mice. *Science* 285: 248–251.

Wang, Y., and Chambers, K. C. (2002). Cooling lesions of the lateral parabrachial nucleus during LiCl activation block acquisition of conditioned taste avoidance in male rats. *Brain Res.* 934: 7–22.

Wang. Y., Lavond, D. G., and Chambers, K. C. (1997a). The effects of cooling the area postrema of male rats on conditioned taste aversions induced by LiCl and apomorphine. *Behav. Brain Res.* 82: 149–158.

Wang. Y., Lavond, D. G., and Chambers, K. C. (1997b). Cooling the area postrema induces conditioned taste aversions in male rats and blocks acquisition of LiCl-induced aversions. *Behav. Neurosci.* 111: 768–776.

Weingarten, S., Senn, M., and Langhans, W. (1993). Does a learned taste aversion contribute to the anorectic effect of bacterial lipopolysaccharide? *Physiol. Behav.* 54: 961–966.

Weiskrantz, L., and Cowey, A. (1963). The aetiology of food reward. *Anim. Behav.* 11: 225–234.

Willi, J., and Grossmann, S. (1983). Epidemiology of anorexia nervosa in a defined region of Switzerland. *Amer. J. Psychiatr.* 140: 564–567.

Wing, E. J., and Young, J. B. (1982). Acute starvation protects mice against *Listeria monocytogenes*. *Infect. Immun.* 28: 771–776.

Yamamoto, T. (1993). Neural mechanisms of taste aversion learning. *Neurosci. Res.* 16: 181–185.

Yamamoto, T. (1994). A neural model for taste aversion learning. In *Olfaction and Taste XI*, K. Kurihara, N. Suzuki, and H. Ogawa (Eds.). Springer, Tokyo, pp.471–474.

Yamamoto, T., Matsuo, R., and Kawamura, Y. (1980). Localization of cortical gustatory area in rats and its role in taste discrimination. *J. Neurophysiol.* 44: 440–455.

Yamamoto, T., Shimura, T., Sako, N., et al. (1994). Neural substrates for conditioned taste aversion in the rat. *Behav. Brain Res.* 65: 123–137.

Young, J. K. (1991). Estrogen and the etiology of anorexia nervosa. *Neurosci. Biobehav. Rev.* 15: 327–331.

Yuan, D. L., and Chambers, K. C. (1999). Estradiol accelerates extinction of lithium chloride-induced conditioned taste aversions through its illness-associated properties. *Horm. Behav.* 36: 287–298.

Chapter 39

Clinical Disorders Affecting Taste: An Update

STEVEN M. BROMLEY and RICHARD L. DOTY

39.1 INTRODUCTION

The importance of taste as a chemical sense is typically underappreciated until it is lost or distorted. Unfortunately the medical world is commonly reactive, rather than proactive, in regards to taste dysfunction. Indeed, very few medical schools provide even a single lecture on this important sensory system and the topic of gustation is lacking in the most popular of neurology textbooks. Such lack of interest contrasts with that exhibited by the corporate world's food and flavor industries, where knowledge of taste function is critical for the success of their products (see Chapters 47–49). Both the medical and corporate worlds, however, have a common need to understand the psychological and biological factors that drive consumption. From the industrial perspective, this relates to providing consumers with the highest quality food products at the lowest possible cost. From the medical perspective, this relates to managing patients with taste dysfunction and the development of programs for improving nutrition, health, and weight management of patients and society at large. Impaired taste function in acutely hospitalized older persons is strongly associated with mortality (Solemdal et al., 2014) and, in general, intact taste functioning can be considered a marker for good health. Even minor distortions in taste perception can represent underlying disease, toxin exposure, or changes in normal anatomy and physiology (Doty and Bromley, 2011).

As is apparent from other chapters of the *Handbook*, considerable progress has been made in understanding the mechanisms of taste transduction and perception. This understanding will undoubtedly influence how taste disorders will be treated in the future and how pharmaceuticals will be developed to minimize their adverse effects on taste function (see Chapters 30 and 40). For example, it is well established that diseases associated with chronic inflammation and immune system dysfunction adversely influence taste function. Type II taste receptor cells (see Chapters 29 and 31), which mediate the perception of sweet, bitter, and umami taste sensations, have been found to be specific targets for autoimmune disease in animal models (Kim et al., 2012), providing a potential focus for directing therapeutics to mitigate their adverse effect on taste. Interestingly, we now know that some patients with taste dysfunction have alterations in the expression of members of the T2R taste receptor gene family (for details of taste receptor genetics, see Chapter 31). In hypogeusia (decreased taste function), the frequency of the T2R40 gene is significantly reduced (Onoda et al., 2011), whereas in phantogeusia (taste sensations in the absence of a stimulus), the frequency of this gene appears to be increased (Hirai et al., 2012). Such a basic understanding may, in the future, lead to gene-related treatments for such disorders.

While it has been known for more than two centuries that taste sensations are mediated via taste buds within the lingual gustatory papillae and the palate (Chapter 29), the relatively recent discoveries of taste receptors in the respiratory epithelium, larynx, epiglottis, stomach, pancreas, and colon will undoubtedly influence the future practice of medicine (for history, see Chapter 1). The gut receptors represent a form of gastrointestinal chemosensation that impacts digestion, chemical absorption, insulin release, and ultimately the metabolism of swallowed foods and beverages (e.g., Breer et al., 2012; Reimann et al., 2012). Gut taste receptors explain why the pancreas releases more insulin when glucose is ingested than when directly

---

Handbook of Olfaction and Gustation, Third Edition. Edited by Richard L. Doty.
© 2015 Richard L. Doty. Published 2015 by John Wiley & Sons, Inc.

Table 39.1 Medical specialties associated with the management of taste dysfunction (listed alphabetically).

cardiology
dentistry/dental surgery
dermatology
endocrinology
gastroenterology
general and vascular surgery
gerontology
internal/family medicine
neurology
neurosurgery
nutrition
oncology
otolaryngology: head and neck surgery
psychiatry/psychology

injected into the blood stream (Geraedts et al., 2012). The dysfunction of such receptors may also explain, at least in part, some cases of anorexia and obesity, and have the potential for providing new therapeutic targets for controlling metabolism (Janssen and Depoortere, 2013). Their decrease provides one explanation of why gastric bypass patients have an immediate decline in their underlying insulin resistance (Geraedts et al., 2012) and often find sweet and fatty meals to be less pleasant (Miras and le Roux, 2010). This new knowledge helps to explain why, to maintain optimal health, one must utilize the gut to receive nutrients rather than to rely on parenteral or intravenous means of feeding.

Despite such advances, the fact remains that relatively little is known about how diseases alter taste function. Even though disorders of taste are common in the general population, the spectrum of disorders influencing taste is broad, and patients are often at a loss as to where to turn for appropriate medical care. From a management perspective, a multidisciplinary mindset on the part of the physician or healthcare professional is clearly needed (Table 39.1). For example, which physician should manage an elderly patient who complaints about a persistent "salty" taste, and is known to have hypertension, thyroid disease, poor dentition, and gastroesophageal reflux? Should it be an internist, endocrinologist, dentist, or gastroenterologist? What about the depressed patient with throat or brain cancer, who suffers from seizures or Bell's palsy, continues to lose weight, has recurrent oral lesions, and complains that things "don't taste right"? Such complexity is one reason why many physicians are at a loss as to how to manage such patients and where to refer them.

The goals of this chapter are to (1) explore the numerous ways that patients present to their physician with taste complaints and what those complaints may imply, (2) provide an approach to obtaining a taste-focused medical

history, physical exam, and sensory evaluation, (3) present the gamut of clinical disorders, across multiple specialties, that can be associated with taste disturbance, (4) discuss clinically-relevant anatomy and physiology associated with taste in the context of specific taste disorders, and (5) suggest options for patient management.

39.2 RANGE OF SYMPTOMS AND THEIR IMPLICATIONS

There is tremendous variability in the ways patients present with taste problems. The range of disturbances include: (1) perceived loss or diminution of function (ageusia or hypogeusia), (2) the distortion of everyday flavors (dysgeusia), (3) persistent abnormal taste sensations in the absence of taste stimuli (phantogeusia), (4) inability to recognize a taste sensation, even though gustatory processing, language, and general intellectual functions are essentially intact (gustatory agnosia), and (5) syndromes of pain with or without taste triggers (e.g., burning mouth syndrome). The overlapping and redundant innervation of the lingual taste buds from multiple cranial nerves and their branches (Chapter 29) makes taste disturbance less likely, but not impossible, in cases of selective neural injury. In fact, distortions of normal gustatory perception may occur with isolated injury to any one of the major nerve pathways, although the patient may not always recognize the problem.

One important distinction that should be made is whether the patient has difficulty with the perception of primary tastes versus the overall *flavor* of a substance. Primary taste sensations include *sweet*, *sour*, *salty*, *bitter*, and possibly *metallic* (iron salts), *umami* (monosodium glutamate, disodium gluanylate, disodium inosinate), and *chalky* (calium salts). Chemical compounds can vary tremendously in their ability to stimulate these primary taste sensations. Like primary colors, these sensations blend to create variable "hues" of taste experience, and when they are combined with the sensory input from the olfactory and trigeminal systems, tasted substances acquire an aroma, texture, temperature, and spiciness – allowing for the overall perception of *flavor*. The primary flavor component of many "tastes" is, in fact, mediated by olfaction (e.g., chocolate, coffee, rose, strawberry, etc.). Stimulation of olfaction occurs during deglutition, when the molecules emanating from the food reach the olfactory mucosa from the oral cavity via the so-called retronasal route. Therefore, when a patient complains of distorted flavor perception, it is essential to also consider coexisting or isolated impairment of the olfactory and trigeminal systems (see Chapters 17 and 50).

Impairment of taste can be devastating to a patient since it not only affects the ability to enjoy food products, but can ultimately alter food choices and patterns of consumption, thereby resulting in weight loss or weight gain, and in some cases malnutrition. People rely on taste function to gauge the quality of ingested nutrients; thus, in patients who are diabetic or hypertensive, the loss of taste may lead to dangerous overcompensation for the loss by adding too much sugar or salt to food. Without proper gustatory function, people are at risk for ingesting potential toxins and spoiled foodstuffs (which usually taste "bad"). Importantly, taste is needed to trigger normal ingestive and digestive reflex systems such as the secretion of oral, gastric, pancreatic, and intestinal juices (Chapter 36).

39.3 MEDICAL EVALUATION

39.3.1 Medical History

A focused clinical history for a person complaining of chemosensory dysfunction, particularly taste, begins with an accurate description of the subjective problem, including (a) onset, (b) frequency, (c) timing, (d) severity, (e) exacerbating or relieving foods or products, (f) relation to daily activities, bodily functions, and states of mind, (g) prior treatments and their efficacy, and (h) co-morbid medical conditions. A determination of current medications is essential. Particular note should be made in the history of problems with speech articulation, salivation, chewing, swallowing, oral pain or burning, dryness of the mouth, periodontal disease, foul breath odor, recent dental procedures or surgeries, recent radiation exposure, medications, and bruxism. Also, inquiry into the patient's diet and oral habits may reveal exposure to oral irritants. Questions about hearing, tinnitus, and balance are important since the vestibulocochlear nerve (CN VIII) travels in close proximity to the facial nerve and can be susceptible to similar etiologies of disturbance (e.g., cerebellopontine angle tumor). Considering the occasional effect of gastroesophageal reflux on taste, asking about stomach problems may also be relevant. Asking about constitutional symptoms, such as fever, malaise, headache, body pains, and rashes, may also be important since such symptoms often accompany cancers or systemic inflammatory conditions (e.g., lupus). The possibility of seizure should be entertained, especially if the patient describes having "black-outs" or witnessed abnormal movements or actions.

39.3.2 Physical Examination

In the evaluation of the oral cavity, an examiner should attend to the condition of the teeth and gums. Notably, a dysgeusia may result from exudates that commonly appear with gingivitis or pyorrhea. The nature and integrity of the fillings, bridges, and other dental work should be inspected. Visual inspection of the tongue surface may show signs of scarring, inflammation, or the absence, atrophy or overgrowth of papillae. A whitish plaque on the tongue can be from candidiasis, lichen planus, leukoplakia, or food product. Inflammation of the tongue, or glossitis, may be present, and can reflect local or systemic conditions. The physician should pay particular note to the color and degree of salivation. Palpation of the tongue should be performed to detect masses, neoplastic lesions or collections deep in the tongue's musculature. The neck should also be evaluated for the presence of masses, thyromegaly, and carotid bruits. Neurologically, the integrity of CN VII, IX, and X can be evaluated by screening for deficits in non-gustatory functions of these nerves, such as facial movements and expression, swallow function, gag, salivation, and voice production. To establish whether taste pores or papillae appear normal or are present in adequate numbers, the tongue can be stained with a dye and then photographed using a low power magnification (e.g., Doty et al., 2001; Miller and Reedy, 1990). Although not performed routinely, biopsy of circumvallate or fungiform papillae for detailed microscopic examination can establish whether pathological changes are present in taste bud tissue. Visual inspection of the ears, externally and internally, via an otoscope is essential. Normally, the tympanic membrane is has a pale shiny gray color; erythema, vesicles, bulging, or an air-fluid level behind the membrane suggests infection. Breath odor can be diagnostic of a medical condition that affects taste function, such as fishy ammonia (renal failure), fruity chewing gum or acetone (diabetic ketoacidosis), musty (fetor hepaticus/ liver disease), putrid (oral or lung infection), or stale smoke (cigarette smoking). There are specific taste and flavor disturbances in conditions such as chronic hepatitis C infection, chronic kidney disease, and other metabolic disorders (Musialik et al., 2012; Armstrong et al., 2010; Palouzier-Paulignan et al., 2012). Considering the direct influence that olfaction has on perceptions of flavor, it is also important to examine the nose, nasal cavity, and paranasal sinus structures (Bromley and Doty, 2010).

39.3.3 Diagnostic Testing

Since most complaints of taste loss really reflect smell loss, olfactory testing should be performed, which is now easy to do given the commercial availability of quantitative smell tests (see Chapter 11). Unfortunately, taste testing is less practical, largely because of time considerations (see Chapter 34). While a number of whole-mouth taste tests have been described in the literature, regional taste testing is required to establish the function of each of the nerves innervating specific taste bud fields, as whole-mouth tests are insensitive to even complete dysfunction of one or several of the nerves that innervate the tongue. Two approaches are available to regionally assess taste function psychophysically. In *chemical testing*, liquid stimuli are presented to selected target regions of the tongue or oral cavity via pipettes, special delivery devices, or filter paper disks. In *electrical testing* (also termed electrogustometry), low levels of electrical current are presented to taste bud fields (Frank and Smith, 1991). At low current ranges, gustatory – not somatosensory – afferents are activated. Equipment designed to present such electrical current to the tongue, termed electrogustometers, typically employ batteries to produce the current. It should be emphasized, however, that even the most simplistic of assessments can have tremendous utility in the clinical setting. For example, for a neurologist trying to distinguish a Bell's Palsy from a stroke syndrome causing facial weakness, the application of a sugar slurry (readily made with packets of sugar and a little water) to the affected side of the tongue (anterior 2/3) followed by the unaffected side can identify a unilateral deficit in taste that is typically found in Bell's Palsy alone.

Imaging studies can be helpful in the evaluation of gustatory disturbance. Plain radiographs of the teeth and jaw may reveal focal lesions of the teeth (e.g., fracture, cavity, impaction) or periodontal tissues (e.g., abscess extension) in proximity to gustatory fibers (e.g., chorda tympani-lingual nerve) and the tongue. Computed tomography (CT), however, is of greater utility in visualizing peripheral and central structures of the head and neck that are frequently affected by trauma, surgery, infection, cancer, and systemic disease. For example, fine axial "cuts" through the mastoid portion of the temporal bone can demonstrate subtle bony defects and mass lesions in and around the middle ear. Coronal CT scans are useful to evaluate paranasal anatomy. Noncontrast CT is used to show bony defects and acute blood accumulations, whereas contrast-enhanced CT highlights areas of inflammation or abnormal vascularity (as seen with many tumors). Magnetic resonance imaging (MRI) is the method of choice for soft-tissue anatomy and central nervous system structures. Particularly when sequences such as DWI (diffusion-weighted imaging) and FLAIR (fluid-attenuated inversion recovery) are used, MRI can discern infarcts of variable ages and other changes within central gustatory pathways. Indeed, gustatory dysfunction has been related to MRI-established ischemia, hemorrhage, or demyelinating plaques in the brainstem and higher cortical areas of some persons with chronic dysgeusia (Tsivgoulis et al., 2011; Onoda and Ikeda, 1999; Yabe et al., 1995). Gadolinium-DTPA (diethylenetriamine pentaacetic acid) enhancement is usually necessary to visualize lesions caused be demyelination, tumor, and infection. An electroencephalogram (EEG) should be obtained in all patients suspected for having seizure activity. Brain stem auditory evoked response (BAER) testing is useful to evaluate the continuity and symmetry of the brain stem auditory pathways and nuclei, and therefore, provides a screen for focal subcortical injury in this region. Although restricted in their resolution and general availability, functional imaging of the brain using Positron Emission Tomography (PET), Single Proton Emission Computed Tomography (SPECT), and functional MRI (fMRI) may be helpful in evaluating cortical and sub-cortical structures related to gustatory processing – particularly in the context of cerebrovascular disease, epilepsy, and dementia. Depending on the suspected underlying medical problem affecting taste, there are a number of laboratory studies that are also available (Bromley, 2000).

39.4 CAUSES OF TASTE DYSFUNCTION

A wide range of disorders and interventions has been associated with at least some disturbance of taste function. Mechanisms for the alteration of taste may include: (1) the release of bad-tasting materials from oral medical conditions and appliances (e.g., gingivitis, purulent sialadenitis), (2) transport problems of tastants to the taste buds (e.g., causes by excessive dryness of the oral cavity, damage to taste pores from a burn, candidiasis), (3) destruction or loss of taste buds themselves (e.g., local trauma, invasive tumor), (4) damage to one or more neural pathways innervating the taste buds (e.g., Bell's Palsy, trauma, dental or surgical procedures), (5) involvement of central neural structures (e.g., tumor, epilepsy, stroke), and (6) systemic disturbances of metabolism that presumably affect the taste system via any of the aforementioned mechanisms (e.g., medications, hypothyroidism, diabetes, liver or kidney disease).

Although total loss or markedly diminished wholemouth gustation exists, such conditions are rarely produced by non-metabolic diseases or disorders unassociated with central ischemic or other damage, since regeneration of taste buds can occur and peripheral damage alone would require the involvement of multiple pathways. Thus, while 433 of 585 patients (74%) with verifiable olfactory loss complained to us of both smell and taste disturbance, less than 4% had verifiable whole-mouth gustatory dysfunction, and even that dysfunction was limited (Deems et al., 1991). Regional deficits in taste dysfunction are much more common. For example, in one study, sensitivity to three relatively low concentrations of NaCl was measured on the tongue tip and 3 cm posterior to the tongue tip in 12 young (20–29 years of age) and 12 elderly (70–79 years of age) subjects. On average, the young subjects were more sensitive to NaCl on the tongue tip than on the more posterior stimulation site and exhibited, at both tongue loci, an increase in detection performance as the stimulus concentration increased. The elderly subjects, who would be expected to exhibit, at worse, moderate deficits on whole-mouth testing, performed at near-chance level (Matsuda and Doty, 1995).

39.4 Causes of Taste Dysfunction

Patients with focal regions of taste impairment can, in some instances, perceive an alteration in overall gustatory perception. Thus, in one series, nearly 57% of patients presenting with Bell's palsy complained of dysgeusia (Adour, 1982) and, in a different study, more than 9% of patients with surgically confirmed acoustic neuromas experienced loss of taste as a primary complaint (Harner and Laws, 1981). However, many people are unaware of such deficits. In fact, it is unusual for patients to specifically recognize their loss of taste sensation on half of the anterior tongue following unilateral sectioning of the chorda tympani in middle ear surgery. This lack of awareness stems, in part, from the redundancy of the multiple taste nerves, as well as compensatory mechanisms.

Interestingly, unilateral anaesthetization of the chorda tympani nerve (CN VII) enhances taste perception on the rear of the tongue (CN IX), particularly on the side contralateral to that of the anaesthesia (Kveton and Bartoshuk, 1994; Yanagisawa et al., 1998). Such enhancement – which would be expected if significant regional lingual damage occurs – seems to be accentuated for unpleasant tastants and, moreover, may induce phantom tastes. Yanagisawa et al. (1998) have reported that this procedure increases the perceived intensity of bitter substances, such as quinine, applied to posterior taste fields, whereas the reverse occurs for the salty taste of NaCl. When both chorda tympani nerves were anesthetized, the effect was perceived bilaterally. In about 40% of their subjects, a phantom taste, usually localized to the posterior tongue contralateral to the anesthesia, appeared in the absence of stimulation and disappeared when the region of origin was anesthetized. Such phantoms may reflect central interactions among the taste nerves (e.g., release of inhibition), since the taste pathways do not cross until they reach the thalamus.

39.4.1 Disorders of the Oral Cavity and Related Structures that Affect Taste

39.4.1.1 Oral Mucosa.
The mucous membranes of the mouth, tongue, and oropharynx are directly exposed to the outside environment, and as such, are unique in their ability to protect, adapt, and regenerate in response to continuous chemical, mechanical, and thermal challenges. Digestion begins in the mouth with tastants being solubilized in saliva and exposed to a myriad of enzymes and resident microflora. A testimony to the durability of the oral mucosa is the fact that humans ingest many diverse foods over a broad range of temperatures and textures, often in the context of abrasive oral appliances and harmful habits such as tobacco use. Moreover, this mucosa tolerates large microbial concentrations of the magnitude found elsewhere only in the colon. This durability stems, in part, from antimicrobial properties of saliva, and the high basal turnover rate of oral epithelial cells (Beidler and Smallman, 1965). Microorganisms that bind to surface cells of the mucosa are rapidly sloughed as the surface is renewed. Subsequently, disorders that alter the secretion and contents of saliva, the turnover rate of epithelial cells, and the balance of microorganisms within the epithelium, have the potential to alter proper mucosal function and to distort taste.

By providing moisture and lubrication, saliva facilitates the chewing and swallowing of food, aids in the production of speech, acts as a buffer to acids and bases, and increases the delivery of solubilized chemicals to taste receptors (Chapter 28). Inadequate salivary flow via impaired production or secretion (e.g., numerous medications, sialolithiasis, irradiation of salivary glands) can result in (1) the tendency of oral mucosa to stick together, (2) impairment of eating and speech function, (3) aggravation of oral infections, (4) impaired ability to properly wear dental prostheses, and, in some instances, (5) abnormal flavor sensations. Clinically, this constellation of problems is considered *xerostomia*, and is one means by which many local and systemic diseases may cause taste disturbance. Importantly, saliva likely plays a significant role in taste bud maintenance (Morris-Witman et al., 2000).

Some lesions involving the oral epithelium can be associated with altered taste perception, tactile irritation, and possibly oral pain (Table 39.2), either by direct involvement with taste-related structures or through association with a taste-altering systemic illness (e.g., local or systemic cancer, autoimmune disorder, vitamin deficiency). Most healthy tissues of the oropharynx appear pink because the epithelium and connective tissues of the lamina propria are relatively translucent and allow red light to reflect from the underlying capillary bed. Disruption of the epithelial architecture, from such pathological processes as chemical necrosis of the surface layers, microbial colonies, exudate, and epithelial thickening from continued trauma, can scatter and reflect incident light creating a discolored whitish lesion. In regions of particular inflammation within the mucosa (e.g., glossitis), reddish lesions can predominate. For example, continued exposure to the heat and smoke from burning tobacco can lead to a rough, white, and fissured tongue surface often with associated areas of inflammation – known as nicotinic stomatitis (Rossie and Guggenheimer, 1990). Leukoplakia is a clinical diagnosis reserved for whitish lesions that cannot be scraped off the tongue surface. Approximately 20% of these lesions represent premalignant epithelial dysplasia, carcinoma *in situ*, or frank invasive squamous cell carcinoma (Rodriguez-Perez and Banoczy, 1982; Allen and Blozis, 1998). The apparent overgrowth or the exaggeration of the length and thickness of the filiform papillae (e.g., "hairy tongue") can result from conditions that impair appropriate desquamation either by reducing salivary flow, limiting oral movements,

Table 39.2 Disorders involving structures of the oral cavity that can alter taste perception.

EPITHELIAL ABNORMALITIES
    Autoimmune
        Behcet's syndrome
        Crohn's disease
        Lupus erythematosus
        Pemphigus vulgaris
        Reiter's syndrome
        Scleroderma
        Sjogren's syndrome
        Wegener's granulomatosis
    Discolored lesions
        Acanthosis nigricans
        Chronic ulcerative stomatits
        Diffuse or multiple papillary/ verrucous lesions
          (e.g., Hairy tongue)
        Erythema multiforme
        Leukoplakia
        Lichen planus
        Median rhomboid glossitis (MRG)
        Nicotinic stomatitis
        Reactive papillary hyperplasia
    Metabolic
        Primary amyloidosis (affecting tongue)
    Infection
        Candidiasis
        Ulcerative lesions
          coxsachievirus
          gonorrhea
          herpes simplex virus
          mycoses
          syphilis
          varicella/ zoster
    Local injury
        Burns/ chemical injury
        Radiation/ chemotherapy
        Trauma
    Neoplastic
        Kaposi's sarcoma
        Oral melanoma
        Squamous cell carcinoma
        Tongue cancer (erosive)
        Verrucous carcinoma
ODONTOGENIC DISEASES
    Acute necrotizing ulcerative gingivitis (ANUG or "trench
      mouth")
    AIDS-related peridontitis
    Dental cares/ abscess
    Gingivitis (acute and chronic)
SALIVARY GLAND ABNORMALITIES
    Acute suppurative sialoadenitis from:
        benign lymphoepithelial lesion
        cystic fibrosis
        dehydration
        granulomatous disease (e.g., TB, sarcoid)
        medications

Table 39.2 (*Continued*)

SALIVARY GLAND ABNORMALITIES
        radiation/ chemotherapy
        salivary gland tumor or cyst
        sialolithiasis
        Sjogren's syndrome
        viral infections
    Chronic sialoadenitis from:
        congenital or trauma-related stenosis
        foreign body
        recurrent sialolithiasis
        salivary gland tumor or cyst
PARANASAL SINUS ABNORMALITIES
    Post-nasal drip
    Purulent sinusitis
GASTROESOPHAGEAL REFLUX DISEASE

or causing temporary imbalances in the growth of normal microbial flora. Numerous drug therapies, primarily antibiotics, have been implicated as a cause for this condition (Thompson and Kessler, 2010). Hairy tongue can sometimes be associated with a disagreeable taste or sensations of gagging when swallowing (Thompson and Kessler, 2010).

Squamous cell carcinoma is particularly worrisome, since: (1) it is responsible for more than 90% of oral cancers, (2) has a considerable mortality rate (5 year survival rate of approximately 52%), and (3) may have only subtle affects – if any – on one's taste function (Salisbury, 1997). Arising from mucosal epithelium, squamous cell carcinoma is common on the ventral and lateral surfaces of the tongue and can be invasive, producing swelling and induration (Allen and Blozis, 1998). After several months of growth, the cancer can develop into an exophytic (sometimes endophytic), painful, ulcerated mass that bleeds easily and has a necrotic odor (Salisbury, 1997). The treatment of oral cancer, as well as other cancers of the head and neck, with agents like radiation and chemotherapy is more often associated with impairment of taste function than the cancers themselves. Radiation therapy alters taste bud cell turnover and can affect taste function, although return after radiation treatment has been observed (Conger, 1973). Additionally, radiation therapy fields often include salivary glands (e.g., parotid or submandibular) that are near the site of the primary tumor or lymph nodes that need treatment; this can result in irreversible damage to acinar cells, profound impairment of salivary flow, subsequent dessication of the mucosa, and xerostomia (Salisbury, 1997). Also, susceptibility to opportunistic infections (e.g., dental caries, candidiasis) increases after therapy. Necrosis of the tongue after arterial chemotherapy has been reported, in addition to the known direct affect on basal cell and taste receptor production (Shih et al, 1999; Duhra and Foulds, 1988).

Local injury to the oral cavity from such insults as physical trauma, burns, and toxic chemical exposures can acutely alter flavor perception and create a dysgeusia associated with the region of mucosa involved. Applying a caustic substance to the oral mucosa can produce variable degrees of epithelial necrosis – resulting in whitish lesions, erosions, ulcerations, or blisters. For example, the inappropriate use of aspirin as a topical anesthetic causes a common and severe form of chemical burn to the musoca (Allen and Blozis, 1998). Although most etiologies of local injury produce only transient lesions that last 1 to 2 weeks, repeated exposures (e.g., heavy use of O_2 liberating mouthwashes) or injury (e.g., prolonged intubation with an oral endotracheal tube), especially in the context of underlying xerostomia, can produce more lasting and symptomatic lesions, and, in some patients, taste disturbances. Interestingly, taste function is relatively resistant to the toxic effects of smoking, although morphological changes to surface papillae on the tongue are observed in heavy smokers (Konstantinidis et al., 2010).

Infections are common causes of distorted taste and oral pain. In fact, any local infection in and around the neural pathways or receptor sites associated with taste perception has the potential to disturb or impair taste. The involved mechanisms can be direct inflammation and injury to a taste-related neural structure or the associated release of mucopurulent material that generates a bad flavor. Important anatomical regions include: (1) the surface of the tongue, mouth, and oropharynx, (2) the teeth and peridontal tissue, (3) the salivary glands, and (4) the middle ear. *Candida albicans* is a commensal yeast that normally exists, in small numbers, in the oral flora. However, when the oral immune system is not functioning optimally – as seen in the very young or old, in the HIV-infected, in patients with leukemia, or in those who have recently taken corticosteroids, chemotherapy, or broad-spectum antibiotics – candidiasis of the tongue and oropharynx can ensue. The development of a pseudomembrane overlying an inflamed oral mucosa is called "thrush," and can be treated with antifungal therapy (Allen, 1992). There are other types of infectious diseases causing oral lesions, many of them viral, and most resulting in some form of pharyngitis, stomatitis, and an ulcerative or blistered lesion of the oral mucosa. Typically they are treated with topical or systemic antibiotics.

There are several autoimmune syndromes that affect the oral mucosa, usually by causing xerostomia and/or ulceration that, on rare occasion, can result in taste disturbance (for review, see Campbell et al., 1983). Sjogren's syndrome (SS) is associated with salivary and lacrimal gland dysfunction – resulting in dry eyes and dry mouth (also called sicca syndrome) – and arthritis. If the symptoms of SS occur in the context of another autoimmune disease (e.g., lupus, scleroderma, rheumatoid arthritis), it is said to be *secondary SS*. Systemic lupus erythematosus (SLE) is an multisystem inflammatory disease of small blood vessels that can produce a wide range of clinical symptoms including superficial ulcerative, vesicular, and white keratotic lesions of the tongue and oral mucosa. Behcet's syndrome is also a multisystem disorder consisting of recurrent oral and genital ulcers, inflammatory disease of the eyes and gastrointestinal tract, arthitis, and on rare occasion cranial neuropathies or menigoencephalitis. *Pemphigus vulgaris* is characterized by serum antibodies directed against components of the epidermis and mucosa, forming characteristic bulla that progress to superficial and friable eroded areas most often involving the palate, buccal mucosa, and tongue (Laskaris et al, 1982). One report describes a patient with Giant cell arteritis presenting with ageusia and lower extremity claudication (Mathew et al., 2010).

39.4.1.2 Tongue.
The tongue is a highly mobile muscular structure with multiple functions, including (1) manipulating food during mastication (chewing), (2) squeezing food into the pharynx during deglutition (swallowing), (3) forming words during articulation (speech), (4) oral cleansing, and (5) taste. Embryologically, the anterior 2/3rds of the tongue (oral part) develops from the first brachial arch while the posterior third (pharyngeal part) develops from the second and third brachial arches, producing two distinct geographic regions identifiable by their nerve supply (CN VII and IX, respectively) and lymphatic drainage. The tongue's arterial supply is chiefly through the lingual artery which arises from the external carotid. Within the pharynx, the oral and nasal cavities communicate, allowing volatile chemicals released during chewing to pass into the nasal passages. Disorders of the tongue, such as neoplasms, infarction and swelling, cannot only affect overall lingual function, but distort surface structures and impair taste. In primary amyloidosis, amyloid deposition within the lingual tissues can impair taste and can be an early symptom of the disease (Ujike et al 1987).

39.4.1.3 Gingival/Dental/Periodontal Structures.
Most diseases affecting the periodontal soft tissues (the gingiva) and the soft and hard palates (periodontium) have an infectious component to their pathology. Dental infections are common among patients who neglect regular dental care. A dentoalveolar abscess, usually resulting from dental caries, can erode from the tooth to the adjacent gingival tissues to create a draining gingival fistula. These fistulas can drain bitter and foul-tasting purulent material. A peridontal cellulitis and/or abscess can involve the adjacent bone or perimandibular soft tissues and is usually associated with

significant gingivitis, variable degrees of facial swelling, and fever. Acute necrotizing ulcerative gingivitis (ANUG, also known as *trench mouth* from the outbreaks among the troops in the trenches during World War I) is now not considered to be contagious, but rather, an opportunistic overgrowth of relatively common microbes in the mouth (Melnick, 1988). Predisposing factors include psychological and environmental stress, nutritional inadequacies, poor oral hygiene, and smoking. The clinical presentation is stereotyped, and involves the sudden onset of gingival pain, bleeding, a bad taste in the mouth, and a fetid mouth odor that can be recognized by the patient (Melnick, 1988). A periodontal infection that is strategically located in the area of the chorda tympani-lingual nerve as it passes near the third molars can injure this nerve and affect taste on the ipsilateral side of the tongue.

39.4.1.4 Salivary Gland Disorders.
In addition to producing saliva, the anatomical orientation of the salivary glands is important to taste since CN VII emerges from the stylomastoid foramen to enter the substance of the parotid gland. Therefore, lesions of the parotid (e.g., tumors, abscesses) can eventually involve this nerve and affect taste. Infections of the salivary glands (saladinitis) can be acute or chronic. Bacterial infections of the gland parenchyma are typically retrograde in nature and the result of a mechanical blockage or diminished production of saliva (Berndt et al, 1931; Gayner et al, 1998). However, there are viral infections, such as the mumps or HIV, which can produce localized gland disease (Gayner et al, 1998). Local symptoms of suppurative sialoadenitis include pain, swelling, induration, and suppurative discharge from the duct orifice into the oral cavity yielding a foul taste. Although infections can occur in any salivary gland, the parotid is the one most commonly involved (Gayner et al, 1998). High pH and viscosity of gland saliva likely contribute to calcium phosphate stone formation within the salivary duct, resulting in obstruction, stagnant salivary flow, dry mouth, and a suppurative infection. Contributing factors to infection include poor oral hygiene and compromised host resistence.

39.4.1.5 Gastroesophageal Reflux Disease.
Gastroesophageal reflux disease (GERD) is defined as a recurrent regurgitation of acidic gastric contents up the esophagus as a result of transient or chronic relaxation of the lower esophageal sphincter (Isolauri et al., 1997). GERD symptoms can be obvious, such as recurrent mid-chest discomfort from esophageal irritation (*heartburn*), or they can be rather subtle, only manifesting as chronic cough, halitosis, hoarseness, or a recurrent acid or sour taste (Richter, 1998; Nelson et al., 2000). While it is not clear whether GERD can result in permanent injury to the gustatory mucosa, dysgeusias due to GERD do occur, but typically resolve with proper treatment of the reflux.

39.4.2 Disorders of the Multiple Peripheral Neural Pathways that Affect Taste

Depending on their location, taste receptors within the taste buds are innervated by branches of several cranial nerves, including CN VII, IX, X, which, unlike CN I, are mixed motor and sensory nerves transmitting multiple forms of information. This becomes important when considering clinical syndromes that involve taste dysfunction since variable degrees of either motor or sensory disturbance may be present, depending upon where and to what extent the taste-mediating cranial nerve is injured (see Table 39.3). All the peripheral gustatory fibers enter the brainstem and project to the nucleus of the tractus solitarius (NTS), which begins in the rosterolateral medulla and extends caudally along the ventral border of the vestibular nuclei.

39.4.2.1 The Facial Nerve (CN VII).
The facial nerve (CN VII) supplies the muscles of facial expression, the stapedius muscle of the middle ear (which helps to dampen the movement of the stapes in response to loud auditory stimuli), the sublingual and submandibular salivary glands, the lacrimal glands, and the membranes of the oral and nasal cavities. In addition, this nerve conveys taste sensation from the taste buds found in the fungiform and filiform papillae on the ispilateral anterior two-thirds of the tongue (via the lingual nerve to the chorda tympani) and on the soft palate (via the lesser palatine nerve branch of the greater petrosal nerve). Because of the relative long and winding course of the facial nerve, it is susceptible to peripheral injury in a number of locations. Thus, from the tongue, gustatory fibers travel through the lingual nerve, passing within the soft tissue of the jaw along the medial surface of the mandible in close proximity to the mandibular third molar. At this point they become the chorda tympani nerve, which penetrates the skull via the stylomastoid foramen and which turns within the tympanic cavity of the middle ear along the medial surface of the tympanic membrane between the malleus and the incus. The fibers then synapse with the geniculate ganglion at the bend of the facial nerve within the facial canal, where the projection becomes the nervus intermedius, which travels next to the eighth nerve through the internal auditory meatus into the cerebellopontine angle. This nerve subsequently penetrates the caudal border of the pons near the medulla and enters the brainstem. Thus, tumors anywhere along the nerve path of the facial nerve — from the oral cavity through the skull base, temporal bone and middle ear, to the cerebellopontine angle and the brainstem — can cause a neuropathy of the nerve.

39.4 Causes of Taste Dysfunction

Table 39.3 Disorders of peripheral nerves that may alter taste perception.

CRANIAL NERVE VII (FACIAL NERVE)
    Infection/ polyneruitis
        Bell's palsy
        Botulism
        Guillain-Barré syndrome
        Herpes zoster
        Human immunodeficiency virus
        Kawasaki disease
        Leprosy
        Lyme disease
        Otitis Media
            Acute bacterial
            Chronic bacterial
            Cholesteatoma
        Syphilis
        Tuberculosis
    Neoplastic
        Glomus tumor
        Primary parotid tumor
        Schwannoma / Acoustic neuroma
        Temporal bone cancer / metatases
    Trauma
        Barotrauma
        Birth trauma
        Injection of local anesthesia (e.g., dental procedure)
        Middle ear surgery
        Temporal or mandibular bone fracture
        Third molar extraction
    Neurovascular
        Brain stem stroke (pontine)
    Inflammatory
        Multiple sclerosis
        Sarcoidosis
    Melkersson syndrome

CRANIAL NERVE IX (GLOSSOPHARYNGEAL NERVE)
    Trauma
        Bronchoscopy
        Laryngoscopy
        Penetrating neck trauma
        Skull base fracture
        Tonsillectomy
    Neoplastic
        Glomus Jugulare tumor
        Neurofibroma
        Skull-based metastases
    Neurovascular
        Carotid artery aneurysm or dissection
        Brain stem stroke (pontine, medullary)
    Infection
        Botulism
        Guillain-Barré syndrome
        Herpes zoster of petrosal ganglion (rare)
        Leprosy
        Meningitis
        Otitis media

Table 39.3 (*Continued*)

CRANIAL NERVE IX (GLOSSOPHARYNGEAL NERVE)
        Pharyngeal / retroparotid abscess
    Inflammatory
        Multiple sclerosis
        Sarcoidosis
CRANIAL NERVE X (VAGUS NERVE)
    Trauma
        Neck surgery
        Penetrating trauma
        Skull base fracture
    Infection
        Cytomegalovirus
        Guillain-Barré syndrome
        Leprosy
        Meningitis
        Pharyngeal / retroparotid abscess
        Poliomyelitis
    Neoplastic
        Glomus Jugulare tumor
        Neurofibroma
        Skull-based metastases
    Neurovascular
        Brain stem stroke (medullary)
        Carotid artery aneurysm or dissection
    Degenerative
        Multisystem atrophy
        Progressive bulbar palsy
    Inflammatory
        Multiple sclerosis
        Sarcoidosis

Bell's Palsy (BP) is a common facial nerve injury associated with gustatory loss. Clinical criteria include the sudden weakness of all muscles of one side of the face in the absence of CNS, ear, or cerebellopontine disease (Taverner, 1973). The etiology is still considered idiopathic, although viruses are often involved (Adour et al, 1975; Takahashi et al., 2001). Thus, the herpes simplex virus-1 genome has been isolated from facial nerve endoneural fluid in affected patients (Murakami et al., 1996). BP affects both sexes and all ages. Although it is frequently associated with ipsilateral loss of taste over the anterior two-thirds of the tongue, the patient may or may not recognize this loss. Involvement of taste in BP localizes the lesion to the neural pathways from the pons, along the nervus intermedius portion of the facial nerve, through the geniculate ganglion, to the point where the chorda tympani joins the facial nerve in the facial canal. A more severe form of BP, Ramsy-Hunt syndrome, results from active herpes infection of the geniculate ganglion, and involves pain and vesicles in the external auditory canal or soft palate. Bartoshuk and Miller (1991) describe a patient with this syndrome whose neural involvement included both

CN V and CN IX on the left side. Intensity ratings of the four major taste qualities were obtained at regular intervals across a nearly two-year period for the front, back, and palate regions. The entire left side of the patient was devoid of taste function about three months after the *Herpes zoster oticus* attack. Although all taste qualities were perceived on the left rear and palate regions after about a year, only sweetness was perceived on the left front. Even by the end of study, full taste recovery of the anterior tongue was not evident.

BP usually begins with pain in or behind the ipsilateral ear with subsequent unilateral facial weakness with maximum paralysis occurring within 48–72 hours). In addition to ipsilateral taste loss, hyperacusis (due to weakness of the stapedius muscle) is often present. The taste-salivary reflex also can be compromised. This reflex results in salivation when a gustatory stimulus, such as a weak acid, stimulates the anterior tongue, and reflects a relay from the nucleus of the solitary tract to the parasympathetic fibers innervating the salivary glands. Regeneration of the facial nerve after BP may be abnormal, such that novel innervation to nearby structures can produce synkinesis of the facial musculature when chewing, or tearing of the ipsilateral eye while eating ("crocodile tears"). Melkersson syndrome, a familial disorder of recurrent BP, can result in permanent facial plegia, and is associated with lingua plicata (furrowed or fissured tongue) and recurrent facial edema.

Taste dysfunction resulting from trauma is much less common than post-traumatic smell dysfunction, with solitary ageusia of one or more primary taste modalities occurring in less than 1% of persons with major head injury (Sumner, 1967; Deems et al., 1991). Our understanding of focal nerve injury (and recovery) associated with potential taste disturbance has improved recently, particularly in iatrogenic situations involving surgical interventions of the head and neck (Graff-Radford and Evans, 2003; McManus et al., 2012; Guinand et al., 2010). However, in cases of global trauma to the head and neck, trauma-related taste disturbances are only rarely unaccompanied by smell impairment. The prognosis for post-traumatic ageusia is far better than for post-traumatic anosmia, and recovery from ageusia usually occurs over a period of a few weeks to months. The etiology of post-traumatic taste disturbance varies. Head trauma that results in injury to the middle ear, a finding often encountered with temporal bone fractures, can potentially result in both ipsilateral impairment of taste function *and* injury to the preganglionic parasympathetic fibers, thereby preventing salivary secretion. Approximately 7–10% of temporal bone fractures result in facial nerve dysfunction (Chang and Cass, 1999). Injury to the lingual nerve from trauma in and around the mouth and tongue is another cause of taste impairment. This branch of the mandibular division of the trigeminal nerve (CN V_3)

is the most proximal pathway to the tongue for general somatic sensation (touch) and special visceral sensation of taste (due to coexisting chorda tympani fibers). Thus, injury to the jaw resulting in mandibular fracture (e.g., punch with a fist), or to the inside of the mouth (e.g., difficult orotracheal intubation, aggressive dental procedure), can result in numbness and taste loss over the anterior two-thirds of the ispilateral side of the affected tongue. Compression injury to the chorda tympani-lingual nerve may have a greater effect on taste than on fine-touch sensory components, since the chorda tympani fibers of the conjoined nerve are more superficial and posterolateral, and therefore more susceptible to partial transection or pressure on the side of the nerve (Wantanabe et al., 1995).

In general, the classification of nerve injury pathology defined by Seddon in 1943 continues to apply to situations of head and neck injury to taste-mediating neural structures and to prognosis (Sedden, 1943). *Neuropraxia* is a conduction block through a nerve resulting from mild trauma without axonal damage and recovery of sensory deficits within, at most, days to months. *Axonotmesis* is a more severe injury to the nerve associated with afferent fiber degeneration, preservation of the nerve sheath, possible neuroma formation, incomplete sensory recovery, and, in most cases, significant dysesthesia/sensory distortions. *Neurotmesis* is the most severe injury typically associated with cutting the nerve all together, anesthesia/loss of function in the nerve distribution and no sensory recovery – particularly if the nerve path is within soft tissue. If the nerve path is within bone at the point of neurotmesis, regeneration may occur (Seddon, 1943; Graff-Radford and Evans, 2003).

A number of surgical procedures can influence taste function. Taste disturbances after tonsillectomy are not uncommon, with as many as eight percent of patients having subjective taste problems six months after surgery. One percent of tonsillectomy patients report suffering continued dysgeusia up to 2–12 years after surgery (Heiser et al., 2010, 2012). Ear surgery, including tympanoplasty, mastoidectomy, and stapedectomy, can damage CN VII, CN VIII, and middle ear structures (Moon and Pullen, 1963; Kveton and Bartoshuk, 1994; Guinand et al., 2010). In the case of mastoidectomy, the risk of hearing loss and facial nerve injury must be emphasized in any consent form. However, the chorda tympani nerve fibers are particularly at risk, and in situations of open cavity middle-ear surgery, chorda tympani division is almost inevitable (Kiverniti and Watters, 2012). This nerve is typically stretched, sectioned, or sacrificed during such surgery, often resulting in taste loss or a metallic dygeusia. Such sensory aberrations can last long after the operation. Bull (1965) found that 78% of patients with bilateral section and 32% of patients with unilateral section of the chorda tympani had persistent adverse symptoms. In a recent review, McManus et al.

(2012) found that 15–22% of patients who have undergone middle ear surgery experience symptoms of taste disturbance and/or dryness of the mouth. Most experienced a gradual recovery to normal function, but in some persistent and troublesome complaints continued. It should be noted that the effects of chronic otitis media alone can impair ipsilateral perception of sour, bitter and salty tastes, with subsequent improvement in function one month following middle-ear surgery *when the chorda tympani nerve is preserved* (Huang et al., 2012). If there is intra-operative injury to the chorda tympani nerve, the post-operative gustatory function a month later remains the same as the pre-operative baseline (McManus et al., 2012). With regard to patient symptoms following open cavity mastoidectomy to both ears, a recent survey showed that 24.3% were aware of a taste disturbance immediately after the surgery; only 8.7% had persistent long-term disturbance (Kivernitis and Watters, 2012). Laryngoscopy and endotracheal intubation can have rare complications that result in lingual nerve injury, often with coexisting hypoglossal nerve damage (Gaut and Williams, 2000; Silva et al., 1992; Evers et al., 1999). However, the specific incidence of taste disturbance with and without somatosensory loss following these procedures has not been well defined.

The common dental procedure, third molar (*wisdom tooth*) extraction, is clearly associated with risk of damage to the lingual nerve/chorda tympani fibers (Graff-Radford and Evans, 2003; Valmeseda-Castellon et al., 2000; Blackburn and Bramley, 1989). The reported incidence of this problem varies widely (from 0.2 to 22%), possibly reflecting variability in the surgical techniques that are employed (Robinson et al., 2000; Scrivani et al., 2000). The chorda tympani-lingual nerve lies in close proximity to the mandibular third molars making it susceptible to dental manipulation and injury. This nerve may also be injured by the injection of local anesthesia, either by direct contact with the needle or as a result of neurotoxic effects of the anesthetic compound (typically lidocaine and epinephrine) (Shafer et al., 1999; Nickel, 1993). Shafer et al. (1999) quantitatively evaluated taste function in 17 patients before third molar extraction, and at 1 month and 6 months thereafter. These investigators found that taste deficits can persist for at least 6 months after surgery, even though they do not typically result in patient complaints of taste loss, and patients with the most deeply impacted teeth exhibit the most severe loss. In a retrospective study of patients' perception of taste after lingual nerve injury and repair, Scrivani et al. (2000) found that among 22 patients who experienced a sensory deficit to the lingual nerve, 20 patients perceived a subjective disturbance of whole mouth taste function before surgical repair. While repair of this nerve provided significant improvement in trigeminal somatosensory function (82%), improvement in whole mouth taste was moderate at best (35%).

Infections can also result in taste disturbance. Otitis media is a suppurative infection of the middle ear caused by such organisms as *S. pneumoniae*, *H. influenzae*, *M. catarrhalis*, and group A streptococci. Recurrent and aggressive cases of otitis media can result in unilateral taste loss or dysgeusia due to chorda tympani nerve injury as it passes through the middle ear. One series reports that otitis media accounts for approximately 3.1% of acute facial palsies (Takahashi et al., 1985). CNS infections such as lyme, TB, and occasionally HIV are routine suspects when a patient presents with a possible "Bell's Palsy," often warranting a lumbar puncture for cerebrospinal fluid analysis. Other inflammatory disease states, such as sarcoidoisis or Wegener's granulomatisis, can also cause isolated facial nerve injury and possibly taste disturbance (Bibas et al., 2001).

39.4.2.2 *The Glossopharyngeal Nerve (CN IX)*. The glossopharyngeal nerve (CN IX) innervates the stylopharyngeus muscle (motor), the parotid gland (autonomic), the tympanic cavity and plexus (sensory), the eustachian tube (sensory), the carotid sinus baroreceptors (sensory), and the pharyngeal mucosa (sensory). The lingual-tonsillar branch of CN IX supplies circumvallate and most foliate taste buds within the posterior third of the tongue. The pharyngeal branch of the CN IX innervates taste buds in the nasopharynx. The nerve cell bodies of these gustatory afferent fibers are located immediately outside the jugular foramen – through which they eventually pass to enter the skull – in the petrosal ganglion. Clinically important functions of this nerve include: a) the salivary reflex (a flow of saliva from the parotid gland through Stensen's duct in response to highly seasoned foods on the tongue), b) the gag reflex (an elevation and constriction of the pharyngeal musculature retraction of the tongue in response to stimulating the base of the tongue and posterior pharyngeal mucosa), and c) the carotid sinus reflex (vagally-mediated reflexive parasympathetic slowing of heart rate with mechanical stimulation of the carotid sinus).

CN IX is small, and unlike CN VII, relatively well protected, making isolated lesions of this nerve extraordinarily rare. In fact, disorders that involve this nerve almost invariably affect the nearby vagus, accessory, or hypoglossal nerves. Nevertheless, damage to CN IX can occur as a result of tonsillectomy, bronchoscopy, or laryngoscopy (Donati et al., 1991; Arnhold-Schneider and Bernemann, 1987; Vories, 1999), reflecting the proximity of its lingual branch to the muscle layer of the palatine tonsillar bed. Neoplasms along the course of CN IX only rarely present with isolated impairment of the nerve. Vernet's syndrome – a syndrome consisting of loss of taste in posterior third of tongue, paralysis of the vocal cord, palate, pharynx, trapezius, and sternocleidomastoid muscles – is caused by lesions of the jugular foramen that affect cranial nerves IX, X, XI. Usually, it is of traumatic origin (e.g., skull base fracture); however, it can occur as a result of internal carotid artery aneurysms, tumors, granulomas, or thrombosis of the jugular bulb (Haerer, 1992).

Glossopharyngeal neuralgia is an important syndrome involving this nerve in which eating, swallowing, or talking may bring on a lancinating pain from one side of the throat, along the course of the eustachian tube to the tympanic membrane and ear. Like trigeminal neuralgia, there can be trigger zones for pain, and with the glossopharyngeal nerve they are usually the base of the tongue, pharyngeal wall, or tonsillar regions. While most cases are considered idiopathic, glosspharyngeal neuralgia has been associated with tongue carcinoma (particularly the posterior third), laryngeal carcinoid, a cerebellopontine angle tumor, and neck trauma (Webb et al., 2000).

On rare occassion there can be a disturbance of taste with increased salivation that accompanies diseases of the middle ear (Haerer, 1992). This owes to the fact that the tympanic plexus is a glossopharyngeal-mediated structure with corresponding visceral fibers in both the inferior salivatory and the gustatory nuclei (Carpenter, 1985).

39.4.2.3 The Vagus Nerve (CN X). The vagus (CN X) mediates sensation from the vocal cords, external ear, external auditory canal, and external surface of the tympanic membrane, as well as visceral sensation from the larynx to the gut. It innervates the taste buds on the epiglottis, aryepiglottal folds, and esophagus (via the internal portion of the superior laryngeal branch). Afferents from the superior laryngeal nerve project centrally from their cell bodies in the inferior nodose ganglion. This nerve provides motor function to the smooth muscle of the pharynx, larynx, and viscera, as well as to all striated muscles of the pharynx, larynx, and palate, except the stylopharyngeus (CN IX) and tensor veli palatini (CN V3) muscles.

Clinically important CN X reflexes include: (1) the vomiting reflex (a severe form of the gag reflex involving the vagus nerve), (2) the swallowing reflex (swallowing in response to pharyngeal wall and back of tongue stimulation by a food bolus), (3) the cough reflex (cough in response to stimulation of the pharynx, laryrnx, trachea, bronchial tree, tympanic membrane or external auditory canal), (4) the sucking reflex (sucking by infants in response to tactile stimulation of the lips near the buccal angle), and (5) the carotid sinus reflex (mentioned above, the efferent limb). In terms of localization, it is important to note that pharyngeal branches separate from the vagus nerve high in the neck; thus, the absence of any sensory changes in the pharynx suggests that the neural lesion is below this level.

While a number of disorders can affect the vagus nerve, such as tumors, vascular lesions, and infections, it is unclear whether isolated vagal injury can actually result in a subjective impairment or distortion of taste. This is of particular current interest given our understanding, described earlier in the chapter, that gastrointestinal chemosensation exists, and that chemosensory cells in the gut are known to interact with visceral afferents and the secretion of satiation peptides.

39.4.2.4 Other Nerves Relevant to Taste. The trigeminal nerve (CN V), largely responsible for somatosensory function of orofacial stuctures, is important to consider with other gustatory-mediating cranial nerves since (a) tactile, thermal, and spicy sensations are essential to the overall flavor of a tasted substance, and (b) anatomically, the gustatory fibers of the facial nerve (chorda tympani) travel in conjunction with the mandibular-trigeminal fibers of the lingual nerve for a short distance near the tongue – making both of these nerves susceptible to orofacial causes of injury.

Isolated damage to the auriculotemporal nerve (ATN) can produce a rare clinical syndrome of post-traumatic gustatory neuralgia – a taste-initiated, episodic, paroxysmal lancinating facial pain in the cutaneous distribution of the auriculotemporal nerve (Scrivani et al., 1998; Soria et al., 1990; Truax, 1989; Sharav et al., 1991). Post-traumatic abnormalities involving this nerve were originally described by Lucie Frey in 1923, such as sweating and flushing in this same region of the ATN following a taste stimulus (Frey's Syndrome). Although typically provoked by tastants, one report suggests Frey's syndrome can be elicited by the smell of food or even emotional excitement (Scrivani et al., 1998). Reported causes include head trauma, parotid gland surgery, temporomandibular joint (TMJ) surgery, carotid endarterectomy, orthognathic surgery, oncologic surgery, viral illness, and local infection (Scrivani et al., 1998; Truax, 1989; Sharav et al., 1991). A condition known as Collet-Sicard Syndrome is a neuropathy of CN XII due to damage to this nerve and nearby structures near the jugular foramen. Trauma and structural lesions to the lateral neck are common causes. The clinical features of this syndrome include: (1) ipsilateral paralysis of the trapezius and sternocleidomastoid muscles, (2) vocal cord and pharynx weakness, (3) hemiparalysis of the tongue, (4) taste lost on the posterior of the tongue, and (5) hemianesthesias of the palate, pharynx and larynx.

39.4.3 Disorders Involving Parts of the Central Nervous System that Affect Taste

The anatomy of the taste system is described in detail in Chapters 29, 32, and 33, and is only cursorily mentioned here to remind the reader of the major brain regions where lesions can influence taste function. Initially, the CN VII, IX and X afferents from the taste receptors enter the brain and synapse within the nucleus tractus solitarius (NTS) of the medulla. From this brain region higher order projections are made to the ventroposteromedial thalamus and then to the parietal operculum and adjacent parainsular cortex, regions that make up the "primary gustatory cortex." A secondary "gustatory cortex" is located in the caudomedial/caudolateral orbitofrontal cortex, just rostral to the primary taste cortex. Within the NTS viscerovisceral and viscerosomatic reflexive connections are also made, via the interneurons of the reticular formation, with cranial nuclei that control (1) the muscles of facial expression, (2) taste-related behaviors, including chewing, licking, salivation and swallowing, and (3) preabsorptive insulin release. The orbitofrontal cortex processes information from several sensory pathways (including olfactory, gustatory, and visual), and has been linked to emotion, feeding and social behaviors in both monkeys and humans (Chapters 32 and 46).

Salivary gland secretion is controlled centrally by the autonomic nervous system and involves multiple structures. While CN IX provides secretomotor function to the lingual glands, as well as to the parotid gland, the fibers of CN VII – in connection with the superior salivatory nucleus – innervate the sublingual and submandibular glands, as well as the buccal glands. Salivation can be produced by electrical stimulation of the brainstem between or adjacent to CN VII and IX (Grovum et al., 2000), the lower Rolandic cortex, the lower precentral gyrus, the postcentral gyrus, the insula, and regions within the frontal cortex (Penfield and Jasper, 1954). Seizures involving the lower Rolandic region are commonly associated with hypersalivation (Penfield and Jasper, 1954).

Damage to brainstem structures that subserve taste may produce dysgeusia, although usually not without significant impairment of other cranial nerves or long tracts (see Table 39.4). Notable brainstem areas involved in taste that are susceptible to injury include the tractus solitarius and its nucleus (where injury produces ipsilateral ageusia) and the pontine tegmentum which involves both gustatory lemnisci (where injury produces bilateral ageusia). Bilateral injury to the thalamus can result in ageusia, although unilateral lesions above the brainstem do not usually cause complete loss of function (likely due to the multiple areas involved in processing taste information). An isolated unilateral lesion to the thalamus or the parietal lobe may cause contralateral taste disturbance.

Taste function is clearly at risk by some central nervous system lesions, as seen in cases of stroke (either ischemic or hemorrhagic) or demyelinating disorders such as multiple sclerosis (Dahlslett et al., 2012; Fleiner et al., 2010; Catalanotto et al., 1984; Onoda and Ikeda, 1999;

Table 39.4 Disorders of central structures that can alter taste perception.

Neurovascular (stroke, arteriovenous malformation)
Brain stem
Nucleus of the tractus solitarius
Pontine tegmentum
Insula (Parietal lobe / Frontal lobe)
Thalamus
Seizures
Amygdala
Hippocampus
Parietal operculum
Rolandic operculum
Medial aspect of the right T1 gyrus
Anterior part of the right T2 gyrus
Neoplastic
Brain stem tumor
Parietal lobe tumor
Temporal lobe tumor
CNS Infection (e.g., Herpes encephalitis)
Inflammatory
Multiple sclerosis
Rasmussen's encephalitis

Yabe et al., 1995). In a review of 15 reported cases, Onoda and Ikeda (1999) found that unilateral impairment of taste function occurs with injury to any one of several central structures, including: (1) the pons – involved in 8 cases (53.3%), (2) the thalamus – involved in five cases (33.3%), (3) the midbrain – involved in one case (6.7%), and (4) the internal capsule – involved in one case (6.7%). Interestingly, based on the side of the respective central lesion and its effect on either unilateral or contralateral gustatory function, they surmised that the gustatory pathways ascend ipsilaterally from the medulla, cross over in their course from the pons to the midbrain, and synapse within the contralateral thalamus. It should be noted, however, that while this interpretation supports a primarily contralateral projection of taste information to the cerebral cortex in humans, gustatory-evoked potentials suggest that unilateral taste stimulation can evoke contralateral and bilateral cortical responses (Plattig, 1968; Kida, 1974). More recent data in the form of repeated case reports support the notion that bilateral diminished taste responses can occur in patients with unilateral midbrain and unilateral paramedian thalamic infarctions (Tsivgoulis et al., 2011; Nakajima et al., 2010). Pritchard et al. (1999) performed formal gustatory testing in six stroke patients with unilateral lesions of the insular cortex and six patients with brain damage outside the insula and found that, compared to controls, damage to the *right* insula produced ipsilateral taste deficits in recognition and intensity, whereas damage to the *left* insula caused an ipsilateral deficit in taste intensity but a *bilateral* deficit in taste recognition. This suggested that taste information from both sides of the tongue must pass through the left insula for proper recognition to occur. In a case presenting with taste dysfunction to our center (Figure 39.1), the MRI showed increased signal intensity suggestive of ischemia in a region of the upper medulla near the right nucleus tractus solitarius (note that, in this image, the patient's right side is on the left of the picture). Taste dysfunction, as indicated by our taste identification test and a localized complaint of dysgeusia, was greater on the right than on the left side of the tongue. Both CN VII and CN IX seem to be involved. In a case report of two patients with different stroke presentations – one with a minor stroke in the anterior circulation and one with a minor stroke in the posterior circulation – both patients reported both olfactory and gustatory dysfunction immediately following their strokes and subsequently experienced weight loss, diminished energy and strength, and an overall decreased in their nutritional status (Green et al., 2008).

Small et al. (1997) found that patients with *right,* but not left, temporal lobe resections for control of intractable epilepsy had elevated citric acid *recognition* thresholds, but normal *detection* thresholds. In normal subjects, orally-presented citric acid increased regional cerebral blood flow bilaterally in the caudolateral orbitofrontal cortex, as measured by PET. Relatively more blood flow occurred in the right anteromedial temporal lobe and right caudomedial orbitofrontal cortex, suggesting that damage to the latter areas might explain the higher taste recognition thresholds that were observed in the patients with right temporal lobe resections.

Gustatory perception may also be affected by cortical injury to areas of association cortex outside of direct gustatory processing. The concept of left unilateral neglect, typically resulting from *contralateral* injury to the right hemisphere (e.g., parietal lobe, thalamus, basal ganglia), appears to exist for taste in a manner similar to the visual, auditory, tactile, and olfactory sensory systems (Bellas et al., 1988). In fact, a specific syndrome of buccal hemi-neglect may exist in which patients neglect mouth and food products in the left half of the mouth and are unable to initiate chewing and swallowing when food is in this location (Andre et al., 2000). This syndrome can result in choking or a socially-embarrassing tendency to drool and regurgitate unnoticed food.

Electrochemical events such as migraine and seizure are presumed to influence taste perception on a central level. With migraine headache, the cortical spreading depression (CSD) associated initiation of aura and early phases of migraine is linked to hypersensitivity to visual (photophobia), auditory (phonophobia) and olfactory (osmophobia) stimuli (Sjostrand et al., 2010; Zanchin et al., 2005; Boulloche et al., 2010; Denuelle et al., 2011). However, it is not clear how gustatory processing changes as a result of migraine. While certain tastes may, in fact, induce a migraine (Blau and Solomon, 1985), central processing of taste information during migraine may acutally "mix" with other sensory information in a condition described as synesthesia (Alstadhaug and Benjaminsen, 2010; Beeli et al., 2005). Possibly as a result of CSD temporarily altering networks involved in cross-modal perceptions, taste can be experienced as a co-percept of vision and colors can actually influence the perceptions of odors and tastes (Alstadhaug and Benjaminsen, 2010). For example, there is a well-established association between seeing yellow and the experience of lemon (Osterbauer et al., 2005). Other cross-modal influences on sensory perception include a more recent look at the effects of auditory stimuli on gustation (North, 2012). While auditory stimuli can influence perceptions of freshness of food products, background music may be an influence on taste perception in general, such as with the "tasting" of wine (North, 2012).

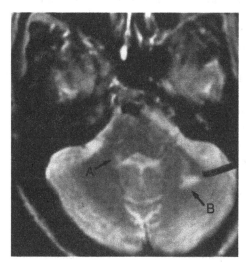
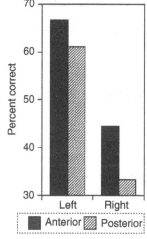

Figure 39.1 L: Axial T2 (2500/90) MR scan through the upper brain stem reveals a hyperintensive infarct lesion (4 x 3 mm) in the right medulla (Arrow A). Note also large infarct (15 x 8 mm) inside the white matter of the left cerebellum (Arrow B). R: Scores on a taste identification test showing a relative decrement on right side of tongue. From a 65-year-old-woman with a history of ministrokes who developed a persistent salty/metallic dysgeusia and soreness on the right side of the tongue following a severe 2-day bout of emesis accompanied by marked dehydration and increased blood pressure but unaccompanied by fever. Copyright © Richard L. Doty, 2001.

With seizure, alteration of taste function through cortical excitation is an area of long-standing interest.

Following in the footsteps of Penfield and Jasper (1954), who explored the function of the brain by observing the effects spontaneous and induced epileptic seizures, Hausser-Hauw and Bancaud (1987) found six specific areas of the brain where gustatory sensations result from electrical stimulation. These areas included the right rolandic operculum, right parietal operculum, right amygdala, right hippocampus, medial aspect of the right T1 gyrus, and the anterior part of the right T2 gyrus. Like olfactory auras, gustatory auras represent seizure activity predominately within the temporal lobes, but they can also be linked on rare occasion to epileptiform activity in the orbitofrontal lobes (Hausser-Hauw and Bancaud, 1987; Acharya et al., 1998; Roper and Gilmore, 1995). Etiologies of the seizures can vary, and identifiable causes include mesial temporal sclerosis, tumors, vascular malformations, infections (e.g., tuberculosis, herpes simplex), and Rasmussen's encephalitis. Examples of taste sensations that have been reported in such cases include "peculiar," "rotten," "sweet," "like a cigarette," "like rotten apples," and "like vomitus" (for review, see West and Doty, 1995). However, some of these "tastes" likely represent smell sensations that are miscategorized as tastes by both the patients and their physicians.

39.4.4 Systemic Disorders and Medications

Excessive dryness of the oral cavity and hyperviscosity of saliva accompany several diseases (e.g., Sjogren's syndrome, xerostomia, pandysautonomia, diabetes mellitus) and are common side-effects of a number of medications (e.g., anticholinergics, antidepressants, antihistamines) (see Chapter 40). Although such conditions likely alter taste when they are chronic, acute effects are less well understood. In general, hyposalivation, per se, appears to have little influence on taste sensitivity in the short run (Christensen et al., 1984).

A number of systemic disorders have been associated with taste dysfunction. Both chronic renal failure (Middleton and Allman-Farinelli, 1999; Armstrong et al., 2010) and end-stage liver disease (Bloomfield et al., 1999; Davidson et al., 1998; Musialik et al., 2012) can produce altered taste function, which, in some cases, seems to improve following transplantation (Bloomfield et al., 1999). Apparently the taste dysfunction associated with liver failure is not necessarily associated with a vitamin deficiency, and may be centrally-mediated (Bergasa, 1998), although it is generally believed that poor nutritional states can be related to taste impairment (Davidson et al., 1998). Avitaminosis, particularly of the B group and C, can result in pellagra, sprue, pernicious anemia, scurvy, or glossitis. Additionally, zinc and copper deficiencies may produce taste disturbances (Mattes and Cowart, 1994). In patients younger than 55 years with uremia, a significant impairment of taste recognition — independent of zinc deficiency—was found by Ciechanover et al. (1980) for salty, sweet, bitter, and sour compared to controls. The impairment for sour and bitter, however, showed improvement immediately after a dialysis session. Interestingly, idiopathic dysgeusia has been reported with blood transfusions (Erick, 1996), and we are familiar with one leukemia patient who could ascertain, by taste, whether transfused blood came from a smoker or a non-smoker.

There is considerable evidence that diabetes is associated with loss of taste function. Thus, there appears to be a progressive loss of taste beginning with glucose and extending to other sweeteners, salty stimuli, and then all stimuli (Hardy et al., 1981). In newly diagnosed non-insulin dependent diabetics (based on plasma glucose concentration and glycosolated hemoglobin percentage), quantifiable taste impairment not only exists, but may be somewhat reversible with correction of the hyperglycemia (Perros et al., 1996). Interestingly, about 10% of first order relatives of late-onset diabetics exhibit decreased suprathreshold taste function that is rather specific to glucose (Settle, 1981). Whether these individuals are more prone to subsequent development of diabetes is not known, but is an intriguing possibility since about 10% of such relatives eventually develop the disease.

Hypothyroidism has been reported to result in significant changes in taste detection thresholds (McConnell et al., 1976); however, the degree of taste dysfunction does not appear to correlate with the severity of the disease (Pittman and Beschi, 1967). Importantly, there is suggestion that replacement of thyroid hormone may normalize threshold sensitivity and suprathreshold performance, although dysgeusias may still be present (Deems et al., 1991).

Pharmacologic agents appear to result in taste disturbances much more frequently than olfactory disturbances (Doty et al., 2008). General anesthesia has been associated with anosmia and hypogeusia, although the mechanisms behind this reported rare complication are unclear (Dhanani and Jiang, 2012). Over 250 medications reportedly alter the sense of taste, including antiproliferative drugs, antirheumatic drugs, antibiotic drugs, psychotrophic drugs and drugs, with sulfhydryl groups, such as penicillamine and captopril (Table 39.5). Angiotensin-converting enzyme (ACE) inhibitors are considered by many authors to be most commonly associated with taste disturbance, such as hypogeusia, or excessive metallic, bitter, or sour dysgeusic tastes (Suliburska et al., 2012; Ackerman and Kasbekar, 1997). Terbinafine, a commonly used antifungal medication, has been linked to taste disturbance that reportedly can last from two days to three years. In the sole study

Table 39.5 Systemic disorders and agents affecting taste.

Medications
  Antibiotics / Antifungals
  Anticonvulsants
  Antidepressants
  Antihistamines and decongestants
  Antihypertensives and cardiac medications
  Anti-inflammatory medications
  Antimanic medications
  Antineoplastics
  Antiparkinson medications
  Antipsychotics
  Antithyroid medications
  Lipid-lowering medications
  Muscle relaxants
Nutritional
  Blood transfusions
  Cancer-related wasting syndrome
  Chronic Renal failure/uremia
  HIV wasting syndrome
  Liver disease (cirrhosis)
  Malnutrition
  Trace metal deficiency (Zinc, Copper)
  Vitamin deficiency (B3, B12, C)
Endocrine
  Adrenocortical insufficiency
  Congenital adrenal hyperplasia
  Cushing's syndrome
  Cretinism
  Diabetes Mellitus
  Hypothyroidism
  Panhypopituitarism
  Pseudohypoparathyroidism
Psychiatric
  Bulimia
  Conversion disorder
  Depression
  Malingering
  Schizophrenia
Genetic
  Familial dysautonomia (Riley-Day)
  Late-onset hereditary spinocerebellar degeneration
  Machado-Joseph disease variant
  Turner's Syndrome / gonadal dysgenesis

beyond a case report that quantitatively examined taste function in patients complaining of terbinafine-related taste loss, all four major qualities were affected. However, smell function was intact (Doty and Haxel, 2005).

As noted earlier, dysgeusia occurs commonly in the context of cancer therapies, with a weighted prevalence from 56–76% depending on the type of cancer treatment (Hovan et al., 2010). In a systematic review of taste disturbances resulting from antiproliferative agents, members of the International Society of Oral Oncology (ISOO)

found that attempts to prevent taste problems from these drugs using prophylactic zinc sulfate or amifostine had minimal benefit and nutritional counseling was helpful in some patients to diminish the symptoms of dygeusia (Hovan et al., 2010). While many antiepileptic medications have been used in the treatment of smell and taste disorders, particularly as they relate to seizures, migraines and other causes of gustatory hallucinosis and dysgeusias, one case report acknowledges that the use of topiramate in a three and a half-year-old boy for his primary generalized seizures resulted in a reversible loss of his ability to detect and recognize tastes and odors during treatment (Ghanizadeh, 2009). Doty and colleagues (2009) ran an independent double-blind study of the influences of eszopiclone (Lunesta™) – a commonly prescribed sleep induction agent – and the influences this medication has on dysgeusia and taste function (Doty et al., 2009). They found that the taste disturbances associated with eszopicolone are more intense and longer lasting in women than in men, stronger in the morning than in the evening, and positively correlated with drug plasma and saliva levels of the same medication (Doty et al., 2009).

In a study of six psychotropic medications, including amytriptyline, clomipramine, desimpramine, imipramine, doxepin, and trifluoperazine, Schiffman et al. (1998) found that these drugs not only have a taste of their own, but they can significantly alter the intensity of other tastants (such as salt and sugar). She also noted that among the medications used in the treatment of HIV and AIDS, the protease inhibitors have a taste themselves, and appear to adversely modify the taste perception of other taste compounds (Schiffman et al., 1999). Repeated oral use of some topical agents, such as hydrogen peroxide or steroids, may affect taste.

It should also be noted that the common antiplatelet therapy — clopidogrel (Plavix™) — used to treat patients with cardiac and stroke-related problems is reported to produce ageusia as a rare side-effect in some cases (Ksouda et al., 2011). Five months after reduction of the clopidogrel dose, researchers note that the ageusia partially improved (Ksouda et al., 2011). Recognizing that this common stroke therapy may result in taste disturbance for some stroke patients needs to be considered when evaluating the effects on gustation in any stroke patient.

39.4.5 Psychiatric Disorders

There are multiple psychiatric conditions known to affect taste function. Patients who are clinically depressed may complain of an unpleasant or diminished taste that is not readily explainable by medications or other causes (Miller and Naylor, 1989); however, the problem with taste can be reversible (Gangdev, 1995). When formally tested, depressed patients demonstrate a higher recognition threshold for sucrose compared to non-depressed controls, a finding that is possible related to a subtle carbohydrate intolerance observed in many depressed patients (Amsterdam et al., 1987; Steiner et al., 1969).

Gustatory hallucinations are associated, on rare occasion, with schizophrenia, cocaine use, and a number of other organic conditions (Carter, 1992; Siegel, 1978). In a study evaluating a large cohort of patients with schizophrenia, schizoaffective disorder, and bipolar I disorder, researchers discovered that tactile, olfactory, and gustatory hallucinations were common among the patients studied and affected as much as 20% of the total study population (Lewandowski et al., 2009). With regard to olfactory or gustatory hallucinations alone, previous reports suggest that 11% of patients with schizophrenia or schizoaffective disorder have chemosensory-specific experiences (Mueser et al., 1990). Of note, tactile, olfactory and gustatory hallucinations are associated with an earlier age of onset of psychotic illness and an earlier age of onset is considered a marker of more severe illness (Lewandowski et al., 2009). While hallucinations appear to be associated with activation of the same brain regions involved in normal sensory processing, there appears to be a component of shared neurobiology underlying many of experiences. This possibly relates to brain regions involved in patients who report synesthesia-like events during migraine attacks and seizures (Weiss and Heckers, 1999; Alstadhaug and Benjaminsen, 2010).

Patients with bulimia nervosa appear to have impaired satiety responses in response to glucose loads; bulimics, unlike controls, do not find sweet tastes less pleasant under such loads (Rodin et al., 1990). Bulimic patients also appear to have a selective loss of taste on the palate, presumably from the local effect of acidic vomitus during a purging episode (Rodin et al., 1990). In a trial evaluating olfactory and gustatory sensitivities among subjects with anorexia nervosa or bulimia nervosa, anorexic patients exhibited lower smell and taste sensitivities than healthy controls (Aschenbrenner et al., 2008; see also Federoff et al., 1995). More importantly, improving the Body Mass Index (BMI) and decreasing eating pathology in these same patients resulted in better olfactory and gustatory performance (Aschenbrenner et al., 2008).

In the context of litigation or identifiable secondary gain, a physician may need to consider malingering on the part of the patient who is complaining of taste loss or distortion. Conversion disorder involving taste is also a possibility in some psychiatric cases.

39.4.6 Genetic Disorders

Familial dysautomonia (also known as Riley-Day syndrome or Hereditary Sensory and Autonomic Neuropathy Type III), an autosomal recessively inherited disease, is a sensory neuropathy that mostly affects Ashkenazi Jewish patients and results in defective autonomic control, insensitivity to pain, hyporeflexia, impaired lacrimation, the absence of fungiform papillae and taste buds on the tongue, and impairment of taste (Gadoth et al., 1997). A recent assessment of the oro-dental-facial features and dysfunction in children with familial dysautonomia at the Barzilai Medical Center in Israel found these children have a normal sense of smell but decreased numbers of lingual fungiform papillae and a specific dysgeusia (Mass, 2012).

Fukutake et al. (1996) described another inherited disease entity in two Japanese siblings which consists of late-onset progressive ataxia, global thermoanalgesia, sensorineural deafness, mild autonomic disturbances, and ageusia in the context of absent fungiform papillae and taste buds. This particular disease entity may represent a variant of late-onset hereditary spinocerebellar degeneration. A few of the same authors reported a case of a family with a possible new variant of Machado-Joseph Disease (confirmed with genetic testing) associated with a sensory dominant axonal neuropathy, autonomic disturbances, and the absence of fungiform papillae (Uchiyama et al., 2001). Gonadal dysgenesis, particularly Turner's Syndrome, has been reportedly linked to a number of sensory abnormalities, including gustatory dysfunction (Henkin, 1967; Valkov et al. 1975).

In addition to altered taste function associated with genetic diseases, it should be noted that taste receptor genes, such as TAS2R38, have been associated, in healthy cohorts, with food and beverage preferences that potentially influence nutrition and health. For example, the TAS2R38 gene is related to vegetable intake (Sacerdote et al., 2007; Duffy et al., 2010; Negri et al., 2012) and variations in TAS2R13 with differences in alcohol consumption in head and neck cancer patients (Dotson et al., 2012).

39.4.7 Special Circumstances

39.4.7.1 Aging. Regional tests of taste function demonstrate marked losses in older persons (Matsuda and Doty, 1995). However, such regional deficits need not translate into marked overall whole mouth loss, given the redundancy of the system. Nonetheless, whole mouth taste acuity does decline with age, and the elderly tend to perceive tastes as being less intense than do younger persons (Weiffenbach et al., 1986). Fortunately, the amount of decline is not as marked as that seen for olfaction. Nonetheless, in conjunction with the loss seen with the sense of smell, decrements in taste function — and most significantly age-related dysgeusias — can be very detrimental to some of the elderly, leading to anorexia, weight loss, malnutrition, impaired immunity, and worsening of medical illnesses.

904   **Chapter 39   Clinical Disorders Affecting Taste: An Update**

39.4.7.2 Pregnancy. Pregnancy proves to be a unique condition with regard to taste function. In one study of 46 pregnant and 41 health women controls, Duffy et al. (1998) noted that sensitivity to quinine hydrochloride (bitter) was highest during the first trimester of pregnancy. The subjects found its taste to be less intense and less disliked as pregnancy progressed. The authors suggest that the increased dislike and intensity of bitter tastes during the first trimester may help to ensure that pregnant women to avoid poisons during a critical phase of fetal development. Similarly, a relative increase in the preference for salt and bitter in the second and third trimesters may support the ingestion of much needed electrolytes to expand fluid volume and support a varied diet (Duffy et al., 1998).

39.4.7.3 Burning Mouth Syndrome. Burning Mouth Syndrome (BMS), also referred to as glossodynia or glossalgia, is the subjective sensation of intense "burning" pain within the mouth without obvious physical cause. Often accompanied by dry mouth, thirst, and altered or dysgeusic taste, BMS usually begins by late morning and is continuous once it has started for the day (Gorsky et al., 1987; Grushka and Sessle, 1991). Interestingly, most of the BMS patients who present for treatment are postmenopausal; however, studies vary in terms of the effectiveness of hormone replacement therapy in reducing oral symptoms. Numerous etiologies suggested by researchers, many of them amenable to treatment, include: (1) nutritional deficiencies (e.g., iron, folic acid, B vitamins, zinc), (2) diabetes mellitus (possibly predisposing to oral candidiasis), (3) denture allergy, (4) mechanical irritation from dentures or oral devices, (5) parafunctional habits of the mouth (e.g., tongue thrusting, teeth grinding, jaw clenching), (6) tongue ischemia as a result of temporal arteritis, (7) oral candidiasis, (8) periodontal disease, (9) reflux esophagitis, and (10) geographic tongue (see Tourne and Fricton, 1992 for review). Psychological factors are clearly important in the diagnosis and management of BMS, since psychological dysfunction (e.g., anxiety, depression) is common in this population (Gorsky et al., 1991). Both neuropathic and psychiatric factors linked to patients with BMS may help to explain why tricyclic antidepressants (e.g., amitriptyline, desipramine, nortriptyline) and benzodiazepines are reportedly useful as therapy in some cases. It is not clear whether the oral burning pain is a result of central or peripheral mechanisms. In a study of 9 men and 24 women, with an average age of 60 years, Formaker et al. (1998) found that the application of a topical anesthetic (dyclonine HCl) reduced the perceptual intensity of salt and sugar, but had variable affects on the burning sensation — such that 12 patients demonstrated increased burning, 14 patients had burning that did not change, and 7 patients experienced decreased burning.

Topical capsaicin may also be helpful, although clinical trials using this substance are limited. A recent review of randomized trials for the treatment of burning mouth syndrome (2012) found a total of 12 relevant articles suggesting that therapies such as capsaicin, α-lipoic acid and clonazepam were more efficacious than some other therapies in reducing BMS symptoms (de Moraes et al., 2012).

39.5 TREATMENT

As with other sensory systems, prognosis for functional recovery of taste is related to the kind of neural damage that is involved. For example, permanent sensory abnormalities are more likely to result from nerve section of a gustatory-mediating cranial nerve than a crush injury (Robinson et al., 2000). Depending on the underlying disorder affecting taste function, spontaneous recovery can occur and some therapeutic strategies can be effective. Taste impairment secondary to nutritional deficiencies may improve with supplementation and a regular diet. When dysfunction is due to viruses, damaged taste afferents often regenerate. Conditions such as Bell's Palsy and radiation-induced xerostomia generally improve over time, a fact that, when known in advance to a patient, is very therapeutic. When identified, infection and inflammation of the mouth surfaces and perioral anatomic structures should be treated with antibiotic or anti-inflammatory medications. In the case of taste loss secondary to hypothyroidism, thyroxin replacement therapy may be beneficial (Mattes and Kare, 1986). Nevertheless, strong evidence for efficacy of most hormonal or pharmacologic treatments reported in the literature is lacking, since double-blind studies are rarely performed.

Dysgeusias and phantogeusias are among the most debilitating taste disorders, often leading to inhibition of eating and loss of weight. Fortunately, spontaneous resolution of most dysgeusias occurs, usually within two years (Deems et al., 1996). In some cases, antifungal and antibiotic treatments have been reported to be useful in resolving phantogeusias and dysgeusias. Antidepressants, such as sertraline, venlafaxine, and mirtazapine, reportedly benefit some patients with dysgeusia (Mizoguchi et al., 2012; Landis et al., 2012; Devere 2012). Since vitamin D deficiency is a cofactor of chemotherapy-induced mucocutaneous toxicity and dysgeusia, adding vitamin D3 1000–2000 units daily may benefit some patients (Fink, 2011). Chlorhexidine mouthwash has been suggested as having possible efficacy for some salty or bitter dysgeusias, possible as a result of its strong positive charge (Helms et al., 1995). One open trial of α-lipoic acid therapy for idiopathic dysgeusia found significant improvements in patients treated with this agent compared to a placebo

(Femiano et al., 2002), leading to the suggestion that the idiopathic form may be a kind of neuropathy "comparable to Burning Mouth Syndrome." Interestingly, something as simple as transient cooling of the mouth in the form of ice cube stimulation can improve some cases of dysgeusia (Fujiyama et al., 2010).

Several other medications may benefit some cases of taste disturbance. In situations where xerostomia and excessive dryness exists, artificial saliva (e.g., Xerolube) may improve comfort and mucosal function. Pilocarpine (Salagen™) is a prescribed medication that stimulates muscarinic cholinergic receptors and subsequently improves salivation for patients suffering from xerostomia related to such conditions as head and neck cancer therapies and Sjogren Syndrome. Cepacol lozenges with benzocaine, used 3–4 times per day, can improve salivation and contribute to anesthesia of mucosal surfaces. Where pain is more evident, lidocaine (Xylocaine™) mouth gel 0.5%–1%, used 3–4 times per day, may also give relief. Amifostine (Ethyol™) is an agent that presumably acts like a free radical scavenger that prevents oxidative damage associated with radiotherapy. Phase III trials have shown that amifostine may reduce xerostomia when given at the time of radiotherapy (Vissink et al., 2010). Oral steroids may be beneficial in some cases of acute and ongoing inflammatory disease (e.g., a prednisone taper from 60 mg to 0 mg over a 7–10 day period). Zinc gluconate around 140 mg per day is potentially helpful in some cases since zinc is a necessary component to regeneration of taste buds and formation of saliva (Devere, 2011, 2012). Additionally, magnesium oxide in doses ranging from 250 mg to 750 mg daily has shown promise in some cases of oral paresthesias as well as migraine and seizures. Some antiepileptic medications are effective, presumably through stabilization of damaged and hyperactive neural fibers transmitting information from oral structures. Among the most commonly used antiepileptic medications to treat chemosensory disturbance is gabapentin, with dose ranges from 300 to 3600 mg per day (Devere, 2012). Another antiepileptic agent that has been reported to show effectiveness for dysgeusia in an epileptic child is lamotrigine (Yamashita et al., 2011).

Trauma and surgery-related injury to peripheral nerves supplying taste (e.g., lingual nerve, facial nerve) may be amenable to surgical repair, particularly in the early months immediately following injury (Robinson et al., 2000; Zuniga et al., 1994). In a series of 53 patients with prior lingual nerve injury who underwent direct repposition of the nerve with epineurial sutures, 41% of the patients could detect some taste solutions postoperatively (Robinson et al., 2000). Recovery of sensory function after repair can varied, light touch and pain functions being most improved. In this study, the study group did not appear to regain completely normal taste sensation or get relief from spontaneous paresthesias or pain syndromes, although many did recover some function and fewer patients were found to bite their tongues by accident (Robinson et al., 2000).

Medication-related changes in taste can, in some instances, be reversed by discontinuance of the offending drug, by employing alternative medications, or by changing drug dosage. The health care professional should be aware of the fact, however, that many pharmacological agents appear to induce long-term alterations in taste that may take months to disappear even after discontinuance of the drug. Although the Physician's Desk Reference (PDR) can be helpful in identifying medications that may result in drug-related taste dysfunction, there is lack of uniformity and control in the acquisition of this information. Many agents are listed with the potential side effects of "altered taste," "ageusia," "dysgeusia," "bad taste," "metallic taste," or "bitter taste" without the benefit of a controlled trial. Therefore, caution should be used in simply attributing a patient's taste disturbance to his or her medication regimen based on PDR classification alone.

Alternative therapies, such as acupuncture, meditation, cognitive-behavioral therapy, and yoga may not only help a patient manage aberrant experiences associated with taste disturbance and oral pain syndromes, but they can be helpful with allowing patients to cope with the psychosocial stressors surrounding their impairment. Specifically, there is some evidence that acupuncture may also stimulate salivary gland secretion and help to alleviate xerostomia (Vissink et al., 2010). Modification of diet and eating habits is also important. Accentuating the other sensory experiences of a meal — such as food texture, aroma, temperature, and color — can optimize the overall eating experience for a patient. In some cases, a flavor enhancer like monosodium glutamate (MSG) can be added to foods to increase palatability and encourage intake (Devere, 2012; Bellisle, 1998; Schiffman, 1997).

Proper oral hygiene and routine dental care are extremely important ways for patients to protect themselves from disorders of the oral cavity that can ultimately result in taste disturbance. Patients should be warned not to overcompensate for their taste loss by adding excessive amounts of sugar or salt. Smoking cessation and the discontinuance of oral tobacco use are essential in the management of any patient with smell and/or taste disturbance and should be repeatedly emphasized.

39.6 CONCLUSIONS

Clinical disorders that influence taste are numerous and may involve multiple organ systems. Because of this, the appropriate diagnosis and management of a patient with gustatory complaints requires a multidisciplinary mind set on the part of the evaluating physician. A disturbance of taste may represent a spectrum of possible problems from local (e.g., oral cavity) to more generalized (e.g., systemic disorders or agents). The ability to taste reflects the proper function of many systems within the body, and as such, may be considered a marker for good health. Conversely, altered taste function, particularly dygeusias that make food distasteful, can influence nutrition and lead to unhealthy states of being — not to mention adversely affecting a person's ability to enjoy food products and overall quality of life. In this chapter, we have presented multiple disorders, across medical specialties, that can influence taste, described anatomy and physiology relating these disorders to the taste system, and provided basic information for the evaluation and management of patients who complain of taste disturbance.

ACKNOWLEDGMENTS

The development of this chapter was supported, in part, by USAMRAA W81XWH-09-1-0467 awarded to R. L. Doty.

REFERENCES

Acharya, V., Acharya, J., and Luders, H. (1998) Olfactory epileptic auras. *Neurology.* 51: 56–61.

Ackerman, B. H., and Kasbekar, N. (1997) Disturbances of taste and smell induced by drugs. *Pharmacotherapy* 17: 482–496.

Adour, K. K. (1982) Current concepts in neurology diagnosis and management of facial paralysis. *NEJM* 307: 348–351.

Adour, K. K., Bell, D. N., and Hilsinger, R. L. Jr., (1975) Herpes simplex virus in idiopathic facial paralysis (Bell's palsy). *JAMA* 233(6): 527–230.

Allen, C. M. (1992) Diagnosing and managing oral candidiasis. *J. Am. Dental Assoc.* 123(1): 77–78, 81–82.

Allen, C. M, and Blozis, G. G. (1998) Oral mucosal lesions. In *Otolaryngology: Head and Neck Surgery*, 3rd edition, Cummings C.W., Fredrickson J.M., Harker L.A., Krause C.J., Richardson M.A., and Schuller D.E. (Eds). St. Louis, Mosby, pp.1527–1545.

Alstadhaug, K. B., and Benjaminsen, E. (2010) Synesthesia and migraine: case report. *BMC Neurol.* 10:121.

Amsterdam, J. D., Settle, R. G., Doty, R. L., et al (1987) Taste and smell perception in depression. *Biol. Psychiatry* 22(12): 1477–1481.

Andre, J. M., Beis, J. M., Morin, N., and Paysant, J. (2000) Buccal hemineglect. *Arch. Neurol.*, 57(12): 1734–1741.

Arnhold-Schneide,r M., and Bernemann, D. (1987) Uber die Haufigkeit von Geschmacksstorungen nach Tonsillektomie. *HNO* 35: 195–198.

Armstrong, J. E., Laing, D. G., Wilkes, F. J. and Kainer, G. (2010) Smell and taste function in children with chronic kidney disease. *Pediatric Nephrol.* 25(8): 1497–1504.

Aschenbrenner, K., Scholze, N., Joraschky, P. and Hummel, T. (2008) Gustatory and olfactory sensitivity in patients with anorexia and bulimia in the course of treatment. *J. Psychiatr. Res.* 43(2): 129–137.

Bartoshuk, L. M., and Miller, I. J. Jr., (1991). Taste perception, taste bud distribution, and spatial relationships. In: T.V. Getchell, R.L. Doty, L.M. Bartoshuk, and J. B. Snow, Jr. (Eds). Smell and Taste in Health and Disease. New York: Raven Press, pp. 205–233.

Beeli, G., Esslen, M., and Jancke, L (2005) When coloured sounds taste sweet. *Nature.* 434: 38.

Beidler, L. M., and Smallman, R. L. (1965) Renewal of cells within taste buds. *J. Cell Biol.* 27: 263–272.

Bellas, D. N., Novelly, R. A., Eskenazi, B., and Wasserstein, J. (1988) The nature of unilateral neglect in the olfactory sensory system. *Neuropsychologia.* 26: 45–52.

Bellisle, F. (1998) Effects of monosodium glutamate on human food palatability. *Ann. N.Y. Acad. Sci.* 855: 438–441.

Bergasa, N. V. (1998) Hypothesis: taste disorders in patients with liver failure may be mediated in the brain. *Am. J. Gastroenterol.* 93(8): 1209–1210.

Berndt, A., Buck, R., and Buxton, R. (1931) The pathogenesis of acute supporative parotitis. *Am. J. Med. Sci.* 182: 639.

Bibas, A., Fahy, C., Sneddon, L., and Bowdler, D. (2001) Facial paralysis in Wegener's granulomatosis of the middle ear. *J. Laryngol. Otol.* 115(4): 304–306.

Blackburn, C. W. and Bramley, P. A. (1989) Lingual nerve damage associated with removal of lower third molars. *Br. Dent. J.* 167: 103–107.

Blau, J. N., and Solomon, F. (1985) Smell and other sensory disturbances in migraine. *J. Neurol.* 232: 275–276.

Bloomfield, R. S., Graham, B. G., Schiffman, S. S., and Killenberg, P. G. (1999) Alterations of chemosensory function in end-stage liver disease. *Physiol. Behav.* 66: 203–207.

Boulloche, N., Denuelle, M., Payoux, P., et al (2010) Photophobia in migraine: an interictal PET study of cortical hyperexcitability and its modulation by pain. *J. Neurol. Neurosurg. Psychiatry.* 81(9): 978–984.

Breer, H., Ederle, J., Frick, C., et al (2012) Gastrointestinal chemosensation: chemosensory cells in the alimentary tract. *Historchem. Cell Biol.* 138(1): 13–24.

Bromley, S. M. (2000) Smell and taste disorders: a primary care approach. *Am. Fam. Phys.* 61: 427–436.

Bromley, S. M. and Doty, R. L. (2010) Olfaction in dentistry. *Oral Dis.* 16(3): 221–232.

Bull, T. R. (1965) Taste and the chorda tympani. *J. Laryngol. Otol.* 79: 479–493.

Campbell, S. M., Montanaro, A., and Bardana, E. J. (1983) Head and neck manifestations of autoimmune disease. *Am. J. Otolaryngol.* 4(3): 187–216.

Carpenter, M. B. (1985) *Core Text of Neuroanatomy*, 3rd edition, Williams & Wilkins: Baltimore, pp.127–128.

Carter, J. L. (1992) Visual, somatosensory, olfactory, and gustatory hallucinations. *Psychiat. Clin. N. Am.* 15(2): 347–358.

Catalanotto, F. A., Dore-Duffy, P., Donaldson, J. O., et al (1984) Quality-specific taste changes in multiple sclerosis. *Ann. Neurol.* 16: 611–615.

Chang, C. Y., and Cass, S. P. (1999) Management of facial nerve injury due to temporal bone trauma. *Am. J. Otol.* 20: 96–114.

Christensen, C. M., Navazesh, M., and Brightman, V. J. (1984) Effects of pharmacologic reductions in salivary flow on taste thresholds in man. *Arch. Oral Biol.* 29: 17–23.

Ciechanover, M., Peresecenschi, G., Aviram, A., and Steiner, J. E. (1980) Malrecogition of taste in uremia. *Nephron* 26(1): 20–22.

Conger, A. D. (1973) Loss and recovery of taste acuity in patients irradiated to the oral cavity. *Radiati. Res.* 53: 338–347.

Davidson, H. I. M., Pattison, R. M., and Richardson, R. A. (1998) Clinical undernutrition states and their influence on taste. *Proc. Nutr. Soc.* 57: 633–638.

Dahlslett, S. B., Goektas, O., Schmidt, F., et al (2012) Psychophysiological and eletrophysiological testing of olfactory and gustatory function in patients with multiple sclerosis. *Eur. Arch. Otorhinolaryngol.* 269(4): 1163–1169.

de Moraes, M., do Amaral Bexerra, B. A., da Rocha Neto, P. C., (2012) Randomized trials for the treatment of burning mouth syndrome: an evidence-based review of the literature. *J. Oral Pathol. Med.* 41(4): 281–287.

Deems, D. A., Doty, R. L., Settle, R. G., et al (1991) Smell and taste disorders, a study of 750 patients from the University of Pennsylvania Smell and Taste Center. *Arch. Otolaryngol. Head Neck Surg.* 117: 519–528.

Deems, D. A., Yen, D. M., Kreshak, A., and Doty, R. L. (1996) Spontaneous resolution of dysgeusia. *Arch. Otolaryngol. Head Neck Surg.* 122: 961–963.

Denuelle, M., Boulloche, N., Pavoux, P., et al (2011) A PET study of photophobia during spontaneous migraine attacks. *Neurology* 76(3): 213–218.

Devere, R. (2011) Dysosmia and dysgeusia: a patient's nightmare. *Chemosense* 12: 2–6.

Devere, R. (2012) Smell and taste in clinical neurology: five new things. *Neurol. Clin. Pract.* 2: 208–214.

Dhanani, N. M. and Jiang, Y. (2012) Anosmia and hypogeusia as a complication of general anesthesia. *J. Clin. Anesth.* 24(3): 231–233.

Donati, F., Pfammatter, J. P., Mauderl,i M., and Vassella, F. (1991) Neurologische Komplikationen nach Tonsillektomie. *Schweiz. Med. Wochenschr.* 121: 1612–1617.

Dotson, C. D., Wallace, M. R., Bartoshuk, L. M. et al. (2012) Variation in the gene TAS2R13 is associated with differences in alcohol consumption in patients with head and neck cancer. *Chem. Senses* 37: 737–744.

Doty, R. L., Bagla, R., Morgenson, M., and Mirza, N. (2001). NaCl thresholds: relationship to anterior tongue locus, area of stimulation, and number of fungiform papillae. *Physiol. Behav.* 72: 373–378.

Doty, R. L and Bromley, S. M. (2011). Disorders of Smell and Taste. In: Longo DL, Fauci AS, Kasper DL et al (eds). *Harrison's Textbook of Internal Medicine*, 18th Edition, Chapter 29, pp. 241–247.

Doty, R. L., and Haxel, B. R. (2005). Objective assessment of terbinafine-induced taste loss. *Laryngoscope* 115: 2035–2037.

Doty, R. L., Shah, M., and Bromley, S. M. (2008) Drug-induced taste disorders. *Drug Saf.* 31(3): 199–215.

Doty, R. L., Treem, J., Tourbier, I., and Mirza, N. (2009) A double-blind study of the influences of eszopiclone on dysgeusia and taste function. *Pharmacol. Biochem. Behav.* 94(2): 312–318.

Duffy, V. B., Bartoshuk, L. M., Striegel-Moore R., and Rodin, J. (1998) Taste changes across pregnancy. *Ann. N.Y. Acad. Sci.* 855: 805–809.

Duffy, V. B., Hayes, J. E., Davidson, A. C. et al. (2010) Vegetable intake in college-aged adults is explained by oral sensory phenotypes and TAS2R38 genotype. *Chemosens. Percept.* 3: 137–143.

Duhra, P., and Foulds, I. S. (1988) Methotrexate–induced impairment of taste acuity. *Clin. Exp. Dermatol.* 13: 126–127.

Erick, M. (1996) Idiopathic dysgeusia associated with blood transfusion: a case report [letter]. *J. Amer. Diet Assoc.* 96: 450.

Evers, K. A., Eindhoven, G. B., and Wierda, J. M. K. H. (1999) Transient nerve damage following intubation for trans-sphenoidal hypophysectomy. *Can. J. Anesth.* 46(12): 1143–1145.

Federoff, I. C., Stoner, S. A., Andersen, A. E. et al. (1995). Olfactory dysfunction in anorexia and bulimia nervosa. *Int. J. Eat. Disord.* 18: 71–77.

Femiano, F., Scully, C., and Gombos, F. (2002) Idiopathic dysguesia: an open trial of alpha lipoic acid (ALA) therapy. *Int. J. Oral Maxillofac. Surg.* 31(6): 625–628.

Fink, M. (2011) Vitamin D deficiency is a cofactor of chemotherapy-induced mucocutaneous toxicity and dysgeusia. *J. Clin. Oncol.* 29(4): e81–82.

Fleiner, F., Dahlslett, S. B., Schmidt, F., et al (2010) Olfactory and gustatory function in patients with multiple sclerosis. *Am. J. Rhinol. Allergy.* 24(5): e93–97.

Formaker, B. K., Mott, A. E., and Frank, M. E. (1998) The effects of topical anesthesia on oral burning in burning mouth syndrome. *Ann. N.Y. Acad. Sci.* 855: 776–780.

Frank, M. E., and Smith, D. V. (1991) Electrogustometry: a simple way to test taste. In *Smell and Taste in Health and Disease*, Getchell, T. V., Doty, R. L., Bartoshuk, L. M. and Snow, J. B. (Eds). Raven Press, New York, pp. 503–514.

Frey, L. (1923) Le syndrome du nerf auriculo-temporal. *Rev. Neurol.* 2: 97.

Fujiyama, R., Ishitobi, S., Honda, K., et al (2010) Ice cube stimulation helps to improve dysgeusia. *Odontology.* 98(1): 82–84.

Fukutake, T., Kita, K., Sakakibara, R., (1996) Late-onset hereditary ataxia with global thermoanalgesia and absence of fungiform papillae on the tongue in a Japanese family. *Brain* 119: 1011–1021.

Gadoth, N., Mass, E., Gordon, C. R., and Steiner, J. E. (1997) Taste and smell in familial dysautonomia. *Dev. Med. Child Neurol.* 39(6): 393–397.

Gangdev, P. S. (1995) Reversible taste impairment in a depressed patient. *S. Afr. Med. J.* 85(11): 1200–1201.

Gaut, A. and Williams, M. (2000) Lingual nerve injury during suspension microlaryngoscopy. *Arch. Otolaryngol. Head Neck Surg.* 126: 669–671.

Gayner, S. M., Kane, W. J., and McCaffrey, T. V. (1998) Infections of the salivary glands. In *Otolaryngology: Head and Neck Surgery*, 3rd edition, Cummings, C. W., Fredrickson, J. M., Harker, L. A., Krause, C. J., Richardson, M. A., and Schuller, D. E. (Eds). Mosby, St. Louis, pp.1234–1246.

Geraedts, M. C., Takahashi, T., Vigues, S., et al (2012) Transformation of postingestive glucose responses after deletion of sweet taste receptor subunits or gastric bypass surgery. *Am. J. Physiol. Endocrinol. Metab.* 303(4): E464–74.

Ghanizadeh, A. (2009) Loss of taste and smell during treatment with topiramate. *Eat. Weight Disord.* 14(2–3): el 37–38.

Gorsky, M., Silverman, S. J., and Chinn, H. (1987) Burning mouth syndrome: a review of 98 cases. *J. Oral Med.* 42: 7–9.

Gorsky, M., Silverman, S. J., and Chinn, H. (1991) Clinical characteristics and management outcome in the burning mouth syndrome. An open study of 130 patients., *Oral Surg., Oral Med. Oral Path.* 72: 192–195.

Graff-Radford, S. B. and Evans, R. W. (2003) Lingual nerve injury. *Headache.* 43(9): 975–983.

Green, T. L., McGregor, L. D., and King, K. M. (2008) Smell and taste dysfunction following minor stroke: a case report. *Can J. Neurosci. Nurs.* 30(2): 10–13.

Grovum, W. L., and Gonzalez, J. S. (2000) Regions in the brainstem and frontal cortex where electrical stimulation elicits parotid and submandibular saliva secretion in sheep. *Brain Res.* 852(1): 1–9.

Grushka, M., and Sessle, B. J. (1991) Burning mouth syndrome. *Dent. Clin. N. Am.* 35(1): 171–184.

Guinand, N., Just, T., Stow, N. W., et al (2010) Cutting the chorda tympani: not just a matter of taste. *J. Laryngol. Otol.* 124(9): 999–1002.

Haerer, A. F. (1992) *DeJong's The neurologic examination*, 5th edition, Lippincott-Raven: Philadelphia, pp. 227–231, 265.

Hardy, S. L., Brennand, C. P., and Wyse, B. W. (1981) Taste thresholds of individuals with diabetes and control subjects. *J. Am. Diet. Assoc.* 79(3): 286–289.

Harner, S. G., and Laws, E. R. Jr., (1981) Diagnosis of acoustic neuroma. *Neurosurgery* 9: 373–379.

Hausser-Hauw, C., and Bancaud, J. (1987) Gustatory hallucinations in epileptic seizures. Electrophysiological, clinical and anatomical correlates. *Brain* 110: 339–359.

Heiser, C., Landis, B. N., Giger, R., et al (2010) Taste disturbance following tonsillectomy – a prospective study. *Laryngoscope.* 120(10): 2119–2124.

Heiser, C., Landis, B. N., Giger, R., (2012) Taste disorders after tonsillectomy: a long-term follow-up. *Laryngoscope* 122(6): 1265–1266.

Helms ,J. A., Della-Fera, M. A., Mott, A. E., and Frank, M. E. (1995) Effects of chlorhexidine on human taste perception. *Arch. Oral Biol.* 40: 913–920.

Henkin ,R. I. (1967) Abnormalities of taste and olfaction in patients with chromatin negative gonadal dysgenesis. *J. Clin. Endocrinol. Metab.* 27(10): 1436–1440.

Hirai, R., Takao, K., Onoda, K., et al (2012) Patients with phantogeusia show increased expression of T2R taste receptor genes in their tongues. *Ann. Otol. Rhinol. Laryngol.* 121(2): 113–118.

Hovan, A. J., Williams, P. M., Stevenson-Moore, P., et al (2010) A systematic review of dysgeusia induced by cancer therapies. *Support Care Cancer.* 18(8): 1081–1087.

Huang, C. C., Lin, C. D., Wang, C. Y., (2012) Gustatory changes in patients with chronic otitis media, before and after middle-ear surgery. *J. Laryngol. Otol.* 126(5): 470–474.

Isolauri, J., Luostarinen, M.,and Isolauri, E. (1997) The natural course of gastroesophageal reflux disease: 17–22 year follow-up of 60 patients. *Am. J. Gastroenterol.* 92: 37–41.

Janssen, S. and Depoortere, I. (2013) Nutrient sensing in the gut: new roads to therapeutics? *Trends Endocrinol. Metab.* 24: 92–100.

Kida, A. (1974) Relations between the taste evoked potential and the background brain wave. *J. Nihon Univ. Med. Assoc.* 34: 43–52.

Kim, A., Feng, P., Ohkuri, T., et al (2012) Defects in the peripheral taste structure and function in the MRL/lpr mouse model of autoimmune disease. *PLoS One.* 7(4): e35588 Epub 2012 Apr 19.

Kiverniti, E. and Watters, G. (2012) Taste disturbance after mastoid surgery: immediate and long-term effects of chorda tympani nerve sacrifice. *J. Laryngol. Otol.* 126(1): 34–37.

Konstantinidis, I., Chatziavramidis, A., Printza, A., et al (2010) Effects of smoking on taste: assessment with contact endoscopy and taste strips. *Laryngoscope.* 120(10): 1958–1963.

Ksouda, K., Affes, H., Hammami, B., et al (2011) Ageusia as a side effect of clopidogrel treatment. *Indian J. Pharmacol.* 43(3): 350–351.

Kveton, J. F., and Bartoshuk, L. M. (1994) The effect of unilateral chorda tympani damage on taste. *Laryngoscope* 104: 25–29.

Laskaris, G., Sklavounou, A., and Stratigos, J. (1982) Bullous pemphigoid, cicatrical pemphigoid, and pemphigous vulgaris. A comparative clinical survey of 278 cases. *Oral Surg.* 54(6): 656–662.

Landis, B. N., Croy, I. and Haehner, A., (2012) Long lasting phantosmia treated with venlafaxine. *Neurocase.* 18(2): 112–114.

Lewandowski, K. E., DePaola, J., Camsari, G. B., et al (2009) Tactile, olfactory, and gustatory hallucinations in psychotic disorders: a descriptive study. *Ann. Acad. Med. Singapore.* 38: 383–387.

Mass, E. (2012) A review of the oro-dento-facial characteristics of hereditary sensory and autonomic neuropathy type III (familial dysautonomia). *Spec. Care Dentist.* 32(1): 15–20.

Mattes, R. D., and Cowart, B. J. (1994) Dietary assessment of patients with chemosensory disorders. *J. Am. Diet Assoc.* 94: 50–56.

Mattes, R. D., and Kare, M. R. (1986) Gustatory sequelae of alimentary disorders. *Digest. Dis.* 4: 129–138.

Mathew, S. D., Bristow, K., and Higgs, J. (2010) Giant cell arteritis presenting as aqeusia and lower extremity claudication. *J. Clin. Rheumatol.* 16(7): 343–344.

Matsuda, T., and Doty, R. L. (1995) Regional taste sensitivity to NaCl: relationship to subject age, tongue locus and area of stimulation. *Chem. Senses* 20: 283–290.

McConnell, R. J., Menendez, C. E., Smith, F. R., et al (1976) Defects of taste and smell in patients with hypothyroidism. *Am. J. Med.* 59: 354–364.

McManus, L. J., Stringer, M. D., and Dawer, P. J. (2012) Iatrogenic injury to the chorda tympani: a systematic review. *J. Laryngol. Otol.* 126(1): 8–14.

Melnick, S. L., Roseman, J. M., Engel, D., and Cogen, R. B. (1988) Epidemiology of acute necrotizing ulcerative gingivitis. *Epidemiol. Rev.* 10: 191–211.

Middleton, R. A., and Allman-Farinelli, M. A. (1999) Taste sensitivity is altered in patients with chronic renal failure receiving continuous ambulatory peritoneal dialysis. *J. Nutr.* 129: 122–125.

Miller, I. J. and Reedy, F. E. (1990). Quantification of fungiform papillae and taste pores in living human beings. *Chem. Senses* 15: 281–294.

Miller, S. L., Mirza, N. and Doty, R. L. (2002). Electrogustometric thresholds: Relationship to anterior tongue locus, area of stimulation, and number of fungiform papillae. *Physiol. Behav.* 75: 753–757.

Miller, S. M., and Naylor, G. J. (1989) Unpleasant taste – a neglected symptoom of depression. *J. Affect. Disord.* 17(3): 291–293.

Miras, A. D. and le Roux, C. W. (2010) Bariatric surgery and taste: novel mechanisms of weight loss. *Curr. Opin. Gastroenterol.* 26(2): 140–145.

Mizoguchi, Y., Monji, A., and Yamada, S. (2012) Dysgeusia successfully treated with sertraline. *J. Neuropsychiat. Clin. Neurosci.* 24(2): e42.

Moon, C. N., and Pullen, E. W. (1963) Effects of chorda tympani section during middle ear surgery. *Laryngoscope* 73: 392–405.

Morris-Witman, J., Sego, R., Brinkley, L., and Dolce, C. (2000) The effects of sialoadenectomy and exogenous EGF on taste bud morphology and maintenance. *Chem. Senses* 25: 9–19.

Murakami, S., Mizobuchi, M., Nakashiro, Y., et al (1996) Bell palsy and herpes simplex virus: identification of viral DNA in endoneurial fluid and muscle. *Ann. Intern. Med.* 124: 27–30.

Musialik, J., Suchecka, W., Klimacka-Nawrot, E., et al (2012) Taste and appetite disorders of chronic hepatitis C patients. *Eur. J. Gastroenterol. Hepatol.* 24(12): 1400–1405.

Mueser, K. T., Bellack, A. S., and Brady, E. U. (1990) Hallucinations in schizophrenia. *Acta. Psychiat. Scand.* 82: 26–29.

Nakajima, M., Ohtsuki, T., and Minematsu, K. (2010) Bilateral hypogeusia in a patient with a unilateral paramedian thalamic infarction. *J. Neurol. Neurosurg. Psychiatry* 81(6): 700–701.

Negri, R., De Feola, M., Di Domenico, S. et al. (2012) Taste perception and food choices. *J. Pediatr. Gastroenterol. Nutr.* 54: 624–629.

Nelson, S. P., Chen, E. H., Syniar, G. M., and Chistoffel, K. K. (2000) Prevalence of symptoms of gastroesophageal reflux during childhood: a pediatric practice-based survey. *Arch. Pediatr. Adolesc. Med.* 154(2): 150–154.

North, A. C. (2012) The effect of background music on the taste of wine. *Br. J. Psychol.* 103(3): 293–301.

Nickel, A. A. (1993) Regional anesthesia. *Oral Maxil. Surg. Clin. N. Am.* 5: 17–24.

Onoda, K., Hirai, R., Takao, K., et al (2011) Patients with Hypogeusia show changes in the expression of the T2R taste receptor genes in their tongues. *Laryngoscope.* 121(12): 2592–2597.

Onoda, K., and Ikeda, M. (1999) Gustatory disturbance due to cerebrovascular disorder. *Laryngoscope* 109: 123–128.

Osterbauer, R. A., Matthews, P. M., Jenkinson, M., et al (2005) Color of scents: Chromatic stimuli modulate odor responses in the human brain. *J. Neurophysiol.* 93: 3434–3441.

Palouzier-Paulignan, B., Lacroix, M. C., Aime, P., 2012) Olfaction under metabolic influences. *Chem. Senses* 37(9): 769–797.

Penfield, W., and Jasper, H. (1954) Epilepsy and the Functional Anatomy of the Human Brain. Boston: Little Brown.

Perros, P., MacFarlane, T. W., Counsell, C., and Frier, B. M. (1996) Altered taste sensation in newly-diagnosed NIDDM. *Diabetes Care* 19: 768–770.

Pittman, J. A., and Beschi, R. J. (1967) Taste thresholds in hyper- and hypothyroidism. *J. Clin. Endocrinol. Metab.* 27: 895–896.

Plattig, K. H. (1968) Über den elektrischen Geschmack, Reizstärkeabhängige evoziere Hirnpotentiale nach elektrischer Reizung der Zunge beim Menschen. *Z. Biol.* 116: 161–211.

Pritchard, T. C., Macaluso, D. A., and Eslinger, P. J. (1999) Taste perception in patients with insular cortex lesions. *Behav. Neurosci.* 113: 663–671.

Reimann, F., Tolhurst, G., and Gribble, F. M. (2012) G-protein-coupled receptors in intestinal chemosensation. *Cell Metab.* 15(4): 421–431.

Richter, J. E. (1998) Extraesophageal manifestations of gastroesophageal reflux disease. *Clin. Perspect.* 1: 28–39.

Robinson, P. P., Loescher, A. R., and Smith, K. G. (2000) A prospective, quantitative study on the clinical outcome of lingual nerve repair. *Brit. J. Oral Maxil. Surg.* 38(4): 255–263.

Rodin, J., Bartoshuk, L., Peterson C., and Schank, D. (1990) Bulimia and taste: possible interactions. *J. Abnorm. Psychol.* 99(1): 32–39.

Rodriguez-Perez, I., and Banoczy, J. (1982) Oral leukoplakia: a histological study. *Acta. Morphol. Acad. Sci. Hung.* 30(3–4): 289–298.

Rossie, K. M, and Guggenheimer, J. (1990) Thermally induced "nicotine" stomatitis. *Oral Surg. Oral Med. Oral Pathol. Oral Radiol. Endod.* 70(5): 597–599.

Roper, S., and Gilmore, R. L. (1995) Orbitofrontal resections for intractable partial seizures. *J. Epilepsy* 8: 186.

Sacerdote, C., Guarrera, S., Smith, G. D. et al. (2007) Lactase persistence and bitter taste response: instrumental variables and mendelian randomization in epidemiologic studies of dietary factors and cancer risk. *Am. J. Epidemiol.* 166: 576–581.

Salisbury, P. L. (1997) Diagnosis and patient management of oral cancer. *Dent. Clin. N. Am.* 41(4): 891–914.

Schiffman, S. S. (1997) Taste and smell losses in normal aging and disease. *JAMA* 278: 1357–1362.

Schiffman, S. S., Graham, B. G., Suggs, M. S., and Sattely-Miller, E. A. (1998) Effect of psychotropic drugs on taste responses in young and elderly persons. *Ann. N.Y. Acad. Sci.* 855: 732–737.

Schiffman, S. S., Zervakis, J., Heffron, S., and Heald, A. E. (1999) Effect of protease inhibitors on the sense of taste. *Nutrition* 15: 767–772.

Scrivani, S. J., Keith, D. A., Kulich, R., et al (1998) Posttraumatic gustatory neuralgia: a clinical model of trigeminal neuropathic pain. *J. Orofac. Pain* 12: 287–292.

Scrivani, S. J., Meghan, M., Donoff, R. B., Kaban, L. B. (2000) Taste perception after lingual nerve repair. *J. Oral Maxil. Surg.* 58: 3–5.

Seddon, H. J. (1943) Three types of nerve injury. *Brain.* 66: 247–288.

Settle, R. G. (1981). Suprathreshold glucose and fructose sensitivity in individuals with different family histories of non-insulin-dependent diabetes mellitus. *Chem. Senses* 6: 435–443.

Shafer, D. M., Frank, M. E., Gent, J. F., and Fischer, M. E. (1999) Gustatory function after third molar extraction. *Oral Surg. Oral Med. Oral Pathol. Oral Radiol. Endodont.* 87: 419–428.

Sharav, Y., Benoliel, R., Schnarch, A., and Greenberg, L. (1991) Idiopathic trigeminal pain associated with gustatory stimuli. *Pain* 44: 171–172.

Shih, C. T., Hao ,S. P., Ng, S. H., and Yen, K. C. (1999) Necrosis of the tongue after arterial chemotherapy. *Otolaryngol. Head Neck Surg.* 121(5): 655–657.

Siegel, R. K. (1978) Cocaine hallucinations. *Am. J. Psychiatry* 135(3): 309–314.

Silva, D. A., Colingo, K. A., and Miller, R. (1992) Lingual nerve repair following laryngoscopy. *Anesthesiol.* 76: 650–651.

Sjostrand, C., Savic, I., Laudon-Meyer, E., et al (2010) Migraine and olfactory stimuli. *Curr. Pain Headache Rep.* 14(3): 244–251.

Small, D. M., Jones-Gotman, M., Zatorre, R. J., et al (1997) A role for the right anterior temporal lobe in taste quality recognition. *J. Neurosci.* 17: 5136–5142.

Smith, W. M., Davidson, T. M., and Murphy, C. (2009) Toxin-induced chemosensory dysfunction: a case series and review. *Am. J. Rhinol. Allergy.* 23(6): 578–581.

Solemdal, K., Møinichen-Berstad, C., Mowe, M. et al. (2014) Impaired taste and increased mortality in acutely hospitalized older people. *Chem. Senses* 39: 263–269.

Soria, E. D., Candaras, M. M., and Truax, B. T. (1990) Impairment of taste in the Guillain-Barré syndrome. *Clin. Neurol. Neurosurg.* 92: 75–79.

Steiner, J. E., Rosenthal-Zifroni, A.,and Edelstein, E. L. (1969) Taste perception in depressive illness. *Israel Ann. Psychiat. Rel. Discip.* 7: 223–232.

Suliburska, J., Duda, G., and Pupek-Musialik, D. (2012) The influence of hypotensive drugs on the taste sensitivity in patients with primary hypertension. *Acta. Pol. Pharm.* 69(1): 121–127.

Sumner, D .(1967) Post-traumatic ageusia. *Brain* 90: 187–202.

Takahashi, H., Hitsumoto, Y., Honda, N., et al (2001) Mouse model of Bell's palsy induced by reactivation of herpes simplex virus type 1. *J. Neuropathol. Exp. Neurol.* 60(6): 621–627.

Takahashi, H., Nakamura, H., Yui, M., and Mori, H. (1985) Analysis of fifty cases of facial palsy due to otitis media. *Arch. Otorhinolaryngol.* 241(2): 163–168.

Taverner, D. (1973) Medical management of idiopathic facial (Bell's) palsy. *Proc. R. Soc. Med.* 66(6): 554–556.

Thompson, D. F. and Kessler, T. L. (2010) Drug-induced Black Hairy Tongue. *Pharmacotherapy.* 30(6): 585–593.

Tourne, L. P., and Fricton, J. R. (1992) Burning mouth syndrome. Critical review and proposed clinical management. *Oral Surg. Oral Med. Oral Pathol. Oral Radiol. Endodont.* 74: 158–167.

Truax, B. T. (1989) Gustatory pain: a complication of carotid endarterectomy. *Neurol.* 39: 1258–1260.

Tsivgoulis, G., Ioannis, H., Vadikolias, K., et al (2011) Bilateral ageusia caused by a unilateral midbrain and thalamic infarction. *J. Neuroimaging* 21(3): 263–265.

Uchiyama, T., Fukutake, T., Arai, K., et al (2001) Machado-Joseph disease associated with an absence of fungiform papillae on the tongue. *Neurology* 56(4): 558–560.

Ujike, H., Yamamoto, M., and Hara, I. (1987) Taste loss as an initial symptom of primary amyloidoisis. *J. Neurol. Neurosurg. Psychiatry.* 50(1): 111–112.

Valkov, I. M., Dokumov, S. I., Genkova, P. I., and Dimov, D. S. (1975) Olfactory, auditory, and gustatory function in patients with gonadal dysgenesis. *Obstetrics Gynecol.* 46(4): 417–418.

Valmeseda-Castellon, E., Berini-Aytes, L., and Gay-Escoda, C. (2000) Lingual nerve damage after third molar extraction. *Oral Surg. Oral Med. Oral Pathol. Oral Radiol. Endod.* 90: 567–573.

Vissink, A., Mitchell, J. B., Baum, B.J., et al (2010) Clinical management of salivary gland hypofunction and xerostomia in head and neck cancer patients: successes and barriers. *Int. J. Radiat. Oncol. Biol. Phys.* 78(4): 983–991.

Vories, A. A. (1999) Dysgeusia associated with tonsillectomy. *Otolaryngol. Head Neck Surg.* 121(3): 303–304.

Wantanabe, K., Tomita, H., and Murkami, G. (1995) Morphometric study of chorda tympani-derived fibers along their course in the lingual nerve. *Nippon Jibiinkoka Gakkai Kaiho [J. Oto-Rhino-Laryngological Soc.Japan]* 98: 80–89.

Webb, C. J, Makura, Z. G. G., and McCormick, M. S. (2000) Glossopharyngeal neuralgia following foreign body impaction in the neck. *J. Laryngol. Otol.* 114: 70–72.

Weiss, A. P. and Heckers, S. (1999) Neuroimaging in hallucinations: a review of the literature. *Psychiat. Res.* 92: 61–74.

West, S. E., and Doty, R. L. (1995). Influence of epilepsy and temporal lobe resection on olfactory function. *Epilepsia* 36: 531–42.

Yabe, I., Andoh, S., Mito, Y., et al (1995) Multiple sclerosis presenting with taste disturbance [in Japanese]. *Neurol. Med.* 43: 383–385.

Yamashita, Y., Ohya, T., Nagamitsu, S., et al (2011) Parageusia in an epileptic child treated with lamotrigine. *Pediatr. Int.* 53(6): 1106–1107.

Yanagisawa, K., Bartoshuk, L. M., Catalanotto, F. A., et al (1998) Anesthesia of the chorda tympani nerve and taste phantoms. *Physiol. Behav.* 63: 329–335.

Zanchin, G., Dainese, F., Mainardi, F., et al (2005) Osmophobia in primary headaches. *J. Headache Pain* 6(4): 213–215.

Zuniga, J. R., Chen, N., and Miller, I. J. (1994) Effects of chorda-lingual injury and repair on human taste. *Chem. Senses* 19: 657–665.

Chapter 40

Influence of Drugs on Taste Function

SUSAN S. SCHIFFMAN

40.1 INTRODUCTION

Medications frequently have objectionable taste-related side-effects that can adversely affect patient compliance with prescribed drug regimens and the successful treatment of disease (al-Shammari et al., 1995; Barbara et al., 1986; Harrison, 1995; Iwai, 1997; Liacouras et al., 1993; Matsui et al., 1996). Hundreds of drugs encompassing all major therapeutic classes have been reported clinically to induce unpleasant and altered taste sensations when administered alone or in combination with other medications (see Table 40.1). These aversive sensations include bitter and metallic tastes, partial or total loss of taste, as well as taste distortions and perversions.

Most of our current knowledge about taste disturbances from medications derives from clinical reports, drug reference books, medication inserts, and chemosensory testing of a small number of patients rather than from comprehensive clinical trials that utilize objective, quantitative tests in conjunction with documentation of medication usage. Thus, neither the incidence nor prevalence of drug-induced taste disturbances in the general population is known (Doty et al., 2008). Estimates of the potential risk of taste disturbances from therapeutic drugs vary widely. An early study of medication use by 4,163 community-dwelling elderly (the Duke Established Populations for Epidemiologic Studies of the Elderly, EPESE) concluded that only 11% of participants took drugs that could alter taste perception (Lewis et al., 1993). A later study of 1,163 persons aged from 32 to 81 years drawn from the San Antonio Longitudinal Study of Aging and the San Antonio Heart study concluded that 389 (or 33%) of the drugs used by this population had the potential to alter the sense of taste (Shinkai et al., 2006). A comparison

of the list of the "Top 200 drugs in the US market by dispensed prescriptions, 2010" (Bartholow, 2011) with the list of drugs in Table 40.1 indicates that 59% of the top 200 drugs have potential taste side-effects. Taste disturbances associated with pharmacological treatments were reported to be the most frequent cause of taste complaints in 2278 patients who visited a Taste and Smell Clinic in Tokyo, Japan during the 10-year period between 1981 and 1990 (Hamada et al., 2002). Drug-induced taste disorders accounted for 21.7% of the diagnoses with a sharp increase in incidence at 50 years of age. In a separate analysis of a portion of the data at the same clinic, Tomita and Yoshikawa (2002) reported that drugs accounted for ~25% of the cases of taste disorders. For specific drugs, the reported incidence of taste disturbances varies widely from less than 5% on average according to drug reference books (PDR, 2005) to as high as 66% for the drug eszopiclone in a clinical study (Doty et al., 2009). The actual incidence and prevalence of drug-related taste disorders for specific drugs awaits quantification with controlled clinical trials because "self-report markedly underestimates chemosensory deficits" (Doty and Haxel, 2005).

Quantification and documentation of taste-related deficits caused by medications present numerous challenges due to several factors. First, standardized taste tests are not routinely utilized by medical practitioners to verify and validate drug-related taste complaints by patients. Second, many taste complaints cannot be classified as standard gustatory sensations such as sweet (sucrose), sour (citric acid), salty (NaCl), bitter (quinine), and umami (monosodium glutamate), or as more controversial tastes such as metallic (iron salts), chalky (calcium salts) or fatty (fatty acids). For example, a patient may complain of a "foul" taste in the mouth that cannot be matched

---

Handbook of Olfaction and Gustation, Third Edition. Edited by Richard L. Doty.
© 2015 Richard L. Doty. Published 2015 by John Wiley & Sons, Inc.

912   **Chapter 40 Influence of Drugs on Taste Function**

Table 40.1 Drugs reported to cause taste disturbances.

Drug Class	Subclass/Indication	
Anesthetics		Amydricaine, Amylocaine, Benzocaine, Cocaine, Dibucaine, Eucaine, Euprocin, Lidocaine, Procaine, Tetracaine, Tropacocaine
Antiepileptics/ anticonvulsants		Acetazolamide, Carbamazepine, Felbamate, Lamotrigine, Phenytoin, Topiramate
Antihistamines and antiallergenic agents	Antihistamines	Azelastine, Chlorpheniramine, Cyproheptadine, Loratadine, Promethazine, Terfenadine
	Steroids	Beclomethasone, Budesonide, Flunisolide, Fluticasone
Antihypertensives and Cardiovascular agents		Adenosine, Amiloride, Amiodarone, Amlodipine, Amrinone, Antithrombin III, Atropine sulfate, Benazepril HCl, Bepridil, Betaxolol, Bisoprolol, Bretylium, Captopril, Chlorthalidone, Cilazapril, Clonidine, Diazoxide, Diltiazem, Dipyridamole, Doxazosin, Enalapril, Esmolol, Ethacrynic Acid, Flecainide, Flosequinan, Fosinopril, Furosemide, Guanfacine, Hydralazine, Hydrochlorothiazide, Inamrinone, Isosorbide Nitrates, Labetalol, Lisinopril, Losartan, Methyldopa, Metolazone, Metoprolol, Mexiletine, Minoxidil, Moricizine, Nifedipine, Nisoldipine, Nitroglycerin, Nylidrin, Oxyfedrine, Pentoxifylline, Perindopril, Phenindione, Procainamide, Propafenone, Propranolol, Quinapril, Quinidine, Ramapril, Spironolactone, Tocainide, Triamterene, Trichlormethiazide
Anti-infectives	Antibacterials	Amoxicillin, Ampicillin, Azithromycin, Aztreonam, Bacampicillin, Carbenicillin, Cefacetrile, Cefadroxil, Cefamandole, Cefodizime, Cefpirome, Cefpodoxime, Ceftriaxone, Cefuroxime, Cephalexin, Chlorhexidine, Cinoxacin, Ciprofloxacin, Clarithromycin, Clindamycin, Clofazimine, Dapsone, Enoxacin, Ethambutol, Ethionamide, Hexetidine, Imipenem, Lincomycin, Lomefloxacin, Mezlocillin, Minocycline, Norfloxacin, Ofloxacin, Penicillins, Piperacillin, Rifabutin, Roxithromycin, Streptomycin, Sulfamethoxazole, Tetracyclines, Ticarcillin, Trimethoprim, Tyrothricin, Vancomycin
	Antivirals	Acyclovir, Amantadine, Foscarnet, Ganciclovir, Idoxuridine, Oseltamivir, Pirodavir, Ribavirin, Rimantadine
	Antifungals	Amphotericin B, Atovaquone, Fluconazole, Griseofulvin, Pentamidine, Terbinafine
	HIV/AIDS drugs	Didanosine, Indinavir, Lamivudine, Nelfinavir, Nevirapine, Ritonavir, Saquinavir, Stavudine, Zalcitabine, Zidovudine
	Antimalarials	Chloroquine, Hydroxychloroquine, Pyrimethamine
	Other	Antimony, Levamisole, Metronidazole, Niclosamide, Niridazole, Stibogluconate
Anti-inflammatory, anti-pyretic, and/or analgesic agents	NSAIDs	Acemetacin, Aspirin, Benoxaprofen, Celecoxib, Choline magnesium trisalicylate, Diclofenac, Etodolac, Fenoprofen, Flurbiprofen, Ibuprofen, Indomethacin, Ketoprofen, Ketorolac, Mefenamic Acid, Nabumetone, Naproxen, Oxaprozin, Phenylbutazone, Piroxicam, Sulindac, Tolmetin
	Opioids	Butorphanol, Codeine, Fentanyl, Hydromorphone, Morphine, Nalbuphine, Pentazocine
	Antirheumatic/ Antiarthritic/ Antigout drugs	Allopurinol, Auranofin, Aurothioglucose, Colchicine, Gold compounds, D-Penicillamine, 5-Thiopyridoxine
	Other	Acetaminophen, Dimethyl Sulfoxide, Tramadol, Hydrocortisone, Dexamethasone
Antilipidemics		Atorvastatin, Cholestyramine, Clofibrate, Fluvastatin, Gemfibrozil, Lovastatin, Pravastatin, Probucol, Simvastatin

40.1 Introduction 913

Table 40.1 (*Continued*)

Drug Class	Subclass/Indication	
Antimigraine drugs		Naratriptan, Rizatriptan, Sumatriptan, Dihydroergotamine mesylate
Antineoplastic and immunosuppressant drugs		Adriamycin, Aldesleukin, Amonafide, Azathioprine, Bleomycin, Busulfan, Carboplatin, Carmustine, Cisplatin, Cyclophosphamide, Dacarbazine, Doxorubicin, 5-Fluorouracil, Lomustine, Methotrexate, Tamoxifen, Tegafur, Vinblastine, Vincristine
CNS drugs/ Sympathomimetics		Amezinium, Amphetamine, Apraclonidine, Benzphetamine, Dexamphetamine, Methamphetamine, Methylphenidate, Phenmetrazine
Dental agents		Mercury, Nickel, Sodium lauryl sulfate, Sodium fluoride
Dermatologic agents		Etretinate, Isotretinoin, Zinc Oxide
Endocrine and diabetes drugs	Antithyroid agents	Carbimazole, Methimazole, Methylthiouracil, Propylthiouracil, Thiouracil
	Insulin/Antidiabetic drugs	Acarbose, Glipizide, Glyburide, Insulin, Metformin, Phenformin, Tolazamide, Tolbutamide
	Endocrine/metabolic bone disorders	Alendronate, Calcitonin, Etidronate, Pamidronate
	Other	Levothyroxine, Protirelin
Gastrointestinal drugs	Anticholinergics/ Antispasmotics	Clindinium, Dicyclomine, Glycopyrrolate, Hyoscyamine, Oxbutynin, Propantheline, Tridihexethyl Chloride
	GERD[1] and antiulcer drugs	Cimetidine, Famotidine, Isopropamide, Lansoprazole, Mesalamine, Methscopolamine, Metoclopramide, Misoprostol, Omeprazole, Ranitidine, Sucralfate, Sulfasalazine
	Appetite suppressants/diet aids	Dinitrophenol, Fenfluramine, Mazindol, Phendimetrazine, Phentermine
	Other	Bismuth, Granisetron, Ursodiol, Monoctanoin, Prochlorperazine
Muscle relaxants and antiparkinsonism drugs		Apomorphine, Baclofen, Benztropine, Bromocriptine, Cyclobenzaprine, Dantrolene, Levodopa, Methocarbamol, Pergolide, Selegiline, Trihexyphenidyl HCl
Nose, throat, and pulmonary agents		Albuterol, Bamifylline, Bitolterol, Cromolyn, Metaproterenol, Nedocromil, Pirbuterol, Pseudoephedrine, Terbutaline
Psychopharmacologic agents	Antidepressants and Bipolar disorder drugs	Amitriptyline, Bupropion, Citalopram, Clomipramine, Desipramine, Doxepin, Fluoxetine, Fluvoxamine, Imipramine, Lithium, Maprotiline, Nortriptyline, Paroxetine, Protriptyline, Sertraline, Tranylcypromine, Trazodone, Trimipramine, Venlafaxine
	Antipsychotics	Clozapine, Fluphenazine, Olanzapine, Pimozide, Prochlorperazine, Risperidone, Trifluoperazine
	Anxiolytics	Alprazolam, Buspirone, Chlormezanone, Clonazepam, Diazepam, Estazolam, Flurazepam, Midazolam, Oxazepam, Triazolam
	Dementia agents	Donepezil, Rivastigmine, Tacrine
	Hypnotics	Eszopiclone, Ethchlorvynol, Propofol, Quazepam, Zolpidem, Zopiclone
Vitamins, minerals, nutrients and related compounds		Calcifediol, Calcitriol, Calcium Salts, Ergocalciferol, Omega Fatty Acids, Phytonadione, Potassium Iodide, Selenium
Miscellaneous	Anti-glaucoma drugs	Dorzolamide, Pilocarpine, Timolol
	Biotechnology drugs	Filgrastim, Interferon-Alpha, Interferon-Gamma, Interleukin-2
	Diagnostics	Gallium, Thallium, Gadodiamide
	Other	Decitabine, Deferoxamine, Dinoprostone, Disulfiram, Leuprolide, Nicotine, Riluzole, Tiopronin

[1] GERD: Gastroesophageal reflux disease

Sources: Micromedex®, 2012; Physicians Desk Reference, 2005; Schiffman, 1983, 1991; Schiffman and Zervakis, 2002; Schiffman et al., 2003; Johnell and Klarin, 2007

914　　Chapter 40　Influence of Drugs on Taste Function

by representative chemosensory stimuli. Third, patients with taste disturbances are frequently taking drugs for a medical condition such as cancer that also causes taste disorders. Thus, the taste disorder may be medication-related, disease-related, or an interaction between the medication and the disease. Fourth, most persons with drug-related taste disorders are taking multiple medications concurrently so it is difficult to determine if a taste disturbance is due to a single medication or to an interaction between several drugs. Fifth, adverse taste effects typically affect only a proportion but not all of the patients who use a drug. Groups that are particularly vulnerable include (1) the elderly who use a disproportionate number of drugs (Qato et al., 2008) and who have reduced drug clearance (Shi and Klotz, 2011), and (2) persons with certain genetic polymorphisms related to bitter taste perception (Hamor and Lafdjian, 1967; Kim et al., 2005; Sandell and Breslin, 2008; Woodling et al., 2004) and the metabolism of drugs (Lee et al., 2005).

Prior reviews on the topic of drug-related taste disorders have summarized information from case reports, quantitative studies of a limited number of persons with taste disturbances, and/or drug reference material supplied by manufacturers (Schiffman, 1983, 1991, 1997; Schiffman and Zervakis, 2002; Schiffman et al., 2003; Ackerman and Kasbekar, 1997; Doty et al., 2008). This review will focus on six topics relevant to understanding the biological bases of drug-induced taste disorders. These include: (1) interaction of drugs with taste receptors on the apical side of the tongue in the oral cavity, (2) genetic differences among individuals that affect taste perception of drugs, (3) taste sensations caused by injectable drugs, (4) drug interactions that result from use of multiple medications, (5) potential biochemical and pharmacological mechanisms by which therapeutic drugs can cause taste disorders, and (6) other factors to consider when assessing the cause of a taste disturbance.

40.2　TASTE COMPLAINTS DUE TO THE SENSORY PROPERTIES OF THE DRUG ITSELF

The majority of active drugs in oral pharmaceutical products are bitter in taste (Szejtli and Szente, 2005), and some chemosensory complaints are directly related to the unpleasant taste properties of the active drugs themselves rather than alteration or distortion of taste sensations. Orally ingested drugs reach taste receptors and channels located on the microvilli of taste buds either during ingestion or alternatively after leaving the general circulation subsequent to absorption from the gut. After exiting the

capillaries adjacent to taste buds, drugs diffuse or are transported across salivary gland membranes into the saliva (Kaufman and Lamster, 2002). Drugs that leave the bloodstream can also permeate paracellular pathways (tight junctions) between taste cells in papillae to reach taste receptors embedded in microvilli (Holland et al., 1991).

40.2.1　Bitter Taste

The fact that drugs typically have bitter tastes is not unexpected because they are xenobiotic compounds with pharmacological and even toxic effects that alter the biochemistry of the body. Many poisonous and potentially harmful compounds in the natural environment of plant and animal origin have bitter, aversive tastes that serve as warning signals to reject and prevent ingestion (Brockhoff et al., 2010; Drewnowski and Gomez-Carneros, 2000; Lindemann, 2001; Meyerhof et al., 2010). Bitter taste sensations are mediated by G-protein coupled taste receptors (Margolskee, 2002) that are encoded by the taste 2 receptor gene family (*TAS2R*) with at least 25 members in humans (Meyerhof et al., 2010, 2011). The human *TAS2R* genes are located at four chromosomal loci including two loci on chromosome 7, a cluster on chromosome 12, and a single gene on chromosome 5 (Adler et al., 2000; Meyerhof et al., 2010; Reed et al., 1999). Tens of thousands of structurally diverse bitter-tasting compounds including drugs are detected by the combinatorial pattern of hTAS2R activation (Meyerhof et al., 2010). A single compound can activate one or multiple TAS2Rs, and conversely a given TAS2R can recognize multiple compounds. Overall, receptors vary in their breath of tuning, and the most broadly tuned receptors (hTAS2R10, hTAS2R14, and hTAS2R46) respond collectively to ~50% of bitter tastants (Meyerhof et al., 2010). The specific subsets of hTAS2R repertoire that encode bitter taste for the vast majority of therapeutic drugs have not yet been characterized. However, the hTAS2R activation patterns for 12 drugs that engender taste complaints are given in Table 40.2; these data show that three drugs activated only one hTAS2R subtype while seven activated multiple subtypes; two drugs (doxepin and metronidazole) did not activate any receptor at the maximal concentrations employed. The finding that some bitter compounds do not activate the hTAS2R subtypes tested suggests that additional (as yet unidentified receptors) may mediate bitterness. Furthermore, mechanisms that do not involve activation of a G-protein coupled taste receptors may also play a role in bitter taste perception of drugs including blockage of potassium channels and permeation of cell membranes with activation of downstream signaling mechanisms (Lindemann, 2001; Naim et al., 2008).

Table 40.2 Activation pattern of hT2R subtypes by drugs associated with taste complaints. NR=Number of hT2R Subtypes Activated.

Drug	hT2R subtype												NR
	3	4	7	10	14	38	39	40	43	44	46	49	
Azathioprine		×		×	×		×				×		5
Chlorhexidine				×									1
Chloroquine	×		×	×			×						4
Chlorpheniramine		×	×	×	×	×	×	×			×		8
Colchicine		×					×				×		3
Cromolyn			×						×			×	3
Dapsone		×		×				×					3
Doxepin													0
Famotidine				×						×			2
Hydrocortisone											×		1
Methimazole					×								1
Metronidazole													0

Source: Meyerhof et al. (2010).

40.2.2 Genetic Differences in Bitter Taste Perception of Drugs

Variation in bitter taste sensitivity as a consequence of genetic polymorphisms in *hTAS2R* genes (Kim et al., 2005; Woodling et al., 2004; Drayna, 2005) is one factor that contributes to inter-individual differences in the perceived aversiveness of drugs. Surveys of 25 *hTAS2R* genes have reported an average of 4.2 single nucleotide polymorphisms (SNPs) per *hTAS2R* gene and at least 151 different haplotypes that occur at a frequency >10% in the population studied (Kim et al., 2005; Drayna, 2005). Widely divergent alleles of the *hTAS2R38* gene (also called the *PTC* gene), for example, govern taste sensitivity to the bitterness of thiourea compounds including the chemical phenylthiocarbamide (PTC); the allele for tasters of PTC differs from the nontaster allele at three locations. "Tasters" are persons who are sensitive to the bitter taste of PTC while "nontasters" are those who find the taste of PTC to be bland. Individuals who were homozygous for the taster form of the *hTAS2R38* gene have been reported to be more sensitive to the bitterness of numerous thiourea compounds including the antithyroid drug methimazole (Hamor and Lafdjian, 1967; Sandell and Breslin, 2008). Genetic polymorphisms of other biochemical components associated with the taste system including gustin and the TRPM5 channel may also contribute to the variation in the perceived aversive taste of the drugs. Gustin (CA6), a zinc metalloprotein secreted by salivary and von Ebner glands, is a taste bud trophic factor (Calò et al., 2011). Polymorphism of the gustin (CA6) gene at rs2274333 (A/G) has been associated with variability in the perception of bitterness of the thiourea compound 6-*n*-propylthiouracil

(PROP) (Padiglia et al., 2010). Genetic polymorphisms in the gene that encodes for a calcium-sensitive, nonselective cation channel TRPM5 expressed in taste buds have been identified (Ketterer et al., 2011); TRPM5 plays a role in downstream signaling of bitter (as well as sweet and umami) taste transduction. Overall, genetic variations that increase the sensitivity to strong bitter tastes are potential contributing factors to taste complaints from drugs.

40.2.3 Metallic Taste

Some therapeutic drugs, such as enoxacin, fenoprofen, and baclofen, produce unpleasant metallic sensations (with or without accompanying bitterness) that occur upon direct contact of the medication with the lingual surface (Schiffman et al., 2002). Artificial sweeteners used as excipients in pharmaceutical products including sodium saccharin and acesulfame-K also have metallic components for some individuals (Schiffman et al., 1979). Medicinal mixtures of traditional herbs such as the hemostatic Ankaferd Blood Stopper (ABS) used during dental procedures produce a metallic taste lasting 3–5 minutes when applied topically (Ercetin et al., 2010). Mineral supplements containing iron, zinc and copper salts also induce metallic tastes upon contact with the lingual surface (Schiffman, 2000). While the transduction mechanisms responsible for metallic sensations induced by drugs are not fully characterized, recent data suggest that the metallic component of artificial sweeteners and mineral salts is mediated in part by TRPV1 (the transient receptor potential cation channel subfamily V member 1) located in sensory nerve endings in the mouth (Riera et al., 2007, 2009). Variants in the *TRPV1* gene that affect its function (Xu et al., 2007) may play a role in

916   Chapter 40   Influence of Drugs on Taste Function

inter-individual differences in the perception of metallic taste.

40.2.4 Intravascular Taste

Taste complaints frequently occur when drugs are administered by the intravascular route. For over 80 years, it has been known that injection of a chemical solution into the peripheral vascular system can produce a taste response within seconds. Weiss et al. (1929) reported that injection of histamine into a vein in the arm produced as sensation on the tongue described as "salty, metallic, electric" in approximately 23 seconds. Fishberg et al. (1933) found that injection of saccharin into an arm vein produced a sweet taste within 16 seconds that rapidly passed from the base to the tip of the tongue. Tarr et al. (1933) reported that injection of sodium dehydrocholate into the arm produced a bitter taste within 16 seconds. Since these early studies, there have been many reports of taste sensations from drugs and nutrients, predominantly metallic and/or bitter tastes, that occur within seconds after injection. Intravenous injections of lidocaine (Owen et al., 2004), ropivacaine (McCartney et al., 2003), iron preparations (Breymann et al., 2001; Macdougall and Roche, 2005), galanin (McDonald et al., 1994), thyrotropin-releasing hormone (TRH) (Callahan et al., 1997), nicotinic acid (Phillips and Lightman, 1981), and arginine (Veldhuis et al., 2006) all elicit metallic taste complaints. The finding that arginine is perceived as metallic when injected but bitter when tasted orally (Schiffman et al., 1981) suggests that the sensory properties of drugs differ depending on whether they contact the apical tongue surface or the basolateral side of taste cells. On the apical surface, drugs directly activate taste receptors on the microvilli. However, activation from the basolateral side must involve downstream signaling mechanisms because G-protein-coupled receptors (GPCRs) that mediate the tastes of bitter, sweet, and umami are presumed to be absent basolaterally. However, amiloride-sensitive sodium channels that serve as receptors for salty taste are present on both the apical and basolateral membranes (Mierson et al., 1996). Some taste disturbances that occur almost instantaneously upon injection are unlikely due to chemical stimulation by a drug but rather a neural response associated with intravascular pressure (Schiffman, 2007).

40.3 MODIFICATION OF NORMAL TASTE SENSATIONS BY DRUGS

Drugs not only have bitter, metallic, or unpleasant tastes of their own but can also modify the sensations of other taste stimuli. While the mechanisms by which medications alter normal taste perception are not well-understood,

modulation or disruption of taste signals can occur at multiple levels including the apical and basolateral side of taste cells, chemosensory neural pathways, and/or the brain. Blood plasma and saliva concentrations also play a role because plasma and salivary levels have been positively correlated with taste distortions (Doty et al., 2009).

40.3.1 Taste Alterations from Interactions of Drugs at the Apical Taste Cell Level

Drugs can modulate taste signals by direct interactions with channels and receptors or by penetrating, accumulating, and/or disrupting taste cell membranes. Antidepressant drugs such as the sertraline have been shown to accumulate in cellular membranes and transform the biochemical properties of the cell (Chen et al., 2012). Permeation of the apical taste cell membrane by drugs to reach the cytosolic of side of the membrane can alter signals by interacting with downsteam signaling mechanisms including G-proteins and TRPM5 (Naim et al., 2008; Gao et al., 2009). Alteration of taste signals may occur within minutes of exposure to drugs at the apical level. For example, application of the diuretic amiloride, a potent inhibitor of sodium transport, to the tongue surface for five minutes reduced the taste intensity of sodium and lithium salts as well as sweeteners including saccharides, glycosides, dipeptides, proteins, and amino acids. However, amiloride had no effect on the perception of potassium or calcium salts, bitter and sour tastes, or amino acids without a sweet or salty component (Schiffman et al., 1983). Adaptation of the tongue to the antifibrillary drug bretylium tosylate was found to potentiate the taste of NaCl and LiCl with no effect on other salts, sweeteners, bitter or sour compounds (Schiffman et al., 1986a). Adaptation of the tongue for several minutes to methyl xanthines (caffeine, theobromine, and theophylline), which are potent antagonists of the adenosine receptor, potentiated the taste of bitter-tasting quinine and artificial sweeteners with a bitter component such as acesulfame-K with little or no effect on sucrose and urea. (Schiffman et al., 1985, 1986b).

Schiffman and colleagues performed a series of experiments to determine if a four-minute lingual application of drugs reported clinically to cause taste disturbances altered the perception of nine suprathreshold oral stimuli (summarized in Schiffman et al., 2002). Table 40.3 is a representative list of the some of the drugs tested along with the taste quality of each drug, taste detection thresholds (DT), and alterations in taste intensity caused by the topical drug exposure. All of these drugs are amphipathic compounds (they contain both hydrophobic and hydrophilic domains) and are membrane permeant. Each of the drugs had sufficient aqueous solubility to determine a taste threshold. The nine stimuli that were evaluated included NaCl, KCl,

CaCl$_2$, sucrose, quinine HCl (QHCl), citric acid, capsaicin (pungent), WS-3 (n-ethyl-p-menthane-3-carboxamide) which has a menthol-like "taste," and FeSO$_4$ (metallic). The main finding was that topical application of a given drug at a level slightly above the taste DT led to some changes in the perceived intensity of the nine test stimuli. (A single arrow indicates that the change in intensity was less than 20% while a double arrow indicates that the change was greater than 20%). Distortions in taste quality of the nine oral stimuli often accompanied these intensity changes. These data show that a short lingual exposure to a drug can alter the taste of subsequent taste stimuli. Alteration of taste perception at the apical surface may be more likely from direct exposure to a drug at the time of ingestion rather than access to apical receptors from the bloodstream because the drug concentration at the apical receptors is typically higher in the former case. It is not yet known if drugs that leave the systemic circulation can accumulate over time in apical taste cell membranes at concentrations high enough to cause the taste alterations reported in Table 40.3.

40.3.2 Taste Alterations Resulting from Drug–Drug Interactions and Polypharmacy

Drug-drug interactions occur when one drug alters the bioavailability and/or pharmacological effect of a co-administered drug (Huang et al., 2008). Polypharmacy increases the risk of toxicity and adverse effects from drug-drug interactions (Johnell and Klarin, 2007) including taste disorders (Schiffman, 2009). Over the last several decades, the rates of prescription drug use and prevalence of polypharmacy have increased significantly in many countries in the world (Rumble and Morgan, 1994; Hovstadius et al., 2010; Hajjar et al., 2007; Gorard, 2006). In a nationally representative sample of community-residing individuals in the United States, 29% of individuals aged 57 through 85 used at least 5 prescription medications concurrently (Qato et al., 2008). A study of elderly cardiovascular patients found that taste losses at the threshold level as well as complaints of altered taste were most prevalent among those who used the greatest number of medications (Schiffman, 2007). Orally administered drugs arrive at the small intestine where they permeate the intestinal membranes and are transported by the hepatic portal vein to the liver and ultimately the systemic bloodstream. During their transit in the intestine and liver, drugs come in contact with metabolic enzymes including cytochromes P450 (CYP) and transporters (e.g., P-glycoprotein) that normally limit the fraction of the dose before it reaches the systemic circulation (van Herwaarden et al., 2009). CYP metabolism promotes the elimination of drugs via chemical reactions that render drugs more polar and water-soluble.

P-glycoprotein (P-gp) is an energy-dependent efflux "pump" that extrudes drugs from the apical surface of intestinal and liver cells back into the intestinal lumen and bile respectively leading to decreased drug accumulation. The combined effect of multiple co-administered drugs on CYP enzymes and transporters can alter normal metabolic processes and trigger significant adverse side effects including taste disorders (Schiffman, 2009). The elderly are at higher risk from drug interactions due to numerous age-related changes in pharmacokinetics including reduced effectiveness of drug metabolism and impaired renal excretion (Corsonello et al., 2010).

Drug-drug interactions that result from inhibition of two CYP isozymes, CYP3A4 and CYP2D6, contribute disproportionately to adverse drug-drug interactions including taste disorders (Schiffman, 2009) because these two isozymes are involved collectively in the metabolism of over 70% of prescription medications (Dantzig et al., 1999; Felmlee et al., 2008; Ingelman-Sundberg, 2005). In Table 40.4, representative drugs from Table 40.1 are presented along with their interactions with CYP3A4, CYP2D6, and/or P-gp (Preissner et al., 2010; Drugbank, 2012; Pubchem, 2012; Flockhart, 2012). The three columns under the label "Substrates" indicate that a drug is either metabolized by CYP3A4 or CYP2D6 or is effluxed by P-gp. The three columns under the label "Inhibitors" indicate that a drug is an inhibitor of CYP3A4 or CYP2D6 metabolism or an inhibitor of P-gp efflux. (Note: a drug can be both a substrate as well as an inhibitor depending upon numerous variables such as the plasma concentration or length of exposure.) Drugs that are substrates of CYP3A4 are frequently substrates of P-gp as well. When a drug that is a substrate of CYP3A4, CYP2D6, or P-gp is coadministered with a drug that is inhibitor of the analogous CYP isozyme or transporter, elevated plasma drug concentrations can occur. Drugs that are strong inhibitors of CYP3A4 such as clarithromycin, indinavir, nelfinavir, ritonavir, and saquinavir, cause a >5-fold increase in the plasma AUC values of co-administered CYP3A4 substrates (Flockhart, 2012). AUC (area under the plasma concentration time curve) has many of important uses in pharmacokinetics. Drugs that are strong inhibitors of CYP2D6 such as bupropion, fluoxetine, paroxetine, and quinidine cause a >5-fold increase in the plasma AUC values of co-administered CYP2D6 substrates. Moderate CYP inhibitors cause a 2-fold to 5-fold increase in the plasma AUC values and weak inhibitors, a >1.25-fold but <2-fold increase. Clarithromycin, a potent inhibitor of CYP3A4 and P-gp, elicited the most taste complaints in a study of over 600,000 elderly patients from the Swedish Prescribed Drug Register (Johnell and Klarin, 2007). The column in Table 40.4 labeled T/N (the ratio of the toxic blood-plasma levels to normal therapeutic blood-plasma levels calculated from Schulz et al., 2012) indicates that

Table 40.3 Taste quality, mean taste detection thresholds (DT), and alterations in perceived intensity of nine oral stimuli after a 4-minute topical exposure of the tongue to drugs.

	Taste Quality	DT (in mM)	NaCl	KCl	CaCl$_2$	Sucrose	QHCl	Citric Acid	Capsaicin	WS-3	FeSO$_4$
Anti-infectives											
Ampicillin	Bitter	1.458	⇓		⇓			⇓		⇑	
Enoxacin	Metallic, Bitter	0.040				↓		⇓			
Ethambutol 2HCl	Bitter	0.247		⇓	↓					⇓	
Levamisole	Bitter	0.449			⇑			↑			
Lomefloxacin HCl	Bitter	0.379						⇑	⇑		
Ofloxacin	Bitter	0.387		⇑				⇑			
Pentamidine isethionate	Bitter	0.062	⇓	⇓	⇓						
Sulfamethoxazole	Sour, Bitter	0.639			↓	↑				↑	
Tetracycline HCl	Sour, Bitter	0.061		↓	↑						
Anti-inflammatory, anti-pyretic, and/or analgesic agents											
Dexamethasone	Bitter	0.0902			⇓						
Diclofenac sodium	Bitter	1.008				↓		⇓	⇓	⇑	
Fenoprofen calcium	Metallic, Sour	1.111									⇓
Antihistamines and antiallergenic agents											
Chlorpheniramine maleate	Bitter	0.085								⇓	
Promethazine	Bitter	0.079	⇓	↓	⇓	⇓		⇑			
Antihypertensives and Cardiovascular agents											
Captopril	Sour, Bitter	0.132		↓		↓					
Diltiazem HCl	Bitter	0.142			⇑	⇑		⇑		⇓	
Enalapril Maleate	Sour, Bitter	0.107						⇓			
Ethacrynic acid	Bitter	0.186						⇑			⇑
Labetalol HCl	Bitter	0.182			↓		↓		⇑		
Mexiletine HCl	Bitter	0.463	⇓		↓	↓	↓		⇓	⇑	⇓
Pentoxifylline	Bitter	0.930	⇑		⇓						
Procainamide HCl	Bitter	0.438		↓	↓				⇑		
Propafenone HCl	Bitter	0.048		⇓			⇓	⇓			
Propranolol HCl	Bitter	0.209							⇑		
Gastrointestinal drugs											
Dicyclomine HCl	Bitter	0.0354		⇓	⇓						
Prochlorperazine	Bitter	0.103						⇑			
HIV/AIDS drugs											
Didanosine	Bitter	24		↓	⇓					⇓	
Indinavir	Bitter	0.0237					⇓	↓	⇑		⇓
Lamivudine	Bitter	4.36		⇓	⇓				↓	↓	
Ritonavir	Bitter	0.0702	↓								
Saquinavir mesylate	Bitter	0.0029	↓				⇓	⇑	⇑		
Stavudine	Bitter	5.99		⇑	⇓						
Zidovudine	Bitter	2.15		⇓					⇓		
Muscle relaxants and antiparkinsonism drugs											
Baclofen	Bitter-Metallic	3.50								↓	
Cyclobenzaprine HCl	Bitter	0.349	⇓	⇓	⇓	⇓			⇑		

40.3 Modification of Normal Taste Sensations by Drugs

Table 40.3 (*Continued*)

	Taste Quality	DT (in mM)	NaCl	KCl	CaCl$_2$	Sucrose	QHCl	Citric Acid	Capsaicin	WS-3	FeSO$_4$
Psychopharmacologic agents											
Amitriptyline HCl	Bitter	0.155	⇓	⇓	⇓	⇓	⇓	⇓	⇓	⇓	
Buspirone HCl	Bitter	0.269			↑	↓	⇓		⇑		
Clomipramine HCl	Bitter	0.122	⇓	⇓	⇓	⇓	⇓		⇓		
Desipramine HCl	Bitter	0.161	↓					⇓	⇓	⇓	⇑
Doxepin HCl	Bitter	0.143	⇓	⇓	⇓	⇓	⇓	⇓	⇓	↓	⇑
Imipramine HCl	Bitter	0.125	⇓	⇓	⇓	↓	↓	⇓	↓	⇓	
Trifluoperazine HC	Bitter	0.069	⇓	⇓	⇓		↓	⇓			⇑

Source: (Modified from Schiffman et al., 2002)

Table 40.4 Representative drugs reported to induce taste alterations: their interactions with CYP450 isozymes and the transporter P-gp, the ratio of the toxic blood-plasma levels to normal therapeutic blood-plasma levels (T/N), and the number targets, enzymes, and transporters with which each drug interacts.

	Substrates			Inhibitors			Ratio (T/N)	Number of known Pharmacokinetic Factors		
Drugs	CYP3A4	CYP2D6	P-gp	CYP3A4	CYP2D6	P-gp		Targets	Enzymes	Transporters
Anesthetics										
Cocaine	×	×		×	×		1.7	8	6	1
Lidocaine	×	×		×	×	×	1.2	4	10	2
Antihistamines										
Chlorpheniramine	×	×			×		64.7	4	4	2
Promethazine		×			×	×	5.0	10	3	1
Terfenadine	×	×		×	×	×	4.0	3	7	1
Antihypertensives, Cardiovascular, and Related Agents										
Amiodarone	×	×		×	×	×	1.3	4	9	1
Amlodipine	×			×	×	×	5.9	9	10	1
Atorvastatin	×			×	×	×		3	8	6
Clonidine		×	×				12.5	3	5	5
Digoxin		×					1.25	1	2	12
Diltiazem	×	×	×	×	×	×	6.2	1	7	1
Dipyridamole						×	2.7	4	0	3
Flecainide		×			×		1.3	2	2	0
Lovastatin	×			×	×	×		3	8	3
Mexiletine	×	×		×			1.1	2	5	0
Nifedipine	×	×	×	×	×	×	1.3	8	11	3
Nisoldipine	×			×		×		5	4	1
Propafenone	×	×			×	×	1.5	2	5	1
Propranolol	×	×	×		×	×	1.7	5	7	2
Quinidine	×		×	×	×	×	1.2	4	9	10
Simvastatin	×	×		×	×	×		2	8	3
Spironolactone						×		2	2	3
Anti-infectives										
Ciprofloxacin			×	×		×	2.9	3	4	1
Clarithromycin	×		×	×		×		1	5	2
Dapsone	×						5.0	2	13	0
Fluconazole				×		×	4.0	1	6	1
Indinavir	×		×	×	×	×	5.0	1	6	6

(*continued*)

Table 40.4 (*Continued*)

Drugs	Substrates			Inhibitors			Ratio (T/N)	Number of known pharmacokinetic factors		
	CYP3A4	CYP2D6	P-gp	CYP3A4	CYP2D6	P-gp		Targets	Enzymes	Transporters
Anti-infectives (Continued)										
Nelfinavir	×		×	×	×	×		1	8	5
Norfloxacin				×				3	6	2
Ritonavir	×	×	×	×	×	×		1	10	6
Saquinavir	×	×	×	×	×	×		1	8	7
Terbinafine	×				×			1	9	0
Anti-inflammatory, anti-pyretic, and/or analgesic agents										
Celecoxib	×				×			2	4	1
Codeine	×	×			×		2.0	3	4	0
Dexamethasone	×		×	×		×		4	17	5
Fentanyl	×			×		×	1.1	3	3	1
Hydrocortisone	×		×	×		×		2	6	3
Tramadol	×	×		×	×		3.3	9	3	0
Antineoplastic and immunosuppressant drugs										
Doxorubicin	×	×	×	×	×	×		2	4	9
Tamoxifen	×	×	×	×	×	×		2	17	3
Vincristine	×		×	×		×		2	3	8
CNS drugs/ Sympathomimetics										
Amphetamine		×			×		2.2	4	2	2
Endocrine and Diabetes Drugs										
Phenformin		×						2	1	2
Gastrointestinal drugs										
Cimetidine			×	×	×	×	7.5	1	12	12
Metoclopramide		×			×			3	4	0
Ranitidine		×	×	×	×	×		1	5	3
Psychopharmacologic Agents										
Alprazolam	×						1.3	20	4	0
Amitriptyline	×	×	×		×	×	1.7	21	9	1
Bupropion	×	×			×		12.0	3	8	0
Buspirone	×	×				×	2.0	2	4	1
Citalopram	×	×			×	×	2.0	1	6	1
Clomipramine	×	×			×	×	1.6	7	4	1
Desipramine		×		×	×	×	2.0	14	8	6
Diazepam	×		×	×			1.2	18	10	1
Donepezil	×	×					1.1	2	3	0
Doxepin		×			×	×	5.0	20	4	1
Fluoxetine	×	×		×	×	×	2.0	2	8	1
Fluvoxamine		×		×	×	×	2.2	1	10	1
Imipramine	×	×	×		×	×	1.4	14	8	5
Midazolam	×		×	×		×	4.0	19	5	2
Nortriptyline	×	×			×		2.0	12	8	0
Paroxetine		×			×	×	2.9	8	4	1
Perphenazine	×	×			×		2.1	3	8	0
Pimozide	×			×	×	×	1.1	4	7	1

40.3 Modification of Normal Taste Sensations by Drugs

Table 40.4 *(Continued)*

Drugs	Substrates			Inhibitors			Ratio (T/N)	Number of known pharmacokinetic factors		
	CYP3A4	CYP2D6	P-gp	CYP3A4	CYP2D6	P-gp		Targets	Enzymes	Transporters
Risperidone	×	×	×	×	×		6.0	14	4	1
Sertraline	×	×		×	×	×	1.2	2	9	1
Thioridazine		×			×		2.0	6	6	0
Trazodone	×	×					1.2	7	4	1
Triazolam	×						2.0	20	5	0
Trimipramine	×	×				×	2.0	11	4	1
Venlafaxine	×	×		×	×	×	2.5	3	5	1
Zolpidem	×	×					2.5	3	8	0

Sources: Preissner et al., 2010; Drugbank, 2012; Pubchem, 2012; Flockhart, 2012; Schulz et al., 2012

many drugs when co-administered with CYP or P-gp inhibitor can be elevated to toxic levels in the blood. For example, the pharmacologic agents diazepam, donepezil, fentanyl, pimozide, sertraline, and trazodone become toxic when blood-plasma levels are only 1.1 to 1.2 times greater than normal therapeutic blood levels. Drug-drug interactions that elevate blood-plasma levels of a medication beyond therapeutic concentrations play a major role in inducing taste disorders (Schiffman, 2009).

An illustration of a toxic effect caused by a drug-drug interaction with a CYP3A4 inhibitor involves the antihistamine terfenadine. In the absence of a CYP3A4 inhibitor, the parent compound terfenadine is nearly completely metabolized presystemically to fexofenadine which is the active form. However, when the parent terfenadine is co-administered with a potent CYP3A4 inhibitor, the parent compound does not undergo metabolism by CYP3A4 to fexofenadine but rather is itself absorbed. Terfenadine, a potent blocker of delayed rectifier potassium channels, can cause cardiac arrhythmias and has been implicated in approximately 125 deaths (Rampe et al., 1993; Flockhart, 1996). Furthermore, terfenadine also causes taste aberrations by a similar action (Schiffman, 2009). Terfenadine was ultimately removed from the United States, Canadian, and UK markets. A second example of a toxic effect caused by a drug-drug interaction involves the P-pg substrate digoxin that inhibits Na/K-ATPase and is used in the treatment of congestive heart failure and atrial fibrillation (Physicians Desk Reference, 2005). Over 30 years ago, Leahey et al. (1978) first reported a drug-drug interaction between digoxin and the antiarrhythmic drug quinidine that is an inhibitor of P-gp. When digoxin and quinidine were co-administered, digoxin levels were more than doubled resulting in ventricular tachycardias and one death. Many other P-gp inhibitors, such as amiodarone and spironolactone, can also elevate digoxin levels beyond therapeutic levels with adverse effects. Taste alterations frequently accompany adverse reactions involving digoxin

and may involve modulation of Na/K-ATPase in the taste system (Schiffman, 2009). Inhibition of CYP2D6 can also cause drug-drug interactions with toxic results, but the degree of the effect is determined by genetic polymorphisms in the CYP2D6 gene that renders a person an "extensive metaboliser" or a "poor metaboliser" of CYP2D6 substrates (Lee et al., 2005). The magnitude of inhibitory drug interactions is greater in extensive metabolisers of CYP2D6 than poor metabolisers.

40.3.3 Pharmacokinetic Factors that may be Causative Factors in Taste Disorders

The potential pharmacokinetic factors that contribute to adverse side effects and toxicity of drugs including taste disorders are very extensive and are in part illustrated by the three columns in Table 40. 4 below the title "Number of known pharmacokinetic factors" and subtitles labeled "Targets," "Enzymes," and "Transporters." A single drug such as the psychopharmacologic agent amitriptyline interacts with 21 targets, 9 enzymes, and 1 transporter including: the sodium-dependent noradrenaline transporter (inhibitor), sodium-dependent serotonin transporter (inhibitor), 5-hydroxytryptamine 2A receptor (antagonist), 5-hydroxytryptamine 1A receptor (pharmacological interaction unknown), delta- and kappa-type opioid receptors (agonist), high affinity nerve growth factor receptor (agonist), brain-derived neurotrophic factor (BDNF)/neurotrophin-3 growth factor receptors (agonist), alpha-1A, 1D, and 2A adrenergic receptors (antagonist), histamine H1 receptor (antagonist), muscarinic acetylcholine receptors M1, M2, M3, M4, and M5 (antagonist), potassium voltage-gated channels subfamily KQT member 2, subfamily A member 1, subfamily D member 2, and subfamily D member 3 (inhibitor), cytochrome P450 2D6 (substrate, inhibitor), cytochrome P450 1A2 (substrate), cytochrome P450 2C19 (substrate, inhibitor), cytochrome P450 2C9 (substrate), cytochrome P450 3A4 (substrate), cytochrome P450 3A5 (substrate), cytochrome P450 2B6 (substrate), cytochrome P450 2C8 (substrate, inhibitor), cytochrome P450 2E1 (substrate, inhibitor), and P-gp/multidrug resistance protein 1 (substrate, inhibitor). Furthermore, amitripyline also interacts with two carriers, serum albumin and alpha-1-acid glycoprotein. It is not yet known which one or combination of these 33 targets, enzymes, transporter, and carriers are responsible for the taste aberrations caused by amitriptyline. The representative drugs in Table 40.4 interact with over different 175 targets, enzymes, and transporters (PubChem, 2012; Drugbank, 2012) which illustrates the magnitude of the difficulty of determining the precise mechanism by which a particular drug or drug combination causes taste aberrations. Furthermore a drug can have toxic consequences that are unrelated to their pharmacokinetic effects involving targets, enzymes, and transporters. Experimental studies are needed to obtain empirical data regarding the mechanisms by which drugs alone and in combination disrupt taste perception.

40.3.4 Reductions in Drug Bioavailability Due to Induction of Drug Metabolism Exacerbate Disease States

Drugs not only serve as substrates and inhibitors but can also be inducers of drug metabolism as well. Increased expression or induction of metabolizing enzymes such as CYP3A4 and transporters such as P-gp by one drug can significantly reduce the plasma concentration of a co-administered drug that is a substrate of CYP3A4 and/or P-gp. Clinical studies have found that expression of CYP3A4 and P-gp is inducible by certain drugs, and that these drugs are frequently associated with taste complaints. Examples of drugs in Table 40.4 that are inducers of CYP3A4 under certain conditions include terfenadine, atorvastatin, nifedipine, quinidine, simvastatin, clarithromycin, ritonavir, terbinafine, dexamethasone, and tamoxifen (Pubchem, 2012). Examples of drugs that are inducers of P-gp include amiodarone, nifedipine, indinavir, nelfinavir, ritonavir, saquinavir, dexamethasone, doxorubicin, tamoxifen, vincristine, cimetidine, midazolam, and trazodone. Dexamethasone has been reported to be an inducer of CYP2D6 (Flockhart, 2012); however, CYP2D6, in general, is difficult to induce. Because induction of CYP3A4 and P-gp by one drug will lead to increased metabolism and efflux respectively of a co-administered drug, the bioavailability of the co-administered drug may be reduced below therapeutic levels. For cancer drugs as an example, induction of CYP3A4 and P-gp can reduce the bioavailability of the CYP3A4 substrate etoposide and the P-gp substrate methotrexate to subtherapeutic levels.

In this case, taste complaints may be a consequence of exacerbation of the disease state (cancer) rather than the CYP3A4 or P-gp inducing drug. Cancer is well-known to cause disturbances in taste (Epstein and Barasch, 2010).

40.3.5 Potential Side-Effects of Drugs Initiated by Interaction with Extra-Oral TAS2Rs

Another potential contributor to taste complaints is the alteration of physiological processes subsequent to interaction of drugs with extra-oral bitter taste receptors. In addition to their presence in the oral cavity, TAS2Rs are expressed in: (1) nasal and laryngeal solitary chemosensory cells (SCCs) (Finger et al., 2003; Tizzano et al., 2011), (2) the respiratory system (Deshpande et al., 2010), (3) the gastrointestinal tract and enteroendocrine STC-1 cells (Wu et al., 2002; Chen et al., 2006), (4) the brain (Singh et al., 2011) and (5) the heart (Foster et al., 2012). Interaction of bitter compounds with TAS2Rs throughout the respiratory system mediates protective airway reflexes (Tizzano et al., 2011) along with the unanticipated effect of relaxation of airway smooth muscle (Deshpande et al., 2010). In the GIT, interaction of bitter compounds with TAS2Rs can impact the motility, GIT hormone release, glucose and insulin homeostasis, and pancreatobiliary secretion (Wu et al., 2002; Dotson et al., 2008). In the brain, activation of TAS2Rs increases intracellular calcium and has been hypothesized to impact the secretion of regulatory peptides involved in the regulation of food intake (such as cholecystokinin) and other biological processes (Singh et al., 2011). The effect of interaction of bitter-tasting compounds with TAS2Rs in the heart is not yet known. Conditioned taste aversions associated with the bitter taste of drugs may in some cases be a consequence of physiological effects produced by interaction of drugs with extra-oral TAS2R receptors; taste aversions serve as an innate response designed to protect an organism against toxic bitter compounds. Conversely, desirable physiological effects produced by interaction of bitter drugs with TAS2Rs, e.g. bronchodilation in asthma, may explain learned tolerance or even preference for some bitter drugs.

40.4 TREATMENT AND PROGNOSIS

At present, there are standard no methods for the treatment of adverse taste disturbances caused by medications. If a taste disturbance is simply due to the taste of the drug during ingestion, numerous taste-masking strategies have been developed to reduce unwanted tastes (Ayenew et al., 2009; Szejtli and Szente, 2005). A frequently employed method is the addition of sweeteners and flavoring agents (Schiffman et al., 1994; Mennella and Beauchamp, 2008). Additional taste masking methods utilize amino acids, salts, microencapsulation, particulate coatings, chemical bitter blockers, and ion exchange resins (Ayenew et al., 2009; Schiffman et al., 2002; Breslin and Beauchamp, 1995; Brockhoff et al., 2011; Lavasanifar et al., 1997; Lorenzo-Lamosa et al., 1997; Ndesendo et al., 1996). Bitter receptor antagonists have recently been identified that reportedly block activation at specific TAS2Rs (Slack et al., 2010; Greene et al., 2011; Brockhoff et al., 2011). The efficacy of taste–masking strategies is influenced by genetic factors. Allelic variation at bitter taste receptors (Kim et al., 2005; Woodling et al., 2004; Drayna, 2005) that elevates bitter sensitivity or at sweet taste receptors (Fushan et al. 2009, 2010) that reduces sweet sensitivity can influence the likelihood of sensory complaints from drugs and affect medication choices (Lipchock et al., 2012; Mennella et al., 2012).

In most (but not all) cases, a taste disorder will resolve within a period of 4 to 10 months after removal of the offending drug or drug combination (Schiffman, 1983; Deems et al., 1996). Taste disorders typically persist for prolonged periods after an offending drug (or drug combination) is withdrawn for numerous reasons. First, the levels and functioning of drug targets, enzymes, and transporters do not return to baseline immediately after cessation of medication use; pharmacokinetic parameters require a period of recovery to return to non-medicated levels after drug withdrawal. Second, most drugs are lipophilic, and can potentially bioaccumulate in tissues. For example, amiodarone, which is deposited and stored in fat, has an elimination half-life of 47 days (Schiffman, 2009). This prolonged half-life is due to the slow release of amiodarone from lipid-rich tissues. The half-life of 47 days specifies the period of time required for the drug level of amiodarone to fall to 50%; at the end of 7 half-lives (329 days), over 99% of amiodarone will be eliminated from the body. Third, some cases of taste loss are due to impaired turnover and replacement of taste cells. Taste cells undergo continuous turnover with a half-life of approximately 10 days (Beidler and Smith, 1991), and disruption of turnover, for example, by chemotherapeutic agents including vinblastine, bleomycin, cisplatin, and methotrexate, are associated with taste loss and alterations.

It is likely that the incidence and prevalence of taste disorders will grow in the coming decades as the number of persons (particularly the elderly) being treated with multiple medications continues to increase globally. Thus, development of effective treatments for taste disorders is necessary to reduce the probability of therapeutic failure due to noncompliance. To achieve this end, experimental studies are required to determine the physiological and pharmacological parameters that are responsible for taste

disturbances caused by specific drugs and/or drug combinations. This knowledge can then be employed in the rational development of treatments for drug-related taste disorders.

REFERENCES

Ackerman, B. H. and Kasbekar N. (1997). Disturbances of taste and smell induced by drugs. *Pharmacotherapy* 17: 482–496.

Adler, E., Hoon, M. A., Mueller, K. L., et al. (2000). A novel family of mammalian taste receptors. *Cell* 100: 693–702.

al-Shammari, S. A., Khoja, T., and al-Yamani, M. J. (1995). Compliance with short-term antibiotic therapy among patients attending primary health centres in Riyadh, Saudi Arabia. *J. Roy. Soc. Health* 115: 231–234.

Ayenew, Z., Puri, V., Kumar, L., and Bansal, A. K. (2009). Trends in pharmaceutical taste masking technologies: A patent review. *Recent Pat. Drug Deliv. Formul.* 3: 26–39.

Barbara, L., Corinaldesi, R., Rea, E., et al. (1986). The role of colloidal bismuth subcitrate in the short-term treatment of duodenal ulcer. *Scand. J. Gastroenterol. Suppl.* 122: 30–34.

Bartholow, M. (2011). Top 200 Drugs of 2010. Published Online: Monday, May 16, 2011. http://www.pharmacytimes.com/print.php?url=/publications/issue/2011/May2011/Top-200-Drugs-of-2010

Beidler, L. M. and Smith, J. C. (1991). Effects of radiation therapy and drugs on cell turnover and taste. In *Smell and Taste in Health and Disease*, T. V. Getchell, R. L. Doty, L. M. Bartoshuk, and J. B. Snow Jr., (Eds.), Raven Press, New York, pp. 753–763.

Breslin, P. A. and Beauchamp, G. K. (1995). Suppression of bitterness by sodium: variation among bitter taste stimuli. *Chem. Senses* 20: 609–623.

Breymann, C., Visca, E., Huch, R., and Huch, A. (2001). Efficacy and safety of intravenously administered iron sucrose with and without adjuvant recombinant human erythropoietin for the treatment of resistant iron-deficiency anemia during pregnancy. *Am. J. Obstet. Gynecol.* 184: 662–667.

Brockhoff, A., Behrens, M., Niv, M. Y., and Meyerhof, W. (2010). Structural requirements of bitter taste receptor activation. *Proc. Natl. Acad. Sci. U S A* 107: 11110–11115.

Brockhoff, A., Behrens, M., Roudnitzky, N., et al. (2011). Receptor agonism and antagonism of dietary bitter compounds. *J. Neurosci.* 31: 14775–14782.

Callahan, A. M., Frye, M. A., Marangell, L. B., et al. (1997). Comparative antidepressant effects of intravenous and intrathecal thyrotropin-releasing hormone: confounding effects of tolerance and implications for therapeutics. *Biol. Psychiatry* 41: 264–272.

Calò, C., Padiglia, A., Zonza, A., et al. (2011). Polymorphisms in TAS2R38 and the taste bud trophic factor, gustin gene co-operate in modulating PROP taste phenotype. *Physiol. Behav.* 104: 1065–1071.

Chen, J., Korostyshevsky, D., Lee, S., and Perlstein, E. O (2012). Accumulation of an antidepressant in vesiculogenic membranes of yeast cells triggers autophagy. *PLoS One* 7(4): e34024. doi:10.1371/journal.pone.0034024

Chen, M. C., Wu, S. V., Reeve, J. R. Jr., and Rozengurt, E. (2006). Bitter stimuli induce Ca^{2+} signaling and CCK release in enteroendocrine STC-1 cells: role of L-type voltage-sensitive Ca^{2+} channels. *Am. J. Physiol. Cell Physiol.* 291: C726–C739.

Corsonello, A., Pedone, C., and Incalzi, R. A. 2010. Age-related pharmacokinetic and pharmacodynamic changes and related risk of adverse drug reactions. *Curr. Med. Chem.* 17: 571–584.

Dantzig, A. H., Shepard, R. L., Law, K. L., et al. (1999). Selectivity of the multidrug resistance modulator, LY335979, for P-glycoprotein and effect on cytochrome P-450 activities. *J. Pharmacol. Exp. Ther.* 290: 854–862.

Deems, D. A., Yen, D. M., Kreshak, A., and Doty, R. L. (1996). Spontaneous resolution of dysgeusia. *Arch. Otolaryngol. Head Neck Surg.* 122: 961–963.

Deshpande, D. A., Wang, W. C., McIlmoyle, E. L., et al. (2010). Bitter taste receptors on airway smooth muscle bronchodilate by localized calcium signaling and reverse obstruction. *Nat. Med.* 16: 1299–1304.

Dotson, C. D., Zhang, L., Xu, H., et al. (2008). Bitter taste receptors influence glucose homeostasis. *PLoS One* 3(12): e3974.

Doty, R. L., and Haxel, B. R. (2005). Objective assessment of terbinafine-induced taste loss. *Laryngoscope* 115: 2035–2037.

Doty, R. L., Shah, M., and Bromley, S. M. (2008). Drug-induced taste disorders. *Drug Saf.* 31: 199–215.

Doty, R. L., Treem, J., Tourbier, I., and Mirza, N. 2009. A double-blind study of the influences of eszopiclone on dysgeusia and taste function. *Pharmacol. Biochem. Behav.* 94: 312–318.

Drayna, D. (2005). Human taste genetics. *Annu. Rev. Genomics Hum. Genet.* 6: 217–235.

Drewnowski, A., and Gomez-Carneros, C. (2000). Bitter taste, phytonutrients, and the consumer: a review. *Am. J. Clin. Nutr.* 72: 1424–1435.

Drugbank. (2012). www.drugbank.ca.

Epstein, J. B., and Barasch, A. (2010). Taste disorders in cancer patients: pathogenesis, and approach to assessment and management. *Oral Oncol.* 46: 77–81.

Ercetin, S., Haznedaroglu, I. C., Kurt, M., et al. (2010). Safety and efficacy of Ankaferd blood stopper in dental surgery. *Int. J. Hematol. Oncol.* 20: 1–5.

Felmlee, M. A., Lon, H. K., Gonzalez, F. J., and Yu, A. M. (2008). Cytochrome P450 expression and regulation in CYP3A4/CYP2D6 double transgenic humanized mice. *Drug Metab. Dispos.* 36: 435–441.

Finger, T. E., Böttger, B., Hansen, A., et al. (2003). Solitary chemoreceptor cells in the nasal cavity serve as sentinels of respiration. *Proc. Natl. Acad. Sci. U S A* 100: 8981–8986.

Fishberg, A. M., Hitzig, W. M., King, F. H. (1933). Measurement of the circulation time with saccharin. *Proc. Soc. Exp. Biol. Med.* 30: 651–652.

Flockhart, D. A. (1996). Drug interactions, cardiac toxicity, and terfenadine: from bench to clinic? *J. Clin. Psychopharmacol.* 16: 101–103.

Flockhart, D. A. (2012). P450 drug interaction table. http://medicine.iupui.edu/clinpharm/ddis/table.aspx

Foster, S. R., Blank, K., Purdue, B., et al. (2012). Screening for potential ligands of cardiac-expressed rat taste receptors. 2012 International Postgraduate Symposium in Biomedical Sciences, September 24–26, 2012, University of Queensland, Australia, p. 68.

Fushan, A. A., Simons, C. T., Slack, J. P., and Drayna, D. (2010). Association between common variation in genes encoding sweet taste signaling components and human sucrose perception. *Chem. Senses* 35: 579–592.

Fushan, A. A., Simons, C. T., Slack, J. P., et al. (2009). Allelic polymorphism within the TAS1R3 promoter is associated with human taste sensitivity to sucrose. *Curr. Biol.* 19: 1288–1293.

Gao, N., Lu, M., Echeverri, F., et al. (2009). Voltage-gated sodium channels in taste bud cells. *BMC Neurosci.* 10: 20.

Gorard, D. A. (2006). Escalating polypharmacy. *QJM* 99: 797–800.

Greene, T. A., Alarcon, S., Thomas, A., et al. (2011). Probenecid inhibits the human bitter taste receptor TAS2R16 and suppresses bitter perception of salicin. *PLoS One* 6(5): e20123.

Hajjar, E. R., Cafiero, A. C., and Hanlon, J. T. (2007). Polypharmacy in elderly patients. *Am. J. Geriatr. Pharmacother.* 5: 345–351.

Hamada, N., Endo, S., and Tomita, H. (2002). Characteristics of 2278 patients visiting the Nihon University Hospital Taste Clinic over a 10-year period with special reference to age and sex distributions. *Acta Otolaryngol. Suppl.* (546): 7–15.

Hamor, G. H., and Lafdjian, A. (1967). Dualistic thiourea moiety taste response of methimazole. *J. Pharm. Sci.* 56: 777–778.

Harrison, C. J. (1995). Rational selection of antimicrobials for pediatric upper respiratory infections. *Pediatr. Infect. Dis. J.* 14: S121–S129.

Holland, V. F., Zampighi, G. A., and Simon, S. A. (1991). Tight junctions in taste buds: possible role in perception of intravascular gustatory stimuli. *Chem. Senses* 16: 69–79.

Hovstadius, B., Hovstadius, K., Astrand, B., and Petersson, G. (2010). Increasing polypharmacy - an individual-based study of the Swedish population 2005–2008. *BMC Clin. Pharmacol.* 10: 16.

Huang, S. M., Strong, J. M., Zhang, L., et al. (2008). New era in drug interaction evaluation: US Food and Drug Administration update on CYP enzymes, transporters, and the guidance process. *J. Clin. Pharmacol.* 48: 662–670.

Ingelman-Sundberg, M. (2005). Genetic polymorphisms of cytochrome P450 2D6 (CYP2D6): clinical consequences, evolutionary aspects and functional diversity. *Pharmacogenomics J.* 5: 6–13.

Iwai, N. (1997). Drug compliance of children and infants with oral antibiotics for pediatric use. *Acta Paediatr. Jpn.* 39: 132–142.

Johnell, K., and Klarin, I. (2007). The relationship between number of drugs and potential drug-drug interactions in the elderly: a study of over 600,000 elderly patients from the Swedish Prescribed Drug Register. *Drug Saf.* 30: 911–918.

Kaufman, E., and Lamster, I. B. (2002). The diagnostic applications of saliva–a review. *Crit. Rev. Oral Biol. Med.* 13: 197–212.

Ketterer, C., Müssig, K., Heni, M., et al. (2011). Genetic variation within the TRPM5 locus associates with prediabetic phenotypes in subjects at increased risk for type 2 diabetes. *Metabolism* 60: 1325–1333.

Kim, U., Wooding, S., Ricci, D., et al. (2005). Worldwide haplotype diversity and coding sequence variation at human bitter taste receptor loci. *Hum. Mutat.* 26: 199–204.

Lavasanifar, A., Ghalandari, R., Ataei, Z., et al. (1997). Microencapsulation of theophylline using ethylcellulose: in vitro drug release and kinetic modelling. *J. Microencapsul.* 14: 91–100.

Leahey, E. B. Jr.,, Reiffel, J. A., Drusin, R. E., et al. (1978). Interaction between quinidine and digoxin. *JAMA* 240: 533–534.

Lee, L. S., Nafziger, A. N., and Bertino, J. S. Jr., (2005). Evaluation of inhibitory drug interactions during drug development: genetic polymorphisms must be considered. *Clin. Pharmacol. Ther.* 78: 1–6.

Lewis, I. K., Hanlon, J. T., Hobbins, M. J., and Beck, J. D. (1993). Use of medications with potential oral adverse drug reactions in community-dwelling elderly. *Spec. Care Dentist.* 13: 171–176.

Liacouras, C. A., Coates, P. M., Gallagher, P. R., and Cortner, J. A. (1993). Use of cholestyramine in the treatment of children with familial combined hyperlipidemia. *J. Pediatr.* 122: 477–482.

Lindemann, B. (2001). Receptors and transduction in taste. *Nature* 413: 219–225.

Lipchock, S. V., Reed, D. R., and Mennella, J. A. (2012). Relationship between bitter-taste receptor genotype and solid medication formulation usage among young children: a retrospective analysis. *Clin. Ther.* 34(3): 728–733.

Lorenzo-Lamosa, M. L., Cuña, M., Vila-Jato, J. L., et al. (1997). Development of a microencapsulated form of cefuroxime axetil using pH-sensitive acrylic polymers. *J. Microencapsul.* 14: 607–616.

Macdougall, I. C., and Roche, A. (2005). Administration of intravenous iron sucrose as a 2-minute push to CKD patients: a prospective evaluation of 2,297 injections. *Am. J. Kidney Dis.* 46: 283–289.

Margolskee, R. F. (2002). Molecular mechanisms of bitter and sweet taste transduction. *J. Biol. Chem.* 277: 1–4.

Matsui, D., Barron, A., and Rieder, M. J. (1996). Assessment of the palatability of antistaphylococcal antibiotics in pediatric volunteers. *Ann. Pharmacother.* 30: 586–588.

McCartney, C. J., Murphy, D. B., Iagounova, A., and Chan, V. W. (2003). Intravenous ropivacaine bolus is a reliable marker of intravascular injection in premedicated healthy volunteers. *Can. J. Anaesth.* 50: 795–800.

McDonald, T. J., Tu, E., Brenner, S., Zabel, P., et al. (1994). Canine, human, and rat plasma insulin responses to galanin administration: species response differences. *Am. J. Physiol.* 266(4 Pt 1): E612–617.

Mennella, J. A., and Beauchamp, G. K. (2008). Optimizing oral medications for children. *Clin. Ther.* 30: 2120–2132.

Mennella, J. A., Finkbeiner, S., and Reed, D. R. (2012). The proof is in the pudding: children prefer lower fat but higher sugar than do mothers. *Int. J. Obes (Lond).* 36: 1285–1291.

Meyerhof, W., Batram, C., Kuhn, C., et al. (2010). The molecular receptive ranges of human TAS2R bitter taste receptors. *Chem. Senses* 35: 157–70.

Meyerhof, W., Born, S., Brockhoff, A., and Behrens, M. (2011). Molecular biology of mammalian bitter taste receptors. A review. *Flavour Fragr J.* 26, 260–268.

Micromedex®. 2012. http://www.thomsonhc.com. MICROMEDEX® 2.0, Truven Health Analytics Inc.

Mierson, S., Olson, M. M., and Tietz, A. E. 1996. Basolateral amiloride-sensitive Na$^+$ transport pathway in rat tongue epithelium. *J. Neurophysiol.* 76: 1297–1309.

Naim, M., Shaul, M. E., Spielman, A. I., et al. (2008). Permeation of amphipathic sweeteners into taste-bud cells and their interactions with post-receptor signaling components: Possible implications for sweet-taste quality. In *Sweetness and sweeteners: Biology, chemistry and psychophysics*, D. K. Weerasinghe and G. E. Dubois (Eds.), USA, Oxford University Press, pp. 241–255.

Ndesendo, V. M., Meixner, W., Korsatko, W., and Korsatko-Wabnegg, B. (1996). Microencapsulation of chloroquine diphosphate by Eudragit RS100. *J. Microencapsul.* 13: 1–8.

Owen, M. D., Gautier, P., and Hood, D. D. (2004). Can ropivacaine and levobupivacaine be used as test doses during regional anesthesia? *Anesthesiology* 100: 922–925.

Padiglia, A., Zonza, A., Atzori, E., et al. (2010). Sensitivity to 6-n-propylthiouracil is associated with gustin (carbonic anhydrase VI) gene polymorphism, salivary zinc, and body mass index in humans. *Am. J. Clin. Nutr.* 92: 539–545.

Phillips, W. S., and Lightman, S. L. (1981). Is cutaneous flushing prostaglandin mediated? *Lancet* 1(8223): 754–756.

Physicians Desk Reference. (2005). *59th ed.* Thomson PDR. Montvale, NJ.

Preissner, S., Kroll, K., Dunkel, M., et al. (2010). SuperCYP: a comprehensive database on Cytochrome P450 enzymes including a tool for analysis of CYP-drug interactions. *Nucleic Acids Res* 38 (Database issue): D237–D243. http://bioinformatics.charite.de/supercyp/index.php?site=drug_search

PubChem. (2012). http://pubchem.ncbi.nlm.nih.gov/

Qato, D. M., Alexander, G. C., Conti, R. M., et al. (2008). Use of prescription and over-the-counter medications and dietary supplements among older adults in the United States. *JAMA* 300: 2867–2878.

Rampe, D., Wible, B., Brown, A. M., and Dage, R. C. (1993). Effects of terfenadine and its metabolites on a delayed rectifier K$^+$ channel cloned from human heart. *Mol. Pharmacol.* 44: 1240–1245.

Reed, D. R., Nanthakumar, E., North, M., et al. (1999). Localization of a gene for bitter-taste perception to human chromosome 5p15. *Am. J. Hum. Genet.* 64: 1478–1480.

Riera, C. E., Vogel, H., Simon S. A.,et al. (2009). Sensory attributes of complex tasting divalent salts are mediated by TRPM5 and TRPV1 channels. *J. Neurosci.* 29: 2654–2662.

Riera, C. E., Vogel, H., Simon, S. A., and le Coutre J. (2007). Artificial sweeteners and salts producing a metallic taste sensation activate TRPV1 receptors. *Am. J. Physiol. Regul. Integr. Comp. Physiol.* 293: R626–R234.

Rumble, R. H., and Morgan, K. (1994). Longitudinal trends in prescribing for elderly patients: two surveys four years apart. *Br. J. Gen. Pract.* 44: 571–575.

Sandell, M., and Breslin, P. (2008). hTAS2R38 receptor genotypes predict sensitivity to bitterness of thiourea compounds in solution and in selected vegetables. In *Expression of Multidisciplinary Flavour Science. Proceedings of the 12th Weurman Symposium.* I. Blank, M. Wüst, and C. Yeretzian (Eds.), Institut of Chemistry and Biological Chemistry, Zürich University of Applied Science. Wädenswil, Switzerland.

Schiffman, S. S. (1983). Taste and smell in disease. *N. Engl. J. Med.* 308: 1275–1279, 1337–1343.

Schiffman, S. S. (1991). Drugs influencing taste and smell perception. In *Smell and taste in health and disease*, T. V. Getchell, R. L. Doty, L. M. Bartoshuk, and J. B. Snow (Eds.), Raven Press, New York, pp. 845–850.

Schiffman, S. S. (1997). Taste and smell losses in normal aging and disease. *JAMA* 278: 1357–1362.

Schiffman, S. S. (2000). Taste quality and neural coding: Implications from psychophysics and neurophysiology. *Physiol. Behav.* 69: 147–159.

Schiffman, S. S. (2007). Critical illness and changes in sensory perception. *Proc. Nutr. Soc.* 66: 331–345.

Schiffman, S. S. (2009). Effects of aging on the human taste system. *Ann. N. Y. Acad. Sci.* 1170: 725–729.

Schiffman, S. S., and Zervakis, J. (2002). Taste and smell perception in the elderly: effect of medications and disease. *Adv. Food Nutr. Res.* 44: 247–346.

Schiffman, S. S., Diaz, C., and Beeker, T. G. (1986a). Caffeine intensifies taste of certain sweeteners: role of adenosine receptor. *Pharmacol. Biochem. Behav.* 24: 429–432.

Schiffman, S. S., Gatlin, L. A., Sattely-Miller, E. A., et al. (1994). The effect of sweeteners on bitter taste in young and elderly subjects. *Brain Res. Bull.* 35: 189–204.

Schiffman, S. S., Gill, J. M., and Diaz, C. (1985). Methyl xanthines enhance taste: evidence for modulation of taste by adenosine receptor. *Pharmacol. Biochem. Behav.* 22: 195–203.

Schiffman, S. S., Lockhead, E., and Maes, F. W. (1983). Amiloride reduces the taste intensity of Na$^+$ and Li$^+$ salts and sweeteners. *Proc. Natl. Acad. Sci. U. S. A.* 80: 6136–6140.

Schiffman, S. S., Reilly, D. A., and Clark, T. B. (1979). Qualitative differences among sweeteners. *Physiol. Behav.* 23: 1–9.

Schiffman, S. S., Rogers, M. O., and Zervakis, J. (2003). Loss of taste, smell, and other senses with age: effects of medication. In *Handbook of Clinical Nutrition and Aging*. C. W. Bales, and C. S. Ritchie. Totowa, NJ: Humana Press, pp. 211–289.

Schiffman, S. S., Sennewald, K., and Gagnon, J. (1981). Comparison of taste qualities and thresholds of D- and L-amino acids. *Physiol. Behav.* 27: 51–59.

Schiffman, S. S., Simon, S. A., Gill, J. M., and Beeker, T. G. (1986b). Bretylium tosylate enhances salt taste. *Physiol. Behav.* 36: 1129–1137.

Schiffman, S. S., Zervakis, J., Graham, B. G., and Westhall, H. L. (2002). Age-related chemosensory losses: effect of medications. In *Chemistry of Taste*, P. Givens, and D. Paredes (Eds.). American Chemical Society, Washington, DC, pp. 94–108.

Schulz, M., Iwersen-Bergmann, S., Andresen, H., and Schmoldt, A. (2012). Therapeutic and toxic blood concentrations of nearly 1,000 drugs and other xenobiotics. *Crit. Care.* 16: R136.

Shi, S., and Klotz, U. (2011). Age-related changes in pharmacokinetics. *Curr. Drug Metab.* 12: 601–610.

Shinkai, R. S., Hatch, J. P., Schmidt, C. B., and Sartori, E. A. (2006). Exposure to the oral side effects of medication in a community-based sample. *Spec. Care Dentist.* 26: 116–120.

Singh, N., Vrontakis, M., Parkinson, F., and Chelikani, P. (2011). Functional bitter taste receptors are expressed in brain cells. *Biochem. Biophys. Res. Commun.* 406: 146–151.

Slack, J. P., Brockhoff, A., Batram, C., et al. (2010). Modulation of bitter taste perception by a small molecule hTAS2R antagonist. *Curr. Biol.* 20: 1104–1109.

Szejtli, J., and Szente, L. (2005). Elimination of bitter, disgusting tastes of drugs and foods by cyclodextrins. *Eur. J. Pharm. Biopharm.* 61: 115–125.

Tarr, L., Oppenheimer, B. S., and Sager, R. V. (1933). The circulation time in various clinical conditions determined by the use of sodium dehydrocholate. *Am. Heart J.* 8: 766–786.

Tizzano, M., Cristofoletti, M., Sbarbati, A., and Finger, T. E. (2011). Expression of taste receptors in solitary chemosensory cells of rodent airways. *BMC Pulm. Med.* 11: 3.

Tomita, H., and Yoshikawa, T. (2002). Drug-related taste disturbances. *Acta Otolaryngol. Suppl.* (546): 116–121.

van Herwaarden, A. E., van Waterschoot, R. A., and Schinkel, A. H. (2009). How important is intestinal cytochrome P450 3A metabolism? *Trends Pharmacol. Sci.* 30: 223–227.

Veldhuis, J. D., Iranmanesh, A., Mielke, K., et al. (2006). Ghrelin potentiates growth hormone secretion driven by putative somatostatin withdrawal and resists inhibition by human corticotropin-releasing hormone. *J. Clin. Endocrinol. Metab.* 91: 2441–2446.

Weiss, S., Robb, G. P., and Blumgart, H. L. (1929). The velocity of blood flow in health and disease as measured by the effect of histamine on the minute vessels. *Am. Heart J.* 4: 664–691.

Wooding, S., Kim, U. K., Bamshad, M. J., et al. (2004). Natural selection and molecular evolution in PTC, a bitter-taste receptor gene. *Am. J. Hum. Genet.* 74: 637–646.

Wu, S. V., Rozengurt, N., Yang, M., et al. (2002). Expression of bitter taste receptors of the T2R family in the gastrointestinal tract and enteroendocrine STC-1 cells. *Proc. Natl. Acad. Sci. U. S. A.* 99: 2392–2397.

Xu, H., Tian, W., Fu, Y., et al. (2007). Functional effects of nonsynonymous polymorphisms in the human TRPV1 gene. *Am. J. Physiol. Renal Physiol.* 293: F1865–F1876.

Part 9

Taste in Nonhuman Species

Chapter 41

Taste Processing in Insects

JOHN I. GLENDINNING

41.1 INTRODUCTION

The sense of taste is essential for the survival of insects. It helps them evaluate the chemical composition of foods (White and Chapman, 1990; Schoonhoven et al., 1992; Glendinning, 2002), select host plants for oviposition (Ma and Schoonhoven, 1973; Städler et al., 1995) and self-medicate themselves against endoparasite infections (Bernays and Singer, 2005). Taste also facilitates digestion by stimulating salivation (Watanabe and Mizunami, 2006) and determining whether ingested food is directed to the mid-gut for digestion or to the crop for storage (Schmidt and Friend, 1991). Below, I review five topics: (a) organization of the taste system, (b) peripheral taste function, (c) taste-mixture interactions, (d) taste processing in the central nervous system, and (d) factors that modulate taste responses. Whenever possible, I highlight significant gaps in our understanding of insect taste. I also emphasize commonalities between the taste systems of insects and mammals.

Before I begin, two caveats are in order. First, the class Insecta contains about 900,000 described species, many of which eat different types of foods. At this point, investigators have studied the taste systems of only a fraction of these species, and have a detailed knowledge of only one species (the fruit fly, *Drosophila melanogaster*). Thus, any generalizations about insect taste must be tempered by our rather myopic view. Second, I cannot possibly cover all of the new information on insect taste in this review. Thus, I have decided to focus on how taste information is processed both peripherally and centrally. For a more detailed review of the molecular and cellular mechanisms of insect taste, I refer the reader to several excellent reviews (Montell, 2009; Isono and Morita, 2010; Kwon et al., 2011).

41.2 ORGANIZATION OF THE TASTE SYSTEM

In contrast to vertebrates, insects can sense foods without actually ingesting them. This was established in the 1920s by Dwight Minnich, who elicited an appetitive response (i.e., extension of the feeding proboscis) in flies and butterflies simply by immersing their "feet" (i.e., tarsi) in a sugar solution (Minnich, 1921; Minnich, 1929). Several years later, Minnich identified the peripheral taste structures that mediated this appetitive response: small hair-like structures protruding from the tarsi (henceforth, taste sensilla). He found that stimulation of a single taste sensillum with sugar water was sufficient to elicit proboscis extension in flies (Minnich, 1931). Remarkably, the significance of this observation–that is, the discovery of the primary organ of taste in insects – was not appreciated until over two decades later (Grabowski and Dethier, 1954; Lewis, 1954).

Subsequent research established that taste sensilla are uniporous structures (Figure 41.1a) that are innervated by the single, unbranched dendrites of 2–6 gustatory receptor neurons (GRNs) (Chapman, 1982). The dendrites terminate near the distal tip of the sensillum (Figure 41.1b). The cell body of each GRN occurs at the base of the taste sensillum, and the axon extends directly to the central nervous system (Dethier, 1976). Because the GRNs within a taste sensillum lie in close proximity to one another (Figure 41.1c), there is the opportunity for activity in one GRN to influence activity in another (see *Taste-Mixture Interactions* below). Taste sensilla are also innervated by a fifth mechanosensory neuron, which provides proprioceptive feedback about when the sensillum is contacting foods (Figure 41.1c).

In some insects, the taste sensilla are restricted to external appendages associated with the oral cavity

---

Handbook of Olfaction and Gustation, Third Edition. Edited by Richard L. Doty.
© 2015 Richard L. Doty. Published 2015 by John Wiley & Sons, Inc.

930　　　　　　　　　　　　　　Chapter 41　Taste Processing in Insects

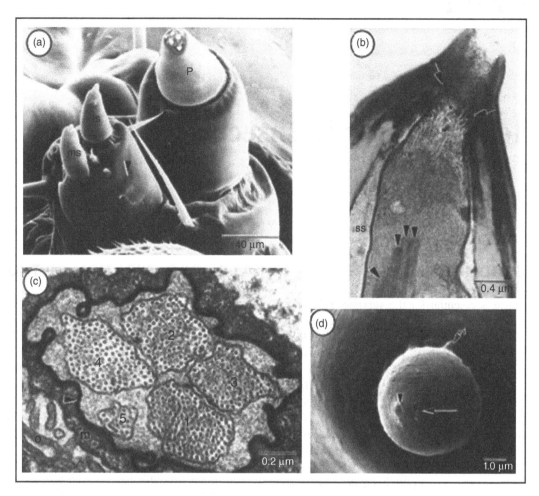

Figure 41.1　SEM micrographs of a taste sensillum from a caterpillar (*Mamestra configurata*, Lepidoptera: Noctuidae). (a) Left maxilla of caterpillar with lateral (*ls*) and medial (*ms*) styloconic sensilla, and a maxillary palp (*p*). (b) Longitudinal section of a lateral styloconic sensillum. At the top of the image, one can see the tip of the pore, which contains a plug of fenestrated fibirils (asterisk), and the dendritic sheath (arrows), which is fused with the cuticular wall of the sensillum. Four distal dendrites, each from a different GRNs (arrowheads), can be seen near the bottom of the image; they are enclosed by the dendritic sheath. *cu*, cuticle; *ss* sensillar sinus. (c) Proximal cross-section of a medial styloconic sensillum, showing GRNs (*1–4*) and a mechanosensory neuron (*5*). All GRNs are enclosed by a dendritic sheath (arrowhead), which is surrounded by a densely granulated intermediate sheath cell (m). *o* outer sheath cell. (d) Apical view of the tip of a lateral styloconic sensillum with the terminal pore (arrowhead). These micrographs are from Shields (1994).

(e.g., see Figure 41.2a), whereas in other insects they are distributed across the body surface (Chapman, 1982). In flies, for example, taste sensilla occur on the proboscis, wing margins, feet and distal tip of the abdomen (Vosshall and Stocker, 2007). The total number of taste sensilla per insect varies greatly across species (i.e., from 30 to over 3000) (Chapman, 1982). Little is known, however, about the functional significance of this variation – for example, whether large numbers of sensilla confer greater sensitivity or discriminatory abilities.

For taste transduction to occur, a chemical stimulus must first diffuse through a mucopolysaccharide substance in the pore at the tip of a taste sensillum and then dissolve into the fluid (or receptor lymph) surrounding the dendrites (Figure 41.1b) (Shields, 1996). Once the chemical stimulus encounters the distal dendritic membrane of a GRN, it is thought to interact with cell-surface receptors and activate transduction mechanisms that lead to the generation of action potentials (Dethier, 1976; Morita, 1992). The nature of the cell-surface gustatory receptors (Grs) was not known until 2000, when Peter Clyne, Coral Warr and John Carlson discovered a family of 19 highly divergent Grs in *Drosophila* (Clyne et al., 2000). Subsequent studies expanded the number of *Drosophila* Grs to 68 (see review in Isono and Morita, 2010) and examined the pattern of expression of the Grs, both within specific GRNs and across the entire population of taste sensilla (see below for details). Although Grs have been discovered in other insect

41.2 Organization of the Taste System

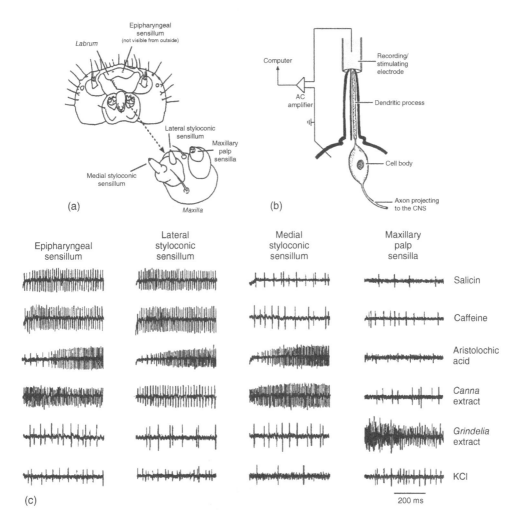

Figure 41.2 (a) Cartoon of the head of a *Manduca sexta* caterpillar, as viewed from below. An enlargement of the maxilla (indicated with an arrow) is provided to clarify the location of the medial and lateral styloconic sensilla. The epipharyngeal sensilla are located underneath the labrum, and thus are not visible in this diagram. This cartoon was adapted from Bernays and Chapman (1994; their Fig. 3.4). (b) Illustration of the tip recording method, which was used to record excitatory responses of individual GRNs located within a taste sensillum. During a tip recording (Hodgson et al., 1955), the tip of a taste sensillum is inserted into the end of a glass recording/stimulating electrode, which is filled with a taste stimulus dissolved in an electrolyte solution (0.1 M KCl in deionized water). The taste stimulus solution diffuses through a pore in the tip of the sensillum and activates transduction mechanism(s) on the distal end of a GRN's dendritic process; the electrode detects the ensuing action potentials. For clarity, only one GRN is indicated. Note that the GRN's axonal process projects directly to the central nervous system without synapsing. (c) Typical excitatory responses of the four different classes of taste sensilla to five aversive taste stimuli: salicin, caffeine, aristolochic acid, *Canna* extract, and *Grindelia* extract (see Glendinning et al., 2002, for exact concentrations of each taste stimulus). The onset of stimulation occurred at the beginning of each trace, and each vertical line reflects the occurrence of an action potential. In most of the traces containing a large number of action potentials, there is a single bitter-sensitive GRN firing regularly, and another GRN (the salt-sensitive taste cell) firing sporadically. The only exception is the multi-unit response of the maxillary palp sensilla to the *Grindelia* extract, which probably contains action potentials from several bitter-sensitive GRNs. Because all aversive taste stimuli were presented in a 0.1 M KCl solution, I also show representative responses of each sensillum to the electrolyte solution alone for comparison (see bottom row of traces). Note that each aversive taste stimulus selectively activates bitter-sensitive GRNs in only a subset of the taste sensilla (e.g., the *Grindelia* extract only stimulated bitter-sensitive GRNs in the maxillary palp). These traces are from Glendinning et al. (2002).

taxa – Lepidoptera (Wanner and Robertson, 2008; Howlett et al., 2012), Hymenoptera (Kent and Roberston, 2009) and Coleoptera (Kent and Roberston, 2009) – progress has been comparatively slow, owing to a paucity of genomic information and molecular tools.

Insect Grs were originally hypothesized to function as G protein-coupled receptors (GPCRs), based on amino acid sequences specifying seven transmembrane domains, a feature typical of GPCRs (Clyne et al., 2000). However, functional studies have produced equivocal support for this hypothesis. On the one hand, there are reports that pharmacological (or genetic) manipulation of metabotropic signaling pathways (i.e., G protein subunits, cAMP, cGMP, calmodulin, protein kinase C, nitric oxide and inositol 1,4,5-trisphosphate) alters taste-mediated responses of insects to chemical stimuli (Amakawa and Ozaki, 1989; Amakawa et al., 1990; Ishimoto et al., 2005; Nakamura et al., 2005; Seno et al., 2005a, 2005b; Kan et al., 2008; Ouyang et al., 2009; Bredendiek et al., 2011; Kan et al., 2011; Vermehren-Schmaedick et al., 2011). On the other hand, there are reports that sugar receptors are ligand-gated ion channels in flies (Murakami and Kijima, 2000) and moths (Sato et al., 2011). At this point, more work is needed to resolve these contradictory findings. It is notable that in insect olfactory receptor neurons (ORNs), odor stimuli are thought to be transduced primarily by ionotropic mechanisms; the metabotropic mechanisms are thought to serve a modulatory role (Silbering and Benton, 2010). This may be the case for insect GRNs as well.

Recent molecular studies in adult *Drosophila* indicate that Grs function as part of heteromeric complexes. For instance, mutation of *Gr66a*, *Gr93a*, or *Gr33a* attenuates responsiveness to caffeine, indicating that the Grs encoded by these genes are necessary for normal caffeine detection (Lee et al., 2009). However, misexpression of two or all three of these Grs in a sugar-sensitive GRN fails to confer caffeine responsiveness, indicating that functional "bitter" receptors consist of four or more Grs (Moon et al., 2009).

Most GRNs terminate in the subesophageal ganglion (SOG; Figure 41.3a) and tritocerebrum; these sites are widely considered to be the primary taste processing centers (Kent and Hildebrand, 1987; Mitchell and Itagaki, 1992; Stocker, 1994; Thorne et al., 2004; Wang et al., 2004; Ignell and Hansson, 2005; Colomb et al., 2007). In some species, GRNs from tarsal taste sensilla terminate in the metathoracic ganglia (Rogers and Newland, 2002). Based on three lines of evidence, Input from GRNs is thought to modulate oromotor feeding circuits in the SOG/tritocerebrum: (a) GRN axons intermingle with (or occur in close apposition to) arborizations from the mouthpart motor neurons (Altman and Kien, 1987); (b) some SOG interneurons have branches that occur in

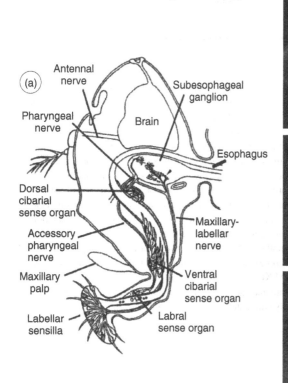
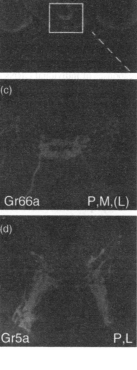

Figure 41.3 (a) Ultrastructure of the head, feeding apparatus and central nervous system of an adult fruit fly *Drosophila melanogaster* (from Stocker and Schorderet, 1981). (b) A whole-mount of the brain of *D. melanogaster*, showing axonal projects from taste cells (labeled with anti-GFP immunohistochemistry). The white box indicates the region of the SOG depicted in panels (c-d). The green axons in panel C are from GRNs expressing the bitter-sensitive Gr66a, and those in panel D are from GRNs expressing the sugar-sensitive Gr5a. It is notable that the projection sites for GRNS expressing Gr66a are spatially distinct from those expressing Gr5a. Panels (b-d) are from Wang et al. (2004). (*See plate section for color version.*)

41.3 Peripheral Taste Function

neuropil with both sensory and motor terminals (Altman and Kien, 1987); and (c) stimulating GRNs with plant extracts modulates the output of the mouthpart motor neurons (Griss et al., 1991). The ability of GRN input to modify oromotor feeding circuits does not appear to involve direct connections between GRNs and mouthpart motor neurons (Gordon and Scott, 2009). Instead, there appear to be intermediary taste circuits within the SOG and/or higher brain centers. One such higher brain center could be the calyces of the mushroom body (a neuropil important for learning and memory), which receives taste input from the SOG via the subesophageal-calycal tract (Schröter and Menzel, 2003). Indeed, there is evidence that disruption of the mushroom body impairs taste-associated learning in adult *Drosophila* (Masek and Scott, 2010).

41.3 PERIPHERAL TASTE FUNCTION

In this section I discuss what is known about peripheral taste responses to chemical stimuli, and how recent studies of *Drosophila* have provided insight into the molecular basis of these responses.

To measure responses of individual GRNs to chemical stimuli, investigators typically make "tip recordings," which involve contacting the pore at the tip of a taste sensillum with a solution containing a taste stimulus dissolved in an electrolyte solution (Hodgson et al., 1955) (Figure 41.2b). Because tip recordings usually contain responses from multiple GRNs, investigators use spike-sorting methods (based on shape, amplitude and/or temporal pattern) to assign spikes to individual GRNs (Schnuch and Hansen, 1990; Smith et al., 1990; Meunier et al., 2003; Glendinning et al., 2006). The GRNs are classified according to their "best" stimulus – that is, the stimulus that generates the strongest excitatory response. Accordingly, a GRN that responds best to sugars is referred to as sugar-sensitive. It should be noted, however, that because most insect GRNs have been tested with only a limited number of taste stimuli, the "best" stimulus for a specific GRN may still be unknown. For the sake of simplicity, I will use best stimulus terminology for labeling GRNs, keeping in mind this caveat.

41.3.1 Aversive Compounds

Many of the compounds that taste bitter to humans also elicit aversive responses (i.e., rapidly inhibit feeding) in insects. Because virtually all naturally occurring poisons taste bitter to humans, the rejection of aversive compounds is thought to represent an evolved mechanism for avoiding poisonous foods (Bate-Smith, 1972; Garcia and Hankins, 1975; Brower, 1984; Brieskorn, 1990). The

effectiveness of this taste-mediated poison detection system is limited by the fact that there is no straightforward relationship between taste and toxicity. Indeed, many aversive compounds are harmless (Bernays and Chapman, 1987; Bernays, 1990; Rouseff, 1990; Bernays, 1991; Glendinning, 1994).

All insects studied to date – for example, Orthopterans (locusts and grasshoppers), Dipterans (flies), Lepidopterans (butterflies and moths), and Coleopterans (beetles) – possess GRNs that respond selectively to aversive and potentially toxic compounds (Schoonhoven and van Loon, 2002; Meunier et al., 2003; Thorne et al., 2004; Marella et al., 2006; Weiss et al., 2011. These GRNs are referred to as "deterrent" or "bitter-sensitive" (Schoonhoven and van Loon, 2002). Importantly, activation of bitter-sensitive GRNs strongly inhibits feeding in insects (Schoonhoven and van Loon, 2002). One study reported that stimulation of as few as two bitter-sensitive GRNs can inhibit feeding (Glendinning et al., 1999b).

The studies cited in the previous paragraph reveal four notable features of bitter-sensitive GRNs. First, the response (i.e., firing rate) increases with stimulus concentration; this indicates that taste intensity is coded, at least in part, in the periphery. Second, many bitter-sensitive GRNs respond to a structurally diverse range of compounds. In *Drosophila*, this capacity stems from the fact that bitter-sensitive GRNs express a diversity of Grs. Third, comparisons across sensilla within the same insect reveal that many GRNs display different molecular receptive ranges (MRRs) (Figure 41.2c). This attribute can be explained in *Drosophila* by the fact that different classes of bitter-sensitive GRNs express different combinations of Grs. Fourth, taste responses to aversive compounds are not mediated exclusively by Grs. For instance, TrpA1 channels are expressed in bitter-sensitive GRNs, and disrupting expression of TrpA1 selectively attenuates the aversive response of *Drosophila* to aristolochic acid (Kim et al., 2010).

41.3.2 Sugars

Most insects show concentration-dependent increases in intake of simple sugars. For instance, the probability that bees, butterflies, and flies will extend their proboscis in response to taste stimulation increases with sugar concentration (Dethier, 1976; Schoonhoven and van Loon, 2002; Omura and Honda, 2003; de Brito Sanchez et al., 2005; Dahanukar et al., 2007; Gordon and Scott, 2009; Zhang et al., 2010). Likewise, the rate at which Lepidopteran caterpillars either bite from disks (Ma, 1972) or swallow solutions (Sasaki and Asaoka, 2006) increases with sugar concentration. *Manduca sexta* caterpillars (Sphingidae; Lepidoptera) are a notable exception to this pattern. They exhibit a strong peripheral taste response to sugars, but nevertheless fail to show immediate appetitive responses to them – for example, they bite at the same rate from disks treated with sugar solutions as from disks treated with water alone (Glendinning et al., 2007).

All insects studied to date possess GRNs that respond robustly to sugars (e.g., glucose, sucrose, fructose, and trehalose) and sugar alcohols (e.g., sorbitol, mannitol, and *myo*-inositol). Whereas Dipterans have sugar-sensitive GRNs that respond to sugars and sugar alcohols, most Lepidopterans have separate GRNs for sugars and sugar alcohols (Dethier, 1976; Schoonhoven and van Loon, 2002; Dahanukar et al., 2007). As in the bitter-sensitive GRNs, the firing rate of carbohydrate-sensitive GRNs increases with stimulus concentration (Dethier, 1976; Glendinning et al., 2007). Second, carbohydrate-sensitive GRNs usually respond to a structurally diverse range of sugars (Dethier, 1976; Schoonhoven and van Loon, 2002; Dahanukar et al., 2007; Glendinning et al., 2007; Jiao et al., 2007). In *Drosophila*, this capacity appears to stem from two factors. The two best characterized sugar receptors – Gr5a and 64a – are not only broadly tuned, but they each respond to a large number of sugars (Dahanukar et al., 2007; Jiao et al., 2007; Slone et al., 2007). It is important to keep in mind that sugars can stimulate feeding by mechanisms other than taste. For instance, *Drosophila* possess post-oral mechanisms for sensing ingested sugars, and that they use input from these mechanisms to learn which foods are most nutritious (Burke and Waddell, 2011; Dus et al., 2011; Fujita and Tanimura, 2011). This post-oral chemosensation may be mediated in part by the cluster of Gr28 receptors in the abdominal multi-dendritic neurons (Thorne and Amrein, 2008).

41.3.3 Salts

All insects studied to date have salt-sensitive GRNs. In flies, monovalent inorganic cations are the most effective stimulants for these GRNs (Dethier, 1976). Cations differ, however, in their stimulatory effectiveness (e.g., in the fleshfly, *Boettcherisca peregrina*, $Li^+ < Na^+ < Cs+ < Rb^+ < K^+$; (den Otter, 1972).

Flies show an appetitive response to low NaCl concentrations (5–10 mM) and aversive response to high NaCl concentrations ($\geq$500 mM; Dethier, 1968; Hiroi et al., 2004). This concentration-dependent change in hedonic response to salts is thought to have evolved because low concentrations of salts satisfy nutritional needs, whereas higher concentrations threaten osmotic equilibrium. To explain the mechanistic basis of this behavior, investigators have silenced the sugar-sensitive GRNs of *Drosophila* that express *Gr5a* (Wang et al., 2004). When these flies were offered a range of NaCl concentrations, they exhibited an attenuated appetitive response to a low (5 mM) concentration, but a normal aversive response to high

concentrations ($\geq$500 mM). These latter findings, together with those of Hiroi *et al.* (2004), indicate that NaCl stimulates two different classes of GRN – low concentrations stimulate Gr5a-expressing (i.e., sugar-sensitive) GRNs, and high concentrations stimulate a different (presumably salt-sensitive) population of GRNs.

In adult *Drosophila*, the aversive response to high NaCl concentrations is thought to be mediated in part by the *dpr* gene (Nakamura et al., 2002). This inference is based on the observation that the proboscis extension reflex was strongly inhibited by high NaCl concentrations in wildtype flies, but not in *dpr*-knockout flies. The *dpr* gene is thought to encode a protein with a single transmembrane domain, and to be expressed in a subset of GRNs (Nakamura et al., 2002).

An important, but unexplained, observation is that some insects can discriminate between different types of salts (Arms et al., 1974; Maes and Bijpost, 1979; Miyakawa, 1981; Miyakawa, 1982). In laboratory rats, the ability to discriminate Na^+ and K^+ salts appears to be mediated by differential input from Na^+-specific (amiloride-sensitive) and Na^+-nonspecific (amiloride-insensitive) cation channels (Spector and Grill, 1992; Smith et al., 2000). The question of whether insects use a similar mechanism is unresolved. One study reported that *Drosophila* expresses two amiloride-sensitive degenerin/epithelial Na+ (DEG/ENaC) channels – Ppk11 and Ppk19 – in GRNs within the gustatory terminal organ of larvae and taste sensilla on the labellum, legs and wing margins of adults (Liu et al., 2003). One concern about the latter study is that 5 mM amiloride was required to inhibit the response of Ppk11 and Ppk19 to NaCl. In mammalian GRNs, one can effectively inhibit the response to Na^+ in mammals with concentrations as low as 30 μM (Vandenbeuch et al., 2008). The need for such a high amiloride concentration in *Drosophila* may reflect a low binding affinity of amiloride for Ppk11 or Ppk19 channels, or non-specific effects of amiloride on insect GRNs.

41.3.4 Amino Acids

Free amino acids stimulate feeding in many insect species, but the behavioral responsiveness to any given amino acid varies considerably across species (Bernays, 1985; Albert and Parisella, 1988; Hirao and Ariai, 1990; Kim and Mullin, 1998; Schoonhoven and van Loon, 2002). Appetitive responses to amino acids are considered adaptive because plant tissues with relatively high protein concentrations should, on average, have correspondingly high concentrations of free amino acids. If so, then amino acid-sensitive GRNs could help insects identify protein-rich plant tissues. Likewise, because some amino acids (e.g., phenylalanine) are considered limiting nutrients for plant feeding insects (Bernays, 1982), amino acid-sensitive GRNs could help the insect maintain nutritional homeostasis.

In caterpillars, amino acids often stimulate specific GRNs (Dethier, 1973; Schoonhoven and van Loon, 2002). For example, two closely related species of *Pieris* caterpillars have an amino acid-sensitive GRN that exhibits concentration-dependent increases in response to ecologically relevant concentrations of 12 L-amino acids (Dethier, 1973; van Loon and van Eeuwijk, 1989). Other caterpillar species (e.g., *Grammia geneura*) detect amino acids with multiple GRNs (Bernays and Chapman, 2001). One of the amino acid sensitive GRNs in *G. geneura* is unusual in that it responds to a diverse array of phagostimulants present in the insect's preferred foodplants, including seven amino acids, three sugars, and a secondary plant compound (catalpol, an iridoid glycoside) (Bernays et al., 2000).

Dipterans use a variety of gustatory mechanisms for detecting amino acids. On the one hand, the tsetse fly (*Flossina fuscipes fuscipes*) has an amino acid-sensitive GRN that responds to 11 out of the 20 L-amino acids present in animal proteins (van der Goes van Naters and den Otter, 1998). On the other hand, the blowfly (*Phormia regina*), fleshfly (*Boettcherisca peregrina*) and hoverfly (*Eristalis tenax*) all lack GRNs that respond to amino acids (Wolbarsht and Hanson, 1965; Shiraishi and Kuwabara, 1970; Goldrich, 1973; Wacht et al., 2000). Instead, these flies detect amino acids in other ways. For instance, the neural responses to 19 amino acids of the labellar sensilla in blowflies and fleshflies fall into one of four categories. The amino acids (1) inhibit the excitatory response of all GRNs (aspartic acid, glutamic acid, histidine, arginine, lysine); (2) stimulate the salt-sensitive GRNs (proline, hydroxyproline); (3) stimulate the sugar-sensitive GRN (valine, leucine, isoleucine, methionine, phenylalanine, tryptophan); or (4) fail to stimulate or inhibit any GRN (glycine, alanine, serine, threonine, cysteine, tyrosine) (Shiraishi and Kuwabara, 1970; Goldrich, 1973). The impact of these four different effects of amino acids on feeding is unclear, however, because none of them elicit proboscis extension in the blowfly (Wolbarsht and Hanson, 1965).

Adult *Drosophila* appear to possess a heterodimeric receptor (Gr8a + Gr66a) which responds to L-canavanine, a naturally occurring a non-proteinogenic amino acid present in leguminous plants (Lee et al., 2012). Activation of this receptor elicits behavioral avoidance. Additional studies are needed to determine whether Gr8a + Gr66a responds to amino acids other than L-canavanine.

41.3.5 Acids

In humans, acids elicit the perception of sourness. Little is known about the behavioral responses of insects to low pH foods. There is one report showing that ascorbic acid stimulates feeding in *Pieris brassicae* caterpillars, but only when it is presented in binary mixture with sucrose (Ma, 1972).

Ascorbic acid occurs in plant tissues at concentrations of 1–10 mM (Schultz and Lechowicz, 1986). The fact that ascorbic acid is an essential vitamin for most herbivorous insects may explain why caterpillars show an appetitive response to ascorbic acid when it co-occurs with sugars. On the other hand, the fact that acidic substances can be toxic and/or depress the pH of the midgut may explain why many insects display aversive responses to highly acidic foods.

Ground beetles (Coleoptera, Carabidae) are the only known taxon of insects with GRNs that that show robust pH-dependent changes in firing rate – the higher the pH, the higher the firing rate (Millius et al., 2006). In other taxa of insects, acids appear to play a modulatory role – that is, alter the responsiveness of specific GRNs to their best stimuli. For example, at low concentrations, acids inhibit the responsiveness of salt-sensitive GRNs, whereas at higher concentrations, they inhibit the responsiveness of sugar- and inositol-sensitive GRNs and stimulate bitter-sensitive GRNs (Dethier and Kuch, 1971; Bernays et al., 1998).

41.3.6 Water

The ability to detect and regulate intake of water is essential, particularly for small terrestrial animals like insects that desiccate easily (Dethier, 1976). In 1957, Myson Wolbarscht discovered a GRN in flies that responds specifically to water (Wolbarsht, 1957); this observation was soon replicated and extended (Evans and Mellon, 1962). Around the same time, investigators demonstrated that stimulation of this water-sensitive GRN stimulates proboscis extension and ingestion in thirsty flies (Dethier and Evans, 1961; Evans, 1961; Dethier, 1976). Subsequent studies have reported similar water-sensitive GRNs in other insects (review in (Schoonhoven and van Loon, 2002). It was not until recently that investigators identified the molecular basis of water taste. Two labs independently reported that Ppk28, a member of the DEG/eNaC family, is the water receptor (Cameron et al., 2010; Chen et al., 2010). There are still questions, however, about the downstream events in water taste transduction. For instance, one study inferred a role of aquaporins in water taste (Solari et al., 2010), and another found evidence for a modulatory role of calmodulin in water taste (Meunier et al., 2009).

41.3.7 Species-Specific Taste Stimuli

Many insect species possess GRNs that are narrowly tuned to chemical stimuli in their diet. Below are a few examples. First, the sugar-sensitive GRNs of blood-feeding insects (e.g., the fleshfly, *Boettcherisca peregrina*) discharge in response to nucleotides in blood (e.g., ADP and GDP) (Furuyama et al., 1999). Second, several herbivorous insects possess GRNs that respond to phytoecdysteroids in plant tissues (Ma, 1969; Tanaka et al., 1994; Marion-Poll and Descoins, 2002; Calas et al., 2007). Stimulation of these GRNs elicits an aversive response. This response is thought to have evolved because phytoecdysteroids are found in about 6% of all land plant species, and their consumption can interfere with ecdysis. Third, several insect species depend on pyrrolizidine alkaloids (PAs) for reproduction and chemical defense against predators. These species possess GRNs, which are narrowly tuned to PAs and stimulate intake of PA-containing foods (Bernays et al., 2002; Bernays et al., 2003). Fourth, some butterflies utilize food sources (e.g., tree sap and rotting fruits) that contain fermentation products like ethanol. Some (e.g., *Vanessa indica*), but not all (*Argyreus hyperbius*) of these butterflies show a stronger peripheral taste response and appetitive feeding response to sugar solutions containing ecologically relevant concentrations of ethanol (Ômura et al., 2008). Fifth, adult *Drosophila* have a population of GRNs that detects CO_2. When activated, they elicit an appetitive feeding response. This CO_2-specific response pathway is thought to help *Drosophila* locate growing yeast, an important food source (Fischler et al., 2007).

41.4 TASTE-MIXTURE INTERACTIONS

41.4.1 Peripheral Taste System

Based on evidence presented above, it is apparent that many insect GRNs are "tuned" to specific classes of taste stimuli (e.g., sugars or salts). These findings were based on studies in which insects were presented with single-component stimuli. What happens when insects encounter real foods, which invariably contain complex mixtures of taste stimuli? Does each class of GRN (e.g., sugar- versus salt-sensitive) function as an independent processing unit (i.e., "silo"), relaying its responses to the central gustatory system with high fidelity? Alternatively, does activation of one GRN influence the response of other GRNs in the same sensillum? If the latter is the case, then this would indicate that taste sensilla serve as a peripheral taste processing unit, much like what has been proposed for vertebrate taste buds (Herness and Zhao, 2009; Chaudhari and Roper, 2010).

It turns out that there is widespread evidence that taste sensilla process the peripheral taste signal. For instance, when two taste stimuli (e.g., S_1 and S_2) that activate different GRNs (e.g., GRN_1 and GRN_2) are presented in a mixture, insect taste sensilla usually generate fewer action potentials than would be predicted based on the sum of the action potentials elicited by each stimulus alone (i.e., they exhibit mixture suppression). Two different types of mixture suppression have been reported. In one type (reciprocal inhibition), S_1 stimulates GRN_1 and S_2 stimulates GRN_2; but when S_1 and S_2 are presented together, the excitatory responses of both GRN_1 and GRN_2 are diminished (Ishikawa, 1963; White et al., 1990; Shields and Mitchell, 1995; Bernays and Chapman, 2001; Glendinning et al., 2007; Wright et al., 2010). In the other type of mixture suppression (non-reciprocal inhibition), S_1 stimulates GRN_1 and S_2 fails to stimulate any GRN; nevertheless, when S_1 and S_2 are presented together, the excitatory response of GRN_1 is diminished (Dethier and Bowdan, 1989; Bernays and Chapman, 2000; Bernays and Chapman, 2001; de Brito Sanchez et al., 2005).

Recent work in my laboratory indicates that peripheral taste processing may be even more complex that hitherto appreciated (Cocco and Glendinning, 2012). The caterpillar *Manduca sexta* (Sphingidae) possesses two classes of sugar-sensitive GRNs – one responds to glucose and inositol (a sugar alcohol) and the other to sucrose. It also possesses a bitter-sensitive GRN, which responds to several aversive compounds including aristolochic acid (AA). While all sugars modulated peripheral taste responses to AA, sucrose had the greatest impact, despite eliciting fewer action potentials per unit time. For instance, when we mixed a single sugar (i.e., sucrose, glucose or inositol) with AA, sucrose (but none of the other sugars) partially blocked the response to AA. When we mixed binary mixtures of sugars with AA, the binary mixtures containing sucrose (i.e., sucrose+inositol or sucrose+glucose) produced the greatest mixture suppression. It follows that activation of the sucrose-sensitive GRN produced greater mixture suppression than activation of the glucose/inositol-sensitive GRN. Additional studies are needed to explain why the taste system of *M. sexta* caterpillars pays particular attention to input from the sucrose-sensitive GRNs.

Little is known about the nature of mixture suppression in insect taste sensilla. While it may involve direct electrical communication via gap junctions (Steinbrecht, 1989; Isidoro et al., 1993) or ephaptic interactions among adjacent GRNs (Jefferys, 1995; Bokil et al., 2001), the most likely explanation is that activation of one GRN elicits paracrine release into the sensillar lumen, which modulates the response of neighboring GRNs. This speculation is based on the existence of extensive paracrine signaling within mammalian taste buds (Herness and Zhao, 2009).

41.4.2 Mixture Interactions in the Central Nervous System

Taste-mixture interactions have also been observed in the central gustatory system. For instance, caffeine normally elicits a taste-mediated aversive response in *Manduca sexta* caterpillars (Glendinning et al., 1999b). However, when caffeine is presented in binary mixture with *myo*-inositol (a sugar alcohol), the aversive response is completely attenuated (Glendinning et al., 2000). Recordings from the peripheral taste system of this insect revealed that the bitter-sensitive GRNs exhibited the same response to caffeine, irrespective of whether the caffeine was presented alone or in binary mixture with *myo*-inositol; likewise, the inositol-sensitive GRNs exhibited the same response to *myo*-inositol, irrespective of whether the myo-inositol was presented alone or in binary mixture with caffeine (Glendinning et al., 2000). The most parsimonious interpretation of these findings is that input from the inositol-sensitive GRNs inhibited a neural circuit in the central nervous system that mediates the aversive response to caffeine.

41.5 TASTE PROCESSING IN THE CENTRAL NERVOUS SYSTEM

In mammals, the brain rapidly processes taste input according to three orthogonal dimensions: intensity, hedonics and quality (Grill and Berridge, 1985; Spector, 2000). Do insects use the same three taste dimensions? With regard to intensity, there are several reports of insects (a) exhibiting concentration-dependent changes in taste-mediated behavioral responsiveness to specific taste stimuli (Dethier, 1968; Bowdan, 1995; Glendinning and Gonzalez, 1995; Glendinning et al., 1999b; Wang et al., 2004), and (b) discriminating different intensities (i.e., concentrations) of taste stimuli (Maes and Bijpost, 1979; Masek and Scott, 2010). With regard to hedonics (a measure of stimulus palatability), there are reports of insects showing taste-mediated preferences or aversions to chemical stimuli (Harley and Thorsteinson, 1967; Dethier, 1968; Dethier, 1976; Bernays and Chapman, 1994; Bowdan, 1995; Glendinning and Gonzalez, 1995; Glendinning et al., 1999b; Wang et al., 2004). Little is known, however, about where in the brain these two taste dimensions are coded (e.g., the SOG or higher brain areas).

The question of whether insects categorize taste stimuli into different qualities (e.g., sweet, salty, sour, bitter and umami) has not been examined. To address this core issue, one cannot simply measure unconditioned behavioral responses (i.e., whether an insect exhibits a preference or aversion to a stimulus). This is because some taste stimuli that are functionally distinct from one another (e.g., high concentrations of NaCl and quinine) elicit similar behavioral responses (i.e., avoidance), but nevertheless stimulate difference classes of GRNs. In the mammal literature, the most commonly used paradigm for determining whether two stimuli elicit distinct taste

qualities is the conditioned taste aversion task (CTA) (Frank and Nowlis, 1989; Wiggins et al., 1989). In brief, one creates a CTA to one taste stimulus (e.g., NaCl), and then asks whether the CTA generalizes to an isostimulatory concentration of another taste stimulus (e.g., quinine). If the animal does *not* generalize the CTA, then one infers that the two taste stimuli have distinct taste qualities. On the other hand, if the animal does generalize the CTA, then one infers that the animal could not distinguish the two taste stimuli because they produce a common taste quality. While this CTA-generalization procedure has been used successfully in mammals, I am unaware of its application to insects. Given that insects are capable of forming CTAs (Gelperin, 1968; Lee and Bernays, 1990; Ayrton and Morgan, 2001, Ayestaran et al., 2010; Salloum et al., 2011), the CTA-generalization paradigm should provide a powerful tool for analyzing taste qualities in insects. Alternatively, one could generate a conditioned taste preference (CTP) to one taste stimulus (Gerber et al., 2004), and then test for generalization of the CTP to another taste stimulus.

The CTA approach was used to determine whether adult *Drosophila* could discriminate among different concentrations and types of sugars (glucose, fructose and maltose), or different types of aversive compounds (caffeine, quinine and berberine). Although the flies readily discriminated stimuli based on sugar intensity (e.g., 100 versus 250 mM fructose), they could not readily discriminate based on sugar type (Masek and Scott, 2010).

41.5.1 Taste Processing in the SOG

There have been several important advances in our understanding of how the SOG processes taste input. One is that GRNs from different gustatory organs and taste sensilla appear to project to spatially distinct regions of the SOG (Kent and Hildebrand, 1987; Shanbhag and Singh, 1992; Stocker, 1994; Pollack and Balakrishnan, 1997; Wang et al., 2004; Jørgensen et al., 2006). Further, these axonal projections are segregated spatially according to GRN type (e.g., whether they respond to sugars, aversive compounds or water) (Figure 41.3) (Thorne et al., 2004; Wang et al., 2004; Inoshita and Tanimura, 2006; Cameron et al., 2010). Accordingly, the spatial pattern of activation of the SOG would provide information to the brain about hedonic features of the taste stimulus (appetitive versus aversive) and its location on the body (mouthparts or legs). It has been proposed that an analogous gustotopic organization exists in the taste cortex of mice (Chen et al., 2011).

Two studies have begun to dissect the circuits controlling ingestive behavior in adult *Drosophila*. One study identified a motor neuron (E49) in the SOG that appears to influence proboscis extension by integrating inputs from bitter-sensitive and sweet-sensitive GRNs (Gordon and Scott, 2009). Another study examined neural control of the cibirial pump, which controls fluid ingestion, via two motor neurons (MN11 and MN12). This study found that MN11 and MN12 exhibit longer patterns of activation in response to taste stimulation with palatable (sucrose) than with unpalatable (caffeine) chemicals (Manzo et al., 2012).

41.5.2 Coding Mechanisms

The nature of the sensory code for insect taste is unresolved. As in mammalian systems, there is debate over the relative importance of three different coding frameworks: labeled-lines, ensemble (i.e., across-fiber) patterns, and temporal patterns of spiking. The labeled-line framework hypothesizes that tastant identity is represented by neural activity in a specific subset of gustatory neurons. According to this model, activation of the sugar-sensitive GRNs would stimulate neural circuits within the SOG that initiate biting, whereas activation of the bitter-sensitive GRNs would inhibit the same neural circuits. Further, the behavioral response to a mixture of sugars and bitter taste stimuli should reflect the algebraic sum of the inputs from the sugar- and bitter-sensitive GRNs, respectively (Schoonhoven and Blom, 1988; Simmonds and Blaney, 1991). One study provided fairly definitive evidence for labeled-line coding in adult *Drosophila* (Marella et al., 2006). The investigators genetically engineered two types of flies (e.g., A and B). In type A flies, they misexpressed a capsaicin receptor (VR1E600K) in the GRNs that express Gr5a (i.e., sugar-sensitive GRNs); whereas in type B flies, they misexpressed VR1E600K in the GRNs that express Gr66a (i.e., bitter-sensitive GRNs). In subsequent behavioral studies, they found that type A flies exhibited an appetitive response to capsaicin, while the type B flies exhibited an aversive response to capsaicin. This result indicates that the hedonic response of an insect is determined by the specific class of GRN (or labeled-line) that is activated. An analogous experiment was performed in mice, and produced similar results (Zhao et al., 2003; Mueller et al., 2005).

The ensemble coding framework hypothesizes that tastant identity is represented by the spatial pattern of activity across large populations of GRNs (Schoonhoven et al., 1992; Smith et al., 2000). Accordingly, discriminable taste stimuli should each elicit different ensemble patterns (i.e., different magnitudes of discharge across the population of GRNs). Ensemble coding was evaluated in a study that sought to identify the framework that best explained how *M. sexta* caterpillars discriminate the taste of three different host plants (Dethier and Crnjar, 1982). After evaluating the three coding frameworks, the authors concluded that ensemble pattern provided the most parsimonious discrimination mechanism. Another study has demonstrated that *M. sexta* uses ensemble coding to discriminate between two aversive taste stimuli (salicin and

Grindelia leaf extract), which activate differences classes of bitter-sensitive GRN (Glendinning et al., 2002).

While the labeled-line and ensemble coding frameworks focus on *which* neurons are activated within the gustatory neuraxis, the temporal coding framework focuses on *how* neurons are activated. This latter framework hypothesizes that tastant identity is represented by the precise temporal pattern of spiking within a neuron or across populations of neurons (Dethier and Crnjar, 1982; Katz et al., 2002; Katz, 2005). Support for this hypothesis comes from a recent study of *Manduca sexta* caterpillars from our laboratory (Glendinning et al., 2006). Several of the bitter-sensitive GRNs in this caterpillar generate a decelerating temporal pattern of spiking when stimulated by caffeine and salicin, and an accelerating temporal pattern of spiking when stimulated by aristolochic acid. We showed that the caterpillars use the distinct temporal patterns of spiking generated by salicin versus aristolochic acid as a basis for discrimination.

It is important to emphasize two caveats associated with the coding frameworks discussed above. First, the three frameworks are not mutually exclusive. In fact, they probably function in a complimentary manner within the same insect (Dethier and Crnjar, 1982). Second, given the profound effects of taste-mixture interactions, physiological state, and experience on responsiveness of individual GRNs to their respective ligands (see below), central processing of the taste input is probably more dynamic and complex than hitherto appreciated.

41.6 FACTORS THAT MODULATE TASTE RESPONSIVENESS

One of the remarkable features of the insect taste system is its plasticity. To illustrate this point, I discuss below several factors known to increase (or decrease) the responsiveness of the gustatory system to specific taste stimuli.

41.6.1 Physical Changes in Taste Sensilla

For a chemical to stimulate a GRN, it must first diffuse through the pore at the tip of a taste sensillum (see Figure 41.1d). Although this terminal pore remains permanently open in many insect species, it opens and closes in Orthopterans (Bernays et al., 1972). The functional significance of terminal pore closure was revealed by a series of experiments in locusts, *Locusta migratoria* (Bernays and Chapman, 1972). When the foregut of the locust becomes distended (e.g., after a meal), stretch receptors in the wall of the foregut are activated and send afferent signals to the brain via the posterior pharyngeal nerves and frontal connectives. These afferent signals stimulate the

41.6 Factors that Modulate Taste Responsiveness

release of a diuretic hormone from the corpora cardiaca, which in turn cause pore closure in many of the taste sensilla. The functional effect of terminal pore closure is that it contributes to meal termination by eliminating phagostimulatory input from the peripheral taste system.

41.6.2 Nutrient Levels in Hemolymph

Several investigators have discovered that changes in the concentration of specific nutrients (e.g., amino acids or trehalose) in the hemolymph directly modulate the responsiveness of GRNs in grasshoppers and caterpillars to the same chemicals (Simmonds et al., 1992; Simpson and Simpson, 1992; Bernays et al., 2004). For example, in locusts, when hemolymph levels of amino acids are low and sugars high, the responsiveness of the GRNs to amino acids is high and that to sugars low (Simpson and Simpson, 1992). On the other hand, when hemolymph levels of amino acids are high and sugars low, the responsiveness of the GRNs to amino acids is low and to that to sugars is high. Because the modulatory effects of hemolymph composition on GRN responsiveness occurred in the absence of neural or humoral links between the central nervous system and GRNs, they were assumed to be mediated peripherally.

41.6.3 Dopaminergic Signaling

There is evidence that dopamine modulates the proboscis extension reflex in adult *Drosophila*, and that the extent of modulation varies as a function of physiological state (i.e., whether the fly is hungry or sated) (Marella et al., 2012). Accordingly, when the fly is hungry, there appears to be an increase in both the activity of its dopaminergic neurons and the probability of proboscis extension. Importantly, this result was observed following taste stimulation with sugars but not aversive compounds or water, implicating a selective role of dopamine in sugar appetite.

41.6.4 Dietary Exposure

When insects are exposed repeatedly to the same taste stimulus (e.g., within and across successive meals), the responsiveness of their gustatory system to these stimuli can increase or decrease. The time-scale over which these changes occur varies from seconds to days. Below, I discuss a four ways that dietary experience modifies taste.

41.6.4.1 Short-Term Sensory Adaptation.
A common feature of all sensory receptor cells is that their firing rate diminishes over time (i.e., adapts) in response to continuous stimulation, but recovers rapidly (i.e., disadapts) upon removal of the stimulus. Adaptation

and disadaptation occur over a period of milliseconds to seconds.

Bitter-sensitive GRNs appear to be more resistant to adaptation than nutrient-sensitive GRNs, a phenomenon that has been reported for a number of species of insects (e.g., Schoonhoven, 1977; Cocco and Glendinning, 2012). From a functional standpoint, this suggests that the amount of afferent input from different GRNs during a meal may not be constant, with the relative amount of input from the bitter-sensitive GRNs increasing over time. The extent to which these adaptation processes contribute to feeding during a meal is unclear, however, because insect taste sensilla do not remain in continuous contact with foods throughout a meal. Some gustatory organs make a series of drumming movements, resulting in intermittent contacts lasting only 100 to 200 ms (Blaney and Duckett, 1975; Devitt and Smith, 1985). These drumming movements could serve to minimize the effects of short-term adaptation in the GRNs, effectively increasing the level of sensory input from GRNs over time (Blaney and Duckett, 1975).

41.6.4.2 Habituation.
When an animal is presented with a stimulus repeatedly across a series of discrete trials, its responsiveness to that stimulus usually wanes (Hinde, 1970). This phenomenon is referred to as habituation if it meets several criteria – for example, it must be mediated centrally, recover spontaneously, generalize to related stimuli, and develop more readily in response to weak than intense stimuli (Thompson and Spencer, 1966). For instance, fruit flies and bees normally exhibit a proboscis extension reflex for 5% sucrose, but this reflex extinguishes following 20 consecutive stimulations within a 5 min period (Duerr and Quinn, 1982; Braun and Bicker, 1992; Scheiner, 2004; Scheiner et al., 2004). This exposure-induced response decrement for sucrose was considered habituation because it met at least two of Thompson and Spencer's habituation criteria.

41.6.4.3 Long-Term Adaptation.
When herbivorous insects are offered a diet containing an unpalatable but harmless compound as their only source of water and nutrients, they will usually sample the diet repeatedly. After ~24 hr of such sampling, the insects eventually adapt their aversive response. This long-term adaptation phenomenon has been reported in several species of herbivorous insect (Jermy et al., 1982; Simmonds and Blaney, 1983; Szentesi and Bernays, 1984; Blaney and Simmonds, 1987; Usher et al., 1988; Glendinning and Gonzalez, 1995; Bernays et al., 2004), but has been examined most extensively in caterpillars of *M. sexta* (Schoonhoven, 1969; Schoonhoven, 1978; Glendinning et al., 1999a, 2001a, 2001b, 2002). For example, 24 hr of dietary exposure to a caffeine diet

eliminates the aversive response to caffeine. The long-term adaptation phenomenon is mediated peripherally (at least in part) because the dietary exposure regime profoundly desensitizes all of the bitter-sensitive GRNs to caffeine. Twenty-four hr of dietary exposure to a salicin diet also eliminates the aversive response to salicin. However, because there is no associated reduction in responsiveness of the bitter-sensitive GRNs, this latter adaptation phenomenon must be mediated centrally. Taken together, these results indicate that the peripheral and central gustatory system of *M. sexta* each have independent mechanisms for overcoming the aversive response to harmless aversive compounds.

Because the long-term adaptation to salicin is mediated by the central gustatory system, and because it generalizes to related taste stimuli (Glendinning et al., 2001b; Glendinning et al., 2002), it can be viewed as a type of habituation. It is important to note, however, that the onset latency for the habituation to salicin (in *M. sexta*) is about 24 hr, whereas that to sucrose (in flies and bees) is ≤5 min (Duerr and Quinn, 1982; Braun and Bicker, 1992; Scheiner, 2004; Scheiner et al., 2004). Further work is needed to determine whether these different onset latencies reflect (a) species differences, (b) greater difficulty habituating to aversive taste stimuli, and/or (c) the existence of multiple habituation mechanisms.

41.6.4.4 Induced Food Preferences.
After feeding on a particular plant species for one to two days, many species of herbivorous insect induce a strong preference for that plant species (review in Szentesi and Jermy, 1990). Despite the widespread occurrence of this phenomenon, the underlying physiological mechanisms are poorly understood. In all likelihood, a variety of mechanisms are activated, depending on the chemical constituents of the plant tissue and the insect species – for example, induction of detoxification enzymes, changes in gustatory and/or olfactory responsiveness, and conditioned preferences (Bernays and Weiss, 1996).

That taste contributes to induced food preferences was established in a study of *M. sexta* caterpillars, which are facultative specialists on plants in the family Solanaceae. When these caterpillars were reared on solanaceous foliage, they induced a strong and exclusive preference for the same foliage (Hanson and Dethier, 1973; Yamamoto, 1974; de Boer and Hanson, 1987; del Campo and Renwick, 2000; Glendinning et al., 2009). A team of investigators discovered that indioside D (a steroidal glycoside) is one of the chemicals in solanaceous foliage to which the caterpillars induce their preference (del Campo et al., 2001). These investigators also presented evidence that GRNs in the caterpillar's styloconic sensilla become "tuned" to indioside D over the course of the induction process, and that

this chemosensory tuning process is sufficient to explain the induced preference for solanaceous leaves (del Campo et al., 2001; del Campo and Miles, 2003). According to the chemosensory tuning hypothesis, the GRNs sensitive to indioside D maintain their normal responsiveness, while those sensitive to other plant chemicals (e.g., sugars, salts and bitter taste stimuli) become less responsive.

41.6.4.5 Health Status.
There is evidence that the peripheral gustatory response of some caterpillar species increases in response to endoparasitic infection (Bernays and Singer, 2005). More specifically, the endoparasitic infections caused specific GRNs to respond more vigorously to specific host plant compounds (i.e., pyrrolizidine alkaloids, PAs). The elevated responsiveness of the PA-sensitive GRNs in turn promoted greater intake of PA-containing plant tissues. This elevated intake was subsequently shown to be adaptive because the PAs were toxic to the endoparasites, and thus helped control (or even eliminate) the infection (Singer et al., 2009).

41.7 CONCLUSION

The discovery of the Grs in *Drosophila* represented a turning point in insect gustation. Investigators were able to use the Grs, together with existing molecular tools, to gain deep insight into how taste input is processed in *Drosophila*. Future work is needed to determine the extent to which the logic of taste processing in *Drosophila* generalizes to other insect species. Given that the genomes of insect species spanning 23 different genera have already been sequenced (http://en.wikipedia.org/wiki/List_of_sequenced_animal_genomes#Insects), it should soon be possible to address fundamental questions about how lifestyle, diet, sex, life stage, and phylogeny all influence taste processing in insects. These advances will not only enrich our understanding of insect feeding biology and evolution, but they may also reveal strategies for controlling insects that transmit disease and cause agricultural damage.

Throughout this review, I have tried to emphasize that despite the many differences between insects and mammals, the taste systems of both taxa appear to share important features. For instance, their taste cells can be grouped into distinct populations based on Gr expression patterns (e.g., some express sugar-sensitive Grs and others bitter-sensitive Grs). Second, both taxa have peripheral taste organs that contain multiple taste receptor cells (taste sensilla in insects and taste buds in mammals), and there is evidence for extensive processing of taste signals within these organs. Third, both taxa appear to display varying degrees of chemotopic organization in their central taste processing areas. By focusing on similarities (rather than differences), we may gain deeper insight into taste processing mechanisms that are universal across species. Such insights should provide valuable clues for future investigations of taste function.

REFERENCES

Albert, P. J. and Parisella, S. (1988). Feeding preferences of eastern spruce budworm larvae in two-choice tests with extracts of mature foliage and with pure amino acids. *J. Chem. Ecol.* 14: 1649–1656.

Altman, J. S. and Kien, J. (1987). Functional organization of the subesophageal ganglion in arthropods. In *Arthropod brain: its evolution, development, structure and functions*, (ed. Gupta, A.P.), 265–301. New York: John-Wiley & Sons.

Amakawa, T. and Ozaki, M. (1989). Protein kinase C-promosted adaptation of the sugar receptor cell of the blowfly, *Phormia regina. J. Insect Physiol.* 35: 233–237.

Amakawa, T., Ozaki, M. and Kawata, K. (1990). Effects of cyclic GMP on the sugar taste receptor of the fly *Phormia regina. J. Insect Physiol.* 36: 281–286.

Arms, K., Feeny, P. and Lederhouse, R. C. (1974). Sodium: stimulus for puddling behavior by tiger swallowtail butterflies, *Papilio glaucus. Science* 185: 372–374.

Ayestaran, A., Martin Giurfa, M. and de Brito Sanchez, G. (2010). Toxic but drank: gustatory aversive compounds induce post-ingestional malaise in harnessed honeybees. *PLoS ONE* 5: e15000.

Ayrton, A. and Morgan, P. (2001). Role of transport proteins in drug absorption, distribution and excretion. *Xenobiot.* 31: 469–497.

Bate-Smith, E. C. (1972). Attractants and repellents in higher animals. In *Phytochemical ecology: Proceedings of the phytochemical society symposium*, (ed. Harborne, J.B.), 45–56. London: Academic Press.

Bernays, E. A. (1982). The insect on the plant–a closer look. In *Proceedings of the 5th international symposium on insect-plant relationships*, eds. Visser, J. H. and Minks, A. K.), 3–17. Wageningen: Pudoc.

Bernays, E. A. (1985). Regulation of feeding behaviour. In *Comprehensive insect physiology, biochemistry and pharmacology. Vol. 4.*, eds. Kerkut, G. A. and Gilbert, L. I.), 1–32. New York: Pergamon Press.

Bernays, E. A. (1990). Plant secondary compounds deterrent but not toxic to the grass specialist acridid *Locusta migratoria*: implications for the evolution of graminivory. *Ent. Exper. Appl.* 54: 53–56.

Bernays, E. A. (1991). Relationship between deterrence and toxicity of plant secondary metabolites for the grasshopper *Schistocerca americana. J. Chem. Ecol.* 17: 2519–2526.

Bernays, E. A., Blaney, W. M. and Chapman, R. F. (1972). Changes in the chemoreceptor sensilla on the maxillary palps of *Locusta migratoria* in relation to feeding. *J. Exp. Biol.* 57: 745–753.

Bernays, E. A. and Chapman, R. F. (1972). The control of changes in peripheral sensilla associated with feeding in *Locusta migratoria* (L.). *J. Exp. Biol.* 57: 755–763.

Bernays, E. A. and Chapman, R. F. (1987). The evolution of deterrent responses in plant-feeding insects. In *Perspectives in chemoreception and behavior*, eds. Chapman, R.F. Bernays, E. A. and Stoffolano, J. G., Jr.), 159–173. New York: Springer-Verlag.

Bernays, E. A. and Chapman, R. F. (1994). *Host-plant selection by phytophagous insects*. New York: Chapman & Hall.

Bernays, E. A. and Chapman, R. F. (2000). A neurophysiological study of sensitivity to a feeding deterrent in two sister species of *Heliothis* with different diet breadths. *J. Insect. Physiol.* 46: 905–912.

Bernays, E. A. and Chapman, R. F. (2001). Electrophysiological responses of taste cells to nutrient mixtures in the polyphagous caterpillar of *Grammia geneura. J. Comp. Physiol. A* 187: 205–213.

Bernays, E. A., Chapman, R. F. and Hartmann, T. (2002). A taste receptor neurone dedicated to the perception of pyrrolizidine alkaloids in the medial galeal sensillum of two polyphagous arctiid caterpillars. *Physiol. Entomol.* 27: 312–321.

Bernays, E. A., Chapman, R. F., Lamunyon, C. W. and Hartmann, T. (2003). Taste receptors for pyrrolizidine alkaloids in a monophagous caterpillar. *J. Chem. Ecol.* 29: 1709–1722.

Bernays, E. A., Chapman, R. F. and Singer, M. S. (2000). Sensitivity to chemically diverse phagostimulants in a single gustatory neuron of a polyphagous caterpillar. *J. Comp. Physiol A* 186: 13–1.9.

Bernays, E. A., Chapman, R. F. and Singer, M. S. (2004). Changes in taste receptor cell sensitivity in a polyphagous caterpillar reflect carbohydrate but not protein imbalance. *J. Comp. Physiol A* 190: 39–48.

Bernays, E. A., Glendinning, J. I. and Chapman, R. F. (1998). Plant acids modulate chemosensory responses in *Manduca sexta* larvae. *Physiol. Entomol.* 23: 193–201.

Bernays, E. A. and Singer, M. S. (2005). Taste alteration and endoparasites. *Nature* 436: 476.

Bernays, E.A. and Weiss, M.R. (1996). Induced food preferences in caterpillars: the need to identify mechanisms. *Ent. Exp. Appl.* 78: 1–8.

Blaney, W. M. and Duckett, A. M. (1975). The significance of palpation by the maxillary palps of *Locusta migratoria* (L.): an electrophysiological and behavioural study. *J. Exp. Biol.* 63: 701–712.

Blaney, W. M. and Simmonds, M. S.J. (1987). Experience: a modifier of neural and behavioural sensitivity. In *Insects-Plants*, eds. Labeyrie, V. Fabres, G. and Lachaise, D.), 237–241. Dordrecht, Netherlands: Dr. W. Junk Publishers.

Bokil, H., Laaris, N., Blinder, K., et al. (2001). Ephaptic interactions in the mammalian olfactory system. *J. Neurosci.* 21: RC173.

Bowdan, E. (1995). The effects of a phagostimulant and a deterrent on the microstructure of feeding by *Manduca sexta* caterpillars. *Entomol. Exper. Appl.* 77: 297–306.

Braun, G. and Bicker, G. (1992). Habituation of an appetitive reflex in the honeybee. *J. Neurophysiol.* 67: 588–598.

Bredendiek, N., Hütte, J., Steingräber, A., et al. (2011). Goalpha is involved in sugar perception in *Drosophila. Chem. Senses* 36: 69–81.

Brieskorn, C. H. (1990). Physiological and therapeutic aspects of bitter compounds. In *Bitterness in foods and beverages*, (ed. Rouseff, R.L.), 15–33. New York: Elsevier.

Brower, L. P. (1984). Chemical defense in butterflies. In *The biology of butterflies*, eds. Vane-Wright, R.I. and Ackery, P.R.), 109–134. London: Academic Press.

Burke, C. J. and Waddell, S. (2011). Remembering nutrient quality of sugar in *Drosophila. Curr. Biol.* 21: 746–750.

Calas, D., Thiéry, D. and Marion-Poll, F. (2007). 20-Hydroxyecdysone deters oviposition and larval feeding in the European grapevine moth, *Lobesia botrana. J. Chem. Ecol.* 32: 2443–2454.

Cameron, P., Hiroi, M., Ngai, J. and Scott, K. (2010). The molecular basis for water taste in Drosophila. *Nature* 465: 91–95.

Chapman, R. F. (1982). Chemoreception: the significance of sensillum numbers. *Adv. Insect Physiol.* 16: 247–356.

Chaudhari, N. and Roper, S. D. (2010). The cell biology of taste. *J. Cell Biol.* 190: 285–296.

Chen, X., Gabitto, M., Peng, Y., et al. (2011). A gustotopic map of taste qualities in the mammalian brain. *Science* 333: 1262–1266.

Chen, Z., Wang, Q. and Wang, Z. (2010). The amiloride-sensitive epithelial Na^+ channel PPK28 Is essential for *Drosophila* gustatory water reception. *J. Neurosci.* 30: 6247– 6252.

Clyne, P. J., Warr, C. G. and Carlson, J. R. (2000). Candidate taste receptors in *Drosophila*. *Science* 287: 1830–1834.

Cocco, N. and Glendinning, J. I. (2012). Not all sugars are created equal: some mask noxious tastes better than others in an herbivorous insect. *J. Exp. Biol.* 215: 1412–1421.

Colomb, J., Grillenzoni, N., Ramaekers, A. and Stocker, R. F. (2007). Architecture of the primary taste center of *Drosophila melanogaster* larvae. *J. Comp. Neurol.* 502: 834–847.

Dahanukar, A., Lei, Y.-T., Kwon, J.Y. and Carlson, J.R. (2007). Two *Gr* genes underlie sugar reception in *Drosophila*. *Neuron* 56: 503–516.

de Boer, G. and Hanson, F. E. (1987). Differentiation of roles of chemosensory organs in food discrimination among host and non-host plants by larvae of the tobacco hornworm, *Manduca sexta*. *Physiol. Entomol.* 12: 387–398.

de Brito Sanchez, G. M., Guirfa, M., de Paula Mota, R. T. and Guauthier, M. (2005). Electrophysiological and behavioural characterization of gustatory responses to antennal "bitter" taste in honeybees. *Eur. J. Neurosci.* 22: 3161–3170.

del Campo, M. L. and Miles, C. I. (2003). Chemosensory tuning to a host recognition cue in the facultative specialist larvae of the moth *Manduca sexta*. *J. Exp. Biol.* 206: 3979–3990.

del Campo, M. L., Miles, C. I., Schroeder, F. C., et al. (2001). Host recognition by the tobacco hornworm is mediated by a host plant compound. *Nature* 411: 186–189.

del Campo, M. L. and Renwick, J. A. A. (2000). Induction of host specificity in larvae of *Manduca sexta*: chemical dependence controlling host recognition and developmental rate. *Chemoecology.* 10: 115–121.

den Otter, C.J. (1972). Differential sensitivity of insect chemoreceptors to alkalai cations. *J. Insect Physiol.* 18: 109–131.

Dethier, V. G. (1968). Chemosensory input and taste discrimination in the blowfly. *Science* 161: 389–391.

Dethier, V. G. (1973). Electrophysiological studies of gustation in Lepidopterous larvae. II. Taste spectra in relation to food-plant discrimination. *J. Comp. Physiol.* 82: 103–134.

Dethier, V. G. (1976). The hungry fly: a physiological study of the behavior associated with feeding. Cambridge, MA: Harvard University Press.

Dethier, V. G. and Bowdan, E. (1989). The effect of alkaloids on the sugar receptors of the blowfly. *Physiol. Entomol.* 14: 127–136.

Dethier, V. G. and Crnjar, R. M. (1982). Candidate codes in the gustatory system of caterpillars. *J. Gen. Physiol.* 79: 549–569.

Dethier, V. G. and Evans, D. R. (1961). The physiological control of water ingestion in the blowfly. *Biol Bull.* 121: 108–116.

Dethier, V. G. and Kuch, J. H. (1971). Electrophysiological studies of gustation in lepidopterous larvae. I. Comparative sensitivity to sugars, amino acids, and glycosides. *Z. Veregl. Physiol.* 72: 343–363.

Devitt, B. D. and Smith, J. J. B. (1985). Action of mouthparts during feeding in the dark-sided cutworm, *Euoxa messoria* (Lepidoptera: Noctuidae). *Can. Entomol.* 117: 343–349.

Duerr, J. S. and Quinn, W. G. (1982). Three *Drosophila* mutations that block associative learning also affect habituation and sensitization. *PNAS* 79: 3646–3650.

Dus, M., Min, S., Keene, A. C., et al. (2011). Taste-independent detection of the caloric content of sugar in *Drosophila*. *PNAS* 108: 11644–11649.

Evans, D. R. (1961). Control of the responsiveness of the blowfly to water. *Nature* 190: 1132–1133.

Evans, D. R. and Mellon, D. (1962). Electrophysiological studies of a water receptor associated with the taste sensilla of the blowfly. *J. Gen. Physiol.* 45: 487–500.

Fischler, W., Kong, P., Marella, S. and Scott, K. (2007). The detection of carbonation by the *Drosophila* gustatory system. *Nature* 448: 1054–1058.

Frank, M. E. and Nowlis, G. H. (1989). Learned aversions and taste qualities in hamsters. *Chem. Senses* 14: 379–394.

Fujita, M. and Tanimura, T. (2011). *Drosophila* evaluates and learns the nutritional value of sugars. *Curr. Biol.* 21: 751–755.

Furuyama, A., Koganezawa, M. and Shimida, I. (1999). Multiple receptor sites for nucleotide reception in the labellar taste receptor cells of the fleshfly *Boettcherisca peregrina*. *J. Insect Physiol.* 45: 249–255.

Garcia, J. and Hankins, W.G. (1975). The evolution of bitter and the acquisition of toxiphobia. In *Olfaction and taste. V. Proceedings of the 5th international symposium in Melbourne, Australia*, eds. Denton, D.A. and Coghlan, J.P.), 39–45. New York: Academic Press.

Gelperin, A. (1968). Feeding behaviour of the praying mantis: a learned modification. *Nature* 217: 399–400.

Gerber, B., Scherer, S., Neuser, K., et al. (2004). Visual learning in individually assayed *Drosophila* larvae. *J. Exp. Biol.* 207: 179–188.

Glendinning, J. I. (1994). Is the bitter rejection response always adaptive? *Phyisol. Behav.* 56: 1217–1227.

Glendinning, J.I. (2002). How do herbivorous insects cope with noxious secondary plant compounds in their diet? *Ent. Exper. Appl.* 104: 15–25.

Glendinning, J.I., Brown, H., Capoor, M., et al. (2001). A peripheral mechanism for behavioral adaptation to specific "bitter" taste stimuli in an insect. *J. Neurosci.* 21: 3688–3696.

Glendinning, J. I., Davis, A. and Rai, M. (2006). Temporal coding mediates discrimination of "bitter" taste stimuli by an insect. *J. Neurosci.* 26: 8900–8908.

Glendinning, J. I., Davis, A. and Ramaswamy, S. (2002). Contribution of different taste cells and signaling pathways to the discrimination of "bitter" taste stimuli by an insect. *J. Neurosci.* 22: 7281–7287.

Glendinning, J. I., Domdom, S. and Long, E. (2001b). Selective adaptation to noxious foods by an insect. *J. Exp. Biol.* 204: 3355–3367.

Glendinning, J. I., Ensslen, S., Eisenberg, M. E. and Weiskopf, P. (1999a). Diet-induced plasticity in the taste system of an insect: localization to a single transduction pathway in an identified taste cell. *J. Exp. Biol.* 202: 2091–2102.

Glendinning, J. I., Foley, C., Loncar, I. and Rai, M. (2009). Induced preference for host plant chemicals in the tobacco hornworm: contribution of olfaction and taste. *J. Comp. Physiol. A* 195: 591–601.

Glendinning, J. I. and Gonzalez, N. A. (1995). Gustatory habituation to deterrent compounds in a grasshopper: concentration and compound specificity. *Anim. Behav.* 50: 915–927.

Glendinning, J. I., Jerud, A. and Weinberg, A. (2007). The hungry caterpillar: an analysis of how carbohydrates stimulate feeding in *Manduca sexta*. *J. Exp. Biol.* 210: 3054–3067.

Glendinning, J. I., Nelson, N. and Bernays, E. A. (2000). How do inositol and glucose modulate feeding in *Manduca sexta* caterpillars? *J. Exp. Biol.* 203: 1299–1315.

Glendinning, J. I., Tarre, M. and Asaoka, K. (1999b). Contribution of different bitter-sensitive taste cells to feeding inhibition in a caterpillar (*Manduca sexta*). *Behav. Neurosci.* 113: 840–854.

Goldrich, N. R. (1973). Behavioral responses of *Phormia regina* (Meigen) to labellar stimulation with amino acids. *J. Gen. Physiol.* 61: 74–88.

Gordon, M. D. and Scott, K. (2009). Motor control in a *Drosophila* taste circuit. *Neuron* 61: 373–384.

Grabowski, C. T. and Dethier, V. G. (1954). The structure of the tarsal chemoreceptors of the blowfly, *Phormia regina* Meigen. *J. Morph.* 94: 1–20.

Grill, H. J. and Berridge, K. C. (1985). Taste reactivity as a measure of the neural control of palatability. *Prog. Psychobiol. Physiol. Psychol.* 11: 1–61.

Griss, C., Simpson, S. J., Rohrbacher, J. and Rowell, C. H. F. (1991). Localization in the central nervous system of larval *Manduca sexta* (Lepidoptera: Sphingidae) of areas responsible for aspects of feeding behavior. *J. Insect Physiol.* 37: 477–482.

Hanson, F. E. and Dethier, V. G. (1973). Role of gustation and olfaction in food plant discrimination in the tobacco hornworm, *Manduca sexta. J. Insect Physiol.* 19: 1019–1034.

Harley, K. L. S. and Thorsteinson, A. J. (1967). The influence of plant chemicals on the feeding behavior, development, and survival of the two-striped grasshopper, *Melanoplus bivittatus* (Say), Acrididae: Orthoptera. *Can. J. Zool.* 45: 305–319.

Hendel, T., Michels, B., Neuser, K., et al. (2005). The carrot, not the stick: appetitive rather than aversive gustatory stimuli support associative olfactory learning in individually assayed *Drosophila* larvae. *J. Comp. Physiol. A* 191: 265–278.

Herness, S. and Zhao, F. (2009). The neuropeptides CCK and NPY and the changing view of cell-to-cell communication in the taste bud. *Physiol. Behav.* 97: 581–591.

Hinde, R. A. (1970). Behavioural habituation. In *Short-term changes in neural activity and behaviour*, eds. Horn, G. and Hinde, R.A.), 3–40. Cambridge: Cambridge University Press.

Hirao, T. and Ariai, N. (1990). Gustatory and feeding responses to amino acids in the silkworm, *Bombyx mori. Jap. J. Appl. Entomol. Zool.* 35: 73–76.

Hiroi, M., Meunier, N., Marion-Poll, F. and Tanimura, T. (2004). Two antagonistic gustatory receptor neurons responding to sweet-salty and bitter taste in *Drosophila. J. Neurobiol.* 61: 333–342.

Hodgson, E. S., Lettvin, J. Y. and Roeder, K. D. (1955). Physiology of a primary chemoreceptor unit. *Science* 122: 417–418.

Howlett, N., Dauber, K., Shukla, A., et al. (2012). Identification of chemosensory receptor genes in *Manduca sexta* and knockdown by RNA interference. *BMC Genomics* 13:211. doi: 10.1186/1471-2164-13-211.

Ignell, R. and Hansson, B. S. (2005). Projection patterns of gustatory neurons in the suboesophageal ganglion and tritocerebrum of mosquitoes. *J. Comp. Neurol.* 492: 214–233.

Inoshita, T. and Tanimura, T. (2006). Cellular identification of water gustatory receptor neurons and their central projection pattern in *Drosophila. PNAS* 103: 1094–1099.

Ishikawa, S. (1963). Responses of maxillary chemoreceptors in the larva of the silkworm, *Bombyx mori*, to stimulation by carbohydrates. *J. Cell. Comp. Physiol.* 61: 99–107.

Ishimoto, H., Takahashi, K., Ueda, R. and Tanimura, T. (2005). G-protein gamma subunit 1 is required for sugar reception in Drosophila. *EMBO J.* 24: 3259–3265.

Isidoro, N., Solinar, M., Baur, R., et al. (1993). Functional morphology of a tarsal sensillum of *Delia radicum* L (Diptera, Anthomyiidae) sensitive to important host-plant compounds. *Int. J. Insect Morph. Embryol.* 39: 275–281.

Isono, K. and Morita, H. (2010). Molecular and cellular designs of insect taste receptor system. *Front. Cell. Neurosci.* 4: 20: 1–16.

Jefferys, J. G. (1995). Nonsynaptic modulation of neuronal activity in the brain: electric currents and extracellular ions. *Physiol. Rev.* 75: 689–723.

Jermy, T., Bernays, E. A. and Szentesi, A. (1982). The effect of repeated exposure to feeding deterrents on their acceptability to phytophagous insects. In *Proceedings of the 5th international symposium on insect-plant relationships*, eds. Visser, H. and Minks, A.), 25–32. Wageningen: PUDOC.

Jiao, Y., Moon, S. J. and Montell, C. (2007). A *Drosophila* gustatory receptor required for the responses to sucrose, glucose, and maltose identified by mRNA tagging. *PNAS* 104: 14110–14115.

Jørgensen, K., Kvello, P., Jørgen Almaas, T. and Mustaparta, H. (2006). Two closely located areas in the subesophageal ganglion and the tritocerebrum receive projections from gustatory receptor neurons located on the antennae and the proboscis in the moth *Heliothis virescens. J. Comp. Neurol.* 496: 121–134.

Kan, H., Kataoka-Shirasugi, N. and Amakawa, T. (2008). Transduction pathways mediated by second messengers including cAMP in the sugar receptor cell of the blow fly: study by the whole cell clamp method. *J. Insect Physiol.* 54: 1028– 1034.

Kan, H., Kataoka-Shirasugi, N. and Amakawa, T. (2011). Multiple pathways from three types of sugar receptor sites to metabotropic transduction pathways of the blowfly: study by the whole cell-clamp experiments. *Comp. Biochem. Physiol. A* 160: 94–99.

Katz, D. B. (2005). The many flavors of temporal coding in the gustatory cortex. *Chem. Senses* 30 (Suppl 1): 80–81.

Katz, D.B., Nicolelis, M.A. and Simon, S.A. (2002). Gustatory processing is dynamic and distributed. *Curr. Opin. Neurobiol.* 12: 448–454.

Kent, K. S. and Hildebrand, J. G. (1987). Cephalic sensory pathways in the central nervous system of larval *Manduca sexta* (Lepidoptera: Sphingidae). *Phil. Trans. Roy. Soc. Lond., B. Biol. Sci.* 315: 1–36.

Kent, L. B. and Roberston, H. M. (2009). Evolution of the sugar receptors in insects. *BMC Evol. Biol.* 18: 41.

Kim, H. S., Lee, Y., Akitake, B., et al. (2010). *Drosophila* TRPA1 channel mediates chemical avoidance in gustatory receptor neurons. *PNAS* 107: 8440–8445.

Kim, J. H. and Mullin, C. A. (1998). Structure-phagostimulatory relationships for amino acids in adult western corn rootworm, *Diabrotica virgifera virgifera. J. Chem. Ecol.* 24: 1499–1511.

Kwon, J., Dahanukar, A., Weiss, L. A. and Carlson, J. R. (2011). Molecular and cellular organization of the taste system in the *Drosophila* larva. *J. Neurosci.* 31: 15300–15309.

Lee, J. C. and Bernays, E. A. (1990). Food tastes and toxic effects: Associative learning by the phytophagous grasshopper *Schistocerca americana* (Drury) (Orthoptera: Acrididae). *Anim. Behav.* 39: 163–173.

Lee, Y., Kang, M.J., Shim, J., et al. (2012). Gustatory receptors required for avoiding the insecticide L-canavanine. *J. Neurosci.* 32: 1429–1435.

Lee, Y., Moon, S. J. and Montell, C. (2009). Multiple gustatory receptors required for the caffeine response in *Drosophila. PNAS* 106: 4495–4500.

Lewis, C. T. (1954). Studies concerning the uptake of contact insecticides. I. The anatomy of the tarsi of certain Diptera of medical importance. *Bull. Ent. Res.* 45: 711–722.

Liu, L., Soren Leonard, A., Price, M. P., et al. (2003). Contribution of *Drosophlia* DEG/ENaC genes to salt taste. *Neuron* 39: 133–149.

Ma, W.-C. (1969). Some properties of gustation in the larvae of *Pieris brassicae. Ent. Exper. Appl.* 12: 584–590.

Ma, W.-C. (1972). Dynamics of feeding responses in *Pieris brassicae* Linn as a function of chemosensory input: a behavioral and electrophysiological study. *Meded. Laudbouwhogeschool Wageningen* 72–11: 1–162.

Ma, W.-C. and Schoonhoven, L. M. (1973). Tarsal contact chemoreception hairs of the large white butterfly *Pieris brassicae* and their possible role in oviposition behavior. *Ent. Exper. Appl.* 16: 343–357.

Maes, F. W. and Bijpost, S. C. A. (1979). Classical conditioning reveals discrimination of salt taste quality in the blowfly *Calliphora vicina. J. Comp. Physiol.* 133: 53–62.

Manzo, A., Silies, M., Gohl, D. M. and K., S. (2012). Motor neurons controlling fluid ingestion in *Drosophila. PNAS* 109: 6307–6312.

Marella, S., Fischler, W., Kong, P., et al. (2006). Imaging taste responses in the fly brain reveals a functional map of taste category and behavior. *Neuron* 49: 285–295.

Marella, S., Mann, K. and Scott, K. (2012). Dopaminergic modulation of sucrose acceptance behavior in *Drosophila. Neuron* 73: 941–950.

Marion-Poll, F. and Descoins, C. (2002). Taste detection of phytoecdysteriods in larvae of *Bombyx mori, Spodoptera littoralis* and *Ostrinia nubilalis, J. Insect Physiol.* 48: 467–476.

Masek, P. and Scott, K. (2010). Limited taste discrimination in *Drosophila. Proc. Natl. Acad. Sci. U.S.A.* 107: 14833–14838.

Meunier, N., Marion-Poll, F. and Lucas, P. (2009). Water taste transduction pathway is calcium dependent in *Drosophila. Chem. Senses* 34: 441–449.

Meunier, N., Marion-Poll, F., Rospars, J. P. and Tanimura, T. (2003). Peripheral coding of bitter taste in *Drosophila. J. Neurobiol.* 56: 139–152.

Millius, M., Merivee, E., Williams, I., et al. (2006). A new method for electrophysiological identification of antennal pH receptor cells in ground beetles: the example of *Pterostichus aethiops* (Panzer, 1796) (Coleoptera, Carabidae). *J. Insect Physiol.* 52: 960–967.

Minnich, D. E. (1921). An experimental study of the tarsal chemoreceptors of two nymphalid butterlies. *J. Exp. Zool.* 33: 172–203.

Minnich, D. E. (1929). The chemical sensitivity of the legs of the blowfly, *Calliphora vomitoria* Linn., to various sugars. *Z. Vergl. Physiol.* 11: 1–55.

Minnich, D. E. (1931). The sensitivity of the oral lobes of the probosscis of the blowfly, Calliphora vomitoria Linn., to various sugars. *J. Exp. Zool.* 60: 121–139.

Mitchell, B. K. and Itagaki, H. (1992). Interneurons of the subesophageal ganglion of *Sarcophaga bullata* responding to gustatory and mechanosensory stimuli. *J. Comp. Physiol.* A 171: 213–230.

Miyakawa, Y. (1981). Bimodal response in a chemotactic behaviour of *Drosophila* larvae to monvalent salts. *J. Insect Physiol.* 27: 387–392.

Miyakawa, Y. (1982). Behavioural evidence for the existence of sugar, salt and amino acid taste receptor cells and some of their properties in *Drosophila* larvae. *J. Insect Physiol.* 28: 405–410.

Montell, C. (2009). A taste of the *Drosophila* gustatory receptors. *Curr. Opin. Neurobiol.* 19: 345–353.

Moon, S. J., Lee, Y., Jiao, Y. and Montell, C. (2009). A *Drosophila* gustatory receptor essential for aversive taste and inhibiting male-to-male courtship. *Current Biol.* 19: 1623–1627.

Morita, H. (1992). Transduction process and impulse initiation in insect contact chemoreceptor. *Zool. Sci.* 9: 1–16.

Mueller, K.L., Hoon, M.A., Erlenback, I., et al. (2005). The receptors and coding logic for bitter taste. *Nature* 434: 225–229.

Murakami, M. and Kijima, H. (2000). Transduction ion channels directly gated by sugars on the insect taste cell. *J. Gen. Physiol.* 15: 455–466.

Nakamura, M., Baldwin, D., Hannaford, S., et al. (2002). Defective proboscus extension response (DPR), a member of the Ig superfamily required for the gustatory response to salt. *J. Neurosci.* 22: 3463–3472.

Nakamura, T., Murata, Y., Mashiko, M., et al. (2005). The nitric oxide–cyclic GMP cascade in sugar receptor cells of the blowfly, *Phormia regina. Chem. Senses* 30 (Suppl.1): i281–i282.

Omura, H. and Honda, K. (2003). Feeding respo nses of adult butterflies, *Nymphalis xanthomelas, Kaniska canace,* and *Vanessa indica,* to components of tree sap and rotting fruits: synergistic effects of ethanol and acetic acid on sugar responsiveness. *J. Insect Physiol.* 49: 1031–1038.

Ômura, H., Honda, K., Asaoka, K. and Inoue, T. A. (2008). Tolerance to fermentation products in sugar reception: gustatory adaptation of adult butterly proboscis for feeding on rotting foods. *J. Comp. Physiol.* A 194: 545–555.

Ouyang, Q., Sato, H., Murata, Y., et al. (2009). Contribution of the inositol 1,4,5-trisphosphate transduction cascade to the detection of "bitter" compounds in blowflies. *Comp. Biochem. Physiol.* A 153: 309–316.

Pollack, G. S. and Balakrishnan, R. (1997). Taste sensilla of flies: function, central neuronal projections, and development. *Microsc. Res. Tech.* 39: 532–546.

Rogers, S. M. and Newland, P. L. (2002). Gustatory processing in thoracic local circuits of locusts. *J. Neurosci.* 22: 8324–8333.

Rouseff, R. L. (1990). *Bitterness in foods and beverages.* New York: Elsevier Science Publishers.

Salloum, A., Violaine Colson, V. and Marion-Poll, F. (2011). Appetitive and aversive learning in *Spodoptera littoralis* larvae. *Chem. Senses* 36: 725–731.

Sasaki, K. and Asaoka, K. (2006). Swallowing motor pattern triggered and modified by sucrose stimulation in the larvae of the silkworm, *Bombyx mori. J. Insect Physiol.* 52: 528–537.

Sato, K., Tanaka, K. and Touhara, K. (2011). Sugar-regulated cation channel formed by an insect gustatory receptor. *PNAS* 108: 11680–11685.

Scheiner, R. (2004). Responsiveness to sucrose and habituation of the proboscis extension response in honey bees. *J. Comp. Physiol.* A 190: 727–733.

Scheiner, R., Sokolowski, M. B. and Erber, J. (2004). Activity of cGMP-dependent protein kinase (PKG) affects sucrose responsiveness and habituation in *Drosophila melanogaster. Learn. Mem.* 11: 303–311.

Schmidt, J. M. and Friend, W. G. (1991). Ingestion and diet destination in the mosquito *Culiseta inornata*: effects of carbohydrate configuration. *J. Insect Physiol.* 37: 817–828.

Schnuch, M. and Hansen, K. (1990). Sugar sensitivity of a labellar salt receptor of the blowfly *Protophormia terraenovae. J. Insect Physiol.* 36: 409–417.

Schoonhoven, L. M. (1969). Sensitivity changes in some insect chemoreceptors and their effect on food selection behavior. *Proc. Kon. Ned. Akad. Wet. Amsterdam, Ser. C* 72: 491–498.

Schoonhoven, L. M. (1977). On the individuality of insect feeding behavior. *Proc. Koninkl. Ned. Akad. Wetensch. (C)* 80: 341–350.

Schoonhoven, L. M. (1978). Long-term sensitivity changes in some insect taste receptors. *Drug Res.* 28: 2367.

Schoonhoven, L.M., Blaney, W. M. and Simmonds, M. S. J. (1992). Sensory coding of feeding deterrents in phytophagous insects. In *Insect-Plant Interactions, Volume IV*, (ed. Bernays, E.A.), 59–79. Boca Raton: CRC Press.

Schoonhoven, L. M. and Blom, F. (1988). Chemoreception and feeding behavior in a caterpillar: towards a model of brain functioning in insects. *Ent. Exper. Appl.* 49: 123–129.

Schoonhoven, L. M. and van Loon, J. J. A. (2002). An inventory of taste in caterpillars: each species is its own key. *Acta Zool. Acad. Sci. Hung.* 48 (Suppl. 1): 215–263.

Schröter, U. and Menzel, R. (2003). A new ascending sensory tract to the calyses of the honeybee mushroom body, the subesophageal-caclcal tract. *J. Comp. Neurol.* 465: 168–178.

Schultz, J. C. and Lechowicz, M. J. (1986). Host-plant, larval age, and feeding behavior influence midgut pH in the gypsy moth (*Lymantria dispar*). *Oecol.* 71: 133–137.

Seno, K., Fujikawa, K., Nakamura, T. and Ozaki, M. (2005a). Gq-alpha subunit mediates receptor site-specific adaptation in the sugar taste receptor cell of the blowfly, Phormia regina. *Neurosci. Lett.* 377: 200–205.

Seno, K., Nakamura, T. and Ozaki, M. (2005b). Biochemical and physiological evidence that calmodulin is involved in the taste response of the sugar receptor cells of the blowfly, *Phormia regina. Chem. Senses* 30: 497–504.

Shanbhag, S. R. and Singh, R. N. (1992). Functional implications of the projections of neurons from individual labellar sensillum of *Drosophila melangaster* as revealed by neuronal-marker horseradish peroxidase. *Cell Tissue Res.* 267: 273–282.

Shields, V. D. (1996). Comparative external ultrastructure and diffusion pathways in styloconic sensilla on the maxillary galea of larval *Mamestra configurata* (Walker) (Lepidoptera: Noctuidae) and five other species. *J. Morph.* 228: 89–105.

Shields, V. D. C. (1994). Ultrastructure of the uniporous sensilla on the galea of larval *Mamestra configurata* (Walker) (Lepidoptera: Noctuidae). *Can. J. Zool.* 72: 2016–2031.

Shields, V. D. C. and Mitchell, B. K. (1995). Responses of maxillary styloconic receptors to stimulation by sinigrin, sucrose and inositol in two crucifer-feeding, polyphagous lepidopterous species. *Phil. Trans. R. Soc. Lond. B* 347: 447–457.

Shiraishi, A. and Kuwabara, M. (1970). The effect of amino acids on the labellar hair chemosensory cells of the fly. *J. Gen. Physiol.* 56: 768–782.

Silbering, A. F. and Benton, R. (2010). Ionotropic and metabotropic mechanisms in chemoreception: 'chance or design'? *EMBO Reports* 11: 173–179.

Simmonds, M. S. J. and Blaney, W. M. (1983). Some neurophysiological effects of azadirachtin on lepidopterous larvae and their feeding response. In *Proceedings of the Second International Neem Conference*, eds. Schmutterer, H. and Ascher, K.R.S.), 163–180. Eschborn: Deutsche Gesellschaft für Technische Zusammernarbeit (GTZ) GmbH.

Simmonds, M. S. J. and Blaney, W. M. (1991). Gustatory codes in lepidopterous larvae. *Symp. Biol. Hung.* 39: 17–27.

Simmonds, M.S.J., Simpson, S.J. and Blaney, W.M. (1992). Dietary selection behaviour in *Spodoptera littoralis*: the effects of conditioning diet and conditioning period on neural responsiveness and selection behavior. *J. Exp. Biol.* 162: 73–90.

Simpson, S. J. and Simpson, C. L. (1992). Mechanisms controlling modulation by haemolymph amino acids of gustatory responsiveness in the locust. *J. Exp. Biol.* 168: 269–287.

Singer, M. S., Mace, K. C. and Bernays, E. A. (2009). Self-medication as adaptive plasticity: increased ingestion of plant toxins by parasitized caterpillars. *PLoS ONE* 4: e4796.

Slone, J., Daniels, J. and Amrein, H. (2007). Sugar receptors in *Drosophila*. *Curr. Biol.* 17: 1809–1816.

Smith, D. V., St. John, S. J. and Boughter, J. D. (2000). Neuronal cell types and taste quality coding. *Physiol. Behav.* 69: 77–85.

Smith, J. J. B., Mitchell, B. K., Rolseth, B. M., et al. (1990). SAPID Tools: microcomputer programs for an analysis of multi-unit nerve recordings. *Chem. Senses* 15: 253–270.

Solari, P., Masala, C., Falchi, A.M., et al. (2010). The sense of water in the blowfly *Protophormia terraenovae*. *J. Insect Physiol.* 56: 1825–1833.

Spector, A.C. (2000). Linking gustatory neurobiology to behavior in vertebrates. *Neurosci. Biobehav. Rev.* 24: 391–416.

Spector, A. C. and Grill, H. J. (1992). Salt taste discrimination after bilateral section of the chorda tympani or glossopharyngeal nerves. *Am. J. Physiol.* 263: R169–R176.

Städler, E., Renwick, J. A. A., Radke, C. D. and Sachdev-Gupta, K. (1995). Tarsal contact chemoreceptor response to glucosinolates and cardenolides mediating ovipositioning in *Pieris rapae*. *Physiol. Entomol.* 20: 175–187.

Steinbrecht, R. A. (1989). The fine structure of thermohygrosensitive sensilla in the silkmoth *Bombyx mori*: receptor membrane structure and sensory cell contacts. *Cell Tissue Res.* 255: 49–57.

Stocker, R. F. (1994). The organization of the chemosensory system in *Drosophila melanogaster*: a review. *Cell Tissue Res.* 275: 3–26.

Stocker, R. F. and Schorderet, M. (1981). Cobalt filling of sensory projections from internal and external mouthparts in *Drosophila*. *Cell Tissue Res.* 216: 513–523.

Szentesi, A. and Bernays, E. A. (1984). A study of behavioural habituation to a feeding deterrent in nymphs of *Schistocerca gregaria*. *Physiol. Entomol.* 9: 329–340.

Szentesi, A. and Jermy, T. (1990). The role of experience in host plant choice by phytophagous insects. In *Insect-plant interactions*, vol. 2 (ed. Bernays, E.A.), 39–74. Boca Raton: CRC Press.

Tanaka, Y., Asaoka, K. and Takeda, S. (1994). Different feeding and gustatory responses to ecdysone and 20-hydroxyecdysone by larvae of the silkworm, *Bombyx mori*. *J. Chem. Ecol.* 20: 125–133.

Thompson, R. F. and Spencer, W. A. (1966). Habituation: a model phenomenon for the study of neuronal substrates of behavior. *Psychol. Rev.* 73: 16–43.

Thorne, N. and Amrein, H. (2008). Atypical expression of *Drosophila* gustatory receptor genes in sensory and central Neurons. *J. Comp. Neurol.* 506: 548–568.

Thorne, N., Chromey, C., Bray, S. and Amrein, H. (2004). Taste perception and coding in *Drosophila*. *Curr. Biol.* 14: 1065–1079.

Usher, B. F., Bernays, E. A. and Barbehenn, R. V. (1988). Antifeedant tests with larvae of *Pseudaletia unipuncta*: variability of behavioral response. *Ent. Exper. Appl.* 48: 203–212.

van der Goes van Naters, W. M. and den Otter, C. J. (1998). Amino acids as taste stimuli for tsetse flies. *Physiol. Entomol.* 23: 278–284.

van Loon, J. J. A. and van Eeuwijk, F. A. (1989). Chemoreception of amino acids in larvae of two species of *Pieris*. *Physiol. Entomol.* 14: 459–469.

Vandenbeuch, A., Clapp, T. R. and Kinnamon, S. C. (2008). Amiloride-sensitive channels in type I fungiform taste cells in mouse. *BMC Neurosci. 2008,* 9: 1.

Vermehren-Schmaedick, A., Scudder, C., Timmermans, W. and Morton, D. B. (2011). Drosophila gustatory preference behaviors require the atypical soluble guanylyl cyclases. *J. Comp. Physiol. A* 197: 717–727.

Vosshall, L.B. and Stocker, R.F. (2007). Molecular architecture of smell and taste in *Drosophila. Annu. Rev. Neurosci.* 30: 505–533.

Wacht, S., Lunau, K. and Hansen, K. (2000). Chemosensory control of pollen ingestion in the hoverfly *Eristalis tenax* by labellar taste hairs. *J. Comp. Physiol. A* 186: 193–203.

Wang, Z., Singhvi, A., Kong, P. and Scott, K. (2004). Taste representations in the *Drosophila* brain. *Cell* 117: 981–991.

Wanner, K. W. and Robertson, H. M. (2008). The gustatory receptor family in the silkworm moth Bombyx mori is characterized by a large expansion of a single lineage of putative bitter receptors. *Insect Molec. Biol.* 17: 621–629.

Watanabe, H. and Mizunami, M. (2006). Classical conditioning of activities of salivary neurons in the cockroach. *J. Exp. Biol.* 209: 766–779.

Weiss, L. A., Dahanukar, A., Kwon, J. Y., et al. (2011). The molecular and cellular basis of bitter taste in *Drosophila. Neuron* 69: 258–272.

White, P. R. and Chapman, R. F. (1990). Tarsal chemoreception in the polyphagous grasshopper *Schistocerca americana*: behavioural assays, sensilla distributions and electrophysiology. *Physiol. Entomol.* 15: 105–121.

White, P. R., Chapman, R. F. and Ascoli-Christensen, A. (1990). Interactions between two neurons in contact chemosensilla of the grasshopper, *Schistocerca americana*. *J. Comp. Physiol. A* 167: 431–436.

Wiggins, L. L., Frank, R. A. and Smith, D. V. (1989). Generalization of learned taste aversions in rabbits: similarities among gustatory stimuli. *Chem. Senses* 14: 103–119.

Wolbarsht, M. L. (1957). Water taste in *Phormia. Science* 125: 1248.

Wolbarsht, M. L. and Hanson, F. E. (1965). Electrical and behavioral responses to amino acid stimulation in the blowfly. In *Olfaction and taste, II*, (ed. Hayashi, T.), 749–760. New York: Oxford.

Wright, G. A., Mustard, J. A., Simcock, N. K., et al. (2010). Parallel reinforcement pathways for conditioned food aversions in the honeybee. *Curr. Biol.* 20: 1–7.

Yamamoto, R. T. (1974). Induction of hostplant specificity in the tobacco hornworm, *Manduca sexta. J. Insect Physiol.* 20: 641–650.

Zhang, Y.-F., van Loon, J. J. A. and Wang, C.-Z. (2010). Tarsal taste neuron activity and proboscis extension reflex in response to sugars and amino acids in *Helicoverpa armigera* (Hubner). *J. Exp. Biol.* 213: 2889–2895.

Zhao, G. Q., Zhang, Y., Hoon, M. A., et al. (2003). The receptors for mammalian sweet and umami taste. *Cell* 115: 255–266.

Chapter 42

Taste in Aquatic Vertebrates

Toshiaki J. Hara

"Our head may look incredibly complicated, but it is built from a simple and elegant blueprint. There is a pattern common to every skull on earth, whether it belongs to a shark, a bony fish, a salamander, or a human."

Neil Shubin (2008)

42.1 INTRODUCTION

Some of the earliest vertebrates, derived from ancestral chordates, were adapted to life as filter feeders in an aquatic environment, probably very much the way that larval lampreys and some jawed fishes survive today. Some fishes suck up mouthfuls of bottom sediments, and some predatory fishes suck in mouthfuls of water carrying the prey into their mouths. All of these methods of feeding require mechanisms for differentiating between edible and the inedible, or between the favorable and unfavorable. These mechanisms are the chemical senses of gustation (taste) and olfaction (smell). The relation of gustation to food intake is manifested in all vertebrates including humans, but many aspects of the complex integration of chemosensory input with antecedent or concomitant physiological activities remain mostly unclear. The evolutionary development of the chemosensory system associated with feeding in vertebrates followed the course in which gustatory perception is responsible for basic food appraisal and bestows the animals with valuable discriminatory power.

Fish, constituting slightly more than one-half of the total number of about 50,000 recognized living vertebrate species, are a pivotal group that has served as a vertebrate model for the study of the gustatory system. Taste buds were first described in fish in early 1800s, and by the turn of the century the peripheral and central gustatory neural organization had been clearly illustrated (Herrick, 1904, 1905). Further elaboration has since made focusing on topographic representation of the gustatory and motor roots in the fish brains (Finger, 1987; Kanwal and Finger, 1992). The gustatory receptor systems in aquatic vertebrates are unique in that, unlike those of the terrestrial, receptors are stimulated by dilute solutions, or "distance chemical receptors". However, there exists a broad line of similarity between both the peripheral and the central gustatory paths in all vertebrate types (Hara, 2007, 2011a), and fishes with enormous hypertrophy/specialization of the gustatory centers may serve as a guide to point the way for researches on the gustatory pathways of higher vertebrates where the system is less accessible. Thus, this chapter is concerned with aspects of the gustatory systems of aquatic vertebrates in perspective of those of all vertebrates, with an emphasis on new insights into the taste bud development, central neural pathways, and physiological as well as behavioral implications.

42.2 STRUCTURAL ORGANIZATION

42.2.1 Peripheral Gustatory Systems

42.2.1.1 Taste Buds. Taste buds, also termed terminal buds, end buds, or cutaneous buds, constitute the structural basis of the peripheral gustatory organ (Figure 42.1). Taste buds occur in virtually all extant vertebrates. Taste (terminal) buds were first described in lampreys; their histology and central connections in larval and adult forms are comparable to those of the gnathostome vertebrates (Baatrup, 1983). The lamprey terminal buds show strong similarity to the taste buds of

---

Handbook of Olfaction and Gustation, Third Edition. Edited by Richard L. Doty.
© 2015 Richard L. Doty. Published 2015 by John Wiley & Sons, Inc.

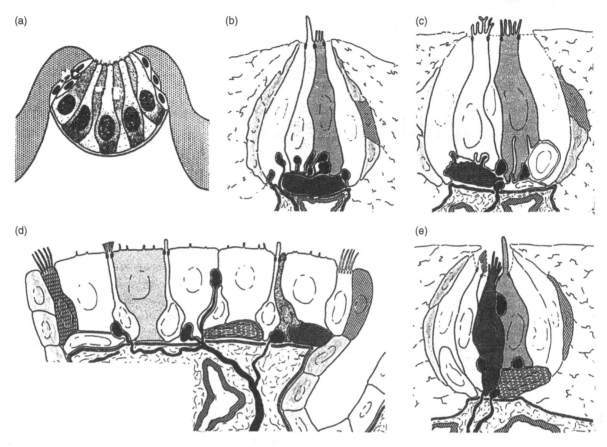

Figure 42.1 Schematic drawings of terminal bud, taste bud, or taste disc, typical of aquatic vertebrates, with the mammalian for comparison. (a) lamprey (b) teleost fish (c) urodelan amphibian and juvenile anuran (d) frog, a salientian amphibian (e) mammal. Modified from Baatrup (1983) and Reutter and Witt (1993).

higher vertebrates, despite the fact that the ciliated cells instead of the microvillar are believed to be the receptor. The presence of cilia may be a specialized character, but is more likely to be another example of the phylogenetic originality of ciliated receptors (Gans and Northcutt, 1983). In teleosts, taste buds are distributed in five subpopulations, reflecting feeding habits, strategies, and habitats of fishes: (1) oral, (2) palatal and laryngeal, (3) branchial (gills), (4) cutaneous, and (5) barbels. Most teleosts lack external taste buds, although some have independently evolved external taste buds on specialized parts of the body, such as barbels, elongated fin rays, and the lips of the mouth. Yellow bullhead catfish (*Ictalurus natalis*), for example, have more than 175,000 taste buds on the entire body surface alone (Atema, 1971). Carp (*Cyprinus carpio*) and goldfish (*Carassius auratus*) have evolved a highly sophisticated food separation system on the roof of the mouth, i.e., palatal organ. Freshwater fish species generally have well-developed taste buds in the oropharyngeal cavity and gill arches/rakers, which is thought to be an adaptation to freshwater habitats where a wider range of physical and chemical changes exist than in sea water (Iwai, 1964; Hara et al., 1993).

Amphibians exhibit the most complex taste bud features of any vertebrates. Juvenile anurans (tadpoles), and larval salamanders (urodeles) have ovoid taste buds which are encased in small processes termed pre-metamorphic papillae resemble the taste buds of fishes. During metamorphosis they undergo transformation to develop large taste organs called taste disks (Figure 42.1d). Despite extensive histological and physiological studies, however, functional significance of the two morphological types developed in the amphibian taste organ is not well understood (Osculati and Sbarbati, 1995).

Generally, a taste bud is buried in the epithelial layers or sits on a dermal papilla without being partitioned by membranes or any other tissues from the surrounding epithelial cells. Thus, the taste bud, while maintaining its entity, is a constituent of the epithelium without partition from the surrounding epithelial cells. In fish, three gustatory cell types constitute a taste bud: (1) rod, or light cell, (2) microvillar, or dark cell, and (3) basal cell. The total number of gustatory cell in a taste bud varies greatly, from 5 to as many as 100 cells. The taste bud evolved while fish took possession of their own distinct ecological niche, thus, resulting in considerable variation in their micro-morphology. The rod cell, most likely a receptor cell, is cylindrical to spindle shaped. Its supra-nuclear portion is characterized by the presence of abundant vesicles. Microvillar cells, often believed to be a supporting cell, but, could well be a receptor cell, are characterized by fine filaments, running through the entire cytoplasm. Junctional complexes exist between microvillar cells and between the rod and microvillar cells. The basal cells, typically one to five per taste bud, are situated at the bottom of the taste bud in depression of the basal lamina. They attach to both receptor and supporting cells by desmosomes (Jakubowski and Whitear, 1990). Basal cells contain many lucent vesicles, rich in serotonin, and possess synaptic connections with both cell types as well with nerve fibers, suggesting that they might act as modulators of gustatory activity (Toyoshima et al., 1984). Unlike those of mammalian taste bud, the basal cells of fish are not the precursors of other cell types within the taste buds.

Gustatory cells of vertebrates, unlike all other type of sensory receptor cells and neurons, are unique in that they originate from local epithelial tissue elements of primarily endoderms, and not from neurogenic ectoderm (neural tube, neural crest or ectodermal placodes). Nevertheless, gustatory cells form synapses and are capable of generating receptor potentials, and even action potentials, and like epithelial cells they have a limited life span and regularly replace themselves (Stone et al., 1995; Vandenbeuch and Kinnamon, 2009). Ontogenetically, taste buds generally develop later than their counterpart the olfactory.

Taste buds are innervated by bundles of unmyelinated nerve fibers that enter from below into their basal portion. Generally, one to three nerve bundles supply each taste bud. Upon entering a taste bud, they further branch out and make an intricate intragemmal nerve plexus, some components of which make synaptic contacts either with the rod cells or with the microvillar cells. However, the synaptic specialization at these nerve-ending receptor junctions is not always clear (Reutter and Witt, 1993).

42.2.1.2 Palatal Organs and Barbels. Although the gustatory system is generally well developed throughout fishes, two fish groups, cyprinids (e. g., goldfishes) and ictalurids (e. g., catfishes), have become highly specialized for the use of sophisticated food-handling apparatuses, i.e., the palatal organ and the barbel. Both groups of fishes are bottom feeders, and have evolved complex mechanisms of food detection and separation. They evolved thousands of extra taste buds, which is a result of the independent evolutions occurred a number of times among teleosts (Northcutt, 2005). The palatal organ, which is a muscular

structure attached to the roof of the mouth, combined with the surface of the gill arches, manipulate to separate palatable food materials from the unpalatable. Barbels, by contrast, densely populated with taste buds, are used for food search and sorting, whereas their intra-oral gustatory system is used for selective food ingestion. Unlike the palatal organs in cyprinids, no muscular structure exists within the barbels (Kiyohara et al., 2002).

42.2.1.3 Facial and Trigeminal Nerve Complication. The facial nerve is closely allied with the trigeminal nerve, and thus strong tactile component exists in the gustatory sensory signal. In vertebrates, the trigeminal nerve generally carries the sensory nerves from the jaw muscles, which is responsible for sensation including pain, temperature, touch, and proprioception. The sensory part of the trigeminal is exclusively of tactile-carrying stimuli from the same region innervated by the facial nerve (Herrick, 1905; Luiten, 1975). Catfish barbels, for example, are heavily innervated by trigeminal nerves mixed with the facial, both of which project to the facial lobe by three branches-ophthalmic, maxillary, and mandibular. Trigeminal fibers are coarser than those of the facial which terminate within the same structural loci, and its input to the primary gustatory complex is restricted to those portions of the nucleus receiving sensory inputs from the face and bsrbels.

Virtually nothing is known about the peripheral termination of the trigeminal nerve, except for the observations that in sea catfish and goatfish some nerve fibers terminate inside taste buds perigemmally without ending on gustatory cells. It thus appears as if the trigeminal fibers have failed to locate their target cells, that is, they become orphaned gustatory nerves. Thus, the relationship between gustatory and tactile sensibility is well recognized but its implications are not fully understood. It seems there may be no pure gustatory area at any level of the gustatory neuraxis (Kiyohara et al., 2002). In rainbow trout, none of single palatine nerve fibers sensitive to tactile stimulation respond to chemical stimulation, and vise versa (Hara, 2011b).

42.3 CENTRAL GUSTATORY PATHWAYS

The gustatory system is the only vertebrate sensory system in which three cranial nerves carry all peripheral gustatory information. Depending upon the location of the taste buds, either the facial (cranial nerve VII), glossopharyngeal (IX), or vagal (X) gustatory nerves innervate the gustatory cells. Generally, all cutaneous taste buds on the body surface and rostral oral regions are innervated by the facial nerve, while taste buds within the posterior

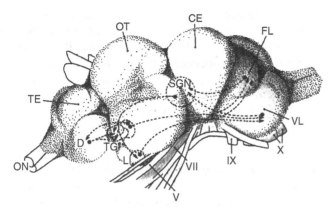

Figure 42.2 Diagram showing dorsolateral view of the brain of a cyprinid fish showing the entrance of trigeminal (V), facial (VII), glossopharyngeal (IX), and vagal (X) nerves, and the main central gustatory connections. CE, cerebellum; D, dorsal telencephalic area; FL, facial lobe; LI, hypothalamic inferior lobe; ON, olfactory nerve; OT, optic tectum; SGN, secondary fustatory nucleus; TE, telencephalon; and TG, tertial gustatory nucleus. Modified from Luiten (1975). (*See plate section for color version.*)

oral cavity and pharyngo-branchial region are innervated by the glossopharyngeal and vagal nerves. Somatotopic organization generally exists in the termination of axons in the primary gustatory nucleus (PGN), the mammalian equivalent of the solitary tract (NST). The gustatory nerves from the different regions of the body thus enter the nucleus in approximately the same order as they are located in the body, that is, axons from more rostral regions of the mouth (or body) enter the more rostral portions of the nucleus (Figure 42.2). In fact, this distribution pattern is common to every skull on earth, whether it belongs to a shark, a bony fish, a salamander, or a human (Shubin, 2008).

As Herrick (1905) pointed out with regard to the major importance of the relation of gustation and olfaction, the secondary center is the *area olfactoria* of the forebrain, whose commissure bears the same relation to the secondary tracts as do those of the secondary gustatory nuclei. The main tertiary tract passes to the inferior lobe, which is, in fishes, the central correlation station for all sensory impression. The olfactory and gustatory tertiary tracts end together throughout the inferior lobe and they have a common descending conduction path, the *tractus lobobulbaris* (Figure 42.2). The ascending fiber connections of the central gustatory system in teleosts are essentially similar to those of mammals. The mammalian SGN projects directly to the amygdala, which receives direct projections from the olfactory bulb.

The inferior lobe and lobobulbar lobe develop extensive descending fiber projections to the primary gustatory centers in the facial (cyprinids) and vagal (catfishes) lobes as well as in the viscerosensory column (salmonids). The vagal lobe projects to motoneurons that innervate the oropharyngeal musculature, which controls swallowing or intraoral food handling in cyprinids. The facial lobe of catfishes, by contrast, projects to somatosensory nuclei in the caudal medulla, which may provide a means to correlate gustatory and somatosensory inputs from a single locus on the body surface.

42.4 FUNCTIONAL PROPERTIES OF THE GUSTATORY SYSTEMS

42.4.1 Signal Transduction and Molecular Basis of Receptors

Chemical stimuli enter the apical portion of the taste bud and bind to receptor molecules located within the membrane of the apical cilia or microvilli of gustatory cells. This interaction leads to a membrane conductance change and depolarization of the cell. Since the discovery that the gustatory cells, but not surface epithelial cells, of salamander *Necturus* are electrically excitable and generate action potentials in response to membrane depolarization (Roper, 1983), gustatory cells of most, if not all, vertebrate species regularly generate action potentials on electrical and chemical stimulation (Takeuchi et al., 2001; Sato et al., 2008). In mammals, three genes encoding the tetrodotoxin (TTX)-sensitive Na^+ currents responsible for the action potentials have been identified (Gao et al., 2009). Gustatory cells are thus only non-neuronal sensory receptor cells to generate action potentials. As mentioned, the taste bud, while maintaining its entity, is a constituent of the epithelium without partition from the surrounding epithelial cells. They replace themselves every few weeks; unlike the olfactory, epithelial cells surrounding the taste buds divide and some of their daughter cells migrate into taste buds to form new gustatory cells. Gustatory cells and epithelial cells arise from a common progenitor, and gustatory receptor cells originate local tissue elements (Stone et al., 1995). Amazingly, how the newly created receptor cells acquire the capability to generate action potentials? The acquisition of receptor phenotype is likely to be the consequence of local tissue interactions.

42.4.2 Responses to Chemical Stimuli

42.4.2.1 Amino Acids. In vertebrates, olfaction is generally recognized as a distance chemical sense with high sensitivity and specificity, while gustation is primarily a contact or close-range sense with moderate sensitivity. However, the gustation in fish is unique in that the receptors are equally sensitive to the same chemical stimuli or even more sensitive than the olfactory, making the distinction between the two sensory modalities blurred.

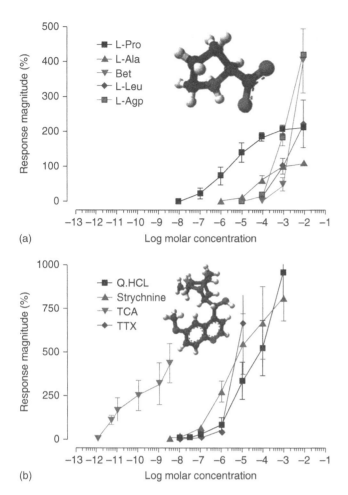

Figure 42.3 Concentration-response (C-R) relationships of gustatory responses recorded from the rainbow trout palatine nerve bundles. L-Pro, L-proline; L-Ala, L-alanine; Bet, betaine; L-Leu, L-leucine; L-Agp, L-α-amino-β-guanidinopropionic acid; Q-HCL, quinine·HCl; TCA, taurocholic acid; and TTX, tetrodotoxin. Insets: molecular structures of L-Pro (a) and Q·HCL (b). Modified from Hara (2011b). (*See plate section for color version.*)

to the background water in studies when the receptors are perfused with water containing varying levels of natural amino acids. Interestingly, natural water is likely to be essential for response generation; replacing natural perfusing water with distilled water eliminates responses to stimulant chemicals, and artificial water only partially restores it. Thus, unknown factor(s) beyond cations in the natural water plays an essential role in maintaining maximal gustatory reception (Kitada and Hara, 1994).

Not all fish are in unison with regard to the chemical response spectra; of the 27 teleost species (comprising 14 families and 29 genera) whose gustatory amino acid specificities have been examined systematically, two general types have been identified: type 1 that are highly selective and respond to only a few amino acids (limited response range), and type 2 whose gustatory systems respond to many amino acids (wide response range)(Hara, 1994a, b). All fish species in type 1, typically salmonids, respond to L-proline (L-Pro), L-alanine (L-Ala), and several other amino acids and related compounds. In rainbow trout, for example, only L-Pro, L-Ala and L-hydroxy-proline (L-Hpr) are stimulatory, with thresholds 10^{-8}–10^{-5} M. Both carp and goldfish may also be placed in this group. A structure–activity study indicates that heterocyclic-imino acids with four, five, or six-membered ring are effective (Marui et al., 1983). The most characteristic feature common to all species of this group is that L-Pro is the dominant stimulant amino acid and it is the only naturally occurring amino acid that is not detected by the counterpart olfactory system (Hara, 2007). These data suggest that the gustatory and olfactory systems detect distinct portions of amino acid spectra, consequently minimizing the redundancy. Further pharmacological characterization of the data indicate that possibly three receptor types are involved in the detection of these amino acids: (1) proline-receptor (L-Pro, L-Hpr, and L-Ala), (2) betaine-receptor (Bet and L-Arg), and (3) leucine-receptor (L-Leu, L-Phe) (Yamashita et al., 2006). Type 2 fishes, by contrast, detect virtually all natural amino acids by gustation, with the sensitivities higher than those of the olfactory counterpart. In channel catfish, for example, responses of the facial nerves innervating extra-oral taste buds to Ala and Arg show thresholds averaging 10^{-11}M, while those of the vagal nerves innervating oropharyngeal taste buds, dominated by the responsiveness to Pro, are less sensitive than the facial (Ogawa and Caprio, 2010). Cross-adaptation results further indicate that all amino acids are detected mainly by L-Ala and L-Arg receptors, supplemented by D-Ala, L-Pro, D-Pro, D-Arg receptors. Japanese eels and some marine fishes are characterized by high sensitivities to Gly, in addition to Ala and Arg (Yoshii et al., 1979; Ishida and Hidaka, 1987).

In larval and adult lampreys, pharyngeal terminal buds are sensitive to several amino acids including Ala and Arg, with the thresholds ranging from 10^{-7} to 10^{-5}M, suggesting

Furthermore, the gustatory system is no longer the sole player in feeding in fish. In fact, recent studies demonstrate that feeding behavior in fish is primarily triggered by olfaction, and complimented by gustation. Electrophysiological studies show that the fish gustatory receptors are specifically stimulated by dilute aqueous solution of low-molecular-weight-chemicals of two main groups: (1) amino acids and (2) primarily aversive chemicals (Figure 42.3; Marui et al., 1983; Hara, 2007, 2011b).

Electrical responses to amino acids are generally characterized by a fast adapting, rapidly returning to the baseline activity even with continuous stimulation, which contrasts with the slow-adapting olfactory response (Hara, 2011b). Threshold concentrations for the more stimulatory amino acids range between the micromolar and nanomolar. Threshold determinations, however, are highly sensitive

the functional relationship between lamprey terminal buds and the taste buds of the gnathostome vertebrates (Baatrup, 1995). Similarly, aquatic toad and frog are sensitive to wide ranges of amino acids and bitter substances with the thresholds ranging from 10^{-9}–10^{-6} M (Yoshii et al., 1982; Gordon and Caprio, 1985).

42.4.2.2 Aversive and/or Toxic Chemicals.

BILE ACIDS To date, only salmonid (*Oncorhynchus* and *Salvelinus* sp.) gustatory systems are known to be extremely sensitive to bile acids. In rainbow trout, both taurine- and glycine-conjugated forms are generally more stimulatory than the free, and taurocholic acid (TCA), one of the most potent chemicals tested, in either the olfactory or gustatory, is detected at 10^{-12}M, with the maximum response magnitude double that of the Pro (cf. Figure 42.3; Yamashita et al., 2006). The response pattern to bile acids is characterized by its slow adapting, tonic response. Contrary to the olfactory, where multiple bile acid receptors exist capable of detecting and discriminating different bile acids (Zhang et al., 2001; Giaquinto and Hara, 2008), gustatory responses to all bile acids are inhibited by pre-treatment (cross-adaptation) with taurolithocholic acid (TLC), suggesting that all bile acids may share the one single receptor/transduction system. The gustatory systems of carp, sea catfish, and tilapia are not sensitive to bile acids up to a concentration of 10^{-4}M.

QUININE, STRYCHNINE, AND TETRODOTOXIN Quinine hydrochloride (Q·HCl), strychnine, and TTX stimulate the salmonid gustatory receptors; rainbow trout and three char species (*Salvelinus* sp.) detect these chemicals at thresholds 10^{-8}–10^{-7} M, with extremely high-response magnitudes (cf. Figure 42.3; Yamamori et al., 1988; Yamashita et al., 1989). Cross-adaptation to strychnine in rainbow trout inhibits responses to Q·HCl and TTX, but without any effect on amino acid and bile acid responses, suggesting that all three groups of chemicals might share the same receptor and transduction mechanisms, or a receptor family with the wide response diversity coexists in a single gustatory cell (Figure 42.4). Aquatic toad *Xenopus* and leopard frog *Rana sphenocephala* also detect Q·HCl, but with much higher thresholds (Yoshii et al., 1982; Gordon and Caprio, 1985).

A paralytic shellfish toxin, saxitoxin (STX), is also potent gustatory stimuli for salmonids. Rainbow trout detect TTX at 2×10^{-7}M, and at 10^{-5}M, it evokes a response magnitude 4 times that of 10^{-3}M Pro (cf. Figure 42.3). These two toxins are vastly different structurally, do not cross-adapt each other, and are therefore unlikely to share the same receptor type. The TTX-sensitive, voltage-dependent Na^+-current is present in most vertebrate gustatory cells (Kinnamon and Cummings, 1992). In frog, the transient inward Na^+-current in a patch-clamp gustatory cell is completely blocked by TTX (Avenet and Lindemann, 1987). However, *in vivo* experiments the generation of a receptor potential in response to gustatory stimuli is not inhibited by TTX (Ozeki and Noma, 1972). Also, in fish perfusion of the palate with TTX had no effect on gustatory responses to amino acids, suggesting that the amino acid-activated TTX-sensitive cation channel(s) is not present in the apical membrane of the gustatory cells, and that TTX does not penetrate the tight junctions known to present at the top of the gustatory cells (Kitada and Hara, 1994). It is noteworthy that either strychnine or bile acid TLCA inhibits gustatory responses to TTX in a competitive fashion (Hara, 2011d).

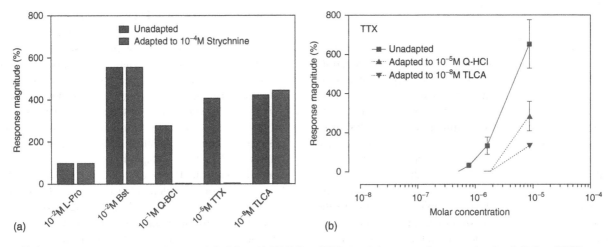

Figure 42.4 Interactions between strychnine, Q·HCl, bile acid (TLCA), snf TTX in rainbow trout, demonstrating that Q·HCl and TTX are inhibited by strychnine, but L-Pro, betaine (Bet) and TLC are unaffected (a); TTX is suppressed by either Q·HCl or TLCA (b). Adapted from Hara (2011d).

CO₂/PH – GUSTATION IS NOT JUST FOR TASTE BUDS The fish gustatory receptors are known to be sensitive to CO_2 (Hara, 2007). In pH-controlled decarbonated natural water, rainbow trout detect CO_2 with a threshold 4×10^{-5} M, which is slightly higher than the CO_2 level found in natural water (1.8×10^{-5} M) equilibrated with air, while the threshold for H^+ is approximately 4×10^{-5} M (Figure 42.5; Yamashita et al., 1989). Responses to CO_2 are independent of pH, and little affected by pretreatment with amino acids or bile acid, and none of the single gustatory fibers responsive to CO_2 responds to other stimuli. The rainbow trout gustatory receptors are thus capable of distinguishing between CO_2 and H^+, and CO_2 receptors are distinct from those that detect other gustatory stimuli. Experimental evidence suggests that a significant CO_2/H^+-driven ventilatory action exists in fishes. The bradycardia and hyperventilation associated with hypercarbia are triggered largely by external CO_2 chemoreceptors on the gills (Gilmour and Perry, 2007).

Similarly, in the gustatory system of the mice, the chorda tympani nerves display robust, dose-dependent, and saturable electrical responses to CO_2 stimulation of the tongue (Chandrashekar et al., (2009). Also in humans, the superior laryngeal nerves detect CO_2 in the larynx (Nishijima and Atoji, 2004). Why do animals need CO_2 sensing? The ancestral fish from which both modern fish and land vertebrates evolved had lungs that enabled them to live in stagnant, poorly aerated water, they may have participated in a series of reflexes that function to modulate respiration or to block the air way.

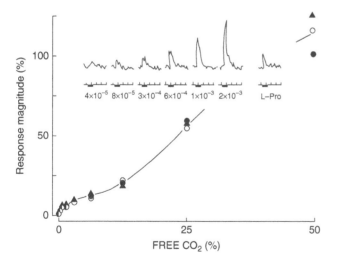

Figure 42.5 Concentration-response curves show the effects of pretreatment with 10^{-3} M Pro (open circles) and 10^{-8} M taurolithocholic acid (solid triangles) on CO_2 responses in rainbow trout. Control responses to CO_2 before adaptation are shown by solid circles. Inset: Typical gustatory responses to CO_2. Modified from Yamashita et al. (1989).

42.4.3 Molecular Basis of Gustatory Receptors and Transduction

Morphological and neurophysiological studies show that the fish gustatory systems are capable of detecting two groups of chemicals via specific receptor mechanisms: (1) generally attractive chemicals represented by amino acids, and (2) generally aversive chemicals including quinine, bile acids and marine toxins. Supporting evidence came from analyses of genomic sequences encoding proteins homologous to mammalian T1Rs and T2Rs, in which two families of G-protein-coupled gustatory receptors identified in several fish species. Two families of G-protein-coupled gustatory receptors fT1Rs and fT2Rs homologous to mammalian T1Rs and T2Rs exist in puffer fish, medaka fish, and zebrafish (Ishimaru et al., 2005). These receptor genes show 60–70% similarities to those of mammals. The former consists of three subtypes, and are expressed in different gustatory cell populations, while the latter, consisting of two to three subtypes, show low degree of similarities to those of mammals, and expressed in different cell groups from those for T1R genes. Calcium imaging analyses of T1Rs in zebrafish and medaka fish using an HEK293T heterologous expression system further demonstrated that both T1R1/3 and a series of T1R2/3 respond to amino acids and denatonium, but not to sugar. However, the kinetics of Ca imaging (ligand concentrations and response time, etc.) is not in accord with values expected of sensory receptors. Thus, inconsistencies between the various aspects of imaging and electrophysiological response characteristics preclude unequivocal classification of the imaging as representing gustatory receptors in question (Oike et al., 2007). In mammals, T1R1 and T1R3 combine to function as a broadly tuned l-amino acid sensors responding to most amino acids, but not to their d-enantiomers (Nelson et al., 2002). Although some inconsistencies exist between receptor activity and gustatory perception, high degrees of T1Rs identity between fish and mammals suggest that the divergence in these gustatory receptor genes occurred prior to that of fish and mammals.

42.4.4 Behavioral Implications

Contrary to the general belief, the gustatory system is no longer the sole player in feeding in fish, and at the same time, not limited to it. Recent studies demonstrate that feeding behavior is initiated primarily by olfaction, and complemented by gustation in many species (Hara, 2007, 2011c). Feeding, one of the two most fundamental processes for the survival of individuals and species, is not operated exclusively by a single sensory system. This is not limited to fishes; as noted by Brillat-Savarin (Fisher, 2011) in 1826, "smell and taste are in fact but a single sense, whose laboratory is the mouth and whose chimney is the nose" in us humans. Identifying the active ingredients in fish food has attracted the interest of many investigators from scientific curiosity or for its potential practical applications. All studies examining different aspects of feeding behaviors elicited by natural and/or food extract in diverse fish species have shown conclusively that amino acids act either singly or in combination to play a major role in stimulating feeding behavior. Upon detection of stimuli, fish trigger appetitive behaviors followed by species-specific search behavior patterns. It is important to note that olfactory receptors, and the gustatory for that matter, are genetically fixed, by which their response capacity is determined. However, subsequent behavioral responses to stimuli may be learned by experience, and stored in temporary memory; consequently the correlation does not always exist between electrophysiological and behavioral observations.

Cysteine (Cys), the most potent olfactory stimulating amino acid, enhances locomotor activity, an initial arousal behavior, in all fish species examined, in exactly the same fashion as do food extracts. The increased locomotor (appetitive) activity is followed by distinct species-specific search behavioral patterns. Importantly, behavioral responses exhibited by each species are stereotypic, eliciting the same behavioral patterns regardless of the type of stimuli, and in a concentration-dependent fashion in all species examined, with thresholds 10^{-9}–10^{-8} M, which approximates the physiological values (Hara, 2006). In channel catfish, by contrast, the entire sequence of feeding behavior patterns in response to food extracts and single amino acids are nearly the same in both the intact and anosmic, implying that gustatory stimuli alone are sufficient to control feeding behavior in this species (Valentinčič and Caprio, 1994).

Q·HCl, strychnine and caffeine induce avoidance behavior and suppress locomotor activities in salmonids and goldfish. When food pellets soaked in Q·HCl solution are sprinkled on the surface of an aquarium, both rainbow trout and goldfish pick off, apparently by vision; the former ingest them immediately, but the latter spit them out after a brief mastication. Salmonids are able to avoid the noxious at a distance, but not once they are taken into the mouth (Jones and Hara, 1985), while, goldfish have the specialized palatal organ that enable them to sort food particles from un-palatable particles. They suck up the mixture, if mixed with gravel, sand, or mud, and manipulate the mix in the mouth, then finally spit out. The well-developed reflex system in the vagal lobes activates the musculature of the palatal organ to effect the sorting operation (Lamb and Finger, 1995; Hara, 2005). Surprisingly, the palatal organs are little sensitive to chemical stimuli.

42.5 CONCLUDING REMARKS

Taste buds, constituting the structural and functional bases of the peripheral gustatory organs, occur in virtually all extant vertebrates. They are onion-shaped clusters of 50–100 neuroepithelial cells including receptor cells. These features prevail throughout the classes of aquatic vertebrates, except for frogs and urodeles that develop large taste organs called taste disks instead. Despite their experimental advantages and extensive investigations, information on their physiological properties and function has been limited. In fish, by contrast, coupled with expansive peripheral receptors, they all have developed elaborate central structures, probably the most complex neural organization found in any vertebrate central nervous system. Contrary to the general understanding in mammals, electrical recording from individual fibers shows that a single gustatory fiber responds best to one type of stimulus (e.g., amino acids, bitter substances, or tactile). Thus, in fish individual sensory fibers may be establishing synaptic contact with only one type of stimulus. Behaviorally, feeding is triggered by single amino acids primarily through olfaction, and complemented or interchangeably by gustation, in naïve fishes. Feeding, one of the two most fundamental processes for the survival of individuals and species, thus is not operated exclusively by a single sensory system. It is also important to note that the behavioral responses are stereotypic, eliciting the same behavioral patterns regardless of the type of stimulus, and concentration-dependent in all species, with thresholds approximating the physiological ones. Pro is the only natural amino acid that is specifically detected by the fish gustatory receptor *ProR*. These amino acid receptors are specifically activated by their primary ligands, recognizing the characteristic structural feature or epitope of an amino acid including (1) length of hydrocarbon chain, (2) difference in functional group, and (3) position of the functional group within the molecule. This is consistent with Ikeda's stereochemical concept of the "UMAMI" taste; "ionic glutamic acid is related first to the characteristics of an amino acid and second to the fact that its amino group occupies the alpha position for the one carboxyl group and the gamma position for the other carboxyl group … the location of every residue is relevant for the taste" (Ikeda, 1909).

REFERENCES

Atema, J. (1971). Structures and functions of the sense of taste in the catfish (*Ictalurus natalis*). *Brain Behav. Evol.* 4: 273–294.

Avenet, P., and Lindemann, B. (1987). Patch-clamp study of isolated taste receptor cells of the frog. *J. Membr. Biol.* 97: 223–240.

References

Baatrup, E. (1983). Ciliated receptors in the pharyngeal terminal buds of larval *Lampetra planeri* (Bloch) (Cyclostomata). *Acta Zool. (Stockh.).* 64: 67–75.

Baatrup, E. (1985). Physiological studies on the terminal buds in the larval brook lamprey, *Lampetra planari* (Bloch). *Chem. Senses.* 4: 549–558.

Chandrashekar, J., Yarmolinsky, D. A., von Buchholtz, L., et al. (2009). The taste of carbonation. *Science.* 326: 443–445.

Finger, T. E. (1987). Gustatory nuclei and pathways in the central nervous system. In *Neurlogy of Taste and Smell.* T. Finger and W. L. Silver (Eds.) John Wiley & Sons, New York, pp. 331–353.

Fisher, M. F. K. (2011). *Jean Anthelme Brillat-Savarin.* The Physiology of Taste. Vintage Books. A Division of Random House, Inc. New York. 446p.

Gans, C., and Northcutt, R. G. (1983). Neural crest and the origin of vertebrates: a new head. *Science* 220: 268–274.

Gao, N., Lu, M., Echeverri, F., et al. (2009). Voltage-gated sodium channels in taste bud cells. *BMC Neurosci.* 10: 20.

Giaquinto, P. C., and Hara, T. J. (2008). Discrimination of bile acids by the rainbow rrout olfactory system: Evidence as potential pheromone. *Biol. Res.* 41: 27–36.

Gilmour, K. M., and Perry, S. F. (2007). Branchial chemoreceptor regulation of cardiorespiratory function. In *Sensory Systems Neuroscience.* T. J. Hara and B. S. Zielinski (Eds) Academic Press, New York, pp. 97–151.

Gordon, K. D., and Caprio, J. (1985). Taste responses to amino acids in the southern leopard frog, *Rana sphenocephala. Comp. Biochem. Physiol.* 81A: 525–530.

Hara, T. J. (1994a). The diversity of chemical stimulation in fish olfaction and gustation. *Rev. Fish Biol. Fish.* 4: 1–35.

Hara, T. J. (1994b). Olfaction and gustation in fish: an overview. *Acta Physiol. Scand.* 152: 207–217.

Hara, T. J. (2005). Olfactory responses to amino acids in rainbow trout: revisited. In *Fish Chemosenses.* K. Reutter and B. G. Kapoor (Eds.) Science Publishers, Inc., Enfield, pp. 31–64.

Hara, T. J. (2006) Feeding behavior in some teleosts is triggered by single amino acid primarily through olfaction. *J. Fish Biol.* 68: 810–825.

Hara, T. J. (2007). Gustation. In *Sensory Systems Neuroscience. Vol. 25. Fish Physiology.* T. J. Hara and B. S. Zielinski (Eds.) Academic Press, New York, pp. 45–96.

Hara, T. J. (2011a). Morphology of the gustatory (taste) system in fishes. In *Encyclopedia of Fish Physiology: From Genome to Environment. Vol.1.* A. P. Farrell (Ed.). Elsevier, San Diego, pp. 187–193.

Hara, T. J. (2011b). Neurophysiology of gustation. In *Encyclopedia of Fish Physiology: From Genome to Environment. Vol.1.* A. P. Farrell (Ed.). Elsevier, San Diego, pp. 218–226.

Hara, T. J. (2011c). Chemosensory behavior. In *Encyclopedia of Fish Physiology: From Genome to Environment. Vol.1.* A. P. Farrell (Ed.). Elsevier, San Diego, pp. 227–235.

Hara, T. J. (2011d). Gustatory detection of tetrodotoxin and saxitoxin, and its competitive inhibition by quinine and strychnine in freshwater fishes. *Mar. Drugs.* 9: 2283–2290.

Hara, T. J., Sveinsson, T., Evans, R. E., and Klaprat, D. A. (1993). Morphological and functional characteristics of the olfactory and gustatory organs of three *Salvelinus* species. *Can. J. Zool.* 71: 414–423.

Herrick, C. J. (1904). The organ and sense of taste in fishes. *Bull. US Comm.* 22: 237–272.

Herrick, C. J. (1905). The central gustatory paths in the brains of bony fishes. *J. Comp. Neurol. Psychol.* 15: 375–456.

Ikeda, K. (1909). New seasonings. *J. Tokyo Chem. Soc.* 30: 820–836.

Ishida, Y., and Hidaka, I. (1987). Gustatory response profiles for amino acid, glycinebetaine, and nucleotides in several marine teleosts. *Nippon Suisan Gakkaishi.* 53: 1391–1398.

Ishimaru, Y., Okada, S., Naito, H., et al. (2005). Two families of candidate taste receptors in fishes. *Mech. Dev.* 122: 1310–1321.

Iwai, T. (1964). A comparative study of the taste buds in gill rakers and gill arches of teleostean fishes, *Bull. Misaki Mar. Biol. Inst., Kyoto Univ.* 7: 19–34.

Jakubowski, M., and Whitear, M. (1990). Comparative morphology and cytology of taste buds. *Z. Mikrosk. Anat. Forsch.* 104: 529–560.

Jones, K. A., and Hara, T. J. (1985). Behavioural responses of fishes to chemical cues: results from a new bioassay. *J. Fish Biol.* 27: 495–504.

Kanwal, J. S., and Finger, T. E. (1992). Central representation and projections of gustatory systems, In *Fish Chemoreception.* T. J. Hara (Ed.). Chapman & Hall, London, pp. 79–102.

Kinnamon, S. C., and Cummings, T. A. (1992). Chemosensory transduction mechanisms in taste. *Annu. Rev. Physiol.* 54: 715–731.

Kitada, Y., and Hara, T. J. (1994). Effects of diluted natural water and altered ionic environments on gustatory responses in rainbow trout (*Oncorhynchus mykiss*). *J. Exp. Biol.* 186: 173–186.

Kiyohara, S., Sakata, Y., Yoshitomi, T., and Tsukahara, J. (2002). The 'goatee' of goatfish. Innervation of taste buds in the barbels and their representation in the brain. *Proc. R. Soc. Lond. B.* 269: 1773–1780.

Lamb, C. F., and Finger, T. E. (1995). Gustatory control of feeding behavior in goldfish. *Physiol. Behav.* 57: 483–488.

Luiten, P. G. M. (1975). The central projections of the trigeminal, facial and anterior lateral line nerves in the carp (*Cyprinus carpio*). *J. Comp. Neurol.* 160: 399–417.

Marui, T., Evans, R. E., Zielinski, B., and Hara, T. J. (1983). Gustatory responses of the rainbow trout (*Salmo gairdneri*) palate to amino acids and derivatives. *J. Comp. Physiol. A.* 153: 423–433.

Nelson, G., Chandrashekar, J., Hoon, M. A., et al. (2002). An amino-acid taste receptor. *Nature* 416: 199–202.

Nishijima, K., and Atoji, Y. (2004). Taste buds and nerve fibers in the rat larynx: an ultrastructural and immunohistochemical study. *Arch. Histol. Cytol.* 67: 195–209.

Northcutt, R. G. (2005). Taste bud development in the channel catfish. *J. Comp. Neurol.* 482: 1–16.

Ogawa, K., and Caprio, J. (2010). Major differences in the proportion of amino acid fiber types transmitting taste information from oral and extraoral regions in the channel catfish. *J. Neurophysiol.* 103: 2062–2073.

Oike, H., Nagai, T., Furuyama, A., et al. (2007). Characterization of ligands for fish taste receptors. *J. Neurosci.* 27: 5584–5592.

Osculati, F., and Sbarbati, A. (1995). The frog taste disc: A prototype of the vertebrate gustatory organ. *Progr. Neurobiol.* 46: 351–399.

Ozeki, M., and Noma, A. (1972). The action of tetrodotoxin, procaine and acetylcholine on gustatory reception in frog and rat. *Jpn. J. Physiol.* 22: 467–475.

Reutter, K., and Witt, M. (1993). Morphology of vertebrate tasteorgans and their nerve supply. In *Mechanisms of Taste Transduction.* Simon, S. A. and Roper, S. D. (Eds.) CRC Press, Boca Raton, pp. 29–82.

Roper, S. (1983). Regenerative impulses in taste cells. *Science* 220: 1311–1312.

Sato, T., Nishishita, K., Okada, Y., and Toda, K. (2008). Electrical properties and gustatory responses of various taste disk cells of frog fungiform papillae. *Chem. Senses.* 33: 371–378.

Shubin, N. (2008). *Your Inner Fish.* Ramdom House, New York, 237p.

Stone, L. M., Finger, T. E., Tam, P. P. L., and Tan, S,-S. (1995). Taste receptor cells arise from local epithelium, not neurogenic ectoderm. *Proc. Natl. Acad. Sci. USA*. 92: 1916–1920.

Takeuchi, H., Tsunenari, T., Kurahashi, T. and Kaneko, A. (2001). Physiology of morphologically identified cells of the bullfrog fungiform papilla. *NeuroReport*. 12: 2957–2962.

Toyoshima, K., Nada, O., and Shimamura, A. (1984). Fine structure of monoamine containing basal cell in the taste buds of the barbells of three species of teleosts. *Cell Tissue Res*. 235: 479–484.

Valentinčič, T., and Caprio, J. (1994).Consummatory feeding behavior to amino acids in intact and anosmic channel catfish *Ictalurus punctatus*. *Physiol. Behav*. 55: 857–863.

Vandenbeuch, A., and Kinnamon, S. C. (2009). Why do taste cells generate action potentials? *J. Biol*. 8: 42.1–5.

Yamamori, K., Nakamura, M., Matsui, T., and Hara, T. J. (1988). Gustatory responses to tetrodotoxin and saxitoxin in fish: a possible mechanism for avoiding marine toxins. *Can. J. Fish. Aquat. Sci*. 45: 2182–2186.

Yamashita, S., Evans, R. E., and Hara, T. J. (1989). Specificity of the gustatory chemoreceptors for CO2 and H+ in rainbow trout (*Oncorhynchus mykiss*). *Can. J. Fish. Aquat. Sci*. 46: 1730–1734.

Yamashita, S., Yamada, T., Hara, T. J. (2006). Gustatory responses to feeding and non-feeding-stimulant chemicals, with special emphasis on amino acids, in rainbow trout. *J. Fish Biol*. 68: 783–800.

Yoshii, K., Kamo, N., Kurihara, K., and Kobatake, Y. (1979). Gustatory responses of eel palatine receptors to amino acids and carboxylic acids. *J. Gen. Physiol*. 74: 301–317.

Yoshii, K., Yoshii, C., Kobatake, Y., and Kurihara, K. (1982). High sensitivity of *Xenopus* gustatory receptors to amino acids and bitter substances. *Am. J. Physiol*. 243: R42–R48.

Zhang, C., Brown, S. B., and Hara, T. J. (2001). Biochemical and physiological evidence that bile acids produced and released by lake char (*Salvelinus namaycush*) function as chemical signals. *J. Comp. Physiol. B* 171: 161–171.

Chapter 43

Comparative Taste Biology with Special Focus on Birds and Reptiles

HANNAH M. ROWLAND, M. ROCKWELL PARKER, PEIHUA JIANG,
DANIELLE R. REED, and GARY K. BEAUCHAMP

Real taste [in] the mouth, according to my theory must be acquired by certain foods being habitual – [and] hence become hereditary.

CHARLES DARWIN'S Notebooks 1836–1844

43.1 INTRODUCTION TO COMPARATIVE TASTE BIOLOGY

Perhaps with the exception of young children, most modern humans are unfamiliar with experimenting with novel and potentially poisonous foods. Thus, our consciously appreciated sense of taste is valuable primarily only inasmuch as it adds pleasure (or displeasure) to dining. We no longer depend upon taste to be the nutrient sensor and early-warning system as it originally served. And if we pause to contemplate the gustatory experience of other animals, we may assume that it is much like our own or perhaps, for some species, like birds, that it is nonexistent. However, this assumption is incorrect. New discoveries from molecular biology and genetics describe the genes and proteins that contribute to taste as a sensory system, and together with mounting evidence from behavioral studies, we have a different story – every organism lives not only in its own dietary niche but, within that niche, also in its own unique world of taste perception and preference. This point was made by Morley Kare more than half a century ago (see Kare, 1970) and is implied by Darwin in the epigraph above.

A striking example of this species variation is the response of animals to sweet fruits. For humans, ripe fruit produces a sensation of pleasure and motivates consumption; it also invokes consumption in nonhuman primates, fruit-eating bats, and rodents, but cats and chickens are indifferent to sweet foods. We ask why – sugar is a source of energy and a signal of nutrients, so why would some animals not like sugar? Cats and chickens both have diets mostly devoid of sugar, which is associated with the evolutionary loss of a gene involved in sweet perception, a gene that humans and most other vertebrates that consume plants still possess. This type of example has inspired the burgeoning integrative field called *comparative taste biology*, which draws on genetics, anatomy, physiology, behavior, and psychology. As comparative taste researchers, we aim to understand the adaptations of animal taste system by examining their feeding behavior, taste preferences, and the underlying molecular and physiological processes that mediate them.

So far, studies of comparative taste biology have focused on traditional animal model species, such as mice and rats, and have often ignored the biological diversity of sensory form and function in most other vertebrates. But recently, research has shown a link between molecular adaptations and a meat-only diet in carnivores (see Box 43.1). The taste biology of birds and reptiles has been particularly neglected. This in part is due to the belief that species in these groups rely more on sight or olfaction than on taste for feeding. This is a notable oversight because both birds and reptiles can provide novel insights into the evolutionary biology of taste perception and food selection. Birds are components of terrestrial food webs and contributors to the evolution of bitter-tasting insect defenses and sweet, sucrose-rich flower nectars, both of

---

Handbook of Olfaction and Gustation, Third Edition. Edited by Richard L. Doty.
© 2015 Richard L. Doty. Published 2015 by John Wiley & Sons, Inc.

958   Chapter 43   Comparative Taste Biology with Special Focus on Birds and Reptiles

BOX 43.1: Examples of Comparative Taste Biology

The study of taste genes is a way to examine how animals adapt to changes in food availability over evolutionary time. New taste receptor genes may be born through genetic duplication and diversification and may proliferate in the population if they are useful, for instance, to warn an animal about new poisons or to provide a cue for a new nutrient. If new mutations result in a new but weak secondary gene function, its duplicated copies can diversify and strengthen that function. If the main function of the original gene becomes less important over time, it can become a pseudogene, while the newly duplicated copies remain functional (Näsvall et al., 2012). This is one model that accounts for the loss of *TAS1R2* (a gene which codes for a part of the sweet receptor) in carnivorous mammals. Below we provide several examples of these types of genetic changes and their relation to taste ecology.

Molecular Changes to the Sweet Taste Receptor

Domestic and wild cats (family Felidae) are obligate carnivores, feeding exclusively on meat. Consistent with their preferences for protein- and fat-rich food but not for carbohydrate-rich food, cats are behaviorally insensitive to sweet carbohydrates or noncaloric sweeteners (Bartoshuk et al., 1975; Beauchamp et al., 1977). This insensitivity is best explained by a lack of a functional T1r2 protein in the sweet taste receptor in cats, caused by the pseudogenization of the *TAS1R2* gene (Li et al., 2005). Apparently this gene has become nonfunctional over time in cats, perhaps as their diet became devoid of gustatory sugars.

Like cats, many other species in the order Carnivora are also exclusive meat eaters – have they also lost the sweet taste receptor? A recent study and found that 7 of the 12 nonfeline species from the order Carnivora – those that feed exclusively on meat or fish – had pseudogenized *TAS1R2* genes, as predicted (Jiang et al., 2012). These seven species belong to different families within Carnivora: Pinnipedia (sea lion, fur seal, Pacific harbor seal), Mustelidae (Asian small-clawed otter), Hyaenidae (spotted hyena), Eupleridae (fossa), and Prionodontidae (banded linsang). These changes represent six separate mutations – only the sea lion and fur seal, two closely related sister species, share the same mutation. These independent pseudogenizing mutations indicate that the loss of *TAS1R2* occurred independently many times during evolution in the order Carnivora, arguing for convergent evolution of this trait. Apparently, the selective pressure for maintaining *TAS1R2* to detect sweet carbohydrates is relaxed in these meat-eating species.

The loss of the sweet receptor suggests that cats are unable to perceive its taste quality under any circumstance, and this supposition has been tested experimentally using two-bowl choice tests. Species with an intact *TAS1R2* gene, such as lesser panda (*Ailurus fulgens*), ferret (*Mustela putorius furo*), genet (*Genetta thierryi*), meerkat (*Suricata suricatta*), yellow mongoose (*Cynictis penicillata*), domestic dog (*Canis lupus familiaris*), and spectacled bear (*Tremarctos ornatus*), show positive responses to sugars (Jiang et al., 2012; Li et al., 2009). Most have a unique acceptance profile when tested with a variety of sugars, perhaps reflecting differences in the amino acid sequences of the T1R2 and T1R3 proteins that make up the sweet taste receptor (Jiang et al., 2012; Li et al., 2009). In contrast, species with a pseudogenized *TAS1R2*, such as the Asian otter (*Amblonyx cinereus*), appear to be insensitive to sweet carbohydrates (Jiang et al., 2012). Other behavioral taste tests showed that the California sea lion (*Zalophus califonianus*) cannot detect sucrose at concentrations up to 2 M (Friedl et al., 1990).

Taken together, these behavioral data agree with genetic data that insensitivity to sweet carbohydrates is due to a lack of a functional sweet taste receptor, which is caused by the pseudogenization of the sweet taste receptor gene *TAS1R2*. We favor the hypothesis that the dietary switch to meat relaxes the selective pressure for maintaining *TAS1R2*, because pseudogenizing events occurred independently during evolution. However, whether the dietary switch to meat caused the pseudogenization of *TAS1R2* or, rather, the pseudogenization of *TAS1R2* gradually drove these species to a meat-based diet remains to be determined.

The bottlenose dolphin (order Cetacea), which displays dietary habitats similar to those of carnivores, also has a pseudogenized *TAS1R2*, based on the draft genome sequence (Jiang et al., 2012). From limited behavioral taste detection data, dolphins show no response or greatly reduced response to sucrose (Friedl et al., 1990; Kuznetsov, 1974), largely agreeing with a lack of a functional T1r2. Together, these findings reinforce the concept of convergent evolution of *TAS1R2* pseudogenization.

Swallowing Foods Whole

Some animals, such as those that swallow their foods without chewing and without sampling its chemical composition, may have a weak or missing taste system. Based on anatomical studies, it is likely that some aquatic mammals, such as sea lions (order Carnivora) and dolphins (Cetacea), have lost some or most taste function (Jiang et al., 2012). Sea lions have taste buds only on the tongue tip and tongue root (Yoshimura et al., 2002). Sea lions use a cranioinertial feeding action: food grasped in the mouth is thrown backward into the throat for swallowing. This feeding technique similar to that of birds and reptiles, so it is perhaps not surprising that their taste buds have a distribution more akin to that of some birds and reptiles (Yoshimura et al., 2002). Dolphins lack conventional taste tissues (fungiform, foliate, vallate papillae) altogether – only a very small number of taste buds are detected at the root of tongue (Yoshimura and Kobayashi, 1997). Thus, both lineages of aquatic mammals appear to have an atrophied taste system.

Morphological changes also appear to correlate with the degree of adaptation to the aquatic environment (Yoshimura et al., 2002): dolphins, which live a completely aquatic life, show drastic loss of taste structure, whereas sea lions, which are amphibious, retain some residual taste structures found in related land mammals (Yoshimura et al., 2002). These morphological changes mirror the genetic and functional changes of their taste system.

In addition to morphological changes that suggest these species have a greatly reduced sense of taste, the genomic data suggest this conclusion, too. In dolphins, all three Tas1r genes are inactivated, and no intact bitter taste receptor genes have been detected in the draft genome sequence of the bottlenose dolphin (Jiang et al., 2012) – the 10 bitter receptors found are all pseudogenes. Although other bitter receptor genes may yet be found with a completed dolphin genome, the fact that 10 out of 10 receptors identified so far are pseudogenes suggests that bitter taste function is either completely lost or greatly compromised in bottlenose dolphins (Jiang et al., 2012). Thus, sea lions and dolphins appear not to rely on taste for their particular food choices.

which rely on a sense of taste. Reptiles represent the first fully terrestrial group in the evolution of land vertebrates and provide a modern window on the taste biology of more ancient sensory systems (see Figure 43.1).

The goal of this chapter is to comprehensively review and consolidate taste research on birds and reptiles. We review what is known about their taste anatomy, physiology, and molecular biology. This provides a basis for understanding taste preferences and feeding behavior. It can also provide insights into broader issues concerning the coevolution of taste and feeding behavior among vertebrates.

We begin by introducing the concept of *taste*, based on research from humans and animal model systems, where information is more complete. We describe the nature of taste as a stimulus, the organization of receptor cells, and how taste perception is evaluated.

43.1.1 Definition and Function

Taste stimuli are structurally diverse chemicals that range from small compounds, such as sodium, to large proteins, such as monellin (a sweet protein found in serendipity berries, *Dioscoreophyllum volkensii*). These stimuli elicit a perceptual response because chemical signals are converted (within special cells) to electrical signals, which travel to brain areas where conscious sensation arise. It is generally agreed by researchers that taste consists of a few distinct qualities: sweet, salty, sour, bitter, and umami (Bachmanov and Beauchamp, 2007); however, other perceptual taste qualities may also exist, such as those associated with calcium (Tordoff et al., 2012) and lipids (Fernelius, 1581).

The best-known biological function of taste is to inform animals about the risks (poisons) and benefits (nutrients) of food. Many plants and animals contain chemical defenses that serve to protect them from consumption (Fahey et al., 2001), and for our ancestors, plants that tasted bitter meant "do not eat" (VanEtten, 1969). Modern humans are inclined to discount taste as a means to find calories. We no longer forage among plants, tasting them and rejecting those that are not nutritious.

Furthermore, sweet taste is no longer a proxy for calories since the rise of high-potency, noncaloric sweeteners like aspartame. But for our ancestors, the sweetness of a piece of fruit was a signal that useful calories were available and the food source should be consumed. And taste can act as a measuring sensor, helping animals to get just the right amount of a nutrient, such as salt, which can kill if ingested at high doses (Boyd, 1973; Ofran et al., 2004). Taste is also a majordomo: once the decision to ingest a food has been made, taste signals other parts of the body (e.g., the digestive system) to prepare for nutrient arrival (Teff, 2000). A final function of taste is to provide pleasure, a feature not much studied in nonhuman animals but currently in the spotlight for humans because of the contribution of hyperpalatable foods to obesity, diabetes, hypertension, and other diseases of excess.

43.1.2 Anatomy

The sensation of taste is initiated by responses of specialized cells in the mouth of mammals that in humans are clustered on the tongue and soft palate, but in many other animals they can be found in other parts of the oral cavity. Chemicals such as capsaicin from hot peppers or menthol from peppermint oil, which stimulate another sensory system, *chemesthesis*, or the common chemical sense, can create a sensation when applied to other areas of the body instead of or in addition to stimulating taste receptors. But for most mammalian species taste perception arises solely from specialized taste cells in the oral cavity. These cells are the reason that sugar tastes sweet only in the mouth and not, for instance, when rubbed on skin. In humans, oral taste cells are organized in *taste buds*, structures that contain many taste receptor cells, as well as cells with other functions. Each taste cell is oriented like a segment of an orange, with microvilli on the surface of the cell pointing inward to the taste pore. Not all cells within the taste bud are of the same type – cells specialize by taste quality and type of taste receptor, and there may be other types of specialization as well (see Chapters 29–31). On the tongue, taste buds can exist either as solitary, fungiform papillae (raised protrusions on the surface of the tongue) or clustered in groups around a cleft, as with foliate and circumvallate papillae, all of which are located in specific subregions of the tongue. Taste cells can also be found in the palate and epiglottis. Taste cells are distinct from solitary chemoreceptors, which are found in some cells in the airways (Finger et al., 2003). The morphology of these solitary cells differs from that of taste cells, although they share some of the same chemical receptors (Finger et al., 2003).

In humans, impulses created from the cells in the taste buds are conveyed to the brain by cranial nerves V, VII, IX, and X. Information from the VII (facial) cranial nerve travels via the chorda tympani to innervate the taste buds

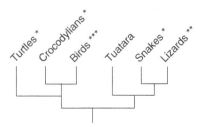

Figure 43.1 Representative phylogeny for the major reptilian lineages. Asterisks represent relative abundances of taste behavior studies in the surveyed literature. Adapted from Chiari et al. 2012.

960 Chapter 43 Comparative Taste Biology with Special Focus on Birds and Reptiles

located on the fungiform papillae of the anterior tongue. The lingual branch of the IX (glossopharyngeal) cranial nerve carries information from the posterior regions of the tongue, including the circumvallate papillae, while the X (vagus) cranial nerve carries information from a patch of the tongue just before the epiglottis. The V (trigeminal) cranial nerve carries information about touch, temperature, and chemesthesis. The primary taste nerves carry information to the nucleus of the solitary tract and thalamus, which radiates to brain areas involved in processing, integration, decision making, and pleasure, such as the primary taste cortex in the anterior insula and orbitofrontal cortex (Rolls, 2012). All of these processes arise from the first step in the taste process: interaction with taste stimuli.

43.1.3 Receptors and Signaling Molecules

Taste perception is initiated when chemicals contact receptors and other membrane-bound channel proteins that are located on the taste cell surface. Bitter taste receptors, the largest group of receptors, are coded by a multigene family identified in humans by two research groups in 2000 (for a recent review, see Boughter and Bachmanov, 2008). This gene family, termed TAS2R (taste type 2 receptor) genes (Adler et al., 2000; Matsunami et al., 2000), includes a group of genes that are each approximately 900 base pairs long. Data-mining approaches against genomes have identified TAS2R gene repertoires that vary greatly in number in a diverse range of taxa, from humans (25 genes) and mice (34 genes) to western clawed frogs (49 genes). Tallies of taste receptor and other genes from different species are available in online databases, such as Ensembl (www.ensembl.org).

A smaller family of genes is involved in sweet and umami reception. This gene family, known as the TAS1R (taste type 1 receptor) family, has just three known members, T1R1, T1R2, and T1R3 (coded by the genes *TAS1R1*, *TAS1R2*, and *TAS1R3*, respectively). These receptors combine to form receptor complexes that sense either sweet (T1R2 + T1R3) or umami (T1R3 + T1R1) in several mammals studied. In addition to these receptor proteins, other proteins previously known for their role in glucose sensing in the pancreas may also be present in taste receptor cells and may provide a secondary system for sweet sensing (Yee et al., 2011) for some species.

Sour and salt receptor(s) are presumed to be ion channels, rather than G-protein-coupled receptors like the T1Rs (sweet) and the T2Rs (bitter). Recent evidence strongly implicates an epithelial sodium channel (ENaC) as one receptor type mediating salt (sodium) taste (Chandrashekar et al., 2010; also see Bosak et al. 2010), but there are likely other salt receptors as well. Sour receptors have not yet been definitively identified. In addition, second messengers that are used for intracellular communication, such as the G-protein α-gustducin, are involved in the cascade of signal transduction events and have well-defined roles in taste perception (for a review, see Breslin and Huang, 2006). As more becomes known about the taste signaling cascade, the list of proteins involved in taste transduction and signaling will expand. This list will certainly include additional transduction components and regulatory genes such as transcription factors. It may also include additional families of receptors not currently described, which may be found in some species and not in others.

43.1.4 Measuring Behavior

Humans have advantages in taste research because they can describe their sensory experience; other animals must demonstrate their perceptions through behavior. One method is to observe what animals eat and how they behave when they encounter foods with particular taste qualities; for example, are they indifferent to it, or do they consume it avidly? Another method is to test taste preferences, typically by offering animals two cups or bowls, one with a fluid or food that contains, for example, sugar or salt, and another with pure water or neutrally flavored food. The amount consumed from each bowl or cup indicates discrimination and preference, but it cannot be used to measure liking unless it is combined with behavioral measures, such as facial expressions (Beauchamp and Mason, 1991; Steiner et al., 2001; for a review of these behavioral methods and their limitations, see Spector, 2003).

Other measurements of taste function involve using electrophysiology to record nerve impulses in response to taste stimuli, or monitoring brain activity patterns by electroencephalography. Metabolic rate within the brain can also be monitored in response to taste stimuli, for example, by using functional magnetic resonance imaging. While these methods are routinely used in humans and laboratory animal models, studies using functional imaging techniques to measure response to taste in birds or reptiles are lacking. However, we expect that this will become possible as the technology becomes adapted for animals other than mammals.

Most of what we know about taste perception and the methods used to study taste behavior come from studies in mammals, particularly humans, mice, and rats. Rodents are models for many human traits due to their commensalism with humans: they eat the same foods and live in the same or similar environments (Crowcroft, 1966; Reed, 2008; Silver, 1995). Mice and rats eat the same types of high-fat, high-sugar foods that humans do (Sclafani and Springer, 1976) and become obese and diabetic in ways similar to humans, which aids our understanding of these metabolic disorders. Further, mice can be genetically engineered to express human proteins that can generate human-typical taste preferences (Mueller et al., 2005; Zhao et al., 2003). While rodents are a useful model system to study human taste, their use has also contributed to the mistaken belief that *all* vertebrate taste systems are similar. In the following sections, we discuss how dietary niches have likely shaped the taste system of birds and reptiles during evolution and bring to light their sensory experience of food and how this may not be easily translated into human experience.

43.2 BIRDS

More than a century ago, Alfred Newton, an English zoologist and ornithologist, wrote in his *Dictionary of Birds*, "The tongue is commonly supposed to be the chief organ of taste; but it is certainly not so in birds, where it is, with a few exceptions, subservient to deglutition [swallowing]" (Newton et al., 1894). Until recently, this point of view has been shared by many researchers working on gustation. Studies showed that birds have fewer taste buds on the tongue than do other vertebrates (Moore and Elliott, 1946) – hardly surprising given how undiscerning birds appear to be when it comes to dietary choice, prioritizing optimal intake over specific taste qualities. Their apparently indiscriminate foraging behavior and limited expressiveness while feeding suggested that birds care little for the taste of the food they eat. However, more recent research has made clear that, although birds do generally have far fewer taste buds than do other vertebrates (Berkhoudt, 1985), they have many more than we previously thought, and in regions of the mouth that previous researchers had never thought to look (Kudo et al., 2008). Similarly, we have discovered that, far from our previous conception of avian feeding behavior as relying more on an acute sense of vision than on a sense of taste, birds can make fine discriminations between different taste stimuli that rival, and sometimes exceed, the abilities of some other vertebrates (Matson et al., 2000; Medina-Tapia et al., 2012).

Birds are adapted to a diverse range of habitats, including tropical rainforests, deserts, scrublands, lowlands, mountain ranges, and the ocean (Hutson, 1991), and they operate within a broad range of dietary niches, consuming fruits and insects (Chaisuriyanun et al., 2011; Stiebel and Bairlein, 2008; Wilkin et al., 2009); lizards and amphibians (Poulin et al., 2001); mammals and birds

(Geer, 1982; Viitala et al., 1995; Wu, 2011); snails and worms (Goss-Custard et al., 1977); fish, squid, and cuttlefish (Croxall and Prince, 1996; Martins et al., 2011; Polito et al., 2011); cereal grains (Perkins et al., 2007); plants (Picozzi et al., 1996; Soininen et al., 2010; Wilson et al., 2006); beeswax (Cronin and Sherman, 1976); sap (Chapman et al., 1999); and nectar (Gartrell, 2000; Lotz and Schondube, 2006). This diversity of life histories broadens the range of potential nutrients and toxins that birds are likely to encounter and demands an equally varied suite of adaptations for acquiring access to potential foodstuffs.

It has been nearly 30 years since Herman Berkhoudt wrote the last comprehensive review of avian taste cells and taste behavior (Berkhoudt, 1985). Since then the field of genomics has revealed the underlying genetics of taste perception in birds, and behavioral studies have expanded beyond chickens and pigeons to include hummingbirds, parrots, and many species of songbirds (see Figure 43.2 for a phylogeny of birds). This work has revealed that the avian tongue, along with the genetic and neurological systems that govern its function, is more complex than Alfred Newton thought possible. Here we examine these advances in avian taste research, including taste cell distribution and frequency, taste behavior, and taste genomics.

43.2.1 Taste System

43.2.1.1 Structure of Avian Taste Buds.
Taste buds have been extensively studied in a variety of mammalian taxa (Miller and Spangler, 1982; Tichy, 1992a, 1992b, 1994), but studies of the taste organs of birds are scarcer and generally limited to the domestic chicken (Ganchrow and Ganchrow, 1985, 1986; Ganchrow et al., 1994, 1995; Gentle, 1971a, 1975, 1978; Lindenmaier and Kare, 1959). The structure of avian taste buds was described by Bath (1906) and later confirmed by studies on chickens and ducks performed by Berkhoudt (1985; for another early example of taste bud identification, see Botezat 1906). In contrast to human taste buds, which have a single structure that can occur within three different types of papillae, avian taste buds are not associated with papillae and are divided into three types based on shape and other morphology. Type I buds, mainly observed in songbirds, chickens, and pigeons, are ovoid structures that comprise a central core of sensory and sustentacular cells (which provide structural support) surrounded by follicular cells (Berkhoudt, 1985). Type II buds are observed in ducks and waders and are narrower and more elongated than type I buds but with the same supporting and follicular

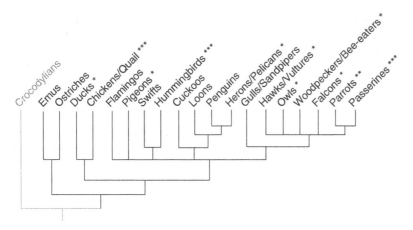

Figure 43.2 Representative phylogeny for the major avian lineages. Asterisks represent relative abundances of taste studies (behavior and anatomy) in the surveyed literature. Adapted from Hackett et al. 2008.

cell types. Type III buds are restricted to parrots; these resemble the buds of mammals but lack follicular cells. Kurosawa et al. (1983) identified a taste canal in the taste buds of chickens, a feature not present in mammalian taste buds; it is a lumen, or tube, allowing communication of the sensory cells with the oral cavity. The diversity of taste bud structures across avian orders suggests divergent evolution similar to the way the beaks of Darwin's finches changed over time and may point toward different taste bud systems adapting to different dietary niches.

43.2.1.2 Distribution and Frequency of Avian Taste Buds.
Avian taste buds are often found in association with salivary ducts (though some also occur freely within the oral mucosa) (Kudo et al., 2008), a feature also observed in reptiles (Ganchrow and Ganchrow, 1985; Gentle, 1971a). Saliva is crucial for the transport of taste stimuli to receptors (Mason and Clark, 1999), as evidenced by the significant reduction in taste ability in chickens with lower salivary flow rate associated with vitamin A and zinc deficiencies (Gentle and Dewar, 1981; Gentle et al., 1981). Birds have higher salivary flow rates than do some mammals (Belman and Kare, 1961), which may compensate for the lower frequency of taste buds. Studies could address this hypothesis by comparing the taste thresholds of species with few versus many taste buds in relation to salivary flow rate.

Unlike mammals, which use their tongues to position food for grinding by the teeth, most birds do not use their tongue during feeding. Instead, they employ a *cranioinertial* feeding action (Berkhoudt, 1985; Gussekloo and Bout, 2005; Zweers et al., 1977): food is initially grasped between the beak tip/anterior mandible and the palatal region of the maxillary glands and then is thrown backward to the posterior surface of the tongue and the esophageal opening. This method of feeding is also seen in many reptiles, such as crocodiles and multiple species of lizards (as well as sea lions; see also Box 43.1).

In mammals, taste buds are predominantly located on the tongue in gustatory papillae, allowing food to come into direct contact with them during feeding. However, in birds we would expect a different distribution of taste buds because of their different feeding action. And, in fact, less than 5% of their taste buds occur on the tongue, generally restricted to the posterior surface (Ganchrow and Ganchrow, 1985; Saito, 1966), where it comes into contact with the food during cranioinertial feeding. This is a low percentage compared with the 73% of lingual taste buds in the hamster (Miller and Smith, 1984) and 79% in the rat (Miller and Spangler, 1982). In contrast, 70% of birds' taste buds occur on the palate, compared with 17% and 15% in the rat and hamster, respectively (Miller and Smith, 1984; Miller and Spangler, 1982), and 25% occur on the lower mandible and the pharyngeal floor (see Figure 43.3a and b). In the emu, structures that resemble taste buds have been reported in the oropharyngeal epithelium, the caudal interramal region, the tongue root, and the laryngo-esophageal junction (see Figure 43.3c). As in mammals, in avians taste buds are innervated by two branches of the facial nerve (VII), the chorda tympani and the palatine (Ganchrow et al., 1986; Gentle, 1983), and the lingual branch of the glossopharyngeal nerve (Halpern, 1962; Mason and Clark, 1999).

The distribution of avian taste buds results in foodstuffs contacting and stimulating each of the taste bud regions sequentially during the feeding action. This is supported by evidence from feeding behavior in ducks and great-crested grebes (*Podiceps cristatus*) (Berkhoudt, 1976, 1985): in ducks, taste buds on the beak tip are used to discriminate between edible and inedible peas (Berkhoudt, 1976), and grebes reject infected fish when held in the mandibular region of the beak, where taste buds are located (Berkhoudt, 1985). Taste buds on the tongue root and esophageal region may act as a final arbiter of food acceptance, after food passes the taste buds in the anterior mouth and those on the palate.

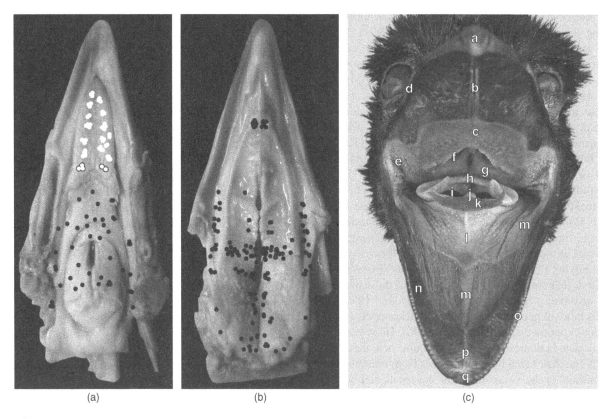

Figure 43.3 The mouth anatomy of the chicken palate (a) and base of mouth (b) and emu palate and base (c). In the chicken (a and b), each dot corresponds to one taste papilla. In the emu (c), the anatomical features are as follows: a, maxillary nail; b, median palatine ridge; c, nonpigmented roof; d, maxillary rhamphotheca; e, rictus; f, left internal nare of choana; g, tongue frenulum; h, laryngeal mound; i, glottis; j, tongue root; k, tongue body; l, nonpigmented floor; m, pigmented floor; n, mandibular rhamphotheca; o, mandibular serrations; p, mandibular plate; q, mandibular nail. Images (a) and (b) courtesy of Kenichi Kudo; (c) courtesy of Martina Crole.

Some studies describe only the occurrence of avian taste buds, without reporting distribution or frequency (Jain, 1977; Moore and Elliott, 1946; Weymouth et al., 1964). More detailed studies, however, have revealed that the frequency of taste buds in birds is small compared with that in mammals: 24 in the blue tit, *Cyanistes caeruleus*, and 46 in the bullfinch, *Pyrrhula pyrrhula* (Gentle 1975), 250–350 in adult chickens (Ganchrow and Ganchrow, 1985; Kudo et al., 2010; Saito, 1966), and 150–375 in ducks (Berkhoudt, 1976; Stornelli et al., 2000), whereas humans have 9000, and rabbits 17,000 (see Gentle, 1975). Studies on the greater rhea (Feder, 1972) and ostrich (Jackowiak and Ludwig, 2008; Tivane, 2008) have not identified any taste buds. Because previous studies are flawed by incomplete sampling of all of the areas in birds where taste buds occur, future studies should encompass the entire oral cavity and more individuals in order to provide accurate measures.

Age-related changes in the number of taste buds in neonatal vertebrates may reflect a difference in gustatory function with increasing age (Yamaguchi et al., 2001). Mammals show an increase in taste bud number from birth to weaning and then a decrease in frequency with age (Shin et al., 2012; Yamaguchi et al., 2001). Increases in taste bud frequency occur with age in chickens, but these studies focused only on the tongue (Lindenmaier and Kare, 1959; Saito, 1966). More recently, Kudo et al. (2008) found no age-related changes when assaying taste bud frequency in the whole oral cavity of chickens up to 140 days in age. Whether there are age-related changes in taste behavior in birds is yet to be studied.

43.2.2 Avian Taste Behavior and Genetics

Studies on taste behavior in birds vary widely in the methods applied to detect sensitivity thresholds and preferences. Some studies use dried foods (e.g., seeds, crumbs, or rice coated with tastants; Alcock, 1970; Rowe and Skelhorn, 2005; Werner et al., 2008), whereas others use two-bottle choice tests with aqueous solutions (Matson et al., 2000). As mentioned above, in free-choice tests such as these, the greater consumption of one item over another is evidence that the animal can discriminate between the two items but

964   **Chapter 43   Comparative Taste Biology with Special Focus on Birds and Reptiles**

should not be considered support for an animal liking or disliking the item (Beauchamp and Mason, 1991). Preference is a difficult concept to measure in birds because, in contrast to humans and other mammals (e.g., rats; Grill and Norgren, 1978), birds exhibit a limited repertoire of behaviors in response to oral stimulation. A characteristic response to aversive stimuli includes persistent tongue and beak movement, head shaking, and beak wiping (Ganchrow et al., 1990; Gentle, 1973; Skelhorn and Rowe, 2009), but there is no analogous repertoire in response to neutral or preferred stimuli (Gentle and Harkin, 1979). Many studies also monitor bird behavior over long periods, which make it difficult to tease apart oral perception from learned associations between taste and postingestional state (e.g., the positive energetic state from ingesting carbohydrates or the emetic effects of ingesting bitter toxins). Thus, studying oral perception requires brief-access tests or methods that circumvent ingestion, such as those used in passive avoidance learning (Marples and Roper, 1997) or methods more akin to those that employ lickometers (used in rodent experiments; Breslin et al., 1996). However, the study of taste hedonics in birds will require new methods to overcome their behavioral differences from mammals.

Contrary to the suggestion that the small number of taste buds reported in birds means that their sense of taste is unimportant (Kassarov, 1999), genomic data and quantitative studies on the physiological, molecular, and behavioral responses of birds to different tastants suggest the opposite. In the following sections, we discuss this evidence for three of the taste qualities for which we have genetic and behavioral data – sweet, umami, and bitter taste (see also Berkhoudt [1985] for an excellent review on the early studies of taste perception).

43.2.3   Sweet Taste

The diets of many avian species include grains rich in starch, and nectars and fruits containing sucrose, glucose, and fructose (Klasing, 1998). Therefore, the gustatory experience of sweet foods by birds was assumed to be much like our own and other vertebrates, with the expectation of a sweet taste preference and the presence of a functional *TAS1R2* (sweet receptor) gene, which codes for one component of the mammalian sweet receptor. However, surveys of the available avian genome sequences have revealed that chicken, turkey, and zebra finch all lack a gene homologous to *TAS1R2* (Lagerstrom et al., 2006; Shi and Zhang, 2006; Zhao et al., 2010), which was likely independently lost during evolution (Zhao et al., 2010) (see also Box 43.1). This result may not have been unexpected since prior research showed that chickens generally lacked behavioral preferences for sugars (see below), though differences in testing methods have given some contradictory results. While this result led to the suggestion that all birds lack the ability to taste sweet, we know that plant nectar, eaten by hummingbirds (family

Trichilidae) and sunbirds (family Necatiniidae), is rich in the disaccharide sucrose and also contains the hexose monosaccharides glucose and fructose and the pentose sugar xylose (Vanwyk and Nicolson, 1995). Fruits are the second most common food category for some birds after insects and are rich in sucrose, fructose, and glucose. Thus, contrary to the suggestion that all birds lack the ability to taste sweetness, we would expect species of birds that exploit nectar and fruit to exhibit behavioral sensitivity and sugar-type preferences and for these species to have a either a functional sweet receptor (i.e., *TAS1R2* gene) or some other molecular mechanisms for detecting sugars (e.g., via glucose transporters). We discuss research on these species after first focusing on those species that lack sweet taste. In 2014 Maude Baldwin, a postgraduate student at Harvard University, cloned the T1R1 and T1R3 taste receptors from the sugar-insensitive chicken, from hummingbirds, and from the hummingbird's closest relative, the insect-eating chimney swift, and tested these in cell culture assay. While the swift and chicken receptors responded only to amino acids, the hummingbird's receptors also fired in response to sweet-tasting sugars, sugar alcohols, and the artificial sweetener sucralose. Baldwin also found that mutations in the birds' T1R1 and T1R3 genes appear to be under positive selection (Baldwin et al., 2014).

43.2.3.1   *Chickens and other Galliforms.*   Wagner and Gunther (1971) conducted one of the only studies showing that chickens (*Gallus domesticus*) prefer sucrose and glucose compared with water. However, testing occurred over a 28- to 30-day period, which does not control for postingestive effects. This is an important oversight, because in long-term tests Kare and Maller (1967) showed that diet can influence taste preferences – birds fed a low-calorie diet selected a 10% sucrose solution over water but had no preference under *ad libitum* feeding. Therefore, long-term tests may reflect dietary inadequacies and responses to postingestive factors rather than actual taste preference behavior.

Some researchers have adapted experimental designs to avoid confounds arising from postingestive effects. Gentle's (1972) study used short presentation periods and showed that chickens rejected 5–30% glucose solutions, rejected or were indifferent to 1–30% fructose solutions, and rejected or were indifferent to 1%, 2.5% 10%, 20%, and 30% sucrose but showed a preference for 5% sucrose. Using electroencephalography, Gentle (1972) found that these same chickens had no responses to glucose concentrations less than 2.5%, rarely responded 10% and 20% fructose, had no response to 1% sucrose, and rarely responded to 10% sucrose. Therefore, many sugar solutions that did not produce behavioral responses did produce nerve responses in the animals (though in other studies electrophysiological responses were low in magnitude [Duncan, 1960; Halpern, 1962]).

In other studies, chickens did not discriminate between water and dextrose or sucrose at concentrations between 2.5% and 25%, and they rejected xylose (Kare and Medway, 1959). Although humans rate saccharin similarly to sucrose at some concentrations (Stromberg and Johnsen, 1990), and many mouse strains show preferences for saccharin (Reed et al., 2004), chickens reject it (Kare and Mason, 1986), perhaps because they cannot perceive its sweet but do perceive its bitter components. Other studies have shown that two other galliforms, bobwhite and Japanese quail, prefer 10% glucose to water (Brindley, 1965; Brindley and Prior, 1968; Hamrum, 1953), and Japanese quail show weak preferences for 1.28% and 2.5% glucose (Urata et al., 1992) and prefer 0.30 M sucrose to distilled water (Harriman and Milner, 1969), although these studies did not rule out postingestive effects. These results for sweet preference are surprising since the amount of sweet food quail encounter in nature is small. We are unaware of any studies on sugar preferences on the other species for which genomic data are available (turkey and zebra finch). However, we expect that their similar diets, which are devoid of sugar, and their lack of *TAS1R2* could also result in behavioral indifference, as is observed in mammalian carnivores (Li et al., 2005). This differs sharply from our expectations regarding nectivorous and frugivorous birds, which have diets high in sugars.

43.2.3.2 *Nectivores.* Baker and Baker (1983) established that nectivorous birds largely fall into two groups, hummingbirds and passerines, and play crucial roles as pollinators for many sugar-producing plants. Most hummingbird-pollinated plants are high in sucrose concentration (0.46–0.81 M) and content (40–46% total sugar) (Baker and Baker, 1983), whereas passerine-pollinated plants are lower in concentration (0.24–0.37 M) and content (0–5% total sugar) (Johnson and Nicolson, 2008). Hummingbirds and other nectar specialists (sunbirds, family Nectariniidae, and honeyeaters, Meliphagidae) are thus thought to prefer high sugar concentrations specific to sucrose, whereas passerine nectivores/generalists should prefer lower concentrations and perhaps nonsucrose sugars.

Initial inspection of the extensive studies on sugar-type preferences in specialist and generalist nectivores (Lotz and Schondube, 2006) seemed to support the hypotheses that specialists exhibit preferences for sucrose over hexose monosaccharide sugars both in the laboratory (Martinez Del Rio, 1990b; Martinez Del Rio et al., 1992; Stiles, 1976) and in the field (Chalcoff et al., 2008), and that generalists have the reverse preference (Brown et al., 2010). However, the story is not so straightforward. Methodology matters when examining sugar preferences, especially in birds, since a species' specific energetic demands can confound interpretation. Different authors control for molarity (Downs, 2000), weight/volume (Blem et al., 2000), or calories (Fleming et al., 2004) in their studies, and as highlighted by Brown et al. (2008), these different methods can yield conflicting results, which do not support the expected dichotomy in preferences. Much is known about the ecology of food choices in birds, but most studies have failed to control for the role of postingestive effects of the compound tested. Because of these issues, we focus only on those studies that have employed equicaloric solutions.

SUGAR-TYPE PREFERENCES Though nectivores are thought to show sucrose preferences, Fleming et al. (2004) found that the broad-tailed hummingbird (*Selasphorus platycercus*) and white-bellied sunbird (*Nectarinia talatala*) show very little discrimination between sucrose and hexose monosaccharide solutions at most concentrations. In another study on nectivores, Fleming et al. (2008) found that the New Holland honeyeater (*Phylidonyris novaehollandiae*) and red wattlebird (*Anthochaera carunculata*) had no preferences for nectar-typical concentrations but did prefer hexose monosaccharides over sucrose solutions less concentrated than natural nectar solutions. They also found that the rainbow lorikeet (*Trichoglossus haematodus*) showed a hexose monosaccharide preference typical of the magnificent hummingbird (*Eugenes fulgens*) and the cinnamon-bellied flowerpiercer (*Diglossa baritula*), a passerine (Schondube and del Rio, 2003). Concentration-dependent preferences for specific sugars may have a physiological root (Lotz and Schondube, 2006). Sucrose must be hydrolyzed to glucose and fructose; thus, a nectivore should prefer hexose monosaccharides at low concentrations to conserve absorption costs (Martinez del Rio, 1990a).

In contrast to specialists, occasional/generalist avian nectivores like the dark-capped bulbul (*Pycnonotus tricolor*) show significant preferences for hexose monosaccharide sugar solutions, irrespective of concentration, when given a choice between hexose monosaccharides and sucrose solutions in equicaloric pairwise choice tests conducted at five different concentrations (5–25%) (Brown et al., 2010). While this suggests different sugar-type preferences in generalist nectivores, this species absorbs hexose monosaccharides more efficiently than sucrose, suggesting absorption efficiency as an explanation, especially since postingestive effects were not controlled here.

CONCENTRATION PREFERENCES Hummingbirds typically prefer stronger nectar concentrations (Blem et al., 2000; Gass et al., 1999) and discriminate between 1% changes in nectar-typical (20%) solutions (Blem et al., 2000). Bacon et al. (2011) hypothesized that the taste of a higher concentration of sucrose would be more "rewarding" than that of lower concentrations, suggesting an avian hedonic aspect to taste. They tested this hypothesis with the rufous hummingbird (*Selasphorus rufus*) foraging at artificial flowers containing either 14% or 20% sucrose, stating that if taste alone determined nectar consumption, the birds should drink more 25% sucrose after first feeding on 14% sucrose but not vice versa. But this was not true – experience, it turns out, can shape their preference,

966   **Chapter 43   Comparative Taste Biology with Special Focus on Birds and Reptiles**

leading the authors to conclude that taste was less important than postingestive feedback in overall preferences. Viscosity of the solutions was not controlled for, nor was its postingestive effect. This highlights the importance of what measures (bout lengths or volumes) should be used to assay taste preferences in studies on birds.

Recently, Medina-Tapia et al. (2012) evaluated the sugar gustatory threshold of the broad-billed hummingbird (*Cynanthus latirostris*) in short taste trials (45 min). They measured bird responses toward sucrose, glucose, fructose, and a 1:1 mixture of glucose-fructose at eight concentrations (0.87–29.2 mM). Chickens and quail exhibit gustatory thresholds for sucrose ranging from 146 to 292 mM (Kare and Medway, 1959), and the cockatiel (*Nymphicus hollandicus*) shows gustatory thresholds of 81.7 and 210.2 mM for glucose and fructose, respectively (Endler and Mappes, 2004). In comparison, the gustatory thresholds of the broad-snouted caiman (*Caiman latiriostris*), a nectivorous reptile, was shown to be three to four orders of magnitude lower (1.31–1.54 mM for sucrose, 1.54–1.75 mM for glucose, 0.87–1.31 mM for fructose, and 1.75–3.5 mM for the 1:1 mixture of glucose–fructose), which is comparable with nectivorous mammals like the bare-tailed woolly opossum (*Caluromys philander*) (Medina-Tapia et al., 2012). Interestingly, Medina-Tapia et al. (2012) suggested that at lower sugar concentrations gustatory thresholds may be more important in determining sugar selection, whereas at intermediate and higher concentrations, assimilation rates and osmotic constraints may be more important factors. This may explain why Bacon et al. (2011), who used concentrations of 14% and 25% sucrose, did not find evidence for taste discrimination in all of their measures of hummingbird taste preference, because these may not be close to gustatory thresholds. Future studies should measure gustatory thresholds before testing for taste versus postingestional feeding cues.

43.2.3.3 Specialist Frugivores.

Fruits differ from nectar in that they are generally considered energetically and nutritionally poor and not adequate to meet a bird's daily metabolic and nutritional demands (Bairlein, 1996). Fruits are generally low in protein and lipid content and high in water and bulk fiber, and they often contain considerable amounts of plant secondary compounds, such as tannins, that bind dietary proteins (making them unavailable for digestion) and negatively affect sugar assimilation (Levey, 1987). Despite this apparent inadequacy, fruits are the second most common food category for birds, after insects (Snow and Snow, 1988).

The hexose monosaccharides fructose and glucose are the most common sugars in fruit pulp (Baker et al., 1998). The sucrose content of passerine-consumed fruit averages only 8% of the total sugar content (Baker et al., 1998). This leads to the hypothesis that frugivorous birds should show a glucose/fructose preference, as opposed to the sucrose preference hypothesized for nectivorous birds. The preferences of some fruit-eating passerines for hexose monosaccharides over sucrose fit this hypothesis well (Brugger and Nelms, 1991; Martinez Del Rio et al., 1988, 1992; Martinez del Rio and Stevens, 1989). The diet of American robins (*Turdus migratorius*) contains 60% fruit and berries (Sallabanks and James, 1999), and they exhibit lower consumption rates of sucrose compared with hexose monosaccharides due to absorption physiology: robins lack the digestive enzyme sucrase, and the birds experience sickness due to sucrose malabsorption (Brugger and Nelms, 1991). Cedar waxwings (*Bombycilla cedrorum*), in contrast, secrete sucrase and can absorb sucrose, but they still show preferences for simple sugars over sucrose (Martinez Del Rio et al., 1989), which has a lower absorption efficiency (61%) than does glucose (92%) or fructose (88%).

As for concentration preferences, Schaefer et al. (2003) found that tanagers (family Thraupidae) preferred a 12% glucose-sucrose mix over an 8% and 2.5% mix and chose more of the 8% food than the 6% and more of the 6% food than the 5% food. Discrimination of sugar content at the 1% level was noted, depending on absolute sugar content in both foods: the birds exhibiting no preference between 12% and 13% foods but a preference for 6% over 5%. Thus, some frugivores prefer stronger concentrations and are able to make accurate discriminations. This finding is similar to the behavior of hummingbirds discussed in the preceding section (Medina-Tapia et al., 2012), where sugar concentration determines preference for sugar type, because of physiology. In contrast, occasional frugivores may have poorer discrimination and weaker preferences because of low reliance on fruits as foods. We discuss the evidence for this hypothesis next.

43.2.3.4 Seasonal Frugivores.

Most birds feed largely on insects and other invertebrates throughout the year but consume large quantities of fruit in the autumn and winter. This includes the thrushes and blackbirds (family Turdidae), starlings (Sturnidae), red-winged blackbirds (Icteridae), and warblers (Sylviidae) (Snow and Snow, 1988). We might expect them to have preferences for hexose monosaccharides over sucrose, similar to specialist frugivores. Tests with equicaloric aqueous solutions of hexose monosaccharides (1:1 mixture of fructose and glucose) and sucrose over four days showed that house finches (*Carpodacus mexicanus*) consume increasing volumes of hexose sugars compared with sucrose at concentrations of 4%, 6%, and 10%, but not 2% (Avery et al., 1999). The observed increase in hexose consumption over sucrose might reflect the same lack of sucrase observed in American robins (Brugger and Nelms, 1991). However, fecal sugar analyses show approximately equal absorption of hexose monosaccharides and sucrose, thus discounting this hypothesis.

Red-winged blackbirds eat more grains, while starlings consume more insects and fruit, leading to the hypothesis that starlings evolved a preference for fructose (Espaillat and Mason, 1990). The responses of both species to aqueous solution of D-fructose (0.1%, 0.5%, 1.0%, and 5.0% w/v) were measured over two-hour test periods and measured consumption at 30-minute intervals. As predicted, red-winged blackbirds drank less fructose per gram of body weight than did starlings. Concentration had no effect on consumption of fructose by blackbirds, whereas consumption by starlings did increase past 0.5% D-fructose. This study demonstrated that the blackbirds' insensitivity to fructose reflects the predominance of grain and insects in their diet and starlings displayed a correlation between fruits and sugar sensitivity. Studies on bird species that rarely consume sugars are uncommon, but for species whose diet is low in sugar, we would hypothesize sugar perception and preference to be weak. However, this is not always the case, as we discuss next.

43.2.3.5 Insectivores, Omnivores, and Carnivores.
The azure-winged magpie (*Cyanopica cyana*) prefers sucrose over glucose and fructose, suggesting that magpies can efficiently absorb sucrose (see Lotz and Schondube 2006), contrary to the expectation that primarily insectivorous species lack sensory adaptations to frugivory (sweet preferences), the great tit (*Parus major*) shows a preference for 7% and 14% glucose solutions (Warren and Vince, 1963). However, each solution was paired with water for eight consecutive 24-hour periods, which did not control for postingestive effects. Ravens feed predominantly on carrion and do not show a preference for five different sugars over water (Harriman and Fry, 1990). Likewise, Rensch and Neunzig (1925) reported rejection or indifference to sugar by the nonfrugivorous owls (family Strigidae) and ducks (Anatidae). That same study showed that carnivorous kestrels (family Falconidae) and herons (order Gressores) accept sugar, although they were offered only a single-bottle stimulus, which assesses only acceptance and not preference. Further comparative work is needed to tease apart sweet taste function and feeding ecology in different avian taxa.

43.2.4 Umami Taste

Birds frequently consume foods containing glutamate and/or amino acids (e.g., meats, insects, grains). Therefore, one might expect that birds should be able to detect and exhibit preferences for free amino acids, and that these species would express at least the components of the umami receptor (T1R1 and T1R3) if not other umami-related genes.

A comparative genome analysis by Shi and Zhang (2006) showed that chickens possess *Tas1r1* and *Tas1r3* required for the umami receptor (as do mouse, cow, dog, cat, and several fish species), and *Tas1r1* has also been documented in turkey, zebra finch, egret, loon, and tubenose seabirds (Zhao et al., 2012). Roura et al. (2008) did a one-by-one comparison of all farm and companion animals for which Tas1r genes are known and found that chicken T1r1 nucleotide homology is lowest, which may equate to qualitative and quantitative differences in taste behavior of chickens compared with other species. A second receptor, a metabotropic G-protein–coupled receptor named mGlu4, which is associated with glutamate taste in rats (Chaudhari et al., 2000), is located on chromosome 26 in the chicken. However, its function, if any, in the chicken taste sensory system is not known.

In humans, the *Tas1r1* and *Tas1r3* genes produce a receptor that is narrowly tuned to L-glutamic acid and L-aspartic acids (Li et al., 2002), whereas in rodents and pigs this receptor has broader tuning, and these species exhibit preferences for many amino acids, among them L-cysteine, L-methionine, and L-alanine (Nelson et al., 2002; Roura et al., 2011). To date, the relationship between umami taste preference and sensitivity and candidate umami receptor genes in birds has not been explored explicitly. However, the available behavioral data on umami/amino acid taste in birds (described below) corroborate the presence of the *Tas1r1* and *Tas1r3* genes in the available genomic data.

43.2.4.1 Galliforms.
Chickens adjust their food intake to compensate for dietary deficiencies in essential amino acids. In feeding-choice scenarios, hatchling broiler chickens show an immediate preference for a diet containing synthetic amino acids rather than one deficient in lysine, methionine, and tryptophan. Cadirci et al. (2009) found that chickens deficient for methionine show a clear preference for methionine-treated water, though color cues on the drinking bottles were necessary and preferences may have been learned following ingestion. Further, Urata et al. (1992) showed that chickens exhibit no preference behavior between 0.01% and 0.64% monosodium glutamate (MSG) in two-bottle choice tests but avoided MSG at concentrations of 1.28–10%. In comparison, Japanese quail showed no preference between water and 0.01% or 5% MSG, but they preferentially consumed water over 10% MSG. Tas1r genes in the chicken may be involved in nutrient sensing and metabolism rather than taste behavior. This is supported by cDNA microarray analysis of

968   **Chapter 43   Comparative Taste Biology with Special Focus on Birds and Reptiles**

gene expression patterns in the hypothalamus, liver, and abdominal fat of chickens, where *Tas1r1* transcripts are differentially expressed between overweight and lean lines of chickens (Byerly et al., 2010). This difference may arise because T1R1 is involved in postingestive nutrient sensing (Ren et al., 2009; Wauson et al., 2012).

43.2.4.2 *Passerines.*

Espaillat and Mason (1990) hypothesized that starlings, because they consume more insects than do red-winged blackbirds, would exhibit greater preference for free amino acids. In short one-bottle choice tests that controlled for postingestive effects, both blackbirds and starlings increased consumption of L-alanine in water with increasing concentration per gram of body mass; male starlings consumed more than did male blackbirds. Another study found that blackbirds strongly prefer rice seed soaked in L-alanine compared to seed soaked in tannic acid (Werner et al., 2008). These results could demonstrate either taste-based preference for amino acids and/or aversion to the potentially sour/bitter sensation of tannic acid.

43.2.5 Bitter Taste

Many birds consume plants and animals that contain defensive toxins (e.g., cardenolides, iridoid glycosides, and cyanoglycosides Brower, 1969; Nishida, 2002), venoms and stings (Schmidt, 1990), urticating hairs (Bowers, 1990; Wyllie, 1981), phenolics (Johnson et al., 2006), and antinutrient factors such as tannins (Dixon et al., 1997) and saponin (Cheeke, 1971). Humans report these compounds as tasting intensely bitter and strongly avoid them in preference tests (for a review, see Bachmanov and Beauchamp, 2007). Toxic cardiac glycosides and cardenolides trigger the vomiting reflex (Brower, 1969) to avoid absorption. Solutions that are rated as bitter by humans (e.g., quinine sulfate) also result in emesis in birds (Alcock, 1970). Bitter tastes (without ingestion) consistently induce nausea in humans, showing that the body not only detects potential toxins but anticipates their ingestion by inducing a prophylactic aversive state (Peyrot des Gachons et al., 2011). Most studies of bitter taste in birds use taste chemicals that are not necessarily part of insects' chemical defense battery (Marples et al., 1989; Rowland et al., 2007, 2010; Skelhorn and Rowe, 2009, 2010; Tullberg et al., 2000; Yang and Kare, 1968). However, the universal trend in these studies is rejection of bitter tastes. Birds respond to these diverse bitter chemicals with specific rejection responses, which include beak wiping and head shaking (Marples and Roper, 1997), vomiting (Alcock, 1970; Brower, 1969), and taste-rejection behavior, where birds attack and release prey on the basis of defense chemicals (Fink and Brower, 1981; Skelhorn and Rowe, 2006a).

These behavioral responses to bitter chemicals are consistent with bioinformatics surveys and direct sequencing, which have shown that turkeys have one bitter taste (Tas2r) gene emus have two (Maehashi et al., submitted), and chickens have three (Hillier et al., 2004). These repertoires are low compared with the functional repertoire of Tas2r genes in other sequenced vertebrate genomes, which ranges from 4 in the platypus to 36 in rats (Dong et al., 2009), up to 64 in frogs (Go, 2006; Shi and Zhang, 2006). The range between birds and other species may suggest a reduced ability to detect and discriminate between bitter substances, or fine-tuning of the receptors to the birds' feeding ecology (specialization), or broad tuning of the receptors allowing many taste chemicals to be perceived by fewer receptors. A study by Davis et al. (2010) on the Tas2r cluster in the white-throated sparrow (*Zonotrichia albicollis*) shows expansion from one to four genes in the passerine lineage since the most recent common ancestor with the chicken (order Galliformes) around 100 million years ago (van Tuinen and Hedges, 2001), followed by a subsequent expansion to 19 genes in the lineage leading to the white-throated sparrow since its most recent common ancestor with the zebra finch (*Taeniopygia guttata*), which has seven genes. The number of intact Tas2r genes in the white-throated sparrow is within the range found in most other vertebrates (i.e., 15–49) (Dong et al., 2009) and may reflect a similar ability to sense and respond to bitter tastants, though this has yet to be confirmed by behavioral assays. What is clear is that avian Tas2r genes have undergone dynamic gene expansion and contraction during the course of vertebrate evolution (Dong et al., 2009; Go, 2006), and species likely differ in responses to bitter compounds. We focus first on those species with small Tas2r repertoires because those with larger repertoires have not yet been studied.

43.2.5.1 *Chickens.*

Chicken bitter taste behavior and neurophysiology have been the most intensively studied because of the chickens' importance as a commercial food source. Though chickens have a small Tas2r repertoire, they display discriminative reactions to bitter taste stimuli as measured by stereotypic head and oral movements in prehatchlings (Vince, 1977) and mature animals (Gentle, 1975) and by tests of preference (Balog and Millar, 1989; Kudo et al., 2010) and discrimination (Skelhorn and Rowe, 2006b, 2007). In electrophysiological tests, chickens exhibit lingual nerve responses to 0.01 M (Halpern, 1962; Kadona et al., 1966) and 0.02 M (Kitchell et al., 1959) quinine hydrochloride (QHCl) and 0.0002 M sucrose octaacetate (Halpern, 1962), and they have chorda tympani responses to 0.1 M and 0.05 M QHCl (Gentle, 1983, 1987).

Results from two-bottle choice trials generally support the results from electrophysiology studies but are affected by the animals' side preferences, testing period length, and dietary deficiencies (Gentle and Dewar, 1981; Gentle et al., 1981). Gentle (1976) found that water deprivation lasting 2–12 hours resulted in increasing acceptance of 0.005 M QHCl and a reduction in the number of aversive responses when the birds were tested for 3 or 20 minutes. In longer-term two-bottle choice tests, chickens show behavioral rejection of QHCl at 0.5 mM and 2.0 mM (Kudo et al., 2010) and at 0.1% (2.77 mM) (Kare et al., 1957). However, while 0.0002 M sucrose octaacetate produces lingual-nerve responses, chickens accept it as well as water (Halpern, 1962). Feed-choice trials suffer from the same confounding factors as two-bottle choice tests, with long test durations indicative of confounding by postingestive effects. Gentle (1971b) found that chickens consume significantly less food coated in 0.5% QHCl within 1 hour; they also consume significantly less food coated in 0.10% quinine hydrobromide compared with a control diet (Balog and Millar, 1989) and reject QHCl concentrations greater than 0.04% over longer periods (Urata et al., 1992).

A useful behavioral assay in chickens is called *passive avoidance training* (reviewed in Rose, 2000). Chicks are allowed to peck at a small, colored bead coated in a distasteful substance (Burne and Rogers, 1997; Marples and Roper, 1997) and are then assayed for avoidance to the colored bead only. This is a pure taste behavior bioassay controlling for postingestive effects. Unfortunately, most passive avoidance tests use methyl anthranilate, which has a strong odor, which complicates disentangling taste from olfaction. However, chicks given bitter taste alone in these assays learn to avoid the visual stimulus, suggesting that denatonium benzoate (Marples and Roper, 1997) and quinine sulfate (Bourne et al., 1991) are aversive to chickens. Similarly, a study on taste-rejection behavior in chicks showed that they more frequently released unpalatable food coated with 4% quinine compared with 1% quinine (Skelhorn and Rowe, 2006a).

In summary, despite their smaller Tas2r repertoire, bitter tastants are perceived as aversive to chickens. Whether their gustatory thresholds or discrimination abilities differ from those of other avian orders has not been directly tested, but in the following sections we describe other avian species bitter taste behavior, which is qualitatively similar.

43.2.5.2 *Specialist Insectivores.* Tropical insectivorous birds, such as the rufous-tailed jacamar *(Galbula ruficauda)* and black-backed oriole *(Icterus abeillei)*, specialize on butterflies (Chai, 1996) that possess chemical defenses and advertise this with warning coloration. These birds are expected to be highly selective in their attacks and consumption of insects, based on their evolutionary history with these prey (Speed et al., 2012), and should

exhibit higher thresholds for bitter stimuli than species naive to such prey, such as chickens. To our knowledge, no studies have assessed gustatory thresholds of specialist insectivores for quinine or denatonium benzoate or even chemicals more akin to those found in their prey. However, studies have reported the foraging behavior of these birds in nature (Fink and Brower, 1981) and in the laboratory (Brower, 1969). For example, Chai (1996) assessed the responses of naive jacamars to poisonous butterflies of the Nymphalidae family and observed that birds attacked prey once or several times but rejected them after tasting, suggesting the butterflies were bitter due to their chemical defenses (though texture, odor, etc., cannot be discounted). Likewise, Fink and Brower (1981) observed orioles attacking and releasing most (75%) monarch butterflies *(Danaus plexippus)* encountered. However, after measuring cardenolide content of the rejected butterflies, the birds did not selectively reject the butterflies high in cardenolides, leading the authors to suggest the birds discriminate based on textural or size cues during capture. Interestingly, once the butterflies were killed, the amount eaten per bird was inversely proportional to cardenolide content, which may reflect the emetic effects of the different body parts as opposed to actual taste cues, because the degree of bitterness may not accurately predict prey toxicity (Glendinning, 1994). Monarchs are high in lipids and proteins (Brower et al., 1988). By not relying solely on taste cues, birds may balance the cost of toxin intake with the nutritional benefits of the butterfly.

43.2.5.3 *Generalist Insectivores.* Many generalist avian foragers in the order Passeriformes learn to associate the visual cue and bitter taste/postingestive effect of prey toxins to avoid repeated consumption (reviewed in Mappes et al., 2005; Schuler and Roper, 1992). The most famous example is the blue jay–monarch butterfly paradigm studied by Lincoln Brower. Monarch butterflies are an interesting case study, because they exhibit a spectrum of palatability. When Brower trained blue jays to eat palatable monarch butterflies, they would quickly eat the first highly toxic one offered. The birds shook their heads and puffed their feathers but had no innate cardenolide-based bitter taste rejection. The perceived bitterness of any food is a function of the genotype of the particular animal under consideration because insects or other foods are not inherently bitter; instead, they contain more or less of certain chemicals that, for some species, engage bitter receptors. As such, bitterness may not accurately predict the potential toxicity of food (Glendinning, 1994), so these generalists may exploit food sources that would otherwise be rejected. In line with this prediction, Skelhorn and Rowe (2010) found that European starlings can associate a bitter taste (denatonium benzoate) with

970   **Chapter 43   Comparative Taste Biology with Special Focus on Birds and Reptiles**

the dose of a second, tasteless toxin (quinine) to optimize their nutrient intake, demonstrating remarkable foraging discrimination.

43.2.5.4 Granivores and Omnivores.

Seeds and grains can contain bitter secondary compounds such as tannins and are often infected by fungi, which increase the concentration of alkaloids (Wolock Madej and Clay, 1991). For parrots and cockatiels, which primarily eat seeds high in tannins and in various stages of ripeness, we might expect these birds to discriminate among seeds as insectivores do among toxic butterflies. Indeed, cockatiels, like humans, are highly sensitive to quinine and even more sensitive than some mammals, which parallels the birds' mammal-typical taste bud structure (Matson et al., 2004). Similar to humans (who overcome tannins in teas and beers), parrots tolerate bitter chemicals in their foods, though this may be explained by their propensity for consuming clay which binds and neutralizes toxins, rather than by taste habituation (Gilardi et al., 1999).

In contrast to granivores, the diets of omnivorous, meat-eating birds such as crows and ravens are often devoid of secondary compounds, and we would expect them to exhibit low thresholds for bitter tastes because bitterness should reliably indicate poisons. This is supported by evidence from common ravens (*Corvus corax*), which reject QHCl at all concentrations and are nearly as averse to drinking a 0.01% solution as they are to 1.6% (Harriman and Fry, 1990).

43.2.6 Summary

Clearly birds can taste, but they do not respond behaviorally the same way to some chemicals that humans do, as illustrated by the dissimilar responses to sweet and bitter compounds. To clarify these issues and enhance our understanding of avian taste perception, shorter test periods are needed to reduce/eliminate postingestive effects. We suggest that future biochemical studies identifying the ligands for the avian Tas2rs would be useful for species conservation and understanding the coevolution of insect defenses and bird responses; functional assays on Tas1r genes in nectivores would elucidate the proximate mechanisms for sugar-type and concentration preferences, and behavioral genetic studies of intra- and interspecies genetic differences could reveal the relationship between dietary niche and taste preferences that could have important implications for our understanding of niche partitioning, habitat choice, dietary changes, and speciation.

43.3 REPTILES

A complementary route for understanding the evolution and diversity of taste behavior comes from comparing studies of birds with those in reptiles, which serve as deep, extant relatives to younger vertebrate classes, such as Aves and Mammalia. Though far less is known about taste in reptiles than in any other vertebrate group, the taste behavior/biology of the squamates (snakes and lizards) is an exception to this, with studies on anatomy, neurophysiology, and behavior that have led to the proposal of squamates as alternative models for taste studies (Young, 1997). Understanding of the reptilian taste system has been hindered by the tendency to focus on other sensory modalities, such as olfaction and vomeronasal systems. We start by discussing the taste system of reptiles, followed by taste genes and behavior.

Class Reptilia comprises four general groups: turtles, crocodilians (crocodiles and alligators), tuatara (a single species, *Sphenodon punctatus*), and squamates (snakes and lizards). Reptiles have a wide geographic range and have radiated into a variety of terrestrial habitats, but they also inhabit marine and freshwater environments. Similar to observations in birds, this diverse range of habitats is mirrored by diets that vary considerably. Turtles are generally considered opportunistic omnivores. For example, the red-eared slider turtle (*Trachemys scripta*) is carnivorous when a juvenile and switches to a more herbivorous diet as it grows (Bjorndal, 1991), whereas the green sea turtle (*Chelonia mydas*) feeds on seagrasses and algae (Garnett et al., 1985). Crocodiles consume aquatic insects, spiders, crustaceans, and fish (Wallace and Leslie, 2008), as well as such terrestrial vertebrates as snakes and mammals (Magnusson et al., 1987). Tuataras feed on a wide range of small animals but predominantly on insects (Wallis, 1981), and squamates feed on mammals, birds, other lizards and snakes, vertebrate eggs, and insects (Vitt et al., 2003; Rodríguez-Robles et al., 1999). The types of foods that birds and reptiles eat overlap somewhat, for instance, insects, nectar, fruit, and spiders, which makes a comparison of evolutionary patterns of taste an interesting avenue to explore.

The avian and reptile literature contains many parallel opinions regarding taste capacity. For instance, it is generally believed that the sense of taste is reduced or completely lost in snakes because taste buds are lacking on the tongue and mucosa of the lower jaw (Auen and Langebartel, 1977; Uchida, 1980; for review, see Young, 1997). However, blind snakes possess taste buds in large numbers (Kroll, 1973). Pit vipers have palatine taste buds that are served by the palatine branch of the facial nerve, and the chorda tympani carries afferents from the oral mucosa of the mandible (Atobe et al., 2004). Similar to birds, snakes have taste buds positioned so that they receive direct stimulation by food particles. Therefore, in the following sections we review the evidence for reptilian taste bud occurrence, taste behavior, and genetics, and similar to the section on birds, we focus on sweet and bitter taste.

43.3.1 Taste System

As is the case for birds, finding the taste cells of reptiles is a matter of knowing where to look. There is a large literature on this topic (reviewed in Schwenk, 2008), summarized in Figure 43.4. Reptilian taste buds are small, flask-shaped organs with the same sensory and support cells as in birds and are considered "intermediate" in form between fish and mammals (Uchida, 1980). Unlike those in birds, however, reptile taste buds generally lack pores, as revealed by ultrastructural examination. Korte (1980) suggested that aquatic reptiles have taste bud structure similar to that in fish and amphibians, which points to convergent evolutionary changes to taste cells in response to a sea versus land environment, and the possibility of underlying differences in the molecular biology of taste cells in aquatic versus terrestrial reptiles.

Taste buds are found in several locations within the oral cavity and pharynx of reptiles and their distribution varies among groups and even among species within groups. The nerves that connect taste cells to the brain are similar to the three cranial nerves in mammals: VII (chorda tympani and palatine branches of the facial nerve), IX (glossopharyngeal), and X (vagus). The palatine branch of the facial nerve carries afferent tracts from palatine taste buds, whereas the chorda tympani afferents from the oral mucosa of the mandible (Atobe et al., 2004). Most turtles, crocodilians, and lizards have taste buds on the tongue and palate, with some exceptions (e.g., Young, 1997). Taste buds in turtles are usually found on the tongue (Uchida, 1980) and may occur in other parts of the mouth and pharynx, but too few taxa have been examined to generalize. Pacific hawksbill (*Eretmochelys imbricata bissa*) and Pacific ridley (*Lepidochelys olivacea*; Cheloniidae) sea turtles appear to lack taste buds (Iwasaki et al., 1996), though buds are present in the red-eared slider turtle (Korte, 1980). Crocodile taste buds were studied by Bath (1906), in the same study that described avian taste buds (see Section 43.2.1.1), and are typically found on the mucosal surfaces of the tongue, palate, and pharynx.

Snakes have no taste buds on the tongue and few or none on the lower jaw; they are instead found mostly on the palate (Auen and Langebartel, 1977; Uchida, 1980; reviewed in Young, 1997). In xenophidian snakes (cobras,

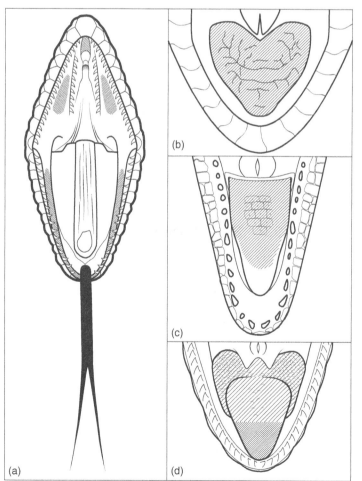

Figure 43.4 Oral cavities of each of the major reptile groups. (a) Dorsal (upper) and ventral (bottom) palates of a generalized snake mouth with its forked tongue in black. The teardrop-shaped cavity in the anterior dorsal palate is the opening of the internal choanae/nares. (b-d) The ventral palate of a turtle (b), crocodilian (c), and lizard (d). Hatched lines in each drawing represent areas where putative taste papillae have been identified histologically. The tongue tip of most lizards, as with most mammals, is highly concentrated with taste papillae, as indicated by denser hatching (d). Images were redrawn from pictures generously provided by K. Schwenk and E. Taylor. Drawing credit: Mary Leonard from the University of Pennsylvania, Philadelphia.

vipers, and sea snakes), taste buds are elevated on small papilla-like projections that also contain mechanoreceptors, which may allow the identification and manipulation of prey in the mouth (Nishida et al., 2000; Schwenk, 2008). This organization is reminiscent of that in ducks and grebes. Although little studied, the rare tuatara, the only extant nonsquamate lepidosaur, possesses numerous filamentous papillae, some of which are purported to be gustatory based on their structure and staining (Schwenk, 1986). The most interesting aspect of the tuatara's taste bud pattern is that the adult structure of the taste papillae is most similar to the papillae of embryonic mammals, squamates, and crocodilians. This finding highlights the research gains that could be made in our understanding of taste developmental and molecular biology by examining the taste physiology and cellular biology of specific groups and/or species of reptiles.

The taste cells, buds, and papillae in reptiles are thus similar in some ways to those of birds and mammals. And as has occurred in some genera of mammals and birds, the function of the reptilian tongue has evolved roles outside of gustation and food manipulation. For example, the tongue can be used to capture and lure prey, to warn predators, to feign death, and to excrete salt. The most pertinent example of an extragustatory role of the reptile tongue is seen in squamate reptiles, where the tongue is essential to their chemosensory system because it captures and transports nonvolatile chemicals to the vomeronasal organ (VNO). An interesting aside is that these same chemicals on the tongue may also make contact with putative taste receptor cells located near the opening of the VNO, emphasizing a possible anatomical convergence that would maximize food chemical detection (Atobe et al., 2004; Kroll, 1973; Schmidt et al., 2010).

When gathering nonvolatile chemicals, lizards lick the food or other item, called a *substrate lick*; when they are gathering volatile chemicals, they lick air, called a *tongue flick* (Greenberg, 1993; Schwenk, 1985). The study of taste behavior in reptiles has therefore been problematic, especially in squamate reptiles, due to a focus on the vomeronasal and olfactory systems for chemosensation. The VNO of most reptiles is exquisitely developed and responds to a remarkable diversity of chemical cues. However, the delivery of chemicals to the VNO requires the involvement of the tongue, especially in squamates (see below). During a tongue flick, a reptile physically transfers chemicals from the external environment to the ducts leading to the chemosensory epithelium of the VNO. It is tempting to postulate that the reliance on the tongue as a chemical cue delivery device in some lizards and snakes may partially explain the complete loss of taste buds on the tongue and oral mucosa in these reptiles. Such a concomitant loss of chemosensory anatomy alongside a suite

of major morphological shifts (mass reduction, lengthening, and forking of the tongue) is not unprecedented in vertebrates. Indeed, porpoises apparently lack an olfactory sense, including loss of their olfactory bulb, which may have arisen alongside the cranial restructuring that was favored in a completely aquatic environment (see Box). Further, the taste gene *Tas1l* in the bottlenose dolphin is considered to be nonfunctional based on a recent study of pseudogenization in carnivore taste genes (Jiang et al., 2012). It would be interesting to know if the carnivore pattern of major taste loss is true in a subset of reptiles that are strict carnivores, such as snakes and crocodilians.

43.3.2 Taste Behavior

Very few studies of reptiles examine taste behavior, due in part to the belief that reptiles have little or no ability to taste, and researchers have a tendency to instead focus on the vomeronasal system. There are also inherent difficulties in testing these species in a controlled laboratory environment or observing their natural behavior, especially because classic gustation research methods like the two-bottle preference test are impractical. Current methods for examining taste in reptiles are to observe whether they reject food adulterated with a bitter chemical or avidly consume those with an added sweet chemical or, in the case of snakes and lizards, whether the tongue behavior changes when it encounters a chemical or food. Anecdotal evidence points to taste behavior in American alligators when food extracts are introduced to water (Schwenk, 2008). They show increased rates of lateral head movements and mouth opening and snapping, and because crocodilians likely do not use olfaction underwater, this suggests a role for taste. What is known, as reviewed below, suggests that reptiles can perceive at least some compounds that humans perceive as sweet or bitter.

43.3.3 Sweet Taste

The study of sweet taste in birds is logical because some birds eat foods that are sweet, such as fruit or nectar, but the motivation to study sweet taste in reptiles is less obvious because the habitual diet for the bulk of reptiles is meat, eggs, and/or insects. That said, some lizards do eat sweet food, the best-known example being fruits and flower nectar (e.g., Liner, 1996; summarized in Olesen and Valido, 2003, 2004). The reaction to sweet taste has been studied directly in the lacertid Boettger's lizard (*Gallotia caesaris*), which exhibits both substrate-licking and tongue-flicking behavior when offered sucrose (Cooper and Pérez-Mellado, 2001). The behavior of the lizard changes with the concentration of the sucrose: as the solution becomes more concentrated, the number of tongue flicks decreases and substrate licks increase. These data suggest that vomeronasal sensing (tongue flicks) is used with lower concentrations of sucrose, whereas taste (substrate licks) is used to assess chemicals at higher sucrose concentration (Cooper and Pérez-Mellado, 2001). Though it is unknown if the VNO can respond to sucrose, the behavioral data suggest that chemical detection strategies shift along the sucrose concentration gradient. It could also be that taste buds in the roof of the mouth, near the opening of the VNO, are sufficient to detect low concentrations of taste chemicals as mentioned above. Although these data suggest that lizards perceive sweet as a taste quality, another study drew the opposite conclusion: Stanger-Hall et al. (2001) found that a sweet mixture (aspartame, with a small amount of dextrose) did not affect food choices in a learning paradigm. However, aspartame is not sweet to many animals (e.g., Li et al., 2011), so it is not the ideal sweet stimulus for these experiments. Overall, these studies suggest that some lizards do respond appetitively to sugars that humans describe as sweet, but given the extreme paucity of data, further study is warranted.

43.3.4 Bitter Taste

As discussed in the section on birds, the ability to sense bitterness may evolve over time and help an animal adapt to the food sources in its environment. As the environment has changed over time, the types of food and toxins that reptiles have found have also changed, sometimes creating an evolutionary mismatch between the presence of a poison (usually a bitter substance) and an animal's ability to taste it (e.g., poisonous cane toads in Australia, highlighted below). Several methods have been used to study the reptilian bitter taste system, typically involving observations of their response to poisonous prey or food that has been tainted with noxious animal secretions (e.g., Barach, 1951; Scott and Weldon, 1989; Weldon and McNease, 1991). When we mention "noxious" or "bitter" compounds, these are, of course, based on human taste perceptions of the same chemicals. Many of the studies below are difficult to interpret because taste is not separated from the other chemosensory systems of reptiles, though investigators do try to make this distinction.

43.3.4.1 Snakes. Burghardt et al. (1973) watched snakes' behavior in response to earthworms that the snakes had been trained to find noxious through pairing with an emetic (lithium chloride, LiCl). Snakes would tongue flick and then accept earthworms following the aversion training; rejection occurred only after the earthworm had been seized in the snake's jaws (and presumably encountered taste cells), suggesting a taste-mediated behavioral response. Stanger-Hall et al. (2001) took an even more direct approach: they reported that lizards avoided contact

with crickets treated with QHCl even after the VNO ducts were sealed, suggesting that taste alone was sufficient to elicit the rejection. Many other studies have drawn the conclusion that reptiles can perceive bitter tastes (e.g., Benes, 1969; Price-Rees et al., 2011, 2012; Shanbhag et al., 2010; Sword, 2001) but typically have not accounted for the contribution of the other chemosensory systems, either through behavioral observation or by occluding the olfactory and vomeronasal systems.

The red-spotted garter snake (*Thamnophis sirtalis concinnus*) feeds on rough-skinned newts (*Taricha granulosa*) (e.g., Brodie et al., 2002) that synthesize and secrete the poison tetrodotoxin from their skin. The garter snakes eat these newts, even the ones with the strongest concentration of the toxin, and sometimes go into cardiac arrest and die. Incredibly, though, some populations of these snakes have adapted to detect the toxin, but newts are rejected only after consumption and not during oral contact, which raises the question of whether it is a simple toxicity reaction during digestion or, less likely, taste receptors in the gut that initiate this response. These two possibilities are not mutually exclusive, but the emphasis here is that we simply do not understand the taste biology of this or any other predator–prey system in reptiles. Garter snakes will reject other types of toxic prey after making oral contact with them (e.g., poison frogs, family Dendrobatidae), which suggests they can perceive the bitterness of toxic prey generally (Brodie and Tumbarello, 1978; Dumbacher et al., 2009) but cannot taste tetrodotoxin. Tetrodotoxin (newts) and alkaloids (poison frogs) elicit different gustatory sensations in humans, with the former having no taste and the latter having bitter taste (J. Daly and R. Saporito, pers. comm.; Daly and Myers, 1967; Saporito et al., 2007). Thus, garter snakes may not be able to taste tetrodotoxin either.

In a different example of reptile bitter taste ecology, invasive cane toads (*Bufo marinus*) in Australia produce bitter bufadienolides that apparently cannot be sensed by many resident species, such as varanid lizards and crocodiles, which routinely eat the cane toads and die as a consequence (e.g., Griffiths and McKay, 2008; Letnic et al., 2008). As with the red-spotted garter snake, it may be that these crocodiles and lizards cannot taste the cane toad poison, though other Australian species do exhibit taste-based recognition of bufadienolides (see below). Such relationships may be examples of an evolutionary lag between the introduction of a new poison in the environment and the ability of a predatory species to perceive and avoid it.

43.3.4.2 Lizards. The most convincing experimental work on bitter taste behavior in lizards was by Stanger-Hall et al. (2001) on green anoles (*Anolis carolinensis*), which

974   **Chapter 43   Comparative Taste Biology with Special Focus on Birds and Reptiles**

learn to avoid contact with crickets treated with QHCl (powder or solution, 1 mM). The crickets were colored with tempera paint, and the anoles exhibited strongest learned avoidance behavior to colored crickets treated with quinine (powder but not QHCl solution) when trained. They found that the avoidance behavior could still be learned by anoles with sealed VNO ducts. This finding strongly suggests that gustation is the mechanism for assessing bitter stimuli in green anoles and that it leads to gustation-based learning.

An interesting study using the agamid eastern garden, or changeable lizard (*Calotes versicolor*) also discovered taste aversion learning. Using prey coated with chloroquine phosphate (a compound that tastes bitter to humans) or sucrose (pastes made with pure powders), Shanbhag et al. (2010) found that the lizards exhibited a strong aversion to chloroquine-coated prey. Further, they demonstrated that the lizards learned to visually associate the bitter prey with the color of the presentation dish. Bitter responses to toxic prey and bitter-compound-coated prey have been shown in other lizard species but not in a learning context (e.g., Benes, 1969). This observation corroborates what is known for the eastern blue-tongued lizard (*Tiliqua scincoides*), which will consume poisonous plants that contain bufadienolides. Populations without experience with the toxic flowers will eat them, but some of these lizards will spit them out before swallowing, suggesting gustation enables toxicity assessment (Price-Rees et al., 2012). There are locally adapted populations of the lizards that, because they can metabolize the bufadienolides, can also survive consuming cane toads that produce the same class of toxic compounds (Price-Rees et al., 2012). This same species exhibits rapid taste learning as well, suggesting it can associate toad-based stimuli with the effects of toxin consumption and thus learn to avoid eating the cane toads, though it is not clear if this is taste or vomeronasally mediated (Price-Rees et al., 2011). Other lizards, such as anoles, will reject grasshoppers raised on toxic diets, suggesting an ecological role for bitter perception that may be widespread across lizard groups (Sword, 2001).

Food aversion learning has also been demonstrated in other lizards (*Basiliscus basiliscus*, *B. vittatus*, *Eumeces schneideri*, and *Mabuya multifasciata*) using LiCl injections, which typically make the animal feel sick (Paradis and Cabanac, 2004). In that same study, the researchers have shown that amphibians (*Bufo paracnemis* and *Pachytriton breviceps*) do not exhibit LiCl-induced food aversion learning, though LiCl may not have the same aversive effects in amphibians as it does in reptiles. The evolution of gustatory-based learning may have been a necessary adaptation as a fully terrestrial vertebrate since the lizards tested span a wide range of terrestrial vertebrates that is quite ancient (Bergmann and Irschick, 2011).

43.3.4.3 *Crocodilians.*
Studies on taste behavior in this group of reptiles are lacking. There are suggestions that some crocodilians will reject poisonous prey or food that has been tainted with noxious animal secretions (e.g., Barach, 1951; Scott and Weldon, 1989; Weldon and McNease, 1991). A fascinating aside to the putative role of gustation in crocodilians is that female crocodilians carry their offspring in their open mouths. It has been hypothesized that gustatory cues (or at least a signal from the oral cavity) must suppress the mother's biting and swallowing reflexes, though there has been no test of this idea (Ferguson, 1981).

43.3.4.4 *Turtles.*
Turtles cannot be trained in operant conditioning using taste alone, but olfactory cues enable conditioning (Manton et al., 1972). Burghardt and Hess (1966) used gustatory cues in food imprinting with neonatal turtles, suggesting that taste plays a role at least in the development of these preferences. However, the turtles in that study had access to olfactory cues from the food items, and it is likely that food choice following conditioning was affected by these cues. Turtles have an exquisitely developed olfactory epithelium and bulb, and much work has demonstrated a central role of olfaction in specific behaviors, such as foraging and open-water homing behavior.

43.3.5 Summary

All indications are that at least some reptiles can taste, but little effort has been focused on this neglected sensory system, in part because of the better-known role that the other specialized chemosensory systems play in their ecology. There is a clear need for behavioral methods to study taste as a sense separate from olfaction and to define the qualities that reptiles can detect and at what concentration. Anatomical studies should be expanded to include molecular methods to characterize taste cells; genomic analysis is just beginning, but the presence of apparently functional taste receptors and other taste signaling molecules (Dong et al., 2009) may encourage further work in this area. The green anole (*Anolis carolinensis*) was the first reptile genome to be sequenced (Alföldi et al., 2011), and it has multiple taste genes (Dong et al., 2009), so this new information may guide new research. Anoles have multiple TAS2R genes that form independent clusters when analyzed alongside other tetrapods and fish (Hillier et al., 2004). Anoles also possess the putative sour receptor PKD1L3 (Chen et al., 2011), as well as multiple G_α subunits (G_q, G_s, G_i, and three forms of G_t; Ohmoto et al., 2011). It would be most informative to conduct sequence alignments for all of the known lizard sequences with their homologues in mammals. We know that the green anole possess several bitter receptors (37), a putative sour receptor (1), and the sweet/umami receptor (3 subunits) (summarized in Dong et al., 2012). Thus, the taste world of at least the green anole may actually be quite similar to that of humans and rodents. Though some research has been conducted on sweet and bitter responses in some lizards, there is a glaring lack of behavioral work concerning any of the other taste modalities. In particular, it would be of fundamental use to know which reptiles can sense umami, salty, and sour compounds and, further, what the detection ranges are for these compounds. Terrestrial species would likely offer the most tractable information to complement human and rodent work on the evolution of taste receptors, especially if taste compound preference experiments were expanded to include nerve recordings.

43.4 DISCUSSION

43.4.1 Challenges and Opportunities in Comparative Taste Biology

When we attempt to study taste from a comparative perspective, the answers even to simple questions, such as whether a given chemical results in taste perception, are not obvious. Most or all chemicals could be tastants for some species. This is a crucial point because often the chemical stimuli chosen for taste research in animals are compounds that have clear tastes to humans and come from the five basic taste qualities. But some species may perceive chemicals and qualities that we cannot, and thus many important tastants are likely unknown and therefore unstudied. Consider the example of artificial sweeteners. Humans detect many compounds as sweet (e.g., aspartame) that apparently do not taste sweet or may even have no taste for other species. This may lead to the conclusion that humans have a "more advanced" sweet taste because it could appear that humans can detect a wider variety of sweeteners. But isn't it also likely that these other species detect other compounds as sweet that to humans are tasteless or at least do not taste sweet? And how would we identify these compounds? Recent research indicates that the giant panda (*Ailuropoda melanoleuca*) has a strong preference for many of the sugars humans describe as sweet. But to humans, bamboo, the major food of the panda, certainly does not taste sweet. Are there compounds in bamboo that are sweet to the panda? This specific example provides a fascinating area for further research, but more important, it illustrates how humans may often be in the dark when examining the taste worlds of other species.

Taste is tied to diet, but beyond this simple idea there is much we do not understand. What we do know

suggests that every species lives in its own taste world and that understanding an animal's diet is a first step toward understanding the taste system. Taste ecology has been hampered by the difficulty of comparing the taste systems of diverse species. Rarely are more than one or two species studied at one time with the same methods, and thus attempts must be made to compare results across studies, as we have done here. What is apparent is that seemingly minor differences in testing methodology may lead to different conclusions; for instance, the choice of tastant concentration is paramount when testing sweet preferences in birds. Largely because of these difficulties, direct cross-species comparisons of taste ecology are an underexplored pathway to discovery.

Studies in animals must move beyond a human-centered view of the taste world. The study of sucrose as the exemplar sweet stimuli may favor some areas of research, such as the wealth of data on sweet perception in birds. This serves other species less well, a point made clearly from the study of carnivores, with their "broken" sweet receptor gene, which do not perceive sweetness at all. As a further example, quinine is unlikely to be an exemplar bitter compound for the snake, which rarely encounters this chemical found in tree bark. Therefore, researchers should be vigilant in finding ways to identify chemicals that are ecologically important to animals, especially those that humans cannot taste.

Likewise, researchers need to reconsider traditional boundaries between taste and other types of chemosensory systems. For instance, the differences between taste and the common chemical sense or chemesthesis (cool, burn, sting, etc.) are clear and straightforward for oral chemosensation because they are mediated by different cranial nerves and have different initial connections in the central nervous system. However, things become more confusing for extraoral taste receptors. In this case these "taste" receptors may communicate directly with trigeminal (chemesthetic) nerves (Finger et al., 2003), thereby confusing these two sensory systems at the anatomical as well as the functional level. We may need to revise what we mean by "taste."

43.4.2 Applied Aspects of Animal Taste

Understanding the taste perceptions of animals has practical implications. An obvious benefit is that understanding what an animal perceives as good or bad taste can help us provide them with more palatable, perhaps more nutritious food, or help us generate repellents for nuisance species. A less obvious but not less important benefit may be the contribution of taste ecology to conservation efforts, especially understanding which chemicals might be most toxic to which species. Each species has its own repertoire of bitter receptors (Conte et al., 2002, 2003; Dong et al., 2009;

976   Chapter 43   Comparative Taste Biology with Special Focus on Birds and Reptiles

Fredriksson and Schioth, 2005; Gloriam et al., 2007; Go, 2006; Lagerstrom et al., 2006; Shi and Zhang, 2006). New bitter receptors emerge or are silenced over time within an animal's genome (Go, 2006; Parry et al., 2004), and some receptors evolve to recognize toxins in the environment (Kim et al., 2005; Wang et al., 2004). Thus, understanding the bitter receptor repertoire in each animal group might point to the chemicals most damaging to them.

43.4.3 Conclusion

Comparative taste biology is an old field with new and exciting opportunities. More studies, especially those that integrate anatomy, behavior, and genotype, are needed to deepen our understanding of the relationship between the sense of taste and feeding ecologies of different species. These studies may also help us understand how our own evolutionary history shaped human taste and diet choice. Such an understanding is necessary if we are to successfully confront a changed food and taste environment that now results in such high rates of obesity, diabetes, and hypertension, all diseases directly relevant to our taste biology.

REFERENCES

Adler, E., Hoon, M. A., Mueller, K. L., et al. (2000). A novel family of mammalian taste receptors. *Cell* 100: 693–702.

Alcock, J. (1970). Punishment levels and the response of black-capped chickadees (*Parus atricapillus*) to three kinds of artificial seeds. *Anim. Behav.* 18, Part 3: 592–599.

Alföldi, J., Di Palma, F., Grabherr, M., et al. (2011). The genome of the green anole lizard and a comparative analysis with birds and mammals. *Nature* 477: 587–591.

Atobe, Y., Nakano, M., Kadota, T., et al. (2004). Medullary efferent and afferent neurons of the facial nerve of the pit viper *Gloydius brevicaudus*. *J. Comp. Neurol.* 472: 345–357.

Auen, E. L., and Langebartel, D. A. (1977). The cranial nerves of the colubrid snakes *Elaphe* and *Thamnophis*. *J. Morphol.* 154: 205–222.

Avery, M. L., Schreiber, C. L., and Decker, D. G. (1999). Fruit sugar preferences of house finches. *Wilson Bull.* 111: 84–88.

Bachmanov, A. A., and Beauchamp, G. K. (2007). Taste receptor genes. *Annu. Rev. Nutr.* 27: 389–414.

Bacon, I., Hurly, T. A., and Healy, S. D. (2011). Hummingbirds choose not to rely on good taste: information use during foraging. *Behav. Ecol.* 22: 471–477.

Bairlein, F. (1996). Fruit-eating in birds and its nutritional consequences. *Comp. Biochem. Physiol. A* 113: 215–224.

Baker, H. G., and Baker, I. (1983). Floral nectar sugar constituents in relation to pollinator type. In *Handbook of Experimental Pollination Biology*, Jones, C. E., and Little, R. J. (Eds). New York: Van Nostrand Reinhold, pp. 117–141.

Baker, H. G., Baker, I., and Hodges, S. A. (1998). Sugar composition of nectars and fruits consumed by birds and bats in the tropics and subtropics. *Biotropica* 30: 559–586.

Baldwin, M. W., Toda, Y., Nakagita, T., O'Connell, M. J., Klasing, K. C., Misaka, T., Edwards, S. V., Liberles, S. D. (2014). Evolution of sweet

taste perception in hummingbirds by transformation of the ancestral umami receptor. *Science*, 345: 929–933.

Balog, J. M., and Millar, R. I. (1989). Influence of the sense of taste on broiler chick feed consumption. *Poultry Sci.* 68: 1519–1526.

Barach, J. P. (1951). The value of the skin secretions of the spotted salamander. *Herpetologica* 7: 58–58.

Bartoshuk, L. M., Jacobs, H. L., Nichols, T. L., et al. (1975). Taste rejection of nonnutritive sweeteners in cats. *J. Comp. Physiol. Psychol.* 89: 971–975.

Bath, W. (1906). Die Geschmacksorgane der Vogel und Krokidile. *Arch. Biontol.* 1: 5–47.

Beauchamp, G. K., Maller, O., and Rogers, J. G. (1977). Flavor preferences in cats (*Felis catus* and *Panthera* sp.). *J. Comp. Physiol. Psychol.* 91: 1118–1127.

Beauchamp, G. K., and Mason, J. R. (1991). Comparative hedonics of taste. In *The Hedonics of Taste*, Bolles, R. C. (Ed). Hillsdale, NJ: Lawrence Erlbaum, pp. 159–181.

Belman, A. L., and Kare, M. R. (1961). Character of salivary flow in the chicken. *Poultry Sci.* 40: 1377.

Benes, E. S. (1969). Behavioral evidence for color discrimination by the whiptail lizard, *Cnemidophorus tigris*. *Copeia* 1969: 707–722.

Bergmann, P. J., and Irschick, D. J. (2011). Vertebral evolution and the diversification of squamate reptiles. *Evolution* 66: 1044–1058.

Berkhoudt, H. (1976). Taste-buds in bill of mallard duck (*Anas-platyrhynchos* L.). *Neth. J. Zool.* 27(3): 310–331

Berkhoudt, H. (1985). Structure and function of avian taste receptors. In *Form and Function in Birds*, King, A. S., and McLelland, J. (Eds). London: Academic Press, pp. 463–498.

Bjorndal, K. A. (1991). Diet mixing: nonadditive interactions of diet items in an omnivorous freshwater turtle. *Ecology* 72: 1234–1241.

Blem, C. R., Blem, L .B., Felix, J., and van Gelder, J. (2000). Rufous hummingbird sucrose preference: precision of selection varies with concentration. *Condor* 102: 235–238.

Bosak, N. P., Inoue, M., Nelson, T., et al. (2010). Epithelial sodium channel (ENaC) is involved in reception of sodium taste: evidence from mice with a tissue-specific conditional targeted mutation of the ENaCα gene. In *AChemS 2010 Annual Meeting Abstracts*. Minneapolis, MN: Association for Chemoreception Sciences, p. 8.

Botezat, E. (1906). Die Nervenendapparate in den Mundteilen der Vogel und die einheitliche Endigungsweise der periohere Nerven bei den Wirbeltieren. *Z. Wiss. Zool.* 84: 205–360.

Boughter, J. D., Jr., and Bachmanov, A. (2008). Genetics and evolution of taste. In *Olfaction and Taste*, Firestein, S., and Beauchamp, G. K. (Eds). San Diego: Academic Press, pp. 371–390.

Bourne, R. C., Davies, D. C., Stewart, M. G., et al. (1991). Cerebral glycoprotein synthesis and long-term memory formation in the chick (*Gallus domesticus*) following passive avoidance training depends on the nature of the aversive stimulus. *Eur. J. Neurosci.* 3: 243–248.

Bowers, M. D. (1990). Recycling plant natural products for insect defense. In *Insect Defenses: Adaptive Mechanisms and Strategies of Prey and Predators*, Evans, D. L., and Schmidt, J. O. (Eds). Albany: State University of New York Press, pp. 353–386.

Boyd, E. M. (1973). *Toxicity of Pure Foods*. Cleveland, OH: CRC Press.

Breslin, P. A., and Huang, L. (2006). Human taste: peripheral anatomy, taste transduction, and coding. *Adv. Otorhinolaryngol.* 63: 152–190.

Breslin, P. A. S., Davis, J. D., and Rosenak, R. (1996). Saccharin increases the effectiveness of glucose in stimulating ingestion in rats but has little effect on negative feedback. *Physiol. Behav.* 60: 411–416.

Brindley, L. D. (1965). Taste discrimination in bobwhite and Japanese quail. *Anim. Behav.* 13: 507–512.

References

Brindley, L. D., and Prior, S. (1968). Effects of age on taste discrimination in bobwhite quail. *Anim. Behav.* 16: 304–307.

Brodie, E. D., and Tumbarello, M. S. (1978). The antipredator functions of *Dendrobates auratus* (Amphibia, Anura, Dendrobatidae) skin secretion in regard to a snake predator (*Thamnophis*). *J. Herpetol.* 12: 264–265.

Brodie, E. D., Jr., Ridenhour, B., and Brodie, E., III (2002). The evolutionary response of predators to dangerous prey: hotspots and coldspots in the geographic mosaic of coevolution between garter snakes and newts. *Evolution* 56: 2067–2082.

Brower, L. P. (1969). Ecological chemistry. *Sci. Am.* 220: 22–29.

Brower, L. P., Nelson, C. J., Seiber, J. N., et al. (1988). Exaptation as an alternative to coevolution in the cardenolide-based chemical defense of monarch butterflies (*Danaus plexippus L.*) against avian predators. In *Chemical Mediation of Coevolution*, Spencer, K. C. (Ed). London: Harcourt Brace Jovanovich, 447–475.

Brown, M., Downs, C. T., and Johnson, S. D. (2008). Sugar preferences of nectar feeding birds – a comparison of experimental techniques. *J. Avian. Biol.* 39: 479–483.

Brown, M., Downs, C. T., and Johnson, S. D. (2010). Sugar preferences and digestive efficiency in an opportunistic avian nectarivore, the dark-capped bulbul *Pycnonotus tricolor*. *J. Ornithol.* 151: 637–643.

Brugger, K. E., and Nelms, C. O. (1991). Sucrose avoidance by American robins (*Turdus-migratorius*) – implications for control of bird damage in fruit crops. *Crop Prot.* 10: 455–460.

Burghardt, G. M., and Hess, E. H. (1966). Food imprinting in the snapping turtle, *Chelydra serpentina*. *Science* 151: 108–109.

Burghardt, G. M., Wilcoxon, H. C., and Czaplicki, J. A. (1973). Conditioning in garter snakes: aversion to palatable prey induced by delayed illness. *Learn. Behav.* 1: 317–320.

Burne, T. H. J., and Rogers, L. J. (1997). Relative importance of odour and taste in the one-trial passive avoidance learning bead task. *Physiol. Behav.* 62: 1299–1302.

Byerly, M. S., Simon, J., Cogburn, L. A., et al. (2010). Transcriptional profiling of hypothalamus during development of adiposity in genetically selected fat and lean chickens. *Physiol. Genom.* 42: 157–167.

Cadirci, S., Smith, W. K., and Mc Devitt, R. M. (2009). Determination of the appetite of laying hens for methionine in drinking water by using colour cue. *Arch. Geflugelkd.* 73: 21–28.

Chai, P. (1996). Butterfly visual characteristics and ontogeny of responses to butterflies by a specialized tropical bird. *Biol. J. Linn. Soc. Lond.* 59: 37–67.

Chaisuriyanun, S., Gale, G. A., Madsri, S., and Poonswad, P. (2011). Food consumed by great hornbill and rhinoceros hornbill in tropical rainforest, Budo Su-Ngai Padi National Park, Thailand. *Raffles Bull. Zool. Suppl.* 24: 123–135.

Chalcoff, V. R., Aizen, M. A., and Galetto, L. (2008). Sugar preferences of the green-backed firecrown hummingbird (*Sephanoides sephaniodes*): a field experiment. *Auk* 125: 60–66.

Chandrashekar, J., Kuhn, C., Oka, Y., et al. C. S. (2010). The cells and peripheral representation of sodium taste in mice. *Nature* 464: 297–301.

Chapman, A., Bradford, M. G., and Hoskin, C. J. (1999). Sap suckers: a novel bird "guild" in wet sclerophyll forests of tropical north Queensland. *Emu* 99: 69–72.

Chiari, Y., Cahais V., Galtier N., and Delsuc F. (2012). Phylogenomic analyses support the position of turtles as the sister group of birds and crocodiles (Archosauria). *BMC BMV Biol*, 10: 65.

Chaudhari, N., Landin, M. A., and Roper, S. D. (2000). A metabotropic glutamate receptor variant functions as a taste receptor. *Nat. Neurosci.* 3: 113–119.

Cheeke, P. R. (1971). Nutritional and physiological implications of saponins – review. *Can. J. Anim. Sci.* 51: 621–632.

Chen, D., Li, P., Guo, W., et al. (2011). Molecular evolution of candidate sour taste receptor gene PKD1L3 in mammals. *Genome* 54: 890–897.

Conte, C., Ebeling, M., Marcuz, A., et al. (2002). Identification and characterization of human taste receptor genes belonging to the TAS2R family. *Cytogenet. Genome Res.* 98: 45–53.

Conte, C., Ebeling, M., Marcuz, A., et al. (2003). Evolutionary relationships of the Tas2r receptor gene families in mouse and human. *Physiol. Genomics* 14: 73–82.

Cooper, W. E., and Pérez-Mellado, V. (2001). Chemosensory responses to sugar and fat by the omnivorous lizard *Gallotia caesaris*: with behavioral evidence suggesting a role for gustation. *Physiol. Behav.* 73: 509–516.

Cronin, E. W. J., and Sherman, P. W. (1976). A resource based mating system the orange-rumped honeyguide. *Living Bird* 15: 5–32.

Crowcroft, P. (1966). *Mice All Over*. Brookfield, IL: Chicago Zoological Society.

Croxall, J. P., and Prince, P. A. (1996). Cephalopods as prey. 1. Seabirds. *Philos. Trans. R. Soc. Lond. B* 351: 1023–1043.

Daly, J. W., and Myers, C. W. (1967). Toxicity of Panamanian poison frogs (Dendrobates): some biological and chemical aspects. *Science* 156: 970–973.

Davis, J. K., Lowman, J. J., Thomas, P. J., et al. (2010). Evolution of a bitter taste receptor gene cluster in a New World sparrow. *Genome Biol. Evol.* 2: 358–370.

Dixon, M. D., Johnson, W. C., and Adkisson, C. S. (1997). Effects of caching on acorn tannin levels and blue jay dietary performance. *Condor* 99: 756–764.

Dong, D., Jin, K., Wu, X., Zhong, Y. (2012) CRDB: Database of Chemosensory Receptor Gene Families in Vertebrate. PLoS ONE 7(2): e31540. doi:10.1371/journal.pone.0031540.

Dong, D., Jones, G., and Zhang, S. (2009). Dynamic evolution of bitter taste receptor genes in vertebrates. *BMC Evol. Biol.* 9: 12.

Downs, C. T. (2000). Ingestion patterns and daily energy intake on a sugary diet: the red lory *Eos borneai* and the malachite sunbird *Nectarinia famosa*. *Ibis* 142: 359–364.

Dumbacher, J. P., Menon, G. K., and Daly, J. W. (2009). Skin as a toxin storage organ in the endemic New Guinean genus *Pitohui*. *Auk* 126: 520–530.

Duncan, C. J. (1960). The sense of taste in birds. *Ann. Appl. Biol.* 48: 409–414.

Endler, J. A, Mappes J. (2004): Predator mixes and the conspicuousness of aposematic signals. *Am. Nat.* 163: 532–547.

Espaillat, J. E., and Mason, J. R. (1990). Differences in taste preference between red-winged blackbirds and European starlings. *Wilson Bull.* 102: 292–299.

Fahey, J. W., Zalcmann, A. T., and Talalay, P. (2001). The chemical diversity and distribution of glucosinolates and isothiocyanates among plants. *Phytochemistry* 56: 5–51.

Feder, F. H. (1972). Contribution to the microscopic anatomy of the digestive system of *Rhea-americana*. *Anat. Anz.* 132: 250–265.

Ferguson, M. W. J. (1981). The structure and development of the palate in *Alligator mississippiensis*. *Arch. Oral Biol.* 26: 427–443.

Fernelius, I. (1581). *Therapeutices Universalis Seu Medendi Rationis, Libri Septem*, Vol. 4. Frankfurt: Andream Wechelum.

Finger, T. E., Bottger, B., Hansen, A., et al. (2003). Solitary chemoreceptor cells in the nasal cavity serve as sentinels of respiration. *Proc. Natl. Acad. Sci. U. S. A.* 100: 8981–8986.

Fink, L. S., and Brower, L. P. (1981). Birds can overcome the cardenolide defense of monarch butterflies in Mexico. *Nature* 291: 67–70.

Fleming, P. A., Bakken, B. H., Lotz, C. N., and Nicolson, S. W. (2004). Concentration and temperature effects on sugar intake and preferences in a sunbird and a hummingbird. *Funct. Ecol.* 18: 223–232.

Chapter 43 Comparative Taste Biology with Special Focus on Birds and Reptiles

Fleming, P. A., Xie, S., Napier, K., et al. (2008). Nectar concentration affects sugar preferences in two Australian honeyeaters and a lorikeet. *Funct. Ecol.* 22: 599–605.

Fredriksson, R., and Schioth, H. B. (2005). The repertoire of G-protein-coupled receptors in fully sequenced genomes. *Mol. Pharmacol.* 67: 1414–1425.

Friedl, W. A., Nachtigall, P. E., Moore, P. W. B., et al. (1990). Taste reception in the Pacific bottlenose dolphin (*Tursiops truncatus gilli*) and the California sea lion (*Zalophus californianus*). In *Sensory Abilities of Cetaceans: Laboratory and Field Evidence*, Thomas, J. A., and Kastelein, R. A. (Eds). New York: Plenum Press, pp. 447–454.

Ganchrow, D., and Ganchrow, J. R. (1985). Number and distribution of taste buds in the oral cavity of hatchling chicks. *Physiol. Behav.* 34: 889–894.

Ganchrow, D., Ganchrow, J. R., Grossisseroff, R., and Kinnamon, J. C. (1995). Taste bud cell generation in the perihatching chick. *Chem. Senses* 20: 19–28.

Ganchrow, D., Ganchrow, J. R., Romano, R., and Kinnamon, J. C. (1994). Ontogeny and taste bud cell turnover in the chicken. 1. Gemmal cell renewal in the hatchling. *J. Comp. Neurol.* 345: 105–114.

Ganchrow, J. R., and Ganchrow, D. (1986). Embryonic-development of taste-buds in the chicken. *Chem. Senses* 11: 602–603.

Ganchrow, J. R., Ganchrow, D., and Oppenheimer, M. (1986). Chorda tympani innervation of anterior mandibular taste buds in the chicken (*Gallus gallus domesticus*). *Anat. Rec.* 216: 434–439.

Ganchrow, J. R., Steiner, J. E., and Bartana, A. (1990). Behavioral reactions to gustatory stimuli in young chicks (*Gallus-gallus domesticus*). *Dev. Psychobiol.* 23: 103–117.

Garnett, S., Price, I., and Scott, F. (1985). The diet of the green turtle, *Chelonia mydas* (L.), in Torres Strait. *Wildl. Res.* 12: 103–112.

Gartrell, B. D. (2000). The nutritional, morphologic, and physiologic bases of nectarivory in Australian birds. *J. Avian Med. Surg.* 14: 85–94.

Gass, C. L., Romich, M. T., and Suarez, R. K. (1999). Energetics of hummingbird foraging at low ambient temperature. *Can. J. Zool.* 77: 314–320.

Geer, T. A. (1982). The selection of tits *Parus* spp by sparrowhawks *Accipiter-nisus*. *Ibis* 124: 159–167.

Gentle, M. J. (1971a). Lingual taste buds of *Gallus domesticus* L. *Br. Poultry Sci.* 12: 245248.

Gentle, M. J. (1971b). Taste and its importance to the domestic chicken. *Br. Poultry Sci.* 12: 77–86.

Gentle, M. J. (1972). Taste preference in the chicken (*Gallus domesticus* L.). *Br. Poultry Sci.* 13: 141–155.

Gentle, M. J. (1973). Diencephalic stimulation and mouth movements in chicken. *Br. Poultry Sci.* 14: 167–171.

Gentle, M. J. (1975). Gustatory behaviour of the chicken and other birds. In *Neural and Endocrine Aspects of Behaviour in Birds*, Wright, P., Caryl, P. G., and Vowle, D. M. (Eds). Amsterdam: Elsevier.

Gentle, M. J. (1976). Quinine hydrochloride acceptability after water deprivation in Gallus-domesticus. *Chem. Senses Flav.* 2: 121–128.

Gentle, M. J. (1978). Extra-lingual chemoreceptors in the chicken (*Gallus domesticus*). *Chem. Senses Flav.* 3: 325–329.

Gentle, M. J. (1983). The chorda tympani nerve and taste in the chicken. *Experientia* 39: 1002–1003.

Gentle, M. J. (1987). Facial nerve sensory responses recorded from the geniculate ganglion of *Gallus gallus* var. *domesticus*. *J. Comp. Physiol. A* 160: 683–691.

Gentle, M. J., and Dewar, W. A. (1981). The effects of vitamin A deficiency on oral gustatory behaviour in the chicken. *J. Comp. Physiol.* 130: 275–279.

Gentle, M. J., Dewar, W. A., and Wight, P. A. L. (1981). The effects of zinc deficiency on oral behavior and taste bud morphology in chicks. *Br. Poultry Sci.* 22: 265–273.

Gentle, M. J., and Harkin, C. (1979). Effect of sweet stimuli on oral behavior in the chicken. *Chem. Senses Flav.* 4: 183–190.

Gilardi, J. D., Duffey, S. S., Munn, C. A., and Tell, L. A. (1999). Biochemical functions of geophagy in parrots: detoxification of dietary toxins and cytoprotective effects. *J. Chem. Ecol.* 25: 897–922.

Glendinning, J. I. (1994). Is the bitter rejection response always adaptive. *Physiol. Behav.* 56: 1217–1227.

Gloriam, D. E., Fredriksson, R., and Schioth, H. B. (2007). The G protein-coupled receptor subset of the rat genome. *BMC Genomics* 8: 338.

Go, Y. (2006). Proceedings of the SMBE Tri-National Young Investigators' Workshop 2005. Lineage-specific expansions and contractions of the bitter taste receptor gene repertoire in vertebrates. *Mol. Biol. Evol.* 23: 964–972.

Goss-Custard, J. D., Jones, R. E., and Newbery, P. E. (1977). The ecology of the wash. I. Distribution and diet of wading birds (Charadrii). *J. Appl. Ecol.* 14: 681–700.

Greenberg, N. (1993). Central and endocrine aspects of tongue-flicking and exploratory behavior in *Anolis carolinensis*. *Brain Behav. Evol.* 41: 210–218.

Griffiths, A. D., and McKay, J. L. (2008). Cane toads reduce the abundance and site occupancy of Merten's water monitor (*Varanus mertensi*). *Wildl. Res.* 34: 609–615.

Grill, M. J., and Norgren, R. (1978). The taste reactivity test. I. Mimetic responses to gustatory stimuli in neurologically intact rats. *Brain Res.* 143: 263–279.

Gussekloo, S. W. S., and Bout, R. G. (2005). The kinematics of feeding and drinking in palaeognathous birds in relation to cranial morphology. *J. Exp. Biol.* 208: 3395–3407.

Hackett S. J., Kimball R. T., Reddy S., et al. (2008). A phylogenomic study of birds reveals their evolutionary history. *Science.* 320: 1763–1768.

Halpern, B. P. (1962). Gustatory nerve responses in the chicken. *Am. J. Physiol.* 203: 541–544.

Hamrum, C. L. (1953). Experiments on the senses of taste and smell in the bob-white quail (*Colinus-virginianus-virginianus*). *Am. Midl. Nat.* 49: 872–877.

Harriman, A. E., and Fry, E. G. (1990). Solution acceptance by common ravens (*Corvus corax*) given two-bottle preference tests. *Psychol. Rep.* 67: 19–26.

Harriman, A. E., and Milner, J. S. (1969). Preference for sucrose solutions by Japanese quail (*Coturnix coturnix japonica*) in 2-bottle drinking tests. *Am. Midl. Nat.* 81: 575–578.

Hillier, L. W., Miller, W., Birney, E., et al. (2004). Sequence and comparative analysis of the chicken genome provide unique perspectives on vertebrate evolution. *Nature* 432: 695–716.

Hutson, A. M. (1991). The daily activity of bird. In *The Cambridge Encyclopedia of Ornithology*, Brooke, M., and Birkhead, T. (Eds). Cambridge: Cambridge University Press, pp. 121–155.

Iwasaki, S. I., Asami, T., and Wanichanon, C. (1996). Fine structure of the dorsal lingual epithelium of the juvenile hawksbill turtle (*Eretmochelys imbricata bissa*). *Anat. Rec.* 244: 437–443.

Jackowiak, H., and Ludwig, M. (2008). Light and scanning electron microscopic study of the structure of the ostrich (*Strutio camelus*) tongue. *Zool. Sci.* 25: 188–194.

Jain, D. K. (1977). Histo-morphological studies on the bucco-pharynx and tongue of frugivorous, carnivorous and omnivorous species of birds. *Pavo* 15: 82–89.

References

Jiang, P., Li, X., Glaser, D., et al. (2012). Major taste loss in carnivorous mammals. *Proc. Natl. Acad. Sci. U. S. A.* 109: 4956–4961.

Johnson, S. D., Hargreaves, A. L., and Brown, M. (2006). Dark, bitter-tasting nectar functions as a filter of flower visitors in a bird-pollinated plant. *Ecology* 87: 2709–2716.

Johnson, S. D., and Nicolson, S. W. (2008). Evolutionary associations between nectar properties and specificity in bird pollination systems. *Biol. Lett.* 4: 49–52.

Kadona, H., Okada, T., and Ohno, K. (1966). Neurophysiological studies on the sense of taste in the chicken. *Res. Bull. Fac. Coll. Agric. Gifu Univ.* 22: 149–159.

Kare, M. R. (1970). *The Chemical Senses of Birds.* Bird Control Seminars Proceedings, Paper 184. http://digitalcommons.unl.edu/icwdmbirdcontrol/184.

Kare, M. R., Black, R., and Allison, E. G. (1957). The sense of taste in the fowl. *Poult. Sci.* 36: 129–138.

Kare, M. R., and Maller, O. (1967). Taste and food intake in domesticated and jungle fowl. *J. Nutr.* 92: 191–196.

Kare, M. R., and Mason, R. M. (1986). The chemical senses in birds. In *Avian Physiology*, Sturke, P. D. (Ed). New York: Springer, pp. 59–73.

Kare, M. R., and Medway, W. (1959). Discrimination between carbohydrates by the fowl. *Poult. Sci.* 38: 1119–1127.

Kassarov, L. (1999). Are birds able to taste and reject butterflies based on "beak mark tasting"? A different point of view. *Behaviour* 136: 965–981.

Kim, U., Wooding, S., Ricci, D., et al. (2005). Worldwide haplotype diversity and coding sequence variation at human bitter taste receptor loci. *Hum. Mutat.* 26: 199–204.

Kitchell, R. L., Ström, L., and Zotterman, Y. (1959). Electrophysiological studies of thermal and taste reception in chickens and pigeons. *Acta Physiol. Scand.* 46: 133–151.

Klasing, K. C. (1998). *Comparative Avian Nutrition.* Wallingford: CAB International.

Korte, G. E. (1980). Ultrastructure of the tastebuds of the red-eared turtle, *Chrysemys scripta elegans. J. Morphol.* 163: 231–252.

Kroll, J. C. (1973). Taste buds in the oral epithelium of the blind snake, *Leptotyphlops dulcis* (Reptilia: Leptotyphlopidae). *Southw. Nat.* 17: 365–370.

Kudo, K., Nishimura, S., and Tabata, S. (2008). Distribution of taste buds in layer-type chickens: scanning electron microscopic observations. *Anim. Sci. J.* 79: 680–685.

Kudo, K., Shiraishi, J., Nishimura, S., et al. (2010). The number of taste buds is related to bitter taste sensitivity in layer and broiler chickens. *Anim. Sci. J.* 81: 240–244.

Kurosawa, T., Niimura, S., Kusuhara, S., and Ishida, K. (1983). Morphological studies of taste buds in chickens. *Jpn. J. Zootech. Sci.* 54: 502–510.

Kuznetsov, V. B. (1974). A method of studying chemoreception in the Black Sea bottlenose dolphin (*Tursiops truncatus*). In *Morfologiya, Fiziologiya I Akustika Morskikh Mlekopitayushchikh*, Sokolov, V. Y. (Ed). Moscow: Nauka.

Lagerstrom, M. C., Hellstrom, A. R., Gloriam, D. E., et al. (2006). The G protein-coupled receptor subset of the chicken genome. *PLoS Comput. Biol.* 2: e54.

Letnic, M., Webb, J. K., and Shine, R. (2008). Invasive cane toads (*Bufo marinus*) cause mass mortality of freshwater crocodiles (*Crocodylus johnstoni*) in tropical Australia. *Biol. Conserv.* 141: 1773–1782.

Levey, D. J. (1987). Sugar-tasting ability and fruit selection in tropical fruit-eating birds. *Auk* 104: 173–179.

Li, X., Bachmanov, A. A., Maehashi, K., et al. (2011). Sweet receptor gene variation and aspartame blindness in primates and other species. *Chem. Senses* 36: 453–475.

Li, X., Glaser, D., Li, W., Johnson, W. E., et al. (2009). Analyses of sweet receptor gene (*Tas1r2*) and preference for sweet stimuli in species of Carnivora. *J. Hered.* 100: S90–S100.

Li, X., Li, W. H., Wang, H., et al. (2005). Pseudogenization of a sweet-receptor gene accounts for cats' indifference toward sugar. *PLoS Genet.* 1: 27–35.

Li, X., Staszewski, L., Xu, H., et al. (2002). Human receptors for sweet and umami taste. *Proc. Nat. Acad. Sci. U. S. A.* 99: 4692–4696.

Lindenmaier, P., and Kare, M. R. (1959). The taste end-organs of the chicken. *Poult. Sci.* 38: 454–550.

Liner, E. (1996). Natural history notes. *Anolis carolinensis* (green anole). Nectar feeding. *Herpetol. Rev.* 27: 78.

Lotz, C. N., and Schondube, J. E. (2006). Sugar preferences in nectar- and fruit-eating birds: behavioral patterns and physiological causes. *Biotropica* 38: 3–15.

Maehashi, K., Fukuda, Y., Matano, M., et al. (submitted). Taste receptor genes of emu: TASR1 and TASR2. GenBank: AB608217.1 and AB630153.1.

Magnusson, W. E., da Silva, E. V., and Lima, A. P. (1987). Diets of Amazonian crocodilians. *J. Herpetol.* 21: 85–95.

Manton, M., Karr, A., and Ehrenfeld, D. W. (1972). An operant method for the study of chemoreception in the green turtle, *Chelonia myda. Brain Behav. Evol.* 5: 188–201.

Mappes, J., Marples, N., and Endler, J. A. (2005). The complex business of survival by aposematism. *Trends Ecol. Evol.* 20: 598–603.

Marples, N. M., Brakefield, P. M., and Cowie, R. J. (1989). Differences between the 7-spot and 2-spot ladybird beetles (Coccinellidae) in their toxic effects on a bird predator. *Ecol. Entomol.* 14: 79.

Marples, N. M., and Roper, T. J. (1997). Response of domestic chicks to methyl anthranilate odour. *Anim. Behav.* 53: 1263–1270.

Martinez del Rio, C. (1990a). Dietary, phylogenetic, and ecological constraints of intestinal sucrase and maltase activity in birds. *Physiol. Zool.* 63: 987–1011.

Martinez Del Rio, C. (1990b). Sugar preferences in hummingbirds the influence of subtle chemical differences on food choice. *Condor* 92: 1022–1030.

Martinez Del Rio, C., Baker, H. G., and Baker, I. (1992). Ecological and evolutionary implications of digestive processes bird preferences and the sugar constituents of floral nectar and fruit pulp. *Experientia (Basel)* 48: 544–551.

Martinez Del Rio, C., Kasarov, W. H., and Levey, D. J. (1989). Physiological basis and ecological consequences of sugar preferences in cedar waxwings. *Auk* 106: 64–71.

Martinez del Rio, C., and Stevens, B. R. (1989). Physiological constraint on feeding behavior: intestinal membrane disaccharidases of the starling. *Science* 243: 794–796.

Martinez Del Rio, C., Stevens, B. R., Daneke, D. E., and Andreadis, P. T. (1988). Physiological correlates of preference and aversion for sugars in three species of birds. *Physiol. Zool.* 61: 222–229.

Martins, S., Freitas, R., Palma, L., and Beja, P. (2011). Diet of breeding ospreys in the Cape Verde archipelago, northwestern Africa. *J. Raptor Res.* 45: 244–251.

Mason, J. R., and Clark, L. (1999). The chemical senses in birds. In *Sturkie's Avian Physiology*, Whittow, G. C. (Ed). Oxford: Elsevier, pp. 39–56.

Matson, K. D., Millam, J. R., and Klasing, K. C. (2000). Taste threshold determination and side-preference in captive cockatiels (*Nymphicus hollandicus*). *Appl. Anim. Behav. Sci.* 69: 313–326.

Matson, K. D., Millam, J. R., and Klasing, K. C. (2004). Cockatiels (*Nymphicus hollandicus*) reject very low levels of plant secondary compounds. *Appl. Anim. Behav. Sci.* 85: 141–156.

980 Chapter 43 Comparative Taste Biology with Special Focus on Birds and Reptiles

Matsunami, H., Montmayeur, J. P., and Buck, L. B. (2000). A family of candidate taste receptors in human and mouse. *Nature* 404: 601–604.

Medina-Tapia, N., Ayala-Berdon, J., Morales-Perez, L., et al. (2012). Do hummingbirds have a sweet-tooth? Gustatory sugar thresholds and sugar selection in the broad-billed hummingbird *Cynanthus latirostris*. *Comp. Biochem. Physiol. A* 161: 307–314.

Miller, I. J., and Smith, D. V. (1984). Quantitative taste bud distribution in the hamster. *Physiol. Behav.* 32: 275–285.

Miller, I. J., and Spangler, K. M. (1982). Taste bud distribution and innervation on the palate of the rat. *Chem. Senses* 7: 99–108.

Moore, C. A., and Elliott, R. (1946). Numerical and regional distribution of taste buds on the tongue of the bird. *J. Comp. Neurol.* 84: 119–131.

Mueller, K. L., Hoon, M. A., Erlenbach, I., et al. (2005). The receptors and coding logic for bitter taste. *Nature* 434: 225–229.

Näsvall, J., Su, L., Roth, J. R., and Andersson, D. I. (2012). Real-time evolution of new genes by innovation, amplification, and divergence. *Science* 338: 384–387.

Nelson, G., Chandrashekar, J., Hoon, M. A., et al. (2002). An amino-acid taste receptor. *Nature* 416: 199–202.

Newton, A., Gadow, H., Lydekker, R., et al. (1894). *Dictionary of Birds*, Vol. 3. London: Adam and Charles Black.

Nishida, R. (2002). Sequestration of defensive substances from plants by Lepidoptera. *Annu. Rev. Entomol.* 47: 57–92.

Nishida, Y., Yoshie, S., and Fujita, T. (2000). Oral sensory papillae, chemo-and mechano-receptors, in the snake, *Elaphe quadrivirgata*. A light and electron microscopic study. *Arch. Histol. Cytol.* 63: 55–70.

Ofran, Y., Lavi, D., Opher, D., et al. (2004). Fatal voluntary salt intake resulting in the highest ever documented sodium plasma level in adults (255 mmol L-1): a disorder linked to female gender and psychiatric disorders. *J. Intern. Med.* 256: 525–528.

Ohmoto, M., Okada, S., Nakamura, S., et al. (2011). Mutually exclusive expression of Galphaia and Galpha14 reveals diversification of taste receptor cells in zebrafish. *J. Comp. Neurol.* 519: 1616–1629.

Olesen, J., and Valido, A. (2003). Lizards as pollinators and seed dispersers: an island phenomenon. *Trends Ecol. Evol.* 18: 177–181.

Olesen, J. M., and Valido, A. (2004). Lizards and birds as generalized pollinators and seed dispersers of island plants. In *Ecología Insular/Island Ecology*, Fernández-Palacios, J., and Morici, C. (Eds). Santa Cruz de la Palma, Spain: Asociación Española de Ecología Terrestre, pp. 229–249.

Paradis, S., and Cabanac, M. (2004). Flavor aversion learning induced by lithium chloride in reptiles but not in amphibians. *Behav. Process.* 67: 11–18.

Parry, C. M., Erkner, A., and le Coutre, J. (2004). Divergence of T2R chemosensory receptor families in humans, bonobos, and chimpanzees. *Proc. Natl. Acad. Sci. U. S. A.* 101: 14830–14834.

Perkins, A. J., Anderson, G., and Wilson, J. D. (2007). Seed food preferences of granivorous farmland passerines. *Bird Stud.* 54: 46–53.

Peyrot des Gachons, C., Beauchamp, G. K., Stern, R. M., et al. (2011). Bitter taste induces nausea. *Curr. Biol.* 21: R247–R248.

Picozzi, N., Moss, R., and Catt, D. C. (1996). Capercaillie habitat, diet and management in a Sitka spruce plantation in central Scotland. *Forestry* 69: 373–388.

Polito, M. J., Trivelpiece, W. Z., Karnovsky, N. J., et al. (2011). Integrating stomach content and stable isotope analyses to quantify the diets of pygoscelid penguins. *PLoS ONE* 6: e26642.

Poulin, B., Lefebvre, G., Ibanez, R., et al. (2001). Avian predation upon lizards and frogs in a neotropical forest understorey. *J. Trop. Ecol.* 17: 21–40.

Price-Rees, S. J., Brown, G. P., and Shine, R. (2012). Interacting impacts of invasive plants and invasive toads on native lizards. *Am. Nat.* 179: 413–422.

Price-Rees, S. J., Webb, J. K., and Shine, R. (2011). School for skinks: can conditioned taste aversion enable bluetongue lizards (*Tiliqua scincoides*) to avoid toxic cane toads (*Rhinella marina*) as prey? *Ethology* 117: 749–757.

Reed, D. R. (2008). Animal models of gene-nutrient interactions. *Obesity (Silver Spring)* 16 Suppl. 3: S23–S27.

Reed, D. R., Li, S., Li, X., et al. (2004). Polymorphisms in the taste receptor gene (Tas1r3) region are associated with saccharin preference in 30 mouse strains. *J. Neurosci.* 24: 938–946.

Ren, X., Zhou, L., Terwilliger, R., et al. (2009). Sweet taste signaling functions as a hypothalamic glucose sensor. *Front Integr. Neurosci.* 3: 12.

Rensch, B., and Neunzig, R. (1925). Experimentelle Untersuchungen über den Geschmackssinn der Vögel *J. Ornithol.*, 73: 633–646.

Rodríguez-Robles, J. A., Bell, C. J., and Greene, H. W. (1999). Gape size and evolution of diet in snakes: feeding ecology of erycine boas. *J. Zool.* 248: 49–58.

Rolls, E. T. (2012). Taste, olfactory and food texture reward processing in the brain and the control of appetite. *Proc. Nutr. Soc.* 71: 488–501.

Rose, S. P. R. (2000). God's organism? The chick as a model system for memory studies. *Learn. Mem.* 7: 1–17.

Roura, E., Humphrey, B., Klasing, K., and Swart, M. (2011). Is the pig a good umami sensing model for humans? A comparative taste receptor study. *Flavour Frag. J.* 26: 282–285.

Roura, E., Humphrey, B., Tedo, G., and Ipharraguerre, I. (2008). Unfolding the codes of short-term feed appetence in farm and companion animals. A comparative oronasal nutrient sensing biology review. *Can. J. Anim. Sci.* 88: 535–558.

Rowe, C., and Skelhorn, J. (2005). Colour biases are a question of taste. *Anim. Behav.* 69: 587–594.

Rowland, H. M., Hoogesteger, T., Ruxton, G. D., et al. (2010). A tale of 2 signals: signal mimicry between aposematic species enhances predator avoidance learning. *Behav. Ecol.* 21: 851–860.

Rowland, H. M., Ihalainen, E., Lindstrom, L., Mappes, J., and Speed, M. P. (2007). Co-mimics have a mutualistic relationship despite unequal defences. *Nature* 448: 64–67.

Saito, I. (1966). Comparative anatomical studies of the oral organs of the poultry. V. Structures and distribution of taste buds of the fowl. *Bull. Fac. Agric. Miyazahi Univ.* 13: 95–102.

Sallabanks, R., and James, F. C. (1999). American robin: *Turdus migratorius*. *Birds North Am.* 462: 1–27.

Saporito, R. A., Donnelly, M. A., Jain, P., et al. (2007). Spatial and temporal patterns of alkaloid variation in the poison frog *Oophaga pumilio* in Costa Rica and Panama over 30 years. *Toxicon* 50: 757–778.

Schaefer, H. M., Schmidt, V., and Bairlein, F. (2003). Discrimination abilities for nutrients: which difference matters for choosy birds and why? *Anim. Behav.* 65: 531–541.

Schmidt, J. O. (1990). Hymenopteran venoms: striving toward the ultimate defense against vertebrates. In *Insect Defenses: Adaptive Mechanisms and Strategies of Prey and Predators*, Evans, D. L., and Schmidt, J. O. (Eds). New York: State University of New York Press, pp. 387–420.

Schmidt, M., Nowack, C., and Wöhrmann-Repenning, A. (2010). On the presence of taste buds close to the vomeronasal organs in Gekkonidae. *Amphib. Rept.* 31: 355–361.

Schondube, J. E., and del Rio, C. M. (2003). Concentration-dependent sugar preferences in nectar-feeding birds: mechanisms and consequences. *Funct. Ecol.* 17: 445–453.

Schuler, W., and Roper, T. J. (1992). Responses to warning colouration in avian predators. *Adv. Stud. Behav.* 21: 111–146.

Schwenk, K. (1985). Occurrence, distribution and functional significance of taste buds in lizards. *Copeia* 1985, 91–101.

References

Schwenk, K. (1986). Morphology of the tongue in the tuatara, *Sphenodon punctatus* (Reptilia: Lepidosauria), with comments on function and phylogeny. *J. Morphol.* 188: 129–156.

Schwenk, K. (2008). Comparative anatomy and physiology of chemical senses in nonavian aquatic reptiles. In *Sensory Evolution on the Threshold: Adaptations in Secondarily Aquatic Vertebrates*, Thewissen, J., and Nummela, S. (Eds). Berkeley: University of California Press, pp. 65–81.

Sclafani, A., and Springer, D. (1976). Dietary obesity in adult rats: similarities to hypothalamic and human obesity syndromes. *Physiol. Behav.* 17: 461–471.

Scott, T. P., and Weldon, P. J. (1989). Chemoreception in the feeding behaviour of adult American alligators, *Alligator mississippiensis. Anim. Behav.* 39: 398–400.

Shanbhag, B. A., Ammanna, V. H. F., and Saidapur, S. K. (2010). Associative learning in hatchlings of the lizard *Calotes versicolor*: taste and colour discrimination. *Amphib. Rept.* 31: 475–481.

Shi, P., and Zhang, J. (2006). Contrasting modes of evolution between vertebrate sweet/umami receptor genes and bitter receptor genes. *Mol. Biol. Evol.* 23: 292–300.

Shin, Y.-K., Cong, W.-n., Cai, H., et al. (2012). Age-related changes in mouse taste bud morphology, hormone expression, and taste responsivity. *J. Gerontol.* 67A: 336–344.

Silver, L. M. (1995). *Mouse Genetics. Concept and Applications.* Oxford: Oxford University Press.

Skelhorn, J., and Rowe, C. (2006a). Avian predators taste-reject aposematic prey on the basis of their chemical defence. *Biol. Lett.* 2: 348–350.

Skelhorn, J., and Rowe, C. (2006b). Prey palatability influences predator learning and memory. *Anim. Behav.* 71: 1111–1118.

Skelhorn, J., and Rowe, C. (2007). Automimic frequency influences the foraging decisions of avian predators on aposematic prey. *Anim. Behav.* 74: 1563–1572.

Skelhorn, J., and Rowe, C. (2009). Distastefulness as an antipredator defence strategy. *Anim. Behav.* 78: 761–766.

Skelhorn, J., and Rowe, C. (2010). Birds learn to use distastefulness as a signal of toxicity. *Proc. R. Soc. B* 277: 1729–1734.

Snow, B., and Snow, D. W. (1988). *Birds and Berries.* Calton, UK: T & AD Poyser.

Soininen, E. M., Hubner, C. E., and Jonsdottir, I. S. (2010). Food selection by barnacle geese (*Branta leucopsis*) in an Arctic pre-breeding area. *Polar Res.* 29: 404–412.

Spector, A. C. (2003). Psychophysical evaluation of taste function in nonhuman mammals. In *Handbook of Olfaction and Gustation*, Doty, R. L. (Ed). New York: Marcel Dekker, pp. 861–879.

Speed, M. P., Ruxton, G. D., Mappes, J., and Sherratt, T. N. (2012). Why are defensive toxins so variable? An evolutionary perspective. *Biol. Rev. Camb. Philos. Soc.* 87: 874–884.

Stanger-Hall, K. F., Zelmer, D. A., Bergren, C., et al. (2001). Taste discrimination in a lizard (*Anolis carolinensis*, Polychrotidae). *Copeia* 20: 490–498.

Steiner, J. E., Glaser, D., Hawilo, M. E., and Berridge, K. C. (2001). Comparative expression of hedonic impact: affective reactions to taste by human infants and other primates. *Neurosci. Biobehav. Rev.* 25: 53–74.

Stiebel, H., and Bairlein, F. (2008). Frugivory in central European birds I: diet selection and foraging. *Vogelwarte* 46: 1–23.

Stiles, F. G. (1976). Taste preferences, color preferences, and flower choice in hummingbirds. *Condor* 78: 10–26.

Stornelli, M. R., Lossi, L., and Giannessi, E. (2000). Localization, morphology and ultrastructure of taste buds in the domestic duck (*Cairina moschata domestica* L.) oral cavity. *Ital. J. Anat. Embryol.* 105: 179–188.

Stromberg, M. R., and Johnsen, P. B. (1990). Hummingbird sweetness preferences – taste or viscosity. *Condor* 92: 606–612.

Sword, G. A. (2001). Tasty on the outside, but toxic in the middle: grasshopper regurgitation and host plant-mediated toxicity to a vertebrate predator. *Oecologia* 128: 416–421.

Teff, K. (2000). Nutritional implications of the cephalic-phase reflexes: endocrine responses. *Appetite* 34: 206–213.

Tichy, F. (1992a). The ultrastructure of taste buds in the newborn lamb. *Acta Vet. Brno* 61: 83–91.

Tichy, F. (1992b). The ultrastructure of taste buds in the newborn pig. *Acta Vet. Brno* 61: 171–177.

Tichy, F. (1994). The ultrastructure of taste buds in the newborn cat. *Acta Vet. Brno* 63: 49–54.

Tivane, C. (2008). A morphological study of the oropharynx and oesophagus of the ostrich (*Struthio camelus*). University of Pretoria.

Tordoff, M. G., Alarcon, L. K., Valmeki, S., and Jiang, P. (2012). T1R3: a human calcium taste receptor. *Sci. Rep.* 2: 496.

Tullberg, B. S., Gamberale-Stille, G., and Solbreck, C. (2000). Effects of food plant and group size on predator defence: differences between two co-occurring aposematic Lygaeinae bugs. *Ecol. Entomol.* 25: 220–225.

Uchida, T. (1980). Ultrastructural and histochemical studies on the taste buds in some reptiles. *Arch. Histol. Jpn.* 43: 459–478.

Urata, K., Manda, M., and Watanabe, S. (1992). Behavioral study on taste responses of hens and female Japanese quail to salty, sour, sweet, bitter and umami solutions. *Anim. Sci. Technol.* 63: 325–331.

VanEtten, C. (1969). Goitrogens. In *Toxic Constituents of Plant Foodstuffs*, Liener, I. E. Food Science and Technology Monographs. New York: Academic Press, pp. 103–142.

van Tuinen, M., and Hedges, S. B. (2001). Calibration of avian molecular clocks. *Mol. Biol. Evol.* 18: 206–213.

Vanwyk, B. E., and Nicolson, S. W. (1995). Xylose is a major nectar sugar in Protea and Faurea. *S. Afr. J. Sci.* 91: 151–153.

Viitala, J., Korpimaki, E., Palokangas, P., and Koivula, M. (1995). Attraction of kestrels to vole scent marks visible in ultraviolet-light. *Nature* 373: 425–427.

Vince, M. A. (1977). Taste sensitivity in the embryo of the domestic fowl. *Anim. Behav.* 25, Part 4: 797–805.

Vitt, L. J., Pianka, E. R., Cooper, W. E., Jr., and Schwenk, K. (2003). History and the global ecology of squamate reptiles. *Am. Nat.* 162: 44–60.

Wagner, M. W., and Gunther, W. C. (1971). Parameters of sugar preference in normal and heat-stressed chicks. *J. Gen. Psychol.* 85: 177–185.

Wallace, K. M., and Leslie, A. J. (2008). Diet of the Nile crocodile (*Crocodylus niloticus*) in the Okavango Delta, Botswana. *J. Herpetol.* 42: 361–368.

Wallis, G. Y. (1981). Feeding ecology of the tuatara (*Sphenodon punctatus*) on Stephens Island, Cook Strait. *N. Z. J. Ecol.* 4: 89–97.

Wang, X., Thomas, S. D., and Zhang, J. (2004). Relaxation of selective constraint and loss of function in the evolution of human bitter taste receptor genes. *Hum. Mol. Genet.* 13: 2671–2678.

Warren, R. P., and Vince, M. A. (1963). Taste discrimination in the great tit (*Parus major*). *J. Comp. Physiol. Psychol.* 56: 910–913.

Wauson, E. M., Zaganjor, E., Lee, A. Y., et al. (2012). The G protein-coupled taste receptor T1R1/T1R3 regulates mTORC1 and autophagy. *Mol. Cell* 47: 851–862.

Weldon, P. J., and McNease, L. (1991). Does the American alligator discriminate between venomous and nonvenomous snake prey? *Herpetologica* 47: 403–406.

Werner, S. J., Kimball, B. A., and Provenza, F. D. (2008). Food color, flavor, and conditioned avoidance among red-winged blackbirds. *Physiol. Behav.* 93: 110–117.

Weymouth, R. D., Lasiewski, R. C., and Berger, A. J. (1964). Tongue apparatus in hummingbirds. *Acta Anat.* 58: 252–270.

Wilkin, T. A., King, L. E., and Sheldon, B. C. (2009). Habitat quality, nestling diet, and provisioning behaviour in great tits Parus major. *J. Avian. Biol.* 40: 135–145.

Wilson, D. J., Grant, A. D., and Parker, N. (2006). Diet of kakapo in breeding and non-breeding years on Codfish Island (Whenua Hou) and Stewart Island. *Notornis* 53: 80–89.

Wolock Madej, C., and Clay, K. (1991). Avian seed preference and weight loss experiments: the effect of fungal endophyte-infected tall fescue seeds. *Oecologia* 88: 296–302.

Wu, Y. (2011). The diet of Saker falcon (*Falco cherrug*) on the eastern fringe of Gurban Tunggut Desert, China. In *Proceedings of the 2011 International Conference on Remote Sensing, Environment and Transportation Engineering (RSETE 2011).* Piscataway, NJ: Institute of Electrical and Electronics Engineers, pp. 8647–8649.

Wyllie, I. (1981). *The Cuckoo.* London: B. T. Batsford.

Yamaguchi, K., Harada, S., Kanemaru, N., and Kasahara, Y. (2001). Age-related alteration of taste bud distribution in the common marmoset. *Chem. Senses* 26: 1–6.

Yang, R. S. H., and Kare, M. R. (1968). Taste response of a bird to constituents of arthropod defensive secretions. *Ann. Entomol. Soc. Am.* 61: 781–782.

Yee, K. K., Sukumaran, S. K., Kotha, R., et al. (2011). Glucose transporters and ATP-gated K+ (KATP) metabolic sensors are present in type 1 taste receptor 3 (T1r3)-expressing taste cells. *Proc. Natl. Acad. Sci. U. S. A.* 108: 5431–5436.

Yoshimura, K., and Kobayashi, K. (1997). A comparative morphological study on the tongue and the lingual papillae of some marine mammals particularly of four species of Odontoceti and Zalophus. *Odontology* 85: 385–507.

Yoshimura, K., Shindoh, J., and Kobayashi, K. (2002). Scanning electron microscopy study of the tongue and lingual papillae of the California sea lion (*Zalophus californianus californianus*). *Anat. Rec.* 267: 146–153.

Young, B. A. (1997). On the absence of taste buds in monitor lizards (*Varanus*) and snakes. *J. Herpetol.* 31: 130–137.

Zhao, G. Q., Zhang, Y., Hoon, M. A., et al. (2003). The receptors for mammalian sweet and umami taste. *Cell* 115: 255–266.

Zhao, H., Xu, D., Zhang, S., and Zhang, J. (2012). Genomic and genetic evidence for the loss of umami taste in bats. *Genome Biol. Evol.* 4: 73–79.

Zhao, H., Zhou, Y., Pinto, C. M., et al. (2010). Evolution of the sweet taste receptor gene Tas1r2 in bats. *Mol. Biol. Evol.* 27: 2642–2650.

Zweers, G. A., Gerritsen, A. F. C., and van Kranenburg-Voogd, P. J. (1977). Mechanics of feeding in the mallard (*Anas platyrhynchos*). In *Contributions to Vertebrate Evolution*, Hecht, M. K., and Szalay, F. S. (Eds). Basel: Karger.

Chapter 44

Functional Organization of the Gustatory System in Macaques

Thomas C. Pritchard and Thomas R. Scott

44.1 INTRODUCTION

The Primate Order spans a range of animals comprising prosimians, New World (platyrrhini) and Old World (catarrhini) monkeys, greater and lesser apes, and humans (homonids). Simians have been genetically distinct from prosimians for nearly 60 million years, Old from New World monkeys for 36 million years, and apes from Old World monkeys for 25 million years. Given these periods of prolonged genetic isolation, the consistency in gustatory structure and function across the Order is perhaps more surprising than the variations.

Primates offer a faithful neural model for human gustation, so it is a matter of concern that the laboratories on three continents that provided this evidence have now closed. This has been due largely to retirements, and to the demands and expense of maintaining primates, but, on a more positive note, to the increasing prominence of non-invasive imaging techniques in humans, particularly functional magnetic resonance imaging (fMRI) and positron emission tomography (PET). Although non-invasive imaging techniques have brought important insights into complex sensory, cognitive, clinical, and behavioral issues in humans, coding by individual neurons remains well beyond its scope, and due to its invasive nature, these studies must be performed on laboratory animals. Non-invasive imaging techniques also lack the spatial and temporal resolution necessary to address important issues in sensory perception and integration that typically occur on a millisecond time scale. Even with these limitations, data collected with fMRI and PET are relevant to our discussion of the gustatory system in nonhuman primates, and for this reason, imaging data, collected mostly from

human subjects, will be included, as necessary, to expand the picture of taste processing in monkeys.

In this chapter, we will summarize what is known about the anatomy, physiology, and function of five levels of the gustatory system in the primate: (1) peripheral nerves, (2) medulla, (3) thalamus, (4) cortex, and (5) ventral forebrain. The extant taste literature is not uniformly distributed across these areas, so this review, likewise, will emphasize research on the peripheral gustatory system and the cerebral cortex. The content of this review also reflects the broad focus of research in the periphery on both New and Old World monkeys, and the great apes, versus the almost singular use of Old World monkeys and humans for central nervous system research.

44.2 PERIPHERAL ORGANIZATION OF TASTE

44.2.1 Taste Buds

Gustatory transduction occurs when sapid stimuli in the oral cavity bathe the exposed pores of taste buds embedded in the epithelial surface of the tongue, soft palate, epiglottis, pharynx, and larynx (Bradley, 1971). Taste buds are small (50 um diameter), goblet-shaped ensembles of 40–60 receptor, sustentacular, and basal (stem) cells (Miller, 1995). Each receptor cell has an apical tuft of microvilli that communicates with the intraoral cavity via a small opening or taste pore. Taste transduction takes place via receptor proteins and ion channels that line the microvillar membrane. Taste buds on the tongue are encased within larger structures called papillae. Three types of papillae

---

Handbook of Olfaction and Gustation, Third Edition. Edited by Richard L. Doty.
© 2015 Richard L. Doty. Published 2015 by John Wiley & Sons, Inc.

have been identified: fungiform, foliate, and circumvallate; they vary in size and shape, and each has a unique regional distribution on the tongue.

Fungiform papillae, which are located along the perimeter of the anterior two-thirds of the tongue, are the easiest to access, and for that reason, they have received the most attention from sensory physiologists. Sergovia et al. (2002), after counting the number of taste pores on the superficial surface of fungiform papillae in humans, reported that each papilla contains an average of six taste buds (range = 0–28). The foliate papillae are a series of four or five epithelial folds along the lateral edge of the tongue, adjacent to the palatoglossal arch. Because taste buds line the walls of the foliate papillae, sapid stimulation is most effective when the mechanical action of chewing opens the papillar folds. Although taste buds of the foliate papillae are described as vestigial in some textbooks, sapid stimulation of the foliates in monkeys evokes a robust response in neurons within the thalamic taste relay (Pritchard, unpublished observations, 2000) and presumably the rest of the gustatory neuraxis.

In monkeys, the circumvallate papillae are 3–6 bunker-like structures that straddle the midline on the posterior tongue (Bradley et al., 1985; Emura et al., 2002). Each papilla is surrounded by a deep trench or vallum that is filled with the mucous secreted by nearby von Ebner's glands (Sbarbati et al., 1999). Mucous within the vallum helps dissolve tastants and transport them through the narrow cleft and toward the taste receptors (Beidler, 1995). Optimal stimulation of the circumvallate taste buds occurs during mechanical stimulation of the posterior tongue by the hard palate during mastication. In monkeys, taste buds line the papillar surface; in humans, they are also found on the epithelial surface. Even though the circumvallate papillae contain almost half of all taste buds, their relative inaccessibility has limited investigation of their response properties.

All mammals examined to date have taste buds on the soft palate, pharynx, epiglottis, and larynx. Unlike taste buds located on the tongue, those on the extralingual surfaces are not embedded in distinct papillae. In both primate and non-primate species, the extralingual taste buds are found in small clusters within islands of keratinizing, stratified, squamous epithelium (Cleaton-Jones, 1971; Klein and Schroeder, 1979; Klein et al., 1979). Taste buds located on the extralingual epithelium are more likely involved with airway protection than taste perception (Harding et al, 1978; Miyaoka et al., 1998; Sweazey et al., 1994; Jowett and Shrestha, 1998).

44.2.2 Taste Bud Innervation

In primates, the taste buds within the fungiform papillae and the anterior half of the foliate papillae are innervated by the chorda tympani (CT) nerve, a branch of the facial nerve (McManus et al., 2011; see Figure 44.1). The lingual nerve provides non-gustatory innervation to the fungiform and foliate papillae on the anterior two-thirds of the tongue. Taste buds within the circumvallate papillae and the posterior foliate papillae, as well as non-gustatory structures

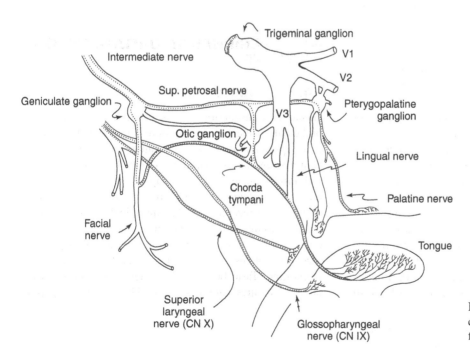

Figure 44.1 Neural innervation of the oral cavity. (Diagram adapted with permission from Oxford University Press).

on the posterior third of the tongue surface, are innervated by the lingual branch of the glossopharyngeal (GP) nerve (Taillibert et al., 1998).

The neural supply of the extralingual taste buds remains uncertain, but research conducted in a variety of primate and nonprimate species implicates branches of the GP and vagus nerves, and just as importantly, excludes other nerves from consideration. Because the GP and vagus nerves carry information from multiple sensory modalities, it is not uncommon for individual gustatory axons to respond to thermal and even tactile stimulation, typically of the same receptive field. For this reason, tracing studies have not been able to isolate gustatory, tactile, and thermal axons from one another, but this design limitation bears more on the central projections of these modalities than the nerves that innervate the peripheral structures.

Based on comparative data, the anterior part of the soft palate is innervated by the greater superficial petrosal nerve (rat, Miller and Spangler, 1982), while the posterior region is supplied by the GP nerve (rat, Hamilton and Norgren, 1984; cat, Yoshida et al., 2000), and possibly the vagus nerve (rat, Hamilton and Norgren, 1984). In monkey and human, the GP nerve innervates the nasopharynx (Kanagasuntheram et al., 1969; Leela et al., 1971) and probably the epiglottis and pharynx (Mu and Sanders, 2000; Toure and Vacher, 2005), but based on data in subprimate species, a role for the superior laryngeal branch of the vagus nerve cannot be ruled out (Hanamori and Smith, 1986; Yoshida et al., 2000). The superior laryngeal nerve, which innervates the larynx in the rat and cat (Norgren, 1984; Stedman et al., 1980; Yoshida et al., 2000), is presumed to supply the larynx in the monkey as well.

44.2.3 Response Properties of Peripheral Taste Nerves

44.2.3.1 *Fiber Types and Labeled Lines.* The first electrophysiological recordings from taste nerve fibers of primates were conducted in Sweden using rhesus macaques more than a half century ago (Gordon et al., 1959). As has become the custom, the authors separated the fibers into types, in this case, those rather specifically responsive to sodium, to acids, or to quinine, and those more broadly tuned to sugars and non-sweet stimuli.

Breadth of sugar sensitivity proved meaningful in the interpretation of gustatory coding strategies. In the late 1960s, whole nerve recordings from the CT nerve of the squirrel monkey established the order of effectiveness of saccharides: fructose > sucrose > glucose (T. Snell, unpublished thesis, 1968). The order of neural effectiveness, however, differed from the order of behavioral preference: sucrose > fructose > glucose (Ganchrow and Fisher, 1968). Pfaffmann (1974) resolved this discrepancy

to his satisfaction upon recording from single fibers in the chorda tympani nerve of the squirrel monkey and discovering that sweet-best fibers reliably showed the sequence that paralleled behavior (sucrose > fructose > glucose) which had been obscured in the whole nerve recordings by sideband sensitivity to fructose in salt-sensitive fibers. Pfaffmann concluded that the perception of sweetness that was driving the monkey's behavioral preferences was carried exclusively by sweet-best fibers and that activity induced in other fibers by sugars was irrelevant to the afferent code, i.e., that sweetness was carried in a labeled line devoted to its detection. Since Pfaffmann had been the author of the competing "across-fiber pattern" theory, in which activity of all fibers was integral to each afferent signal (Pfaffmann, 1941), his conversion, based on monkey data, was a signal moment in the history of gustatory neural coding.

Further recordings from the CT nerve of macaques reaffirmed the existence of four fiber types, although those responding to sweet, salty, and bitter stimuli were more narrowly tuned than those sensitive to acids, which usually had sideband sensitivity to salts (Sato et al., 1975).

This theme of specificity is repeated through the succeeding decades of recordings from primate peripheral taste nerves. Danilova and colleagues recorded single fiber activity from both the CT and GP nerves of the marmoset (a New World monkey), and reported three prevalent fiber types: sweet-best, bitter-best, and sour-best, with a lesser representation of salt (Danilova et al., 2002; Danilova and Hellekant, 2004). Stimuli that activated narrowly tuned fibers elicited the expected preference or avoidance behavior in two-bottle preferences tests, implying that each fiber type serves as a labeled line for its associated taste quality.

Hellekant et al. (1997) established a clearer distinction between the contributions of the CT and GP nerves when they compared the sensitivities of their respective single fibers in macaques. Salt responsiveness characterized 40% of the CT fibers, but only 15% of those in the GP nerve; sucrose activated 34% of CT fibers versus 24% of those in the GP; acids stimulated 17% of CT fibers, but none in the GP. Thus, three of the presumed basic taste qualities were favored in the CT nerve. By comparison, activity in the GP nerve was heavily skewed toward the fourth basic taste quality: quinine activated 61% of GP axons, but only 9% of those in the CT nerve. Upon application of the sweet inhibitor lactisole, responses to sugars were suppressed in sweet-best fibers but not in other fiber types that had side-band sensitivity to sugars. This finding, once again, implied that the perception of sweetness was carried exclusively by sweet-best fibers (Wang et al., 2009). The themes of specificity and separation of function among peripheral nerve fibers of the macaque were reinforced and then extended to chimpanzees, one of the great apes. Recordings

986   **Chapter 44   Functional Organization of the Gustatory System in Macaques**

from single fibers of the CT nerve in chimpanzees revealed an exceptionally low breadth-of-tuning coefficient of 0.30,[1] indicating that each axon had a high degree of response specificity (Hellekant et al., 1997). Here, in distinction to reports from macaques, fibers were divisible into three classes: sucrose (50%), NaCl (31%), and quinine (19%); none responded to acid. Sodium-responsive fibers were further divisible into subclasses that were (1) specifically responsive to sodium (i.e., salt-best fibers) and suppressed by the sodium-channel blocker amiloride, (2) broadly sensitive to sodium and potassium, but not affected by amiloride, and (3) responsive to sodium and monosodium glutamate, a stimulus that presumably carries the umami quality.

Hellekant et al. (1998) then examined whether labeled lines were responsible for coding specific taste qualities, as had been implied by the reports above. They reasoned that if a defined set of fibers were responsible for coding a basic taste quality, then suppression of activity in that group of fibers would block the associated perception; conversely, activation limited to those fibers would create the signature percept. To this end, Hellekant et al. (1998) used gymnemic acid and miraculin to manipulate the sweetness of sapid stimuli. Gymnemic acid suppresses sweet taste in humans (Frank et al., 1992) and chimpanzees (Hellekant et al., 1996); miraculin, contains a glycoprotein whose sugar component is unable to stimulate sweet receptors until its conformation is altered in the presence of acids (Miller and Bartoshuk, 1991). Hellekant et al. (1998) reported that gymnemic acid suppresses the responses of sucrose-best fibers in the CT nerve, while preserving sucrose side-band sensitivity in sodium-best and quinine-best fibers. Conversely, application of miraculin to the tongue caused sucrose-best fibers in the CT nerve to respond to acids, an effect that was subsequently abolished by the application of gymnemic acid. The authors concluded that sucrose-best fibers satisfied the criteria for a labeled line because (1) they are activated either by sucrose or by miraculin's effect on acids; (2) their activity is positively correlated with sweet perception, as judged by behavioral acceptance; and (3) gymnemic acid, which suppresses CT neural responses to sweet stimuli, also blocks sweet perception.

---

[1] The breadth-of-tuning coefficient is a measure of the degree to which a single cell responds to each of the four prototypical stimuli: sucrose, NaCl, HCl, and quinine (Smith and Travers, 1979).

The equation is: $H = -k \sum p_i \log p_i$

where H = breadth of tuning; k = a scaling constant (here, $k = 1.661$ such that H = 1.0 when the neuron responds equally to all four stimuli) p_i = proportional response to each of the four stimuli; i = 1–4 stimuli.

44.2.3.2 *Variability of Taste Perception Across the Primate Order.*

PERCEPTION OF SALTY AND SOUR STIMULI Salty and sour perception is similar across Primate species because transduction for both qualities is mediated by the widely distributed classes of ionotropic receptors.

PERCEPTION OF SWEET STIMULI Differences among primate species with respect to sweetness are the rule rather than the exception. The most accurate diagnostics to evaluate interspecies differences in sweet perception is relative sensitivity to sugars, amino acids, dipeptides, proteins, etc., and the differential effectiveness of sweet-inducing or sweet-suppressing chemicals such as miraculin and gymnemic acid.

Glaser et al. (1995) provided important insight into the evolution of sweet receptors by testing preference for the sweeteners aspartame and thaumatin in 50 species of Primates including prosimians, New and Old World monkeys, lesser and greater apes, and humans. The two-bottle preference data from this massive survey were consistent: all species of prosimians and New World monkeys were indifferent to aspartame and thaumatin; all species of Old World monkeys, and apes, as well as humans preferred them to water. As a demonstration of how idiosyncratic these binding interactions can be, the sweetener alitame, a dipeptide derived from aspartame, was preferred by all primate species. Thus, the binding site for alitame is older than the 60 million year division between prosimians and simians, while the binding site for aspartame and thaumatin is more recent than the 36 million year split between New and Old World monkeys (Kappelman, 1992). On a finer time scale, miraculin induces sweet taste in the presence of acids in simians (New and Old World monkeys), apes, and humans, but not in the more primitive prosimians (Hellekant et al., 1981).

The basis for species differences in sweet perception have been traced to a specific region of the sweet receptor. The genes Tas1R2 and Tas1R3 produce the associated G-protein-coupled receptors T1R2 and T1R3, which together form the heterodimer that detects sweet stimuli. While the docking site on the T1R2 receptor that is stimulated by monosaccharides (e.g., glucose, fructose) and the disaccharide sucrose is universal across the Order, the effectiveness of other molecules appears to vary across primate species.

With the discovery of the T1R family of receptors for sweet stimuli, Nelson et al. (2001) and Li et al. (2011) sequenced the genes between aspartame tasters and non-tasters in three New World monkeys (tamarin, marmoset, squirrel monkey), three Old World monkeys (macaque, patas monkey, baboon), three apes (chimpanzee, gorilla, orangutan), and humans. These studies revealed nine variant sites along the Tas1R2 gene, and 32 variants along the Tas1R3 gene in New and Old World monkeys. Li et al. (2011) concluded that the most likely binding location for aspartame was an allosteric site on the T1R2 receptor that could cause a conformational change in the protein. Based upon the work of Glaser et al. (1995), it appears that prosimians and New World monkeys lack this allosteric site.

The similarity of Old World monkeys, apes, and humans in the taste preference experiments conducted by Glaser et al. (1995) suggest that these species represent an homogenous group. However, use of a different diagnostic test – suppression of sweet taste by gymnemic acid – shows that sweet perception is not identical in these primates. Gymnemic acid suppresses sweet taste in humans, chimpanzees, and gibbons (Hellekant et al., 1990), but not in macaques (Hellekant et al., 1996). Studies of the kinetics of gymnemic acid action on taste imply that it does not compete with the sweet molecule, but rather binds to a site on the receptor that renders the normal sweet receptor sites inoperable (Kennedy, 1989). These results, considered in light of the Glaser et al. (1995) data, suggest that the sweet receptors of Old World monkeys are similar to, but not identical to those of humans and apes. Whether humans and apes (chimpanzees, gibbons) perceive sweet in an identical fashion is an empirical question.

PERCEPTION OF BITTER STIMULI Primates deploy a large family of genes which produce the proteins that, collectively, detect most harmful chemicals (Chandrashekar et al., 2000). Despite the prevalence of harmful, bitter substances in the environment, few studies have investigated bitter perception in primates. The demonstration by Behrens and Meyerhof (2008) that the Tas2R bitter gene family is identical in humans and chimpanzees indicates that the genes and pseudogenes of bitterness have been conserved over the six million years of separation between these two species.

44.3 CENTRAL TASTE PATHWAYS

44.3.1 Nucleus of the Solitary Tract

44.3.1.1 Anatomy. Gustatory axons of the facial, GP, and vagus nerves enter the rostral medulla and coalesce into the solitary tract, which divides its namesake, the nucleus of the solitary tract (NST), into lateral and medial segments (Beckstead and Norgren, 1979). The facial and GP fibers terminate in the lateral division, while axons of the vagus nerve terminate within the medial division; the GP nerve terminates in both (Beckstead and Norgren, 1979; Satoda et al., 1996; Rhoton, 1968; Figure 44.2). These anatomical data are complemented by numerous electrophysiological studies in the rat (Norgren, 1978; Chen and DiLorenzo, 2008), rabbit (Schwartzbaum and DiLorenzo, 1982), hamster (Travers, 1988; Cho and Li, 2008), and monkey (Beckstead et al., 1980; Scott et al., 1986a) which confirm that second-order gustatory neurons are located within the rostral NST while viscerosensory neurons are located further caudally. Based upon data collected in the rat, neurons that respond to taste stimulation of the anterior tongue are located rostrally to those with receptive fields on the posterior tongue (Travers and Norgren, 1995).

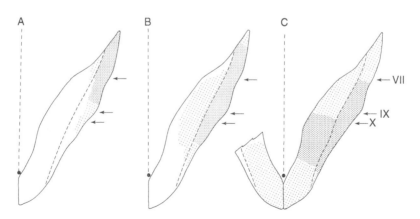

Figure 44.2 Central distributions of cranial nerves VII, IX, and X (Panels A, B, and C, respectively) within the NST of the macaque. The three panels show outlines of the NST in the horizontal plane, with rostral to the top of the page and lateral to the right, viewed from above. In each panel, the vertical dotted line indicates midline, the obex is marked by a filled circle, and the entry level of cranial nerves VII, IX, and X into the NST are marked by arrows. The dotted line coursing through the NST separates the medial and lateral subdivisions of the nucleus. Autoradiographic label density within the NST following injections of cranial nerve VII, IX, and X ganglia is indicated by two stippling densities. Only the vagus projects to the contralateral side of the NST. (Diagram constructed with data originally summarized in Beckstead and Norgren, 1979. Used with permission from Lippincott, Williams, and Wilkins).

44.3.1.2 Physiology.
In Old World monkeys, there is substantial overlap of facial, GP, and vagal terminals within the gustatory region of the NST (Pritchard et al., 2000), yet this rostral-to-caudal (gustatory-to-viscerosensory) organization is preserved (Beckstead et al., 1980). In humans, Topolevec et al. (2004) and Komisaruk et al. (2002) used fMRI to demonstrate that sapid stimulation of the tongue activates the rostral NST whereas tasks designed to modulate blood pressure and heart rate (forced maximal inspiration, sustained isometric muscle contraction, and the Valsalva maneuver) activate the medulla more caudally. Thus, based upon a variety of complementary anatomical, electrophysiological, and imaging studies, the organization of the NST into a rostrolateral, gustatory region and a caudomedial, visceral region is highly conserved across species. The proximity of these two systems in the medulla may contribute to brainstem control of feeding and digestion; further rostrally, however, the organization of the gustatory and visceral afferent systems differ significantly from each other, both within and across species.

Whereas gustatory data from the peripheral nerves of Primates have accumulated over a half century from laboratories around the world, information on the NST began and ended in Oxford in 1984. Two reports emerged from that effort.

In the first, Scott et al. (1986a) reported that NST taste neurons respond at threshold concentrations and have concentration-response functions matching those from human psychophysical studies (Collings, 1974). In distinction to data from peripheral nerves, NST cells are broadly tuned across the prototypical taste stimuli, with a mean breadth-of-tuning coefficient of 0.87. The majority of NST cells respond maximally to sweet (40%) and salty (35%) stimuli more often than to bitter (17%) or sour (8%) fluids.

The second report had its roots in previous rodent research, specifically studies by Scott and colleagues that suggested that the taste system was malleable according to the physiological state of the animal, altering its sensitivity to direct the animal's food choices (Chang and Scott, 1984; Giza and Scott, 1983; Jacobs et al., 1988). Rolls and colleagues countered that human gustatory perception remained unchanged during the progression from hungry to satiated during a meal, a finding that would not permit such malleability at the hindbrain level (Rolls et al., 1981, 1982). In a series of studies, beginning in the NST, Scott and Rolls addressed this issue by cutting through the three levels of ambiguity inherent in the literature (rodent vs primate, anesthetized vs awake, NST vs a presumed cortical perception).

Yaxley et al. (1985), by recording from individual NST neurons of awake macaques over the course of a meal, determined that feeding to satiety has no effect on taste responses to the prototypical stimuli. These data countered the predictions based on previous rodent experiments, but more importantly, they were consistent with data from human psychophysical studies (Rolls et al. 1981). Subsequent studies at the cortical level on this issue are described below.

Isolating taste cells from the NST of awake macaques is a daunting task. Because the NST is a narrow, elongated nucleus consisting of small, tightly-packed neurons, deep in the posterior cranial fossa where brain movements are common, it is unlikely that the literature on the macaque NST will expand in the near term.

44.3.1.3 Projections.
Axons originating in the rostral, gustatory region of the NST ascend through the brainstem within the central tegmental tract and terminate in the parvicellular division of the ventroposteromedial nucleus of the thalamus (VPMpc; Beckstead et al., 1980; see Figure 44.3). The location of the taste relay within the most medial aspect of the ventrobasal complex completes the image of the sensory homunculus that stretches across the ventral nucleus of the thalamus. As Figure 44.4

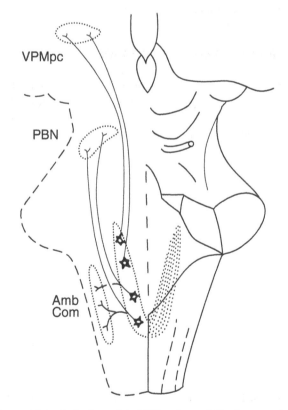

Figure 44.3 Projections of the NST to the caudal brainstem and diencephalon of the Old World monkey. Projections to the PBN originate in the caudal NST, which receives significant afferent input from the vagus nerve. Axons originating in the rostral, primarily gustatory part of the NST, bypass the PBN and terminate in the caudal aspect VPMpc. (Used with permission from John Wiley & Sons, Inc.).

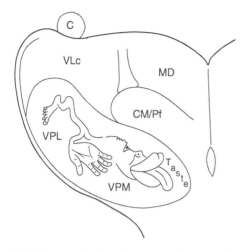

Figure 44.4 Diagram showing the topographic organization of the somatosensory and gustatory systems within the ventrobasal thalamus. The head and face representations are located within the ventroposteromedial nucleus (VPM), which is located medially to the body representation within the ventroposterolateral nucleus (VPL). The thalamic taste area, the most medial tenant of the ventrobasal thalamus, is located with the parvicellular division of the ventroposteromedial nucleus (VPMpc). Abbreviations: CM/Pf=centromedian/parafascicular nucleus; CN=caudate nucleus; IC=internal capsule; LD=lateral dorsal nucleus; VLc=ventral lateral nucleus, pars caudalis; VPI=ventroposteroinferior nucleus.

shows, somesthetic sensation from the torso/limbs, and head/mouth is represented within the ventroposterolateral nucleus (VPL) and VPM nuclei, respectively, while taste and visceral sensation are located most medially within the adjacent VPMpc.

This direct NST-VPMpc projection in nonhuman primates represents a significant departure from the ascending taste pathway rats, hamsters, mice, and cats. In these subprimate species, gustatory neurons in the NST project to the pontine parabrachial nucleus (PBN), which in turn, sends axons to VPMpc (Loewy and Burton, 1978; Norgren, 1978; Norgren and Leonard, 1973; Tokita et al., 2009; Halsell, 1992; Nomura et al., 1979). There is ample electrophysiological evidence showing that the PBN in these species contains gustatory neurons (Halsell and Travers, 1997; Halsell and Frank, 1991, 1992; Cho and Li, 2008, DiLorenzo and Schwartzbaum, 1982), but there is no evidence in any primate species that the PBN neurons respond to taste (Pritchard et al., 2000). Rather than gustation, the PBN in the monkey is an important relay for general visceral and nociceptive information bound for the forebrain, and accordingly, it receives direct projections from the caudal viscerosensory NST (Beckstead et al., 1980) and lamina I of the spinal cord (Westlund and Craig, 1996). A minor, but direct, projection from the PBN to VPMpc exists in monkeys, but based upon the afferent connectivity of the PBN, this presumably relates to visceral sensation. It is also significant that the visceral projections from the PBN terminate in the rostral half of VPMpc rather in than the thalamic taste area which is located further caudally (Pritchard et al., 2000). Thus, VPMpc and NST both have gustatory and visceral subdivisions, but unlike the NST, in the thalamus, the taste area is located within the caudal half of the nucleus (Pritchard et al., 2000).

44.3.2 Thalamus

44.3.2.1 Anatomy. In every species studied to date, electrophysiological studies have confirmed that the thalamic gustatory relay is located in VPMpc (primate: Benjamin, 1963; Pritchard et al., 1986, 1989; Reilly and Pritchard, 1995; subprimate: Scott and Erickson, 1971; Verhagen et al., 2003). As discussed above, gustatory neurons are located only within the caudal half of VPMpc, and further sequestered within the medial half of the nucleus; the lateral half of VPMpc is populated by neurons that respond to oral somatosensory or visceral stimulation (Burton and Benjamin, 1971; Pritchard et al., 1986).

44.3.2.2 Physiology. Like neurons in the NST, the coding properties of cells in the thalamic taste relay have received little attention from the taste community. One reason may be that the sensory thalamus, in general, has been an enigma and often dismissed as a passive relay to the cerebral cortex where conscious perceptions are probably generated. Although clearly not true, when the techniques of chronic recording in awake, behaving monkeys finally reached the gustatory system, little was accomplished in the NST or the thalamus before the attention shifted to the cerebral cortex.

A significant finding reported by Pritchard et al. (1989) was the presence of ipsilateral gustatory receptive fields in VPMpc. Although this result was expected based upon previous anatomical studies in monkeys, the confused clinical literature on this issue has called into question whether taste ascends from tongue to cortex ipsilaterally (Pritchard et al., 1989; Pritchard, 2012). Other electrophysiological evidence that the taste system in primates is organized ipsilaterally was provided by Lenz et al. (1997). In these studies, the posterior thalamus in the vicinity of the ventroposterolateral nucleus, pars oralis and caudalis (VPLo, VPLc, respectively) and VPMpc was electrically stimulated in three awake thalamotomy patients. Lenz's patients reported a variety of intra- and perioral sensations, including a sour taste that localized to the side of the tongue ipsilateral to the locus of stimulation.

Pritchard et al. (1989) reported that gustatory-evoked responses are less sluggish in the thalamus than reported for the NST by Scott et al. (1986a); sugar and salt remain the most effective stimuli. Although the minor decrease in breadth of tuning from 0.87 in the NST to 0.73 in the VPMpc suggests that neurons in the thalamus are almost as broadly tuned as those in the medulla, this appearance primarily reflects the nonlinearity of the entropy statistic (see Pritchard et al., 1989 for a discussion of this issue). In fact, NST neurons are considerably more broadly tuned than taste neurons in the thalamus. As in all other levels of the gustatory neuraxis, VPMpc is an admixture of sensory modalities (Benjamin, 1963; Pritchard et al., 1989). Taste-responsive neurons constitute 36% of those in the nucleus, but tactile (11%) and thermal-responses (<1%) neurons are present, as are non-responsive cells (18%) and neurons that respond prior to stimulus delivery (35%).

44.3.3 Insular-Opercular Cortex

44.3.3.1 Anatomy.
Third-order gustatory neurons in VPMpc project to a small crescent of granular neocortex at the junction of the rostrodorsal insula and the inner operculum (Benjamin and Burton, 1968; Benjamin et al., 1968; Pritchard et al., 1986; see Figure 44.5). This terminal area is densest within the fundus of the superior limiting sulcus and tapers gradually into the dysgranular cortices of the body of the insula and the inner operculum. Because these thalamocortical projections originate in the medial, gustatory portion of VPMpc, this terminal area is, by anatomical definition, primary taste cortex.

Taste responses have been recorded from three areas that either receive direct projections from VPMpc or from primary taste cortex (1) insular-orbital cortex; (2) the inner and outer frontal opercula; and (3) the posterior insula. The latter two contain predominantly, if not exclusively, higher-order cortical taste neurons.

1. The most significant projections from primary taste cortex terminate in the dysgranular and agranular cortices that stretch from the body of the insula ventrally and rostrally across the posterior orbital cortex, and finally into the posterior orbitofrontal cortex (OFC; Carmichael and Price, 1995). The agranular insula also receives direct projections from the dysgranular insula (see Figure 44.6; Mesulam and Mufson, 1982; Mufson and Mesulam, 1982; Morecraft et al., 1992; Barbas, 1993; Carmichael and Price, 1995). Based upon its connectivity to primary taste cortex, the dysgranular and agranular insulae represent higher-order taste cortices. Given the complexity and redundancy of this multi-neuron network, it is not possible to specify or even separate the admixture of secondary from tertiary (from quaternary and beyond) taste cortices in these areas. For this reason, the body of the insula as well as the orbital and orbitofrontal cortices should be designated only as higher-order taste cortices.

2. A minor projection from primary taste cortex to the inner operculum (Brodmann areas (BAs) 1 and 2) and the precentral extension of BA 3b on the lateral convexity (Pritchard, unpublished observations, 1991) probably accounts for the few taste-responsive neurons found in these areas (Sudakov et al., 1971; Ogawa et al., 1989). These are the same areas of

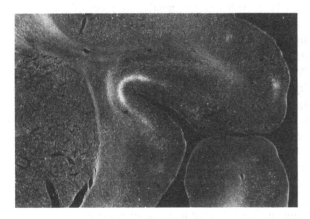

Figure 44.5 Darkfield photomicrograph of terminal label in insulo-opercular cortex after a tritiated amino acid injection into the taste-responsive area of VPMpc.

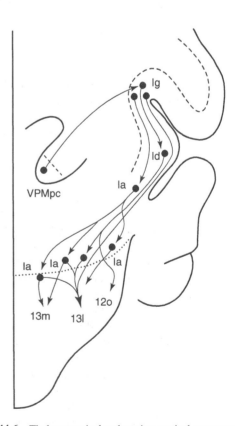

Figure 44.6 Thalamocortical and cortico-cortical gustatory projections of the Old World monkey. This diagram shows the thalamic and insular gustatory relays in the coronal plane appended to the ventral surface of the OFC. Abbreviations: BA12o, BA13l, BA13m; Ia=agranular insula; Id=dysgranular insula; Ig=granular insula.

cortex where evoked potentials have been recorded following electrical stimulation of the chorda tympani and lingual-tonsillar branch of the glossopharyngeal nerve (Benjamin and Burton, 1968; Benjamin et al., 1968). The precentral extension of BA 3b also receives a direct projection from the oral somatosensory area located at the VPMpc/VPM border (Pritchard et al., 1986).

3. There are no direct projections from the gustatory thalamus to the middle or posterior insula, where taste responses have been observed in humans using magnetoencephalography (MEG; Wakita et al. (2009)), fMRI (Del Parigi et al., 2005), and PET (Kobayashi et al., 2002), and reported during psychophysical testing of stroke patients (Pritchard et al., 1999). The most plausible route for taste information to reach the middle and posterior insula is a minor projection that originates in the body of the rostral insula (Mesulam and Mufson, 1982).

44.3.3.2 Physiology.
Sudakov et al. (1971) provided the first description of taste-responsive neurons in the insula, inner operculum, and the lateral convexity of alert monkeys. Perhaps the most salient feature of this and subsequent studies is that taste cortex contains so few taste neurons. This study, the only one conducted in New World monkeys to date, showed that only 3.2% (14 of 437) of the neurons located in primary taste cortex respond to sapid stimulation of the tongue. Although these taste neurons are interspersed with oral, tactile neurons, convergent taste/tactile responses were not encountered. Within the adjacent frontal operculum, gustatory neurons account for only 3.7% of the population. Ogawa et al. (1989) examined these areas in anesthetized monkeys and reported that about half of the taste-responsive neurons in their sample are located within the granular cortex adjacent to the crest of the superior circular sulcus. Only 3.1% of the neurons within the inner and outer opercula respond to sapid stimulation of the tongue, and in the dysgranular cortex of the insula, neurons responsive to mechanical stimulation of the oral cavity outnumber taste neurons 8 to1. Ito and Ogawa (1994), using alert monkeys and working with a larger sample of neurons, reported that most of the taste neurons they sampled were located within the inner operculum. Together, these reports show that somatosensory neurons and other neurons for which the adequate stimulus cannot be identified greatly outnumber gustatory neurons within the insula and frontal operculum.

Detailed electrophysiological experiments have built upon the early anatomical and physiological studies of the insula and operculum (Scott et al., 1986b; Ogawa et al., 1989; Yaxley et al., 1988; Scott et al., 1991; Ito and Ogawa, 1994; Rolls and Baylis, 1994). These latter reports estimate that the taste-responsive area within the rostrodorsal insula

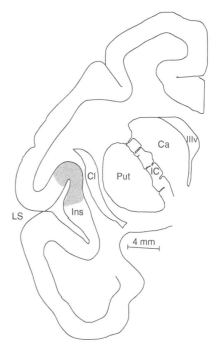

Figure 44.7 The shaded area within the insula shows the taste-responsive region where many of the taste neurons described by Scott and Plata-Salaman (1999) were recorded. Ca=caudate; Cl=claustrum; IC=internal capsule; Ins= insula; LS=lateral sulcus; Put=putamen; TL=temporal lobe; IIIv=lateral ventricle.

occupies 50 mm^3 (Scott and Plata-Salaman, 1999), which is slightly larger than the area defined by less sensitive, anatomical methods. These studies examined both insular and opercular regions but many of the taste-responsive neurons they recorded were located within the dysgranular cortex that lies outside the boundaries of primary taste cortex (Figure 44.7). The minor projection from the thalamus to the dysgranular insula is overshadowed by input from the granular cortex at the insular-opercular junction, which means many if not most of the neurons within the body of the insula are second, or even third-order cortical taste neurons. Consistent with previous studies, Scott and Plata-Salaman (1999) and Lara et al. (2009) report that only 6.5% of insular neurons respond to gustatory stimulation.

One of the main goals of these electrophysiological investigations was to determine whether the Old World monkey could serve as a neural surrogate for human taste perception. To that end, a series of experiments compared electrophysiological activity from the monkey with human psychophysical data with regard to: (1) thresholds and intensity-response functions (Scott et al., 1991), (2) coding of quality for each of four basic tastes (Smith-Swintosky et al., 1991): sweetness (Plata-Salamán et al., 1993), saltiness (Scott et al., 1994), sourness (Plata-Salamán et al., 1995), and bitterness (Scott et al., 1999), (3) coding of

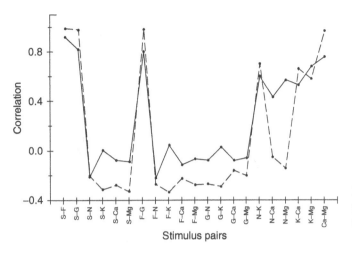

Figure 44.8 Taste quality correlations between human psychophysical data (dotted lines; Kuznicki and Ashbaugh, 1979) and monkey electrophysiological data (solid lines; Smith-Swintosky et al., 1991).

amino acids (Plata-Salamán et al., 1992), and (4) taste mixtures (Plata-Salamán et al., 1996). In each case, the analysis of the macaque neural responses, carried out with sophisticated, quantitative tools associated with patterning theory (Erickson, 1967), implied that neural coding of taste quality in nonhuman primates matches the psychophysical reports of humans. In one analysis, however, patterning theory was wanting. The patterning theory assumes that the gustatory code is carried by all neurons, so whether a neuron is highly responsive or unresponsive to a stimulus, each data point contributes to the response pattern representing that taste. The competing labeled-line theory assumes that the quality code is restricted to the most responsive channel, and that lesser activity from other cells is irrelevant. If perceived taste intensity and stimulus concentration rise in tandem, then the most responsive channel – the putative labeled line – should provide a match between macaque and human. The data from monkeys support the labeled-line theory, and thus, reinforce the conclusions drawn from experiments in the gustatory periphery showing that the taste system encompasses discrete channels, each devoted to coding a basic quality (Scott and Giza, 2000). This discrepancy notwithstanding, patterns of neural activity elicited from the macaque insula (Smith-Swintosky et al., 1991) correlate +0.91 with psychophysical measures of taste similarity in humans (Kuznicki and Ashbaugh, 1979; see Figure 44.8). This high correlation between the macaque neural responses and human psychophysical reports implies that coding of taste quality is complete either before the signal reaches the insula or within the insula itself.

Just as in the NST (Yaxley et al., 1985), the responses of taste neurons in the insular-opercular cortex remained stable over the course of a meal, even as the hedonic value of the sugar with which the animals were satiated shifted from positive to negative (Rolls et al., 1988; Yaxley et al., 1988). This conclusion was confirmed by two more recent reports that used NaCl and sucrose as cues and rewards (Ifuku et al., 2003; Ogawa, et al., 2005). In these experiments, regardless of the hedonic tone conferred on the tastants during a GO/NOGO behavioral task, the evoked responses remained unchanged. Similarly, de Araujo et al. (2003), in an fMRI study of humans, observed that water is an effective stimulus for insular neurons, and that the magnitude of activation elicited does not change as subjects drink to satiety and hedonic value wanes. Contradictory evidence, however, was reported by Small et al. (2001a) who used PET to demonstrate in humans that chocolate activates the insula when the stimulus is appealing, but not after satiety develops. Similarly, Tataranni et al. (1999), Gautier (2001), and Del Parigi et al. (2002), also using PET, reported that activation of the insula is greater in hungry than in sated humans. Thus, a firm conclusion on whether neurons in the insula signal hedonic appeal is obfuscated by species differences and varied experimental approaches.

Taste perception is but one aspect of ingestive behavior that has been localized to the insula. Scott and Plata-Salaman (1999) reported that non-gustatory insular neurons also respond during mouth movements (23.8%), tactile stimulation of the oral cavity (3.7%), approach of the stimulus delivery syringe (1.2%), tongue extension (0.3%) and olfactory stimulation. Insular neurons also respond to the viscosity (53%), temperature (35%), fat content (8%), grittiness (8%), and astringency (8%) of sapid stimuli, as well as to capsaicin, the active ingredient in chili peppers (Critchley and Rolls, 1996; Rolls, 2005). Verhagen et al. (2004) reported that half of all insular neurons respond to only one of these components;[2] half are bi- or multimodal. Collectively, these data speak to the insula's involvement in the perception of flavor, the amalgam of gustatory, olfactory, and somesthetic (touch, temperature, nociception), information.

---

[2] Verhagen et al. (2004) tested all of the stimulus modalities above except astringency.

44.3.3.3 Imaging. It is a curious twist of the discipline and its techniques that the monkey model was developed as a surrogate for human gustation; that after having established itself as such, electrophysiology went into decline, partly to be replaced by neural imaging techniques suitable for use in human subjects; and that the data from imaging is now being used to help complete the picture of taste processing in the monkey. Functional MRI studies using humans have confirmed the original observations in nonhuman primates that that the insula is activated by water (de Araujo et al., 2003), stimulus viscosity (de Araujo and Rolls, 2004), temperature (Guest et al., 2007), and texture (de Araujo and Simon, 2009).

Perhaps unexpectedly, photographs of food activate the anterior insula while those of non-food objects do not (Simmons et al., 2005), reminiscent of the observation in the macaque by Scott and Plata-Salamán (1999) that insular neurons respond selectively to the sight of the feeding syringe. Many studies have shown that activation of the insula is not limited to sensory stimulation originating in the external environment. For example, activation of the insula accompanies stimulation of the esophagus (Aziz et al., 1995; Weusten et al., 1994) and the throat (Roper et al., 1993), as well as general visceral arousal induced by phobic anxiety (Rauch et al., 1995). PET scans reveal insular activation while subjects are instructed to move their mouths and tongues, as well as during speech (Raichle, 1991), implying a role for the insula in oromotor functions. Insular-based seizures (Fiol et al., 1988) and electrical stimulation of the insula induce vomiting (Oppenheimer et al., 1992). Thus, insular cortex is involved in multiple aspects of feeding, including the appearance (vision) and perhaps detection (smell) of food, as well as contact and engagement of the tongue with food, and perhaps even the digestive process.

Beyond the sensory and motor realms, the insula also has been implicated in various forms of cognitive behavior. For example, the insula is active when subjects imagine sweetness or saltiness (Levy et al., 1999). Subjects who are misled and told that the subsequent taste would not be bitter show less activation than normal when they are given quinine (Nitschke et al., 2006). When subjects are instructed to detect a taste that is, in fact, absent, their taste-related attention elicits an insular response (Veldhuizen et al., 2007). Viewing facial expressions of disgust also activates the insula (Phillips et al., 1997; Wicker et al., 2003; Jabbi et al., 2008). Finally, both when anticipating and consuming food, obese subjects show larger activation in the insula than lean subjects (Stice et al. (2008)). Thus, imagination, expectation, attention, emotion, and anticipation all contribute to the set of features that can affect activity in the insula.

Because most of these electrophysiological and imaging studies focused on the dysgranular and agranular insulae that constitute higher-order cortical taste areas, it is not surprising that many of their findings expand the role of the insula beyond sensory analysis of sapid stimuli. If the insula marks the transition from sensory analysis to utilization (i.e., cognition and behavior), how should the mission of the insula be defined? Craig (2009) has suggested that the insula's primary function, broadly defined, is homeostatic control, interoception, and self-awareness (see also Evrard et al., 2012). If one accepts this grand view of the insula, its mission may encompass expansive issues such as nourishment, risk-taking (Ishii et al., 2012), and ultimately protection of the organism. The sense of taste, with its charge to identify and accept nourishing foods while rejecting toxins, fits well into this broad view of insular function.

44.3.4 Orbital Cortex

44.3.4.1 Anatomy. Taste neurons have been recorded across a broad swath of the OFC stretching from the medial orbital sulcus as far laterally as the precentral operculum (area PrCO). Except for the medial orbitofrontal cortex (mOFC) and the caudolateral orbitofrontal cortex (clOFC), the taste-responsive neurons in the OFC are too widely scattered to constitute an "area." Gustatory projections to the mOFC and clOFC follow multiple, multisynaptic pathways through the ventral insula and the posterior orbital cortex (Carmichael and Price, 1995), but divergence, convergence, and overlap of these pathways obscures the circuitry and, ultimately, leaves us with an incomplete understanding of how taste information reaches the OFC. That said, we know that the last leg of the afferent pathway to the OFC originates in the higher-order taste areas within the dysgranular and agranular insulae, and the posterior orbital cortex rather than primary taste cortex (Carmichael and Price, 1995, 1996; Mufson and Mesulam, 1982; Barbas, 1993; Mesulam and Mufson, 1985; Baylis et al., 1995; Pritchard, unpublished observations, 2005). Thus, like other sensory systems that project to the prefrontal cortex, the gustatory afferent neurons that terminate in the OFC are located in sensory association cortices rather than in primary taste cortex (Carmichael and Price, 1995; Barbas, 1993, 2007). Through these detailed descriptions of the local circuitry within the OFC, we also know that the clOFC is synaptically more remote from the insular taste areas than the mOFC.

The highest concentration of taste neurons in the OFC lies in the mOFC, the narrow isthmus of cortex between the medial and lateral orbital sulci (Pritchard et al., 2005). In this region, which consists primarily of BAs 13m and 13a, almost 20% of the neurons respond to gustatory stimulation; only 8% of the neurons in the 1mm perimeter area

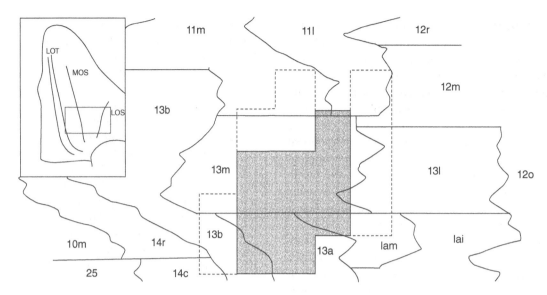

Figure 44.9 The core and perimeter taste areas within the mOFC projected onto an unfolded BA map of the OFC of the macaque. Inset shows the location of this partial OFC map relative to the ventral surface of the frontal lobe. (Adapted from Pritchard et al., 2005).

(BAs 13b, 13l, 12m, 11l, and 11m, and the medial agranular insula) respond to taste (see Figure 44.9[3]). A subsequent study by Lara et al. (2009) reported that 11% of the neurons in the mOFC respond to taste stimulation. The relatively high proportion of taste cells in the mOFC stands in marked contrast to the rest of the OFC where, on average, only 3.8% (range=1.6 to 7.3%;) of the population respond to sapid stimuli (Thorpe et al., 1983; Rolls et al., 1990; Rolls and Baylis, 1994). The clOFC, although not well defined, is located along the lateral edge of the OFC, adjacent to the border with the posterior orbital cortex. The clOFC is within, but is not synonymous with BA47/12o. About 2% of the neurons in the clOFC respond to sapid stimulation of the tongue (Rolls et al., 1990).

44.3.4.2 Physiology.
Although some sensory analysis of taste certainly takes place within the gustatory association cortices in the dysgranular and agranular insulae, as described above, these areas are noteworthy for their role in taste-mediated guidance of feeding behavior and nourishment. Based upon several detailed electrophysiological experiments and a more extensive imaging literature, it appears that the progression from sensory analysis to sensory support for behavior that was the hallmark of the insula, continues in the OFC.

Taste-responsive neurons are sparsely distributed across much of the posterior OFC, but detailed descriptions are only available for two regions: the mOFC (Pritchard et al., 2005; Lara et al., 2009) and the clOFC (Rolls et al., 1990). Tract tracing experiments have demonstrated that the mOFC is one of several entry points to the OFC for projections that originate in the dysgranular and agranular insulae (Carmichael and Price, 1995).

The response properties of mOFC gustatory neurons resemble those of taste neurons in the insula in some ways and the clOFC in others. Like the insula, mOFC neurons are broadly tuned to the four basic taste qualities (mOFC H = 0.79; insula H = 0.70), with sensitivity to each in the 50–60% range. This contrasts sharply with the clOFC where 82% of the neurons respond to glucose, with salty, sour, and bitter sensitivity accounting for less than 12% each. Such narrow tuning is reflected in the extremely low entropy coefficient of clOFC neurons (H=0.39). Surprisingly, the narrow response breadth of clOFC taste neurons is accompanied by significantly more convergent visual, olfactory, auditory, and somatosensory information than observed in the mOFC, which has only a modest number of multimodal neurons; few taste neurons in the insula respond to other sensory modalities (Rolls and Baylis, 1994; Critchley and Rolls, 1996; Rolls, 2004; Thorpe et al., 1983; Scott and Plata-Salaman, 1999; Pritchard et al., 2005). It remains to be seen whether the multimodal character of clOFC neurons provides a platform for the type of sensory integration that defines flavor perception.

The OFC, long recognized for playing a role in emotion, is now considered a vital area of the brain for evaluating reward and risk in the often complicated process of decision-making (Tremblay and Schultz, 1999; O'Neill and Schultz, 2010; Rempel-Clower, 2007; Summerfield and Tsetsos, 2012). Rudebeck and Murray (2011), who have used the term "balkanized" to describe the anatomical

---

[3] Because BA 13l receives direct projections from primary taste cortex, it is possible the percentage of taste neurons reported for this area has been underestimated due to limited sampling.

and multifunctional specialization of the OFC, find support for their view in both the animal and human literatures.

Electrophysiological experiments in the OFC of monkeys that emphasize cue-guided behavior and reward contingencies instead of passive taste properties (e.g., Pritchard et al., 2005) engage a much higher percentage of the cells in the OFC. For example, Hosokawa et al. (2007), using a delayed color matching test that included both positive and negative reinforcement, showed that the responses of 31% of clOFC neurons code the relative preference of the reward. Schultz et al. (1998), using a GO-NOGO paradigm, reported that neurons in BAs 11, 13, and 14 respond to cues that predict reward in proportion to the value of the reward. When the reward contingencies are changed, the responses of some neurons increase, others decrease, and some neurons change from non-responders to responders (Tremblay and Schultz, 2000a, 2000b). The ability of OFC neurons to track the monkey's behavioral adjustments to the fluctuating relationship between cue relevance and reward availability has also been reported by Baeg et al. (2009). The responses of other OFC neurons follow reward magnitude or the taste of the reward (Padoa-Schioppa and Assad, 2006; Padoa-Schioppa, 2007).

The response magnitude of OFC neurons is also modulated by postingestive factors that decrease the rewarding value of food, the subject's incentive to eat or drink, and internal need states. For example, when a specific food is consumed to the point of satiety (sensory specific satiety), the rewarding value of that food declines, sometimes to the point that it loses its ability to serve as a reinforcer. Using this paradigm, Scott et al. (1986a) and Yaxley et al. (1985) concluded that satiety does not affect the magnitude of taste responses in the NST and insula, respectively. By comparison, Rolls et al., (1989) demonstrated that the taste-evoked responses of clOFC neurons decrease as a linear function of satiety. In other words, response magnitude decreased in parallel with reward devaluation. In the mOFC, the relationship between satiety-induced reward devaluation and response magnitude is more complex and not as decisive. For most neurons, response magnitude decreased as satiety developed, although in a more graded fashion than reported for the clOFC. For a small cohort of neurons, response magnitude increased with satiety. Perhaps most interesting, some mOFC neurons that initially did not respond to sapid stimulation, began responding to taste after the monkeys reached satiety (Pritchard et al., 2008). Collectively, this research drives home the message that the OFC is not a higher-order taste area in the traditional sense, but rather an area where taste information is used to support reward-evaluation, risk assessment, and decision-making. Damage to the OFC would be expected to undermine these processes, and predictably, monkeys with OFC lesions show normal satiety, but no evidence of satiety-induced reward devaluation (Izquierdo et al., 2004; Murray and Izquierdo, 2007).

44.3.4.3 *Imaging.*

Functional imaging techniques are now in their adolescence and like their human counterparts, they have a voracious appetite for exploration and adventure. The field of cognitive neuroscience, which historically has struggled against the brain's almost infinite complexity, seemed like a worthy target for these nascent technologies. Attention quickly focused on the OFC, which is probably the crown jewel of cognitive neuroscience, and was directed toward the well-established role of the OFC in emotion, reward, and their derivative behaviors. Multiple studies have shown that the OFC is involved with behaviors driven by both primary (taste) and secondary (money) reinforcers (Bechara et al., 1994; Knutson et al., 2000; O'Doherty et al., 2001).

The imagining literature confirms the general findings of the electrophysiological literature that the OFC is more likely to use sensory information, including taste, to fuel decisions related to reward comparison and decision-making rather than the more mundane business of quality or intensity identification. For example, Fitzgerald et al. (2009) used fMRI to demonstrate that the mOFC encodes the relative subjective value of rewards, both positive and negative, rather than the absolute reward value of the choices available to the subjects. These findings, like the neurophysiological data of Hosokawa et al. (2007), show that the OFC encodes relative rather than absolute preference and aversion for cue-associated consequences. The ability to focus on the relative value of rewards creates a common currency for the daily decisions that people make about real life issues such as the diet versus hot fudge sundae dilemma. Just as importantly, the ability to normalize rewards is a labile process that supports adaptation to an ever-changing environment. Damage to the OFC impairs reversal learning in both monkeys and humans (Iverson and Mishkin, 1970; Fellows and Farah, 2003; Rolls (1994); patients with OFC damage have impaired decision-making (Fellows, 2011).

Imaging studies have confirmed electrophysiological demonstrations that taste-evoked activity in the OFC reflects internal states such as hunger and satiety (Pritchard et al., 2008; Rolls et al., 1989). For example, Kringelbach et al. (2003) have shown that sensory specific satiety decreases fMRI activity in the OFC during consumption of the satiating liquid, but not when the subject drinks a novel liquid. Goldstone et al. (2009), using fMRI, showed that fasting increases activation in the mOFC and clOFC when the subjects view photographs of high caloric foods. Fasting also enhances subjective ratings of these foods, which is consistent with earlier psychophysical reports (Cabanac, 1971; Cameron et al., 2008). Other studies have shown that gustatory-evoked responses in the OFC are modulated by body composition (lean vs obese) and memory (Del Parigi et al., 2002, 2005). The differential emphasis of the OFC and the insula on reward-driven decision-making and sensory processing, respectively, is illustrated by a study conducted by Grabenhorst and Rolls (2008). In this fMRI experiment, activation shifted from the insula to the OFC when the subject was instructed to attend to the pleasantness of the stimulus instead of its intensity.

Taken together, the insula and the OFC clearly run the gamut from advanced and perhaps final processing of the sensory signal to taste-guided cognitive behavior. If a precise transition point for this functional shift exists, it is not currently known. At present, this functional shift in taste processing appears to lie in the immediate vicinity of the dysgranular and agranular insula. Identifying this transition point awaits a better understanding of what each area contributes to regulation of the internal milieu, risk vs reward decisions, and sensory-guided executive function.

44.4 DESCENDING GUSTATORY PROJECTIONS

Two subcortical areas have been implicated in feeding and gustation: the hypothalamus and the amygdala. Gustatory information in the insula and OFC probably reaches the hypothalamus, both directly and indirectly, through the amygdala (Aggleton et al., 1980, Fudge et al., 2005; Mufson et al., 1981; Price and Amaral, 1981).

44.4.1 Amygdala

In the amygdala, overlap of the representations of the internal and external worlds permits needs to be assessed relative to the external resources available to fulfill them (Turner et al., 1980). Because the amygdala is able to guide need-driven behavior, it is well-positioned to influence emotion, motivation, and hedonic tone (Aggleton et al., 1980; LeDoux, 1987). The ability of the amygdala to integrate visceral and exteroceptive information enables it to play an active role in initiation and guidance of feeding (Lénárd and Hahn, 1982; Nakano et al., 1986). Accordingly, normal feeding and taste perception are disrupted by amygdala lesions in monkeys (Jones and Mishkin, 1972; Klüver and Bucy, 1939) and humans (Kim and Umback, 1973; Narabayashi, 1977; Small et al., 2001b, 2001c).

Taste neurons have been recorded in the amygdala of rats (Sadacca et al., 2012), rabbits (Schwartzbaum and Morse, 1978), and macaques (Scott et al., 1993; Yan and Scott, 1996; Kadohisa et al., 2005). Although each study had a different thrust, collectively they showed that the response properties of neurons in the amygdala are similar in some respects to those in the anterior insula from which they presumably receive their input. For example, Scott et al. (1993) showed that taste neurons in the amygdala are distributed over a volume of about 76 mm^3 (vs 50 mm^3 in the insula) and they compose approximately 7% (vs 6.5%) of the cell sample. Amygdalar and insular neurons are similarly tuned across the four prototypical taste stimuli, with a mean breadth-of-tuning coefficient of 0.82 and 0.70, respectively.

The key difference between the amygdala and the insula was in their respective response characteristics. Taste neurons in the amygdala have a higher spontaneous rate than cells in the insula (8 vs 3 spikes/s.), which makes response inhibition a more credible option. In the insula, only 2% of taste responses met the criteria for inhibition (Scott and Plata-Salamán, 1999), while in amygdala, fully 17% did so (Scott et al., 1993). Of greater import, unlike the insula, taste-evoked activity in the amygdala lack the resolution to match the degree of taste quality and intensity discrimination that both monkeys and humans demonstrate. Instead of taste quality or intensity, the taste-evoked responses of amygdalar neurons reflect the hedonic value of the stimuli. For example, response magnitude in the amygdala decreased 58% as the hedonic value of the sweet satiating stimulus shifted from positive to negative (Yan and Scott, 1996). This hedonic-related decrement in taste responsivity stands in marked contrast to the NST and the insula where neurons are unaffected by satiety-induced decrements in hedonic value. The impact of satiety on taste responses in the amygdala appears to complement the finding by Sanghera et al. (1979) that the responses of visual neurons in the amygdala decrease to the sight of food that monkeys are fed to satiety.

A mechanism for such a reduction may be the sensitivity that some amygdala neurons show for vascular and interstitial glucose levels. About 17% of neurons in the amygdala respond to the iontophoretic application of glucose, and the impact is nearly always inhibitory (Karádi et al., 1998). As blood glucose levels rise with feeding, circulating glucose may suppress the responses that would previously have been evoked by taste stimuli. This effect is seen yet more clearly in the hypothalamus, as described below.

In summary, the amygdala focuses on the emotive aspects of taste and feeding rather than the precise coding of taste quality or concentration, which are likely fully processed within the insula. The amygdala, in conjunction with the OFC, the hypothalamus, and perhaps the ventral striatum and anterior cingulate gyrus, may add context to the sensory aspects of taste perception by integrating input from other senses for the appreciation of flavor, and by coordinating the animal's needs with the hedonic aspects of taste.

44.4.2 Hypothalamus

Early lesion-behavioral studies demonstrated that the hypothalamus plays a critical role in feeding (Anand and Brobeck, 1951; Hetherington and Ranson, 1940). The subsequent discovery of presumed gustatory (Mesulam and Mufson, 1982) and olfactory (Takagi, 1986) projections to the hypothalamus in the macaque have made it a prime target to investigate the relationship between the chemical senses and feeding. A role for the hypothalamus in feeding was further bolstered when Oomura et al. (1964) demonstrated that a subset of hypothalamic neurons in the rat respond to fluctuating blood glucose levels. Oomura et al. (1974) then used microiontophoretic injections of glucose into the hypothalamus to reveal two sets of neurons: those whose activity is suppressed by the presence of glucose ("glucose-sensitive" (GS) cells), and those that are unaffected ("glucose-insensitive" (GIS) cells; Aou et al., 1984, Nishino et al., 1988, Oomura et al., 1974; 1991).

Karádi et al. (1991, 1992) showed that GS and GIS cells comprise 27% and 73% of the neurons in the macaque hypothalamus, respectively. GS neurons respond primarily to chemosensory stimuli with olfactory-responsive neurons (86%) more common than taste-responsive neurons (66%); 67% of all GS cells respond to both modalities (Karádi et al., 1989, 1992). GIS cells, by comparison, were significantly less likely to respond to gustatory and olfactory stimuli. Rather, GIS neurons respond to external cues related to the acquisition of food such as a cue light that signals the availability of an operant task to obtain food, or a cue tone that indicates the task has been successfully completed. Thus, Oomura's multifaceted approach revealed two types of hypothalamic cells: GIS cells that respond to food-related, environmental, non-chemical cues, and GS cells that respond to the more intimate chemosensory signals associated with ingestion and digestion.

Burton et al. (1976), recording from individual neurons in the lateral hypothalamic area and substantia innominata, reported that 2.5% of the cells respond to taste, and 9.3% respond to the sight of food. As in other gustatory areas, taste-related neurons constitute only a small fraction of the population, and as in the amygdala, a robust spontaneous rate of approximately 10 spikes/s made response inhibition a relatively common observation. As in the OFC and the amygdala, both excitatory and inhibitory responses to the taste and sight of food are attenuated as the monkeys are fed to satiety with a particular food; responses elicited by other foods are largely unaffected. Unlike the amygdala where satiety decreases taste responses by 58% (Yan and Scott, 1996), Burton et al. (1976) reported response suppression is complete in the hypothalamus. Thus, in the macaque, the amygdala appears to be functionally intermediate between the insula, where satiety is reported to have no effect on taste-evoked activity (Yaxley et al., 1988),

and the hypothalamus, where it causes total suppression. As the authors point out, these hypothalamic data may reflect the neural mechanism underlying psychophysical reports of sensory-specific satiety in humans (Rolls et al., 1981; 1982). Whether these satiety-related neurons are GS cells is not clear.

44.5 CONCLUSION

Gustation, which emerged in coelenterates some 500 million years ago, is among the most ancient of our senses, yet it has been well conserved across phyla. Within the Mammalian Class, taste is consistent in its receptor distribution, peripheral innervation, and, with the notable exception of the parabrachial nucleus, its central pathways.

In the Primate Order, the accumulated data imply that identification of broad taste categories (i.e., qualities) is addressed at the peripheral level as separate lines of communication (labeled lines) transmit information to the central nervous system with a high degree of specificity. Second-order neurons in the nucleus of the solitary tract show greater breadth and this is preserved more centrally. Qualitative specificity is restored in the caudolateral orbitofrontal cortex, where multisensory interactions, learning, and hedonics dominate.

Electrophysiological research in nonhuman primates since the mid-1980s has addressed each level of the gustatory neuraxis. Unfortunately, coverage has been uneven, and despite the precision of the data, the picture that has emerged is so grainy that trying to understand central taste processing has been like assembling a photograph from its randomly acquired pixels. More recently, non-invasive imaging technology has provided optical images of neural activity, typically across large areas of cortex. The imaging data have brought ensemble coding to the fore, fostered a more integrated view of brain activity, and provided at least partial support for the following conclusions.

Functionally, the nucleus of the solitary tract is most likely involved in orchestrating gustatory reflexes associated with ingestion or rejection of gustatory stimuli. The parabrachial nucleus does not appear to play a role in Primate gustation, unlike its involvement in assessing taste quality and hedonic value in rodents (Yamamoto et al., 1994). The contribution of the thalamic relay in Primate taste processing has not yet been determined, but based upon studies in the rat, it is unlikely to be simply a passive relay for gustatory information bound for the cerebral cortex (e.g., Schroy et al., 2005). Primary taste cortex in the granular insula is the likely locus where quality and intensity identification is completed. Neurons here offer the closest match between macaque electrophysiological and human psychophysical reports. Beyond primary taste cortex, gustatory information participates in complex processes that involve multiple cortical and subcortical areas, other sensory modalities, and feedback from internal receptors. Accordingly, taste information contributes to multisensory evaluation of flavor, and reward comparison, as well as food selection and hedonic assessment guided by internal variables such as blood glucose levels. As a central character in this varied and complex mixture, taste helps ensure that diet selection serves the manifold and changing nutritional requirements of Primates.

REFERENCES

Aggleton, J. P., Burton, M. J., and Passingham, R. E. (1980). Cortical and subcortical afferents to the amygdala of the rhesus monkey (*Macaca mulatta*). *Brain Res.* 190: 347–368.

Anand, B. K. and Brobeck, J. R. (1951). Hypothalamic control of food intake in rats and cats. *Yale J. Biol. Med.* 24: 123–140.

Aou, S., Oomura, Y., Lénárd, L., et al. (1984). Behavioral significance of monkey hypothalamic glucose-sensitive neurons. *Brain Res.* 302: 69–74.

Aziz, Q., Furlong, P. L., Barlow, J., et al. (1995). Topographic mapping of cortical potentials evoked by distension of the human proximal and distal oesophagus. *Electroencephalogr. Clin. Neurophysiol.* 96: 219–229.

Baeg, E. H., Jackson, M. E., Jedema, H. P., Bradberry, C. W. (2009). Orbitofrontal and anterior cingulate cortex neurons selectively process cocaine-associated environmental cues in the rhesus monkey. *J. Neurosci.* 29: 11619–11627.

Barbas, H. (1993). Organization of cortical afferent input to orbitofrontal areas in the rhesus monkey. *Neurosci.* 56: 841–864.

Barbas, H. (2007). Specialized elements of orbitofrontal cortex in primates. *Ann. N.Y. Acad. Sci.* 1121: 10–32.

Baylis, L. L., Rolls, E. T., and Baylis, G. C. (1995). Afferent connections of the caudolateral orbitofrontal cortex taste area of the primate. *Neurosci.* 64: 801–812.

Bechara, A., Damasio, A. R, Damasio, H., and Anderson, S. W. (1994). Insensitivity to future consequences following damage to human prefrontal cortex. *Cognition* 50: 7–15.

Beckstead, R. M., Morse, J. R., and Norgren, R. (1980). The nucleus of the solitary tract in the monkey: projections to the thalamus and brain stem nuclei. *J. Comp. Neurol.* 190: 259–282.

Beckstead, R. M., and Norgren, R. (1979). Central distribution of the trigeminal, facial glossopharyngeal and vagus nerves in the monkey. *J. Comp. Neurol.* 184: 455–472.

Behrens, M. and Meyerhof, W. (2008). Bitter taste sensitivity in humans and chimpanzees. Online library DOI: 10.1002/9780470015902.a0020778.

Beidler, L. M. (1995). Saliva. Its functions and disorders. In *Handbook of olfaction and gustation*. Doty, R.L. (Ed.) New York: Marcel Dekker, Inc. pp. 503–520.

Benjamin, R. M. (1963). Some thalamic and cortical mechanisms of taste. In *Olfaction and Taste*, Zotterman Y. (Ed.) New York: Pergamum. vol. 1: pp. 309–329.

Benjamin, R. M., and Burton, H. (1968). Projection of taste nerve afferents to anterior opercular-insular cortex in squirrel monkey (*Saimiri sciureus*). *Brain Res.* 7: 221–231.

Benjamin, R. M., Emmers, R., and Blomquist, A. J. (1968). Projection of tongue nerve afferents to somatic sensory area I in squirrel monkey (*Saimiri sciureus*). *Brain Res.* 7: 208–220.

Bradley, R. M. (1971). Tongue topography. In: *Handbook of sensory physiology*. Beidler, L. M. (Ed.) IV, Berlin: Springer-Verlag pp. 1–30.

Bradley, R. M., Stedman, H. M., and Mistretta, C. M. (1985). A quantitative study of lingual taste buds and papillae in the aging rhesus monkey. In: *Behavior and pathology of aging in rhesus monkeys*. Davis, R.T. and Leathers, C.W. (Eds.) New York: Alan R. Liss, Inc. pp. 187–199.

Burton, M. J., Rolls, E. T., and Mora, F. (1976). Effects of hunger on the responses of neurons in the lateral hypothalamus to the sight and taste of food. *Exp. Neurol.* 51: 668–677.

Burton, H., and Benjamin, R. M. (1971). Handbook of sensory physiology, chemical senses. In *Taste*. Beidler, L. M. (Ed). New York: Springer-Verlag, pt 2, vol. IV, pp. 148–164.

Cabanac, M. (1971). Physiological role of pleasure. *Science* 173: 1103–1107.

Cameron, J. D., Goldfield, G. S., Cyr, M. J. and Doucet, E. (2008). The effects of prolonged caloric restriction leading to weight-loss on food hedonics and reinforcement. *Physiol. Behav.* 94: 474–480.

Carmichael, S. T., and Price, J. L. (1995). Sensory and premotor connections of the orbital and medial prefrontal cortex in macaque monkeys. *J. Comp. Neurol.* 363: 642–664.

Carmichael, S. T., and Price, J. L. (1996). Connectional networks within the orbital and medial prefrontal cortex of macaque monkeys. *J. Comp. Neurol.* 371: 179–207.

Chandrashekar, J., Mueller, K. L., Hoon, M.A., et al. (2000). T2Rs function as bitter taste receptors. *Cell* 100: 703–711.

Chang, F-C. T., and Scott, T. R. (1984). Conditioned taste aversions modify neural responses in the rat nucleus tractus solitarius. *J. Neurosci.* 4: 1850–1862.

Chen J.-Y., and DiLorenzo, P. M. (2008). Responses to binary taste mixtures in the nucleus of the solitary tract: Neural coding with firing rate. *J. Neurophysiol.* 99: 2144–2157.

Cho, Y.K., and Li, C.-S. (2008). Gustatory neural circuitry in the hamster brain stem. *J. Neurophysiol.* 100: 1007–1019.

Cleaton-Jones, P. (1971). Histological observations in the soft palate of the albino rat. *J. Anat.* 110: 39–47.

Collings, V.B. (1974). Human taste response as a function of locus of stimulation on the tongue and soft palate. *Percept. Psychophys.* 16: 169–174.

Craig, A. D. (Bud). (2009). How do you feel–now? The anterior insula and human awareness. *Nat. Rev. Neurosci.* 10: 59–70.

Critchley, H. D., and Rolls, E. T. (1996). Hunger and satiety modify the responses of olfactory and visual neurons in the primate orbitofrontal cortex. *J. Neurophysiol.* 75: 1673–1686.

Danilova, V., Danilova, Y., Roberts, T., et al. (2002). The sense of taste in a New World monkey, the common marmoset: recordings from the chorda tympani and glossopharyngeal nerves. *J. Neurophysiol.* 88: 579–594.

Danilova, V. and Hellekant, G. (2004). Sense of taste in a New World monkey, the common marmoset. II. Link between behavior and nerve activity. *J. Neurophysiol.* 92: 1067–1076.

de Araujo, I. E. T., Kringelbach, M. L., Rolls, E. T., and McGlone, F. (2003). Human cortical responses to water in the mouth, and the effects of thirst. *J. Neurophysiol.* 90: 1865–1876.

de Araujo, I. E. T. and Rolls, E. T. (2004). Representation in the human brain of food texture and oral fat. *J. Neurosci.* 24: 3086–3093.

de Araujo, I. E. and Simon, S. A. (2009). The gustatory cortex and multisensory integration. *Int. J. Obes.* 33: 534–543.

Del Parigi, A., Chen, K., Salbe, A.D., et al. (2005). Sensory experience of food and obesity: a positron emission tomography study of the brain regions affected by tasting a liquid meal after a prolonged fast. *NeuroImage* 24: 436–443.

References

Del Parigi, A., Gautier, J.-F., Chen, K., et al. (2002). Neuroimaging and obesity. mapping the brain responses to hunger and satiation in humans using positron emission tomography. *Ann. N.Y. Acad. Sci.* 967: 389–397.

DiLorenzo, P. M., and Schwartzbaum, J. S. (1982). Coding of gustatory information in the pontine parabrachial nuclei of the rabbit: Magnitude of neural response. *Brain Res.* 251: 229–244.

Emura, S., Hayakawa, D., Chen, H., and Shoumura, S. (2002). Morphology of the dorsal lingual papillae in the Japanese macaque. *Anat. Histol. Embryol.* 31: 313–316.

Erickson (1967). Neural coding of taste quality. In *The Chemical Senses and Nutrition*. M. Kare and O. Maller (Eds.) Baltimore: Johns Hopkins Press, pp. 313–327.

Evrard, H. C., Forro, T., and Logothetis, N. K.(2012). Von Economo neurons in the anterior insula of the macaque monkey. *Neuron* 74: 482–489.

Fellows, L. K. (2011). Orbitofrontal contributions to value-based decision making: evidence from humans with frontal lobe damage. *Ann. N.Y. Acad.* 1239: 51–58.

Fellows, L. K., and Farah, M. J. (2003). Ventromedial frontal cortex mediates affective shifting in humans: evidence from a reversal learning paradigm. *Brain* 126 (pt 8): 1830–1837.

Fiol, M.E., Leppik, I.E., Mireles, R., and Maxwell, R. (1988). Ictus emeticus and the insular cortex. *Epilepsy Res.* 2: 127–131.

Fitzgerald, T. H. B., Seymour, B., and Dolan, R. J. (2009). The role of human orbitofrontal cortex in value comparison for incommensurable objects. *J. Neurosci.* 29: 8388–8395.

Frank, R. A., Mize, S. J. S., Kennedy, L. M., et al. (1992). The effect of *Gymnema syvestre* extracts on the sweetness of eight sweeteners. *Chem. Senses* 17: 461–479.

Fudge J. L., Breitbart M. A., Danish M., and Pannoni V. (2005). Insular and gustatory inputs to the caudal ventral striatum in primates. *J Comp Neurol.* 490: 101–118.

Ganchrow, J., and Fisher, G. L. (1968). Two behavioral measures of the squirrel monkey's (*Saimiri scieureus*) taste for four concentrations of five sugars. *Psychol. Rep.* 22: 503–511.

Gautier, J.-F., Del Parigi, A., Chen, K., et al. (2001). Effect of satiation on brain activity in obese and lean women. *Obesity Res.* 9: 676–684.

Giza, B. K., and Scott, T. R. (1983). Blood glucose levels selectively affect gustatory sensitivity in the rat nucleus tractus solitarius. *Physiol. Behav.* 31: 643–650.

Glaser, D., Tinti, J.M., and Nofre, C. (1995). Evolution of the sweetness receptor in primates. I. Why does alitame taste sweet in all prosimians and simians, and aspartame only in Old World simians? *Chem. Senses* 20: 573–584.

Goldstone, A. P., de Hernandez, C. G., Beaver, J. D., et al., (2009). Fasting biases brain reward systems towards high-calorie foods. *Eur. J. Neurosci.* 30: 1625–1635.

Gordon, G., Kitchell, R., Ström, L., and Zotterman, Y. (1959). The response pattern of taste fibers in the chorda tympani of the monkey. *Acta Physiol. Scand.* 46: 119–132.

Grabenhorst, F., and Rolls, E. T. (2008). Selective attention to affective value alters how the brain processes taste stimuli. *Eur. J. Neurosci.* 27: 723–729.

Guest, S., Grabenhorst, F., Essick, G., et al. (2007). Human cortical representation of oral temperature. *Physiol. Behav.* 92: 975–984.

Halsell, C. B. (1992). Organization of parabrachial nucleus efferents to the thalamus and amygdala in the golden hamster. *J. Comp. Neurol.* 317: 57–78.

Halsell, C. B., and Frank, M. E. (1991). Mapping study of the parabrachial taste-responsive area for the anterior tongue in the golden hamster. *J. Comp. Neurol.* 306: 708–722.

Halsell, C. B., and Travers, S. P. (1997). Anterior and posterior oral cavity responsive neurons are differentially distributed among parabrachial subnuclei in rat. *J Neurophysiol.* 78: 920–938.

Hamilton, R., and Norgren, R. (1984). Central projections of gustatory nerves in the rat. *J. Comp. Neurol.* 222: 560–577.

Hanamori, T., and Smith, D. V. (1986). Central projections of the hamster superior laryngeal nerve. *Brain Res. Bull.* 16: 271–279.

Harding, R., Johnson, P., and McClelland, M. E. (1978). Liquid-sensitive laryngeal receptors in the developing sheep, cat, and monkey. *J. Physiol.* 277: 409–422.

Hellekant, G., Glaser, D., Brouwer, J., and van der Wel, H. (1981). Gustatory responses in three prosimian and two simian primate species to six sweeteners and miraculin and their phylogenetic implications. *Chem. Senses* 6: 165–173.

Hellekant, G., DuBois, G., Geissmann, T., et al. (1990). Taste responses of the chorda tympani proper nerve in the white-handed gibbon (*Hylobates lar*). In *ISOT X: Proceedings of the Tenth International Symposium on Olfaction and Taste*. K.B. Døving (Ed).Oslo: CGS Press, pp. 115–131.

Hellekant, G., Ninomiya, Y., DuBois, G. E., et al. (1996). Taste in chimpanzees. I. The summated response to sweeteners and the effect of gymnemic acid. *Physiol. Behav.* 60: 469–479.

Hellekant, G., Danilova, V., and Ninomiya, Y. (1997). Primate sense of taste: Behavioral and single chorda tympani and glossopharyngeal nerve fiber recordings in the rhesus monkey, *Macaca mulatta. J. Neurophysiol.* 77: 978–993.

Hellekant, G., Ninomiya, Y., and Danilova, V. (1998). Taste in chimpanzees. III. Labeled-line coding in sweet taste. *Physiol. Behav.* 65: 191–200.

Hetherington, A. W. and Ranson, S. W. (1940). Hypothalamic lesions and adiposity in the rat. *Anat. Rec.* 78: 149–172.

Hosokawa, T., Kato, K., Inoue, M., and Mikami, A. (2007). Neurons in the macaque orbitofrontal cortex code relative preference of both rewarding and aversive outcomes. *Neurosci. Res.* 57: 434–445.

Ifuku, H., Hirata, S.I., Nakamura, T., and Ogawa, H. (2003). Neuronal activities in the monkey primary and higher-order gustatory cortices during a taste discrimination delayed GO/NOGO task and after reversal. *Neurosci. Res.* 2: 161–175.

Ishii, H., Ohara, S., Tobler, P.N., et al. (2012). Inactivating anterior insular cortex reduces risk taking. *J. Neurosci.* 32: 16031–16039.

Ito, S., and Ogawa, H. (1994). Neural activities in the frontal-opercular cortex of macaque monkeys during tasting and mastication. *Jap. J. Physiol.* 44: 141–156.

Iversen, S. D., and Mishkin, M. (1970). Perseverative interference in monkeys following selective lesions of the inferior prefrontal convexity. *Exp. Brain. Res.* 11: 376–386.

Izquierdo, A., Suda, R. K., and Murray, E. A. (2004). Bilateral orbital prefrontal cortex lesions in rhesus monkeys disrupt choices guided by both reward value and reward contingency. *J. Neurosci.* 24: 7540–7548.

Jabbi, M., Bastiaansen, J., and Keysers, C. (2008). A common anterior insula representation of disgust observation, experience and imagination shows divergent functional connectivity pathways. *Pub. Lib. Sci.—ONE (PLoS ONE)*, 3: e2939.

Jacobs, K. M., Mark, G. P., and Scott, T. R. (1988). Taste responses in the nucleus tractus solitarius of sodium-deprived rats. *J. Physiol.* 406: 393–410.

Jones, B., and Mishkin, M. (1972). Limbic lesions and the problems of stimulus-reinforcement associations. *Exp. Neurol.* 36: 362–377.

Jowett, A., and Shrestha, R. (1998). Mucosa and taste buds of the human epiglottis. *J. Anat.* 193: 617–618.

Kadohisa, M., Rolls, E. T., and Verhagen, J. V. (2005). Neuronal representations of stimuli in the mouth: the primate insular taste cortex, orbitofrontal cortex and amygdala. *Chem. Senses* 30: 401–419.

Kanagasuntheram, R., Wong, W. C., and Chan, H. L. (1969). Some observations on the innervation of the human nasopharynx. *J. Anat.* 194: 361–376.

Kappelman, J. (1992). The age of the Fayum primates as determined by paleomagnetic reversal stratigraphy. *J. Hum. Evol.* 22: 495–503.

Karádi, Z., Oomura, Y., Nishino, H., and Aou, S. (1989). Olfactory coding in the monkey lateral hypothalamus: behavioral and neurochemical properties of odor-responding neurons. *Physiol. Behav.* 45: 1249–1257.

Karádi, Z., Oomura, Y., Nishino, H., et al. (1991). Complex attributes of lateral hypothalamic neurons in the regulation of feeding of alert rhesus monkeys. *Brain Res. Bull.*, 25: 933–939.

Karádi, Z., Oomura, Y., Nisho, H., et al. (1992). Responses of lateral hypothalamic glucose-sensitive and glucose-insensitive neurons to chemical stimuli in behaving rhesus monkeys. *J. Neurophysiol.* 67: 389–400.

Karádi, Z., Scott, T. R., Oomura, Y., et al. (1998). Complex functional attributes of amygdaloid gustatory neurons in the rhesus monkey. *Ann. N.Y. Acad. Sci.* 855: 488–492.

Kennedy, L. M. (1989). Gymnemic acids: specificity and competitive inhibition. *Chem. Senses* 14: 853–858.

Kim, Y. K., and Umback, W. (1973). Combined stereotaxic lesions for treatment of behavioral disorders and severe pain. In *Surgical Approaches to Psychiatry*. Laitinen, L.V. and Livingston, K.E. (Eds.) Baltimore: Springer Verlag, pp. 182–188.

Klein, P. B., and Schroeder, H. E. (1979). Epithelial differentiation and taste buds in the soft palate of the monkey, *Macaca irus. Cell Tissue Res.* 196: 181–188.

Klein, P. B., Weilemann, W. A., and Schroeder, H. E. (1979). Structure of the soft palate and composition of the oral mucous membrane in monkeys. *Anat. Embryol.* 156: 197–215.

Klüver, H. and Bucy, P. C. (1939). Preliminary analysis of functions of the temporal lobes in monkeys. *Arch. Neurol. Psychiatry* 42: 979–1000.

Knutson, B., Westdorp, A., Kaiser, E., and Hommer, D. (2000). FMRI visualization of brain activity during a monetary incentive delay task. *NeuroImage* 12: 20–27.

Kobayashi, M., Sasabe, T., Takeda, M., et al. (2002). Functional anatomy of chemical senses in the alert monkey revealed by positron emission tomography. *Eur. J. Neurosci.* 16: 975–980.

Komisaruk, B. R., Mosier, K. M., Liu, W.-C., et al. (2002). Functional localization of brainstem and cervical spinal cord nuclei in human with fMRI. *AJNR Am. J. Neuroradiol.* 23: 609–617.

Kringelbach, M. L., O'Doherty, J., Rolls, E. T., and Andrews, C. (2003). Activation of the human orbitofrontal cortex to a liquid food stimulus is correlated with its subjective pleasantness. *Cereb. Cortex* 13: 1064–1071.

Kuznicki, J., and Ashbaught, N. (1979). Taste quality differences within the sweet and salty taste categories. *Sens. Proces.* 3: 157–182.

Lara, A. H., Kennerley, S. W., and Wallis, J. D. (2009). Encoding of gustatory memory by orbitofrontal neurons. *J. Neurosci.* 29: 765–774.

LeDoux, J. E. (1987). Emotion. In *Handbook of Physiology. The nervous system*. Bethesda: American Physiological Society pp. 419–459.

Leela, K., Kanagasuntheram, R., and Ahmed, M. M. (1971). Innervation of the nasopharynx in *Macaca fascicularis. J. Anat.* 110: 49–56.

Lénárd, L., and Hahn, Z. (1982). Amygdalar noradrenergic and dopaminergic mechanisms in the regulation of hunger and thirst-motivated behavior. *Brain Res.* 233: 115–132.

Lenz, F. A., Gracely, R. H., Zirh, T. A., et al. (1997). Human thalamic nucleus mediating taste and multiple other sensations related to ingestive behavior. *J. Neurophysiol.* 77: 3406–3409.

Levy, L. M., Henkin, R. I., Lin, C. S., et al. (1999). Taste memory induces brain activation as revealed by functional MRI. *J. Comput Assist. Tomog.* 23: 499–505.

Li, X., Bachmanov, A. A., Maehashi, K., et al. (2011). Sweet taste receptor gene variation and aspartame taste in primates and other species. *Chem. Senses* 36: 453–475.

Loewy, A. D., and Burton, H. (1978). Nuclei of the solitary tract: Efferent projections to the lower brain stem and spinal cord of the cat. *J. Comp. Neurol.* 181: 421–450.

McManus, L. J., Dawes, P. J. D., and Stringer, M. D. (2011). Clinical anatomy of the chorda tympani: a systematic review. *J. Larygol. Otol.* 125: 1101–1108.

Mesulam, M.-M., and Mufson, E. J. (1982). Insula of the old world monkey. III: Efferent cortical output and comments on function. *J. Comp. Neurol.* 212: 38–52.

Mesulam, M.-M., and Mufson, E. J. (1985). The insula of Reil in man and monkey. *In Cerebral Cortex*, Peters, A. and Jones, E. G. (Eds.) 4, pp. 179–226.

Miller, I. J. (1995). Anatomy of the peripheral taste system. In: *Handbook of Olfaction and Gustation*. Doty, R.L. Ed). New York: Marcel Dekker, Inc. pp. 521–547.

Miller, I. J., Jr., and Bartoshuk, L. M. (1991). Taste perception, taste bud distribution, and spatial relationships. In *Smell and Taste in Health and Disease*. Getchell, T.V., Doty, R.L., Bartoshuk, L.M., and Snow, J.B. Jr. (Eds.) New York: Raven Press, pp. 205–233.

Miller, I. J., and Spangler, K. M. (1982). Taste bud distribution and innervation on the palate of the rat. *Chem. Senses* 7: 99–108.

Miyaoka, Y., Shingai, T., Yoshihiro, T., et al. (1998). Responses of neurons in the parabrachial region of the rat to electrical stimulation of the superior laryngeal nerve and chemical stimulation of the larynx. *Brain Res. Bull.* 45: 95–100.

Morecraft, R. J., Geula, C., and Mesulam, M.-M. (1992). Cytoarchitecture and neural afferents of orbitofrontal cortex in the brain of the monkey. *J. Comp. Neurol.* 323: 341–358.

Mu, L., and Sanders, I. (2000). Sensory nerve supply of the human oro- and laryngopharynx: a preliminary study. *Anat. Rec.* 258: 406–420.

Mufson, E. J., Mesulam M-M., and Pandya D. N. (1981). Insular interconnections with the amygdala in the rhesus monkey. *Neurosci.* 6: 1231–1248.

Mufson, E. J., and Mesulam, M.-M. (1982). Insula of the Old World monkey. II: Afferent cortical input and comments on the claustrum. *J. Comp. Neurol.* 212: 23–37.

Murray, E. A., and Izquierdo, A. (2007). Orbitofrontal cortex and amygdala contributions to affect and action in primates. *Ann. N.Y. Acad. Sci.* 1121: 273–296.

Nakano, Y., Oomura, Y., Lénárd, L., et al. (1986). Feeding-related activity of glucose- and morphine-sensitive neurons in the monkey amygdala. *Brain Res.*, 399: 167–172.

Narabayashi, H. (1977). Stereotaxic amygdalectomy for epileptic hyperactivity—long range results in children. In *Topics in Child Neurology*. Blaw, M.E., Rapin, I., and Kinsbourne, M. (Eds.) New York: Spectrum, pp. 319–331.

Nelson, G., Hoon, M. A., Chandrashekar, J., et al. (2001). Mammalian sweet taste receptors. *Cell* 106: 381–390.

Nishino, H., Oomura, Y., Karádi, Z., et al. (1988). Internal and external information processing by lateral hypothalamic glucose-sensitive and insensitive neurons during bar press feeding in the monkey. *Brain Res. Bull.* 20: 839–845.

References

Nitschke, J. B., Dixon, G. E., Sarinopoulos, I., et al. (2006). Altering expectancy dampens neural response to aversive taste in primary taste cortex. *Nat. Neurosci.* 9: 435–442.

Nomura, S., Mizuno, N., Itoh, K., et al. (1979). Localization of parabrachial nucleus neurons projecting to the thalamus or the amygdala in the cat using horseradish peroxidase. *Exp. Neurol.* 64: 375–385.

Norgren, R. (1978). Projections from the nucleus of the solitary tract in the rat. *Neurosci.* 3: 207–218.

Norgren, R. (1984). Central neural mechanisms of taste. *In Handbook of Physiology–The Nervous System III, Sensory Processes.* Darien-Smith, I. (Ed.) (Section Eds., Brookhart, J., and Mountcastle, V. B.) Washington, DC: American Physiological Society. Pp. 1087–1128.

Norgren, R., and Leonard, C. M. (1973). Ascending central gustatory pathways. *J. Comp. Neurol.* 150: 217–237.

O'Doherty J., Kringelbach, M. L., Hornak, J., et al. (2001). Abstract reward and punishment representations in the human orbitofrontal cortex. *Nat. Neurosci.* 4: 95–102.

Ogawa, H., Ito, S. I., and Nomura, T. (1989). Oral cavity representation at the frontal operculum of macaque monkeys. *Neurosci. Res.* 6: 283–298.

Ogawa, H., Ifuku, H., Nakamura, T., and Hirata, S. (2005). Possible changes in information from the primary to higher-order gustatory cortices studied by recording neural activities during a taste discrimination GO/NOGO task in monkeys. *Chem. Senses* 30: i78–i79.

O'Neill, M. and Schultz, W. (2010). Coding of reward risk distinct from reward value by orbitofrontal neurons. *Neuron* 68: 789–800.

Oomura, Y., Ooyama, H., Sugimori, M., et al. (1964). Glucose inhibition of the glucose-sensitive neurone in the rat lateral hypothalamus. *Nature* 247: 254–256.

Oomura, Y., Ooyama, H., Sugimori, M., et al. (1974). Glucose inhibition of the glucose-sensitive neurons in the lateral hypothalamus. *Nature* 247: 284–286.

Oomura, Y., Nishino, H., Karadi, Z., et al. (1991). Taste and olfactory modulation of feeding related neurons in behaving monkey. *Physiol. Behav.* 49: 943–950.

Oppenheimer, S. M., Gelb, A., Girvin, J. P., and Hachinski, V. C. (1992). Cardiovascular effects of human insular cortical stimulation. *Neurol.* 42: 1927–1932.

Padoa-Schioppa, C. (2007). Orbitofrontal cortex and the computation of economic value. *Ann. N.Y. Acad. Sci.* 1121: 232–253.

Padoa-Schioppa, C., and Assad, J. A. (2006). Neurons in orbitofrontal cortex encode economic value. *Nature* 441: 223–226.

Pfaffmann, C. (1941). Gustatory afferent impulses. *J. Comp. Physiol. Psych.* 17: 243–258.

Pfaffmann, C. (1974). Specificity of the sweet receptors of the squirrel monkey. *Chem. Senses Flavor* 1: 61–67.

Phillips, M. L., Young, A. W., Senior, C., et al. (1997). A specific neural substrate for perceiving facial expressions of disgust. *Nature* 389: 495–498.

Plata-Salamán, C. R., Scott, T. R. and Smith, V. L. (1992). Gustatory neural coding in the monkey cortex: L-amino acids. *J. Neurophysiol.* 67: 1552–1561.

Plata-Salamán, C.R., Scott, T. R., and Smith-Swintosky, V. L. (1995). Gustatory neural coding in the monkey cortex: Acid stimuli. *J. Neurophysiol.* 74: 556–564.

Plata-Salamán, C. R., Scott, T. R., and Smith-Swintosky, V. L. (1993). Gustatory neural coding in the monkey cortex: The quality of sweetness. *J. Neurophysiol.* 60: 482–493.

Plata-Salamán, C. R., Smith-Swintosky, V. L., and Scott, T. R. (1996). Gustatory neural coding in the monkey cortex: Mixtures. *J. Neurophysiol.* 75: 2369–2379.

Price J. L, and Amaral D. G. (1981). An autoradiographic study of the projections of the central nucleus of the monkey amygdala. *J. Neurosci.* 1: 1242–1259.

Pritchard, T. C. (2012). Gustatory system. In *The Human Nervous System.* Paxinos, G. and Mai, J. (Eds.) 3rd edition, New York: Elsevier, pp. 1187–1218.

Pritchard, T. C., Edwards, E. M., Smith, C. A., et al. (2005). Gustatory neural responses in the medial orbitofrontal cortex of the Old World monkey. *J. Neurosci.* 25: 6047–6056.

Pritchard, T. C., Hamilton, R., Morse, J., and Norgren, R. (1986). Projections from thalamic gustatory and lingual areas in the monkey, *Macaca fascicularis. J. Comp. Neurol.* 244: 213–228.

Pritchard, T. C., Hamilton, R. B., and Norgren, R. (1989). Neural coding of gustatory information in the thalamus of *Macaca mulatta. J. Neurophysiol.* 61: 1–14.

Pritchard, T. C., Hamilton, R. B., and Norgren, R. (2000). Projections of the parabrachial nucleus in the Old World monkey. *Exp. Neurol.* 165: 101–117.

Pritchard, T. C., Macaluso, D. A., and Eslinger, P. J. (1999). Taste perception in patients with insular cortex lesions. *Behav. Neurosci.* 113: 663–671.

Pritchard, T. C., Nedderman, E. N., Edwards, E. M., et al. (2008). Satiety-responsive neurons in medial orbitofrontal cortex of the macaque. *Behav. Neurosci.* 122: 174–182.

Raichle, M. E. (1991). Memory mechanisms in the processing of words and word-like symbols. Exploring brain functional anatomy with positron tomography. *Ciba Foundat. Symp.* 163: 198–204.

Rauch, S. L., Savage, C. R., Alpert, N. M., et al. (1995). A positron emission tomographic study of simple phobic symptom provocation. *Arch. Gen. Psychiat.* 52: 20–28.

Reilly, S., and Pritchard, T. C. (1995). The influence of thalamic lesions on taste preference behavior of Old World monkeys. *Exp. Neurol.* 135: 56–66.

Rempel-Clower, N, L. (2007). Role of orbitofrontal cortex connections in emotion. *Ann. N.Y. Acad. Sci.* 1121: 72–86.

Rhoton, A. L. Jr. (1968). Afferent connections of the facial nerve. *J. Comp. Neurol.* 133: 89–100.

Rolls, B. J., Rolls, E. T., Rowe, E. A., and Sweeney, K. I. (1981). Sensory specific satiety in man. *Physiol. Behav.* 27: 137–142.

Rolls, B. J., Rowe, E. A., and Rolls, E. T. (1982). How sensory properties of foods affect human feeding behavior. *Physiol. Behav.* 29: 409–417.

Rolls, E. T. (2004). Convergence of sensory systems in the orbitofrontal cortex of primates and brain design for emotion. *Anat. Rec. A Discov. Mol. Cell Evol. Biol.* 281: 1212–1225.

Rolls, E. T. (2005). Taste and related systems in primates including humans. *Chem. Senses* 30: i76–i77.

Rolls, E. T., and Baylis, L. L. (1994). Gustatory, olfactory, and visual convergence within the primate orbitofrontal cortex. *J. Neurosci.* 14: 5437–5452.

Rolls, E. T., Scott, T. R., Sienkiewicz, Z. J., and Yaxley, S. (1988). The responsiveness of neurons in the frontal opercular gustatory cortex of the macaque monkey is independent of hunger. *J. Physiol.* 397: 1–12.

Rolls, E. T., Sienkiewicz, Z. J., and Yaxley, S. (1989). Hunger modulates the responses to gustatory stimuli of single neurons in the caudolateral orbitofrontal cortex of the macaque monkey. *Eur. J. Neurosci.* 1: 53–60.

Rolls, E. T., Yaxley, S., and Sienkiewicz, Z. J. (1990). Gustatory responses of single neurons in the caudolateral orbitofrontal cortex of the macaque monkey. *J. Neurophysiol.* 64: 1055–1066.

Roper, S. N., Lévesque, M. F., Sutherling, W. W., and Engel, J. Jr. (1993). Surgical treatment of partial epilepsy arising from the insular cortex. *J. Neurosurg.* 79: 226–229.

Rudebeck, P. H., and Murray, E. A. (2011). Balkanizing the primate orbitofrontal cortex: distinct subregions for comparing and contrasting values. *Ann N.Y. Acad Sci.* 1239: 1–13.

Sadacca, B. F., Rothwax, J. T., and Katz, D. B. (2012). Sodium concentration coding gives way to evaluative coding in cortex and amygdala. *J. Neurosci.* 32: 9999–10011.

Sanghera, M. K., Rolls, E. T., and Roper-Hall, A. (1979). Visual responses of neurons in the dorsolateral amygdala of the alert monkey. *Exp. Neurol.* 63: 610–626.

Sato, M., Ogawa, H., and Yamashita, S. (1975). Response properties of macaque monkey chorda tympani fibers. *J. Gen. Physiol.* 66: 781–810.

Satoda, T., Takahashi, O., Murakami, C., et al. (1996). The sites of origin and termination of afferent and efferent components in the lingual and pharyngeal branches of the glossopharyngeal nerve in the Japanese monkey (*Macaca fuscata*). *Neurosci. Res.* 24: 385–392.

Sbarbati, A., Crescimanno, C., and Osculati, F. (1999). The anatomy and functional role of the circumvallate papilla/von Ebner gland complex. *Med. Hypotheses* 53: 40–44.

Schroy, P. L., Wheeler, R. L., Davidson, C., et al. (2005). Role of gustatory thalamus in anticipation and comparison of rewards over time in rats. *Am J Physiol Regul. Integr. Comp. Physiol.* 288: R966–R980.

Schultz, W., Tremblay, L., and Hollerman, J. R. (1998). Reward prediction in primate basal ganglia and frontal cortex. *Neuropharmacol.* 37: 421–429.

Scott, T. R., and Erickson, R. P. (1971). Synaptic processing of taste quality information in thalamus of the rat. *J. Neurophysiol.* 34: 868–884.

Scott, T. R., Giza, B. K., and Yan, J (1999). Gustatory neural coding in the cortex of the alert cynomolgus macaque: The quality of bitterness. *J. Neurophysiol.* 81: 60–71.

Scott, T. R., and Giza, B. K. (2000). Issues of gustatory neural coding: Where they stand today. *Physiol. Behav.* 69: 265–276.

Scott, T. R., Karádi, Z., Oomura, Y., et al. (1993). Gustatory neural coding in the amygdala of the alert macaque monkey. *J. Neurophysiol.* 69: 1810–1820.

Scott, T. R., Plata-Salamán, C. R., Smith, V. L., and Giza, B. K. (1991). Gustatory neural coding in the monkey cortex: Stimulus intensity. *J. Neurophysiol.* 65: 76–86.

Scott, T. R., and Plata-Salamán, C. R. (1999). Taste in the monkey cortex. *Physiol. Behav.* 67: 489–511.

Scott, T. R., Plata-Salamán, C. R., and Smith-Swintosky, V. L. (1994). Gustatory coding in the monkey cortex: The quality of saltiness. *J. Neurophysiol.* 71: 1692–1701.

Scott, T. R., Yaxley, S., Sienkiewicz, Z. J., and Rolls, E. T. (1986a). Gustatory responses in the nucleus tractus solitarius of the alert cynomolgus monkey. *J. Neurophysiol.* 55: 182–200.

Scott, T. R., Yaxley, S., Sienkiewicz, Z. J., and Rolls, E. T. (1986b). Gustatory responses in the frontal operculum of the alert cynomolgus monkey. *J. Neurophysiol.* 56: 876–890.

Schwartzbaum, J. S., and DiLorenzo, P. M. (1982). Gustatory functions of the nucleus tractus solitarius in the rabbit. *Brain Res. Bull.* 81: 285–292.

Schwartzbaum, J. S., and Morse, J. R. (1978). Taste responsivity of amygdaloid units in behaving rabbit: A methodological report. *Brain Res. Bull.* 3: 131–141.

Sergovia, C., Hutchinson, I., Laing, D. G., and Jinks, A. L. (2002). A quantitative study of fungiform papillae and taste pore density in adults and children. *Dev. Brain Res.* 138: 135–146.

Simmons, W. K., Martin, A., and Barsalou, L. W. (2005). Pictures of appetizing foods activate gustatory cortices for taste and reward. *Cereb. Cort.* 15: 1602–1608.

Small, D. M., Zatorre, R. J., Dagher, A., et al. (2001a). Changes in brain activity related to eating chocolate: from pleasure to aversion. *Brain*, 124: 1720–1733.

Small, D. M., Zatorre, R. J., and Jones-Gorman, M. (2001b). Changes in taste intensity perception following anterior temporal lobe removal in humans. *Chem. Senses* 26: 425–432.

Small, D. M., Zatorre, R. J., and Jones-Gotman, M. (2001c). Increased intensity perception of aversive taste following right anteromedial temporal lobe removal in humans. *Brain*, 124: 1566–1575.

Smith, D. V., and Travers, J. B. (1979). A metric for the breadth of tuning of gustatory neurons. *Chem. Senses* 4: 215–229.

Smith-Swintosky, V. L., Plata-Salamán, C. R., and Scott, T. R. (1991). Gustatory neural coding in the monkey cortex: Stimulus quality. *J. Neurophysiol.* 66: 1156–1165.

Stedman, H. M., Bradley, R. M., Mistretta, C. M., and Bradley, B. E. (1980). Chemosensitive responses from the cat epiglottis. *Chem. Senses* 5: 223–245.

Stice, E., Spoor, S., Bohon, C., et al. (2008). Relation of reward from food intake and anticipated food intake to obesity: A functional magnetic imaging study. *J. Abnorm. Psychol.* 117: 924–935.

Sudakov, K., MacLean, P. D., Reeves, A., and Marino, R. (1971). Unit study of exteroceptive inputs to claustrocortex in awake, sitting, squirrel monkey. *Brain Res.* 28: 19–34.

Summerfield, C., and Tsetsos, K. (2012). Building bridges between perceptual and economic decision-making: neural and computational mechanisms. *Front. Neurosci.* 6: 70. doi:10.3389/fnins.2012.00070.

Sweazey, R. D., Edwards, C. A., and Kapp, B. M. (1994). Fine structure of taste buds located on the lamb epiglottis. *Anat. Rec.* 238: 517–527.

Taillibert, S., Bazin, B., and Pierrot-Deseilligny, C. J. (1998). *Neurol. Neurosurg. Psychiat.* 64: 691–692.

Takagi, S. F. (1986). Studies on the olfactory nervous system of the old world monkey. *Prog. Neurobiol.*, 27: 195–250.

Tataranni, P. A., Gautier, J.-F., Chen, K., et al. (1999). Neuroanatomical correlates of hunger and satiation in humans using positron emission tomography. *Proc. Nat. Acad. Sci.* 96: 4569–4574.

Thorpe, S. J., Rolls, E. T., and Maddison, S. (1983). The orbitofrontal cortex: Neuronal activity in the behaving monkey. *Exp. Brain Res.* 49: 93–115.

Tokita, K., Inoue, T., and Boughter, Jr., J. D. (2009). Afferent connections of the parabrachial nucleus in C57BL/6J mice. *Neurosci.* 161: 475–488.

Topolovec, J. C., Gati, J. S., Menon, R. S., et al. (2004). Human cardiovascular and gustatory brainstem sites observed by functional magnetic resonance imaging. *J. Comp. Neurol.* 471: 446–461.

Toure, G., and Vacher, C. (2005). The epiglottis, a glosso-laryngeal structure: an anatomic study of its innervation. *Morpholgie.* 89: 117–120.

Travers, J. B. (1988). Efferent projections from the anterior nucleus of the solitary tract of the hamster. *Brain Res.* 457: 1–11.

Travers, S. P., and Norgren, R. (1995). Organization of orosensory responses in the nucleus of the solitary tract of the rat. *J. Neurophysiol.* 73: 2144–2162.

Tremblay, L., and Schultz, W. (1999). Relative reward preference in primate orbitofrontal cortex. *Nature* 398: 704–708.

Tremblay, L., and Schultz, W. (2000a). Reward-related neuronal activity during go-no-go task performance in primate orbitofrontal cortex. *J. Neurophysiol.* 83: 1864–1876.

Tremblay, L., and Schultz, W. (2000b). Modifications of reward expectation-related neuronal activity during learning in primate orbitofrontal cortex. *J. Neurophysiol.* 83: 1877–1885.

References

Turner, B. H., Mishkin, M., and Knapp, M. (1980). Organization of the amygdalopetal projections from modality-specific cortical association areas in the monkey. *J. Comp. Neurol.* 191: 515–543.

Veldhuizen, M. G., Bendor, G., Constable, R. T., and Small, D. M. (2007). Trying to detect taste in a tasteless solution: Modulation of early gustatory cortex by attention to taste. *Chem. Senses* 32: 569–581.

Verhagen, J. V., Giza, B. K., and Scott, T. R. (2003). Responses to taste stimulation in the ventroposteromedial nucleus of the thalamus in rats. *J. Neurophysiol.* 89: 265–275.

Verhagen, J. V., Kadohisa, M., and Rolls, E. T. (2004). The primate insular/opercular taste cortex: Neuronal representation of the viscosity, fat, texture, grittiness, temperature, and taste of foods. *J. Neurophysiol.* 92: 1685–1699.

Wakita, M., Kobayakawa, T., Saito, S., et al. (2009). Handedness: dependent asymmetrical location of the human primary gustatory area, area G. *Neuroreport* 20: 450–455.

Wang, Y., Danilova, V., Cragin, T., et al. (2009). The sweet taste quality is linked to a cluster of taste fibers in primates: lactisole diminishes preference and responses to sweet in S (sweet best) chorda tympani fibers of *M. fascicularis* monkey. *BMC Physiology* 9: 1 doi:10.1186/1472-6793-9-1

Westlund, K. N., and Craig, A. D. (1996). Association of spinal lamina I projections with brainstem catecholamine neurons in the monkey. *Exp. Brain Res.* 110: 151–162.

Weusten, B. L. A. M., Fransson, H., Wieneke, G. H., and Smout, A. J. P. M. (1994). Multichannel recording of cerebral potentials evoked by esophageal balloon distension in humans. *Digest. Dis. Sci.* 39: 2074–2083.

Wicker, B., Keysers, C., Plailly, J., et al., (2003). Both of us disgusted in my insula: the common neural basis of seeing and feeling disgust. *Neuron* 40: 655–664.

Yamamoto, T., Shimura, T., Sakai, N., and Ozaki, N. (1994). Representation of hedonics and quality of taste stimuli in the parabrachial nucleus of the rat. *Physiol. Behav.* 56: 1197–1202.

Yan, J., and Scott, T. R. (1996). The effect of satiety on responses of gustatory neurons in the amygdala of alert cynomolgus macaques. *Brain Res.* 740: 193–200.

Yaxley, S., Rolls, E. T., Sienkiewicz, Z. J., and Scott, T. R. (1985). Satiety does not affect gustatory activity in the nucleus of the solitary tract of the alert monkey. *Brain Res.*, 347: 85–93.

Yaxley, S., Rolls, E. T., and Sienkiewicz, Z. (1988). The responsiveness of neurones in the insular gustatory cortex of the macaque monkey is independent of hunger. *Physiol. Behav.* 42: 223–229.

Yoshida, Y., Tanaka, Y., Hirano, M., and Nakashima, T. (2000). Sensory innervation of the pharynx and larynx. *Am J Med.* 108: 51S–61S.

Part 10

Central Integration
of Olfaction, Taste,
and the Other Senses

Chapter 45

Chemosensory Integration and the Perception of Flavor

JOHN PRESCOTT and RICHARD STEVENSON

45.1 INTRODUCTION

The use of the word flavour/flavor in English is thought to date from the late 14th or early 15th century. Originally it had a predominantly olfactory meaning, in that it referred to a fragrance or aroma (The New Shorter Oxford English Dictionary, 1993). In contrast, a meaning that is more consistent with our current view but which (in modern terms) undervalues the role of olfaction – "a quality perceived by taste (aided by smell)" – emerged somewhat later, in the late 17th century. Rozin (1982) notes that not all languages carry this latter, multisensory meaning, in that they do not distinguish well between primary tastes qualities and the overall flavor (taste plus smell) of a food in the mouth. In other words, not all languages (not even all European languages) acknowledge the role of olfaction in flavor. This may reflect widespread ignorance until relatively recently of the importance of both taste and smell to flavor. It is also consistent with the everyday, non-technical use of the terms "taste" and "flavor" as interchangeable.

The then existing general confusion between taste qualities and overall flavor was remarked on by a well-known English physician of the early 19th century, William Prout (Brock, 1967). In an essay dating from 1810, Prout explicitly recognized the olfactory component of flavor, noting that "substances in general have the strongest flavor that are volatizable or partly soluble in air as well as water," and further that "it is probable that the major part of our sense of flavor comes from the stimulation of the olfactory nerve by odorous air which is ejected when, during the act of swallowing food, the epiglottis

closes the trachea." Prout's definition of flavor as "that sensation which is produced when substances under certain circumstances are introduced into the mouth, the nostrils being at the same time open" was in fact influential (despite its lack of detail), in that it was adopted in physiology texts of the day.

Writing around the same time from a less scientific perspective, the French gastronomic pioneer, Brillat-Savarin was "tempted to believe that smell and taste are in fact but a single sense, whose laboratory is the mouth and whose chimney is the nose" (Brillat-Savarin, 1825). He thus not only explicitly valued the contribution of both smell and taste in flavor perception, but also anticipated J. J. Gibson's ecological approach to perception (Gibson, 1966). Gibson argued that the primary purpose of perception is to seek out objects in our environment, particularly those that are biologically important. As such, the physiological origin of sensory information is less important than that the information can be used in object identification. The key to effective perception is that sensory information is interpreted as qualities that belong to the object itself. Within this context, flavor can be seen as a functionally distinct sense that is cognitively "constructed" from the integration of distinct physiologically defined sensory systems (primarily olfaction and gustation) in order to identify and respond to objects that are important to our survival, namely foods.

The recognition that our experiences of food flavors reflect combinations of taste, olfactory, and other sensory (e.g., somatosensory) information is therefore not recent. Nonetheless, how these inputs are combined, and the

---

Handbook of Olfaction and Gustation, Third Edition. Edited by Richard L. Doty.
© 2015 Richard L. Doty. Published 2015 by John Wiley & Sons, Inc.

1007

1008   **Chapter 45   Chemosensory Integration and the Perception of Flavor**

resultant interactions between the senses involved in flavor perception, have recently enjoyed increased interest. This has been born, to some extent, out of a realization that in our everyday food experiences we respond, perceptually and hedonically, not to discrete tastes, odors, and tactile sensations, but to flavors constructed from a synthesis of these sensory signals (Prescott, 2004b). In addition, recent neuroimaging studies (see Chapter 35) have provided evidence that flavor is not merely a shorthand description for an aggregation of different senses involved in food perception but rather involves brain structures working as a distinct, functional neural network.

Understanding flavor perception has also benefitted from the emergence of multimodal sensory perception as a distinct field of study (see for example Driver and Spence, 2000). In contrast to earlier emphases on perceptual discrimination – essentially how we resolve William James's "blooming, buzzing confusion" (James, 1890) – multimodal perceptual studies address how information from physiologically distinct sensory systems is combined to provide more accurate and perceptually salient information. This has led to questions regarding binding, which initially referred to the combination of features especially in visual perception (Treisman, 1998), but is now posed to help understand how information is combined to facilitate a single percept from multiple senses (sometimes referred to as the *unity assumption*) (see Section 45.4 below).

45.2 SENSORY INTERACTIONS WITHIN FLAVORS

Cross-modal sensory integration is frequently inferred from the influence of one modality on responses to another. Commonly, this is an enhanced (sometimes supra-additive) response to information from one sensory system due to concurrent input from another modality. For example, in a noisy environment, speech comprehension is improved if we see the speaker's lip movements (Sumby and Polack, 1954). There has similarly been considerable interest in determining the cause of sensory interactions with flavor, as some have significant practical implications. For example, certain interactions can allow a reduction in the quantity of sugar, fat or salt needed to make a product acceptable (e.g., Nasri et al., 2011) (see Section 45.5.2 below). Two basic causes of interactions have been established. One involves events at the periphery, including, for example, chemical reactions between food ingredients, bulk effects of food that alter volatile or tastant release, and receptor competition between the chemicals in food. The other involves events in the brain, and it is this type of interaction that forms the main focus of this section. As the interaction literature is extensive, the interested reader is directed to

Delwiche (2004), Keast et al., (2004), or Stevenson (2009), for detailed treatments.

45.2.1 Interactions with a Peripheral Cause

Interactions between oral irritation and taste, taste and temperature, taste and touch, irritation and temperature, odor and temperature, odor and oral irritation, and oral irritation and touch, all seem to have peripheral causes – with the caveat that the evidence is often suggestive rather than definitive. In this section, these interactions and their probable cause are summarized.

45.2.1.1 Irritation and Taste. Capsaicin, the pungent principal of the chili pepper, can suppress sweetness and bitterness, but not saltiness and sourness (Simons et al., 2002; Prescott et al., 1993). This may occur because capsaicin affects the protein-based taste receptors that detect sweetness and bitterness, but not the ion based channels that detect saltiness and sourness. Sweetness has been shown to be reduced by the mechanical oral irritation produced by the calcium oxalate crystals present in some fruits (Walker and Prescott, 2003). The effect is probably secondary to increased viscosity associated with increased concentration of these crystals in solution. Certain sweet tastants (e.g., sucrose) can also suppress the irritation induced by capsaicin (Prescott et al., 1993) and by carbonic acid (the agent that generates the pungency associated with carbonation; (Yau and McDaniel, 1991). In this case the effect may arise via sucrose's capacity to promote salivation, which then acts as a dilutent. Conversely, irritation can be shown to be increased in the presence of acids (Walker and Prescott, 2003) and salt (Prescott et al., 1993), consistent with the irritant properties of both of these tastants (Gilmore and Green, 1993).

45.2.1.2 Irritation and Touch. Viscous foods appear to reduce capsaicin's burn intensity (e.g., Nasrawi and Pangborn, 1989). However, this may owe more to the capacity of fats to capture capsaicin, preventing receptor binding, than any specific viscosity related effect (Lawless et al., 2000). Irritation may also be reduced by concurrent oral mechanical stimulation although the locus of this effect has yet to be established (Green, 1990).

45.2.1.3 Irritation and Temperature. The common terminology applied to sensations driven by chemical irritants – burning and cooling – and that from thermal stimuli suggests some commonality in processing. Higher temperature (37°C vs 21°C) capsaicin solutions have been shown to produce more burn, although this may have been

45.2 Sensory Interactions within Flavors 1009

through burn suppression in the lower temperature solutions (Prescott et al., 1993). Albin et al. (2008) examined irritant/temperature commonality by applying an irritant to one side of the tongue tip and then placing the tongue tip on a hot or cold thermode. Participants were then asked to judge which side of the tongue was hotter. Whenever one side of the tongue was treated with capsaicin, mustard oil or cinnamaldehyde, and placed on the hot thermode, the treated side was always selected as "hotter." In contrast, when one side of the tongue was treated with menthol and placed against a cold thermode, the treated side was judged as cooler. These effects are probably peripheral in origin, as identical outcomes were obtained after pretreating the tongue with the irritant, waiting until all irritant sensation had dissipated, and then conducting the thermode test.

45.2.1.4 Irritation and Olfaction.
When an irritant with no olfactory component is sniffed at the same time as a relatively pure olfactory stimulus, the irritant exerts a far more potent suppressive effect on odor perception than odor perception exerts on irritant perception (Murphy and Cain, 1980). This effect is almost identical in magnitude to that obtained when the odor and the irritant are applied to different nostrils, indicating that for orthonasal stimuli there is a significant central component to irritant suppression of orthonasal olfactory perception (Cain and Murphy, 1980). However, when an irritant is placed in the mouth, alongside an odor (i.e. odor-irritant mixture), there does not seem to be any centrally mediated effect paralleling that seen for orthonasal olfaction (Prescott and Stevenson, 1995).

45.2.1.5 Temperature and Taste.
Higher temperatures have been consistently found to enhance the sweetness of sucrose – but not other tastants – with cooling having the reverse effect (Bartoshuk et al., 1982). This seems to occur as a consequence of the vehicle either warming or cooling the tongue (Green and Frankmann, 1987). Indeed, simply warming the tongue can generate illusory sensations of sweetness, with cooling generating sensations of sourness and saltiness (Cruz and Green, 2000). These illusory taste effects driven by tongue warming or cooling may account for the changes in sweetness perception observed with higher or lower temperature. They also suggest that thermal stimuli can directly activate certain classes of taste receptor.

More recently Green and Nachtigal (2012) observed a difficulty in perceiving sweet tastes – for example, while licking a lollipop – when the tongue remained outside of the mouth. Once the tongue was retracted into the mouth, however, the sweetness was obvious, raising the possibility that the higher internal mouth temperature was responsible for this effect. By comparing the rate of adaption to sweetness at different temperatures, these investigators were able to show that warming the tongue outside of the mouth by dipping it into a solution with the same temperature as inside the mouth (37°C) produced the same effect on sweetness as withdrawing the tongue into the mouth. An intriguing question is why the same effects did not occur when the study was carried out using the bitter taste of quinine. Why there are particular interactions between sweetness and temperature and not other tastes and temperature is unknown. Such findings are consistent with those of earlier studies that have been unable to consistently show that tastes at different temperatures vary in intensity.

45.2.1.6 Temperature and Olfaction.
Food temperature exerts a relatively straightforward effect on retronasal olfaction. Greater temperature initially increases volatile chemical release, while cooler temperatures initially reduce it, leading to greater intensity of the same odor when presented in a warmer vehicle than in a cooler vehicle (Pangborn et al., 1978).

45.2.1.7 Taste and Touch.
The principal focus in studying tactile/taste interactions has been on the impact of viscosity on taste. Initial studies found that increasing viscosity reduced the intensity of all tastants (Moscowitz and Arabie, 1970). Later studies revealed a more complex picture, showing that the particular type of agent used to generate viscosity, its concentration, and the type of tastant used all effect whether taste intensity suppression occurs (e.g., Christensen, 1980; Cook et al., 2002; Walker and Prescott, 2000). This suggests that the effects of viscosity on taste perception are mediated by the individual chemical properties of the viscous agent, with each varying in the capacity to impede the movement of different tastants to the taste receptor. The impact of taste on viscosity perception has not yielded consistent results (see Theunissen and Kroeze, 1995).

Green and Nachtigal (2012), in a study of how tastes and touch sensations interact, compared the effects of passive to active tasting of sweet, salty, sour, and umami (mono-sodium glutamate, or MSG taste) tastes. In the former condition, tastants were applied with a swab to different parts of tongue, but the tongue remained immobile. In the active condition, the participants were asked to touch the tongue to the roof of the mouth and also swallow. Hence, this condition more closely resembled normal eating. Active tasting had a strong impact on taste intensity, but only for the taste of MSG. Of all the tastes, *umami* tastes seem to have a "mouthfilling" quality. Moreover, as these researchers point out, active food manipulation including chewing is needed to break down the food's physical structure, releasing the glutamate and other amino acids responsible for *umami* tastes. This process may also be behind the fact that, in this study, MSG intensity in

1010　　　**Chapter 45　Chemosensory Integration and the Perception of Flavor**

either condition was highest at the rear of the tongue. No such "tongue geography" effects were seen for the other tastes.

45.2.2 Interactions with a Central Cause

Three types of centrally based interaction are described in this section. Two of these, somatosensory-olfactory interactions involving texture and interactions between taste and smell, occur within the mouth, while others such as the impact of external visual and auditory cues on flavor perception do not.

45.2.2.1 Olfaction and Touch.
It is well established that textural variables such as viscosity and hardness can affect the intensity of odorants (e.g., Lundgren et al., 1986). A centrally-based cause for such interactions has recently been established. Weel et al., (2002) measured volatile release *in vivo* for a set of odorized gels that varied in hardness. While increasing hardness progressively reduced odor intensity, the volatile release profile was the same irrespective of hardness. A similarly compelling finding in regard to viscosity was made by Bult et al., (2007). In this study participants were nasally catheterized. One catheter then delivered odorants to the anterior nares and the other to the posterior nares. Odorants were delivered during the time period in which participants sampled by mouth either plain or thickened milk. Odor intensity was significantly reduced in the presence of thickened milk irrespective of whether the odor was presented to the anterior or posterior nares. Stevenson and Mahmut (2011a) observed the same effect when participants were asked to sniff an odor while holding a viscous fluid in their mouth. The viscous fluid was associated with reduced odor intensity judgments relative to holding plain water in the mouth. While these interactions are robust and centrally mediated, their function and causes remain obscure.

A smaller body of work has suggested that odors consistently associated with creamy or viscous textures can act to enhance participants' judgments of these textural properties (e.g., Weenen et al., 2005). These effects are also centrally mediated, as the addition of the target odorant to the viscous mixture does not alter the physical properties of the stimulus.

45.2.2.2 Effects of Vision and Audition on Flavor Perception.
A food's appearance gives many clues to its flavor, including its texture, freshness, and fattiness, but only food color has been explored in any depth. Color does not seem to exert any consistent effect on taste or odor perception in the mouth (e.g., Frank et al., 1989), in contrast to its well-established effect on odor intensity

judgments in orthonasal perception (e.g., Zampini et al., 2007). Color does, however, exert a significant influence on qualitative aspects of flavor perception, biasing the perceived identity of a flavor in a manner consistent with the color (e.g., DuBose et al., 1980). This effect has been most dramatically demonstrated by Morot et al. (2001), who found that changing the color of white wine to red using an odorless and tasteless colorant, acted to shift the flavor profile from that of a white wine to that of a red. Finally, one interesting study examined the effect of visual texture on actual texture perception (De Wijk et al., 2004). Here, participants could see a custard, which they *believed* they were sampling via a straw, when in fact the straw entered a separate lower compartment containing the target custard. Visual texture in the upper compartment significantly affected participants' judgments of the texture of the target custard in the lower compartment, even though this remained unchanged throughout the experiment. All of these effects would seem to be driven by learning, with visual cues presumably generating expectancies, which in turn act to bias participants perception and/or judgment of the target flavor.

Auditory cues can also influence flavor perception. Vickers and Bourne (1976) originally suggested that a significant part of crispness perception might be driven by auditory cues. Zampini and Spence (2004) directly manipulated this by having participants receive false auditory feedback via headphones, while sampling the same potato chips (crisps). High frequency amplification enhanced judgments of crispness, while attenuation of overall loudness, and the high frequency component in particular, reduced crispness judgments. While finding no parallel effect for judgments of carbonation (Zampini and Spence, 2005), their data indicate that auditory cues can affect textural perception in certain situations.

45.2.2.3 Olfaction and Taste.
Of all forms of flavor interaction, that between taste and smell is the most extensively studied. Pioneering experiments by Murphy, Cain, and Bartoshuk (1977) and Murphy and Cain (1980), in addition to several follow-up studies by other authors, indicated no interaction between these two modalities. However, as Lawless and Schledgel (1984) illustrated, this depends upon what is measured. Lawless and Schledgel (1984) had participants sample a series of odor-taste mixtures, judging on some trials the intensity of the taste, on others the intensity of the olfactory component, and on others still the overall intensity of the mixtures. This rating approach produced results entirely consistent with the preceding studies, demonstrating independence between taste and smell (Murphy et al., 1977; Murphy and Cain, 1980). However, when participants were asked to discriminate between all possible odor-taste mixtures

used in the experiment (e.g., weak sucrose and strong citral vs weak sucrose and weak citral), there was substantial evidence of taste–smell interactions, with some mixtures being much harder to discriminate between than others. Lawless and Schledgel (1984) concluded that the attentional strategy which participants adopted for each of these tasks – an analytical focus on one aspect of the stimulus for intensity judgments, and global comparison for discrimination – were responsible for these apparently discrepant outcomes. This work was instrumental in creating a particular way of thinking about how the task adopted by the experimenter influences the way in which participants evaluate, and perhaps experience, odor-taste mixtures (e.g., Frank et al., 1993) (see also Section 45.4.2 below).

Two types of taste–smell interaction have been documented. The first can be termed taste-induced odor enhancement. The early literature observed several examples (e.g., von Sydow et al., 1974) where the presence of a particular taste would increase intensity judgments of an olfactory attribute. For example, in a sucrose-strawberry mixture, the presence of sucrose would increase the judged intensity of the strawberry odor. The most interesting demonstration of this effect was made by Davidson, Linforth, Hollowood, and Taylor (1999), using chewing gum flavored with sucrose and menthone. First, they established the release rate of the sucrose and menthone by physically measuring these properties *in vivo*. Sucrose was lost rapidly upon chewing, while release of the menthone remained fairly constant over the 5 minutes of recording. Second, they had an experienced sensory evaluation panel chew the same type of gum in the same manner, while making judgments of its mint flavor. The fascinating finding was that variation in mint flavor judgment was *wholly* explained by variation in sucrose concentration. So even though the menthone concentration remained largely constant, reductions in mint flavor tracked reductions in sucrose (see Figure 45.1). More recent findings have confirmed this same general effect, with vanilla flavor judged far stronger in the presence of a stronger sweet taste than a weaker one (Green et al., 2012). These findings suggest that tastes, under appropriate circumstances, can act to enhance or reduce the magnitude of purely olfactory qualities.

The second type of effect, which has received more attention, concerns the effect of odorants on taste perception. Contemporary interest in this form of interaction was triggered by Frank and Byram's (1988) observation that strawberry odor, but not peanut butter odor, could increase judgments of sweetness when added to a sweet base. Subsequently, odor-induced taste enhancement has been obtained in many different settings using diverse odorants and tastants (e.g., Stevenson et al., 1999; Nasri et al., 2011; Prescott, 1999).

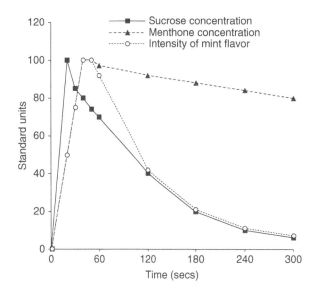

Figure 45.1 Changes in physical release of sucrose and menthone in the mouth during gum chewing, alongside ratings of the intensity of the mint flavor across time (adapted from Davidson, Linforth, Hollowood & Taylor, *J. Agric. Food Chem.*, 47, 4336–4340, 1999).

Sufficient information about odor-induced taste enhancement effects, and taste-induced odor enhancement effects, is available to provide a fairly comprehensive account of their cause. One explanation, which can now be dismissed, is that they emerge from some form of physiochemical interaction (e.g., the tastant "salts out" volatiles in the mixture, thereby increasing its intensity). While salting can occur, it typically does not at tastant concentrations used in most psychophysical experiments (e.g., von Sydow et al., 1974). Another possibility, pertinent to odor-induced taste enhancement, is the ability of certain odorants to stimulate taste receptors. Again, although some odorants have this property, careful testing that has evaluated taste with the odorant in water and the nose pinched (to eliminate retronasal olfaction) finds this not to be the case (Stevenson et al., 2000).

Both taste-induced odor enhancement and odor-induced taste enhancement seem to depend upon some form of psychological congruency between the taste and smell (e.g., Frank and Byram, 1988). Although, at least theoretically, such congruency could result from the odor and taste being hedonically similar (Schifferstein and Verlegh, 1996), this does not seem to be the case as some pleasant odors (e.g., peanut butter) will not enhance pleasant tastes (e.g., sucrose). A more promising possibility concerns prior experience with a particular taste–smell combination. Stevenson, Prescott and Boakes (1999) observed that sniffed odor sweetness, that is, the sweetness that becomes attributed to an odor as a result of its prior history of co-occurrence with sweet tastes, was the single best predictor of the degree to which that odor would

1012   **Chapter 45 Chemosensory Integration and the Perception of Flavor**

enhance the sweetness of sucrose in the mouth. They also reported examples of sweetness suppression by odors that were very low in smelled sweetness, as well as the suppression of sourness by a sweet-smelling odor. Clearly, taste–smell interactions depend to a large degree upon learning (Stevenson et al., 1995).

Enhancement can also be observed under experimental conditions that do not rely on judgments of intensity. Dalton et al., (2000) measured detection thresholds for the sweet almond smelling odorant benzaldehyde against a background of saccharin, and in later experiments against monosodium glutamate (MSG). Participants were able to detect the saccharin-benzaldehyde compound at a significantly lower concentration than either benzaldehyde or saccharin alone. This was not the case for the saccharin-MSG compound, presumably because it was incongruent. One concern with these types of findings is probability summation. Detecting the presence of one stimulus when only one is present is harder than detecting the presence of one stimulus when two are present (Treisman, 1998). When estimates of probability summation are made, subthreshold integration effects between taste and smell disappear (Delwiche and Heffelfinger, 2005). However, a more recent study using reaction times to detect weak tastes, odors and taste-odor combinations indicates that reaction times are quicker even taking into account probability summation (Veldhuizen et al., 2010). This suggests that odor-taste interactions can occur even under conditions that do not favour more cognitive response based strategies.

45.3 FLAVOR LEARNING

The mouth affords plenty of opportunity for peripheral interactions to arise within and between food ingredients and receptors, and several different types of such interactions have been identified. Centrally mediated interactions have also been observed, most notably between taste and smell, and smell, viscosity and hardness, as well as between vision, audition, and flavor. While some of these centrally mediated effects may have pre-attentive (that is, automatic) causes, most seem to rely upon learning.

The ability to identify nutritious food is an essential prerequisite for survival. This requires learning the many and varied signals for safe and nutritious food, as well as those for energetically impoverished and poisonous food. Gibson (1966) suggested that the sensory components with flavors are "functionally united when anatomically separated" (p.137) in order to increase the chances of successful dietary selection. There is considerable evidence that learning underlies this process in humans.

Human flavor learning can be divided into two basic categories, depending upon whether its outcome involves

a perceptual or an affective change (note this is principally an organizational scheme and it does not imply any commonality of mechanism). Three types of flavor learning result in perceptual change: (1) a form of learning in which an odor associated with a previously encountered flavour comes to recover the original flavor (termed redintegration flavor learning); (2) flavor perceptual learning, in which experience with a particular class of flavor leads to an improved ability to describe, categorise, identify and discriminate between members of that class; and (3) flavor recognition learning, in which implicit perceptual/semantic templates of previously consumed foods form the point of comparison for new foods of the same general type (e.g., comparing a shop bought apple pie to that of your Grandma's apple pie). A change of affect towards a flavor, or one of its components, occurs in several forms of flavor learning, including: flavor-flavor, flavor-calorie, flavor-drug and flavor-aversion learning, observational learning, and learned satiety. Each of these different forms of human flavor learning is reviewed below.

45.3.1 Learning that Results in Perceptual Change

45.3.1.1 Redintegrative Flavor Learning. The observation that certain odors are reported as smelling sweet, and that such odors are typically those that occur in sweet tasting foods, prompted speculation that this might be caused by learning (Frank and Byram, 1988). Stevenson et al., (1995) explored this possibility by having participants swill and spit samples composed of one unfamiliar odor in sucrose solution, and another in citric acid solution. On a later sniffing test, it was found that odor-sucrose pairings led to that odor being judged as sweeter smelling, while odor-citric acid pairings led to the odor being judged as sourer smelling. This effect has now been obtained under several different conditions, which varied number of pairings, presentation format, and selection of odors and tastes (e.g., Prescott et al., 2004; Stevenson et al., 1998; Yeomans and Mobini, 2006). The effect has also been observed in rats (Gautam and Verhagen, 2010).

Redintegrative flavor learning has unusual properties relative to other forms of human associative learning (c.f. Shanks, 2010). It occurs rapidly (Prescott et al., 2004), it is resistant to interference (Stevenson et al., 2000), and acquisition seems to require only minimal conscious awareness of the odor-taste contingencies (Stevenson et al., 1998). Redintegrative flavor learning can also occur independently of the affective change that typically accompanies pairing an odor with a pleasant or unpleasant taste (see Flavor-flavor learning below). While odor-sucrose pairings, for example, generate *both* affective and sensory learning, only affective learning is sensitive to participants' internal state (i.e., hungry/full) and to the extent to which the taste is hedonically valenced (i.e., like/dislike; Yeomans et al., 2006; Yeomans and Mobini, 2006; Yeomans et al., 2009).

The perceptual consequences of this type of learning – the acquisition of a taste-like quality – seems to share many properties in common with taste sensations generated by gustatory stimuli. Perceptual similarity has been suggested by priming effects, in which sweet smelling odors facilitate identification of sweet tastes, and sour smells facilitate the identification of sour tastes (White and Prescott, 2007). The same conclusion emerges from other studies where participants smell several sweet odors and taste several sweet tastes, followed by a surprise frequency test, revealing that participants conflate the number of "real" sweet tastes with odor induced "sweet" tastes (Stevenson and Oaten, 2011a). Tastes and odor-induced tastes also seem to share some neurophysiological correlates. Damage to brain areas involved in taste perception is associated with impaired odor induced taste perception, suggesting some commonality of neural processing (Stevenson et al., 2008). This is also suggested by a study that demonstrated increased pain tolerance after sniffing sweet smells, but not pleasant non-sweet smells, an effect akin to that observed with sweet tastes (Prescott and Wilkie, 2007).

Whether somatosensory properties present within a flavor compound can be acquired by odors in a manner akin to taste has also been examined. Sundqvist et al. (2006) paired odors with milks of varying fat and sugar content. Odors paired with high fat milks were later found to enhance the perceived fat content of the same test milk relative to odors paired with low fat milks. Unfortunately, this experiment could not determine whether this effect was driven by somatosensory, gustatory or olfactory cues to fat content. A more recent study (Stevenson and Mahmut, 2010) confirmed that odors can acquire purely tactile properties. Odors were either paired with water, with the thickener carboxymethyl cellulose (CMC) or with CMC and sucrose. The odor paired with the sweetened CMC increased viscosity judgments of a partially thickened solution of CMC to a greater extent than odors that had been paired with water or CMC alone. This suggests that tastes may need to be present to support odor-viscosity learning.

One explanatory model for these effects proposes that each experience of an odor always invokes a search of memory for prior encounters with that odor. If, in the initial experience of the odor, it was paired with a taste, a cross-modal configural stimulus – that is, a flavor – is encoded in memory. Subsequently sniffing the odor alone will evoke the most similar odor memory – the flavor – that will include both the odor and the taste component. Thus, for example, sniffing caramel odor activates memorial representations of caramel flavors, which includes a sweet taste component.

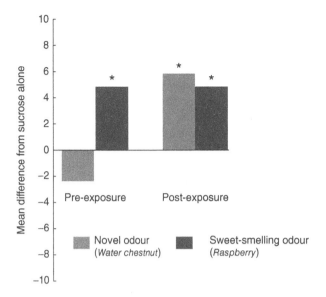

Figure 45.2 Mean ratings of the tasted sweetness of the odor/sucrose mixture relative to sucrose alone, prior to and following repeated pairing of the odor and sucrose, for an already sweet-smelling odor (*raspberry*) and a novel odor (*waterchestnut*). *indicates significantly different (5%) from sucrose alone. Reprinted with modification from *Food Quality and Preference*, 10, J. Prescott, Flavour as a psychological construct: implications for perceiving and measuring the sensory qualities of foods, pp. 349–356, Copyright (1999), with permission from Elsevier.

This results either in perceptions of smelled taste properties such as sweetness or, in the case of a mixture, a perceptual combination of the memorial odor representation with the physically present taste in solution (Stevenson and Boakes, 2004; Stevenson et al., 1998; see Figure 45.2).

45.3.1.2 Flavor Perceptual Learning.
Flavor perceptual learning has mainly been studied by exploring the difference between experts – for wine, beer and food – and novices. Two general points are worth making before looking in more detail at the findings from this literature. First, *perceptual* expertise in the flavor domain seems to be passively acquired (Melcher and Schooler, 1996; Valentin et al., 2007) and it is unclear if training programs can improve upon this. Second, because much of the research focus has been on comparing experts to novices, any observed performance difference could be attributable to selection bias rather than expertise. For example, anosmic people are unlikely to gravitate to wine tasting, so it may be that flavor expertise reflects a bias towards those with naturally better sensory ability or sensitivity. Although this point cannot be definitively settled, it would seem unlikely. One study of chefs (Hirsch, 1990), for example, revealed little difference in chemosensory ability relative to controls. Others have found at best

1014   **Chapter 45   Chemosensory Integration and the Perception of Flavor**

only small differences in general chemosensory ability favouring the expert (Tempere et al., 2011).

Flavor expertise seems to involve two discrete components, which then interact extensively during deployment of the special skill (Frost and Noble, 2002). The first component is perceptual expertise, which as noted above seems to be acquired passively via experience (e.g., Melcher and Schooler, 1996; Valentin et al., 2007). Melcher and Schooler (1996) found that regular wine drinkers and wine tasters were similar in discriminative ability when the test involved only discrimination, as did Valentin et al. (2007) in a comparison of expert and regular beer drinkers. Such findings point to the importance of choosing an appropriate control group when examining perceptual expertise. For example, Solomon (1990) found that expert wine tasters were significantly better at discriminating a set of white wines than participants who had rarely drunk wine before. Such differences tend to disappear when untrained but regular consumers are used instead of naïve controls, suggesting the likely importance of passive exposure. Unfortunately, we do not currently know what exposure parameters are optimal – a significant issue when considering training a sensory panel.

The second component of expertise is the extensive semantic knowledge that is accrued about the product (e.g., Solomon, 1990). Accruing such knowledge has several important benefits. One of these is the ability to verbally communicate about the product in a reliable manner. This has been demonstrated in several wine tasting studies, whereby expert wine tasters have been asked to match descriptions of wine they made on an earlier occasion, to the same wines on a later occasion. They are better at this than novices, but not dramatically so (Lawless, 1995). A more precise matching between language and sensory experience can also improve the reliability of product profiling, such that ratings made at one point in time are more closely aligned to those made at a later time, relative to those from untrained participants (Roberts and Vickers, 1994; Moskowitz, 1996).

A further benefit of semantic knowledge is the ability to organize information into meaningful categories. This is most evident in wine expertise, where knowledge of varietal features aids categorization of wines and hence their identification (Hughson and Boakes, 2002). The benefits of this type of knowledge can be seen in a study reported by Ballester et al. (2008). Here wine tasters and regular wine drinkers were initially asked to judge the similarity of several wines drawn from two varieties. Both groups produced nearly identical sets of similarity judgments, arguably reflecting common underlying perceptual differentiation. However, the participants were then asked to judge the typicality of each wine, that is the degree to which each wine reflects the features typical of its respective variety. Not surprisingly there was little agreement

amongst the novices, but substantial agreement among the experts, a consequence of their well-developed knowledge of varietal features.

45.3.1.3   *Flavor Recognition Learning.* One of the least explored aspects of flavor learning and memory is our ability to retain generalized "flavor templates" for a particular food or beverage that we have previously experienced (Barker, 1982; Mojet and Koster, 2005; Koster et al., 2004). These templates seem to be important in identifying deviations from expectation and do not appear to be directly experienced. They serve an important function in assisting the detection of spoiled, poisonous or otherwise atypical food. Studies suggest that the degree of deviation between a food and its template may need to be moderately large before participants actually notice the difference (e.g., Zellner et al., 2004). However, obvious deviations such as when a sample *looks* like strawberry ice cream but turns out to be savoury salmon mousse are readily detected – and rejected (Yeomans et al., 2008a). At least some of this ability must rely upon quite abstract flavor representations, as patients with semantic dementia demonstrate an impaired ability to identity both congruent and incongruent flavors, while showing a spared ability to discriminate one flavor from another (Piwnica-Worms et al., 2010).

45.3.2   Learning that Results in Affective Change

Likes and dislikes naturally arise from the integrated perception of flavor, since we are responding to substances that we have learned to recognize as foods and that are therefore biologically, culturally, and socially valued. Initial ("gut") responses to foods are almost always hedonic and this naturally precedes accepting or rejecting the food. As with perceptual changes in odors, hedonic properties of flavors arise from associative learning.

45.3.2.1   *Flavor-Flavor Learning.* Pairing an odor with a taste can also manifest as a change in liking for that odor, one that mirrors the participants liking or disliking for the taste. In the first demonstration of this effect, Zellner et al. (1983) had participants sample one tea flavor with sucrose and another flavor in water. Across several experiments they obtained evidence that the sweet-paired tea came to be liked more than the tea experienced in water. Similar changes in odor liking based upon pairing with pleasant or unpleasant tastes have now been obtained in several different paradigms (e.g., Baeyens et al., 1990; Dickinson and Brown, 2007; Wardle et al., 2007). As noted above, these affective changes are independent of alterations in an odor's perceptual taste-like properties (Yeomans and Mobini, 2006). As with changes in an odor's

45.4 Factors Influencing Binding of Sensory Signals in Flavors

perceptual properties, it was originally suggested that odor affective changes occurred independently of awareness (Baeyens et al., 1990). Recent work has cast doubt on this assertion and suggests that as with many other forms of human learning, awareness of the contingencies is necessary for conditioning to occur (Wardle et al., 2007).

45.3.2.2 Flavor-Calorie Learning. Through repeated pairing, flavors become associated with post-ingestional consequences, including those related to the intake of nutrients (e.g., glutamate) or energy in the form of fats or carbohydrates (Mobini et al., 2007; Kern et al, 1993; Prescott, 2004a; Yeomans et al., 2008b). Flavor-calorie learning can occur independently of flavor-flavor learning. Capaldi and Privitera (2007) had participants consume either a high-fat bitter tasting food with a distinct odor or a low-fat bitter tasting food with a different distinct odor. The odor used in training was then tested in a non-bitter test food. The group which had experienced the odor paired with the high-fat food liked the test food flavor more than the group that had experienced their odor in the low-fat food. Clearly, the only difference between these groups was the energy associated with the odor, indicating the capacity of the foods energy content to change liking. Although flavor-calorie learning is likely to be a significant contributor to food preferences, and perhaps to the amount of each food that we eat, it has not been well explored in humans, so many basic questions remain unanswered (e.g., does the type of nutrient matter? How is the effect exerted? Is awareness necessary?).

45.3.2.3 Flavor-Drug Learning. Although most orally consumed psychoactive drugs could in theory become associated with their vehicle's flavor (but see Pliner et al., 1985), this has only been examined in any depth for caffeine. With caffeine, participants can form associations between the presumed pleasure obtained from relieving a deprivation state and the odorous component of the flavor in which it was embedded. As this would suggest, participants need to be deprived of caffeine both to learn such associations and to demonstrate them (e.g., Yeomans et al., 2007).

45.3.2.4 Flavor-Aversion Learning. Experiencing a flavor contingent with nausea commonly results in a long-lasting aversion to that flavor. Survey research suggests that many North Americans have one or more aversions, many of these being formed on the basis of a single flavor-nausea pairing (Logue et al., 1981). Flavor aversion learning can also be demonstrated in the laboratory using nausea induced either by chemicals or motion. Either can produce reliable aversions to flavors consumed

prior to nausea onset (e.g., Cannon et al., 1983). Apart from the rapidity of this form of learning it is especially interesting because it can occur even when a person knows that the nausea results from a cause (e.g., chemotherapy, radiotherapy, flu) other than food (e.g., Bernstein, 1978). Most recently, a form of flavor aversion arising from pairing a flavor with exercise, but without nausea, has also been reported in humans (Havermans et al., 2009).

45.3.2.5 Observational Learning. Much learning about food and flavor occurs during childhood. One component of this is likely to involve observing parental facial and vocal reactions to food, and the formation of attitudes on that basis. There is certainly evidence consistent with this, particularly in younger children (e.g., Marinho, 1942; Harper and Sanders, 1975). Whether observational learning exerts an effect in older children, adolescents, and adults is more contentious, and has not received consistent support (e.g., contrast Baeyens et al., 1996, with Baeyens et al., 2001).

45.3.2.6 Learned Satiety. A final form of flavor learning concerns the ability of prior experience with a food to regulate future intake of that food. An energy dense food will produce more rapid satiety, and it is argued that this information is learned and then used to modulate intake when the food is encountered again (e.g., Booth et al., 1982). This effect has not been consistently demonstrated and its importance in governing energy intake is not currently known (e.g., Brunstrom and Mitchell, 2007).

45.4 FACTORS INFLUENCING BINDING OF SENSORY SIGNALS IN FLAVORS

45.4.1 Spatial and Temporal Contiguity

Studies of interactions between visual, auditory, and somatosensory systems have demonstrated the importance of spatial and/or temporal contiguity in facilitating cross-modal sensory integration (Driver and Spence, 2000). In flavors, the different stimulus elements are associated temporally. However, while both gustatory and somatosensory receptors are spatially located in the mouth, olfactory receptors are not. The question then arises of how odors become bound to taste and touch. Central to this process is the *olfactory location illusion*, in which the odor components of a food appear to originate in the mouth (Rozin, 1982). Thus, we *never* have a sense that the *oranginess* of tasted orange juice is being perceived within the nose, even if we are aware that it is an odor. This illusion is both strong and pervasive, despite the fact that we are frequently presented with evidence of the importance of the olfactory component in flavors, e.g., through a blocked nose during a head cold. One probable manifestation of this phenomenon is the interchangeability of chemosensory terms such as *flavor* and *taste* in common usage – that is, we routinely fail to make a distinction between olfactory and taste qualities within flavors.

How crucial are spatial and temporal contiguity of the discrete sensory inputs to the olfactory location illusion and to odor-taste binding? Von Bekesy (1964) pointed to the likely importance of temporal factors as potential determinants of odor/taste integration by showing that the perceived location of an odor (mouth vs nose) and the extent to which an odor and taste were perceived as one sensation or two could be manipulated by varying the time delay between the presentation of the odor and taste. With a time delay of zero (simultaneous presentation), the apparent locus of the odor was the back of the mouth and the odor/taste mixture was perceived as a single entity. When the odor preceded the taste, the sensation was perceived as originating in the nose. While this report is consistent with models of binding across other sensory modalities, insufficient details were provided to judge the reliability of the conclusions.

Few subsequent studies have addressed this issue. A key role for temporal synchrony in facilitating integration is suggested by a demonstration that odor-induced taste enhancement can occur regardless of whether the odor is presented orthonasally or retronasally, providing that the odor and taste are presented simultaneously (Sakai et al., 2001). Pfieffer et al. (2005) manipulated both spatial and temporal contiguity for the odor and taste while assessing the threshold for benzaldehyde odor (almond/cherry) in the presence of a subthreshold sweet taste, but failed to find convincing evidence of their manipulations. However, a recent study, albeit preliminary, suggests that synchronicity judgments of odor and taste may be less sensitive to onset discrepancies than other multimodal stimulus pairs, including audio-visual stimuli and odors and tastes that are individually paired with visual stimuli (Kobayakawa et al., 2009). One interpretation of such a finding, if confirmed, would be that odor-taste binding operates under less stringent requirements for spatio-temporal synchrony than the multisensory integration of other sensory systems. Binding under conditions in which there is a tolerance for asynchrony might reflect the high adaptive significance of chemosensory binding or the fact that in normal eating much of the retronasal olfactory stimulus occurs subsequent to swallowing, following initial tastes by seconds. Alternatively, at least in the case of temporal asynchrony, congruency between the odor and taste may be crucial. Thus, judgments of audio-visual asynchrony are more difficult when the different modalities are bound by a common origin (Spence and Parise, 2010).

It has been a matter of conjecture as to whether concurrent tastes or oral tactile sensations, or both, are chiefly responsible for the capture and referral of olfactory information to the oral cavity and to binding the sensory inputs to one another and to a physical stimulus such as a food (Murphy and Cain, 1980). Prima facie, somatosensory input is strongly implicated since it provides more detailed spatial information than does taste (Lim and Green, 2008). Thus, tactile stimulation is able to capture taste, presumably by providing superior spatial information and enhancing localization (Lim and Green, 2008; Todrank and Bartoshuk, 1991). Moreover, in neuroimaging studies, retronasal odors have been shown to activate the mouth area of primary somatosensory cortex, whereas the same odors presented via the nose do not (Small et al., 2005). This distinction, which occurs even when subjects are unaware of route of stimulation, suggests a likely neural correlate of the binding process, and supports a role for somatosensory input in the process.

However, several recent studies suggest that the key role in odor localization is played by concurrent tastes. Stevenson et al. (2011) confirmed that the presence of a taste in the mouth while an odor was sniffed was sufficient to localize that odor sensation to the mouth. Further, in a series of studies that compared the presence of tastes with tasteless somatosensory stimuli in the mouth (including the addition of mouth movements), localization was found to be primarily a function of the presence of tastes (see Figure 45.3). Much the same finding was reported by Lim

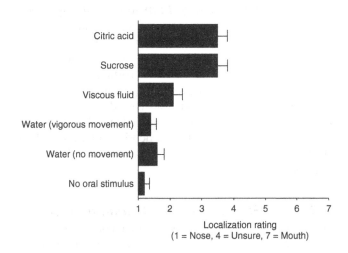

Figure 45.3 The relative importance of tastes in localization of odors to the mouth. Mean (+SEM) judgments of where participants perceive an odor to be arising, when either no stimulus, water under various conditions, a viscous fluid (carboxy methylcellulose) or tastants are present in the mouth (adapted from Stevenson, Oaten and Mahmut (2011b), *Q.J. Exp. Psychol.*, 64, 224–240, 2011).

45.4 Factors Influencing Binding of Sensory Signals in Flavors

and Johnson (2011) who showed that odors experienced retronasally, while poorly localized to the mouth in the absence of another stimulus, increased in the presence of tastants but not tasteless water. Moreover, they showed the effect was strongly a function of odor/taste congruency. Thus, referral was only evident when sweetness was paired with vanilla odor, and saltiness with soy sauce odor, in contrast to these odors with non-congruent tastes. This finding was subsequently replicated using sets of different congruent and incongruent odor/taste combinations presented within a solid, food-like object made from gelatin (Lim and Johnson, 2012). In contrast, however, Stevenson et al. (2011b) using a different measure of odor localization, found little evidence that odor/taste congruency was important, showing that even non-congruent tastes clearly improved odor localization over water alone. Overall, these recent studies strongly support the role of tastes in odor referral over somatosensory input and they also suggest that the effect may at least partly depend on prior learning.

The binding of odors to tastes and tactile stimuli may also rely on processing information about the origins of odor stimulation. Orthonasally presented odors are more readily identified and have lower thresholds than the same odors presented retronasally via the mouth (Pierce and Halpern, 1996; Voirol and Daget, 1986), and there is a strong suggestion that the two routes of stimulation are processed with some independence. Thus, neuroimaging studies show different activation patterns in cortical olfactory areas as a result of route of administration (Small et al., 2005). From an adaptive point of view, this makes sense. Olfaction has been described (Rozin, 1982) as the only dual sense because it functions both to detect volatile chemicals in the air (orthonasal sniffing) and to classify objects in the mouth as foods or not, and each of these roles has unique adaptive significance. Since the mouth acts as the gateway to the gut, our chemical senses can be seen as part of a defense system to protect our internal environment – once something is placed in the mouth, there is high survival value in deciding whether or not consumption is appropriate. Sensory qualities (tastes, retronasal odors, tactile qualities) occurring together in the mouth are therefore bound into a single perception, which identifies a substance as a food.

45.4.2 Cognitive Processes in Flavor Binding

Even though an odor's sniffed "taste" qualities and its ability to enhance that taste in solution are highly correlated (Stevenson et al., 1999), the enhancement of tastes in solution by congruent odors appears to operate under some constraints. This became evident from findings that whether or not an odor enhances taste was dependent on task requirements. Thus, while strawberry odor was found to enhance the sweetness of sucrose in solution when the subjects were asked to judge only sweetness, the enhancement was not evident when other sensory qualities of these mixtures, such as sourness and fruitiness, were rated as well (Frank et al., 1993). Similarly, significantly less sweetness enhancement was found when subjects rated the odor as well as taste intensity of flavors (sweetness plus strawberry or vanilla) than when they rated sweetness alone (Clark and Lawless, 1994).

In attempting to explain such effects, Frank and colleagues (Frank, 2003; Frank et al., 1993; van der Klaauw and Frank, 1996) suggested that, given perceptual similarity between an odor and taste, the conceptual "boundaries" that the subject sets for a given complex stimulus will reflect the task requirements. In the case of an odor/taste mixture in which the elements share a similar quality, combining those elements is essentially optional. If so, then the apparent influence of the number of rating scales on odor/taste interactions may result from the impact of these scales on how attention is directed towards the odor and taste. In keeping with this view, van der Klaauw and Frank (1996) were able to eliminate taste enhancement by directing subjects' attention to the appropriate attributes in a taste/odor mixture, even when they were only required to *rate* sweetness.

These attentional effects appear to correspond to the differing modes of interaction that occur *within* sensory modalities. The blending of odors to form entirely new odors is a commonplace occurrence in flavor chemistry and perfumery (at least for complex odor mixtures), and hence is referred to as *synthetic* interaction (analogous to the blending of light wavelengths). By contrast, the mixing of tastes is typically seen as an *analytic* process, because individual taste qualities do not fuse to form new qualities and, like simultaneous auditory tones, can be distinguished from one another in mixtures. A further category of interaction, namely *fusion* – the notion of sensations combined to form a single percept, rather than combining synthetically to form a new sensation – was proposed and applied to flavor perception by McBurney (1986).

The notion of fusion in flavor perception implies that the percept remains analyzable into its constituent elements even when otherwise perceived as a whole. While our initial response to *apple flavor* is an effortless combining of all of its sensory qualities into a single percept, we can, if required, switch between a synthetic approach to flavor and an analysis of the flavor elements. Hence, apple flavor can be both a synthetic percept and, with minimal effort, a collection of tastes (sweet; sour), textures (crisp; juicy) and odor notes (*lemony*; *acetone-like*; *honey*). A more precise way of conceptualizing flavor therefore is that cross-modal sensory signals are combined to produce a percept, rather than combining – in the way that odors themselves do – to form a new sensation. During normal food consumption,

1018　　**Chapter 45　Chemosensory Integration and the Perception of Flavor**

we typically respond to flavors synthetically – an approach reinforced by the olfactory illusion and by the extent to which flavor components are congruent. As noted earlier, this implies a sharing of perceptual qualities, e.g., sweetness of a taste and of an odor, derived from prior experience of these qualities together.

Conversely, analytic approaches to complex food or other flavor stimuli (e.g., wines) are often used by trained assessors to provide a descriptive profile of discrete sensory qualities, as distinct from an assessment of the overall flavor. Asking assessors to become analytical appears to produce inhibitory effects on odor-taste interactions. In one study using both trained descriptive panellists and untrained consumers (Bingham et al., 1990), solutions of the sweet-smelling odorant *maltol* plus sucrose were rated as sweeter than a solution of sucrose alone by the untrained consumers. In contrast, no such enhancement was found in the ratings of those trained to adopt an analytical approach to the sensory properties of this mixture.

A configural account of odor/taste perceptual learning (Stevenson and Boakes, 2004; see Section 45.3.1.1) implies that when attention is directed towards a flavor, it is attended to as a single compound or configuration, rather than a collection of elements. A configural explanation for the ability of an odor to later summate with the taste to produce enhanced sweetness implies an attentional approach that combines the odor and taste, rather than identifying them as separate elements in the flavor. In other words, for a complete binding of flavor features via configural learning, synthesis of the elements via attending to the whole flavor is critical. The limited evidence that exists suggests that the binding and joint encoding of odors, tastes and tactile sensations is automatic. This is indicated both by the finding that perceptual changes in odors following pairing with tastes appears not to require conscious awareness on the part of the subject of the particular odor-taste contingencies (Stevenson et al., 1998) and data suggesting that a single co-exposure of an odor and taste can result in transfer of the taste properties to the odor (Prescott et al., 2004).

Such learning should be sensitive to manipulations in which attention is directed either towards the identity of their constituent elements (an analytical strategy) or to the configural as a whole (a synthetic strategy) during the pairing of odors and tastes, or in subsequent judgments of the odor/taste mixtures. An analytical strategy should inhibit increases in the taste properties of the odor, and the subsequent ability of the odor to influence tastes in solution. In contrast, treating the elements as a synthetic whole is likely to encourage the blurring of the perceptual boundaries, fostering subsequent odor/taste interactions.

Consistent with this, pre-exposure of the elements of the specific odor-taste flavor compounds that were later repeatedly associated – in Pavlovian terms, US or CS pre-exposure – significantly reduced any change in odor perceptual qualities following the pairing with tastes (Stevenson and Case, 2003). Similarly, prior analytical training in which attention was explicitly directed towards the individual elements in an odor and sweet taste mixture was shown to inhibit the development of a sweet-smelling odor (Prescott and Murphy, 2009). The fact that the training used different odor/taste combinations than were later used in the conditioning procedure suggests that an attentional focus (analytical or synthetic) induced during training was then applied to *new* odor/taste combinations during conditioning. This study also showed identical effects of such a focus on the hedonic value of a flavor – only those odors paired under synthetic conditions showed conditioned liking. A similar inhibition of changes in smelled sweetness of novel odors was also reported by Labbe and Martin (2009) following sensory profiling of the odors in mixtures with sweet tastes during exposure (analytical condition), whereas increases in odor sweetness were found in a condition that used a synthetic mock-discrimination task similar to that of Prescott and Murphy (2009).

The findings from these studies have important theoretical implications in that their results are clearly consistent with configural accounts of perceptual odor-taste learning and flavor representation (Stevenson and Boakes, 2004; Stevenson et al., 1998). Under conditions where attention is directed towards individual stimulus elements during conditioning, the separate representation of these elements may be incompatible with learning of a configural representation that includes both taste and odor properties (see Figure 45.4). This explanation is supported by the demonstration that an analytical approach also acted to inhibit a sweet-smelling odor's ability to enhance a sweet taste when the odor/taste combination was evaluated in solution following repeated pairing (Prescott et al., 2004). In other words, an analytical attentional strategy can be shown to interfere with either the development of a flavor configuration resulting from associative learning, or the subsequent ability of this configuration to combine with a physically present tastant. While contrasting analytical and synthetic attentional approaches therefore appear to support a configurational explanation of flavor formation, it should be nevertheless noted that the presence of negative findings following such manipulations on odor-taste learning (Stevenson and Mahmut, 2011b) means that our understanding of the role of experimental tasks in inducing different attentional approaches remains limited.

Analytical approaches to perception appear to be inhibitory to liking even once that liking has been established. Commercial tea drinks were evaluated by consumers in one of two contexts. One group was asked only to rate their liking for the drink once it had been tasted;

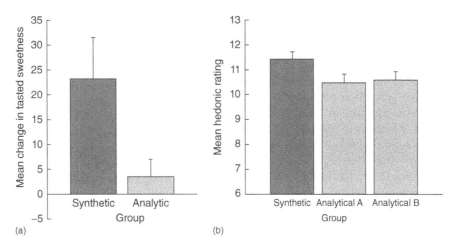

Figure 45.4 The impact of synthetic and analytical approaches to configural stimuli. (a) Mean (+SEM) ratings of the sweetness of a flavor comprised of sucrose in solution together with an odor that has previously been conditioned with this taste so that it smells sweet. Despite this, the enhancement is evident only in a group that treated the elements synthetically during their association. Figure adapted from data reported in Prescott, J., Johnstone, V. and Francis, J. (2004) Odor/taste interactions: Effects of different attentional strategies during exposure. *Chemical Senses*, 29, 331–334. (b) Mean (+SEM) ratings of overall flavour liking for a beverage sample in the synthetic (overall liking only) and the two analytical (overall liking plus attributes) groups. Reprinted with modification from *Food Quality and Preference*, 22, J. Prescott, S. M. Lee, K-O, Kim, Analytic approaches to evaluation modify hedonic responses, pp. 391–393, Copyright (2011), with permission from Elsevier.

two other groups were given this same task but their rating of liking was followed by a series of ratings of different aspects of the drink's flavor (Prescott et al., 2011). This suggests that an analytical attentional strategy was induced by knowledge that the flavor was to be perceptually analyzed, inhibiting the configuration process responsible for the transfer of hedonic properties (see Figure 45.4).

45.5 CLINICAL AND APPLIED ISSUES IN FLAVOR PERCEPTION

45.5.1 Perceptual Loss

Several studies have now examined the impact on eating behavior from damaging one of the three senses involved in oral flavor perception. Impacts have been well documented for olfaction (Aschenbrenner et al., 2008) and somewhat less so for taste (Welge-Lussen et al., 2011), but there is little information on damage to oral somatosensation. Olfactory loss is usually reported at first as a "loss of taste" with the person presenting with both taste and olfactory impairment (e.g., Deems et al., 1991). Clinical testing usually confirms that the loss is restricted to olfaction. The functional impact can be profound, with a loss of enjoyment of food, changes in food preferences, and alteration of body weight (Aschenbrenner et al., 2007). Body weight changes vary, and this may depend upon whether the anosmia is accompanied by parosmia (unpleasant olfactory distortion), which appears to be associated with weight loss (Mattes et al., 1990).

Complete loss of taste is rare, although significant general impairment can occur as a consequence of chemotherapy, radiotherapy, and some drugs (e.g., Zabernigg et al., 2010). This has a similar impact to anosmia, altering food preferences and body weight (Porter et al., 2010). Where taste loss is restricted to just one area of the tongue, this may not be noticed because of the redundancy of the taste afferents and the somatosensory referral of gustatory information across the "missing" taste region (Todrank and Bartoshuk, 1991). In addition, if the loss is restricted to one side, disinhibition may occur on the other, increasing sensitivity that further compensates for partial loss. For somatosensory loss, there are no documented reports other than temporary loss produced by local anaesthesia. This can be sufficiently profound, however, to prevent a person from recognizing the flavor of a common food (Kutter et al., 2011).

The most clinically significant manifestation of sensory loss in flavor is in the elderly, as they may experience independent reductions in their capacity to detect, identify and discriminate both tastes and smells (Hickson, 2005). This can be of major significance in the frail elderly, as poor appetite driven in part by sensory loss, may contribute to reduced food intake, weight loss, and increased risk of dying (Hickson, 2005). A similar albeit less dramatic relationship holds true in the healthy elderly, where reduced appetite may contribute to suboptimal nutrition. This was investigated by Schiffman (1992) who found that augmenting the olfactory and the gustatory components of food boosted intake and enjoyment in this population.

45.5.2 Flavor, Appetite, and Obesity

Shepherd (2006) has argued that a greater understanding of flavor perception can help tackle the obesity epidemic. To date four obesity-flavor perception relationships have emerged. The first concerns changes in flavor hedonics across the course of meal, which act to reduce food intake prior to the onset of physiological signals – sensory specific satiety (Hetherington, 1996; Rolls, Rowe, and Rolls, 1982). Normally the hedonic response to a particular food flavor will decline while consuming that food (e.g., Rolls and Rolls, 1997). One obvious possibility is that this process may be less efficient in obese relative to lean individuals leading them to eat more. However, in the two studies that have investigated this, neither found any significant influences of body mass on sensory specific satiety (Brondel et al., 2007; Snoek et al., 2004).

The second flavor-obesity link concerns preference for high-energy foods. Interest in this arose from a study by Drewnowski et al. (1985) that investigated fat/sugar preferences using real fat/sugar mixes in lean, obese and recovered obese female participants. They found that obese and recovered obese women had much higher fat preferences than lean women, suggesting this might contribute to poorer nutritional choices and weight gain, although sweetness differences were less evident. A more recent study of African-American adults examined whether differences in fat perception might predict food preferences and dietary choice (Liang et al., 2012). Those participants with the poorest ability to discriminate fat content expressed greater preference for fatty foods, and had more abdominal fat, than participants best able to discriminate the fat content of a set of test foods. Whether the differences in fat perception in these studies are a product or a consequence of weight gain is not currently known.

The third flavor-obesity relationship concerns taste. Bartoshuk et al. (2006) found that participants with more taste buds tend to be leaner. In a similar vein obese individuals appear to find the same sweet taste less intense than do lean persons, and they seem to enjoy increases in sweetness intensity in a way that lean participants do not. These differences in perception may again contribute to greater preferences for, and intake of, energy dense food.

The final flavor-obesity link emerged from the claim that patients who have undergone gastric bypass surgery for weight loss seem to experience long-lasting reductions in preference for fatty and sweet foods (e.g., Miras and le Roux, 2010). Animal models (receiving gastric bypass vs. placebo treatment) confirmed that changes in fat and sugar preference do occur, but tests in humans who have undergone gastric bypass surgery reveal only modest differences in sensory and hedonic perception of flavor stimuli (Bueter et al., 2011; le Roux et al., 2011). If these preference shifts are of sufficient magnitude to drive a reduction in energy intake, then they are of great interest, especially if the mechanism can be induced without the need for surgery.

Understanding how flavor perceptions and preferences are formed may help to provide insight into behaviors that promote obesity. As noted earlier, odor-taste pairings can produce learned flavor preferences when the pairing is accompanied by ingestion of valued nutrients. The distinction between this form of learning and flavor-flavor learning (in which a liked stimulus, such as a sweet taste, may not be accompanied by calories) can be seen in terms of the distinction between, respectively, conditioned wanting versus conditioned liking (Winkielman and Berridge, 2003). In the former, the effects of the learning can be measured in appetitive terms using measures of intake and hunger, rather than simply measures of liking. Hence, it has been recently demonstrated in humans that a novel soup flavor paired with ingested monosodium glutamate (MSG) not only increased in rated liking, even when tested without added MSG, but also, relative to a non-MSG control, produced behavioral changes including increases in *ad libitum* food intake and rated hunger following an initial tasting of the flavor (Yeomans et al., 2008b). In other words, the soup flavor became a cue that motivated the desire or want for consumption of the soup (see Figure 45.5).

This research points to an origin for the triggers to unrestrained eating that represents a risk for obesity. Via this same process seen with MSG conditioning, flavors not only become liked by pairing with energy, but also become cues that engage wanting for foods, in particular palatable (often high sugar, high fat) foods, since these are both common and effective sources of energy. For the obese, the sensory cues for the presence of energy may be especially powerful in triggering the desire to eat (Hofmann et al., 2010).

There have been some recent attempts to explore other practical implications of odor-taste learning, opening opportunities to perhaps exploit its consequences for healthier eating. It has been shown, for example, that the enhancement of tastes by congruent odors seen in model systems (that is, solutions) also occurs in foods, with bitter- and sweet-smelling odors enhancing their respective congruent tastes in milk drinks (Labbe et al., 2006). Most recently, an examination of the potential for odors from a range of salty foods to enhance saltiness in solution (Lawrence et al., 2009) and model foods (Batenburg and van der Velden, 2011) raise the possibility that such odors could be used to effectively reduce the sodium content of foods, without the typical concurrent loss of acceptability that occurs (Girgis et al., 2003). Similarly, the finding that odors can take on fat-like properties following associative pairing with fats (Sunqvist et al., 2006) might allow odors

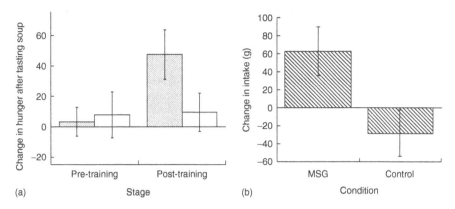

Figure 45.5 Mean (+/− SEM) change in (a) hunger ratings and (b) intake of soup (pre- and post-pairing data shown), following tasting a soup that had previously been repeatedly paired with monosodium glutamate (MSG; cross-hatched bar in (a)) or a control soup that had previously had no added MSG. Reprinted from *Physiology & Behavior*, 93, M. R. Yeomans, N. J. Gould, S. Mobini, J. Prescott, Acquired flavor acceptance and intake facilitated by monosodium glutamate in humans, pp. 958–966, Copyright (2008b), with permission from Elsevier.

to partially substitute for actual fat content in foods. These studies therefore point to an exciting prospect, in which research aimed at understanding multisensory processes in flavor perception may lead to applications that ultimately have important public health consequences.

45.5.3 Implications for the Measurement of Sensory Qualities

A configural view of flavor perception generates important implications for the measurement of sensory qualities. The differing task requirements, which appear to disrupt the fusion of distinct sensory inputs, map very well on to real distinctions that exist between consumer and trained or "expert" perceptions of food and beverage flavors. Without motivation to do otherwise, consumers will naturally integrate sensory information, thereby producing a response (typically hedonic) to the whole food or beverage. On the other hand, trained analytical panels and other experts (wine judges, and so on) are not only trained to view complex flavors as a set of individual elements, but the tasks typically required of them also involve assessment of multiple sensory attributes, each assumed to be independent.

As has been discussed in this chapter, many such attributes are not independent, even though they may be derived from distinct physiological sources. As noted earlier, Bingham et al (1990) showed that consumers, but not trained panels, showed sweetness enhancement effects by maltol odor. Prescott et al., (2011) provided evidence that analytical approaches to judging products can inhibit liking in consumers. Beyond that, there are few "real-world" data addressing this issue.

It is clear that the type of questions that are asked of consumers or subjects determine how perceptually similar flavor qualities are combined or separated. This has obvious consequences for the psychophysical and hedonic evaluation of sensory qualities, and raises the question of what qualities *should* be measured in evaluations of food/beverage flavors. Thus, in a complex mixture such as a sweet food, which part is the "real sweetness"? Clearly, it depends on whether it is important to know about the sweetness of the flavor, or the sweetness associated with the odor or produced by the taste. However, the distinction between odor sweetness and taste sweetness may be artificial if one is evaluating flavors, since the sweetness of the flavor components may be represented cognitively, and even perhaps neurally, as equivalent. They may be functionally equivalent, as well. As noted earlier, odor sweetness can suppress tasted sourness, just as a sweetener such as sucrose does (Stevenson et al., 1999). Frank et al. (1993) suggest that the question might be an empirical one, especially for consumers. That is, what form of sweetness best predicts overall responses to foods? In this regard, overall flavor sweetness may be the most ecologically valid measure. In the case of trained panels, the whole point of descriptive analysis techniques is to deconstruct compound stimuli into their constituent parts. The question might arise therefore of whether information is being lost by relying on the physiological origin of sensory qualities (e.g., taste qualities perceived via taste receptors), rather than on their functional perceptual representations (e.g, taste as a cognitive representation). Moreover, it assumes that a particular stimulus is driving what is perceived. As shown by Davidson et al. (1999) in their study of the interaction of menthone and sweetness in chewing gum, it was the taste that determined overall flavor rather than the odor. These issues may be especially pertinent if trained panel data are being used to interpret and explain consumer responses.

45.6 CONCLUSION: BETTER LIVING THROUGH CHEMOSENSORY SCIENCE!

Interest in the nature and origin of flavor perceptions has recently migrated from the laboratory into the culinary world. The rise of molecular or "modernist" cuisine has been based to a large extent on understanding and utilizing the experience-dependent interactions that have been discussed in this chapter (Prescott, 2012). Chefs practicing this form of cuisine make use of the ability of sensory signals to enhance one another. For example, diners are presented with auditory and visual stimuli that are congruent with the food they are eating, presumably in an attempt to make the flavor experience more immersive. One famous example is the use of the sound of waves on the shore (via an iPod hidden in a conch shell) to accentuate the visual presentation of seafood on a tapioca-based "sandy beach," with a vegetable and seafood based foam as the rolling waves. Other dishes take advantage of well-known perceptual biases based on visual primacy, as well as our poor ability at olfactory recognition. Hence, apparently simple tastes such as an orange flavored jelly coloured a deep red paired with a beetroot flavored jelly coloured orange can arouse surprise, if not confusion in the diner, because of the conflict between what is seen and what is tasted. Perhaps most interesting, however, are attempts to isolate the experience of flavor for the diner. The British chef Heston Blumenthal creates a dish that consists of a tactile stimulus, hardened, whipped egg white minimally-flavored with powdered green tea, but accompanied by a fine mist of lime odor sprayed into the air above the diner. The hidden surprise in this dish is the effortless way in which the textural and taste sensory properties of the "mousse" are combined with the lime odor to produce an integrated flavor. The diner receives a dramatic demonstration of a simple, but genuine, multisensory experience. And while the overall effect is also presented for its hedonic properties (that is, it is meant to be delicious), it has no purpose beyond this and no real nutritional value. It is as close as most diners have ever come to experiencing a flavor without a food. Whether such "experiments" with flavor and foods become merely a culinary footnote remains to be seen, but they are evidence that research on flavor is seeping into the "real world".

REFERENCES

Albin, K. C., Carstens, M. I. and Carstens, E. (2008). Modulation of oral heat and cold pain by irritant chemicals. *Chem. Sens.* 33: 3–15.

Aschenbrenner, K., Hummel, C., Teszmer, K., et al. (2008). The influence of olfactory loss on dietary behaviors, *Laryngoscope*, 118: 135–144.

Baeyens, F., Eelen, P., Van den Bergh, O. and Crombez, G. (1990). Flavor-flavor and color-flavor conditioning in humans. *Learn. Motiv.* 21: 434–455.

Baeyens, F., Eelen, P., Crombez, G. and De Houwer, J. (2001). On the role of beliefs in observational flavor conditioning. *Cur. Psychol.: Dev. Learn. Pers.* 20: 183–203.

Baeyens, F., Vansteenwegen, D., De Houwer, J. and Crombez, G. (1996). Observational conditioning of food valence in humans. *Appetite* 27: 235–250.

Ballester, J., Patris, B., Symoneaux, R. and Valentin, D. (2008). Conceptual vs perceptual wine spaces: does expertise matter? *Food Qual. Pref.* 19: 267–276.

Barker, L. M. (1982). Building memories for food. In: L.M. Barker (Ed.), *The Psychobiology of Human Food Selection*. Westport: AVI Publishing Company, pp. 85–99.

Bartoshuk, L. M., Duffy, V. B., Hayes, J. E., et al. (2006). Psychophysics of sweet and fat perception in obesity: Problems, solutions and new perspectives. *Phil. Trans. R. Soc. B.* 361: 1137–1148.

Bartoshuk, L. M., Rennert, K., Rodin, J. and Stevens, J. C. (1982). Effects of temperature on the perceived sweetness of sucrose. *Physiol. Behav.* 28: 905–910.

Batenburg, M. and van der Velden, R. (2011). Saltiness enhancement by savory aroma compounds. *J. Food Sci.* 76: S280–S288.

Bernstein, I. L. (1978). Learned taste aversions in children receiving chemotherapy. *Science* 200: 1302–1303.

Bingham, A. F., Birch, G. G., De Graaf, et al. (1990). Sensory studies with sucrose–maltol mixtures. *Chem. Senses* 15: 447–456.

Booth, D. A., Mather, P. and Fuller, J. (1982). Starch content of ordinary foods associatively conditions human appetite and satiation, indexed by intake and eating pleasantness of starch-paired flavors. *Appetite* 3: 163–184.

Brillat-Savarin, J.-A. (1825). The Physiology of Taste (1994 ed.). London: Penguin Books.

Brock, W. H. (1967). William Prout on taste, smell, and flavor *J. Hist. Med.* (April): 184–187.

Brondel, L., Romer, M., Van Wymelbeke, et al. (2007). Sensory-specific satiety with simple foods in humans: No influence of BMI? *Int. J. Ob.* 31: 987–995.

Brunstrom, J. M. and Mitchell, G. (2007). Flavor-nutrient learning in restrained and unrestrained eaters. *Physiol. Behav.* 90: 133–141.

Bueter, M., Miras, A., Chichger, H., et al. (2011). Alterations of sucrose preference after Roux-en-Y gastric bypass. *Physiol. Behav.* 104: 709–721.

Bult, J. H. F., de Wijk, R. A. and Hummel, T. (2007). Investigations on multimodal sensory integration: texture, taste, and ortho- and retronasal olfactory stimuli in concert. *Neurosci. Lett.* 411: 6–10.

Cain, W. S. and Murphy, C. L. (1980). Interaction between chemoreceptive modalities of odor and irritation. *Nature* 284: 255–257.

Cannon, D. S., Best, M. R., Batson, J. D. and Feldman, M. (1983). Taste familiarity and apomorphine-induced taste aversions in humans. *Behav. Res. Therapy* 21: 669–673.

Capaldi, E. D. and Privitera, G. J. (2007). Flavor-nutrient learning independent of flavor-taste learning with college students. *Appetite* 49: 712–715.

Christensen, C. M. (1980). Effects of solution viscosity on perceived saltiness and sweetness. *Percept. Psychophys.* 28: 347–353.

Clark, C. C. and Lawless, H. T. (1994). Limiting response alternatives in time-intensity scaling: an examination of the halo-dumping effect. *Chem. Sens.* 19: 583–594.

Cook, D. J., Hollowood, T. A., Linforth, R. S. T. and Taylor, A. J. (2002). Perception of taste intensity in solution of random-coil polysaccharides above and below c. *Food Qual. Pref.* 13: 473–480.

References

Cruz, A. and Green, B. G. (2000). Thermal stimulation of taste. *Nature* 403: 889–892.

Dalton, P., Doolittle, N., Nagata, H. and Breslin, P. A. S. (2000). The merging of the senses: integration of subthreshold taste and smell. *Nature Neurosci.* 3: 431–432.

Davidson, J. M., Linforth, R. S. T., Hollowood, T. A. and Taylor, A. J. (1999). Effect of sucrose on the perceived flavor intensity of chewing gum. *J. Agric. Food Chem.* 47: 4336 –4340.

Deems, D. A., Doty, R. L., Settle, R. G., et al. (1991). Smell and taste disorders, a study of 750 patients from the University of Pennsylvania smell and taste center. *Arch. Otolaryngol. Head Neck Surg.* 117: 519–528.

Delwiche, J. (2004). The impact of perceptual interactions on perceived flavor. *Food Qual. Pref.* 15: 137–146.

Delwiche, J. and Heffelfinger, A. L. (2005). Cross-modal additivity of taste and smell. *J. Sens. Stud.* 20: 512–525.

de Wijk, R. A., Prinz, J. F., Polet, I. A. and van Doorn, R. M. (2004). Amount of ingested custard as affected by its colour, smell, and texture. *Physiol. Behav.* 82: 397–403.

Drewnowski, A., Brunzell, J. D., Sande, K., et al. (1985). Sweet tooth reconsidered: Taste responsiveness in human obesity. *Physiol. Behav.* 35: 617–622.

Dickinson, A. and Brown, K. J. (2007). Flavor-evaluative conditioning is unaffected by contingency knowledge during training with color-flavor compounds. *Learn. Behav.* 35: 36–42.

Driver, J., and Spence, C. (2000). Multisensory perception: Beyond modularity and convergence. *Curr. Biol.* 10, R731–R735.

DuBose, C. N., Cardello, A. V. and Maller, O. (1980). Effects of colorants and flavorants on identification, perceived intensity, and hedonic quality of fruit-flavored beverages and cake. *J. Food Sci.* 45: 1393–1415.

Frank, R.A. (2003). Response context affects judgments of flavor components in foods and beverages. *Food Qual. Pref.* 14: 139–145.

Frank, R. A. and Byram, J. (1988). Taste–smell interactions are tastant and odorant dependent. *Chem. Sens.* 13: 445–455.

Frank, R. A., Ducheny, K. and Mize, S. J. S. (1989). Strawberry odor, but not red color, enhances the sweetness of sucrose solutions. *Chem. Sens.* 14: 371–377.

Frank, R. A., van der Klaauw, N. J. and Schifferstein, H. N. J. (1993). Both perceptual and conceptual factors influence taste-odor and taste-taste interactions. *Percept. Psychophys.* 54: 343–354.

Frost, A.C. & Noble, A.C. (2002). Preliminary study of the effect of knowledge and sensory expertise on liking red wines. *American Journal of Oenology and Viticulture*, 53, 275–284.

Frost, A.C. & Noble, A.C. (2002). Preliminary study of the effect of knowledge and sensory expertise on liking red wines. *American Journal of Oenology and Viticulture*, 53, 275–284.

Gautam, S. and Verhagen, J. (2010). Evidence that the sweetness of odors depends on experience in rats. *Chem. Sens.* 35: 767–776.

Gibson, J. J. (1966). *The Senses Consideered as Perceptual Systems.* Boston: Houghton Mifflin Company.

Gilmore, M. M. and Green. B. G. (1993). Sensory irritation and taste produced by NaCl and citric acid: effects of capsaicin desensitization. *Chem. Senses* 18: 257–272.

Girgis, S., Neal, B., Prescott, J., et al. (2003). A one-quarter reduction in the salt content of bread can be made without detection. *Eur. J. Clin. Nutrit.* 57: 616–620.

Green, B. G. (1990). Effects of thermal, mechanical, and chemical stimulation on the perception of oral irritation. In: B. G. Green, J. R. Mason and M. R. Kare (Eds.), Chemical Senses: Irritation, *Volume 2.* New York: Marcel Dekker, pp. 171–192.

Green, B. G. and Frankman, S. P. (1987). The effects of cooling the tongue on the perceived intensity of taste. *Chem. Sens.* 12: 609–619.

Green, B. G. and Nachtigal, D. (2012). Somatosensory factors in taste perception: Effects of active tasting and solution temperature. *Physiol. Behav.* 12: 15–31.

Green, B. G., Nachtigal, D., Hammond, S. and Lim, J. (2012). Enhancement of retronasal odors by taste. *Chem. Sens.* 37: 77–86.

Harper, L. V. and Sanders, K. M. (1975). The effect of adults' eating on young children's acceptance of unfamiliar foods. *J. Exp. Child Psychol.* 20: 206–214.

Havermans, R.C., Salvy, S-J., and Jansen, A. (2009). Single-trial exercise-induced taste and odor aversion learning in humans. *Appetite* 53: 442–445.

Hetherington, M. M. (1996). Sensory-specific satiety and its importance in meal termination. *Neurosci. Biobehav. Rev.* 20: 113–117.

Hickson, M. (2005). Malnutrition and ageing. *Postgrad. Med. J.* 82: 2–8.

Hirsch, A. R. (1990). Smell and taste: how the culinary experts compare to the rest of us. *Food Tech. September Issue:* 96–101.

Hofmann, W., van Koningsbruggen, G.M., Stroebe, et al. (2010). As pleasure unfolds: Hedonic responses to tempting food. *Psychol. Sci.* 21: 1863–1870.

Hughson, A. L. and Boakes, R. A. (2002). The knowing nose: the role of knowledge in wine expertise. *Food Qual. Pref.* 13: 463–472.

James, W. (1890). *The principles of psychology.* (Vol. 1). New York: Holt.

Keast, R. S. J., Dalton, P. H. and Breslin, P. A. S. (2004). Flavor interactions at the sensory level. In: A. Taylor and D. Roberts (Eds.), *Flavor Perception.* Oxford, Blackwell Publishing Ltd, pp. 228–255.

Kern, D. L., McPhee, L., Fisher, J., et al. (1993). The postingestive consequences of fat condition preferences for flavors associated with high dietary fat. *Physiol. Behav.* 54: 71–76.

Kobayakawa, T., Toda, H., and Gotow, N. (2009). *Synchronicity judgement of gustation and olfaction.* Paper presented at the Association for Chemoreception Sciences, Sarasota, FL.

Koster, M. A., Prescott, J., and Koster, E. P. (2004). Incidental learning and memory for three basic tastes in food. *Chem. Senses,* 29: 441–453.

Kutter, A., Hanesch, C., Rauh, C. and Delgado, A. (2011). Impact of proprioception and tactile sensations in the mouth on the perceived thickness of semi-solid food. *Food Qual. Pref.* 22: 193–197.

Labbe, D., Damevin, L., Vaccher, C., et al. (2006). Modulation of perceived taste by olfaction in familiar and unfamiliar beverages. *Food Qual. Pref.* 17: 582–589.

Labbe, D. and Martin, N. (2009). Impact of novel olfactory stimuli at supra and subthreshold concentrations on the perceived sweetness of sucrose after associative learning. *Chem. Senses* 34: 645–651.

Lawless, H. T. (1995). Flavor. In: M. Friedman and E. Carterette (Eds.), *Handbook of Perception and Cognition. Volume 16, Cognitive Ecology.* San Diego, Academic Press, pp. 325–380.

Lawless, H. T., Hartono, C. and Hernandez, S. (2000). Thresholds and suprathreshold intensity functions for capsaicin in oil and aqueous based carriers. *J. Sens. Stud.* 15: 437–447.

Lawless, H. T. and Schlegel, M. (1984). Direct and indirect scaling of sensory differences in simple taste and odor mixtures. *J. Food Sci.* 49: 44–51.

Lawrence, G., Salles, C. , Septier, C., et al. (2009). Odour–taste interactions: A way to enhance saltiness in low-salt content solutions. *Food Qual. Pref.* 20: 241–248.

le Roux, C., Bueter, M., Theis, et al. (2011). Gastric bypass reduces fat intake and preference. *Am. J. Physiol-Reg. Integrat. Comp. Physiol.* 301: R1057–R1066.

Liang, L., Sakimura, J., May, D., et al. (2012). Fat discrimination: A phenotype with potential implications for studying fat intake behaviors and obesity. *Physiol. Behav.* 105: 470–475.

Lim, J., and Green, B. G. (2008). Tactile interaction with taste localization: Influence of gustatory quality and intensity. *Chem. Senses,* 33: 137–143.

Lim, J., and Johnson, M. B. (2011). Potential mechanisms of retronasal odor referral to the mouth. *Chem. Senses*, 36: 283–289.

Lim, J., and Johnson, M. B. (2012). The role of congruency in retronasal odor referral to the mouth. *Chem. Senses*, 37: 515–521.

Logue, A. W., Ophir, I. and Strauss, K. E. (1981). The acquisition of taste aversions in humans. *Behav. Res. Ther.* 19: 319–333.

Lundgren, B., Pangborn, R. M. and Daget, N. (1986). An interlaboratory study of firmness, aroma and taste. *Lebensm.-Wiss. u.-Technol.* 19: 66–76.

Marinho, H. (1942). Social influence in the formation of enduring preferences. *J. Ab. Soc. Psychol.* 37: 448–468.

Mattes, R. D., Cowart, B., Schiavo, M., et al. (1990). Dietary evaluation of patients with smell and/or taste disorders. *Am. J. Clin. Nutrit.* 51: 233–240.

McBurney, D. H. (1986). Taste, smell, and flavor terminology: Taking the confusion out of fusion. In H. L. Meiselman and R. S. Rivkin (Eds.), Clinical Measurement of Taste and Smell, New York: MacMillan (pp. 117–125).

Melcher, J. M. and Schooler, J. W. (1996). The misremembrance of wines past: verbal and perceptual expertise differentially mediate verbal overshadowing of taste memory. *J. Mem. Lang.* 35: 231–245.

Miras, A. D. and le Roux, C. W. (2010). Bariatric surgery and taste: novel mechanisms of weight loss. *Curr. Opin. Gastroenterol.* 26: 140–145.

Mobini, S., Chambers, L. C. and Yeomans, M. R. (2007). Effects of hunger state on flavor pleasantness conditioning at home: Flavor-nutrient learning vs. flavor-flavor learning. *Appetite* 48: 20–28.

Mojet, J. and Koster, E. P. (2005). Sensory memory and food texture. *Food Qual. Pref.* 16: 251–266.

Morrot, G., Brochet, F. and Dubourdieu, D. (2001). The color of odors. *Brain Lang.* 79: 309–320.

Moskowitz, H. R. (1996). Experts versus consumers: a comparison. *J. Sen. Stud.* 11: 19–37.

Moskowitz, H. R. and Arabie, P. (1970). Taste intensity as a function of stimulus concentration and solvent viscosity. *J. Text. Stud.* 1: 502–510.

Murphy, C., Cain, W. S. and Bartoshuk, L. M. (1977). Mutual action of taste and olfaction. *Sen. Proc.* 1: 204–211.

Murphy, C. and Cain, W. S. (1980). Taste and olfaction: independence vs interaction. *Physiol. Behav.* 24: 601–605.

Nasri, N., Beno, N., Septier, C., et al. (2011). Cross-modal interactions between taste and smell: odor-induced saltiness enhancement depends on salt levels. *Food Qual. Pref.* 22: 678–682.

Nasrawi, C. W. and Pangborn, R. M. (1989). The influence of tastants on oral irritation by capsaicin. *J. Sen. Stud.* 3: 287–294.

The New Shorter Oxford English Dictionary on Historical Principles. Editor: L. Brown, 1993, Vol. 1, Oxford: Clarendon Press.

Pangborn, R. M., Gibbs, Z. M. and Tassan, C. (1978). Effect of hydrocolloids on apparent viscosity and sensory properties of selected beverages. *J. Text. Stud.* 9: 415–436.

Piwnica-Worms, K., Omar, R., Hailstone, J. and Warren, J. (2010). Flavor processing in semantic dementia. *Cortex* 46: 761–768.

Pfieffer, J. C., Hollowood, T. A., Hort, J. and Taylor, A. J. (2005). Temporal synchrony and integration of sub-threshold taste and smell signals. *Chem. Senses* 30: 539–545.

Pierce, J., and Halpern, B. (1996). Orthonasal and retronasal odorant identification based upon vapor phase input from common substances. *Chem. Senses* 21: 529–543.

Pliner, P., Rozin, P., Cooper, M. and Woody, G. (1985). Roles of specific postingestional effects and medicinal context in the acquisition of liking for tastes. *Appetite* 6: 243–252.

Porter, S. R., Fedele, S. and Habbab, K. M. (2010). Taste dysfunction in head and neck malignancy. *Oral Oncol.* 46: 457–459.

Prescott, J. (1999). Flavour as a psychological construct: implications for perceiving and measuring the sensory qualities of foods. *Food Qual. Pref.* 10: 349–356.

Prescott, J. (2004a). Effects of added glutamate on liking for novel food flavors. *Appetite* 42: 143–150.

Prescott, J. (2004b). Psychological processes in flavour perception. In A.J. Taylor and D. Roberts (Eds.), *Flavour Perception* (pp. 256–277). London: Blackwell Publishing.

Prescott, J. (2012). *Taste Matters. Why we like the foods we do.* London: Reaktion Books.

Prescott, J., Allen, S. and Stephens, L. (1993). Interactions between oral chemical irritation, taste and temperature. *Chem. Sens.* 18: 389–404.

Prescott, J., Johnstone, V. and Francis, J. (2004). Odor-taste interactions: effects of attentional strategies during exposure. *Chem. Sens.* 29: 331–340.

Prescott, J., Lee, S. M., and Kim, K. O. (2011). Analytic approaches to evaluation modify hedonic responses. *Food Qual. Pref.* 22: 391–393.

Prescott, J., and Murphy, S. (2009). Inhibition of evaluative and perceptual odour-taste learning by attention to the stimulus elements. *Q. J. Exp. Psychol.* 62: 2133–2140.

Prescott, J. and Stevenson, R. J. (1995). Effects of oral chemical irritation on tastes and flavors in frequent and infrequent users of chili. *Physiol. Behav.* 58: 1117–1127.

Prescott, J. and Wilkie, J. (2007). Pain tolerance selectively increased by a sweet-smelling odor. *Psychol. Sci.* 18: 308–311.

Roberts, A. K. and Vickers, Z. M. (1994). A comparison of trained and untrained judges evaluation of sensory attribute intensities and liking of cheddar cheeses. *J. Sen. Stud.* 9: 1–20.

Rolls, B. J., Rowe, E.A. and Rolls, E.T. (1982). How sensory properties of foods affect human feeding behavior. *Physiol. Behav.* 29: 409–417.

Rolls, E. T. and Rolls, J. H. (1997). Olfactory sensory-specific satiety in humans. *Physiol. Behav.* 61: 461–473.

Rozin, P. (1982). "Taste–smell confusions" and the duality of the olfactory sense. *Percept. Psychophys.* 31: 397–401.

Sakai, N., Kobayakawa, T., Gotow, N., et al. (2001). Enhancement of sweetness ratings of aspartame by a vanilla odor presented either by orthonasal or retronasal routes. *Percept. Mot. Skills* 92: 1002–1008.

Schifferstein, H. N. J. and Verlegh, P. W. J. (1996). The role of congruency and pleasantness in odor-induced taste enhancement. *Acta Psychol.* 94: 87–105.

Schiffman, S. S. (1992). Olfaction in aging and medical disorders. In: M. J. Serby and K. L. Chobor, *Science of olfaction*. New York, Springer-Verlag, pp. 500–525.

Shanks, D.R. (2010). Learning: From association to cognition. *Ann. Rev. Psychol.* 61: 273–301.

Shepherd, G. M. (2006). Smell images and the flavor system in the human brain. *Nature* 444: 316–321.

Simons, C. T., O'Mahony, M. and Carstens, E. (2002). Taste suppression following lingual capsaicin pre-treatment in humans. *Chem. Sens.* 27: 353–365.

Small, D.M., Gerber, J.C., Mak, Y.E., and Hummel, T. (2005). Differential neural responses evoked by orthonasal versus retronasal odorant perception in humans. *Neuron*, 47: 593–605.

Snoek, H. M., Huntjens, L., van Gemart, L. J., et al. (2004). Sensory-specific satiety in obese and normal-weight women. *Am. J. Clin. Nutrit.* 80: 823–831.

Solomon, G. E. A. (1990). Psychology of novice and expert wine talk. *Am. J. Psychol.* 103: 495–517.

References

Spence, C. and Parise, C. (2010). Prior-entry: A review. *Conscious. Cogn.* 19: 364–379.

Stevenson, R. J. (2009). The psychology of flavor. Oxford University Press, Oxford.

Stevenson, R. J. and Boakes, R. A. (2004). Sweet and sour smells: the acquisition of taste-like qualities by odors. In G. Calvert, C.B. Spence and B. Stein (Eds.), Handbook of Multisensory Processes (pp. 69–83). Cambridge: MIT Press.

Stevenson, R. J., Boakes, R. A. and Prescott, J. (1998). Changes in odor sweetness resulting from implicit learning of a simultaneous odor-sweetness association: an example of learned synesthesia. *Learn. Motiv.* 29: 113–132.

Stevenson, R. J., Boakes, R. A. and Wilson, J. P. (2000). Resistance to extinction of conditioned odor perceptions: evaluative conditioning is not unique. *J. Exp. Psychol.: Lean. Mem. Cognit.* 26: 423–440.

Stevenson, R. J., and Case, T. I. (2003). Preexposure to the stimulus elements, but not training to detect them, retards human odor-taste learning. *Behav. Proc.* 61: 13–25.

Stevenson, R. J. and Mahmut, M. K. (2011a). Experience dependent changes in odor-viscosity perception. *Acta Psychol.* 136: 60–66.

Stevenson, R. J. and Mahmut, M. K. (2011b). Discriminating the stimulus elements during human odor–taste learning: A successful analytic stance does not eliminate learning. *J Exp Psychol Anim Behav. Proc.* 37: 477–482

Stevenson, R. J., Mahmut, M. K. and Oaten, M. J. (2011a). The role of attention in the localization of odors to the mouth. *Attn. Percept. Psychophys.* 73: 247–258.

Stevenson, R. J., Miller, L. A. and Thayer, Z. C. (2008) Impairments in the perception of odor-induced tastes and their relationship to impairments in taste perception. *J. Exp. Psychol.: Hum. Percept. Perf.* 34: 1183–1197.

Stevenson, R. J. Oaten, M. J. and Mahmut, M. (2011b) The role of taste and oral somatosensation in olfactory localisation. *Q. J. Exp. Psychol.* 64: 224–240.

Stevenson, R. J. and Oaten, M. (2010). Sweet odours and sweet tastes are conflated in memory. *Acta Psychol.* 134: 105–109.

Stevenson, R. J., Prescott, J. and Boakes, R. A. (1995). The acquisition of taste properties by odors. *Learn. Motiv.* 26: 433–455.

Stevenson, R. J., Prescott, J. and Boakes, R. A. (1999). Confusing taste and smells: how odours can influence the perception of sweet and sour tastes. *Chem. Sens.* 24: 627–635.

Sundqvist, N. C., Stevenson, R. J. and Bishop, I. R. J. (2006). Can odours acquire fat-like properties? *Appetite* 47: 91–99.

Sumby, W.H. and Polack, I. (1954). Visual contribution to speech intelligibility in noise. *J. Acoust. Soc. Am.* 26: 212–215.

Tempere, S., Cuzange, E., Malak, J., et al. (2011). The training level of experts influences their detection thresholds for key wine components. *Chem. Percept.* 4: 99–115.

Theunissen, M. J. M. and Kroeze, J. (1995). The effect of sweeteners on perceived viscosity. *Chem. Sens.* 20: 441–450.

Todrank, J. and Bartoshuk, L. M. (1991). A taste illusion: Taste sensation localized by touch. *Physiol. Behav.* 50: 1027–1031.

Treisman, A. (1998). Feature binding, attention and object perception. *Phil. Trans. R. Soc. Lond.* B 353: 1295–1306.

Valentin, D., Chollet, S., Beal, S. and Patris, B. (2007). Expertise and memory for beers and beer olfactory compounds. *Food Qual. Pref.* 18: 776–785.

van der Klaauw, N.J. and Frank, R.A. (1996). Scaling component intensities of complex stimuli: The influence of response alternatives. *Environ. Int.* 22: 21–31.

Veldhuizen, M., Shephard, T. G. Wang, M-F. and Marks, L. E. (2010). Coactivation of gustatory and olfactory signals in flavor perception. *Chem. Sens.* 35: 121–133.

Vickers, Z. M. and Bourne, M. C. (1976). Crispness in foods-a review. *J. Food Sci.* 41: 1153–1157.

Voirol, E., and Daget, N. (1986). Comparative study of nasal and retronasal olfactory perception. *Lebensm.-Wiss. u.-Technol.* 19: 316–319.

von Bekesy, G. (1964). Olfactory analogue to directional hearing. *J. Appl. Physiol.* 19: 369–373.

von Sydow, E., Moskowitz, H., Jacobs, H. and Meiselman, H. (1974). Odor-taste interaction in fruit juices. *Lebensm.-Wiss. u.-Technol.* 7: 18–24.

Wardle, S. G., Mitchell, C. and Lovibond, P. (2007). Flavor evaluative conditioning and contingency awareness. *Learn. Behav.* 35: 233–241.

Walker, S. and Prescott, J. (2000). The influence of solution viscosity and different viscosifying agents on apple juice flavor. *J. Sens. Stud.* 15: 285–307.

Walker, S. and Prescott, J. (2003). Psychophysical properties of mechanical oral irritation. *J. Sens. Stud.* 18: 325–345.

Weel, K. G. C., Boelrijk, A. E. M. and Alting, A.C. (2002). Flavor release and perception of flavored whey protein gels: perception is determined by texture rather than release. *J. Agric. Food Chem.* 50: 5149–5155.

Weenen, H., Jellema, R. H. and de Wijk, R. A. (2005). Sensory sub-attributes of creamy mouth feel in commercial mayonnaises, custard desserts and sauces. *Food Qual. Pref.* 16: 163–170.

Welge-Lussen, A., Dorig, P., Wolfensberger, M., et al. (2011). A study about the frequency of taste disorders. *J. Neurol.* 258: 386–392.

White, T. L. and Prescott, J. (2007). Chemosensory cross-modal Stroop effects: congruent odors facilitate taste identification. *Chem. Sens.* 32: 337–341.

Winkielman, P. and Berridge, K. C. (2003). Irrational wanting and subrational liking: How rudimentary motivational and affective processes shape preferences and choices. *Polit. Psychol.* 24: 657–679.

Yau, N. J. N. and McDaniel, M. R. (1991). Carbonation interaction with sweetness and sourness. *J. Food Sci.* 57: 1412–1416.

Yeomans, M. R., Chambers, L., Blumenthal, H. and Blake, A. (2008). The role of expectancy in sensory and hedonic evaluation: The case of smoked salmon ice-cream. *Food Qual. Pref.* 19: 565–573.

Yeomans, M. R., Gould, N., Mobini, S. and Prescott, J. (2008). Acquired flavor acceptance and intake facilitated by monosodium glutamate in humans. *Physiol. Behav.* 93: 958–966.

Yeomans, M. R. and Mobini, S. (2006). Hunger alters the expression of acquired hedonic but not sensory qualities of food-paired odors in humans. *J. Exp. Psychol.: Anim. Behav. Proc.* 32: 460–466.

Yeomans, M. R., Mobini, S. and Chambers, L. (2007). Additive effects of flavour-caffeine and flavour-flavour pairings on liking for the smell and flavour of a novel drink. *Physiol. Behav.* 92: 831–839.

Yeomans, M. R., Mobini, S., Elliman, T. D., et al. (2006). Hedonic and sensory characteristics of odors conditioned by pairing with tastants in humans. *J. Exp. Psychol.: Anim. Behav. Proc.* 32: 215–228.

Yeomans, M. R., Prescott, J. and Gould, N. J. (2009). Acquired hedonic and sensory characteristics of odors: Influence of sweet liker and propylthiouracil taster status. *Q. J. Exp. Psychol.* 62: 1648–1664.

Zabernigg, A., Gamper, E-M., Giesinger, J., et al. (2010). Taste alterations in cancer patients receiving chemotherapy: A neglected side-effect? *The Oncologist* 15: 913–920.

Zampini, M., Sanabria, D., Phillips, N. and Spence, C. (2007). The multisensory perception of flavor: assessing the influence of color cues on flavor discrimination responses. *Food Qual. Pref.* 18: 975–984.

Zampini, M. and Spence, C. (2004). The role of auditory cues in modulating the perceived crispness and staleness of potato chips. *J. Sens. Stud.* 19: 347–363.

Zampini, M. and Spence, C. (2005). Modifying the multisensory perception of a carbonated beverage using auditory cues. *Food Qual. Pref.* 16: 632–641.

Zellner, D. A., Rozin, P., Aron, P. and Kulish, C. (1983). Conditioned enhancement of human's liking for flavor by pairing with sweetness. *Learn. Motiv.* 14: 338–350.

Zellner, D. A., Strickhouser, D. and Tornow, C. E. (2004). Disconfirmed hedonic expectations produce perceptual contrast, not assimilation. *Am. J. Psychol.* 117: 363–387.

Chapter 46

Neural Integration of Taste, Smell, Oral Texture, and Visual Modalities

EDMUND T. ROLLS

46.1 INTRODUCTION

46.1.1 Introduction and Overview

The aims of this chapter are to describe how taste, olfactory, visual, oral sensory, and other sensory inputs are combined in the brain, how a representation of reward value is produced, and how cognition and selective attention influence this processing.

Complementary neuronal recordings in primates, and functional neuroimaging in humans, show that the primary taste cortex in the anterior insula provides separate and combined representations of the taste, temperature, and texture (including fat texture) of food in the mouth independently of hunger and thus of reward value and pleasantness. One synapse on, in the orbitofrontal cortex, these sensory inputs are for some neurons combined by associative learning with olfactory and visual inputs, and these neurons encode food reward in that they only respond to food when hungry, and in that activations correlate with subjective pleasantness. Cognitive factors, including word-level descriptions, and selective attention to affective value, modulate the representation of the reward value of taste and olfactory stimuli in the orbitofrontal cortex and a region to which it projects, the anterior cingulate cortex, a tertiary taste cortical area. The food reward representations formed in this way play an important role in the control of appetite, food intake. Individual differences in these reward representations may contribute to obesity.

46.1.2 Food Reward and Appetite

A reason why it is important to understand the brain systems for food reward is that the reward value of food (i.e., whether we will work for a food), measures our appetite for a food, and whether we will eat a food. Thus normally we want food (will work for it, and will eat it) when we like it. "We want because we like": the goal value, the food reward value, makes us want it. For example, neurons in the orbitofrontal cortex and lateral hypothalamus described below respond to the reward value of a food when it is, for example, shown, and these neuronal responses predict whether that food will be eaten (Rolls, 1981, 2005a, 2014; Rolls et al., 1986, 1989). Similarly, in a whole series of studies on sensory-specific satiety in humans based on these discoveries, the reported pleasantness in humans of a food is closely correlated with whether it will then be eaten, and even with how much is eaten (Rolls et al., 1981b, 1983b, 1984). The situation when it has been suggested that wanting is not a result of liking (Berridge et al., 2009), is when behavior becomes a habit. A habit is a stimulus-response type of behavior that is no longer under control of the goal, but is under the control of an overlearned conditioned stimulus (Rolls, 2005a, 2014). The concept here is that food reward normally drives appetite and eating, and it is therefore important to understand the brain mechanisms involved in food reward, in order to understand the control of appetite and food intake.

Handbook of Olfaction and Gustation, Third Edition. Edited by Richard L. Doty.
© 2015 Richard L. Doty. Published 2015 by John Wiley & Sons, Inc.

46.1.3 Investigations in Primates including Humans

The focus of the approach taken here is on complementary neurophysiological investigations in macaques and functional neuroimaging in humans. There are a number of reasons for this focus.

First, there are major anatomical differences in the neural processing of taste in rodents and primates (Rolls and Scott, 2003; Scott and Small, 2009; Small and Scott, 2009). In rodents (and also in primates) taste information is conveyed by cranial nerves 7, 9, and 10 to the rostral part of the nucleus of the solitary tract (NTS) (Norgren and Leonard, 1971; Norgren and Leonard, 1973; Norgren, 1990) (see Figure 46.1). However, although in primates the NTS projects to the taste thalamus and thus to the cortex (Figure 46.1), in rodents the majority of NTS taste neurons responding to stimulation of the taste receptors of the anterior tongue project to the ipsilateral medial aspect of the pontine parabrachial nucleus (PbN), the rodent "pontine taste area" (Cho et al., 2002; Small and Scott, 2009). The remainder project to adjacent regions of the medulla. From the PbN the rodent gustatory pathway bifurcates into two pathways: (1) a ventral "affective" projection to the hypothalamus, central gray, ventral striatum, bed nucleus of the stria terminalis and amygdala; and (2) a dorsal "sensory" pathway, which first synapses in the thalamus and then the agranular and dysgranular insular gustatory cortex (Norgren and Leonard, 1971; Norgren, 1974, 1976, 1990). These regions, in turn, project back to the PbN in rodents to sculpt the gustatory code and guide complex feeding behaviors (Norgren, 1976; Di Lorenzo, 1990; Norgren, 1990; Li et al., 2002; Lundy and Norgren, 2004).

In contrast, in primates (including humans) there is strong evidence to indicate that the PbN gustatory relay is absent (Small and Scott, 2009): (1) Second-order gustatory projections that arise from rostral NTS appear not to synapse in the PbN and instead join the central tegmental tract and project directly to the taste thalamus in primates (Beckstead et al., 1980; Pritchard et al., 1989); (2) Despite several attempts, no one has successfully isolated taste responses in the monkey PbN (Norgren, 1990; Small and Scott, 2009) (the latter cite Ralph Norgren, personal communication and Tom Pritchard, personal communication); (3) In monkeys the projection arising from the PbN does not terminate in the region of ventral basal thalamus that contains gustatory responsive neurons (Pritchard et al., 1989).

Second, a functional difference of rodent taste processing from that of primates is that physical and chemical signals of satiety have been shown to reduce the taste responsiveness of neurons in the nucleus in the solitary tract, and the pontine taste area, of the rat, with decreases in the order of 30%, as follows (Rolls and Scott, 2003; Scott and Small, 2009). Gastric distension by air or with 0.3 M NaCl suppress responses in the NTS, with the greatest effect on glucose (Gleen and Erickson, 1976). Intravenous infusions of 0.5 g/kg glucose (Giza and Scott, 1983), 0.5 U/kg insulin (Giza and Scott, 1987b), and 40 μg/kg glucagon (Giza et al., 1993), all cause reductions in taste responsiveness to glucose in the NTS. The intraduodenal infusion of lipids causes a decline in taste responsiveness in the PbN, with the bulk of the suppression borne by glucose cells (Hajnal et al., 1999). The loss of signal that would otherwise be evoked by hedonically positive tastes implies that the reward value that sustains feeding is reduced at the brainstem level, making termination of a meal more likely (Giza et al., 1992). Further, if taste activity in NTS is affected by the rat's nutritional state, then intensity judgements in rats should change with satiety. There is evidence that they do. Rats with conditioned aversions to 1.0 M glucose show decreasing acceptance of glucose solutions as their concentrations approach 1.0 M. This acceptance gradient can be compared between euglycemic rats and those made hyperglycemic through intravenous injections (Scott and Giza, 1987). Hyperglycemic rats showed greater acceptance at all concentrations from 0.6 to 2.0 M glucose, indicating that they perceived these stimuli to be less intense than did conditioned rats with no glucose load (Giza and Scott, 1987a).

In contrast, in primates, the reward value of taste is represented in the orbitofrontal cortex in that the responses of orbitofrontal taste neurons are modulated by hunger in just the same way as is the reward value or palatability of a taste. In particular, it has been shown that orbitofrontal cortex taste neurons stop responding to the taste of a food with which a monkey is fed to satiety, and that this parallels the decline in the acceptability of the food (Rolls et al., 1989; Critchley and Rolls, 1996c). In contrast, the representation of taste in the primary taste cortex of primates (Scott et al., 1986; Yaxley et al., 1990) is not modulated by hunger (Rolls et al., 1988; Yaxley et al., 1988). Thus in the primary taste cortex of primates (and at earlier stages of taste processing including the nucleus of the solitary tract (Yaxley et al., 1985)), the reward value of taste is not represented, and instead the identity of the taste is represented (Rolls, 2014).

The importance of cortical processing of taste in primates, first for identity and intensity in the primary taste cortex, and then for reward value in the orbitofrontal cortex, is that both types of representation need to be interfaced to visual and other processing that requires cortical computation. For example, it may have adaptive value to be able to represent exactly what taste is present, and to link it by learning to the sight and location of the source of the taste, even when hunger and reward is not being produced, so that the source of that taste can be found in future, when it may have reward value. In line with cortical processing to dominate the processing of taste in primates, there is no

46.1 Introduction

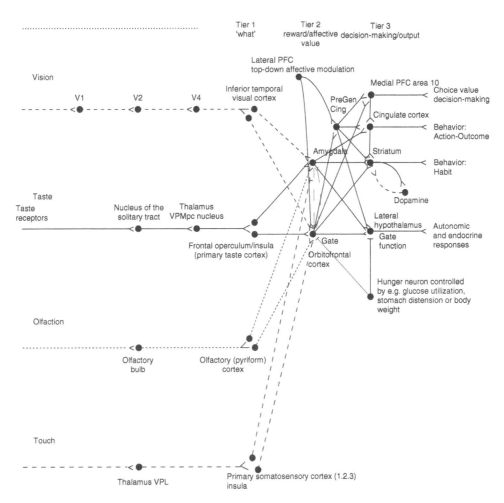

Figure 46.1 Schematic diagram showing some of the gustatory, olfactory, visual and somatosensory pathways to the orbitofrontal cortex, and some of the outputs of the orbitofrontal cortex, in primates. The secondary taste cortex, and the secondary olfactory cortex, are within the orbitofrontal cortex. V1 - primary visual cortex. V4 - visual cortical area V4. PreGen Cing – pregenual cingulate cortex. "Gate" refers to the finding that inputs such as the taste, smell, and sight of food in some brain regions only produce effects when hunger is present (Rolls, 2005a). Tier 1: the column of brain regions including and below the inferior temporal visual cortex represents brain regions in which "what" stimulus is present is made explicit in the neuronal representation, but not its reward or affective value which are represented in the next tier of brain regions (Tier 2), the orbitofrontal cortex and amygdala, and in the anterior cingulate cortex. In Tier 3 areas beyond these such as medial prefrontal cortex area 10, choices or decisions about reward value are taken, with the mechanisms described elsewhere (Rolls, 2008a; Rolls and Deco, 2010; Rolls, 2014). Top-down control of affective response systems by cognition and by selective attention from the dorsolateral prefrontal cortex is also indicated. Medial PFC area 10 – medial prefrontal cortex area 10; VPMpc – ventralposteromedial thalamic nucleus.

modulation of taste responsiveness at or before the primary taste cortex, and the pathways for taste are directly from the nucleus of the solitary tract in the brainstem to the taste thalamus and then to the taste cortex (Figure 46.1) (Rolls, 2014).

The implication is that taste, and the closely related olfactory and visual processing that contribute to food reward, are much more difficult to understand in rodents than in primates, partly because there is less segregation of "what" (identity and intensity) from hedonic processing in rodents, and partly because of the more serial hierarchical processing in primates (Figure 46.1).

Third, the prefrontal cortex (and for that matter the temporal lobe visual cortical areas) have also undergone great development in primates, and one part of the prefrontal cortex, the orbitofrontal cortex, is very little developed in rodents, yet is one of the major brain areas involved in taste and olfactory processing, and emotion and motivation, in primates including humans. Indeed, it has been argued (on the basis of cytoarchitecture, connections, and functions) that the granular prefrontal cortex is a primate innovation, and the implication of the argument is that any areas that might be termed orbitofrontal cortex in rats (Schoenbaum et al., 2009) are homologous only

1030 Chapter 46 Neural Integration of Taste, Smell, Oral Texture, and Visual Modalities

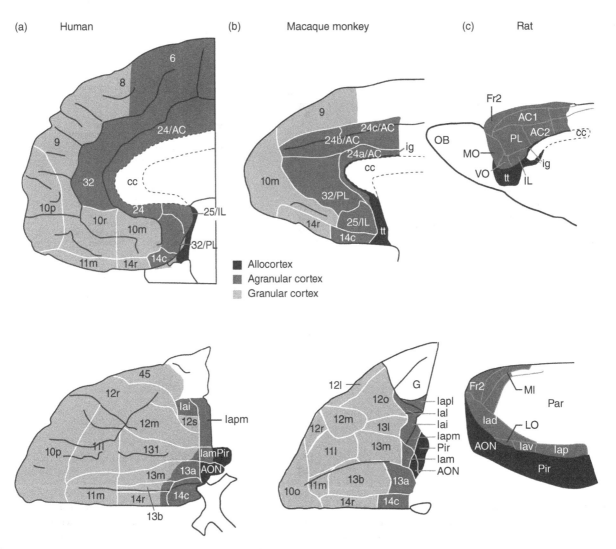

Figure 46.2 Comparison of the orbitofrontal (below) and medial prefrontal (above) cortical areas in humans, macaque monkeys, and rats. (a) Medial (top) and orbital (bottom) areas of the human frontal codex (Öngür et al. (2003)). (b) Medial (top) and orbital (bottom) areas of the macaque frontal cortex (Carmichael and Price (1994)). (c) Medial (top) and lateral (bottom) areas of rat frontal cortex (Palomero-Gallagher and Zilles (2004)). Rostral is to the left in all drawings. Top row: dorsal is up in all drawings. Bottom row: in (a) and (b), lateral is up; in (c), dorsal is up. Not to scale. Abbreviations: AC, anterior cingulate cortex; AON, anterior olfactory 'nucleus'; cc, corpus callosum; Fr2 second frontal area; Ia, agranular insular cortex; ig, induseum griseum; IL, infralimbic cortex; LO, lateral orbital cortex; MO, medial orbital cortex: OB, olfactory bulb; Pr, piriform (olfactory) cortex; PL, prelimbic cortex; tt, tenia tecta; VO, ventral orbital cortex; Subdivisions of areas are labelled caudal (c); inferior (i), lateral (l), medial (m); orbital (o), posterior or polar (p), rostral (r), or by arbitrary designation (a, b). Reproduced with permission from Passingham and Wise (2012). The Neurobiology of the Prefrontal Cortex. Oxford University Press: Oxford.

to the agranular parts of the primate orbitofrontal cortex (shaded mid gray in Figure 46.2), that is to areas 13a, 14c, and the agranular insular areas labelled Ia in Figure 46.2 (Passingham and Wise, 2012). It follows from that argument that for most areas of the orbitofrontal and medial prefrontal cortex in humans and macaques (those shaded light gray in Figure 46.2), special consideration must be given to research in macaques and humans. As shown in Figure 46.2, there may be no cortical area in rodents that is homologous to most of the primate including human orbitofrontal cortex (Preuss, 1995; Wise, 2008; Passingham and Wise, 2012).

46.2 TASTE PROCESSING IN THE PRIMATE BRAIN

46.2.1 Pathways

A diagram of the taste and related olfactory, somatosensory, and visual pathways in primates is shown in Figure 46.1.

The multimodal convergence that enables single neurons to respond to different combinations of taste, olfactory, texture, temperature, and visual inputs to represent different flavors produced often by new combinations of sensory input is a theme of that will be addressed.

46.2.2 The Insular Primary Taste Cortex

Rolls and colleagues have shown that the primary taste cortex in the primate anterior insula and adjoining frontal operculum contains not only taste neurons tuned to sweet, salt, bitter, sour (Scott et al., 1986; Yaxley et al., 1990; Scott and Plata-Salaman, 1999; Rolls and Scott, 2003), and umami as exemplified by monosodium glutamate (Baylis and Rolls, 1991; Rolls et al., 1996b), but also other neurons that encode oral somatosensory stimuli including viscosity, fat texture, temperature, and capsaicin (Verhagen et al., 2004). Some neurons in the primary taste cortex respond to particular combinations of taste and oral texture stimuli, but do not respond to olfactory stimuli or visual stimuli such as the sight of food (Verhagen et al., 2004). Neurons in the primary taste cortex do not represent the reward value of taste, that is the appetite for a food, in that their firing is not decreased to zero by feeding the taste to satiety (Rolls et al., 1988; Yaxley et al., 1988).

Parts of the insula can be activated by visual stimuli related to disgust, such as a face expression of disgust (Phillips et al., 2004), and this could reflect the fact that parts of the insula are part of the visceral efferent system involved in autonomic responses (Critchley, 2005) (and may even overlap partly with the taste-responsive areas (Simmons et al., 2013)).

46.2.3 The Secondary Taste Cortex in the Orbitofrontal Cortex

A secondary cortical taste area in primates was discovered by Rolls and colleagues (Thorpe et al., 1983; Rolls et al., 1989; Rolls et al., 1990) in the orbitofrontal cortex, extending several mm in front of the primary taste cortex. This is defined as a secondary cortical taste area, for it receives direct inputs from the primary taste cortex, as shown by a combined neurophysiological and anatomical pathway tracing investigation (Baylis et al., 1995). Different neurons in this region respond not only to each of the four classical prototypical tastes sweet, salt, bitter, and sour (Rolls et al., 1990, 2003a; Rolls, 1997; Verhagen et al., 2003; Kadohisa et al., 2005b), but also to umami tastants such as glutamate (which is present in many natural foods such as tomatoes, mushrooms, and milk) (Baylis and Rolls, 1991) and inosine monophosphate (which is present in meat and some fish such as tuna) (Rolls et al., 1996b). This evidence, taken together with the identification of glutamate taste receptors (Zhao et al., 2003; Maruyama et al., 2006), leads to the view that there are five prototypical types of taste information channels, with umami contributing, often in combination with corresponding olfactory inputs (Rolls et al., 1998; McCabe and Rolls, 2007; Rolls, 2009a), to the flavor of protein. In addition, other neurons respond to water (Rolls et al., 1990), and others to somatosensory stimuli including astringency as exemplified by tannic acid (Critchley and Rolls, 1996a), and capsaicin (Rolls et al., 2003a; Kadohisa et al., 2004).

Some of the coding principles are illustrated by the two neurons shown in Figure 46.3. The two neurons each

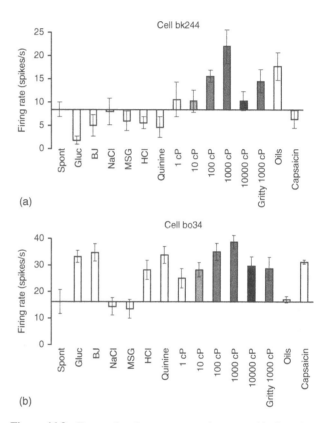

Figure 46.3 Taste and oral somatosensory inputs to orbitofrontal cortex neurons. (a) Firing rates (mean ± sem) of viscosity-sensitive neuron bk244 which did not have taste responses, in that it did not respond differentially to the different taste stimuli. The firing rates are shown to the viscosity series (carboxymethylcellulose 1 – 10,000 centiPoise, to the gritty stimulus (1,000 cP carboxymethylcellulose with Fillite microspheres), to the taste stimuli 1 M glucose (Gluc), 0.1 M NaCl, 0.1 M MSG , 0.01 M HCl and 0.001 M QuinineHCl, and to fruit juice (BJ). Spont = spontaneous firing rate. (b) Firing rates (mean ± sem) of viscosity-sensitive neuron bo34 which had no response to the oils (mineral oil, vegetable oil, safflower oil and coconut oil, which have viscosities that are all close to 50 cP). The neuron did not respond to the gritty stimulus in a way that was unexpected given the viscosity of the stimulus, was taste tuned, and did respond to capsaicin. (After Rolls, Verhagen and Kadohisa, 2003).

1032   Chapter 46   Neural Integration of Taste, Smell, Oral Texture, and Visual Modalities

have their independent tuning to the set of stimuli. It is this independent tuning or coding that underlies the ability of the brain to represent the exact nature of a stimulus or event, and this applies to taste in addition to other sensory modalities (Rolls et al., 2010d; Rolls and Treves, 2011). The encoding of information in the brain can of course not be captured by functional neuroimaging, because that takes an average of the activity of tens of thousands of neurons, whereas each neuron conveys information that is to a considerable extent independent of the information encoded by other neurons (Rolls et al., 2009; Rolls et al., 2010d; Rolls and Treves, 2011).

Taste responses are found in a large mediolateral extent of the orbitofrontal cortex (Critchley and Rolls, 1996a; Pritchard et al., 2005; Rolls, 2008b; Rolls and Grabenhorst, 2008). Indeed, taste neurons have been shown to extend throughout area 13 in a region that is approximately 7–12 mm from the midline (Rolls and Baylis, 1994; Critchley and Rolls, 1996a; Rolls et al., 1996b; Rolls, 2008b), the exact area in which Pritchard et al. (2005) also found a population of taste neurons (see Figure 46.3 with cytoarchitectonic areas indicated after Carmichael and Price, 1994; Öngür and Price, 2000; Petrides and Pandya, 2001; Öngür et al., 2003). We showed in our previous studies that these taste neurons extend from approximately 4 mm anterior to the clinoid process of the sphenoid bone to 12 mm anterior. Pritchard et al. (2005) focused their investigation on a region 5–9 mm anterior to the sphenoid.) Although Pritchard et al. (2005) commented that in their study there was a good proportion of taste neurons in this area, we, in comparing the proportions of taste neurons in different parts of the orbitofrontal cortex extending out laterally through area 12, find similar proportions of taste neurons throughout this mediolateral extent (from 7 mm to 20 mm lateral) (Rolls et al., 1989, 1990, 1996b, 2003a; Rolls and Baylis, 1994; Critchley and Rolls, 1996a; Verhagen et al., 2003; Kadohisa et al., 2004, 2005b). Moreover, even in area 13m, in the region 7–12 mm lateral where Pritchard et al., (2005) found taste neurons, we know that many other properties are represented, including oral texture as exemplified by astringency and fat texture (Critchley and Rolls, 1996a; Rolls et al., 1999); and olfactory properties (Critchley and Rolls, 1996b, c; Rolls et al., 1996) which can become associated by learning with taste stimuli (Rolls et al., 1996c). Thus area 13m contains taste, oral texture, and olfactory representations. Some of these cells are multimodal in these modalities (Rolls and Baylis, 1994; Critchley and Rolls, 1996a; Rolls et al., 1996c), and the majority of these neurons have their responses to taste and/or olfactory stimuli modulated by hunger (Critchley and Rolls, 1996c). In a more recent investigation, Rolls,

Verhagen, Gabbott, and Kadohisa measured the responses of 1753 neurons in rhesus macaques, and found taste neurons in the mid and medial orbitofrontal cortex region extending to within approximately 7 mm of the midline in area 13m, but very few in the more medial areas 10, 14, and 25, as illustrated in Figure 46.4 (Rolls, 2008b).

46.2.4 The Anterior Cingulate Cortex: A Tertiary Taste Cortical Area

The orbitofrontal cortex, including the extensive areas where taste neurons noted above are found, projects to the pregenual cingulate cortex area 32 (Carmichael and Price, 1996). Human imaging studies have shown that reward-related stimuli, such as the taste of sucrose and the texture of oral fat, activate the pregenual cingulate cortex (de Araujo and Rolls, 2004; Rolls, 2005a; Rolls and Grabenhorst, 2008; Rolls, 2009b; Grabenhorst and Rolls, 2011). However, little is known at the neuronal level of whether the responses of single neurons in the pregenual cingulate cortex are tuned to taste stimuli and respond differentially to different taste stimuli. Rolls, Gabbott, Verhagen, and Kadohisa therefore recorded from single neurons in the macaque pregenual cingulate cortex, in order to obtain evidence on these issues (Rolls et al., 2008b).

The responses of a pregenual cingulate cortex neuron with taste responses are shown in Figure 46.5. The neuron increased its firing rate primarily to glucose, fruit juice and cream, with some response to the oily texture of silicone oil, to monosodium glutamate, and to quinine. When the macaque was fed to satiety with glucose, the neuron showed a sensory-specific decrease in its response to the taste of glucose (Rolls, 2008b).

The data for the pregenual cingulate cortex and adjacent areas were obtained by Rolls, Gabbott, Verhagen, and Kadohisa (Rolls, 2008b) in two rhesus macaques in recordings that extended from approximately 10 mm anterior with respect to the sphenoid reference to approximately 13 mm anterior, with the recording sites of the neurons shown in Figure 46.6 (Rolls, 2008b). As shown, most of these neurons were in area 32, with one taste neuron in area 10. Although a small proportion of the neurons were classified as responding to taste, this proportion is not out of line with the proportion of taste neurons recorded with identical techniques in the same laboratory in the primary taste cortex in the macaque anterior insula and adjoining frontal opercular cortex. Of the 12 responsive neurons in the medial wall cortex, 11 had best responses to sweet stimuli (glucose and/or fruit juice, as illustrated in Figure 46.5), and one had best responses to quinine and NaCl. The spontaneous firing rates of neurons in the

46.2 Taste Processing in the Primate Brain

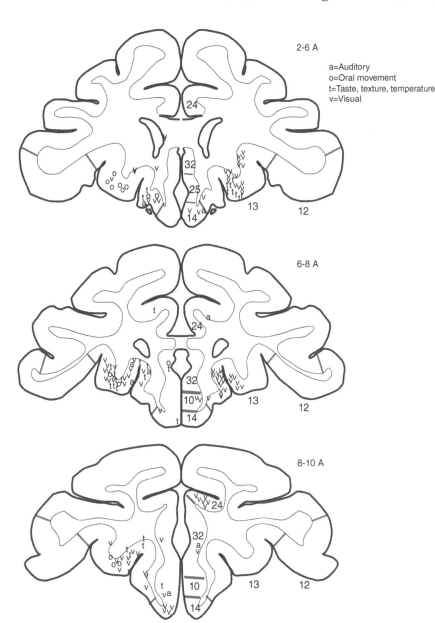

Figure 46.4 The reconstructed positions of the neurons in the medial orbitofrontal cortex with different types of response, with the cytoarchitectonic boundaries determined after Carmichael and Price (1994). The neurons within different planes at distances in mm anterior (a) to the sphenoid reference point are shown on the coronal sections. (Data from Rolls, Verhagen, Gabbott and Kadohisa, 2008; see Rolls, 2008b).

pregenual cingulate cortex were typically in the range 0–5 spikes/s, which increased significantly to 20–30 spikes/s when the neurons were responding selectively to specific taste stimuli.

The presence of a neuronal representation of a primary (unlearned) reinforcer, taste, in the pregenual cingulate cortex is of importance for understanding the functions more generally of the anterior cingulate cortex in complex reward-related learning (Amiez et al., 2006), and in action selection (Grabenhorst and Rolls, 2011; Rushworth et al., 2011), for this evidence shows that primary rewards are represented in at least one part of the anterior cingulate cortex – the pregenual cingulate cortex area 32 (Rolls, 2008b; Rolls, 2009b). Neurons responding to fruit juice used as a reinforcer in saccade countermanding have been found in the dorsal part of the anterior cingulate sulcus area 24c (Ito et al., 2003), and the pregenual cingulate cortex provides a source of inputs to area 24 (Carmichael and Price, 1996). Indeed, establishing that the pregenual cingulate cortex contains a representation of a primary reinforcer, in this case a taste, is of importance more generally in relation to understanding the functions of the pregenual cingulate cortex in emotion (Rolls and Grabenhorst, 2008; Rolls, 2009b; Grabenhorst and Rolls, 2011), in that for example some disorders of emotion in humans produced by anterior cingulate damage include deficits in responding to what are probably other primary reinforcers, face and voice expression (Hornak et al., 2003; Rolls, 2005a, 2014).

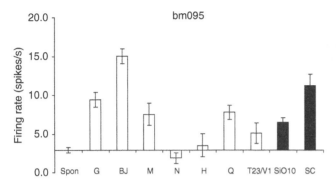

Figure 46.5 Responses of a pregenual cingulate cortex neuron (bm095) with differential responses to tastes and oral fat texture stimuli. The mean (±sem) firing rate responses to each stimulus calculated in a 5 s period over several trials are shown. The spontaneous (Spon) firing rate of 3 spikes/s is shown by the horizontal line, with the responses indicated relative to this line. The taste stimuli were 1 M glucose (G), blackcurrant fruit juice (BJ), 0.1 M NaCl (N), 0.1 M MSG (M), 0.01 M HCl (H) and 0.001 M QuinineHCl (Q); water (T23/V1); single cream (SC); and silicone oil with a viscosity of 10 cP (SiO10). The neuron had significantly different responses to the different stimuli as shown by a one-way ANOVA (F[9,46]=17.7, $p<10^{-10}$). (Data from Rolls, Gabbott, Verhagen, and Kadohisa; see Rolls, 2008b).

46.2.5 The Pleasantness of the Taste of Food, Sensory-Specific Satiety, and the Effects of Variety on Food Intake

The modulation of the reward value of a sensory stimulus such as the taste of food by motivational state, for example hunger, is one important way in which motivational behavior is controlled (Rolls, 2005a, 2007, 2014). The subjective correlate of this modulation is that food tastes pleasant when hungry, and tastes hedonically neutral when it has been eaten to satiety. Following Edmund Rolls' discovery of sensory-specific satiety revealed by the selective reduction in the responses of lateral hypothalamic neurons to a food eaten to satiety (Rolls, 1981; Rolls et al., 1986), it has been shown that this is implemented by neurons in a region that projects to the hypothalamus, the orbitofrontal (secondary taste) cortex, for the taste, odor, and sight of food (Rolls et al., 1989; Critchley and Rolls, 1996c) (Figure 46.7).

This evidence shows that the reduced acceptance of food that occurs when food is eaten to satiety, the reduction in the pleasantness of its taste and flavor, and the effects of variety to increase food intake (Cabanac, 1971; Rolls and Hetherington, 1989; Rolls and Rolls, 1977, 1982, 1997; Rolls et al., 1981a, b, 1982, 1983a, b, 1984; Hetherington, 2007), are produced in the primate orbitofrontal cortex, but not at earlier stages of processing

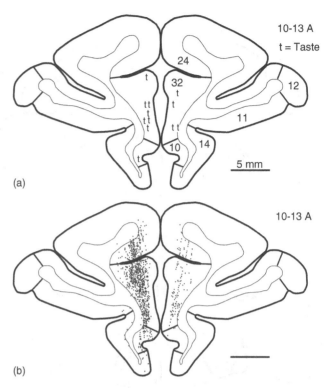

Figure 46.6 The reconstructed positions of the anterior cingulate neurons with taste (t) responses, together with the cytoarchitectonic boundaries determined by Carmichael and Price (1994). Most (11/12) of the taste neurons were in the pregenual cingulate cortex (area 32), as shown. The neurons are shown on a coronal section at 12 mm anterior (a) to the sphenoid reference point. The recording sites were reconstructed using X-radiographs made on every track and subsequent histology using the methods described by Rolls et al., (2003a). (b). The locations of all the 749 neurons recorded in the anterior cingulate region in this study are indicated to show the regions sampled. (Data from Rolls, Gabbott, Verhagen, and Kadohisa; see Rolls, 2008b).

including the insular-opercular primary taste cortex (Rolls et al., 1988; Yaxley et al., 1988) and the nucleus of the solitary tract (Yaxley et al., 1985), where the responses reflect factors such as the intensity of the taste, which is little affected by satiety (Rolls et al., 1983c; Rolls and Grabenhorst, 2008). In addition to providing an implementation of sensory-specific satiety (probably by adaptation of the synaptic afferents to orbitofrontal neurons with a time course of the order of the length of a course of a meal), it is likely that visceral and other satiety-related signals reach the orbitofrontal cortex (as indicated in Figure 46.1) (from the nucleus of the solitary tract, via thalamic and possibly hypothalamic nuclei) and there modulate the representation of food, resulting in an output that reflects the reward (or appetitive) value of each food (Rolls, 2014).

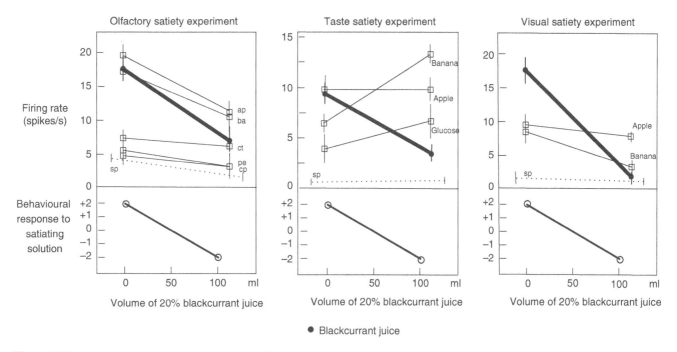

Figure 46.7 Orbitofrontal cortex neuron with visual, olfactory and taste responses, showing the responses before and after feeding to satiety with blackcurrant juice. The solid circles show the responses to blackcurrant juice. The olfactory stimuli included apple (ap), banana (ba), citral (ct), phenylethanol (pe), and caprylic acid (cp). The spontaneous firing rate of the neuron is shown (sp). The taste panel is for the flavor of food in the mouth. Below the neuronal response data for each experiment, the behavioral measure of the acceptance or rejection of the solution on a scale from +2 to -2 is shown. The values shown are the mean firing rate and its s.e. (After Critchley and Rolls, 1996c).

46.2.6 The Representation of Flavor: Convergence of Olfactory, Taste, and Visual Inputs in the Orbitofrontal Cortex and Related Areas Including the Amygdala nd Hippocampus

Taste and olfactory pathways are brought together in the orbitofrontal cortex where flavor is formed by learned associations at the neuronal level between these inputs (see Figure 46.1) (Rolls and Baylis, 1994; Critchley and Rolls, 1996b; Rolls et al., 1996). Visual inputs also become associated by learning in the orbitofrontal cortex with the taste of food to represent the sight of food and contribute to flavor (Thorpe et al., 1983; Rolls et al., 1996c). The visual and olfactory as well as the taste inputs represent the reward value of the food, as shown by sensory-specific satiety effects (Critchley and Rolls, 1996c). Olfactory-to-taste associative learning by these orbitofrontal cortex neurons may take 30–40 trials to reverse in an olfactory-to-taste discrimination task, and this may help to make a flavor stable (Rolls et al., 1996c). Olfactory neurons are found in a considerable anterior-posterior extent of the primate orbitofrontal cortex, extending far into areas 11 and 14 (Rolls and Baylis, 1994; Critchley and Rolls, 1996b, c; Rolls et al., 1996c, 1996), and are not restricted to a posterior region as some have thought (Gottfried and Zald, 2005).

Visual-to-taste association learning and its reversal by neurons in the orbitofrontal cortex can take place in as little as one trial (Thorpe et al., 1983; Rolls et al., 1996c; Deco and Rolls, 2005a). This has clear adaptive value in enabling particular foods with a good or bad taste to be learned and recognized quickly, important in foraging and in food selection for ingestion. The visual inputs reach the orbitofrontal cortex from the inferior temporal visual cortex, where neurons respond to objects independently of their reward value (e.g., taste), as shown by satiety and reversal learning tests (Rolls et al., 1977; Rolls, 2008a, 2012c). The visual-to-taste associations are thus learned in the orbitofrontal cortex (Rolls, 2014). These visual–taste neurons thus respond to expected value (and in humans different orbitofrontal cortex neurons signal expected monetary value (Rolls et al., 2008a)).

Different neurons in the orbitofrontal cortex respond when a visually signalled expected taste reward is not obtained, that is, to negative reward prediction error (Thorpe et al., 1983; Rolls and Grabenhorst, 2008). There is evidence that dopamine neurons in the ventral tegmentum respond to positive reward prediction error (Schultz, 2007), and as such, they do not respond to taste reward (Rolls,

2014). The inputs to the dopamine neurons may originate from structures such as the orbitofrontal cortex, where expected value, reward outcome (e.g., taste), and negative reward prediction error are represented (Rolls, 2014).

The amygdala also contains neurons that respond to taste and oral texture (Sanghera et al., 1979; Scott et al., 1993; Kadohisa et al., 2005b, 2005a). Some neurons in the primate amygdala respond to visual stimuli associated with reinforcers such as taste, but do not reflect the reinforcing properties very specifically, do not rapidly learn and reverse visual-to-taste associations, and are much less affected by reward devaluation by feeding to satiety than are orbitofrontal cortex neurons (Sanghera et al., 1979; Yan and Scott, 1996; Kadohisa et al., 2005a, b; Wilson and Rolls, 2005; Rolls, 2014). The primate orbitofrontal cortex appears to be much more closely involved in flexible (rapidly learned, and affected by reward devaluation) representations than is the primate amygdala (Rolls, 2014).

The primate hippocampus contains neurons that respond to the sight of locations in spatial scenes where taste rewards are found, but not to the sight of objects associated with taste reward (Rolls and Xiang, 2005), and this is part of the evidence for understanding the functions of the hippocampus in episodic memory (Rolls, 2008a, 2010b).

46.2.7 The Texture of Food, Including Fat Texture

46.2.7.1 Viscosity, Particulate Quality, and Astringency.
Some orbitofrontal cortex neurons have oral texture-related responses that encode parametrically the viscosity of food in the mouth (shown using a methyl cellulose series in the range 1–10,000 centiPoise), and others independently encode the particulate quality of food in the mouth, produced quantitatively, for example by adding 20–100 μm microspheres to methyl cellulose (Rolls et al., 2003a) (see Figure 46.3). Somatosensory signals that transmit information about capsaicin (chilli) and astringency are also reflected in neuronal activity in these cortical areas (Critchley and Rolls, 1996a; Kadohisa et al., 2004, 2005b).

46.2.7.2 Oral Fat Texture.
Texture in the mouth is an important indicator of whether *fat* is present in a food, which is important not only as a high value energy source, but also as a potential source of essential fatty acids. In the orbitofrontal cortex, Rolls, Critchley et al., (1999) have found a population of neurons that responds when fat is in the mouth. The fat-related responses of these neurons are produced at least in part by the texture of the food rather than by chemical receptors sensitive to certain chemicals, in that such neurons typically respond not only to foods such as cream and milk containing fat, but also to paraffin oil (which is a pure hydrocarbon) and to silicone oil [$Si(CH_3)_2O)_n$]. Moreover, the texture channels through which these fat-sensitive neurons are activated are separate from viscosity sensitive channels, in that the responses of these neurons cannot be predicted by the viscosity of the oral stimuli, as illustrated in Figure 46.8 (Verhagen et al., 2003; Rolls, 2011a). The responses of these oral fat-encoding neurons are not related to free fatty acids such as linoleic or lauric acid (Verhagen et al., 2003; Kadohisa et al., 2005b; Rolls, 2011a), and the fat responsiveness of these primate orbitofrontal cortex

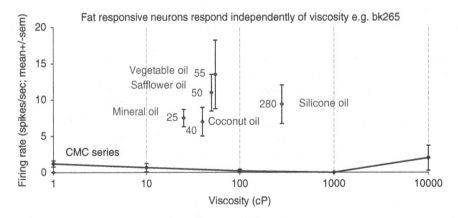

Figure 46.8 A neuron in the primate orbitofrontal cortex responding to the texture of fat in the mouth independently of viscosity. The cell (bk265) increased its firing rate to a range of fats and oils (the viscosity of which is shown in centipoise). The information that reaches this type of neuron is independent of a viscosity sensing channel, in that the neuron did not respond to the methyl cellulose (CMC) viscosity series. The neuron responded to the texture rather than the chemical structure of the fat in that it also responded to silicone oil ($Si(CH_3)_2O)_n$) and paraffin (mineral) oil (hydrocarbon). Some of these neurons have taste inputs. (After Verhagen, Rolls and Kadohisa, 2003).

neurons is therefore not related to fatty acid sensing (Gilbertson et al., 1997; Gilbertson, 1998), but instead to oral texture sensing (Rolls, 2011a; Rolls, 2012b). This has I believe very important implications for the development of foods with the mouth feel of fat, but low energy content (Rolls, 2011a, 2012b). A few neurons do have responses to linoleic and/or lauric acid, but these neurons do not respond to fat in the mouth, and may reflect the bad taste that rancid fats may have (Verhagen et al., 2003; Rolls, 2011a). Some of the fat texture-related orbitofrontal cortex neurons do though have convergent inputs from the chemical senses, in that in addition to taste inputs, some of these neurons respond to the odour associated with a fat, such as the odour of cream (Rolls et al., 1999). Feeding to satiety with fat (e.g., cream) decreases the responses of these neurons to zero on the food eaten to satiety, but if the neuron receives a taste input from for example glucose taste, that is not decreased by feeding to satiety with cream (Rolls et al., 1999). Thus there is a representation of the macronutrient fat in this brain area, and the activation produced by fat is reduced by eating fat to satiety.

Fat texture, oral viscosity, and temperature, for some neurons in combination with taste, are represented in the macaque primary taste cortex in the rostral insula and adjoining frontal operculum (Verhagen et al., 2004), which provides a route for this information to reach the orbitofrontal cortex.

These oral sensory properties of food, and also the sight and smell of food, are also represented in the primate amygdala (Rolls, 2000; Rolls and Scott, 2003; Kadohisa et al., 2005a, 2005b). Interestingly, the responses of these amygdala neurons do not correlate well with the preferences of the macaques for the oral stimuli (Kadohisa et al., 2005b), and feeding to satiety does not produce the large reduction in the responses of amygdala neurons to food (Rolls, 2000; Rolls and Scott, 2003) that is typical of orbitofrontal cortex neurons.

46.2.7.3 Oral Temperature.
In addition, we have shown that some neurons in the insular cortex, orbitofrontal cortex, and amygdala reflect the temperature of substances in the mouth, and that this temperature information is represented independently of other sensory inputs by some neurons, and in combination with taste or texture by other neurons (Kadohisa et al., 2004, 2005a, b; Verhagen et al., 2004). Somatosensory signals that transmit information about capsaicin (chili) are also reflected in neuronal activity in these brain areas (Kadohisa et al., 2004, 2005b). Activations in the human orbitofrontal and insular taste cortex also reflect oral temperature (Guest et al., 2007).

46.3 IMAGING STUDIES IN HUMANS

46.3.1 Taste

In humans it has been shown in neuroimaging studies using functional Magnetic Resonance Imaging (fMRI) that taste activates an area of the anterior insula/frontal operculum, which is probably the primary taste cortex (O'Doherty et al., 2001; de Araujo et al., 2003b; Small, 2010), and part of the orbitofrontal cortex, which is probably the secondary taste cortex (Francis et al., 1999; O'Doherty et al., 2001; de Araujo et al., 2003b; Rolls, 2005b, 2008b). We pioneered the use of a tasteless control with the same ionic constituents as saliva (O'Doherty et al., 2001; de Araujo et al., 2003b), as water can activate some neurons in cortical taste areas (Rolls et al., 1990) and can activate the taste cortex (de Araujo et al., 2003b). Within individual subjects separate areas of the orbitofrontal cortex are activated by sweet (pleasant) and by salt (unpleasant) tastes (O'Doherty et al., 2001).

The primary taste cortex in the anterior insula of humans represents the identity and intensity of taste in that activations there correlate with the subjective intensity of the taste, and the orbitofrontal and anterior cingulate cortex represents the reward value of taste, in that activations there correlate with the subjective pleasantness of taste (Grabenhorst and Rolls, 2008; Grabenhorst et al., 2008a) (Figure 46.9).

We also found activation of the human amygdala by the taste of glucose (Francis et al., 1999). Extending this study, O'Doherty et al., (2001) showed that the human amygdala was as much activated by the affectively pleasant taste of glucose as by the affectively negative taste of NaCl, and thus provided evidence that the human amygdala is not especially involved in processing aversive as compared to rewarding stimuli. Zald et al. (1998, 2002) also showed that the amygdala, as well as the orbitofrontal cortex, respond to aversive (e.g., quinine) and to sucrose taste stimuli.

Umami taste stimuli, of which an exemplar is monosodium glutamate (MSG) and which capture what is described as the taste of protein, activate the insular (primary), orbitofrontal (secondary), and anterior cingulate (tertiary (Rolls, 2008b)) taste cortical areas (de Araujo et al., 2003a). When the nucleotide 0.005 M inosine 5'-monophosphate (IMP) was added to MSG (0.05 M), the BOLD (blood oxygenation-level dependent) signal in an anterior part of the orbitofrontal cortex showed supralinear additivity, and this may reflect the subjective enhancement of umami taste that has been described when IMP is added to MSG (Rolls, 2009a). (The supra-linear additivity refers to a greater activation to the combined

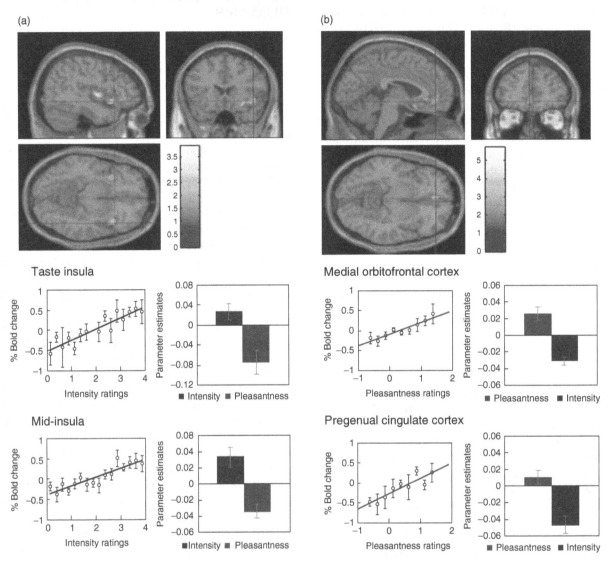

Figure 46.9 Effect of paying attention to the pleasantness vs the intensity of a taste stimulus. (a) Top: A significant difference related to the taste period was found in the taste insula at [42 18−14] $z=2.42$ $p<0.05$ (indicated by the cursor) and in the mid insula at [40−2 4] $z=3.03$ $p<0.025$. Middle: Taste Insula. Right: The parameter estimates (mean ± sem across subjects) for the activation at the specified coordinate for the conditions of paying attention to pleasantness or to intensity. The parameter estimates were significantly different for the taste insula $t=4.5$, df=10, $p=0.001$. Left: The correlation between the intensity ratings and the activation (% BOLD change) at the specified coordinate ($r=0.91$, df=14, $p\ll0.001$). Bottom: Mid Insula. Right: The parameter estimates (mean ± sem across subjects) for the activation at the specified coordinate for the conditions of paying attention to pleasantness or to intensity. The parameter estimates were significantly different for the mid insula $t=5.02$, df=10, $p=0.001$. Left: The correlation between the intensity ratings and the activation (% BOLD change) at the specified coordinate ($r=0.89$, df=15, $p\ll0.001$). The taste stimulus, monosodium glutamate, was identical on all trials. (b) Top: A significant difference related to the taste period was found in the medial orbitofrontal cortex at [−6 14−20] $z=3.81$, $p<0.003$ (towards the back of the area of activation shown) and in the pregenual cingulate cortex at [−4 46−8] $z=2.90$, $p<0.04$ (at the cursor). Middle: Medial orbitofrontal cortex. Right: The parameter estimates (mean ± sem across subjects) for the activation at the specified coordinate for the conditions of paying attention to pleasantness or to intensity. The parameter estimates were significantly different for the orbitofrontal cortex $t=7.27$, df=11, $p<10^{-4}$. Left: The correlation between the pleasantness ratings and the activation (% BOLD change) at the specified coordinate ($r=0.94$, df=8, $p\ll0.001$). Bottom: Pregenual cingulate cortex. Conventions as above. Right: The parameter estimates were significantly different for the pregenual cingulate cortex $t=8.70$, df=11, $p<10^{-5}$. Left: The correlation between the pleasantness ratings and the activation (% BOLD change) at the specified coordinate ($r=0.89$, df=8, $p=0.001$). The taste stimulus, 0.1 M monosodium glutamate, was identical on all trials. (After Grabenhorst and Rolls, 2008). (*See plate section for color version.*)

stimulus MSG+IMP than to the sum of the activations to MSG and IMP presented separately. This evidence that the effect of the combination is greater than the sum of its parts indicates an interaction between the parts to form in this case an especially potent taste of umami, which is part of what can make a food taste delicious (Rolls, 2009a).) Overall, these results illustrate that the responses of the brain can reflect inputs produced by particular combinations of sensory stimuli with supralinear activations, and that the combination of sensory stimuli may be especially represented in particular brain regions, and may help to make the food pleasant.

46.3.2 Odor

In humans, in addition to activation of the pyriform (olfactory) cortex (Zald and Pardo, 1997; Sobel et al., 2000; Poellinger et al., 2001), there is strong and consistent activation of the orbitofrontal cortex by olfactory stimuli (Zatorre et al., 1992; Francis et al., 1999; Rolls et al., 2003b). This region appears to represent the pleasantness of odor, as shown by a sensory-specific satiety experiment with banana vs vanilla odor (O'Doherty et al., 2000), and this has been confirmed by Gottfried et al. (personal communication, see Gottfried, 2013), who also showed that activations in the pyriform (primary olfactory) cortex were not decreased by odor devaluation by satiety. Further, pleasant odors tend to activate the medial, and unpleasant odors the more lateral, orbitofrontal cortex (Rolls et al., 2003b), adding to the evidence that it is a principle that there is a hedonic map in the orbitofrontal cortex, and also in the anterior cingulate cortex, which receives inputs from the orbitofrontal cortex (Rolls and Grabenhorst, 2008; Grabenhorst and Rolls, 2011). The primary olfactory (pyriform) cortex represents the identity and intensity of odor in that activations there correlate with the subjective intensity of the odor, and the orbitofrontal and anterior cingulate cortex represents the reward value of odor, in that activations there correlate with the subjective pleasantness (medially) or unpleasantness (laterally) of odor (Rolls et al., 2003b, 2008c, 2009; Grabenhorst et al., 2007; Rolls and Grabenhorst, 2008; Rolls et al., 2008c; Rolls et al., 2009; Grabenhorst and Rolls, 2011).

46.3.3 Olfactory-Taste Convergence to Represent Flavor, and the Influence of Satiety on Flavor Representations

Taste and olfactory conjunction analyses, and the measurement of supradditive effects indicating convergence and interactions, showed convergence for taste (sucrose) and odor (strawberry) in the orbitofrontal and anterior cingulate cortex, and activations in these regions were correlated with the pleasantness ratings given by the participants (de Araujo et al., 2003c; Small et al., 2004; Small and Prescott, 2005). These results provide evidence on the neural substrate for the convergence of taste and olfactory stimuli to produce flavor in humans, and where the pleasantness of flavor is represented in the human brain. The first region where the effects of this convergence are found is in an agranular part of what cytoarchitecturally is the insula (Ia, at Y=15) that is topologically found in the posterior orbitofrontal cortex, though it is anterior to the insular taste cortex, and posterior to the granular orbitofrontal cortex (see Figure 46.2) (de Araujo et al., 2003c).

McCabe and Rolls (2007) have shown that the convergence of taste and olfactory information appears to be important for the delicious flavor of umami. They showed that when glutamate is given in combination with a consonant, savory, odor (vegetable), the resulting flavor can be much more pleasant than the glutamate taste or vegetable odour alone, and that this reflected activations in the pregenual cingulate cortex and medial orbitofrontal cortex. The principle is that certain sensory combinations can produce very pleasant food stimuli, which may of course be important in driving food intake; and that these combinations are formed in the brain far beyond the taste or olfactory receptors (Rolls, 2009a).

To assess how satiety influences the brain activations to a whole food which produces taste, olfactory, and texture stimulation, we measured brain activation by whole foods before and after the food is eaten to satiety. The foods eaten to satiety were either chocolate milk, or tomato juice. A decrease in activation by the food eaten to satiety relative to the other food was found in the orbitofrontal cortex (Kringelbach et al., 2003) but not in the primary taste cortex. This study provided evidence that the pleasantness of the flavor of food, and sensory-specific satiety, are represented in the orbitofrontal cortex.

46.3.4 Oral Viscosity and Fat Texture

The viscosity of food in the mouth is represented in the human primary taste cortex (in the anterior insula), and also in a mid-insular area that is not taste cortex, but which represents oral somatosensory stimuli (de Araujo and Rolls, 2004). Oral viscosity is also represented in the human orbitofrontal and perigenual cingulate cortices, and it is notable that the perigenual cingulate cortex, an area in which many pleasant stimuli are represented, is strongly activated by the texture of fat in the mouth and also by oral sucrose (de Araujo and Rolls, 2004). We have shown that the pleasantness and reward value of fat texture is represented in the mid-orbitofrontal and anterior cingulate cortex, where activations are correlated with the subjective pleasantness of oral fat texture (Rolls, 2009a; Grabenhorst

1040 Chapter 46 Neural Integration of Taste, Smell, Oral Texture, and Visual Modalities

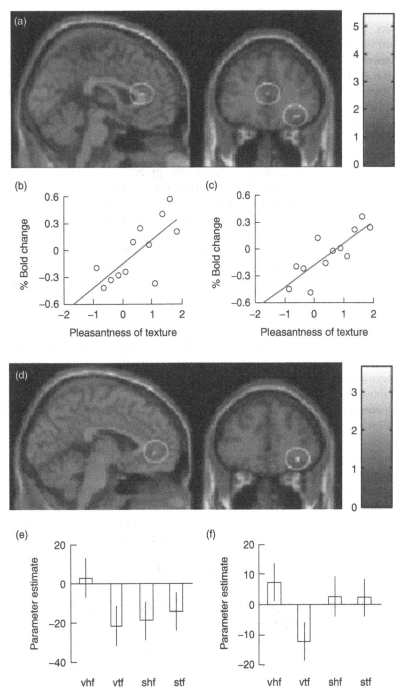

Figure 46.10 Brain regions in which the activations were correlated with the subjective pleasantness of fat texture: Mid-orbitofrontal cortex ([32 34 –14] z=3.38 p=0.013) (a, yellow circle, c showing the relation between the % change in the BOLD signal and the rating of the pleasantness of the texture) and anterior cingulate cortex ([2 30 14] z=3.22, p=0.016) (a, pink circles, and b). (After Grabenhorst et al., 2010). (*See plate section for color version.*)

et al., 2010; Rolls, 2010a) (Figure 46.10). This provides a foundation for studies of whether activations in the fat reward system are heightened in people who tend to become obese (Rolls, 2012a). Interestingly, high fat stimuli with a pleasant flavor increase the coupling of activations between the orbitofrontal cortex and somatosensory cortex, suggesting a role for the somatosensory cortex in processing the sensory properties of food in the mouth (Grabenhorst and Rolls, 2013).

46.3.5 The Sight of Food

O'Doherty et al., (2002) showed that visual stimuli associated with the taste of glucose activated the orbitofrontal cortex and some connected areas, consistent with the primate neurophysiology. Simmons, Martin and Barsalou (2005) found that showing pictures of foods, compared to pictures of places, can also activate the orbitofrontal cortex. Similarly, the orbitofrontal cortex and connected areas were also found to be activated after presentation of food stimuli to food-deprived subjects (Wang et al., 2004).

46.3.6 Top-Down Cognitive Effects on Taste, Olfactory, and Flavor Processing

To what extent does cognition influence the hedonics of food-related stimuli, and how far down into the sensory system does the cognitive influence reach? To address this, we performed an fMRI investigation in which the delivery of a standard test odor (isovaleric acid combined with cheddar cheese odour, presented orthonasally using an olfactometer) was paired with a descriptor word on a screen, which on different trials was "Cheddar cheese" or "Body odor." Participants rated the affective value of the test odor as significantly more pleasant when labeled "Cheddar Cheese" than when labeled "Body odour," and these effects reflected activations in the medial orbitofrontal cortex (OFC) / rostral anterior cingulate cortex (ACC) that had correlations with the pleasantness ratings (de Araujo et al., 2005). The implication is that cognitive factors can have profound effects on our responses to the hedonic and sensory properties of food, in that these effects are manifest quite far down into sensory and hedonic processing (in the orbitofrontal cortex, see Figure 46.1), so that hedonic representations of odors are affected (de Araujo et al., 2005).

Similar cognitive effects and mechanisms have now been found for the taste and flavor of food, where the cognitive word level descriptor was for example "rich delicious flavor" and activations to flavor were increased in the orbitofrontal cortex and regions to which it projects including the pregenual cingulate cortex and ventral striatum, but were not influenced in the insular primary taste cortex where activations reflected the intensity (concentration) of the stimuli (Grabenhorst et al., 2008a) (see Figure 46.11).

46.3.7 Effects of Selective Attention to Affective Value Versus Intensity on Representations of Taste, Olfactory, and Flavor Processing

We have found that with taste, flavor, and olfactory food-related stimuli, selective attention to pleasantness modulates representations in the orbitofrontal cortex (see Figure 46.9), whereas selective attention to intensity modulates activations in areas such as the primary taste cortex (Grabenhorst and Rolls, 2008; Rolls et al., 2008c). Thus, depending on the context in which tastes and odors are presented and whether affect is relevant, the brain responds to a taste, odor or flavor differently. These findings show that when attention is paid to affective value, the brain systems engaged to represent the stimulus are different from those engaged when attention is directed to the physical properties of a stimulus such as its intensity.

The source of the top-down modulation by attention of the orbitofrontal cortex appears to be the lateral prefrontal cortex, as shown by PPI (psychophysiological interaction) analyses (Grabenhorst and Rolls, 2010), and by Granger causality analyses (Ge et al., 2012; Luo et al., 2013). The mechanism probably involves a weak top-down biased competition effect on the taste and olfactory processing (Desimone and Duncan, 1995; Deco and Rolls, 2005b; Rolls, 2008a). Because whole streams of cortical processing are influenced (orbitofrontal and cingulate cortex, and even their coupling to the primary taste cortex, by pleasantness-related processing; and insular taste cortex and the mid-insula by intensity-related processing (Grabenhorst and Rolls, 2010; Luo et al., 2013), the process has been described as a biased activation model of attention (Grabenhorst and Rolls, 2010).

This differential biasing by prefrontal cortex attentional mechanisms (Grabenhorst and Rolls, 2010; Ge et al., 2012) of brain regions engaged in processing a sensory stimulus depending on whether the cognitive demand is for affect-related vs more sensory-related processing may be an important aspect of cognition and attention which have implications for how strongly the reward system is driven by food, and thus for eating and the control of appetite (Grabenhorst and Rolls, 2008; Rolls et al., 2008c; Grabenhorst and Rolls, 2011; Rolls, 2012a). The top-down modulations of processing have many implications for investigations of taste, olfactory, and other sensory processing, and for the development of new food and perfumery products.

46.3.8 Beyond Reward Value to Decision Making

Representations of the reward value of food, and their subjective correlate the pleasantness of food, are fundamental in determining appetite and processes such as economic decision making (Rolls, 2005a; Padoa-Schioppa, 2011; Padoa-Schioppa and Cai, 2011; Rolls, 2014). But after the reward evaluation, a decision has to be made about whether to seek for and consume the reward. We are now starting to understand how the brain takes decisions as described in *The Noisy Brain* (Rolls and Deco, 2010), and this has implications for whether a reward of a particular value will be selected (Rolls, 2008a; Rolls and Grabenhorst, 2008; Rolls and Deco, 2010; Grabenhorst and Rolls, 2011; Rolls, 2011b; Deco et al., 2013; Rolls, 2014).

A tier of processing beyond the orbitofrontal cortex, in the medial prefrontal cortex area 10, becomes engaged when choices are made between odor stimuli based on their pleasantness (Grabenhorst et al., 2008b; Rolls et al., 2010b, a; Rolls et al., 2010c) (tier 3 in Figure 46.1).

1042 Chapter 46 Neural Integration of Taste, Smell, Oral Texture, and Visual Modalities

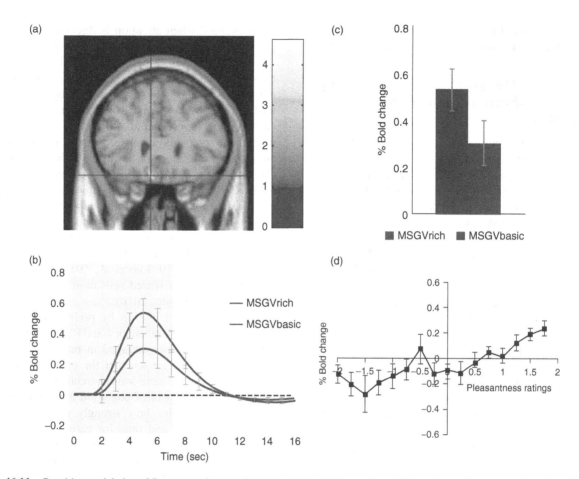

Figure 46.11 Cognitive modulation of flavor reward processing in the brain. (a). The medial orbitofrontal cortex was more strongly activated when a flavor stimulus was labelled "rich and delicious flavor" (MSGVrich) than when it was labelled "boiled vegetable water" (MSGVbasic) ([−8 28 −20]). (The flavor stimulus, MSGV, was the taste 0.1 M MSG + 0.005 M inosine 5'monophosphate combined with a consonant 0.4% vegetable odor.) (b). The timecourse of the BOLD signals for the two conditions. (c). The peak values of the BOLD signal (mean across subjects ± sem) were significantly different ($t=3.06$, df=11, $p=0.01$). (d). The BOLD signal in the medial orbitofrontal cortex was correlated with the subjective pleasantness ratings of taste and flavor, as shown by the SPM analysis, and as illustrated (mean across subjects ±sem, $r=0.86$, $p<0.001$). (After Grabenhorst, Rolls and Bilderbeck, 2008 (Grabenhorst et al., 2008a)). (*See plate section for color version.*)

The choices are made by a local attractor network in which the winning attractor represents the decision, with each possible attractor representing a different choice, and each attractor receiving inputs that reflect the evidence for that choice. (The attractor network is formed in a part of the cerebral cortex by strengthening of the recurrent collateral excitatory synapses between nearby pyramidal cells. One group of neurons with strengthened synapses between its members can form a stable attractor with high firing rates, which competes through inhibitory interneurons with other possible attractors formed by other groups of excitatory neurons (Rolls, 2008a, 2010c). The word attractor refers to the fact that inexact inputs are attracted to one of the states of high firing that are specified by the synaptic connections between the different groups of neurons. The result in this non-linear system is that one attractor wins, and this implements a mechanism for decision making with one winner (Rolls, 2008a; Wang, 2008; Rolls and Deco, 2010; Deco et al., 2013).) The decisions are probabilistic as they reflect the noise in the competitive non-linear decision-making process that is introduced by the random spiking times of neurons for a given mean rate that reflect a Poisson process (Rolls and Deco, 2010; Rolls et al., 2010a). The costs of each reward need to be subtracted from the value of each reward to produce a net reward value for each available reward before the decision is taken (Rolls, 2008a; Rolls and Grabenhorst, 2008; Grabenhorst and Rolls, 2011). The reasoning or rational system with its long-term goals (introducing evidence such as "scientific studies have shown that fish oils rich in omega 3 may reduce the probability of Alzheimer's disease") then competes with the rewards such as the pleasant flavor of food (which are gene-specified (Rolls, 2005a, 2014), though subject to conditioned effects

(Booth, 1985; Rolls, 2005a)) in a further decision process which may itself be subject to noise (Rolls, 2005a, 2008a; Rolls and Deco, 2010). This can be described as a choice between the selfish phene (standing for phenotype) and the selfish gene (Rolls, 2011b, 2012d, 2014). In this context, the findings described in this paper that the cognitive system can have a top-down influence on the reward system are important advances in our understanding of how these decisions are reached.

46.4 SYNTHESIS

These investigations show that a principle of brain function is that representations of the reward / hedonic value and pleasantness of sensory including food-related stimuli are formed separately from representations of what the stimuli are. The pleasantness / reward value is represented in areas such as the orbitofrontal cortex and pregenual cingulate cortex, and it is here that hunger/satiety signals modulate the representations of food to make them implement reward. The satiety signals that help in this modulation may reach the orbitofrontal cortex from the hypothalamus, and in turn, the orbitofrontal cortex projects to the hypothalamus where neurons are found that respond to the sight, smell, and taste of food if hunger is present (Rolls and Grabenhorst, 2008; Rolls, 2014). We have seen above some of the principles that help to make the food pleasant, including particular combinations of taste, olfactory, texture, visual, and cognitive inputs.

A hypothesis is developed elsewhere that obesity is associated in part with overstimulation of these reward systems by very rewarding combinations of taste, odor, texture, visual, and cognitive inputs (Rolls, 2005a, 2011c, 2012a, 2014).

ACKNOWLEDGMENTS

This research was supported by the Medical Research Council. The participation of many colleagues in the studies cited is sincerely acknowledged. They include Ivan de Araujo, Gordon Baylis, Leslie Baylis, Hugo Critchley, Paul Gabbott, Fabian Grabenhorst, Mikiko Kadohisa, Morten Kringelbach, Christian Margot, Francis McGlone, John O'Doherty, Barbara Rolls, Juliet Rolls, Thomas Scott, Zenon Sienkiewicz, Simon Thorpe, Maria Ines Velazco, Justus Verhagen, and Simon Yaxley.

REFERENCES

Amiez, C., Joseph, J. P., and Procyk, E. (2006). Reward encoding in the monkey anterior cingulate cortex. *Cereb. Cortex* 16: 1040–1055.

Baylis, L. L., and Rolls, E. T. (1991). Responses of neurons in the primate taste cortex to glutamate. *Physiol. Behav.* 49: 973–979.

Baylis, L. L., Rolls, E. T. and Baylis, G, C. (1995). Afferent connections of the orbitofrontal cortex taste area of the primate. *Neuroscience* 64: 801–812.

Beckstead, R. M., Morse, J. R. and Norgren, R. (1980). The nucleus of the solitary tract in the monkey: projections to the thalamus and brainstem nuclei. *J. Comp. Neurol.* 190: 259–282.

Berridge, K. C., Robinson, T. E. and Aldridge, J. W. (2009). Dissecting components of reward: 'liking', 'wanting', and learning. *Curr. Opin. Pharmacol.* 9: 65–73.

Booth, D. A. (1985). Food-conditioned eating preferences and aversions with interoceptive elements: learned appetites and satieties. *Ann. N. Y. Acad. Sci.* 443: 22–37.

Cabanac, M. (1971). Physiological role of pleasure. *Science* 173: 1103–1107.

Carmichael, S. T. and Price, J. L. (1994). Architectonic subdivision of the orbital and medial prefrontal cortex in the macaque monkey. *J. Comp. Neurol.* 346: 366–402.

Carmichael, S. T. and Price, J. L. (1996). Connectional networks within the orbital and medial prefrontal cortex of macaque monkeys. *J. Comp. Neurol.* 371: 179–207.

Cho, Y. K., Li, C. S. and Smith, D. V. (2002). Gustatory projections from the nucleus of the solitary tract to the parabrachial nuclei in the hamster. *Chem. Senses* 27: 81–90.

Critchley, H. D. (2005). Neural mechanisms of autonomic, affective, and cognitive integration. *J. Comp. Neurol.* 493: 154–166.

Critchley, H. D. and Rolls, E. T. (1996a). Responses of primate taste cortex neurons to the astringent tastant tannic acid. *Chem. Senses* 21: 135–145.

Critchley, H. D. and Rolls, E. T. (1996b). Olfactory neuronal responses in the primate orbitofrontal cortex: analysis in an olfactory discrimination task. *J. Neurophysiol.* 75: 1659–1672.

Critchley, H. D. and Rolls, E. T. (1996c). Hunger and satiety modify the responses of olfactory and visual neurons in the primate orbitofrontal cortex. *J. Neurophysiol.* 75: 1673–1686.

de Araujo, I. E. T. and Rolls, E. T. (2004). The representation in the human brain of food texture and oral fat. *J. Neurosci.* 24: 3086–3093.

de Araujo, I. E. T., Kringelbach, M. L., Rolls, E. T. and Hobden, P. (2003a). The representation of umami taste in the human brain. *J. Neurophysiol.* 90: 313–319.

de Araujo, I. E. T., Kringelbach, M. L., Rolls, E. T. and McGlone, F. (2003b). Human cortical responses to water in the mouth, and the effects of thirst. *J. Neurophysiol.* 90: 1865–1876.

de Araujo, I. E. T., Rolls, E. T., Kringelbach, M. L., et al. (2003c). Taste-olfactory convergence, and the representation of the pleasantness of flavour, in the human brain. *Eur. J. Neurosci.* 18: 2059–2068.

de Araujo, I. E. T., Rolls, E. T., Velazco, M. I., et al. (2005). Cognitive modulation of olfactory processing. *Neuron* 46: 671–679.

Deco, G. and Rolls, E. T. (2005a). Synaptic and spiking dynamics underlying reward reversal in orbitofrontal cortex. *Cereb. Cortex* 15: 15–30.

Deco, G. and Rolls, E. T. (2005b). Neurodynamics of biased competition and co-operation for attention: a model with spiking neurons. *J. Neurophysiol.* 94: 295–313.

Deco, G., Rolls, E. T., Albantakis, L. and Romo, R. (2013). Brain mechanisms for perceptual and reward-related decision-making. *Prog. Neurobiol.* 103: 194–213.

Desimone, R. and Duncan, J. (1995). Neural mechanisms of selective visual attention. *Ann. Rev. Neurosci.* 18: 193–222.

Di Lorenzo, P. M. (1990). Corticofugal influence on taste responses in the parabrachial pons of the rat. *Brain Res.* 530: 73–84.

Francis, S., Rolls, E. T., Bowtell, R., et al. (1999). The representation of pleasant touch in the brain and its relationship with taste and olfactory areas. *Neuroreport* 10: 453–459.

Ge, T., Feng, J., Grabenhorst, F. and Rolls, E. T. (2012). Componential Granger causality, and its application to identifying the source and mechanisms of the top-down biased activation that controls attention to affective vs sensory processing. *NeuroImage* 59: 1846–1858.

Gilbertson, T. A. (1998). Gustatory mechanisms for the detection of fat. *Curr. Opin. Neurobiol.* 8: 447–452.

Gilbertson, T. A., Fontenot, D. T., Liu, L., et al. (1997). Fatty acid modulation of K+ channels in taste receptor cells: gustatory cues for dietary fat. *Am. J. Physiol.* 272: C1203–1210.

Giza, B. K. and Scott, T. R. (1983). Blood glucose selectively affects taste-evoked activity in rat nucleus tractus solitarius. *Physiol. Behav.* 31: 643–650.

Giza, B. K. and Scott, T. R. (1987a). Blood glucose level affects perceived sweetness intensity in rats. *Physiol. Behav.* 41: 459–464.

Giza, B. K. and Scott, T. R. (1987b). Intravenous insulin infusions in rats decrease gustatory-evoked responses to sugars. *Am. J. Physiol.* 252: R994–1002.

Giza, B. K., Scott, T. R. and Vanderweele, D. A. (1992). Administration of satiety factors and gustatory responsiveness in the nucleus tractus solitarius of the rat. *Brain Res. Bull.* 28: 637–639.

Giza, B. K., Deems, R. O., Vanderweele, D. A. and Scott, T. R. (1993). Pancreatic glucagon suppresses gustatory responsiveness to glucose. *Am. J. Physiol.* 265: R1231–1237.

Gleen, J. F. and Erickson, R. P. (1976). Gastric modulation of gustatory afferent activity. *Physiol. Behav.* 16: 561–568.

Gottfried, J. A. 2015 Structural and functional imaging of the human olfactory system. In: *Handbook of Olfaction and Gustation*, 3rd Edition (Doty R. L., ed). pp. 281–308. New York: Wiley Liss.

Gottfried, J. A. and Zald, D. H. (2005). On the scent of human olfactory orbitofrontal cortex: meta-analysis and comparison to non-human primates. *Brain Res.* 50: 287–304.

Grabenhorst, F. and Rolls, E. T. (2008). Selective attention to affective value alters how the brain processes taste stimuli. *Eur. J. Neurosci.* 27: 723–729.

Grabenhorst, F. and Rolls, E. T. (2010). Attentional modulation of affective vs sensory processing: functional connectivity and a top-down biased activation theory of selective attention. *J. Neurophysiol.* 104: 1649–1660.

Grabenhorst, F. and Rolls, E. T. (2011). Value, pleasure, and choice in the ventral prefrontal cortex. *Trends Cogn. Sci.* 15: 56–67.

Grabenhorst, F. and Rolls, E. T. (2014). The representation of oral fat texture in the human somatosensory cortex. *Hum. Brain Mapp.* 35: 2521–2530.

Grabenhorst, F., Rolls, E. T. and Bilderbeck, A. (2008a). How cognition modulates affective responses to taste and flavor: top down influences on the orbitofrontal and pregenual cingulate cortices. *Cereb. Cortex* 18: 1549–1559.

Grabenhorst, F., Rolls, E. T. and Parris, B. A. (2008b). From affective value to decision-making in the prefrontal cortex. *Eur. J. Neurosci.* 28: 1930–1939.

Grabenhorst, F., Rolls, E. T., Parris, B. A. and D'Souza, A. (2010). How the brain represents the reward value of fat in the mouth. *Cereb. Cortex* 20: 1082–1091.

Grabenhorst, F., Rolls, E. T., Margot, C., et al. (2007). How pleasant and unpleasant stimuli combine in different brain regions: odor mixtures. *J. Neurosci.* 27: 13532–13540.

Guest, S., Grabenhorst, F., Essick, G., et al. (2007). Human cortical representation of oral temperature. *Physiol. Behav.* 92: 975–984.

Hajnal, A., Takenouchi, K. and Norgren, R. (1999). Effect of intraduodenal lipid on parabrachial gustatory coding in awake rats. *J. Neurosci.* 19: 7182–7190.

Hetherington, M. M. (2007). Cues to overeat: psychological factors influencing overconsumption. *Proc. Nutrit. Soc.* 66: 113–123.

Hornak, J., Bramham, J., Rolls, E. T., et al. (2003). Changes in emotion after circumscribed surgical lesions of the orbitofrontal and cingulate cortices. *Brain* 126: 1691–1712.

Ito, S., Stuphorn, V., Brown, J. W. and Schall, J. D. (2003). Performance monitoring by the anterior cingulate cortex during saccade countermanding. *Science* 302: 120–122.

Kadohisa, M., Rolls, E. T. and Verhagen, J. V. (2004). Orbitofrontal cortex neuronal representation of temperature and capsaicin in the mouth. *Neuroscience* 127: 207–221.

Kadohisa, M., Rolls, E. T. and Verhagen, J. V. (2005a). The primate amygdala: neuronal representations of the viscosity, fat texture, temperature, grittiness and taste of foods. *Neuroscience* 132: 33–48.

Kadohisa, M., Rolls, E. T. and Verhagen, J. V. (2005b). Neuronal representations of stimuli in the mouth: the primate insular taste cortex, orbitofrontal cortex, and amygdala. *Chem. Senses* 30: 401–419.

Kringelbach, M. L., O'Doherty, J., Rolls, E. T. and Andrews, C. (2003). Activation of the human orbitofrontal cortex to a liquid food stimulus is correlated with its subjective pleasantness. *Cereb. Cortex* 13: 1064–1071.

Li, C. S., Cho, Y. K. and Smith, D. V. (2002). Taste responses of neurons in the hamster solitary nucleus are modulated by the central nucleus of the amygdala. *J. Neurophysiol.* 88: 2979–2992.

Lundy, R. F. Jr. and Norgren, R. (2004). Activity in the hypothalamus, amygdala, and cortex generates bilateral and convergent modulation of pontine gustatory neurons. *J. Neurophysiol.* 91: 1143–1157.

Luo, Q., Ge, T., Grabenhorst, F. et al. (2013). Attention-dependent modulation of cortical taste circuits revealed by Granger causality with signal-dependent noise *PLoS Comp. Biol.*. 9:e1003265.

Maruyama, Y., Pereira, E., Margolskee, R. F. et al. (2006). Umami responses in mouse taste cells indicate more than one receptor. *J. Neurosci.* 26: 2227–2234.

McCabe, C. and Rolls, E. T. (2007). Umami: a delicious flavor formed by convergence of taste and olfactory pathways in the human brain. *Eur. J. Neurosci.* 25: 1855–1864.

Norgren, R. (1974). Gustatory afferents to ventral forebrain. *Brain Res.* 81: 285–295.

Norgren, R. (1976). Taste pathways to hypothalamus and amygdala. *J. Comp. Neurol.* 166: 17–30.

Norgren, R. (1990). Gustatory system. In: *The Human Nervous System* (Paxinos G, ed), pp. 845—861. San Diego: Academic.

Norgren, R. and Leonard, C. M. (1971). Taste pathways in rat brainstem. *Science* 173: 1136–1139.

Norgren, R. and Leonard, C. M. (1973). Ascending central gustatory pathways. *J. Comp. Neurol.* 150: 217–238.

O'Doherty, J., Rolls, E. T., Francis, S., et al. (2001). The representation of pleasant and aversive taste in the human brain. *J. Neurophysiol.* 85: 1315–1321.

O'Doherty, J., Rolls, E. T., Francis, S., et al. (2000). Sensory-specific satiety related olfactory activation of the human orbitofrontal cortex. *Neuroreport* 11: 893–897.

O'Doherty, J. P., Deichmann, R., Critchley, H. D. and Dolan, R. J. (2002). Neural responses during anticipation of a primary taste reward. *Neuron* 33: 815–826.

Öngür, D. and Price, J. L. (2000). The organisation of networks within the orbital and medial prefrontal cortex of rats, monkeys and humans. *Cereb. Cortex* 10: 206–219.

References

Öngür, D., Ferry, A. T. and Price, J. L. (2003). Architectonic division of the human orbital and medial prefrontal cortex. *J. Comp. Neurol.* 460: 425–449.

Padoa-Schioppa, C. (2011). Neurobiology of economic choice: a good-based model. *Annu. Rev. Neurosci.* 34: 333–359.

Padoa-Schioppa, C. and Cai, X. (2011). The orbitofrontal cortex and the computation of subjective value: consolidated concepts and new perspectives. *Ann. N. Y. Acad. Sci.* 1239: 130–137.

Palomero-Gallagher, N. and Zilles, K. (2004). Isocortex. In: *The Rat Nervous System* (Paxinos G, ed), pp. 729–757. San Diego, CA: Elsevier Academic Press.

Passingham, R. E. P. and Wise, S. P. (2012). *The Neurobiology of the Prefrontal Cortex.* Oxford: Oxford University Press.

Petrides, M. and Pandya, D. N. (2001). Comparative cytoarchitectonic analysis of the human and the macaque ventrolateral prefrontal cortex and corticocortical connection patterns in the monkey. *Eur. J. Neurosci.* 16: 291–310.

Phillips, M. L., Williams, L. M., Heining, M., et al. (2004). Differential neural responses to overt and covert presentations of facial expressions of fear and disgust. *NeuroImage* 21: 1484–1496.

Poellinger, A., Thomas, R., Lio, P., et al. (2001). Activation and habituation in olfaction- an fMRI study. *NeuroImage* 13: 547–560.

Preuss, T. M. (1995). Do rats have prefrontal cortex? The Rose-Woolsey-Akert program reconsidered. *J. Cogn. Neurosci.* 7: 1–24.

Pritchard, T. C., Hamilton, R. B. and Norgren, R. (1989). Neural coding of gustatory information in the thalamus of Macaca mulatta. *J. Neurophysiol.* 61: 1–14.

Pritchard, T. C., Edwards, E. M., Smith, C. A., et al. (2005). Gustatory neural responses in the medial orbitofrontal cortex of the old world monkey. *J. Neurosci.* 25: 6047–6056.

Rolls, B. J. and Hetherington, M. (1989). The role of variety in eating and body weight regulation. In: *Handbook of the Psychophysiology of Human Eating* (Shepherd R, ed), pp. 57–84. Chichester: Wiley.

Rolls, B. J., Rowe, E. A. and Rolls, E. T. (1982). How sensory properties of foods affect human feeding behavior. *Physiol. Behav.* 29: 409–417.

Rolls, B. J., Rolls, E. T. and Rowe, E. A. (1983a). Body fat control and obesity. *Behav. Brain Sci.* 4: 744–745.

Rolls, B. J., Van Duijenvoorde, P. M. and Rowe, E. A. (1983b). Variety in the diet enhances intake in a meal and contributes to the development of obesity in the rat. *Physiol. Behav.* 31: 21–27.

Rolls, B. J., Van Duijvenvoorde, P. M. and Rolls, E. T. (1984). Pleasantness changes and food intake in a varied four-course meal. *Appetite* 5: 337–348.

Rolls, B. J., Rolls, E. T., Rowe, E. A. and Sweeney, K. (1981a). Sensory specific satiety in man. *Physiol. Behav.* 27: 137–142.

Rolls, B. J., Rowe, E. A., Rolls, E. T., et al. (1981b). Variety in a meal enhances food intake in man. *Physiol. Behav.* 26: 215–221.

Rolls, E. T. (1981). Central nervous mechanisms related to feeding and appetite. *Br. Med. J* 37: 131–134.

Rolls, E. T. (1997). Taste and olfactory processing in the brain and its relation to the control of eating. *Crit. Rev. Neurobiol.* 11: 263–287.

Rolls, E. T. (2000). Neurophysiology and functions of the primate amygdala, and the neural basis of emotion. In: *The Amygdala: A Functional Analysis,* Second Edition (Aggleton, J. P., ed), pp. 447–478. Oxford: Oxford University Press.

Rolls, E. T. (2005a). *Emotion Explained.* Oxford: Oxford University Press.

Rolls, E. T. (2005b). Taste, olfactory, and food texture processing in the brain, and the control of food intake. *Physiol. Behav.* 85: 45–56.

Rolls, E. T. (2007). Sensory processing in the brain related to the control of food intake. *Proc. Nutr. Soc.* 66: 96–112.

Rolls, E. T. (2008a). Memory, Attention, and Decision-Making: *A Unifying Computational Neuroscience Approach.* Oxford: Oxford University Press.

Rolls, E. T. (2008b). Functions of the orbitofrontal and pregenual cingulate cortex in taste, olfaction, appetite and emotion. *Acta Physiologica Hungarica* 95: 131–164.

Rolls, E. T. (2009a). Functional neuroimaging of umami taste: what makes umami pleasant. *Am. J. Clin. Nutr.* 90: 803S–814S.

Rolls, E. T. (2009b). The anterior and midcingulate cortices and reward. In: *Cingulate Neurobiology and Disease* (Vogt BA, ed), pp. 191–206. Oxford: Oxford University Press.

Rolls, E. T. (2010a). Neural representation of fat texture in the mouth. In: *Fat Detection: Taste, Texture, and Postingestive Effects* (Montmayeur, J-P., Coutre, I. J., eds), pp. 197–223. Boca Raton, FL.: CRC Press.

Rolls, E. T. (2010b). A computational theory of episodic memory formation in the hippocampus. *Behav. Brain Res.* 205: 180–196.

Rolls, E. T. (2010c). Attractor networks. *WIREs Cogn. Sci.* 1: 119–134.

Rolls, E. T. (2011a). The neural representation of oral texture including fat texture. *J. Texture Stud.* 42: 137–156.

Rolls, E. T. (2011b). Consciousness, decision-making, and neural computation. In: *Perception-Action Cycle: Models, Algorithms and Systems* (V. Cutsuridis, A. Hussain, J. G. Taylor, eds), pp. 287–333. Berlin: Springer.

Rolls, E. T. (2011c). Taste, olfactory, and food texture reward processing in the brain and obesity. *Int. J. Obesity.* 35: 550–561.

Rolls, E. T. (2012a). Taste, olfactory, and food texture reward processing in the brain and the control of appetite. *Proc. Nutr. Soc.* 71: 488–501.

Rolls, E. T. (2012b). Mechanisms for sensing fat in food in the mouth. *J. Food Sci.* 77: S140–S142.

Rolls, E. T. (2012c). Invariant visual object and face recognition: neural and computational bases, and a model, VisNet. *Front. Compu. Neurosci.* 6: 35 (31–70).

Rolls, E. T. (2012d). *Neuroculture. On the Implications of Brain Science.* Oxford: Oxford University Press.

Rolls, E. T. (2014). *Emotion and Decision-Making Explained.* Oxford: Oxford University Press.

Rolls, E. T. and Rolls, B. J. (1977). Activity of neurones in sensory, hypothalamic and motor areas during feeding in the monkey. In: *Food Intake and Chemical Senses* (Katsuki, Y., Sato, M., Takagi, S., Oomura, Y., eds), pp. 525–549. Tokyo: University of Tokyo Press.

Rolls, E. T. and Rolls, B. J. (1982). Brain mechanisms involved in feeding. In: *Psychobiology of Human Food Selection* (Barker LM, ed), pp. 33–62. Westport, CN: AVI Publishing Company.

Rolls, E. T. and Baylis, L. L. (1994). Gustatory, olfactory, and visual convergence within the primate orbitofrontal cortex. *J. Neurosci.* 14: 5437–5452.

Rolls, E. T. and Rolls J. H. (1997). Olfactory sensory-specific satiety in humans. *Physiol. Behav.* 61: 461–473.

Rolls, E. T.and Scott, T. R. (2003). Central taste anatomy and neurophysiology. In: *Handbook of Olfaction and Gustation*, 2nd Edition (Doty RL, ed), pp. 679–705. New York: Dekker.

Rolls, E. T. and Xiang, J. -Z. (2005). Reward-spatial view representations and learning in the hippocampus. *J. Neurosci.* 25: 6167– 6174.

Rolls, E. T. and Grabenhorst, F. (2008). The orbitofrontal cortex and beyond: from affect to decision-making. *Prog. Neurobiol.* 86: 216–244.

Rolls, E. T., and Deco, G. (2010). *The Noisy Brain: Stochastic Dynamics as a Principle of Brain Function.* Oxford: Oxford University Press.

Rolls, E. T. and Treves, A. (2011). The neuronal encoding of information in the brain. *Prog. Neurobiol.* 95: 448–490.

Rolls, E. T., Judge, S. J. and Sanghera, M. (1977). Activity of neurones in the inferotemporal cortex of the alert monkey. *Brain Res.* 130: 229–238.

Rolls, E. T., Rolls, B. J. and Rowe, E. A. (1983c). Sensory-specific and motivation-specific satiety for the sight and taste of food and water in man. *Physiol. Behav.* 30: 185–192.

Rolls, E. T., Sienkiewicz, Z. J. and Yaxley, S. (1989). Hunger modulates the responses to gustatory stimuli of single neurons in the caudolateral orbitofrontal cortex of the macaque monkey. *Eur. J. Neurosci.* 1: 53–60.

Rolls, E. T., Yaxley, S. and Sienkiewicz, Z. J. (1990). Gustatory responses of single neurons in the caudolateral orbitofrontal cortex of the macaque monkey. *J. Neurophysiol.* 64: 1055–1066.

Rolls, E. T., Critchley, H. D. and Treves, A. (1996a) The representation of olfactory information in the primate orbitofrontal cortex. *J. Neurophysiol.* 75: 1982–1996.

Rolls, E. T., Verhagen, J. V. and Kadohisa, M. (2003a). Representations of the texture of food in the primate orbitofrontal cortex: neurons responding to viscosity, grittiness and capsaicin. *J. Neurophysiol.* 90: 3711–3724.

Rolls, E. T., Kringelbach, M. L. and de Araujo, I. E. T. (2003b). Different representations of pleasant and unpleasant odors in the human brain. *Eur. J. Neurosci.* 18: 695–703.

Rolls, E. T., McCabe, C. and Redoute, J. (2008a). Expected value, reward outcome, and temporal difference error representations in a probabilistic decision task. *Cereb. Cortex* 18: 652–663.

Rolls, E. T., Grabenhorst, F. and Franco, L. (2009). Prediction of subjective affective state from brain activations. *J. Neurophysiol.* 101: 1294–1308.

Rolls, E. T., Grabenhorst, F. and Deco, G. (2010a). Decision-making, errors, and confidence in the brain. *J. Neurophysiol.* 104: 2359–2374.

Rolls, E. T., Grabenhorst, F. and Deco, G. (2010b). Choice, difficulty, and confidence in the brain. *NeuroImage* 53: 694–706.

Rolls, E. T., Grabenhorst, F. and Parris, B. A. (2010c). Neural systems underlying decisions about affective odors. *J. Cog. Neurosci.* 22: 1069–1082.

Rolls, E. T., Scott, T. R., Sienkiewicz, Z. J. and Yaxley, S. (1988). The responsiveness of neurones in the frontal opercular gustatory cortex of the macaque monkey is independent of hunger. *J. Physiol.* 397: 1–12.

Rolls, E. T., Critchley, H., Wakeman, E. A. and Mason, R. (1996b). Responses of neurons in the primate taste cortex to the glutamate ion and to inosine 5'-monophosphate. *Physiol. Behav.* 59: 991–1000.

Rolls, E. T., Critchley, H. D., Mason, R. and Wakeman, E. A. (1996c). Orbitofrontal cortex neurons: role in olfactory and visual association learning. *J. Neurophysiol.* 75: 1970–1981.

Rolls, E. T., Critchley, H. D., Browning, A. and Hernadi, I. (1998). The neurophysiology of taste and olfaction in primates, and umami flavor. *Ann. N.Y. Acad. Sci.* 855: 426–437.

Rolls, E. T., Verhagen, J. V., Gabbott, P. L. and Kadohisa, M. (2008b). Taste and texture representations in the primate pregenual cingulate and medial orbitofrontal cortices. In preparation.

Rolls, E. T., Critchley, H. D., Verhagen, J. V. and Kadohisa, M. (2010d). The representation of information about taste and odor in the orbitofrontal cortex. *Chemosens. Percept.* 3: 16–33.

Rolls, E. T., Murzi, E., Yaxley, S., et al. (1986). Sensory-specific satiety: food-specific reduction in responsiveness of ventral forebrain neurons after feeding in the monkey. *Brain Res.* 368: 79–86.

Rolls, E. T., Critchley, H. D., Browning, A. S., et al. (1999). Responses to the sensory properties of fat of neurons in the primate orbitofrontal cortex. *J. Neurosci.* 19: 1532–1540.

Rolls, E. T., Grabenhorst, F., Margot, C., et al. (2008c). Selective attention to affective value alters how the brain processes olfactory stimuli. *J. Cogn. Neurosci.* 20: 1815–1826.

Rushworth, M. F., Noonan, M. P., Boorman, E. D., et al. (2011). Frontal cortex and reward-guided learning and decision-making. *Neuron* 70: 1054–1069.

Sanghera, M. K., Rolls, E. T. and Roper-Hall, A. (1979). Visual responses of neurons in the dorsolateral amygdala of the alert monkey. *Exp. Neurol.* 63: 610–626.

Schoenbaum, G., Roesch, M. R., Stalnaker, T. A. and Takahashi, Y. K. (2009). A new perspective on the role of the orbitofrontal cortex in adaptive behaviour. *Nat. Rev. Neurosci.* 10: 885–892.

Schultz, W. (2007). Multiple dopamine functions at different time courses. *Ann. Rev. Neurosci.* 30: 259–288.

Scott, T. R. and Giza, B. K. (1987). A measure of taste intensity discrimination in the rat through conditioned taste aversions. *Physiol. Behav.* 41: 315–320.

Scott, T. R. and Plata-Salaman, C. R. (1999). Taste in the monkey cortex. *Physiol. Behav.* 67: 489–511.

Scott, T. R. and Small, D. M. (2009). The role of the parabrachial nucleus in taste processing and feeding. *Ann. N. Y. Acad. Sci.* 1170: 372–377.

Scott, T. R., Yaxley, S., Sienkiewicz, Z. J. and Rolls, E. T. (1986). Gustatory responses in the frontal opercular cortex of the alert cynomolgus monkey. *J. Neurophysiol.* 56: 876–890.

Scott, T. R., Karadi, Z., Oomura, Y., et al. (1993). Gustatory neural coding in the amygdala of the alert macaque monkey. *J. Neurophysiol.* 69: 1810–1820.

Simmons, W. K., Martin, A. and Barsalou, L. W. (2005). Pictures of appetizing foods activate gustatory cortices for taste and reward. *Cereb. Cortex* 15: 1602–1608.

Simmons, W. K., Avery, J. A., Barcalow, J. C., et al. (2013). Keeping the body in mind: Insula functional organization and functional connectivity integrate interoceptive, exteroceptive, and emotional awareness. *Hum. Brain Mapp.* 34: 2944–2958.

Small, D. M. (2010). Taste representation in the human insula. *Brain Struct. Funct.* 214: 551–561.

Small, D. M. and Prescott, J. (2005). Odor/taste integration and the perception of flavor. *Exp. Brain Res.* 166: 345–357.

Small, D. M. and Scott, T. R. (2009). Symposium overview: What Happens to the pontine processing? Repercussions of interspecies differences in pontine taste representation for tasting and feeding. *Ann. N. Y. Acad. Sci.* 1170: 343–346.

Small, D. M., Voss, J., Mak, Y. E., et al. (2004). Experience-dependent neural integration of taste and smell in the human brain. *J. Neurophysiol.* 92: 1892–1903.

Sobel, N., Prabhakaran, V., Zhao, Z., et al. (2000). Time course of odorant-induced activation in the human primary olfactory cortex. *J. Neurophysiol.* 83: 537–551.

Thorpe, S. J., Rolls, E. T. and Maddison, S. (1983). Neuronal activity in the orbitofrontal cortex of the behaving monkey. *Exp. Brain Res.* 49: 93–115.

Verhagen, J. V., Rolls, E. T. and Kadohisa, M. (2003). Neurons in the primate orbitofrontal cortex respond to fat texture independently of viscosity. *J. Neurophysiol.* 90: 1514–1525.

Verhagen, J. V., Kadohisa, M. and Rolls, E. T. (2004). The primate insular/opercular taste cortex: neuronal representations of the viscosity, fat texture, grittiness, temperature and taste of foods. *J. Neurophysiol.* 92: 1685–1699.

Wang, G. J., Volkow, N. D., Telang, F., et al. (2004). Exposure to appetitive food stimuli markedly activates the human brain. *NeuroImage* 21: 1790–1797.

References

Wang, X. J. (2008). Decision making in recurrent neuronal circuits. *Neuron* 60: 215–234.

Wilson, F. A. W. and Rolls, E. T. (2005). The primate amygdala and reinforcement: a dissociation between rule-based and associatively-mediated memory revealed in amygdala neuronal activity. *Neuroscience* 133: 1061–1072.

Wise, S. P. (2008). Forward frontal fields: phylogeny and fundamental function. *Trends Neurosci.* 31: 599–608.

Yan, J. and Scott, T. R. (1996). The effect of satiety on responses of gustatory neurons in the amygdala of alert cynomolgus macaques. *Brain Res.* 740: 193–200.

Yaxley S, Rolls ET, Sienkiewicz ZJ (1988). The responsiveness of neurons in the insular gustatory cortex of the macaque monkey is independent of hunger. *Physiol. Behav.* 42: 223–229.

Yaxley, S., Rolls, E. T. and Sienkiewicz, Z. J. (1990). Gustatory responses of single neurons in the insula of the macaque monkey. *J. Neurophysiol.* 63: 689–700.

Yaxley, S., Rolls, E. T., Sienkiewicz, Z. J. and Scott, T. R. (1985). Satiety does not affect gustatory activity in the nucleus of the solitary tract of the alert monkey. *Brain Res.* 347: 85–93.

Zald, D. H. and Pardo, J. V. (1997). Emotion, olfaction, and the human amygdala: amygdala activation during aversive olfactory stimulation. *Proc. Natl. Acad. Sci. U. S. A.* 94: 4119–4124.

Zald, D. H., Hagen, M. C. and Pardo, J. V. (2002). Neural correlates of tasting concentrated quinine and sugar solutions. *J. Neurophysiol.* 87: 1068–1075.

Zald, D. H., Lee, J. T., Fluegel, K. W. and Pardo, J. V. (1998). Aversive gustatory stimulation activates limbic circuits in humans. *Brain* 121: 1143–1154.

Zatorre, R. J., Jones-Gotman, M., Evans, A. C. and Meyer, E. (1992). Functional localization of human olfactory cortex. *Nature* 360: 339–340.

Zhao, G. Q., Zhang, Y., Hoon, M. A., et al. (2003). The receptors for mammalian sweet and umami taste. *Cell* 115: 255–266.

Part 11

Industrial Applications and Perspectives

Chapter 47

Olfaction and Taste in the Food and Beverage Industries

GRAHAM A. BELL and WENDY V. PARR

47.1 INTRODUCTION

Chemosensory research related to food manufacturing and consumer acceptance began in circumstances hardly recognizable today, when the emphasis was on selling what you made, rather than making what would sell (Moskowitz et al., 2011). Modern chemosensory research largely stemmed from the development of the need for massive amounts of ready-made food products by the US military in World War II (Jones et al., 1955; Kamen, 1989). Applications of knowledge gained in this period were initially focused on niche markets such as institutional catering, airline food services, emergency rationing, and provisions for athletes, hikers, and campers (Lyon et al., 1992). During this early period, further developments were made in product design, quality assurance, shelf-life testing, product matching, and screening for bacteria and other pathogens.

Today commercially prepared food and beverages are largely marketed to mass consumers in developed countries, dwarfing these earlier niche markets. With consolidation of commercial markets in Europe in the past decades (Lyon, 1999), and several other major international free trade agreements, a global commercial imperative, which is measured in trillions of dollars (Deloitte, 2010), now drives the interest of the food and beverage industries in the science of olfaction and gustation. For this reason, large food companies commit millions of dollars to chemosensory science (Weller, 1999) and smaller ones maintain a close watch on the new developments of others.

What has not changed in the past century is the fact that sensory processes underpin the acceptance of food and beverage products. When questioned, consumers report that "taste," meaning the experience of flavor and associated sensations in the mouth, is of paramount importance in purchase decisions, particularly for fruit, dairy, and beverage items (Easton et al., 1997). Not surprisingly, smell, taste, and associated sensory modalities are regarded widely in the food and beverage industries as the major determinants of the choices people make when buying or trading for food. Added to this "core belief" is an appreciation of the sheer size of the global food and beverage markets and the involved competitive pressures.

In this chapter we address how sensory evaluation has become a key element in defining the direction modern industry takes to tailor its products to optimize success in the market place. We provide an overview of how technology and such basic factors as age and gender are becoming important to provide new products focused on specific sectors of the general population.

47.2 WHERE IS CHEMOSENSORY SCIENCE FOUND IN THE FOOD AND BEVERAGE INDUSTRIES?

Much of modern chemosensory science is comprised of "basic" science, including central and peripheral nervous system physiology, the molecular biology of chemoreceptors, and the chemistry of odorants and tastants. Discoveries in basic chemosensory science have led to new technologies with applications in the food and beverage industries and related industries upon which they depend (Bell, 1996). For example, it is now commonplace to identify the types of volatiles that arise from foods using gas chromatography and mass spectrometry so as to isolate

---

Handbook of Olfaction and Gustation, Third Edition. Edited by Richard L. Doty.
© 2015 Richard L. Doty. Published 2015 by John Wiley & Sons, Inc.

1051

1052 **Chapter 47 Olfaction and Taste in the Food and Beverage Industries**

the specific chemicals critical for their flavor. All but the smallest of contemporary food and beverage companies now employ scientists and technicians specifically trained in chemistry and sensory evaluation.

Such sophistication was not always the case. In the not-too-distant past, judgment on important matters of product quality and product development was applied in a haphazard way and was often dominated by company management who would hold "organoleptic tester" sessions. Such sessions frequently consisted solely of management personnel tasting and discussing the products around a table (Jellinek, 1985). The first systematic scientific methods applied to the sensory evaluation of products were developed by such pioneers as Amerine et al., (1965), Helm and Trolle (1946), and Peryam and Swartz (1950). Rigorous scientific methodology and statistical approaches soon followed, as exemplified by studies performed by Moskowitz (1983) and Stone and Sidel (1993), and a statistics textbook focused primarily on the sensory evaluation of food (O'Mahony, 1986). Practical sensory evaluation guide books of value to in-house food technologists appeared in the 1990's, including those of Lyon et al. (1992) and Lawless and Heymann (1998; 2010).

Many food companies outsource the specific development of flavor chemicals that are added to their foods. At the heart of the global food and beverage industries are the flavor and fragrance houses, such as International Flavors and Fragrances (IFF), Givaudan, Firmenich, Takasago, Symrise, and Quest International. These companies play an important role in the development of ingredients related to the sensory quality of food and beverages, in addition to producing fragrances for non-food purposes, including soaps, cosmetics, air fresheners and candles (Herman, 2005). They commonly make large intramural and extramural investments in analytical chemistry and in fundamental science of olfaction and gustation. As is common in private enterprise, confidentiality surrounds their in-house research. Nonetheless, an outsider can gain insight into what these companies consider important by reading the published work of institutions they support, such as the Monell Chemical Senses Center in the United States (www.monell.org/research) and the Institut du Gout in France, which is funded in part by the European Union (EU) (www.institutdugout.fr). From time to time, books appear that reveal the work of in-house flavorists and perfumers, few of whom operate in the public domain (Rowe, 2005b).

In addition to such companies, many countries have publically funded organizations in which fundamental science is conducted on questions derived from, or of importance to, olfaction and gustation. Examples of these are the Commonwealth Scientific and Industrial Research Organisation (CSIRO) in Australia and the Institute of Plant and Food Research in New Zealand. However, the trend in the latter institutions over the past two decades has been to publish less in the public domain and more under a veil of confidentiality for fee-paying clients or sponsors. The result is that a greater dividend from public investment passes to the "private sector partners" and the results of research remain obscured at least long enough for a commercial advantage (patents, licenses or transfer of know-how) to be gained by the fee-paying party. By comparison, knowledge transfers readily from the transparent public domain to the opaque and competitive private sector.

It is important to note that food and beverage companies are keenly interested in what is happening in public domain science. In addition to sending their R&D people to key conferences, large food companies regularly bring together leading chemosensory scientists from around the globe for seminars and conferences, where the scientists explain their latest work and discuss, under confidentiality, the possible import of the new knowledge to the inviting company. Occasionally proceedings are published (e.g., Bell and Watson, 1999). More typically, research done in-house is taken up directly by the company and seldom appears in the public domain.

47.3 DRIVERS OF CONSUMER BEHAVIOR

The chemical senses, along with vision and oral somatosensation, can be considered the gatekeepers for food intake (Tuorila and Monteleone, 2009), protecting us from the ingestion of inappropriate or even lethal food items and orienting our eating behaviors towards healthy and safe food items. Research aimed at providing information about how people perceive, conceptualize, judge, and otherwise respond to food and beverages has increased steadily since the 1950s. Human physiological processes, genetic differences (Bartoshuk, 1993; Prescott, 1998), and psychological processes including learning and memory (e.g., Parr et al. 2002), have received much attention. Behavioral psychology, psychometrics, and cognitive psychology have provided methodologies to quantify human responses to multiple dimensions of foods and beverages (e.g., Labbe, et al., 2006). As noted below, in recent years the sensory evaluation field has been integrated with marketing issues, including market segmentation, packaging (e.g., wine labels or bottle shape), cultural and philosophical perspectives (e.g., importance of familiarity, perceived value, geographical location, and "green" issues), and emotional responses to foods and beverages (Lundahl, 2012).

Movements in the latter directions stem from the fact that consumers rarely have the opportunity to taste a food or beverage at the point of selection or purchase. For this reason, factors other than what a food tastes or smells like play a large part in food-choice behavior. Extrinsic factors include packaging, price, and advertisement details (Aaron et al., 1994; Lockshin et al., 2006), nutritional needs (Scott, 2001), health concerns (van Herpen et al., 2012), age- and gender-related factors (Bell and Strauss, 2007), and memories of past consuming behavior (Labbe et al., 2006). In food choice behavior, expectations are an assumed mediating variable (Deliza and MacFie, 1996).

Food and beverage companies, aiming to meet consumer expectations and demand, and to improve their market share, profit, or competitive position, have become dependent on the tools of multidisciplinary product evaluation, employing both consumers and trained panelists to make analytical judgments regarding product acceptability and other factors. The relatively recent focus on individual differences in the perception of, and preferences for, various sensory qualities involved in foods and beverages has led to the development of diversifying product qualities to cater to specific segments of the market, such as specific age groups, a process called market segmentation (Moskowitz et al., 2011).

47.3.1 Product Differentiation, Competitive Edge, and Quality Assurance

Manufacturers often compete in markets where success depends on creating perceived differences in the sensory quality of their products. For every market leader there are usually competitive products vying for market share either by producing a better product (in the mind of the consumer) or one that is impossible to tell apart from the leader: a "me too" (and potentially less expensive) competitor. As a product's success in the marketplace grows, manufacturers are faced with a number of dilemmas. For example, do they tailor products for markets into which they intend to expand or do they try to achieve a universal product that overrides cultural and contextual factors in new market territories? Are the available raw materials across different locals sufficiently consistent in quality to guarantee a final product good enough to sustain demand for it and to uphold the reputation of the brand?

To address the sensory issues, panels of consumers are often recruited to undertake blind assessments of whether or not a difference can be detected between products kept in stock as "gold standards" and those containing changed ingredients or formulations. Protocols for "difference testing" have been published and are available from national standards bodies. In some cases the objective is to establish or maintain a difference between a product and its rival and in others the objective is to ensure that there is no perceivable difference between two products.

47.3.2 The Chemical Basis of Food Acceptance

Working closely with product developers in the food and beverage industry are various types of chemists including synthetic, analytical, organic and inorganic chemists. Their general aim is to understand everything to do with the molecules that underpin the quality of food during its transformation from raw materials to an acceptable product for consumers, and which will return a profit through that transformation in the process known as "value adding." There is a continuing process of introducing new flavors to processed food to "extend the product line," while removing others. Many companies are constantly motivated to find a new flavor or "concept" for a product that will increase sales that will be "the next big thing" (Moskowitz et al., 2011). Importantly, consumers demand greater degrees of safety and truth in product label information than ever before. In response, industry retains an ongoing interest in taking flavorants out of products rather than adding them. For example, manufacturers of traditionally salty products continue to be concerned with formulating reductions in components such as sodium without noticeably diminishing acceptability of the product (e.g., Liem et al., 2011; Prescott and Khu, 1995).

Bottled beverages have appeared on world markets that are clear and colourless, yet taste remarkably like familiar fruits (e.g., strawberry, grape, etc.). These illustrate the art of flavoring a product using invisible flavouring agents and the flavor chemists' quest for "nature identical" chemical formulations. This implies the exact reconstruction of the components, often numbering in the hundreds, of the flavor compounds in the original object. However, exact replication of a chemical flavor profile is rarely needed and economic considerations come into play. Rather, one or more "key" odorant molecules are identified, by reference to chemical analysis of the original object and by methods such as split-stream gas chromatography (GC) and mass spectrometry (MS). In split-stream GC, a panel member smells each chemical component that arises from the vapors of an object sent through an absorbent column to define those chemicals that give the object its characteristic aroma or flavor. The key compounds are then purified or synthesised and, in some cases, mixed in proportions found in the sampled objects. The resulting flavor mixture is further refined until it is deemed by in-house sensory assessment to be "near enough" (for commercial purposes) to the original. The agent is then introduced into a product matrix, such as a liquid beverage or an appropriate food system, such as flavorless vegetable material. It is of interest that two Nobel Prizes have been awarded to chemists who have produced the means by which flavor or fragrance molecules can be identified (Ruzicka and Pauling, in Rowe, 2005a).

1054   Chapter 47   Olfaction and Taste in the Food and Beverage Industries

The process of identifying, isolating, purifying, and synthesizing aroma chemicals is long and hard, even before testing for safety and whether the molecule is 'true" to the intended or recognizable flavor. Stringent regulations, imposed recently by the European Union (EU) and driven by consumer concerns about "ingesting toxicants," have made it harder for a number of chemicals to be allowed into food products. In addition, limitations are now imposed by regulatory bodies of many countries as to what solvents might be used to carry the aroma chemical into the product. A typical dispersion medium for key flavor molecules into a fruit juice beverage is 70% propylene glycol, 20% ethanol or isopropanol and 10% water. However, no alcohol is permitted in Halal-compliant countries and so other solutions need to be found from a very limited list of permitted solvents (Baines and Knights, 2005).

Modern chemistry has described over 20 million chemical compounds (Zviely, 2005) of which a significant, though undefined, proportion have sensory properties. The potential palette for the flavor chemist is therefore practically infinite. Molecules producing olfactory sensations are generally of low molecular weight (45–300 Da; Taylor and Hort, 2004; Turin and Yoshii, 2003) and enter the nose through the nostrils (orthonasal smelling). They are also released from food by the action of mastication and swallowing, making their way by the interior passages to the olfactory receptors through the back of the mouth and throat: retronasal smelling (Burdach and Doty, 1987; Halpern 2004, 2008). The combination of retronasal olfaction and taste perception achieved through the taste receptors in the tongue and mouth are crucial inputs to a "gestalt" or integration process that creates an entirely different flavor perception which gives a person the impression that it is happening in the mouth (Prescott, 1999, 2004, 2012; Chapters 45 and 46). Knowing this, we can expect the palette of the creative flavor chemist to be expanded beyond its present large size, particularly when odorants released by chewing become identified.

Surprisingly little is known about the structure or function of flavor molecules that dictates why they smell the way they do. This largely reflects the fact that many molecules participate in a given flavor sensation. The reader is encouraged to explore the cornucopia of chemistry described in Rowe (2005c) to appreciate just how many species of molecules contribute fragrance to the flavor of foods: alcohols, acids, esters, lactones, aldehydes, acetals, nitriles, ketones, carotenoids, hydrocarbons, and heterocycles. In addition there are the sulphur compounds, many of which have extremely low odor thresholds (such as 10^{-7} for 4-methyl-4-mercaptopentan-2-one, a characteristic odor of the Sauvignon grape must; Jameson, 2005), and which include the thiols, thioesters, acyclic sulphides, polysulfides, and saturated heterocyclic sulphur compounds. There are many more molecular species and it

is typical for hundreds to make up a common odor such as the smell of brewed coffee (Rowe, 2005c).

The relationships between molecular structure and the perceived quality of fragrances have proven to be very complicated. Molecules with similar shapes can have distinctly different perceived qualities, while others with different shapes can have similar qualities (Youngentob 1999; Polak 2000; Turin and Yoshii, 2003). Some molecules inhibit or otherwise alter the activity of other molecules at the level of the olfactory receptors (i.e., in pharmacological terms are antagonists or agonists) or as a result of patterns of activity set up in higher brain regions, most notably the olfactory bulb. The discovery of the gene family that encodes olfactory membrane receptors by Buck and Axel (1991), and the demonstration of combinatorial receptor codes for the odorant reception mechanism (Malnic, et al., 1999), has encouraged the concept that knowledge of the odorant and odorant receptors may be of value in guiding the synthesis of new flavor molecules. However, this has proved to be difficult and progress has been slow. A case is made by Turin (2005) and Turin and Yoshii (2003) for the use of information on the vibrational characteristics of molecules (upon which Turin developed his spectroscopic theory of olfactory reception: Turin, 1996) to aid the discovery of useful molecules, such as an acid-stable lemon fragrance (Turin, 2005). Industry awaits further technological developments from discoveries of the mechanisms of receptor-ligand interactions. These are likely to include "designer molecules" for flavor and so-called "flavor modulators" that block the olfactory action of other molecules (Bell, 1996).

47.4 UNDERSTANDING WHAT THE CONSUMER EXPERIENCES

47.4.1 Cuisine versus the Global Product

The development of a global product that adheres to a single formula and can be made consistently at globally distributed plants is considered a desirable goal by many food and beverage companies. However, despite the spread of branded and identically packaged products, close examination of many "global products" reveals that a good deal of manipulation of the product takes place to meet the demands of local markets. Such factors as climate, familiarity with limited local cuisines, and even socioeconomic status, for example, can define the sweetness considered by the local consumers as "just right." Based on sensory evaluations, many products have to be reformulated from the sensory core to the packaging, brand presentation, size of portion, and even recommendations as to when the product should be consumed. Publication of why products fail in international markets is rare, despite the very useful practice of "flop analysis" which largely falls within the boundary of company know-how (Köster and Mojet, 2012).

However, attempts at such development can have unexpected consequences. For example, in many Asian countries relatively few food products were imported before the reduction of tariff barriers. When new and fresh products, such as fine butter was exported to Indonesia (as refrigeration became more widely available), consumers were puzzled that it didn't have a rancid taste and a pineapple flavor which are the main sensory attributes of Indonesian canned butter: the only product that had been previously available (Easton et al.,1997). Consumer resistance to novelty and to branding or packaging which declares the product to be "foreign" was a frequent finding in the CSIRO's Japan Project (Prescott et al, 1992, 1997; Laing et al., 1993, 1994, Prescott and Bell, 1995), a government supported project that included a review of the food culture of Japan (Bell et al., 1992), a public domain quarterly bulletin on food information and consumer behavior (*Food Japan* and *Food Japan and Asia*, 1991–1994), and confidential field work for sponsoring companies. When novelty and other identifiers of novelty were not made available to sensory panel members, resistance and cultural biases disappeared (Prescott and Bell, 1995).

47.4.2 What is Flavor?

This question is fundamental to chemosensory science and is dealt with in detail in several other chapters of this volume. Flavor, in effect, is a unified "oral" perceptual composite of several sensory modalities: a "gestalt" determined entirely by higher order brain processes from a confluence of independent sensory inputs comprised of separate nerves, physical stimuli, receptors, and genes (Prescott, 1999). The visual analogy is binocular depth perception. With each eye tested separately the world looks quite flat, but with both eyes open it "leaps" into the familiar and unique experience of depth. Depth perception has a spectacular effect on a person's experience and ability to perform visuo-motor tasks, as trying to reach for an item on a table with one eye closed will demonstrate. Similarly, combining smell, taste, and tactile sensations into the unified experience of flavor evokes an entirely different experience compared with each sensory component alone.

Valuable research on the chemical senses that relates to the food and beverage industries comes from consideration of the act of eating and its influences on flavor. Examples include studies of retronasal smelling (e.g., Burdach and Doty, 1987; Halpern, 2004, 2008), release of flavorants and odorants by the act of chewing (e.g., Taylor, 2005), the role of experience in perception of orthonasal and retronasal odor (e.g, Wilson and Stevenson, 2006), and

the oral perception of umami (e.g., Ninomiya, 2003), fat (e.g., Song, 2000, Gilbertson, 2003, Keast, 2012), texture (e.g., Brandt et al., 1963; Moskowitz, 1987; Szczesniak et al., 1975), astringency (e.g., Noble,1999), "tingle" (e.g., Dewis, 2005), and the "burning" or "cooling" effects of oral trigeminal nerve stimulation (e.g., Alimohammadi and Silver, 2002; Green, 2004).

Surprisingly, an appreciation of the multiple sensory processes involved in flavor perception has been largely ignored by industrial scientists until relatively recently (Prescott, 1999, 2006, 2012). This reflects, in part, the delay in the development or application of sophisticated analytical tools that deal with such complexity, including conjoint analysis (Moskowitz, et al, 2001), data mining, and various forms of multivariate analysis (Lawless and Heymann, 2010).

47.5 ACADEMIC CHEMOSENSORY SCIENCE INFORMING INDUSTRY

Basic information from academia concerning the development and regression of chemosensory systems and their function has aided in the creation of products focused on both young and old segments of the population. Research on the influences of age on chemosensation has led food companies to the realization that they can safely abandon the "one size fits all" approach of the 1950s and 1960s and can profitably exploit niche markets or "strata" with targeted products (Moskowitz et al., 2011). Information derived from brain scans and, to a large degree, "artificial noses" has also made its way into the food and beverage industries from the public sector and other industries, providing novel means for assessing the physiological impact of products on the brain and for addressing basic issues of product quality control and differentiation.

47.5.1 Age and Aging

The question of flavor perception is of vital importance when considering how best to formulate food for the very young and for the older person. Attempts at such stratified formulations have been challenging but remain a focus of research of many food and beverage companies.

47.5.1.1 The Child. Consideration of what should be fed to babies and young children has long been a food industry concern, with the focus being mainly on health and nutrition, rather than palatability. For generations diet supplements, such as cod liver oil, were served to children with scant regard to their tastes, which often were disgusting. In fact, the food preferences of children under the age of five are largely dictated by sweetness and familiarity

1056 **Chapter 47 Olfaction and Taste in the Food and Beverage Industries**

(Birch, 1979). Importantly, preferences established early in a child's life can last into later life (Mennella, et al., 2005). Relative to adults, children appear to prefer higher concentrations of sweet, salty, and sour tastants (Popper and Kroll, 2007), although the notion that children are more sensitive than adults to tastes and smells has received only scant support (Guinard, 2001).

"Fussy eating" by children is something most parents know well. Foods that the child will eat without complaint are often given acquiescence by busy and frustrated parents, and diets can become unbalanced, resulting in obesity and other unwanted health effects. Both the child and its parent become a market of interest to the food industry, since the parent is the shopper, while the child is the consumer. In a recent review, Mustonen and Tourila (2009) examined the degree to which chemosensory factors and neophobia (avoidance of novelty) determine flavor preferences and eating habits in children ranging in age from 7 to 13 years. They concluded that neophobia is by far the stronger influence and that children can be educated to dispel their reluctance to try novel foods and thereby broaden and improve nutritional balance in their diets. Understanding the contributions of sensory and non-sensory factors in food selection by children is an area which has yet to be properly explored, and in which the food industry can work with sensory scientists to develop products and promotions to stimulate variety and nutritional balance in the child's diet.

47.5.1.2 The Older Consumer. People in developed countries who are over the age of 65 years have previously been of little interest to the food and beverage industries, owing mainly to their low economic power. This is no longer the case. They have emerged as a consumer force with significant disposable income, in retirement. Their resources, by which their choices can be felt in food sales and catering, has made the chemosensory determinants of their food choices commercially relevant.

The function of the primary sensory modalities that contribute to flavor perception begin to fail in many elderly (Doty et al., 1984; Doty and Kamath, 2014; Matsuda and Doty, 1995; Mojet et al., 2001, 2003). Over the age of 65 years, the perception of flavor becomes attenuated or radically altered, often with negative consequences for general safety as well as for good nutrition and health maintenance (DeVere, 2003; Reiter and Costanzo, 2003). The enjoyment and palatability of foods declines in many people after age 65, although only those with severe deficits report experiencing significant changes in food flavor (Kremer et al., in Mustonen and Tuorila, 2007). The onset of age-related chemosensory loss is gradual for some individuals, but rapid for others. Such loss can reflect the influences of cumulative lifetime damage from viral and other xenobiotic insults to the olfactory epithelium (Deems et al., 1991), as well as neurological changes secondary to neurodegenerative diseases such as Alzheimer's disease, Parkinson's disease, cancer, medication usage, and multiple medical interventions, including treatments for hypertension, diabetes, and cancer (Doty, 2012; Schiffman, 2000; Schiffman and Zervakis, 2002; see also Chapters 18 and 38). Health consequences are not trivial. Chemosensory deficits can alter the motivation to eat and the kinds of food that are chosen. Such deficits can exacerbate medical conditions, impair nutritional status, and result in loss of weight and immunity to infections (Deems et al. 1991; DeVere, 2003).

A challenge for chemosensory researchers is to determine what can be done about improving foods for older consumers. Judicious use of flavor enhancement remains an option and has been shown to have measureable health benefits for some elderly (Schiffman, 2000, 2002). However, Mustonen and Tuorila (2007) have pointed out weaknesses in flavor enhancement strategies (reviewed by Bell and Strauss, 2007), including the heterogeneity of sensory deficits in the older population, their adaptation to deficits, and the failure to obtain positive health outcomes in experiments where foods were experimentally altered. That being said, commercially successful foods have emerged which have focused primarily on members of this age group.

47.5.2 Brain Scanning and Chemosensory Science

Scanning technology for recording human brain metabolism and cerebral blood-flow, such as positron emission tomography (PET) and functional magnetic resonance imaging (fMRI), has been employed for decades in clinical settings. In recent years, chemosensory scientists working in or associated with hospitals exploited this technology and important studies on this topic began to emerge in the scientific literature in the late 1990s (Levy et al., 1997; Ohla et al., 2012b; Savic, 2002; Chapters 13 and 35).

Of immediate interest to scientists and the food industry were the findings of enhanced neural activity in identified brain areas when olfactory and taste stimuli were presented together, but not singly (Small et al, 1997, 2004). This finding suggested that specific brain regions could be identified that may be particularly important for understanding and quantifying flavor processing. In addition, the relatively early and landmark PET study by Small et al., (2001) revealed most compellingly the visualization of brain regions that were engaged during the enjoyment of chocolate: brain regions which diminished and gave way to other regions once satiety had been reached. More research is needed, however, to establish reliability, sensitivity and specificity of such measures. Other studies have found brain areas that are relatively specific to food cravings (Pelchat et al., 2004), food texture and fat (Araujo and Rolls, 2004), taste enhancing effects of umami-related compounds (McCabe and Rolls, 2007), trigeminal nerve stimulation by pungent odorants (Billot et al.,2011; Hummel et al., 2005), visual food cues for pleasant tastes (Ohla et al., 2012 a), selective attention to chemosensory stimuli (Veldhuisen and Small, 2011), and enhancement of tastes by smells (Seo et al., 2011). Somewhat separate brain loci were reported for sweet and salty tasting stimuli (Spetter et al., 2010), for single and binary odor mixtures (Boyle et al., 2009), for pleasant and unpleasant odors (Henkin and Levy, 2001), and variations in odor-induced brain activity attributable to gender (Yousem et al., 1999a; Bengtsson et al, 2001) and age (Yousem et al., 1999b).

The question that has been asked by the food and beverage industries is how they can use such information from this new era of chemosensory science? Can the scanners be substituted for human sensory evaluation panels? The current cost of fMRI scanners and both the cost and need to supply medical isotopes necessary to use PET scanners, suggests that this equipment will most likely stay within medical and research institutions, where industry can, nonetheless, benefit from the funding of research employing such technology. However, the development of standardized protocols and instruments for routine brain scanning of any kind within the food industry is a long way off, both from the perspective of reducing costs for PET and fMRI, and in producing reliable and valid data.

Specific attempts have been made by the food industry to harness surface brain voltages measured through electrodes placed on the scalp (electroencephalography: EEG). Like functional imaging, such studies have the potential for providing useful information about attitudes and liking of people for food products and for minimizing the number of subjects required to obtain such information (Owen and Patterson, 2002; Owen et al., 2002). However, sensitivity and specificity issues are present and, despite fiscal support from the Australian food industry, the experimental results and technology have yet to be adopted by this industry.

47.5.3 Machine Sensing (Artificial Noses)

Another area of interest to industry, much of which has been developed in the public sector, is the invention of specialized detectors which can potentially replace human noses in quality control, product development, and other facets of food technology. Can machines replace people in food manufacturing? In some sectors of the industry the answer is yes. For example, at a number of large chocolate, candy, and cookie manufacturing facilities, visitors can be taken to a viewing point above a vast floor of machines and process lines. Finding the humans on the floor is quite

a task. A handful of floor operators will attend to hectares of production machinery. Engineering solutions govern oven temperatures, stress sensors control line speeds and roller pressures, and machine vision is now common in sorting products for optically discerned faults, dangerous inclusions, and so on. In each case a human has been replaced by a device with some form of reliable and valid decision-making ability.

In general, quality control of the final product still requires human intervention. Large costs are attached to every batch of product, and one of the most expensive events, dreaded by plant managers, is when a fault is discovered in the product after it reaches the consumer. An example of what can occur is that lines that add flavor during the process become temporarily blocked or otherwise fail, resulting in large and costly quantities of cookies arriving at the market that taste terrible.

Nevertheless, machine solutions are being sought to reduce the cost of running and training human judges. Chemical sensors and electronic noses have long been envisaged as playing a useful part in food production (Barnett, 1999; Hibbert, 1999; Levy and Naidoo, 1999; Mackay-Sim, 1999; Persaud, 2003). Although earlier and cumbersome devices were designed to detect a wide range of chemicals, the new trend is to tailor the device for the intended job, creating a sensor array with performance characteristics tuned for one or a few chemicals (Bell and Wu, 2006). The challenge for e-nose technology remains the achievement of useful and timely recognition of vapors in an operational context, in addition to the practicalities concerning sensor lifespan and reliability. An approach to data processing in real time using a Bayesian probability model may bring the technology closer to widespread use in the food and beverage industries (Hibbert and Bell, 2011).

47.6 HUMAN SENSORY EVALUATION

As noted in the introduction, the field of sensory evaluation or "sensory analysis" had its beginnings in the aftermath of WWII when the limitations of having one or a few persons make sensory decisions that could impact the future prosperity of a company became clear (Stone et al., 1974; Piggot, 1984; Jellinek, 1985). Eventually, the replacement of the individual expert taster of old with a group of tasters (members of a panel) which was not under any influence to deliver a replica of the manager's opinion, became the standard. These panels replaced the heavily biased and haphazard managerial discussion groups notorious for influencing each other and delivering what the boss wanted, as delightfully illustrated in Jellinek (1985).

Today, a basic in-house panel or "group taster" is comprised of 10 to 15 healthy non-smoking employees

1058   **Chapter 47   Olfaction and Taste in the Food and Beverage Industries**

who volunteer for this task, often from a diverse set of departments. Their combined judgments are more objective (i.e., reliable and valid statistically) than that of a single individual. Consumer panels, which are hired from outside the company are usually comprised of larger groups. Such panels allow for better sampling the population in general. Today, new information technology is used to reach larger numbers of consumers in ways never possible previously, with interesting implications for how companies will develop and distribute products for newly identifiable consumer subgroups on a massive scale (Moskowitz et al., 2011). Nevertheless, the role of the "expert" has endured in many quarters, particularly in the wine industry where a single individual's "expert" opinion, usually an established wine critic, carries enormous weight and influence.

47.6.1 Influence of Prior Experience on Liking and Choice

Each person brings their unique world of established knowledge and experience to the table when selecting and evaluating food products. Familiarity or expertise with a product, gained through learning, memory, and associated emotions, forms part of this established knowledge. Experientially-gained knowledge can be individual, hence leading to idiosyncratic responding by a person (Parr, 2003), but is often shared within a social or cultural group, as demonstrated by the culture-bound popularity of kimchi in Korea and Vegemite in Australia. In the latter case, the knowledge is often considered in terms of social representations (Dubois, 2000) which can be considered as "shared thoughts" within a group. Since food or beverage selection is a choice behavior, involving discrimination, the features of a product are subjected to scrutiny and recognition processes (Freeman et al., 1993). It is not surprising that familiarity or prior experience is an important determinant of food selection.

Scientific studies demonstrating such influences are widespread. Stevenson and Prescott (1994), for example, found that prior experience with chili, or long-term memory of chili, influenced intensity ratings of perceived burn by capsaicin solutions. Distel et al. (1999) showed that product familiarity determines the liking and perceived strength of an odorant. Ayabe-Kanamura et al. (1998), in a study of the responses of Japanese and German participants to odorants considered typical or untypical for each culture, also found that familiarity influenced participant responses. As noted by Dijksterhuis et al. (2006), prior experience with food or beverages "inarguably has an effect on current liking."

Foods and beverages can be familiar in terms of their intrinsic (e.g., flavor characteristics) and extrinsic (e.g., packaging) characteristics. Labbe et al. (2006) reported a study of the influence of beverage familiarity

on the interaction between olfactory and taste perception. Cocoa-flavored beverages comprised the familiar item, and caffeinated milk comprised the non-familiar item. They found that familiarity or experience modulated taste perception even in participants trained in sensory panel work. Adding vanilla to the familiar cocoa beverage induced a significant sweet enhancement. However, addition of vanilla flavoring to the non-familiar beverage, the caffeinated milk, enhanced perceived sweetness as with the familiar cocoa beverage, but unexpectedly also enhanced perception of bitterness. In another study, van Herpen et al. (2012) found that familiarity with a product's label influenced evaluation of the product (e.g., rating of trust in the label), but did not influence product choice based on perceived health benefits, emphasizing the fact that the influence of familiarity on product selection is a complex issue.

47.6.2 Expertise: Influence of Expert Opinion on Consumer Behavior with Examples from the Wine Industry

Assignment of grades, scores, or awards to foods and beverages to reflect quality is a long-established practice with its roots in industry rather than in science. Ratings of perceived wine quality go back at least as far as the mid nineteenth century with events such as the ranking of the Bordeaux chateaux (1855) and the first Australian wine show in 1845 (Walsh, 2002). Whereas the aim of grading or rating product quality was purportedly to improve products and grow the relevant industry (Walsh, 2002), more recently the award of medals or high grades to food and beverage products has been a major tool of marketing. For some wine companies, use of information about their medals and awards has served as the dominant factor in their marketing strategies, and it would appear that gold medals do sell wines (Murphy, 2002).

Today, there is a large number of forums for evaluation of food and beverage products such as formal shows where quality is judged by designated experts within a field (e.g., olive oil, cheese, or wine awards), and media for disseminating the results of such evaluations (e.g., specialist magazines such as *Wine Spectator* in the USA, and internet sites). It is not surprising that consumers take notice of expert opinion when selecting foods and beverages that they cannot taste at the point of purchase, especially for relatively complex and expensive products that can vary from product to product and carry a high level of perceived risk in a consumer's mind (Hayes and Pickering, 2012). Researchers have shown expert opinion to influence a range of relevant product aspects including choice of wine by consumers (Murphy, 2002), liking of wine (Siegrist and Cousin, 2009), price of wine (Dubois

47.6 Human Sensory Evaluation 1059

and Nauges, 2010; Hadj Ali, et al., 2008) and quantity of wine consumed (Hilger et al., 2011).

The wine industry recently has made attempts to formalize training and selection systems for their judges (O.I.V., 1994). This is aimed at remedying a situation where relative absence of guiding principles based on scientific knowledge leaves those deemed to have expertise within the food and beverage industries relying primarily on their own experiences and anecdotal evidence to achieve their tasks. Further research involving industry and science can only benefit all players, not least the consumer who it would appear from research to date is influenced by expert opinion when selecting, purchasing and consuming food and beverage products.

47.6.3 Genetic Factors

47.6.3.1 Taste Genes. By far the most active area of research into the genetics of human taste perception has been concerned with differences amongst people in perception of bitterness. Taste receptors of the TAS2R family detect bitter compounds (Newcomb et al., 2010). Ability to detect bitter compounds has been, from an evolutionary perspective, deemed an adaptive development that serves as a mechanism for avoidance of toxic compounds. Individual differences in human responsiveness to the bitter tastant known as PROP (6-n-propylthiouracil) have been studied extensively; see reviews by several authors in Prescott and Tepper (2004) and Reed et al. (2006).

That individual differences in sensitivity to a bitter compound can be linked scientifically with responsiveness to oral sensations such as astringency (Pickering et al., 2004) and with self-reported alcoholic beverage intake (Duffy et al., 2004) suggests a genetic basis, at least in part, for some food preferences. Nonetheless, whilst several studies do support a link between a person's PROP status and food choice (Tepper, 2008), the research area is plagued with methodological issues, not least the influence of a myriad of confounding variables that exemplify the difficulty in empirically separating out genetic influences from effects of environmental variables on a person's food preferences. Genetic heterogeneity for PROP sensitivity was demonstrated across a number of cultures (Bell and Song, 2004), demonstrating the degree to which the genes for this taste characteristic may influence behavior between populations of consumers in various countries.

Individual differences in perception of sweetness, sourness, saltiness, and of umami have been less frequently reported. An area where industry has for some years now had information linking genetic variation with differences in taste perception is that of bitterness perception. This has led to an understanding by companies that produce and market artificial sweeteners, that the bitter after-taste perceived by a proportion of the consuming population

(Roudnitzky et al., 2009) likely has a genetic basis. This knowledge allows a company to innovate by, say, choosing to produce an alternate product for that segment of the consuming population that is particularly sensitive to the source compound (e.g., a sweetener that cannot bind bitter taste receptors). The challenge remains for the industries to identify the appropriate subgroups and modify their products accordingly. We might expect the beer industry to take up this approach in future.

There is evidence for marked variation in human ability to detect sweetness (Reed et al., 2006), as well for variation in the sweet-related receptor genes TAS1R1 and TAS1R3. However, actual linkage of the genetic differences with behavioral differences related to perception of sweetness awaits further research (Newcomb et al., 2010). In the case of perception of umami, and the detection of sourness and saltiness, there are few data available that associate individual differences in perception with genetic differences. Nonetheless, as the state of knowledge concerning genetic differences in odor and taste perception improves, the food and beverage industries will have opportunity to consider the relevance of the data, and feasibility of application of results, to their particular products.

47.6.3.2 Smell Genes. As with taste, odor qualities contribute in large part to the desirability and selection of numerous foods and beverages. Perception of odor is arguably more complex to understand than that of taste given that genes for odorant receptors comprise one of the largest gene families in the human genome (Hasin-Brumshtein et al., 2009). What is clear is that marked individual differences occur amongst people in terms of both sensitivity and odor-quality perception (Ayabe-Kanamura et al., 1998). Twin studies have provided an important source of data that suggest that individual differences in olfactory perception have a heritable basis (Finkel et al., 2001), although the influences of heritability appear to be swamped by environmental factors in later life (Doty et al., 2011). Recently, Jaeger et al. (2010), using a genome-wide association approach, demonstrated differences in ability to detect the compound cis-3-hexen-1-ol, a compound that is considered an important source of flavor in fruit (e.g., green kiwifruit), vegetables (e.g., tomatoes), and in some white wines. Further research is underway (Jaeger et al., 2010) to determine whether genetic variation within the relevant area of the genome is able to be associated with behavioral data such as liking of specific foods and beverages containing cis-3-hexen-1-ol. This work has potential to serve as a model for other researchers interested in linking identified genetic variation with human behaviors, including those involved in the preference, selection, and purchasing of food and beverages.

Whether, and in what ways, the food and beverage industries will use the information from the emerging

1060 **Chapter 47 Olfaction and Taste in the Food and Beverage Industries**

science linking human genetic variation with behaviors associated with taste and odor perception remains to be seen. Newcomb et al. (2010) discuss what they perceive as opportunities for the food and beverage industries arising from the new knowledge about genetic variation, as well as some potential pitfalls. The idea that the marketplace could be segmented into groups of individuals more or less likely to prefer and purchase a particular product on the basis of their genetics, must be appealing and likely to be of increasing importance for the industries. They suggest that it will be in new flavor development that the biggest advantage to industry will probably be found. That is, costs of development of a new flavor may be reduced if the probability that it will appeal to a particular market can be accurately predicted on the basis of established genetic diversity in the consumer population. One obvious difficulty with such an approach is that the world is becoming "smaller"; many societies are ethnically multi-cultural, such that identifying individuals' genetic makeup would rely on markers of genetic differences between individuals rather than on readily found criteria such as nominated nationality, visual appearance, and so forth. One way around this would be to genotype children at birth and aim to identify their likely flavor preferences. Clearly, not only would this be an expensive operation, but it raises major ethical issues in terms of to whom such information would be made available.

47.6.3.3 Nutritional Genes. A recent development is the research area of personalized nutrition, with its basis in molecular nutrition studies. It aims to identify the nutritional needs of individuals based on their genetic background (Williams et al., 2008). If it can be understood how nutrients influence humans with different genetic backgrounds, it may be possible to give dietary advice and develop food and beverage products that will improve quality and quantity of life, with such products tailored to meet the nutritional needs of particular groups in society (Newcomb et al., 2010). Clearly, the ethical issues associated with use of individuals' genetic information to guide product development are enormous but will need to be addressed. We can expect the food and beverage industries to pay cautious attention to these genetic approaches.

47.7 SOCIETAL INFLUENCES AND FASHION

That the culture within which one exists markedly influences one's food and beverage choices is not in question. Consumers from geographically different world regions have been shown to have different expectations and preferences regarding food and beverage products (e.g., Do et al.,

2009). Until recently, the majority of research addressing this topic involved comparisons of consumer preference and consumption behavior as a function of culture, where culture was defined in terms of geographic location. Over the last few decades, two new cultural influences on food and beverage choice have emerged: (i) the increased interest of consumers in perceived health benefits and risks pertaining to food and beverage products; and (ii) what can loosely be termed "green issues."

47.7.1 Perceived Health Benefits

Consumers have become increasingly interested in the health benefits or perceived risks related to the food and beverage products that they consume. Informed choice in what one eats and drinks is now a common and accepted expectation for many of today's consumers. Taking wine as an example, the notion that red wine confers a health benefit on the consumer, known as the "French paradox," coincided with increased red wine consumption over white wine consumption, and more recently with an interest in red wines known to be high in compounds believed to be healthy, such as wines made from the grape varietal Tannat (Renaud and de Lorgeril, 1992; Lippi et al., 2010).

As well as favoring foods and beverages with agents perceived to offer health benefits, consumers have shown, over recent decades, an increased interest in avoiding what they perceive to be risky produce in terms of perceived healthiness, to the point of tolerating less palatable foods such as bitter vegetables. However, products that sacrifice pleasant flavor in the interest of being healthy seldom survive in the market; indeed, health benefits tend not to over-ride the sensory hedonic quality of foods, such as low fat or low salt products in determining choice, and is one reason why most diets fail (Prescott, 2012).

47.7.2 Green Choices: Organic and Biodynamic Food

Another influence that is usually short-lived in over-riding chemosensory determinants of food choice is so-called green issues, which are political or ideological in nature. These include individuals' concerns for "ecologically sustainable agricultural practices" such as "organic-" and "biodynamically-" grown produce. Consumer advocates for preferring such products stress absence of pesticides in production, absence of animal-unfriendly methods (free-ranged versus caged chickens), truth in labeling (to reveal inclusion of preservatives and flavor chemicals, including monosodium glutamate), and preferred sources of origin such as locally-made over imported products which have generated "more carbon" in their transport.

Marketing of organic and biodynamic foods and beverages includes notions such as "purity" and "freshness," terms which imply a sensory quality, seldom confirmed by sensory measurement. Further, products produced by biodynamic agriculture typically command a higher price than more conventionally-produced products (Smith and Barquin, 2007, Kramer, 2006). However, research supporting superior sensory or nutritive value from organic or biodynamic produce is lacking. One of the few peer-reviewed studies comparing biodynamic and organic viticulture reported no consistent differences in any of the physical, chemical, and biological parameters measured (Reeve et al., 2005).

From these two social movements, we can deduce that the importance of sensory drivers of choice will act in opposition to the forces driving consumer interest in products with perceived health benefits, and "green value." To survive in the market, such products will need to have a sufficient amount of sensory quality.

47.7.3 Cognitive Processes

Human experience includes imaging (i.e., remembering the sensory qualities such as odors, tastes, tactile sensations), judgment (making a decision), and language (Cowan, 1988; Parr, 2008). Food evaluations and choice involve many cognitive factors in the decision process. These factors can be idiosyncratic (Parr, 2003), not easily modifiable, and can include cognition, emotions, ideas, desires, motives, and context.

O'Mahony and colleagues have shown how cognitive judgment influences a person's food and beverage choices (O'Mahony, 1992): perception of sensory differences will depend on a person's propensity or bias toward saying that something is "different" when they are unsure (O'Mahony, 1991; 1992). Such "response bias" is a cognitive factor central to smell and taste judgment. When two products are easily distinguishable, response bias is not an issue: a consumer will detect the difference every time and select according to established preferences, but when subtle change in a product may exist, difference testing needs to take into account the response bias (O'Mahony, 1991; 1992).

47.7.3.1 Expectations. Expectations generated by both intrinsic (sensory) cues and extrinsic cues (non-sensory stimuli) can influence chemosensory perception leading to biased judgments (Deliza and MacFie, 1996). Coloring a white wine red influences the aromatic description of the wine (Pangborn et al., 1963; Morrot et al., 2001), in wine professionals, as well as in consumers (Parr et al., 2003). Similar effects have been demonstrated with extrinsic cues. Brochet and Morrot (1999) reported that information on a wine's label influenced sensory evaluation of the wine. Siegrist and Cousin (2009) reported

that participants' wine evaluations were influenced by whether negative or positive information (wine critic Parker's ratings) was given to participants prior to their tasting of the wines. They found that favorable sensory expectations were generated by a high Parker score. Similarly, Piqueras-Fiszman and Spence (2012) weighed over 500 wine bottles in a wine retail store and demonstrated that the weight of a wine bottle correlated positively and significantly with wine price.

Several models for integrating the information a consumer receives during their sensory evaluations of a product are assessed by Deliza and MacFie (1996). Such models are based to some degree on Festinger's (1957) notion of cognitive dissonance. In this context, cognitive dissonance refers to mental conflict between what we are currently experiencing in a food or beverage and what we expected to experience. Research shows that the information already in a consumer's head, known as "top-down" or idea-based input (Reisberg, 1997; Dalton, 2000), can over-ride the sensorial input, biasing a person to "perceive" and report sensory experiences that are unrealistic given the nature of the product being experienced. Further, this can be accompanied by marked confidence that the erroneous judgment is in fact correct (Cain and Potts, 1996). This phenomenon renders the wine industry open to damaging fraud by false labeling, as reported in several countries (USA; New Zealand) in 2012.

Many factors influence the discrepancy between expected and experienced liking of food products. These include health and nutrition information (Tuorila et al., 1994) and information about product origin (Caporale and Monteleone, 2001). The rating of liking is influenced by factors such as food labeling, packaging, price, information, advertising, as well as intrinsic factors such as aroma, taste, and textural qualities. However, whether such factors modify the actual sensory experience or whether they merely create a cognitive status that modifies the overall judgment of the product, is not clear. In a study with wine, Siegrist and Cousin (2009) separated these two factors by providing either positive or negative information about a wine critic's ratings of the wines at differing times, namely before the wine tasting or after the wine tasting but prior to the wine evaluation by participants. Results showed that information given prior to tasting the wines influenced participants' ratings, but information given after tasting but before making their judgments did not influence participants' ratings. This suggests that once a person has had an opportunity to taste a product unencumbered with prior information, their judgment thereafter is based on that sensory experience and not on intervening information, be it positive or negative.

In the food and beverage industries expectations of consumers are continually being manipulated by the marketplace: fancy and expensive packaging, label design,

1062   **Chapter 47   Olfaction and Taste in the Food and Beverage Industries**

price, and so forth. What is important in marketing is that the expectation is met. Un-met expectations can result in delayed reactions as a person tries to deal with the discrepancy, and difficulty in forming preferences for the object (Wilson et al., 1989).

47.7.3.2 The Dynamic Nature of Liking.

There are many factors determining the hedonic quality of what we eat and drink, and this is the subject of a recent book (Prescott, 2012). Understanding why people like a particular food, is often a challenging and changing objective. Some factors involved are discussed here.

REPEATED EXPOSURE A consumer's perception and hedonic judgment of a product can change with repeated exposure to the product. Changes in liking as a function of repeated exposure can have positive health benefits and be commercially exploited. Increased liking can be achieved for food with reduced sodium chloride, sucrose or dietary fat, simply by sufficient repeated exposure to the "healthier" product (Sullivan and Birch, 1990). Understanding the parameters of food novelty, and therefore variables such as complexity, which are likely to interact with repeated exposure of food items to influence sensory perception and liking, is a current research area of interest (Köster and Mojet, 2007; Medel, 2011).

COGNITION AND EMOTION Influence of emotions and prior memories often occurs outside conscious awareness: a particular smell can influence a person's behavior (e.g., the kind of judgment they make), even though the person has no awareness of this, and may not even be conscious of the odor's presence (Herz, 1997). The influence of knowledge about food and beverage products that has been gained unintentionally (i.e., without explicit attention or learning) on memory for food items and on food preferences has been studied (Haller et al., 1999; Degel and Köster, 1999; Mojet and Köster, 2002). Use of emotive odor, such as freshly-brewed coffee or freshly-baked bread, is now commonplace in many commercial settings. Research on non-intentional learning and memory, involving odor, suggests that not only do such techniques have a reasonable probability of being successful, but that a consumer may well have no awareness of the ambient stimuli (e.g., pleasant odors) contributing to their decision making, thus raising ethical issues for consideration.

47.8 CONCLUSIONS

Chemosensory science of olfaction and taste and the knowledge it produces is essential to the modern food and beverage industries. Understanding the sensory and non-sensory determinants of consumer behavior involving products is crucial to the economic success of companies within these industries.

Recent advances in food production and processing technology have expanded the choices for, and availability of, food and beverages to increasingly affluent consumer populations. As populous nations such as China, India, Indonesia, and Brazil increase in affluence, global economic growth is likely to continue, and with it chemosensory science. So long as consumers are free to choose what they buy, competitive companies will need to be informed about chemosensory science and apply it innovatively. Understanding why people select the products that they do, whether a function of age, gender, culture, contemporary fashion, genetic predisposition, political persuasion, or experience and familiarity, will require expertise and a constructive interaction between chemosensory scientists both within and outside of commercial food and beverage enterprises. They stand to learn from each other and to influence the direction, aims and methods of research activities, such as those examining the influence of aroma on the perception of sweetness intensity and how to develop low salt and low fat products with acceptable flavors. Understanding the fundamentals of chemoreception at the molecular level will provide flavor chemists with a rational basis for designing the next generation of economically and experientially compelling flavor molecules. Brain scanning by PET, fMRI and other methods offer another level of analysis by which insights into human chemosensory perception will, in future, yield useful knowledge for industrial applications. Developments in sensors, arrays, and algorithms for non-human judgments of product quality will find a greater place in production and quality control fields, and reduce costs of, and dependence on, human judges. Nevertheless, the consumer is the final judge of products in the all-important marketplace and it is their chemosensory systems that will remain an important meeting point for industrialists and chemosensory scientists.

REFERENCES

Aaron, J. I., Mela, D. J. and Evans, R. E. (1994) The influence of attitudes, beliefs and label information on perceptions of reduced-fat spread. *Appetite*, 22(1): 25–38.

Alimohammadi, H., and Silver, W. S. (2002). Chemesthesis: Hot and cold mechanisms. *ChemoSense*, 4(2): 1–2, 5–6, 9.

Amerine, M. A., Pangborn, R. M., and Roessler, E. B. (1965). *Principles of Sensory Evaluation of Food*, New York: Academic.

Araujo de, I. E. and Rolls, E. T. (2004). Representation in the human brain of food texture and oral fat. *J. Neurosci.* 24(12): 3086–3093.

Ayabe-Kanamura, S., Schicker, L., Laska, M., et al. (1998). Differences in perception of everyday odors: A Japanese-German cross-cultural study. *Chem. Senses*, 23: 31–38.

References

Baines, D. and Knights, J. (2005). Applications I: Flavors. In *Chemistry and Technology of Flavors and Fragrances*, Rowe, D. J. (Ed). Oxford UK: Blackwell.

Barnett, D. (1999) Probabilities and possibilities: On-line sensors for food processing. In *Taste and Aromas: The Chemical Senses in Science and Industry*, Bell, G. A., and Watson, A. J. (Eds). London: Blackwell Science.

Bartoshuk, L. M. (1993). The biological basis of food perception and acceptance. *Food Qual. Pref.* 4: 21–32.

Bell, G. A. (1996) Molecular mechanisms of olfactory perception: Their potential for future technologies. *Trends Food Sci. Technol.*, 7, 425 – 431.

Bell, G. A., Ng, F., Waring, J., and Vereker, M. (1992). *Exporting to Japan: A Guide to the Japanese Food Market and Distribution System*. Sydney: CSIRO.

Bell, G. A., and Watson, A. J. (Eds) (1999). *Taste and Aromas: The Chemical Senses in Science and Industry*, London: Blackwell Science.

Bell, G. A., and Song, H.-J. (2004). Genetic Basis for 6-n-Propylthiouracil Taster and Supertaster Status Determined Across Cultures. In *Genetic Variation in Taste Sensitivity*, Prescott, J., and Tepper, B. J. (Eds). New York: Marcel Dekker.

Bell, G. A., and Wu, W. (2006). The industrial e-nose: Protecting people and profits. *ChemoSense*, 8(2): 1–5.

Bell, G. A., and Strauss, S. C. (2007). Dynamics of food flavor perception in the over 60s: New implications for food design and diet for older people. *ChemoSense*, 9(2): 1–10.

Bengtsson, S., Berglund, H., Gulyas, B., et al. (2001). Brain activation during odor perception in males and females. *Neuroreport*, 12(9): 2027–2033.

Billot, P.-E., Comte, A., Galliot, E., et al. (2011). Time course of odorant- and trigeminal-induced activation in the human brain: an event-related functional magnetic resonance imaging study. *Neurosci.* 189: 370–376.

Birch, L. L. (1979). Preschool children's food preferences and consumption patterns. *J. Nutrit. Educ.*, 11: 189–192.

Boyle, J. A., Djordjevic, J., Olsson, M.J., et al. (2009). The human brain distinguishes between single odorants and binary mixtures. *Cerebral Cortex*, 19: 66–71.

Brandt, M. A., Skinner, E. Z. and Coleman, J. A. (1963). Texture profile method. *J. Food Sci.* 28: 404–409.

Buck, L., and Axel, R. (1991). A novel multigene family may encode odorant receptors: a molecular basis for odor recognition. *Cell*, 65(1): 175–187.

Burdach, K. and Doty, R. L. (1987). Retronasal flavor perception: Influences of mouth movements, swallowing and spitting. *Physiol. Behav.* 41: 353–356.

Brochet, F. and Morrot, G. (1999). Influence of the context on the perception of wine: Cognitive and methodological implications. *J. Internat. Sci. Vigne Vin: Special Issue Wine Tasting*, 33: 187–192.

Cain, W. S. and Potts, B. C. (1996). Switch and bait: Probing the discriminative basis of odor identification via recognition memory. *Chem. Senses*, 21: 35–44.

Caporale, G. and Monteleone, E. (2001). Effect of expectations induced by information on origin and its guarantee on the acceptability of a traditional food: Olive oil. *Sci. Aliments*, 21: 243–254.

Cowan, N. (1988). Evolving conceptions of memory storage, selective attention, and their mutual constraints within the human information-processing system. *Psych. Bull.* 104: 163–191.

Dalton, P. (2000). Fragrance perception: from the nose to the brain. *J. Cosmetic Sci.*, 51: 141–150.

Degel, J. and Köster, E. P. (1999). Odors: Implicit memory and performance effects. *Chem. Senses*, 24: 317–325.

Deems, D. A., Doty, R. L., Settle, R. G., Moore-Gillon, V., Shaman, P., Mester, A. F., Kimmelman, C. P., Brightman, V. J. and Snow, J. B. (1991). Smell and taste disorders: A study of 750 patients from the University of Pennsylvania Smell and Taste Center. *Arch. Otolaryngol. Head Neck Surg.* 17: 519–528.

Deliza, R., and MacFie, H. (1996). The generation of sensory expectation by external cues and its effect on sensory perception and hedonic ratings: A review. *J. Sensory Stud.* 11: 103–128.

Deloitte (2010). World Food Processing Overview. www.reportlinker.com/source=d&idreport=11330801

DeVere, R. (2003). Taste and smell disorders: a view from clinical practice, *ChemoSense*, 6(1): 6–7.

Dewis, M.L. (2005). Molecules of taste and sensation. In *Chemistry and Technology of Flavors and Fragrances*, Rowe, D. J. (Ed). Oxford UK: Blackwell.

Dijksterhuis, G., Mojet, J., Köster, E. P., et al. (2006). Workshop summary: The role of memory in food choice and liking. *Food Qual. Pref.* 17: 650–657.

Distel, H., Ayabe-Kanamura, S., Matinez-Gomez, M., et al. (1999). Perception of everyday odors: Correlation between intensity, familiarity and strength of hedonic judgment. *Chem. Senses*, 24: 191–199.

Do, V.-B., Patris, B. and Valentin, D. (2009). Opinions on wine in a new consumer country: A comparative study of Vietnam and France. *J. Wine Res.* 20, 227–245.

Doty, R. L. (2012). Olfaction in Parkinson's disease and related disorders. *Neurobiol. Dis.* 46: 527–552.

Doty, R. L. and Kamath, V. (2014). The influences of age on olfaction: a review. *Front. Psychol. Sci.* 5:20. doi:10.3389/fpsyg.2014.00020.

Doty, R. L., Petersen, I., Menseh, N. and Christensen, K. (2011). Genetic and demographic influences on odor identification ability in the very old. *Psychol. Aging.* 26: 864–871.

Doty, R. L., Shaman, P., Applebaum, S. L., et al. (1984). Smell identification ability: Changes with age. *Science* 226: 1441–1443.

Dubois, D. (2000). Categories as acts of meaning: The case of categories in olfaction and audition. *Cognit. Sci. Qtly.* 1: 33–66.

Dubois, P. and Nauges, C. (2010). Identifying the effect of unobserved quality and expert reviews in the pricing of experience goods: empirical application on Bordeaux wine. *Internat. J. Ind. Org.* 28: 205–221.

Duffy, V. B., Peterson, J. and Bartoshuk, L. M. (2004)Associations between taste genetics, oral sensations and alcohol intake. *Physiol. Behav.* 82: 435–445.

Easton, K., Bell, G. A. and Ng, F. (1997). *Exporting Food to Indonesia. A Guide for Australian Small to Medium Enterprises*. Sydney: CSIRO.

Festinger, L. (1957). *A theory of cognitive dissonance*. Illinois: Row, Peterson.

Finkel, D., Pedersen, N. L. and Larsson, M. (2001). Olfactory functioning and cognitive abilities: A twin study. J. Gerontol.: Psychol. *Sci.* 56B: 226–233.

Freeman, R. P. J., Richardson, N. J., Kendal-Reed, M. S., and Booth, D. A. (1993). Bases of a cognitive technology for food quality. *Brit. Food J.*, 95: 37–44.

Gilbertson, T. A. (2003) The lure of dietary fat: Are taste buds partly to blame? *ChemoSense*, 5(2): 1–6.

Green, B. G. (2004) Oral chemesthesis: An integral component of flavor. *Flavor Perception*, Taylor, A. J. and Roberts, D. D. (Eds). Oxford, UK: Blackwell.

Guinard J.-X. (2001) Sensory and consumer testing with children. *Trends Food Sci. Technol.* 11: 273–283.

Hadj Ali, H., Lecocq, S. and Visser, M. (2008). The impact of gurus: Parker grades and en primeur wine prices. *Econom. J.*, 118: F158–F173.

Chapter 47 Olfaction and Taste in the Food and Beverage Industries

Haller, R., Rummel, C., Henneberg, S., Pollmer, U. and Köster, E. P. (1999). The influence of early experience with vanillin on food preference later in life. *Chem. Senses*, 24: 465–467.

Halpern, B. P. (2004). Retronasal and orthonasal smelling. *ChemoSense*, 6(3): 1–8.

Halpern, B. P. (2008). Mechanisms and consequences of retronasal smelling: computational fluid dynamic observations and psychophysical measures. *ChemoSense*, 10(3): 1–8.

Hasin-Brumshtein, Y., Lancet, D. and Olender, T. (2009). Human olfaction: from genomic variation to phenotypic diversity. *Trends in Genetics*, 25: 178–184.

Hayes, J. E. and Pickering, G. J. (2012). Wine expertise predicts taste phenotype. *Am. J. Enol. Vitic.*, 63: 80–84.

Helm, E. and Trolle, B. (1946). Selection of a taste panel. *Wallerstein Laboratory Communications*, 9: 181–194.

Henkin, R. I. and Levy, L. L. (2001). Lateralization of brain activation to imagination and smell of odors using functional magnetic resonance imaging (fMRI): Left hemisphere localization of pleasant and right hemisphere localization of unpleasant odors. *J. Comp. Assist. Tomog.* 25(4): 493–514.

Herman, S. J. (2005). Applications II: Fragrance. In *Chemistry and Technology of Flavors and Fragrances*, Rowe, D. J., (Ed). Oxford UK: Blackwell.

Herz, R. S. (1997). Emotion experienced during encoding enhances odor retrieval cue effectiveness. *Am. J. Psych.*, 110: 489–506.

Hibbert, D. B. (1999). Electronic noses for sensing and analyzing industrial chemicals. In *Taste and Aromas: The Chemical Senses in Science and Industry*, Bell, G. A., and Watson, A. J. (Eds). London: Blackwell Science.

Hibbert, D. B. and Bell, G. A. (2011). A Bayesian approach to odour recognition. *ChemoSense*, 13(1): 1–6.

Hilger, J., Rafert, G. and Villas-Boas, S. (2011). Expert opinion and the demand for experience goods: an experimental approach in the retail wine market. *Rev. Econ. Stat.* 93(4): 1289–1296.

Hummel, T., Doty, R. L. and Yousem, D. M. (2005). Functional MRI of intranasal chemosensory trigeminal activation. *Chem. Senses*, 30 (suppl 1): i205–i206.

Jaeger, S. R., MacRae, J. F., Salzman, Y., et al. (2010). A preliminary investigation into a genetic basis for cis-3-hexen-1-ol odour perception: a genome-wide association approach. *Food Qual. Pref.* 21: 121–131.

Jameson, S. B. (2005) Aroma Chemicals III: Sulfur Compounds. In *Chemistry and Technology of Flavors and Fragrances*, D. J. Rowe (Ed). Oxford UK: Blackwell.

Jellinek, G. (1985). *Sensory evaluation of food: theory and practice*. London UK: Ellis Horwood.

Jones, L. V., Peryam, D. R., and Thurstone, L. L. (1955). Development of a scale for measuring soldiers' food preferences. *Food Res.* 20: 512–520.

Kamen, J. (1989). Observations, reminiscences and chatter. Sensory evaluation. In *Celebration of Our Beginnings*. Committee E-18 on Sensory Evaluation of Materials and Products. Philadelphia: ASTM, pp. 118–122.

Keast, R. S. J. (2012) Emerging evidence supporting fatty acid taste in humans. *ChemoSense*, 14(3): 1–8.

Köster, E. P., and Mojet, J. (2007). Boredom and the reasons why some new products fail. In *Consumer-led food product development*, MacFie, H. J. H. (Ed.), Cambridge, UK: Woodhead Publishing, pp. 262–280.

Köster, E.P. and Mojet, J. (2012) Flop analysis: a useful tool for future innovations (Part 1). *Agrofood Industry Hitech*, 23: 1, 6–8.

Kramer, M. (2006). *Why I buy bio*. Wine Spectator, October 31.

Labbe, D., Damevin, L., Vaccher, C., et al. (2006). Modulation of perceived taste by olfaction in familiar and unfamiliar beverages. *Food Qual. Pref.*, 17: 582–589.

Laing, D. G., Prescott, J., Bell, G. A., et al. (1993). A cross-cultural study of taste discrimination with Australians and Japanese. *Chem. Senses*, 18(2): 161–168.

Laing, D. G., Prescott, J., Bell, G. A., et al. (1994). Responses of Japanese and Australians to sweetness in the context of different foods. *J. Sensory Stud.* 9(2): 131–155.

Lawless, H. T. and Heymann, H. (1998). *Sensory Evaluation of Food: Principles and Practices*, 1st Edition. New York: Chapman and Hall.

Lawless, H. T., Heymann, H. (2010). *Sensory Evaluation of Food: Principles and Practices*, 2nd Edition. New York: Springer.

Levy, D. C. and Naidoo, B. (1999). How machines can understand smells and tastes: Controlling your product quality with neural networks. In *Taste and Aromas: The Chemical Senses in Science and Industry*, Bell, G. A., and Watson, A. J. (Eds). London: Blackwell Science.

Levy, L. M., Henkin, R. I., Hutter, A., et al. (1997). Functional MRI of human olfaction. *J. Comp. Assist. Tomog.* 21(6): 849–856.

Liem, D. G., Miremadi, F. and Keast, R. S. J. (2011). Reducing sodium in foods: The effect on flavor. *Nutrients*, 3: 694–711.

Lippi, G., Franchini, M., and Guidi, G. C. (2010). Red wine and cardiovascular health: the "French Paradox" revisited. *Internat. J. Wine Res.*, 2:1–7.

Lockshin, L., Jarvis, W., d'Hauteville, F., and Perrouty, J. P. (2006). Using simulations from discrete choice experiments to measure consumer sensitivity to brand, region, price, and awards in wine choice. *Food Qual. Pref.* 17: 166–178.

Lundahl, D. (2012). *Breakthrough Food Product Innovation through Emotions Research*. London: Elsevier.

Lyon, D. H. (1999). Meeting industry needs: The European perspective. In *Taste and Aromas: The Chemical Senses in Science and Industry*, Bell, G. A., and Watson, A. J. (Eds). London: Blackwell Science.

Lyon, D. H., Francombe, M. A., Hasdell, T. A. and Lawson, K. (1992). *Guidelines for Sensory Analysis in Food Product Development and Quality Control*, London: Chapman and Hall.

Mackay-Sim, A. (1999). Electronic sensor technologies for the food and allied industries. In *Taste and Aromas: The Chemical Senses in Science and Industry*, Bell, G. A., and Watson, A. J. (Eds). London: Blackwell Science.

Malnic, B., Hirono, J., Sato, T., and Buck, L. B. (1999). Combinatorial receptor codes for odors. *Cell*, 96: 713–723.

McCabe, C. and Rolls, E. T. (2007). Umami: A delicious flavour formed by convergence of taste and olfactory pathways in the human brain. *Europ. J. Neurosci.* 25(6): 1855–1864.

Matsuda, T. and Doty, R. L. (1995). Age-related taste sensitivity to NaCl: Relationship to tongue locus and stimulation area. *Chem. Senses*, 20: 283–290.

Medel, M. (2011). Perception de la qualité du vin par les consommateurs. Unpublished Ph.D. thesis, University of Burgundy, France. http://www.sudoc.abes.fr/.

Mennella, J. A., Pepino, M. Y. and Reed, D. R. (2005). Genetic and environmental determinants of bitter perception and sweet preferences. *Pediatics*, 115: e216–e222.

Mojet, J., Christ-Hazelhof, F. and Heidema, J. (2001). Taste perception with age: generic or specific losses in threshold sensitivity to the five basic tastes. *Chem. Senses*, 26: 845–860.

Mojet, J. and Köster, E. P. (2002). Texture and flavor memory in foods: An incidental learning experiment. *Appetite*, 38: 110–117.

Mojet, J., Heidema, J. and Christ-Hazelhof, F. (2003). Taste perception with age: generic or specific losses in supra-threshold intensities of five taste qualities. *Chem. Senses*, 28: 394–413.

Morrot, G., Brochet, F. and Dubourdieu, D. (2001). The color of odors. *Brain and Language*, 79: 309–320.

Moskowitz, H. R. (1983). *Product Testing and Sensory Evaluation of Foods*. Westport CT: Food and Nutrition Press.

Moskowitz, H. R. (Ed.) (1987). *Food Texture*. New York: Marcel Dekker.

Moskowitz, H. R., Gofman, A., Itty, B., et al. (2001). Rapid, inexpensive, actionable concept generation and optimization – the use and promise of self-authoring conjoint analysis for the food service industry. *Food Serv. Technol.* 1: 149–168.

Moskowitz, H. R., Krieger, B. and Ettinger Lieberman, L. (2011). The 21st Century development of products: where consumer guidance is taking us. In Kaden, B., Linda G., and Price M. (Eds) *Leading Edge Marketing Research: 21st Century Tools and Practices*. Thousand Oaks Ca.: Sage Publishing, pp. 71–88.

Murphy, P. (2002). Stakeholder presentation – retailer. ASVO Proceedings: Who's running this show? Future directions for the Australian wine show system, Adelaide Aust.: ASVO, pp. 31–32.

Mustonen, S., and Tuorila, H. (2007) The unfulfilled promise of flavour enhancement. *ChemoSense*, 9(3): 7–9.

Mustonen, S., and Tuorila, H. (2009) Sensory education of school children: What can be achieved? *ChemoSense*, 12(1): 1–6.

Newcomb, R. D., McRae, J., Ingram, J., et al. (2010). Genetic variation in taste and odour perception: an emerging science to guide new product development. In *Consumer-driven innovation in food and personal care products*, Jaeger, S.R., and MacFie, H. (Eds). Oxford, U.K: Woodhead Publishing.

Ninomiya, K. (2003). Umami: An oriental or a universal taste? *ChemoSense*, 5(3): 1–8.

Noble, A. C. (1999). Sensory evaluation and the wine industry. *ChemoSense*, 1(2): 1–4.

O. I. V. (1994). *Standard on International Wine Competitions*. France: Office International de la Vigne et du Vin.

Ohla, K., Toepel, U., Le Coutre, J. and Hudry, J. (2012a). Visual-gustatory interaction: orbitofrontal and insular cortices mediate the effect of high-calorie visual food cues on taste pleasantness. *PLoS One 7*, e32434.dx.plos.org.

Ohla, K., Busch, N. A. and Lundstrom, J. N. (2012b). Time for taste– a review of the early cerebral processing of gustatory perception. *Chemosens. Percep.* 5(1): 87–99.

O'Mahony, M. (1986). *Sensory Evaluation of Food*. New York: Marcel Dekker.

O'Mahony, M. (1991). Descriptive analysis and concept alignment. In *Sensory Science Theory and Application in Foods*, Lawless, H. T., and Klein, B. P. (Eds). pp. 223–267, New York: Marcel Dekker.

O'Mahony, M. (1992). Understanding discrimination tests: A user-friendly treatment of response bias, rating and ranking R-index tests and their relationship to signal detection. *J. Sensory Stud.* 7: 1–47.

Owen, C. M. and Patterson, J. (2002). Odour-liking physiological indices: a correlation of sensory and electrophysiological responses to odour. *Food Qual. Pref.* 13(5): 307–316.

Owen, C. M., Patterson, J. and Silberstein, R. B. (2002). Olfactory modulation of steady-state visual evoked potential topography in comparison with differences in odor sensitivity. *J. Psychophysiol.*, 16(2): 71–78.

Pangborn, R. M., Berg H., and Hansen, B. (1963). The influence of color on discrimination of sweetness in dry table wine. *Am. J. Psychol.* 76: 492–495.

Parr, W. V. (2003). The ambiguous nature of our sense of smell. *The Australian & New Zealand Grapegrower & Winemaker*, Annual Technical Issue, pp. 114–116.

Parr, W. V. (2008). Application of cognitive psychology to advance understanding of wine sensory evaluation and wine expertise. In *Applied Psychology Research. Trends*, Kiefer, K. H. (Ed). Nova Science Publishers, pp. 55–76.

Parr, W. V., Heatherbell, D. A., and White, K. G. (2002). Demystifying wine expertise: Olfactory threshold, perceptual skill and semantic memory in expert and novice wine judges. *Chem. Senses*, 27: 747–756.

Parr, W. V., White, K. G. and Heatherbell, D. (2003). The nose knows: Influence of colour on perception of wine aroma. *J .Wine Res.* 14: 79–102.

Pelchat, M. L., Johnson, A., Chan, R., et al. (2004). Images of desire: Food-craving activation during fMRI. *NeuroImages*, 23(4): 1486–1493.

Persaud, K. C. (2003). Olfactory system cybernetics: Artificial noses. In *Handbook of Olfaction and Gustation*, Doty, R. L. (Ed). New York: Marcel Dekker, Inc.

Peryam, D. R. and Swartz, V. W. (1950). Measurement of sensory differences. *Food Technol.* 4: 390–395.

Pickering, G. J., Simunkova, K., and DiBattista, D. (2004). Intensity of taste and astringency sensations elicited by red wines is associated with sensitivity to PROP (6-n-propythiouracil). *Food Qual. Pref.* 15: 147–154.

Piggott, J. R. (Ed.) (1984) *Sensory Analysis of Foods*. London: Elsevier.

Piqueras-Fiszman, B. and Spence, C. (2012). The weight of a bottle as a possible extrinsic cue with which to estimate the price (and quality) of a wine? *Observed correlations. Food Qual. Pref.* 25: 41–45.

Polak, E. H. (2000) Understanding Odour: Vertebrate Olfactory Receptors. *ChemoSense*, 2(2): 1–7.

Popper, R. and Kroll, J. J. (2007) Consumer testing of products using children. In *Consumer-led Food Product Development*, MacFie, H. J. H. (Ed)., Cambridge UK: Woodhead.

Prescott, J. (1998) Comparison of taste perceptions and preferences of Japanese and Australian consumers: Overview and implications for cross-cultural research. *Food Quality and Preference*, 9, 393–402.

Prescott, J. (1999) Flavor as a psychological construct: Implications for perceiving and measuring the sensory qualities of foods. *Food Quality and Preference*, 10, 349–356.

Prescott, J. (2004). Psychological processes in flavour perception. In A.J. Taylor & D.D. Roberts (Eds.) *Flavour Perception*. Blackwell Publishing: Oxford, UK.

Prescott, J. (2006). Understanding consumer responses to food: The role of genetic differences in taste sensitivity. *The Food Technologist*, 28, 174–177.

Prescott, J. (2012) *Taste Matters: Why We Like the Foods We Do*. London: Reaktion Books.

Prescott, J., Laing, D., Bell, G. A., et al. (1992). Hedonic responses to taste solutions: a cross-cultural study of Japanese and Australians. *Chem. Senses*, 17(6): 801–809.

Prescott, J. and Bell, G. A. (1995). Cross-cultural determinants of food acceptability: Recent research on sensory perceptions and preferences. *Trends in Food Sci. Technol.* 6: 201–205.

Prescott, J. and Khu, B. (1995). Changes in preference for saltiness within soup as a function of exposure. *Appetite*, 24: 302.

Prescott, J., Bell, G. A., Gillmore, R., et al. (1997) Cross-cultural comparisons of Japanese and Australian responses to manipulations of sweetness in foods. *Food Qual. Pref.* 8(1): 44–55.

Prescott, J. and Tepper, B. J. (Eds) (2004) *Genetic Variation in Taste Sensitivity*. New York: Marcel Dekker.

Reed, D. R., Tanaka, T. and McDaniel, A. H. (2006). Diverse tastes: genetics of sweet and bitter perception. *Physiol. Behav.* 88: 215–226.

Reisberg, D. (1997). *Cognition: Exploring the Science of the Mind*. New York: W. W. Norton.

Reiter, E. R. and Constanzo, R. M. (2003) The overlooked impact of olfactory loss: Safety, quality of life and disability issues. *Chemo Sense*, 6(1): 1–4.

Renaud, S. and de Lorgeril, M. (1992). Wine, alcohol, platelets, and the French paradox for coronary heart disease. *The Lancet*, 339: 1523–1526.

Reeve, J., Carpenter-Boggs, L., Reganold, J., et al. (2005). Soil and wine grape quality in biodynamically and organically managed vineyards. *Am. J. Enol. Viticult.* 56: 367–376.

Rowe, D. J. (2005a) Introduction. In *Chemistry and Technology of Flavors and Fragrances*, Rowe, D. J. (Ed.). Oxford UK: Blackwell.

Rowe, D. J. (Ed) (2005b) *Chemistry and Technology of Flavors and Fragrances*, Oxford UK: Blackwell.

Rowe, D. J. (2005c) Aroma Chemicals I: C, H, O Compounds. In *Chemistry and Technology of Flavors and Fragrances*, Rowe, D. J. (Ed). Oxford UK: Blackwell.

Roudnitzky, N., Bufe, B., Wooding, S., et al. (2009). Phenotype-genotype correlation in individuals' sensitivities to the bitter off-taste of sulfonyl amide sweetners. *Pangborn Abstracts*, O3.1.

Savic, I. (2002). Imaging of brain activation by odorants in humans. *Curr. Opin. Neurobiol.* 12(4): 455–461.

Schiffman, S. S. (2000). Intensification of sensory properties of foods for the elderly. *J. Nutrition*, 130: 927S–930S.

Schiffman, S. S. (2002) Flavor enhancement and its positive health benefits, *Aroma-Chology Rev.*, 10(2): 1–5.

Schiffman, S. S. & Zervakis, J. (2002). Taste and smell perception in the elderly: effect of medications and disease. *Adv. Food Nutrition*, 44: New York: Academic Press, p. 247.

Scott, T. R. (2001) The sense of taste. *ChemoSense*, 4(1): 1–9.

Seo, H.-S., Iannilli, E., Hummel, C., et al. (2011) A salty-congruent odor enhances saltiness: Functional magnetic resonance imaging study. *Hum. Brian. Mapp.* Doi:1010.1002/hbm.21414.

Siegrist, M. and Cousin, M.-E. (2009). Expectations influence sensory experience in a wine tasting. *Appetite*, 52: 762–765.

Small, D. M., Jones-Gotman, M., Zatorre, R. J., et al. 1997). Flavor processing: more than the sum of its parts. *Neuroreport*, 8(18): 3913–3917.

Small, D. M., Zatorre, R. J., Dagher, A., et al. (2001). Changes in brain activity related to eating chocolate. From pleasure to aversion. *Brain*, 124: 1720–1733.

Small, D. M., Voss, J., Mak, Y. E., et al. (2004). Experience-dependent neural integration of taste and smell in the human brain. *J. Neurophysiol.* 92: 1892–1903.

Smith, D. and Barquin, J. (2007). Biodynamics in the wine bottle. *Skeptical Inquirer*, 31.6.

Song, H.-J. (2000). The gustatory role of fat. *Chemo Sense*, 3(1): 1–3.

Spetter, M. S., Smeets, P. A. M., de Graaf, C. and Viergever, M. A. (2010) Representation of sweet and salty taste intensity in the brain. *Chem. Senses*, 35: 831–840.

Stevenson, R. J. and Prescott, J. (1994). The effects of prior experience with capsaicin on ratings of its burn. *Chem. Senses*, 19: 651–656.

Stone, H. and Sidel, J. (1993). *Sensory evaluation practices*, 2nd ed. San Diego: Academic.

Stone, H., Sidel, J., Oliver, S., et al. (1974). Sensory evaluation by quantitative descriptive analysis. *Food Technol.*, 28: 24.

Sullivan, S. A. and Birch, L. L. (1990). Pass the sugar, pass the salt: Experience dictates preference. *Dev. Psychol.* 26: 546–551.

Szczesniak, A. S., Loew, B. J. and Skinner, E. Z. (1975). Consumer texture profile technique. *J. Food Sci.*, 40: 1253–1257.

Taylor, A. (2005). Flavor release and flavor perception. *ChemoSense*, 7(4): 1–6.

Taylor, A. J. and Hort, J. (2004). Measuring proximal stimuli involved in flavour perception. In *Flavour Perception*, Tayor. A. J. and Roberts, D. D. (Eds). Oxford, UK: Blackwell.

Tepper, B. J. (2008). Nutritional implications of genetic taste variation. The role of PROP sensitivity and other taste phenotypes. *Ann. Rev. Nutrition*, 28: 367–388.

Tuorila, H., Cardello, A. V. and Lesher, L. L. (1994). Antecedents and consequences of expectations related to fat-free and regular-fat foods. *Appetite*, 23: 247–264.

Tuorila, H. and Monteleone, E. (2009). Sensory food science in the changing society: Opportunities, needs, and challenges. *Trends Food Sci. Technol.* 20: 54–62.

Turin, L. (1996). A spectroscopic mechanism for primary olfactory reception. *Chem. Senses*, 21(6): 773–791.

Turin, L. (2005). Rational Odorant Design. In *Chemistry and Technology of Flavors and Fragrances*, Rowe, D. J. (Ed) Oxford UK: Blackwell.

Turin, L., and Yoshii, F. (2003). Structure-odor relationships: A modern perspective. In *Handbook of Olfaction and Gustation*, Doty, R. L. (Ed). New York: Marcel Dekker, Inc.

Van Herpen, E., Seiss, E. and van Trijp, H. C. M. (2012). The role of familiarity in front-of-pack label evaluation and use: A comparison between the United Kingdom and the Netherlands. *Food Qual. Pref.* 26: 22–34.

Veldhuizen, M. G. and Small, D. M. (2011). Modality-specific neural effects of selective attention to taste and odor. *Chem. Senses*, 36(8): 747–760.

Walsh, B. (2002). Stakeholder presentation – wine committee, Royal Adelaide Wine Show. ASVO Proceedings: Who's running this show? *Future Directions for the Australian Wine Show System*, Adelaide: ASVO, 10–12.

Weller, J. (1999). The value of sensory science to the food product developer and food processor. In *Taste and Aromas: The Chemical Senses in Science and Industry*, Bell, G. A., and Watson, A. J. (Eds). London: Blackwell Science.

Williams, C. M., Ordovas, J. M., Lairon, D., et al. (2008). The challenges for molecular nutrition research 1: linking genotype to healthy nutrition. *Genes and Nutrition*, 3: 41–49.

Wilson, T. D., Kraft, D. and Dunn, D. S. (1989). The disruptive effects of explaining attitudes: The moderating effect of knowledge about the attitude object. *J. Exp.Soc. Psych.* 25: 379–400.

Wilson, D. A. and Stevenson, R. J. (2006). *Learning to Smell: Olfactory Perception from Neurobiology to Behavior*. Baltimore: Johns Hopkins University Press.

Youngentob, S. L. (1999). Introduction to the sense of smell: Understanding odours from the study of human and animal behavior. In *Tastes and Aromas: The Chemical Senses in Science and Industry*, Bell, G. A. and Watson, A. J. (Eds). London: Blackwell Science.

Yousem, D. M., Maldjian, J. A., Siddiqi, F., et al. (1999a). Gender effects on odor-stimulated functional magnetic resonance imaging. *Brain Res.* 818: 480–487.

Yousem, D. M., Maldjian, J. A., Hummel, T., et al. (1999b). The effect of age on odor-stimulated functional magnetic resonance imaging. *Am. J. Neuroradiol.* 20: 600–608.

Zvively, M. (2005) Aroma Chemicals II: Heterocycles. In *Chemistry and Technology of Flavors and Fragrances*, Rowe, D. J. (Ed) Oxford UK: Blackwell.

Chapter 48

Olfaction and Gustation in the Flavor and Fragrance Industries

BENJAMIN MATTEI, ARNAUD MONTET, and MATTHIAS H. TABERT

48.1 INTRODUCTION

In the preceding chapters different aspects of the olfactory and gustatory sensory systems have been discussed at length. In this chapter, we discuss how chemosensory perception is used in the flavor and fragrance industry to design flavors and fragrances. We first provide an overview of this industry, including its history. We then explore how this industry employs *chemosensory*, *affective*, and *conceptual* measurement in the development and assessment of products.

48.2 THE FLAVOR AND FRAGRANCE INDUSTRY AND ITS HISTORY

The primary mission of the modern flavor and fragrance industry is to create and blend raw materials into flavors and fragrances that meet market demands. While this industry started as an artisanal activity, it is now a multi-billion dollar chemical industry in which raw materials are either purchased, transformed from existing materials, synthesized *de novo*, or extracted from natural products. Flavorists and perfumers blend these ingredients into proprietary formulas which are then used in the development of the final fragrance or flavor.

The roots of creating flavors and fragrances date back over 5000 years ago to the early Egyptians. Herbs, spices, and flowers were used to flavor food and to make perfumes through distillation. While the more systematic use of distillations by pharmacies to make medicinal remedies

developed in the 16th and 17th centuries, the first aromatic chemicals were synthesized in the 19th century (see Chapter 1). Since then, thousands of aromatic compounds have been extracted or synthesized, and the production of flavors and fragrances has reached a global scale. As a result of such industrialization, the necessity arose to define and apply quality standards in manufacturing and distribution to ensure consistency in the chemosensory properties of the final products.

Although the original motivation behind flavor creation was to enhance the taste of food, other functions came into being as processed foods became popular. These included recreating the taste of original foods, masking off-notes in product formulations, and increasing shelf life. Similarly, the development of fragrances was initially focused on giving the human body, animals, objects, and living spaces a pleasant scent, as well as masking unwanted odors, such as sweat, smoke, dirt, and excrement. Subsequently fragrances were added to a wide range of products and some came to symbolize functions of such products. For example, lemon odor came symbolize cleanliness itself, reflecting the application of lemon scent to a wide range of cleaning agents.

As the industry matured and became more competitive, optimizing the quality of flavors and fragrances became a major priority. To achieve this end, considerable capital investment was made in research as to how to best create fragrances and flavors that meet the demands of increasingly sophisticated consumers from a variety of cultures. To capitalize on such advances, companies that market flavored or odorized products approach flavor and fragrance companies with "briefs" that specify the chemosensory

---

Handbook of Olfaction and Gustation, Third Edition. Edited by Richard L. Doty.
© 2015 Richard L. Doty. Published 2015 by John Wiley & Sons, Inc.

requirements needed in the end product and its desired effects on consumers. Each solicited company then develops a product to meet the specifications outlined in the brief with the goal of winning a contract. Briefs typically contain key performance indicators, such as a requirement to match and replace a current flavor or fragrance because of technical, regulatory, or economic reasons, or to create a new flavor or fragrance for introduction into the marketplace.

48.3 DESIGNING FLAVORS AND FRAGRANCES

Professional artisans, namely perfumers and flavorists, are essential in the creating of new flavors and fragrances. Such artisans employ chemical libraries that consist of successful final formulations, flavor or fragrance keys composed of molecules that illustrate a sensory direction, and single-molecule chemicals (Figure 48.1). The successful formulations are generally organized according to sensory profiles, regulatory information, price, and other factors. These often serve as a starting point to elaborate on and obtain the desired end-result in new creation. Flavor and fragrance keys further facilitate and accelerate the creative process. For example, the use of a flavor key on a bland orange base could steer the overall flavor profile in a blood orange, ripe, peely, or candy direction. Typically, flavor key libraries are specific to a given artisan and define his or her olfactory personality or signature. One of the main tasks of a trainee perfumer or flavorist is to start building his or her own library of keys.

Nowadays, such artisans often work in conjunction with equipment that helps to identify or isolate volatile and non-volatile elements from natural products (e.g., the nectar of a flower). In the case of volatiles, separation is commonly performed using gas chromatography (GC), such that different volatiles appear at different times as they pass through a heated column within the device after the product is injected into it. In split-stream GC-analyses, the various "peaks" that occur over time represent different chemicals – chemicals which can be simultaneously sent to a nozzle where they can be smelled by the perfumer or flavorist. Further chemical identification of such peaks can be made using mass spectrometry (MS). This process makes it possible for perfumers and flavorists to discover combinations of molecules within natural products that guide them in better mimicking nature and inspiring the creative process. Importantly, the merging of sensory and analytical procedures makes it possible to understand and correct for seasonal and other variations in the chemical constituents of natural materials, such as flowers and herbs, thereby providing more consistent products.

48.4 MEASURING HUMAN SMELL AND TASTE PERCEPTION IN THE INDUSTRIAL CONTEXT

Aside from marketplace acceptance, consumer perception is the ultimate performance measure of a flavor or fragrance. In this context, perception is defined as the organization, identification, and interpretation of sensory information. At IFF, we focus on understanding consumer perceptions at three general levels: sensory, affective, and conceptual. The *sensory level* refers to the perceived sensory qualities of a product or its ingredients, such as the identification of its flavor or fragrance notes and their relative intensities. The *affective level* refers to the emotional or hedonic influences of a product on consumers. The *conceptual level* refers to more cognitive aspects of perception. For example, each time a person is exposed to and perceives an odor or taste, a *percept* or mental impression of the agent and its source is formed or reinforced. Different percepts contribute to the formation of *concepts*, or mental representations. For example, concepts like "authentic," "natural," "healthy," or "feminine" may apply to a strawberry flavor in yoghurt or to the fragrance

Figure 48.1 Image from an IFF perfumer evaluating a key in front of his fragrance library.

48.4 Measuring Human Smell and Taste Perception in the Industrial Context

of shea butter in a skin care product. A major challenge of the flavor and fragrance industry is to decipher how such concepts are formed and their influence on the buying habits of consumers.

48.4.1 Assessing Sensory Aspects of Flavors and Fragrances

48.4.1.1 Sensory Panels.
Ultimately, the sensations induced by flavors and fragrances are evaluated by groups of people comprised of sensory panels. In an ideal world, such panels would consist of a large group of extra-sensitive people that are highly trained and able to describe the sensory elements of products with the utmost precision and reliability. In reality, sensory panels are usually composed of around ten to thirty people who are screened for their sensitivity (detection and saturation thresholds as well as their ability to discriminate between closely perceived intensities) and cognitive skills (ability to describe, identify, memorize and focus) (Stone and Sidel, 2004). The performance of sensory panelists are also carefully monitored over time to ensure repeatability, discrimination and consistency. However, less extensive variants are common. For example, in some elements of the fragrance industry, only one or a few extremely sensitive and highly trained panelists, termed evaluators, are used. This is risky, however, since it precludes averaging out possible mistakes or inconsistencies over a number of people.

Reflecting the need for less expensive, more flexible, and more holistic assessments of consumer perceptions, many sensory panels are now comprised of groups of largely untrained consumers (Varela and Areas, 2012). When the objective is to assess the similarity of sensations evoked by two different products, or preferences among different formulations, sensory panelists do not need to be specifically trained to detect flavor or fragrance notes. They just need to be sensitive and reliable enough to perceive the products and to detect possible sensory differences between them. To accommodate cost, resource, logistics, and experimental design considerations, different panels for different objectives are frequently employed. For example, separate panels may be used to assess the sensory proximity of two products, to generate an overview of the sensory space covered in a given marketplace, or to guide product development.

Although members of classic sensory panels are often screened for sensitivity using traditional procedures (see Chapters 11 and 34) and are thoroughly trained to learn and distinguish subtle differences between aromatic notes, this cannot make up for a lifetime of sensory experience in a cultural setting. Hence, the employment of panels within a culture where a product is to be marketed can sometimes be invaluable. This is exemplified by some people in India who are able to distinguish dozens of different curries, much like the Inuit peoples of the Arctic North whose language distinguishes among dozens of different states of ice and snow. Importantly, repeated use of a product can also enhance sensitivity specifically for that product. For instance, some consumers who have been using a product several times a day for years are able to detect the slightest changes in its sensory profile. Moreover, people with an aversion to certain chemicals, which are often learned, may be more sensitive than others regarding the object of aversion. Finally, on a shorter timescale, scale, hunger tends to increase sensitivity and satiety tends to decrease sensitivity for at least some chemicals (Zverev, 2004).

48.4.1.2 Sensory Panels vs Analytical Instruments.
While GCs and MS are extremely useful in detecting and identifying volatiles responsible for producing certain scents or flavors, sensory panels are preferred over such instruments in evaluating products. There are several reasons why this is the case. First, the human nose is more sensitive to some chemicals than such devices, including the more focused "electronic noses" and "electronic tongues." Second, information from different sensory modalities (e.g., smell, taste, and texture) is integrated in the human brain through neural networks, using associative memory and other processes. Mimicking this integration with instruments and electronic neural networks, while theoretically possible, has not yet been achieved and will require much more future research and development. Third, target products often are influenced by the environment with which they interact in ways that are difficult for instruments to measure. This includes, for example, reactions of a perfume on the skin and the influences of a complex oral salivary and bacterial environment on chewed food. In order to reproduce the sensations that a person perceives while using or consuming such products, one would need to reproduce artificially the same processes with instruments, which is an enormous challenge. Finally, what instruments detect is not necessarily relevant for the consumer's ultimate perception of the product.

48.4.1.3 Sensory Profiling Methods Using a Common Sensory Language.
A primary goal of sensory profiling by panels is to identify the different flavor and fragrance notes that can be perceived in an evaluated product and to establish their relative intensities using scaling methods such as rating scales (Lawless and Heymann, 2010; Chapters 11 and 34). The language is said to be common because all the panelists share the same understanding and knowledge of each note. It is therefore easy to analyze the results with very simple statistical tests. The two most well-known methods are the Quantitative Descriptive Analysis (QDA®) (Stone and Sidel, 2004) and Spectrum Descriptive Analysis (Spectrum™) (Civille

and Lyon, 1996). However, these approaches are quite different and most sensory scientists actually use adapted or sometimes hybrid versions of them. Without going too much into detail, during the QDA® the panel will generate a common sensory language composed of descriptors that best represent the different sensations perceived from the products. This phase of language generation is supported by product tasting and/or smelling and discussion moderated by a panel leader who acts as a communication facilitator. The panelists then evaluate the perceived intensities of each of the descriptors and individually rate their perceptions on a scale. They are free to use the scale as they wish and, as a result, the differences in the intensities of the notes between different products will always be relative. The two main differences between QDA® and Spectrum™ is that with Spectrum™ the sensory language will be predefined and the panelists will be calibrated on the use of the intensity scale, both points supported by physical standards called references. Thus, Spectrum™ produces an absolute evaluation of perceived intensities. Both methods have their advantages and do not serve the exact same purpose. QDA® focuses on the sensory discrimination between products whereas Spectrum™ focuses on the absolute measure of perception of the products.

Another method worth mentioning is consensus profiling, a method formalized by the Arthur D. Little corporation in the late 1940s (Caul, 1957; Keane, 1992). It is one of the oldest methods of sensory profiling and is generally performed by a traditional sensory panel. Instead of evaluating the products individually, panelists discuss their perceptions in a group under the supervision of a panel leader until agreement is reached on a list of descriptors and on a score for the perceived intensity of each descriptor. Unfortunately, this archaic method introduces bias due to the panel discussion used to reach consensual intensity scores. Moreover, reliability of the results cannot be measured through statistical analyses. Due to the fact that it is very easy to implement and that it does not require much sensory expertise, this method is still in wide use today.

All of the aforementioned methods generate static sensory profiles that are generally represented as spider graphs (Figure 48.2). However, static sensory profiles do not account for the time component of the perceived sensations. When we first smell a perfume, for example, we do not perceive all the aromatic notes at the same time: first come the top notes generated by highly volatile molecules, then come the heart notes which appear from two minutes to one hour after application, and finally come the base notes, some of which can still be perceived after twenty-four hours.

To address the dynamic nature of such time-related perceived sensations, we can, as a first approximation, assess static sensory profiles sequentially. For instance, while eating a potato chip, we can generate a static sensory profile when the chip is being put on the tongue, followed by another static profile when it has been chewed for ten seconds, and a final one after it has been swallowed. The temporal definition of the description may be enough in most cases, but not always. Therefore, other methods have been created such as Time-Intensity Profiling (Figure 48.3). This dynamic sensory profiling method, which stemmed from early studies in which taste intensity was tracked longitudinally (e.g., Holway and Hurvich, 1937; Sjostrom, 1954; Jellinek, 1964), follows similar rules as the aforementioned static sensory profiling methods, except that the panelists continuously evaluate the intensity of each descriptor over a period of time, one after the other. Although it gives a much higher level of precision time-wise, it is impractical to perform Time-Intensity Profiling on the whole sensory profile of a complex product when dozens of descriptors may apply. For this reason, Time-Intensity Profiling usually employs a limited number of descriptors focused on the dynamic release of a few specific flavor or fragrance notes.

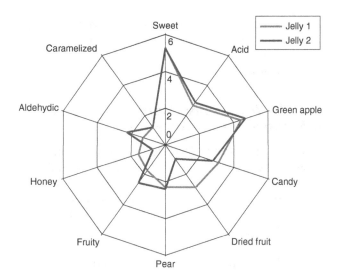

Figure 48.2 Spider graph representing the sensory profiles of two jellies. Here for instance, both Jellies are showing a similar intensity for the descriptor Sweet whereas the Dried Fruit aromatic note is perceived as more intense for Jelly 1 compared to Jelly 2.

It is therefore critical to not only be able to measure the intensities of the different notes detected in the product but also to see how these intensities evolve with time in order to understand the dynamic aspect of product perception. Two very common examples can illustrate this point. First, it is possible to encapsulate flavors in order to better control their release over time. This is how chewing-gums can sometimes still have flavor even after having been chewed for thirty minutes. Second, the perceived intensity of sweetness generated by sugar varies over time according to a specific pattern and that the same pattern is not always reproduced by artificial sweeteners.

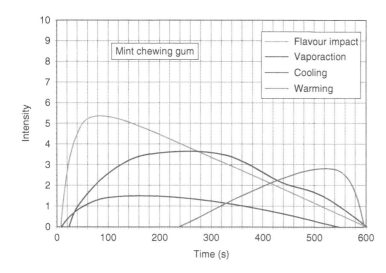

Figure 48.3 Time-Intensity profile of a mint chewing-gum. Here for instance, the overall flavor has strong initial impact when the chewing-gum is put in the mouth whereas the warming effect is maximum after 500 sec of chewing. (*See plate section for color version.*)

Therefore, it is perfectly possible to have the same overall perceived intensity of sweetness in two products, but to see some differences of sweetness intensity over time between the two products. Hence, the sensory perception has to be measured not only qualitatively (notes) and quantitatively (perceived intensities), but also temporally. In short, taste perception is dynamic.

Another approach to dynamic sensory profiling is the Temporal Dominance of Sensations (TDS) method (Le Reverend et al., 2008). Instead of evaluating each descriptor separately, as in the time-intensity profiling method, the panelists consider all the different notes perceived in the product under evaluation and select the dominant one over time (Figure 48.4). This principle is very seductive since it keeps the same temporal definition as the time-intensity profiling but makes it possible to consider a greater variety of flavor or fragrance notes at the same time. One limitation of TDS is that it does not give a clear indication of the perceived intensities at all times and only focuses on the most dominant notes. Another is that it does not provide an operational definition of dominance. A descriptor is traditionally said to be dominant when it represents the most striking perception at a given time. Under which circumstances is a given note perceived as the most dominant compared to the others at a certain time? Experience shows that this can happen in the following situations: when a note is the most intense and partly masks the others, when a new note appears over time while the other ones stay put or decrease in intensity, and when a note is perceived as incongruent with the others. Unfortunately, when the definition of dominant becomes ambiguous, panelists may change their use of definitions over time. This makes interpretation of the results difficult since what is being measured can be inconsistent and idiosyncratic among panelists. For instance, a panelist who is not familiar with cinnamon in chocolate might consider the cinnamon note as dominant whereas another panelist who is used to the association might focus on something else.

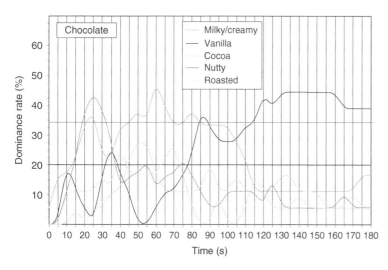

Figure 48.4 TDS profile on a chocolate. The nutty and milky/creamy notes are co-dominant at the beginning. As the chocolate starts to melt down, the milky/creamy note becomes clearly dominant followed by the vanilla note that stays in the mouth even after swallowing. (*See plate section for color version.*)

Although an alternative approach would be to have consumers perform TDS to see what is most relevant for them, this would not be sensory profiling, per se, since it would include parameters such as familiarity, hedonic value, and so on. In order to make the results more objective, one can significantly increase the number of panelists beyond that employed in a typical sensory panel. In this way, the biases induced by individual panelists are averaged. Another way to overcome this problem is to restrain and fix the definition of dominance. A commonly accepted approach is to define as dominant the main changes in the sensory profile over time.

Whatever method is used, quality standards of the sensory language must be implemented to ensure the reliability of results. Each descriptor must be unique, independent from the other descriptors, consensual, understandable, discriminating, stable over time, and reproducible. Many of these requirements can be easily met if a physical standard (reference) is associated with each descriptor. These references can be essential oils, single molecules, mixes of few molecules or even finalized flavors or fragrances. In order to be as precise and specific as possible however, the references should be as unidimensional as possible, meaning that they should ideally evoke a sensation that is unitary. For example, if a reference induces several sensations, the target sensation induced by the reference may vary between panelists. This is why using finished products, where perception is generally very complex, is ill advised in such assessments.

48.4.1.4 Other Sensory Profiling Methods.

The sensory profiling methods presented above require the sensory panelists to be trained thoroughly on a common language and are quite time consuming. This is one reason why many alternative methods have been developed in the past twenty years (for review, see Varela and Ares, 2012). The objective here is not to list and detail all of them, but rather to give the reader an idea of their general principles. Three families of methods that do not require language alignment and training of panelists are described in this section: verbal-based methods, similarity-based methods, and reference-based methods (Valentin et al., 2012).

Two verbal-based methods are Free Choice profiling (FCP) (Heymann, 1994) and Flash profiling (FP) (Dairou and Sieffermann, 2002). In both cases, the panelists evaluate the products individually with their own sensory language. In FCP, an intensity score is given to each descriptor whereas in FP, a ranking of products is performed for each descriptor. Therefore the main difference between the two is that FCP focuses on describing products and FP focuses on discriminating products. In both cases, though, the sensory language used to measure the perception of the products is not common to all panelists. While one can still position the products on a sensory map to gain insights into the nature of product differences (Figure 48.5), clear sensory profiles such as those represented with spider graphs cannot be obtained.

Free Sorting (FS) is a similarity-based method that is simple to implement (Faye et al., 2004). The principle

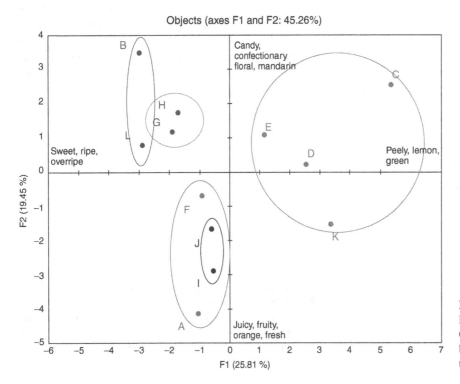

Figure 48.5 Sensory map generated from Free Choice profiles of 12 orange drinks. Clusters of products have been identified here but we can see that the sensory description of the products is not very precise.

48.4 Measuring Human Smell and Taste Perception in the Industrial Context 1073

behind this method is to group products according to their similarities. Because variability exists in similarities among products as well as among panel members, it is possible to compute sensory distances among them. Let us consider a simple example with three products: a can of cola flavored soda, a can of apple juice, and an apple. Here there are two obvious groupings that will account for the similarities of the products. One way is to place the two cans in one group because they are both beverages and the other group will consist of the apple because it is a fruit. The second way is to place the can of apple juice and the apple in the same group because they are both apple flavored and the can of cola in the other group. Both are acceptable and if we were to perform this test with a panel, we would probably observe both ways of grouping the products. It is unlikely that panelists will place the can of cola flavored soda and the apple in the same group. When processing the answers statistically, we will therefore obtain a continuum in which the cola flavored soda is at one end and the apple at the other with the can of apple juice at the midpoint. From this example we can see that while FS allows us to measure sensory similarities and differences between products, it does not provide a reliable measure of the underlying factors that determine the sensory differences.

Finally, for reference-based methods, the recently developed Polarized Sensory Positioning (PSP) is worth mentioning (Teillet et al., 2010; Saldamando et al., 2013). Similar to FS, the basic idea behind the PSP is to measure the sensory distances (similarity or dissimilarity) between each product under evaluation with the addition of a few predefined benchmarks, or anchored products, that serve as references. One of the advantages of this method is that it can be performed with a small number of products, unlike the FS which generally requires many products to give relevant results. However, the choice of the reference products is critical and generally requires a good initial knowledge of the product space, meaning the main types of notes that will help to discriminate between products.

All these alternative methods are much faster and easier to execute than the traditional methods (Varela and Ares, 2012). However, there are some downsides as well. Indeed, appropriate data processing requires more advanced statistical analyses and the interpretation is not always as straightforward as with the traditional methods. Moreover, the information provided by these alternative methods is not precise and specific enough to efficiently guide the development of fragrances and flavors. They are therefore mainly used to get a quick overview of the structure of the sensory space covered by a group of products (market products or prototypes). For instance, performing a sorting on twenty orange beverages in Germany might allow us to identify three main clusters of products, and therefore three main flavor directions present in this market. It will not precisely identify the nature of the flavor

directions, but it will make it possible to select and run a traditional sensory profile on three products representing each of the three clusters. This is much more efficient than profiling the 20 orange beverages right from the start, an approach that may generate a level of detail not required at this initial stage of assessment.

48.4.1.5 Sensory Discrimination Methods.

In the case of a replace/match scenario, we have seen that being able to describe in detail the sensory differences between different samples is not always required. Sometimes, we just need to quantify them in order to select the best match or to validate that a prototype is an acceptable match. This is the situation in which discrimination methods are mainly used in the flavor and fragrance industry.

There are dozens of discrimination methods to evaluate the sensory similarity or difference between products but only a few of them are commonly used. The most famous one is the triangle test (i.e., odd-man-out paradigm), which is widely employed and generally recognized as the standard for discrimination tests (Helm and Trolle, 1946; Peryam and Swartz, 1950). It has been used for many years, is easy to implement, and its results are easy to analyze. Its principle is simple: on each trial panelists evaluate three samples, two of which are the same (either the target or the match proposal) and the third is different. Panelists identify which product is different from the other two. This approach does not require prior knowledge about the products and keeps the total number of samples under evaluation reasonably low (three). The triangle test is not, however, the most sensitive discrimination method.

The two-alternative forced choice method (2-AFC) is another commonly used method. Here only two samples are presented: the target product and the match proposal. Panelists determine which of the two is perceived as stronger for a given descriptor. Although apparently less sensitive from a statistical point of view (the probability of choosing the right answer by chance is 1/2, compared to 1/3 for the Triangle test), the 2-AFC is actually much more powerful than the triangle test (Ennis and Jesionka, 2011). One reasons for this is that the panelists focus on a single specific aspect of the product (i.e., intensity of a given descriptor) instead of having to deal with the overall perception of the product. Also, the panelists are evaluating two instead of three samples, limiting sensory fatigue when carry-over effects are present. However, the 2-AFC not only assumes that the experimenter has prior knowledge of the nature of the sensory difference between the target and the match proposal, but also that the panelists have a common understanding of the descriptor in order to evaluate the same sensation. Unfortunately this is often not the case, particularly in the testing of flavors and fragrances. However, a variant of the 2-AFC can be considered in which the two products (target and match proposal) are presented twice. The first time, panelists are asked to describe the main difference(s) between the two. The second time, each panelist is asked to find back these difference(s). For example, during the first product presentation, one panelist may find the target as sweeter whereas a second panelist may find it more intense. When presented with the products for the second time, the first panelist is then asked to identify the sample that is sweeter whereas the second panelist is asked to identify which sample has the stronger flavor intensity.

48.4.1.6 Measures of Sensory Distances.
In a number of instances, the perceptual distances between chemosensory elements of a product are useful to obtain. As noted in Chapter 11, this can be done via signal detection approaches as well as via more elaborate procedures, such as multidimensional scaling. In the case of replace/match scenarios, sensory profiles often do not need to be described, but an evaluation of the sensory distance between a target product and a match proposal is needed. If this distance is small, differences between two products will most likely not be discriminated by consumers. On the other hand, if it is large, consumers will most likely note differences. The perceived sensory distance may be the result of different intensities of several notes and their evolution over time, or may be restricted to a specific note, depending upon the context and the purpose of the assessment.

48.4.2 Assessing Affective Aspects of Flavors and Fragrances

Flavors and fragrances are universally recognized for their aesthetic value and, as such, hedonics continues to be the primary driver of product development (Montet, Warrenburg & Glazman, 2012). More than any other sensory system, olfaction drives perceptive and cognitive processes closely linked to human emotions. For this reason, and the fact that flavors and fragrances can trigger emotional responses, industry has sought to identify links between flavors, fragrances, and emotions. Products which elicit the strongest emotional responses, such as a stimulating flavor for an energy drink, a relaxing smell for a night cream, and a sensual fragrance for social occasions are more likely to have success in the marketplace.

Emotions give rise to internal effects (physiological and cognitive), as well as external ones (expressive or behavioral). Emotions are commonly measured according to three main components: physiological, expressive/behavioral (non-verbal), and cognitive/subjective (verbal) (Figure 48.6). The main advantage of physiological measures (e.g., skin conductance, blood pressure, and

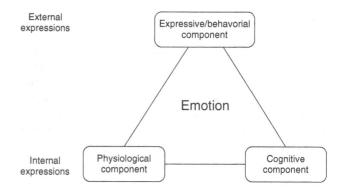

Figure 48.6 The three components of emotional measurement.

brain activity) is to avoid the subjective bias inherent in self-report. However, while they may differentiate two stimuli of opposed valence, they do not offer the required level of sensitivity needed to measure different emotions in finished products at the same level of liking. Moreover, the experimental setup required by this approach is not conducive for consumer testing and is far removed from consumption conditions. Expressive responses are preferable to behavioral responses, since flavors and fragrances do not typically modify behavior directly but trigger feelings. A popular means for such measurement is to assess basic spontaneous facial expressions in response to a flavor or fragrance. According to Darwin (1872), these innate expressions serve a highly adaptive communication function among social group members and include sadness, happiness, anger, fear, surprise, and disgust. Unfortunately, experience has shown that these basic expressions do not add much to the product evaluation beyond a standard liking measure.

Verbal self-report offers the most sensitive probe for assessing emotional responses to chemical stimuli. Using this approach, emotional effects can be computed accurately and reliably given an appropriate experimental design that includes pre- and post-exposure surveys and appropriate control conditions in which the target fragrance or flavor is lacking. A basic affective dimension that can be easily verbalized is that of liking. One can simply ask the consumer, "How much do you like this product," and have the consumer rate the degree of liking on a rating scale. In addition to comparing such ratings across products, relative preferences can be assessed using ranking procedures. For example, a consumer may be asked, "Rank these products from the one you like the most to the one you like the least." Such rankings can be subjected to statistical assessments and derivations of relative distances between products using scaling procedures such as that derived by Thurstone (1927).

Theory largely drives research in this area, and two theoretical models of emotion have served as the basis for a number of tests. The first model is categorical and assumes that such emotions as happiness, sadness, fear, disgust, and anger are innate, universal, and independent of one another (Ekman, 1992; Izard, 1977; Kemper, 1981; Plutchik, 1980). The second model is dimensional and assumes such emotions result from combinations of underlying psychological and psychophysiological dimensions such as valence (pleasant/unpleasant) and arousal (Russell, 1980).

Regardless of their theoretical basis, most measurements of emotion can be assessed using questionnaires and ratings scales. At IFF, raw ingredients have been evaluated on a selection of 44 attributes and over 2000 answers have been collected over the course of several decades (Warren and Warrenburg, 1993; Warrenburg, 2002). From the results of factorial mapping, a circular organization of emotions consistent with eight categories of moods has resulted from the combination of the valence and arousal dimensions (Figure 48.7).

48.4.3 Assessing Cognitive and Conceptual Aspects of Flavors and Fragrances

As is evident throughout this chapter, consumers respond not only to the smell or taste of products, but to what the product means to them based on prior experience or social influences. Hundreds of non-chemosensory factors dictate, alone or in combination with chemosensory factors, the nature of a consumer's ultimate perception of a product. For example, the type of packaging and brand image directly influence perceived quality of a product and whether it is purchased by the end user. Such concepts as "authenticity," "naturalness," "uniqueness" or "newness" are increasingly being considered by the food industry as critical to the success of a product. When a consumer expresses the desire to purchase a "natural" product, it does not necessarily only imply that the ingredients should be natural from a regulatory standpoint. Rather, it means that the expectation of the overall perception of the product be natural. In fact, "natural" changes over time. For example, during a recent test on vanilla ice-cream, consumers were exposed to a vanillin flavor (a molecule that is broadly used in the food industry to mimic vanilla taste, although its sensory profile is quite different from the actual vanilla) versus vanilla extract (which renders more faithfully the taste of the actual vanilla). Interestingly the consumers found the taste of the product with vanillin very authentic and natural whereas they found the taste of the product with the vanilla extract very artificial. This underscores the fact that the concept of naturalness and authenticity do not necessarily correspond to what we may think of as being truly natural or authentic.

48.4.4 Dealing with Stimulus Complexity

It is important to note that an understanding of the perceived odor or taste of individual components of added flavors need not predict the ultimate chemosensory experience elicited by the product to which they are added. In fact, added flavors typically represent a very small percentage of the chemical ingredients of most food products (e.g., less than one percent). Thus, the ultimate flavor of a food product depends upon a combination of ingredients. In chicken soup, for instance, there are dozens of ingredients. The added flavors may enhance or complement aspects of the taste of the soup, but their perceived attributes, by themselves, may not represent the ultimate flavor of the soup.

There are numerous examples of how chemicals interact in complex ways producing the final percept (see Chapter 11). For example, anethole by itself smells like anise (licorice-like) and cis-3-hexenol by itself smells like freshly cut green grass. When you put the mixture of these two chemicals in water, you will smell a mix of anise and green grass, with a clear dominance of anise. However, if you put the same mixture in whole milk, it will smell mostly of green grass with a very slight hint of anise. This is because anethole, unlike cis-3-hexenol, is lipid soluble and will be partly taken up by fat globules contained in the milk, decreasing its release into the air in a milk medium.

Importantly, both the chemical and non-chemical sensory systems influence each other in dictating the flavor and ultimate enjoyment of flavors and fragrances and should not be considered in isolation. Hebb's classic theory of associative learning (Hebb, 1949) provides a theoretical framework of how sensory modalities can interact. The basic idea is that cells which fire together wire together. In other words, if two sensations are perceived

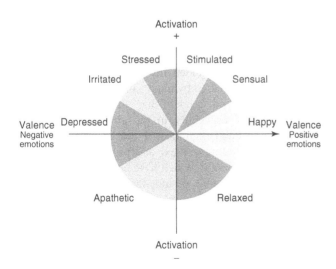

Figure 48.7 Structure of the emotional categories resulting from the organization of 44 emotional attributes.

1076   **Chapter 48   Olfaction and Gustation in the Flavor and Fragrance Industries**

together repeatedly over time, they will tend to reinforce each other, to the point that when only one of these two sensations is generated by an external stimulus, the other sensation will still be perceived, albeit to a lesser degree (see Chapters 45 and 46). For instance, if someone is repeatedly consuming seaweeds, exposure will be made simultaneously to aromatic notes specific to seaweeds as well as to saltiness. Therefore the neurons responsible for the olfactory perception of seaweed aromatic notes will wire with the neurons responsible for the perception of saltiness and will reinforce each other. If the subject is subsequently exposed only to seaweed aromatic notes, it is likely that saltiness will be perceived even though no salt is actually present. This overall process, whose physiological correlates have been documented in the orbitofrontal cortex (see Chapters 35 and 46), is complex since multiple sensations and perceptions become interconnected as a result of experience.

Among such factors are those of context and interaction with non-chemosensory modalities. In regard to the latter, an orange drink that is colored orange or yellow will be perceived, respectively, as having a more orange or a more lemony taste. Also, the way the product is used or consumed plays a role. For instance, depending on the quantity of detergent that is put into a washing machine, the smell of the laundry will be more or less intense and certain fragrance notes may become masked by others. Therefore not only the intensity of the sensation but also the sensory profile will be impacted. Frequently a "magic moment" occurs when a fragrance is most salient and its benefit is most expected. For example, such a moment can occur when the detergent is poured, when the laundry is removed from the machine, when the laundry is hung, or when the clothes are worn. Such moments technically refer to different "life-stages" of the fragrance and correspond to different notes being released.

Idiosyncratic factors further complicate the matter. For example, skin type, diet, medications, pigmentation of the skin, and even mood can influence how a scent will react with body chemistry, thereby impacting its sensory performance. Such complexity is challenging for the chemosensory scientist and has led the flavor industry in particular to focus on sensations beyond the strictly olfactory or gustatory. The need for such a focus became clear when developing formulations that eliminated or greatly reduced the amount of sugar, salt, fat, and monosodium glutamate (MSG) in the final product. For example, removing sugar not only influences the perception of sweetness, but can modify the flavor profile, its overall intensity, and the mouth-feel perception. Sugar generates salivation and mouth-filling sensations that must be added back in a sugar-free product. Also, most non-sugar sweeteners can generate, at least in some people, "off-notes," that is, undesired aromatic sensations, mouth feels, and distorted

taste sensations. Aspartame, for example, can generate bitterness, metallic taste, and a mouth-drying sensation.

48.5 MEETING MARKET ASSESSMENTS OF CONSUMER WANTS AND NEEDS

Before starting the development process for creating a new flavor or fragrance, it is often of great strategic value to characterize the current market landscape for a product category by describing how products are being used and perceived by consumers. Indeed, it helps to know what kind of scented or flavored products consumers are exposed to, most familiar with, and accept – factors that help to identify opportunities for new products rooted in chemosensory trends and directions. As a result of such insight, a consumer goods company may initiate a me-too strategy into an existing market or develop an entirely new product and new market.

Such "preliminary research" helps to understand critical benchmarks in the consumer's mind, such as the collection of percepts that have been generated over time and that are associated to concepts. We know for instance that consumers from different cultural backgrounds and personal experiences will not expect the same type of flavor from a chicken bouillon. In some countries, for example, consumers expect and prefer a boiled chicken taste enriched with spices, whereas in others their expectation and focus is on a dominant roasted chicken taste with an overlay of herbs. This is due partly to culinary history and partly to the types of products present in their marketplace. Therefore, a possible way to approach the internal benchmark of consumers is to look not only at local market products but also at the products to which consumers are culturally and traditionally exposed. The same applies to fragranced products, given that smell is firmly anchored in culture. From an olfactory standpoint, childhood concepts are driven by baby care products with different culturally specific references and percepts. In the United States, these evocations are triggered by floral aldehydic powdery notes while in France they are triggered by musky orange blossom ones.

Clearly, an end product has to be aligned with consumer needs, especially in the affective dimension. Different flavors and fragrances will evoke different feelings and are therefore context and product specific. For instance, lavender notes, which have been shown to have a calming effect on people, should, *a priori*, not be dominant for a product that is positioned as being stimulating and exciting. Although some associations between affect and flavor or fragrance notes are known, most have yet to be discovered and are being explored proactively by industry. This second aspect of creation is definitely the main one that is given consideration when a perfume is being developed. Indeed, briefs provided to perfumers by cosmetic companies are generally not defined in terms of sensory characteristics, but rather in affective and conceptual terms. For instance, a perfume brief may include requirements for "femininity and elegance, echoes with the fashion, food and life style from Northern Italy, and must appeal to modern and sparkling women". This description is stated at the affective and conceptual levels to guide the types of notes that will be most likely to meet customer expectations. Finally, measuring the olfactory and taste perceptual elements of a finished product over time not only ensures that this product meets current consumer expectations, but also provides a perspective for how long the product will continue to be viable in the market.

In flavors more than in fragrances, consumers generally have sensory expectations linked to natural products or existing food benchmarks on the market. It is important for the industry to make sure that the consumers recognize the chemosensory character of the product that is marketed. In other words, the scent of a lavender air freshener should be recognized by the consumer as lavender. But here again the cultural background and the internal benchmarks have to be taken into account as olfactory references for lavender can vary from country to country. Before the industrial revolution, we can assume that the flavor and fragrance concepts were being formed through the exposure to natural and artisanal products. However, with the development of the flavor and fragrance industry, consumers are increasingly exposed to processed products that differ in aroma and flavor from the original products. Therefore, their exposure to such products alters their concepts of what a product should smell or taste like. For example, grenadine syrup, very famous in France, was originally made of pomegranate's pulp. However, with its industrialization, its flavor slowly drifted away from the original flavor and is now a mix of red fruits (often raspberry) and vanilla. Although it is still called grenadine syrup, its flavor is nothing like it was decades ago and today's consumers have an expectation of what an "original" grenadine syrup should taste like that is different from yesterday's consumers. Interestingly, the recent trend of consumers to cook at home, supported by the numerous television shows on cooking, may once again shift flavor concepts towards more natural products.

48.6 CONCLUSIONS

A major mission of flavor and fragrance houses is to design novel scents and tastes to satisfy consumer demands. It is therefore critical to understand consumer expectations and experiences. In this chapter we have shown how artisans create or duplicate flavors and fragrances, and have described a number of procedures by which product formulations and end-stage are compared and evaluated.

We have pointed out that there is much more to the art of understanding the influences of chemosensory creations on the behavior of customers than simply smells or tastes. Although traditionally hedonic factors dominated the focus of our industry, modern research is moving towards a more holistic perspective on product development. Thus, in addition to understanding hedonics, understanding emotional and conceptual processes of the consumer are now viewed as critical for the strategic positioning of products in the marketplace.

REFERENCES

Caul, J. F. (1957). The profile method of flavor analysis. *Adv. Food Res.*, 7: 1–40.

Civille, G. V. and Lyon, B. (1996). ASTM lexicon vocabulary for descriptive analysis. *American Society for Testing and Materials*, Philadelphia.

Dairou, V. and Sieffermann, J. M. (2002). A comparison of 14 jams characterized by conventional profile and a quick original method, the flash profile. *J. Food Sci.*, 67: 826–834.

Darwin, C. (1872). *The Expression of Emotions in Man and Animals.* 1st ed. John Murray, London.

Ekman, P. (1992). Are there basic emotions? *Psychol. Rev.*, 99: 550–553.

Ennis, J. M. and Jesionka, V. (2011). The power of sensory discrimination methods revisited. *J. Sens. Stud.*, 26: 371–382.

Faye, P., Bremaud, D., Daubin, M. D., et al. (2004). Perceptive free sorting and verbalization tasks with naïve subjects: an alternative to descriptive mappings. *Food Qual. Pref.*, 15: 781–791.

Hebb, D. O. (1949). *The Organization of Behavior.* New York: Wiley & Sons.

Helm, E. and Trolle, B. (1946). Selection of a taste panel. *Wallerstein Lab. Communications*, 9: 181–194.

Heymann, H. (1994). A comparison of free choice profiling and multidimensional scaling of vanilla samples. *J. Sens. Stud.*, 9: 445–453.

Holway, A. H., and Hurvich, L. M. (1937). Differential gustatory sensitivity to salt. *Am. J. Psychol.*, 49: 37–48.

Izard, C. E. (1977). *Human Emotions.* Plenum Press, New York, 495 p.

Jellinek, G. (1964). Introduction to and critical review of modern methods of sensory analysis (odor, taste, and flavor evaluation) with special emphasis on descriptive analysis. *J. Nutr. Diet.*, 1: 219–260.

Keane, P. (1992). The flavor profile. In: R. C. Hootman (Ed.), *ASTM Manual on Descriptive Analysis Testing.* Philadelphia: ASTM.

Kemper, T. D. (1981). Social constructionist and positivist approaches to the sociology of emotions. *Am. J. Sociol.*, 87: 336–362.

Lawless, H. T. and Heymann H. (2010). *Sensory Evaluation of Food: Principles and Practices*, 2nd ed. New York, NY. Chapman & Hall.

Le Reverend, F. M, Hidrio, C., Fernandes, A. and Aubry, V. (2008). Comparison between temporal dominance of sensations and time intensity results. *Food Qual. Pref.*, 19: 174–178.

Montet, A., Warrenburg, S. and Glazman, L. (2012). 9.5. Parfums, cosmétiques et arômes alimentaires: enjeux industriels de la mesure des émotions. *Odorat et goût – De la neurobiologie des sens chimiques aux applications*, QUAE, 550 p.

Peryam, D. R. and Swartz, V. W. (1950). Measurement of sensory differences. *Food Technol.*, 4: 390–395.

Plutchik, R. (1980). *Emotion: A Psychoevolutionary Synthesis.* Harper, New York, 440 p.

Russell, J. A. (1980). A circumplex model of affect. *J. Pers. Soc. Psychol.*, 39: 1161–1178.

Saldamando, L. de (2013). Polarized sensory positioning: Do conclusions depend on the poles? *Food Qual. Pref.*, 29: 25–32.

Sjostrom, L. B. (1954). The descriptive analysis of flavor. In: *Food Acceptance Testing Methodology.* Quartermaster Food and Container Institute, Chicago, pp. 4–20.

Stone, H. and Sidel, J. L. (2004). *Sensory Evaluation Practices*, 3rd ed. San Diego, CA. Elsevier Academic Press.

Teillet, E., Schlich, P., Urbano, C., et al. (2010). Sensory methodologies and the taste of water. *Food Qual. Pref.*, 21: 967–976.

Thurstone, L. L. (1927). A law of comparative judgment. *Psychol. Review*, 54: 273–286.

Valentin, D., Chollet S., Lelièvre M. and Abdi H. (2012). Quick and dirty but still pretty good: a review of new descriptive methods in food science. *Int. J. Food Sci. Technol.*, 47: 1563–1578.

Varela, P. and Ares, G. (2012). Sensory profiling, the blurred line between sensory and consumer science. A review of novel methods for product characterization. *Food Res. Int.*, 48: 893–908.

Warren, C. and Warrenburg S. (1993). Mood benefits of fragrances. *Perfum. Flav.*, 18: 9–16.

Warrenburg, S. (2002). Measurement of emotions in olfactory research. *Chemistry of Taste*, 19: 243–259.

Zverev, Y. P. (2004). Effects of caloric deprivation and satiety on sensitivity of the gustatory system. *BMC Neurosci.*, 5: 5.

Chapter 49

The Smell and Taste of Public Drinking Water

GARY A. BURLINGAME and RICHARD L. DOTY

49.1 INTRODUCTION

The use of the human senses to distinguish good water from bad, such as by taste, smell, and appearance, is a basic means of self preservation. Aside from drinking water, natural water sources can also become aesthetically unpleasant, limiting their value in supporting recreation and development. Indeed, the whole water cycle is impacted by the sensory characteristics of the water.

Surprisingly, it was not until the late 1800s that serious scientific inquiry was made as to associations between human sensory observations and the safety of water (Corbin, 1986). While it is true that bad smelling and tasting water can be related to deadly germs or chemical contaminants, it should be pointed out that odorless and good tasting water can be teaming with pathogens or dangerous chemicals. Hence, the senses are not always a failsafe marker for the safety of water, as will be pointed out later in this chapter.

As chemical and biological tests of the constituents of water were being applied during this early period, human sensory techniques were also being developed to categorize the pollution found in rivers and waterways. Smells emanating from such sources of water were assessed and categorized under both hot and cold testing conditions (Whipple, Fair, and Whipple, 1948). Lists of odor descriptors were developed and included, among others, the following descriptors: aromatic, cucumber, chlorinous, earthy, fishy, grassy, geranium, medicinal, moldy, musty, nasturtium, oily, pigpen, sweetish, hydrogen sulfide, vegetable, and violet. It was during this early period that it was discovered that while some odors were found to be specifically produced by anaerobic conditions and waste

discharges, others came from natural organic matter and nuisance algal growth.

As the Germ Theory of Disease fought its way into acceptance in the late 1800s, the origins of waterborne diseases such as typhoid and cholera were being discovered and water treatment movements began in earnest. Filtration and chlorination became the standard means for treating water for potable use. Indeed, the use of chlorine was hailed as one of the greatest public health benefits of the 20th century (Friedman, 1998). Whereas filtration had a neutral or positive effect on the flavor of treated water, chlorination eliminated some odors, made others worse, such as in the formation of chlorophenols, and contributed to the formation of a distinct flavor of public drinking water with its chlorine smell.

In modern times, a safe water supply is provided through a multiple barrier concept – a series of strategies for treating and protecting the quality of the water from the source of origin (reservoir, river, spring, well), through treatment, and on to the consumer's point of use. The goal of treatment is to eliminate microbial pathogens, control nuisances, and produce biologically and chemically stable potable water. Most drinking water systems obtain their water supply from ground water which often needs little if any treatment. When water treatment is needed, such as for turning river or reservoir water into drinking water, several processes are typically used to achieve these ends. These include (a) pre-sedimentation or the settling out of sediment, (b) pre-oxidation for minimizing algae, iron, and manganese contaminants by the use, for example, of chlorine, chlorine dioxide, and potassium permanganate, (c) oxidation to produce chemically stable water, (d) disinfection to remove pathogens, (e) coagulation, flocculation,

---

Handbook of Olfaction and Gustation, Third Edition. Edited by Richard L. Doty.
© 2015 Richard L. Doty. Published 2015 by John Wiley & Sons, Inc.

and sedimentation to remove suspended matter using such additives as ferric chloride, alum, polymers, and lime, (f) fine particulate removal by filtration using such media as sand and anthracite, and (g) post-treatment or the addition of chemicals such as fluoride, corrosion-control phosphates, and ammonia to form a chloramine residual. Advances to oxidation and disinfection have included exposure to ozone and ultraviolet light with or without the addition of hydrogen peroxide. Sophisticated membrane technology is now available for the physical removal of microorganisms and other contaminants, and granular activated carbon filters are employed to remove organic chemical contaminants (Burlingame, 2011; Mackey, 2012).

The fallibility of relying solely on one's senses in accepting the safety of drinking water is illustrated by a well-studied waterborne disease outbreak in Walkerton, Ontario (Parr, 2005). During this outbreak in a groundwater system, seven people died from *E. coli* O157:H7 and half the town's population experienced varying levels of illness. The basis of this problem was contamination of the groundwater by surface runoff from a nearby farm. Although a number of factors contributed to this disaster, the flavor of the water was an important player. The town's people had resisted adding chlorine to their public water supply because they did not want to taste a chlorine residual. Since their groundwater looked and tasted good, it was erroneously assumed to be safe to drink. In North America, most public water systems are small in size and rely on groundwater sources. In the United States, as of 2009, these systems are now regulated under the Environmental Protection Agency's (EPA) Ground Water Rule to prevent microbial contamination of the drinking water.

The impact of post-treatment processes and materials to which water becomes exposed, such as pipes and linings, have been a concern (Khiari, 2002). Treated drinking water has a shelf life and its quality is affected by the materials to which it has been exposed (e.g., iron, copper, plastic, epoxy and bitumen) and by ongoing chemical reactions and biological activity as the water travels from the treatment plant to the consumer's faucet (Khiari et al., 2002; Malleret et al., 2002; Andreas and Gunten, 2009). A common customer complaint of public drinking water comes from iron release or rusty water from systems that use iron water mains, reflecting soluble iron that produces a metallic flavor (Mirlohi et al., 2011). Complaints also arise, although less commonly, from other types of pipes through which water travels including copper, which can occur in drinking water in a reduced or oxidized state in the form of cuprous or cupric ions. Much of the United States' indoor plumbing consists of copper tubing or pipe. Concern about the health effects of too much copper exposure led to a corrosion-based standard for copper by the EPA at 1.3 mg/L. The standard for preventing staining and off tastes was set slightly lower at 1 mg/L.

It is now apparent that copper can produce a copper-like odor or metallic flavor via chemical reactions in the mouth and nasal cavity. For example, the reduced form of copper ions in the presence of oxygen causes lipid oxidation reactions that produce such odorous compounds as carbonyls, resulting in metallic-like sensations (Dietrich et al., 2008). As a result, although individual sensitivities vary, retronasal detection of off flavors can occur for copper at levels below the USEPA standard of 1 mg/L.

In light of such issues, modern water utility customers have choices. They can avoid tap water altogether and use bottled water for drinking. However, the cost of bottled water over tap water is very significant. Customers can also install point-of-use and point-of-entry treatment systems to provide additional treatment to remove chemicals and bacteria, as well as to polish off the flavor of the water. Systems are certified and tested, so most consumers can become aware of their capabilities for filtering and improving the water's taste (see Consumer Reports, 2012). The attitudes that drive customers to make such choices have been investigated in numerous scientific studies, a number by public utilities that need to maintain the public's trust and support (Mackey et al., 2003). Included in such attitudes is the sensory experience of bathing in tap water, as off odors can also affect that experience (Omur-Ozbek et al., 2011).

Worldwide, the taste and smell of drinking water remains an important issue affecting the public's acceptance of water and the understanding of its safety. The World Health Organization's guideline states that an acceptable drinking water should be free of tastes and odors that would be objectionable to the majority of consumers who rely on its use (World Health Organization, 2008). As developing countries establish public water systems and safe water supplies, there will be increased demand for acceptable aesthetic water quality. This is a logical progression that has occurred throughout history, as exemplified by the Romans and other great civilizations.

49.2 DEVELOPING THE MEANS TO DESCRIBE THE TASTE AND ODOR OF WATER

In the 1980s, a resurgence of research brought better science to the assessment of drinking water flavor (Suffet et al, 1988; Desrochers, 2008). Flavor Profile Analysis (FPA) (American Public Health Association, 2007b), which was developed as the Flavor Profile Method in the 1940s by Arthur D. Little, Inc. for application in the commercial food industry, was adapted for assessing drinking water, as were other techniques from this industry (Meilgaard et al., 1987; Lawless and Heymann, 1998). Among tests already developed within the water industry were the

49.3 The Global Impact of Algae

Threshold Odor Number and the Flavor Threshold Test (American Public Health Association, 2007b). During this exciting period, trained sensory panels with outstanding chemosensory capabilities (see Smith et al., 1993) were formed and reliably catalogued the flavors of water and the chemicals that give rise to off tastes and odors (Bartels et al., 1986). A fundamental understanding was made as to how flavor quality and its descriptors vary as odorant concentration is altered logarithmically according to the Weber-Fechner law (Suffet et al., 1995; Rashash et al., 1997). Attempts were made to develop the sniffing of gas chromatography effluents as a method for identifying the chemical compounds that give rise to odors in water (Khiari et al., 1992). By the year 2000, most or all possible tastes and smells in public drinking water had been identified by sensory and chemical means, and the causes explained (Burlingame, 2011). Flavor wheels for water (Suffet et al., 1999), as well as wastewater (Burlingame et al., 2004), were developed and refined, as has been the practice in the wine and beer industries. Such descriptors have made it possible to develop internationally useful taste and odor profiles. Importantly, sensory panels were developed within a number of companies to identify and classify olfactory and taste contaminants from sectors of water distribution systems. In some cases, trained panel members had been shown to outperform even blind persons on standardized taste and smell tests (Smith et al., 1993).

In the United States and many other countries, a measureable residual of chlorine or chloramine is required for the safety of the drinking water. EPA regulations require that the total chlorine levels range between 0.2 mg/L and 4 mg/L. As a result, public drinking water can have a "chlorinous" flavor or smell. The impact of the chlorine residual on public perceptions and attitudes has been the focus of a number of research studies (Mackey et al., 2004). In both the United States and in numerous countries in Europe, the flavor of public drinking water can now be described by common terms, including such descriptors as chlorinous, earthy, and musty with possible off flavors such as metallic, medicinal, plastic, decaying vegetation, sulfidy, grassy, fishy, cucumber, chemical, and solvency. The odor intensity of these descriptions can be quantified by standardized intensity scales. For most waters the ratings range from very weak to moderate in strength. Most recently, the influences of retronasal smell on the flavor experiences of water consumers have become more apparent and undoubtedly will continue to receive continued scientific attention (Dietrich, 2009; Omur-Ozbek and Dietrich, 2011).

Historically, scientists within the water industry largely ignored actual tastes in water, being more focused on its mineral content, as measured by total dissolved solids (Bruvold and Daniels, 1990). This reflected, in part, the fact that people rarely complained about the sweetness, sourness, or bitterness of water. The exception was salty taste, which often arose in geographical areas where salt had intruded into wells or where source water contained high levels of total dissolved solids. Nevertheless, a comprehensive review of the taste of water was published in 2007 (Burlingame et al., 2007).

49.3 THE GLOBAL IMPACT OF ALGAE

During the 1700s and 1800s in Europe, major urban rivers such as the Thames and the Seine smelled quite bad (Corbin, 1986). During the late 1800s and early 1900s, rivers in America followed suit. Water supplies for cities, such as Philadelphia, were selected in part based on how they smelled. Many people preferred their private wells because the wells produced clear and cool water even though the wells were often, especially in populated cities, contaminated by cesspools. The impact of source water characteristics on drinking water quality is still a major focus of research (Suffet et al., 1995; Taylor et al., 2006). Even today, many countries are struggling with the pollution of their surface and ground water supplies. In China, for example, source water pollution and surface runoff have created significant taste and odor problems for drinking water supplies (Yu et al., 2009). Earthy, musty, septic, swampy odors have been reported to make up the smell of the water due to such odorous chemicals as heptanal, benzaldehyde, dimethyl trisulfide, 1-octen-3-ol, benzene acetaldehyde, bis(2-chloroisopropyl) ether, 2-methylisoborneol, 1,10-dimethyl-9-decalol, and β–ionone.

The two most significant odorous chemicals in surface waters worldwide are metabolic byproducts of cyanobacteria, also known as blue-green algae: 2-methylisoborneol (MIB) and 4,8a-dimethyl-decahydronaphthalen-4a-ol, which is commonly known as geosmin (Jüttner and Watson, 2007). These two chemicals impart an earthy, musty smell to water that is difficult to remove. At the levels typically found in water supplies, these agents have no known adverse health effects. They can be detected by olfaction at levels as low as 2–5 ng/L. Moreover, their enantiomers can have different sensory properties. Geosmin, like MIB, can significantly impact a drinking water utility and its customers (Burlingame et al., 1986).

The same odorous chemicals produced by cyanobacteria can be produced by common inhabitants of the soil and aquatic environment known as actinomycetes (Zaitlin and Watson, 2006). The actinomycetes can be cultured and shown to be producers of the odors. They can be isolated from the water environment, but it has been difficult to correlate their detection by culture methods with the presence of earthy-musty odors.

Algal blooms are one cause of surface scum and offensive odors that negatively affect the recreational

1082   **Chapter 49   The Smell and Taste of Public Drinking Water**

uses of water. Aquaculture or the growing of fish for food in ponds can be greatly affected by algal products, including geosmin and MIB, which can taint the taste of fish. The potential for algae to contribute odorous chemicals to water is significant; algae can produce odorous compounds during their growth phase and during their death phase. Algal products can be emitted to the surrounding water or contained within the cells until the cells are lysed, as occurs during water treatment. The list of possible chemicals other than geosmin and MIB that algae can produce is enormous, and include dimethylsulfide, dimethyldisulfide, dimethyltrisulfide, hydrogen sulfide, heptanal, hexanal, pentanal, trans-2-nonenal, trans-hexenal, cis-3-hexen-1-ol, trans-2-cis-6-nonadienal, trimethylamine, β-ionone, 3-methyl butanal, limonene, β-cyclocitral, 3-methoxy-2-isopropylpyrazine and 2-isobutyl-3-methoxypyrazine (Watson, 2010). Reports of taste-and-odor episodes in public water systems have increased over the past 20 years. One possible cause is climate change and an increase in droughts. Even in the aftermath of wildfires, there have been taste-and-odor problems. However, there is no global system for tracking and reporting taste-and-odor episodes in public water supplies.

Water utility companies can no longer claim that algae-caused tastes and odors in drinking water have no associated health risks. In 2010, the United States Geological Survey (USGS), a scientific agency of the federal government, issued a statement indicating that "If Water Looks and Smells Bad, It May Be Toxic" (Graham et al., 2010). They went on to say, "Earthy or musty odors, along with visual evidence of blue-green algae, also known as cyanobacteria, may serve as a warning that harmful cyanotoxins are present in lakes or reservoirs." Importantly, the same cyanobacteria that produce earthy and musty smelling compounds can produce non-odorous toxins. Thus, risk communication has become a rapidly growing field of study to help water utilities better communicate to the public the issue of drinking water safety. As a result, water systems have had to develop better public communication procedures in this age of social networking (Pollard et al., 2009). Information and mis-information gets out quite rapidly and extensively through blogs, twitter, text messaging and the internet. The means by which modern water utilities reach out to their customers include web sites, call centers, press releases, media interviews, newspaper notices, direct mailings, and consumer confidence reports.

49.4 PUBLIC ACCEPTABILITY OF DRINKING WATER FLAVOR

Taste and odor problems in public water supplies are often short-lived and episodic in nature, such as from an industrial spill or from a seasonal bloom of algae in a water supply reservoir. This occurrence characteristic, combined with a lack of regulatory standards, has allowed water utilities to defer applying long term solutions. As a result, customer complaints about the taste and smell of tap water have often been the main motivator for water utilities to take action to treat water to remove nuisance tastes and odors. Customers are sensitive to changes in the flavor of their tap water, but they are not reliable in their descriptions of those changes. Customers tend to make logical associations rather than relying on their sensory judgments, which can make it more difficult to track and respond to customer complaints. Since public perception and attitudes are not always accurately displayed through customer complaints, water companies have developed tools to better track and understand the attitudes of their customers (Burlingame and Mackey, 2007). This allows them to be certain that they are optimizing the selection and funding of taste and odor control strategies.

One approach to control the taste and smell of drinking water is through a quality control program that assesses the processes and controls at a water utility. Benchmarks are in place by which public water systems can evaluate their capabilities to control drinking water taste and odor (McGuire et al., 2004). Such quality programs have been gaining broader acceptance in the drinking water industry worldwide, as embodied in water safety plans (World Health Organization, 2008). Water safety plans utilize a comprehensive risk assessment and risk management approach that covers all the steps involved in providing a high quality drinking water supply.

Since taste- and-odor causing chemicals at the levels found in drinking water are not often directly linked to public health effects, there are no federal requirements for controlling taste and odor in public water systems. In the 1970s, Secondary Maximum Contaminant Levels (SMCLs) were established by the USEPA trans-2-cis-6-nonadienal, to assist public water systems in managing aesthetic water quality (USEPA, 1992). These guidelines cover the nuisance effects from taste, smell, color, and staining, as well as foaming and scale development. They also include guidelines for chloride, copper, corrosivity, foaming agents, iron, manganese, pH, sulfate, total dissolved solids and zinc. The smell of water is addressed by a Threshold Odor Number (TON) – a measure of the dilution needed to give water a just perceptible smell. However, the widely employed TON method is often misapplied at water utilities, such as by abbreviating the technique to make it simpler to perform (Dietrich et al., 2004). The SMCL guidelines that were developed during the 1970s are outdated and clearly in need of revision (Dietrich and Burlingame, 2015).

49.5 MTBE: A CASE STUDY OF ORGANOLEPTIC LIMITATIONS

Failed attempts to set national standards based on population-based odor thresholds are highlighted by the ongoing legal battle over methyl tertiary butyl ether (MTBE) contamination of groundwater. MTBE was added to gasoline in the United States beginning in the 1970s as a replacement for lead (Carter et al., 2006). Leaking underground gasoline storage tanks have since contaminated private and public ground water supplies with MTBE, an agent which leaches through soil and is not easily contained. MTBE is often said to have a sweet chemical turpentine-like odor, although a range of other odor qualities have been reported, including "chemical," "cleaning fluid," "ether," "gas," "gasoline," "paint," "paint thinner," "shellac," "stale garbage," "varnish" and, in the case of taste, "bitter" (e.g., Young et al., 1996). Attempts have been made to establish a population's odor threshold concentration as a standard for treatment and for legal cases (Stocking et al., 2001). Unfortunately, the odor threshold measurement methodology that has been generally employed, most notably ASTM E679 (ASTM, 1997, 2004), has significant limitations when used in attempts to set regulatory standards for MTBE.

As noted in Chapter 11, most psychophysical studies seek to determine individual odor threshold values. Clinically, such values are generally compared to normative data to establish whether a person's ability is normal or not. Based on a limited number of determinations of individual odor thresholds, some of which are methodologically flawed (e.g., failures to employ forced-choice procedures and epidemiological sampling procedures), the water industry has attempted to establish "group thresholds," i.e., concentrations of a contaminant that can be detected by some proportion of a target population of consumers that can be used for regulatory purposes. Unfortunately, practical and economic considerations often make it impossible to totally eliminate off flavors in water and some subset of the population can detect contaminants that are indiscernible to most people. The fly in the ointment is determining the contamination level that protects most members of the consuming public from perceiving off smells in water, yet does so at a cost that minimizes customer costs or, in some cases, taxpayer burden. There are advocates within the water industry for using a 50% group threshold, that is, a threshold value that can be detected by half of the population (e.g., Stocking et al., 2001). However, the acceptance of such a threshold would imply, by definition, that half of the consuming population would be able to discern the contaminant in their water at some level of intensity. It would seem prudent to set a more conservative standard in order to minimize objections of consumers to the acceptability and safety of their drinking water.

In the largest study of its type to date, Stocking et al. (2001) sought to determine the 50% population olfactory threshold for MTBE in water for "consumers," i.e., untrained panel members. Individual thresholds were computed for 57 subjects using a concentration series that employed an odorant and two blanks on each trial. The task was to discern which of the samples was contaminated. Eight concentrations of MTBE were evaluated, beginning with the lowest concentration (ascending method of limits procedure; Chapter 11). This approach has been codified as the ASTM E679 procedure (ASTM, 1997, 2004) and was used by Stocking et al. to argue that an appropriate standard for regulating MTBE should be 15 µg/L (15 ppb).

Analysis of the raw data of this study (Table 49.1) shows that nearly 20% of the subjects, that is, subject #'s 13, 14, 19, 21, 22, 24, 37, 41, 52 and 53, could detect the lowest MTBE concentration that was presented, that is, 2 µg/L. Hence, their actual individual odor thresholds fell below this low concentration. The ASTM E679 procedure requires that the stimuli to be presented fall below the lowest threshold of the subjects, so this procedure was not strictly followed in this study. On the other end of the scale, 14% of the subjects (#'s 12, 16, 20, 28, 31, 34, 35, and 47) could not detect the highest concentration that was presented, again resulting in inaccurate threshold estimates and violating the requirements of the ASTM E679 procedure.

Aside from these study-specific issues is the fact that thresholds based upon single ascending series trials are highly unreliable, throwing into question the accuracy of the ASTM E679 threshold determinations in the first place (Chapter 11). In the three-choice paradigm, correct responses can occur by chance a third of the time on a given trial. Two successive correct responses, which were used as a primary criterion for determining the detection thresholds of the Stocking et al. (2001) study, would be expected to occur by chance 1/9th of the time (i.e., 1/3 x 1/3).

A preferred approach for assessing group sensitivity from data such as that collected by Stocking et al. would be one that takes into account all of the data. For example, the percent of the group that reports detecting the stimulus can be calculated at each concentration, rather than relying on the unreliable single point estimates of each participant's threshold. A correction for guessing can also be applied, such as the application of Abbott's formula: $P_{adjusted} = (P_{observed} - P_{chance})/1 - P_{chance}$. When this approach is taken, ~15% of the panelists of the Stocking et al. study were able to detect the 2 µg/L (2 ppb) and ~25% the 3.5 µg/L concentration of MTBE. The 6 µg/L and the 10 µg/L MTBE concentrations were discernable by about 40% of the study group.

Table 49.1 Raw data presented as Table 2 in Stocking et al. (2001). Shaded areas represent the transition points from detecting (x) and not detecting (o) the vapor phase of the MTBE laden aqueous solution. The trials began at the lowest concentration (2 μg/L) and increased in concentration to the highest concentration (100 μg/L). Gray boxes indicate data used by Stocking et al. to calculate threshold values.

S#	Concentration 2 μg/L	3.5 μg/L	6 μg/L	10 μg/L	18 μg/L	30 μg/L	60 μg/L	100 μg/L	S#	Concentration 2 μg/L	3.5 μg/L	6 μg/L	10 μg/L	18 μg/L	30 μg/L	60 μg/L	100 μg/L
1	x	o	x	x	x	x	x	x	30	o	x	x	o	x	x	x	x
2	o	o	o	x	x	x	x	x	31	x	o	o	x	o	o	x	o
3	o	x	o	o	o	o	x	x	32	x	x	x	x	o	o	o	x
4	o	o	o	o	o	o	x	x	33	x	x	x	o	x	o	x	x
5	x	o	x	o	x	x	o	x	34	o	o	o	o	o	o	x	o
6	o	o	x	x	x	x	x	x	35	o	o	o	x	o	x	x	o
7	o	x	x	o	x	x	x	x	36	o	o	o	x	o	x	x	x
8	x	x	o	x	x	x	x	X	37	x	x	x	x	x	x	x	x
9	o	o	x	o	x	x	x	x	38	x	o	x	x	x	x	x	x
10	o	o	o	o	o	o	o	x	39	x	x	o	o	x	x	x	x
11	x	o	x	x	x	x	x	x	40	o	o	x	o	x	x	x	x
12	o	o	o	o	x	o	x	o	41	x	x	x	x	x	x	x	x
13	x	x	x	x	x	x	x	x	42	o	x	o	x	x	x	x	x
14	x	x	x	x	x	x	x	x	43	o	o	o	o	o	o	x	x
15	o	x	x	o	x	x	x	x	44	o	x	x	o	x	o	x	x
16	o	o	x	o	o	o	x	o	45	o	x	x	o	x	x	x	x
17	x	o	x	x	x	x	x	x	46	x	x	x	x	x	o	x	x
18	o	o	x	x	x	x	x	x	47	o	x	o	o	o	x	x	o
19	x	x	x	x	x	x	x	x	48	o	x	o	x	o	x	x	x
20	x	o	x	x	o	x	o	o	49	o	o	o	x	o	x	x	x
21	x	x	x	x	x	x	x	x	50	o	o	x	x	x	x	x	x
22	x	x	x	x	x	x	x	x	51	o	o	o	x	x	x	x	x
23	x	x	o	o	x	x	x	x	52	x	x	x	x	x	x	x	x
24	x	x	x	x	x	x	x	x	53	x	x	x	x	x	x	x	x
25	o	x	x	x	o	x	x	x	54	x	o	o	o	o	x	x	x
26	o	x	x	x	o	x	x	x	55	o	o	x	x	x	x	x	x
27	x	o	o	o	x	o	o	x	56	o	o	o	o	o	x	x	x
28	O	o	x	x	x	x	x	o	57	o	o	x	x	o	x	o	x
29	O	x	o	o	x	x	x	x									

	2 μg/L	3.5 μg/L	6 μg/L	10 μg/L	18 μg/L	30 μg/L	60 μg/L	100 μg/L
Uncorrected %	43.85	49.12	61.40	59.64	66.67	77.19	89.47	85.96
Abbott Corrected %	15.78	23.68	42.10	39.46	50.00	65.79	84.21	78.94

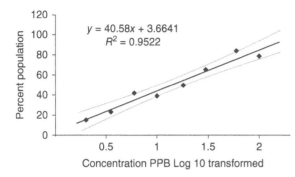

Figure 49.1 Linear least squares regression analysis of percent of population detecting MTBE across concentrations presented in this study. Note excellent goodness-of-fit R^2 value of 0.95.

Table 49.2 Combined data from Stocking et al. 2001 and Campden 2003 replication study.

Combined Stocking 2001 & Campden 2003 Data	2 µg/L	3.5 µg/L	6 µg/L	10 µg/L	18 µg/L	30 µg/L	60 µg/L	100 µg/L
Study 1	25/57	28/57	35/57	34/57	38/57	44/57	51/57	49/57
Study 2	22/55	22/55	23/55	24/55	31/55	30/55	30/55	33/55
Total of both Studies	47/112	50/112	58/112	58/112	69/112	64/112	81/112	82/112
% Correct	41.96	44.64	51.79	51.79	61.61	57.14	72.32	73.21
Abbott's Correction	12.94	16.96	27.69	27.69	42.42	35.71	58.48	59.82

Since the relationship between the ability to detect chemical stimuli and their concentration is often well described by a log linear function, the second author of this chapter and a colleague modeled the Stocking et al. data of Table 49.1 using linear regression. A plot of $\log_{10}$ concentration vs. the percent of subjects detecting MTBE (chance corrected by Abbott's formula) of these data is shown in Figure 49.1 (similar findings were found when the MTBE concentration is expressed in natural logs or when logistic regression is employed). Note the strong linear association between these measures, as indicted by the high R^2 value, an index of the amount of variance that is accounted for in the measurements. The antilog of the concentration values provides the concentration in µg/L (ppb). In these analyses, 10% of the study population is estimated to be able to detect MTBE at levels ranging from 1–2 µg/L; 15% at 2–3 µg/L; and 50% at 13–14 µg/L.

This approach results in a different conclusion than arrived at by Stocking et al. who found 15 µg/L to be a reasonable value for "establishing the federal SMCL for MTBE and other organic chemicals in drinking water" (p. 95). However, is it acceptable that 50% of a population can still detect a contaminant in their tap water when they are told that the water is safe to drink? Some might argue that while an individual can detect the presence of something in the water by smell, they cannot determine whether the sensation is unpleasant or alarming. From this perspective, the detection threshold may be overly conservative as an index of consumer acceptability. Such questions have not received adequate study.

A subsequent study that followed the same procedures as the Stocking et al. study was performed in England in an effort to replicate the 15 µg/L threshold value in a different group of consumers (Campden, 2003). With minor exceptions, the same general conditions applied; that is, a similar number of men and women, three general age groups, and the testing of consumers rather than trained panel members. If one accepts the proposition that the Stocking et al. (2001) and the Campden et al. (2003) studies are two samples of the same general population, then their data can be combined into a single sample of 112 subjects. A summary of the combined data is shown in Table 49.2. As apparent from this table, the same conclusions are apparent when the sample size is essentially doubled. Thus, ~13% of the combined study group detected 2 µg/L of MTBE. When linear regression was applied to the chance corrected and combined data (Figure 49.2), 10% of the study population detected MTBE in the 1–2 µg/L range.

The aforementioned studies represent the largest and most technically sophisticated studies to examine MTBE

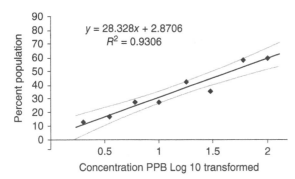

Figure 49.2 Least squares regression analysis of percent of population detecting MTBE across concentrations presented in the combined data of Stocking et al. (2001) and Campden et al. (2003). Note goodness-of-fit R^2 value of 0.93.

thresholds in the general population. Despite this fact, however, methodological considerations limit the degree to which these can be relied upon in establishing solid regulatory standards. Importantly, while Stocking et al. have concluded that 15 µg/L is an appropriate value for establishing a regulatory threshold for MTBE, the fact that their own data suggest that 10–15% of their subjects were able to detect MTBE by smell at 2 µg/L or less make this conclusion debatable. This is especially true when the odor quality of MTBE (e.g., paint thinner) would denote to the consumer that the water is unsafe and contains chemical contamination.

It is unfortunate that measuring taste and odor thresholds in the general population are expensive and time consuming. Importantly, the impact of the background matrix on the odor threshold has not been adequately explored. For example, the odor threshold can be influenced by the presence or absence of free chlorine residual in the water as well as the water's temperature. Considering the difficulty with setting a standard for MTBE, as well as other chemicals that can impact the acceptability of drinking water, the search continues for standards for the taste and odor quality of water (AWWA Water Quality Division Taste and Odor Committee, 2002). To date, no standard test has been developed by which to regulate whether drinking water has an acceptable taste or odor quality or meets public acceptance. Public investment in well-performed chemosensory studies employing larger groups of subjects who are representative of the population at large would avert much debate and cost in the litigation arena, serving the general public at large.

49.6 CONCLUSIONS

The human senses of taste and smell have been successfully used to improve the quality of drinking water. The first author has used his own sense of smell to track an algae-produced cucumber-like odor over 200 miles upriver to its source in a reservoir (Burlingame et al., 1992), to track an off flavor to an illegal, waste cutting oil dump site, to determine that a pipe joint lubricant was causing customer complaints (Burlingame et al., 1994), to determine that the leaching of metals in household plumbing was causing one particular customer to persistently complain about her tap water, and to find the cause of a canned-corn-like malodor at a wastewater treatment plant that could not be determined by chemical testing (Burlingame, 2009a; Cheng et al., 2009). This would not have been possible without a good understanding of the instrument of detection (the human sensory system) and the data it produces. This understanding comes from the many diverse fields of study on the human senses and their application such as provided in this book. For example, the University of Pennsylvania Smell Identification Test (UPSIT; see Chapter 11), which is used for medical diagnoses and research, has been used to identify panelists who have a reduced smell ability. Such persons may explain anomalies in research data during the testing of water samples and some human failures in detecting off flavors that require treatment.

Unlike foods and non-water beverages, drinking water should not have any noticeable taste or smell (Burlingame, 2009b). It should produce a refreshing and pure sensation. However, good tasting water is not chemically pure water, but water that contains a careful blend of minerals at a neutral pH and cool temperature. Nonetheless, in many places where a public drinking water is provided, the consumers become accustomed to a chlorinous flavor in the water because the safety of the water supply is maintained by the addition of chlorine.

Off flavors can have a significant political and economic impact on the public's acceptability of public drinking water. While most of the chemical and biological causes of off flavors in drinking water have been identified, and while sensory and analytical tests are being used to monitor the flavor quality of drinking water, there remains a need for national and international standards for chemicals that cause off flavors in water or for the proper management of the aesthetic quality of drinking water. This was recently highlighted by the January 2014 spill of crude methylcyclohexanemethanol (MCHM) into the Elk River just upstream of a drinking water treatment plant in West Virginia. This spill resulted in a do-not-use order for about 300,000 water customers, a state-of-emergency, and the President declaring the area a federal disaster (Rosen et al., 2014). When analytical testing of the drinking water produced test results that allowed the do-not-use order to be lifted, the customers could still detect the sharp licorice smell of MCHM in their tap water. Sensory testing revealed that the odor threshold was below the analytical detection limit and the limit of health concern (McGuire et al., 2014). The contaminant's odor became the leading factor for guaranteeing the water's safety. How could the public trust the pronoucements that the water was safe when they still could detect the MCHM telltale licorice smell in their homes? As a result, public and private utilities that supply drinking water need to effectively communicate with their customers about the taste and odor quality of the tap water. Customers are the most effective, though not highly accurate, detectors of changes in tap water taste and smell. Customers provide a valuable service to a water utility company in not only ensuring that water system problems are quickly responded to when they inevitably arise, but in helping to develop strategies to be proactive in preventing problems before they occur (Welton et al., 2007).

REFERENCES

American Public Health Association (2007a). Section 2170, Flavor Profile Analysis. In *Standard Methods for the Examination of Water and Wastewater*, 21st edition. Washington, D.C., American Public Health Association, American Water Works Association and Water Environment Federation, pp. 2–19.

American Public Health Association (2007b). *Standard Methods for the Examination of Water and Wastewater*, 21st edition. Washington, D.C., American Public Health Association, American Water Works Association and Water Environment Federation.

Andreas, P. and Gunten, U. V. (2009). Taste and odour problems generated in distribution systems: a case study on the formation of 2,4,6-trichloroanisole. *J. Wat. Supply: Res. Tech. – AQUA* 58(6): 386–394.

ASTM (1997, 2004). Standard Practice for Determination of Odor and Taste Thresholds by a Forced-Choice Ascending Concentration Series Method of Limits, (E679-97 & E679-04). In *Annual Book of Standards*. West Conshohocken, American Society for Testing and Materials.

Bartels, J. H. M., Burlingame, G. A. and Suffet, I. H. (1986). Flavor profile analysis: taste and odor control of the future. *J. A.W.W.A.* 78(3): 50–55.

Bruvold, W. H. and Daniels, J. I. (1990). Standards for mineral content in drinking water. *J. A. W. W. A.* 82(2): 59–65.

Burlingame, G. A., Dann, R. M., and Brock, G. L. (1986). A case study of geosmin in Philadelphia's water. *J. A.W.W.A.* 78(3): 56–61.

Burlingame, G. A., Muldowney, J. J., and Maddrey, R. E. (1992). Cucumber flavor in Philadelphia's drinking water. *J. A.W.W.A.* 84(8): 92–97.

Burlingame, G. A., Choi, J., Fadel, M., et al. (1994). Sniff new mains … before customers complain. *Opflow* 20(10)3.

Burlingame, G. A., Suffet, I. H., Khiari, D. and Bruchet A. L. (2004). Development of an odor wheel classification scheme for wastewater. *Wat. Sci.Tech.* 49(9): 201–209.

Burlingame, G. A., Dietrich, A. M. and Whelton, A. J. (2007). Understanding the basics of tap water taste. *J. A.W.W.A.* 99(5): 100–111.

Burlingame, G. A. and Mackey, E. (2007). Philadelphia obtains useful information from its customers about taste and odour quality. *Wat. Sci. Tech.* 55(5): 257–263.

Burlingame, G. A. (2009). A practical framework using odor survey data to prioritize nuisance odors. *Wat. Sci. Tech.* 59(3): 595–602.

Burlingame, G. A. (2009). *Taste at the Tap*. Denver: American Water Works Association.

Burlingame, G. A., Booth, S. D. J., Bruchet, A., et al. (2011). *Diagnosing Taste and Odor Problems Field Guide*. Denver: American Water Works Association.

Campden Food and Drink Research Association. (2003). *Consumer Odour Threshold of Methyl Tertiary Butyl Ether (MTBE) in Water*. Work commissioned by MHE Corporation. Report, 12th November, 2003.

Carter, J. M., Grady, S. J., Delzer, G. C., et al. (2006). Occurrence of MTBE and other gasoline oxygenates in CWS source waters. *J. A. W. W. A.* 98(4): 91–104.

Cheng, X., Wodarczyk, M., Lendzinski, R., et al. (2009). Control of DMSO in wastewater to prevent DMS nuisance odors. *Wat. Res.* 43: 2989–2998.

Consumer Reports. (2012). Water filters. *Consumer Reports* 77(2): 44–45.

Corbin, A. (1986). *The Foul and the Fragrant – Odor and the French Social Imagination*. Cambridge: Harvard University Press.

Desrochers, R. (2008). Sensory analysis in the water industry. *J. A.W. W. A.* 100(10): 50–54.

Dietrich, A. M., and Burlingame, G. A. (2015). Critical review and rethinking of USEPA secondary standards for maintaining organoleptic quality of drinking water. *Environ. Sci. Technol.* 49: 708–720.

Dietrich, A. M., Hoehn, R. C., Burlingame, G. A., and Gittelman, T. (2004). *Practical Taste-and-Odor Methods for Routine Operations: Decision Tree*. Denver: Water Research Foundation.

Dietrich, A. M., Cuppert, J. D. and Duncan, S. (2008). How much copper is too much? *Opflow* 34(9): 28– 30.

Dietrich, A. (2009). The sense of smell: contributions of orthonasal and retronasal perception applied to metallic flavor of drinking water. *J. Wat. Supply: Res. Tech. – AQUA* 58(8): 562–570.

Friedman, R. (1998). *The Life Millenium: The 100 Most Important Events and People of the Past 1,000 Years*. Darby: Diane Publishing Co.

Graham, J. L., Loftin, K. A., Meyer, Michael T. et al. (2010). Cyanotoxin mixtures and taste-and-odor compounds in cyanobacterial blooms from the midwestern United States. *Environ. Sci. Tech.* 44(19): 7361–7368.

Jüttner, F. and Watson, S. B. (2007). Biochemical and ecological control of geosmin and 2-methylisoborneol in source waters. *Appl. Environ. Micro.* 73(14): 4395–4406.

Khiari, D., Brenner, L., Burlingame, G. A. and Suffet, I. H. (1992). Sensory gas chromatography for evaluation of taste and odor events in drinking water. *Wat. Sci. Tech.* 25(2): 97–104.

Khiari, D., Barrett, S., Chinn, R., et al. (2002). *Distribution Generated Taste-and-Odor Phenomena*. Denver: Water Research Foundation.

Lawless, H. T. and Heymann, H. (1998). *Sensory Evaluation of Food: Principles and Practices*. New York: International Thomson Publishing.

Mackey, E. D., Davis, J., Boulos, L., et al. (2003). *Customer Perceptions of Tap Water, Bottled Water, and Filtration Devices*. Denver: Water Research Foundation.

Mackey, E. D., Baribeau, H., Fonesca, A. C., et al. . (2004). *Public Perception of Tap Water Chlorinous Flavor*. Denver: Water Research Foundation.

Mackey, E. D. (2012). Decision tool aids musty, earthy taste-and-odor assessment. *Opflow* 38(6): 18–20.

Malleret, L. and Bruchet, A. (2002). A taste and odor episode caused by 2,4,6-tribromoanisole. *J. A. W. W. A.* 94(7): 84–94.

McGuire, M., Graziano, N., Sullivan, L., et al. (2004). *Water Utility Self-Assessment for the Management of Aesthetic Issues*. Denver: Water Research Foundation.

McGuire, M. J., Rosen, J., Whelton, A. J. and Suffet, I. H. (2014). An unwanted licorice odor in a West Virginia water supply. *J. A. W. W. A.* 106(6): 72–82.

Meilgaard, M., Civille, G. V. and Carr, B. T. (1987). *Sensory Evaluation Techniques*, volumes 1 and 2. Boca Raton: CRC Press.

Mirlohi, S., Dietrich, A. M. and Duncan, S. E. (2011). Age-associated variation in sensory perception of iron in drinking water and the potential for overexposure in the human population. *Environ. Sci. Tech.* 45: 6575–6583.

Omur-Ozbek, P. and Dietrich, A. (2011). Retronasal perception and flavor thresholds of iron and copper in drinking water. *J. Wat. Health* 9(1): 1–9.

Omur-Ozbek, P., Gallagher, D. L., and Dietrich, A. M. (2011). Determining human exposure and sensory detection of odorous compounds released during showering. *Environ. Sci. Tech.* 45: 468–473.

Parr, J. (2005). Local water diversely known: Walkerton Ontario, 2000 and after. *Environ. Planning D: Soc. and Space* 23(2): 251–271.

Pollard, S., Bradshaw, R., Tranfield, D., et al. (2009). *Developing a Risk Management Culture – 'Mindfulness' in the International Water Utility Sector*. Denver: Water Research Foundation.

Rashash, D. M. C., Dietrich, A. M. and Hoehn, R. C. (1997). FPA of selected odorous compounds. *J. A. W. W. A.* 89(2): 131–141.

Rosen, J. S., Whelton, A. J. and McGuire, M. J. (2014). The crude MCHM chemical spill in Charleston, W. Va. *J. A. W. W. A.* 106(9): 65–74.

Stocking, A. J., Suffet, I. H., McGuire, M. J. and Kavanaugh, M. C. (2001). Implications of an MTBE odor study for setting drinking water standards. *J. A. W. W. A.* 93(3): 95–105.

Smith, R. S., Doty, R. L., Burlingame, G. K. and McKeown, D. A. (1993). Smell and taste function in the visually impaired. *Percept. Psychophys.* 54: 649–655.

Suffet, I. H., Brady, B., Burlingame, G. A., et al. (1988). Development of the flavor profile method into a standard method for sensory analysis of water. *Wat. Sci. Tech.* 20(8): 1–9.

Suffet, I. H., Mallevialle, J. and Kawczynski, E. (Eds). (1995). *Advances in Taste-and-Odor Treatment and Control*. Denver: Water Research Foundation.

Suffet, I. H., Khiari, D. and Bruchet, A. (1999). The drinking water taste and odor wheel for the millenium: beyond geosmin and 2-methylisoborneol. *Wat. Sci. Tech.* 40(6): 1–13.

Taylor, W. D. Losee, R. F., Torobin, M., et al. (2006). *Early Warning and Management of Surface Water Taste-and-Odor Events*. Denver: Water Research Foundation.

USEPA. (1992). *Secondary Drinking Water Regulations: Guidance for Nuisance Chemicals* (EPA 810/K-92-001). Washington, D.C.: United States Environmental Protection Agency.

Water Quality Division Taste and Odor Committee. (2002). Options for a taste and odor standard. *J. A. W. W. A.* 94(6): 80–87.

Watson, S. B. (2010). Algal taste and odor. Chapter 15 in *Manual M57- Algae: Source to Treatment*. Denver: American Water Works Association, pp. 329–375.

Whelton, A. J., Dietrich, A. M., Gallagher, D. L. and Roberson, J. A. (2007). Using customer feedback for improved water quality and infrastructure monitoring. *J. A. W. W. A.* 99(11): 62–76.

Whipple, G. C., Fair, G. M. and Whipple, M. C. (1948). *The Microscopy of Drinking Water*, 4th ed. New York: John Wiley and Sons, Inc.

WHO. (2008). *Guidelines for Drinking-water Quality*, 3rd ed. Geneva: World Health Organization.

Young, W. F., Horth, H., Crane, R. and Arnott, M. (1996). Taste and odour concentrations of potential potable water contaminants. *Wat. Res.* 30: 331–340.

Yu, J. W., Khao, Y. M., Yang, M., et al. (2009). Occurrence of odour-causing compounds in different source waters of China. *J. Wat. Supply: Res. Tech. –AQUA* 58(8): 587–594.

Zaitlin, B. and Watson, S. B. (2006). Actinomycetes in relation to taste and odour in drinking water: myths, tenets and truths. *Wat. Res.* 40: 1741–1753.

Part 12

Other Chemosensory Systems

Chapter 50

Trigeminal Chemesthesis

J. ENRIQUE COMETTO-MUÑIZ and CHRISTOPHER SIMONS

50.1 INTRODUCTION

In addition to the classical modalities of smell (olfaction) and taste (gustation), all body mucosae can respond directly to chemicals in our environment. This broadly distributed mucosal chemosensitivity was originally labeled the "common chemical sense" (Parker, 1912). In fact, such chemical sensitivity is also present in the skin, under the epidermis (Keele, 1962). The name remained until not long ago (Cain, 1981; Silver, 1987) and it is still occasionally used. More recently, this chemosensory modality is referred to as "chemesthesis" or "chemesthesia," in analogy to "somesthesis," since one can think of it as "chemically-induced somesthesis" (Green and Lawless, 1991; Green et al., 1990b). More simply, chemesthesis conveys the concept of chemical "feel" for sensations that are neither odors nor tastes (Bryant and Silver, 2000; Green et al., 1990a). Mucosal chemesthetic sensations are usually sharp and/or pungent. They include: prickling, piquancy, stinging, irritation, tingling, freshness, coolness, burning, and the like.

Although, as mentioned, all body mucosae possess chemosensitivity, those most exposed to environmental chemicals are the face mucosae: nasal, ocular, and oral. Early studies already had established that the trigeminal nerve (cranial nerve V, CN V), which provides chemical sensitivity to these three mucosae, is responsible for a variety of physiological responses to chemical irritants (Allen, 1928; 1929a; b; Dawson, 1962; Stone et al., 1966; Stone and Rebert, 1970; Stone et al., 1968; Ulrich et al., 1972). Thus, it is quite appropriate that this chemosensory system is now often referred to as "trigeminal chemoreception" (Cometto-Muñiz et al., 2010; Doty and Cometto-Muñiz, 2003; Doty et al., 2004; Silver and Finger, 2009; Sliver and

Finger, 1991). Previous reviews have described in detail the anatomy and physiology of the trigeminal chemosensory system (Doty and Cometto-Muñiz, 2003; Doty et al., 2004). Briefly, the three branches of the trigeminal nerve – the ophthalmic, the maxillary, and the mandibular – provide the bulk of chemesthetic sensitivity to the ocular, nasal, and oral mucosae. The glossopharyngeal (CN IX) and vagus (CN X) nerves also contribute to chemesthesis in the mouth (posterior tongue) and the nasopharyngeal and pharyngeal (throat) areas. Studies in mice and rats have shown that nasal solitary chemoreceptor cells (SCCs) can contribute to nasal chemesthesis (Finger et al., 2003; Silver and Finger, 2009; Tizzano et al., 2010), although their development and survival does not seem to be dependent on intact trigeminal innervation (Gulbransen et al., 2008a). SCCs express receptor proteins involved in gustatory chemoreception (T1Rs, T2Rs) and, thus, are also related to the sense of taste (Gulbransen et al., 2008b; Ohmoto et al., 2008; Sbarbati et al., 2004; Tizzano et al., 2011). However, expression of these receptors is thought to enable identification of potentially noxious irritants or pathogenic secretions (Tizzano et al., 2010). SCCs, which have also been described in several organs from the digestive and respiratory systems, have been proposed to be part of a diffuse chemosensory system (Osculati et al., 2007; Sbarbati et al., 2010).

In the present chapter we will summarize some important functional properties of the human trigeminal chemosensory system in the nasal, ocular, and oral mucosae. Among others, we will address issues related to candidate molecular receptors for trigeminal chemesthesis, chemesthetic sensitivity (i.e., detection thresholds), structure-activity relationships, detection of chemical mixtures, and temporal properties of trigeminal chemesthetic sensations.

Handbook of Olfaction and Gustation, Third Edition. Edited by Richard L. Doty.
© 2015 Richard L. Doty. Published 2015 by John Wiley & Sons, Inc.

50.2 MOLECULAR RECEPTORS FOR TRIGEMINAL CHEMESTHESIS

Chemesthetic sensitivity in the mucosae of the human face (nasal, ocular, and oral) is mediated by C and A_{delta} fibers primarily from the trigeminal nerve, although in the case of oral chemesthesis, the glossopharyngeal and vagus nerves also contribute (see reviews in Doty and Cometto-Muñiz, 2003; Doty et al., 2004; Rentmeister-Bryant and Green, 1997). The ion-channels and receptors involved are expressed by a class of peripheral neurons called polymodal nociceptors since they can respond to chemical, thermal, and mechanical stimuli (Belmonte et al., 2004). Some of the channels and receptors found in trigeminal and other sensory neurons are particularly sensitive to one or another prototypical irritant. This is the case, for example, for nicotine (Alimohammadi and Silver, 2000; Thuerauf et al., 2006), capsaicin (Tominaga and Tominaga, 2005), and menthol (Xing et al., 2006). Nevertheless, the ion-channels responding to the last two chemicals are also thermoreceptors that respond to warm/hot and cool/cold temperatures, respectively. In fact, the capsaicin channel not only responds to chemically-related vanilloids (Szallasi and Blumberg, 1999), but also to unrelated VOCs (Silver et al., 2006; Trevisani et al., 2002), to other pungent compounds (Macpherson et al., 2005; McNamara et al., 2005) and even to inorganic volatiles (Trevisani et al., 2005). It has also been shown that menthol (Macpherson et al., 2006) and other pungent substances (Bautista et al., 2005; Jordt et al., 2004) interact with more than one type of channel. Even the nicotinic receptor is modulated by VOCs such as homologous alcohols (Godden et al., 2001). In turn, VOCs can stimulate trigeminal neurons that are insensitive to cooling and capsaicin, suggesting that they can activate other mechanism(s) and receptors (Inoue and Bryant, 2005). VOCs that are reactive towards tissue can produce nociception indirectly by damaging cells and producing the release of intracellular mediators like K^+, H^+, ATP, and glutamate, among others, which, in turn, would activate nociceptors (Garle and Fry, 2003; Lee et al., 2005; Moalem et al., 2005; Vaughan et al., 2006; Wood and Docherty, 1997).

Among the large family of transient receptor potential (TRP) ion-channels that includes the capsaicin (TRPV1) and menthol (TRPM8) receptors (Appendino et al., 2008; Inoue, 2005; Kim and Baraniuk, 2007; Woodard et al., 2007), the TRPA1 nociceptor has been implicated in the chemesthetic response to environmental irritants (Bautista et al., 2006; Macpherson et al., 2007b; McNamara et al., 2007) and even to CO_2 (Wang et al., 2010), as well as to weak organic acids (Wang et al., 2011). The TRPM5 channel has also been implicated in responses to irritants (Lin et al., 2008). At least two mechanisms have been suggested for the activation of TRPA1 by ligands (Peterlin

et al., 2007). Under one mechanism, compounds such as eugenol, carvacrol, and methyl salicylate would act via a classical binding pocket (Xu et al., 2006). Under a second mechanism, electrophilic ligands such as acrolein, trans-2-pentanal, and trans-cinnamaldehyde would act by a covalent modification of thiol (cysteine) and, less likely (LoPachin et al., 2008), amine groups in the ion-channel protein (Hinman et al., 2006; Macpherson et al., 2007a). A recent review has described the molecular mechanisms of such reactions (LoPachin et al., 2008). In fact, the TRPV1 nociceptor has also been found to be activated by covalent modification of cysteine residues (Salazar et al., 2008). These studies explored the molecular/cellular and animal behavioral aspects of chemesthesis. The general picture emerging from them reflects a complex interplay of many factors and variables (acidity, temperature, release of endogenous mediators) modulating the response of any single TRP channel. Added to this is the generally broad chemical spectrum of ligands and modulators acting on any given receptor. As the chemesthetic neural message travels towards higher levels of the nervous system, convergence of input from different TRP channels (Zanotto et al., 2007) and other modulatory influences (Omote et al., 1998; Waters and Lumb, 2008) are likely to occur. In view of such complex array of factors, structure-activity approaches to trigeminal chemesthesis that rely on selective physicochemical properties of irritants governing their transport from the gas phase to a receptor biophase (or area) constitute very useful tools to describe, model, and predict human irritation potency from VOCs as measured psychophysically (Abraham et al., 2010b; Abraham et al., 2010c).

50.3 NASAL AND OCULAR CHEMESTHESIS

50.3.1 Separation of Nasal Trigeminal Chemosensitivity from Olfactory Sensitivity

The great majority of airborne chemicals that we detect with our noses are able to produce both an olfactory and a trigeminal response (Cain, 1974; 1976). That is, they can elicit both an odor and a chemesthetic (i.e., pungent or irritative) sensation (Doty et al., 1978). As a rule, virtually all these volatiles are initially detected by smell and, if their vapor concentration keeps increasing, there is a level where they also begin to evoke nasal chemesthesis (Doty, 1975). This presents a challenge if one wishes to measure nasal pungency thresholds devoid of olfactory biases, for example by using force-choice procedures against blanks. In other words, for most chemicals, a force-choice response between a stimulus (irritant) and a blank (plain

50.3 Nasal and Ocular Chemesthesis

air) will be biased since the participant will easily detect the stimulus by its odor, and will do so at concentrations well below those necessary to barely elicit any trigeminal chemesthetic sensation (i.e., to reach the vapor's pungency threshold). To address this situation, investigators have employed two main strategies. The first is to measure nasal pungency (i.e., chemesthetic) thresholds in subjects lacking a sense of smell (called anosmics) but otherwise healthy, thus avoiding odor biases (Cometto-Muñiz and Cain, 1990). The second is to measure nasal localization (or lateralization) thresholds, rather than detection thresholds, where subjects seek to localize which nostril (left or right) received a puff of irritant vapor when the other nostril simultaneously receives a puff of plain air (Cometto-Muñiz and Cain, 1998; Wysocki et al., 1997). Findings obtained using these two strategies are discussed in the next section.

50.3.2 Nasal Trigeminal Chemosensitivity as Reflected by Nasal Pungency Thresholds in Anosmics

Since anosmic subjects lack functional olfaction, any vapor that they can detect when presented to the nose has to be detected by means of the trigeminal system. Thus, one possibility to measure nasal pungency thresholds unbiased by odors is to measure them in anosmic participants. Early studies suggested that simple physicochemical properties might predict the chemesthetic potency of vapors (Doty, 1975; Doty et al., 1978). A number of physicochemical properties change orderly and systematically along homologous chemical series. These observations prompted a series of investigations that, using carbon chain length as a practical unit of chemical change, measured nasal pungency thresholds in anosmics and odor detection thresholds in normosmics (i.e., subjects with normal olfaction) towards selected members of homologous alcohols (Cometto-Muñiz and Cain, 1990), acetate esters (Cometto-Muñiz and Cain, 1991), 2-ketones and secondary and tertiary alcohols and acetates (Cometto-Muñiz and Cain, 1993), n-alkylbenzenes (Cometto-Muñiz and Cain, 1994), aliphatic aldehydes and carboxylic acids (Cometto-Muñiz et al., 1998a), and terpenes (Cometto-Muñiz et al., 1998b). They all employed a uniform methodology that included delivery of vapors via squeeze bottles, a two-alternative forced-choice procedure against blanks (mineral oil), an ascending concentration approach, a fixed-criterion of five correct choices in a row, and measurement of headspace vapor concentration in the bottles via gas chromatography (Cometto-Muñiz and Cain, 1990).

The outcome from these studies revealed the following (Figure 50.1): (1) In all cases, nasal pungency thresholds (NPTs) were higher than odor detection thresholds (ODTs) by a factor ranging from 1 to 5 orders of magnitude, depending on the chemical. In other words, trigeminal chemesthetic sensitivity was always lower than olfactory sensitivity by the factor mentioned. (2) NPTs and ODTs tend to decrease with carbon chain length within homologous series; that is, trigeminal and olfactory chemosensitivity increase with carbon chain length within the series. (3) Trigeminal chemosensitivity experiences a "cut-off" effect along homologous series, meaning that a certain homolog is reached that fails to elicit an NPT even at saturated vapor concentration for room temperature (see section "50.3.7 Cut-off effect…" further below). Once this cut-off homolog is reached, all larger homologs fail to evoke pungency as well.

50.3.3 Nasal Trigeminal Chemosensitivity and Anosmia

The measurement of NPTs in anosmics as a method to assess trigeminal chemesthetic thresholds in normosmics, assumes that the absence of smell does not interfere in a significant way with nasal trigeminal chemosensitivity. The outcome of psychophysical and electrophysiological studies addressing this assumption have been mixed, as described below, but they lead to the conclusion that nasal trigeminal chemosensitivity is not largely different in normosmics and anosmics.

As noted earlier, psychophysical studies of nasal trigeminal chemesthetic sensitivity in normosmics need to control for odor biases since the great majority of chemicals evoke both odor and irritation but olfactory sensitivity is higher than nasal trigeminal sensitivity (i.e., odor thresholds are lower than chemesthetic thresholds for any given chemical). Unfortunately, effective control for odor biases in normosmics cannot be established in suprathreshold intensity procedures (e.g., (Kendal-Reed et al., 1998), and is sometimes not included even in detection threshold measurement procedures, (e.g. (Gudziol et al., 2001). This could explain the apparent outcome in such studies of higher chemesthetic sensitivity in normosmics compared to anosmics. Nasal detection thresholds for the strong irritant chloroacetyl phenone were not significantly different between congenital anosmics and age- and gender-matched normosmics (Cui and Evans, 1997). The virtually odorless irritant carbon dioxide (CO_2), see (Cain and Murphy, 1980; Cometto-Muñiz and Cain, 1982; Dunn et al., 1982; García Medina and Cain, 1982), elicited lower intensity ratings in congenital anosmics than in normosmics (Frasnelli et al., 2007b) but, in contrast, it elicited intensity ratings no different in anosmics after upper respiratory tract infection or after head trauma than in normosmics (Frasnelli et al., 2007a). In terms of CO_2 detection thresholds, anosmics after upper respiratory tract infection and those after head

1094 Chapter 50 Trigeminal Chemesthesis

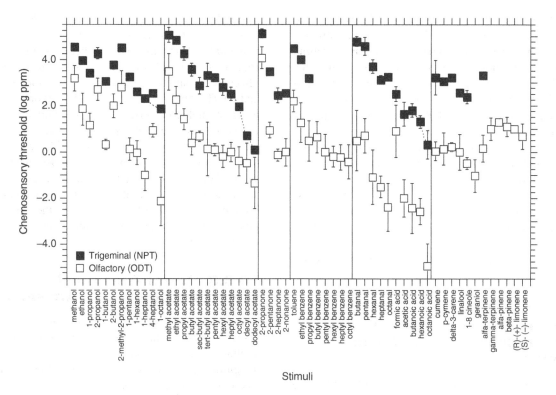

Figure 50.1 Showing trends in trigeminal (chemesthetic) and olfactory thresholds along lineal members of homologous series (plus some branched homologs), including alcohols, acetate esters, 2-ketones, alkylbenzenes, aliphatic aldehydes, and carboxylic acids. Also shown are both type of thresholds for cumene and selected terpenes. Trigeminal data are nasal pungency thresholds (NPTs) measured in anosmic subjects, and olfactory data are odor detection thresholds (ODTs) measured in normosmic subjects. A cut-off effect in chemesthesis detection (i.e., the saturated chemical vapor fails to elicit irritation) begins to develop for compounds whose NPTs are joined by a dashed line and fully develops for compounds whose NPTs are not plotted (see text). Bars indicate standard deviation.

trauma, but not hyposmics, had higher thresholds than normosmics (Frasnelli et al., 2006). A follow-up study found a similar outcome but also included congenital anosmics (Frasnelli et al., 2010). Unfortunately, CO_2 thresholds in these studies were obtained under a "yes–no" response approach without blanks, a method quite susceptible to criterion-based biases, rather than obtained under a more robust forced-choice procedure against blanks.

Electrophysiological studies on human nasal chemesthesis have explored peripheral responses, for example, the negative mucosal potential (NMP) (Kobal, 1985; Thürauf et al., 1993), and central responses generated in the cortex, e.g., trigeminal event-related potentials (tERP) (Kobal and Hummel, 1988; Kobal and Hummel, 1998). Peripheral responses to CO_2 stimulation in congenital anosmics and in anosmics from acquired etiologies showed a larger activation than in normosmics (Frasnelli et al., 2007a; b). In contrast, central responses to CO_2 stimulation in congenital anosmics showed no significant differences with those in normosmics (Frasnelli et al., 2007b), whereas anosmics from acquired etiologies (and hyposmics) showed a significantly smaller activation than normosmics (Frasnelli et al., 2007a; Hummel et al., 1996a). The specific origin for this smaller central activation in certain anosmics remains unclear: in some cases the smaller activation was observed for the peak-to-peak amplitude of the early P1N1 wave (Hummel et al., 1996a) but in other cases it was observed for the base-to-peak amplitude P2 and peak-to-peak amplitude N1P2 (Frasnelli et al., 2007a).

The conflicting results discussed above led us to conclude that any advantage in nasal trigeminal chemosensitivity that normosmics might have over anosmics, if indeed real, appears to be relatively small.

50.3.4 Nasal Trigeminal Chemosensitivity as Reflected by Nasal Localization (i.e., Lateralization) Thresholds in Normosmics and Anosmics

A pioneer investigation concluded that nasal localization (i.e., lateralization) of chemical vapors entering the nose through the right or the left nostril cannot be achieved via olfactory stimulation (odor) but can only be achieved when the vapors additionally activate the nasal trigeminal nerve (chemesthesis) (von Skramlik, 1925). A later paper argued in favor of nasal localization via olfaction (von Békesy, 1964), but it is likely that the odorants presented were at concentrations that activated trigeminal chemesthesis. Further studies concluded that for short chemical vapor presentations, and with the head in a fixed position, nasal localization only occurs via trigeminal chemesthesis, not via smell (Kobal et al., 1989; Schneider and Schmidt, 1967). This finding provided the opportunity to: (1) assess nasal trigeminal chemesthetic sensitivity directly in normosmics via nasal localization thresholds (NLTs), without olfactory interference, and under a robust forced-choice procedure, and (2) use NLTs to directly compare nasal trigeminal chemesthetic sensitivity in normosmics and anosmics, using a common method.

Some studies have measured nasal localization thresholds in normosmics and in anosmics for selected homologous n-alcohols, terpenes and cumene, employing a "squeeze-bottle" delivery system, an ascending concentration approach, and quantification of vapor concentrations using gas chromatography (Cometto-Muñiz and Cain, 1998; Cometto-Muñiz et al., 1998b). Their outcome suggested slightly lower NLTs for normosmics but the difference failed to achieve statistical significance. In fact, the three estimates of trigeminal chemesthetic sensitivity: NLTs in normosmics, NLTs in anosmics, and NPTs (as defined, always in anosmics) produced quite similar results (Figure 50.2). In two studies using a single concentration (neat vapor) of eucalyptol, with no chemical-analytical quantification, congenital anosmics (Frasnelli et al., 2007b) and anosmics from acquired etiologies (Frasnelli et al., 2007a) did not differ significantly from normosmics in their ability to lateralize that vapor. In contrast, a previous study had found the lateralization scores of normosmics to be significantly higher than those of subjects with olfactory dysfunction (both anosmics and hyposmics) when using the neat vapors from eucalyptol and from benzaldehyde, with no analytical quantification of either vapor (Hummel et al., 2003). Another investigation found that a high stimulus volume (21 ml) enhances the ability of subjects to localize neat vapors, compared to a low volume (11 ml) (Frasnelli et al., 2011b).

In conclusion, now in terms of nasal localization, the comparison of nasal trigeminal chemesthetic sensitivity between normosmics and anosmics has produced mixed results, again indicating that any sensitivity advantage for normosmics, if indeed real, seems relatively small.

50.3.5 Ocular Trigeminal Chemosensitivity as Reflected by Eye Irritation Thresholds in Normosmics and Anosmics

As noted, the trigeminal nerve also innervates the eyes, so vapors reaching the ocular mucosa can give rise to eye irritation, a key symptom invariably mentioned in studies of indoor air pollution (Korpi et al., 2009; Wolkoff et al., 2006; Wolkoff et al., 2003). Eye irritation thresholds (EITs) were measured in normosmics for homologous n-alcohols, acetate esters, 2-ketones, n-alkylbenzenes, for various terpenes, and for cumene, using a methodology analogous to that described above for NPTs, including vapor quantification via gas chromatography (Cometto-Muñiz and Cain, 1991, 1995, 1998; Cometto-Muñiz et al., 1998b).

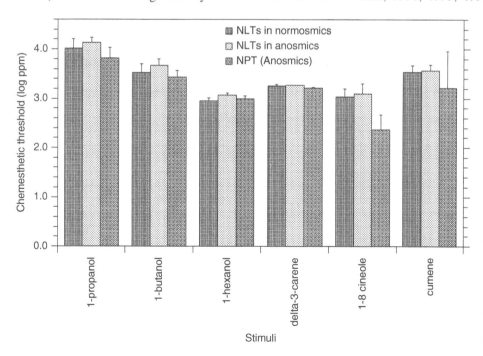

Figure 50.2 Illustrating the similarity between nasal localization thresholds (NLTs) in normosmics and in anosmics for selected homologous n-alcohols, terpenes, and cumene (Cometto-Muñiz and Cain, 1998; Cometto-Muñiz et al., 1998b). Also shown, for comparison, are the nasal pungency thresholds (NPTs) (by definition, always obtained in anosmics) for the same chemicals. Bars indicate standard deviation.

The overall outcome showed the following (Figure 50.3): (1) Both EITs and NPTs tend to decline with carbon chain length along homologous series. (2) In general, EITs are very close to NPTs for any given vapor, although in particular instances the EIT can be higher (e.g., ethanol) or lower (e.g., butyl acetate) than the NPT. (3) In some cases, the saturated vapor of a chemical fails to reach an NPT but does reach an EIT (e.g., 1-octanol, octyl acetate, and geraniol).

Measurement of EITs provided another opportunity to compare the trigeminal chemosensitivity of normosmics and anosmics. The outcome for 10 vapors tested revealed similar EITs across the two groups of subjects (Figure 50.4), again indicating quite comparable trigeminal chemesthetic sensitivity between normosmics and anosmics, now in the ocular mucosa (Cometto-Muñiz and Cain, 1998; Cometto-Muñiz et al., 1998b).

50.3.6 Structure-Activity Relationships in Nasal and Ocular Trigeminal Chemesthesis

A key element for understanding how a chemosensory system functions is to establish the relationship between the structural and physicochemical properties of the stimulating chemicals and the elicited chemosensory responses, including sensitivity, intensity, and quality. In this section we discuss such relationships in terms of the initial step of a sensory response, namely, stimulus detection, that is, thresholds, as an indicator of stimulus potency.

The wide chemical and structural variety of pungent (i.e., irritating) vapors suggests that, in many but not all cases, these irritants exert their effect via "selective" rather than "specific" processes. *Selective* processes rest on the physicochemical transfer of the irritant from the air into the nasal mucus or tear film, and its further transfer through successive biophases (i.e., biological compartments or environments) until reaching the receptor sites responsible for chemesthetic activity. In turn, *specific* effects are those where activity is mostly dependent on a narrowly defined key characteristic or property of the irritant stimulus that allows it to interact effectively with the receptor site. For example, to be active the chemical might need to be a strong electrophile, or to possess a restricted molecular configuration (e.g., only one of two enantiomers is active), a certain molecular size (e.g., molecules above that size are

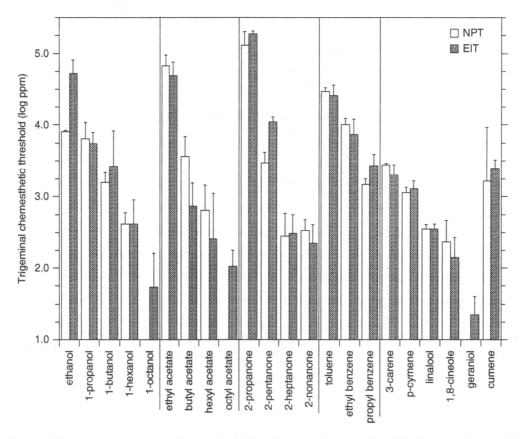

Figure 50.3 Comparability between nasal pungency thresholds (NPTs) and eye irritation thresholds (EITs, in normosmics) for 22 VOCs from four homologous series, selected terpenes, and cumene (Cometto-Muñiz and Cain, 1991, 1995, 1998, Cometto-Muñiz et al., 1998b). Bars indicate standard deviation.

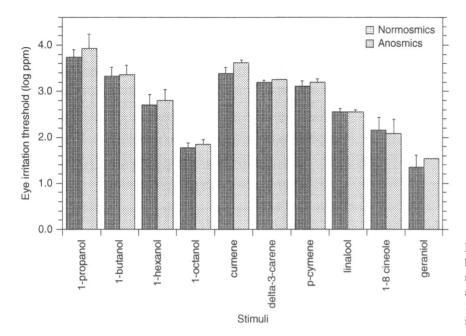

Figure 50.4 Depicting the similarity between eye irritation thresholds (EITs) measured in normosmics and in anosmics across 10 VOCs (Cometto-Muñiz and Cain, 1998; Cometto-Muñiz et al., 1998b). Bars indicate standard deviation.

inactive), or a particular chemical functional group (e.g., only thiols are active).

A number of studies have attempted to correlate trigeminal chemesthetic potency with a variety of structural and physicochemical properties, including: molecular weight, water solubility, molecular geometry, saturated vapor pressure, boiling point, adjusted boiling point, Ostwald solubility coefficient, other partition coefficients, and combinations of these and other parameters, experimental and/or calculated, (Doty, 1975; Doty et al., 1978; Hau et al., 1999; Muller and Greff, 1984; Nielsen et al., 1990, 1992; Roberts, 1986). As pointed out by (Abraham et al., 2010c), many of these attempts lacked a general mechanistic or chemical interpretation of the basis for the chemesthetic potency of pungent vapors. In contrast, a quantitative structure-activity relationship (QSAR) based on a solvation equation (Abraham, 1993a; b) has the advantage of including a mechanistic and chemical interpretation of the properties (or "descriptors") selected to model and predict human trigeminal chemesthetic sensitivity towards airborne compounds (Abraham et al., 2010c). This solvation-based QSAR takes the following general form:

$$SP = c + e.E + s.S + a.A + b.B + l.L \quad (50.1)$$

In the context of chemesthesis, SP (the dependent variable) is a sensory property reflecting chemesthetic potency of a series of irritants (i.e., solutes) towards humans, when the irritants reach, penetrate, diffuse, and trigger stimulation of the target mucosa (nasal or ocular). For example, SP is the logarithm of the reciprocal of nasal pungency thresholds, log (1/NPTs), or of eye irritation thresholds, log (1/EITs). (We used reciprocals so that the more potent is the irritant, the larger numerically is 1/NPT or 1/EIT.) On the other side of the equation, **E**, **S**, **A**, **B**, and **L** are physicochemical properties or descriptors (independent variables) of the series of irritants. These descriptors are defined as follows: **E** is the excess molar refraction of the irritant that can be determined from the refractive index of the compound; it represents the tendency of the irritant to interact with a receptor phase through π- or n-electron pairs. **S** is the dipolarity/polarizability of the irritant. **A** is the overall or effective hydrogen bond acidity of the irritant. **B** is the overall or effective hydrogen bond basicity of the irritant. **L** represents log L^{16} where L^{16} is the gas-hexadecane partition coefficient of the irritant at 298 °K, and it is a measure of the lipophilicity (i.e., lipid solubility) of the irritant. The constant c, and the coefficients e, s, a, b, and l are obtained by multiple linear regression analysis, but these are not merely fitted coefficients because they define the complementary physicochemical properties that characterize the receptor biophase with which the irritants interact. That is, they provide a physicochemical characterization of the environment or phase associated with the chemesthetic receptors. They do so in the following way: "e" gives the propensity of the receptor phase to interact with the irritant's π- and n-electron pairs; "s" quantifies the receptor phase dipolarity/polarizability because a dipolar irritant will interact with a dipolar phase and a polarizable irritant will interact with a polarizable phase; "a" reflects the receptor phase basicity since an irritant that is a hydrogen-bond acid will interact with a basic phase; "b" reflects the receptor phase acidity since an irritant that is a hydrogen-bond

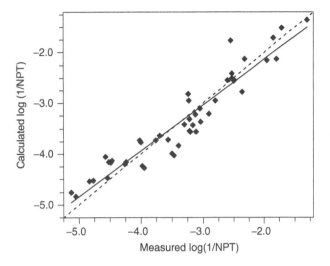

Figure 50.5 Relationship between observed (i.e., experimentally measured) and calculated (from equation (2)) nasal pungency thresholds (NPTs). Both types of thresholds are plotted as log (1/NPT), with NPTs expressed in units of parts per million (ppm) by volume. The dashed line represents the line of identity.

base will interact with an acidic phase; and, finally, "l" measures the lipophilicity of the receptor phase area.

The solvation equation (50.1) has been applied to describe and model experimentally-measured human NPTs from up to 48 volatile organic compounds (VOCs), including alcohols, acetate esters, ketones, alkylbenzenes, aldehydes, carboxylic acids, terpenes and other chemicals (Abraham et al., 1996; Abraham et al., 2001; Abraham et al., 2000; Abraham et al., 1998b; Abraham et al., 1998d; Abraham et al., 2007; Abraham et al., 2010b; Cometto-Muñiz et al., 1998a; 2005a). The solvation equation was found to account for 90 to 95% of the variability in measured NPTs. There is a strong correlation between the measured and the calculated NPTs (Figure 50.5). The latest version of the equation, as applied to NPTs, reads as follows (Abraham et al., 2010c):

$$\text{Log}(1/\text{NPT}) = -7.770 + 1.543\,\mathbf{S} + 3.296\,\mathbf{A} + 0.876\,\mathbf{B} + 0.816\,\mathbf{L} \quad (50.2)$$

$N = 47$, $R^2 = 0.901$, $SD = 0.312$, $F = 95$, $Q^2 = 0.874$, PRESS $= 5.1901$, PSD $= 0.351$

where $\mathbf{S}$, $\mathbf{A}$, $\mathbf{B}$, and $\mathbf{L}$ (descriptor $\mathbf{E}$ was not significant) are as defined after equation (50.1); N is the number of data points (i.e., VOCs), R is the correlation coefficient, SD is the standard deviation, F is the F-statistic, Q and PRESS are the leave-one-out statistics, and PSD is the predictive standard deviation.

The same solvation equation (50.1) has also been successfully applied to describe and model experimentally-measured human EITs and, quite interestingly, also Draize eye scores in rabbits (Abraham et al., 1998a, c, 2000, 2001, 2003, 2009, 2010b). In fact, it was shown that the two data sets assessing ocular chemesthesis: the one on human EITs and the one on Draize scores in rabbits can be made compatible and actually be combined into one single QSAR including up to 91 compounds (Abraham et al., 1998c, 2003). The equation combining the two types of measurement reads as follows (Abraham et al., 2003):

$$\text{SP} = -7.892 - 0.379\,\mathbf{E} + 1.872\,\mathbf{S} + 3.776\,\mathbf{A} + 1.169\,\mathbf{B} + 0.785\,\mathbf{L} + 0.568\,\mathbf{I} \quad (50.3)$$

$N = 91$, $R^2 = 0.936$, $SD = 0.433$, $F = 204.5$

where SP = log (1/EIT) or SP = log (MMAS/P°), i.e., a modified rabbit Draize score as previously defined (Abraham et al., 2003), and $\mathbf{I}$ is an indicator variable taking the value $I = 0$ for human data and $I = 1$ for Draize rabbit data. All other symbols are exactly as defined for equations (50.1) and (50.2).

The success of the solvation model to describe NPTs and EITs indicates that the VOCs included exert their chemesthetic effect mainly via selective, rather than specific, processes as defined at the beginning of this section. Instead, vapors that act mainly via specific processes might depart from the values predicted by the solvation equation. Typically, this departure is in the direction of being more potent than predicted, that is, having a lower NPT or EIT than predicted. We can say that the solvation equation calculates the general chemesthetic potency of any vapor with certain physicochemical properties, but, if in addition, the vapor activates one or more receptors in a strongly specific manner, its potency will be greater than that calculated by the equation, and, in consequence, its NPT or EIT will be lower than calculated. Examples of pungent compounds acting specifically on certain receptors are capsaicin (Tominaga and Tominaga, 2005), nicotine (Alimohammadi and Silver, 2000) and acrolein (Bautista et al., 2006). However, some of them, for example, capsaicin, have very little volatility. There are also compounds for which the solvation model predicts some level of chemesthetic potency but, when tested, fail to elicit chemesthesis. These cases were discovered when studying chemesthetic potency within homologous chemical series. The phenomenon, called the "cut-off" effect, is discussed in the next section.

50.3.7 Cut-Off Effect for Eliciting Trigeminal Chemesthesis along Vapors from Homologous Chemical Series

As illustrated in Figures 50.1, 50.2, 50.3, and 50.4, trigeminal chemesthetic thresholds tend to decrease with carbon chain length along homologous series (Cometto-Muñiz, 2001). Nevertheless, this trend often ends abruptly when a certain homolog is reached for which no threshold can be measured since even the saturated vapor of the compound at room temperature ($\approx 23°C$) fails to elicit chemesthetic detection reliably (Cometto-Muñiz and Cain, 1991, 1994, 1998; Cometto-Muñiz et al., 1998a). We labeled this phenomenon a "cut-off" effect for the detection of trigeminal chemesthesis, in analogy with a similar effect described for anesthesia (Franks and Lieb, 1985; 1990; Horishita and Harris, 2008). In turn, we labeled the smallest homolog failing to elicit chemesthetic detection the "cut-off homolog."

At least two possible mechanisms could explain the occurrence of the cut-off effect: (1) the saturated vapor concentration, at room temperature (23°C), of the cut-off homolog (and all ensuing homologs) falls below the necessary threshold, and (2) the cut-off homolog (and all ensuing homologs) lacks a key feature to trigger chemesthesis. We looked into these two possibilities by following three approaches: (1) calculating a predicted chemesthetic threshold via the solvation QSAR model described just above and comparing it with the saturated vapor concentration (at 23°C) of the cut-off homolog; (2) increasing the saturated vapor concentration of the cut-off homolog by heating it to 37°C and testing if this would precipitate detection; and (3) measuring complete concentration-detection functions (rather than just a threshold point) for the cut-off compound and for its immediate homologous neighbors (Cometto-Muñiz and Abraham, 2008; Cometto-Muñiz et al., 1998a, 2005a, b, 2006, 2007a; b). The combined outcome from these approaches indicated that the cut-off effect often results from a restriction other than an insufficient vapor concentration; for example, it could result from reaching a large enough homolog that exceeds a critical molecular dimension to be able to elicit chemesthesis (Cometto-Muñiz et al., 2010). If so, this molecular size restriction would constitute a specific effect (as defined at the beginning of the previous section) and, in consequence, it would not be predicted by the solvation model represented by equation (50.1).

50.3.8 Chemical Mixtures and Chemesthesis

An interesting functional aspect of trigeminal chemesthesis relates to the stimulation with chemical mixtures, both at threshold (i.e., detection) and suprathreshold (i.e., perceived intensity or magnitude) levels. When probing the rules by which trigeminal chemesthesis detects chemical mixtures, as compared to the detection of the individual constituents, it is important to rule out physicochemical interactions and associations among components that might occur before the mixture even reaches the receptor sites. This consideration is usually not addressed in mixture studies. A recent investigation concluded that, at concentrations near chemesthetic detection thresholds, the association between non-reactive VOCs in mixtures upon reaching their biological site of action would only amount to less than 5% of their total concentration (Abraham et al., 2010a).

In terms of threshold chemesthetic detection of chemical mixtures, an intensive study of nine individual VOCs from a number of homologous series and five of their mixtures including 3, 6, and 9-components, found various degrees of agonism (i.e., additive effects) among the constituents (Cometto-Muñiz et al., 1997). The investigation included two chemesthetic endpoints: NPTs and EITs, and an olfactory endpoint: ODTs. Chemesthetic thresholds showed a larger degree of agonism than olfactory ones, with eye irritation having stronger agonism than nasal pungency. The extent of chemosensory agonism became larger as the number and the lipophilicity (i.e., lipid solubility) of the mixture components increased. Later studies followed a more detailed approach by focusing in binary mixtures and by measuring full concentration-detection (i.e., psychometric or detectability) functions, (see (Cometto-Muñiz et al., 2002), rather than just threshold values. As before, components of the mixtures were chosen from homologous series, allowing us to address the role of structure-activity relationships in the chemesthetic detection of binary mixtures (Cometto-Muñiz et al., 1999, 2001, 2004a). The results were analyzed in terms of response-addition (i.e., sum of detectabilities) and of dose-addition (sum of doses within a selected psychometric function). The outcome pointed towards relative agreement between these two ways of quantifying addition, and indicated a larger degree of addition at low than at high levels of detectability (Cometto-Muñiz et al., 2004b). In these latter studies, which are based on concentration-detection functions, nasal pungency showed a larger degree of addition than eye irritation for mixtures presented at moderate and high detectability levels.

In terms of perceived intensity of chemical mixtures (i.e., suprathreshold range), a study of mixtures of the pungent odorants ammonia and formaldehyde revealed that, as the mixtures increased in concentration, the total nasal perceived intensity switched from being significantly lower to being significantly higher than the sum of the perceived intensities of its components (Cometto-Muñiz et al., 1989). A follow-up investigation confirmed that this effect was due to a gain in the relative contribution of nasal chemesthesis over olfaction as the concentration of the mixtures increased (Cometto-Muñiz and Hernández, 1990). In another study, the ocular mucosa seemed more responsive than the nasal mucosa regarding the

1100 Chapter 50 Trigeminal Chemesthesis

chemesthetic intensity elicited by the complex mixture environmental tobacco smoke (Cain et al., 1987). Under a dose–response approach, the eye irritation intensity produced by a mixture of eight VOCs and two mixtures of four VOCs indicated that the components acted in a simple additive way to produce ocular irritation (Hempel-Jorgensen et al., 1999). A recent paper employed nasal lateralization scores and intensity ratings to compare three single chemicals (menthol, eucalyptol, and allyl isothiocyanate) and two mixtures (menthol/eucalyptol and menthol/allyl isothiocyanate) (Frasnelli et al., 2011a). Intensity ratings probably included odor plus chemesthesis (i.e., total nasal intensity), although it is not specifically stated. Individual stimuli were presented at a single liquid dilution (found to be iso-intense by a different group of experienced subjects) and the mixtures were presented at 50%, probably volume/volume (v/v), but no analytical quantification of the vapor concentrations from either single or mixed stimuli was presented. The results suggested suppression within the mixture activating a common receptor (i.e., menthol/eucalyptol) given that this mixture produced lower intensity ratings than the other mixture, and slightly lower lateralization scores than the other mixture (albeit only under a one-sided t-test since the main effect "stimulus" failed to reach significance for lateralization scores). This outcome should be taken with the caveat that many chemesthetic stimuli that are mainly known to activate a certain receptor type can also activate or inhibit other types (Macpherson et al., 2006) with more or less potency and/or efficacy (Lambert, 2004).

Real world chemical exposures often involve mixtures of at least dozens or hundreds of components (Feron et al., 2002). A number of investigations have looked into complex mixtures of VOCs to establish their ability to elicit nasal and/or ocular trigeminal chemesthesis in humans, at threshold and suprathreshold levels, among other diverse effects. Examples of such complex stimuli include: environmental tobacco smoke (Cain et al., 1987), fragranced products (Millqvist et al., 1999; Opiekun et al., 2003), indoor air (Hudnell et al., 1992; Laumbach et al., 2005; Meininghaus et al., 2003; Otto et al., 1990), and outdoor air surrounding composting facilities (Müller et al., 2004), animal production facilities (Schiffman et al., 2000; Schiffman and Williams, 2005), and other industries (Dahlgren et al., 2003). In such studies, the typical goal was to assess the effect of the mixture (diluted or neat) on different groups of subjects (e.g., unexposed controls, exposed subjects, asthmatics, etc.). Since, in most cases, these very complex mixtures were not fully defined from a chemical standpoint, no attempts were made to relate the chemesthetic impact of the mixtures to that of the individual constituents.

50.3.9 Temporal Properties of Nasal and Ocular Chemesthesis

Since nasal and ocular trigeminal chemesthetic responses often act as a warning signal against the continued exposure to potentially deleterious airborne compounds, these sensations tend to increase with time of exposure before reaching a plateau or declining thereafter (Shusterman et al., 2006; Wise et al., 2009b). At near-threshold levels, an increase in stimulation time produces lower detection thresholds, at least within the first few seconds ($\approx$4–6 sec) of stimulation. This temporal integration of threshold nasal chemesthesis, whether measured with a trigeminally-induced reflex apnea (Dunn et al., 1982) or with the nasal localization approach, has been observed for a variety of irritants including ammonia (Cometto-Muñiz and Cain, 1984; Wise et al., 2005), CO_2 (Wise et al., 2004), ethanol (Wise et al., 2006), homologous alcohols (Wise et al., 2007), and homologous propionates (Wise et al., 2009a). Within short stimulation times (up to 6 sec), the relationship between exposure-time and concentration falls close but often is short of a perfect trade off; that is, to achieve the chemesthetic threshold, it takes somewhat more than doubling the stimulus duration to compensate for a twofold decrease in concentration.

At supra-threshold levels, the perceived intensity of nasal chemesthesis increases with stimulation time for at least the first 4 sec, as shown with ammonia (Cometto-Muñiz and Cain, 1984; Wise et al., 2005) and CO_2 (Wise et al., 2003). Under a protocol testing homologous alcohols with alternative cycles of 3 sec of chemical stimulation and 3 sec of clean air for a 1 min trial, the intensity of nasal chemesthesis showed adaptation, that is, it decreased for recurrent pulses of the alcohol vapor (Wise et al., 2010). Using fixed suprathreshold levels of CO_2 as the nasal chemesthetic stimulus, a series of studies have looked into the effect of repetitive stimulation with CO_2 at different inter-stimulus intervals (ranging from 2 to 90 sec) in subsequent ratings of pain intensity. They concluded that, for intervals ranging from 10 to 90 sec, pain ratings for repetitive CO_2 presentations were lowest after a 10-sec interval and largest after a 90-sec interval (Hummel and Kobal, 1999). For very short intervals (2 to 8 sec), pain ratings increased after 2-sec intervals but decreased after 6- and 8-sec intervals (Hummel et al., 1994; Hummel et al., 1996b). In contrast, the decrease in amplitude to repetitive CO_2 stimulations of both the peripheral negative mucosal potential (NMP) and the central chemo-somatosensory event-related potential (CSSERP) were more pronounced with the shortest inter-stimulus interval, that is, 2 sec. (Hummel et al., 1994; Hummel et al., 1996b). Nevertheless, a more recent report found that the amplitude of chemosensory event-related potentials to repetitive CO_2 nasal stimulations increased when the inter-stimulus interval was shortened from 30 to 10 sec, concluding that the effect is due to trigeminal sensitization during repeated stimulation (Kassab et al., 2009). In another study, a lineal relationship emerged between the stimulus duration of a steady CO_2 pulse increasing from 100 to 300 msec and the amplitude of the P3 component of the trigeminal event-related potential, suggesting that the later components of this potential encode not only for stimulus concentration but also for stimulus duration (Frasnelli et al., 2003).

Other studies on the effect of exposure time on nasal chemesthesis explored much longer durations, in the range of minutes to hours, and under whole body exposures. Chamber studies testing formaldehyde at various fixed concentrations revealed that perceived irritation (nasal and ocular) increased with time during a 30-min exposure (Cain et al., 1986). In realistic daily home or occupational exposures to a chemical vapor that might be causing mild irritation, the nasal chemesthetic threshold for that chemical increases, revealing desensitization, but the effect does not appear to generalize to other irritants (Dalton et al., 2006; Smeets and Dalton, 2002; Wysocki et al., 1997). Exposures of two to three hours to mixtures of VOCs (Hudnell et al., 1992) and to the complex stimulus environmental tobacco smoke (Cain et al., 1987) revealed that the perceived intensity of chemesthesis clearly increases with time.

50.4 ORAL CHEMESTHESIS

50.4.1 Separation of Oral Trigeminal Chemosensitivity from Gustatory Sensitivity

Typically, the detection of water soluble chemicals occurs in the oral cavity and is critical to the survival of organisms. In the case of gustation, chemosensitivity provides a mechanism by which foods can be instantaneously evaluated prior to ingestion. This ability has a profound impact on an organism's survival because, in the case of sweet sensitivity, it allows for the identification of potentially nutritious food sources, whereas for bitter and sour sensitivity, provides the ability to detect and avoid poisonous or spoiled foods (for review, see Brody et al., 2012). Although there are similarities, gustation is differentiated from oral chemesthesis in a number of important ways. Prototypical chemesthetic stimuli include many plant-based molecules such as (a) capsaicinoids (capsaicin from chili peppers} (Szolcsányi and Jancsó-Gábor, 1975; Stevens and Lawless, 1987; Green, 1989; Caterina et al., 1997; piperine from black pepper (Stevens and Lawless, 1987; Green, 1996; Dessirier et al., 1999)), (b) isothiocyantes (allyl isothiocyanate from mustard oil) (Simons et al., 2003; Jordt et al., 2004; Bandell et al., 2004) and allicin from

garlic (Bautista et al., 2005) and (c) the alkaloid nicotine (Jansco et al., 1961; Dessirier et al., 1997; 1998), found in the nightshade family (Solanaceae) of plants. Ingestion of these compounds at low levels elicits a predominantly pungent or burning sensation from the oral mucosa (Keele, 1962; Lawless and Stevens, 1988; Rentmeister-Bryant and Green, 1997) and vermilion border of the lips (Lawless, 1984; Lawless and Stevens, 1988), but have minor or no discernible taste qualities on their own (except for nicotine which has been described as bitter) (Pfaffman, 1959). Recently, oleocanthal from olive oil has been identified which elicits an irritant sensation sensed primarily from the pharynx (Beauchamp et al., 2005; Peyrot des Gachons et al., 2011). Other chemesthetic sensations include (a) pungency from ethanol (Green, 1988; Prescott and Swain-Campbell, 2000), cinnamic aldehyde (Prescott and Swain-Campbell, 2000; Klein et al., 2011) and zingerone (Prescott and Stevenson, 1996a; 1996b), (b) cooling from menthol (Eccles, 1994; Cliff and Green, 1994; 1996) and other compounds including icillin (Wei and Seid, 1983), WS-12 (Sherkheli et al., 2010), WS-23 (Serkheli et al., 2010) and a recently identified high-potency synthetic compound that elicits a strong cooling sensation of long duration (Furrer et al., 2008), (c) warming from camphor (Green, 1990; Moqrich, 2005) and (d) tingling from alpha-hydroxy-sanshool (Bryant and Mezine, 1999; Albin and Simons, 2010) or spilanthal (Ley et al., 2006), unique compounds derived from the sechuan peppercorn and jambu fruit, respectively. These compounds are in contrast to gustatory stimuli that are usually detected at much higher levels (mM) and evoke specific taste sensations including sweet, sour, salty, bitter, umami and possibly fat (for review on fat taste, see Mattes, 2011). Importantly, it should be noted that salts and acids, which evoke sensations of saltiness and sourness, respectively, can elicit pungent sensations from the oral cavity (Simons et al., 1999; Dessirier et al., 2000b; 2001) when used at high concentrations. Thus, unlike many chemesthetic stimuli, acids and salts have both gustatory and irritant properties.

Chemesthetic and gustatory sensitivity are underpinned by different receptor families and different neuronal pathways (for detailed discussion see Section 50.1) and, in some cases, evoke different behaviors. Oral chemesthetic stimuli typically interact with one of three main families of receptors including members of the Transient Receptor Potential (TRP) family (for review, see Clapham, 2003 and Talavera et al., 2008), nicotinic cholinergic receptors (nAChRs; Dessirier et al., 1998; Simons et al., 2003b; Thuerauf et al., 2006), and outward-rectifying 2-pore potassium channels including KCNK3, KCNK9 and KCNK18 (Bautista et al., 2008). Most irritants interact with one of several ionotropic TRP channels including TRPA1, TRPM8, TRPV1, and TRPV2. Nicotine and sanshool, a compound that induces tingling sensations, activate the nAChRs, TRPA1 (see Talavera et al., 2009), and 2-pore potassium channels (Bautista et al., 2008). These receptors are expressed by polymodal nociceptors or, in the case of the 2-pore potassium channels, a mechanosensitive subpopulation of sensory neurons (Bautista et al., 2008; Lennertz et al., 2010; Patapoutian et al., 2009). Ligand-receptor interaction evokes direct excitation of the neuron whose signal is conveyed centrally to the trigeminal nucleus (Sessle, 1989; Carstens et al., 1998). From there, neural projections to the thalamus and, subsequently, primary somatosensory cortex, allow for higher processing of the nociceptive information (Sessle, 1989). When ingested, chemesthetic stimuli evoke reflex pathways and aversive behaviors intended to minimize intake. Salivation (Ley and Simchen, 2007) dilutes the irritant and behavioral responses reduce additional exposure. However, one central feature specific to oral chemesthesis is its contribution to flavor. The sensations associated with chili peppers, wasabi, mustard, and carbonation, while pungent, are often sought after, and contribute to the flavor profile of many cuisines. Not surprisingly, therefore, chemesthetic compounds have a long history in culinary use. Several hypotheses have been forwarded to explain the use of these aversive compounds by humans ranging from thrill-seeking (Rozin and Schiller, 1980) to proposed health benefits. Interestingly, research has identified a range of health-related benefits of various chemesthetic stimuli including antibacterial activity (Ozçelik et al., 2011), cardiovascular protection (Peng and Li, 2010), and anti-obesity effects (Hursel and Westerterp-Plantenga, 2010).

These findings are in contrast to those ascribed to gustation. Identified in the early 2000s, gustatory receptors include the 7-transmembrane, g-protein coupled T1R (Nelson et al., 2001; 2002) and T2R (Adler et al., 2000; Chandrashekar et al., 2000) subfamilies, which are involved in the transduction of sweet and umami and bitter tastes, respectively (see Chapter 31). Although still unidentified in humans, it is believed that ionotropic channels transduce salty (Yarmolinsky et al., 2009) and sour (Yarmolinsky et al., 2009) sensations. Recently, in addition to texture and olfaction, fat has been proposed to have a gustatory component (Chalé-Rush et al., 2007) that involves activation of CD36 (Laugerette et al., 2005), the g-protein coupled receptors GPR40 (Matsumura et al., 2007; Cartoni et al., 2010) and GPR120 (Matsumura et al., 2007; Cartoni et al., 2010), and/or the inactivation of outward rectifying K+ channels (Gilbertson et al., 1997). These receptors are expressed in taste receptor cells which are derived from local epithelium (Stone et al., 1995) and innervated by primary neurons of the chorda tympani, glossophayngeal, and vagus nerves (for review, see Whitehead, 1988). The central projection of these neurons is to specific gustatory centers in the brainstem, thalamus and insular cortex (Whitehead, 1988). When ingested, tastant stimuli largely evoke innate stereotyped behaviors that can often be modified through learning. Sweet, umami, and salt (low concentrations) stimuli elicit appetitive behaviors whereas sour and bitter tastants evoke aversive behaviors which limit intake (for review, see Brody et al., 2012).

50.4.2 Structure-Activity Relationships in Oral Trigeminal Chemesthesis

To date, the vast majority of psychophysical investigation linking oral irritancy to chemical structure-activity relationships (SAR) has been related to capsaicinoids. Several capsaicinoid compounds have been utilized to probe the human irritant response and identify the key structural attributes responsible for the pungency of this class of molecules. Capsaicin, the prototypical chemesthetic irritant, consists of an alkyl chain joined to a vanillyl moiety through an acyl amide (Figure 50.6a). This compound has been studied extensively and has been shown to elicit an irritant sensation in humans at levels as low as 0.2 ppm (Szolcsányi, 1990). Three critical structural features of capsaicin have been identified that impact the perceived pungency of this molecule. The vanillyl moiety is important and contributes greatly to the perceived pungency (Szolcsanyi and Jancso-Gabor, 1975). However, if the ring is opened, as in the case of undecenoyl-3-amino-propanol, pungency can still be elicited albeit at a significantly reduced potency (Szolcsanyi and Jancso-Gabor, 1975). The length of the alkyl chain also impacts pungency with perceived irritation peaking when carbon chain length reaches approximately ten carbons (Szolcsanyi and Jancso-Gabor, 1975). Finally, the presence of an acyl amide group contributes to perceived irritancy. Conversion of the acyl amide into an acyl ester significantly reduces pungency (Szolcsanyi and Jancso-Gabor, 1975).

Structure-activity relationships for other irritants have not been as extensively studied although a key structural feature of many irritants activating TRPA1 is their ability to covalently modify specific cysteine residues within the N-terminal domain of the channel (Hinman et al., 2006).

A substantial body of knowledge generated from the flavor and fragrance industry has provided important insight into the central features of cooling compounds. Using a pharmacophore hypothesis, key structural elements were identified for menthane carboxamides (Figure 50.6b), the most widely used and investigated family of cooling compounds (Furrer et al., 2008). The primary binding interaction between this class of compounds and TRPM8, its cognate receptor, is believed to be hydrophobic in nature and mediated via the menthane core (Furrer et al., 2008). Additionally, the affinity of menthane carboxamides can

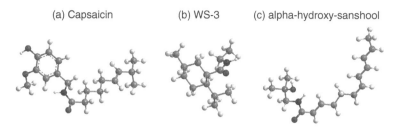

Figure 50.6 Chemical structures of three chemesthetic stimuli. (a) The TRPV1 agonist capsaicin. (b) The TRPM8 agonist WS-3, a simple menthane carboxamide. (c) The KCNK, two-pore potassium channel blocker, alpha-hydroxy-sanshool. (*See plate section for color version.*)

be modulated by the amidic moiety. Finally, the importance of two hydrogen bond acceptors in proximity to the compact menthane core was found to improve potency of these molecules (Watson et al., 1978, Furrer et al., 2008). However, it should be noted that other chemical cores have the capacity to evoke cooling (see, for example, Furrer et al., 2008). For instance, icillin, a compound having little structural similarity to menthol, is a potent activator of TRPM8 (McKemy et al., 2002; Chuang et al., 2004; Dhaka et al., 2007).

Finally, recent structure-activity relationships have been investigated for compounds that elicit a tingle sensation via blockade of 2-pore potassium channels. Sanshools (Figure 50.6c) are long-chained polyenamides and the prototypical stimulus to elicit the tingle sensation. Four key structural elements associated with sanshools have been identified which are necessary for the perceived sensation (Galopin et al., 2002). In particular, a N-isobutylcarboxamide motif is required as well as the presence of a cis-double bond in the fatty chain. Furthermore, the efficacy of these compounds can be enhanced by the presence of a long fatty chain, saturation of the alpha-, beta-, gamma- and delta- carbons, and a hydroxyl on the N-alkyl group (Galopin et al., 2002).

50.4.3 Temporal Properties of Oral Chemesthesis

A unique property of most oral chemesthetic irritants is their ability to elicit temporally dynamic patterns of irritation with repeated application. Psychophysical methods have been described that allow investigators to track oral irritation intensity across time (Lawless and Stevens, 1988; Stevens and Lawless, 1987; Green, 1988; 1989; Dessirier et al., 1999, 2001; Prescott, 1999; Prescott and Stevenson, 1996a; 1996b). Using these techniques, several distinctive patterns (see Figure 50.7) have been identified including sensitization, desensitization, cross-desensitization, and stimulus-induced recovery (SIR). Interestingly, these same patterns of response have been observed electrophysiologically in single-unit recordings of trigeminal cells in Vc (Dessirier et al., 2000c; Sudo et al., 2002). Sensitization is described as the increase in perceived intensity of a given sensation upon repeated exposure. In the oral cavity, the occurrence of sensitization is dependent upon the inter-stimulus intervals with which chemesthetic agents are applied (Green, 1989; Green and Rentmeister-Bryant, 1998; Prescott, 1999). With short inter-stimulus intervals, typically on the order of three minutes or less, the perceived irritation grows until it reaches a plateau (Green, 1991). Many, but not all, chemesthetic compounds elicit this pattern of irritation including the capsaicinoids capsaicin (Lawless, 1984; Stevens and Lawless, 1987; Green, 1989; Prescott, 1999) and piperine (Stevens and Lawless, 1987; Green, 1996; Rentmeister-Bryant and Green, 1997), high concentrations of acids (Green, 1996; Dessirier et al., 2000b) and salts (Dessirier et al., 2001), and menthol (Cliff and Green, 1994, 1996). Interestingly, mustard oil (allyl isothiocyante) was found to elicit a sensitizing pattern of irritation when delivered to the nasal cavity (Brand and Jacquot 2002) but not when delivered to the oral cavity (Simons et al., 2003a). Other irritants, including zingerone (Prescott and Stevenson, 1996a, 1996b) and nicotine (Dessirier et al., 1997) do not evoke sensitization.

Several mechanisms underlying this phenomenon have been proposed (see Carstens et al., 2002) including the peripheral and central sensitization of nociceptive pathways that typically accompany inflammation (Willis, 2009). Spatial recruitment has also been proposed as a mechanism underpinning sensitization (Carstens et al., 2002). In this model, irritant chemicals applied to the lingual surface activate chemosensitive neurons. With repeated application, the irritant compounds diffuse into the lingual epithelium where they encounter and excite previously unactivated neurons. The increasing number of activated nociceptors results in a progressive increase in the perceived intensity of the irritation sensation. Eventually, the rate of diffusion through the epithelium is equaled by the clearance rate of the compound. At this point, spatial recruitment cannot continue and the intensity of the perceived irritant sensation plateaus (see Figure 50.7).

When the repeated application of irritant chemicals occurs after a hiatus or at longer inter-stimulus intervals, typically greater than 5 min, self- or cross-desensitization occurs (Green, 1991; Prescott and Stevenson, 1996b; see Figure 50.7). Self-desensitization, like adaptation, is described as a decrease in perceived sensation intensity with repeated or continued application of a given chemesthetic compound. Cross-desensitization is the term given wherein application of one chemoirritant causes reduced

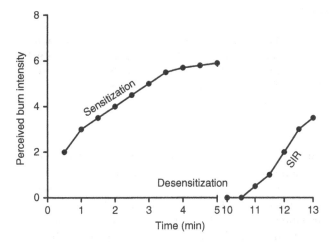

Figure 50.7 Temporal patterns of oral chemesthesis. Repeated application of a chemesthetic irritant at short interstimulus intervals (e.g. 30 sec), elicits a sensitizing pattern of irritation in which the perceived sensation grows more intense until it reaches a plateau. When the repeated application of irritant chemicals occurs after a hiatus or at longer inter-stimulus intervals (greater than 5 min) desensitization occurs. However, desensitization can be overcome by recurrent stimulation at short inter-stimulus intervals. With continued stimulation, stimulus induced recovery (SIR) occurs and the perceived irritant sensation grows more intense.

sensitivity to subsequent applications of a second irritant. Cross-desensitization is often used as evidence supporting the view that two compounds share the same receptor and/or neural mechanisms (e.g., Cliff and Green, 1996; Dessirier et al., 2000a; Green, 1996). To date, many of the chemorirritants tested in psychophysical protocols evoke desensitizing patterns of oral irritation including capsaicin (e.g., Green, 1989, 1991; Szolcsányi, 1990; Prescott, 1999), piperine (Green, 1996; Dessirier et al., 1999), ethanol (Prescott and Swain-Campbell, 2000), acids (Dessirier et al., 2000b), salts (Dessirier et al., 2001), nicotine (Dessirier et al., 1997), allyl isothiocyanate (Simons et al., 2003a), cinnamic aldehyde (Prescott and Swain-Campbell, 2000), zingerone (Prescott and Stevenson, 1996a, 1996b), and menthol (Cliff and Green, 1994; 1996).

Desensitization is likely due to peripheral mechanisms occurring at the receptor or neuronal level. In whole-cell patch-clamp experiments of trigeminal and dorsal root ganglion neurons, repeated application of capsaicin resulted in successively smaller inward currents (Liu and Simon, 1996, 1998; Su et al., 1999; Yao and Qin, 2009). This reduction could be mitigated by the removal of extra-cellular Ca^{+2}, suggesting the importance of this ion in regulating the process (Liu and Simon, 1998). Interestingly, other capsaicinoids, including zingerone and olvanil, elicit cellular desensitization however, the mechanism appears to be Ca^{+2}-independent (Liu and Simon, 1998). An additional mechanism contributing to desensitization could be the result of some irritants ability to inhibit voltage-sensitive Na^+ channels as is the case for capsaicin (Su et al., 1999 and Liu et al., 2001). Clearly, this could also serve as the root cause of cross-desensitization.

An interesting aspect of lingual desensitization is that it can be overcome with subsequent, recurrent stimulation (Green, 1996). This phenomenon, termed Stimulus-Induced Recovery (SIR; Figure 50.7), is both concentration- and inter-stimulus interval-dependent. Initiating and driving SIR to completion requires stimulus concentrations equal to or greater than the concentrations used to elicit the desensitized state. For instance, in studies using capsaicin, a 33 μM solution could elicit SIR only when the desensitizing capsaicin concentration was 33 μM or lower, but not when a 99 μM or 330 μM solution was used (Green and Rentmeister-Bryant, 1998). Additionally, like sensitization, initiation of SIR requires recurrent stimulus applications with short inter-stimulus intervals – the shorter the inter-stimulus interval, the greater the incidence and magnitude of SIR (Green and Rentmeister-Bryant, 1998). Studies investigating the temporal dependence of SIR showed that an inter-stimulus interval greater than 60-sec was inefficient at reversing capsaicin desensitization (Green and Rentmeister-Bryant, 1998). Although still speculative, the mechanism underlying SIR is thought to be the same as that underlying sensitization – namely, peripheral and/or central sensitization, and spatial recruitment.

Many chemesthetic compounds activate multiple receptor sub-types. Additionally, the kinetics of sensitization, desensitization, and adaptation for each of the sub-qualities can differ (Lawless and Stevens, 1990). As a consequence, many chemesthetic compounds evoke unique and complex sensory profiles that are temporally dynamic. An excellent example of this occurs with lingual application of alkylamide compounds such as alpha-hydroxy sanshool (Figure 50.8). Sanshool has been shown to activate 2-pore potassium channels (Bautista et al., 2008), TRPV1 (Sugai et al., 2005; Koo et al., 2007; Menozzi-Smarrito et al., 2009; Riera et al., 2009), and TRPA1 (Koo et al., 2007; Menozzi-Smarrito et al., 2009; Riera et al., 2009). Additionally, lingual nerve recordings in rat have shown that sanshool activates low and high threshold cold-sensitive fibers as well as low threshold mechanosensitive fibers (Bryant and Mezine, 1999). As expected from the diverse receptor and neuronal types that are activated by these compounds, the psychophysical results are equally complex. Lingual application of a synthetic alkylamide elicits a sensation that is described as initially tingling and pungent but after approximately 15 minutes as cooling and numbing (Figure 50.8).

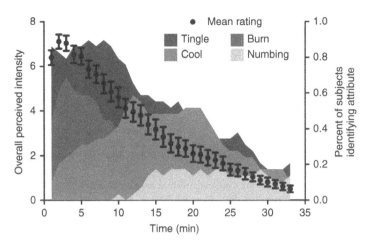

Figure 50.8 Temporal profile and attribute identification following lingual application of a synthetic alkylamide. The figure shows the mean overall intensity perceived over time and the proportion of panelists selecting tingle, burning, cooling or numbing at each time point. Error bars indicate SEM. Figure adapted from Albin and Simons, 2010.

50.4.4 Threshold for Oral Chemesthesis

Testing oral chemesthetic thresholds is difficult due to the presence of sensitization, desensitization and SIR. Typical bias-free methods used to establish threshold sensitivity, such as forced-choice, ascending and descending staircase approaches, cannot be used because they require repeated stimulation. As discussed above, a short inter-stimulus interval can result in sensitization leading to confounded threshold determinations. Conversely, with a longer inter-stimulus interval, desensitization can result leading to detection thresholds that are artificially high. Allowing time for recovery from desensitization is potentially possible, however, in some cases this phenomenon has been shown to last for days (Carstens et al., 2007; Green, 1996; Karrer and Bartoshuk, 1991). Moreover, long inter-stimulus intervals require subjects to remember preceding stimulus intensities and memory effects have been shown to worsen performance on discrimination tasks (Avancini de Almeida, 1999; Cubero et al., 1995).

The limitations associated with investigating chemesthetic compounds restrict the options available for sound psychophysical testing. One potential solution that has been developed is the use of "half-tongue" methodologies (e.g., Dessirier et al., 1997; Simons et al., 1999) in which forced-choice, discrimination tasks are accomplished by placing a compound of interest on one side of a subject's tongue and the control solution on the other. Subjects can compare the sensation on both sides of the tongue simultaneously to identify any intensity differences. Previous investigations have indicated no differences in sensitivity across both sides of the tongue as long as the same dorsal lingual surface is being stimulated bilaterally (Lawless and Stevens, 1988; 1990). Over several days, multiple stimulus concentrations can be compared to identify threshold sensitivity and, as the stimuli being compared are delivered simultaneously, memory effects are obviated.

ACKNOWLEDGMENTS

Preparation of this chapter was supported in part by grants number R01 DC 002741 and R01 DC 005003 from the National Institute on Deafness and Other Communication Disorders (NIDCD), National Institutes of Health (NIH).

REFERENCES

Abraham, M. H. (1993a). Application of solvation equations to chemical and biochemical processes. *Pure Appl. Chem.* 65: 2503–2512.

Abraham, M. H. (1993b). Scales of solute hydrogen-bonding: their construction and application to physicochemical and biochemical processes. *Chem. Soc. Rev.* 22: 73–83.

Abraham, M. H., Acree, W. E. and Cometto-Muñiz, J. E. (2009). Partition of compounds from water and from air into amides. *New J. Chem.* 33: 2034–2043.

Abraham, M. H., Andonian-Haftvan, J., Cometto-Muñiz, J. E. and Cain, W. S. (1996). An analysis of nasal Irritation thresholds using a new solvation equation. *Fundam. Appl. Toxicol.* 31: 71–76.

Abraham, M. H., Gola, J. M. R., Cometto-Muñiz, J. E. and Acree, W. E. (2010a). Hydrogen Bonding between Solutes in Solvents Octan-1-ol and Water. *J. Org. Chem.* 75: 7651–7658.

Abraham, M. H., Gola, J. M. R., Cometto-Muñiz, J. E. and Cain, W. S. (2001). The correlation and prediction of VOC thresholds for nasal pungency, eye irritation and odour in humans. *Indoor Built Environ.* 10: 252–257.

Abraham, M. H., Gola, J. M. R., Kumarsingh, R., et al. (2000). Connection between chromatographic data and biological data. *J. Chromatogr. B-Analyt. Technolog. Biomed. Life Sci.* 745: 103–115.

Abraham, M. H., Hassanisadi, M., Jalali-Heravi, M., et al. (2003). Draize rabbit eye test compatibility with eye irritation thresholds in humans: a quantitative structure-activity relationship analysis. *Toxicol. Sci.* 76: 384–391.

Abraham, M. H., Kumarsingh, R., Cometto-Muñiz, J. E. and Cain, W. S. (1998a). A quantitative structure-activity relationship (QSAR) for a draize eye irritation database. *Toxicol In Vitro* 12: 201–207.

Abraham, M. H., Kumarsingh, R., Cometto-Muñiz, J. E. and Cain, W. S. (1998b). An algorithm for nasal pungency thresholds in man. *Arch. Toxicol.* 72: 227–232.

Abraham, M. H., Kumarsingh, R., Cometto-Muñiz, J. E. and Cain, W. S. (1998c). Draize Eye Scores and Eye Irritation Thresholds in Man Combined into one Quantitative Structure-Activity Relationship. *Toxicol. In Vitro* 12: 403–408.

Abraham, M. H., Kumarsingh, R., Cometto-Muñiz, J. E., et al. (1998d). The determination of solvation descriptors for terpenes, and the prediction of nasal pungency thresholds. *J. Chem. Soc., Perkin Trans.* 2 2405–2411.

Abraham, M. H., Sánchez-Moreno, R., Cometto-Muñiz, J. E. and Cain, W. S. (2007). A quantitative structure-activity analysis on the relative sensitivity of the olfactory and the nasal trigeminal chemosensory systems. *Chem. Senses* 32: 711–719.

Abraham, M. H., Sánchez-Moreno, R., Gil-Lostes, J., et al. (2010b). The biological and toxicological activity of gases and vapors. *Toxicol. in Vitro* 24: 357–362.

Abraham, M. H., Sánchez-Moreno, R., Gil-Lostes, J., et al. (2010c). Physicochemical modeling of sensory irritation in humans and experimental animals. In: *Toxicology of the Nose and Upper Airways*, Morris, J.B. and Shusterman, D.J. (Eds.). New York: Informa Healthcare USA, pp. 376–389.

Adler, E., Hoon, M. A., Mueller, K. L., et al. (2000). A novel family of mammalian taste receptors. *Cell* 100: 693–702.

Albin, K. C. and Simons, C. T. (2010). Psychophysical evaluation of a sanshool derivative (alkylamide) and the elucidation of mechanisms subserving tingle. *PLoS One* 5: e9520.

Alimohammadi, H. and Silver, W. L. (2000). Evidence for nicotinic acetylcholine receptors on nasal trigeminal nerve endings of the rat. *Chem. Senses* 25: 61–66.

Allen, W. F. (1928). Effect on respiration, blood pressure, and carotid pulse of various inhaled and insufflated vapors when stimulating one cranial nerve and various combinations of cranial nerves I. Branches of the trigeminus affected by these stimulants. *Am. J. Physiol.* 87: 319–325.

Allen, W. F. (1929a). Effect of various inhaled vapors on respiration and blood pressure in anesthetized, unanesthetized, sleeping and anosmic subjects. *Am. J. Physiol.* 88: 620–632.

Allen, W. F. (1929b). Effect on respiration, blood pressure and carotid pulse of various inhaled and insufflated vapors when stimulating one cranial nerve and various combinations of cranial nerves III. Olfactory and trigeminals stimulated. *Am. J. Physiol.* 88: 117–129.

Appendino, G., Minassi, A., Pagani, A. and Ech-Chahad, A. (2008). The role of natural products in the ligand deorphanization of TRP channels. *Curr. Pharm. Des.* 14: 2–17.

Avancini de Almeida, T. C., Cubero, E. and O'Mahony, M. (1999). Same-different discrimination tests with interstimulus delays up to one day. *J. Sens. Studies* 14: 1–18.

Bandell, M., Story, G. M., Hwang, S. W., et al. (2004). Noxious cold ion channel TRPA1 is activated by pungent compounds and bradykinin. *Neuron* 41: 849–857.

Bautista, D. M., Jordt, S. E., Nikai, T., et al. (2006). TRPA1 mediates the inflammatory actions of environmental irritants and proalgesic agents. *Cell* 124: 1269–1282.

Bautista, D. M., Movahed, P., Hinman, A., et al. (2005). Pungent products from garlic activate the sensory ion channel TRPA1. *Proc. Natl. Acad. Sci. U. S. A.* 102: 12248–12252.

Bautista, D. M., Sigal, Y. M., Milstein, A. D., et al. (2008). Pungent agents from Szechuan peppers excite sensory neurons by inhibiting two-pore potassium channels. *Nat. Neurosci.* 11: 772–779.

Beauchamp, G. K., Keast, R. S., Morel, D., et al. (2005). Phytochemistry: ibuprofen-like activity in extra-virgin olive oil. *Nature* 437: 45–46.

Belmonte, C., Acosta, M. C. and Gallar, J. (2004). Neural basis of sensation in intact and injured corneas. *Exp. Eye Res.* 78: 513–525.

Brand, G. and Jacquot, L. (2002). Sensitization and desensitization to allyl isothiocyanate (mustard oil) in the nasal cavity. *Chem. Senses* 27: 593–598.

Brody, H, Grayson, M., Scully,T., and Haines,N. eds. *Nature Outlook Taste.* New York: Nature Publishing Group, 2012.

Bryant, B. P. and Mezine, I. (1999). Alkylamides that produce tingling paresthesia activate tactile and thermal trigeminal neurons. *Brain Res.* 842: 452–460.

Bryant, B. and Silver, W. L. (2000). Chemesthesis: The Common Chemical Sense. In: *The Neurobiology of Taste and Smell. 2nd Edition*, Finger, T.E., Silver, W.L. and Restrepo, D. (Eds.). New York: Wiley-Liss, pp. 73–100.

Cain, W. S. (1974). Contribution of the trigeminal nerve to perceived odor magnitude. *Ann. N. Y. Acad. Sci.* 237: 28–34.

Cain, W. S. (1976). Olfaction and the common chemical sense: some psychophysical contrasts. *Sens. Processes* 1: 57–67.

Cain, W. S. (1981). Olfaction and the common chemical sense: Similarities, differences, and interactions. In: *Odor Quality and Chemical Structure (ACS Symposium Series No. 148)*, Moskowitz, H. R. and Warren, C. B. (Eds.). Washington, DC: American Chemical Society, pp. 109–121.

Cain, W. S. and Murphy, C. L. (1980). Interaction between chemoreceptive modalities of odour and irritation. *Nature* 284: 255–257.

Cain, W. S., See, L. C. and Tosun, T. (1986). Irritation and odor from formaldehyde: Chamber studies. In: *IAQ'86. Managing Indoor Air for Health and Energy Conservation.* Atlanta, Georgia, USA: American Society of Heating, Refrigerating and Air-Conditioning Engineers, Inc., pp. 126–137.

Cain, W. S., Tosun, T., See, L. C. and Leaderer, B. (1987). Environmental tobacco smoke - Sensory reactions of occupants. *Atmos. Environ.* 21: 347–353.

Carstens, E., Albin, K. C., Simons, C. T. and Carstens, M. I. (2007). Time course of self-desensitization of oral irritation by nicotine and capsaicin. *Chem. Senses* 32: 811–816.

Carstens, E., Iodi Carstens, M., Dessirier, J. M., et al. (2002). It hurts so good: oral irritation by spices and carbonated drinks and the underlying neural mechanisms. *Food Qual. Pref.* 13: 431–443.

Carstens, E., Kuenzler, N. and Handwerker, H. O. (1998). Activation of neurons in rat trigeminal subnucleus caudalis by different irritant chemicals applied to oral or ocular mucosa. *J. Neurophysiol.* 80: 465–492.

Cartoni, C., Yasumatsu, K., Ohkuri, T., et al. (2010). Taste preference for fatty acids is mediated by GPR40 and GPR120. *J. Neurosci.* 30: 8376–8382.

Caterina, M. J., Schumacher, M. A., Tominaga, M., et al. (1997).The capsaicin receptor: a heat-activated ion channel in the pain pathway. *Nature* 389: 816–824.

Chalé-Rush, A., Burgess, J. R. and Mattes, R. D. (2007). Evidence for human orosensory (taste?) sensitivity to free fatty acids. *Chem. Senses* 32: 423–431.

Chandrashekar, J., Mueller, K. L., Hoon, M. A., et al. (2000). T2Rs function as bitter taste receptors. *Cell* 100: 703–711.

Chuang, H. H., Neuhausser, W. M. and Julius, D. (2004). The super-cooling agent icilin reveals a mechanism of coincidence detection by a temperature-sensitive TRP channel. *Neuron* 43: 859–869.

Clapham, D. E. (2003). TRP channels as cellular sensors. *Nature* 426: 517–524.

Cliff, M. A. and Green, B. G. (1994). Sensory irritation and coolness produced by menthol: evidence for selective desensitization of irritation. *Physiol. Behav.* 56: 1021–1029.

Cliff, M. A. and Green, B. G. (1996). Sensitization and desensitization to capsaicin and menthol in the oral cavity: interactions and individual differences. *Physiol. Behav.* 59: 487–494.

References

Cometto-Muñiz, J. E. (2001). Physicochemical basis for odor and irritation potency of VOCs. In: *Indoor Air Quality Handbook*, Spengler, J.D., Samet, J. and McCarthy, J.F. (Eds.). New York: McGraw-Hill, pp. 20.21–20.21.

Cometto-Muñiz, J. E. and Abraham, M. H. (2008). A cut-off in ocular chemesthesis from vapors of homologous alkylbenzenes and 2-ketones as revealed by concentration-detection functions. *Toxicol. Appl. Pharmacol.* 230: 298–303.

Cometto-Muñiz, J. E. and Cain, W. S. (1982). Perception of nasal pungency in smokers and nonsmokers. *Physiol. Behav.* 29: 727–731.

Cometto-Muñiz, J. E. and Cain, W. S. (1984). Temporal integration of pungency. Chem. Senses 8: 315–327.

Cometto-Muñiz, J. E. and Cain, W. S. (1990). Thresholds for odor and nasal pungency. *Physiol. Behav.* 48: 719–725.

Cometto-Muñiz, J. E. and Cain, W. S. (1991). Nasal pungency, odor, and eye irritation thresholds for homologous acetates. *Pharmacol. Biochem. Behav.* 39: 983–989.

Cometto-Muñiz, J. E. and Cain, W. S. (1993). Efficacy of volatile organic compounds in evoking nasal pungency and odor. *Arch. Environ. Health* 48: 309–314.

Cometto-Muñiz, J. E. and Cain, W. S. (1994). Sensory reactions of nasal pungency and odor to volatile organic compounds: the alkylbenzenes. *Am. Ind. Hyg. Assoc. J.* 55: 811–817.

Cometto-Muñiz, J. E. and Cain, W. S. (1995). Relative sensitivity of the ocular trigeminal, nasal trigeminal and olfactory systems to airborne chemicals. *Chem. Senses* 20: 191–198.

Cometto-Muñiz, J. E. and Cain, W. S. (1998). Trigeminal and olfactory sensitivity: comparison of modalities and methods of measurement. *Int. Arch. Occup. Environ. Health* 71: 105–110.

Cometto-Muñiz, J. E., Cain, W. S. and Abraham, M. H. (1998a). Nasal pungency and odor of homologous aldehydes and carboxylic acids. *Exp. Brain Res.* 118: 180–188.

Cometto-Muñiz, J. E., Cain, W. S. and Abraham, M. H. (2004a). Chemosensory additivity in trigeminal chemoreception as reflected by detection of mixtures. *Exp. Brain Res.* 158: 196–206.

Cometto-Muñiz, J. E., Cain, W. S. and Abraham, M. H. (2004b). Detection of single and mixed VOCs by smell and by sensory irritation. *Indoor Air* 14 Suppl 8: 108–117.

Cometto-Muñiz, J. E., Cain, W. S. and Abraham, M. H. (2005a). Determinants for nasal trigeminal detection of volatile organic compounds. *Chem. Senses* 30: 627–642.

Cometto-Muñiz, J. E., Cain, W. S. and Abraham, M. H. (2005b). Molecular restrictions for human eye irritation by chemical vapors. *Toxicol. Appl. Pharmacol.* 207: 232–243.

Cometto-Muñiz, J. E., Cain, W. S., Abraham, M. H. and Gola, J. M. (1999). Chemosensory detectability of 1-butanol and 2-heptanone singly and in binary mixtures. *Physiol. Behav.* 67: 269–276.

Cometto-Muñiz, J. E., Cain, W. S., Abraham, M. H. and Gola, J. M. (2001). Ocular and nasal trigeminal detection of butyl acetate and toluene presented singly and in mixtures. *Toxicol. Sci.* 63: 233–244.

Cometto-Muñiz, J. E., Cain, W. S., Abraham, M. H. and Gola, J. M. (2002). Psychometric functions for the olfactory and trigeminal detectability of butyl acetate and toluene. *J. Appl. Toxicol.* 22: 25–30.

Cometto-Muñiz, J. E., Cain, W. S., Abraham, M. H. and Kumarsingh, R. (1998b). Trigeminal and olfactory chemosensory impact of selected terpenes. *Pharmacol. Biochem. Behav.* 60: 765–770.

Cometto-Muñiz, J. E., Cain, W. S., Abraham, M. H. and Sánchez-Moreno, R. (2006). Chemical boundaries for detection of eye irritation in humans from homologous vapors. *Toxicol. Sci.* 91: 600–609.

Cometto-Muñiz, J. E., Cain, W. S., Abraham, M. H. and Sánchez-Moreno, R. (2007a). Concentration-detection functions for eye irritation evoked by homologous n-alcohols and acetates approaching a cut-off point. *Exp. Brain Res.* 182: 71–79.

Cometto-Muñiz, J. E., Cain, W. S., Abraham, M. H. and Sánchez-Moreno, R. (2007b). Cutoff in detection of eye irritation from vapors of homologous carboxylic acids and aliphatic aldehydes. *Neuroscience* 145: 1130–1137.

Cometto-Muñiz, J. E., Cain, W. S., Abraham, M. H., et al. J. (2010). Nasal chemosensory irritation in humans. In: *Toxicology of the Nose and Upper Airways*, Morris, J.B. and Shusterman, D.J. (Eds.). New York: Informa Healthcare, pp. 187–202.

Cometto-Muñiz, J. E., Cain, W. S. and Hudnell, H. K. (1997). Agonistic sensory effects of airborne chemicals in mixtures: odor, nasal pungency, and eye irritation. *Percept. Psychophys.* 59: 665–674.

Cometto-Muñiz, J. E., García-Medina, M. R. and Calviño, A. M. (1989). Perception of pungent odorants alone and in binary mixtures. *Chem. Senses* 14: 163–173.

Cometto-Muñiz, J. E. and Hernández, S. M. (1990). Odorous and pungent attributes of mixed and unmixed odorants. *Percept. Psychophys.* 47: 391–399.

Cubero, E., Avancini de Almeida, T. C. and O'Mahony, M. (1995). Cognitive aspects of difference testing: Memory and interstimulus delay. *J. Sens. Studies* 10: 307–324.

Cui, L. and Evans, W. J. (1997). Olfactory event-related potentials to amyl acetate in congenital anosmia. *Electroencephalogr. Clin. Neurophysiol.* 102: 303–306.

Dahlgren, J., Warshaw, R., Thornton, J., et al. (2003). Health effects on nearby residents of a wood treatment plant. *Environ. Res.* 92: 92–98.

Dalton, P., Dilks, D. and Hummel, T. (2006). Effects of long-term exposure to volatile irritants on sensory thresholds, negative mucosal potentials, and event-related potentials. *Behav. Neurosci.* 120: 180–187.

Dawson, W. W. (1962). Chemical stimulation of peripheral trigeminal nerve. *Nature* 196: 341.

Dessirier, J. M., Chang, H. K., O'Mahony, M. and Carstens, E. (2000a). Cross-desensitization of capsaicin-evoked oral irritation by high but not low concentrations of nicotine in human subjects. *Neurosci. Lett.* 290: 133–136.

Dessirier, J. M., Nguyen, N., Sieffermann, J. M., et al. (1999). Oral irritant properties of piperine and nicotine: psychophysical evidence for asymmetrical desensitization effects. *Chem. Senses* 24: 405–413.

Dessirier, J. M., O'Mahony, M. and Carstens, E. (1997). Oral irritant effects of nicotine: psychophysical evidence for decreased sensation following repeated application and lack of cross-desensitization to capsaicin. *Chem. Senses* 22: 483–492.

Dessirier, J. M., O'Mahony, M., Iodi-Carstens, M. and Carstens, E. (2000b). Sensory properties of citric acid: psychophysical evidence for sensitization, self-desensitization, cross-desensitization and cross-stimulus-induced recovery following capsaicin. *Chem. Senses* 25: 769–780.

Dessirier, J. M., O'Mahony, M., Iodi-Carstens, M., et al. (2001). Oral irritation by sodium chloride: sensitization, self-desensitization, and cross-sensitization to capsaicin. *Physiol. Behav.* 72: 317–324.

Dessirier, J. M., O'Mahony, M., Sieffermann, J. M. and Carstens, E. (1998). Mecamylamine inhibits nicotine but not capsaicin irritation on the tongue: psychophysical evidence that nicotine and capsaicin activate separate molecular receptors. *Neurosci. Lett.* 240: 65–68.

Dessirier, J. M., Simons, C. T., Sudo, M., et al. (2000c). Sensitization, desensitization and stimulus-induced recovery of trigeminal neuronal responses to oral capsaicin and nicotine. *J. Neurophysiol.* 84: 1851–1862.

Dhaka, A., Murray, A. N., Mathur, J., et al. (2007). TRPM8 is required for cold sensation in mice. *Neuron* 54: 371–378.

Doty, R. L. (1975). Intranasal trigeminal detection of chemical vapors by humans. *Physiol. Behav.* 14: 855–859.

Doty, R. L., Brugger, W. E., Jurs, P. C., et al. (1978). Intranasal trigeminal stimulation from odorous volatiles: psychometric responses from anosmic and normal humans. *Physiol. Behav.* 20: 175–185.

Doty, R. L. and Cometto-Muñiz, J. E. (2003). Trigeminal Chemosensation. In: *Handbook of Olfaction and Gustation. 2nd Edition*, Doty, R.L. (Ed.). New York: Marcel Dekker, pp. 981–1000.

Doty, R. L., Cometto-Muñiz, J. E., Jalowayski, A. A., et al. (2004). Assessment of upper respiratory tract and ocular irritative effects of volatile chemicals in humans. *Crit. Rev. Toxicol.* 34: 85–142.

Dunn, J. D., Cometto-Muñiz, J. E. and Cain, W. S. (1982). Nasal reflexes: Reduced sensitivity to CO2 irritation in cigarette smokers. *J. Appl. Toxicol.* 2: 176–178.

Eccles, R. (1994). Menthol and related cooling compounds. *J. Pharm. Pharmacol.* 46: 618–630.

Feron, V. J., Cassee, F. R., Groten, J. P., et al. (2002). International issues on human health effects of exposure to chemical mixtures. *Environ. Health Perspect.* 110 Suppl 6: 893–899.

Finger, T. E., Bottger, B., Hansen, A., et al. (2003). Solitary chemoreceptor cells in the nasal cavity serve as sentinels of respiration. *Proc. Natl. Acad. Sci. U. S. A.* 100: 8981–8986.

Franks, N. P. and Lieb, W. R. (1985). Mapping of general anaesthetic target sites provides a molecular basis for cutoff effects. *Nature* 316: 349–351.

Franks, N. P. and Lieb, W. R. (1990). Mechanisms of general anesthesia. *Environ. Health Perspect.* 87: 199–205.

Frasnelli, J., Albrecht, J., Bryant, B. and Lundstrom, J. N. (2011a). Perception of specific trigeminal chemosensory agonists. *Neuroscience* 189: 377–383.

Frasnelli, J., Hummel, T., Berg, J., et al. (2011b). Intranasal localizability of odorants: influence of stimulus volume. *Chem. Senses* 36: 405–410.

Frasnelli, J., Lotsch, J. and Hummel, T. (2003). Event-related potentials to intranasal trigeminal stimuli change in relation to stimulus concentration and stimulus duration. *J. Clin. Neurophysiol.* 20: 80–86.

Frasnelli, J., Schuster, B. and Hummel, T. (2007a). Interactions between olfaction and the trigeminal system: what can be learned from olfactory loss. *Cereb. Cortex* 17: 2268–2275.

Frasnelli, J., Schuster, B. and Hummel, T. (2007b). Subjects with congenital anosmia have larger peripheral but similar central trigeminal responses. *Cereb. Cortex* 17: 370–377.

Frasnelli, J., Schuster, B. and Hummel, T. (2010). Olfactory dysfunction affects thresholds to trigeminal chemosensory sensations. *Neurosci. Lett.* 468: 259–263.

Frasnelli, J., Schuster, B., Zahnert, T. and Hummel, T. (2006). Chemosensory specific reduction of trigeminal sensitivity in subjects with olfactory dysfunction. *Neuroscience* 142: 541–546.

Furrer, S., Slack, J., McCluskey, S. et al. (2008). New Developments in the chemistry of cooling compounds. *Chemosensory Perception* 1: 119–126

Galopin, C. C., Furrer, S. M., and Goeke, A. (2002). Pungent and tingling compounds in Asian cuisine. In: *ACS Symposium Series: Challenges in Taste Chemistry and Biology, Volume 867*, Hofmann T., Ho C.T., and Pickenhagen W. (Eds.). Washington DC: American Chemical Society, pp. 139–152.

García Medina, M. R. and Cain, W. S. (1982). Bilateral integration in the common chemical sense. *Physiol. Behav.* 29: 349–353.

Garle, M. J. and Fry, J. R. (2003). Sensory nerves, neurogenic inflammation and pain: missing components of alternative irritation strategies? A review and a potential strategy. *Altern. Lab. Anim.* 31: 295–316.

Gilbertson, T. A., Fontenot, D. T., Liu, L., et al. (1997). Fatty acid modulation of K+ channels in taste receptor cells: gustatory cues for dietary fat. *Am. J. Physiol.* 272: C1203–C1210.

Godden, E. L., Harris, R. A. and Dunwiddie, T. V. (2001). Correlation between molecular volume and effects of n-alcohols on human neuronal nicotinic acetylcholine receptors expressed in Xenopus oocytes. *J. Pharmacol. Exp. Ther.* 296: 716–722.

Green, B. G. (1988). Spatial and temporal factors in the perception of ethanol irritation on the tongue. *Percept. Psychophys.* 44: 108–116.

Green, B. G. (1989). Capsaicin sensitization and desensitization on the tongue produced by brief exposures to a low concentration. *Neurosci. Lett.* 107: 173–178.

Green, B. G. (1990). Sensory characteristics of camphor. *J. Invest. Dermatol.* 94: 662–666.

Green, B. G. (1991). Temporal characteristics of capsaicin sensitization and desensitization on the tongue. *Physiol. Behav.* 49: 501–505.

Green, B. G. (1996). Rapid recovery from capsaicin desensitization during recurrent stimulation. *Pain* 68: 245–253.

Green, B., Mason, J. and Kare, M., (eds) (1990a). *Chemical Senses. Volume 2. Irritation.* Marcel Dekker, Inc., New York.

Green, B. G. and Lawless, H. T. (1991). The psychophysics of somatosensory chemoreception in the nose and mouth. In: *Smell and Taste in Health and Disease*, Getchell, T.V., Doty R.L., Bartoshuk, L.M. and Snow Jr.,, J.B. (Eds.). New York: Raven Press, pp. 235–253.

Green, B. G., Mason, J. R. and Kare, M. R. (1990b). Preface. In: *Chemical Senses, Volume 2: Irritation*, Green, B. G., Mason, J. R. and Kare, M. R. (Eds.). New York: Marcel Dekker, Inc., pp. v–vii.

Green, B. G. and Rentmeister-Bryant, H. (1998). Temporal characteristics of capsaicin desensitization and stimulus-induced recovery in the oral cavity. *Physiol. Behav.* 65: 141–149.

Gudziol, H., Schubert, M. and Hummel, T. (2001). Decreased trigeminal sensitivity in anosmia. *ORL. J. Otorhinolaryngol. Relat. Spec.* 63: 72–75.

Gulbransen, B., Silver, W. and Finger, T. E. (2008a). Solitary chemoreceptor cell survival is independent of intact trigeminal innervation. *J. Comp. Neurol.* 508: 62–71.

Gulbransen, B. D., Clapp, T. R., Finger, T. E. and Kinnamon, S. C. (2008b). Nasal solitary chemoreceptor cell responses to bitter and trigeminal stimulants in vitro. *J. Neurophysiol.* 99: 2929–2937.

Hau, K. M., Connell, D. W. and Richardson, B. J. (1999). Quantitative structure-activity relationships for nasal pungency thresholds of volatile organic compounds. *Toxicol. Sci.* 47: 93–98.

Hempel-Jorgensen, A., Kjaergaard, S. K., Molhave, L. and Hudnell, K. H. (1999). Sensory eye irritation in humans exposed to mixtures of volatile organic compounds. *Arch. Environ. Health* 54: 416–424.

Hinman, A., Chuang, H. H., Bautista, D. M. and Julius, D. (2006). TRP channel activation by reversible covalent modification. *Proc. Natl. Acad. Sci. U. S. A.* 103: 19564–19568.

Horishita, T. and Harris, R. A. (2008). n-Alcohols inhibit voltage-gated Na+ channels expressed in Xenopus oocytes. *J. Pharmacol. Exp. Ther.* 326: 270–277.

Hudnell, H. K., Otto, D. A., House, D. E. and Molhave, L. (1992). Exposure of humans to a volatile organic mixture II. Sensory.. *Arch. Environ. Health* 47: 31–38.

Hummel, T., Barz, S., Lotsch, J., et al. (1996a). Loss of olfactory function leads to a decrease of trigeminal sensitivity. *Chem. Senses* 21: 75–79.

Hummel, T., Futschik, T., Frasnelli, J. and Huttenbrink, K. B. (2003). Effects of olfactory function, age, and gender on trigeminally mediated sensations: a study based on the lateralization of chemosensory stimuli. *Toxicol. Lett.* 140–141: 273–280.

Hummel, T., Gruber, M., Pauli, E. and Kobal, G. (1994). Chemo-somatosensory event-related potentials in response to repetitive painful chemical stimulation of the nasal mucosa. *Electroencephalogr. Clin. Neurophysiol.* 92: 426–432.

Hummel, T. and Kobal, G. (1999). Chemosensory event-related potentials to trigeminal stimuli change in relation to the interval between repetitive stimulation of the nasal mucosa. *Eur. Arch. Otorhinolaryngol.* 256: 16–21.

Hummel, T., Schiessl, C., Wendler, J. and Kobal, G. (1996b). Peripheral electrophysiological responses decrease in response to repetitive painful stimulation of the human nasal mucosa. *Neurosci. Lett.* 212: 37–40.

Hursel, R. and Westerterp-Plantenga, M. S. (2010). Thermogenic ingredients and body weight regulation. *Int. J. Obes (Lond).* 34: 659–669.

Inoue, R. (2005). TRP channels as a newly emerging non-voltage-gated CA2+ entry channel superfamily. *Curr. Pharm. Des.* 11: 1899–1914.

Inoue, T. and Bryant, B. P. (2005). Multiple types of sensory neurons respond to irritating volatile organic compounds (VOCs): calcium fluorimetry of trigeminal ganglion neurons. *Pain* 117: 193–203.

Jansco, N., Jansco-Gabor, A. and Takats, I. (1961). Pain and inflammation induced by nicotine, acetylcholine and structurally related compounds and their prevention by desensitizing agents. *Acta. Physiol.* 19: 113–132.

Jordt, S. E., Bautista, D. M., Chuang, H. H., et al. (2004). Mustard oils and cannabinoids excite sensory nerve fibres through the TRP channel ANKTM1. *Nature* 427: 260–265.

Karrer, T. and Bartoshuk, L. (1991). Capsaicin desensitization and recovery on the human tongue. *Physiol. Behav.* 49: 757–764.

Kassab, A., Schaub, F., Vent, J., et al. (2009). Effects of short inter-stimulus intervals on olfactory and trigeminal event-related potentials. *Acta Otolaryngol.* 129: 1250–1256.

Keele, C. A. (1962). The common chemical sense and its receptors. *Arch. Int. Pharmacodyn. Ther.* 139: 547–557.

Kendal-Reed, M., Walker, J. C., Morgan, W. T., et al. (1998). Human responses to propionic acid. I. Quantification of within- and between-participant variation in perception by normosmics and anosmics. *Chem. Senses* 23: 71–82.

Kim, D. and Baraniuk, J. N. (2007). Sensing the air around us: the voltage-gated-like ion channel family. *Curr. Allergy Asthma Rep.* 7: 85–92.

Klein, A. H., Carstens, M. I., Zanotto, K. L., et al. (2011). Self- and cross-desensitization of oral irritation by menthol and cinnamaldehyde (CA) via peripheral interactions at trigeminal sensory neurons. *Chem Senses* 36: 199–208.

Kobal, G. (1985). Pain-related electrical potentials of the human nasal mucosa elicited by chemical stimulation. *Pain* 22: 151–163.

Kobal, G. and Hummel, C. (1988). Cerebral chemosensory evoked potentials elicited by chemical stimulation of the human olfactory and respiratory nasal mucosa. *Electroencephalogr. Clin. Neurophysiol.* 71: 241–250.

Kobal, G. and Hummel, T. (1998). Olfactory and intranasal trigeminal event-related potentials in anosmic patients. *Laryngoscope* 108: 1033–1035.

Kobal, G., Van Toller, S. and Hummel, T. (1989). Is there directional smelling? *Experientia* 45: 130–132.

Koo, J. Y., Jang, Y., Cho, H., et al. (2007). Hydroxy-alpha-sanshool activates TRPV1 and TRPA1 in sensory neurons. *Eur. J. Neurosci.* 26: 1139–1147.

Korpi, A., Jarnberg, J. and Pasanen, A. L. (2009). Microbial volatile organic compounds. *Crit. Rev. Toxicol.* 39: 139–193.

Lambert, D. (2004). Drugs and receptors. *Contin. Educ. Anaesth. Crit. Care Pain* 4: 181–184.

Laumbach, R. J., Fiedler, N., Gardner, C. R., et al. (2005). Nasal effects of a mixture of volatile organic compounds and their ozone oxidation products. *J. Occup. Environ. Med.* 47: 1182–1189.

Laugerette, F., Passilly-Degrace, P., Patris, B., et al. (2005). CD36 involvement in orosensory detection of dietary lipids, spontaneous fat preference, and digestive secretions. *J. Clin. Invest.* 115: 3177–3184.

Lawless, H. (1984). Oral chemical irritation: Psychophysical properties. *Chem. Senses* 9: 143–155.

Lawless, H. T. and Stevens, D. A. (1988). Responses by humans to oral chemical irritants as a function of locus of stimulation. *Percep. Psychophys.* 43: 72–78.

Lawless, H. T. and Stevens, D. A. (1990). Differences between and interactions of oral irritants. In: *Chemical Senses, Volume 2: Irritation*, Green, B.G., Mason, J.R. and Kare, M.R. (Eds.). New York: Marcel Dekker, Inc., pp. 197–216

Lee, Y., Lee, C. H. and Oh, U. (2005). Painful channels in sensory neurons. *Mol. Cells* 20: 315–324.

Lennertz, R. C., Tsunozaki, M., Bautista, D. M. and Stucky, C. L. (2010). Physiological basis of tingling paresthesia evoked by hydroxy-alpha-sanshool. *J. Neurosci.* 30: 4353–4361.

Ley, J. P., Krammer, G., Looft, J., et al. (2006). Structure-activity relationships of trigeminal effects for artificial and naturally occurring alkamides related to spilanthol. *Dev. Food Sci.* 43: 21–24.

Ley, J. P and Simchen, U. (2007). Quantification of the saliva-inducing properties of pellitorine and spilanthole. *Recent Highlights Flavor Chem. Biol.* 365–368.

Lin, W., Ogura, T., Margolskee, R. F., Finger, T. E. and Restrepo, D. (2008). TRPM5-expressing solitary chemosensory cells respond to odorous irritants. *J. Neurophysiol.* 99: 1451–1460.

Liu, L., Oortgiesen, M., Li, L. and Simon, S. A. (2001). Capsaicin inhibits activation of voltage-gated sodium currents in capsaicin-sensitive trigeminal ganglion neurons. *J. Neurophysiol.* 85: 745–758.

Liu, L. and Simon, S. A. (1996). Similarities and differences in the currents activated by capsaicin, piperine, and zingerone in rat trigeminal ganglion cells. *J Neurophysiol.* 76: 1858–1869.

Liu, L. and Simon, S. A. (1998). The influence of removing extracellular Ca2+ in the desensitization responses to capsaicin, zingerone and olvanil in rat trigeminal ganglion neurons. *Brain Res.* 809: 246–252.

LoPachin, R. M., Barber, D. S. and Gavin, T. (2008). Molecular mechanisms of the conjugated alpha,beta-unsaturated carbonyl derivatives: relevance to neurotoxicity and neurodegenerative diseases. *Toxicol. Sci.* 104: 235–249.

Macpherson, L. J., Dubin, A. E., Evans, M. J., et al. (2007a). Noxious compounds activate TRPA1 ion channels through covalent modification of cysteines. *Nature* 445: 541–545.

Macpherson, L. J., Geierstanger, B. H., Viswanath, V., et al. (2005). The pungency of garlic: activation of TRPA1 and TRPV1 in response to allicin. *Curr. Biol.* 15: 929–934.

Macpherson, L. J., Hwang, S. W., Miyamoto, T., et al. (2006). More than cool: promiscuous relationships of menthol and other sensory compounds. *Mol. Cell. Neurosci.* 32: 335–343.

Macpherson, L. J., Xiao, B., Kwan, K. Y., et al. (2007b). An ion channel essential for sensing chemical damage. *J. Neurosci.* 27: 11412–11415.

Matsumura, S., Mizushige, T., Yoneda, T., et al. (2007). GPR expression in the rat taste bud relating to fatty acid sensing. *Biomed. Res.* 28: 49–55.

Mattes, R. D. (2011). Accumulating evidence supports a taste component for free fatty acids in humans. *Physiol. Behav.* 104: 624–631.

McKemy, D. D., Neuhausser, W. M. and Julius, D. (2002). Identification of a cold receptor reveals a general role for TRP channels in thermosensation. *Nature* 416: 52–58.

McNamara, C. R., Mandel-Brehm, J., Bautista, D. M., et al. (2007). TRPA1 mediates formalin-induced pain. *Proc. Natl. Acad. Sci. U. S. A.* 104: 13525–13530.

McNamara, F. N., Randall, A. and Gunthorpe, M. J. (2005). Effects of piperine, the pungent component of black pepper, at the human vanilloid receptor (TRPV1). *Br. J. Pharmacol.* 144: 781–790.

Meininghaus, R., Kouniali, A., Mandin, C. and Cicolella, A. (2003). Risk assessment of sensory irritants in indoor air--a case study in a French school. *Environ. Int.* 28: 553–557.

Menozzi-Smarrito, C., Riera, C. E., Munari, C., et al. (2009). Synthesis and evaluation of new alkylamides derived from alpha-hydroxysanshool, the pungent molecule in szechuan pepper. *J. Agric. Food. Chem.* 11: 1982–1989.

Millqvist, E., Bengtsson, U. and Lowhagen, O. (1999). Provocations with perfume in the eyes induce airway symptoms in patients with sensory hyperreactivity. *Allergy* 54: 495–499.

Moalem, G., Grafe, P. and Tracey, D. J. (2005). Chemical mediators enhance the excitability of unmyelinated sensory axons in normal and injured peripheral nerve of the rat. *Neuroscience* 134: 1399–1411.

Moqrich, A., Hwang, S. W., Earley, T. J., et al. (2005). Impaired thermosensation in mice lacking TRPV3, a heat and camphor sensor in the skin. *Science* 307: 1468–1472.

Muller, J. and Greff, G. (1984). Recherche de relations entre toxicite de molecules d'interet industriel et proprietes physico-chimiques: test d'irritation des voies aeriennes superieures applique a quatre familles chimiques. *Food Chem. Toxicol.* 22: 661–664.

Müller, T., Thissen, R., Braun, S., Dott, W. and Fische,r G. (2004). (M)VOC and composting facilities. Part 2: (M)VOC dispersal in the environment. *Environ. Sci. Pollut. Res. Int.* 11: 152–157.

Nelson, G., Chandrashekar, J., Hoon, M. A., et al. (2002). An amino-acid taste receptor. *Nature* 416: 199–202.

Nelson, G., Hoon, M. A., Chandrashekar, J., et al. (2001). Mammalian sweet taste receptors. *Cell* 106: 381–390.

Nielsen, G. D., Hansen, L. F. and Alarie, Y. (1992). Irritation of the upper airways. Mechanisms and structure-activity relationships. In: *Chemical, microbiological, health and comfort aspects of indoor air quality — State of the art in SBS*, Knöppel, H. and Wolkoff, P. (Eds.). Dordrecht: Kluwer Academic Publishers, pp. 99–114.

Nielsen, G. D., Thomsen, E. S. and Alarie, Y. (1990). Sensory irritant receptor compartment properties. Equipotent vapour concentrations related to saturated vapour concentrations, octanol-water, and octanol-gas partition coefficients. *Acta Pharm. Nord.* 2: 31–44.

Ohmoto, M., Matsumoto, I., Yasuoka, A., et al. (2008). Genetic tracing of the gustatory and trigeminal neural pathways originating from T1R3-expressing taste receptor cells and solitary chemoreceptor cells. *Mol. Cell. Neurosci.* 38: 505–517.

Omote, K., Kawamata, T., Kawamata, M. and Namiki, A. (1998). Formalin-induced nociception activates a monoaminergic descending inhibitory system. *Brain Res.* 814: 194–198.

Opiekun, R. E., Smeets, M., Sulewski, M., et al. (2003). Assessment of ocular and nasal irritation in asthmatics resulting from fragrance exposure. *Clin. Exp. Allergy* 33: 1256–1265.

Osculati, F., Bentivoglio, M., Castellucci, M., et al. (2007). The solitary chemosensory cells and the diffuse chemosensory system of the airway. *Eur. J. Histochem.* 51 Suppl 1: 65–72.

Otto, D., Molhave, L., Rose, G., et al. (1990). Neurobehavioral and sensory irritant effects of controlled exposure to a complex mixture of volatile organic compounds. *Neurotoxicol. Teratol.* 12: 649–652.

Ozçelik, B., Kartal, M. and Orhan, I. (2011). Cytotoxicity, antiviral and antimicrobial activities of alkaloids, flavonoids, and phenolic acids. *Pharm. Biol.* 49: 396–402.

Parker, G. H. (1912). The relation of smell, taste, and the common chemical sense in vertebrates. *J. Acad. Nat. Sci. Phila.* 15: 219–234.

Patapoutian, A., Tate, S. and Woolf, C. J. (2009). Transient receptor potential channels: targeting pain at the source. *Nat. Rev. Drug. Discov.* 8: 55–68.

Peng, J. and Li, Y. J. (2010). The vanilloid receptor TRPV1: role in cardiovascular and gastrointestinal protection. *Eur. J. Pharmacol.* 627: 1–7.

Peyrot des Gachons, C., Uchida, K., Bryant, B., et al. (2011). Unusual pungency from extra-virgin olive oil is attributable to restricted spatial expression of the receptor of oleocanthal. *J. Neurosci.* 31: 999–1009.

Peterlin, Z., Chesler, A. and Firestein, S. (2007). A painful trp can be a bonding experience. *Neuron* 53: 635–638.

Pfaffmann, C. (1959). The sense of taste. *Handbook of Physiology: Neurophysiology.* Field J., Magoun H. W., Hall V. E. (eds). (American Physiological Society, Washington, DC), Vol 1, pp 507–533.

Prescott, J. (1999). The generalizability of capsaicin sensitization and desensitization. *Physiol. Behav.* 66: 741–749.

Prescott J. and Stevenson, R. J. (1996a). Psychophysical responses to single and multiple presentations of the oral irritant zingerone: relationship to frequency of chili consumption. *Physiol. Behav.* 60: 617–624.

Prescott, J. and Stevenson, R. J. (1996b). Desensitization to oral zingerone irritation: effects of stimulus parameters. *Physiol. Behav.* 60: 1473–1480.

Prescott, J. and Swain-Campbell, N. (2000). Responses to repeated oral irritation by capsaicin, cinnamaldehyde and ethanol in PROP tasters and non-tasters. *Chem. Senses* 25: 239–246.

Rentmeister-Bryant, H. and Green, B. G. (1997). Perceived irritation during ingestion of capsaicin or piperine: comparison of trigeminal and non-trigeminal areas. *Chem. Senses* 22: 257–266.

Riera, C. E., Menozzi-Smarrito, C., Affolter, M., et al. (2009). Compounds from Sichuan and Melegueta peppers activate, covalently and non-covalently, TRPA1 and TRPV1 channels. *Br. J. Pharmacol.* 157: 1398–1409.

Roberts, D. W. (1986). QSAR for upper-respiratory tract irritation. *Chem. Biol. Interactions* 57: 325–345.

Rozin, P. and Schiller, D. (1980). The nature and acquisition of a preference for chili pepper by humans. *Motivation and Emotion* 4: 77–101.

Salazar, H., Llorente, I., Jara-Oseguera, A., et al. (2008). A single N-terminal cysteine in TRPV1 determines activation by pungent compounds from onion and garlic. *Nat. Neurosci.* 11: 255–261.

Sbarbati, A., Bramanti, P., Benati, D. and Merigo, F. (2010). The diffuse chemosensory system: exploring the iceberg toward the definition of functional roles. *Prog. Neurobiol.* 91: 77–89.

Sbarbati, A., Merigo, F., Benati, D., et al. (2004). Laryngeal chemosensory clusters. *Chem. Senses* 29: 683–692.

Schiffman, S. S., Walker, J. M., Dalton, P., et al. (2000). Potential health effects of odor from animal operations, wastewater treatment, and recycling of byproducts. *Journal of Agromedicine* 7: 7–81.

Schiffman, S. S. and Williams, C. M. (2005). Science of odor as a potential health issue. *J. Environ. Qual.* 34: 129–138.

Schneider, R. A. and Schmidt, C. E. (1967). Dependency of olfactory localization on non-olfactory cues. *Physiol. Behav.* 2: 305–309.

Sessle, B. J. (1989). Neural mechanisms of oral and facial pain. *Otolaryngol. Clin. North Am.* 22: 1059–1072.

Sherkheli, M. A., Vogt-Eisele, A. K., Bura, D., et al. (2010). Characterization of selective TRPM8 ligands and their structure activity response (S.A.R) relationship. *J. Pharm. Pharm. Sci.* 13: 242–253.

Shusterman, D., Matovinovic, E. and Salmon, A. (2006). Does Haber's law apply to human sensory irritation? *Inhal. Toxicol.* 18: 457–471.

Silver, W. L. (1987). The common chemical sense. In: *Neurobiology of Taste and Smell*, Finger, T.E. and Silver, W.L. (Eds.). New York: Wiley, pp. 65–87.

References

Silver, W. L., Clapp, T. R., Stone, L. M. and Kinnamon, S. C. (2006). TRPV1 Receptors and Nasal Trigeminal Chemesthesis. *Chem. Senses* 31: 807–812.

Silver, W. L. and Finger, T. E. (2009). The anatomical and electrophysiological basis of peripheral nasal trigeminal chemoreception. *Ann. N. Y. Acad. Sci.* 1170: 202–205.

Silver, W. L. and Finger, T. E. (1991). The trigeminal system. In: *Smell and Taste in Health and Disease*, Getchell, T.V., Doty, R.L., Bartoshuk, L.M. and Snow Jr., J.B. (Eds.). New York: Raven Press, pp. 97–108.

Simons, C. T., Carstens, M. I. and Carstens, E. (2003a). Oral irritation by mustard oil: self-desensitization and cross-desensitization with capsaicin. *Chem Senses* 28: 459–465.

Simons, C. T., Dessirier, J. M., Carstens, M. I., et al. (1999). Neurobiological and psychophysical mechanisms underlying the oral sensation produced by carbonated water. *J. Neurosci.* 19: 8134–8144.

Simons, C. T., Sudo, S., Sudo, M. and Carstens, E. (2003b). Mecamylamine reduces nicotine cross-desensitization of trigeminal caudalis neuronal responses to oral chemical irritation. *Brain Res.* 991: 249–253.

Smeets, M. and Dalton, P. (2002). Perceived odor and irritation of isopropanol: a comparison between naive controls and occupationally exposed workers. *Int. Arch. Occup. Environ. Health* 75: 541–548.

Stevens, D. A. and Lawless, H. T. (1987). Enhancement of responses to sequential presentation of oral chemical irritants. *Physiol. Behav.* 39: 63–65.

Stone, H., Carregal, E. J. and Williams, B. (1966). The olfactory trigeminal response to odorants. *Life Sci.* 5: 2195–2201.

Stone, H. and Rebert, C. S. (1970). Observations on trigeminal olfactory interactions. *Brain Res.* 21: 138–142.

Stone, H., Williams, B. and Carregal, E. J. (1968). The role of the trigeminal nerve in olfaction. *Exp. Neurol.* 21: 11–19.

Stone, L. M., Finger, T. E., Tam, P. P. and Tan, S. S. (1995). Taste receptor cells arise from local epithelium, not neurogenic ectoderm. *Proc Natl Acad Sci U S A.* 92: 1916–1920.

Su, X., Wachtel, R. E. and Gebhart, G. F. (1999). Capsaicin sensitivity and voltage-gated sodium currents in colon sensory neurons from rat dorsal root ganglia. *Am. J. Physiol.* 277: G1180–G1188.

Sudo, S., Sudo, M., Simons, C. T., et al. (2002). Sensitization of trigeminal caudalis neuronal responses to intraoral acid and salt stimuli and desensitization by nicotine. *Pain* 98: 277–286.

Sugai, E., Morimitsu, Y., Iwasaki, Y., et al. (2005). Pungent qualities of sanshool-related compounds evaluated by a sensory test and activation of rat TRPV1. *Biosci. Biotechnol. Biochem.* 69: 1951–1957.

Szolcsányi, J (1990). Capsaicin, irritation and desensitization. Neurophysiological basis and future perspectives. In: BR Green, JR Mason, MR Kare (eds): *Chemical senses, Vol. 2. Irritation.* Marcel Dekker, New York, 141–168

Szolcsányi, J and Jancsó-Gábor, A. (1975). Sensory effects of capsaicin congeners I. Relationship between chemical structure and pain-producing potency of pungent agents. *Arzneimittelforschung* 25: 1877–1881.

Szallasi, A. and Blumberg, P. M. (1999). Vanilloid (Capsaicin) receptors and mechanisms. *Pharmacol. Rev.* 51: 159–212.

Talavera, K., Nilius, B. and Voets, T. (2008). Neuronal TRP channels: thermometers, pathfinders and lifesavers. *Trends Neurosci.* 31: 287–295.

Talavera, K., Gees, M., Karashima, Y., et al. (2009). Nicotine activates the chemosensory cation channel TRPA1. *Nat. Neurosci.* 12: 1293–1299.

Thuerauf, N., Markovic, K., Braun, G., et al. (2006). The influence of mecamylamine on trigeminal and olfactory chemoreception of nicotine. *Neuropsychopharmacology* 31: 450–461.

Thürauf, N., Hummel, T., Kettenmann, B. and Kobal, G. (1993). Nociceptive and reflexive responses recorded from the human nasal mucosa. *Brain Res.* 629: 293–299.

Tizzano, M., Cristofoletti, M., Sbarbati, A. and Finger, T. E. (2011). Expression of taste receptors in solitary chemosensory cells of rodent airways. *BMC Pulm. Med.* 11: 3.

Tizzano, M., Gulbransen, B. D., Vandenbeuch, A., et al. (2010). Nasal chemosensory cells use bitter taste signaling to detect irritants and bacterial signals. *Proc. Natl. Acad. Sci. U. S. A.* 107: 3210–3215.

Tominaga, M. and Tominaga, T. (2005). Structure and function of TRPV1. *Pflugers Arch.* 451: 143–150.

Trevisani, M., Patacchini, R., Nicoletti, P., et al. (2005). Hydrogen sulfide causes vanilloid receptor 1-mediated neurogenic inflammation in the airways. *Br. J. Pharmacol.* 145: 1123–1131.

Trevisani, M., Smart, D., Gunthorpe, M. J., et al. (2002). Ethanol elicits and potentiates nociceptor responses via the vanilloid receptor-1. *Nat. Neurosci.* 5: 546–551.

Ulrich, C. E., Alarie, Y. and Haddock, M. P. (1972). Airborne Chemical Irritants - Role of trigeminal nerve. *Arch. Environ. Health* 24: 37.

Vaughan, R. P., Szewczyk, M. T., Jr., Lanosa, M. J., et al. (2006). Adenosine sensory transduction pathways contribute to activation of the sensory irritation response to inspired irritant vapors. *Toxicol. Sci.* 93: 411–421.

von Békesy, G. (1964). Olfactory analogue to directional hearing. *J. Appl. Physiol.* 19: 369–373.

von Skramlik, E. (1925). Über die Lokalisation der Empfindungen bei den niederen Sinnen. *Zeitschr. Sinnesphysiol. (II. Abteilung)* 56: 69–140.

Wang, Y. Y., Chang, R. B., Allgood, S. D., et al. (2011). A TRPA1-dependent mechanism for the pungent sensation of weak acids. *J. Gen. Physiol.* 137: 493–505.

Wang, Y. Y., Chang, R. B. and Liman, E. R. (2010). TRPA1 is a component of the nociceptive response to CO_2. *J. Neurosci.* 30: 12958–12963.

Waters, A. J. and Lumb, B. M. (2008). Descending control of spinal nociception from the periaqueductal grey distinguishes between neurons with and without C-fibre inputs. *Pain* 134: 32–40.

Watson, H. R., Hems, R., Rowsell, D. G. and Spring, D. J. (1978). New compounds with the menthol cooling effect. *J. Soc. Cosmet. Chem.* 29:185–200

Wei, E. T. and Seid, D. A. (1983). AG-3-5a chemical producing sensations of cold. *J. Pharm. Pharmacol.* 35: 110–112.

Whitehead, M. C. (1988). Neuroanatomy of the gustatory system. *Gerodontics* 4: 239–243.

Willis, W. D. Jr., (2009). The role of TRPV1 receptors in pain evoked by noxious thermal and chemical stimuli. *Exp. Brain Res.* 196: 5–11.

Wise, P. M., Canty, T. M. and Wysocki, C. J. (2005). Temporal integration of nasal irritation from ammonia at threshold and supra-threshold levels. *Toxicol. Sci.* 87: 223–231.

Wise, P. M., Canty, T. M. and Wysocki, C. J. (2006). Temporal integration in nasal lateralization of ethanol. *Chem. Senses* 31: 227–235.

Wise, P. M., Radil, T. and Wysocki, C.J. (2004). Temporal integration in nasal lateralization and nasal detection of carbon dioxide. *Chem. Senses* 29: 137–142.

Wise, P. M., Toczydlowski, S. E. and Wysocki, C.J. (2007). Temporal integration in nasal lateralization of homologous alcohols. *Toxicol. Sci.* 99: 254–259.

Wise, P. M., Toczydlowski, S. E., Zhao, K. and Wysocki, C. J. (2009a). Temporal integration in nasal lateralization of homologous propionates. *Inhal. Toxicol.* 21: 819–827.

Wise, P. M., Wysocki, C. J. and Radil, T. (2003). Time-intensity ratings of nasal irritation from carbon dioxide. *Chem. Senses* 28: 751–760.

Wise, P. M., Zhao, K. and Wysocki, C. J. (2009b). Dynamics of nasal chemesthesis. *Ann. N. Y. Acad. Sci.* 1170: 206–214.

Wise P. M., Zhao K. and Wysocki C.J. (2010). Dynamics of nasal irritation from pulsed homologous alcohols. *Chem. Senses* 35: 823–829.

Wolkoff, P., Nojgaard, J. K., Franck, C. and Skov, P. (2006). The modern office environment desiccates the eyes? *Indoor Air* 16: 258–265.

Wolkoff, P., Skov, P., Franck, C. and Petersen, L. N. (2003). Eye irritation and environmental factors in the office environment--hypotheses, causes and a physiological model. *Scand. J. Work. Environ. Health* 29: 411–430.

Wood, J. N. and Docherty, R. (1997). Chemical activators of sensory neurons. *Annu. Rev. Physiol.* 59: 457–482.

Woodard, G. E., Sage, S. O. and Rosado, J. A. (2007). Transient receptor potential channels and intracellular signaling. *Int. Rev. Cytol.* 256: 35–67.

Wysocki, C. J., Dalton, P., Brody, M. J. and Lawley, H. J. (1997). Acetone odor and irritation thresholds obtained from acetone-exposed factory workers and from control (occupationally unexposed) subjects. *Am. Ind. Hyg. Assoc. J.* 58: 704–712.

Xing, H., Ling, J., Chen, M. and Gu, J. G. (2006). Chemical and cold sensitivity of two distinct populations of TRPM8-expressing somatosensory neurons. *J. Neurophysiol.* 95: 1221–1230.

Xu, H., Delling, M., Jun, J. C. and Clapham, D. E. (2006). Oregano, thyme and clove-derived flavors and skin sensitizers activate specific TRP channels. *Nat. Neurosci.* 9: 628–635.

Yao, J. and Qin, F. (2009). Interaction with phosphoinositides confers adaptation onto the TRPV1 pain receptor. *PLoS Biol.* 7: e46.

Yarmolinsky, D. A., Zuker, C. S. and Ryba, N. J. (2009). Common sense about taste: from mammals to insects. *Cell* 139: 234–244.

Zanotto, K. L., Merrill, A. W., Carstens, M. I. and Carstens, E. (2007). Neurons in superficial trigeminal subnucleus caudalis responsive to oral cooling, menthol, and other irritant stimuli. *J. Neurophysiol.* 97: 966–978.

Chapter 51

The Vomeronasal Organ

LISA STOWERS and MARC SPEHR

51.1 INTRODUCTION

In 1813, the Danish anatomist Ludvig L. Jacobson (1783–1843) described a richly innervated mammalian organ that "is located in the foremost part of the nasal cavity, in close contact with the nasal cartilage (septum), on palatal elongations of the intermaxillary bone" (Jacobson et al., 1998). From his comparative investigations in different domesticated and wild mammalian species, Jacobson concluded that "the organ exists in all mammals" and likely functions as a secretory and / or "sensory organ which may be of assistance to the sense of smell" (Jacobson et al., 1998). Eighty-two years after its discovery, the "organ of Jacobson" was (re)named "organon vomeronasale (Jacobsoni)" or vomeronasal organ.

While most mammals have a vomeronasal organ (VNO) as the peripheral sensory structure of an accessory olfactory system (AOS), its enormous morphological diversity complicates extrapolation of findings between species (Salazar and Sánchez-Quinteiro, 2009; Salazar et al., 2007). In the following, we therefore focus on the rodent AOS with particular emphasis on mouse models. Mice have emerged as the best-studied mammalian VNO model system and their genetic amenability allows integrated investigations that span the molecular, cellular and systems level (Chamero et al., 2012).

The VNO is a paired cylindrical, bilaterally symmetrical tubular structure that is located at the base of the anterior nasal septum just above the palate (Halpern and Martinez-Marcos, 2003; Meredith, 1991). The entire blind-ended organ is enclosed in a bony capsule and, depending on the species, opens anteriorly into either the nasal or the oral cavity via the vomeronasal or nasopalatine duct, respectively. Viscous secretions of lateral glands fill the VNO lumen with approximately

0.5 mm^3 mucus (Leinders-Zufall et al., 2009). On its medial side, a crescent-shaped pseudo-stratified neuroepithelium is mainly composed of three cell types – basal cells, sustentacular cells, and mature vomeronasal sensory neurons (VSNs) (Figure 51.1A). Each small bipolar VSN extends a single unbranched apical dendrite that terminates in a microvillous knob-like structure at the epithelial surface. While the dendritic VSN endings float in the lumen's mucus, single long unbranched axons gather into vomeronasal nerve bundles that project dorsally under the septal epithelium through openings of the cribriform plate along the medial side of the olfactory bulb to the glomerular layer of the AOB (Meredith, 1991). In total, the mouse VNO harbors a few hundred thousand VSNs (Rodriguez, 2004) that gain structural and metabolic support from a band of microvillous supporting cells in the most superficial epithelial layer. Remarkably, VSNs are short-lived neurons that are continuously replaced from a local stem cell reservoir even in aged individuals (Brann and Firestein, 2010). This robust regenerative capacity is ensured by a population of adult pluripotent cells – basal cells – that are located both along the basal epithelial membrane and in the marginal zone (Halpern and Martinez-Marcos, 2003).

On the lateral side of the VNO, highly vascularized cavernous tissue harbors a large blood vessel that is extensively innervated by sympathetic nerves from the superior cervical ganglion that enter the posterior VNO along nasopalatine fibers (Figure 51.1a) (Ben-Shaul et al., 2010; Meredith and O'Connell, 1979). Sympathetic activity triggers adrenergic release and, consequently, vasoconstriction of the vascular tissue that results in negative intraluminal pressure. This peristaltic vascular pumping mechanism causes massive fluid entry into the VNO in situations of stress and / or novelty-induced arousal. This way, the VNO can take up relatively non-volatile stimuli such as peptides

---

Handbook of Olfaction and Gustation, Third Edition. Edited by Richard L. Doty.
© 2015 Richard L. Doty. Published 2015 by John Wiley & Sons, Inc.

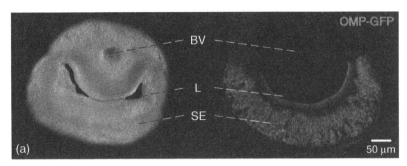

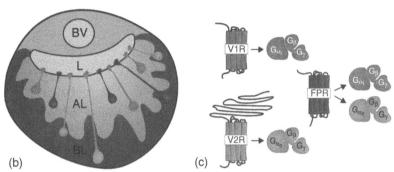

Figure 51.1 The mouse VNO – structure, epithelial dichotomy, and chemosensory GPCRs. (a) Coronal sections reveal the typical VNO anatomy. In both acute vibratome sections (left) and fixed tissue cryosections (right), the crescent shaped sensory epithelium (SE), the lumen (L) and the large lateral blood vessel (BV), that drives the vascular pump, become apparent. (b) schematic diagram illustrating the molecularly distinct apical (AL; yellow) and basal (BL; blue) layers of the sensory neuroepithelium. Color code of individual VSNs indicates chemoreceptor expression as in C. (c) Three classes of vomeronasal chemosensory GPCRs. (*See plate section for color version.*)

or proteins that dissolve in the ventral nasal mucus following direct contact with urine deposits, vaginal secretions, facial gland secretions or saliva (Luo et al., 2003; Wysocki et al., 1980).

Confirmation of Ludvig Jacobson's original hypothesis that the VNO serves a sensory / olfactory function emerged in the late 1960s and early 1970s (Raisman, 1972; Winans and Scalia, 1970) when the neuroanatomical description of segregated parallel projections of the main and accessory olfactory systems to different telencephalic and diencephalic nuclei founded the dual olfactory hypothesis (Scalia and Winans, 1975). Accordingly, the main and accessory systems have traditionally been viewed as both anatomically and functionally separate pathways that detect distinct sets of olfactory cues and trigger different behavioral responses. Today, it has become evident that a clear-cut distinction – the AOS detecting nonvolatile pheromones, while the main system functions as a nonselective molecular analyzer of volatile odorants – is far from true (Spehr et al., 2006b).

AOS ontogeny and postnatal development are best described in the rat and mouse. The rat vomeronasal neuroepithelium is embryologically derived from an evagination of the olfactory placode that can be observed between E12. and E13 (Cuschieri and Bannister, 1975). Based on expression of the olfactory marker protein (OMP), mouse neurons are first identified by E14 (Tarozzo et al., 1998). At E18, the lateral vascular pattern is complete (Szabó and Mendoza, 1988) and the vomeronasal nerve appears fully developed. Thus, all structural VNO components are present at birth and the organ reaches adult size shortly before puberty (Giacobini et al., 2000). Whether the VNO is functional in neonates is less clear. Contrary to previous observations (Coppola and O'Connell, 1989), Hovis and coworkers (Hovis et al., 2012) report that the vomeronasal duct is open and thus allows stimulus access to the sensory neuroepithelium at birth. During the first postnatal week, VSNs form complete microvilli (Mucignat-Caretta, 2010) and release glutamate from their axon terminals (Hovis et al., 2012). It is therefore still subject to debate if the rodent AOS functions in juveniles (Mucignat-Caretta, 2010).

Regulation of VSN development is still poorly understood. Currently, two transcription factors – Mash1 and Bcl11b/Ctip2 – have been attributed critical functions in VSN differentiation (Enomoto et al., 2011; Murray et al., 2003). Genetic ablation of *Mash1* in vomeronasal progenitor cells results in a profound reduction of VSN proliferation in late embryonic and postnatal stages, whereas loss of Bcl11b/Ctip2 function seems to primarily affect postmitotic VSNs. In such newly differentiated neurons, Bcl11b/Ctip2 apparently regulates the fate choice between molecularly distinct VSN types (Enomoto et al., 2011).

51.2 SIGNALING MECHANISMS IN VOMERONASAL NEURONS

51.2.1 Vomeronasal Chemoreceptors

In many, though not all, mammalian species the VNO displays a structural and functional dichotomy (Dulac and

51.2 Signaling Mechanisms in Vomeronasal Neurons

Torello, 2003; Halpern, 1987; Mucignat-Caretta, 2010). Two topographically segregated VSN subpopulations express distinct repertoires of receptors and other putative signaling molecules. VSNs located in the apical layer of the sensory neuroepithelium, closest to the lumen, express the G protein α-subunit $G_{\alpha i2}$ in concert with one member of a multigene family that encodes ~150 intact G protein-coupled receptors (GPCRs) – the V1Rs (Figure 51.1b-c) (Dulac and Axel, 1995; Rodriguez et al., 2002).

While initial attempts to identify putative vomeronasal chemoreceptor genes based on odorant receptor homology were not successful, Dulac and Axel differentially screened cDNA libraries from single rat VSNs. They uncovered a large multigene family of class-A (i.e., rhodopsin-like) GPCRs that are exclusively expressed in vomeronasal tissue (Dulac and Axel, 1995). In mice, this *V1r* family of genes contains more than 150 potentially functional members as well as an approximately equal number of likely pseudogenes (Rodriguez et al., 2002; Roppolo et al., 2007). Antisense probe labeling revealed a punctate, non-overlapping transcription pattern in VSNs restricted to the apical $G_{\alpha i2}$-/PDE4A-positive layer of the sensory epithelium (Berghard and Buck, 1996; Dulac and Axel, 1995; Halpern et al., 1995). *V1r* genes are unusually divergent and polymorphic across alleles. They share intron-free coding regions and are generally clustered on most chromosomes (Capello et al., 2009; Roppolo et al., 2007). Classified into 12 relatively isolated families that each contain between one and 30 members (Rodriguez et al., 2002; Zhang et al., 2004), *V1r* gene choice in a given VSN is tightly controlled. Monogenic, indeed monoallelic, expression ensures that individual VSNs display a single V1R chemodetector "morph" (Rodriguez et al., 1999) and, thus, obtain a distinct functional identity. Deletion of a single 600 kb *V1r* gene cluster on mouse chromosome 6, containing all (but one) *V1r*a and *V1r*b family members, has clearly shown the essential requirement for some V1Rs in mediating VSN chemoresponsivity (Chamero et al., 2012; Del Punta et al., 2002a). In the fifteen years since the discovery of the *V1r* genes the majority of vertebrate V1Rs still remain orphan receptors, so it is technically remarkable that the Dulac group recently succeeded in determining the specificity of 56 individual V1Rs to a range of ethologically relevant cues (Isogai et al., 2011). Using a high-throughput molecular readout of VNO activity mapping, Isogai and coworkers concluded that individual receptor subfamilies have evolved towards the recognition of specific groups of animals or behaviorally relevant chemical structures.

While strict monogenic / monoallelic gene choice is a well-documented hallmark of *V1r* expression (Belluscio et al., 1999; Rodriguez et al., 1999; Roppolo et al., 2007), the underlying molecular mechanisms that ensure only one receptor is expressed by each sensory neuron are far less clear. *V1r* clusters are typically not interrupted by other genes, an arrangement that allows regulatory mechanisms to function at the cluster level. Thus, cluster organization may not just reflect the result of multiple gene-duplication events, but rather play an additional role in expression regulation (Roppolo et al., 2007). Similar to monoallelic and mutually exclusive odorant receptor (OR) gene choice (Lewcock and Reed, 2004; Serizawa et al., 2003), transcription of a nonfunctional *V1r* mutant allele mediates coexpression of a second, functional *V1r* gene. In turn, transcription of this functional *V1r* gene then drives an unknown negative feedback mechanism that maintains monogenic expression. Remarkably, mutant gene transcription silences the *V1r* gene cluster in *cis* relative to the nonfunctional allele, yet allows alternative choice of a functional receptor gene located on the homologous cluster in *trans* (Roppolo et al., 2007). These findings contradict a concept of a molecular "tag" that marks either parental chromosomes / cluster for gene expression. Notably, the negative feedback signal that ensures gene exclusion is also maintained by exogenous expression of an OR gene from a *V1r* locus (Capello et al., 2009), strongly suggesting a common mechanistic basis of monogenic / monoallelic transcription in chemosensory neurons.

Between the apical layer and the basement membrane is a molecularly distinct zone of neurons which form the basal layer. Basal neurons are $G_{\alpha o}$-positive and express members of an unrelated GPCR family – the V2Rs (Figure 51.1b-c) (Herrada and Dulac, 1997; Matsunami and Buck, 1997; Ryba and Tirindelli, 1997). In the mouse genome, about 120 *V2r* genes have apparently intact coding regions, whereas an additional ~160 seem to be pseudogenes (Munger et al., 2009; Young and Trask, 2007). As typical class-C GPCRs, members of the V2R family share a large hydrophobic amino (N)-terminal extracellular domain that is believed to form the primary site of ligand binding (Mombaerts, 2004b; Spehr and Munger, 2009). The *V2r* genes share no significant sequence homology with the *V1r* family, but are distantly related to metabotropic glutamate receptors, Ca^{2+}-sensing receptors, and sweet/umami-sensing *T1r* taste receptors (Dulac and Torello, 2003; Mombaerts, 2004b). Similar to *V1r* genes, *V2r*s are usually found in clusters on most chromosomes and, according to sequence homology, are grouped into four distinct families: family-A, -B, -D, and -C (also known as *V2r2*) (Silvotti et al., 2007; Silvotti et al., 2011; Young and Trask, 2007). While family-A genes comprise the vast majority of *V2r*s with more than 100 members, only 4 genes constitute family-D. Like *V1r* genes, but with the notable exception of family-C receptors, *V2r* genes are also characterized by mutually exclusive monoallelic transcription that leads to a punctate expression pattern restricted to the more basal $G_{\alpha o}$-positive layer of the VNO sensory epithelium. While family-A, B and D receptors, thus, "obey" the one neuron – one receptor rule for chemosensory cells (Mombaerts, 2004a; Munger et al., 2009), those seven highly homologous (>80 %) family-C receptors are found in most, if not all, $G_{\alpha o}$-positive VSNs (Martini et al., 2001), effectively allowing for combinatorial V2R coexpression patterns in basal VSNs. The unusual expression of family-C $V2r$ genes is somewhat reminiscent of the atypical insect olfactory coreceptor Orco (Vosshall and Hansson, 2011) which appears functionally orthologous across species and serves as a chaperoning dimerization partner for the odor-specific subunits of the insect olfactory receptor gene family (Larsson et al., 2004; Nakagawa and Vosshall, 2009). Whether family-C V2Rs serve an analogous function in vomeronasal chemoreception remains to be determined (Spehr and Munger, 2009).

In 2003, two groups discovered that V2R-positive basal VSNs also express members of a family of nine non-classical class Ib major histocompatibility complex (MHC) genes, known as $H2\text{-}Mv$ or M10 (Ishii et al., 2003; Loconto et al., 2003). H2-Mv proteins were initially proposed to associate with V2Rs and, in complex with β2-microglobulin, serve a critical chaperone function for proper V2R trafficking and surface expression (Dulac and Torello, 2003; Loconto et al., 2003). Later, Ishii and Mombaerts reported that a substantial fraction of V2R-positive VSNs does not express any of the nine $H2\text{-}Mv$ genes while other neurons express multiple $H2\text{-}Mv$ family members in a combinatorial manner (Ishii and Mombaerts, 2008), questioning the general requirement of H2-Mv proteins as V2R escort molecules. Moreover, while the $H2\text{-}Mv$ gene family is exclusively found in rodents, other mammalian species such as the opossum express putatively functional $V2r$ genes without intact $H2\text{-}Mv$ open reading frames (Shi and Zhang, 2007; Young and Trask, 2007).

Direct proof that at least some V2Rs mediate VSN chemoresponsivity has been obtained from gene deletion studies. Mice deficient for $Vmn2r26$ ($V2r1b$) or $Vmn2r116$ ($V2rp5$) exhibit dramatically reduced sensitivity to genotype-specific MHC class I peptides (Leinders-Zufall et al., 2009) or male-specific exocrine gland-secreting peptide 1 (ESP1 (Haga et al., 2010)), respectively. In wildtype mice, however, these macromolecules elicit robust and highly sensitive responses with VSN activation thresholds in the picomolar range (Leinders-Zufall et al., 2009).

Contrary to current concepts of isolated $V1r$ gene expression in apical VSNs (see above), combinatorial coexpression of three multigene families in non-random combinations – family-A/B/D $V2r$s (115 genes), family-C $V2r$s (7 genes), and $H2\text{-}Mv$s (9 genes) – appears to bestow a functional identity on a given basal VSNs (Chamero et al., 2012). So far, efforts to deorphanize V2Rs (as well as V1Rs) in recombinant protein expression systems have

largely failed because of a general lack in understanding the regulatory mechanisms that control V1R/V2R cell-surface expression. However, some recent promising advances in heterologous V2R plasma membrane targeting (Dey and Matsunami, 2011) raise expectations that more detailed knowledge on vomeronasal receptor function could soon become available.

A third family of putative vomeronasal chemoreceptors has recently been identified by two groups (Liberles et al., 2009; Rivière et al., 2009). Parallel large scale screening for yet unidentified sensory cell populations and corresponding chemosensory GPCRs identified five candidate non-$V1/2r$ chemoreceptor genes that share many of the structural and functional hallmarks of previously identified olfactory GPCRs. All five genes ($Fpr\text{-}rs1$, $rs3$, $rs4$, $rs6$ and $rs7$) are members of the formyl peptide receptor (FPR)-like gene family, show a putative seven-transmembrane topology, display a selective and punctate transcription pattern in the VNO neuroepithelium, and are monogenically expressed with respect to both their own and other chemoreceptor gene families (Liberles et al., 2009; Rivière et al., 2009). Immunocytochemical localization studies of one particular family member, FPR-rs3, revealed protein translocation to the microvillous dendritic endings of VSNs (Rivière et al., 2009, Ackels et al., 2014). With the exception of $Fpr\text{-}rs1$ which is coexpressed with $G_{\alpha o}$, the remaining vomeronasal $Fpr\text{-}rs$ genes all coexpress with $G_{\alpha i2}$ in the apical epithelial layer (Figure 51.1b–c).

Immune cells such as neutrophils and monocytes express FPR1, the founding family member, as well as the related receptor FPR-rs2 (Le et al., 2002). Both receptors play important roles in host defense against pathogens (Soehnlein and Lindbom, 2010). Immune system FPRs recognize a broad range of chemoattractants such as the prototypical formylated bacterial peptide N-formylmethionine-leucine-phenylalanine (fMLF), mitochondrially encoded peptides or various antimicrobial / inflammatory modulators (Kolaczkowska and Kubes, 2013). These ligand spectra are therefore defined by immunological function than by structural properties. Upon activation, this unusually broad FPR tuning profile enables granulocytes and macrophages to use both pathogen- and host-derived compounds as chemotactic guideposts towards sites of infection or tissue damage. However, neither of the "classical" immune system FPRs is transcribed in VSNs (Liberles et al., 2009; Rivière et al., 2009).

Neither immune system nor vomeronasal Fprs share significant sequence similarity with other chemosensory GPCRs such as V1/2Rs, ORs or trace amine-associated receptors (TAARs; Liberles and Buck, 2006). Moreover, phylogenetic analysis suggests that vomeronasal Fprs, which form a single gene cluster directly adjacent to a stretch of 39 $V1/2r$ genes, selectively evolved from several

51.2 Signaling Mechanisms in Vomeronasal Neurons

recent gene duplications and positive Darwinian selection in the rodent lineage (Liberles et al., 2009), thus, arguing for neofunctionalization of VNO *Fprs* (Bufe et al., 2012). Yet, vomeronasal neurons are activated *in situ* by fMLF as well as the mitochondria-derived formylated peptides ND1-6T or ND1-6I (Chamero et al., 2011) and heterologously expressed FPR-rs1-7 proteins retain agonist spectra similar to immune system FPRs (Rivière et al., 2009). Thus, the exact biological function of vomeronasal FPRs remains to be determined.

51.2.2 Signaling Cascade(s) and General VSN Physiology

The signal transduction mechanisms following detection of ligands by V1R, V2R, or FPR-rs receptors are only partly understood and many important aspects of general VNO physiology still remain elusive. It is clear, however, that VSNs display membrane depolarization, an increased action potential firing rate and, consequently, a transient increase in cytosolic Ca^{2+} levels upon exposure to natural stimuli (Chamero et al., 2007; 2011; Del Punta et al., 2002a; Haga et al., 2010; Inamura and Kashiwayanagi, 2000; Inamura et al., 1997; 1999; Kim et al., 2011; Kimoto et al., 2007; Leinders-Zufall et al., 2000; 2004; Lucas et al., 2003; Nodari et al., 2008; Papes et al., 2010; Spehr et al., 2002). Given their extraordinarily high input resistance (typically several gigaohms (Hagendorf et al., 2009; Liman and Corey, 1996; Shimazaki et al., 2006; Ukhanov et al., 2007)), VSNs are exquisitely sensitive to electrical stimulation. Thus, at resting membrane potentials of between −60 and −75 mV (Liman and Corey, 1996; Ukhanov et al., 2007), depolarizing current injections of only a few picoamperes produce repetitive firing (Munger et al., 2009).

In addition to "standard" Hodgkin-Huxley type voltage-gated conductances such as those of TTX-sensitive sodium and delayed rectifier potassium currents (Bean, 2007), VSNs express several additional ion channels that shape their specific electrophysiological input–output relationship. L-type and T-type Ca_v^{2+} currents (Liman and Corey, 1996) generate low-threshold Ca^{2+} spikes that partly drive action potential discharge (Ukhanov et al., 2007). These Ca_v^{2+} currents are functionally coupled to large-conductance Ca^{2+}-sensitive K^+ (BK) channels in the VSN soma that were proposed to either maintain persistent rhythmic firing (Ukhanov et al., 2007) or, quite contrary, to mediate sensory adaptation via arachidonic acid-dependent BK channel recruitment during stimulation (Zhang et al., 2008). A homeostatic function is served by layer-specific and activity-dependent expression of ether-à-go-go-related (ERG) K^+ channels that control sensory output of basal V2R-positive VSNs. ERG channel expression extends the dynamic range of the VSN stimulus–response function

and is thus ideally suited to adjust VNO neurons to a target output range in a use-dependent manner (Hagendorf et al., 2009). Likely mediated by HCN2 and/or HCN4 channel subunits, hyperpolarization-dependent I_h currents represent another voltage-gated conductance that directly controls VSN excitability (Dibattista et al., 2008).

Layer-specific coexpression of the G-protein α-subunits $G_{\alpha i2}$ and $G_{\alpha o}$ in microvilli and dendritic tips of V1R- and V2R-expressing VSNs, respectively (Berghard and Buck, 1996; Halpern et al., 1995; Matsuoka et al., 2001), had early on raised speculation about a functional role of $G_{\alpha i2}$ and $G_{\alpha o}$ in apical and basal signaling pathways. Conclusive proof of this attractive model, however, came only recently from conditional olfactory neuron-specific gene deletion studies that demonstrated an essential requirement of $G_{\alpha o}$ in VSN responses to MHC I antigens, MUPs, as well as ESP1 (Chamero et al., 2011). Moreover, $G_{\alpha o}$ was found to be critical for sensing mitochondrially encoded FPR-rs1 ligands (Chamero et al., 2011), whereas neuronal responses to fMLF, a stimulus shown to activate at least some of the four $G_{\alpha i2}$-coupled FPRs (Rivière et al., 2009), were not significantly altered in $G_{\alpha o}^{-/-}$ mice (Chamero et al., 2011). With respect to $G_{\alpha i2}$, functional evidence supporting a role of this subunit in V1R-mediated signaling is still lacking. Global deletion of $G_{\alpha i2}$ did not result in an unambiguously VNO-dependent phenotype (Norlin et al., 2003) and functional recordings from defined neurons in those $G_{\alpha i2}$ null mice were not performed. As also true for a previous global $G_{\alpha o}$ deletion model (Tanaka et al., 1999), unconditional knockout of abundantly expressed and highly promiscuous signaling proteins such as $G_{\alpha i2}$ and $G_{\alpha o}$ is rather expected to induce a wide variety of defects (Chamero et al., 2011).

There is more consensus on a critical role of the β2 isoform of phospholipase C (PLCβ2) in VSN signal transduction (Spehr et al., 2006b), although corresponding genetic deletion studies are also lacking. Activation of this enzyme by non-GTP-binding $G_{\beta/\gamma}$ complexes promotes cleavage of phosphatidylinositol-4,5-bisphosphate into the soluble messenger molecule inositol-1,4,5-trisphosphate (IP_3) and the membrane-bound lipid diacylglycerol (DAG). IP_3 is usually coupled to massive Ca^{2+} release from intracellular storage organelles, whereas DAG exerts direct signaling functions at target membrane proteins or is metabolized to polyunsaturated fatty acids. Different models for the actions of either or all products of PLC-dependent lipid turnover have been proposed (Dibattista et al., 2012; Kim et al., 2011; Lucas et al., 2003; Spehr et al., 2002; Yang and Delay, 2010). The common denominator of basically all models is a central role of Ca^{2+} and an important, though maybe not indispensable function of the VNO-specific transient receptor potential (TRP) channel TRPC2 (Liman et al., 1999). The TRPC2 protein is highly localized to VNO sensory microvilli and *TrpC2*$^{-/-}$ mice show severe defects in a variety of

1118 Chapter 51 The Vomeronasal Organ

social and sexual behaviors (Leypold et al., 2002; Stowers et al., 2002). Yet, there are some different phenotypes induced by *TrpC2* deletion *versus* surgical VNO ablation (Kelliher et al., 2006; Pankevich et al., 2004) and recent reports of (residual) urine-evoked vomeronasal sensory activity in *TrpC2*$^{-/-}$ VSNs (Kelliher et al., 2006; Kim et al., 2011; Yang and Delay, 2010) have added another layer of complexity to our current understanding of vomeronasal transduction.

Elevated cytosolic Ca^{2+} - from TRPC2-dependent influx (Lucas et al., 2003) and maybe also IP_3-mediated store depletion (Kim et al., 2011; Yang and Delay, 2010) – appears to affect a variety of downstream cascade proteins. Increased Ca^{2+} was shown to exert both negative and positive feedback regulation in VSNs (Chamero et al., 2012). Down-regulation of TRPC2 by Ca^{2+}/calmodulin functions in vomeronasal sensory adaptation and gain control (Spehr et al., 2009). By contrast, a significant portion of urine-evoked VSN activity appears to be carried by a Ca^{2+}-activated chloride current (I_{Cl}) (Dibattista et al., 2012; Kim et al., 2011; Yang and Delay, 2010) that, similar to transduction mechanisms in OSNs (Pifferi et al., 2009; Stephan et al., 2009), appears to be mediated by members of the recently identified anoctamin family of Ca^{2+}-activated chloride channels (Caputo et al., 2008; Schroeder et al., 2008; Yang et al., 2012). Different groups have shown that both anoctamin1 and anoctamin2 are present in VSNs and largely colocalize with TRPC2 (Billig et al., 2011; Dauner et al., 2012; Dibattista et al., 2012; Rasche et al., 2010). However, whether I_{Cl} contributes to depolarization (analogous to olfactory neurons) or leads to membrane hyperpolarization is directly and only dependent on the *in vivo* chloride equilibrium potential at the VSN microvillar membrane – a physiological parameter that is currently unknown. A third ion channel target of elevated Ca^{2+} conducts a Ca^{2+}-activated nonselective cation current (I_{CAN}) found in both hamster (Liman, 2003) and mouse VSNs (Spehr et al., 2009). I_{CAN}, however, requires much higher Ca^{2+} concentration levels than I_{Cl} for activation (Tirindelli et al., 2009).

51.3 VOMERONASAL CHEMOSENSORY CUES

VNO neurons are responsive to ethologically relevant odors emitted from other animals including so-called pheromones and kairomones, as well as a subset of odors of unknown biological function that are not emitted by other animals (Chamero et al., 2012; Sam et al., 2001; Trinh and Storm, 2003). Pheromones (pherin = to transfer and hormone = to excite) have been commonly assumed to be chemosensory ligands that elicit a specialized behavior or neuroendocrine response upon detection by members of

the same species (Karlson and Luscher, 1959; see Doty, 2010). Kairomones are ligands emitted by other species that upon detection lead to a behavioral advantage in the receiver and disadvantage the signaler. Current knowledge of VNO stimuli has been obtained through the study of the effects of native odor sources such as urine, saliva, and soiled bedding on conspecifics (Ben-Shaul et al., 2010; He et al., 2008; Isogai et al., 2011). Electrophysiology and behavioral observations have guided the isolation and identification of VNO ligands based on three major assumptions. First, it is presumed that relevant VNO ligands are secreted or excreted from an individual into the environment and their emission and detection is likely to be species-specific. Larger molecules that are not a common byproduct of general metabolism are more likely to meet this requirement (although not so complex that synthesis becomes costly) (Wilson, 1963). Second, if the ligand promotes a sexually dimorphic behavior, then its production and emission is likely to be confined to one gender. Third, since the VNO lumen is mucus filled and located at the base of the nasal cavity it is therefore not directly available to traditional volatile odorants. Relevant stimuli must efficiently diffuse or undergo transport to access the sensory receptors (Meredith and O'Connell, 1979).

Six categories of VNO ligands have now been identified that meet the above predictions: exocrine gland-secreting peptides (ESPs), major histocompatibility complex (MHC) peptide ligands, major urinary proteins (MUPs), volatile small molecules, sulfated steroids, and formyl-peptides. These known ligands fall into three groups. ESPs, MHCs and MUPs are all peptides or small proteins that have undergone species-specific gene duplications in rodents (Kimoto et al., 2007; Logan et al., 2008). The majority of these protein ligands have been shown to activate V2R expressing neurons (Chamero et al., 2007; Kimoto et al., 2007; Leinders-Zufall et al., 2004). VNO-stimulating volatile small molecules and sulfated steroids (also small molecules) are commonly emitted by many animal species and largely function to signal gender- or state-specific internal physiology (Nodari et al., 2008; Novotny et al., 1990; Novotny et al., 1985; Novotny et al., 1999b). The majority of these small molecules are detected by V1R expressing neurons (Leinders-Zufall et al., 2000; Meeks et al., 2010). Formylated peptides are produced by bacteria and mitochondria. These ligands activate FPR receptors (Liberles et al., 2009; Rivière et al., 2009).

The 38 mouse ESPs are secreted from glandular tissues including lacrimal, Harderian, and submaxillary (Kimoto et al., 2005). They are differentially expressed across development and between genders (Kimoto et al., 2007, Ferrero et al., 2013), potentially providing a wealth of state-specific cues to guide social behavior. Detailed study of one ESP, ESP1, found sex-specific expression in male tears (Kimoto et al., 2005). When detected by females, ESP1 specifically activates a single V2R whose action is necessary to stimulate female mating behavior: lordosis (Haga et al., 2010). Another recently identified ESP (ESP22) acts as a a juvenile semiochemical that blocks sexual behaviour (Ferrero et al., 2013).

MHC class I peptides, which function in self-recognition in the immune system, have also been shown to function as VNO sensory stimuli (Leinders-Zufall et al., 2004). They are detected by V2R-expressing neurons in the VNO as well as currently undefined neurons in the main olfactory epithelium (Leinders-Zufall et al., 2004, 2009; Spehr et al., 2006a). MHC peptides are thought to be excreted in urine throughout the life of both males and females, providing a stable representation of individuality. Indeed, they have been shown to be required for a female to imprint the memory of her mating partner (Leinders-Zufall et al., 2004).

Mouse MUPs are a family of 21 species-specific genes (Logan et al., 2008; Mudge et al., 2008). They are differentially regulated by thyroxine, growth hormone, and testosterone (Knopf et al., 1983; Szoka and Paigen, 1978) and therefore exhibit age- and gender-specific expression. MUPs are primarily synthesized in the liver and excreted in urine, but they are also expressed by several secretory tissues including the submaxillary, lacrimal, and mammary glands (Finlayson et al., 1965; Shaw et al., 1983). MUPs fold into a β-barrel structure creating an internal ligand binding pocket (Flower, 1996). This structure efficiently binds small molecules found in urine which themselves activate VNO neurons (Sharrow et al., 2002; Timm et al., 2001). MUPs have been proposed to function as pheromone carrier proteins serving to transmit the small volatile ligands found in urine into the mucus-filled VNO, and as stabilizers of these volatile ligands to extend their signaling potency by delaying their release after excretion into the environment (Beynon and Hurst, 2003, 2004; Hurst et al., 1998). Of the 21 MUP genes in the mouse genome, each individual only expresses 4–12 urine MUPs (Hurst et al., 2001; Robertson et al., 1996, 1997). This personally-tailored protein excretion has been proposed to differentially stabilize small molecules in order to facilitate individual recognition (Armstrong et al., 2005; Cheetham et al., 2009; Hurst and Beynon, 2004; Hurst et al., 2001). In addition, MUPs alone, without the associated small molecule ligands, have been shown to stimulate V2R-expressing neurons (Chamero et al., 2007; Papes et al., 2010). Their detection promotes gender-specific behavior, including territorial aggression when detected by other males and attraction when detected by females (Chamero et al., 2007; Roberts et al., 2010), as well as gender-neutral behavior, acting to facilitate conditioned place preference (Roberts et al., 2012). While strain-specific MUPs additionally constitute a context-dependent combinatorial code that governs multiple social behaviors in mice (Kaur et al., 2014), MUP orthologs emitted from heterospecifics are sufficient to signal VNO-mediated innate fear (Papes et al., 2010).

To identify gender-specific bioactive ligands, subtractive gas chromatography–mass spectrometry (GCMS) methods were used to screen for compounds enriched in urine from intact males compared to androgen-negative castrated male urine. This approach identified several volatile VNO-activating ligands including isoamylamine (IAA), isobutylamine (IBA), 3,4-dehydro-exo-brevicomin (DHB), 2-sec-butyl-4,5-dihydrothiazole (SBT), 6-hydroxy-6-methyl-3-heptanone (HMH), and α/β-farnesene (Nishimura et al., 1989; Novotny et al., 1999b; Schwende et al., 1986). All of these small molecules bind MUPs which likely act as carriers to transport them into the VNO lumen (Novotny et al., 2007). Unbound, they have all been shown to directly activate V1R expressing neurons, some at as little as 10^{-11} M concentration (Boschat et al., 2002; Leinders-Zufall et al., 2000). SBT, α/β-farnesene, and HMH are excreted in male urine and have been shown to promote female puberty acceleration (Jemiolo et al., 1986; Novotny et al., 1999a; Novotny et al., 1999b, although see Flanagan et al., 2011) while DHB and 2-heptanone are emitted by females to delay estrus or puberty (Jemiolo et al., 1989; Jemiolo et al., 1986; Novotny et al., 1999b). DHB, α/β-farnesene, and 2-heptanone have been shown to direct social behavior in several evolutionarily diverse species, including insects (Goodwin et al., 2006; Papachristoforou et al., 2012; Schulz et al., 2003; Silk et al., 2010).

Biochemically fractionating urine and assaying corresponding VNO neural response led to the purification of sulfated steroids as a potentially large family of VNO ligands whose activity could account for eighty-percent of the total activity of female urine (Nodari et al., 2008). The ligands primarily activate apical (V1R-expressing) neurons (Isogai et al., 2011; Celsi et al., 2012). Steroid hormones are sulfated to facilitate their clearance and therefore reflect the dynamic physiological state of an individual (Hanson et al., 2004).These ligands include androstan-, androsten-, cholestan-, estratrien-, pregnan-, and pregnen-families of sulfated steroids which collectively have the potential to transmit many qualities about an individual's physiological state (Nodari et al., 2008). A blend of three sulfated steroids are said to act as a migratory pheromone in the lamprey, however it is unknown if sulfated steroids function individually or as a blend in the mouse (Fine and Sorensen, 2008; Sorensen et al., 2005).

The identification of FPRs as exclusively expressed in VNO sensory neurons initiated the discovery of formylated-peptides and inflammation-related ligands as chemosensory cues (Liberles et al., 2009; Rivière et al., 2009). Currently, little is known about the function of these

1120   Chapter 51   The Vomeronasal Organ

diverse ligands, but they are likely to signal the pathogenic state of individuals (Arakawa et al., 2011). The mouse VNO expresses approximately over 250 different sensory receptors (V1Rs, V2Rs, and FPRs) and the known stimuli have been shown to be detected by a single, highly-tuned receptor (Haga et al., 2010; Leinders-Zufall et al., 2000; Nodari et al., 2008). If each receptor functions to detect a single salient cue then the majority of VNO ligands are still largely unknown.

51.4 CIRCUITS AND INFORMATION PROCESSING

Relative to other sensory systems, little is known about either the anatomy or function of VNO-responsive neural circuits. In mammals, sensory information is primarily processed in the cortex by neurons that enable associative learning and a plastic response to stimuli, both within and across individuals. In contrast, the behavioral response to semiochemicals is relatively invariant and highly conserved among individuals. Aligning with these behavioral observations, the circuits activated by the VNO are not traditional cortical structures. Instead, they primarily activate processing networks in the limbic system which utilize more hardwired mechanisms as compared to the cortex (Meredith, 1991; Scalia and Winans, 1975). Further, the circuits activated by VNO stimuli are thought to be relatively "simple" consisting of three primary processing relays between sensory input and output command. The processing centers include the accessory olfactory bulb (AOB), amygdala, and hypothalamus (Figure 51.2) (Lo and Anderson, 2011). Recent knowledge, however, is revealing that the circuit processing of VNO stimuli is likely to be more complex than was previously assumed, as it is influenced by both modulator and feedback mechanisms that enables variable, state specific responses, to VNO stimuli. Further identification and study of VNO stimulated circuits will be necessary to determine how the chemical environment is transformed into a biologically meaningful code that generates emotional behavior and neuroendocrine changes.

51.4.1 The Accessory Olfactory Bulb

Forming the vomeronasal nerve, VSN axon bundles pass through the cribriform plate and target a specialized region at the dorsal/caudal end of the olfactory bulb; the accessory olfactory bulb (AOB; Fig. 51.2). Each of the functional subsets of VSNs remains segregated in two distinct regions of the AOB (Chamero et al., 2012). V1R expressing neurons co-express OCAM and synapse on OCAM-negative expressing mitral cells located in the rostral half of the

AOB, while the V2R expressing cells lack discernible OCAM expression and synapse with OCAM-positive second-order output neurons, mitral cells, found in the caudal half of the AOB (Jia and Halpern, 1996; Mori et al., 2000). VSNs expressing a single type of VR converge with dendrites of mitral cells in 6–10 glomeruli, distributed in broad regions of the AOB (Del Punta et al., 2002b). Some glomeruli are exclusively innervated by neurons expressing the same VR (Del Punta et al., 2002b), while others coalesce with different, but likely closely related, VSNs (Wagner et al., 2006). This glomerular organization of VSN-to-mitral cell connectivity in the AOB likely serves to integrate incoming sensory information (Belluscio et al., 1999; Rodriguez et al., 1999).

The mechanisms underlying VSN axonal pathfinding to select AOB glomeruli are not well understood. Clearly, both the formation of an AOB glomerular map and VSN survival depend on VR expression (Belluscio et al., 1999). In addition, three of the four prominent and highly conserved families of attractive and / or repulsive axon guidance cues – netrins, semaphorins, ephrins, and slits (Bashaw and Klein, 2010; Yu and Bargmann, 2001) – have been implicated VSN axon navigation. Attractive interaction between EphA6 receptor tyrosine kinases on mitral cell dendrites and cell surface-associated ephrin-A5 ligands on V1R-expressing VSNs mediate axon guidance to the anterior AOB, a mechanism opposite to the frequently observed repulsive interactions between Eph receptors and ligands (Knöll et al., 2001). Chemorepulsive mechanisms, however, also function in segregated VSN pathfinding. Secreted semaphorins act on neuropilin-2 (Npn-2) receptors expressed in apical VSNs (Cloutier et al., 2002) and are required for axon fasciculation and segregation, whereas another secreted chemorepellent, Slit-1, activates Robo-2 receptors that mediate basal VSN axon targeting to the posterior half of the AOB (Prince et al., 2009). Notably, Npn-2 and V1Rs play distinct and complementary roles in promoting AOB wiring as the phenotypic defects in $Npn\text{-}2^{-/-}$ mice and animals lacking individual $V1r$s do not match (Cloutier et al., 2002).

In mice and rats, a morphologically distinct AOB primordium is first detected at E15 and E16, respectively (Knöll et al., 2001; Marchand and Bélanger, 1991) and prenatal formation of the AOB vomeronasal nerve layer has been reported (Knöll et al., 2001). The critical period for fine-tuning of AOB wiring and for formation of well-defined glomeruli, however, occurs after axonal coalescence during the first four to six postnatal days (Hovis et al., 2012; Salazar et al., 2006). During this period, development of precise connectivity appears strongly modulated by early VSN activity (Hovis et al., 2012).

As pointed out by Dulac and Wagner, the basic biology of the AOB is almost unexplored and any functional analogy with neurons / circuits of the main olfactory

51.4 Circuits and Information Processing

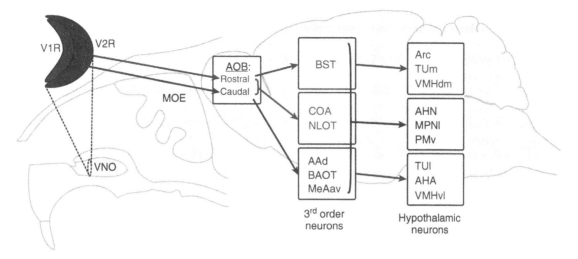

Figure 51.2 Circuits implicated in processing VNO stimuli. Ligands activate a subset of sensory neurons in either the apical, blue, or basal, red, zone of the VNO. The axons from each zone project to glomeruli in spatially distinct regions of the AOB. Some second order mitral cells maintain the apical/basal dichotomy as they project to amygdalar regions, red and blue text, while other amygdalar regions receive information from both zones of the AOB, purple text. The amygdala acts as an information processing hub, which can alter activity throughout the brain, however the hypothalamus has been shown to be a necessary third order target. Regions within the hypothalamus including the Arc, TU, VMH, and AHA maintain the apical and basal segregation established at the periphery suggesting a functional dichotomy within these nuclei. AA, anterior amygdala; AHA, amygdalo-hyppocampal area; AHN, anterior hypothalamic nucleus; AOB, accessory olfactory bulb; Arc, arcuate nucleus; BAOT, bed nucleus of the accessory olfactory tract; BST, bed nuclei stria terminalis; COA, cortical nucleus amygdala; MeA, medial amygdala; MPN, medial preoptic nucleus; NLOT, nucleus lateral olfactory tract; PM, pre mammillary nucleus; TU, tuberal nucleus hypothalamus; VMH, ventromedial hypothalamus; VNO, vomeronasal organ; d, dorsal; m, medial; a, anterior; v, ventral; l, lateral. (*See plate section for color version.*)

bulb is mostly speculative (Dulac and Wagner, 2006). As a result, premature extrapolation of anatomical organization and physiological principles from the main to the accessory bulb has seriously hampered an unbiased assessment of AOB neurobiology. Some fundamental differences between both regions were already pointed out in 1901 by Ramon y Cajal (Larriva-Sahd, 2008): (1) the AOB glomerular layer is relatively thick and composed of smaller and less defined confluent glomeruli; (2) AOB projecting neurons share some morphological analogies to the main bulb's mitral / tufted (M/T) cells, but their somata are rarely mitral shaped; and (3) the AOB shows only a rudimentary external plexifom layer. Indeed, AOB glomeruli are surrounded by only a few periglomerular cells, tightly clustered, and highly variable in size (10–30 μm diameter (Tirindelli et al., 2009)). Since "conventional" plexifom layers are largely missing (Salazar et al., 2006), Larriva-Sahd suggested to refer to an external (ECL) and internal (ICL) cellular layer separated by the complex fiber bundles of the lateral olfactory tract (LOT) which conducts axons from both main and accessory bulb projection neurons (Larriva-Sahd, 2008). In each glomerulus, a few hundred VSN axons terminate and release the excitatory neurotransmitter glutamate upon stimulation (Dudley and Moss, 1995). Two basic models for M/T cell glomerular connectivity have been proposed; homo- and heterotypic connectivity. According to the homotypic wiring scheme,

M/T cell primary dendrites exclusively target glomeruli that are activated by VSN populations that share the same chemoreceptor identity (Del Punta et al., 2002b; Hovis et al., 2012). By contrast, the heterotypic connectivity model predicts that M/T cells connect specifically to glomeruli innervated by neurons expressing receptors of the same subfamily (Wagner et al., 2006), leading to a glomerular map of multiple subfamily-dependent domains (Dulac and Wagner, 2006).

M/T cells in the AOB lack an extensive lateral dendritic tree. Instead, they elaborate extensively branched apical / primary dendrites that terminate as multiple tufts in as many as twelve different glomeruli within the homonymous AOB half (Ben-Shaul et al., 2010; Meeks et al., 2010; Takami and Graziadei, 1991; Urban and Castro, 2005; Wagner et al., 2006). Dendritic tuft size and shape are also polymorphic. Dulac and Wagner described elaborated "ball of yarn"– like tufts, basket-like and bush-like tufts, as well as tufts containing only few fine branches, and tufts with a "dead" end (Dulac and Wagner, 2006). Within such glomerular tufts, local regenerative events – "tuft spikes" – occur independently of somatic spiking, suggesting that M/T cell tufts can accommodate local nonlinear synaptic input integration (Urban and Castro, 2005). Interestingly, both subthreshold and suprathreshold excitation was shown to trigger local dendritic transmitter release from AOB M/T cells (Castro and Urban, 2009).

In addition to local dendritic computation, active back-propagation of sodium spikes from the soma to glomerular tufts can synchronize the activity of "sister" M/T cells that share glomerular projections (Ma and Lowe, 2004). With maximum average firing frequencies <50 Hz, AOB projection neurons are relatively slow, yet divergent in their discharge properties (Zibman et al., 2011). Moreover, both transient and persistent firing responses have been recorded from these neurons (Shpak et al., 2012). As M/T cell axon collaterals contact both adjacent projection neurons as well as interneurons in both the anterior and posterior AOB (Larriva-Sahd, 2008), stimulus-locked as well as sustained firing would be transmitted throughout the AOB.

M/T cells form synapses with inhibitory interneurons both at the glomerular level and along the more proximal parts of their primary dendritic trees (Brennan and Kendrick, 2006). While thoroughly investigated in the main bulb (Price and Powell, 1970; Schoppa and Urban, 2003), the AOB dendrodendritic reciprocal synapse between M/T and granule cells – the main class of AOB interneurons (Keverne and Brennan, 1996) – has received less attention, even though it has been attributed a similarly essential role in information processing (Hayashi et al., 1993; Jia et al., 1999; Taniguchi and Kaba, 2001). Strategically located along the basal segments of M/T cell apical dendrites, AOB dendrodendritic synapses appear to primarily provide GABAergic recursive self-inhibition, rather than spreading lateral inhibition (Luo and Katz, 2004). Such local inhibitory feedback was shown to sharpen M/T cell response selectivity (Hendrickson et al., 2008) and generate γ-frequency field potential oscillations in the AOB (Schoppa, 2006; Sugai et al., 1999). Several groups have suggested that a relatively simple relay-like recursive self-inhibition plays a sensory "gating" role that underlies olfactory learning and memory formation (Brennan and Keverne, 1997; Castro et al., 2007; Hayashi et al., 1993; Kaba et al., 1994).

While local field potential oscillations at γ-frequencies are thought to reflect recursive inhibition at the AOB dendrodendritic synapse, slower oscillations at γ-frequencies have been attributed to feedback interactions with central brain areas (Neville and Haberly, 2003). The AOB receives centrifugal inputs from the amygdala, locus coeruleus, and raphe nucleus (Broadwell and Jacobowitz, 1976, Mohedano-Moriano et al., 2012) as well as the bed nucleus of the stria terminalis the horizontal limb of the diagonal band of Broca (Fan and Luo, 2009; Smith and Araneda, 2010). These feedback afferents enter the AOB via the LOT as well as the bulbar core white matter (Larriva-Sahd, 2008) and they were shown to be critical for olfactory memory formation (Keverne and Brennan, 1996). Initially, attention had focused on both noradrenergic and glutamatergic feedback from the locus coeruleus and corticomedial amygdala, respectively. Based on vaginocervical

stimulation during mating, AOB noradrenaline levels remain increased for ~4 hours (Brennan et al., 1995). During this critical period, noradrenaline release mediates plastic changes in synaptic strength at the dendrodendritic synapse (Brennan and Keverne, 1997; Brennan and Keverne, 2004). Mechanistically, early results pointed to noradrenaline-dependent disinhibition of M/T cells via α2-receptor-mediated suppression of granule cells (Brennan, 2004; Otsuka et al., 2001). More recent findings suggest, however, that an α1-dependent increase in granule cell GABA release inhibits M/T cell firing (Araneda and Firestein, 2006; Smith et al., 2009). Similarly controversial results were obtained when metabotropic receptor targets of glutamatergic feedback were investigated. The jury is still out on many questions about the exact functional consequences of feedback projections, but it seems clear that afferent centrifugal modulation of AOB processing plays an important physiological role in AOS biology.

51.4.2 The Amygdala

Second order neurons project from the AOB to the amygdala, an ancient processing system present in lungfish and tetrapods (Kass, 2007; Medina et al., 2011). The amygdala is highly responsive to unexpected stimuli and therefore signals arousal and attention, especially of biologically relevant signals (Pessoa and Adolphs, 2010; Whalen, 1998). In addition, it is thought to serve as a vigilance system, to determine where incoming sensory information falls on the continuum of positive to negative relevancy to an individual's fitness or survival, and then signals this assessment to the rest of the brain (LeDoux, 2000, 2012; Pessoa and Adolphs, 2010; Salzman and Fusi, 2010).

The amygdala is composed of a heterogeneous group of approximately ten distinct nuclei. Anatomically, these collective nuclei are highly interconnected with each other, the cortex, and the Hypothalamus-Pituitary-Adrenal, HPA, axis. Together they are positioned to act as a "hub" to coordinate the overall brain state (Newman, 1999; Pessoa and Adolphs, 2010). Amygdalar feedback to the cortex is thought to modulate the "upstream" incoming deluge of sensory information in order to focus attention onto significant information. "Downstream" circuits activate the HPA axis to elicit the hormone and neuroendocrine release associated with emotional behavior. These known characteristics of amygdala information-processing generate underlying features observed during VNO-mediated social and survival behavior.

From the AOB, subsets of second order M/T cells directly project to multiple amygdala nuclei including the MeA, COA, BST, NLOT, BAOT, and the AA (Figure 51.2) (Gutierrez-Castellanos et al., 2010; Fig. 51.2), several of which have been functionally implicated in mediating VNO signaling (Bressler and Baum, 1996; Butler et al.,

2011; Kang et al., 2006; Meredith and Westberry, 2004). Notably, while the M/T neurons stimulated by either V1R or V2R sensory neurons both equally project to most of these amygdalar nuclei (Martinez-Marcos and Halpern, 1999; von Campenhausen and Mori, 2000), anatomical and functional evidence suggest that the V1R/V2R segregation established in the AOB is maintained in the anterior MeA and BST (Figure 51.2) (Mohedano-Moriano et al., 2007, 2008). Specifically, cells from the anterior AOB project to the medial-posterior-medial BST, with no cells from the posterior part projecting to this sub nucleus, and cells from the posterior AOB project to the dorsal AA and deep layers of the BAOT and anterio-ventral MeA, with no cells from the anterior AOB projecting to these regions (Mohedano-Moriano et al., 2007; Pardo-Bellver et al., 2012). One clue which may help solve the puzzle of this anatomical distinction may come from functional analysis revealing that sub-nuclei of the MeA are activated by distinct stimuli. Odors from the same species activate both the anterior and posterior MeA while odors from different species have been found to activate only the anterior MeA (Meredith and Westberry, 2004). Consistent with mediating biologically-relevant stimuli, anterior MeA inactivation results in deficits in mating, social memory, and social chemoinvestigative behavior (Ferguson et al., 2001; Kondo, 1992; Kondo and Sachs, 2002; Lehman and Winans, 1982; Lehman et al., 1980; Liu et al., 1997). Most amygdalar neurons are molecularly heterogeneous and little is known about their underlying physiology, however, characterization of the MeA cells show them to have anatomical and physiological properties similar to pyramidal cells in the piriform cortex (Bian et al., 2008). Notably, the MeA output neurons are capable of sustained bursting, a characteristic that may underlie prolonged behavioral responses or neuroendocrine release (Bian et al., 2008).

51.4.3 The Hypothalamus

The hypothalamus is thought to be the third, and final, processing module for VNO stimuli prior to circuits that mainly subserve the generation of motor patterns or neuroendocrine release. The hypothalamus is composed of multiple nuclei that function to maintain homeostasis (such as hunger, thirst, temperature, or blood pressure) by releasing hormones to direct the actions of the pituitary (Besser and Mortimer, 1974). It also functions to release the motor patterns that underlie behavioral responses including fear, aggression, mating, and sleep (Adams, 2006; Canteras et al., 1997; LeDoux, 2012; Lin et al., 2011; Motta et al., 2009). Neural activity from the MeA can directly engage the hypothalamus (Adams, 2006; Canteras et al., 1997; Choi et al., 2005; LeDoux, 2012; Lin et al., 2011; Lo and Anderson, 2011; Motta et al., 2009; Pardo-Bellver et al.,

2012; Petrovich et al., 2001). Alternatively, the signal can be further processed by either the BST, a region which is thought to integrate homeostatic and sensory information (Dong et al., 2001; Dumont, 2009) or by the hippocampus, which may serve to prioritize behavioral response to multiple, behaviorally conflicting, VNO stimuli (Petrovich et al., 2001). Notably, the different circuits established by V1R/V2R activity are maintained both anatomically (Mohedano-Moriano et al., 2008; Pardo-Bellver et al., 2012; Petrovich et al., 2001) and functionally in some hypothalamic nuclei (Fig. 51.2) (Bian et al., 2008; Lin et al., 2011).

Neurons in both the amygdala and the hypothalamus express receptors that detect and respond to peptides and other neural modulators. These include steroid hormones (estrogen, testosterone) (Blake and Meredith, 2011; Merchenthaler et al., 2004; Mitra et al., 2003), peptide hormones (NPY, substance P, oxytocin, GNRH, ACTH) (Ferguson et al., 2001; Jennes and Conn, 1994; Van Pett et al., 2000; Wood and Swann, 2005), and small molecule neurotransmitters (dopamine, serotonin) (Fremeau et al., 1991; Lemberger et al., 2007; Schiller et al., 2006). It is known that these receptors enable VNO activated circuits to modulate hard-wired aspects of their response to account for an individual's state (dominance, gender, hunger, and stress levels). However, precisely how neuromodulators alter circuit activity is currently unknown.

It is well established that the behavior elicited by VNO signaling varies depending on the gender of the receiving animal (Baum, 2009; Segovia and Guillamon, 1993; Stowers and Logan, 2010). For example, male-emitted cues elicit aggression when detected by other males but not females. Consistent with these observations, several nuclei within the amygdala and hypothalamus are sexually dimorphic in volume, usually larger in males (based on differences in absolute number, density, and size of individual neurons) (Dulac and Kimchi, 2007; Guillamon and Segovia, 1997; Kawata, 1995). VNO activated circuits have also been demonstrated to be functionally dimorphic (Blake and Meredith, 2011; Halem et al., 2001; Kang et al., 2006; Kimchi et al., 2007; Wu and Shah, 2011). Current studies have begun to determine the molecular mechanisms that underlie gender-dimorphic differences in VNO-activated circuits, though mechanisms remain unknown (Boehm et al., 2005; Gregg et al., 2010; Kimchi et al., 2007; Xu et al., 2012).

51.5 EVOLUTION AND DECAY OF THE VNO

The VNO is an evolutionarily ancient system, established prior to taste sensing (Grus and Zhang, 2009). Genetic analysis of the essential components of VNO signal transduction in living genomes indicates their likely expression in the ancestor of extant vertebrates. *V1rs* and *TrpC2* are expressed in olfactory tissues of the sea lamprey genome indicating they evolved before the divergence of jawless fish and cartilaginous fish. *V2rs* are expressed by the elephant shark and therefore evolved before the divergence of cartilaginous fish and bony vertebrates (Grus and Zhang, 2009). The segregation of these sensory components into a distinct sensory organ, the AOB, occurred with the evolution of tetrapods (Meisami and Bhatnagar, 1998). Expression of VNO signaling elements has been described in all vertebrate lineages except *sirenia*, which has a rudimentary main olfactory system, as well as birds and *cetacea* which also lack main olfactory system (Meisami and Bhatnagar, 1998). Some lineages, such as bats and primates, consist of species that contain a well-developed VNO and other species which lack evidence for VNO function.

Strikingly, the size and anatomy of the VNO and AOB vary dramatically within lineages (Meisami and Bhatnagar, 1998). This variation largely correlates with the magnitudes of the VR gene repertoire (Grus et al., 2005; Young et al., 2010). Indeed, there is a 56-fold variation in *Vr* gene number out of the 37 mammalian species that currently have sufficient genome sequence coverage for analysis. The platypus, mouse, mouse lemur, and rabbit have the largest *V1r* repertoire, each expressing between 160–280 *V1rs* (Grus et al., 2005, 2007; Young et al., 2010); while many diverse genomes contain 60–110 *V1rs* including the other rodents and lagomorphs, bushbaby, treeshrew, shrew, hedgehog, wallaby, and opossum. However, dolphin, little brown bat, and flying fox have no intact *Vrs*, or *TrpC2*, in their genome (Young et al., 2010). Interestingly, although dogs appear to readily sample and utilize olfactory information, their genome only contains 9 *V1rs*. It has been proposed that domestication imposed selective pressure to favor *Vr* inactivating mutations. However, the genome of wild wolves appears similar to the dog indicating that loss of VNO receptors occurred prior to domestication (Young et al., 2010).

While the human genome contains five *V1rs*, they are not likely to function since *TrpC2* underwent inactivating mutations in Old World primates (Liman and Innan, 2003; Zhang and Webb, 2003). This has led to the hypothesis that the evolution of trichromatic vision functioned to eliminate selective pressures on VNO signaling elements (Liman and Innan, 2003). However, analysis of additional primate genomes indicate that *V1r* loss began in the common ancestor of both New and Old World primates, occurring independently of the gain of color vision (Young et al., 2010). Based on known ethological characteristics, gain or loss of *Vrs* has not been found to correlate with body size, nocturnality, diet, sociality, or mating system (Young et al., 2010).

Currently, the selective pressures that result in extreme VR gene duplication or loss remain unknown.

Moreover, the identity of VR expression does not vary predictably across evolution. The known ratio of V1R/V2R receptors is strongly skewed towards the V2Rs in the zebrafish (6/46) and frog (21/330) (Brykczynska et al., 2013). In contrast, most other sequenced genomes are biased to express V1Rs including platypus (283/15), cow (40/0), and mouse (239/121) (Brykczynska et al., 2013; Grus et al., 2007). It has been noted that the strategy of receptor bias switched from V2R to V1R upon water to land habitation, however, more recent analysis of the VR ratios of terrestrial lizard (1/37), and snake (4/>116) do not support this hypothesis and the underlying logic of receptor bias remains unknown (Brykczynska et al., 2013; Shi and Zhang, 2007).

In addition to the variance of their absolute numbers, the identity of the *Vr* genes in sequenced genomes has undergone rapid and substantial changes between species (Brykczynska et al., 2013; Grus and Zhang, 2004; Young et al., 2010). Eighty percent of analyzed *V1rs* have a more similar gene in the same genome (paralogs) than in any other species (orthologs) (Young et al., 2010). Even in closely related species such as mouse and rat in which 86-94% of rat genes have mouse orthologs, only 11% of the rat *V1rs* have mouse orthologs (Grus and Zhang, 2004, 2008). Analysis of these changes shows weak purifying selection and random drift, with only about 5% of analyzed *V1rs* having undergone positive selection (Park et al., 2011). Frequent gene duplication and pseudogenization has been found to occur rapidly following divergence of two species (Grus and Zhang, 2004). These characteristics result in a "semi-private" repertoire of species-specific VNO receptors (Young et al., 2010) and support the hypothesis that the VNO functions to detect and convey information about species-relevant ligands.

51.6 FUNCTIONAL SIGNIFICANCE OF THE VNO

Since its discovery, the functional relevance of the VNO has posed a mystery. It is not known why most terrestrial vertebrates have at least two distinct olfactory systems; the VNO and the MOE. Species such as humans, that do not possess a functional VNO, demonstrate that the MOE alone is sufficiently sensitive and complex to generate olfactory perceptions and guide behavior. Currently, the functional significance of a secondary olfactory system, the VNO, remains unknown.

Comparison of the two systems shows MOE sensory neurons primarily excite cortical centers, while those from the VNO activate limbic centers (Lo and Anderson, 2011; Stettler and Axel, 2009). This striking difference in the flow of neural activity led Cajal to first speculate that (the AOB) "is a special center, perhaps differentiated for receiving impressions of some particular kind of olfactory excitation" (Cajal, 1901). Early investigators additionally noted that most terrestrial vertebrates detect and respond to specialized chemosignals that transmit ethologically relevant information among and between species. When considered together, the circuit anatomy of the VNO appeared well suited to mediate such responses. The formal test of this hypothesis was enabled by molecular genetics. Ablation of the primary signal transduction channel of the VNO, *TrpC2*, prevents VNO sensory neurons from responding to many ligand sources. Male $TrpC2^{-/-}$ mutants have no defects in courting and mating females, which is thought by some to be facilitated by so-called pheromonal cues (see Doty, 2010). However, mutants are unable to initiate territorial aggression or establish dominance hierarchies; behaviors dependent on semiochemicals (Leypold et al., 2002; Stowers et al., 2002). Mutant females show various defects in mating and maternal behavior (Kimchi et al., 2007). Similarly, the MOE can be silenced by genetic ablation of the primary signal transduction channel, *Cnga2* (Brunet et al., 1996). When assayed for putative pheromone-mediated behavior, $Cnga2^{-/-}$ mutant mice are unable to initiate aggression, both males and females fail to mate, and newborn pups die of dehydration, being unable to detect maternal olfactory cues necessary to initiate suckling (Brunet et al., 1996; Logan et al., 2012; Mandiyan et al., 2005). Together, these studies show that while the VNO is capable of mediating an innate response to ethologically relevant odor cues, the MOE serves this function as well. Therefore, the VNO is not uniquely specialized for chemicals some consider to be "pheromones."

Another striking difference between the two olfactory systems comes from sequence analysis of VNO receptors. Unlike the MOE, where there are orthologous receptors across divergent species, the receptors of the VNO appear to rapidly evolve between species, creating VNO receptor repertoires that are largely specific to one species (Grus and Zhang, 2004, 2008). It is likely that this customized receptor expression enables each species to recognize chemical cues that are of particular relevance to their survival and fitness.

Stimulation of the mouse VNO can have an immediate "releasing" effect to generate behavior (Bronson 1979, Dulac and Torello, 2003). Detection of cues emitted by strange male mice release aggression and countermarking behavior with a variable latency, on the order of seconds-minutes. Likewise, in the female, VNO stimulation can result in heterosexual attraction and lordosis, a receptive mating posture. In addition, the VNO detects cues emitted by other mice that can act over days to slowly change a recipient's neuroendocrine physiology (Bronson,

1979, Dulac and Torello, 2003; see, however, Doty, 2010). No so-called priming cues have been described to impact male behavior, but the detection of male-emitted cues may prime a female to synchronize and accelerate the estrus cycle, and advance the onset of estrus in juveniles, while detection of female-emitted cues may prolong diestrus thereby delaying entry into estrus. Females emit volatile chemicals that promote receiving females to slow or even abort the estrus cycle. When colony size is optimal, males emit volatiles that may counter their action, ensuring short estrus cycles and regular fecundity. However, when the colony becomes too large, females in each social group greatly outnumber males and the estrus delaying agents may become effective, slowing the rate of further colony expansion.

The mouse VNO may also play an important role in detecting kairomones, which are ligands emitted by one species that elicit a behavior to advantage the receiver of another species (Ben-Shaul et al. 2010; Isogai et al., 2011; Papes et al., 2010). Indeed, out of 88 receptors found to be stimulated by ethologically relevant odors, only 17 are activated by cues emitted by other mice while 71 have been found to be activated by scents emitted by other species (Isogai et al. 2011). The behavioral significance of the detection of other species cues varies depending on whether the species is a predator, prey, or neutral. The mouse VNO is known to detect cues from likely predators. Upon detection, they function to elicit defensive behaviors such as avoidance and an increase in ACTH stress hormones (Papes et al. 2010, Isogai et al. 2011).

It is tempting to speculate that the chemosignals mediating essential behaviors common to many different species, such as mammalian newborn suckling or the general act of mating, are detected by the dominant MOE system. The VNO, on the other hand, may detect and transmit cues that elicit behaviors that are not common to many species. Instead it may function to promote species specific behavior such as those to facilitate mating, for example urine marking and lordosis, which are sexual behaviors common to the mouse but not essential to successful mating in all vertebrates, or to respond to other species as appropriate predator or prey, such as a mouse's particular need to avoid a cat. In this hypothesis, the functions of the VNO would vary between species according to their specific ethological needs. Currently, the functional significance of the evolutionarily ancient and genetically costly sensory system remains to be identified.

REFERENCES

Ackels, T., von der Weid, B., Rodriguez, I. and Spehr, M. (2014). Physiological characterization of formyl peptide receptor expressing cells in the mouse vomeronasal organ. Frontiers in Neuroanatomy, 8 (November), 134. doi:10.3389/fnana.2014.00134.

Adams, D. B. (2006). Brain mechanisms of aggressive behavior: an updated review. *Neurosci. Biobehav. Rev.* 30: 304–318.

Arakawa, H., Cruz, S. and Deak, T. (2011). From models to mechanisms: odorant communication as a key determinant of social behavior in rodents during illness-associated states. *Neurosci. Biobehav. Rev.* 35: 1916–1928.

Araneda, R. C. and Firestein, S. (2006). Adrenergic enhancement of inhibitory transmission in the accessory olfactory bulb. *J. Neurosci.* 26: 3292–3298.

Armstrong, S. D., Robertson, D. H., Cheetham, S. A., et al. (2005). Structural and functional differences in isoforms of mouse major urinary proteins: a male-specific protein that preferentially binds a male pheromone. *Biochem. J.* 391: 343–350.

Bashaw, G. J. and Klein, R. (2010). Signaling from axon guidance receptors. *Cold Spring Harbor perspectives in biology* 2, a001941.

Baum, M. J. (2009). Sexual differentiation of pheromone processing: links to male-typical mating behavior and partner preference. *Horm. Behav.* 55: 579–588.

Bean, B. P. (2007). The action potential in mammalian central neurons. *Nat. Rev. Neurosci.* 8, 9579–9967.

Belluscio, L., Koentges, G., Axel, R. and Dulac, C. (1999). A map of pheromone receptor activation in the mammalian brain. *Cell* 97: 209–220.

Ben-Shaul, Y., Katz, L. C., Mooney, R. and Dulac, C. (2010). In vivo vomeronasal stimulation reveals sensory encoding of conspecific and allospecific cues by the mouse accessory olfactory bulb. *Proc. Nat. Acad. Sci. USA* 107: 5172–5177.

Berghard, A. and Buck, L. B. (1996). Sensory transduction in vomeronasal neurons: evidence for G alpha o, G alpha i2, and adenylyl cyclase II as major components of a pheromone signaling cascade. *J. Neurosci.* 16: 909–918.

Besser, G. M. and Mortimer, C. H. (1974). Hypothalamic regulatory hormones: a review. *J. Clin. Path.* 27: 173–184.

Beynon, R. J. and Hurst, J. L. (2003). Multiple roles of major urinary proteins in the house mouse, Mus domesticus. *Biochem. Soc. Trans.* 31: 142–146.

Beynon, R. J. and Hurst, J. L. (2004). Urinary proteins and the modulation of chemical scents in mice and rats. *Peptides* 25: 1553–1563.

Bian, X., Yanagawa, Y., Chen, W. R. and Luo, M. (2008). Cortical-like functional organization of the pheromone-processing circuits in the medial amygdala. *J. Neurophysiol.* 99: 77–86.

Billig, G. M., Pál, B., Fidzinski, P. and Jentsch, T. J. (2011). Ca^{2+}-activated Cl^- currents are dispensable for olfaction. *Nat. Neurosci.* 14: 763–769.

Blake, C. B. and Meredith, M. (2011). Change in number and activation of androgen receptor-immunoreactive cells in the medial amygdala in response to chemosensory input. *Neuroscience* 190: 228–238.

Boehm, U., Zou, Z. and Buck, L. B. (2005). Feedback loops link odor and pheromone signaling with reproduction. *Cell* 123: 683–695.

Boschat, C., Pelofi, C., Randin, O., et al. (2002). Pheromone detection mediated by a V1R vomeronasal receptor. *Nat. Neurosci.* 5: 1261–1262.

Brann, J. H. and Firestein, S. (2010). Regeneration of new neurons is preserved in aged vomeronasal epithelia. *J. Neurosci.* 30: 15686–15694.

Brennan, P. A. (2004). The nose knows who's who: chemosensory individuality and mate recognition in mice. *Hormones and Behavior* 46: 231–240.

Brennan, P. A. and Kendrick, K. M. (2006). Mammalian social odours: attraction and individual recognition. *Phil. Transact. Royal Soc. London. Series B, Biological sciences* 361: 2061–2078.

Brennan, P. A. and Keverne, E. B. (1997). Neural mechanisms of mammalian olfactory learning. *Prog. Neurobiol.* 51: 457–481.

Brennan, P. A. and Keverne, E. B. (2004). Something in the Air? New Insights into Mammalian Pheromones. *Curr. Biol.* 14: R81–R89.

Brennan, P. A., Kendrick, K. M. and Keverne, E. B. (1995). Neurotransmitter release in the accessory olfactory bulb during and after the formation of an olfactory memory in mice. *Neuroscience* 69: 1075–1086.

Bressler, S. C. and Baum, M. J. (1996). Sex comparison of neuronal Fos immunoreactivity in the rat vomeronasal projection circuit after chemosensory stimulation. *Neuroscience* 71: 1063–1072.

Broadwell, R. D. and Jacobowitz, D. M. (1976). Olfactory Relationships of the Telencephalon and Diencephalon in the Rabbit. III. The ipsilateral centrifugal fibers to the olfactory bulbar andretrobulbar formations. *J. Comp. Neurol.* 170: 321–345.

Bronson, F. H. (1979). The reproductive ecology of the house mouse. *Q. Rev. Biol.* 54: 265–299.

Brunet, L. J., Gold, G. H. and Ngai, J. (1996). General Anosmia Caused by a Targeted Disruption of the Mouse Olfactory Cyclic Nucleotide-Gated Cation Channel. *Neuron* 17: 681–693.

Brykczynska, U., Tzika, A. C., Rodriguez, I. and Milinkovitch, M. C. (2013). Contrasted evolution of the vomeronasal receptor repertoires in mammals and squamate reptiles. *Gen. Biol. Evol.* 5: 389–401.

Bufe, B., Schumann, T. and Zufall, F. (2012). Formyl peptide receptors from immune and vomeronasal system exhibit distinct agonist properties. *J. Biol. Chem.* 287: 33644–33655.

Butler, R. K., Sharko, A. C., Oliver, E. M., et al. (2011). Activation of phenotypically-distinct neuronal subpopulations of the rat amygdala following exposure to predator odor. *Neuroscience* 175: 133–144.

Cajal, R. (1901). Textura del lóbulo olfativo accesorio. *Trab. Lab. Invest. Biol.* 1: 141–149.

Canteras, N. S., Chiavegatto, S., Ribeiro do Valle, L. E. and Swanson, L. W. (1997). Severe reduction of rat defensive behavior to a predator by discrete hypothalamic chemical lesions. *Brain Res. Bull.* 44: 297–305.

Capello, L., Roppolo, D., Jungo, V. P., et al. (2009). A common gene exclusion mechanism used by two chemosensory systems. *Eur. J. Neurosci.* 29, 671–678.

Caputo, A., Caci, E., Ferrera, L., et al. (2008). TMEM16A, a membrane protein associated with calcium-dependent chloride channel activity. *Science* 322: 590–594.

Castro, J. B. and Urban, N. N. (2009). Subthreshold glutamate release from mitral cell dendrites. *J. Neurosci.* 29: 7023–7030.

Castro, J. B., Hovis, K. R. and Urban, N. N. (2007). Recurrent dendrodendritic inhibition of accessory olfactory bulb mitral cells requires activation of group I metabotropic glutamate receptors. *J. Neurosci.* 27: 5664–5671.

Celsi, F., D'Errico, A. and Menini, A. (2012). Responses to sulfated steroids of female mouse vomeronasal sensory neurons. *Chem. Senses* 37: 849–858.

Chamero, P., Katsoulidou, V., Hendrix, P., et al. (2011). G protein G(alpha)o is essential for vomeronasal function and aggressive behavior in mice. *Proc. Nat. Acad. Sci. USA* 108: 12898–12903.

Chamero, P., Leinders-Zufall, T. and Zufall, F. (2012). From genes to social communication: molecular sensing by the vomeronasal organ. *Trends Neurosci.* 35 : 597–606.

Chamero, P., Marton, T. F., Logan, D. W., et al. (2007). Identification of protein pheromones that promote aggressive behaviour. *Nature* 450: 899–902.

Cheetham, S. A., Smith, A. L., Armstrong, S. D., et al. (2009). Limited variation in the major urinary proteins of laboratory mice. *Physiol. Behav.* 96: 253–261.

Choi, G. B., Dong, H. W., Murphy, A. J., et al. (2005). Lhx6 delineates a pathway mediating innate reproductive behaviors from the amygdala to the hypothalamus. *Neuron* 46: 647–660.

Cloutier, J.-F., Giger, R. J., Koentges, G., et al. (2002). Neuropilin-2 Mediates Axonal Fasciculation, Zonal Segregation, but Not Axonal Convergence, of Primary Accessory Olfactory Neurons. *Neuron* 33: 877–892.

Coppola, D. M. and O'Connell, R. J. (1989). Stimulus access to olfactory and vomeronasal receptors in utero. *Neurosci. Lett.* 106: 241–248.

Cuschieri, A. and Bannister, L. H. (1975). The development of the olfactory mucosa in the mouse: light microscopy. *J. Anat.* 119: 277–286.

Dauner, K., Lißmann, J., Jeridi, S., et al. (2012). Expression patterns of anoctamin 1 and anoctamin 2 chloride channels in the mammalian nose. *Cell Tissue Res.* 347: 327–341.

Del Punta, K., Leinders-Zufall, T., Rodriguez, I., et al. (2002a). Deficient pheromone responses in mice lacking a cluster of vomeronasal receptor genes. *Nature* 419 : 70–74.

Del Punta, K., Puche, A. C., Adams, N. C., et al. (2002b). A divergent pattern of sensory axonal projections is rendered convergent by second-order neurons in the accessory olfactory bulb. *Neuron* 35: 1057–1066.

Dey, S. and Matsunami, H. (2011). Calreticulin chaperones regulate functional expression of vomeronasal type 2 pheromone receptors. *Proc. Nat. Acad. Sci. USA* 108: 16651–16656.

Dibattista, M., Amjad, A., Maurya, D. K., et al. (2012). Calcium-activated chloride channels in the apical region of mouse vomeronasal sensory neurons. *J. Gen. Physiol.* 140: 3–15.

Dibattista, M., Mazzatenta, A., Grassi, F., et al. (2008). Hyperpolarization-Activated Cyclic Nucleotide-Gated Channels in Mouse Vomeronasal Sensory Neurons. *J. Neurophysiol.* 100: 576–586.

Dong, H. W., Petrovich, G. D. and Swanson, L. W. (2001). Topography of projections from amygdala to bed nuclei of the stria terminalis. *Brain. Res. Rev.* 38: 192–246.

Doty, R. L. (2010). *The Great Pheromone Myth*. Baltimore: Johns Hopkins University Press.

Dudley, C. A. and Moss, R. L. (1995). Electrophysiological evidence for glutamate as a vomeronasal receptor cell neurotransmitter. *Brain Res.* 675: 208–214.

Dulac, C. and Axel, R. (1995). A novel family of genes encoding putative pheromone receptors in mammals. *Cell* 83: 195–206.

Dulac, C. and Torello, A. T. (2003). Molecular detection of pheromone signals in mammals: from genes to behaviour. *Nat. Rev. Neurosci.* 4: 551–562.

Dulac, C. and Wagner, S. (2006). Genetic analysis of brain circuits underlying pheromone signaling. *Annu. Rev. Genet.* 40: 449–467.

Dulac, C. and Kimchi, T. (2007). Neural mechanisms underlying sex-specific behaviors in vertebrates. *Curr. Opin. Neurobiol.* 17: 675–683.

Dumont, E. C. (2009). What is the bed nucleus of the stria terminalis? *Prog. Neuropsychopharmacol. Biol. Psychiatry.* 33: 1289–1290.

Enomoto, T., Ohmoto, M., Iwata, T., et al. (2011). Bcl11b/Ctip2 controls the differentiation of vomeronasal sensory neurons in mice. *J. Neurosci.* 31: 10159–10173.

Fan, S. and Luo, M. (2009). The organization of feedback projections in a pathway important for processing pheromonal signals. *Neuroscience* 161, 489–500.

Ferguson, J. N., Aldag, J. M., Insel, T. R. and Young, L. J. (2001). Oxytocin in the medial amygdala is essential for social recognition in the mouse. *J. Neurosci.* 21: 8278–8285.

Ferrero, D. M., Moeller, L. M., Osakada, T., Horio, N., Li, Q., Roy, D. S., … Liberles, S. D. (2013). A juvenile mouse pheromone inhibits sexual behaviour through the vomeronasal system. *Nature.* doi:10.1038/nature12579

Fine, J. M. and Sorensen, P. W. (2008). Isolation and biological activity of the multi-component sea lamprey migratory pheromone. *J. Chem. Ecol.* 34: 1259–1267.

Finlayson, J. S., Asofsky, R., Potter, M. and Runner, C. C. (1965). Major urinary protein complex of normal mice: origin. *Science* 149: 981–982.

Flanagan, K. A., Webb, W. and Stowers, L. (2011). Analysis of male pheromones that accelerate female reproductive organ development. *PloS one* 6: e16660.

Flower, D. R. (1996). The lipocalin protein family: structure and function. *Biochem. J.* 318 (Pt 1): 1–14.

Fremeau, R. T., Jr., Duncan, G. E., Fornaretto, M. G., et al. (1991). Localization of D1 dopamine receptor mRNA in brain supports a role in cognitive, affective, and neuroendocrine aspects of dopaminergic neurotransmission. *Proc. Nat. Acad. Sci. USA* 88: 3772–3776.

Giacobini, P., Benedetto, A., Tirindelli, R. and Fasolo, A. (2000). Proliferation and migration of receptor neurons in the vomeronasal organ of the adult mouse. *Brain Res. Devel. Brain Res.* 123: 33–40.

Goodwin, T. E., Eggert, M. S., House, S. J., et al. (2006). Insect pheromones and precursors in female African elephant urine. *J. Chem. Ecol.* 32: 1849–1853.

Gregg, C., Zhang, J., Butler, J. E., et al. (2010). Sex-specific parent-of-origin allelic expression in the mouse brain. *Science* 329: 682–685.

Grus, W. E. and Zhang, J. (2004). Rapid turnover and species-specificity of vomeronasal pheromone receptor genes in mice and rats. *Gene* 340: 303–312.

Grus, W. E. and Zhang, J. (2008). Distinct evolutionary patterns between chemoreceptors of 2 vertebrate olfactory systems and the differential tuning hypothesis. *Mol. Biol. Evol.* 25: 1593–1601.

Grus, W. E. and Zhang, J. (2009). Origin of the genetic components of the vomeronasal system in the common ancestor of all extant vertebrates. *Mol. Biol. Evol.* 26: 407–419.

Grus, W. E., Shi, P. and Zhang, J. (2007). Largest vertebrate vomeronasal type 1 receptor gene repertoire in the semiaquatic platypus. *Mol. Biol. Evol.* 24: 2153–2157.

Grus, W. E., Shi, P., Zhang, Y. P. and Zhang, J. (2005). Dramatic variation of the vomeronasal pheromone receptor gene repertoire among five orders of placental and marsupial mammals. *Proc. Nat. Acad. Sci. USA* 102: 5767–5772.

Guillamon, A. and Segovia, S. (1997). Sex differences in the vomeronasal system. *Brain Res. Bull.* 44, 377–382.

Gutierrez-Castellanos, N., Martinez-Marcos, A., Martinez-Garcia, F. and Lanuza, E. (2010). Chemosensory function of the amygdala. *Vitam. Horm.* 83: 165–196.

Haga, S., Hattori, T., Sato, T., et al. (2010). The male mouse pheromone ESP1 enhances female sexual receptive behaviour through a specific vomeronasal receptor. *Nature* 466: 118–122.

Hagendorf, S., Fluegge, D., Engelhardt, C. H. and Spehr, M. (2009). Homeostatic control of sensory output in basal vomeronasal neurons: activity-dependent expression of ether-à-go-go-related gene potassium channels. *J. Neurosci.* 29: 206–221.

Halem, H. A., Baum, M. J. and Cherry, J. A. (2001). Sex difference and steroid modulation of pheromone-induced immediate early genes in the two zones of the mouse accessory olfactory system. *J. Neurosci.* 21: 2474–2480.

Halpern, M. (1987). The organization and function of the vomeronasal organ. *Ann. Rev. Neurosci.* 10: 325–362.

Halpern, M. and Martinez-Marcos, A. (2003). Structure and function of the vomeronasal system: an update. *Prog. Neurobiol.* 70: 245–318.

Halpern, M., Shapiro, L. S. and Jia, C. (1995). Differential localization of G proteins in the opossum vomeronasal system. *Brain Res.* 677: 157–161.

Hanson, S. R., Best, M. D. and Wong, C. H. (2004). Sulfatases: structure, mechanism, biological activity, inhibition, and synthetic utility. *Angew. Chem. Int. Ed. Engl.* 43: 5736–5763.

Hayashi, Y., Momiyama, A., Takahashi, T., et al. (1993). Role of metabotropic glutamate receptors in synaptic modulation in the accessory olfactory bulb. *Nature* 366: 687–690.

He, J., Ma, L., Kim, S., et al. (2008). Encoding gender and individual information in the mouse vomeronasal organ. *Science* 320: 535–538.

Hendrickson, R. C., Krauthamer, S., Essenberg, J. M. and Holy, T. E. (2008). Inhibition shapes sex selectivity in the mouse accessory olfactory bulb. *J. Neurosci.* 28: 12523–12534.

Herrada, G. and Dulac, C. (1997). A novel family of putative pheromone receptors in mammals with a topographically organized and sexually dimorphic distribution. *Cell* 90: 763–773.

Hovis, K. R., Ramnath, R., Dahlen, J. E., et al. (2012). Activity Regulates Functional Connectivity from the Vomeronasal Organ to the Accessory Olfactory Bulb. *J. Neurosci.* 32: 7907–7916.

Hurst, J. L. and Beynon, R. J. (2004). Scent wars: the chemobiology of competitive signalling in mice. *BioEssays* 26: 1288–1298.

Hurst, J. L., Payne, C. E., Nevison, C. M., Marie, A. D., et al. (2001). Individual recognition in mice mediated by major urinary proteins. *Nature* 414: 631–634.

Hurst, J. L., Robertson, D. H. L., Tolladay, U. and Beynon, R. J. (1998). Proteins in urine scent marks of male house mice extend the longevity of olfactory signals. *Anim. Behav.* 55: 1289–1297.

Inamura, K. and Kashiwayanagi, M. (2000). Inward current responses to urinary substances in rat vomeronasal sensory neurons. *Eur. J. Neurosci.* 12: 3529–3536.

Inamura, K., Kashiwayanagi, M. and Kurihara, K. (1997). Inositol-1,4,5-trisphosphate Induces Responses in Receptor Neurons in Rat Vomeronasal Sensory Slices. *Chem. Senses* 22: 93–103.

Inamura, K., Matsumoto, Y., Kashiwayanagi, M. and Kurihara, K. (1999). Laminar distribution of pheromone-receptive neurons in rat vomeronasal epithelium. *J. Physiol.* 517 Pt 3: 731–719.

Ishii, T. and Mombaerts, P. (2008). Expression of nonclassical class I major histocompatibility genes defines a tripartite organization of the mouse vomeronasal system. *J. Neurosci.* 28: 2332–2341.

Ishii, T., Hirota, J. and Mombaerts, P. (2003). Combinatorial coexpression of neural and immune multigene families in mouse vomeronasal sensory neurons. *Curr. Biol.* 13: 394–400.

Isogai, Y., Si, S., Pont-Lezica, L., et al. (2011). Molecular organization of vomeronasal chemoreception. *Nature* 478: 241–245.

Jacobson, L., Trotier, D. and Døving, K. B. (1998). Anatomical description of a new organ in the nose of domesticated animals by Ludvig Jacobson (1813). *Chem. Senses* 23: 743–754.

Jemiolo, B. and Reolini, F., Xie, T. M., Wiesler, D. and Novotny, M. (1989). Puberty-affecting synthetic analogs of urinary chemosignals in the house mouse, Mus domesticus. *Physiol. Behav.* 46: 293–298.

Jemiolo, B., Harvey, S. and Novotny, M. (1986). Promotion of the Whitten effect in female mice by synthetic analogs of male urinary constituents. *Proc. Nat. Acad. Sci. USA* 83: 4576–4579.

Jennes, L. and Conn, P. M. (1994). Gonadotropin-releasing hormone and its receptors in rat brain. *Front. Neuroendocrin.* 15: 51–77.

Jia, C. and Halpern, M. (1996). Subclasses of vomeronasal receptor neurons: differential expression of G proteins (Gi alpha 2 and G(o alpha)) and segregated projections to the accessory olfactory bulb. *Brain Res.* 719: 117–128.

Jia, C., Chen, W. R. and Shepherd, G. M. (1999). Synaptic Organization and Neurotransmitters in the Rat Accessory Olfactory Bulb. *J. Neurophysiol.* 81: 345–355.

Kaba, H., Hayashi, Y. and Nakanishi, S. (1994). Induction of an Olfactory Memory by the Activation of a Metabotropic Glutamate Receptor. *Science* 265: 262–264.

Kang, N., Janes, A., Baum, M. J. and Cherry, J. A. (2006). Sex difference in Fos induced by male urine in medial amygdala-projecting accessory olfactory bulb mitral cells of mice. *Neurosci. lett.* 398: 59–62.

Karlson, P. and Luscher, M. (1959). Pheromones': a new term for a class of biologically active substances. *Nature* 183: 55–56.

Kass, J. H. and Bullock, T. H. (2007). Evolution of nerous systems: A comprehensive reference, Vol. 2: *Non-mammalian vertebrates.*, Vol 2 (Academic Press).

Kaur, A. W., Ackels, T., Kuo, T.-H., Cichy, A., Dey, S., Hays, C., … Stowers, L. (2014). Murine Pheromone Proteins Constitute a Context-Dependent Combinatorial Code Governing Multiple Social Behaviors. *Cell*, 157(3), 676–688. doi:10.1016/j.cell.2014.02.025

Kawata, M. (1995). Roles of steroid hormones and their receptors in structural organization in the nervous system. *Neurosci. Res.* 24 : 1–46.

Kelliher, K. R., Spehr, M., Li, X.-H., et al. (2006). Pheromonal recognition memory induced by TRPC2-independent vomeronasal sensing. *Eur. J. Neurosci.* 23: 3385–3390.

Keverne, E. B. and Brennan, P. A. (1996). Olfactory recognition memory. *J. Physiol.* 90: 503–508.

Kim, S., Ma, L. and Yu, C. R. (2011). Requirement of calcium-activated chloride channels in the activation of mouse vomeronasal neurons. *Nat. Commun.* 2: 365.

Kimchi, T., Xu, J. and Dulac, C. (2007). A functional circuit underlying male sexual behaviour in the female mouse brain. *Nature* 448: 1009–1014.

Kimoto, H., Haga, S., Sato, K. and Touhara, K. (2005). Sex-specific peptides from exocrine glands stimulate mouse vomeronasal sensory neurons. *Nature* 437: 898–901.

Kimoto, H., Sato, K., Nodari, F., et al. (2007). Sex- and strain-specific expression and vomeronasal activity of mouse ESP family peptides. *Curr. Biol.* 17: 1879–1884.

Knöll, B., Zarbalis, K., Wurst, W. and Drescher, U. (2001). A role for the EphA family in the topographic targeting of vomeronasal axons. *Development (Cambridge, England)* 128: 895–906.

Knopf, J. L., Gallagher, J. F. and Held, W. A. (1983). Differential, multihormonal regulation of the mouse major urinary protein gene family in the liver. *Mol. Cell. Biol.* 3: 2232–2240.

Kolaczkowska, E. and Kubes, P. (2013). Neutrophil recruitment and function in health and inflammation. *Nat. Rev. Immunol.* 13: 159–175.

Kondo, Y. (1992). Lesions of the medial amygdala produce severe impairment of copulatory behavior in sexually inexperienced male rats. *Physiol. Behav.* 51: 939–943.

Kondo, Y. and Sachs, B. D. (2002). Disparate effects of small medial amygdala lesions on noncontact erection, copulation, and partner preference. *Physiol. Behav.* 76: 443–447.

Larriva-Sahd, J. (2008). The accessory olfactory bulb in the adult rat: a cytological study of its cell types, neuropil, neuronal modules, and interactions with the main olfactory system. *J. Comp. Neurol.* 510 : 309–350.

Larsson, M. C., Domingos, A. I., Jones, W. D., et al. (2004). Or83b encodes a broadly expressed odorant receptor essential for Drosophila olfaction. *Neuron* 43: 703–714.

Le, Y., Murphy, P. M. and Wang, J. M. (2002). Formyl-peptide receptors revisited. *Trends Immunol.* 23: 541–548.

LeDoux, J. E. (2012). Rethinking the emotional brain. *Neuron* 73: 653–676.

LeDoux, J. E. (2000). Emotion circuits in the brain. *Ann. Rev. Neurosci.* 23: 155–184.

References

Lehman, M. N. and Winans, S. S. (1982). Vomeronasal and olfactory pathways to the amygdala controlling male hamster sexual behavior: autoradiographic and behavioral analyses. *Brain Res.* 240: 27–41.

Lehman, M. N., Winans, S. S. and Powers, J. B. (1980). Medial nucleus of the amygdala mediates chemosensory control of male hamster sexual behavior. *Science* 210: 557–560.

Leinders-Zufall, T., Brennan, P., Widmayer, P. S P, C., et al. (2004). MHC class I peptides as chemosensory signals in the vomeronasal organ. *Science* 306: 1033–1037.

Leinders-Zufall, T., Ishii, T., Mombaerts, P., et al. (2009). Structural requirements for the activation of vomeronasal sensory neurons by MHC peptides. *Nat. Neurosci.* 12 : 1551–1558.

Leinders-Zufall, T., Lane, A. P., Puche, A. C., et al. (2000). Ultrasensitive pheromone detection by mammalian vomeronasal neurons. *Nature* 405: 792–796.

Lemberger, T., Parlato, R., Dassesse, D., et al. (2007). Expression of Cre recombinase in dopaminoceptive neurons. *BMC Neurosci.* 8: 4.

Lewcock, J. W. and Reed, R. R. (2004). A feedback mechanism regulates monoallelic odorant receptor expression. *Proc. Nat. Acad. Sci. USA* 101: 1069–1074.

Leypold, B. G., Yu, C. R., Leinders-Zufall, T., et al. (2002). Altered sexual and social behaviors in trp2 mutant mice. *Proc. Nat. Acad. Sci. USA* 99: 6376–6381.

Liberles, S. D. and Buck, L. B. (2006). A second class of chemosensory receptors in the olfactory epithelium. *Nature* 442: 645–650.

Liberles, S. D., Horowitz, L. F., Kuang, D., et al. (2009). Formyl peptide receptors are candidate chemosensory receptors in the vomeronasal organ. *Proc. Nat. Acad. Sci. USA* 106: 9842–9847.

Liman, E. R. (2003). Regulation by voltage and adenine nucleotides of a Ca^{2+}-activated cation channel from hamster vomeronasal sensory neurons. *J. Physiol.* 548: 777–787.

Liman, E. R. and Corey, D. P. (1996). Electrophysiological characterization of chemosensory neurons from the mouse vomeronasal organ. *J. Neurosci.* 16: 4625–4637.

Liman, E. R., Corey, D. P. and Dulac, C. (1999). TRP2: A candidate transduction channel for mammalian pheromone sensory signaling. *Proc. Nat. Acad. Sci. USA* 96: 5791–5796.

Liman, E.R. and Innan, H. (2003). Relaxed selective pressure on an essential component of pheromone transduction in primate evolution. *Proc. Nat. Acad. Sci. USA* 100: 3328–3332.

Lin, D., Boyle, M. P., Dollar, P., et al. (2011). Functional identification of an aggression locus in the mouse hypothalamus. *Nature* 470: 221–226.

Liu, Y. C., Salamone, J. D. and Sachs, B. D. (1997). Lesions in medial preoptic area and bed nucleus of stria terminalis: differential effects on copulatory behavior and noncontact erection in male rats. *J. Neurosci.* 17: 5245–5253.

Lo, L. and Anderson, D. J. (2011). A cre-dependent, anterograde transsynaptic viral tracer for mapping output pathways of genetically marked neurons. *Neuron* 72: 938–950.

Loconto, J., Papes, F., Chang, E., et al. (2003). Functional expression of murine V2R pheromone receptors involves selective association with the M10 and M1 families of MHC class Ib molecules. *Cell* 112 : 607–618.

Logan, D. W., Brunet, L. J., Webb, W. R., et al. (2012). Learned recognition of maternal signature odors mediates the first suckling episode in mice. *Curr. Biol.* 22: 1998–2007.

Logan, D. W., Marton, T. F. and Stowers, L. (2008). Species specificity in major urinary proteins by parallel evolution. *PloS one* 3: e3280.

Lucas, P., Ukhanov, K., Leinders-Zufall, T. and Zufall, F. (2003). A diacylglycerol-gated cation channel in vomeronasal neuron dendrites is impaired in TRPC2 mutant mice: mechanism of pheromone transduction. *Neuron* 40: 551–561.

Luo, M. and Katz, L. C. (2004). Encoding pheromonal signals in the mammalian vomeronasal system. *Curr. Opin. Neurobiol.* 14: 428–434.

Luo, M., Fee, M. S. and Katz, L. C. (2003). Encoding Pheromonal Signals in the Accessory Olfactory Bulb of Behaving Mice. *Science* 299: 1196–1201.

Ma, J. and Lowe, G. (2004). Action potential backpropagation and multiglomerular signaling in the rat vomeronasal system. *J. Neurosci.* 24: 9341–9352.

Mandiyan, V. S., Coats, J. K. and Shah, N. M. (2005). Deficits in sexual and aggressive behaviors in Cnga2 mutant mice. *Nat. Neurosci.* 8: 1660–1662.

Marchand, R. and Bélanger, M. C. (1991). Ontogenesis of the axonal circuitry associated with the olfactory system of the rat embryo. *Neurosci. Letts.* 129: 285–290.

Martinez-Marcos, A. and Halpern, M. (1999). Differential centrifugal afferents to the anterior and posterior accessory olfactory bulb. *Neuroreport* 10: 2011–2015.

Martini, S., Silvotti, L., Shirazi, A., et al. (2001). Co-expression of putative pheromone receptors in the sensory neurons of the vomeronasal organ. *J. Neurosci.* 21: 843–848.

Matsunami, H. and Buck, L. B. (1997). A multigene family encoding a diverse array of putative pheromone receptors in mammals. *Cell* 90: 775–784.

Matsuoka, M., Yoshida-Matsuoka, J., Iwasaki, N., et al. (2001). Immunocytochemical study of Gi2alpha and Goalpha on the epithelium surface of the rat vomeronasal organ. *Chem. Senses* 26: 161–166.

Medina, L., Bupesh, M. and Abellan, A. (2011). Contribution of genoarchitecture to understanding forebrain evolution and development, with particular emphasis on the amygdala. *Brain Behav. Evol.* 78: 216–236.

Meeks, J. P., Arnson, H. A. and Holy, T. E. (2010). Representation and transformation of sensory information in the mouse accessory olfactory system. *Nat. Neurosci.* 13: 723–730.

Meisami, E. and Bhatnagar, K. P. (1998). Structure and diversity in mammalian accessory olfactory bulb. *Microsc. Re. Techniq* 43: 476–499.

Merchenthaler, I., Lane, M. V., Numan, S. and Dellovade, T. L. (2004). Distribution of estrogen receptor alpha and beta in the mouse central nervous system: in vivo autoradiographic and immunocytochemical analyses. *J. Comp. Neurol.* 473: 270–291.

Meredith, M. (1991). Sensory processing in the main and accessory olfactory systems: comparisons and contrasts. *J. Steroid Biochem.* 39: 601–614.

Meredith, M. and O'Connell, R. J. (1979). Efferent control of stimulus access to the hamster vomeronsasl organ. *J. Physiol.* 286: 301–316.

Meredith, M. and Westberry, J. M. (2004). Distinctive responses in the medial amygdala to same-species and different-species pheromones. *J. Neurosci.* 24 : 5719–5725.

Mitra, S. W., Hoskin, E., Yudkovitz, J., et al. et al(2003). Immunolocalization of estrogen receptor beta in the mouse brain: comparison with estrogen receptor alpha. *Endocrinology* 144 : 2055–2067.

Mohedano-Moriano, A., de la Rosa-Prieto, C., Saiz-Sanchez, D., et al. (2012). Centrifugal telencephalic afferent connections to the main and accessory olfactory bulbs. *Front. Neuroanat.* 6: 19.

Mohedano-Moriano, A., Pro-Sistiaga, P., Ubeda-Banon, I., et al. (2007). Segregated pathways to the vomeronasal amygdala: differential projections from the anterior and posterior divisions of the accessory olfactory bulb. *Eur. J. Neurosci.* 25: 2065–2080.

Mohedano-Moriano, A., Pro-Sistiaga, P., Ubeda-Banon, I., et al. (2008). V1R and V2R segregated vomeronasal pathways to the hypothalamus. *Neuroreport* 19: 1623–1626.

Mombaerts, P. (2004a). Odorant receptor gene choice in olfactory sensory neurons: the one receptor-one neuron hypothesis revisited. *Curr. Opin. Neurobiol.* 14: 31–36.

Mombaerts, P. (2004b). Genes and ligands for odorant, vomeronasal and taste receptors. *Nat. Rev. Neurosci.* 5: 263–278.

Mori, K., von Campenhause, H. and Yoshihara, Y. (2000). Zonal organization of the mammalian main and accessory olfactory systems. *Phil. Transact. Royal Soc. London B, Biol. Sci.* 355: 1801–1812.

Motta, S. C., Goto, M., Gouveia, F. V., et al. (2009). Dissecting the brain's fear system reveals the hypothalamus is critical for responding in subordinate conspecific intruders. *Proc. Nat. Acad. Sci. USA* 106: 4870–4875.

Mucignat-Caretta, C. (2010). The rodent accessory olfactory system. *J. Comp. Physiol.* 196: 767–777.

Mudge, J. M., Armstrong, S. D., McLaren, K., et al. (2008). Dynamic instability of the major urinary protein gene family revealed by genomic and phenotypic comparisons between C57 and 129 strain mice. *Genome Biol.* 9: R91.

Munger, S. D., Leinders-Zufall, T. and Zufall, F. (2009). Subsystem organization of the mammalian sense of smell. *Ann. Rev. Physiol.* 71: 115–140.

Murray, R. C., Navi, D., Fesenko, J., et al. (2003). Widespread defects in the primary olfactory pathway caused by loss of Mash1 function. *J. Neurosci.* 23: 1769–1780.

Nakagawa, T. and Vosshall, L. B. (2009). Controversy and consensus: noncanonical signaling mechanisms in the insect olfactory system. *Curr. Opin. Neurobiol.* 19: 284–292.

Neville, K. R. and Haberly, L. B. (2003). Beta and gamma oscillations in the olfactory system of the urethane-anesthetized rat. *J. Neurophysiol.* 90: 3921–3930.

Newman, S. W. (1999). The medial extended amygdala in male reproductive behavior. A node in the mammalian social behavior network. *Ann. NY Acad. Sci.* 877: 242–257.

Nishimura, K., Utsumi, K., Yuhara, M., et al. (1989). Identification of puberty-accelerating pheromones in male mouse urine. *J. Exp. Zool.* 251: 300–305.

Nodari, F., Hsu, F. F., Fu, X., et al. (2008). Sulfated steroids as natural ligands of mouse pheromone-sensing neurons. *J. Neurosci.* 28: 6407–6418.

Norlin, E. M., Gussing, F. and Berghard, A. (2003). Vomeronasal Phenotype and Behavioral Alterations in Gai2 Mutant Mice. *Curr. Biol.* 13: 1214–1219.

Novotny, M., Harvey, S. and Jemiolo, B. (1990). Chemistry of male dominance in the house mouse, Mus domesticus. *Experientia* 46: 109–113.

Novotny, M., Harvey, S., Jemiolo, B. and Alberts, J. (1985). Synthetic pheromones that promote inter-male aggression in mice. *Proc. Nat. Acad. Sci. USA* 82 : 2059–2061.

Novotny, M. V., Jemiolo, B., Wiesler, D., et al. (1999a). A unique urinary constituent, 6-hydroxy-6-methyl-3-heptanone, is a pheromone that accelerates puberty in female mice. *Chem. Biol.* 6: 377–383.

Novotny, M. V., Ma, W., Wiesler, D. and Zidek, L. (1999b). Positive identification of the puberty-accelerating pheromone of the house mouse: the volatile ligands associating with the major urinary protein. *Proc. R. Soc. Lond. B Biol. Sci.* 266 : 2017–2022.

Novotny, M. V., Soini, H. A., Koyama, S., et al. (2007). Chemical identification of MHC-influenced volatile compounds in mouse urine. I: Quantitative Proportions of Major Chemosignals. *J. Chem. Ecol.* 33: 417–434.

Otsuka, T., Ishii, K., Osako, Y., et al. (2001). Modulation of dendrodendritic interactions and mitral cell excitability in the mouse accessory olfactory bulb by vaginocervical stimulation. *Eur. J. Neurosci.* 13: 1833–1838.

Pankevich, D. E., Baum, M. J. and Cherry, J. A. (2004). Olfactory sex discrimination persists, whereas the preference for urinary odorants from estrous females disappears in male mice after vomeronasal organ removal. *J. Neurosci.* 24 : 9451–9457.

Papachristoforou, A., Kagiava, A., Papaefthimiou, C., et al. (2012). The bite of the honeybee: 2-heptanone secreted from honeybee mandibles during a bite acts as a local anaesthetic in insects and mammals. *PloS one* 7: e47432.

Papes, F., Logan, D. W. and Stowers, L. (2010). The vomeronasal organ mediates interspecies defensive behaviors through detection of protein pheromone homologs. *Cell* 141: 692–703.

Pardo-Bellver, C., Cadiz-Moretti, B., Novejarque, A., et al. (2012). Differential efferent projections of the anterior, posteroventral, and posterodorsal subdivisions of the medial amygdala in mice. *Front. Neuroanat.* 6: 33.

Park, S. H., Podlaha, O., Grus, W. E. and Zhang, J. (2011). The microevolution of V1R vomeronasal receptor genes in mice. *Gen. Biol. Evol.* 3: 401–412.

Pessoa, L. and Adolphs, R. (2010). Emotion processing and the amygdala: from a "low road" to "many roads" of evaluating biological significance. *Nat. Rev. Neurosci.* 11: 773–783.

Petrovich, G. D., Canteras, N. S. and Swanson, L. W. (2001). Combinatorial amygdalar inputs to hippocampal domains and hypothalamic behavior systems. *Brain Res. Brain Res. Rev.* 38 : 247–289.

Pifferi, S., Dibattista, M., Sagheddu, C., et al. 2009). Calcium-activated chloride currents in olfactory sensory neurons from mice lacking bestrophin-2. *J. Physiol.* 587: 4265–4279.

Price, J. L. and Powell, T. P. (1970). The synaptology of the granule cells of the olfactory bulb. *J. Cell Sci.* 7: 125–155.

Prince, J. E. A., Cho, J. H., Dumontier, E. et al. (2009). Robo-2 controls the segregation of a portion of basal vomeronasal sensory neuron axons to the posterior region of the accessory olfactory bulb. *J. Neurosci.: the official journal of the Society for Neuroscience* 29: 14211–14222.

Raisman, G. (1972). An experimental study of the projection of the amygdala to the accessory olfactory bulb and its relationship to the concept of a dual olfactory system. *Experimental Brain Res. Experimentelle Hirnforschung. Expérimentation cérébrale* 14: 395–408.

Rasche, S., Toetter, B., Adler, J., et al. (2010). Tmem16b is specifically expressed in the cilia of olfactory sensory neurons. *Chem. Senses* 35 : 239–245.

Rivière, S., Challet, L., Fluegge, D., et al. (2009). Formyl peptide receptor-like proteins are a novel family of vomeronasal chemosensors. *Nature* 459: 574–577.

Roberts, S. A., Davidson, A. J., McLean, L., et al. (2012). Pheromonal induction of spatial learning in mice. *Science* 338: 1462–1465.

Roberts, S. A., Simpson, D. M., Armstrong, S. D., et al. (2010). Darcin: a male pheromone that stimulates female memory and sexual attraction to an individual male's odour. *BMC Biology* 8: 75.

Robertson, D. H., Cox, K. A., Gaskell, S. J., et al. (1996). Molecular heterogeneity in the Major Urinary Proteins of the house mouse Mus musculus. *Biochem. J.* 316 (Pt 1): 265–272.

Robertson, D. H., Hurst, J. L., Bolgar, M. S., et al. (1997). Molecular heterogeneity of urinary proteins in wild house mouse populations. *Rapid Commun Mass Spectrom* 11: 786–790.

Rodriguez, I. (2004). Pheromone receptors in mammals. *Hormones and Behavior* 46: 219–230.

Rodriguez, I., Del Punta, K., Rothman, A., et al. (2002). Multiple new and isolated families within the mouse superfamily of *V1r* vomeronasal receptors. *Nat. Neurosci.* 5: 134–140.

Rodriguez, I., Feinstein, P. and Mombaerts, P. (1999). Variable patterns of axonal projections of sensory neurons in the mouse vomeronasal system. *Cell* 97: 199–208.

Roppolo, D., Vollery, S., Kan, C.-D., et al. (2007). *Gene* cluster lock after pheromone receptor gene choice. *The EMBO Journal* 26: 3423–3430.

References

Ryba, N. J. P. and Tirindelli, R. (1997). A new multigene family of putative pheromone receptors. *Neuron* 19: 371–379.

Salazar, I. and Sánchez-Quinteiro, P. (2009). The risk of extrapolation in neuroanatomy: the case of the Mammalian vomeronasal system. *Front. Neuroanat.* 3 : 22.

Salazar, I., Quinteiro, P. S., Aleman, N., et al. (2007). Diversity of the Vomeronasal System in Mammals: The Singularities of the Sheep Model. *Microsc. Re. Techniq* 70 : 752–762.

Salazar, I., Sanchez-Quinteiro, P., Cifuentes, J. M. and Fernandez De Troconiz, P. (2006). General organization of the perinatal and adult accessory olfactory bulb in mice. The anatomical record. *Part A, Discoveries in Molecular, Cellular, and Evolutionary Biology* 288: 1009–1025.

Salzman, C. D. and Fusi, S. (2010). Emotion, cognition, and mental state representation in amygdala and prefrontal cortex. *Ann. Rev. Neurosci.* 33: 173–202.

Sam, M., Vora, S., Malnic, B., (2001). Neuropharmacology. Odorants may arouse instinctive behaviours. *Nature* 412: 142.

Scalia, F. and Winans, S. S. (1975). The differential projections of the olfactory bulb and accessory olfactory bulb in mammals. *J. Comp. Neurol.* 161: 31–55.

Schiller, L., Donix, M., Jahkel, M. and Oehler, J. (2006). Serotonin 1A and 2A receptor densities, neurochemical and behavioural characteristics in two closely related mice strains after long-term isolation. *Prog. Neuro-Psychopha.* 30: 492–503.

Schoppa, N. E. (2006). Synchronization of olfactory bulb mitral cells by precisely timed inhibitory inputs. *Neuron* 49: 271–283.

Schoppa, N. E. and Urban, N. N. (2003). Dendritic processing within olfactory bulb circuits. *Trends Neurosci.* 26: 501–506.

Schroeder, B. C., Cheng, T., Jan, Y. N. and Jan, L. Y. (2008). Expression cloning of TMEM16A as a calcium-activated chloride channel subunit. *Cell* 134: 1019–1029.

Schulz, S., Kruckert, K. and Weldon, P. J. (2003). New terpene hydrocarbons from the alligatoridae (crocodylia, reptilia). *Journal of Natural Products* 66: 34–38.

Schwende, F. J., Wiesler, D., Jorgenson, J. W., et al. (1986). Urinary volatile constituents of the house mouse, Mus musculus, and their endocrine dependency. *J. Chem. Ecol.* 12: 277–296.

Segovia, S. and Guillamon, A. (1993). Sexual dimorphism in the vomeronasal pathway and sex differences in reproductive behaviors. *Brain Res. Brain Res. Rev.* 18 : 51–74.

Serizawa, S., Miyamichi, K., Nakatani, H., et al. (2003). Negative feedback regulation ensures the one receptor-one olfactory neuron rule in mouse. *Science* 302: 2088–2094.

Sharrow, S. D., Vaughn, J. L., Zidek, L., et al. (2002). Pheromone binding by polymorphic mouse major urinary proteins. *Protein Sci.* 11: 2247–2256.

Shaw, P. H., Held, W. A. and Hastie, N. D. (1983). The gene family for major urinary proteins: expression in several secretory tissues of the mouse. *Cell* 32: 755–761.

Shi, P. and Zhang, J. (2007). Comparative genomic analysis identifies an evolutionary shift of vomeronasal receptor gene repertoires in the vertebrate transition from water to land. *Genome Research* 17: 166–174.

Shimazaki, R., Boccaccio, A., Mazzatenta, A., et al. (2006). Electrophysiological properties and modeling of murine vomeronasal sensory neurons in acute slice preparations. *Chem. Senses* 31: 425–435.

Shpak, G., Zylbertal, A., Yarom, Y. and Wagner, S. (2012). Calcium-Activated Sustained Firing Responses Distinguish Accessory from Main Olfactory Bulb Mitral Cells. *J. Neurosci.* 32 : 6251–6262.

Silk, P. J., Lemay, M. A., LeClair, G., et al. (2010). Behavioral and electrophysiological responses of Tetropium fuscum (Coleoptera: Cerambycidae) to pheromone and spruce volatiles. *Environ. Entomol.* 39 : 1997–2005.

Silvotti, L., Cavalca, E., Gatti, R., et al. (2011). A recent class of chemosensory neurons developed in mouse and rat. *PloS one* 6: e24462.

Silvotti, L., Moiani, A., Gatti, R. and Tirindelli, R. (2007). Combinatorial co-expression of pheromone receptors, V2Rs. *J. Neurochem.* 103: 1753–1763.

Smith, R. S. and Araneda, R. C. (2010). Cholinergic modulation of neuronal excitability in the accessory olfactory bulb. *J. Neurophysiol.* 104: 2963–2974.

Smith, R. S., Weitz, C. J. and Araneda, R. C. (2009). Excitatory Actions of Noradrenaline and Metabotropic Glutamate Receptor Activation in Granule Cells of the Accessory Olfactory Bulb. *J. Neurophysiol.* 102: 1103–1114.

Soehnlein, O. and Lindbom, L. (2010). Phagocyte partnership during the onset and resolution of inflammation. *Nat. Rev. Immunol.* 10: 427–439.

Sorensen, P. W., Fine, J. M., Dvornikovs, V., et al. (2005). Mixture of new sulfated steroids functions as a migratory pheromone in the sea lamprey. *Nat. Chem. Biol.* 1: 324–328.

Spehr, J., Hagendorf, S., Weiss, J., et al. (2009). Ca^{2+} -calmodulin feedback mediates sensory adaptation and inhibits pheromone-sensitive ion channels in the vomeronasal organ. *J. Neurosci.* 29: 2125–2135.

Spehr, M. and Munger, S. D. (2009). Olfactory receptors: G protein-coupled receptors and beyond. *J. Neurochem.* 109: 1570–1583.

Spehr, M., Hatt, H. and Wetzel, C. H. (2002). Arachidonic acid plays a role in rat vomeronasal signal transduction. *J. Neurosci.* 22: 8429–8437.

Spehr, M., Kelliher, K. R., Li, X. H., et al. (2006a). Essential role of the main olfactory system in social recognition of major histocompatibility complex peptide ligands. *J. Neurosci.* 26: 1961–1970.

Spehr, M., Spehr, J., Ukhanov, K., et al. (2006b). Parallel processing of social signals by the mammalian main and accessory olfactory systems. *Cell. Mol. Life Sci.* 63: 1476–1484.

Stephan, A. B., Shum, E. Y., Hirsh, S., et al. (2009). ANO2 is the cilial calcium-activated chloride channel that may mediate olfactory amplification. *Proc. Nat. Acad. Sci. USA* 106: 11776–11781.

Stettler, D. D. and Axel, R. (2009). Representations of odor in the piriform cortex. *Neuron* 63: 854–864.

Stowers, L. and Logan, D. W. (2010). Sexual dimorphism in olfactory signaling. *Curr. Opin. Neurobiol.* 20: 770–775.

Stowers, L., Holy, T. E., Meister, M., et al. (2002). Loss of Sex Discrimination and Male-Male Aggression in Mice Deficient for TRP2. *Science* 295: 1493–1500.

Sugai, T., Sugitani, M. and Onoda, N. (1999). Effects of GABAergic agonists and antagonists on oscillatory signal propagation in the guinea-pig accessory olfactory bulb slice revealed by optical recording. *Eur. J. Neurosci.* 11: 2773–2782.

Szabó, K. and Mendoza, A. S. (1988). Developmental studies on the rat vomeronasal organ: vascular pattern and neuroepithelial differentiation. *I. Light microscopy. Brain Res.* 467: 253–258.

Szoka, P. R. and Paigen, K. (1978). Regulation of mouse major urinary protein production by the Mup-A gene. *Genetics* 90: 597–612.

Takami, S. and Graziadei, P. P. (1991). Light microscopic Golgi study of mitral/tufted cells in the accessory olfactory bulb of the adult rat. *J. Comp. Neurol.* 311: 65–83.

Tanaka, M., Treloar, H. B., Kalb, R. G., et al. (1999). G(o) protein-dependent survival of primary accessory olfactory neurons. *Proc. Nat. Acad. Sci. USA* 96: 14106–14111.

Taniguchi, M. and Kaba, H. (2001). Properties of reciprocal synapses in the mouse accessory olfactory bulb. *Neuroscience* 108: 365–370.

Tarozzo, G., Cappello, P., De Andrea, M., et al. (1998). Prenatal differentiation of mouse vomeronasal neurones. *Eur. J. Neurosci.* 10 : 392–396.

Timm, D. E., Baker, L. J., Mueller, H., et al. (2001). Structural basis of pheromone binding to mouse major urinary protein (MUP-I). *Prot. Sci.* 10: 997–1004.

Tirindelli, R., Dibattista, M., Pifferi, S. and Menini, A. (2009). From pheromones to behavior. *Physiol. Rev.* 89: 921–956.

Trinh, K. and Storm, D. R. (2003). Vomeronasal organ detects odorants in absence of signaling through main olfactory epithelium. *Nat. Neurosci.* 6: 519–525.

Ukhanov, K., Leinders-Zufall, T. and Zufall, F. (2007). Patch-clamp analysis of gene-targeted vomeronasal neurons expressing a defined *V1r* or *V2r* receptor: ionic mechanisms underlying persistent firing. *J. Neurophysiol.* 98: 2357–2369.

Urban, N. N. and Castro, J. B. (2005). Tuft calcium spikes in accessory olfactory bulb mitral cells. *J. Neurosci.* 25 : 5024–5028.

Van Pett, K., Viau, V., Bittencourt, J. C., et al. (2000). Distribution of mRNAs encoding CRF receptors in brain and pituitary of rat and mouse. *J. Comp. Neurol.* 428: 191–212.

von Campenhausen, H. and Mori, K. (2000). Convergence of segregated pheromonal pathways from the accessory olfactory bulb to the cortex in the mouse. *Eur. J. Neurosci.* 12: 33–46.

Vosshall, L. B. and Hansson, B. S. (2011). A unified nomenclature system for the insect olfactory coreceptor. *Chem. Senses* 36: 497–498.

Wagner, S., Gresser, A. L., Torello, A. T. and Dulac, C. (2006). A multireceptor genetic approach uncovers an ordered integration of VNO sensory inputs in the accessory olfactory bulb. *Neuron* 50: 697–709.

Whalen, P. J. (1998). Fear, vigilance, and ambiguity: initial neuroimaging studies of the human amygdala. *Curr. Dir. Psychol. Sci.* 7: 177–188.

Wilson, E. O. (1963). Pheromones. *Scientific America* 208: 100–114.

Winans, S. S. and Scalia, F. (1970). Amygdaloid nucleus: new afferent input from the vomeronasal organ. *Science* 170: 330–332.

Wood, R. I. and Swann, J. M. (2005). The bed nucleus of the stria terminalis in the Syrian hamster: subnuclei and connections of the posterior division. *Neuroscience* 135: 155–179.

Wu, M. V. and Shah, N. M. (2011). Control of masculinization of the brain and behavior. *Curr. Opin. Neurobiol.* 21: 116–123.

Wysocki, C. J., Wellington, J. L. and Beauchamp, G. K. (1980). Access of urinary nonvolatiles to the mammalian vomeronasal organ. *Science* 207: 781–783.

Xu, X., Coats, J. K., Yang, C. F., et al. (2012). Modular genetic control of sexually dimorphic behaviors. *Cell* 148: 596–607.

Yang, C. and Delay, R. J. (2010). Calcium-activated chloride current amplifies the response to urine in mouse vomeronasal sensory neurons. *J. Gen. Physiol.* 135 : 3–13.

Yang, H., Kim, A., David, T., et al. (2012). TMEM16F Forms a Ca(2+)-activated cation channel required for lipid scrambling in platelets during blood coagulation. *Cell* 151: 111–122.

Young, J. M. and Trask, B. J. (2007). V2R gene families degenerated in primates, dog and cow, but expanded in opossum. *Trends Genet.* 23: 209–212.

Young, J. M., Massa, H. F., Hsu, L. and Trask, B. J. (2010). Extreme variability among mammalian V1R gene families. *Genome Res.* 20: 10–18.

Yu, T. W. and Bargmann, C. I. (2001). Dynamic regulation of axon guidance. *Nat. Neurosci.* 4 Suppl: 1169–1176.

Zhang, J. and Webb, D. M. (2003). Evolutionary deterioration of the vomeronasal pheromone transduction pathway in catarrhine primates. *Proc. Nat. Acad. Sci. USA* 100: 8337–8341.

Zhang, P., Yang, C. and Delay, R. J. (2008). Urine stimulation activates BK channels in mouse vomeronasal neurons. *J. Neurophysiol.* 100: 1824–1834.

Zhang, X., Rodriguez, I., Mombaerts, P. and Firestein, S. (2004). Odorant and vomeronasal receptor genes in two mouse genome assemblies. *Genomics* 83: 802–811.

Zibman, S., Shpak, G. and Wagner, S. (2011). Distinct intrinsic membrane properties determine differential information processing between main and accessory olfactory bulb mitral cells. *Neuroscience* 189: 51–67.

Chapter 52

The Septal Organ, Grueneberg Ganglion, and Terminal Nerve

MINGHONG MA, JOERG FLEISCHER, HEINZ BREER, and HEATHER EISTHEN

52.1 INTRODUCTION

All vertebrate animals develop complicated chemosensory systems to detect a myriad of chemical cues signaling potential food, mates, and danger. In most non-human animals, the nose – a seemingly unitary organ – contains distinct receptors for multiple chemosensory systems. Besides the main olfactory epithelium (MOE), afferents from the trigeminal nerve (CNV), and the vomeronasal organ (VNO), many mammalian species possess two spatially segregated clusters of sensory cells in the nasal cavity forming the septal organ of Masera (SO) and the Grueneberg ganglion (GG) (Figure 52.1). Each of these chemosensory subsystems detects distinct but overlapping olfactory cues. Interestingly, the SO and GG neurons are polymodal; they convey other sensory modalities mediated by mechanical and thermal stimuli, respectively. The key features of the SO and GG will be discussed in Section 52.2 and 52.3, respectively, including their chemoreceptors, signal transduction cascades, central projections, and functional roles.

In addition, animals constantly adjust their sensory responses according to their internal status and external environment, partly achieved via centrifugal modulatory pathways. Activity in the olfactory epithelium is modulated by the terminal nerve (TN), which releases peptides, including gonadotropin-releasing hormone (GnRH; Figure 52.1). Section 52.4 of this chapter reviews the modulatory function of the TN system.

52.2 THE SEPTAL ORGAN (SO)

The SO is a small island of olfactory epithelium lying bilaterally at the ventral nasal septum near the entrance of nasopharynx (Figure 52.1). It was first described in newborn mice by Broman (1921) and subsequently characterized in detail by Rodolfo-Masera (1943). The SO has not been demonstrated in primates, including humans, but is found in many mammals, including rabbit, rat, mouse, hamster, deer mouse, guinea pig, opossum, bandicoot, and koala (Rodolfo-Masera, 1943; Adams and McFarland, 1971; Bojsen-Moller, 1975; Katz and Merzel, 1977; Breipohl et al., 1983, 1989; Kratzing, 1984a,b; Taniguchi et al., 1993), but not in cat or ferret (Breipohl et al., 1983; Weiler and Farbman, 2003). The SO resembles the MOE in the cellular composition; it is composed of ciliated olfactory sensory neurons (OSNs), microvillar cells, supporting cells, and Bowman's glands (Graziadei, 1977; Miragall et al., 1984; Kratzing, 1984a,b; Breipohl et al., 1989; Adams, 1992; Taniguchi et al., 1993; Giannetti et al., 1995a).

During development, the SO first appears around embryonic day E16 in both rat and mouse (Giannetti et al., 1995a; Oikawa et al., 2001; Tian and Ma, 2008b). At this stage, the MOE is separated from the respiratory epithelium with a well-defined boundary and some olfactory marker protein (OMP)-positive cells start to cluster in the presumptive SO region. From E18 to postnatal day 7 (P7), OMP-positive OSNs continue to accumulate in the SO, accompanied by the gradual disappearance of the OMP-GFP cells in the transitional zone between the SO and the MOE (Tian and Ma, 2008b). The SO continues to develop postnatally and reaches its peak in size between 2 to 4 months (Weiler and Farbman, 2003). Recent years have witnessed significant advances in our understanding of the molecular and functional organization of this enigmatic organ.

Handbook of Olfaction and Gustation, Third Edition. Edited by Richard L. Doty.
© 2015 Richard L. Doty. Published 2015 by John Wiley & Sons, Inc.

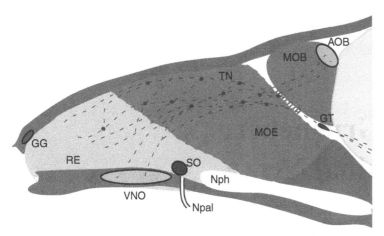

Figure 52.1 Schematic drawing of the mid-sagittal view of the nasal cavity illustrates multiple chemosensory organs in the rodent nose. MOE = main olfactory epithelium; MOB = main olfactory bulb; VNO = vomeronasal organ; AOB = accessory olfactory bulb; SO = septal organ; RE = respiratory epithelium; GG = Grueneberg ganglion (GG); Npal = nasopalatine duct; Nph = nasopharynx; TN = terminal system; GT = ganglion terminale. The trigeminal system is not shown.

52.2.1 Odorant Receptors and Signal Transduction

Surprisingly, the septal organ predominantly expresses only a handful of odorant receptors (ORs) out of a repertoire of ~1200 in mouse. Using the cDNA cloning approach, two groups identified more than 120 candidate OR genes from the mouse SO (Kaluza et al., 2004; Tian and Ma, 2004). However, the expression levels of individual OR genes vary dramatically, confirmed by a customized Affymetrix genechip covering all the mouse OR genes (Zhang et al., 2004) and *in situ* hybridization. The most abundant OR, SR1 (also termed MOR256-3 or Olfr124), is expressed in nearly 50% of the SO neurons. Each of the other eight ORs (MOR267-16, 122–1, 160–5, 236–1, 256–17, 0–2, 235–1 and 244–3 from low- to high-abundance) is expressed approximately in 3.7% (MOR267-16) to 13.3% (MOR244-3) of the SO neurons (Figure 52.2A). Together, these nine OR types account for >90% of the SO neurons (Figure 52.2B). All nine abundant septal organ ORs are also expressed in the most ventrolateral zone of the MOE, even though the relative abundance does not match that in the septal organ (Tian and Ma, 2004).

The unusually high density of SR1 neurons in the SO raises the question whether a single cell expresses a single OR type. Double *in situ* hybridization experiments with combined OR probes reiterate the one cell-one receptor tenet with few exceptions (Tian and Ma, 2004). Interestingly, the coexpression frequency of SR1 vs. a mixture of the remaining eight ORs in single neurons is nearly ten times higher in newborn mice (2.0% at P0 vs. 0.2% at P28), suggesting a reduction of the sensory neurons with multiple ORs during postnatal development. In addition, such reduction is prevented by four-week sensory deprivation or impaired apoptosis, indicating that activity induced by sensory inputs plays a role in ensuring the one cell–one receptor rule in a subset of OSNs in the nose (Tian and Ma, 2008a).

Most SO neurons express the olfactory G protein (G_{olf}) and type III adenylyl cyclase (ACIII), revealed by antibody staining (Ma et al., 2003). This suggests that signal transduction in the SO is mediated by the canonical cAMP pathway, which is further supported by patch clamp recordings from individual OSNs in this region. Odorant responses were mimicked by an adenylyl cyclase activator and a phosphodiesterase inhibitor, blocked by an adenylyl cyclase inhibitor, and eliminated in cyclic nucleotide-gated (CNG) channel knockout mice (Ma et al., 2003; Grosmaitre et al., 2007). Thus, the SO resembles the MOE in signal transduction, but represents a much simpler system by predominantly expressing a small fraction (~1%) of the OR repertoire.

52.2.2 Central Projections

The SO neurons give rise to a few axon bundles that travel across the cribriform plate and terminate in the main olfactory bulb (MOB). In rats, focal injection of horseradish peroxidase (HRP) in the MOB results in labeled cells in the septal organ (Pedersen and Benson, 1986), while injection of HRP in the SO traces the axons to the posterior, ventromedial olfactory bulb with a few densely labeled and many lightly labeled glomeruli (Astic and Saucier, 1988; Giannetti et al., 1992). More recently, these findings were confirmed by placing lipophilic dyes (such as DiI) in the SO (Levai and Strotmann, 2003; Ma et al., 2003). The densely labeled glomeruli (so called "septal" glomeruli) receive inputs mainly (if not exclusively) from the SO, while the lightly labeled glomeruli receive mixed inputs from the SO and the MOE. The projection pattern of the SO is consistent with the fact that this region expresses a few ORs at high densities and many other ORs sparsely. The cortical regions that receive inputs from the mitral/tufted cells innervating the "septal" glomeruli remain to be determined.

52.2 The Septal Organ (SO)

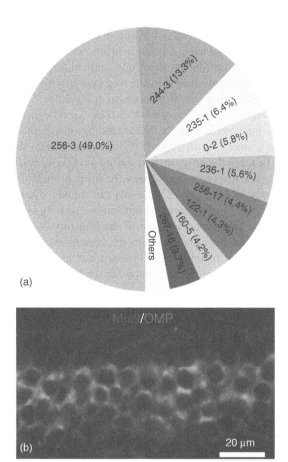

Figure 52.2 The septal organ predominantly expresses nine odorant receptors. (a) The relative contribution of these nine odorant receptors in the septal organ of one-month old mice. The percentage for MOR256-3 (SR1) and MOR244-3 cells was determined by double in situ hybridization against the OMP probe and the percentage for other receptor cells was calculated by normalizing their cell densities to that of MOR244-3 cells (Tian & Ma, 2004; 2008b). (b) The mixed probes from these nine odorant receptors labeled >90% of OMP-positive neurons in the septal organ section. Adapted from (Tian and Ma, 2008b). (*See plate section for color version.*)

52.2.3 Odorant Response Properties

Although the SO primarily covers a small fraction of the OR repertoire, most SO neurons (~70%) are surprisingly responsive to diverse chemicals with different size, shape, and functional groups, as revealed in patch clamp recordings (Grosmaitre et al., 2007; Grosmaitre et al., 2009). These neurons are extremely sensitive to some odorants with a nanomolar threshold and a wide dynamic range covering 3~4 log units of concentration from threshold to saturation (Figure 52.3). This is consistent with a previous EOG study in which the SO was found to respond to many chemicals with a lower threshold than the MOE (Marshall and Maruniak, 1986).

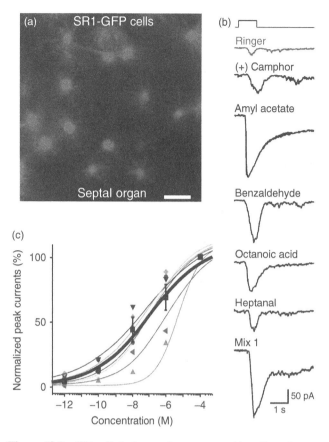

Figure 52.3 SR1 cells in the septal organ respond broadly and over a wide concentration range. (a) The image was taken under fluorescent illumination of the septal organ from an SR1–IRES–tauGFP mouse. Scale bar, 5 μm. (b) A single SR1 cell responded to all five odorants at 1 μM, and to Mix 1 at 1 μM, under voltage-clamp mode. The gray trace (Ringer) was the mechanical response induced by a Ringer puff that was delivered with the same pressure as the odorant puffs. (c) The dose–response curve for amyl acetate was averaged from six SR1 cells. The holding potential was −65 mV for all neurons. Each colored line (and the corresponding symbols) represents the data from a single cell, and the thick black line represents the averaged curve from all six neurons. Error bars indicate SEM. Adapted from (Grosmaitre et al., 2009). (*See plate section for color version.*)

Because SR1 is expressed in 50% of the SO neurons, it is plausible that SR1 confers broad tuning to its host OSNs. This concept is supported by three pieces of evidence. First, all SR1 neurons (genetically labeled for identification, Figure 52.3a), regardless of their location in the SO or in the MOE, are broadly responsive and sensitive (Figure 52.3b,c). Second, genetic deletion of SR1 results in OSNs that are mostly narrowly-tuned to odorants. Third, when expressed in a heterologous cell line, SR1 shows broad response spectrum (Grosmaitre et al., 2009). However, SR1 is not the only broadly-tuned OR in the SO, because more than half of the non-SR1 neurons are also broadly-responsive (Grosmaitre et al., 2009). A recent study confirms that another SO receptor, MOR256-17, is broadly-tuned in a heterologous system (Li et al., 2012). The SO also contains OSNs that are highly selective. For instance, MOR244-3 (the second most abundant OR in the SO), when expressed in heterologous cells, responds to (methylthio)methanethiol (MTMT), a semiochemical in the male urine that appears to attract female mice. As expected, a significant portion of the non-SR1 neurons in the SO responded to MTMT in patch clamp recordings (Duan et al., 2012). It appears that mammalian ORs and their host OSNs show diverse tuning properties, ranging from highly selective (e.g., MOR244-3) to broadly responsive (e.g., SR1 and MOR256-17).

Situated in the direct air path and packed with broadly-responsive OSNs, the SO may serve as a general odor detector and/or a sensor of the total odor concentration to "alert" the organism, a function originally proposed by Rodolfo-Masera (1943). Nevertheless, the potential alerting role was not verified in a behavioral study with surgical removal of the septal organ (Giannetti et al., 1995b), which could be due to the existence of broadly tuned OSNs in the MOE. An alternative and complementary hypothesis suggests that the septal organ may function as a "mini-nose" in surveying food odors as well as social cues (Breer et al., 2006). The finding that the SO contains MOR244-3 neurons which respond to MTMT supports its role in detecting social cues. It is plausible that this enigmatic organ serves multiple chemosensory roles.

52.2.4 Mechanosensitivity of Olfactory Sensory Neurons

Another novel finding arising from the SO is that many OSNs respond not only to odorants, but also to mechanical stimuli delivered by pressure ejections of odorant-free Ringer's solution (Grosmaitre et al., 2007). The mechanical responses directly correlate with the pressure intensity and similar mechanosensitivity also exists in a subset of OSNs in the MOE. Several lines of evidence support that odorant receptors are required in mechanosensitivity of mouse OSNs. First, genetic ablation of key signaling proteins in odor transduction completely eliminates mechanical responses in OSNs. Second, mechanosensitivity is associated with certain odorant receptor types. OSNs expressing SR1, I7 or M71 are mechanosensitive while those expressing MOR23 and mOR-EG are not. Third, loss-of-function mutation of the I7 receptor or genetic ablation of the SR1 receptor abolishes or significantly reduces mechanical responses in the host OSNs (Connelly et al., 2015). It remains to be determined whether odorant receptors serve as mechanical sensors as well as odor detectors.

The mechanosensitivity found in the OSNs is particularly interesting because these neurons could provide airflow information during breathing and sniffing, which is known to be required by the olfactory system when actively sampling external stimuli (Buonviso et al., 2006; Schaefer and Margrie, 2007; Wachowiak, 2011). An odorant signal can be interpreted differently, depending on whether it is due to a strong or weak sniff. For instance, when the air flows faster in the nose, such as during a powerful sniff, the mechanical response can enhance the firing probability and frequency of individual OSNs that are weakly stimulated by odorants. Furthermore, the mechanosensitivity of OSNs may drive the synchronization of the olfactory bulb activity (theta-band oscillation) with the breathing cycle, even in the absence of odorants (Grosmaitre et al., 2007). The SO can thus serve as an airflow sensor in addition to its chemosensory roles.

52.3 THE GRUENEBERG GANGLION (GG)

Mature sensory neurons in the MOE and VNO are characterized by the expression of olfactory marker protein (OMP). Recently GG neurons were also found to express OMP (Fuss et al., 2005; Koos and Fraser, 2005; Fleischer et al., 2006a; Roppolo et al., 2006; Storan and Key, 2006). The GG, which was named after Hans Grüneberg who first described it (Grüneberg, 1973), has been studied almost exclusively in mice and it is yet unclear whether it exists in other mammals.

52.3.1 Morphology of the GG

In the murine nose, the GG is situated bilaterally in the nasal vestibule, adjacent to the opening of the naris (Figure 52.4a,b). The GG has an extent of a few hundred micrometers along the rostral/caudal axis; it resides between the levels of the orifices of the lateral nasal gland (glandula nasalis lateralis or Steno's gland) and the glandula nasalis medialis. It is surrounded by the nasal roof, the septum, and a thin epithelial layer lining the lumen of the nasal cavity (Grüneberg, 1973; Fuss et al., 2005; Koos and Fraser, 2005; Fleischer et al., 2006a; Roppolo et al., 2006; Storan and Key, 2006) (Figure 52.4c).

Two distinct cell types are present in the GG. The first type, hereinafter designated as GG neurons or GG cells, has a round to oval or elliptical shape (Fuss et al., 2005; Koos and Fraser, 2005; Roppolo et al., 2006; Storan and Key, 2006) and expresses OMP (Figure 52.4d) as well as the neuronal marker βIII tubulin (Fleischer et al., 2006a; Brechbuhl et al., 2008). Each GG neuron harbors about 30 to 40 cilia with a length of approximately 15 μm and a diameter of about 0.2 μm. The cilia are arranged into

52.3 The Grueneberg Ganglion (GG)

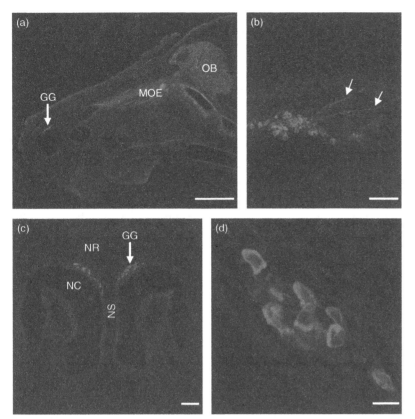

Figure 52.4 Morphological arrangement of GG neurons. (**a**) Sagittal section through the head of a neonatal mouse stained with an OMP-specific antibody. At a position far anterior from the main olfactory epithelium (MOE), OMP-expressing cells in the GG are marked by the arrow (OB: olfactory bulb). (**b**) Higher magnification image of a sagittal section through the very anterior part of the nose harboring the GG. Axons emanating from OMP-positive GG neurons are indicated by arrows. (**c**) Coronal section through the anterior nasal region of a newborn mouse stained with an OMP antiserum reveals the bilaterally symmetric arrangement of the GG between the nasal septum (NS), the nasal roof (NR) and the nasal cavity (NC). (**d**) High magnification of a small group of GG neurons stained with an OMP-specific antibody. Sections are counterstained with DAPI or Toto-3. Scale bars: **A** = 500 μm; **B, C** = 100 μm; **D** = 10 μm. (data from Fleischer et al., 2006b; with kind permission of John Wiley and Sons) (*See plate section for color version.*)

three to four bundles localized to discrete cellular regions. The structure of these cilia is quite exceptional. Their base is located deep in the cell body; consequently, a substantial portion of the cilia is invaginated into the cytoplasm of the soma (Tachibana et al., 1990; Brechbuhl et al., 2008). The cytoskeletal structure of the ciliary axoneme is also unusual: the basal body area comprises 9 triplets of microtubules, the proximal portion of the cilia is composed of a (9+0) microtuble doublet and the more distal regions harbor an 8+1 doublet (Brechbühl et al., 2008). GG neurons and their cilia seem to be largely enveloped by so-called satellite or ensheathing cells, which comprise the second type of cells in the GG. The expression of molecular markers such as the glial fibrillary acidic protein (GFAP) suggests that these cells have a glia-like phenotype (Tachibana et al., 1990; Brechbuhl et al., 2008). In addition, GG neurons are embedded in connective tissue that is superimposed by a keratinized epithelial layer (Grüneberg, 1973; Fuss et al., 2005; Koos and Fraser, 2005; Fleischer et al., 2006a; Roppolo et al., 2006; Storan and Key, 2006). Thus, in contrast to other OMP-positive neurons in the nose, GG cells apparently lack any direct access to the nasal lumen and are not part of the nasal epithelium.

Throughout pre- and postnatal development, the morphology of the GG undergoes considerable changes. In perinatal stages of mice, GG cells are arranged in smaller groups or larger clusters surrounded by the connective tissue of the anterior nasal region. This arrangement becomes evident for the first time at embryonic days 15 and 16 (E15/E16). However, the GG neurons are not born in this connective tissue, but appear to originate in stage E14 from a relatively thick epithelium in the anterior nasal region from which they seem to bud off before penetrating the subjacent connective tissue. In juvenile and adult mice, the ganglion-like structure is transformed into a filiform arrangement of the GG neurons (Fleischer et al., 2006a). Moreover, the approximately 800 GG neurons in perinatal stages decline to less than 700 in adults (Fleischer et al., 2007).

Each GG neuron is endowed with a single axon (Figure 52.4B) that projects to the brain (Fuss et al., 2005; Koos and Fraser, 2005; Fleischer et al., 2006a; Roppolo et al., 2006; Storan and Key, 2006). The first GG axons emerge at about E16 (Fuss et al., 2005). Lesion of the axonal process leads to degeneration of GG neurons. In contrast to OMP-positive neurons in other nasal compartments, GG neurons do not regenerate after axotomy. Consistent with this observation, in experiments using the thymidine analogue BrdU, which labels newly generated cells, only very few stained GG cells were found (Roppolo et al., 2006).

52.3.2 Spontaneous Electrical Activity of GG Neurons

Based on their spontaneous firing pattern in extracellular recording experiments, GG neurons have been grouped into three categories (Liu et al., 2012). The first group (~38% of the cells) comprises neurons with a continuous discharge of action potentials with firing frequencies of 10 to 25 Hz. The second group (~29% of the cells) is characterized by bursts of action potentials followed by electrical quiescence, i.e., long inter-burst intervals (0.3 to 2 seconds). The third group (~33% of the cells) shows only sporadic firing of actions potentials. The molecular basis and the functional relevance of these three distinct spontaneous firing patterns is yet unknown.

52.3.3 Axonal Projection Patterns of GG Neurons

The GG was initially considered to be part of the terminal nerve (TN) (Grüneberg, 1973) (see Section 52.4). The recent finding that GG neurons express OMP favors the view that the GG might, in fact, belong to the olfactory system. In general, OMP-expressing neurons in the nose project their axons to the olfactory bulb (OB) where they synapse on dendritic processes in round structures called glomeruli. The axons of GG neurons coalesce and form a single or a few tracts that run ipsilaterally along the dorsal roof of the nose, penetrate the cribriform plate and approach the MOB. The axon bundles then project along the medial side of the MOB and diverge into two (or sometimes more) smaller bundles that innervate a small group of glomeruli (up to 10) in the posterior region of the MOB (Figure 52.5). These glomeruli are arranged as a chain-like structure surrounding the rostral part of the accessory olfactory bulb (AOB) which resides on the dorsal and posterior region of the MOB.

To reach the relevant glomeruli, one portion of GG fibers takes a dorsal to lateral route, thus innervating glomeruli in the dorsal and in the lateral aspect of the MOB. The other fibers follow a medial to ventral trajectory and innervate glomeruli in these regions of the MOB. Finally, both bundles reach the most ventral region of the MOB (Fuss et al., 2005; Koos and Fraser, 2005; Fleischer et al., 2006a; Roppolo et al., 2006; Storan and Key, 2006; Matsuo et al., 2012). This axonal projection pattern strongly suggests that the GG does not belong to the terminal nerve system but is part of the olfactory system. Further characterization of the glomerular structures innervated by GG fibers revealed that the presynaptic marker synaptotagmin is already present at birth. This finding indicates an early maturation of the GG circuitry and suggests that the GG may already be functional in newborn animals (Matsuo et al., 2012).

52.3.4 Olfactory Signaling Proteins in GG Neurons

Olfactory receptors are necessary to render OSNs responsive to odorants and should be expressed in GG neurons if they also have an olfactory function. Nevertheless, odorant receptors (ORs) are only expressed in very few GG neurons and only at prenatal stages; instead, most GG neurons express the V2r83 receptor (also called Vmn2r1 or V2R2), which belongs to the V2R family of olfactory receptors (Fleischer et al., 2006b) typically expressed in the VNO (Martini et al., 2001). The small subpopulation of V2r83-negative GG neurons expresses members of the trace amine-associated receptors (TAARs); however, only one subtype is expressed in each cell (Fleischer et al., 2007). Thus, similar to OSNs in the MOE (reviewed in Mombaerts, 2004), each GG neuron seems to be endowed with one receptor type only. Expression of V2r83 and TAARs appears to be highest in perinatal stages and diminishes in juveniles and adults (Fleischer et al., 2007).

52.3.5 Olfactory Responsiveness of GG Neurons

Since GG neurons share a variety of characteristic features with OSNs in other nasal compartments, including expression of OMP and olfactory receptors as well as axonal projection to the OB, GG neurons might be expected to respond to chemical stimuli. The first evidence for this hypothesis was provided by calcium-imaging experiments indicating that murine GG neurons are activated by so-called alarm pheromones that have been reported to elicit freezing behavior in conspecifics (Brechbuhl et al., 2008). However, the chemical nature of alarm pheromones is so far unknown.

As a more direct approach to identify chemical substances capable of activating GG neurons, a larger array of odorous compounds was tested. Most do not activate GG neurons. By contrast, some pyrazine derivatives strongly activate GG neurons *in vivo*, as monitored by the expression of the activity-dependent gene c-Fos (Mamasuew et al., 2011a). In addition, electrophysiological recordings demonstrate that the GG-activating pyrazine derivatives evoke electrical responses in GG neurons (W. Hanke and J. Fleischer, unpublished data). More detailed analyses revealed that the three pyrazine derivatives 2,3-dimethylpyrazine (2,3-DMP), 2,5-dimethylpyrazine and 2,3,5-trimethylpyrazine are capable of activating GG neurons. Interestingly, even closely related chemical compounds, such as pyrazine, 2-methylpyrazine, 2,3-diethylpyrazine and o-xylene fail to stimulate the GG (Mamasuew et al., 2011a). These findings suggest that GG neurons are activated by only a few distinct compounds. Moreover, only the V2r83-expressing GG neurons respond to given pyrazine derivatives (Mamasuew et al., 2011a).

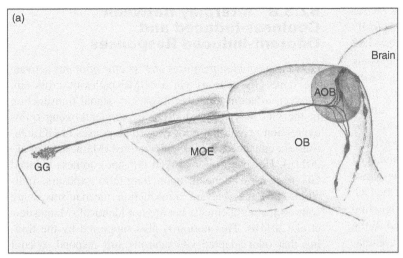
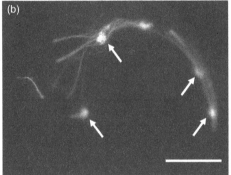
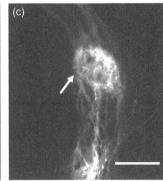

Figure 52.5 Axonal projection of GG neurons. (a) Schematic representation of a saggital section through the head of mouse. Axons emerging from the GG follow a trajectory along the nasal septum and project to the OB. In the posterior region of the OB, at the level of the accessory olfactory bulb (AOB), GG axon bundles separate into two branches to reach the relevant glomeruli in the dorsal, lateral and medial area of the OB. While one branch takes a dorsal trajectory going round the AOB, the other fibers remain on the medial side and project in a ventral direction. (b)-(c) Projection of GG axons in the OB (coronal sections) visualized by staining with the lipophilic tracer DiI (medial is to the left; dorsal is to the top). (b) On their route, GG axons innervate a number of interconnected necklace glomeruli (arrows). (c) High magnification image of a necklace glomerulus (arrow) innervated by GG axons. Scale bars: **B** = 500 μm; **C** = 100 μm. (this figure was kindly provided by Sabrina Stebe)

The functional relevance of the GG activation by pyrazine derivatives is so far unclear. Interestingly, strong activation of GG cells by these compounds was only observed in neonates but not in adults (Mamasuew et al., 2011a), suggesting that the responsiveness may have particular implications for pups. Two important natural sources for dimethylpyrazines have been identified. First, dimethylpyrazines are present in the urine of some predators of mice, such as ferrets (Zhang et al., 2005); accordingly, these compounds may have an alerting function. Second, dimethylpyrazines are present in the urine of grouped females that are neither pregnant nor lactating (Jemiolo et al., 1989). In this context, the relevance of these substances for pups is yet unclear.

52.3.6 Responsiveness to Coolness

In addition to their chemosensory responsiveness, GG neurons are also activated by cool temperatures (Mamasuew et al., 2008; Schmid et al., 2010). Studies of c-Fos expression *in vivo* suggest that the responsiveness of GG neurons to cool temperatures appears to be more intense in neonates than in adults (Mamasuew et al., 2008).

About 75% of all GG neurons respond to cool stimuli (Mamasuew et al., 2010; Schmid et al., 2010). However, only the V2r83-expressing GG cells are activated by cool temperatures; the TAAR-positive GG neurons do not respond to coolness (Mamasuew et al., 2008). Thus, the V2r83-positive GG neurons can be considered to be dual sensory cells responding to both chemical and thermal stimuli. Besides the GG, receptor V2r83 is also expressed in a considerable portion of the sensory neurons in the VNO (Martini et al., 2001), but vomeronasal neurons do not respond to cool stimuli (Mamasuew et al., 2008; Schmid et al., 2010).

52.3.7 Molecular Signaling Cascades

Recent attempts to characterize the molecular mechanisms underlying the responsiveness of GG neurons to chemical and thermal stimuli have led to the identification of distinct signaling molecules. GG cells abundantly express G proteins, especially the G protein α-subunits G_i and G_o (Fleischer et al., 2006b), both of which are also present in vomeronasal neurons (Berghard and Buck, 1996; Jia and Halpern, 1996). In line with the notion that cyclic nucleotide-gated (CNG) ion channels are crucial for olfactory signaling (Brunet et al., 1996), the CNG channel subunit CNGA3 is expressed in numerous GG neurons (Liu

1140 **Chapter 52 The Septal Organ, Grueneberg Ganglion, and Terminal Nerve**

et al., 2009; Mamasuew et al., 2010). However, CNGA3 is different from the subtype CNGA2, which is typically expressed in OSNs of the MOE. Specifically, CNGA3 is poorly responsive to cyclic adenosine monophosphate (cAMP) but is strongly activated by cyclic guanosine monophosphate (cGMP) (Kaupp and Seifert, 2002), leading to the notion that cGMP may be involved in sensory signaling of GG neurons. This concept was supported by the discovery that the transmembrane guanylyl cyclase subtype GC-G is expressed in the GG. In addition, GG neurons express the cGMP-hydrolyzing phosphodiesterase PDE2A (Fleischer et al., 2009; Liu et al., 2009). In the GG, expression of all these three cGMP-associated signaling elements (CNGA3, GC-G and PDE2A) is confined to the V2r83-positive neurons (Fleischer et al., 2009; Mamasuew et al., 2010; Matsuo et al., 2012). Because only these GG neurons are activated by the above mentioned dimethylpyrazine ligands (Mamasuew et al., 2011a), it is conceivable that cGMP signaling is involved in mediating the responsiveness to these odorants. Indeed, in c-Fos experiments (Mamasuew et al., 2011a) and in electrophysiological approaches (Hanke and Fleischer, unpublished results), responses to odorants were reduced in mice deficient for CNGA3 or GC-G compared to wild-type conspecifics.

The transmembrane guanylyl cyclase GC-G has been reported to be activated by bicarbonate, an observation leading to speculations that the GG might respond to CO_2 (Chao et al., 2010). However, CO_2 failed to activate GG neurons in calcium-imaging (Brechbuhl et al., 2008) as well as in c-Fos (Fleischer and Breer, unpublished results) experiments.

In search for the mechanisms underlying the coolness-evoked GG responses, evidence for a novel thermosensory pathway in mammals has accumulated. The coolness-sensitive TRP channel TRPM8, which is crucial for the responsiveness to coolness in neurons of the dorsal root and trigeminal ganglia (Bautista et al., 2007; Dhaka et al., 2007), is absent from GG neurons (Fleischer et al., 2009). Previous studies have shown that in the nematode *Caenorhabditis elegans*, so-called AWC neurons are capable of responding to both chemical and thermal stimuli, thus resembling GG neurons. Interestingly, in AWC neurons, chemo- and thermosensory transduction requires transmembrane guanylyl cyclases and CNG channels (Coburn and Bargmann, 1996; Komatsu et al., 1996; Inada et al., 2006). Based on these findings, it was hypothesized that the GG neurons may employ a similar transduction pathway to respond to cool temperatures, and c-Fos experiments using mice deficient for CNGA3 or GC-G provide results that support this view (Mamasuew et al., 2010; Fleischer and Breer, unpublished results).

52.3.8 Interplay Between Coolness-Induced and Odorant-Induced Responses

Given that cool temperatures and specific odorants activate the same GG neurons via a cGMP pathway, cross-talk between olfactory and thermosensory signal transduction in the GG may occur. Indeed, experiments using c-Fos expression reveal that odor-evoked responses in GG neurons are enhanced at cool temperatures (Mamasuew et al., 2011a). However, odor-evoked but not coolness-induced GG responses attenuate upon long-term exposure, indicating that the relevant transduction mechanisms share some signaling elements but are not identical (Mamasuew et al., 2011b). This notion is also supported by the finding that odor-adapted GG neurons still respond to cool temperatures (Mamasuew et al., 2011b).

The observation that odor-induced activation is abolished at very warm temperatures (35°C) (Fleischer and Breer, unpublished results) suggests that temperature-sensitive ion channels may be also important for chemosensory signaling in the GG. In this context, the thermosensitive potassium channel TREK-1 appears to be an interesting candidate. At cool temperatures, TREK-1 is closed, causing a membrane depolarization; at higher temperatures, TREK-1 is opened and the membrane hyperpolarizes (Maingret et al., 2000). Interestingly, TREK1 is also inhibited by G protein-coupled signaling (Enyedi and Czirjak, 2010). Most recent studies have shown that TREK-1 is expressed in numerous GG neurons (J. Fleischer and H. Breer, unpublished observations). Thus, TREK-1 could function as an important molecular element regarding sensory transduction mechanisms in GG neurons (Figure 52.6).

52.3.9 Is the GG Part of the "Necklace" Olfactory System?

GG neurons project their axons to the caudal MOB where they converge in a small number of interconnected glomeruli surrounding the anterior part of the AOB (Fuss et al., 2005; Koos and Fraser, 2005; Roppolo et al., 2006; Storan and Key, 2006). These glomeruli are arrayed like "beads on a string" and have therefore been designated as "necklace" glomeruli (Shinoda et al., 1989). "Necklace" glomeruli are also innervated by axons of the so-called GC-D neurons from the MOE (Juilfs et al., 1997; Leinders-Zufall et al., 2007; Walz et al., 2007). The question whether axons from the GG and the GC-D neurons may project to the same glomeruli was unequivocally answered in a recent study demonstrating that the two axonal populations do not converge (Matsuo et al., 2012). Therefore, the "necklace" system is composed of two different subsets of glomeruli: one exclusively innervated

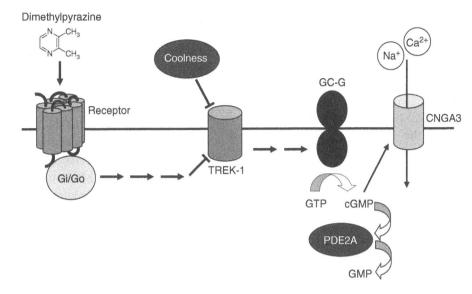

Figure 52.6 Putative chemo- and thermosensory signaling pathways in GG neurons. Activation of GG neurons by cool temperatures and given odorants appears to be mediated by cGMP-associated signaling elements, most notably the transmembrane guanylyl cyclase GC-G and the cGMP-activated ion channel CNGA3. It is unknown how these elements are activated by coolness or odorants, respectively. Presumably, cool temperatures directly inhibit the thermosensitive potassium channel TREK-1. Similarly, TREK-1 could be also closed by appropriate odorants via a G protein-mediated cascade. It is yet elusive how inactivation of TREK-1 might affect cGMP signaling in GG neurons. Finally, cGMP could be hydrolyzed by the phosphodiesterase PDE2A.

by neurons from the GG, the other by GC-D neurons from the MOE. The distinct axonal projection pattern of GG neurons and GC-D neurons is also mirrored by the relevance of different guidance factors: while proper targeting of GC-D axonal processes depends on neuropilin-2 (Walz et al., 2007), the GG uses neuropilin-1 to establish accurate wiring (Matsuo et al., 2012).

52.4 THE TERMINAL NERVE (TN)

The terminal nerve (TN), which is also termed the nervus terminalis and Cranial Nerve 0 (CN 0), is the most anterior of the vertebrate cranial nerves, extending between the nasal cavity and basal forebrain. It is currently thought to function in modulating activity in the olfactory epithelium and, in teleost fishes, in the retina. We include the TN in this chapter in part because of its role as a centrifugal modulatory olfactory subsystem and in part because the neurons of the TN ganglion develop from the nasal placode, as do those of the olfactory and vomeronasal epithelia (Schwanzel-Fukuda and Pfaff, 1989; Wray et al., 1989).

The TN was first discovered in sharks by Fritsch, who referred to it simply as "a supernumerary nerve" (1878). Pinkus (1894) described the TN in lungfish, leading some subsequent investigators to refer to the TN as "Pinkus's nerve". In studies of another lungfish species, Sewertzoff (1902) described part of the nerve that enters the brain at the level of the preoptic area, leading him to suggest that it be called "nervus praeopticus". Locy (1905) later described the TN in a variety of sharks and rays, and named it the "nervus terminalis" because in these animals the nerve enters the brain through the lamina terminalis. Although the Anglicized version of this name, "terminal nerve", is now commonly used, alternative names are still found in the older literature and in some newer studies of non-mammalian vertebrates.

52.4.1 Anatomy of the Terminal Nerve

The anatomy of the TN has been examined in every class of vertebrates, but most studies have focused on cartilaginous fishes, teleost fishes, and amphibians. These studies indicate some anatomical differences among groups, although the functional consequences are unclear. For example, the centrifugal neurites of the bipolar TN ganglion cells sometimes comprise a separate nerve, but the fibers of the TN sometimes run within the olfactory or vomeronasal nerve. The location of the ganglion cell bodies also varies among species, with somata found in one or more ganglia within the TN proper or along the TN pathway through the olfactory bulb or anterior forebrain. For example, in most cartilaginous fishes, a separate TN runs medial to the olfactory nerve, and includes a visible ganglion located medial and external to the nasal cavity. In some shark species, the TN possesses additional ganglia located along the nerve itself (Locy, 1905). In contrast, in teleost fishes the TN runs within the olfactory nerve and is not visible externally. In most teleosts the TN ganglion consists of a cluster of neurons located between the olfactory bulb and the ventral telencephalon, with fibers that extend peripherally to both the olfactory epithelium and the retina (Brookover and Jackson, 1911; Rossi et al., 1972; Springer, 1983; Oka et al., 1986). Because of this association with both the olfactory system and retina, the TN ganglion in teleosts is also frequently called the "nucleus olfacto-retinalis" (Munz et al., 1981).

Based largely on its anterior location, the Grueneberg ganglion was thought by its discoverer to be part of the TN system (Grüneberg, 1973). We now know that in mice the TN ganglion lies near the ventrocaudal portion of the MOB, in the subarachnoid space; additional TN ganglion cells are scattered along the path of the TN fibers in the nasal cavity and along the ventral surface of the MOB and AOB (Jennes, 1986) (Figure 52.1). Rostrally, the TN fibers merge with the olfactory fibers and are distributed widely throughout the nasal cavity (Jennes, 1986). Similarly, in adult humans the TN consists of many small bundles in the nasal cavity, coursing with olfactory fibers along septum and with the ethmoid branch of the trigeminal nerve more anteriorly (McCotter, 1915). Inside the braincase, the TN is separate from the olfactory nerve, and in some cases is large enough to be seen with the naked eye (Brookover, 1914; Johnston, 1914). The TN ganglion is medial to the MOB, on the inside surface of the dura mater near the cribriform plate (Brookover, 1914). Proximal to the ganglion, the fibers run over the surface of the gyrus rectus medial to the olfactory peduncles, entering the brain anterior to the optic chiasm (Johnston, 1914).

In many vertebrates, the cells and fibers of the TN contain GnRH (reviewed in Wirsig-Wiechmann et al., 2002). The TN can also be labeled with antisera directed against neuropeptide Y (NPY) and the molluscan cardioexcitatory neuropeptide FMRFamide. These two peptides have a similar C-terminal and many such antisera cross-react. Although the presence of NPY has been established it is not clear whether the TN also contains FMRFamide-like peptides (Chiba, 2000; Mousley et al., 2006). In any case, modern studies often rely on antisera directed against GnRH or FMRFamide to identify or localize TN cells and fibers.

Immunocytochemical data suggest that the TN may not be present in all vertebrates. In some reptiles, either the immunoreactive characteristics of the TN differ from those in other jawed vertebrates, perhaps a way that varies with developmental stage, or the nerve is not present at all developmental stages. For example, among lizards, GnRH-immunoreactive cells and fibers are present in the TN of adult *Eumeces laticeps* and *Sceloporus undulatus*, but not in adult *Anolis carolinensis* (Rosen et al., 1997). In embryonic and adult *Podarcis sicula*, GnRH-immunoreactive cells and fibers cannot be detected in a nerve, nor in the MOE or MOB (Masucci et al., 1992; D'Aniello et al., 1994). FMRFamide-immunoreactive cells and fibers are present in the TN during development in *Chalcides chalcides*, but not in adults (D'Aniello et al., 2001). Similarly, the TN of the spectacled caiman (*Caiman crocodilus*) displays FMRFamide-like immunoreactivity in embryos but not adults (D'Aniello et al., 1999, 2001). The development of the TN in turtles has been described by Johnston (1913) and Larsell (1917), who were able to visualize the nerve as it courses over the surface of the olfactory nerve due to its several conspicuous ganglia. The histochemical characteristics of the nerve have not been examined in detail in turtles, although FMRFamide-immunoreactive cells and fibers are not present in or around peripheral olfactory structures in adult *Trachemys scripta* (D'Aniello et al., 1999, 2001). On the other hand, a GnRH-immunoreactive TN is present in adult garter snakes (*Thamnophis sirtalis*), with a discrete ganglion at the ventral border between the MOB and rostral telencephalon (Smith et al., 1997). Medina and colleagues (2005) report that Nile crocodiles (*Crocodylus niloticus*) possess a GnRH-immunoreactive TN, but a detailed description of the pathway was not provided.

The presence of a TN in lampreys is also questionable. Based on conventional histology, Heier (1948) first suggested that lampreys lack a TN. Antisera directed against GnRH, NPY, and FMRFamide fail to label any structures reminiscent of the TN in larval or adults of four genera of lampreys (Crim et al., 1979a, b; Meyer et al., 1987; King et al., 1988; Ohtomi et al., 1989; Wright et al., 1994; Tobet et al., 1995; Eisthen and Northcutt, 1996; Chiba, 1999). Consistent with this idea, GnRH neurons in lampreys appear to develop from the diencephalon rather than from the olfactory placode (Tobet et al., 1996). Nevertheless, fibers that project from the nasal sac through olfactory bulb to the ventral forebrain have been labeled with injections of tracers into the nasal sac of larval or adult *Lampetra planeri* and *Ichthyomyzon unicuspis* (Meyer et al., 1987; von Bartheld et al., 1987; Northcutt and Puzdrowski, 1988; von Bartheld and Meyer, 1988). Although this projection was originally interpreted as a TN it may be the extra-bulbar olfactory pathway, the existence of which was unknown at the time these studies were conducted (Eisthen and Northcutt, 1996).

52.4.2 Function of the Terminal Nerve

Because the anterior fibers of the TN are found in the nasal cavity and the posterior fibers extend toward the hypothalamic/preoptic area, it was originally thought to serve a sensory function, perhaps related to reproduction (e.g., Rossi et al., 1972; Demski and Northcutt, 1983). Three lines of evidence support the newer idea that the TN serves a neuromodulatory function: studies have failed to find evidence of sensory activity in the TN; the TN ganglion cells are intrinsically active and release GnRH; and TN peptides modulate activity in the olfactory epithelium and, in teleost fishes, in the retina. We will briefly review these studies below.

Electrophysiological recordings have failed to detect sensory responses in the TN. For example, Bullock and Northcutt (1984) used both hook and suction electrodes to record from the TN in a spiny dogfish (*Squalus acanthias*). The authors detected no response to food, conspecific odorants, or amino acids. They also found no response to mechanical stimuli in the nasal sac or on the body surface nor to auditory, visual, or electrical stimuli. Similarly, another study found that exposure to conspecific sex-related and other odorants did not alter the firing rate of TN ganglion cells in goldfish, *Carassius auratus*. In addition, the TN did not respond to visual, auditory, magnetic, or thermal stimuli (Fujita et al., 1991).

Several researchers have examined the effects of TN lesions on reproductive behaviors and found effects that are subtle and difficult to interpret. For example, lesions of the TN impair initiation of nest-building behavior, but do not abolish reproduction, in male dwarf gouramis, *Colisa lalia* (Yamamoto et al., 1997). In female hamsters (*Mesocricetus auratus*), TN lesions lead to a slight decrease in the latency to exhibit lordosis behavior (Wirsig-Wiechmann, 1997). Male hamsters with TN lesions spent less time investigating female vaginal odors and increased their latency to ejaculate during mating, but lesions had no effect on testosterone surges that result from exposure to female vaginal secretions (Wirsig and Leonard, 1987; Wirsig-Wiechmann, 1993).

The physiological properties of the TN have been the subject of much study in teleost fishes, particularly in dwarf gouramis (reviewed in Oka, 1992, 2002). Electrophysiological experiments demonstrate that the majority of the ganglion cells fire spontaneous action potentials at a constant frequency. Excised TN ganglion cells maintain their pacemaker activity in a dish, demonstrating that the activity is endogenous (Oka, 2002). This rhythmic activity is brought about by the interaction of a tetrodotoxin-resistant persistent sodium current that depolarizes the cell and a persistent potassium current that activates near the resting potential of the cell, initiating repolarization (Oka, 1992, 1996; Abe and Oka, 2000, 2002). The firing frequency of the cells is modulated by the same form of GnRH that is present in the nerve, causing an initial decrease in firing rate, followed by a later increase (Abe and Oka, 2000, 2002). This modulation of firing rate by GnRH may function to synchronize firing (Abe and Oka, 2000). Experiments with bonnethead sharks (*Sphyrna tiburo*) suggest that tonic activity of TN ganglion cells is not unique to teleosts (White and Meredith, 1987).

Few studies have sought to identify anatomical inputs to the TN, perhaps because it was long thought to serve a sensory function. One detailed tract-tracing study with dwarf gouramis and tilapia (*Oreochromis niloticus*) indicates that in teleost fishes the TN receives input from olfactory areas in the forebrain as well as from the nucleus tegmento-olfactorius, a midbrain region that receives input from the reticular formation as well as visual and somatosensory areas (Yamamoto and Ito, 2000). In contrast, in two species of shark projections to the TN ganglion arise from the medial pallium, septum, and preoptic area, and inputs from primarily olfactory areas of the forebrain were not observed (Yáñez et al., 2011). Activity of TN ganglion cells in bonnethead sharks is modulated by acetylcholine and norepinephrine, which appear to be of central origin (White and Meredith, 1993, 1995). Thus, accumulating evidence suggests that TN ganglion cells are spontaneously active, and that this activity may be regulated by centrifugal inputs that differ among species.

Electrical stimulation of the TN ganglion in Atlantic stingrays (*Dasyatis sabina*) leads to an increase in GnRH levels in cerebrospinal fluid (Moeller and Meredith, 1998). While suggestive, this result does not demonstrate that the TN releases GnRH under normal physiological conditions. More direct evidence comes from studies with dwarf gouramis in which slices of brain containing TN ganglion cells have been shown to release GnRH. Importantly, increasing doses of glutamate increase both the rate of pacemaker potentials in TN ganglion cells and the release of GnRH (Oka, 2002; Ishizaki et al., 2004). This result strongly suggests that spontaneous activity of TN ganglion cells leads to the release of peptides.

If the TN releases peptides, where do the peptides go? The anatomical connection between the terminations of the TN fibers and the nasal sensory epithelia remains obscure because the TN fibers taper and are difficult to identify in electronic micrographs. One study indicates that in mice the TN fibers approach the lamina propria underlying the MOE, although it is not clear whether or not the fibers penetrate into the epithelium (Jennes, 1986). TN fibers also wrap around and take up tracers injected into fenestrated capillaries underneath the MOE, suggesting that the fibers could also release compounds into the microcirculation around the MOE (Jennes, 1986). Finally, some have suggested that TN fibers could release compounds into mucus

1144   **Chapter 52   The Septal Organ, Grueneberg Ganglion, and Terminal Nerve**

overlying the MOE through association with subepithelial glands (e.g., Wirsig-Wiechmann et al., 2002).

If the TN releases peptides in or near the nasal epithelia, peptide receptors are required for a response. Receptors for at least one TN-derived peptide, GnRH, are present in nasal sensory epithelia. Specifically, a radiolabeled GnRH receptor antagonist labels the MOE in tiger salamanders, *Ambystoma tigrinum*, and the VNO in prairie voles, *Microtus ochrogaster* (Wirsig-Wiechmann and Jennes, 1993; Wirsig-Wiechmann and Wiechmann, 2001). An antiserum directed against GnRH receptors labels the MOE in another salamander, the mudpuppy *Necturus maculosus* (Zhang and Delay, 2007). Finally, Southern blots confirm the presence of GnRH receptors in both the MOE and VNO in prairie voles (Wirsig-Wiechmann and Wiechmann, 2001). Thus, if the TN were to release GnRH in or near the nasal sensory epithelia, the peptide would be expected to evoke a response.

Indeed, electrophysiological studies with amphibians demonstrate that TN derived peptides alter both the odorant sensitivity and excitability of OSNs. In mudpuppies, GnRH briefly suppresses then tonically enhances the magnitude of the voltage-dependent, tetrodotoxin (TTX)-sensitive sodium current in OSNs (Eisthen et al., 2000; Zhang and Delay, 2007). This effect depends on both PKA and PKG (Zhang and Delay, 2007), but the GnRH receptor subtype and full biochemical pathway involved have not been identified. Interestingly, the neurons are more responsive to GnRH during the breeding season, suggesting that GnRH may play a role in regulating responses to odorants in a way that promotes reproduction (Eisthen et al., 2000; Zhang and Delay, 2007). Lending further support to the idea that the GnRH-containing fibers of the TN play a role in reproductive behavior, estrogen affects GnRH concentrations in the TN in the African clawed frog *Xenopus laevis* (Wirsig-Wiechmann and Lee, 1999). In addition, GnRH concentrations are higher in courted female rough-skinned newts (*Taricha granulosa*) than in uncourted conspecifics (Propper and Moore, 1991). Finally, experiments with another aquatic salamander, the axolotl *Ambystoma mexicanum*, demonstrate that GnRH suppresses responses evoked by amino acids, which serve as feeding cues in aquatic vertebrates (Park and Eisthen, 2003). Given that some amphibians cease feeding during the breeding season it is tempting to interpret this result as further evidence of the involvement of TN-derived GnRH in reproduction. Nevertheless, studies with mudpuppies show that even responses to behaviorally meaningless odorants are suppressed by GnRH (Zhang and Delay, 2007).

Although most functional studies of the TN have focused on GnRH, experiments reveal that other TN peptides serve a modulatory function. The TN in axolotls is NPY-immunoreactive. As with GnRH, application of synthetic axolotl NPY enhances the magnitude of the TTX-sensitive sodium current in OSNs; intriguingly, neurons from food-deprived animals are much more likely to respond to GnRH. Although GnRH suppresses odorant responses evoked by amino acids, NPY enhances them, and the magnitude of the increase is greater in food-deprived animals. These data suggest that TN-derived NPY modulates activity in the MOE with respect to the animal's nutritional state (Mousley et al., 2006). Separate studies with axolotls demonstrate that FMRFamide also modulates both odorant responses and the TTX-sensitive sodium current in OSNs, although it is not yet clear whether the TN contains FMRFamide-related compounds (Park and Eisthen, 2003).

In teleost fishes, an additional branch of the TN innervates the retina, where it likely serves a modulatory function. *In situ* hybridization experiments with the cichlid *Astatotilapia burtoni* indicate that two different subtypes of GnRH receptors are expressed in the amacrine cell layer, probably on interplexiform cells, and in the ganglion cell layer (Grens et al., 2005). Separate studies demonstrate that TN-derived peptides modulate activity of both populations of cells in teleosts. In white perch (*Roccus americana*), GnRH- and FMRFamide-immunoreactive TN fibers synapse on dopaminergic interplexiform cells in the amacrine cell layer (Zucker and Dowling, 1987). GnRH causes release of dopamine from interplexiform cells, reducing the receptive field size and responses to full-field illumination in horizontal cells. FMRFamide alone does not appear to affect activity of horizontal cells, but in some cases suppresses the effects of GnRH (Umino and Dowling, 1991).

In goldfish, GnRH-immunoreactive TN fibers extend into the inner plexiform and amacrine cell layers of the retina. Extracellular recordings from excised retinas demonstrate that GnRH, FMRFamide, and a variety of compounds with some sequence similarity to FMRFamide, such as met-enkephalin, modulate activity of retinal ganglion cells. The effects of these compounds are more pronounced during the winter and spring, again suggesting a relationship between TN activity and reproduction (Walker and Stell, 1986; Stell et al., 1987). Further, the release of FMRFamide-like peptides in the retina can be induced by exposure to light and is temporarily decreased by lesioning the olfactory nerve, presumably eliminating odorant input to the TN ganglion (Fischer and Stell, 1997). *In vivo* experiments with zebrafish (*Danio rerio*) demonstrate that olfactory stimulation with an amino acids enhances light sensitivity and increases firing rate in retinal ganglion cells. This effect appears to be due to TN-stimulated release of dopamine from interplexiform cells (Huang et al., 2005). Finally, exposure of zebrafish to amino acid odorants increases behavioral sensitivity to a looming visual stimulus as well as electroretinogram responses to a pulse of light. These effects are blocked by administration of a D_2 receptor antagonist or co-injection of 6-hydroxydopamine and pargyline, which selectively eliminates the interplexiform cell population (Maaswinkel and Li, 2003). These data demonstrate that the TN modulates retinal activity in teleosts in response to odorants by stimulating dopamine release from interplexiform cells, altering activity of both horizontal cells and retinal ganglion cells.

Taken together, the available data strongly suggest that the TN functions to modulate activity in the nasal sensory epithelia and, in teleost fishes, in the retina, to maximize sensitivity to stimuli most relevant to the animal's condition. Important questions remain to be answered, including identification of the full suite of peptides found in the TN, the route by which TN peptides travel to olfactory and vomeronasal receptor neurons, and the mechanisms underlying changing responsiveness to TN peptides in OSNs.

52.5 CONCLUDING REMARKS

Organization of the peripheral olfactory system is more complicated than previously appreciated. More subsystems may still emerge with new molecular markers and more detailed anatomical and functional analysis. What are the advantages of having multiple chemosensory and modulatory subsystems? First, different receptors expressed in these subsystems can expand the overall detection capacity of the olfactory system for chemicals and other stimuli. Second, critical information can be processed in parallel by multiple subsystems, which send signals to different brain regions for further processing and integration. Third, the chemosensory responses can be modulated based on the internal status and the external environment. These chemosensory systems allow the host organisms to detect the olfactory cues and respond appropriately by adjusting their behaviors, emotions, and hormones. Future studies will elucidate the specific information conveyed by each subsystem and shed light on how the chemosensory information is processed and integrated by the brain.

REFERENCES

Abe, H. and Oka, Y. (2000). Modulation of pacemaker activity by salmon gonadotropin-releasing hormone (sGnRH) in terminal nerve (TN)-GnRH neurons. *J. Neurophysiol.* 83: 3196–3200.

Abe, H. and Oka, Y. (2002). Mechanisms of the modulation of pacemaker activity by GnRH peptides in the terminal nerve-GnRH neurons. *Zool. sci.* 19: 111–128.

Adams, D. R. (1992). Fine structure of the vomeronasal and septal olfactory epithelia and of glandular structures. *Microsc. Res. Tech.* 23: 86–97.

Adams, D. R. and McFarland, L. Z. (1971). Septal olfactory organ in Peromyscus. *Comp. Biochem. Physiol. A* 40: 971–974.

Astic, L. and Saucier, D. (1988). Topographical projection of the septal organ to the main olfactory bulb in rats: ontogenetic study. *Brain Res.* 470: 297–303.

Bautista, D. M., Siemens, J., Glazer, J. M., et al. (2007). The menthol receptor TRPM8 is the principal detector of environmental cold. *Nature* 448: 204–208.

Berghard, A. and Buck, L. B. (1996). Sensory transduction in vomeronasal neurons: evidence for G alpha o, G alpha i2, and adenylyl cyclase II as major components of a pheromone signaling cascade. *J. Neurosci.* 16: 909–918.

Bojsen-Moller, F. (1975). Demonstration of terminalis, olfactory, trigeminal and perivascular nerves in the rat nasal septum. *J. Comp. Neurol.* 159: 245–256.

Brechbuhl, J., Klaey, M. and Broillet, M. C. (2008). Grueneberg ganglion cells mediate alarm pheromone detection in mice. *Science* 321: 1092–1095.

Breer, H., Fleischer, J. and Strotmann, J. (2006). The sense of smell: multiple olfactory subsystems. *Cell. Mol. Life Sci.* 63: 1465–1475.

Breipohl, W., Naguro, T. and Miragall, F. (1983). Morphology of the Masera organ in NMRI mice (comined morphometric, freeze-fracture, light- and scanning electron microscopic investigations. *Verh. Anat. Ges.* 77: 741–743.

Breipohl, W., Naguro, T. and Walker, D. G. (1989). The postnatal development of Maresa's organ in the rat. *Chem. Senses* 14: 649–662.

Broman, I. (1921). Über die Entwickelung der konstanten grösseren Nasenhöhlendrüsen der Nagetiere. *Anat. Entw. Gesch.* 60: 439–586.

Brookover, C. (1914). The nervus terminalis in adult man. *J. Comp. Neurol.* 24: 131–135.

Brookover, C. and Jackson, T. S. (1911). The olfactory nerve and the nervus terminalis of Ameiurus. *J. Comp. Neurol.* 22: 237–259.

Brunet, L. J., Gold, G. H. and Ngai, J. (1996). General anosmia caused by a targeted disruption of the mouse olfactory cyclic nucleotide-gated cation channel. *Neuron* 17: 681–693.

Bullock, T. H. and Northcutt, R. G. (1984). Nervus terminalis in dogfish (Squalus acanthias, Elasmobranchii) carries tonic efferent impulses. *Neurosci. Lett.* 44: 155–160.

Buonviso, N., Amat, C. and Litaudon, P. (2006). Respiratory modulation of olfactory neurons in the rodent brain. *Chem. Senses* 31: 145–154.

Chao, Y. C., Cheng, C. J., Hsieh, H. T., et al. (2010). Guanylate cyclase-G, expressed in the Grueneberg ganglion olfactory subsystem, is activated by bicarbonate. *Biochem. J.* 432: 267–273.

Chiba, A. (1999). Immunohistochemical distribution of neuropeptide Y-related substance in the brain and hypophysis of the arctic lamprey, *Lethenteron japonica. Brain Behav. Evol.* 53: 102–109.

Chiba, A. (2000). Immunohistochemical cell types in the terminal nerve ganglion of the cloudy dogfish, Scyliorhinus torazame, with special regard to neuropeptide Y/FMRFamide-immunoreactive cells. *Neurosci. Lett.* 286: 195–198.

Coburn, C. M. and Bargmann, C. I. (1996). A putative cyclic nucleotide-gated channel is required for sensory development and function in C. elegans. *Neuron* 17: 695–706.

Connelly, T., Yiqun, Yu., Grosmaitre, X., Wang, J., Santarelli, L. C., Savigner, A., Qiao, X., Wang, Z., Storm, D. R., Ma, M. (2015). G Protein-coupled odorant receptors underlie mechanosensitivity in mammalian olfactory sensory neurons. *Proc Natl Acad Sci U S A,* 112: 590–595.

Crim, J. W., Urano, A. and Gorbman, A. (1979a). Immunocytochemical studies of luteinizing hormone-releasing hormone in brains of agnathan fishes. I. Comparisons of adult Pacific lamprey (Entosphenus tridentata) and the Pacific hagfish (Eptatretus stouti). *Gen. Comp. Endocrinol.* 37: 294–305.

Crim, J. W., Urano, A. and Gorbman, A. (1979b). Immunocytochemical studies of luteinizing hormone-releasing hormone in brains of agnathan fishes. II. Patterns of immunoreactivity in larval and maturing Western brook lamprey (Lampetra richardsoni). *Gen. Comp. Endocrinol.* 38: 290–299.

D'Aniello, B., Fiorentin, M., Pinelli, C., et al. (2001). Localization of FMRFamide-like immunoreactivity in the brain of the viviparous skink (Chalcides chalcides). *Brain Behav. Evol.* 57: 18–32.

D'Aniello, B., Pinelli, C., Jadhao, A. G., et al. (1999). Comparative analysis of FMRFamide-like immunoreactivity in caiman (Caiman crocodilus) and turtle (Trachemys scripta elegans) brains. *Cell Tissue Res.* 298: 549–559.

D'Aniello, B., Pinelli, C., King, J. A. and Rastogi, R. K. (1994). Neuroanatomical organization of GnRH neuronal systems in the lizard (Podarcis s. sicula) brain during development. *Brain Res.* 657: 221–226.

Demski, L. S. and Northcutt, R. G. (1983). The terminal nerve: a new chemosensory system in vertebrates? *Science* 220: 435–437.

Dhaka, A., Murray, A. N., Mathur, J., et al. (2007). TRPM8 is required for cold sensation in mice. *Neuron* 54: 371–378.

Duan, X., Block, E., Li, Z., et al. (2012). Crucial role of copper in detection of metal-coordinating odorants. *Proc. Natl. Acad. Sci. U.S.A.* 109: 3492–3497.

Eisthen, H. L., Delay, R. J., Wirsig-Wiechmann, C. R. and Dionne, V. E. (2000). Neuromodulatory effects of gonadotropin releasing hormone on olfactory receptor neurons. *J. Neurosci.* 20: 3947–3955.

Eisthen, H. L. and Northcutt, R. G. (1996). Silver lampreys (Ichthyomyzon unicuspis) lack a gonadotropin-releasing hormone- and FMRFamide-immunoreactive terminal nerve. *J. Comp. Neurol.* 370: 159–172.

Enyedi, P. and Czirjak, G. (2010). Molecular background of leak K+ currents: two-pore domain potassium channels. *Physiol. Rev.* 90: 559–605.

Fischer, A. J. and Stell, W. K. (1997). Light-modulated release of RFamide-like neuropeptides from nervus terminalis axon terminals in the retina of goldfish. *Neuroscience* 77: 585–597.

Fleischer, J., Hass, N., Schwarzenbacher, K., et al. (2006a). A novel population of neuronal cells expressing the olfactory marker protein (OMP) in the anterior/dorsal region of the nasal cavity. *Histochem. Cell Biol.* 125: 337–349.

Fleischer, J., Mamasuew, K. and Breer, H. (2009). Expression of cGMP signaling elements in the Grueneberg ganglion. *Histochem. Cell Biol.* 131: 75–88.

Fleischer, J., Schwarzenbacher, K., Besser, S., et al. (2006b). Olfactory receptors and signalling elements in the Grueneberg ganglion. *J. Neurochem.* 98: 543–554.

Fleischer, J., Schwarzenbacher, K. and Breer, H. (2007). Expression of trace amine-associated receptors in the Grueneberg ganglion. *Chem. Senses* 32: 623–631.

Fritsch, G. (1878). *Untersuchungen über den feineren Bau des Fischgehirns mit besonderer Berücksichtigung der Homologien bei anderen Wirbelthierklassen.* Berlin: Verlag der Gutmann'schen Buchhandlung.

Fujita, I., Sorensen, P. W., Stacey, N. E. and Hara, T. J. (1991). The olfactory system, not the terminal nerve, functions as the primary chemosensory pathway mediating responses to sex pheromones in male goldfish. *Brain Behav. Evol.* 38: 313–321.

Fuss, S. H., Omura, M. and Mombaerts, P. (2005). The Grueneberg ganglion of the mouse projects axons to glomeruli in the olfactory bulb. *Eur. J. Neurosci.* 22: 2649–2654.

Giannetti, N., Pellier, V., Oestreicher, A. B. and Astic, L. (1995a). Immunocytochemical study of the differentiation process of the septal organ of Masera in developing rats. *Brain Res. Dev. Brain Res.* 84: 287–293.

Giannetti, N., Saucier, D. and Astic, L. (1992). Organization of the septal organ projection to the main olfactory bulb in adult and newborn rats. *J. Comp. Neurol.* 323: 288–298.

Giannetti, N., Saucier, D. and Astic, L. (1995b). Analysis of the possible altering function of the septal organ in rats: a lesional and behavioral study. *Physiol. Behav.* 58: 837–845.

Graziadei, P. P. C. (1977). Functional anatomy of the mammalian chemoreceptor system. In Chemical Signals in Vertebrates, Muller-Schwarze D. and Mozell M.M. (Eds.), Plenum, New York, pp.435–454.

Grens, K. E., Greenwood, A. K. and Fernald, R. D. (2005). Two visual processing pathways are targeted by gonadotropin-releasing hormone in the retina. *Brain Behav. Evol.* 66: 1–9.

Grosmaitre, X., Fuss, S. H., Lee, A. C., et al. (2009). SR1, a mouse odorant receptor with an unusually broad response profile. *J. Neurosci.* 29: 14545–14552.

Grosmaitre, X., Santarelli, L. C., Tan, J., et al. (2007). Dual functions of mammalian olfactory sensory neurons as odor detectors and mechanical sensors. *Nat. Neurosci.* 10: 348–354.

Grüneberg, H. (1973). A ganglion probably belonging to the N. terminalis system in the nasal mucosa of the mouse. *Z. Anat. Entwicklungsgesch* 140: 39–52.

Heier, P. (1948). Fundamental principles in the structure of the brain: A study of the brain of Petromyzon fluviatilis. *Acta Anat.* 5: 1–213.

Huang, L., Maaswinkel, H. and Li, L. (2005). Olfactoretinal centrifugal input modulates zebrafish retinal ganglion cell activity: a possible role for dopamine-mediated Ca2+ signalling pathways. *J. Physiol.* 569: 939–948.

Inada, H., Ito, H., Satterlee, J., et al. (2006). Identification of guanylyl cyclases that function in thermosensory neurons of Caenorhabditis elegans. *Genetics* 172: 2239–2252.

Ishizaki, M., Iigo, M., Yamamoto, N. and Oka, Y. (2004). Different modes of gonadotropin-releasing hormone (GnRH) release from multiple GnRH systems as revealed by radioimmunoassay using brain slices of a teleost, the dwarf gourami (Colisa lalia). *Endocrinology* 145: 2092–2103.

Jemiolo, B., Andreolini, F., Xie, T. M., et al. (1989). Puberty-affecting synthetic analogs of urinary chemosignals in the house mouse, Mus domesticus. *Physiol. Behav.* 46: 293–298.

Jennes, L. (1986). The olfactory gonadotropin-releasing hormone immunoreactive system in mouse. *Brain Res.* 386: 351–363.

Jia, C. and Halpern, M. (1996). Subclasses of vomeronasal receptor neurons: differential expression of G proteins (Gi alpha 2 and G(o alpha)) and segregated projections to the accessory olfactory bulb. *Brain Res.* 719: 117–128.

Johnston, J. B. (1913). Nervus terminalis in reptiles and mammals. *J. Comp. Neurol.* 23: 97–120.

Johnston, J. B. (1914). The nervus terminalis in man and mammals. *Anat. Rec.* 8: 185–198.

Juilfs, D. M., Fulle, H. J., Zhao, A. Z., et al. (1997). A subset of olfactory neurons that selectively express cGMP-stimulated phosphodiesterase (PDE2) and guanylyl cyclase-D define a unique olfactory signal transduction pathway. *Proc. Natl. Acad. Sci. U. S. A.* 94: 3388–3395.

Kaluza, J. F., Gussing, F., Bohm, S., et al. (2004). Olfactory receptors in the mouse septal organ. *J. Neurosci. Res.* 76: 442–452.

Katz, S. and Merzel, J. (1977). Distribution of epithelia and glands of the nasal septum mucosa in the rat. *Acta Anat. (Basel)* 99: 58–66.

Kaupp, U. B. and Seifert, R. (2002). Cyclic nucleotide-gated ion channels. *Physiol. Rev.* 82: 769–824.

King, J. C., Sower, S. A. and Anthony, E. L. (1988). Neuronal systems immunoreactive with antiserum to lamprey gonadotropin-releasing hormone in the brain of Petromyzon marinus. *Cell Tissue Res.* 253: 1–8.

References

Komatsu, H., Mori, I., Rhee, J. S., et al. (1996). Mutations in a cyclic nucleotide-gated channel lead to abnormal thermosensation and chemosensation in C. elegans. *Neuron* 17: 707–718.

Koos, D. S. and Fraser, S. E. (2005). The Grueneberg ganglion projects to the olfactory bulb. *Neuroreport* 16: 1929–1932.

Kratzing, J. E. (1984a). The structure and distribution of nasal glands in four marsupial species. *J. Anat.* 139: 553–564.

Kratzing, J. E. (1984b). The anatomy and histology of the nasal cavity of the koala (Phascolarctos cinereus). *J. Anat.* 138: 55–65.

Larsell, O. (1917). Studies on the nervus terminalis: turtle. *J. Comp. Neurol.* 30: 423–443.

Leinders-Zufall, T., Cockerham, R. E., Michalakis, S., et al. (2007). Contribution of the receptor guanylyl cyclase GC-D to chemosensory function in the olfactory epithelium. *Proc. Natl. Acad. Sci. U. S. A.* 104: 14507–14512.

Levai, O. and Strotmann, J. (2003). Projection pattern of nerve fibers from the septal organ: DiI-tracing studies with transgenic OMP mice. *Histochem. Cell Biol.* 120: 483–492.

Li, J., Haddad, R., Chen, S., Santos, V. and Luetje, C. W. (2012). A broadly tuned mouse odorant receptor that detects nitrotoluenes. *J. Neurochem.* 121: 881–890.

Liu, C. Y., Fraser, S. E. and Koos, D. S. (2009). Grueneberg ganglion olfactory subsystem employs a cGMP signaling pathway. *J. Comp. Neurol.* 516: 36–48.

Liu, C. Y., Xiao, C., Fraser, S. E., et al. (2012). Electrophysiological characterization of Grueneberg ganglion olfactory neurons: spontaneous firing, sodium conductance, and hyperpolarization-activated currents. *J. Neurophysiol.* 108: 1318–1334.

Locy, W. A. (1905). On a newly recognized nerve connected with the fore-brain of selachians. *Anat. Anz.* 26: 111–123.

Ma, M., Grosmaitre, X., Iwema, C. L., et al. (2003). Olfactory signal transduction in the mouse septal organ. *J. Neurosci.* 23: 317–324.

Maaswinkel, H. and Li, L. (2003). Olfactory input increases visual sensitivity in zebrafish: a possible function for the terminal nerve and dopaminergic interplexiform cells. *J. Exp. Biol.* 206: 2201–2209.

Maingret, F., Lauritzen, I., Patel, A. J., et al. (2000). TREK-1 is a heat-activated background K(+) channel. *Embo J.* 19: 2483–2491.

Mamasuew, K., Breer, H. and Fleischer, J. (2008). Grueneberg ganglion neurons respond to cool ambient temperatures. *Eur. J. Neurosci.* 28: 1775–1785.

Mamasuew, K., Hofmann, N., Breer, H. and Fleischer, J. (2011a). Grueneberg ganglion neurons are activated by a defined set of odorants. *Chem. Senses* 36: 271–282.

Mamasuew, K., Hofmann, N., Kretzschmann, V., et al. (2011b). Chemo- and thermosensory responsiveness of Grueneberg ganglion neurons relies on cyclic guanosine monophosphate signaling elements. *Neurosignals* 19: 198–209.

Mamasuew, K., Michalakis, S., Breer, H., et al. (2010). The cyclic nucleotide-gated ion channel CNGA3 contributes to coolness-induced responses of Grueneberg ganglion neurons. *Cell. Mol. Life Sci.* 67: 1859–1869.

Marshall, D. A. and Maruniak, J. A. (1986). Masera's organ responds to odorants. *Brain Res.* 366: 329–332.

Martini, S., Silvotti, L., Shirazi, A., et al. (2001). Co-expression of putative pheromone receptors in the sensory neurons of the vomeronasal organ. *J. Neurosci.* 21: 843–848.

Masucci, M., D'Aniello, B., Iela, L., et al. (1992). Immunohistochemical demonstration of the presence and localization of diverse molecular forms of gonadotropin-releasing hormone in the lizard (Podarcis s. sicula) brain. *Gen. Comp. Endocrinol.* 86: 81–89.

Matsuo, T., Rossier, D. A., Kan, C. and Rodriguez, I. (2012). The wiring of Grueneberg ganglion axons is dependent on neuropilin 1. *Development* 139: 2783–2791.

McCotter, R.E. (1915). A note on the course and distribution of the nervus terminalis in man. *Anat. Rec.* 9: 243–246.

Medina, M., Reperant, J., Miceli, D., et al. (2005). GnRH-immunoreactive centrifugal visual fibers in the Nile crocodile (Crocodylus niloticus). *Brain Res.* 1052: 112–117.

Meyer, D. L., von Bartheld, C. S. and Lindörfer, H. W. (1987). Evidence for the existence of a terminal nerve in lampreys and in birds. In *The Terminal Nerve (Nervus Terminalis): Structure, Function, and Evolution,* Demski L. S. and Schwanzel-Fukuda M. (Eds). New York: New York Academy of Sciences, pp. 385–391.

Miragall, F., Breipohl, W., Naguro, T. and Voss-Wermbter, G. (1984). Freeze-fracture study of the plasma membranes of the septal olfactory organ of Masera. *J. Neurocytol.* 13: 111–125.

Moeller, J. F. and Meredith, M. (1998). Increase in gonadotropin-releasing hormone (GnRH) levels in CSF after stimulation of the nervus terminalis in Atlantic stingray, *Dasyatis sabina. Brain Res.* 806: 104–107.

Mombaerts, P. (2004). Genes and ligands for odorant, vomeronasal and taste receptors. *Nat. Rev. Neurosci.* 5: 263–278.

Mousley, A., Polese, G., Marks, N. J. and Eisthen, H. L. (2006). Terminal nerve-derived neuropeptide y modulates physiological responses in the olfactory epithelium of hungry axolotls (Ambystoma mexicanum). *J. Neurosci.* 26: 7707–7717.

Munz, H., Stumpf, W. E. and Jennes, L. (1981). LHRH systems in brain of platyfish. *Brain Res.* 221: 1–13.

Northcutt, R. G. and Puzdrowski, R. L. (1988). Projections of the olfactory bulb and nervus terminalis in the silver lamprey. *Brain Behav. Evol.* 32: 96–107.

Ohtomi, M., Fujii, K. and Kobayashi, H. (1989). Distribution of FMRFamide-like immunoreactivity in the brain and neurohypophysis of the lamprey, Lampetra japonica. *Cell Tissue Res.* 256: 581–584.

Oikawa, T., Saito, H., Taniguchi, K. and Taniguchi, K. (2001). Immunohistochemical studies on the differential maturation of three types of olfactory organs in the rats. *J. Vet. Med. Sci.* 63: 759–765.

Oka, Y. (1992). Gonadotropin-releasing hormone (GnRH) cells of the terminal nerve as a model neuromodulator system. *Neurosci. Lett.* 142: 119–122.

Oka, Y. (1996). Characterization of TTX-resistant persistent Na^+ current underlying pacemaker potentials of fish gonadotropin-releasing hormone (GnRH) neurons. *J. Neurophysiol.* 75: 2397–2404.

Oka, Y. (2002). Physiology and release activity of GnRH neurons. *Prog. Brain Res.* 141: 259–281.

Oka, Y., Munro, A. D. and Lam, T. J. (1986). Retinopetal projections from a subpopulation of ganglion cells of the nervus terminalis in the dwarf gourami (Colisa lalia). *Brain Res.* 367: 341–345.

Park, D. and Eisthen, H. L. (2003). Gonadotropin releasing hormone (GnRH) modulates odorant responses in the peripheral olfactory system of axolotls. *J. Neurophysiol.* 90: 731–738.

Pedersen, P. E. and Benson, T. E. (1986). Projection of septal organ receptor neurons to the main olfactory bulb in rats. *J. Comp. Neurol.* 252: 555–562.

Pinkus, F. (1894). Über einen noch nicht beschriebenen Hirnnerven des Protopterus annectens. *Anat. Anz.* 9: 562–566.

Propper, C. R. and Moore, F. L. (1991). Effects of courtship on brain gonadotropin hormone-releasing hormone and plasma steroid concentrations in a female amphibian (Taricha granulosa). *Gen. Comp. Endocrinol.* 81: 304–312.

Rodolfo-Masera, T. (1943). Su l'esistenza di un particolare organo olfactivo nel setto nasale della cavia e di altri roditori. *Arch. Ital. Anat. Embryol.* 48: 157–212.

Roppolo, D., Ribaud, V., Jungo, V. P., et al. (2006). Projection of the Gruneberg ganglion to the mouse olfactory bulb. *Eur. J. Neurosci.* 23: 2887–2894.

Rosen, G., Sherwood, N. and King, J. A. (1997). Immunoreactive gonadotropin-releasing hormone (GnRHir) is associated with vestibular structures in the green anole (Anolis carolinensis). *Brain Behav. Evol.* 50: 129–138.

Rossi, A., Basile, A. and Palombi, F. (1972). On the functional nature of the nervus terminalis system of teleosts. *Rivista di biologia* 65: 385–409.

Schaefer, A. T. and Margrie, T. W. (2007). Spatiotemporal representations in the olfactory system. *Trends Neurosci.* 30: 92–100.

Schmid, A., Pyrski, M., Biel, M., et al. (2010). Grueneberg ganglion neurons are finely tuned cold sensors. *J. Neurosci.* 30: 7563–7568.

Schwanzel-Fukuda, M. and Pfaff, D. W. (1989). Origin of luteinizing hormone-releasing hormone neurons. *Nature* 338: 161–164.

Shinoda, K., Shiotani, Y. and Osawa, Y. (1989). "Necklace olfactory glomeruli" form unique components of the rat primary olfactory system. *J. Comp. Neurol.* 284: 362–373.

Smith, M. T., Moore, F. L. and Mason, R. T. (1997). Neuroanatomical distribution of chicken-I gonadotropin-releasing hormone (cGnRH-I) in the brain of the male red-sided garter snake. *Brain, behavior and evolution* 49: 137–148.

Springer, A. D. (1983). Centrifugal innervation of Goldfish retina from ganglion cells of the Nervus Terminalis. *J. Comp. Neurol.* 214: 404–415.

Stell, W. K., Walker, S. E. and Ball, A. K. (1987). Functional-anatomical studies on the terminal nerve projection to the retina of bony fishes. *Ann. N. Y. Acad. Sci.* 519: 80–96.

Storan, M. J. and Key, B. (2006). Septal organ of Gruneberg is part of the olfactory system. *J. Comp. Neurol.* 494: 834–844.

Sewertzoff, A. N. (1902). Zur Entwicklungsgeschichte des Ceratodus forsteri. *Anat. Anz.* 21: 593–608.

Tachibana, T., Fujiwara, N. and Nawa, T. (1990). The ultrastructure of the ganglionated nerve plexus in the nasal vestibular mucosa of the musk shrew (Suncus murinus, insectivora). *Arch. Histol. Cytol.* 53: 147–156.

Taniguchi, K., Arai, T. and Ogawa, K. (1993). Fine structure of the septal olfactory organ of Masera and its associated gland in the golden hamster. *J. Vet. Med. Sci.* 55: 107–116.

Tian, H. and Ma, M. (2004). Molecular organization of the olfactory septal organ. *J Neurosci* 24: 8383–8390.

Tian, H. and Ma, M. (2008a). Activity plays a role in eliminating olfactory sensory neurons expressing multiple odorant receptors in the mouse septal organ. *Mol. Cell. Neurosci.* 38: 484–488.

Tian, H. and Ma, M. (2008b). Differential development of odorant receptor expression patterns in the olfactory epithelium: a quantitative analysis in the mouse septal organ. *Dev. Neurobiol.* 68: 476–486.

Tobet, S. A., Chickering, T. W. and Sower, S. A. (1996). Relationship of gonadotropin-releasing hormone (GnRH) neurons to the olfactory system in developing lamprey (Petromyzon marinus). *J. Comp. Neurol.* 376: 97–111.

Tobet, S. A., Nozaki, M., Youson, J. H. and Sower, S. A. (1995). Distribution of lamprey gonadotropin-releasing hormone-III (GnRH-III) in brains of larval lampreys (Petromyzon marinus) *Cell Tissue Res.* 279: 261–270.

Umino, O. and Dowling, J. E. (1991). Dopamine release from interplexiform cells in the retina: effects of GnRH, FMRFamide, bicuculline, and enkephalin on horizontal cell activity. *J. Neurosci.* 11: 3034–3046.

von Bartheld, C. S., Lindorfer, H.W. and Meyer, D.L. (1987). The nervus terminalis also exists in cyclostomes and birds. *Cell Tissue Res.* 250: 431–434.

von Bartheld, C.S. and Meyer, D. L. (1988). Central projections of the nervus terminalis in lampreys, lungfishes, and bichirs. *Brain Behav. Evol.* 32: 151–159.

Wachowiak, M. (2011). All in a sniff: olfaction as a model for active sensing. *Neuron* 71: 962–973.

Walker, S. E. and Stell, W. K. (1986). Gonadotropin-releasing hormone (GnRF), molluscan cardioexcitatory peptide (FMRFamide), enkephalin and related neuropeptides affect goldfish retinal ganglion cell activity. *Brain Res.* 384: 262–273.

Walz, A., Feinstein, P., Khan, M. and Mombaerts, P. (2007). Axonal wiring of guanylate cyclase-D-expressing olfactory neurons is dependent on neuropilin 2 and semaphorin 3F. *Development* 134: 4063–4072.

Weiler, E. and Farbman, A. I. (2003). The septal organ of the rat during postnatal development. *Chem. Senses* 28: 581–593.

White, J. and Meredith, M. (1987). The nervus terminalis of the shark: the effect of efferent impulses on ganglion cell activity. *Brain Res.* 400: 159–164.

White, J. and Meredith, M. (1993). Spectral analysis and modelling of ACh and NE effects on shark nervus terminalis activity. *Brain Res. Bull.* 31: 369–374.

White, J. and Meredith, M. (1995). Nervus terminalis ganglion of the bonnethead shark (Sphyrna tiburo): evidence for cholinergic and catecholaminergic influence on two cell types distinguished by peptide immunocytochemistry. *J. Comp. Neurol.* 351: 385–403.

Wirsig-Wiechmann, C. R. (1993). Nervus terminalis lesions: I. No effect on pheromonally induced testosterone surges in the male hamster. *Physiol. Behav.* 53: 251–255.

Wirsig-Wiechmann, C. R. (1997). Nervus terminalis lesions: II. Enhancement of lordosis induced by tactile stimulation in the hamster. *Physiol. Behav.* 61: 867–871.

Wirsig-Wiechmann, C. R. and Jennes, L. (1993). Gonadotropin-releasing hormone agonist binding in tiger salamander nasal cavity. *Neurosci. Lett.* 160: 201–204.

Wirsig-Wiechmann, C. R. and Lee, C. E. (1999). Estrogen regulates gonadotropin-releasing hormone in the nervus terminalis of Xenopus laevis. *Gen. Comp. Endocrinol.* 115: 301–308.

Wirsig-Wiechmann, C. R. and Wiechmann, A. F. (2001). The prairie vole vomeronasal organ is a target for gonadotropin-releasing hormone. *Chem. Senses* 26: 1193–1202.

Wirsig-Wiechmann, C. R., Wiechmann, A. F. and Eisthen, H. L. (2002). What defines the nervus terminalis? Neurochemical, developmental, and anatomical criteria. *Prog. Brain Res.* 141: 45–58.

Wirsig, C. R. and Leonard, C. M. (1987). Terminal nerve damage impairs the mating behavior of the male hamster. *Brain Res.* 417: 293–303.

Wray, S., Grant, P. and Gainer, H. (1989). Evidence that cells expressing luteinizing hormone-releasing hormone mRNA in the mouse are derived from progenitor cells in the olfactory placode. *Proc. Natl. Acad. Sci. U. S. A.* 86: 8132–8136.

Wright, G. M., McBurney, K. M., Youson, J. H. and Sower, S. A. (1994). Distribution of lamprey gonadotropin-releasing hormone in the brain and pituitary gland of larval, metamorphic, and adult sea lampreys, Petromyzon marinus. *Can. J. Zool.* 72: 48–53.

Yamamoto, N. and Ito, H. (2000). Afferent sources to the ganglion of the terminal nerve in teleosts. *J. Comp. Neurol.* 428: 355–375.

Yamamoto, N., Oka, Y. and Kawashima, S. (1997). Lesions of gonadotropin-releasing hormone-immunoreactive terminal nerve cells: effects on the reproductive behavior of male dwarf gouramis. *Neuroendocrinology* 65: 403–412.

References

Yáñez, J., Folgueira, M., Köhler, E., et al. (2011). Connections of the terminal nerve and the olfactory system in two galeomorph sharks: an experimental study using a carbocyanine dye. *J. Comp. Neurol.* 519: 3202–3217.

Zhang, J. X., Soini, H. A., Bruce, K. E., et al. (2005). Putative chemosignals of the ferret (Mustela furo) associated with individual and gender recognition. *Chem. Senses* 30: 727–737.

Zhang, W. and Delay, R. J. (2007). Gonadotropin-releasing hormone modulates voltage-activated sodium current and odor responses in Necturus maculosus olfactory sensory neurons. *Journal of neuroscience research* 85: 1656–1667.

Zhang, X., Rogers, M., Tian, H., et al. (2004). High-throughput microarray detection of olfactory receptor gene expression in the mouse. *Proc. Natl. Acad. Sci. U. S. A.* 101: 14168–14173.

Zucker, C. L. and Dowling, J. E. (1987). Centrifugal fibres synapse on dopaminergic interplexiform cells in the teleost retina. *Nature* 330: 166–168.

Author Index

Aaron, J. I., 1053
Abad, F., 805, 806
Abaffy, T., 110
Abbasi, A. A., 848
Abbé, E., 638
Abdallah, L., 839
Abe, H., 1143
Abel, R., 541
Abele, M., 427, 436
Abitol, M. L., 595
Ables, J. L., 190
Ablimit, A., 66
Abolmaali, N. D., 284
Abraham, A., 237, 437
Abraham, M. H., 1092, 1097, 1098, 1099
Abraham, N. M., 169
Abrams, M. T., 149
Abu-Hamdan, D. K., 846
Accolla, R., 784, 785
Acebes, A., 538
Aceto, A., 70
Acharya, A. B., 805
Acharya, V., 901
Ache, B. W., 250, 531
Ackerman, B. H., 901, 914
Ackroff, K., 868
Acree, T. E., 688
Adachi, K., 190
Adamek, G. D., 170
Adams, D. B., 1123
Adams, D. R., 17, 47, 69, 74, 1133
Adams, R. G., 385, 497
Adams, S., 627
Adamson, J. E., 357
Aden, E., 414, 421
Ader, R., 873, 874
Adler, C. H., 438
Adler, E., 690, 693, 914, 960, 1102
Adler, J., 513, 762
Adolphs, R., 783 , 1122

Adour, K. K., 891, 895
Adrian (1950), 367
Adrian (1953), 16
Adrian (1991), 358
Adrian, E. D., 168, 169, 209
Aggerbeck, H., 459
Aggleton, J. P., 345, 996
Agnello, D., 872
Agster, K. L., 217
Agu, R. U., 71
Agüero, A., 876, 877
Aguirre, A., 193, 197
Ahlenius, H., 197
Ahlskog, J. E., 416, 428
Ahlstrom, R., 385, 497
Ahmed, M., 735
Ahn, S., 190
AhPin, P., 733
Ahren, B., 839
Aiba, T., 141
Ait-Omar, A., 842
Aitasalo, K., 384
Aitken, M. L., 392
Aitken, R. C. B., 758
Akabas, M. H., 693
Akaishi, T., 840
Akay, C., 499
Akbarian, S., 718
Akerblom, M., 189, 192
Akerlund, A., 380, 382
Akers, K. G., 499
Akeson, R., 465
Akeson. R. A., 465
Akhtar, M. S., 385
Akirav, I., 879
Akpan, N., 458
Aksoy, H., 412
AlAïn, S., 319, 320
al-Shammari, S. A., 911
Alade, M. O., 317

Albert, P. J., 934
Alberti, S., 189
Alberts, J. R., 305, 307, 315, 316, 320, 321
Albin, K. C., 1009, 1101
Albinus, 642, 644
Albone, E. S., 20, 594
Albrecht, J., 246, 832
Alcalay, R. N., 421, 424
Alcock, B., 653, 654
Alcock, J., 963, 968
Alden, M., 708
Alekseyenko, O. V., 76
Alexander, B. K. (1967), 22
Alexander, T. H., 499, 500
Alexson, T. O., 190, 193
Alfano, M., 765
Alföldi, J., 974
Algom, D., 761
Ali, J., 487
Alimohammadi, H., 1055, 1092, 1098
Allavena, R. E., 487
Allen, C. M., 891, 892, 893
Allen, E., 19
Allen, H., 44, 57
Allen, J. J., 596
Allen, W. K., 465
Allen, Z. J., 196
Allen, W. F., 1091
Alleva, A. M., 832, 841
Allison, A. C., 280, 281
Allison, T., 261, 265
Allman-Farinelli, M. A., 848, 901
Almkvist, O., 405
Alonso, M., 184, 198
Alstadhaug, K. B., 900, 903
Altar, C. A., 466, 473
Althaus, J., 12, 19, 183, 385
Altman, J., 19, 183, 738
Altman, J. S., 932, 933
Altman, K. W., 246, 385, 486, 496

Handbook of Olfaction and Gustation, Third Edition. Edited by Richard L. Doty.
© 2015 Richard L. Doty. Published 2015 by John Wiley & Sons, Inc.

Altner, H., 557
Altundag, A., 245
Alvarez, P., 346
Alvarez-Buylla, A., 183, 184, 186, 189, 193, 195, 197
Alves, J. A., 185
Amakawa, T., 932
Amaral, D. G., 996
American Public Health Association, 1081
Amerine, M. A., 1052
Amieva, M. R., 146
Amiez, C., 1033
Amit, Z., 867
Amo, L., 547, 567, 572
Amoore, J. E., 228, 242, 365, 388
Amrein, H., 934
Amsterdam, J. D., 438, 903
Anagnostides, A., 838
Anand, B. K., 997
Anand-Kumar, T. C., 73
Andersen, H. B., 840
Anderson, A. K., 290, 292, 782
Anderson, D. J., 1120, 1123, 1124
Anderson, G. H., 834
Anderson, J., 243
Anderson, J. O., 489, 490
Anderson, N. H., 227, 235
Anderson, S. A., 196
Anderson, V., 386
Andersson, M. N., 539
Andl, T., 731
Andrade, J., 338, 339
Andre, J. M., 415, 900
Andre, R. F., 358, 367, 368
Andreas, P., 1080
Andres, K., 134
Andres, K. H., 18, 646
Andreu-Agullo, C., 188, 193
Andrew, R. J., 611
Angers, S., 731
Angot, E., 190
Anholt, R. R. H., 15
Anisko, J. J., 21
Ankel-Simons, F., 55, 606
Anliker, J. A., 802, 803, 813, 814
Anon., 848
Ansari, K. A., 391, 403, 415, 498
Antal, M., 159
Anton, E. S., 187, 191
Anton, F., 487
Antonini, J. M., 497
Antunes, M. B., 385, 498
Anzinger, A., 393
Aou, S., 997
Aoyama, S., 317
Apfelbach, R., 310, 456
Appendino, G., 1092
Appleton, K. M., 833
Apter, A. J., 381, 382

Arabie, P., 1009
Arakawa, H., 581, 584, 1120
Araneda, R. C., 170, 174, 251, 1122
Araneda, S., 466
Araujo de, I. E., 1057
Arcaro, K. F., 325
Arendt, T., 411, 412
Areas, G., 1069, 1072, 1073
Arevalo, R., 553
Arevian, A. C., 166, 168, 169
Arey, L., 642
Ariai, N., 934
Arias, C., 835
Aristotle, 637
Armelagos, G., 3
Armitage, J. P., 513, 516
Arms, K., 934
Armstrong, C., 17, 461
Armstrong, J. B., 513
Armstrong, J. E., 243, 309, 313, 801, 807, 889, 901
Armstrong, R. N., 70
Armstrong, S. D., 1119
Arnhold-Schneider, M., 897
Arnold, F., 644
Arnold, J., 596
Arnold, S. A., 194
Arnold, S. E., 148
Aroniadou-Anderjaska, V., 160, 161, 162, 165
Aronsohn, E., 247, 248
Arvidson, K., 643
Arvidsson, A., 198
Asa, C. S., 594
Asaoka, K., 933
Asarian L., 869, 870, 872
Aschenbrenner, K., 390, 843, 844, 903, 1019
Aschner, M., 497
Ashbaught, N., 714, 716, 992
Ashe, J. H., 876, 878
Ashwell, K. W., 50, 741
Aspen, V. A., 837, 838
Assad, J. A., 717, 995
Assaf, Y., 286
Asson-Batres, M. A., 143, 144, 145, 146, 147
Astbäck, J., 650
Astic, L., 128, 455, 462, 1134
ASTM, 230, 796, 1083
Atema, J., 645, 948
Atighechi, S., 386
Atobe, Y., 970, 971, 972
Atoji, Y., 643, 953
Au, C., 18, 97, 147
Au, E., 97, 147
Au, W. W., 146
Aubry, L., 521
Auen, E. L., 970, 971

Aujard, F., 617
Aungst, B. J., 458, 467
Aungst, J. L., 159, 163
Avancini de Almeida, T. C., 1105
Avenet, P., 15, 952
Avery, M. L., 966
Axel, R., 15, 16, 95, 109, 123, 171, 210, 214, 367, 465, 553, 554, 593, 1054, 1115, 1124
Axelrod, J., 556
Ayabe-Kanamura, S., 1058, 1059
Ayenew, Z., 922, 923
Ayer–Lelievre, C., 142
Ayestaran, A., 937
Ayrton, A., 937
Azen, E. A., 626, 644
Azevedo, E., 763
Aziz, Q., 993
Azmitia, E. C., 194
Azoulay, R., 306
Azzali, G., 642, 647, 650, 651

Baatrup, E., 947, 948, 952
Baba, T., 423, 498
Bachmanov, A. A., 685, 686, 688, 689, 796, 835, 836, 959, 960, 968
Bacon, A. W., 23, 405, 409, 410
Bacon, I., 965, 966
Bacon-Moore, A. S., 405
Baddeley, A., 252
Baddeley, A. D., 338, 340
Badonnel, K., 66
Baeg, E. H., 717, 995
Baeyens, F., 1014, 1015
Baggio, L. L., 842
Bagshaw, M. H., 776
Bahar, A., 879
Bahar-Fuchs, A., 408, 410, 411
Bahrami, F., 457, 464
Bahrick, L., 322
Bailey, K., 617
Bailey, K. C., 8
Bailey-Hill, K., 111
Baines, D., 1054
Baird, J. C., 236
Bairlein, F., 961, 966
Bajaj, S., 377, 390
Bakardjiev, A., 145
Baker, C. V., 737
Baker, H., 455, 456, 461, 465, 466, 473
Baker, H. G., 965, 966
Baker, I., 965
Baker, K. B., 429
Baker, T. A., 456, 464, 465
Baker, T. C., 533, 539
Bakhit, C., 466, 473
Bakker, K., 386
Bakker, J. M., 429

Author Index

Balakrishnan, R., 937
Balali-Mood, M., 489
Balaraman, S., 499
Balboni, G. C., 76
Baldaccini, N. E., 465
Baldwin, D. H., 557
Baldwin, M. W., 964
Baldwin, R. M., 69
Balin, B. J., 455, 456, 461, 464, 465, 466
Ballard, D. H., 295
Ballester, J., 1014
Balmer, C. W., 94
Balog, J. M., 968, 969
Balogh, R. D., 308
Balordi, F., 190, 191, 193
Balter, M., 517
Balthazart, J., 869
Baltussen, E., 574
Balu, R., 164, 166
Bamberger, C., 651
Bammler, T. K., 70
Banasr, M., 190, 194
Bancaud, J., 765, 776, 901
Bandell, M., 1101
Bandettini, P., 293
Bandoh, H., 557
Bandyopadhyay, S. K., 552
Banger, K. K., 70, 79, 498
Banks, E. M., 325
Banks, M. A., 554
Banks, P. B., 586, 587
Bannister, L. H., 58, 93, 146, 465, 1114
Banoczy, J., 891
Bansal, S., 838
Bao, Z. P., 69
Baraban, S. C., 557
Barach, J. P., 973, 974
Baraniuk, J., 488
Baraniuk, J. N., 1092
Barasch, A., 922
Barbara, L., 911
Barbas, H., 715, 990, 993
Barbe, C. D., 248, 249, 250
Barber, P. C., 174, 465
Barbour, J., 71
Barceloux, D. G., 464
Bardy, C., 195
Bargmann, C. I., 1120, 1140
Barik, S., 458
Barkai, E., 214, 216, 344
Barker, L. M., 1014
Barlow, L. A., 651, 655, 728, 729, 730, 734, 735, 736
Barnea, G., 128
Barnes, D. C., 214, 215
Barnett, D., 1057
Barnett, E. M., 455, 462
Barnett, S. A., 869
Barnett, S. C., 146

Barnham, A. L., 313
Barnicot, N. A., 755
Baron, G., 55
Baron, J., 69, 78
Barquin, J., 1061
Barr, R. G., 800, 806, 807
Barragan, E., 473
Barraud, P., 139
Barreiro-Iglesias, A., 646
Barrett, J., 612, 617
Barron, S., 499
Barrow, A., 458, 467
Barrows, W. M., 20
Barry, M. A., 776, 781
Bartel, D. L., 648, 649
Bartels, J. H. M., 1081
Bartelt, R. J., 532
Bartholow, M., 911
Bartocci, M., 307, 308, 317
Barton, R. A., 607
Bartoshuk, L. M., 3, 9, 10, 235, 346, 501, 630, 642, 645, 669, 672, 751, 753, 754, 755, 756, 757, 758, 759, 760, 761, 762, 763, 764, 765, 766, 782, 836, 891, 895, 896, 958, 986, 1009, 1010, 1016, 1019, 1020, 1052, 1105
Bartter, F. C., 388
Barylko-Pikielna, N., 846
Barz, S., 271, 416, 423
Basak, O., 190, 193
Bashaw, G. J., 1120
Bassler, B. L., 518, 519, 520
Basson, M. D., 756
Bastien-Dionne, P. O., 196
Bate-Smith, E. C., 933
Batenburg, M., 1020
Bath, K. G., 189
Bath, W., 961, 971
Bathellier, B., 168
Batista, C. M., 198
Batista-Brito, R., 196
Batoni, G., 100
Batsell, W. R., 326
Battista, D., 188
Bauer, C., 872
Bauer, S., 142, 144
Baum, B. J., 643
Baum, M. J., 174, 1122, 1123
Baumer, D., 412
Bautista, D. M., 1092, 1098, 1101, 1102, 1104, 1140
Bavier, M., 501, 872
Baxi, K. N., 47
Baxter, N. J., 632
Bayer, S., 738
Bayer, S. A., 183
Baylis, L. L., 714, 715, 716, 775, 784, 991, 993, 994, 1031, 1032, 1035
Bazaes, A., 554

Bazer, G. T., 455, 457, 460
Beach, F. A., 21, 583, 595
Bealer, S. L., 675
Bean, B. P., 1117
Bean, N. J., 21
Beard, M. D., 147, 389
Beatus, P., 189, 193
Beauchamp, G. K., 3, 73, 246, 305, 308, 310, 313, 314, 319, 325, 326, 344, 345, 583, 584, 670, 672, 685, 797, 798, 799, 800, 802, 807, 808, 809, 815, 816, 817, 821, 822, 823, 833, 834, 836, 837, 923, 958, 959, 960, 964, 968, 1101
Bechara, A., 995
Bechgaard, E., 458
Becker, E., 269, 429
Becker, Y., 456, 462
Beckman, M. E., 705
Beckstead, R. M., 706, 710, 711, 780, 987, 988, 989, 1028
Becquemin, M. H., 485, 486
Becques, A., 314
Bedford, E. A., 93, 98
Bedlack, R. S., 409
Bedwell, J. S., 435
Beeli, G., 900
Begbie, J., 98, 737
Begum, M., 390
Behrens, M., 649, 671, 687, 689, 690, 691, 695, 696, 987
Beidler, L. M., 3, 11, 14, 17, 57, 456, 463, 626, 629, 633, 650, 665, 763, 891, 923, 984
Beikirch, R. J., 870
Beites, C. L., 93, 96, 730, 731, 732, 734, 736
Bejsovec, A., 731
Bekkers, J. M., 209, 213, 214
Bekoff, M., 595
Bélanger, M. C., 1120
Belanger, R., 550, 551
Belanger, R. M., 547, 553
Belcher, A. M., 611
Belinsky, S., 457
Belinsky, S. A., 76, 78
Belitz, H.-D., 690
Bell (1803), 10
Bell, G. A., 250, 1051, 1052, 1053, 1054, 1055, 1056, 1057, 1059
Bell, H., 216
Bell, W., 524
Bell, W. E., 525
Bellas, D. N., 900
Bellieni, C. V., 806
Bellingeri, 1818, 10
Bellini, L., 9, 637, 638
Bellisle, F., 839, 905
Bello, N. T., 838
Belluscio, L., 112, 163, 167, 172, 1115, 1120

Belluzzi, O., 159, 170, 194
Belman, A. L., 962
Belmonte, C., 1092
Belvindrah, R., 187, 191
Belzer, L. M., 848
Bemis, W. E., 553
Ben-Arie, N., 70
Ben-Shaul, Y., 173, 1113, 1118, 1121, 1125
Benali, A., 48, 58
Bender, G., 227, 779, 786
Benedict, C., 463
Benes, E. S., 973, 974
Bengtsson, S., 1057
Benignus, V. A., 265
Benjamin, R. M., 14, 710, 711, 776, 779,
    780, 989, 990, 991
Benjaminsen, E., 900, 903
Bennett, B. D., 116, 117
Bennick, A., 632
Benoit, S. C., 874
Benoliel, R., 369
Benowitz, N. L., 459, 468
Bensafi, M., 237, 296, 310
Benson, T. E., 473, 1134
Benton, R., 533, 932
Benuck, I., 585
Benvenuti, S., 547
Beraldo, K. E. A., 833
Bereiter, D. A., 708
Berendse, H. W., 421, 423, 877
Beréziat, J. C., 79
Berg, B. G., 535
Berg, D., 426
Berg, H. C., 513
Berg, L., 263
Bergamasco, N. H. P., 833
Bergasa, N. V., 849, 901
Berger, G., 66
Berger, H., 261
Berger-Sweeney, J., 412
Berggren, B., 838
Berghard, A., 1115, 1117, 1139
Berglund, B., 228, 235, 248, 249
Berglund, H., 298
Bergman, J. E., 390
Bergman, U., 80
Bergmann, O., 197
Bergmann, P. J., 974
Bergquist, C., 459, 469
Bergstrom, U., 76
Berk, L. E., 308
Berkhoudt, H., 647, 961, 962, 963, 964
Berkowicz, D. A., 162
Berlucchi, G., 780
Berman, D. E., 879
Bernabeu, R., 217, 218
Bernard, C., 11, 625
Bernard, J. F., 708
Bernard, R. A., 846

Bernays, E. A., 929, 931, 933, 934, 935,
    936, 937, 938, 939, 940
Berndt, A., 894
Berne, C., 840
Bernemann, D., 897
Bernhardson, B. M., 500
Berns, G. S., 785
Bernstein, I. L., 346, 473, 501, 815, 837,
    865, 868, 869, 871, 872, 873, 876, 877,
    878, 1015
Berntson, G. G., 782
Berridge, K. C., 655, 785, 796, 799, 865,
    867, 868, 869, 875, 937, 1020, 1027
Berry, H., 758
Berteretche, M. V., 501, 782, 785
Berthoud, H. R., 837, 839
Bertino, M., 832, 837
Bertollo, D. N., 435
Beschi, R. J., 389, 846, 901
Beshel, J., 168, 169, 344
Besnard, P., 835
Bessen, R. A., 487
Besser, G. M., 1123
Best, A. R., 216
Bester, H., 708
Betbeder, D., 471
Betchen, S. A., 245
Bettini, S., 552
Betts, K., 496
Beukelaers, P., 187, 190
Bevc, I., 324
Beyer, C., 14
Beynon, R. J., 568, 581, 583, 1119
Bezard, E., 426
Bezencon, C., 695, 841
Bhalla, U. S., 168
Bhandawat, V., 111, 113
Bharathi, V. P., 499
Bhatia, S., 846
Bhatnagar, K. P., 21, 42, 43, 47, 48, 54, 55,
    57, 608, 1124
Bhatt, J. P., 555
Bhinder G., 557
Biaggioni, I., 427, 429
Bian, X., 1123
Bianco, J. I., 147
Biarnes, X., 692
Bibas, A., 897
Bicker, G., 939, 940
Bieber, I. B., 324
Bielavska, E., 877
Biernat, M., 759
Bijani, C., 470
Bijpost, S. C. A., 934, 937
Bilberg, K., 102
Bilko, A., 314, 318, 326
Billig, G. M., 1118
Billing, J., 7
Billot, P.-E., 1057

Bingham, A. F., 1018, 1021
Bingham, P., 317
Bingham, P. M., 317
Bingley, C. A., 667
Birch, L. L., 795, 804, 809, 817, 823, 868,
    869, 1056, 1062
Bird, D., 57
Birnbaum, M. H., 759
Bishop, C. L., 193
Bishop, E. R., Jr., 390
Bitanihirwe, B. K. Y., 101
Bizzini, L., 404
Bjorkman, S., 458, 467
Bjorndal, K. A., 970
Black, A. H., 868
Blackburn, C. W., 897
Blackburn, T. P., 556
Blackwell, H. R., 230
Blais, C. A., 837
Blake, C. B., 1123
Blakeslee, A. F., 387, 755, 756, 762
Blanchard, K. T., 80
Blanck, H. M., 795
Blanco, J. C., 591
Blaney, W. M., 938, 939
Blass, E. M., 305, 315, 316, 319, 320, 322,
    325, 585, 800, 805, 806, 807
Blau, J. N., 900
Blaustein, J., 869
Blaylock, R. L., 497
Blednov, Y. A., 836
Blem, C. R., 965
Blentic, A., 737
Blessing, W. W., 462
Blitz, A. M., 381, 383
Blizard, R. A., 595
Block (1946), 19
Block, M. L., 490
Blom, F., 938
Blomster, L. V., 142, 146, 147
Bloom, G., 93, 280
Bloom, S. J., 308, 313
Bloomfield, R. S., 849, 901
Blossfeld, I., 804, 819, 822
Blozis, G. G., 891, 892, 893
Blumberg, P. M., 1092
Blumenfeld, R. S., 340
Blundell, J. E., 832, 833
Bluthe, R. M., 340, 876
Bo, X., 694
Boakes, R. A., 340, 1011, 1013, 1014, 1018
Bobillier, P., 466
Bocca, E., 365
Bodian, D., 17, 455, 462
Bodner, L., 838
Bodyak, N., 344
Boehm, N., 306
Boehm, U., 115, 1123
Boer, C. C., 845

Author Index

Boesveldt, S., 418, 419, 423
Bogan, N., 164
Bogdanffy, M. S., 65, 72, 457
Bogucka-Bonikowska, A., 836
Bohm, S., 71
Bohnen, N. I., 417, 419, 420, 423, 489, 498
Boistel, J., 21
Boitani, L., 596
Bojsen-Moller, F., 57, 364, 1133
Bokil, H., 936
Bolen, R. H., 610
Bolles, R. C., 868
Bolteus, A. J., 190, 193
Bolze, M. S., 845
Bom, D. C., 667
Bomback, A. S., 501
Bonadonna, F., 565, 567, 572
Bonanni, E., 264
Bond, D. S., 837, 838
Bond, J. A., 64, 78, 79
Bondier, J. R., 497
Bondy, C., 142
Bonfanti, L., 183
Bonfils, P., 242
Bongaerts, J. H. H., 627
Bonhoeffer, T., 167
Bonini, N. M., 525
Bonnet-Font, C., 466
Booker, S. M., 497
Boone, N., 149
Boonstra, R., 587
Boorstin, D. J., 6
Boot, L. M., 21
Booth, D. A., 248, 313, 834, 836, 867, 868, 1015, 1043
Borders, A. S., 146
Bordeu, T. D., 11
Bordey, A., 190, 193
Borg, G., 235, 654, 758, 759
Borg-Neczak, K., 455
Boring, E. G., 3, 12, 764
Borison, H. L., 876
Bornovalova, M., 877
Bornstein, W. S., 776
Borowsky, B., 115, 553
Borror, M. J., 298
Borta, A., 190, 194
Boruch, A. V., 147
Bosak, N. P., 960
Boschat, C., 554, 1119
Bossola, M., 848
Bossy, J., 306
Bostantjopoulou, S., 415, 424
Botchway, C. A., 626
Botezat, E., 961
Bouchard, M., 497, 498
Boucher, Y., 779, 780, 781
Boudriot, F., 646
Boughter, J. D., 649, 703, 796, 960

Bouhlal, S., 836
Boulet, M., 612
Boulloche, N., 900
Bourgery, J. M., 639, 654
Bourne, M. C., 1010
Bourne, R. C., 969
Bourne, G. H., 67
Bout, R. G., 962
Bouter, N., 667
Bouton, M., 879
Bovetti, S., 186, 187, 191, 195, 196
Bovi, T., 420, 426
Bovio, G., 845
Bowdan, E., 936, 937
Bowen, W. H., 627
Bower, J. M., 168, 209
Bowers, J. M., 22
Bowers, M. D., 968
Bowler, R. M., 497, 498
Bowman, N. E., 294
Bowman, W., 11
Boyd, E. M., 959
Boyden, E. S., 20
Boyett-Anderson, J. M., 609
Boyle, J. A., 294, 1057
Bozoyan, L., 190
Bozza, T., 126
Braak, E., 218, 411
Braak, H., 217, 218, 411, 473, 491, 502
Bradford, D., 186
Bradley, E. A., 429, 430
Bradley, R. M., 11, 625, 626, 628, 633, 643, 644, 650, 651, 705, 706, 717, 765, 766, 796, 983, 984
Bradway, S. D., 627
Brady, J. D., 112
Braga-Neto, P., 436
Bragulat, V., 499
Brainerd, H. G., 412
Bramley, P. A., 897
Branchek, T. A., 556
Brand, G., 245, 1103
Brand, J. G., 686, 837
Brandt, I., 454, 457, 464
Brandt, M. A., 1055
Brann, J. H., 1113
Brannon, P. M., 838
Brasser, S. M., 836
Brauchli, P., 263
Braun, G., 939, 940
Braun, J. J., 22, 876, 877, 878
Braune, P., 611
Braus, H., 642
Bray, D., 515
Bray, S. J., 735
Brazel, C. Y., 190, 194
Brechbühl, J., 586, 1136, 1137, 1138, 1140
Bredendiek, N., 932
Breer, H., 64, 112, 887, 1136, 1140

Breipohl, W., 93, 135, 139, 141, 1133
Brenker, C., 118
Brennan, P., 580
Brennan, P. A., 173, 1122
Brenneman, K. A., 455
Breslin, P., 670, 672, 674, 870, 914, 915, 923, 960, 964
Bressler, S. C., 1122
Breton-Provencher, V., 198
Brewer, W. J., 311, 431, 432, 433, 434, 435
Breymann, C., 916
Brezinski, D. A., 79
Brezun, J. M., 194
Brierley, T., 367
Brieskorn, C. H., 933
Briggs, M. H., 392
Brill, M. S., 196
Brillat-Savarin, J.-A., 1007
Brindley, G. S., 752
Brindley, L. D., 965
Briner, H. R., 384
Brinner, L. J., 846
Brittebo, E. B., 69, 73, 76, 79, 80, 454, 456, 457, 463
Britton, R. A., 520
Britton, S. J., 667
Broadwell, R. D., 173, 174, 455, 461, 465, 466, 1122
Brobeck, J. R., 997
Broberg, D. J., 501, 873
Broca, P., 12, 19
Brochet, F., 1061
Brock, W. H., 1007
Brockhoff, A., 670, 691, 692, 914, 923
Brockmann, K., 427
Brodie, E. D., 973
Brodoehl, S., 377, 423
Brody, D., 438
Brody, H., 1101, 1102
Brody, J. A., 428
Bröer, S., 466
Broggio, E., 406
Broman, I., 1133
Bromley, S. M., 243, 245, 500, 501, 751, 765, 887, 889, 890
Brondel, L., 839, 1020
Bronner-Fraser, M., 736, 737
Bronshtein, A. A., 15
Bronson, F. H., 582, 1125
Brook, R. D., 490
Brookover, C., 1142
Brooks, S. E., 598
Broome, B. M., 539
Brosseau, L., 867
Brosvic, G. M., 22, 388, 389, 846
Broughan, C., 313
Brousseau, K., 412
Brouwer, J. N., 629
Brower, L. P., 933, 968, 969

Browman, H. I., 553
Brown, D., 68
Brown, D. S., 595
Brown, G. E., 547, 555, 557
Brown, J., 233
Brown, J. W., 306
Brown, K. J., 1014
Brown, K. S., 230
Brown, M., 965
Brown, P. L., 584
Brown, R. E., 21, 325
Brown, R. J., 842
Brownlie, J., 101
Bruce, D. G., 839
Bruce, H. M., 21
Bruch, R. C., 15
Bruchet, A., 1080
Bruera, E., 845
Brugger, K. E., 966, 967
Brune, N. E. I., 671
Brunet, L. J., 112, 115, 473, 1125, 1139
Brunjes, P. C., 163, 210, 298, 473, 552, 553
Brunner, H. L., 58
Bruno, R. M., 710
Brunstrom, J. M., 838, 1015
Bruvold, W. H., 1081
Bryant, B., 1091
Bryant, B. P., 630, 1092, 1101, 1104
Bryant, R. W., 667
Brykczynska, U., 1124
Buchanan-Smith, H. M., 611
Bucher, H. U., 805, 806
Buchholz, J. A., 141, 142, 143, 147
Buchner, K., 456, 462, 463, 465, 466
Buchsbaum, M. S., 405, 409
Buck, L., 15, 16, 95, 109, 123, 171, 367,
    465, 593, 1054
Buck, L. B., 16, 115, 554, 1115, 1116, 1117,
    1139
Bucket (1970)., 4
Buckingham, M. E., 729
Buckland, M. E., 141, 142
Buckner, R. L., 288, 292
Bucy, P., 783
Bucy, P. C., 996
Buddenberg, T. E., 463
Budge, E. A., 4
Budgett, H. M., 598
Bueter, M., 1020
Bufe, B., 670, 685, 691, 756, 1117
Buiakova, O. I., 93, 99
Bujas, Z., 754
Bulfone, A., 18
Bull, D. F., 865, 872
Bull, T. R., 654, 763, 764, 765, 896
Bullock, T. H., 1122, 1143
Bulsing, P. J., 269
Bult, J. H. F., 1010
Bunce, C., 311

Bunsey, M., 346, 585
Buonviso, N., 164, 167, 168, 169, 1136
Burch, R. E., 848
Burd, G. D., 456, 463
Burdach, K. J., 23, 227, 246, 376, 1054,
    1055
Burdette, D. L., 517
Bures, J., 875, 876, 877, 879
Buresova, O., 875, 876, 879
Burge, J. C., 847, 848
Burge, K. G., 583, 584
Burger, B. V., 567, 572
Burger, K. S., 785
Burghardt, G. M., 973, 974
Burk, K., 195
Burke, C. J., 934
Burkl, W., 643
Burlingame, G. A., 1080, 1081, 1082, 1086
Burmeister, H. P., 284, 285
Burne, R. H., 548, 549 552, 551
Burne, T. H. J., 568, 969
Burnet, F. M., 17
Burnham, M., 320
Burns, W. A., 732
Burrow, L., 846
Burt, G., 875
Burton, H., 14, 706, 711, 776, 779, 780,
    989, 990, 991
Burton, M. J., 997
Burton, N. C., 497, 498
Burwell, R. D., 217
Buschhuter, D., 283, 286
Busenbark, K. L., 413, 423
Bush, C. F., 112
Bushdid, C., 114
Buss (2008), 323
Buss, C., 840
Busse, K., 422
Butenandt, A., 532
Butera, P. C., 870
Butler, R., 459, 469
Butler, R. K., 1122
Butters, N., 233, 236, 287
Buttery, R. G., 365
Büttner, A., 317
Buxton, L. H. D., 355
Buxton, R. B., 281
Byerly, M. S., 968
Bylsma, F. W., 413
Byram, J., 766, 1011, 1012
Byrd, C. A., 552, 553

Cabanac, M., 786, 839, 847, 869, 974, 995,
    1034
Cablk, M. E., 572
Caccappolo, E., 247, 375
Cachat, J., 557
Cacioppo, J. T., 782
Cadirci, S., 967

Cador, M., 874
Cagan, R. H., 15
Caggiano, M., 96, 135, 136
Cahusac, P. M. B., 345
Cai, X., 1041
Cai, Z., 216
Caicedo, A., 650, 651
Cain, W. S., 3, 228, 230, 237, 238, 242, 244,
    246, 247, 248, 249, 309, 310, 311, 339,
    340, 343, 346, 753, 1009, 1010, 1016,
    1061, 1092, 1093, 1095, 1096, 1097,
    1099, 1100, 1101
Cajal, R., 1124
Cajal, R. S., 11
Calas, D., 936
Calderón-Garcidueñas, L., 385, 411, 490,
    493, 496, 502
Callahan, A. M., 916
Callaway, E. M., 213
Calles-Escandon, J., 839
Calò, C., 757, 915
Calof, A. L., 18, 135, 137, 736
Calu, D. J., 344
Cambier, J., 428
Cameron, E. L., 228, 243, 246, 247, 269,
    309, 414, 434
Cameron, H. A., 183
Cameron, J. D., 246, 995
Cameron, P., 935, 937
Camp, C. A., 807
Campagna, S., 565, 569, 570
Campanella, G., 437
Campbell, I. M., 430
Campbell, K. H., 22
Campbell, L. M., 239
Campbell, S. M., 893
Campden et al., 1085, 1086
Campden Food and Drink Research
    Association, 1085
Campeau, S., 587
Campfield, L. A., 839
Cancalon, P., 463
Canciani, M., 392
Canessa, C. M., 686
Cang, J., 161, 164, 165, 166, 167, 168, 633
Cannon, D. S., 870, 1015
Cano, J., 838
Canteras, N. S., 1123
Cao, Y., 553, 649, 694
Capaldi, E. D., 1015
Capdevila, J. H., 76
Capello, L., 1115
Capobianco, A. J., 193
Caporale, G., 1061
Capretta, P. J., 318, 869
Caprio, J., 556, 557, 951, 952, 954
Capsoni, S., 471
Caputo, A., 1118
Carbajal, R., 805, 806

Cardé, R. T., 532, 598
Cardello, A. V., 759, 760, 761, 767
Cardenas, M., 627
Cardesin, A., 242
Carey, R. M., 167, 168
Carleton, A., 166, 194, 785
Carlson, G. C., 161, 162, 169
Carlson, G. P., 69
Carlson, J. R., 16
Carlsson, M. A., 538
Carmichael, M., 595
Carmichael, S. T., 217, 218, 609, 715, 716,
    990, 993, 994, 1030, 1032, 1033, 1034
Carpenter, D. O., 876
Carpenter, G. H., 632
Carpenter, M. B., 898
Carr, B. A., 69
Carr, V. M., 135, 141, 737, 738
Carr, W. J., 21, 22, 343
Carrai, M., 691
Carrasco, M., 237
Carroll, B., 437
Carson, J. A., 845
Carson, K. A., 101
Carstens, E., 1102, 1103
Carter, D., 126
Carter, J. L., 903
Carter, J. M., 1083
Cartmill, M., 50, 54, 55
Cartoni, C., 692, 835, 1102
Cartwright, R. A., 755
Carus, C. G., 643
Carvalho, A. de O., 99
Case, T. I., 1018
Cashdan, E., 323
Cass, S. P., 896
Cassady, B. A., 832
Cassell, M. D., 708, 711
Casserius, 9, 637
Cassileth, P. A., 874
Castella, P., 193
Castellucci, V., 338
Castiglione, M., 194
Castillo, P. E., 170
Castor, S. M., 326, 821
Castro, J. B., 165, 166, 1121, 1122
Catalanotto, F. A., 633, 765, 899
Caterina, M. J., 369, 1101
Caton, R., 261
Catroppa, C., 386
Cattarelli, M., 214, 715
Cau, E., 137, 138, 736
Caul, J. F., 1070
Caul, W. F., 21
Cavanagh, D. G., 66
Cave, A. J. E., 46, 52, 55, 57
Cayre, M., 191, 198
CDC, 831
Cechetto, D. F., 711

Cecil, J. E., 838
Cedervall, T., 490
Celsi, F., 1119
Cenquizca L. A., 174
Centers for Disease Control and Prevention,
    496
Cerda-Reverter, J. M., 553
Cereda, C., 778, 780
Cerf, B., 775, 776
Cerf-Ducastel, B., 297, 776, 779, 781, 782
Cernoch, J. M., 321
Cervantes, S., 731
Chabaud, P., 217
Chaen, T., 66+
Chai, P., 969
Chai, Y., 729, 730
Chaisuriyanun, S., 961
Chalansonnet, M., 164
Chalcoff, V. R., 965
Chalé-Rush, A., 632, 834, 1102
Chalke, H. D., 246
Chalouhi, C., 309
Chamberlain, M. P., 80
Chambers, E., 801
Chambers, K. C., 868, 869, 870, 871, 872,
    874, 877, 878
Chamero, P., 1113, 1115, 1116, 1117, 1118,
    1119, 1120
Chan, A., 406
Chan, P., 111
Chan, S., 835
Chandrashekar, J., 677, 678, 686, 688, 689,
    690, 691, 693, 953, 960, 987, 1102
Chang, C. Y., 896
Chang, F. C., 784
Chang, F.-C. T., 703, 878, 988
Chang, R. B., 16, 676, 687
Chang, T. L., 100
Chang, Waters, and Liman (2010)., 16
Chang, Y. F., 497
Chanquoy, L., 310
Chao, L. L., 496
Chao, O. Y., 456, 458, 464
Chao, Y. C., 1140
Chapin, J. K., 710
Chapman, A., 961
Chapman, I. M., 839
Chapman, R. F., 929, 930, 931, 933, 935,
    936, 937, 938
Chapouton, P., 183
Chapuis, J., 215, 216, 217, 346
Chaput, M., 164
Chaput, M. A., 164
Char, H., 458, 468
Charland-Verville, V., 386
Charles, P. C., 454, 456, 462
Charpak, S., 164
Charpentier, M. J. E., 612
Charrier, I., 320

Chaudhari, N., 17, 525, 645, 647, 650, 690,
    693, 694, 728, 936, 967
Chaudhury, D., 170, 338, 339
Chavanpatil, M. D., 468, 470
Chavez-Jauregui, R. N., 839
Chazal, G., 188, 191, 435
Cheal, M., 305
Checke, S., 867
Cheeke, P. R., 968
Cheeseman, G. H., 228
Cheetham, S. A., 1119
Chehrehasa, F., 147
Chemuturi, N. V., 71, 466
Chen, A. W., 803
Chen, B., 735
Chen, C. F., 216
Chen, D., 298, 974
Chen, J., 136, 137, 143, 916
Chen, J. H., 188
Chen, J. Y., 702, 703, 987
Chen, M., 840
Chen, M. C., 841, 922
Chen, M. J., 393
Chen, S. C., 472
Chen, T. M., 486
Chen, V., 227
Chen, W., 422
Chen, W. R., 160, 164
Chen, X., 96, 136, 137, 784, 785, 937
Chen, X. B., 360
Chen, Y., 65, 69
Chen, Z., 935
Chen, Z.-h., 522
Chen, W. R., 165, 166
Cheng, L. C., 189, 192
Cheng, L. H., 642, 643
Cheng, S. J., 644
Cheng, X., 1086
Cheng, Y. S., 365
Cherninskii, A. A., 263
Chess, A., 126
Chhabra, H. S., 148
Chiari, Y., 959
Chiba, A., 1142
Chiba, Y., 71
Chien, Y. C., 835
Chikaraishi, D. M., 18, 135, 137
Chinkers, M., 117
Cho, H. J., 459, 470, 471
Cho, J. H., 128, 368
Cho, J. Y., 473
Cho, Y. K., 706, 707, 709, 718, 987, 989,
    1028
Choi, G. B., 20, 214, 1123
Choi, S., 841
Chojnacki, A., 193
Chopra, A., 311
Chorover, S. L., 531
Chotro G., 326

Chotro, M. G., 835
Chou, K. L., 419
Chou, Y. H., 537
Choudhury, E. S., 237, 243
Christensen, C. M., 630, 838, 901, 1009
Christensen, M. K., 101
Christensen, S., 526
Christensen, T. A., 541
Christiansen, F., 535
Christie, J. M., 161, 163, 164, 166, 169
Christopoulos, A., 669
Chu, S., 305, 323, 337
Chuah, M. I., 18, 97, 98, 141, 142, 143, 147, 306, 307
Chuang, H. H., 1103
Chung, R. S., 148
Chung, S. K., 359
Chung-Davidson, Y.-W., 557
Chuong, C. M., 730
Church, S. M., 429
Churcher, A. M., 553
Churchill, A., 345
Ciechanover, M., 901
Ciombor, K. J., 161, 164, 170
Civille, G. V., 1070
Clair, A. L., 431
Clancy, A. N., 51
Clapham, D. E., 1101
Clapp, T. R., 649, 693
Clark, A., 568
Clark, A. A., 691, 696
Clark, C., 430, 434, 435, 436
Clark, C. C., 1017
Clark, K. D., 525
Clark, K. E., 319
Clark, L., 962
Clark, M. M., 869
Clark, P. G., 311
Clark, W. E. L., 17, 461
Clarke, J. L., 11
Clarke, M., 519, 520
Clarke, P. M. R., 611
Clarke, R. W., 356, 369
Clay, K., 970
Cleaton-Jones, P., 984
Clee, M. D., 846
Cleland, T. A., 56, 162, 168, 169, 170, 338, 344, 368
Clement, P. A., 358
Clerico, D. M., 39, 47
Cliff, M. A., 1101, 1103, 1104
Cloquet, H., 3, 9
Cloutier, J.-F., 1120
Clyne, P. J., 930, 932
Cobey, S., 344
Cobos, I., 192
Coburn, C. M., 1140
Cocco, N., 936, 939
Cohen, L. B., 167

Cohen, Y., 219
Colasante, G., 192
Coldwell, S. E., 803, 808, 811, 822, 833
Cole, A. M., 100
Cole, K. S., 547, 555
Cole, P., 357, 393
Cole, T. B., 547
Coleman, E., 434
Coleman, J., 679, 835
Coles, M. G. H., 269
Collet, S., 377, 386, 423
Collings, V. B., 708, 752, 753, 988
Collins, S. A., 581
Colliver, J. A., 233
Colman, A. M., 235
Colomb, J., 932
Combarros, O., 780
Comert, A., 384
Connelly, T., 436, 437
Connes, R., 646
Connors, K. A., 103
Conover, J. C., 190, 197, 198, 487, 738
Conrad, P., 501, 848
Conrad, W. A., 357
Constantinescu, C. S., 391
Constantinidis, J., 415
Constanzo, R. M., 1056
Consumer Reports, 1080
Conte, C., 975
Conti, M. Z., 408
Contreras, C. M., 319
Contreras, R. J., 388, 703, 865, 878
Cook, A. L., 149
Cook, B. L., 114
Cook, D. J., 1009
Cook, M. H., 728
Coon, M. J., 63, 68, 69, 73, 74, 75, 76, 79
Cooper, D. S., 763
Cooper, G. P., 845
Cooper, J. G., 48
Cooper, R. L., 874
Cooper, W. E., 972, 973
Cooper, W. E., Jr., 47
Coopersmith, R., 393
Copeland, M., 549, 552
Coppola, D. M., 1114
Coquelin, A., 21
Coraboeuf, E., 21
Corbin, A., 3, 7, 1079, 1081
Corby, K., 271, 410
Corcoran, C., 432, 434
Corey, D. P., 1117
Cornsweet, T. N., 230, 752
Cornu, J-N., 599
Coronas, V., 142, 145
Corps, K. N., 66
Correa, M., 785
Correia, G. L., 424
Correia, S., 387

Corson, S. L., 742
Corsonello, A., 917
Cortina, J. M., 240
Corwin, J., 404, 409, 415, 848
Coryell, C. D., 19
Coskun, V., 187
Costa, M., 498
Costanzo, R. M., 93, 95, 96, 97, 240, 280, 386
Cotsarelis, G., 136
Cottler Fox, M., 647
Couley, G. F., 98
Coulter, D. M., 846
Councill, J. H., 194
Couper, J., 497
Coureaud, G., 319, 320
Court, M. H., 70
Courtes, S., 188, 191
Courtiol, E., 168, 169
Cousin, M.-E., 1058, 1061
Couto, A., 533
Covington, D. K., 840
Covington, J. W., 269
Cowan, N., 1061
Cowan, P. W., 763
Cowart, B. J., 362, 381, 382, 753, 765, 795, 796, 798, 802, 816, 817, 823, 843, 901
Cowey, A., 869
Cox, J. P. L., 547, 549
Cox, T. A., 428
Crabtree, N., 385, 497
Craft, S., 458, 487
Craig, A. D., 706, 779, 989, 993
Craigie, E. H., 548
Cramer, C. K., 420
Cramer, P. E., 217
Crammer, B., 669
Crapper McLachlan, D. R., 456, 457, 459, 473
Crastnopol, B., 245, 377
Cravedi, J. P., 556
Craven, B. A., 43, 51, 57, 58, 593
Creelman, C., 233
Cremer, H., 188
Crespo-Facorro, B., 429, 435, 436
Crim, J. W., 1142
Crino, P. B., 148
Crisostomo, E. A., 436
Criswell, S. R., 497
Critchley, H. D., 218, 609, 715, 716, 786, 992, 994, 1028, 1031, 1032, 1034, 1035, 1036
Crnjar, R. M., 938
Crofton, K. M., 457
Cronin, E. W. J., 961
Crook, C., 305
Crook, C. K., 797, 806, 807, 833, 836
Cross, C. E., 66
Crowcroft, P., 960

Author Index

Crowder, R. G., 339
Croxall, J. P., 961
Croy, I., 227, 377
Cruz, A., 655, 1009
Cruzan, G., 69
Crysdale, W. S., 381
Crystal, S. R., 815, 835, 837, 839
Cubero, E., 1105
Cui, L., 265, 269, 1093
Cullen, M. M., 362
Cumming, A. G., 433
Cummings, D. M., 167
Cummings, T. A., 15, 646, 952
Cunliffe, R., 99
Cunningham, A. M., 141, 142
Cunningham, C. L., 870
Cunningham, G. B., 568
Cunningham, M. G., 101
Cupchik, G. C., 337
Curran, A. M., 571, 572, 598
Curtis, M. A., 183, 186, 197, 198, 487
Cury, K. M., 161, 168
Cuschieri, A., 146, 1114
Cuschieru, A., 93
Czaja, J. A., 869
Czerniawska, A., 455, 460
Czirjak, G., 1140

D'Agostino, J., 69, 70, 80
D'Amico-Martel, A., 98, 728
D'Angelo, L. L., 669, 670
D'Aniello, B., 1142
D'Autreaux, F., 737, 738, 742
da Fonseca, C. O., 458, 469, 474
da Silva, F. H. L., 262, 272
Dabdoub, A., 736
Dade, L. A., 297
Daget, N., 1017
Dagher, A., 785
Dahanukar, A., 933, 934
Dahl, A. R., 63, 67, 68, 72, 73, 76, 78, 79,
     454, 463, 464, 465, 466
Dahlen, J. E., 195
Dahlgren, J., 1100
Dahlof, C. G., 458
Dahlslett, S. B., 391, 899
Dairou, V., 1072
Dal Monte, M., 465
Dale, A. M., 288
Dalham, T., 17
Dallas, L. J., 553
Dalton, P., 56, 228, 246, 247, 297, 309, 339,
     369, 381, 385, 393, 496, 1012, 1061,
     1101
Dalton, P. H., 246, 385, 496
Daly, J. W., 973
Daly, K. C., 344
Daly, M., 324
Damak, S., 629, 689, 694

Damholdt, M. F., 421
Damm, M., 362, 363, 364, 367, 384
Danciger, E., 133
Dando, R., 647, 649, 667, 694
Dandy, W. E., 654, 762
Danielian, P. S., 729
Daniels, C., 272, 385
Daniels, J. I., 1081
Daniher, A., 667
Danilova, V., 985
Danowski, S., 836
Danscher, G., 101
Dantas, Z. N., 459
Dantzer, R., 340, 874
Dantzig, A. H., 917
Dapporto, L., 611
Darch, S. E., 520
Darcy, D. P., 193
Daremberg, C., 316
Darwin, C., 12, 316, 957, 1074
Das, G. D., 19
Das, S., 538
DasGupta, R., 731
Dastidar, P., 358
Daszuta, A., 194
Datiche, F., 346
Datta, N. C., 552
Dauger, S., 738, 742
Daum, R. F., 416, 423
Dauner, K., 1118
DaVanzo, J. P., 21
David, F. O., 167, 169
Davidar, E. R. C., 596
Davidson, H. I. M., 901
Davidson, J. M., 1011, 1021
Davidson, R. J., 797, 806
Davidson, T. L., 833
Davidson, T. M., 228, 242, 309, 461, 499,
     500, 843
Davies, B. W., 517
Davies, G. A., 627
Davis, B., 145
Davis, B. J., 164, 174, 705, 707
Davis, C., 643
Davis, D. P., 525
Davis, H., 345
Davis, J. K., 968
Davis, L. B., 308
Davis, R. L., 216, 535
Davison, I. G., 161, 164, 167, 214
Dawes, C., 627, 628
Daws, C., 837
Dawson, W. W., 1091
de Ajuriaguerra, J., 415
de Araujo, I. E., 297, 702, 713, 717, 776,
     781, 783, 785, 993
de Araujo, I. E. T., 992, 993, 1032, 1037,
     1039, 1041
de Beer, G. R., 44, 45

De Beun, R., 870
de Boer, G., 940
de Brito Sanchez, G. M., 933, 936
de Bruyne, M., 533, 537, 540
de Gabory, L., 486
de Graaf, C., 803, 804, 809, 820, 823, 832,
     833
de Graaf, R., 11
de Haan, L., 429
De Jong, F. H., 869
De la Cruz, O., 118
De la Rosa-Prieto, C., 174
De Lorenzo, A. J., 15, 460, 465, 466
de Lorgeril, M., 1060
De Marchis, S., 186, 193, 196
De Michele G., 437
de Moraes, M., 904
de Ondarza, J., 525
De Panfilis, C., 288
De Saint Jan, D., 160, 161, 169
de Souza Silva, M. A., 456, 458, 463
de Toledo, L., 868
De Wied, D., 876
de Wijk and Cain, 1994:, 310
de Wijk, R. A., 1010
De, S., 500
Deacon, R. M. J., 339
Deamer, N. J., 80, 457, 464
Dean, M. K., 103
Debarbieux, F., 164, 169
Debat, H., 71
DeBorde, D. C., 469
DeBose, J. L., 565
Deco, G., 1029, 1035, 1041, 1042, 1043
DeCoursey, T. E., 676
Deeb, J., 417, 420
Deecke, L., 244, 405, 416
Deems, D. A., 23, 246, 375, 376, 377, 380,
     386, 389, 439, 485, 500, 843, 844, 846,
     849, 890, 896, 901, 904, 923, 1019,
     1056
Deems, R. O., 765, 849
DeFazio, R. A., 648, 649
Degel, J., 343, 345, 346, 1062
DeHamer, M. K., 18, 142, 143, 147
DeHan, R. S., 646
Deichmann, R., 288
Deisig, N., 538
Dekker, M. H. A., 567, 570, 572, 574
del Campo, M. L., 940
Del Parigi, A., 714, 991, 992, 996
Del Punta, K., 172, 173, 1115, 1117, 1120,
     1121
del Rio, C. M., 965
Del Tredici, K., 491, 502
Delank, K. W., 381, 384
Delaunay-El Allam, M., 308, 313, 316, 321,
     325, 326

Delay, R. J., 646, 647, 650, 1117, 1118, 1144
DeLeon, V. B., 58
Deliza, R., 1053, 1061
Dell'Orco, D., 490
Dellacorte, C., 556
Delmaire, C., 413
Deloitte, 1051
DeLorenzo, 1970, 455
Delorme, B., 139, 148
Delplanque, S., 311
Delwiche, J., 629, 1008, 1012
Delwiche, J. F., 763, 764
Demattè, M. L., 345
Demski, L. S., 1143
den Otter, C. J., 934, 935
Dennis, J. C., 58, 592
Denuelle, M., 900
Depoortere, I., 888
Derby, C. D., 250
Derjean, D., 553, 557
Desage, M., 318
Desai, H., 753
Descoins, C., 936
Desgranges, J. C., 646
Deshmukh, L. S., 805
Deshmukh, S. S., 217
Deshpande, D. A., 922
Deshpande, V. S., 68
DeSimone, J. A., 665, 675, 676, 677, 678, 686, 753, 754
Desimone, R., 1041
Desmaisons, D., 164, 169
DeSnoo, K., 797, 805, 811, 833
Desor, J. A., 246, 797, 798, 807, 808, 809, 811, 815, 816, 819, 833, 836
Desrochers, R., 1080
Dessirier, J. M., 1101, 1103, 1104, 1105
DeStefano-Shields, C., 71
Dethier, V. G., 21, 929, 930, 933, 934, 935, 936, 937, 938, 940
Detje, C. N., 487
Detsch, O., 710
Deutsch, J. A., 834
Devanand, D. P., 23, 218, 407, 408, 409, 410, 423
DeVere, R., 904, 905, 1056
DeVito, L. M., 339, 346
Devitt, B. D., 939
Devreotes, P. N., 520, 521
Dewan, A., 116
Dewar, W. A., 962, 969
deWijk, R. A., 247, 309, 632
Dewis M. L., 675, 679, 1055
Dewit, H., 836
DeWys, W. D., 501, 845, 872
Dexter, D. T., 459
Dey, S., 1116
Dhaka, A., 1103, 1140

Dhal, A. R., 486
Dhanani, N. M., 901
Dhawale, A. K., 161, 167
Dhong, H. J., 23
Di Costanzo, A., 458, 468
Di Giorgi Gerevini, V. D., 190, 194
Di Lorenzo, P. M., 702, 703, 705, 706, 707, 708, 709, 711, 718, 779, 784, 1028
Di Scala, G., 218
Diamant, H., 835
Diamond, I., 731
Diamond, P., 839
Diaz-Arnold, A. M., 850
DiBattista, D., 834
Dibattista, M., 1117, 1118
Dicke, M., 532
Dickinson, A., 1014
Dickson, D. W., 489, 498
Dielenberg, R. A., 587
Dietrich, A. M., 1080, 1081, 1082
Dietz, J. R., 116
Dietz, S. B., 165, 166
Dieulafe, L., 41, 44, 47
DiGiuseppe, G., 527
Dijksterhuis, G., 1058
Dillon, T. E., 739
DiLorenzo, P. M., 987, 989
DiMeglio, D., 835, 836
Dimov, D., 437
Dinehart, M. E., 756
Ding, X., 63, 67, 68, 69, 70, 73, 74, 75, 76, 79, 486, 487
Dirschnabel, A. J., 848
Distel, H., 1058
Ditraglia, G. M., 836
Ditzen, M., 537, 541
Divine, K. K., 497
Dixon, M. D., 968
Djeridane, Y., 581
Djordjevic, J., 296, 407, 547
Djurisic, M., 164
Do, V.-B., 1060
Dobell, E., 848
Dobell, E., 848
Docherty, R., 1092
Dockray, G. J., 842
Dodd, G., 58
Dodd, G. H., 593
Dodson, H. C., 58, 465
Doe, C. Q., 736
Doetsch, F., 184, 185, 187, 189, 191, 193
Doetsch, G. S., 703, 707
Dolan, R. J., 210, 289, 297, 345, 346
Dollken, V., 499
Dominguez, H. D., 835
Dominguez, R. A., 391
Domjan, M., 869
Donaldson, L., 338, 339
Donati, F., 897
Donchin, E., 269

Dong, D., 609, 968, 974, 975
Dong, H-W., 160, 162, 1123
Donnelly, A., 467, 469
Donovan, M. D., 466
Doop, M. L., 432
Dorman, D. C., 455, 485, 497, 500
Dorrego, F., 565
Dorries, K. M., 73, 310, 311, 324
Dostert, C., 491
Dotson, C. D., 691, 922
Doty, R. L., 3, 7, 8, 15, 20, 21, 22, 23, 47, 149, 162, 199, 217, 218, 227, 228, 230, 231, 232, 233, 234, 235, 237, 238, 239, 241, 242, 243, 244, 245, 246, 247, 262, 263, 264, 265, 269, 286, 298, 305, 309, 310, 324, 345, 364, 375, 376, 377, 380, 381, 385, 386, 387, 388, 389, 391, 392, 393, 404, 405, 409, 410, 412, 413, 414, 415, 416, 423, 424, 426, 427, 428, 429, 434, 437, 438, 473, 474, 489, 491, 496, 497, 498, 500, 501, 532, 580, 581, 582, 583, 584, 587, 594, 595, 600, 654, 751, 753, 763, 765, 843, 846, 870, 887, 889, 890, 901, 902, 903, 911, 914, 916, 1054, 1055, 1059, 1091, 1092, 1093, 1097, 1114, 1118, 1125
Double, K. L., 417, 423
Doucet, S., 316, 317, 318, 319, 320, 321, 322
Doucette, J. R., 18, 141, 146
Doucette, R., 98, 146, 148
Doucette, W., 170, 171, 344
Douek, E., 3, 23, 380
Dougherty, R. W., 319
Douglas, A. J., 873
Douglas, H. D., 567, 570, 572, 573, 574
Døving, K. B., 16, 549, 550, 551, 552, 557
Dowling, J. E., 1144
Downes, J. J., 305, 337
Downey, L. L., 381, 384
Downs, C. T., 965
Doyle, K. L., 143, 145
Draghia, R., 455
Drake, B., 234
Dravnieks, A., 248
Dray, S. M., 874
Drayna, D., 756, 915, 923
Drenckhahn, D., 651
Dreosti, I. E., 101
Drewett, R. F., 835
Drewnowski, A., 690, 756, 757, 761, 763, 832, 835, 847, 871, 914, 1020
Drexler, M., 248
Drickamer, L. C., 584
Driver, J., 1008, 1015
Drover, V. A., 840
Drucker, D. J., 842
Drugbank, 917, 921, 922
DSM-IV-TR, 871

Author Index

Duan, X., 114, 1136
Dubois, D., 1058
DuBois, G. E., 666, 667, 668, 669, 670, 671, 672, 673, 675, 690
Dubois, P., 1058
DuBose, C. N., 1010
Duckett, A. M., 939
Duda, J. E., 424
Duda, T., 117
Dudai, Y., 879
Dudley, C. A., 1121
Duerr, J. S., 939, 940
Duester, G., 76
Duff, K., 243, 406, 438
Duffy, V. B., 756, 836, 903, 904, 1059
Duggan, C. D., 126
Duhra, P., 893
Dulac, C., 171, 553, 554, 1114, 1115, 1116, 1120, 1121, 1123, 1125
Dumbacher, J. P., 568, 973
Dumont, E. C., 1123
Dunbar, I. A., 594, 595
Duncan, C. J., 686, 964
Duncan, H. J., 362, 386
Duncan, J., 1041
Duncan, R. B., 392
Duncan-Lewis, C. A., 499
Dunn, J. D., 1093, 1100
Dunn, L. M., 412
Dunn, L. T., 876, 877
Dunn, T. P., 429, 430
Duprez, T. P., 284
Durand, K., 306, 308, 321, 322, 323
Durand-Lagarde, M., 269
Dus, M., 934
Dusek, J. A., 346
Dvoryanchikov, G., 648, 649, 694
Dwyer, D. M., 865
Dyer, J., 17
Dyer, M. A., 736

Eade, A. M., 499
Easton, K., 1051, 1055
Eaton, D. L., 70
Eayrs, J. T., 22
Ebbell, B., 355
Ebner, V., 639, 644, 654
Eccles, R., 356, 357, 364, 369, 1101
Eckard, C., 11
Eckel, L. A., 870, 872, 876
Ecker, A., 11
Eckert, R., 524
Eckstein, A., 797, 814, 819
Eddington, N. D., 470
Edwards, D. A., 583, 584
Eeckman, F. H., 168, 169
Egan, J., 525
Egan, J. M., 795
Eggan, K., 126

Egger, V., 165, 166
Eggers, C., 426
Eggers, G., 383
Eglitis, J., 642, 643
Egner, T., 295
Ehlers, M. D., 214
Ehrenberg, C. G., 638
Ehrlich, P., 11, 638
Eibenstein, A., 407
Eichel, B. S., 384
Eichenbaum, H., 218, 219, 231, 233, 296, 337, 339, 340, 343, 344, 345, 346, 585
Eisthen, H. L., 547, 552, 1142, 1144
Eitel, G., 359
Eiting, T. P., 51
Eklund, A., 325
Ekman, G., 230
Ekman, P., 312, 799, 800, 801, 821, 1075
Ekstrand, J. J., 218
Ekstrom, A. D., 345
el Naggar M., 384
El-Deiry, A., 765
El-Sharaby, A., 728, 733
Elaagouby, A., 170
Elad, D., 359
Elder, A., 455
Elder, W. H., 567
Elian, M., 412
Elkins, R. L., 873
Elliot Smith, G., 50, 51, 55
Elliot, P. B., 233
Elliott, R., 961, 963
Elsaesser, R., 552
Elsberg, C. A., 228, 385
Elsherif, H. S., 380
Elson, A. E., 694, 695
Elson, A. E. T., 832
Elsworth, J. D., 427
Elwany, S., 361
Emes, R. D., 74
Emmers, R., 14
Emsley, J. G., 186, 187, 189, 191, 193
Emura, S., 984
Endler, J. A., 966
Endoh-Yamagami, S., 187, 191
Endres, R. G., 515
Engelen, L., 777, 837
Engelman, K., 839, 846
Engelmann, T. W., 646
Engen, E. A., 310, 311, 313, 345
Engen, T., 246, 308, 309, 310, 311, 312, 313, 337, 339, 340, 345, 346, 752
Engström, H., 15, 93, 280, 647
Ennis, J. M., 1073
Ennis, M., 157, 158, 159, 160, 161, 163, 164, 166, 170, 171
Enns, J. T., 251
Enomoto, T., 1114
Ensoli, F., 143

Environmental Protection Agency, 457
Enyedi, P., 1140
Epple, G., 51, 54, 55, 323, 611, 617
Epstein, J. B., 845, 922
Epstein, L. H., 801, 833, 838
Eraly, S. A., 466
Ercetin, S., 915
Erick, M., 901
Erickson, 992
Erickson, J. T., 738
Erickson, R. P., 703, 707, 709, 710, 711, 714, 989, 1028
Erickson, Y., 629, 631
Erikson, K. M., 497, 498
Eriksson, C., 80, 456
Eriksson, P. S., 133
Erlanger, J., 14
Erles, K., 101
Escanilla, O., 162, 171
Esiri, M. M., 73, 455, 462, 473
Eskandar, E. N., 252
Eskenazi, B., 287, 437
Espaillat, J. E., 967, 968
Essick, G. K., 763
Estes, R. D., 58
Ettenberg, A., 876
Eudy, J. D., 210
Evans, C. S., 617
Evans, D. R., 935
Evans, J., 460
Evans, J. E., 455, 460, 461, 498
Evans, R. E., 548, 557
Evans, R. W., 896, 897
Evans, S., 617
Evans, W. J., 264, 265, 266, 269, 1093
Evans-Pennimpede, J. E., 671
Everitt, B. J., 876, 877
Evers, K. A., 897
Evidente, V. G., 428
Evrard, H. C., 993
Exner, S., 11
Eylam, S., 686
Eyman, R. K., 236
Eyre, M. D., 165
Ezeh, P. I., 163, 389
Ezzeddine, Y., 338

Faas, A. E., 314, 835
Fabian, N., 558
Fagius, J., 840
Fahey, J. W., 959
Fahlgren, A., 100
Fahrenkrug, J., 364
Fährmann, W., 646
Falkner, B., 846
Faller, T. H., 71
Fan, S., 174, 1122
Fan, Y., 149
Fantana, A. L., 167

Fantino, M., 869
Farah, M. J., 995
Farb, P., 3
Farbman, A. I., 3, 15, 18, 99, 135, 140, 141,
    142, 143, 147, 501, 646, 647, 648, 650,
    655, 731, 739, 1133
Farinas, I., 738
Farnsworth, D., 755
Farovik, A., 342
Fasano, C. A., 188
Fast, K., 761, 762
Faulcon, P., 438
Faulk, M. E., Jr., 14, 776
Faulkes, C. G., 586
Faure, F., 486, 500
Faurion, A., 688, 775, 776, 779, 780, 781,
    784, 785
Fay, T., 762
Faye, P., 1072
Fazakerley, J. K., 454, 456, 462
Fazeli, P. K., 871
Fazil, M., 472
Fechner, G., 752
Fechner, G. T., 227
Feder, F. H., 963
Federico, G., 66, 141, 142, 145
Federle, M. J., 518
Federspil, P. A., 384
Fedoroff, I. C., 390
Feinstein, P., 98, 127, 128
Fekete, É. M., 874
Felberbaum, R. A., 308
Feldman, M., 837, 838
Feldmesser, E., 118
Fellermann, K., 100
Fellous, S., 526
Fellows, L. K., 995
Felmlee, M. A., 917
Felsted, J. A., 785
Feltrin, K. L., 840
Femiano, F., 904
Feneis, H., 654
Feng, P. C., 454
Fentress, J. C., 655
Ferdenzi, C., 310, 311, 322, 324
Ferguson, D. B., 626
Ferguson, E., 496
Ferguson, J. N., 1123
Ferguson, M. W. J., 974
Ferguson-Segall, M., 21, 388, 393
Ferkin, B. D., 582
Ferkin, M. H., 583
Fernandez, M., 308, 322, 800, 806
Fernandez, P. C., 537, 538
Fernández-Irigoyen, J., 493
Fernandez-Urrusuno, R., 459, 469
Fernberger, S. H., 234
Fernelius, I., 959
Ferner, H., 648

Fernstrom, A., 848
Fernstrom, J. D., 695
Féron, F., 18, 142, 143, 145, 147, 148, 381
Féron, V. J., 486, 1100
Ferrand, N., 118
Ferrando, S., 552, 553, 646
Ferrante, M. A., 388
Ferraris, A., 426
Ferraris, N., 101
Ferreira, J. J., 418
Ferrell, F., 629
Ferrero, D. M., 115, 116
Ferreya-Moyano, H., 473
Ferri, A. L., 189
Ferriday, D., 838
Ferrier, D., 38
Ferris (1964), 355
Ferris, A. M., 843, 844
Ferry, B., 218, 346
Festinger, L., 1061
Festring, D., 673
Fetting, J. H., 765, 766
Fiedler, N., 496
Field, L. J., 116
Field, T. M., 320
Fielder, C. P., 381
Fillion, T. J., 325
Fine, J. M., 1119
Finelli, P. F., 384
Finger, T. E., 48, 552, 644, 645, 648, 649,
    651, 694, 922, 947, 954, 959, 975,
    1091
Fink, L. S., 968, 969
Fink, M., 849, 904
Finkel, D., 1059
Finkenzeller, P., 261, 265
Finlayson, J. S., 1119
Fiol, M. E., 993
Firestein, S., 14, 16, 110, 123, 125, 174,
    547, 548, 1113, 1122
Firtel, R., 521
Fischer, A. J., 1144
Fischer, E., 12
Fischer, Jde S., 469, 474
Fischer, R., 755
Fischer, U., 839
Fischer-Tenhage, C., 599
Fischler, W., 936
Fishberg, A. M., 916
Fishell, G., 190, 191, 193
Fisher, A., 470
Fisher, G. L., 985
Fisher, M. F. K., 953
Fishilevich, E., 533
Fitzgerald, T. H. B., 995
Fixsen, C., 637, 646
Flaherty, C. F., 867
Flamme, E., 667
Flanagan, D., 535, 537

Flanagan, K. A., 1119
Flanagan-Cato, L. M., 870
Fleiner, F., 391, 899
Fleischer, J., 110, 1136, 1137, 1138, 1139,
    1140
Fleischl von Marxow, E., 261
Fleischmann, A., 127
Fleming, A. S., 585
Fleming, P. A., 965
Fletcher, M. L., 167, 169, 213, 216
Fletcher, R. B., 137
Flockhart, D. A., 917, 921, 922
Floemer, F., 307
Flood, P. F., 594
Flor-Henry, P., 431
Florindo, H. F., 472
Flower, D. R., 1119
Flynn, M. R., 497
Fode, C., 737
Foerster, O., 14
Folstein, J. R., 310
Fong, K. J., 66, 76
Forbes, W. B., 139
Forestell, C. A., 323, 799, 800, 834
Formaker, B. K., 631, 904
Forni, P. E., 96, 97, 98, 139
Forster, S., 408
Fort, A., 626
Fortin, N. J., 339, 341, 342, 346
Fosmire, G. J., 849
Fossey, E., 311
Foster, J. D., 465
Foster, N. R., 555
Foster, S. R., 922
Fotiadis, D., 103
Foulds, I. S., 893
Fourcroy, A. F., 7
Fox, A. L., 755, 762
Fox, M. W., 594, 597
Fox, N. A., 797, 806
Fox, P. C., 391
Fox, P. T., 20
Foy, J. W. D., 80
Frame, L. H., 597
Franceschini, I. A., 146
Francis, G. W., 249
Francis, J., 1019
Francis, S., 1037, 1039
Franco, B., 390
Frank, D., 600
Frank, G. K., 785
Frank, M., 702
Frank, M. E., 3, 14, 247, 703, 704, 705, 753,
    754, 762, 937, 989
Frank, R. A., 228, 239, 240, 309, 310, 346,
    766, 986, 1010, 1011, 1012, 1017,
    1021
Frankman, S. P., 1009
Franks, K. M., 214

Author Index

Franks, N. P., 1099
Franzen, A., 76
Fraser, S. E., 1136, 1137, 1138, 1140
Frasnelli, J., 245, 298, 1093, 1094, 1095, 1100, 1101
Frasnelli, J. A., 501, 848
Frazier, C. R. M., 834
Frazier, L. I., 473
Frazier, L. L., 163
Frecka, J. M., 842
Frederickson, C. J., 101
Frederique, F., 715
Fredman, S. M., 460
Fredriksson, R., 975
Freeman, N. J., 611
Freeman, R. P. J., 1058
Freeman, W. J., 168, 169, 295
Freiherr, J., 243
Freisen, W. V., 799, 801, 821
Freitag, J., 125, 557
Fremeau, R. T., 1123
Freud, S., 323
Freundlieb, N., 194
Frey, L., 898
Frey, P. W., 233
Frey, S., 776
Frey, W. H., 458
Freyer, D., 616
Freyer, L., 96
Freygang, W. H., 19
Friberg, U., 643
Fricton, J. R., 904
Friden, M., 343
Fried, H. U., 167
Friedl, W. A., 958
Friedman, B., 101
Friedman, D., 161, 169
Friedman, M., 384
Friedman, R., 1079
Friedrich, R., 172
Friedrich, R. W., 168
Friend, W. G., 929
Friesen, W., 312
Frijters, J. E. R., 237, 248, 250, 672, 847
Friston, K., 295
Fritsch, G., 1141
Fritzsch, B., 729, 738
Fröhlich, E., 490
Fromentin, G., 834
Froriep, R., 651
Frost, A. C., 1014
Frost and Noble, (2002), 1014
Fry, E. G., 967, 970
Fry, J. R., 1092
Fry, T. L., 765
Frye, R. E., 18, 79, 246, 364, 500
Fryhle, C. B., 569
Fuchs, E., 731
Fudge J. L., 996

Fuentes, R. A., 164
Fujii, H., 76
Fujii, M., 386
Fujikura, T., 66
Fujimoto, S., 648, 649, 650
Fujita, I., 1143
Fujita, M., 934
Fujita, Y., 842
Fujiyama, R., 850, 904
Fukami, T., 69
Fukami, H., 644
Fukuda, N., 66, 118
Fukutake, T., 903
Fulford, A. J., 873
Fulle, H. J., 116
Fuller, J. L., 597
Fulton, J. F., 19
Funakoshi, M., 14, 765
Füri, S., 325
Furrer, S., 1101, 1102, 1103
Furstenberg, A. C., 384
Furukawa, M., 245
Furuya, N., 263
Furuyama, A., 936
Fusari, C., 308
Fusetti, M., 409
Fushan, A. A., 757, 923
Fushiki, T., 632
Fusi, S., 1122
Fuss, S. H., 127, 1136, 1137, 1138, 1140

Gabassi, P. G., 339
Gabbott, P. L., 1031, 1032, 1033, 1034
Gachomo, E. W., 99
Gadoth, N., 903
Gaffan, D., 339, 340
Gaillard, D., 838
Gairns, F. W., 763
Gal, S., 460
Galan, R. F., 541
Galanter, E. H., 755, 758
Galef, B. G., 117, 318, 344, 585, 869
Galen (Claudius Galenus), 637
Galindo, M. M., 692
Galindo-Cuspinera, V., 670, 674, 760
Galizia, C. G., 344, 535, 536, 537, 538, 539, 541
Gall, C. M., 141, 147, 159
Gallagher, M., 571
Gallagher, R. M., 458, 467
Galli, R., 188
Gallo, M., 876, 877
Galopin, C. C., 1103
Ganchrow, D., 643, 647, 651, 709, 796, 961, 962, 963
Ganchrow, J., 985
Ganchrow, J. R., 643, 647, 795, 796, 799, 806, 811, 819, 868, 961, 962, 963, 964
Ganeshina, O., 535

Gangdev, P. S., 902
Gans, C., 948
Gansler, D. A., 438
Ganz, T., 99, 100
Gao, L., 142, 145
Gao, N., 916, 950
Gao, Y., 217, 458
Garb, J. L., 326, 865, 873
Garbers, D. L., 117
Garcia, G. J., 360, 486
Garcia, J., 20, 22, 346, 865, 867, 868, 870, 875, 933
Garcia, P. L., 323
Garcia-Gonzalez, D., 187
García Medina, M. R., 1093
Gardner, L. E. Jr., 325
Garland, E. M., 427, 429
Garle, M. J., 1092
Garnett, S., 970
Garrett-Laster, M., 848
Garten, S., 14
Gartrell, B. D., 961
Garzaro, M., 363
Garzotto, D., 189
Gascon, E., 190, 193
Gaskell, B. A., 76
Gass, C. L., 965
Gasser, B., 306
Gasser, H. S., 14
Gaultier, C., 322
Gaut, A., 897
Gautam, S., 1012
Gautam, S. H., 498
Gautier, J.-F., 714, 992
Gautier-Hion, A., 611
Gayner, S. M., 894
Gayoso, J., 552
Gazit, I., 598
Gdynia, H. J., 849
Ge, T., 783, 784, 1041
Geary, N., 869, 870
Gebhardt, L. P., 17, 73, 461
Gebruers, E. M., 840
Geddes, J., 429, 430
Gedikli, O., 765
Geer, T. A., 961
Geha, P., 785
Geier, M., 532, 540
Geisler, M. W., 269
Geissler, A. H., 849
Geissmann, T., 611
Geist, N. R., 41
Gelperin, A., 168, 937
Gelstein, S., 298
Gemberling, G. A., 869
Gennings, J. N., 74
Gent, J. F., 3, 244
Genter, M. B., 68, 69, 71, 72, 76, 78, 79, 80, 455, 457, 464, 466, 474, 486, 487

Gentle, M. J., 961, 962, 963, 964, 968, 969
Genzel, C., 308, 321
George, P., 757
Geraedts, M. C., 695, 887, 888
Geran, L. C., 686
Gerber, B., 344, 937
Gerlach, J., 638
Gerlai, R., 557
Gersemann, M., 100
Gerspach, A. C., 841
Gervais, R., 167, 212
Gervasi, P. G., 67, 68, 466
Gescheider, G. A., 230, 233
Gese, E. M., 596
Gesteland, R. C., 14, 307
Getchell, M. L., 64, 68, 69, 73, 142, 143, 144, 145, 146, 406, 465
Getchell, T. V., 14, 64, 66, 73, 96, 142, 143, 144, 145, 146, 251, 465, 646
Getschman, E., 647, 649
Ghanbari, H. A., 148
Ghanizadeh, A., 902
Ghantous, H., 76, 456, 464
Ghashghaei, H. T., 187
Ghatpande, A. S., 170
Gheusi, G., 19, 166
Ghorbanian, S. N., 361, 381, 382
Ghosh, S., 213
Giachetti, I., 779
Giachino, C., 195
Giacobini, P., 1114
Gialamas, D. M., 598
Giannetti, N., 1133, 1134, 1136
Gianutsos, G., 455
Giaquinto, P. C., 952
Giattina, J. D., 557
Gibbins, S., 805
Gibson, E., 311
Gibson, E. L., 834
Gibson, J. J., 779, 1007, 1012
Gibson, R. S., 849
Gibson, W. P., 754
Giduck, S. A., 765
Gietzen, D. W., 868
Giger, R., 739
Gil-Perotin, S., 187
Gilad, Y., 51, 55, 609
Gilardi, J. D., 970
Gilbert, A. N., 243
Gilbert, P. E., 407, 409
Gilbertson, T. A., 678, 702, 834, 835, 1037, 1055, 1102
Gill, F. B., 566
Gilmore, M. M., 1008
Gilmore, R. L., 901
Gilmore, R. W., 21, 595
Gilmour, K. M., 953
Ginane, C., 629
Giorgi, P. P., 465

Giovanni, M. E., 834
Giraldeau, L. A., 344
Giraldo, J., 103
Girard, S. D., 486
Girardin, M., 358
Girardot, N. F., 758
Gire, D. H., 161, 162, 168, 169
Girgis, S., 1020
Gittleman, J. L., 50
Giugno, L., 876
Giza, B., 703
Giza, B. K., 703, 707, 711, 988, 992, 1028
Glanzman, D. L., 338
Glasbrenner, B., 839
Glaser, D., 612, 629, 986, 987
Glatz, R., 111
Glazman, L., 1074
Gleen, J. F., 1028
Glemarec, A., 250
Glendenning, K. K., 608
Glendinning, J. I., 314, 345, 499, 693, 796, 799, 833, 841, 929, 931, 933, 934, 936, 937, 938, 939, 940, 969
Globus, J. H., 19
Gloriam, D. E., 115, 556, 975
Glover, G. H., 288
Glusman, G., 16, 123, 125
Go, Y., 610, 968, 975, 976
Gobba, F., 497
Godden, E. L., 1092
Godfrey, J., 21
Godfrey, P. A., 123, 125
Goehler, L. E., 865, 872
Goense, J. B., 288
Goetzl, F. R., 832
Goff, W. R., 22, 261, 265
Goker-Alpan, O., 427
Gokoffski, K. K., 144
Gold, G. H., 14, 15, 109, 112
Goldberg, A. N., 845
Golding, J., 814
Golding-Wood, D. G., 382, 383
Goldman, J., 324
Goldman, J. E., 184, 186, 191, 197
Goldman, M. J., 100
Goldman, R., 324
Goldman, S. A., 183
Goldman, W. P., 337
Goldrich, N. R., 935
Goldstein, B. J., 76, 133, 135, 136, 137, 141, 142, 143
Goldstein, D. S., 418, 423, 427, 429
Goldstone, A. P., 995
Golé, L., 520
Golgi, C., 638
Golik, A., 846
Gomer, R. H., 521
Gomes, V. M., 99
Gomez-Carneros, C., 914

Goncalves, M. B., 193
Gondivkar, S. M., 848
Gong, Q., 99, 141, 146
Gonyou, H. W., 319
Gonzalez, J. S., 899
Gonzalez, N. A., 937, 939
González-Pérez, O., 189, 193, 197, 486, 488
Good, J. M., 11, 13
Good, K. P., 245, 414, 431, 433, 434, 435, 438
Good, P. F., 455, 459
Goode, R. L., 357
Goodlin-Jones, B. L., 320
Goodrich-Hunsaker, N. J., 345
Goodspeed, R. B., 843, 844
Goodwin, K. M., 599
Goodwin, T. E., 1119
Gorard, D. A., 917
Gordon, G., 985
Gordon, K. D., 952
Gordon, M. D., 730, 937
Gordon, M. K., 137
Gorell, J. M., 461, 473
Gormican, A., 845
Gorrie, C. A., 148
Gorski, R. A., 870
Gorsky, M., 904
Gosepath, J., 71, 364
Gosling, L. M., 581, 584
Gosnell, B. A., 874
Goss-Custard, J. D., 961
Goto, N., 765, 780
Gottfried, J. A., 209, 210, 216, 217, 218, 289, 290, 292, 293, 297, 337, 345, 346, 1035, 1039
Gottofrey, J., 455, 460
Gotz, M., 184
Goubet, N., 307, 308
Goudsmit, N., 432
Gould, E., 19, 183
Gould, N. R., 100
Goullé, J. P., 497
Gourlay, S. G., 459, 468
Gowers, W. R., 38
Goy, R. W., 869
Goyert, H. F., 247
Gozes, I., 458, 468
Grabauskas, G., 705, 706
Grabenhorst, F., 292, 293, 782, 786, 835, 996, 1032, 1033, 1034, 1035, 1037, 1038, 1039, 1040, 1041, 1042, 1043
Graberg, J., 651
Grabowski, C. T., 929
Gracely, R. H., 759
Gracia-Llanes, F. J., 171
Gradin, M., 806
Gradon, L., 365
Graf, M. V., 320
Graff-Radford, S. B., 896, 897

Author Index 1165

Graham, A., 98, 737
Graham, J. L., 1082
Graham, L., 600
Graillon, A., 800, 806, 811
Granger, N., 98
Grant, A. J., 540
Grant, E. C., 584
Grant, O., 359, 360
Grant, R., 754
Graves, A. B., 23, 410
Gravitz, L., 666
Gray, A. J., 406, 438
Gray, C. M., 169
Graziadei, G. A., 99, 733
Graziadei, P. P., 10, 18, 64, 76, 93, 95, 96,
    134, 135, 139, 140, 141, 173, 646, 733,
    1121
Graziadei, P. P. C., 93, 1133
Grebe, T. W., 514
Green, B. G., 235, 236, 655, 757, 759, 760,
    764, 779, 831, 833, 1008, 1009, 1011,
    1016, 1055, 1091, 1101, 1103, 1104,
    1105
Green, D. M., 233, 752
Green, E., 785, 832, 833
Green, J. E., 404, 409
Green, K. F., 874
Green, N., 70
Green, P., 242
Green, S. M., 610
Green, T., 76, 78
Green, T. L., 900
Greenberg, C. S., 806
Greenberg, M. I., 485
Greenberg, N., 972
Greene, L. S., 808, 816
Greene, T. A., 671, 692, 923
Greenwald, B. D., 468
Greer, C. A., 164, 194, 195, 196, 250
Greff, G., 1097
Greger, J. L., 849
Gregg, C., 1123
Gregorian, C., 187
Gregory, S., 5
Gregson, R. A. M., 237, 239, 430
Grens, K. E., 1144
Gresham, L. S., 79
Gridley, T., 734
Griep, M. I., 501, 848
Grier, J. B., 233
Grieve, F. G., 835
Griffin, C. A., 118
Griffin, J. W., 364
Griffith, L. P., 245
Griffiths, A. D., 973
Grigson, P. S., 718, 867, 876, 878
Grill, H. J., 708, 796, 799, 865, 867, 868,
    869, 875, 878, 934, 937
Grill, M. J., 964

Grillo, M., 456, 463
Grimm, I., 193
Grimm, N., 616
Grinker, J., 847
Grinvald, A., 167
Griss, C., 933
Gritti, A., 184
Groenewegen, H. J., 877
Grolli, S., 66
Grosjean, Y., 533
Grosmaitre, X., 1134, 1135, 1136
Gross, C. G., 19
Gross, G. W., 456, 463
Gross, J. B., 740
Gross-Isseroff, R., 383
Gross-Isseroff, R. G., 429, 430
Grossman, M. I., 246, 832
Grossmann, S., 871
Grosveld, F., 126
Grover Johnson, N., 646
Groves, J., 754
Grovum, W. L., 899
Grubb, B. R., 392
Grubb, M. S., 194
Gruest, N., 314, 868
Grüneberg, H., 1136, 1137, 1138, 1142
Grunewald, B., 535
Grus, W. E., 554, 1123, 1124, 1125
Grushka, M., 754, 766, 904
Gryllus, L., 9
Gsell, A., 599
Gu, J., 65, 68, 70, 74, 76, 79, 80, 457
Guadagni, D. G., 228
Guarneros, M., 385, 411
Gudziol, H., 1093
Gudziol, V., 244, 245, 284
Guengerich, F. P., 79
Guenther, P. M., 804
Guerin, D., 170
Guerin, M. C., 498
Guerout, N., 147
Guerrero-Cazares, H., 184, 487
Guerrieri, F., 539
Guest, S., 993, 1037
Guggenheimer, J., 891
Guilarte, T. R., 497, 498
Guilemany, J. M., 362, 363
Guilford, J. P., 230, 235, 240, 312
Guillamon, A., 76, 1123
Guillemot, F., 137
Guillery, R. W., 709
Guillory, A. W., 308
Guilmette, R. A., 365
Guinand, N., 896
Guinard, J. X., 313, 803, 838, 1056
Gulbransen, B. D., 1091
Guldin, W. O., 711
Gungor, A., 364
Gunten, U. V., 1080

Gunther, W. C., 964
Güntherschulze, J., 608
Guo, A., 345
Guo, J., 345, 731
Guo, X., 418
Guo, Y., 166
Guo, Z., 133, 136, 137, 138
Gupta, A. K., 429
Gupta, D., 384
Gupta, N., 535, 541
Gurkan, S., 625, 628, 644, 650
Guss, J., 382
Gussekloo, S. W. S., 962
Gussing, F., 71
Gustafsson, T., 581
Gustavson, C. R., 870, 871
Gustavson, J. C., 870, 871
Guth, L., 741
Guthrie, K. M., 147, 298, 463, 465
Gutierrez, R., 712, 713, 716
Gutierrez-Castellanos, N., 1122
Guxens, M., 490
Guzzo, A. C., 73

Haak, R., 598
Haase, L., 783, 786
Haberly, L. B., 169, 209, 210, 214, 215,
    289, 292, 295, 1122
Habermehl, K. H., 642, 651
Hack, I., 188, 191
Hack, M. A., 184, 197
Hackett S. J., 962
Hadj Ali, H., 1059
Hadley, W. M., 63, 67, 68, 72, 78, 79, 463,
    464, 465, 466
Haehner, A., 244, 265, 283, 419, 423, 437
Haerer, A. F., 898
Haga, S., 172, 1116, 1117, 1119, 1120
Hagedorn, L., 729
Hagelin, J. C., 565, 567, 570, 572
Hagendorf, S., 1117
Hagg, T., 187, 189, 191, 193, 194
Hagglund, M., 76, 145, 146
Hagmann, P., 286
Hague, C., 112
Hague, K., 429
Hahn Berg, I. C., 627
Hahn, C. G., 67, 141
Hahn, I., 67, 358, 359, 365
Hajjar, E., 66
Hajjar, E. R., 917
Hajnal, A., 1028
Hakanen, J., 188
Halabisky, B., 165, 166
Halász, N., 145, 159
Haldane, E. S., 10
Halem, H. A., 1123
Halgren, E., 269
Hall, B. A., 514

Hall, J. M., 728, 729, 730, 731, 732, 734
Hall, M. J., 756
Hall, R. A., 489
Hall, W. G., 838
Hallem, E. A., 537
Haller, A. V., 11, 642, 643
Haller, R., 326, 1062
Hallock, R. M., 703
Hallschmid, M., 458, 469
Halpern, B., 1017
Halpern, B. P., 227, 262, 629, 713, 765, 962, 964, 968, 969, 1054, 1055
Halpern, M., 47, 48, 171, 553, 1113, 1115, 1117, 1120, 1123, 1139
Halpin, Z. T., 343
Halsell, C. B., 701, 704, 705, 707, 708, 989
Halthur, T. J., 627
Hamada, N., 911
Hamasaki, K., 650
Hamasaki, T., 525
Hamdani, E. H., 552
Hamilton, J. M., 413
Hamilton, K. A., 163
Hamilton, R., 704, 985
Hamilton, R. B., 875
Hammes, B., 871
Hamor, G. H., 914, 915
Hamosh, M., 732
Hamrum, C. L., 965
Hanamori, T., 711, 876, 985
Handley, O. J., 409
Haneberg, B., 459
Hanig, D. P., 764
Hanke, W., 1138
Hankins, W. G., 933
Hannover, A., 638
Hansel, D. E., 71
Hansen, A., 547, 552, 556, 557
Hansen, J. A., 557
Hansen, K., 933
Hanson, F. E., 935, 940
Hanson, L. R., 456, 470
Hanson, S. R., 1119
Hansson, B. S., 533, 932, 1116
Hao, H. N., 100
Hara, S., 763
Hara, T. J., 547, 548, 549, 550, 551, 552, 553, 555, 557, 947, 948, 949, 951, 952, 953, 954
Harada, H., 263
Harada, N., 731
Harada, S., 655
Harberts, E., 487, 488
Harbuz, M. S., 873
Harder, D. B., 763
Harder, J., 100
Harding, J. W., 135, 140
Harding, R., 984
Hardy, A., 171

Hardy, S. L., 901
Hare, T. A., 785
Hari, R., 267
Harkema, J. R., 47, 55, 63, 64, 66, 67
Harkin, C., 964
Harkins, S. W., 237
Harley, C. W., 171
Harley, K. L. S., 937
Harlow, D. E., 736, 738
Harner, S. G., 891
Harper, L. V., 1015
Harper, R., 3
Harriman, A. E., 22, 965, 967, 970
Harrington, A., 3
Harrington, F., 594
Harrington, F. H., 596
Harris, G., 799, 815, 836
Harris, H., 753, 755, 757
Harris, H. E., 384
Harris, J. A., 145
Harris, R. A., 1099
Harris, W., 654
Harrison, C. J., 911
Harrison, D., 800
Harrison, R., 361
Harrison, W. T., 17, 461
Hart, B. L., 596, 872
Hartmann, C., 319
Hartmann, G., 763
Harvey, P. D., 429
Harvey, S., 582
Hasagawa, S., 23
Haschek, W. M., 72, 76, 464
Haschke, M., 471
Hashiguchi, Y., 16, 47, 548, 554, 556
Hashimoto, Y., 243
Hashkes, P. J., 839
Hasin-Brumshtein, Y., 1059
Hasler, A. D., 548
Hass, N., 841
Hasselmo, M., 162, 168, 169
Hasselmo, M. E., 214, 292, 295
Hastings, L., 385, 455, 460
Hatton, E., 6
Hau, K. M., 1097
Hauser, G., 20
Hauser, G. J., 314
Hausner, H., 314, 319, 345
Hausser-Hauw, C., 765, 776, 901
Havermans, R. C., 1015
Havlicek, J., 325
Hawkes, C. H., 265, 266, 271, 391, 405, 410, 412, 414, 416, 423
Haxel, B. R., 386, 902, 911
Hay, D. I., 627
Hayama, T., 765
Hayar, A., 158, 159, 160, 161, 162, 166, 169, 170, 171
Hayashi, R., 429

Hayashi, T., 837
Hayashi, Y., 1122
Hayek, R, 455, 459
Hayes, C. L., 75
Hayes, J. E., 228, 756, 757, 1058
Hayes, R. B., 464
Hayes, U. L., 870, 874
Hayter, D., 598
Haywood, G. J., 596
Haywood, M. W., 596
Hazelbauer, G. L., 513, 514, 515
He, F. J., 833
He, J., 1118
He, W., 693
He, X. L., 117
He, X. Y., 69
Heath, R. B., 839
Heath, T. P., 753
Hebb, D. O., 1075
Hecht, G. S., 703
Heck, G. L., 15, 675, 686
Heckers, S., 903
Hector, M. P., 625
Hedges, S. B., 968
Hedner, M., 241
Heffelfinger, A. L., 1012
Hegg, C. C., 145
Hegoburu, C., 216
Heidel, A., 4
Heier, P., 1142
Heilmann, S., 227, 392
Heimer, L., 211
Heinbockel, T., 164, 165, 166, 540
Heinrichs, C., 874
Heise, S. R., 583
Heisenberg, M., 535
Heisenberg, W., 272
Heiser, C., 896
Hellekant, G., 644, 655, 665, 985, 986, 987
Hellfritsch, C., 671
Hellman, T. J., 651
Helm, E., 227, 1052, 1073
Helmholtz, 638
Helms, J. A., 904
Helson, H., 236
Hemdal, P., 412
Hempel-Jorgensen, A., 1100
Hempstead, J. L., 133, 465
Hendel, T., 937
Henderson, P. W., 318, 869
Henderson, R. G., 498
Hendrickson, R., 172, 173
Hendrickson, R. C., 1122
Hendriks, A. P., 244
Hendrix, L., 324
Hengge, R., 516, 517
Henkin, R., 501

Author Index

Henkin, R. I., 66, 376, 386, 388, 389, 391, 392, 393, 633, 651, 753, 783, 844, 845, 849, 903, 1057
Hennessy, J. W., 873, 874
Henning, S. J., 732
Henriksson, J., 73, 455, 460, 461, 473, 497, 498
Henry, J. D., 596
Hensiek, A. E., 426
Henton, W. W., 22
Hepper, P., 55
Hepper, P. G., 305, 314, 319, 320, 324, 326, 595, 598
Herbert, H., 708, 875, 877
Herman, S. J., 1052
Hermann, F., 651
Hermann, G. E., 877
Hernández, S. M., 1099
Herness, M. S., 650, 754
Herness, S., 649, 666, 936
Herold, S., 195
Herrada, G., 171, 1115
Herrick, C. J., 209, 638, 947, 949, 950
Hersch, M., 651, 796
Hershkovich, O., 626
Herting, B., 418, 423
Hertz, J., 392, 753
Herz, R., 323
Herz, R. S., 296, 297, 337, 345, 1062
Hess, E. H., 974
Hessler, G., 113
Hester, S. D., 71
Hetherington, A. W., 997
Hetherington, M., 1034
Hetherington, M. M., 758, 832, 1020, 1034
Hevezi, P., 17
Heyanka, D. J., 438
Heydel, J. M., 70, 73, 486, 487
Heymann, E. W., 51, 54, 55, 611, 612
Heymann, H., 672, 751, 752, 753, 758, 767, 1052, 1055, 1069, 1072, 1080
Heyward, P. M., 164
Heywood, P. G., 240
Hibbert, D. B., 1057
Hickson, M., 1019
Hidaka, I., 951
Hidalgo, J., 231, 408
Higashisaka, K., 455
Hilberg, O., 383
Hildebrand, J. G., 535, 932, 937
Hilger, J., 1059
Hill, D. L., 704, 705, 741, 742, 836
Hill, J. M., 390
Hill, W. C. O., 44
Hillenius, W. J., 41, 58
Hillier, L., 16
Hillier, L. W., 968, 974
Hilsted, J., 459, 469
Hinde, R. A., 939

Hinds, J. W., 18, 135, 183
Hinman, A., 1092, 1102
Hinrichs, A. L., 691
Hinrichsen, K. V., 651, 652
Hinriksdottir, I., 383
Hintiryan, H., 869, 870
Hirai, R., 887
Hiramatsu, C., 610
Hirao, T., 934
Hiraoka, T., 838
Hirasawa, A., 840
Hirata, K., 648, 649
Hirata, Y., 165, 646
Hiroi, M., 934
Hirota, J., 125, 126
Hirsch, A. R., 1013
Hirsch, B. T., 547
Hirsch, E. C., 459
His, W., 12
Hisamitsu, M., 385
Hisaoka, T., 146
Hitch, G. J., 338, 340
Hitoshi, S., 190, 193, 194
Hjermstad, M. J., 235
Ho, R. J., 471
Ho, W. K., 384
Hodder, K., 339
Hodgson, E. S., 14, 931, 933
Hodos, W., 233
Hoekman, J. D., 471
Hoeppner, T. J., 456, 463
Hof, P. R., 412
Höfer, D., 17, 651, 695, 730
Hofer, H., 643
Hoffman, C. K., 11
Hoffman, E. J., 19
Hoffman, H. J., 375, 765
Hoffmann, A., 640, 642
Hoffmeyer, L. B., 800, 806
Hofmann, M. H., 557
Hofmann, T., 694
Hofmann, W., 1020
Hoglinger, G. U., 190, 194, 198
Hokfelt, T., 146, 648
Holbrook, E. H., 64, 96, 133, 134, 141, 142, 143, 148, 149, 298, 380, 486
Holcomb, J. D., 141, 145
Hold, B., 324, 343
Holden-Wiltse, J., 847
Holen, T., 66
Holl, A., 456, 462
Holland, A. J., 412
Holland, G., 690
Holland, P. C., 868
Holland, V. F., 15, 914
Holland, Y., 456, 463, 466
Holley, A., 164
Hollingworth, H. L., 766
Holloway, W. R., 339, 340

Holmberg, J., 190
Holmberg, K., 557
Holmes, W. G., 320
Holmes, W. M., 549, 551
Holst, B., 667
Holst, J. J., 839
Holt, G. R., 385
Holt, J. P., 357
Holt, S., 871
Holway, A. H., 1070
Holy, T. E., 172
Honda, K., 933
Honegger, K. S., 540
Hong, J. H., 632, 845, 850
Hönigschmied, J. von, 10, 642, 653
Honore, A., 147
Hoon, M. A., 649, 688
Hooper, L., 517
Hopfield, J. J., 209, 538
Horio, N., 675, 688
Horishita, T., 1099
Hormuzdi, S. G., 169
Horn, W., 10
Hornak, J., 1033
Hornung, D. E., 358, 364, 365, 434
Horowitz, L. F., 455, 465
Horschler, I., 359
Hort, J., 1054
Horvát, S., 470
Horvath, G., 599
Horwitz, A. F., 188
Hoseman, W., 384
Hoshino, N., 739
Hosley, M. A., 655, 733, 741
Hosokawa, T., 717, 995
Hosoya, Y., 14, 708
Hossain, M. B., 497
Hou, Y. P., 498
Hou-Jensen, H., 642
Houlihan, D. J., 430, 434
Hounsfield, G. N., 19
Houpt, T. A., 878, 879
House, H. P., 764
Hovan, A. J., 845, 902
Hovemann, B. T., 68
Hovis, K. R., 1114, 1120, 1121
Hovstadius, B., 917
Howard, J. D., 292, 294
Howe, H. A., 17, 455, 462
Howell, G. A., 101
Howlett, N., 932
Hozumi, S., 409
Hreash, R., 876
Hsia, A. Y., 162
Hsu, J. J., 288
Hsu, P., 141, 142
Hu, B., 384
Hu, H., 188, 191
Hu, J., 116, 117, 411

Author Index

Hu, S., 801
Hu, Z. L., 644
Hu, Y., 390
Hua, Z., 68, 69, 73
Huang, A. L., 666, 675, 688
Huang, C., 499
Huang, C. C., 897
Huang, G. Z., 174
Huang, J., 534
Huang, J. H., 836
Huang, L., 693, 960, 1144
Huang, S. M., 917
Huang, S. S., 100
Huang, T., 738, 739, 741
Huang, Y. A., 647, 648, 688, 694
Huang, Y.-J., 648, 649, 666, 670, 672
Huard, J. M., 96, 135, 137, 138
Huard, J. M. T., 64
Huart, C., 271, 273, 274
Hubbard, P. C., 547, 553, 555
Hübener, F., 616
Huck, U. W., 325
Hudnell, H. K., 1100, 1101
Hudry, J., 217, 417, 429, 432
Hudson, R., 249, 320, 326, 385, 613, 616
Huertas, M., 553, 555
Huettel, S. A., 281, 288
Hughes, N. K., 586, 587
Hughson, A. L., 1014
Hull, C. L., 879
Hull, E. M., 584
Hulshoff Pol, H. E., 429
Hummel, C., 14, 265, 1094
Hummel, T., 23, 228, 236, 237, 243, 244,
    245, 265, 269, 270, 271, 284, 309, 311,
    358, 362, 363, 370, 377, 380, 382, 384,
    392, 393, 417, 420, 423, 496, 762,
    1057, 1094, 1095, 1100
Humphrey, T., 306, 307
Huneycutt, B. S., 462, 487
Hunt, J. R., 654
Hunt, T., 867
Hunter, A. J., 338
Hunter, J., 11
Huntley, A. C., 50
Huque, T., 675
Hursel, R., 1102
Hurst, E. W., 17, 461
Hurst, J. L., 568, 581, 582, 583, 1119
Hurtado-Chong, A., 189
Hurvich, L. M., 1070
Hurwitz, T., 430, 434
Huskisson, E. C., 758
Hussain, A., 553, 556
Hussain, A. A., 458, 463, 467
Hussain, M. A., 458, 467
Hutson, A. M., 961
Hutton, J. L., 500, 845
Huttunen, J., 267

Hvastja, L., 310
Hwang, P. M., 693

Iannacchione, V. G., 496
Iannilli, E., 781, 783
Ibarretxe-Bilbao, N., 286
Ichimura, A., 840
Idler, D. R., 547, 555
IFICF, 831
Ifuku, H., 714, 992
Igarashi, K. M., 211, 213, 217
Ignell, R., 932
Iguchi, H., 188
Ihrie, R. A., 186, 196
Iida, M., 643
Iijima, M., 418, 421
Ikan, R., 669
Ikeda, K., 16, 228, 242, 363, 386, 389, 954
Ikeda, M., 780, 890, 899
Ikemoto, S., 211
Ilegems, E., 691
Ileri-Gurel, E., 753
Illig, K. R., 210, 218, 292
Imada, T., 631
Imai, T., 127, 128
Imaizumi, T., 148
Imamura, F., 211
Imamura, K., 213, 251
Imayoshi, I., 195, 196, 198
Inada, H., 688, 1140
Inamura, K., 1117
Incardona, J. P., 730
Ingelman-Sundberg, M., 79, 917
Ingham, P. W., 730
Inglis, S. R., 595
Ingvar, G. H., 19
Innan, H., 51, 1124
Inomata, K., 633
Inoshita, T., 937
Inoue, R., 557
Inoue, T., 1092
Insausti, R., 217
Insley, S. J., 591
Inta, D., 186, 194
Inthavong, K., 486
Ionescu, E., 837
Iravani, J., 67
Irschick, D. J., 974
Irwin, M. T., 610
Irzhanskaia, K. N., 308
Isaacson, J. S., 161, 164, 165, 166, 167, 168,
    193, 209, 214
Isaka, H., 457
Iscan, M., 487
Isezuo, S. A., 846
Ishida, Y., 951
Ishige, T., 71
Ishihara, L., 426
Ishii, H., 993

Ishii, R., 834
Ishii, S., 688
Ishii, T., 126, 171, 1116
Ishikawa, S., 359, 936
Ishimaru, T., 384
Ishimaru, Y., 675, 688, 953
Ishimoto, H., 932
Ishizaki, M., 1143
Ishizuka, K., 433
Isidoro, N., 936
Isles, A. R., 582
Isogai, Y., 1115, 1118, 1119, 1125
Isolauri, J., 894
Isono, K., 930
Itagaki, H., 932
Itani, J., 869
Itaya, S. K., 455, 461
Ito, A., 651, 729, 738
Ito, H., 1143
Ito, I., 541
Ito, S., 714, 991, 1033
Itoh, N., 732
Ivankovic, S., 457
Iverson, G., 242
Iverson, S. D., 995
Iwai, N., 64, 136, 911
Iwai, T., 948
Iwasaki, S., 647
Iwasaki, S. I., 971
Iwatsuki, K., 730, 731, 733
Iwema, C. L., 126
Izard, C. E., 1075
Izquierdo, A., 717, 995

Jabbi, M., 788, 993
Jablonska, B., 187
Jackman, A. H., 243
Jackowiak, H., 963
Jackowski, A., 165
Jackson, E. L., 189, 193, 197
Jackson, J. A., 423
Jackson, J. H., 385
Jackson, R. T., 10, 453, 457
Jackson, T. S., 1142
Jacob, E. G., 386
Jacob, J., 567
Jacob, N. H., 639, 654
Jacob, S., 325
Jacob, T. J. C., 247
Jacobowitz, D. M., 173, 174, 1122
Jacobs, C., 525
Jacobs, K. M., 988
Jacobs, W. W., 799
Jacobson, J., 459
Jacobson, L., 1113
Jacoby, E., 113
Jacquot, L., 1103
Jaeger (1976), 647
Jaeger, S. R., 1059

Author Index

Jafek, B. W., 23, 380, 381, 386, 390
Jaffe, R., 732
Jägel (1991)., 654
Jagodowicz, M., 365, 367
Jahan-Pawar, B., 460
Jahovic, N., 633
Jahr, C. E., 162
Jaillardon, E., 67
Jain, A. K., 471
Jain, B., 380
Jain, D. K., 963
Jain, S., 649
Jakubowski, M., 646, 647, 949
Jalas, J. R., 69
Jameela, H., 592
Jamel, H. A., 803, 808
James, C. E., 802, 804, 810, 833
James, F. C., 966
James, G. A., 784
James, W., 1008
Jameson, S. B., 1054
Jäncke, L., 313
Jancsó-Gábor, A., 1101, 1102
Jang, W., 64
Jang, H. J., 17, 695, 839, 841, 842
Jang, T., 705
Jang, W., 133
Jankovic (1989), 428
Janota, M., 642
Janowitz, H. D., 246
Janowitz, H. D., 832, 838
Jansco, N., 1101
Janssen, N., 390
Janssen, S., 841, 842, 888
Jarolim, K. L., 456
Jaskoll, T., 734
Jasman, A. J. M., 633
Jasmin, L., 715
Jasper, H., 899, 901
Jaspers, V. L. B., 570
Jayaraman, A., 518, 519, 520
Jeannet, P. Y., 458, 467
Jeanrenaud, B., 839
Jedlitschky, G., 70
Jefferis, G. S., 535, 540
Jeffery, E. H., 72, 76, 464
Jeffery, R. W., 837
Jeffery, W. R., 646
Jefferys, J. G., 936
Jeffrey, A. M., 63, 457
Jehl, C., 297, 310, 337
Jellinek, G., 1052, 1057, 1070
Jellinek, J. S., 250
Jellinger, K., 502
Jellinger, K. A., 491, 502
Jemiolo, B., 1119, 1139
Jenal, U., 516, 517
Jenkins, G. N., 837
Jenkins, R., 339

Jenner, M. S., 3, 6
Jennes, L., 1123, 1142, 1143, 1144
Jennings, H. S., 3
Jensen, J. L., 627
Jensen, R. K., 457
Jermy, T., 939, 940
Jesionka, V., 1073
Jesteadt, W., 752
Jezierski, T., 597
Jezzard, P., 281
Jia, C., 145, 147, 171, 173, 497, 498, 1120, 1122, 1139
Jia, Z., 667
Jiang, D. H., 460
Jiang, H. J., 16
Jiang, J. B., 357, 367
Jiang, M. R., 161, 164, 170
Jiang, P., 629, 669, 689, 958, 972
Jiang, R. S., 384
Jiang, X., 729
Jiang, Y., 901
Jiao, Y., 934
Jimbo, D., 408, 409
Jin, K., 190, 198
Jinich, S., 412
Jinks, A., 250, 252
Jintapattanakit, A., 459
Jirowetz, L., 319
Jitpukdeebodintra, S., 733
Jo, S. M., 101
Joerges, J., 250
Johansen, J. P., 216
Johansen, K. K., 422, 426
Johansson, L., 384
Johansson, M. L., 554
John, S. W., 116
Johnell, K., 913, 917
Johns, A., 345
Johns, E. E., 339
Johnsen, P. B., 965
Johnson, B. A., 16
Johnson, A., 48
Johnson, A., 403, 415, 498
Johnson, A. J., 339
Johnson, B. A., 167, 213, 368
Johnson, B. N., 228
Johnson, C. C., 460
Johnson, D. M., 218
Johnson, F., 311
Johnson, F., Jr., 246
Johnson, M. A., 115
Johnson, M. B., 1017
Johnson, M. E., 501
Johnson, N. S., 548, 557
Johnson, R. E., 343
Johnson, S. D., 965, 968
Johnson, W. G., 847
Johnston, D. W., 567
Johnston, J. B., 728, 1142

Johnston, J. W. Jr., 3, 20
Johnston, R. E., 20, 583, 595
Johnston, S., 467
Johnstone, V., 1019
Joint, I., 526
Joly, M., 617
Joly, S., 100
Jones, A. C., 595
Jones, A. S., 356, 357, 368, 369
Jones, B., 996
Jones, B. P., 239, 836
Jones, C. B., 611
Jones, D. J., 871
Jones, D. T., 15, 465
Jones, E. G., 718
Jones, F. N., 3, 235, 236, 242, 248
Jones, I. L., 565, 570
Jones, K. A., 555, 954
Jones, L. V., 758, 1051
Jones, M. H., 3
Jones, N. S., 469
Jones, R. B., 568, 581
Jones, R. L., 459, 469
Jones, R. O., 765
Jones, S. V., 167
Jones, W. D., 540
Jones-Gotman, M., 218, 239, 287, 434, 1056
Jonides, J., 339
Jonsson, A. B., 100
Jönsson, F., 338
Jonz, M. G., 646
Jordan, H., 837
Jordt, S. E., 1092, 1101
Jørgensen, K., 937
Jorissen, M., 67
Jornot, L., 488
Jortner, R. A., 535, 541
Josephs, O., 288
Joshi, A. S., 472
Joshi, D., 616
Jourdan, F., 162
Jowett, A., 642, 984
Joyner, A. L., 190
Juilfs, D. M., 116, 1140
Julis, R. A., 832
Juncos, J. L., 413
Juneja, T., 412
Jung, H. S., 731, 732, 734
Jung, T. M., 384
Jung, Y. G., 363
Jungblut, L. D., 558
Jungwirth, S., 408
Jurisch, A., 637, 642
Just, T., 643, 839
Justus, K. A., 532, 598
Jüttner, F., 1081
Jyotaki, M., 832

Kaba, H., 174, 1122
Kachele, D. L., 705, 707
Kadohisa, M., 218, 776, 996, 1031, 1032, 1033, 1034, 1036, 1037
Kadona, H., 968
Kagan, B. L., 100
Kageyama, R., 138
Kahn, R. S., 429
Kaissling. K. E., 21, 539
Kajiura, S. M., 549, 553
Kajuira, H., 797, 798, 811
Kakita, A., 186, 191
Kalat, J. W., 873
Kalcheim, C., 728
Kaler, G., 466
Kaliner, M., 488
Kaliner, M. A., 66
Kalinke, U., 487
Kallas, H. E., 459, 469
Kallen, F. C., 42, 43, 47, 48, 57
Kallmann, F. J., 389
Kalmus, H., 598, 753, 755, 757
Kalogerakis, M. G., 324
Kaluza, J. F., 1134
Kam, M., 186, 197
Kamami, Y. V., 358
Kamath, V., 246, 433, 434, 435
Kamel, U. F., 391
Kamen, J., 1051
Kamen, J. M., 758
Kaminski, L. C., 763
Kaminsky, L. S., 63
Kampov-Polevoy, A. B., 835, 836
Kan, H., 932
Kanagasuntheram, R., 762, 985
Kandel, E. R., 338
Kanekar, S., 143, 145
Kaneko, N., 186, 188, 191, 296
Kang, J., 556
Kang, N., 174, 1123
Kang, P., 422
Kang, Y., 708, 709, 711
Kant, E., 12
Kanter, E. D., 214
Kanwal, J. S., 947
Kanwisher, N., 292
Kao, H. D., 458
Kaplan, J. N., 322, 612
Kaplan, M. S., 183
Kapoor, A. S., 521
Kappel, P., 611
Kappeler, P., 611
Kappelman, J., 986
Kapsimali, M., 732, 734, 735, 736
Karachunski, P. I., 458
Karádi, Z., 996, 997
Karasulu, H. Y., 470
Kare, M. R., 957, 961, 962, 963, 964, 965, 966, 968, 969

Kareken, D. A., 406, 409
Karimnamazi, H., 708
Karlson, P., 22, 532, 1118
Karlsson, A. C., 567, 572
Karmowski, A., 319
Karn, R. C., 74
Karnekull, S. C., 247, 375
Karniol, I. G., 875
Karnup, S. V., 161
Karrer, T., 1105
Kart, R., 581
Kasbekar N., 901, 914
Kase, N., 390
Kashiwadani, H., 167, 169
Kashiwayanagi, M., 73, 754, 1117
Kashlan, O. B., 686
Kaslin, J., 183
Kasowski, H. J., 159, 160
Kasper, M., 644, 650, 651
Kasper-Sonnenberg, M., 498
Kasperbauer, J. L., 357
Kass, J. H., 1122
Kassab, A., 1101
Kassarov, L., 964
Kastin, A. J., 320
Kastner, S., 710, 718
Katada, S., 110, 111, 113, 114
Katagiri, H., 194, 195
Katakowski, M., 190
Katare, Y. K., 551
Kato, A., 114
Kato, S., 428
Kato, T., 195
Katoh, H., 97, 139
Katotomichelakis, M., 384
Katschinski, M., 838, 839
Katsumata, T., 674, 679
Katsuragi, Y., 672
Katz, D. B., 655, 702, 703, 712, 713, 783, 784, 938
Katz, H., 22
Katz, H. I., 312
Katz, H. M., 344
Katz, L., 164, 165, 167, 168, 213
Katz, L. C., 159, 161, 164, 167, 213, 1122
Katz, S., 1133
Katzav, A., 391
Katzenschlager, R., 417, 428
Kau, A. L., 517
Kauer, J. S., 14, 168, 250, 367
Kauffman, D. L., 632
Kaufman, E., 914
Kaufmann, H., 427, 429
Kaufmann, N., 313
Kauhanen, E., 598
Kaupp, U. B., 1140
Kautz, M. A., 210, 345
Kavoi, B. M., 592
Kawai, T., 632

Kawakami, K., 320
Kawakami, M., 729
Kawamura, Y., 14
Kawana, E., 466
Kawata, M., 1123
Kawauchi, S., 137, 144
Kay, L. M., 164, 167, 168, 169, 210, 344, 346
Kay, R. B., 210
Kay, S. R., 435
Kaya, N., 649
Kayser, J., 266, 268, 271
Kayser, R., 355, 357
Keane, P., 1070
Keast, R. S. J., 670, 672, 757, 847, 1008, 1055
Keedy, M., 526
Keele, C. A., 1091, 1101
Keenan, C. M., 18
Keesey, J. C., 21
Kehoe, P., 868
Keil, W., 317, 345
Kelber, C., 535
Keller, A., 114, 160, 162
Keller, K. L., 757, 802, 812
Keller, M., 580, 583, 584, 585
Keller, P. J., 632
Kelliher, K. R., 1118
Kelling, S. T., 629
Kelly, J. T., 358, 359, 365
Kelsch, W., 194, 195, 196
Kemmer, T., 838
Kemper, T. D., 1075
Kendal-Reed, M., 309, 1093
Kendler, K., 874
Kendrick, K. M., 1122
Kennakin, T., 669
Kennedy, C., 19
Kennedy, C. J., 102
Kennedy, L. M., 987
Kenneth, J. H., 9
Kent, K. S., 932, 937
Kent, L. B., 932
Kent, S., 872
Kent, W. D. T., 865
Kentner, D., 515
Kepecs, A., 168, 169
Kerley, L. L., 599
Kern, D. L., 1015
Kern, E. B., 357
Kern, R. C., 362, 381, 386
Kernie, S. G., 198
Kerr, D. S., 111
Kerr, K. M., 217
Kertelge, L., 420, 424, 426
Keshavan, M. S., 434
Kessen, W., 806
Kesslak, J. P., 340, 404, 405, 415
Kessler, T. L., 892

Author Index

Kettenmann, B., 268
Ketterer, C., 915
Kety, S. S., 19
Keverne, E. B., 173, 553, 1122
Kevetter, G. A., 174
Key, B., 143, 147, 465, 1136, 1137, 1138, 1140
Keyhani, K., 356, 359, 366 369
Keymer, J. E., 515
Khalifah, R. G., 116
Khan, N. A., 835
Khan, N. L., 417, 424, 426
Khan, R. M., 263, 289
Khan, M., 126, 129
Khiari, D., 1080, 1081
Khu, B., 1053
Kiaune, L., 497
Kida, A., 899
Kiefer, S. W., 836, 877
Kielan-Jaworowska, Z., 49
Kien, J., 932, 933
Kiernan, M. C., 501
Kiesow, F., 10, 12
Kijima, H., 932
Kikuta, S., 210
Kim, A., 887
Kim, D., 1092
Kim, D. J., 649, 650
Kim, D. W., 384
Kim, G. H., 816, 818
Kim, H. G., 390
Kim, H. S., 933
Kim, I. D., 468
Kim, J. B., 765
Kim, J. H., 934
Kim, J. K., 66
Kim, J. Y., 377, 418, 730, 731, 733, 734
Kim, K-O, 1019
Kim, M. R., 650, 691, 693
Kim, S., 1117, 1118
Kim, S. K., 359
Kim, S. W., 386
Kim, T. H., 66, 470
Kim, U., 690, 914, 915, 923, 976
Kim, U.-K., 756, 757
Kim, Y., 497
Kim, Y. H., 421
Kim, Y. K., 996
Kimbell, J. S., 51, 67, 367, 485, 486
Kimchi, T., 1123, 1125
Kimmel, A. R., 521
Kimmelman, C. P., 355, 384
Kimoto, H., 554, 556, 1117, 1118
Kimoto, M., 67
Kimura, K., 14
Kincaid, A. E., 487
King, A. J., 573
King, A. S., 567
King, C. T., 707, 742

King, J. C., 1142
King, K. R., 832
Kinnamon, J. C., 647, 648, 650
Kinnamon, S. C., 14, 15, 643, 649, 675, 693, 949, 952
Kinnunen, I., 384
Kinomura, S., 775
Kinoshita, Y., 455, 460, 461
Kiovula, M., 586
Kippin, T. E., 187, 190
Kirkby, H. M., 228
Kirkes. W. S., 11
Kishi, M., 650
Kishi, T., 871
Kisich, K. O., 100
Kissel, P., 415
Kita, H., 709
Kitada, Y., 951, 952
Kitagawa, M., 688
Kitamura, F., 497
Kitchell, R. L., 968
Kittel, P., 135, 136, 140
Kivernitis, E., 896, 897
Kiyohara, S., 949
Kiyokage, E., 159, 160, 163
Kjeldmand L., 610
Kjelvik, G., 297, 407
Klabunde, T., 113
Klajner, F., 838
Klarin, I., 913, 917
Klasing, K. C., 964
Klatzo, I., 499
Kleene, S. J., 103, 112
Kleifield, E. I., 847
Kleiman, D., 581
Klein, A. H., 1101
Klein, P. B., 984
Klein, R., 1120
Klein, S. L., 93
Kleineidam, C., 540
Kleineidam, C. J., 532, 539
Klemm, W. R., 263
Klemola, T., 586
Kleyman, T. R., 686
Klimacka-Nawrot, E., 849
Klimek, L., 362, 383, 384
Klingmuller, D., 390
Klopman, R. B., 611
Kloskowski, A., 574
Klossek, J. M., 384
Klotman, M. E., 100
Klotz, U., 914
Kluver, H., 783, 996
Knaden, M., 539, 540
Knafo, S., 214
Knai, C., 795
Kniep, H. H., 312
Knierim, J. J., 217
Knights, J., 1054

Knöll, B., 1120
Knopf, J. L., 1119
Knopfel, T., 160
Knopp, U., 384
Knupfer, L., 404, 409, 437
Knutson, B., 995
Kobal, G., 14, 23, 228, 243, 262, 265, 266, 269, 270, 1094, 1095, 1100
Kobayakawa, K., 124
Kobayakawa, T., 776, 777, 778, 779, 782, 783, 1016
Kobayashi, K., 958
Kobayashi, M., 243, 609, 991
Koce, A., 557
Kochli, A., 837
Kock, K., 631
Koehl, M. A., 531
Koelega, H., 310, 311
Koelega, H. S., 242
Koelliker, R. A., 638
Koelling, R. A., 346, 867
Koenigkam-Santos, M., 377
Kogan, J. H., 340
Koger, S. M., 340
Koh, L., 487
Koh, M. T., 876, 877
Kohidai, L., 527
Kohl, Z., 198
Kohler, C. G., 432
Kohlrausch, O., 41
Kohwi, M., 196
Koide, T., 549, 552
Koizumi, A., 689
Koizumi, H., 187, 191
Kokovay, E., 184, 187, 188, 191
Kokrashvili, Z., 730, 831, 840
Kokubo, Y., 428
Kolaczkowska, E., 1116
Kolattukudy, P. E., 567, 572
Kolb, B., 340
Kolbinger, A., 520
Kollmann, M., 44, 51, 54
Kolunie, J. M., 584
Komar, D., 598
Komatsu, H., 1140
Komisaruk, B. R., 14, 988
Kondo, H., 382
Kondo, Y., 1123
Konstantinidis, I., 376, 382, 384, 501, 893
Kontos, A. P., 468
Konturek, J. W., 837, 838
Konturek, S. J., 837, 838
Koo, J. Y., 1104
Koob, G. F., 874
Koos, D. S., 1136, 1137, 1138, 1140
Kopala, L., 429, 430, 431, 434
Kopala, L. C., 429, 430, 431, 432, 434, 438, 439
Koppe, T., 41, 45

Koppelhus, U., 527
Korpi, A., 1095
Korpimaki, E., 586
Korsching, S., 172
Korsching, S. I., 556, 646
Korte, G. E., 647, 971
Korten, J. J., 415
Kosaka, K., 159, 160, 196, 553
Kosaka, T., 159, 196, 553
Kosar, E. M., 711
Koshland, D. E., Jr., 3
Koskela, S., 68
Koss, E., 286, 404, 438
Kosten, T., 388, 865, 878
Koster, E. P., 230, 238, 247, 250, 310, 311, 337, 339, 343, 345, 346, 1014, 1055, 1062
Koster, M. A., 1014
Koster, N. L., 141, 142, 145
Kostina, G. N., 568
Kostopanagiotou, G., 501
Kotrschal, K., 644
Koujitani, T., 457
Kourtzi, Z., 292
Kovacs, T., 411, 427
Kowalewski, J., 410
Kowalewsky, S., 591
Kowall, N. W., 459
Koyama, N., 15
Kraegen, E. W., 839
Kraepelin, E., 429
Krahn, D., 836
Kramer, M., 1056, 1061
Krames, L., 20
Krammer, G., 667
Krantz, E. M., 243
Kranzler, H. R., 757, 836
Krarup, B., 754
Kratskin, I. L., 78
Kratzing, J. E., 1133
Krauel, K., 266, 269
Krause, K. F. T., 280
Krause, R., 638
Krause, W., 11, 12, 280
Krautwurst, D., 16, 111
Kream, R. M., 473
Krell, T., 514
Kremer, B., 242
Krestel, D., 592
Kreutzberg, G. W., 463
Kriegeskorte, N., 293
Krimm, R. F., 729, 738, 739, 741, 742
Kringelbach, M. L., 786, 796, 799, 995, 1039
Krishan, M., 470
Krishna, N. S., 66, 70, 141, 142, 143
Krishnan, A. V., 501
Krispin, S., 737
Kristensson, K., 462, 464, 465, 483

Kroeze, J., 1009
Kroeze, J. H. A., 763
Krohn, M., 667
Krolewski, R. C., 138
Kroll, J. C., 970, 972
Kroll, J. J., 796, 1056
Kroner, T., 322
Krout, K. E., 708
Kruger, R., 424
Ksouda, K., 902
Kubes, P., 1116
Kubie, J. L., 14
Kuch, J. H., 935
Kudo, H., 71, 551
Kudo, K., 647, 649, 961, 962, 963, 968, 969
Kuebler, L. S., 538
Kuhlenbeck, H., 19
Kuhn, C., 670, 691
Kuhn, M., 116, 117
Kulkarni, J., de, 434
Kumazawa, T., 631
Kung, C., 524, 525
Kunze, W. A. A., 170
Kuo, H. W., 497
Kuo, Y.-L., 766
Kurahashi, M., 633
Kurahashi, T., 14, 112, 250
Kurihara, K., 15, 631, 672
Kurosaki, M., 71
Kurosawa, T., 962
Kurtz (2004), 366
Kurtz, D. B., 238, 243
Kurtz, M. M., 488
Kusaba, T., 846, 848
Kushkuley, J., 499
Kussmaul, A., 308
Kutas, M., 269
Kutter, A., 1019
Kuwabara, M., 935
Kux, J., 551
Kuznetsov, V. B., 958
Kuznicki, J., 714, 716, 992
Kuznicki, J. T., 251
Kveton, J. F., 765, 782, 891, 896
Kwapil, T. R., 429
Kwon, H. J., 99
Kwon, J., 929
Kwon, J. Y., 540
Kwong, K., 146, 147
Kyriazis, G. A., 695, 840

Laaris, N., 166
Labbe, D., 1018, 1020, 1052, 1053, 1058
Lacar, B., 194
Lacher, V., 540
Ladher, R. K., 736
Ladowsky, R. L., 876
Laeng, B., 832
Laethem, R. M., 79

Lafay, F., 455, 462
Lafdjian, A., 914, 915
Laffort, P., 248
Laframboise, A. J., 552
Lagace, D. C., 195, 196
Lagerstrom, M. C., 964, 976
Lagier, S., 168, 169
Lai, D., 417, 423
Laidre, M. E., 611
Laing, D. G., 243, 248, 249, 250, 251, 252, 309, 311, 1055
Laissue, P. P., 535
Laitman, C. J., 871
Lakard, S., 146, 147
Lakkakula, A. P., 795
Lalonde, E., 642, 643
Lam, D. J., 368
Lam, Y.-W., 709
LaMantia, A. S., 76, 94
Lamb, C. F., 954
Lambers, D. S., 319
Lambert, D., 1100
Laming, D., 752
Lamster, I. B., 914
Lan, J., 460
Lancet, D., 15, 70
Lancisi, J. M., 7
Land, L. J., 456, 463, 466
Land, R., 710
Landacre, F. L., 728
Landau, W. H., 19
Landgraf, R., 469
Landin, A. M., 650
Landis, B. N., 362, 364, 414, 419, 780, 848, 904
Lane, A., 311
Lane, A. P., 383
Lane, R. J., 458
Lang, C. J., 429
Lange, R., 406
Langebartel, D. A., 970, 971
Langendijk, P., 318
Langford, H. G., 846
Langhans, W., 872
Langston, J. W., 473
Lanier, S. A., 836
Lanza, D. C., 148
Lanza, D. L., 69
Lara, A. H., 716, 991, 994
Larjola, K., 310
Larochette, A. C., 801
Laroia, H., 413
Larriva-Sahd, J., 171, 172, 173, 1121, 1122
Larsell, O., 1142
Larsson, M., 305, 337, 405, 413
Larsson, M. C., 533, 1116
Larsson, P., 68, 73, 80, 456
Lasagna, L., 758
Lascano, A. M., 266, 267, 268

Author Index

Lasiter, P. S., 705, 707, 708, 742
Laska, M., 22, 51, 54, 55, 249, 251, 344, 591, 610, 611, 612, 613, 616
Laskaris, G., 893
Lassen, N. A., 19
Lasseter, A. E., 598
Lastein, S., 557
Latz, E., 493
Laudien, J. H., 269
Laudien, M., 66, 100, 362
Laugerette, F., 835, 838, 840, 1102
Laughlin, S. B., 831
Laukaitis, C. M., 74
Laumbach, R. J., 1100
Laurent, G., 164, 167, 168, 169, 344, 346, 541
Lauterbur, P., 19
Lavasanifar, A., 923
Lavenex, P., 345
Laveran, C. L. A., 7
Lavi, E., 455, 466
Lavin, M. J., 869
Lawless, H., 1101, 1103, 1104, 1105
Lawless, H. T., 230, 236, 339, 340, 345, 672, 751, 752, 753, 755, 756, 758, 763, 767, 1008, 1010, 1011, 1014, 1017, 1052, 1055, 1069, 1080, 1091, 1101, 1103
Lawrence, G., 1020
Laws, E. R. Jr., 891
Lawson, M. J., 368, 593
Lawton, D. M., 648, 649
Layman, L. C., 390
Laywell, E. D., 184
Lazard, D., 70, 73, 465
Lazarini, F., 166, 198
Lazic, S. E., 413
Lazzara, P., 344
Le Douarin, N. M., 98, 728
Le Floch, J. P., 390, 754, 848
Le Gros Clark, W. E., 16, 50, 52
Le Magnen, J., 21, 311, 318
Le Reverend, F. M, 1071
le Roux, C. W., 847, 888, 1020
Le, G. G., 434
Le, Y., 1116
Leahey, E. B. Jr., 921
Leahy, M., 599
Leavens, T. L., 497
LeBlanc, J., 839
Lebrun, C., 876
Lecanuet, J. P., 306
Lechowicz, M. J., 935
Leclaire, S., 567, 572
Lederer, F. L., 8
Ledikwe, J. H., 835
LeDoux, J. E., 996, 1122, 1123
Lee, A. C., 93, 99, 307
Lee, C. E., 1144

Lee, C. H., 66, 669
Lee, H. M., 816, 818
Lee, H. P., 359, 360
Lee, J., 668, 669
Lee, J. C., 937
Lee, J. H., 148
Lee, J. M., 822
Lee, J. S., 438
Lee, J. T., 66
Lee, K., 15
Lee, K. H., 143
Lee, L. S., 914, 921
Lee, M. G., 76, 77
Lee, M. J., 733, 734
Lee, P. H., 417, 418, 423, 426, 428
Lee, R. J., 17, 380
Lee, S. H., 381
Lee, S. H. A. G., 100
Lee, S. R., 187
Lee, S. van der, 21
Lee, S. M., 1019
Lee, W. H., 142
Lee, Y., 932, 935, 1092
Leela, K., 985
Lefkowitz, R. J., 127
Legg, J. W., 12, 385
LeGoff, D. B., 837
Lehman, C. D., 23, 754, 765
Lehman, M. N., 1123
Lehmann, D., 266
Lehn, H., 297
Lehrer, R. I., 99
Lehrner, J., 244, 405, 408, 416
Lehrner, J. P., 309, 405
Lei, H., 210
Leick, V., 527
Leigh, A. D., 386
Leinders-Zufall, T., 15, 112, 116, 117, 547, 554, 556, 1113, 1116, 1117, 1118, 1119, 1120, 1140
Lemasson, M., 195, 196
Lemay, A., 459
Lemberger, T., 1123
Lemery, N., 7
Lemke, G., 128
Lemmens, S. G., 842
Lemon, C. H., 703, 705, 706
Lénárd, L., 996
Lennertz, R. C., 1102
Lenochova, P., 323
Lenz, F. A., 711, 989
Leon, M., 16, 20, 167, 213, 305, 315, 368, 580, 585
Leon-Sarmiento, F. E., 391, 409, 496
Leonard, B. C., 101
Leonard, C. M., 706, 989, 1028, 1143
Leonard, G., 384
Leonard, N. L., 705
Leong, S. Y., 191

Leopold, D. A., 47, 298, 362, 364
Leprohon, C. E., 834
Lesage, S., 424, 426
Lesham, M., 798, 808, 816, 837
Leslie, A. J., 970
Lesniak, A., 593
Letnic, M., 973
Leung, C. T., 64
Leung, C. T., 18, 136, 137
Leung, H. T., 879
Leung, P. M. B., 834
Levai, O., 1134
Levenson, C. W., 460
Levesque, S., 490
Leveteau, J., 14
Levey, D. J., 966
Levin, H. S., 386
Levine, S., 873, 874
Levine, S. C., 358
Levitt, H., 752, 753
Levy, D. C., 1057
Lévy, F., 585
Levy, L., 228, 501
Levy, L. L., 1057
Levy, L. M., 288, 376, 392, 993, 1056
Levy, R. L., 458, 467
Lewandowski, K. E., 429, 903
Lewcock, J. W., 127, 1115
Lewey, F. H., 776
Lewis, C. T., 929
Lewis, D., 654, 762
Lewis, I. K., 911
Lewis, J., 487
Lewis, J. L., 65, 67, 68, 72, 73, 80, 454
Lewis, R. J., 51
Lewis, R. W., 568
Lewis, V. R., 598
Lewitt, M. S., 389, 846
Lewkowitz-Shpuntoff, H. M., 376, 389, 390, 438
Lewkowski, M. D., 801
Ley, J., 667
Ley, J. P., 667, 692, 1101, 1102
Leydig, F., 9, 637
Leypold, B. G., 1118, 1125
Li, B., 496
Li, C.-S., 705, 706, 708, 709, 987, 989, 1028
Li, H., 346
Li, J., 126, 627, 1136
Li, J. H., 647
Li, L., 68, 69, 77, 1145
Li, M., 515
Li, W., 216, 217, 218, 297, 344, 555
Li, X., 16, 17, 187, 458, 459, 496, 629, 666, 667, 671, 673, 689, 690, 841, 958, 965, 967, 973, 986, 987
Li, X. J., 644
Li, Y., 18, 148
Li, Y. J., 1102

Li, Y. R., 112
Liacouras, C. A., 911
Liang, L., 1020
Liang, L. C., 835, 847
Liapi, C., 498
Liberini, P., 414
Liberles, S. D., 16, 115, 171, 554, 1116, 1117, 1118, 1119
Licht, T., 190, 195
Lichtenfels, J. R., 58
Lieb, W. R., 1099
Liebetanz, D., 609
Liebl, D. J., 729, 738, 741
Liem, D. G., 800, 803, 804, 809, 810, 820, 822, 823, 832, 833, 1053
Liggett, S. B., 113
Lightman, S. L., 916
Likert, R., 758
Liley, A. W., 796, 797, 805, 811, 833, 835
Lim, D. A., 197
Lim, J., 757, 761, 1016
Lim, J. H., 461, 500
Lima, C., 148
Liman, E. R., 12, 51, 694, 1117, 1118, 1124
Limebeer, C. L., 875
Lin, C., 731
Lin, C. H., 426
Lin, C. S., 710
Lin, D., 1123
Lin, D. M., 18
Lin, H-M., 599
Lin, S. C., 497
Lin, W., 649, 1092
Lindbom, L., 1116
Lindemann, B., 15, 17, 643, 647, 650, 914, 952
Lindemose, S., 527
Linden, R. W., 642
Linden, R. W. A., 625
Lindenmaier, P., 961, 963
Lindig, J., 392
Lindley, M. G., 669
Lindquist, N. G., 456, 463
Liner, E., 972
Ling, G., 70
Linnaeus, C., 7
Linschoten, M. R., 752, 763
Linster, C., 162, 167, 168, 169, 170, 216, 338, 344, 580
Linzmeier, R. M., 100
Liou, A. P., 840
Lipchock, S. V., 822, 923
Lippa, C. F., 409, 410, 414
Lippi, G., 1060
Lipsitt, L. P., 308, 797, 806, 807
Lipson, A. C., 147
Lipton, R. B., 458, 467
Lisberg, A. E., 596
Little, A., 518

Little, A. C., 846
Little, T. J., 841, 842
Litvack, J. R., 382, 384
Liu, C., 75
Liu, C. Y., 1138, 1139 1142
Liu, D., 694
Liu, F., 731, 732, 733
Liu, H., 705
Liu, H. C., 242
Liu, H. K., 189
Liu, H. X., 655, 728, 729, 730, 731, 732, 733, 756
Liu, L., 655, 675, 934, 1104
Liu, M. A., 469
Liu, N., 145
Liu, P., 489
Liu, S., 158, 159, 171
Liu, W. L., 163
Liu, X., 190, 193, 738
Liu, X. F., 456, 463
Liu, Y., 117, 455
Liu, Y. C., 1123
Liu, Y. S., 742
Liu, Z., 183
Livermore, A., 248, 249, 250
Livneh, Y., 195
Lledo, P. M., 166, 168, 183, 195, 196, 198, 217
Lo, L., 1120, 1123, 1124
Lobasso, S., 64
Lobo, M. V., 840
Lockshin, L., 1053
Loconto, J., 171, 1116
Locy, W. A., 1141, 1142
Lodovichi, C., 163
Loewy, A. D., 462, 706, 708, 989
Logan, C., 737
Logan, D. W., 319, 585, 1118, 1119, 1123, 1125
Logothetis, N. K., 288
Logue, A. W., 865, 873, 1015
Loireau, J. N., 611
Lois, C., 19, 183, 185, 195
Lojkowska, W., 410
Lombardi, J. R., 582
Lomvardas, S., 16, 127
Lonai, P., 143
London, B., 386, 392, 485
Long, R. A., 599
Longo, V., 68, 73, 78, 79, 80
Loo (1973), 45
Loo, A. T., 380
Loo, C. K., 393
Loo, S. K., 51, 606
Looy, H., 832
LoPachin, R. M., 1092
Lopez, E., 142, 143, 144, 147
Lopez, G. F., 739
Lopez-Mascaraque, L., 146

Lopez-Rios, J., 193
LopezJimenez, N. D., 688
Lorentzen, M., 839
Lorenzo-Lamosa, M. L., 923
Lorig, T. S., 228, 262, 263, 265, 273, 289
Lorry, D., 7
Lotsch, J., 270, 418
Løtvedt, P. K., 616
Lotz, C. N., 961, 965, 967
Loudon, C., 531
Louilot, A., 344
Louis, E. D., 413, 418
Louon, A., 458
Lovén, C., 9, 10, 638, 640
Lowe, G., 15, 112, 163, 164, 169, 1122
Lowe, M. R., 847
Lower, R., 11
Lu, J., 18, 148
Lu, M., 111, 678
Lu, W., 99
Lucas, A. M., 566, 568
Lucas, F., 839
Lucas, G. A., 867
Lucas, P., 1117, 1118
Lucchina, L. A., 758
Lucchini, R., 385, 498
Lucchini, R. G., 460
Luce, R. D., 343, 760
Lucero, M. T., 141, 142, 145
Luciano, L., 17
Ludwig, C., 11, 625
Ludwig, K., 234
Ludwig, M., 963
Lugaz, O., 630, 781
Luiten, P. G. M., 949, 950
Lumb, B. M., 1092
Lumeng, J., 310
Lumia, A. R., 581, 584
Lumley, L. A., 581
Lumpkin, E. A., 369
Lund, V. J., 358, 362, 364, 384
Lundahl, D., 1052
Lundgren, B., 1010
Lundh, B., 465
Lundström, J. N., 298, 343
Lundy, R. F., 704, 706, 708, 709, 710, 711, 718, 1028
Lung, M. A., 355
Luo, G., 75
Luo, J., 197
Luo, M., 164, 165, 167, 168, 173, 174, 213, 1114, 1122
Luo, Q., 1041
Lupo, D., 78
Luporini, P., 527
Luppi, B., 472, 473
Lüscher, M., 22, 532, 1118
Luscombe-Marsh, N. D., 834
Lush, D., 17

Lush, I. E., 690
Luskin, M. B., 19, 183, 185, 187, 196
Lussana, F., 653
Luxan, M. P., 565
Luzzi, S., 407, 426
Ly, A., 763
Lyall, V., 675, 676, 677, 678, 679, 686, 687, 688
Lyckholm, L., 392
Lydell, K., 583
Lygonis, C. S., 388
Lyman, B. J., 346
Lynch, A. S., 520
Lynch, G., 209
Lyon, B., 1070
Lyon, D. H., 1051, 1052
Lyons, D. B., 128, 129

Ma, C., 466
Ma, H., 647
Ma, J., 163, 164, 169, 842, 1122
Ma, L., 739, 740
Ma, M., 48, 127, 1133, 1134, 1135
Ma, W.-C., 929, 933, 935, 936
Maassab, H. F., 469
Maaswinkel, H., 1145
Macakova, L., 627
Macas, J., 198
Macchi, G., 307
Maccioni, P., 499
MacDonald, C. J., 702, 713
Macdonald, D. W., 21, 596
MacDonald, K. P., 143
MacDonald, K. P. A., 18
Macdougall, I. C., 916
Mace, O. J., 841, 842
Macfarlane, A., 316, 321
MacFie, H., 1053, 1061
Machemer, H., 524, 525
Machold, R., 190
Maciel, S. M., 803, 808
Mackay-Sim, A., 14, 18, 96, 98, 101, 134, 135, 136, 140, 141, 143, 147, 148, 149, 250, 367, 389, 846, 1057
Mackenzie, I. C. K., 754
Mackenzie, P. I., 70
Mackey, E. D., 1080, 1081, 1082
Mackintosh, J. H., 584
Macknin, J. B., 217
MacLean, A., 11
MacLean, P. D., 12
MacLeod, K., 541
MacLeod, P., 14, 779
MacLeod, R. B., 22
Macmillan, N., 233
MacPherson, E. E., 317
Macpherson, L. J., 1092, 1100
Macpherson, L. M. D., 628
Macrides, F., 145, 168, 531

Macrini, T. E., 42, 44, 45, 46, 50, 58
Maeda, N., 626
Maehashi, K., 968
Maejima, M., 597
Maes, F. W., 934, 937
Magendie, F., 11, 638, 643, 653
Magklara, A., 128, 129
Magnusson, W. E., 970
Magoun, H. W., 263
Magrassi, L., 141
Mahajan, S. K., 848
Mahalik, T. J., 135, 141
Mahanthappa, N. K., 18, 136, 143, 147
Maher, B. J., 162
Mahmut, M. K., 1010, 1013, 1018
Maier, W., 42, 44, 45, 46, 50, 55, 56
Mainardi, D., 20, 325
Mainardi, M., 318
Mainen, Z. F., 167
Maingret, F., 1140
Mainland, J., 274
Mainland, J. D., 244, 297
Mainwaring, G., 71
Mair, R. G., 340, 384, 438, 499
Maisetta, G., 100
Mak, G. K., 198
Mak, Y. E., 782
Makin, J. W., 246, 317
Makita, K., 76
Makowska, I., 243, 408
Malaspina, D., 217, 431, 433, 434, 435
Malcolm, R., 847
Malfertheiner, P., 838
Malik, S., 785
Maller, O., 797, 811, 964
Malleret, L., 1080
Mallet, P., 323
Malmberg, H., 362, 382
Malmierca, E., 718
Malnic, B., 16, 111, 123, 125, 1054
Malone, G. T., 236
Malone, J., 516, 517
Malpighi, M., 9, 10, 637, 638
Maltais, F., 355
Mamasuew, K., 1138, 1139, 1140
Manabe, H., 217
Manach, C., 632
Manahan, C., 521
Mandairon, N., 167, 170, 171, 195, 196, 198, 338, 344
Mandiyan, V. S., 1125
Manestar, D., 362
Mangiapane, M. L., 876
Manglapus, G. L., 137, 138
Mangold, J. E., 742
Mania-Farnell, B. L., 145
Manis, M., 759, 760
Mann, N. M., 386
Mann, S., 632

Mansour, A., 465
Manton, M., 974
Mantovani, G., 872
Manzo, A., 938
Maone, T. R., 797, 805, 833
Mappes, J., 966, 969
Mar, J. C., 149
Maras, P. M., 339
Marchand, F., 1120
Marchand, R., 1120
Marchlewska-Koj, A., 20
Marcinkowska, U., 323
Marcus, D. A., 599
Marcus, J., 22
Mardon, J., 567, 572
Marek, C. A., 850
Marella, S., 933, 938, 939
Maremmani, C., 422
Margalit, T., 70
Margolis, F. L., 15, 93, 99, 456, 463
Margolskee, R., 525
Margolskee, R. F., 17, 695, 795, 914
Margrie, T. W., 161, 164, 167, 168, 169, 1136
Marin-Padilla, M., 146
Marinaro, C., 190
Marinho, H., 1015
Marini, S., 68, 70, 73, 79
Marion-Poll, F., 936
Mark, G. P., 703
Markley, E. J., 844, 845
Markman, M., 874
Markon, K. E., 235
Markowitsch, H. J., 711
Marks, L. E., 230, 235, 236, 755, 756
Marlier, L., 307, 317, 318, 319, 320, 321, 322
Marlin, D. W., 869
Marom, R., 320
Maronpot, R. R., 64
Maroon, J., 497
Marotta, A., 437
Marples, N. M., 964, 968, 969
Marr, D., 214
Marr, J. N., 325
Marras, C., 417
Mars, M., 832
Marshall, C. A., 197
Marshall, D. A., 592, 1135
Marston, M., 595
Martel, K. L., 174
Marti-Fabregas, J., 198
Martin, C., 169, 216, 344, 835
Martin, G. M., 874
Martin, G. N., 263
Martin, H. N., 11
Martin, J. R., 876
Martin, L. J., 184
Martin, N., 1018

Author Index

Martin, R. D., 44, 55
Martin, S., 835
Martinez del Rio, C., 965, 966
Martinez, B. A., 437
Martinez-Marcos, A., 47, 1113, 1123
Martinez-Molina, N., 186
Martini, R., 650, 651
Martini, S., 1116, 1138, 1139
Martinon, F., 491
Martins, S., 961
Marttin, E., 458, 467, 469
Martzke, J. S., 387
Marui, T., 951
Maruniak, J., 456, 464, 465
Maruniak, J. A., 141, 229, 298, 473, 583, 1135
Maruyama, Y., 692, 1031
Marxreiter, F., 198
Mascagni, P., 497
Maseeh, A., 849
Masek, P., 937
Masieri, S., 358
Masing, H., 358
Mason, J. R., 960, 962, 964, 967, 968
Mason, R. M., 965
Mass, E., 903
Massolt, E. T., 838
Mast, T. E., 388
Mastella, G., 392
Masterton, R. B., 608
Masucci, M., 1142
Mateo, J. M., 320, 321, 583
Matha, K. V., 237, 437
Mathew, S. D., 893
Mathison, S., 453, 466, 469
Mathuru, A. S., 547
Matigian, N., 149
Matisoo-Smith, E., 803, 813, 814
Matochik, J. A., 583
Matson, A., 633
Matson, K. D., 961, 963, 970
Matsuda, T., 23, 753, 890, 893, 1056
Matsui, A., 609, 610
Matsui, D., 911
Matsumiya, K., 459
Matsumoto, H., 160
Matsumoto, I., 736
Matsumura, S., 1102
Matsunami, H., 111, 112, 171, 554, 648, 649, 690, 960, 1115, 1116
Matsuo, A. Y. O., 556
Matsuo, R., 11, 626, 628, 629, 630, 631, 633, 838
Matsuo, T., 1138, 1140, 1141
Matsuoka, M., 1117
Matsushita, M., 708
Matsuyama, A., 141, 147
Matsuyama, T., 470
Matt, G. E., 500

Mattes, R. D., 389, 501, 632, 765, 832, 833, 834, 835, 836, 837, 838, 842, 843, 846, 847, 849, 868, 873, 901, 904, 1019, 1101
Mattes-Kulig, D. A., 844
Matthews (1999), 100
Maurage, P., 499, 836
Max, M., 688
May, O. L., 704, 705, 742
Mayer, A. D., 585
Mayer, U., 71
Maynard, M., 822
Mazziotta, J. C., 281
Mbiene, J. P., 655, 731, 739
McAbee, G., 429
McAlonan, K., 219
McArthur, C., 632
McBride, K., 500
McBurney, D. H., 3, 11, 345, 625, 629, 690, 752, 753, 754, 755, 762, 1017
McCabe, B. F., 765
McCabe, C., 1031, 1039, 1057
McCabe, K. L., 736
McCaffrey, R. J., 23, 406, 409
McCaffrey, T. V., 355
McCarthy, D. O., 872
McCarthy, V. P., 392
McCartney, C. J., 916
McCartney, W., 3
McCarty, J. H., 187, 191
McCaughey, S. A., 675, 676, 879
McClintock, M. K., 580
McClintock, T. S., 80, 111
McClure, S. M., 785, 788
McConnell, R. J., 389, 846, 901
McCormack, L. J., 384
McCotter, R. E., 593, 1142
McCrory, M. A., 795
McCulloch, M., 599
McCurdy, R. D., 145, 147, 148
McDaniel, M. A., 346, 1008
McDonald, D. G., 338
McDonald, T. J., 916
McFarland, L. Z., 1133
McGann, J. P., 162
McGaughy, J., 214
McGeer, P. L., 428, 459
McGinnis, M. Y., 581
McGuire, B., 596
McGuire, M. J., 1082, 1087
McIlwain, H., 209
McIntosh, T. K., 581
McIntyre, J. C., 126
McKay, H. V., 584
McKay, J. L., 973
McKemy, D. D., 369, 1103
McKenna, P. J., 390
McKenzie, S., 346
McKeown, D. A., 412

McKeown-Eyssen, G., 496
McKhann, G. D., 403
McKinnon, J., 420
McLaughlin, N. C., 407
McLaughlin, S. K., 15, 649, 693
McLaughlin, T., 5, 7, 123, 128
McLean, J. H., 73, 170, 171, 455, 462
McMahon, A. P., 730
McMahon, C., 243
McMahon, D. B. T., 753
McManus, L. J., 704, 896, 897, 984
McMartin, C., 453, 466
McNamara, A. M., 216, 338
McNamara, C. R., 1092
McNamara, F. N., 1092
McNease, L., 973, 974
McPheeters, M., 704
McQuiston, A. R., 159
McReynolds, P., 234
McShane, R. H., 406, 414
Meachum, C. L., 865
Meats, P., 429
Mech, L. D., 581, 595, 596
Mech, S. G., 583
Medel, M., 1062
Medina, L., 1122
Medina, M., 1142
Medina-Tapia, N., 961, 966
Medway, W., 965, 966
Meeks, J. P., 172, 173, 1118, 1121
Mehansho, H., 633
Meier, J. J., 839
Meijlink, F., 731
Meilgaard, M., 1080
Meilgaard, M. C., 767
Meilhac, S. M., 729
Meininghaus, R., 1100
Meisami, E., 199, 473, 1124
Meisel, R. L., 584
Meiselman, H., 777
Meiselman, H. L., 833
Meister, M., 167
Mejia-Gervacio, S., 191, 193
Mela, D. J., 847
Melanson, S. W., 458, 467
Melcher, J. M., 1013, 1014
Mellert, T. K., 66, 465
Mellier, D., 320
Mellon, D., 935
Mellon, J. D., 591
Melnick, S. L., 894
Melville, G. N., 67
Menco, B. P., 498, 648
Menco, B. P. M., 15, 48, 93, 95, 650
Méndez-Gallardo, V., 319
Mendez-Gomez, H. R., 188
Mendoza, A. S., 98, 1114
Mendoza-Torreblanca, J. G., 184
Menezes-Filho, J. A., 498

Author Index

Menn, B., 193, 197
Mennella, J. A., 308, 310, 313, 314, 319, 323, 325, 326, 344, 345, 795, 796, 797, 798, 799, 800, 801, 802, 803, 804, 807, 809, 810, 812, 816, 819, 820, 821, 822, 823, 833, 834, 835, 836, 868, 923, 1056
Menon, G. K., 566, 567
Menon, J., 566
Menozzi-Smarrito, C., 1104
Menzel, R., 535, 933
Mercer, A. R., 535, 537
Merchenthaler, I., 1123
Mercier, S., 834
Meredith, M., 20, 21, 171, 298, 456, 464, 580, 617, 1113, 1118, 1120, 1123, 1143
Meredith, T. L., 553
Merigo et al. (2009), 631
Merigo, F., 842
Merkel, F., 10, 646
Merkle, F. T., 183, 184, 196
Merson, T. D., 189
Mertl-Millhollen, A. S., 51
Merwin, A. A., 870
Merzel, J., 1133
Mese, H., 626
Mesholam, R. I., 217, 409, 414, 423
Mesulam, M. M., 715, 778, 990, 991, 993, 997
Metcalf, J. E., 18
Metzker, K., 611
Meulstee, J., 415
Meunier, N., 933, 935
Meurman, J. H., 626, 633
Meusel, T., 356, 420
Meyer, D. L., 557, 1142
Meyer, E. A., 210
Meyer, F. F., 344
Meyer, M. R., 115, 116
Meyerhof, W., 687, 689, 690, 691, 695, 696, 914, 915, 987
Meynert, T., 280
Meyrick, B., 17
Meza, C. V., 317
Mezine, I., 1101, 1104
Mezzanotte, W. S., 357
Michaloski, J. S., 126
Michel, W. C., 553
Merigo, F., 631
Mickley, G. A., 879
Micromedex®, 913
Middleton, F. A., 499, 500
Middleton, R. A., 848, 901
Midkiff, E. E., 865
Mierson, S., 916
Mikkila, V., 822
Miklavc, P., 547, 555
Miklosi, A., 593

Mikulka, P. J., 870
Milanese, M., 100
Miles, C., 339
Miles, C. I., 940
Milinski, M., 323
Millar, J. S., 587
Millar, R. I., 968, 969
Millar, S. E., 731
Miller, A., 801, 807
Miller, A. M., 317
Miller, I. J. Jr., 3, 629, 640, 642, 643, 645, 650, 654, 733, 763, 895, 961, 962, 983, 985, 986
Miller, J. L., 796
Miller, L. D., 516
Miller, L. W., 320
Miller, M. B., 519
Miller, S. L., 23
Miller, S. M., 902
Millery, J., 547, 558
Milligan, G., 103
Millius, M., 935
Millqvist, E., 1100
Mills, T. M., 458, 467
Milner, J. S., 965
Milton, K., 611
Min, Y. G., 67, 384
Ming, G. L., 183
Minn, A. L., 68
Minnich, D. E., 20, 21, 929
Minor, A. V., 15
Minovi, A., 141, 384
Miquilena-Colina, M. E., 840
Miragall, F., 1133
Miras, A. D., 847, 888, 1020
Mirlohi, S., 1080
Mirza, R. S., 557
Mirzadeh, Z., 184
Miselis, R. R., 877
Mishkin, M., 339, 995, 996
Mishra, A., 381, 392
Mishra, S. K., 190
Mistretta, C. M., 643, 655, 729, 730, 739, 741, 796, 836
Mitchell, B. K., 932, 936
Mitchell, G., 1015
Mitchell, J. E., 871
Mitra, S. W., 1123
Mitsui, S., 213
Miura, H., 650, 728
Miwa, T., 141, 142, 843
Miyakawa, Y., 934
Miyamichi, K., 51, 126, 128, 213
Miyamoto, T., 413, 419
Miyano, M., 669
Miyaoka, Y., 984
Miyawaki, A., 71, 80
Mizoguchi, C., 777
Mizoguchi, Y., 904

Mizrahi, A., 195
Mizunami, M., 929
Mizuno, K., 316, 317, 321
Mlynski, G., 359
Moalem, G., 1092
Mobbs, P. G., 535
Moberg, P. J., 404, 405, 409, 413, 429, 431, 432, 434, 435, 439
Mobini, S., 1012, 1013, 1014, 1015, 1021
Mobley, A. K., 187, 191
Mochizuki, Y., 642
Moe, K. E., 868
Moeller, J. F., 1143
Moessnang, C., 421
Moffett, A. J., 392
Mohedano-Moriano, A., 174, 1123
Mohr, C., 435
Mojet, J., 75, 343, 1014, 1055, 1056, 1062
Mokdad, A. H., 795
Molina, J. C., 316, 326
Molina, L. T., 490
Moll, B., 383
Moll, R., 644
Møller, P., 239, 343, 344, 346
Moller, R., 388
Molliere, D., 386
Molofsky, A. V., 187, 188, 197
Mombaerts, P., 16, 95, 98, 112, 123, 126, 127, 128, 146, 157, 171, 553, 1115, 1116, 1138
Monath, T. P., 456, 462, 466
Moncomble, A. S., 317
Moncrieff, R. W., 248, 311, 312, 313
Monds, R., 520
Monod, G., 556
Monod, J., 670
Monroe, D., 520
Monroe, S., 703, 706, 707, 708, 709, 711, 718, 779
Montagna, W., 317
Montalti, D., 567, 568
Montavon, P., 650
Monteleone, E., 1052, 1061
Monteleone, P., 838
Montell, C., 929
Montet, A., 1074
Montgomery, E. B., Jr., 416, 423, 429
Montgomery, W. F., 317
Monti-Graziadei, G. A., 64, 76, 93, 96, 134, 135, 139, 141
Monticello, T. M., 72
Montmayeur, J. P., 688
Moon, C., 144
Moon, C. N., 764, 896
Moon, R. T., 731
Moon, S. J., 932
Mooney, K. M., 865
Moor, J. G., 838
Moore, C. A., 961, 963

Moore, C. H., 102, 103, 499
Moore, F. L., 1144
Moore, G. F., 369
Moore, J. D., 322, 343
Moore, M., 836
Moore, P. A., 630
Moore, W. J., 41, 44, 45, 47
Moqrich, A., 1101
Morales, J., 6
Morales, J. A., 455, 462
Moran, D. T., 17, 23, 93, 95, 552
Moran, M., 797, 798, 807, 823, 834, 836, 837
Morecraft, R. J., 715, 990
Moreno, M. M., 166
Moretti, M., 71
Morfit, C., 4
Morgan, B. A., 730
Morgan, C. D., 217, 269, 271, 405, 406
Morgan, G. C., 553
Morgan, J. I., 133, 465
Morgan, K., 917
Morgan, K. T., 66, 67, 72, 358
Morgan, P., 937
Mori, I., 100, 456, 462
Mori, K., 14, 123, 125, 157, 164, 167, 195, 196, 217, 465, 1120, 1123
Mori, S., 286
Moriceau, S., 216, 217
Morikawa, Y., 146
Morin-Audebrand, L., 343
Morita, H., 930
Morita, Y., 552
Morizumi, T., 171
Morley, J. F., 423
Morris, E. T., 6, 7
Morris, J. B., 66, 67, 80
Morris, P., 834
Morris, R., 876, 877
Morris-Wiman, J., 633, 655, 891
Morrison, E. E., 48, 93, 95, 96, 97, 280
Morrison, G. R., 22
Morrot, G., 1010, 1061
Morruzzi, G., 263
Morse, J. R., 996
Mortimer, C. H., 1123
Morton, R., 7
Moscovich, M., 437
Mosel, D. D., 850
Moshkin, M. P., 324
Moskowitz, H., 236
Moskowitz, H. R., 248, 249, 250, 752, 759, 760, 832, 1009, 1014, 1051, 1052, 1053, 1055, 1058
Moss, R. L., 1121
Mossa-Basha, M., 381, 383
Mossman, K. L., 845
Mosso, A., 19
Mostafa, B. E., 368

Motoki, D., 838
Motokizawa, F., 14, 263
Motomura, N., 407
Mott, A. E., 382
Motta, G., 776
Motta, S. C., 1123
Moulton, D., 54, 55, 134, 135, 139
Moulton, D. G., 3, 14, 18, 22, 57, 367, 392, 592
Moulton, R., 871
Mouly, A. M., 214, 218
Mousley, A., 1142, 1144
Moyer, B. R., 567
Mozell, M. M., 14, 23, 246, 337, 365, 366, 367
Mroueh, A., 390
Mu, L., 985
Mucignat-Caretta, C., 1114, 1115
Mudd, S., 458
Mudge, J. M., 1119
Mudo, G., 189
Mueller, A., 283, 423
Mueller, B. K., 98
Mueller, C., 762, 764
Mueller, C. A., 392
Mueller, K. L., 685, 938, 961
Mueser, K. T., 429, 903
Mufson, E. J., 715, 778, 990, 991, 993, 996, 997
Mujica-Parodi, L. R., 298
Mukamel, R., 288
Mukherjee, B. P., 876
Mulder, N. H., 845
Muliol, J., 387
Muller, A., 417, 427, 850
Muller, B., 421
Müller, F., 57, 306
Muller, G. W., 669
Muller, J., 1097
Müller, T., 1100
Müller-Schwarze, C., 325
Müller-Schwarze, D., 325
Mullin, C. A., 934
Mullol, J., 66
Mumm, J. S., 141, 143
Mummalaneni, S., 678
Münch, F., 642
Mundinano, I. C., 412, 414, 423
Muneer, P. M., 499
Mungarndee, S. S., 718
Munger, B. L., 762
Munger, S. D., 116, 117, 118, 1115, 1116, 1117
Munz, H., 1142
Murakami, M., 169, 932
Murakami, S., 895
Murase, S., 187, 188
Murlis, J., 538
Murofushi, T., 415

Murphy, C., 217, 218, 228, 242, 246, 266, 269, 271, 286, 297, 310, 375, 405, 406, 409, 410, 412, 752, 754, 766, 785, 832, 833, 834, 1009, 1010, 1016
Murphy, C. L., 1009, 1093
Murphy, G. J., 160, 161, 162
Murphy, M. A., 369
Murphy, M. R., 21
Murphy, P., 1058
Murphy, S., 1018
Murphy, S. E., 78
Murphy, S. P., 833
Murray, A., 647, 648, 872
Murray, A. B., 872
Murray, E. A., 218, 717, 994, 995
Murray, F. S., 310
Murray, J., 872
Murray, M. J., 872
Murray, R. C., 736, 1114
Murray, R. G., 647, 648, 650
Murray, T. K., 338
Murrell, W., 18, 133, 139, 143, 148, 149
Murthy, S. N., 471
Murthy, V. N., 165, 166
Musialik, J., 889, 901
Muskavitch, M. A. T., 734
Mustonen, S., 1056
Myers, A. L., 236
Myers, C. W., 973
Myers, M. D., 838
Mykytowycz, R., 3, 20
Mystakidou, K., 458

Na, L., 471
Nachman, M., 876, 878
Nachtigal, D., 1009
Nada, O., 648, 649
Naessen, R., 141, 280
Nagahara, Y., 18, 134, 140
Nagai, T., 649, 741
Nagai, Y., 111
Nagao, Y., 838, 840, 848
Nagashima, A., 73
Nagato, T., 644, 741
Nagawa, F., 124, 126
Nagayama, S., 163, 167, 211, 213
Nagel, 1905, 3
Nagel, W. A., 248
Nageris, B., 246, 500
Nagler, R. M., 626
Nai, Q., 170
Naidoo, B., 1057
Naim, M., 914, 916
Nair, R., 849
Nait-Oumesmar, B., 197
Naito, Y., 524
Najac, M., 157, 161, 162, 168
Naka, A., 390, 848
Nakagawa, T., 533, 1116

Nakagawa, Y., 17
Nakajima, M., 69, 899
Nakajima, T., 69
Nakajima, Y., 780
Nakamura, H., 143
Nakamura, K., 630
Nakamura, M., 934
Nakamura, T., 14, 15, 109, 112, 713, 932
Nakamura, Y., 783
Nakano, T., 357
Nakano, Y., 996
Nakaoka, Y., 525
Nakash, J., 598
Nakashima, A., 128
Nakashima, M., 877
Nakashima, T., 141, 280
Nakatani, K., 557
Nakayama, A., 735, 736
Nakazato, M., 754
Nam, S. C., 186
Nan, B., 144
Nanda, R., 633
Nannelli, A., 79
Narabayashi, H., 996
Narayanan, C. H., 728
Narayanan, Y., 728
Narens, L., 760
Narla, V. A., 344
Nash, S., 871
Nasrawi, C. W., 1008
Nasri, N., 1008, 1011
Nassi, J. J., 213
Näsvall, J., 958
Nathan, C. F., 872
National Cancer Institute, 457
Nauges, C., 1059
Nauli, A. M., 840
Navazesh, M., 838
Nawrot, M. P., 541
Nawroth, J. C., 831
Nayak, A., 632
Naylor, G. J., 902
Nazarenko, Y., 486
Ndesendo, V. M., 923
Neal, H. V., 728
Nebert, D. W., 76
Nederkoorn, C., 840
Nee, L. E., 409, 410, 427
Neely, G., 235
Nef, P., 68, 69, 465
Negoias, S., 286
Negrin, N. S., 412
Negus, V., 41, 44, 45, 46, 49, 54, 55, 57, 367
Nei, M., 110, 115, 553, 554, 610
Neiditch, M. B., 518
Nelms, C. O., 966, 967
Nelson, D. R., 68, 69
Nelson, G., 16, 498, 672, 688, 689, 690,
    841, 953, 967, 986, 1102

Nelson, G. M., 649
Nelson, L. M., 765
Nelson, S. P., 894
Nelson, T. M., 688
Nemetschek-Ganssler, H., 648
Nerad, L., 876
Neuhaus, E. M., 103, 118
Neuhaus, W., 592
Neunzig, R., 967
Neuwelt, E. A., 490
Neville, K. R., 169, 214, 1122
Nevison, C. M., 581
Nevitt, G. A., 547, 551, 565, 566, 568
Newcomb, R. D., 1059, 1060
Newland, P. L., 932
Newman, K. S., 343
Newman, M., 18
Newman, M. P., 135, 143, 147
Newman, S., 584
Newman, S. W., 1122
Newton, A., 961
Ng, B. A., 364
Ng, K. L., 188, 191
Ngai, J., 125
Nguyen, A. D., 377, 435
Nguyen, L., 190, 193
Nguyen, M. Q., 127
Nguyen-Ba-Charvet, K. T., 128, 188, 191
Nica, R., 164
Niccoli-Waller, C. A., 406
Nicholson, J. K., 517
Nickel, A. A., 897
Nickell, W. T., 170
Nicolaidis, S., 837
Nicolay, D. J., 96
Nicoleau, C., 189
Nicolelis, M. A. L., 4, 57
Nicoll, R. A., 14
Nicolson, S. W., 964, 965
Nie, X., 732, 734
Nie, Y., 689
Niedermeier, W., 846
Niehus, J. S., 870
Nielsen, G. D., 1097
Nielsen, S. W., 594
Nieto-Sampedro, M., 18, 148
Niewalda, T., 539
Nigrosh, B. J., 344
Niimura, Y., 16, 110, 553, 554, 556, 558,
    610
Nijland, M. J., 796
Niki, M., 666, 695
Nikiforovich, G. V., 113
Nikonov, A. A., 556, 557
Nikula, K. J., 79
Nilsson, G., 839
Ninkovic, J., 195
Ninomiya, K., 1055
Ninomiya, Y., 17, 765

Nir, Y., 462
Nishida, M., 16, 548, 554, 556
Nishida, R., 968
Nishida, Y., 972
Nishihira, S., 355
Nishiike, S., 718
Nishijima, K., 643, 953
Nishijo, H., 707
Nishikawa, T., 143, 147
Nishimura, K., 1119
Nishino, H., 997
Nishino, J., 188, 192
Nishioka, K., 424
Nishitani, S., 320
Nishizumi, H., 126, 127
Nissant, A., 217
Nitschke, J. B., 782, 785, 786, 993
Nitta, E., 780
Nityanandam, A., 189
Niven, J. E., 831
Noble, A. C., 767, 835, 836, 1014, 1055
Noble, J., 586
Nodari, F., 1117, 1118, 1119, 1120
Noden, D. M., 98
Noguchi, M., 525
Noguchi, S., 633
Noll, R. B., 310
Nolte, C., 650, 651
Nolte, D. L., 314
Noma, A., 952
Nomura, S., 647, 989
Nomura, T., 708, 710, 711
Nonaka, N., 470, 472
Nordeng, H., 555
Nordin, S., 217, 228, 237, 243, 247, 266,
    269, 375, 381, 405, 409, 413, 843
Norgren, R., 14, 630, 654, 703, 704, 705,
    706, 707, 708, 709, 710, 711, 762, 765,
    780, 796, 799, 867, 875, 878, 964, 985,
    987, 989, 1028
Norlin, E. M., 126, 1117
Normanno, N., 732
Norrison, W., 22
North, A. C., 900
Northcutt, R. G., 553, 655, 728, 729, 730,
    734, 948, 949, 1142, 1143
Nosrat, C. A., 142, 651, 655, 729, 738, 739,
    741
Nosrat, I. V., 651, 655
Notta, A., 12, 385
Nottebohm, F., 183
Novacek, M. J., 45
Novis, B. H., 838
Novoselov, S. V., 66
Novotny, M. V., 583, 1118, 1119
Nowak, R., 320
Nowell, N. W., 581
Nowlis, G. H., 806, 937
Nuñez, A., 718

Nunez-Parra, A., 71
Nusse, R., 730
Nusser, Z., 169
Nysenbaum, A. N., 835

O'Connell, R., 171
O'Connell, R. J., 456, 464, 1113, 1114, 1118
O'Connor, D. T., 846
O'Connor, E. F., 876
O'Doherty, J., 218, 292, 995, 1037, 1039
O'Doherty, J. P., 776, 783, 785, 786, 1040
O'Dwyer, T. W., 568
O'Keeffe, G. C., 190, 194, 198
O'Leary, D. D., 123, 128
O'Mahony, M., 233, 236, 630, 1052, 1061
O'Neill, M., 717, 994
O'Rahilly, R., 57, 306
O'Rourke, N. A., 19
O'Sullivan, R. L., 391
O'Toole, G., 520
O. I. V., 1059
Oakley, B., 633, 642, 644, 651, 655, 729, 732, 733, 741
Oami, K., 526
Oaten, M., 323, 1013
Obata, H., 649, 694
Oberdorster, G., 365, 486
Ofran, Y., 959
Ogawa, H., 14, 117, 355, 631, 701, 706, 707, 708, 710, 711, 713, 714, 765, 775, 776, 777, 779, 990, 991, 992
Ogawa, K., 951
Ogawa, S., 19
Ogden, C. L., 822
Ogden, G. R., 763
Ogiwara, Y., 17
Ogle, W., 13, 385
Ogura, T., 47, 649, 694
Ohara, I., 838
Ohishi, Y., 15
Ohkubo, K., 66
Ohkuri, T., 694
Ohla, K., 778, 1056, 1057
Ohm, T., 473
Ohmoto, M., 974, 1091
Ohmura, S., 648
Öhrwall, H., 10, 12
Ohshima, H., 644
Ohsu, T., 692
Ohtomi, M., 1142
Ohtsuka, T., 138
Ohtubo, Y., 728
Oikawa, T., 1133
Oike, H., 649, 689, 953
Ojha, P. P., 521
Ojima, K., 644
Oka, H., 420, 423

Oka, Y., 113, 114, 124, 168, 554, 556, 646, 687, 1142, 1143
Okan, F., 805
Okano, M., 17, 95
Okubo, K., 66
Okubo, T., 735
Olender, T., 593
Olesen, J. M., 972
Olichney, J. M., 217, 414
Oliveira-Maia, A. J., 671
Oliver, C., 412
Oliver, K. R., 454, 456, 462
Ollman, B. G., 228, 242
Olmsted, J. M. D., 655, 729
Olofsson, J. K., 269, 295
Olsen, K. H., 547, 557
Olsen, S. R., 534, 535, 538
Olsewski, P. K., 873
Olson, G. M., 320, 321
Olson, J. M., 755
Olson, L., 655, 739
Olsson, M. J., 248, 343
Olsson, P., 384
Olsson, S. B., 323
Olsson, Y., 456, 462, 464, 465
Olszewska, E., 363
Omote, K., 1092
Omur-Ozbek, P., 1080, 1081
Omura, H., 933, 936
Ondo, W. G., 417, 423
Öngür, D., 715, 1030, 1032
Ono, K., 188, 191, 633
Onoda, K., 780, 781, 887, 890, 899
Oomura, Y., 709, 997
Oostindjier, M., 314
Öpalski, A., 19
Ophir, D., 361, 845
Opiekun, R. E., 1100
Oppenheimer, S. M., 993
Ordal, G., 513, 515, 516
Orgeur, P., 307, 319
Ormsby, C. E., 878
Ornitz, D. M., 732
Orona, E., 163, 164, 196
Orr-Urtreger, A., 143
Örs, R., 320
Ortenzi, C., 527
Ortmann, R., 306
Osaki, T., 633
Osculati, F., 647, 948, 1091
Ossebaard, C. A., 678
Ossenkopp, K.-P., 870, 872, 876
Oster, H., 799, 800, 801, 804, 806, 811, 814, 819, 835, 836
Osterbauer, R. A., 297, 900
Osterkamp, T., 598
Ota, M. S., 736
Otaki, J. M., 118
Otsuka, T., 174, 1122

Ottersen, O. P., 877
Otto, D., 1100
Otto, T., 339, 340
Ottoson, D., 3, 14, 548, 557
Ouyang, Q., 932
Overbosch, P., 235
Ovesen, L., 501, 754
Owen, C., 264
Owen, C. M., 1057
Owen, M. D., 916
Owens, J. M., 71
Ozaki, M., 932
Ozçelik, B., 1102
Ozdener, H., 650
Ozeck, M., 693
Ozeki, M., 952
Ozenberger, J. M., 364
Ozlugedik, S., 360
Oztokatli, H., 143, 146

Pabinger, S., 848
Pace, U., 15, 109
Pacifico, R., 115
Packard, A., 133, 137, 138, 139
Pade, J., 384
Padiglia, A., 915
Padoa-Schioppa, C., 717, 995, 1041
Pagano, S. F., 184
Paigen, K., 1119
Paik, S. I., 390
Paiva, K. B., 731
Paivio, A., 346
Pal, S. K., 596
Palagi, E., 611
Palladius, R. T. A., 7
Pallotto, M., 195
Palma, V., 190
Palomero-Gallagher, N., 1030
Palouzier-Paulignan, B., 162, 889
Paluch, R. A., 801
Pandipati, S., 170
Pandya, D. N., 1032
Pangborn, R. M., 3, 230, 672, 752, 832, 834, 835, 837, 838, 843, 1008, 1009, 1061
Panizza, B., 637, 653
Pankevich, D. E., 1118
Pantazopoulos, H., 435
Panzanelli, P., 194, 195, 196
Papachristoforou, A., 1119
Papadopoulou, M., 535, 541
Paparo, B. S., 382
Papes, F., 1117, 1119, 1125
Papin, L., 44, 51, 54
Papino, M. Y., 868
Papp, M. I., 427
Pappas, G. D., 435
Paquette, C., 525
Paradis, S., 974
Paran, N., 647

Author Index

Parat, M., 117
Paratcha, G., 189
Pardelli, C., 308
Pardini, M., 438
Pardo, J. V., 290, 291, 292, 776, 1039
Pardo-Bellver, C., 1123
Parent, C. A., 520, 521
Parent, J. M., 197, 198
Parfet, K. A. R., 319
Parise, C., 1016
Parisella, S., 934
Park, D., 1144
Park, S., 432, 435
Park, S. H., 1124
Parker, 1872, 42
Parker, G. H., 3, 654, 655, 1091
Parker, K. J., 463
Parker, L. A., 867, 875, 876
Parker, M. A., 729, 730
Parmentier, M., 16, 117
Parr, J., 1080
Parr, L. W., 755
Parr, W. V, 1052, 1058, 1061
Parra, K. V., 547
Parrao, T., 422, 423
Parrish, T. B., 787
Parrish-Aungst, S., 163, 167, 196
Parrott, D. V. M., 21
Parry, C. M., 976
Parsons, T. S., 41
Partridge, B. L., 548
Pasquet, P., 832
Pass, D. A., 566
Passamonti, L., 785
Passingham, R. E. P., 1030
Passy, J., 12
Pasternak, O., 286
Patapoutian, A., 1102
Patel, A. V., 738, 739, 741
Patel, S., 458
Patel, S. J., 499
Patel, U., 135, 143
Pathania, M., 195
Patierno, S. R., 498
Paton, J. A., 183
Patrick, M. E., 836
Patris, S. M., 309, 310
Patte, R., 248
Patterso, D. S., 390
Patterson, J., 264, 1057
Patterson, M. Q., 249
Patton, H. D., 776
Paudel, K. S., 470
Paula, J., 599
Pauling, L., 19
Paulli, S., 44, 50, 57
Paulraj, S., 596
Paunescu, T. G., 71
Pause, B. M., 237, 242, 265, 266, 269, 271

Pavlidis, P., 754
Pavlov, I. P., 20, 879
Paysan, J., 552
Payton, C. A., 211, 213
Peabody, C. A., 404
Pearsall, M. D., 598
Pearson, A. A., 98
Pearson, P., 797, 798, 821, 834
Pech, L. L., 525
Pedersen, A. M., 837
Pedersen, P. A., 315
Pedersen, P. E., 585, 1134
Pedziwiatr, 1972, 44
Peet, M. M., 17, 461
Peier, A. M., 369
Pelchat, M., 875
Pelchat, M. L., 836, 1057
Pellerin, C., 497
Pelosi, P., 66
Peluso, C. E., 144, 145, 146
Pelz, C., 344
Pelz, D., 538
Pembroke, A., 597, 599
Penalver, A., 572
Pencea, V., 185
Penfield, W., 14, 776, 899, 901
Peng, H. M., 68
Peng, J., 1102
Penn, D., 325
Pennartz, C. M. A., 713
Pentzek, M., 407, 409, 411
Penzoldt, F., 12
Pepino, M. Y., 798, 801, 802, 807, 810, 823, 833, 836
Peppel, P., 487
Peretto, P., 183, 185, 186
Perez, C. A., 649
Perez-Lopez, M., 556
Pérez-Mellado, V., 972, 973
Perez-Orive, J., 535, 540, 541
Perez-Polo, J. R., 146
Perkins, A. J., 961
Perkins, O. M., 317
Perl and Doty, 428
Perl, E., 405
Perl, P. T., 455, 459
Perlman, S., 455, 462
Perras, B., 458, 468
Perros, P., 847, 901
Perroteau, I., 142
Perry, G., 148
Perry, G. C., 595
Perry, G. H., 633
Perry, S. F., 953
Persaud, K. C., 1057
Persson, E., 76, 455, 497
Persuad, K., 58
Peryam, D. R., 758, 1052, 1073
Peryman, D. R., 227

Pesonen, M., 591
Pessoa, L., 1122
Peterlin, Z., 114, 1092
Peters, J. M., 406
Peters, R. P., 581, 595, 596
Petersen, C. C., 128
Petersen, C. I., 732, 733, 734
Peterson, F., 307, 312, 806, 811, 814, 819
Peto, E., 312, 323
Petreanu, L., 195
Petrides, M., 776, 1032
Petrovich, G. D., 1123
Petrulis, A., 337, 338, 339, 340, 343, 346
Pettigrew, R., 593.
Petzold, G. C., 171
Pevsner, J., 15, 103, 465
Pevzner, R. A., 647
Peyrot des Gachons, C., 968, 1101
Pfaar, O., 364, 384
Pfaff, D. W., 3, 98, 1141
Pfaffmann, C., 11, 14, 22, 625, 629, 675, 705, 753, 764, 765, 985, 1101
Pfeiffer, W., 551
Pfieffer, J. C., 1016
Pfiester, M., 599
Pfister, P., 554
Pfurtscheller, G., 262, 272
Phamdelegue, M. H., 539
Phillips, M. E., 164, 166
Phillips, M. L., 993, 1031
Phillips, P. A., 876
Phillips, W., 198
Phillips, W. S., 916
Philpot, B. D., 164
Philpott, C. M., 228
Physicians Desk Reference (PDR), 911, 913
Pickering, G. J., 1058, 1059
Picozzi, N., 961
Pierce, J., 1017
Pierce, J. D., Jr., 230
Piersma, T., 567
Pierson, M. E., 458, 469
Piesse, G. W. H., 3, 4, 5, 6
Pietrowsky, R., 458, 468
Pifferi, S., 1118
Piggott, J. R., 1057
Pignatelli, A., 159, 170
Pihet, S., 307
Pihlström, H., 57, 608
Pimentel, D. O., 161, 169
Pinching, A. J., 16, 157, 158, 250, 391
Pinkus, F., 1141
Pino, D., 195
Pinto, S., 76
Piqueras-Fiszman, B., 1061
Piras, E., 69, 80
Pírez, N., 162
Pirogovsky, E., 413
Pispa, J., 729

Pitcher, B. J., 591
Pitteloud, N., 390
Pittman, D. W., 835
Pittman, J. A., 389, 846, 901
Pivk, U., 627
Piwnica-Worms, K., 1014
Pixley, S. K., 18, 142, 145, 146
Plachez, C., 193
Plailly, J., 218, 219, 295, 296, 429, 436
Plassmann, H., 776
Plata-Salamán, C. R., 712, 714, 716, 775, 776, 779, 784, 991, 992, 993, 994, 996, 1031
Platel, J. C., 186, 190, 194
Plato, C. C., 428
Platt, R., 100
Plattig, K. H., 14, 228, 265, 899
Pless, D., 360
Pliner, P., 344, 1015
Plopper, C. G., 457
Plutchik, R., 1075
Pluznick, J. L., 118
Pobozy, E., 631
Poellinger, A., 289, 1039
Poffenberger, A. T., 766
Polack, I., 1008
Polak, E. H., 465, 1054
Poldrack, R. A., 281
Polich, J., 269
Polito, M. J., 961
Pollard, S., 1082
Pollock, G. S., 143, 147, 937
Pollock, M. S., 557
Polonskaia, E. L., 499
Poncelet, J., 327
Ponsen, M. M., 423
Ponzo, M., 643
Poo, C., 209, 214
Poon, H. F., 71
Popper, R., 796, 1056
Popping, A., 874
Porter, J., 210
Porter, R. H., 51, 54, 55, 246, 308, 316, 317, 320, 321, 322, 343
Porter, S. L., 513, 516
Porter, S. R., 1019
Porter, T. D., 68
Posner, M. I., 343
Postolache, T. T., 390, 438
Postuma, R. B., 413, 419
Potten, C. S., 136
Potter, H., 233, 236, 287
Potter, L. R., 116
Potter, S. M., 157
Potts, B. C., 346, 1061
Potts, W., 325
Poulin, B., 961
Poulton, E. B., 867
Powell, G. F., 392

Powell, T. P. S., 157, 158, 164, 165, 217, 250, 1122
Power, M. L., 837
Powers, J. B., 21, 47, 58
Powers, K. M., 460
Powley, M. W., 69
Powley, T. L., 837
Prada and Nevitt, 568
Prada, P. A., 568, 571, 572
Prah, J. D., 228, 265
Prakash, I., 668, 669
Pramanik, R., 627, 628
Prechtl, (1958), 316
Prediger, R. D., 427, 456, 468, 474
Prediger, R. D. S., 486, 487, 491, 496, 497
Prehn-Kristensen, A., 298
Preissner, S., 917, 921
Prescott, J., 310, 346, 756, 761, 781, 783, 834, 1008, 1009, 1011, 1012, 1013, 1015, 1018, 1019, 1021, 1022, 1039, 1052, 1053, 1054, 1055, 1058, 1059, 1060, 1062, 1101, 1103, 1104
Pressler, R. T., 165, 166, 170
Preston, C. C., 235
Preston, R., 522, 524
Preston, R. R., 525
Preti, G., 298
Preuss, T. M., 1030
Preussman, R., 78
Preyer, W., 308, 322
Pribram, K. H., 776
Price, J. L., 101, 164, 165, 217, 218, 295, 715, 716, 990, 993, 994, 996, 1030, 1032, 1033, 1034, 1122
Price, S., 73, 499
Price-Rees, S. J., 973, 974
Prichard, T. C., 654, 655
Prince, J. E. A., 1120
Prince, P. A., 568, 961
Principato, J. J., 364
Prior, S., 965
Pritchard, J. A., 796
Pritchard, T. C., 642, 654, 706, 707, 710, 711, 712, 714, 715, 716, 717, 718, 762, 775, 776, 778, 779, 782, 783, 875, 899, 988, 989, 990, 991, 993, 994, 995, 1028, 1032
Privitera, G. J., 1015
Proctor, D. F., 67, 358
Proctor, G. B., 627, 632, 633
Proetz, A. W., 13, 358
Pronin, A. N., 671, 690, 691, 692
Propper, C. R., 1144
Prosser, H. M., 188, 191
Proust, M., 337
Prout, W., 1810–1007
Prutkin, J. M., 757, 763
Pryse-Phillips, W., 390
Przekop, P. R., 742

Ptak, J. E., 22
PubChem, 917, 921, 922
Puche, A. C., 193
Puig, A., 391
Pullen, E. W., 764, 896
Pum, M. E., 464
Pumplin, D. W., 647, 649, 650
Punter, P. H., 228, 240
Puopolo, M., 159
Purdon, S. E., 431
Putcha, L., 459, 469
Puverel, S., 188, 191
Puzdrowski, R. L., 1142
Pydi, S. P., 692

Qato, D. M., 914, 917
Qiang, F., 470
Qin, F., 1104
Qu, Q., 189
Quadagno, D. M., 325
Quagliato, L. B., 418
Quarcoo, D., 71
Quay, W. B., 567
Quignon, P., 50, 593
Quina, L. A., 736, 737, 738
Quinn, N. P., 415, 423
Quinn, T. P., 555
Quinn, W. G., 939, 940
Quinones-Hinojosa, A., 183, 184, 197
Quint, C., 392
Quinton, R., 389, 390

Rabin, B. M., 870, 874, 878
Rabin, M. D., 246, 249, 297, 308, 340, 346
Rabin, R. D., 238, 244
Rabl, H., 643
Racette, B. A., 497, 498
Rader, D. J., 517
Rae, T. C., 41, 45
Raff, A. C., 501, 502, 848
Rafols, J. A., 96
Rafter, J. J., 73
Raichle, M. E., 20, 993
Raimbault, C., 317
Rainey, L. H., 806, 811, 814, 819
Raisman, G., 174, 1114
Rajan, R., 210
Rajchard, J., 566
Rak, K., 517
Rake, G., 456
Rakover, S. S., 251
Raliou, M., 757
Rall, W., 14, 165
Ramakekers, F., 874
Ramanathan, R., 459, 469
Ramenghi, L. A., 800, 806
Raming, K., 16
Ramirez, I., 838
Ramjit, A. L., 420

Ramoino, P., 525
Ramón y Cajal, S., 93, 171, 172
Ramon-Cueto, A., 18, 148
Rampe, D., 921
Rampin, O., 211
Ramus, S. J., 340
Randall, H. W., 64
Randall, J. E., 357
Randol, M., 273
Ranganath, C., 340
Rankin, K. M., 757
Ranson, S. W., 997
Rantala, M. J., 323
Rantonen, P. J. F., 626, 633
Ranvier, L., 653
Rao, R. P., 295
Rao, S., 513, 515, 516
Rapps, N., 390
Rasch, B., 297
Rasche, S., 1118
Rashash, D. M. C., 1081
Rasmussen, K., 343
Rasmussen, L. E. L., 324
Rasmussen-Conrad, E. L., 847
Rath, L., 537
Rathi, A., 521
Ratnasuriya, R. H., 871
Rattaz, C., 320
Rauch, S. L., 993
Raudenbush, B., 364
Rausch, R., 287
Ravasco, P., 845
Rawlins, J. N. P., 339
Rawls, A., 318
Rawson, N. E., 76, 94, 143, 145, 146, 217,
    251, 390
Ray, J. P., 295
Rayney, L. H., 307, 312
Raynor, H. A., 833
Razafimanalina, R., 835, 836
Read, E. A., 47, 56
Reale, E., 17
Reber, M., 128
Rebert, C. S., 1091
Reckmeyer, N. M., 835
Reddy, V. G., 458
Reden, J., 386, 392
Redican, W. K., 322
Redl, B., 631
Reed, C. J., 68, 71, 457
Reed, D., 428
Reed, D. M., 428
Reed, D. R., 753, 756, 757, 914, 960, 965,
    1059
Reed, M. K., 262
Reed, P., 339
Reed, R. R., 15, 127, 465, 1115
Reedy, F. E., 763
Reedy, F. E. J., 640, 642, 643

Reep, R. L., 715
Reese, E. A., 556
Reeve, J., 1061
Reger, M. A., 463
Rehn, B., 135
Rehn, T., 370
Rehnberg, B. G., 557, 630, 631
Rehorek, S. J., 58
Reichert, F. L., 762
Reid, L., 17
Reiger, I., 597
Reilly, S., 718, 876, 877, 878, 989
Reimann, F., 887
Reisberg, D., 1061
Reisert, J., 112
Reiss, C. S., 487
Reiter, E. R., 386, 1056
Reitz, M., 469
Reivich, M., 19
Relman, D. A., 517
Remak, R., 654
Remick, A. K., 833
Rempel-Clower, N. L., 717, 994
Remy, N., 5
Ren, X., 549, 552, 968
Ren, Y., 649
Renaud, S., 1060
Renault, V. M., 188
Reneerkens, J., 567, 569, 570, 572, 573,
    574, 575
Renehan, W. E., 705
Rennaker, R. L., 210, 214
Renner, B., 242
Renner, M., 21
Renqvist, Y., 15
Rensch, B., 967
Rentmeister-Bryant, H., 1092, 1101, 1103
Renwick, J. A. A., 940
Repacholi, B., 323, 324
Ressler, K. J., 16, 124, 126, 128, 157, 367,
    465
Restrepo, D., 115, 217, 344
Resurreccion, A. V., 803
Retzius, G., 638, 640, 646
Reutter, K., 645, 646, 647, 649, 650, 651,
    653, 728, 796, 948, 949
Revial, M. F., 172
Revington, M., 66
Revusky, S., 874
Reyher, C. K., 164, 165
Rezek, D. L., 404
Reznik, G., 64
Rhee, C. S., 363
Rhein, L. D., 15
Rhodin, J., 17
Rhoton, A. L. Jr., 706, 987
Rhyu et al., 674
Ribak, C. E., 164
Ricard, J., 190

Ricci, N. R., 527
Rice, J. C., 764
Rich, T. J., 583
Richard, J., 404
Richards, K. M., 599
Richardson, B. E., 847
Richardson, C. T., 837, 838
Richardson, P. R. K., 591
Richman, R. A., 309, 310
Richter, C. P., 11, 20, 22
Richter, J. E., 894
Richter, T. A., 675
Rico, A. G., 871
Riddiford, L. M., 73
Riddle, M. S., 459
Ridout, J. B., 237
Riera, C. E., 915, 1104
Riffell, J. A., 538, 540
Righard, L., 317
Rigter, H., 874
Riley, A., 20
Riley, E. P., 499
Rinberg, D., 164, 167, 168
Rinck, F., 310
Ringstedt, T., 739
Ripamonti, C., 845
Riquelme, P. A., 186
Risberg, J., 19
Risser, J. M., 162
Riti, L., 9
Ritter, S., 876
Riva, A., 644
Rivera, H. M., 870
Rivest, S., 490
Rivière, S., 171, 554, 1116, 1117, 1118,
    1119
Rivlin, R. S., 846
Roalf, D. R., 432, 434
Robbins, D. C., 839
Roberston, H. M., 932
Robert-Koch-Institut, 843
Roberts, A. K., 1014
Roberts, D. W., 1097
Roberts, E., 457, 459, 473
Roberts, E. S., 71, 75, 76
Roberts, I. M., 732
Roberts, J. D., 655, 731
Roberts, R., 171
Roberts, S. A., 1119
Roberts, S. C., 325, 581, 582, 586, 587
Robertson, D. H., 1119
Robertson, G. T., 520
Robertson, H. M., 932
Robertson, M. D., 839
Robin, S., 593
Robinson, A. M., 66
Robinson, P. P., 642, 643, 763, 897, 904, 905
Robinson, S. R., 307, 314, 319
Robinson, T. E., 785

Roblegg, E., 490
Robottom-Ferreira, A. B., 79
Robson, A. K., 242
Roche, A., 916
Rochefort, C., 19, 166, 195, 339
Rochlin, M. W., 739
Rodgers, L. F., 525, 526
Rodin, J., 839, 847, 903
Rodolfo-Masera, T., 1133, 1136
Rodriguez, I., 172, 554, 1113, 1115, 1120
Rodriguez-Echandia, E. L., 633, 838
Rodriguez-Perez, I., 891
Rodríguez-Robles, J. A., 970
Rodriguez-Violante, M., 421
Roelink, H., 730
Roelofs, R. F., 187, 191
Roesch, M. R., 344
Roessner, V., 390
Rogers, J., 838
Rogers, L. J., 568, 969
Rogers, M. E., 70
Rogers, P. J., 841
Rogers, Q. R., 834
Rogers, R. C., 877
Rogers, S. M., 932
Roisen, F. J., 148
Roithmann, R., 364
Rojo, A. I., 468
Roland, W. S. U., 671
Roldan, G., 876, 877
Rolen, S. H., 547, 553, 555
Rolheiser, T. M., 421
Rolland, R. M., 599
Rolls, B. J., 344, 714, 758, 786, 832, 833,
  988, 997, 1034
Rolls, E. T., 218, 290, 292, 338, 344, 345,
  346, 609, 714, 715, 716, 717, 775, 782,
  784, 786, 832, 960, 991, 992, 993, 994,
  995, 996, 1020, 1027, 1028, 1029,
  1031, 1032, 1033, 1034, 1035, 1036,
  1037, 1038, 1039, 1040, 1041, 1042,
  1043, 1057
Rolls, J. H., 1020
Roman, C., 877
Roman, F., 214
Romanov, R. A., 694
Romanovitch, M., 525
Romantshik, O., 316
Rombaux, P., 270, 271, 283, 284, 377, 386,
  423
Ronca, A. E., 307, 315, 316
Ronchini, C., 193
Ronnett, G. V., 148
Rooney, N. J., 597
Röösli, M., 500
Root, C. M., 534, 538
Roper, S. D., 15, 17, 629, 631, 645, 646,
  647, 649, 650, 694, 728, 901, 936, 950
Roper, S. N., 993

Roper, T. J., 566, 568, 964, 968, 969
Roppolo, D., 127, 1115, 1136, 1137, 1138,
  1140
Ros, C., 388
Rosario, V., 3
Rose, C. S., 497
Rose, S. P. R., 969
Rosen, A. M., 702, 703, 706
Rosen, G., 1142
Rosen, J. B., 345, 587
Rosen, J. S., 1087
Rosenblatt, J. S., 305, 320, 321, 322, 585
Rosenblum, P. M., 74
Rosenstein, D., 799, 801, 804, 806, 811,
  814, 819, 835, 836
Roskams, A. J., 97, 133, 141, 142, 144, 145,
  146, 147
Rospars, J. P., 535
Ross, B. M., 246, 337, 340, 346
Ross, C. F., 54
Ross, G. R. T., 10
Ross, G. W., 414, 417, 418
Ross, M. G., 796
Ross, R., 7
Ross, W., 423
Rossetti, D., 632
Rossi, A., 1142, 1143
Rossie, J., 605, 606
Rossie, J. B., 41, 42, 44, 45, 46, 47, 50, 54,
  55, 56, 58
Rossie, K. M., 891
Rossler, P., 693
Rotch, T. M., 386
Roth, J., 423
Rothacker, D., 832
Rothkrug, L., 5
Rothman, A., 126
Rothman, R. J., 596
Rothova, M., 734
Rouadi, P., 355
Rouby, C., 310
Roudnitzky, N., 690, 691, 1059
Rouquier, S., 608, 610
Roura, E., 967
Rouseff, R. L., 933
Rousseaux, M., 438
Roussin, A. T., 702, 703
Rovee, C. K., 234
Rovee-Collier, C. K., 308, 310
Rowe, C., 963, 964, 968, 969
Rowe, D. J., 1052, 1053, 1054
Rowe, F. A., 585
Rowe, T., 45, 47, 50, 57, 58
Rowland, H. M., 968
Roy, C. S., 19
Roy, G., 670
Royer, S. M., 647, 648, 650
Royet, J. P., 218, 290, 292, 297, 393, 406,
  409

Rozánska-Kudelska, M., 362
Rozenfeld, F. E., 583
Rozengurt, E., 841
Rozengurt, N., 841
Rozhnov, V. V., 591
Rozhnov, Y. V., 591
Rozin, P., 323, 762, 766, 867, 868, 869, 873,
  1007, 1015, 1017, 1102
Ruan, Y., 458
Rub, U., 436
Rubin, B. D., 213
Rubin, G. B., 309
Ruch, T. C., 776
Rückerl, R., 490
Rudebeck, P. H., 218, 994
Rudenga, K., 785
Rudenga, K. J., 783, 785, 787
Rudmik, L., 384
Ruf, I., 55, 56
Ruff, J. S., 325
Ruff, R. L., 596
Ruijschop, R. M., 847
Ruis, H., 459, 469
Ruitenberg, M. J., 142, 146
Ruiz, C., 687
Ruiz-Avila, L., 649, 650
Ruiz-Martinez, J., 421
Rumble, R. H., 917
Rumsey, T. S., 871
Runge, E. M., 739
Ruocco, L. A., 463
Rupp, C. I., 217, 286, 429, 432, 836
Ruppert, B., 42
Rush, R. A., 141, 142
Rushworth, M. F., 1033
Russchen, F. T., 295
Russell, J. A., 1075
Russell, M. J., 317, 320, 321, 465
Russo, A. F., 3
Rutishauser, U., 188
Ruusila, V., 591
Ryan, R. P., 516, 517
Ryba, N., 171
Ryba, N. J. P., 554, 1115
Rydberg, E., 19
Rydzewski, B., 382
Rytzner, C., 15, 647
Ryzhikov, A. B., 456
Rzoska, J., 867

Saad, M. J. A., 839
Saalmann, Y. B., 710, 718
Saar, D., 214, 344
Saarikoski, S. T., 69
Saavedra, J. M., 556
Sachs, B. D., 1123
Sachs, C. J., 458, 467
Sachse, S., 535, 537, 538, 539, 541
Sadacca, B. F., 996

Author Index

Sadacca, B. P., 713
Sadathosseini, A. S., 320
Saghatelyan, A., 188, 191, 195
Saglio, P., 557
Sahakian, B. J., 838
Sahay, A., 215
Sahl, H. G., 100
Sahler, C. S., 468
Saimi, Y., 524
Saino-Saito, S., 196
Sainz, E., 688
Saito, H., 111, 114
Saito, I., 647, 962, 963
Sajjadian, A., 412
Sajwan, M. S., 555
Sakai, N., 876, 877, 1016
Sakane, T., 458, 467, 468
Sakano, H., 123, 125, 128
Sakmann, B., 166
Sakuma, K., 268
Sakuma, S., 471
Sakurai, S., 868
Sakurai, T., 692
Salamone, J. D., 785
Salata, J. A., 754
Salazar, H., 1092
Salazar, I., 592, 1113, 1120, 1121
Salazar, J., 460
Salbe, A. D., 847
Saldamando, L. de, 1073
Saleh, T. M., 869
Salehi-Ashtiani, K., 141, 142
Salemme, F. R., 667
Salerno-Kennedy, R., 409
Salibian, A., 567, 568
Salin, P. A., 161
Salinukul, N., 566
Salisbury, P. L., 892
Salkina, G. P., 599
Sallabanks, R., 966
Sallaz, M., 162
Salloum, A., 937
Salvinelli, F., 501
Salzman, C. D., 1122
Sam, M., 1118
Samanen, D. W., 139
Sammeta, N., 80
Samoylov, A. M., 101
Sampson, P. D., 499
Sams, J. M., 462
San Gabriel, A., 692
San Miguel, R., 673
Sanai, N., 183, 184, 186, 197
Sànchez, C. L., 320
Sanchez-Juan, P., 780
Sánchez-Quinteiro, P., 1113
Sandell, M., 914, 915
Sanders, I., 985
Sanders, K. M., 1015

Sandford, A. A., 309
Sandilands, V., 567, 572
Sandow, P. L., 845
Sanematsu, K., 832
Sanghera, M. K., 996, 1036
Sanico, A., 488
Santin, R., 420
Santorelli, G., 437
Santos, D. V., 375
Sanz Ortiz, J., 845
Saoud, M., 431
Saper, C. B., 473, 708, 711
Sapolsky, R. M., 873
Saporito, R. A., 973
Sarin, S., 488
Sarinopoulos, I., 786
Sarkar, M. A., 466
Sarles, H., 838
Sarnat, H. B., 307
Sartor, F., 834
Sas, R., 627, 628
Sasagawa, T., 71
Sasaki, K., 933
Sasner, J., 522
Sassoe-Pognetto, M., 157
Sato, H., 71
Sato, K., 103, 533, 780, 932
Sato, M., 985
Sato, T., 950
Sato, Y., 552, 553, 556
Satoda, T., 706, 987
Satoh, M., 136, 143, 144, 553
Satya-Murti, S., 436
Saucier, D., 556, 1134
Sauer, K., 520
Saunders-Pullman, R., 421, 426, 427
Sauvage, M. M., 341, 342
Saveliev, N. A., 14
Savic, I., 292, 297, 298, 1056
Savidge, J. A., 599
Sawada, M., 167, 196
Sawamoto, K., 184, 185
Sawaya, W. N., 567, 572
Sawchenko, P. E., 873
Sbarbati, A., 644, 647, 733, 948, 984, 1091
Scadding, G. K., 243, 362, 384
Scalera, G., 501, 872
Scalia, F., 1114, 1120
Scalzi, H. A., 648
Scarborough, P. E., 75
Scarpa, A., 10
Schaal, B., 51, 54, 55, 305, 306, 307, 309,
    310, 312, 313, 314, 315, 316, 317, 318,
    319, 320, 321, 322, 323, 344, 345, 834
Schaap, P., 522
Schab, F. R., 339
Schachter, M., 869
Schaefer, A. T., 161, 164, 167, 168, 1136
Schaefer, H. M., 966

Schaefer, M. L., 211
Schaffer, J. P., 306
Schaller, G. E., 514
Schank, J. C., 22
Schatz, R. A., 80
Schaumburg, H., 849
Schaumlöffel, D., 498
Schechter, P. J., 386
Schecklmann, M., 311
Scheibe, M., 356
Scheiner, R., 939, 940
Schellinger, D., 283
Schenk, F., 345
Scherer, P. W., 358
Scherfler, C., 286
Scherr, S., 832
Schieferstein, G. J., 457
Schiestl, F. P., 539
Schiet, F. T., 248, 250
Schiff, J. M., 11
Schifferstein, H. N. J., 672, 1011
Schiffman (1997), 905
Schiffman, S. S., 23, 237, 289, 405, 409,
    464, 501, 631, 667, 686, 753, 784, 835,
    846, 847, 848, 849, 902, 913, 914, 915,
    916, 917, 919, 921, 923, 1019, 1056,
    1100
Schiller, D., 868, 1102
Schiller, F., 4, 12
Schiller, L., 1123
Schilling, A., 617
Schinkele, O., 643
Schioth, H. B., 975
Schirmer, T., 516
Schlegel, M., 1011
Schleidt, M., 308, 321, 324, 343
Schliebs, R., 411, 412
Schlosser, G., 736
Schmachtenberg, O., 554
Schmale, H., 15, 631, 644, 651
Schmid, A., 1139
Schmidt, C. E., 1095
Schmidt, F. A., 377, 423
Schmidt, H. J., 305, 308, 313, 802
Schmidt, J. M., 929
Schmidt, J. O., 968
Schmidt, M., 972
Schmidt, R., 228
Schnabolk, G. W., 466
Schneider, C. V., 11
Schneider, D., 14, 21
Schneider, F., 436
Schneider, G. E., 21
Schneider, J. F., 307
Schneider, J. S., 498
Schneider, L., 527
Schneider, M. A., 324
Schneider, N. G., 459
Schneider, R. A., 1095

Schneider, S. P., 163
Schneider, W., 295
Schneyer, C. A., 629
Schnuch, M., 933
Schoenbaum, G., 218, 219, 344, 347, 715, 1029
Schoenfeld, M. A., 783, 784
Schoenfeld, T. A., 56, 163, 368
Schofield, P. W., 408
Scholz, N. L., 557
Schondube, J. E., 961, 965, 967
Schooler, J. W., 337, 1013, 1014
Schoon, A., 598
Schoon, G. A. A., 598
Schoonhoven, L. M., 929, 933, 934, 935, 938, 939
Schoppa, N. E., 161, 162, 165, 166, 168, 169, 1122
Schoppe, S., 435
Schorderet, M., 932
Schou, M., 870, 875
Schroeder, B. C., 1118
Schroeder, B. O., 100
Schroeder, H. E., 984
Schroers, M., 309
Schroeter, J. D., 486, 497
Schröter, U., 933
Schroy, P. L., 997
Schubert, C. R., 394, 409, 843
Schuett, V. E., 800
Schul, R., 218
Schuler, W., 969
Schulkin, J., 837
Schultz, E., 134
Schultz, E. W., 17, 18, 73, 461
Schultz, J. C., 935
Schultz, J. E., 525
Schultz, W., 717, 994, 995, 1035
Schultze, M., 11
Schulz, M., 917, 921
Schulz, S., 1119
Schulze, F. E., 637
Schussler, P., 842
Schutz, H. G., 759, 760, 761
Schwab, A., 527
Schwabe, (1867), 10
Schwalbe, G., 9
Schwalbe, (1868), 638, 640, 647
Schwann, T., 638
Schwanzel-Fukuda, M., 98, 390, 1141
Schwarting, G. A., 18, 136, 143, 147
Schwartz Levey, M., 136
Schwartz, B. S., 240, 246, 385
Schwartz, C., 811, 834, 836
Schwartz, D. N., 370
Schwartz, G. E., 263
Schwartz, H. G., 654
Schwartz, S. R., 763
Schwartz, T. W., 667, 839

Schwartzbaum, J. S., 987, 989, 996
Schwarzbauer, C., 288
Schwende, F. J., 1119
Schwenk, K., 971, 972
Schwob, J. E., 80, 96, 133, 135, 136, 140, 217, 390
Scinska, A., 835, 836
Sclafani, A., 632, 695, 795, 833, 835, 840, 841, 868, 961
Scoggin, J. A., 458, 467
Scott, A. E., 381, 392
Scott, J. P., 595, 597
Scott, J. W., 14, 58, 163, 164, 165, 168, 367
Scott, K., 849, 937
Scott, T. P., 973, 974
Scott, T. R., 703, 707, 708, 710, 711, 712, 714, 716, 717, 718, 775, 776, 779, 782, 784, 832, 878, 987, 988, 989, 991, 992, 993, 994, 995, 996, 997, 1028, 1031, 1036, 1037, 1053
Scott-Johnson, P., 58
Scranton, R. A., 487
Scrivani, S. J., 897, 898
Seal, E. C. J., 263
Seamon, J. G., 337
Seaworld, 591
Secchi, A., 839
Seddon, H. J., 896
Sedig, L., 421
Segal, M., 194
Segato, F. N., 868
Segovia, C., 643, 804
Segovia, S., 76, 1123
Seibt, A., 616
Seid, D. A., 1101
Seiden, A. M., 361, 362, 384, 386
Seidman, L. J., 414, 430, 431
Seidman, M. D., 500
Seifert, R., 1140
Seju, U., 458, 471
Seki, M., 456, 462
Seki, T., 458, 466, 467, 471
Sekler, I., 466
Sela, L., 296
Self, P. A., 308
Selliseth, N. J., 644
Selset, R., 552
Selvig, K. A., 644
Semb, G., 228, 232
Semke, E., 315
Seno, K., 932
Seo, H.-S., 1057
Sepehrnia, A., 384
Serby, M., 404, 405, 409, 410, 415, 429, 430
Sergovia, C., 984
Series, F., 355
Serizawa, S., 126, 127, 128, 129, 556, 1115
Servant, G., 667, 689, 695
Sessle, B. J., 766, 904, 1102

Seta, K., 127
Seta, Y., 650, 735, 736
Setchell, J. M., 611
Sethupathy, P., 168
Settle, R. G., 836, 847, 901
Sevelinges, Y., 216, 344
Sevy, S., 435
Sewell, L., 429
Sewertzoff, (1902), 1141
Shabadash, S. A., 594
Shafer, D. M., 897
Shaffer, S. E., 833
Shah, A., 325
Shah, A. S., 380
Shah, M., 413, 419, 423
Shah, N. M., 1123
Shahbake, M., 643, 763
Shallenberger, R. S., 670, 688
Shamchuk, A. L., 557
Shamushaki, V. A. J., 555
Shanbhag, B. A., 973, 974
Shanbhag, S. R., 937
Shang, Y., 534, 535
Shanks, D. R., 1012
Shannon, I. L., 626
Shao, Z., 159, 160, 161, 162, 168
Shapera, M. R., 848
Shapiro, R. E., 877
Sharav, Y., 898
Sharer, J. D., 423
Shargorodsky, J., 497
Sharma, J. C., 423
Sharma, T., 429
Sharp, F. R., 19, 20
Sharpe, P. T., 730
Sharpe, S. T., 587
Sharrow, S. D., 582, 1119
Shatzman, A. R., 651
Shaw, A., 11
Shaw, C. A., 410
Shaw, C. L., 567, 572, 573
Shaw, P. H., 1119
Shawkey, M. D., 567
Shea, S. D., 170
Shear, P. K., 438
Shehin-Johnson, S. E., 68
Sheikh, B. N., 189
Shen, Q., 184, 185, 187, 191, 193
Shen, T., 649, 650, 694
Sheng, J., 67, 68
Shenoy, S. K., 127
Shephard, B. C., 405, 416
Shepherd, G., 123
Shepherd, G. M., 14, 101, 102, 103, 157, 160, 164, 191, 210, 250, 368, 456, 463, 466, 1020
Shepherd, R., 848
Sherkheli, M. A., 1101
Sherman, A. H., 392

Author Index

Sherman, P. W., 7, 961
Sherman, S. M., 210, 709, 710
Sherrington, C. S., 19
Shi, C. J., 708, 711
Shi, H., 357
Shi, P., 554, 556, 689, 691, 964, 967, 968, 976, 1116, 1124
Shi, S., 914
Shiau, C. E., 737
Shibata, M., 192
Shibuya, T., 14, 557
Shields, V. D., 930
Shields, V. D. C., 936
Shigematsu, K., 459
Shigemura, N., 757
Shigemura, R., 674
Shih, C. T., 893
Shimamura, A., 646, 650
Shimazaki, R., 1117
Shimazaki, T., 189, 193
Shimoda, M., 317
Shimoda, N., 459, 467
Shimozaki, K., 192
Shin, S. H., 362, 364, 367
Shin, Y. K., 694, 695, 963
Shindo, Y., 693
Shingai, T., 763
Shingo, T., 198
Shinkai, R. S., 911
Shinoda, K., 1140
Shionoya, K., 346
Ship, J. A., 766
Shipley, M. T., 73, 99, 157, 158, 159, 162, 163, 166, 167, 170, 171, 455, 461, 463, 465
Shiraishi, A., 935
Shirasu, M., 114
Shirazki, A., 837
Shirley, S., 465
Shoenfeld, N., 391
Shoham, Z., 869
Shook, B. A., 197, 198, 487
Shore, D. I., 251
Shou, J., 143, 144, 147, 736
Shoulson, I., 413
Shpak, G., 1122
Shrestha, R., 642, 984
Shu, C. H., 392
Shu, H. J., 172
Shubin, N., 947, 950
Shusterman, D., 369, 1100
Shusterman, R., 164, 168, 368
Shykind, B. M., 127
Sidel, J., 1052
Sidel, J. L., 1069
Siderowf, A., 217, 417, 422, 423
Sidky, M., 11
Sieffermann, J. M., 1072
Siegel, R. E., 637

Siegel, R. K., 903
Siegrist, M., 1058, 1061
Siemers, B. M., 610
Sienkiewicz, Z. J., 775
Sikora, E. R., 489
Silas, J., 393
Silbering, A. F., 344, 533, 535, 536, 537, 538, 932
Silk, P. J., 1119
Silkoff, (1999), 357
Sillero-Zubiri, C., 596
Silletti, E., 627
Silva, A. M., 391
Silva, D. A., 897
Silveira, P. P., 834
Silveira-Moriyama, L., 419, 420, 421, 423, 426, 429
Silver, L. M., 960
Silver, W. L., 1091, 1092, 1098
Silver, W. S., 1055
Silverman, I., 324
Silverman, S. L., 459, 469
Silvotti, L., 1115
Simard, A. R., 490
Simbayi, L. C., 876, 878
Simchen, U., 1102
Simitzis, P. E., 314
Simmen, (1999), 359
Simmen, D., 228, 237, 243
Simmonds, M. S. J., 938, 939
Simmons, W. K., 993, 1031, 1040
Simo, S., 188, 191
Simoes, T., 71
Simola, M., 362, 382
Simon, C., 837
Simon, S. A., 4, 57, 655, 675, 695, 993, 1104
Simon, T. W., 250
Simons, C. T., 1008, 1101, 1103, 1104, 1105
Simons, P. J., 840
Simpson, C. L., 939
Simpson, P. J., 145
Simpson, S. B., Jr., 737
Simpson, S. J., 939
Singer, M. S., 929, 940
Singh, N., 388, 692, 922
Singh, R. N., 937
Singhasemanon, N., 497
Sinno, M. H., 872
Sirota, P., 431, 439
Sivam, A., 363, 383
Sjoberg, L., 230
Sjölinder, H., 100
Sjostrand, C., 900
Sjostrom, L. B., 1070
Skelhorn, J., 963, 964, 968, 969
Skinner, B. F., 20
Skinner, J. E., 169
Sklar, P. B., 15

Skoog, D. A., 570, 571
Skoog, K. M., 876
Skrandies, W., 266
Slack, J. P., 667, 671, 695, 923
Sleight, S. D., 457
Sliwa, A., 591
Slone, J., 934
Slosson, E. E., 232
Slot, W. B., 459, 469
Slotnick, B. M., 20, 22, 248, 295, 296, 344
Small, D. M., 20, 217, 218, 297, 344, 499, 500, 580, 714, 718, 766, 776, 777, 778, 779, 780, 781, 782, 783, 785, 786, 787, 789, 832, 900, 992, 996, 1016, 1017, 1028, 1037, 1038, 1039, 1056, 1057
Smallman, R. L., 17, 650, 891
Smallwood, P. M., 125, 127
Smart, I. H., 134
Smart, J. L., 835
Smeets, A. J., 839
Smeets, M., 1101
Smeets, P. A., 786
Smeets, P. A. M., 831
Smith, B. A., 805, 807
Smith, B. H., 344
Smith, C. G., 134
Smith, D. A. R., 239
Smith, D. B., 265
Smith, D. M., 317
Smith, D. V., 4, 361, 384, 643, 650, 651, 655, 675, 678, 703, 705, 707, 709, 733, 754, 763, 765, 782, 934, 938, 962, 985, 986, 1061
Smith, F. J., 839
Smith, F. R., 848
Smith, J., 20
Smith, J. C., 835, 923
Smith, J. J., 216
Smith, J. J. B., 933, 939
Smith, M. T., 1142
Smith, N., 390
Smith, R. S., 170, 174, 1081, 1122
Smith, T. C., 162
Smith, T. D., 21, 22, 41, 42, 43, 44, 45, 46, 47, 48, 50, 51, 54, 55, 56, 57, 58, 605, 606, 608, 617
Smith, T. L., 384
Smith, W. B., 144
Smith, W. M., 461, 499
Smith-Swintosky, V. L., 716, 782, 783, 784, 991, 992
Smotherman, W. P., 22, 306, 307, 314, 326, 874, 875
Smutzer, G. S., 410, 762
Snapyan, M., 185
Sneddon, H., 568
Snell, T., 985
Snoek, H. M., 1020
Snow, B., 966

Snow, D. W., 966
Snow, J. B., 765
Snowdon, C. T., 596
Snyder, D. J., 756, 757, 759, 762, 764, 765, 766
Snyder, L. H., 755, 756
Snyder, S. H., 74, 644
Sobel, E. C., 164
Sobel, N., 20, 217, 228, 274, 287, 288, 289, 298, 338, 357, 370, 416, 423, 436, 1039, 1056
Sobin, C., 435
Sobotta, J., 654
Soehnlein, O., 1116
Sohn, W. J., 733
Sohrabi, H. R., 409
Soini, H. A., 567, 570, 574, 575, 591
Soininen, E. M., 961
Sokoloff, L., 19
Solari, P., 935
Solbu, E. H., 310
Solbu, T. T., 66
Soler, Z. M., 362, 384
Sollars, S. I., 742
Solomon, F., 900
Solomon, G. E. A., 1014
Solomon, G. S., 23, 243, 405, 409
Solomons, T. W. G., 569
Sömmering, S. T., 11, 279, 280, 639, 640, 642, 644
Song, H., 183
Song, H.-J., 1055, 1059
Song, K. H., 470
Sonoda, H., 599
Soranzo, N., 691
Sorensen, L. B., 833
Sorensen, P. W., 555, 1119
Soria, E. D., 898
Soria, J. M., 189
Sorokin, S. P., 64
Sosulski, D. L., 210, 213, 217
Soucy, J., 839
Sorter, A., 500
Souder, W., 565
Soumier, A., 190, 194
Sourjik, V., 515
Soussignan, R., 308, 312, 313, 316, 317, 319
Spangler, K. M., 654, 961, 962, 985
Sparing, R., 780
Sparkes, R. S., 390
Sparks, D. L., 407
Speca, D. J., 553, 554, 556
Spector, A. C., 22, 686, 782, 796, 799, 865, 876, 877, 934, 937, 960
Speed, M. P., 969
Speedie, N., 557
Spehr, J., 1118
Spehr, M., 115, 116, 118, 1114, 1115, 1116, 1117, 1119

Spence, C., 1008, 1010, 1015, 1016, 1061
Spencer, R. F., 455, 461, 465, 466
Spencer, W. A., 939
Spetter, M. S., 782, 783, 785, 832, 1057
Spiegel, R., 404, 409, 437
Spielman, A. I., 11, 626, 644
Spielman, J. S., 600
Spiess, C. K., 490
Spilianakis, C. G., 126
Spitzer, L., 847
Spoor, A., 357
Spors, H., 167, 168
Springer, A. D., 1142
Springer, D., 961
Springer, K., 228
Sprissler, C., 647
Squire, L., 340
Squire, L. R., 269
Squires, A., 141, 142, 145
Squirrel, D. J., 593
Sreebny, L. M., 625
Sreenivasan, K. V., 430
Srinivas, N. R., 458, 467
Srinivasan, M. V., 345
Sriram, K., 497
Srur, E., 641, 643
St John, J. A., 98, 148
St. Andre, J., 876, 877
Stacey, N., 547, 553, 557
Stacey, N. E., 547, 555
Städler, E., 929
Stafford, L. D., 832
Stager, K. E., 565
Stamler, R., 837
Stamps, J. J., 408
Stanger-Hall, K. F., 973
Stapleton, J. R., 702, 703, 713, 714
Starcevic, S. L., 68, 70, 80, 556, 557
Starostik, M. R., 649
Starr, A., 265
Staubli, U., 218, 296
Steck, K., 345
Steckler, T., 873
Stedman, H. M., 985
Stedman, T. J., 431
Steele, J. C., 428
Steen, J. B., 598
Steffens, A. B., 837
Stein, D. J., 391
Stein, L. J., 798, 815, 836, 837
Stein, M., 312, 323
Stein, N., 795, 799, 804
Steinbach, S., 408, 486, 500
Steinbrecht, R. A., 936
Steiner, J. E., 312, 313, 316, 796, 798, 799, 801, 804, 806, 811, 814, 819, 821, 867, 868, 903, 960
Steinert, R. E., 840, 841
Steiniger, von, F., 867

Stell, W. K., 1144
Stenman, J., 183, 196
Stenonis, N., 11
Stephan, A. B., 112, 1118
Stephan, H., 607, 618
Stephani, C., 781
Stephen, K. W., 765, 766
Sterling, D., 677
Stern, I. B., 651, 796
Stern, J. M., 584
Stern, M. B., 414, 416
Sternini, C., 730
Stettenheim, P. R., 566
Stettler, D. D., 210, 214, 1124
Stevens, B., 800, 805
Stevens, C. N., 361, 384
Stevens, D. A., 20, 22, 1101, 1103, 1104, 1105
Stevens, J. C., 231, 240, 246, 753, 755, 756
Stevens, M. H., 361, 384
Stevens, S. S., 227, 230, 235, 236, 755, 758, 759
Stevenson, R. J., 216, 295, 310, 323, 324, 337, 339, 340, 429, 761, 766, 781, 1008, 1009, 1010, 1011, 1012, 1013, 1016, 1017, 1018, 1021, 1055, 1058, 1101, 1103, 1104
Stewart, J. E., 632, 835, 847
Stewart, R. A., 837
Stewart, R. B., 835
Stewart, R. E., 665
Stewart, R. R., 193
Stewart, W. B., 20, 250, 455, 456, 461, 464, 585
Steyn, D., 472
Stice, E., 785, 993
Stickrod, G., 314
Stiebel, H., 961
Stierle J. S., 538, 539
Stiles, F. G., 965
Stillman, J. A., 754
Stirnimann, F., 307, 308, 312
Stjarne, P., 384
Stock, C., 527
Stock, J., 514
Stock, M. J., 839
Stocker, R., 520
Stocker, R. F., 930, 932, 937
Stocking, A. J., 1083, 1084, 1085, 1086
Stoddart, D. M., 592
Stokes, J. R., 627
Stoll, W., 384
Stone, D. M., 473
Stone, F., 832
Stone, H., 1052, 1057, 1069, 1091
Stone, L. M., 633, 651, 655, 728, 949, 950, 1102
Stone, L. S., 655, 728
Stopfer, M., 535, 538, 539, 541

Author Index

Storan, M. J., 1136, 1137, 1138, 1140
Storch, V., 646
Storm, D. R., 1118
Stornelli, M. R., 647, 963
Stößel, A., 43, 57
Stott, W. T., 66
Stowers, L., 584, 1118, 1123, 1125
Strack, A. M., 462
Strassmann, B. I., 22
Stratton, G. M., 8
Strausfeld, N. J., 535
Strauss, G. P., 433, 434
Strauss, S. C., 1053, 1056
Stricker, E. M., 874
Stricker, S., 184
Strickland, M., 313
Strohl, K. P., 357
Stromberg, M. R., 965
Strong, V., 599
Stroop, W. G., 380, 455, 461, 462
Strotmann, J., 64, 124, 126, 128, 367, 1134
Strowbridge, B. W., 161, 164, 165, 166, 169, 217
Stuck, B. A., 269
Stuiver, M., 247, 358
Stunkard, A. J., 865, 873
Stunkart, A. J., 326
Stürckow, B., 20
Sturz, G. R., 676
Su, C. Y., 538
Su, T., 67, 68, 69, 79, 80, 486
Su, X., 1104
Subramaniam, R. P., 359
Sudakov, K., 712, 990, 991
Sudhakaran, I. P., 534
Sudo, S., 1103
Suffet, I. H., 1080, 1081
Suga, Y., 500
Sugai, E., 1104
Sugai, T., 1122
Suh, G. S. B., 540
Suh, Y., 188, 192
Suliburska, J., 901
Sulkowski, W. J., 497
Sullivan, R. M., 162, 209, 216, 308, 315, 320, 321, 584
Sullivan, S. A., 804, 809, 817, 823, 868, 1062
Sullivan, S. L., 124, 465
Sulmont-Rosse, C., 346
Sultan, B., 362, 488
Sultan, S., 167, 198
Sulz, L., 143, 144, 147
Sulzer, M., 10
Sumby, W. H., 1008
Summerfield, C., 295, 717, 994
Summers, M., 5
Sumner, D., 237, 386, 896
Sun, G., 192

Sun, H., 732
Sun, J., 193
Sun, J. D., 76
Sun, L., 117, 193
Sun, W., 191
Sun, X. J., 537
Sun, Y., 737
Sunderland, E., 755
Sündermann, D., 611
Sundermann, E., 434
Sundermann, F. W. Jr., 497
Sundqvist, N. C., 1013, 1020
Sung, C. H., 111
Sung, Y. K., 67
Sung, Y. W., 364
Sunyer, J., 490
Susi, P., 497
Suss, C., 309
Sutterlin, A., 548, 558
Suttisri, R., 669
Suwabe, T., 741, 742
Suyama, S., 190
Suzuki, J., 96
Suzuki, M., 283, 421
Suzuki, N., 209, 213, 214, 456, 462
Suzuki, S. O., 186, 197
Suzuki, W., 340
Suzuki, Y., 407, 501, 653, 735
Svejda, J., 642
Svensson, C. K., 319
Swain, R., 459, 469
Swain-Campbell, N., 1101, 1104
Swank, M. W., 873, 876, 878, 879
Swann, J. M., 1123
Swanson, L. W., 174, 873
Swartz, V. W., 227, 1052, 1073
Sweazey, R. D., 765, 994
Sweeney, R. J., 567, 572
Swets, J. A., 230, 232, 233, 752
Swift, D. L., 358
Swithers, S. E., 831, 833, 838, 841
Sword, G. A., 973, 974
Syed, A. S., 47
Szabó, K., 1114
Szallasi, A., 1092
Szczesniak, A. S., 1055
Szejtli, J., 914, 922
Szente, L., 914, 922
Szentesi, A., 939, 940
Szentesi, A. S., 939, 940
Szeszko, P. R., 432, 434
Szoka, P. R., 1119
Szolcsányi, J., 1101, 1102, 1104
Szucs, E., 358
Szyszka, P., 344, 535, 538, 540, 541

Tabaton, M., 473
Tabert, M. H., 407, 493
Tachdjian, C., 667, 673
Tachibana, T., 1137

Taillibert, S., 704, 985
Takagi, S. F., 3, 4, 6, 14, 229, 243, 387, 557, 608, 997
Takahashi, H., 128, 895, 897
Takahashi, K., 642
Takala, A., 458, 467
Takami, S., 173, 1121
Takano-Maruyama, M., 737
Takaoka, T., 849
Takeda, A., 101
Takeuchi, H., 128, 129, 950
Takeuchi, M., 136, 143
Takeuchi, S., 780
Takigami, S., 171
Talamo, B. R., 148, 473, 474
Talavera, K., 1101, 1102
Talhout, R., 500
Talley, H. M., 74
Tallkvist, J., 461, 492, 497
Tallon, S., 318
Tam, S., 384
Tamano, H., 101
Tamayo, R., 516, 517
Tamura, H., 71
Tan, E. S., 115
Tan, J., 159, 161, 165, 167, 168
Tanabe, T., 14, 218, 608, 609
Tanaka, M., 1117
Tanaka, N. K., 541
Tanaka, Y., 936
Tanapat, P., 19
Tandler, B., 65, 644, 648
Tang, L., 521
Tang, T., 490
Taniguchi, K., 840, 1133
Taniguchi, M., 1122
Tanimura, T., 934, 937
Tank, D. W., 164
Tanner, C. M., 424, 473
Tanner, W. P., Jr., 230, 232
Tarozzo, G., 1114
Tarr, L., 916
Tarttelin, M. F., 870
Tarun, A. S., 71
Taruno A., 694
Tataranni, P. A., 714, 992
Tateda, H., 14
Tateyama, T., 269
Tattersfield, A. S., 198
Tatzer, E., 797, 805, 833
Tautz, J., 540
Tavazoie, M., 185
Taverner, D., 895
Taylor, (1990), 876
Taylor, A., 1055
Taylor, A. J., 1054
Taylor, A. N., 874
Taylor, J. S., 553
Taylor, K., 100

Taylor, N. W., 688
Taylor, W. D., 1081
Teague, R., 141, 142, 143, 147
Teegarden, S. L., 835
Teeter, J. H., 15
Teff, K., 959
Teff, K. L., 835, 839
Teghtsoonian, R., 759
Teicher, M. H., 20, 319, 320, 585
Teichner, W. H., 20
Teijeiro-Osorio, D., 471
Teillet, E., 1073
Temmel, A. F., 363, 843, 848
Tempere, S., 1014
Temple, E. C., 804, 811, 833
Temussi, P., 4
Temussi, P. A., 834
Tengamnuay, P., 469
Tennen, H., 765
Teow, B. H., 837
Tepper, B. J., 756, 762, 763, 833, 834, 848, 1059
Ter-Pogossian, M. M., 19
Teubner, P., 616
Teucher, B., 251
Teuscher, E., 308
Thaler, B. I., 837
Tham, W. W., 296
Theisen, B., 549, 552
Thesen, T., 266
Thesleff, I., 729, 730, 838
Theunissen, M. J. M., 1009
Thewissen, J. G. M., 50
Thiebaud, N., 73, 79, 486
Thiessen, D., 586
Thirlby, R. C., 838
Thirumangalathu, S., 651, 728, 729
Thisse, B., 732
Thisse, C., 732
Thomann, P. A., 286, 377, 423
Thomas, A. M., 310
Thomas, D. A., 584
Thomas, J. E., 742
Thomas, J. E., Hill, D. L., 742
Thomas, R. H., 567, 572
Thomas-Danguin, T., 242
Thompson, D. F., 892
Thompson, K., 71, 466
Thompson, R. F., 939
Thor, D. H., 339, 340
Thored, P., 198
Thornburn, C. C. (1972), 755
Thorne, N., 932, 933, 934, 937
Thorne, R. G., 355, 461
Thornhill, R. A., 18, 134, 549
Thornton-Manning, J. R., 63, 69, 464, 465, 466
Thorpe, S. J., 715, 716, 994, 1031, 1035
Thorsteinson, A. J., 937

Threadgill, D. W., 732
Thuerauf, N., 1092, 1101
Thürauf, N., 1094
Thurstone, L. L., 227, 1074
Tian, H., 127, 1133, 1134, 1135
Tian, L., 468
Tichansky, D. S., 847
Tichy, F., 961
Tie, K., 765
Tieman, D., 767
Tierney, K. B., 497, 549, 551, 557, 558
Tijero, B., 424
Tikhonova, N. A., 647
Tillett, T., 500
Tilton, F. A., 497
Timm, D. E., 1119
Tinklenberg, J. R., 404
Tirindelli, R., 171, 554, 1115, 1118, 1121
Tisay, K. T., 147
Tisdall, F. F., 17, 461, 500
Tissingh, G., 409, 416
Tivane, C., 963
Tizzano, M., 644, 655, 693, 695, 922, 1091
Tjälve, H., 73, 80, 455, 456, 457, 460, 461, 473, 485, 497, 498
Tobet, S. A., 1142
Tobin, V., 842
Tod, E., 600
Todd, R. B., 11
Todd, W. A., 323
Todrank, J., 314, 315, 343, 345, 764, 1016, 1019
Toga, A. W., 281
Toida, K., 159, 160
Tokita, K., 708, 989
Tolhuis, B., 126
Tolhurst, G., 840, 841
Tolis, G., 459
Tomchik, S. M., 647, 648
Tome, M., 139
Tomiczek, C., 781
Tomie, J. A., 345
Tominaga, M., 1092, 1098
Tominaga, T., 1092, 1098
Tomita, H., 754, 765, 846, 911
Tomita, N. E., 803, 808
Tomlinson, A. H., 73, 455, 462
Tomonari, H., 650
Tomota, Y., 407
Toms, M., 253
Tonoike, M., 268
Topolovec, J. C., 779, 780, 988
Tordoff, M. G., 832, 841, 959
Torello, A. T., 1115, 1116, 1125
Torreilles, J., 498
Torrey, T. W., 729
Toto, P. D., 644
Toubas, P., 320, 321
Toubeau, G., 647

Touhara, K., 16, 73, 111, 114
Toulouse, E., 13
Toulouse, N., 309
Tourbier, I. A., 239
Toure, G., 985
Tourne, L. P., 904
Townsend, M. J., 228
Toyono, T., 840
Toyoshim, K., 949
Toyoshima, K., 646, 647, 648, 650
Toyota, B., 228
Trabue, I. M., 3
Tran, A. H., 488
Trant, A. S., 845
Trapido-Rosenthal, H. G., 68, 73
Trask, B. J., 1115, 1116
Trasobares, E., 846
Travers, J. B., 705, 707, 708, 867, 875, 986, 987
Travers, S. P., 703, 704, 707, 708, 799, 987, 989
Treesukosol, Y., 678, 679, 687
Treisman, A., 1008, 1012
Treisman, M., 339
Treloar, H. B., 157
Tremblay, L., 717, 994, 995
Treves, A., 1032
Trevino, R. J., 385
Trevisani, M., 1092
Tribollet, E., 876
Tricas, T., 553, 557
Tricker, A. R., 78
Trifunovic, R., 876
Trimble, M., 591
Trinh, K., 1118
Trojanowski, J. Q., 411
Trolle, B., 227, 1052, 1073
Trombley, P. Q., 101, 102, 103, 162
Tropepe, V., 189, 192, 197
Trotier, D., 48, 298, 383
Truax, B. T., 898
Tsetsos, K., 717, 994
Tsivgoulis, G., 890, 899
Tsuboi, A., 124, 125, 126, 128
Tsuboi, Y., 414, 426, 428
Tsukatani, T., 230, 239
Tsuno, Y., 169, 170
Tsurugizawa, T., 525
Tsuruoka, S., 846
Tucker, A. S., 730
Tucker, D., 14
Tucker, R. M., 835, 847
Tuckerman, F., 651
Tullberg, B. S., 968
Tumbarello, M. S., 973
Tuorila, H., 1052, 1056, 1061
Turaga, D., 172
Turetsky, B. I., 271, 286, 377, 429, 432, 435, 439

Author Index

Turin, L., 104, 1054
Turk, M. A., 457
Turk, M. A. M., 72, 78
Turker, S., 487
Turnbull, B., 803, 813, 814
Turner, B. H., 996
Turner, C. P., 146
Turner, G. C., 535, 541
Turner, L. S., 251
Turner, W., 12, 55
Turnley, A. M., 191
Turton, J., 423
Tyden, E., 68

Ubink, R., 146
Uchenna Agu R., 473
Uchida, N., 161, 168
Uchida, T., 647, 970, 971
Uchiyama, T., 903
Udager, A., 730
Udani, R. H., 805
Ueda, A., 317
Ueda, H., 548
Ueda, K., 649
Ueno, Y., 237, 610, 616
Uesaka, Y., 780
Ugolini, G., 462
Ugur, T., 432, 434
Ujike, H., 893
Ukhanov, K., 112, 1117
Ullrich, N. V., 756
Ulrich, C. E., 1091
Umabiki, M., 847
Umback, W., 996
Umino, O., 1144
Ungureanu, I. M., 671
Uno, Y., 68
Upadhyay, A., 320
Uraih, L. C., 64
Uranagase, A., 142
Urata, K., 965, 967, 969
Urban, N. N., 166, 168, 1121, 1122
Urbick, B., 822
Urweyler, S., 667
USDA, 831
USEPA, 1082
Usher, B. F., 939
Usherwood, P. N., 525
Uvnas-Wallensten, K., 839

Vacher, C., 985
Vagell, M. E., 581
Vahl, T. P., 838
Vaidyanathan, A., 80
Vaidyanathan, S., 363, 392
Vaka, S. R., 471
Valensi, P., 459
Valentinčič, T., 547, 555, 557, 954
Valentin, D., 310, 1013, 1014, 1072

Valentin, G., 12
Valentine, M., 522, 524, 525, 526
Valenzuela, C. F., 499
Valido, A., 972
Valkenburgh, B., 50, 51–4
Valkov, I. M., 903
Valldeoriola, F., 422
Vallesi, A., 527
Valley, M. T., 198
Valmeseda-Castellon, E., 897
Valverde, F., 98, 146
Van Belle, S., 617
Van de Moortele, P.-F. C. B, 775
van den Berge, S. A., 198
van der Goes van Naters, W. M., 935
van der Klaauw, N. J., 1017
van der Veen, M., 6
van der Velden, R., 1020
van Eeuwijk, F. A., 935
Van Gilse, P. H. G., 45
Van Haastert, P., 521
Van Hausen, G. W., 461
Van Herpen, E., 1053, 1058
van Herwaarden, A. E., 917
Van Houten, J. L., 3, 522, 524, 525, 526
van Lennep, E. W., 629
van Loon, J. J. A., 933, 934, 935
Van Ommeren, E., 671
Van Pett, K., 1123
Van Reekum, C. M., 345
Van Toller, S., 262, 309
van Tuinen, M., 968
Van Valkenburgh, B., 46, 50, 57, 58
van Wimersma Greidanus, Tj. B., 874
Vandenbergh, J. G., 20, 21, 582
Vandenbeuch, A., 649, 675, 934, 949
Vander, M. W., 835
Vanderluit, J. L., 187
VanEtten, C., 959
vanHeerden, J., 594
Vankeerberghen, A., 100
Vannelli, G. B., 76
Vanwyk, B. E., 964
Varatharasan, N., 646
Vardhanabhuti, B., 632
Varela, P., 1069, 1072, 1073
Varendi, H., 316, 317, 319
Vargas, G., 145
Varro, M. T., 7
Vaschide, N., 13
Vaschide, S., 309
Vassalli, A., 126
Vassar, R., 124, 126, 128, 157, 465
Vasterling, J. J., 496
Vatner, S. F., 840
Vaughan, R. P., 1092
Vavia, P. R., 468, 470
Vawter, M. P., 874
Vayssettes-Courchay, C., 784

Vazquez, M., 798, 821, 834
Vearrier, D., 485
Velayudhan, L., 408
Velazquez-Perez, L., 436, 437
Veldhuis, J. D., 916
Veldhuizen, M. G., 777, 778, 779, 785, 787, 993, 1012, 1057
Velez, Z., 552
Velicu, I., 66
Vent, J., 386, 393, 499, 500
Vento, J. A., 386
Ventura, A. K., 765, 833
Ventura, R. E., 184
Verbaan, D., 419, 423
Verbalis, J. G., 873, 874, 876
Verbruggen, H., 598
Verendeev, A. L., 20
Verhagen, J. V., 133, 168, 368, 703, 707, 710, 711, 715, 777, 989, 992, 1012, 1031, 1032, 1033, 1034, 1036, 1037
Verleger, R., 269
Verlegh, P. W. J., 1011
Verma, A., 69
Verma, P., 818
Vermehren-Schmaedick, A., 932
Verplancke, G., 585, 587
Verrill, A. H., 6
Verron, H., 322
Verschueren, K., 143
Verson, E., 643
Vesalius, Andrea, 9
Vet, L. E. M., 532
Veyrac, A., 170, 547
Veyseller, B., 377
Vicario-Abejon, C., 188
Vickers, N. J., 532, 540
Vickers, Z. M., 1010, 1014
Vickland, H., 146
Victor, J. D., 702, 703, 784
Vidic, B., 41
Vigues, S., 688, 689
Viitala, J., 961
Vilbig, R., 739, 740
Vince, M. A., 320, 967, 968
Vincent, A. J., 146
Vingerhoeds, M. H., 627
Vinnikova, A. K., 676
Vinolo, M. A., 840
Vintschgau, 1879, 4
Vintschgau, M., 10
Visser, J., 802, 813
Vissink, A., 850, 905
Viswaprakash, N., 102, 499
Vitt, L. J., 970
Vitten, H., 165, 166
Vivino, A. E., 6
Vodyanoy, V., 101, 103, 104
Voets, T., 369
Vogt, R. G., 15, 73

Voigt, J. M., 69, 79
Voirol, E., 1017
Vollmecke, T., 244
Volokhov, A. A., 595
Volta, A., 10
von Bartheld, C. S., 1142
von Baumgarten, R., 134
von Bekesy, G., 377, 1016, 1095
von Brunn, A., 11, 12
von Campenhausen, H., 1123
von Clef, J., 297
von Dannecker, L. E., 111
von Düring, M. V., 646
von Frisch, K., 20, 21
von Haller, A., 7, 39
von Skramlik, E., 4, 1095
von Sydow, E., 1011
von Vintschgau, M., 653
von Weymarn, L. B., 80
Vories, A. A., 897
Vosshall, L. B., 533, 535, 930, 1116
Vroon, A., 427
Vucini', D., 162
Vukovic, J., 142

Wachowiak, M., 158, 162, 167, 168, 169, 580, 1136
Wacht, S., 935
Waclaw, R. R., 189
Waddell, S., 934
Wade, F., 103
Wade, G., 869
Wade, G. N., 869
Wadhams, G. H., 513, 516
Wagner, A., 785
Wagner, M. W., 964
Wagner, R., 654
Wagner, S., 172, 173, 1120, 1121
Waite, P. M. E., 455, 459
Wake, 1999, 55
Wakisaka, S., 651, 733
Wakita, M., 991
Wako, K., 55
Walbourn, L., 865
Waldton, S., 403
Walk, H. A., 339
Walker, D. B., 592
Walker, J. C., 228
Walker, S., 1008, 1009
Walker, S. E., 1144
Wallace, K. J., 345, 587
Wallace, K. M., 970
Wallace, S. J., 458
Wallén, H., 616
Waller, A., 637, 646
Wallis, G. Y., 970
Wallrabenstein, I., 115
Wallraff, H. G., 566
Walsh, T. D., 845

Walt, D. R., 58
Walter, B. A., 487
Walters, E., 80
Walz, A., 1140
Wan, R., 874
Wang, C., 183, 184, 186, 197
Wang, C. X., 470, 471
Wang, D. D., 193
Wang, F., 128, 157
Wang, G. J., 1041
Wang, H., 362, 872
Wang, H. B., 79
Wang, H. L., 426
Wang, J., 408, 411, 422, 423, 472
Wang, J. C., 355
Wang, J. W., 540
Wang, L., 187, 269, 705
Wang, M., 468
Wang, Q., 68, 69, 70
Wang, Q. S., 406, 409
Wang, S., 742
Wang, S. S., 126
Wang, T. W., 189
Wang, X., 976
Wang, X. J., 1042
Wang, Y., 127, 455, 485, 667, 841, 874, 876, 877, 878, 985
Wang, Y. Y., 1092
Wang, Y. Z., 143, 145, 147
Wang, Z., 932, 934, 937
Wang, W., 412
Wanner, K. W., 932
Wansleeben, C., 731
Wantanabe, K., 896
Ward, C. d., 413, 415, 424
Wardlaw, S. A., 79
Wardle, S. G., 1014, 1015
Warner, M. D., 404, 412, 430
Warren, C., 1075
Warren, R. P., 967
Warrenburg, S., 1074, 1075
Waruszewski, B. A., 72
Warwick, Z. S., 835, 847
Washausen, S., 738
Waskett, L., 345
Wasser, S. K., 595
Watanabe, H., 929
Watanabe, K., 466
Watelet, J. B., 63
Water Quality Division Taste and Odor Committee, 1086
Waters, A. J., 1092
Waters, C. M., 518, 519, 520
Waters, H., 16
Watson, A. J., 1052
Watson, H. R., 1103
Watson, S. B., 1081, 1082
Watson, W. L., 810, 817
Watt, L. B., 806

Wattendorf, E., 217, 286, 420
Watters, G., 896, 897
Wauson, E. M., 968
Waxman, D. J., 75
Webb, C. J., 898
Webb, D. M., 1124
Weber, E. H., 227, 231
Webster, M. M., 501, 873
Weddell, G., 654
Wedekind, C., 323, 325
Weech, M., 80, 556
Weel, K. G. C., 1010
Weems, J. M., 69
Weenen, H., 1010
Weeraratne, S., 525
Wehkamp, J., 99, 100
Wehr, M., 168
Wei, C. J., 162
Wei, E. T., 1101
Wei, Y., 69
Weichert, C. K., 41
Weickert, C. S., 184, 197
Weidemann, J., 780
Weidenmuller, A., 540
Weierstall, R., 237, 242
Weiffenbach, J. M., 391, 392, 633, 753
Weigel, M. T., 388
Weiler, E., 48, 58, 140, 1133
Weinandy, F., 196
Weinberg, A., 99
Weinberger, D. R., 429
Weinberger, M. H., 677
Weingarten, H. P., 832
Weingarten, S., 872
Weinhold, I., 359
Weinshilboum, R. M., 70
Weinstock, R. S., 389, 390, 848
Weisfeld, G. E., 322, 324
Weiskopf, N., 288
Weiskrantz, L., 869
Weismann, K., 848
Weiss, A. P., 903
Weiss, D. G., 456, 462, 463, 465, 466
Weiss, L. A., 933
Weiss, M. R., 940
Weiss, P., 456, 463, 466
Weiss, S., 836, 916
Weiss, T., 249
Wekesa, K. S., 15
Welbeck, K., 832
Welch, J., 599
Weldon, P. J., 973, 974
Welge-Lussen, A., 245, 266, 272, 363, 384, 386, 1019
Welker, W. I., 168
Weller, J., 1051
Weller, M. P., 429, 430
Wellington, J. L., 583
Wellis, D. P., 163, 165

Author Index 1193

Wells, D. L., 595, 597, 598, 599, 600
Wells, J. M., 730
Wells, K. L., 734
Wells, M. A., 845
Welsch, U., 646
Weltzien, F.-A., 552
Wen, Z., 458, 461, 473
Wenner, M., 678
Wenning, G. K., 416, 423, 426, 427, 429
Wensley, C. H., 465
Wenzel, B., 228
Wenzel, B. M., 566
Werblin, F., 14
Wercinski, S. A., 572
Werner, S. J., 963, 968
Wessjohann, L., 667
Wesson, D. W., 168, 169, 210, 211, 217,
    218, 368, 581
West, A. K., 98
West, S. A., 518, 520
West, S. E., 8, 901
Westberry, J. M., 1123
Westbrook, G. L., 160, 161, 162, 164, 165,
    166, 169
Westerman, R. A., 134
Westermarck, E., 324
Westerterp-Plantenga, M. S., 1102
Westervelt, H. J., 407, 408, 414
Westlund, K. N., 706, 989
Westrate, J. A., 839
Wetherill, G. B., 752, 753
Wetherton, B. M., 705
Wetter, S., 271, 409, 410, 412, 413
Wetzel, C. H., 16
Wetzig, A., 139
Weusten, B. L. A. M., 993
Wevitavidanalage, M., 393
Wewetzer, K., 147
Wexler, 2005, 360
Weymouth, R. D., 963
Whalen, P. J., 1122
Wharton, T., 11
Wheatcroft, P. E. J., 755
Whelan, R. J., 567, 569, 572, 573, 575
Whelton, A. J., 1087
Whipple, G. C., 1079
Whishaw, I. Q., 340, 345
Whissell-Buechy, D., 755, 756
Whitby-Logan, G. K., 70
White, B. D., 834
White, E. L., 250
White, J., 1143
White, K. D., 837
White, P. R., 929, 936
White, R., 565
White, T. L., 338, 339, 340, 346, 1013
Whitear, M., 644, 646, 949
Whitehead, M. C., 650, 704, 705, 708, 1102
Whitesides, J., 146, 147

Whitman, M. C., 185, 194, 195, 196
Whitney, G., 763
Whittaker, D. J., 567, 574
Whitten, C. F., 837
Whitten, W. K., 21, 582
Whittle, C. E., 7
Wible, J. R., 52, 57
Wicher, D., 533
Wichterle, H., 186
Wicker, B., 993
Wickham, R. S., 845
Widmayer, P., 840, 842
Widström, A. M., 317
Wiechmann, A. F., 1144
Wiersma, A., 629
Wieser, H., 690
Wiggins, L. L., 937
Wiki, 591
Wilcock, G., 473
Wiley, R., 455, 461
Wilkie, J., 1013
Wilkin, T. A., 961
Willaime-Morawek, S., 183, 196
Willander, J., 305, 337
Willatt, D. J., 368
Willcox, M. E., 250
Willhite, D. C., 158
Willi, J., 871
Williams, C. M., 1060, 1100
Williams, C. R., 567, 572
Williams, D. L., 838
Williams, D. R., 428
Williams, H., 597, 599
Williams, H. L., 355, 357
Williams, M., 897
Williams, P. L., 651
Williams, R., 141, 142
Williams, S. M., 516
Williams, S. S., 414
Willis, C. M., 599
Willis, M. A., 532
Willis, T., 10
Willis, W. D. Jr., 1103
Wills, C., 755
Wills, G. D., 579
Wilson, C. W. M., 835
Wilson, D. A., 162, 167, 170, 209, 210, 211,
    214, 215, 216, 217, 289, 295, 315, 337,
    338, 340, 580, 1055
Wilson, D. J., 961
Wilson, E., 598
Wilson, E. M., 117
Wilson, E. O., 1118
Wilson, F. A. W., 1036
Wilson, H. C., 22
Wilson, J. G., 643
Wilson, M., 324
Wilson, M. A., 209
Wilson, R. I., 167, 534, 538

Wilson, R. S., 375, 407, 409, 411, 414, 419
Wilson, T. D., 1062
Wiltrout, C., 344
Winans, S. S., 21, 47, 58, 174, 715, 1114,
    1120, 1123
Winberg, S., 557
Windus, L. C., 147
Winer, J. A., 718
Wing, E. J., 872
Wingreen, N. S., 515
Winkielman, P., 1020
Winne, G. I., 840
Winner, B., 195, 198
Winston, J. S., 291, 292, 782
Wirant, S. C., 596
Wirsig, C. R., 1143
Wirsig-Wiechmann, C. R., 1142, 1143, 1144
Wirth, S., 218
Wisby, W. J., 548
Wise, J. B., 386
Wise, P. M., 230, 237, 837, 1100
Wise, S. P., 718, 1030
Wisen, O., 838
Wisniewski, L., 838
Wisnivesky, J. P., 496
Withee, J., 834
Witherly, S. A., 847
Witmer, L. M., 42
Witschi, J. C., 837
Witt, M., 48, 58, 298, 629, 644, 645, 646,
    647, 648, 649, 650, 651, 653, 729, 796,
    948, 949
Wittko, I. M., 190
Wolbarsht, M. L., 935
Wolf, D. A., 472
Wolf, G., 708, 711
Wolfensberger, M., 362
Wolkoff, P., 1095
Wolock Madej, C., 970
Wolovich, C. K., 617
Wolozin, B., 18, 133, 148
Wolstenholme, J. T., 836
Won, S. J., 468
Wonders, C. P., 196
Wong, G. T., 650, 693
Wong, H. L., 67, 68, 78
Wong, S. T., 112, 115
Wood, E. R., 340, 345
Wood, J. B., 237
Wood, J. N., 1092
Wood, R., 584
Wood, R. I., 1123
Wood, T. K., 518, 519, 520
Woodard, G. E., 1092
Woodhall, E., 142, 147, 148
Wooding, S., 756, 914, 915, 923
Woods, S. C., 831
Wooley, O. W., 837
Wooley, S. C., 837

Woolf, T. B., 165
Woolridge, M. W., 835
World Health Organization, 1080, 1082
Woskow, M. H., 236, 248
Wotman, S., 818
Wray, S., 98, 1141
Wren, A. M., 842
Wrench, R., 566
Wright (1914), 39
Wright, G. A., 936, 937
Wright, G. M., 1142
Wright, H. N., 237, 238, 243
Wright, J., 4, 8, 10–11
Wright, J. von, 310
Wrobel, B. B., 369
Wroblewski, K., 632
Wronski, M., 836
Wu, G., 731
Wu, H., 70
Wu, H. H., 144, 147
Wu, J., 430, 434, 435, 436, 455
Wu, K. N., 298
Wu, L., 69, 111, 112
Wu, M. V., 1123
Wu, S. V., 17, 841, 922
Wu, T., 842
Wu, W., 188, 191, 1057
Wu, X., 422
Wu, Y., 961
Wu, Z., 499
Wuensch, K. L., 318
Wundt (1897), 782
Wustenberg, E. G., 390
Wyllie, I., 968
Wysocki, C. J., 20, 21, 47, 246, 297, 298,
    308, 339, 369, 388, 393, 1093, 1101,
    1114
Wyss, H. v., 642
Wyss, J. M., 217

Xiang, J. -Z., 1036
Xiao, M., 149
Xie, F., 76, 77, 487
Xie, J., 502
Xie, Q., 79
Xing, H., 1092
Xiong, W., 164
Xu, A., 496
Xu, F., 368
Xu, H., 674, 689, 915, 1092
Xu, X., 1123
Xu,W., 217

Yabe, I., 890, 899
Yacoob, S. Y., 553
Yadon, C. A., 216
Yagi, S., 500
Yajima, T., 458, 466, 467
Yaksi, E., 534

Yalowitz, M. S., 710, 711
Yamada, Y., 765
Yamagata, N., 538, 541
Yamagishi, M., 380
Yamaguchi, K., 963
Yamaguchi, M., 195
Yamaguchi, S., 631, 672
Yamamori, K., 952
Yamamoto, C., 209, 776
Yamamoto, N., 1143
Yamamoto, R. T., 940
Yamamoto, T., 630, 631, 707, 708, 709,
    711, 712, 838, 876, 877, 878, 997
Yamamoto, Y., 547, 548, 555
Yamasaki, F., 642
Yamasaki, H., 630
Yamashita, H., 845
Yamashita, J., 581
Yamashita, R., 570
Yamashita, S., 951, 952, 953
Yamashita,Y., 905
Yamauchi, Y., 729
Yamazaki, F., 557
Yamazaki, K., 325, 583
Yambe, H., 557
Yan, J., 996, 997, 1036
Yan, Z., 210
Yanagi, S., 556
Yanagisawa, K., 23, 754, 765, 782, 891
Yáñez, J., 1143
Yang, C., 1117, 1118
Yang, C. C., 367
Yang, G. C., 51, 367, 368
Yang, H., 1118
Yang, P., 194
Yang, Q. X., 288
Yang, R., 649
Yang, R. S. H., 968
Yang, W., 525
Yang, X., 667, 736, 737
Yano, J., 525
Yao, J., 1104
Yarali, A., 537
Yarita, H., 608
Yarmolinsky, D. A., 685, 686, 688, 693,
    694, 696, 702, 1102
Yasui, M., 459
Yasumatsu, K., 309, 317, 690
Yau, K. W., 112
Yau, N. J. N., 1008
Yaxley, S., 708, 714, 717, 775, 776, 779,
    784, 988, 991, 992, 995, 997, 1037,
    1933, 1936
Ye, Q., 629
Yee, C., 649, 655
Yee, C. L., 649, 650, 741
Yee, K. K., 143, 145, 146, 362, 381, 666,
    960
Yeh, C. K., 633

Yeomans, M. R., 1012, 1013, 1014, 1015,
    1020, 1021
Yeshurun, Y., 297, 339, 340
Yeung, D. L., 837
Yiin, Y. M., 833
Yildiz, A., 317
Yin, Q., 517
Yokoi, M., 167, 168, 172, 173
Yokosuka, M., 171, 172
Yokota, S., 455
Yokoyama, T. K., 167, 195
Yonekura, J., 172, 173
Yonelinas, A. P., 340, 341
Yonkers, J. D., 646
Yoon, H., 115
Yoritaka, A., 422, 424
Yoshida, H., 14
Yoshida, I., 217
Yoshida, M., 230
Yoshida, S., 96
Yoshida, T., 144
Yoshida, Y., 985
Yoshie, S., 650
Yoshihara, S., 188, 192, 195, 196
Yoshihara, Y., 125, 552, 554, 556
Yoshii, F., 1054
Yoshii, K., 951, 952
Yoshikawa, K., 111, 114
Yoshikawa, T., 846, 911
Yoshimura, K., 958
Youdim, M. B., 460
Young, B. A., 970, 971
Young, J. A., 629
Young, J. B., 872
Young, J. K., 871
Young, J. M., 16, 123, 124, 125, 531, 593,
    1115, 1116, 1124
Young, K. M., 183, 196
Young, P. T., 11
Young, S. H., 841
Young, S. Z., 193
Young, V. B., 520
Young, W. F., 1083
Youngentob, S. L., 51, 93, 99, 314, 315,
    340, 345, 368, 499, 835, 1054, 1057
Yousem, D. M., 20, 38, 283, 284, 288, 376,
    377, 388, 390, 423, 436, 1057
Yu, C. P., 365
Yu, G-Z., 164
Yu, G., 365
Yu, J. W., 1081
Yu, T. W., 1120
Yuan, D. L., 870
Yuan, J., 515
Yuan, Q., 160, 171
Yun, J. S., 317

Zabernigg, A., 845, 1019
Zaborszky, L., 170

Author Index

Zachar, P. C., 646
Zafra, M. A., 837
Zagotta,W. N., 112
Zahm, D. S., 762
Zahn, R., 527
Zaidi, F. N., 705
Zaitlin, B., 1081
Zajonc, R. B., 344
Zakharov, A., 487
Zalaquett, C., 586
Zald, D. H., 290, 291, 292, 776, 783, 785, 786, 1035, 1037, 1039
Zampighi, G. A., 15
Zampini, M., 1010
Zancanaro, C., 647
Zanchin, G., 900
Zandstra, E. H., 803, 804, 809, 833
Zanger, P., 100
Zanotto, K. L., 1092
Zanusso, G., 100
Zanuttini, L., 310, 339
Zasloff, M., 100
Zatorre, R. J., 20, 218, 239, 287, 434, 1039, 1056
Zatta, P., 455, 459
Zeifman, D., 807
Zeinstra, G., 313
Zeiske, E., 552
Zelano, C., 212, 295, 296, 338, 346
Zelikina, T. I., 594
Zelles, T., 165
Zellner, D. A., 210, 345, 1014
Zencak, D., 188
Zeng, Q., 633
Zervakis, J., 913, 914, 1056
Zhainazarov, A. B., 112
Zhan, F., 631
Zhang, C., 552, 553, 644, 651, 952
Zhang, C. X., 644

Zhang, D., 143
Zhang, F., 525, 667, 673, 689, 690, 695
Zhang, G. H., 728, 733
Zhang, H., 190, 678
Zhang, J., 115, 286, 554, 689, 691, 693, 964, 967, 968, 976, 1116, 1123, 1124, 1125
Zhang, J. H., 69
Zhang, J. X., 567, 572, 1139
Zhang, J. Z., 554, 556
Zhang, K., 422
Zhang, L., 455
Zhang, L. L., 741
Zhang, P., 1117
Zhang, Q.-Y., 68, 69, 75, 76, 79, 143, 145
Zhang, R. L., 198
Zhang, S. Y., 78
Zhang, W., 1144
Zhang, X., 16, 70, 110, 118, 123, 125, 548, 1115, 1134
Zhang, X. G., 471
Zhang, X. L., 68, 70
Zhang, Y., 693, 694
Zhang, Y. J., 472
Zhang, Y.-F., 933
Zhao, C., 188, 189, 192
Zhao, F., 936
Zhao, F. L., 649, 666, 694
Zhao, G. Q., 689, 690, 938, 961, 1031
Zhao, H., 16, 111, 455, 689, 964, 967
Zhao, J., 498
Zhao, K., 51, 56, 356, 357, 359, 360, 361, 363, 366, 367, 368, 369
Zhao, L., 763
Zhao, X., 77
Zhao, Y., 471
Zhen, Z., 502
Zheng, D. R., 306, 307
Zheng, J., 112

Zhou, W., 298
Zhou, X., 66, 69, 74, 75, 76
Zhou, Y., 731, 732, 734
Zhuang, H., 111
Zhuo, X. L., 69, 74, 77
Zibman, S., 1122
Ziegler, D., 458
Zielinski, B., 547, 552
Zielinski, B. S., 64, 68, 70, 80, 549, 550, 551, 552, 556, 557
Zigova, T., 196
Zijlmans, J. C. M., 428
Zilles, K., 1030
Zimmermann, K. W., 644
Zinke, H., 667
Zinner, S. H., 836
Zippel, H. P., 4, 11, 12, 553
Ziswiler, V., 567
Zivadinov, R., 391
Zola-Morgan, S., 340
Zorn, A. M., 730
Zorzon, M., 391
Zotterman, Y., 14, 782
Zou, Z., 172, 557
Zozulya, S., 16
Zub, K., 596
Zubare-Samuelov, M., 668
Zucco, G., 417
Zucco, G. M., 247, 337, 339, 412, 415, 848
Zucker, C. L., 1144
Zufall, F., 116
Zumkley, H., 846
Zuniga, J. R., 23, 763, 905
Zupko, K., 69, 465
Zuwała, K., 647
Zverev, Y. P., 1069
Zviely, M., 1054
Zwaardemaker, H., 12, 228, 248
Zweers, G. A., 962

Subject Index

Note: Page references in *italics* refer to Figures; those in **bold** refer to Tables

[^{14}C]2-deoxy-D-glucose (2-DG) autoradiographic method, 19
[^{18}F] flurorodeoxyglucose method, 19
1-alcohols, 616
1-butanol, 309, 382
1-methyl-4-phenyl-1,2,3,6-tetra-hydropyridine (MPTP), 23–4, 423–4, 438, 460, 468, 474
1-octen-3-ol, 1081
1,10-dimethyl-9-decalol, 1081
2,3,7,8-tetrachlorodibenzo-*p*-dioxin, 78, 79
2,6-dichlorobenzonitrile *see* DCBN
2,6-dichlorophenyl methylsulfone, 76
2,6-dichlorothiobenzamide, 80
2-butanone, 364
2-deoxyglucose, 268
2-heptanone, 554
2-isobutyl-3-[3H]methoxypyrazine, 103
2-ketones, 616
2-methyl-2-propanethiol, 114
2-methyl-but-2-enal (2MB2), 317
2-methylisoborneol (MIB), 1081
2-phenylethylamine, 115–16
3-aminobenzamide, 80
3-drop technique, 753
3-isobutyl-1-methylxanthine (IBMX), 103
3-methylcholanthrene, 78, 79
3-methylindole (3-MI), 69, 77, *77*
3-trifluoromethylpyridine, 76
3'-5'-cyclic adenosine monophosphate (cAMP), 678
4-(2,2,3-trimethylcyclopentyl) butanoic acid, 692
4-(methylnitrosamino)-1-(3-pyridyl)-1-butanone (NNK), 69
4,4'diisothiocyanostilbene-2,2'-disulfonic acid (DIDS), 677

4,8a-dimethyl-decahydronaphthalen-4a-ol (geosmin), 1083
4-aminobiphenyl, 69
4-dihydroxybenzoic acid, 667
5α-androst-16-en-3-one, 73, 114
5-α-androstenone, 73–4, 79, 114, 311, 319, 388, **430–1**
6-hydroxy-6-methyl-3-heptanone, 583
6-hydroxydopamine (6-OHDA), 460, 468
6-*n*-propylthiouracil *see* PROP
7-transmembrane proteins, 533
7-tridecanone, 574
8-cup technique, 753
17α,20β-dihydroxy-4-pregnen-3-one, 74
α-amino-3-hydroxy-5-methyl-4-isoxazolepropionic acid (AMPA), 160–1
α-defensin, 99–100
α-gustducin, 15, 17, 693, 695, 960
α-lipoic acid, 392, 904
α-naphthoflavone, 79
α-synuclein, 198, 414, 424–5, 428, 468, 490–4, 502
α-transducin, 693
β-amyloid, 148, 217–8, 411–2, 468, 490–5, 502, 893
β-arrestin, 127
β-catenin, 731
β-cyclodextrins, methylated, 469
β-defensins, 99–101, *100*, *101*
β-galactosidase, 115, 136
β-glucopyranosides, 685
β-ionone, 79, 1081
β-lactoglobulin, 672
β-naphthoflavone, 78
β-phenylethylmethylethylcarbinol, 231
β,β*ı*-iminodipropionitrile (IDPN), 72

accessory olfactory bulb (AOB), 48, 99, 157, 171–3, 306, 592, 593, 608, 617, **618**, 1120, 1121
accessory olfactory system (AOS), 115, 157, 171, 174, 464, 579, 617, 1113–14
acesulfame-K, 667, 915, 916
acetaminophen (APAP), 72, 76, 464
toxicity, *75*
acetochlor, 78
acetonitrile, 72
acetyl salicylic acid, 467
acetyltransferases, 71
acoustic neuromas, 891
acoustic rhinometry, 357–8
acquired immune deficiency syndrome *see* HIV/AIDS
acrolein, 1092
across-fiber pattern theory, 985
actinomycetes, 1081
acupuncture, 393, 905
acute necrotizing ulcerative gingivitis (ANUG) (trench mouth), 894
acyclovir, 468
Addison's disease, **378**, 388
adaptation, olfactory, 248–249
adaptation, taste, 703
adenoid hypertrophy, 361, 368
adenoidectomy, 361, 369, 381
adenoids, hypertrophied, 381–2
adenylyl cyclase, 15, 112, 128, 145, 389, 523, 525, 526
adolescence, olfaction in, 309–11
adrenocortical insufficiency, *see* Addison's disease
adrenomedullin, 66
'adult' stem cells, 135
Aedes albopictus, 533

Handbook of Olfaction and Gustation, Third Edition. Edited by Richard L. Doty.
© 2015 Richard L. Doty. Published 2015 by John Wiley & Sons, Inc.

1197

Aedes japonicus, 533
Aedes sp., 532
Aethia psittacula (crested auklet), 570, *573, 574*
aflatoxin B1, 69
ageusia, 780, 888, 896, 902–905
aging
   and medication use, taste dysfunction, 911
   olfactory function and, 23, 199, 234, 237 238, 245, 246, 312, 313, 375, 377, 392, 485, 1056
   taste function and, 642, 797, 804, 963
agoenhancers, 667
Ailuropoda melanoleuca (giant panda), 975
Ailurus fulgens (lesser panda), 958
air pollution
   exposure to, 385
   and olfaction, 489–96, 502
air-dilution olfactometers, laboratory-based, 228
alachlor, 78, 464
alanine, olfactory system in transport of, 463
alarm pheromones, 586
Albright hereditary osteodystrophy (AHO), 389
alcohol, 499–500
alcohol abuse, 23, 438, 836
alcohol dehydrogenase, 464
Alcohol Sniff test, 309
alcohols, 569
aldehyde dehydrogenases (ALDHs), 68, 70, 464, 486
aldehyde oxidase homologue, 3 71
aldehydes, 570
algal blooms, 1081–2
aliphatic acetic esters, 616
alkaline phosphatase, 64, 67
allergic rhinitis, 361, 362, 369
   adrenomedullin in, 66
allesthesia, 786, 869
allyl N-geranyl carbamate, 675
aloin, 671
ALS/parkinsonism-dementia Complex of Guam (ALS/PDG), 428
aluminum, 499
   olfactory system in transport of, 459
   salts, as deodorant, 8
Alzheimer's disease (AD), xi, xii, 19, 23, 101, 148, 217, 219, 272, **379, 404–408**, *409*, 410–12, 463, **472**, 492, 549, 1042, 1056
   alpha-synuclein in, 491
   aluminum in, 459
   EEG, 263
   entorhinal cortex in, 218
   metals in, 457
   olfactory bulb volume in, 377
   olfactory CSERPs, 271

olfactory structural imaging in, 286
olfactory system as route for pathogens to brain, 456
single staircase (SS) procedure in, 230
transport in, 473
Amblonyx cinereus (Asian otter), 958
Ambystoma (axolotyl), 646
Ambystoma mexicanum (axolotl), 1144
Ambystoma tigrinum (tiger salamander), 1144
amides, 570
amifostine (EthyolTM), 902, 905
amiloride, 631, 665, 916
amino acids, transport of, 462–3
aminopeptidases, 66, 71
amiodarone, 921, 922, 923
amitriptyline, 921–2
amphibians
   nasal sac in, 41
   taste in, 948
amygdala, 174, 210, *211, 212*, 217, 218, 268, 269, 272, 280, 286, 289, 290, 291, 293, *294*, 296–9, 307, 411, 435, 488, 708, 714, 716, 779, 782–3, 785–7, 789, 870, 875, 877–9, *899*, 901, 950, 996–7, 1028, *1029*, 1035–7, 1120–3
amyl acetate, 246, 309, 310, 312, 313, **422, 432**, 592, 1135
amylase, 627
amyloidosis, 893
amyotrophic lateral sclerosis (ALS), 23, 101, 412
amytriptyline, 902
analgesics, administration via nasal cavity, 467
androstadienone (AND), 298, 311
androstenone, 73–4, 79, 114, 311, 319, 388, **430–1**
anesthetics, 501
aneurysms
   of anterior communicating bifurcation, 384
   of internal carotid artery, 384, 385
angiotensin-converting enzyme (ACE), 66, 901
angiotensin-converting enzyme (ACE) inhibitors, 376, 846
Anguilla anguilla (eel), *462*
Ankaferd Blood Stopper (ABS), 915
ankyrin, 651
anoctamin, 2 112
Anolis carolinensis (green anole), 973–4, 1142
anorexia nervosa, 390, 903
   chronic disease state cancer, 872
   estradiol and, 871–2
   systemic infections and, 872
   toxicity and, 872
anosmia, 8, 12, 843, 1094–5

complete, 375
congenital, 376, 387–8
due to head injury, 13
history, 8
MRI of, 283, 284
nasal trigeminal chemosensitivity in, 1093–4
ocular trigeminal chemosensitivity in, 1095–6
partial, 375
ANP/AFP theory of taste quality coding, 703
anterior cingulate cortex, 1032–3
anterior cortical nucleus, development of, 307
anterior ethmoidectomy, 384
anterior olfactory nucleus (AON), 210
   development of, 307
anterior piriform cortex (APC), 289
   encoded structure, 292
anterior rhinomanometer, 355
Anthochaera carunculata (red wattlebird), 965
antibiotics, administration via nasal cavity, 467–8
antihistamines, in sinus disease, 363
antimigraine medicines, 467
antivirals, administration via nasal cavity, 467–8
Apis mellifera (honeybee), 531–2, *532*
apolipoprotein E, 409
apomorphine, 867
apoptosis, 135
appetite, 832
   brain mechanisms and, 1027
   suppression, 468–9
aquatic vertebrates, olfaction in, 547–58
   anatomy, 549–53
      lateral olfactory tract (LOT), 552
      lateral-medial olfactory tract (LMOT), 552
      medial-medial olfactory tract(MMOT), 552
      nares, 549
      nasal flow, 550–1
      olfactory bulb, 552–3
      olfactory chambers, 549
      olfactory epithelium, 551–2
      olfactory tracts and connections, 552
   applications and perspectives, 558
   behavioral responses, **555**
   fish vs mammalian olfaction, 558
   functional assessment, 557
   genetics and physiology, 553–7
      biotransformation, 556–7
      OR receptors, 553–4, **554**
      signal transduction, 556
      TAAR receptors, 556
      VR receptors, 554–6

Subject Index

history of study, 548–9
terrestrial olfactory adaptation, 557–8
aquatic vertebrates, taste in, 947–54
central gustatory pathways, 949–50
facial and trigeminal nerve complication, 949
functional properties of gustatory systems, 950–4
molecular basis of gustatory receptors and transduction, 953
palatal organs and barbels, 949
peripheral gustatory systems, 947–9
responses to chemical stimuli, 950–4
amino acids, 950–2
aversive and/or toxic chemicals, 952–3
behavioural implications, 953–4
bile acids, 952
carbon dioxide/pH, 953
quinine, strychnine and tetrodotoxin, 952, *952*
signal transduction and molecular basis of receptors, 950
structural organization, 947–9
taste buds, 947–9, *948*
arginine, 916
Argyreus hyperbius, 936
Ariopsis felis(sea catfish), 553
aristolochic acid (AA), 936
Arochlor, 1254 79
Arothron nigropunctatus (pufferfish), *550*
artificial (electronic) noses, 58, 1057
asafoetida, 308, 312
asbestos, 491
ascending method of limits (AML), 230
aspartame (APM), 671, 666, 667
appetite and, 832
assistance neighborhoods, 516
associative learning, 297, 1075
Astatotilapia burton, 1144
Astynax jordani, 646
Astynax mexicanus, 646
Ateles geoffroyi (spider monkey), *613*
atrophic rhinitis syndrome, 369
Atta sexdens, 539
Atta spp., 532
Atta vollenweideri, 532, 539
attention-deficit-hyperactivity disorder (ADHD), 438, 463
auriculotemporal nerve (ATN), disorders of, 898
auto Inducers (AIs), 518
autoimmune diseases, olfactory dysfunction in, 391
Avicenna, 6
avitaminosis, 901
axonotmesis, 896

Bacillus subtilis, chemotaxis in, 516
baclofen, 915

bacterial chemotaxis, 513–18
barbels in aquatic vertebrates, 949
barley lectin, 465
basal cells, 95–6, **140**
globose, 137–8, *138*
Basiliscus basiliscus, 974
Basiliscus vittatus, 974
batrachotoxins, 568
bats
nasal anatomy in, 41
nasal mucosae in, *49*
bears, olfactory abilities in, 591–2
Beck Depression Inventory, 387
behavioral functions of olfaction, early, 314–24
birth and rapid learning of odors, 315–16, *316*
fetal chemoreception, 314–15, *315*
odor communication in nursing niche, 316–20
odor of breast milk, 317
odor of lactating breast, 316–17
olfaction and energy conservation, 319–20
transnatal chemosensory continuity, 317–19
development of social cognition, 320–4
social diversification, 320–4
individual recognition, 322–3
from social discrimination to social recognition, 320–1
social preferences, 323–4
Behcet's syndrome, 893
Bell's palsy, 890, 891, 895–6, 897, 904
Belone belone (garfish), 549, *550*, 552
benzaldehyde, 1081
benzalkonium chloride, 473
benzene, 69
olfactory system in transport of, 464
benzene acetaldehyde, 1081
benzo(a)pyrene, 78, 79
benzocaine, 905
biased random walks, 513
Bible, 8
bifrontal craniotomy, 384
biofilms, 520, *520*
Biolfa test, 309
bipolar disorder, 148, 903
bird odors, 565–75
chemicals, classification, **567**
chemicals, functional groups, *571*
feathers, 566
preen gland, 566–8
skin, 566
sources of chemicals, 566–8
stomach oil, 568
volatile organic compounds (VOCs), 569–74

volatility and vapor pressure classification, **569**
birds, gustation in, 961–70
bitter taste, 968–70
chickens, 968
generalist insectivores, 969–70
granivores and omnivores, 970
specialist insectivores, 969
genetics, 963–4
sweet taste, 964–7
chickens and galliforms, 964–5
insectivores, omnivores, and carnivores, 967
nectivores, 965–6
seasonal frugivores, 966–7
specialist frugivores, 966
taste behavior, 961, 963–4
taste buds
distribution and frequency of, 962–3
structure, 961–2
taste capacity, 970
taste system, 961–2
umami taste, 967–8
galliforms, 967–8
passerines, 968
bis(2-chloroisopropyl) ether, 1081
bitter taste, 690–2
birds and, 968–70
children and, 822
drugs and, 914–15
orofacial expressions in infants, 799
preference in infants, 798, 823
in primates, 986, 987
receptor genes, **687**
receptors, 841
reptiles and, 973–4
taste modulators, 670–2
transduction, 693–4
bladder cancer, 599
Blandin-Nuhn glands, 644
blast injection, 228, *228*, 385
blindness, early, olfactory bulb volume in, 377
blood oxygen level-dependent [BOLD] contrast, 20, 280, 282, 288, 776
Boettcherisca peregrina (fleshfly), 934, 935
Bombycilla cedrorum (Cedar waxwing), 966
Bombyx mori, 532, *532*
olfactory sensory neurons in, *533*
Bone Morphogenetic Proteins (Bmps), 734
Borna disease virus, 462
bottlenose dolphin, 958
botulinum toxin, 831
brain injury, 468
brain stem auditory evoked response (BAER) testing, 890
brain tumor therapy, administration via nasal cavity, 469
breadth-of-tuning coefficient, 986

breast cancer, 599
breast milk, odor of, 317
Breathe Right strip, 364
bretylium tosylate, 916
budesonide, 383
bufadienolides, 973, 974
Bufo americanus (toad), 463
Bufo marinus (cane toad), 973
Bufo paracnemis, 974
bulimia nervosa, 903
bupropion, 917
burning mouth syndrome (BMS), 765–6, 888, 904
burns, oral cavity, 893
buserelin, 469
butanol olfactory threshold, 362
butorphanol, 467
butyric acid, 312

cadmium, 497
  olfactory system in transport of, 460
Caenorhabditis elegans, 1140
caffeine bitterness, 672
Caiman crocodilus (spectacled caiman), 1142
Caiman latiriostris (broad-snouted caiman), 966
calcium oxalate, 1008
Calidris canutus (red knot), anti-predator behavior in, 573–4
Callithrix jacchus (common marmoset), lateral nasal wall in, *52*
Callorhinchus milii (elephant shark), 554
Calotes versicolor (eastern garden or changeable lizard), 974
Caluromys philander (bare-tailed woolly opossum), 966
camera lucida, 279
cancer
  bladder, 599
  breast, 599
  chemosensory abnormalities in, 844–5
  colon, 756
  colorectal, 599
  detection by dogs, 599
  lung, 599
  nasal, 78
  ovarian, 599
  prostate, 469, 599
Candida albicans, 893
candidiasis, 890
  of the tongue, 893
Canidae, olfaction in, 591–600
  glands, 594
  olfactory epithelium, 593
  olfactory-guided behavior, 595–7
    early development of, 595
    scent marking, 595–7
    scent rolling, 597

social recognition, 595
  olfactory receptor genes, 593
  olfactory sensitivity, 592
  olfactory sensory structures, 592–3
  olfactory systems, 592–4
  sniffing, 593
Canis familiaris (domestic dog), 591, 592, 595, 958
  conservation applications, 599
  economic applications, 598–9
  forensic applications, 597–8
    cadaver detection, 598
    scent identification line-ups, 598
    tracking, trailing and air scenting, 598
  health and medical applications, 599
  training, 597
  uses of nose, 597–9
capsaicin, 467, 679, 917, 1008, 1092
capsazepine, 679
captopril, 901
Carassius auratus (goldfish), 548, 948, 1143
carbimazole, 464
carbohydrate, taste qualities of, 733–4
carbon disulfide, 117
carbon tetrachloride, 80
carbonic anhydrase, 67
carbonic anhydrase II, 116
carbonic anhydrase VI, 67
carboxylesterase, 67, 68, 71, 464
carboxylic acids, 569
carboxymethyl cellulose (CMC), 1013
carboxypeptidases, 66
carboydrate sweetener taste, 666–7, 668
cardiolipin, 64
Carnivora, olfaction in, 591–600
carnosine (β-alanyl-L-histidine), 463
carotid sinus reflex, 897, 898
caroverine, 392
Carpodacus mexicanus (house finch), 966
carvacrol, 1092
catalase, 71
catarrhines, 605, *606*
Cathartes aura (turkey vulture), *565*
caudal brainstem
  primate, 706, 708
  rodent, 704–6. 707–8
Cebuella pygmaea, nasal mucosa in, *56*
central insular cortex, 876–7
central nervous system, disorders affecting taste, 898–901, **899**
Cepacol lozenges, 905
cephalexin, 468
cephalometric nasal index, 355
cerebral autosomal dominant arteriopathy with subcortical infarcts and leukoencephalopathy (CADASIL), 438
cetylpyridinium chloride, 679
Chalcides chalcides, 1142
chalky taste, 888

channel calcium modulated homeostasis 1 (CALHM1), 694
CHARGE syndrome, 390
Chelonia mydas (green sea turtle), 970
chemesthesis, 959, 1091
chemical taste thresholds, 752, 753–4
chemoperception, airflow dynamics and, 369–70
chemosensation
  cephalic phase cardiovascular response, 839–40
  cephalic phase gastrointestinal motor and secretory response, 838
  cephalic phase pancreatic exocrine reponse, 838
  cephalic phase renal response, 840
  cephalic phase salivary response, 837–8
  cephalic phase thermogenic response, 839
  influences on nutrient utilization, 837–40
chemosensory disorders, 23, 388, 482
  dietary management in, 849–50
chemosensory event-related potentials (CSERPs), 264–72
  effects of brain disease, 271–2
  functional significance, 268–9
  measurement, 264–5, *266*
  neural sources, 267–8
  in olfactory clinic, 269–71
  representation, 265–7
chemo-somatosensory event-related potential (CSSERP), 1100
chemotherapy, 500–1
  effect on taste, 845
child obesity, 822, 838
children, taste sensitivity and preference in, 800–22
  behavioral tests, 802–4
    paired-comparison and/or forced-choice procedures, 802
    rank order, 803–4
    scaling techniques, 803
    stimulus presentation methods, 804
  early, olfaction in, 308–9
  food and beverage industries and, 1055–6
  mid, olfaction in, 309–11
  reflex-like behaviors, 801
    analgesic responses, 801
    orofacial responses, 801
  sweet taste preference, 833–4
chlorhexidine mouthwash, 904
chloride deficiency, 837
chlormethiazole, 80
choanae (posterior nares), 39
cholecystokinin (CCK), 838, 842
cholecystokinin (CCK) agonists, 469
cholera, 1079
choline acetyl transferase, 650
cholinergic input, 170
chorda tympani, 654

Subject Index

Christianity, perfumes in, 5–6
chromatin negative gonadal dysgenesis *see*
    Turner's Syndrome
chromium, 497
chronic kidney disease, chemosensory
    abnormalities in, 848
chronic obstructive pulmonary disease, 23
chymotrypsin, 838
ciclesonide, 71
ciliated protozoa, 522–7
cinnamaldehyde, 1009
Ciona intestinalis (sea squirt), 553
circumvallate papillae, 727, 728, 732–4,
    959
    in macaques, 984
cirrhosis of the liver, 848
cis-3-hexen-1-ol, 1059
cis-3-hexenol, 1075
citric acid, 20, **470**, 709, 762, 764, 786, 797,
    798, 803, 804, **807, 809, 819, 820**, 822,
    823, 832, 900, 911, 917, 1012, *1016*
clarithromycin, 917
classical conditioning, 20, 289, 868
Clethrionomys (bank voles), 21
Clethrionomys glareolus (bank vole), 581
clomipramine, 902
clopidogrel (PlavixTM), 902
closed head injury (CHI), 387
Clostridium difficile, 520
    cyclic di-GMP in, 517
cobalt, olfactory system in transport of, 460
cobalt protoporphyrin IX, 80
cocaine, 463, 903
coefficient of stability, 240
coffee, odor of, 6, 346
cognitive-behavioral therapy, 905
Colisa lalia (dwarf gourami), 1143
Collet-Sicard syndrome, 898
colon cancer, 756
colorectal cancer, 599
columella, 39
common cold, 380
comparative taste biology, 957–76
    applied aspects of animal taste, 975–6
    challenges and opportunities in, 975
    molecular changes to sweet taste receptor,
        958
    swallowing foods whole, 958
    see also birds, gustation in; reptiles,
        gustation in
computational fluid dynamics (CFD),
    359–60, *360*, 367
computed axial tomography (CAT scan),
    279
computed tomography (CT), 19, 280
    of human olfactory bulb, 283
    of taste disturbance, 890
concanavalin A, 461
conditioned aversion response, 20, 867

conditioned avoidance response, 865
conditioned illness response, 865
conditioned responses (CR), 865
conditioned stimulus (CS), 289, 865, 875–9
conditioned taste aversion, 501, 703, *866*
    basolateral amygdala, 877
    central insular cortex, 876–7
    conditioned responses (CR), 865
    conditioned stimulus (CS), 865, 875–9
    description, 865–7
    estradiol and, 870
    feeding system and, 868–72
    illness systems and, 872–3
    integration of taste and toxicity, 876–8
    learning theories, 867–8
    neural mechanisms, 875–9
    parabrachial nucleus, 877–8
    plasticity of preferences and aversions,
        868–9
    stress hormones, 873–5
    stress systems and, 873–5
    taste stimulus detection, 875
    toxicity stimulus detection, 875–6
    unconditioned responses (URs), 865
    unconditioned stimulus (US), 867, 875–9
Configurational Hypothesis of Facial
    Recognition, 251
Configurational Hypothesis of Olfaction,
    251
congenital smell loss, idiopathic, olfactory
    bulb volume in, 377
constructive interference in steady state
    (CISS), 284
continuous time-frequency analysis, 272
continuous wavelet transform (CWT), 272
Controlled Oral Word Association Test
    (COWAT), 438
Coolidge effect, 343
copper deficiency, 901
cortical spreading depression (CSD), 900
corticobasal degeneration (CBD), 426
corticosteroids
    olfactory function and, 392
    in sinus disease, 363
cough reflex, 898
coumarin, 76, 79
cranioinertial feeding action, 962
cribiform plate, 8, 10, 11, 13, 39, 40, 43, 57,
    279, 281, 284, 356, 364, **379**, 381,
    383–8, 390, 394, 454, 460, 485, 608,
    1113, 1120, 1134, 1138, 1142,
crocodile tears, 896
crocodilians, taste in, 974
Crocodylus niloticus (Nile crocodiles), 1142
Crohn's disease, 100
Cronbach's coefficient alpha, 240
cross-modal magnitude matching, 236
current source density (CSD), 266
Cushing's syndrome, 388–9

cyanide, 79
Cyanistes caeruleus (blue tit), 963
cyanobacteria, 1081–2
Cyanopica cyana (azure-winged magpie),
    967
cyclamate, 667
cyclic di-GMP synthesis pathway, 516–17
cyclodextrins, 469, 473
cyclooxygenases, 71
Cymbopogon, 7
Cynanthus latirostris (broad-billed
    hummingbird), 966
Cynictis penicillata (yellow mongoose), 958
Cynops pyrrhogaster (Japanese newt), 557
Cyprinus carpio (carp), 948
cystatin S, 627
cystic fibrosis (CF; mucoviscidosis), 24, 71
    olfactory dysfunction in, 391–2
cytochrome P450, 54, 64, 68–70, 72–80,
    486, 556, 921
cytokeratin, 96, 644, 651, 686, 731

D4T (2′,3′-didehydro-3′-deoxythymidine),
    467
Danaus plexippus (monarch butterfly), 969
Danio rerio (zebrafish), 549, 551, 1144
Dasyatis sabina (Atlantic stingray), 1143
DCBN, 71, 76, 77, 80, 464
defensins, 96
delayed match and non-match to sample
    tasks (DNMS), 339–40
dementia, 376
    dementia with Lewy bodies, 414
    frontotemporal, 438
    vascular, 437–8
    vasculature, 263
denatonium benzoate, 969
dendritic glutamate spillover, 160–1, *161*
dendrodentitic transmission, 165–6
Dengue fever, 531
dental disorders, 893–4
dentoalveolar abscess, 893
deoxycorticosterone acetate (DOCA), 388
depression, 101
    olfactory dysfunction in, 438
desimpramine, 902
detection threshold, *see* threshold,
    psychophysical
developmental odor hedonics, 311–14
dexamethasone cipeclate, 71
dextromethorphan, 468
diabetes, 7, 901
    chemosensory abnormalities, 847–8
    dog detection of, 599
    gestational, 834
    history of diagnosis, 7–8
    loss of taste function, 901
    non-insulin-dependent, 841
    olfactory dysfunction and, 390

diacylglycerol, 693
dialysis, 501–2
diazepam, 921
Dicentrarchus labrax (sea bass), 553
dichlobenil (2,6-dichlorobenzonitrile), 464
Dicrostonyx groenlandicus (lemming), 587
Dictyostelium, di-cGMP in, 528
Dictyostelium discoideum, chemotaxis in, 513, 520–2, *523*
diethyldithiocarbamate, 80
diethylpyrocarbonate, 678
diethylstilbestrol (DES), 871
difference threshold, 231–2, 752
diffusion tensor imaging (DTI), 286–7
diffusion-weighted imaging, 890
Diglossa baritula (cinnamon-bellied flowerpiercer), 965
digoxin, 921
dimethyl trisulfide, 1081
dimethyl-β-cyclodextran (DM βCD), 469
dipeptidyldipeptidase, 71
diphtheria, 469
discrimination learning, 344
disulfiram, 80
divalent metal transporters (DMTs), 71
divinylbenzene (DVB), 572
dog *see Canis familiaris*
dog appeasing pheromone (DAP), 600
dolphins, 958
donepezil, 923
Down syndrome, **379**, 409, 496
    olfactory dysfunction in, 403, 412–13
doxepin, 904, 916
draw tube olfactometer of Zwaardemaker, 228, *228*
Drosophila ampilophila (pomice fly), 21
Drosophila melanogaster (fruit fly), 532, *532*, 929
    antennal lobe, *536*
    glomeruli in, 535
    Kenyon cells in, 535
    projection neurons (PNs) in, 535
Drosophila
    gustatory receptors in, 930, 940
    carbon dioxide detection, 936
    carbon dioxide receptor, 540
    discrimination between sugar types, 937
    dopaminergic signaling, 939
    labeled line coding in, 938
    olfactory lobe, 539
    olfactory receptors in, 16
    OSNs in, 533
    peripheral taste function, 933
    receptors to amino acids, 935
    salt, aversive response to, 934
    sugar, taste response to, 934
drug-induced Parkinson's disease (DPD), 426–7
dry-plate photography, 279

DT-diaphorase, 71
dual affective architecture model of affect, 782
dual olfactory hypothesis, 1114
Dumetella carolinensis (gray catbirds), 573
dyes, transport of, 462
dysgeusia, 23, 765, 844, 850, 888, 889, 891
dysosmia (parosmia), 24, 375, 385, 844, 850

E-cadherin, 133
echolocation, 41
echo-planar imaging (EPI), 283
ecoturbinals, 42
Edswarsiella ictaluri, 100
Egypt, ancient, use of perfumes in, 4
electro-olfactogram (EOG), 367, 377, 548
electroencephalography (EEG)
    chemosensory event-related potentials (CSERPs), 264–72
    continuous time-frequency analysis, 272, 273–4, *274*
    discrete frequency and continuous time-frequency measures, 272–3
    event-related potentials (ERPs), 262
    frequency domain, 262
    origins, 261
    oscillatory changes, 262–4
        clinical and hedonic significance, 263–4
        coherence, 263
        early research, 262–3
        neuronal mechanisms, 262
        temporal resolution, 264
        theta activity, 262–3
    in electroencephalogram (EEG) , taste disturbance, 890
    time domain, 262
    time-frequency, 262
electrogustatory stimulation, 776
electrogustometry EGM), 752, 754–5, 776
electromyography (EMG), facial responses to odors and, 313
electron microscopy (EM), 280
electronic noses, 1069
electronic tongues, 1069
electrophysiological studies, 14, 245
    measurement of olfactory function, 261–74
empty nose syndrome, 369
encephalitis, arboviral, 380
endocasts, 57
endocrine disorders, olfactory dysfunction and, 388–9
endometriosis, 469
endopeptidases, 66
endoscopic nasal sinus surgery, 360, 361, 362–3
endoturbinals, 42
enoxacin, 915

Enteromorpha, 526
entorhinal cortex, 209–12, 217–219, 286, 290, 307, 343, 409–412, 711
Entosphenus (lamphrey), 462
Environmental Protection Agency (EPA) Ground Water Rule, 1080, 1081
ephedrine, 363
epilepsy, 8, 24, 101, 239, 271, **380**, 429, 437, 438, **458**, 467, 599, 890, 900–902
epithelial sodium channels (ENaCS) taste, 686
epoxide hydrolases (EPHXs), 68, 71, 79, 464, 486
equilibrative nucleoside transporters [ENT], 466
Eretmochelys imbricata bissa (Pacific hawksbill), 971
Eristalis tenax (hoverfly), 935
ERM model, 248
Escherichia coli 0157:H7, 520
    chemotaxis, 513, 514–16, *514*
    lipopolysaccaride (LPS), 584
    quorum sensing in, 520
Esox lucius (pike), 460, 463
essential oils, fungitoxic properties, 7
essential tremor (ET), 413, 423
estra-1,3,5(10),16-tetraen-3-ol (EST), 298
estradiol, 73
    and anorexia nervosa, 871–2
    and conditioned taste aversion, 870
    and food intake, 869–70
estrogen hypersensitivity hypothesis, 871
eszopiclone (Lunesta™), 902
ethanol, 679
    taste qualities of, 835–6
ethmoid, 39
ethmoturbinals, 42, 46, 47, 51, 605, 606, *606*
ethyl hexanoate, 538
ethylene glycol, modified, 573
ethyleneamine diamine tetraacetic acid (EDTA), 469
Eugenes fulgens (hummingbird), 965
eugenol, 1092
eukaryotic microbes, 520–2, *522*
Eumeces laticeps, 1142
Eumeces schneideri, 974
Euplotes, 527, *527*
Euplotes crassus, 527
Euplotes octocarinatus, 527
Euplotes raikovi, 527
Euplotes vannus, 527
Evans blue, olfactory system in transport of, 462
Event-Related Potentials (ERPs), 264
extensively hydrolyzed protein infant formulas (ePHF), 823
external cellular layer (ECL), 172–3
external plexiform layer (EPL), 157

Subject Index

external tufted cells (ETCs), 158–9
ezrin, 651

Facial Action Coding System (FACS), 312, 799
facial nerve, disorders of, 894–7
facial responses to odors, 312
familial dysautomonia, 149, 903
Fast Fourier Transform (FFT), 272
fast-spin-echo two-dimensional sequence, 284
fat taste qualities of, 834–5
fear sweat, 298
Felis catus (cat), 587
fenoprofen, 915
fentanyl, 467, 921
ferrocene, 76
fetal alcohol spectrum disorder (FASD), 499
fetus
  chemoreception, 314–15, *315*
  development of taste, 796
  olfaction in, 307–8
  taste sensitivity in, 797
fexofenadine, 921
Fibroblast Growth Factors (FGFs), 734
filiform papillae, 639, 640, *641*, 643
fish
  facial and trigeminal nerve, 949, *950*
  nasal sac in, 41
  olfaction vs mammalian olfaction, 558
  palatal organs and barbels in, 949
  responses to amino acids, 951
  sensitivity to carbon dioxide, 953
  taste buds in, 947–8
  terminal nerve in, 1143
FLAIR (fluid-attenuated inversion recovery), 890
flash profiling (FP), 1072
flavin-containing monooxygenases, 68
flavonol sulfotransferase (FST) gene families, 70
flavor
  appetite and obesity, 1020–1
  binding of sensory signals, 1015–19
    cognitive processes in, 1017–18
    fusion in flavor perception, 1017–19
    spatial and temporal contiguity, 1015–17
  definition, 796, 888, 1007, 1055
  history, 3, 6, 9, 23
  interactions with central cause, 1010–12
    effects of vision and audition on flavor perception, 1010
    olfaction and taste, 1002–12
    olfaction and touch, 1010
  interactions with peripheral cause, 1008–10
    irritation and olfaction, 1009
    irritation and taste, 1008

irritation and temperature, 1008–9
irritation and touch, 1008
taste and touch, 1009–10
temperature and olfaction, 1009
temperature and taste, 1009
  measurement of sensory qualities, 1021
  perceptual loss, 1019
  sensory interactions within, 1008
  vs taste, 1007
flavor amplification of foods, 849
flavor and fragrance industry, 1067–77
  affective aspects, 1074–5
  analytical instruments, 1069
  cognitive and conceptual aspects, 1075
  consumer wants and needs, 1076–7
  designing flavors and fragrances, 1068
  history of, 1067–8
  measurement, 1068–76
  sensory aspects, 1069–74
  sensory discrimination methods, 1073–4
  sensory distances, 1074
  sensory panels, 1069
  sensory profiling, 1069–73
  stimulus complexity, 1075–6
flavor-aversion learning, 1015
flavor-calorie learning, 1015
flavor-drug learning, 1015
flavor-flavor learning, 1014–15, 1020
flavor learning, 1012–15
flavor preceptual learning, 1013–14
Flavor Profile Analysis (FPA), 1080
Flavor Profile Method, 1080
flavor recognition learning, 1014
Flavor Threshold Test, 1080–1
flavor wheels, 1081
Flehmen-like facial expressions, 617
Flossina fuscipes fuscipes (tsetse fly), 935
flunarizine, 426
fluoro-2-deoxy-D-glucose (FDG), 287
fluoxetine, 917
foliate papillae, 639, 640, *641*, 642, 727, 728
  in macaques, 984
folic acid, 463
follistatin (Fst), 734
food and beverage industries, 1051–62
  age and aging, 1055–6
  brain scanning, 1056–7
  chemical basis of food acceptance, 1053–4
  chemosensory science and, 1051–2, 1055–6
  children and, 1055–6
  cognitive processes, 1061–2
    cognition and emotion, 1062
    dynamic nature of liking, 1062
    expectations, 1061–2
    repeated exposure, 1062
  consumer behavior, 1052–3
  cuisine, 1054–5

global products, 1054–5
human sensory evaluation, 1057–60
  genetic factors, 1059–60
  nutritional genes, 1060
  prior experience on liking and choice, 1058
  smell genes, 1059–60
  taste genes, 1059
  wine industry, 1058–9
machine sensing (artificial noses), 1057
older consumers, 1056
organic and biodynamic food, 1060–1
perceived health benefits, 1060
product differentiation, competitive edge and quality assurance, 1053
societal influences and fashion, 1060–2
food choice behavior, human, 1052–3
food hedonics, 832–3
food intake, chemical senses and, 833
food preference, odor memory and, 344–5
food reward of taste, brain mechanisms and, 1027, 1028
food testing, 767
food texture, primates and, 1036–7
foraging, 610–11
forced-choice discrimination, 752, 753
forced-choice odor identification tests, 240
forced-choice procedures, 230
forced-choice threshold test, 241
forebrain, neural coding in, 710
Fragile X mental retardation, 149
Fragile X mental retardation protein (FMRP), 412
Free Choice profiling (FCP), 1072, *1073*
free modulus method, 235
Free Sorting (FS), 1072–3
frequency analysis, 272
Frey's syndrome, 898
Friedreich's ataxia (FRDA), 436
frontotemporal dementia, 438
frontoturbinals, 47
functional Magnetic Resonance Imaging (fMRI), 20
  cross-adaptation, 292
  in encoding odor objects, 292–3, *293*
  in encoding odor valence, 290–2, *291*
  event-related era, 289–90
  initial olfactory studies, 288–9
  multivariate pattern analysis, 293–4, *294*
  of olfactory memory, 297
  taste in humans, 1037
  taste disturbance, 890
  taste perception, 775
fungiform papillae, 640, *641*, 642–3, 651, 727–8, 733–4, 959, 984

G protein-coupled receptors (GPCRs), 15–17, 95, 103–4, 109–113, 118, 127–28, 306, 389, 521–5, 533, 556,

Subject Index

G protein-coupled receptors (*continued*)
629, 631, 633, 646, 648, 650, 666, 668,
686, 688, 693, 695, 840, 914, 932, 953,
960, 967, 986, 1102, 1115–17,
1134–41
gag reflex, 897, 898
Gaidropsarus (rockling), 644
galanin, 916
Galbula ruficauda (rufous-tailed jacamar),
969
Galen (Claudius Galenus), 8
gallamine triethiodide (Flaxedil), 710
Gallotia caesaris (Boettger's lizard), 972
Gallus domesticus (chicken), 964–5
gamma glutamyl peptidase, 64
gas chromatography (GC), 1068, 1081
gas chromatography/mass spectrometry
(GC/MS), 574
gastroesophageal reflux disease (GERD),
894
general Labeled Magnitude Scale (gLMS),
235–6, 760
generalization, 344
genetic disorders, taste dysfunction and, 903
Genetta thierryi (genet), 958
Germ Theory of Disease, 1079
ghrelin, 838, 842
gingiva, disorders of, 893–4
gingival fistula, 893
gingivitis, 889, 890
glands
Blandin-Nuhn, 644
Bowman's, 48, 64–71, 79–80
in Canidae, 594
parotid, 625, 626, 628
preen, 566–8
salivary, 625, 643–4
scent, 611
serous, 626
sublingual, 625, 626
submandibular, 625, 626
ventral septal, 48
vomeronasal, 48
Von Ebner's, 626, 628, 732–3
glial cytoplasmic inclusions (GCIs), 427
glioblastoma (GBM), 469, 474
glioma, frontal lobe, 384
Global Field Power (GFP), 266
globose basal cells (GBCs), 94, 95, 96, *96*,
133–4, 135–7
glomerular layer (GL), 157, 171–2
glomerular neurons, 158–9, 172
adult-born neurons and
sensory/experience-dependent
plasticity, 166–7
excitatory intraglomerular processing,
160
inhibitory intraglomerular processing,
161–2

interglomerular processing, 162–3
glossitis, 891
glossopharyngeal nerve, 625, 654
disorders of, 897–8
glossopharyngeal neuralgia, 898
glucagon-like peptide-1 (GLP-1), 832, 838,
839, 842
glucocerebrosidase-related parkinsonism,
427
glutathione (GSH), 66
glutathione peroxidase, 71
glutathione-S-transferases (GSTs), 68, 464,
486
glycyrrhizic acid (GA), 669
gold, transportation of, 460, 465, 466
gonadal dysgenesis, 903
Grammia geneura, 935
granule cell layer (GCL), 157, 164–5
granule cells (GCs), 164–5
greater petrosal nerve, 654
Greece, ancient, use of perfumes in, 4–5
Grindelia leaf extract, 938
Grueneberg ganglion (GG), 305, 1133,
1136–41
axonal projection patterns of neurons,
1138
cell types in, 1136–7
coolness-induced vs odorant-induced
responses, 1140
molecular signaling cascades, 1139–40
morphology, 1136–7, *1137*
olfactory responsiveness of neurons,
1138–9
olfactory signaling proteins in neurons,
1138
as part of the 'necklace' olfactory system,
1140–1
responsiveness to coolness, 1139
spontaneous electrical activity of neurons,
1138
Guam ALS/PD complex, 24
guanosine 5′-monophosphate (GMP), 631
guanylyl cyclase D, 16, 116–17
guanylin, 116
Gulf War syndrome, 496–7
gustatory agnosia, 888
gustatory axon projections to the CNS,
development of, 741–2
gustatory cortex, 715–17
human, 775–81
primate, 711–12, 714–15, *714*, 715–17
rodent, 711, 712–14, 715, 716
gustatory hallucinations, 903
gustatory neural coding, 701
excitatory response intervals and sensory
confounds, 701–2
quality coding, with or across neurons,
702–3
stimulus selection, 702

temporal and ensemble coding, 702
gustatory papillae, 639–42, *641*
gustory receptor neurons (GRNs) in insects,
929, 932–3
bitter-sensitive, 933
deterrent, 933
gustatory reflex, 625
gustatory sensory ganglia, formation of,
736–8, *737*
gustducin, 650, 666, 670, 672
gustin (CA6), 915
gut peptides, 842
gut taste receptors, 887–8
flavor preference and appetite, 841
gastric emptying and, 841–2
glucose homeostasis and, 842
gymnemic acid, 665, 669, 986

habituation paradigms, 20
Haemophilus influenzae, 897
Hagen–Poiseuille equation, 356
hairy tongue, 891–2
hallucinatory psychoses, chronic, 390
haloperidone, 426
haplorhines, 605, 606
head trauma, 24
olfactory bulb volume in, 377
olfactory dysfunction in, 385–7, 438
taste dysfunction and, 896
heartburn, 894
Heliothis, 540
Heliothis virescens, glomeruli in, 535
hemiageusia, 780
heparanase, 71
hepatitis, viral, 848
heptanal, 1081
Hereditary Sensory and Autonomic
Neuropathy Type III) *see* familial
dysautonomia
herpes simplex virus, 462
Herpes zoster oticus, 896
Heterocephalus glaber (naked mole rats),
586
high mobility group box-1 (HMGB1), 468
high-potency (HP) sweeteners, 666–7, 668
hippocampus, explicit memory and, 342
Hippocrates, 8
histatins, 627, 632
HIV/AIDS, 380, 438, 454,467, 893, 894,
897, 902
hodulcin, 669
Homeobox A2 (Hoxa2), 737
homobatrachotoxin, 568
Hoplopagrus guentherii (barred snapper),
551
horizontal basal cells (HBCs), 94, 95–6,
133–4, 135–7, 138–9
hormonal stress response, 873

Subject Index

hormone analogues, administration via nasal cavity, 469

horseradish peroxidase (HRP) , transport of, 464

hot wire anemometer, 358

human herpes virus 6 (HHV-6), 488

human immunodeficiency syndrome *see* HIV/AIDS

Human Leucocyte Antigen (HLA), odor memory and, 325–6

hunger, 832–3

Huntington's disease, 24, 198, 413
    olfactory dysfunction in, 403

hydrocephalus, 384

hydroxysteroid sulfotransferase (HSST) gene families, 70

hygiene, odors and, 7

hyperosmia, 375

hypertension, chemosensory abnormalities in, 845–6

hypogeusia, 844, 887, 888

hypogonadopropic hypogonadism, olfactory dysfunction in, 438

hyposalivation, 901

hyposmia, 375, 843–4
    MRI of, 283

hypothalamus in macaques, 997

hypothyroidism, 389
    chemosensory abnormalities in, 846–7
    taste dysfunction in, 901

Ichthyomyzon unicuspis, 1142

Ictalurus natalis (brown bullhead), *462*

Ictalurus natalis yellow bullhead catfish), 948

Ictalurus nebulosus (catfish), *462*

Icterus abeillei (black-backed oriole), 969

idiopathic hypogonadotropic hypogonadism (IHH), 389

imipramine, 902

Imitrex®, 467

immediate neuronal precursor, 135

immunoglobulins, transport of, 464

Implicit Association Test, 345

in utero chemosensory plasticity, 499

in vitro staining, 57

incest avoidance, 324

indinavir, 917

infants, taste preference in, 797–9
    analgesic responses, 800
    ethanol and, 835–6
    fats and, 835
    orofacial expressions, 799–800
    salt and, 836–7
    sucking and consummatory responses, 797–9
    sweet, 833–4

inferior turbinate reduction, 360

inferior turbinectomy, 361

inflammatory bowel disease, 100

influenza, 380, 469

inhibitory avoidance behavior, 874

inosine 5′-monophosphate (IMP), 631

inosine 5′-phosphate, 834

inositol 1,4,5-trisphosphate (IP3), 693

inositol trisphosphate (IP3), 666

inositol trisphosphate receptor (IP3R), 666

insect olfaction, 531–40
    carbon dioxide, 540
    combinatorial odor coding, 535–7
    concentration coding and gain control, 537–8
    contrast enhancement, 537
    discrimination and generalization, 539
    insect response to odors, 531–2
    odor coding dynamics, 541
    odor representation transformed between antennal lobe and mushroom bodies, 540–1
    odor-mixture processing, 538
    olfactory pathway, 532–41, *534*
    processing of concurrent odors from multiple sources, 538–9
    sexual pheromones, 539–40
    signal to noise improvement, 537

insects, taste processing in, 929–41
    dietary exposure, 939–40
        habituation, 939
        health status, 940
        induced food preferences, 940
        long-term adaptation, 939–40
        short-term sensory adaptation, 939
    dopaminergic signaling, 939
    organization of taste system, 929–33
    peripheral taste function, 933–6
        acids, 935
        amino acids, 934–5
        aversive compounds, 933
        salts, 934
        species-specific taste stimuli, 935–6
        sugars, 933–4
        water, 935
    nutrient levels in hemolymph, 939
    taste-mixture interactions, 936–7
        in CNS, 936–7
        peripheral taste system, 936
    taste processing in CNS, 937–8
        coding mechanisms, 938
        in SOG, 937–8
    taste sensilla in, 533, 929–30, *930*, 938–9

insular-opecular cortex in macaques, 990–3, *990*

insulin, 469

insulin-like growth factor-I, 66

internal carotid aneurysms, 384, 385

internal plexiform layer (IPL), 157

interneurons, 157

interturbinal, 47

intravascular taste, drugs and, 916

iron, olfactory system in transport of, 459–60

irritation, 891, 894, 904, 1008, 1009, 1091–1104

ischaemia, SVZ neurogenesis in, 198

Junco hyemalis (dark eyed juncos), 570, 573, 574

just noticeable difference (JND), 231

juvenile recognition of odors, 340

juxtaglomerular cells (JGCs), 157, 158

kallikrein, 66

Kallmann syndrome, 24, 376, 389–90
    odor memory tests in, 239
    olfactory bulb volume in, 377

Kenyon cells, 535

keratin, 133, 136, 144, 566, 627, 733, 741, 984, 1137

ketones, 570

kin recognition, 343

kinocilia, 47

Kirrel2 and Kirrel3, 128

klinotaxis, 549

Kluver-Bucy syndrome, 783

kokumi, 674, 692–3

Korsakoff psychosis, 24, 438
    odor memory tests in, 239

Kovat's retention indices, 571

L-α-phosphatidylcholine, 469

L-buthionine sulphoximine, 80

L-menthol, 369

Labeled Affective Magnitude Scale, 761

Labeled Hedonic Scale, 761

labeled line, 539–40

Labeled Magnitude Scale, 803

labeled scales, 235, 236, 758–61, *758*
    invalid comparisons, 759–60, *760*
    properties of, 758–9

lactating breast, odor of, 316–17

lamotrigine, 905

Lampetra planeri, 1142

lampreys
    responses to amino acids, 951–2
    taste buds in, 947
    terminal nerve in, 1142

Larmor frequency, 282

laryngectomy, 362
    total, olfactory bulb volume in, 377

lateral olfactory tract (LOT), 164, 173, 210-11, 307, 552, 1121

Latimeria chalumnae (coelacanth), 553

Lattice Boltzmann simulations, 359

laureth-9, 469
learned satiety, 1015
learning, 344–6
lectins, transport of, 461
Lepidochelys olivacea (Pacific ridley), 971
Lepisosteus osseus (garfish), 463
leucine, olfactory system in transport of, 462–3
leukoplakia, 891
leukotriene A4 hydrolase, 71
levocetirizine, 363
levosulpiride, 426
Lewy body disease (LBD) (diffuse Lewy body disease), 414
lidocaine, 363, 467, 905, 916
light microscopy, 279
lipoxygenases, 71
lithium chloride (LiCl), 865, 870, 872
   toxicity stimulus detection and, 876
liver diseases, chemosensory abnormalities in, 848–9
liver transplantation, 848–9
lizards, taste in, 973–4
Locusta migratoria, 938
loouping ill virus, 17
loss of consciousness (LOC), duration of, 386–7
Lou Gehrig's disease *see* amyotrophic lateral sclerosis (ALS)
low density lipoprotein receptor, 64
lucifer, olfactory system in transport of, 462
lung cancer, 599
lyme disease, 897
Lyon Clinical Olfactory Test, 309

Mabuya multifasciata, 974
macaques, gustatory system in, 983–98
   central taste pathways, 987–96
      nucleus of solitary tract, 987–9
      thalamus, 989–90
      insular-opecular cortex, 990–3, *990*
      orbital cortex, 993–6
   descending gustatory projections, 996–7
      amygdala, 996
      hypothalamus, 997
   peripheral organization, 983–5
      circumvallate papillae, 984
      foliate papillae, 984
      fungiform papillae, 984
      neural innervation of oral cavity, *984*
      taste bud innervation, 984–5
      taste buds, 983–5
      taste pores, 984
   peripheral taste nerves, 985–7
   see also primates
Machado-Joseph disease (SCA3/MJD), 436, 903
macrosmatic species, 12, 22, 51, 54, 55, 57
magnetic resonance imaging *see* MRI

magnetoencephalography (MEG), 267
   spatio-temporal pattern of activation, 779–80
   taste pathways, 776–9
   taste perception, 775
magnitude estimation, 235–6, 755–6, 759
Maillard reaction, 674
main olfactory bulb (MOB), 11, 14, 16–21, 24, 43, 51, 63, 70–4, 76, 80–1, 98, 99, 101, 109, 123–4, 128, 135–7, 140–1, 144–8, 157–174, 183–199, 209–219, 245, 250, 261, 279, 280–6, 295, 299, 306–7, 339, 344, 346, 365, 368, 376–8, 380, 384, 385–8, 390, 393, 411–4, 423–8, 434–5, 453–4, 459–66, 473, 486–7, 490–1, 493–6, 499, 502, 549, 552, 556, 584–5, 592, 606–8, 617–18, 625, 711, 846, 950, 972, 1029–30, 1054, 1113
major urinary proteins (MUPs), 582
malaria, 531
malignant melanoma, 599
malingering, 376
Mamestra configurata, taste sensillum in, *930*
mammalian olfaction vs fish olfaction, 558
mammals, taste bud cells types in, *645, 647–50, 648*
   molecular markers, 650–1
   substances associated with, **649**
   type I (glia-like cells), 648
   type II (receptor cells), 648
   type III (presynaptics cells), 648
   type IV (basal cells), 648–50
   type V (marginal cells), 650
mammary pheromone, 317
Manduca sexta, 70, 532, *532, 533*
   aversion adaptive response, 939
   caffeine and, 936–7
   central gustatory system, 940
   ensemble coding in, 938
   food preferences, 940
   glomeruli in, 535
   sugar, taste response to, 933–4
   sugar-sensitive GRNs, 936
   taste sensilla of, *931*
manganese, 460–1, 497–498
manganese occupational neurotoxicity, 497
manganese-superoxide dismutase, 468
manganism (manganese toxicity), 460, 498
marginoturbinal, 46
mass spectrometry (MS), 1068
masticatory reflex, 625
mastoidectomy, taste dysfunction and, 896
match-to-sample odor identification test, 309
matrix metalloproteinase (MMP), 9 500
maxilloturbinals, 42, 46
measure of response bias (beta), 233

meatuses, 39
medial orbitofrontal cortex, 411
medial-medial olfactory tract(MMOT), 552
medications, taste dysfunction and, 901–2
mediodorsal thalamus (MDT), in coding olfactory prediction error, 295–6
meditation, 905
Megaderma lyra (bat), nasal mucosae in, 49
Megaerops ecaudatus (bat) nasal mucosae in, *49*
megalin, 64
Melkersson syndrome, 896
memory
   amygdala in, 342
   in perception of odorant mixtures, 252–3
   working, 297, 338–40
   see also olfactory memory
memory tests, 238–9
meningiomas, 384
meningoencephalitis, 893
menstrual synchrony, 22
menthol, 1092
mercury, olfactory system in transport of, 461
mere odor familiarization, 308
Mesocricetus auratus (hamster), 1143
metabotropic glutamate receptor 1 (mGluR1) receptors, 161
metaiodobenzylguanidine (MIBG), 423, 428
metal transporters, 486
metallic taste, 888, 915–16
metals
   olfaction and, 497, 502–3
   transport of, 457–61
methimazole, 76, 80, 464, 915
methyl anthranilate, 969
methyl bromide, 80
methyl-accepting proteins (MCPs), 514
methyl cyclohexanemethanol (MCHM), 1087
methyl-cyclopentenolone, 263
methyl methacrylate, 71
methyl salicylate, 1092
methyl tertiary butyl ether (MTBE) in drinking water, 1083–6
methyl xanthines, 916
metoclopramide, 426
metronidazole, 914
metyrapone, 79, 80
microbial chemical sensing, 513–28
   bacterial chemotaxis, 513–18
   biofilms, 520, *520*
   ciliated protozoa, 522–7
   eukaryotic microbes, 520–2, *522*
   quorum sensing, 518–20
Microcebus murinus (lesser mouse lemur), 54, *55*
   lateral nasal wall in, *52*
   nasal mucosa in, *56*

Subject Index

respiratory and olfactory epithelia in, *48*
microsmatic animals, 12
microsmia, 375
microsomal epoxide hydrolase, 71
Microtus ochrogaster (prairie vole), 581, 1144
Microtus pennsylvanicus (meadow vole), 582
microvillar cells, 95, *96*
midazolam, 467
Middle Ages
    perfumery during, 6
    stenches during, 7
middle turbinate medialization, 384
migraine, 900
mild cognitive impairment (MCI), 411
Mini-Mental Status Examination, 410, 438
miraculin, 986
mirtazapine, 904
mitral and tufted cells (M/TCs), 157
mitral cell layer (MCL), 157, 186, *285*, *306*, *424*
mitral cells (MCs), 140, 157, *157*–174, 196, 199, 211, 212, 215–217, 250, 252, *280*, *285*, 307, 338, 339, 453, *454*, 460, 461, 490, 552, 553, *1120*, *1121*
mixture suppression, 672
mogroside V, 668
mometasone, 363
monellin, 669, 959
monoamine oxidase (MAO), 71
monooxygenases, 464
monosodium glutamate (MSG), 672, 674, 702, 783, 834, 905
    neuroimaging and, 1037
    touch sensation and, 1009–10
Moraxella catarrhalis, 897
Morganucodon, 49
motivational matching, 319
mouse hepatitis virus, 462
MPTP, 23, 24, 426–7, 468, 474
    MPTP-induced parkinsonism (MPTP-P), 23, 423, 426, *427*, 438
MRI, 20, 279, 280, *281*
    functional, 283
    functioning of, 281–2
    of human olfactory bulb, 283–6, *284*, *285*
    signal generation, 282
    in taste disturbance, 890
    transformation of MR signal into spatial image, 282–3
MRI-compatible olfactometers, 289
MSA-Cerebellar (MSA-C), 427
MSA-Parkinsonism (MSA-P), 427
mucins, 626, 627
mucus, nasal, components, 66
multidimensional scaling (MDS), 236–7
multidrug resistance P-glycoproteins (MDR), 486

multidrug resistance-related proteins (MRP), 71, 486
multiple chemical hypersensitivity, 375
multiple chemical sensitivity (MCS), 496
multiple-choice identification tests, 237
multiple-item odor identification tests, 240
multiple sclerosis, 23, 899
    olfactory bulb volume in, 377
    olfactory dysfunction in, 391, 403
multiple system atrophy (MSA), 423, 427–8
multipotent progenitor cell, 135
mumps, 894
murine coronavirus, 462
Mus musculus (mouse)
    olfactory system, 123
    scent marking in, 581, *582*
muscarinic acetylcholine receptor (M3-R), 112
mustard oil, 1009
Mustela putorius furo (ferret), 958
myasthenia gravis (MG), 24, 468, 496
    olfactory dysfunction in, 391
Myxine glutinosa (hagfish), 557
Myxococcus xanthus, chemotaxis in, 516

n-aldehydes, 616
n-amyl acetate, 592
N-cadherin, 737
n-carboxylic acids, 616
N-cyclopropyl E2, Z6-nonadienamide, 675
n-decanal, 574
N-geranyl 2-methylbutanamide, 675
N-geranyl isobutanamide, 675
N-geranylcyclopropylcarboximide, 679
n-hexanal, 570
N-(2-hydroxyethyl)-4-methyl-2-((4-methyl-1H-indol-3-yl)thio)pentanamide, 678
N-(3-methoxyphenyl)-4-chlorocinnamide, 679
N-methyl-D-asparate (NMDA) receptors, 161
N-nitrosonornicotine, 78, 79
N-nitrosopiperidine, 78
N-nitrosopyrrolidine, 78
n-octanal, 570, 574
NADPH:quinone oxidoreductase, 71
NADPH-cytochrome P450 reductase (CPR), 68
nafarelin, 469
naphthalene, 69, 76
NAR, 357
nares, 39, 549
nasal airflow
    animal studies, 367–8
    measurement of, 357–60
    influence on olfactory and trigeminal senses, 356
    in nasal pathology and after clinical treatment, 361–4

odorant deposition and uptake, 365–6
olfactory function in humans and, 360–1, *361*
pharmacological treatment of, 363
nasal airway resistance (NAR), 356
nasal anatomy
    human, 355
    in dog, *43*
    in mammals, 39–41, *40*, 42–51, *49*
    microanatomy, 47–9
    phylogenetic variation, 49–50
    primate internal nasal anatomy, 51–6, *49*, *52–3*
    primates, olfactory reduction in, 54–6
    techniques for studying, 57–8
    in terrestrial vertebrates, 41–2
nasal biotransformation enzymes, 68–72
    detoxication of inhaled toxicants, 72
    functions of, 72–8
    metabolic activation and xenobiotic toxicity, 76–8
    modification of olfactory stimuli, 73–4
    modulation of endogenous signaling molecules, 75–6
    protection from inhaled toxicants, 72–3
nasal cancer, 78
nasal cavity
    anatomy, 64
    drugs/toxicants administed via, for CNS and systemic effects, **458–9**
    functional physiology, 355
    mammalian, development of, 41–2
    substance transported to brain, **454–5**
    therapeutic drug administration via, 466–73, **470–2**
nasal cavity volume (NCV), 357, 364
nasal chemoreceptive (nasochemoreception) functioning, 306–7
nasal chemoreceptive structures, 305–6
nasal cilia, 66–7
nasal conchae, 39
nasal cycle, 41, 364–5
nasal decongestants in sinus disease, 363
nasal dilators, 363–4
nasal endoscopy, 376
nasal epithelium, 47, 64
nasal flow in aquatic vertebrates, 550–1
nasal fluid dynamics, 356–7
nasal fossa, 39, 42, *43*
    terminology, **44–5**
nasal localization thresholds (NLTs), 1095
nasal obstruction index, *382*
nasal patency, 368–9
    potential mechanisms of nasal flow perception, 368–9
nasal polyps, 362–3
nasal resistance, 361, 363
nasal sac in fish, 41
    in amphibians, 41

nasal septal cartilage, 39
nasal septum, 39, 42
nasal sinus disease, 361, 364, 366, 368–370, 381
nasal surgery, 383–4
nasal trigeminal chemosensitivity
  and nasal localization thresholds in normosmics and osmotics, 1094–5
  nasal pungency thresholds in anosmia, 1093–4
  vs olfactory sensitivity, 1092–3
  structure-activity relations in, 1096–198
  temporal properties of, 1100–1
nasopharyngeal carcinoma, 384
nasosinus disease, see nasal sinus disease
nasoturbinals, 39–41, 42, 46, 51, 54, 605
Navier–Stokes equation, 359
Near Infrared Spectroscopy (NIRS), 307, 308
Nectarinia talatala (white-bellied sunbird), 965
Necturus (mudpuppy), 646, 647, 950
Necturus maculosus (mudpuppy), 1144
negative allosteric modulators (NAMs), 667
negative mucosal potential (NMP), 1094
negativity bias, 782
Neisseria meningitidis, 469
nelfinavir, 917
Neogobius melanostomus (round goby), 550, 551
neonates
  facial reactivity to odors, 312, 315
  olfaction in, 308
neophobia, 344, 869
neostigmine, 468
neural bud, 739
neural crest derived cells (NCDC), 96
neurodegenerative diseases, 468
neuroepithelium, 47
neurofibrillary tangles (NFTs), 411
neurogenic inflammation, 488–9
neuroimaging in humans, 1037–43
  affective value vs intensity, 1041
  odor, 1039
  oral viscosity and fat texture, 1039–40
  reward and decision making, 1041–3
  sight of food, 1040–1
  taste, 1037–9
  top-down cognitive effects on taste, olfactory and flavor processing, 1041
neuromodulatory inputs, 169–1
neuron-specific enolase (NSE), 650
neuropeptide Y, 469
neuropeptides, 694–5
neurophysiology
  human, 1028–30
  primate, 608–9
Neuropilin-1, 128
neuropraxia, 896

neurotensin, 838
neurotmesis, 896
neurotransmitters, 694–5
nickel, 498
  olfactory system in transport of, 461
nicotine, 69, 679
  nasal sprays, 468–9
nicotinic acid, 916
nicotinic stomatitis, 891
NIH Toolbox Pediatric Odor Identification Test, 309
Nile virus, 462
nitrobenzene, 388
NNK, 72, 78, 464
no-slip, 356
nonallergic rhinitis, 362
noncontact infrared thermometry, 368
non-esterified fatty acids (NEFA), 847
nongustatory papillae, 638–9
nonparametric R index, 233
norepinephrine (NE) input, 170–1
nose as respiratory organ, history of, 355
nose-brain barrier, 73
novel-odor-recognition, 340
nuclear magnetic resonance (NMR), 281–2
nucleus of the lateral olfactory tract, development of, 307
nucleus of the solitary tract (NST), 727, 1028
  macaques, 987–9
  primate, 706–7
  rodent, 706
Nymphicus hollandicus (cockatiel), 966

obesity
  chemosensory abnormalities in, 847
  child, 822, 838
  flavor, appetite and, 1020–1
  ghrelin in, 842
observational learning, 1015
obstructive sleep apnea, 364
ocular trigeminal chemosensitivity
  and eye irritation thresholds in normoscmics and anosmics, 1095–6
  structure-activity relations in, 1096–198
  temporal properties of, 1100–1
odor deprivation, 298
odor discrimination, 298
odor event-related potential (OERP), 377
odor feature differentiation, 297
odor identification tests, 237–8
odor memory, 19, 214, 218, 238–9, 241, 289, 296, 337–47, 388, **404–5**, 413, **415–6**, **430**, 434, 437, 1013
odor naming tests, 237
odor objects, 337
odor perceptual representation, 297
odor responses, 167–8
odor toxicosis association, 346

odorant deposition, 365–6
odorant flow rate, 265
odorant flux, 366
odorant mixtures, perception of, 247–53
odorant receptor gene regulation, 123–9
  class I odorant receptors, 125
  class II odorant receptors, 126
  negative feedback regulation - one receptor rule, 127–8
  odorant receptor gene choice, 126
  organization and structure of, 123–9
  regulation of odorant receptor-instructed axonal projection, 128–9
  single odorant receptor gene expression, 126–8
  zonal expression, 124–6
odorant receptors, molecular structure, 110, *110*
odorant sorption, 365–6
odorant transport model, 365
odorant-binding proteins (OBP), 64, 66
odor-evoked heart-rate reflex and active investigation of odorants, 338
odor-induced taste enhancement, 1011, 1016
odor-nursing association, 325
odor-odor associations, 346
odor-odor learning, 337
odor-tactile associations, 345–6
odor-taste association, 346
odor-taste learning, 337
odor-verbal associations, 346
odor-visual stimuli associations, 345
oesophageal salivary reflexes, 625
olfacto-facial reflex model, 312
olfactometer, 12, 21
olfactory nerve, 80, 98, 146–7, 149, 157, 264–6, 279, 280–5, 306, 365–6, 376, 385, 394, 424, 453–4, 460–73, 487, 490, 549, 553, 595, 950, 1007, 1141–2, 1144
olfactory agnosia, 375
olfactory bulb *see* main olfactory bulb (MOB); *see also* accessory olfactory bulb (AOB)
olfactory bulbectomy, 21, 80, 136, 138, 144, 146, 147, 184
olfactory chambers in aquatic vertebrates, 549
olfactory cleft, 71, 358–367, 369, 370, 383
olfactory conditioning, 612, *613*
olfactory cortex
  anatomy and physiology, 209–19
  in nonhuman primates, 609
  structures and function, 210–19, *211*, *212*
olfactory detection thresholds, *see* thresholds, psychophysical
olfactory discrimination, 616–17
olfactory disorders

Subject Index

causes, 378–92, **378–80**
classification, 375–6
classification of, 13
clinical evaluation, 376–8
incidence, 375
medical history, 376
quantitative olfactory testing and MR imaging, 376–8
treatment, 392–4
olfactory ensheathing cells (OECs), 97–8, *97*, 146–7
olfactory epithelial pathway, 453, 466
olfactory epithelium, 14, 16–9, 47–9, 63–71, 93–104, 109–10, 115, 123, 125, 128–9, 133–149, 158, 212, 215, 216, 246, 280–1, 356–7, 360, 363, 366–8, 376–7, 380, 385, 388, 390, 392, 453–474, 485–6, 488–92, 500–502, 547–558, 593, 606, 608, 735–6, 846, 974, 1056, 1119
in aquatic vertebrates, 551–2
in Canidae, 593
in nonhuman primates, 608
basal cells, 11, 12, 18, 19, 47, 63–67, 93–96, 133–149, 280, 380, 394, 464, 490, 491, 606
cell types, 64
development, 93–4
impact of autonomic modulation on, 488–9
innate immunity in, 99–101
microvillar cells, 95, *96*
migrating cells, 98
olfactory ensheathing cells, 97–8, *97*
olfactory marker protein (OMP), 99, *99*
olfactory nerve and olfactory bulb formation, 98–9, *98–9*
olfactory receptor neuron, 95, *95*
supporting cells, 96, *97*
xenobiotic damage, 486
zinc nanoparticles and olfactory receptor neuron signaling, 101–4, *104*
olfactory epithelium, neurogenesis in adult, 133–9
cell types in, 133–4, *134*
ectomesenchymal stem cell in, 139
globose basal cell lineage in, 137–8, *138*
multipotent stem cells, 135–7
replacement of sensory neurons in, 134–5
basal cell proliferation, **140**
clinical applications, 148–9
brain diseases, 148–9
transplantation repair of spinal cord, 148
molecular regulation of, 141–8
growth factor functions in, 141–6, *144*, **147**

growth factors and receptors, 141, **142–3**
macrophages, 146
olfactory ensheathing cells (OECs) and, 146–7
regulation of, 139–41, *140*
olfactory event-related potentials (OERPs), 412, 413
olfactory evoked response potentials (OERPs), 309
in schizophrenia, 435
olfactory hallucination *see* phantosmia
olfactory imaging
functional imaging, 287–98, *287*, 299
attending to odors, 294–6
encoding odor objects, 292–3, *293*
encoding odor valence, 289–92, *290*
event-related era of fMRI, 289–90
human olfactory first study, 287–8
initial fMRI studies, 288–9
odor imagery, 296
odor memory, 296–7
odor mixture discrimination and disambiguation, 293–4
odors as putative chemosignals, 298
olfactory learning, plasticity and experience, 297–8
history of, 279–81
structural imaging, 283–7, 299
in clinical diagnostics, 286–7
of human olfactory bulb, 283–6
olfactory labeling, 462
olfactory landscape, 539
olfactory learning, in nonhuman primates, 616
olfactory location illusion, 1015–16
olfactory marker protein (OMP), 15, 48, 99, *99*, 133, 307
olfactory memory, 296–7, 337–47
definitions, 337
explicit memory, 340–2
habituation/discrimination, 338–9
implicit memory, 342–4
implicit vs explicit memory, 343–4
interference, 339
long-term memory, 340–4
novelty and change detection, 342–3
persistence of, 324–7
recognition memory, 338–44
semantics-free odor memory, 343
serial memory, 339, *339*
short-term memory (working), 338–40
in squirrel monkey, 616
olfactory mucosa, 63–81
agents causing damage in rodents, **457**
blood vessels, 65–6
cellular anatomy, *65*
composition, 64–7
cytochrome P450 68–70

enzymes, 70–1
epithelial cell types, 64
interspecies differences, 67
lymphatic vessels, 65–6
nasal biotransformation enzymes, 68–72
secretions, 66–7
subepithelial structure, 65–6
olfactory nerve layer (ONL), 157
olfactory nerves
development of, 306
discovery of, 10–11
formation, 98–9, *98–9*
pathway, 453
olfactory neuroblastomas, 148
olfactory neuroepithelium, early studies of, 11–12
olfactory neuron
dysfunction, 362
in insects, 533
olfactory pathways, nose–brain, studies, 17–18
olfactory Pavlovian conditioning, 289
olfactory percept, 262
olfactory predictive coding, 295–6, *296*
olfactory receptor cells, regeneration of, 18–19
olfactory receptor genes, 110, 111, 595
in primates, 609–10, **609**, **610**
olfactory receptor neurons (ORNs), 64, 76, 93, 95, *95*
signaling, 101–4, *104*
olfactory receptors, 109–18
functional assay systems, 111–12
ligand binding mode and G protein coupling, 112–13, *113*
ligand selectivity, 114
in non-olfactory tissues, 117–18
odorant receptor multigene family, 109–11
signal transduction and modulation of activation, 112
vertebrate odorant receptors, 109–14
vertebrate olfactory receptors in olfactory epithelium, 115–17
olfactory recess, 42
olfactory reference syndrome, 390
olfactory reflex, 625
olfactory sensory neurons (OSN), 109, 307, 533–5, *534*
development of, 306
olfactory social communication, 611
olfactory stop response, 264
olfactory toxicants delivered via blooodstream, 464
olfactory tracts in aquatic vertebrates, 552
olfactory tubercle (OT), 210, 211–12
development of, 307
olfactory vector hypothesis, 496
olfactory-to-taste associative learning, 1035

Oncorhynchus, 952
Oncorhynchus keta (chum salmon), 551
Oncorhynchus tshawytscha (king salmon), 460
operant conditioning, 21, 612, 616
optogenetics, 20
oral cavity disorders affecting taste perception, **892**
oral chemesthesis, 1101–5
  half-tongue methodologies, 1105
  oral trigeminal vs gustatory sensitivity, 1101
  structure-activity relationships (SARs) in, 1102–3
  temporal properties, 1103–4, *1104*, 1105
    cross-desensitization, 1103
    desensitization, 1103, 1104
    self-desensitization, 1103
    sensitization, 1103
    stimulus-induced recovery (SIR), 1103, 1104
  threshold, 1105
oral mucosa, disorders of, 891–3
oral pain phantoms, 765–6
orbital cortex in macaques, 993–6
orbitofrontal cortex (OFC), 218–19, 775
Orco, 533
Oreochromis niloticus (tilapia), 1143
organic and biodynamic food, 1060–1
ORNs, 157
oscillations in odor response, 168–9
osmophobia, 900
otitis media, 897
ovarian cancer, 599
oxycodone, 467
oxymetazoline, 363
oxymorphone, 467
oxytocin nasal sprays, 469

p-cresidine, 464
P450 isozymes, 464
Pachytriton breviceps, 974
Pacific sockeye salmon, 548
pair-bonding, 343
palatal organs in aquatic vertebrates, 949
Pan troglodytes (chimpanzee), lateral nasal wall in, *52*
pandysautonomia, 901
paneth cells, 99
panic disorder, olfactory dysfunction in, 438
papillae, 637, *638*
  embryonic origins, 728–9
  see also under types
parabrachial nucleus (PbN), 877–8, 1028
  in rat, 706–7, 708
  taste relay, 780
paracellular transport, 453, 466
Paramecium, 522
Paramecium caudatum, 525, 526

Paramecium tetraurelia, motility, 522, 524–5, *524*
  chemoresponse pathways, 525, *526*
  response to quinidine, 526
paranasal recesses, 42
paranasal sinus, 42
paranasal spaces, 42
paraquat, 468
parasagital meningliomas, 385
paraseptal cartilages, 42
Parkinson's disease (PD), 23, 149, 217
  alpha-synuclein in, 491
  aluminum in, 459
  autonomic system in, 489
  diffusion tensor imaging (DTI), in, 286–7
  drug modeling in lab animals, 468
  EEG, 263
  familial, 424–6, **425**
  iron in, 459–60
  manganese and, 460–1
  metals in, 457
  olfactory bulb volume in, 377
  olfactory CSERPs, 271
  olfactory dysfunction in, 391, 403, 414–24, **415–22**
  olfactory system as route for pathogens to brain, 456
  SVZ neurogenesis in, 198
  transport in, 473, 474
  volumetric brain mapping (VBM) in, 286
Parkinson-Dementia Complex of Guam (PDG), 403, 423, 428, 496
Parkinsonism, forms of, 426–8
parosmia *see* dysosmia
parotid gland, 625, 626, 628
paroxetine, 917
partial anosmia, 375
particle image velocimetry (PIV), 358–9
particulate matter associated with lipopolysaccharides (PM-LPS), 490
Parus major (great tit), 967
passive avoidance learning, 964, 969
pathogen associated molecular patterns (PAMPs), 100
Pavlovian conditioning, 868
  aversive, 297
Peabody Picture Vocabulary Test-Revised (PPVT-R), 412
pellagra, 901
Pemphigus vulgaris, 893
penicillamine, 901
pentadecan-2-one, 570
pepper, 6
peptidylglycine alpha-amidating monooxygenase, 71
perfume use
  in early medicine, 6–8
  history of, 4–6
Perfumer's strip, 228, *228*

periamygdaloid cortex, development of, 307
periglomerular cells (PGCs) cells, 158, 159
perillyl alcohol (POH), 469
perinatal chemosensory matching, 319
periodontal structures, disorders of, 893–4
peripheral nerve
  disorders of, **895**
  trauma, 905
pernicious anemia, 901
Peromyscus leucopus noveboracensis (white footed mice), 587
Peromyscus maniculatus bairdi (deer mice), 582, 587
peroxidase, 66
peroxiredoxins, 66
perphenazine, 426
Persia, use of perfumes in, 5
Petromyzon marinus (lamprey), 548, 557
Pfeiffer Functional Activities Questionnaire (FAQ), 410
phalloidin, 676
phantogeusia, 393, 887, 888
phantom tastes, 891
phantosmia (olfactory hallucination), 24, 375, 384, 393
pharmaceuticals, taste dysfunction and, 911–23, **912–13**
  due to drug sensory properties, 914–16, **915**
    bitter taste, 914–15
    intravascular taste, 916
    metallic taste, 915–16
  modification of normal taste sensations, 916–22
    apical taste cell level, 916–17
    drug bioavailability reduction, 922
    drug-drug interactions and polypharmacy, 917–18
    pharmacokinetic factors, 921–2
    side-effects and TAS2Rs, 922
  treatment and prognosis, 922–3
pharmaceuticals, transport of, 463–4
phenacetin, 464
phenobarbital, 78, 79
phenol sulfotransferase (PST) gene families, 70
phenoxyalkanoic acid salt **5** (lactisole), 669
phenyl ethyl alcohol (PEA), 264, 365
  detection, 381, *382*
phenylthiocarbamide (PTC), 755, 836, 915
phenylthiocarbamide (PTC)/6-propyl-2-thiouracil (PROP), 671
pheromonal cues, 1125
pheromones (chemosignals), 22, 298, 532
  alarm, 586
  animal welfare and, 600
  mammary, 317
  sexual, 539–40

Subject Index

phonophobia, 900
Phormia regina (blowfly), 935
phorone, 80
phosphatidic acid (PA), 672
phosphatidylethanolamine, 64
phosphatidylinositol 4, 5-bisphosphate (PIP2), 679, 693
phosphoinositide-3-kinase (PI3K), 112
phospholipase A2, 71
phospholipase Cβ2 (PLCβ2), 666, 693
photophobia, 900
Phylidonyris novaehollandiae (New Holland honeyeater), 965
Picture Identification Test, 241
Pieris, 532, 935
Pieris brassicae, 935
pilocarpine, 905
pimozide, 921
pinnipeds, olfaction in, 591
piperonyl butoxide, 79
piriform aperture, 39
piriform cortex, 210, 212–17, *213*, 295
    development of, 307
Pithecia, maxilloturbinal in, *52*
pituitary adenomas, 385
plague, 7, *8*
Plato, 8
platyrrhines, 605,*606*
Plexin-A1, 128
Pliny, 8
Plotosus lineatus (striped eel catfish), 549, 552
Plunc, 67
pneumoencephalography, 279
pneumonia, 380
Podarcis sicula, 1142
Podiceps cristatus (great-crested grebe), 962
Poiseuille resistance (*R*), 356
poisoned-partner effect, 869
polar bears, 591
Polarized Sensory Positioning (PSP), 1073
poliomyelitis, 17, 18, 461–2, 473
polyacrylate (PA), 572
polydimethylsiloxane (PDMS), 572, 573
polyinosinic:polycytidylic acid, 80
polymodal nociceptors, 1092
polymyositis (PM), 391
polypectomy, 366, 384
polyposis, 361
polyps
    nasal, 8
    turbinate hypotrophy, 355
positive allosteric modulators (PAMs), 666–8, *667*
Positive and Negative Syndrome Scale (PANSS), 435
positron emission tomography (PET), 19, 20, 279
    in encoding odor objects, 292

human olfactory processing, 287–8
of olfactory memory, 297
in taste disturbance, 890
in taste perception, 775
using Pittsburgh Compound B (PiB), 411
post-traumatic gustatory neuralgia, 898
posterior ethmoidectomy, 384
posterior piriform cortex (PPC), 289
    PPC encoded quality, 292
predictive coding, 295
prednisolone, 363
preen gland, 566–8
preference paradigms, 20, 21
pregnancy, taste function in, 904
presbyosmia, 375
primary olfactory cortex, 166, 411
primates
    caudal brainstem, 706, 707–8
    gustatory cortex, 711–12, 714–15, *714*, 715–17
    nucleus tractus solitarius (NTS), 706–7
    taste bud innervation in, 704
    thalamus, 710, 711
primates, behavior, 610–12
    foraging, 610–11
    sniffing at mouths, 611
    olfactory cues, 610–11
    olfactory social communication, 611
    scent glands, 611
    urine washing, 611
    scent-marking, *612*
primates, food texture and, 1036–7
    oral fat texture, 1036–7
    oral temperature, 1037
    viscosity, particulate quality and astringency, 1036
primates, neurophysiology, 1028–30
primates, olfaction in, 600–18
    anatomy, 605–6
    cribiform plate in, 608, **608**
    genetics, 609–10
    neuroanatomy, 606–8
    neurophysiology, 608–9
    olfactory bulbs in, 606–8, **607**
    olfactory epithelium in, 608
    olfactory receptor (OR) genes in, 609–10, **609**, **610**
    physiology, 612–17
    olfactory detection thresholds, **614–16**
    vomeronasal (VNO) system, 617–18
primates, taste perception in, 986–7
primates, taste processing in, 1030–7
    anterior cingulate cortex, 1032–3
    insular primary taste cortex, 1031
    olfactory, taste and visual representation of flavor, 1035–6
    pathways, 1030–1
    pleasantness of taste, 1034

secondary taste cortex in orbitofrontal cortex, 1031–2
priming cues, 1125
Principal Component Analysis (PCA), 266, *268*
pristane (2, 6, 10, 14-Tetramethylpentadecane), 568
probenecid, 671
procion, olfactory system in transport of, 462
progesterone, 73
progressive supranuclear palsy (PSP), 23, 24, 410, **420**, 423, 428–9
    olfactory dysfunction in, 428–9
    PSP-P, 428–9
proline-rich proteins (PRPs), 626, 632
PROP, 755, 802, 915, 1059
    ethanol sensitvity and, 836
propofol, 501
prostate cancer, 469, 599
protein, taste qualities of, 834
Proterorhinus marmoratus (tube nose goby), 549
Proustian effects, 337
pseudohypoparathyroidism (PHP), 24, 389
Pseudomonas aeruginosa
    biofilm, *520*
    chemotaxis in, 517
Pseudomonas fluorescens
    biofilm, 520
    chemotaxis in, 517
pseudopseudohypoparathyroidism (PPHP), 389
psoriasin, 66
psychiatric disorders
    olfactory disfunction in, 390–1
    taste dysfunction in, 902–3
psychopathy, 24
psychophysical measures of human oral sensation, 751–67
    direct scaling of suprathreshold intensity, 755–8
    in food science, 767
    genetic factors in oral sensory variation, 756–7
    global intensity scales, 761–2
    individual differences in oral sensation, 755–6
    labeled scales, 758–61, *758*
    localised taste loss, 765
    magnitude estimation, 755
    magnitude matching, 755, 756
    oral anesthesia, 766
    oral disinhibition, 765–6
    oral sensory anatomy: videomicroscopy of the tongue, 763–4, *764*
    regional vs whole-mouth sensation, 762–7
    retronasal olfaction, 766–7

psychophysical measures of human oral
sensation (*continued*)
   spatial taste testing, 764–6
   taster status classification, 757–8
   threshold procedures, 752–3
   threshold vs. intensity, 751–2
   whole-mouth oral sensation, 762–3
psychophysical olfactory tests, 230–40
   adaptation, 246–7
   detection of malingering, 245
   perception of odorant mixtures, 247–53
   reliability, 240–1
   responsivity vs sensitivity, 247
   stimulus control and presentation, 228–9
   subject variables, 245–6
   subject variables, 245–6
   test comparison, 241–4, **242–4**
   unilateral vs bilateral, 244–5
Ptilocercus lowii, lateral nasal wall in, *52*
public health, odors and, 7
pure autonomic failure (PAF), 429
purinergic receptor P2Y, 112
Pycnonotus tricolor (dark-capped bulbul),
   965
pyorrhea, 889
pyrazole, 79
pyridine, 388
Pyrrhula pyrrhula (bullfinch), 963
pyrrolizidine alkaloids (PAs), 936, 940

quality discrimination tests, 236–7
quality recognition tests, 237
Quantitative Descriptive Analysis (QDA),
   1069, 1070
quantitative structure-activity relationship
   (QSAR), 1097
quinidine, 917
quinine, 672, 867, 891
quinine HCl (QHCl), 917
quinine sulfate, 969
quorum sensing, 518–20

R-(+)-carvone, 310
rabies virus, 462
radiation therapy, effect on taste, 845, 892
radiography, taste disturbance and, 890
rainbow trout
   sensitivity to carbon dioxide, 953
   TTX detection, 952
Ramsay-Hunt syndrome, 895
Rana (grass frog), 462
Rana sphenocephala (leopard frog), 952
Rana temporaria (common frog), 548
rapid eye movement (REM) sleep behavior
   disorder, 413–14, 427
Rasmussen's encephalitis, 901
Rattus norvegicus, 21
rebaudioside A (REBA), 668, *668*, 669, 671

receiver operating characteristic (ROC)
   curves of odor recognition memory,
   340–1, *341*, 342
receptor expression enhancing proteins
   (REEPs), 111
receptor function, studies of, 14–15
receptor transporting proteins (RTPs), 111
recessus cupularis, 55
recognition memory, 297
recognition threshold, 752
redintegrative flavor learning, 1012–13
regional cerebral blood flow (rCBF), 287
release-of-inhibition phantoms, 765
Renaissance, 9–13
repetition priming, 342–3
reptiles
   bitter taste, 973–4
   gustation in, 970–5
   nasal anatomy in, 41
   substrate lick, 972
   sweet taste, 972–3
   taste behavior, 972
   taste buds, 971–2
   taste capacity, 970
   taste system, 970–2
   tongue flick, 972
resiniferatoxin, 679
resistant to inhibitors of cholinesterase 8
   homologue B (Ric-8B), 111
response criterion, 232, 249
response latency approach, 237
restless leg syndrome, 24, 438
retinoic acid (RA) signaling, 93–4
retronasal odor perception, 227
retrovirally-derived vector (RRV), 135
Rett's syndrome, 148
Reynolds averaging, 359
Reynolds number, 356, 358, 369
rhinarium, 605
rhinomanometry, 357–8, 360, 361, 362
rhinometry, 357–8
Rhinopoma (mouse-tailed bat), 46
rhinoscopes, 376
rhinosinusitis, 66, 283, 298, 369, 380, 381
   acute and chronic, 382–3, *383*
   acute viral-related, 382
   and olfactory loss, 362
Rhizobium, chemotaxis in, 516
rhodanese, 67, 68, 72, 79, 464
Rhodobacter sphaeroides, chemotaxis in,
   516
Richardson syndrome, 428
Riley-Day syndrome *see* familial
   dysautomonia
risperidone, 426
ritonavir, 917
Roccus americana (white perch), 1144
rodents
   brainstem, 708–9

   caudal brainstem, 704–6, 707–8
   gustatory cortex, 711, 712–14, 715, 716
   nucleus tractus solitarius (NTS), 706
   parabrachial nucleus (PBN), 706–7, 708
   pontine taste area, 1028
   taste bud innervation in, 704
   thalamus, 710–11
rodents, olfactory communication in,
   579–87
   active intraspecific, 580–3, 586–7
   aggression, 584
   food foraging and preference, 585–6, 587
   intra- vs interspecific, 580, *580*
   maternal behavior, 584–5
   nesting and interspecific
      preference/avoidance, 587
   passive intraspecific, 583–6, 587
   scent marking, 580–3
   sociosexual behaviors, 583–4
   warning signals, 586
roentgenograms (x-ray), 279
Roman Empire, use of perfumes in, 4, 6
ropivacaine, 916
rose water, 6
rotenone, 468

saccharin, 666, 667, 671, *671*
Saccharomyces cerevisiae, 103
Saguinus
   ethmoturbinal in, *53*
   maxilloturbinal in, *52*
Saimiri (squirrel monkey), 322
Saimiri sciureus (squirrel monkey), *613*, 616
salicin, 671
saliva, 625–33
   amylase, taste produced by, 631–2
   artificial, 905
   as solvent, 629
   bitter, 631
   composition, 626–7, **626**
   effects of impaired flow, 891
   ligase, taste produced by, 632
   oral senses and, 632–3
   production, 625
   proteins **627**
   role in astringency, 632
   role in taste, 629–32
   mucosal membrane, 627–8
   salt, 629–30
   sensory receptors, 627
   sour, 630
   sweet, 630–1
   taste buds, 628, 629
   taste pores, 628–9
   umami, 631
   unstimulated (resting), 625–6
salivary androgen binding protein (sABP),
   66
salivary glands, 625

Subject Index 1213

disorders of, 894
tongue and, 643–4
salivary proteins, 632
salivary reflex, 11, 625, 897
Salmonella typhimurium, chemotaxis in, 513
salt taste, 233, 686–7, *see also* sodium chloride
modulators, 677–9
orofacial expressions in infants, 799
preference in infants, 798–9, 823
primates and, 986
qualities of, 836–44
saliva and, 629–30
Salvelinus sp., 952
San Diego Odor Identification Test, 309
saquinavir, 917
sarcoidoisis, 896
saxitoxin, 952
scatol, 263
Sceloporus undulatus, 1142
scent glands, 611
scent marking, 595–7, 612
scent rolling, 597
Schistocerca gregaria, 532, *532*
schizoaffective disorder, 903
schizophrenia, 19, 24, 148, 149, 217, 903
odor memory tests in, 239
olfactory bulb volume in, 377
olfactory CSERPs, 271
olfactory dysfunction in, 390, 403, 429–36, **430–3**
olfactory structural imaging in, 286
scratch and sniff odorized strips, 228, *228*
scurvy, 901
Scyliorhinus canicula (small-spotted catshark), 553
sea lions, 958
seasonal affective disorder, 24, 390, 438
seasonal allergic rhinitis, 363
sebokeratinocytes, 566
Secondary Maximum Contaminant Levels (SMCLs), 1082
secondary olfactory cortex (medial orbitofrontal cortex), 411
secretory leukoprotease inhibitor, 66
sedatives, administration via nasal cavity, 467
Selasphorus platycercus (broad-tailed hummingbird), 965
Selasphorus rufus (rufous hummingbird), 965
Selective Reminding Test (SRT), 410
self-administered computerized olfactory test system (SCOTS), *229*
Semiliki virus, 462
sendai virus, 462
sensilla, taste in insects, 533, 929–30, *930*, 938–9

sensory-dietary synergy, 841–3
sensory specific satiety, 786
septal deviation, 360, 364
septal olfactory organ (SOO), 42, 48
septal organ of Masera (SO), 305, 1133–6
central projections, 1134–5
mechanosensitivity of olfactory sensory neurons, 1136
odorant receptors and signal transduction, 1134
odorant response properties, 1135–6
septal perforation, 360
septoplasty, 361, 363, 369
serotonergic (5-HT) input, 171
serous glands, 626
sertraline, 904, 921
sevoflurane, 501
sexual sweat, 298
short axon cells (SACs), 158, 159, 165
Shy-Drager syndrome, 427
sialadenitis, 890, 894
sialolithiasis, 891
signal detection theory (SDT), 232–3, 752
silica, 491
Siluridae, 645
silver staining, 279
single photon emission computed tomography (SPECT), 279, 890
single staircase (SS) procedures, 230, *231*
sinus, normal anatomy, *381*
Sjögren's syndrome, 24, 633, 893, 901, 905
olfactory dysfunction in, 391
smallpox, 279
Smell Identification Test, *see* University of Pennsylvania Smell Identification Test
Smell Wheel, *229*
smoking
cessation, 468–9
tongue and, 891
snakes, taste in, 973
Snap & Sniff™ wand, *229*
sniff bottles, 228, *228*, 229, 230
Sniff Magnitude Test, 239–40, *239*, 245, 309
sniff rate analysis paradigms, 21
Sniffin' Sticks test, **243**, 244, 270, 309, **408**, **422**, **433**
sniffing, Canidae, 593
sniffing at mouths, primates, 611
social recognition of body odors, 343
socially transmitted food preference (STFP), 117
sodium chloride (NaCl), 11, 22, 233, 388–9, 625, 629–30, 665, 669–70, 672, 674–5, 678–9, 686–7, 702–3, 705, 707–9, 711, 713–4, 716, 753–6, 762, 764. 778, 783, 786, 798, **807**, **815**, 832, 836, 840, 890–1, 911, 917, **918**,

934, 938, 986, 993, 1028, 1031–2, 1034, 1037 1062, *see also* salt taste
sodium cyclamate, 670, 672
sodium dehydrocholate, 916
sodium depletion, 837
sodium glycocholate, 469
sodium-hydrogen exchanger isoform 1 (NHE1), 676
sodium lauryl sulfate, 753
sodium saccharin, 915
sodium taste, 686
sodium tauro-24,25-dihydrofusidate (STDHF), 469
Solea senegalensis (sole), 552
solid phase microextraction (SPME), 572–3, *572*, 574
solvents, transport of, 464
somatostatin, 838
Sonic Hedgehog (Shh), 733–4
sour taste modulators, 675–7, 687–8
acid entry/exit balance, 676
calcium as, 675
carbonic anhydrase and, 677
cell shrinkage and, 675–6
mannitol and cytochalasin B in, 676
NADPH oxidase-and cAMP-linked membrane proton channels, 676–7
stimulus flow rate and, 675
taste cell subset in sour transduction, 675
zinc and, 677
sour taste
in infants, 823
in primates, 986
saliva and, 630
spatial processing, 250–1
Spectrum Descriptive Analysis, 1069, 1070
sperm, count, low, 469
Spermophilus beldingi (ground squirrel), 583
Sphenodon punctatus, 970
sphenoidectomy, 384
Spheroides maculatus (puffer), 549
Sphyrna lewini (scalloped hammerhead shark), 549, *550*, 557
Sphyrna tiburo (bonnethead sharks), 1143
spice use
in early medicine, 6–8
history of, 4–6
spinocerebellar ataxias (SCAs), 436–7
spironolactone, 921
split-half reliability coefficient, 240
sporadic olivopontocerebellary atrophy, 427
sprue, 901
Squalus acanthias (spiny dogfish), 1143
squamous cell carcinoma, 892
squeeze bottle, 228, *228*
SQUID (Superconducting QUantum Interference Device) sensors, 776
stapedectomy, taste dysfunction and, 896

Subject Index

Staphylococcus aureus, 100
statherin, 627
Steele, Richardson, and Olszewski
    syndrome *see* progressive supranuclear
    palsy (PSP), 23–4, 410, **420**, 423, 428,
    429
step-through shock avoidance, 874
stereo-olfaction, 549
steviol glycoside (SG) sweetener, 669
stevioside, 666, 669
stimulus-induced recovery (SIR), 1103,
    1104
stimulus matching task, 237
stimulus-stimulus association, 345–6
stir bar sorptive extraction (SBSE)
    (Twister), 573, 574
strepsirrhines, 605
*Streptococcus pneumonia*e, 897
striatonigral degeneration, 427
stroke, 438, 890, 899
    SVZ neurogenesis and, 198
Strongest Component Model, 248
structure-activity relationships (SARs) in
    oral chemesthesis, 1102–3
styrene, 69, 76
    olfactory system in transport of, 464
sublingual glands, 625, 626
submandibular gland, 625, 626
substance P, 66
subventricular zone (SVZ)
    cytoarchitecture of, 183–4, *185*
    neuroblast migration, 184–6
    phenotype specification and maintenance,
        196–7
    proliferation and migration.control of,
        186–94, **187–90**
        cell cycle regulators, 190–1
        cytoskeletal proteins, 191
        extra-cellular matric and guidance
            molecules, 191
        growth factors, 192–3
        intercellular signaling pathways, 193
        neurotransmitters, 193–4
        transcription factors and miRNA, 192,
            192
'sucking' papillae, 642
sucking reflex, 898
sucralose, 670
sucrose, 388–9, 630–1, 643, 666–9,
    688–9, 702–3, 707–9, 711, 714, 751,
    753–4, 762–4, 767, 780, 783, *784*,
    786–7, 797, 800–4, **805–820**, 833–4,
    836, 838, 841–2, 847, 865, 867,
    870–1, 874, 876, 878, 903, 911,
    916–7, 934–6, 938–40, 957–8,
    964–9, 972–5, 986, 992, 1008–9,
    1011–4, *1015*, 1017–9, 1021, 1032,
    1037, 1040, 1062
sucrose taste simulation, 666

sulcus terminalis, 639
sulfoglycosphingolipids, 64
sulfotransferases (SULTs), 70, 486
sumatriptan, 467
Suncus murinus (insectivore shrew), 867
superoxide dismutase, 71
suprasellar meningiomas, 385
suprathreshold scaling procedures, 233–6,
    751–767
Suricata suricatta (meerkat), 958
sustentacular cells, 64
swallowing reflex, 898
sweet taste, 688–90
    birds and, 964–7
    children and, 822, 823
    orofacial expressions in infants, 799
    preference for in children and infants,
        798, 833–4
    primates and, 986–7
    reptiles and, 972–3
    saliva and, 630–1
sweet taste receptors, 689, 840
sweet taste, modulators of, 666–72
    antagonists as, 669–70
    osmolytes as, 668–9
    positive allosteric modulators (PAMs),
        666–8, *667*
sweet taste stimuli, transduction, 693–4
sweet water aftertaste (SWA), 669–70, *670*
Sydney Children's Hospital Odor
    Identification Test, 309
synchrony in odor response, 169
synergistic sweeteners, 667
synthetic interaction, 1017
systemic disorders, taste dysfunction and,
    901, **902**
systemic lupus erythematosus (SLE), 893

T&T Olfactometer, 230
TAAR receptors, 16, *110*, 115–6, 118, 553,
    554, 556, 1116, 1138–9
T2R16, 685
Taeniopygia guttata (zebra finch), 968
tannins, 632
Taricha granulosa (rough-skinned newt),
    973, 1144
taste
    anatomy, 959–60
    receptors and signalling molecules, 960
    measuring behavior, 960
    definition, 665, 796, 959
taste, function of, 781–8, 959
    affective coding of, 785–8
    familiarity coding of, 785
    intensity coding of, 781, 782–3
    quality coding of, 783–5
taste adaptation, 703
taste blindness, 755

taste buds, 9, 10, 14, 15, 17, 18, 23, 375,
    380, 625–633, 637–655, 666, 672,
    676, 680, 686–688, 694, 695, 704, 708,
    718, 727–742, 754–757, 762–764,
    796, 804, 845, 850, 887–892, 897,
    898, 903, 905, 914, 915, 936, 940,
    947–954, 958–976, 983–985, 1020
    anatomy, 959
    density, 643
    development of, 651–3, *653*
    distribution, **642**
    embryonic epithelium and mesenchyme
        signaling, 729–30
    embryonic origins, 728–9
    extralingual, 643, 644
    gustatory nerves, 729
    history of identification, 637, *640*
    innervation of, 653–4, *654*
    lingual, distribution of, 640–1
    molecular regulation of anterior
        development, 730–2
    molecular regulation of cell lineage,
        734–6
    Notch signaling, 734, 735
    regeneration of, 18–19
    saliva and, 628, 629
    structure, 947–9, *948*
    transcription factors and taste bud
        development, 735–6
    vertebrate, cell types of, 645–50
taste buds, species variation in
    in amphibians, *645*, 646–7
    in birds, 647, 961–3
    in fish, 645–6, *645*, 948–9
    in macaques, 983–5
    in mammals, *645*, 647–50, *648*
    in reptiles, 647, 971–2
taste detection thresholds (DT), 916–17,
    918
taste disturbance
    causes, 890–1
    chemical testing, 889–90
    diagnostic testing, 889–90
    electrical testing (electrogustometry), 890
    imaging, 890
    medical evaluation, 889–90
    medical history, 889
    physical examination, 889
    regional testing, 889
    symptoms, 888–9
    treatment, 904–5
    whole-mouth testing, 889
taste epithelium, anatomy of, 727–8
taste evoked probability map, 777
taste-induced odor enhancement, 1011
taste perception
    ontogeny of, 796
    research approaches and methdologies,
        796–7

research findings of basic tastes, **805–21**
taste phantoms, 765
taste pore, 651, 796, 959
    in macaques, 984
    saliva and, 628–9
taste potentiated odor aversion, 346
taste projections, laterality of, 780–1
taste quality, drugs and, **918–21**
taste receptors, 17, 840–1
    amino acid, 841
    fat, 840–1
    G protein-coupled, 688–93
    in gastrointestinal tract, 840
    ion channels as, 686–8
    type 1 taste receptors (T1RS), 688–9
    type 2 taste receptors (T2RS), 690–2
taste reward prediction error, 1035
taste system, development of, 651–3, *652*,
    727–42
taste transduction, molecular basis, 685–96
taurine, olfactory system in transport of, 463
TDI, *see* Sniffin' Sticks test
Temporal Dominance of Sensations (TDS),
    1071, *1071*
temporal lobe epilepsy, 437, **437, 438**
temporal processing, 251–2
tenia tecta, 210
Tenrec ecaudatus, *55*
terbinafine, 901
terfenadine, 383, 921
terminal nerve (nervus terminalis), 21, 43,
    48, 57, 1141–5
    anatomy, 1141–2
    function, 1143–5
test-retest reliability coefficient, 240
testosterone, 73, 74, *74*
tetanus, immunization, 469
tetradecan-2-one, 570
Tetrahymena, 526–7
teviol glycoside (SG) sweeteners, 671
thalamus, 210, 212, 218, 219, 295, 298, 346,
    428, 435, 436, 609, 706, 707,
    710–712, 714, 715, 718
    in macaques, 989–90
    primate, 710, 711
    rodent, 710–11
thallium, olfactory system in transport of,
    461
Thamnophis sirtalis (garter snakes), 1142
Thamnophis sirtalis concinnus (red-spotted
    garter snake), 973
thaumatin, 669
theophylline, 392–3
thiamine propyldisulfide (Alinamin), 229
thiazide diuretics, taste distrubance and, 846
thiophene, 388
Threshold Odor Number (TON), 1080, 1082
threshold, psychophysical, 11–2, 22–3, 66,
    76, 227, 230–1,240, **242–4**, 244–8,

269–70, 283–4, 286, 308–310, 339,
    341, 362–9, 375, 380–5, 388–9, 391,
    403, **404–8**, 409, 413–4, **415–22**, 427,
    429, **430–3**, 435–40, 461, 501, 538,
    592–3, 608–9, 612–4, 616, 629, 630,
    632–3, 670, 674, 686, 708, 714,
    751–5, 757, 762, 765, 767, 783,
    796–802, **807**, **810**, **813**, **818**, 832,
    834–6, 845–8, 900, 903, 916–18,
    953–4, 962, 966, 969, 970, 991, 1012,
    1016–17, 1054, 1081–7, 1091–1105
thrush, 893
thyrotropin-releasing hormone (TRH), 916
thyroxin replacement therapy, 904
tidal water pump, 359
Tiliqua scincoides (eastern blue-tongued
    lizard), 974
Time-Intensity Profiling, 1070, *1071*
Tinca tinca (tench), *550*
'tissue-resident' stem cells, 135
tobacco, 246, 500
    second-hand smoke (SHS) exposure, 500
    third-hand smoke (THS) exposure, 500
toluene, 69
    olfactory system in transport of, 464
tongue
    anatomy of, 637–44
    blood supply to gustatory papillae, 644
    development of, 651–3, *652*
    disorders of, 893
    history, 637–8, *640*
    innervation of, 653–4, 739–41, *740*
    necrosis after chemotherapy, 892
    salivary glands, 643–4
    solitary chemosensory cells (SCC), 644
    see also papillae
tongue map, 764
tonsillectomy, taste dysfunction and, 896
topiramate, 902
trace amine-associated receptors (TAARs),
    16, 115–16, 553
Trachemys scripta (red-eared slider turtle),
    980, 1142
trans-cinnamaldehyde, 1092
trans-2-cis-6-nonadienal, 1082
trans-2-pentanal, 1092
trans-stilbene oxide, 79
transcellular transport, 453
transcranial magnetic stimulation (TMS),
    393, 394
    transcription factor Phox2b, 737–8
    transcription factors, including Islet 1 and
        Tlx3, 737
transcytosis, 453
Transient Receptor Potential (TRP) family,
    1101
    TRPM5 666, 694, 695
transient receptor potential polycystic
    (TRPP) marker proteins, 675

transit amplifying cell, 135
transnatal chemosensory continuity, 318
transneuronal transport, 466
transport of agents by olfactory system
    to brain and circulation, 453–64
    CNS transport, 466
    consequences of, 473–4
    mechanisms, 464–6
    transneuronal transport, 465–6
    transporters in nasal epithelia, 466
    uptake, 464–5
transverse lamina, 42
traumatic brain injury (TBI), 468
trazodone, 921
Tremarctos ornatus (spectacled bear), 958
trench mouth, 894
triacylglycerols, 834
Trichoglossus haematodus (rainbow
    lorikeet), 965
tridecan-2-one, 570
triethylamine, 252
trifluoperazine, 902
trigeminal chemesthesis, 1091–1105
    chemical mixtures and, 1099–100
    cut-off effect, along vapors from
        homologous chemical series,
        1098–109
    molecular receptors for, 1092–3
trigeminal event-related potentials (tERP),
    1094
trigeminal nerve, 305–6, *306*, 655
    development of, 306
    disorders of, 898
trigeminal neuralgia, 780, 898
trimethylamine, 310, 312
Triturus (newt), *646*
tropotaxis, 549
Trypan blue, olfactory system in transport
    of, 462
tuberculosis, 897
tufted cells (TCs), 157, 163
tumors
    corpus callosum, 385
    intranasal and intracranial, 384
Tupaia glis (tree shrews), 45, *46*, 54
turbinals, 43
    anatomical and physiological roles of,
        46–7
    in mammals, 41–2
    phylogenetic variation, 50
    mucosal properties, 50–1
turbinate hypertrophy, 360
turbinectomy, 360
turbinoplasty, 363, 369
Turdus migratorius (American robin), 966
Turner's syndrome, 388, 903
turtles, taste in, 974
two-alternative forced choice method
    (2-AFC), 753, 1073–4

Subject Index

two-bottle taste test, 22
tympanoplasty, taste dysfunction and, 896
typhoid, 1079
tyrosine hydroxylase, 468

UDP glucuronosyltransferase (UDPGT or
    UGT), 68, 70, 464, 486
umami, 17, 525, 626, 631, 648, **649**, 665
    birds and, 967–8
    neuroimaging and, 1037
    orofacial expressions in infants, 799
    preference in infants, 798
    saliva and, 631
    taste receptor, 634
    touch sensation and, 1009
    transduction, 693–4
umami, taste modulators, 672–5
    agoenhancers as, 673–4, *674*
    antagonists as, 674–5
    glycoconjugated peptides as, 674
    N-geranyl cyclopropylcarboxamide
        (NGCC) as, 674–5
    pure PAMs as, 672–3, *673*
uncertainty principle, 272
uncinectomy, 384
unconditioned responses (URs), 865
unconditioned stimulus (UCS), 289
unconditioned stimulus (US), 865, 875–9
unity assumption, 1008
University of Pennsylvania Smell
    Identification Test (UPSIT), 237, *238*,
        241, **244**, 245, 247, 309, 381, 383,
        385–391, **404–8**, 409–14, **415–22**,
        429, **430–3**, 434–8, 493, 496, 502,
        1086
upper respiratory infections, 380–1
    smell loss and, 378
urea bitterness suppression, 672
urine washing, 611
urocortin peptide, 473
urodilatin, 116
uroguanylin, 116

vaccines, administration via nasal cavity,
    469
vagus nerve, 654–5
    disorders of, 898
valeric acid (VA), 264
vallate papillae, 639, 640, *641*, 642, 651
value adding, 1053
Vanessa indica, 936
vascular dementia, 437–8
vascular parkinsonism (VP), 428
vasculature dementia (VaSD), EEG, 263
vasoactive intestinal peptide (VIP), 66,
    468,650

vasopressin, 468
Venezuelan equine encephalitis virus, 462
venlafaxine, 904
ventral septal glands, 48
Vernet's syndrome, 897–8
vesicular stomatitis virus (VSV), 487, 462
Vibrio cholerae
    cyclic di-GMP in, 517
    biofilm matrix synthesis by cyclic
        di-GMP in, 516, *517*
Vibrio fischeri, quorum sensing in, 518
Vibrio harveyi
    quorum sensing in, 518–19, *519*
    autoinducers in, *518*
viral-related smell dysfunction
    olfactory bulb volume in, 377
viruses, transport of, 461–2
visual analog scale (VAS), 758
visual-to-taste association learning, 1035
vitamin A, 392
vitamin B complex, 392, 393
vitamin B12 derivatives, administration via
    nasal cavity, 469
vitamin D deficiency, 849
volatile organic compounds (VOCs)
    in humans, trigeminal sensitivity to,
        1098–1105
    in birds, 569, 571, 574
        analytical techniques, 570–2
        coatings for, 572, **573**
        solid phase microextraction (SPME),
            572–3, *572*, 574
        solvent extraction, 572
        stir bar sorptive extraction (SBSE)
            (Twister), 573, 574
volumetric brain mapping (VBM), 286
vomer bone, 39
vomeromodulin, 66
vomeronasal glands, 48
vomeronasal organ (VNO), 16, 17, 20–2,
    40, 42–4, 47–8, 56–8, 69, 95,
    100–101, 110, 115, 157, 171–74, 298,
    305–6, 464, 547, 553–6, 593, 617–8,
    972–4, 1113–25
    accessory olfactory bulb, 48, 99, 157,
        171–3, 306, 592–3, 608, 617, **618**,
        1120–1
    amygdala, 1122–3
    circuits and information processing,
        1120–3, *1121*
    evolution and decay of, 1123–4
    functional significance and, 1124–5
    hypothalamus, 1123
    pheromones and, 298
    in primates, 617–18
    removal, 22

signaling cascades, 1117–18
signaling mechanisms, 1114–17
structure and function, 1113, *1114*
vomeronasal chemosensory cues,
    1118–22
vomeronasal sensory neurons (VSNs), 1113
    physiology and signaling cascades,
        1117–18
    signaling mechanisms, 1114–17
    vomeronasal chemoreceptors, 1114–17
vomeronasal systems, human, development
    of, 305–6, *306*
Von Ebner's Gland Complex (CVP/VEG),
    development of, 732–3
von Ebner's glands, 626, 628

water, drinking, 1079–87
    description of taste and odor, 1080–1
    global impact of algae in, 1081–2
    methyl tertiary butyl ether (MTBE) in,
        1083–6
    public acceptability of flavor, 1082
wavelet analysis, 272
Weber-Fechner law, 1081
Weber's glands, 644
Weber's law, 231
Wegener's granulomatosis, 896
Wernicke-Korsakoff syndrome, 499
Westermarck effect, 324
wheatgerm agglutinin (WGA), 461
    WGA conjugated to horseradish
        peroxidase (WGA-HRP), 461, 466
Wheel Smell Identification Test, 309
wine industry, 1058–9
wisdom tooth extraction, 897
witchcraft, 5
Wnt planar cell polarity (PCP) pathway, 733
Wnt/ß-catenin, 733
working memory, 297, 338–40
World Trade Center (9/11), olfaction and,
    496, 502
WS-3 (n-ethyl-p-menthane-3-carboxamide),
    917
Wsp chemosensory pathway, *517*

X-linked Recessive Dystonia-parkinsonism,
    also termed 'Lubag', I, 428
xenobiotics, 453, 485–8
    metabolism, olfactory, modification of,
        78–81
Xenopus, 557, 952
Xenopus laevis (African clawed frog),
    1144
xerostomia, 626, 850, 891, 892, 901, 904
xylene, olfactory system in transport of,
    464

Subject Index

Y-maze, 548, 557
yes/no identification tests, 237
yoga, 905

Z-4-decenal, 574
(Z)-5-tetradecen-1-ol, 114
Zalophus califonianus (California sea lion),
958

Zicam (zinc gluconate), 461, 499
zidovudine (AZT;
3′-azido-2′,3′-dideoxythymidine),
467
zinc, 499
nanoparticles, 101–4, *104*
olfactory system in transport of, 461
zinc deficiency, 846, 849, 901

zinc-enriched neurons (ZENs), 101
zinc gluconate, 905
zinc sulfate, 392, 902
ziziphins, 669
zoniporide, 676
Zonotrichia albicollis (white-throated
sparrow), 968